FUNDAMENTAL WORLD OF QUANTUM CHEMISTRY

Fundamental World of Quantum Chemistry

A Tribute to the Memory of Per-Olov Löwdin

Volume I

Edited by

Erkki J. Brändas
Uppsala University,
Uppsala, Sweden

and

Eugene S. Kryachko
Bogoliubov Institute for Theoretical Physics, Kiev, Ukraine
and University of Leuven, Leuven, Belgium

KLUWER ACADEMIC PUBLISHERS
DORDRECHT / BOSTON / LONDON

A C.I.P. Catalogue record for this book is available from the Library of Congress.

ISBN 978-1-4020-1290-7 (Vol. I)
ISBN 978-1-4020-1291-4 (Vol. II)
ISBN 978-1-4020-1286-0 (Set)

Published by Kluwer Academic Publishers,
P.O. Box 17, 3300 AA Dordrecht, The Netherlands.

Sold and distributed in North, Central and South America
by Kluwer Academic Publishers,
101 Philip Drive, Norwell, MA 02061, U.S.A.

In all other countries, sold and distributed
by Kluwer Academic Publishers,
P.O. Box 322, 3300 AH Dordrecht, The Netherlands.

Printed on acid-free paper

Cover Picture:
Löwdin's asymmetrical wave function Ψ in love with Hamiltonian

Photo: Per-Olov Löwdin (Michael Hehenberger.
Property of the Uppsala Quantum Chemistry Group)

FOREWORD TO VOLUME I

It is with respect and devotion that we point out Per-Olov Löwdin's pioneering and limitless efforts to found and establish the science of Quantum Chemistry. With his extraordinary capacity, Per-Olov Löwdin made basic contributions to the development, provided the conceptual framework and introduced fundamental concepts, and added the rigour to the field. Let us remind the reader of Löwdin's definition of a molecule as a Coulombic system of electrons and nuclei whose Hamiltonian with the center of mass translational degrees of freedom removed, has at least a bound state (in Molecules in Physics, Chemistry and Biology II, J. Maruani (Ed.), Kluwer, Dordrecht, 1988). But let us also recall his critique, i. e. of the impossibility to derive the Coulomb Hamiltonian from the first principles and the non-unique existence of molecular structures due to the general incompatibility between the number of nuclear positions and the number of internuclear distances. One of the most urgent problems of modern quantum chemistry is to treat the motions of the atomic nuclei and the electrons on a more or less equivalent basis.

Per-Olov Löwdin also established, developed and strengthened the links between physics, chemistry and biology. In addition, by organizing a staggering amount of regular summer and winter schools, institutes and symposia, by founding the International Journal of Quantum Chemistry and the Advances of Quantum Chemistry, he commended undisputable leadership at the highest national and international levels.

The clarity of expression, the vision of the possible and the optimism for the future are beautifully reflected in his "*Present Situation of Quantum Chemistry*" published in J. Phys. Chem. 61, 55-68 (1957), where, with eagerness of reaching a wide audience and capability of attracting many scientists, Per-Olov Löwdin introduced and defined the field of Quantum Chemistry by formulating its goals, i.e.

(i) to be able to predict accurately the properties of a hypothetical polyatomic molecule before it is synthezised in the laboratories;

(ii) to obtain such knowledge of the electronic structure of matter that one can construct new substances of particular value to mankind;

(iii) to learn to think in terms of electrons and their quantum mechanical behaviour.

In his revelations "*Quantum Theory of Many Particle Systems I-III*" (Phys. Rev. 97, 1474-1520 (1955)), Per-Olov Löwdin made a rigorous and ground-breaking formulation of the all important correlation problem in quantum chemical applications, from which emanated fundamental concepts like the correlation energy, the natural spin orbital, the general configuration interaction (CI) method, the reduced density matrix and the charge and bond order matrix. The constraints imposed by symmetry considerations in various independent particle models termed the "*symmetry dilemma*" (Rev. Mod. Phys.

35, 496 (1963)) have further led to important developments in solid state physics, chemical physics and physical chemistry.

Well-known is Löwdin's series of "*Studies in Perturbation Theory I-XIV*", a framework for the analysis of intermolecular forces as well as the technical development of very large CI expansions. The impact of his work on Quantum Biology are to be found in new applications and developments, i.e. in possible conformation work on prions, proton transcorrelations in DNA, macromolecular recognitions and specific motifs, in cancer research, etc. where there is a highly relevant quantum mechanical component to be identified. These accomplishments together with close to 70 summer and winter schools and Sanibel Symposia have contributed to the education of several generations of quantum chemists and solid state physicists.

It is a great priviledge and honor to have participated in this edification. Looking at the picture of Per-Olov Löwdin, climbing in the mountains, we recollect the famous song by the Prophetes Deborah after the victory over the Philistines: "The mountains flowed before the Lord." This is precisely whom he was, is and will be - Per-Olov Löwdin or Pelle for all of us. Indeed, these years, when he was with us, were the most colourful, exciting and fascinating period in the history of quantum chemistry. It was the most enjoyable time for all of us because we all worked in quantum chemistry "just for FUN", as he always maintained. He will always be with us, the whole family of Quantum Chemists - including his own family with the "mother of the Uppsala Quantum Chemistry Group", Karin Löwdin, as prima motor and inner connection, and he is particularly close to all who have contributed to this tribute volume to his memory appearing as Volumes 4 and 5 in the Series "Conceptual Trends in Quantum Chemistry", already included Volume 1, "Conceptual Trends in Quantum Chemistry" (1994), Volume 2, "Structure and Dynamics of Atoms and Molecules: Conceptual Trends" (1995), and Volume 3, "Conceptual Perspectives in Quantum Chemistry" (1997), edited by Jean-Louis Calais and Eugene Kryachko. We deeply thank all authors for their wonderful and profound contributions.

Erkki J. Brändas

Uppsala University
Uppsala

Eugene S. Kryachko

Bogoliubov Institute for Theoretical Physics,
Kiev, Ukraine
and University of Leuven, Leuven, Belgium

Contents of Volume I

Contents of Volume II

PER-OLOV LÖWDIN

A Scientific and Personal Appreciation

Harrison Shull

Lives of great men all remind us
We can make our lives sublime,
And, departing, leave behind us,
Footprints in the sands of time.
Longfellow

Per-Olov Löwdin was a veritable giant in the scientific universe, a dynamo of energy, full of *joie de vivre*, and a kindly mentor to us all. We rejoice in the life he shared so willingly with each of us. We can only hope that our own 'footprints in the sands of time' will reflect in some small measure the indelible strides he contributed scientifically and personally all around the world.

I have been fortunate, very fortunate, that my life has been so entwined with that of Per. My first contact with Pelle (an affectionate name for Per) was one of those coincidences which, one realizes afterwards, changed the direction of one's life.

Per received his doctorate in theoretical physics in Uppsala, Sweden in 1948, the same year I received mine in physical chemistry at Berkeley. Following a year with Robert Mulliken in Chicago, I joined the faculty at Iowa State College. The first post-World War II scientific conference I was privileged to attend was a gathering at Shelter Island at the eastern tip of Long Island.

The Shelter Island Conference was an exciting and stimulating occasion. Attendance was limited to 25, and I was very fortunate to have been included. For that I can thank Robert Mulliken with whom I had had a recent post-Doctoral appointment. It was an adventure just to get to Shelter Island.

E.J. Brändas and E.S. Kryachko (eds.),
Fundamental World of Quatum Chemistry, Vol. I, 1–14.

First was, of course, travel to New York, followed by a long bumpy creaky ride on the Long Island railroad. That ride seemed very much shorter than it really was, however, because I had a chance to talk science with George Kimball all the way. The end of the line was followed by a ferry ride to Shelter Island and a taxi ride to the Ram's Head Inn.

Not long ago, I had the opportunity to stop in at the Ram's Head Inn. It seemed very much the same from the outside, but the plush living room furnished with over-stuffed chairs where we gave our talks and held our spirited discussions was now occupied by a dining room. A plaque on the wall commemorates the first theoretical physics conference there in, I believe, 1947, but there is no history or indication of any other meetings there.

The younger scientists in our 1951 conference were assigned two to a room in the inn. Luckily, Pelle was my a room-mate, and we spent hours in the late evenings talking about science, Sweden, the future, and everything in between. Needless to say, I was charmed by Pelle, impressed with his enthusiasm, his broad education, the range of his intellect, and his scientific acumen. It was clear that his background, the breadth and depth of his education, and his mathematical skills were exceptional, far exceeding my own.

Each of the participants gave short presentations of their current work. Per talked of his calculations on the problem of the solid state. I talked briefly about the calculations I had done in Mulliken's laboratory. The atmosphere was very informal, friendly, with lots of fun and banter. Sir John Lennard-Jones made some notes on his pad using a magazine he had picked up as backing support. When he jumped up to make a point, he was photographed waving the magazine - Playboy, or something even more out of character. There was great fun in one session held in a local schoolhouse when the Inn was otherwise occupied. The session was continued in a 3rd grade classroom, and Charles Coulson was the butt of a lot of laughter as he tried to fit his long legs under the desk.

This stimulating initial contact with Pelle at Shelter Island encouraged me to think of a sabbatical in Sweden. I applied for a Guggenheim Fellowship for the 1954-1955 academic year. The Foundation awarded me a fellowship but wrote they didn't have enough money to grant me the minimum I needed. They hoped their grant would help. Iowa State had no sabbatical program, so I requested a leave of absence without pay for the year. Initially, the Chairman of the Department said he wouldn't give me that leave, to which I immediately replied that I would resign. He relented, but it was a significant event that influenced my later decision mot to return to Iowa State after my leave. With insufficient funds from my grant, I went into debt, securing a personal loan from my brother, so my family and I could go to Sweden, and I could work with Pelle.

We arrived in Uppsala in September 1954, and were warmly welcomed by Per who had found a convenient apartment for us. It was within walking distance of Professor Waller's Institute of Mechanics and Mathematical Physics where Per had become Docent. There was no office available for me in the Institute, so Per arranged for a small table in the Institute's library for me to use as a 'desk'.

I consulted with Per as to what scientific work I should undertake. I had done my thesis on the triplet spectroscopy of benzene, so I suggested I might do some theoretical calculations on that molecule. In our second meeting, Per persuaded me to scale back that altogether too ambitious program, and he suggested we look at ethylene as perhaps a simpler, more amenable, but interesting, molecule. And a day later, we both realized that two electrons as a model system was all we could handle with the primitive computing facilities available then. I soon became known as the "two-electron chemist".

One day, doing some literature work on the calculations of Professor Hylleraas of Norway, I ran across an old calculation of his on Helium using a basis set of Laguerre functions with a single exponent. These results from decades earlier were significantly better than a similar more recent calculation by Taylor and Parr, using a basis of the H atom Legendre functions. Per was delighted with this discovery, suggested immediately an improvement in the basis set used by Hylleraas, and set Klaus Appel to help me complete new calculations. These were very instructive because they demonstrated the importance of including the continuum of the H- atom basis in doing configuration interaction calculations.

Per and I jointly published a letter in the Journal of Chemical Physics recommending the use of the new basis set in preference to the H atom set which was much more slowly convergent. A rather amusing exchange followed. Linus Pauling wrote us, gently pointing out that these functions were already well known, and in fact, he wrote, they were described in the widely used text of Pauling and Wilson. To our surprise and my discomfort, the functions were indeed described in Pauling and Wilson although they seemed to me to be presented somewhat obscurely. Chagrined, Per responded courteously to Pauling, thanking him for leading us to that earlier work.

Not long thereafter, I was studying some other older papers in the library, and, as I frequently did, I looked at adjacent articles as well. I chanced upon a paper by Pauling in which he called the reader's attention to the superior properties of this very same orbital basis set. I was fascinated to find, however, that Pauling included a footnote, apparently added in proof, about the history of these Laguerre functions. He reported that a referee had informed him that this was not a new discovery, but that these functions were well known and had been described in the literature in the late 1920's. So does science evolve! Per and I had a good laugh about this sequence of events.

Coincidentally the Nobel Prize in Chemistry that year was awarded to Linus Pauling. Pelle used his very best efforts and succeeded in securing an invitation for me to attend the ceremonies. It was exciting to meet Pauling, to see the ceremonies, and to attend the banquet even though, as one towards the bottom of the rigid Swedish protocol list, I was relegated to an anteroom off the main banquet area. It shows, however, how hard Per worked to be sure his visitors always could take advantage of all that went on in Sweden. That never changed throughout his life.

In the fall of 1954, Per proposed that we have a small conference in Uppsala. Besides those of us in Sweden, there would be invited perhaps as many as four other scientists, and, for that purpose, he secured a small grant from the *King Gustav VI Adolfs 70-years Fund for Swedish Culture*. As I recall, the initial invitees were Mulliken, Coulson, Heitler, Daudel, and Pullman. Each scientist to whom Per wrote, however, not only eagerly accepted, but asked if he could bring some colleagues with him. Per enlisted the administrative support of Inga Fischer-Hjalmars in Stockholm and searched far and wide for further financial support. This he obtained not only from the Swedish Natural Science Research Council but also from five Swedish Corporations. In record time a conference of some sixty individuals took place in Stockholm and Uppsala in March. At the end of World War II, scientists everywhere were starved for opportunities to meet other scientists to talk science. Per felt that need and did something about it. It was an early example of Pelle's drive and willingness to go to great lengths to accomplish an immediate worthwhile goal.

The conference was a rousing success. Beside the original invitees except for Pauling, those attending included Pople, Longuet-Higgins, Hall,

Boys, and McWeeny from England, both Alberte and Bernard Pullman from France, Professor Hylleraas from Norway, Bill Lipscomb and Donald Hornig from the USA, Prof. Araki from Japan, and both Professors Waller and Otto Klein from Sweden. The talks were multi-lingual, two in German, all the many French participants spoke in French, and the remainder used English. Most of the English-speaking participants other than the Swedes, however, understood little French or German, and the scientists giving papers in those languages were dismayed at their inability to communicate with the English speaking attendees. It is interesting to note that at the next such multi-national conference just a few years later, all the French and German presenters were using British English and probably shook their heads at the strange pronunciation of the attending American scholars who still couldn't converse in either French or German.

Today, we have become so used to the ready availability of computers, we tend to forget how recently they became widespread. Despite the development and use of computers in WWII, in the early post-war years calculations were done by hand on mechanical calculators - the Marchant and Frieden in the United states, and the highly capable Facit in Sweden. Robert Mulliken in the Physics Department at the University of Chicago had a computing assistant who had worked with him for a number of years doing nothing but such mechanical calculations. The first transition to something more suited to heavier computation for the average scientist was the adaptation of an IBM accounting machine (the 402A, I believe) by hand-wiring its control board to do simple algebraic processes according to the 'program' on the punch cards fed into the machine. An iterative program was established by running the cards through the machine repetitively, feeding the stack of output cards back into the input stack by hand until one was satisfied with the level of convergence of the calculation.. It is impossible to express today how thrilled I was the first time I succeeded in diagonalizing a 5X5 secular determinantal equation. It seemed (and was!) a miracle!

For one unused to the length and difficulty of thesis work in Sweden, the depth and difficulty of Pelle's thesis work was astonishing to me. The computational complexity of the solid state system he investigated was overwhelming, and his ability to generate meaningful results in these pre-computer days was extraordinary. All the work of Per's solid state computations was completed using mechanical calculating machines. He did have a precise, thorough, and very helpful computing accomplice in Klaus Appel, who also helped me a great deal when I worked in Sweden

Per's dissertation itself was not widely available, but in 1956 Per published in the Philosophical Magazine Supplement a thorough 172 page review of the "Quantum Theory of Cohesive Properties of Solids". This was based on his dissertation at the University of Uppsala where his thesis work was completed under the supervision of Professor Ivar Waller, a very well-known solid-state theoretical physicist. In one footnote in that comprehensive paper, Per says: "The author is greatly indebted to Professor I. Waller, Uppsala, for suggesting to him in 1946 the classical question of the failure of the Cauchy relations as a suitable thesis problem ...". It was clear to me, however, that Per's relationship with Waller was not a warm one.

Following the conferring of his Doctorate, Pelle worked in Professor John Slater's group at MIT, and the influence of that contact is readily apparent in this same Philosophical Magazine review. The close relationship that Per established then with Slater continued throughout Slater's life, and Slater spent his post-retirement years with Per in Gainesville.

The appearance of the Phil. Mag. review was preceded, however, by the publication of Pelle's seminal work on the "Quantum Theory of Many-Particle Systems" in three papers in the Physical Review - papers that became the heart of the work of his group and of quantum scientists all over the world for many years.

In the first paper, Per explored the Density Matrix, Natural Spin-Orbitals, and the Configuration Interaction methods of finding approximate solutions to Schrödinger's equation. He discussed the practical methods of using configuration interaction to obtain approximate wave functions for interesting systems, and advocated these techniques, followed by transformation to natural orbitals, as a way to speed the convergence of the calculations. Furthermore, the eigenvalues of the natural expansion allowed useful interpretation as occupation numbers of the respective orbitals.

In the second paper, he explored the Hartree-Fock approximation within this same framework, again using projection operators, showing that in the Hartree-Fock representation, the first order density matrix contains all of the information about the system. He established relations between the eigenvalues of two different states in the Hartree-Fock representation and excitation energies, and showed the validity of an extension of Koopman's Theorem. He discussed Slater's 'average exchange potential' and related the work to the charge and bond order matrices of Mulliken and others. All this in clear elegant mathematical formalism.

The third paper took up the spin problem with the necessity for introducing more than a single determinant to represent the system including spin. Here he introduced projection operators as a neat way to handle spin degeneracies, and he tied together the work of physicists on the solid state and the work of chemists and physicists on molecular orbital and valence bond methods of molecular calculations.

This brief discussion of the contents of these papers does not do justice to the richness of subjects they contain. They set a high standard for their lucid presentations of complicated material, their orderly organization, and their thoughtful discussion. Such an elegant presentation became a hallmark of all of Löwdin's later papers. This early set became almost a 'Bible' for the group of people working with Pelle, and in the wider scientific community as well. Per soon ran out of reprints, but he arranged to have additional copies printed in Sweden. As a matter of curiosity, I recently looked to see how many references to Per's work, as recorded in the Science Citation Index, referred to these three papers. The total number of citations to Per's work has regularly exceeded 300 each year for decades on end, and of these citations, consistently about 20%, even as recently as five years ago, were references to this trilogy in the Physical Review.

Upon my return to the States in 1955, I joined the faculty at Indiana University. When I arrived, the University had already begun the process of securing University-wide computing facilities. Our first was the IBM accounting machine which I have already mentioned earlier. This was followed by an all-electronic 604, a "very fast" machine with ten memory locations and about as much power as our current $10 hand calculators. The limited 'fast' memory required careful programming to make optimum use of its speed for only the most highly repetitious parts of computations. We acquired 'ice-boxes' of mechanical ten-digit relays for back-up slow or archival storage. We made a regular habit of running to the appropriate one of these, lifting the lid to be sure the desired number really did get into storage, and, on occasion, kicking the box to encourage a sticking counter to move into position. That machine was followed by an IBM 650 drum storage machine, and eventually by a larger IBM 709. What a thrill! Continuing the work Per and I had started in Uppsala, my group at Indiana became the largest users of computing facilities there.

By this time, Pelle had finally become independent of Professor Waller, forming his own "Quantum Chemistry Group" located in new buildings on Rundelsgränd, just near the main University building. The review article in Phil. Mag. published in 1956 is reported by Per in his acknowledgements as the "first contribution from the 'Quantum Chemistry Group' which has been newly established at Uppsala University for research in atomic, molecular, and solid-state theory." He then expresses his "most sincere gratitude to His Majesty, the Swedish King, for his personal support ... and for his great personal understanding of the importance of basic research".

In the meantime, the United States defense establishment, following the example of its Office of Naval Research, established research offices in each service, offering basic research grants to scientists to continue the interaction between the military and the basic research community, an interaction that had contributed so significantly to winning WWII. As part of the principle behind the Marshall Plan, these grants were also extended to non-US scientists. Among the grantees of the Air Force Office of Scientific Research was Per-Olov Löwdin. Following receipt of such a grant, he wrote me to ask if I would be interested in coming back to Sweden as Assistant Director of the "Quantum Chemistry Group" in Uppsala supported by his Air Force grant. Although I had been at Indiana only two years, I asked for a leave of absence to take this unusual opportunity, and my Department administration enthusiastically supported this request. So early in 1958, my family and I were once again on a trans-Atlantic ship bound for Sweden, staying there until the summer of 1959.

As I suspected, being Assistant Director of the group was an administrative position in title only. Per was very much ***the*** Director, and he really had no need for an Assistant Director. I had a fine office next door to Jan Linderberg and Anders Fröman, and near Jan Nordling, Ludde Calais and others, who were students at that time, working towards their doctor's degree with, however, no visible mechanism for them to receive it. They trusted that Per would find a way before they reached the point of needing an academic degree process. Of course, they were right to place such confidence in Per's ability to find a solution to their problem.

I resumed the work I had carried on in Uppsala during my first visit. The Quantum Chemistry Group now had a small computer, mostly programmed by Klaus Appel, and with that we could make faster progress. I suggested it would be interesting to see what an actual Natural Orbital looked like and proceeded to do some Helium calculations to that end. With Klaus Appel's help and the new Laguerre functions as basis, we did extensive calculations on the radial functions of Helium, extracting the natural orbitals. This work led to several joint publications with Per, detailing the calculations.

Shortly after my arrival in Sweden, Pelle discussed with me his ideas of a summer institute or school somewhere in Sweden. Per chose Vålådalen in central Sweden as the location of the first such summer session. I helped in getting out some of the advertsing, and Per worked on the funding and the participants. A wonderful group of students and guests assembled in that mountainous location in the summer of 1958. A number of special guests, including both Mulliken and Pauling, joined the school for a symposium in its last week. It was both scientifically exciting and great fun to observe their lively interactions. The summer school was a great success in every way and enriched Per's personal life as well.

A beautiful young Swedish woman named Karin was there providing support for the school and its participants, and soon thereafter she became Mrs. Per-Olov Löwdin. She was a wonderful complement to Per and served as his helpmate and gracious partner for the rest of his life. They moved in to their permanent home just above the offices of the Group. It was a wonderful location, especially convenient for Pelle, as he could work either at his home or at the office downstairs, whichever was most convenient, with no more than two minutes transit time between them.

At Vålådalen, Per also exhorted us to exercise our bodies as well as our minds, and he set an outstanding example, leading participants on runs, hikes, soccer games, Swedish dancing, and more. He continued such activities at all the subsequent schools. He personally trained regularly, and he made it easy and inviting for all his colleagues and students to do so too.

Quite apart from our scientific work, Per and I had many sessions discussing a variety of other problems and concerns. A potentially relatively easy solution to the problem of providing for Per's advancement to Full Professor was for the University to promote him to a "Personal" Chair, and I believe Uppsala University's top administration supported that solution wholeheartedly. It is my understanding, however, that Professor Waller did not approve of Per's advancement and was in a position to block that solution. It was discouraging to Per, but he used his resourcefulness and energy to pursue an alternate more difficult path. This required getting the Swedish Riksdag (or national legislature) to establish and provide funding for a new Institute at the University, an Institute in Quantum Chemistry. Per's connections with the Swedish scientific establishment, with the Wallenberg Foundation, and with the King were impeccable, and in due time the Institute was formalized in that way. Of course, Per could not automatically become the Chair Professor of the Institute. He was required to submit a formal application for the new Chair in competition with other applicants from all over the world, as is the Swedish custom. The result was a foregone conclusion. In retrospect, one can reflect on how fortunate it was that he didn't get the Personal Chair. In that case the continuity of the Institute in perpetuity would not have been assured, for it would have ceased operation on Per's retirement. With the formation of the Institute in Quantum Chemistry, and with Per as the first occupant of its Chair, the Institute continues its formal existence, carrying on in Per's tradition. Of course, with the Institute's formation, the problem of obtaining formal Doctor's degrees for Per's students was solved.

Per also discussed with me at length on a number of occasions the idea of his spending part of each year in the United States. He had been negotiating with the University of Florida, especially with Professor Harry Sisler and Dean Grinter. Pelle wanted to know what I thought about the University of Florida, the characteristics of American Universities in general, and of Florida in particular, what he should ask for in negotiating with Florida, and on and on. So, in a way, Florida had a silent American partner unknown to them in its negotiations with Per.

The great success at Vålådalen was followed the next summer by the second such summer school at Skogshem on Lidingö in Stockholm. I attended that summer school for a brief time on my way home to the United States. These schools were outstanding successes, and Per continued the herculean efforts needed to keep them going every year, to find funds for student and guest stipends, and to adapt to changing conditions around the world. He made special efforts to acquire financial support for students and participants from third world countries. He especially supported and nurtured those from countries struggling with development, and he encouraged the formation of centers for Quantum Chemistry all around the world. He was a supporter of world peace, the equality of all mankind, fairness and justice, and he worked behind the scenes to further such causes.

When Pelle came to Florida, he started right away to look for a place to establish a similar winter school, and happily settled on Sanibel Island, then a lovely, remote, relatively unspoiled locale. Thus began the long series of winter "Sanibel" conferences, so-called even for those schools located elsewhere when Sanibel became too up-scale, too accessible, and too expensive. I dare say there are few Quantum Chemists today who have not attended one or more of these schools. They were marked not only by outstanding lectures, up to date symposia, and tough problem-solving sessions, but also by a relaxed atmosphere. Early afternoons were free for individual choice activities, games, swimming, relaxing by the pool. Interactions among changing mixes of participants on all kinds of subjects were evident. Wide-ranging discussions continued late at night over pretzels and beer long after the last lecture. And Karin was everywhere, ensuring all were comfortable, had plentiful food and drink, and that every one's needs were attended to.

Ten years passed before I returned again to Uppsala, thanks this time to a National Science Foundation Senior Fellowship. Pelle and Karin were as welcoming as ever. Per's energy continued unabated, and Karin's support of him was legendary. At 'Midsommar' as always, they opened up their country home to visitors to celebrate the season, take a sauna and jump in the river, dance around the 'Majstång', and enjoy the midnight light. Lucia celebrations in the dark of winter were always a part of the Group activities, enjoyed especially by foreigners who had no similar holidays. The lovely Löwdin apartment just above the Group's location was regularly filled with visitors. They could not help but enjoy the Löwdin's eclectic art collection, Per's piano artistry, and Karin's gracious and welcoming hospitality.

At the Group, the number of visitors and students at the Institute had expanded to fill all available spaces at Rundelsgränd, even including an expansion into new space. I was fortunate to be able to share an office with Sten Lunell. There were numerous scientists from Latin America, scientists from the Soviet Union and Eastern Europe, as well as the usual contingent from the more developed countries. Preparations were under way for a move of the Institute to a more permanent home near Kemikum and Fysikum on the Uppsala main campus.

Six years later, I had a final opportunity to spend a full academic year with Pelle when I resigned from my administrative posts at Indiana University and received a fully paid leave that permitted us to return once more. My scientific contributions had diminished during the time I had been away from science, but the welcome by both the Löwdins was as warm as ever. I made good use of the opportunity to bring my scientific reading up to date. Hospitality to all was the watchword of Per's running of his Institute. He continued to give elegant lectures and to contribute significantly to each of his visitors and of his students.

I have seen Per and Karin on many other occasions - in Japan, in Sweden at the Nobel ceremonies, in Uppsala as a lecturer supported by Per's 70th birthday fund. We have enjoyed the company of Per and Karin at Sanibel and St. Augustine winter schools, and Per has lectured a number of times at my home institution. On one occasion I had arranged a lecture for Pelle at Indiana. He was due late in the afternoon, so we invited a large group to a party in our home to welcome him. At the appointed hour, he called from Indianapolis that there was no way he could get to Bloomington that evening because of the bad weather. In the background, he could hear the sounds of the great party that he was missing. He promised, however, to come down on a bus the next morning to give a seminar and meet my group. At the appointed hour, he got to the bus depot in good time, and, having a few minutes to spare, he asked the driver where he could post a letter before he got on the bus. The bus driver pointed to a postbox across the street, and, while Per was posting his letter, the driver drove off without him. Per was astonished, called apologetically, and unfortunately for us had to fly on without a visit.

In my life as a University administrator, I have watched the introduction of new subjects of study, and their subsequent abandonment over the years. I have seen the formation of only two or three genuine new disciplines such as computer science and linguistics, to name two. Per single-handedly started, nurtured, set the standards, and breathed and lived the new discipline of Quantum Chemistry. I can remember Bob Parr's concern about where the jobs would be for the increasing numbers of students in our respective groups once each research university had hired a quantum chemist. Neither he nor I anticipated that the universities would then hire a second, a third, nor that there would be such world-wide growth in the field, or that industrial companies would hire these students in large numbers. But Per did anticipate that. He foresaw early the inevitable tie between quantum chemistry and molecular biology, and he welcomed students and researchers in these new areas to his summer and winter schools. He started and nurtured the International Journal of Quantum Chemistry to include not only the traditional atomic and molecular calculations, but also the new computational advances, and the advances in quantum biology.

Throughout it all, he honored others willingly, he set an example of excellence in all he did, and he demanded it of others, raising the level of achievement wherever he touched science and scientists. He loved the discussions of science, experiencing the intermix of nature, physical exertion, and the mental discipline of the lecture, of learning, and of research. He was thoughtful, kind, gracious, and yet energetic, efficient, and effective. He understood the politics of science and how universities worked, and he used that knowledge effectively for furthering the things he most believed in. One of my last contacts with him was a request for help in supporting a candidate for a university administrative position. He left no stone unturned in his constant quest for the betterment of quantum chemistry and for the advancement of those who tilled in that vineyard.

Per lives on through his disciples, through his scientific contributions, through his building of solid institutional organizations supporting the discipline, through his conscious active involvement with scientists from all over the world without concern for their countries' politics, or their countries' status. Only the unceasing drive for individual excellence mattered. And Per lives on through his family, particularly Karin, who was at his side and a vital support through all this great saga of accomplishment. We mourn his death, but we celebrate his achievements, achievements which place him among the scientific great of all time.

I end with a poem written by my wife as her personal tribute to Per-Olov Löwdin. To my mind, it conveys my own feelings far better than I am able to express them myself, not only for Per's scientific genius, but his organizational ability, his humanity, his friendship, and his love for all peoples.

The Swedish Doorman

Wil Shull
February 2001

He didn't wear a uniform, but handsome he was, and always on duty,
Standing at relaxed attention, with a smile and twinkling blue eyes,
Extending his hand in friendship with welcome warm, robust and ready
To lend a helping hand, to give directions, to lift and carry some of one's
heaviest luggage, and he was
Always there to open doors:
Doors of thought, doors to the mind, doors to science, doors of investigation
and
Exploration, doors of contemplation, doors to new paths, new ways, new tries.

Doors even to the unthought and the unseen,
And doors long shut, bolted, and sealed,
Doors to exchange of information, to sharing and cooperation,
Doors to equality, opportunity, and education,
Doors to international progress, productive organization, and
effectiveness,
Doors to family, friendship, and fun, doors of companionship and
competitions,
Doors to progress, hope and life itself -- doors to the heart.

Though he has left his post, passing through the final doorway of
***here and now*,**
The doors he has opened will evermore remain ajar, as will those of our love,
respect,
Appreciation, and admiration for the magnificent **Swedish Doorman**

THE KIND AND PERSONAL INFLUENCE OF PER-OLOV LÖWDIN

Jerome Karle
Laboratory for the Structure of Matter, Naval Research Laboratory, Washington, D.C. 20375-5341

An Introduction to Sanibel Symposia

In 1977, the University of Uppsala celebrated its five hundredth anniversary. As part of the celebration, Per-Olov Löwdin and colleagues arranged a Symposium on "Quantum Chemistry – A Scientific Melting Pot." My wife, Isabella, and I received invitations to participate in this symposium and we were very pleased to accept for several reasons. Among them was the opportunity to meet Per-Olov and a large number of the members of the quantum chemistry and related communities. It was a fine symposium and our meeting with Per-Olov and his charming wife, Kärin, led to a continuing friendship.

By 1977, we had much research experience in the determination of molecular and crystal structures by diffraction methods, an area that Per-Olov appreciated as a useful source of information for relating structure to function and for facilitating studies in quantum chemistry. This was, evidently, quite consistent with the chosen title of the 1977 Symposium. From this point on, up to the present, we have made many visits to the Sanibel Symposia, which have involved our active participation in both structural chemistry and quantum chemistry.

When we first attended the Sanibel Symposia, they were no longer held on Sanibel Island. Palm Coast was the venue, an area somewhat north of Daytona and on the shore of the Atlantic Ocean. After that, the Sanibel Symposia were moved further north and held in an oceanographic facility belonging to the University of Florida. Across the road from this facility was a large park that housed a variety of sea creatures. Among them were a number of dolphins who enjoyed playing ball with scientists (and anyone else who would play with them). This was pleasant recreation if one were prepared to catch and throw a rather wet ball. After a few years at the oceanographic facility,

E.J. Brändas and E.S. Kryachko (eds.), Fundamental World of Quatum Chemistry, Vol. I, 15–20.

the Sanibel Symposia moved to the current meeting place, the Ponce de Leon hotel in St. Augustine. The Symposia always seemed to have a good mix of scientific information and relaxation in an atmosphere of friendliness and it is very nice that the Symposia continue in the same pattern, one of the many valuable legacies of Per-Olov.

Extending Scientific Interests

As time went on, I found the opportunity to engage in studies that could bring structural chemistry and quantum chemistry together. This thought was motivated by the expectation that information obtainable from each of these research areas would be enhanced as a consequence of the combination. This was certainly not an original thought, but it seemed that there may be some new opportunities because of a number of advances in theory and practice, including computational facilities. The possibility of making some contributions to theory and pactice as well as obtaining interesting information added to my interest.

The opportunity to pursue these thoughts arose when Lou Massa, of Hunter College and the City University of New York, and I started a collaboration which was extended some time later to include one of several of Lou Massa's fine students, Lulu Huang. Since that time, there were a number of problems of interest in which we all collaborated. In several instances, the work and thoughts of Per-Olov played a valuable role. In the next section, I shall describe some of his contributions, at times indispensable, always clarifying.

Some contributions from the Work of Per-Olov Löwdin to Our Research Program

An Application of Projector Matrices

In an article entitled "Quantum Crystallography and the Use of Kernel Projector Matrices [1], we stated, in the first part of the Introduction, the following:

"We outline here a general method for combining X-ray diffraction data for crystals and quantum mechanics with the objective of obtaining accurate wave functions that are consistent with the X-ray data. The name that we have chosen for this type of investigation is "quantum crystallography."

"The approach to quantum crystallography that has been taken throughout this manuscript is to define the electron density in a crystal in terms of a single determinant of orbitals. Of course, N-representable representations of the electron density distribution need not be restricted to a single determinant of orbitals. In the description of the manner in which the single-determinant approach is to be achieved, it will be seen that a main feature is a projector matrix that occurs in the quantum mechanical definition of the electron density distribution. Besides a description of quantum crystallography, key considerations in this article concern the initiation of the calculations for quantum crystallography with a good quality projector matrix and how such a matrix may be obtained for large structures."

In a second article entitled "Kernel Projector Matrices for Leu[1]-Zervamicin," [2] we stated in part of the Introduction the following:

"In this study, we present purely quantum mechanical calculations. The only use of crystallography is to provide atomic coordinates. The purpose of the calculations is to illustrate how fragment calculations, using standard programs, can be used to obtain parts of projector matrices. These parts can be added together to form the projector matrix for a large molecule, and the calculations increase linearly with increase in the number of atoms in the structure. When the calculation is programmed for high-performance parallel computers, the increase in computing time should be very small. The details of the computation of the individual parts of the projector matrix from the fragments and how the parts are combined are discussed in detail. Once the projector matrix is completed, wave functions follow by standard methods of quantum mechanics."

Per-Olov Löwdin's Contributions Concerning Density Matrices

1. Natural orbitals and natural geminals

Löwdin is a pioneer in the application of density matrices to the fields of quantum chemistry and quantum biology. He invented the concept of natural orbitals and natural geminals. These have turned out to be fundamental and useful tools applied world wide. Natural orbitals are fundamental to the obtaining of electron density matrices and electron density distributions from theoretical calculations and experimental data [3-8]. The natural geminals are two-body wave functions which form the basis for our understanding of correlation energies [9-12]. These are applications of considerable importance to chemistry on the atomic and molecular level.

2. Projectors and idempotency

His work using the concept of projectors is at least two-fold. One example is the concept of projecting, out of an arbitrary space of orbitals, a particular subspace of orbitals from which one could obtain a rigorous solution to the Schrödinger equation. This is expressed in the form of a matrix which is characterized by the mathematical property of idempotency, namely, the square of the matrix is equal to the original matrix. Such projectors have become the foundation for many applications, for example, the numerical implementation of Hartree-Fock theory, the most widely used model in quantum chemistry and materials science. The projectors have also found important use in the development of the field of quantum crystallography, in which the agreement between quantum chemical calculations and X-ray data is optimized. A second application of these projectors has been the projecting out of a wave function a particular spin component, an eigenfunction of the spin operator for a physical system. This is valuable for the calculation of spectra and an understanding of certain types of chemical bonds [12-17].

3. N-representability

Löwdin was one of the first people to comprehend the importance of the matter of N-representability, namely the need to establish a criterion that would ensure that quantum mechanical calculations would result in a valid and chemically realistic solution to the Schrödinger equation. He also made valuable contributions to the algebra and the mathematical structure of the problem. N-representability has become universally recognized as a condition for the valid use of density matrices [6, 7, 8].

4. Löwdin orthogonalization

Orbitals that are orthogonal provide a special simplicity to the mathematical formulation of the theory of quantum mechanics and its applications. The most widely used practical method for transforming nonorthogonal orbitals to orthogonal ones was invented by Löwdin and it bears his name, viz. "Löwdin orthogonalization" [18-22].

There must be thousands of examples, similar to the ones that I have mentioned, in which progress in quantum chemistry was stimulated and facilitated by Per-Olov Löwdin's brilliant contributions to theory and his efforts to make it understandable. The many facets of his contributions, in addition to his broad theoretical studies, may be seen in his status reports and his reviews, his leadership in establishing and editing the International Journal of Quantum Chemistry and in bringing people together at the Sanibel Symposia, from a broad variety of disciplines, whose interests represent the theme of the 1977 meeting in Uppsala, "Quantum Chemistry – A Scientific Melting Pot."

One of the aspects of the Sanibel Symposia that had been a particular pleasure for my wife and me was the kind hospitality of Kärin and Per-Olov Löwdin.

References

1. Quantum Crystallography and the Use of Kernel Projector Matrices. L. Massa, L. Huang and Karle, J. Int. J. Quant. Chem., Quantum Chem. Symp. 1995, 29, 371.
2. Kernel Projector Matrices for Leu[1]-Zervamicin. L. Huang, L. Massa and J. Karle, Int. J. Quant. Chem., Quantum Chemistry Symp. 1996, 30, 1691.
3. Quantum Theory of Many-Particle Systems. I. Physical Interpretation by Means of Density Matrices, Natural Spin-Orbitals, and Convergence Problems in the Method of Configurational Interaction. Per-Olov Löwdin, Phys. Rev. 1955, 97, 1474.
4. Quantum Theory of Many-Particle Systems. II. Study of the Ordinary Hartree-Fock Approximation. Per-Olov Löwdin, Phys. Rev. 1955, 97, 1490.
5. Quantum Theory of Many-Particle Systems. III. Extension of the Hartree-Fock Scheme to Include Degerate Systems and Correlation Effects. Per-Olov Löwdin, Phys. Rev. 1955, 97, 1509.
6. Calculation of Lower Bounds to Energy Eigenvalues by Reduced Density Matrices and the Representability Problem. Per-Olov Löwdin and T. K. Lim, Int. J. Quant. Chem. 1969, 38, 697.
7. Some Aspects on the Development of the Theory of Reduced Density Matrices and the Representability Problem, in "Density Matrices and Density Functionals". Per-Olov Löwdin, Eds. R.

Erdahl and V. H. Smith, D. Reidel Publishing Company, Dordrecht, The Netherlands, 1987.

8. Remarks about the Wavefunction Representability of the First-Order Reduced Density Matrices. Per-Olov Löwdin, Int. J. Quant. Chem., Symp. 1986, QC20, 769.
9. Natural Spin-Orbitals for Helium. H. Shull and Per-Olov Löwdin, J. Chem. Phys. 1955, 23, 1565.
10. Superposition of Configurations and Natural Spin Orbitals. Applications to the He Problem. H. Shull and Per-Olov Löwdin, J. Chem. Phys. 1959, 30, 617.
11. The Long-Range Interaction of Two 1s Hydrogen Atoms Expressed in Terms of Natural Spin Orbitals. J. O. Hirschfelder and Per-Olov Löwdin, Mol. Phys. 1959, 2, 229.
12. Long-Range Interaction of Two 1s-Hydrogen Atoms Expressed in Terms of Natural Spin-Orbitals. J. O. Hirschfelder and Per-Olov Löwdin, Mol. Phys. 1965, 9, 491.
13. The Normal Constants of Motion in Quantum-Mechanics Treated by Projection Technique. Per-Olov Löwdin, Revs. Mod. Phys. 1962, 34, 520.
14. Angular Momentum Wavefunctions Constructed By Projector Operators. Per-Olov Löwdin, Revs. Mod. Phys. 1964, 36, 966.
15. The Projection Operator for a Space Spanned by a Linearly Dependent Set. Per-Olov Löwdin, Int. J. Quant. Chem. 1969, 3, 767.
16. Some Properties of Inner Projections. Per-Olov Löwdin, Int. J. Quant. Chem. 1971, 4S, 231.
17. On the Sanibel Coefficients in the Expansion of Spin-Projected Slater Determinants.
Per-Olov Löwdin, Int. J. Quant. Chem. 1983, 24, 729.
18. On the Non-Orthogonality Problem Connected with the Use of Atomic Wave Functions in the Theory of Molecules and Crystals. Per-Olov Löwdin, J. Chem. Phys. 1950, 18, 365.
19. On the Non-Orthogonality Problem. Per-Olov Löwdin, Adv. Quant. Chem. 1970, 5, 185.
20. Some Comments on the Construction of an Orthonormal Set of LCAO Basis Functions for Crystals. T. Ahlenius , J.-L. Calais and Per-Olov Löwdin, J. Phys. Chem. 1973, 6, 1896.
21. On the Historical Development of the Valence Bond Method and the Non-Orthogonality Problem. Per-Olov Löwdin, Theochem. 1991, 229, 1.
22. Some Remarks on the Resemblance Theorems Associated with Various Orthonomalization Procedures. Per-Olov Löwdin, Int. J. Quant. Chem. 1993, 48, 225.

P.-O. LÖWDIN AND THE QUANTUM MECHANICS OF MOLECULES

R. G. Woolley
Dept. of Chemistry and Physics,
Nottingham Trent University,
Nottingham NG11 8NS, UK.
guy.woolley@ntu.ac.uk

B. T. Sutcliffe
Laboratoire de Chimie Physique Moléculaire,
Université Libre de Bruxelles,
B-1050 Bruxelles, Belgium
bsutclif@ulb.ac.be

Abstract It is surely no exaggeration to say that all Per-Olov Löwdin's career was devoted to the quantum mechanics of molecules. But late in his career he began detailed consideration of a formal definition of the term *molecule* within quantum mechanics. The authors first heard him talk of such matters at a meeting held in honour of Prof. Raymond Daudel in 1986. The paper arising from this talk appeared in 1988 and was followed during the next six or so years by others in the same vein. In tribute to Per-Olov we should like here to examine the work in these papers.

1. Introduction

Towards the end of his scientific career, Per-Olov Löwdin actively devoted time to foundational questions of quantum chemistry. He gave many talks at conferences and workshops in which he surveyed the whole range of theoretical techniques employed in quantum chemistry. He also drew attention to what he saw as significant problems that his generation had not succeeded in resolving, and expressed the hope that the next

E.J. Brändas and E.S. Kryachko (eds.),
Fundamental World of Quatum Chemistry, Vol. I, 21–65.

generation of leaders in the field of quantum chemistry would take up the challenge; some of these talks were written up and published. In the following we refer only to the published work [1]–[6] that we have located [7], and concentrate on his attempts to give a quantum definition of a molecule.

His approach is presented fairly fully in just two of the papers [1, 2] which have much common material that we describe below, whereas the other papers [3]–[6] give only summary discussion, and draw attention to some of the unsolved problems. We have felt it appropriate to summarize Per-Olov's views (Section 3) very much in his own words which we have taken freely from these papers. We start with a summary account of the historical perspective as a background to Löwdin's concerns. The development of quantum chemistry is a well-known story. It is generally accepted that because of the great difference in the masses of electrons and nuclei it is a good approximation to treat the electronic and nuclear motions as being essentially independent. Thus the electronic states can be determined at different nuclear geometries by treating the nuclei as fixed sources of a potential for the electronic motion. The charge distribution of the electrons is then a function of the nuclear positions and determines the nuclear motion [8, 9]. Our main concern is to emphasize how its procedures require identical nuclei to be treated as distinguishable particles. We take this feature to be the defining characteristic of quantum chemistry.

It is clear that Löwdin was not satisfied with the traditional account. In [2] he starts with the famous Dirac quotation about the theoretical basis for the whole of chemistry being known with the completion of quantum mechanics (see Section 2 below), but reminds us that the situation in quantum mechanics is just the same as with Newton's laws; these laws regulate the planetary motion but do not describe the planetary system itself - the various planets, their masses, the actual initial conditions appropriate to the solar system. He notes that "quantum chemistry is to a large extent still based on a great deal of chemical and physical insight of mostly experimental nature", and that there is a very long road from the statement that a molecule is a collection of electrons and nuclei, subject to the laws of quantum mechanics, to an account of chemistry, longer perhaps than Dirac ever imagined. The first steps on this road involve an examination of the quantum mechanics of a molecule that is not based on the Born-Oppenheimer approximation. At the WATOC90 conference in Toronto, Löwdin declared "One of the most urgent problems of modern quantum chemistry is to treat the motions of the atomic nuclei and the electrons on a more or less equivalent basis" [5]. The motivation for such a step had been identified in his earlier lecture in Paris [1]: "In this way, we may have a somewhat better chance to try to eliminate - or at least minimize - the elements of chemical insight derived from experimental experience which are now so essential as starting points of the theory."

His programme (Section 3) is to start with the non-relativistic Hamiltonian[1], H, for a collection of electrons and nuclei interacting via Coulombic forces, and to use the results of modern mathematical analysis to see how far he can go with the characterization of the properties of H. The relevant branch of mathematics is the functional analysis for operators on a Hilbert space; this subject is now very extensive and there is a wealth of results, as well as recognized open problems, relevant to the quantum mechanics of the Coulomb Hamiltonian [10, 11]. We attempt here to draw this information together as a development of the approach Löwdin sketched in his late papers [1]–[6].

2. Quantum chemistry: the historical perspective

In 1929 Dirac wrote famously [12]

> The fundamental laws necessary for the mathematical treatment of large parts of physics and the whole of chemistry are thus fully known, and the difficulty lies only in the fact that application of these laws leads to equations that are too complex to be solved.

Whatever might have been felt by the Chemists of the day about Dirac's claim (and about this, see Simões [13]), there can be no doubt that quantum mechanical ideas began to be very influential in chemistry from the early 1930s onwards. Not only did Pauling's quantum mechanically based, if still qualitative, valence bond ideas begin to supplement the traditional Lewis and Langmuir account of bonding but also the work of Hückel, Hund and Mulliken led to the idea of molecular orbitals. The molecular orbital approach led to simplifications in the complex equations of quantum theory in such a way as to enable certain kinds of model calculations to be made on molecules. The results of such calculations could be correlated with experimental observations, and quantitative developments along these lines continued throughout the 1940s. With the development of electronic computers, however, the complexity of the equations began to seem rather less of a problem and from the early 1950s onwards there were attempts to understand some of the molecular aspects of chemistry by means of quantum mechanical electronic structure calculations. These were performed within the clamped nu-

[1]The *Coulomb Hamiltonian*

clei scheme using linear combinations of atomic orbitals and aimed to evaluate computationally the molecular integrals that arise when this is done.

In this context it is interesting to recall what Coulson said in his Tilden lecture given in October 1951 [14]. He quotes Dirac as above but then goes on to say

> Yet the true contribution of wave mechanics is not here, as I have to remind myself when I am being urged to start immense schemes for the numerical evaluation of molecular integrals. But if it is not here, where is it? The answer is simple. Wave mechanics has shown us what is going on, and at the deepest possible level. We shall find, $\cdots$, that it has taken the concepts of the experimental chemist–the imaginative perception that those who had lived in their laboratories and allowed their minds to dwell creatively upon the facts that they found–and it has shown how they all fit together; how, if you wish, they have one single rationale; and how this hidden relationship to each other can be brought out.

What Coulson seems to be saying here is that Dirac was wrong in supposing that all that there was to chemistry could be comprehended if only we could solve the equations of wave mechanics. Coulson sees chemistry as a much more complicated and imaginative enterprise than Dirac apparently does; instead he suggests that wave mechanics has provided a rationalising underpinning to many chemical ideas. He did not, in 1951 anyway, believe that the most important contribution that quantum mechanics could make to chemistry was by means of computation. Nevertheless, although computational quantum chemistry comprised and continues to comprise only a small part of chemistry, it has grown in influence and utility, as was recognized explicitly in the award in 1998 of the Nobel Prize for Chemistry to Kohn and Pople. It is the citation for Pople that is of particular relevance here:

> John Pople is rewarded for developing computational methods making possible the theoretical study of molecules, their properties and how they act together in chemical reactions. These methods are based on the fundamental laws of quantum mechanics as defined by, among others, the physicist E. Schrödinger. A computer is fed with particulars of a molecule or a chemical reaction and the output is a description of the properties of that molecule or how a chemical reaction may take place. The result is often used to illustrate or explain the results of different kinds of experiment. Pople made his computational techniques easily accessible to researchers by designing the GAUSSIAN computer program. The first version was published in 1970. The program has since been developed and is now used by thousands of chemists in universities and commercial companies the world over.

As this citation makes clear, many aspects of chemical behaviour can be rationalized in terms of molecular electronic structure calculations made in the clamped nuclei approach.

In the clamped nuclei approach, the nuclei form a geometrical figure whose shape and size are disposable. The nuclei generate an attractive potential for electronic motion, and this potential is invariant under all uniform translations and all orthogonal transformations of the figure. The Coulomb Hamiltonian for electronic motion is solved with this potential to yield the electronic wavefunction and electronic energy for particular choices of the nuclear geometry. The classical molecular structure is taken to be that nuclear geometry at which the sum of electronic energy and the classically calculated nuclear repulsion energy is a minimum. The vibration-rotation spectrum can be understood in terms of the structure so calculated, freely rotating while the nuclei perform small vibrations. Technically this outcome is achieved by treating the sum of the electronic energy and the nuclear repulsion energy as a potential with a minimum at the classical structure and solving the quantum mechanical vibration-rotation problem for nuclear motion about this minimum. This is a development of the approach first proposed by Eckart [15]. By constructing this potential as the nuclei are separated in various ways, nuclear collision calculations can be made, in terms of which many kinds of simple chemical reactions can be understood. The basis of the scheme is found in the work of Born and Oppenheimer [16] who gave a quantum mechanical justification for the treatment of the nuclei, in first approximation, simply as providers of a potential for electronic motion.

Eckart's work is in the context of classical mechanics and so makes no reference to the work of Born and Oppenheimer; however they share the same classical viewpoint, namely that in a stable molecule the nuclei undergo only small oscillations about an equilibrium configuration. Eckart proposes a coordinate frame fixed in the molecule which is such that the equilibrium positions of the nuclei are A constant vectors $\mathbf{a}_i$. The potential energy for the nuclei may be expressed in terms of displacement coordinates

$$\rho_i(\mathbf{z}_i) = \mathbf{z}_i - \mathbf{a}_i$$

which vanish when the molecule is at its equilibrium configuration; here the $\mathbf{z}_i$ are nuclear coordinates expressed relative to the centre of nuclear mass in the frame fixed in the molecule.

It is possible, by perfectly standard methods, to transform a classical Hamiltonian into a quantum mechanical one. Equally, Eckart's Hamiltonian may be derived directly from the Schrödinger equation for a system of A nuclei moving in the specified potential [17]. The resulting quantum mechanical Hamiltonian must, if it is to be regarded as arising from the Coulomb Hamiltonian, be invariant under any permutation of identical particles. Consider, however, the changes induced in the displacement coordinates by the permutation of identical nuclei. If a particular permutation induces the transposition $\mathbf{z}_i \leftrightarrow \mathbf{z}_j$ then the displacement coordinates transform as

$$\rho_i(\mathbf{z}_i) \to \rho_i(\mathbf{z}_j), \quad \rho_j(\mathbf{z}_j) \to \rho_j(\mathbf{z}_i)$$

and are usually different from the initial ones. So in general one cannot expect the Eckart Hamiltonian to be invariant under the permutation of identical nuclei. In fact the Eckart Hamiltonian is invariant only under that sub-group of permutations that can be realised by the operations comprising the point group of the equilibrium framework. An extreme example of the difference between the Coulomb Hamiltonian, and the Eckart model Hamiltonian is afforded by a chiral molecule for which the only symmetry operation is the identity (point group C_1). Most chiral molecules contain some identical nuclei of course.

Historically it was found that the Eckart approach in classical form worked pretty well to describe molecular spectra and that it worked well in quantum mechanical form too, provided that the symmetry of the problem was restricted to that of the point group of the equilibrium framework, that is, to the point group of the classical molecular geometry. It thus appears that the quantum mechanical form of the Eckart Hamiltonian is appropriate if the displacement coordinates can be regarded as describing distinguishable particles. The more extended kinds of potential energy surface needed to describe simple chemical reactions have to be similarly symmetry restricted.

It would thus seem necessary for the success of the standard approaches that the nuclei be treated as identifiable particles, just as they are in classical chemical structure theories. In the original work by Born and Oppenheimer both the electrons and the nuclei are assumed at the outset to be treated as quantum mechanical particles. In Section 3 of their paper [16] one reads (in a translation by S. M. Blinder [18]):

> An arbitrary configuration of electrons and nuclei cannot always be treated by a general approximation procedure. We will here consider only states which correspond to a stable molecule. We will begin with the following question:
>
> Is there a system of values of the relative nuclear coordinates ξ_i such that the eigenfunctions ψ_n of the energy operator (6), in so far as they depend on the ξ_i, have values significantly different from zero only in a small neighborhood of this set?
>
> This wave-mechanical requirement corresponds obviously to the classical condition that the nuclei undergo only small oscillations about the equilibrium configuration; the $|\psi_n|^2$ is the probability of finding a certain configuration of given energy.

The energy operator referred to here as (6) is just the full Coulomb Hamiltonian in which the A nuclear coordinates are assumed expressible in terms of the $3A-6$ relative nuclear coordinates ξ_i and 6 more coordinates which denote the position and orientation of the relative nuclear configuration. In the discussion to establish an answer to the question that they pose in the second paragraph above, they do not require the wave functions to have any particular permutational symmetry. But if the nuclei and the electrons were to be treated on an equal footing in a Coulomb Hamiltonian for the whole molecular system, which of course was the starting point of the work of Born and Oppenheimer, any identical nuclei would be indistinguishable and any wavefunction involving them would have to reflect the symmetries required by such indistinguishability. Thus in using the clamped nuclei Hamiltonian to construct a potential, by holding the nuclei in fixed relative positions and hence identifying them, we are discarding a fundamental symmetry of the full problem. This point appears to have been first recognised in the literature by Berry [19].

3. Löwdin's definition of a molecule

Although Löwdin fully accepted the tremendous range and utility of modern quantum chemistry it is evident in his late writings that he was sensitive to the fact that its foundation, particularly as regards 'molecular symmetry', relies substantially on practical chemical experience. He hoped that a fundamental analysis of the Coulomb Hamiltonian for a molecule would help reduce this dependence on chemical insights, and initiated such a study in the paper [1]. In the following we review the problem taking Löwdin's study as our starting point.

The Hamiltonian operator for a system of N electrons and A atomic nuclei may be written as

$$\mathsf{H} = \sum_{g}^{A} \frac{\mathsf{p}_g^2}{2m_g} + e^2 \sum_{g<h}^{A} \frac{Z_g Z_h}{r_{gh}} + \sum_{i}^{N} \left(\frac{\mathsf{p}_i^2}{2m} - e^2 \sum_{g}^{A} \frac{Z_g}{r_{ig}} \right) + e^2 \sum_{i<j}^{N} \frac{1}{r_{ij}} \quad (1.1)$$

where the individual terms have obvious physical interpretations. The most basic operator property of H is that it is *essentially self-adjoint*, as famously proved by Kato in 1951. The first, second and last terms on the rhs of (1.1) are positive, and the crude estimate [1]

$$\sum_{i}^{N} \left(\frac{\mathsf{p}_i^2}{2m} - e^2 \sum_{g}^{A} \frac{Z_g}{r_{ig}} \right) \geq -\frac{me^4 AN}{2\hbar^2} \sum_{g}^{A} {Z_g}^2 \quad (1.2)$$

confirms that H is *bounded from below*; thus the spectrum has a lowest eigenvalue E_0, so that

$$\mathsf{H} \geq E_0 1 \tag{1.3}$$

For normalized wavefunctions in the L^2–Hilbert space one then has the variation principle,

$$<\mathsf{H}>= \frac{<\Phi|\mathsf{H}|\Phi>}{<\Phi|\Phi>} \geq E_0 \tag{1.4}$$

and by using a scaling argument originally due to Hylleraas, it can be shown that

$$<\mathsf{H}>\leq 0 \tag{1.5}$$

i.e. E_0 is always negative. Löwdin notes that a density matrix (or as he prefers to say 'system operator') formulation is equally possible with

$$<\mathsf{H}>= Tr[\mathsf{H}\Gamma], \quad Tr[\Gamma] = 1, \quad \text{etc.} \tag{1.6}$$

The lowest eigenvalue E_0, however, lies at the *bottom of a continuum*, and so the associated eigenfunction describes a *scattering* state. The reason for this is easily found by examining the symmetries of H. The Coulomb Hamiltonian has the following constants of the motion [1, 2]:

i) The total momentum[2]

$$\mathsf{P} = \sum_k \mathsf{p}_k \tag{1.7}$$

ii) The total angular momentum

$$\mathsf{L} = \sum_k \mathsf{r}_k \times \mathsf{p}_k \tag{1.8}$$

iii) The permutations $\mathcal{P}$ of the electronic coordinates $\mathbf{x}_i$ which lead with the Pauli Principle [12] to the antisymmetry property of the wavefunctions

$$\mathcal{P}\Psi = (-1)^p \Psi \tag{1.9}$$

iv) The permutations of such nuclear coordinates as are associated with nuclei of the same mass.

As a consequence of i) the centre-of-mass of the whole system, with coordinate vector

$$\boldsymbol{\xi} = \frac{1}{M}\sum_k m_k \mathbf{r}_k, \quad M = \sum_k m_k \tag{1.10}$$

[2] The sums are now taken over *all* the particles in the system.

moves like a *free particle* having an energy spectrum covering the interval $[0, \infty]$, and so E_0 must be at the bottom of a continuum. Thus it is desirable to separate out the centre-of-mass motion by writing

$$\mathsf{H} = \frac{\mathsf{P}_\xi^2}{2M} + \mathsf{H}' = \mathsf{T}_\xi + \mathsf{H}' \tag{1.11}$$

and shift the focus of the investigation to the internal Hamiltonian H' which may have both discrete and continuous eigenvalues. There are infinitely many possible choices of internal coordinates, so that the form of H' is not determined, but whatever coordinates are chosen the essential point is that H' is the *same* operator specified by the decomposition (1.11).

Since the total angular momentum is constant, and can be written in the form

$$\mathsf{L} = \mathsf{L}_\xi + \mathsf{L}_{int} \tag{1.12}$$

where both terms are separately conserved, one could further transform H' by making its rotational invariance explicit. Löwdin however does not pursue this further, and contents himself with the remark that the angular momentum operators have discrete spectra, and so do not alter the analysis. At this point in the discussion, in both [1] and [2], he observes that there are in any case other difficulties that - for the moment at least - must be given higher priority. He writes [2] "The Coulombic Hamiltonian H' does not provide much obvious information or guidance, since there is [sic] no specific assignments of the electrons occurring in the systems to the atomic nuclei involved - hence there are no atoms, isomers, conformations etc. In particular one sees no molecular symmetry, and one may even wonder where it comes from. Still it is evident that all this information must be contained somehow in the Coulombic Hamiltonian. In this situation, it is possible to proceed in two completely different ways." Firstly, there is the traditional approach based on the Born-Oppenheimer argument which emerged from physical and chemical insights of essentially experimental nature as to the existence of molecules, their shapes, symmetry etc. Most quantum chemistry is of this type and is hence still highly dependent on empirical experience. One may also use mathematical analysis; in order to show the existence of a molecule in its ground state, one has to show that the lowest eigenvalue E_0 of the associated Hamiltonian H' is discrete and corresponds to a *closed state*.

That this statement can be true follows from two fundamental results. Firstly, a classical theorem asserts that, given a self-adjoint operator such as H' and a trial wavefunction that yields an expectation value $< \mathsf{H}' >$ which is below the bottom of the essential spectrum, H' has at least one discrete negative eigenvalue. Secondly, the location of the bottom of the essential spectrum is defined by the famous Hunzicker-Van Winter-Zhislin (HVZ) theorem. It is obtained by looking at the lowest energy of all possible separated clusters. Taken together these two theorems confirm the validity of a *conjecture* on which quantum chemistry was originally founded; namely if one can find an expectation value $< \mathsf{H}' >$ which is a lower energy than the lowest energy of all possible separated clusters then there exists a bound ground state. Löwdin cites the study by Hylleraas of the energy of the He atom, the study of the H_2 molecule by James and Coolidge and other examples [1, 2].

There are still formidable problems however, of which the most severe perhaps is the fact that the Coulomb Hamiltonian (1.1) is one and the same for all possible isomers associated with a given chemical formula. Löwdin proposes a general energy surface that may provide a description of isomerism [1, 2]; the energy surface is constructed by writing

$$\mathfrak{E}(\boldsymbol{\alpha} < \mathsf{H}' >) = \frac{< \Phi|\mathsf{H}'|\Phi >}{< \Phi|\Phi >} \tag{1.13}$$

where $\Phi = \Phi(\mathbf{X}, \mathbf{Q}, \boldsymbol{\alpha})$; $\mathbf{X}$ and $\mathbf{Q}$ stand for all the electronic and nuclear variables and are integrated out in the evaluation of $< \mathsf{H}' >$. Then $\mathfrak{E}$ depends on a set of parameters $\boldsymbol{\alpha} = \alpha_1, \alpha_2, \dots \alpha_p$, and its minima might be associated with stable isomers in a way familiar from the standard classical discussion of potential energy surfaces; this calculation can also be formulated in a density matrix framework. He concludes that the calculation of the hypersurface $\mathfrak{E}(\boldsymbol{\alpha})$ is a computational effort much bigger than anything yet undertaken in quantum chemistry, and that a new computer architecture might be required. On the other hand molecular structure itself is problematic in this setting [1]. Suppose the molecule contains A nuclei which, to simplify the discussion, are taken to be distinguishable, and that one nuclear coordinate is fixed in the process of elimination of the centre-of-mass motion. There are then $(3A - 3)$ coordinates to be determined, whereas there are $1/2A(A - 1)$ internuclear distances in a 'structure', and so for $A > 6$ the system is over-determined. There must therefore be further constraints to make the additional internuclear distances compatible with the others. Hence the existence of specific 'molecular structures' cannot be taken for granted[3].

Finally, in Per-Olov's last published paper devoted to these questions [6], the fundamental problem for a quantum theory of molecules is identified as: *where does molecular symmetry come from ?* The tenor is

[3]If rotation is also separated off [5] there are $(3A - 6)$ coordinates and the condition for over-determination becomes $A > 4$.

very much of handing over unsolved problems to a new generation: in his words "It is definitely not easy to *go beyond* the conventional Born-Oppenheimer approximation and - even if some scientists have tackled the problem with a certain amount of success - it seems evident that some new fundamental ideas may be needed."

4. The spectrum of the Coulomb Hamiltonian

In this section we develop further the ideas that Löwdin introduced in his discussion of the bound states of the Coulomb Hamiltonian and their identification as molecules. Nothing that is said here is at all new but it is perhaps not yet well represented in the theoretical chemistry literature so a brief summary of the fundamental results may be helpful. In [1] Löwdin acknowledges the help he received from correspondence with Prof. Barry Simon and draws attention to the book [20] *The Analysis of Operators* by Reed and Simon. Chapter XIII of that book contains proofs of many of the results cited by Löwdin. In [20] some necessary results arising from the spectral theory of self-adjoint operators are simply quoted; details may be found in earlier volumes by the same authors, particularly in Section VII [21] and Section X of [22]. An alternative source of similarly relevant material is the book by Thirring [23], who also provides a useful summary, expressed in accessible language and without too much detail, in his contribution to the Schrödinger centenary volume [24].

In 1951 Kato [25] established that the Coulomb Hamiltonian, H, is essentially self-adjoint. This property, which is stronger[4] than Hermiticity, guarantees that the time evolution

$$\Psi(t) = \exp(-i\mathsf{H}t/\hbar)\Psi(0)$$

of a Schrödinger wavefunction is unitary, and so conserves probability. This is not true for operators that are Hermitian but not self-adjoint. It is easy enough to construct examples of such operators; an example given by Thirring [24] is of the radial momentum operator $-i\hbar\partial/\partial r$ acting on functions $\phi(r)$, $\phi(0) = 0$ with $0 \leq r < \infty$. He comments

> Having been confronted with these facts of life one might wonder whether Schrödinger by writing down a formal Hamiltonian without specifying its domain has actually predicted anything about atoms and molecules. That he has actually done so comes from the fact that the Coulomb potential, though unbounded, is small compared with the kinetic energy.

It is the smallness of the Coulomb potential that is used to establish that H is bounded from below, see (1.3).

[4] They are the same for finite-dimensional operators

There are various ways in which the spectrum $\sigma(\mathsf{A})$ of a self-adjoint operator A may classified. ¿From the point of view of measure theory the natural decomposition is into *pure point*, *absolutely continuous* and *singular continuous* parts. The sets are closed but need not be disjoint. The *discrete* spectrum $\sigma_d(\mathsf{A})$ is the subset of the pure point spectrum that consists of isolated eigenvalues of finite multiplicity. The *essential* spectrum $\sigma_{ess}(\mathsf{A})$ is the complement of the discrete spectrum and is the infinite dimensional part.

$$\sigma_{ess}(\mathsf{A}) = \sigma(\mathsf{A}) \backslash \sigma_d(\mathsf{A})$$

The discrete spectrum and the essential spectrum are, by definition, disjoint; however, although the essential spectrum is always closed, the discrete spectrum need not be. $\sigma_{ess}(\mathsf{A})$ includes the absolutely continuous spectrum $\sigma_{ac}(\mathsf{A})$ and the singular continuous spectrum, $\sigma_{sc}(\mathsf{A})$ this last consisting of infinitely degenerate eigenfunctions. As we shall see, for the Coulomb Hamiltonian the singular continuous spectrum is empty, $\sigma_{sc}(\mathsf{H}) = \emptyset$.

The absolutely continuous spectrum describes states in which the particles leave a bounded region in a finite time; these are the scattering states of the system. The discrete spectrum describes bound states in which the particles stay infinitely long in a bounded region. This is the part of the spectrum that is most usefully investigated in the search for the molecule. If one considers the spectrum of the Coulomb Hamiltonian for the Helium atom then the essential spectrum clearly begins at the first ionisation energy, an energy E somewhat below zero. Between this beginning and zero there are an infinite number of states corresponding to those of the hydrogen-like system remaining after the ionisation, together with a free electron. Such states are part of the essential and not of the discrete spectrum. At energies above zero the system consists of a freely moving nucleus and two free electrons. Thus in the case of the Helium atom $\sigma_d(\mathsf{H}') = [E_0, \mathrm{E})$ and $\sigma_{ess}(\mathsf{H}') = [\mathrm{E}, \infty)$. This sort of description can be generalised to the formal Hamiltonian appropriate to any molecular formula, as we now describe.

An extremely powerful technique for the investigation of eigenvalues (the discrete spectrum) of an operator is the *min-max* principle pioneered by Weyl and Courant [26]. This characterizes the (exact) eigenvalues in terms of an extremum property of the operator. A modern formulation is Theorem XIII.1 in [20]; a self-adjoint operator A that is bounded from below, that is $\mathsf{A} \geq c\mathsf{I}$ for some finite c, has eigenvalues $\lambda_n(\mathsf{A})$ that can be expressed as

$$\lambda_n(\mathsf{A}) = \sup U_A(\phi_1, \phi_2, \ldots \phi_{n-1}) \tag{1.14}$$

where

$$U_A(\phi_1, \phi_2, \ldots \phi_m) = \inf < \psi|\mathsf{A}|\psi > \tag{1.15}$$

and $< \psi|\psi >= 1$, $\psi \in [\phi_1, \ldots \phi_m]^\perp$

$$[\phi_1, \ldots \phi_m]^\perp = \{\psi| < \psi|\phi_i >= 0,\ i = 1, 2, \ldots m\} \tag{1.16}$$

Then for each fixed n, either

a) there are n eigenvalues below $\sigma_{ess}(\mathsf{A})$, and λ_n is the nth eigenvalue or

b) λ_n is the bottom of $\sigma_{ess}(\mathsf{A})$, and there are at most $(n-1)$ eigenvalues below $\sigma_{ess}(\mathsf{A})$. The significance of this result is that if $\sigma_{ess}(\mathsf{A}) = [a, \infty)$ for some a and we know that $\lambda_n < a$, there must be at least n eigenvalues [20].

On the other hand if the variation in (1.14)-(1.16) is restricted to a n-dimensional subspace of the domain of the operator A described by an orthogonal projection P, then the min-max principle can be used to characterize the eigenvalues $\hat{\lambda}_n$ of PAP, and this leads to

$$\lambda_n \leq \hat{\lambda}_n \tag{1.17}$$

which is the *Rayleigh-Ritz* principle. This is essentially the result that Löwdin [1, 2] attributes to Weyl[5], for if a variational calculation with some trial function yields an expectation value $< \mathsf{A} >$ below the bottom of $\sigma_{ess}(\mathsf{A})$, there must be at least one eigenvalue below $< \mathsf{A} >$, by (1.17).

We noted in Section 1 that the operator H_∞ obtained from (1.1) by taking the limit $\{m_g \to \infty, \forall g\}$ is the basis of modern quantum chemistry; it is of course the familiar electronic Hamiltonian. Suppose we have found the energy eigenvalues $U_m(\mathbf{x}^n), \mathbf{x}^n = (\mathbf{x}_1^n, \mathbf{x}_2^n, \ldots \mathbf{x}_A^n)$ and corresponding eigenfunctions $\Psi_m(\mathbf{x}^e, \mathbf{x}^n)$ of H_∞. The energies are to be used as effective potentials in the Schrödinger equation for the nuclear motion,

$$\left[\sum_g^A \frac{\mathsf{p}_g^2}{2m_g} + U_m(\mathbf{x}^n)\right] \Phi(\mathbf{x}^n)_i = \mathcal{E}_{mi}\Phi(\mathbf{x}^n)_i \tag{1.18}$$

to yield molecular energies $\{\mathcal{E}_{mi}\}$. The approximate ground-state energy $\mathcal{E}_{11}$ found in this way provides a *lower bound* to the true ground state energy E_0, according to the operator inequality

$$\mathsf{H} \geq U_1(\mathbf{x}^n) + \sum_g^A \frac{\mathsf{p}_g^2}{2m_g} \geq \mathcal{E}_{11} \tag{1.19}$$

[5] But Löwdin was mistaken in his attribution to Weyl's paper [27], which contains fundamental results about σ_{ess} for bounded self-adjoint operators.

On the other hand, if the product wavefunction $\Psi(\mathbf{x}^e, \mathbf{x}^n)_1 \Phi(\mathbf{x}^n)_1$ is used in a Rayleigh-Ritz calculation, the energy so obtained is an upper bound to E_0, so we have the estimates

$$\mathcal{E}_{11} \leq E_0 \leq \mathcal{E}_{11} + \sum_g^A \frac{c_g^2}{2m_g} \tag{1.20}$$

where the constants c_g are independent of the nuclear masses [24, 28, 29, 30].

In order to use these results as information about the ground-state however we need quantitative information also about the bottom of $\sigma_{ess}(\mathsf{A})$; this is much harder for the general $N-$particle Coulomb Hamiltonian. A theorem of Weyl, [27] shows that if A is self-adjoint and an operator B differs from it by only a compact perturbation, $\sigma_{ess}(\mathsf{A}) = \sigma_{ess}(\mathsf{B})$. The essential spectrum of the kinetic energy operator, $-\nabla^2$, is easily seen to be $\sigma_{ess}(-\nabla^2) = [0, \infty]$ (e.g. consider its Fourier transform); Weyl's theorem suffices to prove that $\sigma_{ess}(-\nabla^2 + V) = \sigma_{ess}(-\nabla^2) = [0, \infty)$ for a large class of two-body Schrödinger operators (after removal of the centre-of-mass contribution) with $V \to 0$ at infinity, including the case of the Coulomb interaction $1/r$. Suppose we write

$$\mathsf{H} = \mathsf{H}_0 + V \tag{1.21}$$

and define the resolvents of H and H_0 by

$$R(E) = \frac{1}{\mathsf{H} - E} \qquad R_0(E) = \frac{1}{\mathsf{H}_0 - E} \tag{1.22}$$

Then by standard manipulations [31]

$$R(E) = R_0(E) - R_0(E) V R(E) \tag{1.23}$$

In the two-body Coulombic case where $\mathsf{H}_0 = -\nabla^2$, and $V = 1/r$, the crucial result is that $R_0(E)V$ is a compact operator, and so is bounded [30].

This property fails in the $N-$body case where V is a sum over two-particle interactions

$$V = \sum_{i<j} V_{ij}(\mathbf{x}_i - \mathbf{x}_j)$$

which are such that $V_{ij}(\mathbf{X}) \to 0$ as $X \to \infty$, where X is the ij relative distance coordinate. Then even though the two-body potential is well-behaved at infinity, V fails to go to zero at infinity in tubes where

$$\sum_{i=1}^{n-1} |\mathbf{x}_i - \mathbf{x}_n|^2 \to \infty$$

while some $|\mathbf{x}_i - \mathbf{x}_j|$ remain finite. As a result, $R_0(E)V$ is not compact, and Weyl's theorem is no longer applicable. During the 1960's and 1970's a different decomposition of the resolvent was discovered and utilized in scattering theory, and this culminated finally in the proof of the so-called HVZ theorem (Theorem XIII.17 in [20]) which demonstrates that the essential spectrum of the internal Coulombic Hamiltonian H' can be written as $\sigma_{ess}(\mathsf{H}') = [\Sigma, \infty)$ where Σ is the energy of the lowest two-body[6] cluster decomposition of the $N-$particle system. Subsequently a rather different method of proof based on geometrical ideas was developed; a valuable discussion of the HVZ theorem from this point of view can be found in [30].

The Coulomb interaction belongs to a special class of two-body potentials that are called *dilatation analytic.* Indeed, suppose that $V_{ij}(\mathbf{X})$ has the property that the operator

$$\mathsf{J}(\theta) = V(e^{\theta}\mathbf{X})(-\nabla^2 + 1)^{-1}$$

is compact and has an analytic continuation from real θ to $\{\theta, |Im\ \theta| < \epsilon\}$ for some $\epsilon > 0$. This suffices to guarantee that any eigenvalues of the internal Hamiltonian H' are of finite multiplicity and can only accumulate at thresholds [32]; physically, the thresholds correspond to fragmentation energies. The exploitation of the dilatation analyticity of the Coulombic interaction is commonly referred to as the complex coordinate rotation technique [32, 33, 34]. It allows a particularly vivid visualization of the entire spectrum of H', and generalizes the description of the Helium atom we gave earlier. The continuous spectrum of H' lies above a certain (possibly infinite) number of discrete states which are bound stationary states of the $N-$particle cluster (atom or molecule). The singular continuous spectrum of H' is empty, $\sigma_{sc}(\mathsf{H}') = \emptyset$. In the complex energy plane the continuous spectrum is a branch cut on the real E axis starting at some value Σ and extending to $+\infty$; it is rich in structure having numerous sets of other bound states embedded in it. These are associated with the fragments that the $N-$particle cluster can break up into. A new cluster fragment does not appear until the energy has exceeded a characteristic threshold value. This description of the spectrum is what one would expect intuitively from the facts of mass spectrometry.

Let $\mathbf{x}$ and $\mathbf{p}$ stand collectively for all the coordinate and momentum operators of the N particles described by H, and write a position space wavefunction as $\psi^N(\mathbf{x})$ with $\mathbf{x} = (\mathbf{x}_1, \mathbf{x}_2, \ldots \mathbf{x}_N)$, where as discussed

[6] It is necessary to consider the two-cluster threshold only, since the three-cluster threshold can be shown (see result (20b) of XIII.5 of [20]) to be above the two-cluster one and so on.

previously, Section 3, only $(N-1)$ of the $\mathbf{x}_i$ are independent of the centre-of-mass coordinate $\boldsymbol{\xi}$. Now introduce a dilatation transformation with a unitary operator parameterized by an angle θ such that

$$\mathsf{U}(\theta)\psi^N(\mathbf{x}) = e^{i3(N-1)\theta/2}\psi^N(\mathbf{x}e^{i\theta})$$

The effect of $\mathsf{U}(\theta)$ is simply to transform $\mathbf{x}$ and $\mathbf{p}$ as

$$\mathbf{x} \to \mathbf{x}e^{i\theta} \qquad \mathbf{p} \to \mathbf{p}e^{i\theta}$$

Then we can construct the unitary transform of the molecular Hamiltonian H under $\mathsf{U}(\theta)$

$$\mathcal{H} = \mathsf{U}(\theta)\mathsf{H}\mathsf{U}(\theta)^{-1}$$

With the assumption of Coulombic interactions between the particles the following conclusions can then be stated [34]:

i Bound-state eigenvalues of $\mathcal{H}$ are independent of θ, and identical to those of H, for $|\theta| < \pi/2$;

ii scattering thresholds corresponding to the possibility of fragmentation into all possible cluster fragments in different states of excitation are also independent of θ for $|\theta| < \pi/2$;

iii the segments of continua beginning at each scattering threshold rotate by an angle 2θ into the lower half plane $(\theta \geq 0)$, each about its own threshold,

iv in addition, discrete eigenvalues of $\mathcal{H}$ may appear in the sector $0 > \arg(z-\Sigma) > -2\theta$ where z is the complex energy and Σ is the bottom of $\sigma_{ess}(\mathsf{H})$ - these are associated with bound states with finite lifetimes (resonances).

5. Symmetry in the Coulomb Hamiltonian

It would seem pertinent before we consider molecular symmetry further, to analyse more fully the consequences of the symmetries of the Coulomb Hamiltonian identified in Section 3.

Let us initially look at the translational symmetry and let the position variables in the problem be designated as $\mathbf{x}_i$ which will be treated as a column matrix of three cartesian components $x_{\alpha i}$, $\alpha = x, y, z$. We regard the $\mathbf{x}_i$ collectively as the 3 by N_T matrix $\mathbf{x}$. When it is necessary to distinguish between electrons and nuclei, the variables may be split up into two sets, one set consisting of N variables, $\mathbf{x}_i^e$, describing the electrons and the other set of A variables, $\mathbf{x}_i^n$, describing the nuclei with $N_T = N + A$. Each of the particles has mass m_i and charge $Z_i e$. The charge-numbers Z_i are positive integers for a nucleus and minus one for an electron. In a neutral system the charge-numbers sum to zero.

All that is needed to remove the centre of mass motion from the full molecule Hamiltonian is a coordinate transformation symbolised by

$$(\mathbf{t}\,\boldsymbol{\xi}) = \mathbf{x}\,\mathbf{V} \tag{1.24}$$

In (1.24) $\mathbf{t}$ is a 3 by $N_T - 1$ matrix and $\boldsymbol{\xi}$ is a 3 by 1 matrix, so that the combined (bracketed) matrix on the left of (1.24) is 3 by N_T. $\mathbf{V}$ is an N_T by N_T matrix which, from the structure of the left side of (1.24), has a special last column whose elements are

$$V_{iN_T} = {M_T}^{-1} m_i, \qquad M_T = \sum_{i=1}^{N_T} m_i \tag{1.25}$$

Hence $\boldsymbol{\xi}$ is the standard centre-of-mass coordinate

$$\boldsymbol{\xi} = {M_T}^{-1} \sum_{i=1}^{N_T} m_i \mathbf{x}_i \tag{1.26}$$

As the coordinates $\mathbf{t}_j, j = 1, 2, ...N_T - 1$ are to be translationally invariant, we require

$$\sum_{i=1}^{N_T} V_{ij} = 0, \qquad j = 1, 2, ...N_T - 1 \tag{1.27}$$

on each remaining column of $\mathbf{V}$ and it is easy to see that (1.27) forces $\mathbf{t}_j \rightarrow \mathbf{t}_j$ as $\mathbf{x}_i \rightarrow \mathbf{x}_i + \mathbf{a}$, all i.

The $\mathbf{t}_i$ are independent if the inverse transformation

$$\mathbf{x} = (\mathbf{t}\,\boldsymbol{\xi})\mathbf{V}^{-1} \tag{1.28}$$

exists. The structure of the right side of (1.28) shows that the bottom row of $\mathbf{V}^{-1}$ is special and, without loss of generality, we may require its elements to be

$$(\mathbf{V}^{-1})_{N_T i} = 1, \qquad i = 1, 2, ...N_T \tag{1.29}$$

The inverse requirement on the remainder of $\mathbf{V}^{-1}$ implies that

$$\sum_{i=1}^{N_T} (\mathbf{V}^{-1})_{ji} m_i = 0, \qquad j = 1, 2, ...N_T - 1 \tag{1.30}$$

When we write the column matrix of the cartesian components of the partial derivative operator as $\partial/\partial\mathbf{x}_i$, the coordinate change (1.24) gives

$$\frac{\partial}{\partial \mathbf{x}_i} = \sum_{j=1}^{N_T-1} V_{ij}\frac{\partial}{\partial \mathbf{t}_j} + m_i {M_T}^{-1}\frac{\partial}{\partial \boldsymbol{\xi}} \tag{1.31}$$

and when it seems more convenient this column matrix of derivative operators will also be denoted as the vector grad operator $\vec{\nabla}(\mathbf{x}_i)$.

If a second set $\mathbf{t}'$ of translationally invariant coordinates is constructed it may be related to the original set by

$$(\mathbf{t}'\,\boldsymbol{\xi}) = (\mathbf{t}\,\boldsymbol{\xi})\overline{\mathbf{V}}, \quad \overline{\mathbf{V}} = \mathbf{V}^{-1}\mathbf{V}' \tag{1.32}$$

where $\mathbf{V}'$ is the transformation matrix that defines $\mathbf{t}'$. The matrix $\overline{\mathbf{V}}$ is

$$\begin{pmatrix} \mathbf{G} & \mathbf{0} \\ \mathbf{0} & 1 \end{pmatrix}, \quad G_{ij} = (\overline{\mathbf{V}})_{ij}, \quad i,j = 1,2,\ldots A-1 \tag{1.33}$$

It is easily seen that the form of (1.31) is preserved under a change from $\mathbf{t}$ to $\mathbf{t}'$. It is thus the case that any set of translationally invariant coordinates can be related to any other set by means of a linear transformation.

The Hamiltonian (1.1) in the new coordinates becomes

$$\begin{aligned} \mathsf{H}(\mathbf{t},\boldsymbol{\xi}) &= -\frac{\hbar^2}{2}\sum_{i,j=1}^{N_T-1}\frac{1}{\mu_{ij}}\vec{\nabla}(\mathbf{t}_i).\vec{\nabla}(\mathbf{t}_j) + \frac{e^2}{8\pi\epsilon_o}\sum_{i,j=1}^{N_T}{}'\frac{Z_iZ_j}{r_{ij}(\mathbf{t})} - \frac{\hbar^2}{2M_T}\nabla^2(\boldsymbol{\xi}) \\ &= \mathsf{H}'(\mathbf{t}) - \frac{\hbar^2}{2M_T}\nabla^2(\boldsymbol{\xi}) \end{aligned} \tag{1.34}$$

Here

$$1/\mu_{ij} = \sum_{k=1}^{N_T} {m_k}^{-1}V_{ki}V_{kj}, \qquad i,j = 1,2,\ldots N_T-1 \tag{1.35}$$

The $N_T - 1$ dimensional square matrix composed of all the $1/\mu_{ij}$ is denoted as $\boldsymbol{\mu}^{-1}$. The operator r_{ij} is the interparticle distance operator expressed as a function of $\mathbf{t}_i$. Thus

$$r_{ij}(\mathbf{t}) = \left(\sum_\alpha\Big(\sum_{k=1}^{N_T-1}((\mathbf{V}^{-1})_{kj} - (\mathbf{V}^{-1})_{ki})t_{\alpha k}\Big)^2\right)^{1/2} \tag{1.36}$$

In (1.34) the $\vec{\nabla}(\mathbf{t}_i)$ are grad operators expressed in the cartesian components of $\mathbf{t}_i$ and the last term represents the centre-of-mass kinetic energy. Since the centre-of-mass variable does not enter the potential energy term, the centre-of-mass motion may be separated off completely so that the eigenfunctions of H are of the form

$$T(\boldsymbol{\xi})\Psi(\mathbf{t}) \tag{1.37}$$

where $\Psi(\mathbf{t})$ is a wavefunction for the Hamiltonian $\mathsf{H}'(\mathbf{t})$ (1.34) in (1.11) which we will refer to as the *translationally invariant* Hamiltonian.

It should be emphasised that the $\mathbf{x}_i$ simply specify field points, and cannot actually be particle coordinates because of the indistinguishability of sets of identical particles. Weyl [35] and later Mackey [36] both stress that, in the case of sets of identical particles, in addition to supporting the canonical quantum conditions, the space on which quantum mechanical operators act must be confined to a sub-space of the full Hilbert space of definite permutational symmetry. This means that the effect of any operator on a function in this sub-space must be to produce another function in the subspace. Multiplication of a properly symmetrised function by a single coordinate variable produces a new function which is not in the symmetrised sub-space. Thus only operators symmetric in all the coordinates of identical particles can properly be deployed in the calculation of expectation values that represent observables. Weyl discusses this in Section C 9 of Chapter IV of [35]. He says of the two particle case:

> Physical quantities have only an objective significance if they depend *symmetrically* on the two individuals.

and he then goes on to generalise this conclusion to the symmetrical form for the quantities constructed from the variables of N identical particles. He closes his discussion by looking at the two electron problem. He says that although it might be supposed that the electrons as a pair of twins could be named "Mike" and "Ike"

> it is impossible for either of these individuals to retain his identity so that one of them will always be able to say "I'm Mike" and the other "I'm Ike". Even in principle one cannot demand an alibi of an electron! In this way the Leibnizian principle of *coincidentia indiscernibilium* holds in quantum mechanics.

This discussion holds for identical particles of any kind that are to be described by quantum mechanics and it precludes the specification of, for example, the expected value of a particular coordinate chosen from a set describing many identical particles. To avoid periphrasis we shall, however, continue to speak of the variables as if they were particle coordinates.

It is straightforward to enumerate the other symmetries of the translationally invariant Hamiltonian. The general permutation of identical particles can be written as

$$\mathcal{P}(\mathbf{x}^e\mathbf{x}^n) = (\mathbf{x}^e\mathbf{x}^n)\begin{pmatrix} \mathbf{P}^e & \mathbf{0} \\ \mathbf{0} & \mathbf{P}^n \end{pmatrix} \equiv \mathbf{x}\mathbf{P} \tag{1.38}$$

where $\mathbf{P}^e$ and $\mathbf{P}^n$ are standard permutation matrices. They are orthogonal with determinant ± 1 according to whether the permutation is of even or odd parity. The matrix $\mathbf{P}^n$ will have non-zero entries only for each group of identical atomic nuclei and is most conveniently visualised as having block diagonal form, one block for each group of identical nuclei. The effect of this permutation on the translationally invariant coordinates is:

$$\mathcal{P}(\mathbf{t}\,\boldsymbol{\xi}) = (\mathbf{t}\,\boldsymbol{\xi})\begin{pmatrix} \mathbf{H} & \mathbf{0} \\ \mathbf{0} & 1 \end{pmatrix} \tag{1.39}$$

where

$$(\mathbf{H})_{ij} = (\mathbf{V}^{-1}\mathbf{P}\mathbf{V})_{ij}, \qquad i,j = 1,2,.....N_T - 1 \tag{1.40}$$

The matrix $\mathbf{H}$ is not necessarily in standard permutational form neither is it orthogonal, even though it has determinant ± 1 according to the sign of $|\mathbf{P}|$. Thus under any permutation of like particles, the translationally invariant coordinates will transform into linear combinations of themselves. Any chosen transformed coordinate will, generally, involve both $\mathbf{P}^e$ and $\mathbf{P}^n$ in its definition. If it is desired to identify electrons with a particular set of translationally invariant coordinates, specialised coordinate choices must be made to avoid $\mathbf{P}^n$ becoming involved in the definition of their transformed forms, and even more specialised choices must be made to ensure that members of the chosen set transform only into each other under $\mathbf{P}^e$.

Because the translationally invariant Hamiltonian is also invariant under all rotations and rotation-reflections of the translationally invariant coordinates, it will have eigenfunctions which provide a basis for irreducible representations (irreps) of the orthogonal group in three dimensions O(3). Thus the eigenfunctions are expected to be of two kinds classified by their parity; each kind consists of eigenfunction sets, each with degeneracy $2J + 1$, according to the irrep $J = 0, 1, 2, \ldots$ of SO(3) to which the eigenfunctions belong. The representations of O(3) are distinct for each parity, and so there is no group theoretical reason to expect eigenfunctions with different parity to be degenerate.

Simultaneously the eigenfunctions will provide irreps for the permutation group $\mathcal{S}$ of the system. This group comprises the direct product of the permutation group $\mathcal{S}_N$ for the electrons with the permutation groups $\mathcal{S}_{A_i}$ for each set of identical nuclei i comprising A_i members. The physically realisable irreps of this group are restricted by the requirement that, when spin is properly incorporated into the eigenfunctions, the eigenfunctions form a basis only for the totally symmetric representation, if bosons (spin 0, 1, 2 etc) or of the antisymmetric representation, if fermions (spin 1/2, 3/2, 5/2 etc). Both of these representations are one-dimensional. We shall speak of irreps of the translationally invariant Hamiltonian which correspond to physically realisable states as *permutationally allowed.* In general such irreps will be many dimensional and so we would expect to have to deal with degenerate sets of eigenfunctions in attempting to identify a molecule in the solutions to the translationally invariant problem.[7]

Since all possible sets of translationally invariant coordinates can be related by linear transformations, the spectrum of the translationally invariant Hamiltonian is quite independent of the choice of translationally invariant coordinates made. It is, however, clear from the form of (1.36) that the asymptotic form of any particular $\mathsf{H}'(\mathbf{t})$ for a given interparticle configuration, will not in general be expressible at all simply in terms of the chosen set of the $\mathbf{t}_i$. Further, any linear transformations of the $\mathbf{t}_i$ that may be induced by permutations according to (1.39), will complicate further the expression of asymptotic behaviour. Thus extreme care must be exercised in any attempt to describe cluster decomposition for any given Coulomb Hamiltonian, and so situate the bottom of its essential spectrum. These problems have been addressed by Balslev[8] [37] who was able to show that the essential spectrum of the translationally invariant Coulomb Hamiltonian started at the lowest two-cluster threshold with compatible symmetry. In particular he showed that for the atomic Coulomb Hamiltonian there were infinite sequences of discrete eigenvalues for each symmetry type. It is the case then, that the HVZ theorem that Löwdin [1, 2] referred to, still holds when all symmetries are considered. However, actually to describe a given cluster decomposition it is, in general, necessary to perform linear transformations of the translationally invariant coordinates so that the interparticle asymptotes may be properly represented and also preserve permutational symmetry. This last is important because there may be particles of the same kind in each of the clusters.

5.1 Symmetry and asymptotic behaviour

As an example of what is involved in cluster decomposition consider a neutral system with two nuclei and N electrons. Let the two nuclei have

[7]See the discussion of Theorem XIII.46 in [20].
[8]see also Notes to Section XIII.5 in [20]

masses m_b and m_c, and charges Z_b and Z_c and be located by coordinates $\mathbf{x}_b$ and $\mathbf{x}_c$. Then if one is investigating this system as if it were a diatomic molecule, it is customary to define a relative nuclear variable as

$$\mathbf{t}^n = \mathbf{x}_b - \mathbf{x}_c \tag{1.41}$$

A set of translationally invariant electronic variables may be written as

$$\mathbf{t}_i^e \equiv \mathbf{t}_i = \mathbf{x}_i^e + v_b\mathbf{x}_b + v_c\mathbf{x}_c, \quad i = 1, 2, \ldots N, \quad v_b + v_c = -1 \tag{1.42}$$

The inverse transformation is

$$\mathbf{x}_b = m_c M^{-1}\mathbf{t}^n + \mathbf{X}$$

$$\mathbf{x}_c = -m_b M^{-1}\mathbf{t}^n + \mathbf{X}$$

$$\mathbf{x}_i^e = \mathbf{t}_i + \frac{(m_b v_c - m_c v_b)}{M}\mathbf{t}^n + \mathbf{X} \tag{1.43}$$

where $M = m_b + m_c$ and $\mathbf{X} = M^{-1}(m_b\mathbf{x}_b + m_c\mathbf{x}_c)$ is the usual centre-of-nuclear mass coordinate. It is seen that generally the definition of $\mathbf{t}_i$ will be changed if $\mathbf{x}_b$ and $\mathbf{x}_c$ are interchanged, so if permutations of the nuclear coordinates are to have no effect on the translationally invariant electronic coordinates, v_b and v_c must be chosen to be the same for identical nuclei. This is just the sort of coordinate choice that seems to have been made by Born and Oppenheimer [16] in Part VIII of their paper where they offer an independent treatment of the diatomic molecule[9].

The nuclear-electronic distances have to be supplied from (1.43) according to the particular choices made for v_b and v_c. A natural choice for a homonuclear diatomic, and a possible one for a heteronuclear one too, is to choose both v_b and v_c to be $-1/2$, giving

$$\mathbf{t}_i = \mathbf{x}_i^e - \frac{1}{2}(\mathbf{x}_b + \mathbf{x}_c) \tag{1.44}$$

$$\mathbf{x}_i^e = \mathbf{t}_i + \frac{(m_c - m_b)}{2M}\mathbf{t}^n + \mathbf{X} \tag{1.45}$$

so that in the homonuclear case we get

$$(\mathbf{x}_b - \mathbf{x}_j^e) = (\frac{\mathbf{t}^n}{2} - \mathbf{t}_j), \quad (\mathbf{x}_c - \mathbf{x}_j^e) = (\frac{\mathbf{t}^n}{2} + \mathbf{t}_j)$$

[9]It is not quite clear from the paper precisely how the electronic coordinates were chosen.

Let us consider the asymptotic behaviour of the electron-nucleus potential terms here. As $\mathbf{t}^n$ becomes very large, the nuclear repulsion term becomes very small and, for any fixed values for the $\mathbf{t}_i$, the electron-nucleus attraction terms become small too, while the electronic repulsion terms remain constant. Thus the asymptotes of the problem so formulated are a pair of bare nuclei with the ionised electrons referred to a point midway between them. This is clearly not a very good candidate for the lowest two-cluster decomposition.

A much better choice for the lowest two-cluster decomposition would be into a pair of neutral atoms. In these circumstances it would be reasonable to define

$$\mathbf{t}'_i = \mathbf{x}^e_i - \mathbf{x}_b, \quad i = 1, 2, \ldots Z_b$$

$$\mathbf{t}'_i = \mathbf{x}^e_i - \mathbf{x}_c, \quad i = Z_b + 1, \ldots N \tag{1.46}$$

Notice that with this choice, the definition of the translationally invariant electronic coordinates is changed, if the nuclei are permuted. The inverse of this transformation for the nuclei is as above in (1.43) while that for the electrons is

$$\mathbf{x}^e_i = \mathbf{t}'_1 + \frac{m_c}{M}\mathbf{t}^n + \mathbf{X}, \quad i = 1, 2, \ldots Z_b$$

$$\mathbf{x}^e_i = \mathbf{t}'_2 - \frac{m_b}{M}\mathbf{t}^n + \mathbf{X}, \quad i = Z_b + 1, \ldots N$$

If i and j are in different clusters then

$$(\mathbf{x}^e_i - \mathbf{x}^e_j) = (\mathbf{t}_i - \mathbf{t}_j \pm \mathbf{t}^n)$$

so that the inter-electronic part of the potential has a dependence upon the translationally invariant nuclear coordinates. The coordinates of the electron-nucleus attraction terms take the form:

$$(\mathbf{x}^e_i - \mathbf{x}_b) = (\mathbf{t}_i), \quad (\mathbf{x}^e_i - \mathbf{x}_c) = (\mathbf{t}_i + \mathbf{t}^n)$$

if i is in the first cluster and

$$(\mathbf{x}^e_i - \mathbf{x}_c) = (\mathbf{t}_i), \quad (\mathbf{x}^e_i - \mathbf{x}_b) = (\mathbf{t}_i - \mathbf{t}^n) \tag{1.47}$$

if i is in the second cluster. It is easily seen that this choice does allow the dissociation into a pair of separate atoms but at the cost of making the electronic repulsion term dependent upon what is formally, a nuclear variable and also of considerably complicating the description of the permutational symmetry of the electrons.

The two sets (1.44) and (1.46) of translationally invariant coordinates are related by

$$\mathbf{t}'_i = \mathbf{t}_i - \frac{1}{2}\mathbf{t}^n, \quad i = 1, 2, \ldots Z_b$$

$$\mathbf{t}'_i = \mathbf{t}_i + \frac{1}{2}\mathbf{t}^n \quad i = Z_b + 1, \ldots N \tag{1.48}$$

and obviously, any wave function obtained in one set of coordinates can be re-expressed, though perhaps not at all conveniently, in the other. This establishes that the exact solutions to the problem expressed in either set of coordinates yield the same energies, provided that the same boundary conditions are imposed, in spite of having different looking wave functions; it is this fact that makes it possible to establish the HVZ theorem while restricting the permutational symmetry of the solutions.

Clearly a somewhat different approach is necessary when considering how rotation-reflection symmetry can be maintained asymptotically. The translationally invariant coordinates are not immediately adapted to describe such symmetry. Descriptions of this kind of symmetry are usually offered in terms of three angular coordinates and a parity specification, together with $3N_T - 6$ *internal* coordinates, which are invariant under all rotation-reflections. There are many rather delicate topological problems raised by such a choice. The space defined is a quotient manifold that is only locally Euclidean and so coordinates can be defined only locally. In fact it always takes more than one coordinate system to cover the quotient manifold, and this complicates greatly any explicit account of the rotation-reflection symmetry at the asymptotes. We shall follow Löwdin in his discussion of angular momentum (rotational symmetry) and take the view that there are other difficulties that for the moment must be given higher priority.

5.2 Symmetry and particle identity

In view of the above discussion and the arguments advanced by Born and Oppenheimer [16], it may be helpful to express $\mathsf{H}'(\mathbf{t})$ in terms of two sets of coordinates. One set consists of $A - 1$ translationally invariant coordinates $\mathbf{t}_i^n$ expressed entirely in terms of the original $\mathbf{x}_i^n$:

$$\mathbf{t}_i^n = \sum_{j=1}^{A} \mathbf{x}_j^n V_{ji}^n, \qquad i = 1, 2, \ldots, A - 1 \tag{1.49}$$

here $\mathbf{V}^n$ is a non-singular matrix whose last column is special, with elements

$$V_{iA}^n = M^{-1} m_i, \qquad\qquad M = \sum_{i=1}^{A} m_i \tag{1.50}$$

so that the coordinate $\mathbf{X}$, defined by its last column, is the coordinate of the centre-of-nuclear mass. The elements in the first $A-1$ columns of $\mathbf{V}^n$ each sum to zero, exactly as in the general case, to ensure translational invariance. The other set comprises N translationally invariant coordinates of the form

$$\mathbf{t}_i^e = \mathbf{x}_i^e + \sum_{j=1}^{A} v_j \mathbf{x}_j^n \tag{1.51}$$

For definiteness we shall choose $v_i = -m_i/M$ so that the translationally invariant electronic coordinates are the original electronic coordinates expressed relative to the centre-of-nuclear-mass. For generality[10] the systems that we shall consider must have $A \geq 4$. The inverse relations are

$$\mathbf{x}_i^e = \mathbf{X} + \mathbf{t}_i^e \tag{1.52}$$

$$\mathbf{x}_i^n = \mathbf{X} + \sum_{j=1}^{A-1} \mathbf{t}_j^n ((\mathbf{V}^n)^{-1})_{ji} \tag{1.53}$$

with

$$((\mathbf{V}^n)^{-1})_{Ai} = 1, \qquad i = 1, 2, ...A \tag{1.54}$$

while the inverse requirement on the remaining rows gives

$$\sum_{i=1}^{A} ((\mathbf{V}^n)^{-1})_{ji} m_i = 0 \qquad j = 1, 2, ...A-1 \tag{1.55}$$

The centre-of-mass and the centre-of-nuclear-mass are related by

$$\boldsymbol{\xi} = \mathbf{X} + \frac{m}{M_T} \sum_{i=1}^{N} \mathbf{t}_i^e \tag{1.56}$$

The translationally invariant Coulomb Hamiltonian now takes the form:

$$\mathsf{H}'(\mathbf{t}) \to \mathsf{H}^e(\mathbf{t}^e) + \mathsf{H}^n(\mathbf{t}^n) + \mathsf{H}^{en}(\mathbf{t}^n, \mathbf{t}^e) \tag{1.57}$$

The part of the Hamiltonian which can be associated with electronic motion is

$$\mathsf{H}^e(\mathbf{t}^e) = -\frac{\hbar^2}{2\mu} \sum_{i=1}^{N} \nabla^2(\mathbf{t}_i^e) - \frac{\hbar^2}{2M} \sum_{i,j=1}^{N}{}' \vec{\nabla}(\mathbf{t}_i^e) \cdot \vec{\nabla}(\mathbf{t}_j^e) + \frac{e^2}{8\pi\epsilon_0} \sum_{i,j=1}^{N}{}' \frac{1}{|\mathbf{t}_j^e - \mathbf{t}_i^e|} \tag{1.58}$$

[10]The cases $A = 1$, $A = 2$ and $A = 3$ (the nuclear configurations that define, respectively, a point, a line and a plane) must be given special attention and to consider them here would deflect us from the main thrust of the argument.

with

$$1/\mu = 1/m + 1/M \tag{1.59}$$

while the part that can be associated with nuclear motion is

$$\mathsf{H}^n(\mathbf{t}^n) = -\frac{\hbar^2}{2}\sum_{i,j=1}^{A-1}\frac{1}{\mu^n_{ij}}\vec{\nabla}(\mathbf{t}^n_i).\vec{\nabla}(\mathbf{t}^n_j) + \frac{e^2}{8\pi\epsilon_0}\sum_{i,j=1}^{A}{}' \frac{Z_i Z_j}{r_{ij}(\mathbf{t}^n)} \tag{1.60}$$

where $r_{ij}(\mathbf{t}^n)$ is defined just as in the general case (1.36) but using the $\mathbf{t}^n_i$ and $(\mathbf{V}^n)^{-1}$, that is

$$r_{ij}(\mathbf{t}^n) = \left(\sum_{\alpha}(\sum_{k=1}^{A-1}((\mathbf{V}^{n-1})_{kj} - (\mathbf{V}^{n-1})_{ki})t^n_{\alpha k})^2\right)^{1/2} \tag{1.61}$$

and the inverse mass matrix is similarly specialised as

$$1/\mu^n_{ij} = \sum_{k=1}^{A} m_k{}^{-1} V^n_{ki} V^n_{kj}, \qquad i,j = 1,2,...A-1 \tag{1.62}$$

The electronic and nuclear motion are coupled only via a potential term:

$$\mathsf{H}^{en}(\mathbf{t}^n,\mathbf{t}^e) = -\frac{e^2}{4\pi\epsilon_0}\sum_{i=1}^{A}\sum_{j=1}^{N}\frac{Z_i}{r'_{ij}(\mathbf{t}^n,\mathbf{t}^e)} \tag{1.63}$$

and the electron-nucleus distance expression becomes

$$|\mathbf{x}^n_i - \mathbf{x}^e_j| \equiv r'_{ij} = \left|\sum_{k=1}^{A-1}\mathbf{t}^n_k(\mathbf{V}^n)^{-1}_{ki} - \mathbf{t}^e_j\right| \tag{1.64}$$

It follows from (1.39) that this choice of translationally invariant coordinates is such that the $\mathbf{t}^e_i$ transform under a permutation of the $\mathbf{x}^e_i$ exactly as do the $\mathbf{x}^e_i$ and remain unchanged under any permutation of the $\mathbf{x}^n_i$. The $\mathbf{t}^n_i$ are invariant under any permutation of the $\mathbf{x}^e_i$ while a permutation of the $\mathbf{x}^n_i$ produces a transformation among the $\mathbf{t}^n_i$ just as in (1.39) but with the variables restricted to the $\mathbf{t}^n_i$, the permutations only among the identical nuclei and with transformation matrices involving only the nuclear variables. This choice makes $\mathsf{H}^e(\mathbf{t}^e)$ trivially invariant under permutations of the original electronic coordinates and independent of any particular choice of translationally invariant nuclear coordinates. Similarly $\mathsf{H}^n(\mathbf{t}^n)$ is independent of any particular choice of translationally invariant electronic coordinates and is also invariant under any permutation of the original coordinates of identical nuclei. The interaction operator $\mathsf{H}^{en}(\mathbf{t}^n, \mathbf{t}^e)$ is obviously invariant under a permutation of the original electronic coordinates and is also invariant under a permutation of the original coordinates of identical nuclei.

Although, for a given choice of boundary conditions, the spectrum of the translationally invariant Hamiltonian is precisely the same with this choice of coordinates as it would be for any other choice, we might hope to recognise in these coordinates an eigenfunction corresponding to a molecule, rather more easily than we would in some other choices. We might hope too, that if we find a suitably localised eigenfunction $\Psi(\mathbf{t}^n, \mathbf{t}^e)$ then it might be possible to write it in good approximation in the electron-nuclear product form $\Phi(\mathbf{t}^n)\psi(\mathbf{t}^e, r_{ij}(\mathbf{t}^n))$ in the spirit of the Born and Oppenheimer analysis. It should be noted explicitly that the choice of a product form to approximate the exact eigenfunction does not mean that any permutational symmetry of the problem is being neglected here or will in future be neglected in this context.

Let us consider now the behaviour of such functions under identical particle permutations. Let a particular irreducible representation of the symmetric group of the electronic coordinates be denoted as $[\boldsymbol{\lambda}]^N$ and let the conjugate representation be denoted as $[\tilde{\boldsymbol{\lambda}}]^N$. For electrons (or any spin 1/2 particles) the representation of the symmetric group carried by the spin-eigenfunctions $\Theta_{S,M_S,i}$ must be one described by a no more than two-rowed Young diagram, that is $[\boldsymbol{\lambda}]^N \equiv [\lambda_1, \lambda_2]$ where

$$\lambda_1 = N/2 + S, \qquad \lambda_2 = N/2 - S$$

The representations are independent of the choice of M_S and i labels the rows (columns) of the representation. The dimension of the representation is given by the Wigner number

$$f_S^N = \frac{(2S+1)N!}{(N/2+S+1)!(N/2-S)!}$$

Assuming that the translationally invariant part of the Coulomb Hamiltonian for the chosen system has eigenfunctions in the discrete spectrum then, among them, there will be a degenerate set that provides a basis for the representation conjugate to that for the chosen spin-eigenfunctions. The representation and the conjugate representation have the same dimension and a basis of space-spin products can be formed which belongs to the antisymmetric representation of the symmetric group and hence satisfies the Pauli Principle. For example, suppose that a 10 electron system, such as ammonia, was being considered and it was hoped to identify a singlet state. In this case (ignoring for the moment the nuclear variables) one would be looking for a set of 42 degenerate eigenfunctions of the Coulomb Hamiltonian which provided a basis for the irrep $\widetilde{[5,5]}$ under permutations of electronic variables. These would be functions of the kind earlier called *permutationally allowed.*

One could provide precisely similar arguments to deal with protons and so to describe ammonia, for example, one might look for a nuclear spin doublet arising from the protons and then $[\boldsymbol{\lambda}]^A$ would just be $[2,1]$ and one would be looking for a pair of degenerate functions to provide a basis for the irrep $\widetilde{[2,1]}$ under permutation of the protons. Thus to describe ammonia in a singlet electronic spin state with a doublet nuclear spin state one would have to find a degenerate set of 84 eigenfunctions among the eigenfunctions of the Coulomb Hamiltonian to provide a basis for the permutationally allowed irrep of $\mathcal{S}_N(10) \times \mathcal{S}_A(3)$.

This sort of argument could be extended to particles with spins other than 1/2 and with Bose rather than Fermi statistics. To do so is, however, much more difficult and, so far as the authors know, has never been systematically attempted. The difficulties arise because it is much harder in the general case than it is with particles of spin 1/2, to associate the spin functions with their space parts to produce functions of appropriate symmetry. Although it is true that particles of spin s can provide a basis for representations of the symmetric group corresponding to Young diagrams with, at most, only $2s+1$ rows, it is not in general possible to determine the lengths of these rows simply from the total S and N values in the problem. One cannot then make any simple association between particular spin-eigenfunctions and a matched spatial function. To achieve the required results it is necessary to deploy the formal machinery of the symmetric group in constructing a Clebsch-Gordan series from the full set of space-spin products in order to isolate in the direct product space either the antisymmetric irreps for fermions, or symmetric irreps for bosons. For the present, it will have to be sufficient to know that it could be done.

If one moves from ammonia to a larger molecule, for example, the simple hydrocarbon with empirical formula C_8H_8, then a host of other problems begin to emerge. Consider first the 56 electrons. It is easy, though tedious, to show that the dimension of the permutationally allowed representation $\widetilde{[28,28]}$ for the singlet state, given by the Wigner formula f_S^N with $N = 56$ and $S = 0$, is

$$53 \times 47 \times 44 \times 43 \times 41 \times 37 \times 35 \times 34 \times 31 \simeq 2.6\ 10^{14}$$

so that for this rather simple system one can expect the eigenfunctions, if any, in the discrete spectrum of $\mathsf{H}'(\mathbf{t})$ to be very extensively degenerate even without considering any degeneracies arising from the nuclear variables.

Interestingly enough such a possibility may have troubled Born and Oppenheimer [16]. In their discussion of equation (15) in Part I of the paper they write V_n as the sum of the electronic and nuclear repulsion energy for the nth electronic state and say [18]:

> Moreover we assume that V_n is a non-degenerate eigenvalue. As a matter of fact, this is never the case, since, because of the indistinguishability of the electrons the resonance degeneracy, discovered by Heisenberg and Dirac, enters; .... But since we are concerned here only with the systematics of the approximation procedure, we will not consider these degeneracies. Their consideration would require higher approximations in the secular equation.

The work of Dirac referred to in this quotation is from Section 3 of [12] but nowhere is the sheer extent of the likely degeneracy at all discussed.

Our primary concern here is not with the extent of the degeneracy, but rather the occurence of *isomers*. The Coulomb Hamiltonian $\mathsf{H}'(\mathbf{t})$ for C_8H_8 is the molecular Hamiltonian for cubane, cyclooctatetrene, vinylbenzene and many other compounds too. Indeed it is even the molecular Hamiltonian for a system with optical isomers, 3-vinyl hexa-1,4-diyne, which has the molecular formula

$$\begin{array}{ccccccccc} & & & & \mathrm{H} & & & & \\ & & & & | & & & & \\ \mathrm{CH_2} & = & \mathrm{CH} & - & \mathrm{C} & - & \mathrm{C} & \equiv & \mathrm{CH} \\ & & & & | & & & & \\ & & & & \mathrm{C} & \equiv & \mathrm{C} & - & \mathrm{CH_3} \end{array}$$

and in which the central carbon is clearly chiral by the conventional rules. For a chiral molecule the only symmetry operation is the identity (point group C_1).

This might not seem too serious a matter for it might be argued that the different isomers corresponded simply to different eigenstates of the same Hamiltonian, H'. But in classical structural chemistry, different isomers mean different geometries and it is the idea of a distinct geometry that is problematic in quantum mechanics. If we write the variables corresponding to the carbon nuclei in C_8H_8 as $\mathbf{x}_j^n$, $j = 1, \ldots 8$ and those corresponding to the protons as $\mathbf{x}_{i+8}^n$, $i = 1, \ldots 8$ then a particular CH interparticle distance is

$$x_{ij}^{CH} = |\mathbf{x}_{i+8}^n - \mathbf{x}_j^n|$$

One might be tempted to suppose that the calculation of the expected values of such interparticle distances with a particular eigenfunction of H' would determine the geometry. However we have seen in Section 5 that

x_{ij}^{CH} is not a proper observable. The only possible operator incorporating these distances is the symmetrical sum

$$\sum_{i,j=1}^{8} x_{ij}^{CH}$$

and all that can be inferred from its expectation value is that, on average, all the CH interparticle distances are the same. This is not to suppose that this average value is the same for all the eigenfunctions of H' that might be investigated in a search for isomers, it is simply that what differences there might be, cannot support the detailed geometrical interpretation which is characteristic of classical chemical structure theory.

However, one might hope to get somewhere with the idea of a geometrical structure by considering the development of the definition to deal with a system described in a frame fixed in the body. Since we are interested only in a localised region of configuration space in which we hope to identify the molecule, it is reasonable to start from the Hamiltonian (1.57). The idea here is to somehow fix a coordinate frame in the system and to define its orientation by means of 3 Eulerian angles, ϕ_m defined entirely in terms of the $A-1$ translationally invariant $\mathbf{t}_i^n$. The remaining motions of the system are then described in terms of $3A-6$ internal coordinates q_k which are so defined as to be invariant to any orthogonal transformations of the $\mathbf{t}_i^n$, together with the $\mathbf{z}_i$ which are the $\mathbf{t}_i^e$ expressed in the frame fixed in the body.

In such a coordinate system the Hamiltonian takes the general form

$$\mathsf{H} = \mathsf{H}_{rot} + \mathsf{H}_{vib} + \mathsf{H}_{rv}$$

where the particular forms of each term depend on how the choices are made for the Eulerian angles and the internal coordinates. But if we can make choices here such that the rotational part of this Hamiltonian looks something like

$$\mathsf{H}_{rot} = \frac{\mathsf{L}_x^2}{2I_{xx}^0} + \frac{\mathsf{L}_y^2}{2I_{yy}^0} + \frac{\mathsf{L}_z^2}{2I_{zz}^0}$$

where the L_α are the components of the total angular momentum operator and $I_{\alpha\alpha}^0$ are the nuclear principal moments of inertia then we are some way towards identifying a chemical molecule. One thinks this because in traditional molecular spectroscopy, the molecule is identified by its rotational characteristics, even if vibrational motion is also involved, and its rotational characteristics are explained in terms of moments of inertia of a geometrical figure composed of point masses.

Starting from the Coulomb Hamiltonian, to arrive at a Hamiltonian which had an angular part of this form, a neutral and natural choice would seem to be to choose the Eulerian angles to define an orthogonal matrix **C** that diagonalises the nuclear inertia tensor. This yields moments of inertia and puts the Hamiltonian in principal axis form, just the form appropriate to describe a rigid rotator in classical mechanics. The principal axis approach to the molecule was attempted in classical mechanics by Eckart [38] in 1934, shortly before the approach [15] that we have referred to in Section 2. It was tried too, almost simultaneously, by Hirschfelder and Wigner [39] using quantum mechanics. It has been used on many subsequent occasions, even in nuclear structure theory [40]. It leads to a rotational Hamiltonian of the form

$$\mathsf{H}_{rot} = \frac{I_{xx}\mathsf{L}_x^2}{2(I_{zz}-I_{yy})^2} + \frac{I_{yy}\mathsf{L}_y^2}{2(I_{xx}-I_{zz})^2} + \frac{I_{zz}\mathsf{L}_z^2}{2(I_{yy}-I_{xx})^2}$$

This is not at all like the rigid rotor operator given above. Here the operator is divergent whenever two nuclear moments of inertia are the same. It is thus quite impossible to describe a symmetric top molecule in this formulation. It seemed to pose such a severe problem that Eckart observed in the abstract of his paper [38] that:

> The ordinary moments of inertia appear in the Lagrangian kinetic energy but these are replaced by other functions of the radii of gyration in the Hamiltonian. This throws doubt upon all molecular configurations assigned on the basis of empirical values of moments of inertia.

Indeed it turns out more generally, inspite of some heroic efforts by van Vleck [41], that the Hamiltonian so derived is largely ineffective in describing molecules in terms of their traditional geometrical structures and so it found no use in the elucidation of molecular spectra. As we have seen, Eckart [15], got round this difficulty by defining a frame fixed in the body so that in that frame, a particular chosen geometrical shape, the reference geometry for the molecule, could be composed from constant vectors. This choice led to the rotational part of his Hamiltonian having exactly the classically expected form when the nuclei occupied their equilibrium positions. Furthermore Eckart was able to show that when this happened, the interaction part H_{rv} of his Hamiltonian, vanished. Of course, Eckart did not derive his form from the Coulomb Hamiltonian, he simply used what he knew about classical molecular structure to build an appropriate potential and it is unsurprising that the results of using that Hamiltonian confirm the assumptions used in its construction.

It should be noticed that in the principal axis form, no permutational symmetry has been disregarded and the Hamiltonian remains invariant under any permutation of identical particle coordinates. In the later

Eckart form, some permutational symmetry has been disregarded. It is often assumed that this disregard can be justified if the potential energy surface that arises from making the Born Oppenheimer approximation to solutions of the Coulomb Hamiltonian, has certain features. This view originates in some work [42] of Longuet-Higgins in the early 1960s but it has been much extended and developed since then. It is to an examination of this view that we turn in the next section.

6. The Born Oppenheimer approximation and the potential energy surface.

As was pointed out by us in 1976, [43], the general arguments made by Born and Oppenheimer in [16] are really quite formal. This is equally true of the later work by Born which is described in appendix VIII of his book with Huang [44]. To make the arguments mathematically precise and to estimate errors, it is necessary to find a way of dealing with the mathematical problems raised by the invariance of the Hamiltonian to uniform translations, orthogonal transformations and permutations of the variables, while preserving a plausible distinction between electronic and nuclear variables. A suitable translationally invariant Hamiltonian has been given in (1.57) but as we have seen, it is not an appropriate one for the description of the atomic dissociation asymptotes of the problem. This Hamiltonian is relevant for the purposes of describing electrons and nuclei separately, only in respect of eigenfunctions which are suitably localised, and, as we shall see, it is the Hamiltonian that has been used to provide a mathematical underpinning for the Born-Oppenheimer approach. So far as is known at present, it cannot be used to provide a similar underpinning for the later approach of Born [44] and we shall not consider this further here. It is recognised, of course, that it is this later Born approach which provides the basis for the standard descriptions of molecular dissociation and other reactions.

The full electronic Hamiltonian arising from (1.57) is obtained as

$$\mathsf{H}^{elec} = \mathsf{H}^{e}(\mathbf{t}^{e}) - \frac{e^2}{4\pi\epsilon_0}\sum_{i=1}^{A}\sum_{j=1}^{N}\frac{Z_i}{r'_{ij}(\mathbf{t}^{n},\mathbf{t}^{e})} + \frac{e^2}{8\pi\epsilon_0}\sum_{i,j=1}^{A}{}'\,\frac{Z_iZ_j}{r_{ij}(\mathbf{t}^{n})} \qquad (1.65)$$

and this is the Hamiltonian used by Klein *et al* [45] in their consideration of the precise formulation of the Born-Oppenheimer approximation for polyatomic systems[11]. This Hamiltonian is invariant under all orthogonal transformations of coordinates so that its eigenfunctions, if any, are

[11] A detailed consideration of the diatomic system can be found in [30, 46].

angular momentum eigenstates with definite parity. It is also invariant under the permutation of the variables of any set of identical particles.

In [45] it is assumed that (1.65) has a discrete eigenvalue which has a minimum as a function of the $\mathbf{t}_i^n$ in the neighborhood of some values $\mathbf{t}_i^n = \mathbf{a}_i$. Because of the rotation-inversion invariance, such a minimum exists on a three dimensional sub-manifold for all $\mathbf{a}_i$ such that

$$\mathbf{a}_i \to \mathbf{R}\mathbf{a}_i, \quad \mathbf{R} \in \mathrm{O}(3)$$

The $\mathbf{a}_i$ therefore define the geometrical shape of the minimum in the usual way. If the minimum figure is a plane then the potential well is diffeomorphic to SO(3) while if it is non-planar then it is diffeomorphic to O(3) and so the well is actually a symmetric double well. In either case, Klein *et al* show that the eigenvalues and eigenfunctions of the full problem can be obtained as WKB-type expansions to all orders of the expansion parameter, the square root of the ratio of the electronic to a typical nuclear mass. Because of the way the Hamiltonian is formulated the invariance of the Hamiltonian under permutations of the electronic variables is readily considered and the electronic wavefunction can easily be chosen in permutationally allowed form, no matter what the nuclear geometry happens to be. However it is rather less easy to consider permutational invariance when some of the nuclei are identical and, in fact, such permutations are not considered in [45].

In our view it would be very desirable to attempt to extend the arguments of Klein *et al* to include invariance under the permutation of identical nuclear variables in a way analogous to that in which rotational invariance is considered. It would have to be assumed that the required minimum exists on an appropriate sub-manifold for all $\mathbf{a}_i$ such that

$$\mathbf{a}_i \to \mathcal{P}\mathbf{a}_i, \quad \forall \mathcal{P} \in \mathcal{S}_A$$

This requirement would, presumably, result is a multi-well problem to be solved. Quantum mechanical problems of this form have been studied, at least in the semi-classical limit by Helffer and Sjöstrand [47] for example. The eigenfunctions of the full Hamiltonian with such a multi-well potential would, again presumably, be viewable as superpositions of various rotationally invariant forms associated with particular nuclear geometries. But, on average, there would be only one interparticle distance for like particles. It should be stressed however, that this is speculative and we have no firm mathematical results to support such a view. However that may be, such a multiple minimum view is not the one usually taken in spectroscopic calculations. Rather, as we have seen in our discussion of the Eckart Hamiltonian, a single specific assignment of molecular geometry is made in which the nuclei are identified. In the context of the present discussion this might be considered as a choice of geometry that defines just one of the permutationally equivalent minima and, as we have also seen, this is to neglect some of the permutational symmetry of the problem. The work by Longuet-Higgins [42] offered a justification of this in terms of the idea of a *feasible* permutation. If, within the chosen well on the potential energy surface, a permutation of identical nuclei could be described by a point group operation on the nuclear framework then such a permutation was *feasible*. If such a permutation could be described only by an energetically demanding dismantling and reassembling of the the molecular model, then such a permutation would not be a feasible one. What permutations are feasible depends upon the energy range being considered. Such non-point group operations as rotation about a single bond are often considered feasible operations in this context and there is nowadays a pretty complete theory of the symmetry feasible operations, usually called the theory of the the Nuclear Permutation Inversion Group. A critical account of this theory and an exposition of others in the same vein, can be found in the monograph by Ezra [48].

If our analysis of the problem is correct, however, these approaches cannot be considered fully satisfactory. The arguments given above imply that in order to construct the potential well in terms of which feasible operations can be defined, it is necessary largely to ignore the permutational symmetry that feasible operations are invoked to restore at least in part. There is also something of a logical difficulty in such approaches. The permutations of the variables of identical particles are simply mathematical operations in the quantum theory: they do not correspond to physical operations. However, once the idea of feasibility is associated with a permutation then some physical effect seems inevitably to be implied. So it is not clear if the idea of a feasible permutation is equivalent to the more abstract mathematical idea in the underlying theory.

Our position here can be summarised by saying that we believe that the work of Klein *et al* [45] and of others, for example Hagedorn [49], has put the Born-Oppenheimer approximation on a firm foundation but without considering fully the permutational invariances of the problem. We think that such invariances might well be included without seriously affecting the arguments used and that the basic intuition of Born and Oppenheimer, that wave functions of electron-nuclear product form are adequate approximations in the circumstances specified, will remain a correct one. If we are right however, there will be a major difference between the solutions to the traditional and our extended formulation of the Born-Openheimer problem. The solutions to the extended formulation will provide basis functions for irreps of $\mathcal{S}$, the full permutation group of the system. This is because all the permutational symmetry has been included in posing the extended problem. But with such solutions, as we have seen, the idea of molecular structure ceases to have its usual meaning and any definition of the term 'molecule' becomes problematic.

7. Discussion

In the preceding sections we have attempted to describe some of the basic properties of the Coulomb Hamiltonian for a collection of electrons and nuclei. Per-Olov Löwdin sought to found a mathematical definition of a molecule on such a discussion, but recognized that the theory was far from completion. We may mention several aspects of the quantum mechanics of molecules that he left as problems for the future (Section 3); firstly his suggestion of an 'energy surface' as an account of isomerism has an *ad hoc* character, for the nature of the parameters $\boldsymbol{\alpha}$ was left entirely obscure from the point of view of the fundamental theory. Then, what is the mathematical nature of the constraints required to make the $\frac{1}{2}A(A-1)$ interatomic distances in a system containing A nuclei consistent with a definite molecular structure ? Finally we recall that we noted earlier that, from the perspective of the theory of the Coulomb Hamiltonian for a system of electrons and nuclei, Löwdin found the occurrence of isomerism and molecular symmetry a mystery whose unravelling would require new ideas. Recently, Simon has presented a list of open problems in mathematical physics [11], among which Problem 12 is of relevance here:

> **Problem 12**: Is there a mathematical sense in which one can justify from first principles current techniques for determining molecular configurations ?

This problem, although stated in mathematically vague terms, should be viewed as asking for some precise way to go from fundamental quantum theory to configurations of molecules; evidently Simon does not see *ab initio* electronic structure theory as a complete answer.

The characteristic feature of any molecular structure is the identification of particular atoms with specific locations in the molecule. Typically this entails that identical nuclei are treated as distinguishable entities ('methyl hydrogens', 'hydroxyl hydrogens' etc.). How then can identical particles be regarded as distinguishable in a quantum theory ? It is clear that no continuous approximation can lead to a change in the symmetry group of any given Hamiltonian, since symmetries are either present or absent. It is also clear that a *linear* Schrödinger equation for a *finite* system cannot generate states that exhibit spontaneously broken symmetries. Symmetry breaking is a familiar occurrence in quantum field theories [50] and we leave open the possibilities that either the system of electrons and nuclei can be considered to be macroscopic (N the number of particles $\to \infty$) with molecules appearing as *quasi-particles*, or, if finite, is coupled to a quantized field that represents its "environment" as in discussions of decoherence; however we do not wish to pursue this line here beyond noting that the properties of the Coulomb Hamiltonian are likely to be important in such discussions [56, 57, 58].

Let us, with Löwdin [2], confine attention to the Coulomb Hamiltonian for a molecule, equation (1.1), on the assumption that somehow all the information about molecular symmetry is contained within it. Several observations seem pertinent. We know that the treatment of identical particles is a matter of fundamental difference between classical and quantum mechanics. In classical mechanics we describe motion by specifying the orbits of the individual particles under the given classical dynamical law. Even though the particles may be identical, if the initial conditions have been set, it makes perfect sense to say that at a given instant in time, t, particle 1 is moving along orbit 1, particle 2 is moving along orbit 2 and so on. Any permutation of the identical particles will also be a solution of the dynamical problem, since the classical equations of motion are invariant under such permutations, but this solution belongs to *another set of initial conditions* i.e to a distinct physical situation [31]. So a classical description of a molecular structure involving identical nuclei in terms of an association between a specified particle and a specified potential well in which it is localized is perfectly natural.

The characteristic features of the quantum mechanical treatment of permutation symmetry can be seen in the simple case of two identical particles. Given a set of single-particle states $\{\phi_k(\mathbf{x})\}$, which we may take to be orthonormal, a general state of a two-particle system can be constructed as the superposition

$$\Psi(1,2) = \sum_{k,k'} c_{k,k'}\ \phi_k(\mathbf{x}_1)\ \phi_{k'}(\mathbf{x}_2) \tag{1.66}$$

according to the usual rules of quantum mechanics the coefficients $\{c_{k,k'}\}$ have the interpretation that $|c_{k,k'}|^2$ is the probability that particle 1 is in state k, while particle 2 is in state k$'$. Let P_{12} be the linear operator that permutes the variables of particles 1 and 2, such that acting on Ψ(1,2) it gives the state

$$\Phi(1,2) \equiv P_{12}\ \Psi(1,2) = \sum_{k,k'} c_{k,k'}\ \phi_k(\mathbf{x}_2)\ \phi_{k'}(\mathbf{x}_1) \tag{1.67}$$

Now the probability that particle 1 is in state k while particle 2 is in state k′is $|c_{k',k}|^2$ which in general is different from $|c_{k,k'}|^2$, so that the state $\Phi(1,2)$ resulting from the action of P_{12} is a *different* vector in the Hilbert space.

However if the two particles are *identical*, $\Phi(1,2)$ and $\Psi(1,2)$ should be states representing the *same* physical situation, and hence can differ at most by a sign; equivalently the probabilities $|c_{k,k'}|^2$ and $|c_{k',k}|^2$ must be equal and so [31]

$$c_{k,k'} \;=\; \pm\; c_{k',k}$$

The + sign here means that the state Ψ is *symmetric* in the particle variables

$$\Psi(1,2) \;=\; \Psi(2,1)$$

while the − sign is associated with an *antisymmetric* state

$$\Psi(1,2) \;=\; -\;\Psi(2,1)$$

in turn, these are eigenstates of the permutation operator P_{12} with eigenvalues +1 and −1 respectively. As shown by Dirac [12] the permutation P_{12} can be considered to be a dynamical variable in quantum mechanics; P_{12}, considered as a dynamical variable, has no classical limit, and there is no analogous quantity in classical mechanics.

If the particles are identical all physical observables Ω_i must be symmetric functions of the variables, and so will commute with the permutation operator

$$[P_{12},\; \Omega_i] \;=\; 0$$

one of the Ω_i will be the Hamiltonian H of course, and thus P_{12} is a constant of the motion

$$\dot{P}_{12} \;=\; \frac{i}{\hbar}[\mathsf{H},\; P_{12}] \;=\; 0$$

The eigenstates of H will be simultaneous eigenstates of P_{12}. Thus the dynamical evolution of the system cannot cause transitions from one subspace to the other i.e. they are incoherent spaces and are never mixed in superpositions. The restriction of the possible states of the system to lie in either the symmetric subspace or the antisymmetric subspace is therefore consistent with the equations of motion. It is also reasonable to suppose that if the two particles are perturbed by some interaction V it cannot disturb the permutation symmetry of the system,

$$[P_{12}, \mathsf{V}] = 0$$

since, following Weyl's whimsical example referred to in Section 5, no perturbation can distinguish between "Mike" and "Ike" [35]. The textbook discussion of the states of the Helium atom is a paradigmatic example of these ideas, but they are of course quite generally applicable to any identical particles (in any numbers) in a quantum mechanical system. The exchange degeneracy characteristic of identical particles in quantum mechanics is associated with the occurrence of superpositions such as (1.66). Thus it seems essential to find reasons why such superpositions can be suppressed, so that the state reduces to something like a Hartree product

$$\Psi(1,2) = \psi(1,2) \approx \alpha_k(\mathbf{x}_1)\, \alpha_{k'}(\mathbf{x}_2)$$

A clue to the nature of the functions $\alpha_k(\mathbf{x})$ may perhaps be gleaned from a remark attributed to Pauli: "No description of the motion of an electron in an atom in space and time is given......this is clear from the fact that, outside the domain of validity of geometrical optics it is impossible to construct 'rays' that can be considered as orbits of particles." [51] In the context of a quantum mechanical theory, geometrical optics corresponds to the semiclassical limit ($\hbar \to 0$) i.e. the WKB approximation. Of course $\hbar$ has a definite value depending on the system of units used, and has the dimensions of action (or angular momentum), and $\hbar = 0$ *cannot* be realized in the physical world. What we mean by the semiclassical limit is the situation where, if S a typical classical action in the system under consideration, there is a parameter $\delta = (\hbar/S)$ that is negligibly small.

The quantum mechanical Hamiltonian for a N-particle system in a coordinate representation is

$$\mathsf{H} = -\hbar^2 \sum_i \frac{\nabla_i^2}{2m_i} + \mathsf{V}(\mathbf{x})$$

with Schrödinger equation

$$\mathsf{H}\, \psi(\mathbf{x}) = E\, \psi(\mathbf{x})$$

The limits $\hbar \to 0$ or $m_i \to \infty$ for some i are closely related; they lead to *singular* perturbation problems since in the limit the coefficient of the *highest-order* differential operator in the Schrödinger equation is put to zero. Thus in quantum mechanics, $\hbar = 0$ is a point of singularity; if it were not, the semiclassical limit would be trivial for we could solve the quantum mechanical problem with $\hbar$ as a parameter and then set $\hbar = 0$ at the end. The discrete spectrum of the Hamiltonian is associated with infinitely long-lived states, and the semi-classical limit is further complicated by the conjunction of the two limits $t \to \infty$, $\hbar \to 0$ which do *not* commute; on the other hand, it is precisely this fact that leads to a remarkable richness of physically relevant phenomena [52, 53]. This suggests that a possibly crucial contribution to Löwdin's programme is the detailed exploration of the asymptotic behaviour of the Schrödinger equation for the Coulomb Hamiltonian, with the consequences of the occurrence of identical nuclei being fully accounted for.

Suppose that the quantum mechanical Hamiltonian is written

$$\mathsf{H} = \mathsf{H}(\mathbf{x}, \mathbf{p})$$

and consider its transform with an unitary operator U that is purely a function of the coordinates,

$$\mathsf{U} = e^{i\,S(\mathbf{x})/\hbar}$$

so that

$$\mathsf{H}(\mathbf{x}', \mathbf{p}') = \mathsf{U}\,\mathsf{H}\,\mathsf{U}^{-1} = H\,(\mathbf{x}, \mathbf{p} + \boldsymbol{\nabla} S) \tag{1.68}$$

The traditional derivation of the semiclassical limit [54] is that since $\mathbf{p} \to -i\hbar\nabla$, the momentum drops out of the Hamiltonian in the limit $\hbar \to 0$ and the Schrödinger equation reduces to the classical Hamilton-Jacobi (HJ) equation

$$H(\mathbf{x}, \boldsymbol{\nabla} S) = E$$

The solutions of the HJ equation determine the dynamics of the classical system described by H; the WKB approximation says that for a partially (or completely) integrable system there is a *regular spectrum* of energy levels that can be obtained by imposing the Einstein-Brillouin-Keller or Maslov conditions on the action integrals over closed curves C_k on invariant N-tori, [55]

$$\oint_{C_k} \mathbf{p}.d\mathbf{x} = (n_k + \gamma)\,h, \qquad k = 1, ...N \tag{1.69}$$

A limitation of this formulation, for our purposes, is that it is based on picking out particular classical solutions according to the EBK quantization condition (1.69), that is, it approximates the quantum solution from the *classical side* of the singular limit $\hbar \to 0$, for which there is no permutation operator as a dynamical variable. However the semiclassical limit *can* be studied in a Hilbert space setting; for example the *microscope transformation* used by Combes and Seiler [30] to give a rigorous version of the Born-Oppenheimer theory of a diatomic molecule is essentially a semiclassical theory. It is applicable if there is a minimum in the potential $V_{min} = V(\mathbf{x}_0)$ associated with a particular configuration of the particles[12] that is deep enough for the lowest energy eigenstates to be localized about $\mathbf{x}_0$. One can look at these states with a 'microscope' with a certain resolving power that depends on Planck's constant - loosely speaking, this means that it is possible to study the limit $\hbar \to 0$ from the *quantum side* of the problem, and it is possible that the consequences for permutation symmetry can be studied in this setting. As yet this has not been done, as far as we are aware.

The microscope transformation is based on a unitary transformation of the form

$$\mathsf{U}_\lambda = e^{i\, S_\lambda(\mathbf{x},\mathbf{p})/\hbar}, \quad S_\lambda^* = S_\lambda$$

where λ is a parameter to be determined. The usual rules yield

$$\Psi_\lambda = \mathsf{U}_\lambda^{-1}\,\psi, \quad \mathsf{H}_\lambda = \mathsf{U}_\lambda^{-1}\,\mathsf{H}\,\mathsf{U}_\lambda$$

$$\mathbf{x}_\lambda = \mathbf{x} - \frac{\partial S_\lambda}{\partial \mathbf{p}}, \quad \mathbf{p}_\lambda = \mathbf{p} + \frac{\partial S_\lambda}{\partial \mathbf{x}}$$

The Hermitian operator S_λ is chosen so that the transformation produces a translation to make $\mathbf{x}_0$ the origin of the coordinates, and a dilation (scale transformation)

$$\frac{\partial S_\lambda}{\partial \mathbf{p}} = (1 - \lambda)\,(\mathbf{x} - \mathbf{x}_0), \quad \frac{\partial S_\lambda}{\partial \mathbf{x}} = \frac{(1 - \lambda)}{\lambda}\,\mathbf{p} \tag{1.70}$$

These relations show that S_λ is a *non-integrable* function of the phase-space variables $(\mathbf{x}, \mathbf{p})$, that is, it doesn't have a definite value at $(\mathbf{x}, \mathbf{p})$, but it does have well-defined derivatives given by equation (1.70) *except* at $\lambda = 0$; at this point the transformation equations have a singularity.

Under this transformation a Hamiltonian of the form

$$\mathsf{H} = \sum_i \frac{\mathsf{p}_i^2}{2m_i} + \mathsf{V}(\mathbf{x})$$

is transformed to

[12] The multiminima case can also be treated in this way.

$$\mathsf{H}_\lambda = \mathsf{V}(\mathbf{x}_0) + \lambda^2 \mathsf{K}(\lambda)$$

where

$$\mathsf{K}(\lambda) = -\frac{\hbar^2}{\lambda^4}\sum_i \frac{\nabla_i^2}{2m_i} + \frac{1}{\lambda^2}\left(\mathsf{V}(\mathbf{x}_0 + \lambda(\mathbf{x} - \mathbf{x}_0)) - \mathsf{V}(\mathbf{x}_0)\right)$$

We now put $\lambda = \sqrt{\hbar}$ so as to eliminate λ from the kinetic energy term in $\mathsf{K}(\lambda)$; with this choice for λ, unitary equivalence of the spectrum implies that the eigenvalues of the original Hamiltonian H are related to those of $\mathsf{K}(\lambda)$ by

$$E_n = \mathsf{V}(\mathbf{x}_0) + \hbar\, \mu_n(\lambda)$$

Provided V is analytic in λ it can be expanded about $\lambda = 0$, and this puts $\mathsf{K}(\lambda)$, in lowest order, into the form of a sum of coupled oscillators so that the first approximation for the eigenvalue function μ_n is

$$\mu_n \approx \sum_k \omega_k \left(n_k + \frac{1}{2}\right)$$

In the Born-Oppenheimer calculation for the diatomic molecule [30], analyticity of V in λ could be proven, and the role of $\sqrt{\hbar}$ was taken by the usual BO expansion parameter $\kappa = (m_e/M_{Nu})^{\frac{1}{4}}$.

The singular nature of the transformation is demonstrated by the modification of the spectrum associated with the limit $\kappa \to 0$. The spectrum of the Coulomb Hamiltonian for a molecule was discussed in Section 4; for the diatomic molecule, $\sigma_{ess}(\mathsf{H})$ starts at the lowest two-body threshold $\Sigma = \lambda_A(m_A) + \lambda_B(m_B)$ given by the minimal value of the sums of pairs of binding energies for atoms A and B with finite masses m_A and m_B respectively. On the other hand the spectrum of the electronic Hamiltonian, H_∞, is purely continuous, $\sigma(\mathsf{H}_\infty) = [\mathsf{V}_{min}, \infty)$. In the limit $m_A, m_B \to \infty$, Σ does not generally converge to V_{min}; instead the missing part of the continuous spectrum $[\mathsf{V}_{min}, \lambda_A(\infty) + \lambda_B(\infty)]$ is provided by an accumulation of bound states in this interval [59].

While the microscope transformation is formally applicable to the polyatomic case, it may not be sufficient to control the asymptotic behaviour. The work of Klein et al. was based on a powerful symbolic operator method, the *pseudodifferential calculus* [45, 60]. Interestingly, these investigations utilize a formalism closely associated with Löwdin, namely the partitioning technique. The complement of the adiabatic subspace can be used to fold the non-adiabatic terms into a modified

Hamiltonian acting on the adiabatic subspace, as an energy dependent operator. This turns out to be a useful procedure because bounds can be established for its contribution. Whether these techniques are sufficient to investigate the consequences of treating identical nuclei explicitly remains to be seen however. In our view the intrinsically quantal nature of indistinguishability of particles is the major hurdle that must be cleared if Löwdin's programme is to be developed further.

8. Dedication

This paper is dedicated to the memory of Per-Olov Löwdin.

References

[1] P-O. Löwdin in *Molecules in Physics, Chemistry and Biology II*, Ed. J. Maruani, Kluwer, Dordrecht, 1988, p 3.

[2] P-O. Löwdin, 1989, Pure and Applied Chemistry, **61**, 2065.

[3] P-O. Löwdin, Studies in Physical and Theoretical Chemistry, **62**, Elsevier, Amsterdam, 1989, p.605.

[4] P-O. Löwdin, Int. J. Quant. Chem., **38**, 741 (1990).

[5] P-O. Löwdin, J. Mol. Struct. (Theochem), **230**, 13 (1991).

[6] P-O. Löwdin, Int J. Quant Chem., **51**, 473 (1994).

[7] P-O. Löwdin, List of published papers, provided by Prof. E. Brandas, Institute for Quantum Chemistry, Uppsala University, Sweden.

[8] B. H. Bransden and C. J. Joachain, *Physics of Atoms and Molecules*, Longmans, London 1983.

[9] A. Messiah, *Quantum Mechanics*, Vol. II North-Holland, Amsterdam, 1960.

[10] B. Simon, J. Math. Phys. **41**, 3523 (2000).

[11] B. Simon, in *Mathematical Physics 2000*, Eds. A. Fokas, A. Grigoryan, T. Kibble and B. Zegarlinski, Imperial College Press, London, 2000, p.283.

[12] P. A. M. Dirac, Proc. Roy. Soc. **A 123**, 714 (1929).

[13] A. Simões, Physics in Perspective, in press, Sept 2002.

[14] C. A. Coulson, J. Chem. Soc, 2069 (1955).

[15] C. Eckart, Phys. Rev., **47**, 552 (1935).

[16] M. Born and J. R. Oppenheimer, Ann. der Phys., **84**, 457 (1927).

[17] J. C. Louck, J. Mol. Spec.,**61**, 107 (1976).

[18] S. M. Blinder at www.ulb.ac.be/cpm/people/scientists/bsutclif/welcom

[19] R. S. Berry, Rev. Mod. Phys., **32**, 447 (1960).

[20] M. Reed and B. Simon, *Methods of Modern Mathematical Physics, IV, Analysis of Operators*, Academic Press, New York, 1978.

[21] M. Reed and B. Simon, *Methods of Modern Mathematical Physics, I, Functional Analysis*, revised and enlarged edition, Academic Press, New York, 1980.

[22] M. Reed and B. Simon, *Methods of Modern Mathematical Physics, II, Fourier Analysis, Self-Adjointness*, Academic Press, New York, 1975.

[23] W. Thirring, *A Course in Mathematical Physics, 3, Quantum Mechanics of Atoms and Molecules*, Tr. E. M. Harrell, Springer-Verlag, New York, 1981.

[24] W. Thirring in *Schrödinger, centenary celebrations of a polymath*, Ed. C. W. Kilminster CUP, Cambridge, 1987, p 65.

[25] T. Kato, Trans. Am. Math. Soc., **70**, 212 (1951).

[26] R. Courant and D. Hilbert, *Methods of Mathematical Physics*, Vol. 1, Interscience Publishers, Inc. New York, 1953.

[27] H. Weyl, Rend. Circ. Mat. Palermo,**27**, 373 (1909).

[28] V. F. Brattsev, Soviet Physics - Doklady, 10, 44 (1965).

[29] S. Epstein, J. Chem. Phys., **44**, 836 and 4062 (1966).

[30] J.-M. Combes and R. Seiler, in *Quantum Dynamics of Molecules*, Ed. R. G. Woolley, NATO ASI B57, Plenum Press, New York, 1980, p.435.

[31] P. Roman, *Advanced Quantum Theory*, Addison-Wesley Publishing Co., Reading, Mass., 1965.

[32] E. Balslev and J.-M. Combes, Commun. Math. Phys. **22**, 280 (1973).

[33] J. Aguilar and J.-M. Combes, Commun. Math.Phys. **22**, 269 (1973).

[34] W. P. Reinhardt, Ann. Rev. Phys. Chem. **33**, 223 (1982).

[35] H. Weyl, *The Theory of Groups and Quantum Mechanics*, 2nd Edition, Tr. H. P. Robertson, Dover, New York, 1931.

[36] G. Mackey, *The Mathematical Foundations of Quantum Mechanics*, Benjamin, New York, 1963.

[37] E. Balslev, Ann. Phys.(N.Y.), **73**, 49 (1972).

[38] C. Eckart, Phys. Rev., **46**, 384 (1934).

[39] J. O. Hirschfelder and E. Wigner, Proc. Nat. Acad. Sci., **21**, 113 (1935).

[40] B. Buck, L. C. Biedenharn and R. Y. Cusson, Nucl. Phys, **A 317**, 215 (1979).

[41] J. H. van Vleck, Phys. Rev., **47**, 487 (1935).

[42] H. C. Longuet-Higgins, Mol. Phys., **6**, 445 (1963).

[43] R. G. Woolley and B. T. Sutcliffe, Chem. Phys. Lett., **45**, 393 (1976).

[44] M. Born and K. Huang, *Dynamical Theory of Crystal Lattices* , Oxford University Press, Oxford, 1955.

[45] M. Klein, A. Martinez, R. Seiler, and X. P. Wang, Commun. Math. Phys., **143**, 607 (1992).

[46] J.-M. Combes, P. Duclos and R. Seiler in *Rigorous atomic and molecular physics*, Eds. G. Velo and A. Wightman, Plenum, New York, 1981, p. 185.

[47] B. Helffer and J. Sjöstrand, Commun. Partial Differ. Equations **9**, 337 (1984).

[48] G. Ezra, *Symmetry properties of molecules*, Lecture Notes in Chemistry **28**, Springer, Berlin, 1982.

[49] G. Hagedorn, Commun. Math. Phys., **77**, 1 (1980).

[50] P. W. Anderson, *Basic Notions in Condensed Matter Physics*, Benjamin/Cummings/Addison-Wesley, New York (1984).

[51] W. Pauli, quoted by B. L. van der Waerden, in *The Physicist's Conception of Nature*, Ed. J. Mehra, Reidel, Dordrecht, 1973, p.282.

[52] M. V. Berry, in *Les Houches Lecture series 52*, Eds. M. J. Giannoni, A. Voros and J. Zinn-Justin, North-Holland, Amsterdam, 1991.

[53] M. V. Berry, in *Quantum Mechanics: Scientific perspectives on divine action* Eds: R. J. Russell, P. Clayton, K. Wegter-McNelly and J. Polkinghorne, Vatican Observatory CTNS publications, Rome, 2001, p.41.

[54] P. A. M. Dirac, *The Principles of Quantum Mechanics*, 4th Edition revised, Oxford University Press, Oxford, 1967.

[55] I. C. Percival, J. Phys. B Atom. Molec. Phys. **6**, L229, (1973).

[56] R. G. Woolley, Israel. J. Chem. **19**, 30 (1980).

[57] R. G. Woolley, Structure and Bonding, **52**, 1 (1982).

[58] R. G. Woolley, J. Mol. Struct. (Theochem), **230**, 17 (1991).

[59] J.-M. Combes, Acta Phys. Austr. **17**, Suppl., 139 (1977)

[60] C. L. Feffermann, Bull. Am. Math. Soc., **9**, 129 (1983)

SYMMETRY BREAKING IN THE INDEPENDENT PARTICLE MODEL

J. L. Stuber and J. Paldus*
Department of Applied Mathematics, University of Waterloo
Waterloo, Ontario, Canada N2L 3G1

Eulogy

Felix qui potuit rerum cognoscere causas.

It is with great respect and admiration that we dedicate this paper to the memory of Professor Per-Olov Löwdin. His pioneering contributions to many aspects of quantum sciences, and his selfless dedication to their promotion and dissemination, have left an indelible mark on our field of endeavour and brought a lasting benefit and solid base for its future development. He introduced numerous basic concepts which we constantly rely on, as is also testified[1–4] – in a small way – by this paper, built a remarkable following by establishing both the Uppsala and Gainesville Quantum Theory Groups, tirelessly organized innumerable Schools and Symposia, and founded and edited an influential journal and a monograph series which will continue his legacy to future generations of quantum chemists. With his untimely departure, we have not only lost a great scientist, an exemplary educator and organizer, but above all, a most wonderful and kind human being. For all of us who had the privilege to interact with Per-Olov, his charisma, friendliness, generosity, encouragement, and congenial nature, as well as his engaging smile and charming wit, will never be forgotten.

Ars longa, vita brevis.
Atque in perpetuum, frater, ave atque vale.

*Also at Department of Chemistry, and Guelph-Waterloo Center for Graduate Work in Chemistry, University of Waterloo, Waterloo, Ontario, Canada N2L 3G1.

E.J. Brändas and E.S. Kryachko (eds.),
Fundamental World of Quatum Chemistry, Vol. I, 67–139.

1. Introduction

Looking at the literature, one finds that, in atomic and molecular physics, comparatively little interest has so far been devoted to the GHF-scheme, and the purpose of this paper is to try to focus the interests of some quantum chemists to this rather interesting problem of looking for the "absolute minimum" of the original Hartree-Fock method.

(P.-O. Löwdin and I. Mayer, Ref. 5, p. 93.)

In view of the complexity of quantum mechanical equations of motion describing atomic, molecular, and solid state phenomena – even when modelling very small systems at the nonrelativistic level within the Born-Oppenheimer approximation – one is obliged to introduce rather drastic approximations in order to proceed. The standard approach in the investigation of the molecular electronic structure is then to rely on the so-called *ab initio* electronic model Hamiltonian $\hat{\mathrm{H}}$, which is defined on a suitable finite-dimensional subspace $\mathcal{W}_N$ of the infinite-dimensional N-electron component of the Fock space $\mathcal{F}$, and to look for the solutions of the time-independent Schrödinger equation associated with this model Hamiltonian. The space $\mathcal{W}_N$ then represents a vector space of N-linear alternating forms, built as the antisymmetrized tensor product of the one-electron space $\mathcal{W}_1$, which in turn is spanned by a chosen set of nonorthogonal atomic spin orbitals (ASOs) $|\chi_\Lambda\rangle \equiv |\Lambda\rangle_{\Lambda=1,\cdots,M}$, defining a given *ab initio* model.

Even this drastically simplified problem is not easy to solve exactly, except for sufficiently small values of N and M, in which case a brute force diagonalization of $\hat{\mathrm{H}}$ within $\mathcal{W}_N$ is feasible. The key difficulty arises due to the presence of the two-body terms in the Hamiltonian, which represent the long-range Coulomb interaction between electrons and thus prevent the separation of the problem into the lower dimensional ones. Thus, an obvious first step is to approximate the two-body component of $\hat{\mathrm{H}}$ by a separable, one-electron potential, which then enables the reduction of the N-body problem into the effective one-body problems. This in turn implies that the pertinent wave function must belong to the manifold of separable wave functions, whose members consist of single antisymmetrized products (or decomposable N-forms) of one-electron functions or molecular spin orbitals (MSOs), characterizing the so-called *independent particle model* (IPM). In view of the spin independence of $\hat{\mathrm{H}}$, the MSOs $|\psi_i\rangle \equiv |I\rangle$ are almost invariably assumed to be given by the product of a molecular orbital (MO) $|\phi_i\rangle \equiv |i\rangle$ and pure-spin eigenstate $|\sigma\rangle$ of $\hat{\mathrm{S}}_z$, $\sigma = \pm\frac{1}{2}$ (or $\sigma = \alpha, \beta$). We refer to these as canonical MSOs, in contrast to the general ones that involve an arbitrary linear combination of spin-up ($|\frac{1}{2}\rangle \equiv |\alpha\rangle$) and spin-down ($| -\frac{1}{2}\rangle \equiv |\beta\rangle$)

states. The (canonical) atomic spin orbitals (ASOs) are similarly given as $|\Lambda\rangle = |\lambda\rangle|\sigma\rangle$, $\sigma = \alpha, \beta$. In general, we shall label the ASOs and AOs by Greek upper and lower case symbols, respectively (reserving the symbols σ,α, and β for the spin functions), while the MSOs and MOs will be correspondingly labelled by the upper and lower case letters of the Latin alphabet.

Although one can employ different criteria to define the 'best' IPM (say, by requiring the maximum overlap with the exact wave function, leading to the so-called Brueckner orbitals), the most useful and practical approach is to rely on the variation principle and to search for the IPM wave function $|\Phi_0\rangle$ which yields the best possible upper bound to the exact energy. This leads to the Hartree-Fock (HF) equations and HF MSOs or MOs[6–10] (for an overview of various aspects of HF theory, see Ref. 11).

Without restricting the generality, one can assume the MSOs (or, correspondingly, MOs) form an orthonormal basis of $\mathcal{W}_1$, $\mathcal{W}_1 = \mathrm{Span}(\{|\Lambda\rangle\})$, $\dim \mathcal{W}_1 = M$. Thus, expressing the MSOs $|I\rangle$ as linear combinations of the ASOs $|\Lambda\rangle$ (the LCAO approximation)

$$|I\rangle = \sum_{\Lambda=1}^{M} C_{\Lambda I}|\Lambda\rangle\,, \tag{1}$$

(or, similarly, the MOs $|i\rangle$ as linear combinations of the AOs $|\lambda\rangle$), it is the coefficients $C_{\Lambda I}$ (or $c_{\lambda i}$) that play the role of variational parameters (subject to the orthonormality requirement $\langle I|J\rangle = \delta_{IJ}$). Then, designating a single antisymmetrized product wave function by $|\Phi\rangle$, the variational parameters $C_{\Lambda I}$ are determined so as to minimize the energy expectation value $\mathcal{E}[\Phi]$,

$$\mathcal{E}[\Phi] = \langle\Phi|\hat{\mathrm{H}}|\Phi\rangle/\langle\Phi|\Phi\rangle\,, \tag{2}$$

where $|\Phi\rangle = |\{I_1 I_2 \cdots I_N\}\rangle$ represents a N-particle IPM wave function built from a subset of MSOs $|I_\mu\rangle$ (usually chosen on the basis of the so-called Aufbau Principle). Here we use braces to indicate the antisymmetric character of $|\Phi\rangle$ (for details of the notation employed, see e.g. Refs. 12–15).

The necessary condition determining the HF M(S)Os is given by the requirement that the first variation of the energy functional (2) vanishes or, alternatively, that

$$\begin{aligned}\delta^{(1)}\mathcal{E}[\Phi_0] &= \delta^{(1)}\langle\Phi|\hat{\mathrm{H}}|\Phi\rangle|_{\Phi=\Phi_0} \\ &= 0\,,\end{aligned} \tag{3}$$

subject to the normalization constraint

$$\langle \Phi_0 | \Phi_0 \rangle = 1 \,, \tag{4}$$

with $|\Phi_0\rangle$ designating the HF wave function. This necessary condition is usually stated in the form of HF equations, which are one-electron pseudo-eigenvalue equations for the MSOs $|I\rangle$ (or MOs $|i\rangle$), or equivalently in the form given by Brillouin's theorem. In view of the variational condition (3), the HF wave function $|\Phi_0\rangle$ then represent a stationary point on the energy hypersurface, defined by Eq. (2), over a single antisymmetrized (or single determinantal) manifold represented by the coefficients $C_{\Lambda I}$, subject to the orthonormality requirement

$$\langle I | J \rangle = \delta_{IJ} \,. \tag{5}$$

Clearly, this variational manifold will generally exhibit a number of stationary points of various kinds. Since the HF equations represent a pseudo-eigenvalue problem,[7,9,10]

$$\hat{\mathrm{F}} | I \rangle = \epsilon_I | I \rangle \,, \tag{6}$$

where the Fock operator $\hat{\mathrm{F}} \equiv \hat{\mathrm{F}}(\{|J\rangle\})$ depends in a highly nonlinear fashion on all the MSOs defining $|\Phi\rangle$ (or $|\Phi_0\rangle$), they generally possess a number of distinct solutions, which are associated with different stationary points of the variational-energy hypersurface just mentioned.

The usual way of solving the HF problem is based on the iterative *self-consistent field* (SCF) procedure, starting with a suitable initial guess for the MSOs $|I\rangle$ (or MOs $|i\rangle$). Although we are interested in the solution corresponding to the lowest possible energy, there is no guarantee that the SCF procedure will take us to this desired end. In fact, even when relying on the so-called *direct energy minimization* (DEM) algorithms (cf., e.g. Ref. 16), the resulting solution will generally depend on the initial approximation. Indeed, only when we start from a point lying in the same 'catchment region' of the desired minimum of the energy hypersurface shall we reach the solution associated with this local minimum via the iterative DEM process.

In most actual applications, when one studies the electronic structure for closed-shell (CS) ground states of molecules in or near their equilibrium geometry, there are typically no difficulties in reaching the desired solution, which in fact corresponds to the absolute minimum on the energy hypersurface. However, in many situations when we are away from the equilibrium geometry or when we explore quasidegenerate systems, this is not the case. In fact, in many such instances the usual SCF procedure either does not converge at all, converges extremely slowly, or can even oscillate between two 'stationary' solutions.

An even more challenging situation arises when handling the open-shell (OS) systems, in which case the HF potential energy surface may possess a highly singular behavior and several candidates for 'ground state' solutions may coexist in the same region of geometries.[17–19] Furthermore, when breaking the chemical bonds and dissociating a system (even a CS one) into the OS fragments, the HF solution invariably tends to a wrong dissociation channel (the so-called *size-inconsistency problem*). A standard procedure in this case is to relax certain symmetry requirements on the wave function that are implied by the symmetry group of the Hamiltonian and to employ a suitable broken symmetry solution in the dissociated limit [usually one foregoes the total spin conservation by giving up the double occupancy and employs the *unrestricted HF* (UHF) solution, as will be discussed in greater detail below]. Yet, in the OS cases, even this procedure may lead to wrong dissociation products.[19]

In all such cases it is useful to obtain more information about a given solution that is implied by the vanishing (3) of the first variation of the energy functional by examining the character of the pertinent stationary point. An obvious first step in this direction is to examine the second variation $\delta^{(2)}\mathcal{E}$ of the energy functional at that stationary point, as was first considered by Thouless.[20] The positive definiteness of $\delta^{(2)}\mathcal{E}$ guarantees that the stationary point represents a *local* minimum, rather than some kind of a saddle point or even a local maximum. Such a solution is referred to as *stable*. This local stability does not prevent the existence of other solution(s) with lower energy, although such solutions, if they do exist, will be separated from a given one by a potential barrier. However, when the solution is not stable (i.e., $\delta^{(2)}\mathcal{E}$ is not positive semi-definite), there certainly exists another solution with lower energy that is not separated from a given one by a potential barrier. Such an *unstable solution* will generally violate certain symmetries of the Hamiltonian and we refer to it as a *broken symmetry* (bs) solution.

Historically, ever since the early days of HF theory[21] , the first explorations of the stability problem were carried out heuristically, without examining the second variation (or the Hessian) of the energy functional. Let us mention here, for example, the work of Sawada and Fukuda[22] who examined this problem in extended systems, namely the symmetry adapted plane wave solution for the three-dimensional Fermi gas interacting via a delta-function-type potential and the BCS-Bogoliubov Hamiltonian. In the molecular context, we mention several studies[23–29] that explored the existence of bs IPM solutions without examining the second variation. In fact, long before a systematic formulation of stability conditions and their exploration, various broken symmetry solutions have been employed (particularly those breaking the total spin symmetry) in order to achieve the size consistency and account to for a significant part of the correlation energy, particularly for systems far away from the equilibrium geometry.

As already mentioned, the formulation of the stability problem based on the second variation of the energy functional or Hessian was initiated by Thouless[20] in the context of nuclear physics. In this formulation it is very important to carefully specify the variational manifold considered. Specifically, when the studied Hamiltonian is invariant with respect to certain symmetry group operations, we can formulate corresponding *symmetry restricted* stability conditions, depending on the variational space or manifold at hand. This is closely related with the so-called *Löwdin's symmetry dilemma*,[3,4] alluded to at the outset.

As is well known, for the Hamiltonian $\hat{\mathrm{H}}$ that possesses symmetries (constants of motion, invariance or Casimir operators, etc.) $\hat{\Lambda}_a$, associated with the symmetry group $\mathfrak{G}$ of $\hat{\mathrm{H}}$, so that for

$$\begin{aligned}[\hat{\mathrm{H}}, \hat{\Lambda}_a] &\equiv \hat{\mathrm{H}}\hat{\Lambda}_a - \hat{\Lambda}_a\hat{\mathrm{H}}\,, \qquad (a = 1, 2, \ldots) \\ &= 0\,,\end{aligned} \tag{7}$$

with

$$[\hat{\Lambda}_a, \hat{\Lambda}_b] = 0\,, \qquad (a, b = 1, 2, \ldots) \tag{8}$$

there exists a solution $|\Psi\rangle$ of the time independent Schrödinger equation

$$\hat{\mathrm{H}}|\Psi\rangle = E|\Psi\rangle\,, \tag{9}$$

that is automatically *symmetry adapted* (sa) and thus represents a simultaneous eigenstate of the operators $\hat{\Lambda}_a$,

$$\hat{\Lambda}_a|\Psi\rangle = \lambda_a|\Psi\rangle\,, \qquad (a = 1, 2, \ldots)\,. \tag{10}$$

However, when searching for an approximate wave function $|\Phi\rangle$ by relying on the variation principle, in which case $|\Phi\rangle$ belongs to a proper submanifold that is embedded in the full N-electron space $\mathcal{W}_N$ to which the exact wave function $|\Psi\rangle$ belongs, the imposition of the symmetry requirements (10) on $|\Phi\rangle$, i.e., requiring that $|\Phi\rangle$ is sa, namely that

$$\hat{\Lambda}_a|\Phi\rangle = \lambda_a|\Phi\rangle\,, \qquad (a = 1, 2, \ldots) \tag{11}$$

represents additional constraints on the variational problem at hand. Of course, such additional constraints can only raise the energy, since they further reduce the flexibility of the variational manifold. Consequently, disposing with one or more constraints (11) leads to a bs solution $|\Phi'\rangle$ with lower energy than that associated with the sa solution $|\Phi\rangle$.

Although Löwdin's symmetry dilemma[3,4] addresses the general problem of the global minimum for variationally based functionals, it has clear implications for the HF or, more generally, the IPM stability problem. Yet, while the local instability implies the existence of another solution with lower energy, local stability is no guarantee that a given solution represents a global minimum on the variational energy hypersurface. Unfortunately, it is very difficult to assess the global character of solutions.

The *ab initio* model Hamiltonians that are routinely used in molecular electronic structure studies are spin-free and thus commute with both the $\hat{S}_z$ and $\hat{S}^2$ total spin operators. As long as we employ pure or canonical spin orbitals $|I\rangle$, $|I\rangle = |i\rangle|\sigma\rangle$, with $|\sigma\rangle = |\alpha\rangle$ or $|\beta\rangle$, the single antisymmetrized product wave function $|\Phi\rangle$, which characterizes the IPM, is automatically an eigenstate of $\hat{S}_z$. In the CS case, when the MOs are doubly occupied, it is also an eigenstate of $\hat{S}^2$. We then refer to the corresponding $|\Phi\rangle$ as the *restricted HF* (RHF) wave function. Even in the OS case, $|\Phi\rangle$ will be an eigenstate of $\hat{S}^2$ as long as all OS MOs (i.e., those that are singly occupied) have the same spin component, say $|\alpha\rangle$. This maximum spin-projection (or high-spin) wave function is referred to as the *restricted open-shell HF* (ROHF) wave function.

Since RHF-type wave functions do not properly dissociate when breaking chemical bonds, one often enlarges the variational space by relaxing the double occupancy requirement and associates different orbitals with different spins (DODS approximation), leading to the *unrestricted HF* (UHF) wave functions of the DODS type. These UHF wave functions are no longer eigenstates of $\hat{S}^2$ (while still eigenstates of $\hat{S}_z$) and thus belong to the class of broken spin-symmetry wave functions. While for the OS systems the UHF wave functions always yield a lower-energy solution than ROHF, this is not necessarily the case for CS ground states. In fact, the RHF wave functions are usually stable in the neighborhood of the equilibrium geometry and break the spin symmetry only when one or more bonds are sufficiently stretched. Although UHF wave functions will usually describe the correct dissociation channel, as already mentioned, any resulting potential energy surface or curve will be nonanalytic at the boundary between regions of different symmetry, in this case between the UHF and RHF solutions, if such a boundary exists. Needless to say that the DODS-type wave function does not represent the most general spin-symmetry breaking, since we can use MSOs that are not pure-spin eigenstates, as we shall see later on. Such IPM-type wave functions are referred to as the *generalized HF* (GHF) wave functions (cf. Refs. 4, 5).

In the RHF case, Thouless' stability conditions[20] may be spin adapted, leading to the *singlet* and *nonsinglet* (*triplet*) *stability conditions.*[30] In the former case the singlet character of the wave function is always preserved so that when the instability does occur, the resulting bs solution violates a symmetry other than spin symmetry, usually breaking the spatial one. The triplet instability, on the other hand, leads to the DODS-type UHF solutions. These latter solutions have been known and used for a long time,[21,27,31] as already mentioned, while the singlet unstable space-symmetry breaking wave functions were more surprising.[30,32–35] In both cases, however, the stability conditions provided precise conditions for the existence of these bs solutions.

The singlet instabilities, as well as the ensuing bs solutions displaying both the diagonal and off-diagonal (i.e., respectively, atomic-charge and bond-order alternating) charge density waves, have been investigated for various π-electron systems, modelled via the Pariser-Parr-Pople (PPP) semiempirical Hamiltonian.[32–38] Later on *ab-initio* studies of such instabilities have been carried out.[39–42] In a similar way, stability conditions can be formulated for Brueckner-type solutions,[43] as well as for simple open-shell ROHF solutions.[44,45] The latter instabilities have been investigated at both the semi-empirical[46] and *ab initio*[47,48] levels. Recently these conditions were explored in connection with the highly singular behavior of the ROHF solutions,[18,19] first described by Mulliken.[17]

Following the initial multiplet classification[30] of Thouless' HF instabilities for CS ground states, a much more general classification was developed by Fukutome,[49–52] who also considered the $\hat{S}_z$ symmetry breaking (leading to various *torsional* spin-current or density waves) and time-reversal symmetry. This original classification based on the subgroups of the SU(2)$\otimes\mathfrak{T}$ group, where $\mathfrak{T} = \{\hat{1}, -\hat{1}, \hat{\Theta}, -\hat{\Theta}\}$ with $\hat{\Theta}$ designating the time-reversal operation, was later extended to the SU(2)$\otimes\mathfrak{T}\otimes\mathfrak{P}$ group by Ozaki and Fukutome[53–57] for certain point groups $\mathfrak{P}$. An alternate, simplified derivation of some of these results was later published by Löwdin *et al.*[58,59] Extensive exploitation of this theory, primarily at the semi-empirical level, has been carried out both by Fukutome *et al.*[60–63] and by Yamaguchi and collaborators[64–70] (see also Ref. 71 and *loc. cit.*). At the *ab initio* level various GHF-type solutions were explored by Löwdin and Mayer.[5]

The general classification of Fukutome *et al.*,[50,51,63] as well as various *ad hoc* explorations of the GHF-type solutions,[5] have always been restricted to CS-type wave functions. The main purpose of this paper is thus to extend this classification to a much wider class of wave functions, including those going beyond the IPM (say, MC SCF wave functions and the like). In view of the above mentioned convergence problems of the standard SCF procedure, we rely on the DEM formulation which guarantees, at least in principle, the convergence to the desired GHF solutions.

In closing this introduction we would be amiss not to mention at least the most important reviews[63,71–73] and books[74] on this topic. On the general topic of HF solutions we refer to the monograph edited by Carbó and Klobukowski[11] , which also contains several topics on, or related to, the stability problem.

2. HF equations and Thouless' stability conditions

In order to introduce the basic notation and to delineate the problem, we first formulate the HF equations in a form suitable for the direct energy minimization (DEM) algorithm. This formulation avoids the introduction of Lagrange multipliers that are usually employed to enforce the orthonormality (5) of the MSOs (or MOs), and thus the normalization of the IPM wave function, Eq. (4), also having a natural link to the classification approach developed in Sec. 3. Although the exploitation of Lagrange multipliers is straightforward in the general spin-orbital or CS cases, their definition in the OS case is far from being unambiguous (cf. Refs. 10, 75–80). For this and other reasons (e.g., the poor convergence or even non-convergence of the standard SCF algorithm in some cases) our formulation is based on the DEM algorithm. We then state general Thouless' stability conditions in terms of this formalism.

2.1 Independent particle model

We represent the IPM wave function $|\Phi\rangle$, involving the MSOs $|I_\mu\rangle$, $\mu = 1, \ldots, N$, in the second quantized form (for the notation employed, see Refs. 12–15)

$$|\Phi\rangle \equiv |\{I_1 I_2 \cdots I_N\}\rangle = \hat{\mathrm{X}}^\dagger_{I_1} \hat{\mathrm{X}}^\dagger_{I_2} \cdots \hat{\mathrm{X}}^\dagger_{I_N} |0\rangle \,, \tag{12}$$

where $\hat{\mathrm{X}}^\dagger_I$ ($\hat{\mathrm{X}}_I$) are the creation (annihilation) operators that are associated with the orthonormal set of MSOs $|I\rangle$, while $|0\rangle$ represents the normalized physical vacuum state, $\langle 0|0\rangle = 1$, so that $|\Phi\rangle$ is also normalized. The fact that $|\Phi\rangle$ is totally antisymmetric is indicated by enclosing the list of MSO labels in braces.

The electronic Hamiltonian $\hat{\mathrm{H}}$ involves at most two-body interactions,

$$\hat{\mathrm{H}} = \hat{\mathrm{Z}} + \hat{\mathrm{V}} \,, \tag{13}$$

and its one- and two-body components $\hat{\mathrm{Z}}$ and $\hat{\mathrm{V}}$, respectively, have the following general form

$$\hat{\mathrm{Z}} = \sum_{IJ} Z_{IJ} \hat{\mathrm{X}}_I^\dagger \hat{\mathrm{X}}_J , \qquad \hat{\mathrm{V}} = \tfrac{1}{4} \sum_{IJKL} V_{IJ;KL}^{(\mathcal{A})} \hat{\mathrm{X}}_I^\dagger \hat{\mathrm{X}}_J^\dagger \hat{\mathrm{X}}_L \hat{\mathrm{X}}_K , \tag{14}$$

where the corresponding matrix elements are defined as

$$\begin{aligned} Z_{IJ} &= \langle I | \hat{\mathrm{z}} | J \rangle , \\ V_{IJ;KL}^{(\mathcal{A})} &= V_{IJ;KL} - V_{IJ;LK} , \quad V_{IJ;KL} = \langle IJ | \hat{\mathrm{v}} | KL \rangle , \end{aligned} \tag{15}$$

and $\hat{\mathrm{z}}$ and $\hat{\mathrm{v}} = r_{12}^{-1}$ represent the corresponding one- and two-electron operators in their first quantized form. The energy expectation value $\mathcal{E}[\Phi]$, Eq. (2), then takes the form

$$\mathcal{E}[\Phi] = \sum_I Z_{II} n_I + \tfrac{1}{2} \sum_{IJ} V_{IJ;IJ}^{(\mathcal{A})} n_I n_J , \tag{16}$$

where the spin-orbital occupation numbers n_I are given by the diagonal elements of the occupation matrix $\mathbf{N} = (N_{IJ})$,

$$N_{IJ} = \langle \Phi | \hat{\mathrm{X}}_I^\dagger \hat{\mathrm{X}}_J | \Phi \rangle = n_I \delta_{IJ} . \tag{17}$$

As mentioned already in Sec. 1, the *ab initio* models are defined in terms of the AO (or ASO) basis, which spans the one-electron space $\mathcal{W}_1$, but, generally, is not orthogonal. Recalling that we label the ASOs with Greek capitals (e.g., $|\Lambda\rangle, |\Gamma\rangle$, etc.), we define the overlap matrix $\mathbf{S} = (S_{\Lambda\Gamma})$,

$$S_{\Lambda\Gamma} = \langle \Lambda | \Gamma \rangle , \tag{18}$$

which is positive definite with unit diagonal, $S_{\Lambda\Lambda} = 1$, and $0 < |S_{\Lambda\Gamma}| < 1$ when $\Lambda \neq \Gamma$. Clearly, $\mathbf{S}$ represents the metric in the chosen one-electron space $\mathcal{W}_1$. In the LCAO approximation, Eq. (1), the orthonormality of the MSOs takes the form

$$\mathbf{C}^\dagger \mathbf{S} \mathbf{C} = \mathbf{I} \equiv (\delta_{IJ}) , \tag{19}$$

where $\mathbf{C} = (C_{\Lambda I})$ represents the LCAO coefficient matrix.

Introducing the Fock-Dirac first-order density matrix in the ASO basis as

$$\mathbf{P} \equiv (P_{\Lambda\Gamma}) = \mathbf{C} \mathbf{N} \mathbf{C}^\dagger , \tag{20}$$

the energy functional (16) takes the form

$$\mathcal{E}[\Phi] \equiv \mathcal{E}[\mathbf{P}] = \sum_{\Lambda\Gamma} Z_{\Lambda\Gamma} P_{\Gamma\Lambda} + \tfrac{1}{2} \sum_{\Lambda\Gamma\Xi\Omega} V_{\Lambda\Omega;\Gamma\Xi}^{(\mathcal{A})} P_{\Xi\Omega} P_{\Lambda\Gamma} , \tag{21}$$

where the one- and two-electron matrix elements are defined analogously as in Eqs. (15). In fact, designating the representation of $\hat{\mathrm{Z}}$ and $\hat{\mathrm{V}}$ operators in the ASO basis by $\mathbf{Z}$ and $\mathbf{V}$ (or $\mathbf{V}^{(\mathcal{A})}$ for the antisymmetrized version), respectively, we can write Eq. (21) in the following compact form

$$\mathcal{E}[\mathbf{P}] = \tfrac{1}{2}\mathrm{Tr}[(\mathbf{Z}+\mathbf{F})\mathbf{P}]\,, \tag{22}$$

where $\mathbf{F} \equiv (F_{\Lambda\Gamma})$ has matrix elements

$$F_{\Lambda\Gamma} = Z_{\Lambda\Gamma} + \sum_{\Xi\Omega} V^{(\mathcal{A})}_{\Lambda\Omega;\Gamma\Xi} P_{\Xi\Omega}\,. \tag{23}$$

We emphasize that the superscript $(\mathcal{A})$ indicates the antisymmetrized version of the two-electron matrix elements [cf. Eq. (15)],

$$V^{(\mathcal{A})}_{\Lambda\Omega;\Gamma\Xi} = \langle\Lambda\Omega|\hat{\mathrm{v}}|\Gamma\Xi\rangle - \langle\Lambda\Omega|\hat{\mathrm{v}}|\Xi\Gamma\rangle\,, \tag{24}$$

while

$$V^{(\mathcal{A})}_{IJ;KL} = \sum_{\Lambda\Gamma\Xi\Omega} V^{(\mathcal{A})}_{\Lambda\Gamma;\Xi\Omega} C^*_{\Lambda I} C^*_{\Gamma J} C_{\Omega L} C_{\Xi K}\,, \tag{25}$$

and similarly for Z_{IJ}.

Using the LCAO approximation, Eq. (1), the coefficients $C_{\Lambda I}$ play the role of variational parameters in Eq. (22), subject to the constraints (19). Thus, computing the variation $\delta\mathcal{E}$ of the energy functional (21) or (22),

$$\delta\mathcal{E} \equiv \delta\mathcal{E}[\Phi] = \delta^{(1)}\mathcal{E} + \delta^{(2)}\mathcal{E} + \cdots\,, \tag{26}$$

the critical points of the energy hypersurface are determined by requiring that the first variation $\delta^{(1)}\mathcal{E}$ vanishes,

$$\delta^{(1)}\mathcal{E} = 0\,, \tag{27}$$

subject to the constraints (19). Introducing the Lagrange multipliers $\mathbf{E} \equiv (E_{IJ})$, one arrives at the standard LCAO form of HF equations (often called Hartree-Fock-Roothaan equations[9]), namely

$$\mathbf{F}(\mathbf{C})\mathbf{C} = \mathbf{SCE}\,, \tag{28}$$

where the Fock operator matrix $\mathbf{F}$ depends on the LCAO coefficients $C_{\Lambda I}$, Eq. (23). As is well known, these equations can be brought to the canonical form in which $\mathbf{E}$ is diagonal, yielding the ASO representation of Eq. (6), with the diagonal elements of $\mathbf{E}$ representing the (spin-)orbital energies ϵ_I.

2.2 Thouless' representation

When we wish to avoid the explicit use of Lagrange multipliers, it is best to rely on the DEM or steepest descent algorithm (cf., e.g. Ref. 16). For this purpose, it is convenient to employ a natural parameterization of the energy functional as implied by Thouless' theorem. This theorem states that any two single antisymmetrized product (or single determinant) wave functions $|\Phi\rangle$ and $|\widetilde{\Phi}\rangle$ that are not mutually orthogonal can be related via the transformation

$$|\widetilde{\Phi}\rangle = e^{\hat{\mathrm{T}}_1}|\Phi\rangle\,, \tag{29}$$

where $\hat{\mathrm{T}}_1$ is a particle-number-preserving, one-body operator

$$\hat{\mathrm{T}}_1 = \sum_{IJ} K_{IJ}\hat{\mathrm{X}}_I^\dagger\hat{\mathrm{X}}_J\,, \tag{30}$$

with $|\Phi\rangle$ given by Eq. (12) and $|\widetilde{\Phi}\rangle$ analogously defined in terms of another MSO basis $\{|\widetilde{I}_\mu\rangle\}$, with which we associate the transformed creation and annihilation operators $\hat{\mathrm{Y}}_{\widetilde{I}}^\dagger$ and $\hat{\mathrm{Y}}_{\widetilde{I}}$, respectively, so that

$$|\widetilde{\Phi}\rangle = |\{\widetilde{I}_1\widetilde{I}_2\cdots\widetilde{I}_N\}\rangle = \hat{\mathrm{Y}}_{\widetilde{I}_1}^\dagger\hat{\mathrm{Y}}_{\widetilde{I}_2}^\dagger\cdots\hat{\mathrm{Y}}_{\widetilde{I}_N}^\dagger|0\rangle\,. \tag{31}$$

We thus assume that the MSOs $|I\rangle$ represent an intermediate guess in the numerical procedure, while $|\widetilde{I}\rangle$ are the yet undetermined HF MSOs. The search for the desired IPM wave function then reduces to the search for the appropriate parameters K_{IJ} defining $\hat{\mathrm{T}}_1$. These parameters represent the elements of the so-called *mixing matrix* $\mathbf{K} = (K_{IJ})$. Although, in principle, $\mathbf{K}$ is arbitrary, it is desirable to preserve the orthonormality of the MSOs (and thus the normalization of $|\widetilde{\Phi}\rangle$ provided that the initial reference is normalized). This is achieved via a unitary transformation, which requires $\hat{\mathrm{T}}_1$ to be skew-Hermitian, i.e.,

$$\hat{\mathrm{T}}_1^\dagger = -\hat{\mathrm{T}}_1\,, \tag{32}$$

which in turn implies that

$$\mathbf{K}^\dagger = -\mathbf{K}\,. \tag{33}$$

To determine the form of the energy functional (22) using Thouless' parameterization, we first relate the orthonormal basis sets $\{|I\rangle\}$ and $\{|\widetilde{I}\rangle\}$ via the similarity transformation implied by Eq. (29) (note that

$\hat{T}_1|0\rangle = 0$), namely (see Appendix A)

$$|\widetilde{I}\rangle = \hat{Y}^\dagger_{\widetilde{I}}|0\rangle = e^{\hat{T}_1}\hat{X}^\dagger_{\widetilde{I}}e^{-\hat{T}_1}|0\rangle = \sum_{m=0}^{\infty}\frac{1}{m!}\left[\hat{T}_1, \hat{X}^\dagger_{\widetilde{I}}\right]^{(m)}|0\rangle$$
$$= \sum_{m=0}^{\infty}\frac{1}{m!}\sum_J \hat{X}^\dagger_J(\mathbf{K}^m)_{J\widetilde{I}}|0\rangle = \sum_J (e^{\mathbf{K}})_{J\widetilde{I}}|J\rangle\,, \tag{34}$$

where the superscript (m) implies a multiple commutator,

$$[\hat{A},\hat{B}]^{(m)} = \underbrace{[\hat{A},[\hat{A},[\cdots,[\hat{A}}_{m\ \hat{A}\text{'s}},\hat{B}]\cdots]]]\,, \qquad [\hat{A},\hat{B}]^{(0)} = \hat{B}\,, \tag{35}$$

and

$$\mathbf{K}^0 = \mathbf{I}\,. \tag{36}$$

Assuming the MSO basis $\{|I\rangle\}$ in the LCAO form, Eq. (1), with the coefficient matrix $\mathbf{C}$ satisfying the constraint (19), we employ a similar expansion for the new MSOs $|\widetilde{I}\rangle$,

$$|\widetilde{I}\rangle = \sum_\Lambda \widetilde{C}_{\Lambda\widetilde{I}}|\Lambda\rangle\,, \tag{37}$$

where the new coefficient matrix $\widetilde{\mathbf{C}}$,

$$\widetilde{\mathbf{C}} = \mathbf{C}e^{\mathbf{K}} \tag{38}$$

satisfies again the constraint (19). Exploiting this form of the coefficient matrix $\widetilde{\mathbf{C}}$, Eq. (38), we obtain for the energy functional $\mathcal{E}[\mathbf{P}] \equiv \mathcal{E}[\mathbf{K}]$, Eq. (22), an analogous expression

$$\mathcal{E}[\mathbf{K}] = \tfrac{1}{2}\mathrm{Tr}[(\mathbf{Z} + \mathbf{F}(\mathbf{K}))\mathbf{P}(\mathbf{K})]\,, \tag{39}$$

where now

$$\mathbf{P}(\mathbf{K}) = e^{\mathbf{K}}\mathbf{N}e^{-\mathbf{K}} \tag{40}$$

and

$$(\mathbf{F}(\mathbf{K}))_{IJ} = Z_{IJ} + \sum_{KL} V^{(A)}_{IK;JL}(\mathbf{P}(\mathbf{K}))_{LK} \tag{41}$$

are the transformed density and Fock matrices in the reference basis set $\{|I\rangle\}$, Eq. (1).

The advantage of Eq. (39) is that any variation with respect to $\mathbf{K}$ is unconstrained, except for requiring the skew-Hermiticity, Eq. (33), which, however, represents a simple linear constraint. In this way we can avoid the explicit use of Lagrange multipliers.

2.3 Variation of the energy expectation value

Relying on Eq. (39) we can write for the m-th variation $\delta^{(m)}\mathcal{E}$, Eq. (26), that

$$\delta^{(m)}\mathcal{E} = \tfrac{1}{2}\mathrm{Tr}\left(\mathbf{Z}\,\delta^{(m)}\mathbf{P} + \sum_{\ell=0}^{m} \delta^{(\ell)}\mathbf{F}\,\delta^{(m-\ell)}\mathbf{P}\right). \tag{42}$$

Using again the commutator expansion (A.1) for $\mathbf{P}$, Eq. (40), and notation of Eq. (35), we get

$$\mathbf{P}(\mathbf{K}) = e^{\mathbf{K}}\mathbf{N}e^{-\mathbf{K}} = \sum_{m=0}^{\infty} \frac{1}{m!}[\mathbf{K},\mathbf{N}]^{(m)}\,, \tag{43}$$

implying that the m-th variation $\delta^{(m)}\mathbf{P}$ is given by

$$\delta^{(m)}\mathbf{P} = \frac{1}{m!}[\mathbf{K},\mathbf{N}]^{(m)}\,. \tag{44}$$

Similarly, the variation of the Fock matrix is given in each order through that of the density matrix, i.e.,

$$(\delta^{(m)}\mathbf{F})_{IJ} = \sum_{KL} V^{(\mathcal{A})}_{IK;JL}(\delta^{(m)}\mathbf{P})_{LK}\,, \qquad m > 0 \tag{45}$$

while

$$(\delta^{(0)}\mathbf{F})_{IJ} \equiv F_{IJ} = Z_{IJ} + \sum_{KL} V^{(\mathcal{A})}_{IK;JL} N_{LK}\,. \tag{46}$$

Finally, using the identity

$$\mathrm{Tr}(\mathbf{F}\,\delta^{(m)}\mathbf{P}) = \mathrm{Tr}[\mathbf{Z}\,\delta^{(m)}\mathbf{P} + (\delta^{(m)}\mathbf{F})\,\mathbf{N}]\,, \tag{47}$$

we can further simplify the expression for $\delta^{(m)}E$, Eq. (42), so that for the first and second variations of interest we get

$$\delta^{(1)}\mathcal{E} = \mathrm{Tr}(\mathbf{F}\,\delta^{(1)}\mathbf{P}) = \mathrm{Tr}(\mathbf{F}[\mathbf{K},\mathbf{N}])\,, \tag{48}$$

$$\begin{aligned}\delta^{(2)}\mathcal{E} &= \mathrm{Tr}(\mathbf{F}\,\delta^{(2)}\mathbf{P}) + \tfrac{1}{2}\mathrm{Tr}(\delta^{(1)}\mathbf{F}\,\delta^{(1)}\mathbf{P}) \\ &= \tfrac{1}{2}\mathrm{Tr}(\mathbf{F}[\mathbf{K},\mathbf{N}]^{(2)}) + \tfrac{1}{2}\mathrm{Tr}(\delta^{(1)}\mathbf{F}\,[\mathbf{K},\mathbf{N}])\,.\end{aligned} \tag{49}$$

2.4 DEM form of HF equations

Setting the first variation $\delta^{(1)}\mathcal{E}$, Eq. (48), to zero, we find the required HF equations determining the critical points on the energy hypersurface $\mathcal{E}[\Phi]$. Using the cyclic property of the trace, we see that

$$\begin{aligned}\delta^{(1)}\mathcal{E} &= \mathrm{Tr}(\mathbf{F}[\mathbf{K},\mathbf{N}]) = \mathrm{Tr}(\mathbf{FKN} - \mathbf{FNK}) = \mathrm{Tr}([\mathbf{N},\mathbf{F}]\mathbf{K}) \\ &= \sum_{IJ}([\mathbf{N},\mathbf{F}])_{IJ}K_{JI} = -\sum_{IJ}([\mathbf{N},\mathbf{F}])_{IJ}K^{*}_{IJ}\,.\end{aligned} \tag{50}$$

The last expression on the righthand side of Eq. (50) may be viewed as a directional derivative in the 'K-space', interpreting (IJ) as a multi-index. Thus, $\delta^{(1)}\mathcal{E}$ has a form

$$\delta^{(1)}\mathcal{E} = -\vec{a}\cdot\vec{v}^*\,, \tag{51}$$

where $\vec{a}$ is some fixed vector, while $\vec{v}$ can vary. The direction of the steepest descent then corresponds to the choice $\vec{v} = \vec{a}$, yielding

$$\delta^{(1)}\mathcal{E} = -\vec{a}\cdot\vec{a}^* = -|\vec{a}|^2\,. \tag{52}$$

By analogy, the direction of steepest descent in our case is given by

$$\mathbf{K} = [\mathbf{N},\mathbf{F}]\,. \tag{53}$$

Now, when $\delta^{(1)}\mathcal{E}$ will vanish for all admissible directions $\mathbf{K}$, both sides of the expression (53) must vanish as well, so that

$$([\mathbf{N},\mathbf{F}])_{IJ} = \sum_K (N_{IK}F_{KJ} - F_{IK}N_{KJ}) = F_{IJ}(n_I - n_J) = 0\,, \tag{54}$$

where we used the expression (17) for the orbital occupancies.

The last Eq. (54) represents in fact Brillouin's theorem, which is equivalent to the standard form of HF equations. Indeed, at convergence, Eq. (54) implies that the Fock matrix $\mathbf{F}$ is block diagonal with vanishing entries between the occupied and unoccupied orbitals.

2.5 DEM algorithm

The directional derivative (53) offers an alternative approach for finding of the HF solutions, namely through the direct energy minimization. The 'vector' $\mathbf{K}$ gives the direction which, at least locally, most effectively lowers the energy. We thus have to determine how far it is worthwhile to follow this direction (see a similar approach used with the standard SCF procedure in Refs. 30, 32, 33, 39, 41, 46). This can be done by parameterizing the orbital coefficient matrix with a real scalar parameter ν as follows [cf. Eq. (38)]

$$\widetilde{\mathbf{C}}(\nu) = \mathbf{C}e^{\nu\mathbf{K}}\,, \tag{55}$$

and by considering the energy functional (39) as a function of ν. Performing a one-dimensional minimization yields an optimal value $\nu_c > 0$ which, in turn, determines a new coefficient matrix $\widetilde{\mathbf{C}}(\nu_c)$, Eq. (55), and the corresponding new Fock matrix $\mathbf{F}(\widetilde{\mathbf{C}}(\nu_c))$. The latter can be used to obtain a new direction 'vector' $\mathbf{K}$ and the process repeated until the Brillouin condition (54),

$$[\mathbf{N},\mathbf{F}] = 0\,, \tag{56}$$

is satisfied to the prescribed tolerance.

The above outlined algorithm defines the DEM procedure. Unlike the standard SCF approach, this procedure does not require the introduction of Lagrange multipliers that ensure the orthonormality of the MSOs via Eq. (19). More importantly, this procedure always achieves, at least in principle, numerical convergence, unlike the standard SCF procedure.

2.6 Thouless' stability conditions

We must explore the conditions for the second variation $\delta^{(2)}\mathcal{E}$ to be positive definite with respect to all admissible variations,

$$\delta^{(2)}\mathcal{E} > 0\,, \tag{57}$$

guaranteeing the local stability (i.e., a local minimum).

Recalling Eq. (49), we first examine each term on the right hand side separately. For the first term we find

$$\begin{aligned} \tfrac{1}{2}\mathrm{Tr}(\mathbf{F}[\mathbf{K},\mathbf{N}]^{(2)}) &= \tfrac{1}{2}\mathrm{Tr}(\mathbf{F}[\mathbf{K},[\mathbf{K},\mathbf{N}]]) \\ &= \tfrac{1}{2}\mathrm{Tr}(\mathbf{F}(\mathbf{K}^2\mathbf{N} - 2\mathbf{KNK} + \mathbf{NK}^2)) \\ &= \tfrac{1}{2}\mathrm{Tr}(\mathbf{K}\{\mathbf{N},\mathbf{F}\}\mathbf{K}) - \mathrm{Tr}(\mathbf{KFKN})\,, \end{aligned} \tag{58}$$

where $\{A, B\}$ designates the anticommutator of A and B,

$$\{A, B\} = AB + BA\,. \tag{59}$$

Expanding, next, the right hand side as an explicit sum, we find

$$\begin{aligned} \tfrac{1}{2}\mathrm{Tr}(\mathbf{F}[\mathbf{K},\mathbf{N}]^{(2)}) &= \tfrac{1}{2}\sum_{IJKL} K_{IJ}K_{KL}\Big((\{\mathbf{N},\mathbf{F}\})_{JK}\delta_{LI} - 2F_{JK}N_{LI}\Big) \\ &= \tfrac{1}{2}\sum_{IJKL} K_{IJ}K_{KL}(n_J + n_K - 2n_I)\delta_{LI}F_{JK}\,. \end{aligned} \tag{60}$$

Symmetrizing, finally, the right hand side with respect to the interchange of summations over I, J and K, L, we finally get

$$\begin{aligned} \tfrac{1}{2}\mathrm{Tr}(\mathbf{F}[\mathbf{K},\mathbf{N}]^{(2)}) = \tfrac{1}{2}\sum_{IJKL} & K_{IJ}K_{KL}[\tfrac{1}{2}(n_J + n_K - 2n_I)\delta_{LI}F_{JK} \\ & + \tfrac{1}{2}(n_L + n_I - 2n_K)\delta_{JK}F_{LI}]\,. \end{aligned} \tag{61}$$

Turning now to the second term on the right hand side of Eq. (49) and using the expression for the first variation of the Fock matrix $\delta^{(1)}\mathbf{F}$,

Eq. (45) with $m = 1$, we get

$$\begin{aligned}\tfrac{1}{2}\mathrm{Tr}(\delta^{(1)}\mathbf{F}\,[\mathbf{K},\mathbf{N}]) &= \tfrac{1}{2}\sum_{IJKL} V^{(\mathcal{A})}_{JL;IK}([\mathbf{K},\mathbf{N}])_{IJ}([\mathbf{K},\mathbf{N}])_{KL} \\ &= \tfrac{1}{2}\sum_{IJKL} K_{IJ}K_{KL}V^{(\mathcal{A})}_{JL;IK}(n_I - n_J)(n_K - n_L)\,,\end{aligned} \tag{62}$$

where in the last step we again used the explicit expansion for the commutator expressions. In this case, no symmetrization is required, since the result is manifestly invariant under the exchange of I, J and K, L summation labels.

Combining both expressions, Eqs. (61) and (62), the second variation of the energy $\delta^{(2)}\mathcal{E}$ becomes

$$\begin{aligned}\delta^{(2)}\mathcal{E} = \tfrac{1}{2}\sum_{IJKL} K_{IJ}K_{KL}\,\Big[&\tfrac{1}{2}(n_J + n_K - 2n_I)\delta_{LI}F_{JK} \\ &+\tfrac{1}{2}(n_L + n_I - 2n_K)\delta_{JK}F_{LI} + V^{(\mathcal{A})}_{JL;IK}(n_I - n_J)(n_K - n_L)\Big]\,.\end{aligned} \tag{63}$$

This expression is the key to the formulation of the stability problem. We recall that in view of the skew-Hermitian character of the mixing matrix $\mathbf{K}$, only the off-diagonal matrix elements K_{IJ} for $I < J$ are required, since

$$K_{JI} = -K^*_{IJ}\,. \tag{64}$$

Moreover, without restricting the generality of our considerations, we can set the diagonal elements K_{II} (which, in principle, must be purely imaginary) to zero, since whenever $I = J$ or $K = L$, the right hand side of Eq. (63) vanishes as long as the MSOs satisfy the Brillouin condition (54). Thus, we can cast the expression (63) into the matrix form

$$\delta^{(2)}\mathcal{E} = \tfrac{1}{2}\mathbf{v}^\dagger \cdot \mathbf{M} \cdot \mathbf{v}\,, \tag{65}$$

where

$$\mathbf{v} = \begin{pmatrix} \mathbf{k} \\ \mathbf{k}^* \end{pmatrix} \tag{66}$$

is the *augmented vector* involving the column vector $\mathbf{k}$ and its complex conjugate $\mathbf{k}^*$, whose components consist of the entries K_{IJ} ($I < J$) of the mixing matrix. The *stability matrix* $\mathbf{M}$ then assumes the form

$$\mathbf{M} = \begin{pmatrix} \mathbf{A} & \mathbf{B} \\ \mathbf{B}^* & \mathbf{A}^* \end{pmatrix}, \tag{67}$$

where

$$A_{IJ;KL} = \tfrac{1}{2}(n_I + n_K - 2n_J)\delta_{LJ}F_{IK} - \tfrac{1}{2}(n_L + n_J - 2n_K)\delta_{IK}F_{LJ} + V^{(\mathcal{A})}_{IL;JK}(n_I - n_J)(n_K - n_L)\,, \tag{68}$$

and

$$B_{IJ;KL} = \tfrac{1}{2}(n_I + n_L - 2n_J)\delta_{KJ}F_{IL} + \tfrac{1}{2}(n_K + n_J - 2n_L)\delta_{IL}F_{KJ} + V^{(\mathcal{A})}_{IK;JL}(n_I - n_J)(n_K - n_L)\,. \tag{69}$$

The matrix $\mathbf{M}$ is Hermitian, since $\mathbf{A}^\dagger = \mathbf{A}$ and $\mathbf{B}^\dagger = \mathbf{B}^*$, and consequently all eigenvalues are real.

We thus find that in order for $\delta^{(2)}\mathcal{E}$ to be strictly positive definite, Eq. (57), for all nontrivial admissible vectors $\mathbf{v}$, the matrix $\mathbf{M}$ must be strictly positive definite. Since $\mathbf{M}$ is Hermitian, it is diagonalizable, and we can write

$$\mathbf{M}\mathbf{v} = \mu\mathbf{v}\,. \tag{70}$$

Thus, the positive definiteness of $\mathbf{M}$, which guarantees that a given HF solution is *stable*, requires that all the eigenvalues μ are positive. On the other hand, if one or more eigenvalues μ are negative, the solution is said to be *unstable*, indicating the existence of another solution with lower energy. This new solution can be obtained by proceeding in the local steepest descent direction implied by the eigenvector $\mathbf{v}$ that is associated with the largest (in absolute value) negative eigenvalue. Since the entries of $\mathbf{v}$ consist of the matrix elements K_{IJ} ($I < J$), it is an easy task to reconstruct the full matrix $\mathbf{K}$, which is then employed in the DEM procedure, Eq. (55).

We note, finally, that the stability problem (70) can be factorized further when the stability matrix $\mathbf{M}$ is real valued, which occurs when the MSOs and the Fock matrix are real. In this case, we can split the stability problem into its real and imaginary parts. The real subproblem then corresponds to the stability problem within the space of real-valued solutions, while considering both the real and imaginary subproblems we obtain the stability conditions associated with the space of complex-valued HF solutions. These subproblems result when we decompose the eigenvector $\mathbf{v}$ into its real and imaginary components,

$$\mathbf{v} = \begin{pmatrix} \mathbf{k}^{(\mathbf{R})} \\ \mathbf{k}^{(\mathbf{R})} \end{pmatrix} + \mathrm{i} \begin{pmatrix} \mathbf{k}^{(\mathbf{I})} \\ -\mathbf{k}^{(\mathbf{I})} \end{pmatrix}, \tag{71}$$

and substitute into the eigenvalue problem (70). Since $\mathbf{A}$ and $\mathbf{B}$ are assumed to be real-valued, we obtain

$$(\mathbf{A} + \mathbf{B})\mathbf{k}^{(\mathbf{R})} = \mu^{(\mathbf{R})}\mathbf{k}^{(\mathbf{R})}\,, \tag{72}$$

for the real subproblem, and

$$(\mathbf{A} - \mathbf{B})\mathbf{k}^{(\mathbf{I})} = \mu^{(\mathbf{I})}\mathbf{k}^{(\mathbf{I})}, \qquad (73)$$

for the imaginary subproblem. If the eigenvalues of both the real and imaginary subproblems are positive, then the solution is said to be *complex stable*, and if only the real subproblem has positive eigenvalues, the solution is *real stable.*

3. Classification of broken-symmetry solutions

Classification of broken symmetry wave functions is accomplished by investigating their transformation properties under the action of the invariance operators $\hat{\mathrm{g}}$ constituting the symmetry (or invariance) group $\mathfrak{G}$ of the Hamiltonian, i.e.

$$[\hat{\mathrm{g}}, \hat{\mathrm{H}}] = 0, \qquad (\forall \hat{\mathrm{g}} \in \mathfrak{G}) \qquad (74)$$

or, equivalently,

$$\hat{\mathrm{g}}\hat{\mathrm{H}}\hat{\mathrm{g}}^{-1} = \hat{\mathrm{H}}, \qquad (\forall \hat{\mathrm{g}} \in \mathfrak{G}) \qquad (75)$$

implying the invariance of $\hat{\mathrm{H}}$ under the similarity transformation by $\hat{\mathrm{g}}$'s. Generally, when $|\Psi\rangle$ is an eigenstate of $\hat{\mathrm{H}}$, then $\hat{\mathrm{g}}|\Psi\rangle$ is also an eigenstate belonging to the same eigenvalue as $|\Psi\rangle$. Thus the space spanned by the set $\{\hat{\mathrm{g}}|\Psi\rangle\}$ is a carrier space of a representation of $\mathfrak{G}$. Further, the irreducible representations of $\mathfrak{G}$ may be uniquely labeled by the eigenvalues (or quantum numbers) of pertinent Casimir operators $\hat{\Lambda}_a$.

Now, the *exact* eigenstates of the Hamiltonian $\hat{\mathrm{H}}$ can be so chosen as to be simultaneously the eigenstates of the symmetries $\hat{\Lambda}_a$ associated with the invariance group $\mathfrak{G}$, while for the variationally determined *approximate* wave functions, the symmetry requirements (11) represent additional constraints. Those wave functions which transform according to some irreducible representation of $\mathfrak{G}$ are referred to as *symmetry adapted* (sa) or invariant, while those leaving invariant only some proper subgroup $\mathfrak{H}$ of $\mathfrak{G}$ are called *broken symmetry* (bs) wave functions. The latter can thus be classified by the subgroups of the full invariance group $\mathfrak{G}$.

Rather than exploring directly the transformation properties of the wave function, it is advantageous to examine the transformations that link sa and bs solutions. Given a sa reference state $|\Phi_0\rangle$, any bs wave function $|\Phi\rangle$ can be related to $|\Phi_0\rangle$ via some nonsingular transformation $\hat{\mathrm{U}}$,

$$|\Phi\rangle = \hat{\mathrm{U}}|\Phi_0\rangle. \qquad (76)$$

In the case of a single-determinantal IPM wave function, this transformation is generally achieved via the unitary transformation of Thouless' theorem (29), in which case

$$\hat{\mathrm{U}} = e^{\hat{\mathrm{T}}_1}\,, \tag{77}$$

with $\hat{\mathrm{T}}_1$ having the form given by Eq. (30). Generally the relationship (76) is preserved under the action of an invariant subgroup $\mathfrak{H}$ of $\mathfrak{G}$,

$$\hat{\mathrm{h}}|\Phi\rangle = \hat{\mathrm{h}}\hat{\mathrm{U}}|\Phi_0\rangle = (\hat{\mathrm{h}}\hat{\mathrm{U}}\hat{\mathrm{h}}^{-1})\hat{\mathrm{h}}|\Phi_0\rangle = \hat{\mathrm{U}}(\hat{\mathrm{h}}|\Phi_0\rangle)\,, \tag{78}$$

provided that

$$\hat{\mathrm{h}}\hat{\mathrm{U}}\hat{\mathrm{h}}^{-1} = \hat{\mathrm{U}} \qquad (\forall\,\hat{\mathrm{h}} \in \mathfrak{H} \subset \mathfrak{G})\,. \tag{79}$$

For a nondegenerate $|\Phi_0\rangle$ that transforms according to a one-dimensional irreducible representation of $\mathfrak{G}$, $|\Phi_0\rangle$ remains invariant up to a complex phase, so that

$$\hat{\mathrm{g}}|\Phi_0\rangle = e^{\mathrm{i}\delta(\hat{\mathrm{g}})}|\Phi_0\rangle\,, \qquad (\forall\,\hat{\mathrm{g}} \in \mathfrak{G})\,, \tag{80}$$

in which case $|\Phi\rangle$ also transforms via the corresponding subduced representation

$$\hat{\mathrm{h}}|\Phi\rangle = \hat{\mathrm{U}}\hat{\mathrm{h}}|\Phi_0\rangle = \hat{\mathrm{U}}e^{\mathrm{i}\delta(\hat{\mathrm{h}})}|\Phi_0\rangle = e^{\mathrm{i}\delta(\hat{\mathrm{h}})}|\Phi\rangle\,, \qquad (\hat{\mathrm{h}} \in \mathfrak{H} \subset \mathfrak{G})\,, \tag{81}$$

assuming that Eq. (79), and thus Eq. (78), hold. We may, therefore, conclude that for a transformation $\hat{\mathrm{U}}$, satisfying Eq. (79), the bs wave function $|\Phi\rangle$ possesses the transformation properties and any associated constants of the motion $\hat{\Lambda}_a$ (under the subgroup $\mathfrak{H}$, $\mathfrak{H} \subset \mathfrak{G}$) inherited from the sa reference $|\Phi_0\rangle$. This constrains the possible transformations linking the sa and bs wave functions and, ultimately, the bs wave functions themselves. Once the full invariance group $\mathfrak{G}$ is known, it only remains to elucidate its subgroup structure and to investigate the corresponding classes of bs solutions.

3.1 Fukutome's subgroup classification

In general, the invariance group of a standard, spin-free, electronic Hamiltonian $\hat{\mathrm{H}}$ involves the spin rotation group $\mathfrak{S} \equiv \mathrm{SU}(2)$ and the time-reversal group $\mathfrak{T}$, i.e. $\mathfrak{G} = \mathfrak{S} \otimes \mathfrak{T}$. While $\hat{\mathrm{H}}$ is also particle-number preserving, this type of symmetry is trivially satisfied by all the wave functions employed in the studies of the molecular electronic structure [cf., however, the Bardeen-Cooper-Schrieffer (BCS) type wave functions[81,82]]. In many cases, $\hat{\mathrm{H}}$ is also invariant with respect to a spatial point group symmetry $\mathfrak{P}$, which is generally violated in the case that the instability arises from the stability problem associated with the full invariance group $\mathfrak{G} = \mathfrak{S} \otimes \mathfrak{T}$, although it may occur simultaneously with other symmetry breakings. The exploration of the full $\mathfrak{S} \otimes \mathfrak{T} \otimes \mathfrak{P}$ subgroup structure requires a separate consideration for each specific point group $\mathfrak{P}$ (cf., e.g., Refs. 54, 56, 57). On the other hand, in each specific case, the subgroups of $\mathfrak{P}$ are well known and the corresponding symmetry breaking is not difficult to interpret (see eg. Refs. 30, 33, 46). For this reason, we focus on the spin-rotation and time-reversal group $\mathfrak{G} = \mathfrak{S} \otimes \mathfrak{T}$, which is admitted by all spin-free, electronic Hamiltonians.

We recall that the defining representation for the spin-rotation group $\mathfrak{S}$ can be given in terms of the spin operators $\hat{\mathrm{S}}_a\,(a = 1, 2, 3)$ as follows

$$\mathfrak{S} = \{\hat{\mathrm{U}}_{\mathrm{R}}(\vec{n}, \theta) \,:\, \vec{n} \in \mathbb{R}^3,\, \|\, \vec{n}\, \| = 1,\, \theta \in (-2\pi, 2\pi]\}\,, \tag{82}$$

where

$$\hat{\mathrm{U}}_{\mathrm{R}}(\vec{n}, \theta) = \exp\left(\mathrm{i}\theta \sum_{a=1}^{3} \hat{\mathrm{S}}_a n_a\right), \quad \vec{n} = (n_1, n_2, n_3) \tag{83}$$

and

$$\hat{\mathrm{S}}_a = \tfrac{1}{2}\hat{\sigma}_a = \tfrac{1}{2} \sum_{\sigma_1 \sigma_2} (\sigma_a)_{\sigma_1 \sigma_2} \left(\sum_{i=1}^{N} \hat{\mathrm{E}}^{i\sigma_1}_{i\sigma_2} \right), \tag{84}$$

where $\sigma_a\,(a = 1, 2, 3)$ are the well-known Pauli matrices and

$$\hat{\mathrm{E}}^{i\sigma_1}_{j\sigma_2} = \hat{\mathrm{X}}^{\dagger}_{i\sigma_1} \hat{\mathrm{X}}_{j\sigma_2} \tag{85}$$

are the unitary group $\mathrm{U}(2n)$ generators (cf., e.g. Refs. 83–85 and *loc. cit.*) defined in terms of the second quantization creation and annihilation operators on a canonical spin-orbital basis

$$|I\rangle \equiv |i\sigma\rangle := |i\rangle|\sigma\rangle \equiv |i\rangle \otimes |\sigma\rangle\,, \quad \sigma = \pm\tfrac{1}{2}\,. \tag{86}$$

Using the explicit form of Pauli matrices,

$$\sigma_1 = \begin{pmatrix} 0 & 1 \\ 1 & 0 \end{pmatrix}, \quad \sigma_2 = \begin{pmatrix} 0 & -\mathrm{i} \\ \mathrm{i} & 0 \end{pmatrix}, \quad \sigma_3 = \begin{pmatrix} 1 & 0 \\ 0 & -1 \end{pmatrix}, \tag{87}$$

we can thus express the spin operators $\hat{\mathrm{S}}_a\,(a = 1, 2, 3)$ as follows

$$\begin{aligned} \hat{\mathrm{S}}_1 &= \tfrac{1}{2} \sum_{i\sigma} \hat{\mathrm{E}}^{i\sigma}_{i\bar{\sigma}}\,, \\ \hat{\mathrm{S}}_2 &= -\tfrac{\mathrm{i}}{2} \sum_{i\sigma} (-1)^{\sigma - 1/2} \hat{\mathrm{E}}^{i\sigma}_{i\bar{\sigma}}\,, \\ \hat{\mathrm{S}}_3 &= \tfrac{1}{2} \sum_{i\sigma} (-1)^{\sigma - 1/2} \hat{\mathrm{E}}^{i\sigma}_{i\sigma}\,, \end{aligned} \tag{88}$$

where $|\bar{\sigma}\rangle = |-\frac{1}{2}\rangle \equiv |\beta\rangle = |\bar{\alpha}\rangle$ if $|\sigma\rangle = |\frac{1}{2}\rangle \equiv |\alpha\rangle$, and vice versa.

We note that our definition, Eq. (82), accounts for the double-covering of the rotation group by incorporating both rotations and their inverses in the defining representation by virtue of the fact that $\theta \in (-2\pi, 2\pi]$, and we adopt the phase convention

$$\hat{\mathrm{U}}_{\mathrm{R}}(\vec{n}, \theta + 2\pi) = -\hat{\mathrm{U}}_{\mathrm{R}}(\vec{n}, \theta) \tag{89}$$

for $\theta \notin (-2\pi, 2\pi]$. Although this convention is only valid when $\hat{\mathrm{U}}_{\mathrm{R}}$ acts on single electron states $|I\rangle$, no problem arises in our subsequent derivations where only similarity transformations arise.

The time-reversal group $\mathfrak{T}$ for a many-electron system is given by

$$\mathfrak{T} = \{\pm\hat{1}, \pm\hat{\Theta}\}\,, \tag{90}$$

where $\hat{\Theta}$ is skew-Hermitian and $\hat{\Theta}^2 = -\hat{1}$. Unlike in the case of scalar wave functions, $\hat{\Theta}$ is not simply equivalent with the complex conjugation operator $\hat{\mathrm{K}}$, but, in general, is defined by the expression

$$\hat{\Theta} = \hat{\mathrm{U}}_{\mathrm{R}}(\vec{e}_2, -\pi)\hat{\mathrm{K}} = \exp(-\mathrm{i}\pi\hat{\mathrm{S}}_2)\hat{\mathrm{K}}\,. \tag{91}$$

Using the standard phase convention[82,86] this reduces to

$$\hat{\Theta} = -\mathrm{i}\hat{\sigma}_2\hat{\mathrm{K}}\,, \tag{92}$$

when acting on single two-component spinors. The inverse of $\hat{\Theta}$, using the identification (89), is then given by

$$\hat{\Theta}^{-1} = -\hat{\Theta} = \exp(\mathrm{i}\pi\hat{\mathrm{S}}_2)\hat{\mathrm{K}}\,. \tag{93}$$

The full $\mathfrak{G} = \mathfrak{S} \otimes \mathfrak{T}$ group may thus be characterized as follows

$$\begin{aligned} \mathfrak{G} &\equiv \mathfrak{S} \otimes \mathfrak{T} \\ &= \{\hat{\mathrm{U}}_R(\vec{n}, \theta), \hat{\mathrm{U}}_R(\vec{n}, \theta)\hat{\mathrm{K}} \,:\, \vec{n} \in \mathbb{R}^3, \|\, \vec{n} \,\| = 1,\, \theta \in (-2\pi, 2\pi]\}\,, \end{aligned} \tag{94}$$

since the spin rotation involved in $\hat{\Theta}$, Eq. (91), can be absorbed in the spin component $\mathfrak{S}$ of $\mathfrak{G}$. The resulting group $\mathfrak{G}$ possesses eight subgroups (including trivial ones), which are not difficult to identify by considering the subgroups of $\mathfrak{S}$ and $\mathfrak{T}$ (cf. Refs. 50, 63). Here we assume, without restricting the generality, a specific choice of coordinate axes, since other choices simply differ via a fixed spin rotation.

The trivial subgroups are the full group $\mathfrak{G} \equiv \mathfrak{S} \otimes \mathfrak{T}$ and the identity group $\mathfrak{E} = \{\hat{1}\}$, while the constituting component groups $\mathfrak{S}$ and $\mathfrak{T}$ are proper subgroups of $\mathfrak{G}$. The only non-trivial subgroup of $\mathfrak{S}$ is the axial group $\mathfrak{A}$ generated by $\hat{\mathrm{S}}_3$,

$$\mathfrak{A} = \{\hat{\mathrm{U}}_{\mathrm{R}}(\vec{e}_3, \theta) = \exp(\mathrm{i}\hat{\mathrm{S}}_3\theta) : \theta \in (-2\pi, 2\pi]\}\,, \tag{95}$$

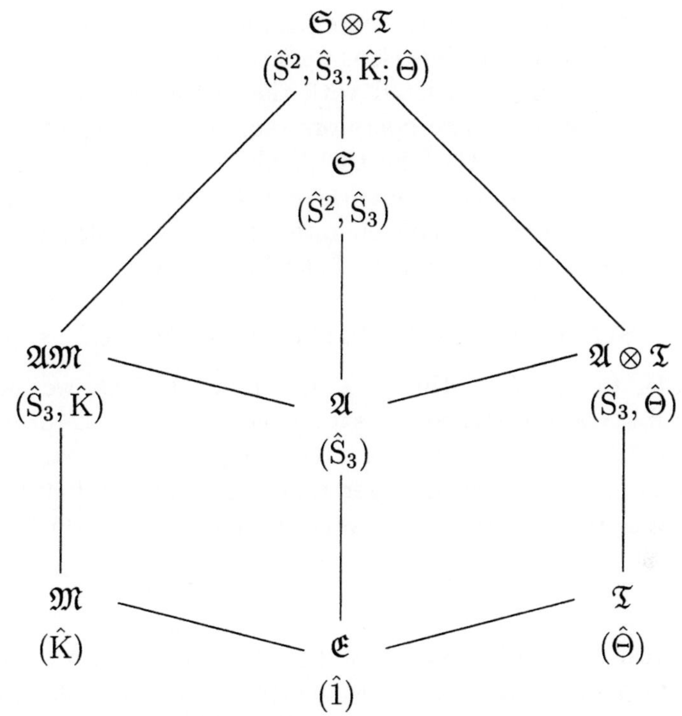

Figure 1. Subgroup inclusion diagram displaying the subgroup structure for the $\mathfrak{S} \otimes \mathfrak{T}$ group placed at the top of the figure and the trivial subgroup $\mathfrak{E} \equiv \{\hat{1}\}$ at the bottom. The relevant Casimir or invariance operators for each subgroup, whose eigenvalues label the corresponding broken symmetry solution classes, are given in parentheses.

which in turn generates the subgroups $\mathfrak{A} \otimes \mathfrak{T}$ and $\mathfrak{A}\mathfrak{M}$, where[50,63]

$$\mathfrak{A} \otimes \mathfrak{T} = \{\hat{\mathrm{U}}_{\mathrm{R}}(\vec{e}_3, \theta), \hat{\mathrm{U}}_{\mathrm{R}}(\vec{e}_3, \theta)\hat{\Theta} : \theta \in (-2\pi, 2\pi]\}\,, \tag{96}$$

$$\mathfrak{A}\mathfrak{M} = \{\hat{\mathrm{U}}_{\mathrm{R}}(\vec{e}_3, \theta), \hat{\mathrm{U}}_{\mathrm{R}}(\vec{e}_3, \theta)\hat{\mathrm{K}} : \theta \in (-2\pi, 2\pi]\}\,. \tag{97}$$

The only remaining subgroup is then the complex conjugation group $\mathfrak{M}$,

$$\mathfrak{M} = \{\hat{1}, \hat{\mathrm{K}}\}\,, \tag{98}$$

sometimes misleadingly referred to as the 'magnetic' group. The subgroup structure is immediately obvious from the above definitions and is illustrated in Fig. 1.

The bs wave functions can now be characterized by these subgroups, representing their invariance groups. Clearly, the trivial subgroup $\mathfrak{E}$

characterizes the most general wave functions that break all the symmetries. Each class of wave functions possesses one or more good quantum numbers, which are associated with the operators $\hat{\Lambda}_a$ listed for each subgroup in Fig. 1. These operators represent a *complete set of commuting operators* (CSCO)[87] for each (sub)group and the corresponding quantum numbers uniquely label the relevant eigenstates. The transformation properties of invariant operators and the classification of the bs wave functions are studied in the following sections.

3.2 Structure of invariant operators

In order to determine the structure of bs solutions, we have to find the possible transformations $\hat{\text{U}}$, satisfying the invariance property (79), that link the bs and sa solutions. For this purpose we examine a more general problem of finding the operators $\hat{\text{A}}$ that are invariant with respect to similarity transformations by various subgroups $\mathfrak{H} \subset \mathfrak{G}$, i.e.,

$$\hat{\text{h}}\hat{\text{A}}\hat{\text{h}}^{-1} = \hat{\text{A}}\,, \qquad (\forall\,\hat{\text{h}} \in \mathfrak{H} \subset \mathfrak{G})\,. \tag{99}$$

This is most easily achieved by mapping a given operator $\hat{\text{A}}$ into a set of invariant operators having the desired properties. For this purpose we use the *invariant mean* map (cf., e.g. Ref. 88)

$$\mathcal{M}_{\mathfrak{H}}(\hat{\text{A}}) = \frac{1}{|\mathfrak{H}|}\sum_{\hat{\text{h}}\in\mathfrak{H}} \hat{\text{h}}\hat{\text{A}}\hat{\text{h}}^{-1}\,, \tag{100}$$

into the space of invariant operators, since it is easily seen that

$$\hat{\text{h}}\mathcal{M}_{\mathfrak{H}}(\hat{\text{A}})\hat{\text{h}}^{-1} = \mathcal{M}_{\mathfrak{H}}(\hat{\text{A}})\,, \tag{101}$$

and the mapping $\mathcal{M}_{\mathfrak{H}}$ is a projection,

$$\mathcal{M}_{\mathfrak{H}}(\mathcal{M}_{\mathfrak{H}}(\hat{\text{A}})) = \mathcal{M}_{\mathfrak{H}}(\hat{\text{A}})\,. \tag{102}$$

Here we assumed that $\mathfrak{H}$ has a finite order $|\mathfrak{H}|$. Clearly, for the one-parameter continuous groups with $\delta \in (-2\pi, 2\pi]$, the invariant mean is defined analogously by

$$\mathcal{M}_{\mathfrak{H}}(\hat{\text{A}}) = \frac{1}{4\pi}\int_{-2\pi}^{2\pi} \hat{\text{h}}(\theta)\hat{\text{A}}\hat{\text{h}}^{-1}(\theta)d\theta\,, \tag{103}$$

and similarly for multiparameter compact Lie groups such as $\mathfrak{S}$ or $\mathfrak{S} \otimes \mathfrak{T}$, albeit with differing weighting depending on the parameterization employed.

For one body operators $\hat{\mathrm{A}}_1$,

$$\hat{\mathrm{A}}_1 = \sum_{ij}\sum_{\sigma_1\sigma_2} A_{i\sigma_1,j\sigma_2}\hat{\mathrm{E}}^{i\sigma_1}_{j\sigma_2}\,, \qquad (A_{i\sigma_1,j\sigma_2} \in \mathbb{C}) \tag{104}$$

which include the operator $\hat{\mathrm{T}}_1$ that is relevant for our purposes, we only need to know the transformation properties of the U$(2n)$ generators $\hat{\mathrm{E}}^{i\sigma_1}_{j\sigma_2}$, Eq. (85), which are given in Appendix A.

Considering, first, the finite subgroup $\mathfrak{M}$, Eq. (98), we find for the invariant mean, Eq. (100), of the operator $\hat{\mathrm{A}}_1$, Eq. (104), the expression

$$\mathcal{M}_{\mathfrak{M}}(\hat{\mathrm{A}}_1) = \tfrac{1}{2}(\hat{\mathrm{A}}_1 + \hat{\mathrm{K}}\hat{\mathrm{A}}_1\hat{\mathrm{K}}) = \sum_{ij}\sum_{\sigma_1\sigma_2} A^{(\mathfrak{M})}_{i\sigma_1,j\sigma_2}\hat{\mathrm{E}}^{i\sigma_1}_{j\sigma_2}\,, \tag{105}$$

where

$$A^{(\mathfrak{M})}_{i\sigma_1,j\sigma_2} = \tfrac{1}{2}(A_{i\sigma_1,j\sigma_2} + A^*_{i\sigma_1,j\sigma_2})\,. \tag{106}$$

If $\hat{\mathrm{A}}_1$ is already invariant, i.e., $\mathcal{M}_{\mathfrak{M}}(\hat{\mathrm{A}}_1) = \hat{\mathrm{A}}_1$, we can conclude that its matrix elements are real, i.e. $A_{i\sigma_1,j\sigma_2} \in \mathbb{R}$.

Turning next to the subgroup $\mathfrak{T}$, Eq. (90), and realizing that $-\hat{1}$ and $-\hat{\Theta}$ yield the same similarity transform as $\hat{1}$ and $\hat{\Theta}$, respectively, we get

$$\mathcal{M}_{\mathfrak{T}}(\hat{\mathrm{A}}_1) = \tfrac{1}{2}(\hat{\mathrm{A}}_1 + \hat{\Theta}\hat{\mathrm{A}}_1\hat{\Theta}^{-1}) = \sum_{ij}\sum_{\sigma_1\sigma_2} A^{(\mathfrak{T})}_{i\sigma_1,j\sigma_2}\hat{\mathrm{E}}^{i\sigma_1}_{j\sigma_2}\,, \tag{107}$$

where now

$$A^{(\mathfrak{T})}_{i\sigma_1,j\sigma_2} = \tfrac{1}{2}(A_{i\sigma_1,j\sigma_2} + (-1)^{\bar{\sigma}_1+\bar{\sigma}_2-1}A^*_{i\bar{\sigma}_1;j\bar{\sigma}_2})\,. \tag{108}$$

Again, $\hat{\mathrm{A}}_1$ is already invariant under the action of $\mathfrak{T}$ if

$$A_{i\bar{\sigma}_1,j\bar{\sigma}_2} = (-1)^{\sigma_1+\sigma_2-1}A^*_{i\sigma_1,j\sigma_2}\,, \tag{109}$$

i.e., when

$$A_{i\bar{\sigma},j\bar{\sigma}} = A^*_{i\sigma,j\sigma} \quad \text{and} \quad A_{i\bar{\sigma},j\sigma} = -A^*_{i\sigma,j\bar{\sigma}}\,. \tag{110}$$

For the continuous subgroups $\mathfrak{A}$ and $\mathfrak{S}$ we have to employ the integral form for the invariant mean, Eq. (103). For the axial spin rotation group $\mathfrak{A}$ this gives using Eq. (A.9)

$$\begin{aligned}\mathcal{M}_{\mathfrak{A}}(\hat{\mathrm{A}}_1) &= \frac{1}{4\pi}\int_{-2\pi}^{2\pi} e^{\mathrm{i}\hat{\mathrm{S}}_3\theta}\hat{\mathrm{A}}_1 e^{-\mathrm{i}S_3\theta}d\theta \\ &= \sum_{ij}\sum_{\sigma_1\sigma_2}\delta_{\sigma_1,\sigma_2}A_{i\sigma_1,j\sigma_2}\hat{\mathrm{E}}^{i\sigma_1}_{j\sigma_2} \\ &= \sum_{ij}\sum_{\sigma_1\sigma_2} A^{(\mathfrak{A})}_{i\sigma_1,j\sigma_2}\hat{\mathrm{E}}^{i\sigma_1}_{j\sigma_2}\,,\end{aligned} \tag{111}$$

with

$$A^{(\mathfrak{A})}_{i\sigma_1,j\sigma_2} = \delta_{\sigma_1\sigma_2} A_{i\sigma_1,j\sigma_2} \,. \tag{112}$$

Thus, if all the elements of $\hat{\mathrm{A}}_1$ in the off-diagonal spin blocks vanish, $\hat{\mathrm{A}}_1$ is already invariant under $\mathfrak{A}$.

For the full spin rotation group $\mathfrak{S}$, we have to average over the whole group. In the case of one-body operators,* this can be achieved via two-step projection, first by $\hat{\mathrm{S}}_3$, followed by $\hat{\mathrm{S}}_1$, yielding

$$\begin{aligned} \mathcal{M}_{\mathfrak{S}}(\hat{\mathrm{A}}_1) &= \frac{1}{4\pi}\int_{-2\pi}^{2\pi} e^{\mathrm{i}\hat{\mathrm{S}}_1\theta}\mathcal{M}_{\mathfrak{A}}(\hat{\mathrm{A}}_1)e^{-\mathrm{i}\hat{\mathrm{S}}_1\theta}d\theta \\ &= \sum_{ij}\sum_{\sigma_1\sigma_2} A^{(\mathfrak{S})}_{i\sigma_1,j\sigma_2}\hat{\mathrm{E}}^{i\sigma_1}_{j\sigma_2}\,, \end{aligned} \tag{113}$$

where now

$$A^{(\mathfrak{S})}_{i\sigma_1,j\sigma_2} = \tfrac{1}{2}\delta_{\sigma_1\sigma_2}(A_{i\sigma_1,j\sigma_2} + A_{i\bar{\sigma}_1,j\bar{\sigma}_2})\,. \tag{114}$$

Thus, $\hat{\mathrm{A}}_1$ is invariant under the spin rotation group if all off-diagonal blocks vanish and the remaining spin blocks are identical, i.e.

$$A_{i\bar{\sigma},j\bar{\sigma}} = A_{i\sigma,j\sigma}\,. \tag{115}$$

The above derived results for the coefficients of the projected operators and the conditions for their invariance with respect to the relevant subgroups are summarized in Tables 1 and 2.

4. Symmetry restricted HF equations and stability conditions

Exploiting the classification of invariant one-body operators given in the preceding section, we can characterize the structure of possible IPM bs solutions and derive explicit stability conditions for each class of solutions. Using the standard canonical MSO basis $\{|I\rangle\}$ with $|I\rangle = |i\sigma\rangle \equiv |i\rangle \otimes |\sigma\rangle$, Eq. (86), the operator $\hat{\mathrm{T}}_1$, Eq. (30), takes the form

$$\hat{\mathrm{T}}_1 = \sum_{ij}\sum_{\sigma_1\sigma_2} K_{i\sigma_1,j\sigma_2}\hat{\mathrm{E}}^{i\sigma_1}_{j\sigma_2}\,, \tag{116}$$

so that we can describe the desired structure of bs solutions in terms of the mixing matrix $\mathbf{K}$, $\mathbf{K} = (K_{i\sigma_1,j\sigma_2})$, or the corresponding coefficient matrix $\widetilde{\mathbf{C}}$, Eq. (38).

*For general many-body operators, the invariant mean for the spin rotation group is not obtained by the simple prescription described here. Instead it requires a suitable parameterization and weighting of the group elements to ensure the entire group is covered exactly once.

group	matrix elements
$\mathfrak{S}\otimes\mathfrak{T}$	$A^{(\mathfrak{S}\otimes\mathfrak{T})}_{i\sigma_1,j\sigma_2} = \frac{1}{4}\delta_{\sigma_1\sigma_2}(A_{i\sigma_1,j\sigma_2} + A^*_{i\sigma_1,j\sigma_2} + A_{i\bar{\sigma}_1,j\bar{\sigma}_2} + A^*_{i\bar{\sigma}_1,j\bar{\sigma}_2})$
$\mathfrak{S}$	$A^{(\mathfrak{S})}_{i\sigma_1,j\sigma_2} = \frac{1}{2}\delta_{\sigma_1\sigma_2}(A_{i\sigma_1,j\sigma_2} + A_{i\bar{\sigma}_1,j\bar{\sigma}_2})$
$\mathfrak{A}\mathfrak{M}$	$A^{(\mathfrak{A}\mathfrak{M})}_{i\sigma_1,j\sigma_2} = \frac{1}{2}\delta_{\sigma_1\sigma_2}(A_{i\sigma_1,j\sigma_2} + A^*_{i\sigma_1,j\sigma_2})$
$\mathfrak{A}\otimes\mathfrak{T}$	$A^{(\mathfrak{A}\otimes\mathfrak{T})}_{i\sigma_1,j\sigma_2} = \frac{1}{2}\delta_{\sigma_1\sigma_2}(A_{i\sigma_1,j\sigma_2} + (-1)^{\bar{\sigma}_1+\bar{\sigma}_2-1}A^*_{i\bar{\sigma}_1,j\bar{\sigma}_2})$
$\mathfrak{A}$	$A^{(\mathfrak{A})}_{i\sigma_1,j\sigma_2} = \delta_{\sigma_1\sigma_2}A_{i\sigma_1,j\sigma_2}$
$\mathfrak{M}$	$A^{(\mathfrak{M})}_{i\sigma_1,j\sigma_2} = \frac{1}{2}(A_{i\sigma_1,j\sigma_2} + A^*_{i\sigma_1,j\sigma_2})$
$\mathfrak{T}$	$A^{(\mathfrak{T})}_{i\sigma_1,j\sigma_2} = \frac{1}{2}(A_{i\sigma_1,j\sigma_2} + (-1)^{\bar{\sigma}_1+\bar{\sigma}_2-1}A^*_{i\bar{\sigma}_1,j\bar{\sigma}_2})$
$\mathfrak{E}$	$A^{(\mathfrak{E})}_{i\sigma_1,j\sigma_2} = A_{i\sigma_1,j\sigma_2}$

Table 1. Matrix elements of invariant one-body operator Â, Eq. (104), for the $\mathfrak{S}\otimes\mathfrak{T}$ group and its subgroups (cf. Fig. 1)

group	matrix elements
$\mathfrak{S}\otimes\mathfrak{T}$	$A_{i\bar{\sigma},j\bar{\sigma}} = A_{i\sigma,j\sigma} \in \mathbb{R}$, $A_{i\sigma,j\bar{\sigma}} = 0$
$\mathfrak{S}$	$A_{i\bar{\sigma},j\bar{\sigma}} = A_{i\sigma,j\sigma}$, $A_{i\sigma,j\bar{\sigma}} = 0$
$\mathfrak{A}\mathfrak{M}$	$A_{i\sigma,j\sigma} \in \mathbb{R}$, $A_{i\sigma,j\bar{\sigma}} = 0$
$\mathfrak{A}\otimes\mathfrak{T}$	$A_{i\bar{\sigma},j\bar{\sigma}} = A^*_{i\sigma,j\sigma}$, $A_{i\sigma,j\bar{\sigma}} = 0$
$\mathfrak{A}$	$A_{i\sigma,j\bar{\sigma}} = 0$
$\mathfrak{M}$	$A_{i\sigma_1,j\sigma_2} \in \mathbb{R}$
$\mathfrak{T}$	$A_{i\bar{\sigma},j\bar{\sigma}} = A^*_{i\sigma,j\sigma}$, $A_{i\bar{\sigma},j\sigma} = -A^*_{i\sigma,j\bar{\sigma}}$
$\mathfrak{E}$	

Table 2. Structure of the matrix elements of the invariant one-body operator Â, Eq. (104), for the $\mathfrak{S}\otimes\mathfrak{T}$ group and its subgroups. All matrix elements are complex valued unless indicated otherwise.

Relying on the results summarized in Table 2, and taking into account the skew-symmetry restrictions imposed on $\hat{T}_1$, Eq. (32), or on the corresponding mixing matrix **K**, Eq. (33), we find for the structure of the latter the results that are summarized in Table 3. Since the coefficient matrix $\widetilde{\mathbf{C}}$ is obtained by exponentiation, its structure is analogous to

group	matrix structure	restrictions
$\mathfrak{S}\otimes\mathfrak{T}$	$\mathbf{K}=\begin{pmatrix}\mathbf{K}_{\mathrm{R}} & 0\\ 0 & \mathbf{K}_{\mathrm{R}}\end{pmatrix}$	$(\mathbf{K}_{\mathrm{R}})_{ij}\in\mathbb{R}$
$\mathfrak{S}$	$\mathbf{K}=\begin{pmatrix}\mathbf{K}_{\mathrm{R}} & 0\\ 0 & \mathbf{K}_{\mathrm{R}}\end{pmatrix}$	
$\mathfrak{AM}$	$\mathbf{K}=\begin{pmatrix}\mathbf{K}_{\alpha} & 0\\ 0 & \mathbf{K}_{\beta}\end{pmatrix}$	$(\mathbf{K}_{\alpha})_{ij}, (\mathbf{K}_{\beta})_{ij}\in\mathbb{R}$
$\mathfrak{A}\otimes\mathfrak{T}$	$\mathbf{K}=\begin{pmatrix}\mathbf{K}_{\alpha} & 0\\ 0 & \mathbf{K}_{\alpha}^{*}\end{pmatrix}$	
$\mathfrak{A}$	$\mathbf{K}=\begin{pmatrix}\mathbf{K}_{\alpha} & 0\\ 0 & \mathbf{K}_{\beta}\end{pmatrix}$	
$\mathfrak{M}$	$\mathbf{K}=\begin{pmatrix}\mathbf{K}_{\alpha\alpha} & \mathbf{K}_{\alpha\beta}\\ -\mathbf{K}_{\alpha\beta}^{t} & \mathbf{K}_{\beta\beta}\end{pmatrix}$	$(\mathbf{K}_{\sigma_1\sigma_2})_{ij}\in\mathbb{R}$
$\mathfrak{T}$	$\mathbf{K}=\begin{pmatrix}\mathbf{K}_{\alpha\alpha} & \mathbf{K}_{\alpha\beta}\\ -\mathbf{K}_{\alpha\beta}^{*} & \mathbf{K}_{\alpha\alpha}^{*}\end{pmatrix}$	$(\mathbf{K}_{\alpha\beta})^{\mathrm{t}}=\mathbf{K}_{\alpha\beta}$
$\mathfrak{E}$	$\mathbf{K}=\begin{pmatrix}\mathbf{K}_{\alpha\alpha} & \mathbf{K}_{\alpha\beta}\\ -\mathbf{K}_{\alpha\beta}^{\dagger} & \mathbf{K}_{\beta\beta}\end{pmatrix}$	

Table 3. Structure of the invariant mixing matrix **K** for various solution classes. Entries are complex valued unless indicated otherwise.

that of the **K** matrix, as shown in Table 4 (from now on, we drop the tilde in order to simplify the notation).

It is important to emphasize at this stage that the classification given in Table 4 is identical with that given by Fukutome,[50,63] except that our formulation[89] applies to a more general class of wave functions, including various OS situations. For example, the earlier formulated doublet stability conditions[44–46] for simple OS ROHF solutions go beyond the scope of Fukutome's formulation, while they easily fit within the above exposed formalism.

In the past, several designations have been employed for various bs solutions. The most comprehensive classification, based on the eight possible subgroups of the spin-rotation and time-reversal group $\mathfrak{S}\otimes\mathfrak{T}$, was suggested by Fukutome.[50,63] This nomenclature, together with the

group	matrix structure	restrictions
$\mathfrak{S}\otimes\mathfrak{T}$	$\mathbf{C}=\begin{pmatrix}\mathbf{C}_\mathrm{R} & 0\\ 0 & \mathbf{C}_\mathrm{R}\end{pmatrix}$	$(\mathbf{C}_\mathrm{R})_{ij}\in\mathbb{R}$
$\mathfrak{S}$	$\mathbf{C}=\begin{pmatrix}\mathbf{C}_\mathrm{R} & 0\\ 0 & \mathbf{C}_\mathrm{R}\end{pmatrix}$	
$\mathfrak{AM}$	$\mathbf{C}=\begin{pmatrix}\mathbf{C}_\alpha & 0\\ 0 & \mathbf{C}_\beta\end{pmatrix}$	$(\mathbf{C}_\alpha)_{ij},(\mathbf{C}_\beta)_{ij}\in\mathbb{R}$
$\mathfrak{A}\otimes\mathfrak{T}$	$\mathbf{C}=\begin{pmatrix}\mathbf{C}_\alpha & 0\\ 0 & \mathbf{C}^*_\alpha\end{pmatrix}$	
$\mathfrak{A}$	$\mathbf{C}=\begin{pmatrix}\mathbf{C}_\alpha & 0\\ 0 & \mathbf{C}_\beta\end{pmatrix}$	
$\mathfrak{M}$	$\mathbf{C}=\begin{pmatrix}\mathbf{C}_{\alpha\alpha} & \mathbf{C}_{\alpha\beta}\\ \mathbf{C}_{\beta\alpha} & \mathbf{C}_{\beta\beta}\end{pmatrix}$	$(\mathbf{C}_{\sigma_1\sigma_2})_{ij}\in\mathbb{R}$
$\mathfrak{T}$	$\mathbf{C}=\begin{pmatrix}\mathbf{C}_{\alpha\alpha} & \mathbf{C}_{\alpha\beta}\\ -\mathbf{C}^*_{\alpha\beta} & \mathbf{C}^*_{\alpha\alpha}\end{pmatrix}$	
$\mathfrak{C}$	$\mathbf{C}=\begin{pmatrix}\mathbf{C}_{\alpha\alpha} & \mathbf{C}_{\alpha\beta}\\ \mathbf{C}_{\beta\alpha} & \mathbf{C}_{\beta\beta}\end{pmatrix}$	

Table 4. Structure of the orbital coefficient matrix **C** for various solution classes. Entries are complex valued unless indicated otherwise.

standard, generally employed terminology (which, however, is often ambiguous), is summarized in Table 5, together with the proposed simplified classification that will be employed in the following text. Thus, the wave functions that are invariant with respect to the $\mathfrak{S}\otimes\mathfrak{T}, \mathfrak{AM}$, and $\mathfrak{M}$ groups will be referred to as the *real RHF*, *real UHF*, and *real GHF* solutions, respectively. Similarly, the terms *complex RHF*, *complex UHF*, and *complex GHF* will be associated with solutions that are invariant under the $\mathfrak{S}, \mathfrak{A}$, and $\mathfrak{C}$ groups, respectively. These terms clearly indicate the structure of the pertinent solutions, while being sufficiently concise and exact. The remaining classes associated with the $\mathfrak{A}\otimes\mathfrak{T}$ and $\mathfrak{T}$ groups are then referred to as the *paired UHF* and *paired GHF* types. In Fukutome's terminology, implied by the character of the corresponding solutions for cyclic or infinite linear polyenes, these latter

group	Fukutome's designation	suggested designation	standard designation
$\mathfrak{S}\otimes\mathfrak{T}$	TICS time invariant closed shells	**real RHF**	RHF, ROHF
$\mathfrak{S}$	CCW charge current waves	**complex RHF**	restricted (open shell) HF
$\mathfrak{A}\mathfrak{M}$	ASDW axial spin density waves	**real UHF**	UHF (DODS UHF)
$\mathfrak{A}\otimes\mathfrak{T}$	ASCW axial spin current waves	**paired UHF**	unrestricted HF, different orbitals for different spins UHF
$\mathfrak{A}$	ASW axial spin waves	**complex UHF**	
$\mathfrak{M}$	TSDW torsional spin density waves	**real GHF**	GHF
$\mathfrak{T}$	TSCW torsional spin current waves	**paired GHF**	generalized HF
$\mathfrak{E}$	TSW torsional spin waves	**complex GHF**	

Table 5. Nomenclature for Hartree-Fock solution classes. This list provides a translation table between the subgroup classification, Fukutome's designation[50] , and the one adopted in this work. The standard designation is also provided as a reference.

types are referred to as solutions possessing the *current* rather than *density waves*, specifically the axial-spin-current and torsional-spin-current waves, respectively (cf. Table 5).

4.1 Symmetry restricted HF equations

When relying on the DEM procedure, the general expressions for the direction of the steepest descent, Eq. (53), must be appropriately adjusted for each symmetry class of solutions by giving the desired form to the mixing matrix **K**, and thus to the first variation of the energy functional $\delta^{(1)}\mathcal{E}$. In this subsection we present such a formulation for the real and complex GHF, UHF, and RHF solution classes.

4.1.1 GHF equations. For the real and complex GHF solution classes, which correspond to the complex conjugation group $\mathfrak{M}$ and to the trivial group $\mathfrak{E}$, respectively, the direction of steepest descent is essentially given by the general expression (53), where the matrix elements of **K** are real and complex valued for the respective classes, with the Fock matrix given by Eq. (46).

4.1.2 UHF equations. The real and complex UHF solution classes that are associated, respectively, with the subgroups $\mathfrak{A}\mathfrak{M}$ and $\mathfrak{A}$, are characterized by the block diagonal mixing matrix

$$\mathbf{K} = \begin{pmatrix} \mathbf{K}_\alpha & 0 \\ 0 & \mathbf{K}_\beta \end{pmatrix}, \tag{117}$$

assuming that we employ canonical MSOs. The Fock matrix is blocked correspondingly,

$$\mathbf{F} = \begin{pmatrix} \mathbf{F}_\alpha & 0 \\ 0 & \mathbf{F}_\beta \end{pmatrix}, \tag{118}$$

with the spin blocks

$$(\mathbf{F}_\sigma)_{ij} = F_{i\sigma,j\sigma}\,. \tag{119}$$

The explicit expressions for these matrix elements are given in Appendix B.

The direction of the steepest descent is then implied by the expression for the first variation of the energy functional $\delta^{(1)}\mathcal{E}$, Eq. (48), which in this case has the form

$$\delta^{(1)}\mathcal{E} = \mathrm{Tr}([\mathbf{N},\mathbf{F}]\mathbf{K}) = \mathrm{Tr}\left\{\begin{pmatrix} [\mathbf{N}_\alpha,\mathbf{F}_\alpha] & 0 \\ 0 & [\mathbf{N}_\beta,\mathbf{F}_\beta] \end{pmatrix}\begin{pmatrix} \mathbf{K}_\alpha & 0 \\ 0 & \mathbf{K}_\beta \end{pmatrix}\right\}, \tag{120}$$

where $\mathbf{N}_\alpha$ and $\mathbf{N}_\beta$ are diagonal matrices of the MSO occupation numbers $n_{i\sigma}$, $(\sigma = \alpha, \beta)$, $n_{i\sigma} = 0$ or 1. Using the same argument as in Sec. 2.4, we thus find that

$$\mathbf{K}_\sigma = [\mathbf{N}_\sigma, \mathbf{F}_\sigma]\,, \qquad \sigma = \alpha, \beta\,. \tag{121}$$

At first sight, the expression corresponding to the α and β spins seem to be independent of one-another. However, they are coupled through the expression for the Fock matrix, Eq. (B.12).

4.1.3 RHF equations. For the real and complex RHF solutions that correspond, respectively, to the subgroups $\mathfrak{S}\otimes\mathfrak{T}$ and $\mathfrak{S}$, the mixing matrix has a simple spin-independent, block-diagonal form

$$\mathbf{K} = \begin{pmatrix} \mathbf{K}_{\mathrm{R}} & 0 \\ 0 & \mathbf{K}_{\mathrm{R}} \end{pmatrix} . \tag{122}$$

Although the corresponding coefficient matrix $\mathbf{C}$ is also spin independent, the Fock matrix has identical spin blocks $\mathbf{F}_\alpha$ and $\mathbf{F}_\beta$ only in the CS case, while in the ROHF case the $\mathbf{N}_\alpha$ and $\mathbf{N}_\beta$ occupancy matrices are different and thus the $\mathbf{F}_\alpha$ and $\mathbf{F}_\beta$ blocks are also not the same. Thus, for high-spin OS HF equations, the Fock matrix is spin dependent despite the fact that the orbital coefficients are spin independent.

Indeed, employing the expression (120) for $\delta^{(1)}\mathcal{E}$ with $\mathbf{K}_\sigma = \mathbf{K}_{\mathrm{R}}$, $(\sigma = \alpha, \beta)$, and taking a trace over the spin index, we get

$$\delta^{(1)}\mathcal{E} = \mathrm{Tr}\{([\mathbf{N}_\alpha, \mathbf{F}_\alpha] + [\mathbf{N}_\beta, \mathbf{F}_\beta])\mathbf{K}_{\mathrm{R}}\} , \tag{123}$$

which implies that

$$\mathbf{K}_{\mathrm{R}} = [\mathbf{N}_\alpha, \mathbf{F}_\alpha] + [\mathbf{N}_\beta, \mathbf{F}_\beta] . \tag{124}$$

If the α and β occupancies are identical, as in the CS singlet case, the problem greatly simplifies, since all the spin-dependence disappears from the Fock matrix $\mathbf{F}$ and

$$\mathbf{K}_{\mathrm{R}} = 2[\mathbf{N}_{\mathrm{R}}, \mathbf{F}_{\mathrm{R}}] . \tag{125}$$

However, when $\mathbf{N}_\alpha \neq \mathbf{N}_\beta$ we deal with conditions that correspond to Roothaan's restricted OS (ROHF) equations.

At this point it is important to observe that the spin dependence or independence of the Fock matrix for the RHF class of solutions, which distinguishes, respectively, the cases characterized by Eqs. (124) and (125), does not arise from the classification of broken symmetry solutions, but instead from the so-called self-consistent symmetries of HF equations that are discussed next.

4.1.4 Self-consistent symmetries. Generally speaking, the self-consistent symmetries do not arise from the invariance properties of the Hamiltonian, but are the consequence of the specific approaches determining the 'best' IPM wave function on the basis of the variation principle. In the case of HF equations, these symmetries may be ascribed to the interaction of the so-called *mean field* with itself that characterizes the self-consistent process of solving the HF equations.

Let $\hat{\mathrm{g}}$ represent an invariance operator of our Hamiltonian $\hat{\mathrm{H}}$, Eq. (74), (75), that transforms a given IPM wave function $|\Phi\rangle$ into $|\widetilde{\Phi}\rangle$,

$$|\widetilde{\Phi}\rangle = \hat{\mathrm{g}}|\Phi\rangle\,. \tag{126}$$

The corresponding density matrix $\mathbf{P}$, Eq. (20), is similarly transformed,

$$\widetilde{\mathbf{P}} = \mathbf{\Gamma}\mathbf{P}\mathbf{\Gamma}^{-1}\,, \tag{127}$$

with $\mathbf{\Gamma} \equiv \mathbf{\Gamma}(\hat{\mathrm{g}})$ designating a suitable matrix representation of $\hat{\mathrm{g}}$. Without restricting generality, we can assume the representation $\mathbf{\Gamma}$ to be unitary, so that

$$\begin{aligned}\mathcal{E}[\widetilde{\Phi}] &= \langle\widetilde{\Phi}|\hat{\mathrm{H}}|\widetilde{\Phi}\rangle = \langle\Phi|\hat{\mathrm{g}}^{\dagger}\hat{\mathrm{H}}\hat{\mathrm{g}}|\Phi\rangle = \langle\Phi|\hat{\mathrm{g}}^{-1}\hat{\mathrm{H}}\hat{\mathrm{g}}|\Phi\rangle \\ &= \langle\Phi|\hat{\mathrm{H}}|\Phi\rangle = \mathcal{E}[\Phi]\,,\end{aligned} \tag{128}$$

where we used the fact that $\hat{\mathrm{H}}$ is invariant with respect to $\hat{\mathrm{g}}$. However, this does not imply that $|\widetilde{\Phi}\rangle = \hat{\mathrm{g}}|\Phi\rangle$ and $|\Phi\rangle$ are identical, or even that the Fock-Dirac density matrices $\widetilde{\mathbf{P}}$ and $\mathbf{P}$ coincide. In the event when $\widetilde{\mathbf{P}} \neq \mathbf{P}$, and hence $\mathbf{F}(\widetilde{\mathbf{P}}) \neq \mathbf{F}(\mathbf{P})$, the solutions $|\widetilde{\Phi}\rangle$ and $|\Phi\rangle$ are distinct but energetically degenerate, which can be verified through Eq. (128). If, instead, the action of $\mathbf{\Gamma}$ yields $\widetilde{\mathbf{P}} = \mathbf{P}$ (and consequently $\mathbf{F}(\widetilde{\mathbf{P}}) = \mathbf{F}(\mathbf{P})$), the solutions are then identical. In such a case the operator $\hat{\mathrm{g}}$ represents the *self-consistent symmetry* of the Fock matrix.

Recall that $\mathbf{P} = \mathbf{C}\mathbf{N}\mathbf{C}^{\dagger}$, Eq. (20), and $|I\rangle = \sum_{\Lambda} C_{\Lambda I}|\Lambda\rangle$, Eq. (1). Arranging the MSOs and ASOs as the entries of the corresponding row matrices $\mathbf{\Phi}$ and $\mathbf{\Lambda}$, we can cast Eq. (1) into the matrix form

$$\mathbf{\Phi} = \mathbf{\Lambda}\mathbf{C}\,. \tag{129}$$

Since only $\phi_I(\vec{x}) = \langle\vec{x}|I\rangle$ and $\chi_{\Lambda}(\vec{x}) = \langle\vec{x}|\Lambda\rangle$, $\vec{x} \equiv (\vec{r}, s)$, depend on the spatial ($\vec{r}$) and spin (s) variables, the action of $\hat{\mathrm{g}}$ or $\mathbf{\Gamma}$ leaves matrix $\mathbf{C}$ unchanged, i.e. $\hat{\mathrm{g}}|I\rangle = \sum_{\Lambda} C_{\Lambda I}\hat{\mathrm{g}}|\Lambda\rangle$, so that $\mathbf{C}$ is invariant with respect to $\hat{\mathrm{g}}$,

$$\widetilde{\mathbf{C}} = \mathbf{\Gamma}\mathbf{C}\mathbf{\Gamma}^{-1} = \mathbf{C}\,. \tag{130}$$

We thus find for $\widetilde{\mathbf{P}}$ that

$$\widetilde{\mathbf{P}} = \mathbf{\Gamma}\mathbf{P}\mathbf{\Gamma}^{-1} = \mathbf{\Gamma}\mathbf{C}\mathbf{N}\mathbf{C}^{\dagger}\mathbf{\Gamma}^{-1} = \widetilde{\mathbf{C}}\widetilde{\mathbf{N}}\widetilde{\mathbf{C}}^{\dagger} = \mathbf{C}\widetilde{\mathbf{N}}\mathbf{C}^{\dagger}\,, \tag{131}$$

so that, naturally, the equality $\widetilde{\mathbf{P}} = \mathbf{P}$ requires that $\widetilde{\mathbf{N}} = \mathbf{N}$.

Now, a wave function (or the corresponding density matrix) that transforms according to some irreducible representation $\mathbf{\Gamma}$ of the invariance group $\mathfrak{G}$ (or equivalently, belongs to the symmetry species $\mathbf{\Gamma}$) may be characterized by an appropriate set of quantum numbers. Since $\mathbf{\Gamma}$ is also a representation for any subgroup $\mathfrak{H}$ of $\mathfrak{G}$, $\mathfrak{H} \subset \mathfrak{G}$ (though not necessarily irreducible), the same quantum numbers are preserved under the action of $\mathfrak{H}$. In principle, this invariance should also be preserved by the solutions of HF equations. Specifically, a solution of the symmetry adapted HF equations corresponding to the group $\mathfrak{G}$ should also be a solution of HF equations that are symmetry adapted relative to the subgroup $\mathfrak{H}$ of $\mathfrak{G}$, i.e., it should also represent a critical point on the energy hypersurface associated with a larger variational space associated with the subgroup $\mathfrak{H}$.

For example, a real valued IPM wave function that is invariant under the action of the group $\mathfrak{AM}$ must be automatically invariant with respect to its proper subgroup $\mathfrak{M}$. This invariance property is inherited by the corresponding HF equations, and any UHF solution will also represent a critical point, albeit not necessarily a minimum, of the GHF functional. However, this inheritance property is not generally preserved for certain classes of symmetry restricted HF equations. Thus, while the CS RHF solutions also represent critical points of the UHF energy functional, this is not the case for the OS ROHF solutions, which never satisfy the UHF equations. (*Mutatis mutandis*, the ROHF simple-open-shell solutions are *always* doublet unstable[44–46]). This implies that the 'best' IPM solutions, i.e., the HF solutions, do not necessarily inherit some underlying symmetries, as is easily understood in the case of the latter example in terms of the above derived result. Indeed, in the case of CS singlet RHF solutions, the occupation numbers (or the density matrix) are clearly invariant with respect to any rotation in the spin space, so that $\widetilde{\mathbf{N}} = \mathbf{N}$ and $\widetilde{\mathbf{P}} = \mathbf{P}$, thus preserving the energy. However, in the OS case, the self-consistent symmetry is absent, so that $\widetilde{\mathbf{N}} \neq \mathbf{N}$, and the ROHF solution does not represent a stationary point on the UHF energy hypersurface, the latter solution always having a lower energy than the ROHF one due to the spin polarization.

We thus conclude that solutions of the $\mathfrak{S} \otimes \mathfrak{T}$, $\mathfrak{S}$, $\mathfrak{A} \otimes \mathfrak{T}$, and $\mathfrak{T}$ type possess a self-consistent symmetry only when $\mathbf{N}_\alpha = \mathbf{N}_\beta$. Since the real and complex valued high-spin ROHF solutions, associated with the groups $\mathfrak{S} \otimes \mathfrak{T}$ and $\mathfrak{S}$, respectively, do not possess the self-consistent symmetry, as $\mathbf{N}_\alpha \neq \mathbf{N}_\beta$, they are not solutions of the UHF equations. This fact leads to the distinction between the cases characterized by Eqs. (124) and (125) for the solutions of the RHF classes. Consequently, the identical ($\mathbf{N}_\alpha = \mathbf{N}_\beta$) and distinct ($\mathbf{N}_\alpha \neq \mathbf{N}_\beta$) occupancy cases for the RHF solutions must be handled separately (cf., e.g. Refs. 30, 45) despite belonging to the same class of bs solutions.

4.2 Symmetry restricted stability conditions

For each symmetry class of Sec. 3 we can now employ the structure of the mixing matrix **K** (Table 3) to determine the corresponding form of the second variation $\delta^{(2)}\mathcal{E}$, Eq. (49), and thus of the stability matrix **M**, Eq. (67). In doing so, it is important to remove certain trivial transformations involving, separately, the occupied and virtual orbitals (and doubly and singly occupied orbitals in the ROHF case). Such transformations only rotate a given solution into an equivalent one, and correspond to zero eigenvalues of the stability matrix. Once such transformations are eliminated, a zero eigenvalue of **M** indicates a possible boundary between the stable and unstable regimes.

In the following, we only address the real and complex RHF, UHF, and GHF solutions, omitting the paired-solution classes.

4.2.1 GHF stability conditions. The general form of matrix elements of the stability matrix **M**, Eq. (67), or of its constituting submatrices **A** and **B**, is given by Eqs. (68) and (69). In the GHF case, the above mentioned trivial transformations are those rotations involving the MSOs having the same occupancies, i.e., between the occupied ($n_I = 1$) and between the unoccupied ($n_I = 0$) MSOs. Labelling, thus, the occupied and virtual MSOs with the upper case letters from, respectively, the beginning $(A, B, \cdots)$ and the end $(R, S, \cdots)$ of the Latin alphabet, so that $n_A = n_B = \cdots = 1$ and $n_R = n_S = \cdots = 0$, we arrive at the following form for the matrix elements of the stability matrix

$$A_{AR;BS} = \delta_{AB}F_{SR} - \delta_{SR}F_{AB} + V^{(\mathcal{A})}_{AS;RB} \tag{132}$$

$$B_{AR;BS} = V^{(\mathcal{A})}_{AB;RS} \,. \tag{133}$$

Clearly, the column vector **k** defining the augmented vector **v**, Eq. (66), involves only the entries K_{AR}. We thus recover the general spin-orbital form of Thouless' stability conditions.[20]

4.2.2 UHF stability conditions. To find the stability conditions for the real and complex UHF solutions, we employ the canonical spin-orbital basis, writing $|I\rangle = |i\sigma\rangle \equiv |i\rangle \otimes |\sigma\rangle$, etc., and take into account the partial spin dependence of the matrix elements of $\hat{\mathrm{V}}$, Eq. (B.9) and of $\hat{\mathrm{F}}$, Eq. (B.12) and (B.13). This brings the general expressions (68) and (69) to the form

$$A_{i\sigma_1 j\sigma_2;k\sigma_3\ell\sigma_4} = -\tfrac{1}{2}(n_{i\sigma_1} + n_{k\sigma_1} - 2n_{j\sigma_2})\delta_{\ell j}\delta_{\sigma_1\sigma_3}\delta_{\sigma_2\sigma_4}(\mathbf{F}_{\sigma_1})_{ik} \\ - \tfrac{1}{2}(n_{\ell\sigma_2} + n_{j\sigma_2} - 2n_{k\sigma_1})\delta_{ik}\delta_{\sigma_1\sigma_3}\delta_{\sigma_2\sigma_4}(\mathbf{F}_{\sigma_2})_{\ell j} \\ + (\delta_{\sigma_1\sigma_2}\delta_{\sigma_3\sigma_4}V_{i\sigma_1\ell\sigma_3;j\sigma_1k\sigma_3} \\ - \delta_{\sigma_1\sigma_3}\delta_{\sigma_2\sigma_4}V_{i\sigma_1\ell\sigma_2;k\sigma_1j\sigma_2})(n_{i\sigma_1} - n_{j\sigma_2})(n_{k\sigma_3} - n_{\ell\sigma_4}) , \tag{134}$$

$$B_{i\sigma_1 j\sigma_2;k\sigma_3\ell\sigma_4} = \tfrac{1}{2}(n_{i\sigma_1} + n_{\ell\sigma_1} - 2n_{j\sigma_2})\delta_{kj}\delta_{\sigma_1\sigma_4}\delta_{\sigma_2\sigma_3}(\mathbf{F}_{\sigma_1})_{i\ell} \\ + \tfrac{1}{2}(n_{k\sigma_2} + n_{j\sigma_2} - 2n_{\ell\sigma_1})\delta_{i\ell}\delta_{\sigma_1\sigma_4}\delta_{\sigma_2\sigma_3}(\mathbf{F}_{\sigma_2})_{kj} \\ + (\delta_{\sigma_1\sigma_2}\delta_{\sigma_3\sigma_4}V_{i\sigma_1k\sigma_3;j\sigma_1\ell\sigma_3} \\ - \delta_{\sigma_1\sigma_4}\delta_{\sigma_2\sigma_3}V_{i\sigma_1k\sigma_2;\ell\sigma_1j\sigma_2})(n_{i\sigma_1} - n_{j\sigma_2})(n_{k\sigma_3} - n_{\ell\sigma_4}) . \tag{135}$$

The **K**-matrix associated with real and complex UHF solutions has the structure given in Table 3, i.e.

$$K_{i\sigma_1,j\sigma_2} = \delta_{\sigma_1\sigma_2}(\mathbf{K}_{\sigma_1})_{ij} , \tag{136}$$

implying that only the entries of the type $W_{i\sigma j\sigma;k\sigma'\ell\sigma'}$, $W = A, B$, are admissible, so that Eqs. (134) and (135) become

$$A_{i\sigma j\sigma;k\sigma'\ell\sigma'} = -\tfrac{1}{2}(n_{i\sigma} + n_{k\sigma} - 2n_{j\sigma})\delta_{\ell j}\delta_{\sigma\sigma'}(\mathbf{F}_{\sigma})_{ik} \\ - \tfrac{1}{2}(n_{\ell\sigma} + n_{j\sigma} - 2n_{k\sigma})\delta_{ik}\delta_{\sigma\sigma'}(\mathbf{F}_{\sigma})_{\ell j} \\ + (V_{i\sigma\ell\sigma';j\sigma k\sigma'} - \delta_{\sigma\sigma'}V_{i\sigma\ell\sigma;k\sigma j\sigma})(n_{i\sigma} - n_{j\sigma})(n_{k\sigma'} - n_{\ell\sigma'}) , \tag{137}$$

$$B_{i\sigma j\sigma;k\sigma'\ell\sigma'} = -\tfrac{1}{2}(n_{i\sigma} + n_{\ell\sigma} - 2n_{j\sigma})\delta_{kj}\delta_{\sigma\sigma'}(\mathbf{F}_{\sigma})_{i\ell} \\ + \tfrac{1}{2}(n_{k\sigma} + n_{j\sigma} - 2n_{\ell\sigma})\delta_{i\ell}\delta_{\sigma\sigma'}(\mathbf{F}_{\sigma})_{kj} \\ + (V_{i\sigma k\sigma';j\sigma\ell\sigma'} - \delta_{\sigma\sigma'}V_{i\sigma k\sigma;\ell\sigma j\sigma})(n_{i\sigma} - n_{j\sigma})(n_{k\sigma'} - n_{\ell\sigma'}) . \tag{138}$$

Eliminating, finally, the trivial terms corresponding to mixing between the occupied and, separately, between the unoccupied MSOs of the same spin, and labelling the occupied MSOs associated with spin σ by $a(\sigma), b(\sigma)$ and the unoccupied ones by $s(\sigma), r(\sigma)$, we finally obtain for the relevant matrix elements

$$A_{a(\sigma)r(\sigma);b(\sigma)s(\sigma)} = -\delta_{r(\sigma)s(\sigma)}(\mathbf{F}_{\sigma})_{a(\sigma)b(\sigma)} \\ + \delta_{a(\sigma)b(\sigma)}(\mathbf{F}_{\sigma})_{s(\sigma)r(\sigma)} + V^{(\mathcal{A})}_{a(\sigma)s(\sigma);r(\sigma)b(\sigma)} , \\ A_{a(\sigma)r(\sigma);b(\bar{\sigma})s(\bar{\sigma})} = V_{a(\sigma)s(\bar{\sigma});r(\sigma)b(\bar{\sigma})} , \tag{139}$$

$$B_{a(\sigma)r(\sigma);b(\sigma)s(\sigma)} = V^{(\mathcal{A})}_{a(\sigma)b(\sigma);r(\sigma)s(\sigma)} , \\ B_{a(\sigma)r(\sigma);b(\bar{\sigma})s(\bar{\sigma})} = V_{a(\sigma)b(\bar{\sigma});r(\sigma)s(\bar{\sigma})} , \tag{140}$$

where again $\bar{\sigma} = -\sigma$ (or $\bar{\alpha} = \beta, \bar{\beta} = \alpha$). In implementing these stability conditions, attention must be paid to the fact that the number of occupied MSOs is, generally, different for each spin σ, and each of the orbital labels a, b and r, s is associated with the occupancy of a particular spin label.

In the case of identical occupancies $\mathbf{N}_\alpha = \mathbf{N}_\beta$, characterizing the CS case, the stability of the RHF solutions within the space of real or complex UHF solutions reduces to two subproblems. This is similar to the real and imaginary subproblems $(\mathbf{A}+\mathbf{B})$ and $(\mathbf{A}-\mathbf{B})$, Eqs. (72) and (73), respectively. However, in the case of the RHF solutions, this leads to the so-called *singlet-triplet stability conditions*, first formulated by Čížek and Paldus.[30,32–35] The singlet problem relates to the stability within the space of proper RHF solutions and is identical with the stability conditions formulated in the next subsection 4.2.3. However, the triplet (or non-singlet) instability implies the existence of a UHF solution that breaks the spin symmetry, and has a lower energy than the original RHF solution.

In the case of distinct occupancies $\mathbf{N}_\alpha \neq \mathbf{N}_\beta$, characterizing the OS case, the resulting RHF (or, rather, ROHF) solutions are always unstable within the UHF solution space due to the lack of self-consistent symmetry, as explained earlier (cf. Sec. 4.1.4). Typically, such ROHF solutions do not correspond to a critical point on the energy hypersurface within the variational space allowing for a spin-symmetry breakdown.

4.2.3 RHF stability conditions. For the real and complex RHF solution classes, the matrix elements of $\hat{\mathrm{Z}}$ and $\hat{\mathrm{V}}$ operators are spin-independent, Eqs. (B.8) and (B.9), i.e.,

$$Z_{i\sigma_1,j\sigma_2} = Z_{ij}\delta_{\sigma_1\sigma_2}, \qquad V_{i\sigma_1 j\sigma_2;k\sigma_3\ell\sigma_4} = V_{ij;k\ell}\delta_{\sigma_1\sigma_3}\delta_{\sigma_2\sigma_4}\,, \tag{141}$$

and the block-diagonal mixing matrix $\mathbf{K}$ has identical spin blocks

$$K_{i\sigma_1,j\sigma_2} = (\mathbf{K}_{\mathrm{R}})_{ij}\delta_{\sigma_1\sigma_2}\,, \tag{142}$$

although the Fock matrix, depending on the orbital occupancies, may still be spin dependent. Consequently, the RHF stability matrices $\mathbf{A}$ and $\mathbf{B}$ involve spin-free matrix elements, namely

$$(\mathbf{A}_{\mathrm{R}})_{ij;k\ell} = \sum_{\sigma\sigma'} A_{i\sigma j\sigma;k\sigma'\ell\sigma'}\,, \tag{143}$$

and similarly for $(\mathbf{B}_{\mathrm{R}})_{ij;k\ell}$, with $A_{i\sigma j\sigma;k\sigma'\ell\sigma'}$ and $B_{i\sigma j\sigma;k\sigma'\ell\sigma'}$ given by the UHF expressions (137) and (138), respectively, albeit with spin independent two-electron integrals. Thus, for the RHF stability matrices we get

$$(\mathbf{A}_{\rm R})_{ij;k\ell} = -\tfrac{1}{2}\sum_{\sigma}(n_{i\sigma} + n_{k\sigma} - 2n_{j\sigma})\delta_{\ell j}(\mathbf{F}_\sigma)_{ik} - \tfrac{1}{2}\sum_{\sigma}(n_{\ell\sigma} + n_{j\sigma} - 2n_{k\sigma})\delta_{ik}(\mathbf{F}_\sigma)_{\ell j} + V_{i\ell;jk}\sum_{\sigma\sigma'}(n_{i\sigma} - n_{j\sigma})(n_{k\sigma'} - n_{\ell\sigma'}) - V_{i\ell;kj}\sum_{\sigma}(n_{i\sigma} - n_{j\sigma})(n_{k\sigma} - n_{\ell\sigma})\,, \tag{144}$$

$$(\mathbf{B}_{\rm R})_{ij;k\ell} = \tfrac{1}{2}\sum_{\sigma}(n_{i\sigma} + n_{\ell\sigma} - 2n_{j\sigma})\delta_{kj}(\mathbf{F}_\sigma)_{i\ell} + \tfrac{1}{2}\sum_{\sigma}(n_{k\sigma} + n_{j\sigma} - 2n_{\ell\sigma})\delta_{i\ell}(\mathbf{F}_\sigma)_{kj} + V_{ik;j\ell}\sum_{\sigma\sigma'}(n_{i\sigma} - n_{j\sigma})(n_{k\sigma'} - n_{\ell\sigma'}) - V_{ik;\ell j}\sum_{\sigma}(n_{i\sigma} - n_{j\sigma})(n_{k\sigma} - n_{\ell\sigma})\,. \tag{145}$$

Unlike the GHF and UHF cases, the presence of the self-consistent symmetry when $\mathbf{N}_\alpha = \mathbf{N}_\beta$ requires that we differentiate between the stability conditions for identical ($\mathbf{N}_\alpha = \mathbf{N}_\beta$) and distinct ($\mathbf{N}_\alpha \neq \mathbf{N}_\beta$) occupancy classes.

Singlet stability. The RHF stability conditions for the case of identical spin-up and spin-down occupancies, $\mathbf{N}_\alpha = \mathbf{N}_\beta$, correspond to singlet RHF solutions, when

$$\mathbf{F}_\alpha = \mathbf{F}_\beta = \mathbf{F}_{\rm R}\,. \tag{146}$$

Labelling the doubly occupied orbitals by a and b, and the unoccupied ones by r and s, the nonvanishing matrix elements (144) and (145) become

$$\begin{aligned}(\mathbf{A}_{\rm R})_{ar;bs} &= 2\delta_{ab}(\mathbf{F}_{\rm R})_{sr} - 2\delta_{rs}(\mathbf{F}_{\rm R})_{ab} + 4V_{as;rb} - 2V_{as;br}\\ (\mathbf{B}_{\rm R})_{ar;bs} &= 4V_{ab;rs} - 2V_{ab;sr}\,.\end{aligned} \tag{147}$$

These are precisely the CS singlet RHF stability conditions of Čížek and Paldus.[30,35]

High-spin stability. For the single-determinantal IPM wave functions, the only other possibility to generate a total spin eigenstate is to

$$(\mathbf{A}_\mathrm{R})_{ap;bq} = -\delta_{pq}\Big(\sum_\sigma (\mathbf{F}_\sigma)_{ab} - (\mathbf{F}_\alpha)_{ab}\Big) + \delta_{ab}\Big(\textstyle\sum_\sigma (\mathbf{F}_\sigma)_{qp} - (\mathbf{F}_\alpha)_{qp}\Big) + V_{aq;pb} - V_{aq;bp}$$
$$(\mathbf{A}_\mathrm{R})_{ap;bs} = \delta_{ab}\Big(\sum_\sigma (\mathbf{F}_\sigma)_{sp} - \tfrac{1}{2}(\mathbf{F}_\alpha)_{sp}\Big) + 2V_{as;pb} - V_{as;bp}$$
$$(\mathbf{A}_\mathrm{R})_{ap;qs} = V_{as;pq}$$

$$(\mathbf{A}_\mathrm{R})_{ar;bq} = \delta_{ab}\Big(\sum_\sigma (\mathbf{F}_\sigma)_{qr} - \tfrac{1}{2}(\mathbf{F}_\alpha)_{qr}\Big) + 2V_{aq;rb} - V_{aq;br}$$
$$(\mathbf{A}_\mathrm{R})_{ar;bs} = -\delta_{rs}\sum_\sigma (\mathbf{F}_\sigma)_{ab} + \delta_{ab}\sum_\sigma (\mathbf{F}_\sigma)_{sr} + 4V_{as;rb} - 2V_{as;br}$$
$$(\mathbf{A}_\mathrm{R})_{ar;qs} = -\tfrac{1}{2}\delta_{rs}\Big(\sum_\sigma (\mathbf{F}_\sigma)_{aq} + (\mathbf{F}_\alpha)_{aq}\Big) + 2V_{as;rq} - V_{as;qr}$$

$$(\mathbf{A}_\mathrm{R})_{pr;bq} = V_{pq;rb}$$
$$(\mathbf{A}_\mathrm{R})_{pr;bs} = -\tfrac{1}{2}\delta_{rs}\Big(\sum_\sigma (\mathbf{F}_\sigma)_{pb} + (\mathbf{F}_\alpha)_{pb}\Big) + 2V_{ps;rb} - V_{ps;br}$$
$$(\mathbf{A}_\mathrm{R})_{pr;qs} = -\delta_{rs}(\mathbf{F}_\alpha)_{pq} + \delta_{pq}(\mathbf{F}_\alpha)_{sr} + V_{ps;rq} - V_{ps;qr}$$

$$(\mathbf{B}_\mathrm{R})_{ap;bq} = V_{ab;pq} - V_{ab;qp}$$
$$(\mathbf{B}_\mathrm{R})_{ap;bs} = 2V_{ab;ps} - V_{ab;sp}$$
$$(\mathbf{B}_\mathrm{R})_{ap;qs} = \tfrac{1}{2}\delta_{pq}\Big(\sum_\sigma (\mathbf{F}_\sigma)_{as} - 2(\mathbf{F}_\alpha)_{as}\Big) + V_{aq;ps}$$

$$(\mathbf{B}_\mathrm{R})_{ar;bq} = 2V_{ab;rq} - V_{ab;qr}$$
$$(\mathbf{B}_\mathrm{R})_{ar;bs} = 4V_{ab;rs} - 2V_{ab;sr}$$
$$(\mathbf{B}_\mathrm{R})_{ar;qs} = 2V_{aq;rs} - V_{aq;sr}$$

$$(\mathbf{B}_\mathrm{R})_{pr;bq} = \tfrac{1}{2}\delta_{pq}\Big(\sum_\sigma (\mathbf{F}_\sigma)_{br} - 2(\mathbf{F}_\alpha)_{br}\Big) + V_{pb;rq}$$
$$(\mathbf{B}_\mathrm{R})_{pr;bs} = 2V_{pb;rs} - V_{pb;sr}$$
$$(\mathbf{B}_\mathrm{R})_{pr;qs} = V_{pq;rs} - V_{pq;sr}$$

Table 6. **Entries of the high-spin, real and complex RHF (or ROHF), stability matrix, Eq. (67). See the text for the notation convention used.**

associate all singly occupied orbitals with the same spin component, say the spin-up (or α) one, in which case the total spin quantum number S equals the total axial spin quantum number M, i.e. $S = M$. Thus, in this high-spin case we have a certain number of doubly occupied orbitals, which we again label by a and b, and $2S$ singly occupied orbitals, labelled by p and q. The remaining unoccupied spatial orbitals are again labelled by r and s. The HF equations for this case were first formulated by Roothaan[10] and the stability conditions for the simple OS doublet case by Paldus and Čížek.[44,45] For the general high-spin case the stability conditions were first derived by Stuber[89] and the resulting formulas are summarized in Table 6. Note that the only nontrivial excitations are of the types *ap*, *ar*, and *pr*.

In the simplest doublet case, these conditions may be shown to be equivalent to those given earlier.[44,45] This has been also verified by actual applications, which are illustrated on a few examples in the following section.

5. Applications

In this section we present several examples that illustrate the exploitation of the above presented theory in the investigations of the existence and character of various HF solutions. For this purpose we have chosen systems for which a similar investigation was carried out earlier using independent and, sometimes, *ad hoc* approaches. In this way we can verify both the correctness of our formalism as well as to evaluate its effectiveness.

The most complete search for the *ab initio* UHF and, in particular, GHF solutions of various types has been undertaken in the past by Löwdin and Mayer,[5] who employed the standard STO-6G and 4-31G basis sets (cf., e.g. Ref. 90) in their exploration of bs solutions for the beryllium and carbon atoms and for the boron hydride molecule. In their work, these authors[5] relied on the standard SCF software[90] when searching for various HF solutions and they often encountered serious convergence difficulties in this task.

In addition to the CS systems just mentioned (specifically Be and BH), we also re-examine the stability problem for a genuine OS case of the beryllium hydride (BeH) molecule. This system was examined[18] in connection with a highly singular behavior of the ROHF potential energy curve (PEC), first described by Mulliken,[17] since the ROHF solution represents a natural point of departure in spin-adapted post-Hartree-Fock approaches, namely when applying the unitary group approach (UGA)[83–85] to the coupled-cluster (CC) formalism for OS systems.[91,92] In view of a new, general formulation of these stability conditions for any high-spin OS case, it is useful to verify that in the simple doublet case we obtain the same result as with the original, specialized version.[18,45]

All the above mentioned developments deal primarily with the character of a given HF solution and only in the presence of the instability give us the prescription how to reach a lower-energy bs solution. In general, we can only be assured that our solution represents a local minimum and we have no criterion to check whether it represents the absolute minimum or, in fact, whether additional solutions associated with other local minima do exist or not.

As already mentioned earlier, in the case of CS ground states for systems in or near their equilibrium geometry, we can be reasonably confident that the self-consistent field solutions obtained in a standard way are, in general, the desired absolute-energy-minimum solutions. However, this is much less certain in the OS cases, including the CS systems away from the equilibrium geometry. Particularly in OS cases we often encounter a highly singular behavior of the potential energy curves (PECs) or surfaces when we rely on the ROHF formalism, as will be illustrated in the case of the BeH molecule.

In some cases, the corresponding UHF solution yields a realistic and smooth PEC, which, however, does not properly dissociate, as witnessed, for example, by the asymmetric dissociation of the $^3A''$ state of the CH_2 radical.[19] In this case, we in fact encounter the instability in both the OS singlet ($^1A''$) and triplet ($^3A''$) states, resulting in a highly singular ROHF PEC for both states. The singularity is again closely linked with the stability of these solutions.[19] We also mention that one can often generate a multitude of distinct UHF solutions once one or more chemical bonds are significantly stretched. These solutions may differ very little in their energy, but possess widely different character and spin properties.[19] Often a very slight change in the initial approximation will lead to another such solution. This will also be seen in Sec. 5.2 dealing with the BH molecule.

Finally, we wish to mention a pioneering study by Kowalski and Jankowski,[93,94] who have undertaken to explore *all* possible solutions of HF equations.† Although they were able to do so only for a very simple model system,[97] this does not diminish the significance of their undertaking. Below, we shall also briefly discuss their results in the light of the stability problem.

5.1 Be atom

The ground state of Be seems to be the simplest system possessing a GHF solution. Treated at the STO-6G level, Löwdin and Mayer[5] obtained a GHF solution having only a slightly lower energy than the $M = 0$ UHF solution (−14.505190 *vs.* −14.505074 a.u.). Both the UHF and GHF solutions break not only the spin, but also the spatial symmetry.

†Note that the primary goal of these authors was to explore all possible solutions of nonlinear systems of algebraic equations arising in various coupled-cluster approaches, see Refs. 95,96.

Energy (a.u.)	STO-3G	4-31G
RHF	−14.503 361	−14.556 871
UHF	−14.505 074	−14.557 743
GHF	−14.505 190	−14.557 798
^{1}FCI	−14.556 089	−14.605 415

	STO-3G		4-31G	
Averages	UHF	GHF	UHF	GHF
$\langle \hat{S}_x^2 \rangle$	0.12489	0.06891	0.09654	0.05266
$\langle \hat{S}_y^2 \rangle$	0.12489	0.13270	0.09654	0.10237
$\langle \hat{S}_z^2 \rangle$	0.00000	0.06889	0.00000	0.05263
$\langle \hat{S}^2 \rangle$	0.24978	0.27050	0.19308	0.20766
$\langle \hat{L}_x^2 \rangle$	0.13384	0.13270	0.10171	0.10237
$\langle \hat{L}_y^2 \rangle$	0.13384	0.07145	0.10171	0.05413
$\langle \hat{L}_z^2 \rangle$	0.00000	0.07147	0.00000	0.05410
$\langle \hat{L}^2 \rangle$	0.26769	0.27561	0.20343	0.21059

Table 7. The RHF, UHF, GHF, and FCI energies for the STO-6G and 4-31G models of the Be ground state, together with the mean values of spin and angular momentum operators. All mean values vanish in the RHF and FCI cases, and thus are not listed. For all solutions we also have that $\langle \hat{L} \rangle = \langle \hat{S} \rangle = \langle \hat{L} \cdot \hat{S} \rangle = 0$, with the exception of the 4-31G GHF solution when $\langle \hat{L} \cdot \hat{S} \rangle = -0.0007$.

We have employed our system of codes[98] and both the STO-6G and 4-31G basis sets[99] to explore the real RHF, UHF, and GHF solutions. A comparison with the corresponding full CI (FCI) results, representing the exact solution for a given model, is also made. The total energies associated with these solutions, together with the average values of spin and angular momentum operators, are given in Table 7. For the STO-6G basis set that was employed by Löwdin and Mayer[5] we find a complete agreement with their results (the only difference being that the role of p_x and p_y orbitals in the GHF solutions has been replaced by p_x and p_z orbitals).

lowest root	RHF	UHF	GHF
real RHF	+0.555403		
	(+0.641265)		
imag RHF	+0.270367		
	(+0.295903)		
real UHF	−0.017139	$+3.2 \cdot 10^{-7}$	
	(−0.025599)	$(+5.2 \cdot 10^{-8})$	
imag UHF	+0.135184	+0.142465	
	(+0.147951)	(+0.158646)	
real GHF	−0.017139	−0.002064	$+4.8 \cdot 10^{-9}$
	(−0.025599)	(−0.003247)	$(-1.4 \cdot 10^{-7})$
imag GHF	−0.017139	−0.002064	−0.001098
	(−0.025599)	(−0.003247)	(−0.001733)

Table 8. The lowest root of the pertinent stability problems for the 4-31G Be model. The corresponding values for the STO-6G model are enclosed in parentheses.

In Table 8 we also provide the lowest roots in the relevant symmetry-restricted stability problems. We see that the RHF solution is stable within the space of real and complex RHF solutions (i.e., there are no broken space-symmetry solutions as, for example, in the case of the O^{2-} ion[41]). However, this RHF solution is unstable relative to the real UHF solutions. Following the corresponding eigenvector of the stability problem (using our DEM procedure) we arrive at the UHF solution. The UHF stability problem has a zero eigenvalue, indicating a trivial mode associated with the spatial degeneracy and shows the instability relative to the real GHF solutions. Again, the resulting real GHF solution has a zero eigenvalue (within the imposed tolerance on the accuracy). Interestingly enough, the GHF solution is unstable in the space of complex GHF solutions. Unfortunately, the complex GHF class of solutions has not yet been implemented in our system of codes.

The orbital coefficients for the $M = 0$ RHF and UHF, as well as for the GHF occupied orbitals, are given in Table 9. Although the occupied spin orbitals are rather similar in all cases, significant differences occur for the highest occupied ones. Again, these results are consistent with those provided by Löwdin and Mayer,[5] except for the trivial interchange of coordinate axes. It is also worthwhile to point out a very specific

orbital	ϕ_1(RHF)	ϕ_1(UHF)	ϕ_1(GHF)	ϕ_2(RHF)	ϕ_2(UHF)	ϕ_2(GHF)
1s(α)	+0.995 521	+0.995 423	+0.995 416			
2s(α)	+0.017 048	+0.017 410	+0.017 435			
2p$_x$(α)						+0.000 003
2p$_y$(α)		−0.000 004	+0.000 520			
2p$_z$(α)		−0.000 714				+0.000 520
1s(β)				+0.995 521	+0.995 423	+0.995 416
2s(β)				+0.017 048	+0.017 410	+0.017 435
2p$_x$(β)			+0.000 003			
2p$_y$(β)					+0.000 004	−0.000 520
2p$_z$(β)			+0.000 520		+0.000 714	

(a)

orbital	ϕ_3(RHF)	ϕ_3(UHF)	ϕ_3(GHF)	ϕ_4(RHF)	ϕ_4(UHF)	ϕ_4(GHF)
1s(α)	−0.279 907	−0.270 900	−0.270 276			
2s(α)	+1.033 982	+0.998 776	+0.996 347			
2p$_x$(α)						+0.001 049
2p$_y$(α)		−0.001 381	+0.189 031			
2p$_z$(α)		−0.258 690				+0.188 999
1s(β)				−0.279 907	−0.270 900	−0.270 276
2s(β)				+1.033 982	+0.998 776	+0.996 347
2p$_x$(β)			+0.001 105			
2p$_y$(β)					+0.001 381	−0.189 031
2p$_z$(β)			+0.188 999		+0.258 690	

(b)

Table 9. The (a) core RHF, UHF, and GHF MSOs ϕ_1 and ϕ_2 and (b) highest occupied RHF, UHF, and GHF MSOs ϕ_3 and ϕ_4 for the Be STO-6G model.

structure of the GHF solution, having $\langle \hat{\mathrm{S}} \rangle = 0$, yet not being an eigenstate of $\hat{\mathrm{S}}_3$. In Tables 10 and 11 we then provide the orbital energies for both occupied and virtual orbitals. The spatial symmetry implied

orbitals	RHF	UHF	GHF
ϕ_1	−4.541 835	−4.541 005	−4.540 950
ϕ_2	−4.541 835	−4.541 005	−4.540 950
ϕ_3	−0.252 096	−0.263 392	−0.264 157
ϕ_4	−0.252 096	−0.263 392	−0.264 157
ϕ_5	+0.219 697	+0.222 707	+0.222 911
ϕ_6	+0.219 697	+0.222 707	+0.222 911
ϕ_7	+0.219 697	+0.222 707	+0.224 644
ϕ_8	+0.219 697	+0.222 707	+0.224 644
ϕ_9	+0.219 697	+0.233 990	+0.233 224
ϕ_{10}	+0.219 697	+0.233 990	+0.233 224

Table 10. Be STO-6G orbital energies.

by the orbital coefficients (Table 9) reflects the character of the pertinent class of HF solutions, and may be understood when considered in the light of the spin and time reversal symmetries. Consequently, the orbital energies occur in identical pairs, although only the RHF spin orbitals are identical, up to the spin component, while for the UHF and GHF solutions each spin orbital is different.

5.2 BH molecule

We next examine a six-electron system, namely the boron hydride (BH) molecule using the 4-31G basis set.[99] This system was also explored by Löwdin and Mayer[5] and was shown to exhibit several UHF and GHF class solutions. We thus use these results as a benchmark to test our independent approach to bs solutions. On the whole, we were able to reproduce Löwdin and Mayer's results, as well as to extend them.

The exact ground state of the BH molecule, as described by the 4-31G model, is a singlet and the pertinent FCI PEC is shown in Fig. 2 together with the corresponding lowest lying triplet state. The RHF solution describing this state is not size consistent and approaches the incorrect dissociation limit involving the mixture of covalent and ionic valence-bond structures, as usual. The high-spin ROHF solution, whose PEC is also shown in Fig. 2, is size consistent, but displays a singular behavior in the interval $3 \leq R \leq 4$ a.u.

Both singlet RHF and triplet ROHF solutions are stable within their respective solution spaces, but are unstable relative to the UHF solutions

orbitals	RHF	UHF	GHF
ϕ_1	−4.674 903	−4.676 606	−4.676 710
ϕ_2	−4.674 903	−4.676 606	−4.676 710
ϕ_3	−0.291 995	−0.300 192	−0.300 717
ϕ_4	−0.291 995	−0.300 192	−0.300 717
ϕ_5	+0.105 159	+0.106 038	+0.106 095
ϕ_6	+0.105 159	+0.106 038	+0.106 095
ϕ_7	+0.105 159	+0.106 038	+0.107 119
ϕ_8	+0.105 159	+0.106 038	+0.107 119
ϕ_9	+0.105 159	+0.113 888	+0.113 425
ϕ_{10}	+0.105 159	+0.113 888	+0.113 425
ϕ_{11}	+0.596 879	+0.594 378	+0.592 933
ϕ_{12}	+0.596 879	+0.594 378	+0.592 933
ϕ_{13}	+0.596 879	+0.596 164	+0.596 121
ϕ_{14}	+0.596 879	+0.596 164	+0.596 121
ϕ_{15}	+0.596 879	+0.596 164	+0.597 427
ϕ_{16}	+0.596 879	+0.596 164	+0.597 427
ϕ_{17}	+0.805 437	+0.808 405	+0.808 577
ϕ_{18}	+0.805 437	+0.808 405	+0.808 577

Table 11. Be 4-31G orbital energies.

for all internuclear separations R considered, as shown in Fig. 3. As the triplet ROHF solution lacks the self-consistent symmetry, only the real stability roots for the singlet ground state are shown in Fig. 3. For the internuclear separations $R > 3.388$ a.u. there are two negative roots, implying the existence of at least two distinct UHF solutions. In fact, the UHF solution that is associated with the negative root appearing at the critical internuclear separation $R_c = 3.388$ a.u. is the 'standard' UHF solution (referred to as UHF/1 by Löwdin and Mayer[5]) that splits off the RHF solution at the critical point R_c and correctly dissociates as $R \to \infty$ to the same limit as the triplet ROHF solution.[‡] We shall thus

[‡]Nonetheless, it is worth mentioning that in order to obtain this solution with standard software packages, e.g. GAMESS,[99] one has to employ a suitable starting approximation. For example, one can use the triplet ROHF orbitals for larger R values and to continue the solution towards the critical point R_c. With other choices, particularly in the vicinity of R_c,

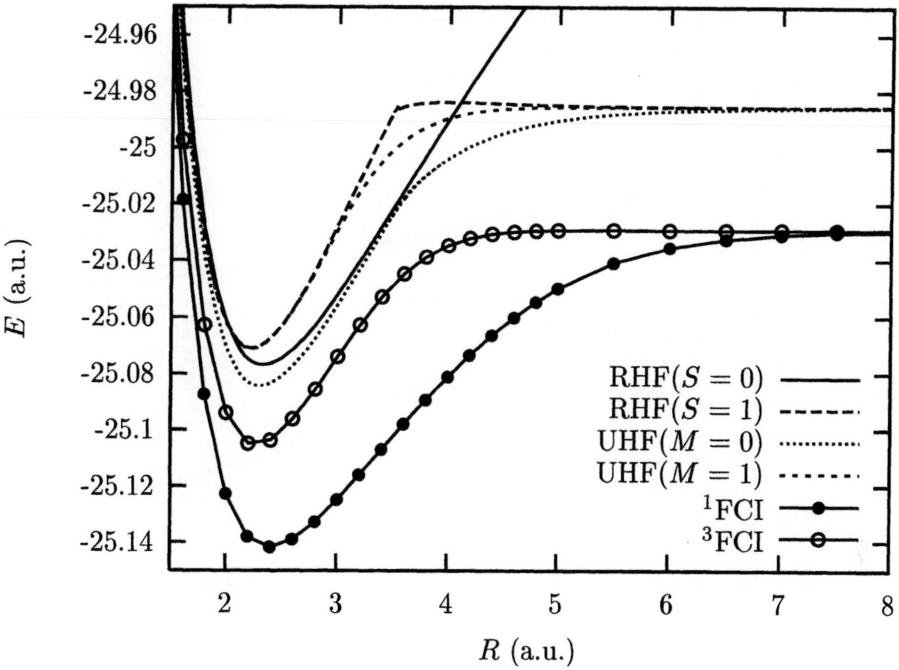

Figure 2. Potential energy curves for the BH 4-31G model obtained with various approximations. See the text for details.

concentrate on the 'nonstandard' UHF (and GHF) solutions that are associated with the instability that persists for all internuclear separations R considered (cf. Fig. 3).

The $M = 0$ UHF solutions are very close in energy as shown in Fig. 4. For greater clarity we have subtracted the energy (in a.u.) given by the polynomial

$$E(R) = A + B(R - b) + C(R - c)^2 , \tag{148}$$

with $A = -25.026$, $B = 0.0393$, $C = -0.035$, $b = 3.40$, and $c = 3.75$, and shown by the thin dashed line in Fig. 4, labelled 'scale', yielding the plot shown in Fig. 5. From this figure we see that the UHF solution that exists in the entire region $3.44 \leq R \leq 3.94$ a.u. connects the other two UHF solutions, namely the left hand side (lhs) solution that splits off at $R \approx 3.44$ a.u. and the 'standard' right hand side (rhs) UHF

---

the standard SCF scheme converges either to the RHF solution or to some higher energy triplet-like solutions.

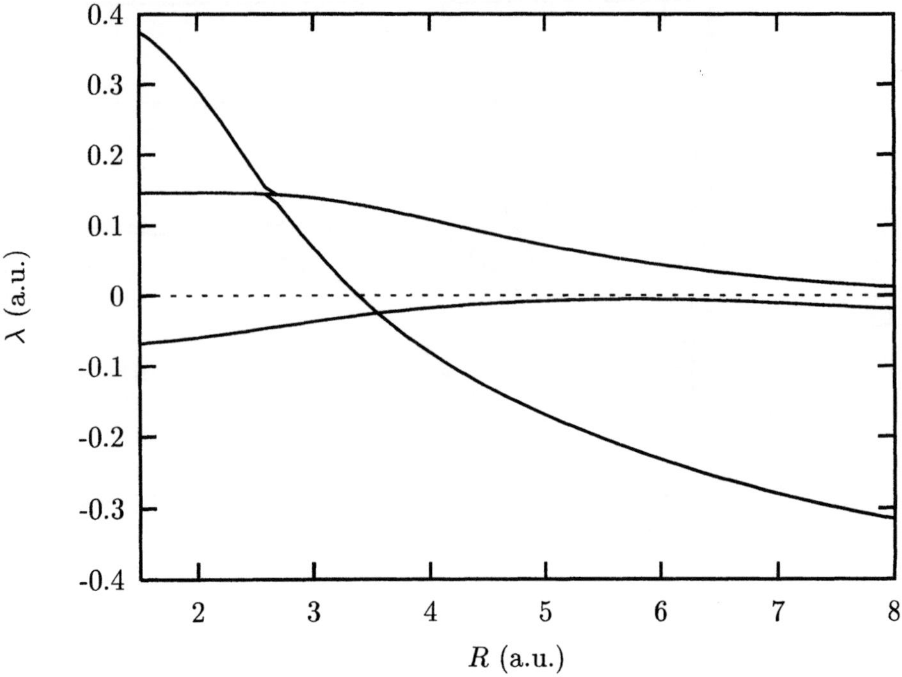

Figure 3. The three lowest lying roots of the RHF stability problem (relative to UHF) for the BH 4-31G model.

solution discussed above. In the intermediate region $3.55 < R < 3.64$ a.u. these lhs and rhs UHF solutions are unstable as implied by the corresponding roots of the stability problem shown in Fig. 6 and in detail in Fig. 7. Comparing with the results of Löwdin and Mayer,[5] we can identify our persistent and lhs UHF solutions with their UHF/2 and UHF/3 solutions, respectively.

The UHF solutions just discussed are all unstable relative to the GHF solutions as shown in Fig. 7. The pertinent GHF solutions arise for $R < 4.10$ a.u. and interconnect with each other and the lhs UHF solution at $R = 3.44$ a.u., similarly as the UHF solutions. Two of the GHF solutions correspond to those of Löwdin and Mayer,[5] while the additional GHF solution, which links with the UHF solution at $R = 3.44$ a.u., is only found in this work. All GHF solutions, within their respective stability problems, yield only zero eigenvalues that are associated with trivial modes, and no lower energy solutions can be obtained using the corresponding eigenvectors.

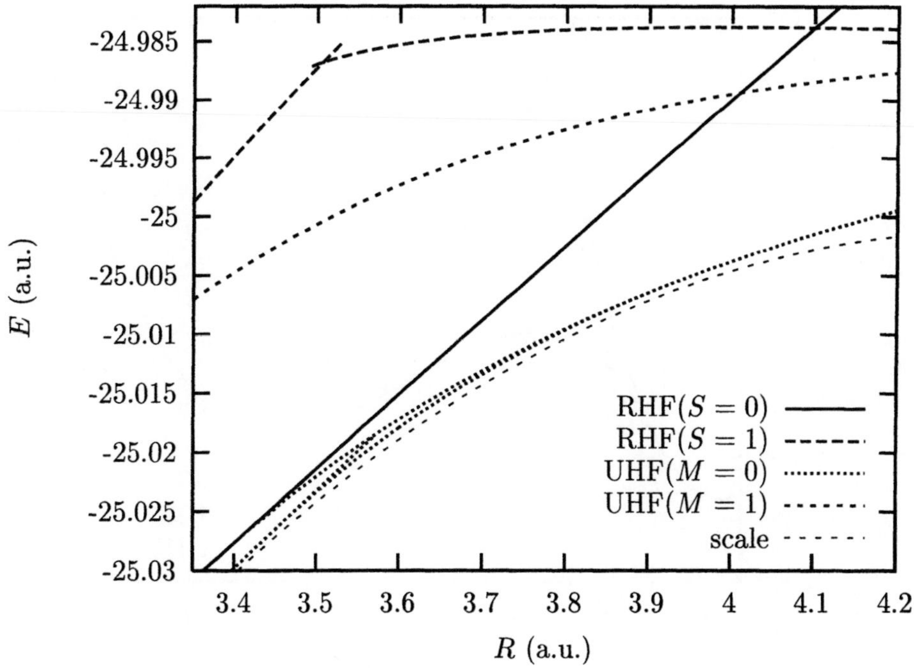

Figure 4. Details of the scaled RHF and UHF potential energy curves for the BH 4-31G model. See the text for details.

The inter-relationship between the UHF and GHF solutions can be better understood in terms of the average spin and spin-deviation quantities shown in Fig. 8 and Fig. 9. The GHF solutions, in contrast to the preceding example, have nonzero average spin. The Fig. 9 also shows the transition structure of the persistent UHF solution that connects with rhs and lhs solutions at $R = 3.44$ a.u. and $R = 3.94$ a.u. The three GHF solutions split off at $R = 3.44$, 3.72, and 4.10 a.u., and interconnect the rhs UHF and lhs UHF and GHF solutions. Fig. 8 and Fig. 9 indicate the usefulness of the spin characteristics both as a diagnostic and interpretative tool.

5.3 BeH molecule

We next consider an example of a five-electron OS system which clearly demonstrates the singular behavior of the ground state PECs obtained at the spin-adapted ROHF level and its relationship with the doublet instability of the corresponding solutions.[18] We can also expect

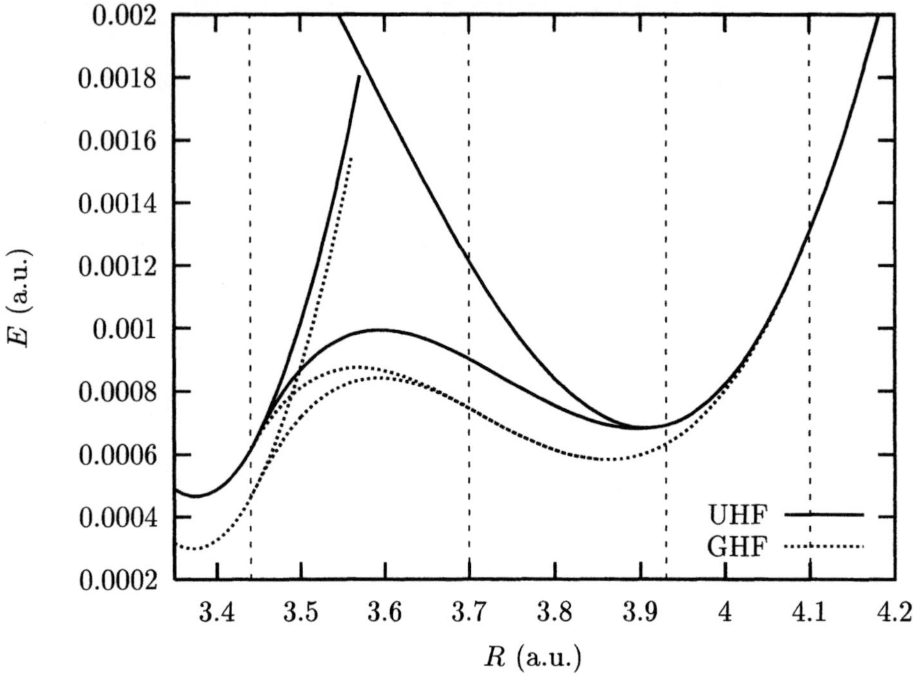

Figure 5. Details of the scaled UHF and GHF potential energy curves for the BH 4-31G model. The vertical dashed lines indicate the critical internuclear separations R_c for the instability onset (cf. Fig. 6 and Fig. 7) and thus for the occurence of a new bs solutions with lower energy. See the text for details.

that this molecule admits a GHF solution, at least for large internuclear separations R, when this solution must merge with that found for the Be atom. To verify this claim, we shall employ again the 4-31G basis set.[99]

The exact FCI PECs for three low-lying doublet states (^{2}FCI) and for a quartet state (^{4}FCI) having the same dissociation limit are shown together with the ROHF and UHF ground state solutions in Fig. 10. The low-lying doublet excited state PECs belong to different symmetry species and cross in the vicinity of $R = 3.8$ a.u. The singular character of the ROHF solution is easily understood in view of the fact that in the bonding region the hydrogen 1s orbital and the Be 2s orbital form a doubly occupied σ-bonding orbital, while the remaining electron populates the corresponding antibonding σ-orbital. In the dissociation limit, however, we have a CS Be atom and an OS H atom. Clearly, the ROHF

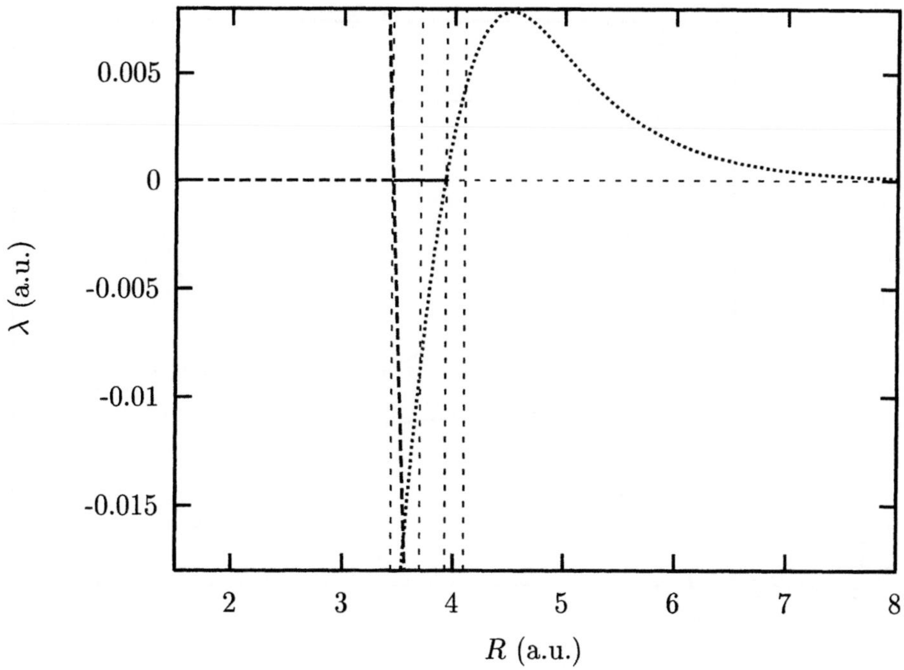

Figure 6. The lowest energy UHF stability roots for the BH 4-31G model. The style of the curves corresponds to that in Fig. 4 and Fig. 5 and the vertical dashed lines have the same meaning as in Fig. 5.

wave function is unable to describe this transition in a smooth manner, both types of limiting wave functions having a mutually incompatible character. Interestingly enough, both types of these solutions can co-exist in a finite region of internuclear separations $4.0 < R < 4.6$ a.u., but eventually cease to exist. This behavior is reflected in that of the lowest stability root, shown in Fig. 11, which in each case approaches rapidly zero. Qualitatively, this indicates the disappearance of the local minimum and the emergence of a local maximum or a saddle point.

In view of the absence of self-consistent symmetry for non-singlet RHF (or ROHF) solutions, the UHF solution exist in the entire range of internuclear separations. The corresponding PEC is also plotted in Fig. 10. Strictly speaking, it lies below the ROHF PECs and smoothly interpolates between the bonding and dissociated regimes. This UHF solution is stable within the space of both real and complex UHF solutions as

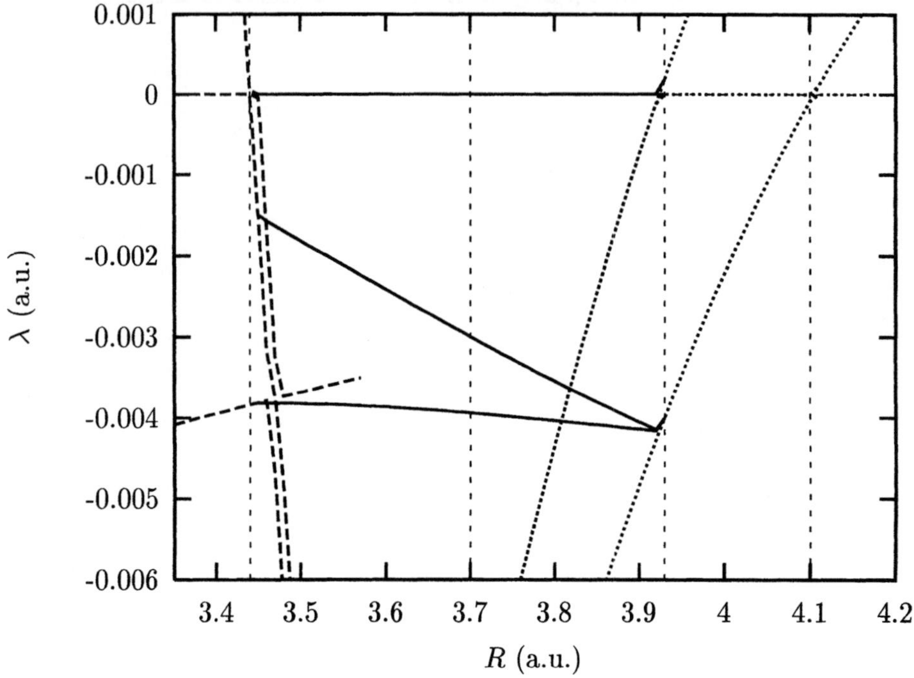

Figure 7. Details of the lowest energy UHF stability roots for the BH 4-31G model. The style of the curves corresponds to that in Fig. 4 and Fig. 5 and the vertical dashed lines have the same meaning as in Fig. 5.

shown in Fig. 12. However, it is not stable in the space of GHF solutions, where we find a negative root for $R > 6.90$ a.u. (cf. Fig. 13).

As already mentioned at the outset, this behavior may have been expected in view of the Be results, Sec. 5.1. The pertinent real GHF solution is reached by following the eigenvector of the lowest real GHF stability root, indicating the local direction of steepest descent. Similarly as in the case of the Be atom, the resulting GHF solution has only slightly lower energy than the UHF solution, as shown in Fig. 14, where we also indicate the limiting $R \to \infty$ values for both UHF and GHF solutions by dashed horizontal lines. The GHF solution is stable relative to the real GHF solutions but, as in the case of the Be atom, is unstable relative to the complex GHF solutions.

The inter-relationship of the RHF, UHF, and GHF solutions is again best elucidated by considering the average spin deviations (Fig. 15). For each of these solutions the average spin is $\langle \hat{\mathrm{S}} \rangle = (0, 0, \frac{1}{2})$. This does not

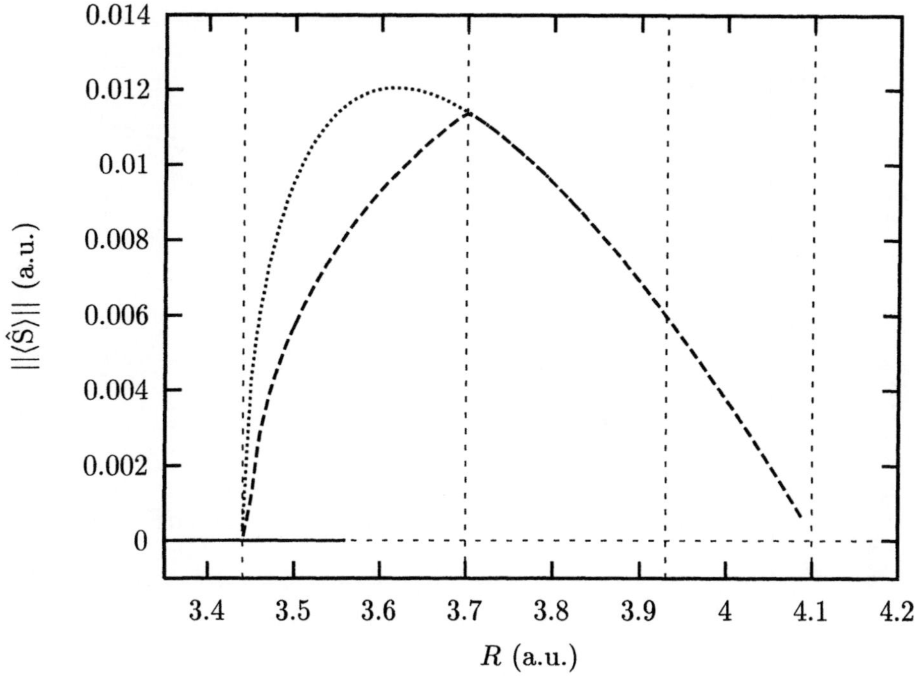

Figure 8. Average value of the total spin $\langle \hat{S} \rangle$ associated with different GHF solutions of the BH 4-31G model. The vertical dashed lines have the same meaning as in Fig. 5.

imply that the UHF or GHF solution is an eigenstate with $M = \frac{1}{2}$, but only the fact that the spin-deviations occur symmetrically and average out for the x and y components. Clearly, for the high-spin RHF solution, when $S = M = \frac{1}{2}$, we have that

$$\langle \hat{S}^2 \rangle - \langle \hat{S} \rangle^2 = S(S+1) - M^2 = \tfrac{1}{2}, \tag{149}$$

represented by the horizontal dashed line in Fig. 15. The corresponding spin deviation for the UHF solution (represented by the solid line in Fig. 15) stays at more or less the RHF value for small R values, then sharply rises around $R \approx 4$ a.u. and finally approaches the value of about 0.7 in the dissociation limit. This is compatible with the fact that the spin deviation $\langle \hat{S}^2 \rangle - \langle \hat{S} \rangle^2$ for the Be and H atoms are, respectively, 0.193 and 0.5, i.e., in total amount to 0.693. The corresponding value for GHF solution starts to deviate at $R = 6.90$ a.u. as shown in Fig. 15,

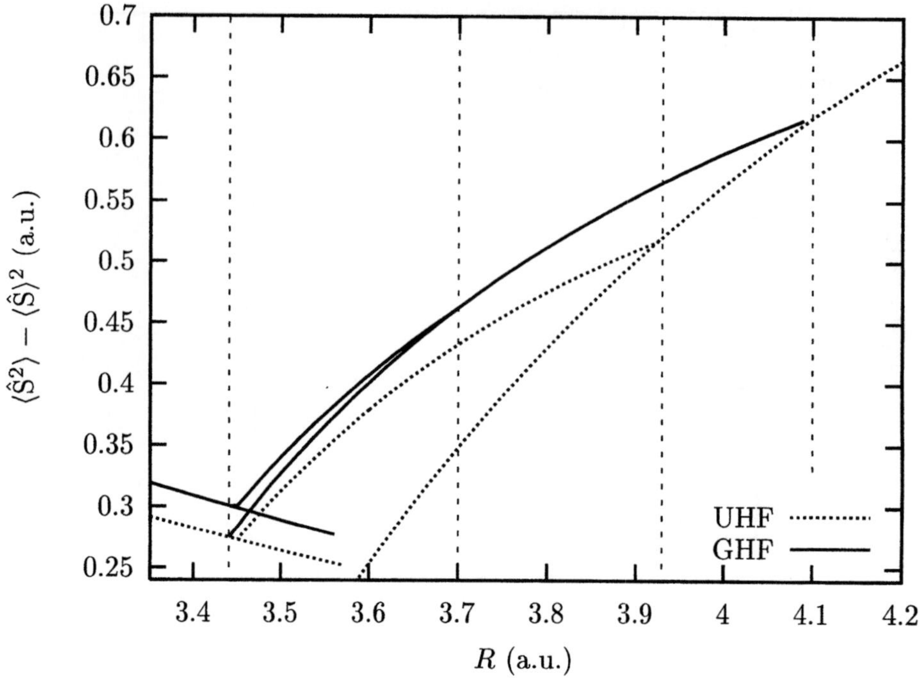

Figure 9. Spin deviation $\langle \hat{S}^2 \rangle - \langle \hat{S} \rangle^2$ associated with different UHF and GHF solutions of the BH 4-31G model (cf. Fig. 4). The vertical dashed lines have the same meaning as in Fig. 5.

being slightly above the UHF curve, which is again consistent with the results for the Be atom.

5.4 H4 model

The widely studied minimum basis set[100] H4 model[97] consists of two, slightly stretched, hydrogen molecules in the trapezoidal configuration, with three identical internuclear separations $R_{ij} \equiv R(\mathrm{H}_i\text{–}\mathrm{H}_j)$ and C_{2v} point group symmetry (i.e., $R_{12} = R_{23} = R_{34} = 2$ a.u. and $R_{14} > 2$ a.u., with a sequential numbering of atomic sites along the perimeter). The geometry is thus fully determined by a single parameter $\alpha \equiv \angle \mathrm{H}_1\mathrm{H}_2\mathrm{H}_3 - \pi/2$. The interest in this model[101–118] stems from the fact that it enables a continuous variation of the degree of quasidegeneracy from the fully degenerate square configuration ($\alpha = 0$) to a nondegenerate linear one ($\alpha = \pi/2$). The reason for employing slightly stretched H_2 molecules

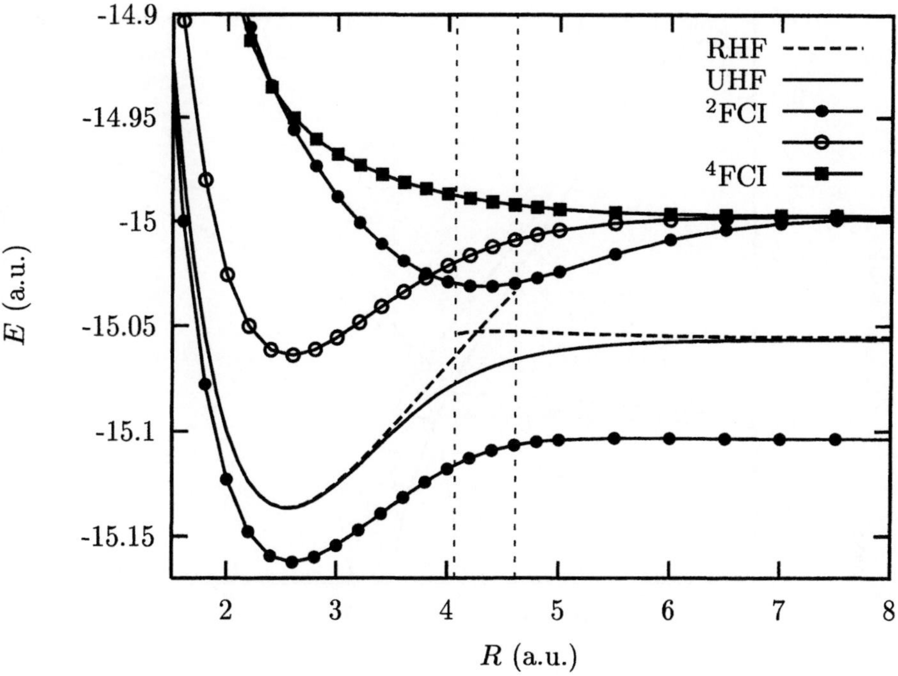

Figure 10. The RHF, UHF, and FCI potential energy curves for the doublet ground state of the BeH 4-31G model. The FCI potentials for the lowest doublet and quartet excited states are also shown for the sake of reference. The range of coexistence of both ROHF solutions is indicated by the vertical dashed lines.

is to enhance the role of correlation effects and the range of singular behavior of certain post-HF approaches (cf. Ref. 111).

Clearly, the most challenging situation for the IPM approaches arises in the vicinity of the degenerate limit ($\alpha = 0$), where the correlation energy reaches its maximum value. This is likely the reason why the authors[93, 94] had chosen an almost square configuration ($\alpha = 0.005$) for their study. By exploiting recently developed homotopy methods[119] and the powerful software package HOMPAC,[120] they were able to find twelve distinct RHF solutions, the same number as in the dimension of the corresponding symmetry adapted FCI space. The overlap S of these RHF solutions with the exact FCI states never exceeded the value of $S = 3/4$, except for the lowest energy bs solution ($S = 0.96$, solution No. 3 of Ref. 93) with the first excited state and for the highest energy solution ($S = 0.95$, solution No. 12 of Ref. 93) with the corresponding

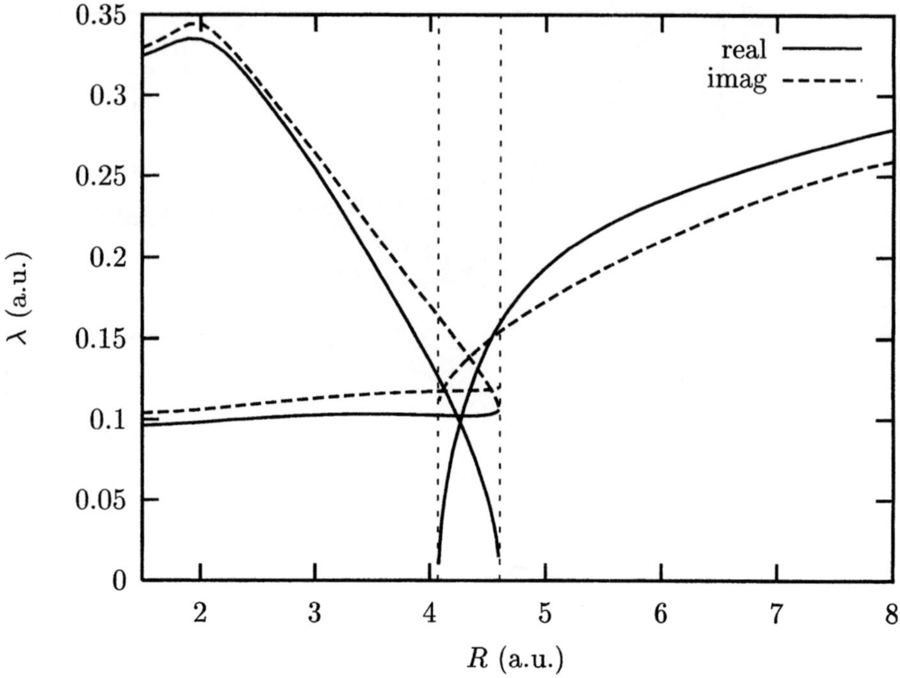

Figure 11. The lowest real and imaginary RHF stability roots for the BeH 4-31G model. The vertical dashed lines have the same meaning as in Fig. 10.

highest energy FCI solution. Seven of these singlet RHF solutions are spatially sa, while the remaining five break spatial symmetry.

Looking at the stability problem within the space of real RHF solutions, we find that only the lowest two of these solutions (solutions No. 1 and 2 of Ref. 93) correspond to a local minimum. These solutions represent the ground and $(1a_1)^2(2a_1)^2$ biexcited states. Interestingly enough, it is the lowest lying bs solution (solution No. 3 of Ref. 93) which has the largest overlap with the lowest lying $^1A_1$ excited state (i.e., the second FCI state from below). This solution, however, is unstable within the space of real RHF solutions. In fact, all other solutions are unstable, the number and absolute value of negative roots roughly increasing with the increasing energy of these solutions.

The two solutions with the highest energy (solutions No. 11 and 12 of Ref. 93) correspond in fact to a local maximum, since all the roots

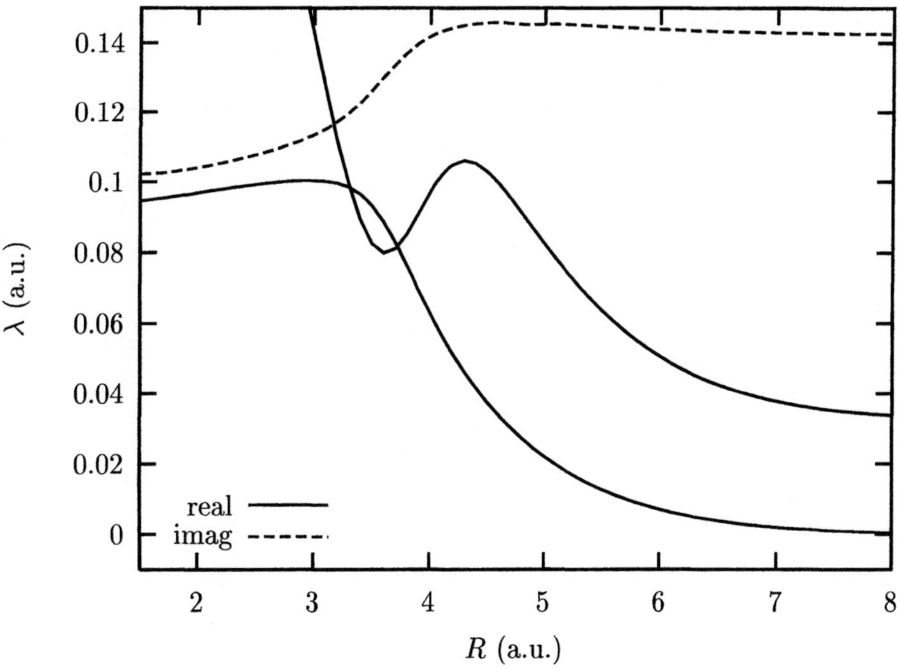

Figure 12. The low lying UHF stability roots relative to the UHF solution space.

of the real RHF stability problem are negative.§ Not surprisingly, the highest $^1A_1$ state (solution No. 12), corresponding essentially to the single quadruply excited configuration $(2a_1)^2(2b_2)^2$ of this model, has one of the largest overlaps ($S = 0.95$) with the corresponding highest energy FCI solution. We thus observe the hole-particle symmetry, characterizing the minimum basis set model, being reflected in the qualitative nature of these results. Of course, this symmetry is partially obscured by the asymmetry in the orbital occupancies in the Fock operator.

6. Concluding remarks

In this paper we have presented a novel approach to the stability problem of HF solutions that is based on the direct energy minimization

§This is also the case for the solution No. 9 when we use the MOs given in Table 1 of Ref. 93, even though the highest root is only -0.97×10^{-4}. However, when we properly converge this solution, the highest root becomes slightly positive, namely 0.18×10^{-3}, implying the saddle point character.

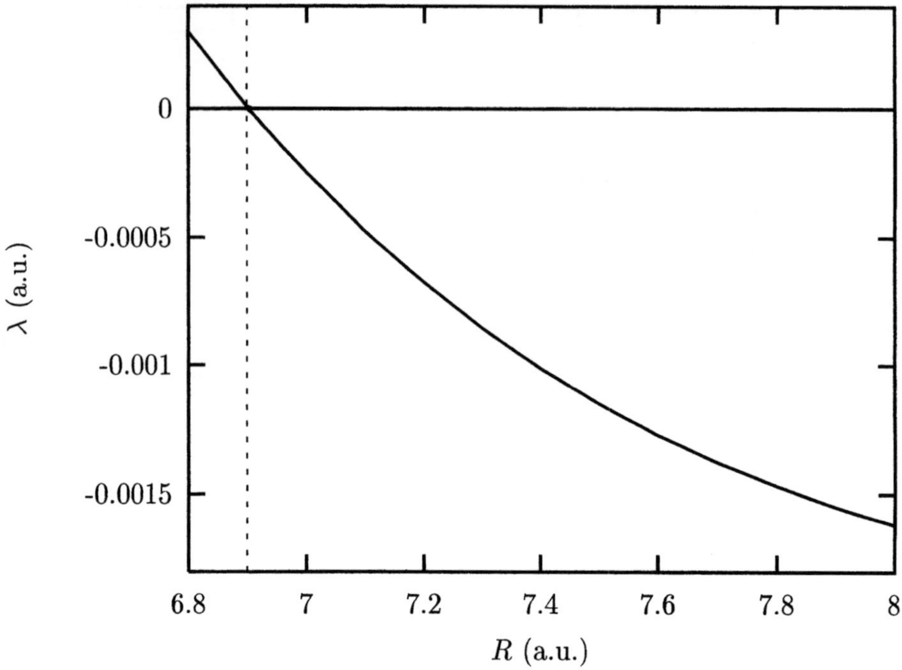

Figure 13. The lowest UHF stability root relative to the GHF solution space.

algorithm (cf., e.g. Ref. 16). The advantage of this approach is at least two-fold: It accounts for the constraints implied by the orthonormality of the MSOs or MOs without introducing the Lagrange multipliers and it enables the exploitation of the energy optimization techniques based on the steepest descent algorithms. The first feature enables an extension of Fukutome's general classification of instabilities to open-shell systems or, in fact, to a rather general class of variational wave functions, while the second innovation enables the design of a more efficient and reliable software for finding of HF solutions than is the standard self-consistent iterative process. The latter feature is particularly important when searching for various broken-symmetry solutions that are often associated with very shallow local minima on the mean energy hypersurface. While at present our software is capable of handling only real-type RHF, UHF, and GHF solutions, the work is in progress on the extension to the complex domain, as well as on modifications that will enable a better handling of solutions that are associated with various saddle points.

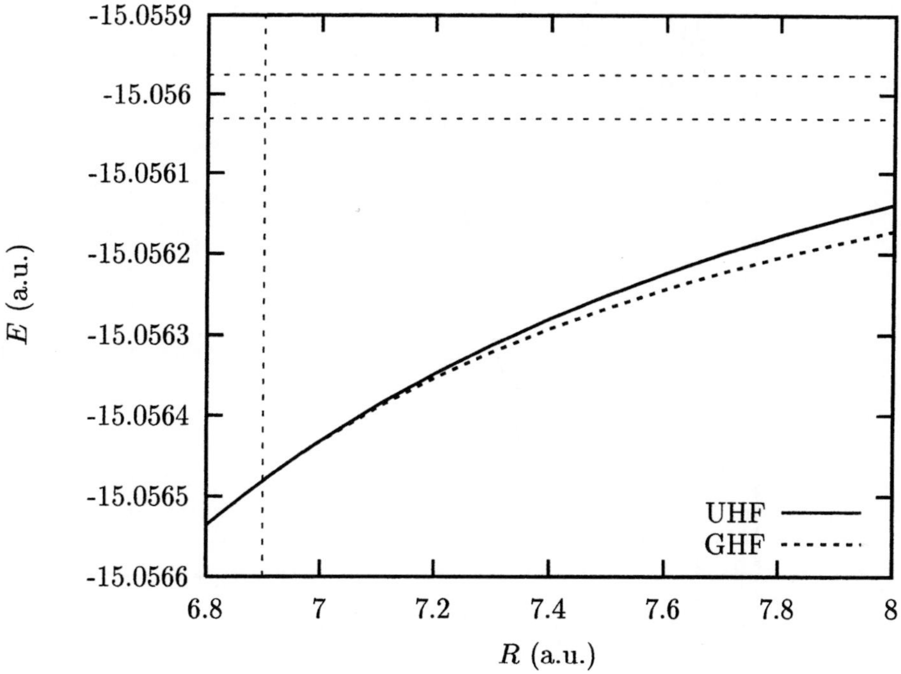

Figure 14. Details of the UHF and GHF potential energy curves for $R > 6.80$ a.u. for the BeH 4-31G model. The dissociation limit for each solution is indicated by the horizontal dashed line. The critical internuclear distance R_c at which the GHF solution splits off the UHF one ($R_c = 6.90$ a.u.) is indicated by the vertical dashed line.

The theoretical developments that we have just outlined have been illustrated in the preceding section by considering several characteristic examples for which the various UHF and GHF solutions were (at least partially) known from the literature. However, these previously determined solutions were obtained in a more or less *ad hoc* manner via standard quantum chemical software. Thus, for example, in the case of the boron hydride molecule we were able to present a more complete picture by including an additional GHF solution that has escaped the conventional computational scheme.

The generality of the present formalism has been also illustrated on the case of the beryllium hydride doublet ground state. This system was again examined earlier[18] using the doublet stability conditions for the ROHF solutions involving a single open-shell MSO.[44,45] This study was undertaken in order to explain a highly singular nature of the ROHF

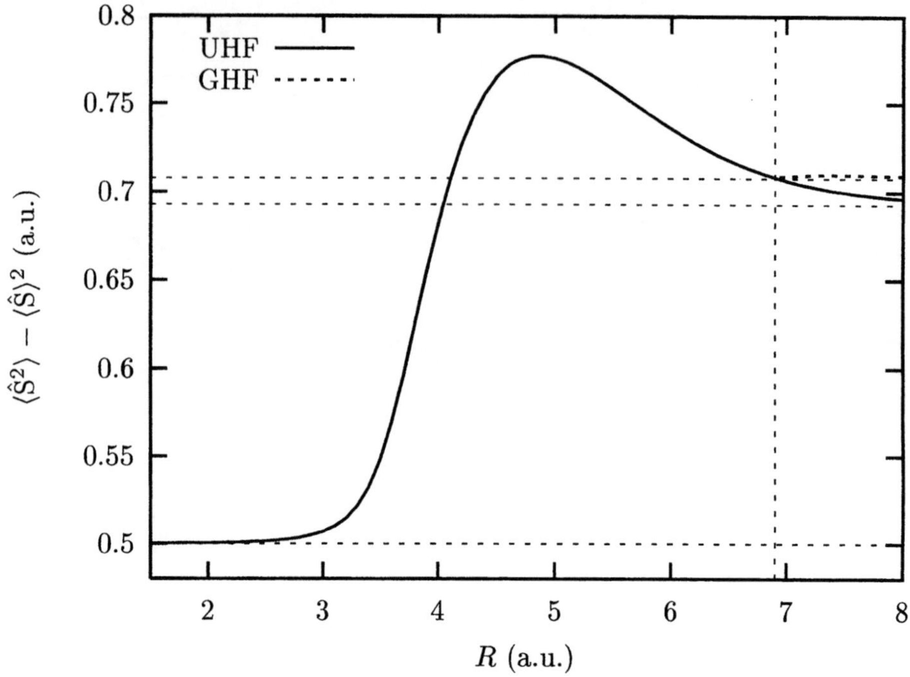

Figure 15. Average spin deviation $\langle \hat{S}^2 \rangle - \langle \hat{S} \rangle^2$ for the UHF and GHF solutions of the BeH 4-31G model. At $R = 6.90$ a.u., the GHF solution splits from the UHF curve. Both curves differ from the value of 1/2 of the exact spin eigenstate, indicated by the dashed horizontal line. The vertical dashed line indicates the onset of the instability (cf. Fig. 14).

potential energy curve. Although the new general formulation that is applicable to an arbitrary high-spin ROHF solution is formally rather distinct from the existing doublet stability version,[44,45] we obtain the same results for the critical values of the internuclear separations that delimit the singularity region and explain the 'hysteresis'-type behavior of the ROHF potentials.

Although in this chapter we have primarily focused on formal mathematical aspects of the stability problem and on reliable algorithms for the determination of the implied broken-symmetry solutions and their classification, we would like to make at least a couple of brief remarks concerning the practical aspects and possible physical meaning of these phenomena. For a more comprehensive discussions of these topics we refer the reader to the existing reviews[71–74] and to the papers they quote.

Perhaps the most relevant aspect of the HF stability problem is associated with the fact that the HF - or more generally the IPM - solutions represent the point of departure or reference in most approaches that account for the many-electron correlation effects, generally referred to as post-HF methods. The key importance of correlation in most quantum chemical applications is well known and need not be emphasized. These effects are particularly felt when exploring the entire potential energy surfaces, since the RHF method is not size consistent, and thus is not capable to correctly describe the dissociation channels involving open-shell fragments. The exploitation of the DODS-type UHF solutions can often overcome this size-consistency problem, but leads to spin-nonadapted description, which has its own shortcomings and problems (cf., e.g. Ref. 15). Also the multitude of various UHF solutions, often with closely lying energies and limited ranges of existence, which in turn imply the nonanalyticity of the pertinent potential energy curves or surfaces (cf., e.g. Refs. 34,48 etc.), as also the examples given in Sec. 5 demonstrate, represent an obstacle in the search for a proper reference state for post-HF methods.

In general, it is difficult to give a physical significance to various broken symmetry solutions, since their existence may be connected with the limitations of the *ab initio* model employed or with the inadequacy of the IPM. Nonetheless, we can state that these solution generally imply the importance and the nature of the correlation effects that are missing in the symmetry adapted RHF-type solutions.

Yet, the presence of the symmetry breaking, be it of spatial, spin, or of other nature, especially if occurring for a sufficiently accurate *ab initio* model (i.e., for models employing sufficiently large and flexible basis sets), may indeed imply the actual tendency of the system. As an example, let us mention the spatial symmetry breaking found for the linear polyenic systems with cyclic boundary conditions[33] that can be used to model linear metallic-like systems. Here the presence of broken symmetry solutions possessing the off-diagonal charge-density waves[30,33,38] indeed implies the symmetry breaking[121,122] leading to the bondlength alternating structures, thus modelling the real physical phenomenon found in polyacetylene. This symmetry breaking does survive even when the correlation effects are accounted for[123–126] and represents a real physical effect whose origin is not difficult to understand (cf. Ref. 39). The spin-density waves, associated with various UHF and GHF solutions, play a similar role in magnetic and other phenomena.

Undoubtedly, it is for the extended systems where the IPM model and the broken-symmetry solutions are most often exploited. Let us mention, for example, recent studies of tunable conductivity or superconductivity and magnetism by Yamaguchi and collaborators in exotic materials in the region of metal-insulator transitions[127,128] or of anomalous phases of photodoped systems.[129,130] In fact, the analogues of the UHF and GHF methods have been also introduced in the density functional theory (DFT), in particular the so-called generalized spin-density functional theory (see, e.g. Ref. 131), and exploited in the study of high-spin dendritic or polymeric ferromagnetic ground states for various organic ferromagnets.[130] Even when the HF or DFT broken-symmetry solutions do not provide a reliable description of the studied phenomena, they nonetheless represent a useful first step.

In closing, let us note that the broken-symmetry solutions continue to play a useful role in studies of finite systems as well. For example, the GHF-type solutions have been recently used to handle spin-frustrated systems, such as closed chains having an odd number of spin sites (e.g., the H_3 and H_5 molecules),[132–134] while the UHF-type solutions proved their usefulness in handling of unstable metal-metal bonds[135,136] and generalized DFT spin orbitals in the studies of polyradical systems[137] and radical reactions.[138]

Finally, we must mention a new type of approaches, proposed by Yamaguchi's group, that combine the standard quantum chemical methods, such as the HF or coupled-cluster methods, with quantum electrodynamics (QED), by treating the N-electron system and the Dirac sea on an equal footing. In particular, for the so-called QED-SCF method, these authors carried out Fukutome's type classification of pertinent solutions.[139]

A. Transformation properties of the unitary group generators

The similarity transformations of the unitary group U($2n$) generators $\hat{\mathrm{E}}^{i\sigma_1}_{j\sigma_2}$, Eq. (85), by various group elements of $\mathfrak{G} = \mathfrak{S} \otimes \mathfrak{T}$, which are required in computation of invariant means, Eq. (100) or (103), are easily evaluated using the multicommutator expansion

$$e^{\hat{\mathrm{A}}}\hat{\mathrm{B}}e^{-\hat{\mathrm{A}}} = \sum_{m=0}^{\infty} \frac{1}{m!}[\hat{\mathrm{A}}, \hat{\mathrm{B}}]^{(m)}, \tag{A.1}$$

where $[\hat{\mathrm{A}}, \hat{\mathrm{B}}]^{(m)}$ is defined by Eq. (35). Although we can obtain the desired expressions directly by relying on the commutation relations for the unitary group generators (cf., e.g. Refs. 83–85), it is both easier and more transparent to proceed stepwise by considering first the transformation of individual creation and annihilation operators.

For transformations involving spin operators, we first observe that

$$\begin{aligned}\left[\hat{S}_1, \hat{X}^\dagger_{k\sigma}\right] &= \tfrac{1}{2} X^\dagger_{k\bar{\sigma}}\,,\\ \left[\hat{S}_2, \hat{X}^\dagger_{k\sigma}\right] &= \tfrac{i}{2}(-1)^{\sigma-(1/2)}\hat{X}^\dagger_{k\bar{\sigma}}\,,\\ \left[\hat{S}_3, \hat{X}^\dagger_{k\sigma}\right] &= \tfrac{1}{2}(-1)^{\sigma-(1/2)}\hat{X}^\dagger_{k\sigma}\,,\end{aligned} \tag{A.2}$$

and similar ones for the corresponding annihilation operator $\hat{X}_{k\sigma}$ that result by Hermitian conjugation. These relationships follow immediately by relying on the explicit form of the spin operators $\hat{S}_a$, $a = 1, 2, 3$ that are given by Eqs. (88) and the commutator relations

$$\left[\hat{E}^I_J, \hat{X}^\dagger_K\right] = \hat{X}^\dagger_I \delta_{JK}\,, \qquad \left[\hat{E}^I_J, \hat{X}_K\right] = -\hat{X}_J \delta_{IK}\,, \tag{A.3}$$

as well as on the fact that

$$(\hat{E}^I_J)^\dagger = \hat{E}^J_I \quad \text{and} \quad [\hat{A}, \hat{B}]^\dagger = [B^\dagger, A^\dagger]\,. \tag{A.4}$$

Using the above results we easily find for the spin rotation of the creation operator $\hat{X}^\dagger_{k\sigma}$ that

$$\begin{aligned} e^{i\hat{S}_1\theta}\hat{X}^\dagger_{k\sigma}e^{-i\hat{S}_1\theta} &= \sum_{m=0}^{\infty}\frac{(-1)^m\theta^{2m}}{(2m)!}[\hat{S}_1, \hat{X}^\dagger_{k\sigma}]^{(2m)} \\ &+ i\sum_{m=0}^{\infty}\frac{(-1)^m\theta^{2m+1}}{(2m+1)!}[\hat{S}_1, \hat{X}^\dagger_{k\sigma}]^{(2m+1)} \\ &= \cos(\theta/2)\,\hat{X}^\dagger_{k\sigma} + i\sin(\theta/2)\,\hat{X}^\dagger_{k\bar{\sigma}}\,,\end{aligned} \tag{A.5}$$

and, similarly

$$e^{i\hat{S}_2\theta}\hat{X}^\dagger_{k\sigma}e^{-i\hat{S}_2\theta} = \cos(\theta/2)\,\hat{X}^\dagger_{k\sigma} + (-1)^{\bar{\sigma}-(1/2)}\sin(\theta/2)\,\hat{X}^\dagger_{k\bar{\sigma}}\,, \tag{A.6}$$

$$e^{i\hat{S}_3\theta}\hat{X}^\dagger_{k\sigma}e^{-i\hat{S}_3\theta} = \cos(\theta/2)\,\hat{X}^\dagger_{k\sigma} + i(-1)^{\sigma-(1/2)}\sin(\theta/2)\,\hat{X}^\dagger_{k\sigma}\,, \tag{A.7}$$

Again, analogous relationships for the annihilation operator $\hat{X}_{k\sigma}$ are easily obtained by Hermitian conjugation.

Finally, for the similarity transformed U(2n) generators we find

$$\begin{aligned} e^{i\hat{S}_1\theta}\hat{E}^{i\sigma_1}_{j\sigma_2}e^{-i\hat{S}_1\theta} &= (\cos(\tfrac{\theta}{2})\,\hat{X}^\dagger_{i\sigma_1} + i\sin(\tfrac{\theta}{2})\hat{X}^\dagger_{i\bar{\sigma}_1})(\cos(\tfrac{\theta}{2})\hat{X}_{j\sigma_2} - i\sin(\tfrac{\theta}{2})\hat{X}_{j\bar{\sigma}_2}) \\ &= \cos^2(\theta/2)\,\hat{E}^{i\sigma_1}_{j\sigma_2} + \sin^2(\theta/2)\,\hat{E}^{i\bar{\sigma}_1}_{j\bar{\sigma}_2} \\ &+ i\sin(\theta/2)\,\cos(\theta/2)\left(\hat{E}^{i\bar{\sigma}_1}_{j\sigma_2} - \hat{E}^{i\sigma_1}_{j\bar{\sigma}_2}\right) \\ &= \tfrac{1}{2}\left(\hat{E}^{i\sigma_1}_{j\sigma_2} + \hat{E}^{i\bar{\sigma}_1}_{j\bar{\sigma}_2}\right) + \tfrac{1}{2}\cos(\theta)\left(\hat{E}^{i\sigma_1}_{j\sigma_2} - \hat{E}^{i\bar{\sigma}_1}_{j\bar{\sigma}_2}\right) \\ &+ \tfrac{i}{2}\sin(\theta)\left(\hat{E}^{i\bar{\sigma}_1}_{j\sigma_2} - \hat{E}^{i\sigma_1}_{j\bar{\sigma}_2}\right)\,,\end{aligned} \tag{A.8}$$

and, similarly

$$e^{\mathrm{i}\hat{\mathrm{S}}_3\theta}\hat{\mathrm{E}}^{i\sigma_1}_{j\sigma_2}e^{-\mathrm{i}\hat{\mathrm{S}}_3\theta} = \delta_{\sigma_1\sigma_2}\hat{\mathrm{E}}^{i\sigma_1}_{j\sigma_2} + \delta_{\sigma_1\bar{\sigma}_2}e^{\mathrm{i}\theta\mathrm{sign}(\sigma_1)}\hat{\mathrm{E}}^{i\sigma_1}_{j\sigma_2}\,. \tag{A.9}$$

Turning now to the time-reversal operator, we first note that under the complex conjugation operation $\hat{\mathrm{K}}$, the operator $c\hat{\mathrm{X}}^{\dagger}_{k\sigma}$, where $c \in \mathbb{C}$, transforms as

$$\hat{\mathrm{K}}(c\hat{\mathrm{X}}^{\dagger}_{k\sigma})\hat{\mathrm{K}} = c^*\hat{\mathrm{X}}^{\dagger}_{k\sigma}\,, \tag{A.10}$$

while the time reversal operation gives

$$\begin{aligned}\hat{\Theta}(c\hat{\mathrm{X}}^{\dagger}_{k\sigma})\hat{\Theta}^{-1} &= e^{-\mathrm{i}\pi\hat{\mathrm{S}}_2}\hat{\mathrm{K}}(c\hat{\mathrm{X}}^{\dagger}_{k\sigma})e^{\mathrm{i}\pi\hat{\mathrm{S}}_2}\hat{\mathrm{K}} \\ &= c^*\cos(-\pi/2)\,\hat{\mathrm{X}}^{\dagger}_{k\sigma} + c^*(-1)^{\bar{\sigma}-(1/2)}\sin(-\pi/2)\,\hat{\mathrm{X}}^{\dagger}_{k\bar{\sigma}} \\ &= c^*(-1)^{\sigma-(1/2)}\hat{\mathrm{X}}^{\dagger}_{k\bar{\sigma}}\,, \end{aligned} \tag{A.11}$$

where we used Eq. (93) for $\hat{\Theta}^{-1}$. Thus, for the U($2n$) generators we get

$$\hat{\mathrm{K}}(c\hat{\mathrm{E}}^{i\sigma_1}_{j\sigma_2})\hat{\mathrm{K}} = c^*\hat{\mathrm{E}}^{i\sigma_1}_{j\sigma_2}\,, \tag{A.12}$$

and

$$\hat{\Theta}(c\hat{\mathrm{E}}^{i\sigma_1}_{j\sigma_2})\hat{\Theta}^{-1} = c^*(-1)^{\sigma_1+\sigma_2-1}\hat{\mathrm{E}}^{i\bar{\sigma}_1}_{j\bar{\sigma}_2}\,. \tag{A.13}$$

B. Spin-independent matrix elements

The matrix elements of a spin-free, electronic Hamiltonian $\hat{\mathrm{H}}$, Eq. (13), that is invariant with respect to the action of the group $\mathfrak{S} \otimes \mathfrak{T}$, are spin independent. In the canonical MSO basis $\{|I\rangle\}$, $|I\rangle = |i\sigma\rangle \equiv |i\rangle \otimes |\sigma\rangle$, we thus have

$$Z_{IJ} = Z_{i\sigma_1,j\sigma_2} = Z_{ij}\delta_{\sigma_1\sigma_2}\,, \tag{B.1}$$

$$V_{IJ;KL} = V_{i\sigma_1,j\sigma_2;k\sigma_3,\ell\sigma_4} = V_{ij;k\ell}\delta_{\sigma_1\sigma_3}\delta_{\sigma_2\sigma_4}\,. \tag{B.2}$$

In terms of the integrals over the ASOs, when

$$Z_{IJ} = \sum_{\Lambda\Xi} C^*_{\Lambda I}Z_{\Lambda\Xi}C_{\Xi J}\,, \tag{B.3}$$

we find when employing the canonical ASO basis (using the notation mirroring the one used for the MSOs by replacing Latin symbols with Greek ones, while reserving the symbol $|\sigma\rangle$ for pure-spin functions in both cases), i.e., writing

$$|\Lambda\rangle = |\lambda\sigma\rangle \equiv |\lambda\rangle \otimes |\sigma\rangle\,, \tag{B.4}$$

that

$$Z_{IJ} \equiv Z_{i\sigma_1,j\sigma_2} = \sum_{\lambda\xi} \sum_{\sigma_3\sigma_4} C^*_{\lambda\sigma_3,i\sigma_1} Z_{\lambda\xi} \delta_{\sigma_3\sigma_4} C_{\xi\sigma_4,j\sigma_2}$$
$$= \sum_{\mu\nu} Z_{\mu\nu} \sum_{\sigma} C^*_{\mu\sigma,i\sigma_1} C_{\nu\sigma,j\sigma_2} \,. \tag{B.5}$$

This expression can be further simplified for each specific case of the chosen symmetry class as we shall see. For the two-electron matrix elements we find similarly that

$$V_{IJ;KL} \equiv V_{i\sigma_1,j\sigma_2;k\sigma_3,\ell\sigma_4}$$
$$= \sum_{\mu\nu\rho\tau} V_{\mu\nu;\rho\tau} \sum_{\sigma\sigma'} C^*_{\mu\sigma,i\sigma_1} C^*_{\nu\sigma',j\sigma_2} C_{\tau\sigma',\ell\sigma_4} C_{\rho\sigma,k\sigma_3} \,. \tag{B.6}$$

Again, this expression further reduces for specific broken symmetry classes.

In the case of the real or complex UHF solutions, the coefficient matrix **C** is block diagonal, each of the two submatrices belonging to one spin component, i.e.,

$$C_{\mu\sigma,i\sigma'} = C_{\mu\sigma,i\sigma} \delta_{\sigma\sigma'} \equiv \delta_{\sigma\sigma'} (\mathbf{C}_\sigma)_{\mu i} \,, \tag{B.7}$$

so that (B.5) becomes

$$Z_{i\sigma_1,j\sigma_2} = \delta_{\sigma_1\sigma_2} \sum_{\mu\nu} Z_{\mu\nu} (\mathbf{C}_{\sigma_1})_{\mu i} (\mathbf{C}_{\sigma_1})_{\nu j} \,, \tag{B.8}$$

and, similarly, for (B.6) we get

$$V_{i\sigma_1,j\sigma_2;k\sigma_3,\ell\sigma_4} = \tag{B.9}$$
$$\delta_{\sigma_1\sigma_3} \delta_{\sigma_2\sigma_4} \sum_{\mu,\nu,\rho,\tau} V_{\mu\nu;\rho\tau} (\mathbf{C}^*_{\sigma_1})_{\mu i} (\mathbf{C}_{\sigma_1})_{\rho k} (\mathbf{C}^*_{\sigma_2})_{\nu j} (\mathbf{C}_{\sigma_2})_{\tau\ell} \,.$$

In the case of the real or complex RHF CS solutions, a further simplification results due to the fact that both spin blocks are identical, namely

$$\mathbf{C}_\sigma = \mathbf{C}_\mathrm{R} \,, \qquad \sigma = \pm\tfrac{1}{2} \ (\text{or}\, \sigma = \alpha, \beta) \,. \tag{B.10}$$

The corresponding Fock matrix in terms of canonical spin orbitals has the form

$$F_{i\sigma_1,j\sigma_2} = Z_{i\sigma_1,j\sigma_2} + \sum_{k\sigma} V^{(\mathcal{A})}_{i\sigma_1,k\sigma;j\sigma_2,k\sigma} n_{k\sigma} \,, \tag{B.11}$$

which, in the real or complex UHF case becomes

$$F_{i\sigma_1,j\sigma_2} = \delta_{\sigma_1\sigma_2}\Bigg[\sum_{\mu\nu} Z_{\mu\nu}(\mathbf{C}^*_{\sigma_1})_{\mu i}(\mathbf{C}_{\sigma_1})_{\nu j} + \sum_{\mu\nu\rho\tau} V_{\mu\nu;\rho,\tau}\Big((\mathbf{C}^*_{\sigma_1})_{\mu i}(\mathbf{C}_{\sigma_1})_{\rho j}\sum_k\sum_\sigma(\mathbf{C}^*_\sigma)_{\nu k}(\mathbf{C}_\sigma)_{\tau k}n_{k\sigma} - (\mathbf{C}^*_{\sigma_1})_{\mu i}(\mathbf{C}_{\sigma_1})_{\tau j}\sum_k(\mathbf{C}^*_{\sigma_1})_{\nu k}(\mathbf{C}_{\sigma_1})_{\rho k}n_{k\sigma_1}\Big)\Bigg], \tag{B.12}$$

so that $\mathbf{F}$ has a spin-block structure with diagonal blocks

$$(\mathbf{F}_\sigma)_{ij} = F_{i\sigma,j\sigma}\,. \tag{B.13}$$

For real and complex RHF solutions, these blocks are again spin independent in the CS case thanks to Eq. (B.10), while for the high-spin OS cases (i.e., for the ROHF solutions) there will be different spin-orbital occupancies $n_{k\sigma_1}$ in the last term of (B.12).

Acknowledgements

The continued support by the National Science and Engineering Research Council of Canada (J.P.) is gratefully acknowledged. We are also indebted to Professor Karol Jankowski for sharing with us useful information concerning the homotopy methods and to Dr. Karol Kowalski for providing us with a copy of his Ph.D. Thesis on this topic. We also thank Professor Kizashi Yamaguchi and Dr. Shusuke Yamanaka for numerous pre- and re-prints of papers on various aspects of broken symmetry solutions and their exploitation. Finally, we are indebted to Dr. Xiangzhu Li for his kind help with some computations.

References

[1] Löwdin, P.-O. (1953) in: *Proceedings of the Nikko Symposium on Molecular Physics*, Maruzen, Tokyo, p. 13.

[2] Löwdin, P.-O. (1955) *Phys. Rev.* **97**, 1509.

[3] Löwdin, P.-O. (1963) *Discussion of the Hartree-Fock Approximation*, in: P. Lykos and G.W. Pratt, *Rev. Mod. Phys.* **35**, 496.

[4] Löwdin, P.-O. (1969) *Adv. Chem. Phys.* **14**, 283.

[5] Löwdin, P.-O. and Mayer, I. (1992) *Adv. Quantum Chem.* **24**, 79.

[6] Hartree, D.R. (1928) *Proc. Cambridge Phil. Soc.* **24**, 89,111,426.

[7] Fock, V. (1930) *Z. Phys.* **61**,126.

[8] Slater, J.C. (1930) *Phys. Rev.* **35**, 210.

[9] Roothaan, C.C.J. (1951) *Rev. Mod. Phys.* **23**, 69.

[10] Roothaan, C.C.J. (1960) *Rev. Mod. Phys.* **32**, 179.

[11] Carbó, R. and Klobukowski, M. (Eds) (1990) *Self-Consistent Field, Theory and Applications*, Elsevier, Amsterdam.

[12] Paldus, J. and Čížek, J. (1975) *Adv. Quantum Chem.* **9**, 105.

[13] Paldus, J. (1981) *Diagrammatic Methods for Many-Fermion Systems*, Katholieke Universiteit, Nijmegen, The Netherlands.

[14] Paldus, J. (1992) in: *Methods in Computational Molecular Physics*, NATO ASI Series B: Physics, Vol. 293, S. Wilson and G.H.F. Diercksen, Eds., Plenum Press, New York, pp. 99–194.

[15] Paldus, J. and Li, X. (1999) *Adv. Chem. Phys.* **110**, 1.

[16] Fernández Rico, J. (1990) in Ref. 11, pp. 220–248.

[17] Mulliken, R.S. (1971) *Int. J. Quantum Chem., Symp.* **5**, 95.

[18] Li, X. and Paldus, J. (1995) *J. Chem. Phys.* **102**, 2013.

[19] Li, X. and Paldus, J. (1995) *J. Chem. Phys.* **103**, 6536.

[20] Thouless, D.J. (1960) *Nucl. Phys.* **21**, 225; also in: *The Quantum Mechanics of Many-Body Systems*, Academic, New York, 1961.

[21] For a general overview, see Löwdin, P.-O. (1966) in: *Quantum Theory of Atoms, Molecules, and the Solid State, A Tribute to John C. Slater*, P.-O. Löwdin, Ed., Academic, New York, p. 601.

[22] Sawada, K. and Fukuda, N. (1961) *Progr. Theoret. Phys. (Kyoto)* **25**, 653.

[23] Coulson, C.A., in Ref. 21, p. 97.

[24] Coulson, C.A. and Fischer, I. (1949) *Phil.Mag.* **40**, 386.

[25] Ooshika, Y. (1957) *J. Phys. Soc. Japan* **12**, 1246.

[26] Adams, W.H. (1962) *Phys. Rev.* **127**, 1650.

[27] Koutecký, J. (1967) *J. Chem. Phys.* **46**, 2443.

[28] Misurkin, I.A. and Ovchinnikov, A.A. (1967) *Zh. Eksp. Teoret. Fiz.* **3**, 431.

[29] Lefebvre, R. and Smeyers, Y.G. (1967) *Int. J. Quantum Chem.* **1**, 403.

[30] Čížek, J. and Paldus, J. (1967) *J. Chem. Phys.* **47**, 3976.

[31] Pauncz, R. (1967) *Alternant Molecular Orbital Method*, Saunders, Philadelphia.

[32] Čížek, J. and Paldus, J. (1970) *J. Chem. Phys.* **53**, 821.

[33] Paldus, J. and Čížek, J. (1970) *Phys. Rev. A* **2**, 2268.

[34] Čížek, J. and Paldus, J. (1971) *Phys. Rev. A* **3**, 525.

[35] Paldus, J. and Čížek, J. (1970) *J. Polymer Sci. C* **29**, 199.

[36] Fukutome, H. (1968) *Progr. Theoret. Phys. (Kyoto)* **40**, 998.

[37] Fukutome, H. (1968) *Progr. Theoret. Phys. (Kyoto)* **40**, 1227.

[38] Paldus, J. and Čížek, J. (1969) *Progr. Theoret. Phys. (Kyoto)* **42**, 769.

[39] Bénard, M. and Paldus, J. (1980) *J. Chem. Phys.* **72**, 6546.

[40] Bénard, M. (1979) *J. Chem. Phys.* **71**, 2546.

[41] Paldus, J. and Čížek, J. (1985) *Can. J. Chem.* **63**, 1803.

[42] Bénard, M., Laidlaw, W.G., and Paldus, J. (1986) *Chem. Phys.* **103**, 43.

[43] Paldus, J., Čížek, J., and Keating, B. (1973) *Phys. Rev. A* **8**, 640.

[44] Paldus, J. and Čížek, J. (1969) *Chem. Phys. Lett.* **3**, 1.

[45] Paldus, J. and Čížek, J. (1970) *J. Chem. Phys.* **52**, 2919.

[46] Paldus, J. and Čížek, J. (1971) *J. Chem. Phys.* **54**, 2293.

[47] Paldus, J. and Veillard, A. (1977) *Chem. Phys. Lett.* **50**, 6.

[48] Paldus, J. and Veillard, A. (1978) *Mol. Phys.* **35**, 445.

[49] Fukutome, H. (1971) *Progr. Theoret. Phys. (Kyoto)* **45**, 1382.

[50] Fukutome, H. (1974) *Progr. Theoret. Phys. (Kyoto)* **52**, 115.

[51] Fukutome, H. (1974) *Progr. Theoret. Phys. (Kyoto)* **52**, 1766.

[52] Fukutome, H. (1975) *Progr. Theoret. Phys. (Kyoto)* **53**, 1320.

[53] Ozaki, M. and Fukutome, H. (1978) *Progr. Theoret. Phys. (Kyoto)* **60**, 1322.

[54] Ozaki, M. (1979) *Progr. Theoret. Phys. (Kyoto)* **62**, 1183.

[55] Ozaki, M. (1980) *Progr. Theoret. Phys. (Kyoto)* **63**, 84.

[56] Ozaki, M. (1982) *Progr. Theoret. Phys. (Kyoto)* **67**, 83.

[57] Ozaki, M. (1982) *Progr. Theoret. Phys. (Kyoto)* **67**, 415.

[58] Löwdin, P.-O., Calais, J.-L., and Calazans, J. (1981) *Int. J. Quantum Chem.* **20**, 1201.

[59] Löwdin, P.-O. (1986) *Proc. Ind. Acad. Sci. (Chem. Sci.)* **96**, 121.

[60] Fukutome, H. (1972) *Progr. Theoret. Phys. (Kyoto)* **47**, 1156.

[61] Fukutome, H. (1973) *Progr. Theoret. Phys. (Kyoto)* **49**, 22.

[62] Fukutome, H. (1973) *Progr. Theoret. Phys. (Kyoto)* **50**, 1433.

[63] Fukutome, H. (1981) *Int. J. Quantum Chem.* **20**, 955.

[64] Yamaguchi K., Fueno, T., and Fukutome, H. (1973) *Chem. Phys. Lett.* **22**, 461.

[65] Fukutome, H., Takahashi, M., and Takabe, T. (1975) *Progr. Theoret. Phys. (Kyoto)* **53**, 1580.

[66] Yamaguchi K. and Fukutome, H. (1975) *Progr. Theoret. Phys. (Kyoto)* **54**, 1599.

[67] Yoshioka, Y., Yamaguchi K., and Fueno, T. (1978) *Mol. Phys.* **35**, 33.

[68] Takabe, T. and Fukutome, H. (1976) *Progr. Theoret. Phys. (Kyoto)* **56**, 689.

[69] Takahashi, M. and Fukutome, H. (1978) *Progr. Theoret. Phys. (Kyoto)* **59**, 1787.

[70] Igawa, A. and Fukutome, H. (1980) *Progr. Theoret. Phys. (Kyoto)* **64**, 491.

[71] Yamaguchi K. (1990) in Ref. 11, pp. 727–823.

[72] Paldus, J. (1990) in Ref. 11, pp. 1–45.

[73] Mestechkin, M.M. (1990) in Ref. 11, pp. 312–385.

[74] Mestechkin, M.M. (1977) *Density Matrix Method in Molecular Theory*, Naukova Dumka, Kiev (in Russian).

[75] Davidson, E.R. (1973) *Chem. Phys. Lett.* **21**, 565.

[76] Binkley, J.S., Pople, J.A., and Dobosh, P.A. (1974) *Mol. Phys.* **28**, 1423.

[77] McWeeny, R. and Diercksen, G. (1968) *J. Chem. Phys.* **49**, 4852.

[78] Guest, M.F. and Saunders, V.R. (1974) *Mol. Phys.* **28**, 819.

[79] Faegri, K. and Manne, R. (1976) *Mol. Phys.* **31**, 1037.

[80] Krebs, S. (1999) *Comput. Phys. Commun.* **116**, 137.

[81] Bardeen, J., Cooper, L.N., and Schrieffer, J.R. (1957) *Phys. Rev.* **108**, 1175.

[82] Blaizot, J.P. and Ripka, G. (1986) *Quantum Theory of Finite Systems*, MIT Press, Cambridge.

[83] Paldus, J. (1976) in: *Theoretical Chemistry: Advances and perspectives*, Vol. 2, D. Henderson and H. Eyring, Eds., Academic Press, New York, pp. 131–290.

[84] Paldus, J. (1994) in: *Contemporary Mathematics*, Vol. 160, American Mathematical Society, Providence, R.I., pp. 209–236.

[85] Paldus, J. (1996) in: *Atomic, Molecular, and Optical Physics Handbook*, Chap. 4, *Dynamical Groups*, G.W.F. Drake, Ed., American Institute of Physics, Woodbury, N.Y., pp. 76–87.

[86] Greiner, W. and Müller, B. (1994) *Quantum Mechanics: Symmetries*, Springer-Verlag, Berlin.

[87] Chen, J.-Q. (1989) *Group Representation Theory for Physicists*, World Scientific, Singapore.

[88] Naimark, M.A. and Štern, A.I. (1982) *Theory of Group Representations*, Springer-Verlag, Berlin.

[89] Stuber, J. L. (2002) Ph.D Thesis, University of Waterloo, Waterloo, Ontario, Canada.

[90] The GAUSSIAN 92 system of programs, REVISION E.2, Frisch, M.J., Trucks, G.W., Head-Gordon, M., Gill, P.M.W., Wong, M.W., Foresman, J.B., Johnson, B.G., Schlegel, H.B., Robb, M.A., Replogle, E.S., Gomperts, R., Andres, J.L., Raghavachari, K., Binkley, J.S., Gonzalez, C., Martin, R.L., Fox, D.J., Defrees, D.J., Baker, J., Stewart, J.J.P., and Pople, J.A. (1992), Gaussian, Inc., Pittsburgh PA.

[91] Li, X. and Paldus, J. (1994) *J. Chem. Phys.* **101**, 8812.

[92] Jeziorski, B., Paldus, J., and Jankowski, P. (1995) *Int. J. Quantum Chem.* **56**, 129.

[93] Kowalski, K. and Jankowski, K. (1998) *Phys. Rev. Lett.* **81**, 1195.

[94] Kowalski, K. (1999) Ph.D Thesis, Nicholas Copernicus University, Toruń, Poland (in Polish).

[95] Kowalski, K. and Jankowski, K. (1998) *Chem. Phys. Lett.* **290**, 180.

[96] Kowalski, K. and Piecuch, P. (2000) *Phys. Rev. A* **61**, 52506.

[97] Jankowski, K. and Paldus, J. (1980) *Int. J. Quantum Chem.* **18**, 1243.

[98] Stuber, J. L. (2002), The Virgil v1.1C software package. Software developed for Ph.D. thesis. Available on-line from Mathematical Sciences Inc., <http://www.mathematicalsciences.com/virgil/>

[99] Schmidt, M.W., Baldridge, K.K., Boatz, J.A., Elbert, S.T., Gordon, M.S., Jensen, J.H., Koseki, S., Matsunaga, N., Nguyen, K.A., Su, S.J., Windus, T.L., Dupuis, M., and Montgomery, J.A. (1993) *J. Comput. Chem.* **14**, 1347.

[100] Huzinaga, S. (1965) *J. Chem. Phys.* **42**, 1293.

[101] Wilson, S., Jankowski, K., and Paldus, J. (1983) *Int. J. Quantum Chem.* **23**, 1781.

[102] Kaldor, U. (1985) *Int. J. Quantum Chem.* **28**, 103.

[103] Wilson, S., Jankowski, K., and Paldus, J. (1985) *Int. J. Quantum Chem.* **28**, 525.

[104] Iijima, N. and Saika, A. (1985) *Int. J. Quantum Chem.* **27**, 481.

[105] Paldus, J., Wormer, P.E.S., and Bénard, M. (1988) *Coll. Czech. Chem. Commun.* **53**, 1919.

[106] Zarrabian, S. and Paldus, J. (1990) *Int. J. Quantum Chem.* **38**, 761.

[107] Jankowski, K., Paldus, J., and Wasilewski, J. (1991) *J. Chem. Phys.* **95**, 3549.

[108] Kucharski, S.A., Balková, A., and Bartlett, R.J. (1991) *Theor. Chim. Acta* **80**, 321.

[109] Balková, A., Kucharski, S.A., Meissner, L., and Bartlett, R.J. (1991) *Theor. Chim. Acta* **80**, 335.

[110] Balková, A., Kucharski, S.A., and Bartlett, R.J. (1991) *Chem. Phys. Lett.* **182**, 511.

[111] Paldus, J., Piecuch, P., Pylypow, L., and Jeziorski, B. (1993) *Phys. Rev. A* **47**, 2738.

[112] Piecuch, P., Toboła, R., and Paldus, J. (1993) *Chem. Phys. Lett.* **210**, 243 and *loc. cit.*

[113] Jankowski, K., Paldus, J., Grabowski, I., and Kowalski, K. (1994) *J. Chem. Phys.* **101**, 3085 and *loc. cit.*

[114] Piecuch, P. and Paldus, J. (1994) *Phys. Rev. A* **49**, 3479.

[115] Finley, J.P., Chaudhuri, R.K., and Freed, K.F. (1995) *J. Chem. Phys.* **103**, 4990.

[116] Li, X. and Paldus, J. (1997) *J. Chem. Phys.* **107**, 6257.

[117] Mahapatra, U.S., Datta, B., and Mukherjee, D. (1999) *J. Chem. Phys.* **110**, 6171.

[118] Meissner, H. and Paldus, J. (2000) *J. Chem. Phys.* **113**, 2612.

[119] Morgan, A.P. (1987) *Solving Polynomial Systems Using Continuation for Engineering and Scientific Problems*, Prentice Hall, Englewood Cliffs, NJ.

[120] Morgan, A.P., Sommese, A.J., and Watson, L.T. (1989) *ACM Trans. Math. Softw.* **15**, 93.

[121] Paldus, J. and Chin, E. (1983) *Int. J. Quantum Chem.* **24**, 373.

[122] Paldus, J., Chin, E., and Grey, M.G. (1983) *Int. J. Quantum Chem.* **24**, 395.

[123] Pauncz, R. and Paldus, J. (1983) *Int. J. Quantum Chem.* **24**, 411.

[124] Paldus, J. and Takahashi, M. (1984) *Int. J. Quantum Chem.* **25**, 423.

[125] Takahashi, M. and Paldus, J. (1984) *Int. J. Quantum Chem.* **26**, 349.

[126] Takahashi, M. and Paldus, J. (19854) *Int. J. Quantum Chem.* **28**, 459.

[127] Nagao, H., Nishino, M., Mitani, M., Yoshioka, Y., and Yamaguchi, K. (1997) *Int. J. Quantum Chem.* **65**, 947.

[128] Nagao, H., Kawabe, H.,Nishino, M., Shigeta, Y., Mitani, M., and Yamaguchi, K. (1999) *Synth. Met.* **103**, 2651.

[129] Nagao, H., Nishino, M., Shigeta, Y., Yoshioka, Y., and Yamaguchi, K. (2000) *J. Chem. Phys.* **113**, 11237.

[130] Mitani, M., Mori, H., Takano, Y., Yamaki, D., Yoshioka, Y., and Yamaguchi, K. (2000) *J. Chem. Phys.* **113**, 4035.

[131] Yamanaka, S., Yamaki, D., Kiribayashi, S., and Yamaguchi, K. (2001) *Int. J. Quantum Chem.* **85**, 421.

[132] Yamaki, D., Shigeta, Y., Yamanaka, S., Nagao, H., and Yamaguchi, K. (2001) *Int. J. Quantum Chem.* **84**, 546.

[133] Yamanaka, S., Shigeta, Y., Ohta, Y., Yamaki, D., Nagao, H., and Yamaguchi, K. (2001) *Int. J. Quantum Chem.* **84**, 369.

[134] Yamanaka, S., Yamaki, D., Shigeta, Y., Nagao, H., and Yamaguchi, K. (2001) *Int. J. Quantum Chem.* **84**, 670.

[135] Yamaguchi, K. (1979) *Chem. Phys. Lett.* **66**, 395.

[136] Yamaguchi, K. (1979) *Chem. Phys. Lett.* **68**, 477.

[137] Yamanaka, S., Takano, Y., and Yamaguchi, K. (2002) *Generalized Spin Density Functional Study of Multicenter Polyradical Systems*, (preprint).

[138] Yamanaka, S., Ohsaku, T., Yamaki, D., and Yamaguchi, K. (2002) *Generalized Spin Density Functional Study of Chemical Reactions*, (preprint).

[139] Ohsaku, T. and Yamaguchi, K. (2001) *Int. J. Quantum Chem.* **85**, 272.

The Hueckel-Hubbard Hamiltonian and its Gel'fand Basis

F. A. Matsen
The University of Texas, Austin TX 78712

Abstract: It is a privilege to offer this article to the Tribute Volume in the memory of my dear friend, Per-Olov Löwdin. He encouraged me to develop Hueckel-Hubbard theory in its spin-free unitary-group realization. In the present article I review this theory. I note first that the electron is a fermion, which is conventionally represented by the tensor product of a freeon (spin free) orbital and a pure spin orbital. In my formulation I employ only the freeon orbitals, which I believe expose more clearly the basic dynamics of the interelectronic spin-free coulomb interaction in the many-electron problem. The Hilbert space of the Unitary-Group Hamiltonian is spanned by the Gel'fand basis. This basis is uniquely represented by Gel'fand diagrams whose construction is described below. After reviewing the theory I present the results of numerical calculations for the linear four-site system. This system has twenty singlet states, fifteen triplet states and one quintet state. I diagonalize the Hueckel-Hubbard Hamiltonian in the Gel'fand basis. Since this basis spans the Hilbert space, its eigenvalues and eigenvectors are exact except for numerical errors introduced in the diagonalization.

Key words: spin-free theory, Hueckel-Hubbard theory, Gel'fand diagrams, molecular orbitals, site orbitals, freeon orbitals, unitary group Hamiltonian.

1. MY HAMILTONIAN

I employ the unitary-group form of the Hueckel-Hubbard Hamiltonian,

$$\mathbf{H} = -t\sum_{<nn>}\left(\mathbf{E}_{rs} + \mathbf{E}_{sr}\right) + U\mathbf{d}$$

E.J. Brändas and E.S. Kryachko (eds.),
Fundamental World of Quatum Chemistry, Vol. I, 141–153.

Here $\mathbf{E}_{rs}$ is an infinitesimal generator of the unitary group U(n), where n is the number of site, and the number of freeon (spin-free) orbitals, t is the electron-transfer (hopping) integral (the negative of the Hueckel beta.) U is the repulsion energy for two electrons on the same site and

$$\mathbf{d} = \left(\mathbf{E}_{rr}^2 - \mathbf{E}_{rr}\right)/2$$

counts the number of doubly-occupied site orbitals. A one-parameter Hamiltonian is obtained by dividing by $U + t$.

$$H = (1-x)\sum_{<nn>}(\mathbf{E}_{rs} + \mathbf{E}_{sr}) + x\mathbf{d},$$

where $x = U/(t + U)$ is called the "correlation index". It monitors the transition from a conductor to an insulator (i.e., the Mott transition). The Hilbert space of **H** is spanned by a set of orthonormal Gel'fand states, which are uniquely represented by Gel'fand diagrams. The Gel'fand diagrams are constructed on a Young diagram, which for an N-electron system is composed of N boxes arranged in two columns of lengths $L_1 \geq L_2 \geq 0$ with the freeon spin quantum number given by

$$S = (L_1 - L_2)/2$$

See Table 1.

Table 1. Young diagrams and the attendant spins.

N = 1	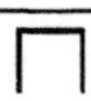	S = (1 – 0)/2 = ½
N = 2	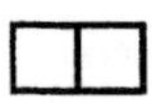	S = (1 – 1)/2 = 0
N = 2		S = (2 – 0)/2 = 1
N = 3	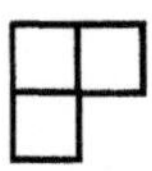	S = (2 – 1)/2 = ½
N = 3	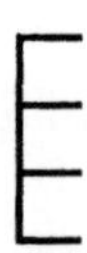	S = (3 – 0)/2 = 3/2
N = 4	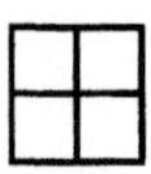	S = (2 – 2)/2 = 0

N = 4	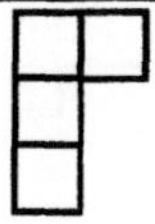	S = (3 – 1)/2 = 1
N = 4	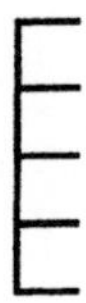	S = (4 – 0)/2 = 2

2. THE SITE-ORBITAL AND THE MOLECULAR-ORBITAL, SINGLET GEL'FAND STATES

For x = 1 the site-orbital Gel'fand states are the eigenvectors. Their diagrams are constructed by adding freeon (spin-free) site orbitals to the appropriate Young diagram in non-decreasing order (a < b < c < d) along rows and in ascending order down columns. The twenty singlet site-orbital Gel'fand states are listed in Table 2 together with an integer label and the dimensionless eigenvalues (d).

For x = 0 the molecular orbital Gel'fand states are the eigenvectors. These are listed in Table 2 as well (in the order of increasing energy, E), where they have been assigned a numerical index. Also listed are the dimensionless eigenvalues, which are the simple sums of the orbital energies ($\varepsilon_k = -2\cos(k\pi/(n+1))$), where n is the number of sites; i.e., $\varepsilon_1 = -1.6180$, $\varepsilon_2 = -0.6180$, $\varepsilon_3 = 0.6180$, $\varepsilon_4 = 1.6180$. Note that for molecular-orbital states we have E(j) + E(21-j) = 0.

Table 2. The site-orbital and the molecular-orbital singlet Gel'fand states.

	site-orbital diagrams	**E**	molecular-orbital diagrams	**E**
|1>	a b c d	0.0	1 1 2 2	- 4.472
|2>	a c b d	0.0	1 1 2 3	-3.263

	site-orbital diagrams	E	molecular-orbital diagrams	E
|3>	a a b c	1.0	1 1 2 4	-2.236
|4>	a a b d	1.0	1 2 2 3	-2.236
|5>	a a c d	1.0	1 1 3 3	-2.0
|6>	a b b c	1.0	1 2 3 3	-1.0
|7>	a b b d	1.0	1 1 3 4	-1.0
|8>	b b c d	1.0	1 2 2 4	-1.0
|9>	a b c c	1.0	1 1 4 4	0.0
|10>	a c c d	1.0	1 2 3 4	0.0
|11>	b c c d	1.0	1 3 2 4	0.0
|12>	a b d d	1.0	2 2 3 3	0.0
|13>	a c d d	1.0	1 2 4 4	1.0
|14>	b c d d	1.0	1 3 3 4	1.0

	site-orbital diagrams	E	molecular-orbital diagrams	E
|15>	a a b b	2.0	2 2 3 4	1.0
|16>	a a c c	2.0	2 2 4 4	2.0
|17>	a a d d	2.0	1 3 4 4	+2.236
|18>	b b c c	2.0	2 3 3 4	+2.236
|19>	b b d d	2.0	2 3 4 4	+3.236
|20>	c c d d	2.0	3 3 4 4	+4.472

3. TRIPLET STATES

Table 3. The site-orbital and molecular triplet Gel'fand states arranged in the order of increasing energy.

	site-orbital diagrams	E	molecular-orbital diagrams	E
|1>	a b c d	0.0	1 1 2 3	-3.236
|2>	a c b d	0.0	1 1 2 4	-2.326

	site-orbital diagrams	**E**	molecular-orbital diagrams	**E**
|3>	a d b c	0.0	1 2 2 3	-2.326
|4>	a a b c	1.0	1 2 2 4	-1.236
|5>	a a b d	1.0	1 3 2 3	1.0
|6>	a a c d	1.0	1 1 3 4	1.0
|7>	a b b c	1.0	1 2 3 4	0.0
|8>	a b b d	1.0	1 3 2 4	0.0
|9>	b b c d	1.0	1 4 2 3	0.0
|10>	a c b c	1.0	1 4 2 4	1.0
|11>	a c c d	1.0	2 2 3 4	1.0

	site-orbital diagrams	E	molecular-orbital diagrams	E
|12>	b c c d	1.0	1 3 3 4	+1.236
|13>	a d b d	1.0	2 3 3 4	+2.326
|14>	a d c d	1.0	1 4 3 4	+2.362
|15>	b d c d	1.0	2 4 3 4	+3.236

Finally, there is only one quintet state:

Table 4. The freeon quintet state diagrams.

	site-orbital diagrams	E	molecular-orbital diagrams	E
|1>	a b c d	0.0	1 2 3 4	0.0

4. NUMERICAL CALCULATIONS

For $0 \leq x \leq 1$ I calculate numerically the eigenvalues and eigenvectors in the site-orbital basis. The matrix elements of the Hueckel-Hubbard Hamiltonian in this basis can be written as follows:

$$
\begin{aligned}
<p\,|\,\mathrm{H}\,|\,q> &= (1-x)\sum_{<nn>} <p\,|\,(\mathrm{E}_{rs}+\mathrm{E}_{sr})\,|\,q> + xU <p\,|\,d\,|\,q> \\
&= (1-x)\sum_{<nn>} <p\,|\,\mathrm{E}_{rs}\,|\,q> + <p\,|\,\mathrm{E}_{sr}\,|\,q> + xdU\delta(p,q)
\end{aligned}
$$

To evaluate the first term I employ the formulas given in the appendix. The diagonalization was carried out by means of Mathematica software.

5. RESULTS

The singlet eigenvalues are plotted in Figure 1, the triplet eigenvalues are plotted in Figure 2 and the quintet eigenvalue is plotted in Figure 3. The eigenvectors in the site-orbital basis for the ground state are given in Table 5.

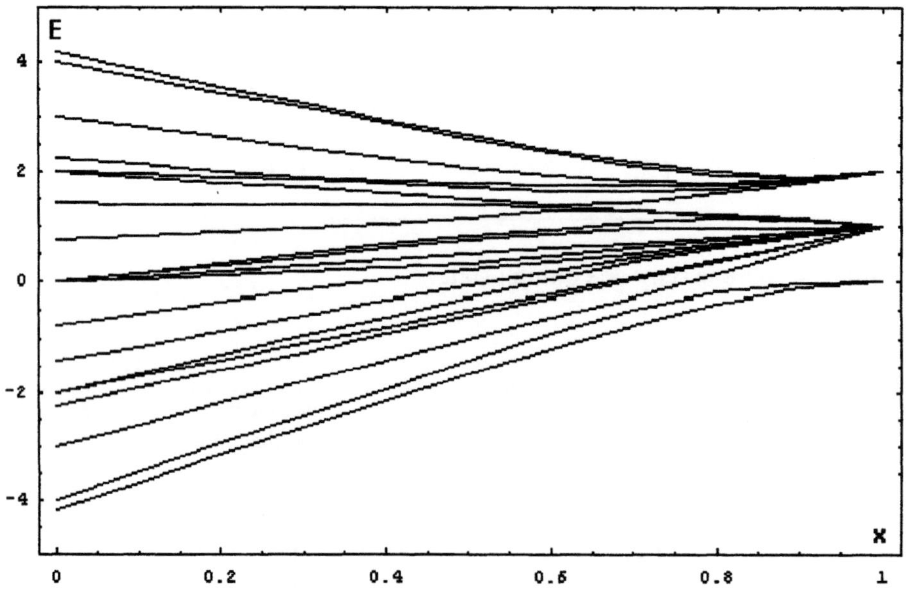

Figure 1. The singlet eigenvalues.

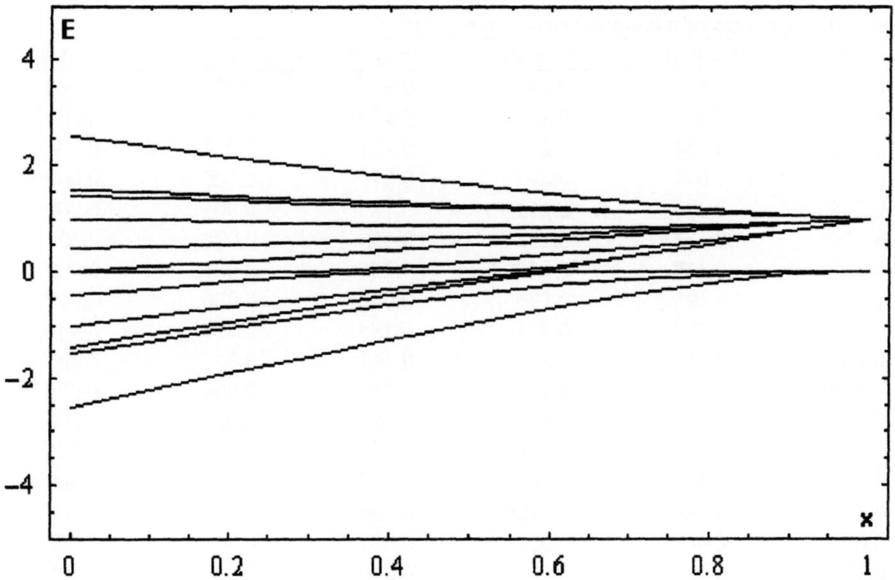

Figure 2. The triplet eigenvalues.

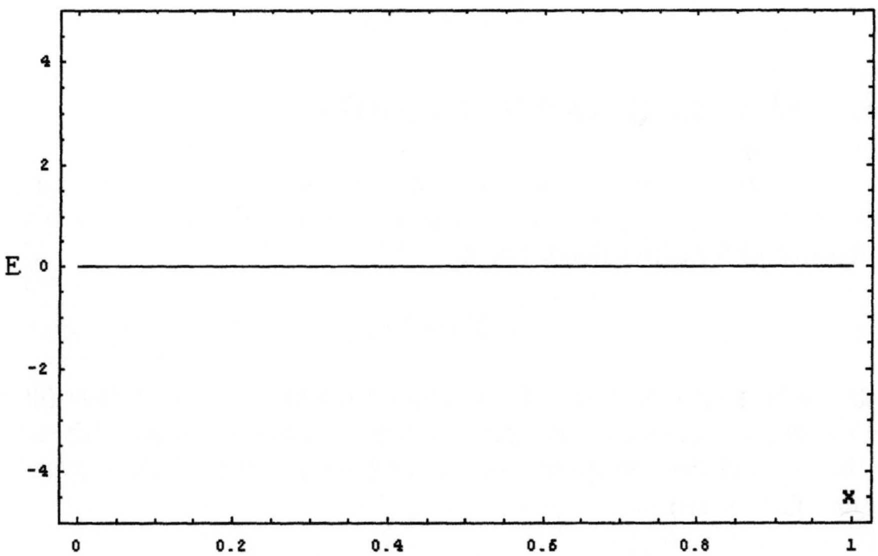

Figure 3. The quintet eigenvalue.

Table 5. Components of the ground state eigenvector

x = 0.0	x = 0.25	x = 0.50	x = 0.75	x = 0.999	x = 1.0
0.373	0.414	0.487	0.627	0.799	1.000
0.285	0.336	0.405	0.500	0.601	0.000
0.103	0.154	0.203	0.211	0.001	0.000
0.097	0.074	0.043	0.010	0.000	0.000
0.248	0.230	0.213	0.185	0.001	0.000
0.179	0.154	0.114	0.053	0.000	0.000
0.382	0.377	0.349	0.258	0.001	0.000
0.225	0.193	0.136	0.048	0.000	0.000
0.245	0.229	0.213	0.185	0.001	0.000
0.103	0.154	0.203	0.211	0.001	0.000
0.179	0.154	0.114	0.053	0.000	0.000
0.122	0.144	0.172	0.180	0.001	0.000
0.097	0.074	0.043	0.010	0.000	0.000
0.382	0.377	0.349	0.258	0.001	0.000
0.164	0.166	0.146	0.080	0.000	0.000
0.097	0.047	0.005	0.009	0.000	0.000
0.125	0.117	0.102	0.062	0.000	0.000
0.076	0.060	0.036	0.008	0.000	0.000
0.335	0.296	0.221	0.096	0.000	0.000
0.164	0.166	0.146	0.080	0.000	0.000

6. SUMMARY AND DISCUSSION

We have computed the spectrum of the four-site Hueckel-Hubbard Hamiltonian (twenty singlet states, fifteen triplet states and one quintet state as a function of the correlation index, x:

$$x = U/(U + t),$$

which monitors the transition of the system from a metal to an insulator (the Mott transition). We employ the unitary group form of the Hueckel-Hubbard Hamiltonian and the orthonormal set of Gel'fand of Gel'fand states, which span its Hilbert space.

The Gutzwiller's ansatz

$$|\varphi\rangle \propto \prod_i (1 - x\mathbf{d}) | \varphi_0\rangle = P(x) | \varphi_0\rangle$$

tells us that for x close to 1, the ground state wave function should contain no terms with d greater than 0 (no ionic states). An examination of the eigenvectors (Table 5) shows that in our framework this result comes out naturally: at x=1 the only non-zero projections are $\langle I|1\rangle = \langle I|2\rangle = 1$. Recalling the definition of the states |1> and |2>, one can see that those are the states with no double-occupancy.

7. ACKNOWLEDGEMENTS.

I am happy to acknowledge the technical assistance of Konstantin Mardanov, and the financial assistance of the Robert Welch Foundation of Houston, Texas and the Theoretical Chemistry Institute of the University of Texas at Austin.

8. REFERENCES

Gel'fand M. and Zetlin, M. L., *Dokl..Akad Nauk SSSR* **71**, 825 (1950)

Gel'fand M. and Graev M. I., *Dokl..Akad Nauk SSSR* **29**, 1329 (1965); English translation ser. **2** 64 116 (1967)

Gutzwiller, M.C., Phys.Rev. **137** (1965) 1726Rassetti, Mario (ed), *The Hubbard Model*, World Scientific (1991)

Matsen F. A., *Freeon Dynamics,* Adv. Quantum Chem.,**27**.1(1996)

Matsen F. A. and Pauncz R., *The Unitary Group in Quantum Chemistry*. Elsevier (1988)

APPENDIX

To evaluate the matrix elements of H, we need only the matrix elements of the form

$$\langle G|E_{rs}|G'\rangle = \langle G'|E_{sr}|G\rangle,$$

where G is a Gel'fand tableau. For $r = s$, we have

$$\langle G|E_{rr}|G'\rangle = \delta(G,G')g_r,$$

where g_r is the occupation number of the r-th orbital in the G-th Gel'fand state. For $r \neq s$ we convert each Gel'fand tableau into a Gel'fand symbol denoted $G = [R_{i,\rho},\ldots,R_{\rho,\rho};R_{i,\rho-1},\ldots,R_{\rho-1,\rho-1};\ldots;R_{11}]$, where R_{ij} is the number of squares in the i-th row of that Young diagram which remains after squares containing highest $(j-i)$ orbitals have been removed from G.

In the formulas to follow we indicate only those R's, which do not appear twice inside the brackets:

$$\langle R_{11}-1|E_{21}|R_{11}\rangle=\sqrt{(R_{11}-R_{12}+1)(R_{11}-R_{22})}$$

$$\langle R_{11}-1,R_{12}-1|E_{31}|R_{11},R_{12}\rangle=\sqrt{\frac{(R_{12}-R_{13}-1)(R_{11}-R_{22})(R_{23}-R_{12})(R_{12}-R_{33}+1)}{(R_{12}-R_{22})(R_{12}-R_{22}+1)}}$$

$$\langle R_{11}-1,R_{22}-1|E_{31}|R_{11},R_{22}\rangle=-\sqrt{\frac{(1-R_{11}+R_{12})(R_{13}-R_{22}+2)(R_{22}-R_{23}-1)(R_{33}-R_{22})}{(R_{12}-R_{22}+1)(R_{12}-R_{22}+2)}}$$

$$\langle R_{11}-1,R_{12}-1|E_{32}|R_{11},R_{12}\rangle=\sqrt{\frac{(R_{12}-R_{11})(R_{12}-R_{13}-1)(R_{23}-R_{12})(R_{12}-R_{33}+1)}{(R_{12}-R_{32})(R_{12}-R_{22}+1)}}$$

$$\langle R_{22}-1|E_{32}|R_{22}\rangle=-\sqrt{\frac{(R_{11}-R_{22}+1)(R_{22}-R_{13}-2)(R_{22}-R_{23}-1)(R_{22}-R_{33})}{(R_{12}-R_{22}+1)(R_{12}-R_{22}+2)}}$$

$$\langle R_{11}-1,R_{12}-1,R_{13}-1|E_{41}|R_{11},R_{12},R_{13}\rangle=$$
$$=-\sqrt{\frac{(R_{11}-R_{22})(R_{12}-R_{23})(R_{12}-R_{33}+1)(R_{22}-R_{13}-1)(R_{13}-R_{14}-1)(R_{13}-R_{24})(R_{13}-R_{34}+1)(R_{13}-R_{44}+2)}{(R_{12}-R_{22}+1)(R_{12}-R_{22})(R_{13}-R_{23})(R_{13}-R_{23}+1)(R_{13}-R_{33}+1)(R_{13}-R_{33}+2)}}$$

$$\langle R_{11}-1,R_{12}-1,R_{23}-1|E_{41}|R_{11},R_{12},R_{23}\rangle=$$
$$=-\sqrt{\frac{(R_{11}-R_{22})(R_{12}-R_{13}-1)(R_{12}-R_{33}+1)(R_{23}-R_{22})(R_{23}-R_{14}-2)(R_{23}-R_{24}-1)(R_{34}-R_{23})(R_{23}-R_{44}+1)}{(R_{12}-R_{22})(R_{12}-R_{22}+1)(R_{23}-R_{13}-1)(R_{23}-R_{13}-2)(R_{23}-R_{33})(R_{23}-R_{33}+1)}}$$

$$\langle R_{11}-1,R_{12}-1,R_{33}-1|E_{41}|R_{11},R_{12},R_{33}\rangle=$$
$$=-\sqrt{\frac{(R_{11}-R_{22})(R_{12}-R_{13}-1)(R_{12}-R_{23})(R_{33}-R_{22}-1)(R_{33}-R_{14}-3)(R_{23}-R_{24}-2)(R_{33}-R_{34}-1)(R_{44}-R_{33})}{(R_{12}-R_{22})(R_{12}-R_{22}+1)(R_{33}-R_{13}-2)(R_{33}-R_{13}-3)(R_{33}-R_{23}-1)(R_{33}-R_{23}-2)}}$$

$$\langle R_{11}-1,R_{22}-1,R_{13}-1|E_{41}|R_{11},R_{22},R_{13}\rangle=$$
$$=-\sqrt{\frac{(R_{11}-R_{12}-1)(R_{22}-R_{23}-1)(R_{22}-R_{33})(R_{12}-R_{13})(R_{13}-R_{14}-1)(R_{13}-R_{24})(R_{13}-R_{34}+1)(R_{13}-R_{44}+2)}{(R_{22}-R_{12}-1)(R_{22}-R_{12}-2)(R_{13}-R_{23}+1)(R_{13}-R_{23})(R_{13}-R_{33}+1)(R_{13}-R_{33}+2)}}$$

$$\langle R_{11}-1,R_{22}-1,R_{23}-1|E_{41}|R_{11},R_{22},R_{23}\rangle=$$
$$=-\sqrt{\frac{(R_{22}-R_{13}-2)(R_{11}-R_{12}-1)(R_{22}-R_{33})(R_{23}-R_{12}-1)(R_{23}-R_{14}-2)(R_{23}-R_{24}-1)(R_{34}-R_{23})(R_{23}-R_{44}+1)}{(R_{22}-R_{12}-1)(R_{22}-R_{12}-2)(R_{23}-R_{13}-1)(R_{23}-R_{13}-2)(R_{23}-R_{33}+1)(R_{23}-R_{33})}}$$

$$\langle R_{11}-1,R_{22}-1,R_{33}-1|E_{41}|R_{11},R_{22},R_{33}\rangle=$$
$$=\sqrt{\frac{(R_{11}-R_{12}-1)(R_{22}-R_{13}-2)(R_{22}-R_{23}-1)(R_{33}-R_{12}-2)(R_{33}-R_{14}-3)(R_{33}-R_{24}-2)(R_{33}-R_{34}-1)(R_{44}-R_{33})}{(R_{22}-R_{12}-1)(R_{22}-R_{12}-2)(R_{33}-R_{13}-2)(R_{33}-R_{13}-3)(R_{33}-R_{23}-1)(R_{33}-R_{23}-2)}}$$

$$\langle R_{12}-1,R_{13}-1|E_{42}|R_{12},R_{13}\rangle =$$
$$=\sqrt{\frac{(R_{12}-R_{11})(R_{12}-R_{23})(R_{12}-R_{33}+1)(R_{13}-R_{22}+1)(R_{13}-R_{14}-1)(R_{24}-R_{13})(R_{13}-R_{34}+1)(R_{13}-R_{44}+2)}{(R_{12}-R_{22})(R_{12}-R_{22}+1)(R_{13}-R_{23})(R_{13}-R_{23}+1)(R_{13}-R_{33}+1)(R_{13}-R_{33}+2)}}$$

$$\langle R_{12}-1,R_{23}-1|E_{42}|R_{12},R_{23}\rangle =$$
$$=-\sqrt{\frac{(R_{12}-R_{11})(R_{12}-R_{13}-1)(R_{12}-R_{33}+1)(R_{22}-R_{23})(R_{23}-R_{14}-2)(R_{23}-R_{24}-1)(R_{23}-R_{34})(R_{23}-R_{44}+1)}{(R_{12}-R_{22})(R_{12}-R_{22}+1)(R_{23}-R_{13}-1)(R_{23}-R_{13}-2)(R_{23}-R_{33})(R_{23}-R_{33}+1)}}$$

$$\langle R_{12}-1,R_{33}-1|E_{42}|R_{12},R_{33}\rangle =$$
$$=-\sqrt{\frac{(R_{12}-R_{11})(R_{12}-R_{13}-1)(R_{12}-R_{23})(R_{33}-R_{22}-1)(R_{33}-R_{14}-3)(R_{33}-R_{24}-2)(R_{33}-R_{34}-1)(R_{44}-R_{33})}{(R_{12}-R_{22})(R_{12}-R_{22}+1)(R_{33}-R_{13}-2)(R_{33}-R_{13}-3)(R_{33}-R_{23}-1)(R_{33}-R_{23}-2)}}$$

$$\langle R_{22}-1,R_{13}-1|E_{42}|R_{22},R_{13}\rangle =$$
$$=\sqrt{\frac{(R_{22}-R_{11}-1)(R_{22}-R_{23}-1)(R_{22}-R_{33})(R_{12}-R_{13})(R_{13}-R_{14}-1)(R_{13}-R_{24})(R_{13}-R_{34}+1)(R_{13}-R_{44}+2)}{(R_{22}-R_{12}-1)(R_{22}-R_{12}-2)(R_{13}-R_{23})(R_{13}-R_{23}+1)(R_{13}-R_{33}+1)(R_{13}-R_{33}+2)}}$$

$$\langle R_{22}-1,R_{23}-1|E_{42}|R_{22},R_{23}\rangle =$$
$$=\sqrt{\frac{(R_{22}-R_{11}-1)(R_{22}-R_{13}-2)(R_{22}-R_{33})(R_{23}-R_{12}-1)(R_{23}-R_{14}-2)(R_{23}-R_{24}-1)(R_{34}-R_{23})(R_{23}-R_{44}+1)}{(R_{22}-R_{12}-1)(R_{22}-R_{12}-2)(R_{23}-R_{13}-1)(R_{23}-R_{13}-2)(R_{23}-R_{33})(R_{23}-R_{33}+1)}}$$

$$\langle R_{22}-1,R_{33}-1|E_{42}|R_{22},R_{33}\rangle =$$
$$=-\sqrt{\frac{(R_{22}-R_{11}-1)(R_{22}-R_{13}-2)(R_{22}-R_{23}-1)(R_{33}-R_{12}-2)(R_{33}-R_{14}-3)(R_{33}-R_{24}-2)(R_{33}-R_{34}-1)(R_{44}-R_{33})}{(R_{22}-R_{12}-1)(R_{22}-R_{12}-2)(R_{33}-R_{13}-2)(R_{33}-R_{13}-3)(R_{33}-R_{23}-1)(R_{33}-R_{23}-2)}}$$

$$\langle R_{13}-1|E_{43}|R_{13}\rangle =\sqrt{\frac{(R_{13}-R_{12})(R_{13}-R_{22}+1)(R_{13}-R_{14}-1)(R_{24}-R_{13})(R_{13}-R_{34}+1)(R_{13}-R_{44}+2)}{(R_{13}-R_{23})(R_{13}-R_{23}+1)(R_{13}-R_{33}+1)(R_{13}-R_{33}+2)}}$$

$$\langle R_{23}-1|E_{43}|R_{23}\rangle =\sqrt{\frac{(R_{23}-R_{12}-1)(R_{23}-R_{22})(R_{23}-R_{14}-2)(R_{23}-R_{24}-1)(R_{34}-R_{23})(R_{23}-R_{44}+1)}{(R_{23}-R_{13}-1)(R_{23}-R_{13}-2)(R_{23}-R_{33})(R_{23}-R_{33}+1)}}$$

$$\langle R_{33}-1|E_{43}|R_{33}\rangle =\sqrt{\frac{(R_{33}-R_{12}-2)(R_{33}-R_{22}-1)(R_{33}-R_{14}-3)(R_{33}-R_{24}-2)(R_{33}-R_{34}-1)(R_{44}-R_{33})}{(R_{33}-R_{13}-2)(R_{33}-R_{13}-3)(R_{33}-R_{23}-1)(R_{33}-R_{23}-2)}}$$

DIFFERENT ORBITALS FOR DIFFERENT SPINS, LÖWDIN'S IDEA

RUBEN PAUNCZ

Department of Chemistry, Technion
Israel Institute of Technology
Haifa, Israel

1. Introduction

I met Per-Olov Löwdin for the first time in 1958 in Vålådalen. We have been in touch earlier, and he invited me to come to the first Summer Institute. After the Summer Institute he invited me to come to Uppsala for a year. He suggested that we work on the alternant molecular orbital method, this is a special version of the different orbitals for different spins method . This year had a great influence on my scientific work, Löwdin's wonderful ideas have been an inspiration and a great deal of my work was connected with three ideas suggested by him:

a) The different orbitals for different spins method.
b) The pairing theorem.
c) The spin projection operator.

In this review devoted to his memory I would like to present an overview of these three ideas.

2. The different orbitals for different spins method

2.1 THE SPIN-FREE HAMILTONIAN

In most molecular calculations one uses the Born-Oppenheimer (1927) approximation according to which one considers the solution of the electronic Hamiltonian when the nuclei are at fixed positions. For most problems one uses the following Hamiltonian (atomic units are used):

$$\mathcal{H} = \sum_{i=1}^{N}[-\frac{1}{2}\nabla_i^2 - \sum_{\alpha}\frac{Z_\alpha}{r_{\alpha i}}] + \sum_{i<j}^{N}\frac{1}{r_{ij}} = \sum_{i=1}^{N}h(i) + \sum_{i<j}^{N}g(i,j) \tag{1}$$

Here we have neglected small terms which correspond to spin-orbit, spin-spin etc interaction and we have also neglected relativistic terms. This

E.J. Brändas and E.S. Kryachko (eds.),
Fundamental World of Quatum Chemistry, Vol. I, 155–181.

Hamiltonian does not contain the spin variables, so the Hamiltonian operator commutes with the operators $\mathbf{S}^2$ and $\mathbf{S}_z$, so the eigenvalues of the latter operators are constants of motions.

2.2 THE ANTISYMMETRY PRINCIPLE

One of the basic principles of quantum mechanics states (Dirac 1926) that the wavefunctions corresponding to fermions should satisfy the antisymmetry principle:

$$P\Psi = (-1)^p \Psi \tag{2}$$

The wavefunction should change sign if we interchange the coordinates (spatial and spin) of two electrons.

If a trial function Φ does not fulfill the antisymmetry condition, we can project from it its antisymmetric component by the use of the *antisymmetrizer* $\mathcal{A}$:

$$\mathcal{A} = \frac{1}{\sqrt{N!}} \sum_P (-1)^p P \tag{3}$$

The antisymmetrizer is a projection operator, with the present normalization it is essentially idempotent.

2.3 THE MOLECULAR ORBITAL METHOD

The molecular orbital method in its most general formulation chooses the trial variational wavefunction in the form of an antisymmetrized product of N functions, each of which depends on the space and spin variables of only one electron.

$$\Psi = \mathcal{A}\psi_1(1)...\psi_N(N) \tag{4}$$

The individual functions (called *orbitals*) depend on the spatial and spin variables. In its simplest variation ψ_i is a product of a spatial function ϕ_i and a spin function, the latter is an eigenfunction of the S_z operator and corresponds to the eigenvalue $\frac{1}{2}$ or $-\frac{1}{2}$, respectively (denoted by α and β).

$$\psi_i = \phi_i \left\{ \begin{array}{c} \alpha \\ \beta \end{array} \right. \tag{5}$$

In the most general case the orbital is a linear combination of two spatial orbitals u_i and v_i multiplied by α and β, respectively (*general spin orbital*).

$$\psi_i = u_i\alpha + v_i\beta \tag{6}$$

where the orbital components u_i and v_i could be complex functions.

It is obvious from the form of the total Hamiltonian, which contains terms involving the relative coordinates of two electrons, that this formulation cannot yield the exact wave function. However, it serves as a very good starting point in the investigation of the general properties of atoms and molecules. It gives a natural link to the older concept of orbit in describing the structure of atom. In the case of molecules, it gives a simple description of the excitation and ionization processes, while the correlation with chemical properties is best achieved by means of a unitary transformation leading to *localized orbitals*, (Boys (1960), Edmiston and Ruedenberg (1963)) which correspond to inner shells, lone pairs and bond orbitals, respectively.

2.4 THE RESTRICTED HARTREE-FOCK METHOD

Let us consider the case when the number of electrons is even $N = 2n$. Let us choose the following trial function:

$$\Psi = \mathcal{A}a_1(1)\alpha(1)...a_n(n)\alpha(n)a_1(n+1)\beta(n+1)...a_n(2n)\beta(2n) \tag{7}$$

This single determinant corresponds to the picture that n spatial orbitals are doubly occupied, each occurs with α and β spins. By minimizing the expectation value of the Hamiltonian, one arrives at the *restricted Hartree-Fock equations.* The most often used form is when the molecular orbitals are themselves linear combinations of basic orbitals (*atomic orbitals*), and then the Hartree-Fock equations are given in matrix form (Roothaan (1951), Hall (1951)).

The physical picture of the Hartree-Fock (*self consistent field method, SCF*) is that each electron moves in the average field given by the rest of the electrons. The total wavefunction takes into account part of the electron correlation: in an antisymmetric wavefunction the probability of finding two electrons with the same spin in the same neighborhood is zero. Each electron is surrended by a *Fermi hole.* The method is called *resctricted Hartree-Fock, RHF* because of the restriction of each spatial orbital being doubly occupied by two electrons with opposite spins.

The single Slater determinant with doubly occupied orbitals is an eigenfunction of the operators $\mathbf{S}^2$ and $\mathbf{S}_z$, it belongs to the eigenvalue 0 (singlet state).

The RHF method gives good results for the total energy, bond lengths, and some other properties of the molecules. The small error in the total energy (0.5%) is still too large when we would like to calculate transition energies, dissociation energies, and so on. It is well known that the simple molecular orbital picture gives the wrong answer for the dissocation of the simplest molecule (H_2). Löwdin (1959) defined the difference between the SCF energy and the best energy obtained using the nonrelativistic Hamiltonian as the *correlation energy.*

2.5 GENERAL HARTREE-FOCK METHOD

Let us take a determinantal wavefunction where the spin orbitals are of the form given in Eq. 6 (general spin orbitals). The minimization of the total energy leads to the general Hartree-Fock method (GHF). This notation was introduced first by Sykja and Calais (1982) when studying pairing and general spin orbitals.

The general Hartree-Fock equations are quite complicated integro-differential equations of a nonlinear nature. Lieb and Simon (1977) proved the *mathematical existence* of the solutions. A great step forward in the studying of these equations was made by Fukutome, he found that it is possible to classify the solutions of the GHF equations according to their symmetry properties with respect spin-rotation and time-reversal. Fukutome (1981) summarized their achievement in a review paper. Calais (1985) gave a transparent presentation of Fukutome's method when discussing gap equations and extended systems. He showed that the spinor forms of the spin orbitals corresponding to the eight Fukutome symmetry classes can be displayed in the way given in Table 1.

Calais and Sykja (1982) discussed pairing and general spin obitals for the electron gas. Löwdin and Mayer (1992), Mayer and Löwdin (1993) discussed various aspects of the GHF method, their papers contain a comprehensive list of papers on the subject. A detailed discussion of the method is outside the scope if this review. In the following we shall consider a realization corresponding to the Fukutome class ASDW.

2.6 DIFFERENT ORBITALS FOR DIFFERENT SPINS

The idea of different orbitals for different spins starts with Slater (1951). He pointed out that in the Li atom the $1s\alpha$ funtion is influenced by a different exchange potential than the $1s\beta$ function, since exchange occurs only between electrons with parallel spins, so one can expect different orbitals

Table 1: Spinor forms of the spin orbitals corresponding to the eight Fukutome symmetry classes, Calais (1985)

1. Time reversal invariant closed shells (TICS)

$$|\Phi>= \begin{bmatrix} u_1, & 0, & u_2, & 0, & \dots & ,u_{\frac{N}{2}}, & 0 \\ 0, & u_1, & 0, & u_2, & \dots & ,0, & u_{\frac{N}{2}} \end{bmatrix}; \quad u_i \text{ real}$$

2. Charge Current Waves (CCW)
   Same as for TICS but with complex orbitals u_i

3. Axial Spin Current Waves (ASCW)

$$|\Phi>= \begin{bmatrix} u_1, & 0, & u_2, & 0, & \dots & ,u_{\frac{N}{2}}, & 0 \\ 0, & u_1^*, & 0, & u_2^*, & \dots & ,0, & u_{\frac{N}{2}}^* \end{bmatrix};$$

4. Axial Spin Density Waves (ASDW)

$$|\Phi>= \begin{bmatrix} u_1, & 0, & u_2, & 0, & \dots & ,u_{\frac{N}{2}}, & 0 \\ 0, & v_1, & 0, & v_2, & \dots & ,0, & v_{\frac{N}{2}} \end{bmatrix}; \quad u_i, v_i \text{ real}$$

5. Axial Spin Waves (ASW)
   Same as (4) but with complex orbitals u_i and v_i

6. Torsional Spin Current Waves (TSCW)

$$|\Phi>= \begin{bmatrix} u_1, & w_1, & u_2, & w_2, & \dots, & u_{\frac{N}{2}}, & w_{\frac{N}{2}}, \\ -w_1^*, & u_1^*, & -w_2^*, & u_2^* & \dots, & -w_{\frac{N}{2}}^*, & u_{\frac{N}{2}}^* \end{bmatrix}$$

7. Torsional Spin Density Waves (TSDW)

$$|\Phi>= \begin{bmatrix} u_1, & w_1, & u_2, & w_2, & \dots, & u_{\frac{N}{2}}, & w_{\frac{N}{2}}, \\ t_1, & v_1, & t_2, & v_2 & \dots & ,t_{\frac{N}{2}}, & v_{\frac{N}{2}} \end{bmatrix}; u_i, w_i, t_i, v_i \text{ real}$$

8. Torsional Spin Waves (TSW)
   Same as (7) but with complex orbitals

associated with α and β functions.

The idea has been generalized by Löwdin (1955). One of the simplest ways of improving the description of the electron correlation between electrons with antiparallel spins is to assign different spatial orbitals for the α and β spinfunctions:

$$\Psi = \mathcal{A}\, a_1(1)\alpha(1)...a_n(n)\alpha(n)\, b_1(n+1)\beta(n+1)...b_n(2n)\beta(2n) \qquad (8)$$

Such a wavefunction is still a single determinant. Without any loss of generality we can assume that both the a_i's and the b_j's form an orthonormal set of orbitals. The determinant remains essentially invariant if we form linear combinations of columns, this degree of freedom can be used in orthogonalizing the set of a's and the b's.

There is one more degree of freedom in the choice of a proper set of orbitals. Löwdin (1961), Amos and Hall (1961) found a *pairing theorem* according to which the set of orbitals a's and b's can be subjected to orthogonal transformations (without any essential change in the single determinant) in such a way that the new orbitals should satisfy the following relations:

$$< a_i | a_j >= \delta_{ij} \qquad < b_k | b_l >= \delta_{kl} \qquad < a_i | b_j >= \lambda_i \delta_{ij} \qquad (9)$$

The pairing theorem has many important consequences, we shall present a detailed discussion of it in the second part of the review.

There is one important difference between the new determinant and the one corresponding to the RHF scheme: the new determinant is not a proper eigenfunction of $\mathbf{S}^2$ (it is still an eigenfunction of $\mathbf{S}_z$ with the eigenvalue 0). Löwdin (1958) suggested to use the spin projection operator method in order to obtain a proper spin eigenfunction.

$$\Psi_S = \mathcal{O}_S \Psi \qquad (10)$$

We shall discuss this method in more detail in the third part of the review. We shall see that the use of the projection operator has important consequences on the way the wavefunction describes the electron correlation.

2.7 THE ALTERNANT MOLECULAR ORBITAL METHOD

Löwdin (1953) introduced a simple realization for the DODS method. It is based on some properties of alternant systems. Coulson and Rusbroke (1940) introduced the concept of *alternant hydrocarbons*, they are conjugated systems in which we can divide the atoms into two sets (denoted by stars and circles) in such a way that no two atoms belonging to the same set are neighbors to each other. Benzene, naphthalene are alternant hydrocarbons, azulene, fulvene are not alternants.

Coulson and Rushbroke (1940) proved that the molecular orbitals of alternant systems (obtained in the simple Hückel method) have interesting properties. To each orbital in the occupied set corresponds an orbital in the empty set with the following properties:

$$\epsilon_i = \alpha + x_i\beta \qquad \epsilon_i' = \alpha - x_i\beta \tag{11}$$

α and β are the Hückel parameters. The corresponding molecular orbitals have the same coefficients on the starred atoms and differ in sign for the unstarred atoms.

$$\begin{aligned} \psi_i &= \sum_\mu^* \chi_\mu c_{\mu i} + \sum_\nu^o \chi_\nu c_{\nu i} \\ \psi_i' &= \sum_\mu^* \chi_\mu c_{\mu i'} - \sum_\nu^o \chi_\nu c_{\nu i'} \end{aligned} \tag{12}$$

Löwdin (1953) suggested to use these properties and to form the a_i's and b_j's by the following construction:

$$\begin{aligned} a_i &= \cos\theta_i\ \psi_i + \sin\theta_i\ \psi_i' \\ b_i &= \cos\theta_i\ \psi_i - \sin\theta_i\ \psi_i' \end{aligned} \tag{13}$$

One can easily verify that the orbitals satisfy the following equations:

$$< a_i|a_j >= \delta_{ij} \qquad < b_k|b_l >= \delta_{kl} \qquad < a_i|b_j >= \cos 2\theta_i\ \delta_{ij} \tag{14}$$

Substituting Eq. (12) into Eq. (13) we have the following result:

$$\begin{aligned} a_i &= u_i \sum_\mu^* \chi_\mu c_{\mu i} + v_i \sum_\nu^o \chi_\nu c_{\nu i} \\ b_i &= v_i \sum_\mu^* \chi_\mu c_{\mu i} + u_i \sum_\nu^o \chi_\nu c_{\nu i} \end{aligned} \tag{15}$$

where

$$u_i = \cos\theta_i + \sin\theta_i \qquad v_i = \cos\theta_i - \sin\theta_i \tag{16}$$

Eq. (16) shows that for $0 \leq \theta_i \leq \frac{\pi}{4}$ the a_i's have larger amplitudes than the b_i's on the first set (starred atoms), whereas in the second set (unstarred atoms) the situation is reversed. For $\theta_i = \frac{\pi}{4}$ this localization is maximal, so that the *alternant orbitals* are reduced to the one or to the other set, respectively. One more attractive feature of the method is that for $\theta_i = 0, i = 1, ..., n$ both alternant orbitals yield the same occupied orbitals and we are back at the doubly occupied one-determinantal representation.

Löwdin (1953) presented the idea of the alternant molecular orbitals at the Nikko conference in Japan, and very soon Ito and Yoshizumi (1955) made an application of the method for the benzene molecule with promising result.

2.8 THE ENERGY EXPRESSION IN THE AMO METHOD

As I mentioned in the introduction, we started to work on the energy expression with Löwdin and de Heer in 1960 in Uppsala. We derived first the expression for the simplest case when all the θ's are the same (one-parameter version), (Pauncz, de Heer, Löwdin (1962)), de Heer (1962) derived the expression for different parameters, and de Heer and Pauncz (1963) obtained the expression for arbitrary multiplicities. The expression was made more transparent later by Pauncz, Kirtman and Palke (1987). We present here the latest version.

The energy is a function of the nonlinear parameters, $\theta_i, i = 1, ..n$. These parameters enter the energy expression primarily through the variables: $\lambda_i = \cos 2\theta_i$. There are three important functions of the λ's that appear repeatedly. They are defined as follows:

$$\Lambda_q = \sum_{k=0}^{n-q} (-1)^k C(S,k)\, S_k(x_1, ..., x_n) \quad (q = 0, 1, 2) \tag{17}$$

In this expression $x_k = \lambda_k^2$, and $S_k(x_1, ..., x_n)$ is the kth symmetric sum formed from $x_1, ..., x_n$.

$S_0 = 1 \quad S_1 = x_1 + \ldots + x_n \quad S_2 = x_1 x_2 + \ldots + x_{n-1} x_n \ldots S_n = x_1 x_2 \ldots x_n$

The coeffcients *C(S,k)* arise from the use of the spin projection operator, they are called *Sanibel coefficients.* For the singlet state they are given as follows:

$$C(0,k) = \frac{(-1)^k}{n+1} \begin{pmatrix} n \\ k \end{pmatrix}^{-1} \tag{18}$$

Λ_0 is the normalization integral.

We shall also need expressions which involve the first and the second derivatives of $\Lambda_k, k = 0, 1, 2$ with respect to the x_i's. They are defined as follows:

$$L_k^i = \frac{\partial}{\partial x_i}\Lambda_k \qquad L_k^{ij} = \frac{\partial^2}{\partial x_i \partial x_j}\Lambda_k \tag{19}$$

Let us divide the energy expression into two parts: the first corresponds to the one-electron operators in the Hamiltonian *h(i)* and the second to the electron repulsion *g(i,j)*. The one-electron part is :

$$E_1 = \sum_{i=1}^{n} w_i - \frac{1}{\Lambda_0}\sum_{i=1}^{n} \lambda_i \Delta w_i \left(L_0^i + L_1^i\right) \tag{20}$$

Here

$$w_i = h_i + h'_i \qquad \Delta w_i = h'_i - h_i \tag{21}$$

.

h_i, and h'_i are the diagonal matrix elements of the one-electron Hamiltonian over the occupied orbitals and their virtual pairs.

$$h_i = \int \psi_i \, h \, \psi_i dv \qquad h'_i = \int \psi'_i \, h \, \psi'_i dv \tag{22}$$

The two-electron part reads as follows:

$$E_2 = \frac{1}{\Lambda_0}\left(E_{2a} + E_{2b} + E_{2c}\right) \tag{23}$$

where

$$E_{2a} = \sum_{i=1}^{n}[A_i\gamma_{ii} + A_{i'}\gamma_{i'i'} - B_i\gamma_{ii'} - C_i\delta_{ii'}] \tag{24}$$

The basic two-electron integrals γ_{ij}, δ_{ij} and ζ_{ij} are respectively the coulomb, exchange and a new kind of integral which is characteristic of the AMO method:

$$\gamma_{ij} = (ii|jj), \qquad \delta_{ij} = (ij|ji), \qquad \zeta_{ij} = (ij|i'j') \tag{25}$$

The two-electron integrals are defined as follows:

$$(ij|kl) = \int\int \psi_i(1)^*\psi_j(1)g(1,2)\psi_k(2)^*\psi_l(2)dv_1dv_2 \tag{26}$$

The $\lambda(\theta)$ dependent factors in Eq. (24) are given as follows:

$$A_i = \frac{1}{4}(1+\lambda_i)^2(L_1^i + L_0^i) \qquad A_{i'} = \frac{1}{4}(1-\lambda_i)^2(L_1^i + L_0^i)$$
$$B_i = \frac{1}{2}(1-\lambda_i)^2(L_1^i - L_0^i) \quad C_i = (1-\lambda_i)^2 L_1^i \tag{27}$$

E_{2b} and E_{2c} are defined as follows:

$$E_{2b} = \sum_{i<j}^{n}[\Gamma(i,j) - \Delta(i,j)], \qquad E_{2c} = -\sum_{i<j}^{n} Z(i,j) \tag{28}$$

where

$$\Gamma(i,j) = \Gamma_{ij}\gamma_{ij} + \Gamma_{ij'}\gamma_{ij'} + \Gamma_{i'j}\gamma_{i'j} + \Gamma_{i'j'}\gamma_{i'j'}$$
$$\Delta(ij) = \Delta_{ij}\delta_{ij} + \Delta_{ij'}\delta_{ij'} + \Delta_{i'j}\delta_{i'j} + \Delta_{i'j'}\delta_{i'j'} \tag{29}$$
$$Z(ij) = Z_{ij}\zeta_{ij} + Z_{i'j'}\zeta_{i'j'}$$

The definition of the functions involving the λ's is the following:

$$\Gamma_{ij} = (1+\lambda_i)\,(1+\lambda_j)(L_2^{ij} + [\lambda_i + \lambda_j]L_1^{ij} + \lambda_i\lambda_j L_0^{ij})$$
$$\Delta_{ij} = \frac{1}{2}\,(1+\lambda_i)\,(1+\lambda_j)(L_2^{ij} + [2(\lambda_i+\lambda_j) - 1 - \lambda_i\lambda_j]L_1^{ij} + \lambda_i\lambda_j L_0^{ij})$$
$$Z_{ij} = F_{ij} + G_{ij} \qquad Z_{ij'} = F_{ij} - G_{ij} \tag{30}$$
$$F_{ij} = \sin 2\theta_i \sin 2\theta_j\,(L_2^{ij} + L_1^{ij}) \quad G_{ij} = -\frac{1}{4}\sin 4\theta_i \sin 4\theta_j\,(L_1^{ij} + L_0^{ij})$$

From the expressions of Γ_{ij} and Δ_{ij} one obtains the corresponding factors involving *i'* and/or *j'* by reversing the sign of the λ_i and/or λ_j.

2.9 THE SPIN-PROJECTED EXTENDED HARTREE-FOCK METHOD

Following Löwdin (1953) let us construct the following variational wavefunction:

$$\Psi_S = \mathcal{O}_S \mathcal{A} \prod_{i=1}^{n} a_i(i)\alpha(i) b_i(n+i)\beta(n+i) \tag{31}$$

where the a_i's and b_j's satisfy the pairing relations (Eq. 9). Löwdin suggested to call the method using this variational wavefunction *spin-projected extended Hartree-Fock method.* The best orbitals a_i's and b_i's and the pairing parameters λ_i's are to be determined from the minimization of the total energy calculated with this trial function.

The most important contributions to this method have been made by Mayer (1974) and coworkers. He derived the total energy expression, the latter has a similar structure to the one calculated from the AMO method, with the basic difference that in the integrals the a_i's and b_j's appear instead of the ψ_i's and ψ_i''s but the λ dependent functions are essentially the same. Mayer presented different methods for the determination of the best orbitals using a generalized Brillouin theorem.

Mayer (1980) has summarized their achievements in an excellent and comprehensive review in which he presented the basic equations, the different methods for solving the variational equations and comparison with similar methods (Goddard's GF method, Goddard (1968 a,b)). He covers the literature up to 1980.

In the present review we shall consider two developments which took place after 1980 : a) Karadakov's method b) The paired orbital method.

2.10 KARADAKOV'S METHOD

Karadakov (1986) considers a unitary transformation of the reference DODS Slater determinant. The energy expectation value corresponding to the transformed DODS Slater determinant is expanded in a Taylor series about the reference Slater determinant, including second order terms. Karadakov proposes several direct iterative approaches to the optimization of the energy, for example, implementations of steepest decent and conjugate gradient techniques, Newton-Raphson, Fletcher and limited configuration interaction approaches. He also considers the stability conditions for spin-projected extended Hartree-Fock methods.

2.11 PAIRED ORBITAL METHOD

Pauncz and coworkers (1987) considered another approach to the spin-projected extended Hartree-Fock method. They retain the AMO energy expression but they intend to find the best paired orbitals used in the construction of the alternant molecular orbitals. For alternant systems the Coulson- Rushbroke theorem provided a natural starting point for choosing the orbitals ψ_i and ψ_i', but for a general system one has to find the "best pairs". If we have an m-dimensional basis and n orbitals are double occupied in the restricted Hartree- Fock scheme, then the space spanned by the virtual orbitals has the dimension m-n. In this virtual space we have to find an n dimensional subspace which contains the best pairs for each ψ_i. In the first paper these orbitals are determined using the somewhat arbitrary criterion that the sum of repulsion integrals between ψ_i and ψ_i' should be maximum.

In a later paper Pauncz (1992) considered orthogonal transformation in the m-n dimensional virtual space in order to find the best paired orbitals, which lead to the minimum of the energy. The paired orbital method corresponds to Mayer's (1980) observation that any DODS wavefunction can be considered as a special one of the AMO type, with a particular choice of the starting MO's used in the AMO orbital mixing procedure.

2.12 APPLICATION OF THE DODS METHOD

In the early applications the AMO method was used either in the one-parameter version, or in the many parameter version. Applications to alternant hydrocarbons yielded excellent result, for example for the benzene the two-parameter AMO function yielded about 90 % of the correlation energy as compared with a full CI calculation in the same basis. Calais (1964a,b, 1965) extended the use of AMO to band calculations with encouraging results. The summary of these applications are found in Pauncz's monograph (1967) and in Mayer's review (1980).

The paired orbital method (Pauncz (1992)) as applied to the water molecule yielded only 27 % for the equilibrium distance, the result was better when the OH distance was stretched by the factors 1.5 and 2. In the last case it yielded 62 % of the correlation energy. The comparison was with the full CI calculation of Harrison and Handy (1983), the latter used 256,473 configurations,

During the last decade there was a decline in the applications probably because of the disappointing results in the water calculations.

3. The pairing theorem

The pairing theorem was presented as a conjecture by Löwdin (1961), it was proved by Amos and Hall (1961), they named the orbitals satisfying the pairing theorem *corresponding orbitals.* King et al (1967) pointed out several possibilites for the use of corresponding orbitals in quantum mechanics. Löwdin used it in the DODS method in the AMO version. Mayer (1986) gave a new proof of the theorem, this is more simple than the original proof.

Karadakov (1985) extended the pairing theorem, Mayer (1986) presented a constructive proof of Karadakov's theorem. Löwdin (1991) summarized the pairing theorem and its extension using Karadakov's and Mayer's derivations. Recently Hamza and Mayer (2001 a,b) used the pairing theorem in the derivation of an explicit analytic expression for the Heitler-London interaction energy between two "unperturbed" closed-shell molecules. In the following we shall follow closely Löwdin's (1991) presentation.

3.1 THE MIRROR THEOREM

Löwdin (1991) pointed out that the pairing theorem is a special case of a much more general theorem usually called *the mirror theorem.* Let **R** and **S** be two linear operators, then the mirror theorem says that the products **RS** and **SR** has the same nonvanishing eigenvalues.

Let us assume that the operator **RS** has the eigenfunction u_k which belongs to the nonvanishing eigenvalue μ_k, while the operator **SR** has the eigenfunction v_l belonging to the nonvanishing eigenvalue ν_l:

$$\mathbf{RS}\, u_k = u_k\, \mu_k, \qquad \mathbf{SR}\, v_l = v_l\, \nu_l, \qquad \mu_k \neq 0, \quad \nu_l \neq 0 \tag{32}$$

For the proof let us consider the function $v'_k = \mathbf{S}u_k$. This function cannot be zero, since $u_k" = \mathbf{R}v'_k = \mathbf{RS}u_k = u_k\mu_k \neq 0$. Let us apply the operator product **SR** to this function: $\mathbf{SR}\, v'_k = \mathbf{SRS}\, u_k = \mathbf{S}\,(u_k\, \mu_k) = v'_k\, \mu_k$. This shows that μ_k is also an eigenvalue to **SR**.

In the next step consider the function $u'_l = \mathbf{R}v_l$. This function cannot be zero since $v_l" = \mathbf{S}u'_l = \mathbf{SR}\, v_l \neq 0$. Now let us form: $\mathbf{RS}\, u'_l = \mathbf{RSR}\, v_l = \mathbf{R}(v_l\nu_l) = u'_l\nu_l$. This shows that ν_l is also an eigenvalue to the operator **RS**. Hence the operators **RS** and **SR** have the same nonvanishing eigenvalues.

3.2 THE PAIRING THEOREM

Let us consider two linear spaces $\mathcal{A}$ and $\mathcal{B}$ spanned by the elements $\{a_1, a_2, ..., a_m\}$ and $\{b_1, b_2, ..., b_n\}$. We can denote these sets by the row vectors $\mathbf{a} = \{a_k\}$, and $\mathbf{b} = \{b_k\}$, respectively. As an example we can take the set of molecular orbitals $\psi_1, .., \psi_m$ and $\phi_1, ..., \phi_n$. Let us assume that $m \geq n$. We shall assume that the two subspaces are partly overlapping and that $\mathcal{B}$ is not fully contained in $\mathcal{A}$. We assume that the basis vectors in both spaces are orthogonalized and normalized.

We can form two projection operators:

$$\mathbf{P_a} = |\mathbf{a}><\mathbf{a}| = \sum_{k=1}^{m} |a_k><a_k| \quad \mathbf{P_b} = |\mathbf{b}><\mathbf{b}| = \sum_{k=1}^{n} |b_k><b_k| \quad (33)$$

Here and in the following we are using Dirac's notation. These projectors are self-adjoint, idempotent and invariant under unitary transformations of **a** and **b**, respectively. With the use of these operators we can divide the spaces $\mathcal{A}$ and $\mathcal{B}$ into two orthogonal subspaces each:

$$\mathcal{A}' = \mathbf{P_b}\mathcal{A}, \qquad \mathcal{A}" = (1 - \mathbf{P_b})\mathcal{A}, \qquad (34)$$

$$\mathcal{B}' = \mathbf{P_a}\mathcal{B}, \qquad \mathcal{B}" = (1 - \mathbf{P_a})\mathcal{B} \qquad (35)$$

One can easily show that $\mathbf{P_b}\mathcal{A}" = 0$ (using $\mathbf{P_b^2} = \mathbf{P_b}$), and that the entire subspace $\mathcal{A}"$ is orthogonal to the space $\mathcal{B}$. In the same way, the entire subspace $\mathcal{B}"$ is orthogonal to the space $\mathcal{A}$.

Let us introduce the two rectangular matrices:

$$\Delta =<\mathbf{a}|\mathbf{b}> \quad \Delta^{\dagger} =<\mathbf{b}|\mathbf{a}> \qquad (36)$$

These matrices are of the order $m \times n$, and $n \times m$, respectively. In the following we shall deal with the subspaces:

$$\mathcal{A}' = \mathbf{P_b}\mathcal{A} = \mathbf{b} <\mathbf{b}|\mathbf{a}> \quad \text{and} \quad \mathcal{B}' = \mathbf{P_a}\mathcal{B} = \mathbf{a} <\mathbf{a}|\mathbf{b}>.$$

The metric matrices for these two subspaces are given as follows:

$$\begin{aligned}\Omega'_a &=<\mathcal{A}'|\mathcal{A}'>=<\mathbf{P_b}\mathcal{A}|\mathbf{P_b}\mathcal{A}>=<\mathcal{A}|\mathbf{P_b}\mathcal{A}> \\ &=<\mathbf{a}|\mathbf{b}><\mathbf{b}|\mathbf{a}> = \Delta\Delta^{\dagger}\end{aligned} \qquad (37)$$

$$\begin{aligned}\Omega'_b &=<\mathcal{B}'|\mathcal{B}'>=<\mathbf{P_a}\mathcal{B}|\mathbf{P_a}\mathcal{B}>=<\mathcal{B}|\mathbf{P_a}\mathcal{B}> \\ &=<\mathbf{b}|\mathbf{a}><\mathbf{a}|\mathbf{b}> = \Delta^{\dagger}\Delta\end{aligned} \qquad (38)$$

The dimensions of the two subspaces are given by the numbers of nonvanishing eigenvalues of these metric matrices. The products $\Delta\Delta^{\dagger}$ and $\Delta^{\dagger}\Delta$ are quadratic matrices of order m and n, respectively. They have the same nonvanishing eigenvalues (mirror theorem). If $m > n$ the product $\Delta\Delta^{\dagger}$ has at least $(m - n)$ eigenvalues equal to zero. If the number of nonvanishing eigenvalues is $p \leq n$ then this number gives the order of the two subspaces $\mathcal{A}'$ and $\mathcal{B}'$.

Since the two products $\Delta\Delta^\dagger$ and $\Delta^\dagger\Delta$ are Hermitian matrices they can be diagonalized by the unitary matrices **U** and **V**, respectively.

$$\mathbf{U}^\dagger\boldsymbol{\Delta}\boldsymbol{\Delta}^\dagger\mathbf{U} = \mu_{\mathbf{a}} \quad \mathbf{V}^\dagger\boldsymbol{\Delta}^\dagger\boldsymbol{\Delta}\mathbf{V} = \mu_{\mathbf{b}} \tag{39}$$

Since the two matrices $\Omega_a = \Delta\Delta^\dagger$ and $\Omega_b = \Delta^\dagger\Delta$ are semipositive definite, each of the eigenvalues is greater or equal to zero. Because of the mirror theorem they contain the same nonvanishing eigenvalues $\mu_1, ..., \mu_p$. Let us collect the nonvanishing eigenvalues into a diagonal matrix μ of order p.

Löwdin (1991) shows that one can partition the matrix **U** into two parts: $\mathbf{U} = (\mathbf{U}', \mathbf{U}'')$ where the first part of order $m \times p$ is associated with the nonvanishing eigenvalues μ.

$$\Delta\Delta^\dagger\mathbf{U}' = \mathbf{U}'\mu \tag{40}$$

He introduces a new matrix **V'** of order $n \times p$ through the relation:

$$\mathbf{V}' = \Delta^\dagger\mathbf{U}'\mu^{-1/2} \tag{41}$$

Using these matrices one can introduce the following sets:

$$\mathbf{a}' = \mathbf{a}\mathbf{U}', \quad \mathbf{b}' = \mathbf{b}\mathbf{V}' \tag{42}$$

Löwdin shows that the orthonormalized sets **a'** and **b'** have the overlap matrix:

$$\Delta' =< \mathbf{a}'|\mathbf{b}' >= (\mathbf{U}')^\dagger < \mathbf{a}|\mathbf{b} > \mathbf{V}' = (\mathbf{U}')^\dagger\boldsymbol{\Delta}\mathbf{V}' = \mu^{1/2} \tag{43}$$

This is the *pairing theorem*. Putting $\mu_k^{1/2} = \lambda_k$ one obtains the special form:

$$< a'_k|b'_l >= \lambda_k\delta_{kl}, \quad 0 < \lambda_k \leq 1 \tag{44}$$

This is the form used in AMO theory. One can distinguish between two cases: a) If $\lambda_k = 1$ then the orbitals are common for the two spaces $\mathcal{A}$ and $\mathcal{B}$, they would correspond to the set of *doubly occupied orbitals.* The remaining pairs of orbitals are referred to as *corresponding orbitals.* They have fundamental importance in the treatment of the correlation problem. Löwdin's treatment is very similar to the transparent proof of the pairing theorem given by Mayer (1997).

3.3 EXTENDED PAIRING THEOREM (KARADAKOV)

Karadakov (1985) pointed out that one can extend the pairing theorem by including the space spanned by the *virtual orbitals.* Mayer (1986) gave a constructive proof of the theorem. Löwdin's (1991) presentation follows closely the ideas developed by Karadakov and Mayer.

Löwdin divides the *occupied* orbitals into two groups: 1) The *doubly occupied* orbitals $a'_k = b'_k$ with $\lambda_k = 1$, they span the intersection $\mathcal{A}'_{int} = \mathcal{B}'_{int}$ of order q.

2) The pairs of *corresponding* orbitals a'_j and b'_j for which $0 < \lambda < 1$, they span the subspaces $\mathcal{A}'_{corr}$ and $\mathcal{B}'_{corr}$, both subspaces are of the order $(p-q)$.

$$\mathbf{A}' = (\mathbf{A}'_{\mathbf{int}}, \mathbf{A}'_{\mathbf{corr}}) \quad \mathbf{B}' = (\mathbf{B}'_{\mathbf{int}}, \mathbf{B}'_{\mathbf{corr}}) \tag{45}$$

Following Löwdin (1991) let us define the following subspace:

$$\mathcal{C} = (\mathbf{1} - \mathbf{P}_b)\mathcal{B}' = (\mathbf{1} - \mathbf{P}_b)\mathbf{P}_a\mathcal{B} \tag{46}$$

In terms of the column vectors we can express this relation as follows:

$$\mathbf{c} = (\mathbf{a} < \mathbf{a}|\mathbf{b} > -\mathbf{b} < \mathbf{b}|\mathbf{a} >< \mathbf{a}|\mathbf{b} >) = (\mathbf{a} - \mathbf{b}\Delta^{\dagger})\Delta \tag{47}$$

The metric matrix of this subspace can be obtained in the following way:

$$< \mathbf{c}|\mathbf{c} > = < (\mathbf{a} - \mathbf{b}\Delta^{\dagger})\Delta|(\mathbf{a} - \mathbf{b}\Delta^{\dagger})\Delta >$$

$$= \Delta^{\dagger}(\mathbf{1} - \Delta\Delta^{\dagger})\Delta = \Omega_b(\mathbf{1} - \Omega_b) \tag{48}$$

This matrix may be brought to the diagonal form $\lambda(\mathbf{1} - \lambda))$, it has $(p - q)$ nonvanishing eigenvalues. We observe that the subspace $\mathcal{B}_{int}$ has no component in $\mathcal{C}$. The nonvanishing component comes from the subspace $\mathcal{B}'_{corr}$. Let us define:

$$\mathcal{C}_{virt} = (\mathbf{1} - \mathbf{P_b})\mathcal{B}'_{corr} \tag{49}$$

Löwdin defines the subspace $\mathcal{D}$ as:

$$\mathcal{D} = (\mathbf{1} - \mathbf{P}_a)\mathcal{A}' = (\mathbf{1} - \mathbf{P}_a)\mathbf{P}_b\mathcal{A} \tag{50}$$

In terms of orbitals this relation can be written as:

$$\mathbf{d} = (\mathbf{b} < \mathbf{b}|\mathbf{a} > -\mathbf{a} < \mathbf{a}|\mathbf{b} >< \mathbf{b}|\mathbf{a} >) = (\mathbf{b} - \mathbf{a}\Delta)\Delta^{\dagger} \tag{51}$$

The subspace has the metric matrix:

$$< \mathbf{d}|\mathbf{d} >=< (\mathbf{b} - \mathbf{a}\Delta)\Delta^{\dagger}|(\mathbf{b} - \mathbf{a}\Delta)\Delta^{\dagger} >$$

$$= \Delta(\mathbf{1} - \Delta^{\dagger}\Delta)\Delta^{\dagger} = \Omega_a(\mathbf{1} - \Omega_a) \tag{52}$$

This matrix has also $(p-q)$ nonvanishing eigenvalues. The subspace $\mathcal{A}'_{int}$ is orthogonal to this subspace and we can define the space $\mathcal{D}_{virt}$ by the relation:

$$\mathcal{D}_{virt} = (\mathbf{1} - \mathbf{P_a})\mathcal{A}'_{corr} \tag{53}$$

The orbitals belonging to this subspace are orthogonal to the entire space $\mathcal{A}$ but have an overlap with the space $\mathcal{A}'_{corr}$. Löwdin shows the following relations:

$$< \mathbf{a}'|\mathbf{c}_{virt} >= (1 - \lambda^2)^{1/2} \qquad < \mathbf{b}'|\mathbf{d}_{virt} >= (1 - \lambda^2)^{1/2} \tag{54}$$

The final result is that the overlap matrix between the two types of virtual orbitals is diagonal:

$$< \mathbf{c}_{virt}|\mathbf{d}_{virt} >= -\lambda \tag{55}$$

The right-hand side is a diagonal matrix of order $(p-q)$. This is the extended pairing theorem of Karadakov.

4. The spin projection operator

Pratt (1953) was the first one to construct an operator which, when applied to a product of spin functions of n α's and n β's yields an eigenfunction of S^2 with the eigenvalue 0. Berencz and Pauncz (1958) generalized Pratt's method. In the Symposium in Paris 1958 Löwdin presented a very elegant construction of projected spinfunctions. Harris (1967a, b) derived the matrix elements of the Hamiltonian between spin-projected functions. The properties of the spin projection coefficients have been discussed at the first Sanibel conference, they were later called *Sanibel coefficients* (Manne (1966)). The construction of spin-eigenfunctions was discussed in the monographs of the author (Pauncz (1979, 2000)). In this overview I would like to present Löwdin's method.

4.1 LÖWDIN'S SPIN-PROJECTION OPERATOR

Let us start with a primitive spin function $\theta(\mu, \nu)$ which is a product of $\mu = n_\alpha$ α's and $\nu = n_\beta$ β's. Such a function is an eigenfunction of the $\mathbf{S}_z$ operator and it belongs to the eigenvalue $s = \frac{1}{2}(n_\alpha - n_\beta)$, but it is not an eigenfunction of $\mathbf{S}^2$. It is a mixture of different spin-eigenfunctions.

Löwdin's idea was to use an operator which annihilates all the contributions except the one which comes from a given eigenvalue of $\mathbf{S}^2$. This operator can be written in the following form:

$$\mathcal{O}_K = \prod_S \frac{[\mathbf{S}^2 - S(S+1)\mathbf{1}]}{[K(K+1) - S(S+1)]} \quad S \neq K \tag{56}$$

The operator is a projection operator, it is idempotent.

Löwdin (1958) defined the square bracket of spin functions $[\alpha^p \beta^q]$ as the sum of all primitive spinfunctions which have p α's and q β's. The number of terms in the square bracket is

$$\binom{p+q}{p} = \binom{p+q}{q}$$

Löwdin showed that the square brackets fulfill the following relations:

$$\begin{aligned} \mathbf{S}_+^k[\alpha^p \beta^q] &= \frac{(p+k)!}{p!}[\alpha^{p+k} \beta^{q-k}] \\ \mathbf{S}_-^k[\alpha^p \beta^q] &= \frac{(q+k)!}{q!}[\alpha^{p-k} \beta^{q+k}] \end{aligned} \tag{57}$$

Here $\mathbf{S}_+$ and $\mathbf{S}_-$ are the *step-up* and *step-down* spin operators.

$$\begin{aligned} \mathbf{S}_+\alpha = 0 \qquad \mathbf{S}_+\beta = \alpha \\ \mathbf{S}_-\alpha = \beta \qquad \mathbf{S}_-\beta = 0 \end{aligned} \tag{58}$$

Using these expressions Löwdin transformed the projection operator into the following form:

$$\mathcal{O}_{lm} = (2l+1)\frac{(l+m)!}{(l-m)!}\sum_{p=0}^{\frac{N}{2}-l}(-1)^p \frac{\mathbf{S}_-^{l-m+p}\,\mathbf{S}_+^{l-m+p}}{p!(p+2l+1)!} \tag{59}$$

For the *principal case* $m=l$ this expression reduces to the following one:

$$\mathcal{O}_{ll} = (2l+1)! \sum_{p=0}^{\frac{N}{2}-l} (-1)^p \frac{\mathbf{S}_-^p \mathbf{S}_+^p}{p!(p+2l+1)!} \tag{60}$$

The final result for the projected spin function read as follows:

$$\mathcal{O}_{ll}\theta_1(\mu,\nu) = \mathcal{O}_{ll}[\alpha^\mu][\beta^\nu] = \frac{2l+1}{\mu+1} \sum_{p=0}^{\nu} (-1)^p \begin{pmatrix} \mu \\ p \end{pmatrix}^{-1} [\alpha^{\mu-p}\beta^p]\,[\alpha^p\beta^{\nu-p}] \tag{61}$$

In this expression the first square bracket refers to electrons $1...\mu$, and the second one to electrons $\mu+1\ ...\mu+\nu$. The primitive spinfunction θ_1 has the following form:

$$\theta_1 = \alpha(1)...\alpha(\mu)\,\beta(\mu+1)...\beta(\mu+\nu) \tag{62}$$

4.2 LÖWDIN'S PATH DIAGRAM

Let us have another primitive spin function θ_k which belongs to the same S_z eigenvalue, i.e. it has μ α's and ν β's. This function can be obtained from θ_1 by an appropriate permutation of the variables:

$$P_k\theta_1 = \theta_k \tag{63}$$

The projection operator commutes with any permutation (it is symmetric in the coordinates of the electrons), so the projection of θ_k can be obtained as follows:

$$\mathcal{O}_S\theta_k = \mathcal{O}_S\mathbf{P}_k\theta_1 = \mathbf{P}_k\mathcal{O}_S\theta_1 = \mathbf{P}_k\Theta_1 \tag{64}$$

The number of primitive spin functions is much larger than the independent spin eigenfunctions, so not all the projections are linearly independent.
In order to find the linearly independent projected functions, Löwdin introduced the *path diagram.* With every primitive spin function we can associate a graph, in which to an α corresponds a line pointing in 45 degrees, while to a β a line pointing in -45 degrees.

Löwdin (1958) proposed to find the independent projections by taking those primitive spinfunctions whose path diagram is completely above the axis. In Fig. 1 this means θ_1 and θ_2. The number is certainly correct because formally these graphs are identical to the graphs in the *branching diagram*

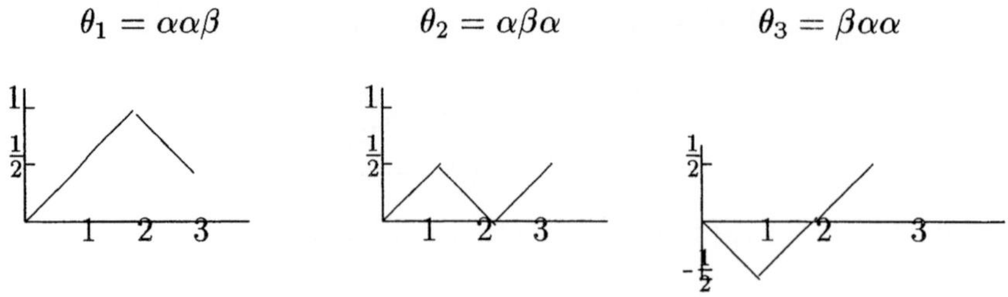

Figure 1: Path diagrams for N=3, $S_z = 1/2$

which shows in a graphical way the linearly independent spin eigenfunctions. Löwdin's conjecture was proved by Gershgorn (1968), a simpler proof was given later by the author (Pauncz 1976).

4.3 ALTERNATIVE DERIVATION OF THE PROJECTED FUNCTIONS

Löwdin (1956) presented an alternative method for the projected wavefunctions. Following Löwdin let us define first the following quantities:

$$T_k = [\alpha^{\mu-k}\beta^k][\alpha^k\beta^{\nu-k}] \quad \mu+\nu = N = 2n \quad \mu-\nu = 2M \tag{65}$$

The first square bracket in T_k refers to the first μ electrons, while the second square bracket corresponds to the rest of the electrons.

One can show that the effect of operator $\mathbf{S}^2$ on T_k yields the following result:

$$\begin{aligned} \mathbf{S}^2 T_k &= [2k(n-k)+n+M^2]T_k \\ &+ [(n-k+1)^2 - M^2]\, T_{k-1} + (k+1)^2 T_{k+1} \end{aligned} \tag{66}$$

In applying the projection operator to $T_0 \equiv \theta_1$ we apply successively the operator $\mathbf{S}^2$. The final result can be cast in the following simple form:

$$\mathcal{O}_S\theta_1 = \mathcal{O}_S T_0 = \sum_{k=0}^{\nu} C(S,k)T_k \tag{67}$$

Löwdin (1956) obtained simple recursion formulas for the coefficients $C(S,k)$. These formulas are obtained if we write down the condition that the projected function is a spin eigenfunction with the quantum number S and use Eq. (66):

$$\begin{aligned}\mathbf{S}^2 \sum_{k=0}^{\nu} C(S,k)T_k &= \sum_{k=0}^{\nu} C(S,k)[(2k(n-k)+n+M^2)T_k \\ &\quad +(k+1)^2T_{k+1} + ((n-k+1)^2 - M^2)T_{k-1}] \\ &= S(S+1)\sum_{k=0}^{\nu} C(S,k)T_k \end{aligned} \tag{68}$$

We can write this equation in the form:

$$\sum_{k=0}^{\nu} d_k T_k = 0 \tag{69}$$

The different T_k's are orthogonal to each other, so they are linearly independent. Then from Eq. (69) follows that all the d_k's are zeroes, and this yields the following recursion formula for the coefficients:

$$\begin{aligned} d_k = [2k(n-k)+n+M^2 - S(S+1)]C(S,k) \\ +((n-k)^2 - M^2)C(S,k+1) + k^2C(S,k-1) = 0 \end{aligned} \tag{70}$$

From this equation the coefficients can be calculated starting from $C(S,0)$. The spin-projection coefficients received much attention, they have been extensively discussed in the early Sanibel conferences and they are often called "Sanibel coefficients". The calculation of these coefficients is quite simple using the Löwdin recursion formula. Manne (1966) has prepared a table containing the Sanibel coefficients up to 10 electrons. There are different formulae for these coefficients, an extensive review of the different formula can be found in Pauncz's book (1979). The most systematic treatment is given by Smith and Harris (1969) and they also show relations to alternative formulae. For a quite recent treatment of the Sanibel coefficients see Löwdin (1983), Pauncz (1989).

4.4 THE PHYSICAL MEANING OF THE PROJECTED FUNCTION

Let us start with the square bracket $[\alpha^k\beta^l]$, which according to its definition is the sum of all possible spinfunctions with k α's and l β's. The

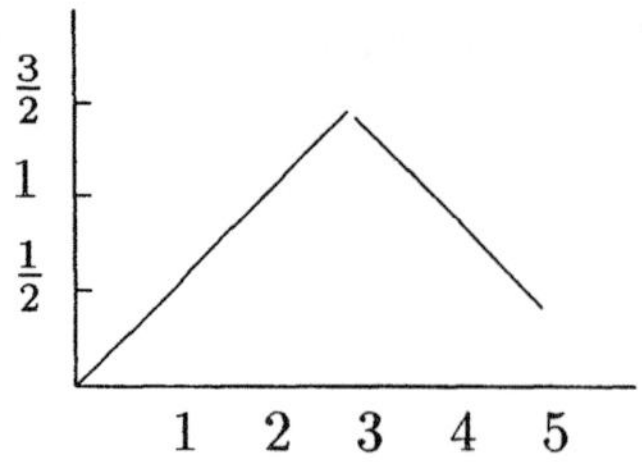

Figure 2: Diagram of the last spin eigenfunction for $N = 5,\ S = 1/2$.

square bracket is an eigenfunction of $\mathbf{S}_z$ operator and it belongs to the eigenvalue: $\frac{1}{2}(k-l)$. One can also show that the square bracket is also an eigenfunction of the $\mathbf{S}^2$ operator:

$$\mathbf{S}^2[\alpha^k\beta^l] = s(s+1)[\alpha^k\beta^l] \quad s = \frac{1}{2}(k+l) \tag{71}$$

This is the state of highest multiplicity for $k+l$ electrons, this state is non-degenerate, i.e., this is the only state which belongs to the quantum number $s = \frac{1}{2}(k+l)$.

Now we can see the physical meaning of the projected spin function. It is the spin function which is obtained by coupling together two resultant spins $\frac{1}{2}\mu$ and $\frac{1}{2}\nu$ to a resultant spin $\frac{1}{2}(\mu-\nu)$. This meaning was already pointed out by Pratt (1953). In the *branching diagram* this is the last spin eigenfunction, it can be represented by the following diagram (Fig. 2):
In the starting picture the orbitals $a_1, \ldots, a_\mu$ are associated with spins α. According to this picture these spins are coupled together with the resultant maximum spin $\frac{\mu}{2}$. This subsystem of a highest multiplicity gives a very strong correlation between the corresponding electrons. In a similar way the orbitals $b_1, \ldots, b_\nu$ are associated with β spins, they are coupled together to a resultant maximum spin $\frac{\nu}{2}$. Again we have a very strong correlation between the electrons in this subsystem. Finally the two subsystems are coupled together to form the resultant spin $\frac{1}{2}(\mu-\nu)$. The correlation between the two subsytems is introduced through the pairing of the corresponding orbitals.

From this picture it is quite clear why was the method successful for conjugated systems (and solids), but was not so successful when applied to other systems.

Acknowledgement

The author is grateful to Profs. E. Brändas and E. Kryachko for their kind invitation to contribute to the volume dedicated to Per-Olov Löwdin's memory. The author expresses his sincere gratitude to Prof. I. Mayer for many helpful discussions on the subjects covered in this review. These discussions took place during the author's visit at Prof. Mayer's research laboratory and during Prof. Mayer's visit to the author's laboratory. As it is obvious from the review Prof. Mayer made many very important contributions to this field.

A tribute to the memory of Per-Olov Löwdin

I was privileged to work with Per-Olov Löwdin on some of the subjects covered in this review. He was a great scientist with a firm grasp of the essential basic physical facts,he had a fine intuition and he was a very refined practitioner of difficult mathematical techniques. I was also privileged to work with him as a lecturer in the different Summer Schools and Winter Institutes arranged in Upsala (Sweden) and Gainesville (Florida). He was a wonderful lecturer with great enthusiasm and he always captivated his audience. He was also great in mountaineering, I was happy to follow his footsteps in the Norwegian mountains in the extracurricular activities in the Summer Schools. In the review I emphasized his contribution, this was only a small part of his very wide contribution to quantum molecular science. I shall always treasure his wonderful memory.

References

Amos, A.T. and Hall G.G. (1961) Single determinant wave functions, *Proc. Royal Society (London)* **A 263**, 483-493.

Berencz, F. and Pauncz R. (1958) Construction of $\mathbf{S}^2$ eigenfunctions by the method of spin operators I. General theory, *Proc. Phys. Soc.* **71**, 145-151.

Born, M. and Oppenheimer R. (1927) Quantum theory of molecules, *Ann. Physik* **84**, 457-484.

Boys, S.F. (1960) Construction of some molecular orbitals to be approximately invariant for changes from one molecule to another, *Rev. Mod. Phys.* **32**, 296-299.

Calais, J.L. (1964a) Different bands for different spins I. The cohesive en–

ergy of an alkali metal, *Ark. Fys.* **28**, 479-498.

——— (1964b) Different bands for different spins II Application to a linear chain of hydrogen atoms, *Ark. Fys.* **28**, 511-527.

——— (1965) Different bands for different spins III Solid atomic hydrogen, *Ark. Fys.* **29**, 255-273.

——— (1985) Gap equations and instabilities for extended systems, *Adv. Quantum Chem.* **17**, 225-250.

Coulson, C. A., and G. S. Rushbroke (1940) The method of molecular orbitals, *Proc. Camb. Phil. Soc.* **36**, 193-200.

de Heer, J. (1962) Studies in the alternant molecular orbital method III. A many-parameter energy expression for systems with closed shell structure, *J. Chem. Phys.* **37**, 2080-2083.

——— (1962) A refined alternant molecular orbital treatment of the ground state of benzene, *J. Phys. Chem* **66**, 2288-2293.

——— (1963) The method of different orbitals for different spins and its application to alternant hydrocarbons, *Rev. Mod. Phys.* **35**, 631-637.

——— and Pauncz R. (1963) Studies in the alternant molecular orbital method V. A many-parameter energy expression for states with different multiplicities; Application to benzene, *J. Chem. Phys.* **39**, 2314-2320.

Dirac, P.A.M. (1926) On the theory of quantum mechanics, *Proc Roy Soc. (London)* **112**, 661.

Edmiston, S. and Ruedenberg, K. (1963) Localized atomic and molecular orbitals, *Rev. Mod. Phys.* **35**, 457-465.

Fukutome, H. (1981) Unrestricted Hartree-Fock theory and its applications to molecules and chemical reactions, *Int. J. Quantum Chem.* **20**, 955-1065.

Gershgorn, Z. (1968) Proof of the linear independence of properly selected projected spin eigenfunctions, *Int. J. Quantum Chem.* **2**, 341-348.

Goddard, W. A. III (1968) Improved quantum theory of many-electron systems III. The GF method, *J. Chem. Phys.* **48**, 450-461.

——— (1968) Improved quantum theory of many-electron systems. IV. Properties of the GF wavefunctions, *J. Chem. Phys.* **48**, 5337-5347.

Hall, G.G. (1951) A method of calculating ionization potentials, *Proc. Roy. Soc.* **A 205**, 541-552.

Hamza A. and Mayer I. (2001a) Overlap repulsion with Löwdin's pairing theorem I, *Int. J. Quantum Chem.* **82**, 53-59.

——— ——— (2001b) Overlap repulsion with Löwdin's pairing theorem II, *Int. J. Quantum Chem.* **82**, 105-112.

Harris, F. E. (1967) Molecular orbital theory, *Adv. Quantum Chem.* **3**, 61-127.

————,(1967) Open shell molecular orbital theory, *J. Chem. Phys.* **46**, 2769-2776.

Harrison, R.J. and Handy, N.C. (1983) Full CI calculations on BH H_2O NH_3, and HF, *Chem. Phys. Letters* **95**, 386-391.

Ito T. and Yoshizumi H. (1955) Application of the alternant orbital method to benzene, *J. Phys. Soc. Japan* **10**, 201-207.

Karadakov, P. (1985) An extension of the pairing theorem, *Int. J. Quantum Chem.* **27**, 699-707.

——— (1986) The spin-projected Hartree-Fock method: Direct optimization schemes and stability conditions, *Int. J. Quantum Chem.* **30**, 239-264.

King, H.F., Stanton, R.E., Kim, H., Wyatt, R.E, and Parr, R.G. (1967) Corresponding orbitals and the nonorthogonality problem in molecular quantum mechanics, *J. Chem. Phys.* **47**, 1936-1941.

Löwdin, P.-O. (1953) A method of alternant molecular orbitals, *Symposium on Molecular Physics at Nikko, Japan.* Maruzen, Tokyo, 13.

——— (1955) Quantum theory of many-particle system. III. Extension of the Hartee-Fock scheme to include degenerate systems and correlation effects, *Phys. Rev.* **97**, 1509-1520.

——— (1956) Quantum theory of cohesive properties of solids, *Adv. in Phys.* **5**, 1-172.

——— (1958) Nature des fonctions de la mésomerie, in: *Calcul des Foncons d'Onde Moléculaire*, Centre National de la Recherche Scientifique, Paris , pp. 23-37.

——— (1959) Correlation problem in many-electron quantum mechanics I. Review of different approaches and discussion of some current ideas, *Adv. Chem. Phys.* **2**, 207-325.

——— (1964) Angular momentum wavefunctions constructed by projection operators, *Rev. Mod. Phys.* **36**, 966-976.

——— (1983) On the Sanibel coefficients in the expansion of spin-projected Slater determinants, *Int. J. Quantum Chem.* **24**, 729-745.

——— (1991) On the pairing theorem and its extension, *Israel J. of Chem.* **31**, 297-302.

Löwdin P.-O. and Mayer I. (1992) Some studies of the general Hartree-Fock method, *Adv. Quantum Chem.* **24**, 79-114.

Manne, R. (1966), Table of Sanibel (Spin projection) coefficients, *Theor. Chim. Acta* **6**, 116-121.

Mayer, I. (1974) On the derivation of the extended Hartree-Fock equations I., *Acta Phys. Acad. Sci. Hung.* **37**, 39-52.

——— (1980) The spin-projected extended Hartree-Fock method, *Adv. Quant. Chem.* **12**, 189-262.

——— (1997) Simple proof of the pairing theorem, *Int. J. Quantum Chem.* **63**, 31-33.

——— (1986) Simple constructive proof of Karakadov's extended pairing theorem, *Int. J. Quantum Chem.* **29** , 31-34.

——— and Löwdin P.-O. (1993) Some comments on the general Hartree-Fock method, *Chem. Phys. Letters* **202**, 1-6.

Pauncz, R.,(1967) *Alternant Molecular Orbital Method*, W. B. Saunders Co. Philadelphia.

——— (1976) Investigations into the properties of projected spin functions, in: *Quantum Science: Methods and Structure; A Tribute to Per-Olov Löwdin* Eds. J.L. Calais, O. Goscinski, J. Linderberg and Y. Öhrn, Plenum Publ. Co. New York, 357-365.

——— (1979) *Spin Eigenfunctions: Construction and Use*, Plenum Publishing Co. New York.

——— (1987) B.Kirtman, and W.E.Palke Studies in the paired orbital method.I, *Int. J. Quantum Chem.* 1987 **21S**, 533-538.

——— (1988) Studies in the paired orbital method II. The energy expression and its derivatives with respect to the nonlinear parameters, *Theor. Chim. Acta* **73**, 383-391.

——— (1988) Studies in the paired orbital method III. The structure of the PO wavefunction, in: *The Structure of Small Molecules and Ions* Eds. R. Naaman and Z. Vager, Plenum Press, New York, 237-242.

——— (1989) Sanibel coefficients revisited, *J. Mol. Structure (Theochem)* **199**, 257-263.

——— (1992) Studies in the paired orbital method IV. Orthogonal transformations in the virtual space, *Int. J. Quantum Chem.* **26S**, 161-169.

——— (2000) *The construction of spin eigenfunctions An exercise book*, Kluwer Academic/ Plenum Publishers, New York.

Pratt, G.W.Jr. (1953) Eigenfunctions of $\mathbf{S}^2$ by a spin operator method, *Phys. Rev.* **92**, 278-288.

Roothaan, C.C.J. (1951) New developments in molecular orbital theory,

Rev. Mod. Phys. **23**, 69-89.

Slater, J.C. (1951) A simplification of the Hartree-Fock method, *Phys. Rev.* **81**, 385-390.

Smith V.H. Jr. and Harris, F.E. (1969) Projection of exact spin eigenfunctions, *J. Math. Phys* **10**, 771-778.

THE PAULI EXCLUSION PRINCIPLE, SPIN-STATISTICS CONNECTION, AND PERMUTATION SYMMETRY OF MANY-PARTICLE WAVE FUNCTIONS

I.G. Kaplan*

Instituto de Investigaciones en Materiales, UNAM, Apdo. Postal 70-360, 04510 Mexico, D.F., MEXICO and Department of Chemistry, Kansas State University, Manhattan, KS 66506-3701

Abstract. The Pauli exclusion principle, spin-statistics theorem, unusual statistics (parastatistics, fractional statistics, etc.), and their connection with the permutation symmetry of many-particle wave functions are discussed. For composite particles, the spin-statistics connection can be violated; e.g., the two-fermion Cooper pairs are described by the symmetric wave functions but have the fermion occupation numbers of one-particle states. Although, all elementary particles known at present are fermions or bosons; as was shown by author in 1976 , the quasiparticles in a periodical lattice, e.g. the Frenkel excitons or magnons, obey the modified parafermi statistics. In 1982 Wilczek revealed 2D objects (anyons) obeying the so-called fractional statistics.

It is demonstrated that the indistinguishability principle is insensitive to the permutation symmetry of the wave function and cannot be used as a criterion for the verification of the Pauli exclusion principle. Heuristic arguments are given in favor that the existence in Nature only the non-degenerate permutation representations (symmetric and antisymmetric) is not occasional. As follows from our analysis of possible scenarios, the permission of degenerate permutation representations leads to contradictions with the concept of particle identity and their independence. Thus, the prohibition of degenerate permutation states by the Pauli exclusion principle follows from the general physical assumptions underlaying quantum theory, but the problem of the spin-statistics connection is still open. It is pointed out that the Pauli exclusion principle and the Jahn-Teller effect have some similar features.

* Electronic mail: kaplan@fisica.unam.mx

E.J. Brändas and E.S. Kryachko (eds.),
Fundamental World of Quatum Chemistry, Vol. I, 183–220.

1. HISTORICAL SURVEY; SPIN-STATISTICS CONNECTION

Wolfgang Pauli formulated his principle before the creation of quantum mechanics (1925-1927). He arrived at formulation of this principle trying to explain the regularities in the classification of atomic spectral terms in a strong magnetic field. In paper submitted for publication in January 1925, Pauli formulated his principle as follows [1]:

> "*In an atom there cannot be two or more equivalent electrons for which the values of all four quantum numbers coincide. If an electron exists in an atom for which all of these numbers have definite values, then this state is* 'occupied'."

At that time, the fourth quantum number was not described by any model. Pauli called the property associated with it as the "characteristic two-valuedness of the quantum properties of the electron which cannot be described classically" [2].

This non-classical two-valued nature of electron is now called *spin*. In anticipating the quantum nature of the magnetic moment of electron before the creation of quantum mechanics, Pauli exhibited a striking intuition. It is interesting to note that this intuition together with Pauli's inherent rigor of thought did not allow him at once to admit the hypothesis of the spin that had been proposed by Kroning and independently by Uhlenbeck and Goudsmit to explain the origin of the fourth quantum number of electron. Pauli's objections resulted from the fact that their spin hypothesis was based on the classical concept of rotation of the electron about its own axis. And though Uhlenbeck and Goudsmit [3] explained the doublet splitting in the spectra of alkali atoms by assuming that the ratio of the intrinsic moment of the rotating electron to mechanical moment was twice as great as in the orbital motion, Pauli was very skeptical inclined toward this hypothesis. Upon meeting with Bohr who had fallen under the influence of explanation of the doublet splitting into the favor on the rotating-electron hypothesis, Pauli expressed the regret that a new ``heresy" had arisen in atomic physics, as van der Waerden wrote in his recollections [4].

It is now clear that Pauli was right in not agreeing with the classical interpretation of the fourth degree of freedom. The spin cannot in principle be described by classical physics. Pauli recalled [5]:

> "Although at first I strongly doubted the correctness of this idea because of its classical mechanical character, I was finally converted to it by Thomas [6] calculations on the magnitude of doublet splitting. On the other hand, my earlier doubts as well as the cautious expression 'classically non-describable two-valuedness' experienced a certain verification during later developments, as Bohr was able to show on the basis of wave mechanics that the electron spin cannot be measured by classically describable experiments (as, for instance, deflection of molecular beams in external electromagnetic fields) and must therefore be considered as an essentially quantum mechanical property of the electron."

The first studies devoted to applying the newborn quantum mechanics to many-particle systems were those by Heisenberg [7] and Dirac [8]. In these studies, the Pauli principle, formulated as the prohibition for two electrons to occupy the same quantum state, was derived as consequence of the antisymmetry of the wave function of the system of electrons. Dirac [8] came to the conclusion that the light quanta must be described by the symmetric wave functions. He specially noted that a system of electrons cannot be described by the symmetric wave functions since the latter allow any number of electrons to occupy a quantum state.

Thus, with the creation of quantum mechanics, the prohibition on the occupation numbers of electron system states was supplemented by the prohibition of all types of permutation symmetry of electron wave functions except for antisymmetric one. Later on, an analysis of experimental data has permitted to formulate the Pauli exclusion principle for all known elementary particles. Namely:

The only possible states of a system of identical particles possessing spin s are those for which the total wave function transforms upon interchange of any two particles as

$$P_{ij}\Psi(1,\ldots,i,\ldots j,\ldots,N) = (-1)^{2s}\Psi(1,\ldots,i,\ldots j,\ldots,N) \qquad (1)$$

That is, it is symmetric for integer values of s and antisymmetric for half-integer s.

The exclusion principle also holds for the permutation symmetry of composite-particle wave functions, e.g. for nuclei. The latter consist of nucleons: protons and neutrons which are fermions because have s=1/2. Depending on the value of the total nuclear spin, one can speak of boson nuclei and fermion nuclei. The nuclei with even number of nucleons have an integer value of the total spin and belong to the boson particles. The nuclei with odd number of nucleons have a half-integer value of the total spin and belong to the fermion particles. The same is true for an arbitrary many-fermion system. The group-theoretical approach for finding quantum states of arbitrary systems allowed by the Pauli exclusion principle was elaborated in Refs. [9-11], see also book [12].

The well-known example is the $^{16}O_2$ molecule. The nucleus ^{16}O is a boson composite particle, so the total wave function of the $^{16}O_2$ molecule must be symmetric under the permutations of nuclei. From this follows that in the ground electronic state, which is antisymmetric, the nuclear wave function should be also antisymmetric, so, only odd values of the rotational angular momentum are permitted. This was revealed in spectroscopic measurements many year ago [13]. Recent tests of possible violation of the Pauli exclusion principle for the $^{16}O_2$ molecule [14,15] give the probability of such violation $\leq$ (5-8)x 10^{-7}.

All experimental data known to date agree with the Pauli exclusion principle. Several types of experiments on the search for possible small violations of the Pauli principle have been suggested in the literature [16-18], a comprehensive discussion was presented at the conference [19], see reports by Greenberg [20] and Gillaspy [21]. A test of Pauli principle was made earlier by searching for γ quanta with energies around 20 MeV, which should be produced with transitions of nucleons in the ^{12}C nucleus from the 2p shell to the occupied 1s shell. This search has given a lower limit for the formation time of a "non Pauli" nucleus $\tau \geq 2 \times 10^{20}$ years [22]. The search for X-ray radiation produced by the transition of an electron in the germanium detector to the occupied $(1s)^2$ shell, either in the case of the hypothetical spontaneous decay of an electron in this shell or in the case of violation of the Pauli principle, gave an even greater lower limit $\tau \geq 1.5 \times 10^{25}$ years [23]. One of the last experimental test of the Pauli exclusion principle: the search for the Pauli forbidden $1s^4$ state of the Be atom was published in Ref. [24].

The exclusion principle, first discovered by Pauli for electrons and later spread to all particles, was based on the analysis of experimental data. Pauli himself was never satisfied by this. In his Nobel Prize lecture [5] Pauli said:

> "Already in my initial paper, I especially emphasized the fact that I could not find a logical substantiation for the exclusion principle nor derive it from more general assumptions. I always had a feeling, which remains until this day, that this is the fault of some flaw in the theory."

Let us stress that this was said in 1946, or after the Pauli famous theorem [25] of the relation between spin and statistics. In this theorem, Pauli did not give a direct proof. He showed that due to some physical contradictions, the second quantization operators for particles with integral spins cannot obey the fermion commutation relations; while for particles with half-integral spins, the second quantization operators cannot obey the boson commutation relations. From this Pauli concluded that particles with integral spin have to obey the Bose-Einstein statistics, while those with half-integral spin have to obey the Fermi-Dirac statistics.

Thus, according to the Pauli theorem, the connection between the value of spin and the permutation symmetry of many-particle wave function, Eq. (1), follows if we assume that particles can obey only two types of commutation relations: boson or fermion relations. At that time it was believed that it is really so. However in 1953, Green [26] (and independently Volkov [27]) showed that the more general, paraboson and parafermion trilinear commutation relations, satisfying all physical requirements and containing the boson and fermion commutation relations as particular cases, can be introduced. A corresponding parastatistics is classified by its rank p. For $p = 1$ the parafermi statistics becomes identical to the Fermi-Dirac statistics, while the parabose statistics is reduced to the Bose-Einstein statistics [28], see also the Isihara book [29].

As was shown by Greenberg and Messiah [28], all elementary particles then known are bosons or fermions. This conclusion does not spread on the quasiparticle case. In 1976, Kaplan [30] revealed that the parafermi statistics is realized for quasiparticles in a crystal lattice, e.g. for the Frenkel excitons or magnons, but due to a periodical crystal field, the Green trilinear commutation relations are modified by the quasi-impulse conservation law. In next section, we present the parastatistics commutation relations and discuss in details the case of quasi-particles in a periodical lattice.

For elementary particles, the permutation symmetry of wave function is directly connected with its statistics: particles with integer spin obey the Bose-Einstein statistics, and particles with half-integer spin obey the Fermi-Dirac statistics. The situation becomes complicated in the case of composite particles. Although, the wave function of composite particles has the boson (fermion) permutation symmetry, it is not true for its second quantization operators [31-34]. The simplest example is the Cooper pair operators [35]. They are constructed as a product of two fermion operators

$$b_{\mathbf{k}}^{+} = c_{\mathbf{k}\alpha}^{+} c_{-\mathbf{k}\beta}^{+}, \qquad b_{\mathbf{k}} = c_{-\mathbf{k}\beta} c_{\mathbf{k}\alpha} \ . \tag{2}$$

Direct calculation results in

$$[b_{\mathbf{k}}, b_{\mathbf{k}'}^{+}]_{-} = [b_{\mathbf{k}}^{+}, b_{\mathbf{k}'}^{+}]_{-} = [b_{\mathbf{k}}, b_{\mathbf{k}'}]_{-} = 0, \qquad \mathbf{k} \neq \mathbf{k}' \tag{3}$$

$$[b_{\mathbf{k}}, b_{\mathbf{k}}^{+}]_{-} = 1 - (\hat{n}_{\mathbf{k}} + \hat{n}_{-\mathbf{k}}) \tag{4}$$

where $\hat{n}_{\mathbf{k}} = c_{\mathbf{k}\alpha}^{+} c_{\mathbf{k}\alpha}$ is the electron number operator. It is easy to verify that

$$(b_{\mathbf{k}}^{+})^2 = (b_{\mathbf{k}})^2 = 0. \tag{5}$$

Thus, for **k** ≠ **k**´ Cooper´s pairs are bosons, for **k** = **k**´ they obey the Pauli principle and have the fermion occupation numbers for one-particle states. This is an additional (to Green´s parastatistics) example that the variety of allowed commutation relations is not exhausted by the boson and fermion ones. Note: the reason that the fermion-type occupation numbers for Cooper's pairs does not prevent the formation of the superconducting condensate is in the following. The condensate, in which all pairs are in the same state with the total impulse **K**= 0, is formed by pairs having different vectors **k**. Hence, the system of such pairs possesses the boson properties.

For boson and fermion elementary particles there is the one-to-one correspondence between the symmetry of the wave function and the commutation relations for the field operators, whereas in the case of parastatistics this is not so. As noted in Ref.[28], the problem of correspondence between the parastatistics commutation relations and the permutation symmetry of wave function, if any, is open. In subsequent studies [36-38] was shown that the parafermi (parabose) statistics of order p can be connected with the permutation symmetry characterized by all Young diagrams with N boxes with the first row (column) no longer than p (in Section 3 we will return to this connection and show that it relates only to a part of the factorized total wave function which does not contain some internal degrees of freedom). Since, there are no prohibitions on the existence of elementary particles obeying the parastatistics commutation relations, the proof of the spin-statistical theorem [25] loses its base.

As we noted above, the situation with the spin-statistical theorem becomes quite serious in the case of composite particles. For instance, the composite particles with even number of fermions are described by symmetric wave functions, but the field operators for such particles do not obey the boson commutation relations. The Cooper pairs are a good example of this kind.

In dynamical interactions, the internal structure of composite particles reveals itself at distances between interacting particles at which the overlap of their wave functions becomes appreciable. This leads to the well known in atomic and molecular physics exchange interaction stipulated by the requirement of antisymmetry of many-electron wave functions. Hence, the exchange interaction is a consequence of the Pauli exclusion principle. The same origin has an effective repulsion at short distances between two composite particles consisting of identical fermions [39-41].

After 1940, numerous proofs of the spin-statistics theorem were published. All these proofs contain some explicit (or implicit) assumptions, see book by Duck and Sudarshan [42], review on it by Wightman [43], and Proceedings [19]. As emphasized by Berry and Robbins [44], the relation between spin and statistics "cries out for understanding".

2. PARASTATISTICS AND STATISTICS OF QUASIPARTICLES IN A PERIODICAL LATTICE

2.1 Short review of parastatistics

As was discussed in previous section, Green [26] introduced in 1953 a generalized method of field quantization with trilinear commutation relations, which include the boson and fermion commutation relations as particular cases,

$$[[a_k^+, a_{k'}]_\pm, a_{k''}]_- = -2\delta_{kk''}\, a_{k'} \tag{6}$$

$$[[a_k^+, a_{k'}]_\pm, a_{k''}^+]_- = 2\delta_{k'k''}\, a_k^+ \tag{7}$$

$$[[a_k^+, a_{k'}^+]_\pm, a_{k''}^+]_- = [[a_k, a_{k'}]_\pm, a_{k''}]_- = 0\,, \tag{8}$$

upper sign for *parabosons* (bosons satisfy these equations also), lower sign for *parafermions* (and fermions).

Green found an infinite set of solutions, labeled by the integers $p\,(p = 1,2,...,\infty)$,

$$a_k = \sum_{\alpha=1}^{p} b_k^{(\alpha)}. \tag{9}$$

It is the *Green Ansatz*. For parabosons:

$$\alpha = \beta \quad \left[b_k^{(\alpha)}, b_\ell^{(\alpha)+}\right]_- = \delta_{k\ell}\,, \quad \left[b_k^{(\alpha)+}, b_\ell^{(\alpha)+}\right]_- = \left[b_k^{(\alpha)}, b_\ell^{(\alpha)}\right]_- = 0, \tag{10}$$

as for bosons.

$$\alpha \neq \beta \quad \left[b_k^{(\alpha)}, b_\ell^{(\beta)+}\right]_+ = \left[b_k^{(\alpha)+}, b_\ell^{(\beta)+}\right]_+ = \left[b_k^{(\alpha)}, b_\ell^{(\beta)}\right]_+ = 0, \tag{11}$$

as for fermions.

For parafermions:

$$\alpha=\beta \quad \left[b_k^{(\alpha)}, b_\ell^{(\alpha)+}\right]_+ = \delta_{k1}, \quad \left[b_k^{(\alpha)+}, b_\ell^{(\alpha)+}\right]_+ = \left[b_k^{(\alpha)}, b_\ell^{(\alpha)}\right]_+ = 0, \tag{12}$$

as for fermions,

$$\alpha\neq\beta \quad \left[b_k^{(\alpha)}, b_\ell^{(\beta)+}\right]_- = \left[b_k^{(\alpha)+}, b_\ell^{(\beta)+}\right]_- = \left[b_k^{(\alpha)}, b_\ell^{(\beta)}\right]_- = 0, \tag{13}$$

as for bosons.

As follows from Eqs.(10) and (12), for $p=1$, the solution of the Green Ansatz reduces to the usual boson and fermion operators in paraboson and parafermion cases, respectively. It is worth-while to note that such mixed boson-fermion behavior of the operators $b_k^{(\alpha)}$ is well known in solid state physics and called as *paulionic*, see the book by Davydov [45]. The value of p in the Green Ansatz is called the *rank of parastatistics*. The application of operators (9) to the vacuum state is easily found using Eqs.(10)-(13). So,

$$a_k|0\rangle = 0, \quad a_k a_\ell^+|0\rangle = p\delta_{k\ell}|0\rangle \qquad \text{for all } k,\ell \tag{14}$$

For the parafermi statistics, the maximum occupation numbers of one state is equal to the rank of parastatistics p

$$\left(a_k^+\right)^N|0\rangle \neq 0 \qquad \text{for } N\leq p, \tag{15}$$

$$\left(a_k^+\right)^{p+1}|0\rangle = 0. \tag{16}$$

At $p=1$, the parafermi statistics is reduced to the Fermi-Dirac statistics. For parabose statistics there is no restriction in the occupation numbers, as in the Bose-Einstein statistics.

The particle number operator in parastatistics is defined as

$$\hat{N}_k = \tfrac{1}{2}\left[a_k^+, a_k\right]_\pm \mp \tfrac{1}{2}p. \tag{17}$$

It satisfies the general property of the particle number operators

$$\left[\hat{N}_k, a_\ell^+\right]_- = \delta_{k\ell} a_\ell^+ \tag{18}$$

and in the Bose (Fermi) case turns to

$$\hat{N}_k = a_k^+ a_k . \tag{19}$$

As was shown in Refs.[46,47], the parafermion and paraboson algebras are the Lie algebras of the orthogonal and symplectic groups, respectively. For example, the algebra for ν parafermion operators a_r and their adjonts $a_r^+ (r = 1,2,..., \nu)$, which constitutes a parafermi ring, is the Lie algebra of the orthogonal group $SO_{2\nu+1}$ in $2\nu+1$ dimensions; more details about the parastatistics is presented in the book by Ohnuki and Kamefuchi [48].

In spite of numerous studies of all known elementary particles, see Refs.[19,28], the elementary particles obeying the parastatistics were not revealed. On the other hand, as discussed in Refs.[49-51], the ordinary fermions, which differ by some internal quantum numbers but are similar dynamically, can be described by the parafermi statistics. In this case, fermions with different internal quantum numbers are considered as non-identical distinguishable particles. The parafermi statistics of rank p describes systems with p different types of fermions. As a result, *quarks* with 3 colors obey the parafermi statistics of rank $p = 3$; *nucleons* in nuclei (isotope spin $\frac{1}{2}$) obey the parafermi statistics of rank $p = 2$.

Let us stress, the fermions characterized by the same value of internal quantum number obey the Fermi-Dirac statistics. It is the system of *p* fermion groups with different internal quantum numbers that obeys the parafermi statistics of rank *p*.

2.2 Statistics of quasiparticles in a periodical lattice

Although no elementary *paraparticles* has been revealed; in the case of *quasiparticles,* the situation can be different. As was shown by Kaplan [30], the parastatistics in the form modified by the quasi-impulse conservation law is realized for quasiparticles in a periodical lattice. Namely: the Frenkel excitons and magnons obey a modified parafermi statistics of rank *M*, where *M* is the number of equivalent lattice sites within the delocalization region of a collective excitation. Later on the results [30] for the Frenkel excitons and magnons were extended to polaritons [52], defectons in quantum crystals [53], and the Wannier-Mott excitons [54]. Recently, we proved that delocalized holes [55] and delocalized coupled hole pairs [56] also obey the modified parafermi statistics. This modified parafermi statistics for a general system of non specified quasiparticles in a periodical lattice [57] is considered below.

Let us study a system of spinless excitations of one type on the sites of a crystal lattice. The nature of excitations is not specified, it can be electron or spin excitations, holes or hole pairs on sites (in the last case it is sites of "superlattice" formed by centers of mass of hole pairs [56]). In the absence of dynamic interaction between excitations, the model Hamiltonian in the site representation can be written as follows:

$$H = \varepsilon_0 \sum_n b_n^+ b_n + \sum_{n,n'} M_{nn'} b_n^+ b_{n'} \qquad (20)$$

where b_n^+, b_n are the creation and annihilation operators of excitation at site n, ε_0 is the creation energy of site excitation in a lattice, $M_{nn'}$ is the so-called *hopping integral*, characterizing the efficiency of excitation transfer from site n to site n′.

The operators b_n^+, b_n acting on different sites are independent and must commute. So, they obey the Bose commutation relations:

$$[b_n, b_{n'}^+]_- = [b_n^+, b_{n'}^+]_- = [b_n, b_{n'}]_- = 0 \qquad \text{for } n \neq n'. \qquad (21)$$

In the Hamiltonian (20), it is assumed that at a site only one excitation and of one type can be created, that is $(b_n^+)^2|0\rangle = 0$, where $|0\rangle$ is the vacuum state. The operators acting on one site can be defined in such manner [45] that they satisfy the Fermi commutation relations:

$$[b_n, b_n^+]_+ = 1, \qquad [b_n^+, b_n^+]_+ = [b_n, b_n]_+ = 0 \,. \qquad (22)$$

Thus, the excitation operators in site representation have to be attributed to the paulion operators. In the case of the Cooper pair operators, only the Pauli relation (5) is valid; Eq.(22) is not valid because Cooper's operators are a product of two fermion operators. Other difference is in the following: unlike the Cooper pair operators, which are defined in impulse space and so are delocalized, the operators of excitation are localized at lattice sites.

The introduced excitations do not belong to stationary states. They can move in crystal and occupy any sites. This is provided by the hopping term in the Hamiltonian (20). The stationary states can be created by some unitary transformation:

$$B_{\mathbf{q}} = \frac{1}{\sqrt{M}} \sum_{n} u_{\mathbf{q}n} b_n \, , \; B_{\mathbf{q}}^{+} = \frac{1}{\sqrt{M}} \sum_{n} u_{\mathbf{q}n}^{*} b_n^{+} \qquad (23)$$

which diagonalizes the Hamiltonian (20) in the quasi-impulse space,

$$H = \sum_{\mathbf{q}} \varepsilon_{\mathbf{q}} B_{\mathbf{q}}^{+} B_{\mathbf{q}} \, . \qquad (24)$$

In the Eq.(23), M is the number of equivalent lattice sites. For simple lattices with one site per cell, the unitary transformation (23) is determined by the translation symmetry and $u_{\mathbf{q}n} = \exp(-i\mathbf{q}r_n)$. The self-energy of the diagonalized Hamiltonian (24) is equal to

$$\varepsilon_{\mathbf{q}} = \varepsilon_0 + \sum_{n'(\neq n)} M_{nn'} \exp[i\mathbf{q}(\mathbf{r}_n - \mathbf{r}_{n'})] \, . \qquad (25)$$

The action of the operator $B_{\mathbf{q}}^{+}$ on the vacuum state $|0\rangle$ creates a quasiparticle in the quasi-impulse space. In the coordinate space, the quasiparticle is delocalized among all equivalent sites. It can be exciton, magnon, delocalized hole or coupled hole pair, etc..

Since the operators b_n^{+}, b_n obey neither the boson nor the fermion commutation relations, the unitary transformation in general case is not canonical; this means that it does not preserve the commutation properties of the transformed operators and the operators (23) do not correspond to the paulion quasiparticles. As was shown by author [30] for an exponential unitary transformation, such operator obey the modified parafermi statistics with trilinear commutation relations:

$$[[B^+_{\mathbf{q}}, B_{\mathbf{q}'}]_-, B_{\mathbf{q}''}]_- = -2M^{-1}B_{\overline{\mathbf{q}}} \qquad \overline{\mathbf{q}} = \mathbf{q}'+\mathbf{q}''-\mathbf{q} \tag{26}$$

$$[[B^+_{\mathbf{q}}, B_{\mathbf{q}'}]_-, B^+_{\mathbf{q}''}]_- = 2M^{-1}B^+_{\overline{\mathbf{q}}} \qquad \overline{\mathbf{q}} = \mathbf{q}-\mathbf{q}'+\mathbf{q}''. \tag{27}$$

Instead of the Kronecker symbols, as in the parafermion relations (6) and (7), the value of $\overline{\mathbf{q}}$ in the right-hand side of Eqs.(26) and (27) is determined by the quasi-momentum conservation law. This is the consequence of the translation symmetry. The absence of the Kronecker symbols in Eqs.(26) and (27) leads to important physical consequences which we discuss below.

It can be shown [30,55] that one state may be occupied by up to *M* quasiparticles:

$$(B^+_{\mathbf{q}})^N|0\rangle \neq 0 \qquad \text{for N} \leq M; \qquad (B^+_{\mathbf{q}})^{M+1}|0\rangle = 0 \tag{28}$$

This means that quasiparticles in the quasi-impulse space obey the parafermi statistics of rank *M* with the modified trilinear commutation relations (26) and (27).

The state with N quasiparticles, each with the same **q**, has to be normalized:

$$|N_{\mathbf{q}}\rangle = C_N(B^+_{\mathbf{q}})^N|0\rangle, \tag{29}$$

$$C_N = [N!(1-\frac{1}{M})(1-\frac{2}{M})\cdots(1-\frac{M-1}{M})]^{-\frac{1}{2}} . \tag{30}$$

The expression for C_N is found by the induction method using the operator equation for shifting the operator $B_{\mathbf{q}}$ to the right, see Refs. [30,55], and taking into account that

$$B_{\mathbf{q}}|0\rangle = 0 \quad \text{and} \quad B_{\mathbf{q}'}B^+_{\mathbf{q}}|0\rangle = \delta_{\mathbf{q}\mathbf{q}'}|0\rangle. \tag{31}$$

The second equation in (31) does not contain the rank of parastatistics, as in Eq.(14), because, in contrary to the Green Ansatz (9), the transformation (23) is normalized.

The result of applying operators $B_{\mathbf{q}}^{+}$ and $B_{\mathbf{q}}$ to the state vector (29) is the following:

$$B_{\mathbf{q}}^{+}\left|N_{\mathbf{q}}\right\rangle=\sqrt{(N_{\mathbf{q}}+1)\left(1-\frac{N_{\mathbf{q}}}{M}\right)}\left|N_{\mathbf{q}}+1\right\rangle, \tag{32}$$

$$B_{\mathbf{q}}\left|N_{\mathbf{q}}\right\rangle=\sqrt{N_{\mathbf{q}}\left(1-\frac{N_{\mathbf{q}}-1}{M}\right)}\left|N_{\mathbf{q}}-1\right\rangle. \tag{33}$$

From Eq.(32) follows that the effect of applying $B_{\mathbf{q}}^{+}$ on a state with a maximum occupation number $N_{\mathbf{q}}=M$ is equal to zero. As $M\to\infty$, relations (30), (32), and (33) turn into the well known relations for bosons.

The result of applying the product $B_{\mathbf{q}}^{+}\ B_{q}$ to the state vector (29) is easily found using Eqs.(32) and (33),

$$B_{\mathbf{q}}^{+}B_{\mathbf{q}}\left|N_{\mathbf{q}}\right\rangle=N_{\mathbf{q}}\left(1-\frac{N_{\mathbf{q}}-1}{M}\right)\left|N_{\mathbf{q}}\right\rangle. \tag{34}$$

So, the operator $B_{\mathbf{q}}^{+}B_{\mathbf{q}}$ is not a particle number operator in a state **q**, as in the case of boson, fermion or paulion operators. It can be proved that for the introduced above modified parafermi statistics, the operator of particle number in a state **q** does not exist, see Ref.[30,55], and independent one-quasiparticle states cannot be defined. The mathematical reason for this is the absence of the Kronecker symbols in the trilinear commutation relations (26) and (27).

From the principal impossibility of introducing the operator of particle number $\hat{N}_{\mathbf{q}}$ for quasiparticles in crystal follows that the ideal gas of such quasiparticles does not exist. The Hamiltonian H_0, in which interaction terms are not included, does not correspond to the Hamiltonian of an ideal quasiparticle system,

$$H_0 = \sum_{\mathbf{q}} \varepsilon_{\mathbf{q}} B_{\mathbf{q}}^{+} B_{\mathbf{q}} \neq \sum_{\mathbf{q}} \varepsilon_{\mathbf{q}} \hat{N}_{\mathbf{q}} \ . \tag{35}$$

Right-hand part of Eq.(35) cannot be written, because $\hat{N}_{\mathbf{q}}$ cannot be defined.

The form of the Hamiltonian H_0 means that quasiparticles do not interact (dynamically) with each other, nevertheless, it is not an ideal system. For study the physical meaning of H_0, let us calculate its expectation value in the vector state $\left|N_{\mathbf{q}}\right\rangle$ with N quasiparticles. We have to calculate

$$E(N_{\mathbf{q}}) = \left\langle N_{\mathbf{q}} \middle| H_0 \middle| N_{\mathbf{q}} \right\rangle = C_N^2 \left\langle 0 \middle| B_{\mathbf{q}}^N \left(\sum_{\mathbf{q}'} \varepsilon_{\mathbf{q}'} B_{\mathbf{q}'}^{+} B_{\mathbf{q}'} \right) (B_{\mathbf{q}}^{+})^N \middle| 0 \right\rangle . \tag{36}$$

Using an operator equation for shifting $B_{\mathbf{q}}$ to the right, after cumbersome but straight- forward calculations, see Ref. [55], we obtain

$$E(N_{\mathbf{q}}) = N\varepsilon_{\mathbf{q}} \left(1 - \frac{N-1}{M} \right) + \frac{N(N-1)}{M(M-1)} \sum_{\mathbf{q}'(\neq \mathbf{q})} \varepsilon_{\mathbf{q}'} = N \left[\varepsilon_{\mathbf{q}} + \frac{N-1}{M} (\bar{\varepsilon} - \varepsilon_{\mathbf{q}}) \right] \tag{37}$$

where $\bar{\varepsilon} = \left(1/_{M} - 1 \right) \sum_{\mathbf{q}'(\neq \mathbf{q})} \varepsilon_{\mathbf{q}'}$ is the mean energy of the quasiparticle band. The second term in the right-hand part of Eq.(37) is called the *kinematic* interaction. It is proportional to the concentration of quasiparticles. At small concentration it approaches to zero and gas of quasiparticles can be considered as a Bose gas. This kind of interaction depending on the deviation of quasiparticle statistics from the Bose (Fermi) behavior was firstly revealed by Dyson [58] in his studies of magnetic systems.

In high T_c ceramics, the maximum T_c is achieved for a hole concentration in the CuO_2 planes equals to 0.2-0.25 per CuO_2 units [59]. The same order of magnitude has to be for the coupled hole-pair concentration [56]. For excitons a much higher concentrations can be achieved in modern laser installations. Thus, the deviations from the Bose-Einstein statistics for the quasiparticle systems are not negligible and have to be taken into account.

The kinematic interaction is larger the larger is the difference between $\varepsilon_{\mathbf{q}}$ and the mean energy of the quasiparticle band. For state $\varepsilon_{\mathbf{q}} < \bar{\varepsilon}$, the kinematic interaction is positive, so, in the ground state, the kinematic interaction leads to repulsion. For higher states with $\varepsilon_{\mathbf{q}} > \bar{\varepsilon}$, the kinematic interaction is attractive.

According to the expression (37) for energy $E(N_{\mathbf{q}})$, the kinematic interaction mixes all states of the quasiparticle band. Therefore, we cannot define quasiparticles in some particular state. The ideal gas of quasiparticles does not exist fundamentally. Hence, the Gentile distribution function [60], derived for an ideal paragas, cannot be applied to a quasiparticle system in the quasi-impulse space.

As is shown above, the quasiparticles in the quasi-impulse space obey the modified parafermi statistics of rank M, so one state can be occupied up to M quasiparticles. The number of quasiparticles is equal to the number of created excitations. The latter cannot exceed the number M of lattice sites at which the excitation can be created. So, all created quasiparticles can occupy one state, for example, the ground state. This means that, in spite of non-boson behavior, there is no statistical prohibition of the Bose-Einstein condensation (BEC) phenomenon in quasiparticle systems. On the other hand, the gas of quasiparticles is always non-ideal. Thus, the BEC phenomenon has to be considered like that in the Bose gas with weak repulsion, as in the ^{4}He liquid case, but using the proper trilinear commutation relations (the kinematic repulsion instead of dynamical repulsion in liquid helium).

3. INDISTINGUISHABILITY OF IDENTICAL PARTICLES AND THE SYMMETRY POSTULATE

In this section we focus on the symmetry restrictions imposed by the Pauli exclusion principle. The point is that the Schrödinger equation is invariant under any permutation of identical particles. The

Hamiltonian of an identical particle system commutes with the permutation operators,

$$[P, H]_- = 0. \tag{38}$$

From this follows that the solutions of the Schrödinger equation may belong to any representation of the permutation group, including the degenerate ones. However, according to the exclusion principle, a system of identical particles can be only in those states that are not degenerate with respect to permutations: in a symmetric or an antisymmetric state; all other types of symmetry are forbidden. The question might be asked: *whether this limitation on the solutions of the Schrödinger equation follows from the fundamental principles of quantum mechanics or it is an independent principle*?

We will discuss possible answers on this question developing some ideas of our previous publications [61-63].

There are two points of view on the problem of independency of the Pauli exclusion principle from other fundamental quantum-mechanical postulates. Some physicists, including one of the founders of quantum mechanics Dirac [64] (see also Shiff [65] and Messiah [66]), believe that there are no laws in Nature that forbid the existence of particles described by wave functions with more complicated permutation symmetry than those of bosons and fermions, and that the existing limitations are only due to the specific properties of the known elementary particles. Messiah [66] has even introduced the term *symmetry postulate* to emphasize the primary nature of the constraint on the allowed types of the wave function permutation symmetry. By using the Schur lemma, Messiah and Greenberg [67] have shown that the existence of permutation degeneracy should not introduce additional uncertainty into the characteristics of a state. This also follows directly from the Wigner-Eckart theorem generalized for the permutation group, see Eq. (4.60) in book [12]. Namely: the matrix element of an operator L, that is symmetric in all the particles, can be presented as

$$\left\langle \Psi_r^{[\lambda]} \middle| \hat{L} \middle| \Psi_{\bar{r}}^{[\lambda]} \right\rangle = \delta_{r\bar{r}} \left\langle \Gamma^{[\lambda]} \middle\| \hat{L} \middle\| \Gamma^{[\lambda]} \right\rangle \tag{39}$$

where index r labels the basic functions of the representation $\Gamma^{[\lambda]}$ of the permutation group. The double vertical line in the right-hand side of this formula means that the matrix element is independent on the basic function index. Thus, the expectation value of operator L is the same for all functions belonging to the degenerate state.

Another point of view is that the symmetry postulate is not an independent principle and can be derived from the fundamental principles of quantum mechanics, in particular, from the principle of indistinguishability of identical particles. The typical argumentation presented in some textbooks and monographs [68-70] is the following.

From the requirement that the states of a system obtained by permutations of identical particles must all be physically equivalent, one concludes that the change in the wave function resulting from the transposition of any two identical particles should only cause multiplication by an insignificant phase factor,

$$P_{12}\Psi(x_1,x_2)=\Psi(x_2,x_1)=e^{i\alpha}\Psi(x_1,x_2) \tag{40}$$

where α is a real constant and x is the set of spatial and spin variables. One more application of the permutation operator P_{12} gives

$$\Psi(x_1,x_2)=e^{i2\alpha}\Psi(x_1,x_2) \tag{41}$$

or

$$e^{2i\alpha}=1 \quad \text{and} \quad e^{i\alpha}=\pm 1. \tag{42}$$

Since all particles are assumed to be identical, the wave function should change in exactly the same way under transposition of any pair of particles, i.e. it should be either totally symmetric or totally antisymmetric.

The incorrectness of this proof is in the following. Equation (40) is valid only for one-dimensional representations. The statement that wave functions describing the same physical state may differ by no more than a phase factor is evidently not true. For instance, according to Eq.(39), the value of the physical quantities characterizing a system of identical particles are the same for all functions belonging to a multi-dimensial representation, and all these different analytical functions describe the same physical state. According to the group theory, the application of a group operation to one of basic functions, belonging to some multi-dimensial representation, transforms it in a linear combination of basic functions,

$$P_{12}\Psi_i = \sum_k \Gamma_{ki}(P_{12})\Psi_k \,. \tag{43}$$

The application of the permutation operator P_{12} to both sides of Eq.(43) leads to the initial function.

$$P_{12}\{P_{12}\Psi_i\} = \sum_\ell \left[\sum_k \Gamma_{\ell k}(P_{12})\Gamma_{ki}(P_{12})\right]\Psi_\ell = \tag{44}$$

$$= \sum_\ell \Gamma_{li}\left(P_{12}{}^2 = I\right)\Psi_\ell = \Psi_i,$$

since $\Gamma_{\ell i}(I) = \delta_{\ell i}$. Hence, Eq.(43) leads to the same result as

$$P_{12}\Psi_i = \pm\Psi_i. \tag{45}$$

By requiring that under permutations the wave function must change by no more than a phase factor, one actually *postulates* that the representation of the permutation group to which the wave function belongs is one-dimensional. Thus, the proof in Refs.[68-70] is based on the initial statement that then is proved as a final result.

In the above proof there is an additional incorrectness: the indistinguishability principle is directly related to the behavior of the wave function. However, since the wave function is not an observable, the indistinguishability principle is related to it only indirectly via the expressions for measurable quantities. A rigorous proof should be based on a rigorous formulation of the indistinguishability principle for identical particles. One possible formulation is the following:

All observable quantities are invariant under the permutations of identical particles, and vice versa, the permutations of identical particles cannot be observed.

Since in quantum mechanics, the physical quantities are expressed as bilinear forms of wave functions, the indistinguishability principle requires the invariance of these bilinear forms and can be formulated as [71]:

$$P\left\langle \Psi \right| \hat{L} \left| \Psi \right\rangle = \left\langle \Psi \right| \hat{L} \left| \Psi \right\rangle. \tag{46}$$

Often, one limits oneself to the requirement that the probability of a given configuration of a system of identical particles must be invariant under permutations [72,73],

$$P\left| \Psi(x_1,\ldots,x_N) \right|^2 = \left| \Psi(x_1,\ldots,x_N) \right|^2. \tag{47}$$

For a function to satisfy Eq. (47), it is sufficient that under permutations it would change as

$$P\,\Psi(x_1,\ldots,x_N) = e^{i\alpha_p(x_1,\ldots,x_N)}\Psi(x_1,\ldots,x_N), \tag{48}$$

i.e. unlike the case of the requirement of condition (40), in the general case, the phase is a function of coordinates and the permutation; and Eq. (41) evidently does not hold.

Most other proofs of the symmetry postulate contain unjustified constraints. A critical survey of such proofs can be found in Refs. [67,71,72]. The proof of the symmetry postulate based on the indistinguishability principle was given by Girardeau [73] and by the author [71]. In Ref. [73], the indistinguishability principle was used in the form (47), in Ref. [71] it was used in the form (46). As was noted later by the author [61,62], these two formulations of the indistinguishability principle are correct only when the system is in a non-degenerate state. In a degenerate state, the system can be described with an equal probability by any one of the basic vectors of the degenerate state. As a result, we can no longer select a pure state (the one that is described by the wave function) and should regard the degenerate state as a mixed one, where each basis vector enters with the same probability. We must sum both sides of Eqs.(46) and (47) over all wave functions that belong to the degenerate state. As showed by von Neumann [74], the diagonal element of the density matrix for a degenerate state has the form

$$D(x_1,\ldots,x_N;x_1,\ldots,x_N) = \frac{1}{f_\lambda}\sum_{r=1}^{f_\lambda} \Psi_r^{[\lambda]}(x_1,\ldots,x_N)^* \, \Psi_r^{[\lambda]}(x_1,\ldots,x_N), \tag{49}$$

where the expression (49) is written for the case of the f_λ-dimensional representation $\Gamma^{[\lambda]}$ of the permutation group π_N. The possibility of expressing the density matrix through only one of the functions implies that the degeneracy with respect to permutations has been eliminated. However, the latter cannot be achieved without violating the identity of the particles.

It is not difficult to check that for every representation $\Gamma^{[\lambda]}$ of the permutation group π_N, the probability density, Eq. (49), is a group invariant:

$$PDP^{-1} = D \qquad \text{for all} \qquad P \in \pi_N . \tag{50}$$

From this follows that the probability density obeys the indistinguishability principle even in the case of multi-dimensional representations of the permutation group. Thus, the indistinguishability principle is insensitive to the symmetry of wave function and cannot be used as a criterion for selecting the correct symmetry.

It is worth-while to note that from the discussion above does not follow that the symmetry of wave function is not significant and one can perform quantum mechanical study using only the density matrix which, as we have shown, does not depend upon the symmetry of wave function. This symmetry controls the atomic and molecular states allowed by the Pauli principle. We demonstrated it above for the rotational states of the $^{16}O_2$ molecule. Another example is the allowed multiplets in atomic spectroscopy. For instance, in the $(np)^2$ electronic shell only ^1S, ^3P, and ^1D states are realized. This follows directly from the antisymmetry of the total electronic wave function. The symmetry of many-particle state and its multiplicity are completely dictated by the symmetry of wave function attributed to this state [12].

Let us stress also the importance in many physical problems of the geometrical phase of wave function studied by Berry [75]. The Berry phase is the geometrical phase, which appears in addition to the familiar dynamical phase in the wave function of a quantum system that has undergo a cyclic adiabatic change. Even though the initial and final states the system will be the same, the phases of initial and final wave functions will be different [75,76]. As was shown by Berry [75], the Aharonov-Bohm effect [77] can be explained as a special case of the geometrical phase factor, see also Refs. [78,79].

Thus, the widespread application of the density functional theory based on the Kohn-Sham equation does not mean that the concept of wave function in quantum mechanics lost its importance.

Although the Pauli exclusion principle cannot be rigorously derived from other quantum-mechanical postulates, there are some heuristic arguments indicating that the description of an identical particle system by the multi-dimensial representations of the permutation group leads to some contradictions with the concept of the particle identity and their independency. In next Section, we discuss these arguments in detail.

4. SOME CONTRADICTIONS WITH THE CONCEPT OF PARTICLE IDENTITY AND THEIR INDEPENDENCE IN THE DEGENERATE PERMUTATION STATES

4.1 Possible scenarios of the allowance of degenerate permutation states

In this section, we discuss the properties of a quantum mechanical system of identical particles that does not obey the symmetrization postulate and can be in states with all possible permutation symmetries. To the best of our knowledge, Steinmann [80] was the first who considered the identical particle system in a degenerate permutation state. He considered the so-called triangular representation of the permutation group π_3 characterized by the Young diagram $[\lambda] = [21]$[1] and came to the conclusion that particles in the representation $\Gamma^{[21]}$ are distinguishable because if we performed the reduction $\pi_3 \rightarrow \pi_2$, the two-particle state will be a mixture of symmetric and antisymmetric states. The Steinmann arguments were criticized by Hartle and Taylor [81] who showed that, in general, the indistinguishability is preserved in the case of multi-dimensial representations $\Gamma^{[\lambda]}$. They concluded that there are no theoretical reasons against an existence of degenerate permutation states in quantum mechanics although it is disagreeable. The same conclusion

[1] The classification of irreducible representations of permutation groups according to the Young diagrams can be found in Ref.[12].

was made earlier by Casher et al. [82] who named this situation rather unpalatable.

As we showed in previous Section, the indistinguishability principle is insensitive to the permutation symmetry and is satisfied by functions belonging also to the degenerate permutation states described by the multi-dimensial representations of the permutation group. In this sense, the critique of Steinmann arguments [80] by Hartle and Taylor [81] was correct. Nevertheless, the general conclusion [81,82] that there are no theoretical prohibitions on the existence of some unknown identical particle systems in the degenerate permutation states is quite not obvious and it is worthwhile to revise this conclusion. As we will see below, the assumption in Refs [81,82] that the triangular representation for three particles can originate from the state of two particles described by the linear combination of symmetric and antisymmetric functions leads to the distinguishable particles, see Eq. (66).

Let us consider a quantum mechanical system of identical elementary particles without the restrictions imposed by the symmetrization postulate and base our study on the Hartree–Fock one-particle approximation. In this case, the states of a system of identical particles with the number of particles not conserved can be presented as vectors in the Fock space **F** [83]. The latter is a direct sum of spaces $\mathbf{F}^{(N)}$ corresponding to a fixed number of particles N

$$\mathbf{F} \doteq \sum_{N=0}^{\infty} \mathbf{F}^{(N)} . \tag{51}$$

Each of the space $\mathbf{F}^{(N)}$ can be presented as a direct product of one-particle spaces $\mathbf{f}$:

$$\mathbf{F}^{(N)} = \underbrace{\mathbf{f} \otimes \mathbf{f} \otimes ... \otimes \mathbf{f}}_{N} . \tag{52}$$

The basic vectors of $\mathbf{F}^{(N)}$ are the product of one-particle vectors $|v_k(k)\rangle$ belonging to space **f**; k in the parenthesis denotes the set of particle spin and space coordinates,

$$|\xi^{(N)}\rangle = |v_1(1)\rangle |v_2(2)\rangle ... |v_N(N)\rangle . \tag{53}$$

For simplicity, let us consider the case where all one-particle vectors in Eq. (53) are different. There will be no qualitative changes in the results if some of the vectors $|v_k\rangle$ coincide. One can produce N! new many-particle vectors by applying to the vector (53) N! permutations of the particle coordinates. These new vectors also belong to $\mathbf{F}^{(N)}$ and form in it a certain invariant subspace which is reducible. The *N*! basic vectors of the latter, $P|\xi^{(N)}\rangle$, make up the regular representation of the permutation group π_N. As is known, the regular representation is decomposed into irreducible representations, each of which appears a number of times equal to its dimension. The space $\varepsilon^{(N)}$ falls into the direct sum

$$\varepsilon_\xi^{(N)} \doteq \sum_{\lambda_N} f_{\lambda_N} \varepsilon_\xi^{[\lambda_N]}. \tag{54}$$

where $\varepsilon_\xi^{[\lambda_N]}$ is an irreducible subspace of dimension f_λ drawn over thebasic vectors $|[\lambda_N]r\rangle$, and $[\lambda_N]$ is a Young diagram with N boxes. The basic vectors can be constructed of non-symmetrized basic vector $|\xi^{(N)}\rangle$ by using the Young operators $\omega_{rt}^{[\lambda_N]}$ [12],

$$|[\lambda_N]rt\rangle = \omega_{rt}^{[\lambda_N]}|\xi^{(N)}\rangle = \left(\frac{f_\lambda}{N!}\right)^{1/2} \sum_P \Gamma_{rt}^{[\lambda_N]}(P)P|\xi^{(N)}\rangle \tag{55}$$

where $\Gamma_{rt}^{[\lambda_N]}(P)$ are the matrix elements of representation $\Gamma^{[\lambda_N]}$ and index t distinguishes between the bases in accordance with the decomposition of $\varepsilon_\xi^{(N)}$ into f_λ invariant subspaces and describes the symmetry under permutations of the particle vector indices (these permutations are often called as place permutations [36,37] for distinguishing them from the particle permutations P).

Thus, a space with a fixed number of particles can always be divided into irreducible subspaces $\varepsilon_\xi^{[\lambda_N]}$, each of which is characterized by a certain permutation symmetry given by a Young diagram with N boxes. The symmetry postulate demands that the basis vectors of a system of N identical particles belong to one of the two subspaces characterized by irreducible one-dimensional representations, either [N] or $[1^N]$. All other subspaces are "empty". Let us examine the situation that arises when no symmetry constraints are imposed.

As is known, one of the consequences of the different permutation symmetry of state vectors for bosons and fermions is the dependence of the energy of system on the particle statistics. For the same law of dynamic interaction, the so-called exchange terms enter the expression for the energy of fermion and boson system with opposite signs. Let us obtain the expression for the energy of a system of particles belonging to an irreducible subspace $\varepsilon_{\xi}^{[\lambda_N]}$ with an arbitrary Young diagram $[\lambda_N]$.

The energy of the system in a degenerate state is

$$E = Tr(HD) \tag{56}$$

where D is the density operator defined, similarly to Eq. (49), as

$$D_t = \frac{1}{f_\lambda}\sum_{r=1}^{f_\lambda} \left|[\lambda]rt\right\rangle\left\langle[\lambda]rt\right|. \tag{57}$$

We assume that the Hamiltonian includes only one- and two-particle interaction operators:

$$H = \sum_i h_i + \sum_{i<j} g_{ij} \,. \tag{58}$$

The calculation of the trace over the functions with symmetry $[\lambda_N]$ yields

$$E_t^{[\lambda]} = \frac{1}{f_\lambda}\sum_{r=1}^{f_\lambda} \left\langle[\lambda]rt\right|H\left|[\lambda]rt\right\rangle. \tag{59}$$

The matrix element in Eq. (59) has been calculated in Ref. [84] in a general case of non-orthogonal one-particle vectors. In the case where all vectors in Eq. (53) are different and orthogonal one gets

$$E_t^{[\lambda]} = \sum_a \left\langle v_a \right|h\left| v_a\right\rangle + \sum_{a<b}[\left\langle v_a v_b\right|g\left| v_a v_b\right\rangle + \Gamma_{tt}^{[\lambda]}(P_{ab})\left\langle v_a v_b\right|g\left| v_b v_a\right\rangle] \tag{60}$$

where $\Gamma_{tt}^{[\lambda]}(P_{ab})$ is the diagonal matrix element of the transposition of vectors $|v_a\rangle$ and $|v_b\rangle$ in the right-hand part of Eq.(53).

Thus, the energy of system in a state with symmetry corresponding to a Young diagram [λ] depends not only on [λ], but also on a type of the Young tableaus t characterizing the symmetry of the basic functions according to the place permutations of the one-particle vectors, index t. Since the transitions between f_λ irreducible subspaces, characterized by index t, are allowed, there are f_λ split levels with symmetry [λ].

Only exchange terms in Eq.(60) depend upon the symmetry of states. For one-dimensional representations, $\Gamma_{tt}^{[\lambda]}(P_{ab})$ do not depend on the number of particles and P_{ab}: $\Gamma^{[N]}(P_{ab})=1$ and $\Gamma^{[1^N]}(P_{ab})=-1$ for all P_{ab} and N. For multi-dimensional representations, the matrix elements $\Gamma_{tt}^{[\lambda]}$ (P_{ab}) depend on [λ] and P_{ab}; in general, they are different for different pairs of identical particles[2]. The latter is difficult to conform with the identity of particles. But it is not single contradiction with the concept of particle identity in the degenerate permutation states

Taking into account that the transitions between states with different symmetry [λ_N] are strictly forbidden and each state of N particle system with different [λ_N] has a different analytical formula for its energy, we must conclude that each type of symmetry [λ_N] corresponds to a certain kind of particles with statistics determined by this permutation symmetry. On the other hand, the classification of state with respect to the Young diagrams [λ_N] is connected exclusively with identity of particles. Therefore, it must be some additional *inherent particle characteristics* which establishes for the N particle system to be in a state with a definite permutation symmetry, like integer and half-integer values of particle spin for bosons and fermions, and this inherent characteristic has to be different for different [λ_N]. So, the particles belonging to the different types of permutation symmetry [λ_N] are not identical.

Let us trace down the genealogy of irreducible subspaces $\varepsilon_{\xi}^{[\lambda_N]}$. In Fig.1, the genealogy for all irreducible subspaces with N = 2 to 4 is

---

[2] The matrices of transpositions for all irreducible representations of groups π_2 - π_6 are presented in book [12], Appendix 5.

presented. We call the hypothetical particles characterized by the multi-dimensial representations of the permutation group as *intermedions* implying that they obey some intermediate between fermion and boson statistics. For bosons and fermions there are two non-intersecting chains of irreducible representations: [N]→[N+1] and $[1^N]\rightarrow[1^{N+1}]$, respectively; and the energy expression for each type of particles has the same analytical form that does not depend on the number of particles in a system. The situation changes drastically, if we put into consideration the milti-dimensial representations. The number of different statistics depends upon the number of particles in a system and rapidly increases with N. For the multi-dimensial representations we cannot select any non-intersecting chains, as in the fermion and boson cases. According to Fig.1, the intermedion particles with a definite $[\lambda_N]$ in the Nth generation can originate from particles of different kinds $[\lambda_{N-1}]$ in the (N-1)th generation, even from fermions or bosons (in the special case $[\lambda_3]$ =[21], it originates from both $[1^2]$ and [2]). The physical picture in which adding one

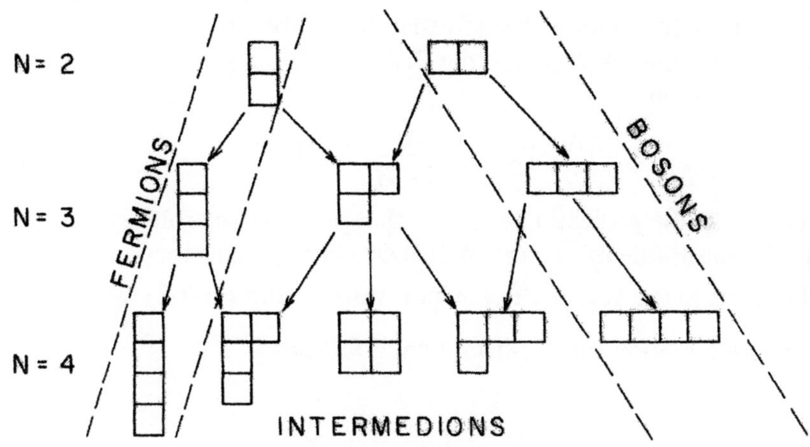

Fig. 1 Symmetry of the irreducible representations of the permutation group for N = 2 to 4 and their genealogy.

particle changes properties of all particles cannot correspond to a system of *independent* identical particles (although, it cannot be excluded for some quasiparticle systems where we have not an independency of quasiparticles, see Section 2.2).

For an ideal gas, it is evident that adding a particle identical to a system of N identical particles cannot change the properties of a new (N+1)-particle system. Let us show that the interaction of identical particles does not change the permutation symmetry of non-interacting particle system [71].

The wave vector $|\Psi\rangle$ of a system of interacting particles characterized by the Hamiltonian

$$H = H_0 + \hat{V} \tag{63}$$

is defined in the same Fock space as the wave vector $|\Psi_0\rangle$ of non-interacting system and it can be generated from the latter by using some unitary transformation

$$|\Psi\rangle = \hat{U}|\Psi_0\rangle. \tag{62}$$

The form of this unitary transformation can be obtained if we use the well-known relationship that the Brillouin-Wigner perturbation theory [85] is based on:

$$|\Psi\rangle = |\Psi_0\rangle + \frac{Q}{E - H_0}\hat{V}\,|\Psi\rangle \tag{63}$$

where Q is the projection operator that projects an arbitrary vector onto the multitude of vectors of Hilbert (Fock) space that are orthogonal to the vector $|\Psi_0\rangle$. Upon substituting Eq.(62) into Eq.(63), we get an equation for $\hat{U}$, from which follows

$$\hat{U}^{-1} = 1 - \frac{Q}{E - H_0}\hat{V}. \tag{64}$$

Since the interaction operator $\hat{V}$ of identical particles is invariant with respect to permutations of identical particles, the operator $\hat{U}$ is also invariant. Hence, according to Eq.(62), the states $|\Psi\rangle$ and $|\Psi_0\rangle$ have the same permutation symmetry.

Thus, the scenario, in which all symmetry types $[\lambda_N]$ are allowed and each of them corresponds to a definite particles statistics, contradicts to the concept of particle identity and their independency from each other.

Let us consider another virtual possibility and begin from the ordinary fermions. As is well known, in quantum mechanics of identical particles in the absence of spin interactions, the total spin S is a good quantum number and labels the energy levels of the system. The symmetry of the coordinate wave function depends on the value of S which cause the dependence on S of the total energy of the system regardless of the dynamical interaction law. The symmetry of the coordinate wave function corresponds to some coordinate Young diagram $[\lambda_N]_{coord}$ (uniquely connected with the spin Young diagram $[\lambda_N]_{spin}$) that can belong to a multi-dimensial representation of the group π_N [12]. Introduction in the case of fermion system an additional quantum numbers, e.g. the isotopic spin, leads to new possible pairs of $[\lambda_N]_{coord}$ and $[\lambda_N]_{spin}$ providing the symmetry of the total wave function $[1^N]$. If particles are characterized by the ordinary and isotopic spin, the Young schemes $[\lambda]_{spin}$ and $[\lambda]_{coord}$ can contain up to four rows and four columns respectively (where $[\lambda]_{spin}$ corresponds to both the ordinary and isotopic spin coordinates).

Hence, the system of fermions can be described by the multi-dimensial permutation representations of the permutation group; however, the latter correspond not to the total wave function but to its factorized parts. The total wave function is completely antisymmetric in accordance with the Pauli exclusion principle.

The analogous situation takes place in the parastatistics case. The establishment in Refs.[36-38] that parafermi (parabose) commutation relations for parastatistics of the order p correspond to the permutation symmetry characterized by Young diagrams with the first row (column) no larger than p does not related to the total wave function, because the additional degrees of freedom characterizing paraparticles are not included in the wave function. Multiplying this wave function belonging to a multi-dimensial permutation representation by the wave function describing the additional degrees of freedom, we can obtain the proper permutation symmetry for the total wave function. In the case of parafermi statistics it will be $[1^N]$, as for a fermion system.

So, we must consider the possibility of realization of the multi-dimensial representations $[\lambda_N]$ for the *total* wave function including *all* degrees of freedom of the particles under consideration. As we showed in the scenario considered first, the assumption that for the total wave function all possible $[\lambda_N]$ can be allowed leads to contradictions with the concept of particle identity and their independency. Therefore, we have to consider the possibility that for some type of intermedions for each fixed N only one type of $[\lambda_N]$ is permitted (as it is in the case of fermions, $[1^N]$, and bosons, [N]). But in this scenario, we meet with a serious problem which is more evident in particular case of a two-intermedion system.

Let us stress that the permutation group can be always applied to the Hilbert space of independent particles (the latter is not true for quasiparticles, see next subsection 4.2). For N=2, the permutation group has only two irreducible representations: symmetric (boson) and antisymmetric (fermion) one. We have no other option but to assume that the symmetry of two-intermedion system coincides with the two-fermion or two-boson system. The point is that on the contrary to the statement in Refs.[81,82], in general case, the wave function of two identical particles cannot be described by some superposition

$$\Psi_n(x_1,x_2) = c_1\Psi^{[2]}(x_1,x_2) + c_2\Psi^{[1^2]}(x_1,x_2)\,. \tag{65}$$

This superposition describes distinguishable particles. In fact,

$$P_{12}\left|\Psi_n(x_1,x_2)\right|^2 = \left|c_1\Psi^{[2]}(x_1,x_2) - c_2\Psi^{[1^2]}(x_1,x_2)\right|^2 \neq \left|\Psi_n(x_1,x_2)\right|^2. \tag{66}$$

Certainly, at large distances in the absence of overlap of one-particle wave functions, the electrons are distinguishable, and the two-electron wave function can be constructed as a simple unsymmetrized product. In this limit, the physical consequences for antisymmetrized and unsymmetrized states will be the same. The paradox of the requirement of definite symmetry for two electrons in different galaxies discussed by Mirman [86] does not exist in reality. However, in general case, electrons can be located in one region of space with overlapping wave functions. The total wave function must be antisymmetrized. It belongs to the irreducible representation with symmetry $[1^N]$ and corresponds to the indistinguishable electrons. The same arguments are valid for N > 2, a mixture of several $[\lambda_N]$ corresponds to distinguishable particles.

The situations labeled in Refs. [81,82] as "disagreeable" or "unpalatable" is physically forbidden if we consider a system of identical independent particles (but not quasiparticles, as anyons, see below). So, the two independent identical particles must be considered only in the pure fermion or boson state. The addition of the third particle identical to the two others changes the fermion (or boson) statistics on the intermedion statistics with $[\lambda_3] = [21]$ and this takes place even in an ideal gas of intermedions. Again, we obtained a contradiction with the concept of particle identity and their independence.

4.2 Anyons and fractional statistics

In all discussions presented above, we did not consider the dimension of space in which the particles are moving. As was shown by Leinaas and Myrheim [87], see also presentation in Ref.[88], from the topological properties of one- and two-dimensional spaces follows that the behavior of an arbitrary two-particle system under their exchange can possess a continuum of possible intermediate cases between fermion and boson cases. This follows from the topological differences between 3D space and space with less dimension: in 3D space, two consequent exchanges are equivalent to zero exchange because each path may be continuously deformed into the point, whereas in space of less dimensions two exchanges are not topologically equivalent to zero exchange , and it is possible to assign any value to the phase due to exchange.

These ideas were developed by Wilczek [89] who considered charge particles with charge q orbiting around a tube with magnetic flux Φ and demonstrated that the total change of the orbital angular momentum is proportional to $q\Phi$ and, hence, it can accept arbitrary fractional values. This result is valid for a two-dimensional orbital motion and, in general, should not be surprising if one consider an essential difference between the rotation symmetries in 2D and in 3D spaces. In a more common (in quantum textbooks) 3D case, the non-abelian nature of the rotation group implies that its representations are labeled by a discrete index, the integer and half-integer angular momentum. The latter is directly follows from the commutation relations for the operators of angular momentum [65,69]. In 2D space, the rotation group is abelian and has a continuum of representations, corresponding to a continuum of allowed angular momenta.

The exchange of the two identical flux-tube-particle composits leads to a phase factor $e^{i\theta}$, which is not necessarily equal to ±1. Hence, the flux-tube-particle composites, considered by Wilczek, have unusual statistics, interpolating between bosons and fermions. In so far as the exchange of two such identical objects can give *any* phase, Wilczek [89] named them *anyons*. This name has since generalized to any 2D objects with so called *fractional statistics*. It was accepted that a system of identical objects obeys the fractional statistics if the wave function changes by a complex phase factor (not by ±1) when two objects are interchanged. As was shown by Laughlin [90], see also Refs. [91,92], the observed anomalous quantum Hall effect with fractional charge [93] origins from the fractionally charged excitations of the 2D electronic gas. These fractionally charged excitations can be considered as quasiparticles obeying the fractional statistics. Let us remark that it is a rather conventional term, since the result of interchange of two anyons was found only for a two-anyon system without taking into account the influence of other anyons; while the behavior of a multi-anyon system is drastically differs from the fermion and boson systems, for which the result of two-particle exchange .does not depend upon the number of particles in a system.

In subsequent studies [94], it was revealed that the anyons in principle cannot be considered as independent objects. The movement of each anyon depends upon location of all others. The phase due to exchange of two identical anyons is path-dependent. As a result, the permutation group cannot be applied to the Hilbert space of anyons. One has to distinguish topologically non-equivalent trajectories leading to the same permutations of anyons. The Hilbert space of identical anyons induced the representations of so-called *braid group* [95,96].

Once the permutation group is replaced by the braid group, the simple construction of many-particle wave functions passing from the solution of one-particle problems to the solution of many-particle problems does not work any more. The statistical distribution function of anyon system cannot be constructed. The anyons are not independent objects, and even in the absence of dynamical interactions, an ideal gas of anyons does not exist. Anyons can be considered as fermions with an additional attractive interaction [95] or, in some cases, as bosons with a weak repulsion of statistics origin.

This resembles the kinematic interaction in the quasiparticle systems in a periodical lattice, see Section 2.2 and Refs.[30,55].

And what can be said in this connection about the Pauli exclusive principle? We cannot apply it to anyons because the permutation group cannot be applied to the Hilbert space of anyon systems. The discussion of degenerate representations of the permutation group loses its sense. This situation is specific for quasiparticle systems.

Returning to a system of identical particles, we resume that all contradictions of possible scenarios discussed above are resolved if only the one-dimensional irreducible representations of the permutation group are permitted, as follows from the Pauli exclusive principle. Thus, the existence in Nature of only symmetric and antisymmetric types of permutation symmetry is not occasional. It is intimately connected with the identity of particles and their independence.

5. CONCLUDING REMARKS

In spite of more than 75 years studies of the Pauli exclusion principle and spin-statistics connection, we still have not a rigorous theoretical ground for it. In the case of composite particles, for which the Pauli exclusion principle is also satisfied, the spin-statistics connection for some systems is even violated. The latter takes place for two-fermion particles (e.g. the Cooper pairs) which are described by the symmetrical wave functions but have the fermion occupation numbers of one-particle states.

As we showed in Section 3, the indistinguishability principle is insensitive to the permutation symmetry of wave function and is satisfied also by wave functions belonging to the multi-dimensial representations of the permutation group characterized by the Young diagrams $[\lambda_N]$ of general type. So, the indistinguishability principle cannot be used for the verification of the Pauli exclusion principle. Experimental data and checking known to date confirm the Pauli exclusion principle; all elementary particles belong only to one of two statistics: fermion or boson statistics (for quasiparticles, it is not true).

Although, the Pauli exclusion principle has no rigorous theoretical proof and follows from experiment, the existence in Nature only the non-degenerate permutation states is not occasional. For different virtual scenarios discussed in Section 4, the permission for an identical particle system to be in a degenerate permutation state leads to contradictions with the concept of particle identity and their independence. It is important to stress that the introduction in the parastatistics theory the permutation symmetries, corresponding to the multi-dimensial representations, relates to the wave functions that do not include the internal degrees of freedom characterizing paraparticles. It does not contradict that the total wave function possesses the proper permutation symmetry in accordance with the Pauli exclusion principle.

On the other hand, for quasiparticle systems the Pauli principle in most cases cannot be applied. The system of two anyons possesses the symmetry of states intermediate between fermion and boson symmetries. While nothing can be said about the permutation symmetry of the multi-anyon wave function because a result of the interchange of two anyons depends upon location and trajectories of all other anyons. The complex situation takes place in the case of quasiparticles in a periodical lattice. They can be considered as independent objects only in the boson limit.

If we reduce the Pauli exclusion principle only to the prohibition of all types of the permutation symmetry, except the completely symmetric and antisymmetric one, it can be considered as a consequence of general physical assumptions underlaying quantum theory. However, the problem of spin-statistics connection is still open.

The final remark. I would like to point out at an interesting similarity between the Pauli exclusion principle and the Jahn-Teller effect [97]. According to the latter, see Refs. [98,99], molecules (non-linear) and crystals are not stable if the ground electronic state is degenerate. Due to the vibronic interactions, the symmetry of the ground state conformation lowers down to a symmetry for which the system has the non-degenerate ground state. It belongs to some one-dimensional representation of a point (space) group of symmetry. The same follows from the Pauli exclusion principle for the permutation group: only the one-dimensional representations of the permutation group are realized. This similarity in these two quite different, at first glance, physical phenomena is worth-while for the special study.

REFERENCES

1. Pauli W. 1925 *Zs. Phys.* **31,** 373.
2. Pauli W. 1925 *Zs. Phys.* **31,** 765.
3. Uhlenbeck G.E. and Goudsmit S. 1925 *Naturwissenschaften* **13**, 953.
4. van der Waerden B.L. in 1960 *Theoretical Physics in the Twentieth Century* Eds. M. Fierz and V.F. Weisskopf (Cambridge, MA).
5. Pauli W. *Nobel Lecture* in [4].
6. Tomas W. 1926 *Nature* **117**, 514.
7. Heisenberg W. 1926 *Zs. Phys.* **38,** 411; *Ibid.* **39,** 499.
8. Dirac P.A.M. 1926 *Proc. Roy .Soc.*London A **112,** 661.
9. Kaplan I.G. 1967 *Sov. Phys. JETP* **24**, 114.
10. Kaplan I.G. and Rodimova O.B. 1974 *Sov. Phys. JETP* **39**, 1764.
11. Kaplan I.G. and Rodimova O.B. 1976 *Int. J. Quant. Chem.* **10**, 690.
12. Kaplan I.G. 1975 *Symmetry of Many-Electron System* (Academic Press: New York).
13. Dicke H.G. and Babcock H.D. 1927 *Proc. Natl. Acad. Sci. U.S.A.* **13,** 670.
14. Angelis M., Gagliardi G., Gianfrani L. and Tino G.M. 1996 *Phys Rev. Lett.* **76,** 2840.
15. Hilborn R.C. and Yuca C.L. 1996 *Phys. Rev. Lett.* **76,** 2844.
16. Ignatiev A.Yu. and Kuzmin V.A. 1987 *Sov. Phys. Nucl. Phys.* **46,** 444
17. Greenberg D.W. and Mohapatra R.N. 1987 *Phys. Rev. Lett.* **59,** 250.
18. Okun L.B. 1989 *Sov. Phys. Uspekhi* **32,** 543.
19. Eds. Hilborn R.C. and Tino G.M. 2000 *Spin-Statistics Connection and Commutation Relations* (AIP Conf. Proc. No.545, AIP, Melville, New York).
20. Greenberg O.W. in [19] p.113.
21. Gillaspy J.D. in [19] p.241.
22. Logan B.A. and Ljubicic A. 1979 *Phys. Rev. C.* **20,** 1957.
23. Avignone F.T., III, Brodzinski R.L., Hensley W.K., Miles H.S., and Reeves J.N.1986 *Phys. Rev.* D **34,** 97.
24. Javorsek D., II, et al. 2000 Phys. Rev. Lett. **85,** 2701.
25. Pauli W. 1940 P*hys. Rev*. **58,** 716.

26. Green H.S. 1953 *Phys. Rev.* **90,** 270.
27. Volkov D.V. 1959 *Sov. Phys. JETP* **9,** 1107; 1960 *Ibid.* **11,** 375.
28. Greenberg O.W. and Messiah A.M. 1965 *Phys. Rev.* **138,** B1155.
29. Isihara A. 1971 *Statistical Physics* (Academic Press: New York).
30. Kaplan I.G. 1976 *Theor. Math. Phys.* **27,** 466.
31. Girardeau M.D. 1963 *J.Math. Phys.* **4**, 1096.
32. Girardeau M.D. 1971 *Phys. Rev. Lett.* **27,** 1416.
33. Gilbert J.D. 1977 *J. Math. Phys.* **18,** 791.
34. Kaplan I.G. 1980 *Problems in Theor. and Nucl. Phys.* (Saratov State University Press, Saratov, in Russian) No.7 p.66.
35. Schrieffer J.R. 1988 *Theory of Superconductivity* (Addison-Wesley: Redwood City,CA).
36. Landshoff P.V. and Stapp H.P. 1967 *Ann. Phys. (N.Y.)* **43,** 72.
37. Ohnuki Y. and Kamefuchi S. 1969 *Ann. Phys.* **51,** 337.
38. Stolt R.H. and Taylor J.R. 1970 *Phys. Rev.* D **8,** 2226; 1970 Nucl. Phys. B **9,** 1.
39. Zel'dovich Ya.B. 1960 *Sov. Phys.-JETP* **10**, 403.
40. Baz' A.I. 1971 *Sov. Phys.-JETP Lett.* **14**, 423.
41. Zel'dovich Ya.B. 1974 *Sov. Phys.-JETP Lett.* **20**, 93.
42. Duck I. and Sudarshan E.C.G. 1997 *Pauli and the Spin-Statistics Theorem* (World Scientific: Singapore).).
43. Wightman A.S. 1999 *Amer. J. Phys.* **67,** 742.
44. Berry M.and Robbins J. in [19] p.3.
45. Davydov A.S. 1962 *Theory of Molecular Excitons* (McGraw-Hill: New York).
46. Kamefuchi S. and Takahachi Y. 1962 *Nucl. Phys.* **36**, 177.
47. Rayan C. and Sudarshan E.C.G. 1963 *Nucl. Phys.* **47**, 207.
48. Ohnuki Y. and Kamefuchi S. 1982 *Quantum Field Theory and Parastatistics* (Springer: Berlin).
49. Chernikov N.A.1962 *Acta Physica Polonica* **21,** 52.
50. Greenberg O.W. 1964 *Phys. Rev. Lett.* **13,** 598.
51. Govorkov A.B. 1983 *Sov. J. Part. Nucl.* **14,** 520.
52. Avdyugin A.N., Zavorotnev Yu.D., and Ovander L.N. 1983 *Sov. Phys. Solid State* **25,**1437.
53. Pushkarov D.I. 1986 *Phys. Status Solidi* b **133,** 525.

54. Nguen A. and Hoang N.C. 1990 *J. Phys.: Condens. Matter* **2,** 4127.
55. Kaplan I.G. and Navarro O. 1999 *J. Phys.: Condens. Matter* **11,** 6187.
56. Kaplan I.G. and Navarro O. 2000 *Physica* C : *Superconductivity* **341-348,** 217.
57. Kaplan I.G. in [19] p.72.
58. Dyson F. 1956 *Phys. Rev.* **102**, 1217.
59. Zhang H. and Sato H. 1993 *Phys. Rev. Lett.* **70**, 1697.
60. Gentile G. 1940 *Nuovo Cimento* **17**, 493.
61. Kaplan I.G. in 1980 *Group Theoretical Methods in Physics*, Ed. V.I. Man'ko (Nauka: Moscow) v. 1, p.175.
62. Kaplan I.G. 1992 *J. Mol. Struct.* **272,** 187.
63. Kaplan I.G. 2002 Int. J. Quant. Chem. (in Press).
64. Dirac P.A.M. 1958 *The Principles of Quantum Mechanics* (Clarendon Press: Oxford).
65. Schiff L.I. 1955 *Quantum Mechanics* (Mc Graw-Hill: New York).
66. Messiah A.M. 1962 *Quantum Mechanics* (North-Holland: Amsterdam).
67. Messiah A.M. and Greenberg O.W.1964 *Phys. Rev.* **136,** B248.
68. Corson E.M. 1951 *Perturbation Methods in Quantum Mechanics of Electron Systems* (University Press: Glasgow).
69. Landau L.D. and Lifschitz E.M. 1965 *Quantum Mechanics* (Addison-Wesley: Massachusetts).
70. Blokhintzev D.I. 1964 *Principles of Quantum Mechanics* (Allyn and Bacon: Boston, MA).
71. Kaplan I.G. 1976 *Sov. Phys. Uspekhi* **18,** 988.
72. Girardeau M.D. 1965 *Phys. Rev.* **139,** B560.
73. Girardeau M.D. 1969 *J. Math. Phys.*, **10,** 1302.
74. von Neumann J.V. 1932 *Mathematische Grundlagen der Quantenmechanik* (Springer: Berlin).
75. Berry M.V. 1984 *Proc. Roy. Soc. London* A **392,** 45.
76. Berry M.V. 1988 *Scientific American,* December, 26.
77. Aharonov Y. and Bohm D. 1959 Phys. Rev. **115**, 485.
78. Mead C.A. and Truhlar D.G. 1979 J. Chem. Phys. **70**, 2284.
79. Mead C.A. 1980 *Chem. Phys.* **49**, 23; *Ibid.* **49**, 33.
80. Steinmann O. 1966 *Nuovo Cimento* **44,** 755.
81. Hartle J.B. and Taylor J.R. 1969 *Phys. Rev.* **178,** 2043.

82. Casher A, Frieder G, Gluck M., and Peres A 1965 *Nucl. Phys.* **66,** 632.
83. Schweber S.S. 1961 *An Introduction to Relativistic Quantum Field Theory* (Row Peterson: New York).
84. Kaplan I.G. and Rodimova O.B. 1973 *Int. J. Quant. Chem.* **7,** 1203.
85. March N.H., Young W.H., and Sampanthar S. 1967 *The Many-Body Problem in Quantum Mechanics* (Cambridge University Press: Cambridge).
86. Mirman R. 1973 *Nuovo Cimento* **18B,** 110.
87. Leinaas J.M. and Myrheim J. 1977 *Nuovo Cimento* **37B**, 1.
88. Canright G.S. and Girvin S.M. 1990 *Science* **247,** 1197.
89. Wilczek F. 1982 *Phys. Rev. Lett.***48**, 1144; *Ibid.* **49**, 957.
90. Laughlin R.B. 1983 *Phys. Rev. Lett.* **50**, 1395; 1988 *Ibid.* **60**, 2677.
91. Haldane F.D.M. 1983 *Phys. Rev. Lett.* **51**, 605.
92. Halperin B.I. 1984 *Phys. Rev. Lett.* **52**, 1583.
93. Tsui D.C., Störmer H.L., and Gossard A.C. 1982 *Phys. Rev. Lett.* **48**, 1559.
94. Chen Yi-Hong, Wilczek F., Witten E., and Halperin B.I. 1989 *Int. J. Mod. Phys.* B **3,** 1001.
95. Moram S. 1983 *The Mathematical Theory of Knots and Braids* (North-Holland:Amsterdam).
96. Jones V.F.R. 1985 *Bull. Amer. Math. Soc.* **12**, 103; 1987 *Ann. Math.* **126**, 335.
97. Jahn H.A. and Teller E. 1937 *Proc. Roy. Soc*. (London) A **161,** 220.
98. Englman R. 1972 *The Jahn-Teller Effect in Molecules and Crystals* (Wiley: London).
99. Bersuker I.B. and Pollinger V.Z. 1989 *Vibronic Interactions in Molecules and Crystals* (Springer: Berlin).

THE LENGTH AND BREADTH OF LÖWDIN-ORTHOGONALIZATIONS

Vipin Srivastava*
Cavendish Laboratory, University of Cambridge, Cambridge CB3 0HE, U.K.
and
School of Physics, University of Hyderabad, Hyderabad 500 046, India

Abstract: A geometrical view of Löwdin's orthogonalization procedures is taken to understand their distinctive features. Vital new information are found to be hidden in these procedures, which open them up for novel applications e.g., to understand cognitive phenomena like 'discrimination', and organizing large data sets such as gene expression profiles, etc. in biology. In some of these applications the restriction of linear-independence of the given set of vectors that are to be orthogonalized comes in the way. Ways of getting round this restriction are also discussed. In this process it is found that Löwdin's Canonical orthogonalization is exactly equivalent to the well known Principal component analysis.

1. Introduction

The problem of conversion of a given set of linearly-independent vectors into a set of mutually orthogonal vectors has been studied for a long time. It finds wide applications in mathematics, physics and chemistry. The methods involve a straightforward and simple mathematics and differ in special conditions they satisfy. We present a simple analysis of the orthogonalization methods in terms of the projections of the given vectors on the orthogonalized bases vectors and find that the methods possess some powerful properties hitherto unknown. The discovered properties are not only curious but also have algorithmic value in that they enable geometrical constructions of the orthogonal basis sets and the eigenvalues of the metric matrix of the given vector set. These properties also underline the scenarios where the different methods should be

E.J. Brändas and E.S. Kryachko (eds.),
Fundamental World of Quatum Chemistry, Vol. I, 221–235.

applicable. We expect that the present insight should also provide clues to understand certain aspects of cognitive memory, e.g. along the lines of [1].

2. Orthogonalization methods

The Gram-Schmidt orthogonalization procedure, the oldest and commonly used in mathematics (also applied recently to spin-glass like neural networks [1]), gives an orthonormal set depending on the *sequence* in which the given vectors are picked. The work on *sequence-independent* orthogonalization, in a particular context of physical problems, was started by Landshoff [2] and Wannier [3], and later developed comprehensively by Löwdin [4]. Two notable methods of this class are Löwdin's *Symmetric* [5] and *Canonical* [6] orthogonalizations. The Symmetric orthogonalization contains the results of Landshoff and Wannier [4]. The Canonical orthogonalization is generally known in the physics literature (see [7] for references) after Schweinler and Wigner [8] (hereafter referred to as S-W) who in fact introduced a versatile parameter "m" that highlighted an *extremal* property of Löwdin's Canonical orthogonalization. Earlier Löwdin [9] had talked about an *optimum-principle* obeyed by the Canonical orthogonalization which we find in a sense resembles the S-W extremal property. Recently Chaturvedi et al. [7] have given a new orthonormal basis that has another extremal property in terms of the S-W parameter m. It turns out that this basis is the same as Löwdin's Symmetric orthogonalization if the vectors in the given set are normalized.

2.1 General framework:

Let $\mathbf{V}$ represent a set of linearly independent vectors, $\vec{v}_1, \vec{v}_2, \vec{v}_3, \cdots, \vec{v}_N$, in a N- dimensional space which can in general be a complex vector space. We can define a general non-singular linear transformation $\mathbf{A}$ for the basis $\mathbf{V}$ to go to a new basis $\mathbf{Z}$:

$$\mathbf{Z} = \mathbf{VA}\,. \tag{1}$$

The set $\mathbf{Z}(\equiv \{\vec{z}_\kappa\})$ will be orthonormal if

$$<\mathbf{Z}|\mathbf{Z}> = <\mathbf{VA}|\mathbf{VA}> = \mathbf{A}^\dagger <\mathbf{V}|\mathbf{V}> \mathbf{A} = \mathbf{A}^\dagger\mathbf{MA} = \mathbf{I}\,, \tag{2}$$

where $\mathbf{M}$ is a Hermitian metric matrix of the given basis $\mathbf{V}$. The substitution

$$\mathbf{A} = \mathbf{M}^{-1/2}\mathbf{B}\,, \tag{3}$$

where $\mathbf{B}$ is an arbitrary unitary matrix, leads to the general solution of the orthogonalization problem. The specific choice $\mathbf{B} = \mathbf{I}$ gives the *Symmetric Orthogonalization*, $\mathbf{Z} \equiv \mathbf{\Phi} = \mathbf{V}\mathbf{M}^{-1/2}$, while $\mathbf{B} = \mathbf{U}$, where $\mathbf{U}$ diagonalizes $\mathbf{M}$,

$$\mathbf{U}^{\dagger}\mathbf{M}\mathbf{U} = \mathbf{d}\,, \tag{4}$$

gives the *Canonical Orthogonalization*, $\mathbf{Z} \equiv \mathbf{\Lambda} = \mathbf{V}\mathbf{U}\mathbf{d}^{-1/2}$. The orthonormal basis due to Chaturvedi et al., denoted here by $\mathbf{\Gamma}$, is related to $\mathbf{\Lambda}$ [7], and coincides with $\mathbf{\Phi}$ in a particular circumstance indicated below.

2.2 Projections of V on Z:

Taking our cue from Schweinler and Wigner [8] we consider the matrix

$$\begin{pmatrix} |(\vec{v}_1, \vec{z}_1)|^2 & |(\vec{v}_1, \vec{z}_2)|^2 & \dots & |(\vec{v}_1, \vec{z}_N)|^2 \\ |(\vec{v}_2, \vec{z}_1)|^2 & |(\vec{v}_2, \vec{z}_2)|^2 & \dots & |(\vec{v}_2, \vec{z}_N)|^2 \\ \vdots & \vdots & & \vdots \\ |(\vec{v}_N, \vec{z}_1)|^2 & |(\vec{v}_N, \vec{z}_2)|^2 & \dots & |(\vec{v}_N, \vec{z}_N)|^2 \end{pmatrix}$$

of projection-squares of the given vectors $\{\vec{v}_k\}$ on a basis set $\{\vec{z}_\kappa\}$. The elements in a row corresponding to a particular $\vec{v}_k$ add up to $|\vec{v}_k|^2$:

$$\sum_{\kappa} |(\vec{v}_k, \vec{z}_\kappa)|^2 = |\vec{v}_k|^2\,, \quad k = 1, ..., N\,, \tag{5}$$

and the elements in a column for a particular $\vec{z}_\kappa$ add up to a real positive number c_κ:

$$\sum_{k} |(\vec{v}_k, \vec{z}_\kappa)|^2 = (\mathbf{AMMA}^{\dagger})_{\kappa\kappa} = (\mathbf{BMB}^{\dagger})_{\kappa\kappa} = c_\kappa\,, \quad \kappa = 1, ..., N\,. \tag{6}$$

Note that we have the identity

$$\sum_{\kappa=1}^{N} c_\kappa = \sum_{k=1}^{N} |\vec{v}_k|^2\,, a\ constant\ for\ a\ given\ set\ \mathbf{V}\,. \tag{7a}$$

Further, we have

$$\sum_{\kappa} {c_\kappa}^2 = m, \quad the\ S - W\ parameter. \tag{7b}$$

A basis set $\mathbf{Z}$ will satisfy the set of simultaneous equations (6) with the positive real numbers c_κ's obeying the identity (7a) [10]. Specific bases will satisfy additional conditions on the values of c_κ's either through (7b) or otherwise. For instance, if $c_\kappa = |\vec{v}_k|^2$ with $\kappa = k$ (i.e. $\mathbf{B}$=$\mathbf{I}$ in (6)), we will get the *Symmetric* basis $\mathbf{Z} = \mathbf{\Phi}$; $m = m_{max}$, which will arise for the maximally lop-sided distribution of the c_κ's (satisfying (7a)), will give the *Canonical* basis $\mathbf{Z} = \mathbf{\Lambda}$; and $m = m_{min}$, which will correspond to an average distribution, $c_1 = c_2 = ... = c_N = (c_1 + c_2 + ... + c_N)/N$, will give the basis $\mathbf{Z} = \mathbf{\Gamma}$ of Chaturvedi et al. [7]. For normalized $\vec{v}_k$'s the basis $\mathbf{\Gamma}$ and the Symmetric basis $\mathbf{\Phi}$ become the same.

2.3 New information:

Following useful information is embedded in the above identification: In the Symmetric case, where $\mathbf{Z} = \mathbf{\Phi}$, since the sum of squared-projections of all the $\vec{v_k}$'s on a $\vec{\phi_\kappa}$, say $\vec{\phi_l}$, is equal to the sum of squared-projections of all the $\vec{\phi_\kappa}$'s on the $\vec{v_k}$ with $k = l$, the symmetry properties of $\mathbf{V}$, if any, are preserved in $\mathbf{\Phi}$. This feature also ensures that $\mathbf{\Phi}$ resembles the original set $\mathbf{V}$ in that Löwdin's resemblance measure [4], $\langle \mathbf{Z} - \mathbf{V} | \mathbf{Z} - \mathbf{V} \rangle$, has its smallest value when $\mathbf{B}$=$\mathbf{I}$ or $\mathbf{Z} = \mathbf{\Phi}$. If seen slightly differently, the above symmetry interestingly implies that the squared-projections of all the $\vec{v_k}$'s on a $\vec{\phi_l}$ add up to the squared length of $\vec{v_l}$, the feature that can be used to geometrically generate the Symmetric basis set.

The last property above turns into a stricter condition in the case of $m = m_{min}$, where $\mathbf{Z} = \mathbf{\Gamma}$: the basis vectors $\vec{\gamma}_\kappa$ are arranged such that the sum of squared-projections of all the $\vec{v}_k$'s on each $\vec{\gamma}_\kappa$ is the *same* – equal to the average of $|\vec{v}_k|^2$ – irrespective of how the $\vec{v}_k$'s are arranged. In effect, the set $\mathbf{\Gamma}$ is arranged so as to cancel the effects of inhomogeneity in the distribution of $\vec{v}_k$'s.

On the other hand in the $m = m_{max}$ case, with $\mathbf{Z} = \mathbf{\Lambda}$, the basis vectors $\vec{\lambda}_\kappa$'s must be oriented such that they sample those directions in which bunches of $\vec{v}_k$'s tend to be oriented. In order to attain $m = m_{max}$ the Canonical basis set is arranged in such an optimal fashion that the sum of squared-projections of all the $\vec{v}_k$'s on one of the $\vec{\lambda}_\kappa$'s, say $\vec{\lambda}_1$ – *i.e.* c_1 – is the largest. After the $\vec{\lambda}_1$, is fixed, the rest of the set, orthogonal to $\vec{\lambda}_1$, is oriented such that another $\vec{\lambda}_\kappa$, say $\vec{\lambda}_2$, is able to maximise the *total* squared-projection of all the $\vec{v}_k$'s on it. This will be c_2 and will be smaller than c_1. All the $\vec{\lambda}_\kappa$'s are arranged according to this optimum-principle that ensures for the given set $\{\vec{v}_k\}$ the most lop-sided distribution of c_κ's with $c_1 > c_2 > ... > c_N$. This particular set of c_κ's in fact comprises the eigenvalues of $\mathbf{M}$ in the basis $\mathbf{\Lambda}$ (take $\mathbf{B}$=$\mathbf{U}$ in eqn.(6)) which are generically non-degenerate. The arrangement of $\vec{\lambda}_\kappa$'s that yields the S-W condition of $m = m_{max}$ manifests Löwdin's optimal property [4,9] of the Canonical orthogonalization. Note that just as the c_κ gains its largest values $(\mathbf{d})_{\kappa\kappa}$ for $\mathbf{B}$=$\mathbf{U}$, the quantity

$$\sum_\alpha |A_{\alpha\kappa}|^2 = (\mathbf{A}^\dagger \mathbf{A})_{\kappa\kappa} = (\mathbf{B}^\dagger \mathbf{M}^{-1} \mathbf{B})_{\kappa\kappa}\,, \tag{8}$$

has its smallest values for $\mathbf{B}=\mathbf{U}$, in which case it is $(\mathbf{d}^{-1})_{\kappa\kappa}$. If d_1^{-1} is the smallest value and the associated basis vector is $\vec{\lambda}_1$, Löwdin [9] showed that for all vectors orthogonal to $\vec{\lambda}_1$ the sum $\sum_\alpha |A_{\alpha\kappa}|^2$ has the smallest value $d_2^{-1}(> d_1^{-1})$ associated with $\vec{\lambda}_2$, and that one can go on in this manner to find the smallest d_κ^{-1}'s associated with the respective $\vec{\lambda}_\kappa$'s.

It should be noted that the value of m_{max} depends on the distribution of orientations of $\vec{v}_k$'s relative to each other, whereas the value of m_{min} is independent of this. There can be any number of $\mathbf{V}$'s satisfying a particular identity (7a) but differing in distributions of orientations of the vectors $\vec{v}_k$'s. Each of these will have a different m_{max} but the same m_{min}. The inhomogeneity in the distribution of directions of a given set of vectors will decide the value of m_{max} – the larger the inhomogeneity, the larger will be the value of m_{max}.

Now for the sake of completeness we will examine the Gram-Schmidt orthogonalization (represented by $\mathbf{Z}=\mathbf{\Omega}$) in the framework of eqns. (6). The bases $\mathbf{\Omega}$ $(= \{\vec{\omega}_\kappa\})$, where for a predetermined sequence of $\vec{v}_k$'s a sequence of $\vec{\omega}_\kappa$'s is generated such that a $\vec{v}_k$ has non-zero projections only on those $\vec{\omega}_\kappa$'s for which $\kappa \leq k$, satisfy the modified set of equations (6) with $|(\vec{v}_k, \vec{z}_\kappa)|^2=0$ for all $\kappa > k$ together with the identity (7a). It is clear at the outset that none of the $\mathbf{\Omega}$'s ($N!$ in number) can coincide with the Symmetric basis $\mathbf{\Phi}$, since for any $\mathbf{\Omega}$ at least c_1 is necessarily greater than $|\vec{v}_1|^2$. But, an $\mathbf{\Omega}$ corresponding to a certain sequence of $\vec{v}_k$'s can accidently coincide with the bases $\mathbf{\Gamma}$ or $\mathbf{\Lambda}$. The possibility of the coincidence with $\mathbf{\Lambda}$ will however be precluded if the $\vec{v}_k$'s are normalized for then $\mathbf{\Lambda}$ coincides with $\mathbf{\Phi}$ as seen above.

3. Applications to cognitive and biological phenomena

Recently we have proposed [1] that Schmidt's orthogonalization is a plausible cognitive function the brain might be employing to discriminate an object from another. An object is represented by an N-dimensional vector in an N-dimensional feature space represented by a neural network of N neurons. The information about the object is encoded in the brain as the synaptic efficacies are modified by it in accordance with the Hebb's hypothesis [1]. Our proposal is that an information about a new object is orthogonalized with respect to the information already in store before it is taken in by the synapses. In real terms the process of orthogonalization amounts to comparing the new information with those in the store, identifying its similarities and differences with the old information and then encoding these similarities and differences in the synapses. This process is not only intuitively appealing but is very efficient too in the sense that the storage of the information is completely free from noise, which inevitably creeps in due to the correlations among the information coming in for storage, consequently the retrieval of the stored information is exact and prompt.

However the above scheme is constrained by the orthogonalization procedure's "*sequence-dependence*" which makes it applicable to phenomena like our thought process and use of language etc. that are sequential in time. However, there are a host of cognitive phenomena that are not sequential in time. For instance, detection of changes in a visual array, categorization according to numerosities, etc. [11]. It is expected that the understanding developed here of the sequence-independent orthogonalizations will prove valuable in exploring such phenomena.

4. Problem of linear-dependence

The canonical orthogonalization (CO) can be useful also in organising large volumes of data, e.g. gene-expression profiles [12]. However, the requirement that the given vectors ought to be linearly-independent puts a serious limitation as there can be only n linearly-independent vectors in an n-dimensional space. We show here how the restriction of linear-independence can be circumvented to apply the CO to any number of vectors in n-dimensions. In this process we also find that the canonical orthogonalization procedure generalized to handle linearly-dependent vectors is in fact the same as the well known Principal Component Analysis (PCA) [13] as applied to data that has *not* been centred about the mean.

Let $\boldsymbol{V} = \{\vec{v}^1, \vec{v}^2, \cdots, \vec{v}^N\}$ possess N vectors with $N > n$, the dimensionality of each vector. Consequently the vectors will not be linearly-independent and we will not be able to use canonical orthogonalization to sample those directions in which the given vectors have a hierarchy of maximal projections.

4.1 The ways out:

To overcome the above problem we propose to view N vectors in n-dimension as n vectors in N-dimension. That is, consider n vectors, $\underset{\to}{v}_1, \underset{\to}{v}_2, \cdots, \underset{\to}{v}_n$, such that $\underset{\to}{v}_1$ comprises the first components of $\vec{v}^1$ through $\vec{v}^N$, (i.e. $\underset{\to}{v}_1 = (v_1^1, v_1^2, \cdots, v_1^N)$), $\underset{\to}{v}_2$ comprises the second components of $\vec{v}^1$ through $\vec{v}^N$, and so on. Now construct the $n \times n$ metric matrix $\boldsymbol{M}$ from the N-dimensional vectors $\underset{\to}{v}_1, \underset{\to}{v}_2, \cdots, \underset{\to}{v}_n$.

In other words, while under normal circumstances of $N \leq n$ one constructs $\vec{M}$ by taking V^TV (V, is an $n \times N$ matrix).

$$\vec{M} = V^TV = \begin{pmatrix} \vec{v}^1.\vec{v}^1 & \vec{v}^1.\vec{v}^2 & \cdots & \vec{v}^1.\vec{v}^N \\ \vec{v}^2.\vec{v}^1 & \vec{v}^2.\vec{v}^2 & \cdots & \vec{v}^2.\vec{v}^N \\ \vdots & \vdots & & \vdots \\ \vec{v}^N.\vec{v}^1 & \vec{v}^N.\vec{v}^2 & \cdots & \vec{v}^N.\vec{v}^N \end{pmatrix}, \tag{9}$$

for $N > n$ we propose to construct $\underset{\rightarrow}{M}$ by taking VV^T,which can be represented in terms of $\underset{\rightarrow}{v_i}$'s as,

$$\underset{\rightarrow}{M} = VV^T = \begin{pmatrix} \underset{\rightarrow}{v_1}.\underset{\rightarrow}{v_1} & \underset{\rightarrow}{v_1}.\underset{\rightarrow}{v_2} & \cdots & \underset{\rightarrow}{v_1}.\underset{\rightarrow}{v_n} \\ \underset{\rightarrow}{v_2}.\underset{\rightarrow}{v_1} & \underset{\rightarrow}{v_2}.\underset{\rightarrow}{v_2} & \cdots & \underset{\rightarrow}{v_2}.\underset{\rightarrow}{v_n} \\ \vdots & \vdots & & \vdots \\ \underset{\rightarrow}{v_n}.\underset{\rightarrow}{v_1} & \underset{\rightarrow}{v_n}.\underset{\rightarrow}{v_2} & \cdots & \underset{\rightarrow}{v_n}.\underset{\rightarrow}{v_n} \end{pmatrix}. \tag{10}$$

The matrix $\underset{\rightarrow}{M}$ is some times referred to as 'sum of squares and cross products' (SSCP) matrix in the literature [14]. It is in fact the so called Covariance matrix [13] without subtracting the mean. In Covariance matrix one constructs $\underset{\rightarrow}{v_m}$ as $((v_m^1 - \overline{v}_m), (v_m^2 - \overline{v}_m), \cdots, (v_m^N - \overline{v}_m))$, where $\overline{v}_m$ is the average of the mth component of all the N vectors, and is ignored in constructing the SSCP matrix. The PCA is typically done using $\underset{\rightarrow}{M}$. While the i-j element of $\vec{M}$ gives the overlap of ith and j^{th} vectors, i.e. overlap of all the features of vectors, $\vec{v}^i$ and $\vec{v}^j$, the p-q element in $\underset{\rightarrow}{M}$ represents the overlap of pth and q^{th} features (or components) of all the given N vectors.

When $N > n$, $\vec{M}$ can be constructed but CO cannot be applied as the given vectors will not be linearly-independent. However, we can have the condition of (no. of vectors) $\leq$ (dimensionality) satisfied by interchanging N and n and can then apply CO. This amounts to applying CO to $\underset{\rightarrow}{M}$ [15]. We find numerically that the CO applied through $\underset{\rightarrow}{M}$ for n *normalized* N-dimensional vectors is equivalent to the PCA applied through $\underset{\rightarrow}{M}$ constructed from N *normalized* n-dimensional vectors.

We also attempted the problem of CO for $N > n$ vectors through the usual $\vec{M}$-route. We find that we can proceed in a hierarchical manner. Divide N vectors into N/n (+1, if N is not a multiple of n) groups. Apply CO to each group and isolate the CO basis vectors which get the maximum projection in each group. Make groups from these vectors with $\leq n$ vectors in each group. Apply CO to these groups and continue until we are left with one group of $\leq n$ vectors. The CO applied to this group will yield the final vector that captures the maximum projections of the given N vectors. Computationally this method is more economical than the CO or PCA applied through $\underset{\rightarrow}{M}$.

5. Computations

Before coming to the situation of main interest, namely $N > n$, we will study the case of $N = n$ where the canonical orthogonalization holds in normal course. We considered a number of random examples of 3 vectors in 3-dimensions ($n = 3$), a sample of which is listed in Table I. Metric matrices $\vec{\boldsymbol{M}}$ and $\underset{\rightarrow}{\boldsymbol{M}}$ were constructed as in (9) and (10). We find that $\vec{\boldsymbol{M}}$ and $\underset{\rightarrow}{\boldsymbol{M}}$ always have the same eigenvalues but their eigenvectors are different. However, the eigenvectors of $\underset{\rightarrow}{\boldsymbol{M}}$ coincide with the canonically orthogonalized set $\boldsymbol{\Lambda} = \boldsymbol{VUd}^{-1/2}$ obtained from $\vec{\boldsymbol{M}}$.

The eigenvectors of $\underset{\rightarrow}{\boldsymbol{M}}$ represent the so called 'principal components' [13], or the uncorrelated vectors which are ordered so that the first few retain most of the variation present in the original vectors. We use the common criterion of sum of the projection-squares of the given vectors on the principal components [16]. It is maximum along the first principal component. Thus the principal component analysis (PCA) based on $\underset{\rightarrow}{\boldsymbol{M}}$ is the same as the canonical orthogonalization (CO) based on $\vec{\boldsymbol{M}}$ in the case of 3 vectors in 3-dimensions (3-d). This should hold in general for n vectors in n-dimensions (n-d).

Note that $\underset{\rightarrow}{\boldsymbol{M}}$ has information on overlap of pairs of features of the given vectors. Often one wants to look at the features in the background of the features possessed by all the vectors (e.g. feature no. 1 in the background of this feature in all the vectors). Then one considers the correlation matrix $\underset{\rightarrow}{\boldsymbol{C}}$ obtained from $\underset{\rightarrow}{\boldsymbol{M}}$ by using normalised $\underset{\rightarrow}{v_i}$'s ($= \underset{\rightarrow}{v_i} \,/\, |\, \underset{\rightarrow}{v_i} \,|$) in place of $\underset{\rightarrow}{v_i}$'s. This amounts to starting with $\boldsymbol{V^T}$ and normalizing each of its columns. Call it $\tilde{\boldsymbol{V}}$. Apply canonical orthogonalization to $\tilde{\boldsymbol{V}}^{\boldsymbol{T}}\tilde{\boldsymbol{V}}$ and the principal component analysis to $\tilde{\boldsymbol{V}}\tilde{\boldsymbol{V}}^{\boldsymbol{T}}$. The eigenvectors of $\tilde{\boldsymbol{V}}\tilde{\boldsymbol{V}}^{\boldsymbol{T}}$ coincide with the vectors in the set $\tilde{\boldsymbol{V}}\tilde{\boldsymbol{U}}\tilde{\boldsymbol{d}}^{-1/2}$. However two of the three eigenvalues change a little bit as compared to the earlier values corresponding to $\boldsymbol{V^T V}$ and $\boldsymbol{V V^T}$.

We will now turn to the general case of $N > n$ to study how CO can be applied in this case and to further analyse the equivalence of PCA and the CO. Consider a set of 24 vectors (i.e. $N = 24$) in 3-dimensions (i.e. $n = 3$). An arbitrary set $\boldsymbol{W}$ of normalized vectors is listed in Table II. The SSCP matrix $\boldsymbol{WW^T} \equiv \underset{\rightarrow}{\boldsymbol{M}}$ is constructed. The eigenvectors of $\underset{\rightarrow}{\boldsymbol{M}}$ represent the mutually orthogonal principal components. The sum of projection squares of $\vec{v}^i$'s maximises along one of the eigenvectors and is also equal to the largest eigenvalue of $\underset{\rightarrow}{\boldsymbol{M}}$. This eigenvector represents the first principal component.

Clearly the canonical orthogonalization can not be applied to the given 24 vectors in 3-d. Taking our cue from the example of 3 vectors in 3-d we will consider three vectors $\underset{\to}{v_i}$'s in 24-dimensions, i.e. the 24×3 matrix $\tilde{\boldsymbol{W}}^{\boldsymbol{T}}$. Note that the $\underset{\to}{v_i}$'s have to be normalized as 24-d vectors while $\vec{v}^i$'s are normalized as 3-d vectors. Such normalized vectors are $\underset{\to}{\tilde{w}_i}$'s and $\vec{w}_i$'s respectively; the $\tilde{}$ represents the difference in the mode of normalization. Thus, $\tilde{\boldsymbol{W}}^{\boldsymbol{T}}$ is not the simple transpose of $\boldsymbol{W}$. Since now the number of $\underset{\to}{\tilde{w}_i}$'s is less than the dimensionality of the space, namely 24, we will compute $\vec{\boldsymbol{M}}$ as $(\tilde{\boldsymbol{W}}^{\boldsymbol{T}})^{\boldsymbol{T}}(\tilde{\boldsymbol{W}}^{\boldsymbol{T}})$, which turns out to be $\tilde{\boldsymbol{W}}\tilde{\boldsymbol{W}}^{\boldsymbol{T}}$ and therefore represented as $\underset{\to}{\tilde{M}}$– this will be a 3×3 matrix. Compute the eigenvalues and eigenvectors of $\underset{\to}{\tilde{M}}$ and then the canonical orthonormal basis set $\boldsymbol{\Lambda} = \tilde{\boldsymbol{W}}^{\boldsymbol{T}}\boldsymbol{U}\boldsymbol{D}^{-1/2}$. The projection-squares of $\underset{\to}{\tilde{w}_i}$'s on $\underset{\to}{\lambda_i}$'s add up to the eigenvalues of $\underset{\to}{\tilde{M}}$, with the maximum being along $\underset{\to}{\lambda_1}$.

The eigenvectors of $\underset{\to}{M}$, which characterise the directions of the principal components, and the eigenvectors of $\underset{\to}{\tilde{M}}$, which mark the directions of the canonically orthogonalized bases, are aligned parallel to each other within an accuracy of less than 4%. Consequently, the PCA as applied to $\underset{\to}{M}$, is equivalent to the CO as applied to $\underset{\to}{\tilde{M}}$.

We tried the hierarchical way of circumventing the problem of linear dependence of the same set of given vectors $\boldsymbol{W}$ in order to apply the CO scheme. The 24 vectors in 3-d were divided into 8 groups of linearly-independent vectors. The CO was applied to each one of them. From each group we picked out that basis vector along which the projection-squares maximised in that group. We again divided these 8 vectors into three sets comprising 3, 3 and 2 vectors and applied CO to these three groups and picked out 3 main basis vectors along which the projection-squares maximised in each group. Finally we applied the CO to these three vectors and came up with the final canonically orthogonalized vector along which the sum of projection-squares of these three vectors is the maximum. The projection-squares of the 24 given vectors on this final vector add up to 22.4576 which is very close to 22.4703, the largest eigenvalue of $\underset{\to}{M}$ constructed from the given 24 vectors for doing the PCA. The results of this scheme are summarised in Table III. Thus the hierarchical scheme of CO works successfully in determining the direction along which a given set of vectors projects maximally even though it may be a set of linearly-dependent vectors.

6. Conclusion

In sum we find that it is easy to visualise all the orthogonalization methods in terms of the projections of the given linearly-independent vectors on the orthogonal bases. This exercise elucidates interesting geometrical features of the orthogonalized bases hidden in their optimal, extremal and symmetry properties and also shows how these basis sets as well as the eigenvalues of the metric matrix of the given **V** can be constructed purely from geometrical considerations.

We further show that it is possible to circumvent the restriction that a given set of vectors should be linearly-independent in order to apply Löwdin's canonical orthogonalization to find a direction in which the given vectors project maximally. Two proposals are made and tested numerically. It is also found that when the CO is used in these fashions, it is equivalent to the PCA, but the hierarchical scheme of CO proposed here is computationally more economical than the PCA.

Acknowledgements

The author enjoyed extensive discussions with Prof. P.-O. Löwdin and also his encouragement during the course of this work. Thanks are due to A. Ramesh Naidu for help in computations. This work was supported by a grant from the Ministry of Information Technology, Government of India, and the Leverhulme Foundation (U. K.).

References

*E mail: vs265@cam.ac.uk and vpssp@uohyd.ernet.in

1. Vipin Srivastava and S.F.Edwards, Physica A **276**, 352 (2000).
2. R. Landshoff, Z. Physik **102**, 201 (1936).
3. G. H. Wannier, Phys. Rev. **52**, 191 (1937).
4. See P. O. Löwdin, Advn. Quant. Chem. no. 5, 185 (1990) for a review.
5. P.- O. Löwdin, Arkiv Mat. Astr.Fysik **35A**, 9 (1947).
6. P.- O. Löwdon, Advn. Phys. **5**, 1 (1956).
7. S. Chaturvedi, A.K. Kapoor and V. Srinivasan, J. Phys. **A 31**, L367 (1998).
8. H. C. Schweinler and E. P. Wigner, J. Math. Phys. **97**, 1693 (1970).
9. P.-O. Löwdin, Phys. Rev. **97**, 1474 (1955).
10. The set of eqns. (6) will have many more solutions than those that form orthonormal bases. Many of these may not even be bases. A $\boldsymbol{N}$-dimensional orthonormal basis set will require $\boldsymbol{N(N-1)/2}$ conditions to be fulfilled.
11. B. Butterworth, Science **284**, 928 (1999); also S. Dehaene et al. *ibid* 970 (1999).
12. Vipin Srivastava et al. (2002), to be published.
13. Jolliffe I T, *Principal Component Analysis* (Springer-Verlag 1986).
14. Searle S R, *Matrix Algebra Useful for Statistics* (John Wiley & Sons, 1982), See Chapter 13
15. When N and n are interchanged, the n vectors in N-dimension will be represented by $\boldsymbol{V^T}$. And $\vec{\boldsymbol{M}}$ for $\boldsymbol{V^T}$ will be $\boldsymbol{(V^T)^T(V^T)}$, i.e. $\boldsymbol{VV^T}$, i.e. $\underset{\rightarrow}{\mathrm{M}}$
16. Rojas R, *Neural Networks* (Springer, 1996)

Table I: Canonical Orthogonalization and Principal Component Analysis of 3 vectors in 3-dimension

Input matrix of vectors (each column represents a vector):

$$\boldsymbol{V} = \begin{pmatrix} 5 & 7 & 3 \\ 2 & 1 & 8 \\ 4 & 6 & 9 \end{pmatrix} \xrightarrow{Normalization} \begin{pmatrix} .7454 & .7548 & .2417 \\ .2981 & .1078 & .64478 \\ .5963 & .6470 & .7252 \end{pmatrix}$$

$$\boldsymbol{V^T V}$$

$$\vec{\boldsymbol{M}} = \begin{pmatrix} 1 & .9806 & .8048 \\ .9806 & 1 & .7212 \\ .8048 & .7212 & 1 \end{pmatrix}$$

Eigenvalues Eigenvectors

$$\boldsymbol{d} = \begin{pmatrix} 2.6759 \\ .3134 \\ .0107 \end{pmatrix}; \quad \boldsymbol{U} = \begin{pmatrix} .6029 & .2611 & -.7539 \\ .5858 & .4966 & .6405 \\ .5416 & -.8278 & .1464 \end{pmatrix}$$

$$\boldsymbol{VUd^{-1/2}}$$

$$\boldsymbol{\Lambda} = \begin{pmatrix} .6251 & .6598 & -.4171 \\ .3619 & -.7185 & -.5940 \\ .6916 & -.2203 & .6879 \end{pmatrix}$$

$$\downarrow$$

$$(\boldsymbol{V^T \Lambda})_{max} = \mathbf{2.6759} = \boldsymbol{d_{max}}$$

$$\boldsymbol{V V^T}$$

$$\underset{\rightarrow}{\boldsymbol{M}} = \begin{pmatrix} 1.1838 & .4595 & 1.1081 \\ .4595 & .5161 & .7151 \\ 1.1081 & .7151 & 1.3001 \end{pmatrix}$$

Eigenvalues Eigenvectors

$$\boldsymbol{d} = \begin{pmatrix} 2.6759 \\ .3134 \\ .0107 \end{pmatrix}; \quad \boldsymbol{U'} = \begin{pmatrix} .6251 & -.6598 & -.4171 \\ .3619 & .7185 & -.5940 \\ .6916 & .2203 & .6879 \end{pmatrix}$$

$$\downarrow$$

$$(\boldsymbol{V^T U'})_{max} = \mathbf{2.6759} = \boldsymbol{d_{max}}$$

Table II (a): The given vectors (each column represents a vector)

<u>24 vectors in 3-dimensions</u>:

$$
\boldsymbol{V} = \left(\begin{matrix} 8 & 20 & 45 & 10 & 19 & 39 & 7 & 31 & 36 & 8 & 23 & 53 \\ 16 & 24 & 44 & 20 & 26 & 42 & 14 & 30 & 34 & 16 & 26 & 50 \\ 40 & 44 & 54 & 50 & 53 & 61 & 35 & 43 & 45 & 40 & 45 & 57 \end{matrix} \right.
$$

$$
\left. \begin{matrix} 7 & 34 & 44 & 6 & 18 & 58 & 2 & 17 & 47 & 13 & 18 & 48 \\ 14 & 32 & 40 & 12 & 20 & 52 & 4 & 14 & 38 & 6 & 16 & 40 \\ 35 & 44 & 48 & 30 & 44 & 50 & 10 & 15 & 27 & 15 & 20 & 32 \end{matrix} \right)
$$

$$
\downarrow \; Normalization(columnwise)
$$

$$
\mathbf{W} = \left(\begin{matrix} .1826 & .3706 & .5426 & .1826 & .3064 & .4659 & .1826 & .5089 & .5380 & .1826 \\ .3651 & .4447 & .5306 & .3651 & .4192 & .5018 & .3651 & .4925 & .5081 & .3651 \\ .9129 & .8154 & .6512 & .9129 & .8546 & .7288 & .9129 & .7060 & .6725 & .9129 \end{matrix} \right.
$$

$$
\begin{matrix} .4047 & .5729 & .1826 & .5300 & .5758 & .1826 & .3490 & .6266 & .1826 & .6380 \\ .4575 & .5405 & .3651 & .4988 & .5234 & .3651 & .3878 & .5618 & .3651 & .5254 \\ .7918 & .6162 & .9129 & .6858 & .6281 & .9129 & .8531 & .5402 & .9129 & .5629 \end{matrix}
$$

$$
\left. \begin{matrix} .7100 & .6269 & .5750 & .6838 \\ .5740 & .2893 & .5111 & .5698 \\ .4079 & .7234 & .6389 & .4558 \end{matrix} \right)
$$

<u>3 vectors in 24-dimensions</u>:

$$
\boldsymbol{V^T} = \begin{pmatrix} 8 & 16 & 40 \\ 20 & 24 & 44 \\ 45 & 44 & 54 \\ 10 & 20 & 50 \\ 19 & 26 & 53 \\ 39 & 42 & 61 \\ 7 & 14 & 35 \\ 31 & 30 & 43 \\ 36 & 34 & 45 \\ 8 & 16 & 40 \\ 23 & 26 & 45 \\ 53 & 50 & 57 \\ 7 & 14 & 35 \\ 34 & 32 & 44 \\ 44 & 40 & 48 \\ 6 & 12 & 30 \\ 18 & 20 & 44 \\ 58 & 52 & 50 \\ 2 & 4 & 10 \\ 17 & 14 & 15 \\ 47 & 38 & 27 \\ 13 & 6 & 15 \\ 18 & 16 & 20 \\ 48 & 40 & 32 \end{pmatrix} \xrightarrow{Normalization(columnwise)} \boldsymbol{\tilde{W}^T} = \begin{pmatrix} .0535 & .1107 & .1976 \\ .1338 & .1660 & .2174 \\ .3011 & .3043 & .2668 \\ .0669 & .1383 & .2471 \\ .1271 & .1798 & .2619 \\ .2609 & .2905 & .3014 \\ .0468 & .0968 & .1729 \\ .2074 & .2075 & .2125 \\ .2408 & .2351 & .2224 \\ .0535 & .1107 & .1976 \\ .1539 & .1798 & .2224 \\ .3546 & .3458 & .2816 \\ .0468 & .0968 & .1729 \\ .2275 & .2213 & .2174 \\ .2944 & .2766 & .2372 \\ .0401 & .0830 & .1482 \\ .1204 & .1383 & .2174 \\ .3880 & .3596 & .2471 \\ .0134 & .0277 & .0494 \\ .1137 & .0968 & .0741 \\ .3144 & .2628 & .1334 \\ .0870 & .0415 & .0741 \\ .1204 & .1107 & .0988 \\ .3211 & .2766 & .1581 \end{pmatrix}
$$

Table II (b):

Canonical Orthogonalization

$$(\tilde{\boldsymbol{W}}^{\boldsymbol{T}})^{\boldsymbol{T}}(\tilde{\boldsymbol{W}}^{\boldsymbol{T}}) = \tilde{\boldsymbol{W}}\tilde{\boldsymbol{W}}^{\boldsymbol{T}} \equiv \underset{\rightarrow}{\tilde{M}} = \begin{pmatrix} 1 & .9836 & .8743 \\ .9836 & 1 & .9433 \\ .8743 & .9433 & 1 \end{pmatrix}$$

$$\text{Eigenvalues } \boldsymbol{D} = \begin{pmatrix} 2.8683 \\ .1298 \\ .0091 \end{pmatrix} \text{ \& Eigenvectors } \boldsymbol{U} = \begin{pmatrix} .5756 & -.6155 & .5384 \\ .5893 & -.1443 & -.7949 \\ .5669 & .7748 & .2797 \end{pmatrix}$$

$$(\tilde{\boldsymbol{W}}\boldsymbol{\Lambda})^2 = \left(\tilde{\boldsymbol{W}}\left[\tilde{\boldsymbol{W}}^{\boldsymbol{T}}\boldsymbol{U}\boldsymbol{D}^{-1/2}\right]\right)^2 = \begin{pmatrix} .9503 & .0492 & .0005 \\ .9961 & .0027 & .0012 \\ .9219 & .0780 & .0001 \end{pmatrix}$$

Sum of projection-squares of 3 vectors (in 24-d) on $\boldsymbol{\Lambda}$

$$= \text{sum of column elements in } (\tilde{\boldsymbol{W}}\boldsymbol{\Lambda})^2 = \begin{pmatrix} 2.8683 \\ .1298 \\ .0091 \end{pmatrix}$$

Principal Component Analysis

$$\boldsymbol{W}\boldsymbol{W}^{\boldsymbol{T}} \equiv \underset{\rightarrow}{M} = \begin{pmatrix} 5.2457 & 4.9612 & 6.9608 \\ 4.9612 & 5.1063 & 7.7808 \\ 6.9608 & 7.7808 & 13.6479 \end{pmatrix}$$

$$\text{Eigenvalues } \boldsymbol{d} = \begin{pmatrix} 22.4703 \\ 1.4671 \\ .0625 \end{pmatrix} \text{ \& Eigenvectors } \boldsymbol{E} = \begin{pmatrix} .4437 & -.7413 & .5036 \\ .4690 & -.2868 & -.8354 \\ .7637 & .6068 & .2204 \end{pmatrix}$$

Sum of projection-squares of the given 24-vectors (in 3-d) on $\boldsymbol{E}$

$$= \text{sum of column elements in } (\boldsymbol{W}^{\boldsymbol{T}}\boldsymbol{E})^2 = \begin{pmatrix} 22.4703 \\ 1.4671 \\ .0625 \end{pmatrix}$$

Table III: Hierarchical scheme of canonical orthogonalization of 24 vectors of Table II (a):

$$\boldsymbol{W} \equiv \boldsymbol{W}^{(0)} =$$

$$\left[\underbrace{\begin{matrix} .1826 & .3706 & .5426 \\ .3651 & .4447 & .5306 \\ .9129 & .8154 & .6512 \end{matrix}}_{1} \;\; \underbrace{\begin{matrix} .1826 & .3064 & .4659 \\ .3651 & .4192 & .5018 \\ .9129 & .8546 & .7288 \end{matrix}}_{2} \;\; \underbrace{\begin{matrix} .1826 & .5089 & .5380 \\ .3651 & .4925 & .5081 \\ .9129 & .7060 & .6725 \end{matrix}}_{3} \right.$$

group# 1 2 3

$$\underbrace{\begin{matrix} .1826 & .4047 & .5729 \\ .3651 & .4575 & .5405 \\ .9129 & .7918 & .6162 \end{matrix}}_{4} \;\; \underbrace{\begin{matrix} .1826 & .5300 & .5758 \\ .3651 & .4988 & .5234 \\ .9129 & .6858 & .6281 \end{matrix}}_{5} \;\; \underbrace{\begin{matrix} .1826 & .3490 & .6266 \\ .3651 & .3878 & .5618 \\ .9129 & .8531 & .5402 \end{matrix}}_{6}$$

group# 4 5 6

$$\left. \underbrace{\begin{matrix} .1826 & .6380 & .7100 \\ .3651 & .5254 & .5740 \\ .9129 & .5629 & .4079 \end{matrix}}_{7} \;\; \underbrace{\begin{matrix} .6269 & .5750 & .6838 \\ .2893 & .5111 & .5698 \\ .7234 & .6389 & .4558 \end{matrix}}_{8} \right]_{3\times 24}$$

group# 7 8

$\lambda_i^{(1)}$'s that maximise the sum of projection-squares in each group at stage (1)

↓

$$\boldsymbol{W}^{(1)} = \left[\underbrace{\begin{matrix} .3722 & .3216 & .4210 \\ .4554 & .4334 & .4659 \\ .8088 & .8418 & .7783 \end{matrix}} \;\; \underbrace{\begin{matrix} .3960 & .4439 & .3964 \\ .4650 & .4758 & .4517 \\ .7918 & .7593 & .7993 \end{matrix}} \;\; \underbrace{\begin{matrix} .5488 & .6375 \\ .5192 & .4639 \\ .6552 & .6151 \end{matrix}} \right]_{3\times 8}$$

$\lambda_i^{(2)}$'s that maximise the sum of projection-squares in each group at stage (2)

↓

$$\boldsymbol{W}^{(2)} = \begin{bmatrix} .3721 & .4123 & .5941 \\ .4521 & .4644 & .4923 \\ .8106 & .7838 & .6361 \end{bmatrix}_{3\times 3}$$

$\lambda_i^{(3)}$ that maximises the sum of projection-squares in this group at stage (3)

↓

$$\lambda_1^{(3)} = \begin{bmatrix} .4626 \\ .4732 \\ .7497 \end{bmatrix}$$

Sum of projection-squares of 24 vectors in 3-dimension on $\lambda_1^{(3)}$
$= (\boldsymbol{W}^{\boldsymbol{T}} \, \lambda_1^{(3)})^2 = 22.4576$

Correlation Between Position and Momentum. A Phase-Space View

Jens Peder Dahl
Chemical Physics, Department of Chemistry
Technical University of Denmark
DTU 207, DK-2800 Kongens Lyngby, Denmark
and
Abteilung für Quantenphysik, Universität Ulm
D-89069 Ulm, Germany
e-mail: jpd@kemi.dtu.dk

Abstract

In the wave-function picture, the correlation between a particle's position and momentum is hidden in the Fourier transformation between the position and momentum wave functions, but in a phase-space picture it is contained in one and the same phase-space function. Focusing on the Weyl-Wigner representation, I discuss the correlation between position and momentum as a function of time for some very simple examples, and introduce the quantitative concept of radial and angular kinetic characters of a wave packet. As a basis for the discussion, I present the salient features of the Weyl-Wigner representation. As a by-product, the well-known expression for the probability current density is derived via the phase-space picture.

1 Introduction

I first met Per-Olov Löwdin in the beautiful summer of 1959, at the summer school on quantum chemistry which he was organizing at Skogshem outside Stockholm. This was a very successful arrangement, with first-class teachers and a most stimulating atmosphere. I possessed only a rudimentary knowledge of quantum mechanics when I arrived, so the school was a big challenge to me, and neither before nor after have I absorbed so much new material in such a short time. To attend Löwdin's lectures was, in particular, like a revelation. Here he was, still a young

E.J. Brändas and E.S. Kryachko (eds.),
Fundamental World of Quatum Chemistry, Vol. I, 237–260.

and strong man, with a charming appearance and a beautiful handwriting, explaining the nature of quantum mechanics and its underlying linear algebra with a natural ease. He had his important articles on many-particle systems, density matrices, and the correlation problem behind him [1, 2], and he enjoyed explaining it all to us, not by hand-waving arguments, but always in exact terms.

During his whole life, Per-Olov Löwdin retained his deep interest in the foundation of quantum mechanics, and in the development of theoretical methods for determining the structure and properties of atoms, molecules and solids. He never got tired of communicating his insight and knowledge to the big audiences that met him all over the world. His intellectual and moral influence on a large number of quantum chemists, including myself, has been remarkable.

To contribute to a volume consisting of tributes to the memory of such a man is not an easy task. What I have chosen to do is to present some simple aspects of a special representation of quantum mechanics to which I have paid much attention for several years, namely, the phase-space representation of quantum mechanics. In this representation, one views the position and momentum characteristics of a quantum state in one and the same picture. In his lectures at the summer school of 1959, Per-Olov Löwdin emphasized, that while the modulus of a position-space wave function gives rise to a probability density in position space, the phase of the same wave function provides the connection to momentum space. A most characteristic feature of the phase-space representation of quantum mechanics is that it treats the position and momentum spaces symmetrically.

The phase-space representation of quantum mechanics that I refer to in the present paper is the so-called Weyl-Wigner representation. Other phase-space representations have been considered in the literature, but the Weyl-Wigner representation has come to play the role of a canonical phase-space representation, because of its simplicity [3, 4].

In the Weyl-Wigner representation, one works with Wigner functions instead of wave functions and with dynamical phase-space functions instead of operators. It is a representation in its own right. Hence, it is relevant to apply physical intuition to the form and behavior of phase-space functions, just as physical intuition may be applied to the form and behavior of wave functions and operators. When we do so, we discover that our understanding of quantum states becomes enlarged, because the two types of intuition may work differently and, therefore, supplement each other. It is a purpose of the present work to illustrate this with some very simple quantum states as examples. Focus is, in particular, put on the problem of visualizing the correlation between a particle's position and momentum, from a phase-space point of view.

In the following, I first recall the salient aspects of the Weyl-Wigner representation, including a discussion of the equation of motion for the Wigner function. I then draw attention to an important aspect of a phase-space representation, namely, its potential for generating well-defined position densities for any operator. I give reference to some of our earlier work on the subject, in which we concluded that the Weyl-Wigner representation also in this respect is the natural representation to use. As a further accentuation of this conclusion, I now show that the well-known expression for the probability current density may be identified with the velocity density as derived from the Weyl-Wigner representation. Subsequently, I specialize to free-particle motion and the correlation between position and momentum of a free particle. This concludes in the introduction of two quantitative measures, referred to as the radial and angular kinetic characters of a wave packet. In the discussion part, I point to another measure of a similar character.

But before proceeding, it is quite in the spirit of Per-Olov Löwdin to remind ourselves, that even if there are several equally valid representations of quantum mechanics, there is only one physical reality. Löwdin made many reflections concerning this situation. He himself introduced many valuable and practically useful concepts in quantum mechanics. However, he also made important contributions at the abstract level. In this context, let me quote the following passage from his article *Quantum Theory as a Trace Algebra* [5]:

> If one has obtained a valid theory, it may be tempting to try to interpret the auxiliary quantities physically as well in order to get a physical model of the system. Even if such an attempt is quite successful, it remains worthwhile to remember that, e. g, in von Neumann's abstract formulation of quantum theory, there are many auxiliary quantities which are undefined and contentless, and that there may be several different realizations leading to entirely different physical models even though the true *physical results* are the same. For this reason, it continues to be worthwhile to try to avoid specific "physical" realizations and to keep the theory on an abstract level.

While I fully agree as to the general validity of these remarks, an objective of the present paper is to introduce several auxiliary quantities and interpret them in "physical" terms.

2 The Weyl-Wigner Representation

In this section, we introduce the most basic properties of the Weyl-Wigner representation. In so doing, we consider a single spinless particle allowed to move in D spatial dimensions, with $\boldsymbol{r}$ being the D-dimensional position vector, and $\boldsymbol{p}$ the corresponding linear momentum vector. The Cartesian coordinates of $\boldsymbol{r}$ and $\boldsymbol{p}$ will be denoted x_i and p_i, respectively. Thus $\boldsymbol{r} = (x_1, x_2, \ldots, x_D)$ and $\boldsymbol{p} = (p_1, p_2, \ldots, p_D)$. Note, however, that we here use the word particle in a generalized sense. The D coordinates could, for instance, stand for the total set of Cartesian coordinates of N identical particles, instead of the D coordinates of a single individual particle.

With the particle in the quantum state $|\psi\rangle$, the position-space wave function $\psi(\boldsymbol{r}) = \langle \boldsymbol{r}|\psi\rangle$ and the momentum-space wave function $\varphi(\boldsymbol{p}) = \langle \boldsymbol{p}|\psi\rangle$ are connected by a Fourier transformation,

$$\psi(\boldsymbol{r}) = \left(\frac{1}{2\pi\hbar}\right)^{D/2} \int \varphi(\boldsymbol{p}) e^{i\boldsymbol{p}\cdot\boldsymbol{r}/\hbar} d\boldsymbol{p}, \tag{1}$$

$$\varphi(\boldsymbol{p}) = \left(\frac{1}{2\pi\hbar}\right)^{D/2} \int \psi(\boldsymbol{r}) e^{-i\boldsymbol{p}\cdot\boldsymbol{r}/\hbar} d\boldsymbol{r}. \tag{2}$$

Here, and in the following, the integrations are understood to be from $-\infty$ to ∞ for each Cartesian position or momentum coordinate. The state $|\psi\rangle$ is assumed to be normalized to unity. Thus,

$$\langle\psi|\psi\rangle = \int \psi^*(\boldsymbol{r})\psi(\boldsymbol{r}) d\boldsymbol{r} = \int \varphi^*(\boldsymbol{p})\varphi(\boldsymbol{p}) d\boldsymbol{p} = 1. \tag{3}$$

We have suppressed the possible time dependence of the state $|\psi\rangle$. It may be easily added when needed.

The Wigner function is now defined as follows [6, 8, 9, 10, 3, 11],

$$W(\boldsymbol{r}, \boldsymbol{p}) = \left(\frac{1}{2\pi\hbar}\right)^{D} \int \psi^*(\boldsymbol{r} - \boldsymbol{r}'/2)\psi(\boldsymbol{r} + \boldsymbol{r}'/2) e^{-i\boldsymbol{p}\cdot\boldsymbol{r}'/\hbar} d\boldsymbol{r}'. \tag{4}$$

The connection between the position and momentum wave functions, Eqs. (1) and (2), shows that the Wigner function may be written in the equivalent form,

$$W(\boldsymbol{r},\boldsymbol{p}) = \left(\frac{1}{2\pi\hbar}\right)^D \int \varphi^*(\boldsymbol{p}-\boldsymbol{p}'/2)\varphi(\boldsymbol{p}+\boldsymbol{p}'/2)e^{i\boldsymbol{r}\cdot\boldsymbol{p}'/\hbar}d\boldsymbol{p}'. \quad (5)$$

The expressions (4) and (5) display the symmetric way in which $W(\boldsymbol{r},\boldsymbol{p})$ depends on $\boldsymbol{r}$ and $\boldsymbol{p}$. Integrations give

$$\int W(\boldsymbol{r},\boldsymbol{p})d\boldsymbol{p} = \psi^*(\boldsymbol{r})\psi(\boldsymbol{r}) = \rho(\boldsymbol{r}), \quad (6)$$

$$\int W(\boldsymbol{r},\boldsymbol{p})d\boldsymbol{r} = \varphi^*(\boldsymbol{p})\varphi(\boldsymbol{p}) = \pi(\boldsymbol{p}), \quad (7)$$

and

$$\int\int W(\boldsymbol{r},\boldsymbol{p})d\boldsymbol{r}d\boldsymbol{p} = 1. \quad (8)$$

$\rho(\boldsymbol{r})$ is the position-space particle density, $\pi(\boldsymbol{p})$ the momentum-space particle density. Thus, the Wigner function has the particle densities in position space and momentum space as proper marginal densities.

Next, let $\hat{A}$ be some operator acting on $|\psi\rangle$. The expectation value of $\hat{A}$ in the state $|\psi\rangle$ is then

$$\langle\hat{A}\rangle = \langle\psi|\hat{A}|\psi\rangle. \quad (9)$$

If $\hat{A}$ is Hermitian, $\langle\hat{A}\rangle$ is real. In the phase-space representation, we have the expression

$$\langle\hat{A}\rangle = \int\int a(\boldsymbol{r},\boldsymbol{p})W(\boldsymbol{r},\boldsymbol{p})d\boldsymbol{r}d\boldsymbol{p} \quad (10)$$

where $a(\boldsymbol{r},\boldsymbol{p})$ is the dynamical phase-space function corresponding to the operator $\hat{A}$. For a Hermitian operator $a(\boldsymbol{r},\boldsymbol{p})$ is real. We have furthermore, that if $\hat{A}$ is of the form $F(\hat{\boldsymbol{r}}) + G(\hat{\boldsymbol{p}})$, then $a(\boldsymbol{r},\boldsymbol{p})$ will be simply $F(\boldsymbol{r}) + G(\boldsymbol{p})$. Otherwise, the noncommutativity between $\hat{\boldsymbol{r}}$ and $\hat{\boldsymbol{p}}$ will come into play. The transformation involved is the Weyl transformation [7, 8, 9, 10, 3, 11]. It may, for instance, be represented in the following form

$$a(\boldsymbol{r},\boldsymbol{p}) = \int \langle\boldsymbol{r}+\boldsymbol{r}'/2|\hat{A}|\boldsymbol{r}-\boldsymbol{r}'/2\rangle e^{-\boldsymbol{p}\cdot\boldsymbol{r}'/\hbar}d\boldsymbol{r}', \quad (11)$$

in which the matrix element of $\hat{A}$ is defined with respect to two eigenstates of the position operator $\hat{\boldsymbol{r}}$, corresponding to the eigenvalues $\boldsymbol{r} + \boldsymbol{r}'/2$ and $\boldsymbol{r} - \boldsymbol{r}'/2$, respectively. The dynamical phase-space function $a(\boldsymbol{r}, \boldsymbol{p})$ is called the Weyl transform of the operator $\hat{A}$.

Let $\hat{C} = \hat{A}\hat{B}$, and let $a(\boldsymbol{r}, \boldsymbol{p})$ and $b(\boldsymbol{r}, \boldsymbol{p})$ be the dynamical phase-space functions corresponding to the operators $\hat{A}$ and $\hat{B}$, respectively. The dynamical phase-space function corresponding to the operator $\hat{C}$ is then given by the star product

$$c(\boldsymbol{r}, \boldsymbol{p}) = a(\boldsymbol{r}, \boldsymbol{p}) * b(\boldsymbol{r}, \boldsymbol{p}), \tag{12}$$

where

$$c(\boldsymbol{r}, \boldsymbol{p}) = \exp\left[\frac{i\hbar}{2}\left(\frac{\partial}{\partial \boldsymbol{r}_1} \cdot \frac{\partial}{\partial \boldsymbol{p}_2} - \frac{\partial}{\partial \boldsymbol{p}_1} \cdot \frac{\partial}{\partial \boldsymbol{r}_2}\right)\right] a(\boldsymbol{r}, \boldsymbol{p}) b(\boldsymbol{r}, \boldsymbol{p}). \tag{13}$$

Here, the subscript 1 on a differential operator indicates that this operator acts only on the first function in the product $a(\boldsymbol{r}, \boldsymbol{p})b(\boldsymbol{r}, \boldsymbol{p})$. Similarly, the subscript 2 is used with operators that only act on the second function in the product.

The relations (6), (7), (8), and (10) show that, operationally, $W(\boldsymbol{r}, \boldsymbol{p})$ plays the role of a phase-space probability density, and we shall find it useful to refer to it in that way. Doing so implies, however, a conceptual adaptation, for although $W(\boldsymbol{r}, \boldsymbol{p})$ is real-valued it is not restricted to non-negative values, as a true probability density would be. In accordance with this, some authors prefer the alternative designation quasiprobability density.

In any case, we may say that $W(\boldsymbol{r}, \boldsymbol{p})$ is a measure of the way the phase-space point $(\boldsymbol{r}, \boldsymbol{p})$ supports the given quantum state. Another way to express this support is to note that we may also write

$$W(\boldsymbol{r}, \boldsymbol{p}) = \left(\frac{1}{\pi\hbar}\right)^D \langle\psi|\hat{\Pi}(\boldsymbol{r}, \boldsymbol{p})|\psi\rangle \tag{14}$$

where $\hat{\Pi}(\boldsymbol{r}, \boldsymbol{p})$ is the *inversion operator* with respect to the phase-space point $(\boldsymbol{r}, \boldsymbol{p})$ [12, 13, 14, 15]. It has the form

$$\hat{\Pi}(\boldsymbol{r}, \boldsymbol{p}) = \left(\frac{1}{4\pi\hbar}\right)^D \int\int e^{i(\boldsymbol{r}\cdot\boldsymbol{u}+\boldsymbol{p}\cdot\boldsymbol{v})/\hbar} e^{-i(\hat{\boldsymbol{r}}\cdot\boldsymbol{u}+\hat{\boldsymbol{p}}\cdot\boldsymbol{v})/\hbar} d\boldsymbol{u} d\boldsymbol{v}. \tag{15}$$

It is Hermitian and unitary, i. e.,

$$\hat{\Pi}(\boldsymbol{r}, \boldsymbol{p})^\dagger = \hat{\Pi}(\boldsymbol{r}, \boldsymbol{p}), \qquad \hat{\Pi}(\boldsymbol{r}, \boldsymbol{p})^2 = 1. \tag{16}$$

It is, in fact, an observable in Dirac's sense [15, 16] because it satisfies an algebraic equation, $\hat{\Pi}(\boldsymbol{r},\boldsymbol{p})^2 = 1$.

Assume now, that we have an orthornormal set of states $|\psi_n\rangle$,

$$\langle\psi_n|\psi_m\rangle = \delta_{nm}. \tag{17}$$

We may then introduce the more general phase-space densities

$$\begin{aligned} W_{nm}(\boldsymbol{r},\boldsymbol{p}) &= \left(\frac{1}{2\pi\hbar}\right)^D \int \psi_n^*(\boldsymbol{r}-\boldsymbol{r}'/2)\psi_m(\boldsymbol{r}+\boldsymbol{r}'/2)e^{-i\boldsymbol{p}\cdot\boldsymbol{r}'/\hbar}d\boldsymbol{r}' \\ &= \left(\frac{1}{\pi\hbar}\right)^D \langle\psi_n|\hat{\Pi}(\boldsymbol{r},\boldsymbol{p})|\psi_m\rangle \end{aligned} \tag{18}$$

which are probability densities for $n = m$, transition densities for $n \neq m$. Eq. (8) generalizes to

$$\int\int W_{nm}(\boldsymbol{r},\boldsymbol{p})d\boldsymbol{r}d\boldsymbol{p} = \delta_{nm}, \tag{19}$$

and Eq. (10) becomes replaced by

$$\langle\psi_n|\hat{A}|\psi_m\rangle = \int\int a(\boldsymbol{r},\boldsymbol{p})W_{nm}(\boldsymbol{r},\boldsymbol{p})d\boldsymbol{r}d\boldsymbol{p}. \tag{20}$$

Let us finally note that the function $W_{nm}(\boldsymbol{r},\boldsymbol{p})$ may be considered the Weyl transform of the operator $(1/2\pi\hbar)^D|\psi_m\rangle\langle\psi_n|$. This is readily seen by inserting this operator into the definition (11) for the Weyl transformation. The Wigner function $W(\boldsymbol{r},\boldsymbol{p})$ of Eq. (4) is, in particular, the Weyl transform of the operator $(1/2\pi\hbar)^D\hat{\rho}$, where

$$\hat{\rho} = |\psi\rangle\langle\psi| \tag{21}$$

is the particle density operator. The position-space representative of $\hat{\rho}$ is the position-space density matrix

$$\rho(\boldsymbol{r},\boldsymbol{r}') = \langle\boldsymbol{r}|\hat{\rho}|\boldsymbol{r}'\rangle = \psi(\boldsymbol{r})\psi^*(\boldsymbol{r}'), \tag{22}$$

whose diagonal element is the position-space particle density. Similarly, we have the momentum-space density matrix

$$\pi(\boldsymbol{p},\boldsymbol{p}') = \langle\boldsymbol{p}|\hat{\rho}|\boldsymbol{p}'\rangle = \varphi(\boldsymbol{p})\varphi^*(\boldsymbol{p}'), \tag{23}$$

the diagonal element of which is the momentum-space particle density $\pi(\boldsymbol{p})$.

3 Phase-Space Dynamics

Let us now add the time dependence to our description. The density operator (21) becomes then

$$\hat{\rho}(t) = |\psi(t)\rangle\langle\psi(t)|, \tag{24}$$

and the Wigner function (4) reads

$$W(\boldsymbol{r},\boldsymbol{p},t) = \left(\frac{1}{2\pi\hbar}\right)^D \int \psi^*(\boldsymbol{r}-\boldsymbol{r}'/2,t)\psi(\boldsymbol{r}+\boldsymbol{r}'/2,t)e^{-i\boldsymbol{p}\cdot\boldsymbol{r}'/\hbar}d\boldsymbol{r}'. \tag{25}$$

The time evolution of the state vector $|\psi(t)\rangle$ is determined by the time-dependent Schrödinger equation

$$i\hbar\frac{\partial}{\partial t}|\psi(t)\rangle = \hat{H}|\psi(t)\rangle \tag{26}$$

where $\hat{H}$ is the Hamiltonian. The time evolution of the density operator is given by the von Neumann equation

$$\frac{\partial\hat{\rho}(t)}{\partial t} = \frac{1}{i\hbar}[\hat{H},\hat{\rho}(t)] \tag{27}$$

where $[,]$ is the usual commutator. From this equation, we obtain the equation of motion for the Wigner function $W(\boldsymbol{r},\boldsymbol{p},t)$, by recalling that $W(\boldsymbol{r},\boldsymbol{p},t)$ is the Weyl transform of the operator $(1/2\pi\hbar)^D\hat{\rho}(t)$ and exploiting the expression (13) for the star product. The result is

$$\frac{\partial}{\partial t}W(\boldsymbol{r},\boldsymbol{p},t) = \{h(\boldsymbol{r},\boldsymbol{p}),W(\boldsymbol{r},\boldsymbol{p},t)\}_M. \tag{28}$$

Here, $h(\boldsymbol{r},\boldsymbol{p})$ is the Weyl transform of the quantum-mechanical Hamiltonian $\hat{H}$. In most cases, $\hat{H}$ will be of the form

$$\hat{H} = \frac{\hat{\boldsymbol{p}}^2}{2M} + V(\hat{\boldsymbol{r}}), \tag{29}$$

with $\hat{\boldsymbol{p}}^2/2M$ representing the kinetic energy and $V(\hat{\boldsymbol{r}})$ the potential energy. The phase-space Hamiltonian is then

$$h(\boldsymbol{r},\boldsymbol{p}) = \frac{\boldsymbol{p}^2}{2M} + V(\boldsymbol{r}). \tag{30}$$

The symbol $\{,\}_M$ is the Moyal bracket whose general form is

$$\{a(\boldsymbol{r},\boldsymbol{p}),b(\boldsymbol{r},\boldsymbol{p})\}_M = \frac{2}{\hbar}\sin\left[\frac{\hbar}{2}\left(\frac{\partial}{\partial \boldsymbol{r}_1}\cdot\frac{\partial}{\partial \boldsymbol{p}_2}-\frac{\partial}{\partial \boldsymbol{p}_1}\cdot\frac{\partial}{\partial \boldsymbol{r}_2}\right)\right]a(\boldsymbol{r},\boldsymbol{p})b(\boldsymbol{r},\boldsymbol{p}). \quad (31)$$

Retaining only the first term in the expansion of the sin- operator gives the Poisson bracket of classical mechanics [17], i. e.,

$$\{a(\boldsymbol{r},\boldsymbol{p}),b(\boldsymbol{r},\boldsymbol{p})\}_P = \frac{\partial a(\boldsymbol{r},\boldsymbol{p})}{\partial \boldsymbol{r}}\cdot\frac{\partial b(\boldsymbol{r},\boldsymbol{p})}{\partial \boldsymbol{p}}-\frac{\partial a(\boldsymbol{r},\boldsymbol{p})}{\partial \boldsymbol{p}}\cdot\frac{\partial b(\boldsymbol{r},\boldsymbol{p})}{\partial \boldsymbol{r}}. \quad (32)$$

With a Hamiltonian of the form (30) we get

$$\{h(\boldsymbol{r},\boldsymbol{p}),W(\boldsymbol{r},\boldsymbol{p},t)\}_P = \frac{\partial V(\boldsymbol{r})}{\partial \boldsymbol{r}}\cdot\frac{\partial W(\boldsymbol{r},\boldsymbol{p},t)}{\partial \boldsymbol{p}}-\frac{\boldsymbol{p}}{M}\cdot\frac{\partial W(\boldsymbol{r},\boldsymbol{p},t)}{\partial \boldsymbol{r}}. \quad (33)$$

It is easy to verify that in the special, but important case where $V(\boldsymbol{r})$ is a polynomial of at most second degree in the components of $\boldsymbol{r}$, the Moyal bracket $\{h(\boldsymbol{r},\boldsymbol{p}),W(\boldsymbol{r},\boldsymbol{p},t)\}_M$ simplifies to the Poisson bracket $\{h(\boldsymbol{r},\boldsymbol{p}),W(\boldsymbol{r},\boldsymbol{p},t)\}_P$. The equation of motion for the Wigner function, Eq. (28), becomes then the classical Liouville equation [17],

$$\frac{\partial}{\partial t}W(\boldsymbol{r},\boldsymbol{p},t) = \{h(\boldsymbol{r},\boldsymbol{p}),W(\boldsymbol{r},\boldsymbol{p},t)\}_P, \quad (34)$$

and hence the Wigner function develops in time in the same way as a classical phase-space distribution. In the special case of a free particle, where the potential energy vanishes, we get

$$\frac{\partial}{\partial t}W(\boldsymbol{r},\boldsymbol{p},t) = -\frac{\boldsymbol{p}}{M}\cdot\frac{\partial W(\boldsymbol{r},\boldsymbol{p},t)}{\partial \boldsymbol{r}}. \quad (35)$$

This differential equation has the solution

$$W(\boldsymbol{r},\boldsymbol{p},t) = W(\boldsymbol{r}-\boldsymbol{p}t/M,\boldsymbol{p}), \quad (36)$$

where $W(\boldsymbol{r},\boldsymbol{p})$ is the Wigner function at time zero. In the following, we shall study the spreading of a free-particle wave packet by means of this equation. But first, we shall rest for a while in position space.

4 Generalized Position-Space Densities

In the previous section's Eqs. (6) and (7), we noted that the particle densities in position space and momentum space could be obtained by integrating the Wigner function over momentum space and position space, respectively. The position-space particle density $\rho(\boldsymbol{r})$ and the momentum-space particle density $\pi(\boldsymbol{p})$ are both quantities that can be measured. We shall now consider some more general densities which usually cannot be measured. They are, nevertheless, of a considerable conceptual value.

The densities in question may be obtained from expressions like that of Eq. (10) for the expectation value of an operator $\hat{A}$. Thus, we define the position space density associated with the operator $\hat{A}$ and the state described by the Wigner function $W(\boldsymbol{r},\boldsymbol{p})$ as the quantity

$$\alpha(\boldsymbol{r}) = \int a(\boldsymbol{r},\boldsymbol{p})W(\boldsymbol{r},\boldsymbol{p})d\boldsymbol{p}. \tag{37}$$

It is a density because simple integration over $\boldsymbol{r}$ yields the expectatation value

$$\langle\hat{A}\rangle = \int \alpha(\boldsymbol{r})d\boldsymbol{r}. \tag{38}$$

We may, in general, assume that $\hat{A}$ is a Hermitian operator. Hence, $a(\boldsymbol{r},\boldsymbol{p})$ is a real-valued quantity, and so is the density $\alpha(\boldsymbol{r})$. But we can of course not assume that it is a non-negative quantity.

Expressions like that of Eq. (10) also occur in other phase-space representations. But the functions corresponding to $a(\boldsymbol{r},\boldsymbol{p})$ and $W(\boldsymbol{r},\boldsymbol{p})$ are then different, and so are the functions corresponding to $\alpha(\boldsymbol{r})$. In previous work [18, 19, 20, 21], we have considered various density functions for one- and many-electron atoms, namely kinetic energy densities, potential-energy densities and exchange-energy densities. For the kinetic-energy densities, we made comparisons between the density functions derived from different representations, and concluded that the function derived from the Weyl-Wigner representation had some particularly appealing features. This represents yet another reason for adapting the Weyl-Wigner representation as the canonical phase-space representation. Still another reason is represented by the following section, in which we show that the well-known expression for the probability current density follows naturally from the Weyl-Wigner representation, as the density function corresponding to the velocity operator.

5 Probability Current Density

In this section we shall consider the density function corresponding to the velocity operator

$$\hat{\boldsymbol{v}} = \frac{\hat{\boldsymbol{p}}}{M}, \tag{39}$$

with M being the mass of our particle. The Weyl transform of $\hat{\boldsymbol{v}}$ is simply

$$\boldsymbol{v} = \frac{\boldsymbol{p}}{M}. \tag{40}$$

We shall consider a general time-dependent state, and hence work with the time-dependent Wigner function $W(\boldsymbol{r}, \boldsymbol{p}, t)$, Eq. (25). We denote the density function defined by the dynamical function $\boldsymbol{v}$ and the Wigner function $W(\boldsymbol{r}, \boldsymbol{p}, t)$ by $\boldsymbol{S}(\boldsymbol{r}, t)$. Thus,

$$\boldsymbol{S}(\boldsymbol{r}, t) = \frac{1}{M} \int \boldsymbol{p} W(\boldsymbol{r}, \boldsymbol{p}, t) d\boldsymbol{p}. \tag{41}$$

With the form of the Wigner function available, $\boldsymbol{S}(\boldsymbol{r}, t)$ may be readily evaluated. It may, however, also be evaluated directly from the wave function, and we shall now show that $\boldsymbol{S}(\boldsymbol{r}, t)$ is identical with the familiar probability current density.

Let us first treat the one-dimensional case, $D = 1$. The expression (25) becomes then

$$W(x, p, t) = \frac{1}{2\pi\hbar} \int \psi^*(x - x'/2, t) \psi(x + x'/2, t) e^{-ipx'/\hbar} dx', \tag{42}$$

where x and p denote the one-dimensional position and momentum variables, respectively, both ranging from $-\infty$ to ∞. Eq. (41) becomes

$$S(x, t) = \frac{1}{M} \frac{1}{2\pi\hbar} \int \int p \psi^*(x - x'/2, t) \psi(x + x'/2, t) e^{-ipx'/\hbar} dx' dp. \tag{43}$$

To proceed, we recall the following representations of the δ-function and its derivative

$$\delta(y) = \left(\frac{1}{2\pi\hbar}\right) \int_{-\infty}^{\infty} e^{-ipy/\hbar} dp, \tag{44}$$

$$\delta'(y) = \left(\frac{-i}{2\pi\hbar^2}\right) \int_{-\infty}^{\infty} p e^{-ipy/\hbar} dp. \tag{45}$$

They allow us to write Eq. (43) in the form

$$S(x,t) = \frac{i\hbar}{M} \int \delta'(x') \psi^*(x - x'/2, t) \psi(x + x'/2, t) dx'. \tag{46}$$

We now perform a partial integration and exploit that $\psi(x)$ vanishes at $\pm\infty$. This gives

$$S(x,t) = -\frac{i\hbar}{M} \int \delta(x') \frac{d}{dx'} \left[\psi^*(x - x'/2, t) \psi(x + x'/2, t)\right] dx', \tag{47}$$

or,

$$S(x,t) = -\frac{i\hbar}{2M} \int \delta(x') \left\{ - \left[\frac{d}{dx} \psi^*(x - x'/2, t)\right] \psi(x + x'/2, t) \right.$$
$$\left. + \psi^*(x - x'/2, t) \left[\frac{d}{dx} \psi(x + x'/2, t)\right] \right\}, \tag{48}$$

and finally,

$$S(x,t) = \frac{\hbar}{2iM} \left\{ \psi^*(x,t) \frac{d}{dx} \psi(x,t) - \left[\frac{d}{dx} \psi^*(x,t)\right] \psi(x,t) \right\}. \tag{49}$$

Let us now return to the D-dimensional case of Eq. (41). This is a vector equation. Consider the equation defined by the first component p_1 of $\boldsymbol{p}$, i. e.,

$$S_1(\boldsymbol{r},t) = \frac{1}{M} \int p_1 W(\boldsymbol{r},\boldsymbol{p},t) d\boldsymbol{p}. \tag{50}$$

This equation may be treated in a similar way as Eq. (43), by repeated use of Eqs. (44) and (45). The equation defined by p_2 may be similarly treated, and we finally arrive at the following expression for $\boldsymbol{S}(\boldsymbol{r},t)$:

$$\boldsymbol{S}(\boldsymbol{r},t) = \frac{\hbar}{2iM} \left\{ \psi^*(\boldsymbol{r},t) \boldsymbol{\nabla} \psi(\boldsymbol{r},t) - \left[\boldsymbol{\nabla} \psi^*(\boldsymbol{r},t)\right] \psi(\boldsymbol{r},t) \right\}. \tag{51}$$

This is the standard expression for the usual probability current density [22], as foreseen. Together with the position-space particle density $\rho(\boldsymbol{r},t)$, it satisfies the continuity equation

$$\frac{\partial}{\partial t} \rho(\boldsymbol{r},t) + \boldsymbol{\nabla} \cdot S(\boldsymbol{r},t) = 0. \tag{52}$$

As an application of the expression (41), we show in Fig. 2 the probability current density for a Gaussian wave packet to be discussed in Sec. 7, after a general section on units and scaling.

6 Free-Particle States. Units and Scaling

The concept of scaling plays a significant role in physics. This was often emphasized by Löwdin, and his article on its use in atomic and molecular physics [23] has been much recognized.

In the following, we shall consider some free-particle states. However, we realize that for a free particle there is no natural length scale. This is in contrast to a particle moving in a potential like, for instance, a harmonic potential or a Coulomb potential. In a harmonic-oscillator potential, the characteristic length is $\sqrt{\hbar/M\omega}$, where ω is the angular frequency of the oscillator and M is the particle mass, in the Coulomb problem the characteristic length is the Bohr radius $a_0 = (\hbar^2/m_e)(4\pi\epsilon_0/e^2)$, where m_e is the electronic mass, e is the electronic charge and ϵ_0 is the so-called permittivity of vacuum. But for a free particle there is no way of combining the parameters of the Hamiltonian into a quantity of dimension length. This implies that the choice of a proper length unit is arbitrary. If, however, some choice has been made, then natural units of other physical quantities offer themselves. The best we can do is, therefore, to specify the natural set of units as one in which

- The unit of length is $1/\alpha_0$ where α_0 is a free parameter of dimension inverse length.
- The unit of linear momentum is $\alpha_0\hbar$.
- The unit of mass is the particle mass M.
- The unit of energy is $\alpha_0^2\hbar^2/M$.
- The unit of time is $M/\alpha_0^2\hbar$.

With the physical quantities measured in these units, the Hamiltonian (29) simply becomes

$$\hat{H} = \frac{1}{2}\boldsymbol{p}^2, \tag{53}$$

and the position-space representation of the Schrödinger equation (26) reads

$$i\frac{\partial}{\partial t}\psi(\boldsymbol{r},t) = -\frac{1}{2}\boldsymbol{\nabla}^2\psi(\boldsymbol{r},t). \tag{54}$$

The time development of the Wigner function is given by Eq. (36) with $M = 1$. Hence, Eq. (36) becomes

$$W(\boldsymbol{r}, \boldsymbol{p}, t) = W(\boldsymbol{r} - \boldsymbol{p}t, \boldsymbol{p}), \tag{55}$$

Assume that the position-space wave function for a chosen free-particle state depends on $\boldsymbol{r}$ through the dimensionless quantity $\alpha\boldsymbol{r}$. It may then be natural to choose $a_0 = 1/\alpha$. This is tantamount to putting $\alpha = 1$ in the expression for the wave function, whereby the wave function is transformed into a standard form. Thus, the role of α becomes that of a scale parameter.

In the remainder of this paper, we shall assume that all physical quantities have been expressed in the units described in the present section.

7 Gaussian Wave Packet in One Dimension

In this section, we shall analyze the time development of a simple one-dimensional state on the basis of the concepts introduced in the previous sections. The state we shall consider is that described by a Gaussian wave packet centered at $x = 0$ at time $t = 0$. The initial position-space wave function, which we denote $\psi_0(x)$, reads

$$\psi_0(x) = \left(\frac{\alpha^2}{\pi}\right)^{\frac{1}{4}} e^{-\frac{1}{2}\alpha^2 x^2}. \tag{56}$$

But in accordance with the remarks of the previous section, we may put $\alpha = 1$, and hence restrict the analysis to the wave function

$$\psi_0(x) = \left(\frac{1}{\pi}\right)^{\frac{1}{4}} e^{-\frac{1}{2}x^2}. \tag{57}$$

The corresponding momentum-space wave function is

$$\varphi_0(p) = \left(\frac{1}{\pi}\right)^{\frac{1}{4}} e^{-\frac{1}{2}p^2}. \tag{58}$$

Thus, there is complete symmetry between x and p at time zero. This symmetry is, however, not preserved in time.

To determine the form of ψ_0 for $t > 0$ (or $t < 0$), one must solve the one-dimensional form of the time-dependent Schrödinger equation (54). For such a simple initial wave function as that of Eq. (57), this is a fairly straightforward problem. One possibility is to use the free-particle propagator (see e. g. [11]). Another possibility is to first solve the equation in momentum space and then transform the result to position space by a Fourier transformation (see e. g. [24]). In any case, the result is

$$\psi_0(x,t) = \left(\frac{1}{\pi(1+it)^2}\right)^{\frac{1}{4}} e^{-\frac{1}{2}x^2/(1+it)}. \tag{59}$$

The position-space particle density is obtained by taking the absolute square of this result,

$$\rho_0(x,t) = \left(\frac{1}{\pi(1+t^2)}\right)^{\frac{1}{2}} e^{-x^2/(1+t^2)}. \tag{60}$$

The momentum-space particle density is independent of t,

$$\pi_0(p,t) = \left(\frac{1}{\pi}\right)^{\frac{1}{2}} e^{-p^2}. \tag{61}$$

Let us now instead look at the problem in phase space. We then begin by determining the Wigner function (4) corresponding to the wave function (57). The result is well known and takes the form

$$W_0(x,p) = \frac{1}{\pi}e^{-x^2-p^2}. \tag{62}$$

This Wigner function is simply the product of $\rho_0(x,0)$ and $\pi_0(p,0)$. The Wigner function at a later time may, of course, be obtained from the $\psi_0(x,t)$ of Eq. (59), but it is much easier and more instructive to use Eq. (55), which immediately gives

$$\begin{aligned} W_0(x,p,t) &= W_0(x-pt,p) \\ &= \frac{1}{\pi}e^{-(x-pt)^2-p^2}. \end{aligned} \tag{63}$$

Integration over p and x, respectively, is readily seen to give the expressions (60) and (61), as it should. For the probability current density (41) we get

$$S_0(x,t) = \left(\frac{1}{\pi(1+t^2)^3}\right)^{\frac{1}{2}} txe^{-x^2/(1+t^2)}. \tag{64}$$

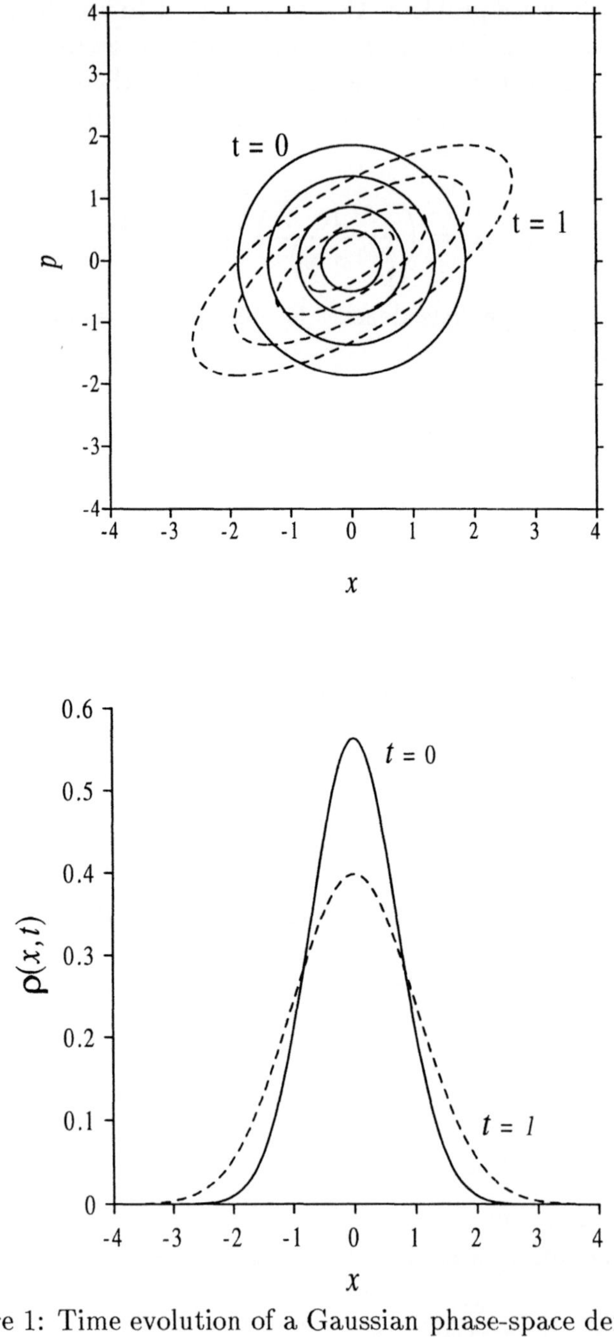

Figure 1: Time evolution of a Gaussian phase-space density (top) and its associated position-space particle density (bottom). The phase-space function is represented by its contour curves.

Figure 1 shows the Wigner function (63) at $t = 0$ and at $t = 1$, together with the corresponding position-space particle density $\rho_0(x,t)$ at the same instants of time. $\rho_0(x,t)$ broadens with time. It is this behavior that is referred to as the spreading of the wave packet. It makes predictions about the position of the particle increasingly imprecise as time goes on. The momentum-space particle density (61) is, as already emphasized, independent of time.

It is often contended that the spreading of the wave packet is a purely quantum-mechanical effect. Figure 1 allows us to elaborate this statement. For it illustrates how the phase-space probability density evolves in time in exactly the same manner as a classical phase-space density does. In this picture, the spreading of the wave-packet comes about, because phase-space points with large p values move faster than points with low p values. Quantum mechanics is important when constructing the initial distribution. It must cover a suitable region of phase space to honor the uncertainty principle. But once a quantum-mechanically proper initial distribution has been set up at time zero, it evolves from there on in the same way as a classical distribution. As already emphasized in connection with Eq. (34), this is true even in the presence of a potential, provided the potential is a polynomial of at most second degree in x.

This discussion illustrates the kind of intuition one may build around the phase-space representation of quantum mechanics. It also shows how the wave-function picture and the phase-space picture supplement each other, in accordance with our remarks in the Introduction. In the wave-function picture, there is no simple way of visualizing the correlation between position and momentum. It is hidden in the Fourier transformations (1) and (2) that connect the wave functions in position space and momentum space. In the phase-space picture, on the other hand, the correlation may be visualized by means of a contour plot like that in Fig 1, even as a function of time. The fact that $W_0(x,p,t)$ is not a true probability density needs not disturb us.

In the following section, we shall see more examples of the mutual supplementation of the wave-function picture and the phase-space picture, as we leave the one-dimensional world.

We close this section by presenting Fig. 2. It shows the probability current density derived from the Wigner function (63) by means of the expression (41), again at time $t = 0$ and $t = 1$. The current density at $t = 0$ is zero. This is in accordance with the fact that $\psi_0(x,0)$ is real. For, as Eq. (49) shows, the current density vanishes for any real-valued

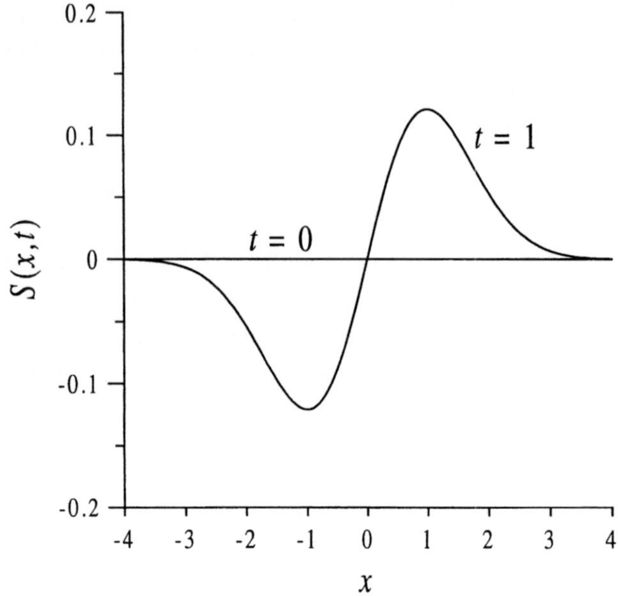

Figure 2: Probability current density for a Gaussian wave packet.

wave function.

8 Gaussian Wave Packet in Two Dimensions

In this section we move to two dimensions and consider a Gaussian wave packet $|\psi_a\rangle$ centered on the origin of position space at time $t = 0$:

$$\begin{aligned} \psi_a(\boldsymbol{r}) &= \frac{1}{\sqrt{\pi}} e^{-\frac{1}{2}(x^2+y^2)} \\ &= \frac{1}{\sqrt{\pi}} e^{-\frac{1}{2}r^2}. \end{aligned} \tag{65}$$

Here $\boldsymbol{r} = (x, y)$ and $r = \sqrt{x^2 + y^2}$. The momentum-space wave function is similarly

$$\begin{aligned} \varphi_a(\boldsymbol{p}) &= \frac{1}{\sqrt{\pi}} e^{-\frac{1}{2}(p_x^2+p_y^2)} \\ &= \frac{1}{\sqrt{\pi}} e^{-\frac{1}{2}p^2}, \end{aligned} \tag{66}$$

with $\boldsymbol{p} = (p_x, p_y)$ and $p = \sqrt{p_x^2 + p_y^2}$. These wave functions are simple products of wave functions of the type (57) and (58) for motion in the

x and y directions.

The Wigner function is also the product of the Wigner functions corresponding to motion along x and y. Thus, the initial Wigner function has the form

$$W_a(\boldsymbol{r}, \boldsymbol{p}) = \frac{1}{\pi^2} e^{-(r^2+p^2)}. \tag{67}$$

The Wigner function at a later time is

$$\begin{aligned} W_a(\boldsymbol{r}, \boldsymbol{p}, t) &= W_a(\boldsymbol{r} - \boldsymbol{p}t, \boldsymbol{p}) \\ &= \frac{1}{\pi^2} e^{-r^2-(1+t^2)p^2+2t\boldsymbol{r}\cdot\boldsymbol{p}}, \end{aligned} \tag{68}$$

or

$$W_a(\boldsymbol{r}, \boldsymbol{p}, t) = \frac{1}{\pi^2} e^{-r^2-(1+t^2)p^2+2trp\cos u}, \tag{69}$$

where u is the angle between the vectors $\boldsymbol{r}$ and $\boldsymbol{p}$.

The particle densities in position space and momentum space are

$$\rho_a(\boldsymbol{r}, t) = \frac{1}{\pi(1+t^2)} e^{-r^2/(1+t^2)} \tag{70}$$

and

$$\pi_a(\boldsymbol{p}, t) = \frac{1}{\pi} e^{-p^2}, \tag{71}$$

respectively. These expressions allow us, for instance, to calculate the expectation values of $\hat{r}$ and $\hat{p}$ as

$$\langle \hat{r} \rangle = 2\pi \int_0^\infty r\rho_a(\boldsymbol{r}, t) r dr = \frac{\sqrt{\pi(1+t^2)}}{2} \tag{72}$$

and

$$\langle \hat{p} \rangle = 2\pi \int_0^\infty p\pi_a(\boldsymbol{p}, t) p dp = \frac{\sqrt{\pi}}{2}. \tag{73}$$

In the previous section, we discussed the correlation between x and p with reference to Fig. 1. The Wigner function of the present section depends on three variables, namely, r, p and u. One way to visualize such a function is to draw a (r, p) contour diagram like that of Fig. 1 for each value of u. We have used this procedure in our earlier work on the Wigner function of the hydrogen atom [18]. We shall not pursue it here, but instead take an independent look at the u dependence of the Wigner function. This dependence describes the correlation between the directions of the position vector $\boldsymbol{r}$ and the linear-momentum vector $\boldsymbol{p}$. In the wave-function picture, this correlation is again hidden in the Fourier transformations (1) and (2), but in the phase-space picture it stands out clearly. We shall now construct a graphical representation of this correlation.

To ease the discussion, let us introduce the symbols $\langle \hat{A} \rangle_q$ and $\langle a \rangle_c$, where $\langle \hat{A} \rangle_q = \langle \hat{A} \rangle$, while $\langle a \rangle_c$ denotes the average of the dynamical phase-space function $a(\boldsymbol{r}, \boldsymbol{p}, t)$ with a Wigner function. With the state vector and the Wigner function corresponding to each other, we have of course that

$$\langle F(\hat{\boldsymbol{r}}) \rangle_q = \langle F(\boldsymbol{r}) \rangle_c, \tag{74}$$

and similarly

$$\langle G(\hat{\boldsymbol{p}}) \rangle_q = \langle G(\boldsymbol{p}) \rangle_c, \tag{75}$$

where F and G are arbitrary functions, as in the text following Eq. (10). In general, $\langle \hat{A} \rangle_q$ and $\langle a \rangle_c$ will only be identical when $a(\boldsymbol{r}, \boldsymbol{p}, t)$ is the Weyl transform of $\hat{A}$.

The motivation for working with two different symbols for "expectation values" is the fact that the Wigner representation is a representation in its own right. In accordance with our remarks in the Introduction, this implies that we may apply physical intuition to a dynamical phase-space function without first asking about its corresponding operator. In particular, it is of interest to consider the dynamical functions $\boldsymbol{r} \cdot \boldsymbol{p} = x p_x + y p_y = r \cos u$ and $\boldsymbol{r}_\perp \cdot \boldsymbol{p} = x p_y - y p_x = r \sin u$, where $\boldsymbol{r} = (x, y)$ as usual and $\boldsymbol{r}_\perp = (-y, x)$. We may also refer to $\boldsymbol{r}_\perp \cdot \boldsymbol{p}$ as the angular momentum ℓ. It is the Weyl transform of the operator $\hat{\ell} = \hat{x}\hat{p}_y - \hat{y}\hat{p}_x$. The function $\boldsymbol{r} \cdot \boldsymbol{p}$ is, on the other hand, the Weyl transform of the operator $\frac{1}{2}(\hat{x}\hat{p}_x + \hat{p}_x\hat{x} + \hat{y}\hat{p}_y + \hat{p}_y\hat{y})$.

Using the Wigner function (68) we find that

$$\langle \boldsymbol{r} \cdot \boldsymbol{p} \rangle_c = \langle r p \cos u \rangle_c = t \tag{76}$$

and

$$\langle \boldsymbol{r}_\perp \cdot \boldsymbol{p} \rangle_c = \langle r p \sin u \rangle_c = 0. \tag{77}$$

We also find that

$$\langle (\boldsymbol{r} \cdot \boldsymbol{p})^2 \rangle_c = \langle r^2 p^2 \cos^2 u \rangle_c = \frac{1 + 4t^2}{2} \tag{78}$$

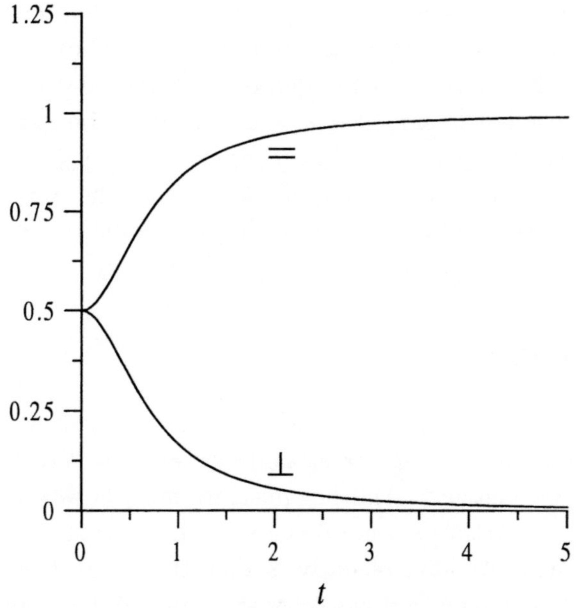

Figure 3: Radial (=) and angular (⊥) kinetic character of a Gaussian wave packet in two dimensions.

and

$$\langle(\boldsymbol{r}_\perp \cdot \boldsymbol{p})^2\rangle_c = \langle r^2 p^2 \sin^2 u\rangle_c = \frac{1}{2}. \tag{79}$$

Normalizing by exploiting that

$$\langle r^2 p^2\rangle_c = 2t^2 + 1, \tag{80}$$

we get that

$$\chi_{(=)} = \frac{\langle r^2 p^2 \cos^2 u\rangle_c}{\langle r^2 p^2\rangle_c} = \frac{4t^2+1}{2(2t^2+1)} \tag{81}$$

and

$$\chi_{(\perp)} = \frac{\langle r^2 p^2 \sin^2 u\rangle_c}{\langle r^2 p^2\rangle_c} = \frac{1}{2(2t^2+1)}. \tag{82}$$

We may refer to these expressions, which add up to 1 at any instant, as the radial and angular kinetic characters of our wave packet. They are represented graphically in Fig. 3.

$\chi_{(=)}$ and $\chi_{(\perp)}$ are measures of the correlation between the directions of $\boldsymbol{r}$ and $\boldsymbol{p}$ as functions of time. They are both equal to one half at

$t = 0$. This is in accordance with the fact that the Wigner function, Eq. (67), is a product of an r distribution and a p distribution at $t = 0$. All angles between $\boldsymbol{r}$ and $\boldsymbol{p}$ are then equally probable, and the chance that $\boldsymbol{p}$ be parallel to $\boldsymbol{r}$ is the same as the chance that $\boldsymbol{p}$ be perpendicular to $\boldsymbol{r}$. But as the time-dependent Wigner function of Eq. (69) shows, small values of u will be favored for larger values of t, and $\chi_{(=)}$ will become the dominant part, tending to one, while $\chi_{(\perp)}$ tends to zero. This is, in fact, the behavior of $\chi_{(=)}$ and $\chi_{(\perp)}$ shown in Fig. 3.

9 Discussion

The quantities $\chi_{(=)}$ and $\chi_{(\perp)}$ that we have introduced above are probably the simplest measures of the correlation between the directions of $\boldsymbol{r}$ and $\boldsymbol{p}$ that one may construct. Another measure may, however, be obtained by focusing on a particle's kinetic energy and its separation into radial and angular parts. We have recently studied this separation for arbitrary dimension D, and shown that the separation runs differently in the wave-function picture and the phase-space picture [25]. For a free particle, the time dependence of the separation in the phase-space picture follows a similar trend as that shown in Fig. 3. In fact, the radial kinetic energy behaves like $\chi_{(=)}$, while the angular kinetic energy behaves like $\chi_{(\perp)}$. For the time-independent ground state of the hydrogen atom we found the remarkable result that the kinetic energy is purely radial in the wave-function picture and purely angular in the phase-space picture. This finding, which relates to arbitrary dimension D, accentuates the difference in intuition that one may apply within the two pictures.

10 Acknowledgments

It is a pleasure to acknowledge the support of the Alexander von Humboldt-Stiftung and the great hospitality enjoyed at the Abteilung für Quantenphysik. Wolfgang P. Schleich is, in particular, thanked for many fruitful discussions.

References

[1] P.-O. Löwdin, Phys. Rev. **97**, 1474, 1490, 1509 (1955).

[2] P.-O. Löwdin, Adv. Chem. Phys. **2**, 207 (1959).

[3] J. P. Dahl, in *Conceptual Trends in Quantum Chemistry*, p. 199 (Eds. E. S. Kryachko and J. L. Calais, Kluwer, Dordrecht, 1994).

[4] J. J. Wlodarz, Phys. Lett. A **286**, 97 (2001).

[5] P.-O. Löwdin, Int. J. Quant. Chem. **12**, Suppl. 1, 197 (1977).

[6] E. Wigner, Phys. Rev. **40**, 749 (1931).

[7] H. Weyl, *The Theory of Groups and Quantum Mechanics* (Methuen, London, 1931).

[8] H. J. Groenewold, Physica **12**, 405 (1946).

[9] J. E. Moyal, Proc. Cambridge Phil. Soc. **45**, 99 (1949).

[10] M. Hillery, R.F. O'Connell, M. O. Scully, and E. P. Wigner, Physics Reports **106**, 121 (1984).

[11] W. P. Schleich, *Quantum Optics in Phase Space* (Wiley-VCH, Berlin, 2001).

[12] A. Grossmann, Commun. Math. Phys. **48**, 191 (1976).

[13] A. Royer, Phys. Rev. A **15**, 449 (1977).

[14] J. P. Dahl, Phys. Scripta **25**, 499 (1982).

[15] J. P. Dahl, Physica **114A**, 439 (1982).

[16] P. A. M. Dirac, *The Principles of Quantum Mechanics*, 4th ed. (Oxford University Press, Oxford, 1958).

[17] H. Goldstein, *Classical Mechanics*, 2nd ed. (Addison-Wesley, Reading, Massachusetts, 1980).

[18] J. P. Dahl and M. Springborg, Mol. Phys. **47**, 1001 (1982).

[19] M. Springborg and J. P. Dahl, in *Local Density Approximations in Quantum Chemistry and Solid-State Physics*, p. 381 (Eds. J. P. Dahl and J. Avery, Plenum, New York, 1984).

[20] J. P. Dahl and B. Amstrup, J. Mol. Struct. (Theochem) **261**, 345 (1992).

[21] M. Springborg and J. P. Dahl, J. Chem. Phys. **110**, 9360 (1999).

[22] L. I. Schiff, *Quantum Mechanics*, 3rd ed. (McGraw-Hill, New York, 1968).

[23] P.-O. Löwdin, Mol. Spectroscopy **3**, 46 (1959).

[24] J. P. Dahl, *Introduction to the Quantum World of Atoms and Molecules* (World Scientific, Singapore, 2001).

[25] J. P. Dahl and W. P. Schleich, Phys. Rev. A **65**, 022109 (2002).

Harmonic Polynomials, Hyperspherical Harmonics, and Sturmians

John Avery
H.C. Ørsted Institute
University of Copenhagen

1 Introduction

Per-Olov Löwdin was one of the great pioneers of quantum chemistry; and his work initiated an astonishing number of fruitful lines of research. For example, one can think of his pioneering papers on angular momentum projection operators, on Sturmian basis functions, and on scaling in quantum theory [1-4]. In this chapter, which is dedicated to Professor Löwdin's memory, I would like discuss the connections between angular momentum theory (in d dimensions), Sturmians, generalized Sturmians and scaling.

2 Homogeneous and harmonic polynomials

The theory of angular momentum in d-dimansional spaces is most easily approached by considering the properties of harmonic polynomials; and in order to understand harmonic polynomials, we can begin by discussing first monomials and then homogeneous polynomials.

2.1 Monomials

Let $x_1, x_2, x_3, ..., x_d$ be the Cartesian coordinates of a d-dimensional space. A *monomial* of order n in these coordinates is defined as a product of the form

$$m_n = x_1^{n_1} x_2^{n_2} x_3^{n_3} ... x_d^{n_d} \tag{1}$$

E.J. Brändas and E.S. Kryachko (eds.),
Fundamental World of Quatum Chemistry, Vol. I, 261–296.

where

$$n_1 + n_2 + \ldots + n_d = n \tag{2}$$

It is easy to see from this definition that

$$\frac{\partial m_n}{\partial x_j} = n_j x_j^{-1} m_n \tag{3}$$

and therefore from equations (2) and (3),

$$\sum_{j=1}^{d} x_j \frac{\partial m_n}{\partial x_j} = n m_n \tag{4}$$

For example,

$$m_4 = x_1^2 x_2 x_3 \tag{5}$$

is a monomial of order 4 in the coordinates x_1, x_2, and x_3, since

$$n_1 + n_2 + n_3 = 2 + 1 + 1 = 4 \tag{6}$$

and

$$\sum_{j=1}^{3} x_j \frac{\partial m_4}{\partial x_j} = x_1(2x_1 x_2 x_3) + x_2(x_1^2 x_3) + x_3(x_1^2 x_2) = 4x_1^2 x_2 x_3 = m_4 \tag{7}$$

2.2 Homogeneous polynomials

A *homogeneous* polynomial of order n is defined as a linear combination of monomials of order n. For example,

$$f_4 = 2x_1^2 x_2 x_3 + 3x_4^4 + 4x_5^2 x_6^2 \tag{8}$$

is a homogeneous polynomial of order 4. Since equation (4) holds for every monomial term in a homogeneous polynomial f_n, it follows that

$$\sum_{j=1}^{d} x_j \frac{\partial f_n}{\partial x_j} = n f_n \tag{9}$$

We now introduce the *generalized Laplacian operator*

$$\Delta \equiv \sum_{j=1}^{d} \frac{\partial^2}{\partial x_j^2} \tag{10}$$

and the *hyperradius*

$$r^2 \equiv \sum_{j=1}^{d} x_j^2 \tag{11}$$

From equations (9), (10), and (11), it follows (with a little work) that if f_α is a homogeneous polynomial of order α and if r is the hyperradius, then

$$\Delta\left(r^\beta f_\alpha\right) = \beta(\beta + d + 2\alpha - 2)r^{\beta-2} f_\alpha + r^\beta \Delta f_\alpha \tag{12}$$

Equation (12) is an important result, which is of great utility in the theory of harmonic polynomials.

2.3 Harmonic polynomials

A *harmonic* polynomial h_n of order n is defined as a homogeneous polynomial of order n which also obeys the generalized Laplace equation:

$$\Delta h_n = 0 \tag{13}$$

For example

$$h_4 = 3x_4^2 x_5 x_6 - x_5^3 x_6 \tag{14}$$

is a harmonic polynomial of order 4. In the case of harmonic polynomials the relationship shown in equation (12) takes on a simpler form. Because of equation (13), the last term in (12) vanishes if the polynomial is harmonic, and we have

$$\Delta\left(r^\beta h_\alpha\right) = \beta(\beta + d + 2\alpha - 2)r^{\beta-2} h_\alpha \tag{15}$$

2.4 Canonical decomposition of a homogeneous polynomial into a series of harmonic polynomials

For every homogeneous polynomial f_n it is possible to find harmonic polynomials h_n, h_{n-2} h_{n-4} such that

$$f_n = h_n + r^2 h_{n-2} + r^4 h_{n-4} + ... \tag{16}$$

In other words, every homogeneous polynomial can be decomposed into a series of harmonic polynomials multiplied by appropriate powers of the hyperradius. To see that this is the case, we can apply the generalized Laplacian operator Δ repeatedly to both sides of equation (16), making use of equation (15). We thus obtain the relations:

$$\begin{aligned}\Delta f_n &= 2(d+2n-4)h_{n-2} + 4(d+2n-6)r^2 h_{n-4} + ... \\ \Delta^2 f_n &= 8(d+2n-6)(d+2n-8)h_{n-4} + ...\end{aligned} \tag{17}$$

and in general

$$\Delta^\nu f_n = \sum_{k=\nu}^{\left[\frac{1}{2}n\right]} \frac{(2k)!!}{(2k-2\nu)!!} \frac{(d+2n-2k-2)!!}{(d+2n-2k-2\nu-2)!!} r^{2k-2\nu} h_{n-2k} \tag{18}$$

where $\left[\frac{1}{2}n\right]$ means "the largest integer in $n/2$", and where

$$j!! \equiv \begin{cases} j(j-2)(j-4)...4\times 2 & j = \text{even} \\ j(j-2)(j-4)...3\times 1 & j = \text{odd} \end{cases} \tag{19}$$

In this way, we obtain a set of simultaneous equations which can be solved for the harmonic polynomials h_n, h_{n-2}...etc. For example, when $n=2$, equation (16) becomes

$$f_2 = h_2 + r^2 h_0 \tag{20}$$

If we apply Δ to both sides of this equation, we obtain:

$$\Delta f_2 = \Delta(r^2 h_0) = 2dh_0 \tag{21}$$

Equations (20) and (21) form a set of simultaneous equations which can be solved for the two unknowns, h_2 and h_0. We thus obtain:

$$\begin{aligned}h_0 &= \frac{1}{2d}\Delta f_2 \\ h_2 &= f_2 - \frac{r^2}{2d}\Delta f_2\end{aligned} \tag{22}$$

It can be seen from this example, and from the more general case where there are a larger number of simultaneous equations, that the harmonic polynomials in the series (16) can be expressed in terms of various powers of r^2 and of Δ applied to f_n, and that it is possible to rewrite (16) in the form:

$$h_n = f_n + a_2 r^2 \Delta f_n + a_4 r^4 \Delta^2 f_n + a_6 r^6 \Delta^3 f_n + ... \tag{23}$$

where a_2, a_4, a_6, etc. are constants. In order to determine these constants, we can apply Δ to both sides of equation (23). The result is

$$0 = \Delta f_n + a_2 \Delta(r^2 \Delta f_n) + a_4 \Delta(r^4 \Delta^2 f_n) + ... \tag{24}$$

If we now make use of equation (15) and collect terms in $\Delta^\nu f_n$, we obtain the simultaneous equations

$$\begin{aligned} 1 + 2(d + 2n - 4)a_2 &= 0 \\ a_2 + 4(d + 2n - 6)a_4 &= 0 \\ a_4 + 6(d + 2n - 8)a_6 &= 0 \\ \vdots &= \vdots \end{aligned} \tag{25}$$

From the set of equations (25) it follows that

$$\begin{aligned} h_n &= f_n - \frac{r^2}{2(d+2n-4)} \Delta f_n + ... \\ &= \sum_{j=0}^{[n/2]} \frac{(-1)^j (d+2n-2j-4)!!}{(2j)!!(d+2n-4)!!} r^{2j} \Delta^j f_n \end{aligned} \tag{26}$$

This gives us an expression for the harmonic polynomial of highest order in the canonical decomposition of the homogeneous polynomial f_n, equation (16). Using similar arguments, one can show that the harmonic polynomials of lower order in the decomposition of equation (16) are given by:

$$\begin{aligned} h_{n-2\nu} &= \frac{(d+2n-4\nu-2)!!}{(2\nu)!!(d+2n-2\nu-2)!!} \\ &\times \sum_{j=0}^{[\frac{n}{2}-\nu]} \frac{(-1)^j (d+2n-4\nu-2j-4)!!}{(2j)!!(d+2n-4\nu-4)!!} r^{2j} \Delta^{j+\nu} f_n \end{aligned} \tag{27}$$

A particularly important special case of equation (27) is the case where n is even and where $n - 2\nu = 0$. In that case, the only term in the series in equation (27) is the $j = 0$ term, and the equation reduces to

$$h_0 = \frac{(d-2)!!}{n!!(d+n-2)!!}\Delta^{\frac{1}{2}n} f_n \tag{28}$$

h_0 is just a constant (or zero), since Δ has been applied to f_n so many times that no dependence on the coordinates remains. For example, if we apply the generalized Laplacian operator Δ once to the 4th order homogeneous polynomial shown in equation (8), we obtain

$$\Delta f_4 = 2x_2x_3 + 36x_4^2 + 8x_5^2 + 8x_6^2 \tag{29}$$

A second application of Δ yields

$$\Delta^2 f_4 = 104 \tag{30}$$

so that in this example

$$h_0 = \frac{104(d-2)!!}{4!!(d+4-2)!!} = \frac{13}{d(d+2)} \tag{31}$$

where d is the dimension of the space. This example illustrates the fact that if h_0 occurs at all in the canonical decomposition of a homogeneous polynomial, it is just a constant. We can extend our illustrative example and calculate all of the harmonic polynomials in the canonical decomposition of f_4. From equation (27) it follows that for any 4th order homogeneous polynomial f_4, the harmonic polynomials in the canonical decomposition

$$f_4 = h_4 + r^2 h_2 + r^4 h_0 \tag{32}$$

are given by

$$\begin{aligned}
h_0 &= \frac{1}{8d(d+2)}\Delta^2 f_4 \\
h_2 &= \frac{1}{2(d+4)}\Delta f_4 - \frac{r^2}{4d(d+4)}\Delta^2 f_4 \\
h_4 &= f_4 - \frac{r^2}{2(d+4)}\Delta f_4 + \frac{r^4}{8(d+2)(d+4)}\Delta^2 f_4
\end{aligned} \tag{33}$$

One can verify that when the harmonic polynomials shown in equation (33) are substituted into (32), equation (32) is an identity. Furthermore, by applying Δ to the polynomials in (33), it is easy to show that they are indeed harmonic. When f_4 is the particular 4th order homogeneous polynomial shown in equation (8), substitution of (29) and (30) into (33) yields

$$h_2 = \frac{1}{d+4}\left[x_2x_3 + 18x_4^2 + 4x_5^2 + 4x_6^2\right] - \frac{26r^2}{d(d+4)}$$

$$h_4 = f_4 - \frac{r^2}{d+4}\left[x_2x_3 + 18x_4^2 + 4x_5^2 + 4x_6^2\right] + \frac{13r^4}{(d+2)(d+4)} \qquad (34)$$

2.5 Generalized angular momentum

The canonical decomposition of a homogeneous polynomial f_n into a series of harmonic polynomials multiplied by appropriate powers of the hyperradius, equation (16), can be viewed as a decomposition of f_n into eigenfunctions of the *generalized angular momentum operator* Λ^2 (sometimes called the grand angular momentum operator):

$$\Lambda^2 \equiv -\sum_{i>j}^{d}\sum_{j=1}^{d}\left(x_i\frac{\partial}{\partial x_j} - x_j\frac{\partial}{\partial x_i}\right)^2 \qquad (35)$$

When $d = 3$, this operator can be seen to reduce to the familiar angular momentum operator

$$L^2 = L_1^2 + L_2^2 + L_3^2 \qquad (36)$$

where

$$L_1 = \frac{1}{i}\left(x_2\frac{\partial}{\partial x_3} - x_3\frac{\partial}{\partial x_2}\right) \qquad (37)$$

L_2 and L_3 being similarly defined with cyclic permutations of the coordinates. By expanding the square in equation (35), it is possible to rewrite the generalized angular momentum operator in the form

$$\Lambda^2 = -r^2\Delta + \sum_{i,j=1}^{d} x_ix_j\frac{\partial^2}{\partial x_i\partial x_j} + (d-1)\sum_{i=1}^{d} x_i\frac{\partial}{\partial x_i} \qquad (38)$$

We now let Λ^2 act on a homogeneous polynomial f_n, making use of equation (9). This yields

$$\Lambda^2 f_n = -r^2 \Delta f_n + n(d-1) f_n + \sum_{i,j=1}^{d} x_i x_j \frac{\partial^2 f_n}{\partial x_i \partial x_j} \tag{39}$$

The last term in equation (39) can be simplified by means of the relationship

$$\sum_{i,j=1}^{d} x_i x_j \frac{\partial^2 f_n}{\partial x_i \partial x_j} = n(n-1) f_n \tag{40}$$

the proof of which is similar to the proof of equation (9). Combining (39) and (40), we obtain

$$\Lambda^2 f_n = -r^2 \Delta f_n + n(n+d-2) f_n \tag{41}$$

From (41) it follows that if the generalized angular momentum operator is applied to a harmonic polynomial of order λ, the result is

$$\Lambda^2 h_\lambda = \lambda(\lambda + d - 2) h_\lambda \tag{42}$$

where λ is zero or a positive integer. When $d = 3$ (and $\lambda = l$), this reduces to

$$L^2 h_l = l(l+1) h_l \tag{43}$$

Thus we can see that in a 3-dimensional space, a harmonic polynomial of order l is an eigenfunction of angular momentum corresponding to the eigenvalue $l(l+1)$, while in the general case of a d-dimensional space, a harmonic polynomial of order λ is an eigenfunction of generalized angular momentum with the eigenvalue $\lambda(\lambda + d - 2)$.

2.6 Hyperangular integrations

In a 3-dimensional space, the volume element can be expressed in the form

$$dx_1 dx_2 dx_3 = r^2 dr\, d\Omega \tag{44}$$

where $d\Omega$ is the element of solid angle. This is a particular case of the more general relationship

$$dx_1 dx_2 ... dx_d = r^{d-1} dr\, d\Omega \tag{45}$$

where r is the hyperradius and where $d\Omega$ is the generalized solid angle element. Eigenfunctions of the generalized angular momentum operator Λ^2 belonging to different eigenvalues are orthogonal to each other when integrated over generalized solid angle, so that

$$\int d\Omega\, h_\lambda^* h_{\lambda'} = 0 \quad \text{if } \lambda \neq \lambda' \tag{46}$$

We saw above that h_0 is just a constant, and therefore when $\lambda' = 0$ we can write

$$\int d\Omega\, h_\lambda^* h_0 = h_0 \int d\Omega\, h_\lambda^* = 0 \quad \text{if } \lambda \neq 0 \tag{47}$$

from which it follows that

$$\int d\Omega\, h_\lambda = 0 \quad \text{if } \lambda \neq 0 \tag{48}$$

Since the hyperradius is unaffected by an integration over generalized solid angle, we can combine (48) and (16) to obtain the result

$$\int d\Omega\, f_n = \begin{cases} r^n h_0 \int d\Omega & n = \text{even} \\ 0 & n = \text{odd} \end{cases} \tag{49}$$

where f_n is an arbitrary homogeneous polynomial of order n. We already have a formula for h_0, equation (28), but in order to complete the angular integration, we need to evaluate the total solid angle $\int d\Omega$. We can do this by multiplying both sides of equation (44) by e^{-r^2} and integrating:

$$\int_0^\infty dr\, r^{d-1} e^{-r^2} \int d\Omega = \prod_{j=1}^{d} \int_{-\infty}^\infty dx_j e^{-x_j^2} \tag{50}$$

The integral over the hyperradius on the left side of equation (50) can be expressed in terms of the gamma function

$$\int_0^\infty dr\, r^{d-1} e^{-r^2} = \frac{\Gamma(d/2)}{2} \tag{51}$$

and the gamma function can also be used to evaluate the integrals over the Cartesian coordinates on the right hand side of (50):

$$\int_{-\infty}^\infty dx_j e^{-x_j^2} = \Gamma(1/2) = \pi^{\frac{1}{2}} \tag{52}$$

Combining (50), (51) and (52) we obtain an expression for the total solid angle in a d-dimensional hyperspace:

$$\int d\Omega = \frac{2\pi^{\frac{d}{2}}}{\Gamma\left(\frac{d}{2}\right)} \tag{53}$$

Thus, finally, we obtain a powerful angular integration formula which allows us to find the integral over the solid angle of any homogeneous polynomial of order n: If n is even,

$$\int d\Omega\, f_n = \frac{2\pi^{d/2} r^n (d-2)!!}{\Gamma(d/2) n!!(d+n-2)!!} \Delta^{\frac{1}{2}n} f_n \tag{54}$$

If n is odd the integral vanishes. It is interesting to notice that in equation (54), we integrate by differentiating!

It is possible to generalize equation (54) and to construct a formula which will allow us to calculate the integral over solid angle of any function whatever of $x_1, x_2, ..., x_d$, provided that the function can be expanded about the origin in a convergent series of polynomials in the coordinates: Suppose that such an expansion is possible for the function $F(\mathbf{x})$. Then we can write:

$$F(\mathbf{x}) = \sum_{n=0}^{\infty} f_n(\mathbf{x}) \tag{55}$$

where the f_n's are homogeneous polynomials of order n. Then from equation (54), we have:

$$\int d\Omega F(\mathbf{x}) = \frac{(d-2)!!2\pi^{d/2}}{\Gamma\left(\frac{d}{2}\right)} \sum_{n=0,2,..}^{\infty} \frac{r^n}{n!!(n+d-2)!!} \Delta^{n/2} f_n(\mathbf{x}) \tag{56}$$

The terms for which n is odd have been omitted from the sum because all of them are zero. If we apply the generalized Laplacian operator $n/2$ times to $F(\mathbf{x})$ and set $\mathbf{x} = 0$ in the resulting expression, the only term which survives is the constant:

$$\left[\Delta^{n/2} F(\mathbf{x})\right]_{\mathbf{x}=0} = \Delta^{n/2} f_n(\mathbf{x}) \tag{57}$$

Combining (56) and (57) and letting $\nu \equiv n/2$, we obtain

$$\int d\Omega F(\mathbf{x}) = \frac{(d-2)!!2\pi^{d/2}}{\Gamma\left(\frac{d}{2}\right)} \sum_{\nu=0}^{\infty} \frac{r^{2\nu}}{(2\nu)!!(d+2\nu-2)!!} \left[\Delta^{\nu} F(\mathbf{x})\right]_{\mathbf{x}=0} \tag{58}$$

When $d = 3$, this reduces to

$$\int d\Omega F(\mathbf{x}) = 4\pi \sum_{\nu=0}^{\infty} \frac{r^{2\nu}}{(2\nu+1)!} \left[\Delta^{\nu} F(\mathbf{x})\right]_{\mathbf{x}=0} \tag{59}$$

It is easy to implement the angular integration formulas (54), (58), and (59) using programming systems like Mathematica or Maple, where differentiation is a rapidly-performed operation. When $F(\mathbf{x})$ is a polynomial, the series in (58) and (59) terminate.

3 Hyperspherical harmonics

3.1 The relationship between harmonic polynomials and hyperspherical harmonics

In 3-dimensional space, the Laplacian operator can be written in terms of radial differentiations and the angular momentum operator:

$$\Delta = \frac{1}{r^2}\frac{\partial}{\partial r} r^2 \frac{\partial}{\partial r} - \frac{L^2}{r^2} \tag{60}$$

The d-dimensional generalization of this relation is:

$$\Delta = \sum_{j=1}^{d} \frac{\partial^2}{\partial x_j^2} = \frac{1}{r^{d-1}}\frac{\partial}{\partial r} r^{d-1} \frac{\partial}{\partial r} - \frac{\Lambda^2}{r^2} \tag{61}$$

where r is the hyperradius and where Λ^2 is the generalized angular momentum operator. Now suppose that $h_\lambda(\mathbf{x})$ is a harmonic polynomial of order λ. We can let

$$Y_\lambda(\mathbf{u}) \equiv r^{-\lambda} h_\lambda(\mathbf{x}) \tag{62}$$

where the unit vector $\mathbf{u}$ is defined by

$$\mathbf{u} \equiv \frac{1}{r}\{x_1, x_2, ..., x_d\} \tag{63}$$

In other words, if we divide a harmonic polynomial of order λ by r^λ, we obtain a function $Y_\lambda(\mathbf{u})$ which is independent of the hyperradius, and which depends only on the hyperangles. Such a function is called a *hyperspherical harmonic* [8-10]. We would like to show that a hyperspherical harmonic, generated in this way from a harmonic polynomial h_λ, is an eigenfunction of generalized angular momentum corresponding to the eigenvalue $\lambda(\lambda + d - 2)$. To do this we let Δ act on h_λ, making use of equations (61) and (62):

$$\begin{aligned} \Delta h_\lambda &= \left[\frac{1}{r^{d-1}}\frac{\partial}{\partial r}r^{d-1}\frac{\partial}{\partial r} - \frac{\Lambda^2}{r^2}\right] r^\lambda Y_\lambda(\mathbf{u}) \\ &= \\ &= \left[\lambda(\lambda + d - 2) - \Lambda^2\right] r^{\lambda-2} Y_\lambda(\mathbf{u}) = 0 \end{aligned} \tag{64}$$

from which it follows that

$$\Lambda^2 Y_\lambda(\mathbf{u}) = \lambda(\lambda + d - 2) Y_\lambda(\mathbf{u}) \tag{65}$$

since r commutes with Λ^2. It is interesting to notice that equation (65) can also be derived in a different way, starting with (42) and dividing both sides by r^λ. Notice also that when $d = 3$, (and $\lambda = l$), (65) reduces to

$$L^2 Y_l(\mathbf{u}) = l(l+1) Y_l(\mathbf{u}) \tag{66}$$

The familiar spherical harmonics of a 3-dimensional space are, in fact, a special case of hyperspherical harmonics. Just as the familiar spherical harmonics require another index, m, to distinguish between the various linearly independent eigenfunctions of L^2 belonging to a particular value of l, so the hyperspherical harmonics in spaces of higher dimension require $d-2$ other indices to distinguish between the various eigenfunctions of Λ^2 belonging to a particular value of λ. In mathematical discussions of hyperspherical harmonics, it is usual to organize this set of additional indices according to the chain of subgroups:

$$SO(d) \supset SO(d-1) \supset SO(d-2) .... \supset SO(2) \tag{67}$$

For example, the familiar spherical harmonics in a 3-dimensional space are eigenfunctions of both L^2 and L_z. The index l labels the different eigenfunctions of L^2, and it can also be thought of as a label for the class 1 irreducible representations of the group $SO(3)$. The index m of the familiar spherical harmonics labels the eigenfunctions of L_z, and it is also a label for the class 1 irreducible representations of $SO(2)$. Similarly, in a 4-dimensional space, the index λ labels the eigenfunctions of Λ^2 (with d=4), and since L^2 and L_z commute with each other and with Λ^2, it is possible to construct a set of hyperspherical harmonics which are simultaneous eigenfunctions of all three operators. These 4-dimensional hyperspherical harmonics will be labeled by 3 indices, λ, l, and m, corresponding respectively to the irreducible representations of $SO(4)$, $SO(3)$, and $SO(2)$. The chain of subgroups shown in equation (67) is often called the *standard tree.* Although the sets of hyperspherical harmonics generated by organizing the minor indices according to the standard tree are convenient in mathematical discussions, they are not necessarily the most convenient for use in physical problems. One can alternatively generate sets of hyperspherical harmonics where the minor indices are organized according to nonstandard trees. General nonstandard trees for many-dimensional spaces have been studied in detail by Aquilanti and coworkers at the University of Perugia, and by a number of Russian authors; but in this article we will consider only standard trees.

3.2 The standard tree

Suppose that we have generated a harmonic polynomial h_μ of order μ in the $d-1$ dimensional space of the coordinates $x_1, x_2, ..., x_{d-1}$. If we multiply this harmonic polynomial by $x_d^{\lambda-\mu}$, the result will be a homogeneous polynomial of order λ in the d-dimensional space of the coordinates $x_1, x_2, ..., x_d$:

$$f_{\lambda,\mu}(x_1, ..., x_d) = x_d^{\lambda-\mu} h_\mu(x_1, ..., x_{d-1}) \tag{68}$$

From equation (26) it follows that if we apply the operator

$$O_\lambda = \sum_{k=0}^{[\lambda/2]} \frac{(-1)^k (d+2\lambda-2k-4)!!}{(2k)!!(d+2\lambda-4)!!} r^{2k} \Delta^k \tag{69}$$

to this homogeneous polynomial, the result will be a harmonic polynomial of order λ. The effect of the d-dimensional Laplacian operator on the homogeneous polynomial is easy to calculate: Since

$$\Delta h_\mu(x_1, ..., x_{d-1}) = 0 \tag{70}$$

it follows that

$$\Delta^k [x_d^{\lambda-\mu} h_\mu(x_1, ..., x_{d-1})] = \frac{(\lambda-\mu)!}{(\lambda-\mu-2k)!} x_d^{\lambda-\mu-2k} h_\mu(x_1, ..., x_{d-1}) \tag{71}$$

Thus we obtain:

$$\begin{aligned} &O_\lambda \left[x_d^{\lambda-\mu} h_\mu(x_1, ..., x_{d-1}\right] \\ &= r^{\lambda-\mu} h_\mu(x_1, ..., x_{d-1}) \frac{(\lambda-\mu)!}{(d+2\lambda-4)!!} \\ &\times \sum_{k=0}^{[(\lambda-\mu)/2]} \frac{(-1)^k (d+2\lambda-2k-4)!!}{(2k)!!(\lambda-\mu-2k)!} \left(\frac{x_d}{r}\right)^{\lambda-\mu-2k} \end{aligned} \tag{72}$$

In other words, the result of the projection is the harmonic polynomial $h_\mu(x_1, ..., x_{d-1})$ in the $d-1$-dimensional subspace, multiplied by $r^{\lambda-\mu}$ and by a finite series in x_d/r, where r is the hyperradius in the d-dimensional space. Some examples of equation (72) for various values of $\lambda - \mu$ are shown in Table 1. Table 2 illustrates the same relationships as Table 1, but Table 2 shows the special case where $d = 3$, $\lambda = l$, $\mu = m$ and $h_\mu(x_1, ..., x_{d-1}) = (x_1 + ix_2)^m$. It is easy to see that the mth order polynomial $(x_1 \pm ix_2)^m$ is harmonic, since

$$\frac{\partial^2}{\partial x_1^2}(x_1 \pm ix_2)^m = m(m-1)(x_1 \pm ix_2)^m \tag{73}$$

and

$$\frac{\partial^2}{\partial x_2^2}(x_1 \pm ix_2)^m = -m(m-1)(x_1 \pm ix_2)^m \tag{74}$$

so that

$$\Delta(x_1 \pm ix_2)^m = 0 \tag{75}$$

Starting with this mth order harmonic polynomial in the space of x_1 and x_2, we can construct lth order harmonic polynomials in the 3-dimensional space x_1, x_2, x_3 by multiplying by x_3^{l-m} and projecting out the harmonic polynomial of largest order in the canonical decomposition, as shown in Table 2. If we then divide by $r_{(3)}^l$, where

$$r_{(3)} \equiv \left(x_1^2 + x_2^2 + x_3^2\right)^{1/2} \tag{76}$$

Table 1: Equation (72) for various values of $\lambda - \mu$.

$\lambda - \mu$	$O_\lambda \left[x_d^{\lambda-\mu} h_\mu(x_1, ..., x_{d-1})\right]$
0	$h_\mu(x_1, ..., x_{d-1})$
1	$x_d h_\mu(x_1, ..., x_{d-1})$
2	$\left(x_d^2 - \frac{r^2}{d+2\lambda-4}\right) h_\mu(x_1, ..., x_{d-1})$
3	$\left(x_d^3 - \frac{3x_d r^2}{d+2\lambda-4}\right) h_\mu(x_1, ..., x_{d-1})$

Table 2: Equation (72) (Table 1) in the special case where $d = 3$, $\lambda = l$, $\mu = m$, and $h_\mu(x_1, .., x_{d-1}) = (x_1 + ix_2)^m$. The familiar 3-dimensional spherical harmonics (unnormalized) are shown in the right-hand column.

$l-m$	$O_l\left[x_3^{l-m}(x_1+ix_2)^m\right]$	$Y_{l,m}(\theta,\phi)$
0	$(x_1+ix_2)^m$	$\sin^m\theta e^{im\phi}$
1	$x_3(x_1+ix_2)^m$	$\cos\theta\sin^m\theta e^{im\phi}$
2	$\left(x_3^2 - \frac{r_{(3)}^2}{2l+1}\right)(x_1+ix_2)^m$	$\left(\cos^2\theta - \frac{1}{2l+1}\right)\sin^m\theta e^{im\phi}$
3	$\left(x_3^3 - \frac{3x_3 r_{(3)}^2}{2l+1}\right)(x_1+ix_2)^m$	$\left(\cos^3\theta - \frac{3\cos\theta}{2l+1}\right)\sin^m\theta e^{im\phi}$

is the radius in the 3-dimensional space, we obtain functions which, apart from a normalization factor, are just the familiar 3-dimensional spherical harmonics $Y_{l,m}$. Having found 3-dimensional harmonic polynomials labeled by l and by m, the indices of the irreducible representations of $SO(3)$ and $SO(2)$, we can use these as a starting point for constructing 4-dimensional harmonic polynomials in the hyperspace x_1, x_2, x_3, x_4, as shown in Table 3. This table is the same as Table 1, except that it illustrates the special case where $d = 4$, $\mu = l$ and $h_\mu(x_1, ..., x_{d-1}) = h_{l,m}(x_1, x_2, x_3)$, the harmonic polynomials $h_{l,m}$ being those shown in Table 2. In Table 3, the index λ labels the irreducible representations of $SO(4)$. Thus the indices λ, l, m are organized according to the chain of subgroups

Table 3: Equation (72) (Table 1) in the special case where $d = 4$, $\mu = l$ and $h_\mu(x_1, ..., x_{d-1}) = h_{l,m}(x_1, x_2, x_3)$. The (unnormalized) 4-dimensional hyperspherical harmonics are shown in the right-hand column.

$\lambda - l$	$O_\lambda\left[x_4^{\lambda-l} h_{l,m}(x_1, x_2, x_3)\right]$	$Y_{\lambda,l,m}(\chi, \theta, \phi)$
0	$h_{l,m}$	$\sin^l \chi Y_{l,m}(\theta, \phi)$
1	$x_4 h_{l,m}$	$\cos\chi \sin^l \chi Y_{l,m}(\theta, \phi)$
2	$\left(x_4^2 - \frac{r_{(4)}^2}{2\lambda}\right) h_{l,m}$	$\left(\cos^2 \chi - \frac{1}{2\lambda}\right) \sin^l \chi Y_{l,m}(\theta, \phi)$
3	$\left(x_4^3 - \frac{3x_4 r_{(4)}^2}{2\lambda}\right) h_{l,m}$	$\left(\cos^3 \chi - \frac{3\cos\chi}{2\lambda}\right) \sin^l \chi Y_{l,m}(\theta, \phi)$

shown in equation (67), d being equal to 4. If we divide these polynomials by $r_{(4)}^{\lambda}$ where

$$r_{(4)} \equiv \left(x_1^2 + x_2^2 + x_3^2 + x_4^2\right)^{1/2} \tag{77}$$

and if we normalize the resulting functions, we obtain the set of 4-dimensional hyperspherical harmonics shown in Table 4. In Table 4, $u_1, ..., u_4$ are components of a 4-dimensional unit vector related to the Cartesian coordinates $x_1, ..., x_4$ by equation (63). The components of this unit vector can also be expressed in terms of angular coordinates:

$$\begin{aligned} u_1 &= \sin\chi \sin\theta \cos\phi \\ u_2 &= \sin\chi \sin\theta \sin\phi \\ u_3 &= \sin\chi \cos\theta \\ u_4 &= \cos\chi \end{aligned} \tag{78}$$

In equation (78), θ and ϕ are the usual angles of the spherical polar coordinates in a 3-dimensional space, while χ is defined by

$$\sin\chi \equiv \frac{r_{(3)}}{r_{(4)}} \tag{79}$$

from which it follows that

$$\cos\chi = \frac{x_4}{r_{(4)}} \tag{80}$$

The hyperspherical harmonics in a 4-dimensional spece can be shown to have the general form:

$$Y_{\lambda,l,m}(\mathbf{u}) = \mathcal{N}_{\lambda,l} C_{\lambda-l}^{l+1}(\cos\chi) \sin^l \chi Y_{l,m}(\theta, \phi) \tag{81}$$

Where $\mathcal{N}_{\lambda,l}$ is a normalization factor

$$\mathcal{N}_{\lambda,l} = (-1)^{\lambda} i^l (2l)!! \sqrt{\frac{2(\lambda+1)(\lambda-l)!}{\pi(\lambda+l+1)!}} \tag{82}$$

and where $C_{\lambda}^{\alpha}(\cos\chi)$ is a Gegenbauer polynomial

$$C_{\lambda}^{\alpha}(\cos\chi) = \sum_{t=0}^{[\lambda/2]} \frac{(-1)^t \Gamma(\lambda+\alpha-t)}{t!(\lambda-2t)!\Gamma(\alpha)} (2\cos\chi)^{\lambda-2t} \tag{83}$$

Table 4: 4-dimensional hyperspherical harmonics. The unit vectors $u_1, ..., u_4$ are related to the anngles and coordinates through equations (78) and (63).

λ	l	m	$\sqrt{2}\pi\ Y_{\lambda,l,m}(\mathbf{u})$
0	0	0	1
1	1	1	$i\sqrt{2}(u_1 + iu_2)$
1	1	0	$-i2u_3$
1	1	−1	$-i\sqrt{2}(u_1 - iu_2)$
1	0	0	$-2u_4$
2	2	2	$-\sqrt{3}(u_1 + iu_2)^2$
2	2	1	$2\sqrt{3}u_3(u_1 + iu_2)$
2	2	0	$-\sqrt{2}(2u_3^2 - u_1^2 - u_2^2)$
2	2	−1	$-2\sqrt{3}u_3(u_1 - iu_2)$
2	2	−2	$-\sqrt{3}(u_1 - iu_2)^2$
2	1	1	$-i2\sqrt{3}\ u_4(u_1 + iu_2)$
2	1	0	$2i\sqrt{6}\ u_4u_3$
2	1	−1	$2i\sqrt{3}\ u_4(u_1 - iu_2)$
2	0	0	$4u_4^2 - 1$

The interested reader may enjoy verifying that equation (81) leads to the 4-dimensional hyperspherical harmonics shown in Tables 3 and 4. The angular integration formula (54) may also be used to verify that the 4-dimensional hyperspherical harmonics shown in Table 4 are properly normalized and that they obey the orthonormality relation:

$$\int d\Omega\, Y^*_{\lambda',l',m'} Y_{\lambda,l,m} = \delta_{\lambda',\lambda}\delta_{l',l}\delta_{m',m} \tag{84}$$

3.3 Gegenbauer polynomials

Gegenbauer polynomials, equation (83), play a role in the theory of hyperspherical harmonics which is analogous to the role played by Legendre polynomials in the theory of the familiar 3-dimensional spherical harmonics. Indeed, the Legendre polynomials are a special case of Gegenbauer polynomials corresponding to $d = 3$ and $\alpha = 1/2$. All of the beautiful theorems which hold for 3-dimensional spherical harmonics have d-dimansional generalizations in which Legendre polynomials are replaced by Gegenbauer polynomials. For example, the familiar expansion of the Green's function of the Laplacian operator in terms of Legendre polynomials,

$$\frac{1}{|\mathbf{x}-\mathbf{x}'|} = \frac{1}{r_>(1+\epsilon^2-2\epsilon\mathbf{u}\cdot\mathbf{u}')^{1/2}} = \frac{1}{r_>}\sum_{l=0}^{\infty}\left(\frac{r_<}{r_>}\right)^l P_l(\mathbf{u}\cdot\mathbf{u}') \tag{85}$$

has the d-dimensional generalization:

$$\frac{1}{|\mathbf{x}-\mathbf{x}'|^{d-2}} = \frac{1}{r_>^{d-2}(1+\epsilon^2-2\epsilon\mathbf{u}\cdot\mathbf{u}')^{\alpha}} = \frac{1}{r_>^{d-2}}\sum_{\lambda=0}^{\infty}\left(\frac{r_<}{r_>}\right)^\lambda C_\lambda^\alpha(\mathbf{u}\cdot\mathbf{u}') \tag{86}$$

where $\mathbf{x}$ and $\mathbf{u}$ are d-dimensional vectors and where

$$\epsilon \equiv \frac{r_<}{r_>} \tag{87}$$

and

$$\alpha \equiv \frac{d-2}{2} \tag{88}$$

Similarly, the sum rule for spherical harmonics,

$$\frac{4\pi}{2l+1}\sum_m Y^*_{lm}(\mathbf{u}')Y_{lm}(\mathbf{u}) = P_l(\mathbf{u}\cdot\mathbf{u}') \tag{89}$$

has the d-dimensional generalization

$$\frac{2\pi^{\frac{d}{2}}(d-2)}{\Gamma\left(d/2\right)\left(2\lambda+d-2\right)}\sum_{\{\mu\}} Y^*_{\lambda\{\mu\}}(\mathbf{u}')Y_{\lambda\{\mu\}}(\mathbf{u}) = C^\alpha_\lambda(\mathbf{u}\cdot\mathbf{u}') \tag{90}$$

while the expansion of a plane wave in terms of Legendre polynomials,

$$e^{i\mathbf{p}\cdot\mathbf{x}} = \sum_{l=0}^{\infty} i^l(2l+1)j_l(pr)P_l(\mathbf{u}_p\cdot\mathbf{u}) \tag{91}$$

has the generalization

$$e^{i\mathbf{p}\cdot\mathbf{x}} = \sum_{\lambda=0}^{\infty} i^\lambda(2\lambda+d-2)(d-4)!!j^d_\lambda(pr)C^\alpha_\lambda(\mathbf{u}_p\cdot\mathbf{u}) \tag{92}$$

where $\mathbf{u}_p$ is a d-dimensional unit vector in the direction of $\mathbf{p}$ and where

$$j^d_\lambda(pr) \equiv \sum_{t=0}^{\infty}\frac{(-1)^t(pr)^{2t+\lambda}}{(2t)!!(d+2t+2\lambda-2)!!} \tag{93}$$

is what might be called a "hyperspherical Bessel function". Proofs of these relations (and a more complete discussion) may be found in references [9] and [10].

4 Sturmians

4.1 Fock's mapping and hydrogenlike Sturmians in momentum space

In 1935, V. Fock published a remarkable paper [17-19] in which he solved the Schrödinger equation of a hydrogenlike atom in momentum space. Fock demonstrated that if 3-dimensional momentum space is projected onto the surface of a 4-dimensional hypersphere by the mapping

$$\begin{aligned} u_1 &= \frac{2k_\mu p_1}{k_\mu^2+p^2} = \sin\chi_p\sin\theta_p\cos\phi_p \\ u_2 &= \frac{2k_\mu p_2}{k_\mu^2+p^2} = \sin\chi_p\sin\theta_p\sin\phi_p \end{aligned}$$

$$u_3 = \frac{2k_\mu p_3}{k_\mu^2 + p^2} = \sin\chi_p \cos\theta_p$$
$$u_4 = \frac{k_\mu^2 - p^2}{k_\mu^2 + p^2} = \cos\chi_p \tag{94}$$

then the Fourier transforms of the hydrogenlike orbitals can be expressed very simply in terms of the 4-dimensional hyperspherical harmonics through the relationship

$$\chi^t_{n,l,m}(\mathbf{p}) = M(p) Y_{n-1,l,m}(\mathbf{u}) \tag{95}$$

where

$$M(p) \equiv \frac{4k_\mu^{5/2}}{(k_\mu^2 + p^2)^2} \tag{96}$$

If one lets $k_\mu = Z/n$, then these momentum-space solutions are the Fourier transforms of the familiar hydrogenlike orbitals. On the other hand, if k_μ is held constant for all values of the quantum numbers n, l, and m, then the functions defined by equations (95) and (96) are the Fourier transforms of the Shull-Löwdin hydrogenlike Sturmians [3-6], as shown in Table 5. It can be seen that substitution of the hyperspherical harmonics shown in Table 4 into equation (95), with $u_1, ..., u_4$ defined by (94), yields the functions shown in Table 5.

In direct space, these hydrogenlike Sturmians are defined by the relations

$$\chi_{nlm}(\mathbf{x}) = R_{nl}(r) Y_{lm}(\theta, \phi) \tag{97}$$

and

$$R_{nl}(r) = \mathcal{N}_{nl}(2k_\mu r)^l e^{-k_\mu r} F(l + 1 - n|2l + 2|2k_\mu r) \tag{98}$$

where

$$\mathcal{N}_{nl} = \frac{2k_\mu^{3/2}}{(2l+1)!}\sqrt{\frac{(l+n)!}{n(n-l-1)!}} \tag{99}$$

and where

$$F(a|b|x) = 1 + \frac{ax}{b} + \frac{a(a+1)x^2}{b(b+1)2!} + ... \tag{100}$$

is a confluent hypergeometric function. It is an interesting exercise to evaluate the Fourier-Bessel transforms of $\chi_{1,0,0}(\mathbf{x})$, $\chi_{2,0,0}(\mathbf{x})$, and

Table 5: Hydrogenlike Sturmians in momentum space

$$\chi^t_{1,0,0}(p) = \frac{4k_\mu^{5/2}}{\sqrt{2}\pi(p^2+k_\mu^2)^2}$$

$$\chi^t_{2,0,0}(p) = -2\frac{4k_\mu^{5/2}}{\sqrt{2}\pi(p^2+k_\mu^2)^2}\frac{k_\mu^2-p^2}{p^2+k_\mu^2}$$

$$\chi^t_{2,1,0}(p) = -2i\frac{4k_\mu^{5/2}}{\sqrt{2}\pi(p^2+k_\mu^2)^2}\frac{2k_\mu p_3}{p^2+k_\mu^2}$$

$$\chi^t_{3,0,0}(p) = \frac{4k_\mu^{5/2}}{\sqrt{2}\pi(p^2+k_\mu^2)^2}\left[4\left(\frac{k_\mu^2-p^2}{p^2+k_\mu^2}\right)^2-1\right]$$

$$\chi^t_{3,1,0}(p) = 2i\sqrt{6}\frac{4k_\mu^{5/2}}{\sqrt{2}\pi(p^2+k_\mu^2)^2}\left(\frac{k_\mu^2-p^2}{p^2+k_\mu^2}\right)\left(\frac{2k_\mu p_3}{p^2+k_\mu^2}\right)$$

$\chi_{2,1,0}(\mathbf{x})$, and to show that the transforms are those given in Table 5. The first few radial functions are

$$\begin{aligned} R_{10}(r) &= 2k_\mu^{3/2}e^{-k_\mu r} \\ R_{20}(r) &= 2k_\mu^{3/2}(1-k_\mu r)e^{-k_\mu r} \\ R_{21}(r) &= \frac{2k_\mu^{5/2}}{\sqrt{3}}\, r\, e^{-k_\mu r} \end{aligned} \tag{101}$$

If k_μ is set equal to Z/n, these correspond to the familiar hydrogenlike radial functions. On the other hand, when k_μ is held constant, the one-electron basis set defined by equations (98)-(100) are the Sturmian basis functions introduced by Shull and Löwdin, who were able to show that such a basis set is complete without the inclusion of the continuum. One might ask why completeness is so important, since in practice one always truncates a basis set. The answer is that the usual hydrogenlike radial functions, which are polynomials multiplied by $e^{-Zr/n}$, have nodes at larger and larger values of r as n increases. Thus the radial functions corresponding to large values of n are not very useful for constructing an approximate series representation of a function whose nodes occur at small values of r. Calculational experience has shown that a series of Sturmians is able to give a very accurate and rapidly convergent representation of an atomic or molecular orbital, and therefore Sturmian basis sets are becoming more and more widely used in the quantum mechanical treatment of many-electron systems. The Shull-Löwdin Sturmians are solutions of the differential equation

$$\left[-\frac{1}{2}\nabla^2 + \frac{1}{2}k_\mu^2 - \frac{nk_\mu}{r}\right]\chi_{nlm}(\mathbf{x}) = 0 \tag{102}$$

Notice that if k_μ were replaced by Z/n, this would be the differential equation obeyed by the familiar hydrogenlike orbitals; but for the Sturmians, the energy $-k_\mu^2/2$ is the same for all the members of the basis set, regardless of the quantum numbers. Another (complex conjugated) member of the basis set, with quantum numbers n', l', and m', will obey

$$\left[-\frac{1}{2}\nabla^2 + \frac{1}{2}k_\mu^2 - \frac{n'k_\mu}{r}\right]\chi^*_{n'l'm'}(\mathbf{x}) = 0 \tag{103}$$

We can use equations (102) and (103) to show that the Sturmians obey a potential-weighted orthogonality relation in direct spece: If we multiply the equations from the left respectively by $\chi^*_{n'l'm'}(\mathbf{x})$ and $\chi_{nlm}(\mathbf{x})$, and if we afterwards integrate over coordinates and take the difference between the two equations, making use of Hermiticity, we obtain

$$(n - n') \int d^3x \chi^*_{n'l'm'}(\mathbf{x}) \frac{1}{r} \chi_{nlm}(\mathbf{x}) = 0 \tag{104}$$

Therefore the potential-weighted integral in equation (104) must vanish if $n \neq n'$; and from the orthogonality of the spherical harmonics, we know that the integral must also vanish if $l \neq l'$ and/or $m \neq m'$. With the normalization shown in equation (99), the potential-weighted orthonormality relation obeyed by the Sturmian basis set is

$$\int d^3x \; \chi^*_{n'l'm'}(\mathbf{x}) \chi_{nlm}(\mathbf{x}) \frac{1}{r} = \frac{k_\mu}{n} \delta_{n'n} \delta_{l'l} \delta_{m'm} \tag{105}$$

The orthonormality of the 4-dimensional hyperspherical harmonics can be used to show that hydrogenlike Sturmian basis sets also obey a weighted orthonormality relation in reciprocal space: From Fock's mapping, equation (94), it follows that

$$d\Omega = \sin^2 \chi_p \sin \theta_p \; d\chi_p d\theta_p d\phi_p = \left(\frac{2k_\mu}{k_\mu^2 + p^2} \right)^3 d^3p \tag{106}$$

Combining equations (106), (95) and (96) with the orthonormality relations for the 4-dimensional hyperspherical harmonics, we obtain:

$$\int d^3p \left(\frac{k_\mu^2 + p^2}{2k_\mu^2} \right) \chi^{t*}_{n',l'm'}(\mathbf{p}) \chi^t_{n,l,m}(\mathbf{p}) = \delta_{n'n} \delta_{l'l} \delta_{m'm} \tag{107}$$

from which it is possible to derive (105). The function space of a basis set which obeys this type of momentum-space orthonormality relation is called a Sobolev space [44]. One can use (107) to derive the Sturmian expansion of a plane wave: Since the basis set $\chi^{t*}_{n'l'm'}(\mathbf{p})$ is complete, we can use it to expand any function of $\mathbf{p}$; and therefore we can write:

$$e^{i\mathbf{p}\cdot\mathbf{x}} = \left(\frac{k_\mu^2 + p^2}{2k_\mu^2} \right) \sum_{n'l'm'} \chi^{t*}_{n'l'm'}(\mathbf{p}) a_{n'l'm'} \tag{108}$$

The coefficients $a_{n'l'm'}$ can be determined by multiplying both sides of (108) by $d^3p\ \chi^t_{nlm}(\mathbf{p})$ and integrating:

$$\int d^3p\ \chi^t_{nlm}(\mathbf{p})e^{i\mathbf{p}\cdot\mathbf{x}} = \sum_{nlm} a_{nlm}\int d^3p\left(\frac{k_\mu^2+p^2}{2k_\mu^2}\right)\chi^{t*}_{nlm}(\mathbf{p})\chi^t_{n'l'm'}(\mathbf{p}) \tag{109}$$

Making use of the weighted orthonormality relation, we obtain:

$$a_{nlm} = (2\pi)^{3/2}\chi_{nlm}(\mathbf{x}) \tag{110}$$

$$e^{i\mathbf{p}\cdot\mathbf{x}} = (2\pi)^{3/2}\left(\frac{k_\mu^2+p^2}{2k_\mu^2}\right)\sum_{nlm}\chi^{t*}_{nlm}(\mathbf{p})\chi_{nlm}(\mathbf{x}) \tag{111}$$

4.2 The many-center one-electron problem

Suppose that we wish to solve the Schrödinger equation

$$\left[-\frac{1}{2}\nabla^2 + v(\mathbf{x}) - \epsilon\right]\varphi_\mu(\mathbf{x}) = 0 \tag{112}$$

for an electron moving in the many-center potential

$$v(\mathbf{x}) = -\sum_a \frac{Z_a}{|\mathbf{x}-\mathbf{X}_a|} \tag{113}$$

produced by a number of nuclei with charges Z_a and positions $\mathbf{X}_a$. We can try to represent the solution by a superposition of Sturmian basis functions localized on the various nuclei. If we let τ stand for the set of indices $\{anlm\}$, then we can write

$$\varphi_\mu(\mathbf{x}) = \sum_\tau \chi_\tau(\mathbf{x})C_{\tau,\mu} \tag{114}$$

where

$$\chi_\tau(\mathbf{x}) \equiv \chi_{nlm}(\mathbf{x}-\mathbf{X}_a) \tag{115}$$

One finds that convergence is most rapid if the exponent k_μ characterizing the basis set is linked to the energy ϵ by

$$\epsilon \equiv -\frac{1}{2}k_\mu^2 \tag{116}$$

If one makes this substitution as well as (114), the Schrödinger equation, (112), becomes

$$\sum_{\tau} \left[-\frac{1}{2}\nabla^2 + \frac{1}{2}k_\mu^2 + v(\mathbf{x})\right] \chi_\tau(\mathbf{x}) C_{\tau,\mu} = 0 \tag{117}$$

If we express a member of the Sturmian basis set in terms of its Fourier transform

$$\chi_\tau(\mathbf{x}) = \frac{1}{(2\pi)^{3/2}} \int d^3p \; e^{i\mathbf{p}\cdot\mathbf{x}} \chi_\tau^t(\mathbf{p}) \tag{118}$$

we can obtain the relationship

$$\left[-\frac{1}{2}\nabla^2 + \frac{1}{2}k_\mu^2\right] \chi_\tau(\mathbf{x}) = \frac{1}{(2\pi)^{3/2}} \int d^3p \; e^{i\mathbf{p}\cdot\mathbf{x}} \left(\frac{k_\mu^2 + p^2}{2}\right) \chi_\tau^t(\mathbf{p}) \tag{119}$$

from which it follows that

$$\begin{aligned}
&\int d^3x \; \chi_{\tau'}^*(\mathbf{x}) \left[-\frac{1}{2}\nabla^2 + \frac{1}{2}k_\mu^2\right] \chi_\tau(\mathbf{x}) \\
&= \int d^3p \left(\frac{k_\mu^2 + p^2}{2}\right) \left(\frac{1}{(2\pi)^{3/2}} \int d^3x \; e^{-i\mathbf{p}\cdot\mathbf{x}} \chi_{\tau'}^t(\mathbf{x})\right)^* \chi_\tau^t(\mathbf{p}) \\
&= \int d^3p \left(\frac{k_\mu^2 + p^2}{2}\right) \chi_{\tau'}^{t*}(\mathbf{p}) \chi_\tau^t(\mathbf{p}) \equiv k_\mu^2 \mathcal{S}_{\tau',\tau}
\end{aligned} \tag{120}$$

From (115), (118) and (95), we have

$$\chi_\tau^t(\mathbf{p}) = e^{-i\mathbf{p}\cdot\mathbf{X}_a} \; \chi_{nlm}^t(\mathbf{p}) = e^{-i\mathbf{p}\cdot\mathbf{X}_a} \frac{4k_\mu^{5/2}}{(k_\mu^2 + p^2)^2} Y_{n-1,l,m}(\mathbf{u}) \tag{121}$$

With the help of (96) and (106), substitution of (121) into (120) yields

$$\mathcal{S}_{\tau',\tau} = \int d\Omega \; e^{i\mathbf{p}\cdot(\mathbf{X}_{a'} - \mathbf{X}_a)} Y_{n'-1,l',m'}^*(\mathbf{u}) Y_{n-1,l,m}(\mathbf{u}) \tag{122}$$

where we must remember that τ stands for the set of indices $\{anlm\}$, and where

$$\mathcal{S}_{\tau',\tau} \equiv \frac{1}{k_\mu^2} \int d^3x \; \chi_{\tau'}^*(\mathbf{x}) \left[-\frac{1}{2}\nabla^2 + \frac{1}{2}k_\mu^2\right] \chi_\tau(\mathbf{x}) \tag{123}$$

The integrals shown in equation (122) were first introduced by T. Shibuya and C.E. Wulfman [22-24], who extended the momentum-space methods of Fock to the many-center one-electron problem. In order to evaluate these integrals, Shibuya and Wulfman derived the Sturmian expansion of a plane wave, equation (111), which they rewrote in the form

$$e^{i\mathbf{p}\cdot\mathbf{R}} = \left(\frac{2\pi}{k_\mu}\right)^{3/2} (1+u_4) \sum_{nlm} \chi_{nlm}(\mathbf{R}) Y^*_{n-1,l,m}(\mathbf{u}) \tag{124}$$

Substitution of this expansion into (122) yields

$$\begin{aligned} \mathcal{S}_{\tau',\tau} &= \left(\frac{2\pi}{k_\mu}\right)^{3/2} \sum_{n''l''m''} \chi_{n''l''m''}(\mathbf{X}_{a'} - \mathbf{X}_a) \\ &\times \int d\Omega \; (1+u_4) Y^*_{n''-1,l'',m''}(\mathbf{u}) Y^*_{n'-1,l',m'}(\mathbf{u}) Y_{n-1,l,m}(\mathbf{u}) \end{aligned} \tag{125}$$

Shibuya and Wulfman evaluated the hyperangular integrals shown in equation (125) by means of the coupling coefficients for 4-dimensional hyperspherical harmonics, which they derived by means of group theory. Alternatively, one can make use of the hyperangular integration formulas (54) or (58). One can show that when $d = 4$, these formulas imply that

$$\int d\Omega \prod_{j=1}^{4} u_j^{n_j} = \frac{4\pi^2}{(n_1+n_2+n_3+n_4+2)!!} \prod_{j=1}^{4} (n_j - 1)!! \tag{126}$$

The selection rules become more transparent if the Shibuya-Wulfman integrals are evaluated in two steps: From (124) it follows that

$$\begin{aligned} &\int d\Omega e^{i\mathbf{p}\cdot\mathbf{R}} Y_{n-1,l,m}(\mathbf{u}) \\ &= \left(\frac{2\pi}{k_\mu}\right)^{3/2} \sum_{n'l'm'} \chi_{n'l'm'}(\mathbf{R}) \int d\Omega (1+u_4) Y^*_{n'-1,l',m'}(\mathbf{u}) Y_{n-1,l,m}(\mathbf{u}) \end{aligned} \tag{127}$$

The hyperangular integral in equation (127) vanishes unless $l' = l$, $m' = m$, and $n' = n, n \pm 1$:

$$\int d\Omega (1+u_4) Y^*_{n'-1,l',m'}(\mathbf{u}) Y_{n-1,l,m}(\mathbf{u})$$

$$= \delta_{l'l}\delta_{m'm}\delta_{n'n} - \delta_{l'l}\delta_{m'm}\delta_{n',n+1}\frac{1}{2}\sqrt{\frac{(n-l)(n+l+1)}{n(n+1)}}$$

$$-\delta_{l'l}\delta_{m'm}\delta_{n',n-1}\frac{1}{2}\sqrt{\frac{(n+l)(n-l-1)}{n(n-1)}} \tag{128}$$

so that

$$\left(\frac{k_\mu}{2\pi}\right)^{3/2}\int d\Omega e^{i\mathbf{p}\cdot\mathbf{R}} Y_{n-1,l,m}(\mathbf{u})$$

$$= \chi_{nlm}(\mathbf{R}) - \frac{1}{2}\sqrt{\frac{(n-l)(n+l+1)}{n(n+1)}}\chi_{n+1,l,m}(\mathbf{R})$$

$$-\frac{1}{2}\sqrt{\frac{(n+l)(n-l-1)}{n(n-1)}}\chi_{n-1,l,m}(\mathbf{R}) \tag{129}$$

The Shibuya-Wulfman integrals may then be evaluated by resolving the product $Y^*_{n'-1,l',m'}(\mathbf{u})Y_{n-1,l,m}(\mathbf{u})$ into a superposition of hyperspherical harmonics, afterwards making use of (129). With the help of Table 4, one can show that

$$\begin{aligned} \mathcal{S}_{a'100,a100} &= (1+s)e^{-s} \\ \mathcal{S}_{a'100,a200} &= -2s^2e^{-s}/3 \\ \mathcal{S}_{a'200,a200} &= (3+3s-2s^2+s^3)e^{-s}/3 \end{aligned} \tag{130}$$

and so on, where

$$\mathbf{s} \equiv k_\mu(\mathbf{X}_{a'} - \mathbf{X}_a) \tag{131}$$

An alternative method for evaluating the Shibuya-Wulfman integrals is discussed in reference [10]. From equation (131) it can be seen that the Shibuya-Wulfman integrals are functions of the dimensionless parameters $\mathbf{s}$, defined by equation (128). Likewise it turns out that the matrix elements of the potential can be expressed in terms of this dimensionless perameter. If we let

$$\int d^3x\ \chi^*_{\tau'}(\mathbf{x})v(\mathbf{x})\chi^*_{\tau'}(\mathbf{x}) \equiv -k_\mu\mathcal{V}_{\tau',\tau} \tag{132}$$

then $\mathcal{V}_{\tau',\tau}$ is a function only of $\mathbf{s}$. For example, in the case of diatomics one finds that

$$\mathcal{V}_{a'100,a100} = \frac{1}{s} - \frac{e^{-2s}}{s}(1+s)$$

$$\mathcal{V}_{a'100,a200} = -\frac{1}{2s} + \frac{e^{-2s}}{2s}(1 + 2s + 2s^2)$$
$$\mathcal{V}_{a'200,a200} = \frac{1}{s} - \frac{e^{-2s}}{2s}(2 + 3s + 2s^2 + 2s^3) \quad (133)$$

The matrix elements of the potential can be shown to be related to the Shibuya-Wulfman integrals through a sum rule. If we let

$$K_{\tau',\tau} \equiv \sqrt{\frac{Z_{a'} Z_a}{n'n}} \mathcal{S}_{\tau',\tau} \quad (134)$$

then one can show [10] that

$$\sum_{\tau''} K_{\tau',\tau''} K_{\tau'',\tau} = \sqrt{\frac{Z_{a'} Z_a}{n'n}} \mathcal{V}_{\tau',\tau} \quad (135)$$

When the basis set is truncated, this sum rule is only approximate, but in the limit of a complete set, it becomes exact.

If we multiply equation (117) from the left by a complex conjugate function from the Sturmian basis set and integrate over volume, we obtain:

$$\sum_{\tau} \int d^3x \chi^*_{\tau'}(\mathbf{x}) \left[-\frac{1}{2}\nabla^2 + \frac{1}{2}k_\mu^2 + v(\mathbf{x}) \right] \chi_\tau(\mathbf{x}) C_{\tau,\mu} = 0 \quad (136)$$

Expressing this in terms of the matrices $\mathcal{V}_{\tau',\tau}$ and $\mathcal{S}_{\tau',\tau}$ and dividing through by k_μ, we obtain the Sturmian secular equation for the one-electron many-center problem:

$$\sum_{\tau} \left[\mathcal{V}_{\tau',\tau} - k_\mu \mathcal{S}_{\tau',\tau} \right] C_{\tau,\mu} = 0 \quad (137)$$

As mentioned above, both $\mathcal{V}_{\tau',\tau}$ and $\mathcal{S}_{\tau',\tau}$ do not depend explicitly on k_μ, but only on the dimensionless parameters $\mathbf{s}$ defined by equation (131). Thus the procedure for solving the one-electron many-center problem is as follows: We choose a particular molecular geometry which is of interest, but with an unknown size factor - i.e., the shape of the molecule is known at the outset, but the scale is unknown. We then solve the Sturmian secular equations, (136), and obtain a spectrum of optimum k_μ values. The largest of these values corresponds to the ground state, the next largest to the first excited state, and so on. Substituting the k_μ values into equation (131) we at last find the scale of the molecule for which the solution is appropriate, the energy being given by (116). By repeating this procedure, scaling the **s** parameters to larger and smaller values, keeping the internuclear angles constant, one can obtain energies and wavefunctions as functions of the size of the molecule, for the ground state and for the excited states. It is interesting to notice that that the eigenvalues obtained by solving the secular equation (136) are not energies but values of the Slater exponents which characterize the Sturmian basis set. In a certain sense, one does not know what basis set one is using until after the secular equation has been solved. One knows the form of the basis set, but not the k_μ value which characterizes all the functions in the set. After the secular equation has been solved, one knows both the basis set and the size of the molecule for which the solution is appropriate. For a given molecular size, the ground state solution corresponds to the largest k_μ value, the first excited state to the next largest, and so on, as we would expect, since the solutions for progressively higher energies become progressively more diffuse. Figure 1 shows some of the $m = 0$ excited state obtained in this way for the HHe^{2+} ion as functions of the internuclear separation.

4.3 Generalized Sturmians

In 1968, Goscinski completed an important paper in which he generalized the concept of Sturmian basis sets [7]. He regarded a Sturmian basis set as a set of solutions of the Schrödinger equation

$$\left[-\frac{1}{2}\Delta + \beta_\nu V_0(\mathbf{x}) - E\right] \phi_\nu(\mathbf{x}) = 0 \tag{138}$$

where $V_0(\mathbf{x})$ is a zeroth-order potential, and where the constants β_ν are chosen in such a way that all of the solutions to equation (138) correspond to the same value of the energy E, regardless of their quantum numbers ν.

$$\Delta \equiv \sum_{j=1}^{N} \left(\frac{\partial^2}{\partial x_j^2} + \frac{\partial^2}{\partial y_j^2} + \frac{\partial^2}{\partial z_j^2} \right) \equiv \sum_{j=1}^{N} \Delta_j \tag{139}$$

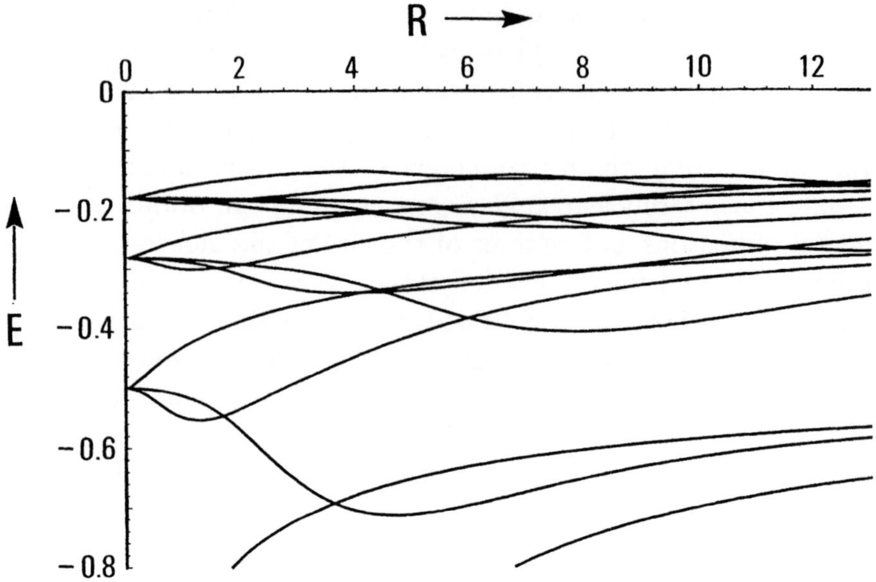

Figure 1: This figure shows some $m = 0$ excited states of a molecular ion with one electron and two nuclei. The nuclear charges are $Z_1 = 1$ and $Z_2 = 2$. Energies are given in Hartrees, and internuclear distances in Bohrs. In the united atom limit, the excited states shown in the figure correspond to $n = 2$, $n = 3$, $n = 4$ and $n = 5$. The calculation was made by means of the method described in equations (112)-(137).

Table 6: Singlet excited state energies (in Hartrees) for the 2-electron isoelectronic series, calculated by James Avery. The basis set used consisted of 36 generalized Sturmians of the Goscinski type. Experimental values are taken from the NIST tables (http://physics.nist.gov/asd). Discrepencies between calculated and experimental energies for the ions may be due to experimental inaccuracies, since, for an isoelectronic series, the accuracy of the generalized Sturmian method increases with increasing atomic number.

	He	Li^+	Be^{2+}	B^{3+}	C^{4+}
$2s\ ^1S$	−2.1429	−5.0329	−9.1730	−14.564	−21.206
expt.	−2.1458	−5.0410	−9.1860	−14.582	−21.230
$3s\ ^1S$	−2.0603	−4.7297	−8.5099	−13.402	−19.406
expt.	−2.0611	−4.7339	−8.5183	−13.415	−19.425
$4s\ ^1S$	−2.0332	−4.6276	−8.2837	−13.003	−18.785
expt.	−2.0334	−4.6299	−8.2891		
$5s\ ^1S$	−2.0210	−4.5811	−8.1806	−12.820	−18.500
expt.	−2.0210	−4.5825			
$6s\ ^1S$	−2.0144	−4.5562	−8.1250	−12.721	−18.346
expt.	−2.0144	−4.5571			
$7s\ ^1S$	−2.0105	−4.5412	−8.0917	−12.662	−18.253
expt.	−2.0104	−4.5418			
$8s\ ^1S$	−2.0080	−4.5315	−8.0701	−12.624	−18.194
expt.	−2.0079				
$9s\ ^1S$	−2.0063	−4.5248	−8.0554	−12.598	−18.153
expt.	−2.0062				
$10s\ ^1S$	−2.0051	−4.5201	−8.0449	−12.579	−18.124
expt.	−2.0050				
$11s\ ^1S$	−2.0042	−4.5166	−8.0371	−12.566	−18.102
expt.	−2.0041				
$12s\ ^1S$	−2.0034	−4.5140	−8.0312	−12.555	−18.086
expt.	−2.0034				

$$\mathbf{x} = \{\mathbf{x}_1, \mathbf{x}_2, ..., \mathbf{x}_N\} \tag{140}$$

$$\mathbf{x}_j \equiv \{x_j, y_j, z_j\} \qquad j = 1, 2, ..., N \tag{141}$$

$$V_0(\mathbf{x}) = -\sum_{j=1}^{N} \frac{Z}{r_j} \tag{142}$$

$$\phi_\nu(\mathbf{x}) = |\chi_\mu \chi_{\mu'} \chi_{\mu''} ...| \tag{143}$$

4.4 References

1. P.O. Löwdin, Phys Rev 1955, 97, 1474.
2. P.O. Löwdin, Appl Phys Supl 1962, 33, 251.
3. Shull, H. and Löwdin, P.O., J Chem Phys 1959, 30, 617.
4. Löwdin, P.O., J Mol Spect 1959, 3, 46-66.
5. Rotenberg, Manuel, Ann Phys (New York) 1962, 19, 62.
6. Rotenberg, Manuel, Adv At Mol Phys 1970, 6, 233-68.
7. Goscinski, O., *Preliminary Research Report No. 217*, Quantum Chemistry Group, Uppsala University, 1968.
8. Vilenkin, N.K., **Special Functions and the Theory of Group Representations**, American Mathematical Society, Providence R.I., 1968.
9. Avery, J., **Hyperspherical Harmonics; Applictions in Quantum Theory**, Kluwer Academic Publishers, Dordrecht, 1989, 255 pages.
10. Avery, J., **Hyperspherical Harmonics and Generalized Sturmians**, Kluwer Academic Publishers, Dordrecht, Netherlands, 2000.
11. Avery, J., *Hyperspherical Harmonics, Some Properties and Applications*, in , **Conceptual Trends in Quantum Chemistry, Volume 1**, Kryachko E.S. and Calais, J.L. editors, Kluwer Academic Publishers, Dordrecht, Netherlands, 1994, pages 135-169.
12. Avery, J., *Hyperspherical Harmonics in Quantum Theory*, in **Problemes Non Lineaires Appliques, 1990-1991; Calcul Scientifique et Chimie Quantique**, Écoles CEA-EDF-INRIA, Rocquencourt, France, 1991, pages 1-28.

13. Wen, Z.-Y. and Avery, J., J Math Phys 1985, 26, 396.
14. Aquilanti, V., Cavalli, S., De Fazio, D, and Grossi, G. *Hyperangular Momentum: Applications to Atomic and Molecular Science*, in **New Methods in Quantum Theory**, Tsipis, C.A., Popov, V.S., Herschbach, D.R., and Avery, J.S., Eds., Kluwer, Dordrecht, 1996.
15. Aquilanti, V., Cavalli, S., Coletti, C. and Grossi, G., Chem Phys 1996, 209, 405.
16. Aquilanti, V., Cavalli, S. and Coletti, C., Chem Phys 1997, 214, 1.
17. Fock, V.A., Z Phys 1935, 98, 145.
18. Fock, V.A., Kgl Norske Videnskab Forh 1958, 31, 138.
19. Bandar, M. and Itzyksen, C., Rev Mod Phys 1966, 38, 330 and 346.
20. Avery, J. and Herschbach, D.R., Int J Quantum Chem 1992, 41, 673.
21. Avery, J., *D-Dimensional Hydrogenlike Orbitals*, in **Dimensional Scaling in Chemical Physics**, Herschbach, D.R., Avery, J. and Goscinski, O. editors, Kluwer Academic Publishers, Dordricht, Netherlands, (1992), pages 139-164.
22. Shibuya, T. and Wulfman, C.E., Proc Roy Soc A 1965, 286, 376.
23. Wulfman, C.E., *Dynamical groups in atomic and molecular physics*, in **Group Theory and its Applications**,
24. Judd, B.R., **Angular Momentum Theory for Diatomic Molecules**, Academic Press, New York, 1975.
25. Monkhorst, H.J. and Jeziorski, B., J Chem Phys 1979, 71, 5268-9.
26. Navasa, J. and Tsoucaris, G., Phys Rev A 1981, 24, 683.
27. Koga, T. and Matsuhashi, T., J Chem Phys 1987, 87, 1677-80, 4696-9.
28. Duchon, C; Dumont-Lepage, M.C.; and Gazeau, J.P., J Chem Phys 1982, 76, 445-7.
29. Avery, J., *Hyperspherical Sturmian Basis Functions in Reciprocal Space*, in **New Methods in Quantum Theory**, Tsipis, C.A., Popov, V. Herschbach, D.R. and Avery J. editors., Kluwer Academic Publishers, Dordrecht, (1995).
30. Avery, J., J Phys Chem 1993, 97, 2406.

31. Avery, J. and Antonsen, F., Theor Chim Acta 1993, 85, 33.
32. Avery, J., J Math Chem 1994, 15, 233.
33. Avery, J., Børsen Hansen, T., Wang, M. and Antonsen, F., Int J Quantum Chem 1996, 57, 401-411.
34. Avery, J., and Børsen Hansen, T., Int J Quantum Chem 1996, 60, 201-211.
35. Aquilanti, V. and Avery, J., Chem Phys Letters 1997, 267, 1-8.
36. Avery, J., J Math Chem 1997, 21, 285-304.
37. Avery , J. and Antonsen, F., J Math Chem 1998, 24, 175-185.
38. Avery, J., Adv Quantum Chem 1999, 31, 201-229.
39. Avery, J., J Mol Struct (Theochem) 1999, 458, 1-9.
40. Avery, J., J Math Chem 1998, 24, 169-174.
41. Avery, J. and Sauer, S., *Many-Electron Sturmians Applied to Molecules*, in **Quantum Systems in Chemistry and Physics, Volume 1**, Hernández-Laguna, A., Maruani, J., McWeeney, R. and Wilson, S. editors, Kluwer Academic Publishers, (2000).
42. Avery, J. and Shim, R., Int J Quantum Chem 2000, 79, 1-7.
43. Aquilanti, V., and Avery, J., Adv. Quantum Chem. 2001, 39, 71-102.
44. Weniger, E.J., J Math Phys 1985, 26, 276.

STURMIAN ORBITALS IN QUANTUM CHEMISTRY: AN INTRODUCTION

Vincenzo Aquilanti and Andrea Caligiana
Dipartimento di Chimica, Università di Perugia, I-06123 Perugia, Italy

Abstract

The Sturmian basis set approach to the solution of quantum chemical problems is the configuration space counterpart of hyperspherical analysis in momentum space and leads to promising algorithms alternative to the usual SCF-plus-configuration interaction hierarchies. Some of the underlying concepts are akin to those which led Shull and Löwdin to introduce "natural spin" orbitals in 1959. This paper is an account of implementations aimed at the solution of the simplest problems of atomic and molecular structure, in order to exhibit the most characteristic features. We review applications to one-electron multicenter (e.g. H_2^+) and multielectron one-center (e.g. atoms) problems. We also introduce the basic steps of a computational scheme for molecular structure problems, where Sturmian orbitals appear as building blocks in a valence-bond type of approach. The example is the hydrogen molecule. The mathematical ingredients are considered, particularly the possibility of exploiting relationships with Slater-type orbitals in order to circumvent the bottleneck of calculation of matrix elements.

1 General

Being the Schrödinger equation (SE) for the non-relativistic hydrogen atom analytically soluble in a closed form, the use of hydrogenoid orbitals as basis sets to approximate the wave functions of atoms and molecules has been extensively considered since the early days of quantum mechanics. Slater-type orbitals (STO) and other variants have been proposed and amply tested as alternatives. In order to tackle the obstacles encountered for the efficient computation of matrix elements (specifically the bielectronic integrals), preference has been given to the Gaussian orbitals (GTO), particularly convenient for such purposes.

E.J. Brändas and E.S. Kryachko (eds.),
Fundamental World of Quatum Chemistry, Vol. I, 297–316.

The current software available for *ab initio* calculations of atomic and molecular properties, typically exploiting Gaussian basis set expansions, almost invariably initiate with Self Consistent Field (SCF) or Multiconfiguration Self Consistent Field (MC-SCF) calculations yielding molecular orbitals as linear combinations of GTO centered on the various nuclei of the molecule, and then proceed with post-Hartree-Fock techniques (Configuration Interaction, Coupled Clusters, ...). Among the disadvantages of such a procedure, we mention the difficulties encountered in the calculation of long-range interactions and of the properties of excited states, and the slow convergence in reproducing the cusps in the electronic densities at nuclei.

The *Sturmian Orbitals* are a set of function closely related to hydrogenoid orbitals. They solve a differential eigenvalue equation of the Sturm-Liouville type (hence the name "Sturmians")

$$\left[-\frac{1}{2}\nabla^2 + \frac{1}{2}k_\mu^2 - \frac{nk_\mu}{r}\right]\chi_{nlm}(\mathbf{x}) = 0 \tag{1}$$

which is very similar to the SE for the one-electron atom: the difference is that in Eq.(1) the energy here written as $-k_\mu^2/2$ will be treated as an arbitrary parameter, while the "nuclear charge" nk_μ is allowed to vary. It is interesting to note that a basis set of functions as χ_{nlm} were termed "natural orbital" by Shull and Löwdin [1] and observed to enjoy the completeness property with no need of inclusion of the continuum, as required for variational calculations using the constant energy orbitals defined by the same Eq.(1).

The Sturmian orbitals can be straightforwardly expressed as linear combinations of Slater-type orbitals. The use of the latter has been superseded by the GTO, for which the slower convergence is often amply balanced by the much simpler algorithm for the calculation of integrals. Apart from this, the employed techniques are identical. The approach involving Sturmian orbitals will be shown to proceed along a route fully alternative to modern *ab initio* quantum chemistry: it will be seen that there is no need of iterative Hartree-Fock calculations, because the introduction of the Sturmian expansion of the electronic wave function leads after algebraic manipulations to a generalized eigenvalue problem (GEP) where the matrix elements contain mono- and bi-electronic integrals among Sturmian orbitals and the eigenvalue is not the electronic energy E but the momentum wave number p_0 such that $E = -p_0^2/2$ (atomic units are implied). The typical integrals of quantum chemistry depend explicitly on the nuclear coordinates R_i, while in the Sturmian approach parameters $s_i = p_0 R_i$ appear. The basis feature is therefore that the Born-Oppenheimer separation is not settled before the calculations. The s_i parameters are instead fixed and only after the solution of the GEP from the relationships $R_i = s_i/p_0$ one finds out what is the point in the potential energy surface which corresponds to an energy $-p_0^2/2$. Another unusual property of this approach is that the eigenvalues obtained by solving the GEP come out as correct solutions not only for ground states, but also for the excited states corresponding to the same symmetry.

The role of the momentum wave number p_0 emerges from the momentum space perspective. Indeed, the Fourier transform of the Sturmian orbitals are those which can be directly related to the hyperspherical harmonics of the 4-dimensional space through the Fock projection [2]:

$$\chi^t_{n,l,m}(\mathbf{p}) = \frac{4k_\mu^{5/2}}{(k_\mu^2 + p^2)^2} \cdot Y_{n,l,m}(\chi, \vartheta, \varphi) \tag{2}$$

(see discussions in the modern Sturmian language perspective in [7, 8]). This perspective attributes to Bohr's principal quantum number n the intriguing role of a hyperangular momentum quantum number. Working in momentum space, before the configuration space counterpart employing Sturmian orbitals were conceived, Shibuya and Wulfman studied the one-electron multicenter problem [9], stimulating generalizations and extensions leading to equivalent formulations in either of the two dual spaces [6, 13, 17, 18, 19, 20, 21, 33, 34, 35].

Illustrating the approach to atomic structure as a many-electron one-nucleus problem, we discuss here the procedure of building a polyelectronic basis set as a product of "monoelectronic" Sturmian orbitals [4, 11, 12, 16, 27, 28, 36]: an alternative method, which defines N-electronic Sturmians as solutions of the $3N$-dimensional hydrogen atom can be elegantly formulated and also has a momentum space counterpart in terms of hyperspherical harmonics [15, 14, 31], but still lacks of developments in the algebraic and computational implementation in order to be computationally competitive.

We next consider the application of the Sturmian orbital procedure to the *ab initio* quantum chemistry of molecules as many-electron many-nuclear systems. The recent proposal, as reviewed in Avery's book [10, 41], involves a MO-LCAO approach where one-electron molecular orbitals are computed first, and then their products are employed to build-up N-electron configurations. Such an approach is currently being implemented, one difficulty being the recovery of the Born-Oppenheimer separation, because, as indicated above, the matching of s_i parameters and R_i coordinates occurs through scaling by the eigenvalues p_0.

This paper presents an account of a promising method for the application of the Sturmian orbital approach to molecular structure calculations. The outline is as follows. In the next section we give explicit expressions and main properties of Sturmian orbitals. In the presentation of the one-electron many-center system, the example provided was inspired by a suggestion by Shull and Löwdin [3], and employs a one-center expansion (Sec.2.1). The generalization to polyelectronic Sturmians as products of one electron functions (Sec. 2.2) is illustrated by a discussion of the accurate calculation of the ground and of some excited levels ^1S of the helium atom. The new proposal (Sec.2.3) for the extension of generalized Sturmian sets to molecular structure is founded on concepts and formalisms previously elaborated. In the framework of quantum chemistry jargon, it will be recognized as a valence-bond type of approach: the calculation of the potential energy curve for the H_2 molecule provides encouraging results already at the level of minimal basis sets. Sec.2.4 is devoted to a discussion of the practical obstacles to the implementation of this approach, the calculation of integrals, which can be in part circumvented through the angular momentum algebra applicable in the momentum space perspective, and in part tackled by exploiting the connections with STO and the existence of programs dealing with integrals involving the latter. Conclusion are in Sec.3.

2 Sturmian basis sets for atomic and molecular calculations

There are four different coordinate systems (polar, parabolic, spheroidal and spheroelliptical) which allow separation and solution of the partial differential equation (1): to each of them we can associate a Sturmian basis set corresponding to alternative symmetry properties [29]. For definiteness, here we will only deal with the most common one of these sets, which employs polar coordinates (r, θ, ϕ). In Eq.(1) the "nuclear charge" nk_μ is allowed to be different for each orbital, at variance with the situation which occurs when defining the hydrogenoid orbitals, and the explicit expression for the eigenfunctions $\chi_{nlm}(r, \theta, \phi)$ is:

$$\chi_{nlm}(r,\theta,\phi) = (2k_\mu)^{3/2} \left[\frac{(n-l-1)!}{2n[(n+l)!]^3}\right]^{1/2} \times \tag{3}$$

$$\times e^{-k_\mu r}(2k_\mu r)^l L_{n-l-1}^{2l+1}(2k_\mu r)\, Y_{lm}(\theta,\phi)$$

In this equation all ingredients are familiar: the L_{n-l-1}^{2l+1} are Laguerre polynomials as in [32] and Y_{lm} is a spherical harmonic. A peculiarity of such a basis set is that the "energy" $-k_\mu^2/2$ can not be positive, and so the set is complete without any continuum states. The Sturmian orbitals obey a potential weighted orthonormality relation:

$$\frac{1}{k_\mu}\int \mathrm{dx}\ \chi^*_{n'l'm'}(\mathbf{x})\frac{1}{r}\chi_{nlm}(\mathbf{x}) = \frac{1}{n}\delta_{n'n}\delta_{l'l}\delta_{m'm} \tag{4}$$

and are normalized to the unity:

$$\int \mathrm{dx}\ \chi^*_{nlm}(\mathbf{x})\chi_{nlm}(\mathbf{x}) = 1 \tag{5}$$

Alternatives for variable separations in momentum space and constructions of the corresponding Sturmian basis sets lead to an even richer phenomenology [30]; these have been classified in Ref. [30], explicit properties and connections being discussed there. They will not be considered at all in this account.

2.1 One-electron many-center systems

If we wish to treat a quantum system consistent of one electron moving in the Coulomb field of some fixed nuclei, the electronic wavefunction can be expanded as a linear combination of Sturmians orbitals centered on the nuclei. We are going to solve the Schrödinger equation (SE)

$$\left[-\frac{1}{2}\nabla^2 + v(\mathbf{x}) - E_\mu\right]\varphi_\mu(\mathbf{x}) = 0 \tag{6}$$

where the potential is

$$v(\mathbf{x}) = -\sum_a \frac{Z_a}{|\mathbf{x}-\mathbf{X}_a|} \tag{7}$$

The index a labels the nuclei: Z_a's are the nuclear charges and vectors $\mathbf{X}_a$'s correspond to positions of nuclei. The symbol $\mathbf{x}$ stands for the set of spacial electron coordinates. The eigenfunction $\varphi_\mu(\mathbf{x})$ is expressed as

$$\varphi_\mu(\mathbf{x}) = \sum_\tau c_{\tau,\mu}\ \chi_\tau(\mathbf{x}) \tag{8}$$

where τ stands for the set of indices $\{anlm\}$ and

$$\chi_\tau(\mathbf{x}) \equiv \chi_{nlm}(\mathbf{x}-\mathbf{X}_a) \tag{9}$$

We demand that the parameter k_μ, appearing in the definition of our basis set (3), be related to the unknown value of the electronic energy according to the relationship

$$E_\mu = -\frac{k_\mu^2}{2} \tag{10}$$

Substitution of Eq.(8) into Eq.(6), multiplication on the left side by $\chi^*_{\tau'}(\mathbf{x})$ and integration over electronic coordinates yields:

$$\sum_{\tau} c_{\tau,\mu} \int \mathrm{d}\mathbf{x}\ \chi^*_{\tau'}(\mathbf{x}) \left[-\frac{1}{2}\nabla^2 + v(\mathbf{x}) + \frac{k_\mu^2}{2}\right] \chi_\tau(\mathbf{x}) = 0 \qquad (11)$$

By considering Eq.(1), we can simplify Eq.(11) as:

$$\sum_{\tau} c_{\tau,\mu} \int \mathrm{d}\mathbf{x}\ \chi^*_{\tau'}(\mathbf{x}) \left[v(\mathbf{x}) + \frac{nk_\mu}{|\mathbf{x}-\mathbf{X}_a|}\right] \chi_\tau(\mathbf{x}) = 0 \qquad (12)$$

Now, dividing by $-k_\mu$, we obtain a generalized eigenvalue problem (GEP)

$$\sum_{\tau} c_{\tau,\mu} \left[\left(\frac{1}{k_\mu}\int \mathrm{d}\mathbf{x}\ \chi^*_{\tau'}(\mathbf{x}) v(\mathbf{x}) \chi_\tau(\mathbf{x})\right)\right. \qquad (13)$$

$$\left. -k_\mu \left(\frac{1}{k_\mu}\int \mathrm{d}\mathbf{x}\ \chi^*_{\tau'}(\mathbf{x}) \frac{n}{|\mathbf{x}-\mathbf{X}_a|} \chi_\tau(\mathbf{x})\right)\right] = 0$$

whose eigenvalue k_μ is related to the energy of the electron by Eq.(10): this is a general feature of the Sturmian approach. The matrix elements between brackets can be calculated provided we are able to obtain nuclear attraction integrals (NAI's) involving Sturmian orbitals. Apparently Eq.(13) implies an iterative procedure since the eigenvalue k_μ is also inside the matrix elements, but actually it is not difficult to show that the value of such NAI's does not depend on k_μ. To apply this method to a two-nuclei monoelectronic molecule like H_2^+, whose internuclear distance is $\mathbf{R}$, one has to evaluate three types of NAI's. One of them is trivial, coinciding with Eq.(4). A more difficult one is:

$$S^{nlm}_{n'l'm'}(\mathbf{R}) = \frac{1}{k_\mu}\int \mathrm{d}\mathbf{x}\ \chi^*_{n'l'm'}(\mathbf{x}) \frac{1}{|\mathbf{x}-\mathbf{R}|} \chi_{nlm}(\mathbf{x}-\mathbf{R}) \qquad (14)$$

representing the interaction between the overlap of two orbitals centered on different nuclei and one of the two nuclei. This integral can be evaluated analytically in momentum space, with the aid of angular momentum theory as proposed by Shibuya and Wulfman [9]; the expression they derived contains 3-j and 9-j coefficients which are fundamental ingredients of angular momentum algebra and can be efficiently evaluated. More recently, Avery proposed an alternative way for this calculation, exploiting an interesting angular integration theorem [10]. The third type of NAI is:

$$W^{nlm}_{n'l'm'}(\mathbf{R}) = \frac{1}{k_\mu}\int \mathrm{d}\mathbf{x}\ \chi^*_{n'l'm'}(\mathbf{x}) \frac{1}{|\mathbf{x}-\mathbf{R}|} \chi_{nlm}(\mathbf{x}) \qquad (15)$$

representing the interaction between the overlap of two orbitals centered on the same nucleus and another nucleus (this integral is denoted $\mathcal{V}_{n'l'm'}^{nlm}$ in [42]). It is possible to evaluate it analytically in configuration space, exploiting the well known expansion:

$$\frac{1}{|\mathbf{x}-\mathbf{x}'|} = \frac{1}{r_>}\sum_{l=0}^{\infty}\left(\frac{r_<}{r_>}\right)^l \frac{4\pi}{2l+1}\sum_m Y_{lm}^*(\hat{\mathbf{x}}')Y_{lm}(\hat{\mathbf{x}}) \tag{16}$$

Both integrals (14) and (15) were introduced in quantum mechanics by Shibuya and Wulfman in their pioneering work. They depend on a parameter $s = k_\mu R$, and this change of variable leads to the disappearance of variable R from the integrals. For example

$$\mathcal{S}_{100}^{100} = (1+s)e^{-s} \tag{17}$$

$$W_{100}^{100} = \frac{1}{s} - \frac{e^{-2s}}{s}(1+s) \tag{18}$$

This means that one has to pick up a value for s, calculate the matrix elements, obtain k_μ solving the GEP, and eventually obtain the internuclear distance R.

The Sturmian method was exploited by Koga et al. for a high-accuracy calculation of the H_2^+ electronic ground state [34]. The authors did not mention how they managed the overcompleteness problem, which arises when a large basis set is used, as Avery pointed out in Ref. [35]. The most direct way for avoiding such a problem was proposed by Löwdin and others [3, 5], who suggested to proceed with an expansion of the molecular orbital in terms of a complete set centered on a single point in the space, which will be the midpoint in the H_2^+ case. It is interesting to explore the suggestion, inserting such a single-point expansion in Eq.(6): one obtains the secular problem:

$$\sum_{nl}\left[2W_{nlm}^{n'l'm}(s/2) - k_\mu \delta_{nl}^{n'l'}\right] c_{nlm,\mu} = 0 \tag{19}$$

where m is $0, 1, \ldots$ for states $\sigma, \pi, \ldots$. The secular equation turns out to be split in two blocks because of the symmetry of H_2^+: indeed, the sum involves only even l's for gerade states, and only odd l's for ungerade states. Results obtained by solving Eq.(19) are given in Fig.(1). As the internuclear distance increases, a larger and larger basis set would be necessary to approximate correctly the H_2^+ wavefunction, and one should face the problem of computing $W_{nlm}^{n'l'm'}$ matrix elements for high quantum numbers, but the accuracy obtained at such an elementary level is most encouraging.

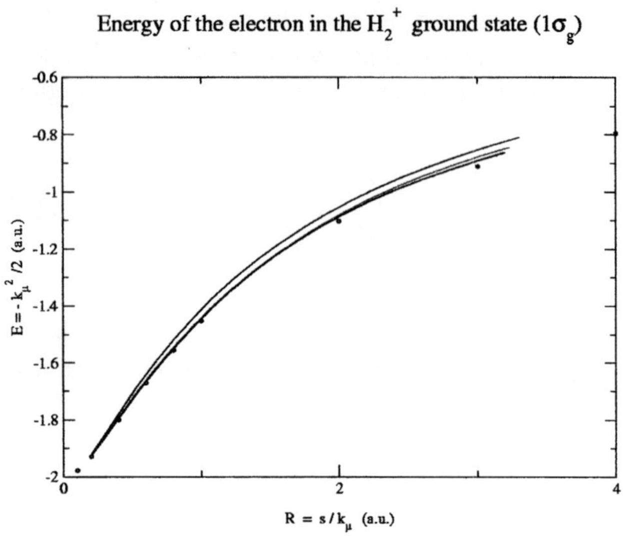

Figure 1: The circles stand for the 10-figure accuracy data by Koga et al. [34]. The three curves represents the results of Eq.(19) for three different basis sets: $\{n_{max}, l_{max}\} = \{3,2\}, \{6,5\}, \{8,6\}$.

When H_2^+ is treated in momentum space, one obtains a secular problem that apparently is different from Eq.(13):

$$\sum_{\tau} c_{\tau,\mu} \left[S_{\tau'}^{\tau} - k_\mu \delta_{\tau,\tau'} \right] = 0 \tag{20}$$

However, as shown e.g. in Ref. [10], there is a relationship between different types of NAI

$$-\int d\mathbf{x}\ \chi_{\tau'}^{*}(\mathbf{x})\ v(\mathbf{x})\ \chi_{\tau}(\mathbf{x}) = \tag{21}$$

$$= \sum_{\tau''} \left[S_{n''l''m''}^{n'l'm'}(\mathbf{X}_{a''} - \mathbf{X}_{a'}) \right] \frac{Z_{a''}}{n''} \left[S_{n''l''m''}^{nlm}(\mathbf{X}_{a''} - \mathbf{X}_{a}) \right]$$

As a consequence, one can convince himself that Eq.(13) is just an iteration of Eq.(20), and the two equations would be are absolutely equivalent in the limit of an infinite basis set [10]. Formally, Eq. (13) is a Generalized Eigenvalue Problem of the particular type:

$$\left(\mathbf{A}^2 - \lambda \mathbf{A} \right) \mathbf{c} = 0 \tag{22}$$

where $\mathbf{A}$ is a generic Hermitian matrix while Eq.(20) corresponds to

$$\left(\mathbf{A} - \lambda \mathbf{I} \right) \mathbf{c} = 0 \tag{23}$$

where **I** is the unit matrix and thus it is an ordinary eigenvalue problem. The two equations are therefore formally equivalent, but when $\mathbf{A}^2$ can be computed directly the practical implementation of algorithms for calculations, basis sets of the same dimension lead to a better variational convergence for Eq.(13) than for Eq.(20), although computing time is higher because (13) is a GEP. Relative advantages and disadvantages are such that use of either can be preferable in different situations. Their study is of relevance because their solutions are of use in the procedure for building up molecular orbitals suggested in [10].

2.2 Many-electron one-center systems

The SE for a N-electronic non-relativistic atom is:

$$\left[-\frac{1}{2}\nabla^2 + V_0(\mathbf{x}) + V'(\mathbf{x}) + \frac{p_0^2}{2}\right]\Psi(\mathbf{x}) = 0 \tag{24}$$

where

$$\nabla^2 = \sum_{j=1}^{N} \nabla_j^2 \tag{25}$$

$$V_0(\mathbf{x}) = -\sum_{j=1}^{N} \frac{Z}{r_j} \tag{26}$$

and

$$V'(\mathbf{x}) = \sum_{i>j}^{N} \frac{1}{r_{ij}} \tag{27}$$

The energy of the system is indicated as $-p_0^2/2$, and $\mathbf{x}$ is the set of all the electronic coordinates $\{\mathbf{x}_1, \mathbf{x}_2, \ldots, \mathbf{x}_N\}$. A convenient N-electronic Sturmian basis set $\Phi_\nu(\mathbf{x})$ is defined by the equation [4, 28, 12, 10, 36]:

$$\left[-\frac{1}{2}\nabla^2 + \beta_\nu V_0(\mathbf{x}) + \frac{p_0^2}{2}\right]\Phi_\nu(\mathbf{x}) = 0 \tag{28}$$

Again, in the spirit of the Sturmian approach for the definition of the basis sets, the unknown energy of the system we are actually looking for is seen to play the role of a constant parameter, while the "nuclear charge" β_ν is the eigenvalue. The solutions of Eq.(28) enjoy a potential weighted orthonormality relationship [36, 10]:

$$\int d\mathbf{x}\Phi_{\nu'}^*(\mathbf{x})V_0(\mathbf{x})\Phi_\nu(\mathbf{x}) = \delta_{\nu',\nu}\frac{2E}{\beta_\nu} \tag{29}$$

The $\Phi_\nu(\mathbf{x})$'s can be written as products of Sturmian atomic orbitals:

$$\Phi_\nu(\mathbf{x}) = \chi_\mu(\mathbf{x}_1)\ \chi_{\mu'}(\mathbf{x}_2)\cdots\chi_{\mu''}(\mathbf{x}_N) \tag{30}$$

provided that two subsidiary conditions are satisfied:

$$k_\mu^2 + k_{\mu'}^2 + k_{\mu''}^2 + ... = p_0^2 \tag{31}$$

and

$$nk_\mu = n'k_{\mu'} = n''k_{\mu''} = ... = Z\beta_\nu \tag{32}$$

Expansion of the atomic wavefunction in Sturmian functions:

$$\Psi(\mathbf{x}) = \sum_\nu C_\nu \Phi_\nu(\mathbf{x}) \tag{33}$$

and insertion of Eq.(33) in Eq.(28) leads after some manipulations involving Eqs.(28) and (29) to the secular equation [10, 11, 12, 36] :

$$\sum_\nu \left[T'_{\nu',\nu} + T^0_{\nu',\nu} - p_0\delta_{\nu',\nu}\right] C_\nu = 0 \tag{34}$$

where

$$T^0_{\nu',\nu} \equiv -\frac{1}{p_0}\int \mathrm{d}\mathbf{x}\ \Phi^*_{\nu'}(\mathbf{x})V_0(\mathbf{x})\Phi_\nu(\mathbf{x}) = Z\left(\sum_{\mu\subset\nu}\frac{1}{n^2}\right)\delta_{\nu',\nu} \tag{35}$$

and

$$T'_{\nu',\nu} \equiv -\frac{1}{p_0}\int \mathrm{d}\mathbf{x}\Phi^*_{\nu'}(\mathbf{x})V'(\mathbf{x})\Phi_\nu(\mathbf{x}) \tag{36}$$

The latter integral can be computed analytically, since all the orbitals involved are centered on the same nucleus.

When calculations are performed using this method the convergence rate toward the ground state energy is good, and moreover the upper eigenvalues give very good approximations to the excited state energies. We performed a calculation over Helium (^1S states), using a basis set made of ^1S bielectronic Sturmian functions with quantum numbers up to $n_{1,max} = 4, l_{1,max} = 2, n_{2,max} = 10, l_{2,max} = 2$. Accurate ^1S states for helium are available since long time [22], and continue to be a benchmark for alternative basis sets in variational calculations [23, 24]. The ground state energy we have obtained is 2.902690 a.u., to be compared with the best variational value 2.903724377 a.u. (error = 0.036%). Considering also upper eigenvalues, we have been able to reproduce carefully by our *ab initio* calculation the Helium ^1S electronic transition energies between the ground state and several excited states (Table 1). In the

Table 1: The ^1S levels of the helium atom.

^1S **Levels**	**Calculated**$^{(a)}$ (a.u)	**Transition energies** $1s\ ns - 1s^2$ (cm^{-1})	**Experimental**$^{(b)}$
$1s^2$	-2.902690	-	-
$1s\ 2s$	-2.147417	165763	166277.547
$1s\ 3s$	-2.061402	184641	184641.369
$1s\ 4s$	-2.033607	190742	190741.633
$1s\ 5s$	-2.021181	193469	193468.726
$1s\ 6s$	-2.014546	194925	194924.965
$1s\ 7s$	-2.010585	195794	195794.328

(a)This work.

(b)NIST Atomic Spectra Database (*http://physics.nist.gov/cgi-bin/AtData/main_asd*).

literature there are some recent examples of accurate calculations on Helium excited states [25, 26] performed with various methods, but they seem to be not suitable for extensions to more complex atoms.

The capability of the Sturmian method to reproduce atomic electronic transition energies can be shown to be a general feature, and was tested also on atoms having more than two electrons [10, 36].

2.3 Many-center many-electron systems

In this section we propose an extension of the Sturmian formalism for the study of systems having many nuclei and electrons. The SE for a molecule coincides with Eq.(24), provided that we define $V_0(\mathbf{x})$ according to

$$V_0(\mathbf{x}) = \sum_a V_{0,a}(\mathbf{x}) = \sum_a \sum_{i=1}^{N} \frac{Z_a}{|\mathbf{x}_i - \mathbf{X}_a|} \tag{37}$$

We expand the electronic wavefunction as in Eq.(33), and the basis functions $\Phi_\nu(\mathbf{x})$ are defined by an equation analogous to Eq.(28):

$$\left[-\frac{1}{2}\nabla^2 + \beta_\nu V_p^\ddagger(\mathbf{x}) + \frac{p_0^2}{2}\right] \Phi_\nu(\mathbf{x}) = 0 \tag{38}$$

The potential $V_p^\ddagger(\mathbf{x})$ is the sum of N contributions, each of them representing the interaction of one electron with one of the nuclei:

$$V_p^\ddagger(\mathbf{x}) = -\sum_{j=1}^{N} \frac{Z_{P,j}}{|\mathbf{x}_j - \mathbf{X}_{p,j}|} \tag{39}$$

This means that actually there is not only one equation defining the basis set, because there are many choices for $V_p^\ddagger(\mathbf{x})$: the basis set can be divided into several subsets, each of them associated to an equation like (38) having a different $V_p^\ddagger(\mathbf{x})$. A function $\Phi_\nu(\mathbf{x})$ is a product of Sturmian orbitals $\chi_\mu(\mathbf{x}-\mathbf{X}_a)$ like Eq.(30), provided that two subsidiary are satisfied: the first one is Eq.(31), and the second is a modification of Eq.(32) taking into account that in general Sturmian orbitals appearing in Eq.(30) are centered on the different nuclei of the molecule:

$$\frac{nk_\mu}{Z_{a'}} = \frac{n'k_{\mu'}}{Z_{a'}} = \frac{n''k_{\mu''}}{Z_{a''}} = \cdots = \beta_\nu \tag{40}$$

A combination of Eqs.(40) and (31) leads to the relationship:

$$\beta_\nu = \frac{p_0}{\sqrt{(\frac{Z_a}{n})^2 + (\frac{Z_{a'}}{n'})^2 + (\frac{Z_{a''}}{n''})^2 + \cdots}} \tag{41}$$

This approach can be interpreted as a Valence Bond (VB) method, since we perform CI over configurations which are products of non-orthogonal atomic orbitals. In order to clarify the concepts exposed and show how this method works in practice, we will show an application to the simplest molecule: H_2. The crudest approximation to the ground state $^1\Sigma_g$ wavefunction is:

$$\Psi_{^1\Sigma_g}(\mathbf{x}_1,\mathbf{x}_2) = \frac{1}{\sqrt{2}}\left[1s_A(1)1s_B(2) + 1s_B(1)1s_A(2)\right] \times \tag{42}$$

$$\times \frac{1}{\sqrt{2}}\left[\alpha(1)\beta(2) - \beta(1)\alpha(2)\right]$$

In this case the basis set is constituted by two bielectronic functions; $\Phi_1(\mathbf{x}) = 1s_A(1)1s_B(2)$ corresponds to the potential

$$V_1^\ddagger(\mathbf{x}) = -\frac{1}{|\mathbf{x}_1 - \mathbf{X}_A|} - \frac{1}{|\mathbf{x}_2 - \mathbf{X}_B|} \tag{43}$$

in Eq.(38), and $\Phi_2(\mathbf{x}) = 1s_B(1)1s_A(2)$ corresponds to

$$V_2^\ddagger(\mathbf{x}) = -\frac{1}{|\mathbf{x}_1 - \mathbf{X}_B|} - \frac{1}{|\mathbf{x}_2 - \mathbf{X}_A|} \tag{44}$$

We can express the overall interaction between electrons and nuclei as:

$$V_0(\mathbf{x}) = V_1^\ddagger(\mathbf{x}) + V_2^\ddagger(\mathbf{x}) \tag{45}$$

According to Eq.(3), $1s$ Sturmian orbitals are defined as

$$1s(1) = \left(\frac{k_\mu^3}{\pi}\right)^{1/2} e^{-k_\mu r_1} \tag{46}$$

The exponent k_μ can be computed combining Eqs.(40) and (41): it is $1/\sqrt{2}$. In general, the exponent of an orbital will depend on the principal quantum numbers of all the orbitals forming the configuration $\Phi_\nu(\mathbf{x})$ in which it appears.

Now we will calculate an approximation to H_2 electronic energy exploiting the wave function (42). Entering Eq.(42) inside (24) and writing the potential $V_0(\mathbf{x})$ as in Eq.(45) one can obtain the 1x1 secular problem:

$$\int [1s_A(1)1s_B(2) + 1s_B(1)1s_A(2)] \tag{47}$$

$$\left[-\frac{1}{2}\nabla^2 + V_1^\ddagger(\mathbf{x}) + V_2^\ddagger(\mathbf{x}) + \frac{1}{r_{12}} + \frac{p_0^2}{2}\right] 1s_A(1)1s_B(2)d1d2 = 0$$

The latter expression can be simplified with the aid of Eq.(38):

$$\int [1s_A(1)1s_B(2) + 1s_B(1)1s_A(2)] \tag{48}$$

$$\left[(1-\beta_1)V_1^\ddagger(\mathbf{x}) + V_2^\ddagger(\mathbf{x}) + \frac{1}{r_{12}}\right] 1s_A(1)1s_B(2)d1d2 = 0$$

Taking into account that the basis set functions associated to the same operator $V_p^\ddagger(\mathbf{x})$ enjoy potential weighted orthonormality relationships

$$\int d\mathbf{x}\phi_{\nu'}^*(\mathbf{x})V_p^\ddagger(\mathbf{x})\phi_\nu(\mathbf{x}) = \delta_{\nu',\nu}\frac{2E}{\beta_\nu} \tag{49}$$

we can modify Eq.(48) as follows:

$$\frac{2E}{\beta_1} - 2E + \int 1s_A(1)1s_B(2)\left[V_2^\ddagger(\mathbf{x}) + \frac{1}{r_{12}}\right] 1s_A(1)1s_B(2)d1d2 \tag{50}$$

$$+\int 1s_B(1)1s_A(2)\left[(1-\beta_1)V_1^\ddagger(\mathbf{x}) + V_2^\ddagger(\mathbf{x}) + \frac{1}{r_{12}}\right] 1s_A(1)1s_B(2)d\mathbf{x} = 0$$

If we write down all the integrals explicitly and exploit Eq.(41), we can rewrite Eq.(50) as:

$$-\sqrt{2}p_0 + p_0 - \int 1s_B^2(2)d2 \int \frac{1s_A^2(1)}{|\mathbf{x}_1 - \mathbf{X}_B|}d1 - \int 1s_A^2(1)d1 \int \frac{1s_B^2(2)}{|\mathbf{x}_2 - \mathbf{X}_A|}d2 \tag{51}$$

$$+\int \frac{1s_A^2(1)1s_B^2(2)}{r_{12}} d1d2$$

$$+(1-\frac{p_0}{\sqrt{2}})\left[-\int 1s_A(2)1s_B(2)d2 \int \frac{1s_A(1)1s_B(1)}{|\mathbf{x}_1-\mathbf{X}_A|}d1\right.$$

$$\left.-\int 1s_A(1)1s_B(1)d1 \int \frac{1s_A(2)1s_B(2)}{|\mathbf{x}_2-\mathbf{X}_B|}d2\right]+$$

$$\left[-\int 1s_A(2)1s_B(2)d2 \int \frac{1s_A(1)1s_B(1)}{|\mathbf{x}_1-\mathbf{X}_B|}d1\right.$$

$$\left.-\int 1s_A(1)1s_B(1)d1 \int \frac{1s_A(2)1s_B(2)}{|\mathbf{x}_2-\mathbf{X}_A|}d2\right]$$

$$+\int \frac{1s_A(1)1s_B(1)1s_A(2)1s_B(2)}{r_{12}}d1d2 = 0$$

Eventually, we divide the latter by p_0 and write the NAI's in a more compact form, exploiting the definitions (14) and (15). The expression for the eigenvalue is:

$$p_0 = \left[\sqrt{2}+\sqrt{2}W_{100}^{100}(s)+\frac{1}{p_0}\int \frac{1s_A(1)1s_B(1)1s_A(2)1s_B(2)}{r_{12}}d1d2+\right. \tag{52}$$

$$\left.\frac{1}{p_0}\int \frac{1s_A^2(1)1s_B^2(2)}{r_{12}}d1d2 - 2\sqrt{2}\mathcal{M}_{100}^{100}(s)\mathcal{S}_{100}^{100}(s)\right] / \left[1+\mathcal{M}_{100}^{100}(s)\mathcal{S}_{100}^{100}(s)\right]$$

The $\mathcal{M}$'s are two-center overlap integrals between Sturmians:

$$\mathcal{M}_{nlm}^{n'l'm'} \equiv \int \chi_{n'l'm'}^*(\mathbf{x})\chi_{nlm}(\mathbf{x})d\mathbf{x} \tag{53}$$

For example,

$$\mathcal{M}_{100}^{100} = \frac{1}{3}e^{-s}(3+3s+s^2) \tag{54}$$

We can show that these $\mathcal{M}$ integrals can be evaluated in momentum space exploiting angular momentum algebra, analogously to the case of the $\mathcal{S}$ integrals encountered previously.

Mono- and bi-electronic integrals in Eq.(52) actually do not depend on the value of p_0 value: rather, they depend on the product $s = k_\mu \cdot R$ (R being the internuclear distance). Therefore, we pick up a value for s, compute the matrix elements and then we can get the eigenvalue p_0; eventually, the physical internuclear distance is obtained as $R = \sqrt{2}s/p_0$.

Results on the potential energy curve for the ground state of H_2, calculated by this procedure, are reported in Fig.(2). Even if really minimal, the basis set of Eq.(42) gives a good qualitative picture of the

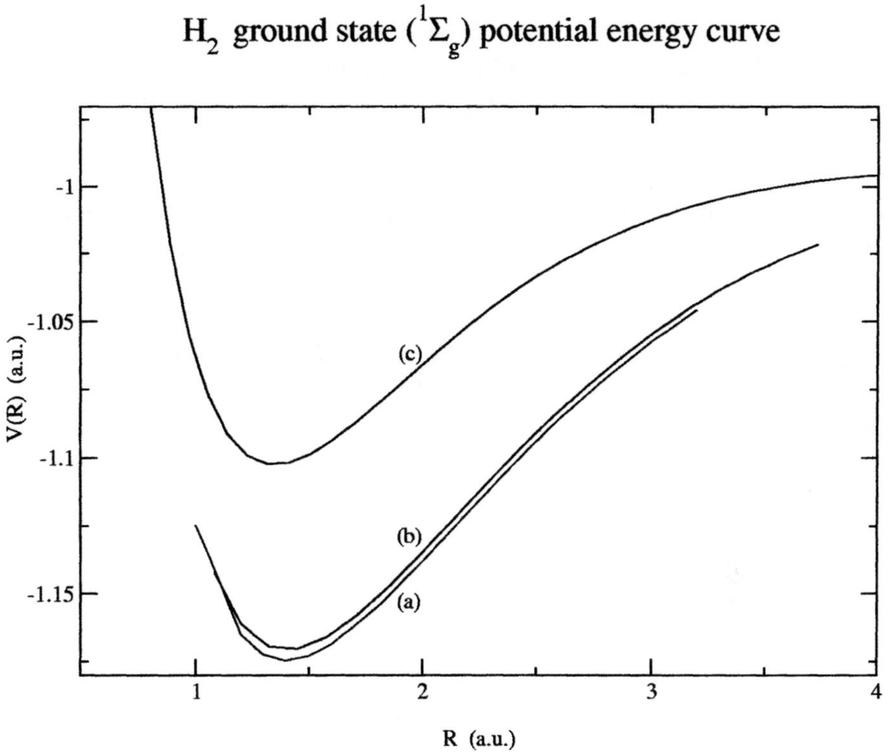

Figure 2: Comparison between the benchmark *ab initio* potential energy curve for H_2 denoted as (a), and the Sturmian approach. Curve (c) corresponds to the single configuration function in Eq.(42) and curve (b) has been obtained including in the expansion all configurations from all Sturmian orbitals up to $3d$. In the latter case, the variance in the energy minimum with the one given in [37] is +0.3%.

chemical bond: the equilibrium distance is close to the real one, and the asymptotic behavior is correct. A more elaborate calculation is also shown in Fig.(2). We have tested that even modest size calculations employing larger basis sets yield accurate results also for excited states.

2.4 The integrals. Relationships with Slater-type orbitals

The evaluation of integrals is always an important issue of quantum chemical methods, because it can involve large efforts in terms of computer times and therefore may limit the concrete possibility of extending the calculations to cover complex systems. This discussion will pertain specifically to the mono- and bi-electronic integrals encountered in Sec.2.

Generally, the use of a basis set as defined in Eq.(38) leads to monoelectronic integrals $\mathcal{S}, W$ and $\mathcal{M}$ containing two orbitals which can differ in the exponents appearing in Eq.(3). This represents a limit for the application of momentum space harmonic analysis, at least according to its present developments. To circumvent this bottleneck and yet be able to test the perspective power of the method, we explored the possibility of making recourse to the relationship between Sturmian orbitals and STO.

In practice, since Rico and collaborators have proposed, implemented and made available an efficient program for the calculation of two-center monoelectronic integrals involving STO, based on the recurrence relationships which they enjoy [38, 39], and since the Sturmian orbitals can be given as simple linear combinations of the STO (explicitly, indicating the latter as ψ_{nlm}:

$$\begin{aligned}
\chi_{1,0,0} &= \psi_{1,0,0} \\
\chi_{2,0,0} &= \psi_{1,0,0} - \sqrt{3}\ \psi_{2,0,0} \\
\chi_{2,1,m} &= \psi_{2,1,m} \\
\chi_{3,0,0} &= \psi_{1,0,0} - 2\sqrt{3}\ \psi_{2,0,0} + \sqrt{10}\ \psi_{3,0,0} \\
\chi_{3,1,m} &= 2\sqrt{\tfrac{2}{3}}\ \psi_{2,1,m} - \sqrt{5}\ \psi_{3,1,m} \\
\chi_{3,2,m} &= \psi_{3,2,m} \\
\chi_{4,0,0} &= \psi_{1,0,0} - 3\sqrt{3}\ \psi_{2,0,0} + 3\sqrt{10}\ \psi_{3,0,0} - \sqrt{35}\ \psi_{4,0,0} \\
\chi_{4,1,m} &= \sqrt{5}\ \psi_{2,1,m} - 5\sqrt{\tfrac{3}{2}}\ \psi_{3,1,m} + \sqrt{21}\ \psi_{4,1,m} \\
\chi_{4,2,m} &= \tfrac{3}{\sqrt{2}}\ \psi_{3,2,m} - \sqrt{7}\ \psi_{4,2,m}
\end{aligned} \tag{55}$$

etc.) we found it expedient to use the program in order to calculate the bicenter monoelectronic integrals $\mathcal{S}$ and $\mathcal{M}$. The W integrals can instead be calculated analytically even though the exponents differ.

However, this program doesn't calculate the three-center monoelectronic integrals

$$\int d\mathbf{x}\ \chi_{nlm}(\mathbf{x}) \frac{1}{|\mathbf{x} - \mathbf{X}_a|} \chi_{n'l'm'}(\mathbf{x} - \mathbf{X}_{a'}) \tag{56}$$

which can occur in molecules with three or more nuclei.

The most demanding effort for our valence-bond Sturmian approach appears, as usual in quantum chemistry, when the calculation of the bielectronic multicenter integrals is needed:

$$\int\int d\mathbf{x}d\mathbf{x}'\ \chi_{\tau_A}(\mathbf{x})\chi_{\tau_B}(\mathbf{x}) \frac{1}{|\mathbf{x} - \mathbf{x}'|} \chi_{\tau_C}(\mathbf{x}')\chi_{\tau_D}(\mathbf{x}') \tag{57}$$

Our test cases presented in Sec.2.3 were tackled in a rather tortuous manner, making recourse to another program due to Rico and coworkers

[40] which in turn in order to obtain integrals for STO expands them as linear combinations of GTO. The resulting Gaussian integrals are finally calculated using available techniques. The ensuing disadvantage of this inelegant procedure is that, since a satisfactory fit of an STO requires about 10 Gaussians, an integral involving four STO's requires the calculation of 10^4 integrals among GTO. Moreover our integrals such as (57) are linear combinations of integrals among STO: for example an integral involving four $3s$-Sturmians requires a sum over 3^4 integrals of the STO program, and therefore $3^4 \times 10^4$ integrals over Gaussians. Therefore this program, which recommends itself for the accuracy which can be adjusted due to flexibility in the choice of the number Gaussians required to approximate a STO and satisfactorily covers the range of permitted quantum numbers ($n_{max} = 6, l_{max} = 3$), although appropriate for test calculations, must be superseded by more direct approaches to tackle integrals such as (57). Alternatives should also be explored, involving for example elliptic or other suitable coordinates.

3 Further remarks and conclusions

The account presented in this paper on developments regarding the use of Sturmians orbitals in quantum chemistry demonstrates how fruitful was the idea put forward long ago by Shull and Löwdin of "natural spin orbitals". The first steps of the proposed approach show perspectives of advantages over conventional methods, particularly for excited states. Not considered in this paper have been the alternatives which open by exploiting the various ways available to separate variables for Eq.(1): they have been recently classified both in configuration [29] and in momentum space [30]. Since the computation of integrals is a feature which will eventually decide the crucial potentialities and scope of the Sturmian approach, the alternative basis sets and the relationships among them are an important goal to further study; some progress have been made for the parabolic set [13, 14].

Needless to say, these orbitals might find a specific role also whenever in quantum chemistry there is a need for an expansion in a complete basis set. Further extensions which are of relevance for the application to heavier systems are those involving the inclusion of relativistic effects: see Chapter 8 in Ref. [10] and recent work by Smitkowski [43, 44, 45, 46]. For the particular scheme introduced in Sec.(2.3) the numerical results on H_2 indicate that the development of a direct algorithm for the calculation of bielectronic multicenter integrals would greatly contribute to practical implementations to systems of increasing complexity. Our current effort on this problem shows the importance of using momentum space techniques, which allow the exploit of powerful angular and hyperangular momentum algebra.

Acknowledgments

We thank Simonetta Cavalli, Cecilia Coletti and Gaia Grossi for their cooperation on various aspects of this work, and John Avery and Francesco Tarantelli for several inspiring discussions. This research is supported by the Italian National Research Council (CNR), the Ministero dell' Istruzione, dell' Università e della Ricerca (MIUR), the Ente Nazionale per l'Energia Alternativa (ENEA), the Agenzia Spaziale Italiana (ASI) and by European Contract "Theoretical Studies of Electronic and Dynamical Processes in Molecules and Clusters" [Contract no. 1-1999-00121], and COST Actions D9 and D23.

References

[1] H. Shull and P.-O. Löwdin, *J. Chem. Phys.*, **30**, 617 (1959).

[2] V. Fock, *Z. Phys.*, **98**,145 (1935).

[3] H. Shull and P.-O. Löwdin, *Svensk Kem. Tidskr.*, **67**, 373 (1955).

[4] O. Goscinski, Preliminary Research Report No. 217, Quantum Chemistry group, Uppsala University (1968).

[5] K.M. Howell and H. Shull, *J. Chem. Phys.* , **30**, 627 (1955).

[6] B. R. Judd, *Angular Momentum Theory for Diatomic Molecules.* Academic Press, New York (1975).

[7] V. Aquilanti, , S. Cavalli, C. Coletti, D. Di Domenico, and G. Grossi. *Quantum Systems in Chemistry and Physics. Vol I: Basic Problems and Model Systems*, page 289 (2000).

[8] J. Avery, *Hyperspherical Harmonics, Applications in Quantum Theory.* Kluwer Academic Publishers, Dordrecht, The Netherlands (1989).

[9] T. Shibuya and C.E. Wulfman, *Proc. Roy. Soc.*, **A286**, 376 (1965).

[10] J. Avery, *Hyperspherical Harmonics and Generalized Sturmians.* Kluwer Academic Publishers, Dordrecht, The Netherlands (2000).

[11] J. Avery, *J. Mol. Struct. (Theochem)*, **458**, 1 (1999).

[12] J. Avery, *Adv. Quantum Chem.* , **31**, 201 (1999).

[13] V. Aquilanti, S. Cavalli, C. Coletti, and G. Grossi, *Chem. Phys.*, **209**, 405 (1996).

[14] V. Aquilanti, S. Cavalli, and C. Coletti, *Chem. Phys.*, **214**, 1 (1997).

[15] J. Avery and D.R. Herschbach, *Int. J. Quantum Chem.*, **41**, 673 (1992).

[16] J.P. Gazeau and A. Maquet, *J. Chem. Phys.*, **73**, 5147 (1980).

[17] B.K. Novosadov *Opt. Spectrosc.*, **41**, 490 (1976).

[18] Brett I. Dunlap *Chem. Phys. Lett.*, **30**, 39 (1975).

[19] Ch. Duchon, M.Cl. Dumont-Lepage and J.P. Gazeau, *J. Phys. A*, **15**, 1227 (1982).

[20] Ch. Duchon, M.Cl. Dumont-Lepage and J.P. Gazeau, *J. Chem. Phys.*, **76**, 445 (1982).

[21] H.J. Monkhorst and B. Jeriorski, *J. Chem. Phys.*, **71**, 5268 (1979).

[22] C.L. Pekeris, *Phys. Rev.*, **126**, 1470 (1962).

[23] A.J. Thakkar and T. Koga, *Phys. Rev. A*, **50**, 854 (1994).

[24] S.P. Goldman, *Phys. Rev. A*, **57**, 677 (1998).

[25] G.W.F Drake and Zong-Chao Yan, *Chem. Phys. Lett.*, **229**, 486 (1994).

[26] J.J. De Groote, M. Masili and J.E. Hornos *J. Phys. B*, **31**, 4755 (1998).

[27] J. Avery, *J. Math. Chem.*, **21**, 285 (1997).

[28] V. Aquilanti and J. Avery, *Chem. Phys. Lett.*, **267**, 1 (1997).

[29] V. Aquilanti, A. Caligiana and S. Cavalli, *Int. J. Quantum. Chem.* (2002), in press.

[30] V. Aquilanti, A. Caligiana, S. Cavalli and C. Coletti, *Int. J. Quantum. Chem.* (2002), in press.

[31] V. Aquilanti, S. Cavalli, C. Coletti, D. Di Domenico and G. Grossi, *Int. Rev. Phys. Chem.*, **20**, 673 (2001).

[32] P. M. Morse and H. Feshbach, *Methods of Theoretical Physics.* McGraw-Hill, New York (1953).

[33] T. Koga and T. Matsuhashi, *J. Chem. Phys.*, **87**, 1677 (1987).

[34] T. Koga and T. Matsuhashi, *J. Chem. Phys.*, **89**, 983 (1988).

[35] J. Avery and R. Shim, *Int. J. Quantum Chem.*, **83**, 1 (2001).

[36] V. Aquilanti and J. Avery, *Adv. Quantum Chem.*, **39**, 71 (2001).

[37] W. Kolos and L. Wolniewicz, *J. Chem. Phys.*, **49**, 404 (1968).

[38] J.F. Rico, R. López and G. Ramírez, *J. Comp. Chem.*, **9**, 790 (1988).

[39] J.F. Rico, R. López, M. Paniagua and G. Ramírez, *Comp. Phys. Commun.*, **64**, 329 (1991).

[40] J.F. Rico, R. López, I. Ema and G. Ramírez, *Comp. Phys. Commun.*, **105**, 216 (1997).

[41] J. Avery and S. Sauer, in *Quantum Systems in Chemistry and Physics, Vol. 1*, Kluwer Academic Publishers (1998).

[42] J. Avery, *this book, and private communication* (2002).

[43] R. Szmytkowski, *J. Phys. B*, **30**, 825 (1997).

[44] R. Szmytkowski, *J. Phys. A*, **31**, 4963, (1998).

[45] R. Szmytkowski, *Phys. Rev. A*, **65**, 012503 (2001).

[46] R. Szmytkowski, *J. Phys. B* , **35**, 1379 (2002).

MOLECULAR ORBITAL CALCULATIONS WITH NUMERICALLY OPTIMIZED ATOMIC ORBITAL BASES

James D. Talman and Raphael Yan

Department of Applied Mathematics and Centre for Chemical Physics,
University of Western Ontario
London, Ontario, Canada N6A 5B7

Abstract

The problem of variationally optimizing the atomic orbitals that form the bases used in constructing molecular orbitals is discussed. The basis functions are angular momentum functions centered on the various nuclei with radial factors that are defined numerically on a radial mesh. The variational principle leads to an ordinary differential equation similar to the radial Schrödinger equation, but with an inhomogeneous term. An iterative scheme for determining a parameter in the equation, akin to an eigenvalue, required to maintain the normalization of the orbital is described. A possible numerical problem stemming from the fact that the nuclear Coulomb potential is not differentiable at the distance to the other nuclei is investigated. Results of the variational optimization for a number of small molecules are given and compared with results using corresponding bases of Gaussian-type orbitals.

E.J. Brändas and E.S. Kryachko (eds.),
Fundamental World of Quatum Chemistry, Vol. I, 317–334.

1 Introduction

Basis functions of atomic orbitals used in constructing single-particle molecular orbitals are almost invariably taken to be superpositions, or contractions, of Gaussian-type orbitals (GTOs). The reason for this is of course that a GTO centered at one nucleus, or at some other spatial point, is readily expanded in terms of GTOs at another point, and similarly, that multicenter integrals involving GTOs centered at different nuclei are very readily calculated. Although the use of contracted GTOs has enabled almost the entire field of computational molecular structure, an intrinsic limitation is that it is essentially unfeasible to optimize such an orbital in a particular molecular calculation. In practice, such orbitals are usually optimized in an atomic calculation, and then scaled in some way to account for the molecular environment. Alternatively, as in the case of tempered distributions of the nonlinear parameters, they may depend on one or two parameters that can be chosen variationally.

If one wishes to go beyond bases of GTOs, it is necessary to confront the problem of translation of angular momentum wave functions, i.e., of expanding an angular momentum wave function centered at one point in terms of wave functions centered at another point. The solution of this problem solves, in principle, the problem of computing the various multicenter integrals required in a molecular calculation. An early extensive investigation of this problem, as part of a much larger study on the cohesive properties of solids, was made by P-O Löwdin [1], who introduced the term *α-function* for the expansion functions. In the case of an angular momentum wave function for which the radial factor is defined numerically, the α-function can be constructed using a numerical technique for computing spherical Bessel, or Hankel, transforms [2, 3]. These methods for computing spherical Bessel transforms and α-functions can then be applied to molecular calculations in which the radial factors in the atomic basis orbitals are defined purely numerically [4].

This possibility of using numerically defined orbitals raises the interesting problem of the variational optimization of the basis functions, as is done in numerical atomic Hartree-Fock calculations [5]. The use of such optimized basis orbitals can then permit the accurate determination of molecular orbitals using only a minimal number of basis functions. The resulting orbitals should also be more physically realistic since they can describe properly the cusp behavior of electronic wave functions at the nuclei. The problem of variational optimization of numerical orbitals has been investigated in a recent article [6] in which the basic equations are presented, and the results of a preliminary application to the methane molecule are described.

The purpose of this article is to discuss more fully the theory presented in [6], to describe in some detail the numerical methods employed, to assess the approximations involved and to present results for a number of small molecules. In particular, a rapidly convergent method for finding the eigenvalue of the ordinary differential equation for the optimized orbitals will be described, and a problem associated with the non-differentiability of the nuclear potential seen by the orbitals will be discussed in detail.

It may be remarked that, because the angular momentum summations for the molecular orbitals are truncated at small l values, this program does not completely solve the Hartree-Fock problem for molecular electronic structure, which is a difficult three-dimensional problem. However, the approach and resulting equations are closely similar to the Hartree-Fock method. In the limit of large l summations, the Hartree-Fock limit would be attained, but in practice going beyond quite small l values will probably be attended by severe numerical difficulties.

The problem of numerical optimization of atomic orbitals has also been discussed recently by Andrae [7]. The optimization procedure has been applied recently to the H_2 and H_3^+ molecules [8].

2 The Mathematical Problem

We consider a system of N_n nuclei located at fixed positions $\mathbf{R}_n$ and N_o orbitals (AOs) $\chi_i(\mathbf{r})$ of the form $f_{l_i m_i}(\mathbf{r} - \mathbf{R}_i)$ centered at the position $\mathbf{R_i}$. The $f_{lm}(\mathbf{r})$ are angular momentum functions of the form

$$\chi(\mathbf{r}) = \chi_{lm}(r) Y_{lm}(\theta, \phi). \quad (1)$$

(In practice, the real and imaginary parts of Y_{lm} are used.) The Gram-Schmidt orthonormalization procedure is used to form an orthonormal molecular basis of molecular orbitals (MOs) ϕ_i:

$$\phi_i(\mathbf{r}) = \sum_{k\leq i} d_{ki}\chi_k(\mathbf{r}). \tag{2}$$

The coefficients d_{ki} are determined by forming the overlap matrix, Ω,

$$\omega_{ij} = \int \chi_i(\mathbf{r})\chi_j(\mathbf{r})d\mathbf{r}, \tag{3}$$

and finding its Cholesky factorization, $\Omega = R^T R$. The elements d_{ki} are then the elements of $D = R^{-1}$. This differs from the common practice of diagonalizing the overlap matrix to find its square root. This method is used in order to take into account the variational derivative of the overlap matrix with respect to the atomic orbitals as discussed in Ref. 1.

Linear combinations (MOs) of the ϕ_i,

$$\begin{aligned}\psi_i(\mathbf{r}) &= \sum_j u_{ji}\phi_j(\mathbf{r}) \\ &= \sum_{kj} d_{kj}u_{ji}\chi_k(\mathbf{r}) \\ &= \sum_k c_{ki}\chi_k(\mathbf{r})\end{aligned} \tag{4}$$

are formed by solving the Roothaan-Hartree-Fock equations [9],[10]. Explicitly, the coefficients c_{ki} are the elements of the product $C = DU$ and the elements of U are obtained from the eigenvectors of the equation

$$\hat{H}\mathbf{u}^i = \lambda_i\mathbf{u}^i; \tag{5}$$

$$u_{ji} = \mathbf{u}^i_j. \tag{6}$$

The Roothaan-Hartree-Fock equation in the AO basis is

$$\hat{H}_0\mathbf{v}^i = \lambda_i\Omega\mathbf{v}^i \tag{7}$$

where

$$\hat{h}^0_{kk'} =< k|H_{\rm sp}|k' > + \sum_{ll'} w_{ll'}[< kl|v|k'l' > -\frac{1}{2} < lk|v|k'l' >] \tag{8}$$

and $\hat{H} = D^T \hat{H}_0 D$, $\mathbf{v}^i = D\mathbf{u}^i$. In the restricted Hartree-Fock approach being followed here, the $w_{kk'}$ are the elements of the density matrix P given in the MO basis by

$$P = 2\sum{}' \mathbf{u}^i \mathbf{u}^{iT} \tag{9}$$

and in the AO basis by $W = DPD^T$. The sum in Eq. (9) is on the occupied orbitals.

The electronic energy is given, in the AO basis, by

$$\begin{aligned} E &= \sum_{kk'} w_{kk'} < k|H_{\rm sp}|k' > \\ &+ \frac{1}{2} \sum_{klk'l'} [w_{kk'} w_{ll'} - \frac{1}{2} w_{lk'} w_{kl'}] < kl|v|k'l' > . \end{aligned} \tag{10}$$

Here

$$H_{\rm sp} = -\frac{1}{2}\nabla^2 + \sum_{i=1}^{N_n} \frac{Z_i}{|\mathbf{r} - \mathbf{R}_i|} \tag{11}$$

and $< kl|v|k'l' >$ are the two-particle matrix elements of the electron- electron repulsion.

The variational equation for the χ_k is

$$\frac{\delta E}{\delta \chi_m(\mathbf{r})} = 0. \tag{12}$$

As discussed in [6], E depends on χ_k directly through the matrix elements of $\hat{H}_0$ and indirectly through the elements of W, which depends on the overlap matrix Ω. The resulting equation is

$$\sum_n \left\{ w_{mn} H_{\rm sp} + \sum_{ll'} \left[w_{mn} w_{ll'} - \frac{1}{2} w_{ln} w_{ml'} \right] v_{ll'}(\mathbf{r}) - \kappa_{mn} \right\} \chi_n(\mathbf{r}) = 0. \tag{13}$$

where

$$v_{ll'}(\mathbf{r}) = \int \frac{1}{|\mathbf{r} - \mathbf{r}'|} \chi_l(\mathbf{r}') \chi_{l'}(\mathbf{r}') d\mathbf{r}'. \tag{14}$$

The elements κ_{mn} stem from the transformation from the AO basis to the orthonormal MO basis [6]. They are given by

$$\kappa_{mn} = \frac{1}{2} {\rm Tr}[\hat{H}(Y_{mn} P + P Y_{mn}^T)], \tag{15}$$

$$[Y_{mn}]_{ij} = d_{mi}d_{nj} + d_{ni}d_{mj},\ j > i,$$
$$= d_{mi}d_{nj},\ j = i,$$
$$= 0,\ j < i. \tag{16}$$

The equations are solved iteratively since the $w_{kk'}$ and $v_{ll'}(\mathbf{r})$ depend on the χ_i.

The system of equations (13) cannot be uncoupled since, in general, the density matrix with elements w_{mn} is singular. Therefore, the problem is solved by separating out the diagonal terms of the equation, and treating the off-diagonal terms as obtained in the previous iteration as an inhomogeneous term. This yields a Schrödinger-like equation for an orbital χ_m with an inhomogeneous term. At the same time, if the procedure is to be meaningful, it is necessary to maintain the normalization of χ_m. This normalization condition is dealt with by introducing a term in the energy with a Lagrange multiplier times the normalization integral. A similar situation arises in atomic Hartree-Fock calculations in which the equations for the orbitals are also coupled [5].

The radial factor in the orbital is obtained by projecting the equation onto the (lm) angular momentum subspace centered at the orbital center. This yields the equation

$$-\frac{1}{r^2}\frac{d}{dr}r^2\frac{d\chi_m}{dr} + \frac{l_m(l_m+1)}{r^2}\chi_m(r) + 2[V_d(r) - E_m]\chi_m(r) + 2Z_m(r) = 0 \tag{17}$$

where E_m is the Lagrange multiplier and $Z_m(r)$ represents the inhomogeneous terms. The potential $V_d(r)$ contains terms from the electron-nuclear attraction energy, and from the direct and exchange electron-electron repulsion energy. Since these terms are projected onto an angular momentum subspace, the electron-nuclear attraction energy contributes terms of the form r^l/R^{l+1}, $r < R$, or R^l/r^{l+1}, $r > R$, where R is the distance to one of the other nuclei. This behavior produces a mild singularity, a discontinuous third derivative, in the resulting orbitals, which may cause numerical difficulties.

3 Numerical Solution

The differential equation (17) is solved on a numerical mesh of points equally spaced in the variable suggested by Chernyshiva et al. [11]

$$\rho = \ln r + \alpha r \tag{18}$$

at mesh points $\rho_i = \rho_{\min} + (i-1)\delta\rho$. With this change of variable, it is necessary to include a weight factor

$$w(r) = \frac{r}{1+\alpha r} \tag{19}$$

in radial integrals. Eq. (17) becomes, if $\chi(r)$ is replaced by $\bar{\chi}(r) = r\chi(r)$, the reduced radial equation

$$-\frac{d^2\bar{\chi}_m}{d\rho^2} + 2w(r)^2[\bar{V} - E_m]\bar{\chi}_m + 2w(r)^2 r Z_m = 0. \tag{20}$$

where

$$\bar{V} = V_d(r) + \frac{l(l+1)}{r^2} + \frac{1/4 + \alpha r}{r^2(1+\alpha r)^2}. \tag{21}$$

The Numerov method [5] is used to find the numerical solution.

In order to make the projection of the the equation onto an angular momentum subspace centered at a particular nucleus, it is necessary to make the Löwdin α-function expansion of all the orbitals in terms of angular momentum functions centered at that nucleus. The basic expansion is of the form

$$f_{lm}(\mathbf{r} - \mathbf{a}) = \sum_{LM} \alpha_{LM}[r, \mathbf{a}, f_{lm}] Y_{LM}(\hat{r}). \tag{22}$$

The product of two such expansions can be expressed in the same form, using the familiar expansion for the product of spherical harmonics in terms of spherical harmonics.

The principal source of error in the calculation is the truncation of the angular momentum sums that arise in the expansion of a function centered at one point in terms of angular momentum functions centered at another point. A variety of upper limits for these summations are used in the calculation. The total energy calculation and the orbital optimization procedure are carried out totally independently. Therefore, the degree to which the self-consistent iterations for the orbitals yields a minimum gives an indication of the overall accuracy of the calculations.

The total energy calculation requires the evaluation of the nuclear attraction integrals and the electron-electron repulsive (generally four-center) integrals. The nuclear attraction summations are terminated at a value $L_{N,max} = 12$. The four-center integrals are computed by expanding each factor in a product about the mid-point of the line joining their centers and multiplying the two expansions. The expansion of each factor in the product is truncated at a value $L_{max} = 14$. The product expansion is truncated at a smaller value $\Lambda^{(1)}_{max} = 10$. The four-center integrals can be expected to be very small if the two orbitals in each product are on different centers, but the number of terms in the sum is very large. In these cases, the sums are truncated at $\Lambda^{(2)}_{max} = 6$. The results obtained with these limits appear to be accurate to $\approx 10^{-4}$hartree. The method is described more fully in Ref [12].

The orbital optimization requires the expansion of each orbital and each nuclear attraction potential about each center, followed by multiplication of the various expansions. The expansions of the orbitals and potentials are truncated at $L_{opt,1} = 18$. The overall accuracy seems to be most sensitive to this truncation; with this choice the results appear to be accurate to about a mhartree, apart from the case of N_2 discussed below. The nuclear attraction contributions to the equation require the multiplication of two such expansions. This is carried out over $L_{opt,1}$ terms in each expansion. The product is required only up to l_{max}, the largest of the basis angular momenta.

The electron-electron repulsion contribution requires the calculation of potential operators formed from 'charge densities' that are products of orbitals. The products are formed using $L_{opt,1}$ terms in the sum, but are truncated at a smaller limit $L_{opt,2} = 10$. The final terms require the product of these potentials with the orbital expansions; these are carried out over $L_{opt,2}$ terms, but the product is only required up to l_{max} terms.

The numerical methods for the expansion of an orbital about a different center and the calculation of multicenter integrals use Fourier transform methods [2, 3] that require the use of a logarithmic mesh, i.e., $\alpha = 0$ in Eq. (18). Cubic interpolation is used to transform orbitals and other functions from one mesh to another. Tests of a number of choices indicate that a mesh as defined in Eq. (18) with $\rho_{\min} = -7.0$, $\delta\rho = 0.1$ and $\alpha = 0.5$ with approximately 175 mesh points gives satisfactory numerical stability. The logarithmic mesh employed in the Fourier transform technique uses mesh parameters $\rho_{\min} \approx -14.0$, $\delta\rho \approx 0.065$ and 256 mesh points.

4 Eigenvalue Determination

The reduced equation (20) has the form

$$-\frac{d^2\phi}{d\rho^2} + 2w(r)^2\left[\frac{l(l+1)}{2r^2} + V_d(r) - E\right]\phi(\rho) = Y(\rho). \tag{23}$$

where $Y(\rho) = -2r(\rho)w(r(\rho))^2 Z(r(\rho))$.

The solution of Eq. (23) can be expressed in the well-known form

$$\phi(\rho) = f_1(\rho)\int_\rho^\infty f_2(\rho')Y(\rho')d\rho' + f_2(\rho)\int_0^\rho f_1(\rho')Y(\rho')d\rho' \tag{24}$$

where $f_1(\rho)$ and $f_2(\rho)$ are solutions of the corresponding homogeneous equation with $f_1(0) = 0$, $f_2(\rho) \to 0, \rho \to \infty$ and the Wronskian $f_2(\rho)f_1'(\rho) - f_1(\rho)f_2'(\rho) = 1$.

The solution can also be expressed formally in terms of the eigenfunctions and eigenvalues of the homogeneous reduced equation

$$-\frac{d^2\eta_n}{d\rho^2} + 2w(\rho)^2\left[\frac{l(l+1)}{2r^2}V_d(r) - \lambda_n\right]\eta_n(\rho) = 0 \tag{25}$$

as

$$\phi(\rho) = \sum_n \frac{<\eta_n|Y>}{\lambda_n - E}\eta_n(\rho) \tag{26}$$

$$<f|g>= \int_0^\infty w(\rho)^2 f(\rho)g(\rho)d\rho \tag{27}$$

The eigenvalue condition is

$$\begin{aligned} S(E) &= <\phi|\phi> \\ &= \sum_n \left[\frac{<\eta_n|Y>}{\lambda_n - E}\right]^2 \\ &= 1 \end{aligned} \tag{28}$$

In the usual situation, $Y(\rho)$ is small, and $\phi(\rho)$ is approximately one of the eigenfunctions, say $\eta_k(\rho)$. This permits an initial estimate for E of

$$E_0 = \lambda_k \pm < \eta_k|Y > . \qquad (29)$$

This result suggests that there is no unique solution to the eigenvalue problem; in fact, Eq. (28) indicates that there is a pair of roots for any k for which

$$\sum_{n \neq k} \left[\frac{< \eta_n|Y >}{\lambda_n - \lambda_k} \right]^2 < 1. \qquad (30)$$

The ambiguity in the square root in Eq. (29) can cause a serious problem with the convergence of the Hartree-Fock iterative process in that the χ_k may 'turn over' between iterations. Therefore, to maintain consistency between iterations, the sign is chosen to be negative, so that the initial ϕ is η_k.

Solution of Eq. (23) with E_0 yields a value for $S(E_0)$. It is also possible to determine $S'(E_0)$:

$$S'(E_0) = 2 < \phi|\phi_E > . \qquad (31)$$

The function ϕ_E, the derivative of ϕ with respect to E, satisfies the inhomogeneous equation

$$-\frac{d^2\phi_E}{d\rho^2} + 2w(r)^2 \left[\frac{l(l+1)}{2r^2} + V_d(r) - E \right] \phi_E(\rho) = 2\phi(\rho). \qquad (32)$$

and can therefore be solved using Eq. (24).

We assume that near the desired solution $S(E)$ can be approximated by a single pole:

$$S(E) = \frac{A^2}{(\mu - E)^2}. \qquad (33)$$

Then

$$S(E_0) = \frac{A^2}{(\mu - E_0)^2}. \qquad (34)$$

$$S'(E_0) = \frac{2A^2}{(\mu - E_0)^3}. \qquad (35)$$

$$\mu = E_0 + \frac{2S(E_0)}{S'(E_0)}. \tag{36}$$

$$A = \pm\frac{2S(E_0)^{3/2}}{S'(E_0)}. \tag{37}$$

Substituting into the equation $S(E) = 1$ yields a new estimate for E:

$$E = E_0 + 2S(E_0)[1 - S(E_0)^{1/2}]/S'(E_0). \tag{38}$$

If $S(E_0) \approx 1$ this reduces to the usual Newton-Raphson correction.

In the calculations described below, this procedure yielded eigenvalues accurate to 10^{-5} within 2-4 iterations in all the minimum basis set cases.

5 Discontinuous Derivative Problem

It was remarked above that, because of nuclei at other centers than the particular orbital center, the potential in the orbital equation has discontinuous derivatives at r values corresponding to the distance to the other nuclei. This singularity occurs in both the potential and the inhomogeneous term in Eq. (17). The latter dependence occurs because of the product of the potential with the other orbitals. In order to assess the effect of this problem, the equations were also solved on a segmented mesh, in which the mesh was constructed at points uniformly spaced between the various internuclear separations with the wave functions and their derivatives constructed to be continuous through the end points.

In Table 1 we show the energy after each iteration for the systems N_2, H_2O, CH_4 and NH_3. These were obtained at the geometries defined in the next section. The interesting observation is that at the first iteration there is a significant advantage to using the mesh that takes into account the singularities, but that this advantage does not persist to the final iteration. Similar results were observed in H_2 and CO; an exception was HF in which the results of the uniform mesh were lower than on the segmented mesh.

The explanation for this observation appears to be that the behavior of a molecular orbital at a particular nucleus is dominated

Table 1: Energy values at successive iterations for a uniform mesh, and a segmented mesh.

N_2		H_2O	
uniform	segmented	uniform	segmented
−108.70740	−108.70740	−75.91372	−75.91372
−108.73641	−108.75479	−75.97223	−75.97678
−108.79692	−108.79904	−76.00848	−75.00810
−108.79886	−108.80030	−76.01024	−76.01028
−108.79945	−108.80073	−76.01072	−76.01080
−108.79954	−108.80075	−76.01061	−76.01073
CH_4		NH_3	
uniform	segmented	uniform	segmented
−40.03810	−40.03810	−56.05479	−56.05479
−40.09205	−40.10819	−56.10096	−56.10654
−40.18710	−40.18677	−56.16877	−56.16888
−40.18819	−40.18867	−56.17036	−56.17046
−40.18881	−40.18904	−56.17115	−56.17119
−40.18879	−40.18906	−56.17129	−56.17135

by the atomic orbitals sited at that nucleus. These play the dominant role in representing the cusp behavior at the nucleus. As self-consistency is approached, orbitals situated at off-nucleus sites apparently are not important for the cusp behavior. This is probably manifested in a cancellation of the potential discontinuity between the potential term and the inhomogeneous term.

It should be commented that these results differ somewhat from those presented in the next section, which were obtained with a number of improvements in the numerical methods.

6 Results

In this section, results for a number of diatomic molecules and ten-electron molecules using minimal bases are presented and compared with corresponding results applying GTOs.

A summary of the results is presented in Table 2. The molecules

Table 2: Minimal basis calculations results for small molecules.

molecule	STO-3G	4-31G	numerical	iterations
H_2	−1.117	−1.127	−1.12854	10
N_2	−107.496	−108.754	−108.81379	6
CO	−111.225	−112.552	−112.63332	4
HF	−98.571	−99.887	−100.03674	10
H_2O	−74.963	−75.907	−76.01476	11
NH_3	−55.454	−56.102	−56.17682	13
CH_4	−39.727	−40.140	−40.19046	5

chosen are those for which detailed results using GTOs are given in the text of Szabo and Ostlund [13]. The molecules and bond lengths are: H_2, (1.400), N_2 (2.074), CO (2.132), CH_4 (2.05), NH_3 (1.913), H_2O (1.809), and HF (1.733). In the NH_3 and H_2O cases the H-H bond angles are 106.67° and 104.52° respectively and in the methane case the angles are equal. The minimal basis consist of five functions, $1s$, $2s$, $2p_{x,y,z}$ for the first row elements, and a single $1s$ function for hydrogen.

The Gaussian basis labelled STO-3G consists of the same basis functions, each a superposition of three GTOs constructed to approximate Slater functions. There is no optimization of the orbitals beyond the introduction of a screening parameter to simulate the contraction of an atom when bound in a molecule. The STO-4-31G basis consists of nine basis functions, with 'inner' and 'outer' $2s$ and $2p_{x,y,z}$, orbitals. The $1s$ orbital is a superposition of four GTOs, the inner $2s$ and $2p$ orbitals are superpositions of three GTOs and the outer orbitals are a single GTO. The orbital parameters are optimized for each *atomic* wave function and scaling factors are introduced for the valence (inner and outer) orbitals.

It is observed that the numerical optimization of the orbitals produces a very substantial improvement on the minimal basis STO-3G basis and and improvements of the order of 50-150 mhartree on the STO-4-31G split valence basis.

This result for N_2 is in significant disagreement with the result of −108.90836 obtained by Moncrieff and Wilson [14] using a basis of 14 s and 7 p orbitals at a bond length of 2.068 Bohr. The minimal basis at this bond length gives an energy of -108.81344. In this particular case, the minimal basis is evidently inadequate to describe the occupied MOs. It is found that including the $3s$ orbital in the basis makes a substantial improvement in the energy, lowering it to -108.89031. Including also the $3p_z$ orbital gives a further lowering to -108.90402. It appears that the 3 s and $2p_z$ orbitals are not quite sufficient to describe the five molecularσ orbitals. It is somewhat surprising that the lowest single-particle state is consistently found to be $1\sigma_u$. It may be conjectured that the electron density at the bond center favours a node in the σ orbital. This effect is apparently not subject to experimental verification. Adding the $3s$ orbitals in the CO calculation lowers the energy to -112.69117.

A calculation for CH_4 with $7s$ orbitals and 4 p orbitals on C and 4 s orbitals on each proton at a bond length of 2.05 Bohr has been made by Thakkar [15], yielding an energy of -40.19265. This is slightly below the present result, but the discrepancy can perhaps again be ascribed to the truncation of the angular momentum sums.

To go beyond the minimal basis calculation, it is necessary to include 'polarization' orbitals in the basis. In the standard terminology this leads to the introduction of the 6-31G* and 6-31G** bases for H and the first row elements. The 6-31G* basis includes the addition of $3d$ orbitals on the first row elements, and the improvement of the $1s$ inner shell orbital to be a contraction of six GTOs. There are in fact six $3d$ orbitals, one of which is essentially an s orbital with $r^2e^{-\zeta r}$ radial dependence. In the results given in [13], there is no optimization of the $3d$ nonlinear parameter. The 6-31G** basis includes the addition of $2p$ orbitals on the H atoms. A single GTO is used, again with no optimization of the nonlinear parameter. There is of course no 6-31G* basis for H_2 and no 6-31G** basis for N_2 and CO.

Results of calculations using polarization orbitals computed in this way are shown in Table 3 and compared with corresponding results using GTOs as given in [13]. The improvement in energies is considerable, but is less marked than in the minimal basis calculations, reflecting the difficulty in making improvements as the wave functions become successively more accurate. Table 4 shows a comparison of the optimized energy results with estimates of the exact restricted Hartree-Fock energies as tabulated by Hariharan

Table 3: Energies of small molecules including polarization orbitals.

molecule	6-31G*	optimized	6-31G**	optimized
H_2			−1.131	−1.13306
N_2	−108.942	−108.98031		
CO	−112.737	−112.77560		
HF	−100.003	−100.05886	−100.011	−100.06159
H_2O	−76.011	−76.04796	−76.023	−76.05369
NH_3	−56.184	−56.20813	−56.195	−56.21217
CH_4	−40.195	−40.20643	−40.202	−40.21494

and Pople [16]. In the N_2 and CO cases, the numerical results include also $3s$ and $3p_z$ orbitals on each nuclei. As discussed above, these have a substantial effect when included in the minimal basis. When included with the $3d$ orbitals, they decrease the energy by 47 mhartree in the N_2 case and 38 mhartree in the CO case.

Table 4: Comparison of the calculated energies with the estimated Hartree-Fock limit.

molecule	optimized	HF limit
H_2	−1.133	−1.134
N_2	−108.980	−108.997
CO	−112.776	−112.791
HF	−100.062	−100.071
H_2O	−76.054	−76.065
NH_3	−56.212	−56.225
CH_4	−40.215	−40.225

There is a variety of sources of possible errors in these calculations. These are primarily the finite difference approximation, truncation of the angular momentum sums, and possible uncertainties arising from the termination of the self-consistency iterations. Rather than imposing a convergence criterion on the orbitals, the

Table 5: N_2 energies at various bond lengths and their deviations δ from a least squares fit.

R	energy	δ
1.908	−108.76757	0.00000
1.958	−108.78883	0.00020
2.008	−108.80448	−0.00056
2.058	−108.80448	0.00039
2.108	−108.81712	0.00020
2.158	−108.81795	−0.00046
2.208	−108.81394	0.00047
2.258	−108.80928	−0.00037
2.308	−108.80171	0.00012

iterations have been terminated when the total energy stopped decreasing; however, this may cause the result to be slightly sensitive to the initial choice of orbitals and may introduce a somewhat random error. The largest source of error appears to be the truncation of the angular momentum summations. However, in the practical problem of obtaining potential energy surfaces the angular momentum truncations probably do not introduce significant errors, but the uncertainty in the termination of the energy iteration process may introduce random errors. To investigate the dependence of the computational errors on the geometry, the N_2 energy has been computed at various bond lengths, as shown in Table 5. The bond lengths were chosen to be at the radial mesh points and half-way between mesh points to indicate any possible effect of the discontinuous derivative problem. The deviations of a least squares fit to the data are also shown. The results indicate that the errors stemming from inconsistency in terminating the iterations are substantially less than one mhartree, and they do not show any systematic error relating to the position of the bond length relative to the mesh points. The least squares fit yields an energy minimun of −108.81787 at a bond length of 2.134 Bohr.

7 Discussion

We have found that with a judicious choice of the radial mesh, the differential equations for the optimized orbitals can be solved in a reliable way, and that the eigenvalue associated with the maintenance of the normalization can be found in a very small number of iterations. The possible numerical problem arising from the discontinuous derivative of the potential appears to be secondary when self-consistency of the molecular orbitals is achieved, since the cusp behaviour of the orbitals at a nucleus is mainly described by the orbitals centered at that nucleus.

The results show that numerical optimization of the basis orbitals makes a very substantial improvement on energies computed in the minimal GTO basis. The improvement of the contribution of polarization orbitals is also significant, but less marked. The calculations using the minimal basis sets are robust and give very consistent results. However, including polarization orbitals is quite vexatious. As well as being much more time consuming, satisfactory convergence of the iterative procedure tends to be sensitive to the choice of the mesh. Undoubtedly, this problem arises because the total energy is much less sensitive to the polarization orbitals.

Future work in this project will include the calculation of the derivatives and second derivatives of the energy with respect to the nuclear coordinates. The feasibility of extending the calculations to considerably larger systems will also be studied to establish whether these methods can provide a useful addition to the standard methods of quantum chemistry to which P-O Löwdin has made so many contributions. This is a valid goal; as is well-known the processing time for an MO-LCAO Hartree-Fock calculation is proportional to N_o^5 so that minimizing N_o is an important objective. Another area of interest being pursued is the application of these methods to density functional calculations.

Acknowledgment

This research has been supported by the Natural Sciences and Engineering Research Council of Canada.

References

[1] Löwdin, P-O, Adv. Phys **5**, 1 (1956).

[2] J.D. Talman, J. Comput. Phys. **29**, 35 (1978).

[3] J.D. Talman, in it ETO Multicenter Integrals , C.A. Weatherford and H.W. Jones Eds. Reidel (Dordrecht, 1982).

[4] J.D. Talman, Int. J. Quantum Chem.: Quantum Chemistry Symp. **27**, 321 (1993).

[5] C. Froese Fischer, *The Hartree-Fock Method for Atoms*, Wiley-Interscience (New York, 1977).

[6] J.D. Talman, Phys. Rev. Lett. **84**, 855 (2000).

[7] D. Andrae, Mol. Phys. **99**, 327 (2001).

[8] J.D. Talman, Can. J. Phys. **79** 673, (2001).

[9] C.C.J. Roothaan, Rev. Mod. Phys. **23**, 69 (1951).

[10] G.G. Hall, Proc. Roy. Soc. Lond. **A205**, 541 (1951).

[11] L.V. Chernyshev, N.A. Cherepkov and V. Radojevic, Comp. Phys. Commun. **11**, 270 (1976).

[12] J.D. Talman, J. Chem. Phys. **80**, 2000 (1984).

[13] A. Szabo and N.S. Ostlund, *MODERN QUANTUM CHEMISTRY Introduction to Advanced Electronic Structure Theory*, McGraw-Hill (New York, 1989)

[14] D. Moncrieff and S. Wilson, J. Phys. B: At. Mol. Opt. Phys. **26** 1605, (1993).

[15] A.J. Thakkar, private communication.

[16] P.C. Hariharan and J.A. Pople, Theoret. Chim. Acta **28**, 213 (1973).

ABOUT OVERLAP AND CHEMICAL REALITY

Giuseppe Del Re[1]
Department of Applied Mathematics,
University of Waterloo, Waterloo, Ontario, Canada
Giuseppe.Delre@unina.it

Introduction: matrices and the physics of orbitals

This paper is intended as a presentation of research along the lines of Per-Olov Löwdin's scientific work, and is dedicated to his memory on two accounts: the scientific importance of his ideas for the future of the theory of molecules and his generous efforts to make his collaborators familiar with the mathematical techniques best suited for processing computational output in view of understanding molecular reality.[1] I wish to express here my special gratitude to Per-Olov for all I learned from him on matrices as tools of theoretical chemistry.

I shall briefly review and complete with some new results a theme that might be generally described as "orthogonalization techniques and the physics of orbitals." I hope to show convincingly that proper use of the matrix methods advocated and developed by Per-Olov Löwdin can reach far beyond the familiar Löwdin orthogonalization procedure; and that, among other things, those methods can yield a quantitative representation of chemical bonds from molecular computations only designed to determine energies and geometries.[2] Valuable papers on orthogonality, albeit in a more pragmatistic spirit, have been published by Walter Thiel and his collaborators. A recent review by that group may be considered as a fairly complete study of the orthogonality problem in connection

[1]Permanent address: Departmei emistry, University of Naples "Federico II," Campus Monte S. Angelo, Via Cinthia, 80126 Napoli, Italy.

E.J. Brändas and E.S. Kryachko (eds.),
Fundamental World of Quatum Chemistry, Vol. I, 335–370.

with predictive computations on molecules; it also contains references to work on that topic made during the last few decades.[3]

Per-Olov Löwdin's research may seem at first sight essentially mathematical, for he placed emphasis on projection operators, density matrices and orthogonalization as mathematical tools. This, however, does not diminish the role of genuine physics in his work; for his presentation of important results had as its background the scientific climate of the years 1950 to 1970, when the pioneering work of Slater, Pauling and Mulliken was still quite actual, and it was easy to appreciate the physical significance of mathematical treatments even when it was not explicitly pointed out. As a proof of this claim, suffice it to mention that, in his fundamental study of 1956,[4] on introducing the nonorthogonality problem, Löwdin spoke of a "most naturally occurring basis set," obviously with reference to atomic orbitals and the physical significance of overlap, which Pauling in particular had proved in terms of numerical correlations.[5] Among the various conceptual difficulties related to overlap was that the Hückel model was successful for understanding conjugation in aromatic molecules, and yet was based, in contrast to Pauling's findings, on a set of $2p\pi$ orbitals whose overlap could be neglected without loss of explanatory and predictive power. Among other things, Löwdin solved the mathematical problem of making the two approaches compatible. This was a specific merit of his famous orthogonalization procedure.

The purpose of this paper is, (i), to emphasize the general variational nature of Löwdin's orthogonalization, (ii), to illustrate its various facets by showing how it can be used to transform the raw output of current *ab initio* computations into physically significant MO-LCAO descriptions of molecules. We shall present the procedure in question step by step. Our hope is that in this way many researchers will be persuaded that when one reads from spectroscopists that "there are no such *things* as orbitals,"[6] the message to be retained is actually that orbitals are meaningless unless they are uniquely defined and shown to describe objective features of molecular structure and properties at a well defined level of approximation. If statements of that sort were to be taken literally, then the only criterion of acceptance of a notion in science would be accurate prediction of observed quantities, and most of physics would have to be rejected, beginning with Kepler's planetary orbits and ending with quarks. The creation-annihilation operator scheme (N-representation) of theoretical physics, which depends on physical single-particle states that are filled or emptied, would be disqualified except for noninteracting particles; since such a disqualification would be unreasonable, then also orbitals, which play the same role in quantum chemistry, should be accepted as ingredients of a description of molecular reality, albeit within the limits of validity of an appropriate independent particle model.

1 Construction of a physical atomic orbital basis

1.1 Löwdin orthogonalization as an optimization problem

The special optimization problem solved by Löwdin consists in looking for a linear transformation $\mathbf{T}$ of a nonorthogonal basis $|\boldsymbol{\chi}>$ such that the new basis

$$|\boldsymbol{\chi}^{(L)}>= |\boldsymbol{\chi}> \mathbf{T} \tag{1.1}$$

be orthogonal and as close as possible to $|\boldsymbol{\chi}>$. We have the simple condition

$$\begin{aligned}&\delta(<\boldsymbol{\chi}^{(L)}|\boldsymbol{\chi}> + <\boldsymbol{\chi}|\boldsymbol{\chi}^{(L)}> + \\ &\quad -\boldsymbol{\lambda} <\boldsymbol{\chi}^{(L)}|\boldsymbol{\chi}^{(L)}> - <\boldsymbol{\chi}^{(L)}|\boldsymbol{\chi}^{(L)}> \boldsymbol{\lambda} = 0.\end{aligned} \tag{1.2}$$

where $\boldsymbol{\lambda}$ is a Hermitean matrix of Lagrange multipliers. In matrix form, introducing the new Hermitean Lagrange matrix

$$\boldsymbol{\Lambda} = \mathbf{M} + \mathbf{M}^{\dagger} \text{ with } \mathbf{M} = \mathbf{S}\mathbf{T}\boldsymbol{\lambda}\mathbf{T}^{-1},$$

the above can be written as

$$\delta(\mathbf{T}^{\dagger}\mathbf{S} + \mathbf{S}\mathbf{T} - \mathbf{T}^{\dagger}\boldsymbol{\Lambda}\mathbf{T}) = 0. \tag{1.3}$$

This reduces to

$$\mathbf{S} - \boldsymbol{\Lambda}\mathbf{T} = 0 \tag{1.4}$$

and its Hermitean conjugate, with the condition that $<\boldsymbol{\chi}^{(L)}|\boldsymbol{\chi}^{(L)}>= \mathbf{I}$. The solution of (1.4) is

$$\mathbf{T} = \boldsymbol{\Lambda}^{-1}\mathbf{S}, \quad \mathbf{S}\boldsymbol{\Lambda}^{-1}\mathbf{S}\boldsymbol{\Lambda}^{-1}\mathbf{S} = \mathbf{I}, \quad \mathbf{T} = \mathbf{S}^{-\frac{1}{2}}, \tag{1.5}$$

QED.

This is the well known Löwdin orthogonalization. It is characterized by the close similarity of the orthogonalized basis set to the original one. It is perhaps useful to recall the application of equation (1.5) to a Hückel basis of six $2p\pi$ atomic orbitals. With an SCF atomic orbital basis,[7] the Löwdin atomic orbitals are the linear combinations whose coefficients are shown in the histograms on the left of figure 1.

Figure 1: $2p\pi$ orbitals of benzene orthogonalized by a standard Löwdin procedure (left) and by the modified Löwdin procedure (right). as explained in the text.

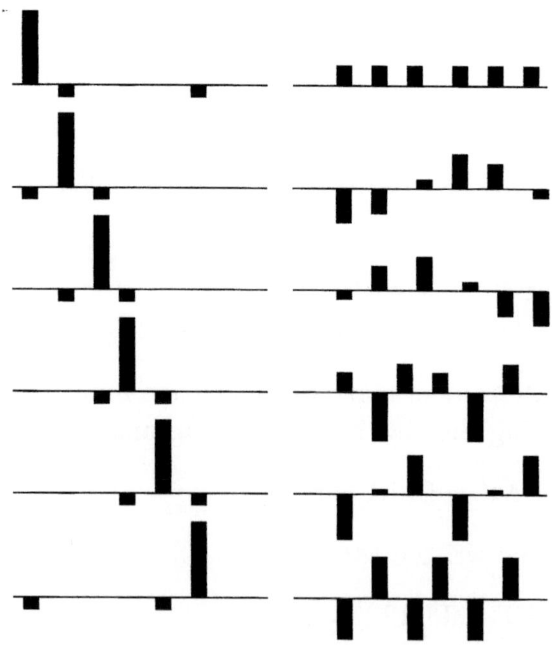

The closeness of these pseudo-atomic orbitals to the true ones is evident; however, the current opinion that they are in fact more localized that the original ones is somewhat vague, for it depends on what is meant by "localized". For one thing, because of quantum interference effects, the squares of the coefficients given above cannot be interpreted as the probabilities of finding an electron prepared in φ_j in one of the $2p\pi$ orbitals.

Concerning basis orthogonalization in general, there are problems which require that it be performed according to a criterion other than minimum deviation from the original basis. A case in point is when it is required that the elements of the orthogonal basis be such that the diagonal elements of a function of the overlap matrix should become as different as possible. This is done again by applying Löwdin's orthogonalization, but a special unitary transformation of the orthogonalized basis should follow. Denoting by $\bar{\mathbf{S}}$ the diagonal form of $\mathbf{S}$ and by $\mathbf{Q}$ its eigenvector matrix, the complete transformation $\mathbf{T}$ is obtained by letting $\mathbf{W} = \mathbf{Q}$ in

$$\mathbf{T} = \mathbf{Q}\,\bar{\mathbf{S}}^{-1/2}\,\mathbf{Q}^{\dagger}\,\mathbf{W} = \mathbf{S}^{-1/2}\mathbf{Q}. \tag{1.6}$$

In the new basis derived from eq.n (1.1) with $\mathbf{T}$ defined as in eq.n 1.6), any analytical function of overlap, e.g. $-\mathbf{I} + 3.5 * (\mathbf{S} - \mathbf{I})$, is diagonal. We shall call "modified Löwdin orthogonalization" the orthogonalization procedure under consideration. Application of equation 1.6 to the benzene π system yields orbitals represented by the right hand histograms of figure 1.

It will be shown below, on discussing the physical significance of bases used for *ab initio* computations, that an appropriate choice of orthogonalization criteria can transform a set of numbers at best useful for detecting computational errors into an interesting description of molecular reality.

1.2 Physical significance of an atomic orbital basis

Some researchers tend to consider Löwdin's orthogonalization a mere trick to be used if overlap causes computational difficulties. This is certainly a limitation of its originally intended scope. As mentioned, the interest of Löwdin's orthogonalization of atomic orbital bases was due, among other things, to its usefulness in reconciling the established parallelism between overlap and chemical strength with successful MO-LCAO models neglecting overlap; but actually Löwdin provided tools that can help to substantiate a reflection on why and how orbitals can be treated as the building

Table 1: Construction and *in situ* energies of physical s carbon atomic orbitals from a DZ-631G computation.

Data from *ab initio* output

C s subshell overlap			C s subshell Fock matrix		
1.0000	0.2191	0.1843	-11.2803	-2.6411	-2.1907
0.2191	1.0000	0.8123	-2.6411	-1.1415	-1.2551
0.1843	0.8123	1.0000	-2.1907	-1.2551	-1.1077

orthogonalizing basis transformations

standard Löwdin for $2s$			modified Löwdin for $2s$		
1.0000	-0.1939	-0.1137	1.0000	-0.2175	0.0567
0	1.5509	-0.7601	0	0.5592	-1.6341
0	-0.7915	1.5209	0	0.5158	1.6351

new Fock-matrix blocks

standard Löwdin for $2s$			modified Löwdin for $2s$		
-11.2803	-0.1749	-0.0413	-11.2803	-0.1529	0.0945
-0.1749	0.1336	-0.7599	-0.1529	-0.7753	-0.1491
-0.0413	-0.7599	-0.1646	0.0945	-0.1491	0.7444

Notes. 1. Schmidt $1s$-$2s$ orthogonalization. 2. The rows and columns correspond to $1s$, $2s_I$, $2s_O$, in this order. 3. The Fock-matrix blocks are for carbon in the CH_2 biradical. 4. The $2s_I$-$2s_O$ overlap after Schmidt orthogonalization to $1s$ is 0.8190. 5. Energies in hartrees.

blocks of a theory of structure-property relations in molecules. On this point there was once a general consensus of many great quantum chemists, particularly Robert S. Mulliken.[8] By contrast, from a perusal of more recent literature, it would seem that computationalists have missed the significance of facts such as that *ab initio* computations use basis sets constructed as approximations of free atom orbitals in terms of Gaussian functions. The success of this recipe provides strong support for the claim that free-atom AO's have a definite physical status, and are therefore important in explanations of why the structure-property relations in molecules are what they are and not otherwise.

How can orthogonalization be used to bear out the physically significant atomic orbitals and their role in bonding? We shall answer this question, pending further illustrations, on the example of the inner-outer $2s$ orbitals of carbon in the well known DZ-631G basis used in many *ab initio* programs for molecular energies.[9] For the carbon $1s$, $2s_I$ and $2s_O$ in the biradical CH_2 (with CH distances 1.089 Å and angle 109°) that basis gives the overlap and Fock-matrix blocks shown in the first section of table 1.

There are several strange things there. The original overlap matrix shows that the secondary $2s$ is on the average quite close to the primary one, the overlap being larger than 0.8; it is therefore clear that it is a bad choice not only as the physics of the problem is concerned, but also computationally, because the high overlap contributes to making certain matrices poorly conditioned, whence the danger of loss of accuracy.

The Fock matrix obtained from the basis used in the computations has high $1s2s$ couplings, and the inner and outer components of $2s$ are almost equivalent in energy and strongly interacting. How much of this is due to the molecular environment (and therefore physically significant), how much to the high intra-atomic overlap values due to the criteria that have guided the choice of the basis set? For the $1s2s$ couplings there is nothing obviously unphysical, because their values are roughly proportional to the corresponding overlap values and the proportionality constants are expected to be high, for they should be roughly proportional to the mean of the corresponding *in situ* orbital energies times the overlap value

(Mulliken approximation). For the $2s_I 2s_O$ Fock matrix block, on the other hand, there are evident grounds for doubt. The proportionality constant found for the $1s2s$ pairs is too high and the two energies are very close, as expected from the overlap value (cf. sec. 1.3). Neither feature seems to make sense from a physical point of view. Of course, the fact that we are working with *in situ* orbitals, and the Fock matrix elements contain therefore contributions from the hydrogen atoms, might be the cause of the strange results; but how can we separate those external effects from the effect of having such a large intra-subshell overlap?

It appears, therefore, that some sort of orthogonalization is required; but it is not obvious which, even ignoring a number of difficulties possibly resulting from two-electron orbitals (see note below). A rational orthogonalization procedure appears to consist of two steps:
1. Schmidt orthogonalization of the $2s$ orbitals to the $1s$ orbital, justified by the fact that only one $1s$ orbital is present in the basis;
2. Modified Löwdin orthogonalization of the $2s$ orbitals, which should yield a valence $2s$ and a pseudo-excited $2s'$ AO, the two being as separate in energy as possible.

Note 1.2.1 – We ignore here possible difficulties arising from the fact that two-electron integrals depend twice on a basis transformation. The linear dependence of electronegativities on hybridization found empirically by Hinze and Jaffé [10] is a partial justification of this simplification; but of course an ad hoc study should be performed.

As an illustration of the difference between the standard Löwdin orthogonalization and the modified one also the results of the former are reported in table 1. Note that the AO's found are not environment dependent (although the corresponding Fock matrix elements are), and therefore their physical significance is the merit of basis orthogonalization.

That in either case the off-diagonal elements of the new Fock matrices should be in general smaller than those of the original one is a well known effect of orthogonalization, and makes the coupling elements more realistic. The new inner and outer orbitals are little apart and strongly coupled after a standard Löwdin transformation, but they are widely separated and weakly coupled after the modified Löwdin transformation defined in equation (1.6). GAMESS computations with the same basis for the ^1D state of carbon give for the $1s$ and the $2s$ orbital energies -11.3485 and -0.7096 hartrees; these values are fairly close to those given in table 2 which result from a modified Löwdin orthogonalization of the basis.

These results are of great import in connection with an idea published by Mulliken late in his life,[11] which was practically ignored because of shift in interest to computational work, namely the suggestion that molecules should be described in terms of MVAO'S — M*odified in situ* V*alence* A*tomic* O*rbitals.* If quantum chemistry is ever to resume its role of a theoretical reflection on molecules, that idea will be a precious starting point, the more so as wise use of computational tools and of the matrix formalism advocated by Per-Olov Löwdin would make the indispensable processing of numerical data comparatively easy and greatly effective.

1.3 Overlap as a measure of AO coupling

That the overlap of atomic orbitals should be a quantity associated with certain features of molecular structure was established by the pioneers of quantum chemistry and extensively confirmed by the very fact that approximations to the Hartree-Fock orbitals of free atoms provide good bases for *ab initio* computations. A discussion of the various quantum mechanical approaches to the chemical bond was given in F. Hund's classical paper on LCAO bond orbitals.[12] There it was shown that the MO-LCAO approach accounts for bonds and their properties, and is quite reasonable in that it represents single electron states in a bond A-B as direct products of spin functions by bond orbitals represented as (weighted) *sums* of two atomic orbitals $|\chi>$ and $|\varphi>$ (wavefunctions χ and φ) whose product is significant in the bond region between A and B. This point has been lost during the last few decades of the twentieth century because of the general shift in interest from understanding to predicting, and of the continuing fragmentation of science.[13] In particular, the idea has been lost that overlap values allow an estimate of the strength of a chemical bond.[14] On the other hand, even the homolythic dissociation limit can be accounted for in an MO-LCAO frame, provided the somewhat forced identification of the MO-LCAO scheme with a variational procedure be relaxed.[15] Without pausing too long on this, we recall the essential arguments related to the central role assigned to overlap in the theory of the chemical bond by defining a function $s \equiv s_{AB}$ in the one-electron configuration space $\mathcal{S}$, such that $ds = \chi\varphi dxdydz$, $dxdydz$ being the volume element of $\mathcal{S}$. As is well known, the potential energy terms appearing in the total *in situ* energy of an electron in the bond orbital $c_A\chi + c_B\varphi$ can be classified into two types, "resonance" R and "electrostatic" C. They have the general forms

$$S = \int_{-\infty}^{\infty} ds,\ R = \int_{-\infty}^{\infty} f ds,\ C = \int_{-\infty}^{\infty} (g_1 dq_A + g_2 dq_B). \qquad (1.7)$$

Here f is a function of the coordinates in $\mathcal{S}$, and is expected to behave as the inverse of a distance; $dq_A = |\chi|^2 dxdydz$ and $dq_B = |\varphi|^2 dxdydz$ are the space distribution differentials for the AO's associated with A and B, respectively. The integrals of type C are exemplified by the attraction of the core of A on the density cloud associated with B. It is also well known that the terms of type C include long range electrostatic terms tending to become proportional to the inverse of the distance of A from B (when, say, g_1 is proportional to inverse distance of an electron of B from the nucleus of A, and g_2 is zero).

A well known analysis of this and other diatomic cases was presented by Mulliken;[16] he showed among other things that the integrals of type R defined in equation (1.7] are either approximately proportional to overlap or are multiplied by overlap in the formulas for energy. In fact, the famous "Mulliken approximation" consists in making $ds \approx S(dq_A + dq_B)/2$ for integration purposes. As far as this approximate equality in the mean is valid, overlap becomes a ubiquitous representative of binding. However, it is meaningless both when complete orthogonalization of a basis is performed and when overlap is an artefact of computational choices. The following question has therefore to be posed: is Mulliken's approximation (and related ones, such as Wolfsberg-Helmholz's rule)[17] justified in numerical results that may be associated with real features of molecules? A systematic answer would be out of the scope of this article, but an instructive illustration is given in table 3, once again for the case of the electron-deficient CH_2 molecule.

Since the work of Pauling it has been well known that the energy separation $2\Delta\varepsilon$ of the bonding and antibonding levels of H_2^+ is

$$2\Delta\varepsilon \propto \frac{(R - SC)}{1 - S^2}, \tag{1.8}$$

which shows that the final level separation of this fundamental model is of the order of $S/(1 - S^2)$ times a constant as long as the series expansion in S of the resonance integral β does not include any constant term, which has to be the case. This means that the level separation vanishes when $S = 0$, i.e. that there is no bonding if there is no overlap.

The result of equation 1.8 can be generalized for a general independent-electron bond model based on two AOs, one per center. In fact, the eigenvalue equation for a Hamiltonian matrix $\mathbf{H}$ with eigenvectors $\mathbf{C}$, eigenvalue matrix $\mathbf{E}$, and overlap matrix $\mathbf{S}$ can be written in orthogonalized form with a diagonal overlap matrix $\bar{\mathbf{S}} = \mathbf{U}^\dagger\mathbf{S}\mathbf{U}$ as:

$$\bar{\mathbf{H}}\bar{\mathbf{C}} = \bar{\mathbf{C}}\mathbf{E}, \text{ with } \bar{\mathbf{C}} = \mathbf{U}\mathbf{C},\ \bar{\mathbf{H}} = \bar{\mathbf{S}}^{-1/2}\mathbf{U}^\dagger\mathbf{H}\mathbf{U}\bar{\mathbf{S}}^{-1/2}. \tag{1.9}$$

The expression of the general element of the transformed matrix $\bar{\mathbf{H}}$ is better suitable for studying the role of overlap in the final MOs and in their energies. We have (assuming $\mathbf{U}$ real)

$$\bar{H}_{jk} = \sum_{p,q} \frac{U_{pj}H_{pq}U_{qk}}{\sqrt{\theta_{jj}\theta_{kk}}} \tag{1.10}$$

where θ_{jj} is the j-th eigenvalue of $\mathbf{S}$.

The above transformation is comparatively simple when the matrix $\bar{\mathbf{H}}$ has dominant 2×2 blocks. This happens, as will be shown later, for Fock matrices associated with saturated molecules. Then, if the individual bond blocks are diagonalized, Hadamard's theorem makes it possible to determine the dependence of the eigenvalue distribution on bond-bond overlaps and couplings with inclusion of minor terms. For this reason we are going to pause on the explicit formulas for a single 2×2 problem, even if the algebra involved is elementary. The overlap block is a 2×2 symmetric matrix with equal

Table 2: *In situ* energies (hartrees) of pure physical AO's.

1	2	3	4	5	6	7
0.889	0.989	1.089	1.189	1.289	1.589	∞
-11.2575	-11.2679	-11.2803	-11.2935	-11.3067	-11.3425	-11.3460
-0.7976	-0.7851	-0.7753	-0.7685	-0.7641	-0.7624	-0.7077
-0.4282	-0.4225	-0.4099	-0.3910	-0.3679	-0.2962	-0.3387
-0.4050	-0.3872	-0.3750	-0.3693	-0.3684	-0.3784	0.0730
0.0844	0.0835	0.0816	0.0793	0.0764	0.0662	0.0730
0.7545	0.7510	0.7444	0.7370	0.7306	0.7210	0.7830
0.7294	0.7312	0.7291	0.7239	0.7171	0.7041	0.7552
0.7502	0.7567	0.7559	0.7516	0.7463	0.7311	0.7834
0.8035	0.8018	0.7989	0.7952	0.7913	0.7789	0.7834

Note. The first two rows give the numbering of the different cases and the C-H_2 distances d in Å; ∞ corresponds to C alone (^1D state). DZ-631G results.

Table 3: Overlap and β values (hartrees) and their renormalized ratios* for CH_2 with the first C-H bond of length 1.089 Å and the second C-H bond of length d (Å).

d	S_1	β_1	r_1	S_2	β_2	r_2
0.889	0.7094	-0.6618	0.9912	0.8057	-0.7837	1.0335
0.989	0.7114	-0.6666	0.9955	0.7616	-0.7282	1.0159
1.089	0.7130	-0.6711	1.0000	0.7130	-0.6711	1.0000
1.189	0.7142	-0.6756	1.0050	0.6617	-0.6143	0.9864
1.289	0.7151	-0.6798	1.0101	0.6092	-0.5593	0.9754
1.589	0.7163	-0.6916	1.0257	0.4559	-0.4135	0.9636

(*). The renormalized ratios are the β/S ratios multiplied by a factor such that the symmetric case ($d = 1.089$) yield 1. The S_1 values increase with increasing lengths of the C-H_2 bonds because the hybrid atomic orbital to which it corresponds is the closer to the maximum bond-overlap one the lower the the value of S_2. DZ-631G results.

diagonal elements, and therefore is diagonalized by a 45° rotation. Taking such a rotation for **U** and considering that the eigenvalues of **S** are $1+S$, $1-S$ (S being the off-diagonal element of **S**), we also have:

$$\bar{\mathbf{H}} = \begin{vmatrix} \frac{\bar{\alpha}+\beta}{1+S} & \frac{\Delta\alpha}{\sqrt{(1-S^2)}} \\ \frac{\Delta\alpha}{(\sqrt{(1-S^2)}} & \frac{\bar{\alpha}-\beta}{1-S} \end{vmatrix} = \begin{vmatrix} \tau-\delta & \gamma \\ \gamma & \tau+\delta \end{vmatrix}, \qquad (1.11)$$

where we have adopted the traditional notation $\alpha_1 = H_{11}$, $\alpha_2 = H_{22}$, $\beta = H_{12}$; $\bar{\alpha}$ is $(\alpha_2+\alpha_1)/2$ and $\Delta\alpha$ is $(\alpha_2-\alpha_1)/2$. The notation in the second matrix of eq.n (1.11) allows us to rewrite the above results in a more legible form. The half-separation of the two final diagonal elements has the form of a deviation of a standard coupling term with respect to the actual β-value:

$$\delta = \frac{1}{1-S^2}(\bar{\alpha}S-\beta). \qquad (1.12)$$

The midpoint of the diagonal elements (to be called "characteristic energy midpoint") of the final matrix has the form

$$\tau = \frac{\bar{\alpha}-\beta S}{1-S^2} = \bar{\alpha}+\delta S. \qquad (1.13)$$

The off-diagonal element of **H** referred to the orthogonalized basis is

$$\gamma = \frac{\Delta\alpha}{\sqrt{1-S^2}}, \qquad (1.14)$$

which becomes the original energy separation when S is negligible. The two eigenvalues are

$$\begin{aligned} \varepsilon_1 &= \tau - \sqrt{\delta^2+\gamma^2}, \\ \varepsilon_2 &= \tau + \sqrt{\delta^2+\gamma^2}. \end{aligned} \qquad (1.15)$$

The role of overlap is very clear here. Were $S=0$, the characteristicic energy midpoint τ of $\bar{\mathbf{H}}$ would be the same as the mean separated-atom energy zero (characteristic midpoint of **H**), but actually it rises from $\bar{\alpha}$ to ∞ as S tends to one. If $\beta = kS = k_0S + k'g(S)$ with $g(S)$ a quantity of order ≥ 2 in S, then expression (1.13) takes the asymptotic form

$$\tau = \bar{\alpha}_0 + (\bar{\alpha}_0 - k_0)\frac{S^2}{1 - S^2} + \mathcal{O}(S^3), \tag{1.16}$$

where the subscript 0 refers to the limit for $S \to 0$.

As to the final level separation, it is twice the quantity

$$\Delta\varepsilon = \sqrt{\delta^2 + \gamma^2} = \frac{\sqrt{[(\bar{\alpha} - k)^2 - (\Delta\alpha)^2]S^2 + (\Delta\alpha)^2}}{1 - S^2}, \tag{1.17}$$

with k as defined in the preceding paragraph. When compared with equation (1.8), expression (1.17) is characterized by three main points:

i. it gives the same result as (1.8) if, as is the case for the latter expression, $\Delta\alpha$ is zero;

ii. when $\Delta\alpha$ and $\bar{\alpha} - k$ are equal $\Delta\varepsilon = \Delta\alpha/(1 - S^2)$;

iii. when $\Delta\alpha$ is dominant, $\Delta\varepsilon \to \Delta\alpha$, no matter what the value of S is.

The general conclusion is that, when the energy separation of the overlapping orbitals yielding the initial matrix **H** is small, the separation of the final energy levels diverges as $S/(1-S^2)$ as $S \to 1$; but the divergence decreases as the separation of the uncoupled levels increases.

We have given here these rather simple derivations because, although they were undoubtedly published by the pioneers of quantum chemistry, they may be difficult to find.[18]

Table 4 contains a numerical example of the changes occurring in the 2×2 matrices associated with the inner-outer 2s orbitals of carbon in CH_2 (cf. table 1). The importance of the above derivations is greater, however, if applied to a two-center-two-AOs description of a bond, because then it shows not only the importance of choosing an appropriate physical basis (which will not be orthogonal), but the way in which overlap combines with the AO coupling β to determine the behavior of the eigenvalues when overlap changes as a result of changes in the bond length. We shall pause on the significance of the above expression in this case when discussing the application of Löwdin's ideas to the detection of bond orbitals in *ab initio* SCF computations.

Table 4: Numerical illustration of modified Löwdin orthogonalization for the carbon $2s$ orbitals in the DZ-631G basis for carbon ^1D. Intra-subshell F-matrices for carbon $2s$ with transformed bases

	original		mod. Löwdin		eigenorb.s	
	-0.5302	-0.7145	-0.7077	-0.0742	-0.7113	0
	-0.7145	-0.6153	-0.0742	0.7830	0	0.7867
τ	-0.5727		0.0377		0.0377	
λ	-0.0426		0.7454		0.7490	

Notes. 1. τ stands for the midpoint of the diagonal elements, λ for their separation. 2. The overlap value is 0.8190, the energies are in hartrees 3. For the procedure see eq.ns (1.12),(1.13), (1.14) and for additional numerical details see table 1.

Note 1.3.1 – Hadamard's theorem states that the j-th eigenvalue of a matrix $\mathbf{A}$ of order n is located in a circle D_j with center A_{jj} and radius equal to the sum of the absolute values of the off diagonal elements of a row, $|A_{jk}|$. This theorem holds when the overlap matrix associated to $\mathbf{A}$ is the unit matrix. Otherwise, a way to apply it is *via* Löwdin's orthogonalization. In the case of Fock matrices close to block-diagonal ones, Löwdin orthogonalization of the part $\mathbf{A}^{(ext)}$ of such a matrix that lies outside the blocks makes it possible to apply Hadamard's theorem to estimate the maximum correction to the ε's of eq.ns (1.15) due to the fact that $\mathbf{A}^{(ext)}$ is not exactly zero. Application of a perturbation expansion is not as reliable as a rigorous theorem, but gives more reasonable estimates. As usual, in highly symmetric molecules the role of small overlaps may be numerically important, but it should be kept in mind that symmetry is removed by the slightest perturbation; now, perturbations are always present in real situations; therefore the isolated-molecule model is anyway unreliable below a certain threshold.

2 Molecular structure and matrix manipulation

2.1 The role of overlap

We have used several times already the expression "physical significance" and similar ones, but we have kept our considerations nevertheless on the numerical side, merely showing that optimum orthogonalization provides results that make sense in familiar interpretation schemes. We will now dwell a little longer on the notion of "physical significance" in connection with atomic orbitals and overlap — more precisely, with bases for LCAO descriptions of molecular structure and the associated overlap.

In this connection, a distinction should be made among three cases.

- There are bases with high overlap values only chosen on grounds of computational simplicity; the above expressions (and the numerical illustration in table 2) show how wary people should be in interpreting what their bases mean; attempts at using those orbitals for explaining facts are unjustified.
- There are problems where overlap is physically significant, but has a more or less standard value (ordinary σ bonds at equilibrium); in that case certain correlations (e.g. β values with bond energies) should not be extrapolated to situations other than equilibrium ones, because they have been derived on a limited range of overlap values, and might not hold for non-standard bonds or for internuclear distances markedly different than equilibrium ones.
- Finally, one may be interested in following the evolution of a bond from a very short length to the dissociation limit using Hund's two-state independent electron model. Then explicit account of different functional dependencies on the energy amplitudes and on overlap might be critical (cf. ref. 13). In those more general cases three points in particular should be kept in mind:
    i. the physical significance of the energy difference of the intervening states should be discussed with account of the inflation resulting for large values of overlap (cf. eq.n (1.17));
    ii. if the β of the original $\mathbf{H}$ has a physical meaning and S is quite large, then that meaning should be given in terms of the difference between β and its Mulliken approximation with account of the inflation due to S;

iii. the zero-point to which reference is made in physical interpretations should be carefully discussed, because the mean energy of the levels obtained by diagonalization changes with overlap dramatically.

From the numerical point of view, the application of the equations for the two-AO problem to the carbon $2s$ DZ 631G orbitals discussed above is very instructive. In connection with physical significance, it has to be applied to AO's belonging to different atoms. Before going over to that side of the story, we add the following commentary.

When a physicist is investigating the physical nature (static properties and becoming) of a complicated system, he usually begins by looking for simplified descriptions that can serve as foundations of his study. Those descriptions usually have two facets: a picture in terms of familiar analogies or model systems (say, H_2^+ or H_2) and a very simple mathematical treatment. Many of today's scientists forget this point because they work on specialized developments of existing theories, and they often accept those theories at their face value; on the other hand, our claim is found in the writings of many great physicists (Einstein, Fermi, Feynman and others). For example, in the famous paper of 1932 in which he solved the problem of the spontaneous emission of an excited atom, Fermi referred only to a simple one-electron atom,[19] and yet no one doubted that his results were a solution of the physical problem under study, even though attempts to get accurate quantitative predictions of spontaneous emission are still hindered by difficulties. Feynman, on the other hand, showed that a simple mathematical model borrowed from a two-state description of the inversion vibration of ammonia could be used to introduce the main concepts and equations of quantum electrodynamics.[20] In either case it is evident that the physical and mathematical descriptions used have a physical significance, in that they show how the general theory can produce precisely a behavior as that which has to be explained, even if they are far from capable of giving accurate quantitative predictions.

In the case of theoretical chemistry, a deviation from this basic approach to science began when attempts were made to adjust the parameters of the Hückel method in the hope that they could yield more that rough estimates of π-electron contributions to molecular properties. This, and the great interest the spectroscopists had in precise excitation energies, caused emphasis to be shifted to numerical accuracy, and the question of physical meaning was altogether lost.

This was precisely what Löwdin's mathematical tools were intended to prevent: as mentioned already, their aim was to use matrix calculus to establish a correspondence between concepts derived from analogies or simple reference systems and mathematical quantities that appeared to be characteristic of molecular structure. Pauling had given genial intuitive instances of this line of thought; it remained for others, particularly Mulliken, Slater and Löwdin, to begin the work of giving it a formalized and logically unquestionable form. We emphasize once again that work is still waiting to be carried out in that connection, and it is now high time that somebody should devote time to it, the more so as reliable numerical results are easily available.

2.2 *Ab initio* bond orbitals

With the above premises, we shall now give a systematic illustration, on a simple example, of how matrix manipulation and critical use of overlap can extract from computational outputs the physics behind them. The matrix methods advocated by Löwdin allow to show in particular that bonds, in the form of bond orbitals, are intrinsically contained ab-initio SCF-MO computations.

The example is that of ethene, treated *ab initio* by a minimum STO3G basis. The C=C bond length is 1.337 Å, a C-H bond length is 1.071 Å, and the angle of the CH_2 group is 110°. (The optimum STO3G-SCF geometry is different — 1.318 and 1.089 Å, 115° —, but, since the procedure to be illustrated starts from energies and eigenvectors, only a reasonable geometry is needed here.)

We start from the SCF eigenvalues $\mathbf{E}$ and the SCF eigenvectors $\mathbf{C}^{(ini)}$. The initial overlap and Fock matrices, $\mathbf{S}^{(ini)}$ and $\mathbf{F}^{(ini)}$, can be computed, if necessary, from $\mathbf{C}^{(ini)}$ and $\mathbf{E}$. The steps of our study are listed below.

1. The physical atomic orbitals (in the adopted Gaussian approximation) are determined by the procedure described above, which here reduces to a Schmidt orthogonalization of the $2s$ orbital of carbon to the corresponding $1s$ orbital. We call $\boldsymbol{\chi}^{(pao)}$ the pure-AO basis thus obtained, to be treated formally as a row matrix. The resulting overlap and Fock matrices over $\boldsymbol{\chi}^{(pao)}$ are denoted by $\mathbf{S}^{(pao)}$ and $\mathbf{F}^{(pao)}$.

2. The maximum-localization hybrids of carbon for the inner (and, in the case of a double-ζ basis, the outer) components of the physical valence shell are determined; these yield a block-diagonal unitary transformation $\mathbf{T}^{(hyb)}$ to be applied to the basis $\boldsymbol{\chi}^{(pao)}$ to provide a new basis $\boldsymbol{\chi}^{(hyb)}$ of H(ybrid) A(tomic) O(rbitals) HAO. The new basis will also be reordered so as to give successive numbers to HAO's participating in the same bond; in this way the blocks of the new overlap and Fock matrices $\mathbf{S}^{(hyb)}$ and $\mathbf{F}^{(hyb)}$, obtained after transformation with $\mathbf{T}^{(hyb)}$, will have nonintersecting diagonal blocks corresponding to bonds.

We shall not report here in detail the procedure for determining optimum hybrids, but it it is in the spirit of this paper to point out that also in this case orthogonalization is a very important step, having significant consequences on the physical content of the results. The orthogonalization step arises as follows. Consider for simplicity a saturated molecule. For each bond the ideal hybrids will obey the maximum overlap principle, and they are easily found by diagonalizing the formal square of the pertinent overlap block. Unfortunately, the hybrids thus found for a given atom A will not be mutually orthogonal, a property the final hybrids must have if they have to represent actual atomic orbitals. To overcome this difficulty the best single-bond hybrids are replaced by the columns of the unitary "hybridization matrix" $\mathbf{U}^{(A)}$ whose columns have the highest possible scalar products with the corresponding columns of the matrix $\mathbf{P}$ formed by the maximum-overlap hybrids. The condition thus specified is expressed by an equation similar to (1.3):

$$\delta(\mathbf{U}^{(A)\dagger}\mathbf{P} + \mathbf{P}^{\dagger}\mathbf{U}^{(A)} - \mathbf{U}^{(A)\dagger}\boldsymbol{\Lambda}\mathbf{U}^{(A)}) = 0, \tag{2.18}$$

the novelty being that $\mathbf{P}$ might be singular. Differentiation with respect to $\mathbf{U}^{(A)}$ and to $\mathbf{U}^{(A)\dagger}$ yields the already found equation

$$\mathbf{P} = \boldsymbol{\Lambda}\mathbf{U}^{(A)},$$

but straightforward multiplication by $\boldsymbol{\Lambda}^{-1}$ is not general enough a solution in this case, precisely because $\mathbf{P}$ might be singular. The correct procedure is as follows.[21] Letting $\mathbf{P} = \mathbf{VmW}^\dagger$, with $\mathbf{m}$ diagonal, we take $\boldsymbol{\Lambda} = \mathbf{VmV}^\dagger$ and $\mathbf{U}^{(A)} = \mathbf{VW}^\dagger$. If $\mathbf{V}$ and $\mathbf{W}$ are unitary, then

$$\boldsymbol{\Lambda}\mathbf{U}^{(A)} = \mathbf{VmV}^\dagger\mathbf{VW}^\dagger = \mathbf{VmW}^\dagger = \mathbf{P}, \tag{2.19}$$

q.e.d. This proves that, if the unitary matrices $\mathbf{V}$ and $\mathbf{W}$ can be found, our *Ansatz* for $\mathbf{U}^{(A)}$ is correct and avoids any inversion. Now, as can be proven by direct substitution of the expression adopted for $\mathbf{P}$, $\mathbf{W}$ and $\mathbf{V}$ are the respective eigenvector matrices, and $\mathbf{m}$ the common eigenvalue matrix of the products $\mathbf{P}^\dagger\mathbf{P}$ and $\mathbf{P}^\dagger\mathbf{P}$. An example of $\mathbf{U}^{(A)}$ is given in table 5. It is particularly instructive that the adoption of a basis close to physically significant atomic orbitals and the construction of hybrids hints at the fact that the HCH angle adopted is too narrow. The straight-bond principle (which is the LCAO translation of Bayer's strain theory) is sufficient to suggest the correct geometry prior to any energy computation, as far as directions go.[22] An analogous principle for bond distances is still under investigation.

3. The above procedure allows construction of a basis of "maximum-localization" hybrid atomic orbitals $\boldsymbol{\chi}^{(hyb)}$, which will be obtained from $\boldsymbol{\chi}$ by a block-diagonal transformation $\mathbf{U}$ whose blocks are associated with the various atoms of the given molecules as follows. For each atom A we specify the bonds it is expected to form, and we find the linear combinations of its $2s$ and $2p$ atomic orbitals (as given by $\boldsymbol{\chi}$) that will yield the maximum overlap $S^{(max)}$ for each of the bonds in question. We then orthogonalize the linear combinations thus obtained so as to keep those with the higher $S^{(max)}$ closer to their form before orthogonalization. As mentioned, for a given atom A the final result will be a set of column vectors that form an orthogonal transformation $\mathbf{U}^{(A)}$ of the original atomic orbitals of A, to be called the "hybridization matrix of A". This will be the block of $\mathbf{U}$ associated with A.

Table 5: Hybrids of carbon in ethene and their properties.

coefficients of $2s$, $2pz$,	0.5567	0.5567	0.6165	0
$2px$, $2py$, as columns:	-0.4360	-0.4360	0.7873	0
	0.7071	-0.7071	-0.0000	0
	0	0	0	1.0000
p-characters (n in sp^n):	2.2264	2.2264	1.6308	∞
electronegativities:	2.7077	2.7077	2.9245	1.7500
centroid distances in Å:	0.1831	0.1831	0.1922	0
unit vectors of the hybrids:	-0.5248	-0.5248	1.0000	0
	0.8512	-0.8512	-0.0000	0
	0	0	0	1.0000
angles with the axes:	121.6550	121.6550	0.0000	90.0000
	31.6550	148.3450	90.0000	90.0000
	90.0000	90.0000	90.0000	0.0000
angles between hybrids:	0	116.6900	121.6550	90.0000
	116.6900	0	121.6550	90.0000
	121.6550	121.6550	0	90.0000
	90.0000	90.0000	90.0000	0

Note. Angles in degrees: the angle between the hybrids directed towards the hydrogens is 117°, larger than the assumed methylene angle, in agreement with the fact that in the assumed geometry that angle is smaller than the optimum one.

The procedure should be repeated for each many-orbital atom, whereas the single diagonal element 1 is assigned to the atoms participating in the binding with just one atomic orbital. Eventually all the intra-atomic blocks of **U** are obtained. **U** will be called the "hybridization matrix" for the given molecule.

In the case of the valence electrons of ethene, if the hydrogen orbitals are numbered 1,2,11,12 and the valence orbitals of the two carbon atoms are numbered 3,4,5,6 and 7,8,9,10, **U** can be described as a 12×12 unit matrix in which the 3-6 and 7-10 diagonal blocks are replaced by the hybridization matrices of the two carbond atoms.

Note 2.2.1 – In order to avoid results that are confusing, although correct, care must be taken that the diagonal matrix **m**, which is obtained as the square root of an eigenvalue matrix, have all elements positive and in the right order.

Note 2.2.2 – The same procedure can be applied without any problem to a basis set including d orbitals, as in the case of iron.[23]

Note 2.2.3 – The identification of potential bonds could be made automatic, but this would make the procedure far less flexible, in contrast to its nature of an intepretive, not a predictive tool. The rule of valency are clearcut enough to guide the choice in ordinary circumstances, and a preprogrammed choice could never allow the study of exceptional cases, e.g. the evolution of a bond when the interatomic distance increases.

Note 2.2.4 – The treatment of nonbonding hybrids can be made by assigning to them a "weight" equivalent to the square of an overlap, and making them pure-s before orthogonalization. The π orbitals are separated and must be treated as orbitals with weight zero.

2.3 Localized and quasi-localized orbitals

The further steps of the procedure we are describing deal with the question: is it possible to determine a unitary transformation of the SCF-LCAO MOs such that they correspond to bonds?

4. The hybrid basis and the overlap and Fock matrices referred to it are constructed. They are given by

$$\begin{aligned}\boldsymbol{\chi} \equiv \boldsymbol{\chi}^{(hyb)} &= \boldsymbol{\chi}\mathbf{U},\\ \mathbf{S} \equiv \mathbf{S}^{(hyb)} &= \mathbf{U}^{\dagger}\,\mathbf{S}^{(pao)}\,\mathbf{U},\\ \mathbf{F} \equiv \mathbf{F}^{(hyb)} &= \mathbf{U}\,\mathbf{F}^{(pao)}\,\mathbf{U},\end{aligned} \tag{2.20}$$

respectively.

In saturated molecules $\mathbf{S}$ has small inter-bond elements, so that it becomes a block-diagonal matrix $\mathbf{s}$ if all elements lower than a certain threshold value are neglected and the columns rearranged. If all elements that have been set to zero in $\mathbf{S}$ are also neglected in $\mathbf{F}$, a block-diagonal matrix $\mathbf{f}$ will result from $\mathbf{F}^{(pao)}$ *via* $\mathbf{F}$. The eigenvectors of $\mathbf{f}$ with overlap $\mathbf{s}$ will be a "bond-orbital matrix" $\mathbf{b}$

which is not obtained by a linear transformation of the original SCF eigenvector matrix, but will be used as a reference matrix. The matrix **S** with the blocks that are not ignored in **s** is shown in table 6 for the valence orbitals. The corresponding Fock matrix for the valence shell is shown in table 7.

As can be seen, the overlap values for normal σ bonds are > 0.65, whereas, as expected, they are much smaller for the π orbitals. The reason why localization is quite effective is a comparatively difficult problem in numerical analysis, but its plausibility can be seen by thinking of applying the Jacobi diagonalization procedure (or in general a perturbation technique) to the matrix **S**. After the steps involving diagonalization of the main blocks the remaining off-diagonal elements will be of the order of the elements originally outside those blocks, and therefore their contribution to the final eigenvalues will be at most of that order.[24] On the example of table 6 the actual maximum off-diagonal element is 0.15. From equation (1.16) it appears that the overlap factor corresponding to this term is approximately 0.025. This, and the close proportionality of β to overlap also discussed above, makes it understandable that the bond blocks of **S** over hybrids really describe most of the physics contained in the overlap and Fock matrices.

2.4 Derivation of quasi-bond SCF molecular orbitals [25]

We shall now illustrate the derivation of quasi-bond SCF orbitals. We shall continue to refer explicitly to the example of a minimal-basis GTO-SCF computation on ethene. The procedure for obtaining the optimum hybrids outlined above provides a block-diagonal unitary transformation **U** of the basis which only mixes basis elements of the same atom and minimizes overlaps of atomic orbitals (AO) not participating in the same bond.

Let the Fock, overlap, and eigenvector matrices on hybrids be **F**, **S**, and **C**. **C**, of course, is symmetry adapted, so that no correlation between orbitals and bonds can be detected. As mentioned, these three matrices are given in tables 6, 7, and in the top subtables of 9, 10. As mentioned, in the same tables the matrices **s** and **f** are highlighted in boldface.

Table 6: overlap matrix $\mathbf{S}^{(hyb)}$ referred to hybrid AO's and ordered according to bonds (valence orbitals only). The boldface blocks are the only nonvanishing ones of s, which is by definition the overlap matrix corresponding to full localization.

HC		HC		CCσ	
1.0000	**0.6628**	0.1450	0.1034	0.1062	0.1505
0.6628	**1.0000**	0.1034	0	0	0.0968
0.1450	0.1034	**1.0000**	**0.6628**	0.1062	0.1505
0.1034	0.0000	**0.6628**	**1.0000**	0.0000	0.0968
0.1062	0.0000	0.1062	0.0000	**1.0000**	**0.7678**
0.1505	0.0968	0.1505	0.0968	**0.7678**	**1.0000**
0.0409	0.1059	-0.0400	-0.1426	0.0968	0.0000
0.0400	0.0409	0.0091	-0.0400	0.1505	0.1062
-0.0400	-0.1426	0.0409	0.1059	0.0968	0.0000
0.0091	-0.0400	0.0400	0.0409	0.1505	0.1062
0	0	0	0	0	0
0	0	0	0	0	0

CH		CH		CCπ	
0.0409	0.0400	-0.0400	0.0091	0	0
0.1059	0.0409	-0.1426	-0.0400	0	0
-0.0400	0.0091	0.0409	0.0400	0	0
-0.1426	-0.0400	0.1059	0.0409	0	0
0.0968	0.1505	0.0968	0.1505	0	0
0.0000	0.1062	0.0000	0.1062	0	0
1.0000	**0.6628**	0.0000	0.1034	0	0
0.6628	**1.0000**	0.1034	0.1450	0	0
0.0000	0.1034	**1.0000**	**0.6628**	0	0
0.1034	0.1450	**0.6628**	**1.0000**	0	0
0	0	0	0	**1.0000**	**0.2485**
0	0	0	0	**0.2485**	**1.0000**

Table 7: Fock matrix $\mathbf{F}^{(hyb)}$ transformed to hybrids and ordered according to bonds. The boldface blocks are the only nonvanishing ones of **f**, which is by definition the effective Fock matrix with full bond localization.

HC		HC		CCσ	
-12.7590	**-18.3980**	-6.5272	-4.0121	-4.7812	-8.0195
-18.3980	**-11.3508**	-4.0121	-3.9157	-3.9924	-4.3424
-6.5272	-4.0121	**-12.7590**	**-18.3980**	-4.7812	-8.0195
-4.0121	-3.9157	**-18.3980**	**-11.3508**	-3.9924	-4.3424
-4.7812	-3.9924	-4.7812	-3.9924	**-19.3895**	**-24.0782**
-8.0195	-4.3424	-8.0195	-4.3424	**-24.0782**	**-19.3895**
-1.4841	-2.7235	1.2835	3.3007	-4.3424	-3.9924
-1.6066	-1.4841	-0.7952	1.2835	-8.0195	-4.7812
1.2835	3.3007	-1.4841	-2.7235	-4.3424	-3.9924
-0.7952	1.2835	-1.6066	-1.4841	-8.0195	-4.7812
0	0	0	0	0	0
0	0	0	0	0	0

HC		HC		CCπ	
-1.4841	-1.6066	1.2835	-0.7952	0	0
-2.7235	-1.4841	3.3007	1.2835	0	0
1.2835	-0.7952	-1.4841	-1.6066	0	0
3.3007	1.2835	-2.7235	-1.4841	0	0
-4.3424	-8.0195	-4.3424	-8.0195	0	0
-3.9924	-4.7812	-3.9924	-4.7812	0	0
-11.3508	**-18.3980**	-3.9157	-4.0121	0	0
-18.3980	**-12.7590**	-4.0121	-6.5272	0	0
-3.9157	-4.0121	**-11.3508**	**-18.3980**	0	0
-4.0121	-6.5272	**-18.3980**	**-12.7590**	0	0
0	0	0	0	**-1.7454**	**-8.9759**
0	0	0	0	**-8.9759**	**-1.7454**

Note 2.4.1 – The orbitals whose coefficients are given in the matrix $\mathbf{b}$ only form an orthogonal set in the basis $\chi^{(hyb)}$ to the extent to which the inter-bond overlaps can be neglected. How far this neglect is acceptable depends on the problem at hand, as mentioned in note 1.3.1; there is anyway no problem as long as we use $\mathbf{b}$ only in order to obtain a unitary transformation of $\mathbf{C}$.

5. A procedure is then applied that yields a unitary transformation of $\mathbf{C}$ such that it be as close as possible to $\mathbf{b}$, yielding a new set of MOs $\mathbf{C}^{(QL)}$ which differ from the SCF ones of tables 9 and 10 by a unitary transformation. The construction of the quasi-localized SCF orbitals can be carried out separately for the occupied and the virtual orbitals.

The procedure in question requires an application of conditional extremization, once again in the spirit of Löwdin's orthogonalization. We are looking for a unitary transformation $\mathbf{T}$ such that $\mathbf{CT}$ be as close as possible to a matrix $\mathbf{V}$ congruent with $\mathbf{C}$. We start again from an equation like (1.3) or (2.18) and write:

$$\delta(\mathbf{V}^{\dagger}\mathbf{CT} + \mathbf{T}^{\dagger}\mathbf{C}^{\dagger}\mathbf{V} - \mathbf{T}^{\dagger}\boldsymbol{\Lambda}\mathbf{T}) = 0 \tag{2.21}$$

Explicit differentiation gives

$$\mathbf{T} = (\mathbf{RR}^{\dagger})^{-1/2}\mathbf{R}, \text{ with } \mathbf{R} = \mathbf{C}^{\dagger}\mathbf{V}. \tag{2.22}$$

$\mathbf{T}$ is evidently unitary. If $\mathbf{V}$ is replaced by $\mathbf{b}$, this equation gives the transformation needed to construct the QLFOs (Q(uasi) L(ocalized) F(ock) O(rbital)s $\mathbf{C}^{(QL)}$ we are looking for. The quasi-diagonal matrix

$$\mathbf{E}^{(QL)} = \mathbf{C}^{(QL)\dagger}\mathbf{F}\mathbf{C}^{(QL)} \tag{2.23}$$

corresponds to the eigenvalue matrix $\mathbf{E}$ obtained from the complete *ab initio* computation. Its diagonal elements are shown in table 8 together with the canonical-orbital and the bond-orbital energies. The eigenvectors obtained by diagonalizing $\mathbf{F}$, the bond orbitals, and the quasi-localized SCF MOs are shown in tables 9 and 10.

Note 2.4.1 – We have looked for QLFOs obtained from the SCF-MOs by multiplication by $\mathbf{T}$ on the right. The reason is obviously that what

Table 8: Comparison of energies of canonical orbitals (1st line), bond orbitals (2nd line), quasi-bond orbitals QLFO (3rd line). All energies in eV.

-289.694	-289.650	-26.211	-19.716	-16.054	-14.299	-12.035
-288.954	-288.948	-24.589	-18.338	-18.338	-18.338	-18.338
-288.125	-288.125	-22.290	-17.132	-17.132	-17.132	-17.132
-8.587	9.621	17.850	19.180	19.390	27.112	30.271
-8.587	9.621	18.838	18.838	18.838	18.838	20.189
-8.587	9.621	20.964	20.964	20.964	20.964	29.354

we want is to replace the columns of **C** by linear combinations of the same, and this requires precisely that **T** should be on the right. It is possible, however, to define formally a transformation **Q** of **C** that yields the same result by multiplication on the left: it is sufficient to write:

$$\mathbf{C}^{(QL)} = \mathbf{C}\mathbf{T}\mathbf{C}^{-1}\,\mathbf{C} = \mathbf{Q}\,\mathbf{C}.$$

This allows association of the transformation **T** with a linear transformation of the basis.

Note 2.4.2 – An interesting basis transformation suggested by Löwdin's orthogonalization technique is the construction leading from the HAO basis to quasi-atomic orbitals such that each is orthogonal to every other orbital except its partner in the bond it forms. Starting from a basis $\boldsymbol{\chi}$ with overlap **S** we want a basis $\boldsymbol{\chi}^{(loc)}$ such that the overlap matrix is **s** and otherwise equivalent to $\boldsymbol{\chi}$. We Löwdin-orthogonalize χ with respect to **s**. The transformation is

$$\boldsymbol{\chi}^{(loc)} = \boldsymbol{\chi}\mathbf{T}, \quad \mathbf{T} = \mathbf{S}^{-\frac{1}{2}}\mathbf{s}^{\frac{1}{2}}; \tag{2.24}$$

the procedure evidently consists of two steps:
1. total orthogonalization,
2. restoration of **s**.

This implies that the hybrid AO's shown above and the $1s$ orbitals of the hydrogen atoms are allowed to have tails on atoms other than their partners in the bonds to which they belong. In our example, those tails are extremely small (coefficients smaller that 0.025 in absolute value): which is another confirmation that localization has been very effective.

Table 9: SCF, fully localized (FL) and SCF quasi-localized (QL) orbitals for ethene. Valence shells only. First part.

SCF bonding σ orbitals					*CC* π_+
-0.2641	-0.2320	0.1219	0.2195	0.3534	0
-0.2758	-0.2967	0.1976	0.2061	0.2713	0
0.2641	-0.2320	0.1219	0.2195	-0.3534	0
0.2758	-0.2967	0.1976	0.2061	-0.2713	0
0.0000	-0.0822	0.3699	-0.4032	-0.0000	0
0.0000	0.0822	0.3699	-0.4032	0.0000	0
-0.2758	0.2967	0.1976	0.2061	-0.2713	0
-0.2641	0.2320	0.1219	0.2195	-0.3534	0
0.2758	0.2967	0.1976	0.2061	0.2713	0
0.2641	0.2320	0.1219	0.2195	0.3534	0
0	0	0	0	0	0.6328
0	0	0	0	0	0.6328
FL bonding σ orbitals					*CC* π_+
0.5790	0	0	0	0	0
0.5174	0	0	0	0	0
0	0.5790	0	0	0	0
0	0.5174	0	0	0	0
0	0	0.5318	0	0	0
0	0	0.5318	0	0	0
0	0	0	0.5174	0	0
0	0	0	0.5790	0	0
0	0	0	0	0.5174	0
0	0	0	0	0.5790	0
0	0	0	0	0	0.6328
0	0	0	0	0	0.6328
QL bonding σ orbitals					*CC* π_+
0.5872	-0.0455	-0.0443	-0.0090	0.0048	-0.0000
0.5268	-0.0158	-0.0057	-0.0296	0.0367	0.0000
-0.0455	**0.5872**	-0.0443	0.0048	-0.0090	0.0000
-0.0158	**0.5268**	-0.0057	0.0367	-0.0296	-0.0000
-0.0090	-0.0090	**0.5431**	-0.0448	-0.0448	-0.0000
-0.0448	-0.0448	**0.5431**	-0.0090	-0.0090	-0.0000
-0.0296	0.0367	-0.0057	**0.5268**	-0.0158	-0.0000
-0.0090	0.0048	-0.0443	**0.5872**	-0.0455	0.0000
0.0367	-0.0296	-0.0057	-0.0158	**0.5268**	0.0000
0.0048	-0.0090	-0.0443	-0.0455	**0.5872**	-0.0000
-0.0000	0.0000	-0.0000	0.0000	-0.0000	**0.6328**
0.0000	-0.0000	0.0000	-0.0000	0.0000	**0.6328**

Table 10: SCF, fully localized (FL) and SCF quasi-localized (QL) orbitals for ethene. Valence shells only. Second part.

SCF anti-bonding σ orbitals					CC π_-
-0.6154	0.6406	0.1416	-0.5971	0.6330	0
0.4921	-0.6739	-0.0550	0.6740	-0.6786	0
0.6154	0.6406	0.1416	0.5971	0.6330	0
-0.4921	-0.6739	-0.0550	-0.6740	-0.6786	0
-0.0000	-0.5811	1.4548	-0.0000	-0.0771	0
-0.0000	0.5811	-1.4548	0.0000	-0.0771	0
0.4921	0.6739	0.0550	-0.6740	-0.6786	0
-0.6154	-0.6406	-0.1416	0.5971	0.6330	0
-0.4921	0.6739	0.0550	0.6740	-0.6786	0
0.6154	-0.6406	-0.1416	-0.5971	0.6330	0
0	0	0	0	0	-0.8157
0	0	0	0	0	0.8157
FL anti-bonding σ orbitals					*CC* π_-
-1.2035	0	0	0	0	0
1.2313	0	0	0	0	0
0	-1.2035	0	0	0	0
0	1.2313	0	0	0	0
0	0	1.4673	0	0	0
0	0	-1.4673	0	0	0
0	0	0	1.2313	0	0
0	0	0	-1.2035	0	0
0	0	0	0	1.2313	0
0	0	0	0	-1.2035	0
0	0	0	0	0	-0.8157
0	0	0	0	0	0.8157
QL anti-bonding σ orbitals					*CC* π_-
-1.2352	-0.0274	-0.0380	0.0175	-0.0026	-0.0000
1.2700	0.0990	0.1439	-0.0692	0.0791	0.0000
-0.0274	**-1.2352**	-0.0380	-0.0026	0.0175	0.0000
0.0990	**1.2700**	0.1439	0.0791	-0.0692	0.0000
0.1195	0.1195	**1.5608**	-0.0324	-0.0324	0.0000
-0.0324	-0.0324	**-1.5608**	0.1195	0.1195	0.0000
-0.0692	0.0791	-0.1439	**1.2700**	0.0990	-0.0000
0.0175	-0.0026	0.0380	**-1.2352**	-0.0274	-0.0000
0.0791	-0.0692	-0.1439	0.0990	**1.2700**	0.0000
-0.0026	0.0175	0.0380	-0.0274	**-1.2352**	0.0000
0.0000	-0.0000	-0.0000	-0.0000	0.0000	**-0.8157**
0.0000	-0.0000	0.0000	-0.0000	0.0000	**0.8157**

2.5 The physical significance of AOs and BOs, again

The above analysis appears to be clear evidence that the myths according to which MO-LCAO schemes have nothing to do with molecular reality are false. For one thing, despite the unphysical appearance of a sophisticated Gaussian basis, matrix manipulations in the spirit of Per-Olov Löwdin lead to a picture where the AOs have the properties one would expect of real atomic orbitals. For another, the fact that the resulting basis itself produces maximum localization hybrids whose angles are close to those obtained with geometry optimization based on energy — as in the present example and in many other examples we have examined — is suggestive of a physical significance of AOs themselves that goes in the direction of the already cited, fundamental paper by F. Hund. For a third thing, HAOs make it possible to show that bonds appear quite naturally in *ab initio* SCF computations, without artificial transformations such as those currently used. The latter have induced highly qualified scientists to reject the whole orbital picture, at the same time as work by Roald Hoffmann [26] and others showed that the organic chemists had a lot to gain from the use of orbitals as interpretive tools.

A few remarks on the physics of the whole orbital picture may be useful here. First: what should one really mean, when he says that atomic orbitals have a measure of reality in a molecule? This seems to be in contrast with the acknowledged fact that AO overlap somehow makes them spatially undistinguishable. The answer is that what makes an atomic orbital atomic is that it is (albeit approximately) a stationary state of an electron in the atom to which it belongs, so that whatever "happens" to that atom in particular affects that atomic orbital and only indirectly the others.

To make the matter more explicit, it should be considered that — as already recalled — physical argumentations are typically based on idealized systems and situations. Quantum mechanics has two powerful mathematical tools in this connection: the adiabatic approximation and the sudden approximation.[27] In the case of a chemical bond, the sudden approximation would apply to an ideal process in which the two atoms are moved almost instantaneously from a distance for which they form no bond to one where they form the bond under consideration. Under the sudden approximation the electrons would still occupy, at the end, states that were stationary at the beginning and a relaxation process would then begin. By contrast, under the adiabatic approximation the system would pass through a continuous succession of stationary states. Hund's picture could be developed and put in a more elaborate form by a wise use of these two schemes. For example, the process of hybridization, which corresponds to symmetry breaking due to a weak field, could be described in terms of the adiabatic approximation, and the subsequent orbital mixing could be described in terms of the sudden approximation. This would result in a back-and-forth oscillation of the electrons occupying the two AOs;[28] in fact, as is well known, the name "resonance integral" for the AO coupling in MO-LCAO schemes corresponds precisely to that picture. Of course, the interchange rate would be so fast (of the order of *1/100* fs) that the situation would not really differ from a stationary state; the more so, as stationary states are themselves idealizations from the point of view of quantum statistical mechanics.

Note 2.5.3 – The Bohr picture, still accepted at face value by many quantum chemists, limits physical significance to stationary states. As a matter of fact, the latter are only approximate states of molecules even in a gas at equilibrium, as was shown by von Neumann in the context of quantum statistical mechanics; the physical justification of Bohr's picture is given in the classic textbook by Kemble.[29]

Another perplexing feature of recent discussions on orbitals in general is worth a brief commentary.[30] Already in a meeting held near Sorrento in 1963 it was pointed out that the quantum chemists, when they speak of 'observability', seem not to realize that only the fundamental measurements (length, time and mass in the macroscopic world) are direct observations; observations on submicroscopic objects are heavily "theory-laden", which is why such great men of science as Wilhelm Ostwald and Ernst Mach refused to believe in the reality of molecules until Perrin's experiments (1913) made the fluctuations of molecular motion evident at the microscope. To make a long story short, observability in the context of atomic physics is one-to-one correspondence between the readings on instruments and submicroscopic properties to be observed. As long as a property fulfills this condition in virtue of a non ambiguous definition, the only other condition to accept it as physically significant is that it should correspond to a model that is at least a limiting case of a real situation. This scheme can be applied to planetary orbits in the solar system: the actual motion of a planet is actually far more complicated than just an ellipse, but there is no ambiguity in the assignment of a particular ellipse to a particular planet. By the same token, if orbitals are defined unambiguously (as has been shown to be possible in this paper), and if they correspond to a limiting situation under the adiabatic or the sudden approximation, there is no objection in principle to considering them — *more precisely, their properties* — observable. In practice, it might happen that there are ambiguities due, for example, to the fact that correlation corrections are of the same order of magnitude as the orbital energy separation, and then, of course, one should speak of the "breakdown" of the orbital assignment just as is currently done when potential energy curves of molecular states cross, and breakdown of the adiabatic approximation ensues.

A second point concerns in particular the physical significance of bond orbitals. As mentioned, tests with configuration interaction on quasi-localized SCF MOs show that the number of configurations required for convergence is greatly reduced, and excitations from occupied to virtual orbitals associated with the same bond are the most important ones. This is in agreement with findings from PCILO computations.[31] This point, to be confirmed by further tests, suggests that a bond is recognized as a physically important unit by the fundamental equation of quantum chemistry, the stationary-state Schrödinger equation.

The above remarks are intended to confirm the claim made at the beginning of this paper, that Per-Olov Löwdin, with his attention to matrices in quantum chemistry, paved the road that will make it possible for theoretical chemistry to take advantage of the advances made in computational methods, its progress towards explaining the structure-property relations of polyatomic molecules, so far only understood in terms of empirical rules, in terms of quantum mechanics.

ACKNOWLEDGMENT. This study was essentially carried out in the fall of 2001, when the author was a visiting scientist at the Department of Applied Mathematics of the University of Waterloo, Ontario, Canada, whose hospitality is gratefully acknowledged here. The author expresses in particular his deep appreciation to Professor Joseph Paldus, FRSC, for having made that visit possible and for the discussions that took place on that occasion.

REFERENCES

1 – Many important contributions in this connection were published by Löwdin as technical reports of the Quantum Chemistry Group of Uppsala University between 1957 and 1962.

2 – Among the papers by P.-O. Löwdin, we have in mind mainly his fundamental review cited below. The background and initial phase of his interest in orthogonality is well presented by John C. Slater in his famous treatise on the *Quantum Theory of Molecules and Solids* (New York: McGraw-Hill 1963)), where he refers in particular to the seminal paper of Löwdin's master G. Wannier, "Structure and Electronic Excitation Levels in Insulating Crystals," *Phys. Rev.* 52(1937), 191, and to Löwdin's papers "On the Nonorthogonality Problem connected with the use of Atomic Wave Functions in the Theory of Molecules and Crystals," *J. Chem. Phys.* 18(1950), 365 and "Note on Orthogonal Atomic Orbitals," *J. Chem. Phys.* 19(1951), 220.

3 – Wolfgang Weber, Walter Thiel: "Orthogonalization corrections for semiempirical methods," *Theor. Chem. Acc.* 103(2000), 495-506. Most of the recent work on orthogonality was cited in Weber and Thiel's paper.

4 – Per-Olov Löwdin: "Quantum theory of cohesive properties of solids," *Phil. Mag. Suppl.* 5(1956), 1-172.

5 – L. Pauling, *The Nature of the Chemical Bond* (Ithaca, N.Y.: Cornell U. Press 1960), p. 91ff.

6 – J. F. Ogilvie, "The nature of the chemical bond 1993," in: E. S. Kryachko and J. L. Calais, *Conceptual Trends in Quantum*

Chemistry (Dordrecht: Kluwer 1994). For the relation of models to reality in quantum chemistry see: G. Del Re, "The Ontological Status of Molecular Structure", *Hyle, An International Journal for the Philosophy of Chemistry (Karlsruhe)* 4(1998), 1-23; "Models and Analogies in Science," ibid. 6(2000), 10-15.

7 – E. Clementi and C. Roetti, *Atomic Data and Nuclear Data Tables* 14(1974), 11-478.

8 – R. S. Mulliken, "Electronic population analysis...," *J. Chem. Phys.* 23(1955), 1833, 1841, 2338, 2343; "The Rydberg States of Molecules," *J. Am. Chem. Soc.* 86(1964), 3183-3197; idem, ibidem 88(1966), 1849-61.

9 – GAMESS package by M. W. Schmidt, K. K. Baldridge, J. A. Boatz, S. T. Elbert, M. S. Gordon, J. H. Jensen, S. Koseki, N. Matsunaga, K. A. Nguyen, S. J. Su, T. L. Windus, together with M. Dupuis, J. A.Montgomery *J. Comput. Chem.* 14(1993), 1347-1363.

10 – J. Hinze and H. H. Jaffé: "Electronegativity. I. Orbital Electronegativity of Neutral Atoms." *J. Am. Chem. Soc.* 84(1962), 540-546.

11 – R. S. Mulliken 1966, cit.

12 – F. Hund: "Zur Frage der Chemischen Bindung – On the Problem of the Chemical Bond", *Zs. f. Physik,* 73(1931), 1-34.

13 – René Thom: *Prédire n'est pas expliquer* (Paris: Eshel 1991, Champs-Flammarion 1993); Frederick Seitz, "Decline of the generalist," *Nature* 403(2000) pp. 483-4.

14 – There is a wealth of literature on this and related points going back to the decades before the 1970's. Among Mulliken's contributions one is especially interesting for the subject of this paper: R. S. Mulliken: "Overlap and Binding Power of 2*s*, 2*p* Hybrid Orbitals", *J. Chem. Phys.* 19(1951), 900-911.

15 – G. Del Re: "A heuristic MO-LCAO model for standard and nonstandard chemical bonds" in; K. D. Sen (ed.), *Reviews in Mod-*

ern Quantum Chemistry: a celebration of the contributions of Robert Parr, (Singapore: World Scientific 2002), pp.180-221.

16 – R. S. Mulliken, “Quelques aspects de la théorie des orbitales moléculaires” *J. chimie phys.* 46(1949), 497ff, particularly pp. 530-540.

17 – M. Wolfsberg, L. Helmholz: “The spectra and electronic structure of the tetrahedral ions MnO_4^{--}. CrO_4^{--},” *J. Chem. Phys.* 20(1952), 837-843.

18 – But see J. Slater, op. cit., and in particular his summary of C. Coulson and I. Fischer’s “Notes on the Molecular Orbital Treatment of the Hydrogen Molecule,” *Phil. Mag.* 40(1949), 386-393.

19 – E. Fermi, “Quantum Theory of Radiation,” *Revs. Mod. Phys.* 4(1932), 87-132, sec. 4.

20 – Richard P. Feynman: *Lectures on Physics* (Redwood City, Calif.: Addison-Wesley 1989), ch. 8, sec. 1ff.

21 – G.Del Re: “Hybridization and Localization in the Tight-Binding Approximation,” *Theoret. Chim. Acta (Berl.)* 1(1963), 188-195; G. Del Re: “On the general theoretical foundations of methods for optimum hybrids,” *J. Mol. Str. (Theochem)* 169(1988), 487-508; G. Del Re, F. Zuccarello: Pauling’s hybridization as a tool of computational chemistry. *Croat. Chem. Acta,* 64(1991), 449-468.

22 – cf. F. Zuccarello, G. Del Re: “Hybrid orbital evolution and reaction path in addition reactions,” *J. Mol. Structure (Theochem)* 260(1992), 249-257, and refs. therein.

23 – G. Del Re, A. Peluso: “Chemical effects and surface properties: the nature of an adsorbed complex,” *Int. J. Quant. Chem.* 49(1994), 429-448.

24 – Cf. E. Durand: *Solutions numériques des équations algébriques* (Paris:Masson 1961), Tome II, p. 93-94.

25 – The scheme of this procedure and an application including CI appeared in a paper by C. Angeli, G. Del Re, M. Persico: “Quasi-bond orbitals from maximum-localization hybrids for ab-initio CI

calculations," *Chem. Phys. Lett.* 233(1995), 102-110.

26 – Cf. e.g. R. Hoffmann, "Interaction of orbitals through space and through bonds," *Accts. Chem. Res.* 4(1971), 1-30.

27 – Albert Messiah: *Mécanique Quantique* (Paris: Dunod 1960; Engl. transl. exists), ch. 17, sec. 7-14.

28 – Feynman, op. cit., l.c.

29 – Edwin C. Kemble: *The Fundamental Principles of Quantum Mechanics* (New York: Dover 1937), ch. 12, sec. 53d, esp. pp. 439.

30 – The discussions we are referring to were concluded with a response by J. C. H. Spence, M. O'Keeffe and J. M. Zuo, J. Chem. Ed. 78(2001), 877 to E. Scerri's contention (ibid. 77(2000), 1492) that to speak of the observability of orbitals is nonsense. Unless I have missed something, the fact that an atomic orbital is a quantum state conceptually equivalent to what in classical physics is a planetary orbit was ignored in that discussion.

31 – S. Diner, J. P. Malrieu, P. Claverie, *Theor. Chim. Acta* 13(1969), 1-17; ibid. 13(1969), 18-28.

Quantum Mechanics of Many-Electrons Systems and the Theories of Chemical Bond

Marco Antonio Chaer Nascimento and André Gustavo Horta Barbosa
Instituto de Química da Universidade Federal do Rio de Janeiro
Cidade Universitária, CT Bloco A sala 412
Rio de Janeiro, RJ 21949-900 . Brazil
(chaer@iq.ufrj.br)

Abstract In this paper we briefly review the basic requirements that must be satisfied by any wavefunction representing many-electron systems. Following that, we examine the conditions under which the classical concepts of molecular structure, chemical structure and chemical bond can be translated into a quantum-mechanical language. Essential to this aim is the utilization of an independent particle model (IPM) for a many-electron system. In spite of the great popularity of the HF model only Valence-Bond (VB) type wavefunctions with optimized, singly occupied and non-orthogonal (except when symmetry imposes) localized orbitals, can provide a quantum-mechanical translation of the classical concepts of chemical structure and chemical bond. Moreover, the use of the HF model as the reference IPM gives rise to unphysical effects such as non-dynamic correlation energy and delocalization energy. Finally, the concept of resonance is redefined in a physically meaningful way, being related to point-group or "accidental" degeneracy. For the latter case, a set of selection rules to determine the possible symmetries of the resonant structures is discussed and applied to representative cases, for which Generalized Multistructural (GMS) wavefunctions are used to recover the full symmetry of the molecule.

1. INTRODUCTION

Prior to the development of quantum mechanics, chemists would picture a molecule as a three-dimensional arrangement of atoms, linked by chemical bonds. The relative positions of the atoms usually define a **molecular structure** and each particularly way of connecting the atoms defines a different **chemical structure**. The nature of the forces keeping the atoms together was totally unknown. Among the models proposed to describe the formation of a chemical bond, Lewis[1] idea of atoms sharing pairs of electrons was the most successful in explaining how single and multiple bonds could be formed between atoms and why some atoms may present different "valences". Also, with his ideas Lewis was able to extend the concepts of acid and base as substances capable of accepting or donating a pair of electrons.

E.J. Brändas and E.S. Kryachko (eds.),
Fundamental World of Quatum Chemistry, Vol. I, 371–405.

The advent of quantum mechanics forced a complete revision of that picture and as soon as it was realized that Schroedinger's equation could be only exactly solved for very simple systems, the need for approximate methodologies became evident.

The quantum theory of molecular structure has been developed along two distinct models, namely the valence-bond (VB)[2-9] and the molecular orbital (MO)[10-13] models. These models furnish quite different pictures of the molecules. In the VB model, the identity of the constituent atoms of the molecule is to a greater extent preserved and the molecule is regarded as a set of atoms held together by "chemical bonds", with each particular set of bonds joining the atoms defining a different "structure". This is a very appealing model inasmuch as it represents the quantum mechanical translation of the classical ideas to which chemists were so much deep-rooted[1,14-21].

In the MO model the atomic nuclei are viewed simply as centers of charge furnishing a potential field for the motion of all the electrons of the molecule. The properties of the molecules are described in terms of molecular orbitals, which are, in general, delocalized over the entire molecule. Contrary the VB model, the "chemical bond" is not viewed anymore as a result of adjacent atoms pairing their electrons, although molecular orbitals could, in principle, be more or less localized in regions which would correspond to the classical chemical bonds.

Independently of which model one chooses to study molecular structure, it is important to realize that there are certain basic conditions which must be satisfied by **any wavefunction representing a many-electrons system**, either an atom or a molecule. After reviewing these conditions we will show that only wavefunctions satisfying them can be used to translate to quantum mechanics the concepts of chemical structure and chemical bond. Moreover, this translation will allow us to considerably enlarge the classical concept of chemical bond as resulting from the pairing of electrons by adjacent atoms of the molecule.

2. INDEPENDENT PARTICLE MODELS

Independent particle models (IPM) are extremely useful for interpreting and rationalizing the results of quantum mechanical calculations. In fact, most of our understanding about the structure and properties of atoms and molecules derives from calculations based on IPM models.

The first general treatment for many-electron atoms was proposed by Hartree[22], who suggested that electrons in atoms would move independently of each other, the motion of each one of the electrons being influenced by the electrostatic potential of the nucleus and an average field due to all the other electrons of the system. Hartree's model was in fact the first IPM used to describe many-electron systems.

His model was extended by Fock[23] (Hartree-Fock, HF, model) to take into account the Pauli exclusion principle, making use of Slater determinantal wavefunctions[24] which, by construction, generate an antisymmetrized sum of orbital products. Initially proposed for atoms, this model was later extended to molecules, by incorporating Hund[10-11] and Mulliken's idea[12] of molecular orbitals.

Another IPM model, the VB model, normally, but erroneously, restricted to the description of molecules, originated with the work by Heisenberg[25], for the He atom, immediately followed by the Heitler and London[2] (HL) treatment of the H_2 molecule. The success of these authors in explaining the stability of the H_2 molecule, followed by Pauling's very successful extension of the HL treatment to more complex molecules, were probably the reason why the VB model became associated to molecules.

Any attempts at translating to quantum mechanics the classical concepts of chemical bond must be made within the framework of an IPM. This is so because the number of bonds, in the Lewis sense, between two atoms is almost always smaller than the number of electrons in each atom. Therefore, only a small fraction of the electrons (the valence electrons) take part in the bonding process. Thus, unless the one-electron states of an atom are univocally determined, it would be impossible to separate the valence from the non-valence states and to predict how many and what kind of bonds can be formed. On the other hand, since each particular way of binding the atoms of a molecule defines a different structure, the translation to quantum mechanics of the concept of chemical structure also depends on the knowledge of the individual one-electron states.

At this point it would be important to establish which attributes an IPM must have in order to allow a precise quantum mechanical translation of those classical concepts. In the first place, the 1-electron states of each individual particle of the many-electron systems must be univocally determined. This feature would allow us:

a) to associate important properties, such as ionization potentials, electron affinities, etc., to 1-electron states of the system;

b) to interpret electronic spectra as resulting from changes in individual particles' states;
c) to interpret the formation of chemical bonds in terms of individual 1-electron states from the isolated atoms.

Another important requirement is that an IPM should provide approximate wavefunctions presenting all the essential symmetries of the exact ones. This is the only guarantee that the approximate solution will have some overlap with the exact solution or, in another words, that the IPM is really providing an approximation to the state of the system that one wants to describe. Which are these symmetries?

From the expression of the hamiltonian for a many-electrons system (atom or molecule) it is clear that a permutation among identical particles will leave it unchanged. This was first recognized by Heisenberg[26] who expressed this fact by saying that, for quantum systems, **the permutation among identical particles is a fundamental constant of motion**:

$$\mathrm{H}P_{ij} = P_{ij}\mathrm{H}\,,\quad \forall \text{ ij particles}\,.$$

Wigner[27] put this statement in a more rigorous basis by saying that **physically acceptable** wavefunctions for microscopic systems must transform as irreducible representations (IR) of the symmetric (or permutation) group. As a consequence, the wavefunction for quantum systems must be symmetric or antisymmetric under the permutation of any two identical particles of the system. For the case of many-fermion systems (particles with half-integer spin) the Pauli[28] exclusion principle states that the wavefunction must be antisymmetric.

In order to fully express the correct symmetries of the state being approximated, spin must be included in our description. However, since non-relativistic hamiltonians do not contain spin coordinates, the spin is included through an *ad-hoc* procedure, by introducing the two spin functions, α and β, which are simultaneous eigenfunctions of the S^2 and S_z operators:

$$S_z\begin{pmatrix}\alpha\\ \beta\end{pmatrix} = \hbar\begin{pmatrix}+1/2\\ -1/2\end{pmatrix}\begin{pmatrix}\alpha\\ \beta\end{pmatrix}$$

$$S^2\begin{pmatrix}\alpha\\ \beta\end{pmatrix} = \frac{3}{4}\hbar^2\begin{pmatrix}\alpha\\ \beta\end{pmatrix}.$$

Thus, the spin is only an indicator[29], it does not have any direct dynamical role in the electronic Hamiltonian. However its inclusion as an electronic coordinate is essential to properly account for the antisymmetry requirement of the total electronic wavefunction. As a consequence, the **exact** non-relativistic wavefunction for a many-electron system can be expressed as the direct product between spatial, φ_i, and spin χ_r, parts, which span different and mutually disjoint subspaces in the Hilbert space:

$$\Psi_{exact} = \sum_{i,r} C_{ir}\varphi_i\chi_r \qquad (1).$$

3. CONSTRUCTING WAVEFUNCTIONS SIMULTANEOUSLY SATISFYING PERMUTATION AND PAULI SYMMETRIES

It must be stressed that the permutation symmetry inherent of many-particle quantum systems is the most important symmetry requirement that one should impose on electronic many-particle wavefunctions. In the literature a greater emphasis is placed on point group symmetry. However, only a small fraction of the existing molecules present that symmetry. Besides, since point group symmetry is defined by the particular position of the nuclei in a given molecule, it does not refer directly to the electronic wavefunction. Consequently, the point group symmetry requirement is automatically fulfilled (apart from accidental degeneracies) for any well defined electronic wavefunction that satisfies the virial theorem. On the other hand, the permutation symmetry depends directly on the form of the electronic wavefunction to be correctly accounted for. Löwdin was very active in this specific area, translating to "quantum chemical language" existent mathematical formalisms to deal with projection operators to symmetry adapt wavefunctions and operators.

Having established the symmetry requirements for the exact non-relativistic wavefunction of a many-electron system, the next step is to devise a general procedure to generate wavefunctions which exhibit the correct symmetries. In order to do that, a brief review of the symmetric group is presented[30].

Given **n** "objects", the symmetric group S_n is the group formed by all possible permutations among these objects.

The number of irreducible representations (IR) of S_n is equal to the number of **partitions** of **n**, i.e., the number of different ways in which **n** can be partitioned as a sum of integers, λ_t,

$$n = \lambda_1 + \lambda_2 + \lambda_3 + \ldots + \lambda_t \quad ,$$

such that $\lambda_1 \geq \lambda_2 \geq \lambda_3 \ldots \geq \lambda_t$.

Each different set of values $\{\lambda_i\}$ defines a partition and therefore an irreducible representation of S_n :

$$[\lambda_1, \lambda_2, \lambda_3, \ldots, \lambda_t] \qquad \text{partition}$$

For example, the groups S_2 and S_3 will have:

n = 2 [2] , [1,1] ∴ [1^2] two IRs and
n = 3 [3] , [2,1] , [1,1,1] ∴ [1^3] three IRs , respectively.

The dimension of a given IR of S_n will be equal to the number of Young tableaux which can be constructed by filling out the cells of the corresponding Young diagrams with integer numbers (representing the objects of the group), in such a way that the numbers will always increase from left to right, along the rows, and from top to bottom, along the columns, of the Young diagrams. For example:

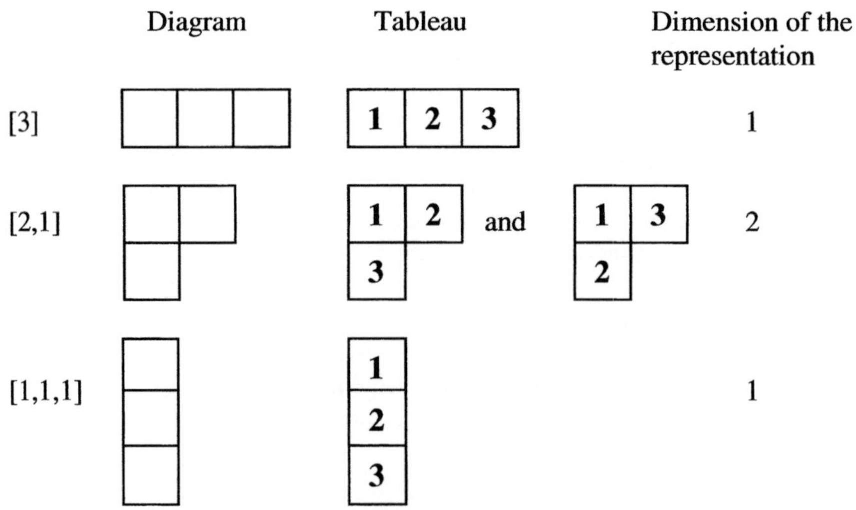

Therefore, the S_3 group has three IRs, two of them unidimensional and one two-dimensional.

Since the spin and spatial parts of the total wavefunction belong to mutually disjoint subspaces, both spin and spatial parts must independently transform like IRs of S_n. Thus, eq (1) can be rewritten as:

$$\Psi_{exact} = \sum_{i,r} c_{ir} \varphi_i^{(\mu)} \chi_r^{(\delta)} \qquad (2),$$

where (μ) and (δ) are any two IRs of S_n.

Although eq. (2) expresses the correct permutation symmetry, one must assure that ψ will also satisfy Pauli principle. As shown by Weyl[31], if one takes (μ) and (δ) to be dual representations, ψ will necessarily be antisymmetric. Dual representations have the same dimension and are the conjugate transpose of each other. In the particular case that φ and χ belong to disjoint subspaces, it suffices that their Young tableaux be the transpose of each other. Therefore, the final **exact** non-relativistic wavefunction which exhibits all the correct symmetries of the system can be written as:

$$\Psi_{exact} = \sum_{i} \Phi_i^{(\mu)} \mathrm{X}_i^{(\bar{\mu})} \qquad (3).$$

4. IPM - WAVEFUNCTIONS WITH THE CORRECT SYMMETRIES

Given **n** independent electrons, how to construct Ψ which conforms to all the symmetries of the problem?

In the absence of terms in the hamiltonian which couple the spin and spatial coordinates of the electrons, S and M_s are good quantum numbers:

$$[H, S^2] = [H, S_z] = 0 \ .$$

Therefore, one could start by constructing the Young tableau corresponding to the spin part of the wavefunction specified by the quantum numbers S and M_S. The Young tableau corresponding to the spatial part will be just the transpose of the spin part tableau.

Starting from a set of orbitals $\{\varphi_i\}$ and spin functions $\{\alpha, \beta\}$ the total wavefunction which exhibits the correct symmetries can be obtained with the help of the appropriate Young operator[30],

$$Y = AS \qquad (4).$$

associated to the tableaux corresponding to the irreducible representations of the spatial and spin wavefunctions. In equation (4), the A and S operators are defined as

$$S=\sum_{p} P_p \quad \text{and} \quad A=\sum_{q}(-1)^q Q \ ,$$

where P and Q are permutation operators which interchanges numbers in the rows and columns of the tableau respectively, q being the parity of the permutation. For the simple case of 2-e singlet, one has:

spin part $\begin{array}{|c|}\hline 1 \\ \hline 2 \\ \hline \end{array}$

$$Y = A(1,2)$$
$$A(1,2)\alpha\beta = [\alpha\beta - \beta\alpha]$$

spatial part $\begin{array}{|c|c|}\hline 1 & 2 \\ \hline \end{array}$

$$Y = S(1,2)\varphi_a\varphi_b = [\varphi_a\varphi_b + \varphi_b\varphi_a]$$

The total wavefunction, with the correct symmetries will be, apart from normalization factors:

$$\Psi = [\varphi_a\varphi_b + \varphi_b\varphi_a][\alpha\beta - \beta\alpha] \qquad (5).$$

If one takes $\varphi_1 = \varphi_2 = \varphi$, this wavefunction can be written as the expansion of a 2x2 determinant:

$$\Psi = \begin{vmatrix} \varphi_a\alpha & \varphi_a\beta \\ \varphi_b\alpha & \varphi_b\beta \end{vmatrix}$$

However, it must be emphasized that taking $\varphi_a = \varphi_b = \varphi$ is a constraint because there is no physical reason to assume that the two electrons should be in the same orbital.

As an illustrative example we can consider the following starting function for a three-electron doublet system:

$$\Psi = \varphi_a\varphi_b\varphi_c\alpha\beta\alpha$$

$$\begin{array}{|c|c|}\hline 1 & 3 \\ \hline 2 \\ \cline{1-1}\end{array} \quad = [E + (13)][E - (12)] = [E - (12) + (13) - (132)]$$

$$[E - (12) + (13) - (132)]\alpha\beta\alpha = \alpha\beta\alpha - \beta\alpha\alpha + \alpha\beta\alpha - \beta\alpha\alpha = \alpha\beta\alpha - \beta\alpha\alpha$$

$$\begin{array}{|c|c|}\hline 1 & 2 \\ \hline 3 \\ \cline{1-1}\end{array} \quad = [E + (12)][E - (13)] = [E - (13) + (12) - (123)]$$

$$[E - (13) + (12) - (123)]\varphi_a\varphi_b\varphi_c = \varphi_a\varphi_b\varphi_c - \varphi_c\varphi_b\varphi_a + \varphi_b\varphi_a\varphi_c - \varphi_c\varphi_a\varphi_b$$

The correct antisymmetric wavefunction for the three electron system is given by the direct product between spatial and spin parts:

$$\Psi = [\varphi_a\varphi_b\varphi_c - \varphi_c\varphi_b\varphi_a + \varphi_b\varphi_a\varphi_c - \varphi_c\varphi_a\varphi_b][\alpha\beta\alpha - \beta\alpha\alpha] \qquad (6).$$

However, when considering the Slater determinant wavefunction for this system we find that only the first, the fourth and the fifth terms are present in the correct antisymmetric expansion:

$$D_1 = \begin{vmatrix} \varphi_a\alpha & \varphi_b\beta & \varphi_c\alpha \\ \varphi_a\alpha & \varphi_b\beta & \varphi_c\alpha \\ \varphi_a\alpha & \varphi_b\beta & \varphi_c\alpha \end{vmatrix} = \begin{cases} \varphi_a\varphi_b\varphi_c\alpha\beta\alpha + \varphi_b\varphi_c\varphi_a\beta\alpha\alpha + \varphi_c\varphi_a\varphi_b\alpha\alpha\beta - \\ \varphi_c\varphi_b\varphi_a\alpha\beta\alpha - \varphi_b\varphi_a\varphi_c\beta\alpha\alpha - \varphi_a\varphi_c\varphi_b\alpha\alpha\beta \end{cases}$$

To restore the permutation symmetry one must superpose the other two determinants generated by column permutations:

$$D_2 = \begin{vmatrix} \varphi_a\beta & \varphi_b\alpha & \varphi_c\alpha \\ \varphi_a\beta & \varphi_b\alpha & \varphi_c\alpha \\ \varphi_a\beta & \varphi_b\alpha & \varphi_c\alpha \end{vmatrix} = \begin{cases} \varphi_a\varphi_b\varphi_c\beta\alpha\alpha + \varphi_b\varphi_c\varphi_a\alpha\alpha\beta + \varphi_c\varphi_a\varphi_b\alpha\beta\alpha - \\ \varphi_c\varphi_b\varphi_a\alpha\alpha\beta - \varphi_b\varphi_a\varphi_c\alpha\beta\alpha - \varphi_a\varphi_c\varphi_b\beta\alpha\alpha \end{cases}$$

$$D_3 = \begin{vmatrix} \varphi_a\alpha & \varphi_b\alpha & \varphi_c\beta \\ \varphi_a\alpha & \varphi_b\alpha & \varphi_c\beta \\ \varphi_a\alpha & \varphi_b\alpha & \varphi_c\beta \end{vmatrix} = \begin{cases} \varphi_a\varphi_b\varphi_c\alpha\alpha\beta + \varphi_b\varphi_c\varphi_a\alpha\beta\alpha + \varphi_c\varphi_a\varphi_b\beta\alpha\alpha - \\ \varphi_c\varphi_b\varphi_a\beta\alpha\alpha - \varphi_b\varphi_a\varphi_c\alpha\alpha\beta - \varphi_a\varphi_c\varphi_b\alpha\beta\alpha \end{cases}$$

In summary, the necessary terms to account for the antisymmetry of the wavefunction come from three distinct determinants, and the correct wavefunction has to be written as a combination of many 3x3 determinants:

$$D_1 + D_2 + D_3 \supset [\varphi_a\varphi_b\varphi_c - \varphi_c\varphi_b\varphi_a + \varphi_b\varphi_a\varphi_c - \varphi_c\varphi_a\varphi_b][\alpha\beta\alpha - \beta\alpha\alpha] \quad (7).$$

This is an important point to be remembered for latter discussion.

4.1 The Hartree-Fock Model

The Hartree-Fock model[32] is almost universally used as the reference IPM in atomic and molecular structure calculations. Starting from Slater determinantal wavefunctions[24] built from a set of orthogonal orbitals, the HF model solves self-consistently for the set of 1-e states which minimizes the energy of the system.

Although its contribution to the study of many-electron systems is unquestionable, it would be interesting to investigate if the HF model is the appropriate IPM to translate to quantum mechanics the classical concepts of molecular structure, chemical structure and chemical bond.

The first point to be noticed is that while Ψ_{HF} is invariant to unitary transformations among the orbitals (1-e states), the kinetic and potential energy operators (T and V) are not, implying that the energy of the 1-e states cannot be univocally determined. Consequently one cannot specify univocally which 1-e states would or would not be involved in forming chemical bonds. And since the connectivity between atoms is not precisely determined, chemical structures cannot be defined. Therefore, for molecules, the HF model allows us to retain at most the concept of molecular structure, within the Born-Oppenheimer approximation.

The second point to be noticed is that Ψ_{HF} is an antisymmetrized sum of spin-orbital products but does not transform like any of the irreducible representations of the symmetric group. The absence of permutation symmetry has several important consequences. First of all, since permutation symmetry is a manifestation of electrons' indistinguishability, strictly speaking, at the HF level the electrons of a many-electron system are not exactly indistinguishable. The antisymmetry requirement is accounted for by mixing the spin and spatial coordinates of the electrons. However, since there is no term in the non-relativistic Hamiltonian to account for spin, this mixing of coordinates is totally artificial. Finally, the lack of permutation symmetry gives rise to unphysical effects, two of which will now be discussed.

The HF wavefunction for the H_2 molecule can be written as

$$\Psi_{HF} = 1\sigma_g(1)1\sigma_g(2)[\alpha\beta - \beta\alpha] \qquad (8).$$

As we pull the hydrogen atoms apart, the energy of the HF wavefunctions shows the typical behavior depicted in figure 1.

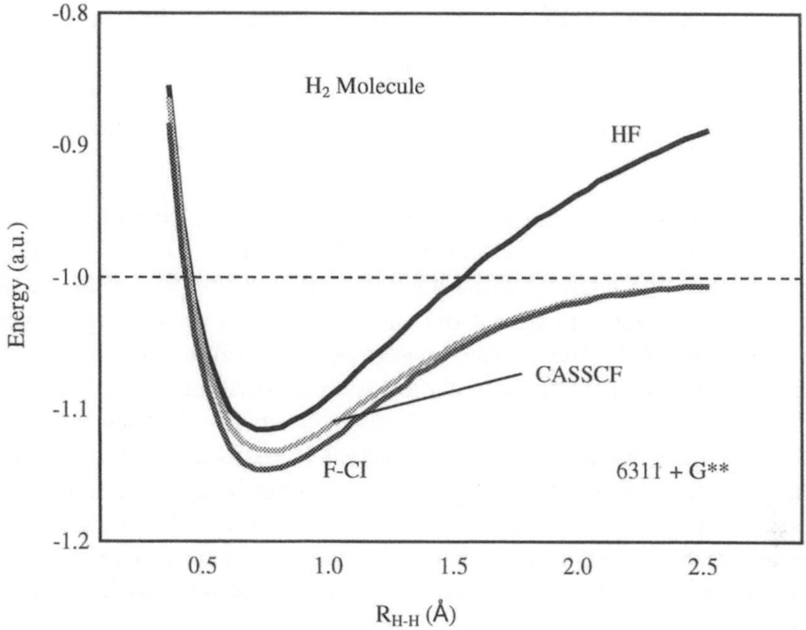

Figure 1. Potential energy surface for H_2 at different levels of calculation.

It is well known that this wrong behavior can be corrected by a simple 2 configurations ($1\sigma_g^2$ and $1\sigma_u^2$) interaction calculation (CI) or by a 2x2 multiconfigurational self-consistent calculation (MCSCF). As the atoms move apart, the true electronic correlation energy (dynamical correlation) should decrease and tend to zero as R approaches infinity. Therefore, the misbehavior of E_{HF} at large R cannot be attributed to the fact that at the HF level the dynamical correlation effects are not taken into account. To distinguish this behavior from the true correlation effects, Sinanoğlu and coworkers[33-36], in the back sixties, coined the term non-dynamical correlation energy to represent, among other spurious effects, the difference between E_{HF} and the sum of the energy of the isolated atoms as R →∍ (figure 1).

The origin of this effect has been extensively discussed[37-40] in the literature and is generally attributed to the degeneracy or near-degeneracy of some molecular orbitals as one dissociates the molecule. Some authors[39-40] also associated non-dynamical correlation with long range and dynamic correlation with short-range electronic effects. In the specific case of the hydrogen molecule, at large R, the orbitals $1\sigma_g$ and $1\sigma_u$ indeed become degenerate. However, this is not the real origin of this effect.

Let us examine the expression of Ψ_{HF} in terms of the atomic orbitals. Taking $1\sigma_g = 1s_a + 1s_b$ as usual, one obtains:

$$\Psi_{HF} = [1s_a 1s_b + 1s_b 1s_a][\alpha\beta - \beta\alpha] + [1s_a 1s_a + 1s_b 1s_b][\alpha\beta - \beta\alpha] \qquad (9).$$

It is clear from eq (9) that the first term has the correct permutation symmetry while the second one does not transform like any irreducible representation of S_2. On the other hand, the effect of the CI (or the MCSCF) calculation is exactly to cancel out the second term of Ψ_{HF} , producing a wavefunction with the correct symmetries (permutation and Pauli). Therefore, if one starts the calculation with a wavefunction exhibiting the correct symmetries, this effect would never be observed. Consequently, the so-called **non-dynamical correlation effect is really not a physical effect** related to some particular property of the system being studied **but just a manifestation of the absence of permutation symmetry**. In conclusion, if the wavefunction chosen to represent the many-electron system exhibits the proper symmetries, **there is no such a thing as non-dynamic correlation effect!**

Many other problems related to the improper choice of the HF model as the IPM can be found in the literature. The HF instability analysis[41-46] can provide an indicator of spurious behavior of the HF wavefunctions in some circumstances. The HF being non-linear, there is no guarantee that the iterative process will converge to the global minimum. In fact, the global minimum of the HF wavefunctions is only attained, in general, if ones disregard the constants of motion associated to the exact Hamiltonian. This is the origin of the so-called "symmetry-dilemma" of the HF methodology, as emphasized by Löwdin[46,47].

As another example of unphysical effect originated from the improper use of HF as reference IPM let us consider the so-called delocalization energy. The HF description of conjugated systems invariably gives rise to very delocalized molecular orbitals, and the greater stability of these conjugated systems, relative to their respective localized counterparts, is attributed to this delocalization effect. The energy gain due to the fact that, at least for some of the electrons, a larger volume of the molecule is accessible, is called delocalization energy.

HF instabilities have been well known since the sixties. However, only very recently a systematic study of the stability of the HF solutions for conjugated systems has been conduced[45]. The results of this study showed that the HF solutions for almost all of the systems considered present instabilities. In another words, the HF solutions do not correspond to the global minimum in the potential energy surfaces (PES). On the other hand, if one starts from a wavefunction which possesses the correct symmetry, one obtains a localized description of the π system, as will be latter discussed. Thus, the delocalization effect observed in conjugated systems at the HF level is simply another manifestation of the absence of permutation symmetry.

To conclude this section, let us summarize the results of our analysis of the HF model. In spite of its great contribution to the study of atomic and molecular structures, it must be recognized that the HF model gives rise to unphysical effects, such as non-dynamic correlation and electronic delocalization, which are just consequences of the fact that Ψ_{HF} does not exhibits the symmetries required for a many-electron wavefunction. From the chemical point of view, the HF model leaves no place for the classical concepts of chemical bond and chemical structure, allowing at most the definition of molecular structure (within the Born-Oppenheimer approximation) and the binding energy of the whole system. Thus, care must be exercised not to push the HF model beyond its conceptual limits.

4.2 An Alternative IPM: The Classical and Modern VB Approaches

An alternative IPM, which takes into account all the symmetries required for a wavefunction representing many-electron systems, has its origin in the work by Heisenberg[25] for the He atom. Having identified that the permutation symmetry of electrons is a constant of motion and taking into account Wigner's symmetry requirements[27], Heisenberg studied the following wavefunctions as candidates to describe the He atom ground state:

$$\Psi_{\pm}(He) = [\varphi_a\varphi_b \pm \varphi_b\varphi_a] \quad (10).$$

From his analysis, Heisenberg concluded that the ground state of He should correspond to the symmetric wavefunction.

One year latter, Heitler and London[2] applied Heisenberg's idea to the H_2 molecule by writing:

$$\Psi_{HL}(H_2) = [1s_a 1s_b + 1s_b 1s_a] \quad (11),$$

where $1s_a$ and $1s_b$ are atomic orbitals centered on the hydrogen atoms H_a and H_b, respectively. Using this function Heitler and London were able to explain the stability of the H_2 molecule. Notice that neither Heisenberg nor

Heitler-London included spin in their wavefunction. The spin part was later included to account for antisymmetry requirement for the total wavefunction.

The great success of this paper attracted Pauling's attention, who recognized that the Heitler-London procedure could not only be extended to larger molecules but also serve as a bridge between Lewis' classical idea of chemical bond and the new quantum theory. In a series of papers[4-6,9], where he introduced the concepts of hybridization, resonance, electronegativity, etc., and also in a famous book[50] that influenced many generations of chemists, Pauling developed his ideas which formed the basis of the classical "Valence-Bond" theory.

The great success achieved by Heitler-London and Pauling ideas obscured Heisenberg original work and led to the wrong conclusion that VB-type wavefunctions could only be written for molecules. This is certainly not the case. VB wavefunctions are generated by projection operators designed to yield an irreducible representation of the symmetric group. These operators act separately on spatial and spin coordinates since this is the only physically meaningful way to proceed. What is special about VB-type wavefunctions is the correct permutation symmetry that they exhibit and which is a requirement for wavefunctions describing either atoms or molecules.

From the qualitative point of view, the VB model is a conceptually correct IPM and provides a framework to translate to quantum mechanics the classical ideas of chemical structure and chemical bond. However, the non-orthogonality of the atomic orbitals posed a great problem in evaluating matrix elements of the hamiltonian. Also, most of the time numerical accuracy could only be attained with the inclusion of highly unrealistic ionic structures (resonance hybrids), causing VB theory to loose its most important characteristics: chemical interpretability and compactness of the wavefunctions. Coulson and Fisher[51] identified the origin of the problem and showed that the need for the ionic structures was associated to the lack of orbital relaxation upon the formation of the bond. However, the solution of this problem in feasible terms required new approaches to calculate VB wavefunctions that only began to appear in the late ninety sixties. In the meantime, unfortunately, the VB theory fell in disuse and the HF-MO approach became the only practical model for molecular structure calculations.

The modern VB approaches were proposed by Goddard[52] and Gerratt[53]. They considered the explicit optimization of the singly-occupied orbitals (therefore implicitly including the effect of the ionic structures), obtaining highly accurate monoconfigurational VB-type wavefunctions. They named their models as GVB (Generalized VB) and SCVB (Spin-Coupled VB) respectively.

These wavefunctions exhibit two extremely important properties not shared by the HF wavefunctions. Besides taking into account all the symmetries required from a multielectronic wavefunction, the resulting optimized orbitals are uniquely determined within a given basis set. Therefore, those unique orbitals allow an IPM interpretation in the sense discussed in section 2.

The orbitals (1e-states) can be associated to properties of the system (atom or molecule). Also, the 1-e states involved in chemical bonding can be univocally identified. Thus, the connectivity among atoms can be precisely identified and chemical structures for the molecules can be defined.

Each different way of connecting the atoms will define a possible chemical structure for the molecule, the best one corresponding to the structure with the lower energy, in the variational sense. This structure will correspond to the best possible representation of the molecule at the IPM level. This structure almost always corresponds to pairing the electrons of the neighbor atoms of the molecule, each pair being pictorially represented by a line connecting the atoms involved. The wavefunction associated to this structure is called GVB-PP (perfect pairing)[54], which means that only one Young tableau is used as spatial projection operator. When more than one tableau is used we have the GVB or SCVB models.

Since the wavefunction defining this best structure has the correct symmetries, the difference between the exact energy of the molecule and that of the chemical structure will correspond to the real correlation effects, as defined either by Wigner[55] and Löwdin[56]. Therefore, GVB-PP or SCVB wavefunctions are the ideal references for computing the real electronic correlation effects. Accordingly, one can now redefine the correlation energy, within a basis set, as:

$$E_{correlation} = E_{exact} - E_{chemical\ structure} \quad (12),$$

or, for atoms:

$$E_{correlation} = E_{exact} - E_{GVB\text{-}PP} \quad (13).$$

When the more appropriate IPM is used, the correlation energy turns out to be a much smaller fraction of the total energy than suggested by the HF calculations. Thus, when GVB or SCVB wavefunctions are used as reference IPM much less computational effort is needed to calculate the real correlation effects.

The use of projection operators to attack the problem of electron correlation is just one more of the important contributions of Löwdin which permeates all the modern VB approaches for calculating electronic structures. In his own words:

"It is my opinion that, at least for electrons, one may remove the essential part of the correlation effects simply by a proper treatment of the symmetry properties, and this leads to a close connection between the shell structure and the constants of motion underlying the independent particle model." (P.O. Löwdin)[57].

4.2.1 The Concept of Resonance

One of the most characteristic concepts of VB theory is resonance[50]. This effect in invoked whenever two or more perfect pairing schemes (chemical structures) have equal or nearly equal energies. In these cases, the molecule is represented by a linear superposition of "chemical structures".

This linear superposition is called resonance and each chemical structure is denominated a resonance hybrid. The benzene molecule is probably the one most used to illustrate this effect, its structure being represented by a superposition of the 2 Kekulé and the 3 Dewar structures:

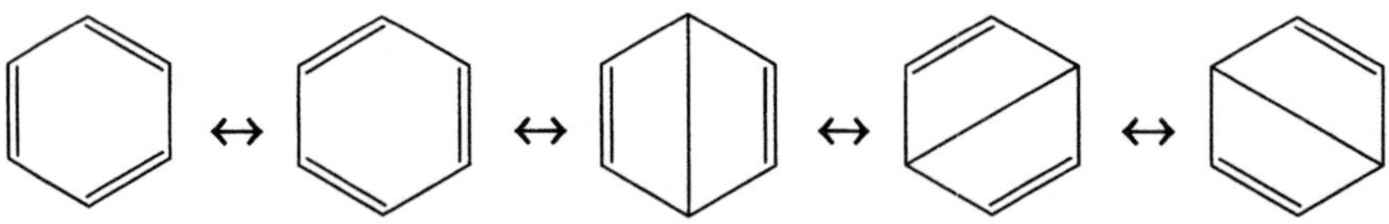

However, if one solves self-consistently for the best orbitals of this molecule, the best VB-type wavefunction will correspond to perfectly localized pairs of electrons in the plane of the molecule, which can be clearly identified with the C-C and C-H bonds, and six mono-occupied degenerated orbitals in the plane perpendicular to the molecule (π orbitals), each one localized at one of the carbon atoms. For the allyl radical, the same situation is observed, with the three "π" electrons localized each on a carbon atom. These results clearly indicate that a much deeper analysis of this concept is required.

Resonance is related to degeneracy or near-degeneracy effects. Degeneracy may arise in molecular quantum mechanics due to the existence of symmetry groups that commute with the molecular hamiltonian[58,59]. It is easily shown that the eigenfunctions of the exact hamiltonian must transform as irreducible representations of the commuting symmetry groups. When a given group has degenerate representations, some or all eigenstates of the hamiltonian of the system will reflect its degeneracy. A k-degenerate eigenvalue induces in the Hilbert space of functions, a k-dimensional subspace spanned by its eigenfunctions. Since they span the same subspace, the eigenfunctions can always be made orthogonal. This is the ordinary case of degenerate representations. However, under certain circumstances, another kind of degeneracy is possible. When the degenerate eigenvalues do not belong to the same subspace one says that there exists an "accidental degeneracy". In this case, the tensorial Hilbert space is factored into irreducible non-overlapping subspaces, each one associated to an eigenvalue. The well-known example of accidental degeneracy is the hydrogen atom. Its eigenfunctions should transform as irreducible representations (S, P, D, F, ...) of the SO(3) group. However the eigenfunctions associated with the same principal quantum number are degenerate: [1s], [2s,2p], [3s,3p,3d] ... In fact, the correct (non-relativistic) group of symmetry of the hydrogen atom is SO(4), which is isomorphic to the rotation group in four spatial dimensions. Accidental degeneracies signal that a given system has more symmetry than it appears to have. From the group theory point of view, it seems natural to relate degenerate resonance structures to point group degeneracy, or accidental degeneracies. This attribution is essential if one wants to relate "resonance" to identifiable physical effects, and to associate a "chemical structure" with the resonance hybrids. The structure of the resultant partitioned Hilbert space will guide us in understanding the fundamental nature of the resonance phenomenon in each case.

It should be clear that in the presence of either point group or "accidental" degeneracies, one VB configuration (chemical structure) is not enough to qualitatively describe the system. It is necessary to consider other VB configurations, leading naturally to the concept of resonance.

In the quantum chemistry literature resonance is sometimes considered to be equivalent to "delocalization". This happened mainly because of the "successful" MO monoconfigurational description of the "pi" system of benzene. However, the Hartree-Fock wave function for the benzene molecule (and for all aromatic molecules) is unstable[35], providing a qualitatively wrong description of its electronic structure. When dealing with pure states, delocalization reflects the failure of a given level of approximation to provide an N-representable wave function[9]. On the other hand, when there is resonance (in the precise sense defined above), we are dealing with mixed states, and the delocalization signals the intrinsic complex instability of degenerate or quasi-degenerate states.

4.2.2 Symmetry Conditions for Resonance Hybrids

It remains to state the symmetry conditions to be obeyed by resonance hybrids. For the total wave function to be able to split into different adiabatic states, it should be decomposable into independent parts. If the nuclear framework has some sort of spatial symmetry, it is easy to know the possible structures of the resonance hybrids. However, we must distinguish between two different situations: point group and accidental degeneracies[59,60].

When a given state belongs to a k-dimensional degenerate irreducible representation, the degenerate eigenfunctions belong to the same tensor subspace, and can always be made orthogonal. Point group degenerate states are always subject to Jahn-Teller distortions. The nuclear framework follows the symmetry descent coordinate until the complete removal of the degeneracy[61]. Consequently, it is not possible to have one PES minimum, or resonance, made by point group degenerate states. However, if these point group degenerate states are quasi-degenerate with a different state, the situation becomes much more complicate, and will not be considered in detail here. In these cases, resonance between these states may be possible, and the symmetries of the resonance hybrids will follow the symmetry descent path of the full point group of the system. An example of this situation recently described in the framework of MO theory is the NO_3 radical[62].

The real Hilbert space is always partitioned into a direct sum of subspaces, each representing a different energy eigenvalue of the spectrum of the hamiltonian operator:

$$\Lambda = \lambda_1 \oplus \lambda_2 \oplus \lambda_3 \ldots$$

When two or more eigenvalues happen to be equal, or nearly equal, we say that there is an "accidental degeneracy". Since the states belong to different subspaces, there is no symmetry descent path to follow. The direct product decomposition is the mathematical tool to analyse the symmetry of the allowed individual adiabatic states. It is related to the "ascent in symmetry" method[63] and justified by the Littlewood-Richardson rules for decomposition of tensor spaces in independent parts[64]. These rules define the only permissible decompositions of a tensor space (in our case, point group space), providing us with the possible symmetries of the resonance hybrids, which reproduce the total symmetry of the system. A simplified statement, suitable for our purposes, is that "***the direct product between the subsystem point group and the group that relates (maps) the subsystems should recover the full symmetry of the system***". Notice that both the subsystems´

point group and the group that maps the subsystems must be invariant subgroups of the full point group. Only invariant subgroups of a larger group can accommodate coherent states. Each chemical structure must be associated with a coherent quantum state. A GVB wavefunction, with fully optimised orbitals will, in general, transform as an irreducible representation of the molecule point group. When this requirement is not fulfilled, the GVB wavefunction will necessarily transform as an irreducible representation associated with an invariant subgroup of the full molecular group.

In Table 1 all point groups are classified according to the possibility of being described by a direct product decomposition[65].

Table 1

Point groups decomposable in direct product forms	*Point groups not decomposable in direct product forms*
C_n, S_n (n = 4k + 2, k=1, 2, ...)	C_n, S_n (n ≠ 4k + 2, k=1, 2, ...)
D_{nd} (n odd)	D_{nd} (n even)
C_{nh}, D_{nh}, T_h, O_h, I_h	C_i, C_s, C_{nv}, D_n, T, T_d, O, I

In table 2 we explicit the possible forms of decomposing the direct product groups.

Table 2

Possible point group decompositions in direct products
$C_n = C_{n/2} \otimes C_2$ (n = 4k + 2, k=1, 2, ...)
$S_n = S_{n/2} \otimes C_i$ (n = 4k + 2, k=1, 2, ...)
$D_{nd} = D_n \otimes C_i$ (n odd)
$C_{nh} = C_n \otimes C_i$, $C_n \otimes C_s$ (n even); $C_n \otimes C_s$ (n odd)
$D_{nh} = D_n \otimes C_i$, $D_n \otimes C_s$, $C_{nv} \otimes C_s$ (n even); $D_n \otimes C_s$, $C_{nv} \otimes C_s$ (n odd)
$D_{\infty h} = C_{\infty v} \otimes C_i$, $C_{\infty v} \otimes C_s$
$T_h = T \otimes C_i$
$O_h = O \otimes C_i$
$I_h = I \otimes C_i$

Good examples are the core hole excited states of homonuclear molecules. When one electron is removed from a core orbital, the original $D_{\infty h}$ symmetry is lowered to $C_{\infty v}$. The $D_{\infty h}$ group can be decomposed into two $C_{\infty v}$ components related by a C_i or C_s operation, so it is fair to consider that the core-hole excited states are described by resonance between the two structures. The adiabatic subsystems have, by definition, zero overlap in the real space. Their interaction is defined only in complex space through the explicit overlap between the many-electron states.

An inspection on Table 2 shows that it is not possible to relate the benzene (D_{6h} symmetry) to Kekulé (D_{3h}) or Dewar (D_{2h}) structures. The ground state of benzene is not degenerate, and there is no theoretical or experimental evidence of a conical intersection with a degenerate state near the ground state geometry[66]. If there is no intersection of degenerate point group state, one cannot follow the symmetry descent path in this case. The only possibility would be that of an accidental degeneracy, but this is ruled out by the impossibility of direct product decomposition. Thus, as already stated before, the ground state of the benzene molecule is *not* described by a resonant mixture of Kekulé and/or Dewar structures.

Similarly, the allyl radical cannot be represented by two resonat hybrids such as:

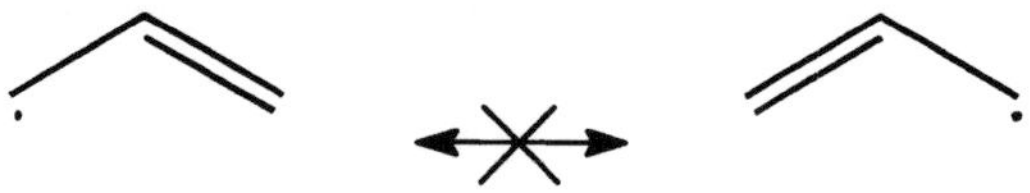

In fact, resonance is not possible for any AB_2 (C_{2V}) molecule, unless the system presents near-degenerate states.

In summary, we have enlarged the concepts of chemical structure and resonance in such a way as to make them conform the more general theories of molecular quantum mechanics. Classical VB concepts have been extremely useful in rationalizing empirical facts but became inadequate in the light of the new theoretical developments. The new concepts presented here are consistent both with the mathematical models of quantum chemistry and with empirical chemical facts, and their formulation recognize the latest research advancements.

5. THE GENERALIZED MULTISTRUCTURAL WAVE FUNCTION (GMS)

The GMS wave function[67-70] combines the advantages of the MO and VB models, preserving the classical chemical structures, but dealing with self-consistently optimised orbitals. From a formal point of view, it is able to reproduce all VB or MO based variational electronic wave functions in its framework. Besides that, it can deal in a straightforward way with the non-adiabatic effects of degenerate or quasi-degenerate states, calculating their interaction and properties.

The GMS wave function can be defined as

$$\Psi_{GMS} = \sum_{I=1}^{N_{struc}} \sum_{i=1}^{N_{sef}} c_i^I \varphi_i^I$$

where φ_i^I represents the ith spin eigenfunction (N_{sef}) of the Ith structure (N_{struc}) and the c_i^I its weight in the expansion. There are no restrictions whatsoever on the form of the wave function φ_i^I. Each of the φ_i^I can be individually optimized at the Hartree-Fock, or multiconfigurational (GVB, CASSCF) level, followed or not by configuration interaction (CI) treatment.

Each one of the φ_i^I is represented in a basis of orthogonal orbitals $\{\Phi_g^I\}$ optimized for the Ith structure. Although the orbitals of a given structure are taken to be orthogonal to each other,

$$\left\langle \Phi_\alpha^I \middle| \Phi_\beta^I \right\rangle = \delta_{\alpha,\beta} \quad ,$$

no such restriction exists for the orbitals belonging to different structures I and II,

$$\left\langle \Phi_\alpha^I \middle| \Phi_\beta^{II} \right\rangle = S_{\alpha,\beta}^{I,II} \quad .$$

The coefficients c_i^I are obtained variationally by solving the equations,

$$\begin{aligned} &\left\langle \delta\Psi_{GMS} \middle| H - E \middle| \Psi_{GMS} \right\rangle = 0 \\ &(H - SE)C = 0 \end{aligned} \quad ,$$

where H and S are the interaction supermatrices containing the diagonal (same structure) and interstructural matrix elements. The matrix elements involving orbitals belonging to different structures are computed using a biorthogonalization procedure.

GMS wavefunctions are particularly suitable to treat systems which exhibit resonance. It also provides a very efficient and convenient way of treating correlation effects, avoiding large CI expansions[67] . In order to illustrate the usage of GMS wavefunctions, we will present two different applications. For a more detailed discussion and several other examples, the reader is referred to the appropriate references[67-70].

As discussed in the previous section, when core electrons of homonuclear diatomic molecules are excited or ionized, the solutions with the localized hole in one of the atoms has a lower ($C_{\infty v}$) symmetry than the full symmetry ($D_{\infty h}$) of the molecule. In these cases, the superposition of two $C_{\infty v}$ structures will recover the full symmetry of the system. Thus, it is correct to describe the core-hole ion or excited state by two hybrids of resonance. The O_2 and N_2 molecules can be used to illustrate this situation. For the O_2 molecule we considered[67] the two ion states resulting from the ionization of one of the 1s core electrons, while for the N_2 molecule we investigated several excited states[71-72]. In both cases the final states were represented by a superposition of two GVB-PP wavefunctions :

$$\Psi_{GMS}^{GVB} = c_1 \Psi_{GVB}(A - A^*) + c_2 \Psi_{GVB}(A^* - A) \ ,$$

where A^* stands for either O^+ or N^o .

Another interesting application has to do with the calculation of gas-phase acidities of carboxylic acids. In this case, accurate values can only be obtained if the stabilization of the carboxylate anion is properly taken into account[73]. This can be accomplished by a GMS wavefunction such as:

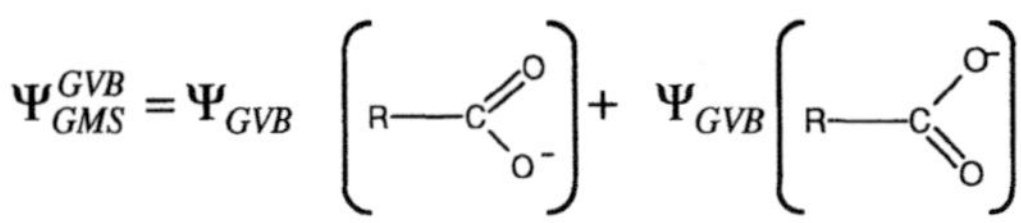

Some of the results of these calculations are summarized in Table 3.

Table 3. Illustrative Examples of GMS Calculations

System	Property	Results	Expt.
O_2	Ionization Potential (eV)	$^2\Sigma^-$ 544.72	544.2[81]
		$^4\Sigma^-$ 543.31	543.3
N_2	Transition Energy (eV) and Optical Oscillator Sthrength	$1\sigma_g \rightarrow 1\pi_g$ 400.96 $1\sigma_u \rightarrow 1\pi_g$ 401.02 $f = 0.192$	401.1[82] unresolved $f = 0.2 \pm 0.02$[83]
H_3CCOOH	Gas-Phase Acidity (kcal/mol)	$\Delta H_{298}^0 = 346.26$	345.29[84]

6. A Quantum-Electrodynamic View of the Chemical Bond.

It still remains the question about the origin of the chemical bond. From the energetic point of view, the problem of the origin of the chemical bond has been analyzed by several authors, using different lines of reasoning and either the VB or MO approaches. In particular, the papers by Ruedenberg[74], Wilson and Goddard[75], and Kutzelnigg[76] present a very detailed analysis of the problem and in spite of the fact that quite different approaches have been used, the same conclusions have been reached about the energetic of the chemical bond formation.

In this section we would like to show that those same conclusions could be reached, in a much simpler way, first by identifying the quantum phenomenon responsible for the chemical bond formation and then trying to analyze how this effect would contribute to the stabilization of the molecule, i.e., to the formation of a chemical bond. In fact, this type of analysis provides a much deeper insight into the nature of the chemical bond and allows us to extend this concept beyond its classical limits.

Quantum electrodynamics is a very successful theory, which can be beautifully formulated in terms of Feynman's space-time diagrams[77]. The reader not familiar with these ideas could refer to the excellent book by Mattuck[78]. However, before starting our analysis, let us refer to the famous double-slit experiment show in figure 2.

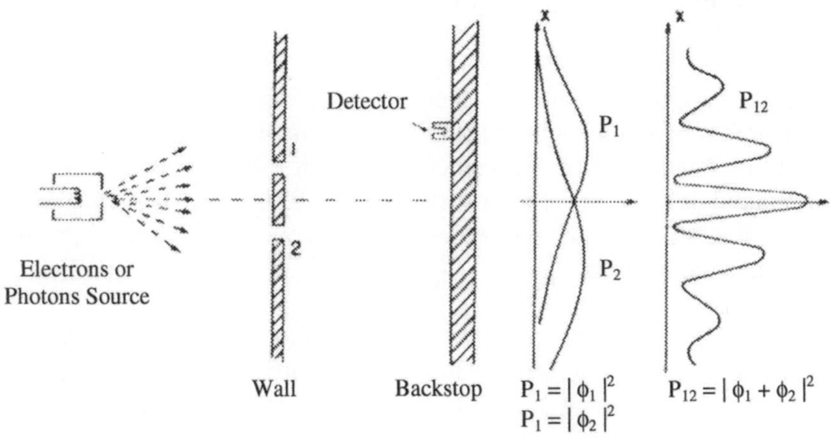

Figure 2. Diffraction of electrons or photons.

This simple experiment reveals one of the most fundamental and intriguing laws governing the quantum world **but very often forgotten**. If photons or electrons are sent through slit 1 (with slit 2 closed), the distribution of intensity for electrons arriving at different points of the backstop is represented by curve P_1. Similarly, curve P_2 represents the same when the quantum particles are sent through slit 2 with slit 1 closed. Classically, if the experiment is repeated with both slits opened, the total intensity would be just the sum $P_1 + P_2$. However, this is **not** what is experimentally observed, as shown on the right of figure 2. On the other hand, the observed distribution P_{12} can be obtained from curves P_1 and P_2 just by adding the amplitudes (ϕ_1 and ϕ_2) for each separate event, and expressing the result as:

$$P_{12} = |\phi_1 + \phi_2|^2 = \underbrace{\phi_1^2 + \phi_2^2}_{\textbf{Classical}} + \underbrace{2\,\phi_1\,\phi_2}_{\textbf{Interference}}$$

The first two terms represent the classical result while the third term is a quantum effect. It is important to emphasize that the closer the frequencies (energies) of the photons or the energy of the electrons, the stronger is the interference effect. Thus, the fundamental law revealed by this experiments is the following:

"When an event can occur in several alternative ways, the amplitude for observing the event is the sum of the amplitudes for each way considered separately."

Returning to the quantum-electrodynamic description of a chemical bond, let us first recall that within the Born-Oppenheimer approximation, the nuclei will be fixed in space. And since the chemical bond concept can only be formulated within the framework of an IPM, only the zero-order Feynman diagrams (those not showing interaction lines between the electrons) need to be used.

Figure 3 shows the diagram corresponding to electron 1 moving from point 1 to point 2 under the influence of proton 1 (i.e., exchanging virtual photons with proton 1), while electron 2 moves from point 3 to 4, exchanging virtual photons with proton 2.

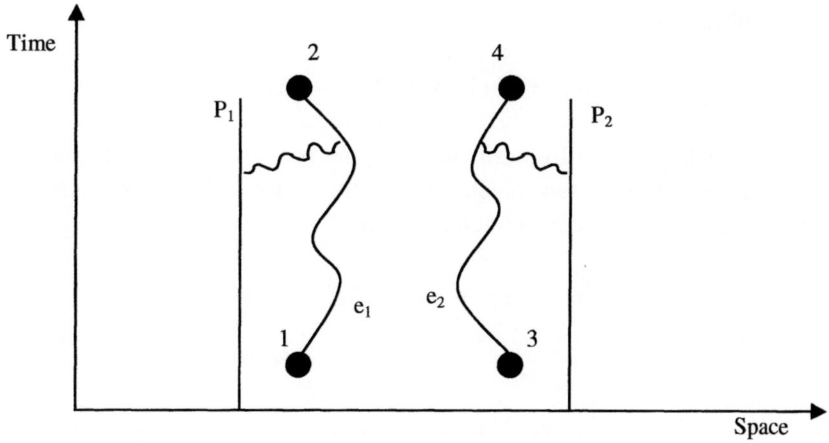

Figure 3

The amplitude for observing this event is just the product of the amplitudes for electron 1 to move from point 1 to 2 and for electron 2 to move from point 3 to 4, since their motions are independent:

$$\mathbf{A_1 = E\,(1 \to 2)\ E\,(3 \to 4)} \quad .$$

But since electrons are indistinguishable, the event can also take place as shown in figure 4:

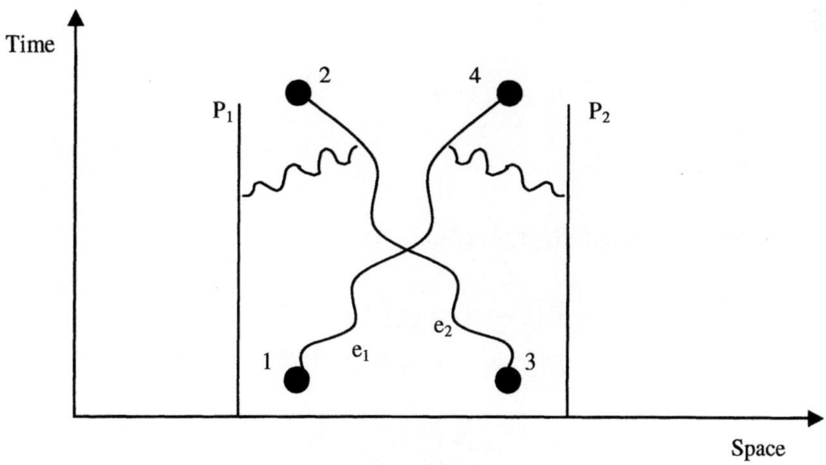

Figure 4

whose amplitude is:

$$\mathbf{A_1 = E\,(1 \rightarrow 4)\;\; E\,(3 \rightarrow 2)}\,.$$

According to our fundamental law, the probability of observing such a system constituted by two electrons moving independently of each other, under the influence of the two nuclei, will be:

$$\mathbf{P = |\,A_1 + A_2\,|^2 = A_1^{\,2} + A_2^{\,2} + \underbrace{2\,A_1\,A_2}_{\text{Interference}}}\,.$$

At this point it is important to mention that considering the electrons independent of each other does not rule out the possibility that they interact through an average field as usually assumed in IPMs. This condition can be simply incorporated in the model used to compute the amplitudes E.

Let us now try to rewrite this result using an independent particle model. To this purpose we define orbitals φ_a and φ_b , centered on the nuclei A and B, respectively, as the best orbitals (obtained in a self-consistent way) to describe the motion of electrons 1 and 2 in the field of the two nuclei. Using these orbitals (amplitudes), for the first diagram (figure 3) one can write:

$$E(1 \rightarrow 2) = \varphi_a(1) e^{-iE_a t/\hbar}$$
$$E(3 \rightarrow 4) = \varphi_b(2) e^{-iEb_b t/\hbar}$$

and

$$A_1 = \varphi_a(1)\varphi_b(2) e^{-iE\;t/\hbar} \tag{14},$$

with $E = E_a + E_b$.

Similarly, for the second diagram (figure 4) one has:

$$E(1 \rightarrow 4) = \varphi_a(2) e^{-iE_a t/\hbar}$$
$$E(3 \rightarrow 2) = \varphi_b(1) e^{-iEb_b t/\hbar}$$

and

$$A_2 = \varphi_a(2)\varphi_b(1) e^{-iEt/\hbar} \tag{15}.$$

Therefore, the total amplitude, i.e., the wavefunction describing the molecule, will be:

$$\Psi = A = A_1 + A_2 = [\varphi_a\varphi_b + \varphi_b\varphi_a]e^{-iEt/\hbar} \qquad (16),$$

and

$$P = \int A^* A dv = 1 + 1 + S^2 \ ,$$

where $S = \int \varphi_a \varphi_b dv$ is the overlap between the orbitals centered on nuclei A and B. Thus, the **interference effect**, in the framework of the IPM used to calculate the amplitudes, appears as an **overlap integral between the orbitals φ_a and φ_b.**

Incidentally, the form of the wavefunction which emerges from this treatment is, apart from the phase factor, identical to VB-type wavefunction for the H_2 molecule, eq (11), i.e., the wavefunction that takes into account the permutation symmetry. In fact, the two wavefunctions are identical because the full VB-type wavefunction also contains the same phase factor, which is never written as it vanishes whenever a property is calculated. Thus VB-type wavefunctions are totally consistent with quantum electrodynamics.

The next point to be examined is how this interference effect contributes to the stabilization of the system (H_2 molecule). In order to answer this question one needs to compute the energy associated to the wavefunction of eq (11), at different values of the internuclear distance:

$$E = \int \Psi^* H \Psi dv = E_A + E_B + \frac{Q + A}{1 + S^2} \qquad (17),$$

where H is the exact non-relativistic Born-Oppenheimer hamiltonian for the H_2 molecule, and $E_A + E_B$ is the total energy of the isolated atoms. According to eq (17), the molecule will be formed if :

$$\frac{Q + A}{1 + S^2} < 0 \qquad \text{or} \qquad Q + A < 0 .$$

But Q is just the equivalent of the classical electrostatic energy and must therefore be positive ($Q > 0$). Thus, the molecule will be formed only if $A < 0$ and $|A| > Q$. The term A, known as the exchange integral in the classical VB theory, is made up of three contributions, two of them proportional to S, and the other exactly equal to the exchange integral as defined in the HF model:

$$A = -S(C_1 + C_2) + K + S^2/R \quad (18).$$

In equation (18), $K \geq 0$; $C_1, C_2 > 0$ and $S^2/R \geq 0$. Therefore if $S = 0$, i.e., in the absence of interference, **no chemical bond is formed**.

Thus, we arrive to the first important conclusion about the origin of the chemical bond: from the quantum-mechanical point of view, the **chemical bond is a consequence of interference effects**.

Having identified the quantum origin of the chemical bond, we are now in a much better position to analyze its energetic. Since the chemical bond is a manifestation of interference effects, it is reasonable to assume that the total energy of the molecule can be written as the sum of two terms, one of them containing the contributions corresponding to the classical terms, the other being related to the stabilization brought about by the interference effect:

$$E_{quantum} = \frac{\int \Psi^* H \Psi dv}{\int \Psi^* \Psi dv} = E_{classical} + E_{interference} \quad (19),$$

with

$$E_{chemical} = E_c(1) + E_c(2) + V_{ee} + V_{nn} + V_{en} \quad (20),$$

where $E_c, V_{ee}, V_{nn}, V_{en}$ are respectively the kinetic energy of electrons, their coulombic repulsion, the nuclei repulsion, and the electron-nuclei attraction terms.

Using the wavefunctions of eq (11) it is indeed possible to write $E_{quantum}$ as the sum of two contributions, with

$$E_{interference} = \frac{E - S^2 E_{classical}}{1 + S^2} \quad (21),$$

where $E = E(S)$ and such that $\lim_{S \to 0} E(S) = 0$.

Notice that in the absence of interference $(S = 0)$, $E_{interference} = 0$ and $E_{quantum} = E_{classical}$, that is, no molecule can be formed. Thus, the energetic of the bond formation is completely defined by $E_{interference}$. It is possible to write $E_{interference}$ as a contribution of kinetic and potential energy terms:

$$E_{interference} = E^T_{interference} + E^V_{interference} \quad (22).$$

How would these two contributions vary with the internuclear distance? The answer to this question, shown in figure 5, is that the stabilization of the molecule is due to a reduction of the kinetic energy of the electrons. This is exactly the conclusion reached by previous works[74-76].

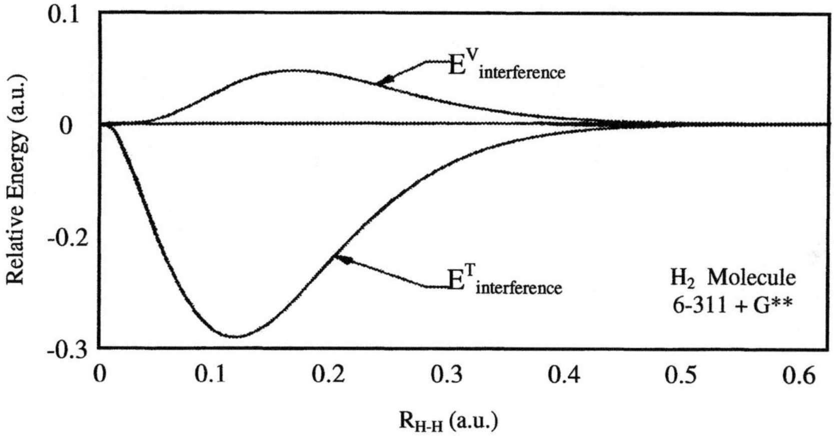

Figure 5. Kinetic and potential contributions to the energy of interference.

The final conclusion of our analysis concerning the origin of the chemical bond can be summarized as follows:

> From the quantum-mechanical point of view, the formation of a chemical bond is a **consequence of interference effects**. From the energetic point of view, the interference responsible for the formation of the chemical bond manifests itself as a **reduction** of the **kinetic energy** of the **electrons** as the bond is formed.

7. EXTENDING THE CONCEPT OF CHEMICAL BOND

We can now reexamine the question of the stability of the conjugated systems, such as benzene. As previously discussed, for many of these systems, the stabilization cannot be attributed to resonance effects. What would then be the source of their stabilization?

Before addressing this question, let us briefly examine the bond in H_2^+. This system is interesting because, having just one electron, it should be the best example of the inadequacy of classical picture of a chemical bond.

In fact, in the early days of quantum mechanics, this molecule was used by Mulliken[7] to illustrate the limitations of the VB theory and the superiority of the MO model.

The space-time diagrams for this molecule are shown in figure 6.

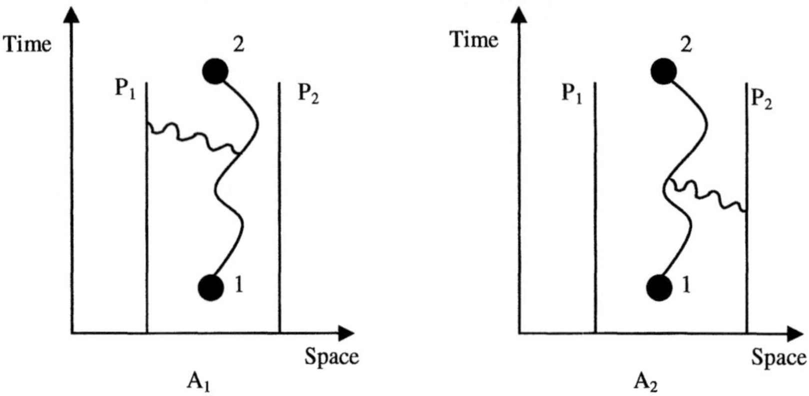

Figure 6

The total amplitude will be $A = A_1 + A_2$. Using the same notation as before:

$$\Psi = \varphi_a(1) + \varphi_b(1)$$

and

$$\Psi\Psi^* = \varphi_a{}^*(1)\,\varphi_a(1) + \varphi_b(1)\varphi_b(1) + \underbrace{2\varphi_a(1)\varphi_b(1)}_{\textbf{Interference}}$$

Thus, in spite of the fact that the molecule has only one electron, the interference term, between the $1e^-$ states φ_a and φ_b , will respond for the formation of the bond. Notice that in this case, the wavefunction takes the form of a molecular orbital. However, the attentive reader will not see this fact as a limitation of the VB model. Much on the contrary, this example exposes the limitations of the MO model. The only reason why Ψ takes the form of an MO is because, for 1-e system, there is no need to take into account the permutation symmetry of the electrons! And for this same reason, there is no need for a VB-type wavefunction.

Figure 7 shows the self-consistently optimized orbitals in the "π" space of benzene, obtained from a CAS-GVB calculation which takes into account the permutation symmetry[80]. Each of these orbitals contains just one electron.

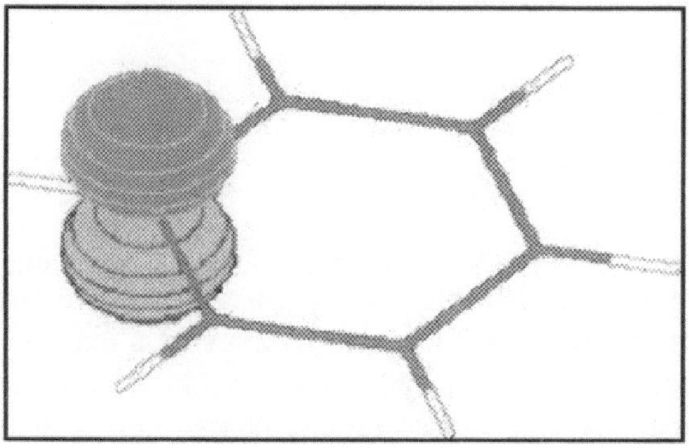

Figure 7. π orbital of benzene. Only one orbital is shown for clarity.

As discussed in section 4.2.2, there is no resonance in this system. On the other hand, we have six 1-e states exactly degenerate, defining a situation of maximum interference. For the ground singlet state of the molecule the spin part of the wavefunction is such that electrons in adjacent carbon atoms have opposite spins and therefore we expect a considerable overlap, that is, interference among them. This is exactly what is obtained from the calculations[80]. Of course, all the six electrons take part in the process and the stabilization results from this strong interference effect. Thus, one could extend the classical concept of chemical bond by saying that the stabilization of benzene is brought about by a six-electron bond[80]. Similarly, the stabilization of the allyl radical is due to a 3-e bond (figure 8).

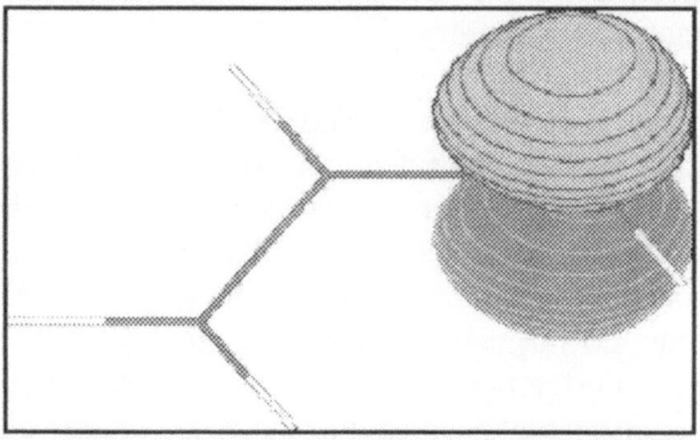

Figure 8. π orbital of allyl radical. Only one orbital is shown for clarity.

However, before being tempted to conclude that the larger the number of electrons in degenerate or near-degenerate 1e$^-$-states the larger the stabilization of the molecule, one must not forget the spin part of the wavefunction which will limit the overlap between electrons with the same spin.

8. CONCLUSIONS

The basic requirements that must be satisfied by any wavefunction representing a non-relativistic many-electron system have been reviewed. It has been shown that in order to properly represent a many-electron system, any wavefunction must respect both the permutation and Pauli symmetries. For non-relavitistic systems, these requirements imply that both the spatial and spin parts of the wavefunction must transform like irreducible representations of the symmetric group while the total wavefunction must be antisymmetric.

Two independent particle models, namely the HF(MO) and the VB models, have been analyzed for these symmetry requirements. From these two models, VB-type wavefunctions are the only ones to satisfy the required symmetries. As a consequence, strictly speaking, only VB-type wavefunctions with optimized, singly occupied and non-orthogonal (except when symmetry imposes) localized orbitals are acceptable wavefunctions to describe many-electron systems at the IPM level. Also, only-VB type wavefunctions can provide a quantum-mechanical translation of the classical concepts of molecular structure, chemical structure and chemical bond. The HF model can at most be used to define molecular structure. Moreover, the use of the HF model as the reference IPM gives rise to unphysical effects such as non-dynamic correlation energy and delocalization energy, which are both consequences of the fact that the HF model does not take into account the permutation symmetry.

The concept of resonance, as derived from the classical VB theory has been also examined and redefined in a more physically meaningful way, being related to point-group or accidental degeneracy. For the later case, a set of selection rules to determine the possible symmetries of the resonant structures have been presented. According to these rules, benzene and other conjugated systems do not exhibit resonance, their stability being derived from the strong interference effect among the degenerate or almost degenerate one-electron π-like states. For the cases were resonance is really operative, the GMS wavefunction provides a very convenient and efficient way of taking this effect into account.

Finally, a quantum electro-dynamical analysis of the chemical bond is presented. This type of analysis clearly reveals that the chemical bond is formed as a result of quantum interference effects. It also reveals that only VB-type wavefucntions are consistent with quantum-electrodynamics and that the benzene (and other non-resonant "conjugated systems") stability can be understood in terms of the interference of six one-electron states, thus allowing us extend the concept of chemical bond as resulting not from pairing of electrons but from the interference of degenerate or near-degenerate one-electron states.

Acknowledgments

The authors acknowledge CNPq, FAPERJ and Instituto do Milenio de Materiais Complexos for financial support.

References

[1]G.N. Lewis, J. Am. Chem. Soc. **38**, 762 (1916)
[2]W. Heitler and F. London, Z. Phys. **44**, 445 (1927)
[3]G. Rumer, Göttingen Nach. 337 (1932)
[4]L. Pauling, Proc. Natl. Acad. Sci. USA **18**, 293 (1932)
[5]L. Pauling and G.W. Wheland, J. Chem. Phys. **1**, 362 (1933)
[6]L. Pauling and J. Sherman, J. Chem. Phys. **1**, 606 (1933)
[7]H. Eyring, J. Am. Chem. Soc. **54**, 3191 (1932)
[8]J. C. Slater, Phys. Rev. **37**, 481 (1931) ; **38**, 1109 (1931)
[9]L. Pauling, J. Am. Chem. Soc. **53**, 1367 (1931) ; **53**, 3225 (1931)
[10]F. Hund, Z. Phys. **51**, 759 (1928)
[11]F. Hund, Angew. Chem. Intl. Ed. Engl. **16**, 87 (1977)
[12]R.S. Mulliken, Phys. Rev. **32**, 761 (1928)
[13]E. Hückel, Z. Phys. **60**, 204 (1931) ; **72**, 310 (1932) ; **76**, 628 (1932)
[14]C.A. Russell, *The History of Valency* (University, Leicester, 1971)
[15]J. Weyer, Angew. Chem. Intl. Ed. Engl. **13**, 591 (1974)
[16]K. Hafner, Angew. Chem. Intl. Ed. Engl. **18**, 641 (1979)
[17]L. Pauling, J. Mol. Sci. **2**, 1 (1981)
[18]E. Frankland, Philos. Trans. **142**, 440 (1852)
[19]A. S. Couper, CR Acad. Sci. **46**, 1157 (1858) ; Philos. Mag. **16**, 104 (1858)
[20] A. M. Butlerov, Z. Chem. **4**, 549 (1861)
[21]A. Kekule, Liebigs Ann. Chem. **106**, 129 (1858)
[22]D. R. Hartree, Proc. Camb. Phil. Soc. **24**, 328 (1928)
[23]V. Fock, Z. Phys. **15**, 126 (1930)
[24]J. C. Slater, Phys. Rev. **35**, 509 (1930) ; **41**, 255 (1932)
[25]W. Heisenberg, Z. Phys. **38**, 411 (1926)

[26]W. Heisenberg, Z. Phys. **39**, 501 (1926)
[27]E. Wigner, Z. Phys. **40**, 883 (1926)
[28]W. Pauli, Z. Phys. **31**, 765 (1925)
[29]J. H. van Vleck and A. Sherman, Rev. Mod. Phys. **7**, 167 (1935)
[30]R. Pauncz, *Spin Eigenfunctions* (Plenum, New York, 1979)
[31]H. Weyl, *The Theory of Groups and Quantum Mechanics* (Dover, New York, 1930)
[32]R. McWeeny, *Methods of Molecular Quantum Mechanics* (Academic Press, San Diego, 1989)
[33]O. Sinanoğlu, Adv. Chem. Phys. **6**, 321 (1964)
[34]H. J. Silverstone and O. Sinanoğlu, J. Chem. Phys. **44**, 1899 (1966)
[35]H. J. Silverstone and O. Sinanoğlu, J. Chem. Phys. **44**, 3608 (1966)
[36]O. Sinanoğlu, Adv. Chem. Phys. **14**, 237 (1969)
[37]B. O. Ross, Adv. Chem. Phys. **69**, 399 (1987)
[38]D. K. W. Mok, R. Neuman and N. C. Handy, J. Phys. Chem. **100**, 6225 (1996)
[39]E. Valderrama, E. V. Ludeña and J. Hinze, J. Chem. Phys. **106**, 9227 (1997)
[40]E. Valderrama, E. V. Ludeña and J. Hinze, J. Chem. Phys. **110**, 2343 (1999)
[41]D. J. Thouless, *The Quantum Mechanics of Many-Body Systems* (Academic Press, New York, 1961)
[42]J. Čízěk and J. Paldus, J. Chem. Phys. **47**, 3976 (1967)
[43]J. Paldus and J. Čízěk, J. Chem. Phys. **52**, 2119 (1970)
[44]R. Seeger and J. A. Pople, J. Chem. Phys. **66**, 3945 (1977)
[45]D. Dehareng and G. Dive, J. Comp. Chem. **21**, 483 (2000)
[46]J. Olsen, D. L. Yeager and P. Jorgensen, Adv. Chem. Phys. **54**, 1 (1983)
[47] P. O. Löwdin, Rev. Mod. Phys. **35**, 496 (1963)
[48]P. O. Löwdin, Adv. Chem. Phys., **14**, 285 (1969)
[49]J. Schütt and M. C. Böhm, J. Am. Chem. Soc. **114**, 7252 (1992)
[50]L. Pauling, *The Nature of the Chemical Bond* (Cornell, Ithaca, 1939)
[51]C. A. Coulson and I. Fischer, Phil. Mag. **40**, 383 (1949)
[52]W. A. Goddard III, Phys. Rev. **157**, 73 (1967); **157**, 81 (1967)
[53]J. Gerratt and W. N. Lipscomb, Proc. Nat. Acad. Sci. **59**, 332 (1968)
[54]W. A. Goddard, T. H. Dunning, W. J. Hunt, J. P. Hay, Acc. Chem. Res. **6**, 368 (1973)
[55]E. Wigner, Phys. Rev. **46**, 1002 (1934)
[56]P. O. Löwdin, Adv. Chem. Phys. **2**, 207 (1959)
[57]P. O. Löwdin in *Quantum Theory of Atoms, Molecules and the Solid State – A Tribute to John C. Slater*, P. O. Löwdin ed., (Academic Press, New York, 1966), p.601
[58]R. McWeeny, *Symmetry* (Pergamon Press, Oxford, 1963)
[59]P. O. Löwdin, Rev. Mod. Phys. **39**, 259 (1967)

[60]C. D. H. Chisholm, *Group Theoretical Techniques in Quantum Chemistry* (Academic Press, London, 1976)
[61]A. D. Liehr, J. Phys. Chem. **67**, 389 (1963)
[62]W. Eisfeld and K. Morokuma, J. Chem. Phys. **113**, 5587 (2000)
[63]D. J. Klein, C. H. Carlisle and F. A. Matsen, Adv. Quantum Chem. **5**, 219 (1970)
[64]X. Li and J. Paldus, J. Math. Chem. **13**, 273 (1993)
[65]S. L. Altmann and P. Herzig, *Point Group Theory Tables* (Clarendon Press, Oxford, 1994)
[66]E. C. Silva, D. L. Cooper, J. Gerratt and M. Raimondi, J. Chem. Phys. **101**, 3866 (1994)
[67]E. Hollauer and M.A.C. Nascimento, Chem. Phys. Lett. **184**, 470 (1991)
[68]E. Hollauer and M.A.C. Nascimento, J. Chem. Phys. **99**, 1207 (1993)
[69]M.A.C. Nascimento in *Quantum Systems in Chemistry and Physics,* R. McWeeny, S. Wilson and J. Marvani (editors) (Kluwer, Amsterdam, 1997), p. 87
[70]A.G.H. Barbosa and M.A.C. Nascimento in *Valence Bond Theory*, D. L. Cooper ed. (Elsevier, Amsterdam, 2002), p. 117
[71]C.E. Bielschowsky, E. Hollauer and M.A.C. Nascimento, Phys. Rev. **A45,** 7942 (1992)
[72]M.P. de Miranda, C.E. Bielschowsky and M.A.C. Nascimento, J. Phys. **B28**, L15 (1995)
[73]J.D. Mottta Neto and M.A.C. Nascimento, J. Phys. Chem. **100**, 15105 (1996)
[74]K. Ruedenderg, Rev. Mod. Phys. **34**, 326 (1962)
[75]W. A. Goddard III and C.W. Wilson Jr., Theor. Chim. Acta **26**, 211 (1972)
[76]W. Kutzelnigg, Theor. Chim. Acta **43**, 1 (1976)
[77]R. P. Feynman, *Quantum Electrodynamics* (Benjamin, Reading, 1973)
[78]R.D. Mattuck, *A Guide to Feynman Diagrams in the Many-Body Problem* (McGraw-Hill, New York, 1967)
[79]R.S. Mulliken, Phys. Rev. **33**, 730 (1929)
[80]A.G.H. Barbosa (Ph.D. Thesis, Instituto de Química da UFRJ, 2002)
[81]K. Siegbahn, *ESCA Applied to Free Molecules* (North-Holland, Amsterdam, 1969)
[82]R.N.S. Sodhi and C.E. Brion, J. Elecron Spectrosc. **34**, 363 (1984)
[83]C. Camilloni, E. Fainelli, G. Petrocelli and G. Stefani, J. Phys. **B20**, 1839 (1987)
[84]J.E. Bartness and R.J. McIver Jr in *Gas Phase Ion Chemistry* (Academic Press, New York, 1979), Vol 2, p.88

BRILLOUIN-WIGNER EXPANSIONS IN QUANTUM CHEMISTRY

A robust approach to the quantum many-body problem in molecules

I. Hubač

Department of Chemical Physics,

Faculty of Mathematics and Physics,

Comenius University,

842 15 Bratislava, Slovakia

hubac@fmph.uniba.sk

S. Wilson

Rutherford Appleton Laboratory,

Chilton, Oxfordshire OX11 0QX, England.

S.Wilson@rl.ac.uk

Abstract The applicability of the Brillouin-Wigner perturbation expansion to many-body systems is critically examined for the both the single reference and the multireference formulations. The partitioning technique is used to gain insight into the Brillouin-Wigner expansion and its relation with the more familiar and more widely used Rayleigh-Schrödinger perturbation theory. From the partitioned hamiltonian matrix, a matrix Lippmann-Schwinger-*like* equation is obtained in Brillouin-Wigner form. By comparing this with the corresponding Rayleigh-Schrödinger-*based* equation, an *a posteriori* correction, which attempts to restore linear scaling of properties with particle number, is defined and analyzed. It is shown that the Brillouin-Wigner expansion for many-body quantum systems forms the basis of a robust approach to the correlation problem in cases where a multireference function is required. The relativistic Brillouin-Wigner expansion is briefly considered.

E.J. Brändas and E.S. Kryachko (eds.),
Fundamental World of Quatum Chemistry, Vol. I, 407–432.

Keywords: Brillouin-Wigner perturbation theory; many-body perturbation theory; Rayleigh-Schrödinger perturbation theory; the partitioning technique; electron correlation; multi-reference functions; Lippmann-Schwinger-*like* equation; *a posteriori* correction of correlation energy

Introduction

Brillouin-Wigner perturbation theory is often dismissed as a tenable approach to the many-body problem that arises in the study of the electronic structure of atoms and molecules (see, for example, the monographs by March, Young and Sampanthar (1) or Wilson (2)). However, even when first proposed in the 1930s by Lennard-Jones (3), by Brillouin (4) and by Wigner (5), it was recognized that Brillouin-Wigner procedure for the approximation of the solutions of the time-independent electronic Schrödinger equation has certain advantages over the alternative Rayleigh-Schrödinger perturbation expansion when treating systems involving some degree of degeneracy.

In more recent years, the difficulties experienced in developing a robust approach to the problem of describing electron correlation effects in atomic and molecular systems which require the use of a multi-configuration reference function have led to a renewal of interest in the application of the Brillouin-Wigner expansion to many-body systems (6) (see also, for example, the short review by Hubač, Mach and Wilson (7) published in 2001). Most many-body theories are based on the Rayleigh-Schrödinger perturbation expansion (1), (2), (8) (9). Indeed, (Rayleigh-Schrödinger) many-body perturbation theory in its Møller-Plesset(10) form was shown to be the most widely used approach to the electron correlation problem in molecules in a review published in 2000 (8). Many-body, multi-reference theories based on the Rayleigh-Schrödinger expansion are plagued by what is known as the intruder state problem - a problem caused by overlap of the set of energy levels in the reference space with those in the complementary space as the perturbation is "switched on". In the Brillouin-Wigner perturbation theory, the reference energy is the "fully-dressed" energy, ensuring a gap between the reference space and it orthogonal complement, and thus rapid convergence of the expansion. However, Brillouin-Wigner perturbation theory, as it is usually formulated cannot be regarded as a many-body theory; that is, a theory in which the energy and other expectation values scale linearly with the number of electrons in the system. This linear scaling with particle number is important in studies carried out within the framework of non-relativistic quantum mechanics and the terms "size extensivity" and "size consistency" have been coined to describe theories satisfying this requirement. Linear scaling with particle number is mandatory in fully relativistic formulations for which even the hydrogen atom constitutes an "infinitely many-bodied"problem (11), (12).

The partitioning technique, which was extensively studied by in a pioneering series of papers (13)-(25) by the late Professor P.-O. Löwdin, provides a basis for the study of perturbation expansions within a unified framework. It provides considerable insight into the relation between the two principal perturbation expansions, the widely used Rayleigh-Schrödinger and the less familiar Brillouin-Wigner theories. The partitioning technique also suggests a variety of generalizations, including the generalized Brillouin-Wigner perturbation expansion (26),(27). In this work, we use the partitioning technique to re-examine the relation between different perturbation expansions paying particular attention to the use of *a posteriori* corrections to the Brillouin-Wigner expansion which attempt to recover a valid many-body formalism and linear scaling with particle number.

In section 1, we consider a number of aspects of the Brillouin-Wigner perturbation expansions which are probably less familiar than the more widely used Rayleigh-Schrödinger expansion. We describe the Brillouin-Wigner perturbation theory in its single reference formulation in subsection 1.1. and multireference formulation in subsection 1.2. The partitioning technique, which was developed extensively by Löwdin (13)-(25) is employed in section 1.3 to provide a different perspective on the Brillouin-Wigner expansion. In subsection 1.4, we make a comparison of the Rayleigh-Schrödinger perturbation expansion and the Brillouin-Wigner theory. In section 2, we turn our attention to Brillouin-Wigner perturbation theory and the many-body problem. Our discussion is divided into two sections:- in subsection 2.1 we consider the application of the Brillouin-Wigner expansion to many-body theories, then, in subsection 2.2, we describe in more detail the development of approximate many-body theories by means of *a posteriori* corrections to Brillouin-Wigner-based expansions. Section 3 contains our concluding remarks.

1. The Brillouin-Wigner perturbation expansion

The theory, which is today called "Brillouin-Wigner perturbation theory", was introduced in three seminal papers, which were published in the 1930s. The first of these was published in 1930 by Lennard-Jones (3) is often overlooked, although some authors, such as, for example, Dalgarno (28) and Wilson (2) refer to the method as "Lennard-Jones-Brillouin-Wigner perturbation theory". Subsequently, Brillouin published his well known contribution in 1932 (4) and Wigner's paper, which emphasized the utility of the method for systems involving degeneracy, appeared three years later (5).

Brillouin-Wigner perturbation theory has a number of advantages over the more widely used Rayleigh-Schrödinger perturbation theory. It also has significant disadvantages which have resulted in its neglect as a foundation for many-body theoryies over the past fifty years. Before describing these advantages and disadvantages let us first recall the basic structure of the Brillouin-Wigner perturbation theory first for a single reference function and then for a multireference function.

1.1 Single-reference Brillouin-Wigner perturbation theory

The time-independent Schrödinger equation can be written

$$\mathcal{H}\Psi_\mu = \mathcal{E}_\mu \Psi_\mu, \quad \mu = 0, 1, 2, ... \tag{1.1}$$

where $\mathcal{H}$ is the hamiltonian operator and Ψ_μ and $\mathcal{E}_\mu$ are the eigenfunctions and eigenvalues, respectively, for the μth state. Putting

$$\mathcal{H} = \mathcal{H}_0 + \lambda \mathcal{H}_1 \tag{1.2}$$

where $\mathcal{H}_0$ is some zero-order hamiltonian, $\mathcal{H}_1$ is a perturbation and λ is a perturbation parameter which allows the perturbation to be "switched on", allows the Schrödinger equation (1.1) to be written in the form

$$\left(\mathcal{E}_\mu - \mathcal{H}_0\right) \Psi_\mu = \lambda \mathcal{H}_1 \Psi_\mu \tag{1.3}$$

which can then be rewritten in the matrix form

$$\left(\mathcal{E}_\mu - \varepsilon_m\right) \langle \Phi_m | \Psi_\mu \rangle = \lambda \langle \Phi_m | \mathcal{H}_1 | \Psi_\mu \rangle \tag{1.4}$$

or

$$\langle \Phi_m | \Psi_\mu \rangle = \frac{\lambda \langle \Phi_m | \mathcal{H}_1 | \Psi_\mu \rangle}{\left(\mathcal{E}_\mu - \varepsilon_m\right)} \tag{1.5}$$

In equations (1.4) and (1.5), the state $|\Phi_m\rangle$ is the unperturbed eigenket of $\mathcal{H}_0$ with eigenenergy ε_m; that is

$$\mathcal{H}_0 |\Phi_m\rangle = \varepsilon_m |\Phi_m\rangle, m = 0, 1, 2, ... \tag{1.6}$$

We find it convenient to put

$$\langle \Phi_\mu | \Psi_\mu \rangle = 1 \tag{1.7}$$

in order to simplify the following formulae. Thus, $|\Psi_\mu\rangle$ is not normalized to unity. Equation (1.7) is termed the "intermediate normalization condition". Setting $\mu = m$ in equation (1.4) and using equation (1.7), the exact energy eigenvalue is given by

$$\mathcal{E}_\mu = \varepsilon_\mu + \lambda \langle \Phi_\mu | \mathcal{H}_1 | \Psi_\mu \rangle \tag{1.8}$$

The eigenstate is constructed using the completeness of the unperturbed basis, $|\Phi_m\rangle$, so that

$$|\Psi_\mu\rangle = \sum_m |\Phi_m\rangle \langle \Phi_m | \Psi_\mu \rangle \tag{1.9}$$

or

$$|\Psi_\mu\rangle = |\Phi_\mu\rangle \langle \Phi_\mu | \Psi_\mu \rangle + \sum_m{}' |\Phi_m\rangle \langle \Phi_m | \Psi_\mu \rangle \tag{1.10}$$

where the prime on the summation excludes the case $m = \mu$. Using equation (1.5) for $\langle \Phi_m | \Psi_\mu \rangle$ and equation (1.7), we can write equation (1.10) as

$$|\Psi_\mu\rangle = |\Phi_\mu\rangle + \lambda \sum_m{}' |\Phi_m\rangle \frac{1}{(\mathcal{E}_\mu - \varepsilon_m)} \langle \Phi_m | \mathcal{H}_1 | \Psi_\mu \rangle \tag{1.11}$$

where the prime again indicates that the case $m = \mu$ is exluded from the summation. Equation (1.11) is the basic formula of the Brillouin Wigner perturbation theory for a single reference function. From equation (1.11) we can develop a series expansion of $|\Psi_\mu\rangle$ in powers of λ with coefficients depending on the perturbed energy $\mathcal{E}_\mu$ rather on ε_m as would be the case in the more familiar Rayleigh-Schrödinger perturbation series. Iterating this basic formula, we find

$$\begin{aligned} |\Psi_\mu\rangle \;=\; & |\Phi_\mu\rangle + \lambda \sum_m{}' |\Phi_m\rangle \frac{1}{(\mathcal{E}_\mu - \varepsilon_m)} \langle \Phi_m | \mathcal{H}_1 | \Phi_\mu \rangle \\ & + \lambda^2 \sum_{j,m}{}' |\Phi_j\rangle \frac{1}{(\mathcal{E}_\mu - \varepsilon_j)} \langle \Phi_j | \mathcal{H}_1 | \Phi_m \rangle \frac{1}{(\mathcal{E}_\mu - \varepsilon_m)} \langle \Phi_m | \mathcal{H}_1 | \Phi_\mu \rangle \\ & + \ldots \end{aligned} \tag{1.12}$$

(The above procedure resembles the Lippmann-Schwinger method (30) in constructing the incoming scattering state.)

By introducing the Brillouin-Wigner type propagator (resolvent)

$$\mathcal{B}_\mu = \sum_{m \neq \mu}{}' |\Phi_m\rangle \frac{1}{(\mathcal{E}_\mu - \varepsilon_m)} \langle \Phi_m | \tag{1.13}$$

equation (1.12) can be written in the form

$$|\Psi_\mu\rangle = \left(1 + \lambda \mathcal{B}_\mu \mathcal{H}_1 + \lambda^2 \mathcal{B}_\mu \mathcal{H}_1 \mathcal{B}_\mu \mathcal{H}_1 + \ldots\right) |\Phi_\mu\rangle \tag{1.14}$$

Substituting the expansion (1.12) for the exact wave function into equation (1.8) for the exact energy eigenvalue, the perturbed energy values are given by

$$\begin{aligned}\mathcal{E}_\mu &= \varepsilon_\mu + \lambda \langle \Phi_\mu | \mathcal{H}_1 | \Phi_\mu \rangle \\ &+\lambda^2 \sum_m{}' \langle \Phi_\mu | \mathcal{H}_1 | \Phi_m \rangle \frac{1}{(\mathcal{E}_\mu - \varepsilon_m)} \langle \Phi_m | \mathcal{H}_1 | \Phi_\mu \rangle \\ &+\lambda^3 \sum_{j,m}{}' \langle \Phi_\mu | \mathcal{H}_1 | \Phi_j \rangle \frac{1}{(\mathcal{E}_\mu - \varepsilon_j)} \langle \Phi_j | \mathcal{H}_1 | \Phi_m \rangle \frac{1}{(\mathcal{E}_\mu - \varepsilon_m)} \\ &\langle \Phi_m | \mathcal{H}_1 | \Phi_\mu \rangle + \ldots \end{aligned} \qquad (1.15)$$

or, in terms of the resolvent (1.13),

$$\begin{aligned}\mathcal{E}_\mu &= \varepsilon_\mu + \lambda \langle \Phi_\mu | \mathcal{H}_1 | \Phi_\mu \rangle + \lambda^2 \langle \Phi_\mu | \mathcal{H}_1 \mathcal{B}_\mu \mathcal{H}_1 | \Phi_\mu \rangle \\ &+\lambda^3 \langle \Phi_\mu | \mathcal{H}_1 \mathcal{B}_\mu \mathcal{H}_1 \mathcal{B}_\mu \mathcal{H}_1 | \Phi_\mu \rangle + \ldots \end{aligned} \qquad (1.16)$$

It should be noted that the expansion (1.16) is not a simple power series expansion in λ since the $\mathcal{E}_\mu$ which appears in the denominators of $\mathcal{B}_\mu$ also depends on λ. However, if the terms $1/(\mathcal{E}_\mu - \varepsilon_m)$ are expanded in powers of λ, then the usual Rayleigh-Schrödinger series for $\mathcal{E}_\mu$ is recovered.

Because of the primed summation, the Brillouin-Wigner series terminates no more than k steps for a finite k-dimensional space. For example, it is easy to see the Brillouin-Wigner expansion to the second order is exact for a two dimensional matrix.

Introducing the wave operator, $\Omega_\mu(\lambda)$, which is defined by the relation

$$|\Psi_\mu\rangle = \hat{\Omega}_\mu(\lambda) |\Phi_\mu\rangle \qquad (1.17)$$

so that the application of the wave operator to the reference function $|\Phi_\mu\rangle$ yields the exact wave function $|\Psi_\mu\rangle$, we can compare equations (1.14) and (1.17) and write

$$\hat{\Omega}_\mu(\lambda) = 1 + \lambda \mathcal{B}_\mu \mathcal{H}_1 + \lambda^2 \mathcal{B}_\mu \mathcal{H}_1 \mathcal{B}_\mu \mathcal{H}_1 + \ldots \qquad (1.18)$$

which can, in turn, be rewritten as the recursion

$$\hat{\Omega}_\mu(\lambda) = 1 + \lambda \mathcal{B}_\mu \mathcal{H}_1 \hat{\Omega}_\mu(\lambda) \qquad (1.19)$$

Equation (1.19) is the Bloch equation (29) in Brillouin-Wigner form.

Equation (1.16) for the exact energy can be rewritten in the form

$$\mathcal{E}_\mu = \varepsilon_\mu + \lambda \left\langle \Phi_\mu \middle| \mathcal{H}_1 \hat{\Omega}_\mu(\lambda) \middle| \Phi_\mu \right\rangle \qquad (1.20)$$

or, defining the reaction operator, $\mathcal{V}_{\mu}(\lambda)$, as

$$\mathcal{V}_{\mu}(\lambda) = \mathcal{H}_1 \hat{\Omega}_{\mu}(\lambda), \tag{1.21}$$

we have

$$\mathcal{E}_{\mu} = \varepsilon_{\mu} + \lambda \langle \Phi_{\mu} | \mathcal{V}_{\mu}(\lambda) | \Phi_{\mu} \rangle . \tag{1.22}$$

Combining the Bloch equation, equation (1.19), with equation (1.21) gives

$$\mathcal{V}_{\mu}(\lambda) = \mathcal{H}_1 + \lambda \mathcal{H}_1 \mathcal{B}_{\mu} \mathcal{V}_{\mu}(\lambda) \tag{1.23}$$

which is a Lippmann-Schwinger-*like* equation (30) in Brillouin-Wigner form. Equations (1.22) and (1.23) together define the exact energy eigenvalue. The solution of these two equations is entirely equivalent to the solution of the Schrödinger equation, equation (1.1).

1.2 Multi-reference Brillouin-Wigner perturbation theory

We turn now to the Brillouin-Wigner perturbation expansion for a system described in zero order by a multireference function. The multireference formalism closely parallels that given in the previous section for the single reference case.

Let

$$\{\Phi_{\nu}; \nu = 0, 1, 2, ..., p-1\} \tag{1.24}$$

be a set of linearly independent functions which are eigenfunctions of the zero-order hamiltonian and which constitute the chosen *reference space*, which we label with $\mathcal{P}$. Here we put aside the problem of how the space $\mathcal{P}$ is actually chosen. This will be determined by the nature of the particular system and the physical processes under investigation. Let P be the projection operator on to this chosen reference space $\mathcal{P}$; that is

$$P = \sum_{\nu \in \mathcal{P}} |\Phi_{\nu}\rangle \langle \Phi_{\nu}| \tag{1.25}$$

and let Q be its orthogonal complement given by

$$Q = \sum_{\nu \notin \mathcal{P}} |\Phi_{\nu}\rangle \langle \Phi_{\nu}| \tag{1.26}$$

The projectors (1.25) and (1.26) satisfy the relations

$$\begin{aligned} P^2 &= P, \\ Q^2 &= Q, \\ PQ &= 0, \\ P + Q &= I. \end{aligned} \tag{1.27}$$

Consider the projection of the exact wave function Ψ_μ onto the reference space $\mathcal{P}$, i.e.

$$\left|\Psi_\mu^{\mathcal{P}}\right\rangle = P\left|\Psi_\mu\right\rangle, \tag{1.28}$$

$\left|\Psi_\mu^{\mathcal{P}}\right\rangle$ is often termed the *model function.* It can be written as a linear combination of the set (1.24)

$$\left|\Psi_\mu^{\mathcal{P}}\right\rangle = \sum_{\nu\in\mathcal{P}} \left|\Phi_\nu\right\rangle C_{\nu\mu} \tag{1.29}$$

The eigenfunctions $\left|\Psi_\mu^{\mathcal{P}}\right\rangle$ are, in general, non-orthogonal, but are assumed to be linearly independent.

As in the single-reference formulation of the theory, the exact eigenstate is constructed using the completeness of the unperturbed basis. For a single state (but a multireference function), equation (1.9) is now written

$$\begin{aligned} \left|\Psi_\mu\right\rangle &= \sum_{\nu\in\mathcal{P}} \left|\Phi_\nu\right\rangle \left\langle\Phi_\nu|\Psi_\mu\right\rangle + \sum_{m\notin\mathcal{P}} \left|\Phi_m\right\rangle \left\langle\Phi_m|\Psi_\mu\right\rangle \\ &= \left|\Psi_\mu^{\mathcal{P}}\right\rangle + \sum_{m\notin\mathcal{P}} \left|\Phi_m\right\rangle \left\langle\Phi_m|\Psi_\mu\right\rangle \end{aligned} \tag{1.30}$$

Using (1.4) and (1.5) we can then write (1.30) in the form

$$\left|\Psi_\mu\right\rangle = \left|\Psi_\mu^{\mathcal{P}}\right\rangle + \lambda \sum_{m\notin\mathcal{P}} \left|\Phi_m\right\rangle \frac{1}{(\mathcal{E}_\mu - \varepsilon_m)} \left\langle\Phi_m\left|\mathcal{H}_1\right|\Psi_\mu\right\rangle \tag{1.31}$$

The exact wave function is then given by the following recursion

$$\left|\Psi_\mu\right\rangle = \left|\Psi_\mu^{\mathcal{P}}\right\rangle + \lambda\mathcal{B}_\mu^{\mathcal{P}}\mathcal{H}_1\left|\Psi_\mu\right\rangle \tag{1.32}$$

in which the resolvent is given by

$$\mathcal{B}_\mu^{\mathcal{P}} = \sum_{m\notin\mathcal{P}} \left|\Phi_m\right\rangle \frac{1}{(\mathcal{E}_\mu - \varepsilon_m)} \left\langle\Phi_m\right| \tag{1.33}$$

From (1.32) we can write

$$\left|\Psi_\mu\right\rangle = \left(1 + \lambda\mathcal{B}_\mu^{\mathcal{P}}\mathcal{H}_1 + \lambda^2\mathcal{B}_\mu^{\mathcal{P}}\mathcal{H}_1\mathcal{B}_\mu^{\mathcal{P}}\mathcal{H}_1 + \ldots\right)\left|\Psi_\mu^{\mathcal{P}}\right\rangle \tag{1.34}$$

Equation (1.34) is analogous to equation (1.14) in the single reference case. Comparing equation (1.34) with the equation defining the wave operator for the multireference case, that is

$$\left|\Psi_\mu\right\rangle = \hat{\Omega}_\mu^{\mathcal{P}}(\lambda)\left|\Psi_\mu^{\mathcal{P}}\right\rangle, \tag{1.35}$$

we can write the wave operator in the form

$$\hat{\Omega}_{\mu}^{\mathcal{P}}(\lambda) = 1 + \lambda \mathcal{B}_{\mu}^{\mathcal{P}} \mathcal{H}_1 + \lambda^2 \mathcal{B}_{\mu}^{\mathcal{P}} \mathcal{H}_1 \mathcal{B}_{\mu}^{\mathcal{P}} \mathcal{H}_1 + ... \qquad (1.36)$$

or

$$\hat{\Omega}_{\mu}^{\mathcal{P}}(\lambda) = 1 + \lambda \mathcal{B}_{\mu}^{\mathcal{P}} \mathcal{H}_1 \hat{\Omega}_{\mu}^{\mathcal{P}}(\lambda) \qquad (1.37)$$

which is the Bloch equation for multireference Brillouin-Wigner theory. The corresponding reaction operator is given by analogy with equation (1.23) by

$$\mathcal{V}_{\mu}^{\mathcal{P}}(\lambda) = \mathcal{H}_1 + \lambda \mathcal{H}_1 \mathcal{B}_{\mu}^{\mathcal{P}} \mathcal{V}_{\mu}^{\mathcal{P}}(\lambda) \qquad (1.38)$$

which is the Lippmann-Schwimger-*like* equation for multireference Brillouin-Wigner theory.

1.3 The Partitioning Technique

The partitioning technique, which was investigated extensively by Löwdin (13)-(25), provides a valuable perspective on the different type of perturbation theory and the relation between the corresponding expansions for the wave function and for the energy. Here we shall concentrate on the approach to the Brillouin-Wigner perturbation theory via the partitioning technique.

We shall proceed immediately to the multi-reference function case since the single reference formalism is just a special case of this. We shall also work with the matrix formulation of the theory since this is the form that will be employed in practical applications.

We divide the zero order functions into the p functions defining the reference space and the complementary space as follows:-

$$\begin{pmatrix} \Phi_m \\ \Phi_n \end{pmatrix} = \begin{pmatrix} \Phi_0 \\ \Phi_1 \\ \cdots \\ \Phi_{p-1} \\ \Phi_p \\ \cdots \\ \Phi_n \end{pmatrix} \qquad (1.39)$$

where the model space is defined by

$$\Phi_m = \begin{pmatrix} \Phi_0 \\ \Phi_1 \\ \cdots \\ \Phi_{p-1} \end{pmatrix} \qquad (1.40)$$

and the complementary space is given by

$$\Phi_n = \begin{pmatrix} \Phi_p \\ \Phi_{p+1} \\ \cdots \\ \Phi_n \end{pmatrix} \tag{1.41}$$

The hamiltonian matrix then takes the form

$$\begin{pmatrix} H_{0,0} & H_{0,1} & \cdots & H_{0,p-1} & H_{0,p} & \cdots & H_{0,n} \\ H_{1,0} & H_{1,1} & \cdots & H_{1,p-1} & H_{1,p} & \cdots & H_{1,n} \\ \cdots & \cdots & \cdots & \cdots & \cdots & \cdots & \cdots \\ H_{p-1,0} & H_{p-1,1} & \cdots & H_{p-1,p-1} & H_{p-1,p} & \cdots & H_{p-1,n} \\ H_{p,0} & H_{p,1} & \cdots & H_{p,p-1} & H_{p,p} & \cdots & H_{p,n} \\ \cdots & \cdots & \cdots & \cdots & \cdots & \cdots & \cdots \\ H_{n,0} & H_{n,1} & \cdots & H_{n,p-1} & H_{n,p} & \cdots & H_{n,n} \end{pmatrix} \tag{1.42}$$

where

$$H_{q,r} = \langle \Phi_q | \mathcal{H} | \Phi_r \rangle \, . \tag{1.43}$$

The square sub-matrix with rows and column labelled by the elements of the reference space is then defined as follows:-

$$\mathbf{H}_{mm} = \begin{pmatrix} H_{0,0} & H_{0,1} & \cdots & H_{0,p-1} \\ H_{1,0} & H_{1,1} & \cdots & H_{1,p-1} \\ \cdots & \cdots & \cdots & \cdots \\ H_{p-1,0} & H_{p-1,1} & \cdots & H_{p-1,p-1} \end{pmatrix} \tag{1.44}$$

The non-square sub-matrix with rows labelled by the elements of the reference space and columns labelled by the elements of the complementary space is defined as follows:-

$$\mathbf{H}_{mn} = \begin{pmatrix} H_{0,p} & \cdots & H_{0,n} \\ H_{1,p} & \cdots & H_{1,n} \\ \cdots & \cdots & \cdots \\ H_{p-1,p} & \cdots & H_{p-1,n} \end{pmatrix} \tag{1.45}$$

whilst that with rows labelled by the elements of the complementary space and columns labelled by the elements of the reference space is defined as follows:-

$$\mathbf{H}_{nm} = \begin{pmatrix} H_{p,0} & H_{p,1} & \cdots & H_{p,p-1} \\ \cdots & \cdots & \cdots & \cdots \\ H_{n,0} & H_{n,1} & \cdots & H_{n,p-1} \end{pmatrix} \tag{1.46}$$

The square sub-matrix with rows and column labelled by the elements of the complementary space is then defined as follows:-

$$\mathbf{H}_{nn} = \begin{pmatrix} H_{p,p} & \dots & H_{p,n} \\ \dots & \dots & \dots \\ H_{n,p} & \dots & H_{n,n} \end{pmatrix} \tag{1.47}$$

The matrix (1.42) may then be written in the form

$$\begin{pmatrix} \mathbf{H}_{mm} & \mathbf{H}_{mn} \\ \mathbf{H}_{nm} & \mathbf{H}_{nn} \end{pmatrix} \tag{1.48}$$

The exact wave function for the μth state can be written as a linear combination of the zero order functions (1.39). The vector of expansion coefficients can be written in the form

$$\begin{pmatrix} \mathbf{C}_m \\ \mathbf{C}_n \end{pmatrix} \tag{1.49}$$

where $\mathbf{C}_m$ corresponds to the reference space and $\mathbf{C}_n$ to the complementary space. The matrix Schrödinger equation for a single state, the μth state, then takes the form

$$\begin{pmatrix} \mathbf{H}_{mm} & \mathbf{H}_{mn} \\ \mathbf{H}_{nm} & \mathbf{H}_{nn} \end{pmatrix} \begin{pmatrix} \mathbf{C}_m \\ \mathbf{C}_n \end{pmatrix} = \mathcal{E}_\mu \begin{pmatrix} \mathbf{C}_m \\ \mathbf{C}_n \end{pmatrix} \tag{1.50}$$

This matrix eigenvalue equation can be written as two simultaneous matrix equations:-

$$\begin{aligned} \mathbf{H}_{mm}\mathbf{C}_m + \mathbf{H}_{mn}\mathbf{C}_n &= \mathcal{E}_\mu \mathbf{C}_m \\ \mathbf{H}_{nm}\mathbf{C}_m + \mathbf{H}_{nn}\mathbf{C}_n &= \mathcal{E}_\mu \mathbf{C}_n \end{aligned}$$

The second of these equations gives

$$\mathbf{H}_{nm}\mathbf{C}_m = (\mathcal{E}_\mu - \mathbf{H}_{nn})\,\mathbf{C}_n \tag{1.51}$$

which can then be rearranged to give

$$\mathbf{C}_n = (\mathcal{E}_\mu - \mathbf{H}_{nn})^{-1}\,\mathbf{H}_{nm}\mathbf{C}_m \tag{1.52}$$

where we have assumed that the inverse matrix exists. Substituting the expression for $\mathbf{C}_n$ given in (1.52) into the first of the simultaneous matrix equations above gives

$$\left[\mathbf{H}_{mm} + \mathbf{H}_{mn}\,(\mathcal{E}_\mu - \mathbf{H}_{nn})^{-1}\,\mathbf{H}_{nm}\right]\mathbf{C}_m = \mathcal{E}_\mu \mathbf{C}_m \tag{1.53}$$

Defining the matrix

$$\mathbf{H}_{mm}^{(\text{effective})} = \mathbf{H}_{mm} + \mathbf{H}_{mn} \left(\mathcal{E}_{\mu} - \mathbf{H}_{nn}\right)^{-1} \mathbf{H}_{nm} \tag{1.54}$$

we can write (1.53) in the form

$$\mathbf{H}_{mm}^{(\text{effective})} \mathbf{C}_{m} = \mathcal{E}_{\mu} \mathbf{C}_{m} \tag{1.55}$$

so that the effective hamiltonian matrix $\mathbf{H}_{mm}^{(\text{effective})}$ with rows and columns labelled by the elements of the reference space nevertheless has the exact energy for the μth state, $\mathcal{E}_{\mu}$, as an eigenvalue. This is the fundamental result of the partitioning technique. The matrix $\mathbf{H}_{mm}^{(\text{effective})}$ has p eigensolutions. The solution with the lowest eigenvalue is taken to correspond to $\mathcal{E}_{\mu}$, the exact energy for the μth state.

Different types of perturbation theory may be obtained by expanding the inverse matrix in (1.54) using the matrix identity

$$(\mathbf{X} - \mathbf{Y})^{-1} = \mathbf{X}^{-1} + \mathbf{X}^{-1}\mathbf{Y}\,(\mathbf{X} - \mathbf{Y})^{-1} \tag{1.56}$$

which can be proved immediately by multiplying from the right by $(\mathbf{X} - \mathbf{Y})$. From (1.56) we can write by iteration

$$(\mathbf{X} - \mathbf{Y})^{-1} = \mathbf{X}^{-1} + \mathbf{X}^{-1}\mathbf{Y}\mathbf{X}^{-1} + \mathbf{X}^{-1}\mathbf{Y}\mathbf{X}^{-1}\mathbf{Y}\mathbf{X}^{-1} + ... \tag{1.57}$$

The particular choice of $\mathbf{X}$ and $\mathbf{Y}$ determines the perturbation expansion.

The Brillouin-Wigner perturbation theory is obtained by writing the inverse matrix in the effective hamiltonian matrix (1.54) in the form

$$\left(\mathcal{E}_{\mu} - \mathbf{H}_{nn}\right)^{-1} = \left(\mathcal{E}_{\mu} - \mathbf{H}_{nn}^{(0)} - \mathbf{H}_{nn}^{(1)}\right)^{-1} \tag{1.58}$$

where $\mathbf{H}_{nn}^{(0)}$ is a matrix of the zero order hamiltonian in the complementary space

$$\langle \Phi_q \,|\mathcal{H}_0|\, \Phi_r \rangle \tag{1.59}$$

and $\mathbf{H}_{nn}^{(1)}$ is the matrix of the perturbation operator in the complementary space

$$\langle \Phi_q \,|\mathcal{H}_1|\, \Phi_r \rangle \tag{1.60}$$

Putting

$$\mathbf{X} = \mathcal{E}_{\mu} - \mathbf{H}_{nn}^{(0)} \tag{1.61}$$

and

$$\mathbf{Y} = \mathbf{H}_{nn}^{(1)} \tag{1.62}$$

equation(1.56) becomes

$$(\mathcal{E}_\mu - \mathbf{H}_{nn})^{-1} = \left(\mathcal{E}_\mu - \mathbf{H}_{nn}^{(0)}\right)^{-1} + \left(\mathcal{E}_\mu - \mathbf{H}_{nn}^{(0)}\right)^{-1} \mathbf{H}_{nn}^{(1)} (\mathcal{E}_\mu - \mathbf{H}_{nn})^{-1} \tag{1.63}$$

and equation (1.57) becomes

$$\begin{aligned}(\mathcal{E}_\mu - \mathbf{H}_{nn})^{-1} &= \left(\mathcal{E}_\mu - \mathbf{H}_{nn}^{(0)}\right)^{-1} + \left(\mathcal{E}_\mu - \mathbf{H}_{nn}^{(0)}\right)^{-1} \mathbf{H}_{nn}^{(1)} \left(\mathcal{E}_\mu - \mathbf{H}_{nn}^{(0)}\right)^{-1} \\ &+ \left(\mathcal{E}_\mu - \mathbf{H}_{nn}^{(0)}\right)^{-1} \mathbf{H}_{nn}^{(1)} \left(\mathcal{E}_\mu - \mathbf{H}_{nn}^{(0)}\right)^{-1} \times \\ &\times \mathbf{H}_{nn}^{(1)} \left(\mathcal{E}_\mu - \mathbf{H}_{nn}^{(0)}\right)^{-1} \\ &+ \ldots\end{aligned} \tag{1.64}$$

The matrix $\mathbf{H}_{nn}^{(0)}$ is diagonal in the basis $|\Phi_q\rangle$. If we put

$$\mathbf{B}_\mu^{\mathcal{P}} = \left(\mathcal{E}_\mu - \mathbf{H}_{nn}^{(0)}\right)^{-1} \tag{1.65}$$

then (1.63) may be written

$$(\mathcal{E}_\mu - \mathbf{H}_{nn})^{-1} = \mathbf{B}_\mu^{\mathcal{P}} + \mathbf{B}_\mu^{\mathcal{P}} \mathbf{H}_{nn}^{(} 1) (\mathcal{E}_\mu - \mathbf{H}_{nn})^{-1} \tag{1.66}$$

and (1.64) takes the form

$$\begin{aligned}(\mathcal{E}_\mu - \mathbf{H}_{nn})^{-1} &= \mathbf{B}_\mu^{\mathcal{P}} + \mathbf{B}_\mu^{\mathcal{P}} \mathbf{H}_{nn}^{(} 1) \mathbf{B}_\mu^{\mathcal{P}} \\ &+ \mathbf{B}_\mu^{\mathcal{P}} \mathbf{H}_{nn}^{(} 1) \mathbf{B}_\mu^{\mathcal{P}} \mathbf{H}_{nn}^{(} 1) \mathbf{B}_\mu^{\mathcal{P}} + \ldots\end{aligned} \tag{1.67}$$

These equations should be compared with equation (1.14) for the single reference function case and (1.35) for the multi-reference function case.

The effective hamiltonian matrix $\mathbf{H}_{mm}^{(\text{effective})}$ defined in equation (1.54) can be written in the form

$$\mathbf{H}_{mm}^{(\text{effective})} = \mathbf{H}_{mm}^{(0)} + \mathbf{V}_{mm} \tag{1.68}$$

where the reaction matrix $\mathbf{V}_{mm}$ is given by

$$\mathbf{V}_{mm} = \mathbf{H}_{mm}^{(1)} + \mathbf{H}_{mn}^{(1)} (\mathcal{E}_\mu - \mathbf{H}_{nn})^{-1} \mathbf{H}_{nm}^{(1)} \tag{1.69}$$

Using (1.66) we can write

$$\mathbf{V}_{mm} = \mathbf{H}_{mm}^{(1)} + \mathbf{H}_{mn}^{(1)} \mathbf{B}_\mu^{\mathcal{P}} \mathbf{V}_{mm} \tag{1.70}$$

which is a Lippmann-Schwinger-*like* equation. The solution of the eigenvalue equation (1.55) with effective hamiltonian matrix given by (1.68) and the reaction matrix given by (1.70) is entirely equivalent to the solution of the Schrödinger equation (1.1).

1.4 Comparison of Brillouin-Wigner perturbation theory and Rayleigh-Schrödinger perturbation theory

We are now in a position to compare the Brillouin-Wigner perturbation expansion with the more familiar Rayleigh-Schrödinger perturbation theory. Brillouin-Wigner perturbation theory has two disadvantages in comparison with the much more widely used perturbation theory of Rayleigh and Schrödinger which are immediately apparent. These are:-

(i) **Brillouin-Wigner perturbation theory is iterative whereas as Rayleigh-Schrödinger perturbation theory is not.**

The Brillouin-Wigner series is iterative, since the exact energy is contained in the denominators arising in the expressions for the energy components and this can only be determined by a self-consistency procedure (unless it is already known!).

Consider, for example, the Brillouin-Wigner perturbation through second order for the ground state ($\mu = 0$):-

$$\mathcal{E} = \langle \Phi_0 | \mathcal{H}_0 | \Phi_0 \rangle + \langle \Phi_0 | \mathcal{H}_1 | \Phi_0 \rangle + \sum_{r \neq 0} \frac{\langle \Phi_0 | \mathcal{H}_1 | \Phi_r \rangle \langle \Phi_r | \mathcal{H}_1 | \Phi_0 \rangle}{\mathcal{E} - E_r} \qquad (1.71)$$

Equation (1.71) is solved iteratively. The right-hand-side of this equation may be written $f^{(2)}(\mathcal{E})$ and equation (1.71) itself can then be put in the form

$$\mathcal{E}_{i+1}^{(2)} = f^{(2)}(\mathcal{E}_i^{(2)}) \qquad (1.72)$$

Taking an initial value $\mathcal{E}_0^{(2)}$ successive application of (1.72) defines a sequence of approximations $\mathcal{E}_i^{(2)}, i = 0, 1, 2, ...$ which, if they are convergent, converge to the second order Brillouin-Wigner energy, which we denote $\mathcal{E}^{(2)}$. In general, of course, $\mathcal{E}^{(2)}$ is not necessarily equal to the exact energy, $\mathcal{E}$.

In third order Brillouin-Wigner perturbation theory, we have

$$\begin{aligned}\mathcal{E} = \langle \Phi_0 | \mathcal{H}_0 | \Phi_0 \rangle + \langle \Phi_0 | \mathcal{H}_1 | \Phi_0 \rangle + \sum_{r \neq 0} \frac{\langle \Phi_0 | \mathcal{H}_1 | \Phi_r \rangle \langle \Phi_r | \mathcal{H}_1 | \Phi_0 \rangle}{\mathcal{E} - E_r} \\ + \sum_{r \neq 0} \sum_{s \neq 0} \frac{\langle \Phi_0 | \mathcal{H}_1 | \Phi_r \rangle \langle \Phi_r | \mathcal{H}_1 | \Phi_s \rangle \langle \Phi_s | \mathcal{H}_1 | \Phi_0 \rangle}{(\mathcal{E} - E_r)(\mathcal{E} - E_s)} \qquad (1.73)\end{aligned}$$

The right-hand-side of this equation may be written $f^{(3)}(\mathcal{E})$ and equation(1.73) itself can then be put in the form

$$\mathcal{E}_{i+1}^{(3)} = f^{(3)}(\mathcal{E}_i^{(3)}) \qquad (1.74)$$

Again, we take an initial value of the energy $\mathcal{E}_0^{(3)}$, say. Successive applications of (1.74) defines a sequence of approximations $\mathcal{E}_i^{(3)}, i = 0, 1, 2, ...$ which, if they are convergent, converge to the third order Brillouin-Wigner energy, which we denote $\mathcal{E}^{(3)}$. Again, in general, $\mathcal{E}^{(3)}$ is not necessarily equal to the exact energy, $\mathcal{E}$.

In contrast to the Brillouin-Wigner approach, the Rayleigh-Schrödinger perturbation theory is manifestly non-iterative since the energy denominators depend only on the eigenvalues of the zero-order hamiltonian.

(ii) **Brillouin-Wigner perturbation theory is not explicitly a many-body theory in that the energy expressions in each order do not scale linearly with particle number.**

This property of the Brillouin-Wigner expansion, which was first made evident in the work of Brueckner (31) in the mid-1950s, is the main reason for the paucity of applications of the method to the atomic and molecular electronic structure problem until recent years. In their treatise on the many-body problem in quantum mechanics, March, Young and Sampanthar write that *"the Brillouin-Wigner form of the [many-body] theory is completely inappropriate"*. They explicitly show that

> *"... the first term $\propto N$ and that all succeeding terms [are] individually negligible compared to the first term, whatever the strength of the perturbation. However, it would be wrong to conclude that the above argument proves that the Brillouin-Wigner series is convergent and that only the first term need by considered. Among other reasons, we know that [the first order term] does not include dynamical correlations between particles and these must be important (on physical grounds) for a strongly interacting system. The situation therefore must be that many ($\sim N$) of the small terms beyond the first are of roughly comparable size, and add up to change the energy/particle by a finite amount. Thus it will be completely misleading to apply many-body perturbation theory in the Brillouin-Wigner form, short of considering an infinite number of terms in the limit of large N."*

In contrast to the Brillouin-Wigner perturbation theory, it is well known that Rayleigh-Schrödinger perturbation theory in its "many-body" form does afford a sound theoretical basis for the description of many-body systems. Again, March, Young and Sampanthar (1) write

> *"the Rayleigh-Schrödinger perturbation theory ... if carefully and systematically used, can yield and energy/particle proportional to N as required, in spite of the appearance of spurious terms proportional to N^2, etc., in any given order. In fact, Brueckner showed that the non-physical terms cancel up to fourth order, and the generalization to all orders was effected by Goldstone."*

They continue

> *"To third order (and to any higher order) one can in fact verify that the terms of the Rayleigh-Schrödinger perturbation expansion either van-*

> *ish or increase as N in the limit N, the non-physical terms proportional to N^2, N^3, etc., all neatly cancelling each other."*

and then conclude

> *"... the Rayleigh-Schrödinger series is more useful than the Brillouin-Wigner form in that it yields an expansion leading to the energy per particle as independent of the size of the system for large N."*

It is for these reasons coupled with the efficiency of the corresponding computer algorithms in low-order that Rayleigh-Schrödinger perturbation theory has been regarded as the method of choice in describing electron correlation effects in atoms and molecules for almost thirty years (8).

Let us now turn to the advantages of the Brillouin-Wigner perturbation theory. Brillouin-Wigner perturbation theory has five main advantages over the Rayleigh-Schrödinger perturbation theory. However, let us first recall that the Rayleigh-Schrödinger perturbation theory for a single reference function ground state corresponds to the following choice for **X** and **Y** in the expansion (1.57):-

$$\mathbf{X} = \varepsilon_0 - \mathbf{H}_{nn}^{(0)} \tag{1.75}$$

and

$$\mathbf{Y} = \mathbf{H}_{nn}^{(1)} - \Delta E \tag{1.76}$$

where

$$\Delta E = \mathcal{E} - \varepsilon_0 \tag{1.77}$$

The Rayleigh-Schrödinger perturbation theory is obtained by expanding ΔE in powers of λ. The advantages of the Brillouin-Wigner perturbation theory are:-

***(i)* Rayleigh-Schrödinger theory can be regarded as a approximation to Brillouin-Wigner theory.**

This is a point made in the original work of Lennard-Jones (3) and emphasized in the review by Dalgarno (28). Consider, for example, the second order energy, $\mathcal{E}^{(2)}$, in the Brillouin-Wigner expansion. This may be written in sum-over-states form as

$$\mathcal{E}^{(2)} = \sum_{r \neq 0} \frac{\langle \Phi_0 | \mathcal{H}_1 | \Phi_r \rangle \langle \Phi_r | \mathcal{H}_1 | \Phi_0 \rangle}{\mathcal{E} - E_r} \tag{1.78}$$

If we make the approximation

$$\mathcal{E} \sim E_0 \tag{1.79}$$

then we are led immediately to the Rayleigh-Schrödinger second order energy component

$$E_0^{(2)} = \sum_{r \neq 0} \frac{\langle \Phi_0 | \mathcal{H}_1 | \Phi_r \rangle \langle \Phi_r | \mathcal{H}_1 | \Phi_0 \rangle}{E_0 - E_r} \tag{1.80}$$

If we write

$$\mathcal{E} = E_0 + \Delta E_0 \tag{1.81}$$

then the denominator in (1.78) becomes

$$\begin{aligned} \frac{1}{\mathcal{E} - E_r} &= \frac{1}{E_0 - E_r + \Delta E_0} \\ &= \frac{1}{E_0 - E_r} + \\ &+ \left[\frac{1}{E_0 - E_r} (-\Delta E_0) \frac{1}{E_0 - E_r + \Delta E_0} \right] \end{aligned} \tag{1.82}$$

and

$$\frac{1}{\mathcal{E} - E_r} \sim \frac{1}{E_0 - E_r} \tag{1.83}$$

if ΔE_0 is small and the second term on the right hand side of (1.82) is negligible.

(ii) **Brillouin-Wigner theory is formally much simpler than the Rayleigh-Schrödinger theory.**

This point is also emphasized in the review by Dalgarno (28) but was also made in the original publication by Wigner (5). The Brillouin-Wigner perturbation series is a simple geometric series whereas in every order beyond second order in the energy the Rayleigh-Schrödinger perturbation theory gives rise to a very much more complicated structure. In every order of the Rayleigh-Schrödinger series there is a principle term of the form

$$\mathcal{E}_{RS}^{(p)} = \langle \Phi_0 | \mathcal{H}_1 (\mathcal{R}\mathcal{H}_1)^{p-1} | \Phi_0 \rangle, \quad p = 1, 2, ... \tag{1.84}$$

where $\mathcal{R}$ is the Rayleigh-Schrödinger resolvent. Equation (1.84) is analogous to the Brillouin-Wigner term $\mathcal{E}^{(p)}$ together with other terms in third and higher orders which are often called 'renormalization' terms. In the Rayleigh-Schrödinger perturbation theory, there is only one 'renormalization' term in third order, four in fourth order, thirteen in fifth order, rising to, for example, 4,861 in tenth order of perturbation (32), (33).

(iii) **Convergence of the Brillouin-Wigner perturbation theory is often more rapid than that of the Rayleigh-Schrödinger theory for a given problem** (3), (28), (4), (34).

It is well known, for example, that energy eigenvalue for a two-state problem is given exactly by second order Brillouin-Wigner perturbation theory. However, the Rayleigh-Schrödinger perturbation expansion, in general, has to be taken to infinite order to solve this simple problem. Specifically, taking a zero order matrix

$$\mathbf{H}_0 = \begin{bmatrix} 0 & 0 \\ 0 & \alpha \end{bmatrix}, \tag{1.85}$$

and a perturbation

$$\mathbf{H}_1 = \begin{bmatrix} 0 & \beta \\ \beta & 0 \end{bmatrix}, \tag{1.86}$$

and then solving the secular problem

$$\begin{vmatrix} -\varepsilon & \beta \\ \beta & \alpha - \varepsilon \end{vmatrix} = 0 \tag{1.87}$$

or

$$\varepsilon^2 - \alpha\varepsilon - \beta^2 = 0 \tag{1.88}$$

gives an exact solution

$$\varepsilon = \frac{1}{2}\left[\alpha \pm \sqrt{\alpha^2 + 4\beta^2}\right]. \tag{1.89}$$

The Brillouin-Wigner perturbation expansion through second order for this problem is

$$\varepsilon = \frac{\beta^2}{\varepsilon - \alpha} \tag{1.90}$$

which is seen to be equivalent to equation (1.88) and thus the exact solution, equation(1.89). For the two-state problem defined above second order Brillouin-Wigner perturbation theory provides an exact result.

The Rayleigh-Schrödinger perturbation series for this problem can be obtained by writing (1.89) in the form

$$\varepsilon = \frac{\alpha}{2}\left[1 \pm \sqrt{1+x}\right], \quad x = \frac{4\beta^2}{\alpha^2} \tag{1.91}$$

and using the identity

$$\sqrt{1+x} = 1 + \frac{1}{2}x - \frac{1}{8}x^2 + \frac{1}{16}x^3 - \frac{5}{128}x^4 + \frac{7}{256}x^5 - \frac{21}{1026}x^6 ... \tag{1.92}$$

The Rayleigh-Schrödinger expansion therefore has the form

$$\varepsilon = -\frac{\beta^2}{\alpha} + \frac{\beta^4}{\alpha^3} - 2\frac{\beta^6}{\alpha^5} + ... \tag{1.93}$$

and summation to all orders is required to obtained the exact energy eigenvalue, assuming convergence.

(iv) **Brillouin-Wigner perturbation theory may converge for problems for which the Rayleigh-Schrödinger theory does not** (28), (5).

An example is again provided by the two-state problem defined by (1.85) and (1.86). The Brillouin-Wigner perturbation expansion leads to an exact solution by second order irrespective of the particular values of the parameters defining the zero-order problem and the perturbation, i.e. α and β. For this problem the Brillouin-Wigner perturbation expansion has an infinite radius of convergence. Let us explicitly introduce the perturbation parameter, λ, and determine the radius of convergence of the Rayleigh-Schrödinger perturbation expansion. We write the hamiltonian matrix for the perturbed problem as

$$\mathbf{H}_0 + \lambda \mathbf{H}_1 = \begin{bmatrix} 0 & 0 \\ 0 & \alpha \end{bmatrix} + \lambda \begin{bmatrix} 0 & \beta \\ \beta & 0 \end{bmatrix} \tag{1.94}$$

which has exact solutions

$$\varepsilon = \frac{1}{2}\left[\alpha \pm \sqrt{(\alpha^2 + 4\beta^2\lambda^2)}\right] \tag{1.95}$$

The Rayleigh-Schrödinger perturbation expansion now takes the form

$$\varepsilon = -\frac{\beta^2}{\alpha}\lambda^2 + \frac{\beta^4}{\alpha^3}\lambda^4 - 2\frac{\beta^6}{\alpha^5}\lambda^6 + ... \tag{1.96}$$

The radius of convergence may be obtained directly by setting the discriminant in (1.24) to zero, i.e. $\alpha^2 + 4\beta^2\lambda^2 = 0$. We are thus led to the result that the Rayleigh-Schrödinger perturbation expansion for the two-state model converges for values of λ satisfying

$$|\lambda| > \frac{1}{2}\left|\frac{\alpha}{\beta}\right| \tag{1.97}$$

For values of λ which do not satisfy (1.97) the expansion (1.96) will diverge. Since $\lambda = 1$ corresponds to the fully perturbed system which is of interest, we have the condition

$$\frac{1}{2}\left|\frac{\alpha}{\beta}\right| < 1 \tag{1.98}$$

However, it is important to note that it is usually possible to obtain valuable estimates of the energy eigenvalue by employing Padé approximants in cases where the Rayleigh-Schrödinger perturbation theory diverges because such approximants are able to handle a wider class of functions than the power series usually assumed in perturbation theory and furthermore certain Padé approximants are known to have invariance properties (see, for example, (35)).

(v) **Brillouin-Wigner perturbation theory is often more convenient to use for degenerate problems.**

This is one of the properties of the method which were emphasized by Wigner (5) in his orginal paper. Indeed, as we have seen in sections 1.1 and 1.2, the Brillouin-Wigner perturbation theory can be formally applied to degenerate problems without modification.

For example, setting $\alpha = 0$ in the two-state problem defined by (1.85) and (1.86) above so as to generate a degenerate problem leads immediately from (1.87) and (1.88) to the exact solution $\varepsilon^2 = \beta^2$. The Brillouin-Wigner second order energy (1.90) becomes $\varepsilon = \beta^2/\varepsilon$ which clearly agrees with the exact solution. On the other hand, the non-degenerate Rayleigh-Schrödinger perturbation theory breaks down because the radius of convergence (1.97) is now zero.

2. Brillouin-Wigner perturbation theory and the many-body problem

Recent years have witnessed (6)-(7) a revival of interest in Brillouin-Wigner perturbation theory and its application to the quantum many-body problem principally because it is seen as a possible remedy to the intruder state problem which often plagues multireference Rayleigh-Schrödinger perturbation theory. Wenzel and Steiner (40) write (p. 4714)

> *"... the reference energy in Brillouin-Wigner perturbation theory is the fully dressed energy ... This feature guarantees the existence of a natural gap and thereby rapid convergence of the perturbation series."*

When we consider the application of multireference Brillouin-Wigner methods to many-body systems, two distinct approaches can be emerge:-

(i) the Brillouin-Wigner perturbation theory can be employed to solve the equations associated with an explicitly many-body method. For

(ii) a posteriori corrections can be developed both for the Brillouin-Wigner perturbation theory itself and for other methods which can be implemented via a Brillouin-Wigner-*based* expansion.

We consider these two approaches in turn. However, let us first mention that, although we are concerned here with the treatment of the non-relativistic many-body problem in molecules using Brillouin-Wigner expansions, the relativistic formalism is entirely analogous. Recently, we have demonstrated that a fully relativistic formulation can be developed by describing a four-component Brillouin-Wigner perturbation theory study of a model problem (55).

2.1 Application to many-body theories

March, Young and Sampanthar (1) write

> *"... it will be completely misleading to apply many-body perturbation theory in the Brillouin-Wigner form, short of considering an infinite number of terms in the limit of large N. "*

The Brillouin-Wigner perturbation theory can be employed to solve the equations associated with an explicitly many-body method. For example, the full configuration interaction method, which although in practice can only be carried out within a small basis set so that basis set truncation errors are unacceptable large, is a valid many-body theory. If the equations of full configuration interaction are solved by means of a Brillouin-Wigner perturbation expansion taken through all orders then it is obvious that a valid many-body description of the target system will be obtained.

More importantly, the Brillouin-Wigner perturbation expansion taken through all orders can be applied to other many-body theories. problem can be solved in this Hubač and his coworkers (36)-(47) have explored the use of Brillouin-Wigner perturbation theory in solving the equations of coupled cluster theory (48)-(51). By adopting an exponential expansion for the wave operator the Brillouin-Wigner coupled cluster theory (36)- (39) is obtained which is entirely equivalent to other many-body formulations for the case of a single reference function since the perturbation expansion is summed through all orders. The application of the Brillouin-Wigner coupled cluster approach to the multi-reference function problem has given some useful insights (52)-(53). In recent work, multireference generalized Brillouin-Wigner coupled cluster theory, in particular, has been shown to have considerable potential (26, 27).

2.2 Approximate many-body theories via *a posteriori* corrections

A posteriori corrections can be obtained to the finite order Brillouin-Wigner perturbation theory and to methods such as limited configuration interaction which are not valid many-body theories. These *a posteriori* corrections are based on a very simple idea which we now outline.

Just as the full configuration interaction problem can be solved by making a Brillouin-Wigner perturbation expansion taken through all orders, so the equations of limited configuration interaction can also be solved by means of an all orders Brillouin-Wigner expansion. The resulting theory will not, of course, constitute a valid many-body theory. However, because the limited configuration interaction expansion can be written as a Brillouin-Wigner expansion, *a posteriori* corrections can be introduced (54).

If we compare the Brillouin-Wigner resolvent with the Rayleigh-Schrödinger resolvent, we see that they differ in their denominators. Using identity relation (56)

$$\frac{1}{(\mathcal{E} - E_k)} = \frac{1}{(E_0 - E_k)} + \frac{1}{(E_0 - E_k)} (-\Delta E) \frac{1}{(\mathcal{E} - E_k)} \tag{1.99}$$

where the exact energy, $\mathcal{E}$, is written as

$$\mathcal{E} = E_0 + \Delta E. \tag{1.100}$$

E_0 is the ground state eigenvalue of $\mathcal{H}_0$ and ΔE is the level shift. The term on the left-hand-side of equation (1.99) is the Brillouin-Wigner denominator. The first term on the right-hand-side of equation (1.99) is the Rayleigh-Schrödinger denominator We can obtain an *a posteriori* correction to *any* Brillouin-Wigner series by rewriting (1.99) in the form

$$\frac{1}{(E_0 - E_k)} = \frac{1}{(\mathcal{E} - E_k)} + \frac{1}{(E_0 - E_k)} (\Delta E) \frac{1}{(\mathcal{E} - E_k)}. \tag{1.101}$$

The Rayleigh-Schrödinger denominator is now on the left-hand-side of our equation and the first term on the right-hand-side is the Brillouin-Wigner denominator. If we recall that the Rayleigh-Schrödinger perturbation theory series employing the linked diagram theorem, which ensures that a valid many-body theory is supported, uses factors of $(E_0 - E_k)^{-1}$ as denominators then the second term on right-hand-side of eq.(1.101) can be viewed as a correction term for Brillouin-Wigner series. In the partitioning approach, the exact energy is an eigenvalue of the effective hamiltonian

$$\mathbf{H}_{mm}^{(\text{effective})} = \mathbf{H}_{mm}^{(0)} + \mathbf{V}_{mm} \tag{1.102}$$

where the reaction matrix $\mathbf{V}_{mm}$ is obtained by solving the Lippmann-Schwinger-*like* equation

$$\mathbf{V}_{mm} = \mathbf{H}_{mm}^{(1)} + \mathbf{H}_{mn}^{(1)} \mathbf{B}_{\mu}^{\mathcal{P}} \mathbf{V}_{mm} \tag{1.103}$$

We have emphasized already that the solution of the eigenvalue equation (1.55) with effective hamiltonian matrix given by (1.102) and the reaction matrix given by (1.103) is entirely equivalent to the solution of the Schrödinger equation (1.1). Equation (1.103) is solved iteratively. Using the relation between the Brillouin-Wigner and Rayleigh-Schrödinger denominators (1.101), we introduce an *a posteriori* correction into (1.103) by writing

$$\mathbf{V}_{mm} = \mathbf{H}_{mm}^{(1)} + \mathbf{H}_{mn}^{(1)} \mathbf{B}_{\mu}^{\mathcal{P}} \mathbf{V}_{mm} + [\mathbf{X}_{mm}] \tag{1.104}$$

where the term $[\mathbf{X}_{mm}]$ is added only after the solution of (1.103) have been obtained. $\mathbf{X}_{mm}$ is obtained by evaluating the term corresponding to the second component of (1.101 for each element of the reaction matrix. This simple idea has been used to find *a posteriori* corrections for the limited configuration interaction method (54) in both the single reference and multireference formulations as well as for state specific multireference Brillouin-Wigner coupled cluster theory (52),(53). This approach is general and can be applied to any quantum chemical method which can be formulated as a Brillouin-Wigner perturbation expansion.

3. Concluding remarks

The need for robust many-body methods capable of describing the molecular electronic structure problem in situations where a multireference function is required has stimulated a renewal of interest in the Brillouin-Wigner perturbation expansion. Brillouin-Wigner-*based* methods do not suffer from the intruder state problem which has plagued multireference Rayleigh-Schrödinger-*based* methods for many years.

Brillouin-Wigner perturbation theory was studied extensively in a series of papers by P.-O. Löwdin using the partitioning technique. In this paper, we have employed the partitioning technique to re-examine the Brillouin-Wigner perturbation theory both in its single reference and multireference forms. We have examined its application to the equations of a valid many-body theory such as full configuration interaction or single reference coupled cluster theory and in the development of *a posteriori* corrections which recover "approximate many-body theories "for methods such as limited configuration interaction.

We mention again in closing the generalized Brillouin-Wigner perturbation theory, originally described by Löwdin, which show considerable promise in solving the equations of coupled cluster theory for the multireference case and in obtaining *a posteriori* corrections for any quantum chemical method for which the basic equations can be caste in the form of a Brillouin-Wigner expansion.

Acknowledgments

This work was carried out under the auspices of the EU COST D23/0001/01 programme. SW acknowledges the support of EPSRC under Research Grant GR/M74627.

References

[1] N.H. March, W.H. Young and S. Sampanthar, *The many-body problem in quantum mechanics*, (Cambridge University Press, 1967) Dover Reprint

[2] S. Wilson, *Electron correlation in molecules*, (Clarendon Press, Oxford, 1984)

[3] J.E. Lennard-Jones, Proc. Roy. Soc. (London) **A129**, 598 (1930)

[4] L. Brillouin, J. Phys. Radium **7**, 373 (1932)

[5] E.P. Wigner, Math. Naturwiss. Anz. Ungar. Akad. Wiss **53**, 475 (1935)

[6] I. Hubač and S. Wilson, J. Phys. B: At. Mol. Opt. Phys. **33**, 365 (2000)

[7] I. Hubač, P. Mach and S. Wilson, Adv. Quantum Chem. **39**, 225 (2001)

[8] S. Wilson, in *Chemical Modelling: Applications and Theory*, Senior Reporter: A. Hinchliffe, Specialist Periodical Reports **1**, 364-452 (The Royal Scoiety of Chemistry, London, 2000)

[9] S. Wilson, in *Chemical Modelling: Applications and Theory*, Senior Reporter: A. Hinchliffe, Specialist Periodical Reports **2**, *in press*, (The Royal Scoiety of Chemistry, London, 2002)

[10] C. Møller and M.S. Plesset, *Phys. Rev.*, **46**, 618 (1934)

[11] H.M. Quiney, I.P. Grant and S. Wilson, in *Many-Body Methods in Quantum Chemistry*, Lecture Notes in Chemistry **52**, 307, ed. U. Kaldor, Springer-Verlag (1989)

[12] S. Wilson, J. Molec. Struct. THEOCHEM **547**, 279 (2001)

[13] P.-O. Löwdin, J. Chem. Phys. **19**, 1396 (1951)

[14] P.-O. Löwdin, J. Molec. Spectros. **10**, 12 (1963)

[15] P.-O. Löwdin, J. Molec. Spectros. **13**, 326 (1964)

[16] P.-O. Löwdin, J. Math. Phys. **3**, 969 (1962)

[17] P.-O. Löwdin, J. Math. Phys. **3**, 1171 (1962)

[18] P.-O. Löwdin, J. Molec. Spectros. **14**, 112 (1964)

[19] P.-O. Löwdin, J. Molec. Spectros. **14**, 119 (1964)

[20] P.-O. Löwdin, J. Molec. Spectros. **14**, 131 (1964)

[21] P.-O. Löwdin, J. Math. Phys. **6**, 1341 (1965)

[22] P.-O. Löwdin, Phys. Rev. A **139**, 357 (1965)

[23] P.-O. Löwdin, in *Perturbation Theory and its Application to Quantum Mechanics*, ed. C.H. Wilcox, John Wiley and Sons, New York (1966)

[24] P.-O. Löwdin, Intern. J. Quantum Chem. **2**, 867 (1968)

[25] P.-O. Löwdin, and O. Goscinski, Intern. J. Quantum Chem. **5**, 685 (1971)

[26] I. Hubač and S. Wilson, Adv. Quantum Chem. **39**, 209 (2001)

[27] I. Hubač and S. Wilson, J. Phys. B: At. Mol. Opt. Phys. **34**, 4259 (2001)

[28] A. Dalgarno, in *Quantum Theory. I. Elements*, edited by D.R. Bates, Vol. 1, Chap. 5 (Academic Press, New York, 1961)

[29] C. Bloch, Nucl. Phys. **6**, 329 (1958)

[30] B.P. Lippmann and J. Schwinger, Phys. Rev. **79**, 469 (1950)

[31] K.A. Brueckner, Phys. Rev. **100**, 36 (1955)

[32] S. Wilson, in *Methods in Computational Molecular Physics*, NATO ASI Series B, Vol. 293, p. 203, (Plenum Press, New York, 1992)

[33] S. Wilson, in *Handbook of Molecular Physics and Quantum Chemistry*, **2**, ed. S. Wilson, John Wiley, Chichester (2002)

[34] P.M. Morse and H. Feshbach, *Methods of Theoretical Physics*, Chap. 9.1 (McGraw-Hill, New York, 1953)

[35] S. Wilson, D.M. Silver and R.A. Farrell, Proc. Roy. Soc. (Lond.) **A356**, 363 (1977)

[36] I. Hubač and P. Neogrady, Phys. Rev. **A50**, 4558 (1994)

[37] I. Hubač, in *New Methods in Quantum Theory*, NATO ASI Series, ed. C.A. Tsipis, V.S. Popov, D.R. Herschbach and J.S. Avery, pp. 183, Kluwer, Dordrecht (1996)

[38] J. Mášik and I. Hubač, Coll. Czech. Chem. Commun. **62**, 829 (1997)

[39] J. Mášik and I. Hubač, in *Quantum Systems in Chemistry and Physics: Trends in Methods and Applications*, ed. R. McWeeny, J. Maruani, Y.G. Smeyers and S. Wilson, pp. 283, Kluwer Academic Publishers, Dordrecht (1997)

[40] W. Wenzel and M.M Steiner, J. Chem. Phys. **108**, 4714 (1998)

[41] W. Wenzel, Intern. J. Quantum Chem. **70**, 613 (1998)

[42] J. Mášik and I. Hubač, Adv. Quantum Chem. **31**, 75 (1998)

[43] J. Mášik, P. Mach and I. Hubač, J. Chem. Phys. **108**, 6571 (1998)

[44] P. Mach, J. Mášik, J. Urban and I. Hubač, Molec. Phys. **94**, 173 (1998)

[45] J. Mášik, P. Mach, J. Urban, M. Polasek, P. Babinec and I. Hubač, Collect. Czech. Chem. Comm. **63**, 1213 (1998)

[46] P. Čársky, V. Hrouda, V. Sychrovsky, I. Hubač, P. Babinec, P. Mach, J. Urban and J. Mášik, Collect. Czech. Chem. Comm. **60**, 1419 (1995)

[47] J. Pittner, P. Nechtigall, P. Čársky, J. Mášik and I. Hubač, J. Chem. Phys. **110**, 10275 (1999)

[48] J. Paldus, in *Methods in Computational Molecular Physics*, NATO ASI Series, edited by S. Wilson and G.H.F. Diercksen, p. 99*ff*, Plenum Press, New York (1992)

[49] J. Paldus, in Relativistic and Correlation Effects in Molecules and Solids, NATO ASI Series, edited by G.L. Malli, p. 207, Plenum Press, New York (1994)

[50] *Recent Advances in Coupled Cluster Methods*, Recent Advances in Computational Chemistry, Vol. 3, edited by R.J. Bartlett, World Scientific, Singapore (1997)

[51] J. Paldus, in *Handbook of Molecular Physics and Quantum Chemistry*, **2**, ed. S. Wilson, John Wiley, Chichester (2002)

[52] I. Hubač, J. Pittner, and P. Čársky, J. Chem. Phys. **112**, 8779 (2000)

[53] J. Sancho-García, J. Pittner, P. Čársky, and I. Hubač, J. Chem. Phys. **112**, 8785 (2000)

[54] I. Hubač, P. Mach and S. Wilson, J. Phys. B: At. Mol. Opt. Phys. **33**, 4735 (2000)

[55] H.M. Quiney, I. Hubač and S. Wilson, J. Phys. B: At. Mol. Opt. Phys. **34**, 4323 (2001)

[56] B.H. Brandow, Rev. Mod. Phys. **39**, 771 (1967)

PARTITIONING TECHNIQUES IN COUPLED-CLUSTER THEORY

Steven R. Gwaltney*, Gregory J. O. Beran, and Martin Head-Gordon
Department of Chemistry
University of California
and Chemical Sciences Division
Lawrence Berkeley National Laboratory
Berkeley, CA 94720-1460

1. Introduction

Partitioning techniques have a long history in quantum chemistry. They were first introduced in 1951 by Per-Olov Löwdin, [1] who then developed these ideas in his famous series of papers entitled "Studies in Perturbation Theory." [2, 3, 4, 5, 6, 7, 8, 9, 10, 11, 12] Of most interest to this chapter is Ref. [5], where Löwdin showed how to derive standard many-body perturbation theory (MBPT) using partitioning techniques.

More recently, these partitioning approaches have been applied to coupled-cluster (CC) theory. [13, 14, 15, 16, 17, 18, 19, 20, 21, 22, 23] In particular, partitioning techniques have been applied to the similarity-transformed Hamiltonian $\bar{H}$ which arises in CC theory. At first the goal was to develop simplified methods [24, 25] to equation-of-motion coupled-cluster (EOM-CC) theory for excited states. [26, 27, 28, 29, 30] EOM-CC for excited states is equivalent to CC linear response [31, 32, 33, 34, 35, 36, 37, 38] and is closely related to symmetry-adapted-cluster configuration interaction (SAC-CI). [39, 40]

Next, Stanton and coworkers extended Löwdin's technique for deriving perturbation theory based on a partitioning approach [5] to derive perturbation corrections to CC theories via a partitioning of $\bar{H}$. They first applied this approach to deriving a triple excitation correction [41, 42] to ionization potential (IP-)EOM-CC theory. [43] Stanton also used this approach to provide a new justification [44] for the extremely popular CCSD(T) [45, 46, 47] method. In ad-

*Present address: Department of Chemistry, Mississippi State University, Mississippi State, MS 39762

E.J. Brändas and E.S. Kryachko (eds.),
Fundamental World of Quatum Chemistry, Vol. I, 433–457.

dition, Crawford and Stanton developed an asymmetric CCSD(T) variant [48], which is equivalent to the CCSD(T)$_{\Lambda}$ method of Kucharski and Bartlett. [49] One notable feature of the work of Stanton and coworkers is that, while they would start with a perturbative expansion of $\bar{H}$, when deriving final working equations, they would fall back to a standard MBPT order expansion. Watts and Bartlett used a similar approach in developing a triple excitation correction to EOM-CC for excited states. [50]

In 1998 this group began studying the question of whether it was possible, within a CC framework, to separate dynamical and non-dynamical correlation. [51, 52] The goal was to mimic CASSCF [53] within some subset of orbitals (for example, the set of all valence orbitals). The approach taken was to start with a coupled-cluster doubles (CCD) [18, 19] wave function within a set of active orbitals and vary the orbital coefficients to minimize the CCD energy. This is the optimized-orbital CCD (OO-CCD or just OD) method. [20, 51, 54, 55] By performing valence-orbital OD (VOD) calculation, [52] we capture an approximation to the correlation energy within the active space.

However, the out-of-valence correlation effects remain to be described. Thus, in analogy to the CASPT2 [56, 57] correction to CASSCF, we developed a perturbative correction for OD. Since this was a true second-order correction to the OD or VOD energy, we named the methods OD(2) [58] and VOD(2). [59] To develop this perturbative correction, we followed Stanton's extension [41] of Löwdin's partitioning technique [5] on $\bar{H}$. However, we consistently kept all terms that appear through second order in this approach, which we call similarity-transformed perturbation theory. Subsequently, this approach has been extended to CCSD(2) [60], which is based on CC singles and doubles (CCSD), [20, 61] and QCCD(2) [62]. The quadratic coupled-cluster doubles (QCCD) method [63] is a CC variant designed to handle difficult problems such as multiple-bond breaking.

Hirata *et al.* [64] also followed this approach we have advocated. However, they chose an alternative definition for their denominators. Later, we will discuss precisely what these differences are and what effects they have.

2. Theory

Here we provide a brief overview of Stanton's generalization [41] of Löwdin's partitioning technique [5]. For a more exhaustive development, please see Ref. [59]. However, before we begin, it will be useful to lay out a few definitions. The place to start is with a set of spin-orbitals, which we divide into occupied and virtual sets, based on their occupation in our chosen reference determinant. Further, we subdivide the occupied and virtual spin-orbitals into inactive occupied, active occupied, active virtual, and inactive virtual sets. We will use the letters $i, j, k, l, \ldots$ to refer to occupied spin-orbitals, with I, J, K,

$L, \ldots$ standing for active occupied spin-orbitals, and $i', j', k', l', \ldots$ standing for inactive occupied spin-orbitals. Similarly, $a, b, c, d, \ldots$ will refer to virtual spin-orbitals, with $A, B, C, D, \ldots$ standing for active virtual spin-orbitals, and $a', b', c', d', \ldots$ standing for inactive virtual spin-orbitals. For spin-orbitals of unspecified occupancy we will use $p, q, r, s, \ldots,$ with capitals and primes carrying their same meanings.

Using this notation a CC wave function can be written as

$$|\Psi_{\text{CC}}\rangle = e^T|0\rangle, \tag{1.1}$$

where $|0\rangle$ is the reference determinant, and

$$T = \sum_n \frac{1}{(n!)^2} t_{I_1 I_2 I_3 \cdots I_n}^{A_1 A_2 A_3 \cdots A_n} \{A_1^\dagger I_1 A_2^\dagger I_2 A_3^\dagger I_3 \cdots A_n^\dagger I_n\}. \tag{1.2}$$

In principle, n can be any number up to the number of electrons. The two cases that concern us here are CCD, with n equal to two, and CCSD, with n equal to one and two. The operator T acts to excite electrons out of active occupied orbitals and into active virtual orbitals.

The primary quantity of interest in this chapter is the similarity-transformed Hamiltonian,

$$\bar{H} \equiv e^{-T} H_N e^T = \left(H_N e^T\right)_c \tag{1.3}$$

Here, $H_N = H - \langle 0|H|0\rangle$ is the normal product Hamiltonian. The subscript c in Eq. 1.3 stands for connected. Each T amplitude must share at least one index in common with H. Because of the similarity transformation, $\bar{H}$ is not Hermitian, and it has different right-hand ($\mathcal{R}$) and left-hand ($\mathcal{L}$) eigenvectors. However, the left- and right-hand eigenvalues are the same, and the eigenvectors form a biorthogonal set

$$\langle 0|\mathcal{L}_i \mathcal{R}_j|0\rangle = C_{ij}\delta_{ij}. \tag{1.4}$$

Spin-orbital equations for $\bar{H}$ can be found elsewhere [29, 59, 65] and so are not repeated here.

The next step is to divide the space of all possible n-electron determinants ($|\mathbf{h}\rangle$) which can be formed from our orbitals into a primary space ($|\mathbf{p}\rangle$) and a secondary space ($|\mathbf{q}\rangle$) such that $|\mathbf{p}\rangle \oplus |\mathbf{q}\rangle = |\mathbf{h}\rangle$. The primary space ($|\mathbf{p}\rangle$) will be further divided into $|0\rangle$ plus ($|\mathbf{g}\rangle$). The space ($|\mathbf{g}\rangle$) is defined as those determinants which can be formed when the operator T acts on $|0\rangle$. For CCD this is the space of double excitations from active occupied to active virtual orbitals, while for CCSD this is the space of single and double excitations from active occupied to active virtual orbitals. Using this notation, then, the CC equations can be expressed as

$$\langle 0|\bar{H}|0\rangle = E_{\text{CC}}, \tag{1.5}$$

and

$$\langle \mathbf{g}|\bar{H}|0\rangle = \mathbf{0}. \tag{1.6}$$

The energy E_{CC} is the CC correlation energy. The total energy is $E = E_{CC} + \langle 0|H|0\rangle$.

However, our interests lie in finding approximations to the exact eigenvalues (within a given basis set) E_{exact} of $\bar{H}$, which are the same as the exact eigenvalues of H_N. Therefore, we write the problem of finding the exact eigenvalues of $\bar{H}$ in the matrix representation

$$\begin{bmatrix} \bar{\mathbf{H}}_{\mathbf{pp}} & \bar{\mathbf{H}}_{\mathbf{pq}} \\ \bar{\mathbf{H}}_{\mathbf{qp}} & \bar{\mathbf{H}}_{\mathbf{qq}} \end{bmatrix} \begin{bmatrix} \mathbf{R}_{\mathbf{p}} \\ \mathbf{R}_{\mathbf{q}} \end{bmatrix} = E_{\text{exact}} \begin{bmatrix} \mathbf{R}_{\mathbf{p}} \\ \mathbf{R}_{\mathbf{q}} \end{bmatrix}, \tag{1.7}$$

where $\mathbf{R}_{\mathbf{p}}$ and $\mathbf{R}_{\mathbf{q}}$ are the projections of $\mathcal{R}$ into the $|\mathbf{p}\rangle$ space and $|\mathbf{q}\rangle$ space, respectively. Expanding Eq. (1.7) gives

$$\bar{\mathbf{H}}_{\mathbf{pp}}\mathbf{R}_{\mathbf{p}} + \bar{\mathbf{H}}_{\mathbf{pq}}\mathbf{R}_{\mathbf{q}} = E_{\text{exact}}\mathbf{R}_{\mathbf{p}}, \tag{1.8}$$

$$\bar{\mathbf{H}}_{\mathbf{qp}}\mathbf{R}_{\mathbf{p}} + \bar{\mathbf{H}}_{\mathbf{qq}}\mathbf{R}_{\mathbf{q}} = E_{\text{exact}}\mathbf{R}_{\mathbf{q}}. \tag{1.9}$$

Solving for $\mathbf{R}_{\mathbf{q}}$ in Eq. (1.9) yields

$$\mathbf{R}_{\mathbf{q}} = \left(E_{\text{exact}}\mathbf{1} - \bar{\mathbf{H}}_{\mathbf{qq}}\right)^{-1} \bar{\mathbf{H}}_{\mathbf{qp}}\mathbf{R}_{\mathbf{p}}, \tag{1.10}$$

where $\mathbf{1}$ is a unit matrix of rank($|\mathbf{q}\rangle$). Substituting Eq. (1.10) back into Eq. (1.8) gives

$$\left\{\bar{\mathbf{H}}_{\mathbf{pp}} + \bar{\mathbf{H}}_{\mathbf{pq}}\left(E_{\text{exact}}\mathbf{1} - \bar{\mathbf{H}}_{\mathbf{qq}}\right)^{-1} \bar{\mathbf{H}}_{\mathbf{qp}}\right\}\mathbf{R}_{\mathbf{p}} = E_{\text{exact}}\mathbf{R}_{\mathbf{p}}. \tag{1.11}$$

This equation contains an energy-dependent effective Hamiltonian which acts only in the $|\mathbf{p}\rangle$ space, but whose eigenvalues are still exact. By multiplying on the left by $\mathbf{L}_{\mathbf{p}}$ (the projection of $\mathcal{L}$ onto the $|\mathbf{p}\rangle$ space) and integrating, we get an expression for the energy

$$\begin{aligned} & E_{\text{exact}}\langle 0|\mathbf{L}_{\mathbf{p}}\mathbf{R}_{\mathbf{p}}|0\rangle \\ &= \langle 0|\mathbf{L}_{\mathbf{p}}\left\{\bar{\mathbf{H}}_{\mathbf{pp}} + \bar{\mathbf{H}}_{\mathbf{pq}}\left(E_{\text{exact}}\mathbf{1} - \bar{\mathbf{H}}_{\mathbf{qq}}\right)^{-1} \bar{\mathbf{H}}_{\mathbf{qp}}\right\}\mathbf{R}_{\mathbf{p}}|0\rangle. \end{aligned} \tag{1.12}$$

Unfortunately, this equation is computationally intractable, since it includes the inverse of $\bar{\mathbf{H}}_{\mathbf{qq}}$, and since the energy appears on both sides of the equation.

To develop a perturbation series from this partitioned equation for the energy and to obtain computationally usable expressions, we first need to split $\bar{H}$ into a zeroth-order part $\bar{H}^{[0]}$ and a first-order part $\bar{H}^{[1]}$. The energy will also be split into a zeroth-order part $E^{[0]}$ and higher order parts ($E^{[1]}$, $E^{[2]}$, $E^{[3]}$, ...) referred to collectively as ΔE. Finally, we need to expand $\mathbf{R}_{\mathbf{p}}$ and $\mathbf{L}_{\mathbf{p}}$ as

$$\mathbf{R}_{\mathbf{p}} = \mathbf{R}_{\mathbf{p}}^{[0]} + \mathbf{R}_{\mathbf{p}}^{[1]} + \mathbf{R}_{\mathbf{p}}^{[2]} + \cdots, \tag{1.13}$$

and

$$\mathbf{L_p} = \mathbf{L_p^{[0]}} + \mathbf{L_p^{[1]}} + \mathbf{L_p^{[2]}} + \cdots. \tag{1.14}$$

Our choice of the zeroth-order term for $\bar{H}$ is based on three criteria. The first is that we want our zeroth-order energy and wave function to be the CC energy and wave function. Second, we want our perturbation expansion to have the same orbital-invariance properties as the CC method upon which it is built. Namely, we want this method to be invariant to rotations within the various orbital subspaces. Third, we want any inverse matrices in the (large) $|\mathbf{q}\rangle$ space to be easily obtainable. Therefore, we choose as the zeroth-order piece of the Hamiltonian

$$\bar{H}^{[0]} = |\mathbf{p}\rangle\langle\mathbf{p}|\bar{H}|\mathbf{p}\rangle\langle\mathbf{p}| + |\mathbf{q}\rangle\langle\mathbf{q}|(E^{[0]} + \bar{F}_{i'j'} + \bar{F}_{IJ} + \bar{F}_{AB} + \bar{F}_{a'b'})|\mathbf{q}\rangle\langle\mathbf{q}|, \tag{1.15}$$

where $\bar{F}_{pq}$ represents the one-electron part of $\bar{H}$. In principle, any one-electron operator could be used in the $|\mathbf{q}\rangle$ space, but by choosing the one which arises naturally from $\bar{H}$, we help break the connection between this method and mean-field theory.

Our one other choice is what to use for $\mathbf{R_p^{[0]}}$ and $\mathbf{L_p^{[0]}}$. Since our goal above was for the CC wave function to be the zeroth-order wave function, we will choose the (biorthonormal) eigenvectors of $\bar{\mathbf{H}}_{\mathbf{pp}}$ which correspond to the CC solution. These are $\mathbf{R_p^{[0]}} = 1$ and $\mathbf{L_p^{[0]}} = 1 + \Lambda$. [29] The operator Λ is a deexcitation operator from CC gradient theory [66] which acts to describe the response of the t amplitudes to an external perturbation. Note that with this partitioning of $\bar{H}$, both $\mathbf{R_p^{[1]}}$ and $\mathbf{L_p^{[1]}}$ are zero, since $\bar{H}^{[1]}$ acting on a $|\mathbf{p}\rangle$ space function gives only a $|\mathbf{q}\rangle$ space contribution.

We are now ready to expand the inverse in Eq. (1.12) to give a perturbation series. Doing so yields

$$\begin{aligned}
(E_{\text{exact}}\mathbf{1} - \bar{\mathbf{H}}_{\mathbf{qq}})^{-1} &= \left(E^{[0]}\mathbf{1} + \Delta E\mathbf{1} - \bar{\mathbf{H}}_{\mathbf{qq}}^{[0]} - \bar{\mathbf{H}}_{\mathbf{qq}}^{[1]}\right)^{-1} \\
&= \left\{\left(E^{[0]}\mathbf{1} - \bar{\mathbf{H}}_{\mathbf{qq}}^{[0]}\right)\left[\mathbf{1} - \left(E^{[0]}\mathbf{1} - \bar{\mathbf{H}}_{\mathbf{qq}}^{[0]}\right)^{-1}\left(\bar{\mathbf{H}}_{\mathbf{qq}}^{[1]} - \Delta E\mathbf{1}\right)\right]\right\}^{-1} \\
&= \left(E^{[0]}\mathbf{1} - \bar{\mathbf{H}}_{\mathbf{qq}}^{[0]}\right)^{-1} \\
&\quad + \left(E^{[0]}\mathbf{1} - \bar{\mathbf{H}}_{\mathbf{qq}}^{[0]}\right)^{-1}\left(\bar{\mathbf{H}}_{\mathbf{qq}}^{[1]} - \Delta E\mathbf{1}\right)\left(E^{[0]}\mathbf{1} - \bar{\mathbf{H}}_{\mathbf{qq}}^{[0]}\right)^{-1} + \cdots.
\end{aligned} \tag{1.16}$$

Finally, by substituting Eqs. (1.13,1.14,1.16) into Eq. (1.12), collecting orders, and using our choices for $\mathbf{R_p^{[0]}}$ and $\mathbf{L_p^{[0]}}$ given above, we arrive at the following expressions for the energy. [58, 59] The zeroth-order energy is

$$E^{[0]} = \langle 0|\,(1 + \Lambda)\,\bar{\mathbf{H}}_{\mathbf{pp}}^{[0]}|0\rangle = E_{\text{CC}}. \tag{1.17}$$

This expression for the energy is equivalent to Eq. (1.5) when the t amplitudes are converged and therefore Eq. (1.6) is satisfied. Recovering the CC energy as the zeroth-order energy is consistent, since the CC wave function was chosen as zeroth order. This also means we have a well-defined total energy, in that any perturbation correction is added to the CC energy.

Because $\mathbf{R_p}$ and $\mathbf{L_p}$ contain no first-order parts and because $\bar{H}^{[1]}$ contains none of $\bar{\mathbf{H}}_{\mathbf{pp}}$,

$$E^{[1]} = 0. \tag{1.18}$$

At second order we get the first new correction,

$$E^{[2]} = \langle 0|(1+\Lambda)\bar{\mathbf{H}}^{[1]}_{\mathbf{pq}} \left(E^{[0]}\mathbf{1} - \bar{\mathbf{H}}^{[0]}_{\mathbf{qq}}\right)^{-1} \bar{\mathbf{H}}^{[1]}_{\mathbf{qp}}|0\rangle. \tag{1.19}$$

The $E^{[2]}$ correction can be broken into six terms. [59] The first two come from the 1 in the $1+\Lambda$ and involve single and double excitations, respectively. These are called the MP2 singles and MP2 doubles, since they resemble these terms from non-Hartree–Fock (non-HF) MBPT(2). [67] In fact, in the limit of $|\mathbf{p}\rangle = |0\rangle$, Eq. (1.19) becomes MBPT(2). The other four terms include Λ and have been named Λ singles, Λ doubles, Λ triples, and Λ quadruples. They involve single excitations, double excitations, semi-internal triple excitations, and semi-internal quadruple excitations, respectively. Specifically, the triple excitations can involve no more than two external virtual orbitals, along with one external occupied orbital, or no more than two external occupied orbitals, along with one external virtual. The quadruple excitation can never involve more than two external orbitals of any kind. Spin-orbital equations for these terms can be found in Ref. [59]. Since the perturbative correction contains only linked diagrams, it is fully size extensive.

If all orbitals are active, the doubles terms (and, when using CCSD, the singles terms) are zero, leaving just the triple and quadruple excitation terms. The triples term has the same form as the triples correction for CCSD(T) [45]. However, the triples terms in CCSD(2) and CCSD(T) have three significant differences, and examining them can be informative. The first is that instead of Λ appearing on the left-hand side of the equation for CCSD(2), CCSD(T) uses $T^{\dagger}$. Using Λ on the left side was first tried by Kucharski and Bartlett as CCSD(T)$_{\Lambda}$ [49] and then independently by Crawford and Stanton as a-CCSD(T). [48] The second difference between CCSD(T) and CCSD(2) is using $\bar{\mathbf{H}}^{[1]}_{\mathbf{pq}}$ and $\bar{\mathbf{H}}^{[1]}_{\mathbf{qp}}$ instead of bare integrals. This, along with the appearance of a quadruples term, is the difference between a-CCSD(T) and the CC(2)PT(2) method of Hirata *et al.* [64] The third difference, which is also the distinction between CCSD(2) and CC(2)PT(2), is in the choice of $\bar{F}$ for use in $\bar{\mathbf{H}}^{[0]}_{\mathbf{qq}}$. The other methods use just bare Fock matrix elements.

The quadruples term of CCSD(2) is related to the quadruples term of CC(2)PT(2) in the same way as the CCSD(2) and CC(2)PT(2) triples terms are related, namely the use of $\bar{F}$ versus F in $\bar{\mathbf{H}}^{[0]}_{\mathbf{qq}}$. In addition, these differ from the $E^{[5]}_{Q}(\Lambda)$ term of Kucharski and Bartlett [49] by the use of $\bar{\mathbf{H}}^{[1]}_{\mathbf{pq}}$ and $\bar{\mathbf{H}}^{[1]}_{\mathbf{qp}}$ instead of bare integrals. As Kucharski and Bartlett discuss, the cost of calculating this term scales as $O_a^4 V_a^3 V_t^2$, where O_a is the number of active occupied spin-orbitals, V_a is the number of active virtual spin-orbitals, and V_t is the total number of active plus inactive spin-orbitals. For all but the smallest test systems, this would be impractical.

This problem also arises in MBPT(5). [68] There, it is solved through a factorization trick. However, the factorization does not apply exactly here. Still, we choose to follow Kucharski and Bartlett's lead [49, 69] and force a factorization approximation. The factorization approximation is

$$\begin{aligned}
&\langle 0|\Lambda_2|\mathbf{D}\rangle\langle\mathbf{D}|\bar{\mathbf{H}}^{[1]}_{\mathbf{pq}}|\mathbf{q}\rangle\langle\mathbf{q}|\left(E^{[0]}\mathbf{1}-\bar{\mathbf{H}}^{[0]}_{\mathbf{qq}}\right)^{-1}|\mathbf{q}\rangle\langle\mathbf{q}|\bar{\mathbf{H}}^{[1]}_{\mathbf{qp}}|0\rangle\\
&\approx\frac{1}{2}\langle 0|\bar{\mathbf{H}}^{[1]}_{\mathbf{pq}}|\mathbf{d}\rangle\langle\mathbf{d}|\left(E^{[0]}\mathbf{1}-\bar{\mathbf{H}}^{[0]}_{\mathbf{qq}}\right)^{-1}|\mathbf{d}\rangle\langle\mathbf{d}|\Lambda_2|\mathbf{q}\rangle\langle\mathbf{q}|\bar{\mathbf{H}}^{[1]}_{\mathbf{qp}}|0\rangle,
\end{aligned}\tag{1.20}$$

where $|\mathbf{D}\rangle$ is the space of double excitations among active orbitals, $|\mathbf{d}\rangle$ is the space of all double excitations, and $|\mathbf{q}\rangle$ is the space of quadruple excitations. Equation (1.20) would hold exactly if $\langle 0|\bar{\mathbf{H}}^{[1]}_{\mathbf{pq}}\left(E^{[0]}\mathbf{1}-\bar{\mathbf{H}}^{[0]}_{\mathbf{qq}}\right)^{-1}|\mathbf{d}\rangle$ equalled $\langle 0|\Lambda_2|\mathbf{D}\rangle$. [49] In fact, $\langle 0|\bar{\mathbf{H}}^{[1]}_{\mathbf{pq}}\left(E^{[0]}\mathbf{1}-\bar{\mathbf{H}}^{[0]}_{\mathbf{qq}}\right)^{-1}|\mathbf{D}\rangle$ is the leading (lowest-order) term in the equation for Λ_2 [70]. In tests [49, 69, 71, 72] the factorization approximation changed the energy at the micro-Hartree level. Therefore, we *define* the (2) method to include this approximation.

After the factorization the cost of the quadruples term is reduced to $V_a^5 V_t$. The triples term scales as $O_t O_a^2 V_a^2 V_t^2$, where O_t is the total number of occupied spin-orbitals. When all orbitals are active, the cost of the (2) triples will be twice that of the (T) triples correction. [48] Depending on the basis set, then, when all orbitals are active, the total cost of the (2) correction will be between twice and a few times that of a (T) calculation.

In order to compare our new approach with other perturbative corrections to CCSD, it is helpful to examine through what order of MBPT CCSD(2) is complete. For both HF and non-HF references, CCSD(2) contains all contributions through fourth order. In addition, it includes a quadruples-quadruples term, which is fifth-order with respect to all references, and the singles-triples term that is fifth-order from a HF reference. What is distinctive about the current approach, other than the use of $\bar{H}$ elements throughout the equations, is that the redefinition of order upon which this approach is based brings in contributions from quadruple excitations along with the triples contribution. Although the triples contribution to the energy is often an order of magnitude larger than the quadruples term, the current analysis suggests the presence of the quads is still important for balance.

One issue which has not been discussed yet is the appropriate choice of orbitals. Earlier, we wrote that two of our criteria for choosing $\bar{H}^{[0]}$ were that we wanted the method to be invariant to rotations within the various orbital subspaces and that we wanted any inverse matrices spanning the $|\mathbf{q}\rangle$ space to be easily obtainable. The invariance criterion was met by choosing $\bar{\mathbf{H}}_{\mathbf{pp}}^{[0]}$ to consist of all of $\bar{H}$ and by choosing $\bar{\mathbf{H}}_{\mathbf{qq}}^{[0]}$ to consist of all of $\bar{F}$ within the various orbital subspaces. Now we use that orbital invariance to satisfy the final criterion. The simplest possible form of $\left(E^{[0]}\mathbf{1} - \bar{\mathbf{H}}_{\mathbf{qq}}^{[0]}\right)^{-1}$ to calculate is when $\bar{\mathbf{H}}_{\mathbf{qq}}^{[0]}$ is diagonal. Making $\bar{\mathbf{H}}_{\mathbf{qq}}^{[0]}$ diagonal requires that $\bar{F}_{i'j'}$, $\bar{F}_{IJ}$, $\bar{F}_{AB}$, and $\bar{F}_{a'b'}$ be made diagonal. Therefore, we exploit the orbital invariance built into the theory to rotate the orbitals within each subset in such a way as to diagonalize these blocks of $\bar{F}$. A complication arises from the fact that $\bar{F}$, like $\bar{H}$, is not Hermitian. Hence, it has different left- and right-hand eigenvectors. To make the various blocks of $\bar{F}$ diagonal, then, requires that the bra and ket indices get different transformation matrices. Because of the tensorial nature of the equations, [73, 74] this poses no fundamental difficulties. However, it does add some to the complexity of the implementation. This treating of the bra and ket indices differently does not add as much extra work and storage as it might seem, though, since in the formation of $\bar{H}$, bra-ket symmetry is already broken. [29]

Now that $\bar{F}_{PQ} = \bar{\varepsilon}_P \delta_{PQ}$, and $\bar{F}_{p'q'} = \bar{\varepsilon}_{p'} \delta_{p'q'}$, can we ascribe any physical significance to these new "orbital eigenvalues"? For $\bar{\varepsilon}_{p'}$ this is straightforward, since $\bar{F}_{p'q'} = F_{p'q'}$, [59] the bare Fock operator for both orbitals inactive. Thus, $\bar{\varepsilon}_{p'}$ is just the Koopmans' Theorem [75] approximation to the ionization potential (IP) or the electron affinity (EA) for that orbital.

For $\bar{F}_{PQ}$ the situation is somewhat more complicated. Here, we must look to IP-EOM-CC [43] and EA-EOM-CC theory for guidance. [76] In a standard IP-EOM-CCSD calculation, the final state with one electron removed is expressed as a linear combination of determinants with one electron removed from the reference determinant and determinants with one electron removed and one electron excited from an occupied to a virtual orbital. In EA-EOM-CCSD the final state with an extra electron added to the reference determinant is formed as a linear combination of determinants with one electron added and determinants with one electron added and one excited. Within the basis of these determinants, $\bar{H}$ is diagonalized, and the eigenvalues are the IP or EA state energies.

However, nothing requires that the diagonalization be over these spaces. In particular, if an IP-EOM-CC calculation is performed over the space of only those determinants with one electron removed, then the part of $\bar{H}$ which gets diagonalized is $\bar{F}_{IJ}$. Similarly, an EA-EOM-CC calculation over the space of determinants with one electron added involves diagonalizing $\bar{F}_{AB}$. Thus, the $\bar{\varepsilon}_P$ elements which appear in the denominators can be interpreted as one particle approximations to the correlated IP's and EA's. Hence, the denominators in the similarity-transformed perturbation theory involve improved (or at least more theoretically complete) energies. The importance of this effect will be demonstrated later.

Finally, some extra care must be taken in defining QCCD(2). To the $(1+\Lambda)$ of Eq. 1.17, QCCD adds a $\frac{1}{2}\Lambda^2$ term. [63] The motivation of QCCD was an analysis of the extended CCD (ECCD) approach of Arponen and coworkers [77, 78] and experience with the variational CCD (VCCD) method. [79] What these two methods have in common is the appearance of an exponential operator in place of the linear $(1+\Lambda)$ of standard CC theory. For model bond-breaking problems in small basis sets, VCCD has proven to be extremely reliable. [79] Unfortunately, its factorial scaling [79] eliminates it from ever being used routinely. The n^{10} scaling of ECCD [80] also effectively eliminates it from normal use. One way to derive QCCD is to replace $(1 + \Lambda)$ with e^{Λ}, expand the exponential in a power series, and then truncate it after the quadratic term. The reason for truncating here is that this is as far as it is possible to go and keep the scaling of the method at n^6, specifically V^6 (as opposed to the O^2V^4 scaling of standard CCD and CCSD). [63] Therefore, since QCCD already includes quadruple excitations iteratively (from $\frac{1}{2}\Lambda^2$), it is necessary to exclude from QCCD(2) any contribution from the fully internal quadruple excitations. [62] This avoids overcounting the quadruples correlation energy.

So far, we have discussed CCSD(2) and variants thereof. However, the theory presented here is not limited to these methods. One approach that has been studied is to perform either a CCSD or an OD calculation within a valence space and then to calculate only the singles and doubles parts of the (2) correction. [81] This attempt to mimic a full-space CCSD or OD calculation is primarily motivated by cost considerations, rather than any special theoretical underpinnings. We need not stop there. Any method for which we can define $\bar{H}$ for all of $|\mathbf{h}\rangle$ is a potential starting point for a (2) correction. Note that this disqualifies the quadratic configuration interaction (QCISD) method, [82] since $\bar{H}$ cannot be defined unambiguously outside of the $\mathbf{pp}$ block. Another approach which may show some promise, though, will be to exploit the exponential parameterization of perfect pairing [83, 84, 85] as the base for a PP(2) method. If successful, PP(2) would take the high accuracy work presented here into the realm of very large molecules.

3. Examples

While an elegant theoretical framework is pleasing, ultimately the success or failure of any electronic structure method depends on its numerical performance. To test these methods' performances, we have chosen a series of simple test cases. As an example of breaking one bond, we present F_2 stretch curves. For breaking two bonds simultaneously, we have the double dissociation of H_2O. And for breaking a triple bond, we dissociate N_2. For water [86] and for diatomic nitrogen [87] full configuration interaction (FCI) results are available. The FCI results are exact solutions of the time-independent Schrödinger equation within the chosen basis set. For all three examples we use Dunning's cc-pVDZ basis. [88, 89] Considering how far this basis set is from the basis set limit, comparisons to experimental results are not warranted. The methods presented here have been implemented within the Q-Chem [90] and PSI [91] suites of programs. All calculations presented here were done with Q-Chem, except for some new CCSDT results, for which ACES II [92] was used.

Figure 1.1. Total energies for the dissociation of F_2 in the cc-pVDZ basis.

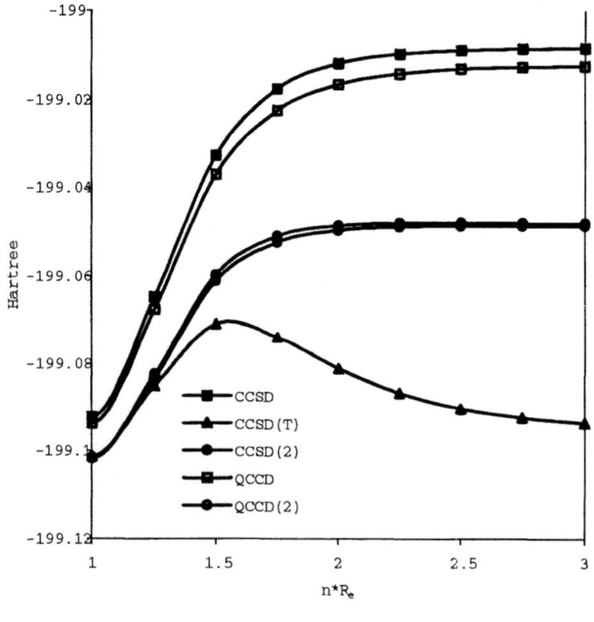

Diatomic fluorine in a cc-pVDZ basis is too large for an FCI calculation to be possible yet. Instead, this example will be used to compare CCSD(2) and QCCD(2) to CCSD(T) in a case where CCSD is qualitatively correct. [93] In Table 1.1 we give the total energies for HF, CCSD, QCCD, CCSD(T), CCSD(2), and QCCD(2) calculated at 1^*R_e, 1.25^*R_e, 1.5^*R_e, 1.75^*R_e, 2^*R_e, 2.25^*R_e, 2.5^*R_e, 2.75^*R_e, and 3^*R_e. For R_e we used 1.41193 Å. [94] The data are plotted in Figure 1.1. Both the CCSD and the QCCD curves are stable, but they overestimate the dissociation energy. The CCSD(T) curve, on the other hand, shows a non-variational collapse by 1.5^*R_e. The CCSD(2) and QCCD(2)

results are quite similar to each other. The largest disagreement between them is 1.4 milliHartrees at 1.5*R_e. Clearly, then, taking the qualitatively correct CC wave function as the starting point for our perturbation correction, instead of using a perturbation correction based on the qualitatively incorrect HF wave function, which is what CCSD(T) does, offers a clear advantage.

Another issue that arises is how to properly handle core orbitals. In Table 1.1 all orbitals were included in the active space. But it is a common practice to exclude core orbitals from the correlation treatment. The typical rationale is that the core orbitals do not change significantly during chemical reactions. However, the core correlation contribution to the total atomization energy of benzene is 7 kcal/mol. [95] Not including this would make it impossible to reach "chemical accuracy", typically defined as 1-2 kcal/mol. Built into the (2) correction is a natural way to account for core correlation effects. If the core orbitals, and, if appropriate, the core virtual orbitals, are treated as inactive, the external and semi-internal parts of (2) correction will automatically provide for some treatment of core-virtual and core-core correlation.

To check the effectiveness of this approach, in Table 1.2 we report the results of CCSD(2) calculations where the core orbitals were excluded entirely from the correlation treatment ("no core") and where the core orbitals were treated as inactive for the (2) calculation ("perturbative core"). Realistically, this is not a terribly demanding example, with only four core electrons and with a very large core-valence separation in the fluorine atom. When the core electrons are completely excluded, the resultant error ranges from 3.54 milliHartrees to 3.67 milliHartrees. Since we are almost always interested only in energy differences, rather than total energies, this result is quite acceptable. Far more impressive is the calculation where the core was treated perturbatively. Here, the errors range from 0.06 milliHartrees to 0.12 milliHartrees. The non-parallelality error (the difference between the smallest and largest errors) is only 0.04 kcal/mol. The extra cost of including the core correlation effects in this manner is quite small. The cost of the most expensive quadruples term, at V^6, does not depend on how the core orbitals are treated. The cost of the most expensive triples term, at $O_t O_a^2 V_a^2 V_t^2$, only grows linearly with the number of core orbitals. Therefore, for routine use it makes the most sense to treat the core orbitals perturbatively in this manner. This even holds for basis sets such as Dunning's cc-pVXZ basis sets [88] that were not designed to treat core-core or core-valence correlation. It has been shown that for the cc-pVQZ basis set, calculations including core correlation give geometries in better agreement with experiment than calculations not including core correlation. [96]

For the water calculations we followed the procedure used by Olsen *et al.* [86] in their FCI study. The geometries are given in Ref. [86], and all electrons were correlated. In these calculations both hydrogens are pulled off simultaneously.

Table 1.1. Total energies for the dissociation of F_2 in the cc-pVDZ basis. Energies are in Hartrees.

	HF	*CCSD*	*QCCD*	*CCSD(T)*	*CCSD(2)*	*QCCD(2)*
1.00*Re	−198.685670	−199.092063	−199.093603	−199.101151	−199.101428	−199.101677
1.25*Re	−198.612171	−199.064661	−199.067480	−199.084940	−199.082309	−199.083135
1.50*Re	−198.527711	−199.032754	−199.036989	−199.070790	−199.059633	−199.061055
1.75*Re	−198.463489	−199.017531	−199.022378	−199.073801	−199.050892	−199.052306
2.00*Re	−198.419839	−199.011839	−199.016554	−199.081058	−199.048526	−199.049610
2.25*Re	−198.391611	−199.009685	−199.014096	−199.086653	−199.047953	−199.048748
2.50*Re	−198.374025	−199.008840	−199.013039	−199.090213	−199.047894	−199.048522
2.75*Re	−198.363063	−199.008508	−199.012604	−199.092300	−199.047969	−199.048518
3.00*Re	−198.355748	−199.008369	−199.012420	−199.093534	−199.048056	−199.048571

Table 1.2. Error relative to all electron CCSD(2) for no core and perturbative core CCSD(2) F_2 calculations. Energy differences are in milliHartrees.

	no core	*perturbative core*
1.00*R_e	3.668	0.120
1.25*R_e	3.620	0.100
1.50*R_e	3.567	0.074
1.75*R_e	3.542	0.062
2.00*R_e	3.540	0.060
2.25*R_e	3.543	0.060
2.50*R_e	3.547	0.062
2.75*R_e	3.549	0.063
3.00*R_e	3.550	0.063

Figure 1.2. Total energies for the double dissociation of H_2O in the cc-pVDZ basis.

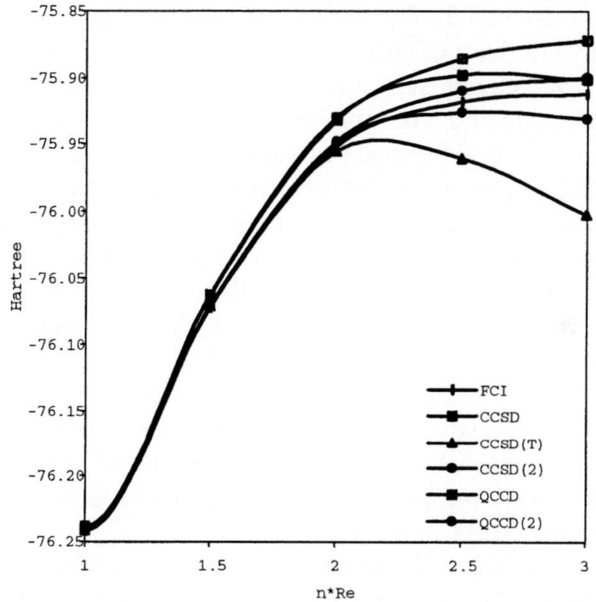

The errors with respect to FCI are given in Table 1.3, and the total energies are shown in Figure 1.2.

The most noticeable result from Table 1.3 is the failure of most of these RHF-based calculations at long bond lengths. The root of this is the incorrect behavior of the RHF reference; it is going to the wrong dissociation limit. This causes CCSD and therefore CC(2)PT(2) and CCSD(2) to break down between 2*R_e and 2.5*R_e. However, CCSD(T), a-CCSD(T), and CCSDT, [97, 98] where the triple excitations are fully included iteratively, begin to break down

Table 1.3. The error with respect to FCI[a] for the double dissociation of water with the cc-pVDZ basis. Errors are in milliHartrees.

	*1.0*Re*	*1.5*Re*	*2.0*Re*	*2.5*Re*	*3.0*Re*
CCSD[a]	3.774	10.043	22.032	20.307	10.849
QCCD[b]	3.202	8.581	19.827	32.696	40.069
CCSD(T)[a]	0.658	1.631	−3.820	−42.564	−90.512
a-CCSD(T)[c]	0.698	2.131	0.698	−21.946	−41.842
CCSDT[a]	0.493	1.423	−1.405	−24.752	−40.126
CC(2)PT(2)	0.269	0.957	1.017	−7.529	−10.584
CCSD(2)[b]	0.401	1.637	3.350	−7.654	−18.640
QCCD(2)[b]	0.349	1.267	3.514	8.490	12.053
CCSDTQ[a]	0.019	0.130	−0.446		

[a] Ref. [86]. [b] Ref. [62]. [c] Ref. [48].

before 2*R_e. Even CCSDTQ [99, 100] appears to be breaking down by 2*R_e. An alternative to these calculations would be to use UHF orbitals with spin and spatial symmetry broken. Size extensive methods based on these orbitals have the advantage of dissociating to the correct energy asymptote, but in the intermediate bond breaking region, around 1.5*R_e to 2*R_e, UHF-based CCSD, CCSDT, CCSD(T), and CCSD(2) actually have a higher energy than their RHF-based brethren. [60, 86] Note that the CC(2)PT(2) results presented here used factorized quadruples, unlike the original definition of the method. [64]

The only two methods which do not suffer non-variational collapse in Table 1.3 are QCCD and QCCD(2). It was precisely this kind of problem that QCCD was designed to handle. [63] For both of these methods the error grows as the bonds lengthen, but neither turns over, which is where the energy starts going down with increased bond distance. What is most encouraging, though, is that the (2) correction stays stable. In fact, in all cases, as long as the underlying CC method is behaving properly, the (2) correction gives a very reasonable approximation to the FCI energy.

Earlier, we discussed the differences between CCSD(2) and the standard CCSD(T). Table 1.4 gives a numerical example of the effects of those differences on the triples contribution to the energy. The concept of a correct triples energy is difficult to define, but as a reference point in the first line we give the difference between CCSDT and CCSD. However, Table 1.3 shows CCSDT becoming much too negative at long bond distances, which implies that this difference should not be the target triples energy.

The triples correction in CCSD(T) grossly overestimates the effects of the triple excitations, especially at long bond distances. The first step in improving the situation is to replace T on the left side of the triples energy expression with Λ, giving a-CCSD(T). [48] This as much as halves the triples estimate,

Table 1.4. Triples and quadruples contributions in the cc-pVDZ double dissociation of water. Energies are in milliHartrees.

	*1.0*Re*	*1.5*Re*	*2.0*Re*	*2.5*Re*	*3.0*Re*
CCSDT triples[a]	−3.281	−8.620	−23.437	−45.059	−50.975
CCSD(T) triples[a]	−3.116	−8.412	−25.852	−62.871	−101.361
a-CCSD(T) triples[b]	−3.076	−7.912	−21.334	−42.253	−52.691
CC(2)PT(2) triples[c]	−2.838	−7.218	−18.227	−36.137	−43.884
CCSD(2) triples[c]	−2.736	−6.741	−16.399	−31.312	−36.826
CC(2)PT(2) quadruples[c]	−0.637	−1.869	−2.788	8.301	22.451
CCSD(2) quadruples[c]	−0.606	−1.666	−2.283	3.351	7.337

[a]Ref. [86]. [b]Ref. [48]. [c]Ref. [60].

returning the behavior to that of CCSDT. Another 8 to 17% reduction in the size of the triples correction comes from replacing the bare integrals in the numerator with $\bar{H}$ elements, giving CC(2)PT(2). [64] This change also necessitates the inclusion of the quadruples term, which is also listed in Table 1.4. The final difference is the inclusion of improved orbital energies in the denominator of CCSD(2), which gives another 4 to 16% decrease in the size of the triples term.

Ironically, at all points the CC(2)PT(2) energy is closer to the FCI energy than CCSD(2) is. The same improved energies in the denominator, which dampen the triples correction, also serve to dampen the quadruples correction, from 5% at 1^*R_e to a whopping 67% at 3^*R_e. Since the quadruples energy becomes positive beyond 2^*R_e, the too-large quadruples correction partially cancels the too-large triples correction.

Our third example, the dissociation of diatomic nitrogen, is one of the more notorious problems for electronic structure theory. For some recent treatments within a CC framework, see, for example, Refs. [101] and [102]. What makes N_2 so difficult is that breaking three bonds should, in principle, require a method that includes simultaneously exciting six electrons. However, the resultant method would scale with the fourteenth power of the system size. Because of the prohibitive cost of such an approach, approximations are introduced. Unfortunately, in all cases except for two mentioned later, this leads to a non-variational collapse. [62, 101, 102]

In Table 1.5 we present results of calculations on N_2. The procedure given in Ref. [87] was followed. The two core orbitals were left entirely uncorrelated, which leads to an insignificant inconsistency between the excluded HF core orbitals in the FCI and CCSD calculations versus the optimized core orbitals in the QCCD calculations. These results are also plotted in Figure 1.3. All of the standard CC methods, including CCSDTQ, which scales as O^4V^6, show a turn-over. Although the CCSDTQ energy never falls below the FCI energy,

Table 1.5. The error with respect to FCI[a] for the dissociation of N_2 with the cc-pVDZ basis. Bond lengths are in Angstroms, and errors are in milliHartrees.

	CCSD[a]	*QCCD*[b]	*CCSD(T)*[c]	*CCSDT*[c]	*CCSD(2)*[b]	*QCCD(2)*[b]	*CCSDTQ*[c]
1.0679	12.522	10.626	1.560[d]	1.431[d]	1.501	0.809	
1.1208	14.469	12.216	1.879	1.759	1.742	0.93	0.229
1.1737	16.682	13.991	2.475	2.351	2.012	1.055	0.323
1.27	21.481	17.744	3.087	3.631	2.645	1.33	0.583
1.4288	31.807	25.514	5.182	6.834	4.738	2.127	1.357
1.5875	45.19	35.27	7.70	10.31	7.32	3.74	2.41
1.7463	60.4	47	7.6	8.1	11.1	6.6	2.6
1.905	72.1	59.5	−6.1[d]	−16.1[d]	10.4	9.8	0.5
2.0638	67.806	71.525	−58.810[d]	−67.686[d]	−7.051	13.927	

[a] Ref. [87] and Ref. [62]. [b] Ref. [62]. [c] Ref. [101]. [d] Present work.

Figure 1.3. Total energies for the dissociation of N_2 in the cc-pVDZ basis.

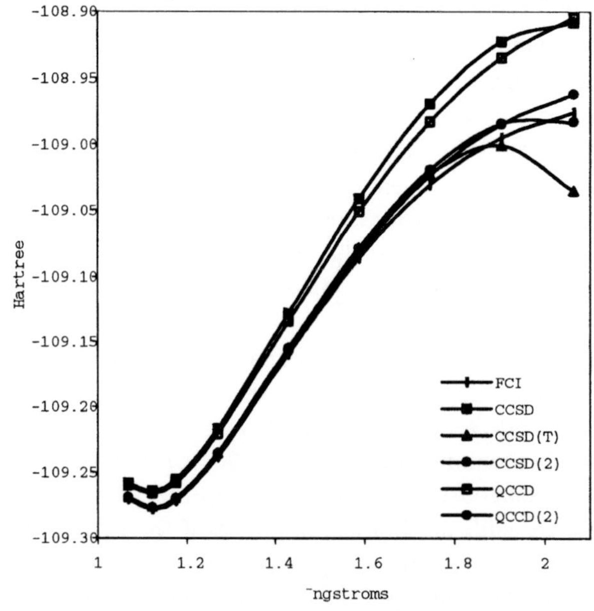

the rapidly decreasing error suggests the non-variational collapse has started. On the other hand, once again, the only two methods which remain stable are QCCD and QCCD(2).

As a final example, we have calculated bond lengths and harmonic vibrational frequencies for a series of radical diatomics. For simple closed-shell molecules, CCSD(T) has proven to give extremely accurate bond lengths. [96] However, in a recent systematic study of radical molecules, on average CCSD(T) did not peform significantly better than CCSD for bond lengths and vibrational frequencies. [103] The cause was several spectacular failures by the triples correction in CCSD(T). For one such example, the NO molecule, we found that, while CCSD(2) corrected much of CCSD(T) failure, the most effective way to fix the calculation was to optimize the orbitals. [60]

Table 1.6. Errors versus experiment for bond lengths and harmonic vibrational frequencies in a series of twelve radical diatomics. Bond length errors are in pm, and vibrational frequency relative errors are in percent.

	CCSD[a]	*OD*[a]	*QCCD*	*CCSD(T)*	*OD(T)*	*CCSD(2)*	*OD(2)*	*QCCD(2)*
errors in bond lengths								
MAD[b]	0.57	0.57	0.43	0.77	0.48	0.28	0.29	0.32
RMS[c]	0.88	0.83	0.66	1.02	0.54	0.39	0.37	0.39
percent errors in harmonic vibrational frequencies								
MAD[b]	4.10	4.36	3.56	4.15	1.01	1.85	1.65	1.22
RMS[c]	4.96	5.32	4.35	6.31	1.20	2.25	2.11	1.61

[a] Ref. [103]. [b] Mean Absolute Deviation [c] RMS Deviation

In Table 1.6 we test the generality of that observation on a fairly difficult set of doublet diatomics. The molecules chosen were CH, OH, FH^+, BO, CN, CO^+, CF, N_2^+, NO, O_2^+, OF, and F_2^+. The experimental results were taken from Ref. [103] and references therein. Five points, spaced at 0.5 pm intervals, were calculated using the cc-pVTZ basis set. [88] Equilibrium bond distances and vibrational frequencies were calculated based on fits using PSI. Core orbitals were not included in the CC calculation. As discussed above, though, the (2) calculations do include a core correction perturbatively. On average, CCSD and OD behave very similarly, with QCCD somewhat superior. On the other hand, OD(T) [58] is vastly more reliable than CCSD(T). This demonstrates the strong dependence of orbital choice on the triples correction in (T). The (2) correction shows the opposite behavior. In fact, the similarity of the CCSD(2) and OD(2) results mirrors the similarity of CCSD and OD. This is an outgrowth of the insensitivity of (2) to the reference chosen. Overall, for bond lengths QCCD(2) gives the best results. For the vibrational frequencies the smallest average error belongs to OD(T). However, this is likely somewhat fortuitous, since the OD(T) bond length error is relatively large.

4. Excited States?

So far we have restricted ourselves to the ground state (or at least the lowest energy state of a given multiplicity and symmetry). However, nowhere in our derivation was it assumed that $\mathcal{R}$ and $\mathcal{L}$ referred to the ground state. Although we eventually inserted $\mathbf{R}_{\mathbf{P}}^{[0]} = 1$ and $\mathbf{L}_{\mathbf{P}}^{[0]} = 1 + \Lambda$, we could have also used the eigenvectors of $\bar{\mathbf{H}}_{\mathbf{pp}}$ representing the kth state, $\mathbf{R}_{p\ k}^{[0]}$ and $\mathbf{L}_{\mathbf{p}\ k}^{[0]}$. These eigenvectors are the EOM-CCSD excited states. [29] The equivalent of Eq. 1.19 for excited states then becomes

$$E_k^{[2]} = \langle 0|\mathbf{L}_{p\ k}^{[0]}\bar{\mathbf{H}}_{\mathbf{pq}}^{[1]}\left(E^{[0]}\mathbf{1} - \bar{\mathbf{H}}_{\mathbf{qq}}^{[0]}\right)^{-1}\bar{\mathbf{H}}_{\mathbf{qp}}^{[1]}\mathbf{R}_{\mathbf{p}\ k}^{[0]}|0\rangle. \tag{1.21}$$

Hirata *et al.* attempted this using their variant of (2) theory. [64] The results were decidedly mixed. For total energies of ground states, we want the energy to scale linearly with system size. This is size extensivity, and (2) for ground states has this feature. For excited states the property of interest is size intensivity. A size intensive method for excited states is one where the excitation energy of an isolated subsystem does not depend on the presence of any other subsystems. EOM-CCSD has this property. In preliminary numerical tests, CC(2)PT(2) for excited states does not appear to be size intensive. [104] To better understand the problem, let us consider a system consisting of two non-interacting subsystems A and B. The Hamiltonian of the system H_{AB} is separable into $H_{AB} = H_A + H_B$. Similarly, $T_{AB} = T_A + T_B$, and $\Lambda_{AB} = \Lambda_A + \Lambda_B$. Plugging into Eq. 1.5 gives

$$\begin{aligned} E_{AB}^{[0]} &= \langle 0|\left(H_{AB}e^{T_{AB}}\right)_c|0\rangle \qquad (1.22)\\ &= \langle 0|\left(H_A e^{T_A}\right)_c|0\rangle + \langle 0|\left(H_A e^{T_A}\right)_c|0\rangle = E_A^{[0]} + E_B^{[0]}. \end{aligned}$$

This is a simple proof of the proper scaling with system size of the CC energy. The proof for a (2) method is

$$\begin{aligned} E_{AB}^{[2]} &= \langle 0|(1+\Lambda_A+\Lambda_B)\bar{\mathbf{H}}_{AB}^{[1]}\left(E^{[0]}\mathbf{1} - \bar{\mathbf{H}}_{AB}^{[0]}\right)^{-1}\bar{\mathbf{H}}_{AB}^{[1]}|0\rangle \\ &= \langle 0|(1+\Lambda_A)\bar{\mathbf{H}}_A^{[1]}\left(E^{[0]}\mathbf{1} - \bar{\mathbf{H}}_A^{[0]}\right)^{-1}\bar{\mathbf{H}}_A^{[1]}|0\rangle \qquad (1.23)\\ &\quad + \langle 0|(1+\Lambda_B)\bar{\mathbf{H}}_B^{[1]}\left(E^{[0]}\mathbf{1} - \bar{\mathbf{H}}_B^{[0]}\right)^{-1}\bar{\mathbf{H}}_B^{[1]}|0\rangle \\ &= E_A^{[2]} + E_B^{[2]}. \end{aligned}$$

The situation for excited states is more complicated. Given an excitation localized on subsystem A, $\mathcal{R}_k^{[0]}$ can be written as $\mathcal{R}_k^{[0]} = r_0 + \mathcal{R}_A$, where r_0 is a numerical constant that is either small or zero, depending on the symmetry of the excited state. Since $\mathcal{L}_k^{[0]}$ has no constant term, it can be written as $\mathcal{L}_k^{[0]} = \mathcal{L}_A$. Inserting these into Eq. 1.21 gives

$$\begin{aligned} E_{AB}^{[2]} &= \langle 0|\mathcal{L}_A\bar{\mathbf{H}}_{AB}^{[1]}\left(E^{[0]}\mathbf{1} - \bar{\mathbf{H}}_{AB}^{[0]}\right)^{-1}\bar{\mathbf{H}}_{AB}^{[1]}(r_0+\mathcal{R}_A)|0\rangle \qquad (1.24)\\ &= \langle 0|\mathcal{L}_A\bar{\mathbf{H}}_A^{[1]}\left(E^{[0]}\mathbf{1} - \bar{\mathbf{H}}_A^{[0]}\right)^{-1}\bar{\mathbf{H}}_A^{[1]}(r_0+\mathcal{R}_A)|0\rangle \\ &= E_A^{[2]}. \end{aligned}$$

If $E_{AB}^{[2]}$ were a correction to just the excitation energy, this would be correct. However, $E_{AB}^{[2]}$ is a correction to the total energy for the excited state. To obey size intensitivy, instead of $E_{AB}^{[2]}$ just including $E_{A}^{[2]}$ for the excited state of A, it would need to also include $E_{B}^{[2]}$ for the *ground state* of subsystem B. Unfortunately, at the current time we see no clear way around this problem.

5. Conclusions

In this chapter we have presented a new approach to developing perturbative corrections to coupled-cluster theory. We started with Stanton's generalization [41] of Löwdin's partitioning approach [5] to deriving perturbation theories. However, we veered from the previous paths by taking $\bar{H}$ as the fundamental quantity and keeping all terms through a consistent order in $\bar{H}$. Previously, people had always fallen back to arguments based on what order terms appeared in standard MBPT when deciding where to truncate their expansion. This brought several aspects of new physics into our models. The first was the inclusion of quadruple excitations along with and as a complement to triple excitation corrections. The second was the use of two-electron integrals that had been dressed by the CC correlation operator T. The other significant piece of new physics was the appearance of correlated "orbital energies" in the denominators of the final spin-orbital equations.

Based on results such as those presented here, we feel that QCCD represents the best performing iterative n^6 method available, and that QCCD(2) is the best method which scales as iterative n^6 plus a non-iterative n^7 component. However, currently the V^6 scaling of QCCD limits its practicality to barely over one hundred basis functions. (In fact, relative to the high cost of the QCCD, the (2) part of QCCD(2) is quite cheap, and therefore in most cases should probably be included any time QCCD is used). For wide-scale use, then, we turn to CCSD(2). Except for situations where CCSD breaks down, such as severe bond-breaking or multireference problems, its performance is on par with that of QCCD(2). And with its cost only a small multiple of the widely-used CCSD(T), we feel that this could be, and possibly should be, the method of choice in the near future for high-accuracy calculations on small to medium sized molecules.

The situation for excited states is much less clear. The analysis above shows where the lack of size intensivity arises in EOM-CCSD(2). However, we feel that there must be some way around this problem. Hopefully someone, possibly inspired by what is written here, will devise a solution.

Acknowledgments

This work was supported by the Director, Office of Energy Research, Office of Basic Energy Sciences, Chemical Sciences Division of the US Department of Energy under Contract No. DE-AC03-76SF00098. MHG is on appointment as a Miller Research Professor in the Miller Institute for Basic Research in Science. One of us (SRG) would like to acknowledge the debt of gratitude owed to Per-Olov Löwdin. Even in his late seventies, when I first met him, his energy and passion was an inspiration for this young scientist.

References

[1] P.-O. Löwdin, J. Chem. Phys. **19**, 1396 (1951).

[2] P.-O. Löwdin, J. Mol. Spectrosc. **10**, 12 (1963).

[3] P.-O. Löwdin, J. Mol. Spectrosc. **13**, 326 (1964).

[4] P.-O. Löwdin, J. Mol. Spectrosc. **13**, 331 (1964).

[5] P.-O. Löwdin, J. Math. Phys. **3**, 969 (1962).

[6] P.-O. Löwdin, J. Math. Phys. **3**, 1171 (1962).

[7] P.-O. Löwdin, J. Mol. Spectrosc. **14**, 112 (1964).

[8] P.-O. Löwdin, J. Mol. Spectrosc. **14**, 119 (1964).

[9] P.-O. Löwdin, J. Mol. Spectrosc. **14**, 131 (1964).

[10] P.-O. Löwdin, J. Math. Phys. **6**, 1341 (1965).

[11] P.-O. Löwdin, Phys. Rev. **139**, A357 (1965).

[12] P.-O. Löwdin, J. Chem. Phys. **43**, S175 (1965).

[13] F. Coester, Nucl. Phys. **1**, 421 (1958).

[14] F. Coester and H. Kümmel, Nucl. Phys. **17**, 477 (1960).

[15] J. Čížek, J. Chem. Phys **45**, 4256 (1966).

[16] J. Čížek, Adv. Chem. Phys. **14**, 35 (1969).

[17] J. Paldus, J. Čížek, and I. Shavitt, Phys. Rev. A **5**, 50 (1972).

[18] J. A. Pople, R. Krishnan, H. B. Schlegel, and J. S. Binkley, Int. J. Quantum Chem. **14**, 545 (1978).

[19] R. J. Bartlett and G. D. Purvis, III, Int. J. Quantum Chem. **14**, 561 (1978).

[20] G. D. Purvis, III and R. J. Bartlett, J. Chem. Phys. **76**, 1910 (1982).

[21] R. J. Bartlett, in *Modern Electronic Structure Theory, Part II*, edited by D. R. Yarkony (World Scientific, Singapore, 1995), pp. 1047–1131.

[22] J. Paldus and X. Li, in *Advances in Chemical Physics*, edited by I. Prigogine and S. A. Rice (Wiley, New York, 1999), Vol. 110, pp. 1–175.

[23] T. D. Crawford and H. F. Schaefer III, in *Reviews in Computational Chemistry*, edited by K. B. Lipkowitz and D. B. Boyd (Wiley-VCH, New York, 2000), Vol. 14, pp. 33–136.

[24] J. Geertsen, M. Rittby, and R. J. Bartlett, Chem. Phys. Lett. **164**, 57 (1989).

[25] S. R. Gwaltney, M. Nooijen, and R. J. Bartlett, Chem. Phys. Lett. **248**, 189 (1996).

[26] K. Emrich, Nucl. Phys. A **351**, 379 (1981).

[27] K. Emrich, Nucl. Phys. A **351**, 397 (1981).

[28] H. Sekino and R. J. Bartlett, Int. J. Quantum Chem. Symp. **18**, 255 (1984).

[29] J. F. Stanton and R. J. Bartlett, J. Chem. Phys. **98**, 7029 (1993).

[30] D. C. Comeau and R. J. Bartlett, Chem. Phys. Lett. **207**, 414 (1993).

[31] H. Monkhorst, Int. J. Quantum Chem. Symp. **11**, 421 (1977).

[32] D. Mukherjee and P. K. Mukherjee, Chem. Phys. **39**, 325 (1979).

[33] S. Ghosh, D. Mukherjee, and S. N. Bhattacharyya, Mol. Phys. **43**, 173 (1981).

[34] E. Dalgaard and H. Monkhorst, Phys. Rev. A **28**, 1217 (1983).

[35] S. Ghosh and D. Mukherjee, Proc. Ind. Acad. Sci. **93**, 947 (1984).

[36] M. Takahashi and J. Paldus, J. Chem. Phys. **85**, 1486 (1986).

[37] H. Koch and P. Jørgensen, J. Chem. Phys. **93**, 3333 (1990).

[38] H. Koch, H. J. A. Jensen, P. Jørgensen, and T. Helgaker, J. Chem. Phys. **93**, 3345 (1990).

[39] H. Nakatsuji and K. Hirao, J. Chem. Phys **68**, 4279 (1978).

[40] H. Nakatsuji, Chem. Phys. Lett. **67**, 329 (1979).

[41] J. F. Stanton and J. Gauss, Theor. Chim. Acta **93**, 303 (1996), **95**, 97 (1997) (E).

[42] J. C. Saeh and J. F. Stanton, J. Chem. Phys. **111**, 8275 (1999).

[43] J. F. Stanton and J. Gauss, J. Chem. Phys. **101**, 8938 (1994).

[44] J. F. Stanton, Chem. Phys. Lett. **281**, 130 (1997).

[45] K. Raghavachari, G. Trucks, J. A. Pople, and M. Head-Gordon, Chem. Phys. Lett. **157**, 479 (1989).

[46] G. E. Scuseria, Chem. Phys. Lett. **176**, 27 (1991).

[47] J. D. Watts, J. Gauss, and R. J. Bartlett, J. Chem. Phys. **98**, 8718 (1993).

[48] T. D. Crawford and J. F. Stanton, Int. J. Quantum Chem. **70**, 601 (1998).

[49] S. A. Kucharski and R. J. Bartlett, J. Chem. Phys. **108**, 5243 (1998).

[50] J. D. Watts and R. J. Bartlett, Chem. Phys. Lett. **258**, 581 (1996).

[51] C. D. Sherrill, A. I. Krylov, E. F. C. Byrd, and M. Head-Gordon, J. Chem. Phys. **109**, 4171 (1998).

[52] A. I. Krylov, C. D. Sherrill, E. F. C. Byrd, and M. Head-Gordon, J. Chem. Phys. **109**, 10669 (1998).

[53] B. O. Roos, in *Ab Initio Methods in Quantum Chemistry, Part II*, Vol. 69 of *Advances in Chemical Physics*, edited by K. P. Lawley (John Wiley & Sons, Chichester, 1987), pp. 399–445.

[54] G. E. Scuseria and H. F. Schaefer III, Chem. Phys. Lett. **142**, 354 (1987).

[55] T. B. Pedersen, H. Koch, and C. Hättig, J. Chem. Phys. **110**, 8318 (1999).

[56] K. Andersson, P.-Å. Malmqvist, B. O. Roos, A. J. Sadlej, and K. Wolinski, J. Phys. Chem. **94**, 5483 (1990).

[57] K. Andersson, P.-Å. Malmqvist, and B. O. Roos, J. Chem. Phys. **96**, 1218 (1992).

[58] S. R. Gwaltney and M. Head-Gordon, Chem. Phys. Lett. **323**, 21 (2000).

[59] S. R. Gwaltney, C. D. Sherrill, M. Head-Gordon, and A. I. Krylov, J. Chem. Phys. **113**, 3548 (2000).

[60] S. R. Gwaltney and M. Head-Gordon, J. Chem. Phys. **115**, 2014 (2001).

[61] J. F. S. J. Gauss, J. D. Watts, and R. J. Bartlett, J. Chem. Phys. **94**, 4334 (1991).

[62] S. R. Gwaltney, E. F. C. Byrd, T. Van Voorhis, and M. Head-Gordon, Chem. Phys. Lett. **353**, 359 (2002).

[63] T. Van Voorhis and M. Head-Gordon, Chem. Phys. Lett. **330**, 585 (2000).

[64] S. Hirata, M. Nooijen, I. Grabowski, and R. J. Bartlett, J. Chem. Phys **114**, 3919 (2001), **115**, 3967 (2001) (E).

[65] P. Piecuch and R. J. Bartlett, in *Advances in Quantum Chemistry*, edited by P.-O. Löwdin (Academic, San Diego, 1999), Vol. 34, pp. 295–380.

[66] R. J. Bartlett, in *Geometrical Derivatives of Energy Surfaces and Molecular Properties*, Vol. 166 of *NATO ASI series C.*, edited by P. Jørgensen and J. Simons (Reidel, Dordrecht, 1986), pp. 35–61.

[67] W. J. Lauderdale, J. F. Stanton, J. Gauss, J. D. Watts, and R. J. Bartlett, Chem. Phys. Lett. **187**, 21 (1991).

[68] S. A. Kucharski and R. J. Bartlett, in *Advances in Quantum Chemistry*, edited by P.-O. Löwdin (Academic, Orlando, 1986), Vol. 18, pp. 281–344.

[69] S. A. Kucharski and R. J. Bartlett, J. Chem. Phys. **108**, 9221 (1998).

[70] J. Gauss, J. F. Stanton, and R. J. Bartlett, J. Chem. Phys. **95**, 2623 (1991).

[71] S. A. Kucharski, J. D. Watts, and R. J. Bartlett, Chem. Phys. Lett. **302**, 295 (1999).

[72] S. A. Kucharski and R. J. Bartlett, J. Chem. Phys. **110**, 8233 (1999).

[73] M. Head-Gordon, P. E. Maslen, and C. A. White, J. Chem. Phys. **108**, 616 (1998).

[74] M. Head-Gordon, M. Lee, P. Maslen, T. Van Voorhis, and S. Gwaltney, in *Modern Methods and Algorithms of Quantum Chemistry Proceedings, Second Edition*, Vol. 3 of *NIC Series*, edited by J. Grotendorst (John von Neumann Institute of Computing, Jülich, 2000), pp. 593–648.

[75] A. Szabo and N. S. Ostlund, *Modern Quantum Chemistry, Introduction to Advanced Electronic Structure Theory* (McGraw-Hill, New York, 1990).

[76] M. Nooijen and R. J. Bartlett, J. Chem. Phys. **102**, 3629 (1995).

[77] J. Arponen, Ann. Phys. **151**, 311 (1983).

[78] J. S. Arponen, R. F. Bishop, and E. Pajanne, Phys. Rev. A **36**, 2519 (1987).

[79] T. Van Voorhis and M. Head-Gordon, J. Chem. Phys. **113**, 8873 (2000).

[80] P. G. Szalay, M. Nooijen, and R. J. Bartlett, J. Chem. Phys. **103**, 281 (1995).

[81] G. J. O. Beran, S. R. Gwaltney, and M. Head-Gordon, J. Chem. Phys. (in press).

[82] J. A. Pople, M. Head-Gordon, and K. Raghavachari, J. Chem. Phys. **87**, 5968 (1987).

[83] I. I. Ukrainskii, Theor. Math. Phys. **32**, 816 (1978).

[84] J. Cullen, Chem. Phys. **202**, 217 (1996).

[85] T. Van Voorhis and M. Head-Gordon, J. Chem. Phys. **112**, 5633 (2000).

[86] J. Olsen, P. Jørgensen, H. Koch, A. Balkova, and R. J. Bartlett, J. Chem. Phys. **104**, 8007 (1996).

[87] H. Larsen, J. Olsen, P. Jørgensen, and O. Christiansen, J. Chem. Phys. **113**, 6677 (2000), **114**, 10985 (2001) (E).

[88] T. H. Dunning, Jr., J. Chem. Phys. **90**, 1007 (1989).

[89] Basis sets were obtained from the Extensible Computational Chemistry Environment Basis Set Database, Version 1.0, as developed and distributed by the Molecular Science Computing Facility, Environmental and Molecular Sciences Laboratory which is part of the Pacific Northwest Laboratory, P.O. Box 999, Richland, Washington 99352, USA, and funded by the U.S. Department of Energy. The Pacific Northwest Laboratory is a multi-program laboratory operated by Battelle Memorial Institute for the U.S. Department of Energy under contract DE-AC06-76RLO 1830. Contact David Feller, Karen Schuchardt, or Don Jones for further information.

[90] J. Kong, C. A. White, A. I. Krylov, D. Sherrill, R. D. Adamson, T. R. Furlani, M. S. Lee, A. M. Lee, S. R. Gwaltney, T. R. Adams, C. Ochsenfeld, A. T. B. Gilbert, G. S. Kedziora, V. A. Rassolov, D. R. Maurice, N. Nair, Y. Shao, N. A. Besley, P. E. Maslen, J. P. Dombroski, H. Daschel, W. Zhang, P. P. Korambath, J. Baker, E. F. C. Byrd, T. Van Voorhis, M. Oumi, S. Hirata, C.-P. Hsu, N. Ishikawa, J. Florian, A. Warshel, B. G. Johnson, P. M. W. Gill, M. Head-Gordon, and J. A. Pople, J. Comp. Chem. **21**, 1532 (2000).

[91] T. D. Crawford, C. D. Sherrill, E. F. Valeev, J. T. Fermann, M. L. Leininger, R. A. King, S. T. Brown, C. L. Janssen, E. T. Seidl, Y. Yamaguchi, W. D. Allen, Y. Xie, G. Vacek, T. P. Hamilton, C. B. Kellogg, R. B. Remington, and H. F. Schaefer III, PSI 3.0, development version, PSITECH, Inc., Watkinsville, GA 30677, U.S.A., 2000.

[92] ACES II program is a product of the Quantum Theory Project, University of Florida Authors: J. F. Stanton, J. Gauss, J. D. Watts, M. Nooijen, N. Oliphant, S. A. Perera, P. G. Szalay, W. J. Lauderdale, S. R. Gwaltney, S. Beck, A. Balková, D. E. Bernholdt, K.-K. Baeck, H. Sekino, P. Rozyczko, C. Huber, and R. J. Bartlett. Integral packages included are VMOL (J. Almlöf and P. R. Taylor), VPROPS (P. R. Taylor), and a modified version of the ABACUS integral derivative package (T. U. Helgaker, H. J. Aa. Jensen, J. Olsen, P. Jørgensen, and P. R. Taylor).

[93] R. J. Bartlett and J. F. Stanton, in *Reviews in Computational Chemistry*, edited by K. B. Lipkowitz and D. B. Boyd (VCH, New York, 1994), Vol. 5, pp. 65–169.

[94] K. P. Huber and G. Herzberg, *Molecular Spetra and Molecular Structure IV. Constants of Diatomic Molecules* (Van Nostrand Reinhold, New York, 1979).

[95] J. M. L. Martin and G. de Oliveira, J. Chem. Phys. **111**, 1843 (1999).

[96] K. L. Bak, J. Gauss, P. Jørgensen, J. Olsen, T. Helgaker, and J. F. Stanton, J. Chem. Phys. **114**, 6548 (2001).

[97] J. Noga and R. J. Bartlett, J. Chem. Phys. **86**, 7041 (1987).

[98] G. E. Scuseria and H. F. Schaefer, III, Chem. Phys. Letters **152**, 382 (1988).

[99] S. A. Kucharski and R. J. Bartlett, Theor. Chim. Acta **80**, 387 (1991).

[100] S. A. Kucharski and R. J. Bartlett, J. Chem. Phys. **97**, 4282 (1992).

[101] J. W. Krogh and J. Olsen, Chem. Phys. Lett. **344**, 578 (2001).

[102] P. Piecuch, S. A. Kucharski, and K. Kowalski, Chem. Phys. Lett. **344**, 176 (2001).

[103] E. F. C. Byrd, C. D. Sherrill, and M. Head-Gordon, J. Phys. Chem. A **105**, 9736 (2001).

[104] S. Hirata, private communication.

Density matrix theory – a retrospective

Roy McWeeny
Università di Pisa
Dipartimento di Chimica e Chimica Industriale
Via Risorgimento 35, 56100 Pisa (Italy)

May 21, 2002

Abstract

Density matrix theory (DMT), especially the theory of *reduced* density matrices, provides a general and powerful tool for the application of quantum mechanics to many-electron systems. Starting from the developments in the mid-1950s, in which Löwdin played an important part, the evolution of the theory during the subsequent half century is reviewed.

The most important one- and two-electron density functions (electron density, spin density, current density, pair density, etc.) are defined and their main properties indicated. These lead to a general physical interpretation of contributions to the total electronic energy of a system, even in the presence of applied fields, and provide a basis for semi-empirical methods such as density functional theory (DFT). The importance of N-representability conditions soon becomes clear.

Illustrative calculations of the density functions for simple systems (in an IPM approximation) reveal the difficulties of 'eliminating' the N-electron wave function in favour of 1- and 2-electron density functions. And this prompts a search for alternative methods of calculating the densities, starting directly from the usual spin-free Schrödinger equation and constructing 'Pauli-compatible' densities which satisfy antisymmetry requirements and are, in principle, exact. In general, such densities still involve an N-electron 'root function' and contain $N!$ terms.

Finally, it is shown that DMT is easily generalized to wave functions of fully relativistic form, in which 2-component spin-orbitals are replaced by 4-component Dirac spinors.

E.J. Brändas and E.S. Kryachko (eds.),
Fundamental World of Quatum Chemistry, Vol. I, 459–485.

1 Early history

Density matrix theory (DMT) originated long ago, in the work of von Neumann [1] and Dirac [2], but was little used before the 1950s except in statistical mechanics. Per-Olov Löwdin was one of the key players in the second phase of development of DMT. I met him in the spring of 1954, when he was visiting John Slater's famous "Solid-State and Molecular Theory Group" (SSMTG) at MIT. At that time I was spending a year with SSMTG, on leave of absence from the Chemistry Department at King's College, Newcastle on Tyne (UK). Our backgrounds were complementary: he was from Ivor Waller's Institute in Uppsala, where he had broken new ground with fully quantum-mechanical calculations of the cohesive energies of alkali halide crystals; I had been working on molecular electronic structure theory, after completing my doctoral thesis[1] with Charles Coulson in Oxford. I had been appointed to the Chemistry Department mainly to teach a course called "Mathematics for Chemists", but my research activity was already centred around molecular applications of quantum mechanics and included the possible use of valence bond theory. As it happened, VB theory was another prevailing interest in SSMTG, where Al Meckler was making *ab initio* calculations on the oxygen molecule. Meckler and I were both using 'Löwdin orthogonalized' orbitals [3] and when, after a brief correspondence, Slater asked if I could join the group for a time I could hardly wait to get started – so I arrived in Boston with my wife and one-year old daughter in July 1953, at the beginning of what was to be a very memorable year.

Per and I never actually collaborated, but we had a common driving interest in DMT. We both appreciated its potential for N-electron quantum mechanics and saw the possibility of deriving density matrices for 1 electron, 2 electrons, and 'clusters' of n electrons, by integrating over 'superfluous' variables. And this led at once to the now well-known expressions for the total electronic energy of a general system in terms of a single two-electron density function – and to the hope of 'eliminating' all reference to the N-electron wave function itself.

I had been fortunate, during my time in Newcastle on Tyne, in attending seminars held by Stanley Rushbrooke on the statistical mechanics of many-body systems in the condensed phase; so I knew about the work of Husimi [4], who had already defined the *reduced* density matrices for few-body clusters, and it was clear that similar quantities could be useful in many-electron

[1] On momentum space solutions of the Schrödinger equation – a topic considered suitable for a physics graduate with little chemical background.

quantum mechanics. Per was looking towards the generalization of Hartree-Fock theory and Slater's introduction of a 'correlation potential' to describe the field 'felt' by an electron in the presence of all the others; while I was thinking about the many-configuration generalizations required in valence bond theory. We became good friends and I remember the three months of Per's visit with great affection. About a year later we both published our density matrix work, and much more; he in his marvellous Physical Review papers [5], and I in Royal Society Proceedings [6] – following an informal presentation in Technical Report No.7 from SSMTG (15 April 1955) – and some years later in Reviews of Modern Physics [7]. Our friendship lasted more than 40 years and included many periods of close association in both Sweden and Florida. I am happy to have the opportunity of dedicating the present survey to Per's memory.

2 Definitions and concepts

The main ideas and procedures of density matrix theory are by now well established. Here they are simply collected, with explanatory comments, first for the 1-electron density matrices and then for their 2-electron analogues – for brevity the 1-DMs and 2-DMs.

2.1 The 1-electron densities

The most familiar of all the density functions is the electron density[2] itself. As everyone knows, the most general definition of the electron density is[3]

$$P(\mathbf{r}) = N \int \Psi(\mathbf{x}, \mathbf{x}_2, \dots \mathbf{x}_N)\Psi^*(\mathbf{x}, \mathbf{x}_2, \dots \mathbf{x}_N) ds d\mathbf{x}_2 \dots d\mathbf{x}_N. \tag{1}$$

This is the probability/unit volume of finding an electron (no matter which) at point $\mathbf{r}$. The wave function Ψ (in principle exact) is assumed antisymmetric and the spin variables are eliminated in the integrations.

$P(\mathbf{r})$ is in fact the 'diagonal element', $P(\mathbf{r}) = P(\mathbf{r};\mathbf{r})$, of the (spinless) 1-electron density matrix $P(\mathbf{r};\mathbf{r}')$, in which the $\mathbf{r}$ of Ψ^* in (1) is replaced by a second variable $\mathbf{r}'$. When spin variables are explicitly included, $P(\mathbf{r};\mathbf{r}')$is

[2] Also called the 'number density' (probable number of electrons per unit volume) or the 'charge density' (probable amount of electric charge per unit volume, in units of $-e$).

[3] We denote space and spin variables collectively by $\mathbf{x} = \mathbf{r}, s$

replaced by

$$\rho(\mathbf{x};\mathbf{x}') = N\int \Psi(\mathbf{x},\mathbf{x}_2,\dots\mathbf{x}_N)\Psi^*(\mathbf{x}',\mathbf{x}_2,\dots\mathbf{x}_N)d\mathbf{x}_2\dots d\mathbf{x}_N, \tag{2}$$

which follows from the product $\Psi\Psi^*$ by integrating over the space-spin coordinates of all electrons but one. More correctly, following Husimi [4], equation (2) defines the 1-electron *reduced* density matrix (RDM), or more briefly the "1-DM". It is to be noted that the spinless 1-DM $P(\mathbf{r};\mathbf{r}')$ is obtained by identifying s', s and integrating:

$$P(\mathbf{r};\mathbf{r}') = \int_{s'=s} \rho(\mathbf{x};\mathbf{x}')ds. \tag{3}$$

A key property of the 1-DM is that the expectation value of any symmetric sum of 1-electron operators, such as the electron-nuclear potential energy ($V_{\rm en}$) and the electronic kinetic energy (T) in the Hamiltonian, can be written directly in terms of the *one*-electron quantity. Thus, with the 'standard' (non-relativistic, clamped-nuclei) Hamiltonian

$$\mathsf{H} = \mathsf{T} + V_{\rm en} + V_{\rm ee} = \sum_{i=1}^{N} \frac{\mathsf{p}^2(i)}{2m} + \sum_{i=1}^{N} V(\mathbf{r}_i) + \sum_{i\neq j=1}^{N} g(i,j) \tag{4}$$

the electron-nuclear potential energy and the kinetic energy terms yield expectation values

$$V_{\rm en} = \int V(\mathbf{r})P(\mathbf{r})d\mathbf{r},\ (a) \qquad \langle \mathsf{T}\rangle = \int_{\mathbf{r}'=\mathbf{r}} \frac{\mathsf{p}^2}{2m}P(\mathbf{r};\mathbf{r}')d\mathbf{r},\ (b) \tag{5}$$

where the off-diagonal quantity is required only for the *kinetic* energy. We note that the expectation value $\langle \mathsf{T}\rangle$ is obtained essentially by a *trace* operation – putting $\mathbf{r}_1' \to \mathbf{r}_1$ (i.e. taking a 'diagonal element') and then integrating over $\mathbf{r}$ (i.e. taking a 'diagonal sum'). Thus, with an alternative notation,

$$\langle \mathsf{T}\rangle = \mathrm{tr}\left(\frac{\mathsf{p}^2}{2m}\right) P(\mathbf{r};\mathbf{r}'), \tag{6}$$

'tr' denoting the 'integral trace'.

Since $P(\mathbf{r};\mathbf{r}')$ arises by spin integration in (3), it is important to know the general form of $\rho(\mathbf{x};\mathbf{x}')$. In fact,

$$\begin{aligned}\rho(\mathbf{x};\mathbf{x}') &= P_{\alpha,\alpha}(\mathbf{r};\mathbf{r}')\alpha(s)\alpha^*(s') + P_{\alpha,\beta}(\mathbf{r};\mathbf{r}')\alpha(s)\beta^*(s')\\ &+ P_{\beta,\alpha}(\mathbf{r};\mathbf{r}')\beta(s)\alpha^*(s') + P_{\beta,\beta}(\mathbf{r};\mathbf{r}')\beta(s)\beta^*(s'),\end{aligned} \tag{7}$$

this being the integral-operator form of the density *operator* ρ, whose effect on a space-spin function is given by $\rho\psi(\mathbf{x}) = \int \rho(\mathbf{x};\mathbf{x}')\psi(\mathbf{x}')d\mathbf{x}'$. For any state of definite spin, however, only the α,α and β,β components are non-zero. On using (3),

$$P(\mathbf{r};\mathbf{r}') = P_{\alpha,\alpha}(\mathbf{r};\mathbf{r}') + P_{\beta,\beta}(\mathbf{r};\mathbf{r}') \tag{8}$$

and on using the analogue of (5b), with spin included, to evaluate the expectation value of the total spin z component, we obtain

$$\langle \mathsf{S}_z \rangle = \int_{\mathbf{x}'=\mathbf{x}} \mathsf{S}_z \rho(\mathbf{x};\mathbf{x}')d\mathbf{x} = \int Q_z(\mathbf{r};\mathbf{r})d\mathbf{r}, \tag{9}$$

where, generally,

$$Q_z(\mathbf{r};\mathbf{r}') = \tfrac{1}{2}[P_{\alpha,\alpha}(\mathbf{r};\mathbf{r}') - P_{\beta,\beta}(\mathbf{r};\mathbf{r}')] \tag{10}$$

and has a diagonal element[4] which is the density of spin angular momentum around the z axis, first introduced by Weissman [8] and McConnell [9]. Thus

$$P(\mathbf{r}) = P_\alpha(\mathbf{r}) + P_\beta(\mathbf{r}), \tag{11}$$

$$Q_z(\mathbf{r}) = \tfrac{1}{2}[P_\alpha(\mathbf{r}) - P_\beta(\mathbf{r})] \tag{12}$$

are, respectively, the electron density and the *spin density*. In fact, however, Q_z is only one component of a *vector* density[5], the field components $Q_x(\mathbf{r}), Q_y(\mathbf{r})$ being defined in strict analogy with (9). Attention is usually focused on Q_z (z being the quantization axis) only because for a pure spin state (quantum numbers S, M_S) the components transverse to the axis are everywhere zero. For such a state, $Q_x = Q_y = 0$ and $Q_z = (M_S/S)Q_z^{\rm st}$, where $Q_z^{\rm st}$ denotes the density for a standard state – the 'top' state ($M_S = S$) of the multiplet. This is not the case in the presence of spin-orbit coupling.

Another spatial density function of great importance is the *current density* $\mathbf{J}(\mathbf{r})$, which is again a vector density. The components $J_x(\mathbf{r}), J_y(\mathbf{r}), J_z(\mathbf{r})$ determine the flux of electron (probability) density out of the volume element $d\mathbf{r}$ at any point $\mathbf{r}$ in space. Since the flux is zero everywhere for a state described by a *real* wavefunction, it is necessary in discussing this density to introduce in the Hamiltonian a magnetic field – however weak – in order to define the quantization axis and to introduce an imaginary component in Ψ. The presence of such a field requires only that the usual kinetic energy term in the Hamiltonian, namely $(1/2m)\sum_i \mathsf{p}^2(i)$, be replaced by

$$\mathsf{T} = (1/2m)\textstyle\sum_i \boldsymbol{\pi}^2(i), \qquad \pi_\lambda = \mathsf{p}_\lambda + eA_\lambda \qquad (\lambda = x,y,z). \tag{13}$$

[4]In referring to diagonal elements we drop both the second variable and the corresponding spin index. The spin density is often normalized to unity, by dividing by M.

[5]More correctly a *pseudo*vector density, being an angular momentum per unit volume.

Here $\boldsymbol{\pi}$ is the 'gauge invariant momentum' and contains the vector potential $\mathbf{A}$ of the field: $\boldsymbol{\pi}$ is in fact the operator associated with the electron *velocity*. The components of the current density may be defined most symmetrically as

$$J_\lambda(\mathbf{r}) = m^{-1}[\tfrac{1}{2}(\pi_\lambda + \pi'^{\dagger}_\lambda)P(\mathbf{r};\mathbf{r}')]_{\mathbf{r}'=\mathbf{r}}, \tag{14}$$

where the adjoint operator $\pi'^{\dagger}_\lambda$ works on the *primed* variables in the density matrix.

The diagonal element $J_\lambda(\mathbf{r})$, has all the properties of a *local* flux density component at point $\mathbf{r}$, as defined in classical physics. In particular, for a wavefunction satisfying the time-dependent Schrödinger equation, it is easy to derive a conservation equation for the electron density:

$$\text{div } \mathbf{J}(\mathbf{r}) = -\frac{\partial P(\mathbf{r})}{\partial t} \tag{15}$$

– the rate of decrease of the probability of finding an electron in volume element $d\mathbf{r}$ at point $\mathbf{r}$ is the divergence of the current density at that point. It follows that for a system in a stationary state the currents must be *circulating* (div $\mathbf{J} = 0$); and that these induced currents (created by the external field) will produce a secondary magnetic field. The components of this field, at point $\mathbf{r}_0$, are found to be

$$B^{\text{ind}}_\lambda(\mathbf{r}_0) = \left(\frac{\mu_0}{4\pi}\right)\int \frac{[(\mathbf{r}-\mathbf{r}_0)\times -e\mathbf{J}(\mathbf{r})]_\lambda}{|\mathbf{r}-\mathbf{r}_0|^3}d\mathbf{r} \tag{16}$$

– just as if (remembering the Biot-Savart law) the electron distribution were a conducting medium containing electric currents of density $-e\mathbf{J}(\mathbf{r})$. Another interesting property of the current density is that the total electronic energy of an arbitrary system in the presence of any external field whatever, with vector potential $\mathbf{A}$, may be written [10]

$$E = E_0 - \int \mathbf{A}\cdot[-e\mathbf{J}_0(\mathbf{r})]d\mathbf{r} - \tfrac{1}{2}\int \mathbf{A}\cdot[-e\mathbf{J}_{\text{ind}}](\mathbf{r})d\mathbf{r}. \tag{17}$$

Here E_0 and $\mathbf{J}_0$ are the energy and current density in the zero-field limit ($\mathbf{J}_0$ vanishing except for systems of high symmetry in states with unquenched angular momentum e.g. paramagnetic atoms), while $\mathbf{J}_{\text{ind}}$ is the field-linear part of the current density and therefore describes the currents *induced* by the applied field. The two field-dependent terms in (17) represent, respectively, the potential energy of a system of 'permanent' electric currents, of density $-e\mathbf{J}_0(\mathbf{r})$, and that of a system of induced currents, of density $-e\mathbf{J}_{\text{ind}}(\mathbf{r})$, in a conducting medium in the presence of a field – as would be calculated from classical physics.

2.2 The 2-electron densities

In order to formulate expressions for the total energy of an N-electron system we need to include the electron repulsion terms in $\frac{1}{2}\sum_{i,j} g(i,j)$. The corresponding expectation value, with an arbitrary 2-body operator $g(i,j)$, follows in terms of the *two*-electron density matrix (spin included) $\pi(\mathbf{x}_1, \mathbf{x}_2; \mathbf{x}'_1, \mathbf{x}'_2)$ and takes the form

$$V_{ee} = \tfrac{1}{2}\,\mathrm{tr}(1)\mathrm{tr}(2) g(1,2)\pi(\mathbf{x}_1, \mathbf{x}_2; \mathbf{x}'_1, \mathbf{x}'_2)$$

– subscripts 1,2 now labelling *two* points in space. Here $\pi(\mathbf{x}_1, \mathbf{x}_2; \mathbf{x}'_1, \mathbf{x}'_2)$ is the analogue of $\rho(\mathbf{x}; \mathbf{x}')$ and is the *pair* density matrix or 2-DM. When $g(1,2)$ involves neither differential nor spin operators (being the usual function of interelectron distance), primed variables may be dropped and we may also integrate immediately over spins. The result is

$$V_{\mathrm{ee}} = \tfrac{1}{2}\int g(1,2)\Pi(\mathbf{r}_1, \mathbf{r}_2) d\mathbf{r}_1 d\mathbf{r}_2, \tag{18}$$

where $\Pi(\mathbf{r}_1, \mathbf{r}_2)$ is the diagonal element, or 'pair density', analogous to the electron density $P(\mathbf{r}_1)$ in the 1-electron case.

The function $\pi(\mathbf{x}_1, \mathbf{x}_2; \mathbf{x}'_1, \mathbf{x}'_2)$ may be written in a form resembling (7), but with 16 components instead of 4: of these, for a state of definite spin, only 6 are non-zero and (integrating over spins) only 4 give contributions to $\Pi(\mathbf{r}_1, \mathbf{r}_2; \mathbf{r}'_1\mathbf{r}'_2)$. This *spinless* pair density has a diagonal element

$$\Pi(\mathbf{r}_1, \mathbf{r}_2) = \Pi_{\alpha\alpha}(\mathbf{r}_1, \mathbf{r}_2) + \Pi_{\alpha\beta}(\mathbf{r}_1, \mathbf{r}_2) + \Pi_{\beta\alpha}(\mathbf{r}_1, \mathbf{r}_2) + \Pi_{\beta\beta}(\mathbf{r}_1, \mathbf{r}_2), \tag{19}$$

whose terms indicate, respectively, the probability that the electrons at points $\mathbf{r}_1$ and $\mathbf{r}_2$ are both 'up-spin', one 'up-spin' and one 'down-spin', or both 'down-spin'.

In spite of the physically appealing form of (19) it does not offer any basis for a 'two-fluid' model, with N_α electrons of one kind and N_β of the other, even though such models are commonly invoked. The more significant interactions between two particles are related to the *coupling* of their spins (e.g. to the spin scalar product $\mathsf{S}(i)\cdot\mathsf{S}(j)$; and the nature of such couplings, for electrons at any two points in space, is determined by various 'spin-coupling densities' [11] which are derived from the pair density matrix $\pi(\mathbf{x}_1, \mathbf{x}_2; \mathbf{x}'_1, \mathbf{x}'_2)$. In fact, the terms 'up-spin' and 'down-spin' electrons are simplistic abstractions.

2.3 The total electronic energy

The primary importance of the density functions is that the total electronic energy may be written, subject only to the use of (4), as[6]

$$E = T + V_{en} + V_{ee}, \tag{20}$$

where the kinetic energy, electron-nuclear potential energy, and electron-electron potential energy are given in (5b), (5a), and (18), respectively. In summary,

$$\begin{aligned} T &= (1/2m)\int_{\mathbf{r}'=\mathbf{r}} \mathbf{p}\cdot\mathbf{p}'^{\dagger}P(\mathbf{r};\mathbf{r}')d\mathbf{r}, \\ V_{en} &= \int V(\mathbf{r})P(\mathbf{r})d\mathbf{r}, \\ V_{ee} &= \tfrac{1}{2}\int g(1,2)\Pi(\mathbf{r}_1,\mathbf{r}_2)d\mathbf{r}_1 d\mathbf{r}_2. \end{aligned} \tag{21}$$

These energy contributions give an expression (20) which became the starting point for many further developments.

Back in the 1950s we were already dreaming of eliminating the N-electron wavefunction altogether and working entirely in terms of the 1-DM and 2-DM: but it soon became clear that formidable problems stood in the way of progress along such lines. Briefly, assuming some approximate form of the 2-DM[7] as a basis for a variational calculation, how can we make sure that it would indeed be compatible with some acceptable N-electron wave function? This so-called 'N-representability problem' provided an intriguing challenge for pure mathematicians and was first taken up by John Coleman – whose work (see for example Ref.[12]) inspired that of many others over the last 50 years and culminated in a recent monograph [13]. In spite of much progress, the problem remains unsolved – except in a somewhat formal sense not lending itself easily to practical application. On the other hand, density *functional* theory (DFT), which has its roots in the same energy expressions (21), has become an important practical tool for making approximate calculations on large many-electron systems. The success of DFT has been achieved, it should be remembered, only at the cost of abandoning any insistence on mathematical rigour and setting up a formalism based on an independent-particle model (IPM), into which semi-empirical features can

[6]Note that T is expressed in a symmetrical form, the vector operators $\mathbf{p}, \mathbf{p}'$ working on the unprimed and primed variables, respectively, and the dagger indicating the adjoint. In the presence of a magnetic field $\mathbf{p}$ should be replaced by $\boldsymbol{\pi}$, its 'gauge invariant' counterpart.

[7]From which the 1-DM follows by a further integration

readily be inserted. Such models were frequently discussed in Slater's group in 1953-54 and Slater's 'Xα' method could indeed be considered an early precursor of DFT: historical detail has been given elsewhere [14]. Here it is not the intention to discuss DFT: reference may be made to the standard sources [15].

In the absence of necessary and sufficient conditions for the N-representability of density matrices the best one can do is to start from any suitable *ansatz*, for example a wave function of some particular concrete form from which the RDMs can be obtained by integration as already indicated. Thus, the 1-DM arising from a pure-state wave function expressed in configuration interaction (CI) form can always be cast in the explicit form [5,16]

$$\rho(\mathbf{x};\mathbf{x}') = \sum_r \lambda_r \psi_r(\mathbf{x})\psi_r^*(\mathbf{x}'), \quad 0 \leq \lambda_r \leq 1, \quad \sum_r \lambda_r = N, \tag{22}$$

where the 'spin-orbitals' (never mentioned until now) are simply the functions that diagonalize the 1-DM and are called the 'natural' spin-orbitals (NSOs). They are eigenfunctions of the (usually approximate) 1-DM and therefore satisfy $\rho\psi_r(\mathbf{x}) = \lambda_r\psi_r(\mathbf{x})$, or more fully

$$\int \rho(\mathbf{x};\mathbf{x}')\psi_r(\mathbf{x}') = \lambda_r\psi_r(\mathbf{x}), \tag{23}$$

where the eigenvalue λ_r is the 'occupation number' of ψ_r in the CI wave function. The expansion (22) is in principle infinite but in practice is always truncated.

The pair density matrix $\pi(\mathbf{x}_1,\mathbf{x}_2;\mathbf{x}_1',\mathbf{x}_2')$ may similarly be expressed in terms of natural spin 'geminals' (NSGs); but again the NSGs must be derived from a wave function and, for various reasons [17], their use has been limited.

In one particular case, an N-representable pair density can be constructed explicitly. This case corresponds to the truncated form of (22) in which

$$\rho(\mathbf{x};\mathbf{x}') = \sum_{r=1}^{N} \psi_r(\mathbf{x})\psi_r^*(\mathbf{x}'), \tag{24}$$

which is the 'Fock-Dirac' density matrix [18,19]. This quantity exhibits the projection operator property of idempotency ($\rho\rho = \rho$) $\int \rho(\mathbf{x};\mathbf{x}'')\rho(\mathbf{x}'';\mathbf{x}')d\mathbf{x}'' = \rho(\mathbf{x};\mathbf{x}')$; and this property itself is sufficient to guarantee that the density derives from a 1-determinant wave function, the spin-orbitals being the N eigenfunctions of ρ with unit eigenvalues. What is more (see Ref.[5]), the corresponding RDMs for any numbers of electrons are all functionals of $\rho(\mathbf{x};\mathbf{x}')$:

for example, the 2-DM is

$$\pi(\mathbf{x}_1, \mathbf{x}_2; \mathbf{x}_1', \mathbf{x}_2') = \begin{vmatrix} \rho(\mathbf{x}_1; \mathbf{x}_1') & \rho(\mathbf{x}_1; \mathbf{x}_2') \\ \rho(\mathbf{x}_2; \mathbf{x}_1') & \rho(\mathbf{x}_2; \mathbf{x}_2') \end{vmatrix} \tag{25}$$

and the n-electron density matrix is an $n \times n$ determinant of the same form. Consequently all the terms in (20) become functionals of $\rho(\mathbf{x}; \mathbf{x}')$ and variational procedures based on the Fock-Dirac density matrix (making no reference to the wavefunction itself) are therefore soundly based and will give correct upper bounds to the energy. Indeed [20], a variationally optimized energy can be found directly by iterative construction of the 1-DM, without ever invoking either a wave function or spin-orbitals. Unfortunately, this is true only at the 1-determinant (Hartree-Fock) level of approximation.

The more general 1-DM (22) is *not* idempotent; and there is no known way of building from it the 2-electron density $\pi(\mathbf{x}_1, \mathbf{x}_2; \mathbf{x}_1', \mathbf{x}_2')$ which determines the electron repulsion energy. In this event, whatever approximate form may be assumed for π, which determines all the terms in (21), variational calculations based on (20) – though perfectly possible – cannot be guaranteed free of variational collapse. Nor can the results of such calculations have any guaranteed physical meaning within the framework of orthodox quantum mechanics. To illustrate the nature of this dilemma, it is sufficient to consider an elementary example.

2.4 An example: density functions for a simple system

To illustrate the practical calculation of the various density functions let us consider the boron atom, using an IPM approximation. The ground-state electron configuration is then assumed to be

$$\mathrm{B}[1s^2 2s^2 2p^1]$$

In the absence of spin-orbit coupling, any state is a simultaneous eigenstate of the mutually commuting operators $\mathsf{H}_0, \mathsf{L}^2, \mathsf{L}_z, \mathsf{S}^2, \mathsf{S}_z$: such states are characterized by E and the quantum numbers L, M_L, S, M_S. The ground state is here 6-fold degenerate, with energy $E = 2\epsilon_{1s} + 2\epsilon_{2s} + \epsilon_{2p}$; and the density matrix contains a closed-shell term common to all states, giving zero contributions to both the spin and current densities, plus a p-electron contribution (which varies according to the values of M_L, M_S). We therefore consider only the p-shell density contributions. The 'states' specified by quantum numbers M_L, M_S are not of course true spectroscopic states: the Hamiltonian is invariant under common space and spin rotations and we should be looking for simultaneous eigenfunctions of $\mathsf{J}^2, \mathsf{J}_z$. In the limit of zero spin-orbit coupling the 'correct zero-order' eigenfunctions can still be written in 1-determinant form; but the p-electron will occupy a 2-component spin-orbital $\psi = a\phi^\alpha\alpha + b\phi^\beta\beta$. The density contributions are easily found: omitting the common closed-shell term, they are as shown in Table 1.

Table 1. Densities for the Boron atom (L-S coupling)

J, M_J	Charge	Spin	Current
$\frac{3}{2}, \frac{3}{2}$	$P = \frac{1}{2}(x^2 + y^2)f^2$ (Torus)	$Q_x = 0$ $Q_y = 0$ $Q_z = \frac{1}{2}P$	$J_x = -yf^2$ $J_y = xf^2$ $J_z = 0$
$\frac{3}{2}, \frac{1}{2}$	$P = \frac{1}{6}(4z^2 + x^2 + y^2)f^2$ (Ellipsoid)	$Q_x = -\frac{2}{3}xzf^2$ $Q_y = -\frac{2}{3}yzf^2$ $Q_z = \frac{1}{12}(4z^2 - x^2 - y^2)f^2$	$J_x = -\frac{1}{3}yf^2$ $J_y = \frac{1}{3}xf^2$ $J_z = 0$
$\frac{1}{2}, \frac{1}{2}$	$P = \frac{1}{3}(x^2 + y^2 + z^2)f^2$ (Sphere)	$Q_x = \frac{2}{3}xzf^2$ $Q_y = \frac{2}{3}yzf^2$ $Q_z = \frac{1}{6}(z^2 - x^2 - y^2)f^2$	$J_x = -\frac{2}{3}yf^2$ $J_y = \frac{2}{3}xf^2$ $J_z = 0$

Only contributions from the p-shell are shown in Table 1: the $1s^2 2s^2$ 'core' contributes a spherically symmetric charge density, but zero spin and current densities. The 2p orbitals (real form) are $xf(r), /yf(r), /zf(r)$. When L-S coupling is assumed J, M_J are the appropriate quantum numbers.

It is to be noted that changing the sign of M_J reverses the direction of the spin polarizaton and the sense of the current flow. In the absence of electron interaction all states would have the same energy, namely $E = 2\epsilon_{1s} + 2\epsilon_{2s} + \epsilon_{2p}$.

The richness of structure in the various density functions is not evident on inspecting only the electron density $P(\mathbf{r})$; for it depends on the eigenvalues of the maximal set of mutually commuting operators required in specifying the state uniquely. This structure disappears if we take an ensemble average (thus renouncing all knowledge of observables other than the energy): the ensemble electron density then becomes spherically symmetrical and the spin and current densities disappear.

3 Symmetry considerations

The Hamiltonian operator most commonly employed in electronic structure calculations, namely (4), does not include spin operators or any terms of relativistic origin – which are usually introduced subsequently (if at all) as small corrections to the non-relativistic energy. In non-relativistic DMT it is in principle quite unnecessary to introduce electron spin explicitly: one might simply assume that the Schrödinger equation $\mathsf{H}\Phi = E\Phi$ *could* be solved exactly and then ask which of the solutions are physically admissible (many of them will not be) and whether they would impose constraints on the RDMs.

Let us suppose then that exact eigenfunctions of the Hamiltonian are available, namely functions $\Phi_\kappa(\mathbf{r}_1, \mathbf{r}_2, \dots \mathbf{r}_N)$, containing only *spatial* variables and satisfying the partial differential equation

$$\mathsf{H}\Phi_\kappa(\mathbf{r}_1, \mathbf{r}_2, \dots \mathbf{r}_N) = E\Phi_\kappa(\mathbf{r}_1, \mathbf{r}_2, \dots \mathbf{r}_N) \tag{26}$$

where, for generality, we may consider a set of g degenerate solutions with a common energy $E_1 = E_2 = \dots = E_g = E$. Such solutions normally exhibit a complicated behaviour under electron permutations, but relatively few of these exact eigenfunctions are physically acceptable: the rest must be thrown away! It is the *symmetry* that determines *which* eigenfunctions can be used in constructing wave functions, with spin included, that will satisfy the Pauli principle. Such functions will be described as 'Pauli-compatible'. In extracting an RDM from any given wavefunction, symmetry will play a crucial role.

Given a degenerate set of g exact eigenfunctions of H, the question of how they should be used in constructing an exact wave function, with spin included, was answered in the early days of quantum mechanics. The relevant theorem [21] is well known but is so often ignored as to be worth stating clearly:

> If $\Phi_1, \Phi_2, \dots \Phi_g$ is a set of g-fold degenerate exact eigenfunctions of the Hamiltonian H, of given energy E, providing an irreducible representation D of the permutation group $\mathcal{S}_N$, then the normalized function
>
> $$\Psi(\mathbf{x}_1, \mathbf{x}_2, \dots \mathbf{x}_N) = g^{-\frac{1}{2}} \sum_{\kappa=1}^{g} \Phi_\kappa(\mathbf{r}_1, \mathbf{r}_2, \dots \mathbf{r}_N)\Theta_\kappa(s_1, s_2, \dots s_N) \tag{27}$$
>
> will be an exact wavefunction *recognizing the existence of spin and satisfying the Pauli (antisymmetry) principle*, provided the spin

eigenfunctions Θ_κ carry the representation, D_S say, *associate*[8] to D. The *only* acceptable space-spin wavefunction of energy E is then the linear combination (27).

The implications of this theorem are far-reaching. If Φ is an exact *non-degenerate* eigenfunction of H $(g = 1)$ the only acceptable wavefunction is the product $\Phi\Theta$ in which Θ carries the identity representation, being invariant under all spin permutations, while Φ is totally antisymmetric under permutation of spatial variables. This particular state, however, is usually of little interest: it is a state of maximum multiplicity in which all N spins are parallel coupled to a resultant $S = \frac{1}{2}N$. For all other physically allowed states, $g > 1$ and is dependent upon N and S. In fact $g = f_S^N$, where 'Wigner's number' [19] is

$$f_S^N = \frac{(2S+1)N!}{(\frac{1}{2}N+S+1)!(\frac{1}{2}N-S)!}. \tag{28}$$

Thus, for a singlet ground state of a 10-electron system $g = 42$ – this being the number of linearly independent spin eigenfunctions with $N = 10, S = 0$, corresponding to the 42 different paths, in the well known Branching Diagram (BD), that lead from the origin to the point $S = 0$ after N steps. This means that the ground-state singlet wavefunction involves 42 degenerate and linearly independent eigenfunctions of the spin-free Hamiltonian H.

The above considerations are independent of any *orbital approximations* that may be employed in the construction of a wavefunction: they apply to exact solutions of the Schrödinger equation with any given spinless Hamiltonian. In particular, they apply to *model* systems in which the full Hamiltonian given in (4) is replaced by

$$\mathsf{H}_0 = \sum_{i=1}^{N} \mathsf{F}(i), \tag{29}$$

where 2-electron operators do not appear (even though F may be a Fock-type operator which implicitly includes electron interaction in a limited way). The above theorem applies equally to both real and model N-particle systems, provided only that the particles are fermions.

The Hamiltonian (29) provides of course the basis of most *orbital* approximations: it has exact eigenfunctions of the form

$$\Phi^0_{ij\ldots p}(\mathbf{r}_1, \mathbf{r}_2, \ldots \mathbf{r}_N) = \phi_i(\mathbf{r}_1)\phi_j(\mathbf{r}_2)\ldots \phi_p(\mathbf{r}_N), \tag{30}$$

[8]The 'associate' or 'dual' $(\check{\mathrm{D}})$ of a representation D contains matrices $\check{\mathbf{D}}(\mathsf{P}) = \epsilon_P\tilde{\mathbf{D}}(\mathsf{P}^{-1})$ where the tilde denotes transposition and $\epsilon_P(=\pm 1)$ is the usual parity factor.

where ϕ_k denotes an eigenfunction of the 1-electron operator F (with eigenvalue ϵ_k, say). The total electronic energy, in IPM approximation, is

$$E^0_{ij\dots\ p} = \epsilon_i + \epsilon_j + \dots\ + \epsilon_p \tag{31}$$

and obviously the state is $N!$-fold degenerate, if the 'occupied' orbitals are all different, since there will be $N!$ permuted products $\mathsf{P}\Phi^0_{ij\dots\ p}$ giving the same $E^0_{ij\dots\ p}$. This permutation degeneracy is resolved by seeking, for a state of given S (which is a *symmetry* label), the f^N_S linearly independent combinations that provide the representation $\breve{\mathrm{D}}_S$. When spin is admitted and these functions are combined to give the unique antisymmetric space-spin function (27), the permutation degeneracy is removed completely – the resultant wavefunction then being Pauli compatible. The 'closed-shell' ground state, fundamental in Hartree-Fock theory, is obtained by filling the lowest-energy orbitals in (30) *in pairs*, adding spin factors α, β within each pair, and antisymmetrizing to obtain a Slater determinant. But this is merely a convenient way of imposing the correct symmetry without using group theory; and it leads at once to the conclusion that *no spin-orbital can be occupied by more than one electron* – the (limited) statement of the Pauli principle that underpins the whole *aufbau* approach in electronic structure theory.

The group theoretical approach is general but less simple. Thus, in the case $N = 10, S = 0$, the irreducible representation D_S is carried by branching diagram functions whose symmetry species may be indicated by Young tableaux (see, for example, Refs.[22,23]) such as

1	3	5	7	9
2	4	6	8	10

– in which electron spin indices appear in the 'boxes'. The tableau is essentially an 'instruction' for generating a symmetry function from a primitive product such as $\alpha\beta\alpha\beta\dots$, namely, "symmetrize with respect to the indices within each row and antisymmetrize with respect to those in the columns". In this example the result is a Weyl-Rumer function of the form

$$(\alpha(s_1)\beta(s_2) - \beta(s_1)\alpha(s_2)) \times (\alpha(s_3)\beta(s_4) - \beta(s_3)\alpha(s_4)) \times \dots,$$

as used in valence bond theory. The other linearly independent spin functions are associated with 'standard' tableaux in which the indices appear in the boxes in a different order, but always increasing along the rows and down the columns: in this example there are 42 such possibilities, corresponding to the 42 standard branching diagram functions. The *spatial* functions that carry the associate representation follow from the spin tableaux by interchanging rows and columns; so the associated orbital function could be obtained by *symmetrizing* with respect to indices (1,2), (3,4), etc. and *antisymmetrizing* with respect to (1,3,5,7,9) and (2,4,6,8,10). On using a primitive product in which the orbitals are identical within pairs (i.e. there is one doubly occupied orbital for each pair) it is easy to show that the sum of products in (27) can be reproduced by antisymmetrizing (in both space and spin variables) a single product

$$\phi_1(\mathbf{r}_1)\phi_1(\mathbf{r}_2)\phi_2(\mathbf{r}_3)...[\alpha(s_1)\beta(s_2) - \beta(s_1)\alpha(s_2)]...$$

– which yields the Slater determinant $|\phi_1\alpha\ \phi_1\beta\phi_2\alpha\ \phi_2\beta....|$.

It is clear from such examples that IPM-type wavefunctions, built up from products of orthogonal spin-orbitals, are most easily handled using Slater determinants. However, it is entirely possible to eliminate the spin factors and to work in terms of spatial functions alone [24], imposing the correct permutation symmetry by means of a projection operator: this spin-free approach (apart from its practical utility in dealing with non-IPM wavefunctions [25] is important in revealing the presence of 'hidden' constraints, which must be observed when we pass from the N-electron wavefunction to the electron density and other density functions by integrating over variables that appear to be redundant.

The electron density (1), usually derived from a wave function expressed in terms of Slater determinants, may be obtained more directly in the spin-free formulation, from the set $\{\Phi_\kappa\}$ in (27), without the need for spin integration; and indeed *any one* of the spinless functions Φ_κ may be used for this purpose. But the definition (1) then needs modification: it is replaced by

$$P(\mathbf{r}) = \sum_{i=1}^{N} \int \Phi_\kappa(\mathbf{r}_1, \mathbf{r}_2, ... \mathbf{r}_N)\delta(\mathbf{r} - \mathbf{r}_i)\Phi_\kappa^*(\mathbf{r}_1, \mathbf{r}_2, ... \mathbf{r}_N)d\mathbf{r}_1 d\mathbf{r}_2 ... d\mathbf{r}_N \quad (32)$$

– where each integration may give a different result, owing to the lack of any simple symmetry in the spatial function. A more symmetrical form may be obtained by writing the space-spin Ψ in the form (27) and then using (1): orthogonality of the spin factors leads to the result (normalized)

$$P(\mathbf{r}) = N \int (f_S^N)^{-1} \sum_{\kappa=1}^{f_S^N} \Phi_\kappa(\mathbf{r}, \mathbf{r}_2, ... \mathbf{r}_N)\Phi_\kappa^*(\mathbf{r}, \mathbf{r}_2, ... \mathbf{r}_N)d\mathbf{r}_2 ... d\mathbf{r}_N. \quad (33)$$

This means that spin may be ignored, the electron density being derived in a form *analogous to* (1), viz.

$$P(\mathbf{r}) = N \int \Phi(\mathbf{r}, \mathbf{r}_2, ... \mathbf{r}_N)\Phi^*(\mathbf{r}, \mathbf{r}_2, ... \mathbf{r}_N)d\mathbf{r}_2 ... d\mathbf{r}_N, \quad (34)$$

provided the pure-state density $\Phi\Phi^*$ is replaced by an *ensemble density*:

$$\Phi\Phi^* \longrightarrow (f_S^N)^{-1}(\Phi_1\Phi_1^* + \Phi_2\Phi_2^* + ...\Phi_\kappa\Phi_\kappa^* + ...), \tag{35}$$

corresponding not to a pure state but rather to a *statistical ensemble* in which every degenerate state appears with the same weight. Consequently, when (as in formal DFT) attention is focused solely on the (spatial) electron density, this density cannot be assumed *pure state* N-representable (i.e. derivable from a single state function of correct symmetry); at best it can be assumed only *ensemble* N-representable, the ensemble containing in general a considerable number of linearly independent spin-free wave functions.

It is easily verified explicitly, taking for example the case of 4 electrons in 4 arbitrary orbitals, that both (32) and (34) – the latter with the ensemble density (35) – yield exactly the same singlet and triplet densities as the conventional approach through (1). But the the real problem in any of these approaches is still that of how to guarantee Pauli compatibility; and it seems that this inevitably involves the imposition of symmetry constraints on the *many-electron state function itself* (pure or mixed).

As long as N-representability conditions for the pair function Π remain unknown, the goal of rigorously eliminating the wavefunction in favour of even the 1- and 2-electron density matrices, let alone the electron density itself, still seems to be unattainable. All that can be done is to impose the proper permutation symmetry conditions on a variational *wavefunction* and then proceed to the required density matrices by integrating over $N-1$ or $N-2$ variables. Even this is not, in general, an easy task – easy as it may be for an IPM 1-determinant wavefunction. Here we note only two aspects of the problem: (i) the *condition* that a spin-free wavefunction must satisfy in order to provide a Pauli-compatible starting point; and (ii) how to *impose* the condition on an arbitrary function.

The necessary and sufficient condition that a spin-free function Φ must satisfy is simply that it belong to an irreducible representation $\breve{D}_S$; and this is equivalent to saying that it must be an eigenfunction of the class operator

$$\mathsf{C} = \sum_{i<j=1}^{N} \mathsf{P}_{ij} \tag{36}$$

with eigenvalue

$$c_S^N = N - \tfrac{1}{4}N^2 - S(S+1), \tag{37}$$

– as follows readily from the Dirac spin-exchange identity. As for the *imposition* of this condition on a suitable trial function, for a ground state of multiplicity $2S+1$, the standard procedure is to take any N-electron function whatever, Ω say, and apply a projection operator

$$\mathsf{O}^{S}_{\kappa\kappa} = \sum_{\mathsf{P}} \breve{D}_S(\mathsf{P})_{\kappa\kappa}\mathsf{P}, \tag{38}$$

for whatever value of the species symbol κ may be convenient, to obtain $\Phi_{S,\kappa} = \mathsf{O}^{S}_{\kappa\kappa}\Omega$. In practice, Ω might be, for example, an orbital product $\Omega_K = \phi_{k_1}\phi_{k_2}\dots\,\phi_{\kappa_N}$; and on taking $\kappa = 1$ (for the first tableau of the dual representation $\breve{\mathsf{D}}_S$) the operator O^{S}_{11} produces a Weyl-Rumer function of the kind used in spin-free forms of valence bond theory [25] Unfortunately, the projected function then contains $N!$ terms and, in the general case where the orbitals are arbitrary (and thus non-orthogonal), the reduction to obtain P and Π is feasible only for few-electron systems.

4 Relativistic generalizations

In any relativistic generalization of the formalism used so far, the usual 2-component spin-orbitals (as encountered in Section 2.4), namely,

$$\psi = \phi^{(\alpha)}\alpha + \phi^{(\beta)}\beta \tag{39}$$

– which are used in constructing general *non*-relativistic wavefunctions – will have to be replaced by the 4-component spinors that occur as solutions of the Dirac equation (see, for example, Ref.[26]). A set of *non-interacting* Dirac electrons would be described using an antisymmetrized product of such spinors; and 'switching on' the interaction would lead to a function expressible as a linear combination of the totality of all possible antisymmetrized products. Instead of building the wavefunction from a complete set of 2-component spin-orbitals, we will therefore use a complete set[9] of 'Dirac spin-orbitals', writing (with e standing for a Dirac basis vector)

$$\psi_k = \phi^{(1)}_k\mathsf{e}_1 + \phi^{(2)}_k\mathsf{e}_2 + \phi^{(3)}_k\mathsf{e}_3 + \phi^{(4)}_k\mathsf{e}_4, \tag{40}$$

thus preserving the analogy with the usual 2-component form (39). A detailed study of the Dirac equation and its solutions will not be required but it is useful to sketch the derivation.

[9] More accurately, the set of *positive energy* solutions of the Dirac equation – which are orthogonal to the negative energy set and separated from them by $\approx 2m_0c^2$, m_0 being the electronic *rest mass*.

4.1 The Dirac equation

The starting point is the relativistic Hamiltonian for an electron moving in the presence of fields $(\mathbf{E}, \mathbf{B})$ arising from the scalar and vector potentials ϕ and $\mathbf{A}$: this is usually written in the form (see, for example, Ref.[27])

$$\frac{(H + e\phi)^2}{c^2} - (\mathbf{p} + e\mathbf{A})^2 = c^2 m_0^2 \tag{41}$$

– showing that the quantity on the left is a *relativistically invariant scalar.*

In the absence of fields, this quantity becomes $(H/c)^2 - \mathbf{p}^2$ and is the squared magnitude of the energy-momentum 4-vector with components

$$p_1 \quad p_2 \quad p_3 \quad p_4 \qquad (p_4 = H/c)$$

in a Minkowski space with metric tensor $g_{11} = g_{22} = g_{33} = -1, g_{44} = +1$. The invariant scalar is then

$$-(p_1^2 + p_2^2 + p_3^2) + p_4^2 = c^2 m_0^2. \tag{42}$$

When fields are applied the components are augmented by the addition of the 4-vector with components

$$eA_1 \quad eA_2 \quad eA_3 \quad e\phi/c$$

to give

$$-(\pi_1^2 + \pi_2^2 + \pi_3^2) + \pi_4^2 = c^2 m_0^2. \tag{43}$$

The passage to quantum theory results on making the usual operator substitutions $(H \to \mathsf{H} = i\hbar\partial/\partial t, \quad p_\mu \to \mathsf{p}_\mu = (\hbar/i)\partial/\partial x_\mu)$ in the components

$$\pi_\mu = p_\mu + eA_\mu \quad (\mu = 1, 2, 3), \quad \pi_4 = (H + e\phi)/c.$$

The operator associated with the left-hand side of (43), acting on an operand ψ, yields the Klein-Gordon equation

$$\left(\pi_4^2 - (\pi_1^2 + \pi_2^2 + \pi_3^2) - c^2 m_0^2\right)\psi = 0, \tag{44}$$

in which, for a stationary state, the operator H in π_4 is equivalent to multiplication by E (the energy parameter in the usual time-dependent phase factor).

The Dirac equation arises when (44) is 'factorized' to obtain a linear form

$$(\pi_4 + \boldsymbol{\alpha} \cdot \boldsymbol{\pi} + \beta c m_0)(\pi_4 - \boldsymbol{\alpha} \cdot \boldsymbol{\pi} - \beta c m_0)\psi = 0 \tag{45}$$

in which $\boldsymbol{\alpha} \cdot \boldsymbol{\pi} = (\alpha_1\pi_1 + \alpha_2\pi_2 + \alpha_3\pi_3)$ and the operators α_μ ($\mu = 1, 2, 3$) and β are assumed to commute with all others. The condition for the equivalence of (44) and (45) is that the operators (putting $\beta = \alpha_4$) satisfy

$$\alpha_\mu\alpha_\nu + \alpha_\nu\alpha_\mu = 2\delta_{\mu\nu} \quad \mu, \nu = 1, 2, 3, 4. \tag{46}$$

The usual procedure at this point is to employ only the factor nearest the operand in (45), obtaining the Dirac equation

$$[\pi_4 - (\alpha_1\pi_1 + \alpha_2\pi_2 + \alpha_3\pi_3) - \beta cm_0]\,\psi = 0, \tag{47}$$

and to pass immediately to a *matrix representation* in which the αs are represented as 4×4 matrices, while ψ becomes a 4-component column. For present purposes, however, it is more convenient to employ a purely symbolic approach to expose the parallel between 2-component (Pauli) and 4-component (Dirac) forms.

To this end, following Temple [28], we define an energy-momentum 4-vector in the form

$$\Pi = \pi_1\mathsf{E}_1 + \pi_2\mathsf{E}_2 + \pi_3\mathsf{E}_3 + i\pi_4\mathsf{E}_4, \tag{48}$$

where the E-symbols will have properties such that the operator associated with the left-hand side of (43) can be written as Π^2. This will in fact be an identity provided

$$\mathsf{E}_\mu\mathsf{E}_\nu + \mathsf{E}_\nu\mathsf{E}_\mu = -2\delta_{\mu\nu} \quad \mu, \nu = 1, 2, 3, 4. \tag{49}$$

which resembles (46) except for a sign reversal – every E-symbol is a square root of -1.

Equation (44) now states that Π^2 has an eigenvalue $c^2m_0^2$ and it follows that

$$\Pi\psi = (\pi_1\mathsf{E}_1 + \pi_2\mathsf{E}_2 + \pi_3\mathsf{E}_3 + i\pi_4\mathsf{E}_4)\,\psi = \lambda\psi, \tag{50}$$

where the eigenvalue λ must be $\lambda = \pm cm_0$. Temple [28] was able to solve the Dirac equation in the form (50), obtaining the bound-state energy levels and discussing the Zeeman effect, by *purely symbolic* arguments – a considerable achievement for a research student in 1929!

4.2 Matrix form of the Dirac equation

To maintain the parallel with 2-component theory, it is expedient to express the 'Dirac space' as a Kronecker product of two spin spaces. To this end, we construct from the Es two operator sets with the same properties as the Pauli operators: these are

$$\mathsf{S}_1 = i\mathsf{E}_2\mathsf{E}_3, \qquad \mathsf{S}_2 = i\mathsf{E}_3\mathsf{E}_1, \qquad \mathsf{S}_3 = i\mathsf{E}_1\mathsf{E}_2 \tag{51}$$

and

$$\mathsf{T}_1 = i\mathsf{E}_4\mathsf{E}_5, \qquad \mathsf{T}_2 = i\mathsf{E}_5, \quad . \quad \mathsf{T}_3 = i\mathsf{E}_4, \tag{52}$$

where $i\mathsf{E}_5 = \mathsf{E}_1\mathsf{E}_2\mathsf{E}_3\mathsf{E}_4$.

The 4 E-symbols, $\mathsf{E}_1, \ldots \mathsf{E}_4$ are characterized simply by their abstract properties (49) and are the generators of a Clifford algebra [29]: any product of such symbols is equivalent to one of 16 independent results

$$i1,\ \mathsf{E}_\mu,\ \mathsf{E}_\mu\mathsf{E}_\nu \qquad (\mu,\nu = 1,..5,\ \mu < \nu) \tag{53}$$

to within a factor ± 1 or $\pm i$. Other sets of quantities, with similar rules of combination, are well known: the Pauli spin operators, for example, $\mathsf{S}_x, \mathsf{S}_y, \mathsf{S}_z$ obey the rules

$$\mathsf{S}_x^2 = \mathsf{S}_y^2 = \mathsf{S}_z^2 = 1, \quad \mathsf{S}_x\mathsf{S}_y = -\mathsf{S}_y\mathsf{S}_x, \quad \mathsf{S}_y\mathsf{S}_z = -\mathsf{S}_z\mathsf{S}_y, \quad \mathsf{S}_z\mathsf{S}_x = -\mathsf{S}_x\mathsf{S}_z. \tag{54}$$

If, as usual, we use indices $1, 2, 3$ for the x, y, z terms and add a factor i, then the symbols

$$\mathsf{A}_1 = i\mathsf{S}_x,\ \mathsf{A}_2 = i\mathsf{S}_y,\ \mathsf{A}_3 = i\mathsf{S}_z,\ \mathsf{A}_4 = i1, \tag{55}$$

– to which a fourth member has been added – are characterized by the properties

$$\mathsf{A}_\mu^2 = -1, \quad \mathsf{A}_\mu\mathsf{A}_\nu + \mathsf{A}_\nu\mathsf{A}_\mu = 0 \quad (\mu \neq \nu), \tag{56}$$

for all values of $\mu, \nu = 1, 2, 3, 4$. The set (56) thus also forms a Clifford algebra.

Another set of symbols, $\{\mathsf{B}_\sigma\}$, assumed to commute with all the A-symbols and to have analogous properties, namely

$$\mathsf{B}_\sigma^2 = -1, \quad \mathsf{B}_\sigma\mathsf{B}_\tau + \mathsf{B}_\tau\mathsf{B}_\sigma = 0 \quad (\sigma \neq \tau), \tag{57}$$

forms a second Clifford algebra, 'conjugate' to the first in the sense $\mathsf{B}_\mu = \mathsf{q}\mathsf{A}_\mu\mathsf{q}^{-1}$, which obviously ensures that each set has the same properties as the other. The element q is easily found to be

$$\mathsf{q} = \tfrac{1}{2}(\mathsf{A}_1\mathsf{B}_1 + \mathsf{A}_2\mathsf{B}_2 + \mathsf{A}_3\mathsf{B}_3 - 1). \tag{58}$$

The two conjugate triads (supplemented by the addition of the imaginary units $\mathsf{A}_4, \mathsf{B}_4$) serve to generate the full set of 16 E-symbols in the form

$$\mathsf{A}_\sigma,\ \mathsf{B}_\tau,\ i\mathsf{A}_\sigma\mathsf{B}_\tau,\ i1, \tag{59}$$

– a factor i being attached where necessary so that the square of any member of the set will be -1. This can be verified explicitly, but at this point it is convenient to pass to a *matrix representation* of the E-symbols – thus making contact with the usual Dirac formulation.

To do so, we associate with each triad a set of 2 x 2 Pauli matrices[10]. Thus,

$$\mathsf{A}_1 \to \mathbf{S}_x = \begin{pmatrix} 0 & 1 \\ 1 & 0 \end{pmatrix}, \quad \mathsf{A}_2 \to \mathbf{S}_y = \begin{pmatrix} 0 & \bar{\imath} \\ i & 0 \end{pmatrix}, \quad \mathsf{A}_3 \to \mathbf{S}_z = \begin{pmatrix} 1 & 0 \\ 0 & \bar{1} \end{pmatrix}, \tag{60}$$

where it is easily confirmed that the 2 x 2 matrices have exactly the same properties (apart from a possible factor i) as the A-symbols in (55). The same three matrices provide a similar representation of the B-symbols; and the matrix to be associated with a product $\mathsf{A}_\sigma \mathsf{B}_\tau$ is the *outer product* of the matrices associated with the separate factors. Thus, with AB we associate the outer product[11] $\mathbf{A} \times \mathbf{B}$ to obtain the 4×4 matrices associated with the symbols in (59).

The first six members of the outer-product set are $1_A \mathsf{B}_\tau, \mathsf{A}_\mu 1_B$, where 1_A and 1_B are unit operators for the A- and B-sets, respectively. The associated outer-product matrices for the first three members will be denoted by[12] $\mathbf{S}_1, \mathbf{S}_2, \mathbf{S}_3$, respectively; and those for the second three by $\mathbf{T}_1, \mathbf{T}_2, \mathbf{T}_3$. In terms of the E-symbols in (53) they provide representations of the two conjugate triads, (51) and (52).

The full matrices are

$$\mathbf{S}_1 = \left(\begin{array}{cc|cc} 0 & 1 & 0 & 0 \\ 1 & 0 & 0 & 0 \\ \hline 0 & 0 & 0 & 1 \\ 0 & 0 & 1 & 0 \end{array}\right), \quad \mathbf{S}_2 = \left(\begin{array}{cc|cc} 0 & \bar{\imath} & 0 & 0 \\ i & 0 & 0 & 0 \\ \hline 0 & 0 & 0 & \bar{\imath} \\ 0 & 0 & i & 0 \end{array}\right), \quad \mathbf{S}_3 = \left(\begin{array}{cc|cc} 1 & 0 & 0 & 0 \\ 0 & \bar{1} & 0 & 0 \\ \hline 0 & 0 & 1 & 0 \\ 0 & 0 & 0 & \bar{1} \end{array}\right) \tag{61}$$

and

$$\mathbf{T}_1 = \left(\begin{array}{cc|cc} 0 & 0 & 1 & 0 \\ 0 & 0 & 0 & 1 \\ \hline 1 & 0 & 0 & 0 \\ 0 & 1 & 0 & 1 \end{array}\right) \quad \mathbf{T}_2 = \left(\begin{array}{cc|cc} 0 & 0 & \bar{\imath} & 0 \\ 0 & 0 & 0 & \bar{\imath} \\ \hline i & 0 & 0 & 0 \\ 0 & i & 0 & 0 \end{array}\right) \quad \mathbf{T}_3 = \left(\begin{array}{cc|cc} 1 & 0 & 0 & 0 \\ 0 & 1 & 0 & 0 \\ \hline 0 & 0 & \bar{1} & 0 \\ 0 & 0 & 0 & \bar{1} \end{array}\right). \tag{62}$$

[10]Factors of i in (55) are here disregarded; and for typographical convenience minus signs are replaced by bars.

[11]The outer product $\mathbf{A} \times \mathbf{B}$ is a matrix of block form, in which the ij-block is $A_{ij}\mathbf{B}$; a unit matrix is associated with a unit operator.

[12]The 4×4 matrices will not be confused with the 2×2 matrices in (60).

The matrices which contain only real elements are symmetric while those with pure imaginary elements are skew. The matrices for the first few E-symbols, in this representation are given by

$$\imath\mathbf{E}_1 = \mathbf{T}_1\mathbf{S}_1, \quad \imath\mathbf{E}_2 = \mathbf{T}_1\mathbf{S}_2, \quad \imath\mathbf{E}_3 = \mathbf{T}_1\mathbf{S}_3, \quad \imath\mathbf{E}_4 = \mathbf{T}_3 \tag{63}$$

and from them all 16 E-matrices may be generated by simple matrix multiplication. Thus, we find

$$\mathbf{E}_1 = \left(\begin{array}{cc|cc} 0 & 0 & 0 & \bar{\imath} \\ 0 & 0 & \bar{\imath} & 0 \\ \hline 0 & \bar{\imath} & 0 & 0 \\ \bar{\imath} & 0 & 0 & 0 \end{array}\right), \quad \mathbf{E}_2 = \left(\begin{array}{cc|cc} 0 & 0 & 0 & \bar{1} \\ 0 & 0 & 1 & 0 \\ \hline 0 & \bar{1} & 0 & 0 \\ 1 & 0 & 0 & 0 \end{array}\right), \quad \mathbf{E}_3 = \left(\begin{array}{cc|cc} 0 & 0 & \bar{\imath} & 0 \\ 0 & 0 & 0 & \imath \\ \hline \bar{\imath} & 0 & 0 & 0 \\ 0 & \imath & 0 & 0 \end{array}\right) \tag{64}$$

while

$$\mathbf{E}_4 = \mathrm{diag}(\bar{\imath}\,\bar{\imath}\,\imath\,\imath). \tag{65}$$

To complete the translation of (50) into matrix form, we need to confirm that ψ can be represented as a column of 4 components $\phi_1, \phi_2, \phi_3, \phi_4$. To this end, we express the basis vectors $\mathsf{e}_1, \dots, \mathsf{e}_4$ as outer products of those that carry the representations of the A- and B-operators in (56) and (57): let us call them $\mathsf{a}_1, \mathsf{a}_2$ and $\mathsf{b}_1, \mathsf{b}_2$, respectively, choosing them as eigenvectors of the ators (dropping the phase factors $\imath$) A_3 (in (55))and B_3 (in the conjugate . Thus

$$\mathsf{A}_3 \left(\mathsf{a}_1 \quad \mathsf{a}_2 \right) = \left(\mathsf{a}_1 \quad \mathsf{a}_2 \right) \begin{pmatrix} 1 & 0 \\ 0 & -1 \end{pmatrix}, \tag{66}$$

$$\mathsf{B}_3 \left(\mathsf{b}_1 \quad \mathsf{b}_2 \right) = \left(\mathsf{b}_1 \quad \mathsf{b}_2 \right) \begin{pmatrix} 1 & 0 \\ 0 & -1 \end{pmatrix}. \tag{67}$$

The Dirac basis can then be expressed as

$$\mathbf{e} = (\mathsf{e}_1\ \mathsf{e}_2\ \mathsf{e}_3\ \mathsf{e}_4) = (\mathsf{a}_1\mathsf{b}_1\ \mathsf{a}_1\mathsf{b}_2\ \mathsf{a}_2\mathsf{b}_1\ \mathsf{a}_2\mathsf{b}_2), \tag{68}$$

where it follows from (66) and (67) that the basis vectors $\mathsf{e}_1, \dots \mathsf{e}_4$ are simultaneous eigenvectors of A_3 and B_3 with eigenvalues $(+1,+1), (+1,-1), (-1,+1), (-1,-1)$, respectively.

The Dirac equation in (50) may now be transcribed into matrix form: the state vector ψ becomes (cf. (40))

$$\psi = \mathbf{e}\boldsymbol{\phi} \tag{69}$$

where $\boldsymbol{\phi}$ is a column (containing the 4 components $\phi_1, \dots \phi_4$), representing ψ in the Dirac basis; while the energy-momentum operator Π is represented by the 4 matrix

$$\mathbf{\Pi} = \pi_1\mathbf{E}_1 + \pi_2\mathbf{E}_2 + \pi_3\mathbf{E}_3 + \imath\pi_4\mathbf{E}_4, \tag{70}$$

where the $\mathbf{E}$-matrices are given in (64) and (65). The matrix form of (50), namely $\mathbf{\Pi}\boldsymbol{\phi} = \lambda\boldsymbol{\phi}$, then yields the four conditions

$$\begin{aligned}
(\pi_4 - m_0c)\phi_1 - \pi_3\phi_3 - (\pi_1 - i\pi_2)\phi_4 &= 0 \\
(\pi_4 - m_0c)\phi_2 - (\pi_1 + i\pi_2)\phi_3 + \pi_3\phi_4 &= 0 \\
(\pi_4 + m_0c)\phi_3 - \pi_3\phi_1 - (\pi_1 - i\pi_2)\phi_2 &= 0 \\
(\pi_4 + m_0c)\phi_4 - (\pi_1 + i\pi_2)\phi_1 + \pi_3\phi_2 &= 0
\end{aligned} \qquad (71)$$

which provide the so-called 'standard form' of the Dirac equations. Of course other forms follow on making any unitary transformation of $\mathbf{\Pi}$ and $\boldsymbol{\phi}$; (71) is most convenient for present purposes. Since $\pi_4 = (\mathsf{H} + e\phi)/c$, there will be two solutions with ϕ_1, ϕ_2 large and ϕ_3, ϕ_4 negligible, corresponding to eigenfunctions of H with energy $\approx m_0c^2$; and two more with the roles of ϕ_1, ϕ_2 and ϕ_3, ϕ_4 reversed, corresponding to energy $\approx -m_0c^2$. When the positive-energy solutions are written in the form (40), namely $\psi = \sum_{\mu=1}^{4} \phi_\mu \mathsf{e}_\mu$, the 'large components' will correspond to $\mu = 1, 2$ and the 'small components' to $\mu = 3, 4$. (Some authors, notably Slater, use the opposite convention.)

4.3 Relativistic density matrices

Our main task is to generalize the density matrices: thus, for example, anticipating the use of determinants of 4-component spin-orbitals, the one-body density will become

$$\rho(\mathbf{x}; \mathbf{x}') = \sum_{i,j} \rho_{ij}\psi_i(\mathbf{x})\psi_j^*(\mathbf{x}'). \qquad (72)$$

An alternative 'natural' expansion may again be found by diagonalizing the density matrix (cf. Ref.[5]) to obtain

$$\rho(\mathbf{x}; \mathbf{x}') = \sum_{k} n_k\psi_k(\mathbf{x})\psi_k^*(\mathbf{x}), \qquad (73)$$

in which ψ_k is now a 'natural' spin-orbital with occupation number n_k.

Let us now introduce formal 'spin variables' in the usual way: for a general 2-component spin-orbital, (39) will become $\psi_k(\mathbf{x}) = \phi_k^\alpha(\mathbf{r})\alpha(s) + \phi_k^\beta(\mathbf{r})\beta(s)$, but for a Dirac spin-orbital (40) we shall have to write

$$\psi_k(\mathbf{x}) = \sum_{\mu=1}^{4} \phi_k^\mu(\mathbf{r})e_\mu(s,t), \qquad (74)$$

where there are now *two* spin variables, so that $\mathsf{a}_\rho \rightarrow a_\rho(s)$, $\mathsf{b}_\sigma \rightarrow b_\sigma(t)$, and a Dirac basis vector will become $\mathsf{e}_\mu \rightarrow e_\mu(s,t)$. The index $\mu = 1,2,3,4$ will then correspond to index *pairs*:

$$\begin{aligned} e_1(s,t) &= a_1(s)b_1(t), \qquad e_2(s,t) = a_1(s)b_2(t), \\ e_3(s,t) &= a_2(s)b_1(t), \qquad e_4(s,t) = a_2(s)b_2(t). \end{aligned} \tag{75}$$

And consequently the 1-DM (7) will take the 16-component form

$$\begin{aligned} \rho(\mathbf{r},s,t:\mathbf{r}',s',t') &= P^{11}(\mathbf{r};\mathbf{r}')a_1(s)b_1(t)a_1^*(s')b_1^*(t') + ... \\ &\quad + ... \\ &\quad + P^{44}(\mathbf{r};\mathbf{r}')a_2(s)b_2(t)a_2^*(s')b_2^*(t'). \end{aligned} \tag{76}$$

When the non-spatial variables are removed by integration, in the usual way, we get the *spinless* density matrix

$$P(\mathbf{r};\mathbf{r}') = \sum_\mu P^{\mu\mu}(\mathbf{r};\mathbf{r}'), \tag{77}$$

where the label $\mu = 1,..4$ corresponds to the Dirac basis labelling and only terms with $\mu = \nu$ have survived the spin integrations. These are

$$P^{\mu\mu}(\mathbf{r};\mathbf{r}') = \sum_k n_k \phi_k^\mu(\mathbf{r})\phi_k^{\mu *}(\mathbf{r}'). \tag{78}$$

The terms for $\mu = 1,2$ are large, for $\mu = 3,4$ they are small, and the spinless density (77) is thus

$$P(\mathbf{r};\mathbf{r}') = P^L(\mathbf{r};\mathbf{r}') + P^S(\mathbf{r};\mathbf{r}'), \tag{79}$$

a sum of the 'Large' and 'Small' terms

$$\begin{aligned} P^L(\mathbf{r};\mathbf{r}') &= \sum_k n_k[\phi_k^{(1)}(\mathbf{r})\phi_k^{(1)*}(\mathbf{r}') + \phi_k^{(2)}(\mathbf{r})\phi_k^{(2)*}(\mathbf{r}')], \\ P^S(\mathbf{r};\mathbf{r}') &= \sum_k n_k[\phi_k^{(3)}(\mathbf{r})\phi_k^{(3)*}(\mathbf{r}') + \phi_k^{(4)}(\mathbf{r})\phi_k^{(4)*}(\mathbf{r}')]. \end{aligned} \tag{80}$$

Here we have used the natural expansion (73), with spin-orbitals written in the form (74). The second term in (80), absent in a Pauli-type approximation, contains the correction arising from the use of a 4-component formulation; it is of order $(2m_0c)^{-2}$ and is usually negligible except at singularities in the potential. As expected, for $N = 1$, (79) reproduces the density obtained from a standard treatment of the Dirac equation; but now there is no restriction on the particle number.

According to (14), the current density should also follow from (79): it should thus be

$$J_x(\mathbf{r}) = J_x^L(\mathbf{r}) + J_x^S(\mathbf{r}), \tag{81}$$

with similar results for J_y, J_z, the 'L' and 'S' terms being defined as in (80). Again, the Small terms are of order $(2m_0c)^{-2}$ relative to the Large. In this case, however, (81) does *not* reproduce the standard result in the case $N = 1$. This is because in the present approach the distributions of charge and spin are treated separately; the former leads to the 'conventional' current resulting from the circulation of point charges, while the latter leads to a 'spin current density' arising as the curl of a magnetization density associated with the spin magnetic dipoles. The conventional current is correctly given by (81), the spin current will be derived separately from the spin density.

The spin density should follow from the density matrix (76), which includes the spin variables. As in (77), $Q_x(\mathbf{x};\mathbf{x}')$ will be a sum of terms containing the various spinor components, summed over all spin-orbitals in the natural expansion. A typical term in the expansion, from Dirac spinor ψ_k, will be

$$Q_x^k(\mathbf{r};\mathbf{r}') = \int \left[\sum_{\mu\nu} \mathsf{S}_x \phi_k^\mu(\mathbf{r}) e_\mu(s,t) \phi_k^{\nu\,*}(\mathbf{r}') e_\nu^*(s',t') \right]_{s'=s,t'=t} dsdt, \tag{82}$$

where the operator works on the unprimed variables. On noting that $\mathbf{S}_1$ in (61) describes the effect of S_1 on the Dirac basis vectors according to $\mathsf{S}_1\mathbf{e} = \mathbf{e}\mathbf{S}_1$, it follows that

$$\mathsf{S}_1(\mathsf{e}_1\ \mathsf{e}_2\ \mathsf{e}_3\ \mathsf{e}_4) = (\mathsf{e}_1\ \mathsf{e}_2\ \mathsf{e}_3\ \mathsf{e}_4)\mathbf{S}_1 = (\mathsf{e}_2\ \mathsf{e}_1\ \mathsf{e}_4\ \mathsf{e}_3). \tag{83}$$

On using these results in the 16-component analogue of (9) we obtain

$$Q_x(\mathbf{r};\mathbf{r}') = Q_x^L(\mathbf{r};\mathbf{r}') + Q_x^S(\mathbf{r};\mathbf{r}'), \tag{84}$$

where the Large and Small terms are given by

$$\begin{aligned} Q_x^L(\mathbf{r};\mathbf{r}') &= \sum_k n_k[\phi_k^{(2)}(\mathbf{r})\phi_k^{(1)\,*}(\mathbf{r}') + \phi_k^{(1)}(\mathbf{r})\phi_k^{(2)\,*}(\mathbf{r}')], \\ Q_x^S(\mathbf{r};\mathbf{r}') &= \sum_k n_k[\phi_k^{(4)}(\mathbf{r})\phi_k^{(3)\,*}(\mathbf{r}') + \phi_k^{(3)}(\mathbf{r})\phi_k^{(4)\,*}(\mathbf{r}')]. \end{aligned} \tag{85}$$

There are corresponding results for Q_y, Q_z and again the 'L' and 'S' terms differ by two orders of magnitude, the large terms reproducing the densities obtained in 2-component Pauli approximation (cf. (9) *et seq.*. When the curl of the corresponding magnetization density is added to (81) we obtain, for the case $N = 1$, the 'Dirac current density' (fully discussed by Slater [30]) which already includes the spin term.

In summary, although (to quote Slater) " ... the Dirac wavefunction has the capability of giving complete information about the orientation of the spin in space... ", the same is true for any number of electrons, provided the wavefunctions are built from 4-component spin-orbitals. There is no need to invoke explicitly the Dirac equation or the expansion of small components in terms of large (see, for example, Ref.[31]). The charge, current and spin densities defined in the conventional (2-component) approach, in terms of the usual density matrices, are immediately generalized to the 'fully relativistic' (4-component) level.

5 References

1. J. von Neumann, Göttinger Nachr. 1927, 245 (1927); see also J. von Neumann, *Mathematische Grundlagen der Quantenmechanik*, Springer, Berlin (1932) (Translation by R.T. Beyer: *Mathematical Foundations of Quantum Mechanics*, Princeton (1955)).

2. P.A.M. Dirac, Proc. Camb. Phil. Soc. **25**, 62 (1928); see also P.A.M. Dirac, *The Principles of Quantum Mechanics 3ed*, Oxford (1947).

3. P.-O. Löwdin, J. Chem. Phys. **18**, 365 (1950).

4. K. Husimi, Proc. Phys.-Math. Soc. Jpn. 22, 264 (1940).

5. P.-O. Löwdin, Phys. Rev. **97**, 1474 (1955); *ibid.*, **97**, 1490 (1955); *ibid.*, 1508 (1955).

6. R. McWeeny, Proc. Roy. Soc. Lond. **A232**, 114 (1955); see also *ibid.*, **A253**, 242 (1959).

7. R. McWeeny, Rev. Mod. Phys. **32**, 335 (1960).

8. S.I. Weissman, J. chem. Phys. **25**, 890 (1956).

9. H.M. McConnell, J. chem. Phys. **28**, 1188 (1958).

10. R. McWeeny, Proc. Ind. Acad. Sci. **96**, 263 (1986).

11. R. McWeeny and Y. Mizuno, Proc. Roy. Soc. Lond. **A259**, 554 (1961).

12. A.J. Coleman, Rev. Mod. Phys. **35**, 668 (1963).

13. A.J. Coleman, and V.I. Yukalov, *Reduced Density Matrices*, Springer,

Berlin (2000).

14. J.C. Slater, *Solid-State and Molecular Theory*, Wiley, New York.

15. E.S. Kryachko and E.V. Ludena, *Energy Density Functional Theory of Many-Electron Systems*, Kluwer, Dordrecht (1990); R.G. Parr and W. Yang, *Density Functional Theory of Atoms and Molecules*, Oxford, New York (1989).

16. E.R. Davidson, *Reduced Density Matrices in Quantum Chemistry*, Academic, New York (1976).

17. R. McWeeny and W. Kutzelnigg, Int. J. Quantum Chem. **2**, 187 (1968).

18. V.A. Fock, Z. Phys. **15**, 126 (1930).

19. P.A.M. Dirac, Proc. Camb. Phil. Soc. **26**, 376 (1930).

20. R. McWeeny, Proc. Roy. Soc. Lond. **A235**, 496 (1956).

21. E.P. Wigner, *Group Theory and its Application to the Quantum Mechanics of Atomic Spectra*, Academic, New York (1959). (Translated from the German edition (1931) by J. J. Griffin.)

22. R. Pauncz, *Spin Eigenfunctions: Construction and Use*, Plenum. New York (1979).

23. R. McWeeny, *Methods of Molecular Quantum Mechanics 2ed*, Academic, London.

24. F.A. Matsen, Adv. in Quantum Chem. **1**, 59 (1964).

25. R. McWeeny, Int. J. of Quantum Chem. **34**, 25 (1988); see also *ibid*, **74**, 87 (1999).

26. R.E. Moss, *Advanced Molecular Quantum Mechanics*, Chapman and Hall, London (1973).

27. W. Rindler, *Introduction to Special Relativity 2ed*, Oxford, New York (1991).

28. G. Temple, Proc. Roy. Soc. Lond. **A127**, 349 (1930); *ibid.* **A128**, 487 (1930).

29. W.K. Clifford, *Elements of Dynamics*, MacMillan, London (1878); see also R. Tucker, *Clifford's Mathematical Papers*, Chelsea, New York (1968), which contains a reprint of the first edition of 1882.

30. J.C. Slater, *Quantum Theory of Atomic Structure*, Volumes 1 and 2, McGraw-Hill, New York (1960). (See especially ch.23)

31. P.-O. Löwdin, J. Mol. Spectrosc. **14**, 131 (1964).

GEOMETRIC ASPECTS OF THE THEORY OF DENSITY MATRICES AND DENSITIES

John E. Harriman
Department of Chemistry and Theoretical Chemistry Institute
University of Wisconsin, Madison, WI 53706, USA
harriman@chem.wisc.edu

1. Introduction

I first learned of reduced density matrices (RDMs) for many-electron systems from Per Olov Löwdin's 1955 papers in *The Physical Review.* [25, 26, 27] Largely as a result of the impression made by those papers and related work, I chose to spend time in Per's Uppsala Quantum Chemistry Group, where I also met John Coleman and learned of the n-representability problem.[4] Among Löwdin's seminal contributions to quantum chemistry was the introduction and use of mathematical techniques not previously common in the field. It is in that spirit that the ideas reported here were developed. One aspect of this presentation will be the attempt to define quantities carefully and to distinguish between closely-related but distinct concepts, and to use notation and terminology preserving these distinctions.

Density matrices. A density matrix provides a more general description of the state of a quantum system than does a wave function or state vector, which are special cases. Introduced initially in quantum statistical mechanics[7, 33] the density matrix describes an ensemble. If the system has probability w_k of being in a state with wave function Ψ_k, the density matrix is

$$\hat{D} = \sum_k w_k |\Psi_k\rangle\langle\Psi_k| \,. \tag{1}$$

This operator can be represented as an integral operator with kernel

$$D(\tau;\tau') = \sum_k w_k \Psi_k(\tau)\Psi_k^*(\tau') \tag{2}$$

E.J. Brändas and E.S. Kryachko (eds.),
Fundamental World of Quatum Chemistry, Vol. I, 487–541.

where τ represents all the variables for the system and τ' represents a similar set with primes added to distinguish them from the variables in the first set. In general a density matrix is time dependent, but the treatment here will concentrate on density matrix properties at one time. If a basis set $\{\Phi_L\}$ is introduced then the density operator, as any other operator, can be represented by a matrix with elements

$$D_{LM} = \int \Phi_L^* \hat{D} \Phi_M \, d\tau \, . \tag{3}$$

Such matrices will be the representation mostly used here.

RDMs for electrons. The systems of primary interest here are electrons in a molecule, treated in the Born-Oppenheimer approximation. Relativistic effects are initially neglected, so the relevant electronic Hamiltonian in atomic units is

$$\hat{H} = -\frac{1}{2}\sum_{j=1}^{n} \nabla_j^2 - \sum_{\nu}\sum_{j=1}^{n} \frac{Z_\nu}{r_{j\nu}} + \sum_{1 \le j < k}^{n} \frac{1}{r_{jk}} + \sum_{\mu < \nu} \frac{Z_\mu Z_\nu}{r_{\mu\nu}} \, , \tag{4}$$

where indices j and k label electrons, μ and ν label nuclei, and Z_ν is the nuclear charge (in units of e). The last term is constant and could be omitted. Nearly all molecular electronic structure calculations involve an orbital basis set, denoted here as $\{\chi_j, \; j = 1, \ldots, b_0\}$. Unless otherwise indicated, it will be assumed for simplicity that the orbitals are orthonormal.[1] These are combined with the spin functions α and β to give a spin orbital basis $\{\phi_j, \; j = 1, \ldots, b = 2b_0\}$, also orthonormal. The introduction of a finite basis set defines a *model problem.* Basis functions for p electrons are normalized, antisymmetrized products

$$\Phi_{J_p}(1, 2, \ldots, p) = \hat{A}_p \phi_{j_1}(1)\phi_{j_2}(2) \cdots \phi_{j_p}(p) \tag{5}$$

where 1, also sometimes written $x_1 = (\vec{r}_1, \xi_1)$, denotes the position and spin coordinates of electron 1, etc., $J_p = (j_1, j_2, \ldots, j_p)$ identifies the spin orbitals involved, and

$$\hat{A}_p = (p!)^{-1/2} \sum_{\pi \in \mathbb{S}_p} \varepsilon(\pi)\hat{\pi} \tag{6}$$

is the normalizing antisymmetrizer in terms of permutations π with parity $\varepsilon(\pi)$. There are $\binom{b}{p}$ antisymmetrized product functions. The "exact"

[1]Initial basis sets in molecular electronic structure calculations are often not orthonormal, but preliminary processing will yield an orthonormal set.

n-electron wave function for the system in the model defined by the spin orbital basis can be written as

$$\Psi = \sum_{J_n} C_{J_n} \Phi_{J_n} \tag{7}$$

with $\sum_{J_n} |C_{J_n}|^2 = 1$, and various approximations to it can also be expressed in this form. The p-electron reduced density matrix (p-RDM or p-matrix) in the kernel representation is then

$$\begin{aligned} D^{(p)}(x_1, \ldots, x_p; x_{1'}, \ldots, x_{p'}) &\stackrel{\text{def}}{=} \\ \int \Psi(x_1, \ldots, x_p, x_{p+1}, \ldots, x_n) &\Psi^*(x'_1, \ldots, x'_p, x_{p+1}, \ldots, x_n) \\ dx_{p+1} \cdots dx_n \,, \end{aligned} \tag{8}$$

normalized to trace 1, or

$$\begin{aligned} \Gamma^{(p)}(x_1, \ldots, x_p; x_{1'}, \ldots, x_{p'}) &\stackrel{\text{def}}{=} \\ \binom{n}{p} \int \Psi(x_1, \ldots, x_p, x_{p+1}, \ldots, x_n) &\Psi^*(x'_1, \ldots, x'_p, x_{p+1}, \ldots, x_n) \\ dx_{p+1} \cdots dx_n \,, \end{aligned} \tag{9}$$

normalized to trace $\binom{n}{p}$.

Since all electrons are equivalent, the expectation value of a p-electron operator is determined by the p-RDM. If

$$\hat{W} = \sum_{j_1 < j_2 < \ldots < j_p} \hat{w}_{j_1, j_2, \ldots, j_p} \tag{10}$$

then

$$\begin{aligned} \left\langle \hat{W} \right\rangle &= \iint \delta(x'_{j_1} - x_{j_1}) \cdots \delta(x'_{j_p} - x_{j_p}) \\ &\times \hat{w}_{1,\ldots,p} \Gamma^{(p)}(x_1, \ldots, x_p; x_{1'}, \ldots, x_{p'}) dx'_{j_1} \cdots dx'_{j_p} dx_{j_1} \cdots dx_{j_p} \,. \end{aligned} \tag{11}$$

Physical properties involve only one or two electrons at a time ($p = 1$ or 2) and it is convenient to have a more compact notation for these cases. We thus define

$$\hat{\gamma} = \hat{\Gamma}^{(1)} \,, \qquad \hat{\Gamma} = \hat{\Gamma}^{(2)} \,. \tag{12}$$

The spectral expansion of a RDM,

$$\Gamma^{(p)}(x_1, \ldots, x_p; x'_1, \ldots, x'_p) = \sum_K \lambda_K^{(p)} \Psi_K^{(p)}(x_1, \ldots, x_p) \Psi_K^{(p)*}(x'_1, \ldots, x'_p), \tag{13}$$

defines the natural p-states. In particular, when $p = 1$, natural spin orbitals are the true one-electron states within the many-electron system. It would obviously be desirable to be able to determine RDMs, especially the two-matrix $\mathbf{\Gamma}$, directly but that has not been possible because the set of such matrices has not been characterized in a useful way. This is the n-representability problem.[4]

Densities. Still greater simplification is possible for operators that are spin-independent and purely multiplicative in the coordinate representation. Their expectation values can be expressed in terms of the electron densities

$$\begin{aligned} \rho(\vec{r}) &= \int \gamma(\vec{r},\xi;\vec{r},\xi)\, d\xi \\ \mathsf{P}(\vec{r}_1,\vec{r}_2) &= \int \Gamma(\vec{r}_1.\xi_1,\vec{r}_2.\xi_2;\vec{r}_1.\xi_1,\vec{r}_2.\xi_2)\, d\xi_1 d\xi_2 \,. \end{aligned} \tag{14}$$

There is no n-representability problem for (one electron) densities; any reasonable $\rho(\vec{r})$ can be obtained from many n-electron wave functions, including single determinants.[9, 12, 24, 34, 35] This makes density functional theory possible. There are symmetric two-electron functions that are nowhere negative and well behaved that are not n-representable two-electron densities,[6] and the n-representable set has not been characterized.

Preview of this chapter. The relationships between density matrices, reduced density matrices, and densities will be discussed in the following sections. Reduced density matrices and densities are considered as elements in vector spaces, and useful meaning can be associated with the components in different orthogonal subspaces. Geometric ideas like distances and angles between vectors are shown to be relevant to the properties of density matrices. The information content of descriptions at different levels can be related using the subspaces to be introduced. Much of the presentation involves "Definitions" and "Theorems." This is intended more as a means of organizing results and providing for convenient reference that as an attempt at mathematical rigor. The proof of a theorem is indicated as ▷ .... ▌ The significance of the results is summarized in the final section.

2. Geometric Formulation

The states of a quantum system are described by vectors in a Hilbert space. The set of Hermitian operators acting on this Hilbert space is itself a real linear space, since a real linear combination of Hermitian operators is a Hermitian operator. A finite spin orbital basis defines a model problem in which the infinite-dimensional Hilbert space is replaced by a finite-dimensional space. For a system of p electrons with b spin orbitals, the dimension of this space is $\binom{b}{p}$. Operators in this model problem are equivalent to finite-sized matrices: a Hermitian operator acting on the p-electron Hilbert space becomes a $\binom{b}{p} \times \binom{b}{p}$ Hermitian matrix.

2.1 Function spaces

The essential feature of a Hilbert space, that it be "complete" in an appropriate sense, is trivial in the case of a finite-dimensional space. The term "Hilbert space" will continue to be used for spaces isomorphic to function spaces spanned by a (finite) basis. Let $\mathcal{H}_1$ be the b-dimensional, one-electron Hilbert space consisting of all linear combinations of the b spin orbital basis functions. Equivalently, the b-component vectors of expansion coefficients could be considered. If the basis is subjected to a unitary transformation the coefficients will be changed but the physical function they define remains the same.

Additional Hilbert spaces could be defined as direct products:

$$\begin{aligned} \mathcal{H}_2^{\otimes} &= \mathcal{H}_1 \otimes \mathcal{H}_1 \\ \mathcal{H}_3^{\otimes} &= \mathcal{H}_1 \otimes \mathcal{H}_1 \otimes \mathcal{H}_1 \\ &\vdots \end{aligned} \tag{15}$$

where $\mathcal{H}_2^{\otimes}$ consists of all linear combinations of products of spin orbitals, $\phi_j(x_1)\phi_k(x_2)$ and so on. Such products will be of some use later, but they are in general not acceptable as spaces of electronic wave functions because of the antisymmetry requirement (the Pauli principle). Antisymmetric spaces can be defined in terms of Grassmann products:

$$\begin{aligned} \mathcal{H}_2 &= \mathcal{H}_1 \wedge \mathcal{H}_1 \\ \mathcal{H}_3 &= \mathcal{H}_1 \wedge \mathcal{H}_1 \wedge \mathcal{H}_1 \\ &\vdots \end{aligned} \tag{16}$$

where the $\wedge$ symbol indicates an antisymmetrized product. Thus for example, $\mathcal{H}_2$ contains linear combinations of antisymmetrized products $[\phi_j(x_1)\phi_k(x_2) - \phi_k(x_1)\phi_j(x_2)]/\sqrt{2}$.

Index sets. It will be convenient to have a compact notation for sets of indices. The symbol J_p will be used for an ordered set of p distinct indices: $J_p = \{j_1, j_2, \ldots, j_p\}$ with $j_1 < j_2 < \ldots < j_p$. The symbol (J_p)

denotes a set of p indices that are not necessarily ordered. The label n will be reserved for the number of electrons in the system of interest. Other numbers of electrons, usually less than n as for reduced density matrices, will be denoted by p, q, etc.

Fock space. The Hilbert spaces for different numbers of electrons can be combined to produce a Fock space

$$\mathcal{H}_F = \mathcal{H}_0 \oplus \mathcal{H}_1 \oplus \mathcal{H}_2 \ldots \oplus \mathcal{H}_b \,. \tag{17}$$

The space $\mathcal{H}_0$ is just the set of complex numbers. With a finite number of spin orbitals available, there are a limited number of Hilbert spaces possible, so the Fock space is also of finite dimension.

$$\dim(\mathcal{H}_p) = \binom{b}{p}\,, \qquad \dim(\mathcal{H}_F) = \sum_{p=0}^{b} \dim(\mathcal{H}_p) = 2^b \,. \tag{18}$$

2.2 Operators, matrices, and tensors

In a formal sense, an operator in quantum mechanics is a map from a Hilbert space to itself, with an appropriate domain specified. In the coordinate representation of the Schrödinger picture commonly used in quantum chemistry, operators are constructed from functions of particle coordinates, derivatives with respect to those coordinates, and spin operators. As in the case of the Hamiltonian in Eq. (4), an operator is defined for a system with a definite number of electrons, n, and acts on functions in $\mathcal{H}_n$.

In a matrix representation, an operator $\hat{F}$ is replaced by a matrix with elements

$$F_{J_n,K_n} = \int \Phi^*_{J_n}(1,\ldots,n)\hat{F}\Phi_{K_n}(1,\ldots,n)\,dx_1\cdots dx_n \,. \tag{19}$$

An alternative is to use tensors with components

$$\mathsf{F}^{(J_n)}_{(K_n)} = \int \phi^*_{j_1}(1)\cdots\phi^*_{j_n}(n)\hat{F}\phi_{k_1}(1)\cdots\phi_{k_n}(n)\,dx_1\cdots dx_n \tag{20}$$

when the indices in each set (superscript or subscript) are distinct and ordered $[(J_n) = J_n,\ (K_n) = K_n]$, and for other index sets by the requirement that the tensor be antisymmetric in the indices in each set.

Creation and annihilation operators. The tensor form is useful when a formulation in terms of creation and annihilation operators is used. These operators are defined to be consistent with the spin orbital basis functions, so that[2]

$$\phi_j \leftrightarrow \check{a}_j^\dagger | \, \rangle \,, \qquad \Phi_{J_p} \leftrightarrow \check{a}_{j_1}^\dagger \check{a}_{j_2}^\dagger \cdots \check{a}_{j_p}^\dagger | \, \rangle \,. \tag{21}$$

In particular, for reduced density matrices for a system in a pure state Ψ

$$\begin{aligned} \gamma_k^j &= \langle \Psi | \check{a}_j^\dagger \check{a}_k | \Psi \rangle \\ \Gamma_{k_1 k_2}^{j_1 j_2} &= \frac{1}{2} \langle \Psi | \check{a}_{j_1}^\dagger \check{a}_{j_2}^\dagger \check{a}_{k_2} \check{a}_{k_1} | \Psi \rangle \,. \end{aligned} \tag{22}$$

The anticommutation relationships of the creation and annihilation operators leads to the antisymmetry of the tensor Γ.

Integral operator kernels. In the basis-set model problem any operator $\hat{F}$ acting on $\mathcal{H}_p$ can also be represented as an integral operator with kernel

$$\begin{aligned} & F(1, \ldots, p; 1', \ldots, p') = \\ & \qquad \sum_{J_p, K_p} F_{J_p, K_p} \Phi_{J_p}(1, \ldots, p) \Phi_{K_p}^*(1', \ldots, p') \\ & = \sum_{(J_p),(K_p)} F_{(K_p)}^{(J_p)} \phi_{j_1}(1) \cdots \phi_{j_p}(p) \phi_{k_1}^*(1') \cdots \phi_{k_p}^*(p') \,. \end{aligned} \tag{23}$$

The first sum on the right is over ordered index sets; that on the second line is in principle over all index sets but the antisymmetry of the coefficients means a term is zero unless the indices in each set (j's or k's) are distinct. The functions Φ_{J_p} are normalized antisymmetrized products, so the product of two such functions will include a factor of $1/p!$ and in the tensor form there are $p!$ terms differing only in sign corresponding to J_p and the same number corresponding to K_p. It follows that

$$F_{(K_p)}^{(J_p)} = \frac{1}{p!} F_{J_p, K_p} \tag{24}$$

when the indices on the left are ordered. Most of the discussion to follow will be in terms of matrices defined with respect to the antisymmetrized product functions $\{\Phi_{J_p}\}$.

---

[2]Fock space operators are indicted here as $\check{f}$; $\hat{f}$ is an operator on a Hilbert space.

2.3 Matrix and related spaces

One of the fundamental ideas used here is that matrices can be considered to be elements in a vector space.[8, 10]

Definition 1 *Let $\mathcal{E}_p$ be the real linear space of $\binom{b}{p} \times \binom{b}{p}$ Hermitian matrices.*[3] *It can be generalized to a space $\mathcal{E}_p^\star$ in which complex coefficients are allowed and the matrices are not necessarily Hermitian. Alternatively, the matrices may be restricted to be real symmetric, in which case the space will be denoted by $\mathcal{E}_p^{rs}$.*

These spaces have subspace relationships[4] $\mathcal{E}_p^{rs} \subset \mathcal{E}_p \subset \mathcal{E}_p^\star$. Define a scalar product on any of them,

$$(\mathbf{F}, \mathbf{G}) = \operatorname{Tr} \mathbf{F}^\dagger \mathbf{G} \; . \tag{25}$$

When attention is restricted to $\mathcal{E}_p$ or $\mathcal{E}_p^{rs}$, the adjoint has no effect and can be omitted. This scalar product defines a norm and distance in the usual way:

$$\begin{aligned} \|\mathbf{F}\| &= (\mathbf{F}, \mathbf{F})^{1/2} \\ d(\mathbf{F}, \mathbf{G}) &= \|\mathbf{F} - \mathbf{G}\| = (\mathbf{F} - \mathbf{G}, \mathbf{F} - \mathbf{G})^{1/2} \; . \end{aligned} \tag{26}$$

Of course scalar products, and thus norms and distances, are unchanged by any unitary transformation of the basis with respect to which the matrices are defined. The many complications that can arise with infinite-dimensional Hilbert spaces and operators acting on them are absent in the model defined by a finite basis set. The matrix spaces $\mathcal{E}_p$, $\mathcal{E}_p^\star$, and $\mathcal{E}_p^{rs}$ are simply Euclidean vector spaces of possibly very large but finite dimension. Their dimension is determined by that of the Hilbert space on which the corresponding operators act. If $B = \binom{b}{p}$ then for real vector spaces

$$\dim(\mathcal{E}_p) = B^2 \; , \qquad \dim(\mathcal{E}_p^{rs}) = B(B+1)/2 \; . \tag{27}$$

The extended space $\mathcal{E}^\star$ can be considered as a complex space of dimension B^2 or as a real space of dimension $2B^2$. Usually only Hermitian matrices will be considered here, so it will be assumed in what follows that "matrix" means "Hermitian matrix" unless otherwise specified.

---

[3]The matrices themselves are not necessarily real, but coefficients in linear combinations must be real.

[4]The notation $\mathcal{F} \subset \mathcal{G}$ (adapted from the corresponding notation for sets) means that the space $\mathcal{F}$ is a subspace of the space $\mathcal{G}$. Matrices (vectors in these spaces) are denoted by bold letters.

Other spaces isomorphic to these spaces of matrices will also be involved.

Definition 2 *Let ${}^{ker}\mathcal{E}_p$ be the space of kernels and $\hat{\mathcal{E}}_p$ the space of operators corresponding to matrices in $\mathcal{E}_p$,*

$$\begin{aligned} A(1,\ldots,p;1',\ldots,p') &= \sum_{J_p,K_p} A_{J_p,K_p}\Phi_{J_p}(1,\ldots,p)\Phi^*_{K_p}(1',\ldots,p') \\ \hat{A} &= \sum_{J_p,K_p} A_{J_p,K_p}|J_p\rangle\langle K_p| \,. \end{aligned} \tag{28}$$

Note that the operator may differ from the operator originally used in the construction of the matrix, which could act on an infinite-dimensional Hilbert space. The operators in $\hat{\mathcal{E}}_p$ have domain restricted to and range projected onto the finite dimensional Hilbert space of the basis set model. Spaces corresponding to $\mathcal{E}^\star_p$ and $\mathcal{E}^{rs}_p$ are defined similarly. In what follows matrices, kernels, or operators will be used as convenient. When their role as elements in a vector space is crucial to the development, any of them will sometimes be called vectors. When symbols A, $\mathbf{A}$, and $\hat{A}$ are used in connection with one another it is to be understood that they are related as in Eqs. (28). Sometimes when meaning is clear from context, $A \in \mathcal{E}$ will be used generically for $\mathbf{A} \in \mathcal{E}$ or $\hat{A} \in \hat{\mathcal{E}}$, as well as for a kernel in ${}^{ker}\mathcal{E}$.

We begin with a treatment of unreduced density matrices, possibly the density matrix $\mathbf{D}^{(n)}$ for an n-electron system, but the results apply equally well to density matrices for other kinds of systems so such density matrices will be denoted by $\mathbf{D}$. Denote the number of states being considered by B, so $\mathbf{D}$ is a $B \times B$ matrix. For the case of an n-electron system $B = \binom{b}{n}$ and $\mathbf{D} = \mathbf{D}^{(n)} = \mathbf{\Gamma}^{(n)}$. For the general case B is the dimension of a relevant Hilbert space and need not be related to b. The matrix space in this case will be denoted by $\mathcal{E}$ (for Hermitian matrices) or $\mathcal{E}^{rs}$ (for real symmetric matrices).

2.4 Basis matrices

Since $\mathcal{E}$ is a vector space it will be convenient to have a basis, preferably orthonormal. One possibility, sometimes known as the canonical basis, is to choose B^2 matrices each having a single nonzero element.[8] Matrices in the basis will be distinguished by superscript labels, leaving the subscript position available for indexing matrix elements. We could label each matrix by a pair of indices, for example $\mathbf{A}^{\kappa\lambda}$ with elements $A^{\kappa\lambda}_{kl} = \delta_{k\kappa}\delta_{l\lambda}$. This choice of basis is useful for general matrices in a real space for real matrices or a complex space for complex matrices.

Restriction to Hermitian or real symmetric matrices will impose restrictions on the coefficients, and it is convenient to build these into the basis. It is also sometimes preferable to replace the pair indices labeling the matrices by single indices.

The diagonal single-nonzero-element matrices can be retained and labeled by the nonzero element:

$$A^{\kappa}_{kl} = \delta_{k\kappa}\delta_{l\kappa} \ . \tag{29}$$

An alternative choice for the diagonal basis matrices will be considered below. The basis for $\mathcal{E}^{rs}$ is completed by adding symmetrized off-diagonal matrices[10] for $B \leq \kappa \leq B(B+1)/2 - 1$,

$$X^{\kappa}_{jk} = 2^{-1/2}\left(\delta_{j,jj(\kappa')}\delta_{k,kk(\kappa')} + \delta_{j,kk(\kappa')}\delta_{k,jj(\kappa')}\right) \tag{30}$$

where $\kappa' = \kappa - B + 1$, and

$$\begin{aligned} jj(\kappa') &= \kappa' - \frac{1}{2}kk(\kappa')^2 + \frac{3}{2}kk(\kappa') - 1 \\ kk(\kappa') &= \left\lfloor \frac{1}{2}\sqrt{8\kappa' - 7)} + \frac{3}{2} \right\rfloor \ . \end{aligned} \tag{31}$$

The symbol $\lfloor x \rfloor$ indicates the "floor" of x, which is the greatest integer not larger than x. This numbering scheme for index pairs corresponds to a matrix of κ' values

$$\begin{pmatrix} - & 1 & 2 & 4 & 7 & \dots \\ 1 & - & 3 & 5 & 8 & \dots \\ 2 & 3 & - & 6 & 9 & \dots \\ 4 & 5 & 6 & - & 10 & \dots \\ 7 & 8 & 9 & 10 & - & \dots \\ \vdots & \vdots & \vdots & \vdots & \vdots & \ddots \end{pmatrix}$$

where the number in the j, k and k, j position is the κ' for that index pair. This scheme has the advantage that when B is increased new $j, k \leftrightarrow \kappa'$ combinations will be added but those already present will not be changed. Of course other ways of ordering index pairs are possible and may have more advantages in some contexts. The basis for $\mathcal{E}$ consists of that for $\mathcal{E}^{rs}$ plus additional imaginary antisymmetric matrices for $B(B+1)/2 \leq \kappa \leq B^2 - 1$ with

$$X^{\kappa}_{jk} = i\, 2^{-1/2}\left(\delta_{j,jj(\kappa'')}\delta_{k,kk(\kappa'')} - \delta_{j,kk(\kappa'')}\delta_{k,jj(\kappa'')}\right) \tag{32}$$

where $\kappa'' = \kappa - B(B+1)/2 + 1$.

2.5 Trace and traceless subspaces

Density matrices have trace 1 or, possibly, for reduced density matrices some other fixed trace. In order to deal with the trace constraint, introduce

Definition 3 *The matrix* $\mathbf{X}$ *is a multiple of the unit matrix with trace one*

$$\mathbf{X} = B^{-1}\mathbf{1}\,, \qquad \mathrm{Tr}\,\mathbf{X} = 1 \tag{33}$$

Then each density matrix can be separated into

$$\mathbf{D} = \mathbf{X} + \overline{\mathbf{D}}\,, \qquad \mathrm{Tr}\,\overline{\mathbf{D}} = 0\,. \tag{34}$$

Of course any matrix in $\mathcal{E}$ can be similarly divided into a multiple of $\mathbf{X}$ and a traceless part, $\mathbf{F} = (\mathrm{Tr}\,\mathbf{F})\mathbf{X} + \overline{\mathbf{F}}$. This suggests a division of $\mathcal{E}$ into two subspaces

Definition 4

$$\mathcal{E} = \mathcal{E}_0 + \overline{\mathcal{E}} \tag{35}$$

where $\mathcal{E}_0$ *is a one-dimensional space consisting of all real multiples of* $\mathbf{X}$ *(or of the* $B \times B$ *unit matrix) and* $\overline{\mathcal{E}}$ *is a space of matrices with trace zero.*

Note for future use that for any $\mathbf{F}$, $\mathbf{FX} = \mathbf{XF} = \mathbf{F}/B$.

Theorem 1 *The subspaces* $\mathcal{E}_0$ *and* $\overline{\mathcal{E}}$ *are orthogonal.*

▷ Two subspaces are orthogonal if any vector in one of them is orthogonal to any vector in the other. From the definitions of these subspaces, any $\mathbf{F}$ in $\mathcal{E}_0$ is of the form $\mathbf{F} = f\mathbf{1}$ and a general element[5] $\overline{\mathbf{G}} \in \overline{\mathcal{E}}$ has $\mathrm{Tr}\,\overline{\mathbf{G}} = 0$. It follows that

$$\left(\mathbf{F}, \overline{\mathbf{G}}\right) = f\mathrm{Tr}\,\overline{\mathbf{G}} = 0\,, \tag{36}$$

which establishes the result. ▮

A similar decomposition into orthogonal subspaces applies to $\mathcal{E}^{rs}$ and $\mathcal{E}_0^{rs} = \mathcal{E}_0$, $\overline{\mathcal{E}^{rs}} \subset \overline{\mathcal{E}}$. In what follows results will be stated in terms of $\mathcal{E}$ and matrices in this space, but unless otherwise noted they apply also to $\mathcal{E}^{rs}$ and matrices in it. The principal difference is that between the dimensions of the spaces.

This decomposition suggests a different choice of the diagonal basis matrices. One basis matrix should be proportional to the unit matrix (or $\mathbf{X}$) but normalized with respect to the trace scalar product.

---

[5]The notation $a \in \mathbb{A}$ means that a is an element of the set $\mathbb{A}$; $\in$ will be used here also to mean that a matrix, kernel, or operator is a vector in the specified vector space.

Definition 5 *A renormalized unit matrix*

$$\mathbf{X}^0 = B^{1/2}\mathbf{X} = B^{-1/2}\mathbf{1} \qquad (37)$$

is to be included in the matrix basis.

Next, construct $B-1$ diagonal matrices that are orthogonal to $\mathbf{X}^0$ and orthonormal. One possible choice is to define, for $1 \leq \kappa \leq B-1$,

$$X^{\kappa}_{jk} = \delta_{jk} \times \begin{cases} [\kappa(\kappa+1)]^{-1/2} & \text{if } 1 \leq j \leq \kappa \\ [\kappa/(\kappa+1)]^{1/2} & \text{if } j = \kappa+1 \\ 0 & \text{if } j > \kappa+1\,. \end{cases} \qquad (38)$$

Note that $\mathbf{X}^0$ is a basis for the one-dimensional space $\mathcal{E}_0$ while the $\mathbf{X}^\kappa$ with $\kappa > 0$ are part of a basis for $\overline{\mathcal{E}^{rs}}$ or $\overline{\mathcal{E}}$.

The set of all trace-one matrices is not a linear space, but it can be thought of as a translated linear subspace, $\overline{\mathcal{E}} + \mathbf{X}$. If $\mathbf{F}$ and $\mathbf{G}$ have traces f and g, respectively, then

$$\begin{aligned} (\overline{\mathbf{F}},\overline{\mathbf{G}}) &= (\mathbf{F} - f\mathbf{X}, \mathbf{G} - g\mathbf{X}) \\ &= (\mathbf{F},\mathbf{G}) - f\,(\mathbf{X},\mathbf{G}) - g\,(\mathbf{F},\mathbf{X}) + fg\,(\mathbf{X},\mathbf{X}) \\ &= (\mathbf{F},\mathbf{G}) - \frac{f}{B}\mathrm{Tr}\,\mathbf{G} - \frac{g}{B}\mathrm{Tr}\,\mathbf{F} + \frac{fg}{B}\mathrm{Tr}\,\mathbf{X}. \\ &= (\mathbf{F},\mathbf{G}) - \frac{fg}{B}\,. \end{aligned} \qquad (39)$$

Also readily established is

Theorem 2 *Distances between density matrices in $\mathcal{E}$ and between their corresponding traceless parts in $\overline{\mathcal{E}}$ are equal.*

▷ If $\mathbf{D}$ and $\mathbf{D}'$ are density matrices then

$$\overline{\mathbf{D}} - \overline{\mathbf{D}'} = (\mathbf{D} - \mathbf{X}) - (\mathbf{D}' - X) = \mathbf{D} - \mathbf{D}' \qquad (40)$$

and distance is the norm of this difference.▌

Suppose that

$$\hat{D} = |\Psi\rangle\langle\Psi|\,, \qquad \hat{D}' = |\Psi'\rangle\langle\Psi'| \qquad (41)$$

where Ψ and Ψ' are orthogonal. It is then obvious that $(\mathbf{D},\mathbf{D}') = 0$ so these pure state density matrices are orthogonal vectors in $\mathcal{E}$. Equation (39) then gives

$$(\overline{\mathbf{D}},\overline{\mathbf{D}'}) = -1/B\,. \qquad (42)$$

Just as in two or three dimensions, an angle between two vectors can be defined by the relation $(\mathbf{F},\mathbf{G}) = \|\mathbf{F}\|\|\mathbf{G}\|\cos\theta$. From the result above we have

Theorem 3 *If $\overline{\mathbf{D}}$ and $\overline{\mathbf{D}'}$ are the vectors in $\overline{\mathcal{E}}$ corresponding to orthogonal pure states then the angle between them is* $\arccos(-1/B) > \pi/2$. *The distance between them is* $d(\overline{\mathbf{D}}, \overline{\mathbf{D}'}) = \sqrt{2}$.

▷ The angle follows from the results above. The distance is the same as that between **D**and $\mathbf{D}'$, which is determined by

$$d(\mathbf{D}, \mathbf{D}')^2 = (\mathbf{D} - \mathbf{D}', \mathbf{D} - \mathbf{D}') = 1 - 0 - 0 + 1 = 2 \tag{43}$$

since each density matrix has norm 1 and they are orthogonal.▌

As B, the number of possible orthogonal states, increases this angle will approach $\pi/2$. The distance is independent of B.

2.6 Hyperspheres

A density matrix describes a *pure state* if and only if it is idempotent,

$$\mathbf{D}^2 = \mathbf{D}\,, \qquad \hat{D} = |\Psi\rangle\langle\Psi| \tag{44}$$

and since $\mathrm{Tr}\,\mathbf{D} = 1$ it follows that for a pure state

$$1 = \mathrm{Tr}\,\mathbf{D}^2 = \|\mathbf{D}\|^2\,. \tag{45}$$

This establishes[10]

Theorem 4 *The pure state density matrices in $\mathcal{E}$ lie on the surface of a hypersphere of radius 1,*

With $\mathbf{F} = \mathbf{G} = \mathbf{D}$ in Eq. (39) the result becomes

Theorem 5 *For any pure state density matrix* $\mathbf{D}$, $\overline{\mathbf{D}}$ *lies on the surface of a hypersphere of radius* $1 - 1/B$ *in* $\overline{\mathcal{E}}$.

The hypersphere in $\overline{\mathcal{E}}$ is the intersection of the hypersphere in $\mathcal{E}$ with the hyperplane perpendicular to the vector $\alpha\mathbf{X}^0$ at $\alpha = 1$. This is analogous to the intersection of a sphere and a plane in a circle. Note that $\|\mathbf{X}\|^2 = B(1/B^2) = B$ and $\|\overline{\mathbf{D}}\|^2 = (B-1)/B$ so $\|\mathbf{X}\|^2 + \|\overline{\mathbf{D}}\|^2 = 1 = \|\mathbf{D}\|^2$. Before leaving this topic, note that

Theorem 6 *If* $\mathbf{D}$ *is a density matrix and is on the hypersphere of radius 1 then it is a pure state.*

▷ Without loss of generality, take the basis to be one in which **D** is diagonal, and denote the diagonal elements (eigenvalues) by $\{\lambda_j, j = 1, \ldots, B\}$. Since **D** is a density matrix, $\lambda_j \geq 0$ for all j and $\mathrm{Tr}\,\mathbf{D} = \sum_j \lambda_j = 1$. Since **D** is on the hypersphere and has the $\{\lambda_j\}$ as its only nonzero elements, $\sum_j \lambda_j^2 = 1$. These conditions can be satisfied only if one λ_j is 1 and the rest are 0, so $\mathbf{D}^2 = \mathbf{D}$.▌

The matrix $\mathbf{X}$ is a density matrix describing an ensemble in which all states are equally populated; the opposite of a pure state. It has been shown that when the number of states, B, is small there is a correlation between the distance of a density matrix $\mathbf{D}$ from $\mathbf{X}$ and the entropy (proportional to $\mathbf{D}\ln\mathbf{D}$) of the ensemble.[18, 30]

2.7 Diagonal matrices and pure states

It is sometimes useful to consider the subspaces $\mathcal{N} \subset \mathcal{E}$ and $\overline{\mathcal{N}} \subset \overline{\mathcal{E}}$ consisting of diagonal matrices. Of course these subspaces are not invariant with respect to unitary transformations of the basis. The number of possible mutually orthogonal pure states Ψ_J is limited to B. They might, for example, be the eigenfunctions of some Hamiltonian. Think of choosing some such set as a basis. From the results above we can conclude that the corresponding $\{\overline{\mathbf{D}}_j\}$, considered as points in $\overline{\mathcal{N}}$, are the vertices of a regular simplex.[6] Of course any pure state can be converted into any other by a unitary transformation, not only the states corresponding to vertices of the simplex but also states that are linear combinations of these particular states and are in $\mathcal{E}$ but not in $\mathcal{N}$. There is a subset of the points on the surface of the hypersphere discussed above corresponding to pure state density matrices. This set is simply connected and any point within it can be linked continuously to any other. It is of a lower "dimension" than is the surface of the hypersphere, dimension used here to mean the number of parameters that must be specified to determine a point in the set.

The dimension of $\mathcal{E}$ is B^2 [for $\mathcal{E}^{rs}$ it is $B(B+1)/2$] and that of $\overline{\mathcal{E}}$ is $B^2 - 1$ [for $\overline{\mathcal{E}^{rs}}$ it is $B(B+1)/2 - 1$]. To fix a point on the surface of a hypersphere of specified radius requires fixing the values of a number of parameters one less than the dimension of the space containing the hypersphere. The number of real parameters required to determine a pure state density matrix is $2B-2$. There are B complex coefficients in the expansion of the wave function with one normalization constraint. In addition, an overall phase factor will cancel out when the density matrix is formed. (It can have no effect on observable properties in any case.) If the coefficients are required to be real, there are B of them and normalization leaves $B-1$ independent parameters.

[6] A regular simplex is the generalization to any number of dimensions of an equilateral triangle, a regular tetrahedron, etc.

2.8 Eigenensembles

Eigenstates and eigenvalues of an operator play an extremely important role in quantum mechanics. This concept can be slightly generalized for density matrices. If $\hat{A}\Psi_a = a\Psi_a$ then expectation values with respect to Ψ satisfy the *zero dispersion* condition $\left\langle \hat{A}^2 \right\rangle = \left\langle \hat{A} \right\rangle^2$. The converse is also true: the zero dispersion condition implies an eigenstate.[7] Suppose that $\hat{A}$ has a degenerate eigenvalue,

$$\hat{A}\Psi_{a,k} = a\Psi_{a,k} \ , \qquad k = 1, \ldots, d_a \tag{46}$$

and that $\hat{D}$ is an ensemble of these eigenstates,

$$\hat{D} = \sum_{k=1}^{d_a} w_k |\Psi_{a,k}\rangle\langle\Psi_{a,k}| \ , \qquad 0 \leq w_k \leq 1 \ , \ \sum_{k=1}^{d_a} w_k = 1 \ . \tag{47}$$

It is then readily shown that

$$\left\langle \hat{A}^2 \right\rangle = \mathrm{Tr}\, \hat{A}^2\hat{D} = \left(\mathrm{Tr}\, \hat{A}\hat{D}\right)^2 = \langle A \rangle^2 \ . \tag{48}$$

A pure state, when one w_k is one and the rest are zero, is a special case. [14]

Definition 6 *A density matrix $\hat{D}$ is an* eigenensemble *of an operator $\hat{A}$ if*

$$\left\langle \hat{A} \right\rangle^2 = \left(\mathrm{Tr}\, \hat{D}\check{A}\right)^2 = \left\langle \hat{A}^2 \right\rangle = \mathrm{Tr}\, \left(\hat{D}\hat{A}^2\right) \tag{49}$$

If $\hat{A}$ is a p-particle operator[8] then $\hat{A}^2$ will be a $2p$-particle operator (assuming $2p \leq n$, the number of electrons in the system). Equation (49) can then be expressed in terms of matrices in $\mathcal{E}_{2p}$, and in fact in terms of matrices in a two-dimensional subspace of $\mathcal{E}_{2p}$ where it defines a parabola. Components of the density matrix orthogonal to this subspace are unconstrained by the condition.[14]

2.9 A special class of matrices

In addition to the pure state density matrices that have a single eigenvalue equal to one and the rest zero, and **X** or other elements of $\mathcal{E}_0$ that

---

[7] This condition has been used by Nakatsuji[29] to establish the equivalence of the "density equation" or contracted Schrödinger equation to the Schrödinger equation.
[8] See Definition 11.

have all eigenvalues equal, there is another class of matrices that will appear significantly in several contexts.[11] They will be called T-matrices and denoted by $\mathbf{T}(t,\tau)$.

Definition 7 *A matrix* $\mathbf{T}(t)$ *is a* T*-matrix if it has* t *eigenvalues equal to* $1/t$ *and the rest zero.*

Pure state density matrices and $\mathbf{X}$ are special cases of T-matrices with $t = 1$ or $t = B$, respectively. Reduced density matrices arising from a single determinant are T-matrices. The index τ is used only when a T-matrix is diagonal in the basis being used. It identifies which diagonal elements are nonzero, and thus has $\binom{B}{t}$ possible values. If $\mathbf{T}(t)$ is a T-matrix then $\overline{\mathbf{T}}(t) = \mathbf{T}(t) - \mathbf{X}$ will have t eigenvalues equal to $1/t - 1/B$ and $B - t$ eigenvalues equal to $-1/B$. As for the pure states discussed previously, any $\mathbf{T}(t)$ can be converted into any other by an appropriate transformation of the basis. The norms of these matrices are determined by

$$\begin{aligned} \|\mathbf{T}(t)\|^2 &= t(1/t)^2 = 1/t \\ \|\overline{\mathbf{T}}(t)\|^2 &= t(1/t - 1/B)^2 + (B-t)(-1/B)^2 = \frac{B-t}{tB}\,. \end{aligned} \tag{50}$$

and since $\|\mathbf{X}\|^2 = 1/B$, $\|\overline{\mathbf{T}}(t)\|^2 + \|\mathbf{X}\|^2 = \|\mathbf{T}(t)\|^2$. When reduced density matrices are considered, this means that the p-RDMs arising from n-electron single determinants are on a hypersphere of radius $\binom{n}{p}^{-1}$ in $\mathcal{E}_p$.[10]

2.10 A simple example

To illustrate some of these general results in a simple case, consider $B = 3$ with real symmetric matrices. The basis matrices are

$$\begin{aligned} &\mathbf{X}^0 = \tfrac{1}{\sqrt{3}}\begin{pmatrix} 1 & 0 & 0 \\ 0 & 1 & 0 \\ 0 & 0 & 1 \end{pmatrix} \quad \mathbf{X}^1 = \tfrac{1}{\sqrt{2}}\begin{pmatrix} 1 & 0 & 0 \\ 0 & -1 & 0 \\ 0 & 0 & 0 \end{pmatrix} \\ &\mathbf{X}^2 = \tfrac{1}{\sqrt{6}}\begin{pmatrix} 1 & 0 & 0 \\ 0 & 1 & 0 \\ 0 & 0 & -2 \end{pmatrix} \quad \mathbf{X}^3 = \tfrac{1}{\sqrt{2}}\begin{pmatrix} 0 & 1 & 0 \\ 1 & 0 & 0 \\ 0 & 0 & 0 \end{pmatrix} \\ &\mathbf{X}^4 = \tfrac{1}{\sqrt{2}}\begin{pmatrix} 0 & 0 & 1 \\ 0 & 0 & 0 \\ 1 & 0 & 0 \end{pmatrix} \quad \mathbf{X}^5 = \tfrac{1}{\sqrt{2}}\begin{pmatrix} 0 & 0 & 0 \\ 0 & 0 & 1 \\ 0 & 1 & 0 \end{pmatrix} \end{aligned} \tag{51}$$

Any point on the hypersphere containing the pure states can be expressed in terms of 4 angles as

$$\begin{aligned}\mathbf{S} &= \frac{1}{\sqrt{3}}\mathbf{X}^0 + \left(\frac{2}{3}\right)^{1/2}\left[\cos\alpha_1\mathbf{X}^1 + \sin\alpha_1\cos\alpha_2\mathbf{X}^2\right. \\ &\quad + \sin\alpha_1\sin\alpha_2\cos\alpha_3\mathbf{X}^3 + \sin\alpha_1\sin\alpha_2\sin\alpha_3\cos\alpha_4\mathbf{X}^4 \\ &\quad \left. + \sin\alpha_1\sin\alpha_2\sin\alpha_3\sin\alpha_4\mathbf{X}^5\right] \end{aligned} \tag{52}$$

and a general normalized wave function expressed as a vector in the three-dimensional Hilbert space is $(\cos\theta,\ \sin\theta\cos\phi,\ \sin\theta\sin\phi)$. The corresponding pure-state density matrix can be expressed in terms of the basis matrices as

$$\mathbf{P} = \sum_{\kappa=0}^{5} C_\kappa \mathbf{X}^\kappa \tag{53}$$

with coefficients

$$\begin{aligned} C_0 &= 1/\sqrt{3} \\ C_1 &= \left[\cos^2\theta - \sin^2\theta\, \cos^2\phi\right]/\sqrt{2} \\ C_2 &= \left[\cos^2\theta + \sin^2\theta\, \left(3\, \cos^2\,\phi - 1\right)/2\right]/\sqrt{6} \\ C_3 &= \sqrt{2}\,\sin\theta\, \cos\theta\, \cos\phi \\ C_4 &= \sqrt{2}\,\sin\theta\, \cos\theta\, \sin\phi \\ C_5 &= \sqrt{2}\,\sin^2\theta\, \sin\phi\, \cos\phi\,. \end{aligned} \tag{54}$$

It is readily verified that $\mathbf{P}^2 = \mathbf{P}$, $\mathrm{Tr}\,\mathbf{P} = 1$ and

$$\|\overline{\mathbf{P}}\|^2 = \sum_{\kappa=1}^{5} C_\kappa^2 = 2/3 = (B-1)/B \text{ for } B = 3 \tag{55}$$

so this matrix is on the hypersphere in $\overline{\mathcal{E}^{rs}}$. It is apparent that $\mathbf{P}(\theta,\phi)$ is less general than $\mathbf{S}(\alpha_1,\ldots,\alpha_4)$ and thus that there must be points on the hypersphere that are not pure state density matrices.

Even in this simple case $\dim(\mathcal{E}^{rs}) = 6$ and $\dim(\overline{\mathcal{E}^{rs}}) = 5$, making it difficult to visualize the entire space. We can, however, look at sections in which the components in the directions of $\mathbf{X}^3$, $\mathbf{X}^4$, and $\mathbf{X}^5$ are zero, equivalent to considering diagonal matrices. In a basis of eigenstates of some operator of interest the pure state density matrices will each have one diagonal matrix element equal to one and all other matrix elements zero, and they span $\mathcal{N}$. The corresponding points in $\overline{\mathcal{N}}$ are the vertices of an equilateral triangle. The midpoint of each side of the triangle is a T-matrix. There are three such.

3. Reduction and invariant subspaces

Attention will now be concentrated on systems of electrons and a set of related matrix spaces introduced.

3.1 Reduction

Starting from an n-electron density matrix $\mathbf{D}^{(n)} = \mathbf{\Gamma}^{(n)}$, often but not necessarily corresponding to a pure state with wave function Ψ, a series of reduced density matrices can be defined for p electrons, with $1 \leq p < n$. It is convenient to extend the range of p to include n, in which case the density matrix is not reduced, and also $p = 0$, in which case the "density matrix" is just a number. Each $\mathbf{D}^{(p)}$ or $\mathbf{\Gamma}^{(p)}$ is a $B_p \times B_p$ matrix, with $B_p = \binom{b}{p}$ the dimension of the p-electron Hilbert space in the model problem defined by a basis set of b spin orbitals.

The reduction process is not limited to density matrices. In the basis-set model any operator $\hat{F}^{(p)}$ on a p-electron Hilbert space is equivalent to an integral operator with kernel

$$F^{(p)}(1,\ldots,p;1',\ldots,p') = \sum_{J_p,K_p} F^{(p)}_{J_p,K_p}\Phi_{J_p}(1,\ldots,n)\Phi^*_{K_p}(1',\ldots,p') \quad (56)$$

where

$$F^{(p)}_{J_p,K_p} = \int \Phi^*_{J_p}(1,\ldots,n)\hat{F}^{(p)}\Phi_{K_p}(1,\ldots,p)\,dx_1\cdots dx_p\,. \quad (57)$$

There is a sequence of matrix spaces $\mathcal{E}_p$ for different values of p, and reduction defines a map from one to another. In terms of the integral kernels, the map ${}^p_q\!\downarrow : \mathcal{E}_p \to \mathcal{E}_q$ is defined by an equation like Eq. (8) or (9). This map is in general many-to-one so a true inverse is not possible, but two inverse-like maps can be defined. One gives a *set* sometimes called the "preimage":

$${}^p_q\!\Uparrow \mathbf{F}^q \stackrel{\text{def}}{=} \left\{F^{(p)} \in \mathcal{E}_p \middle|\, {}^p_q\!\downarrow \mathbf{F}^{(p)} = \mathbf{F}^{(q)}\right\}\,. \quad (58)$$

This map will here be called *expansion*. Note that a density matrix $\hat{D}^{(p)}$ is n-representable if and only if ${}^n_p\!\Uparrow \hat{D}^{(p)}$ *includes* a positive matrix. It is also possible to define a map ${}^p_q\!\uparrow : \mathcal{E}_q \to \mathcal{E}_p$ that yields a specific matrix in the set defined by ${}^p_q\!\Uparrow$. This will be made precise later, and the map will be called *antireduction*.[9]

[9] This map was called "expansion" by Kummer,[22] who introduced it. For a chemist, a natural term for the opposite of reduction would be "oxidation".

3.2 Subspaces

The space $\mathcal{E}_p$ of matrices corresponding to p-electron operators can be divided into subspaces based on reduction behavior.[11] Suppose that p and q satisfy the conditions $0 \leq q < p \leq b/2$. Then

$$\begin{aligned}
\dim(\mathcal{E}_q) &= \binom{b}{q}^2 < \dim(\mathcal{E}_p) = \binom{b}{p}^2 \\
\dim(\mathcal{E}_q^{rs}) &= \frac{1}{2}\binom{b}{q}\left[\binom{b}{q}+1\right] < \dim(\mathcal{E}_p^{rs}) = \frac{1}{2}\binom{b}{p}\left[\binom{b}{p}+1\right]
\end{aligned} \tag{59}$$

The spaces $\mathcal{E}_p$ will be discussed, but with appropriate changes the results obtained apply also to $\mathcal{E}_p^{rs}$. Take any set $\{\mathbf{A}^\kappa,\ \kappa = 1, \ldots, \dim(\mathcal{E}_p)\}$ of linearly independent vectors (matrices) in $\mathcal{E}_p$ and construct the matrices (vectors) $\{{}^p_q\!\downarrow\ \mathbf{A}^\kappa\}$ in $\mathcal{E}_q$. There are more vectors in this set than the dimension of the space, so they must be linearly dependent, *i.e.* there are nonzero coefficients $\{C_\kappa\}$ such that

$$\sum_{\kappa=1}^{\dim(\mathcal{E}_p)} C_\kappa\ \left({}^p_q\!\downarrow\ \mathbf{A}^\kappa\right) = 0. \tag{60}$$

The reduction process is linear, so it follows that

$${}^p_q\!\downarrow \left(\sum_{\kappa=1}^{\dim(\mathcal{E}_p)} C_\kappa \mathbf{A}^\kappa\right) = 0 \tag{61}$$

and there are thus matrices in $\mathcal{E}_p$ that will give zero on reduction from $\mathcal{E}_p$ to $\mathcal{E}_q$. The set of all such matrices is a linear subspace of $\mathcal{E}_p$ which will be denoted by $\mathcal{K}_p^{red}(q)$.

Definition 8 *The null space*[10] *of a reduction map is*

$$\mathcal{K}_p^{red}(q) \stackrel{def}{=} \left\{\mathbf{F}^{(p)} \in \mathcal{E}_p \middle|\ {}^p_q\!\downarrow\ \mathbf{F}^{(p)} = 0\right\} \tag{62}$$

and the orthogonal complement of $\mathcal{K}_p^{red}(q)$ *in* $\mathcal{E}_p$ *is*

$$\begin{aligned}
\mathcal{J}_p^{red}(q) &= \mathcal{E}_p \setminus \mathcal{K}_p^{red}(q) \\
&= \left\{\mathbf{F}^{(p)} \in \mathcal{E}_p \middle| \left(\mathbf{F}^{(p)}, \mathbf{A}^{(p)}\right) = 0\,, \forall \mathbf{A}^{(p)} \in \mathcal{K}_p^{red}(q)\right\}\,.
\end{aligned} \tag{63}$$

[10] Such a space is also known as a kernel. The term kernel will be reserved here for integral operator kernel functions.

Note that if reduction of a matrix from $\mathcal{E}_p$ to $\mathcal{E}_q$ gives zero, then further reduction still gives zero. That is, the null spaces are nested:

$$\mathcal{K}_p^{red}(q') \subset \mathcal{K}_p^{red}(q) , \qquad \text{if } q' < q . \tag{64}$$

A sequence of subspaces $\mathcal{E}_{p,\pi}$ of $\mathcal{E}_p$ can be defined in terms of reduction for $0 < \pi \leq p \leq b/2$, as the intersection of two spaces[11]

Definition 9 *For* $1 \leq \pi \leq p$,

$$\mathcal{E}_{p,\pi} \stackrel{def}{=} \mathcal{K}_p^{red}(\pi - 1) \cap \mathcal{J}_p^{red}(\pi) . \tag{65}$$

The final space $\mathcal{E}_{p,0}$ *is just the set of elements of* $\mathcal{E}_p$ *that are proportional to the unit matrix. If* $\mathbf{A} \in \mathcal{E}_{p,\pi}$ *then* ${}^{p}_{q}\!\downarrow \mathbf{A}$ *is zero if* $q < \pi$ *but nonzero if* $q \geq \pi$.

Theorem 7 *Subspaces* $\mathcal{E}_{p,\pi}$ *and* $\mathcal{E}_{p\pi'}$ *with* $p \leq b/2$ *and* $\pi, \pi' \leq p$, $\pi' \neq \pi$, *are orthogonal.*

▷ Suppose that $\mathbf{A} \in \mathcal{E}_{p,\pi}$ and $\mathbf{B} \in \mathcal{E}_{p\pi'}$ with $\pi' < \pi$. From Eq. (65), $\mathbf{A} \in \mathcal{K}_p^{red}(\pi - 1)$. Either $\pi' = \pi - 1$ so $\mathcal{K}_p^{red}(\pi') = \mathcal{K}_p^{red}(\pi)$ or $\pi' < \pi - 1$ so $\mathcal{K}_p^{red}(\pi') \subset \mathcal{K}_p^{red}(\pi)$ [Eq. (64)], so $\mathbf{A} \in \mathcal{K}_p^{red}(\pi')$. But $\mathbf{B}$ must be in the orthogonal complement of $\mathcal{K}_p^{red}(\pi')$ and is therefore orthogonal to $\mathbf{A}$.▌

It follows that $\mathcal{E}_p = \sum_{\pi=0}^{p} \oplus \mathcal{E}_{p,\pi}$, and a basis for $\mathcal{E}_p$ can be defined consisting of basis matrices for the separate subspaces. The term "reduction index" will be used for π. The subspaces can also be characterized in other ways, to be described below. One is to give a prescription for separating any matrix $\mathbf{F}^{(p)}$ into components in the various $\mathcal{E}_{p,\pi}$. The other is the fact that they are the invariant subspaces with respect to the group of unitary transformations of the spin orbital basis.

Since the subspaces are orthogonal,

$$\dim(\mathcal{E}_p) = \sum_{\pi=0}^{p} \dim(\mathcal{E}_{p,\pi}) \tag{66}$$

and similarly for $\mathcal{E}_p^{rs}$. Consider the effects of reduction by various amounts of elements of $\mathcal{E}_{p,\pi}$ for various values of π. For some $\mathbf{A}^{(p,\pi)} \in \mathcal{E}_{p,\pi}$ with $\pi \leq p \leq b/2$

$${}^{\;\,p}_{p-1}\!\downarrow \mathbf{A}^{(p,\pi)} \qquad \begin{cases} = 0 & \text{if } \pi = p \\ \neq 0 & \text{if } \pi < p \end{cases}$$

[11]For sets $\mathbb{A}$ and $\mathbb{B}$, $\mathbb{A} \cap \mathbb{B}$ is the set of elements that are in both $\mathbb{A}$ and $\mathbb{B}$. The "sets" here are linear spaces.

$$
{}^{\;\,p}_{p-2}\!\downarrow\ \mathbf{A}^{(p,\pi)} \qquad \left\{ \begin{array}{ll} =0 & \text{if } \pi = p \text{ or } p-1 \\ \neq 0 & \text{if } \pi < p-1 \end{array} \right.
$$

$$
\vdots \quad . \tag{67}
$$

By considering the definitions in Eqs. (62)–(65) or by counting the number of possible linearly independent elements in a subspace and considering their reductions, we establish

Theorem 8 *The dimension of* $\mathcal{E}_{p,\pi}$ *is independent of p and is*

$$
\dim(\mathcal{E}_{p,\pi}) = \left\{ \begin{array}{ll} \dim(\mathcal{E}_{\pi}) - \dim(\mathcal{E}_{\pi-1}) & \textit{if } \pi > 0 \\ 1 & \textit{if } \pi = 0 \end{array} \right. \tag{68}
$$

This is also valid if $\mathcal{E}$ is replaced by $\mathcal{E}^{rs}$.

When $p > b/2$ a similar hierarchy of spaces exists, but the dimension of $\mathcal{E}_{p-1}$ is greater than that of $\mathcal{E}_p$. The possible range of the reduction index is always $0 \leq \pi \leq \min(p, b-p)$. It is rarely necessary to consider $p > b/2$ except in connection with so-called "particle-hole" symmetry, which will not be discussed here.

The dimensions of the spaces are

$$
\dim(\mathcal{E}_p) = \binom{b}{p}^2, \qquad \dim(\mathcal{E}_p^{rs}) = \frac{1}{2}\binom{b}{p}\left[\binom{b}{p}+1\right] \tag{69}
$$

leading, with Eqs. (68), to subspace dimensions

$$
\begin{aligned}
\dim(\mathcal{E}_{p,\pi}) &= \binom{b+1}{\pi}^2 \frac{b+1-2\pi}{b+1} \\
\dim(\mathcal{E}_{p,\pi}^{rs}) &= \frac{1}{2}\binom{b+1}{\pi}\left[\binom{b+1}{\pi}+1\right]\frac{b+1-2p}{b+1} .
\end{aligned} \tag{70}
$$

These expressions are valid for $0 \leq \pi \leq \min(p, b-p)$.

Theorem 9 *If* $\mathbf{A} \in \mathcal{E}_{p,\pi}$ *then* ${}^{p}_{q}\!\downarrow \mathbf{A} \in \mathcal{E}_{q,\pi}$*, and the reduction map from* $\mathcal{E}_{p,\pi}$ *to* $\mathcal{E}_{q,\pi}$ *is one-to-one onto.*

$\triangleright$ Let $\{\mathbf{A}^{\kappa},\ \kappa = 1, \ldots, \dim(\mathcal{E}_{p,\pi})\}$ be a basis for $\mathcal{E}_{p,\pi}$. Then the $\{{}^{p}_{q}\!\downarrow \mathbf{A}\}$ provide a basis for $\mathcal{E}_{q,\pi}$. To see this note first that they are linearly independent. If there were a linear combination of them that was zero, then there would be a corresponding linear combination of the $\mathbf{A}^{\kappa}$ in $\mathcal{K}_p^{red}(q)$, but this is not possible for $\mathbf{A} \in \mathcal{E}_{p,\pi}$. Since $\dim(\mathcal{E}_{q,\pi}) = \dim(\mathcal{E}_{p,\pi})$, $\dim(\mathcal{E}_{p,\pi})$ linearly independent vectors constitute a basis. This one-to-one relationship between bases for the two spaces establishes the result

3.3 Reducing bases and antireduction.

It is clearly possible to choose a basis for each $\mathcal{E}_p$ such that each of the basis matrices is entirely in one of the $\mathcal{E}_{p,\pi}$. Further, because of the one-to-one character of the map from $\mathcal{E}_{p,\pi}$ to $\mathcal{E}_{q,\pi}$ with $q < p$, the basis for $\mathcal{E}_{q,\pi}$ can be chosen to be the reduction of that for $\mathcal{E}_{p,\pi}$.

Definition 10 *A reducing basis for the matrix spaces $\mathcal{E}_p$ is the union of bases $\left\{\mathbf{Y}^{(p,\pi)\kappa}\right\}$ for each $\mathcal{E}_{p,\pi}$ with the property that ${}^{p}_{q}\!\downarrow \mathbf{Y}^{(p,\pi)\kappa} \propto \mathbf{Y}^{(q,\pi)\kappa}$.*

Theorem 9 also establishes the existence of the antireduction map, since a one-to-one map can be inverted. An explicit prescription for constructing ${}^{q}_{p} \uparrow \mathbf{A}^{(q)}$ from $\mathbf{A}^{(q)}$ will be given below.

3.4 Components of operators

Any matrix in $\mathcal{E}_p$ can be uniquely resolved into components in the various subspaces $\mathcal{E}_{p,\pi}$. A general procedure and explicit methods for low p will be discussed subsequently. Note that if

$$\mathbf{A}^{(p)} = \sum_{\pi=0}^{p} \mathbf{A}^{(p,\pi)} , \qquad \mathbf{B}^{(p)} = \sum_{\pi=0}^{p} \mathbf{B}^{(p,\pi)} \tag{71}$$

then

$$\left(\mathbf{A}^{(p)}, \mathbf{B}^{(p)}\right) = \sum_{\pi=1}^{p} \left(\mathbf{A}^{(p,\pi)}, \mathbf{B}^{(p,\pi)}\right) . \tag{72}$$

If one of the matrices has nonzero components for only some values of π, then only those same components of the other matrix are required for the evaluation of the scalar product. Recall that the expectation value of an operator can be expressed as its scalar product with the density matrix describing the system state.

An operator corresponding to a physical observable for an n-electron system can be written as a sum of symmetric, p-particle operators acting on the n-electron Hilbert space

$$\hat{F} = \hat{F}^{(0)} + \hat{F}^{(1)} + \ldots + \hat{F}^{(p)} \tag{73}$$

where $\hat{F}^{(0)}$ is a constant, often not present, and

$$\begin{aligned} \hat{F}^{(1)} &= \sum_{j_1=1}^{n} \hat{f}_1(j) \\ \hat{F}^{(2)} &= \sum_{j_2>j_1=1}^{n} \hat{f}_2(j_1, j_2) \end{aligned}$$

$$\vdots$$

$$\hat{F}^{(p)} = \sum_{\substack{j_p > \cdots > j_1 \\ = 1}}^{n} \hat{f}_p(j_1, \ldots, j_p) \,. \tag{74}$$

For ordinary observables $p = 1$ or 2, but higher values can arise for operators like the square of the Hamiltonian.[12] Note the terminology used here: a p-electron operator for an n-electron system involves at most p-electron interactions. When operators are expressed in terms of creation and annihilation operators, n is not included in the specification of the operator but p is still well-defined. As noted in the introduction, the expectation value of a p-electron operator can be evaluated from the p-electron reduced density matrix for the n-electron system.

A term $\hat{F}^{(q)}$ will be called the q-electron *part* of $\hat{F}$. It can be thought of as a q-electron operator and its contribution to the expectation value is determined by the q-electron RDM,

$$\left\langle \hat{F}^{(q)} \right\rangle = \mathrm{Tr}\, \hat{F}^{(q)} \hat{D}^{(n)} = \binom{n}{q} \mathrm{Tr}\, \hat{f}_q(1, \ldots, q) \hat{D}^{(q)} = \mathrm{Tr}\, \hat{f}_q(1, \ldots, q) \hat{\Gamma}^{(q)} \,. \tag{75}$$

It is often convenient to combine parts into an effective, single p-electron part, defining a reduced operator. For the Hamiltonian in Eq. (4), for example,

$$\begin{aligned} \hat{H} &= \sum_{j=1}^{n} \hat{f(j)} + \sum_{1=j<k}^{n} g(j,k) \\ &= \sum_{1=j<k}^{n} \hat{K}(j,k) \end{aligned} \tag{76}$$

where $\hat{f} \leftrightarrow \hat{f}_1$, $\hat{g} \leftrightarrow \hat{f}_2$, the constant nuclear-nuclear Coulomb term $\hat{F}^{(0)}$ has been omitted, and

$$\hat{K}(j,k) = \frac{1}{n-1} \left[\hat{f}(j) + \hat{f}(k) \right] + \hat{g}(j,k) \tag{77}$$

if the expectation values is to be evaluated as $\mathrm{Tr}\, \hat{K}(1,2)\hat{\Gamma}$, and $\hat{K}(1,2)$ is the *reduced Hamiltonian.*[13] The (possibly reduced) p-electron operator

[12] Note that the matrix of an operator product $\hat{A}\hat{B}$ with a finite basis is not in general the same as the product of matrices $\mathbf{AB}$; $\hat{H}^k$ is divergent for $k > 2$, even though $\mathbf{H}^k$ will exist.
[13] The coefficients of $(\hat{f}_j + \hat{f}_k)$ and $\hat{g}_{jk}$ depend on the normalization of the density matrix.

$\hat{f}_p(1,\ldots,p)$ will usually have *components* in subspaces $\hat{\mathcal{E}}_{p,\pi}$ for various π's between 0 and p. These distinctions can be summarized as

Definition 11 *For an operator $\hat{F}$ (in the coordinate representation) that can be written as in Eqs. (73) and (74)*

- *$\hat{F}$ is a p-electron operator, where p is the maximum number of indices occurring in any $\hat{f}(j_1,\ldots,j_p)$*
- *the q-electron part of $\hat{F}$ is the symmetric sum of the operators $\hat{f}(j_1,\ldots,j_q)$ (with $q \leq p$).*
- *the matrix F corresponding to $\hat{F}$ may have components in $\mathcal{E}_{p,\pi}$ for $0 \leq \pi \leq p$; the component $\mathbf{F}^{(p,\pi)}$ in$\mathcal{E}_{p,\pi}$ is the π-electron component of* **F** *(and defines a corresponding π-electron component of $\hat{F}$).*

3.5 Extended matrix spaces

Before explicit expressions for reduction, antireduction, and expansion superoperators are considered it will be convenient to be able to discuss all of the $\mathcal{E}_p$ and the connections between them in a common way. Just as Hilbert spaces can be combined into a Fock space, we can combine matrix spaces for matrices of different sizes. In this spirit, define

$$\mathcal{S} = \mathcal{E}_0 \oplus \mathcal{E}_2 \oplus \ldots \oplus \mathcal{E}_b \ . \tag{78}$$

with a corresponding definition for $\mathcal{S}^{re}$. The elements of these spaces are block-diagonal matrices of dimension $2^b \times 2^b$, and only block-diagonal matrices with the specified structure are included. The dimensions of the spaces are

$$\begin{aligned} \dim(\mathcal{S}) &= \sum_{p=0}^{b} \binom{b}{p}^2 = \binom{2b}{b} \\ \dim(\mathcal{S}^{re}) &= \sum_{p=0}^{b} \frac{1}{2}\binom{b}{p}\left[\binom{b}{p}+1\right] = \frac{1}{2}\binom{2b}{b} + 2^{b-1} \ . \end{aligned} \tag{79}$$

Note that $\mathcal{E}_b$, like $\mathcal{E}_0$, is the one-dimensional space of real numbers. The size of the matrices in $\mathcal{E}_p$ and its dimension increase as p increases from 0 to $b_0 = b/2$ and then decrease again as p is further increased. It is also possible to allow off-diagonal blocks, and the space of matrices not restricted to block-diagonal form will be denoted by $\mathcal{S}^{\mathcal{F}}$. It is of dimension 2^{2b}. The superoperators of interest will correspond to matrices in this extended space (supermatrices).

3.6 Some superoperators and matrices

A supermatrix acts on matrices in a space like $\mathcal{E}_p$ to produce another matrix in the same space or in a different space. A particular form of supermatrix will be of use here. If **A**, **B**, and **F** are matrices such that the matrix product **AFB** is defined, then the supermatrix **A**|**B** is defined on the space containing **F** by

$$(\mathbf{A}|\mathbf{B})\,\mathbf{F} = \mathbf{AFB}\,. \tag{80}$$

This could be expressed as an explicit supermatrix multiplying the "vector" $\underline{F}$:

$$[(\mathbf{A}|\mathbf{B})\mathbf{F}]_{jk} = \sum_{lm} \underbrace{A_{jl}B_{km}}_{[\mathbf{AB}]_{jk,lm}} F_{lm}\,. \tag{81}$$

Corresponding superoperators acting on kernel or operator spaces can also be defined. Note that if $[\mathbf{A},\mathbf{C}] = 0$ and $[\mathbf{B},\mathbf{D}] = 0$ then $[\mathbf{A}|\mathbf{B},\mathbf{C}|\mathbf{D}] = 0$.

Recall that

$$\Phi_{J_p} \leftrightarrow |J_p\rangle = \breve{a}^\dagger_{j_1} \cdots \breve{a}^\dagger_{j_p} |\ \rangle. \tag{82}$$

where the indices are ordered $j_1 < j_2 < \cdots < j_p$. Now define a superoperator[14]

$$\widehat{\Lambda}_- \stackrel{\text{def}}{=} \sum_{j=1}^{b} \breve{a}_j|\breve{a}^\dagger_j \tag{83}$$

and consider its effect on a ket-bra product operator in $\hat{\mathcal{E}}^\star_p$

$$\widehat{\Lambda}_-|K_p\rangle\langle L_p| = \sum_{j=1}^{b} \breve{a}_j|K_p\rangle\langle L_p|\breve{a}^\dagger_j = \breve{a}_j|K_p\rangle\,(\breve{a}_j|L_p\rangle)^\dagger\ . \tag{84}$$

The effect of the annihilation operator is

$$\breve{a}_j|K_p\rangle = \begin{cases} \varepsilon^-(K_p,j)\,|K^-_{p-1}(j)\rangle & \text{if } j \in K_p \\ 0 & \text{if } j \notin K_p \end{cases} \tag{85}$$

where $K^-_p(j)$ is the set $K_p \setminus \{j\}$ obtained by removing j from K_p and $\varepsilon^-(K_p,j) = \pm 1$ is determined by the position of j in K_p.[15] This will be the same as the sign of the term in the expansion of Φ_{K_p} that has $\phi_{j_p}(x_p)$ as one factor and the remaining factors with indices on spin orbitals and

[14] Superoperaters are indicated here by a wide hat.

[15] Note that since K_p is ordered, the set $K_p \setminus \{j\}$ will also be ordered.

their arguments both ordered. A corresponding result applies to $\breve{a}_j|L_p\rangle$. The sum over j will pick out all those j's that are in both K_p and L_p, and for those annihilation is equivalent to integration of $\phi_j(x_p)\phi_j^*(x_p)$ over x_p. This is the same as reduction by 1, apart from a normalization factor:

$$\widehat{\Lambda}_-|K_p\rangle\langle L_p| \leftrightarrow C\int \Phi_{J_p}(x_1,\ldots,x_{p-1},x_p)\Phi^*_{L_p}(x'_1,\ldots,x'_{p-1},x_p)\,dx_p\,. \tag{86}$$

Any operator in $\hat{\mathcal{E}}_p$ can be expressed as a linear combination of such ket-bra products, and we see that $\widehat{\Lambda}_-$ provides a map from $\mathcal{E}_p$ to $\mathcal{E}_{p-1}$. In fact, the definition of $\widehat{\Lambda}_-$ is independent of p. It is thus a map from $\mathcal{S}$ (or $\mathcal{S}^{re}$) to itself with a matrix representation in $\mathcal{S}$ that has nonzero blocks linking $\mathcal{E}_p$ blocks with p's differing by one.

Another superoperator can be defined by

$$\widehat{\Lambda}_+ \stackrel{\text{def}}{=} \sum_{j=1}^{b} \breve{a}_j^\dagger|\breve{a}_j\,. \tag{87}$$

Its effect on a ket-bra operator is

$$\widehat{\Lambda}_+|K_p\rangle\langle L_p| = \sum_{j=1}^{b} \breve{a}_j^\dagger|K_p\rangle\left(\breve{a}_j^\dagger|L_p\rangle\right)^\dagger \tag{88}$$

and

$$\breve{a}_j^\dagger|K_p\rangle = \begin{cases} 0 & \text{if } j\in K_p \\ \varepsilon^+(K_p,j)\pm|K^+_{p+1}(j)\rangle & \text{if } j\notin K_p \end{cases} \tag{89}$$

where $K^+_{p+1}(j)$ is the set $K_p\cup\{j\}$, appropriately reordered, and $\varepsilon^+(K_p,j)$ is a sign factor determined by the permutation required to change j, K_p to $K^+_{p+1}(j)$. One expects that $\widehat{\Lambda}_+$ is the adjoint of $\widehat{\Lambda}_-$, and it is shown in the Appendix that this is the case.

The commutator of these superoperators is readily evaluated from their definitions and the properties of creation and annihilation operators.

$$\widehat{\Lambda}_+\widehat{\Lambda}_- = \sum_{j,k=1}^{b} \breve{a}_j^\dagger\breve{a}_k|\breve{a}_k^\dagger\breve{a}_j \tag{90}$$

and

$$\widehat{\Lambda}_-\widehat{\Lambda}_+ \;=\; \sum_{j,k=1}^{b} \breve{a}_j\breve{a}_k^\dagger|\breve{a}_k\breve{a}_j^\dagger = \sum_{j,k=1}^{b}\left(\delta_{jk}-\breve{a}_k^\dagger\breve{a}_j\right)|\left(\delta_{jk}-\breve{a}_j^\dagger\breve{a}_k\right)$$

$$= \sum_{j,k=1} \left[\delta_{jk}^2 - \delta_{jk}\left(\check{1}|\check{a}_j^\dagger \check{a}_k + \check{a}_k^\dagger \check{a}_j|\check{1}\right) + \check{a}_k^\dagger \check{a}_j|\check{a}_j^\dagger \check{a}_k\right]$$

$$= b\check{1} - \check{1}|\check{N} - \check{N}|\check{1} + \widehat{\Lambda}_+\widehat{\Lambda}_- \tag{91}$$

where $\check{1}$ is the identity operator and $\check{N} = \sum_j \check{a}_j^\dagger \check{a}_j$ is the number operator.[16] Thus

$$\left[\widehat{\Lambda}_+, \widehat{\Lambda}_-\right] = \check{1}|\check{N} + \check{N}|\check{1} - b\check{1} = 2\widehat{\Lambda}_3 , \tag{92}$$

which defines $\widehat{\Lambda}_3$. Note that $\check{a}_j^\dagger \check{a}_k$ and $\check{a}_j \check{a}_k^\dagger$ commute with $\check{N}$.

Consider the effect of $\widehat{\Lambda}_3$ on a ket-bra product. To avoid possible duplication of effort later, generalize to the case in which the ket and bra each have definite particle number, as above, but now allow these numbers to be different.

$$2\widehat{\Lambda}_3|K_p\rangle\langle L_q| = |K_p\rangle\langle L_q|\check{N} + \check{N}|K_p\rangle\langle L_q| - b|K_p\rangle\langle L_q| = (p+q-b)|K_p\rangle\langle L_q| . \tag{93}$$

Such a ket-bra operator is an eigenoperator of the superoperator $\widehat{\Lambda}_3$. A reduced density matrix $\hat{D}^{(p)}$ or other operator on $\mathcal{H}_p$ is expanded in terms of such ket-bra operators with all $p = q$ the same, so it is an eigenoperator of $\widehat{\Lambda}_3$ with eigenvalue $\mu = (2p - b)/2 = p - b_0$. When $b = 2b_0$ for integer b_0, μ is an integer.[17] The range of p is $0 \leq p \leq b$, so $-b_0 \leq \mu \leq b_0$. For the cases of greatest interest $p < b_0$ so μ will be negative.

The superoperators $\widehat{\Lambda}_\pm$ are raising and lowering operators for the eigenvalue of $\widehat{\Lambda}_3$:

$$\begin{aligned}
\widehat{\Lambda}_3\widehat{\Lambda}_+|K_p\rangle\langle L_q| &= \widehat{\Lambda}_3 \sum_j \varepsilon^+(K_p, j)\varepsilon^+(L_q, j)|K_{p+1}^+(j)\rangle\langle L_{q+1}^+(j)| \\
&= \frac{1}{2}[(p+1) + (q+1) - b] \sum_j \varepsilon^+(K_p, j)\varepsilon^+(L_q, j) \\
&\quad \times |K_{p+1}^+(j)\rangle\langle L_{q+1}^+(j)| \\
\widehat{\Lambda}_3\widehat{\Lambda}_-|K_p\rangle\langle L_q| &= \widehat{\Lambda}_3 \sum_j \varepsilon^-(K_p, j)\varepsilon^-(L_q, j)|K_{p-1}^-(j)\rangle\langle L_{q-1}^-(j)| \\
&= \frac{1}{2}[(p-1) + (q-1) - b] \sum_j \varepsilon^-(K_p, j)\varepsilon^-(L_q, j) \\
&\quad \times |K_{p+1}^-(j)\rangle\langle L_{q+1}^-(j)| .
\end{aligned} \tag{94}$$

[16] The summation indices can be redefined in the last term to interchange j and k,

[17] This will be assumed. In the unlikely case that the number of spin orbitals is odd, b_0 can be *defined* to be $b/2$. In this case μ will have half-integer values.

In each case the sum over j is for all $j \in K_p \cup J_q$. It follows that

$$\left[\widehat{\Lambda}_3, \widehat{\Lambda}_\pm\right] = \pm\widehat{\Lambda}_\pm \,. \tag{95}$$

This suggests that an additional superoperator be defined as

$$\widehat{\Lambda^2} \stackrel{\text{def}}{=} \frac{1}{2}\left(\widehat{\Lambda}_+\widehat{\Lambda}_- + \widehat{\Lambda}_-\widehat{\Lambda}_+\right) + \widehat{\Lambda_z^2} \,. \tag{96}$$

Using the same algebraic manipulations as in the case of angular momentum operators, we find that $\widehat{\Lambda^2}$ commutes with $\widehat{\Lambda}_3$, $\widehat{\Lambda}_+$, and $\widehat{\Lambda}_-$ (see Appendix). It can also be shown, following the arguments used in the case of angular momentum operators, that $\widehat{\Lambda^2}$ has eigenvalues $\lambda(\lambda+1)$ where λ is an integer, and that $-\lambda \leq \mu \leq \lambda$. Possible values of λ are $\min(|\mu|) = 0 \leq \lambda \leq \max(|\mu|) = b_0$.

It has been shown that $\widehat{\Lambda}_-$ is essentially equivalent to the reduction map and that the eigenvalue of $\widehat{\Lambda}_3$ is determined by the particle number p. The superoperator $\widehat{\Lambda^2}$ does not change p so it is a map from $\mathcal{E}_p$ to $\mathcal{E}_p$. The vectors (matrices) in $\mathcal{E}_p$ are eigenvectors of $\widehat{\Lambda}_3$, and linear combinations can be can be chosen to be eigenvectors of $\widehat{\Lambda^2}$ as well. Suppose that this has been done and denote the common eigenvectors by $\mathbf{V}^{\lambda,\mu}$. If $\widehat{\Lambda}_-$ is applied to $\mathbf{V}^{\lambda,-\lambda}$ the result must be zero, since μ cannot be reduced below $-\lambda$. Similarly, we have seen that ${}^{\,p}_{p-1}{\downarrow}\ \mathbf{A}^{(p,\pi)} = 0$ for $\mathbf{A}^{(p,\pi)} \in \mathcal{E}_{p,\pi}$ if $\pi = p \leq b_0$. This suggests a relationship $\lambda = b_0 - \pi$.

Suppose that $\left\{\mathbf{Y}^{(p,\pi)\kappa}\right\}$ is an orthonormal basis for $\mathcal{E}_{p,\pi}$ for each allowed p and π. It will be convenient for the current discussion to use $\mu = p - b_0$ and $\lambda = b_0 - \pi$, and let $|\lambda,\mu,\kappa\rangle = \mathbf{Y}^{(p,\pi)\kappa}$. Then

$$\widehat{\Lambda}_3|\lambda,\mu,\kappa\rangle = \mu|\lambda,\mu,\kappa\rangle \qquad \widehat{\Lambda^2}|\lambda,\mu,\kappa\rangle = \lambda(\lambda+1)|\lambda,\mu,\kappa\rangle \,. \tag{97}$$

It is also readily verified from the definitions above that

$$\begin{aligned} \widehat{\Lambda}_+\widehat{\Lambda}_- &= \widehat{\Lambda^2} - \widehat{\Lambda}_3^2 + \widehat{\Lambda}_3 \\ \widehat{\Lambda}_-\widehat{\Lambda}_+ &= \widehat{\Lambda^2} - \widehat{\Lambda}_3^2 - \widehat{\Lambda}_3 \end{aligned} \tag{98}$$

and it follows that

$$\begin{aligned} \widehat{\Lambda}_+\widehat{\Lambda}_-|\lambda,\mu,\kappa\rangle &= \left[\lambda(\lambda+1) - \mu^2 + \mu\right]|\lambda,\mu,\kappa\rangle \\ \widehat{\Lambda}_-\widehat{\Lambda}_+|\lambda,\mu,\kappa\rangle &= \left[\lambda(\lambda+1) - \mu^2 - \mu\right]|\lambda,\mu,\kappa\rangle \,. \end{aligned} \tag{99}$$

Using the adjoint relationship of $\widehat{\Lambda}_+$ and $\widehat{\Lambda}_-$ we find

$$\begin{aligned}
&\left(\widehat{\Lambda}_+|\lambda,\mu,\kappa'\rangle,\widehat{\Lambda}_+|\lambda,\mu,\kappa\rangle\right)\\
&= \langle\lambda,\mu,\kappa'|\widehat{\Lambda}_-\widehat{\Lambda}_+|\lambda,\mu,\kappa\rangle\\
&= [\lambda(\lambda+1)-\mu(\mu+1)]\underbrace{\langle\lambda,\mu,\kappa'|\ \lambda,\mu,\kappa\rangle}_{\delta_{\kappa,\kappa'}}
\end{aligned} \tag{100}$$

from which it can be concluded that

$$|\lambda,\mu+1,\kappa\rangle = [\lambda(\lambda+1)-\mu(\mu+1)]^{-1/2}\,\widehat{\Lambda}_+|\lambda,\mu,\kappa\rangle\ . \tag{101}$$

Similarly

$$|\lambda,\mu-1,\kappa\rangle = [\lambda(\lambda+1)-\mu(\mu-1)]^{-1/2}\,\widehat{\Lambda}_-|\lambda,\mu,\kappa\rangle\ . \tag{102}$$

These relationships allow the construction of an orthonormal basis for $\mathcal{E}_{p,\pi}$ from one for $\mathcal{E}_{p,\pi\pm1}$. Of course, bases for these spaces obtained in other ways will not necessarily be so related.

Powers $\left(\widehat{\Lambda}_\pm\right)^t$ will raise or lower μ and p by t, and for $\mathbf{A}^{p,\pi}\in\mathcal{E}_{p,\pi}$ or $\mathbf{A}^{(q,\pi)}\in\mathcal{E}_{q,\pi}$

$$\begin{aligned}
\left(\widehat{\Lambda}_-\right)^t\mathbf{A}^{(p,\pi)} &= c_-(p,\pi,t)\,{}^{\,p}_{p-t}\!\downarrow\,\mathbf{A}^{(p,\pi)}\in\mathcal{E}_{p-t,\pi}\\
\left(\widehat{\Lambda}_+\right)^t\mathbf{A}^{(q,\pi)} &= c_+(q,\pi,t)\,{}^{q+t}_{\;q}\!\uparrow\,\mathbf{A}^{(q,\pi)}\in\mathcal{E}_{q+t,\pi}
\end{aligned} \tag{103}$$

where the coefficients are given by

$$\begin{aligned}
c_-(p,\pi,t) &= \prod_{s=1}^{t}[(b_0-\pi)(b_0-\pi+1)\\
&\qquad -(p-b_0+s-1)(p-b_0+s-2)]^{1/2}\\
c_+(q,\pi,t) &= \prod_{s=1}^{t}[(b_0-\pi)(b_0-\pi+1)\\
&\qquad -(p-b_0+s)(p-b_0+s-1)]^{1/2}\ .
\end{aligned} \tag{104}$$

This gives an explicit expression for the antireduction map ${}^p_q\!\uparrow$ in terms, ultimately, of creation and annihilation operators.

Projection operators. Any matrix in $\mathcal{E}_p$ is a sum of components in the subspaces $\mathcal{E}_{p,\pi}$,

$$\mathbf{A}^{(p)} = \sum_{\pi=0}^{p}\mathbf{A}^{(p,\pi)}\ . \tag{105}$$

It can be projected onto a particular subspace, and thus the component in that subspace obtained, by use of a projection operator

$$\widehat{O}_{\pi}^{(p)} \stackrel{\text{def}}{=} \prod_{\pi' \neq \pi} \frac{\widehat{\Lambda^2} - \lambda'(\lambda'+1)}{\lambda(\lambda-1) - \lambda'(\lambda'+1)} \tag{106}$$

where $\lambda = b_0 - \pi$, $\lambda' = b_0 - \pi'$, and the product is over π' from 0 to p, excluding $\pi' = \pi$. The expression for $\widehat{\Lambda^2}$ in terms of creation and annihilation operators becomes rather messy because of the presence of $\widehat{\Lambda}_3^2$ in Eq. (96), but if the projection operator is restricted to a particular $\mathcal{E}_p$ then $\mu = p - b_0$ can be substituted for $\widehat{\Lambda}_3$ to give

$$\widehat{\Lambda^2} = (p - b_0)^2 \widehat{1} + \frac{1}{2} \sum_{j,k=1}^{b} \left(\breve{a}_j^\dagger \breve{a}_k | \breve{a}_k^\dagger \breve{a}_j + \breve{a}_j \breve{a}_k^\dagger | \breve{a}_k \breve{a}_j^\dagger \right) . \tag{107}$$

Grassmann product representation. The superoperator $\widehat{\Lambda}_-$ is equivalent to reduction, and $\widehat{\Lambda}_+$ to antireduction, for appropriate proportionality constants. An alternative representation of $\widehat{\Lambda}_+$ will help to clarify its meaning. When acting on a ket-bra operator, $\widehat{\Lambda}_-$ produces a sum of ket-bras each differing from the original one by removal of one spin orbital common to the ket and the bra. The effect of $\widehat{\Lambda}_+$ on a ket bra is also to produce a sum of ket-bras, differing from the original one by the addition of the same spin orbital to each, chosen from among those originally not present in either. There is another way of representing this process.

The Grassmann product of functions has been described above. A Grassmann product of operators is also possible, and can be defined in terms of the integral kernel representation. If

$$\begin{aligned} \hat{A}^{(p)} \quad &\leftrightarrow \quad A^{(p)}(x_1, \ldots, x_p; x_1', \ldots, x_p') \\ \hat{B}^{(q)} \quad &\leftrightarrow \quad B^{(q)}(x_1, \ldots, x_q; x_1', \ldots, x_q') \end{aligned} \tag{108}$$

then $\hat{A}^{(p)} \wedge \hat{B}^{(q)}$ is a $(p+q)$-electron operator with kernel

$$\begin{aligned} &[A \wedge B]^{(p+q)}(x_1, \ldots, x_{p+q}; x_1', \ldots, x_{p+q}') \\ &\quad = \hat{O}_{p+q}^{a} \hat{O}_{p+q}^{a}{}' A^{(p)}(x_1, \ldots, x_p; x_1', \ldots, x_p') \\ &\qquad \times B^{(q)}(x_1, \ldots, x_q; x_1', \ldots, x_q') \end{aligned} \tag{109}$$

where $\hat{O}_{p+q}^{a}$ is the $(p+q)$-electron antisymmetrizing projection operator acting on the unprimed variables and $\hat{O}_{p+q}^{a}{}'$ is the same except that it acts on primed variables.

We next note that in the basis-set model the kernel corresponding to the identity operator on $\mathcal{E}_p$ is

$$I^{(p)}(x_1,\ldots,x_p;x'_1,\ldots,x'_p) = \sum_{K_p} \Phi_{K_p}(x_1,\ldots,x_p)\Phi^*_{K_p}(x'_1,\ldots,x'_p) \quad (110)$$

and in particular, for $p=1$, $I^{(1)}(x;x') = \sum_{k=1}^{b} \phi_k(x)\phi_k^*(x')$. Antisymmetrization will annihilate any product in which the same spin orbital appears more than once, so

$$\begin{aligned} &\widehat{\Lambda}_+ A^{(p)}(x_1,\ldots,x_p;x'_1,\ldots,x'_p) \\ &\propto \; A^{(p)}(x_1,\ldots,x_p;x'_1,\ldots,x'_p) \wedge I^{(1)}(x_{p+1};x'_{p+1}) \,. \end{aligned} \quad (111)$$

Because the antisymmetrizing projection operator does not produce a normalized antisymmetrized product, a proportionality constant will be required.

3.7 Matrix components

Rather than dealing with the subspaces $\mathcal{E}_{p,\pi}$ or bases for them, individual matrices can be divided into their components in the subspaces. This approach was pioneered by Absar and Coleman.[1, 2, 5] With the notation used here,

$$\mathbf{F}^{(1)} = \mathbf{F}^{(1,0)} + \mathbf{F}^{(1,1)} \quad (112)$$

with

$$\mathbf{F}^{(1,0)} = (\operatorname{Tr}\mathbf{F})\,\mathbf{X}^{(1)} \qquad \mathbf{F}^{(1,1)} = \mathbf{F}^{(1)} - \mathbf{F}^{(1,0)} \quad (113)$$

and

$$\mathbf{G}^{(2)} = \mathbf{G}^{(2,0)} + \mathbf{G}^{(2,1)} + \mathbf{G}^{(2,2)} \quad (114)$$

with

$$\begin{aligned} \mathbf{G}^{(2,0)} &= (\operatorname{Tr}\mathbf{G})\,X^{(2)} \\ \mathbf{G}^{(2,1)} &= \frac{4b}{b-2}\left({}^{2}_{1}\!\downarrow\, \mathbf{G}^{(2)}\right) \wedge \mathbf{X}^{(1)} - \frac{2(b-1)}{b-2}(\operatorname{Tr}\mathbf{G})\,\mathbf{X}^{(2)} \\ \mathbf{G}^{(2,2)} &= \mathbf{G}^{(2)} - \frac{4b}{b-2}\left({}^{2}_{1}\!\downarrow\right) \wedge \mathbf{X}^{(1)} + \frac{b}{b-2}(\operatorname{Tr}\mathbf{G})\,\mathbf{X}^{(2)} \end{aligned} \quad (115)$$

where $\mathbf{X}^{(q)}$ is the trace-one multiple of the identity matrix on $\mathcal{E}_q$. While the Coleman-Absar results were based on a group-theoretic analysis they can also be related to the treatment above.

If the notation is generalized to include ${}^{p}_{0}\!\downarrow\, \mathbf{A} = \operatorname{Tr}\mathbf{A}$, $a \wedge \mathbf{B}^p = a\mathbf{B}^p$ when a is a number, and $\mathbf{X}^{(0)} = 1$, then the general form can be expressed as

$$\mathbf{A}^{(p,\pi)} = \sum_{\pi'=0}^{\pi} C^{p}_{\pi\pi'} \left({}^{p}_{\pi'}\!\downarrow \mathbf{A}^{(p)}\right) \wedge \mathbf{X}^{(p-\pi')} . \tag{116}$$

Equations (67) and the requirement that $\sum_{\pi=0}^{p} \mathbf{A}^{(p,\pi)} = \mathbf{A}^{(p)}$ can then be used to evaluate the coefficients.

This expression is linear in $\mathbf{A}^{(p)}$ and can be used to define a projection operator that acts on any matrix in $\mathcal{E}_p$ to give its component in $\mathcal{E}_{p,\pi}$. Alternatively, a basis for $\mathcal{E}_p$ can be chosen to have each element entirely in one subspace, $\mathbf{Y}^{(p,\pi)\kappa} \in \mathcal{E}_{p,\pi}$ for $\kappa = 1, \ldots, \dim(\mathcal{E})_{p,\pi}$. The projection operator can then be written

$$\hat{O}_{p,\pi} \stackrel{\text{def}}{=} \sum_{\kappa=1}^{\dim(\mathcal{E}_{p,\pi})} |\mathbf{Y}^{(p,\pi)\kappa}\rangle\langle\mathbf{Y}^{(p,\pi)\kappa}| . \tag{117}$$

While such matrix bases can be constructed, This is unlikely to be a practical approach except for small p.

3.8 Group theoretic considerations

A unitary transformation of the spin orbital basis does not affect the essential features of the relationships among reduced density matrices. One reason for considering such transformations is that n-representability is preserved. Theoretical methods involving transformation to natural orbitals and MCSCF theory can also be formulated in a way involving such transformations. For any p, $\mathcal{E}_p$ is a carrier space for a representation of the unitary group $\mathbb{U}(b)$. The subspaces $\mathcal{E}_{p,\pi}$ are invariant subspaces [1, 2, 3, 5, 31, 32] and it can be shown that this decomposition provides a complete reduction of the representation.[3] This property can be used to define the components of any matrix in the subspaces or to construct a reducing basis. This approach will not be presented here.

3.9 Explicit reducing basis for small p

As suggested above, one way to construct a reducing basis involves application of the expansion superoperator to map the basis for $\mathcal{E}_{p-1,\pi}$ to one for $\mathcal{E}_{p,\pi}$ followed by orthogonalization of remaining basis matrices to those for all $\mathcal{E}_{p,\pi}$ with $\pi < p$ to give a basis for $\mathcal{E}_{p,\pi=p}$. The particular choice of this final basis for each p is arbitrary within a unitary transformation, but once chosen will fix the bases for that π and higher p's.

Some explicit expressions for the results will be presented here for the case of real symmetric matrices and $p \leq 2$. Imaginary antisymmetric matrices can be added by analogy, and extension to larger p is straightforward in principle, although a bit messy in practice. Basis matrices will be denoted by $\mathbf{Y}^{(p,\pi)\kappa}$, with κ numbered within $\mathcal{E}_p$ starting with $\kappa = 1$ and not starting over when π changes.

The starting point can be taken to be $p = 0$, for which only $\pi = 0$ is possible and the single basis matrix for $\mathcal{E}_{0,0}$ is the one by one "matrix" $\mathbf{Y}^{(0,0)1} = 1$.

By extension of the Grassmann product definition to include the case when one factor is just a number and renormalizing, the basis element for $\mathcal{E}_{1,0}$ is found to be $\mathbf{Y}^{(1,0)1} = b^{-1/2}\mathbf{I}^{(1)}$. This is the matrix $\mathbf{X}^0$ in $\mathcal{E}_1$. The basis for $\mathcal{E}_{1,1}$ consists of other diagonal matrices in $\mathcal{E}_1$ orthogonalized to $\mathbf{Y}^{(1,0)1}$, for example those defined in Eq. (38) and now denoted by $\mathbf{Y}^{(1,1)\kappa}$ with $\kappa = 2, \ldots, b$ All trace zero matrices in $\mathcal{E}_1$ are in $\mathcal{E}_{1,1}$ and a convenient choice of basis matrices is those defined in Eqs. (30) and (31), now denoted by $\mathbf{Y}^{(1,1)\kappa}$ with $\kappa = b+1, \ldots, b(b+1)/2$.

Consider next $p = 2$ and define $B = b(b+1)/2$. The basis for $\mathcal{E}_{p,0}$ consists of $\mathbf{Y}^{(2,0)1} = B^{-1/2}I^{(2)}$. The basis elements for $\mathcal{E}_{2,1}$ are of the form

$$\mathbf{Y}^{(2,1)\kappa} = c\mathbf{Y}^{(1,1)\kappa} \wedge \mathbf{I}^{(1)} \tag{118}$$

where c is chosen so that the result is normalized. Matrices that are of the form $\mathbf{Y}^{(1,1)\kappa} \wedge \mathbf{A}^{jj}$, where A^{jj} has $A^{jj}_{jj} = 1$ as its only nonzero element, are successively orthogonalized to the $\mathbf{Y}^{(2,1)\kappa}$ defined above and to one-another and then renormalized to provide part of the basis for $\mathcal{E}_{2,2}$. Additionally, matrices with nonzero elements only for four distinct indices are in $\mathcal{E}_{2,2}$ and a convenient independent orthonormal set of them can be added to the basis.

Explicit expressions are given in Table 1 for the case when $b = 4$ and $p \leq 2$. While the number $b = 4$ is unreasonably small, these expressions serve to make explicit the general procedures outlined above. It is more compact to give expressions for ket-bra operators rather than the equivalent matrices.

4. Densities

Suppose that $\Gamma^{(p)}$ is the kernel associated with a p-electron reduced density matrix. Integration over spin variables gives the kernel associated with a spinless or charge-density matrix

$$\begin{aligned}\Gamma^{(\bar{p})}&(\vec{r}_1, \ldots, \vec{r}_p; \vec{r}'_1, \ldots, \vec{r}'_p) \\ &= \int \Gamma^{(p)}(\vec{r}_1, \xi_1, \ldots, \vec{r}_p, \xi_p; \vec{1}'\xi_1, \ldots, \vec{p}'\xi_p)\, d\xi_1 \cdots d\xi_p\end{aligned} \tag{119}$$

Table 1. A possible choice of basis matrices when $b = 4$

$p = 0,\ \dim(\mathcal{E}_0) = 1$
$\pi = 0,\ \dim(\mathcal{E}_{0,0}) = 1$

$\hat{Y}_1^{(0,0)} = \hat{1}$

$p = 1,\ \dim(\mathcal{E}_1) = 10$
$\pi = 0,\ \dim(\mathcal{E}_{1,0}) = 1$

$\hat{Y}_1^{(1,0)} = \frac{1}{2}\sum_j |j\rangle\langle j|$

$\pi = 1,\ \dim(\mathcal{E}_{1,1}) = 9$

$\hat{Y}_1^{(1,1)} = \frac{1}{\sqrt{2}}\left(|1\rangle\langle 1| - |2\rangle\langle 2|\right)$
$\hat{Y}_2^{(1,1)} = \frac{1}{\sqrt{6}}\left(|1\rangle\langle 1| + |2\rangle\langle 2| - 2|3\rangle\langle 3|\right)$
$\hat{Y}_3^{(1,1)} = \frac{1}{\sqrt{12}}\left(|1\rangle\langle 1| + |2\rangle\langle 2| + |3\rangle\langle 3| - 3|4\rangle\langle 4|\right)$
$\hat{Y}_{4\ldots 10}^{(1,1)} = \frac{1}{\sqrt{2}}\left(|j\rangle\langle k| + |k\rangle\langle j|\right)\ ,\ 1 \le j < k \le 4$

$p = 2,\ \dim(\mathcal{E}_2) = 21$
$\pi = 0,\ \dim(\mathcal{E}_{2,0}) = 1$

$\hat{Y}_1^{(2,0)} = \frac{1}{\sqrt{6}}\sum_{j<k} |jk\rangle\langle jk|$

$\pi = 1,\ \dim(\mathcal{E}_{2,1}) = 9$

$\hat{Y}_2^{(2,1)} = \frac{1}{2}\left(|1,3\rangle\langle 1,3| + |1,4\rangle\langle 1,4| - |2,3\rangle\langle 2,3| - |2,4\rangle\langle 2,4|\right)$
$\hat{Y}_3^{(2,1)} = \frac{1}{\sqrt{12}}\left(2|1,2\rangle\langle 1,2| - |1,2\rangle\langle 1,5| + |1,4\rangle\langle 1,4| - |2,3\rangle\langle 2,3|\right.$
$\left. + |2,4\rangle\langle 2,4| - 2|3,4\rangle\langle 3,4|\right)$
$\hat{Y}_4^{(2,1)} = \frac{1}{\sqrt{6}}\left(|1,2\rangle\langle 1,2| + |1,3\rangle\langle 1,3| - |1,4\rangle\langle 1,4| + |2,3\rangle\langle 2,3|\right.$
$\left. - |2,4\rangle\langle 2,4| - |3,4\rangle\langle 3,4|\right)$
$\hat{Y}_{5\ldots 10}^{(2,1)} = \frac{1}{2}\left(|j,\ell\rangle\langle j,\ell| + |k,\ell\rangle\langle j,\ell| + |j,m\rangle\langle k,m| + |k,m\rangle\langle j,m|\right)$
$j < k, \ell < m$, 6 distinct choices

$\pi = 2,\ \dim(\mathcal{E})_{2,2} = 11$

$\hat{Y}_{11}^{(2,2)} = \frac{1}{\sqrt{12}}\left(2|1,2\rangle\langle 1,2| - |1,2\rangle\langle 1,5| - |1,4\rangle\langle 1,4| - |2,3\rangle\langle 2,3|\right.$
$\left. - |2,4\rangle\langle 2,4| + 2|3,4\rangle\langle 3,4|\right)$
$\hat{Y}_{12}^{(2,2)} = \frac{1}{2}\left(|1,3\rangle\langle 1,3| - |1,4\rangle\langle 1,4| - |2,3\rangle\langle 2,3| + |2,4\rangle\langle 2,4|\right)$
$\hat{Y}_{13\ldots 18}^{(2,2)} = \frac{1}{2}\left(|j,\ell\rangle\langle j,\ell| + |k,\ell\rangle\langle j,\ell| - |j,m\rangle\langle k,m| - |k,m\rangle\langle j,m|\right)$
$j < k, \ell < m$, 6 distinct choices
$\hat{Y}_{19\ldots 21}^{(2,2)} = \frac{1}{\sqrt{2}}\left(|j,k\rangle\langle \ell,m||\ell,m\rangle\langle j,k|\right)$

and dropping the primes produces a p-electron density

$$\mathsf{P}^{(p)}(\vec{r}_1,\ldots,\vec{r}_p) = \Gamma^{(\bar{p})}(\vec{r}_1,\ldots,\vec{r}_p;\vec{r}_1,\ldots,\vec{r}_p) \tag{120}$$

The commonly interesting cases are $p = 1$ or 2, and the notation is simplified to

$$\Gamma^{(0)} = \Gamma^{(\bar{2})}, \qquad \gamma^{(0)} = \Gamma^{(\bar{1})} \tag{121}$$

and

$$\rho = \mathsf{P}^{(1)}, \qquad \mathsf{P} = \mathsf{P}^{(2)}. \tag{122}$$

Since they are defined here in terms of $\Gamma^{(p)}$ (rather than $D^{(p)}$) they are normalized to

$$\int \rho(\vec{r})\, d\vec{r} = n\,, \qquad \int \mathsf{P}(\vec{r}_1,\vec{r}_2) = n(n-1)/2 \tag{123}$$

where n is the number of electrons in the system of interest. It is sometimes convenient to use

$$\breve{\rho} = n^{-1}\rho\,, \qquad \breve{\mathsf{P}} = \binom{n}{2}^{-1}\mathsf{P} \tag{124}$$

or in the general case $\breve{\mathsf{P}}^{(p)} = \binom{n}{p}^{-1}\mathsf{P}^{(p)}$, integrating to 1.

4.1 Spaces of kernels and of densities

Let ${}^{ker}\mathcal{E}_1^r$ denote the space of one-electron spatial kernels and $\mathcal{E}_1^0$ the corresponding space of $b_0 \times b_0$ matrices, such as those of 1-electron spinless operators with respect to the orbital basis functions. As noted above, it is assumed that the b spin orbitals are products of b_0 orbitals $\{\chi_j\}$ and spin functions α or β, so $b = 2b_0$. Corresponding to a charge density matrix γ^0 is

$$\gamma^{(0)}(\vec{r};\vec{r}\,') = \sum_{j,k=1}^{b_0} \gamma_{j,k}^{(0)}\chi_j(\vec{r})\chi_k^*(\vec{r}\,') \tag{125}$$

in the kernel space.

The (one-electron) density $\rho(\vec{r})$ is an element of a function space spanned by products of the orbitals, but in general these products may be linearly dependent.[18] If the orbitals are real, as is commonly the

[18]In principle, a set of functions either is or is not linearly dependent. In practice there may be near linear dependencies that are hard to deal with. This problem will be ignored in the present, formal treatment.

case for molecular electronic structure calculations, then $b_0(b_0+1)/2$ products can be constructed and the dimension of the space is

$$b_1 = \dim(\mathcal{F}) \leq \frac{1}{2} b_0(b_0+1) \, . \tag{126}$$

Assume that an orthonormal basis $\{f_\kappa(\vec{r}),\ \kappa = 0, \ldots, b_1 - 1\}$ for $\mathcal{F}$ has been constructed. This can be done in such a way that the function f_0 is given by

$$f_0(\vec{r}) = C \sum_{k=1}^{b_0} |\chi_k(\vec{r})|^2 \, , \tag{127}$$

(where C is a normalization constant) is proportional to the density corresponding to a density matrix $\mathbf{X}^0$, and has $\int f_0(\vec{r})\, d\vec{r} > 0$, while

$$\int f_\kappa(\vec{r})\, d\vec{r} = 0 \, , \qquad \kappa > 0 \, . \tag{128}$$

Note that the orthonormality of the basis functions is with respect to the usual L^2 scalar product

$$\int f_\kappa(\vec{r}) f_\lambda(\vec{r})\, d\vec{r} = \delta_{\kappa,\lambda} \tag{129}$$

(functions in the density space are real, even when the orbitals are not necessarily real), but that the trace normalization of the density matrix corresponds to $\int \rho(\vec{r})\, d\vec{r}$. A density can be expanded as

$$\rho(\vec{r}) = \sum_{\kappa=0}^{b_1-1} \rho_\kappa f_\kappa(\vec{r}) \, . \tag{130}$$

The expansion coefficients define a vector in real, b_1-dimensional vector space $\mathcal{F} = \mathcal{F}_1$. Of course the density matrix and density could be replaced by any other kernel and function in the relevant spaces.

Although (as noted in the Introduction) any reasonable density is n representable, when a basis set model is introduced there may be positive functions with appropriate normalization in $\mathcal{F}$ that cannot be obtained from acceptable (positive) density matrices expressible in terms of the given basis.[13]

Densities for $p > 1$ are functions in spaces spanned by symmetrized products of the functions $\{f_\kappa\}$. The permutational symmetry of the kernels $\Gamma^{(p,0)}$ for $p > 1$ is more complicated and attention will initially be concentrated on the one-electron case. It will also be assumed for simplicity that basis orbitals are real and that the matrices of interest are real symmetric, and the subscript $(p = 1)$ will often be omitted.

4.2 The collapse map

The process of obtaining a density from a density matrix kernel defines a map from the kernel space to the density function space. This map will be called "collapse" and denoted by $\hat{\delta}$.[13, 15] The same symbol will be used to denote corresponding map from the matrix space to the vector space of density coefficients, $\hat{\delta} : \mathcal{E}_1 \rightarrow \mathcal{F}$.

As noted above, products of basis orbitals will usually be linearly dependent, at least in a practical sense. Because of this the dimension b_1 of $\mathcal{F}$ will be less than that of ${}^{ker}\mathcal{E}_1^r$. It follows that there are elements in ${}^{ker}\mathcal{E}_1^r$ that give 0 when $\hat{\delta}$ is applied. The null space of this map is denoted by $\mathcal{K}_r$,

$$\mathcal{K}_r \stackrel{\text{def}}{=} \left\{ A \in {}^{ker}\mathcal{E}_1^r \middle| \, \hat{\delta}A = 0 \right\} = \left\{ A(\vec{r};\vec{r}\,') \middle| \, A(\vec{r};\vec{r}) = 0 \right\} \tag{131}$$

The orthogonal complement of $\mathcal{K}_r$ in ${}^{ker}\mathcal{E}_1^r$,

$$\mathcal{L}_r \stackrel{\text{def}}{=} {}^{ker}\mathcal{E}_1^r \setminus \mathcal{K}_r \tag{132}$$

will be called the *local* subspace, for reasons that will be developed below. These same symbols will be used for corresponding subspaces of the matrix space $\mathcal{E}_1$.

It should be noted that the subspace decomposition ${}^{ker}\mathcal{E}^r = \mathcal{K}_r \oplus \mathcal{L}_r$ differs in a fundamental way from the decomposition $\mathcal{E}_1 = \mathcal{E}_{1,0} \oplus \mathcal{E}_{1,1}$. Not only are the dimensions involved different, but the subspace decomposition associated with the collapse map depends on the choice of basis orbitals, via the linear dependency relationships among their products. The subspace decomposition based on reduction maps depends only on the number of orthonormal orbitals; it is independent of their specific nature.

Assume a basis $\left\{ \mathbf{X}^\kappa, \kappa = 0, 1, \ldots, \dim(\mathcal{E}_1^0) - 1 \right\}$ for $\mathcal{E}_1^0$ that is orthonormal with respect to the trace scalar product and includes $\mathbf{X}^0$. The $\mathbf{X}^\kappa$ with $\kappa > 1$ are a basis for $\vec{\mathcal{E}}_1^0$ and have trace zero. Corresponding to each of these basis matrices in $\mathcal{E}_1^0$ is a function in $\mathcal{F}_1$, $\hat{\delta}X_\kappa(\vec{r})$, and in general these functions are linearly dependent. Suppose that there are w independent equations of the form

$$\sum_\kappa C_{j\kappa} \hat{\delta} X_\kappa(\vec{r}) = 0 \, , \qquad j = 1, \ldots, w \, . \tag{133}$$

Integration of the jth equation gives

$$0 = \sum_\kappa \int C_{j\kappa} \hat{\delta} X_\kappa(\vec{r}) \, d\vec{r} = C_{j0} b_0^{1/2} \tag{134}$$

since $\int \hat{\delta} X_\kappa(r)\, d\vec{r} = \mathrm{Tr}\,\mathbf{X}^\kappa$ is zero for $\kappa > 0$, and it follows that $\hat{\delta} X_0$ does not appear in any of these equations. For each j between 1 and w define

$$\mathbf{Z}^j = \sum_\kappa C_{j\kappa} \mathbf{X}^\kappa \,. \tag{135}$$

These all have $\hat{\delta} Z_j = 0$ and are independent. As part of a new basis for $\mathcal{E}_1^0$, construct a set $\{\mathbf{Y}^\kappa,\ \kappa = \dim(\mathcal{E}_1^0) - w, \ldots, \dim(\mathcal{E}_1^0) - 1\}$ by orthonormalizing the $\{\mathbf{Z}^j\}$ with respect to the trace scalar product. Complete this basis by adding $\mathbf{Y}^0 = \mathbf{X}^0$ and $\dim(\mathcal{E}_1^0) - w - 1$ matrices $\mathbf{Y}^\lambda$ as linear combinations of the $\mathbf{X}^\kappa$ normalized and orthogonal to the previously-defined $\mathbf{Y}$'s. As a consequence of these definitions and the fact that $\dim(\mathcal{L}_1) = \dim(\mathcal{F}_1) = b_1$,

$$\begin{aligned} \mathcal{K}_1 &= \mathrm{span}\left\{\mathbf{Z}^j,\ j = 1, \ldots, w\right\} \\ &= \mathrm{span}\left\{\mathbf{Y}^\kappa,\ \kappa = \dim(\mathcal{E}_1^0) - w, \ldots, \dim(\mathcal{E}_1^0) - 1\right\} \\ \mathcal{L}_1 &= \mathrm{span}\left\{\mathbf{Y}^\kappa,\ \kappa = 0, b_1 - 1\right\} \,. \end{aligned} \tag{136}$$

It is possible to choose basis orbitals such that all products are linearly independent. Such a basis is called a LIP basis.[19] With a LIP basis $\mathcal{L}_r = \mathcal{E}_1^0$. Note that while the $\{\mathbf{Y}^\kappa\}$ in $\mathcal{L}_r$ are orthogonal, by construction, with respect to the trace scalar product in $\mathcal{E}_1^0$, the corresponding densities $\hat{\delta}\mathbf{Y}^\kappa$ are not in general orthogonal with respect to the scalar product $(f, g)_\mathcal{F} = \int f^*(\vec{r}) g(\vec{r})\, d\vec{r}$ in $\mathcal{F}$. They are guaranteed to be linearly independent, however, and are a basis for $\mathcal{F}$ so any function in $\mathcal{F}$ has a unique expansion in terms of them.

4.3 Momentum representation

When density matrices are considered, results will be independent of the representation used. In particular, the orbitals $\chi_j(\vec{r})$ in the basis can be replaced by their Fourier transforms $\tilde{\chi}_j(\vec{p})$ to work in the momentum representation. If operators are also used in the momentum representation then the same matrices will result as in the position representation. Scalar products are also unchanged by Fourier transformation, so the entire structure of the matrix spaces discussed above remains unchanged. The effect of the collapse map will change however, because linear dependencies among products of functions are not in general the same as linear dependencies among products of their Fourier transforms.[16]

---

[19] A LIP basis will not be complete and, unless carefully chosen *after* a calculation has been done, is unlikely to be a "good" basis set by usual criteria.[28]

Example: harmonic oscillator functions. As a simple example of linear dependencies of products and of the differences between the position and momentum cases, consider a basis of harmonic oscillator functions in one dimension

$$\psi_n(x) = C_n H_n(x) e^{-x^2/2} \tag{137}$$

where $C_n = (2^n n! \sqrt{\pi})^{-1/2}$ is a normalization constant and H_n is a Hermite polynomial, a polynomial of degree n. A product

$$\psi_m(x)\psi_n(x) = C_m C_n H_m(x) H_n(x) e^{-x^2} \tag{138}$$

will involve a polynomial of degree $m+n$. It can be expressed as a linear combination of harmonic oscillator functions of $\sqrt{2}x$, since the exponential will then be the same as that in the product. The Hermite polynomials are related by

$$H_m(x)H_n(x) = \sum_{\kappa=0}^{\nu} W_{mn\kappa} H_{2\kappa+\alpha}(\sqrt{2}x) \tag{139}$$

where ν and α are determined by the condition $m+n = 2\nu+\alpha$, $\alpha = 0$ or 1. This for reflects the fact that H_n involves only even powers, if n is even, or only odd powers, if n is odd. The coefficients can be shown to be

$$\begin{aligned} W_{mn\kappa} &= 2^{-\alpha/2} \frac{m!n!}{(2\kappa+\alpha)!} \frac{(-1)^{\nu-\kappa}}{2^\nu} \\ &\times \sum_k \frac{(-4)^k (2\nu+\alpha-2k)!}{k!(m-k)!(n-k)!(\nu-\kappa-k)!} \,. \end{aligned} \tag{140}$$

The sum includes all values of k for which all the factorials are defined. Consider products with $0 \le m, n \le M$. There are $(M+1)(M+2)/2$ products possible but the highest power of x to appear will be x^{2M}, and only $2M+1$ functions $\psi_\kappa(\sqrt{2}x)$ are required. For any $M > 1$ there must be linear dependencies. For example, $\psi_0(x)\psi_0(x)$, $\psi_1(x)\psi_1(x)$, and $\psi_0(x)\psi_2(x)$ can all be expressed in terms of $\psi_0(\sqrt{2}x)$ and $\psi_2(\sqrt{2}x)$.

The Fourier transform of a Harmonic oscillator function is[20]

$$\tilde{\Psi}_n(p) = (-i)^n \psi_n(p) \,. \tag{141}$$

[20] Although overall phase factors have no physical significance, the relative phases of a function and its Fourier transform must be preserved. This can be illustrated by the following amusing but fallacious argument. In reduced units the Hamiltonian for a harmonic oscillator can be written in terms of momentum and position operators as $\hat{H} = (\hat{x}^2 + \hat{p})^2/2$, which is the same in either the x or the p representation. This suggests that $\tilde{\psi}_n$ can be taken to be the same as ψ_n. Let $f(x)$ be any reasonable function. It can be expanded as a linear combination of harmonic oscillator functions, $f(x) = \sum_n a_n \psi_n(x)$ and its Fourier transform will thus be $\tilde{f}(p) = \sum_n a_n \psi_n(p)$. Conclusion: any function is equal to its Fourier transform!

Suppose that a linear dependency condition is expressible as

$$\sum_{m,n} a_{m,n}\psi_m(x)\psi_n(x) = 0 \,. \tag{142}$$

There will then be an equation involving momentum-space functions

$$\sum_{m,n} i^{m+n} a_{m,n}\psi_m(p)\psi_n(p) = 0 \,, \tag{143}$$

which involves a different set of coefficients. The number of linear dependencies will be the same in the two representations.

Just as for the position representation, the momentum representation leads to a subspace decomposition based on the collapse map. Let ${}^{ker}\mathcal{E}_1^p$ denote the space of spinless momentum kernels. There will be kernels in this space that give 0 when $\hat{\delta}$ is applied. The null space of this map is denoted by $\mathcal{K}_p$,

$$\mathcal{K}_p \stackrel{\text{def}}{=} \left\{ A \in {}^{ker}\mathcal{E}_1^p \middle|\, \hat{\delta}A = 0 \right\} = \left\{ A(\vec{p};\vec{p}\,') \middle|\, A(\vec{p};\vec{p}) = 0 \right\} \tag{144}$$

The orthogonal complement of $\mathcal{K}_p$ in ${}^{ker}\mathcal{E}_1^p$,

$$\mathcal{L}_p \stackrel{\text{def}}{=} {}^{ker}\mathcal{E}_1^p \setminus \mathcal{K}_p \tag{145}$$

is the "local" subspace in the momentum representation. These same symbols will be used for corresponding subspaces of the matrix space $\mathcal{E}_1^0$.

4.4 Subspaces based on both representations

Let $\hat{\delta}_r$ and $\hat{\delta}_p$ denote the maps from $\mathcal{E}_1^0$ to the density spaces $\mathcal{F}_r$ and $\mathcal{F}_p$, respectively. They lead to two decompositions of the matrix space:

$$\mathcal{E}_1^0 = \mathcal{K}_r \oplus \mathcal{L}_r = \mathcal{K}_p \oplus \mathcal{L}_p \,. \tag{146}$$

Unfortunately, $\mathcal{K}_r$ is not contained within $\mathcal{K}_p$, nor is $\mathcal{K}_p$ contained within $\mathcal{K}_r$, nor are these subspaces orthogonal. The subspaces $\mathcal{L}_r$ and $\mathcal{L}_p$ are also not related in any simple way. A subspace decomposition of $\mathcal{E}_1^0$ based on both $\hat{\delta}_r$ and $\hat{\delta}_p$ can nevertheless be defined:

$$\mathcal{E}_1^0 = \mathcal{K} \oplus \mathcal{L} \oplus \mathcal{KL} \oplus \mathcal{LK} \oplus \mathcal{M} \tag{147}$$

where

$$\mathcal{K} \stackrel{\text{def}}{=} \left\{ \mathbf{A} \in \mathcal{E}_1^0 \middle|\, \hat{\delta}_r\mathbf{A} = 0 \text{ and } \hat{\delta}_p\mathbf{A} = 0 \right\}$$

$$\begin{aligned}
\mathcal{L} &\stackrel{\text{def}}{=} \{\mathbf{A} \in \mathcal{E}_1^0 | \mathbf{A} \perp \mathcal{K}_r \text{ and } \mathbf{A} \perp \mathcal{K}_p\} \\
\mathcal{KL} &\stackrel{\text{def}}{=} \left\{\mathbf{A} \in \mathcal{E}_1^0 \middle| \hat{\delta}_r \mathbf{A} = 0 \text{ and} \mathbf{A} \perp \mathcal{K}_p\right\} \\
\mathcal{LK} &\stackrel{\text{def}}{=} \left\{\mathbf{A} \in \mathcal{E}_1^0 \middle| \hat{\delta}_p \mathbf{A} = 0 \text{ and} \mathbf{A} \perp \mathcal{K}_r\right\} \\
\mathcal{M} &\stackrel{\text{def}}{=} \mathcal{E}_1^0 \setminus (\mathcal{K} \oplus \mathcal{L} \oplus \mathcal{KL} \oplus \mathcal{LK}) \,. \qquad (148)
\end{aligned}$$

It might seem that $\mathcal{M}$ should be empty–that the other subspaces exhaust all of $\mathcal{E}_1^0$, but this is not the case.[16]

4.5 Local and nonlocal components of operators

The concept of a local operator in important in quantum chemistry, especially in density functional theory. An operator is local if it (*i.e.* its Schrödinger picture, coordinate representation) is simply a function of position, like typical Coulombic potential energy operators. The kinetic energy and Hartree-Fock exchange operators are nonlocal. This concept can become ambiguous in the model problem defined by a finite basis set, where two operators having the same matrix representation must be regarded as equivalent. Of course if the momentum representation is used, the definition of locality must be changed. Initially only the traditional meaning associated with the position representation will be considered.

The expectation value of a local operator is determined by the density:

$$\begin{aligned}
\langle f(\vec{r}) \rangle &= \int\!\!\int \delta(x - x') f(\vec{r}) \hat{\gamma}(x; x')\, dx' dx \\
&= \int f(\vec{r}) \gamma^{(0)}(\vec{r}; \vec{r})\, d\vec{r} = \int f(\vec{r}) \rho(\vec{r})\, d\vec{r} \,. \qquad (149)
\end{aligned}$$

Theorem 10 *If the matrix* $\mathbf{G}$ *corresponding to an operator* $\hat{G}$ *is entirely within* $\mathcal{L}_r$ *then there is some function* $g(\vec{r})$ *such that* $\int g(\vec{r})\rho(\vec{r})\, d\vec{r} = \left\langle \hat{G} \right\rangle$ *(within the model defined by the basis).*

▷ Since $\mathbf{G}$ is assumed to be entirely within $\mathcal{L}_r$ it can be expanded as

$$\mathbf{G} = \sum_{\kappa=0}^{b_1 - 1} G_\kappa \mathbf{Y}^\kappa \,, \qquad (150)$$

The density matrix for the state with respect to which the expectation value is to be computed can be expanded as

$$\mathbf{D} = \sum_{\kappa=0}^{b_1 - 1} d_\kappa \mathbf{Y}^\kappa + \sum_{\kappa \geq b_1} d_\kappa \mathbf{Y}^\kappa \qquad (151)$$

and because of the orthonormality of the basis matrices

$$\left\langle \hat{G} \right\rangle = (\mathbf{D}, \mathbf{G}) = \sum_{\kappa=0}^{b_1-1} d_\kappa G_\kappa \, . \tag{152}$$

The density will be given by

$$\rho(\vec{r}) = \sum_{\kappa=0}^{b_1-1} d_\kappa \hat{\delta Y}_\kappa(\vec{r}) \, . \tag{153}$$

Consider a function expanded in terms of the orthonormal basis for $\mathcal{F}_1$ as

$$g(\vec{r}) = \sum_{k=1}^{b_1} g_k f_k(\vec{r}) \, . \tag{154}$$

Then

$$\int g(\vec{r})\rho(\vec{r}) = \sum_{\kappa=0}^{b_1-1} \sum_{k=1}^{b_1} d_\kappa g_k \underbrace{\int f_k(\vec{r}) \hat{\delta Y}_\kappa(\vec{r}) \, d\vec{r}}_{A_{k\kappa}} \, . \tag{155}$$

The matrix $\mathbf{A}$ is not unitary but is nonsingular. The integral of $g\rho$ will be equal to the expectation value of $\hat{G}$ if

$$\sum_k g_k A_{k\kappa} = G_\kappa \, , \qquad g_k = \sum_\kappa \left(\mathbf{A}^{-1}\right)_{k\kappa} G_\kappa \tag{156}$$

so the function required by the theorem exists.▌

Near-locality of kinetic energy. The kinetic energy is a classic example of a nonlocal property, but just how nonlocal is it? We are now in a position to give a quantitative measure of the locality of an operator, within a basis-set model. Of course other measures would be possible. For any operator $\hat{F}$, construct the matrix $\mathbf{F}$ and divide it into components $\mathbf{F}^K \in \mathcal{K}_r$ and $\mathbf{F}^L \in \mathcal{L}_r$. Then the locality of $\mathbf{F}$ is given by

$$\operatorname{loc}(\mathbf{F}) \stackrel{\text{def}}{=} \frac{||\mathbf{F}^L||}{||\mathbf{F}||} \, . \tag{157}$$

From the harmonic oscillator example, and from other cases that have been investigated, we expect that $\dim(\mathcal{L}_r)$ grows linearly with the orbital basis size b_0, while $\dim(\mathcal{E}_1^0)$ grows quadratically. It would thus seem that a randomly selected matrix in $\mathcal{E}_1^0$ will become less and less local as the size of the basis is increased.

The kinetic energy in one dimension, $\hat{T} \propto -d^2/dx^2$, can be treated exactly with a basis of harmonic oscillator functions, and it can be shown that in this case in the limit as $b_0 = n_{max} + 1 \rightarrow \infty$, $\text{loc}(\mathbf{T}) \rightarrow \sqrt{5}/3 \sim 0.75$. For realistic quantum chemical basis sets loc(**T**) can be investigated numerically and is found to be typically > 0.98. The kinetic energy is really quite local, by this measure.[17, 20, 21]

It is not too surprising that the kinetic energy is nearly local. since it involves derivatives evaluated at $\vec{r}\,' = \vec{r}$. The Hartree-Fock exchange operator does not have this property, but it is also found to be very nearly local.[19]

4.6 Two-electron densities

Two electron densities are more difficult to treat for several reasons. The most important is probably that there are well-behaved symmetric functions of $\vec{r}_1$ and $\vec{r}_2$ that are not n representable.[6] This problem will not be addressed here; it is the behavior of the two-electron spinless or charge density matrix that leads to complications now to be discussed.

Theorem 11 *The two-electron charge density matrix is the sum of symmetric and antisymmetric components,*[21]

$$\Gamma^{(0)}(\vec{r}_1, \vec{r}_2; \vec{r}_1', \vec{r}_2') = \Gamma^a(\vec{r}_1, \vec{r}_2; \vec{r}_1', \vec{r}_2') + \Gamma^s(\vec{r}_1, \vec{r}_2; \vec{r}_1', \vec{r}_2'), \tag{158}$$

where Γ^a changes sign if $\vec{r}_1$ and $\vec{r}_2$ are interchanged or $\vec{r}_1'$ and $\vec{r}_2'$ are interchanged, and Γ^s is unchanged by either of these interchanges.

▷ The basis for the two-electron Hilbert space $\mathcal{H}_2$ with respect to which kernels in ${}^{ker}\mathcal{E}_2$ are defined can be chosen to be singlet or triplet spin eigenfunctions, which are products of symmetric geminals and singlet spin functions or of antisymmetric geminals and triply spin functions. In general, Γ will involve cross products between singlet and triplet spin geminals, but the spin integration leading to $\Gamma^{(0)}$ will give zero for them, so only singlet-singlet and triplet-triplet spatial products remain. They have the permutational symmetry claimed in the theorem.▌

As a consequence of this the two-electron spatial kernel space can divided into orthogonal subspaces[22]

$${}^{ker}\mathcal{E}_2^r = {}^{ker}\mathcal{E}_2^a \oplus {}^{ker}\mathcal{E}_2^s\,. \tag{159}$$

[21] This was established by Kutzelnigg[23] for spin eigenstates but is true in general.
[22] Any symmetric geminal is orthogonal to any antisymmetric geminal.

The collapse map from the kernel space to the density space again consists of dropping primes,

$$\hat{\delta}_2 \Gamma^{(0)}(\vec{r}_1, \vec{r}_2; \vec{r}_1', \vec{r}_2') = \Gamma^{(0)}(\vec{r}_1, \vec{r}_2; \vec{r}_1, \vec{r}_2) = \mathsf{P}(\vec{r}_1, \vec{r}_2) \ . \tag{160}$$

This defines a map from ${}^{ker}\mathcal{E}_2^r$ to the two-electron density space $\mathcal{F}_2$, and a decomposition of $\mathcal{F}_2$ into a null space of the collapse map and its orthogonal complement,

$$\mathcal{F}_2 = \mathcal{K}_2 \oplus \mathcal{L}_2 \ . \tag{161}$$

Recall that with a LIP basis $\mathcal{L}_1 = {}^{ker}\mathcal{E}_1^r$ and $\mathcal{K}_1$ is empty. This is not the case when two-electron operators and densities are considered. Even with a LIP basis there are linear dependencies among collapses of distinct elements of ${}^{ker}\mathcal{E}_2^r$.

Take as an orthonormal basis for the one-electron spatial Hilbert space $\mathcal{H}_1^r$ the functions $\{\chi_j(\vec{r}),\ r = 1, \ldots, b_0\}$. To simplify matters, the functions will be assumed to be real and only real combinations will be considered. Additionally, it will be assumed that this is a LIP basis, *i.e.* products, including squares, of the χ's are linearly independent. Define

$$\rho_{jk}(\vec{r}) \stackrel{\text{def}}{=} \chi_j(\vec{r})\chi_k(\vec{r}) \qquad j \leq k \ . \tag{162}$$

Note that although these functions are defined with $j \leq k$, they are also meaningful (but not independent) when $j > k$ since $\chi_j\chi_k = \chi_k\chi_j$.

Permutationally symmetric and antisymmetric two-electron Hilbert spaces and their bases are next defined. For $\mathcal{H}_2^s$,

$$\begin{aligned} \Phi_{jk}^+(\vec{r}_1, \vec{r}_2) &\stackrel{\text{def}}{=} \frac{1}{\sqrt{2}} \left[\chi_j(\vec{r}_1)\chi_k(\vec{r}_2) + \chi_k(\vec{r}_1)\chi_j(\vec{r}_2)\right] \ , \qquad j < k \\ \Phi_{jj}^+(\vec{r}_1, \vec{r}_2) &\stackrel{\text{def}}{=} \chi_j(\vec{r}_1)\chi_k(\vec{r}_2) \end{aligned} \tag{163}$$

and for $\mathcal{H}_2^a$

$$\Phi_{jk}^-(\vec{r}_1, \vec{r}_2) \stackrel{\text{def}}{=} \frac{1}{\sqrt{2}} \left[\chi_j(\vec{r}_1)\chi_k(\vec{r}_2) - \chi_k(\vec{r}_1)\chi_j(\vec{r}_2)\right] \ , \qquad j < k \ . \tag{164}$$

The dimensions of these spaces are included in Table 2.

The one-electron collapse map $\hat{\delta}_1$ with a LIP basis is a one-to-one map from $\mathcal{E}_1$ to $\mathcal{F}_1$. The situation is more complicated for two electrons, however. Two-electron charge density matrices $\mathbf{\Gamma}^0$ are elements of a space

$$\mathcal{E}_2^r = \mathcal{E}_2^a \oplus \mathcal{E}_2^s \tag{165}$$

with dimension

$$\begin{aligned}\dim(\mathcal{E}_2^r) &= \dim(\mathcal{E}_2^a) + \dim(\mathcal{E}_2^s) \\ &= \frac{1}{2}\binom{b_0}{2}\left[\binom{b_0}{2}+1\right] + \frac{1}{2}\binom{b_0+1}{2}\left[\binom{b_0+1}{2}+1\right] \\ &= \frac{1}{4}b_0^2\,(b_0^2+3) \end{aligned} \tag{166}$$

whereas the dimension of the density space is

$$\begin{aligned}\dim(\mathcal{F}_2) &= \frac{1}{2}\binom{b_0+1}{2}\left[\binom{b_0+1}{2}+1\right] \\ &= \frac{1}{8}b(b-1)(b^2-b+2)\,, \end{aligned} \tag{167}$$

the same as that of $\mathcal{E}_2^s$. This means that even with a LIP basis there must be linear dependencies among the collapses of the elements of $\mathcal{E}_2^r$.

An element of the operator space corresponding to $\mathcal{E}_2^r$ can be expressed as

$$G_{jk,lm}^{\pm} = \begin{cases} \frac{1}{\sqrt{2}}\left(|\Phi_{jk}^{\pm}\rangle\langle\Phi_{lm}^{\pm}| + |\Phi_{lm}^{\pm}\rangle\langle\Phi_{jk}^{\pm}|\right) & \text{if } l,m \neq j,k \\ |\Phi_{jk}^{\pm}\rangle\langle\Phi_{jk}^{\pm}| & \text{if } l,m = j,k \end{cases} \tag{168}$$

If the χ's are orthonormal then the G's are orthonormal with respect to the trace scalar product in $\mathcal{E}_2^r$. Note that they are symmetric with respect to interchange of the pair j,k with the pair l,m. In addition G^+ is unchanged by interchange of j with k or of l with m, while G^- changes sign with either such interchange, and is thus unchanged if both are done.

Elements of the two-electron density space are of the form

$$P_{jk,lm}(\vec{r}_1,\vec{r}_2) = \frac{1}{2}\left[\rho_{jk}(\vec{r}_1)\rho_{lm}(\vec{r}_2) + \rho_{lm}(\vec{r}_1)\rho_{jk}(\vec{r}_2)\right] \,. \tag{169}$$

Table 2. Dimensions of various spaces

Space	Dimension	Space	Dimension
$\mathcal{H}_1^r$	b_0	$\mathcal{E}_1^r$	$\binom{b_0+1}{2}$
$\mathcal{F}_1$	$\binom{b_0+1}{2}$		
$\mathcal{H}_2^a$	$\binom{b_0}{2}$	$\mathcal{E}_2^a$	$\frac{1}{2}\binom{b_0}{2}\left[\binom{b_0}{2}+1\right]$
$\mathcal{H}_2^s$	$\binom{b_0+1}{2}$	$\mathcal{E}_2^s$	$\frac{1}{2}\binom{b_0+1}{2}\left[\binom{b_0+1}{2}+1\right]$
$\mathcal{F}_2$	$\frac{1}{2}\binom{b_0+1}{2}\left[\binom{b_0+1}{2}+1\right]$		

Table 3. Independent elements of $\mathcal{E}_2$ and $\mathcal{F}_2$

$\mathcal{E}_2^s$	$\mathcal{E}_2^a$	$\mathcal{F}_2$	number possible
		4 distinct indices	
$G^+_{jk,lm}$	$G^-_{jk,lm}$	$P_{jk,lm}$	$\binom{b_0}{4}$
$G^+_{jl,km}$	$G^-_{jl,km}$	$P_{jl,km}$	$\binom{b_0}{4}$
$G^+_{jm,kl}$	$G^-_{jm,kl}$	$P_{jm,kl}$	$\binom{b_0}{4}$
		3 distinct indices	
$G^+_{jl,kl}$	$G^-_{jl,kl}$	$P_{jl,kl}$	$b_0(b_0-1)$
$G^+_{jk,ll}$		$P_{jk,ll}$	$b_0(b_0-1)$
		2 distinct indices	
$G^+_{jk,jk}$	$G^-_{jk,jk}$	$P_{jk,jk}$	$\binom{b_0}{2}$
$G^+_{jj,kk}$		$P_{jj,kk}$	$\binom{b_0}{2}$
$G^+_{jk,kk}$		$P_{jk,kk}$	$b_0(b_0-1)$
		1 index	
$G^+_{jj,jj}$		$P_{jj,jj}$	b_0

For orthonormal χ's,

$$\int P_{jk,lm}(1,2)\, d\vec{r}_1 d\vec{r}_2 = \delta_{jk}\delta_{lm}\,. \tag{170}$$

The P's are symmetric with respect to interchange of j with k or l with m and interchange of j,k with l,m.

To go farther it is necessary to specify how many of the indices are distinct. Consider first the case in which j,k,l,m are four distinct indices. It is assumed that $j<k<l<m$. The independent elements of $\mathcal{E}_2^\tau$ and $\mathcal{F}_2$ are indicated in Table 3. There is an obvious enumerative correspondence between elements of $\mathcal{E}_2^s$ and those of $\mathcal{F}_2$, but this does *not* mean that $\hat{\delta}_2$ is a one-to-one map from $\mathcal{E}_2^s$ onto $\mathcal{F}_2$.

There are six independent elements of $\mathcal{E}_2$ for each j,k,l,m and only three elements of $\mathcal{F}_2$, so there must be linear dependencies among the $\hat{\delta}_2 G$'s. It is readily determined that

$$\hat{\delta}_2 G^{\pm}_{jk,lm} = \sqrt{2}\,(P_{jl,km} \pm P_{jm,kl}) \tag{171}$$

and association of these indices with j,k,l,m in ways consistent with the assumed orderings gives two more such relationships. Apply an orthogonal transformation to the G's to give

$$K_1^2(jk,lm) \;=\; \frac{1}{2}\left[G^+_{jl,km} - G^+_{jm,lk} + G^-_{jl,km} - G^-_{jm,lk}\right]$$

$$K_2^2(jk,lm) = \frac{1}{2}\left[G^+_{jk,lm} - G^+_{jm,lk} + G^-_{jk,lm} + G^-_{jm,lk}\right]$$
$$K_3^2(jk,lm) = \frac{1}{2}\left[G^+_{jk,lm} - G^+_{jl,km} - G^-_{jk,lm} + G^-_{jl,km}\right] . \quad (172)$$

which have $\hat{\delta}_2 K = 0$ and

$$L_1^2(jk,lm) = \frac{1}{2}\left[G^+_{jl,km} + G^+_{jm,lk} + G^-_{jl,km} + G^-_{jm,lk}\right]$$
$$L_2^2(jk,lm) = \frac{1}{2}\left[G^+_{jk,lm} + G^+_{jm,lk} + G^-_{jk,lm} - G^-_{jm,lk}\right]$$
$$L_3^2(jk,lm) = \frac{1}{2}\left[G^+_{jk,lm} + G^+_{jl,km} - G^-_{jk,lm} - G^-_{jl,km}\right] . \quad (173)$$

for which $\hat{\delta}_2 L$'s are independent elements of $\mathcal{F}_2$. There are $\binom{b_0}{4}$ ways of choosing $j < k < l < m$ and thus elements of this type provide $3\binom{b_0}{4}$ elements in $\mathcal{K}_2$ and the same number in $\mathcal{L}_2$.

When one index is duplicated the indices can be labeled as j,k and $l \neq j,k$. Pairs j,l and k,l are possible for G^+ or G^-, but j,k and l,l is possible only for G^+. In this case

$$\hat{\delta}_2 G^+_{jl,kl} = \sqrt{2}\,(P_{jl,kl} + P_{jk,ll})$$
$$\hat{\delta}_2 G^-_{jl,kl} = \sqrt{2}\,(P_{jl,kl} - P_{jk,ll})$$
$$\hat{\delta}_2 G^+_{jk,ll} = P_{jk,ll} \quad (174)$$

An orthogonal transformation gives

$$K^2(jl,kl) = \frac{1}{2}G^+_{jl,kl} - \frac{1}{\sqrt{2}}G^+_{jk,ll} + \frac{1}{2}G^-_{jl,kl}$$
$$L_1^2(jl,kl) = \frac{1}{2}G^+_{jl,kl} + \frac{1}{\sqrt{2}}G^+_{jk,ll} + \frac{1}{2}G^-_{jl,kl}$$
$$L_2^2(jl,kl) = \frac{1}{\sqrt{2}}\left[G^+_{jl,kl} - G^-_{jl,kl}\right] . \quad (175)$$

with $\hat{\delta}_2 K^2 = 0$ and the $\hat{\delta}_2 L^2$'s independent elements of $\mathcal{F}_2$. There are $b_0\binom{b_0-1}{2}$ choices of j,l and k,l and thus elements of this type provide $b_0\binom{b_0-1}{2}$ elements of $\mathcal{K}_2$ and $2b_0\binom{b_0-1}{2} = b_0(b_0-1)(b_0-2)$ elements of $\mathcal{L}_2$.

Two distinct indices can be combined as jk,jk for G^+ or G^-, or as jj,kk or jk,kk for G^+ only. Application of $\hat{\delta}_2$ gives

$$\hat{\delta}_2 G^+_{jk,jk} = P_{jj,kk} + P_{jk,jk}$$
$$\hat{\delta}_2 G^-_{jk,jk} = P_{jj,kk} - P_{jk,jk}$$

$$\begin{aligned}\hat{\delta}_2 G^+_{jj,kk} &= \sqrt{2} P_{jj,kk} \\ \hat{\delta}_2 G^+_{jk,kk} &= \sqrt{2} P_{jk,kk} \,. \end{aligned} \tag{176}$$

Since $P_{jk,kk}$ does not otherwise occur,

$$L^2(jk,kk) = G^+_{jk,kk} \tag{177}$$

is an element of $\mathcal{L}_2$. There are $b_0(b_0-1)$ elements of this type. The relevant orthogonal transformation of the other kernels in this group is

$$\begin{aligned} K^2(jk,jk) &= \frac{1}{2} G^+_{jk,jk} - \frac{1}{\sqrt{2}} G^+_{jj,kk} + \frac{1}{2} G^-_{jk,jk} \\ L^2_1(jk,jk) &= \frac{1}{2} G^+_{jk,jk} + \frac{1}{\sqrt{2}} G^+_{jj,kk} + \frac{1}{2} G^-_{jk,jk} \\ L^2_2(jk,jk) &= \frac{1}{\sqrt{2}} G^+_{jk,jk} - \frac{1}{\sqrt{2}} G^-_{jk,jk} \,, \end{aligned} \tag{178}$$

with again $\hat{\delta}_2 K^2 = 0$ and the $\hat{\delta}_2 L^2$'s independent elements of $\mathcal{F}_2$. There are $\binom{b_0}{2}$ choices of j,k for these G's, so elements of this type provide $\binom{b_0}{2}$ elements of $\mathcal{K}_2$ and $\binom{b_0}{2}$ elements of $\mathcal{L}_2$.

Finally, all four indices can be the same, in which case only G^+ is possible and the b_0 choices of this index give b_0 elements in $\mathcal{L}_2$.

The total number of independent elements provided for $\mathcal{K}_2$ is

$$3\binom{b_0}{4} + b_0\binom{b_0-1}{2} + \binom{b_0}{2} = \frac{1}{8} b_0(b_0-1)(b_0^2 - b_0 + 2) \tag{179}$$

while the number of independent elements provided for $\mathcal{L}_2$ is

$$3\binom{b_0}{4} + 2b_0\binom{b_0-1}{3} + b_0(b_0-1) + 2\binom{b_0}{2} + b_0 = \frac{1}{8} b_0(b_0-1)(b_0^2 - b_0 + 2) \,. \tag{180}$$

Comparison with the dimensions of the spaces given in Table 2 shows that

$$\dim(\mathcal{K}_2^r) = \dim(\mathcal{E}_2^a) \,, \qquad \dim(\mathcal{L}_2^r) = \dim(\mathcal{E}_2^s) \tag{181}$$

but it is only their dimensions that these spaces have in common. All elements of $\mathcal{K}_2^r$ and most elements of $\mathcal{L}_2^r$ have components in both $\mathcal{E}_2^a$ and $\mathcal{E}_2^s$.

4.7 Density subspaces

The one-electron density is proportional to the integral of the two-electron density over the coordinates of the second electron, and a second integration gives a normalization constant. Recall that $\{f_k,\ k = 1, \ldots, b_1 - 1\}$ is an orthonormal basis for $\mathcal{F}_1$ with the property

$$\int f_k(\vec{r})\, d\vec{r} = 0 \qquad \text{if } k > 0\,. \tag{182}$$

An orthonormal basis for $\mathcal{F}_2$ can be taken to be

$$F_{jk}(\vec{r}_1, \vec{r}_2) = \begin{cases} \frac{1}{\sqrt{2}}\,[f_j(\vec{r}_1)f_k(\vec{r}_2) + f_k(\vec{r}_1)f_j(\vec{r}_2)] & \text{if } j < k \\ f_j(\vec{r}_1)f_j(\vec{r}_2) & \text{if } j = k\,, \end{cases} \tag{183}$$

and these functions have the property

$$\int F_{jk}(\vec{r}_1, \vec{r}_2)\, d\vec{r}_1 = \varepsilon_{jk}\delta_{j,0} f_k(\vec{r}_1) \tag{184}$$

where $\varepsilon_{jk} = 2^{-1/2}$ if $j < k$ or 1 if $j = k$, and

$$\int F_{jk}(\vec{r}_1, \vec{r}_2)\, d\vec{r}_1 d\vec{r}_2 = \delta_{j,0}\delta_{k,0}\,. \tag{185}$$

Define subspaces of the density spaces as

$$\begin{aligned} \mathcal{F}_{1,0} &= \text{span}\,\{f_0\} \\ \mathcal{F}_{1,1} &= \text{span}\,\{f_j,\ j > 0\} \\ \mathcal{F}_{2,0} &= \text{span}\,\{F_{00}\} \\ \mathcal{F}_{2,1} &= \text{span}\,\{F_{0j}\ j > 0\} \\ \mathcal{F}_{2,2} &= \text{span}\,\{F_{jk}\ k \geq j > 0\}\,. \end{aligned} \tag{186}$$

The component of the density in $\mathcal{F}_{1,0}$ is determined by normalization while that in $\mathcal{F}_{1,1}$ integrates to zero. For a two-electron density the component in $\mathcal{F}_{2,0}$ is determined by normalization, that in $\mathcal{F}_{2,1}$ contributes to the one-electron density when integrated over the coordinates of one electron, and the component in $\mathcal{F}_{2,2}$ gives zero when integrated over the coordinates of one electron. If

$$\rho = \sum_j \rho_j f_j\,, \qquad \mathsf{P} = \sum_{j \leq k} \mathsf{P}_{jk} F_{jk} \tag{187}$$

then $\rho_k \propto \mathsf{P}_{0k}$.

5. Summary

It is often useful to think of density matrices or other operators and densities or other functions as elements of vector spaces. This is well known for the case of wave functions as state vectors in an infinite-dimensional Hilbert space. When a finite basis set is introduced all of the relevant spaces are finite-dimensional in the model thereby defined. The process of reduction or its "inverse" links a series of spaces relating to different numbers of electrons. It has been demonstrated that the vector spaces in which density matrices and the matrices of quantum mechanical operators exist can be divided into orthogonal subspaces in a number of ways. These decompositions can be thought of as related to information content. The subspace decomposition based on reduction introduces another integer parameter called the reduction index. The reduction index π is in some ways more meaningful than the particle number p. The reduced Hamiltonian, an element of $\mathcal{E}_2$, determines the full Hamiltonian matrix in $\mathcal{E}_n$, and conversely only the two-electron reduced density matrix is all that is needed to evaluate the expectation value of a Hamiltonian involving no more than two-particle operators. More generally, the matrix of a p-electron operator can be divided into components with $\pi \leq p$ and for any q between π and the maximum fixed by the size of the basis the corresponding matrices in $\mathcal{E}_q$ have π components determined by the π component of the matrix in $\mathcal{E}_p$.

The expectation value of an operator $\hat{A}$ in a state described by density matrix $\hat{D}$ is $\mathrm{Tr}\,\hat{A}\hat{D}$, and in the model problem defined by a finite basis this becomes $\mathrm{Tr}\,\mathbf{AD}$, in terms of matrices. For matrices in any $\mathcal{E}_p$,

$$\mathrm{Tr}\,\mathbf{A}^{(p)}\hat{D}^{(p)} = \sum_{\pi=0}^{p} \mathrm{Tr}\,\mathbf{A}^{(p,\pi)}\hat{D}^{(p,\pi)} \tag{188}$$

because of the orthogonality of the subspaces and of the components of vectors in these subspaces.

The subspaces characterized by particle numbers p and reduction indices π are invariant subspaces for unitary transformations of the spin orbital basis, which further characterizes their significance. The approach use here does not make explicit use of this group-theoretic characterization but instead uses superoperators, conveniently representable in terms of creation and annihilation operators, which are closely analogous to familiar angular momentum operators. An operator $\widehat{\Lambda}_3$ corresponds to $\hat{J}_z$ and its eigenvalue determines the particle number p. An operator $\widehat{\Lambda^2}$ corresponds to $\hat{J}^2$ and its eigenvalue determines the reduction index π. It can be used to construct a projection operator onto the subspace $\mathcal{E}_{p,\pi}$ of a matrix space $\mathcal{E}_p$.

An alternative decomposition of the matrix spaces is linked to the process of obtaining a density matrix from a density. The map involved has been called collapse, and its properties have been investigated for the cases of $p = 1$ and $p = 2$. The null space of this map and its orthogonal complement divide the space of matrices into two or more subspaces. Components of a density matrix in these spaces correspond to the part that determines, and is determined by, the density and the part that does not affect the density. A corresponding separation for operators allows a quantitative measure of how nearly local the operator is.

The decomposition based on collapse differs in an important way from that based on reduction. The decomposition $\mathcal{E}_p = \sum_\pi \oplus \mathcal{E}_{p,\pi}$ is independent of the choice of basis, depending only on the number of basis functions, while the decomposition $\mathcal{E}_1 = \mathcal{K} \oplus \mathcal{L}$ is basis set dependent, even different for a position-space basis and the corresponding Fourier-transformed momentum-space basis. If is also found, in work cited but not presented here, that important operators like the kinetic energy or Hartree-Fock exchange are much more nearly local for realistic quantum chemical basis sets than the simple dimensions of the subspaces would suggest. A lot of physics is built into these basis sets. It is possible to choose basis sets that minimize or eliminate nonlocal character. Such LIP bases (all products linearly independent) are not a good *a priori* choice, but sets consisting of the occupied SCF orbitals have been found to be LIP.[20]

For the two-electron charge density matrix and density it is not possible to eliminate the $\mathcal{K}$ subspace. Analysis of the collapse map is also complicated by the possibility of two permutational symmetries in the two electron charge density matrix. It appears likely that further work can provide more understanding of this problem and the implications for analysis of electron correlation.

The use of geometric ideas, as presented here, supplements the topological ideas often used in reduced density matrix theory and the Hilbert space ideas at the foundations of quantum mechanics. Although the number of dimensions involved is too large for simple pictures, distances and angles as well as orthogonal components can be simply defined and contribute further understanding.

Appendix

1. Adjoint relationship between $\widehat{\Lambda}_+$ and $\widehat{\Lambda}_-$

The adjoint $\hat{A}^\dagger$ of an operator $\hat{A}$ is defined by the requirement that $\left(\hat{A}^\dagger\phi, \psi\right) = \left(\phi, \hat{A}\psi\right)$ for all ϕ in the range of $\hat{A}$ and all ψ in the domain of $\hat{A}$. Consider $\widehat{\Lambda}_-$ initially as a map from $\mathcal{E}_p$ to $\mathcal{E}_{p-1}$ and $\widehat{\Lambda}_+$ as a map from $\mathcal{E}_{p-1}$ to $\mathcal{E}p$. Expand general elements $\hat{A}^{(p)} \in \hat{\mathcal{E}}_p$ and $\hat{B}^{(p-1)} \in \hat{\mathcal{E}}_{p-1}$ as

$$\begin{aligned} \hat{A}^{(p)} &= \sum_{K_p,L_p} A^{(p)}_{K_p,L_p} |K_p\rangle\langle L_p| \\ \hat{B}^{(p-1)} &= \sum_{K'_{p-1},L'_{p-1}} B^{(p-1)}_{K'_{p-1},L'_{p-1}} |K'_{p-1}\rangle\langle L'_{p-1}| \, . \end{aligned} \tag{A1}$$

Then

$$\widehat{\Lambda}_- \hat{A}^{(p)} = \sum_{J_p,K_p} \sum_j A^{(p)}_{J_p,k_p} \breve{a}_j |J_p\rangle\langle K_p| \breve{a}^\dagger_j \tag{A2}$$

and

$$\begin{aligned} \left(\hat{B}^{(p-1)}, \widehat{\Lambda}_- \hat{A}^{(p)}\right) &= \sum_{L_{p-1},M_{p-1}} \sum_{J_p,K_p} \sum_j B^{(p-1)*}_{L_{p-1},M_{p-1}} A^{(p)}_{J_p,K_p} \\ &\quad \times \left(|L_{p-i}\rangle\langle M_{p-1}|, \breve{a}_j |J_p\rangle\langle K_p|\breve{a}^\dagger_j\right) \, . \end{aligned} \tag{A3}$$

The scalar product on the right is equal to

$$\langle M_{p-1}|\breve{a}_j|J_p\rangle\langle K_p|\breve{a}^\dagger_j|L_{p-1}\rangle = \left(\breve{a}^\dagger_j |L_{p-1}\rangle\langle M_{p-1}|\breve{a}_j, |J\rangle_p\langle K_p|\right) \tag{A4}$$

and thus the sum is equal to $\left(\widehat{\Lambda}_+ \hat{B}^{(p-1)}, \hat{A}^{(p)}\right)$.

2. Commutation relations and eigenvalues of $\widehat{\Lambda^2}$

We have the basic commutation relationships

$$\left[\widehat{\Lambda}_-, \widehat{\Lambda}_+\right] = 2\widehat{\Lambda}_3 \, , \qquad \left[\widehat{\Lambda}_3, \widehat{\Lambda}_\pm\right] = \widehat{\Lambda}_\pm \tag{A5}$$

and there is a fundamental, readily verified commutator relationship

$$[\hat{A}, \hat{B}\hat{C}] = \hat{B}[\hat{A}, \hat{C}] + [\hat{A}, \hat{B}]\hat{C} \, . \tag{A6}$$

Using this and Eq. (96) we find

$$\begin{aligned} \left[\widehat{\Lambda}_3, \widehat{\Lambda^2}\right] &= \frac{1}{2}\left[\widehat{\Lambda}_3, \widehat{\Lambda}_+\widehat{\Lambda}_-\right] + \frac{1}{2}\left[\widehat{\Lambda}_3, \widehat{\Lambda}_-\widehat{\Lambda}_+\right] \\ &= \frac{1}{2}\left\{\widehat{\Lambda}_+\left[\widehat{\Lambda}_3, \widehat{\Lambda}_-\right] + \left[\widehat{\Lambda}_3, \widehat{\Lambda}_+\right]\widehat{\Lambda}_- + \widehat{\Lambda}_-\left[\widehat{\Lambda}_3, \widehat{\Lambda}_+\right] + \left[\widehat{\Lambda}_3, \widehat{\Lambda}_-\right]\widehat{\Lambda}_+\right\} \\ &= \frac{1}{2}\left\{\widehat{\Lambda}_+\widehat{\Lambda}_- + \left(-\widehat{\Lambda}_+\right)\widehat{\Lambda}_- + \widehat{\Lambda}_-\left(-\widehat{\Lambda}_+\right) + \widehat{\Lambda}_-\widehat{\Lambda}_+\right\} = 0 \, . \end{aligned} \tag{A7}$$

Next,

$$\begin{aligned} \left[\widehat{\Lambda}_+, \widehat{\Lambda^2}\right] &= \frac{1}{2}\widehat{\Lambda}_+ \underbrace{\left[\widehat{\Lambda}_+, \widehat{\Lambda}_-\right]}_{-2\widehat{\Lambda}_3} + \frac{1}{2}\underbrace{\left[\widehat{\Lambda}_+, \widehat{\Lambda}_-\right]}_{-2\widehat{\Lambda}_3}\widehat{\Lambda}_+ \\ &\quad + \widehat{\Lambda}_3 \underbrace{\widehat{\Lambda}_+, \widehat{\Lambda}_3}_{\widehat{\Lambda}_+} + \underbrace{\widehat{\Lambda}_+, \widehat{\Lambda}_3}_{\widehat{\Lambda}_+}\widehat{\Lambda}_3 = 0 \, . \end{aligned} \tag{A8}$$

A parallel treatment shows that $\left[\widehat{\Lambda}_-, \widehat{\Lambda^2}\right] = 0$.

Since $\widehat{\Lambda^2}$ and $\widehat{\Lambda}_3$ commute they have simultaneous eigenfunctions (or eigenvectors). Suppose that

$$\widehat{\Lambda^2}\hat{V} = a\hat{V}\ , \qquad \widehat{\Lambda}_3\hat{V} = \mu\hat{V}\ . \tag{A9}$$

Then $\widehat{\Lambda}_+\hat{V}, \widehat{\Lambda}_+^2\hat{V}, \ldots$ will be eigenfunctions of $\widehat{\Lambda}_3$ with eigenvalues $\mu+1, \mu+2, \ldots$ and $\widehat{\Lambda}_-\hat{V}, \widehat{\Lambda}_-^2\hat{V}, \ldots$ will be eigenfunctions of $\widehat{\Lambda}_3$ with eigenvalues $\mu-1, \mu-2, \ldots$. In each case they will still be eigenfunctions of $\widehat{\Lambda^2}$ with eigenvalue a. Note, however, that

$$\begin{aligned} a = \left\langle \widehat{\Lambda^2} \right\rangle &= \left(\hat{V}, \widehat{\Lambda^2}\hat{V}\right) = \frac{1}{2}\left(\hat{V}, \widehat{\Lambda}_+\widehat{\Lambda}_-\hat{V}\right) + \frac{1}{2}\left(\hat{V}, \widehat{\Lambda}_-\widehat{\Lambda}_+\hat{V}\right) + \left(\hat{V}, \widehat{\Lambda}_3^2\hat{V}\right) \\ &= \|\widehat{\Lambda}_+\hat{V}\|^2 + \|\widehat{\Lambda}_-\hat{V}\|^2 + \mu^2 \geq \mu^2 \end{aligned} \tag{A10}$$

so the sequences increasing or decreasing m with fixed a must terminate.

These results parallel those for angular momentum operators $\hat{J}_z$, $\hat{J}_\pm$, and $\hat{J}^2$ and the eigenstates $|j,m\rangle$. Following the development of the operator theory of angular momentum in standard texts, these properties lead to the following conclusions:

- The eigenvalues a of $\widehat{\Lambda^2}$ are of the form $\lambda(\lambda+1)$ where λ is nonnegative and is either an integer or half-integer.
- The eigenvalues μ of $\widehat{\Lambda}_3$ occur with integer separation and are integers, if λ is an integer, or half-integer, if λ is half-integer.
- $-\lambda \leq \mu \leq \lambda$.

References

[1] Ilias Absar. Reduced Hamiltonian orbitals. II. Optimal orbital basis sets for the many-electron problem. *Int. J. Quantum Chem.*, 13(6):777–790, 1978.

[2] Ilias Absar and A. J. Coleman. Reduced Hamiltonian orbitals. I. A new approach to the many-electron problem. *Int. J. Quantum Chem. Symp.*, 39(10):319–330, 1976.

[3] Mark E. Casida and John E. Harriman. Geometry of density matrices. VI. Superoperators and unitary invariance. *Int. J. Quantum Chem.*, 30(2):161–212, 1986.

[4] A. J. Coleman. Structure of fermion density matrices. *Rev. Mod. Phys.*, 32(3):668–687, 1963.

[5] A. J. Coleman and I. Absar. Reduced Hamiltonian orbitals. III. Unitary invariant decomposition of Hermitian operators. *Int. J. Quantum Chem. Symp.*, 18:1279–1307, 1980.

[6] Ernest R. Davidson. N-representability of the electron pair density. *Chem. Phys. Lett.*, 246:209–213, 1995.

[7] P. A. M. Dirac. The basis of statistical quantum mechanics. *Proc. Cambridge Phil. Soc.*, 25:62–66, 1929.

[8] U. Fano. Description of states in quantum mechanics by density matrix and operator techniques. *Rev. Mod. Phys.*, 29(1):74–93, 1957.

[9] T. L. Gilbert. Hohenberg-Kohn theorem for nonlocal external potentials. *Phys. Rev. B*, 12(6):2111–2120, 1975.

[10] John E. Harriman. Geometry of density matrices. I. Definitions, N matrices and 1 matrices. *Phys. Rev. A*, 17(4):1249–1256, 1978.

[11] John E. Harriman. Geometry of density matrices. II. Reduced density matrices and N representability. *Phys. Rev. A*, 17(4):1257–1268, 1978.

[12] John E. Harriman. Orthonormal orbitals for the representation of an arbitrary density. *Phys. Rev. A*, 24(2):680–682, 1981.

[13] John E. Harriman. Geometry of density matrices. IV. The relationship between density matrices and densities. *Phys. Rev. A*, 27(3):632–645, 1983.

[14] John E. Harriman. Geometry of density matrices. V. Eigenstates. *Phys. Rev. A*, 30(1):19–29, 1984.

[15] John E. Harriman. Densities, operators, and basis sets. *Phys. Rev. A*, 34(1):29–39, 1986.

[16] John E. Harriman. Electron densities, momentum densities, and density matrices. *Z. Naturforsch.*, 47a:203–210, 1992.

[17] John E. Harriman. Kinetic energy matrices in a basis of equidentiy orbitals. *J. Chem. Phys.*, 104(15):5912–5921, 1996.

[18] John E. Harriman. Distance and entropy for density matrices. *J. Chem. Phys.*, 115(20):9223–9232, 2001.

[19] John E. Harriman and Douglas E. Hoch. Locality of exchange matrices for common gaussian basis sets. *Int. J. Quantum Chem.*, 63(1):111–119, 1997.

[20] Douglas E. Hoch and John E. Harriman. Linear dependencies among basis set products and near locality of some nonlocal operators. *J. Chem. Phys.*, 102(24):9590–9597, 1995.

[21] Douglas E. Hoch and John E. Harriman. Basis set dependence of the locality of the kinetic energy operator. *J. Chem. Phys.*, 104(15):5898–5911, 1996.

[22] Hans Kummer. n-Representability problem for reduced density matrices. *J. Math. Phys.*, 8(10):2063–2081, 1967.

[23] Werner Kutzelnigg. Über die Symmetrie-Eigenschaften der reduzierten Dichtematrizen und der natürlichen Spin-Orbitale und Spin-Geminale (der natürlichen Ein- und Zwei-Elektronen-Functionen). *Z. Naturforsch.*, 18a(10):1058–1064, 1963.

[24] Elliott H. Lieb. Density functionals for Coulomb systems. *Int. J. Quantum Chem.*, 24(3):243–277, 1983.

[25] Per Olov Löwdin. Quantum theory of many particle systems. I. Physical interpretations by means of density matrices, natural spin-orbitals, and convergence problems in the method of configuration interaction. *Phys. Rev.*, 97(4):1474–1489, 1955.

[26] Per Olov Löwdin. Quantum theory of many particle systems. II. Studies of the ordinary Hartree-Fock approximation. *Phys. Rev.*, 97(4):1490–1508, 1955.

[27] Per Olov Löwdin. Quantum theory of many particle systems. III. Extension of the Hartree-Fock scheme to include degenerate systems and correlation effects. *Phys. Rev.*, 97(4):1509–1520, 1955.

[28] Robert C. Morrison. Density and density matrix from optimized linearly independent product basis functions for Be. *Int. J. Quantum Chem. Symp.*, 2:43–49, 1988.

[29] Hiroshi Nakatsuji. Equation for the direct determination of the density matrix. *Phys. Rev. A*, 14(1):41–50, 1976.

[30] Gary T. Smith, Hartmut L. Schmider, and Vedene H. Smith, Jr. Electron correlation and the eigenvalues of the one-matrix. *Phys. Rev. A*, 65:032508, 2002.

[31] Chia-Chung Sun, Xue-Qui Li, and Au-Chin Tang. Lie algebra and unitary invariant decomposition of Hermitian operators. *Int. J. Quantum Chem.*, 25(4):653–666, 1984.

[32] Au-Chin Tang and H. Guo. Characteristic operators and unitary invariant decopmposition of Hermitian operators. *Int. J. Quantum Chem.*, 23(1):217–226, 1983.

[33] John von Neumann. Thermodynamik quantenmechanischer Gesamptheiten. *Göttinger Nachrichten*, 1927:245, 1927.

[34] Gil Zumbach and Klaus Maschke. New approach to the calculations of density functionals. *Phys. Rev. A*, 28(2):544–554, 1983.

[35] Gil Zumbach and Klaus Maschke. Erratum: New approach to the calculations of density functionals. *Phys. Rev. A*, 29(3):1585–1587, 1984.

Density matrix variational theory: Strength of Weinhold-Wilson inequalities

Maho Nakata, Masahiro Ehara and Hiroshi Nakatsuji*
Department of Synthetic Chemistry and Biological Chemistry, Faculty of Engineering, Kyoto University, Kyoto 606-8501, Japan

Abstract. We examine the strength of the Weinhold-Wilson (WW) inequalities for calculating the second-order density matrix(2-RDM) by the density matrix variational theory (DMVT) using the P, Q and G conditions as subsidiary conditions. We calculated the 2-RDM of various molecular electronic states and found that some violations of WW inequalities occur especially for the systems for which the DMVT(PQG) calculations were less accurate. We then developed the DMVT method including further the WW inequalities as the restrictive conditions, DMVT(PQG+WW), and applied it to CH_4, C_2, $CH_2(^1A_1)$ and H_2O. The WW inequalities certainly improved the results, but the improvement was not so remarkable.

1. Introduction

As far as our world involves only up-to-two body elementary operators, this world should be described solely by the second-order reduced density matrix (2-RDM). Professor Löwdin not only did a lot of great works in this and related field(1; 2), but also encouraged many young researchers. HN is one of such researchers and would like to thank Professor Per-Olov Löwdin for his great contributions in the field of quantum molecular science and his encouragement warmly given to him.

Recently, special attentions have been paid to the direct determination of the reduced density matrix (RDM), which is called as density matrix theory(DMT). This approach adopts the 2-RDM as the basic variable of quantum mechanics and there are two formalisms in the DMT. One is based on the density equation(3), which is equivalent to the Schrödinger equation in the necessary and sufficient sense. This approach is called density equation theory (DET). Recent development of the DET(4; 5; 6) has been remarkable and they have been summarized in the review article (7). The other is based on the Ritz variational principle expressed in terms of the 2-RDM and is referred to as the density matrix variational theory (DMVT). The quality of the calculated 2-RDM in the DMVT is dependent on how well we can restrict our variable 2-RDM to be N-representable(8).

* hiroshi@sbchem.kyoto-u.ac.jp

E.J. Brändas and E.S. Kryachko (eds.),
Fundamental World of Quatum Chemistry, Vol. I, 543–557.

The DMVT method was first introduced by Garrod and Percus. They proposed the variational method using P, Q and G conditions and some trivial N-representability conditions(9), and applied it to the ground state of Be(10). Erdahl proposed to use convex programming for the DMVT and applied it to He_2(11). The applications in these early studies were limited to very small systems. Recently, Erdahl and Jin(12) developed the DMVT using 3-RDM as variable and applied it to the model system of one-dimensional periodic lattice of electron pairs. Mazziotti and Erdahl(13) examined the positive semidefinite condition of 3- and 4-RDMs for solving the DMVT combined with the DET and calculated Lipkin model, namely, a boson model of two-energy-level system.

In this series of our DMVT studies(14; 15), we could efficiently implement the DMVT using the semidefinite programming algorithm (SDPA)(16), and successfully calculate the 2-RDMs of the ground states of many different spin-space symmetries for *many atoms and molecules*. We also applied this method to the potential energy surfaces of molecules and reproduced the full-CI curves in good approximation up to the dissociation limit(15); the G condition was found to be very important for describing the dissociation limit.

Another promising approach was initiated by one of the authors(17). Since the exact Ψ is an eigen function of the Hamiltonian that has so simple structure composed of only one- and two-body operators, the Ψ itself should also have a simple structure reflecting this simplicity of the Hamiltonian. Some explicit expressions of the structure of the exact wave function were given and the theories for the ground and excited states was formulated. Applications were given to a simple model system and to atoms and molecules.

For improving the DMVT(PQG) method developed previously(14; 15), we may use some additional N-representability conditions. Some inequalities for the 2-RDM were proposed by Weinhold and Wilson(18), Davidson(19) and McRae and Davidson(20). Since all of these inequalities can be written as linear conditions, it is easy to include these conditions in our formalism. In our previous work(15), we actually examined these inequalities for the resultant 2-RDM of the DMVT(PQG) calculations and found that two linear inequalities were violated in some cases.

In this paper, we examine the Weinhold-Wilson (WW) inequalities for calculating the 2-RDM by the DMVT(PQG) method. Examinations have been done for all the systems that were calculated in the previous studies(14; 15). Then, we propose an efficient formalism for including these conditions as the subsidiary conditions.

2. Theory and calculation

2.1. Definitions and basic algorithm

First and second order reduced density matrices (1-, 2-RDMs), γ and Γ, are defined by:

$$\gamma_j^i = \langle \Psi | a_i^\dagger a_j | \Psi \rangle \tag{1}$$

$$\Gamma_{j_1 j_2}^{i_1 i_2} = \frac{1}{2} \langle \Psi | a_{i_1}^\dagger a_{i_2}^\dagger a_{j_2} a_{j_1} | \Psi \rangle \tag{2}$$

where $a^\dagger$ and a are creation and annihilation operators, respectively. Practical complete N-representability condition is not known for the 2-RDM: we know only some necessary conditions. In the present DMVT, we use P, Q, and G conditions. The P, Q, and G-matrices are defined by,

$$P_{j_1 j_2}^{i_1 i_2} = \langle \Psi | a_{i_1}^\dagger a_{i_2}^\dagger a_{j_2} a_{j_1} | \Psi \rangle \tag{3}$$

$$Q_{j_1 j_2}^{i_1 i_2} = \langle \Psi | a_{i_1} a_{i_2} a_{j_2}^\dagger a_{j_1}^\dagger | \Psi \rangle \tag{4}$$

$$G_{j_1 j_2}^{i_1 i_2} = \langle \Psi | a_{i_1}^\dagger a_{i_2} a_{j_2}^\dagger a_{j_1} | \Psi \rangle \tag{5}$$

respectively. We enforce all of these matrices to be positive semidefinite. We also use seven trivial conditions of 2-RDM, which are antisymmetric condition, hermiticity, trace condition, number of electrons, number of spins, and expectation values of S_z and S^2.

In the DMVT, we take 2-RDM as a variational variable, and minimize the energy within N-representability conditions, namely,

$$E_{\min} = \min_{\Gamma \in \mathcal{P}^{(2)}} \mathrm{Tr} H \Gamma, \tag{6}$$

where H is the Hamiltonian of the system, $\mathcal{P}^{(2)}$ is a set of 2-RDM that satisfy approximate or nearly complete N-representability condition. In the previous papers (14; 15), we performed two types of calculations using the approximate N-representability conditions: the DMVT(PQ) adopts the trivial representability conditions plus P and Q conditions as the approximate conditions, and the DMVT(PQG) further includes G condition. In this work, we also include the Weinhold-Wilson inequalities, and we call the method as the DMVT(PQG+WW) method.

2.2. Weinhold-Wilson inequalities

The Weinhold-Wilson inequalities which are independent from P, Q, G, and 7 trivial conditions are:

Condition IV:

$$1 - \gamma_i^i - \gamma_j^j + 2\Gamma_{ij}^{ij} \geq 0 \tag{7}$$

Condition V:

$$\gamma_i^i - 2\Gamma_{ij}^{ij} \geq 0 \tag{8}$$

Condition VI:

$$\gamma_i^i - 2\Gamma_{ij}^{ij} - 2\Gamma_{ik}^{ik} + 2\Gamma_{jk}^{jk} \geq 0 \tag{9}$$

Condition VII:

$$1 - \gamma_i^i - \gamma_j^j - \gamma_k^k + 2\Gamma_{ij}^{ij} + 2\Gamma_{ik}^{ik} + 2\Gamma_{jk}^{jk} \geq 0 \tag{10}$$

Condition VIII: Positive semidefiniteness of the Ω matrix

$$\Omega = \begin{pmatrix} \gamma_1^1 & 2\Gamma_{12}^{12} & 2\Gamma_{13}^{13} & \cdots & 2\Gamma_{1t}^{1t} & \gamma_1^1 \\ 2\Gamma_{12}^{12} & \gamma_2^2 & 2\Gamma_{23}^{23} & \cdots & 2\Gamma_{2t}^{2t} & \gamma_2^2 \\ 2\Gamma_{13}^{13} & 2\Gamma_{23}^{23} & \gamma_3^3 & \cdots & 2\Gamma_{3t}^{3t} & \gamma_3^3 \\ \vdots & \vdots & \vdots & & \vdots & \vdots \\ 2\Gamma_{1t}^{1t} & 2\Gamma_{2t}^{2t} & 2\Gamma_{3t}^{3t} & \cdots & \gamma_t^t & \gamma_t^t \\ \gamma_1^1 & \gamma_2^2 & \gamma_3^3 & \cdots & \gamma_t^t & 1 \end{pmatrix}. \tag{11}$$

Note that different definition is employed for Γ by factor 2. In this study, we examined the 2-RDM by the DMVT(PQG) method with respect to these five inequalities, and we found that two types of conditions VI and VII were violated for some systems. Therefore, we develop the DMVT method including these conditions.

2.3. DMVT METHOD INCLUDING WEINHOLD-WILSON INEQUALITIES

Since all the Weinhold-Wilson inequalities are linear conditions, it is possible to include all of these inequalities in the SDP formalism simultaneously. However, the number of conditions becomes large since it scales as N^3, and therefore we enforce the inequalities only for those molecules for which the WW inequalities were violated in the preceding DMVT(PQG) calculation. Then, the algorithm of the DMVT (PQG+WW) is as follows:

1. Standard DMVT(PQG) is executed.
2. Weinhold-Wilson inequalities of type IV$\sim$ VIII are examined.
3. DMVT(PQG+WW) is performed with additional WW conditions if some of the WW inequalities are violated.

The procedure of 2 and 3 is repeated until all the inequalities are satisfied. Note that the inequality holds after it is included in the condition. Actually, this iteration is necessary only for one or two times. As noted in section 2.2., the conditions VI and VII were violated for the resultant 2-RDM calculated by the DMVT(PQG), and therefore we developed the method of including only these two WW conditions. In the DMVT method using SDPA, the conditions should be reduced to the standard form and we introduce two types of constraint matrices $\boldsymbol{C}_{ijk}^{\mathrm{VI}}$ and $\boldsymbol{C}_{ijk}^{\mathrm{VII}}$. In the following, we employ the same notations as those in the previous paper(14). In the DMVT(PQG+WW) calculations, we extend the variable matrix $\boldsymbol{Y}$ given by Eq. (3.25) of Ref.(14) as $\boldsymbol{Y}'$,

$$\boldsymbol{Y}' = \begin{pmatrix} \boldsymbol{P} & 0 & 0 & 0 & 0 \\ 0 & \boldsymbol{Q} & 0 & 0 & 0 \\ 0 & 0 & \boldsymbol{G} & 0 & 0 \\ 0 & 0 & 0 & \boldsymbol{W}^{\mathrm{VI}} & 0 \\ 0 & 0 & 0 & 0 & \boldsymbol{W}^{\mathrm{VII}} \end{pmatrix} = \begin{pmatrix} \boldsymbol{Y} & 0 & 0 \\ 0 & \boldsymbol{W}^{\mathrm{VI}} & 0 \\ 0 & 0 & \boldsymbol{W}^{\mathrm{VII}} \end{pmatrix} \tag{12}$$

where $\boldsymbol{W}^{\mathrm{VI}}$, and $\boldsymbol{W}^{\mathrm{VII}}$ are the diagonal matrices and their diagonal elements are given in the right hand sides of Eqs.(9) and (10), respectively.

We define the constraint matrix $\boldsymbol{C}_{ijk}^{\mathrm{VI}}$ by

$$(\boldsymbol{C}_{ijk}^{\mathrm{VI}})_{p_1,p_2,q_1,q_2} = \begin{cases} -\frac{2}{N-1}\delta^i_{p_1}\delta^i_{q_1}\delta^{p_2}_{q_2} + 2\delta^i_{p_1}\delta^j_{p_2}\delta^i_{q_1}\delta^j_{q_2} + 2\delta^i_{p_1}\delta^k_{p_2}\delta^i_{q_1}\delta^k_{q_2} \\ \qquad - 2\delta^j_{p_1}\delta^k_{p_2}\delta^j_{q_1}\delta^k_{q_2}, \\ \qquad \text{for } 1 \le p_1,p_2,q_1,q_2 \le 3n, \\ 1 \qquad \text{for } 3n{+}1 \le p_1,p_2,q_1,q_2 \le 3n{+}n^{\mathrm{VI}}, \\ 0 \qquad \text{otherwise} \end{cases} \tag{13}$$

and C_{ijk}^{VII} by

$$(\boldsymbol{C}_{ijk}^{\mathrm{VII}})_{p_1,p_2,q_1,q_2} = \begin{cases} \frac{2}{N-1}\left(\delta^i_{p_1}\delta^i_{q_1}\delta^{p_2}_{q_2} + \delta^j_{p_1}\delta^j_{q_1}\delta^{p_2}_{q_2} + \delta^k_{p_1}\delta^k_{q_1}\delta^{p_2}_{q_2}\right) \\ \qquad - 2\delta^i_{p_1}\delta^j_{p_2}\delta^i_{q_1}\delta^j_{q_2} + 2\delta^j_{p_1}\delta^k_{p_2}\delta^j_{q_1}\delta^k_{q_2}, \\ \qquad \text{for } 1 \le p_1,p_2,q_1,q_2 \le 3n, \\ 1 \quad \text{for } 3n{+}n^{\mathrm{VI}}{+}1 \le p_1,p_2,q_1,q_2 \le 3n{+}n^{\mathrm{VI}}{+}n^{\mathrm{VII}}, \\ 0 \quad \text{otherwise,} \end{cases} \tag{14}$$

where n^{VI} and n^{VII} are the numbers of the Weinhold-Wilson inequality conditions, which were not satisfied in the previous iteration. The constants for constraints $\boldsymbol{C}^{\mathrm{VI}}$ and $\boldsymbol{C}^{\mathrm{VII}}$ are all 0 and 1, respectively.

Then, these constraints work as

$$\boldsymbol{C}_{ijk}^{\mathrm{VI}} \bullet \boldsymbol{Y}' = \sum_{p_1p_2q_1q_2} (-\frac{2}{N-1}\delta^i_{p_1}\delta^i_{q_1}\delta^{p_2}_{q_2} + 2\delta^i_{p_1}\delta^j_{p_2}\delta^i_{q_1}\delta^j_{q_2}$$

$$
\begin{aligned}
&+2\delta^i_{p_1}\delta^k_{p_2}\delta^i_{q_1}\delta^k_{q_2} - 2\delta^j_{p_1}\delta^k_{p_2}\delta^j_{q_1}\delta^k_{q_2})Y^{p_1p_2}_{q_1q_2} + (\boldsymbol{W}^{\mathrm{VI}})_{ijk} \\
&= -\sum_{p_2}\frac{2}{N-1}\Gamma^{ip_2}_{ip_2} + 2\Gamma^{ij}_{ij} + 2\Gamma^{ik}_{ik} - 2\Gamma^{kj}_{kj} + (\boldsymbol{W}^{\mathrm{VI}})_{ijk} \\
&= -\gamma^i_i + 2\Gamma^{ij}_{ij} + 2\Gamma^{ik}_{ik} - 2\Gamma^{kj}_{kj} + (\boldsymbol{W}^{\mathrm{VI}})_{ijk} \\
&= 0 \qquad (15)
\end{aligned}
$$

and

$$
\begin{aligned}
\boldsymbol{C}^{\mathrm{VII}}_{ijk} \bullet \boldsymbol{Y}' &= \sum_{p_1p_2q_1q_2}(\frac{2}{N-1}\left(\delta^i_{p_1}\delta^i_{q_1}\delta^{p_2}_{q_2} + \delta^j_{p_1}\delta^j_{q_1}\delta^{p_2}_{q_2} + \delta^k_{p_1}\delta^k_{q_1}\delta^{p_2}_{q_2}\right) \\
&\quad - 2\delta^i_{p_1}\delta^j_{p_2}\delta^i_{q_1}\delta^j_{q_2} - 2\delta^i_{p_1}\delta^k_{p_2}\delta^i_{q_1}\delta^k_{q_2} \\
&\quad - 2\delta^j_{p_1}\delta^k_{p_2}\delta^j_{q_1}\delta^k_{q_2})Y^{p_1p_2}_{q_1q_2} + (\boldsymbol{W}^{\mathrm{VII}})_{ijk} \\
&= \gamma^i_i + \gamma^j_j + \gamma^k_k - 2\Gamma^{ij}_{ij} - 2\Gamma^{ik}_{ik} - 2\Gamma^{kj}_{kj} + (\boldsymbol{W}^{\mathrm{VII}})_{ijk} \\
&= 1, \qquad (16)
\end{aligned}
$$

since $(\boldsymbol{W}^{\mathrm{VI}})_{ijk}$ and $(\boldsymbol{W}^{\mathrm{VII}})_{ijk}$ are

$$
\begin{aligned}
(\boldsymbol{W}^{\mathrm{VI}})_{ijk} &= \gamma^i_i - 2\Gamma^{ij}_{ij} - 2\Gamma^{ik}_{ik} + 2\Gamma^{jk}_{jk} \\
(\boldsymbol{W}^{\mathrm{VII}})_{ijk} &= 1 - \gamma^i_i - \gamma^j_j - \gamma^k_k + 2\Gamma^{ij}_{ij} + 2\Gamma^{ik}_{ik} + 2\Gamma^{jk}_{jk}. \qquad (17)
\end{aligned}
$$

We further antisymmetrize the indices $i_1 \leftrightarrow i_2$ and $j_1 \leftrightarrow j_2$, hermitize the indices $(i_1, i_2) \leftrightarrow (j_1, j_2)$ to be suitable for the SDPA formalism, and pack the elements to remove unnecessary variables.

The resultant SDP formalism of the DMVT(PQG+WW) method is now given by,

$$
\left\{
\begin{array}{ll}
\text{Minimize} & \boldsymbol{H} \bullet \boldsymbol{Y}' \\
\text{subjected to} & \boldsymbol{F}_i \bullet \boldsymbol{Y}' = c_i \\
& \tilde{\boldsymbol{E}}^{i_1i_2}_{j_1j_2} \bullet \boldsymbol{Y}' = \delta^{i_1}_{j_1}\delta^{i_2}_{j_2} - \delta^{i_1}_{j_2}\delta^{i_2}_{j_1} \\
& \tilde{\boldsymbol{J}}^{i_1i_2}_{j_1j_2} \bullet \boldsymbol{Y} = 0, \\
& \boldsymbol{C}^{\mathrm{VI}}_{ijk} \bullet \boldsymbol{Y}' = 0, \\
& \boldsymbol{C}^{\mathrm{VII}}_{ijk} \bullet \boldsymbol{Y}' = 1.
\end{array}
\right. \qquad (18)
$$

where the matrices of $\boldsymbol{F}_i$, $\tilde{\boldsymbol{E}}$, and $\tilde{\boldsymbol{J}}$ are defined in ref.(14).

3. Results

First we examined the Weinhold-Wilson inequalities for the 2-RDM obtained by the DMVT(PQG) calculations. We calculated the various electronic states of molecules, which were studied in the previous works(14; 15). For N_2, CO, BH_3 and H_2O, the examinations were performed at both the equilibrium geometry and dissociation limit. In table I, we presented the results of the number of the violated conditions whose absolute errors were lager than 1×10^{-5} together with the DMVT(PQG) and full-CI energies. The violation occurred for the conditions VI and VII in some cases, while the conditions IV and V and VIII held for all the electronic states. For all the other atoms and molecules which were calculated in the previous paper and were not included in table I, all of these inequalities were correctly held. As seen in table I, when the DMVT(PQG) calculations are less accurate, the violation of the inequalities becomes large: the number of violations are large for BH_3, $CH_2(^1A_1)$, C_2 and CH_4.

We also calculated the artificially correlation enhanced system, introduced in the previous study(15): namely, the Hamiltonian is partitioned into F, Fock operator and the rest, V as:

$$H = F + \lambda V, \tag{19}$$

where λ is a real parameter that controls the strength of the electron correlation. Using this Hamiltonian, the DMVT(PQG) calculations were performed for H_2O and the results for some λ were summarized in table II. The violations have occurred for the condition VI and they were large for $\lambda = 1.5$. Note that no violation has occurred for large λ.

We examined the violations in details for C_2, $CH_2(^1A_1)$ and CH_4 and artificially correlation enhanced system with $\lambda = 1.5$ of H_2O. In tables III$\sim$ VI, we listed the violations with the indices and values, whose absolute value is larger than 1×10^{-5} for C_2, CH_4 and H_2O at the equilibrium geometry. As seen in table III, large violations occurred for both conditions VI and VII for C_2: the largest values were in the order of 1×10^{-2}. The violations mainly occurred for the valence orbitals, which are $p\sigma$, $p\pi$ and $p\pi^*$ orbitals, except for some violations with respect to the 1s orbital for condition VI. The violation of conditions VI and VII was also found for the $^1A_1$ state of CH_2. In this case, the violation occurred for the variable for the valence orbitals and not for the 1s core orbitals. In the case of CH_4, the order of the violations was smaller and the absolute errors were within 2.1×10^{-3} as shown in table V. In this case, the violations were also found for the 1s orbital of C. For H_2O with artificially correlation enhanced Hamiltonian of $\lambda = 1.5$, the violations occur only for the condition VI. The errors are calculated to be very small: the order is 1×10^{-4}. As shown in these examples, the absolute values of the violations are related to the accuracy of the DMVT(PQG), which indicates that these conditions may be effective as the necessary conditions for the DMVT method.

We performed, therefore, the DMVT(PQG+WW) calculations including these two WW conditions. The results were presented in table VII. The largest improvement was obtained for C_2 as expected, but it was not so drastic as 7.2×10^{-4} au in total energy. The deviation from the full-CI value is still large. For other systems, $CH_2(^1A_1)$, CH_4 and H_2O, the effect of these conditions was not so prominent and the improved energies were 1.7×10^{-4}, 9.0×10^{-5} and 1.0×10^{-5} au, respectively. These N-representability conditions actually improved the results, however, did not effectively work at least in the combination with the DMVT(PQG) calculation.

4. Conclusion

We examined the Weinhold-Wilson inequalities for the 2-RDM determined by the DMVT(PQG) method. The violations of the conditions VI and VII were found for some electronic states of several molecules, though they were not so large; the largest error was in the order of 1×10^{-2}. We therefore developed the formalism of the DMVT including these inequalities for the SDPA, which is denoted as DMVT(PQG+WW). These conditions certainly improved the results, however, they were not so drastic; the largest improvement was in the order of mhartree. These conditions were not so effective at least for the DMVT method including P, Q and G conditions.

5. Acknowledgment

This research was supported by a Grant-in-Aid for Scientific Research from the Ministry of Education, Science, Culture and Sports.

References

1. P.-O. Löwdin, Phys. Rev. **97**, 1474 (1955); **97**, 1490 (1955); **97**, 1512 (1955).
2. P.-O. Löwdin, Adv. Quantum Chem. **2**, 213 (1965); **3**, 323 (1967); **5**, 185 (1970); **12**, 263 (1980); **17**, 285 (1985); **19**, 87 (1988).
3. H. Nakatsuji, Phys. Rev. A **14**, 41(1976).
4. C. Valdemoro, Phys. Rev. A **45**, 4462 (1992); F. Colmenero C. Pérez del Valle, and C. Valdemoro, Phys. Rev. A **47**, 971 (1993), F. Colmenero and C. Valdemoro Phys. Rev. A **47**, 979 (1993).
5. H. Nakatsuji and K. Yasuda, Phys. Rev. Lett. **76**, 1039(1996), K. Yasuda and H. Nakatsuji, Phys. Rev. A **56**, 2648(1997).
6. D. A. Mazziotti, Phys, Rev. A **57**, 4219(1998); Chem. Phys. Lett. **289**, 419(1998).

7. H. Nakatsuji, in *Many-Electron Densities and Reduced Density Matrices*, ed by J. Cioslowski, Kluwer Academic, New York, 2000.
8. A. J. Coleman, Rev. Mod. Phys. **35**, 668 (1963).
9. C. Garrod, and J. K. Percus, J. Math. Phys., **5**, 1756 (1964).
10. C. Garrod, M. V. Mihailović, and M. Rosina, J. Math. Phys. **16**, 868 (1975), C. Garrod and M. A. Fusco, Int. J. Quantum Chem. **x**, 495 (1976).
11. R. M. Erdahl, Rep. Math. Phys., **15** 147 (1979).
12. R. M. Erdahl and B. Jin, J. Mol. Struct.: THEOCHEM, **527**, 207 (2000); R. M. Erdahl and B. Jin, in *Many-Electron Densities and Reduced Density Matrices*, ed by J. Cioslowski, Kluwer Academic, New York, 2000.
13. D. A. Mazziotti and R. M. Erdahl, Phys. Rev A, **63**, 042113 (2001).
14. M. Nakata, H. Nakatsuji, M. Ehara, M. Fukuda, K. Nakata, K. Fujisawa, J. Chem. Phys., **114**, 8282 (2001).
15. M. Nakata, M. Ehara, and H. Nakatsuji, J. Chem. Phys., **116**, 5432, (2002).
16. K. Fujisawa, M. Kojima, K. Nakata, SDPA (SemiDefinite Programming Algorithm) User's Manual Version 5.00, August 1999.
URL is `http://is-mj.archi.kyoto-u.ac.jp/~fujisawa/software.html`.
17. H. Nakatsuji, J. Chem. Phys., **113**, 2949 (2000); H. Nakatsuji and E. R. Davidson,*ibid*, **115**, 2000 (2001); H. Nakatsuji, *ibid*, **115**, 2465 (2001); *ibid*, **116**, 1811 (2002); H. Nakatsuji, Phys. Rev. A, in press; H. Nakatsuji and M. Ehara, J. Chem. Phys., submitted.
18. F. Weinhold and E. B. Wilson Jr, J. Chem. Phys., **47**, 2298 (1967).
19. E. R. Davidson, J. Math. Phys., **10**, 725 (1969).
20. W. B. McRae and E. R. Davidson, J. Math. Phys., **13**, 1527 (1972).

Table I. Number of violated Weinhold-Wilson inequalities for the 2-RDM determined by DMVT(PQG), whose absolute errors are larger than 1×10^{-5}.

System	State	Violation	DMVT(PQG)	FullCI
BH_3 [a]	$^1A_1$	VI: 12, VII: 12	-26.3926(120)	-26.3822(100)
BH_3 [b]	$^1A_1$	VI: 14, VII: 6	-25.8678(103)	-25.8482(100)
C_2 [a]	$^1\Sigma$	VI: 16, VII: 32	-75.4793(117)	-75.4340(100)
C_2 [b]	$^1\Sigma$	VII: 4	-75.2187(105)	-75.1855(100)
CO [a]	$^1\Sigma$	No	-112.4544(108)	-112.4426(100)
CO [b]	$^1\Sigma$	VI: 12	-112.1156(102)	-112.1096(100)
N_2 [a]	$^1\Sigma_g^+$	VI: 4	-108.7123(108)	-108.7002(100)
N_2 [b]	$^1\Sigma_g^+$	No	-108.4982(100)	-108.4982(100)
H_2O [a]	$^1A_1$	VI: 6	-75.7310(104)	-75.7290(100)
H_2O [b]	$^1A_1$	No	-75.4589(100)	-75.4589(100)
BH_2	$^2A_1$	VII: 6	-25.7089(115)	-25.7032(100)
CH^+	$^1\Sigma^+$	VI: 8	-37.8896(107)	-37.8853(100)
CH^-	$^3\Sigma^-$	VII: 2	-37.9714(99)	-37.9718(100)
CH	$^2\Pi$	VI: 2	-38.1916(111)	-38.1871(100)
CH_2	$^1A_1$	VI: 14, VII: 17	-38.8228(119)	-38.8110(100)
CH_2	$^3B_1$	VI: 2, VII: 4	-38.8556(107)	-38.8534(100)
CH_2(linear)	$^3\Sigma_u^-$	VI: 10, VII: 8	-38.8358(103)	-38.8342(100)
FH_2^+	$^1A_1$	VI: 8	-99.8305(103)	-99.8294(100)
H_2O^+	$^2A_1$	VI: 5	-75.4218(106)	-75.4192(100)
NH_2	$^2A_1$	VI: 4, VII: 8	-55.3570(111)	-55.3525(100)
NH_2	$^2B_1$	VI: 1, VII: 3	-55.4195(108)	-55.4157(100)
CH_4	$^1A_1$	VI: 30, VII:24	-40.2030(124)	-40.1905(100)

[a] At equilibrium geometry
[b] At dissociation limit

Table II. Number of violated Weinhold-Wilson inequalities for the 2-RDM of H_2O determined by DMVT(PQG), whose absolute errors are larger than 1×10^{-5}.

λ	Violation	DMVT(PQG)	FullCI
0.5	VI: 8	-56.5158(102)	-56.5155(100)
1.0	VI: 6	-75.7310(104)	-75.7290(100)
1.5	VI: 8	-95.0064(106)	-94.9978(100)
$\lambda > 1.5$	No	-	-

Table III. Violation of Weinhold-Wilson inequalities for C_2

Condition VI

(i,j,k)	violation	(i,j,k)	violation	(i,j,k)	violation	(i,j,k)	violation
$(5,\overline{4},\overline{7})$	-1.95×10^{-2}	$(\overline{5},4,7)$	-1.94×10^{-2}	$(\overline{5},3,6)$	-1.94×10^{-2}	$(5,\overline{3},\overline{6})$	-1.941×10^{-2}
$(6,4,7)$	-9.29×10^{-3}	$(\overline{6},\overline{4},\overline{7})$	-9.29×10^{-3}	$(\overline{7},\overline{3},\overline{6})$	-9.29×10^{-3}	$(7,3,6)$	-9.28×10^{-3}
$(7,\overline{5},\overline{7})$	-8.01×10^{-3}	$(\overline{7},5,7)$	-8.00×10^{-3}	$(\overline{6},5,6)$	-7.99×10^{-3}	$(6,\overline{5},\overline{6})$	-7.98×10^{-3}
$(\overline{7},\overline{3},6)$	-4.19×10^{-3}	$(7,3,\overline{6})$	-4.17×10^{-3}	$(6,4,\overline{7})$	-4.17×10^{-3}	$(\overline{6},\overline{4},7)$	-4.17×10^{-3}

Condition VII

(i,j,k)	violation	(i,j,k)	violation	(i,j,k)	violation	(i,j,k)	violation
$(2,5,\overline{5})$	-3.09×10^{-2}	$(\overline{2},5,\overline{5})$	-3.09×10^{-2}	$(3,4,\overline{6})$	-1.37×10^{-2}	$(\overline{3},\overline{4},6)$	-1.37×10^{-2}
$(3,4,\overline{7})$	-1.36×10^{-2}	$(\overline{3},\overline{4},7)$	-1.36×10^{-2}	$(3,\overline{3},6)$	-9.03×10^{-3}	$(3,\overline{3},\overline{6})$	-9.02×10^{-3}
$(4,\overline{4},7)$	-9.01×10^{-3}	$(4,\overline{4},\overline{7})$	-9.00×10^{-3}	$(\overline{3},\overline{4},\overline{6})$	-6.50×10^{-3}	$(3,4,6)$	-6.49×10^{-3}
$(3,4,7)$	-6.50×10^{-3}	$(\overline{3},\overline{4},\overline{7})$	-6.50×10^{-3}	$(2,\overline{5},\overline{6})$	-6.40×10^{-3}	$(\overline{2},5,6)$	-6.38×10^{-3}
$(\overline{2},5,7)$	-6.45×10^{-3}	$(2,\overline{5},\overline{7})$	-6.43×10^{-3}	$(2,5,7)$	-2.67×10^{-3}	$(\overline{2},\overline{5},\overline{7})$	-2.67×10^{-3}
$(2,5,6)$	-2.67×10^{-3}	$(\overline{2},\overline{5},\overline{6})$	-2.67×10^{-3}	$(2,4,7)$	-1.49×10^{-3}	$(\overline{2},\overline{4},\overline{7})$	-1.49×10^{-3}
$(2,3,6)$	-1.48×10^{-3}	$(\overline{2},\overline{3},\overline{6})$	-1.47×10^{-3}	$(\overline{3},\overline{5},\overline{6})$	-1.26×10^{-3}	$(3,5,6)$	-1.26×10^{-3}
$(\overline{4},\overline{5},\overline{7})$	-1.25×10^{-3}	$(4,5,7)$	-1.24×10^{-3}	$(1,5,8)$	-1.76×10^{-4}	$(\overline{1},\overline{5},\overline{8})$	-1.69×10^{-4}

Table IV. Violation of Weinhold-Wilson inequalities for $CH_2(^1A_1)$

Condition VI

(i,j,k)	violation	(i,j,k)	violation	(i,j,k)	violation	(i,j,k)	violation
$(5,\bar{4},\bar{5})$	-1.07×10^{-3}	$(\bar{5},4,5)$	-1.07×10^{-3}	$(7,\bar{4},\bar{6})$	-6.48×10^{-4}	$(\bar{7},4,6)$	-6.46×10^{-4}
$(2,\bar{3},\bar{6})$	-3.56×10^{-4}	$(\bar{2},3,6)$	-3.57×10^{-4}	$(6,\bar{6},\bar{7})$	-2.16×10^{-4}	$(\bar{6},6,7)$	-2.17×10^{-4}
$(\bar{6},\bar{2},7)$	-6.98×10^{-5}	$(6,2,\bar{7})$	-6.90×10^{-5}	$(\bar{2},3,\bar{6})$	-4.02×10^{-5}	$(2,\bar{3},6)$	-3.89×10^{-5}
$(2,3,6)$	-3.66×10^{-5}	$(\bar{2},\bar{3},\bar{6})$	-3.61×10^{-5}				

Condition VII

(i,j,k)	violation	(i,j,k)	violation	(i,j,k)	violation	(i,j,k)	violation
$(\bar{2},5,\bar{7})$	-2.39×10^{-3}	$(2,\bar{5},7)$	-2.39×10^{-3}	$(\bar{2},\bar{5},7)$	-1.60×10^{-3}	$(2,5,\bar{7})$	-1.59×10^{-3}
$(2,5,7)$	-1.54×10^{-3}	$(\bar{2},\bar{5},\bar{7})$	-1.54×10^{-3}	$(3,4,6)$	-1.54×10^{-3}	$(\bar{3},\bar{4},\bar{6})$	-1.54×10^{-3}
$(\bar{2},5,7)$	-1.21×10^{-3}	$(2,\bar{5},\bar{7})$	-1.20×10^{-3}	$(3,4,\bar{6})$	-1.14×10^{-3}	$(\bar{3},\bar{4},6)$	-1.14×10^{-3}
$(2,4,5)$	-4.47×10^{-4}	$(\bar{2},\bar{4},\bar{5})$	-4.47×10^{-4}	$(3,6,7)$	-4.46×10^{-4}	$(\bar{3},\bar{6},\bar{7})$	-4.47×10^{-4}
$(\bar{3},6,7)$	-1.87×10^{-4}	$(3,\bar{6},\bar{7})$	-1.86×10^{-4}				

Table V. Violation of Weinhold-Wilson inequalities for CH_4

Condition VI

(i,j,k)	violation	(i,j,k)	violation	(i,j,k)	violation	(i,j,k)	violation
$(3,\bar{1},5)$	-2.06×10^{-3}	$(\bar{3},1,\bar{5})$	-2.06×10^{-3}	$(4,\bar{1},6)$	-2.06×10^{-3}	$(\bar{4},1,\bar{6})$	-2.06×10^{-3}
$(5,3,\bar{8})$	-1.21×10^{-3}	$(\bar{5},\bar{3},8)$	-1.21×10^{-3}	$(6,4,\bar{8})$	-1.21×10^{-3}	$(\bar{6},\bar{4},8)$	-1.21×10^{-3}
$(4,\bar{1},\bar{6})$	-9.15×10^{-4}	$(\bar{4},1,6)$	-9.15×10^{-4}	$(3,\bar{1},\bar{5})$	-9.15×10^{-4}	$(\bar{3},1,5)$	-9.15×10^{-4}
$(4,1,\bar{6})$	-5.49×10^{-4}	$(\bar{4},\bar{1},6)$	-5.49×10^{-4}	$(3,1,\bar{5})$	-5.49×10^{-4}	$(\bar{3},\bar{1},5)$	-5.49×10^{-4}
$(1,3,\bar{4})$	-4.87×10^{-4}	$(1,\bar{3},4)$	-4.87×10^{-4}	$(\bar{1},\bar{3},4)$	-4.87×10^{-4}	$(\bar{1},3,\bar{4})$	-4.86×10^{-4}
$(1,\bar{2},\bar{3})$	-4.76×10^{-4}	$(\bar{1},2,3)$	-4.75×10^{-4}	$(\bar{1},2,4)$	-4.76×10^{-4}	$(1,\bar{2},\bar{4})$	-4.76×10^{-4}
$(1,\bar{3},\bar{4})$	-4.75×10^{-4}	$(\bar{1},3,4)$	-4.75×10^{-4}	$(\bar{5},3,8)$	-2.97×10^{-4}	$(5,\bar{3},\bar{8})$	-2.97×10^{-4}
$(6,\bar{4},\bar{8})$	-2.97×10^{-4}	$(\bar{6},4,8)$	-2.97×10^{-4}				

Condition VII

(i,j,k)	violation	(i,j,k)	violation	(i,j,k)	violation	(i,j,k)	violation
$(1,5,\bar{6})$	-1.80×10^{-3}	$(1,\bar{5},6)$	-1.80×10^{-3}	$(\bar{1},5,\bar{6})$	-1.80×10^{-3}	$(\bar{1},\bar{5},6)$	-1.80×10^{-3}
$(1,\bar{5},\bar{6})$	-1.15×10^{-3}	$(\bar{1},5,6)$	-1.15×10^{-3}	$(1,\bar{5},\bar{7})$	-1.15×10^{-3}	$(\bar{1},5,7)$	-1.15×10^{-3}
$(1,\bar{6},\bar{7})$	-1.15×10^{-3}	$(\bar{1},6,7)$	-1.15×10^{-3}	$(3,\bar{4},8)$	-7.38×10^{-4}	$(3,\bar{4},\bar{8})$	-7.38×10^{-4}
$(\bar{3},4,8)$	-7.38×10^{-4}	$(\bar{3},4,\bar{8})$	-7.38×10^{-4}	$(2,4,5)$	-5.50×10^{-4}	$(\bar{2},\bar{4},\bar{5})$	-5.50×10^{-4}
$(2,3,6)$	-5.50×10^{-4}	$(\bar{2},\bar{3},\bar{6})$	-5.50×10^{-4}	$(2,4,\bar{8})$	-1.27×10^{-4}	$(\bar{2},\bar{4},8)$	-1.27×10^{-4}
$(2,3,\bar{8})$	-1.27×10^{-4}	$(\bar{2},\bar{3},8)$	-1.27×10^{-4}	$(3,4,\bar{8})$	-1.27×10^{-4}	$(\bar{3},\bar{4},8)$	-1.27×10^{-4}

Table VI. Violation of condition VI for H_2O with various λ

$\lambda = 0.50$

(i,j,k)	violation	(i,j,k)	violation
$(5,\bar{4},\bar{5})$	-1.05×10^{-5}	$(\bar{5},4,5)$	-9.78×10^{-6}
$(5,\bar{5},7)$	-6.82×10^{-6}	$(\bar{5},5,\bar{7})$	-6.43×10^{-6}
$(5,2,\bar{5})$	-4.94×10^{-6}	$(\bar{5},\bar{2},5)$	-4.79×10^{-6}
$(5,\bar{5},\bar{6})$	-4.47×10^{-6}	$(\bar{5},5,6)$	-4.12×10^{-6}

$\lambda = 1.00$

(i,j,k)	violation	(i,j,k)	violation
$(5,\bar{4},\bar{5})$	-7.52×10^{-5}	$(\bar{5},4,5)$	-7.46×10^{-5}
$(\bar{5},\bar{2},5)$	-4.25×10^{-5}	$(5,2,\bar{5})$	-4.15×10^{-5}
$(5,\bar{5},6)$	-3.40×10^{-6}	$(\bar{5},5,\bar{6})$	-2.53×10^{-6}

$\lambda = 1.50$

(i,j,k)	violation	(i,j,k)	violation
$(\bar{5},\bar{2},5)$	-3.69×10^{-4}	$(5,2,\bar{5})$	-3.68×10^{-4}
$(\bar{5},3,5)$	-3.00×10^{-4}	$(5,\bar{3},\bar{5})$	-2.99×10^{-4}
$(\bar{5},4,5)$	-1.83×10^{-4}	$(5,\bar{4},\bar{5})$	-1.81×10^{-4}
$(5,\bar{5},6)$	-5.63×10^{-5}	$(\bar{5},5,\bar{6})$	-5.44×10^{-5}

Table VII. Total energy by DMVT(PQG), DMVT(PQG+WW), and FullCI.

	DMVT(PQG)	DMVT(PQG+WW)	ΔE^a	FullCI
CH_4, STO-6G, 1s core	-40.20999	-40.20990	9.0×10^{-5}	-40.19049
C_2, STO-6G, 1s core	-75.47932	-75.47860	7.2×10^{-4}	-75.43398
$H_2O(\lambda = 1.50)$, STO-6G	-95.00638	-95.00637	1.0×10^{-5}	-94.99780
$CH_2(^1A_1)$, STO-6G	-38.82261	-38.82278	1.7×10^{-4}	-38.81099

[a] Improvement by including Weinhold-Wilson conditions.

DENSITY MATRICES FOR ELECTRONS WITH STRONG INTERACTIONS

Robert Erdahl
Department of Mathematics and Statistics, Queen's University
Kingston Ontario, Canada K7L 3N6

It is a pleasure for me to join in honouring Per - Olov since it was at one of the early Sanibel Conferences, over which he presided with such elan, that my interest in reduced density matrices was first engaged. This occured during the lectures of Per - Olov and John Coleman, who were good friends. The lectures of Per - Olov were based on his important 1955 papers [Löw1,Löw2,Löw3], and in these lectures he pronounced his vision that reduced density matrices would provide an accurate description of how correlations are induced in systems of quantum particles, and how these correlations relate to observed phenomena; John's lectures became his classic *The structure of fermion density matrices* [Col1], and led, in 2000, to *Reduced Density Matrices - Coulson's Challenge* [Col2]. Per - Olov's vision provided the common ground over which the Queen's Group of John Coleman, Bob Erdahl and Hans Kummer made contact with many theoretical chemists and physicists, and for the friendships that resulted we are indebted to him. In this chapter I first describe some early work where the correlations in beryllium and a weakly bound pair of helium atoms were given an accurate description, and then describe how the methods used were developed to yield a complete solution of the correlation problem for strongly interacting electrons.

1. INTRODUCTION

The first accurate estimate of a reduced density matrix, without appeal to a wave function, was made by Garrod, Mihailovic and Rosina in their work on the beryllium atom [Ros1]. The entries of the reduced density matrix were treated as variational parameters and adjusted so that the total energy was a minimum. The conditions they imposed on the reduced density matrix were that it be positive semidefinte, and the associated G- and Q-matrices be positive semi-definite, conditions that an N-representable reduced density matrix must satisfy. These conditions

E.J. Brändas and E.S. Kryachko (eds.),
Fundamental World of Quatum Chemistry, Vol. I, 559–585.

did not ensure that the reduced density matrix was N-representable, but were effective. The lower bound was only slightly below the exact ground state energy calculated using a complete configuration interaction treatment, a comparison that was meaningful because the same finite dimensional model Hilbert space was used for both calculations. The lower bound they calculated was accurate to seven figures: the lower bound was -14.60999 atomic units, and the configuration interaction energy was -14.609987. In this paper we refer to the collection of three conditions these authors imposed, the D-, G-, and Q-conditions, as a single one: *the reduced density matrix is required to be 2-positive.*

The second test of this *energy method*, and the first test for a molecule, appeared in 1979 when Erdahl calculated a lower bound to the energy for a pair of weakly bound helium atoms [Erd4]. By requiring that the density matrix be 2-positive a lower bound was calculated that was accurate to five figures when compared to a complete configuration interaction treatment. The Be and He^2 calculations showed that accurate lower bounds could be achieved for atoms and molecules, and in particular, showed the importance of the condition that the reduced density matrix be 2-positive.

Although the work on Be and He^2 showed promise, it was not until this past year that the strength of the 2-positivity condition was thoroughly tested in two lines of research. This work confirmed the earlier observations, that the energy method combined with 2-positivity yields accurate information for atoms and molecules. Nakatsuji and his co-workers [Nak2] did a series of computational experiments where accuracies of between four and five figures were typically achieved. More precisely, they reported the correlation energy as a percentage of the exact correlation energy for a variety of atoms and molecules. They found these percentages ranged between 100%-110% for atoms and diatomic molecules, and between 110%-120% for triatomic molecules; since these percentages are for lower bounds they never go below 100%.

An even more exacting test of 2-positivity appeared in the latest paper of Mazziotti [Maz2] who computed binding energy curves for LiH and H_2O. These computational experiments showed that accurate binding energy curves could be calculated when 2-positivity is invoked; the curves he calculated compared favorably with curves generated by a full configuration interaction treatment. These results were compared with curves generated using second and fourth order perturbation theory, and these deteriorated rapidly with increasing bond length. These experiments showed that lower bounds can successfully track energy shifts over a range of molecular geometries, subtle shifts that cannot be tracked by perturbation theory.

Another significant theme was reported on in the Mazziotti paper. Experiments were made by selectively adding higher order conditions to tighten the lower bounds computed using 2-positivity. These were a portion of the 3-*positivity conditions* selected on the basis of physical intuition about correlations, and experience gained with the lower order conditions. This strategy adds flexibility and strength to the lower bound method in much the way the selection of configurations adds to the configuration interaction method. These computational experiments are first steps in developing a systematic approach to handling the overwhelming array of higher order conditions. This possibility for the lower bounds and the energy method was first discussed in [Erd5,Erd6], under the heading *Dual Configuration Interaction.*

The optimism generated by the Be and He^2 calculations was tempered by another line of investigation in the 70s. Mihailovic and Rosina applied the methods they developed for the beryllium atom to light nuclei [Ros2], and found lower bounds falling below exact values by as much as fifteen percent. For example, the series of nuclei ^{15}O, ^{16}O, ^{17}O, ^{18}O has $3, 4, 5$ and 6 valence nucleons above a ^{12}C core. The greatest errors they observed were for the last in this series, where they computed a lower bound of $-60.06 MeV$, which was $8.55 MeV$ below the configuration interaction energy of $-53.91 MeV$, representing an error of 15.86%. For this series the percentage errors are given respectively by $5.82, 3.76, 14.93$ and 15.86%, showing that errors increase as half-filled structures are approached. The results for ^{15}O and ^{16}O were reasonable, but for ^{17}O, ^{18}O they commented, "new relevant conditions" should be found to supplement 2-positivity. In response to the dominating two-body terms, the nucleons in a shell form highly correlated arrangements that are inadequately characterized by 2-positivity.

That 2-positivity can effectively characterize electron correlations but not nucleon correlations can be explained as follows. The electrons in atoms and molecules configure themselves largely in response to the nuclear attraction terms in the Hamiltonian, but protons and neutrons in the nucleus configure themselves in response to the nucleon-nucleon attraction terms. That is, for atoms and molecules correlations are largely driven by one-body operators, whereas for nuclei correlations are driven by two-body operators. The simpler correlations in atoms and molecules are effectively characterized by 2-positivity, but the more complex nuclear correlations are not.

The main purpose for writing this chapter is to review three recent papers on strongly interacting fermion systems that resolve the difficulties that Mihailovic and Rosina faced. Erdahl and Jin [Erd5,Erd6] studied a lattice model where electrons interact through two-body forces alone, and Mazziotti and Erdahl [Maz3] studied the Lipkin model for electrons where the strength of the two-body forces can be turned on by adjusting a parameter V. There is convincing evidence in [Erd5,Erd6] supporting the comment of Mihailovic and Rosina that "new relevant conditions" are required beyond 2-positivity. The *Figure 4* of these articles is reproduced below, and makes it abundantly clear that 2-positivity cannot effectively characterize the correlations induced by strong 2-body forces. These three papers systematically explore higher order conditions, and show that by requiring the reduced density matrix to be 3-positive the problems faced by Mihailovic and Rosina are completely resolved. Both 3-positivity and 4-positivity were explored, and when the reduced density matrix is required to be 4-positive the lower bound estimates are accurate to ten figures.

The condition that the reduced density matrix be k-positive, for $k = 2, 3, 4, ...$, is a family of conditions that can be used to characterize increasingly complex correlations; the definition of k-positivity is given in Section 2. The most significant result reported in [Erd5,Erd6,Maz3] is that the speed of convergence in the order k is very fast. These authors discovered that: *for small values of* k*, where complexity is still within bounds and calculations feasible, the accuracy of the lower bound method exceeds that of other approximate methods currently employed in many-body theory.*

A secondary purpose in writing this chapter is to point out the significance of the *k-spectrum* of an Hermitian operator, which is a notion of spectrum that fits well when approximations are carried through to order k. The k-spectrum can be used to examine the question of whether the exact reduced density matrix can be calculated without recourse to the wave function, which would amount to a complete solution of the N-representability problem. Another question that can be explored is whether information on excited states can be extracted, and if so, which excited states are accessible. The theory of the k-spectrum is not fully developed at the moment, so only partial answers can be given to these important questions. The k-spectrum is a notion naturally associated with the *dispersion method*, which is the other major method used to estimate reduced density matrices.

Another secondary purpose is to point out the significance of the *Optimization Theorem*, which is discussed in *Section 5.* With this Theorem effective algorithms can be developed for solving the two variational problems associated with reduced density matrix theory, the first being the *Energy Problem:*

$$\min_{p \in \mathfrak{P}_0^k} \quad \langle p, h \rangle.$$

The variation is over the set of k-densities, $\mathfrak{P}_0^k$, which is a kth order approximation of the set of von Neumann density matrices and a variant of the set of kth order reduced density matrices. The special feature of $\mathfrak{P}_0^k$ is that the elements are k-positive. The second variational problem is the *Dispersion Problem:*

$$\min_{p \in \mathfrak{P}_0^k} \quad \langle p, (h - \lambda 1)^2 \rangle.$$

The dispersion operator $(h - \lambda 1)^2$ replaces h in the Energy Problem, but all else is unchanged. This problem is central for the dispersion method for estimating von Neumann densities for quantum states, and for calculating the k-spectrum of an operator. Both the k-density p and the parameter λ are varied in the search for dispersion-free states p_λ so that $\langle p_\lambda, (h - \lambda 1)^2 \rangle = 0$, in which case p_λ approximates the von Neumann density with energy eigenvalue λ.

The main result reported in *Two algorithms for the lower bound method of reduced density matrix theory* [Erd2] was a proof of the Optimization Theorem. This Theorem was then applied to develop second order algorithms to solve the Energy Problem to obtain an energy lower bound for He^2. It is this Theorem that forms the basis for semi-definite programming, which was used by Nakatsuji and his co-workers, and by Mazziotti, in their tests of the strength of 2-positivity. One of the important features of semi-definite programming is that there is a duality theory, much as with linear programming, that can be used to interpret results. In particular, this duality theory can be used to interpret correlations.

The Dispersion Problem is equivalent to the problem of solving the Contracted Schrödinger Equation, an equivalence first established in the 1978 paper of Erdahl and Grudzinski [Erd2], and is discussed below in Section 4. Current research on the Contracted Schrödinger Equation can be characterized as a search for an effective perturbation approach to the Dispersion Problem. It can be inferred from the results reported in [Maz3] that these perturbation approaches do not lead to the same accuracies achieved with an exact solution of the Dispersion Problem.

2. K^{th} ORDER APPROXIMATIONS FOR STATES

An Hermitian operator p is a von Neumann density if it is both positive and has unit trace. More precisely, if $\mathfrak{F}$ is the finite dimensional Fock space for a quantum model where electrons are distributed amongst a finite number of states, then p is a von Neuman density if: *(1)* $\langle p, q^\dagger q \rangle \geq 0$ *for all operators* q *on* $\mathfrak{F}$; *(2) trace*$(p) = 1$. By the formula $\langle p, q^\dagger q \rangle$ we mean the trace scalar product of the operators p and $q^\dagger q$. We will denote by $\mathfrak{P}$ the cone of non-negative operators on $\mathfrak{F}$, and by $\mathfrak{P}_0$ the convex set of all von Neumann densities. The elements of $\mathfrak{P}_0$ are the quantum states for the model.

We will discuss a lattice model where spin-up, spin-down electrons move on a one-dimensional lattice Λ of size $|\Lambda| = r$, so that $\dim \mathfrak{F} = 2^{2r}$. An annihilator for a spin-up electron on lattice site $\mu \in \Lambda$ will be denoted by a_μ, and that for a spin-down electron by b_μ. An arbitrary operator on $\mathfrak{F}$ can be written as a polynomial in the $2r$ annihilation and $2r$ creation operators.

Approximating states by k-densities. By relaxing the condition that a von Neumann density be positive a family of approximations can be constructed. Since an operator can be represented as a polynomial in the annihilation and creation operators it can be assigned a degree; for example, if $q = \sum_\mu q_\mu a_\mu^\dagger + \sum_{\mu,\nu,\rho} q_{\mu\nu\rho} a_\mu^\dagger b_\nu^\dagger b_\rho$, then $\deg q = 3$. We will say that an Hermitian operator p is *k-positive* if it satisfies the condition that $\langle p, q^\dagger q\rangle \geq 0$ *for all operators* q *where* $\deg q \leq k$.

DEFINITION OF K-DENSITY. *An Hermitian operator p is a $k-$density if: (1) $\langle p, q^\dagger q\rangle \geq 0$ for all operators q, where $\deg q \leq k$;(2) trace$(p) = 1$.*

The k-densities are both k-positive and have unit trace; they approximate von Neumann densities to k^{th} order.

We will denote the cone of all k-positive operators by $\mathfrak{P}^k$, and the convex set of all $k-$densities by $\mathfrak{P}_0^k$. The $k-$densities satisfy the relations $\mathfrak{P}_0 \subset ... \subset \mathfrak{P}_0^3 \subset \mathfrak{P}_0^2 \subset \mathfrak{P}_0^1$ and $\bigcap_k \mathfrak{P}_0^k = \mathfrak{P}_0$.

Matrix representations. Let Q^k be the linear space of operators q satisfying the condition $\deg q \leq k$, and let $\{q_1, q_2, ..., q_s\}$ be a basis for Q^k.

DEFINITION OF K-MATRIX. *Suppose that p is a von Neumann density, or a k-density. Then the $k-$matrix for p, relative to the basis $\{q_1, q_2, ..., q_s\}$, is the matrix $\mathbf{P}$ with entries $P_{ij} = \langle p, q_i^\dagger q_j\rangle$.*

The k-matrix is an $s \times s$ matrix representation for p, where $s = dim(\mathcal{Q}^k)$, and is a variant of the kth order reduced density matrix.

The expectation value of an operator x can be evaluated using $k-$matrices as long as the operator has an expansion of the form $x = \sum_{i,j=1}^s x_{ij} q_i^\dagger q_j$. For then $\langle p, x\rangle = \langle p, \sum_{i,j=1}^s x_{ij} q_i^\dagger q_j\rangle = \sum_{i,j=1}^s \langle p, q_i^\dagger q_j\rangle x_{ij}$ $= \sum_{i,j=1}^s P_{ij} x_{ij} = trace(\mathbf{P}^\dagger \mathbf{X}) = trace(\mathbf{PX})$. The matrix representation for x is obtained by taking the coefficients in the expansion to be the matrix elements. In this sequence of equalities, the last follows from the fact that $\mathbf{P}$ is Hermitian. If x is Hermitian the coefficients can be chosen so that $x_{ji} = x_{ij}^*$ ($*$ denotes complex conjugation), in which case the matrix $\mathbf{X}$ is Hermitian.

For the one-dimensional lattice model it is straightforward to choose a basis $\{q_1, q_2, ..., q_s\}$ so that: *(1)* $\langle q_i, q_j \rangle = 2^{2r}\delta_{ij}$, were $r = \mid \Lambda \mid$; *(2)* $\dim(\mathcal{Q}^k)1 = \sum_{i=1}^{s} q_i^\dagger q_i$. *Throughout our discussion we will assume that these conditions on the basis* $\{q_1, q_2, ..., q_s\}$ *are always in force.*

REPRESENTATION THEOREM. *An Hermitian operator p is a $k-$density if and only if the corresponding $k-$matrix $\mathbf{P}$ satisfies the two conditions: (1) $\mathbf{P}$ is positive semi-definite; (2) $trace(\mathbf{P}) = \dim(Q^k)$.*

Proof. (1) Let $q \in Q^k$ be arbitrary, and let $q = \sum_{i=1}^{s} Q_i q_i$ be the expansion in the basis. Then the equalities $\langle p, q^\dagger q \rangle = \sum_{i,j=1}^{s} Q_i^* \langle p, q_i^\dagger q_j \rangle Q_j = \mathbf{Q}^\dagger \mathbf{P} \mathbf{Q}$ show that p is non-negative to k^{th} order if and only if $\mathbf{P}$ is positive semi-definite; $\mathbf{Q}$ is the column vector of coefficients in the expansion.

(2) The formula $\dim(\mathcal{Q}^k)1 = \sum_{i=1}^{s} q_i^\dagger q_i$ establishes that $\frac{1}{\dim(\mathcal{Q}^k)}\mathbf{I}$ is a matrix representation of the identity operator. As a consequence the equalities $1 = \langle p, 1 \rangle = trace(\mathbf{P}\frac{1}{\dim(\mathcal{Q}^k)}\mathbf{I}) = \frac{1}{\dim(\mathcal{Q}^k)} trace(\mathbf{P})$ show that $trace(p) = 1$ if and only if $trace(\mathbf{P}) = \dim(\mathcal{Q}^k)$.

Thus the two conditions on the matrix representation are equivalent to the two conditions for a $k-$density, namely, that p be non-negative to k^{th} order and have unit trace.□

The cone of matrix representations for the elements of $\mathfrak{P}_0^k$ will be denoted by $\mathbb{P}_0^k$. The convex set of $k-$matrices $\mathbb{P}_0^k$ is compact, but the corresponding convex set of $k-$densities $\mathfrak{P}_0^k$ is not.

The Pauli subspace. It follows from the fermion commutation relations that the entries of a $k-$matrix are related by a system of linear equalities. For example, consider the pair transport operator $T_{\mu\nu}^2 = 2(b_\mu^\dagger a_\mu^\dagger a_\nu b_\nu + b_\nu^\dagger a_\nu^\dagger a_\mu b_\mu)$, which moves a spin-up, spin-down pair of electrons between sites μ, ν of Λ. If we define $q_{\pm 0} = a_\mu b_\mu \pm a_\nu b_\nu$, $q_{\pm 1} = a_\mu^\dagger b_\nu \pm b_\mu a_\nu^\dagger$, $q_{\pm 2} = a_\mu b_\nu^\dagger \pm b_\mu^\dagger a_\nu$, $q_{\mp 3} = a_\mu^\dagger a_\nu \mp b_\mu b_\nu^\dagger$, $q_{\mp 4} = a_\mu a_\nu^\dagger \mp b_\mu^\dagger b_\nu$, then, by the commutation relations it follows that

$$\begin{aligned} T_{\mu\nu}^2 &= q_{+i}^\dagger q_{+i} - q_{-i}^\dagger q_{-i}, \qquad & i = 0, 1, 2 \\ &= q_{-i}^\dagger q_{-i} - q_{+i}^\dagger q_{+i}, \qquad & i = 3, 4. \end{aligned}$$

This system of equalities is equivalent to the following linear conditions on the $k-$matrix.

$$\begin{aligned}\mathbf{Q}_{+0}^{\dagger}\mathbf{P}\mathbf{Q}_{+0} - \mathbf{Q}_{-0}^{\dagger}\mathbf{P}\mathbf{Q}_{-0} &= \mathbf{Q}_{+i}^{\dagger}\mathbf{P}\mathbf{Q}_{+i} - \mathbf{Q}_{-i}^{\dagger}\mathbf{P}\mathbf{Q}_{-i}, \quad i = 1,2 \\ &= \mathbf{Q}_{-i}^{\dagger}\mathbf{P}\mathbf{Q}_{-i} - \mathbf{Q}_{+i}^{\dagger}\mathbf{P}\mathbf{Q}_{+i}, \quad i = 3,4\end{aligned}$$

The column vector $\mathbf{Q}_{\pm i}$ is the coordinate vector for $q_{\pm i}$ referred to the fixed basis for Q^k, and the equivalence is established using the identities $\langle p, q_{\pm i}^{\dagger} q_{\pm i}\rangle = \mathbf{Q}_{\pm i}^{\dagger}\mathbf{P}\mathbf{Q}_{\pm i}$. The coordinate vectors $\mathbf{Q}_{\pm i}$ are linearly independent by the independence of the operators $q_{\pm i}$. It follows that these equalities represent four independent conditions on the $k-$matrix $\mathbf{P}$.

The linear conditions on $\mathbf{P}$ can be rewritten as

$$\begin{aligned}\mathbf{P} \quad &\perp \mathbf{Q}_{+0}\mathbf{Q}_{+0}^{\dagger} - \mathbf{Q}_{-0}\mathbf{Q}_{-0}^{\dagger} - \mathbf{Q}_{+i}\mathbf{Q}_{+i}^{\dagger} + \mathbf{Q}_{-i}\mathbf{Q}_{-i}^{\dagger}, \quad i = 1,2, \\ &\perp \mathbf{Q}_{+0}\mathbf{Q}_{+0}^{\dagger} - \mathbf{Q}_{-0}\mathbf{Q}_{-0}^{\dagger} - \mathbf{Q}_{-i}\mathbf{Q}_{-i}^{\dagger} + \mathbf{Q}_{+i}\mathbf{Q}_{+i}^{\dagger}, \quad i = 3,4,\end{aligned}$$

so that $\mathbf{P}$ is orthogonal to a four-dimensional subspace of Hermitian matrices with respect to the trace scalar product. There are many other such linear conditions on $\mathbf{P}$, and taken together they are equivalent to requiring that $\mathbf{P}$ lie in the orthogonal complement of a real linear space of Hermitian matrices which we denote by $\mathcal{S}$. We will call $\mathcal{S}$ the *Pauli subspace* since it encodes the content of the Pauli principle.

Since an arbitrary $k-$matrix is orthogonal to $\mathcal{S}$, the matrix representation for an Hermitian operator is far from unique. Suppose that $\mathbf{H}$ is the matrix representation of some Hamiltonian, and that $\mathbf{S}$ is an arbitrary matrix in $\mathcal{S}$. Then $\mathbf{H}+\mathbf{S}$ is an equally valid representation since the identity $< p, h > = trace\mathbf{PH} = trace\mathbf{P}(\mathbf{H}+\mathbf{S})$ clearly holds for all $k-$matrices $\mathbf{P} \in \mathbb{P}_0^k$. The explanation for such a large number of representations is straight forward: *the operator s corresponding to a matrix $\mathbf{S} \in S$ is equal to the zero operator*. The operator s is constructed by taking the matrix elements to be coefficients in an expansion, but can then be reduced to 0 using the commutation relations.

By the *Representation Theorem* the $k-$matrix representation for the convex set of $k-$densities is given by $\mathbb{P}_0^k = \{\mathbf{P} \in \mathcal{P} \cap \mathcal{S}^{\perp} \mid trace(\mathbf{P}) = \mid \mathcal{Q}^k \mid\}$; in this characterization $\mathcal{P}$ is the cone of positive semi-definite matrices.

Additional properties of matrix representations. To complete this section we list several useful properties for matrix representations. Assume throughout that $r = \mid \Lambda \mid$.

R1. The $k-$matrix for the von Neumann density $\frac{1}{2^{2r}}1$ is equal to the identity matrix $\mathbf{I}$: The entries for the matrix representation of $\frac{1}{2^{2r}}1 \in \mathfrak{P}_0^k$

are given by $P_{ij} = \langle p, q_i^\dagger q_j \rangle = \langle \frac{1}{2^{2r}} 1, q_i^\dagger q_j \rangle = \frac{1}{2^{2r}} \langle q_j, q_i \rangle = \delta_{ij}$, since by assumption $\langle q_j, q_i \rangle = 2^{2r} \delta_{ij}$.

R2. Each Hermitian operator h admitting an expansion of the form $h = \sum_{i,j=1}^s h_{ij} q_i^\dagger q_j$ has a unique representation that is orthogonal to $\mathcal{S}$: If $\mathbf{H}$ is an arbitrary representation, and $\pi_{\mathcal{S}}$ is orthogonal projection onto the Pauli space, then $\mathbf{H} + \pi_{\mathcal{S}}(\mathbf{H}) = \pi_{\mathcal{S}^\perp}(\mathbf{H})$ is an equally valid representation that is orthogonal to $\mathcal{S}$, and uniquely determined by h.

R3. If $\mathbf{S} \in \mathcal{S}$, then trace$(\mathbf{S}) = 0$: Since $\mathbf{I}$ is the k−matrix for the von Neuman density $\frac{1}{2^{2r}} 1$, the elements of the Pauli space must be orthogonal to $\mathbf{I}$ with respect to the trace scalar product, therefore traceless.

R4. The operator 1 is represented by the matrix $\frac{1}{\dim(\mathcal{Q}^k)} \mathbf{I}$: This representation was established in the proof of *Representation Theorem*, and is the unique representation that is orthogonal to $\mathcal{S}$.

R5. If x is an arbitrary operator such that trace$(x) = 0$, then trace$\mathbf{X} = 0$: This follows from the equalities $0 = trace(x) = \langle 1, x \rangle = \langle 1, \sum_{i,j=1}^s x_{ij} q_i^\dagger q_j \rangle = \sum_{i,j=1}^s x_{ij} \langle q_i, q_j \rangle = 2^{2r} \sum_{i,j=1}^s x_{ij} \delta_{ij} = 2^{2r} \sum_{i=1}^s x_{ii} = 2^{2r} trace(\mathbf{X})$.

3. CONVERGENCE IN K

Here we review the three papers [Erd5,Erd6,Maz3] on lower bounds and the energy method, where the difficulties uncovered by Mihailovic and Rosina in their work on light nuclei were resolved; these papers study models where electrons interact through strong two-body forces. Approximations for von Neumann densities were generated by solving the Energy Problem

$$\min_{p \in \mathfrak{P}_0^k} \quad \langle p, h \rangle,$$

for a variety of values of k. The most significant discovery was the rapid convergence of the sequence of approximations for $k = 2, 3, ...$, leading to ten figure accuracy when $k = 4$. This rapid convergence turns the energy method into an effective tool for studying correlations induced by strong two-body forces.

Modeling a one-dimensional superconductor. Erdahl and Jin [Erd3,Erd4,Jin1] examined rate of convergence with order k for a model of a one-dimensional superconductor where electrons interact through two-body forces alone . They showed there is an enormous jump in accuracy in going from second to third order estimates, with the third order estimates being sufficiently accurate to resolve the difficulties uncovered by Garrod, Mihailovic and Rosina in their work on light nuclei. They showed that approximating to third order gives a practical solution for the class of problems where fermions interact through strong two-body forces.

In this superconducting model spin-up, spin-down pairs of electrons wander on a one-dimensional periodic lattice Λ. The *local Hamiltonian*

$$h_{\mu\nu} = \alpha_E{}^2 E_{\mu\nu} + \alpha_T{}^2 T_{\mu\nu},$$

where $\mu, \nu \in \Lambda$ are nearest neighbors, determines how the pairs interact along a nearest neighbor bond and the Hamiltonian is formed by summing these bond contributions:

$$h = \sum_{|\mu-\nu|=1} h_{\mu\nu}.$$

There are an even number of lattice points equi-spaced on a ring, which are represented by the integers $\{1, 2, ..., |\Lambda|\}$. The restriction $\mid \mu - \nu \mid = 1$ limits the summation to nearest neighbor contributions, and since Λ is periodic the identity $\mid \Lambda \mid +1 = 1$ holds.

Along a bond the pairs interact through an electrostatic force $\alpha_E{}^2 E_{\mu\nu}$, the magnitude and sign determined by the coefficient α_E. The pairs also respond to a transport term $\alpha_T{}^2 T_{\mu\nu}$, which moves pairs between adjacent lattice sites. The details of these interactions are given by the expressions

$$\begin{aligned} {}^2E_{\mu\nu} &= \frac{1}{4}(e_{a_\mu} e_{a_\nu} + e_{b_\mu} e_{b_\nu} + e_{a_\mu} e_{b_\nu} + e_{b_\mu} e_{a_\nu}), \\ {}^2T_{\mu\nu} &= 2(b_\mu^\dagger a_\mu^\dagger a_\nu b_\nu + b_\nu^\dagger a_\nu^\dagger a_\mu b_\mu); \end{aligned}$$

the operator a_μ annihilates a spin-up electron at $\mu \in \Lambda$, and the operator b_μ, a spin-down electron; $e_{a_\mu} = a_\mu^\dagger a_\mu - a_\mu a_\mu^\dagger$. The electrostatic operator has expected value $+1$ for states when sites μ, ν are either both occupied or unoccupied, and expected value -1 in the other case where one site is occupied, the other not.

The superscript 2 is added to emphasize that these interactions are *two-body*. That is, *these operators are orthogonal to all scalar and one-body operators with respect to the trace scalar product.* More generally,

DEFINITION OF K-BODY OPERATOR. *A k-body operator is an Hermitian operator, which can be represented as a polynomial of degree 2k in the annihilation and creation operators, is of even degree, and is orthogonal to all (k-1)-body operators, all (k-2)-body operators, ..., and all scalar operators, with respect to the trace scalar product.*

As p varies over the set of von Neumann densities $\mathfrak{P}_0$, the vector of matrix elements $\beta = [\beta_T, \beta_E] = [\langle p,^2 T_{\mu\nu}\rangle, \langle p,^2 E_{\mu\nu}\rangle]$ fills in the *representable region* $\mathcal{R}$ pictured in *Figure 1.* Each von Neumann density $p \in \mathfrak{P}_0$ is represented by a point in $\mathcal{R}$, and each point in $\mathcal{R}$ represents at least one von Neumann density.

Each point $\beta \in \mathcal{R}$ corresponds to a quantum state with total energy given by $\epsilon = \langle p, h\rangle =\mid \Lambda \mid \langle p, h_{\mu\nu}\rangle =\mid \Lambda \mid (\alpha_T\langle p,^2 T_{\mu\nu}\rangle + \alpha_E\langle p,^2 E_{\mu\nu}\rangle) =\mid \Lambda \mid \alpha \cdot \beta$, where $\alpha = [\alpha_E, \alpha_T]$; the number of lattice sites $\mid \Lambda \mid$ is equal to the number of bonds in the ring, and therefore equal to the number of like contributions of bond energy. The minimum energy state for a Hamiltonian with coefficients $\alpha = [0, \alpha_T], \alpha_T < 0$, is represented by the point $\beta_* \in \mathcal{R}$ that is pictured. This follows since β_* is a point of tangency for a line that is tangent to $\mathcal{R}$, and perpendicular to α; the energy is constant along the tangent line, and the vector α points in the direction of increasing energy.

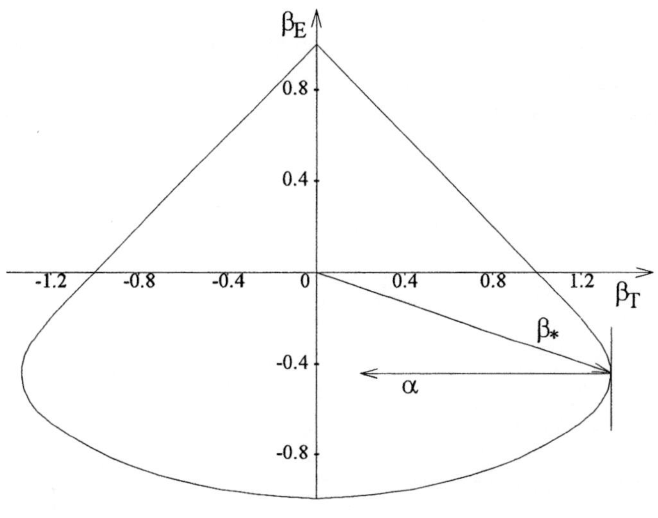

Figure 1. The representable region$\mathcal{R}$

The open circles slightly outside $\mathcal{R}$ in *Figure 2* are third order estimates of boundary points obtained by the lower bound method, and are accurate to three figures. These results are for three spin-up, spin-down pairs of electrons wandering on a ring with six lattice sites.

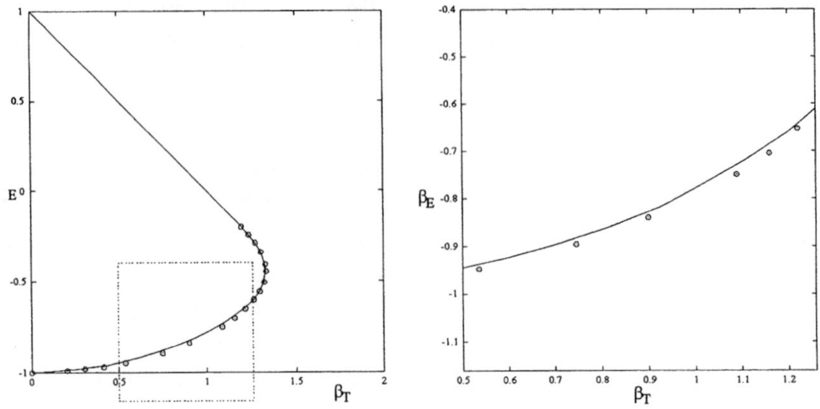

Figure 2. Accuracy of the lower bound method

As the number of lattice sites increases the electrons experience additional correlations, so the representable region shrinks. That is, if $\mathcal{R}_i$ is the representable region for a lattice with $i = |\Lambda|$ sites, then $\mathcal{R}_4 \supset \mathcal{R}_6 \supset \mathcal{R}_8 \supset \mathcal{R}_{10} \supset \ldots$. This phenomenon is accurately tracked by the third order estimates, and *Figure 3* shows that convergence to the limiting case where $|\Lambda| \to \infty$ is rapid.

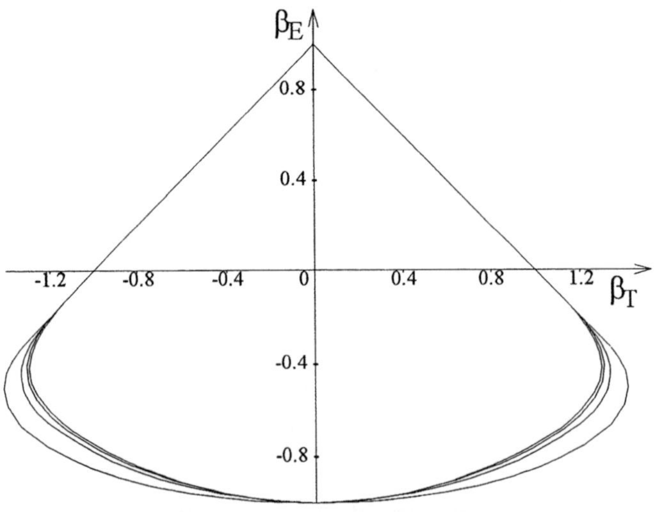

Figure 3. The representable regions $\mathcal{R}_4 \supset \mathcal{R}_6 \supset \mathcal{R}_8 \supset \mathcal{R}_{10}$

Convergence with the order k. Second order estimates of the representable region $\mathcal{R}$ were also made for the cases $|\Lambda| = 4, 6$. These are the *pentagonal* regions $\mathcal{R}_4^2 \subset \mathcal{R}_6^2$ that appear in *Figure 4*.

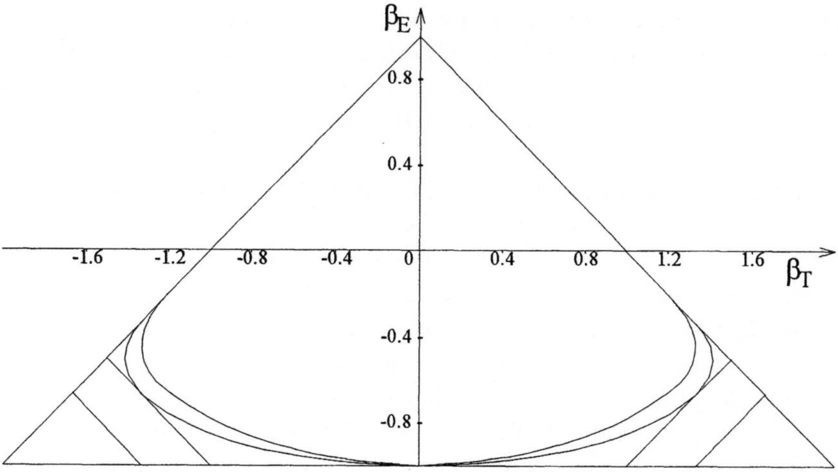

Figure 4. The estimates $\mathcal{R}_4^2 \subset \mathcal{R}_6^2$

Not only do the pentagonal regions poorly represent the corresponding representable regions, they *increase in size* in going from four to six lattice sites rather than shrink! The limiting case, when $|\Lambda| \to \infty$, is the triangle with vertices at $[0,1], [2,-1], [-2,-1]$. Thus the second order estimates completely misrepresent the physical situation and are of little value. The two representable regions $\mathcal{R}_4 \supset \mathcal{R}_6$ have been added for comparison.

These results confirm the observation of Mihailovic and Rosina that 2-positivity cannot characterize the correlations induced by two-body forces, and show that 3-positivity can. The vast difference between the second and third order estimates shows that convergence to the exact representable region is rapid in the order. The third order conditions give an effective characterization of the correlations induced by two-body operators, and it is clear that 4-positivity would give a precise description of these correlations.

The Lipkin model. The extraordinary speed of convergence with k was confirmed in the recent work of Mazziotti and Erdahl [Maz3], where the Lipkin model was studied using the energy method. Second, third and fourth-order approximations were studied for a range of strengths of the two-body forces. The jump in accuracy in going from second to third order approximations was again impressive, and the fourth order estimates effectively removed all remaining errors. The accuracy goes from about four to ten figures when third order estimates are replaced by fourth. A significant conclusion was that fourth order estimates were an order of magnitude more accurate than estimates calculated using other many-body methods, showing that convergence in k is rapid enough to be of enormous practical significance. These results show that 4-positivity gives a solution to the correlation problem for strong two-body forces.

The Lipkin model was introduced to study the strong two-body forces that dominate the dynamics of protons and neutrons in the nucleus. This is a two level system where the energy levels -1 and $+1$ are each N-fold degenerate, and the strength of a two-body term can be adjusted by varying the parameter V. When V is zero all N fermions occupy the ground state, but when V is turned on pairs are promoted to the excited state; the action of the two-body term is to move pairs of fermions between the ground and excited states. For large interaction strengths both levels are nearly equally occupied; there are two half-filled shells, which is the configuration Mihailovic and Rosina found most difficult. In modeling nuclear systems V is chosen to be negative so the two-body forces are attractive, but in the work of Mazziotti and Erdahl V was positive, the two-body forces were repulsive, and an electronic system was modeled.

Convergence in order k was tracked for three different particle numbers, $N = 10, 30, 75$, and for two values of the interaction strength, $V = 0.8, 1.6$. Comparisons were made by recording the percent of the correlation accounted for using second, third, and fourth order approximations, with the results recorded in Table I (taken from [Maz3]).

Table I: Convergence of correlation energy with k

N	V	E_{corr}	$\mathfrak{P}_0^2$	$\mathfrak{P}_0^3$	$\mathfrak{P}_0^4$
10	0.8	-0.0384	731	100.0376	100.000103
30	0.8	-0.0130	2165	100.0103	100.0000670
75	0.8	-0.00526	5238	100.00167	100.0000113
10	1.6	-0.186	478	101.85	100.0958
30	1.6	-0.128	707	102.86	101.90
75	1.6	-0.117	758	100.644	100.428

The third column contains the exact correlation energy computed using a full configuration interaction treatment.

As anticipated convergence in k is astonishing: the accuracy of the fourth order approximations of the energy exceeded ten figures in some instances. These calculations also confirm that second order approximations are of no value, falling below the exact by large margins: the correlation energy for the case $N = 75, V = 0.8$ is over-estimated by a factor greater than 50.

A significant comparison is with other wave-function and many-body perturbation methods currently employed in quantum theory. These comparisons are made in Table II, taken from [Maz3], where correlation energies are tabulated for a variety of methods.

Table II: Comparison with other many-body methods

N	V	E_{corr}	$SDTQCI$	$MP4$	$CCSD$	CSE	$\mathfrak{P}_0^4$
10	0.8	-0.0384	98.85	99.703	101.07	99.9856	100.000103
30	0.8	-0.0130	95.94	96.43	101.08	99.9860	100.0000670
75	0.8	-0.00526	94.25	94.5	100.642	99.9092	100.0000113
10	1.6	-0.186	88.9	100.224	*	100.0336	100.0958
30	1.6	-0.128	47.5	53.9	*	100.486	101.90
75	1.6	-0.117	20.9	23.6	*	99.9657	100.428

Comparisons are made with: (1) a configuration interaction wave calculation with single, double, triple and quadruple excitations ($SDTQCI$); (2) fourth order many-body perturbation theory ($MP4$); (3) solution of the single-double coupled cluster equations ($CCSD$), and; (4) the contacted Schrödinger equation (CSE). Of the comparison methods the single-double coupled cluster treatment gives the best results, but failed to converge in the strong interaction case where $V = 1.6$. The lower bounds computed using the energy method are four orders of magnitude more accurate than those single-double coupled cluster results.

These comparisons show that the rate of convergence of the lower bound method in k is fast enough that when $k = 4$ accuracies are achieved that are superior to other many-body techniques by several orders of magnitude. The correlations induced by two-body interactions are given a precise characterization by 4-positivity.

4. THE K-SPECTRUM

The ground state can also be estimated by solving the *Dispersion Problem*:

$$\min_{p \in \mathfrak{P}_0^k} \langle p, (h - \lambda 1)^2 \rangle.$$

In order that the dispersion problem be well formulated the condition $\deg(h - \lambda 1) \leq k$ must be imposed, which ensures that the values of $\langle p, (h - \lambda 1)^2 \rangle$ are non-negative; this is a direct consequence of the condition $\langle p, q^\dagger q \rangle \geq 0, \deg(q) \leq k$, on k-densities.

A solution of the dispersion problem is a pair (λ, p_λ), where $\lambda \in \mathbb{R}, p_\lambda \in \mathfrak{P}_0^k$ satisfy the equality $\langle p_\lambda, (h - \lambda 1)^2 \rangle = 0$: the k-density p_λ estimates the von Neumann density for the state with energy eigenvalue λ. The search for such dispersion free k-densities requires variation of both λ and p since the expectation value $\langle p, (h - \lambda 1)^2 \rangle$ is typically positive. A complete solution of the dispersion problem requires a list of all pairs (λ, p_λ), where $\lambda \in \mathbb{R}$, and $p_\lambda \in \mathfrak{P}_0^k$ is dispersion free.

It appears that information on both the ground and arbitrary excited states can be extracted by solving the Dispersion Problem, but this is certainly too much to ask from an approximate theory. In fact, we conjecture below that it is only the properties of the ground and first excited states that can effectively be explored; the second and higher excited states require increasing expenditures in computation making these inaccessible. More specifically, we argue that for two-body Hamiltonians where $\deg(h - \lambda 1) = 2$, information on the ground and first excited states can be obtained by solving the fourth order Dispersion Problem, but information on higher excited states requires solution of successively higher order dispersion problems with computations increasingly more difficult.

Definition and properties of the k-spectrum. If $\lambda \in \mathbb{R}$ and $p_\lambda \in \mathfrak{P}_0$ satisfy the equation $\langle p_\lambda, (h - \lambda 1)^2 \rangle = 0$, then λ is a point of the spectrum $\sigma(h)$ of h, and p_λ is the corresponding von Neumann density. There is an analogue statement when p_λ is a k-density.

DEFINITION OF k–SPECTRUM. *If $\lambda \in \mathbb{R}$ and $p_\lambda \in \mathfrak{P}_0^k$ satisfy the equation $\langle p_\lambda, (h - \lambda 1)^2 \rangle = 0$, then λ is a point of the k–spectrum $\sigma^k(h)$ of h, and p_λ is the corresponding k–density.*

A complete solution of the kth order dispersion problem amounts to determining the k-spectrum $\sigma^k(h)$.

THEOREM ON THE k–SPECTRUM. *Assume that* $\deg(h - \lambda 1) \leq k$. *Then: (1) $\sigma^k(h) \supseteq \sigma^{k+1}(h) \supseteq \sigma^{k+2}(h) \supseteq \dots \supseteq \sigma(h)$; (2) if $\lambda \in \sigma(h)$ and $p_\lambda \in \mathfrak{P}_0$ satisfy the equality $\langle p_\lambda, (h - \lambda 1)^2 \rangle = 0$, then (λ, p_λ) is a solution of the kth order dispersion problem.*

Proof. (1) The first assertion is a direct consequence of the inclusions $\mathfrak{P}_0^k \supset \mathfrak{P}_0^{k+1} \supset \mathfrak{P}_0^{k+2} \supset \dots \supset \mathfrak{P}_0$.

(2) By assumption $p_\lambda \in \mathfrak{P}_0$ and $\langle p_\lambda, (h-\lambda 1)^2\rangle = 0$. Since $\mathfrak{P}_0 \subset \mathfrak{P}_0^k$, it follows that $p_\lambda \in \mathfrak{P}_0^k$ so that (λ, p_λ) is a solution of the kth order dispersion problem. □

It is the second assertion of the Theorem that is the striking one since it raises the possibility that exact solutions can be extracted by solving the kth order dispersion problem, *even though this is an optimization problem for k-densities, which are k^{th} order approximations.*

The equivalent matrix formulation. We consider a Hamiltonian h which has at most two–body interactions, but is otherwise arbitrary. By passing to a matrix representation the kth order dispersion problem becomes

$$\min_{\mathbf{P}\in\mathbb{P}_0^k} \quad trace\mathbf{P}(\mathbf{H}^2 - 2\lambda\mathbf{H} + \frac{\lambda^2}{\dim Q^k}\mathbf{I}),$$

where $k \geq 4$. A solution is a pair $(\lambda, \mathbf{P}_\lambda)$ satisfying the equation

$$trace\mathbf{P}_\lambda(\mathbf{H}^2 - 2\lambda\mathbf{H} + \frac{\lambda^2}{\dim Q^k}\mathbf{I}) = 0,$$

where $\lambda \in \mathbb{R}$ and $\mathbf{P}_\lambda \in \mathbb{P}_0^k$; $\mathbf{P}_\lambda$ is a dispersion-free k-matrix. In this formulation $\mathbf{H}^2$ is a matrix representation for h^2, and not the square of $\mathbf{H}$, which is a matrix representation for h.

The second assertion of the *Theorem on the k-spectrum* now takes the form: *if $\lambda \in \sigma(h)$, and $\mathbf{P}_\lambda$ is the k-matrix for the corresponding von Neumann density, then $(\lambda, \mathbf{P}_\lambda)$ is a solution of the matrix k^{th} order dispersion problem.* The content of the second assertion is more readily examined using this matrix formulation.

Consider the sequence of approximating convex sets $\mathfrak{P}_0^4 \supset \mathfrak{P}_0^5 \supset \mathfrak{P}_0^6 \supset \ldots \supset \mathfrak{P}_0$. By passing to representations by 4–matrices we get the strictly decreasing sequence $\mathbb{P}^4(\mathfrak{P}_0^4) \supset \mathbb{P}^4(\mathfrak{P}_0^5) \supset \mathbb{P}^4(\mathfrak{P}_0^6) \supset \ldots \supset \mathbb{P}^4(\mathfrak{P}_0)$ of compact convex sets, which converges to the representable set $\mathbb{P}^4(\mathfrak{P}_0)$ of 4–matrices. With this last sequence we have introduced new notation: $\mathbb{P}^4(\mathfrak{P}_0^k)$ is the convex set of 4–matrices corresponding to the elements of $\mathfrak{P}_0^k$. The first element in this sequence, $\mathbb{P}^4(\mathfrak{P}_0^4)$, is equal to $\mathbb{P}_0^4$.

The function $\varphi(x) = \langle x, (h-\lambda 1)^2\rangle$ is non-negative on the convex sets $\mathfrak{P}_0^k, k = 4, 5, \ldots$, and takes the value 0 on $p_\lambda \in \mathfrak{P}_0 \subset \ldots \subset \mathfrak{P}_0^5 \subset \mathfrak{P}_0^4$. Switching to representations by 4-matrices, the function

$$f(\mathbf{X}) = trace\mathbf{X}(\mathbf{H}^2 - 2\lambda\mathbf{H} + \frac{\lambda^2}{\dim Q^4}\mathbf{I})$$

is non-negative on the compact convex sets $\mathbb{P}^4(\mathfrak{P}_0^4) \supset \mathbb{P}^4(\mathfrak{P}_0^5) \supset \mathbb{P}^4(\mathfrak{P}_0^6) \supset ... \supset \mathbb{P}^4(\mathfrak{P}_0)$, and takes the value 0 the 4-matrix $\mathbf{P}_\lambda \in \mathbb{P}^4(\mathfrak{P}_0) \subset ... \subset \mathbb{P}^4(\mathfrak{P}_0^5) \subset \mathbb{P}^4(\mathfrak{P}_0^4)$, which is the 4-matrix representation of p_λ. If $p \in \mathfrak{P}_0^k$ is a k-density, $\mathbf{P} = \mathbb{P}^4(p)$ is the corresponding 4-matrix, and $\varphi(p) = f(\mathbf{P})$; $\mathbf{P}_\lambda$ is the 4-matrix representation of p_λ, and $\varphi(p_\lambda) = f(\mathbf{P}_\lambda) = 0$. It follows that the hypoerplane with equation $f(\mathbf{X}) = 0$ supports each of the compact convex sets $\mathbb{P}^4(\mathfrak{P}_0^k), k = 4, 5, ...$, and that $\mathbf{P}_\lambda$ is in each support set.

The 2-spectrum - some exact results. The important question raised by the *Theorem on the k-spectrum* is whether exact results can be extracted working at the level of kth order approximations. The most important case is where the Hamiltonian h has two-body terms, and approximations are sought by solving the 4th order dispersion problem. How the 4-spectrum $\sigma^4(h)$ relates to the exact spectrum $\sigma(h)$ is crucial in this case, and is speculated on in this section. These speculations are based on results on the 2-spectrum for a one-body Hamiltonian.

In order that the second order dispersion problem be well-formulated the Hamiltonian must satisfy the condition $\deg(h - \lambda 1) \leq 2$, so that h can include one-body terms, scalar terms, and possibly terms linear in the annihilation and creation operators. We consider the case where $h = ^1 h$ is a particle-conserving, one-body operator, and quote some properties of the 2-spectrum. Although this is the simplest possible choice for Hamiltonian, calculation of the 2-spectrum is far from simple.

In *Spectral properties of cones of approximately representable reduced density matrices,* Erdahl and Grudzinski introduce the notion of k-spectrum and determine some of its properties [5]. For the simple case where $^1h = N$, the number operator, the spectrum is given by $\sigma(N) = \{0, 1, 2, ..., 2r\}$, where $2r$ is the number of one electron states. *The 2-spectrum is:* $\sigma^2(N) = \{0, [1, 2r - 1], 2r\}$. Since $\sigma(N) \subset \sigma^2(N)$ this result is consistent with the first part of the *Theorem on the k-spectrum.*

The 2-spectrum of the number operator has two isolated points, which are the top and bottom eigenvalues. There is also a continuous part, the closed line segment between 1 and $2r - 1$, which are eigenvalues for the first and next-to-top excited states. The 2-spectrum is symmetric under reflection through the point r, a symmetry that can be explained by exchanging particles with holes.

The most important observation is that the structure of $\sigma^2(N)$ permits only four points of the spectrum to be effectively estimated via the second order dispersion problem: the ground and first excited states, and by symmetry, the top and next-to-top excited states. In these cases the estimated spectral elements are exact. Moreover, in these four cases the k-matrix estimates are also exact, and represent von Neumann densities. By contrast, no information can be extracted on intermediate elements of $\sigma(N)$ lying between 1 and $2r-1$, and no information can be extracted on the corresponding von Neumann densities.

Consider a general one body operator ${}^1h$, with spectrum $\sigma(h_1) = \{\lambda_0, \lambda_1, \lambda_2, ..., \lambda_{2r-1}, \lambda_{2r}\}$. The following result on the 2−spectrum was reported as *Proposition 5.1* in [5].

PROPOSITION. $\sigma^2({}^1h) \subseteq \{\lambda_0, [\lambda_1, \lambda_{2r-1}], \lambda_{2r}\}$.

It is a reasonable *conjecture* that the containment can be replaced by equality, in conformity with the results on the number operator. This being the case, λ_1 for the first excited state is the left hand point for the continuous part of the 2−spectrum and it is again only the ground and first excited states that can effectively be studied. Information on the second, third, or higher excited states is inaccessible through the second order dispersion problem.

Based on the results on the 2−spectrum, and on the discussion above, we formulate the following conjectures of the 4−spectrum for a general Hamiltonian ${}^2h$ with two-body interactions.

CONJECTURE. *If* $\sigma({}^2h) = \{\lambda_0, \lambda_1, \lambda_2, ..., \lambda_{s-1}, \lambda_s\}$, *then* $\sigma^4({}^2h) = \{\lambda_0, [\lambda_1, \lambda_{s-1}], \lambda_s\}$.

If this conjecture holds information on the states with energy eigenvalues $\lambda_2, \lambda_3, \lambda_4, ...$ is inaccessible through the fourth order dispersion problem. On the other hand, if for $\lambda = \lambda_0, \lambda_1$ the fourth order dispersion problem

$$\min_{p \in \mathfrak{P}_0^4} \langle p, ({}^2h - \lambda 1)^2 \rangle.$$

has a unique minimizer, the exact 4-matrix can be calculated for the ground and first excited states.

It is reasonable to ask whether information on higher excited states can be extracted through the dispersion problem for higher values of k. For this, the following conjecture applies.

CONJECTURE. *If* $\sigma({}^2h) = \{\lambda_0, \lambda_1, \lambda_2, ..., \lambda_{s-1}, \lambda_s\}$, *then* $\sigma^5({}^2h) = \{\lambda_0, \lambda_1, [\lambda_2, \lambda_{s-2}], \lambda_{s-1}, \lambda_s\}, \sigma^6({}^2h) = \{\lambda_0, \lambda_1, \lambda_2, [\lambda_3, \lambda_{s-3}], \lambda_{s-2}, \lambda_{s-1}, \lambda_s\}, ...$.

Information on the second excited state can be extracted by calculating the 5-spectrum, information on the third by calculating the 6-spectrum, and continuing, so that information on the higher excited states requires increasing expenditures of computational energy, and this quickly becomes infeasible.

In summary, the fundamental question raised by the *Theorem on the k–spectrum* is whether the exact k–matrix for a ground or excited state can be calculated by solving the kth order dispersion problem. In the crucial case, where the interactions are 2–body and approximations carried through to fourth order, we speculate that exact results can be achieved for points $\lambda \in \sigma(h)$ that are isolated in $\sigma^4(h)$, or that lie at endpoints of continuous portions of $\sigma^4(h)$. On the other hand, if $\lambda \in \sigma(h)$ lies interior to a continuous part of $\sigma^4(h)$ the corresponding solution can't be distinguished from nearby solutions which are non-physical.

The link to the Contracted Schrödinger Equation. If Ψ is an eigenfunction of an operator Ω so that $\Omega\Psi = \lambda\Psi$, then this condition can be formulated in three equivalent ways: *(1)* $\langle \Psi, (\Omega - \lambda 1)^\dagger (\Omega - \lambda 1)\Psi \rangle = 0$; *(2)* $\langle \Psi, q^\dagger(\Omega - \lambda 1)\Psi \rangle = 0$ for all operators q on $\mathfrak{F}$; *(3)* $\langle \Psi, \Omega^\dagger \Omega \Psi \rangle = \lambda^* \lambda, \langle \Psi, \Omega\Psi \rangle = \lambda$. The state Ψ, or the corresponding von Neumann density, is *dispersion free* with respect to the operator Ω. These statements hold for arbitrary operators where the eigenvalues are generally complex numbers.

The following theorem shows how these conditions carry over to k^{th} order approximations if a condition is appended on the degree of the operator Ω.

DISPERSION THEOREM. *Let Ω be an operator such that $deg(\Omega) = m \leq k$, and let λ be a complex number. Then the following conditions on a k–density $p \in \mathfrak{P}_0^k$ are equivalent: (1) $\langle p, (\Omega - \lambda 1)^\dagger (\Omega - \lambda 1) \rangle = 0$; (2) $\langle p, q^\dagger (\Omega - \lambda 1) \rangle = 0$ for all operators q satisfying the condition* $\deg(q) \leq$ *$2k - m$; (3) $\langle p, \Omega \rangle = \lambda, \langle p, \Omega^\dagger \Omega \rangle = \lambda^* \lambda$.*

Proof. 1. Assume that condition *(1)* holds and that $\deg(\Omega - \lambda 1) = k$. Let $g = (\Omega - \lambda 1)$. Since $p \in \mathfrak{P}_0^k$, the inequality $0 \leq \langle p, (g + tq)^\dagger (g + tq) \rangle = t^* \langle p, q^\dagger g \rangle + t \langle p, g^\dagger q \rangle + t^* t \langle p, q^\dagger q \rangle$ holds for all $t \in \mathbb{C}$, $q \in Q^k$. It follows that $\langle p, q^\dagger g \rangle = 0$ which is condition *(2)* for this case.

Assume now that condition *(1)* holds and that $\deg(\Omega - \lambda 1) = m < k$. Assume also that $\{s_1, s_2, ..., s_s\}$ is a basis for Q^{k-m} with the property that $\dim(Q^{k-m}) 1 = \sum s_i^\dagger s_i$. Then $0 = \dim(Q^{k-m}) \langle p, (\Omega - \lambda 1)^\dagger (\Omega - \lambda 1) \rangle = \sum \langle p, (\Omega - \lambda 1)^\dagger s_i^\dagger s_i (\Omega - \lambda 1) \rangle$. Since $\deg(s_i(\Omega - \lambda 1)) \leq k$ each term in this sum is non-negative so must be zero. The argument employed for the case $\deg(\Omega - \lambda 1) = k$ can now be used to show that each term $\langle p, q^\dagger s_i (\Omega - \lambda 1) \rangle = 0$ when $\deg(q) \leq k$. Since an arbitrary operator $q^\dagger$ satisfying the condition that $\dim(q) \leq 2k - m$ can be written as $q^\dagger = \sum q_i^\dagger s_i$ where $\deg(q_i) \leq k$, it follows that $\langle p, q^\dagger (\Omega - \lambda 1) \rangle = 0$ which is condition *(2)* for this case.

2. Assume that condition *(2)* holds. By choosing $q = 1$ it follows that $\langle p, (\Omega - \lambda 1)\rangle = \langle p, \Omega\rangle - \lambda = 0$. By choosing $q = \Omega$ it then follows that $\langle p, \Omega^\dagger(\Omega - \lambda 1)\rangle = \langle p, \Omega^\dagger\Omega\rangle - \langle p, \Omega^\dagger\rangle\lambda = \langle p, \Omega^\dagger\Omega\rangle - \lambda^*\lambda = 0$. It follows that condition *(3)* holds.

3. Assume that condition *(3)* holds. Then $\langle p, (\Omega - \lambda 1)^\dagger(\Omega - \lambda 1)\rangle = \langle p, \Omega^\dagger\Omega\rangle - \langle p, \Omega^\dagger\rangle\lambda - \lambda^*\langle p, \Omega\rangle + \lambda^*\lambda = \lambda^*\lambda - 2\lambda^*\lambda + \lambda^*\lambda = 0$, so condition *(1)* holds.□

This is a slight strengthening of *Proposition 3.2* in [Erd2]; see also [Erd1] and [Erd3] for related materials.

In the case where $\Omega = h$ the main content of the *Dispersion Theorem* can be re-stated as follows.

COROLLARY. *Assume that* $\deg(h - \lambda 1) \leq k$. *Then* $(\lambda, p_\lambda), \lambda \in \mathbb{R}, p_\lambda \in \mathfrak{P}_0^k$, *is a solution of the* k^{th} *order dispersion problem if and only if* $\langle p, q^\dagger(h - \lambda 1)\rangle = 0$ *holds for all operators* q *satisfying the condition* $\deg(q) \leq 2k - \deg(h)$.

The equation

$$\langle p, q^\dagger(h - \lambda 1)\rangle = 0$$

is the *Contracted Schrödinger Equation* that appears in the work of Mazziotti [Maz1], Nakatsuji [Nak1] and Valdemoro [Val1,Vql2]. This Corollary establishes the equivalence of solutions of the Dispersion Problem, and solutions obtained via the Contracted Schrödinger Equation.

In the pioneering work of Mazziotti, Nakatsuji and Valdemoro on the Contracted Schrödinger Equation they did not impose the condition that the solution be k-positive. Their common strategy was to estimate the kth order reduced density matrix in terms of the first and second order reduced density matrices. This truncation strategy provided a method for solving the Contracted Schrödinger Equation, and led to impressive results (for some interesting new developments see [Maz4,Maz5]). This strategy for solving the Contracted Schrödinger Equation can be characterized as a perturbation approach to solving the Dispersion Problem.

5. THE FUNDAMENTAL OPTIMIZATION THEOREM

The von Neumann density for a quantum system can be estimated by varying either the energy or the dispersion. Both variations require finding a solution to an optimization problem of the form

$$\min_{p \in \mathfrak{P}_0^k} \quad \langle p, \omega\rangle,$$

where ω is an Hermitian operator, and the variation is over the convex set of k−densities. If ω is the Hamiltonian h the minimizing k−density p approximates the ground state, and if ω is the dispersion operator $(h-\lambda 1)^2$ and the minimum dispersion vanishes, the minimizing k-density approximates the von Neumann density for the state with energy eigenvalue λ.

If ω can be written as $\omega = \sum_{i,j=1}^{s} \omega_{ij} q_i^\dagger q_j$ there is a matrix formulation

$$\min_{\mathbf{P} \in \mathbb{P}_0^k} \quad trace(\mathbf{P}\Omega),$$

where the variation is over the convex set of k−matrices; Ω is the matrix representation of ω with entries $\omega_{ji} = \omega_{ij}^*$, and $\{q_1, q_2, ..., q_s\}$ is a basis for Q^k.

If the condition that $\mathbf{P}$ be orthogonal to the Pauli space $\mathcal{S}$ is removed, this is the variational formulation of the eigenvalue problem for Hermitian matrices: if $\mathbf{v}$ is the eigenvector for the bottom eigenvalue, $\mathbf{v}\mathbf{v}^T$ is the minimizer. The condition $\mathbf{P} \perp \mathcal{S}$ changes things significantly, and defines a new class of problems with a wide range of applications.

Characterizing the minimizer. The minimizing k-matrix can be calculated by applying the following theorem.

OPTIMIZATION THEOREM. $\mathbf{P} \in \mathbb{P}_0^k$ *minimizes* $trace(\mathbf{P}\Omega)$ *if and only if there is a real number* λ_0, *and a matrix* $\mathbf{S} \in \mathcal{S}$ *so that: (1)* $\mathbf{X} = \Omega + \mathbf{S} - \lambda_0 \mathbf{I} \in \mathcal{P}$; *(2)* $\mathbf{XP} = \mathbf{0}$.

The *Euler equation* $\mathbf{XP} = \mathbf{0}$ determines the entries of both $\mathbf{P}$ and $\mathbf{X}$, and therefore characterizes the minimizer.

The quantities λ_0 and $\mathbf{S}$ appearing in the statement of the *Theorem* can be interpreted as follows: λ_0 is the bottom eigenvalue of $\Omega + \mathbf{S}$, and $\mathbf{S}$ is the element of $\mathcal{S}$ that maximizes this bottom eigenvalue. That is, $\mathbf{S} \in \mathcal{S}$ is a solution of the *dual optimization problem*:

$$\max_{\mathbf{S} \in \mathcal{S}} \quad \lambda_0(\Omega + \mathbf{S})$$

It is this formulation which links the basic optimization problem to a wide variety of applications.

This Theorem forms the basis for semi-definite programming, a theory that has developed rapidly in response to the proof of the Optimization Theorem published by Overton in 1988 [Ove1]. In fact, the first proof was published in 1979 in [Erd4], a paper motivated by the need to develop effective algorithms for the lower bound method of reduced density matrix theory. In that paper the theorem was applied to solve the lower bound problem for He^2, two helium atoms forming a weakly bound diatomic molecule.

A symmetric formulation. Let ω be an Hermitian operator which we assume is traceless. Then, by *R2* and *R4*, there is a matrix representation Ω which is both traceless and orthogonal to $\mathcal{S}$. The k-matrix $\mathbf{P}$ minimizing $trace\mathbf{P}\Omega$ must satisfy the inequality $trace\mathbf{P}\Omega < 0$, in which case there is a positive scale factor κ so that $\kappa\mathbf{P}$ has the form $-\Omega + \mathbf{K} - \eta_0\mathbf{I} = \mathbf{Y}$, where $\mathbf{K}$ is a matrix orthogonal to $\mathbf{I}, \Omega, \mathcal{S}$ in the trace scalar product. That is, $\mathbf{K} \in \mathcal{K} = \{\mathbf{I}, \Omega, \mathcal{S}\}^{\perp}$. The Euler equation then assumes the symmetric form

$$\mathbf{XY} = \mathbf{0}, \qquad \mathbf{X} = \Omega + \mathbf{S} - \lambda_0\mathbf{I}, \quad \mathbf{Y} = -\Omega + \mathbf{K} - \eta_0\mathbf{I},$$

where $\mathbf{S} \in \mathcal{S}$, $\mathbf{K} \in \mathcal{K}$. A solution is sought where $\mathbf{X}, \mathbf{Y} \in \mathcal{P}$. The real number η_0 and the matrix $\mathbf{K}$ play the same role as do the quantities $\lambda_0, \mathbf{S}$: η_0 is the bottom eigenvalue of the matrix $-\Omega + \mathbf{K}$, and $\mathbf{K}$ is the matrix in $\mathcal{K}$ that maximizes this bottom eigenvalue.

Second order convergence for algorithms. By introducing bases $\{\mathbf{S}_1, \mathbf{S}_2, ..., \mathbf{S}_s\}$, $\{\mathbf{K}_1, \mathbf{K}_2, ..., \mathbf{K}_t\}$ for $\mathcal{S}, \mathcal{K}$ the matrices $\mathbf{S}, \mathbf{K}$ can be written as $\mathbf{S} = \alpha_1\mathbf{S}_1 + \alpha_2\mathbf{S}_2 + ... + \alpha_s\mathbf{S}_s$, $\mathbf{K} = \beta_1\mathbf{K}_1 + \beta_2\mathbf{K}_2 + ... + \beta_t\mathbf{K}_t$, and the Euler equation is converted to a system of quadratic equations in the parameters $\lambda_0, \alpha_1, \alpha_2, ..., \alpha_s; \eta_0, \beta_1, \beta_2, ..., \beta_t$. There is one equation for each matrix entry, enough to determine these parameters. Since the Euler equation is quadratic in the parameters, algorithms are easily devised that *converge quadratically* to a solution.

The importance of the *Euler Equation* can be illustrated by briefly considering an alternate strategy for solving the basic problem - directly maximizing the objective function $\lambda_0(\Omega + \mathbf{S})$ that appears in the dual formulation. By writing $\mathbf{S} = \alpha_1\mathbf{S}_1 + \alpha_2\mathbf{S}_2 + ... + \alpha_s\mathbf{S}_s$, λ_0 becomes a function of the vector variable $\alpha = [\alpha_1, \alpha_2, ..., \alpha_s] \in \mathbb{R}^s$. However, λ_0 is not a smooth function of α so cannot be well approximated by any of the standard methods. As α moves towards the position of the maximum the bottom eigenvalue is shifted upward, compressing the bottom part of the spectrum, and at the maximum point the bottom eigenvalue is generally highly degenerate. In the vicinity of this maximum λ_0 is the infimum of several analytic functions passing through the position of the maximum, and the non-smooth character of this function is exaggerated. In this region, sampling function values or constructing tangent planes provides little useful information in navigating towards the maximum. One can anticipate that standard algorithms to maximize λ_0 will slow considerably in the neighborhood of the maximum. This was confirmed in the computational work of Rosina and his collaborators [10,15] who used an algorithm based on approximations by tangent planes.

Dual description of correlations. By the *Spectral Theorem* for Hermitian matrices, the optimal matrix $\mathbf{X}$ can be written as

$$\mathbf{X} = \Omega + \mathbf{S} - \lambda_0 \mathbf{I} = \sum \mathbf{G}_i \mathbf{G}_i^\dagger,$$

which by passing to the operator representation becomes

$$\omega - \dim(Q^k)\lambda_0 1 = \sum g_i^\dagger g_i;$$

$\mathbf{G}_i$ is the coordinate vector for the operator g_i referred to the fixed basis for Q^k. The vectors $\mathbf{G}_i$ belong to the kernel of the $k-$matrix $\mathbf{P}$, and the corresponding operators g_i are annihilators for the corresponding approximate von Neuman density: $g_i p = 0$.

The operators g_i characterize the correlations in the minimizing $k-$density, so play an important interpretive role. For example, in the case of the dispersion problem $\omega = (h - \lambda 1)^2$ and

$$(h - \lambda 1)^2 - \dim(Q^k)\lambda_0 1 = \sum g_i^\dagger g_i.$$

The condition that $\lambda \in \sigma^k(h)$ is equivalent to $\lambda_0 = 0$; the minimizing k-density $p \in \mathfrak{P}_0^k$ is dispersion free and satisfies the equation

$$\langle p, (h - \lambda 1)^2 \rangle = \sum \langle p, g_i^\dagger g_i \rangle = 0.$$

This implies that the individual terms $\langle p, g_i^\dagger g_i \rangle = 0$ for each of the operators $g_i \in Q^k$ appearing in the expansion. These "killers," g_i, of the state completely characterize the correlations in the state. If $\lambda \in \sigma^k(h)$, p can be taken to be the exact von Neumann density for the state with energy eigenvalue λ, and the killers g_i give an exact description of the correlations in this quantum state.

6. CONCLUDING REMARKS

It is a pleasure to see the focus now brought to research on reduced density matrices, a theory which is now in full bloom. Many new researchers have joined in the search to find answers to the remaining questions. The successes of the last five years are impressive - clear strategies have have been laid down to do quantum mechanics without wave functions, but by directly calculating the reduced density matrix. This review reports on one of these strategies that has been particulary successful - accuracies have been recorded that are orders of magnitude better that those achievved using other widely used methods.

But many questions remain. There is the technical problem of developing codes that adapt semi-definite programing to the new task of solving the central energy and dispersion problems. Although there was a gap of over twenty years between the first two papers in this direction, it is certain that many papers will follow.

Another important problem is computation of the k-spectrum for a variety of quantum systems. This problem is central to the question of extracting information on excited states. Another problem is the interpretation of the duality theory accompanying semi-definite programing, which we put under the more general heading *dual configuration interaction.* There are many more problems!

I would like to take this opportunity to thank David Mazziotti and John Coleman for their help and advice in writing this chapter.

References

[Col1] A. J. Coleman, *The structure of fermion density matrices,* Rev. Mod. Phys. **35** (1963 668-687.

[Col2] A. J. Coleman and V. I. Yukalov, *Reduced Density Matrices - Coulson's Challenge*, Springer-Verlag, New York, 2000.

[Erd1] R. M. Erdahl and C. Garrod, *Trace relations, one-particle symmetries and the convex structure of reduced density matrices*, published in *Reduced Density Operators with Applications to Physical and Chemical Systems II*, Queen's Papers in Pure and Applied Mathematics **40**, Queen's Press (1974), 22-27.

[Erd2] R. M. Erdahl and H. Grudzinski, *Spectral properties of cones of approximately representable reduced density matrices,* Reports Math. Phys. **14** (1978), 405-418.

[Erd3] R. M. Erdahl, *Representability*, Int. J. Quantum Chem. **13** (1978), 697-718.

[Erd4] R. M. Erdahl, *Two algorithms for the lower bound method of reduced density matrix theory*, Reports Math. Phys. **15** (1979), 147-162.

[Erd5] R. M. Erdahl and B. Jin, *The lower bound method for reduced density matrices*, J. Mol. Struct.:THEOCHEM **527** (2000), 207 - 220.

[Erd6] R. M. Erdahl and B. Jin, *On calculating approximate and exact density matrices*, in Chapter 4 in *Many-electron densities and density matrices*, edited by J. Cioslowski (Kluwer, Boston, 2000).

[Jin1] Beiyan Jin, *Quantum phases for a two-body spin-invariant nearest neighbor interactions*, doctoral disertation Queen's University (1998).

[Löw1] P. - O. Löwdin, *Quantum theory of many particle systems I: Physical interpretations by means of density matrices, natural spin-orbitals, and convergence problems in the method of configuration interaction*, Phys. Rev. **97** (1955) 1474 - 1489.

[Löw2] P. - O. Löwdin, *Quantum theory of many particle systems II: Study of ordinary Hartree-Fock approximations*, Phys. Rev. **97** (1955) 1490 - 1508.

[Löw3] P. - O. Löwdin, *Quantum theory of many particle systems III: Extension of the Hartree-Fock scheme to include degeenerate systems and correlation effects*, Phys. Rev. **97** (1955) 1509 - 1520.

[Maz1] D. A. Mazziotti, *Contracted Schrödinger Equation: determining quantum energies and the two-particle density matrix without wave functions*, Phys. Rev. A **57** (1998), 4219-4234.

[Maz2] D. A. Mazziotti, *Variational minimization of atomic and molecular ground-state energies via the two-particle reduced density matrix*, Phys. Rev. A **65** (2002) 062511.

[Maz3] D. A. Mazziotti and R. M. Erdahl, *Uncertainty relations and reduced density matrices: Mapping many-body quantum mechanics onto four particles*, Phys. Rev. A **63** (2001) 042113.

[Maz4] D. A. Mazziotti, *Variational method for solving the contracted Schrödinger equation through a projection of the N-particle power method onto the two-particle space*, J. Chem. Phys. **116** (2002) 1239-1249.

[Maz5] D. A. Mazziotti, *Purification of correlated reduced density matrices*, Phys. Rev. E **65** (2002), 026704.

[Nak1] K. Yasuda and H. Nakatsuji, *Direct determination of the quantum mechanical density matrix using the density equation II*, Phys. Rev. A **56** (1997), 2648-2657.

[Nak2] M. Nakata, H. Nakatsuji, M. Ehara, M. Fukuda, K. Nakata, K. Fujisawa, *Variational calculations of fermion second-order reduced density matrices by semi-definite programming algorithm*, J. Chem. Phys. **114** (2001), 8282 - 8202.

[Ove1] M. L. Overton, *On minimizing the maximum eigenvalue of a symmetric matrix*, SIAM J. Matrix Anal. Appl. **9** (1988), 256-268.

[Ros1] C. Garrod, M.V. Mihailovic and M. Rosina, *The variational approach to the two-body density matrix*, J. Math. Phys. **16** (1975), 868-874.

[Ros2] M.V. Mihailovic and M. Rosina, *The variational approach to the density matrix for light nuclei*, Nucl. Phys. **A237** (1975), 221-228.

[Val1] F. Colmenero, C. Perez and C. Valdemoro, *Approaching q-order reduced density matrices in terms of the lower order ones I. General relations*, Phys. Rev. A **47** (1993), 971-978.

[Val2] F. Colmenero and C. Valdemoro, *Approaching q-order reduced density matrices in terms of the lower order ones II. Applications*, Phys. Rev. A **47** (1993), 979-985.

Chapter 1

ANALYTICAL HARTREE-FOCK WAVE FUNCTIONS FOR ATOMS AND IONS

Ajit J. Thakkar
Department of Chemistry, University of New Brunswick
Fredericton, New Brunswick, Canada E3B 6E2
ajit@unb.ca
http://www.unb.ca/chem/ajit/

Toshikatsu Koga
Department of Applied Chemistry, Muroran Institute of Technology
Muroran, Hokkaido 050-8585, Japan
koga@mmm.muroran-it.ac.jp

1. Introduction

Hartree-Fock (HF) theory is the fundamental model of atomic structure [1]. Even though it is now commonplace to consider electronic correlation and possibly relativistic effects, there is still room for applications of HF wave functions because they are accurate enough for many purposes. Moreover, they serve as a good starting point for the construction of more accurate wave functions. Although atomic HF wave functions can be obtained in numerical form [2], it is more convenient to tabulate and use wave functions expressed in some analytic form. Early work on analytical wave functions was reviewed by Slater [1].

Analytical approximations to HF wave functions can be obtained by solving the Roothaan-Hall equations [3, 4]. Procedures for obtaining analytical approximations to HF wave functions which are simultaneous eigenfunctions of $\hat{L}^2, \hat{S}^2, \hat{L}_z$ and $\hat{S}_z$ are well-developed for a single open shell [5], two open shells [6], and many open shells [7].

The Roothaan-Hall equations require a basis set to be chosen. Slater-type functions (STFs) are a conventional choice for the basis set in atomic computations even though Gaussian-type functions (GTFs) have become

E.J. Brändas and E.S. Kryachko (eds.),
Fundamental World of Quatum Chemistry, Vol. I, 587–599.

dominant in molecular calculations. The STF χ_{nlm} is defined by

$$\chi_{nlm}(\mathbf{r}; n, \zeta) = R_{nl}(r; n, \zeta) Y_{lm}(\theta, \phi) \tag{1.1}$$

in which $Y_{lm}(\theta, \phi)$ is a spherical harmonic, and the radial function R_{nl} is given by

$$R_{nl}(r; n, \zeta) = [(2n)!]^{-1/2}(2\zeta)^{n+1/2} r^{n-1} \exp(-\zeta r) \tag{1.2}$$

where the parameter ζ is called the exponent and n the principal quantum number. The $\{n\}$ are restricted to integers; non-integer n will not be considered here. Variation of the n and ζ parameters changes the basis set and hence the subspace of $L^2(\mathbf{R}^{3N})$, where N is the number of basis functions, on which the HF equations are projected and solved.

The first comprehensive and widely-used collection of analytical HF wave functions was that of Clementi [8, 9] for atomic species with $Z \leq 36$. This was later superseded by the more accurate and even more comprehensive compilation of Clementi and Roetti [10] which goes up to $Z = 54$. Clementi's tables are very widely used. A search of the ISI Web of Science [11] on May 27, 2002 found 2248 citations to Clementi's tables [8, 9], and 4186 citations to the Clementi-Roetti tables [10].

Many current publications make use of Clementi's tables even though they have been superseded in the last decade. The purpose of this brief chapter is to make the quantum chemistry and atomic physics community aware of more accurate analytical HF wave functions that have been freely available for some time now. Hartree atomic units are used throughout.

2. Single-zeta wave functions

Wave functions in minimal or "single-zeta" STF basis sets are of interest even though they do not give quantitatively accurate results. This is so because they provide physically intuitive quantities such as effective nuclear charges and screening constants. They are also used in many semi-empirical methods based on neglect of differential overlap. For these reasons, single-zeta wave functions are included in Clementi's tables. His single-zeta wave functions are based upon the tacit assumption that the principal quantum number n of an STF is the same as the principal quantum number of the atomic orbital (AO) it is approximating.

We [12] were able to produce significantly improved single-zeta wave functions for Al to Xe by dropping this restriction and treating both n and ζ in each STF as variational parameters. No improvements were found for the atoms from He to Mg. Our computations were made with our corrected version [13] of Pitzer's ATOM program [14] into which we

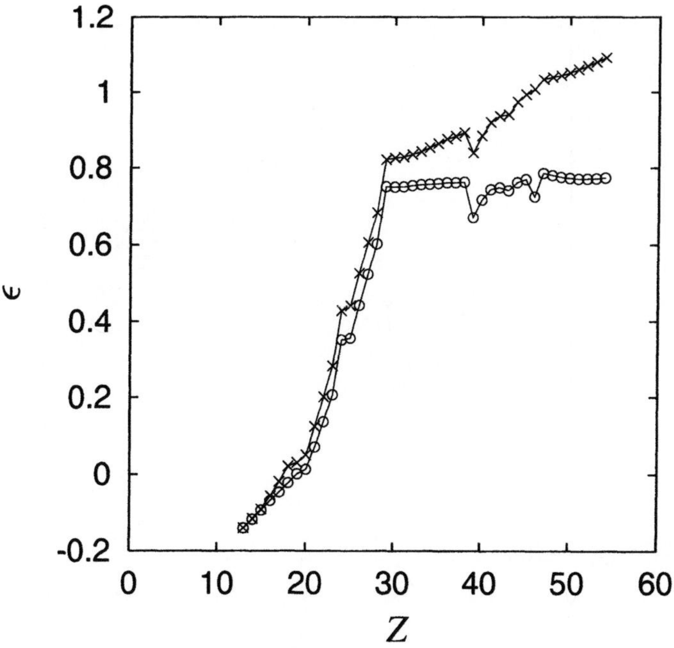

Figure 1.1. Energy errors, $\epsilon = \log_{10}(E - E_{\text{HF}})$, of the conventional ($\times$), and unconventional single-zeta ($\bigcirc$) wave functions for the ground states of the neutral atoms from Al to Xe.

have incorporated Powell's conjugate direction algorithm [15] for optimization of the exponents.

Since energy errors can vary over several orders of magnitude, we use a logarithmic error measure defined by

$$\epsilon = \log_{10}(E - E_{\text{HF}}) \tag{1.3}$$

with respect to the exact Hartree-Fock energy E_{HF} computed [16, 17, 18] with our enhanced version of Froese-Fischer's MCHF program [19].

Figure 1.1 shows that the conventional single-zeta wave functions had energy errors larger than 10 hartrees for $Z \geq 46$ reaching a maximum of 12.35 hartrees for Xe. By contrast, the largest error in our improved functions is 5.94 hartrees for Xe which is 6.41 hartrees less than the corresponding error for its conventional counterpart. We were gratified that the energies of the outermost valence orbitals were also improved in our single-zeta wave functions [12]. Moreover, some, but not all, of the unphysical positive d-orbital energies found for some transition-metal atoms in the conventional single-zeta approximation turned negative in our improved wave functions.

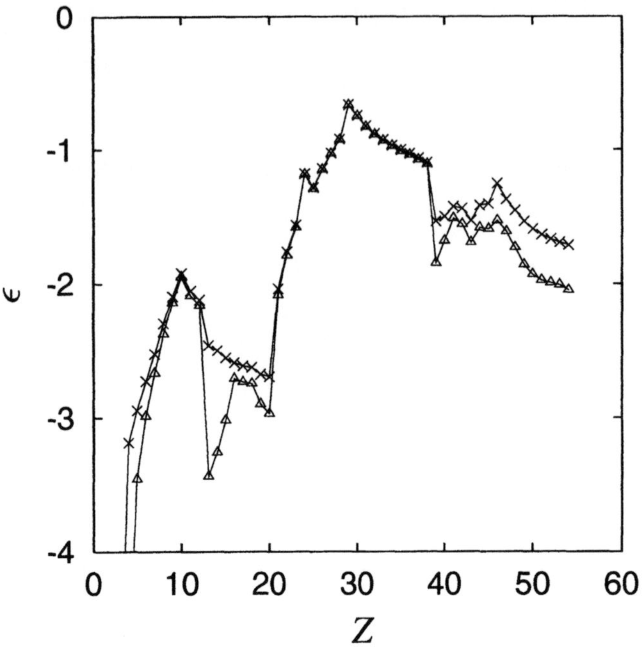

Figure 1.2. Energy errors, $\epsilon = \log_{10}(E - E_{\text{HF}})$, of the conventional ($\times$), and unconventional double-zeta ($\triangle$) wave functions for the ground states of the neutral atoms from He to Xe.

3. Double-zeta wave functions

We [20] then examined double-zeta STF basis sets which are exactly twice the size of single-zeta basis sets. Reoptimization of the Clementi-Roetti wave functions [10] with respect to the exponents brought about energy improvements ranging up to 4×10^{-3} hartrees for the Se atom. These reoptimized functions are hereafter referred to as conventional double-zeta wave functions. This investigation revealed that Clementi and Roetti [10] had used a vector coupling coefficient with the wrong sign for the $[\text{Kr}]5s4d^8(^4F)$ state of the Rh atom.

As in the single-zeta approximation, conventional double-zeta wave functions are constructed under the tacit assumption that both STFs that represent a given AO have their n equal to the principal quantum number of the AO. We [20] then lifted this restriction and optimized both the n and ζ parameters of the STF. We found improved wave functions for all the atoms from Li to Xe, but not for the He atom, in this manner.

The energy errors of the conventional and unconventional double-zeta wave functions with respect to the numerical HF limit are shown in Figure 1.2. The unconventional double-zeta wave functions have smaller

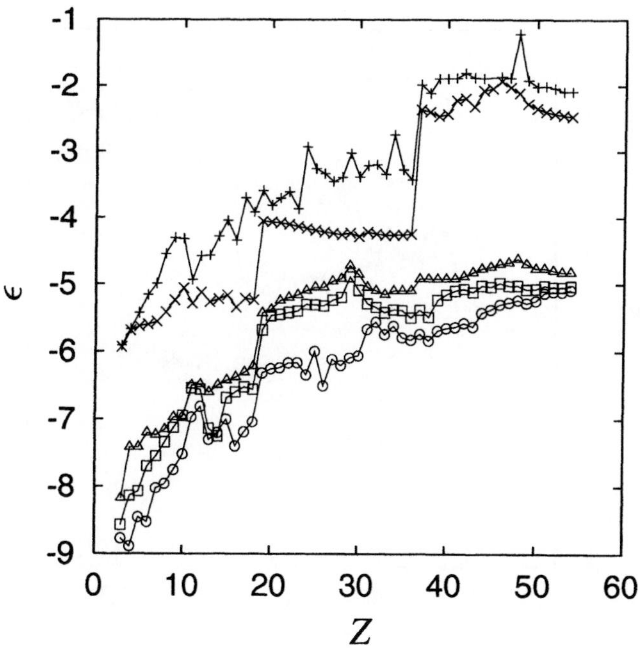

Figure 1.3. Energy errors, $\epsilon = \log_{10}(E - E_{\mathrm{HF}})$, of the Clementi-Roetti (+), K93 (×), Bunge et al. (△), K95 (□) and K98 (○) wave functions for the ground states of the neutral atoms from He to Xe. The Clementi-Roetti result for Rh is omitted because it is in error.

energy errors than the reoptimized conventional ones. The largest improvement was 0.0267 hartrees for Pd(^{1}S). The unconventional double-zeta wave functions also give improved energies for the valence orbitals for most, but not all, atoms. In particular, the valence orbital energies are worsened by the unconventional double-zeta approximation for Si, P, K, Ca, Ge, As and Zr.

4. Near Hartree-Fock wave functions

Analytical HF wave functions in larger STF basis sets are considered next. Improvements to the Clementi-Roetti wave functions for the ground states of the neutral atoms from He to Xe were made at roughly the same time by two groups unaware of each other's work.

Bunge and coworkers [21, 22] constructed analytical HF wave functions more accurate than those of Clementi and Roetti [10] by using larger basis sets, and by partial optimization of the $\{n\}$ quantum numbers of the STF. Koga and coworkers [23] used exactly the same number and type of STF as Clementi and Roetti had done. We then variation-

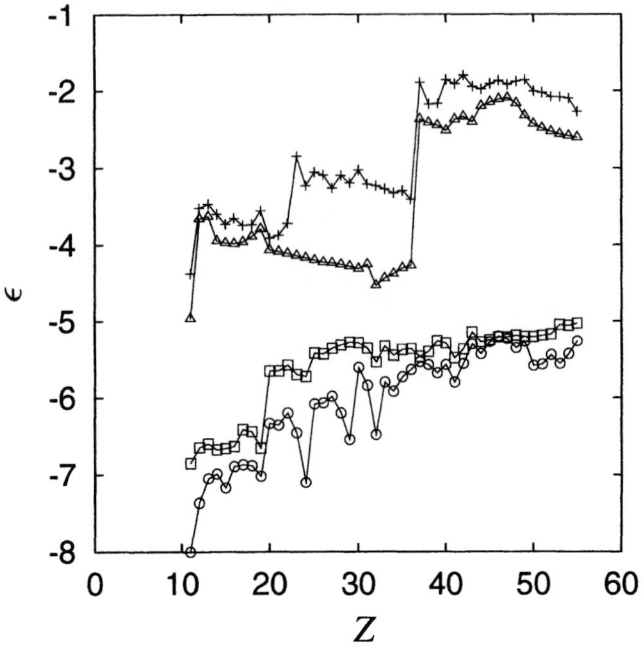

Figure 1.4. Energy errors, $\epsilon = \log_{10}(E - E_{\mathrm{HF}})$, of the Clementi-Roetti (+), K93 (△), K95 (□) and K98 (○) wave functions for the ground states of the cations from Na^+ to Cs^+.

ally reoptimized the $\{\zeta\}$ exponents of the STF while making minimal modifications to the $\{n\}$ quantum numbers of the STF as required to avoid computational linear dependence. We refer to the wave functions of Koga et al [23] as the K93 wave functions.

As in the case of double-zeta wave functions, Koga et al [23] found that Clementi and Roetti [10] had used a vector coupling coefficient with the wrong sign for the [Kr]5s4d^8(^{4}F) state of the Rh atom. Figure 1.3 shows that the energy errors in the K93 wave functions are both smaller and smoother with respect to the nuclear charge Z than those of Clementi and Roetti. Figure 1.3 also shows that the improvements obtained by Bunge and coworkers are greater than those obtained by Koga and coworkers. This is not surprising because the latter used smaller basis sets.

Koga and coworkers [24] also performed a similar reoptimization of the Clementi-Roetti wave functions for the ground states of the cations from Li^+ to Xe^+, and the anions from H^- to I^-. They also reoptimized a wave function for Cs^+ that had been reported by McLean and McLean [25] to cover all singly-charged ions with 54 electrons or less.

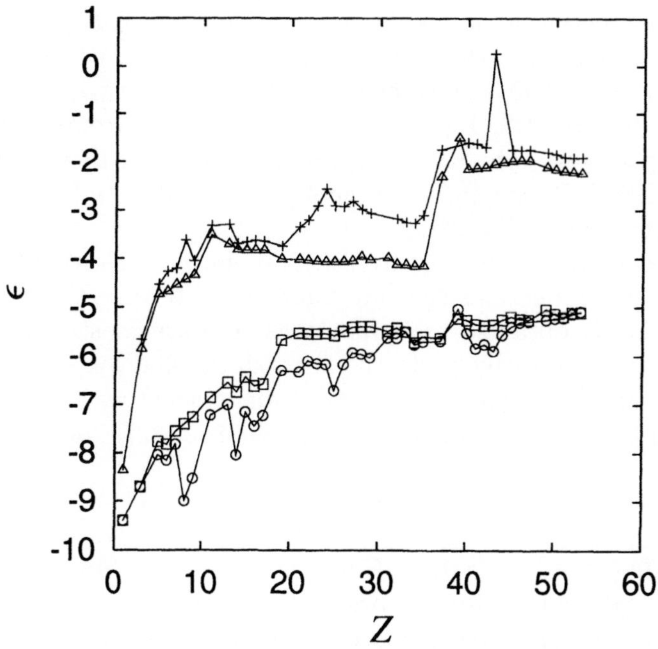

Figure 1.5. Energy errors, $\epsilon = \log_{10}(E - E_{\rm HF})$, of the Clementi-Roetti (+), K93 (△), K95 (□) and K98 (○) wave functions for the ground states of the anions from H^- to I^-. The Clementi-Roetti result for Tc^- is omitted because it is in error.

Figures 1.4 and 1.5 show that energy improvements of a few millihartrees were found for all ions with $Z \geq 37$. The energy improvement for Tc^- was exceptionally large (1.83 hartrees) and was traced to a probable input error by Clementi and Roetti. They [10] had reported a highly improbable exponent of 99.8674 for a 3p STF. If this exponent is reduced by a factor of 10 then a much more reasonable energy is obtained although it is still 0.1 hartrees above the value of Koga and coworkers [24]. A qualitative improvement was found for the stability of Ag^- which was correctly found to have a lower energy than Ag by the reoptimized K93 wave functions [23, 24] whereas the opposite is true for the Clementi-Roetti wave functions.

Two years later, we [16] reported improved wave functions, referred to as the K95 functions, for the ground states of all the neutral atoms from He to Xe, all the cations from Li^+ to Cs^+, and all the stable anions from H^- to I^-. Our wave functions for the neutral atoms use exactly as many STFs as those of Bunge [21, 22] but the $\{n\}$ and $\{\zeta\}$ were reoptimized. We used one extra STF of each symmetry for the anions to describe the diffuse atomic orbital occupied by the added electron and thus obtain anion wave functions of a quality comparable to the cations and neutral atoms. Generally, the K95 basis sets have smaller $\{n\}$ than the Bunge wave functions. The energy errors of these K95 wave functions can be seen in Figures 1.3, 1.4 and 1.5 for the neutral atoms, cations and anions respectively. The K95 energies are lower than those of Clementi and Roetti [10], Bunge and coworkers [21, 22], and the K93 energies of Koga and coworkers [23, 24].

Most recently, we [26] undertook the daunting task of creating analytical HF wave functions that are more accurate than the K95 and all other previously obtained wave functions, and are constrained to agree well, but not necessarily perfectly, with short- and long-range asymptotic conditions. In particular, each radial orbital was constrained to nearly satisfy the electron-nuclear cusp condition [27, 28]:

$$\lim_{r\to 0} \frac{\mathrm{d}f_{nl}(r)}{\mathrm{d}r} = -\frac{Z}{l+1} \lim_{r\to 0} f_{nl}(r) \tag{1.4}$$

in which $f_{nl}(r) = R_{nl}(r)/r^l$. Similarly, the wave function was constrained to satisfy the virial theorem [29, 30] closely.

We used even larger basis sets of STFs. We chose the $\{n, \zeta\}$ by minimizing, in the argot of optimization theory, a suitable "objective function" F instead of the energy. The objective function was defined as

$$F = E + (w_1/n_{\text{occ}}) \sum_{j=1}^{n_{\text{occ}}} |\gamma_j - 1| + w_2|V/T + 2| \tag{1.5}$$

in which E, V and T are the total, potential and kinetic energies, respectively; w_1 and w_2 are empirically adjusted, positive weights; $\gamma_j = -(l+1)f_j'(0)/[Zf_j(0)]$ is the cusp ratio for the jth radial orbital; and n_{occ} is the number of occupied radial orbitals. The second term of F is a "penalty function" that increases F whenever the orbital cusp ratios deviate from their exact values of unity. The third term of F is a penalty function that increases F whenever the virial ratio V/T of the wave function deviates from -2, the exact value predicted by the virial theorem [29, 30]. It is important to understand that the objective function minimization is used only to choose the basis set. The linear coefficients used to expand the atomic orbitals in terms of the basis set are found variationally by solving the restricted HF equations.

Moreover, each orbital was constrained to have good long-range behavior [31, 32, 33, 34] by requiring that:

$$R_{nl}(r) \asymp \exp\left(-\left(-2\varepsilon_{\text{h}}\right)^{1/2} r\right) \qquad \text{as } r \to \infty \tag{1.6}$$

where ε_{h} is the orbital energy of the highest occupied atomic orbital. The constraints on the long-range behavior were handled by simply setting, after each optimization cycle, the smallest STF exponents $\zeta_{\min,l}$ of each symmetry l to $(-2\varepsilon_h)^{1/2}$. The optimization cycles were repeated until the computed ε_h led to insignificant changes in the $\zeta_{\min,l}$.

The energy errors of the resulting wave functions, referred to as K98, can be seen in Figures 1.3–1.5 for the neutral atoms, cations and anions respectively. The K98 energies are only a few microhartrees above the numerical HF limit. They are lower than the Clementi-Roetti, Bunge et al., and K93 energies in all cases. They are also lower than the K95 energies in all but one of the 149 cases. The exception is Y^- where it is too difficult to simultaneously describe the loosely bound 5p orbital, get all the p-orbital cusps right, and get a lower energy than K95 without further enlargement of the basis set. The K98 energies for the cations are more accurate than for the neutral atoms; this is also true for the Clementi-Roetti, K93 and K95 energies. The anion energy errors are comparable to those for the neutral species only for the K95 and K98 wave functions in which slightly larger basis sets were used for the anions to achieve this desirable result.

An extended analysis was given by Koga and coworkers [26] to demonstrate the superiority of the K98 wave functions over the Clementi-Roetti functions, the Bunge et al. functions, and the K95 functions. Their analysis covered total energies, valence orbital energies, deviations from the virial theorem, cusp ratios, the electron density at the nucleus, long-range behavior, and moments of the electron density $\langle r^k \rangle$ for $-2 \leq k \leq 2$.

5. Heavy atoms

There are few compilations of analytic HF wave functions for the heavier atoms with $Z \geq 55$. McLean and McLean [25] reported a set of HF wave functions for the neutral atoms from Cs ($Z = 55$) to U ($Z = 92$). We will denote their functions by MM81. Koga and Thakkar [35] reoptimized the exponents $\{\zeta\}$ and obtained wave functions denoted KT93R with energy improvements averaging 2.8×10^{-3} hartrees as shown in Figure 1.6. They then constructed improved wave functions by adding an extra f-type STF to the basis set for the atoms from Ce ($Z = 58$) to Th ($Z = 90$), choosing the $\{n\}$ differently, and optimizing the exponents. These improved functions, denoted KT93I, led to energies lower than those of McLean and McLean by an average of 26.0×10^{-3} hartrees. The average energy errors with respect to the numerical HF limit [18] decreased from 28.49 millihartrees for the MM81 functions to 26.03 for the KT93R functions to 1.84 for the KT93I functions.

We [36] subsequently constructed an improved set of HF wave functions for the ground states of the atoms from Cs to U, and the first

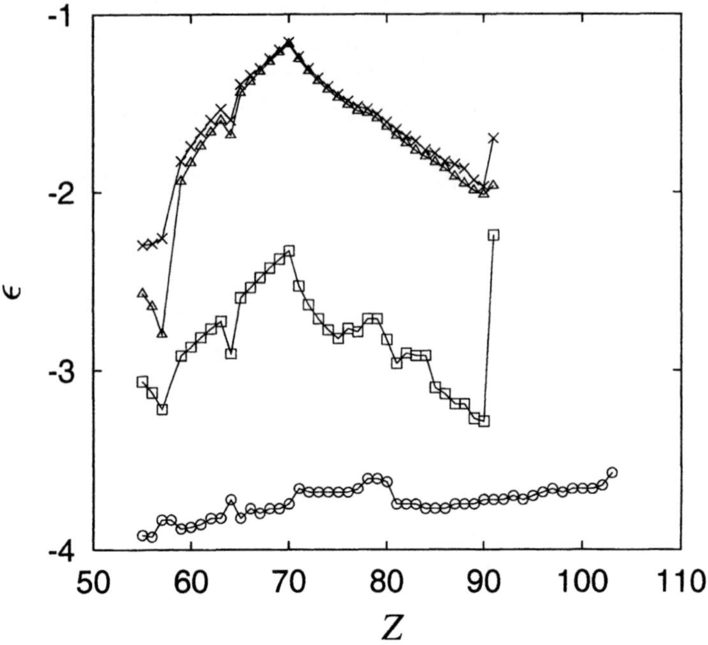

Figure 1.6. Energy errors, $\epsilon = \log_{10}(E - E_{\mathrm{HF}})$, of the MM81 ($\times$), KT93R ($\triangle$), KT93I ($\square$) and K98 ($\bigcirc$) wave functions for the ground states of the neutral atoms from Cs to Lr. Results for Ce ($Z = 58$) and U ($Z = 92$) are shown only for the K98 functions because different states are used in MM81, KT93R, and KT93I.

reported set of HF wave functions for the atoms from Np ($Z = 93$) to Lr ($Z = 103$). We worked with the experimental ground states which meant that states different from the previous work were considered for the Ce and U atoms. The average total energy error was 0.19 millihartrees—an order of magnitude better than the KT93I functions as shown in Figure 1.6. An extensive analysis of total energies, valence orbital energies, moments of the electron density $\langle r^k \rangle$ and moments of the electron momentum density $\langle p^k \rangle$ for $-2 \leq k \leq 4$, the electron density at the nucleus, and the electron momentum density at zero-momentum showed that the K98 functions were the most accurate for all the considered properties except $\langle r^{-2} \rangle$ for which KT93I was slightly more accurate.

6. Other recent work

Clementi and Roetti [10] tabulated HF wave functions for many isoelectronic sequences. Their results have been improved for the isoelectronic series of He, Li, Be, B, C, N, O, F and Ne by Koga and coworkers [37]. Their energy errors do not exceed 1 microhartree as compared with a maximum energy error of 1980 microhartrees for the Clementi-Roetti wave functions. We placed special emphasis on designing the wave functions so that they behave smoothly as a function of Z. Our wave functions are therefore well suited for studies of changes in properties across these isoelectronic series.

Even-tempered wave functions for the neutral atoms of the third and fourth rows were reported by Koga et al. [38]. More accurate doubly even-tempered (DET) wave functions for all the neutral atoms [39] and singly-charged ions [40] with fewer than 55 electrons were also reported. The DET wave functions are roughly of the same accuracy as the K93 functions.

7. Summary

The venerable compilations of Clementi and Roetti [10] and McLean and McLean [25] are now largely outdated. The more recent compilation of Bunge and coworkers [22] has also been superseded. Our HF wave functions are better optimized and available in a variety of basis set sizes, including sizes identical to those used in the Clementi and Roetti, McLean and McLean, and Bunge et al. collections. These improvements are not just minor quantitative ones. For example, the best Clementi and Roetti wave function for Si produces a value of the curvature of the electron momentum density $\Pi''(0)/2$ that is positive (+1.10) [41] whereas the true HF value is negative (−0.51) [42].

All our wave functions are freely available over the Internet [43]. If Internet access is limited, then the wave functions can be obtained by contacting either of the authors of this article.

Acknowledgments

It is an honor to dedicate this article to the memory of Per-Olov Löwdin who contributed so much to the development and teaching of quantum chemistry, including the topic of this article [44, 45, 46]. One of us (AJT) received a rigorous education at the 1972 summer school organized by Professor Löwdin.

Our papers that are reviewed here would not have been possible without the contributions of our coauthors: Takashi Imai, Katsutoshi Kanayama, Masaya Omura, Yousuke Seki, Eri Shibata, Hiroshi Tatewaki, the late Hirohide Teruya, Shinya Watanabe, Tadanori Watanabe and Ryuji Yasuda.

This work was supported in part by the Natural Sciences and Engineering Research Council of Canada, and in part by a Grant-in-Aid for Scientific Research from the Ministry of Education of Japan.

References

[1] J. C. Slater, *Quantum Theory of Atomic Structure* (McGraw Hill, New York, 1960), Vols. I, II.

[2] C. Froese-Fischer, *The Hartree-Fock Method for Atoms* (Wiley, New York, 1977).

[3] C. C. J. Roothaan, Rev. Mod. Phys. **23**, 69 (1951).

[4] G. G. Hall, Proc. Roy. Soc. (Lond.) **A205**, 541 (1951).

[5] C. C. J. Roothaan, Rev. Mod. Phys. **32**, 179 (1960).

[6] S. Huzinaga, Phys. Rev. **122**, 131 (1961).

[7] C. C. J. Roothaan and P. S. Bagus, Meth. Comput. Phys. **2**, 47 (1963).

[8] E. Clementi, Res. Report RJ-256, IBM, San Jose, CA (1963).

[9] E. Clementi, IBM J. Res. Develop. **9**, 2 (1965).

[10] E. Clementi and C. Roetti, At. Data Nucl. Data Tables **14**, 177 (1974).

[11] ISI (Institute for Scientific Information) Web of Science, Thomson Scientific, `http://woscanada.isihost.com/`

[12] T. Koga and A. J. Thakkar, Theor. Chim. Acta **85**, 363 (1993).

[13] T. Koga and A. J. Thakkar, Theor. Chim. Acta **85**, 391 (1993).

[14] R. M. Pitzer, QCPE Bulletin **10**, 14 (1990).

[15] M. J. D. Powell, Comput. J. **7**, 155 (1964)

[16] T. Koga, S. Watanabe, K. Kanayama, R. Yasuda, and A. J. Thakkar, J. Chem. Phys. **103**, 3000 (1995).

[17] T. Koga, H. Tatewaki, and A. J. Thakkar, J. Chem. Phys. **100**, 8140 (1994).

[18] T. Koga, S. Watanabe, and A. J. Thakkar, Int. J. Quantum Chem. **54**, 261 (1995).

[19] C. Froese-Fischer, Comput. Phys. Commun. **4**, 107 (1972).

[20] T. Koga, Y. Seki, and A. J. Thakkar, Bull. Chem. Soc. Jpn. **66**, 3135 (1993).

[21] C. F. Bunge, J. A. Barrientos, A. V. Bunge, and J. A. Cogordan, Phys. Rev. A **46**, 3691 (1992).

[22] C. F. Bunge, J. A. Barrientos, and A. V. Bunge, At. Data Nucl. Data Tables **53**, 113 (1993).

[23] T. Koga, H. Tatewaki, and A. J. Thakkar, Phys. Rev. A **47**, 4510 (1993).

[24] T. Koga, Y. Seki, A. J. Thakkar, and H. Tatewaki, J. Phys. B **26**, 2529 (1993).

[25] A. D. McLean and R. S. McLean, At. Data Nucl. Data Tables **26**, 197 (1981).

[26] T. Koga, K. Kanayama, S. Watanabe, and A. J. Thakkar, Int. J. Quantum Chem. **71**, 491 (1999).

[27] E. Steiner, J. Chem. Phys. **39**, 2365 (1963).

[28] D. P. Chong, J. Chem. Phys. **47**, 4907 (1967).

[29] V. A. Fock, Z. Phys. **63**, 855 (1930).

[30] P.-O. Löwdin, J. Mol. Spectrosc. **3**, 46 (1959).

[31] N. C. Handy, M. T. Marron, and H. J. Silverstone, Phys. Rev. **180**, 45 (1969).

[32] G. S. Handler, D. W. Smith, and H. J. Silverstone, J. Chem. Phys. **73**, 3936 (1980).

[33] C. L. Davis, H. J. A. Jensen, and H. J. Monkhorst, J. Chem. Phys. **80**, 840 (1984).

[34] T. Ishida and K. Ohno, Theor. Chim. Acta **81**, 355 (1992).

[35] T. Koga and A. J. Thakkar, Phys. Rev. A **48**, 4775 (1993); Phys. Rev. A **50**, 891 (1994).

[36] T. Koga, K. Kanayama, T. Watanabe, T. Imai, and A. J. Thakkar, Theor. Chem. Acc. **104**, 411 (2000).

[37] T. Koga, M. Omura, H. Teruya, and A. J. Thakkar, J. Phys. B **28**, 3113 (1995).

[38] T. Koga, H. Tatewaki, and A. J. Thakkar, Theor. Chim. Acta **86**, 477 (1993).

[39] T. Koga, H. Tatewaki, and A. J. Thakkar, Theor. Chim. Acta **88**, 273 (1994).

[40] T. Koga, E. Shibata, and A. J. Thakkar, Theor. Chim. Acta **91**, 47 (1995).

[41] A. J. Thakkar, A. L. Wonfor, and W. A. Pedersen, J. Chem. Phys. **87**, 1212 (1987).

[42] T. Koga and A. J. Thakkar, J. Phys. B **29**, 2973 (1996).

[43] Thakkar home page, `http://www.unb.ca/chem/ajit/` In 2003, the URL may change to `http://www.unbf.ca/chem/ajit/`

[44] P.-O. Löwdin, Phys. Rev. **90**, 120 (1953).

[45] P.-O. Löwdin, Phys. Rev. **94**, 1600 (1954).

[46] P.-O. Löwdin and K. Appel, Phys. Rev. **103**, 1746 (1956).

The Origin of the Molecular Atomization Energy Explained with the HF and HF-CC Models.

E. Clementi
a Alexander von Humboldt-Stiftung contribution
permanent address: Via Carloni, 38, 22100 Como, Italy
e-mail: clementi@fis.unico.it

G. Corongiu
Universita' degli Studi dell' Insubria. Dipartimento di Scienze CC.FF.MM.
Via Lucini 3, I-22100 Como, Italy

Abstract. The Hartree-Fock-Clementi-Corongiu method, HF-CC, is revisited, aiming at an unified formulation for post Hartree-Fock energy computations in atomic and molecular systems. A previous HF-CC *molecular functional*, (Clementi, E.; Corongiu, G. *Theochem* **2001**, *543*, 39) is modified and re-calibrated with a new optimization of the parameters, and tested with a sample of 131 molecules, including radicals, H-bond and van der Waals systems. For this sample of 131 molecules, the computed HF-CC atomization energies have an average standard deviation of 1.89 kcal/mol.
The origin of molecular binding is analyzed with both the HF and the HF-CC models. The atomization energy is decomposed into "the HF classical" energy (the sum of the HF nuclear-electron, HF kinetic and HF Coulomb energies), "the HF exchange" energy and correlation energy; the latter is computed with the HF-CC scaling functional constructed with atomic, covalent, ionic and van der Waals contributions. The correlation energy is first decomposed into the atomic and the molecular components, and these are further decomposed into nuclear-electron, kinetic, coulomb and exchange contributions; these decompositions characterize the HF-CC model and are used to explain the origin of the chemical bond. Computations on van der Waals systems, ranging from He-He to Ar-naphtalene, show the validity of the HF-CC method also for the computations of long range weak interactions.

E.J. Brändas and E.S. Kryachko (eds.),
Fundamental World of Quatum Chemistry, Vol. I, 601–629.

1. Introduction. Theoretical chemistry is more and more tightly connected with computational chemistry; with the former we postulate models, with the latter we verify the model ability at first to reproduce and thereafter to predict chemical observations. A basic quantity, which characterize any molecular system, is its binding energy. Today we have made notable progress in both theoretical and computational chemistry, but we lack a reasonably simple but general description for the binding of the atoms in classical molecules (formed for example with the first and second row atoms), for weaker binding (like the H-bonds), and for even weaker bonds, like those in van der Waals complexes. To accomplish this task we need not only a model, but also a computer program reasonably fast and able to reproduce molecular binding with accuracy and also conceptually and computationally flexible to allow analyses of its components. This takes time to be developed.

Since the days when chemistry had only an approximated theory, mainly the LCAO-MO, and when the Valence Bond theory was still vigorous, Per-Olov Löwdin main research preoccupation was the solution of the Schroedinger equation using rigorous quantum mechanical algorithms. However, the "Gainesville spirit" he created, was always characterized by sympathy and hospitality to shortcuts on the notable mathematical complexity of *ab-initio* models, generally leading to semi-empirical parameters inserted *ad-hoc* at some given level of a theoretical formulation.

The advent and extraordinary development of the computer industry, both hardware and software, made it possible in the early sixties the routine use of the Hartree-Fock approximation, first to atoms [1] and diatomic molecules [2] and then to larger systems either molecules or supermolecules [3]. But the *ab-initio* solution of the correlation energy problem for the very large molecular systems of current interest has not progressed sufficiently. Löwdin in one of his classical papers [4] reminded us that to obtain reliable numerical estimates on the correlation energy for an electron-pair one could consider Wigner proposal [5], a perturbation algorithm, in which the correlation energy is computed as a functional of the Hartree-Fock electronic density. Löwdin's view and Wigner's proposal were taken up by Clementi [1]; however, the correlation energy was not computed as a perturbation on the Hartree-Fock wave-function but was considered an integral part of the Hartree-Fock self consistent field process, adding to the Fock Hamiltonian a functional, representing the Coulomb hole effect, the origin of the correlation energy [4,5]. We recall that the Coulomb hole algorithm [6] makes use of empirical parameters and that the resulting wave function, the Clementi-determinant, is only formally equivalent to the HF determinant. Today, the computational chemistry community makes use of the Kohn-Sham[7] determinant, which differs from the Clementi-determinant in the adoption of a different functional, for example, the B3LYP functional [8].

Clearly, for those chemical systems for which the HF representation is inade-

quate, due to static correlation, the "generalized Hartree-Fock" model [9-12] (i.e. a multi-configuration-scf approach, but with only a few determinants) must be used.
Recently, Clementi and Corongiu departed from these avenues and simply scaled the Hartree-Fock total energy, with explicit scaling factors for the nuclear-electron, kinetic, Coulomb and exchange energies [13-16]; this new approach constitutes the HF-CC approximation.
In this work the different proposals presented in Refs. 13 to 16, are re-examined and modified, yielding a more coherent HF-CC theory for atoms and molecules, including also van der Waals interactions; the theory is tested with a large number of molecular systems, and the computed atomization energy is analyzed in detail.

2. Scaling the Hartree-Fock energy. The HF-CC process is an iterative process which starts at the conclusion of the Hartree-Fock self-consistent-field process. Let us designate with $\phi_{HF}(i)$, $\epsilon_{HF}(i)$, H_{HF}, H, Ψ_{HF}, and Ψ the i-th HF orbital, its orbital energy, the HF Hamiltonian, the Schroedinger Hamiltonian, the HF wave-function and the Schroedinger exact non-relativistic wave-function, respectively. We recall the standard HF equations for the initial trial i-th HF orbital:

$$< \phi^0_{HF}(i)|H_{HF}|\phi^0_{HF}(i) >= \epsilon^0_{HF}(i) \qquad (1)$$

leading, at the end of the optimization of the variational parameters (SCF convergence at the m-th iteration) to

$$< \phi^m_{HF}(i)|H_{HF}|\phi^m_{HF}(i) >= \epsilon^m_{HF}(i) \qquad (2)$$

The HF wave function,Ψ_{HF}, built with the $\phi^m_{HF}(i)$, has the energy:

$$< \Psi_{HF}|H|\Psi_{HF} >= E_{HF} \qquad (3)$$

At this point we start with the HF-CC iterative process. The trial i-th HF-CC orbital $\phi^0_{HFCC}(i)$ is defined as:

$$\phi^0_{HFCC}(i) = \phi^m_{HF}(i) \qquad (4)$$

and the HF Hamiltonian is replaced with H_{HFCC}, a scaled HF Hamiltonian, of the form H_{HFCC}=$H_{HF}(1+S)$, where S is the scaling factor, assumed to be a small perturbation, defined later on as a functional of Ψ_{HF}. An equivalent form of the scaled HF Hamiltonian, explicitly decomposes H_{HF} into its four components H_1, H_2, H_3, H_4, i.e. the nuclear-electron, the kinetic, the coulomb and the exchange components, respectively, and applies scaling factors S_ν (with ν=1,2,3,4) to each component.

The beginning of the HF-CC self-consistent-field process starts with the HF-CC equations

$$< \phi^0_{HFCC}(i)|H_{HF}(1+S)|\phi^0_{HFCC}(i) >= \epsilon^0_{HFCC}(i) \tag{5}$$

After re-optimization of the variational parameters the HF-CC SCF process is concluded at the n-th iteration yielding the HF-CC orbitals

$$< \phi^n_{HFCC}(i)|H_{HF}(1+S)|\phi^n_{HFCC}(i) >= \epsilon^n_{HFCC}(i) \tag{6}$$

With these orbitals one constructs Ψ_{HF-CC}, the HF-CC determinantal wave-function (the HF-CC determinant, for short) with energy E_{HFCC}. Note, that S corresponds to a small perturbation, therefore (1-S) is near to unity, and consequently Ψ_{HFCC} is nearly equal to Ψ_{HF}, and E_{HFCC} is nearly equal to E_{HF}. Let us define the energy ΔE trough the equality

$$\Delta E = E_{HFCC} - E_{HF} \tag{7}$$

Then

$$E_{HFCC} = E_{HF} + \Delta E \tag{8}$$

The HF-CC technique attempts to construct S such that

$$\Delta E = E_{corr} \tag{9}$$

To satisfy Eq. 9, we use a functional of the HF function and empirical parameters discussed in Sections 4 and 5.

3. Analyses of the correlation energy from experiments and HF computations. It seems reasonable that in order to devise a theoretical model for the computation of the correlation energy, we should consider, *firstly*, how large it is relative to the other components of the total energy, both in atoms and molecules, *secondly*, to look for eventual general trends, which might suggest a strategy for the formulation of the model.
This can be done rather readily, as shown below; however, it is convenient first to recall a few simple relations. We indicate as $E_T(a)$, $E_{HF}(a)$, $E_C(a)$, $E_R(a)$, the exact total energy, and the Hartree-Fock, correlation and relativistic energies components of atom "a" in its ground state; for molecular systems the same notation is used, but without the index '(a)'. The following relations are most familiar:

$$E_T(a) = E_{HF}(a) + E_C(a) + E_R(a) \tag{10}$$

$$E_T = E_{HF} + E_C + E_R \tag{11}$$

and

$$E_{HF} = \sum_a E_{HF}(a) + \Delta E_{HF} \tag{12}$$

$$E_C = \sum_a E_C(a) + \Delta E_C \tag{13}$$

$$E_R = \sum_a E_R(a) + \Delta E_R \tag{14}$$

For negligible small ΔE_R one approximates

$$E_R = \sum_a E_R(a) \tag{15}$$

From these relations one obtains:

$$E_T = \sum_a E_T(a) + \Delta E_{HF} + \Delta E_C + \Delta E_R \tag{16}$$

The quantities ΔE_{HF}, ΔE_C, ΔE_R are the HF, the correlation and the relativistic contribution to molecular binding, ΔE_R is often neglected for low Z atoms, since relatively small.
We add one more definition, used later on. For the components of the HF energy we use the notation $E_{HF}(ne)$, $E_{HF}(k)$, $E_{HF}(coul)$, $E_{HF}(ex)$, and $E(nn)$ for the HF nuclear-electron, HF kinetic, HF coulomb, HF exchange energy, and for the nuclear-nuclear repulsion, respectively; in addition we designate as "HF classical energy", $E_{HF}(class)$, the energy sum of the following components

$$E_{HF}(class) = E_{HF}(ne) + E_{HF}(k) + E_{HF}(coul) + E(nn) \tag{17}$$

thus

$$E_{HF} = E_{HF}(class) + E_{HF}(ex) \tag{18}$$

Note that with Eq. 18 we re-state that the HF model is the original "classical Hartree" model plus its Hartree-Fock quantum correction, the exchange energy.
Now we can return to the initial questions of this section. Concerning the first point (how large is the correlation correction) we simply compare the correlation correction with the other components of the total energy (see Eqs. 11 and 12).
In Fig. 1 this is done for *atoms*. The top-left inset reports (in hartrees) the energy difference between Hartree-Fock and Dirac-Fock, i.e. the relativistic

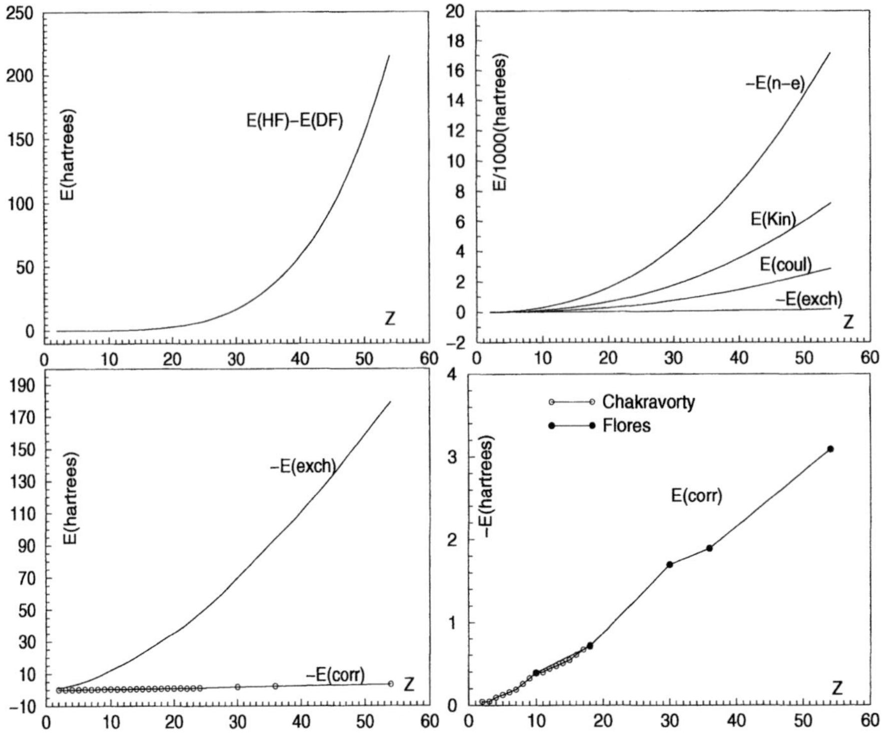

Figure 1. Top left inset: energy difference between Hartree-Fock and Dirac-Fock. Top right inset: $E_{HF}(ne)$, $E_{HF}(k)$,$E_{HF}(c)$, and $E_{HF}(ex)$. Bottom left inset: Hartree-Fock exchange, and correlation energy. Bottom right inset: correlation energy. All energies are in hartrees.

energy; one can readily see that the difference is very large, particularly relative to binding energies in molecules. The top-right inset reports (in hartrees) $E_{HF}(ne)$, $E_{HF}(k)$, $E_{HF}(coul)$, and $E_{HF}(ex)$; note that the latter can hardly be seen with the adopted scale. Therefore we change scale, and in the bottom-left inset we report again the exchange energy and also the correlation energy, not given above since too small. The correlation energies from Z=2 to Z=18 are taken from Chakravorty et al. [17] refining Clementi data of the early sixties [18]; for Z=10, 18, 36 and 52 we report the data from J. R. Flores [19], and for Z=30 unpublished data by Corongiu (see Ref. 13 for details). Since again the scale is too large, we report the correlation data once more on the bottom-right inset. From Fig. 1 one concludes that the *correlation energy is a very small fraction* of the total energy in atoms, the more so the larger is Z; a second conclusion is that, energetically, an atom is mainly a classical electrostatic system with special boundary conditions, the quantization rules; indeed, $E_{HF}(class)$ is by far larger than the sum of $E_{HF}(ex)$ and E_C.

For molecules the total energy is the sum of the atomic energies plus the experimental atomization energy (obtained from the experimental enthalpy of formation at 298 K, corrected for zero point energy and temperature factor). Note that we either compute the relativistic correction or we assume that for light atoms the molecular relativistic correction is reasonably ap proximated by the sum of the relativistic correction of the component atoms at dissociation. As pointed out long ago [20] if we require accuracy up to a few decimals of a kcal/mol this approximation starts to break down for molecules with atoms with atomic number $Z > 13$ (E_R reaches the value of E_C). To simplify the notation, we shall drop the explicit reference to relativistic corrections, since Table 1 contains mostly molecules with low Z atoms.

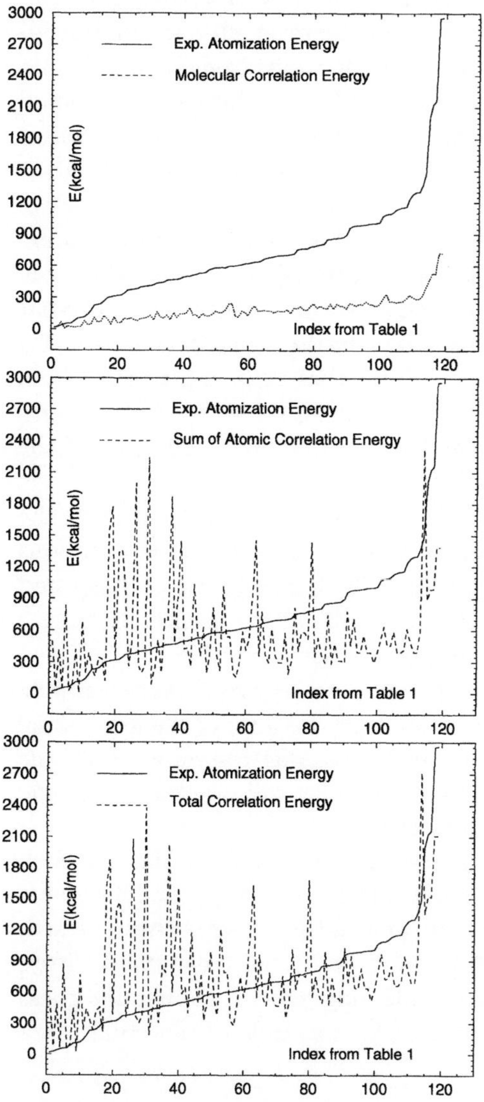

Figure 2. Correlation and atomization energies for the molecules of Table 1. Top inset: atomization energy and ΔE_C. Middle inset: atomization energy and $\sum_a E_C(a)$. Bottom inset: atomization energy and E_C.

In Fig. 2 we compare the correlation energy and the atomization energy for the first 112 molecules of Table 2, ordered for increasing binding energy (atomization energy). Recalling that E_C is the sum of $\sum_a E_C(a)$ and ΔE_C (see Eq. 13) we compare the atomization energy with ΔE_C, top inset, with $\sum_a E_C(a)$, middle inset, and with E_C, bottom inset. For the molecules we have considered $\sum_a E_C(a)$ and E_C are, *grosso modo*, of the same magnitude of the molecular binding, but with a very different and irregular pattern relative to the binding pattern. This "irregularity" is a consequence of the fact that only the valence electrons

participate to molecular binding, but $E_C(a)$ increases proportionally to Z; thus molecules like H_2O and H_2S with similar atomization energies, namely 232.33 and 182.46 kcal/mol, respectively, have notably different $\sum_a E_C(a)$, 161.88 and 380.41 kcal/mol, respectively. Note, however, that we are considering a sample of molecules composed mainly of atoms from Z=1 to Z=10, and only a few with $Z > 10$; a sample of molecules with heavy atoms would show a $\sum_a E_C(a)$ consistently larger than the corresponding binding energy.
From the top inset of Fig. 2 we learn that ΔE_C is nearly a constant fraction of the molecular atomization, as previously noted [15,16]. (In Section 6 we shall return on this decomposition.) Indeed, this interesting result is expected, recalling G.N. Lewis pairing theory: unpaired electrons in the atom are paired in the formation of the molecular bond; further we know that for each new binding pair there is a roughly constant contribution to ΔE_C [5]. One can also deduce from the above inset that ΔE_C must result from "localized effects"; indeed, if the delocalization would be very important, then ΔE_C would increase primarily and proportionally to the size of the molecule, contrary to the data of the inset.
From the middle and bottom insets of Fig. 2 it appears as most evident the reason for the well known difficulties encountered in *ab−initio* computations of E_C, since no distinction is made between the computation of ΔE_C and the computation of $\sum_a E_C(a)$. Indeed, it is difficult to accurately ensure that the computation of ΔE_C does not include some fraction of $\sum_a E_C(a)$ and vice versa; to avoid this one resorts to compute fully E_C, a very complex task, even in moderate size molecules. Note that, because of the impressing advances of the computer industry, including parallel computers [21], the computation of E_C for increasingly large molecules will eventually become feasible even retaining today *ab-initio* methods.
In conclusion, the experimental molecular atomization and the Hartree-Fock energies point strongly to a computational correlation energy model capable to clearly distinguish between $\sum_a E_C(a)$ and ΔE_C. Alternatively stated, we should distinguish between "atomic pair" and "molecular pair". Note that this distinction was used by Clementi long ago to empirically guess the correlation correction for HF binding; ΔE_C was called "molecular extra correlation energy" [22].

4. The scaling factor for atomic systems. In this section we comment on the algorithm for the computation of the scaling factor $S(Z,\Gamma,C)$ for an atom with atomic number Z, state Γ, and configuration C [13,14]. The Coulomb hole effect, the correlation correction origin, is assumed to be proportional to the probability for electron p (with spin α) to "share the space" with electron q (with spin β). Thus, we have assumed [13] that the sum of the product of the Hartree-Fock radial density ρ_p of electron p times the radial density ρ_q of electron q, over all possible pairs pq, is a reasonable starting function to compute the correlation energy. Accordingly, the atomic scaling factor $(1 + S)$ of Eq. 5 is constructed with the functional:

$$S(Z, \Gamma, C)_\nu = \alpha_\nu [S(Z)]^{\beta_\nu} / N^{\gamma_\nu} \tag{19}$$

where α, β, and γ are three semi empirical constants valid for all the neutral atoms of the periodic system, N is the number of electrons, ν=1, 2, 3, 4 for the nuclear-electron, kinetic, coulomb and exchange operator, respectively, and

$$S(Z) = \sum_p \sum_q \int \rho_p \rho_q d\rho_p d\rho_q \tag{20}$$

For positive and negative ions there are equivalent expressions, but with different values for the α, β, and γ semi empirical constants [13]. With these scaling factors the correlation energy for the atoms from Z=1 to Z=54 was computed [13] and found qualitatively close to the correct values; to improve on it was assumed that Eq. 19 is only the first term of a more complex expression. With Eq. 19 the average standard deviation between fitted and the best available [18,19] correlation energies was 10.03 kcal/mol for neutral atoms and 8.45 kcal/mol for singly positive ions and 6.66 kcal/mol for doubly positive ions [13].
Since the accuracy of the fitting is insufficient for molecular computations, we aim to compute atomization energies with an accuracy of 1 to 2 kcal/mol (see later on) we have kept fixed the values of the constants and for each atom we have fitted S(Z) to the best available correlation energies.
As we well know, the HF energy can be decomposed into its components: the nuclear-electron, the kinetic, the coulomb and the exchange energies. The same can be done for the HF-CC energy, and therefore, for the correlation energy. Below we define the components of the correlation energy:

$$\Delta(ne) = \Delta E_{HFCC}(ne) - \Delta E_{HF}(ne), \tag{21}$$

$$\Delta(k) = \Delta E_{HFCC}(k) - \Delta E_{HF}(k), \tag{22}$$

$$\Delta(coul) = \Delta E_{HFCC}(coul) - \Delta E_{HF}(coul) \tag{23}$$

$$\Delta(ex) = \Delta E_{HFCC}(ex) - \Delta E_{HF}(ex). \tag{24}$$

In Fig. 3 we decompose the correlation energy into the sum of the above contributions. *These contributions decrease in absolute value from nuclear-electron, to kinetic to coulomb to exchange, as expected for a scaling model.* The physical interpretation is that the scaling brings about a compression

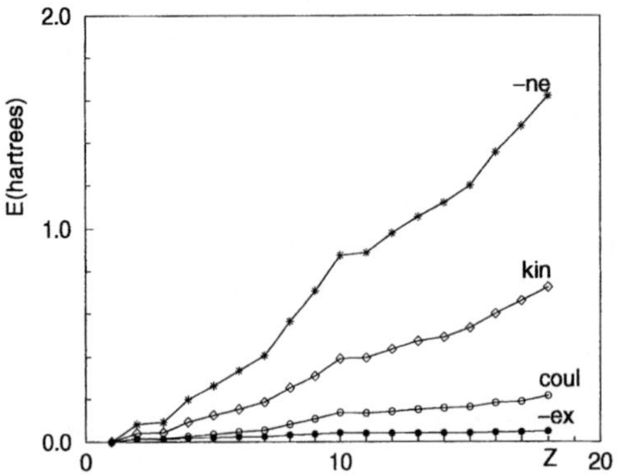

Figure 3. Decomposition of atomic correlation energy into its components.

of the electronic density around the nucleus, thus an increase in the nuclear-electron attraction and a proportional variation in the other components, as demanded by the virial theorem. In Fig. 4 the HF-CC electronic density compression, relative to HF, is exemplified for the Helium, Neon and Argon atoms; we report the difference in $\rho(1s)$, $\rho(2s)$, $\rho(2p)$, $\rho(3s)$, $\rho(3p)$ and $\rho(total)$ between HF-CC and HF-CC. Note that the difference in the density (the compression) has the maximum around the nucleus due to our choice of the functional; indeed we use the same scaling for all electrons, inner shell or valence, i.e. we operate an "average scaling" which will effect preferentially more and more the inner shells the higher the atomic Z. Indeed, for Ar the 3s and 3p density is nearly unchanged.

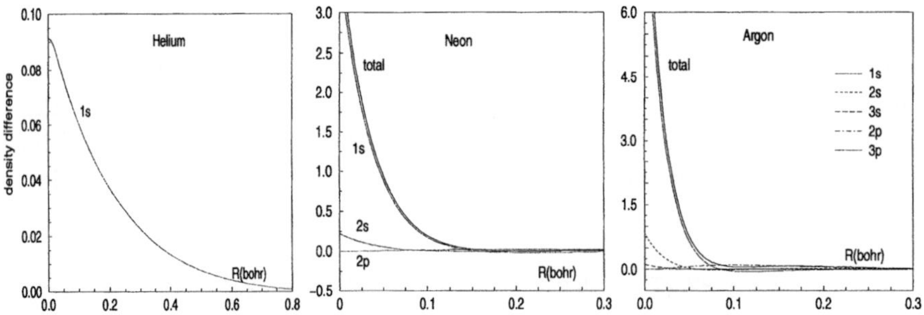

Figure 4. Difference between HF-CC and HF electronic densities for the Helium, Neon and Argon atoms.

5. Scaling factor for an atom in a molecular system. We have learned that *an atom in a molecule* has an electronic structure dissimilar from the one in the *separated atom*; there are ample experimental evidences that the dissimilarity concerns mostly the open shell electrons (for the atom in a molecule), or, equivalently, it mainly concerns the bonding regions. We have also learned that, when energetically advantageous, an electron (or a fraction of electron) can transfer from atom "a" to atom "b". Below we shall label Z'_a the electronic charge of atom "a" in a molecule (at the nuclear

coordinates of interest). Finally, we have learned that atomic orbitals in the molecule can hybridize (from ground state to valence state), and that atomic orbitals of one atom can combine with atomic orbitals of another atom. All this is at the base of the Valence Bond and of the LCAO-MO (today Hartree-Fock) approximations. We propose to use these simple physical chemistry concepts in deriving a functional.
To start with, it is obvious that the correlation energy for "the atom 'a' in a molecule" will not be equal to the correlation energy of the free atom 'a', and consequently its scaling factors will be different from those given in Eq. 19. At the same time the inner shell electrons will have essentially the same correlation as in the separated atom. We propose the following form for the *molecular scaling factor* for atom "a" with nuclear charge Z_a and electronic charge Z'_a in the molecular field M:

$$[1 - \alpha_\nu S_\nu(Z, Z', a, M)^{\beta_\nu}/n^{\gamma_\nu}] \tag{25}$$

Tentatively, Z'_a is equated to Mulliken's gross charge [23] obtained from the HF function. The HF-CC molecular functional is given below:

$$\begin{aligned} S_\nu(Z, Z', a, M) &= S_\nu(Z, a) + (Z'_a - Z_a) \times \Delta S(Z, a) \\ &\quad + k(Z_a) \sum_b B_{a,b} f(r_{ab}) + w(Z_a) \end{aligned} \tag{26}$$

where $S_\nu(Z, a)$ is the atomic functional $S_\nu(Z)$ for atom "a" given in Eq. 19; ΔS designates the difference between the neutral atom value for $S_\nu(Z, a)$ and its corresponding ionic values $S_\nu(Z, a^+)$ or $S_\nu(Z, a^-)$, respectively. Further, in a molecule for each atom pair 'a' and 'b' we define the bond order matrix element, B_{ab} and a function of the inter nuclear separation, $f(r_{ab})$ [16]; these quantities are needed to construct a term for the bonding region of M. Finally, we indicate as $w(Z_a)$ the term for van der Waals interactions, when either atom "a"or atom "b" or both are noble gas element:

$$\begin{aligned} w(Z_a) &= \sum_b \{k_1(Z_a) exp(-r_{ab}) + [k_2(Z_a) B_{a,b}/r_{ab}^6 \\ &\quad + k_3(Z_a) B_{a,b}/r_{ab}^8] \times [I_a I_b \alpha_a \alpha_b/(I_a + I_b)]\} \end{aligned} \tag{27}$$

where I_a, I_b, α_a and α_b are the first ionization potentials and the polarizabilities of atoms 'a' and 'b', respectively. In Eq. 27 we have included the term $1/r_{ab}^8$ suggested, for example, by Dunker and Gordon [24], Duquette *et al.* [25], by Kolos *et al.* [26] and more recently by Aziz [27].
Physically, with the modifications from the atomic scaling factors of Eq. 19 to the molecular scaling factors of Eq. 26, it is recognized that in a molecule there are atoms, thus we include the $S_\nu(Z, a)$ term; however, in a molecule there might be charge transfer, like for ionic bonding. Therefore, there is

need of a correction proportional to the charge transferred (either lost or gained) given by the (Z'-Z) term. In addition the electronic structure of the atom 'a' (in the molecule) is modified, due to bond formation with atom "b", thus the need of a term containing the bond order, assumed to be proportional to the electronic density in the bonding region. Finally, there can be van der Waals forces acting on the atoms; with Eq. 27 we assume that the van der Waals interaction energy is pair-wise, to a first approximation. The present molecular scaling functional differs from the one previously reported [14-16]; note, for example, that a factor $1/r_{ab}$ replaces $1/r_{ab}^3$ given in Ref. 16, thus stressing the effect of the bond order.
The HF-CC *molecular* semi-empirical constants are $k(Z_a)$, for the molecular binding and $k_1(Z_a)$, $k_2(Z_a)$ and $k_3(Z_a)$ for the van der Waals forces. For *approximated computations* yielding computed atomization energies within 4-6 Kcal/mol from experimental values, $k(Z_a)$ is proportional to the number of valence electrons (via a very simple expression requiring a few empirical parameters [16]) ; for *more accurate computations* more parameters are needed [16] to represent $k(Z_a)$.
To complete our definitions of the quantities of the functional, we recall that the bond order B_{ab} for the atoms 'a' and 'b' in a "m" atom molecule, is

$$B_{ab} = \sum_{i \in a} \sum_{j \in b} (P_{ij} S_{ij})(P_{ji} S_{ji}) \tag{28}$$

where a=1,..,m and b=1,..,m and where

$$P_{ij} = 2 \sum_{k} c_{ik} c_{kj} \tag{29}$$

are the density matrix elements, with c_{ik} the SCF coefficients and S_{ij} the overlap integrals.

6. Validation of the molecular scaling functional. The molecules of Tables 1 and 2 have been computed both at the HF and at the HF-CC level, and the corresponding values are given in Table 1. The basis set adopted, a triple zeta with polarization [28], is the same as the one in the previous HF-CC studies, in order to allow comparison. For the systems in Table 2 the standard deviation for the computed atomization energies relative to laboratory data is 1.89 kcal/mol, thus near to the previous [16] standard deviation of 2.25 kcal/mol; note that the present results are somewhat more accurate than those reported for G2 [29] and for DFT with Becke functional [8]. Thus we have once more shown that the HF-CC method can yield very reasonable atomization energies, even for somewhat different forms of the scaling functional [14-16].
The energy difference $E_{HF-CC} - E_{HF}$, should be equal to the total correlation energy, E_C, of the system, previously obtained in Section 3, using HF

data, and experimental bindings. There are, however, very small differences between E_C computed from HF and HF-CC and E_C computed from HF and experimental data, since the E_{HF-CC} energies do not reproduce *exactly* the experimental binding; the difference amounts in average to about 2 kcal/mol, as expected, since the average deviation in the computed atomization energies is 1.89 kcal/mol.

Concerning the basis sets, we have computed the HF energies with the triple zeta plus one d polarization function given in [28] and with the triple zeta plus two d and one f polarization functions given in Turbomole [30]; the latter set yields molecular HF energies not far from the HF limit. Note that the energy difference between the two sets of HF computations has been added to the HF-CC and HF energy data of Table 1, which, therefore, are not far from exact non-relativistic energies.

We have adopted the smaller basis set [28] in order to decrease the computational cost in the very often repeated computations of all the molecules in Table 1, needed to optimize the HF-CC parameters. The parametrisation is basis set dependent, since the functional makes use of basis set dependent quantities, like the gross charges and the bond order. However, being the HF-CC method a post Hartree-Fock method, only basis sets of "triple zeta plus polarization" level are acceptable. For these extended bases the basis set dependency of the HF-CC parameters is not too strong; indeed, the use of the HF-CC parameters optimized for the smaller basis set [28], but applied to HF-CC computations performed with the larger basis set [30], yields nearly the same atomization energies, with 2 to 3 percent differences; this simplifies a parameter re-optimisation.

In Fig. 5 we compare the electronic density from an HF-CC and an HF wave function (same geometry and same basis set); we have considered two examples, H_2 and C_2H_2. For H_2 one can notice the increase in the HF-CC density near the nuclear position and in the bonding region. The same holds for C_2H_2. but the compression of the electronic density is relatively much

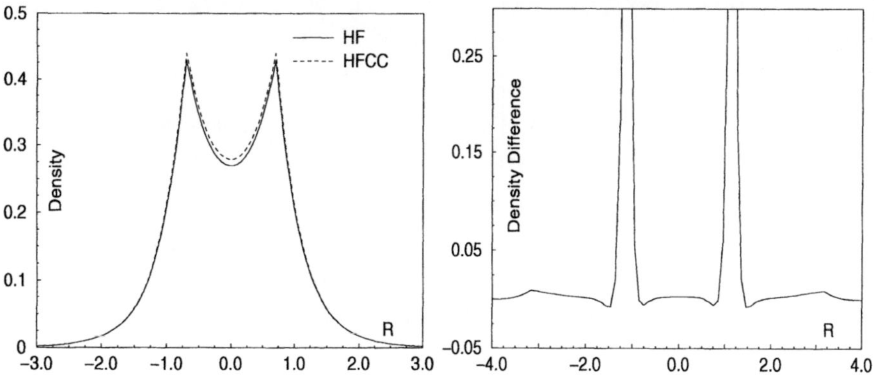

Figure 5. HF-CC and HF electronic densities for H_2 and C_2H_2.

larger around the carbon nuclei than for the hydrogen (recall the discussion of Fig. 4); for this reason for C_2H_2 we plot the *density difference*, very large around the carbon nuclei, much less around the hydrogen nuclei and even less for the carbon-carbon bonding region.

7. The correlation energy from HF-CC and HF computations. We recall that the total correlation energy is given by either the sum $\sum_a E_C(a) + \Delta E_C$ or as the difference $E_{HF-CC} - E_{HF}$. Therefore

$$\Delta E_C = E_{HF-CC} - E_{HF} - \sum_a E_C(a) \tag{30}$$

and we can decompose ΔE_C into its components, namely the nuclear-electron, $\Delta E_C(ne)$, the kinetic, $\Delta E_C(k)$, the coulomb, $\Delta E_C(coul)$, and the exchange, $\Delta E_C(ex)$. These components are reported in Fig. 6. The first information gained from the data of the left inset is that ΔE_C results mainly from $\Delta E_C(ne)$, a result expected since we have scaled the HF total energy, where the nuclear-electron interaction is the largest component. Then follows smaller contributions: $\Delta E_C(k)$, $\Delta E_C(coul)$ and the very small contribution from $\Delta E_C(ex)$. In the right inset, we decompose ΔE_C into $\Delta E_C(ex)$ and $\Delta E_C(class)$, which is the sum of $\Delta E_C(ne)$, $\Delta E_C(k)$, $\Delta E_C(coul)$. As expected, $\Delta E_C(class)$ is the predominant component. Physically, there is a compression in the electronic density, mainly near the nuclei, and, but less, in the bonding regions.

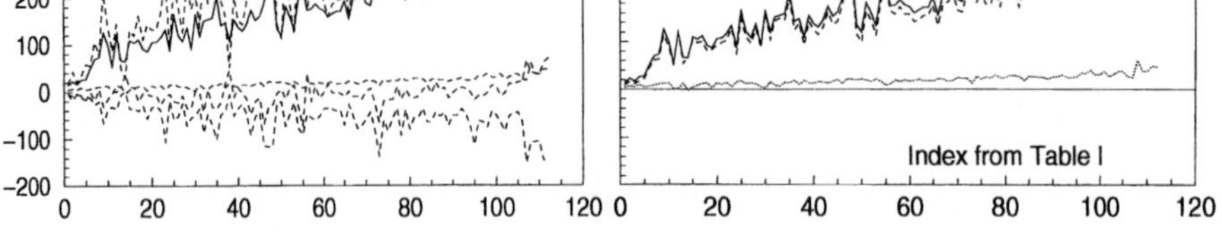

Figure 6. Decomposition of molecular correlation energy, ΔE_C for the molecules of Table 1. Left inset: from top to bottom, $\Delta E_C(ne)$, $\Delta E_C(ex)$, $\Delta E_C(coul)$, and $\Delta E_C(k)$. Right inset: $\Delta E_C(class)$ and $\Delta E_C(ex)$.

Note that this type of decomposition can be done for any post-Hartree-Fock energy, (like CI, coupled cluster, DFT). *Different models, even yielding exactly the same total energy, not necessarily will have exactly the same decomposition.* Therefore, we are of the opinion that this decomposition is very sensitive to compare and to differentiate models.

We return now to the decomposition of the atomization energy, considered also in Fig. 2. We decompose the atomization energy into the molecular correlation energy, ΔE_C, and HF molecular energy, ΔE_{HF}, the latter decomposed into the HF classical molecular energy, $\Delta E_{HF}(class)$, and the HF exchange molecular energy, $\Delta E_{HF}(ex)$. This decomposition is given in Fig. 7. *The new information obtained from this decomposition is that* $\Delta E_{HF}(class)$ *is very close to* ΔE_C. In the top inset of the figure we provide all the components, but to distinguish ΔE_C from $\Delta E_{HF}(class)$, we have provided the middle and bottom insets.

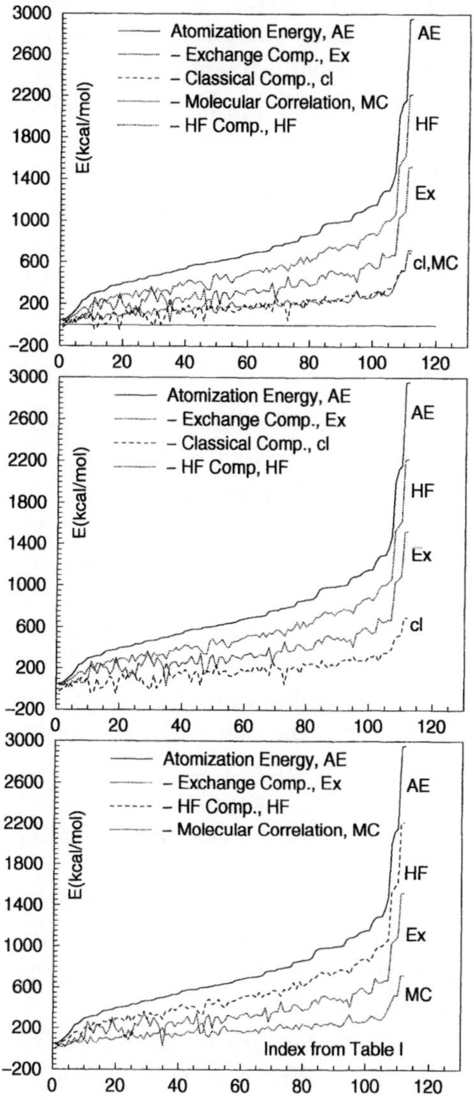

Figure 7. Atomization energy, AE, decomposed into ΔE_{HF}, $\Delta E_{HF}(class)$, $\Delta E_{HF}(ex)$ and ΔE_C, labelled as HF, cl, Ex, MC.

From this analysis we conclude that the atoms in a molecule are held together 1) by the classical electrostatic forces (nuclear-electron, kinetic and Coulomb) of one atom with the others, a result implicit in the Hartree-Fock model; 2) by the molecular correlation energy which concerns mainly the space between bonding atoms, and brings about an atom to atom attraction about as large as the one above; 3) by the exchange energy of one atom with the others, which brings about *the largest attraction.*

The short expression "one atom with the others" stands for "the electron of one atom with the electrons on the other atoms" and is intended to stress the "atoms in the molecule" model implicit in our analyses.

8. Validation of the decomposition $E_C = \sum_a E_C(a) + \Delta E_C$. The decomposition of the total molecular correlation energy into the sum of the correlation energy for the component atoms of the molecule plus a left over, namely the "molecular extra correlation energy", ΔE_C, is by definition mathematically correct. In this section we shall show that the transferability of the atomic functionals to the molecular functional is quantitatively satisfied, namely it adds to the HF molecular energy exactly $\sum_a E_C(a)$; therefore, the decomposition $E_C = \sum_a E_C(a) + \Delta E_C$ is physically meaningful.
The molecular functional of Eq. 26 is constructed with the atomic functional $S(Z, \Gamma, C)_\nu$ (hereafter designated more simply as $S_\nu(Z, a)$) plus additional correction terms to account for the electronic rearrangement from isolated atoms to atoms in the molecule. Therefore, by considering in Eq. 26 only atomic functionals, namely;

$$S_\nu(Z, Z', a, M) = S_\nu(Z, a) \tag{31}$$

we should recover $\sum_a E_C(a)$, if, and only if, the transferability of the atomic functionals to the molecular functional is quantitatively satisfied.
Note, however, that the above functionals in a molecular computation are used to scale both on the one- and many-center integrals. Below we prove that the scaling of the one-center integrals yields quantitatively $\sum_a E_C(a)$ and the scaling of the many-center integrals brings about a relatively small fraction of ΔE_C, since it deals with interactions of one atom with different atoms. The proof is obtained from a computational experimentation.
We have computed the total energy for the molecules of Table 1 with the reduced functional given in Eq. 31, scaling only the one-center integrals: the resulting HF-CC energies are indeed the HF energies plus $\sum_a E_C(a)$, as postulated. It follows that the average error in the computed atomization energy for the molecules of Table 1 is the same as that for the corresponding HF computations, namely 194.38 kcal/mol.
We have also computed the total energy for molecules of Table 1 with the reduced functional of Eq. 31 scaling both the one-center and the many-center integrals; the resulting HF-CC energies are the HF energies plus $\sum_a E_C(a)$ plus a fraction of ΔE_C below determined.
The average deviation in the computed atomization energy for the molecules of Table 1 for this new set of computations is 158.92 kcal/mol: therefore about 20 per cent of ΔE_C is accounted by scaling the many center-integrals with the atomic functional of Eq. 31. The remaining 80 per cent requires the other terms of the molecular functional in Eq. 26.
The above analysis on selected terms in the molecular functional is now extended, and we shall consider not only the atomic functional but also the effect of charge transfer, namely we shall consider a molecular functional limited to the terms given below:

$$S_\nu(Z, Z', a, M) = S_\nu(Z, a) + (Z_a^{'} - Z_a)\Delta S(Z, a) \tag{32}$$

With the above functional we have computed the atomization energy for the molecules of Table 1; the computed atomization energies have an average deviation from the laboratory data of 171.51 kcal/mol; the difference 171.51-158.92 = 12.59 kcal/mol is due to the charge transfer effect. In Fig. 8 we report the atomization energies computed in the HF-CC approximation using only selected terms of the molecular functional, namely Eq. 31 and Eq. 32; we have also added for reference the experimental atomization energy (see Table 2). The two plots (Eq. 31 and Eq. 32) are very close and not easy to distinguish; from this figure we learn that the many-center contribution of the two functionals (Eq. 31 and Eq. 32) yields a non indifferent fraction of ΔE_C.

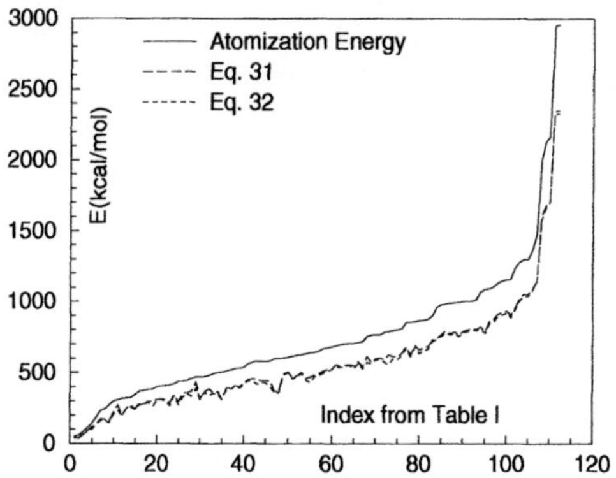

Figure 8. Atomisation energy from atomic functionals of Eq. 31 and Eq. 32.

In conclusion, the decomposition $E_C = \sum_a E_C(a) + \Delta E_C$ not only is algebraically correct, but (1) is suggested by comparing HF atomization energies with laboratory data, (2) can be implemented by transferring the atomic functional to the molecular functional, (3) verified numerically in this section. These results suggest to consider atomic functional different from those in Eq. 31 and Eq. 32 and to retain the rest of the HF-CC approach, if we wish not only accurate correlated energies but also correctly correlated densities.

9. Van der Waals interactions. The van der Waals interactions are very weak but of long range; therefore we expect that these interaction play a critical role in many dynamical processes, for example in biochemical systems, as in molecular docking, protein folding, and solute-solvent interactions. Accurate ab-initio computational investigation on van der Waals interactions have been limited to relatively small systems; large chemical systems are explored essentially by semi-empirical methods. The HF-CC is one of the few techniques which consider van der Waals interactions and correlation effects with a computational cost relatively low, nearly equal to that for HF

computations. The van der Waals interactions for noble gas di-atomics have been the subject of intense studies, and the relative potential energy curves are well determined [24-27, 31-34].
We have computed a number of van der Waals system to explore the validity of our functional, particularly Eq. 27. In Fig. 9 we report the HF-CC potential energy curves for He_2, Ne_2 and Ar_2 obtained with optimized parameters; the HF-CC computed values are indicated with open circles, the experimental determined minima are indicated with a full circle. One can verify that not only the energy minimum, but also the inter nuclear distance are in good agreement with the laboratory data [31-34].

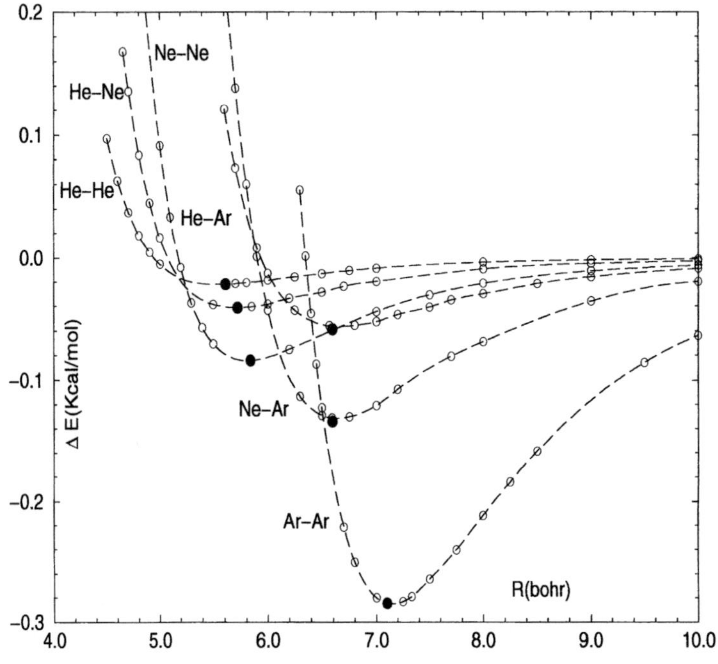

Figure 9. HF-CC van der Waals potential for noble gas diatomics. Open circles for HF-CC computations, full circles for experimental values.

The same parameters are used to compute the van der Waals interaction energies for the mixed systems He-Ne, He-Ar and Ne-Ar; the corresponding potential energy curves are given in Fig. 9. Again we note that the overall agreement between HF-CC computations and laboratory data [33] is satisfactory.
The van der Waals interaction energies of He, Ne and Ar with C_6H_6 have been analyzed in a number of papers [35-42]; the energetically most favorable approach of a noble gas atom to the benzene molecule is found to be along a line perpendicular to the C_6H_6 molecular plane and passing through the

center of the ring. However, differently from the case of the above diatomic systems, the laboratory and computer simulations for the C_6H_6 interactions energies are still under investigation. After parameter optimisation we obtain the potential energy curves labeled A given in Fig. 10; in the figure we report the literature's range of values both for the energy minima (vertical bar) and for the inter nuclear distances (horizontal bar); the open circles indicate the computed HF-CC values. We conclude that the agreement of the HF-CC computation with the literature data is reasonable. The parameters obtained above can now be used to compute van der Waals energies either at different positions on an energy surface or in different systems including large ones. An example of computations at different positions on the energy surface is given by the curves labelled B.

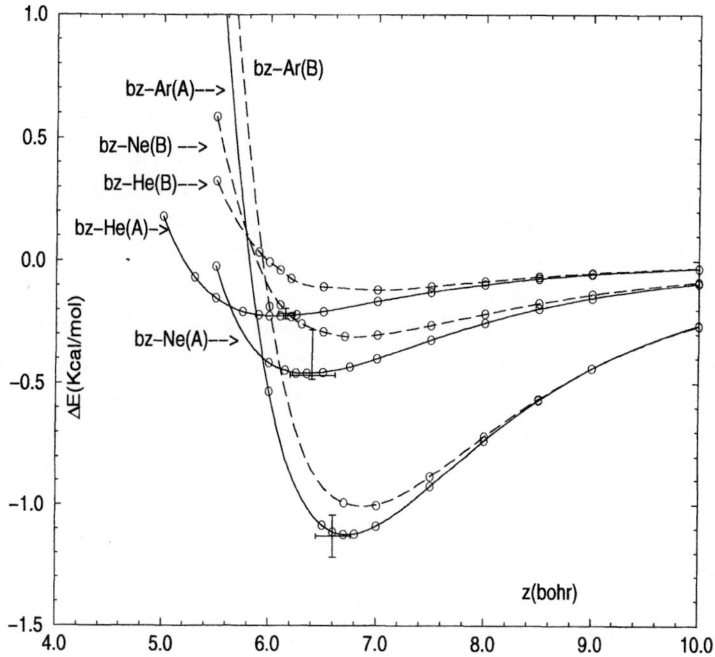

Figure 10. HF-CC van der Waals potential for He, Ne and Ar approaching C_6H_6 along lines perpendicular to the molecular plane. Open circles for HF-CC computations, full circles for experimental values.

For a study of van der Waals interaction energy in a larger system we recall the recent work on naphtalene interacting with He, Ne and Ar [43], included in Table 2 for the equilibrium geometry. Note, however that we are only at the beginning in experimenting with these parameters.

10. Conclusions. In the present version, the HF-CC method does not seek a correlated wave function nor a correlated density, but simply to develop a methodology to scale the energy of the Hartree-Fock model, a very robust model, and obtain near to exact non-relativistic energies for atoms and for molecules, at a computational complexity not exceeding that for the Hartree-Fock model, since we are interested in computations of large molecular systems. The analyses on the atomization energy, leading to the formulation of the HF-CC approximation, has shown the usefulness to model a molecule with three concerted representations (sub models), well familiar to chemistry; first, a general representation, where the electrons are in an average Hartree-Fock sea, (an extension of the LCAO-MO model by R. S. Mulliken); then, a first specific representation to correlate the inner shell electrons, the latter modeled as electrons of a specific atom in a molecule (a concept from the Valence Bond), and finally, a second specific representation to correlate the electrons in the bonding regions with a tight binding and local perspective (as originally suggested by G. N. Lewis).

The set of these three basic sub models appears to be able to accurately describe the molecular binding in strong, medium and weak bonds. The HF-CC method has been tested for a large number of molecules and a variety of somewhat different functionals, and has consistently yielded atomization energies within few kcal/mol (from 1 to 3) from the experimental data; this accuracy compares well with results from the G2 and DFT approximations. The method is computationally as fast as the HF model, thus faster than DFT or G2 approximations.

This study is providing clear evidence that post Hartree-Fock methods can yield accurate energy, if properly parametrized, while retaining electronic densities nearer to the HF then to the correct one. Therefore, this work re-opens the questions on whether the DFT approximation yields a "fully" correlated density, and how much the DFT method is a scaling model relying on empirical parametrisation.

We have also shown how to partition the computed correlation energy and the Hartree-Fock energy in a way which explicitly shows the underlying physics for each one of the above representations. The partitioning provides a proposal on the nature of the chemical bond. Finally, we have also includes a preliminary but promising study on van der Waals interactions.

Acknowledgment. One of us (E. C.) acknowledges a A. von Humboldt Award from the Alexander von Humboldt-Stiftung and Prof. S. Peyerimhoff and Dr. G. Diercksen hospitality and interest during a stay in Bonn and in Garching; discussions with Prof. F. Pirani (Perugia, Italy) on unpublished data are also gratefully acknowledged.

Table 1. E(HF-CC) and E(HF) (hartrees).

	Molecule	Name	E(HF-CC)	E(HF)
1	BeH	beryllium hydride	-15.252550	-15.134849
2	LiH	lithium hydride	-8.069171	-7.986909
3	BH	boron hydride	-25.280226	-25.128504
4	H_2	diatomic hydrogen	-1.173901	-1.132850
5	HF	hydrogen fluoride	-100.441795	-100.069693
6	H_2S	hydrogen sulfide	-399.411527	-398.716085
7	H_2O	water	-76.437074	-76.066035
8	PH_3	phosphine	-343.153180	-342.490330
9	H_2O_2	hydrogen peroxide	-151.567128	-150.845593
10	NH_3	ammonia	-56.572541	-56.223765
11	$AlCL_3$	aluminum trichloride	-1623.308056	-1620.727551
12	CCL_4	carbon tetrachloride	-1878.976530	-1875.929295
13	SiH_4	silane	-291.878664	-291.261288
14	BCL_3	boron trichloride	-1405.637305	-1403.385499
15	$CHCl_3$	chloroform	-1419.366863	-1417.012503
16	CH_2Cl_2	dichloromethane	-959.759541	-958.086494
17	CH_2O	formaldehyde	-114.515721	-113.919526
18	$CH_3S.$	radical (2A')	-438.071854	-437.152155
19	$SiCl_4$	silicon tetrachloride	-2130.600825	-2127.254818
20	$CH_3O.$	radical (2A')	-115.056074	-114.470082
21	C_2H_2	acetylene	-77.339633	-76.853639
22	$CH_2OH.$	radical (2A')	-115.070829	-114.467972
23	$TiCL_4$	titanium tetrachloride	-2690.595934	-2686.760641
24	CH_4	methane	-40.515395	-40.215641
25	N_2H_4	hydrazine	-111.873999	-111.233472
26	CH_2F_2	difluoromethane	-239.022890	-238.016854
27	$C_2H_3.$	radical (2A')	-77.905071	-77.417087
28	CHF_3	trifluoromethane	-338.298518	-336.936574
29	BF_3	boron trifluoride	-324.612604	-323.353767
30	C_2CL_4	tetrachloroethylene	-1917.070172	-1913.797464
31	CH_3SH	methanethiol	-438.724523	-437.762145
32	CF_4	carbon tetrafluoride	-437.566404	-435.852950
33	C_2HCl_3	trichloroethylene	-1457.455412	-1454.872007
34	$HCOOH$	formic acid	-189.784088	-188.854428
35	C_2N_2	cyanogen	-185.669291	-184.648388
36	CH_3OH	methanol	-115.741490	-115.097828
37	$C2H_2Cl_2$	1,1-dichloroethylene	-997.838447	-995.940000
38	Si_2H_6	disilane	-582.568561	-581.371653
39	C_2H_2O	keten	-152.614344	-151.793059
40	C_2H_3Cl	chloroethylene	-538.217165	-537.008450
41	C_2H_4	ethylene	-78.590198	-78.068359
42	C_2H_3F	fluoroethylene	-177.841036	-176.967086
43	C_2HF_3	trifluoroethylene	-376.329771	-374.751240
44	$CH_3CO.$	radical (2A')	-153.178939	-152.360709
45	CH_3NH_2	methylamine	-95.864857	-95.260640

Table 1 continued 1.

	Molecule	Name	E(HF-CC)	E(HF)
46	C_2F_4	tetrafluoroethylene	-475.572887	-473.641017
47	CH_3ONO	methylnitrite	-245.029316	-243.772888
48	CH_3NO_2	nitromethane	-245.033659	-243.771014
49	$C_2H_5.$	radical (2A')	-79.158768	-78.628386
50	B_2H_6	diborane	-53.279035	-52.837742
51	CH_3CN	methylcyanide	-132.772600	-131.977627
52	C_2H_4S	thiirane	-476.804005	-475.618398
53	CH_3SiH_3	methylsilane	-331.207408	-330.325540
54	$C_2H_2O_2$	glyoxal	-227.842308	-226.689708
55	CF_3CN	methylcyanide	-430.521105	-428.668744
56	$C_2H_3Cl_3$	1,1,1-trichloroethane	-1458.680623	-1456.064505
57	C_2H_4O	oxirane	-153.792929	-152.936888
58	CH_3COCl	acetylchloride	-613.489041	-611.939870
59	$CH3CHO$	acetaldehyde	-153.839578	-152.982330
60	C_3H_4	cyclopropene	-116.620718	-115.871881
61	C_2H_5Cl	chloroethane	-539.450227	-538.206330
62	$C_2H_5O.$	radical (2A')	-154.361126	-153.521090
63	C_3H_4	allene	-116.657889	-115.912196
64	CH_3CCH	propine	-116.660375	-115.913952
65	CH_3COF	acetylfluoride	-253.126568	-251.914114
66	C_2H_6	ethane	-79.826971	-79.264125
67	C_2H_5N	aziridine	-133.921819	-133.098869
68	$C_2H_3F_3$	1,1,1-trifluoroethane	-377.614132	-375.994992
69	C_2H_3CN	acrylonitrile	-170.850152	-169.831624
70	$(CH_3)_2S$	dimethylsulfide	-478.034786	-476.802332
71	C_2H_5SH	ethanethiol	-478.036384	-476.812791
72	$C_2H_4O_2$	methylformate	-229.085821	-227.892236
73	C_2F_6	hexafluoroethane	-675.375698	-672.694549
74	$(CH_3)_2O$	dimethylether	-155.034778	-154.129407
75	CH_3COOH	acetic acid	-229.117449	-227.918152
76	C_2H_5OH	ethanol	-155.050451	-154.150661
77	C_3H_6	cyclopropane	-117.899116	-117.108259
78	$(CH_3)_2SO$	dimethylsulfoxide	-553.242852	-551.664265
79	C_3H_6	propylene	-117.910598	-117.123605
80	C_2H_5NO	acetamide	-209.237938	-208.074456
81	$(CH_3)_2NH$	dimethylamine	-135.169102	-134.301902
82	$C_2H_5NH_2$	ethylamine	-135.178699	-134.313503
83	$C_2H_6CH.$	radical (2A')	-118.472043	-117.682122
84	C_4H_4S	thiirane	-553.037839	-551.390892
85	C_3H_6O	acetone	-193.164222	-192.045667
86	C_3H_7Cl	propyl chloride	-578.766669	-577.257444
87	C_4H_6	bicyclobutane	-155.956373	-154.934041
88	C_4H_6	methylencyclopropane	-155.958412	-154.951729
89	C_4H_4O	furan	-230.036013	-228.721657
90	C_4H_6	cyclobutane	-155.979686	-154.962225

Table 1 continued 2.

	Molecule	Name	E(HF-CC)	E(HF)
91	C_4H_6	2-butyne	-155.982635	-154.973031
92	C_3H_8	propane	-119.143186	-118.315055
93	C_4H_6	trans-1,3-butadiene	-155.995259	-154.985746
94	C_4H_5N	pyrrole	-210.182665	-208.896062
95	$C_4H_4N_2$	pyridazine	-264.295633	-262.749182
96	C_3H_8O	methylethylether	-194.357794	-193.193048
97	C_3H_8O	iso-propanol	-194.371106	-193.208655
98	AlC_3H_9	trimethylaluminum	-362.195696	-360.837732
99	C_4H_8	cyclobutane	-157.206390	-156.159257
100	C_4H_8	iso-butene	-157.229471	-156.178017
101	$(CH_3)_3N$	trimthylamine	-174.474209	-173.344834
102	C_5H_5N	pyridine	-248.290495	-246.790044
103	C_5H_8	spiropentane	-195.266371	-193.994818
104	C_4H_{10}	trans-butane	-158.456660	-157.365150
105	C_4H_{10}	isobutane	-158.453587	-157.365584
106	C_6H_6	benzene	-232.258683	-230.791923
107	$C_5H_5Cl_3Ti$	c.pentadienyl$TiCl_3$	-2423.880066	-2419.505062
108	ALC_6H_{15}	trimethyaluminum	-480.122334	-477.957315
109	$C_{10}H_8$	azulene	-385.843187	-383.425790
110	$C_{10}H_8$	naphtalene	-385.914557	-383.495660
111	$C_{14}H_{10}$	anthracene	-539.559514	-536.191220
112	$C_{14}H_{10}$	phenantrene	-539.563831	-536.202420
1	He_2	helium dimer	-5.807452	-5.723188
2	Ne_2	neon dimer	-257.876557	-257.093992
3	Ar_2	argon dimer	-1055.087338	-1053.634048
4	$He-Ne$	helium-neon	-131.841987	-131.408601
5	$He-Ar$	helium-argon	-530.447240	-529.678765
6	$Ne-Ar$	neon-argon	-656.481870	-655.364107
7	$(H_2O)_2$	water dimer	-152.882063	-152.139567
8	$(H_2O)_3$	water trimer	-229.333456	-228.210232
9	H_2O-Li^+	water-lithium ion	-83.774515	-83.361478
10	H_2O-Na^+	water-sodium ion	-238.540546	-237.780843
11	H_2O-K^+	water-potassium ion	-676.212079	-675.097440
12	H_2O-F^-	water-fluorine ion	-176.326261	-175.557085
13	H_2O-Cl^-	water-chlorine ion	-536.730577	-535.652403
14	C_6H_6-He	benzene-helium	-235.162754	-233.653403
15	C_6H_6-Ne	benzene-neon	-361.197625	-359.338607
16	C_6H_6-Ar	benzene-argon	-759.803917	-757.607139
17	$C_{10}H_8-He$	naphtalene-helium	-388.818696	-386.356928
18	$C_{10}H_8-Ne$	naphtalene-neon	-514.853758	-512.042130
19	$C_{10}H_8-Ar$	naphtalene-argon	-913.460493	-910.311083

Table 2. Atomization energy (kcal/mol) computed and experimental.

	Molecule	Name	AE(comp.)	AE(Exp.)
1	BeH	beryllium hydride	53.47	49.76
2	LiH	lithium hydride	57.26	58.11
3	BH	boron hydride	80.67	82.55
4	H_2	diatomic hydrogen	109.54	109.54
5	HF	hydrogen fluoride	128.21	141.13
6	H_2S	hydrogen sulfide	184.25	182.46
7	H_2O	water	232.33	232.33
8	PH_3	phosphine	246.34	241.86
9	H_2O_2	hydrogen peroxide	268.32	268.30
10	NH_3	ammonia	302.93	297.76
11	$AlCL_3$	aluminum trichloride	307.91	306.17
12	CCL_4	carbon tetrachloride	315.73	315.60
13	SiH_4	silane	323.47	321.89
14	BCL_3	boron trichloride	321.71	322.62
15	$CHCl_3$	chloroform	345.28	344.26
16	CH_2Cl_2	dichloromethane	376.30	370.18
17	CH_2O	formaldehyde	376.26	373.49
18	$CH_3S.$	radical (2A')	379.94	380.60
19	$SiCl_4$	silicon tetrachloride	383.16	383.32
20	$CH_3O.$	radical (2A')	401.59	400.50
21	C_2H_2	acetylene	405.97	405.04
22	$CH_2OH.$	radical (2A')	410.84	408.84
23	$TiCL_4$	titanium tetrachloride	409.56	412.15
24	CH_4	methane	419.84	419.86
25	N_2H_4	hydrazine	435.12	437.85
26	CH_2F_2	difluoromethane	439.82	439.34
27	$C_2H_3.$	radical (2A')	447.04	443.96
28	CHF_3	trifluoromethane	463.52	462.01
29	BF_3	boron trifluoride	467.34	469.87
30	C_2CL_4	tetrachloroethylene	470.92	471.28
31	CH_3SH	methanethiol	475.74	473.00
32	CF_4	carbon tetrafluoride	482.38	481.82
33	C_2HCl_3	trichloroethylene	497.26	493.56
34	$HCOOH$	formic acid	500.88	500.73
35	C_2N_2	cyanogen	499.52	502.31
36	CH_3OH	methanol	517.93	512.20
37	$C2H_2Cl_2$	1,1-dichloroethylene	522.24	518.93
38	Si_2H_6	disilane	528.50	529.63
39	C_2H_2O	keten	534.58	531.95
40	C_2H_3Cl	chloroethylene	544.50	538.87
41	C_2H_4	ethylene	563.20	563.21
42	C_2H_3F	fluoroethylene	571.36	571.51
43	C_2HF_3	trifluoroethylene	579.57	580.75
44	$CH_3CO.$	radical (2A')	575.11	580.87
45	CH_3NH_2	methylamine	582.47	581.18

Table 2 continued 1.

	Molecule	Name	AE(comp.)	AE(Exp.)
46	C_2F_4	tetrafluoroethylene	582.88	583.50
47	CH_3ONO	methylnitrite	598.11	598.26
48	CH_3NO_2	nitromethane	600.84	600.70
49	$C_2H_5.$	radical (2A')	606.23	602.25
50	B_2H_6	diborane	608.91	608.62
51	CH_3CN	methylcyanide	621.01	615.08
52	C_2H_4S	thiirane	622.04	622.97
53	CH_3SiH_3	methylsilane	626.19	626.11
54	$C_2H_2O_2$	glyoxal	633.84	633.43
55	CF_3CN	methylcyanide	642.97	639.71
56	$C_2H_3Cl_3$	1,1,1-trichloroethane	638.59	641.57
57	C_2H_4O	oxirane	646.64	649.94
58	CH_3COCl	acetylchloride	671.33	666.91
59	$CH3CHO$	acetaldehyde	675.90	676.42
60	C_3H_4	cyclopropene	678.78	682.03
61	C_2H_5Cl	chloroethane	690.75	690.89
62	$C_2H_5O.$	radical (2A')	689.43	697.12
63	C_3H_4	allene	702.11	702.78
64	CH_3CCH	propine	703.66	704.20
65	CH_3COF	acetylfluoride	706.75	705.43
66	C_2H_6	ethane	711.78	711.79
67	C_2H_5N	aziridine	714.64	719.04
68	$C_2H_3F_3$	1,1,1-trifluoroethane	758.00	757.10
69	C_2H_3CN	acrylonitrile	766.09	763.35
70	$(CH_3)_2S$	dimethylsulfide	766.86	766.78
71	C_2H_5SH	ethanethiol	767.86	767.61
72	$C_2H_4O_2$	methylformate	786.65	784.16
73	C_2F_6	hexafluoroethane	788.16	787.59
74	$(CH_3)_2O$	dimethylether	798.40	796.85
75	CH_3COOH	acetic acid	806.49	802.18
76	C_2H_5OH	ethanol	808.23	809.40
77	C_3H_6	cyclopropane	853.48	851.80
78	$(CH_3)_2SO$	dimethylsulfoxide	853.64	854.17
79	C_3H_6	propylene	860.68	859.22
80	C_2H_5NO	acetamide	869.23	867.18
81	$(CH_3)_2NH$	dimethylamine	869.81	868.82
82	$C_2H_5NH_2$	ethylamine	875.83	875.78
83	$C_2H_6CH.$	radical (2A')	899.24	898.83
84	C_4H_4S	thiirane	961.62	962.07
85	C_3H_6O	acetone	976.05	978.08
86	C_3H_7Cl	propyl chloride	985.75	983.79
87	C_4H_6	bicyclobutane	985.83	985.87
88	C_4H_6	methylencyclopropane	987.11	989.44
89	C_4H_4O	furan	992.02	993.03
90	C_4H_6	cyclobutane	1000.46	1000.46

Table 2 continued 2.

	Molecule	Name	AE(comp.)	AE(Exp.)
91	C_4H_6	2-butyne	1002.31	1002.79
92	C_3H_8	propane	1006.63	1004.63
93	C_4H_6	trans-1,3-butadiene	1010.23	1011.90
94	C_4H_5N	pyrrole	1071.17	1071.09
95	$C_4H_4N_2$	pyridazine	1085.40	1087.85
96	C_3H_8O	methylethylether	1097.51	1093.24
97	C_3H_8O	iso-propanol	1105.87	1106.37
98	AlC_3H_9	trimethylaluminum	1135.38	1134.40
99	C_4H_8	cyclobutane	1142.71	1146.91
100	C_4H_8	iso-butene	1153.33	1156.73
101	$(CH_3)_3N$	trimthylamine	1157.69	1158.62
102	C_5H_5N	pyridine	1235.26	1237.07
103	C_5H_8	spiropentane	1276.78	1281.63
104	C_4H_{10}	trans-butane	1296.75	1298.55
105	C_4H_{10}	isobutane	1294.97	1300.42
106	C_6H_6	benzene	1368.38	1366.95
107	$C_5H_5Cl_3Ti$	c.pentadienyl$TiCl_3$	1482.09	1481.27
108	ALC_6H_{15}	trimethyaluminum	2006.14	2005.55
109	$C_{10}H_8$	azulene	2120.85	2127.05
110	$C_{10}H_8$	naphtalene	2165.64	2160.97
111	$C_{14}H_{10}$	anthracene	2956.05	2951.38
112	$C_{14}H_{10}$	phenantrene	2958.76	2956.71
1	He_2	helium dimer	0.022	0.022
2	Ne_2	neon dimer	0.084	0.084
3	Ar_2	argon dimer	0.284	0.284
4	$He - Ne$	helium-neon	0.042	0.042
5	$He - Ar$	helium-argon	0.056	0.059
6	$Ne - Ar$	neon-argon	0.136	0.134
7	$(H_2O)_2$	water dimer	4.97	(5.)
8	$(H_2O)_3$	water trimer	13.95	(16.)
9	$H_2O - Li^+$	water-lithium ion	36.26	35.33
10	$H_2O - Na^+$	water-sodium ion	25.29	24.89
11	$H_2O - K^+$	water-potassium ion	17.70	18.67
12	$H_2O - F^-$	water-fluorine ion	22.09	25.57
13	$H_2O - Cl^-$	water-chlorine ion	13.68	14.16
14	$C_6H_6 - He$	benzene-helium	0.117	a
15	$C_6H_6 - Ne$	benzene-neon	0.437	b
16	$C_6H_6 - Ar$	benzene-argon	1.197	c
17	$C_{10}H_8 - He$	naphtalene-helium	0.270	–
18	$C_{10}H_8 - Ne$	naphtalene-neon	0.621	–
19	$C_{10}H_8 - Ar$	naphtalene-argon	1.565	–

The experimental range of data is: a) 0.1973-0.2398 kcal/mol; b) 0.4312-0.4866 kcal/mol; c) 1.0446-1.2176 kcal/mol.

References.

1. E. Clementi, *IBM J. Res. and Dev.*, **9**, 2 (1965); see also *IBM J. Res. and Dev. Supplement, "Tables of Atomic Functions"*, **9**, 2 (1965); see also E. Clementi, C. Roetti, *Atomic Data and Nuclear Data Tables,* **14**, 177 (1974).

2. A.D. McLean, M. Yoshimine, *IBM J. Res. and Dev. Supplement, "Tables of Linear Molecule Wave Functions"*, **11**, 15 (1967).

3. See for example E. Clementi, W. von Niessen, *J. Chem. Phys.*, **54**, 521 (1971); see also E. Clementi, *J. Chem. Phys.*, **46**, 3851 (1967).

4. P.O. Löwdin, *Phys. Rev.*, **97**, 1474 (1955).

5. E. Wigner, *Phys. Rev.*, **46**, 1002 (1934).

6. S. Chakravorty, E. Clementi, *Phys. Rev. A*, **39**, 2290 (1989).

7. See for example E.S. Kryachko, E. Ludeña,, *Energy Density Functional Theory of Many-Electron Systems*, Kluwer Academic Publishers, London (1990).

8. A.D. Becke, *J. Chem. Phys.*, **98**, 1372 (1993).

9. D.R. Hartree, W. Hartree, B. Swirles, *Tran. R. Soc.*, **A 238**, 299 (1939).

10. A. Veillard, E. Clementi, *J. Chem. Phys.*, **49**, 2415 (1968).

11. G.C. Lie, E. Clementi, *J. Chem. Phys.*, **60**, 1275 (1974).

12. G.C. Lie, E. Clementi, *J. Chem. Phys.*, **60**, 1288 (1974).

13. E. Clementi, G. Corongiu, *Int. J. Quantum Chem.*, **62**, 571 (1997).

14. E. Clementi, G. Corongiu, *Chem. Phys. Letters*, **282**, 335 (1998).

15. E. Clementi, G. Corongiu, *J. Mol. Struct. (Theochem)*, **493**, 1 (1999).

16. E. Clementi, G. Corongiu, *J. Mol. Struct. (Theochem)*, **543**, 39 (2001).

17. S.J. Chakravorty, R.S. Gwaltney, E.R. Davidson, F.A. Parpia, C.F. Fisher, *Phys. Rev.*, **A 47**, 3649 (1993).

18. E. Clementi, A. Veillard, *J. Chem. Phys.*, **44**, 3050 (1966).

19. J. Flores, *Methods and Techniques in Computational Chemistry, METECC-94*, E. Clementi, Ed., Stef, Cagliari, 133 (1993).

20. E. Clementi, *J. Mol. Spect.*, **12**, 18 (1964).

21. E. Clementi, G. Corongiu, *Parallel Computing*, **25**, 1583 (1999).

22. E. Clementi, *J. Chem. Phys.*, **36**, 33 (1962).

23. R.S. Mulliken, *J. Chem. Phys.*, **23**, 343 (1955); *J. Chem. Phys.*, **23**, 2338 (1955).

24. A.M. Dunker, R.G. Gordon *J. Chem. Phys.*, **64**, 354 (1976).

25. G. Duquette, T.M. Ellis, G. Scoles, R.O. Watts, *J. Chem. Phys.*, **68**, 2544 (1978).

26. W. Kolos, G. Corongiu, E. Clementi, *Int.J.Quantum Chem.*, **17**, 775 (1980).

27. R.A. Aziz, *J. Chem. Phys.*, **99**, 4518 (1993).

28. E. Clementi, G. Corongiu,, *Chem. Phys. Letters*, **90**, 359 (1982).

29. L.A. Curtiss, K. Raghavachari, J.C. Pople, *J. Chem. Phys.*, **106**, 1063 (1997).

30. R. Ahlrichs, M. von Arnim, *Methods and Techniques in Computational Chemistry, METECC-95*, E. Clementi, G. Corongiu, Eds., Stef, Cagliari, 509, (1995).

31. R.A. Aziz, M.J. Slaman, *J. Chem. Phys.*, **94**, 8047 (1991).

32. R.A. Aziz, M.J. Slaman, *Chem. Phys.*, **130**, 187 (1989).

33. M. Keil, L.J. Danielson, P.J. Dunlop, *J. Chem. Phys.*, **94**, 296 (1991).

34. Barrow, D. A.; Aziz, R. A. *J. Chem. Phys.*, **89**, 6189 (1988).

35. Th. Brupbacher, J. Makarewicz, A. Bauder, *J. Chem. Phys.*, **101**, 9736 (1994).

36. H. Koch, B. Fernandez, J. Makarewicz, *J. Chem. Phys.*, **111**, 198 (1999).

37. P. Hobza, O. Bludsky, H.L. Selzle, E.W. Schlag, *J. Chem. Phys.*, **97**, 335 (1992).

38. R.G. Satink, H. Piest, G. von Helden, G. Meijer, *J. Chem. Phys.*, **111**, 10750 (1999).

39. Th. Weber, E. Riedle, H.J. Neusser, E.W. Schlag, *J. Mol. Structure.*, **249**, 69 (1991).

40. T.D. Klots, T. Emilsson, H.S. Gutowsky, *J. Chem. Phys.*, **97**, 5335 (1992).

41. G. Liuti, F. Pirani, *Chem. Phys. Letters*, **122**, 245 (1985).

42. R. Cambi, D. Cappelletti, G. Liuti, F. Pirani, *J. Chem. Phys.*, **95**, 1852 (1991).

43. E. Clementi, G. Corongiu, *J. Phys. Chem.*, **105**, 10379 (2001).

SOME EXACT ENERGY RELATIONSHIPS

Peter Politzer
Department of Chemistry
University of New Orleans
New Orleans, LA 70148
U.S.A

1. ATOMIC ENERGIES

The electrostatic potential V_0 that is produced at the nucleus of an atom by its electrons is,

$$V_0 = -\int \frac{\rho(\mathbf{r})d\mathbf{r}}{r} \tag{1}$$

with $\rho(\mathbf{r})$ being the electronic density. It has been shown by Lowdin [1] and by others [2-4] that V_0 is related to the atom's energy E^{at} by,

$$\left(\frac{\partial E^{at}}{\partial Z}\right)_N = V_0 \tag{2}$$

where N is the number of electrons and Z is the nuclear charge. This relationship can be obtained from the Hellmann-Feynman theorem [5,6].

Eq. (2) indicates that, for a fixed number of electrons, the atomic energy can be expressed in terms of V_0 and the nuclear charge. If the electronic density and hence V_0 are written as functions of a variable nuclear charge, $\rho(\mathbf{r}, Z')$ and $V_0(Z')$, then [2,7],

$$E^{at} = \int_{Z'=0}^{Z} [V_0(Z')]_N dZ' \tag{3}$$

It should be noted that eq. (3) relates the total energy, which includes two-electron interactions, to V_0, a one-electron property. This is clearly a manifestation of the Hohenberg-Kohn theorem [8], and links it to the Hellmann-Feynman.

Integrating eq. (3) by parts yields,

E.J. Brändas and E.S. Kryachko (eds.),
Fundamental World of Quatum Chemistry, Vol. I, 631–638.

$$E^{at} = ZV_0 - \int_{Z'=0}^{Z} \left[Z' \left(\frac{\partial V_0(Z')}{\partial Z'} \right) \right]_N dZ' \tag{4}$$

This is a simpler but still exact version of an atomic energy formula that was presented some time ago [7]:

$$E^{at} = 0.5ZV_0 - 0.5 \int_{Z'=0}^{Z} \left[Z' \left(\frac{\partial V_0(Z')}{\partial Z'} \right) - V_0 \right]_N dZ' \tag{5}$$

Since ZV_0 is the nuclear-electronic interaction energy, V_{ne}, and the virial theorem states that $E^{at} = 0.5(V_{ne} + V_{ee})$, where V_{ee} is the electronic interaction energy, then it follows from eq. (4) that,

$$-\int_{Z'=0}^{Z} \left[Z' \left(\frac{\partial V_0(Z')}{\partial Z'} \right) \right]_N dZ' = 0.5(V_{ee} - V_{ne}) \tag{6}$$

or from eq. (5),

$$-\int_{Z'=0}^{Z} \left[Z' \left(\frac{\partial V_0(Z')}{\partial Z'} \right) - V_0(Z') \right]_N dZ' = V_{ee} \tag{7}$$

Eqs. (6) and (7) explicitly and rigorously relate the two-electron quantity V_{ee} to the one-electron V_0.

2. MOLECULAR ENERGIES

The molecular equivalent of eq. (2), which also comes from the Hellmann-Feynman theorem, is,

$$\left(\frac{\partial E^{mol}}{\partial Z_A} \right)_{N, Z_{B \neq A}, \{\mathbf{R}_i\}} = V_{0,A} \tag{8}$$

in which $V_{0,A}$ is the electrostatic potential at nucleus A due to the electrons and the other nuclei:

$$V_{0,A} = \sum_{B \neq A} \frac{Z_B}{|\mathbf{R}_B - \mathbf{R}_A|} - \int \frac{\rho(\mathbf{r})d\mathbf{r}}{|\mathbf{r} - \mathbf{R}_A|} \tag{9}$$

Analogously to the atomic case, eq. (8) indicates that the molecular energy can be written as a function of the electrostatic potential at one nucleus and the charge of the latter, provided that the number of electrons and all other nuclear charges are held constant. For some purposes, however, it may be preferable to not focus upon a single nucleus but rather to encompass all of them. This can be done by following the example of Wilson [9], and introducing a scaling parameter λ such that the charge on any nucleus i is $Z_i' = \lambda z_i$; in the actual molecule, $\lambda = 1$ and $Z_i' = z_i = Z_i$. This permits all of the nuclear charges to increase in a concerted fashion from zero to their real values.

The Hellmann-Feynman theorem can still be applied; since λ appears only in the nuclear and nuclear-electronic interaction operators in the molecular Hamiltonian, one obtains, as shown by Wilson [9],

$$\left(\frac{\partial E^{mol}}{\partial \lambda}\right)_{N,\{z_i\},\{\mathbf{R}_i\}} = 2\lambda \sum_A \sum_{B>A} \frac{z_A z_B}{|\mathbf{R}_A - \mathbf{R}_B|} - \sum_A z_A \int \frac{\rho(\mathbf{r},\lambda)d\mathbf{r}}{|\mathbf{R}_A - \mathbf{r}|} \qquad (10)$$

(It will be assumed henceforth that the $\{z_i\}$ and $\{\mathbf{R}_i\}$ are kept fixed.) Wilson integrated eq. (10) between $\lambda = 0$ and $\lambda = 1$ to obtain a molecular energy formula [9]:

$$E^{mol} = \sum_A \sum_{B>A} \frac{z_A z_B}{|\mathbf{R}_A - \mathbf{R}_B|} - \sum_A z_A \int_{\lambda=0}^{1} \int \frac{\rho(\mathbf{r},\lambda)d\mathbf{r}}{|\mathbf{R}_A - \mathbf{r}|} d\lambda \qquad (11)$$

This can be rewritten as,

$$E^{mol} = \sum_A z_A \sum_{B>A} \frac{z_B}{|\mathbf{R}_A - \mathbf{R}_B|} - \sum_A z_A \int_{\lambda=0}^{1} \frac{\rho(\mathbf{r},\lambda)d\mathbf{r}}{|\mathbf{R}_A - \mathbf{r}|} d\lambda \qquad (12)$$

or

$$E^{mol} = \sum_A z_A \int_{\lambda=0}^{1} \left[V_{0,A}(\lambda)\right]_N d\lambda \qquad (13)$$

which can also be written in the form,

$$E^{mol} = \sum_A \int_{Z_A'=0}^{Z_A} \left[V_{0,A}(Z_A')\right]_N dZ_A' \qquad (14)$$

Eq. (14) has the nice feature of being the direct molecular analogue of eq. (3), but the notation obscures the key point, explicit in eq. (13), that $\rho(\mathbf{r},\lambda)$ and hence $V_{0,A}$ are functions of the scaling of the *entire* set of nuclear charges, not just Z'_A. Eqs. (13) and (14) fulfill the expectation stemming from the Hellmann-Feynman theorem, eq. (8), but do so while encompassing all of the nuclei.

Integrating eq. (14) by parts gives,

$$E^{mol} = \sum_A Z_A V_{0,A} - \sum_A \int_{Z'_A=0}^{Z_A} \left[Z'_A \left(\frac{\partial V_{0,A}(Z'_A)}{\partial Z'_A} \right) \right]_N dZ'_A \qquad (15)$$

This is the molecular analogue of eq. (4), and again is a simpler but exact version of a formula presented earlier [7]:

$$E^{mol} = 0.5 \sum_A Z_A V_{0,A} - 0.5 \sum_A \int_{Z'_A=0}^{Z_A} \left[Z'_A \left(\frac{\partial V_{0,A}(Z'_A)}{\partial Z'_A} \right) - V_{0,A} \right]_N dZ'_A \qquad (16)$$

It is interesting that the molecular expressions, eqs. (14) - (16), are simply summations over their atomic counterparts, eqs. (3) - (5); there are no mixing or cross-terms. This is only possible, however, when the relationships are all written in terms of the electrostatic potentials at the nuclei. Since $\sum_A Z_A V_{0,A}$ equals $V_{ne} + 2V_{nn}$, where V_{nn} is the nuclear interaction energy, then invoking the virial theorem, $E^{mol} = 0.5(V_{ne} + V_{ee} + V_{nn})$, leads to, from eq. (15),

$$-\sum_A \int_{Z'_A=0}^{Z_A} \left[Z'_A \left(\frac{\partial V_{0,A}(Z'_A)}{\partial Z'_A} \right) \right]_N dZ'_A = 0.5(V_{ee} - V_{ne}) - 1.5V_{nn} \qquad (17)$$

or from eq. (16),

$$-\sum_A \int_{Z'_A=0}^{Z_A} \left[Z'_A \left(\frac{\partial V_{0,A}(Z'_A)}{\partial Z'_A} \right) - V_{0,A}(Z'_A) \right]_N dZ'_A = V_{ee} - V_{nn} \qquad (18)$$

Evaluation of V_{ne}, V_{ee} and V_{nn} for a series of atoms and molecules has shown that the integral terms in eqs. (4), (5), (15) and (16) are only about 15% as large in magnitude as are the first terms [3,10,11]. This suggests that reasonably good approximate atomic and molecular energy formulas might be obtained by simply introducing appropriate scaling factors into the first terms. There has indeed been a great deal of work along these lines (reviewed elsewhere [10-15]), with some success. Average errors as low as 0.10% have been achieved for molecular energies [11]; unfortunately this corresponds to about 77 kcal/mole, which is as large or larger than many bond energies.

3. INTERACTION ENERGIES

Consider the interaction of several atoms to form an N-electron molecule. The interaction energy ΔE is,

$$\Delta E = E^{mol} - \sum_A E_A^{at} \tag{19}$$

These energies will be taken to be functions of a single nuclear charge scaling parameter λ and the respective numbers of electrons. Thus,

$$\Delta E(\lambda, N') = E^{mol}(\lambda, N') - \sum_A E_A^{at}(\lambda, N'_A) \tag{20}$$

Then,

$$d\Delta E(\lambda, N') = \left(\frac{\partial E^{mol}}{\partial \lambda}\right)_{N'=N} d\lambda + \left(\frac{\partial E^{mol}}{\partial N'}\right)_{\lambda} dN'$$

$$-\sum_A \left(\frac{\partial E_A^{at}}{\partial \lambda}\right)_{N'_A=N_A} d\lambda - \sum_A \left(\frac{\partial E_A^{at}}{\partial N'_A}\right)_{\lambda} dN'_A \tag{21}$$

Integrating,

$$\int_{\lambda=0}^{1}\int_{N'=0}^{N} d\Delta E(\lambda, N') d\lambda dN' = \int_{\lambda=0}^{1}\left[\left(\frac{\partial E^{mol}}{\partial \lambda}\right)_{N'=N} - \sum_A \left(\frac{\partial E_A^{at}}{\partial \lambda}\right)_{N'_A=N_A}\right] d\lambda$$

$$+ \int_{N'=0}^{N} \left(\frac{\partial E^{mol}}{\partial N'}\right)_{\lambda=1} dN' - \sum_A \int_{N'_A=0}^{N_A} \left(\frac{\partial E_A^{at}}{\partial N'_A}\right)_{\lambda=1} dN'_A \tag{22}$$

The integrals over λ on the right side of eq. (22) can be treated as in going from eq. (10) to eq. (14). The integrals over N' and N'_A are seen to involve the atomic and molecular chemical potentials [16],

$$\mu_A^{at} = \left(\frac{\partial E_A^{at}}{\partial N_A}\right)_Z \tag{23}$$

and

$$\mu^{mol} = \left(\frac{\partial E^{mol}}{\partial N}\right)_{v(\mathbf{r})} \tag{24}$$

where $v(\mathbf{r})$ is the electrostatic potential at **r** due to the nuclei. Thus, eq. (22) becomes,

$$\Delta E = \sum_A \int_{Z'_A=0}^{Z_A} \left\{ \left[V_{0,A}^{mol}(Z'_A)\right]_N - \left[V_{0,A}^{at}(Z'_A)\right]_{N_A} \right\} dZ'_A$$

$$+ \int_{N'=0}^{N} \mu^{mol} dN' - \sum_A \int_{N'_A=0}^{N_A} \mu_A^{at} dN'_A \tag{25}$$

The first term in Eq. (25) could have been obtained from eqs. (3) and (14). Again, it should be noted that $V_{0,A}^{mol}$ in eq. (25) reflects the scaling of *all* of the nuclear charges.

In the special case of a homonuclear diatomic molecule, for which $N = 2N_A$, eq. (25) takes the form,

$$\Delta E = 2 \int_{Z'_A=0}^{Z_A} \left\{ \left[V_{0,A}^{mol}(Z'_A)\right]_N - \left[V_{0,A}^{at}(Z'_A)\right]_{N_A} \right\} dZ'_A$$

$$+ 2 \int_{N'_A=0}^{N_A} \left(\mu^{mol} - \mu_A^{at}\right) dN'_A \tag{26}$$

4. DISCUSSION AND SUMMARY

Several exact relationships have been presented, for atomic and molecular energies and for the interaction energy of atoms combining to form a molecule. They all come from the Hellmann-Feynman theorem, and focus particularly upon the electrostatic potentials at nuclei.

The evaluation of the integrals in eqs. (3) - (5) and (14) - (16) is complicated by the restriction to isoelectronic pathways. It has been demonstrated [17] that this requirement can be eliminated in eqs. (3) - (5) by assuming that the atomic chemical potential, μ_A^{at}, is equal to zero, and in eqs. (14) - (16) by assuming $\mu_{mol} = 0$. The integrations can then be carried out over more convenient pathways, such as the neutral systems, i.e. $N' = Z'$ in eqs. (3) - (5) and $N' = \sum Z'_A$ in eqs. (14) - (16). This has been tested, at the Hartree-Fock level; the errors were usually about 1% for atoms [7], and 0.5% for diatomic molecules [17].

The chemical potential is rigorously equal to zero for neutral systems within the framework of Thomas-Fermi theory [18-20], which Lieb and Simon have shown to become exact as the sum of the nuclear charges approaches infinity [21]. Komorowski [22] and Ganguly [23,24] have speculated that zero chemical potentials may in general characterize atoms and molecules at equilibrium.

An interesting conceptual point that comes out of eqs. (25) and (26) is that the stabilization of a molecule with respect to the free atoms is related to the changes in the electrostatic potentials at the nuclei and in the chemical potentials. The computed $V_{0,A}^{at}$ and $V_{0,A}^{mol}$ have been compared for groups of atoms and molecules using both Hartree-Fock [12] and density functional procedures [10]; the change that occurs for a given atom upon formation of a molecule can generally be ascribed to electron donation or withdrawal.

The interpretation and evaluation of chemical potentials (or electronegativities [16]) continues to pose problems, as was recently discussed [25]. It has even been suggested, for example, that there should be two scales rather than one [26-28]. The chemical potential is frequently taken to be,

$$\mu = -0.5(I + A) \tag{27}$$

where I and A are the ground-state ionization potential and electron affinity of the system [16.20,29]. However there are difficulties associated with this formulation of the chemical potential [19,25,30,31]; in the case of molecules, for example, eq. (27) does not take into account the dependence upon the nuclear potential that is indicated by eq. (24) [25,32-37]. Thus, the effort to arrive at meaningful values for atomic and molecular chemical potentials continues.

REFERENCES

[1] P.-O. Lowdin, J. Mol. Spectrosc. **3**, 46 (1959).
[2] L. L. Foldy, Phys. Rev. **83**, 397 (1951).
[3] S. Fraga, Theoret. Chim. Acta **2**, 406 (1964).
[4] R. Gaspar, Int. J. Quantum Chem. **1**, 139 (1967).
[5] H. Hellmann, Einfuhrung in die Quantenchemie, Deuticke, Leipzig, 1937.
[6] R. P. Feynman, Phys. Rev. **56**, 340 (1939).
[7] P. Politzer and R. G. Parr, J. Chem. Phys. **61**, 4258 (1974).
[8] P. Hohenberg and W. Kohn, Phys. Rev. B **136**, 864 (1964).
[9] E. B. Wilson, Jr., J. Chem. Phys. **36**, 2232 (1962).
[10] P. Politzer, P. Lane and J. S. Murray, In: Reviews in Modern Quantum Chemistry: A Celebration of the Contributions of R. G. Parr, K. D. Sen, Ed., World Scientific, Singapore, 2002, in press.
[11] P. Politzer, P. Lane and M. C. Concha, Int. J. Quantum Chem., in press.
[12] P. Politzer, Israel J. Chem. **19**, 224 (1980).

[13] P. Politzer, In: Chemical Applications of Atomic and Molecular Electrostatic Potentials, P. Politzer and D. G. Truhlar, Eds., Plenum, New York, 1981, ch. 2.
[14] P. Politzer, In: Single-Particle Density in Physics and Chemistry, N. H. March and B. M. Deb, Eds., Academic, New York, 1987, ch. 3.
[15] N. H. March, Electron Density Theory of Atoms and Molecules, Academic, London, 1992, appendix 6.1.
[16] R. G. Parr, R. A. Donnelly, M. Levy and W. E. Palke, J. Chem. Phys. **68**, 3801 (1978).
[17] P. Politzer, J. Chem. Phys. **80**, 380 (1984).
[18] E. H. Lieb, Rev. Mod. Phys. **53**, 603 (1981).
[19] N. H. March, J. Phys. Chem. **86**, 2262 (1982).
[20] R. G. Parr and W. Yang, Density-Functional Theory of Atoms and Molecules, Oxford University Press, New York, 1989.
[21] E. H. Lieb and B. Simon, Phys. Rev. Lett. **31**, 681 (1973); Adv. Math. **23**, 22 (1977).
[22] L. Komorowski, Chem. Phys. **76**, 31 (1983); Chem. Phys. Lett. **103**, 201 (1983).
[23] P. Ganguly, J. Am. Chem. Soc. **117**, 2655 (1995).
[24] P. Ganguly, J. Phys. Chem. A **104**, 8432 (2000).
[25] P. Politzer, M. E. Grice and J. S. Murray, J. Mol. Struct. (Theochem) **549**, 69 (2001).
[26] L. Komorowski, Chem. Phys. **114**, 55 (1987).
[27] R. G. Pearson, Acc. Chem. Res. **23**, 1 (1990); Coord. Chem. Rev. **100**, 403 (1990).
[28] L. C. Allen, Acc. Chem. Res. **23**, 175 (1990).
[29] R. G. Pearson, Inorg. Chem. **27**, 734 (1988); J. Org. Chem. **54**, 1423 (1989).
[30] J. F. Liebman and J. E. Huheey, Phys. Rev. D **36**, 1559 (1987).
[31] K. S. Lackner and G. Zweig, Phys. Rev. D **36**, 1562 (1987).
[32] P. Politzer and H. Weinstein, J. Chem. Phys. **71**, 4218 (1979).
[33] R. F. Nalewajski, J. Am. Chem. Soc. **106**, 944 (1984).
[34] W. J. Mortier, K. V. Genechten and J. Gasteiger, J. Am Chem. Soc. **107**, 829 (1985).
[35] T. K. Ghanty and S. K. Ghosh, J. Phys. Chem. **95**, 6512 (1991).
[36] P. Politzer, J. E. Huheey, J. S. Murray and M. Grodzicki, J. Mol. Struct. (Theochem) **259**, 99 (1992).
[37] P. K. Chattaraj and S. Nath, Int. J. Quantum Chem. **49**, 705 (1994).

SHELL EFFECTS IN THE RELAXATION ENERGY OF 1*s*-CORE IONIZATION OF ATOMS FROM He THROUGH Xe

J. Maruani[1], A. I. Kuleff[1,2], Ya. I. Delchev[2] and C. Bonnelle[1]

[1] *Laboratoire de Chimie Physique - Matière et rayonnement, CNRS and UPMC, 11, rue Pierre et Marie Curie, 75005 Paris, France - maruani@ccr.jussieu.fr*

[2] *Institute for Nuclear Research and Nuclear Energy, Bulgarian Academy of Sciences, 72, bd Tzarigradsko Chaussee, 1784 Sofia, Bulgaria - akuleff@inrne.bas.bg*

Abstract In this paper we investigate the variation, along the periodic table, of inner-core ionization energies, using both relativistic and non-relativistic calculations. We show that the electron relaxation energy displays variations typical of the shell filling process, while its relativistic contribution shows specifically distorted shell features. The sensitivity of these differential properties to the shell-filling process may be used as a probe of the accuracy of calculations of core excitation or ionization energies.

Keywords Shell effects; relaxation energy; relativistic effects; core ionization

1. INTRODUCTION

Core-electron binding energies, measured by x-ray absorption (XAS), photoemission (XPS) and related spectroscopies, are important for studying the atomic composition and chemical bonding of various systems, due to their sensitivity to the chemical and physical environment of the atoms in the molecular or solid structure [1-3]. The binding energy E_R of an electron in a given atomic orbital $Znlj$ is the difference between the energies of the initial (i) and final (f) states, the initial state being the neutral atom **Z** and the final state the core-hole ion $\mathbf{Z}_+$ with an electron missing in the nlj orbital. In the Koopmans approximation [4] the electron binding energy is simply the SCF energy ε_{nlj} of the corresponding orbital in the neutral atom, or the difference E_K $(i\rightarrow f)$ between the orbitals involved in the atom if one considers an excitation (or ionization) [5]. However the energy E_R $(i\rightarrow f) = E_f - E_i$ actually received by the system to undergo the excitation (or ionization) also involves an energy ΔE_{RK} of relaxation of the electron cloud accompanying the transition, i.e.:

E.J. Brändas and E.S. Kryachko (eds.), Fundamental World of Quatum Chemistry, Vol. I, 639–655.

$$E_R\ (i \rightarrow f) = E_K\ (i \rightarrow f) + \Delta E_{RK}\ (i \rightarrow f), \tag{1}$$

where ΔE_{RK} is negative since in the relaxation process the system restitutes part of the Koopmans energy provided to the system in the excitation.

The computation of core-hole state energies does not follow the standard procedures because straight minimization of the energy normally results in filling the inner levels for reaching the lowest state of desired symmetry (variational collapse). Therefore, specific methods have been devised in order to calculate core-hole states [6-12]. In particular, it has been shown [13] how the amount of relaxation involved in an electronic transition increases with the number of electrons on the atom where the inner level is located and with the energy depth of this level.

Table 1. For elements from He (Z = 2) through Xe (Z = 54), measured inner-core (IC) [23] and outer-valence (OV) [24] ionization energies (in eV).

Elem.	IC [23]	OV [24]	Elem.	IC [23]	OV [24]	Elem.	IC [23]	OV [24]
H	13.598	13.598	K	3607.4	4.341	Rb	15199.7	4.177
He	24.587	24.587	Ca	4038.1	6.113	Sr	16104.6	5.695
Li	54.75	5.392	Sc	4492.8	6.540	Y	17038.4	6.380
Be	111.0	9.322	Ti	4965.3	6.820	Zr	17997.6	6.840
B	188.0	8.298	V	5465.1	6.740	Nb	18985.6	6.880
C	283.8	11.260	Cr	5989.2	6.766	Mo	19999.5	7.099
N	401.6	14.534	Mn	6539.0	7.435	Tc	21044.0	7.280
O	532.0	13.618	Fe	7111.3	7.870	Ru	22117.2	7.370
F	685.4	17.422	Co	7708.9	7.860	Rh	23219.9	7.460
Ne	866.9	21.564	Ni	8332.8	7.635	Pd	24350.3	8.340
Na	1072.1	5.139	Cu	8978.7	7.726	Ag	25514.0	7.576
Mg	1305.0	7.646	Zn	9658.6	9.394	Cd	26711.2	8.993
Al	1559.6	5.986	Ga	10367.1	5.999	In	27939.9	5.786
Si	1838.9	8.185	Ge	11103.1	7.899	Sn	29200.1	7.344
P	2149.0	10.486	As	11866.7	9.810	Sb	30491.2	8.641
S	2472.0	10.360	Se	12657.8	9.752	Te	31813.8	9.009
Cl	2822.4	12.967	Br	13473.7	11.814	I	33169.4	10.451
Ar	3202.9	15.759	Kr	14325.6	13.999	Xe	34561.4	12.130

Due to Heisenberg's uncertainty and Pauli's exclusion principles, the properties of multifermionic systems display periodicities arising from the fermions being grouped into shells and subshells. The shell structure of the one-particle energy spectrum of these systems generates so-called shell effects, at the different hierarchical levels of matter (nuclei, atoms, metal clusters, ...) [14-19]. These effects are related to the variation patterns of the global properties of these systems, revealing the shell-filling process. In atoms, shell effects are manifested as peculiarities in the variations, with atomic number

Z, of such elemental properties as ionization potential, electron affinity, electronegativity or chemical hardness [17-22]. In Table 1, there are given measured values for the first (valence) and last (1*s*-core) ionization potentials of atoms, from He through Xe. As expected, valence energies display much stronger shell effects than core energies (Fig. 1).

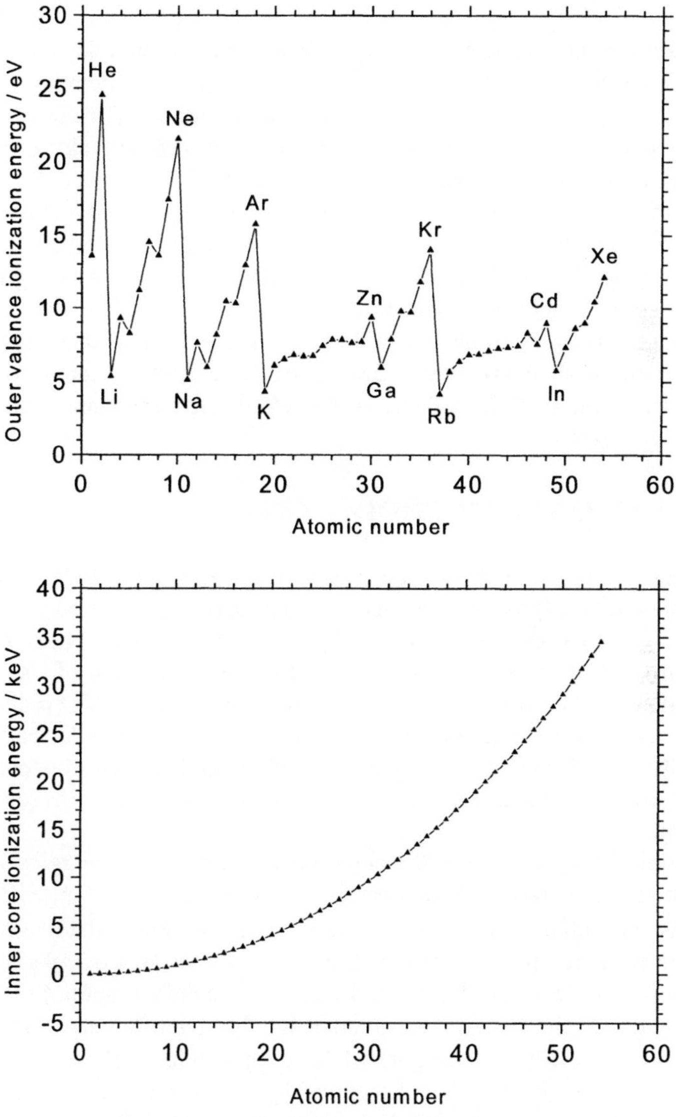

Fig. 1. Comparison between measured inner-core [23] (lower graph) and outer-valence [24] (upper graph) ionization energies for atoms, from He through Xe. The valence energies display an obvious shell structure, whereas this does not appear at first sight for the core energies.

In previous work [20-22] we have shown how the total binding energy of an atom is divided into two contributions: the first one displays a monotonous variation as a function of Z and describes the gross behavior of the system, whereas the second one shows an oscillating variation which involves the information on the shell structure. In particular, we have shown how the oscillating part of the energy properties is determined by the more external electrons [21], which may help define the core-valence separation in the system.

In this paper, we investigate the degree of dependence of the relaxation energy in a core ionization on the shell-filling process. This term expresses the perturbation of the electron cloud by the presence of the core hole and is a function of the whole electron distribution. Therefore, if the relaxation energy is determined with sufficient accuracy, its variation as a function of Z should display shell effects. In other words, the investigation of the variation of this term as a function of Z may help evaluating the accuracy of its determination. The fact that shell effects stem from tiny irregularities in the energy spectrum also suggests that their investigation could help understanding the specific properties of a system under core ionization.

We report a thorough investigation of the dependence of the relaxation energy of 1*s* core ionization on the shell structure in a series of atoms from Helium through Xenon. We also investigate the shell dependence of the relativistic corrections to the ionization and relaxation energies.

2. COMPUTATIONAL METHODOLOGY

We have performed calculations on elements of the periodic table from Helium (Z = 2) through Xenon (Z = 54) for both the neutral atoms and their 1*s* core-hole ions in their ground state. We have used Bruneau's multi-configuration Dirac-Fock (MCDF) *ab-initio* program [25-27], which is based on a numerical resolution of the Breit-Dirac equation for atomic systems corrected for various effects such as QED radiative contributions to the energies and nuclear size and motion terms. The angular part of the calculation is fully vectorized using Racah algebra, allowing consistent treatment of open-shell systems, while the radial part is computed through a numerical resolution of the radial equation.

The program yields, on request, the eigenstates multiplicities, energies and vectors, the relevant atomic integrals, and such properties as the effective charge Z*, the mean radius $\langle 1/r \rangle$ or the spin-orbit coupling parameter λ. In addition to the relativistic (y-rt), non-relativistic (n-rt) multi-configuration Hartree-Fock (MCHF) calculations can also be made, by setting the speed of light (c = 137.036 in the default option) to a very large value (10^6 in our n-rt calculations). The average level, extended average level and optimal level methods [28, 29] are incorporated in the code, which allows the computation of several states in a single run. This is achieved by minimizing the sum (or a weighted sum) of the energies corresponding to those states whose wave-functions are built with a same set of orbitals. This procedure is more appropriate than optimizing a single state and using frozen orbitals to describe the other states because it yields orthogonal wavefunctions for the considered states (which may not be the case in separate calculations). The code also allows using different functions with same quantum numbers and thus adapted descriptions of the orbitals in the initial and final states. These two features are important in a correct treatment of excitation and ionization processes.

Our calculations were performed on atoms and ions in the ground-state, in the basic configuration scheme, optimizing the levels of the lowest spectroscopic multiplet term. For core-hole ions we used the same configuration as for atoms but with a hole in level 1*s*. We made both n-rt and y-rt calculations, including radiative contributions to the energy but neglecting nuclear size and motion terms (since we investigate electronic shell effects and wish to avoid the influence - probably tiny but ambiguous - of nuclear shell structure). Total binding energies were calculated for all atoms in the ground state (E_A) and in the 1*s*-core ionized state (E_I). Both ΔDF (relativistic) and ΔHF (non-relativistic) relaxed ($E_R = E_I - E_A$) and respective Koopmans non-relaxed ($E_K = |\varepsilon_{1s}|$) ionization energies were derived. In both relativistic and non-relativistic calculations, the relaxation energy ΔE_{RK} was defined as $E_R - E_K$. Our aim was to investigate shell effects in the relaxation and relativistic parts of the ionization energy and, eventually, in the relativistic part of relaxation itself.

3. RESULTS AND DISCUSSION

Following the definition (1), the relaxation energy ΔE_{RK} $(i \rightarrow f)$ is obtained as the difference between the relaxed core-hole energy E_R $(i \rightarrow f) = E_I - E_A$ (where E_I and E_A are the total ground-state energies for the core-hole ion and the neutral atom, respectively) and the Koopmans, non-relaxed core ionization energy E_K $(i \rightarrow f) = |\varepsilon_{1s}|$ (where ε_{1s} is the HF or DF energy of orbital 1*s* in the neutral atom). The results from the y-rt (ΔDF) and n-rt (ΔHF) calculations for 1*s*-core ionization of elements from He (Z = 2) through Xe (Z = 54) are given in Table 2. A subscript + in columns 2 and 4 is used to designate the core-hole ions. In column 3 there are recalled the spectroscopic terms on which the optimization was performed for the neutral atom (upper line) and the core-hole ion (lower line), keeping the basic configuration unchanged except for the 1*s* vacancy (column 4).

The values given in columns 5 to 9 are obtained when including (upper line) and not including (lower line) the relativistic effects. For both the atom (column 5) and the ion (column 6), the total binding energy of electrons is lowered when relativity is included. Since there are more electrons in the atom than in the ion, both relaxed (column 7) and Koopmans (column 8) ionization energies are larger in relativistic calculations. Similarly, all electrons being involved in the relaxed but not in the Koopmans process, the relaxation energy (column 9) has a larger magnitude when relativity is included. The relativistic increment to the relaxation energy is given in column 10, where it appears to be larger than experimental accuracy even for light elements with Z > 10.

Table 2. For elements from He (Z = 2) through Xe (Z = 54) and for the spectroscopic term (3rd column) in the basic-configuration scheme (4th column) of the neutral (upper line) and core-ionized (lower line) atoms, calculated relativistic (upper line) and non-relativistic (lower line) energies (in eV). Total binding energies are given for the atom in the ground state (E_A) and in the 1s-core ionized state (E_I). Both ΔDF and ΔHF relaxed ($E_R = E_I - E_A$) and respective Koopmans non-relaxed ($E_K = |\varepsilon_{1s}|$) ionization energies are given. In both relativistic and non-relativistic calculations, the relaxation energy ΔE_{RK} is defined as E_R - E_K. The relativistic correction to the relaxation energy, $\delta\Delta E_{RK}$ (difference between the first and second lines in column 9), is given in column 10.

Z	El.	Term	Config.	E_A	E_I	$E_R = \Delta E_{IA}$	$E_K = -\varepsilon_{1s}$	ΔE_{RK}	$\delta\Delta E_{RK}$
2	He	$^1S_0$	$1s^2$	-77.872	-54.425	23.447	24.980	-1.533	-0.001
	He^+	$^2S_{1/2}$	$1s^1$	-77.870	-54.423	23.447	24.979	-1.532	
3	Li	$^2S_{1/2}$	$(He)2s^1$	-202.267	-139.047	63.220	67.429	-4.209	-0.008
	Li_+	$^3S_1$	$(He^+)2s^1$	-202.255	-139.033	63.222	67.423	-4.201	
4	Be	$^1S_0$	$(He)2s^2$	-396.603	-273.267	123.336	128.805	-5.469	-0.022
	Be_+	$^2S_{1/2}$	$(He^+)2s^2$	-396.552	-273.216	123.336	126.783	-5.447	
5	B	$^2P_{1/2}$	$(He)2s^22p^1$	-667.614	-467.330	200.284	209.463	-9.179	-0.047
	B_+	$^3P_0$	$(He^+)2s^22p^1$	-667.470	-467.201	200.269	209.401	-9.132	
6	C	$^3P_0$	$(He)2s^22p^2$	-1025.891	-730.472	295.419	308.322	-12.903	-0.086
	C_+	$^4P_{1/2}$	$(He^+)2s^22p^2$	-1025.560	-730.194	295.366	308.183	-12.817	
7	N	$^4S_{3/2}$	$(He)2s^22p^3$	-1480.986	-1072.189	408.797	425.561	-16.764	-0.220
	N_+	$^5S_2$	$(He^+)2s^22p^3$	-1480.401	-1071.656	408.745	425.289	-16.544	
8	O	$^3P_2$	$(He)2s^22p^4$	-2036.870	-1493.174	543.696	562.910	-19.214	-0.232
	O_+	$^4P_{5/2}$	$(He^+)2s^22p^4$	-2035.666	-1492.224	543.442	562.424	-18.982	
9	F	$^2P_{3/2}$	$(He)2s^22p^5$	-2707.109	-2009.905	697.204	718.707	-21.503	-0.337
	F_+	$^3P_2$	$(He^+)2s^22p^5$	-2705.067	-2008.321	696.746	717.912	-21.166	
10	Ne	$^1S_0$	$(He)2s^22p^6$	-3501.178	-2631.827	869.351	893.009	-23.658	-0.499
	Ne_+	$^2S_{1/2}$	$(He^+)2s^22p^6$	-3497.946	-2629.321	868.625	891.784	-23.159	
11	Na	$^2S_{1/2}$	$(Ne)3s^1$	-4409.370	-3331.056	1078.314	1103.285	-24.971	-0.690
	Na_+	$^3S_1$	$(Ne_+)3s^1$	-4404.407	-3327.211	1077.196	1101.477	-24.281	
12	Mg	$^1S_0$	$(Ne)3s^2$	-5439.141	-4128.274	1310.867	1336.806	-25.939	-0.924
	Mg_+	$^2S_{1/2}$	$(Ne_+)3s^2$	-5431.793	-4122.586	1309.207	1334.222	-25.015	
13	Al	$^2P_{1/2}$	$(Ne)3s^23p^1$	-6592.339	-5024.293	1568.046	1595.502	-27.456	-1.168
	Al_+	$^3P_0$	$(Ne_+)3s^23p^1$	-6581.803	-5016.196	1565.607	1591.895	-26.288	
14	Si	$^3P_0$	$(Ne)3s^23p^2$	-7874.803	-6026.488	1848.315	1877.298	-28.983	-1.438
	Si_+	$^4P_{1/2}$	$(Ne_+)3s^23p^2$	-7860.130	-6015.192	1844.938	1872.483	-27.545	
15	P	$^4S_{3/2}$	$(Ne)3s^23p^3$	-9291.407	-7139.597	2151.810	2182.651	-30.841	-1.940
	P_+	$^5S_2$	$(Ne_+)3s^23p^3$	-9271.434	-7124.247	2147.187	2176.088	-28.901	
16	S	$^3P_2$	$(Ne)3s^23p^4$	-10843.301	-8363.580	2479.721	2512.175	-32.454	-2.417
	S_+	$^4P_{5/2}$	$(Ne_+)3s^23p^4$	-10816.662	-8343.129	2473.533	2503.570	-30.037	

17	**Cl**	$^2P_{3/2}$	$(Ne)3s^23p^5$	-12538.048	-9706.995	2831.053	2865.134	-34.081	-2.958
	Cl₊	$^3P_2$	$(Ne_+)3s^23p^5$	-12503.149	-9680.220	2822.929	2854.052	-31.123	
18	**Ar**	$^1S_0$	$(Ne)3s^23p^6$	-14380.447	-11174.563	3205.884	3241.609	-35.725	-3.563
	Ar₊	$^2S_{1/2}$	$(Ne_+)3s^23p^6$	-14335.440	-11140.048	3195.392	3227.554	-32.162	
19	**K**	$^2S_{1/2}$	$(Ar)4s^1$	-16361.386	-12747.308	3614.078	3651.219	-37.141	-4.260
	K₊	$^3S_1$	$(Ar_+)4s^1$	-16304.110	-12703.370	3600.740	3633.621	-32.881	
20	**Ca**	$^1S_0$	$(Ar)4s^2$	-18487.527	-14439.754	4047.773	4086.180	-38.407	-5.042
	Ca₊	$^2S_{1/2}$	$(Ar_+)4s^2$	-18415.535	-14384.504	4031.031	4064.396	-33.365	
21	**Sc**	$^2D_{3/2}$	$(Ar)3d^14s^2$	-20762.747	-16262.646	4500.101	4541.178	-41.077	-5.920
	Sc₊	$^3D_1$	$(Ar_+)3d^14s^2$	-20673.470	-16194.258	4479.212	4514.369	-35.157	
22	**Ti**	$^3F_2$	$(Ar)3d^24s^2$	-23195.835	-18219.700	4976.135	5019.770	-43.635	-6.872
	Ti₊	$^4F_{3/2}$	$(Ar_+)3d^24s^2$	-23086.312	-18135.955	4950.357	4987.120	-36.763	
23	**V**	$^4F_{3/2}$	$(Ar)3d^34s^2$	-25790.273	-20313.972	5476.301	5522.601	-46.300	-7.959
	V₊	$^5F_1$	$(Ar_+)3d^34s^2$	-25657.199	-20212.345	5444.854	5483.195	38.341	
24	**Cr**	$^7S_3$	$(Ar)3d^54s^1$	-28551.265	-22557.273	5993.992	6044.410	-48.418	-7.113
	Cr₊	$^8S_{7/2}$	$(Ar_+)3d^54s^1$	-28391.183	-22435.429	5955.761	5997.059	-41.305	
25	**Mn**	$^6S_{5/2}$	$(Ar)3d^54s^2$	-31480.978	-24931.459	6549.519	6601.364	-51.845	-10.391
	Mn₊	$^7S_3$	$(Ar_+)3d^54s^2$	-31289.466	-24785.623	6503.843	6545.297	-41.454	
26	**Fe**	$^5D_4$	$(Ar)3d^64s^2$	-34580.063	-27456.353	7123.710	7178.541	-54.831	-11.810
	Fe₊	$^6D_{9/2}$	$(Ar_+)3d^64s^2$	-34352.853	-27283.515	7069.338	7112.359	-43.021	
27	**Co**	$^4F_{9/2}$	$(Ar)3d^74s^2$	-37857.958	-30135.661	7722.297	7780.237	-57.940	-13.365
	Co₊	$^5F_5$	$(Ar_+)3d^74s^2$	-37590.218	-29932.166	7658.052	7702.627	-44.575	
28	**Ni**	$^3F_4$	$(Ar)3d^84s^2$	-41317.602	-32971.969	8345.633	8406.813	-61.180	-15.059
	Ni₊	$^4F_{9/2}$	$(Ar_+)3d^84s^2$	-41004.061	-32733.851	8270.210	8316.331	-46.121	
29	**Cu**	$^2S_{1/2}$	$(Ar)3d^{10}4s^1$	-44963.138	-35976.844	8986.294	9052.088	-65.794	-16.826
	Cu₊	$^3S_1$	$(Ar_+)3d^{10}4s^1$	-44598.492	-35700.550	8897.942	8946.910	-48.968	
30	**Zn**	$^1S_0$	$(Ar)3d^{10}4s^2$	-48800.519	-39133.575	9666.944	9734.945	-68.001	-18.784
	Zn₊	$^2S_{1/2}$	$(Ar_+)3d^{10}4s^2$	-48377.729	-38813.035	9564.694	9613.911	-49.217	
31	**Ga**	$^2P_{1/2}$	$(Zn)4p^1$	-52822.427	-42446.278	10376.149	10446.955	-70.806	-20.946
	Ga₊	$^3P_0$	$(Zn_+)4p^1$	-52334.617	-42076.297	10258.320	10308.180	-49.860	
32	**Ge**	$^3P_0$	$(Zn)4p^2$	-57033.894	-45921.853	11112.041	11185.704	-73.663	-23.189
	Ge₊	$^4P_{1/2}$	$(Zn_+)4p^2$	-56473.436	-45496.641	10976.795	11027.269	-50.474	
33	**As**	$^4S_{3/2}$	$(Zn)4p^3$	-61438.001	-49563.306	11874.695	11951.475	-76.780	-25.646
	As₊	$^5S_2$	$(Zn_+)4p^3$	-60796.753	-49076.611	11720.142	11771.276	-51.134	
34	**Se**	$^3P_2$	$(Zn)4p^4$	-66034.571	-53369.643	12664.928	12745.034	-80.106	-28.334
	Se₊	$^4P_{5/2}$	$(Zn_+)4p^4$	-65303.747	-52814.671	12489.076	12540.848	-51.772	
35	**Br**	$^2P_{3/2}$	$(Zn)4p^5$	-70829.563	-57347.402	13482.161	13565.740	-83.579	-31.192
	Br₊	$^3P_2$	$(Zn_+)4p^5$	-69999.720	-56716.879	13282.841	13335.228	-52.387	
36	**Kr**	$^1S_0$	$(Zn)4p^6$	-75826.291	-61499.587	14326.704	14413.750	-87.046	-34.059
	Kr₊	$^2S_{1/2}$	$(Zn_+)4p^6$	-74887.258	-60785.814	14101.444	14154.431	-52.987	

37	**Rb**	$^2S_{1/2}$	$(Kr)5s^1$	-81016.004	-65809.989	15206.015	15296.742	-90.727	-37.239
	Rb_+	$^3S_1$	$(Kr_+)5s^1$	-79956.809	-65004.371	14952.438	15005.926	-53.488	
38	**Sr**	$^1S_0$	$(Kr)5s^2$	-86404.793	-70291.078	16113.715	16208.138	-94.423	-40.610
	Sr_+	$^2S_{1/2}$	$(Kr_+)5s^2$	-85213.730	-69384.578	15829.152	15882.965	-53.813	
39	**Y**	$^2D_{3/2}$	$(Kr)4d^15s^2$	-91994.712	-74948.307	17046.405	17145.462	-99.057	-44.216
	Y_+	$^3D_1$	$(Kr_+)4d^15s^2$	-90659.778	-73932.006	16727.772	16782.613	-54.841	
40	**Zr**	$^3F_2$	$(Kr)4d^25s^2$	-97792.782	-79786.197	18006.585	18110.329	-103.744	-48.041
	Zr_+	$^4F_{3/2}$	$(Kr_+)4d^25s^2$	-96300.996	-78650.098	17650.898	17706.601	-55.703	
41	**Nb**	$^6D_{1/2}$	$(Kr)4d^45s^1$	-103801.137	-84811.025	18990.112	19099.623	-109.511	-52.244
	Nb_+	$^7D_1$	$(Kr_+)4d^45s^1$	-102139.751	-83545.492	18594.259	18651.526	-57.267	
42	**Mo**	$^7S_3$	$(Kr)4d^55s^1$	-110025.852	-90019.500	20006.352	20120.579	-114.227	-56.200
	Mo_+	$^8S_{7/2}$	$(Kr_+)4d^55s^1$	-108179.263	-88612.776	19566.487	19624.514	-58.027	
43	**Tc**	$^6S_{5/2}$	$(Kr)4d^55s^2$	-116465.015	-95409.364	21055.651	21174.492	-118.841	-60.639
	Tc_+	$^7S_3$	$(Kr_+)4d^55s^2$	-114417.067	-93848.318	20568.749	20626.951	-58.202	
44	**Ru**	$^5F_5$	$(Kr)4d^75s^1$	-123123.965	-100999.01	22124.955	22250.344	-125.389	-65.711
	Ru_+	$^6F_{11/2}$	$(Kr_+)4d^75s^1$	-120859.267	-99272.428	21586.839	21646.517	-59.678	
45	**Rh**	$^4F_{9/2}$	$(Kr)4d^85s^1$	-130007.499	-106779.827	23227.672	23358.853	-131.181	-70.681
	Rh_+	$^5F_5$	$(Kr_+)4d^85s^1$	-127508.030	-104873.729	22634.301	22694.801	-60.500	
46	**Pd**	$^1S_0$	$(Kr)4d^{10}$	-137118.265	-112762.885	24355.380	24492.770	-137.390	-75.462
	Pd_+	$^2S_{1/2}$	$(Kr_+)4d^{10}$	-134366.225	-110664.654	23701.571	23763.499	-61.928	
47	**Ag**	$^2S_{1/2}$	$(Kr)4d^{10}5s^1$	-144460.380	-118939.021	25521.359	25664.420	-143.061	-80.902
	Ag_+	$^3S_1$	$(Kr_+)4d^{10}5s^1$	-141434.980	-116631.109	24803.871	24866.030	-62.159	
48	**Cd**	$^1S_0$	$(Kr)4d^{10}5s^2$	-152031.915	-125313.707	26718.208	26867.223	-149.015	-86.621
	Cd_+	$^2S_{1/2}$	$(Kr_+)4d^{10}5s^2$	-148712.069	-122779.789	25932.280	25994.674	-62.394	
49	**In**	$^2P_{1/2}$	$(Cd)5p^1$	-159832.083	-131884.746	27947.337	28102.846	-155.509	-92.660
	In_+	$^3P_0$	$(Cd_+)5p^1$	-156195.995	-129108.172	27087.823	27150.672	-62.849	
50	**Sn**	$^3P_0$	$(Cd)5p^2$	-167865.730	-138658.413	29207.317	29369.487	-162.170	-98.926
	Sn_+	$^4P_{1/2}$	$(Cd_+)5p^2$	-163890.145	-135621.211	28268.934	28332.178	-63.244	
51	**Sb**	$^4S_{3/2}$	$(Cd)5p^3$	-176135.867	-145637.366	30498.501	30667.701	-169.200	-105.550
	Sb_+	$^5S_2$	$(Cd_+)5p^3$	-171796.297	-142320.611	29475.686	29539.336	-63.650	
52	**Te**	$^3P_2$	$(Cd)5p^4$	-184642.419	-152820.987	31821.432	31998.190	-176.758	-112.697
	Te_+	$^4P_{5/2}$	$(Cd_+)5p^4$	-179913.150	-149204.554	30708.596	30772.657	-64.061	
53	**I**	$^2P_{3/2}$	$(Cd)5p^5$	-193391.095	-160214.564	33176.531	33360.765	-184.234	-119.781
	I_+	$^3P_2$	$(Cd_+)5p^5$	-188244.989	-156277.821	31967.168	32031.621	-64.453	
54	**Xe**	$^1S_0$	$(Cd)5p^6$	-202384.813	-167821.342	34563.471	34755.530	-192.059	-127.226
	Xe_+	$^2S_{1/2}$	$(Cd_+)5p^6$	-196793.369	-163542.014	33251.355	33316.188	-64.833	

In Figure 2 there are displayed, as a function of atomic number Z, the total binding energy of electrons in neutral atoms: E_A, and in their 1s core-hole ions: E_I, and the 1s-core ionization energy in both Koopmans: E_K, and ΔDF: E_R, approximations, as given by relativistic (y-rt) and non-relativistic (n-rt) calculations. Again we notice that, contrary to valence ionization energies, core ionization energies display no apparent peculiarities, both functions E_K and E_R increasing monotonically with Z. This is because the energy necessary to extract a deep-core electron from an atom with a large number of electrons overwhelms the shell-effect influence from the outer electrons.

In previous work [20, 21] we made use of the discrete second derivative to enhance tiny irregularities in the variations of energies with atomic number Z, that is, to make visible the shell structure and peculiarities stemming from the shell-filling process. In Figure 3 there are shown the second-order finite-difference curves of the quantities displayed in Figure 2. There appears no qualitative difference between the relativistic and non-relativistic curves, except for the Z-increasing drift. The curves now show the saw-tooth-like pattern characteristic of shell effects. The second derivative of the ground-state atomic energy E_A reveals all the peculiarities of the shell structure, all atoms with increased stability due to a full or half filling of the outermost orbital being singled out. The second derivative of the total binding energy E_I for the 1*s* core-hole ion shows the same tendencies, but with effects more clearly pronounced, particularly for large Z, and some distortion in the transition-metal series. The irregularities in the shell-filling process appear very clearly, for instance with the peaks for Cr and Cu and the oscillations in the Nb-Pd sequence (Table 2, column 4). This can be understood from the fact that, due to the core vacancy, the ions have decreased stability, which increases their dependence on the valence electrons and their sensitivity to the shell-filling process.

Also in Figure 3 there are shown the discrete second derivatives of the Koopmans: E_K, and ΔDF: E_R, 1*s*-core ionization energies. The shell-structure effects appear more enhanced in these differential properties. We observe maxima for the atoms with filled shells (Ne, Ar, Kr) and for the atoms with half-filled *p*-subshell (N, P, As, Sb). We also observe sharp maxima / minima for such atoms as Cr / Mn, Cu / Zn, Mo / Tc as well as for Nb, Ru, Pd, where irregularities occur in the filling process of the *d*-subshell (Table 2, column 4). For the atoms with filled *s*-subshell there is an even distribution of maxima (Be, Mg) and minima (Ca, Sr). For the transition-metal series, from Ca through Zn (3*d*) and especially from Sr through Cd (4*d*), there appears regular staggering patterns. These can be understood by looking at the sequences of basic configurations (Table 2), having in mind that the overlap between the valence electrons and the core vacancy varies according to the configuration: $nd^m(n+1)s^1$ or $nd^{m+1}(n+1)s^0$.

In order to investigate the variations of the relativistic component of the considered energies, we display in Figure 4 the relativistic corrections to the ground-state energies of neutral atoms and of their 1*s* core-hole ions, δE_A and δE_I respectively (upper left), and to the Koopmans and ΔDF 1*s*-core ionization energies, δE_K and δE_R (upper right), together with their discrete second derivatives (lower line). As expected from Figure 2, the upper curves decrease (left) or increase (right) monotonically, with no apparent dependence on the shell-filling process.

Fig. 2. Total binding energy for neutral atoms: E_A, and for their 1s core-hole ions: E_I, and 1*s*-core ionization energy in the Koopmans: E_K, and the ΔDF: E_R, approximations, for elements from He through Xe. Calculations were performed with and without including relativistic effects: in the figures, 'rt' stands for relativistic effects and 'y' and 'n' for including or not including them.

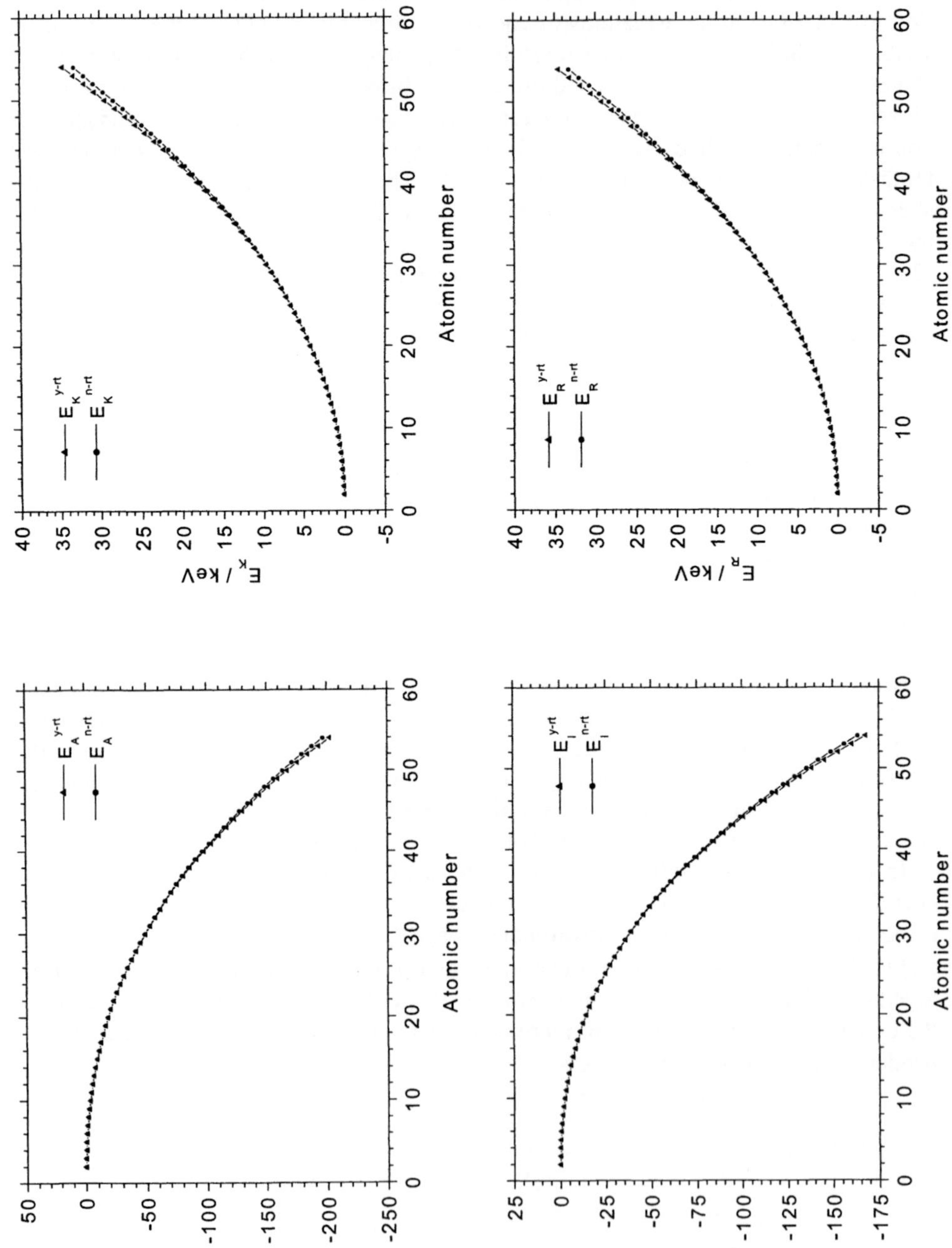

Fig. 3. Discrete second derivatives of the quantities displayed in Fig. 1: neutral atom energy E_A, core-hole ion energy E_I, and 1*s*-core ionization Koopmans (E_K) and ΔDF (E_R) energies, for elements from He through Xe. Calculations were performed with and without including relativistic effects: in the figures, 'rt' stands for relativistic effects and 'y' and 'n' for including or not including them.

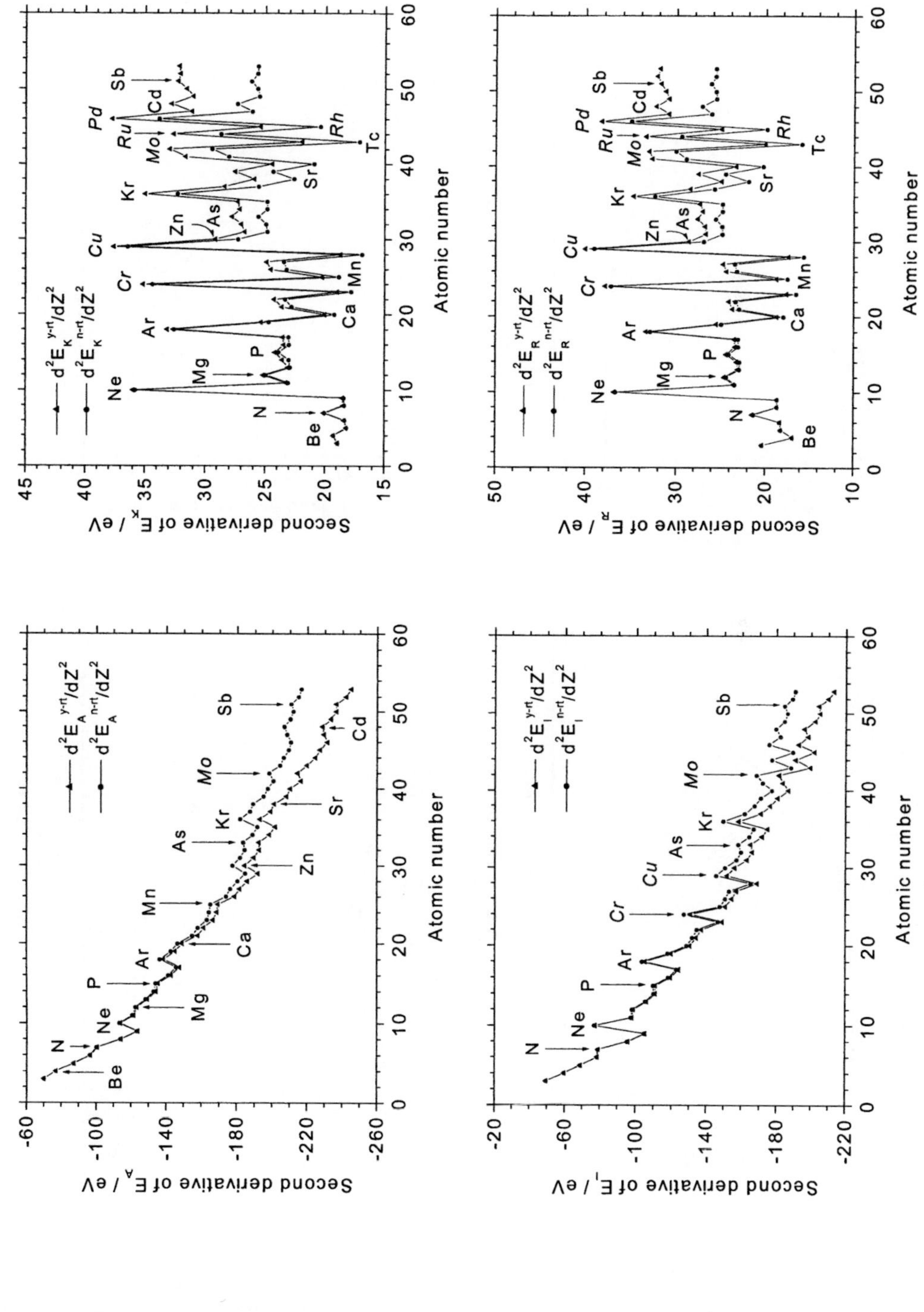

Fig. 4. Relativistic corrections to the ground-state energies of neutral atoms: δE_A, and of their 1s core-hole ions: δE_I (upper left), and to the Koopmans: δE_K, and ΔDF: δE_R, corresponding energies (upper right), together with their discrete second derivatives (lower line).

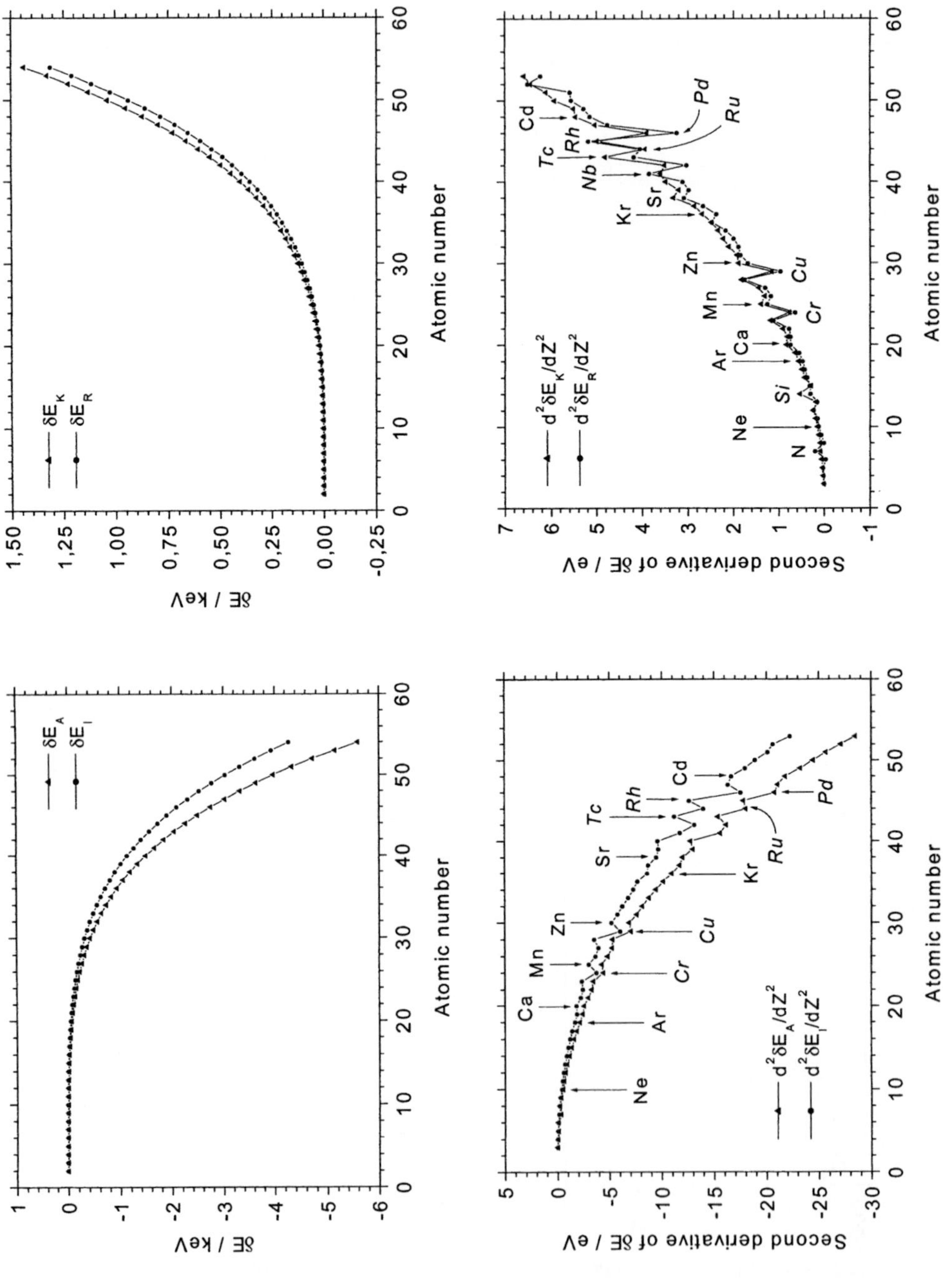

The corresponding second-order finite-difference curves presented below do not exhibit the usual shell-effect pattern (no maxima for rare-gas elements), but they reveal a selective sensitivity to shell-filling irregularities. We clearly see the peculiarities for the Cr / Mn and Cu / Zn couples and again a succession of maxima (Z odd) and minima (Z even) between Nb and Pd. These observations reveal the selective contribution of relativistic effects to shell-filling irregularities.

While the curves shown on the upper and lower lines of Figure 4 present the relativistic contributions extracted from the curves shown on Figs 2 and 3, respectively, the upper-left and lower-left curves of Figure 5 present the y-rt and n-rt relaxation energies and their discrete second derivatives extracted from the curves on the right column of Figs 2 and 3, respectively. As the relaxation energies are already differences between differential properties (the Koopmans and ΔDF ionization energies), they display some structure, though not as resolved as that of the derivatives. The features appearing in their derivatives are essentially the same as those appearing in the derivatives of Koopmans and ΔDF ionization energies (right column of Fig. 3), except that now they are enhanced for the lighter elements and reduced for the heavier. In addition, Be, Mg, Mn and Cd, which have a filled n*s*-subshell, appear clearly as minima, as also do Ca and Sr, and there is a progressive reversal of extrema from N through Sb.

The last differential quantity we have investigated is the relativistic contribution to the relaxation energy. The curves shown on the right-hand side of Figure 5 result from the difference of the y-rt and n-rt curves facing them on the left-hand side. Since these latter are very similar - except for the relativistic drift - their structures cancel out in the difference, which shows no resolution. The discrete second derivative curve shown in the lower-right corner is the difference between the curves shown in the corresponding diagram of Figure 4. Now the rare gases appear, but with a progressive reversal of extrema similar to that of the series N through Sb, which is enhanced. There is also a tiny similar feature arising for the series Mg, Ca, Sr, whereas Zn and Cd as well as Mn and Tc appear clearly as minima. Strong oscillations again occur between Nb and Pd, with a slightly distorted pattern. This curve expresses the selective contribution of relativistic relaxation to the shell-filling irregularities.

4. CONCLUSION

In this paper we have studied the influence of relaxation and relativity on the shell-filling process for elements from He through Xe. For this purpose we have investigated the variations, along the periodic table, of inner-core ionization energies in both Koopmans and ΔDF approximations, using both relativistic and non-relativistic calculations. We have shown that, for these four levels of calculation (y-rt and n-rt, ΔDF and Koopmans), the ionization energies, as well as such differential quantities as their relativistic component, the relaxation energy and, eventually, the relativistic component of the relaxation energy, display variations typical of the shell-filling process, while presenting specific differences which characterize their contribution.

Fig. 5. Relativistic and non-relativistic relaxation energies for 1*s*-core ionization: ΔE_{RK} / y-rt and n-rt (upper left), and relativistic component obtained from their difference: $\delta\Delta E_{RK}$ (upper right), together with their discrete second derivatives (lower line).

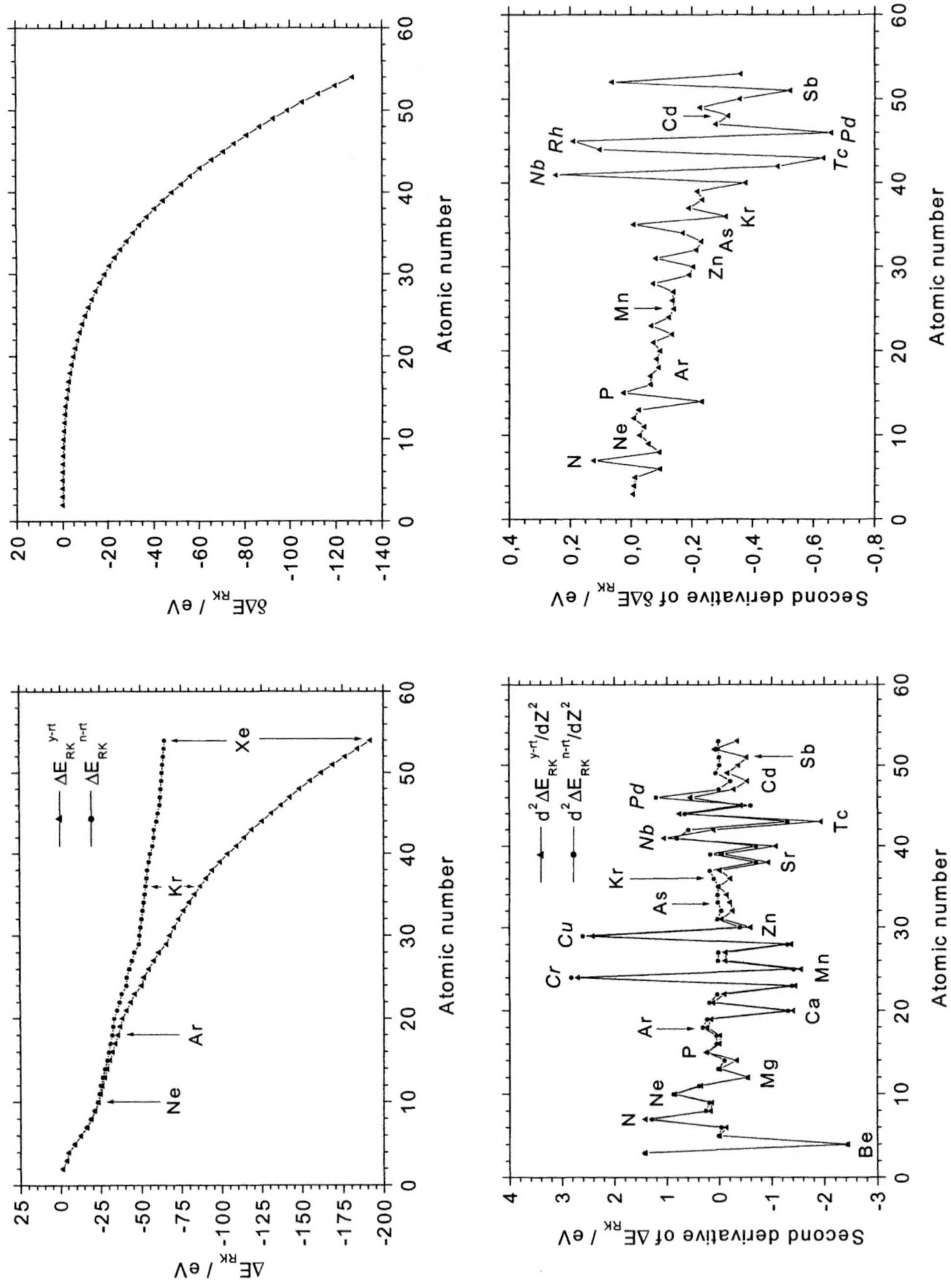

This procedure permits to extract the selective contributions of relaxation, relativity and cross effects to the peculiarities of the shell-filling process. It is remarkable that, after so many subtractions between numbers of decreasing magnitude (Table 2), one can still recognize characteristic shell patterns, while discovering novel shell features. The sensitivity of these differential properties to the shell-filling process could be used as a probe of the accuracy of calculations of core excitation or ionization energies.

It was recalled that, contrary to the valence ionization energies, core ionization energies display no apparent peculiarities. This is related to the specificity of the Coulomb field, which yields a deeply bound $1s$ orbital while the rest of the spectrum has a very weak influence. In future work we will investigate shell effects in the relaxation energy of $2s$ and $2p$-core ionizations of neutral and valence-ionized atoms and in the spin-orbit splitting of the $2p$-levels.

Dedication

As this paper is dedicated to Per-Olov Löwdin, who always showed a broad range of interests, from mathematical physics through quantum chemistry to molecular biology, it may be appropriate at this point to remind his interest for music, not only for piano performing (he had a beautiful grand piano at home) but also for the theory of tone and harmony [30], which stems from the very same root as the quantum theory. As long ago as the 6th century BC, Pythagoras had devised the first diatonic scale by using strings of different lengths, and in the 20th century AD, Louis de Broglie showed that quantum numbers in atomic spectra are related to standing waves of the rotating electrons. Actually, it has even been shown that masses of stable elementary particles are distributed as the note frequencies of the usual chromatic scale [31], and tentatives have been made to devise musical scores from molecular spectra by assigning a note pitch to each frequency and a duration to its intensity [32]. Musical scores have also been built from sequences of nucleotidic bases in nucleic acids [33 a,b] and from sequences of aminoacids in proteins [33 c,d].

The relevance of this paper to physics and music lies in the fact that shell effects show that type of quasi-periodic regularity that is common in art and music. Indeed, musical scores have recently been built from the shell effects appearing in the variations of ionization potential, electron affinity, electronegativity and chemical hardness along the periodic table [34].

Acknowledgments

This work was partly supported by a twinning convention between the Universities Pierre et Marie Curie (Paris) and Saint Kliment Ohridski (Sofia), and by the European project COST-D9/010/98. Doctor Jean Bruneau from CEA is gratefully acknowledged for discussion and advice about the use of his program, and Pr Michel Tronc for a critical reading of the manuscript. One of us (A. I. K.) thanks the French Government for a PhD scholarship.

REFERENCES

1. K. Siegbahn, C. Nordling, G. Johansson *et al.*, *ESCA applied to free molecules*, North-Holland, Amsterdam, 1971.
2. W.L. Jolly, K.D. Bomben and C.J. Eyermann, Atomic Data & Nuclear Data Tables **31**, 433 (1984).
3. S. Hüfner, in M. Cardona, P. Fulde, K. von Klitzing and H.-J. Queisser (eds), *Photoelectron spectroscopy*, Series in solid-state physics **82**, Springer, Berlin, 1995, p. 31.
4. T. Koopmans, Physica **1**, 104 (1933).
5. R. Daudel, G. Leroy, D. Peeters and M. Sana, *Quantum Chemistry,* John Wiley, New York, 1983, Sec. 4.2.
6. E. Clementi and A. Routh, Int. J. Quant. Chem. **6**, 525 (1972); B. Lévy, Ph. Millié, J. Ridard and J. Vinh, J. Elec. Spect. & Rel. Phen. **4**, 13 (1974), and references therein.
7. H. Hsu, E.R. Davidson and R.M. Pitzer, J. Chem. Phys. **65**, 609 (1976); A.R. Rossi and E.R. Davidson, J. Chem. Phys. **96**, 10682 (1992).
8. D. Firsht and R. McWeeny, Molec. Phys. **32**, 1637 (1976); C. Amovilli and R. McWeeny, in M. Defranceschi and Y. Ellinger (eds), *Strategies and Applications in Quantum Chemistry*, Kluwer, Dordrecht, 1996, p. 165.
9. C. Kozmutza, E. Kapuy, M.A. Robb, R. Daudel and I.G. Csizmadia, J. Comp. Chem. **3**, 14 (1982); C. Kozmutza and E. Tfirst, private communication.
10. A. Filipponi, E. Bernieri and S. Mobilio, Phys. Rev. B **38**, 3298 (1988); S. Bodeur, P. Millié and I. Nenner, Phys. Rev. A **41**, 252 (1990).
11. H.J.Aa. Jensen, P. Jorgensen and H. Agren, J. Chem. Phys. **87**, 451 (1987); H.J.Aa. Jensen, H. Agren and J. Olsen, in E. Clementi (edr), *Modern Techniques in Computational Chemistry*, ESCOM, Leiden, 1990, chap. 8. See also: H. Agren and H.J.Aa. Jensen, Chem. Phys. **172**, 45 (1993).
12. D.P. Chong, Chem. Phys. Lett. **232**, 486 (1995); J. Chem. Phys. **103**, 1842 (1995); Can. J. Chem. **74**, 1005 (1996); D.P. Chong, C.H. Hu and P. Duffy, Chem. Phys. Lett. **249**, 491 (1996); C. Bureau and D.P. Chong, *ibid.* **264**, 186 (1997).
13. J. Maruani, M. Tronc and C. Dezarnaud, C. R. Acad. Sci. (Paris) II **318**, 1191 (1994); A. Khoudir, J. Maruani and M. Tronc, in A. Hernandez-Laguna, J. Maruani, R. McWeeny and S. Wilson (eds), *Quantum Systems in Chemistry and Physics*, vol. 2, Kluwer, Dordrecht, 2000, p. 57; J. Maruani, A. Khoudir, A. Kuleff, M. Tronc, G. Giorgi and C. Bonnelle, Adv. Quant. Chem. **39**, 307 (2001).
14. P. Ring and P. Schuck, *The Nuclear Many-Body Problem*, Springer, Berlin, 1980.
15. M. Brack, Rev. Mod. Phys. **65**, 677 (1993).
16. D. Bonatsos, A.I. Kuleff, J. Maruani, P.P. Raychev and P.A. Terziev, Int. J. Quant. Chem. **89**, in press (2002).
17. D.A. Kirzhnits, Yu.E. Lozovik and G.V. Shpatakovskaya, Sov. Phys. Usp. **18**, 649 (1976).
18. B.-G. Englert and J. Schwinger, Phys. Rev. A **32**, 26, 36, 47 (1985).
19. B.-G. Englert, *Lecture Notes in Physics* **300**, Springer, Berlin, 1988.
20. Ya.I. Delchev, R.L. Pavlov, K.A. Pavlova, L.P. Marinova and J. Maruani, Int. J. Quant. Chem. **52**, 1349 (1994).
21. Ya.I. Delchev, A.I. Kuleff, J. Maruani and R.L. Pavlov, Adv. Quant. Chem. **31**, 53 (1999).
22. A.I. Kuleff, Ya.I. Delchev, P.Tz. Yotov, Tz. Mineva and J. Maruani, Int. J. Quant. Chem. **89**, in press (2002).
23. J.A. Bearden and A.F. Burr, Rev. Mod. Phys. **39**, 125 (1967).
24. C.E. Moore, Nat. Bur. Stand. Ref. Data Series **34** (1970).
25. J. Bruneau, J. Phys B **16**, 4135 (1983); CEA / CEL *MCDF Technical Report* (1995).
26. C. Bonnelle, P. Jonnard, C. Barré, G. Giorgi and J. Bruneau, Phys. Rev. A **55**, 3422 (1997).

27. C. Bonnelle, *Electron distributions in heavy elements by x-ray and electron spectroscopies*, in U. Kaldor and S. Wilson (eds), *Theoretical chemistry and physics of heavy and superheavy elements*, Kluwer, Dordrecht, in print.
28. S.J. Rose, N.C. Pyper and I.P. Grant, J. Phys. B **9**, L 307 (1978).
29. I.P. Grant, P.J. McKenzie, P.H. Norrington, D.F. Mayers and N.C. Pyper, Comput. Phys. Commun. **21**, 207 (1980).
30. P.-O. Löwdin, *On the origin of the tone system and the theory of harmony*, Proc. Amer. Phil. Soc. **135**, 533 (1991); Proc. Roy. Acad. Arts and Sciences (Uppsala) **28**, 33 (1991); Proc. Roy. Acad. Mus. (Stockholm), Coll. *Quid est sonus - quid est tonus* (1991).
31. J. Sternheimer: (a) C.R.A.S. (Paris) **297**, 829 (1983); (b) Rev. Biomath. (Paris) **94**, 1 (1986).
32. S. L. Leach and G. G. Englert, J. Elect. Spect. & Rel. Phen. **41**, 181 (1986).
33. (a) K. Hayashi and N. Munakata, Nature **310**, 96 (1984); (b) S. Ohno and M. Jabara, Chem. Scripta **26** B, 43 (1986); (c) Andy Coghlan, New Scient. **142**, 10 (1994); (d) Yoichi Fukagawa, *Protein Music* (Y. F. ed., Tokyo, 1999).
34. J. Maruani, R. Lefebvre and M. Rantanen, *Science and Music, a unifying concept: from the music of the depths to the music of the spheres*, presented at the yearly general meeting of the European Academy in Brussels, Belgium, in September 2000, and at the ICTCP-IV congress in Marly-le-Roi, France, in July 2002.

SECOND ORDER PROPERTIES IN TENSOR PRODUCT SPACE BY CI TECHNIQUES: COMPUTATION OF DISPERSION CONSTANTS.

by Gian Luigi Bendazzoli,
Dipartimento di Chimica Fisica ed Inorganica, Università di Bologna, viale Risorgimento 4, I-40136 Bologna, Italy.

Abstract. In perturbation theory second order properties are given by matrix elements of the reduced resolvent of the zeroth order hamiltonian H_0; in this paper we consider how these quantities can be computed in a direct CI framework when H_0 describes non interacting subsystems. We take the case of dispersion constants in the theory of intermolecular interactions as working example to present and discuss some techniques which have general applicability. They include a Lanczos expansion of the London formula and contour integration techniques derived from Hugenholtz' convolution. Numerical applications to LiH and Be atom are presented and discussed.

Keywords: Lanczos, dispersion constants, direct Full CI, LiH, Be.

E.J. Brändas and E.S. Kryachko (eds.),
Fundamental World of Quatum Chemistry, Vol. I, 657–674.

1 Introduction

Perturbation theory has been extensively studied by Löwdin [1] as a fundamental tool of quantum chemistry to compute molecular properties. The latter can be classified according to perturbation theory as first, second , ... order properties. In this paper we focus on a particular aspect of the computation of second order properties whithin the framework of direct CI method [2]. In a number of important cases we are interested in the properties of systems composed of weakly interacting subsystems $A, B, \ldots$, and at zeroth order we have an unperturbed hamiltonian of the form:

$$H_0 = H_A + H_B + \ldots \tag{1}$$

The most familiar example is the intermolecular interaction [3, 6, 4, 5, 7, 8] in the long-range region where exchange effects between electrons belonging to different molecules become negligible [9]. In this formalism H_A and H_B are the hamiltonians of the separated molecules, the perturbation is the Coulomb interaction between the charged particles of molecules $A, B, \ldots$, the state space of the whole system is the (non antisymmetrized) tensor product of the separated molecular spaces. Another, more formal example of perturbation theory (PT) in (antisymmetrized) tensor product space is the many body PT treatment of electron correlation [10]. Now H_0 of eq. (1) is the Fock operator, which in the canonical MO basis can be viewed as a sum of N commuting terms $\epsilon_k a_k^\dagger a_k$ (ϵ_k, $k = 1, \ldots N$ are the orbital energies); the perturbation is the so called fluctuation potential. Another example of more physical interest is given by the translational, rotational, vibrational and electronic degrees of freedom in a molecule in the so called 'crude Born Oppenheimer' approximation. In all these cases the zeroth order hamiltonian H_0 is a sum of terms concerning separate subsystems and the quantities of interest are given by matrix elements of the reduced resolvent [11] of H_0, an operator in the tensor product space $\mathcal{S}_A \otimes \mathcal{S}_B \otimes, \ldots$. We suppose to have techniques to solve both the eigenstate and the perturbative equations for the separate subsystems; the problem we address is how to compute from these data second order properties in the space $\mathcal{S}_A \otimes \mathcal{S}_B \otimes, \ldots$. This is the case of two interacting molecules A, B, that can be individually treated by direct CI methods. In the following we will focus our attention on this example, keeping in mind that the techniques can be adapted to other situations.

The zeroth order eigenvector of the composite system A, B is simply given by the product of the wavefunctions $\Phi_A \Phi_B$ with energy $E_A^0 + E_B^0$. When the Coulomb interaction between the nuclei and electrons of A and those of B is expanded in a multipolar series [12], we are left with a perturbation operator of product type $V_A V_B$: here V_A and V_B can be e.g. components of the electron dipole operator of molecules A and B. A typical second order quantity to be computed is the following dispersion constant:

$$C^d_{AB} = - < \Phi_A\Phi_B|(E^1_A - V_A)(E^1_B - V_B)|\Phi^d_{AB} > \qquad (2)$$

where $E^1_A =< \Phi_A|V_A|\Phi_A >$ is a first order quantity concerning only molecule A and $|\Phi^d_{AB} >$ is the solution of the following first order equation in the tensor product space:

$$(H_A - E^0_A + H_B - E^0_B)\Phi^d_{AB} = -(E^1_A - V_A)(E^1_B - V_B)\Phi_A\Phi_B \qquad (3)$$

Equation (3) is a linear inhomogeneous equation in a vector space, of type $A|x >= |b >$, where the operator A is singular, i.e. it has a null eigenspace and corresponding eigenvectors $|0_i >$. The condition for the existence of a solution is $0 =< 0_i|A|x >=< 0_i|b >$, so that $|b >$ must be orthogonal to $|0_i >$; this condition is fulfilled by eq. (3). When a solution $|x >$ exists, it is not unique: there is a solution manifold $|x > + \sum_i c_i|0_i >$, where c_i are arbitrary coefficients. In the intermediate normalization [13], one chooses the solution of minimal norm, orthogonal to all $|0_i >$; we denote it by $|x_\perp >$. A convenient tool to treat singular inhomogeneous equations is the reduced resolvent estensively studied by Löwdin [11]. Given a singular operator A, its reduced resolvent is an operator valued function of the complex variable z, here denoted by the symbol $(z - A)^{-1}_\perp$ and defined by [14]:

$$(z - A)^{-1}_\perp = (z - A)^{-1}P_\perp = P_\perp(z - A)^{-1}P_\perp = P_\perp(z - A)^{-1} \qquad (4)$$

where $P_\perp$ is the projection in the orthogonal complement of the null space of A. The reduced resolvent is defined for any value of z not belonging to the spectrum of A; it is also defined for $z = 0$ by analytic continuation. The reduced resolvent projects in the orthogonal complement of the null space of A and coincides there with the ordinary resolvent; for $z = 0$ it provides the solution of minimal norm:

$$-(0 - A)^{-1}_\perp|b >= |x_\perp > \qquad (5)$$

Our purpose now is to compare different ways to compute matrix elements of the reduced resolvent of a sum of operators like $H_A - E_A + H_B - E_B$, using only quantities obtained by direct CI techniques in the spaces of isolated molecules. We remind that in this framework we are essentially restricted to work in a Krylov subspace of the CI space. Indeed, the fundamental operation of direct CI is the following: given a vector $\mathbf{x}$ in CI space compute the vector $\mathbf{y}$ resulting from the application of the hamiltonian $\mathbf{Hx} = \mathbf{y}$ directly from the list of one and two electron integrals. Hamiltonian matrices cannot be completely diagonalized, inverted, nor even fully computed or stored: this is because the dimensions of the CI spaces are usually several millions (even billions [15]). Iterative computation of eigenvectors and iterative solution of inhomogeneous linear equations [2] [16] [17] are all based on this fundamental operation.

We restate the problem as follows. Given two operators O_A and O_B acting in two (finite dimensional) vector spaces $\mathcal{S}_A$ and $\mathcal{S}_B$, find convenient representations of the reduced resolvent $(O_A+O_B)_\perp^{-1}$ and related operators in the tensor product space $\mathcal{S}_A \otimes \mathcal{S}_B$. In our case $O_A = H_A - E_A$,$O_B = H_B - E_B$, and $\mathcal{S}_A$ and $\mathcal{S}_B$ are the Full CI spaces of the separated molecules.

The simplest way to treat operators in tensor product spaces is via the spectral resolution. If P_i^A and P_j^B are eigenprojections of O_A and O_B, associated to the eigenvalues A_i and B_j, one has for the resolvent of O_A+O_B

$$(z - O_A - O_B)^{-1} = \sum_i \sum_j P_i^A P_j^B \frac{1}{z - A_i - B_j} \tag{6}$$

If both O_A and O_B have zero eigenvalues $A_0 = B_0 = 0$, the reduced resolvent is obtained by omitting from the double sum of eq. (6) the terms containing the projections P_0^A, P_0^B. Actually, in order to compute eq. (3) we do not strictly need the reduced resolvent, but, rather:

$$\mathcal{R}_{AB} = \sum_{i>0} \sum_{j>0} P_i^A P_j^B \frac{1}{z - A_i - B_j} \tag{7}$$

Eq. (6) splits into three terms of product form by collecting all terms containing P_0^A or P_0^B in $\mathcal{R}_{A0}$ and $\mathcal{R}_{0B}$ respectively and leaving everithing else in the 'entangled' term $\mathcal{R}_{AB}$:

$$\begin{aligned}(z - O_A - O_B)_\perp^{-1} &= \mathcal{R}_{0A} + \mathcal{R}_{0B} + \mathcal{R}_{AB} \\ &= (z - O_B)_\perp^{-1} P_0^A + (z - O_A)_\perp^{-1} P_0^B + \mathcal{R}_{AB} \end{aligned} \tag{8}$$

Notice that all the $\mathcal{R}$ operators are mutually exclusive i.e. $\mathcal{R}_{A0}\mathcal{R}_{B0} = 0$ etc.; the solution of eq. (3) is given by (5):

$$\Phi_{AB}^d = -\mathcal{R}_{AB}(E_A^1 - V_A)\Phi_A(E_B^1 - V_B)\Phi_B \tag{9}$$

We now consider how the matrix elements of $\mathcal{R}_{AB}$ can be expressed in terms of matrix elements of the reduced resolvents of isolated molecules.

2 Lanczos expansion of the London formula.

The first possibility is obtained from eq. (7): it is the well known London formula for the dispersion constant:

$$C_{AB}^d = - \sum_{i>0,\, j>0} \frac{< 0|V_A|i >< i|V_A|0 >< 0|V_B|j >< j|V_B|0 >}{E_A^i - E_A^0 + E_B^j - E_B^0} \tag{10}$$

where $S_i =< i|V_A|0 >$ is the overlap of the vector $-(E_A^1 - V_A)\Phi_A$ with the excited eigenvector $|i >$ of the Hamiltonian. Since complete diagonalization of hamiltonian matrices is not feasible in direct CI, an obvious approximation to eq. (10) is to truncate the sums including only a limited number of terms. However the number of eigenvectors to be included is in fact limited by considerations of computational cost, more than by effective convergence of the result. Eigenvector computation is performed via the Lanczos [19] or Davidson [21] methods and usually involves several iteration steps. Eigenvectors have been introduced in formula (10) as an expansion basis, but we can use for practical purposes other, smaller expansion sets, provided they span a space suitable for describing the physics of the process: this is the *expansion in pseudostates* [20]. The Lanczos procedure presents some attractive possibilities in this connection. Starting from an initial, normalized, vector $\mathbf{b}_1$ in FCI space of dimension N, Lanczos recursively generates an orthonormal basis $\mathbf{b}_1, \mathbf{b}_2, \ldots \mathbf{b}_n$, in which the Hamiltonian is tridiagonal, the only non-zero matrix elements being:

$$\alpha_i = \mathbf{b}_i^T \mathbf{H} \mathbf{b}_i \; ; \beta_i = \mathbf{b}_{i-1}^T \mathbf{H} \mathbf{b}_i \; . \tag{11}$$

After n recursions, by collecting the vectors $\mathbf{b}_i$ into a $N \times n$ matrix, we can write:

$$\mathbf{H}\mathbf{B}_n = \mathbf{B}_n \mathbf{J}_n + \mathbf{R}_{n+1} \quad n < N \tag{12}$$

where $\mathbf{J}_n$ is the tridiagonal (Jacobi) $n \times n$ matrix with elements given by eq.s (11) and $\mathbf{R}_{n+1}$ is a remainder term (see Appendix 1). By neglecting the latter we get a tridiagonal approximation of the hamiltonian. The Jacobi matrix is easily diagonalized; we get in this way n eigenvalues ϵ_k approximating those of the hamiltonian, and an unitary $n \times n$ matrix $\mathbf{V}_n$ composed of n eigenvectors $\mathbf{v}_k$:

$$\mathbf{J}_n \mathbf{V}_n = \mathbf{V}_n Diag\{\epsilon_k, k = 1, \ldots n\} \tag{13}$$

The $N-$dimensional Ritz vector $\mathbf{B}_n \mathbf{v}_k$ will (in principle) converge to an eigenvector $\mathbf{h}_i$ of the Hamiltonian $\mathbf{H}$, provided the overlap $S_i = \mathbf{h}_i^T \mathbf{b}_0$ is non vanishing. A Krylov-space analysis shows that convergence is faster to the extreme eigenvaues of $\mathbf{H}$. [22], [23]. Therefore, for each molecule, we can approximate the reduced resolvent by its inner projection [24, 25] in the Lanczos space of order n generated from the starting vector $-(E_A^1 - V_A)\Phi_A$. This vector is orthogonal the zeroth order FCI wavefunction Φ_A and so will be the Lanczos vectors generated from it. The tridiagonal Jacobi matrix of each molecule can be diagonalized and the London formula can be rewritten as:

$$C_{AB}^d \approx - \sum_{i=1}^{n} \sum_{j=1}^{n'} \frac{R_{Ai}^2 R_{Bj}^2}{\epsilon_A^i - E_A^0 + \epsilon_B^j - E_B^0} \tag{14}$$

where R_{Ai} is the overlap of the vector $-(E_A^1 - V_A)\Phi_A$ with the $i-th$ Ritz eigenvector of the Jacobi-Lanczos matrix. Thex R_{Ai}'s are known as *residues* in the Lanczos literature [26], and are given by the elements of the first row of the eigenvector matrix of $\mathbf{J}_n$ multiplied by the norm $\| (E_A^1 - V_A)\Phi_A \|$. Eq. (14) involves only matrix elements of operators in the Lanczos space; they can be computed by direct CI techniques. Moreover a practical convergence criterion can be obtained by monitoring the behaviour of eq. (14) for increasing values of n, n', until a stabilization is achieved. The Lanczos recursion is known to develop linear dependencies among the $\mathbf{b}_k$ vectors, leading to the appearance of spurious (or ghost) Ritz eigenvalues or to multiple copies of the same eigenvalue. This problem has been the subject of many investigations (see e.g. [23]) and it is of great relevance in the computation of eigenvalues. However we are not so much interested in the accuracy of *individual eigenvalues,*but rather in the sum (14). Spurious eigenvalues are associated to small residues R_{Ai}^2 and multiple eigenvalues have residues adding up to the correct value [26]. Moreover, practical tests for identifying the spurious eigenvalues have been developed [27] and can be used to assess the reliability of the calculation of eq. (14). As pointed out in a large body of literature [28], the Lanczos recursion is efficient in approximating the spectral distribution of the starting vector $\mathbf{b}_0$ and its moments (see e.g. [29]). It should also be mentioned that Lanczos methods are well established in other fields, from atomic physics [30] to dynamical problems [31].

3 The Casimir-Polder formula.

The other classic approach, the Casimir-Polder formula [32] involves integration of the polarizability at imaginary frequency:

$$C_{AB}^d = \frac{1}{2\pi} \int_0^{+\infty} \alpha_A(i\omega)\alpha_B(i\omega)d\omega \tag{15}$$

There exists a large literature on eq. (15); in this context we limit ourselves to [33] and references therein. One can extract from eq. (15) the following representation of the operator $\mathcal{R}_{AB}$:

$$\mathcal{R}_{AB} = \frac{2}{\pi} \int_0^{+\infty} \frac{H_A - E_A^0}{(H_A - E_A^0)^2 + \omega^2} \frac{H_B - E_B^0}{(H_B - E_B^0)^2 + \omega^2} d\omega \tag{16}$$

Eq. (15) is naturally implemented in a direct CI way: the $\alpha(i\omega)$ are obtained from the FCI solution of the 1st order perturbative equation via the quantities $L(z)$. The latter are defined as:

$$L_A(z) =< \Phi_A^0|(E_A^1 - V_A)(z - H_A + E_A^0)_\perp^{-1}(E_A^1 - V_A)|\Phi_A^0 > \tag{17}$$

The L's are in general complex quantities, fulfilling $L(z^*) = L^*(z)$, and are related to the generalized susceptibility χ [34]. In the simplified notation: $|b> = (E_A^1 - V_A)|\Phi_A^0>$:

$$
\begin{aligned}
L(x+iy) &= <b|\frac{x-H_A+E_A^0}{(x-H_A+E_A^0)^2+y^2}|b> \\
&\quad -i<b|\frac{y}{(x-H_A+E_A^0)^2+y^2}|b> \qquad (18)\\
L(z^*) &= L^*(z) \qquad (19)\\
\chi(z) &= L(z)+L(-z^*)=\chi^*(-z) \qquad (20)
\end{aligned}
$$

The dynamic polarizabilities at real or imaginary frequency are given by:

$$
\begin{aligned}
\alpha(\omega) &= L(\omega)+L(-\omega)=\alpha(-\omega) \qquad (21)\\
\alpha(i\omega) &= L(i\omega)+L(-i\omega)=2\Re\{L(i\omega)\}=\alpha(-i\omega) \qquad (22)
\end{aligned}
$$

The α's are computed in a set of integration points on the imaginary axis and the integral (15) is evaluated numerically. The imaginary part of $L(i\omega)$ is not used. This approach has been used to compute the C_6 dispersion coefficient of LiH [35].

4 Hugenholtz' convolution.

The analogies pointed out in the introduction suggest another possibility for representing the operator $\mathcal{R}_{AB}$. In the context of many body perturbation theory, Hugenholtz [10] introduced a *convolution product* $*$ as a tool to develop his time independent formalism. The convolution product (see Appendix 2) is defined via a contour integral in the complex plane; in our case, the operator $\mathcal{R}_{AB}$ is the convolution product of the reduced resolvents, defined by the following eq. (23):

$$
\begin{aligned}
\mathcal{R}_{AB}(z) &= (z-H_A+E_A^0)_\perp^{-1} * (z-H_B+E_B^0)_\perp^{-1} \\
&= -\frac{1}{2\pi i}\oint (z-H_A+E_A^0-\zeta)_\perp^{-1}(\zeta-H_B+E_B^0)_\perp^{-1}d\zeta \qquad (23)
\end{aligned}
$$

where the contour includes the poles of the second factor and leaves out those of the first. The poles of the second factor are the excitation energies of molecule B (from the ground state) and fall on the positive real axis of the $\zeta=\xi+i\eta$ plane. For $z=0$, the poles of the first factor are all negative and $\zeta=0$ is a regular point. The second order dispersion constant, eq. (2), becomes:

$$
C_{AB}^d = L_A(z) * L_B(z) \ \text{ computed at } \ z=0 \qquad (24)
$$

This eq. (24) can also be implemented by direct CI techniques: one performs separate computations of $L(z)$ in the CI spaces of molecules A and B, and computes the contour integral numerically.

However, before going into the computational details, it is interesting to derive some properties of eq. (24) by exploiting the flexibility of the contour. Consider a set of disjoint circles each of them surrounding a B pole. For each circle the second factor in eq. (23) gives the projection on the corresponding eigenvector(s) [36], and we get a London-like formula of mixed type:

$$\sum_k L_A(E_B^0 - E_B^k) < 0|V_B|k >< k|V_B|0 > \tag{25}$$

The London formula (10) is obtained by expanding the L factor in eigenstates. From eq. (25) one can derive e.g. Lanczos type approximations in the space of molecule B. Eq. (25) requires the solution of the 1st order perturbative equation at real *negative* frequencies $\omega_k = E_B^0 - E_B^k$, i.e. in a region where poles are far away and convergence should be easy.

What is the relation to the Casimir-Polder formula? Consider a semi-circle in the $\Re(\zeta) > 0$ plane and let its radius go to $+\infty$; the contribution of the semi circumference will vanish because the integrand is $O(|\zeta|^{-2})$ as it can be seen from eq. (18). We are left with an integral on the imaginary axis, but we don't recover the Casimir-Polder formula. From eq.s (18,22):

$$C_{AB}^d = -\frac{1}{2\pi i}\int_{+\infty}^{-\infty} L_A(-i\eta)L_B(i\eta)id\eta \tag{26}$$

$$\begin{aligned} &= \frac{1}{4\pi}\int_0^{+\infty} \alpha_A(i\eta)\alpha_B(i\eta)d\eta + \\ &+ \frac{1}{\pi}\int_0^{+\infty} < \frac{\eta}{(H_A - E_A^0)^2 + \eta^2} >< \frac{\eta}{(H_B - E_B^0)^2 + \eta^2} > \end{aligned} \tag{27}$$

Eq (27) is the sum of half the Casimir-Polder term eq. (15) plus an extra term. That means the two terms of eq. (27) must be equal, as it can be derived from the equality:

$$\int_0^{\infty} \frac{ab}{(a^2+\eta^2)(b^2+\eta^2)}d\eta = \int_0^{\infty} \frac{\eta^2}{(a^2+\eta^2)(b^2+\eta^2)}d\eta = \frac{\pi}{2(a+b)} \tag{28}$$

Of course eq. (28) suggests another representation of the $\mathcal{R}_{AB}$ operator very similar to (16) and the following formula for the dispersion constant:

$$C_{AB}^d = \frac{2}{\pi}\int_0^{+\infty} \Im[L_A(i\omega)]\Im[L_B(-i\omega)]d\omega \tag{29}$$

Last, we may observe that formal representations of the reduced resolvent in tensor product space can be obtained from other mathematical identities, like e.g. the Laplace transform, but they are not convenient in the present context of direct CI.

5 Contour integration.

We have at our disposal some independent formulae, eq.s (14,15,24, 25,27,29), that can be used to compute the same quantity and therefore to check each other. As concerns the practical applicability of contour integration, we must carefully consider the computational aspects. For sake of simplicity we consider the case of two identical molecules $A = B$. First of all, it is convenient to choose a contour minimizing the number of points where the integrand should be evaluated. From the property $L(z^*) = L^*(z)$ the optimal contour is symmetrical by reflection on the $x-$axis, and, since the integrand is $L(z)L(-z)$, also symmetrical by inversion around the origin. Therefore we choose the positive imaginary axis $z = iy$, and keep in mind that we have to go backwards from $+\infty$ to 0 and multiply the result by 2. The integrand is only $O(y^{-2})$, so we have to choose carefully the integration points and weights. We adopted the following procedure:

Given the line integral along the unlimited curve Γ in the complex plane, defined by the equation $z = \gamma(t), \quad 0 \leq t < \infty$

$$I = \int_{\Gamma} f(z)dz = \int_{0}^{\infty} f(z)\frac{d\gamma}{dt}dt \tag{30}$$

we perform a change of variable i) to bring the integral to a finite domain, and, ii) to have a new integrand as smooth and regular as possible. In particular, if we choose as new variable:

$$G(z) = \int_{\Gamma}^{z} g(\zeta)d\zeta = \int_{0}^{z} g(\zeta)\gamma' dt \tag{31}$$

where $z \in \Gamma$, our integral transforms to:

$$I = \int_{0}^{G_\infty} \frac{f(z)}{g(z)} dG \tag{32}$$

where G_∞ is the limiting value of $G(z)$ on the curve for $t \to \infty$. If we put $g(z) = f(z)$, the integral is trivially reduced to that of the constant function $= 1$. But we can choose a $g(z)$ similar to $f(z)$, in particular having the same asymptotic behaviour for $t \to \infty$. The integrand will be smooth and if we are able to compute the limit of $f(z)/g(z)$ for $t \to \infty$, we can be confident in the numerical accuracy of the computation. Convenient functions $g(z)$ and their primitives $G(z)$ can be obtained from the Lanczos approximation for the cases here considered, and are displayed in Appendix 3.

Next we divide the interval $0 \leq G \leq G_\infty$ in $M = 2^m$ equal parts of lenght Δ and find, by Newton's method, the points $z_1, z_2, \ldots, z_{m-1}$ lying on the curve and fulfilling $G(z_k) = k\Delta$. These are the integration points: we compute $L(z_k)$ by solving a system of linear equations. The final integration is done by trapezoidal rule followed by Romberg extrapolation [37] to $\Delta \to 0$. Alternatively, we use gaussian integration; the z_k's are obtained by mapping the Gauss-Legendre points in the interval $[0, G_\infty]$.

5.1 Numerical applications and discussion.

We applied our techniques to the computation of the dipole-dipole dispersion constants of LiH-LiH molecules and Be-Be atoms. All the Full CI calculations are very fast and can be easily performed on a PC. In table I we collect the results obtained for LiH. The AO basis set used, suggested by Sadlej and Urban [38], is small but especially designed for dipole polarizability calculations. The internuclear distance is 3.015 bohr and the z axis is parallel to the bond. The form of a matrix element between molecules A and B is given by:

$$\begin{aligned} < aa * bb > &= < \Phi_A^0 | a_A (z - H_A + E_A^0)_\perp^{-1} a_A | \Phi_A^0 > \\ &* < \Phi_B^0 | b_B (z - H_B + E_B^0)_\perp^{-1} b_B | \Phi_B^0 > \end{aligned} \tag{33}$$

where a and b are components of the dipole operator for molecule A and B respectively. In the table we report only the $< xx * xx >$ and $< zz * zz >$ components; they suffice for our discussion. The methods used include the London-Lanczos formula (14), the mixed type eq. (25) and three contour integrations, eq.s (15,26,29). The agreement between the numerical integrations and the London-Lanczos results allows us to consider the latter procedure as capable of providing the Full CI limit. In this case, the computational cost of the London-Lanczos expansion is less than that of the numerical integration. The points of the latter were chosen as explained in the previous section, using a 20 residual Lanczos primitive $G(z)$. The procedure proved to be very effective in integrating slowly decreasing functions as eq. (26) or (29). However, the Casimir-Polder integration still shows a slightly better convergence behaviour over the Hugenholtz and the eq. (29). It should also be remarked that the values of imaginary frequency at which $L(i\omega)$ is computed are different for eq.s (15), (26), 29) and much smaller than those predicted by the conventional Gaussian integration schemes. That means that the Lanczos primitive $G(z)$ takes care very effectively of the asymptotic behaviour of the integrand.

The second system considered is the beryllium atom. In Table 2 we report values obtained by the methods described in this paper (London-Lanczos and various numerical integrations). The dispersion constant in this case is contributed by only one non zero matrix element and the C_6 coefficient is given by $C_6 = 6C^d$. The first AO basis set used is the 9s9p5d defined in ref. [39] and used by Koch *et al.* to compute the C_6 coefficient [17]; we used both the cartesian and spherical variants of the basis. As concerns the numerical integrations, the same comments done for LiH results apply in this case. From the values reported in Table 2, we expect that the Full CI value of C_6 for Be 9s9p5d (cartesian) is 212.502±.001 a.u., to be compared with 212.5 a.u. reported in ref. [17]. Somewhat larger values are obtained by using the spherical version of the 9s9p5d basis (213.13 a.u.) or the 9s9p5d3f basis (214.36 a.u.). The latter was obtained by adding three f orbitals with exponents 0.36, 0.06 and 0.001 to the 9s9p5d basis. The value of the static polarizability obtained with this basis (37.7263 a.u.) is very close to the 'basis set limit' (37.73±0.05 a.u.) computed by Tunega *et el.* using explicitly correlated gaussians [40].

The stability of the London-Lanczos values is achieved after some more iterations than the previous calculation on LiH in spite of the much larger dimension of the Full CI space. However this cannot be considered as an indication that the London-Lanczos expansion converges satisfactorily in all cases. Some insight in the convergence properties of eq. (14) can be gained by the following arguments. The Lanczos procedure is known to converge first to the extreme eigenvalues of the matrix [23]. On the other hand, by inspection of London formula eq. (10), the states most contributing to the dispersion constant have i) low excitation energies and ii) large dipole transition moments with the ground state. They are likely to be found right in the middle of the spectrum of the molecular hamiltonian, so we cannot use CI spaces too rich in higly excited states with small transition dipoles: all the computational effort will be wasted in converging to those states! Such situations arise when we use uncontracted AO bases, especially those containing many s orbitals with high exponents. We constructed such an example by uncontracting the basis of Sadlej and Urban [38] for beryllium: this gave us a Full CI space of only 170,000 det.s, but we could not achieve convergence using as many as 100 Lanczos vectors. A more striking example is provided by the B3 basis of Papadopoulos *et al.*[41]. This is a 16s15p13d uncontracted 126 orbital basis yielding a ≈8 million determinant Full CI space: after 200 Lanczos we are still far from convergence. On the other hand, the 9s9p5d or 9s9p5d3f bases here used are contracted in the s symmetry and gave no problem. The main requirement for the validity of the London-Lanczos expansion seems to be a suitable AO basis, without uncontracted s functions with high exponents; diffuse and polarization functions give no harm. This is indeed a limit of the London-Lanczos procedure in its present form. As concerns possible remedies when using unsuitable AO bases, we limit ourselves to some simple suggestions and remarks. A first possibility is to truncate the orbital basis at the MO level, provided we can properly classify the MO's according to their ability to contribute to the unwanted CI excited states. A second idea to be developed is to introduce a preconditioner in the Lanczos recursion, in such a way to reduce the components of the Lanczos pseudostates on the high excited CI eigenvectors. In this scheme one would loose the orthogonality of the vectors. Both possibilities require further investigation; we will leave it to a future work.

6 Conclusion.

We have presented and discussed some old and new techniques to compute matrix elements of the reduced resolvent in the tensor product of CI spaces in a direct CI framework. The Lanczos expansion of the London formula is capable of providing the Full CI limit of the dispersion constant, at least when the AO basis meets suitable conditions. Alternatives to the Casimir-Polder formula have been derived from Hugenholtz' convolution and practical integration procedures have been proposed and applied.

7 Acknowledgments.

This work has been partly supported by the "Ministero dell'Istruzione, dell'Università e della Ricerca" of Italy (MIUR), and partly by E.C. under contract HPRN-CT-1999-00005. Thanks are due to prof. V. Magnasco (Genua) for a critical reading of the manuscript.

8 Appendix 1. The Lanczos recursion.

Given a starting vector $\mathbf{v}$, the steps of the Lanczos recursion are:

$$\mathbf{q}_0 = \frac{\mathbf{v}}{\| \mathbf{v} \|} \; ; \; \mathbf{q}_{-1} = \mathbf{0} \; ; \; \beta_0 = 0 \tag{34}$$

$$\alpha_k = \left(\mathbf{H}\mathbf{q}_{k-1} - \beta_k \mathbf{q}_{k-2}\right)^T \mathbf{q}_{k-1} \tag{35}$$

$$\mathbf{p}_k = \mathbf{H}\mathbf{q}_{k-1} - \alpha_k \mathbf{q}_{k-1} - \beta_k \mathbf{q}_{k-2} \tag{36}$$

$$\beta_{k+1} = \| \mathbf{p}_k \| \tag{37}$$

$$\mathbf{q}_k = \frac{\mathbf{p}_k}{\beta_{k+1}} \, , k = 1, 2, \ldots \tag{38}$$

The first step is a normalization; the Lanczos vectors referred to in the text are $\mathbf{b}_k = \mathbf{q}_{k-1}$. Only two Lanczos basis vectors need to be kept on disk. The remainder term of eq. (12) is given explicitly in the following equation:

$$\mathbf{H}\mathbf{B}_n = \mathbf{B}_n \mathbf{J}_n + \beta_{n+1} \mathbf{q}_n (\mathbf{e}_n)^T \tag{39}$$

where $\mathbf{e}_n$ is the N-vector with k-th component $= \delta_{kn}$, $k = 1, 2, \ldots, N$. In principle the recursion should stop when $\beta_{n+1} = 0$; in practice, due to roundoff errors this condition is never met.

9 Appendix 2: the Hugenholtz convolution.

We briefly remind the definition and essential properties of Hugenholtz convolution product. Given two meromorphic functions $f(z)$ and $g(z)$, such that $|zf(z)|$ and $|zg(z)|$ are bounded for $|z| \to \infty$, the convolution product $f * g$ is defined as:

$$f * g(z) = -\frac{1}{2\pi i} \oint f(z - \zeta) g(\zeta) d\zeta \tag{40}$$

The integration contour includes all the poles of the second factor and leaves out those of the first. The convolution product is commutative and associative [10]. The classic example is:

$$\frac{1}{a-z} * \frac{1}{b-z} = \frac{1}{a+b-z} \tag{41}$$

The definition can be extended to operator valued functions acting in separate (finite dimensional) vector spaces; it turns out that the convolution product is an operator in the tensor product space. This applies in particular to the resolvent (reduced or not), which is a function of the complex variable z. The simplest way to prove it is via the spectral resolution: one performs the convolution product of the eigenvalues and exchanges summations and integration.

10 Appendix 3. The functions $g(z)$ and their primitives.

- Hugenholtz case. From the Lanczos approximation to $L(z)$:

$$L(z) \approx \sum_{i=1}^{n} \frac{R_i^2}{z - \epsilon_i + E^0} \tag{42}$$

$$g(z) = -\frac{1}{2\pi i} \sum_{i,j} \frac{R_i^2}{z - \epsilon_i + E^0} \frac{R_j^2}{z + \epsilon_j - E^0} \tag{43}$$

$$G(z) = -\sum_{i,j} \frac{R_i^2 R_j^2}{2\pi i} \ln \frac{z + \epsilon_i - E^0}{z_0 + \epsilon_i - E^0} \frac{z - \epsilon_j + E^0}{z_0 - \epsilon_j + E^0} \tag{44}$$

where z_0 is the starting point of the curve (here $z_0 = 0$). The limit $\lim_{y\to\infty} G(z) = G_\infty$ is the London-Lanczos value of eq. (14). The limiting value of the ratio $L(z)L(-z)/g(z)$ is found to be 1 by the following argument. The exact, Full CI function $L(z)L(-z)$ admits an expansion similar to eq. (43), but where the sums are extended to all eigenvectors of the Hamiltonian and the approximate energies are replaced by the exact eigenvalues E_k. Therefore:

$$\frac{L(z)L(-z)}{-2\pi i g(z)} = \frac{\sum_{i,j} \frac{S_i^2}{z-E_i+E^0} \frac{S_j^2}{z+E_j-E^0}}{\sum_{k,l} \frac{R_k^2}{z-\epsilon_k+E^0} \frac{R_l^2}{z+\epsilon_l-E^0}} \tag{45}$$

$$\lim_{|z|\to\infty} \frac{L(z)L(-z)}{-2\pi i g(z)} = \frac{\sum_i S_i^2 \sum_j S_j^2}{\sum_k R_k^2 \sum_l R_l^2} = 1 \tag{46}$$

because

$$\sum_i S_i^2 = \sum_k R_k^2 = \| (E_A^1 - V_A)\Phi_A \|^2$$

- Casimir-Polder. Here $z = iy$; in terms of the real variable y, and setting $A_i = \epsilon_i - E^0$, $B_j = \epsilon_j - E^0$

$$
\begin{aligned}
g(y) &= \frac{2}{\pi}\sum_{i,j}\frac{R_i^2 A_i}{A_i^2+y^2}\frac{R_j^2 B_j}{B_j^2+y^2} \qquad (47)\\
G(y) &= \sum_{i\neq j}\frac{2R_i^2R_j^2}{\pi}\frac{B_j\arctan(y/A_i)-A_i\arctan(y/B_j)}{B_j^2-A_i^2}+\\
&- \sum_i \frac{R_i^2}{\pi}\left(\frac{\arctan(y/A_i)}{A_i}+\frac{y}{A_i^2+y^2}\right) \qquad (48)
\end{aligned}
$$

By an argument similar to eq. (45), it can be shown that:

$$
\lim_{y\to\infty}\frac{\alpha_A(iy)\alpha_B(iy)}{4g(y)} = 1 \qquad (49)
$$

- Similar expressions can be written for the second term eq. (27):

$$
\begin{aligned}
g(y) &= -\frac{2}{\pi}\sum_{i,j}\frac{R_i^2 y}{A_i^2+y^2}\frac{R_j^2 y}{B_j^2+y^2} \qquad (50)\\
G(y) &= -\sum_{i.nej}\frac{2R_i^2R_j^2}{\pi}\frac{A_j\arctan(y/A_i)-B_j\arctan(y/B_j)}{A_i^2-B_j^2}+\\
&- \sum_i \frac{R_i^2}{\pi}\left(\frac{\arctan(y/A_i)}{A_i}-\frac{y}{A_i^2+y^2}\right) \qquad (51)
\end{aligned}
$$

Again, for $y \to \infty$ one finds a limit $= 1$. At $y = 0$ one has $g(0) = 0$, but it is easy to compute the limit for $y \to 0$ of the ratio: for two equal molecules is given by:

$$
\left(\frac{\| (H-E_0)_\perp^{-1}V|0>\|^2}{\sum_i \frac{R_k^2}{A_k^2}}\right)^2
$$

References

[1] P.O. Löwdin and O. Goscinski, Int. J. of Quantum Chem., **5**, 685 (1971) and reference [4] therein.

[2] G.L. Bendazzoli and S. Evangelisti, in *Adv. Quantum Chem.* **39** J.R. Sabin and E. Brändas (Ed.s) (Academic Press 2001), p. 190.

[3] A. Dalgarno in *Adv. Chem. Phys.* **12** J.O. Hirschfelder (Ed.) (J. Wiley & Sons, New York, 1967), p. 143.

[4] P. Arrighini *Intermolecular Forces and Their Evaluation by Perturbation Theory, Lecture Notes in Chemistry n. 25* (Springer Verlag, Berlin Heidelberg New York 1981).

[5] A.D. Buckingham, P.W. Fowler and J.M. Hutson, Chem. Rev. **88**, 963 (1988)

[6] R. McWeeny *Methods of Molecular Quantum Mechanics,* **14**, (Academic Press London (1989), Ch. 14.

[7] V. Magnasco and R. McWeeny, in *Theoretical Models of Chemical Bonding part 4*, Z.B. Maksic (Ed.) (Spinger Verlag, Berlin Heidelberg New York 1991), p. 133.

[8] see e.g. the references reported in the website of K. Szalewicz's group http://www.physics.udel.edu/~mas/TheoManual/SAPTtheo.html

[9] R. Ahlrichs, Theor. Chim. Acta (Berl.), **41** 7 (1975).

[10] N.M. Hugenholtz, Physica, **23**, 481 (1957).

[11] P.O. Löwdin in *Perturbation Theory and Its application in Quantum Mechanics*, Calvin H. Wilcox (Ed.), (John Wiley & Sons, Inc. 1966), p. 255.

[12] H. Margenau, Rev. Mod. Phys., **11**, 1 (1939).

[13] J.O. Hirschfelder, W. Byers Brown and S.T. Epstein in *Adv. Quantum Chem.* **1**, P.O. Löwdin (Ed.) (Academic Press, 1964), p. 255.

[14] P.O. Löwdin, Int. J. Quantum Chem. **II**, 867 (1968).

[15] E. Rossi, G.L. Bendazzoli, S. Evangelisti and D. Maynau, Chem. Phys. Lett., **310** , 530 (1999).

[16] W. Meyer, Chem. Phys., bf 17, 27 (1976);

[17] H. Koch and R.J. Harrison, J. Chem. Phys., **95**, 7479 (1991).

[18] F. London, Z. Phys. Chem., **B11** , 222 (1930).

[19] C. Lanczos, J. Res. Nat. Bureau Stand., **45**, 255 (1950).

[20] V. Magnasco and M. Ottonelli, Trends in Chemical Physics, **7**, 215 (1999).

[21] E.R. Davidson, J. Comp. Phys., **17**, 87 (1975).

[22] Yu V. Vorobyev, *Method of Moments in Applied Mathematics*, (Gordon and Breach, New York London Paris 1965).

[23] G.H. Golub and C.F. van Loan *Matrix Computations*, (North Oxford Academic, Oxford 1983);

[24] P.O. Löwdin, Internat. J. Quantum Chem., **4** 231 (1971); O. Goscinski and E. Brändas, Internat. J. Quantum Chem., **V**, 131 (1971).

[25] H.D. Meyer and S. Pal, J. Chem. Phys., **91**, 6195 (1989).

[26] R.E. Wyatt, in *Adv. Chem. Phys.* **LXXIII** J.O. Hirschfelder (Ed.) (J. Wiley & Sons, New York, 1989), p. 231.

[27] J. Cullum and R.A. Willoughby, J. Comp. Phys., **44**, 329 (1981).

[28] see e.g. *The Recursion Method and its Applications*, D.G. Pettifor and D.L. Wearie (Ed.s), Springer Series in Solid-State Sciences n. 58, (Springer Verlag, Berlin Heidelberg New York Tokio 1985).

[29] H.O. Karlsson, G.L. Bendazzoli, O.Goscinski and S. Evangelisti Internat. J. Quantum Chem., **63**, 719 (1996).

[30] H.O. Karlsson and O.Goscinski J. Phys. B, **27**, 1061 (1994).

[31] R.A. Friesner, J-P. Brunet, R.E. Wyatt, C. Leforestier and S. Binkley, The Internat. J. Superc. Appl., **1**, 9 (1967).

[32] H.B.G. Casimir and D. Polder, Phys. Rev., **73** , 360 (1948).

[33] R. Ahlberg and O.Goscinski, J. Phys. B, **7**, 1194 (1974); R. Ahlberg, Internat. J. Quantum Chem. Symp., **8**, 363 (1974)

[34] see e.g. L. Landau and E. Lifchitz, *Physique Statistique*, (Mir, Moskow 1967), Ch. 12 (the symbol α is used for χ).

[35] G.L. Bendazzoli, V. Magnasco, G. Figari and M. Rui, Chem. Phys. Letters, **300**, 146 (2000).

[36] see e.g. T. Kato *Perturbation Theory for Linear Operators*, (Springer Verlag, Berlin Heidelberg New York 1966).

[37] J. Stoer and R. Bulirsch, *Introduction to Numerical Analysis*, (Springer Verlag, New York Heidelberg Berlin, 1980).

[38] A. J. Sadlej and M. Urban, J. Mol. Struct. (THEOCHEM), **234**, 147 (1991).

[39] R.L. Graham, D.L. Yaeger, J. Olsen, P. Jørgensen, R. Harrison, S. Zarrabian and R. Bartlett, J. Chem. Phys., **85**, 6544 (1986).

[40] D. Tunega, J. Noga and W. Klupper, Chem. Phys. Letters, **269**, 435 (1997).

[41] M.G. Papadopoulos, J. White and A.D. Buckingham, J. Chem. Phys., **102**, 371 (1995).

Table 1. Dipole-dipole dispersion matrix elements for LiH at 3.015 bohr. Basis 10s6p4d/5s3p2d [38], 33 AO's, $\approx$ 70,000 det.s symmetry adapted (C_{2v}). The static polarizability values in this basis are $\alpha_{\parallel} = 26.77$ and $\alpha_{\perp} = 30.00$ a.u. [35].

Method	$< xx * xx >$	Notes
London Lanczos eq.(14)	22.74252	10 residues
London Lanczos eq.(14)	22.74930	20 residues
London Lanczos eq.(14)	22.74931	25 residues, stable
Mixed eq.(25)	22.74931	20 residues
Casimir Polder eq.(15)	22.74931	eq. (32) Romberg 16 pt.s
Casimir Polder eq.(15)	22.74931	eq. (32) Romberg 32 pt.s
Hugenholtz eq. (26)	22.74933	eq. (32) Romberg 16 pt.s
Hugenholtz eq. (26)	22.74934	eq. (32) Romberg 32 pt.s
eq. (29)	22.74936	eq. (32) Romberg 16 pt.s
eq. (29)	22.74933	eq. (32) Romberg 32 pt.s
Method	$< zz * zz >$	Notes
London Lanczos eq.(14)	18.20685	10 residues
London Lanczos eq.(14)	18.24400	20 residues
London Lanczos eq.(14)	18.24413	30 residues, stable
Mixed eq. (25)	18.24406	20 residues
Casimir Polder eq.(15)	18.24424	eq. (32) Romberg 16 pt.s
Casimir Polder eq.(15)	18.24433	eq. (32) Romberg 32 pt.s
Hugenholtz eq.(26)	18.24424	eq. (32) Romberg 16 pt.s
Hugenholtz eq.(26)	18.24433	eq. (32) Romberg 32 pt.s
eq. (29)	18.24517	eq. (32) Romberg 16 pt.s
eq. (29)	18.24455	eq. (32) Romberg 32 pt.s

Table 2. Dispersion constant for Be atom. The 9s9p5d AO basis is taken from [39], the 9s9p5d3f is the previous one augmented by three f orbitals with exponents 0.36, 0.06 and 0.001.

Basis 9s9p5d, cartesian; 66 AO's, $\approx$ 600,000 det.s (D_{2h}) $\alpha = 37.5312$ a.u. [17].

Method	E^d	notes
London Lanczos eq.(14)	35.41335	20 residues
London Lanczos eq.(14)	35.41694	40 residues, stable
Casimir Polder eq.(15)	35.41704	eq. (32) Romberg 32 pt.s
eq.(29)	35.41703	eq. (32) Romberg 32 pt.s
Hugenholtz eq.(26)	35.41697	eq. (32) Romberg 32 pt.s
Casimir Polder eq.(15)	35.41694	eq. (32) Gauss-Legendre 32 pt.s
eq.(29)	35.41694	eq. (32) Gauss-Legendre 32 pt.s

Basis 9s9p5d, spherical; 61 AO's, $\approx$ 400,000 det.s, (D_{2h}) $\alpha = 37.5790$ a.u.

Method	E^d	notes
London Lanczos eq.(14)	35.52120	40 residues, stable

Basis 9s9p5d3f, spherical; 82 AO's, $\approx$ 1,400,000 det.s (D_{2h}); $\alpha = 37.76325$ a.u.

Method	E^d	notes
London Lanczos eq.(14)	35.72629	40 residues, stable

INDEX

FUNDAMENTAL WORLD OF QUANTUM CHEMISTRY

Fundamental World of Quantum Chemistry

A Tribute to the Memory of Per-Olov Löwdin

Volume II

Edited by

Erkki J. Brändas

Uppsala University,
Uppsala, Sweden

and

Eugene S. Kryachko

Bogoliubov Institute for Theoretical Physics, Kiev, Ukraine
and University of Leuven, Leuven, Belgium

KLUWER ACADEMIC PUBLISHERS

DORDRECHT / BOSTON / LONDON

A C.I.P. Catalogue record for this book is available from the Library of Congress.

ISBN 978-1-4020-1290-7 (Vol. I)
ISBN 978-1-4020-1291-4 (Vol. II)
ISBN 978-1-4020-1286-0 (Set)

Published by Kluwer Academic Publishers,
P.O. Box 17, 3300 AA Dordrecht, The Netherlands.

Sold and distributed in North, Central and South America
by Kluwer Academic Publishers,
101 Philip Drive, Norwell, MA 02061, U.S.A.

In all other countries, sold and distributed
by Kluwer Academic Publishers,
P.O. Box 322, 3300 AH Dordrecht, The Netherlands.

Printed on acid-free paper

Cover Picture:
Löwdin's asymmetrical wave function Ψ in love with Hamiltonian

Photo: Per-Olov Löwdin (Michael Hehenberger.
Property of the Uppsala Quantum Chemistry Group)

FOREWORD TO VOLUME II

It is with respect and devotion that we point out Per-Olov Löwdin's pioneering and limitless efforts to found and establish the science of Quantum Chemistry. With his extraordinary capacity, Per-Olov Löwdin made basic contributions to the development, provided the conceptual framework and introduced fundamental concepts, and added the rigour to the field. Let us remind the reader of Löwdin's definition of a molecule as a Coulombic system of electrons and nuclei whose Hamiltonian with the center of mass translational degrees of freedom removed, has at least a bound state (in Molecules in Physics, Chemistry and Biology II, J. Maruani (Ed.), Kluwer, Dordrecht, 1988). But let us also recall his critique, i. e. of the impossibility to derive the Coulomb Hamiltonian from the first principles and the non-unique existence of molecular structures due to the general incompatibility between the number of nuclear positions and the number of internuclear distances. One of the most urgent problems of modern quantum chemistry is to treat the motions of the atomic nuclei and the electrons on a more or less equivalent basis.

Per-Olov Löwdin also established, developed and strengthened the links between physics, chemistry and biology. In addition, by organizing a staggering amount of regular summer and winter schools, institutes and symposia, by founding the International Journal of Quantum Chemistry and the Advances of Quantum Chemistry, he commended undisputable leadership at the highest national and international levels.

The clarity of expression, the vision of the possible and the optimism for the future are beautifully reflected in his "*Present Situation of Quantum Chemistry*" published in J. Phys. Chem. 61, 55-68 (1957), where, with eagerness of reaching a wide audience and capability of attracting many scientists, Per-Olov Löwdin introduced and defined the field of Quantum Chemistry by formulating its goals, i.e.

(i) to be able to predict accurately the properties of a hypothetical polyatomic molecule before it is synthezised in the laboratories;

(ii) to obtain such knowledge of the electronic structure of matter that one can construct new substances of particular value to mankind;

(iii) to learn to think in terms of electrons and their quantum mechanical behaviour.

In his revelations "*Quantum Theory of Many Particle Systems I-III*" (Phys. Rev. 97, 1474-1520 (1955)), Per-Olov Löwdin made a rigorous and ground-breaking formulation of the all important correlation problem in quantum chemical applications, from which emanated fundamental concepts like the correlation energy, the natural spin orbital, the general configuration interaction (CI) method, the reduced density matrix and the charge and bond order matrix. The constraints imposed by symmetry considerations in various independent particle models termed the "*symmetry dilemma*" (Rev. Mod. Phys.

35, 496 (1963)) have further led to important developments in solid state physics, chemical physics and physical chemistry.

Well-known is Löwdin's series of "*Studies in Perturbation Theory I-XIV*", a framework for the analysis of intermolecular forces as well as the technical development of very large CI expansions. The impact of his work on Quantum Biology are to be found in new applications and developments, i.e. in possible conformation work on prions, proton transcorrelations in DNA, macromolecular recognitions and specific motifs, in cancer research, etc. where there is a highly relevant quantum mechanical component to be identified. These accomplishments together with close to 70 summer and winter schools and Sanibel Symposia have contributed to the education of several generations of quantum chemists and solid state physicists.

It is a great privilege and honor to have participated in this edification. Looking at the picture of Per-Olov Löwdin, climbing in the mountains, we recollect the famous song by the Prophetes Deborah after the victory over the Philistines: "The mountains flowed before the Lord." This is precisely whom he was, is and will be - Per-Olov Löwdin or Pelle for all of us. Indeed, these years, when he was with us, were the most colourful, exciting and fascinating period in the history of quantum chemistry. It was the most enjoyable time for all of us because we all worked in quantum chemistry "just for FUN", as he always maintained. He will always be with us, the whole family of Quantum Chemists - including his own family with the "mother of the Uppsala Quantum Chemistry Group", Karin Löwdin, as prima motor and inner connection, and he is particularly close to all who have contributed to this tribute volume to his memory appearing as Volumes 4 and 5 in the Series "Conceptual Trends in Quantum Chemistry", already included Volume 1, "Conceptual Trends in Quantum Chemistry" (1994), Volume 2, "Structure and Dynamics of Atoms and Molecules: Conceptual Trends" (1995), and Volume 3, "Conceptual Perspectives in Quantum Chemistry" (1997), edited by Jean-Louis Calais and Eugene Kryachko. We deeply thank all authors for their wonderful and profound contributions.

John Coleman regrets that he was unable to contribute to this book and wrote that "Per-Olov had an enormous influence on my carrier, enticing me out of the ghetto of pure mathematics into the exciting area of the N-body problem. In my scientific articles there are several allusions to his effect on me. His encouragement and contually renewed invitations to participate in the Sanibel conferences during 10 years permitted me to attempt to spread my ideas and to broaden my, initially, very limited understanding of theoretical physics."

Erkki J. Brändas

Uppsala University
Uppsala

Eugene S. Kryachko

Bogoliubov Institute for Theoretical Physics,
Kiev, Ukraine
and University of Leuven, Leuven, Belgium

Contents of Volume II

Contents of Volume I

Non Hermitian Quantum Mechanics: Formalism and Applications

Edvardas Narevicius and Nimrod Moiseyev
Department of Chemistry and Minerva Center of Nonlinear Physics in Complex Systems Technion – Israel Institute of Technology Haifa 32000, Israel,
e-mail: nimrod@tx.technion.ac.il

Abstract. A brief review of our research in the field of non Hermitian quantum mechanics is presented with an emphasis on the special properties of the non Hermitian Hamiltonians. Specifically, the self orthogonality property and its relevance to physical phenomena which are difficult to interpret by the conventional Hermitian quantum mechanics is addressed. We also show the advantage in using the non Hermitian quantum mechanics to study physical systems where ionization and molecular dissociation occur simultaneously.

Keywords: non Hermitian quantum mechanics, self orthogonality

1. Introduction

Recently there is a growing interest in non Hermitian quantum mechanics. Non Hermitian (NH) quantum mechanical (QM) formalism is being used to study and explain physical phenomena. Sharp structures in electron/molecule collision cross sections and delocalization of bacteria population in drifting aqueous medium are only a few of many examples.

In conventional QM any physically observable quantity is given by the expectation value of a corresponding Hermitian operator. Also the problems studied using the NH-QM can be defined in terms of Hermitian operators only. However this may lead to a system with a very large number of degrees of freedom. For example, consider the dynamics of molecular systems where the electronic and the nuclear coordinates are strongly coupled to one another. In such a case the Born Oppenheimer approach is not applicable and the calculations must involve both nuclear and electronic coordinates. The calculations may be simplified using the non Hermitian quantum mechanics that enables us to take into consideration the coupling between the channels that are open for dissociation and ionization in a simple way(Narevicius and Moiseyev, 1998; Narevicius and Moiseyev, 2000). Another example is a dynamical system coupled to a bath. Often the solution of the full problem is impossible due to the current available computational

E.J. Brändas and E.S. Kryachko (eds.),
Fundamental World of Quantum Chemistry, Vol. II, 1–32.

sources and technology. The calculations become possible by including complex absorbing potential terms (i.e., non Hermitian operators) into the Hamiltonian which introduce the environmental dynamical effects on the studied system(Riss and Meyer, 1993; Moiseyev, 1998).

Finally, the non Hermitian quantum mechanical machinery has to be used to describe classical effects represented by non Hermitian operators. Diffusion effects, spatial fluctuations in inhomogeneous systems(Hatano and Nelson, 1996) both fall in this class of problems.

2. Physical phenomena associated with two types of non Hermitian Hamiltonians

It is well known that the Hermiticity of an operator is defined only within a given space of functions. For example, kinetic energy operator $-d^2/dx^2$ is Hermitian if the functions are in Hilbert space. However if the same operator acts on functions that diverge exponentially at $x \to \pm\infty$ it becomes non Hermitian. On the other hand operator d/dx is not Hermitian even if the functions are in Hilbert space. We will divide the non Hermitian problems into two types. Type I NH - operators are non Hermitian if the functions are not in the Hilbert space (the same operators ARE Hermitian when the functions are in the Hilbert space). Type II NH - operators are not Hermitian although the functions are in the Hilbert space.

The first type of non Hermitian problems is met in nuclear and atomic physics in the study of metastable resonance states. Resonance states represent exponentially decaying systems and are described by a state possessing a complex energy $E - i\Gamma/2$. The real part of the energy stands for the energy position of the resonances state and the complex part is related to the decay probability $w = \Gamma/\hbar$.

The resonance states can be obtained by imposing the outgoing boundary conditions on the general solutions of the Schrödinger equation. We will show here that this approach is equivalent to the definition of a resonance as a state associated with a complex pole of the scattering matrix $S(k)$.

For the sake of simplicity let us assume that V is a short-range one dimensional potential and $V \to \infty$ (i.e. the threshold energy $E_{th} = 0$) as the "reaction" coordinate $r \to \infty$. In such a case a general solution of the time-independent Schrödinger equation at $r \to \infty$ is given by

$$\begin{aligned} \phi(r \to \infty) = A(k)e^{-ikr} \ + \ B(k)e^{+ikr} &\approx e^{-ikr} + S(k)e^{+ikr} \\ E \ = \ (\hbar k)^2/2\mu, & \end{aligned} \tag{1}$$

where $A(k)$ and $B(k)$ are the amplitudes of the incoming and outgoing plane waves respectively. The S-matrix is defined as the ratio between the amplitude of the out-going plane wave and the in-coming wave. The S-matrix has a pole in two cases:

(1) when $B(k)$ has a pole. These are false poles which are not associated with the resonance phenomenon. False poles are independent of the potential, $V = \lambda v(r)$, and exist even when λ approaches zero(Korsch et al., 1984).

(2) when the amplitude, $A(k)$, of the in-coming wave vanishes. It may happen at values of k located on the positive imaginary axis of k, i.e. $k_0 = i|k_0|$. These poles of the S-matrix will be associated with the bound states. This can be easily shown by observing that the regular solution falls of to zero as $\exp ik_0 r = \exp(-|k_0|r)$ at infinity. If the pole of the S-matrix occurs at the lower halfplane the regular wave function has the asymptotic form,

$$\phi(r \to \infty) = e^{irRe(k)+rIm(k)} \tag{2}$$

i.e., it diverges at infinity. This is in a full agreement with the definition of the metastable state as the solution of Schrödinger equation with the outgoing boundary conditions.

The second type non Hermiticity is reported in several fields of physics. For example, the diffusion effects in convective media can be represented by a stationary diffusion equation(Hatano and Nelson, 1996; Hatano and Nelson, 1997; Hatano and Nelson, 1998) that is very similar to the time independent Schrödinger equation. However the stationary diffusion equation contains the non Hermitian derivative term $g * d/dx$ responsible for the convection. It has been shown(Hatano and Nelson, 1996; Hatano and Nelson, 1997; Hatano and Nelson, 1998) that for a critical value of convection velocity $g = g_c$ the bacteria population described by the stationary diffusion equation becomes completely delocalized. We will show below that this phenomenon is associated with the self orthogonality property of the non Hermitian operators.

Also complex terms in the Hamiltonian may arise in the study of systems having several channels for decay. One of the examples is an electron scattering from an H_2 molecule experiment. During the collision the meta stable intermediate state H_2^- is formed. This state can dissociate to $H + H^-$ and at the same time to loose the additional electron to create a neutral H_2 molecule or two hydrogen atoms. The H_2^- metastable state can be described using the complex analogue of the Born-Oppenheimer (BO) approximation. For a fixed $H - H$ distance the H_2^- molecule decays exponentially (due to autoionization) with a rate depending on that distance. This exponential decay can be accounted for by a single resonance state possessing a *complex* electronic "energy". The real part of the complex eigenvalue is the energy position of the resonance state and the imaginary part is proportional to its decay rate (inverse lifetime). Consequently, the states corresponding to the H_2^- intermediate can be obtained by solving the non Hermitian Schrödinger equation with a single complex potential energy surface. In this problem the self orthogonality property was encountered again and helped to explain the sharp resonant structures in the energy dependence of vibrational excitation of H_2 by electron impact.

3. Complex scaling Method

Most of the computational algorithms in quantum mechanics have been developed for Hermitian operators. As discussed above, the physical Hamiltonians are Hermitian only when they operate on bounded functions which get finite values at any point in the coordinate space. Since the functions in the type II non Hermitian problems are also kept in the Hilbert space the same numerical methods can be used to tackle both type II NH and Hermitian problems.

However the functions in type I non Hermitian problems are not in the Hilbert space. They diverge asymptotically and are difficult to deal with from numerical point of view. As will show here, there exists a similarity transformation that transforms the non hermiticity of type I to NH of type II and enables one to use the same well known algorithms (such as variational method) to solve the type II problems.

In the basis of this method stands the similarity transformation $\hat{S}$ which changes the asymptotically diverging resonance functions, ϕ^{res}, to square integrable functions. That is,

$$(\hat{S}\hat{H}\hat{S}^{-1})(\hat{S}\phi_n^{res}) = (\epsilon_n - (i/2)\Gamma_n(\hat{S}\phi_n^{res}) \tag{3}$$

such that,

$$\hat{S}\phi_n^{res} \rightarrow 0 \quad \text{as} \quad \mathrm{r} \rightarrow \infty \tag{4}$$

and $\hat{S}\phi_n^{res}$ are in the Hilbert space although ϕ_n^{res} are not. The complex-scaling operator to be defined below is only one example of a possible similarity transformation for which Eq. 4 is satisfied(Moiseyev and Hirschfelder, 1988).

The complex-scaling operator is given by

$$\hat{S} = e^{i\theta r\partial/\partial r} \tag{5}$$

such that,

$$\hat{S}f(r) = f(re^{i\theta}) \tag{6}$$

for any analytical function f (r). By substituting Eqs. 6 and 2 into Eq. 4 (this is equivalent to scaling r by $\exp(i\theta)$ in Eq. 2), one obtains

$$\hat{S}\phi_n^{res}(r \to \infty) = B(k_n)e^{+i|k_n|exp(i(\theta-\gamma_n)r)} = B(k_n)e^{i\alpha_n r}e^{-\beta_n r}, \quad (7)$$

where,

$$\begin{aligned} k_n &= |k_n|\, e^{-i\gamma_n}, \\ \alpha_n &= a_n(\cos\theta - \tan\gamma_n \sin\theta), \\ \beta_n &= a_n(\sin\theta - \tan\gamma_n \cos\theta), \\ \gamma_n &= \arctan(\Gamma/2\epsilon_n). \end{aligned}$$

One immediately sees from Eq. 7 that for a sufficiently large value of the rotational angle θ the exponential factor β_n gets a positive value and thereby the scaled resonance wave function becomes square integrable. In other words, when

$$\theta \geq \theta_c, \quad (8)$$

where the critical angle θ_c is given by(Balslev and Combes, 1971; Simon, 1973)

$$\theta_c = \gamma_n \equiv \arctan \Gamma_n / 2(\epsilon_n - E_{th}) \quad (9)$$

with ϵ_n, Γ_n and E_{th} being, respectively, the real and imaginary part of the complex energy and the threshold energy (E_{th} was taken to be zero in our studied example) then,

$$\hat{S}\phi_n^{res}(r \to \infty) \to 0. \quad (10)$$

Note, by passing, that the bound states of the unscaled Hamiltonian are a special case of Eq. 7 where $\Gamma_n = 0$ and $a_n = i\,|a_n|$. Therefore, the critical angle θ_c for the bound state is $\theta_c = 0$.

Here we proved (for a cut-off potentials) that by scaling the "reaction" coordinate the resonance wave function becomes square integrable and, consequently, the number of particles in coordinate space is conserved. Therefore, complex scaling has the advantage of associating the resonance phenomenon with the discrete part of the spectrum of the complex-scaled Hamiltonian. Moreover, the resonance state is associated with a single square integrable function, rather than with a collection of continuum eigenstates of the unscaled Hermitian Hamiltonian. Complex scaling may be viewed as a procedure which "compresses" information about the evolution of a resonance state at infinity into a small well-defined part of space. The tail in spatial space of a single, time-independent, square integrable resonance wave function contains all the information about the system, including information on partial width and on the way in which the system evolves as the distance between the "particle" and the "target" increases to infinity. The complex eigenvalues of the complex-scaled Hamiltonian (see Eq. 3) which are associated with the resonance phenomenon are θ-independent. The resonance complex-scaled eigenfunctions are, however, θ-dependent. We may interpret θ as a control parameter which "carries" the information of the decay process from infinity to a finite region in space whose size depends on the value of θ. As θ is increased, the information about the decay process is compressed into a smaller part of the coordinate space. However, θ cannot exceed a problem dependent critical value(Moiseyev, 1998).

4. Generalized definition of the inner product and self orthogonality

The inner product is defined in quantum mechanics as the scalar product, that is,

$$\int_{\text{all-space}} f\hat{H}g\, d\tau = \int_{\text{all-space}} g\hat{H}^{\dagger *} f\, d\tau \qquad (11)$$

where f, g are functions in the Hilbert space. When f and g are exponentially divergent functions - as the resonance Siegert eigenfunctions - these functions are NOT in the Hilbert space, and are NOT contained in the Hermitian domain of Hamiltonian. By scaling the coordinates in the time-independent Schrodinger equation by a complex factor we transform it to become the type II NH problem and the exponentially divergent resonance states become square integrable. However, in spite of the fact that the complex-scaled resonance states are in the generalized Hilbert space, they are not associated with an Hermitian operator ($\hat{H}(xe^{i\theta})$ is non Hermitian) and as we shall show in this section the scalar Hermitian product $\langle f|g\rangle$ should be replaced by a generalized inner product $(f|g)$ (the so-called c-product). See for example Ref. (Brändas and Elander, 1989) about the different possible definitions of a general inner product. It is convenient to define the inner product using the matrix representation of different operators. The matrices representing operators have right (column) eigenvectors, $|i)$, and left (row) eigenvectors, $(j|$. The left eigenvectors are obtained by taking the transpose of right eigenvectors of the transposed matrices. Therefore, when a matrix is complex and symmetric (associated with non Hermitian Hamiltonian) the left eigenvectors are equal (no complex conjugation) to the right ones, $(f|g) \equiv \langle f^*|g\rangle$. The bi-orthogonality condition implies that, $(j|i) = \delta_{i,j}$. Note, that if the Hamiltonian matrix is Hermitian the left j-th state is the complex conjugate of the right j-th state and the inner product is the ordinary metric scalar product, i.e. $(j|i) = \langle j|i\rangle$.

We will show here that the definition of the inner product in NH QM as the c-product gives rise to the eigenstate self-orthogonality phenomenon. The self-orthogonality is associated with the incomplete spectrum of the NH matrix as was shown by Moiseyev and Friedland(Moiseyev and Friedland, 1980). The theorem proved by Moiseyev and Friedland states that if $\hat{A}$ and $\hat{B}$ are two Hermitian operators, one can always find at least one value of a complex parameter λ for which spectrum of $\hat{A}+\lambda\hat{B}$ is incomplete(Moiseyev and Friedland, 1980). The spectrum is incomplete when the number of independent eigenvectors of a matrix is smaller than N. We will explain this phenomena using a following example. Suppose that non Hermitian Hamiltonian, H, which depends on some parameter λ can be represented by the complex symmetric matrix 3×3

$$H(\lambda) = \begin{pmatrix} 6 & e^{i\lambda} & -1 \\ e^{i\lambda} & 5 & 2e^{i\lambda} \\ -1 & 2e^{i\lambda} & 1 \end{pmatrix} \tag{12}$$

The fact that the matrix that represents non Hermitian operator is complex symmetric is the direct result of the use of the generalized inner product. One can easily see that the matrix given in Eq. 12 has three distinctive eigenvalues and eigenvectors for all but one value of parameter λ. When the parameter approaches value of $\lambda = \pi/2$ two of the eigenvalues and eigenfunctions approach each other and finally coalesce, i.e. $E_2 \to E_3$ and $\Psi_2 \to \Psi_3$ as $\lambda \to \pi/2$. For $\lambda = \pi/2$ matrix (Eq. 12) has only two distinct eigenvectors:

$$E_1 = 6, \quad \Psi_1 = \begin{pmatrix} 1 \\ i/3 \\ -1/3 \end{pmatrix} \tag{13}$$

$$E_2 = E_3, \quad \Psi_2 = \Psi_3 = \begin{pmatrix} 0 \\ 1 \\ i \end{pmatrix}$$

Whenever two eigenvectors approach each other they gradually become self orthogonal within the definition of the generalized inner product, i.e. $(\Psi_2|\Psi_2) \to 0$. In our example eigenvector Ψ_2 is self orthogonal for $\lambda = \pi/2$:

$$\begin{pmatrix} 0 & 1 & i \end{pmatrix} \cdot \begin{pmatrix} 0 \\ 1 \\ i \end{pmatrix} = 0 \tag{14}$$

If we will try to normalize the eigenvector Ψ_2 such that $(\Psi_2|\Psi_2) = 1$ we will have to divide its components by 0!

Moiseyev and Friedland assumed that the incomplete spectrum phenomenon which goes along with the formation of a self-orthogonal state is a rare phenomenon and is not associated with any physical measurable quantity. However, recent theoretical studies of the appearance of relatively sharp peaks in the cross section measured in scattering experiments of low energy electrons from Hydrogen molecule have shown that the self-orthogonality phenomenon is neither accidental nor rare(Narevicius and Moiseyev, 1998; Narevicius and Moiseyev, 2000). It was also shown that delocalization phenomenon discovered by Hatano and Nelson is associated with the self orthogonality of an eigenstate of the non Hermitian Hamiltonian(Narevicius and Moiseyev, submitted). Moreover, Michael Berry has shown that there are large set of physical phenomena that happen due to the appearance of self-orthogonal states such as excess noise in unstable lasers (Berry, 2001), propagation near the optic axes of absorbing anisotropic crystals (Berry, 1994), and the diffraction of atomic beams by crystals of light (Berry, 1998).

5. Non Hermitian adiabatic theory

In this section we will provide the formalism for an extension of the Born-Oppenheimer adiabatic approximation that deals with unstable autoionizing molecules and takes into account the coupling between electronic and nuclear motion. The new formalism will be applied in two cases: (i) full collision process of vibrational excitation of H_2 molecule by electron impact(Narevicius and Moiseyev, 2000), (ii) half collision process of interatomic Coulombic decay of electronically excited Ne cationic dimer(Santra et al., 2000; Moiseyev et al., 2001).

(i) *e/H_2 scattering process*

The most usual method that allows the coupling between electronic and vibrational degrees of freedom of molecule is the Born-Oppenheimer (BO) approximation. Within the framework of this approximation electronic energy levels are found for fixed nuclei configuration. The electronic energy obtained as a function of nuclei distance servers later as a potential energy surface for vibrational motion of nuclei.

The BO approximation is applicable whenever the energy distance between the electronic states is larger then the coupling term between them. In such a case nucleus move on a single electronic potential surface and the coupling to the others can be neglected. However for fixed internuclear distances $R < 3.2a.u.$ the H_2^- intermediate formed in the e/H_2 collision process has no bound states. Due to the autoionization process it supports only continuum states. The distance between the electronic states of continuum is zero, hence the usual BO approximation would require dealing with an immense number of electronic potential surfaces strongly coupled with each other.

For a fixed $H - H$ distance the H_2^- molecule decays exponentially with a rate depending on that distance. This exponential decay can be described by a single resonance state possessing a *complex* electronic "energy". The real part of the complex eigenvalue is the energy position of the resonance state and the imaginary part is proportional to its decay rate (inverse lifetime). *From here follows that instead of using many real coupled electronic potential surfaces within the BO approximation we can solve the nuclear Schrödinger equation with a single complex potential energy surface.*

We will derive here the expression for the vibrational excitation cross section of electron/molecule collision using the complex adiabatic approximation. Let us consider a scattering event where the electron represented by $\sqrt{\frac{1}{k_i}}e^{ik_ir}$, collides with a molecule which has N internal electrons. We assume that the electronic state of the molecule is varied adiabatically during the collision. The molecular degrees of freedom are represented by coordinate R. Prior to collision the molecule is in its ν vibrational state, $\chi_\nu(R)$, and in its n_0-th electronic state, $\phi_{n_0}(\{r\}_N; R)$. Assuming adiabaticity, the initial state, $\chi_\nu(R)\Phi_i(\{r\}_{N+1}; R)$, properly antisymmetrized in the electronic coordinates such that

$$\Phi_i(\{r\}_{N+1}; R) = \mathcal{A}(\phi_i(\{r\}_N; R)\sqrt{\frac{1}{k_i}}e^{ik_ir_{N+1}}) \tag{15}$$

is an eigenfunction of the interaction free Hamiltonian

$$H_0 = T(R) + T(\{r\}_{N+1}) + V_{mol}(\{r\}_N, R) \tag{16}$$

where $T(\{r\}_{N+1})$ and $T(R)$ are respectively electronic and nuclear kinetic energy operators and $V_{mol}(\{r\}_N, R)$ is the potential energy of the $N+1$ electrons and nucleus of the molecule. After the inelastic collision the molecule is found in its vibrationally excited state $\chi_{\nu'}(R)$ and the scattered electron is in a state defined by $\sqrt{\frac{1}{k_f}}e^{ik_fr_{N+1}}$. The final state of the non interacting electron-molecule system is $\chi_{\nu'}(R)\Phi_f(\{r\}_{N+1}; R)$ where

$$\Phi_f(\{r\}_{N+1}; R) = \mathcal{A}(\phi_f(\{r\}_N; R)\sqrt{\frac{1}{k_f}}e^{ik_fr_{N+1}}). \tag{17}$$

The energy of the scattered electron is given by the energy conservation law:

$$\frac{|\hbar k_i|^2}{2m_e} + E_\nu = \frac{|\hbar k_f|^2}{2m_e} + E'_\nu \equiv E \tag{18}$$

where m_e is the mass of an electron, E_ν and E'_ν are the energies of the initial and the final states of the molecule. The Hamiltonian during the collision process is given by,

$$H = T(R) + T(\{r\}_{N+1}) + V_{int}(\{r\}_{N+1}, R) \tag{19}$$

where V_{int} is the $N+1$ electron and nucleus interaction potential. V_{int} includes the coordinates of both the colliding electron and the electrons of the molecule. Since the electronic coordinate is much faster than the molecular one the Born-Oppenheimer (BO) approximation is applicable here. The adiabatic BO Hamiltonian is defined by

$$H_{ad}(\{r\}_{N+1}; R) = T(\{r\}_{N+1}) + V_{int}(\{r\}_{N+1}; R) \tag{20}$$

where the nuclear coordinate R is a parameter. The adiabatic Hamiltonian H_{ad} possesses R-dependent eigenenergies $E_n^{ad}(R)$ and eigenfunctions $\phi_n^{ad}(\{r\}_{N+1}; R)$ such that

$$H_{ad}(\{r\}_{N+1}; R)\phi_n^{ad}(\{r\}_{N+1}; R) = E_n^{ad}(R)\phi_n^{ad}(\{r\}_{N+1}; R). \tag{21}$$

The adiabatic energy $E_n^{ad}(R)$ serves as a potential for the motion of the heavy nuclei:

$$(T(R) + E_n^{ad}(R))\chi_{n,\alpha}^{ad}(R) = \epsilon_{n,\alpha}^{ad}\chi_{n,\alpha}^{ad}(R) \tag{22}$$

where $\chi_{n,\alpha}^{ad}(R)$ and $\epsilon_{n,\alpha}^{ad}$ are the nuclear vibrational wave function and energy. Within the Born-Oppenheimer approximation the total wave function is a product of the nuclear wave function and the electronic wave function which depends parametrically on the nuclear coordinate:

$$\Psi_{n,\alpha}^{ad} = \phi_n^{ad}(\{r\}_{N+1}; R)\chi_{n,\alpha}^{ad}(R) \tag{23}$$

According to the Lippman-Schwinger equation the probability to change the vibrational state of the molecule from $\chi_\nu(R)$ to $\chi_{\nu'}(R)$ due to the collision process is given by,

$$\sigma = |\langle\Psi_f|V + VGV|\Psi_i\rangle|^2 \tag{24}$$

where respectively the initial and the final states are given by $\Psi_i = \chi_\nu(R)\Phi_i(\{r\}_{N+1}; R)$ and $\Psi_f = \chi_{\nu'}(R)\Phi_f(\{r\}_{N+1}; R)$. V is defined as $V = V_{int} - V_{mol}$. The resonant part of the excitation probability is

$$\sigma_{res} = \left| \lim_{\epsilon \to 0+} \left\langle \chi_{\nu'}(R)\Phi_i(\{r\}_{N+1}; R) \left| V \frac{1}{E - H + i\epsilon} V \right| \right. \right. \\ \left. \left. \chi_\nu(R)\Phi_f(\{r\}_{N+1}; R) \right\rangle_{r,R} \right|^2. \tag{25}$$

Using the spectral representation of the Green operator $(E - \hat{H})^{-1}$ in Eq. (25) within the Born-Oppenheimer approximation framework one gets:

$$\sigma_{res} = \left| \sum_{n,\alpha} \frac{\left\langle \chi_{\nu'} \left| \left\langle \Phi_i | V | \Psi^{ad}_{n,\alpha} \right\rangle_r \left\langle \Psi^{ad}_{n,\alpha} | V | \Phi_f \right\rangle_r \right| \chi_\nu \right\rangle_R}{E - E^{ad}_{n,\alpha}} \right|^2 \quad (26)$$

The molecular anion, which is formed when an electron collides with a neutral molecule, supports no bound states. However, the electron can be trapped for a finite period of time in the quasibound resonance state. As was mentioned above, there are several methods, such as complex scaling (CS), which allow us to "concentrate" the information about the resonance phenomena into a *single square integrable state* which is associated with a complex energy eigenvalue, $E^{res}_n = \epsilon_n - \frac{i}{2}\Gamma_n$. That is, $\hat{H}_{CS}|n\rangle = E^{res}_n|n\rangle$ where $\hat{H}_{CS}$ is the complex scaled Hamiltonian. The solution space of the complex scaled molecular anion Hamiltonian can be divided into subspace of resonance states and into the subspace of continuum scattering states. Whenever the resonance is isolated (the distance between two consecutive resonances is larger then the width, Γ of each resonance) and has a large lifetime ($\Gamma/\hbar$ is larger than any characteristic frequency of the system) we can make an assumption that a scattering event involves a single resonance state.

A common situation is that, $\Delta E_{vib} \leq \Gamma < \Delta E_{elec}$ where ΔE_{elec} and ΔE_{vib} are respectively the energy spacing between electronic levels and vibrational levels. In such a case scattering event proceeds via a single *electronic* resonance state, n_0. However it involves a number of vibrational states $\{\chi^{ad}_\nu\}$ belonging to the same electronic resonance state, n_0. Consequently the sum over electronic energy levels, n, can be omitted from the sum in the expression for the resonant part of the excitation probability in Eq. 26:

$$\sigma_{res} = \left| \sum_{\alpha}^{res} \frac{\left\langle \chi_{\nu'} \left\langle \Phi_i | V | \phi^{ad}_{n0} \right\rangle_r \chi^{ad}_{n0,\alpha} \right\rangle_R \left\langle \chi^{ad}_{n0,\alpha} \left\langle \phi^{ad}_{n0} | V | \Phi_f \right\rangle_r \chi_\nu \right\rangle_R}{E - E^{ad}_{n,\alpha}} \right|^2 \quad (27)$$

Moiseyev and Peskin(Moiseyev and Peskin, 1990) have shown that the integrals over the electronic coordinates in the Eq. 27 are related to the partial widths. That is,

$$\begin{aligned} \left| \left\langle \Phi_i | V | \phi^{ad}_{n0}(\{r\}_{N+1}; R) \right\rangle_r \right|^2 &= \Gamma^{(i)}_{n0}(R) \\ \left| \left\langle \Phi_f | V | \phi^{ad}_{n0}(\{r\}_{N+1}; R) \right\rangle_r \right|^2 &= \Gamma^{(f)}_{n0}(R) \end{aligned} \quad (28)$$

where $\Gamma^{(i)}_{n_0}(R)$ is the partial width of the electronic resonance state n_0 as function of the nuclear coordinate R. The products of this decay channel are neutral molecule and a free electron associated with a wave number k_i. $\Gamma^{(f)}_{n_0}(R)$ is the partial width for the decay channel that results in a neutral molecule and a scattered electron associated with k_f wave number.

Within the local approximation the energy dependence of the partial widths (via the scattered electron wave number k) is neglected and $\Gamma^{(i)}_{n_0} = \Gamma^{(f)}_{n_0}$. Using this approximation and Eqs. 27 28 one gets that:

$$\sigma_{res} = \left| \sum_{\alpha}^{res} \frac{\left\langle \chi_{\nu'} | \Gamma^{\frac{1}{2}}_{n_0}(R) | \chi^{ad}_{n_0,\alpha} \right\rangle_R \left\langle \chi^{ad}_{n_0,\alpha} | \Gamma^{\frac{1}{2}}_{n_0}(R) | \chi_{\nu} \right\rangle_R}{E - E^{ad}_{n,\alpha}} \right|^2 \tag{29}$$

Since Γ_{n_0} is the width of the electronic resonance state n_0 it is related to the imaginary part of the adiabatic electronic energy $E^{ad}_{n_0}$:

$$\Gamma_{n_0}(R) = -2Im\, E^{ad}_{n_0}(R) \tag{30}$$

Eq. 29 given above not only simplifies the calculation of σ but also provides a simple physical insight to the problem of electron scattering from a neutral molecule. Since the final expression for the vibrational excitation cross section does not depend on the electron coordinates we can say that the electron-neutral molecule collision prepares a nuclear wave packet on the potential energy surface of the molecular anion intermediate. The prepared wave packet is propagated via the nuclear Born-Oppenheimer molecular anion Hamiltonian with the complex potential $E^{ad}_{n_0}(R)$. Finally, the anion undergoes autionization process resulting in the molecule in a vibrationally excited state. The probability amplitude to obtain a specific vibrational state ν of the molecule is given in Eq. 29. This expression is in a complete agreement with the results obtained by Domcke and Cederbaum(Cederbaum and Domcke, 1981) by applying the local approximation to the exact expression for σ.

(ii) *interatomic Coulombic decay of* $\left[Ne_2^+\right]^*$

We will demonstrate here an application of the complex analogue of the Born-Oppenheimer approximation to a half collision process, namely, to the reaction

$$Ne_2 + \hbar\omega \rightarrow \left[Ne_2^+\right]^* + e^- \rightarrow \left[Ne_2^{2+} + e^-(\mathrm{ICD})\right] + e^- \rightarrow 2Ne^+ + 2e^-,$$

where the electronically excited cationic dimer, $\left[Ne_2^+\right]^*$, has a hole in its inner-valence 2s-type molecular orbital $2^2\Sigma_u^+$. We will present the derivation of the expressions for the kinetic energy distributions of the emitted interatomic Coulombic decay electron and of the Ne^+ fragments.

Within the framework of the Born-Oppenheimer approximation the initial state, that is, the ground state of neutral neon dimer, is given by

$$\Psi_0[Ne_2](\{\vec{r}\}_N, R) = \psi^{\text{elec}}[Ne_2](\{\vec{r}\}_N; R)\chi_0[Ne_2](R). \qquad (31)$$

$N = 20$ is the number of electrons, $\{\vec{r}\}_N$ is the notation for the coordinates of the N electrons, $\psi^{\text{elec}}[Ne_2]$ is the adiabatic electronic wave function in the electronic ground state of Ne_2, and $\chi_0[Ne_2]$ is the vibrational eigenfunction of the nuclear Hamiltonian with the electronic ground state potential energy surface.

Utilizing electromagnetic radiation from a synchrotron source, for instance, an inner-valence electron can be ejected and a superposition of autoionizing states of Ne_2^+ in the $2^2\Sigma_u^+$ electronic state is generated. We can approximate these resonance states by the BO solutions only when the non Hermitian approach is taken and the potential energy surface obtained from the electronic structure calculations is complex, $E[2^2\Sigma_u^+ Ne_2^+](R) = \text{Re}E[2^2\Sigma_u^+ Ne_2^+](R) - \frac{i}{2}\Gamma[2^2\Sigma_u^+ Ne_2^+](R)$. Under these circumstances the autoionizing states of $\left[Ne_2^+\right]^*$ can be factorized into an electronic and a nuclear part,

$$\Psi_\nu[2^2\Sigma_u^+ Ne_2^+](\{\vec{r}\}_{N-1}, R) = \psi^{\text{elec}}[2^2\Sigma_u^+ Ne_2^+](\{\vec{r}\}_{N-1}, R)\chi_\nu[2^2\Sigma_u^+ Ne_2^+](R). \qquad (32)$$

Exposing Ne_2 to radiation with photon energy $\hbar\omega$ significantly above the inner-valence ionization threshold can lead to formation of a wave packet of the $(N-1)$-electron system $\left[Ne_2^+\right]^*$,

$$\Phi^{\text{WP}}[2^2\Sigma_u^+ Ne_2^+](\{\vec{r}\}_{N-1}, R) = \sum_\nu \left(\chi_\nu[2^2\Sigma_u^+ Ne_2^+]|\mu(R)|\chi_0[Ne_2]\right)_R \Psi_\nu[2^2\Sigma_u^+ Ne_2^+](\{\vec{r}\}_{N-1}, R) \qquad (33)$$

and a photoelectron in state ϕ_{PE}, which in the sudden approximation does not interact with the Ne_2^+ cation. Thus the dipole transition moment $\mu(R)$ is given by

$$\mu(R) = \left(\mathcal{A}\{\psi^{\text{elec}}[2^2\Sigma_u^+ Ne_2^+]\phi_{\text{PE}}\}|e\sum_{i=1}^{N}\vec{r}_i \cdot \vec{e}_z|\psi^{\text{elec}}[Ne_2]\right)_{\vec{r}}, \qquad (34)$$

where $(\cdot|\cdot)_{\vec{r}}$ indicates integration over the electronic coordinates, e denotes the charge of an electron, and $\vec{e}_z$ is a unit vector along the polarization axis of linearly polarized light. The symbol $\mathcal{A}$ is used to emphasize that the product state of cation and photoelectron must be properly antisymmetrized.

The wave packet $\Phi^{WP}[2^2\Sigma_u^+ Ne_2^+]$ undergoes electronic decay according to the ICD mechanism. The products of this relaxation process are dissociative Ne_2^{2+} in its electronic ground state and a free ICD electron in state ϕ_{ICD} with kinetic energy E_{kin}. Within the framework of the BO approximation the final states can be written as

$$\Psi_f[Ne_2^{2+} + e^-(ICD)](\{\vec{r}\}_{N-1}, R) = \mathcal{A}\{\phi_{ICD}\psi^{elec}[Ne_2^{2+}]\}(\{\vec{r}\}_{N-1}; R)\chi_{E_f}[Ne_2^{2+}](R). \quad (35)$$

The dissociative nuclear wave functions $\chi_{E_f}[Ne_2^{2+}]$ are associated with continuous energies $E_f[Ne_2^{2+}]$. Therefore the energy of the state $\Psi_f[Ne_2^{2+} + e^-(ICD)]$ is $E_{kin} + E_f[Ne_2^{2+}]$.

We treat the coupling between the ICD electron and the dication by means of the Lippmann-Schwinger equation extended to non Hermitian Hamiltonians:

$$\Phi_f[Ne_2^+] = (1 + \hat{G}(E_{kin} + E_f[Ne_2^{2+}])\hat{V})\Psi_f[Ne_2^{2+} + e^-(ICD)]. \quad (36)$$

$\left(\Phi^{WP}[2^2\Sigma_u^+ Ne_2^+] | \Phi_f[Ne_2^+]\right)$ is the probability amplitude that the wave packet $\Phi^{WP}[2^2\Sigma_u^+ Ne_2^+]$ decays into Ne_2^{2+} with energy $E_f[Ne_2^{2+}]$ and an ICD electron with energy E_{kin}. The expression explicitly reads

$$\left(\Phi^{WP}[2^2\Sigma_u^+ Ne_2^+] | \Phi_f[Ne_2^+]\right) = \sum_\nu \left(\chi_0[Ne_2] | \mu(R) | \chi_\nu[2^2\Sigma_u^+ Ne_2^+]\right)_R \cdot \frac{\left(\Psi_\nu[2^2\Sigma_u^+ Ne_2^+] | \hat{V} | \Psi_f[Ne_2^{2+} + e^-(ICD)]\right)_{\vec{r},R}}{E_{kin} + E_f[Ne_2^{2+}] - E_\nu[2^2\Sigma_u^+ Ne_2^+]} \quad (37)$$

What remains to be discussed is the determination of the interaction matrix elements $\left(\Psi_\nu[2^2\Sigma_u^+ Ne_2^+] | \hat{V} | \Psi_f[Ne_2^{2+} + e^-(ICD)]\right)_{\vec{r},R}$. $\hat{V}$ is defined as

$$\hat{V} = \hat{H}[2^2\Sigma_u^+ Ne_2^+] - \hat{H}[Ne_2^{2+} + e^-(ICD)]. \quad (38)$$

where $\hat{H}[2^2\Sigma_u^+ Ne_2^+]$ is a complex, non Hermitian Hamiltonian whereas $\hat{H}[Ne_2^{2+} + e^-(ICD)]$ is Hermitian. By using the BO expressions for $\Psi_\nu[2^2\Sigma_u^+ Ne_2^+]$ (Eq. (32)) and $\Psi_f[Ne_2^{2+} + e^-(ICD)]$ (Eq. (35)) it is straightforward to derive that

$$\left(\Psi_\nu[2^2\Sigma_u^+ Ne_2^+] | \hat{V} | \Psi_f[Ne_2^{2+} + e^-(ICD)]\right)_{\vec{r},R} = \left(\chi_\nu[2^2\Sigma_u^+ Ne_2^+] | d(R) | \chi_{E_f}[Ne_2^{2+}]\right)_R, \quad (39)$$

where

$$d(R) = \left(\psi^{\text{elec}}[2^2\Sigma_u^+\text{Ne}_2^+]|\hat{V}|\mathcal{A}\{\phi_{\text{ICD}}\psi^{\text{elec}}[\text{Ne}_2^{2+}]\}\right)_{\vec{r}}. \tag{40}$$

Moiseyev and co-workers have shown (Narevicius and Moiseyev, 2000; Moiseyev and Peskin, 1990) that, apart from a multiplicative constant, in the local approximation

$$d(R) = e^{i\varphi(R)}\sqrt{\Gamma(R)}. \tag{41}$$

Here $\Gamma(R) = \Gamma[2^2\Sigma_u^+\text{Ne}_2^+](R)$ is the local electronic decay width of inner-valence ionized neon dimer and $\varphi(R)$ is the phase of the resonance width amplitude.

The kinetic energy distribution of the ICD electrons, $\sigma(E_{kin})$, is obtained by adding up all decay probabilities $\left|\left(\Phi^{\text{WP}}[2^2\Sigma_u^+\text{Ne}_2^+]|\Phi_f[\text{Ne}_2^+]\right)\right|^2$ that are associated with an ICD electron of energy E_{kin} in the final state,

$$\sigma(E_{kin}) = \int dE_f[\text{Ne}_2^{2+}]\left|\left(\Phi^{\text{WP}}[2^2\Sigma_u^+\text{Ne}_2^+]|\Phi_f[\text{Ne}_2^+]\right)\right|^2. \tag{42}$$

Assuming $\mu(R) = 1$ and $\varphi(R) = 1$, Eq. (29) follows immediately by inserting Eqs. (37), (39), and (41) into Eq. (42). In a similar fashion the kinetic energy distribution of the Ne^+ fragments, $\tilde{\sigma}(E_f)$, is given by

$$\tilde{\sigma}(E_f) = \int dE_{kin}\left|\left(\Phi^{\text{WP}}[2^2\Sigma_u^+\text{Ne}_2^+]|\Phi_f[\text{Ne}_2^+]\right)\right|^2. \tag{43}$$

6. Physical phenomena explained by non Hermitian quantum mechanics

6.1. Can broad overlapping resonances be observable?

Resonance phenomena is widely observed in atomic, nuclear and particle physics. Yet its definition is not entirely rigorous. Usually, in literature, the term resonance is taken to be exactly synonymous with a pole of S-matrix. However when the pole is located sufficiently far from the real energy axis it is not clear if the pole will produce any physically observable effect. For example consider a one dimensional problem of a particle going through an Eckart potential barrier which is used as a simple model for chemical reactions. It has been shown(Ryaboy and Moiseyev, 1993) that in the case of Eckart potential and other "soft" potential barriers the corresponding S-matrix has a series of poles that start at about the energy of the top of potential barrier and move to the lower energies. Some of the resonances are even located at the energies bellow the top of the potential barrier. However these poles do not induce any observable structure in the transition probability as the function of energy. Consequently we have to classify these poles as unphysical poles and we will not associate them with the resonance phenomena.

Interestingly, the same unphysical poles of 1D Eckart barrier become an approximation for physically observable 2D resonance states due to the trapping of a light particle in between two heavy particles(Narevicius and Moiseyev, 1998). This phenomenon may be observed in the photodissociation studies of ArHCl or ArHBr. When ArHCl/ArHBr is promoted to the electronically excited state hydrogen atom is repulsed from the both heavy atoms, however it can get trapped between the two prior to the dissociation.

In order to present a qualitative analysis of this problem we consider here a model system there the two heavy atoms are represented by two hard disks (with Dirichlet boundary conditions) B and C having the same radius r and the light hydrogen atom A that is allowed to move freely in the plane outside of the two disks as shown in Fig(6.1). Within the framework of the adiabatic approximation the 1D poles of an Eckart-type potential are approximations to the physical 2D resonances associated with a trapping of a light particle.

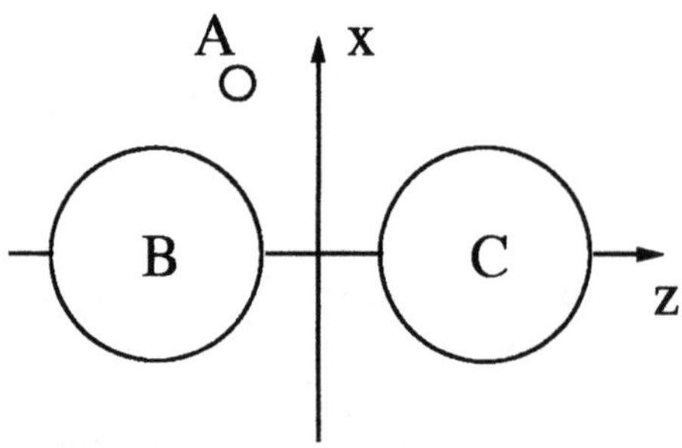

Figure 1. Illustrative picture of the studied model systems where the B and C heavy atoms are described as two disks and the light particle, A, moves freely in the plane, excluding the two disks area; i.e. $V = \infty$, when $x^2 + (z \pm R/2)^2 \leq r^2$ where x, z denotes the coordinates of A.

Usually the adiabatic approximation is taken, as in the BO approximation, when one particle is much heavier than another one. We will use here a non usual adiabatic approach. Here we will treat x as the adiabatic coordinate since the motion of hydrogen along x should be much slower than the motion along the z-axis in order to get a temporarily trapping of the light "particle" in between the two heavy atoms (represented as B and C in Fig.(6.1)). Let us first consider the model where the two heavy atoms are described as two disks with the same radius r and hydrogen atom moving freely in the plane, excluding the area of the two disks. When x is treated as a parameter we reduce the problem from 2D to 1D particle-in-a-box problem with a box length $L(x) = R - 2\sqrt{r^2 - x^2}$, when $-r \leq x \leq +r$, and $L = \mathcal{L}$ elsewhere. $\mathcal{L}$ is the length of the artificial box which has been used in the numerical calculations. The adiabatic potentials are, $E_{n_z}(x) = (\hbar\pi n_z/L(x))^2/(2m)$; $n_z = 1, 2, ...$ In Fig.(6.1) we represent

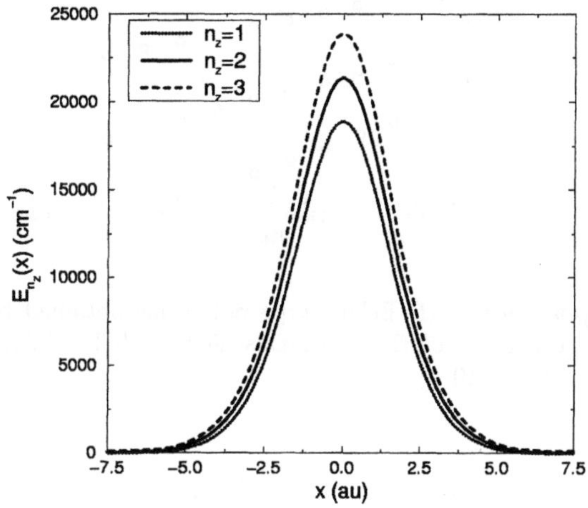

Figure 2. The adiabatic potentials, $E_{n_z}(x)$, obtained when the x-coordinate has been treated as a parameter.

the adiabatic potentials $E_{n_z}(x)$ obtained from our numerical calculations for the *physical* ArHCl potential (soft repulsive potential)and not for the hard-disks/spheres pictures which is used above to clarify the trapping phenomenon of hydrogen in-between two heavy atoms in the absence of attractive force. The shape of the adiabatic 1D potentials, obtained in our numerical calculation is very close to the Eckart potential barriers (i.e inverse of square of cosh function). The coalescence between these two types of potentials is remarkable even on large scale plots.

In the last step of the adiabatic calculations we calculated the $n_x = 0, 1, 2...$ complex eigenvalues of the complex scaled Hamiltonian,

$$H_\theta(x; n_z) = -\exp(-2i\theta)\frac{\hbar^2}{2m}\frac{d^2}{dx^2} + E_{n_z}(x\exp(+i\theta))$$

using the Colbert-Miller discrete variable representation (DVR) expressions (Colbert and Miller, 1992). It is clear from the equation given above that within the framework of the adiabatic approximation the complex poles can be assigned by two good quantum numbers, n_x, n_z. The poles of the 1D Eckart-type potential barriers, $E_{n_z}(x)$, are given in Fig.(6.1). They are in a remarkable agreement with the numerical

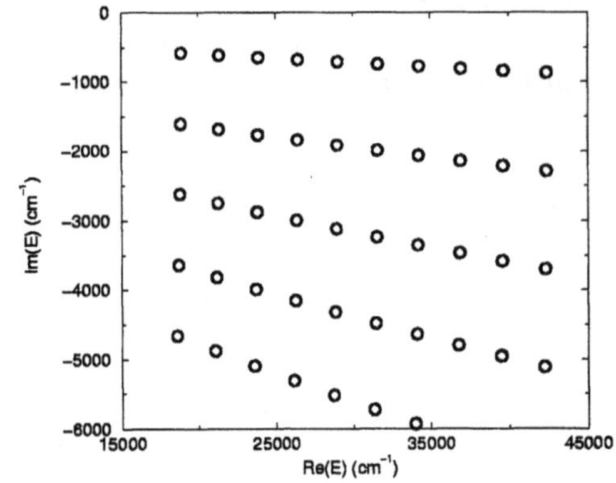

Figure 3. The poles of the 1D Eckart-type potentials obtained by calculating the θ independent eigenvalues of the complex-scaled ArHCl adiabatic Hamiltonians, $H_\theta(x; n_z)$, $n_z = 1, 2, \ldots, 10$.

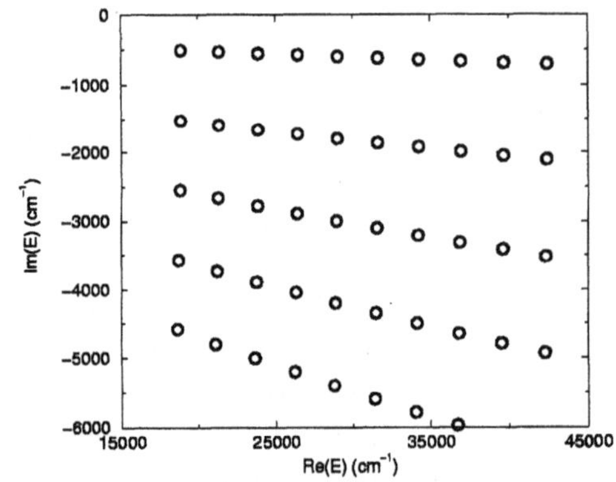

Figure 4. The numerically exact poles of the S-matrix obtained by calculating the eigenvalues of the complex scaled ArHCl 2D Hamiltonian (i.e x$\rightarrow x \exp(i\theta)$). The Ar-Cl distance was kept fixed at the equilibrium distance.

exact results presented in Fig.(6.1). Beside the physical insight we get by the use of the adiabatic approach we also gain in the drastic reduction of the computational effort which is required for calculating the poles. Since the number of operations for calculating the N eigenvalues of a NxN matrix is scaled as N^3, we reduce the length of the calculations by 6 orders of magnitude by using the adiabatic approach

(N=100 has been taken as a representative number of basis functions for the z coordinate).

Here we have shown that the non-resonance poles of 1D Eckart type potential barrier (associated with complex scaled eigenfunctions which are localized in the forbidden classical region in the 2D phase space) are the resonance poles of the 2D problem and are associated with complex-scaled eigenfunctions which are localized in the classically accessible region in the 4D phase space.

6.2. FINGERPRINTS OF OVERLAPPING RESONANCES IN e/H_2 SCATTERING CROSS SECTION MEASUREMENTS

In this section we will propose a mechanism that leads to the structure in the cross section even when the resonances are overlapping i.e. $\Gamma_n > |\epsilon_n - \epsilon_{n\pm1}|$ where ϵ_n is the energy position of the resonance state and Γ_n is the corresponding width. We will describe a case where the *interference* among resonance states results in sharp structure in the scattering cross section σ. This is very different from the known Fano interferences(Fano, 1961) which take place between a given resonance state and a weakly energy dependent background. Fano interferences lead to the distortion of the shape of the resonance peak in the cross section but not to a substantial increase of the lifetime of the trapped particle in the scattering process. In our case the structure is the result of the interference among resonances and consists of peaks where each one of them is much narrower than the width of a single resonance state Γ_n. Since the lifetime of the trapped particle is proportional to the inverse of the width of the peak in the cross section, it implies that the narrowing of the peak in the cross section increases the lifetime of the trapped particle. Moreover, the interference takes place between many resonances which are separated by a large energy interval as compared to the corresponding resonance widths i.e., $|\epsilon_n - \epsilon_m| \gg Max(\Gamma_n, \Gamma_m)$. *Therefore we may consider the new mechanism presented here as a collective coherent resonance phenomenon.* As it will be discussed below, this phenomenon was first observed, without realizing it, in the experiments of inelastic electron scattering from an H_2 molecule(Fano, 1985). The sharp structure in the experimental electron/hydrogen-molecule scattering cross section is associated with the short-lived vibrational states of the autoionizing H_2^- intermediate. We have calculated the positions and the widths of these H_2^- states within the local approximation and to our surprise we obtained that the width of the resonance states were 2-3 times larger than the widths of the peaks in the vibrational excitation cross sections(Narevicius and Moiseyev, 1998). Yet, the mechanism that leads to this phenomenon remained unclear.

We will show here that the nuclear motion of H_2^- induces interference between resonances such that the structures appearing in the scattering cross section are narrower than the width of a single resonance state, Γ_n. First, however, we would like to stress that the interference phenomenon between the resonance states itself is associated with the generalization of the inner product in quantum mechanics. This generalization is required when resonances, metastable states, are associated with complex, rather than real eigenvalues of the Hamiltonian.

The transition probability amplitude, $t(E)$, for a scattering experiment where the initial and the final states are identical, i.e. $|\phi_i\rangle = |\phi_f\rangle = |0\rangle$, is given by the Lippmann-Schwinger equation,

$$t(E) = \langle 0|V + VGV|0\rangle = t_{direct}(E) + t_{res}(E) \tag{44}$$

The resonant term, $t_{res}(E)$, is given by

$$t_{res}(E) = \int d\epsilon\, \rho(\epsilon)a(\epsilon)/(E-\epsilon), \tag{45}$$

where $\hat{H}|\epsilon\rangle = \epsilon|\epsilon\rangle$, $a(\epsilon) = |\langle 0|V|\epsilon\rangle|^2 \geq 0$ and $\rho(\epsilon)$ stands for the density of states. However, when $\hat{H}$ is an Hermitian operator the eigenvalues, ϵ, get real values only and the information about the resonance phenomena is spread over a large number of continuum states. Using the complex scaling (CS) method allows us to "concentrate" the information about the resonance phenomena into a *single square integrable state* which is associated with a complex energy eigenvalue, $E_n^{res} = \epsilon_n - \frac{i}{2}\Gamma_n$. That is, $\hat{H}|n\rangle = E_n^{res}|n\rangle$ where $|n\rangle$ is an eigenfunction which is *not* in the Hermitian domain of the Hamiltonian H. This enables us to replace the integral in the expression for the $t_{res}(E)$ by a sum over the discrete resonance states,

$$t_{res}(E) = \sum_n a_n/(E - \epsilon_n + \frac{i}{2}\Gamma_n). \tag{46}$$

Since the eigenfunction $|n\rangle$ is not in the Hermitian domain of the Hamiltonian the definition of the inner product that we should use should be questioned. If we will keep the usual definition of the scalar product in quantum mechanics the coefficients a_n in Eq. 46 will get real positive values only (as well as $a(\epsilon)$ in Eq. 45) and the possibility of interference among different resonance states which leads to the trapping of an electron due to the molecular vibrations will be eliminated. As was mentioned before the generalized definition of the inner product $(...|...)$ rather than the usual scalar product $\langle ...|...\rangle$ has to be used since the Hamiltonian is not Hermitian(Moiseyev, 1998; Brändas and Elander, 1989). Only the application of the generalized inner product can give rise to the complex coefficients, $a_n = (0|V|n)(n|V|0)$, in Eq. 46. The fact that $\{a_n\}$ can get complex and real negative values is essential to obtain the interference phenomenon which increases the lifetime of H_2^- intermediate in the e^-/H_2 scattering experiments.

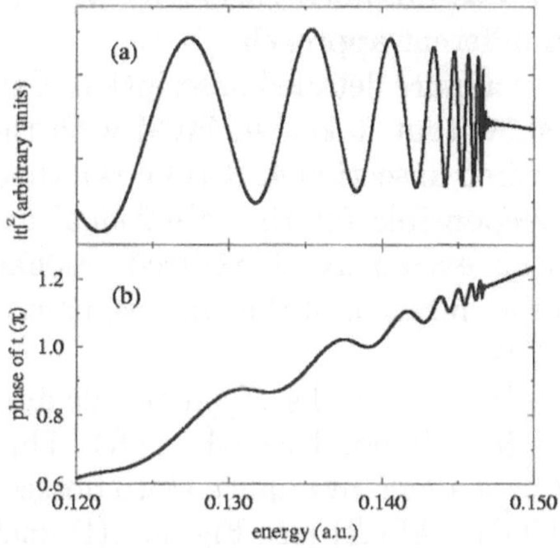

Figure 5. (a) The behavior of the probability to excite H_2 molecule from the ground vibrational $\nu = 0$ to the $\nu = 4$ excited state in e/H_2 collision. (b) The phase of the corresponding probability amplitude.

In Figs 6.2 a,b we represent the electron-H_2 scattering cross section and the phase of the corresponding excitation probability amplitude. The widths of the peaks in Fig 6.2a are $2-3$ times narrower than the width of the H_2^- resonance states. The phase shows an oscillatory behavior similar to the behavior of the transition probability amplitude phase as measured in the experiments of an electron passing through a quantum dot(Schuster et al., 1997). Note that our numerical results

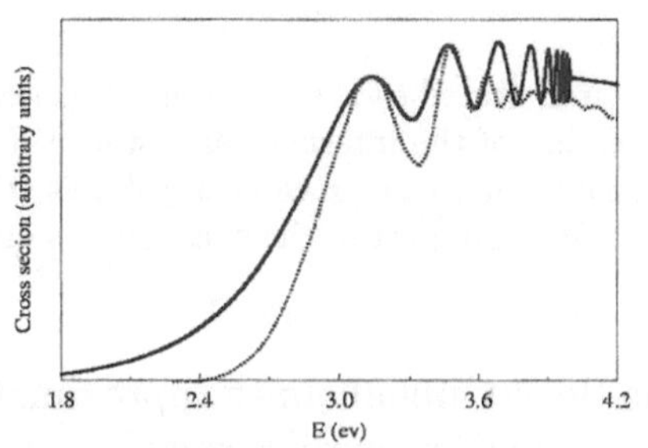

Figure 6. Solid line shows the calculated probability to excite H_2 molecule from the ground vibrational $\nu = 0$ to the $\nu = 4$ excited state in e/H_2 collision. Dotted line represents the shifted experimental results.

presented in Fig. 6.2 are in a remarkable semiquantative agreement with the experimental cross section(Fano, 1985). The experimental cross section was shifted up in energy by 0.45 *ev* to emphasize the agreement between the vibrational structures seen in the spectrum. Similar agreement was obtained earlier by Mündel et. al.(Mündel et al., 1985) using a different approach.

We will now give a more detailed description of the electron trapping phenomena and show that it is associated with the self-orthogonality phenomenon described in section 4. It is known that the ${}^2\Sigma_u^+$ resonance of H_2^- which is responsible for the vibrational excitation in the low energy region, is an extremely short-lived resonance with a lifetime comparable with the duration of the non resonant scattering(Moiseyev and Corcoran, 1979).

For a fixed $H-H$ distance the H_2^- molecule autoionizes with a rate depending on the $H-H$ bond length, $\Gamma(R)$. This exponential decay can be described by a single resonance state possessing a *complex* electronic "energy" $V(R) - \frac{i}{2}\Gamma(R)$ (see Fig. 6.2)(Domcke and Cederbaum, 1980). Therefore instead of using many real coupled electronic potential

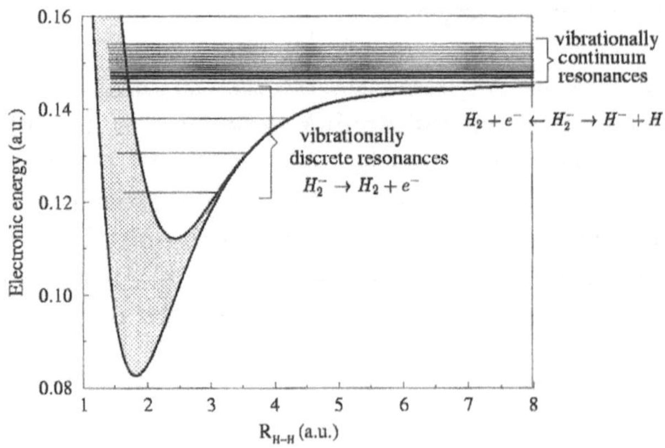

Figure 7. **The shaded area describes the electronic "energy" $\epsilon(R) \pm \Gamma(R)/2$ of autoionizing H_2^- as a function of the internuclear distance R. The horizontal parallel lines stand for the autionization resonance energy levels (the resonance widths are not shown here). The continuum part of the spec trum is quantized due to the finite box approximation.**

surfaces within the Born-Oppenheimer approximation we can solve the nuclear Schrödinger equation with a single complex potential energy surface(Narevicius and Moiseyev, 1998), $\widehat{H} = \widehat{T}_{nuclear} + V(R) - \frac{i}{2}\Gamma(R)$.

By solving the nuclear H_2^- Schrödinger equation with the complex potential surface we obtained two distinctive types of resonance states which are presented in Fig. 6.2. The first one is associated with vibrationally bound motion of autoionizing H_2^- molecule, i.e. $H_2^- \rightarrow H_2+e^-$. We will refer to these states as vibrationally-discrete autoionization-resonance states. The second type of states is associated with the free motion of the nuclei and is referred to here as vibrationally-continuum autoionization-resonances where $H_2(bound) + e^- \overset{A}{\leftarrow} H_2^-(continuum) \overset{B}{\rightarrow} H^- + H$ (i.e. a branch cut of autoionizing states). Note that the autoionization (A) takes place at the inner classical turning point where the width of the complex potential surface is the largest. The dissociation (B) occurs at the outer region where the width of the complex potential is zero. As we will show below the formation of branch cut of resonances due to the nuclear motion plays the key role in the mechanism which is responsible for the enhancement of the electron trapping by the hydrogen molecule.

The initial state populates both types of resonance states of H_2^-. Interference between the vibrational discrete and the vibrational continuum autoionization resonances takes place although the resonance positions of the vibrationally-discrete autoionization-resonance (disc-res) states and the vibrationally-continuum autoionization-resonance states (cont-res) are very "far" from one another (as compared to their width). One may think that when $E \sim \epsilon_n(disc-res)$ there is only one dominant term in the series resonance expansion of $t_{res}(E)$ (see Eq. 46). This is however not the case. The numerators associated with the branch-cut resonances ,$a_n(cont-res)$, get complex values where both the real and the imaginary parts are larger than the corresponding ones of $a_n(disc-res)$ by *several orders of magnitude.* In the numerical calculations the amplitudes of the continuum type resonances were found to be larger by 8-10 orders of magnitude than the discrete type resonance amplitudes. Consequently, $|a_n(cont-res)| >> |a_{n'}(disc-res)|$ but however at $E = E_n(disc-res)$,

$$\left|\frac{a_n(disc-res)}{\frac{i}{2}\Gamma_n(disc-res)}\right| \approx \left|\frac{a_{n'}(cont-res)}{\Delta E + \frac{i}{2}\Gamma_{n'}(cont-res)}\right| \tag{47}$$

where,

$$\Delta E = E_n(disc-res) - E_{n'}(cont-res) \tag{48}$$

even when the difference between the $H_2^-(disc-res) \rightarrow H_2 + e^-$ resonances and $H_2 + e^- \leftarrow H_2^-(cont-res) \rightarrow H^- + H$ resonances is much larger then the corresponding resonance widths,

$$\Delta E \gg Max(\Gamma_n, \Gamma_{n'}).$$

The number of dominant terms in the expression for $t_{res}(E)$ (Eq. 46) determines the number of effective indirect scattering events in the scattering experiment. In our case due to the large amplitudes of the branch-cut (continuum) resonances this number is large and therefore we deal with a multiple scattering process and a collective resonance phenomenon. By collectiveness we mean that even at the resonance energy, $E = \epsilon_n$, a large number of resonances have dominant contributions to the expansion of t_{res} given in Eq. 46.

The large amplitude of the continuum resonance states is a direct result of the non Hermitian properties of H_2^- Hamiltonian (i.e. the resonance eigenfunctions which are associated with complex eigenvalues are not in the Hermitian domain of the molecular Hamiltonian). Let us explain this point in some more detail. As was mentioned above Moiseyev and Friedland(Moiseyev and Friedland, 1980) have proved that if two $N \times N$ real symmetric matrices H_1 and H_2 do not commute, there exists at least one value of parameter $\lambda = \lambda_b$ such that matrix $H_1 + \lambda H_2$ possesses incomplete spectrum. That is at $\lambda \to \lambda_b$ there are at least two specific eigenstates i and j for which $\lim_{\lambda \to \lambda_b}(\epsilon_i - \epsilon_j) = 0$ and also $\lim_{\lambda \to \lambda_b}(\psi_i - \psi_j) = 0$. Since ψ_i and ψ_j are orthogonal (within the general inner product definition i.e., $(\psi_i|\psi_j) \equiv \langle \psi_i^*|\psi_j \rangle = 0$ and not in the usual scalar product definition) upon coalescence their length is reduced to zero $\langle \psi_i^*|\psi_i \rangle = \varepsilon$, $\varepsilon \to 0$. When $\lambda \neq \lambda_b$ but is very close to it we still require normalization of ψ_i, $\langle \psi_i^*|\psi_i \rangle = 1$. Therefore the components of the corresponding eigenvectors are divided by a small number $\epsilon^{1/2}$ and the amplitude of ψ_i becomes enormously large.

We will show here that the phenomenon of spectrum incompleteness is inherent in the case of H_2^-. The Hamiltonian of the intermediate H_2^- within the BO approximation is given by, $H(R) = T(R) + V(R) - \frac{i}{2}\Gamma(R)$ We can rewrite this Hamiltonian introducing an autoionization strength parameter λ in a following way: $H(R) = T(R) + V(R) - \lambda\frac{i}{2}\Gamma(R)$. For $\lambda = 1$ we obtain the original physical Hamiltonian of H_2^-. Using real basis functions or finite grid methods the matrices representing the operators $T(R), V(R)$ and $\Gamma(R)$ are real symmetric ones. The H_2^- Hamiltonian in the matrix form is $\mathbf{H} = \mathbf{T} + \mathbf{V} - \lambda\frac{i}{2}\mathbf{\Gamma} = \mathbf{H_{Re}} - \lambda\frac{i}{2}\mathbf{\Gamma}$. Since matrices $\mathbf{H_{Re}}$ and $\mathbf{\Gamma}$ do not commute it follows from the theorem proved by Moiseyev and Friedland that there exists λ, such that the spectrum of $\mathbf{H}$ is incomplete. In order to find the values of λ where the spectrum is incomplete in the case of H_2^- we have studied the λ dependence of the spectrum. In Fig. 6.2 we present the spectrum of H_2^- obtained while varying λ from 0 to 1 (i.e. increasing the strength of the ionization phenomenon). There are 9 bound vibrational states of H_2^- when the autoionization process is neglected i.e. $\lambda = 0$. They all acquire finite width whenever the value of λ is increased. However the width and the position of each state changes at a very different rate. The most dramatic change occurs to the 8-th bound state ($\lambda = 0$).

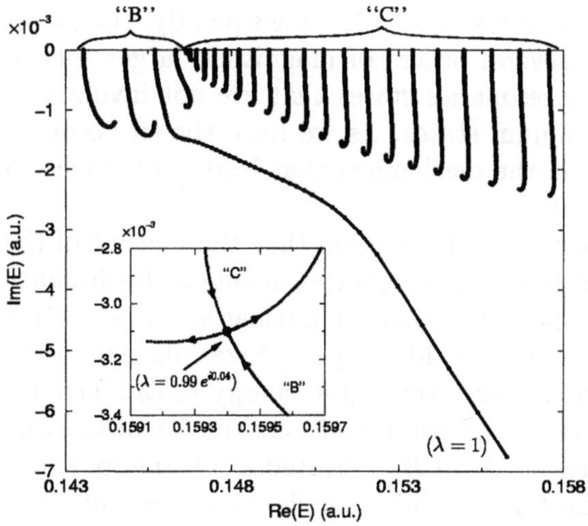

Figure 8. **λ-trajectory calculations of the complex eigenvalues of $Re(\mathbf{H}) - i\lambda Im(\mathbf{H})$ where H is a matrix representation of the nuclear H_2^- Hamiltonian with complex electron potential surface. For $\lambda = 0$ the autoionization is artificially suppressed and the real eigenvalues obtained are divided into vibrationally bound states (denoted by "B" and into dissociative continuum solutions denoted by "C" (note that the discretization is due to the finite box approximation). The inset shows that very close to $\lambda = 1$ (the physical solution) one of the continuum solutions and the 8-th "B" solution coalesce.**

This state is "pushed out" above the energy threshold for the dissociation. Interestingly, whenever the position of the bound state crosses the threshold energy the behavior of the continuum states changes as well. That is, the width of the continuum states starts to decrease with increasing λ. The reason for this behavior is that for some complex values of the parameter λ close to $\lambda = 1$ the bound state which enters the continuum part of the spectrum crosses the continuum states one by one. Since the crossing does not happen at $\lambda = 1$, the picture we see at the physical value of $\lambda = 1$ is that of avoided crossing. It should be stressed that the crossing here implies coalescence of two eigenstates, one eigenstate originating from the bound state at $\lambda = 0$ and the second one from the continuum at $\lambda = 0$. *Indeed, we found several branch points associated with the coalescence of bound and continuum states which are in the vicinity of* $\lambda = 1$. For example in the inset of Fig. 6.2 we show the branch point at $\lambda = 0.99145098 + i0.04006515$. The trajectory approaching from bellow is associated with the 8-th "bound" state and the trajectory approaching from above is associated with the "continuum" state. This branch point is the closest one to $\lambda = 1$ and the two states that cross at this point are very close to be self orthogonal for the physical value of $\lambda = 1$. Consequently, the amplitudes of these states are by several orders of magnitude larger then the amplitudes of the discrete resonance states that are not involved in the crossings with the continuum states. As we have shown above this effect plays a crucial role in the mechanism that leads to the electron trapping by the H_2 molecule.

We have shown in this section that the lifetime of the trapped electron in the electron scattering experiment can be dramatically increased due to the coupling between the nuclear and the electronic motions. It should be stressed that this coupling takes place between molecular autoionizing states over a very large energy range. This unusually strong coupling is due to the existence of a continuum of short lifetime autoionizing resonance states of H_2^-, located at an energy above the threshold energy of dissociation. This branch-cut of resonances is generated by the nuclear motion.

6.3. Critical phenomenon explained by the self orthogonality in non Hermitian quantum mechanics

In this section we will show that the self-orthogonality property of eigenstates of non Hermitian Hamiltonians is responsible for the appearance of critical phenomenon (e. g. sudden delocalization) in different fields of physics. We will study in more details the delocalization phenomenon discovered by Hatano and Nelson(Hatano and Nelson, 1996; Hatano and Nelson, 1997; Hatano and Nelson, 1998) for a class of non-Hermitian quantum-mechanical problems.

The non Hermitian Hamiltonian which is subject of our studies is given in dimensionless units by,

$$\hat{\mathcal{H}} = -\frac{1}{2}\left(\frac{\partial}{\partial x} - g\right)^2 + V(x). \tag{49}$$

We have chosen the potential term, $V(x)$, to be a finite potential well (width equal to 1 and depth equal to -1). The real parameter g is a drift velocity in studies of water flow in aqueous media, winds, etc., or a non-Hermitian external field originating in the transverse magnetic field (Hatano and Nelson, 1996; Hatano and Nelson, 1997; Hatano and Nelson, 1998). Note that for $g = 0$ the Hamiltonian presented above is Hermitian and possesses only one bound state. Nelson and co-workers have shown that a bound state localized inside the potential well turns to be an extended delocalized state as the vector potential parameter g gets a critical value, $g_c = \sqrt{-2E_{bound}(g=0)}$, where $E_{bound}(g=0)$ is the ground state eigenvalue of the Hermitian Hamiltonian $H = \hat{\mathcal{H}}(g = 0)$(Hatano and Nelson, 1996; Hatano and Nelson, 1997; Hatano and

Nelson, 1998; Nelson and Shnerb, 1998). Nelson and Shnerb proposed a simple experiment to study delocalization in inhomogeneous biological systems(Dahmen et al., 2001). A bacteria colony living in illuminated area defined in our model Hamiltonian by $-1/2 \leq x \leq +1/2$ (i.e., inside the potential well $V(x) = -1$). At high wind velocity, $g \geq g_c$ all bacteria are blown into the "desert" area (in our model Hamiltonian where $V(x) = 0$) and the colony dies out. In their experimental setup a solution with photosynthetic bacteria in a circular pipe is illuminated only a in small area covered by $V(x) = -1$, while the rest of the sample is kept dark. The source is moved slowly around the sample to model convective flow. In their model Hamiltonian x is an angle variable whereas L is the radius of the circular pipe. An experiment similar to the proposed one has been carried out recently by Neicu et al. (Neicu et al., 2000). However, the pie geometry used in this experiment is two dimensional as was pointed out by Schnerb (Shnerb, 2001) and the radial symmetry with periodic boundary conditions (in angle variable) is necessary to obtain the sudden delocalization phenomenon.

We have chosen here to impose one dimensional periodic boundary conditions on the Hamiltonian given above. That is, $-\infty \leq x \leq +\infty$ where $V(x) = V(x+L)$. In our studied case even at g=0 all states are delocalized in the sense that they are Bloch states, $\Psi_\alpha = \exp(ikx)\phi_{\alpha,k}(x)$, where $-\pi/2 \leq k \leq +\pi/2$ and $\phi_{\alpha,k}(x) = \phi_{\alpha,k}(x+L)$. However, the $\alpha = 1$ band is constructed from *multiple bound states* which are *localized* inside the periodic potential wells. The amplitudes of $\phi_{\alpha=1,k}(x)$ in between two adjacent potential wells are exponentially small as $L \to \infty$. The second band labeled by $\alpha = 2$ is embedded in the continuum part of the spectrum. The delocalization discussed by Nelson and co-workers is different from the Bloch delocalization mentioned above. Nelson's delocalization implies that the amplitudes of $\phi_{\alpha=1,k}(x)$ in between two adjacent potential wells is as large as inside the periodic potential wells. As we will show bellow Nelson's delocalization is associated with the merging of two different energy bands of non Hermitian Hamiltonian. For some critical value g_{c_j} two different band edges coalesce at $k = (2j-1)\pi/L\, j = 1, 2, ...$ At these branch points (bp) the Bloch wavefunctions, $\phi_{\alpha=1,k=(2j-1)\pi/L}(x) = \phi_{\alpha=2,k=(2j-1)\pi/L}(x) \equiv \phi_j^{bp}$, are self-orthogonal. Nelson's delocalization can be explained in terms of infinite large value of the standard deviation Δx due to the self-orthogonality property of ϕ_j^{bp}. On the basis of this argument it is clear that infinitely large standard variation would be obtained for any dynamical property which is represented by an operator that does not commute with the non Hermitian Hamiltonian.

As it was discussed in the introduction the left eigenfunctions of non Hermitian Hamiltonians are not equal to the right eigenfunctions. The right Bloch states, $\phi^R_{\alpha,k}$ were obtained by calculating the eigenfunctions of the Bloch Hamiltonian, $\hat{\mathcal{H}}_B = \frac{1}{2}\left(-\frac{i\partial}{\partial x} + ig + k\right)^2 + V(x)$. The left Bloch states are eigenstates of $(\hat{\mathcal{H}}^\dagger_B)^* = (\hat{\mathcal{H}}_B(-g))^*$. Consequently the left eigenfunctions are defined as,

$$\phi^L_{\alpha,k} = \phi^{R*}_{\alpha,k}(-g) \tag{50}$$

Bi-orthogonality implies that,

$$(\phi^L_{\alpha'\neq\alpha,k}(g)|\phi^R_{\alpha,k}(g)) \equiv \int_{-\infty}^{+\infty} \phi^L_{\alpha'\neq\alpha,k}(x;g)\phi^R_{\alpha,k}(x;g)dx = 0 \tag{51}$$

However, following the general proof of Moiseyev and Friedland by varying one complex parameter or two real parameters one can find such values that the above equality holds also when $\alpha' = \alpha$. Namely, one should find critical values of k (Bloch wave vector) and g (imaginary vector potential parameter) for which a branch point is obtained, $E_{\alpha=1}(g = g_{bp}, k_{bp}) = E_{\alpha=2}(g = g_{bp}, k_{bp})$. In the vicinity of a branch point the eigenvalue of the non Hermitian Hamiltonian varies as,

$$E \sim [(k - k_{bp}) - i(g - g_{bp})]^{\frac{1}{2}}. \tag{52}$$

The square root results from the fact that the branch point is obtained from coalescence of two eigenvalues. A branch point does not only imply degenerate eigenvalues but also coalescence of the corresponding eigenfunctions, $\phi^{R,L}_{\alpha=1,k_{bp}} - \phi^{R,L}_{\alpha=2,k_{bp}} \rightarrow 0$ as $g \rightarrow g_{bp}$. Due to the bi-orthogonality property at the branch point $\phi^{R,L}_{\alpha=1,k_{bp}} = \phi^{R,L}_{\alpha=2,k_{bp}}$ is a self-orthogonal function. That is, $(\phi^L_{\alpha'=\alpha,k_{bp}}(g_{bp})|\phi^R_{\alpha,k_{bp}}(g_{bp})) = 0$. The results presented in Fig. 6.3 show that the self-orthogonality is obtained for $g = g_{bp} = 0.485$ and $k = k_{bp} = \pi/L$ where $L = 10$ is the "lattice" constant in our model Hamiltonian. As $L \rightarrow \infty$ the branch point value of g, g_{bp}, approaches the critical value obtained by Nelson for the sudden appearance of delocalization when x is an angle variable. As one can see from our results the periodical boundary conditions are crucial to get the self-orthogonality phenomenon. To obtain a branch point in the spectrum of non Hermitian Hamiltonian we need to have two real parameters. The first parameter, g, is the drift velocity and the second parameter, Bloch wave vector k , is provided by the periodic boundary conditions.

In order to show that the self-orthogonality is associated with a branch point which results from the coalescence of the "ground bound" Bloch band with the continuum Bloch band we carried out k-trajectory calculations (from k=0 to k=$2\pi/L$). From the results presented in

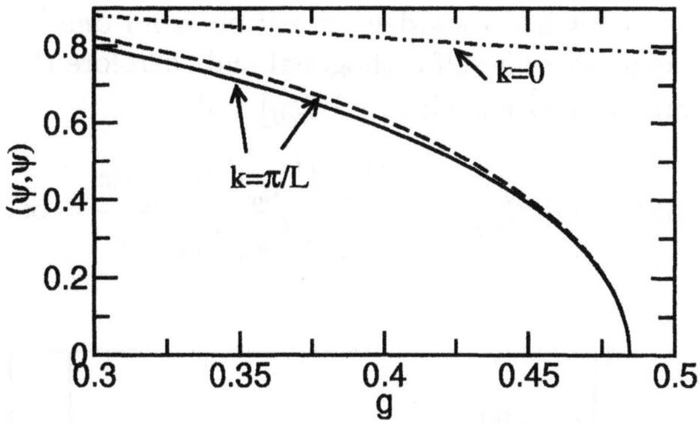

Figure 9. Coalescence of two complex eigenstates at $g = g_{bp}$ when $k = \pi/2$. At the branch point $g = g_{bp}$ the eigenfunction becomes self-orthogonal and $(\Psi|\Psi) = 0$.

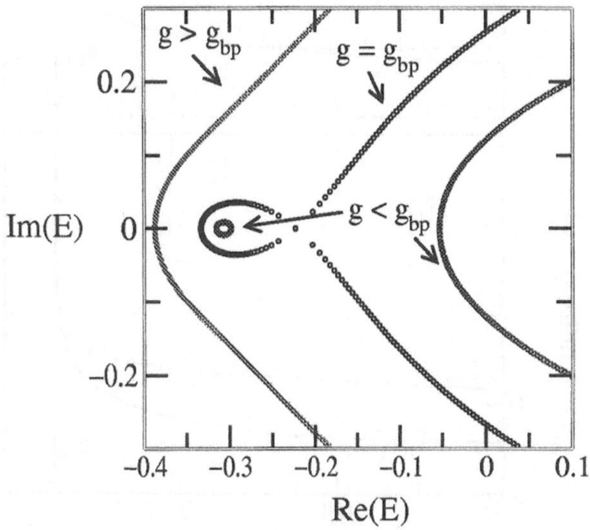

Figure 10. The complex eigenvalue band structure for three different values of g. The Bloch wave vector is varied within $0 \leq k \leq 2\pi/L$.

Fig. 6.3 one can see that when $g \neq g_{bp}$ the Bloch bands are separated in the complex energy plane. The non Hermitian "ground bound" Bloch band is a circle in the complex energy plane whereas the non Hermitian continuum Bloch band is a line. At the critical value of g, g_{bp}, the two bands touch at the branch point obtained when $k = \pi/L$. At this point we obtained the self-orthogonality as shown in Fig. 6.3

Calculation of the real and imaginary parts of the eigenvalues of the Hamiltonian as function of the wave vector k show a series of branch points obtained at $k_{bp} = \pi/L, 3\pi/L, ...$ At each one of these branch points (associated with a sudden transition from one Bloch band to another) the eigenstate is self-orthogonal and therefore the expectation value of an operator $\hat{O}$ for which $[\hat{O}, \hat{\mathcal{H}}_B] \neq 0$,

$$(\phi^L_{\alpha,k_{bp}}|\hat{O}|\phi^R_{\alpha,k_{bp}}) = \frac{\int_{-\infty}^{+\infty} \phi^L_{\alpha,k_{bp}} \hat{O} \phi^R_{\alpha,k_{bp}} dx}{\int_{-\infty}^{+\infty} \phi^L_{\alpha,k_{bp}} \phi^R_{\alpha,k_{bp}} dx} = \infty. \tag{53}$$

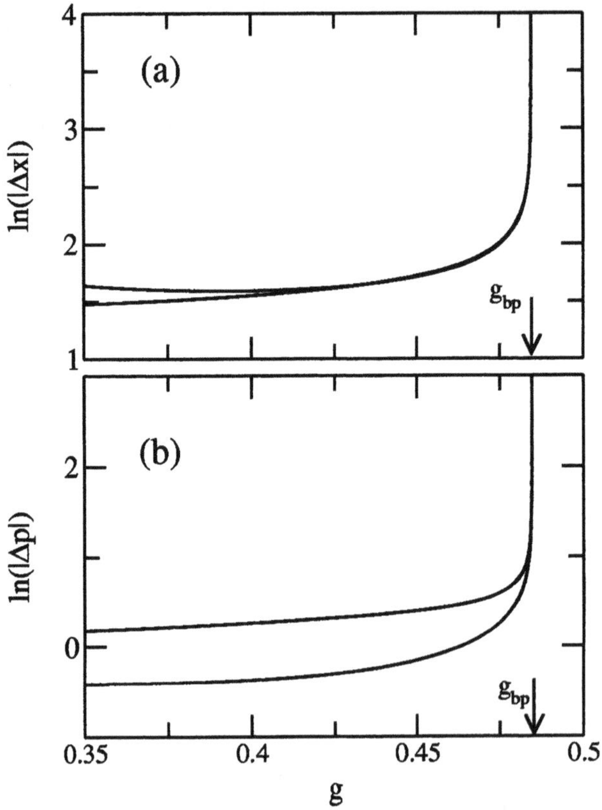

Figure 11. The standard deviations of position (a) and momentum (b) as function of drift velocity parameter g. The standard deviations diverge at $g = g_{bp}$.

The results presented in Fig 6.3 illustrate the sudden delocalization phenomenon when $\hat{O} = \Delta x$ and $\hat{O} = \Delta \hat{p}_x$. In a way this non Hermitian quantum behavior resembles the classical chaotic dynamics. The standard deviation of position and momentum diverge exponentially when the initial conditions are defined as a Gaussian ensemble distribution of particles localized in the chaotic region of the classical phase space.

In this section we addressed a class of problems which are studied by non Hermitian quantum mechanics. We have shown that in the case where periodic boundary conditions can be employed the ground eigenstate becomes self-orthogonal when a physical parameter in the Hamiltonian approaches a critical value. At this critical point the expectation value of *any* operator which does not commute with the Hamiltonian (e.g., position; momentum), gets infinitely large value. Hence, the delocalization discovered by Nelson et. al. is only one example of many possible critical phenomena.

7. Conclusions

In NH-QM the left eigenfunction of Hamiltonian is not equal to the complex conjugate of the right eigenfunction. Consequently, it may happen that the inner product of the function is zero, i.e. $\int \psi^L \psi^R d\tau = 0$. We have shown here that this phenomenon is neither rear nor accidental. It happens in several physical systems we have studied. One of the examples is molecular dissociation accompanied by simultaneous ionization. The numerical advantage of NH-QM was shown by formulating the extension to the BO approximation which allows the solution of the nuclear equation of motion with a complex electronic potential energy surface. Several example were given including e/H_2 collision, photodissocation of $ArHCl$ van der Waals complex and the interatomic Coulombic decay in Ne dimer.

This review of our research in the field of NH-QM is dedicated to Prof Per-Olov Löwdin. Besides being a scientific leader and one of the most influential figures in theoretical chemistry during the last 50 years had the charisma and the charm which attracted many young students and researchers to the field of quantum chemistry. For us who knew him personally (NM spent one month visiting in Uppsala in summer 1981) he became an example of an ideal supervisor and a research group leader who supported his students and created a friendly atmosphere in the group which helped to the satisfy the equation: to do science = FUN.

Acknowledgements

This work was supported in part by the Israel-US Binational Science Foundation and by the Israeli Academy of Sciences.

References

E. Narevicius and N. Moiseyev, Phys. Rev. Lett. **81**, 2221 (1998).
E. Narevicius and N. Moiseyev, Phys. Rev. Lett. **84**, 1681 (2000).
U. V. Riss and H-D. Meyer J. Phys. B: At. Mol. Opt. Phys. **26**, 4503 (1993).
N. Moiseyev, J. Phys. B. **31**, pp. 1431-1441, 1998.
N. Hatano and D. R. Nelson, Phys. Rev. Lett. **77**, 5706 (1996).
N. Hatano and D. R. Nelson, Phys. Rev. B **56**, 8651 (1997).
N. Hatano and D. R. Nelson, Phys. Rev. B **58**, 8384 (1998).
H.J. Korsch, R. Möhlenkamp, H.-D. Meyer, J. Phys. B **17** 2955 (1984) and references therein.
N. Moiseyev, J.O. Hirschfelder, J. Chem. Phys. **88** 1063 (1988).
E. Balslev, J.M. Combes, Commun. Math. Phys. **22** 280 (1971).
B. Simon, Commun. Math. Phys. **27** 1 (1992) ; Ann. Math. **97** 247 (1973).
N. Moiseyev, Physics Reports, **302**, 211 (1998).
The Lertorpet Symposium View on a Generalized Inner Product, edited by E. Brändas and N. Elander, Lecture Notes in Physics, Vol. 325 (Springer, Berlin, 1989).
N. Moiseyev and S. Friedland, Phys. Rev. A. **22**, 618 (1980).
E. Narevicius and N. Moiseyev, submitted.
M. V. Berry, submitted to J. Modern Optics, November 2001.
M. V. Berry, Current Science, **67**, 220, (1994).
M. V.Berry, J. Phys. A **31** 3493 (1998).
E. Narevicius and N. Moiseyev, J. Chem. Phys. **113**, 6088 (2000).
R. Santra, J. Zobeley, L.S. Cederbaum and N. Moiseyev, Phys. Rev. Lett., **85**, 4490 (2000).
N. Moiseyev, R. Santra, J. Zobeley and L.S. Cederbaum, J. Chem. Phys, **114**, 7351 (2001).
N. Moiseyev and U. Peskin, Phys. Rev. A **42**, 255 (1990).
L. S. Cederbaum and W. Domcke, J. Phys. B, **14**, 4665 (1981) and references therein.
V. Ryaboy and N. Moiseyev, J. Chem. Phys. **98** 9618 (1993).
E. Narevicius and N. Moiseyev, Mol. Phys. **94**, 897-903 (1998).
D. T. Colbert and W. H. Miller, J. Chem. Phys. **96**, 1982 (1992).
U. Fano, Phys. Rev. **124**, 1866 (1961).
M. Allan, J. Phys. B, **18**, L451 (1985).
R. Schuster, E. Buks, M. Heiblum, D. Mahalu, V. Umansky, H. Shtrikman, Nature **385**, 417 (1997)
C. Mündel, M. Berman and W. Domcke, Phys. Rev. A **32**, 181 (1985).
N. Moiseyev and C. Corcoran, Phys. Rev. A, **20**, 814, (1979).
W. Domcke and L. S. Cederbaum, J. Phys. B, **13**, 2829 (1980).
D. R. Nelson and N. M. Shnerb, Phys. Rev. E **58**, 1383 (1998).
K. A. Dahmen, D. R. Nelson and N. M. Shnerb, Mathematical Biology **41**, 1 (2001).
T. Neicu, A. Pradhan, D.A. Larochelle and A. Kudrolli, Phys. Rev. E **62**, 1059 (2000).
N. M. Shnerb, Phys. Rev. E **63**, 11906 (2001).

Chapter 2

ON THE GEOMETRIC PHASE EFFECT IN JAHN-TELLER SYSTEMS

A. J. C. Varandas

Departamento de Química
Universidade de Coimbra
3004-535 Coimbra
Portugal.
varandas@qtvs1.qui.uc.pt

Abstract Starting with the hierarchy of approximations involved in the solution of the many-body problem, we examine the case of degeneracies in highly symmetric molecules which are subject to distorting forces in accordance with the Jahn-Teller theorem. The implications of such degeneracies on the single-surface treatment of the nuclear dynamics are then highlighted in relation to the so-called geometric phase effect. The focus will be in the regions close to the degeneracy seam where a linear approximation to the Hamiltonian may be sufficient to understand the basic features of the nuclear dynamics. Being a topological effect, it is expected that the cyclic phases so derived can still be useful for the treatment of the nuclear dynamics problem over the full configuration space. As numerical illustrations, we present calculations of the vibrational cone states for H_3 and several isotopomers involving μ, D, and T. From these results, a phenomenological criterion is suggested which may help to assess the importance of the geometric phase effect in quantum dynamics studies of such triatomic species. Naturally, a greater emphasis will be given to the traditional two-state Jahn-Teller problem, although we examine also the geometric phase effect for arbitrary integral $N-$fold electronic degeneracies. In particular, we discuss a novel interpretation in which the geometric phase effect is viewed as a property inherent to inversion of the adiabatic electronic wave-vectors in the $N-$dimensional space defined by the electronic wave-vector basis, the so-called function space. The relationship between the geometric phase and mixing angles is also discussed. We conclude with a prospective view on future work and pinpoint some continuing challenges.

E.J. Brändas and E.S. Kryachko (eds.),
Fundamental World of Quantum Chemistry, Vol. II, 33–92.

1. Introduction

A complete theoretical description of the structure and properties of molecules and crystals is known to stem from the laws of electronic and nuclear motion. Naturally, the many mathematical difficulties encountered in solving the resulting quantum mechanical Schrödinger equation for such a many-body problem led to introduce several essential approximations. The most celebrated entail the dividing to conquer idea of separating the motions of the rapidly moving electrons from the much slower moving nuclei. They date back to the early days of quantum mechanics [1, 2], with the most important being the adiabatic approximation. In this approximation, the electronic Schrödinger equation is first solved for a set of fixed nuclear arrangements, yielding the adiabatic potential energy surfaces for specified electronic states. Such potential energy surfaces should be invariant under all uniform translations and rigid rotations of the nuclei that form the molecule. Besides enhancing the efficiency of the numerical process in the solution of the Schrödinger equation for the many-body problem, the adiabatic approximation leads to an useful physical significance. In particular, the notion of spatial structure (molecular configuration) would be unconceivable without it.

Deviations to the adiabatic approximation emerged soon after this approximation was introduced. Among the most striking example of coupling of the electronic and nuclear motions is the so-called Jahn-Teller [3] (JT) effect. According to the JT theorem, highly symmetrical molecular configurations in degenerate electronic states are subject to distorting forces acting along certain nontotally symmetric vibrational modes [3–5]. It then implies that if the adiabatic potential energy surfaces of the system, which are a formal solution of the electronic part of the problem, become degenerate[1], then at least one of such surfaces has no extremum at the crossing point. Strictly speaking, the JT theorem[2] does not lead to predictions concerning observable magnitudes but shows that knowledge of the topographical features of the adiabatic potential energy surface is not enough to specify the nuclear behavior.

In addition to the proper JT effect, nonadiabatic effects are expected to play an important role in the case of mixing of close-in-energy electronic states (*e.g.*, when two adiabatic potential energy surfaces become nearly-degenerate at some finite or infinite number of points of

[1]Alternatively, we may address to such intersecting adiabatic potential energy surfaces as sheets of the same degenerate adiabatic potential energy surface at the crossing point.

[2]Exceptions to the JT theorem are molecular configurations which are linear at $_0$, and half-spin systems with strong spin-orbit coupling Kramers degeneracies (Ref. [19], and references therein).

the molecule configurational space) as in the case often referred to as pseudo-Jahn-Teller effect, which together with the Renner-Teller effect (instability of linear molecules) form the so-called vibronic effects. Nonadiabatic processes are therefore ubiquitous and occur in many important fields, which cover from molecular spectroscopy to reaction dynamics and a vast number of other fields both in the gas phase (or in matrices) as well as in crystals [4,5].

Conventionally, degeneracies in potential energy surfaces can be classified as normal, when they are required by the molecular symmetry-point group, or accidental, when they are not. In the present work, all cases belong to the first type, with the simplest one being the E-type intersections occurring in X_3-type JT systems [6] (see section 3.1). In turn, accidental degeneracies include both symmetry-allowed ones (if permitted by the different symmetry of the involved states) and degeneracies involving states of the same symmetry which can only occur by the non-crossing rule [7–10] in triatomic and larger systems. Clearly, the question to ask is how frequently can intersections of adiabatic potential energy surfaces occur. For polyatomic systems with three atoms or more, two adiabatic potential energy surfaces can intersect even if the corresponding states have the same symmetry and spin multiplicity [7–10]. This is due to the well known fact that only two independent relations between three electronic Hamiltonian matrix elements are sufficient for the existence of a doubly-degenerate electronic energy eigenvalue. Thus, their intersection can occur along a $(3N-8)$ [$(3N-7)$ in complex space [11]] hyperline as the energy is plotted against the $(3N–6)$ nuclear coordinates. As a result, electronic degeneracies between states of the same symmetry are common in nature: double-degeneracies for triatomics, triple-degeneracies also for tetratomics, and so on. For the double-degeneracy case, such intersections are for nonlinear arrangements of the atoms of the conical type, having the topology of a double cone. In this case, we may distinguish two geometrical variables defining the so-called $g-h$ plane [12] such that if one were to plot the energy in the subspace of these two variables (bond lengths, angles, etc) the potential energy would have the form of a double cone in the region of degeneracy; in the remaining $(3N-8)$ directions, the energies of ground and excited states are equal. The splitting of the potential energy surfaces in the $g-h$ plane is therefore a linear function away from the intersection point but close to it. A final remark to note that, as for the minima and saddle points (*e.g.*, Ref. [13]), significant progress has been made on the direct location of points on a surface intersection [12,14].

By focusing on JT systems, we may then address the question: can we study the dynamics by using a single surface equation for the nuclear motion (thus using a given sheet of the adiabatic potential energy surface)? An answer to this question must start by recalling the seminal observation made by Herzberg and Longuet-Higgins [9] about four decades ago. They have shown that a real-valued electronic wave function changes sign when the nuclear coordinates traverse a cyclic path around a conical intersection (two-fold electronic degeneracy); for reviews see, *e.g.*: Refs. [15–19]. Such a geometric phase (GP) effect was placed in a wider quantum mechanical context by Berry [20], and hence it also became known as the Berry phase effect[3]. Although research on the GP effect has a long history and is currently a very active field of research [16, 20–39], the issue of cyclic phases at a N-fold electronic degeneracy is a much more recent one [24, 26, 40–42]. In this review, we discuss a novel approach approach toward this goal by focusing on linear JT systems [5, 43], and compare the results with the predictions from other methods.

The Chapter is organized as follows. In section 2, we discuss the hierarchy of approximations to the solution of the many-body problem, and the topological implications of the JT degeneracy in studies of single-surface nuclear dynamics. A numerical illustration which employs a single-surface formalism for the calculation of the vibrational levels for the upper sheet of doublet H_3 and its isotopomers is also presented in this section. Some symmetry aspects of JT systems are then reviewed in section 3, where the linear form of the vibronic Hamiltonian matrix is also discussed. The aim is then use Lie symmetry groups in section 4 to discuss a recent proposition [42] that the GP effect is a property inherent to inversion of the adiabatic electronic wave-vectors in the N−dimensional electronic wave-vector space, the so-called function space. This is done by illustrating the approach using case studies of increasing values of N. Although the material presented is unavoidably biased toward our own work, a fair coverage of other viewpoints is also presented. Section 5 contains the concluding remarks and pinpoints remaining challenges.

2. General principles

Schrödinger's equation describing the complete many-body problem assumes the form

$$\hat{H}_T \Omega(\mathbf{R}', \mathbf{r}') = E \Omega(\mathbf{R}', \mathbf{r}') \tag{2.1}$$

where $\hat{H}_T(\mathbf{R}', \mathbf{r}') = \hat{T}_N(\mathbf{R}') + \hat{H}_e(\mathbf{R}', \mathbf{r}')$ is the total Hamiltonian operator, $\Omega(\mathbf{R}', \mathbf{r}')$ is the total molecular wave function, E is the total energy,

[3]Given the analogy with the effect of a true magnetic field confined to a narrow solenoid [101], Mead [22] suggested the further designation of molecular Aharonov-Bohm effect.

and $\mathbf{R}'$ ($\mathbf{r}'$) is the set of nuclear and electronic coordinates[4] in a coordinate frame fixed in the laboratory (SF frame); for simplicity, the spin coordinates of nuclei and electrons are not indicated [19]. Thus, $\hat{T}_N(\mathbf{R}')$ is the kinetic energy operator for the motion of the nuclei, and

$$\hat{H}_e\left(\mathbf{r}',\mathbf{R}'\right) = \hat{T}_e\left(\mathbf{r}'\right) + \hat{V}_{NN}\left(\mathbf{R}'\right) + \hat{V}_{Ne}\left(\mathbf{r}',\mathbf{R}'\right) + \hat{V}_{ee}\left(\mathbf{r}'\right) \tag{2.2}$$

is the Hamiltonian for the electronic motion: $\hat{T}_e(\mathbf{r}')$ is the kinetic energy operator for the electrons, and $\hat{V}_{NN}(\mathbf{R}')$, $\hat{V}_{Ne}(\mathbf{r}',\mathbf{R}')$, and $\hat{V}_{ee}(\mathbf{r}')$ are the operators for the electrostatic potential energies arising from the nucleus-nucleus, nucleus-electron, and electron-electron interactions[5]. Although a separation of coordinates is generally impossible due to interactions between the nuclear and electronic degrees of freedom, it may partially or completely be done in practice depending on such interactions. For example, due to the uniformity and isotropy of space, the translational and rotational degrees of freedom of an isolated molecule can be regarded as cyclic coordinates (note though that the separation of the rotational degrees of freedom is not trivial [44]). For N nuclei, this leads to $(3N-3)$ translational-free coordinates $\mathbf{R}'$ of which three orientational or external coordinates $\tilde{\mathbf{R}}$, the so-called Euler angles, can be specified by fixing a coordinate frame (BF frame) in the non-rigid assembly of particles. The remaining $(3N-6)$ internal coordinates $\mathbf{R}$ specify the geometry of the system, and define the so-called configuration or shape space. It is formally the quocient space in which a point represents a whole orbit in the translational-reduced $^{3N-3}$ space. The shape coordinates can then be any $3N-6$ independent functions of coordinates on that manifold (known also as internal coordinates). In fact, they can be any $3N-6$ independent invertible functions on the original configuration space that are invariant under both translations and rotations.

For the simplest three-body case (ABC), the shape space is the space of triangles formed by three distinguishable bodies. Many systems of coordinates have been used for this case [45], in particular the Jacobi coordinates $[\vec{r}, \vec{R}, \theta \equiv \cos^{-1}(\vec{r}\cdot\vec{R}/|rR|)$, where $\vec{r}$ is the bond distance vector associated to a given diatomic fragment and $\vec{R}$ the vector connecting the centre-of-mass of such a diatomic fragment with the remaining atom], the valence coordinates $(R_1, R_2, \angle\mathrm{ABC})$, and the three bond distances (R_1, R_2, R_3). Further examples are the sets of linear symmetrized

[4]The nuclear coordinates are often called slow variables, as opposed to the electronic coordinates or fast variables.

[5]Whenever causing no ambiguity, the 'hat' notation employed to denote an operator will be omitted.

combinations of bond distances[6] [46],

$$\begin{pmatrix} Q_1 \\ Q_2 \\ Q_3 \end{pmatrix} = \begin{pmatrix} 1/\sqrt{3} & 1/\sqrt{3} & 1/\sqrt{3} \\ 0 & 1/\sqrt{2} & -1/\sqrt{2} \\ 2/\sqrt{6} & -1/\sqrt{6} & -1/\sqrt{6} \end{pmatrix} \begin{pmatrix} R_1 \\ R_2 \\ R_3 \end{pmatrix} \tag{2.3}$$

or of squared bond distances,

$$\begin{pmatrix} \mathcal{Q} \\ \beta \\ \gamma \end{pmatrix} = \begin{pmatrix} 1 & 1 & 1 \\ 0 & \sqrt{3} & -\sqrt{3} \\ 2 & -1 & -1 \end{pmatrix} \begin{pmatrix} R_1^2 \\ R_2^2 \\ R_3^2 \end{pmatrix} \tag{2.4}$$

these being closely related to the well known hyperspherical coordinates [27,45,47–50] (ρ, θ, φ). Such coordinates are especially convenient for representing the potential energy surface by making a triangular plot for a fixed perimeter (radius) of the molecular triangle (circle). To account for the hidden variable, the coordinates in Eq. (2.3) are often divided by the sum of the three bond distances such that they become unitless; alternatively, those in Eq. (2.4) are scaled by $\mathcal{Q}$. Such coordinates are denoted as in Eq. (2.3) and Eq. (2.4) but carrying a star as superscript. Thus, they become representable in an equilateral triangle of unit height that circumscribes the equilateral triangle and circle shown in Figure 2.1. These represent the physically accessible regions of configuration space within a given representation. To account for the hidden variable, the plotted energy is chosen to be that of the structure whose perimeter (radius) has the lowest energy. This provides the basis for the so-called relaxed triangular plots discussed elsewhere [46]. Using the symmetry coordinates in Eq. (2.4), Figure 2.1 shows the locus of the possible symmetry point groups assuming that the group operations are applied to identical nuclei A, B, and C. Of course, for all ABC molecules with three distinct nuclei, the C_s symmetry is intrinsic to the entire physical circle [equivalent considerations will apply to the physical triangle defined by the linear symmetrized coordinates in Eq. (2.3)]. Also shown are the molecular conformations obtained in a cyclic path around the origin. Thus, in a relaxed triangular plot of the potential surface, the plotted energy is that of the energetically most favorable species of a given symmetry. For example, the locus of all species with D_{3h} symmetry coincides with the C_3 rotation axis, and hence it is the energy of the optimum such a species that will be shown in the relaxed triangular

[6]The corresponding displacements relative to a reference equilateral triangle geometry are illustrated in panels (b) to (d) of Figure 2.6 when discussing the normal coordinates of an X_3-type molecule.

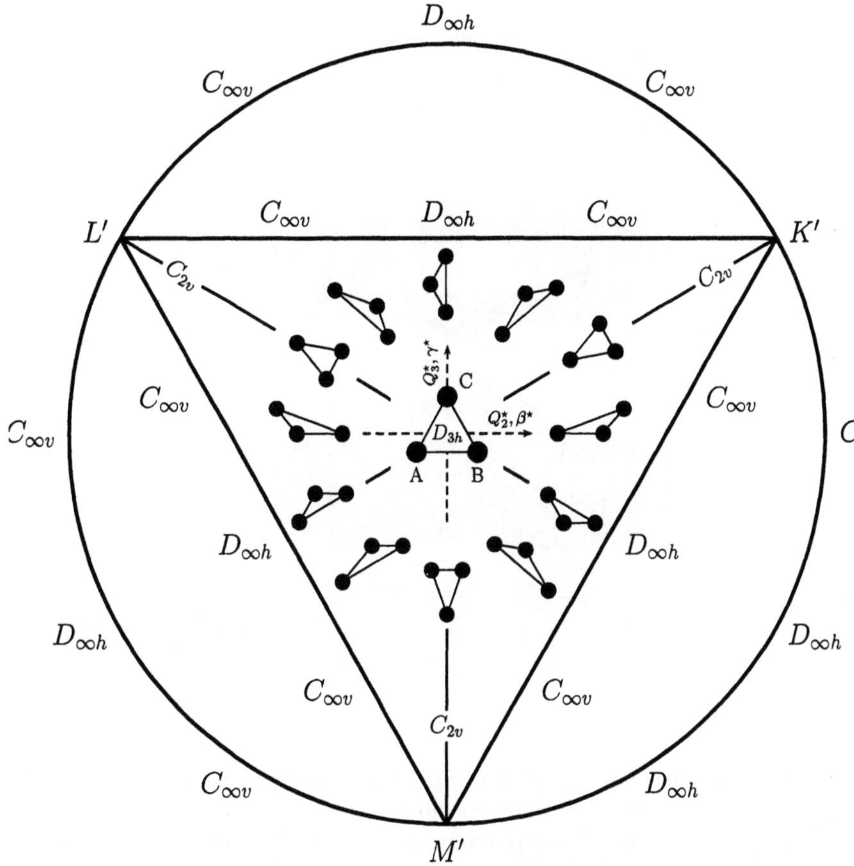

Figure 2.1. Locus of the possible symmetry point groups assuming that the group operations are applied to identical nuclei A, B, and C (which are denoted by the distinctly colored dots). The C_s symmetry is intrinsic to the entire physical circle based on the coordinates in Eq. (2.4) [due to an obvious similarity, equivalent considerations will apply to the physical triangle when using Eq. (2.3)]. Also shown are the molecular conformations obtained by following a closed path around the origin. K', L', and M' indicate the points where the 'physical triangle' and 'physical circle' centered at the origin of the coordinate system touch the circunscribing equilateral triangle employed for the triangular plot [46]. Physically, such points indicate the diatomic species formed by one of the atoms and the united-atom resulting from the remaining diatomic molecule when its atoms collapse into each other for a vanisnhingly small interatomic separation. Note that Q_2 and Q_3 correspond, respectively, to the symmetrized displacements Q_{E_x} and Q_{E_y} used later in the text (an equivalent correspondence applies to the unstarred coordinates β and γ).

plot. A perpective view of such a plot is shown in Figure 2.2 for the popular H_3 double many-body expansion [51–53] (DMBE) potential energy

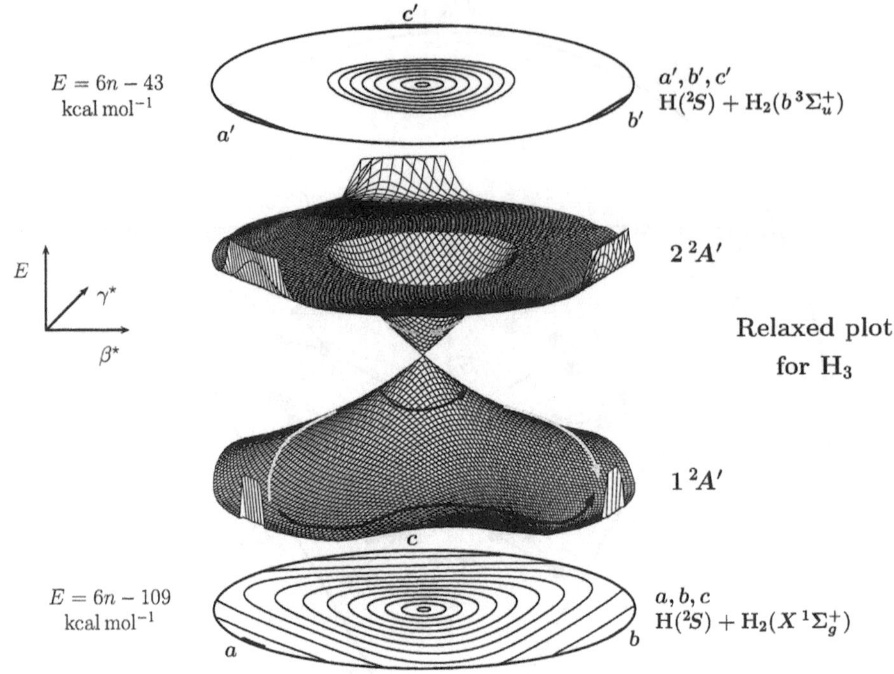

Figure 2.2. Relaxed triangular plot of the double many-body expansion [51–53] (DMBE) potential energy surface [54] for doublet H_3 using the starred coordinates from Eq. (2.4). The lines in red and yellow indicate two distinct paths for the adiabatic exchange of a H atom, while the line in green shows a path on the upper adiabatic sheet encircling the conical intersection. The energies referring to the various contours of both sheets are also indicated in kcal mol^{-1}.

surface [54]. Finally, we observe that the linear geometries lie along the delimiting line of the physical triangle or circle.

2.1 The adiabatic approximation

Within the adiabatic approximation to Eq. (2.1), one first solves the Schrödinger equation for the electronic Hamiltonian operator[7]

$$\hat{H}_e(\mathbf{r}, \mathbf{R})\Psi_I(\mathbf{r}; \mathbf{R}) = V_I(\mathbf{R})\Psi_I(\mathbf{r}; \mathbf{R}) \qquad (2.5)$$

[7]As usual, the semicolon indicates the parametric dependence on **R**.

Since $\{\Psi_I(\mathbf{r};\mathbf{R})\}$ form a complete orthonormal set, the total wave function may be written as the Born-Huang [1, 2] development

$$\Omega(\mathbf{R}',\mathbf{r}) = \sum_I \chi_I(\mathbf{R}')\Psi_I(\mathbf{r};\mathbf{R}) \tag{2.6}$$

where $\{\chi_I(\mathbf{R}')\}$ are the nuclear wave functions, and the summation includes in principle an integration over the continuum [55]. Substitution of Eq. (2.6) in Eq. (2.1) now leads to an infinite set of coupled equations which may be written in matrix form as

$$\left\{-\frac{\hbar^2}{2\mu}\left[\mathbf{I}\nabla^2_{\mathbf{R}'} + 2\mathbf{F}(\mathbf{R}')\cdot\boldsymbol{\nabla}_{\mathbf{R}'} + \mathbf{G}(\mathbf{R}')\right] + \mathbf{V}(\mathbf{R}) - E\mathbf{I}\right\}\boldsymbol{\chi}(\mathbf{R}') = \mathbf{0} \tag{2.7}$$

where $\nabla_{\mathbf{R}'}$ is the gradient with respect to all nuclear degrees of freedom which locate the nuclei in a center-of-mass frame, $\mathbf{I}$ is the appropriate identity matrix, $\boldsymbol{\chi}(\mathbf{R}')$ is a column vector whose components are the nuclear wave functions $\chi_I(\mathbf{R}')$, and the elements of the vector matrix $\mathbf{F}(\mathbf{R}')$ and the scalar matrices $\mathbf{G}(\mathbf{R}')$ and $\mathbf{V}(\mathbf{R})$ are given by[8]

$$\mathbf{F}_{IJ}(\mathbf{R}') = \langle\Psi_I(\mathbf{r};\mathbf{R})|\nabla_{\mathbf{R}'}\Psi_J(\mathbf{r};\mathbf{R})\rangle \tag{2.8}$$

$$G_{IJ}(\mathbf{R}') = \langle\Psi_I(\mathbf{r};\mathbf{R})|\nabla^2_{\mathbf{R}'}\Psi_J(\mathbf{r};\mathbf{R})\rangle \tag{2.9}$$

$$V_{IJ}(\mathbf{R}) = \langle\Psi_I(\mathbf{r};\mathbf{R})|\hat{H}_e|\Psi_J(\mathbf{r};\mathbf{R})\rangle \tag{2.10}$$

where $\mathbf{F}_{IJ}(\mathbf{R}')$ and $G_{IJ}(\mathbf{R}')$ are, respectively, the first- and second-order nonadiabatic matrix coupling elements, which by using the Hellmann-Feynman theorem may be shown to behave hyperbolically near the crossing seam where $V_I = V_J$. Such a behavior is responsible for numerical computational difficulties in the adiabatic approximation. Note that the potential matrix is diagonal in this approximation, $V_{IJ} = V_I\delta_{IJ}$, while

$$\mathbf{C} = -\frac{\hbar^2}{2\mu}\left[2\mathbf{F}(\mathbf{R}')\cdot\boldsymbol{\nabla}_{\mathbf{R}'} + \mathbf{G}(\mathbf{R}')\right] \tag{2.11}$$

is a nondiagonal matrix. Because the off-diagonal operators $\hat{C}_{IJ}$ induce transitions between the electronic states Ψ_I and Ψ_J, they are called operators of nonadiabaticity. In turn, $V_I = V_{II}(\mathbf{R})$ is an adiabatic potential or sheet of the electronically adiabatic manifold. If just the I-th term is used in Eq. (2.6), the motion of the nuclei can lead only to deformations of the electronic distribution, with no electronic transitions being allowed. The distribution of the nuclei is then described by the wave function $\chi_I(\mathbf{R}')$ in the potential $V_{II} + C_{II}$, the so-called proper adiabatic approximation [5].

[8]The bra-ket notation $\langle|\rangle$ implies heretofore integration over the electronic coordinates only.

2.2 The diabatic approximation

In the diabatic approach, the eigenfunctions of the electronic Hamiltonian for the nuclear internal coordinates clamped at $\mathbf{R} = \mathbf{R}_0$,

$$\hat{H}_e(\mathbf{r}, \mathbf{R}_0)\Psi_I^d(\mathbf{r}; \mathbf{R}_0) = V_I(\mathbf{R}_0)\Psi_I^d(\mathbf{r}; \mathbf{R}_0) \tag{2.12}$$

are used as a basis in the expansion of the total wave function

$$\Omega(\mathbf{R}', \mathbf{r}) = \sum_I \chi_I^d(\mathbf{R}')\Psi_I^d(\mathbf{r}; \mathbf{R}_0) \tag{2.13}$$

Substitution of such an ansatz into Eq. (2.1) leads to a system of coupled differential equations similar to Eq. (2.7), although with some notable differences. First, the potential matrix is nondiagonal in the new basis, except at $\mathbf{R} = \mathbf{R}_0$. Second, all matrix elements of the nuclear kinetic energy operator vanish identically since the basis functions do not depend on $\mathbf{R}$. Such an approximation is called the Born-Oppenheimer (BO) approximation [1]. Of course, for an infinite number of expansion functions, the adiabatic and diabatic approaches are equivalent. The two approximations are then related by a unitary transformation, with the eigenvalues of the potential matrix coinciding with the adiabatic potentials $V_I(\mathbf{R})$ in Eq. (2.5). In fact, the BO approximation involves necessarily some power expansion of the potential energy with respect to the nuclear displacements from $\mathbf{R}_0$, and hence it is expected to have convergence properties different from those in the proper adiabatic approximation. Indeed, the BO approximation can break down simply due to the signficant anharmonicity of the potential energy surface. However, it is formally most convenient since the electronic states $\Psi_I(\mathbf{r}) \equiv \Psi_I^d(\mathbf{r}; \mathbf{R}_0)$ form the basis of irreducible representations of the symmetry group appropriate to the nuclear configuration $\mathbf{R}_0$ (Ref. [19], and references therein).

2.3 Generalized Born-Oppenheimer approximation

As first pointed out by Longuet-Higgins [6, 9, 10], the adiabatic approximation breaks down at electronic degeneracies of the conical type, since a real electronic wave function changes sign whenever traversing a nuclear path which encircles the crossing point. This may be stated more formally in the two following theorems [10]:

I. *On going once around any closed path on the surface that contains the conical intersection the sign of the electronic wave function changes sign.*

II. *If a real adiabatic electronic wave function changes sign when a polyatomic traverses a one-dimensional closed loop on a two-dimensional surface in the $(3N-6)$-dimensional nuclear configuration space, then the corresponding electronic state must become discontinuous and degenerate with another one at an odd number of points lying on that surface and within that loop.*

Such theorems provide a criterion to localize a conical intersection, as it has been demonstrated by *ab initio* calculations a few years after their proposition by Varandas, Tennyson and Murrell [56] for LiNaK, and more recently for other systems like O_3 [57]. The sign change in the electronic wave function has important implications. In fact, since the total electronuclear wave function must be continuous and single-valued, the nuclear wave function must also change sign to compensate the sign-change of the electronic wave function. As a result, both $\Psi_I(\mathbf{r};\mathbf{R})$ and $\chi_I(\mathbf{R}')$ are not single-valued when taken as real. However, the wave functions in the Born-Huang [2] expansion can be single-valued once allowed to be complex (see Appendix A of Ref. [53]). One has,

$$\Omega(\mathbf{R}',\mathbf{r}) = \sum_I \chi_I(\mathbf{R}')\, e^{iA_I(\mathbf{R})}\Psi_I(\mathbf{r};\mathbf{R}) \tag{2.14}$$

where $A_I(\mathbf{R})$ is a geometric phase. Two possibilities then arise. One is to associate this phase to the electronic wave function by writing

$$\tilde{\Psi}_I(\mathbf{r};\mathbf{R}) = e^{iA_I(\mathbf{R})}\Psi_I(\mathbf{r};\mathbf{R}) \tag{2.15}$$

where the tilde indicates the complex nature of the wave function. If this is chosen, the generalized adiabatic[9] approximation assumes the form

$$\left\{-\frac{\hbar^2}{2\mu}\left[\nabla_{\mathbf{R}'} + i\nabla_{\mathbf{R}'}A_I(\mathbf{R})\right]^2 + \overline{V}_I(\mathbf{R}) - E\right\}\chi_I(\mathbf{R}') = 0 \tag{2.16}$$

where

$$\overline{V}_I(\mathbf{R}) = V_I(\mathbf{R}) - \frac{\hbar^2}{2\mu}\tilde{G}_{II}(\mathbf{R}') \tag{2.17}$$

with $\tilde{G}_{II}(\mathbf{R}')$ having the form in Eq. (2.9) but with tilded electronic wave functions. Note that

$$\tilde{\mathbf{F}}_{IJ}(\mathbf{R}') = e^{[A_I(\mathbf{R})-A_J(\mathbf{R})]}\mathbf{F}_{IJ}(\mathbf{R}) + i\delta_{IJ}\nabla_{\mathbf{R}'}A_J(\mathbf{R}) \tag{2.18}$$

and, for real-valued electronic orthonormalized wave functions, on then obtains

$$\tilde{\mathbf{F}}_{II}(\mathbf{R}') = i\nabla_{\mathbf{R}'}A_I(\mathbf{R}) \tag{2.19}$$

[9] Also called generalized Born-Oppenheimer approximation [59].

since [58] $\mathbf{F}_{II}(\mathbf{R}) = \langle \Psi_I(\mathbf{r};\mathbf{R}) | \nabla_{\mathbf{R}'} \Psi_I(\mathbf{r};\mathbf{R}) \rangle = 0$.

Conversely, Eq. (2.14) may be written as

$$\Omega(\mathbf{R}',\mathbf{r}) = \sum_I \tilde{\chi}_I(\mathbf{R}')\, \Psi_I(\mathbf{r};\mathbf{R}) \tag{2.20}$$

where the complex nuclear wave functions are now chosen to make $\Omega(\mathbf{R}',\mathbf{r})$ be single-valued,

$$\tilde{\chi}_I(\mathbf{R}') = e^{iA_I(\mathbf{R})} \chi_I(\mathbf{R}') \tag{2.21}$$

The resulting generalized adiabatic approximation (also termed generalized Born-Oppenheimer) assumes now the form [33]

$$\left\{ -\frac{\hbar^2}{2\mu} \left[\nabla^2_{\mathbf{R}'} + G_{II}(\mathbf{R}') \right] + V_I(\mathbf{R}) - E \right\} \tilde{\chi}_I(\mathbf{R}') = 0 \tag{2.22}$$

As noted by Mead and Truhlar [21–23], the ansatz in Eq. (2.14) leads to the appearance of a vector potential in the nuclear Schrödinger equation. Since the $\mathbf{R}$-dependence of $A_I(\mathbf{R})$ must reflect the presence of any conical intersection in accordance with the geometric phase [10,20] condition, the proper phase can only be constructed once the conical intersections have been located. A general approach for determining $A_I(\mathbf{R})$ has been suggested by Kendrick and Mead [59], although it remains a nontrivial task. The alternative ansatz in Eq. (2.20), where the complex phase factor is absorbed in the real nuclear wave functions [60–62], turns out to be especially convenient for X_3-type systems when adopting hyperspherical coordinates since the complex phase factor concerns in principle only the φ hyperangle [27,63]. Such a procedure has been recently been extended to asymmetric cases [33,35,37,39], as it will be discussed further in section 2.3.

The case of a 2*D* Hilbert space in the vicinity of the crossing seam. Consider the nuclear motion restricted to a two-state electronic manifold (*i.e.*, all nonadiabatic couplings to higher electronic states can be ignored). The adiabatic representation assumes the explicit form

$$\left[-\frac{\hbar^2}{2\mu} (\nabla^2_{\mathbf{R}} + \langle \Psi_1 | \nabla^2_{\mathbf{R}} \Psi_1 \rangle) + V_1 - E \right] \chi_1 = \frac{\hbar^2}{2\mu} \left[\langle \Psi_1 | \nabla^2_{\mathbf{R}} \Psi_2 \rangle + 2 \langle \Psi_1 | \nabla_{\mathbf{R}} \Psi_2 \rangle \cdot \nabla_{\mathbf{R}} \right] \chi_2 \tag{2.23}$$

$$\left[-\frac{\hbar^2}{2\mu}\left(\nabla^2_{\mathbf{R}} + \langle\Psi_2|\nabla^2_{\mathbf{R}}\Psi_2\rangle\right) + V_2 - E\right]\chi_2 = \frac{\hbar^2}{2\mu}\left[\langle\Psi_2|\nabla^2_{\mathbf{R}}\Psi_1\rangle + 2\langle\Psi_2|\nabla_{\mathbf{R}}\Psi_1\rangle\cdot\nabla_{\mathbf{R}}\right]\chi_1 \quad (2.24)$$

where we have supressed obvious coordinate dependences on $\{\Psi_I(\mathbf{r};\mathbf{R})\}$ and $\{\chi_I(\mathbf{R})\}$. The adiabatic and diabatic electronic wave functions are then related by the unitary transformation

$$\begin{pmatrix}\Psi_1\\ \Psi_2\end{pmatrix} = \begin{bmatrix}\cos\gamma(\mathbf{R}) & -\sin\gamma(\mathbf{R})\\ \sin\gamma(\mathbf{R}) & \cos\gamma(\mathbf{R})\end{bmatrix}\begin{pmatrix}\Psi_1^d\\ \Psi_2^d\end{pmatrix} \quad (2.25)$$

where $\gamma(\mathbf{R})$ is the mixing angle [33, 34, 64] in the orthogonal transformation that diagonalizes the potential matrix. This varies with $\mathbf{R}$ only, which explains our use of $\nabla_{\mathbf{R}}$.

Since the diabatic basis functions do not depend on $\mathbf{R}$, the two real electronically adiabatic wave functions may then be represented by the corresponding vectors of $\mathbf{R}$-dependent expansion coefficients [33], namely

$$\Psi_1 = \begin{bmatrix}\cos\gamma(\mathbf{R})\\ -\sin\gamma(\mathbf{R})\end{bmatrix}, \quad \Psi_2 = \begin{bmatrix}\sin\gamma(\mathbf{R})\\ \cos\gamma(\mathbf{R})\end{bmatrix} \quad (2.26)$$

Assuming now that the electronic basis is orthonormalized, one has

$$\langle\Psi_1|\nabla_{\mathbf{R}}\Psi_2\rangle = \nabla_{\mathbf{R}}\gamma(\mathbf{R}) \quad (2.27)$$

$$\langle\Psi_2|\nabla_{\mathbf{R}}\Psi_1\rangle = -\langle\Psi_1|\nabla_{\mathbf{R}}\Psi_2\rangle = \nabla_{\mathbf{R}}\gamma(\mathbf{R}) \quad (2.28)$$

Similarly, one gets

$$\langle\Psi_I|\nabla^2_{\mathbf{R}}\Psi_I\rangle = -\langle\nabla_{\mathbf{R}}\Psi_I|\nabla_{\mathbf{R}}\Psi_I\rangle = -\left[\nabla_{\mathbf{R}}\gamma(\mathbf{R})\right]^2 \quad (2.29)$$

while, for $I \neq J$, one obtains $\langle\nabla_{\mathbf{R}}\Psi_I|\nabla_{\mathbf{R}}\Psi_J\rangle = 0$. Moreover, one gets

$$\langle\Psi_1|\nabla^2_{\mathbf{R}}\Psi_2\rangle = \nabla_{\mathbf{R}}\langle\Psi_1|\nabla_{\mathbf{R}}\Psi_2\rangle - \langle\nabla_{\mathbf{R}}\Psi_1|\nabla_{\mathbf{R}}\Psi_2\rangle = \nabla^2_{\mathbf{R}}\gamma(\mathbf{R}) \quad (2.30)$$

and

$$\langle\Psi_2|\nabla^2_{\mathbf{R}}\Psi_1\rangle = -\nabla^2_{\mathbf{R}}\gamma(\mathbf{R}) \quad (2.31)$$

Replacing Eq. (2.27) to Eq.(2.31) in Eq. (2.23) and Eq. (2.24), then yields

$$\left\{-\frac{\hbar^2}{2\mu}\left[\nabla^2_{\mathbf{R}} - \left(\nabla_{\mathbf{R}}\gamma(\mathbf{R})\right)^2\right] + V_1 - E\right\}\chi_1 = \frac{\hbar^2}{2\mu}\left[\nabla^2_{\mathbf{R}}\gamma(\mathbf{R}) + 2\nabla_{\mathbf{R}}\gamma(\mathbf{R})\cdot\nabla_{\mathbf{R}}\right]\chi_2 \quad (2.32)$$

$$\left\{-\frac{\hbar^2}{2\mu}\left[\nabla^2_{\mathbf{R}} - \left(\nabla_{\mathbf{R}}\gamma(\mathbf{R})\right)^2\right] + V_2 - E\right\}\chi_2$$
$$= -\frac{\hbar^2}{2\mu}\left[\nabla^2_{\mathbf{R}}\gamma(\mathbf{R}) + 2\nabla_{\mathbf{R}}\gamma(\mathbf{R})\cdot\nabla_{\mathbf{R}}\right]\chi_1 \quad (2.33)$$

Following Longuet-Higgins [6] and Baer [28], we now consider a complex nuclear wave function of the form

$$\tilde{\chi} = \frac{1}{\sqrt{2}}(\chi_1 + i\chi_2) \quad (2.34)$$

which may also be written as

$$\tilde{\chi} = e^{iA(\mathbf{R})}\chi \quad (2.35)$$

where χ is a real-valued nuclear wave function, and $A(\mathbf{R})$ is some phase that may depend on the nuclear configuration $\mathbf{R}$. After some algebraic manipulation, Eq. (2.23) and Eq. (2.24) lead to the following single-surface equations [33]

$$\left\{\frac{\hbar^2}{2\mu}\left[-\nabla^2_{\mathbf{R}} + \left(\nabla_{\mathbf{R}}\gamma(\mathbf{R})\right)^2 + i\nabla^2_{\mathbf{R}}\gamma(\mathbf{R}) + 2i\nabla_{\mathbf{R}}\gamma(\mathbf{R})\cdot\nabla_{\mathbf{R}}\right]\right.$$
$$\left. + V_1 - E\right\}\tilde{\chi} + \frac{i}{\sqrt{2}}\left(V_2 - V_1\right)\chi_2 = 0 \quad (2.36)$$

$$\left\{\frac{\hbar^2}{2\mu}\left[-\nabla^2_{\mathbf{R}} + \left(\nabla_{\mathbf{R}}\gamma(\mathbf{R})\right)^2 + i\nabla^2_{\mathbf{R}}\gamma(\mathbf{R}) + 2i\nabla_{\mathbf{R}}\gamma(\mathbf{R})\cdot\nabla_{\mathbf{R}}\right]\right.$$
$$\left. + V_2 - E\right\}\tilde{\chi} + \frac{i}{\sqrt{2}}\left(V_1 - V_2\right)\chi_1 = 0 \quad (2.37)$$

and, after neglecting the second term in Eq. (2.36) and Eq. (2.37), to

$$\left\{\frac{\hbar^2}{2\mu}\left[-\nabla^2_{\mathbf{R}} + \left(\nabla_{\mathbf{R}}\gamma(\mathbf{R})\right)^2 + i\nabla^2_{\mathbf{R}}\gamma(\mathbf{R})\right.\right.$$
$$\left.\left. + 2i\nabla_{\mathbf{R}}\gamma(\mathbf{R})\cdot\nabla_{\mathbf{R}}\right] + V_{1,2} - E\right\}\tilde{\chi} = 0 \quad (2.38)$$

which should be valid whenever motion of the nuclei can be confined to the vicinity of the conical intersection ($V_1 \simeq V_2$). Note that different sets of eigenvalues are obtained depending on whether one uses V_1 or V_2. Of course, Eq. (2.38) leads to the corresponding Born-Oppenheimer equations when the derivative coupling terms are constant or zero.

Given the degeneracy of the electronic adiabatic wave functions at the crossing seam, we may also write

$$\tilde{\Psi} = \frac{1}{\sqrt{2}}(\Psi_1 + i\Psi_2) \quad (2.39)$$

In the complex plane $\{|\chi_1\rangle, i|\chi_2\rangle\}$, the complex vector state $|\tilde{\chi}\rangle$ will then be characterized by the argument $A(\mathbf{R}) = \arg\tilde{\chi}$. Similarly, in the complex plane $\{|\Psi_1\rangle, i|\Psi_2\rangle\}$, $|\tilde{\Psi}\rangle$ will satisfy $A(\mathbf{R}) = \arg\tilde{\Psi}$ as it should [*cf.*, Eq. (2.15) and Eq. (2.21)]. Consider now Eq. (2.39) written in the form

$$\tilde{\Psi}(\mathbf{r};\mathbf{R}) = \exp[iB(\mathbf{r},\mathbf{R})]\Psi(\mathbf{r};\mathbf{R}) \tag{2.40}$$

where $B(\mathbf{r},\mathbf{R})$ is a phase which we assume for generality [42] to depend both on the nuclear and electronic coordinates. Using Eq. (2.40), one gets

$$\langle\tilde{\Psi}(\mathbf{r};\mathbf{R})|\nabla_{\mathbf{R}}\tilde{\Psi}(\mathbf{r};\mathbf{R})\rangle = i\nabla_{\mathbf{R}}A(\mathbf{R}) + \langle\Psi(\mathbf{r};\mathbf{R})|\nabla_{\mathbf{R}}\Psi(\mathbf{r};\mathbf{R})\rangle \tag{2.41}$$

where $\nabla_{\mathbf{R}}A(\mathbf{R}) = \langle\tilde{\Psi}|\nabla_{\mathbf{R}}B(\mathbf{r};\mathbf{R})|\tilde{\Psi}\rangle$. In turn, from Eq. (2.39), we obtain [33]

$$\langle\tilde{\Psi}|\nabla_{\mathbf{R}}\tilde{\Psi}\rangle = i\langle\Psi_1|\nabla_{\mathbf{R}}\Psi_2\rangle \tag{2.42}$$

where we have used the fact that $\langle\Psi_I|\nabla_{\mathbf{R}}\Psi_I\rangle = 0$, and $\langle\Psi_1|\nabla_{\mathbf{R}}\Psi_2\rangle = -\langle\Psi_2|\nabla_{\mathbf{R}}\Psi_1\rangle$. A comparison of Eq. (2.42) with Eq. (2.41) then leads to

$$\langle\Psi_1|\nabla_{\mathbf{R}}\Psi_2\rangle = \nabla_{\mathbf{R}}A(\mathbf{R}) \tag{2.43}$$

Furthermore, by using Eq. (2.27), one obtains up to a constant:

$$A(\mathbf{R}) = \gamma(\mathbf{R}) \tag{2.44}$$

Since the mixing angle $\gamma(\mathbf{R})$ has been shown [34, 36] to lead to the proper sign change when looping around a conical intersection (as well as the proper behavior in loops which do not encircle it), we may, for most purposes, identify $A(\mathbf{R})$ with the GP angle; $\Delta\gamma = \pi$ for a cyclic path encircling the conical intersection, otherwise $\Delta\gamma = 0$. Although the geometric phase from Eq. (2.44) depends on the full set of internal coordinates, it shows [34] a behavior similar to the traditional Longuet-Higgins phase of $\phi/2$ close to the seam (ϕ is the pseudorotation angle). Returning to Eq. (2.39), we further note that a development using $\tilde{\Psi} = \frac{1}{\sqrt{2}}(i\Psi_1+\Psi_2)$ would give $A(\mathbf{R}) = -\gamma(\mathbf{R})$. Of course, this sign change has no practical implications, as it simply implies a clockwise measurement of the phase angle if positive phase angles are assumed to be defined counterclockwise in Eq. (2.44).

A final remark about Eq. (2.43) to note that it can in principle be used to define a strictly diabatic two-state electronic basis by solving

$$\mathbf{F} = \nabla_{\mathbf{R}}A(\mathbf{R}) \tag{2.45}$$

where, for the two state case, $A(\mathbf{R})$ would simply be an angle heretofore denoted as $\alpha(\mathbf{R})$, the so-called adiabatic-to-diabatic transformation angle [65–67]. Such an angle would then suffice to transform the adiabatic basis to a diabatic one in which all the derivative couplings vanish identically[10]. In fact, Eq. (2.45) may be solved by integration along paths in the nuclear configuration space [68, 69], but the result may depend both on the initial and final points of integration, as well as on the path chosen:

$$\alpha(s, s_0) = \exp\left(-\int_{s_0}^{s} \mathbf{F}(s) \cdot ds\right) \alpha(s_0) \tag{2.46}$$

where the integral is performed along a line that combines the two points s and s_0, and the dot under the integral sign indicates the scalar product between the vector matrix $\mathbf{F}(s)$ and the differential length element along the integration path ds. The phase accumulated along a closed path, the so-called topological phase [70], will then be given by

$$\xi_\alpha = \oint_C \mathbf{F}(s) \cdot ds \tag{2.47}$$

According to the GP theorems [6,9,10], one should then have [20,69,71]

$$\xi_\alpha = p\pi \tag{2.48}$$

where $p=0$ if C does not enclose any conical intersection, and $p=1$ if it does enclose one such an intersection. In fact, for the title two-state problem, $\alpha(\mathbf{R})$ has been shown to be (up to a constant) identical to the mixing angle $\gamma(\mathbf{R})$ that diagonalizes the potential matrix [34], and (up to a constant) also identical to the geometric phase $A_I(\mathbf{R})$ [33].

Cone states of trihydrogen and its isotopomers. In this section we review some calculations of the cone states of H_3 and its isotopomers [33,35,37,39] using the formalism derived in section 2.3. For an ABC isotopomer of X_3, the crossing seam in hyperspherical coordinates (ρ, θ, φ) is defined at an arbitrary value of the hyperradius ρ by [34]

$$\varphi_s = \tan^{-1}\left[\frac{\cos\chi_{\mathrm{AC}} - t\cos\chi_{\mathrm{AB}} - \left(\frac{d_{\mathrm{A}}}{d_{\mathrm{C}}}\right)^2 + t\left(\frac{d_{\mathrm{A}}}{d_{\mathrm{B}}}\right)^2}{\sin\chi_{\mathrm{AC}} + t\sin\chi_{\mathrm{AB}}}\right] \tag{2.49}$$

$$\theta_s = 2\sin^{-1}\left[\frac{\left(\frac{d_{\mathrm{A}}}{d_{\mathrm{B}}}\right)^2 - 1}{\cos(\varphi_s - \chi_{\mathrm{AB}}) - \left(\frac{d_{\mathrm{A}}}{d_{\mathrm{B}}}\right)^2 \cos\varphi_s}\right] \tag{2.50}$$

[10] Although such a perfect diabatization cannot generally exist for a polyatomic system [68, 102], a method has been suggested for constructing a strictly diabatic basis set for the two-state case by assuming un-normalized basis functions [99]. See also Ref. [18], and references therein.

where

$$t = \left[\left(\frac{d_{\rm A}}{d_{\rm C}}\right)^2 - 1\right]\left[\left(\frac{d_{\rm A}}{d_{\rm B}}\right)^2 - 1\right]^{-1} \tag{2.51}$$

and

$$d_{\rm X}^2 = \frac{m_{\rm X}}{\mu}\left(1 - \frac{m_{\rm X}}{M}\right); \quad \chi_{\rm XY} = 2\tan^{-1}\left(\frac{m_{\rm Z}}{\mu}\right)$$

$$\mu = \sqrt{\frac{m_{\rm A}m_{\rm B}m_{\rm C}}{M}}; \quad M = m_{\rm A} + m_{\rm B} + m_{\rm C} \tag{2.52}$$

with X, Y, and Z standing for atoms A, B, and C with masses m_A, m_B, and m_C. If the isotopomer is of the AB_2 type, the following simplified expression results

$$\theta_s = 2\sin^{-1}\left|\frac{m_{\rm B} - m_{\rm A}}{m_{\rm B} + 2m_{\rm A}}\right| \tag{2.53}$$

while φ_s assumes the value π (0) when $m_{\rm A} > m_{\rm B}$, and zero (π) when $m_{\rm A} < m_{\rm B}$; the value $\varphi_s = 0$ is chosen here for $m_{\rm A} < m_{\rm B}$. Table 2.1 illustrates the values obtained for all isotopomers of H_3, involving μ, D, and T. For example, for HD_2, one has $\theta_s = 0.5048\,\text{rad}$ and $\varphi_s = 0$. Since $\theta_s \neq 0$, only closed paths with $\theta \geq \theta_s$ will enclose the seam for this system, with the accumulated phase change along the path being equal to π. For all other loops with $\theta < \theta_s$, the accumulated phase change is 0. Of course, for homonuclear triatomics, any cycle in φ will enclose the conical intersection since this occurs at $\theta_s = 0$.

The vibrational levels were obtained by solving a generalized adiabatic equation of the form in Eq. (2.22), $\hat{H}\chi = E\chi$, where

$$\hat{H} = \hat{H}_0 + i\hat{H}_1 \tag{2.54}$$

$$\begin{aligned}\hat{H}_0 = &-\frac{\hbar^2}{2\mu}\left\{\frac{\partial^2}{\partial\rho^2} + \frac{16}{\rho^2}\left[\frac{1}{\sin\theta}\frac{\partial}{\partial\theta}\sin\theta\frac{\partial}{\partial\theta} + \frac{1}{4\sin^2(\theta/2)}\frac{\partial^2}{\partial\varphi^2}\right]\right\}\\ &+ \frac{\hat{J}^2 - \hat{J}_z^2}{\mu\rho^2\cos^2(\theta/2)} + \frac{\hat{J}_z^2 + 4i\hbar\hat{J}_z\cos(\theta/2)(\partial/\partial\varphi)}{2\mu\rho^2\sin^2(\theta/2)} + \frac{15\hbar^2}{8\mu\rho^2}\\ &+ \frac{\sin(\theta/2)}{\mu\rho^2\cos^2(\theta/2)}\frac{1}{2}\left[\hat{J}_+^2 + \hat{J}_-^2\right] + V(\rho,\theta,\varphi) + \frac{\hbar^2}{2\mu}\left[\nabla\gamma(\rho,\theta,\varphi)\right]^2\end{aligned} \tag{2.55}$$

$$\hat{H}_1 = -\frac{\hbar^2}{2\mu}\left[\nabla^2\gamma(\rho,\theta,\varphi) + 2\nabla\gamma(\rho,\theta,\varphi)\cdot\nabla\right] \tag{2.56}$$

As usual, $\hat{J}$, $\hat{J}_z$, and $\hat{J}_\pm$ represent the total angular momentum, its component along the z-axis, and the raising/lowering operators in the

Table 2.1. Values of φ_s, θ_s, and $\cos(\theta_s/2)$ for the various isotopomers of tri-hydrogen.

System	φ_s/ rad	θ_s/ rad	$\cos(\theta_s/2)$
μ_3	0	0	1.0000
μ_2H	π	0.8668	0.9075
μ_2D	π	0.9534	0.8885
μ_2T	π	0.9837	0.8815
μH_2	0	1.6192	0.6898
μHD	0.1528	1.8025	0.6206
μHT	0.2133	1.8743	0.5921
μD_2	0	2.0241	0.5301
DT_2	0	0.2854	0.9898
μDT	0.0650	2.1150	0.4910
μT_2	0	2.2164	0.4463
H_3	0	0	1.0000
H_2D	π	0.4023	0.9798
H_2T	π	0.5786	0.9584
HD_2	0	0.5048	0.9683
T_3	0	0	1.0000
HDT	0.3386	0.6667	0.9450
HT_2	0	0.8211	0.9169
D_3	0	0	1.0000
D_2T	π	0.2497	0.9922
DT_2	0	0.2854	0.9898
T_3	0	0	1.0000

body-fixed frame (respectively), which are defined by the set of external coordinates. In turn, V has been taken as the DMBE potential energy surface [54] for such trihydrogenic systems, which may be written in terms of the three internal coordinates (ρ, θ, φ), with the hyperangles θ and φ being related [27] to the usual Smith-Whitten's (Θ, Φ) hyperangles through the relations $\theta = \pi - 4\Theta$ and $\varphi = 2\pi - 2\Phi$. Thus, the range of θ has been extended from 0 to π, allowing to use Legendre (or Jacobi) polynomials in $\cos\theta$ as the finite basis representation (FBR) in θ. Note that the GP effect may be turned-off by ignoring the imaginary operator $i\hat{H}_1$ in the calculations. As pointed out in section 2.3, Eq. (2.55) and Eq. (2.56) are strictly valid in the vicinity of the crossing seam, although their general use may be validated from the fact that the GP effect is determined by the behavior of the potentials in the vicinity of the degeneracy [9, 10, 19, 72, 73].

A mixed grid-basis method has been employed to determine the action of the Hamiltonian on the wave function (Lanczos vector). Uniform grids have been utilized for the coordinates ρ and φ, while the actions

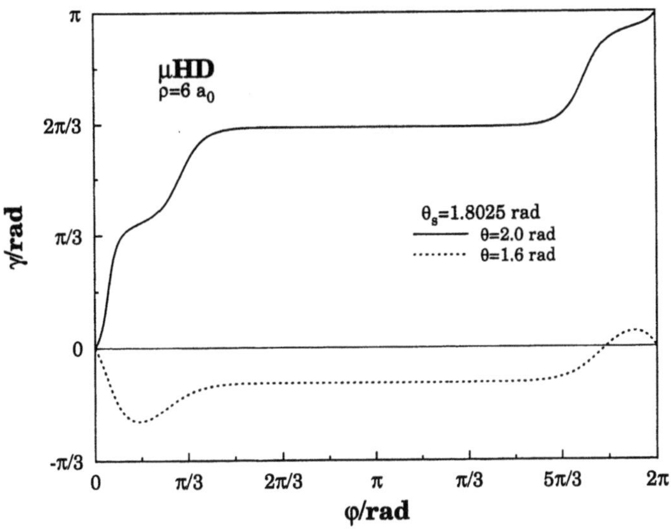

Figure 2.3. Variation of the geometric phase [39] in μHD for two distinct values of θ, one smaller and the other larger than θ_s.

of the associated kinetic energy operators have been performed by using a prime-factor fast Fourier transform (PFFT) technique [27]; the reader being referred to the original papers for details. All calculations [33,35,37,39] have been carried out for $J = 0$ by using a minimum residual filter diagonalization (MFD) technique [74] which, similarly to the simple Lanczos algorithm, has low storage requirements (only two iteration vectors) and utilizes just a single Lanczos recursion for the eigenvalue problem (it can also eliminate spurious and ghost eigenvalues).

Figure 2.3 illustrates for μHD the variation of the geometric phase for two distinct values of θ, one smaller and the other larger than θ_s. Clearly, for values $\theta < \theta_s$, there is no sign change upon completion of a loop as this does not encircle the conical intersection. Conversely, a sign change does occur for $\theta > \theta_s$. In turn, Figure 2.5 compares the results so obtained for H_3, D_3, HD_2, μH_2, and μHD. It is seen that the differences between the NGP and GP results decrease along the sequence $H_3 \sim D_3 > HD_2 > \mu H_2 > \mu HD$, as shown from the root-mean-square deviations for the lowest M calculated levels:

$$\Delta = \sqrt{\frac{\sum_{i=1}^{M} (\mathrm{GP}_i - \mathrm{NGP}_i)^2}{M}} \tag{2.57}$$

No symmetry assignment was judged justified. In fact, such an anal-

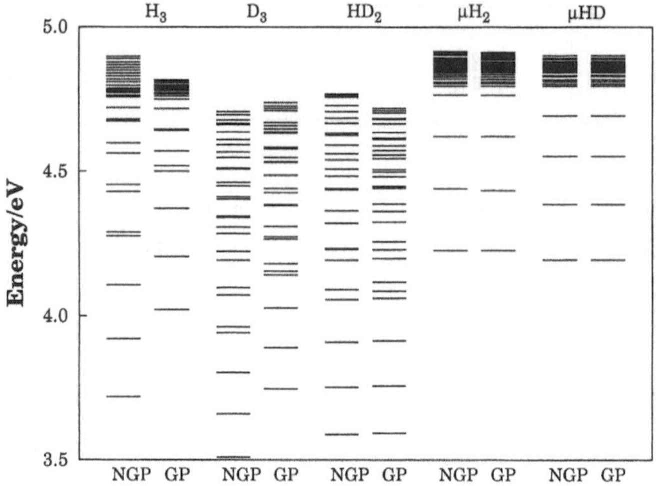

Figure 2.4. First thirty cone states [33, 35, 37, 39] calculated with consideration of the geometric phase effect (GP) and without considering it (NGP) for H_3, D_3, HD_2, μH_2, and μHD.

ysis would be cumbersome and virtually impossible to carry out accurately for the highly excited vibrational states of such floppy systems. Moreover, it is well established for the ground state of the homonuclear trimers [33, 75, 76] that the lowest energy levels for these systems are of A_1 (NGP) and E (GP) symmetries. Thus, their energy difference is a genuine measure of the GP effect, which is exactly what is aimed at. Using a bunch of M levels to define such a phenomenological quantity may have the advantage of taking into account the GP effect on higher vibrational levels.

The logarithms of such deviations are displayed for $M = 30$ are shown in Figure 2.5 as a function of the cosine of the tilting angle of the C_3 axis, which increases from $\theta_s = 0$ for the homonuclears up to $\theta_s = 2.2164\,\text{rad}$ in μT_2 (see Table 2.1). Figure 2.5 shows that the results for the studied isotopomers of trihydrogen may be quantitatively described by the form

$$\Delta = a\mathcal{P}_{\text{loop}}^{m} \tag{2.58}$$

where ($\mathcal{P}_{\text{loop}}$) measures the relative probability of looping around such an axis, namely

$$\mathcal{P}_{\text{loop}} \propto \int_0^{2\pi}\int_{\theta_s}^{\pi} \sin(\theta/2)d\theta d\varphi \Big/ \int_0^{2\pi}\int_0^{\pi} \sin(\theta/2)d\theta d\varphi = \cos(\theta_s/2) \tag{2.59}$$

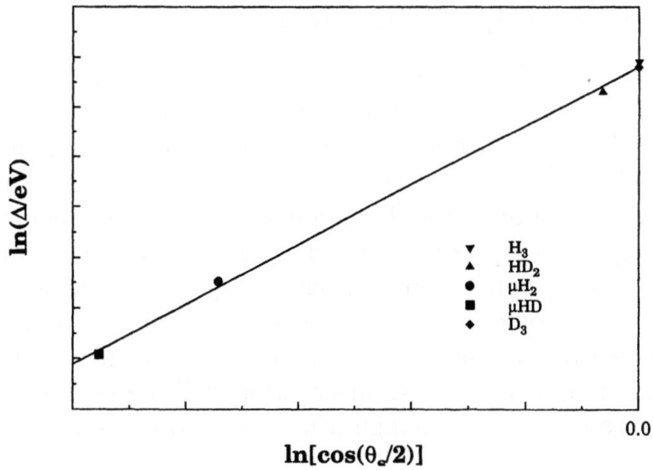

Figure 2.5. Straigth line in Eq. (2.58) showing the logarithms of the root mean square deviation as a function of the cosine of the tilting angle of the C_3 axis.

The optimum least squares parameters are $\ln(a/\mathrm{eV}) = -2.18 \pm 0.07$ and $m = 11.85 \pm 0.26$. The simple correlation in Eq. (2.58) may therefore be used to estimate the role of the GP effect in vibrational calculations of the cone states for other isotopomers of H_3 not considered in the calculations. Note that such a criterion ignores the total energy, which is a significant parameter when dealing with the dynamics of the exchange reactions in the lower sheet of the potential energy surface (the energy must be at least high enough to allow a loop around the conical intersection); see the red and yellow paths shown in the bottom sheet of Figure 2.2. Yet, such a role may not be so decisive when studying the vibrational spectroscopy of the upper cone as this lies above the crossing seam. In summary, the results in Figure 2.5 suggest that, although essential for accurate single-surface vibrational calculations on the homonuclear trihydrogenic systems, the GP effect can be significantly less so for some of their heteronuclear isotopomers, particularly for those involving a single muonium particle. It may be speculated at this point that, for an A+BC elementary reaction, the impact of the GP effect is likely to get further diminished since, in a classical sense, it is not mandatory for a reactive trajectory that it completes an entire loop around the crossing seam before yielding the products; see Figure 2.2. In fact, the probability of doing so should increase with the impact parameter, and hence with the value of J, as it really appears to be the case [77–79] for $H + D_2$.

3. Linear Jahn-Teller systems

In this and following sections we focus in the so-called linear JT systems. Thus, we concentrate on regions close to the crossing seam, and hence can reduce the vibronic JT Hamiltonian to its simplest linear form. Such a simplification has obvious advantages, since the integrals of the motion can be determined without solving any vibronic equations, and the symmetry properties of the adiabatic potentials can be more easily clarified [5]. As we hope to show, such a simplification may also help in establishing the geometric phase, especially in the case of an arbitrary integral N-fold degeneracy. For this, we introduce the so-called representation in electronic function space, which will be examined together with the representation in configuration space in section 3.2.

3.1 Matrix vibronic Hamiltonian

Let us use different letters to distinguish operators which give to an abstract element A of a point group G one of two possible realizations: $\hat{\mathsf{A}}$, to transform the nuclear coordinates only; $\hat{A}$, to act on the electronic degrees of freedom, including those which depend parametrically on the nuclear coordinates. Thus,

$$(\hat{\mathsf{A}}\hat{H}_e)\hat{A}\Psi = \hat{A}\hat{H}_e\Psi \tag{2.60}$$

since Ψ is an eigenfunction of $\hat{H}_e$, and hence Eq. (2.60) just expresses the fact that the transformed eigenfunction is an eigenfunction of the transformed Hamiltonian with the same eigenvalue. A similar result follows for an arbitrary Ψ by expanding this in terms of eigenfunctions of $\hat{H}_e$. Since Ψ is then arbitrary, one obtains

$$(\hat{\mathsf{A}}\hat{H}_e)\hat{A} = \hat{A}\hat{H}_e \tag{2.61}$$

An electronic operator which satisfies Eq. (2.61) for all elements $\hat{A}$ of some point group G will be called an invariant operator[11] [80]. Such an operator can be expanded as

$$\hat{H}_e = \sum_{\mu,s,j} \hat{h}^s_{\mu j} Q^s_{\mu j} \tag{2.62}$$

where $Q^s_{\mu j}$ is a nuclear coordinate transforming as the jth component of the μ-th irreducible representation of G, s is an index that accounts

[11] It can be shown that the sum of two invariant operators is still an invariant operator, and so is the product of two invariant operators. Similarly, the Hermitian conjugate of an invariant operator is also an invariant operator.

for the different occurrences of the same irreducible representation, and $\hat{h}^s_{\mu j}$ is a fixed electronic operator which is independent of the nuclear coordinates. Using Eq. (2.61), one then gets

$$\sum_{j,k,\mu,s} \hat{h}^s_{\mu j} \left[D^{\mu}_{jk}(A) Q^s_{\mu k} \right] \hat{A} = \hat{A} \sum_{k,\mu,s} \hat{h}^s_{\mu k} Q^s_{\mu k} \tag{2.63}$$

which must hold independently of the values of the coordinates, and so be valid for each term $Q^s_{\mu k}$. One has

$$\sum_j \hat{h}^s_{\mu j} \hat{A} D^{\mu}_{jk}(A) = \hat{A} \hat{h}^s_{\mu k} \tag{2.64}$$

or, equivalently,

$$\hat{A} \hat{h}^s_{\mu k} \hat{A}^{-1} = \sum_j \hat{h}^s_{\mu j} D^{\mu}_{jk}(A) \tag{2.65}$$

Thus, $\hat{h}^s_{\mu k}$ transform under $\hat{A} \cdots \hat{A}^{-1}$ as the irreducible representation μ of G.

Let now $\hat{H}_e$ (invariant of the symmetry point group G_0 at the conical intersection) be expanded near the conical intersection as

$$\hat{H}_e(\mathbf{r}, \mathcal{Q}) = \hat{H}_e(\mathbf{r}, \mathcal{Q}_0) + \sum_{\Gamma\gamma} \hat{h}_{\Gamma\gamma} \mathcal{Q}_{\Gamma\gamma} + \frac{1}{2} \sum_{\Gamma_1\gamma_1} \sum_{\Gamma_2\gamma_2} \hat{g}_{\Gamma_1\gamma_1\Gamma_2\gamma_2} \mathcal{Q}_{\Gamma_1\gamma_1} \mathcal{Q}_{\Gamma_2\gamma_2} + \cdots \tag{2.66}$$

where $\mathcal{Q}_{\Gamma\gamma}$ are symmetry adapted displacements, $\hat{h}_{\Gamma\gamma} = \frac{\partial \hat{H}_e(\mathbf{r},\ _0)}{\partial \mathcal{Q}_{\Gamma\gamma}}$, and $\hat{g}_{\Gamma_1\gamma_1\Gamma_2\gamma_2} = \frac{\partial^2 \hat{H}_e(\mathbf{r},\ _0)}{\partial \mathcal{Q}_{\Gamma_1\gamma_1} \partial \mathcal{Q}_{\Gamma_2\gamma_2}}$. Recall now that the irreducible tensors that belong to Γ in the direct product $\Gamma_1 \times \Gamma_2$ assume the form [81]

$$\{\hat{g}\,(\Gamma_1 \times \Gamma_2)\}_{\Gamma\gamma} = \sum_{\gamma_1\gamma_2} \hat{g}_{\Gamma_1\gamma_1\Gamma_2\gamma_2} \langle \Gamma_1\gamma_1\Gamma_2\gamma_2 | \Gamma\gamma \rangle \tag{2.67}$$

$$\{\mathcal{Q}_{\Gamma_1} \times \mathcal{Q}_{\Gamma_2}\}_{\Gamma\gamma} = \sum_{\gamma_1\gamma_2} \mathcal{Q}_{\Gamma_1\gamma_1} \mathcal{Q}_{\Gamma_2\gamma_2} \langle \Gamma_1\gamma_1\Gamma_2\gamma_2 | \Gamma\gamma \rangle \tag{2.68}$$

where $\langle \Gamma_1\gamma_1\Gamma_2\gamma_2 | \Gamma\gamma \rangle$ are Clebsh-Gordan coefficients of the point group G_0. Thus, the quadratic terms in Eq. (2.66) may be written as a convolution of irreducible tensors, yielding [5]

$$\hat{H}_e(\mathbf{r}; \mathcal{Q}) = \hat{H}_e(\mathbf{r}; \mathcal{Q}_0) + \sum_{\Gamma\gamma} \hat{h}_{\Gamma\gamma} \mathcal{Q}_{\Gamma\gamma} + \frac{1}{2} \sum_{\Gamma_1\Gamma_2} \sum_{\Gamma\gamma} \{\hat{g}\,(\Gamma_1 \times \Gamma_2)\}_{\Gamma\gamma} \{\mathcal{Q}_{\Gamma_1} \times \mathcal{Q}_{\Gamma_2}\}_{\Gamma\gamma} \tag{2.69}$$

In the diabatic electronic (BO) basis of states $|\bar{\Gamma}_n\bar{\gamma}_n\rangle$, Eq. (2.69) assumes therefore the matrix form

$$\mathbf{H}_e(\mathcal{Q}) = \mathbf{H}_0 + \sum_{\Gamma\gamma} \mathbf{h}_{\Gamma\gamma}\mathcal{Q}_{\Gamma\gamma} + \frac{1}{2}\sum_{\Gamma_1\Gamma_2}\sum_{\Gamma\gamma} \mathbf{g}(\Gamma_1 \times \Gamma_2)_{\Gamma\gamma}\{\mathcal{Q}_{\Gamma_1} \times \mathcal{Q}_{\Gamma_2}\}_{\Gamma\gamma} \tag{2.70}$$

where $\mathbf{H}_0$ is diagonal, and $\mathbf{h}$ and $\mathbf{g}$ have matrix elements

$$[\mathbf{h}_{\Gamma\gamma}]_{mn} = \langle\bar{\Gamma}_m\bar{\gamma}_m|\hat{h}_{\Gamma\gamma}|\bar{\Gamma}_n\bar{\gamma}_n\rangle \tag{2.71}$$

$$\left[\mathbf{g}\left(\Gamma_1 \times \Gamma_2\right)_{\Gamma\gamma}\right]_{mn} = \langle\bar{\Gamma}_m\bar{\gamma}_m|\hat{g}\left(\Gamma_1 \times \Gamma_2\right)_{\Gamma\gamma}|\bar{\Gamma}_n\bar{\gamma}_n\rangle \tag{2.72}$$

Note that the matrix elements $\langle\bar{\Gamma}_m\bar{\gamma}_m|\hat{h}_{\bar{\Gamma}\bar{\gamma}}|\bar{\Gamma}_n\bar{\gamma}_n\rangle$ are nonzero if, and only if, $\hat{\Gamma} \subset \hat{\Gamma}_m \times \hat{\Gamma}_n$, with the same being true for the matrix elements of $\hat{g}(\Gamma_1 \times \Gamma_2)_{\Gamma\gamma}$. If all the N degenerate states belong to Γ, the Wigner-Eckart theorem [81] allows the potential matrix to be written as

$$\begin{aligned}\mathbf{H}_e(\mathcal{Q}) = \mathbf{H}_0 &+ \sum_{\Gamma\gamma} h_\Gamma \mathcal{Q}_{\Gamma\gamma}\mathbf{C}_{\Gamma\gamma} \\ &+ \frac{1}{2}\sum_{\Gamma_1\Gamma_2}\sum_{\Gamma\gamma} g\left(\Gamma_1 \times \Gamma_2\right)_\Gamma \{\mathcal{Q}_{\Gamma_1} \times \mathcal{Q}_{\Gamma_2}\}_{\Gamma\gamma}\mathbf{C}_{\Gamma\gamma}\end{aligned} \tag{2.73}$$

where h_Γ and $g_\Gamma\left(\Gamma_1 \times \Gamma_2\right)$ are reduced matrix elements of the operators $\hat{h}_{\Gamma\gamma}$ and $\{\hat{g}\left(\Gamma_1 \times \Gamma_2\right)\}_{\Gamma\gamma}$, and $\mathbf{C}_{\Gamma\gamma}$ are matrices composed of appropriate Clebsh-Gordan coefficients in the space of BO electronic states $|\Gamma\gamma\rangle$. Clearly, $\mathbf{H}_e$ reduces to $\mathbf{H}_0$ if $h_\Gamma = g\left(\Gamma_1 \times \Gamma_2\right)_\Gamma = 0$ for $\Gamma \neq A_1$: the motion will be reduced to harmonic oscillations. Thus, the vibronic effects are connected with the last two terms: linear and quadratic vibronic interactions. In the simplest case of the linear JT vibronic Hamiltonian (quadratic and higher-order terms of the vibronic interaction are then ignored), it assumes the explicit matrix form

$$\mathbf{H}_e(\mathcal{Q}) = \frac{1}{2}\sum_{\Gamma\gamma}\left(P_{\Gamma\gamma}^2 + \omega_\Gamma^2 Q_{\Gamma\gamma}^2\right)\mathbf{C}_{A_1} + \sum_{\Gamma\gamma} h_\Gamma Q_{\Gamma\gamma}\mathbf{C}_{\Gamma\gamma} \tag{2.74}$$

with the irreducible representations of JT active modes satisfying $\Gamma \subset [\Gamma_i^2]$ if Γ_i is a normal representation or $\Gamma \subset \{\Gamma_i^2\}$ if Γ_i is a double representation.

H_3: a well studied Jahn-Teller system. Although H_3 (in general, X_3) belongs to the D_{3h} symmetry point group, we may use the C_{3v} subgroup of D_{3h} since no out-of-plane bending is possible for a triatomic system. Following section 3.1, the electronic Hamiltonian $\hat{H}_e$ may then be expanded in terms of the set of coordinates (Q_{E_x}, Q_{E_y}) which belong to the irreducible representations (E_x, E_y) of the C_{3v} symmetry point group since they define the relevant subspace of configuration space (see section 3.2) for the title system. From Eq. (2.62), one gets

$$\hat{H}_e = \left[\hat{h}_{E_x}, \hat{h}_{E_y}\right] \begin{bmatrix} Q_{E_x} \\ Q_{E_y} \end{bmatrix} \tag{2.75}$$

where $\hat{h}_{E_x}$ and $\hat{h}_{E_y}$ are electronic operators that are independent of the nuclear coordinates. Define next complex coordinates and electronic states according to [82]

$$Q_+ = Q_{E_x} + iQ_{E_y} = \rho \exp(i\phi) \tag{2.76}$$
$$Q_- = Q_{E_x} - iQ_{E_y} = \rho \exp(-i\phi) \tag{2.77}$$

and

$$\Psi_+ = \Psi_1 + i\Psi_2 \tag{2.78}$$
$$\Psi_- = \Psi_1 - i\Psi_2 \tag{2.79}$$

Thus, the potential matrix assumes the form

$$\mathbf{W} = \left[\langle E|\hat{h}_{E_x}|E\rangle, \langle E|\hat{h}_{E_y}|E\rangle\right] \begin{bmatrix} Q_{E_x} \\ Q_{E_y} \end{bmatrix} \tag{2.80}$$

Using now the direct product decompositions for $\mathbf{C}_{3v}$, it is known (see Table 57 of Ref. [83], and Appendix II of Ref. [19]) that

$$u_{A_1}v_{A_1} \propto A_1; \quad u_{A_1}v_{A_2} \propto A_2; \quad u_{A_1}v_x \propto x; \quad u_{A_1}v_y \propto y; \tag{2.81}$$
$$u_{A_2}v_{A_2} \propto A_1; \quad u_{A_2}v_x \propto y; \quad u_{A_2}v_y \propto -x; \tag{2.82}$$
$$u_xv_x + u_yv_y \propto A_1; \quad u_xv_y - u_yv_x \propto A_2; \tag{2.83}$$
$$-u_xv_x + u_yv_y \propto x; \quad u_xv_y + u_yv_x \propto y \tag{2.84}$$

where the symbol $\propto$ means "transforms under $\mathbf{C}_{3v}$ as", and v_x and v_y are arbitrary functions with the transformation properties of the corresponding subscripts. Thus,

$$\langle x|\hat{h}_{E_x}|y\rangle = \langle x|u_xv_y\rangle = 0 \tag{2.85}$$
$$\langle x|\hat{h}_{E_x}|x\rangle = \langle x|u_xv_x\rangle = -1 \tag{2.86}$$

leading to

$$\langle E|\hat{h}_{E_x}|E\rangle = -\omega_{E_x}\boldsymbol{\sigma}_x \tag{2.87}$$

$$\langle E|\hat{h}_{E_y}|E\rangle = \omega_{E_y}\boldsymbol{\sigma}_y \tag{2.88}$$

where ω_{E_x} and ω_{E_y} are constants, and $\boldsymbol{\sigma}_x$ and $\boldsymbol{\sigma}_y$ are the Pauli spin matrices[12]. Defining now the scaled coordinates $X_1 = -\omega_{E_x}Q_{E_x}$ and $X_2 = \omega_{E_y}Q_{E_y}$, one has

$$\mathbf{W} = (\boldsymbol{\sigma}_x, \boldsymbol{\sigma}_y)\begin{pmatrix} X_1 \\ X_2 \end{pmatrix} \tag{2.89}$$

which can also be written in matrix form as

$$\mathbf{W} = \boldsymbol{\sigma}\cdot\mathbf{X} \tag{2.90}$$

where $\boldsymbol{\sigma} = (\boldsymbol{\sigma}_x, \boldsymbol{\sigma}_y)$ and $\mathbf{X} = (X_1, X_2)$.

Diagonalization of $\mathbf{W}$ then leads to the electronically adiabatic potentials

$$V_{\pm} = \pm\sqrt{X_1^2 + X_2^2} \tag{2.91}$$

Noting now that

$$\omega_{E_x} = \langle\Psi_+|\hat{h}_{E_x}|\Psi_-\rangle \tag{2.92}$$

$$\omega_{E_y} = \langle\Psi_-|\hat{h}_{E_y}|\Psi_+\rangle \tag{2.93}$$

one obtains $\omega_E = \omega_{E_x} = \omega_{E_y}$, and from Eq. (2.91):

$$V_{\pm} = \pm\omega_E\rho \tag{2.94}$$

Clearly, this double-cone shape (diabolo) is perfectly visible from Figure 2.1 if we focus close to the conical intersection. This result illustrates also the JT theorem by showing that any existing stationary points must lie away from the line of D_{3h} symmetry where the conical intersection occurs.

3.2 Configuration space versus electronic function space

The configuration space forms the natural coordinate frame on which to represent the relevant adiabatic potential energy surface. However, for the purposes of studying the JT effect, we do not need the full $3N-6$ configuration space but only the space defined by the subset of configurational coordinates that are JT active as such a subspace contains all distorted configurations that may be reached by the JT reactive coordinates. Thus, whenever free of ambiguity, we imply heretofore by

[12]For the explicit form of the Pauli matrices, see Eq. (2.103).

configuration space the subspace defined by the JT active coordinates. In the case of a normal representation, they transform as the irreducible coordinate representations[13] Λ in the point group G of the JT origin that belong to the nontotally symmetric part of the symmetrized square of Γ. We may formally write

$$\Lambda = [\Gamma^2] - A_1 \tag{2.95}$$

If we now assume that the system is constrained to a minimum number of JT active vibrational coordinates (*i.e.*, the full vibrational representation contains only one set of normal coordinates for each JT active symmetry), the complete set of such active vibrations subtends a $([\Gamma^2] - 1)$-dimensional real Cartesian space which is known as the configuration space v[14]. For example, since in C_{3v} the symmetric product $[E_1 \times E_1] = A_1 + E_1$, the JT active vibrations are of the e-type, thus spaning a $2D$ configuration space which is a subset of the full $(3D)$ configuration space. This is the case for the trihydrogen prototype studied in section 2.3, where the JT active normal coordinates are Q_{E_x} and Q_{E_y}; see Figure 2.6. In turn, the symmetric product $[T_1 \times T_1] = [T_2 \times T_2] = A_1 + E + T_2$, and hence the JT active vibrations for a three-fold degenerate T state (T_1 or T_2) are of e or t_2 symmetry. Thus, for a tetrahedral molecule such as ML_4, there is one e mode which is of purely bending type and two t_2 modes, one of which is primarily bending while the other is primarily stretching. Since the interference of the two t_2 modes is essentially negligible, and direct experimental observation of bond length deformations is lacking, the analysis of the JT effect can be restricted to the $5D$ space of bending deformations [84] without having to consider the full $9D$ configurational space of the five-atom molecule. Such normal coordinates will be denoted[15] Q_{E_x}, Q_{E_y} (for the e representation) and Q_{T_x}, Q_{T_y}, Q_{T_z} (for the t_2 representation); the coordinate system and two characteristic vibrational coordinates are shown in Figure 2.7.

The possible subgroups of configuration points contained in the configurational space will therefore belong to a subgroup of G, which may be found by using group theoretical methods [84, 85]. Suffice it to say here that the lowest subgroup attainable is denoted the kernel of v in G, and symbolized $K(G, v)$. This contains all symmetry operations that

[13] Using a notation similar to $|\Gamma, \gamma\rangle$ (for an irreducible representation that transforms as the γth component of Γ, the JT active coordinates will receive for clarity symmetry labels $\Lambda\lambda$ such as $Q_{\Lambda\lambda}$.

[14] For the so-called multimode JT effect, see Refs. [5, 103].

[15] Frequently they are labelled Q_θ, Q_ϵ (for the e representation) and Q_ξ, Q_η, Q_ζ (for the t_2 representation).

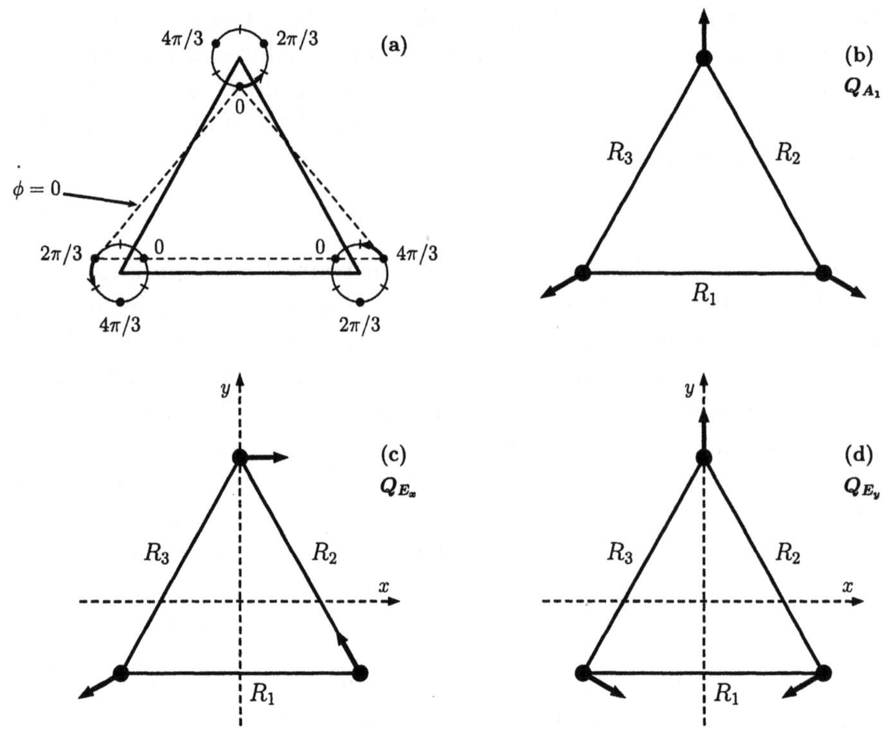

Figure 2.6. Normal mode coordinates for a X_3-type molecule. The arrows in panels (b) to (d) indicate the atomic displacements which are described by a form similar to Eq. (2.3), when the R_i are replaced by displacements with respect to a reference equilateral triangle geometry. Q_{E_x} and Q_{E_y} are the Jahn-Teller active normal coordinates. Also shown in panel (a) are the distortions of the lowest sheet of the adiabatic potential in the linear $E \otimes e$ problem.

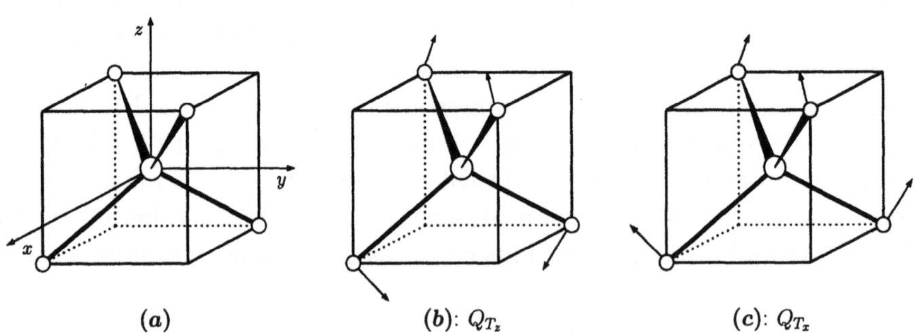

Figure 2.7. Coordinate system and two characteristic vibrational coordinates for a ML_4 system: Q_{T_z} and Q_{T_x}.

leave the entire set of JT coordinates unchanged, *i.e.*, the elements of the group G in the Λ representation that correspond to unit matrices. Maximal subgroups of G which leave only a subset of the JT coordinates unchanged may also be defined, being called epikernels of v and symbolized $E(G,v)$. Clearly, the kernels mark the borderline of the symmetry lowering JT process, while the epikernels describe the allowed intermediate stages. For example, let us limit ourselves to the $2D$ space of the e-vibrations (Q_{E_x} and Q_{E_y}) spaning the e representation of T_d. Since the character equals 2 for the identity operation and for the class of the three C_2 operations, the kernel will be $K(T_d,e)=D_2$. In fact, the e vibrations can destroy all T_d symmetry elements, except those belonging to D_2. Thus, D_2 is the lowest symmetry attainable by e vibrations only, since in D_2, both Q_{E_x} and Q_{E_y} (or any linear combinations thereof) become totally symmetric being unable to reduce further the symmetry. Stated differently, an arbitrary point in the $2D$ (Q_{E_x}, Q_{E_y})-space is characterized by the kernel symmetry D_2. In turn, the intermediate group D_{2d} is an epikernel $E(T_d,e)$ since the totally symmetric representation occurs only once in the $T_d \rightarrow D_{2d}$ descent in symmetry:

$$\begin{array}{ccccc} T_d & \longrightarrow & D_{2d} & \longrightarrow & D_2 \\ (e) & & (a_1+b_1) & & (a+a) \end{array} \tag{2.96}$$

Note that only one of the two e-components is invariant under the D_{2d} operations. If that particular component is activated, the symmetry is lowered to D_{2d} and not further down to D_2. The kernels and epikernels can generally be obtained upon inspection of a descent-in-symmetry sequence of the parent point group. This is illustrated in Figure 4 of Ref. [84], which shows the genealogy of the T_d point group, including the subductional chains for the e and t_2 vibrations [86]. The resulting kernels and epikernels are summarized in Table 2.2 for some illustrative point groups. Also given is the homomorphic image [87] of G in v, $H(g,v)$, which represents the symmetry group of the configuration space v corresponding to the quocient group $G/K(G,v)$. Thus, it gives the number of equivalent configurations within a given subgroup of the parent symmetry group.

We now turn to the electronic function space defined by the set of orthonormal electronic BO states Ψ_I^d or $\left\{|\Gamma\gamma_1\rangle, |\Gamma\gamma_2\rangle, \cdots, |\Gamma\gamma_{|\Gamma|}\rangle\right\}$, which will be used later to discuss the GP effect in JT systems. Such a space has a striking similarity with that suggested by Ceulemans [85], although he used it for analising the shape of the adiabatic potential near the instability point. We may then think of such a set of electronic wave vectors as unit vectors in terms of which an arbitrary adiabatic electronic state

Table 2.2. Configurational space, kernel, epikernels, and homomorphic image for some prototype JT systems [85].

		Configuration space				Function space			
G	Γ	v	$K(G,\Gamma)$	$E(G,\Gamma)$	$H(G,\Gamma)$	Γ	$K(G,f)$	$H(G,f)$	$H'(G,f)$
C_{3v}	E	e	C_1	C_s	C_{3v}	E	C_1	C_{3v}	C_{6v}
D_{3h}	E'	e'	$C_s(\sigma_h)$	C_{2v}	C_{3v}	E'	C_s	C_{3v}	C_{6v}
	E''	e'	$C_s(\sigma_h)$	C_{2v}	C_{3v}	E''	C_1	C_{6v}	C_{6v}
T_d	E	e	D_2	D_{2d}	C_{3v}	E	D_2	C_{3v}	C_{6v}
	T_1, T_2	$e + t_2$	C_1	$D_{2d}D_2C_{3v}C_{2v}C_2C_s$	O	$T1, T_2$	C_1	O	O_h

can be expressed

$$|\Gamma\gamma\rangle = \sum_{i=1,|\Gamma|} c_i|\Gamma\gamma_i\rangle \tag{2.97}$$

with the normalization condition requiring

$$\sum_i c_i^\star c_i = 1 \tag{2.98}$$

The projective function space f so defined can be taken to be real, with allowed transformations involving rotations or reflections of the points on the unit sphere. The full symmetry group will therefore be the full orthogonal group O of dimension $|\Gamma|$, with its elements being represented by orthogonal matrices describing the transformation of the directional cosines c_i. In turn, symmetry elements of the finite group G will induce orthogonal matrix transformations: the Γ matrix representation of G may be therefore a subgroup of $O(|\Gamma|)$ [85]. The effective subgroup of the electronic space, once eliminated any redundant symmetry elements of G that are represented by the same transformation matrix, defines the homomorphic image of G in f, $H(G, f)$. A comparison of homomorphisms in this function projective space f and the configuration space v has been presented in Ref. [85], being some given in Table 2.2. It is important to note that antipodal pairs of eigenfunctions, $|\Gamma\gamma\rangle$ and $-|\Gamma\gamma\rangle$, will refer to the same points in configuration space as a physical state may be described by any wave vector in a ray [88]. Thus, in constructing a cross section of the JT potential energy surface, one should realize that the mapping between the full space and its effective image in function space is two-valued, since the potential is an even function of the electronic kets. The true point group of f space is therefore the extended homomorphic image of G in f, including function space inversion, which is commonly denoted as $H'(G, f)$.

4. Lie symmetry and the geometric phase effect in Jahn-Teller systems

After some general remarks on Lie symmetry groups, a discussion on the Lie symmetry of the vibronic Hamiltonian and its use to elucidate the JT effect are presented following closely some recent work by Varandas and Xu [42]. Naturally, a more detailed presentation of the theory will be given for the simplest JT systems with two-fold and three-fold electronic degeneracies since they are of key importance for understanding the case of an arbitrary N-fold degeneracy. Note that we imply by JT system the so-called JT center [5], *i.e.*, the group of atoms involved in the degenerate electronic states. Thus, even though the X_3 orbital doublet case [19] is a well known prototype for the $E\otimes e$ case, the conclusions drawn should be far more general.

4.1 General considerations on Lie group symmetry

Of interest in this work are the so-called real connected compact Lie groups. Because they are real, this implies that we restrict our attention to Lie groups whose elements are describable by a finite number of real parameters, each of which can have a continuous range of values. Because the range of each of the group parameters is assumed to be closed (and therefore finite), they receive the designation of compact. In turn, they are connected because it is assumed that all of their elements can be reached by proceeding continuously from the identitity, so, by letting the parameters vary continuously from zero. Moreover, they often contain no invariant Abelian subgroups, and hence their designation as simple. An example of such a real connected simple compact Lie group is the rotation group in three dimensions $O(3)$; the parameters are in this case the Euler angles, which are real and have ranges $[0, 2\pi]$ and $[0, \pi]$. An example of a nonsimple Lie group is the group $U(2)$ of all unitary transformations in two dimensions, which contains the group of all matrices of the form $e^{i\zeta}\mathbf{I}$, where $\mathbf{I}$ is the unit 2×2 matrix as an Abelian invariant subgroup.

Any connected compact simple Lie group G has a corresponding simply connected compact Lie group $\tilde{G}$, the so-called covering group, in the sense that $\tilde{G} = G\times Z$ or $G = \tilde{G}/Z$, where Z is an Abelian group of finite order. For each $\tilde{G}$ there are only a finite number of G's. For example, the group $SU(2)$ of all unimodular 2×2 matrices, is the $\tilde{G}$ for itself and for the $3D$ rotational group $O(3)$ [which is isomorphic with $SU(2)/Z_2$, where Z_2 is the two-element group consisting of 1 and -1 times the unit matrix]. Thus the connected simple compact Lie groups fall into sets, each set being characterized by the simply connected member $\tilde{G}$. Such sets can in turn be classified into four main classes. These correspond (modulo Z) to the four sets of classical groups [89,90]: $SU(l+1)$, the group of all unitary unimodular $(l+1)\times(l+1)$ matrices of order $r = (l+1)^2-1$; $O(2l+1)$, the group of all real orthogonal $(2l+1)\times(2l+1)$ matrices of order $r = l(2l+1)$; $USp(2l)$, te group of all unitary simpletic $(2l)\times(2l)$ matrices of order $r = l(2l+1)$; $O(2l)$, the group of all real orthogonal $(2l)\times(2l)$ matrices of order $r = l(2l-1)$. Note that r specifies the number of Lie group parameters, and may also be defined as the number of simultaneously commuting elements in the associated Lie algebra. Note further that for all groups except $O(2l)$, $l = 1, 2, \ldots$. For

$O(2l)$, $l = 3, 4, \ldots$ as $O(2)$ is Abelian and $O(4) = O(3) \times O(3)$. The reader is referred to the specialized literature for further details [81, 90]. In particular, those interested in seeing how the Lie groups may be relevant to the physics of Jahn-Teller systems are addressed to the paper by Judd [91].

4.2 The $N = 2$ case: $E \otimes e$

We begin with some general remarks on electronic degenerate states. Owing to the symmetry of the nuclear configuration (belonging to the point group G), such a degeneracy implies the existence [5] of at least one rotation C_n or roto-reflection S_n symmetry axis of order $n > 2$. When rotated around such an axis by an angle ϕ, the electronic states transform as $\exp(\pm ik\phi)$ with $k = 1, 2, \cdots, n$. Due to time reversal symmetry, the same representation applies to wave functions related by complex conjugation. Thus, the group G containing a n-fold rotational axis may have different double irreducible representations E_k ($k = 1, 2, \cdots, \frac{n-2}{2}$ for n even; $k = 1, 2, \cdots, \frac{n-1}{2}$ for n odd), in addition to the totally symmetric A ($k = 0$) and low-symmetry B ($k = n/2$ for n even) representations. Moreover, the corresponding basis functions $\Psi_{\pm k}$ transform as $\exp(\pm ik\phi)$, with the basis of $[E_k^2]$ being provided by three functions: $\Psi_k^\dagger \Psi_k$, $\Psi_k^\dagger \Psi_{-k}$ and $\Psi_{-k}^\dagger \Psi_k$. The last two possess the transformation properties of $\exp(\pm i2k\phi)$ and belong to E_{2k}, except when n is a multiple of 4 ($k = \frac{n}{4}$); in this case, the two functions belong to $B_1 + B_2$. For electronic doublets, there are then two possible situations. The first occurs when the electronic E_k ($k \neq \frac{n}{2}$) degeneracy is removed by nuclear displacements that transform according to the E_{2k} representation: $E \otimes e$ case. The second is possible only in polyatomic systems with C_n or S_n axes of order $n = 4k$, with the nuclear displacements of B_1 or B_2 symmetry being then active in the linear vibronic interaction: $E \otimes (b_1 + b_2)$ case. We focus here on the more widespread $E \otimes e$ case.

The Jahn-Teller Hamiltonian of the linear $E \otimes e$ problem, written in the basis of electronic states[16] $\{|E_x(\mathbf{r})\rangle, |E_y(\mathbf{r})\rangle\}$ assumes the form [5]

$$\mathbf{H}_T(\mathbf{Q}) = \mathbf{H}_e + \mathbf{H}_v + \mathbf{H}_{ev} \tag{2.99}$$

$$\mathbf{H}_e = E_0 \boldsymbol{\sigma}_0 \tag{2.100}$$

$$\mathbf{H}_v = \frac{1}{2}\left[P_{E_x}^2 + P_{E_y}^2 + \omega_E^2\left(Q_{E_x}^2 + Q_{E_y}^2\right)\right]\boldsymbol{\sigma}_0 \tag{2.101}$$

$$\mathbf{H}_{ev} = V_E(Q_{E_x}\boldsymbol{\sigma}_x + Q_{E_y}\boldsymbol{\sigma}_y) \tag{2.102}$$

[16] We prefer this more explicit notation to Ψ_I^d ($I = 1, 2$) used in presenting the General principles.

where $P_{\Gamma\gamma}$ is the conjugate momentum associated to $Q_{\Gamma\gamma}$, $\mathbf{Q}_0$ a reference nuclear configuration that will be taken as the point of electronic degeneracy, and $|E_x(\mathbf{r})\rangle = |E_x(\mathbf{r},\mathbf{Q}_0)\rangle$ an eigenvector of the Hamiltonian $\mathbf{H}_e(\mathbf{r},\mathbf{Q}_0)$ (a similar definition holds for $|E_x(\mathbf{r})\rangle$). We further observe that E_0 is the energy of the orbital doublet, ω_E the vibrational frequency of the E mode, V_E the so-called linear vibronic interaction constant, and $\boldsymbol{\sigma}_i$ $(i = 0, x, y, z)$ the Pauli matrices:

$$\boldsymbol{\sigma}_0 = \begin{bmatrix} 1 & 0 \\ 0 & 1 \end{bmatrix}, \quad \boldsymbol{\sigma}_x = \begin{bmatrix} 0 & 1 \\ 1 & 0 \end{bmatrix}, \quad \boldsymbol{\sigma}_y = \begin{bmatrix} 0 & -i \\ i & 0 \end{bmatrix}, \quad \boldsymbol{\sigma}_z = \begin{bmatrix} 1 & 0 \\ 0 & -1 \end{bmatrix} \tag{2.103}$$

The symmetry group of the electronic Hamiltonian $\mathbf{H}_e$ is the second-order unitary group $U(2)$. The operators $\hat{I}_{ij} = |\psi_i\rangle\langle\psi_j|$, where $i, j = 1, 2, \cdots, n$, form a manifold of n^2 elements closed with respect to the commutation operation

$$[\hat{I}_{ij}, \hat{I}_{kl}] = \hat{I}_{il}\delta_{kj} - \hat{I}_{kj}\delta_{il} \tag{2.104}$$

and hence an algebra. Note that they all commute with the unit matrix representing the electronic Hamiltonian of the n-fold degenerate electronic term, and hence can be chosen as the generators of the nth-order unitary group. Note also that the eigenfunctions of $\mathbf{H}_e$ may be written as the spinors

$$\Psi_+ = \begin{pmatrix} 1 \\ 0 \end{pmatrix}, \quad \Psi_- = \begin{pmatrix} 0 \\ 1 \end{pmatrix} \tag{2.105}$$

which can formally be regarded as two different states of some particle differing in the value $(\pm 1/2)$ of the projection of the energy spin along the $z-$axis.

Since the electronic Hamiltonian may be written as

$$\mathbf{H}_e = \frac{1}{3} E_0 \left(\boldsymbol{\sigma}_x^2 + \boldsymbol{\sigma}_y^2 + \boldsymbol{\sigma}_z^2 \right) \tag{2.106}$$

the unitary group $U(2)$ must have the rotation group $O(3)$ in $3D$ spin-space as a subgroup. With respect to these rotations the operators $\hat{I}_{ij}$ are components of a second-rank tensor which can be arranged in linear combinations transforming according to the irreducible representations $D^{(1/2)} \times D^{(1/2)} = D^{(0)} + D^{(1)}$ of the group $O(3)$. The components of the corresponding irreducible tensors are [5]

$$\left\{ \psi^{(1/2)} \times \psi^{(1/2)} \right\}_0^{(0)} = \sqrt{2}\boldsymbol{\sigma}_0, \quad \left\{ \psi^{(1/2)} \times \psi^{(1/2)} \right\}^{(1)} = \sqrt{2}\boldsymbol{\sigma} \tag{2.107}$$

where we have used the notation $\{\psi^{(l)} \times \psi^{(l)}\}_M^{(L)}$, with $l = \frac{1}{2}$, $L = 0, 1, \cdots, 2l$, and $M = -L, -L+1, \cdots, L-1, L$. Thus, an arbitrary operator of the group $U(2)$ may be written as

$$\mathbf{G}(\xi_0, \xi_x, \xi_y, \xi_z) = \exp\left[i(\xi_0 \boldsymbol{\sigma}_0 + \xi_x \boldsymbol{\sigma}_x + \xi_y \boldsymbol{\sigma}_y + \xi_z \boldsymbol{\sigma}_z)\right] \tag{2.108}$$

where $\{\xi_i\}$ are the parameters of the group. Excluding the trivial operation associated to ξ_0, one gets the three-parameter subgroup $SU(2)$ of the group $O(3)$, which consists of the unitary matrices

$$\mathbf{G}(\xi_x, \xi_y, \xi_z) = \exp\left[i(\xi_x \boldsymbol{\sigma}_x + \xi_y \boldsymbol{\sigma}_y + \xi_z \boldsymbol{\sigma}_z)\right] = \exp(i\boldsymbol{\xi}\boldsymbol{\sigma}) \tag{2.109}$$

where $\boldsymbol{\xi} = \{\xi_x, \xi_y, \xi_z\}$. The symmetry elements of the electronic Hamiltonian are summarized on the left hand side of Figure 2.8.

In turn, the vibrational Hamiltonian can be rewritten as

$$\mathbf{H}_v = \hbar\omega_E \left(b^\dagger_{E_x} b_{E_x} + b^\dagger_{E_y} b_{E_y} + 1\right) \boldsymbol{\sigma}_0 \tag{2.110}$$

where the creation and annihilation operators

$$b^\dagger_{\Gamma\gamma} = -\frac{i}{\sqrt{2\hbar\omega_\Gamma}}(P_{\Gamma\gamma} + i\omega_\Gamma Q_{\Gamma\gamma}), \quad b_{\Gamma\gamma} = \frac{i}{\sqrt{2\hbar\omega_\Gamma}}(P_{\Gamma\gamma} - i\omega_\Gamma Q_{\Gamma\gamma}) \tag{2.111}$$

obey the commutation relations

$$\left[b_{\Gamma\gamma}, b^\dagger_{\Gamma'\gamma'}\right] = \delta_{\Gamma\Gamma'}\delta_{\gamma\gamma'} \tag{2.112}$$

Such an Hamiltonian is invariant under the group $U(2)$; for a nD oscillator, the appropriate symmetry group will be $U(n)$. The operators $\hat{J}_{ij} = b^\dagger_i b_j$ $(i, j = 1, 2, \cdots, n)$ form therefore a manifold of n^2 elements closed with respect to the commutation operation

$$[\hat{J}_{ij}, \hat{J}_{kl}] = \hat{J}_{il}\delta_{kj} - \hat{J}_{kj}\delta_{il} \tag{2.113}$$

and hence an algebra. Note that they commute with the Hamiltonian of a nD isotropic oscillator, and hence can be chosen as generators of the nth-order unitary group. Similarly, under the rotations of the $O(3)$ group, the operators b_i transform as the components of a nD vector, and the operators $b^\dagger_i b_j$ as second-rank tensors. They can be arranged in linear combinations transforming according to the irreducible representations $D^{(1/2)} \times D^{(1/2)} = D^{(0)} + D^{(1)}$ of the $O(3)$ group. The components of the corresponding irreducible tensors are

$$\left\{b^{(1/2)} \times b^{(1/2)}\right\}^{(0)}_0 = \sqrt{2}(J_{E_xE_x} + J_{E_yE_y}) \tag{2.114}$$

$$\left\{b^{(1/2)} \times b^{(1/2)}\right\}^{(1)}_0 = -i\sqrt{2}(J_{E_xE_y} - J_{E_yE_x}) \tag{2.115}$$

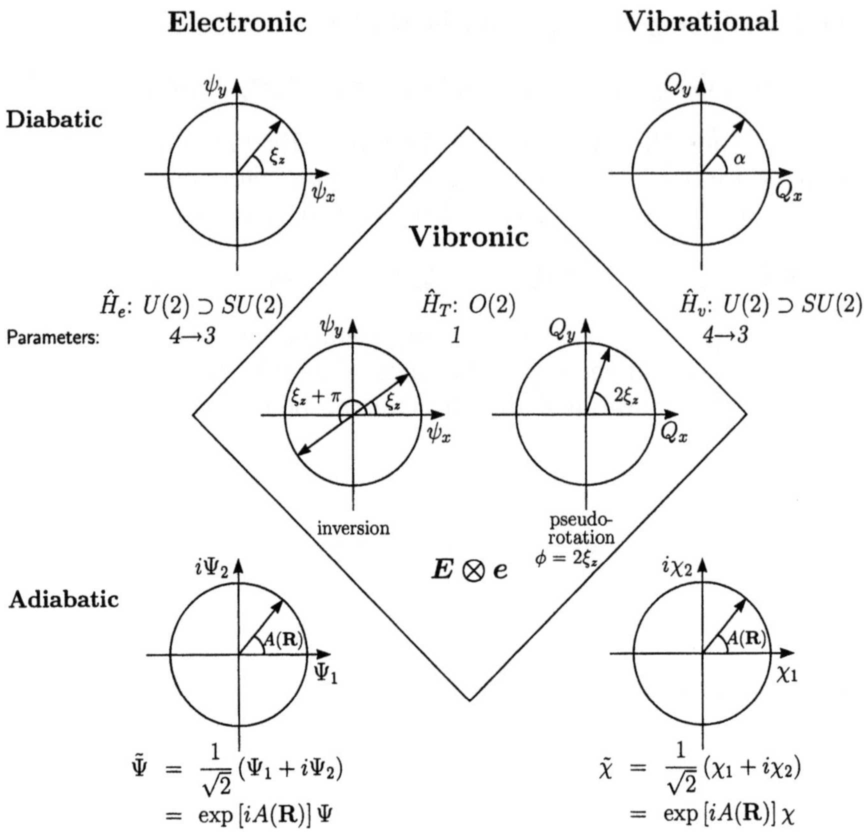

Figure 2.8. Highly schematic illustration of the symmetry elements in the electronic, vibrational, and linear vibronic Hamiltonians for the $E \otimes e$ problem. For simplicity, the notation $\psi_x = |E_x(\mathbf{r})\rangle = |E_x(\mathbf{r}, \mathbf{Q}_0)\rangle$ and $Q_x = Q_{E_x}$ (correspondingly for the y component) is used in this Figure.

$$\left\{b^{(1/2)} \times b^{(1/2)}\right\}^{(1)}_{\pm} = \sqrt{2}\left[(J_{E_xE_x} - J_{E_yE_y}) \pm i(J_{E_xE_y} + J_{E_yE_x})\right] \tag{2.116}$$

For a $2D$ oscillator, it is convenient to take the following operators as generators of the group

$$\hat{L}_0 = b^\dagger_{E_x} b_{E_x} + b^\dagger_{E_y} b_{E_y}, \quad \hat{L}_x = b^\dagger_{E_x} b_{E_x} - b^\dagger_{E_y} b_{E_y}, \tag{2.117}$$

$$\hat{L}_y = b^\dagger_{E_x} b_{E_y} + b^\dagger_{E_y} b_{E_x}, \quad \hat{L}_z = -i(b^\dagger_{E_x} b_{E_y} - b^\dagger_{E_y} b_{E_x}) \tag{2.118}$$

with the operators $\hat{L}_x$, $\hat{L}_y$ and $\hat{L}_z$ having the same commutation relations as the Pauli matrices; thus, they too form an algebra. As a result, any arbitrary operator of the group $U(2)$ describing the symmetry of the vibrational Hamiltonian can be written as

$$\hat{G}(\alpha_x, \alpha_y, \alpha_z) = \exp\left[i(\alpha_x \hat{L}_x + \alpha_y \hat{L}_y + \alpha_z \hat{L}_z)\right] = \exp(i\boldsymbol{\alpha}\mathbf{L}) \qquad (2.119)$$

where $\boldsymbol{\alpha} = \{\alpha_x, \alpha_y, \alpha_z\}$. The symmetry elements of the vibrational Hamiltonian are summarized on the right hand side of Figure 2.8.

In the absence of a vibronic interaction, the symmetry of the $E \otimes e$ JT system is described by at least the group $SU(2) \times SU(2)$. Thus, an arbitrary element of this group will depend on six parameters $\boldsymbol{\xi}$ and $\boldsymbol{\alpha}$,

$$\hat{G} = \exp[i(\boldsymbol{\xi\sigma} + \boldsymbol{\alpha}\mathbf{L})] \qquad (2.120)$$

Taking the vibronic interaction into account lowers the symmetry of the problem. For example, by taking into account the linear vibronic interaction, the symmetry of the system is lowered to the one-parameter (α_z) axial group $O(2)$ with the element being

$$\hat{G} = \exp[i\alpha_z(L_z + \frac{1}{2}\sigma_z)] \qquad (2.121)$$

This is due to the requirement that $[\hat{G}, \hat{H}] = 0$, or

$$[(\boldsymbol{\xi\sigma} + \boldsymbol{\alpha}\mathbf{L}), (Q_{E_x}\boldsymbol{\sigma}_x + Q_{E_y}\boldsymbol{\sigma}_y)] = \mathbf{0} \qquad (2.122)$$

which leads to the following five restrictions on the parameters

$$\alpha_x = \alpha_y = 0 \qquad (2.123)$$

$$\xi_x = \xi_y = 0 \qquad (2.124)$$

$$\alpha_z = \xi_z \qquad (2.125)$$

and hence reduces the six parameters to just one. The infinitesimal operator

$$\hat{J}_z = \hat{L}_z + \frac{1}{2}\hat{\sigma}_z \qquad (2.126)$$

will then be an integral of motion, since it commutes with the Hamiltonian in Eq. (2.99). Such a result is illustrated in the central part of Figure 2.8.

In the absence of vibronic interactions the motions of the electronic subsystem and the nuclear one can then be treated separately. The electronic states $|E_x(\mathbf{r})\rangle$ and $|E_y(\mathbf{r})\rangle$ with a certain value of the projection of the energy spin $\frac{1}{2}\boldsymbol{\sigma}$ on the $z-$axis correspond (in the case of a

X_3 molecule) to the wave of electronic density propagating along the perimeter of the molecule clockwise or counterclockwise, respectively. In turn, the vibrational states with a certain value of L_z correspond to the wave of distortions of an equilateral triangle running clockwise or counterclockwise. Of course, if no vibronic interaction exists, the energy spin $\frac{1}{2}\boldsymbol{\sigma}$ and vibrational angular momentum L are conserved separately. This ceasses to be the case when the vibronic interaction is introduced. The projection of the vibronic angular momentum J onto the $z-$axis is now the good quantum number: the wave of electronic density and the wave of nuclear distortions propagate coherently. Figure 2.6(a) shows how the shape of the molecular triangle evolves during the motion along the trough of the potential energy surface. To see the motions performed by the atoms themselves, one sets all $Q_{\Gamma\gamma}$ equal to zero except Q_{E_x} and Q_{E_y} that are expressed using polar coordinates (ρ, ϕ). Recalling now that $Q_{\Gamma\gamma}$ can be expressed in terms of the Cartesian coordinates of the nuclei [5], one gets a set of relations from which the Cartesian atomic displacements follow as $x_i = \frac{\rho}{\sqrt{3}} \cos\left(\phi - \frac{m_i \pi}{3}\right)$ and $y_i = \frac{\rho}{\sqrt{3}} \sin\left(\phi - \frac{m_i \pi}{3}\right)$, with $m_i = 0, 4, 2$ for $i = 1, 2, 3$, respectively. Figure 2.6(a) shows that the atoms circumscribe circles in a coherent way, a motion usually referred to as pseudo-rotation. The changes occurring during such a motion can then be thought as a wave of distortions propagating around the geometric center and transforming the equilateral triangle into an isosceles one [92]. Analogous waves of distortion can be observed in the case of linear $E \otimes e$ coupling for tetrahedral and cubic systems [92].

By means of Eq. (2.117), Eq. (2.118) and Eq. (2.111), the operator $\hat{L}_z$ assumes in the coordinate representation the form

$$L_z = \frac{1}{\hbar}\left(Q_{E_x} P_{E_y} - Q_{E_y} P_{E_x}\right) \tag{2.127}$$

or, if one adopts the polar coordinates $Q_x = \rho \cos\phi$, $Q_y = \rho \sin\phi$:

$$\hat{L}_z = -i\frac{\partial}{\partial \phi} \tag{2.128}$$

To diagonalize the Hamiltonian $\mathbf{H}_T$, we may use the unitary transformation

$$\mathbf{S}^\dagger \left(Q_{E_x}\boldsymbol{\sigma}_x + Q_{E_y}\sigma_y\right)\mathbf{S} = \rho\boldsymbol{\sigma}_z \tag{2.129}$$

where $\mathbf{S}$ is an unitary matrix defined by

$$\mathbf{S} = \frac{1}{\sqrt{2}}\begin{bmatrix} e^{-i\phi/2} & e^{-i\phi/2} \\ e^{i\phi/2} & -e^{i\phi/2} \end{bmatrix} \tag{2.130}$$

Such a transformation leads to the adiabatic representation, and to the integral of motion

$$\hat{S}^\dagger \hat{J}_z \hat{S} = \hat{L}_z \tag{2.131}$$

Table 2.3. Lie symmetries of the linear JT Hamiltonians.

System	symmetry group [5]	order[a)]
Trigonal, pentagonal, hexagonal and cubic		
$E \otimes e$	$O(2)$	1
Cubic		
$\Gamma_8 \otimes t_2$	$O(3)$	3
$(s+p) \otimes t_{1u}$	$O(3)$	3
$T \otimes (e+t_2)^{b)}$	$SO(3)$	3
$\Gamma_8 \otimes (e+t_2)^{b)}$	$O(5)$	10
$(s+p) \otimes (a_{1g}+t_{1u}+e_g+t_{2g})^{b)}$	$O(4)$	6
Icosahedral		
$T \otimes h$	$SO(3)$	3
$H \otimes (g+2h)^{b)}$	$SO(5)$	10
$H \otimes h$	$SO(3)$	3
$G \otimes (g+h)^{b)}$	$O(4)$	6

[a)] The order r specifies the number of Lie group independent parameters; see section 4.
[b)] For special coupling and frequency conditions.

In other words, the operator $\hat{L}_z$ is an integral of motion in the adiabatic basis.

As a result of the above analysis, the adiabatic electronic wave vectors $\{|\Psi_I(\mathbf{r};\mathbf{Q})\rangle\}$ propagate around the conical intersection as the one-parameter $2D$ rotation $O(2)$ of the initial electronic wave vectors $|E_x(\mathbf{r})\rangle$ and $|E_y(\mathbf{r})\rangle$. As indicated in Table 2.3, note that the symmetry of the Hamiltonian for the $E \otimes e$ is $O(2)$, not $SO(2)$. This is because for arbitrary rotations around the z axis, the Hamiltonian for the $E \otimes e$ case has two more symmetry operations [93]: $\mathbf{P} = \hat{R}_{E_y}\boldsymbol{\sigma}_x$ where $\hat{R} = \exp(i\pi b^\dagger_{E_y} b_{E_y})$ is an operator which changes the sign of Q_{E_y}, and $\boldsymbol{\Theta} = \hat{K}\boldsymbol{\sigma}_x$ where $\hat{K}$ is the complex conjugation operator. Moreover, $\mathbf{P}$ and $\boldsymbol{\Theta}$ anti-commute with $\hat{L}_z$, and hence all the vibronic levels of the linear $E \otimes e$ case are doubly degenerate. It means that such an axial symmetry group includes the improper rotations, and hence the group must be $O(2)$, not $SO(2)$.

In polar coordinates, the parameter $\phi = \xi_z$ is commonly referred to as the pseudo-rotational angle. The adiabatic electronic wave vectors can then be represented as

$$\begin{pmatrix} |\Psi_1(\mathbf{r};\xi_z)\rangle \\ |\Psi_2(\mathbf{r};\xi_z)\rangle \end{pmatrix} = \begin{bmatrix} \cos\xi_z & -\sin\xi_z \\ \sin\xi_z & \cos\xi_z \end{bmatrix} \begin{pmatrix} |E_x(\mathbf{r})\rangle \\ |E_y(\mathbf{r})\rangle \end{pmatrix} \tag{2.132}$$

or in matrix form

$$\boldsymbol{\Psi}(\mathbf{r};\xi_z) = \mathbf{R}_2(\xi_z)\boldsymbol{\psi}(\mathbf{r}) \tag{2.133}$$

It is clear that $\mathbf{R}_2(\xi_z+\pi) = -\mathbf{R}_2(\xi_z)$ under the inversion $\hat{\imath}_2$ defined as $\xi_z \to \xi_z + \pi$. This shows that the adiabatic electronic wave vectors change direction in the electronic wave-vector space $\{\psi_x, \psi_y\}$. Equivalently, we know that a real-valued electronically adiabatic wave function changes sign when parallel-transported in configuration-space along a closed path around the conical intersection, as originally observed by Longuet-Higgins [9,10]. Thus, we have acquired a novel understanding of the GP effect for the orbital doublet degeneracy: the GP effect is a property inherent to the inversion of the adiabatic electronic wave vectors (in the 2D electronic wave-vector space) in the group $O(2)$. We emphasize that an electronically adiabatic wave vector inverts its direction under inversion in the electronic wave-vector space or, equivalently, after a circular propagation in configuration space around the the conical intersection. Since a physical state may be described by any wave vector $|\Psi\rangle$ in a ray [88], $\exp[iA(\mathbf{R})]|\Psi\rangle$, the corresponding wave functions will be double-valued.

Similarly to Eq. (2.132), the general adiabatic electronic wave vectors $\{|\Psi_I(\mathbf{r};\mathbf{R})\rangle\}$ can be represented by a one-parameter 2D transformation of the initial electronic wave vectors $|E_x(\mathbf{r})\rangle$ and $|E_y(\mathbf{r})\rangle$; $\mathbf{R}$ represents the full set of nuclear coordinates. For the two-state problem, one obtains Eq. (2.25) which may be written in matrix form as

$$\boldsymbol{\Psi}(\mathbf{r};\mathbf{R}) = \mathbf{T}_2(\gamma(\mathbf{R}))\boldsymbol{\psi}(\mathbf{r}) \tag{2.134}$$

where $\gamma(\mathbf{R})$ is the mixing angle [33, 34, 64]. By defining the two real electronic wave functions as in Eq. (2.26) one obtains after some algebra the geometric phase [33] in Eq. (2.44). We emphasize [34] that the mixing angle $\gamma(\mathbf{R})$ has the correct sign-change behavior: $\Delta\gamma = \pi$ for a closed path encircling the intersection seam, while $\Delta\gamma = 0$ for a closed path which does not enclose the seam.

4.3 The $N = 3$ case: $T \otimes (e + t_2)$

As it is well known [5], three-fold irreducible representations are inherent to the cubic groups and icosahedral groups. In cubic groups without an inversion center, the symmetric square product $[T_1^2] = [T_2^2] = A_1 + E + T_2$ and, in the presence of inversion, $[T_{1g}^2] = [T_{1u}^2] = [T_{2g}^2] = [T_{2u}^2] = A_{1g} + E_g + T_{2g}$. Thus, for triple degenerate electronic states, the doubly degenerate E-type vibrations and the three-fold degenerate T-type vibrations are JT active: $T \otimes (e + t_2)$ case. The simplest example is an octahedral molecule of type ML_6 which belongs to the point group O_h. In cubic systems with JT active T_2 vibrations, the adiabatic potential in the T_2 subspace has O symmetry for $G = O$, O_h, and T_d, and T-symmetry for $G = T$, and T_h. The same symmetry is inherent to the adiabatic potential in the combined space of E and T_2 vibrations.

The reduction of the spherical group $O(3)$, itself a subgroup of $U(5)$, to the molecular group O_h leads to the resolution $D^{(2)} = E_g + T_{2g}$. Thus, the force constant of the normal vibration in the linear Jahn-Teller problem is split into ω_E^2 and ω_T^2 components, with the Jahn-Teller Hamiltonian of the linear $T \otimes (e + t_2)$ case having the form

$$\mathbf{H}_T(\mathbf{Q}) = \mathbf{H}_e + \mathbf{H}_v + \mathbf{H}_{ev} \tag{2.135}$$

$$\mathbf{H}_e = E_0 \mathbf{C}_A \tag{2.136}$$

$$\mathbf{H}_v = \frac{1}{2}\Big[P_{E_x}^2 + P_{E_y}^2 + P_{T_x}^2 + P_{T_y}^2 + P_{T_z}^2 + \omega_E^2\left(Q_{E_x}^2 + Q_{E_y}^2\right) + \omega_T^2\left(Q_{T_x}^2 + Q_{T_y}^2 + Q_{T_z}^2\right)\Big]\mathbf{C}_A \tag{2.137}$$

$$\mathbf{H}_{ev} = V_E(Q_{E_x}\mathbf{C}_{E_x} + Q_{E_y}\mathbf{C}_{E_y}) + V_T(Q_{T_x}\mathbf{C}_{T_x} + Q_{T_y}\mathbf{C}_{T_y} + Q_{T_z}\mathbf{C}_{T_z}) \tag{2.138}$$

where $\{Q_{\Gamma\gamma}\}$ are the real normal coordinates transforming as the $D^{(2)}$ representation, and $\{P_{\Gamma\gamma}\}$ the corresponding conjugated momenta. The matrices $\{\mathbf{C}_{\Gamma\gamma}\}$ are those of Clebsch-Gordan coefficients determined by the real electronic basis of the T term, $\{|T_x\rangle, |T_y\rangle, |T_z\rangle\}$, which are defined by

$$\mathbf{C}_A = \begin{bmatrix} 1 & 0 & 0 \\ 0 & 1 & 0 \\ 0 & 0 & 1 \end{bmatrix}, \mathbf{C}_{E_x} = \begin{bmatrix} \frac{1}{2} & 0 & 0 \\ 0 & \frac{1}{2} & 0 \\ 0 & 0 & -1 \end{bmatrix}, \mathbf{C}_{E_y} = \begin{bmatrix} -\sqrt{\frac{3}{2}} & 0 & 0 \\ 0 & \sqrt{\frac{3}{2}} & 0 \\ 0 & 0 & 0 \end{bmatrix} \tag{2.139}$$

$$\mathbf{C}_{T_x} = \begin{bmatrix} 0 & 0 & 0 \\ 0 & 0 & -1 \\ 0 & -1 & 0 \end{bmatrix}, \mathbf{C}_{T_y} = \begin{bmatrix} 0 & 0 & -1 \\ 0 & 0 & 0 \\ -1 & 0 & 0 \end{bmatrix}, \mathbf{C}_{T_z} = \begin{bmatrix} 0 & -1 & 0 \\ -1 & 0 & 0 \\ 0 & 0 & 0 \end{bmatrix} \tag{2.140}$$

Note that the symmetry group of the electronic Hamiltonian $\mathbf{H}_e$ is now the third-order unitary group $SU(3)$, being the vibrational Hamiltonian invariant under the group $SU(5)$. Thus, in the absence of vibronic interactions, the symmetry of the total Hamiltonian is described by the group $SU(3) \times SU(5)$. In the case of unequal vibrational frequencies $\omega_E \neq \omega_T$, the symmetry of $\mathbf{H}_v$ will be broken in the absence of vibronic coupling. In this case, the symmetry of $\mathbf{H}_v$ will be $SU(3)$ but this has no implications on the subsequent analysis.

We now take into account the vibronic interaction which will reduce the symmetry of the total Hamiltonian. The eight integrals of motion of the group $SU(3)$ can be split into two sets: one containing three components of the tensor convolution $\{\Psi^{(1)} \times \Psi^{(1)}\}^{(1)}$, and the other containing five components of the tensor convolution $\{\Psi^{(1)} \times \Psi^{(1)}\}^{(2)}$. When the symmetry is reduced from $SU(3)$ to $SO(3)$, the last cease to be good quantum numbers, while the former remain to be integrals of motion; the three components form a vector of the energy spin $S = 1$ according to

$$\mathbf{S}_x = \begin{bmatrix} 0 & 0 & 0 \\ 0 & 0 & -i \\ 0 & i & 0 \end{bmatrix}, \quad \mathbf{S}_y = \begin{bmatrix} 0 & 0 & i \\ 0 & 0 & 0 \\ -i & 0 & 0 \end{bmatrix}, \quad \mathbf{S}_z = \begin{bmatrix} 0 & -i & 0 \\ i & 0 & 0 \\ 0 & 0 & 0 \end{bmatrix} \tag{2.141}$$

The 24 integrals of motion of the group $SU(5)$ can be split into four sets using

$$\{b^{(l)} \times b^{(l)}\}_M^{(L)} = \sum_{m_1,m_2} b^\dagger_{m_1} b_{m_2} (lm_1 lm_2|LM) \tag{2.142}$$

where $l = 2$, $L = 1, \cdots, 2l$; $m_1, m_2 = -l, -l+1, \cdots, l$, and $(lm_1lm_2|LM)$ are the Clebsch-Gordan coefficients of the group $O(5)$. In the reduction from $SU(5)$ to $O(5)$ there remain ten integrals of motion: the $L = 1$, and $L = 3$ components of the tensors in Eq. (2.142). The subsequent reduction of the group $O(5)$ to the group $O(3)$ leaves only the three components of the tensor $L = 1$ as the integrals of motion: these are the components of the angular momentum $\mathbf{L} = (L_x, L_y, L_z)$. Thus, introduction of the vibronic interaction leads to three integrals of motion which are the three components of the total angular momentum $\mathbf{J} = \mathbf{L} + \mathbf{S}$. Since this is an integer, there is no need to consider improper rotations: the symmetry of the Hamiltonian is reduced to $SO(3)$. Similar considerations apply to the $E \otimes t_2$ case illustrated Figure 2.9.

Now, by analogy with the $N = 2$ case, we assign to each chosen parameter a 3×3 unitary planar rotation matrix [94]. Thus, the full matrix assumes the form

$$\mathbf{T}_3 = \mathbf{t}_{12}(\xi_{12})\mathbf{t}_{23}(\xi_{23}) \tag{2.143}$$

where ξ_{ij} are the two relevant angles in the $3D$ polar-coordinate system, and $\mathbf{t}_{nm}$ are 3×3 matrices defined by $[\mathbf{t}_{nm}]_{nn} = [\mathbf{t}_{nm}]_{mm} = \cos(\xi_{nm})$ and $[\mathbf{t}_{nm}]_{nm} = -[\mathbf{t}_{nm}]_{mn} = \sin(\xi_{nm})$, with all other entries satisfying

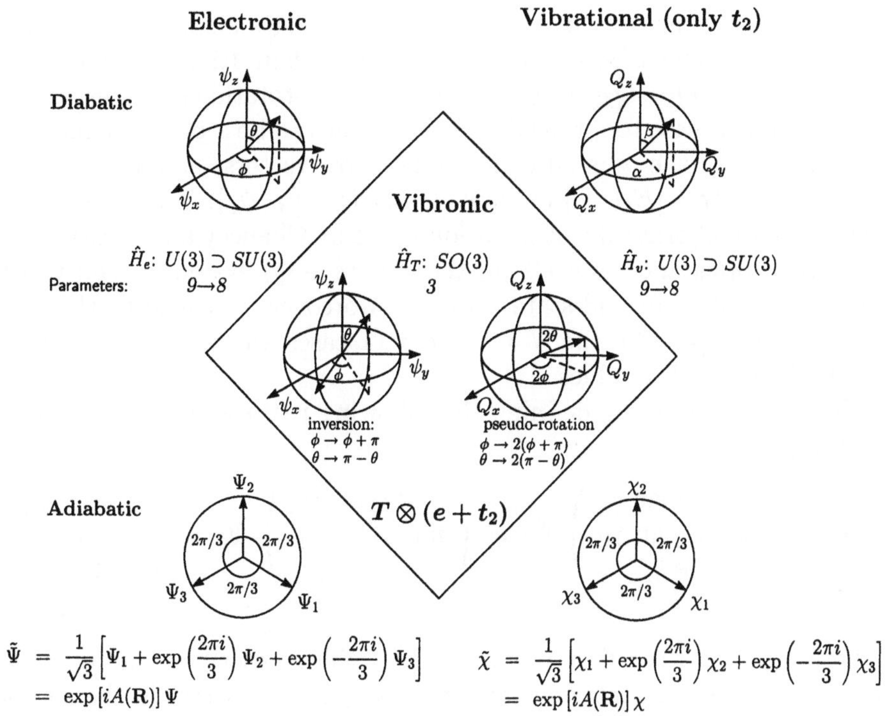

Figure 2.9. Highly schematic illustration of the symmetry elements of the electronic, vibrational, and linear vibronic Hamiltonians for the $T \otimes (e + t_2)$ case. For simplicity, only the t_2 vibrational component is illustrated, while the notation $\psi_x = |E_x(\mathbf{r})\rangle = |E_x(\mathbf{r}, \mathbf{Q}_0)\rangle$ and $Q_x = Q_{T_x}$ (correspondingly for y and z) has been used.

$[\mathbf{t}_{nm}]_{ij} = \delta_{ij}$. The real electronic wave functions will now be given by the three rows of $\mathbf{T}_3$ matrix, namely

$$\begin{pmatrix} |\Psi_1(\mathbf{r}; \xi_{23}, \xi_{12})\rangle \\ |\Psi_2(\mathbf{r}; \xi_{23}, \xi_{12})\rangle \\ |\Psi_3(\mathbf{r}; \xi_{23}, \xi_{12})\rangle \end{pmatrix} = \begin{bmatrix} \cos\xi_{23}\sin\xi_{12} & \sin\xi_{23}\sin\xi_{12} & \cos\xi_{12} \\ \cos\xi_{23}\cos\xi_{12} & \sin\xi_{23}\cos\xi_{12} & -\sin\xi_{12} \\ -\sin\xi_{23} & \cos\xi_{23} & 0 \end{bmatrix} \begin{pmatrix} |E_x(\mathbf{r})\rangle \\ |E_y(\mathbf{r})\rangle \\ |E_z(\mathbf{r})\rangle \end{pmatrix} \tag{2.144}$$

which shows that, under inversion $\hat{i}_3$ in $\{|E_x(\mathbf{r})\rangle, |E_y(\mathbf{r})\rangle, |E_z(\mathbf{r})\rangle\}$ ($\xi_{12} \to \pi - \xi_{12}, \xi_{23} \to \xi_{23} + \pi$), the adiabatic electronic wave vectors $|\Psi_1\rangle$ and $|\Psi_3\rangle$ will change direction (*i.e.*, the corresponding electronic wave functions change sign). In other words, $|\Psi_1\rangle$ and $|\Psi_3\rangle$ will be subject to GP effect while $|\Psi_2\rangle$ is not. Of course, state labelling is arbitrary, and hence the conclusion is that only one pair of states changes sign. We may then think of $2\xi_{12}$ and $2\xi_{23}$ in the $5D$ nuclear configuration space spanned by $(Q_{E_x}, Q_{E_y}, Q_{T_x}, Q_{T_y}, Q_{T_z})$ as being pseudo-rotational angles for the nuclei. In this case too, the GP effect is shown to be a property inherent to inversion of the adiabatic electronic wave vectors, now in the space $\{|E_x(\mathbf{r})\rangle, |E_y(\mathbf{r})\rangle, |E_z(\mathbf{r})\rangle\}$ or equivalently $\{\psi_x, \psi_y, \psi_z\}$. The above result seems to corroborate the observation made by Chancey and O'Brien [26] that the appearance of a GP in the $T \otimes (e + t_2)$ system is tied to the occurrence of a ground-state triplet. Moreover, it suggests that the procedure adopted for the $N = 2$ and 3 JT cases warrants generalization to the higher index values.

We may proceed by defining the following real electronic wave functions:

$$\Psi_1 = \begin{pmatrix} \cos\gamma_{23}\sin\gamma_{12} \\ \sin\gamma_{23}\sin\gamma_{12} \\ \cos\gamma_{12} \end{pmatrix} \quad \Psi_2 = \begin{pmatrix} \cos\gamma_{23}\cos\gamma_{12} \\ \sin\gamma_{23}\cos\gamma_{12} \\ -\sin\gamma_{12} \end{pmatrix} \quad \Psi_3 = \begin{pmatrix} -\sin\gamma_{23} \\ \cos\gamma_{23} \\ 0 \end{pmatrix} \tag{2.145}$$

where $\gamma_{12}(\mathbf{R})$ is the first mixing angle (which describes mixing of Ψ_1 and Ψ_2 to form an intermediate state Ψ_{12}), and $\gamma_{23}(\mathbf{R})$ is the second mixing angle (which describes mixing of Ψ_{12} and Ψ_3). In addition, by analogy with Eq. (2.39), we write the complex wave function $\tilde{\Psi} = \frac{1}{\sqrt{3}}\sum_{n=1}^{3}\exp\left[\frac{2(n-1)\pi i}{3}\right]\Psi_n$ which, upon evaluation using Eq. (2.145), yields

$$\langle\tilde{\Psi}|\nabla\tilde{\Psi}\rangle = -\frac{i}{\sqrt{3}}\left(\nabla\gamma_{12} + \cos\gamma_{12}\nabla\gamma_{23} - \sin\gamma_{12}\nabla\gamma_{23}\right) \tag{2.146}$$

Thus, if we extend Eq. (2.43) to define $A(\mathbf{R})$, one obtains

$$\langle\tilde{\Psi}|\nabla\tilde{\Psi}\rangle = i\nabla A(\mathbf{R}) \tag{2.147}$$

Comparing Eq. (2.146) with Eq. (2.147) then yields

$$\nabla A(\mathbf{R}) = -\frac{1}{\sqrt{3}}\left(\nabla\gamma_{12} + \cos\gamma_{12}\nabla\gamma_{23} - \sin\gamma_{12}\nabla\gamma_{23}\right) \tag{2.148}$$

which provides an explicit relation between the GP angle and the mixing angles $\gamma_{12}(\mathbf{R})$ and $\gamma_{23}(\mathbf{R})$. We emphasize that there are three possible mixing angles, from which we may define the two independent Lie-group parameters (see also the next subsection).

4.4 The $N = 4$ case: $G \otimes (g + h)$

Consider now a four-fold electronic degeneracy, which occurs in Jahn-Teller systems such as $(s+p)\otimes(a_{1g}+t_{1u}+e_g+t_{2g})$ and $G\otimes(g+h)$. The

Lie-group symmetry of the corresponding Jahn-Teller Hamiltonians with inclusion of linear vibronic interactions [5] is in this case $O(4)$. Thus, there will be 6 Lie-group parameters. On the other hand, in the four-fold degeneracy problem, we have 4 electronic wave vectors and hence 6 possible mixing angles (one for each pair). Such electronic wave vectors define a $4D$ space in which an arbitrary adiabatic electronic vector can be characterized by its modulus and three angles. The question then arises of how to relate these three angles to the six parameters predicted from Lie group theory (or, equivalently, to the six mixing angles). In other words, the general $\mathbf{T}_4$ matrix assumes the form

$$\mathbf{T}_4 = \mathbf{t}_{12}(\xi_{12})\mathbf{t}_{23}(\xi_{23})\mathbf{t}_{34}(\xi_{34})\mathbf{t}_{45}(\xi_{45})\mathbf{t}_{56}(\xi_{56})\mathbf{t}_{61}(\xi_{61}) \qquad (2.149)$$

although only three of these 4×4 unitary planar rotation matrices (one for each mixing angle) can be treated independently. In principle, any set of such angles (out of a total of 20 possibilities) will be equally appropriate, but the three corresponding matrices must form an ordered sequence in order to lead to physically meaningful solutions: upon inversion, an arbitrary wave vector must either change sign or remain sign unchanged. Possible selections are therefore $\mathbf{t}_{12}\mathbf{t}_{23}\mathbf{t}_{34}$, $\mathbf{t}_{34}\mathbf{t}_{45}\mathbf{t}_{56}$, and so on, but not, *e.g.*, $\mathbf{t}_{12}\mathbf{t}_{45}\mathbf{t}_{61}$. Of course, since all indices are equivalent, any ordered sequence can be relabelled as $\mathbf{T}_4 = \mathbf{t}_{12}\mathbf{t}_{23}\mathbf{t}_{34}$ which is the one picked for the following analysis.

Following the previous cases, the real electronic wave functions will now be given by the four rows of $\mathbf{T}_4$, namely

$$\Psi_1 = \begin{pmatrix} -S_{12}S_{23}S_{34} \\ S_{12}C_{23} \\ S_{12}S_{23}C_{34} \\ C_{12} \end{pmatrix}, \qquad \Psi_2 = \begin{pmatrix} -C_{12}S_{23}S_{34} \\ C_{12}C_{23} \\ C_{12}S_{23}C_{34} \\ -S_{12} \end{pmatrix}$$
$$\Psi_3 = \begin{pmatrix} -C_{23}S_{34} \\ -S_{23} \\ C_{23}C_{34} \\ 0 \end{pmatrix}, \qquad \Psi_4 = \begin{pmatrix} -C_{34} \\ 0 \\ -S_{34} \\ 0 \end{pmatrix} \qquad (2.150)$$

where $C_{ij} = \cos\xi_{ij}$, and $S_{ij} = \sin\xi_{ij}$. We now require a procedure for the inversion operation in $4D$, $\hat{\imath}_4$. Since we are not aware of such a procedure, it will be based on requiring that the image point obtained upon application of $\hat{\imath}_4$ must show its N coordinates x_i sign-reversed. For a vector of unit length in $4D$, one has

$$x_1 = C_{12}, \quad x_2 = S_{12}C_{23}$$
$$x_3 = S_{12}S_{23}C_{34}, \quad x_4 = S_{12}S_{23}S_{34} \qquad (2.151)$$

and hence it is easy to show that inversion $\hat{i}_4$ corresponds to $\xi_{12} \to \pi - \xi_{12}$, $\xi_{23} \to \pi - \xi_{23}$, and $\xi_{34} \to \pi + \xi_{34}$. Clearly, the wavefunctions Ψ_1 and Ψ_4 in Eq. (2.150) are seen to change sign, while Ψ_2 and Ψ_3 remain sign-unchanged. This corresponds to case (2, 2) of Ref. [41], and has no correspondence in the model Hamiltonian work of Ref. [40].

We proceed by calculating the first derivative coupling terms as a function of the mixing angles. After some algebra, one obtains:

$$\langle \Psi_1 | \nabla \Psi_2 \rangle = -\nabla \gamma_{12}, \quad \langle \Psi_1 | \nabla \Psi_3 \rangle = -s_{12} \nabla \gamma_{23}, \quad \langle \Psi_1 | \nabla \Psi_4 \rangle = -s_{12} s_{23} \nabla \gamma_{34} \tag{2.152}$$

$$\langle \Psi_2 | \nabla \Psi_3 \rangle = -c_{12} \nabla \gamma_{23}, \quad \langle \Psi_2 | \nabla \Psi_4 \rangle = -c_{12} s_{23} \nabla \gamma_{34}, \quad \langle \Psi_3 | \nabla \Psi_4 \rangle = -c_{23} \nabla \gamma_{34} \tag{2.153}$$

where $c_{ij} = \cos \gamma_{ij}$ and $s_{ij} = \sin \gamma_{ij}$. By now writing the complex electronic wave function as $\tilde{\Psi} = \frac{1}{\sqrt{4}} \sum_{n=1}^{4} \exp\left[\frac{2(n-1)\pi i}{4}\right] \Psi_n$, one obtains

$$\langle \tilde{\Psi} | \nabla \tilde{\Psi} \rangle = \frac{i}{\sqrt{4}} \left(\langle \Psi_1 | \nabla \Psi_2 \rangle + \langle \Psi_2 | \nabla \Psi_3 \rangle + \langle \Psi_3 | \nabla \Psi_4 \rangle + \langle \Psi_4 | \nabla \Psi_1 \rangle \right) \tag{2.154}$$

and, by defining the GP angle $A(\mathbf{R})$ as in previous cases, one gets

$$\nabla A(\mathbf{R}) = -\frac{1}{\sqrt{4}} \left(\nabla \gamma_{12} + c_{12} \nabla \gamma_{23} + c_{23} \nabla \gamma_{34} - s_{12} s_{23} \nabla \gamma_{34} \right) \tag{2.155}$$

which provides a novel explicit relation for the GP angle in terms of the mixing angles.

4.5 The arbitrary N-integral case

The Lie symmetries of other JT Hamiltonians are summarized in Table 2.3 [5]. We consider in this subsection the general integral ($N \geq 3$)-fold degeneracy case. The necessary orthogonal transformation matrix assumes the form

$$\mathbf{T}_N = \mathbf{t}_{12}(\xi_{12}) \mathbf{t}_{23}(\xi_{23}) \mathbf{t}_{34}(\xi_{34}) \cdots \mathbf{t}_{N-1,N}(\xi_{N-1,N}) \tag{2.156}$$

where $\mathbf{t}_{nm}$ are $N \times N$ matrices defined in a manner similar to Eq. (2.149). The N real electronic wave functions become now the N rows of $\mathbf{T}_N$, which can be written by induction using the previous cases. To examine the effect of the inversion $\hat{i}_N$, we then require the following generalization of Eq. (2.151):

$$x_1 = C_{12}$$

$$x_k = \left(\prod_{i=1}^{k-1} S_{i,i+1}\right) C_{k,k+1} \qquad (k = 2, 3, \cdots, N-1)$$

$$x_N = \left(\prod_{i=1}^{N-1} S_{i,i+1}\right) \tag{2.157}$$

As it can be seen by induction from previous cases, $\hat{\imath}_N$ implies $\gamma_{i,i+1} \to \pi - \gamma_{i,i+1}$ $(i = 1, 2, \cdots, N-2)$, and $\gamma_{N-1,N} \to \pi + \gamma_{N-1,N}$.

After extensive algebraic manipulation, one obtains for $\langle \Psi_k^{(N)} | \nabla \Psi_i^{(N)} \rangle$ $(i = 1, 2, N)$:

$$\langle \Psi_k^{(N)} | \nabla \Psi_1^{(N)} \rangle = \delta_{2,k} \nabla \gamma_{12} + \sum_{m=1}^{N-2} \prod_{j=1}^{m} s_{j,j+1} \delta_{m+2,k} \nabla \gamma_{m+1,m+2} \tag{2.158}$$

$$\langle \Psi_k^{(N)} | \nabla \Psi_2^{(N)} \rangle = -\delta_{1,k} \nabla \gamma_{12} + c_{12} \delta_{3,k} \nabla \gamma_{23} + c_{12} \sum_{m=2}^{N-2} \prod_{i=2}^{m} s_{i,i+1} \delta_{m+2,k} \nabla \gamma_{m+1,m+2} \tag{2.159}$$

$$\langle \Psi_k^{(N)} | \nabla \Psi_N^{(N)} \rangle = \langle \Psi_k^{(N)} | \frac{\partial \Psi_N^{(N)}}{\partial \gamma_{N-1,N}} \rangle \nabla \gamma_{N-1,N} \tag{2.160}$$

In turn, for $i = N - m \geq 3$, one gets:

$$\begin{aligned} \langle \Psi_k^{(N)} | \nabla \Psi_{N-m}^{(N)} \rangle &= \langle \Psi_k^{(N)} | \frac{\partial \Psi_{N-m}^{(N)}}{\partial \gamma_{N-m-1,N-m}} \rangle \nabla \gamma_{N-m-1,N-m} \\ &+ c_{N-m-1,N-m} \Bigg(\delta_{N-m+1,k} \nabla \gamma_{N-m,N-m+1} \\ &+ \sum_{l=0}^{N-5} \prod_{j=0}^{l} s_{N-m+j,N-m+j+1} \delta_{N-m+2+l,k} \nabla \gamma_{N-m+1+l,N-m+2+l} \Bigg) \end{aligned} \tag{2.161}$$

Note that the only nonzero integrals $\langle \Psi_k^{(N)} | \frac{\partial \Psi_N^{(N)}}{\partial \gamma_{ij}} \rangle$ in Eq. (2.158) to Eq. (2.161) are

$$\langle \Psi_1 | \frac{\partial \Psi_3}{\partial \gamma_{23}} \rangle = -s_{12}, \qquad \langle \Psi_2 | \frac{\partial \Psi_3}{\partial \gamma_{23}} \rangle = -c_{12} \tag{2.162}$$

$$\langle \Psi_1 | \frac{\partial \Psi_4}{\partial \gamma_{34}} \rangle = -s_{12} s_{23}, \qquad \langle \Psi_2 | \frac{\partial \Psi_4}{\partial \gamma_{34}} \rangle = -c_{12} s_{23}, \quad \langle \Psi_3 | \frac{\partial \Psi_4}{\partial \gamma_{34}} \rangle = -c_{23} \tag{2.163}$$

$$\langle\Psi_1|\frac{\partial\Psi_5}{\partial\gamma_{45}}\rangle = -s_{12}s_{23}s_{34}, \quad \langle\Psi_2|\frac{\partial\Psi_5}{\partial\gamma_{45}}\rangle = -c_{12}s_{23}s_{34}$$
$$\langle\Psi_3|\frac{\partial\Psi_5}{\partial\gamma_{45}}\rangle = -c_{23}s_{34}, \quad \langle\Psi_4|\frac{\partial\Psi_5}{\partial\gamma_{45}}\rangle = -c_{34} \tag{2.164}$$

By now writing the N-state complex electronic wave function as

$$\tilde{\Psi} = \frac{1}{\sqrt{N}}\sum_{n=1}^{N}\exp\left[\frac{2(n-1)\pi i}{N}\right]\Psi_n \tag{2.165}$$

we obtain for the derivative coupling term

$$\langle\tilde{\Psi}|\nabla\tilde{\Psi}\rangle = \frac{1}{N}\sum_{m\neq n}^{N}\exp\left[\frac{2(n-m)\pi i}{N}\right]\langle\Psi_m|\nabla\Psi_n\rangle \tag{2.166}$$

or

$$\langle\tilde{\Psi}|\nabla\tilde{\Psi}\rangle = \frac{i}{\sqrt{N}}\left(\langle\Psi_1|\nabla\Psi_2\rangle + \langle\Psi_2|\nabla\Psi_3\rangle + \langle\Psi_3|\nabla\Psi_4\rangle + \cdots + \langle\Psi_N|\nabla\Psi_1\rangle\right) \tag{2.167}$$

If $A(\mathbf{R})$ is then defined similarly to previous cases, one gets

$$\nabla A(\mathbf{R}) = \frac{1}{\sqrt{N}}\left(\langle\Psi_1|\nabla\Psi_2\rangle + \langle\Psi_2|\nabla\Psi_3\rangle + \langle\Psi_3|\nabla\Psi_4\rangle + \cdots + \langle\Psi_N|\nabla\Psi_1\rangle\right) \tag{2.168}$$

which gives the explicit relation between the GP angle and the first derivative coupling terms. Moreover, by using Eq. (2.158) and Eq. (2.159), we obtain the relation between the GP angle and the $N-1$ mixing angles for the general $N \geq 3$ state case:

$$\nabla A(\mathbf{R}) = -\frac{1}{\sqrt{N}}\left(\nabla\gamma_{12} + \sum_{n=1}^{N-2} c_{n,n+1}\nabla\gamma_{n+1,n+2} - \prod_{n=1}^{N-2} s_{n,n+1}\nabla\gamma_{N-1,N}\right) \tag{2.169}$$

A fimal remark concerns the fact that, except for the $E\otimes e$ case, the present approach gives $\nabla A(\mathbf{R})$ rather than $A(\mathbf{R})$ explicitly in terms of the $N-1$ mixing angles. This should not pose any problem as the general single-surface equation, Eq. (2.16), just requires the specification of $\nabla A(\mathbf{R})$.

4.6 A comparison with other approaches

Although work on this topic may be traced back to the papers by Keating and Mead [24] and Chance and O'Brien [26] in the eighties, the most recent attempts to deal generally with this topic have been by Manolopoulos and Child [40] and Baer [41]. It is on these more recent works that we focus on this subsection.

Manolopoulos and Child [40] studied the sets of possible sign changes that can occur when N real quantum states are transported adiabatically around a N-fold degeneracy by considering a $N \times N$ Hamiltonian of the form

$$\mathbf{H}(\theta) = \mathbf{F}\cos\theta + \mathbf{G}\sin\theta \tag{2.170}$$

where $\mathbf{F}$ and $\mathbf{G}$ are real symmetric matrices which are assumed to scale linearly with some radius r to produce the degeneracy at $r = 0$. The only other restriction is that $H(\theta)$ should otherwise be nondegenerate at all θ for $r \neq 0$ in order to allow a unique specification of its adiabatic eigenvectors $|x_i(\theta)\rangle$ for all θ between 0 and 2π. Thus, they solved the eigenvalue equation

$$\hat{H}(\theta)|x_i(\theta)\rangle = E_i(\theta)|x_i(\theta)\rangle \tag{2.171}$$

for $i = 1, 2, \cdots N$, and classified the sets of possible sign changes using as criterion the overlap of the adiabatic wave functions at different atomic configurations, namely

$$\gamma_i = \langle x_i(0)|x_i(2\pi)\rangle \tag{2.172}$$

Thus, their approach closely resembles the Pancharatnam connection [95, 96] theorem utilized by Fukumoto *et al.* [97] to locate conical intersections[17]. Since $\hat{H}(\theta+\pi) = -\hat{H}(\theta)$, which implies $E_{i'}(\theta+\pi) = -E_i(\theta)$ where i and $i' = N+1-i$ represent the ith highest and ith lowest eigenvalues at each value of θ around the cycle, and

$$|x_{i'}(\theta+\pi)\rangle = \eta_i|x_i(\theta)\rangle \tag{2.174}$$

where $\eta_i = \pm 1$. Thus, $|x_{i'}(\pi)\rangle = \eta_i|x_i(0)\rangle$ and $|x_i(2\pi)\rangle = \eta_{i'}|x_{i'}(\pi)\rangle = \eta_{i'}\eta_i|x_i(0)\rangle$, which in turn implies the following pairing theorem for the signs [40]:

$$\gamma_i = \gamma_{i'} = \eta_i\eta_{i'} \tag{2.175}$$

[17]The Pancharatnam connection has been suggested to define the phase of the wave function at a configuration B in relation to the wave function at A by maximizing

$$||\psi(A)\rangle + |\psi(B)\rangle|^2 = 2 + 2\Re\{\langle\psi(A)|\psi(B)\rangle\} \tag{2.173}$$

where $|\psi(A)\rangle$ and $|\psi(B)\rangle$ are normalized electronic wave functions at A and B, respectively. Thus, it amounts to choose the phase for $|\psi(B)\rangle$ such that $\langle\psi(A)|\psi(B)\rangle$ is real and positive. If, as it is the case here, we consider a real Hamiltonian with real wave functions, it just specifies the sign of the wavefunction at B in relation to the sign at A. This turns out to be the phase specification for the electronic wave functions used by Herzberg and Longuet-Higgins [9,10].

Table 2.4. Examples of allowed adiabatic sign changes at a N−fold degeneracy.

N	MC[a]	B[b]	VX[c]
1	$+$	$+$	$+$
2	$--$	$--$	$--$
3	$+++$	$+++$	n.o.[d]
	$-+-$	$-+-$	$-+-$
4	$++++$	$++++$	n.o.
	$----$	$----$	n.o.
	n.o.	$--++$	$--++$

[a] Ref. [40].
[b] Ref. [41].
[c] Ref. [42].
[d] Not observed.

Thus, for odd $N = 2k + 1$ values, the central eigenvector will not change sign: $\gamma_{k+1} = \eta_{k+1}^2 = +1$. For even N, the above pairing theorem can be strenghtened by noting that the matrix $\mathbf{U}(\theta)$ defined by $U_{ij}(\theta) = \langle x_i(0)|x_i(\theta)\rangle$ is orthogonal, $\det \mathbf{U}(\theta) = \pm 1$. Since it is continuous and $\mathbf{U}(0)$ is the identity matrix, it follows that $\det \mathbf{U}(\theta) = +1$. Thus, $\det \mathbf{U}(\pi) = +1$, which taken in conjunction with the pairing theorem, yields

$$\gamma_1\gamma_2\cdots\gamma_k = (-1)^k \qquad \text{for } n = 2k \tag{2.176}$$

$$\gamma_1\gamma_2\cdots\gamma_k\eta_{k+1} = (-1)^k \qquad \text{for } n = 2k+1 \tag{2.177}$$

Thus, there will be 2^{k-1} sign changes (rather than 2^k) for $N = 2k$, while the number of possible sign combinations is 2^k if $N = 2k + 1$, since the sign of η_{k+1} is unrestricted in this case. The second column of Table 2.4 illustrates the allowed adiabatic sign changes around $N = (1-4)$-fold degeneracies predicted from the Manopolous-Child Hamiltonian model in Eq. (2.170).

The approach followed by Baer [41] stems from the adiabatic-to-diabatic transformation matrix defined by

$$\mathbf{A}(s|s_0) = \exp\left(\int_{s_0}^{s} \mathbf{F}\cdot ds\right)\mathbf{A}(s_0) \tag{2.178}$$

from which follows the so-called topological matrix:

$$\mathbf{D} = \mathbf{A}(s_0|s_0)\mathbf{A}^\dagger(s_0) = \exp\left(\oint_C \mathbf{F}\cdot ds\right) \tag{2.179}$$

In Eq. (2.178), the integral is performed along a line that combines the points s and s_0, while the dot under the integral sign stands for the scalar product between the derivative coupling vector-matrix $\mathbf{F}(s)$ and ds, a differential length element along the line of integration. In turn, $\mathcal{C}$ in Eq. (2.179) is any cyclic path. In order that the diabatic potentials be uniquely defined, Baer [41] suggests as a necessary condition that $\mathbf{D}$ is diagonal and has the following terms

$$D_{jk}(\mathcal{C}) = \delta_{jk} e^{i\chi_k(\mathcal{C})} \tag{2.180}$$

where $\chi_k(\mathcal{C})$ is some phase, thence terms of unit norm. In the case of the two-state problem, $\mathbf{A}$ becomes the 2×2 matrix in Eq. (2.25) with $\gamma(\mathbf{R})$ replaced by the diabatic-to-adiabatic transformation angle $\alpha(\mathbf{R})$ (as noted in section 2.3, for the two-state problem, the two angles are identical to each other up to an additive constant [34]).

Despite the debate on the existence of rigorous diabatic potentials (Refs. [30, 38, 98, 99], and references therein) for the general N-state problem, Baer [41] suggests to use the following $N \times N$ adiabatic-to-diabatic transformation matrix

$$\mathbf{A}^{(N)} = \prod_{i=1}^{N-1} \prod_{j>i}^{N} \mathbf{Q}_{ij}^{(N)}(\gamma_{ij}) \tag{2.181}$$

where $\mathbf{Q}_{ij}^{(N)}(\gamma_{ij})$ is a $N \times N$ planar rotation matrix similar to $\mathbf{t}_{ij}(\xi_{ij})$ in Eq. (2.156): in the diagonal two cosine functions and the rest of the $(N-2)$ elements are $(+1)$s; in the off-diagonal positions, one has two sine functions at the relevant positions and the rest are zeros. Defining the values of the angles γ_{ij} at the end point as α_{ij}, the topological angles, Baer [41, 100] concludes that in order that the $\mathbf{D}$ matrix be diagonal such angles must be multiples of π (or zero):

$$\alpha_{ij} = n_{ij}\pi \tag{2.182}$$

where $n_{ij} = n_{ji}$ are integers. The elements of $\mathbf{D}^{(N)}$ assume then the form

$$D_{ij}^{(N)} = \delta_{ij} \prod_{k \neq i}^{N} \cos\alpha_{ik} = \delta_{ij}(-1)^{\sum_{k\neq i}^{N} n_{ik}} \qquad i = 1, 2, \cdots N \tag{2.183}$$

and hence the topological matrix $\mathbf{D}^{(N)}$ becomes a diagonal matrix with diagonal elements -1 or $+1$. From his analysis, Baer [100] notes three types of situations. The first relates to systems (so-defined [41] Jahn-Teller systems) for which all diagonal elements of $\mathbf{D}^{(N)}$ are negative. In relation to these systems, he observes that they can only occur if, and only if, N (the dimension of the relevant Hilbert space, also denoted sub-Hilbert space) is an even number. For example, such a situation cannot be observed for the three-state case. The second situation refers to systems for which all diagonal elements of $\mathbf{D}^{(N)}$ are positive (so-defined [41] Renner-Teller systems), which have been found to exist for all cases. Finally, there are in-between situations, but not all possible sign combinations can occur for a given N. For example, the $\mathbf{D}^{(3)}$ matrix may contain two (-1)s and one $(+1)$ [denoted $(1,2)$ case, where the first digit indicates the number of $(+1)$s and the second the number of (-1)s] but never two $(+1)$s and one (-1) [$(2,1)$ case]. A summary of the allowed adiabatic sign changes around $N = (1-4)$-fold degeneracies predicted from Baer's approach is presented in the third column of Table 2.4.

Shown in the fourth column of Table 2.4 are the results obtained from the approach reported in Ref. 42 which has been very much followed in this section. According to this approach, the adiabatic wave vectors for the general N-state case are related to the diabatic wave vector basis through a $N \times N$ transformation matrix involving as parameters the angles necessary to represent an adiabatic wave vector in the corresponding projective function space. For example, the Lie-group symmetry of the Hamiltonians for the $(s+p)\otimes(a_{1g}+t_{1u}+e_g+t_{2g})$ and $G\otimes(g+h)$ JT systems with inclusion of linear vibronic interactions [5] belong to the symmetry group $O(4)$, and hence there will be 6 Lie-group parameters from which we can select the three angles necessary to characterize such an adiabatic wave vector in function space. As discussed in subsection 4.4, the involved procedure implies the $\mathbf{T}_4$ matrix in Eq. (2.150), which is written as a product of three 4×4 unitary planar rotation matrices (one for each mixing angle) subject to the constraint that one gets an ordered sequence for the matrix product. In fact, only such a choice has been found to yield physically meaningful solutions, namely, upon inversion an arbitrary wave vector must either change sign or remain sign unchanged. By direct inspection of each adiabatic wave vector upon inversion in the function space, it has then been shown that in this situation only the $(2,2)$ case is observed. The notable feature from Table 2.4 is perhaps the fact that all three methods lead to somewhat different predictions, which may not be surprising given the distinct Hamiltonian models that have been employed. In particular, since the systems that have been considered in the present work are JT systems with high symmetry requirements, this may explain why the approach discussed in the present work is the most restrictive one among the three compared in Table 2.4 for the sign-changes of the adiabatic wave functions. A final remark to note that the five-fold electronically degenerate case $H \otimes h$ presents the only situation in Table 2.3 where the number of parameters in the Lie group, $SO(3)$, is smaller than the number of angles required to define the corresponding adiabatic electronic wave vector in function space: 3 versus 4 in that order. The associated $\mathbf{T}_5$ matrix may then be written as a product of three 5×5 unitary planar rotation matrices by fixing one angle and insuring that any arbitrary electronically adiabatic wave vector will either change sign or remain sign unchanged upon inversion in function space. The results agree with thôse for the general case.

5. Concluding remarks

After surveying the hierarchy of approximations involved in the solution of the many-body problem, we focused on the case of highly symmetric systems with electronic degeneracies of the Jahn-Teller type. A generalized adiabatic (Born-Oppenheimer) treatment of the nuclear dynamics was then presented to account for the GP effect. The focus was in regions close to the degeneracy locus where a linear approximation to the Hamiltonian may be sufficient to understand the basic features of the nuclear dynamics. In fact, being a topological effect, it is expected that the cyclic geometric phases so derived can still be useful for the treatment of the full nuclear dynamics problem, irrespective of the molecule configuration space region. Such an approach has been used in the full-dimensional zero total angular momentum calculations of the cone states of trihydrogen and some of its isotopomers summarized in section 2.3. From these results, an attempt was also made toward the prediction of how important the GP effect can be in single-surface quantum dynamics calculations for H_3 and its isotopomers. To achieve this goal, a phenomenological parameter defined as the root mean square deviation (Δ) between the vibrational energies for the cone states calculated with consideration the GP effect and those obtained without it (NGP) has been suggested. Using five distinct isotopomers of H_3, it has been shown that Δ can be expressed as a simple linear relationship in terms of the probability of looping around the C_3 axis which gets tilted due to the mass-dependence of the hyperspherical coordinates used for the calculations. The good linear fit so obtained suggests that it should also be valid for the H_3 isotopomers not studied thus far. It suggests that, although essential for accurate single-surface vibrational calculations on the homonuclear trihydrogenic systems, the GP effect can be significantly less so for some of their heteronuclear isotopomers, particularly for those involving a single muonium particle.

Although a greater emphasis has naturally been given to the traditional two-state JT problem, we examined also the GP effect at other integral $N-$fold electronic degeneracies using the corresponding linear JT Hamiltonians. The major results may be summarized as follows. For an N-fold electronic JT degeneracy, propagation of the adiabatic electronic wave vectors around the point of degeneracy has been represented as a $(N-1)$-parameter rotation in the N-dimensional projective function space defined by the electronic wave-vector basis. For the $N=2$ and $N=3$ JT cases, inversion in such a function space has been shown to lead to the GP effect as predicted from its traditional analysis in configuration space [9, 10, 26]. Specifically, for $N=2$, any adiabatic electronic wave vector can be represented as a one-parameter (*i.e.*, the mixing angle, which is identical up to an initial phase to the GP angle [33, 34]) rotation in the $2D$ wave-vector space. In turn, for $N=3$, such a rotation will take place in the $3D$ function space, with inversion then showing that only two adiabatic electronic wave vectors are subject to the GP effect. The method was then extended to higher degeneracies, with the following predictions. For $N=4$, the GP effect when viewed from the action of the inversion $\hat{i}_4$ in function space leads to the prediction that two electronic wave vectors must change sign (and hence be subject to the GP effect) while the remaining two are not. The same approach may be used for any integer$-N$ value degeneracy by examining the effect of applying $\hat{i}_N$ to the adiabatic electronic wave vectors in the corresponding N-dimensional function space. Not unexpectedely, perhaps, the predictions so obtained did not fully coincide with the results reported [40, 41] for $N=(1-4)$-degeneracies, nor do the latter agree among themselves. This may partly be attributed to the fact that we focused on linear JT systems which have very specific symmetry requirements. For example, Table 2.4 shows that for $N=4$, we predict only the $(2,2)$ sign-change case in agreement with Baer [41]. However, we disagree with both the Manolopoulos-Child [40] and Baer [41] predictions in not observing the $(4,0)$ and $(0,4)$ cases. In an attempt to rationalize such discrepancies, we may conjecture that a high degeneracy can be viewed as the coalescence of (low-N)-fold ones, *e.g.*, a four-fold degeneracy may be seen as the result of two coalescing two-fold degeneracies. In this case, the traditional GP theorems would apply to each of the latter in the vicinity of the degeneracy seam. If these were both of the conical type (*i.e.*, each involving a pair of sign changing states) then the GP theorem would imply that in the vicinity of the coalescence point none of the four states could change sign upon adiabatic transportation of the wave function over a closed loop encircling the degeneracy point (since this actually represents an even number of conical intersection points). Stated differently, only one of the $N=2$ intersections may be conical leading to a single pair of sign changing states. Such a reasoning would then lead to the prediction that the $(0,6)$ case might exist, since an odd number of two-state conical intersections would be involved. However, the $N=6$ case may alternatively be viewed as the coalescence of $N=4$ and $N=2$ degeneracies. If the GP theorem is here too applicable, then only one of these intersections can be of the conical type leading again to just one pair of states subject to the GP effect. Because there is no reason of principle to believe that one of these two merging schemes should prevail over the other, the maximum number of changing states common to both interpretations of the $N=6$ case is two, *i.e.*, one has a $(4,2)$ case. This is indeed the result which is obtained through the analysis of the corresponding $\mathbf{T}_6$ matrix. In fact, by an extension of the above argument, one is led to conclude that only one pair of sign changing states is observable for any arbitrary N-fold degeneracy, which is indeed the result predicted from the approach [42] discussed in the present work. Note that the odd-N state cases poses no problem, since only a single sign-unchanging state needs to be added to the closest even-N state case. Despite assumedely introduced in a naïve form, the ideas reported above raise the intriguing question of whether the GP theorems indirectly apply to the case of an arbitrary integer N-fold degeneracy. Clearly, a proper verification of the above conjectures requires further study. This would be greatly enhanced by employing an algebraic code for the calculation and analysis of the involved orthogonal matrices. A further remark to observe that no electronic degeneracies higher than five-fold are reported in Table 2.3, and hence the issue of higher electronic degeneracies may loose some practical relevance.

In termination, objections to the present simple interpretation will doubtless appear. The road is rocky, and we simply hope to have brought up an issue which deserves careful consideration. Indeed, any molecule possessing a three-fold rotation axis or any higher axis of rotational symmetry will be bound to present highly degenerate electronic states. Unfortunately, to our knowledge, no global surface has yet been reported which shows a three-fold or higher electronic degeneracy. Moreover, even if it was available, it is not clear which should be the best way to carry out the quantum dynamics studies on such a system. The simplest approach would be a generalized adiabatic treatment by accounting for the involved topological effects. In this work, we have reported an expression for the gradient of the GP angle in terms of the relevant mixing parameters or, in the two-state case, an expression for the GP angle itself. However, a generalization of the adiabatic approach similar to that of the two-state problem is still unavailable. This is certainly another challenging issue that warrants investigation.

6. Dedication

This paper is dedicated to the memory of Per-Olov Löwdin.

Acknowledgments

I thank all my coworkers, especially Z.R. Xu and L.P. Viegas for their valuable contributions to the work here reported. This work has the support of Fundação para a Ciência e Tecnologia, Portugal.

References

[1] M. Born and J. R. Oppenheimer, Ann. Phys. **84**, 457 (1927).

[2] M. Born and K. Huang, *Dynamical Theory of Crystal Lattices* (Oxford, London, 1954).

[3] H. A. Jahn and E. Teller, Proc. R. Soc. **161A**, 220 (1937).

[4] R. Englman, *The Jahn-Teller Effect in Molecules and Crystals* (Wiley (Interscience), New York, 1972).

[5] I. B. Bersuker and V. Z. Polinger, *Vibronic Interactions in Molecules and Crystals* (Springer-Verlag, Berlin, 1989).

[6] H. C. Longuet-Higgins, Adv. Spectrosc. **2**, 429 (1961).

[7] J. von Neumann and E. Wigner, Phys. Z. **30**, 467 (1929).

[8] E. Teller, J. Phys. Chem. **41**, 109 (1937).

[9] G. Herzberg and H. C. Longuet-Higgins, Faraday Discuss. Chem. Soc. **35**, 77 (1963).

[10] H. C. Longuet-Higgins, Proc. R. Soc., London **A.344**, 147 (1975).

[11] T. F. George, K. Morokuma, and Y.-W. Lin, Chem. Phys. Lett. **30**, 54 (1975).

[12] D. R. Yarkony, J. Phys. Chem. **100**, 18612 (1996).

[13] Y. Yamaguchi, Y. Osamura, J. D. Goddard, and H. F. Schaeffer, *New Dimension to Quantum Chemistry: Analytic Derivativ Methods in* ab initio *Molecular Electronic Structure Theory* (Oxford University Press, Oxford, 1994).

[14] M. J. Beapark, M. A. Robb, and H. B. Schlegel, Chem. Phys. Lett. **223**, 269 (1994).

[15] C. A. Mead, Rev. Mod. Phys. **64**, 51 (1992).

[16] D. R. Yarkony, Rev. Mod. Phys. **68**, 985 (1996).

[17] A. Kupperman, in *Dynamics of Molecules and Chemical Reactions*, edited by R. E. Wyatt and J. Z. H. Zhang (Marcel, Dekker, Inc., New York, 1996), p. 411.

[18] M. Baer, Phys. Rep. **358**, 75 (2002).

[19] A. J. C. Varandas and Z. R. Xu, in *Adv. Chem. Phys.*, edited by G. D. Billing and M. Baer (Wiley, New York, in press)

[20] M. V. Berry, Proc. R. Soc. **A392**, 45 (1984).

[21] C. A. Mead and D. G. Truhlar, J. Chem. Phys. **70**, 2284 (1979).

[22] C. A. Mead, Chem. Phys. **49**, 23 (1980).

[23] C. A. Mead, J. Chem. Phys. **72**, 3839 (1980).

[24] S. P. Keating and C. A. Mead, J. Chem. Phys. **82**, 5102 (1985).

[25] F. S. Ham, Phys. Rev. Lett. **58**, 725 (1987).

[26] C. C. Chancey and M. C. M. O'Brien, J. Phys. A: Math. Gen. **21**, 3347 (1988).

[27] A. J. C. Varandas and H. G. Yu, J. Chem. Soc. Faraday Trans. **93**, 819 (1997).

[28] M. Baer and R. Englman, Chem. Phys. Lett. **265**, 105 (1997).

[29] M. Baer, J. Chem. Phys. **107**, 2694 (1997).

[30] B. K. Kendrick, C. A. Mead, and D. G. Truhlar, J. Chem. Phys. **110**, 7594 (1999).

[31] A. J. C. Varandas and Z. R. Xu, Chem. Phys. **259**, 173 (2000).

[32] A. J. C. Varandas and Z. R. Xu, Chem. Phys. Lett. **316**, 248 (2000).

[33] A. J. C. Varandas and Z. R. Xu, J. Chem. Phys. **112**, 2121 (2000).

[34] Z. R. Xu, M. Baer, and A. J. C. Varandas, J. Chem. Phys. **112**, 2746 (2000).

[35] Z. R. Xu and A. J. C. Varandas, Int. J. Quantum Chem. **80**, 454 (2000).

[36] Z. R. Xu and A. J. C. Varandas, Int. J. Quantum Chem. **83**, 279 (2001).

[37] Z. R. Xu and A. J. C. Varandas, J. Phys. Chem. **105**, 2246 (2001).

[38] B. Kendrick, C. A. Mead, and D. G. Truhlar, Chem. Phys. Lett. **277**, 31 (2002).

[39] A. J. C. Varandas and L. P. Viegas, Chem. Phys. Lett. (submitted for publication).

[40] D. E. Manolopoulous and M. S. Child, Phys. Rev. Lett. **82**, 2223 (1999).

[41] M. Baer, Chem. Phys. Lett. **322**, 520 (2000).

[42] A. J. C. Varandas and Z. R. Xu, Int. J. Quantum Chem. (submitted for publication).

[43] R. Englman, *The Janh-Teller Effect* (Wiley, New York, 1973).

[44] A. Kiselev, Can. J. Phys. **56**, 615 (1978).

[45] J. Zúñiga, A. Bastida, and A. Requena, J. Chem. Soc. Faraday Trans. **93**, 1681 (1997).

[46] A. J. C. Varandas, Chem. Phys. Lett. **138**, 455 (1987).

[47] R. C. Whitten and F. T. Smith, J. Math. Phys. **9**, 1103 (1968).

[48] B. R. Johnson, J. Chem. Phys. **73**, 5051 (1980).

[49] B. R. Johnson, J. Chem. Phys. **79**, 1906 (1983).

[50] B. R. Johnson, J. Chem. Phys. **79**, 1916 (1983).

[51] A. J. C. Varandas, Adv. Chem. Phys. **74**, 255 (1988).

[52] A. J. C. Varandas, *Lecture Notes in Chemistry*, edited by A. Laganá and A. Riganelli (Springer, Berlin, 2000), vol. 75, p. 33.

[53] A. J. C. Varandas, in *Conical Intersections: Electronic Structure, Dynamics and Spectroscopy*, edited by W. Domcke, D. R. Yarkony and H. Köppel (World Scientific Publishing, Singapore, in press).

[54] A. J. C. Varandas, F. B. Brown, C. A. Mead, D. G. Truhlar, and N. C. Blais, J. Chem. Phys. **86**, 6258 (1987).

[55] B. T. Sutcliffe, in *Lecture Notes in Chemistry* (Springer, Berlin, 1999), vol , pp. 61.

[56] A. J. C. Varandas, J. Tennyson, and J. N. Murrell, Chem. Phys. Lett. **61**, 431 (1979).

[57] S. S. Xantheas, S. T. Elbert, and K. Ruedenberg, J. Chem. Phys. **93**, 7519 (1990).

[58] R. Jackiw, Comments At. Mol. Phys. **21**, 71 (1988).

[59] B. K. Kendrick and C. A. Mead, J. Chem. Phys. **102**, 4160 (1995).

[60] Y.-S. M. Wu, B. Lepetit, and A. Kuppermann, Chem. Phys. Lett. **186**, 319 (1991).

[61] Y.-S. M. Wu and A. Kuppermann, Chem. Phys. Lett. **201**, 178 (1993).

[62] A. Kuppermann and Y.-S. M. Wu, Chem. Phys. Lett. **213**, 636 (1993).

[63] G. D. Billing and N. Marković, J. Chem. Phys. **99**, 2674 (1993).

[64] A. J. C. Varandas, J. Chem. Phys. **107**, 867 (1997).

[65] M. Baer, Chem. Phys. **15**, 49 (1976).

[66] M. Baer, Mol. Phys. **40**, 1011 (1980).

[67] T. Pacher, L. S. Cederbaum, and H. Köppel, Adv. Chem. Phys. **84**, 293 (1993).

[68] M. Baer, Chem. Phys. Lett. **35**, 112 (1975).

[69] D. R. Yarkony, J. Chem. Phys. **105**, 10456 (1996).

[70] R. Englman and M. Baer, J. Phys.: Condens. Matter **11**, 1059 (1999).

[71] D. R. Yarkony, J. Phys. Chem. A **101**, 4263 (1997).

[72] H. C. Longuet-Higgins, U. Öpik, M. H. L. Pryce, and R. A. Sack, Proc. R. Soc. **244A**, 1 (1958).

[73] M. S. Child and H. C. Longuet-Higgins, Phyl. Trans. R. Soc. London A **254**, 259 (1961).

[74] H. G. Yu and S. C. Smith, Ber. Bunsenges. Phys. Chem. **101**, 400 (1997).

[75] A. J. C. Varandas, H. G. Yu, and Z. R. Xu, Mol. Phys. **96**, 1193 (1999).

[76] A. J. C. Varandas and Z. R. Xu, Int. J. Quantum Chem. **75**, 89 (1999).

[77] B. K. Kendrick, J. Chem. Phys. **112**, 5679 (2000).

[78] B. J. Kendrick, J. Chem. Phys. **114**, 4335 (2001).

[79] A. Kuppermann and Y.-S. M. Wu, Is the geometric phase effect important for the $H + D_2 \rightarrow HD + D$ reaction?, *Workshop on Quantum Reaction Dynamics* (Pasadena, U.S.A., 2001).

[80] T. C. Thompson and C. A. Mead, J. Chem. Phys. **82**, 2408 (1985).

[81] C. D. H. Chisholm, *Group Theoretical Techniques in Quantum Chemistry* (Academic Press, 1976).

[82] W. Moffitt and A. D. Liehr, Phys. Rev. **106**, 1195 (1957).

[83] G. Herzberg, *Molecular Spectra and Molecular Structure. III. Electronic Spectra and Electronic Structure of Polyatomic Molecules* (van Nostrand, New York, 1966).

[84] A. Ceulemans, D. Beyens, and L. G. Vanquickenborne, J. Am. Chem. Soc. **106**, 5824 (1984).

[85] A. Ceulemans, J. Chem. Phys. **87**, 5374 (1987).

[86] J. E. B. Wilson, J. C. Decius, and P. C. Cross, *Molecular Vibrations: the Theory of Infrared and Raman Spectra* (McGraw-Hill, New York, 1955).

[87] P. Murray-Rast, H. Bürgi, and J. D. Dunitz, Acta Crystallogr., Sect. A **A35**, 703 (1979).

[88] M. Hammermesh, *Group Theory and its Application to Physical Problems* (Dover, New York, 1989).

[89] L. O'Railfeartaigh, in *Group Theory and Its Applications*, edited by E. M. Loebl (Academic Press, New York, 1968), p. 469.

[90] L. S. Pontryagin, *Topological Groups* (Princeton University Press, Princeton, New Jersey, 1958).

[91] B. R. Judd, Can. J. Phys. **52**, 999 (1974).

[92] A. D. Liehr, J. Phys. Chem. **67**, 389 (1963).

[93] C. H. Leung and W. H. Kleiner, Phys. Rev. B **10**, 4434 (1974).

[94] S. J. Leon, *Linear Algebra with Applications* (MacMillan Publishing Company, New York, 1990).

[95] S. Pancharatnam, Ind. Acad. Sci. A **44**, 247 (1956).

[96] J. Samuel and R. Bhandari, Phys. Rev. Lett. p. 2339 (1988).

[97] Y. Fukumoto, H. Koizumi, and K. Makoshi, Chem. Phys. Lett. **313**, 283 (1999).

[98] M. Baer, A. M. Mebel, and G. D. Billing, J. Phys. Chem. A **106**, 6499 (2002).

[99] E. S. Kryachko and A. J. C. Varandas, Int. J. Quantum Chem., (submitted for publication).

[100] M. Baer and A. Alijah, Chem. Phys. Lett. **319**, 489 (2000).

[101] Y. Aharonov and B. Bohm, Phys. Rev. **115**, 485 (1959).

[102] C. A. Mead and D. G. Truhlar, J. Chem. Phys. **77**, 6090 (1982).

[103] H. Köppel, W. Domcke, and L. S. Cederbaum, Adv. Chem. Phys. **72**, 59 (1984).

Recent developments and applications of the state-specific approach to excited states and their dynamics

CLEANTHES A. NICOLAIDES*
*Physics Department, National Technical University, Athens, Greece, and Theoretical and Physical Chemistry Institute, National Hellenic Research Foundation, 48 Vasileos Constantinou Ave., Athens 11635, Greece. e-mail:can@eie.gr

1. INTRODUCTION

Progress in quantum chemistry requires a practical combination of appropriate formalism with efficient computational methods and physically relevant function spaces, so as to achieve reliable, interpretable and applicable solutions of the time-independent Schrödinger equation (TISE) as well as of the time-dependent Schrödinger equation (TDSE). This necessity becomes much more imperative when the object of study are dynamical processes and properties of excited electronic states, in which case the theory and methods for the calculation of reliable electronic wavefunctions and spectra have to be more sophisticated and more complex than what is normally used for, say, the ground state, especially if the state of interest is in the continuous spectrum, i.e., it is a resonance (unstable) state.

A few years ago [1], I reviewed briefly certain aspects of theoretical and computational work on atomic and molecular excited states that was done in Athens during the previous two decades or so. The paper was entitled " *The state-specific approach to the solution of problems of electronic structure and dynamics involving excited states* ", where the generic name "*state-specific approach*", (SSA), signifies the framework within which formalisms and methods of computation have been, and are

E.J. Brändas and E.S. Kryachko (eds.),
Fundamental World of Quantum Chemistry, Vol. II, 93–106.

being developed and applied for the quantitative understanding of properties of excited electronic states of atoms and of small molecules , and of the interplay between electronic structure and spectroscopy or dynamics. The discussion in [1] was concerned with problems that require the solution of the TISE as well as of the TDSE. Among other things , I pointed to cases of excited states where , if a SSA calculation is possible , convergence to a reliable solution is economic and physically transparent. Furthermore, the possibility of using state-specific , and, according to the situation , well-correlated wavefunctions, allows their systematic incorporation into bigger theoretical and calculational schemes , involving theory of scattering or the solution of the TDSE for the determination of the many-electron time-dependent wavefunction , $\Psi(\mathbf{r},t)$. For example, comparison of the SSA with other methods that are based on the use of a single set of basis functions and diagonalization of real or complex Hamiltonians can be found in [1-3] . In [2] , whose title is " *On the application of conventional quantum chemistry methods of computation to states perturbed by the continuous spectrum*" , I argued that for such states , due to the different mathematical properties of the scattering function space, even a full CI calculation with commonly used single basis sets and a real Hamiltonian, need not lead to accurate results.Instead, theory and calculation must account for the fact that such states are superpositions of bound and unbound components which must be represented by different function spaces when the real Hamiltonian is used.

The invitation to contribute to this volume , dedicated to the memory of a scientist whom I knew personally and whom I respected deeply, Per-Olov Löwdin, offered me the opportunity of continuing the descriptive presentation, along the lines of [1] , of more recent developments and applications of the SSA. The following themes are discussed briefly , leaving the bulk of the material to the original publications:

1) Certain aspects of the SSA and its earlier applications.

2) The time-dependent evolution of atomic polyelectronic nonstationary states [4-6].

3) Bound and resonance states of multiply excited states of atomic negative ions (ANIs) [3,7,8].

4) The interaction of the electromagnetic field with high-n Rydberg states and with scattering states [4,5].

The discussion assumes a nonrelativistic Hamiltonian , i.e. it is concerned with systems of small effective nuclear charge. Therefore, the radial one-electron functions are nonrelativistic. However, even here , relativistic mixing of nonrelativistic configurations may often be crucially important , as for example, in the case of J-dependent relativistic decays of metastable bound excited states of ANIs , where the details of both electron correlation and relativistic corrections play a significant role . For such problems, state-specific calculations with interchannel coupling have shown that the physics of the many-electron problem is described well at the level of the Breit-Pauli Hamiltonian [9].

I note that although the reference list contains mainly SSA , and SSA-related papers, the connection to other theoretical and experimental work can be found in the papers cited therein.

2. ESSENTIALS OF THE SSA FOR THE CALCULATION AND USE OF CORRELATED WAVEFUNCTIONS OF ATOMIC AND MOLECULAR EXCITED STATES

I start by quoting from [1] (page 120) : "*Excited states are characterized by one or more open-shell symmetry-adapted configurations whose relative importance is a function of the nuclear charge , Z, and/or of geometrical parameters in the Born-Oppenheimer Hamiltonian. In general, this leads to complexity as regards the calculation of state-specific self-consistent orbitals. In addition, the bulk of the wavefunction constitution of many excited states is determined by energy near-degeneracies or by strong off-diagonal perturbations, whose ab initio calculation is quite sensitive to the function spaces used. Both of the above features give rise to a multitude of possible correlation corrections. Finally, for excited states embedded in the continuous spectrum, new and fundamental questions arise.....Essential to the SSA is the emphasis on formulating theoretical models which can incorporate physically relevant and well-optimized function spaces* ".

The above comments about correlation effects and function spaces , i.e., about the significance , for the calculation of wavefunctions , of choices of zero-order configurations with self-consistent orbitals and of correlation configurations with virtual orbitals representing one-, two-, three- e.t.c. excitations from occupied zero-order orbitals , were experienced computationally in the early 1970s in work on atomic states with holes in the inner subshells , [10] and references therein. Since then, a more thorough understanding has been developed , both formally and computationally, for states of a variety of structures and energies. Accordingly, some of the SSA features and strategies are:

1) Regardless of where the energy of the excited state is , the choice of the zero-order reference space must be state-specific, or nearly so . If correlation is important, the reference zero-order wavefunction , (the "*Fermi-Sea*"

wavefunction), must be multiconfigurational, with self-consistent orbitals. For atoms and diatomics, these orbitals can be numerical. For their computation, we have been using the programs of Froese-Fischer [11] and of McCullough [12], which were adapted to and incorporated into the overall procedure for the calculation of state-specific correlated wavefunctions.

2) According to the SSA, the N-electron wavefunction consists of parts, (one or more configurations), whose optimal representation is achieved by separately optimized function spaces, i.e., not by the use of the same orbital basis set. For example, such situations characterize zero-order versus correlation spaces, valence versus Rydberg or covalent versus ionic configurations. These spaces can, in principle, be defined and computed separately and the total wavefunction is put together via nonorthonormal configuration-interaction (NONCI). The result is a compact wavefunction, whose level of accuracy depends on the property under consideration. For example, just a 2x2 or a 6x6 NONCI calculation proved sufficient to understand quantitatively the sudden polarization in the zwitterionic states of twisted butadiene and ethylene [13]. As regards spectroscopy of atoms and small molecules, the typical and challenging example is the case of "valence-Rydberg-continuum" (VRC) configurational mixtures. The valence state can be formally separated from the Rydberg-continuum series, and its wavefunction can be optimized separately, with its own correlation vectors, before it is allowed to interact with the series. It thus becomes a "*diabatic*" state, as in the prototypical example of the He_2^+ $1\sigma 2\sigma^2 \leftrightarrow 1\sigma^2 n\sigma$ $^2\Sigma_g^+$ (n = 2,3,...) spectrum [14]. When the mixed states of interest are in the discrete spectrum, they can be treated by diagonalization of the energy matrix, in terms of NONCI. In the general case of VRC, the SSA has been developed by unifying the treatment of the discrete and the continuous spectrum via a combination of CI and K-matrix methods [15].

Finally, the variational convergence to a localized wavepacket with a local energy minimum inside the continuous spectrum, yielding a "diabatic"-type of state by excluding the parts of function space leading to fragmentation, is the basis for the calculation of resonance and autoionizing states in multielectron atoms [e.g., 1-3,8].

3) The combination of compact, state-specific wavefunctions with the concept of nonorthonormality between the sets of orbitals of such wavefunctions representing two states participating in a transition process, leads to the heuristic as well as to the quantitative understanding of the interplay between electronic structure and transition rates and spectra [10,16,17]. As regards the heuristic part, in the review of [17] a set of rules for the easy identification of strong nonrelativistic correlation effects in atomic excited states throughout the Periodic Table were presented. These rules are the result of analysis and accumulated computational experience by

D.R.Beck and the author, on topics of electron correlation and spectroscopy [10,16,17]. For example, in spectroscopies involving the creation of zero order states with holes in inner subshells, the pair correlations with "*symmetric exchange of orbital symmetry*" (SEOS), which are characterized by large off-diagonal matrix elements, have significant influence on the spectra. Such situations arise in the photoionization of the ns subshell below an np subshell, ($p^2 \leftrightarrow$ sd SEOS), or in the 5p photoionization of the actinides, (df $\leftrightarrow$ pg SEOS), or in the nd photoabsorption in the rare earths, ($f^2 \leftrightarrow$ dg SEOS) [17]. As regards the quantitative part, the SSA has allowed the recognition of the contribution, to a good approximation, of the influence of the different self-consistent fields and of real or virtual multielectron correlations in photoabsorption induced by one or by more photons and causing the excitation of one or more electrons [10,16,18,19]. In the case of two electrons in the continuous spectrum, appropriate theory was developed for the treatment of electron correlation using square-integrable complex functions [19].

3. THE TIME-EVOLUTION OF NONSTATIONARY STATES OF POLYELECTRONIC ATOMS

Although it is possible to formulate quantum mechanical theories of many dynamical processes within a time-independent framework, based on which computations can be carried out, (at least in principle), it is also desirable to develop an understanding from a time-dependent point of view, where the requirement is the determination of a polyelectronic time-dependent wavefunction, $\Psi(\mathbf{r},t)$. This requirement implies that the many-electron problem, which is especially difficult for excited states, (as compared with the ordinary ground or low-lying discrete states for which there is no conceptual challenge), is augmented by the necessity of developing and implementing appropriate theory for the computation of $\Psi(\mathbf{r},t)$, satisfying the TDSE for the physical system of interest. Paradigms of processes where knowledge of the electronic $\Psi(\mathbf{r},t)$ is necessary, are the ones resulting from the interaction of short and intense laser pulses with atomic states, leading to ionization, or the time-evolution of an excited state dissipating into a purely continuous spectrum of free particles. In such cases, $\Psi(\mathbf{r},t)$ describes a nonstationary state.

3.1 The state-specific expansion approach (SSEA) for nonstationary systems

The boundary conditions on $\Psi(\mathbf{r},t)$ are that at $t = 0$ it is represented by a square-integrable wavefunction , $\Psi(\mathbf{r},0)$, and that its norm over the whole space of all the coordinates is conserved as it evolves under the influence of the Hamiltonian , H . On the other hand, flux is not conserved if observation concerns only the space defined by $\Psi(\mathbf{r},0)$.

If $\Psi(\mathbf{r},0)$ is an exact square-integrable solution of the TISE , thereby representing a discrete , stationary state , then its nonstationary evolution for $t > 0$ is caused by an interaction operator which is external to the atom . This interaction may be time-independent , as in the case of an external static electric field, or time-dependent, as in the case of dynamic electromagnetic fields. In either case, the atom-field interaction operators can be written explicitly .

On the other hand, there are numerous resonance phenomena that are measured inside the continuous spectrum of the multielectron Hamiltonian , without any external field. If looked at from a time-dependent point of view, they are caused by nonstationary states whose evolution is due to the time-independent nonseparable Hamiltonian itself, be it nonrelativistic or relativistic. Then , $\Psi(\mathbf{r},0)$ ($\equiv \Psi_o(\mathbf{r})$) is not an eigenfunction of H or of an explicit local operator. Instead, it represents a localized wavepacket that must be calculated under appropriate physical constraints . The corresponding assumption is that it is formed at $t = 0$ and has a lifetime that is longer than the duration of excitation . Given the fact that it lies in the continuous spectrum , its construction must be done under constraints that are computationally relevant . The problem is accentuated when the binding potential responsible for the temporary localization is weak. For example, such situations occur with the transient formation of resonance states of ANIs , [1-3, 7,8] and references therein.

If one brings to mind the huge body of work that has been published by numerous workers , over decades , on the problem of solving reliably the Schrödinger equation for just the field-free ground state, it becomes clear that the ab initio calculation and possibility of analysis of $\Psi(\mathbf{r},t)$ for even small atomic and molecular systems is a demanding task .

Regardless of whether H is the time-dependent matter-electromagnetic field interaction or a time-independent polyelectronic Hamiltonian , the approach that has been followed by Mercouris, Komninos and Nicolaides [4-6] is to expand $\Psi(\mathbf{r},t)$ in terms of *state-specific* wavefunctions of the complete set of stationary and quasi-stationary states :

$$\Psi(\mathbf{r},t) = \sum_n a_n(t)\, \psi_n(\mathbf{r}) + \int dE\, b_E(t)\varphi(\mathbf{r},E) \quad . \qquad (1)$$

The $\psi_n(\mathbf{r})$ are square-integrable wavefunctions , representing discrete states or the localized parts of resonance (multiply excited) states, and the $\varphi(\mathbf{r},E)$ are energy-normalized scattering states. The substitution of eq. (1) into the TDSE results in a system of coupled integro-differential equations , involving bound-bound , bound-scattering and scattering-scattering matrix elements with the time-dependent coefficients being the unknowns . Of these matrix elements, special attention must be paid to the field-induced matrix elements involving initial and final states that are either Rydberg or scattering orbitals , since for these not only can there be mathematical questions of singularities but also the electric dipole approximation (EDA) may break down [4,5].

It is immediately clear from eq. (1) that in order to obtain $\Psi(\mathbf{r},t)$ accurately and with good convergence rate, three conditions must be met. These conditions refer to (i) the quality of the wavefunctions in expansion (1) , (ii) the number of the wavefunctions , and (iii) the method of integration of the TDSE over time. These issues have been dealt with in [4-6] and in other applications of the SSEA . One computational advantage is the fact that the state-specific wavefunctions are compact and are handled numerically. From the point of view of interpretation, the SSEA allows the direct determination of the significance of each state of the spectrum to the phenomenon under study. For atom-field interactions, the number of the discrete states in eq. (1) is always relatively small , regardless of the size of the system. It is the continuum that presents the computational difficulty, since , as intensity increases, one has to go to high energies while making the energy step small. On the average, expansion (1) contains about 5000 to 10000 terms, the overwhelming majority being the number of energy-normalized scattering states. Calculations with up to 80000 coupled equations have been carried out , without problems of convergence.

Here I must add that the propagation of $\Psi(\mathbf{r},t)$ in time, starting at t = 0 , implies the possibility of solving the TDSE up to durations that are physically relevant to the problem of interest. In some cases this is possible and in others it is not , since the required integration over time is too long . Two examples for the latter case, can be found in [6,20]. The first such example concerns the ab initio study of the two-photon resonant ionization of the He 1s2s ^1S metastable state [20] . The aim was to compare with observations that were obtained by a weak pulse of duration 5 ns [21] . At the wavelength of 501.7 nm , which is the energy difference between the ^1S and the 1s3p $^1P^o$ states, this is equivalent to $3x10^6$ cycles , and so the time of integration over the whole pulse was computationally prohibitive. Instead, we opted for the understanding of the basic features of the observations via the calculation of a few Rabi oscillations, whereby we determined the dependence of the resonance profile, of the Rabi frequency and of the magnitude of the oscillations, on the frequency and on the peak intensity [20]. In this way, the observed phenomenon [21] of the dependence on intensity of the width of the peak around the resonance frequency emerged naturally from the ab initio solution of the TDSE.

The second example concerns the time evolution of a relativistically autoionizing state, the J = 5/2 level of He^- 1s2s2p $^4P^o$, whose lifetime is very long (about 350 μs [22]) . This level decays into the (He $1s^2$ ^1S + εf) $^2F^o$, J = 5/2 , continuum via spin-spin interactions. The aim was to integrate over many lifetimes so as to obtain the survival probability in the regime where exponential decay breaks down. However, a number of methods for propagating $\Psi(\mathbf{r},t)$ to such long times proved uneconomical and insufficient [6b].

On the other hand, it did prove possible to obtain meaningful $\Psi(\mathbf{r},t)$ and the corresponding survival probability , $P(t) = | \langle\Psi(\mathbf{r},0)|\Psi(\mathbf{r},t)\rangle|^2$, for a few prototypical nonstationary states of Ca , He^- and Li^- , thereby determining the regimes of exponential decay and of deviations from it , for t ≈ 0 and for t >> τ , where τ is the lifetime for exponential decay [6]. Additional discussion and results on this problem are presented in [6]. One interesting result is that for each unstable state , apart from the concept of lifetime it is possible to define and compute a "*stationarity coefficient*" , reflecting the degree of stability around t = 0 , before irreversible exponential decay takes over.

4. THE INTERACTION OF THE ELECTRO-MAGNETIC FIELD WITH HIGH-N RYDBERG STATES AND WITH SCATTERING STATES

During the last 15 years or so, there has been much interest and progress in the experimental and theoretical study of the laser-induced dynamics of excitation of Rydberg levels and wavepackets [23, 24] . One major characteristic of the Rydberg eigenfunctions is their very large extent , since the mean value of their radius grows as n^2 . Similarly, there are laser-induced processes , such as above threshold ionization, that involve the coupling of scattering states , whose energy-normalized wavefunctions extend to infinity .Therefore, for phenomena involving Rydberg-Rydberg or Rydberg-scattering or scattering-scattering state transitions, the analysis of the laser-atom interaction problem within the SSA must take the fact of the large extent , which is specific to such states, into account. This was done recently in [5] , where a new theory for the quantitative understanding of the full laser-atom interaction was formulated and implemented. The justification is presented in the following passage , which I quote from pages 412 and 1440 of [5] :

"Among the fundamental assumptions of atomic, molecular and nuclear spectroscopy is the long-wavelength approximation (LWA) and the concomitant treatment of matter-electromagnetic field interaction in terms of the first few terms of the multipole expansion of the $\vec{A}(\vec{r})$ *operator in the Coulomb gauge, where* $\vec{A}(\vec{r})$ *is the vector potential that is coupled with the momentum* $\vec{p}$ *of the charged particle .The LWA is based on the condition that* $2\pi r/\lambda$ *is much smaller than unity .* λ *is the wavelength of radiation and r signifies the "atomic dimensions" of the two states , (real or virtual), involved in the transition matrix element . Under this condition , the low-order expansion of the vector potential , taken in the transverse gauge, is valid :*

$$\vec{A}(\vec{r}) = \vec{\varepsilon}_{\vec{k}\mu} e^{i\vec{k}\vec{r}} \quad , \quad \vec{\varepsilon}_{\vec{k}\mu}\vec{k} = 0, \quad \mu = 1,2 \tag{2}$$

$$\vec{A}(\vec{r}) = \vec{\varepsilon}_{\vec{k}\mu}[1 + \vec{k}\vec{r} + ...] \qquad \text{(LWA)} \tag{3}$$

$\vec{\varepsilon}_{\vec{k}\mu}$ *is the polarization vector,* μ *numbers the two possible transverse polarizations of the field and* $\vec{k}$ *is the photon wave vector . The dominant term of the interaction based on the multipole expansion is the electric dipole approximation (EDA) , according to which* $\vec{A}(\vec{r}) \approx \vec{A}(0) = 1$. *If symmetry forbids EDA transitions, then the probability of transition between two states is obtained from the matrix elements of the next terms (electric quadrupole, magnetic dipole, etc).*

The validity and wide applicability of the EDA is based on the fact that, in conventional spectroscopic cases, at least one of the two states of the transition matrix element is described by a bound wavefunction with dimensions much smaller than the wavelength , λ *, of the radiation. However, there is no a priori reason for the EDA to hold when both initial-*

and final-state wavefunctions are very extended, having dimensions at least as large as λ. Such cases can appear, even for low-frequency radiation , when the physics involves the coupling of real or virtual states represented by high-n Rydberg or by scattering wavefunctions".

The calculations of [5] on the size of matrix elements obtained within the LWA and from the full interaction showed that there is a physically meaningful difference as the quantum number n increases. In turn , the solution of the TDSE , using expansions of the form of eq.(1) with realistic laser pulses [5b], showed that the time-dependent probability of single-photon excitation from the ground state of Hydrogen to high-n Rydberg levels, implies that the physics of Rydberg wavepacket formation cannot be described reliably by the EDA . This conclusion was drawn from the numerical results of the implementation of the multipolar interaction Hamiltonian [25], which reduces directly to the length form of the EDA , when the LWA is valid.

The multipolar interaction Hamiltonian is given by [25]

$$H_I = e\sum_j\int_0^1 \vec{r}_j\vec{E}_T(\lambda\vec{r}_j)d\lambda - \frac{e}{mc}\sum_j\int_0^1 \lambda\vec{p}_j\vec{r}_j\times\vec{B}_j(\lambda\vec{r}_j)d\lambda + \frac{e^2}{2mc^2}\left\{\sum_j\int_0^1 \lambda\vec{r}_j\times\vec{B}_j(\lambda\vec{r}_j)d\lambda\right\}^2 \quad (4)$$

and involves the electric , $\vec{E}$, and magnetic , $\vec{B}$, fields rather than the vector potential. The λ-integration permits the writing of expressions involving infinite expansions in a compact form. The three terms above are the electric, the paramagnetic and the diamagnetic operators. In reference [5a] it was shown how to reduce eq.(4) to workable expressions for the computation of matrix elements. The calculations of [5b] employed only the full electric operator. However, for very strong fields relative to the states of interest , the magnetic contributions ought to be accounted for as well.

5. MULTIPLY EXCITED BOUND AND RESONANCE STATES OF ATOMIC NEGATIVE IONS

Given an excited state of an N-electron neutral atom, it is possible that the addition of an electron to an outer empty spin-orbital will produce either a resonance (nonstationary) state of the (N+1)-electron ANI or, if energy , symmetry and parity permit it, a bound excited state of the ANI . From the formal point of view, the difference between the two types of states is fundamental, since in the case of resonance states the continuous spectrum has physical significance [2]. On the other hand, theoretical and computational analysis shows that the crucial characteristic for the solution of the corresponding TISE is the possibility of producing a reliable (N+1)-electron square-integrable wavefunction, Ψ_o , of energy $E_o = <\Psi_o|H|\Psi_o>$. If E_o is below the energy of the corresponding N-electron excited state and if symmetry and parity do not allow Coulomb interaction with the continuous spectrum of the N-electron atom, Ψ_o represents a discrete state of the nonrelativistic Hamiltonian.

The zero order description of excited states of ANIs of both categories is achieved by either one main configuration or by a superposition of multiply excited configurations.For example, the addition of a 2p electron to the $2p^2$ ^3P doubly excited state of He, which is a discrete state even though its energy is embedded in the scattering continuum ,59.65 eV above the ^1S He ground state , may lead to a He^- state of $^4S^o$ symmetry, $2p^3$ $^4S^o$, which is below the He $2p^2$ ^3P by about 0.32 eV and therefore it is bound due to symmetry. On the other hand, the $2p^3$ configuration can also give rise to $^2P^o$ symmetry.In this case, configurational interaction is strong, while there is also mixing between bound and open channel scattering components.Therefore,the corresponding state is nonstationary.

For an extensive review on resonance states of ANIs, see [26].For a review of bound excited states of ANIs, until the end of the 1980s, see [27].

5.1 Multiply excited bound states of ANIs having all their electrons with unpaired spins

It turns out that the formation of a discrete state of an ANI can take place for very high energies, even though this appears , at first thought, improbable. This fact was pointed out in [7] , with quantitative predictions based on systematic SSA calculations for states of B^- , C^- , N^- and O^- . Specifically, having recognized certain special electronic structure characteristics of such states, Piangos and Nicolaides [7] first hypothesized that multiply excited states with all their spins unpaired must constitute the ceiling of the discrete spectra of a variety of ANIs , at least in the low-Z part of the Periodic Table. Indeed, after the identification of zero-order configurations according to a particular building-up procedure, correlated wavefunctions and electron affinities were calculated , with inter-and intrashell correlations included. Very interesting results were obtained. For example, the O^- $1s2s2p^33s3p^3$ ^{10}S state lies a full 600.6 eV above the O $1s^22s^22p^4$ ^3P ground state , and yet it is bound!

5.2 Comments on multiply excited resonance states of ANIs

The fundamental reason for the formation of resonance states of ANIs is the capacity of the (N+1)-electron system to come into a nonstationary state whose wavefunction is temporarily (t=0) localized as a wavepacket, Ψ_o , of energy $E_o = <\Psi_o|H|\Psi_o>$, which is a local energy minimum inside the continuous spectrum. This possibility depends on the electronic structure and symmetry of the state and on the result of the competition between the corresponding attractive and repulsive forces . Thus, whereas the localized part of autoionizing states of neutral atoms and of positive ions is bound sufficiently strongly , the formation of resonance states in ANIs , where the nuclear attraction on the outer electrons is weak, involves , in general, subtleties of interelectronic interactions. These facts have consequences as to the degree of difficulty and complexity of computation , as to the possibility of extracting pictures of electronic structure and as to the efficiency of the existing theoretical approaches for identifying resonance states [1-3,8].

Of special interest are electronic structures which , in zero order , are multiply excited and are characterized by strong configurational mixings. Prototypical paradigms are the n =2 and n =3 intrashell triply excited states of He^- [26,8] , such as the well-known $2s^22p\ ^2P^o$ resonance. According to the SSA , given the constraints that are imposed by the nature of such states, the calculation of Ψ_o is achieved optimally in two phases . The first aims at the solution of appropriate multiconfigurational Hartree-Fock (MCHF) equations . (See, however, [3] , and references to earlier work on the resonance spectrum of H^-).

When the effective nuclear attraction on the outer electrons is strong , i.e. in cases of positive ions or even of neutral atoms , experience has shown that , in general, the choice of configurations and the corresponding convergence can be achieved without significant conceptual or numerical difficulty. Once the MCHF solution , Φ_{MCHF} , is known, the remaining part of Ψ_o is obtainable variationally , to an accuracy that is relevant to the needs and significance of the problem.

On the other hand, when it comes to ANIs , the effective nuclear attraction is weakened considerably . The ab initio choice of the important zero order function space and its self-consistent optimization can range from difficult to impossible. Critical to such calculations is the role of "*open channel-like*" (OCL) configurations and the possibility of obtaining convergence to a localized wavepacket in their presence [8] . Furthermore, when going to a configuration interaction level, beyond Φ_{MCHF} , it is often expedient to use nonorthonormal basis functions in order to represent optimally different parts of electron correlation . In fact , high accuracy is often crucial since , given the small energy differences , the density of multiply excited resonance states becomes relatively high even at low excitation levels. Obviously, in such cases one needs reliable positions and , often, widths , for the identification of resonance spectra [8].

Once and E_0 have been calculated, the aim is to obtain the remaining energy due to the interaction with the open channels , the so-called energy shift, and the decay width. The computation of these quantities imply the application of theory which incorporates the effect of the continuous spectrum. This can be achieved by implementing methods that employ either real functions of the scattering type or L^2 complex functions spaces [1-3,8,9,15,19].

6. CONCLUSIONS

The above brief discussion and comments dealt with the implementation of the state-specific approach for the solution of the multielectron time-independent and time-dependent Schrödinger equations , with emphasis on a variety of prototypical problems involving excited electronic states of atoms and small molecules. A gleaning of old as well as of recent developments and applications by the Athens group was done , concerning issues of the theory , analysis and computation of electronic structure, of spectroscopy and of time-dependent dynamics. As a conclusion, I think that the statement from the Abstract of [1] is still relevant: " *If, in conjunction with suitable computational methods, attention is given to the state-specific choice and optimization of function spaces, especially when studying excited states, the accuracy of the description of electronic structure and of the results of computation of properties increases , while the complexity of the many-electron problem is reduced significantly. These facts allow the consistent understanding of the interplay between major features of electronic structure and properties or processes, as well as the practical numerical solution of computationally very demanding problems*".

REFERENCES

1. C.A.Nicolaides, 1996, Int.J.Quantum Chem. **60** , 119.
2. C.A.Nicolaides, 1999, Int.J.Quantum Chem. **71** , 209.
3. M.Bylicki and C.A.Nicolaides , 2002, Phys.Rev. **A 65** , 012504.

4. Th.Mercouris,Y.Komninos, S.Dionissopoulou and C.A.Nicolaides, 1994, Phys.Rev. **A 50** , 4109 and Th.Mercouris, Y.Komninos, S.Dionissopoulou and C.A.Nicolaides, 1997, J.Phys. **B 30** , 2133.
5. Y.Komninos, Th.Mercouris and C.A.Nicolaides, 2002, Phys.Rev. **A 65** , 043412 and Th.Mercouris, Y.Komninos and C.A.Nicolaides, 2002, J.Phys. **B 35** , 1439.
6. C.A.Nicolaides and Th.Mercouris , 1996, J.Phys. **B 29** , 1151, Th.Mercouris and C.A.Nicolaides, 1997, J.Phys. **B 30** , 811 and Th.Mercouris and C.A.Nicolaides, 2002, Phys.Rev. **A 65** , 012112.
7. N.A.Piangos and C.A.Nicolaides, 1998, J.Phys. **B 31** , L147.
8. C.A.Nicolaides and N.A.Piangos, 2001, Phys.Rev. **A 64** , 052505, C.A.Nicolaides and N.A.Piangos, 2001, J.Phys. **B 34** ,99 and N.A.Piangos and C.A.Nicolaides, 2001, J.Phys. **B 34** , L633.
9. G.Miecznik,T.Brage and C.Froese-Fischer, J.Phys. **B 25** ,4217 (1992) and G.Aspromallis, Ch.Sinanis and C.A.Nicolaides, 1996, J.Phys. **B 29** , L1.
10. D.R.Beck and C.A.Nicolaides, 1978. In *Excited States in Quantum Chemistry* (C.A.Nicolaides and D.R.Beck eds.), Reidel, Dordrecht,, pp. 105,143, 329.
11. C.Froese-Fischer, 1978, Comp.Phys.Commun. **14** , 145.
12. E.A.McCullough, 1986, Comp.Phys.Rep. **4**, 265.
13. I.D.Petsalakis, G,Theodoracopoulos, C.A.Nicolaides and R.J.Buenker, 1984, J.Chem.Phys. **81** , 3161 and 5952.
14. C.A.Nicolaides, N.C.Bacalis and Y.Komninos, 1992, Chem.Phys.Lett. **192** , 486.
15. Y.Komninos, G.Aspromallis and C.A.Nicolaides, 1995, J.Phys. **B 28** , 2049 and Y.Komninos, G.Aspromallis and C.A.Nicolaides, 1996, J.Phys. **B 29** , L193.
16. C.A.Nicolaides and D.R.Beck, 1975, Chem.Phys.Lett. **36** , 79.
17. C.A.Nicolaides, 1987. In *Giant resonances in atoms, molecules and solids* (J.P.Connerade, J.M.Esteva and R.C.Karnatak eds.), Plenum, New York, p.213, D.R.Beck and C.A.Nicolaides, 1982, Phys.Rev. **A 26** , 857 and D.R.Beck and C.A.Nicolaides, 1986, Phys.Rev. **A 33** , 3885.
18. C.A.Nicolaides, 1993, J.Phys. **B 26** , L291.
19. Th.Mercouris, C.Haritos and C.A.Nicolaides, 2001, J.Phys. **B 34** , 3789.
20. C.A.Nicolaides, S.Dionissopoulou and Th.Mercouris, J.Phys. 1996, **B 29** , 231.
21. H.Haberland , M.Oschwald and J.T.Broad, 1987, J.Phys. **B 20** , 3367 and H.Haberland and M.Oschwald, 1988, J.Phys. **B 21** ,1183.
22. L.M.Blau, R.Novick and D.Weinflash, 1970, Phys.Rev.Lett. **24** , 1268 and T.Andersen, L.H.Andersen, P.Balling, H.K.Haugen, P.Hvelplund,W.W.Smith and K.Taulbjerg , 1993, Phys.Rev. **A 47** , 890.
23. G.Alber and P.Zoller, 1991, Phys.Rep. **199** , 231.
24. J.J.Jones and L.D.Noordam, 1997, Adv.At.Mol.Phys. **38** , 1.
25. E.A.Power and S.Zienau, 1959, Phil. Trans.R.Soc. **251** , 427, R.G.Wooley, 1975, Adv.Chem.Phys. **33** ,153 and E.A.Power and T.Thirunamachandran, 1980, Proc.R.Soc. **A 372** , 265.
26. S.J.Buckman and C.W.Clark, 1994, Rev.Mod.Phys. **66** , 539.
27. C.A.Nicolaides, G.Aspromallis and D.R.Beck, 1989, J.Mol.Struc.(Theor.) **199** , 283.

ACCURATE DETERMINATION OF PARTIAL RATES FROM A MULTICHANNEL WAVE FUNCTION

R. LEFEBVRE
Laboratoire de Photophysique Moléculaire
Université de Paris-Sud, 91405 Orsay, France
and U.F.R. de Physique Fondamentale et Appliquée,
Université Pierre et Marie Curie, 75231 Paris, France

AND

B. STERN
U.F.R. de Physique Fondamentale et Appliquée, Université Pierre et Marie Curie, 75231 Paris, France

Abstract.

The partial rates of a decay process can be derived from the asymptotic fluxes of the multichannel wave function of the associated resonance. A propagation scheme of the coupled equations combining a variable step with exterior complex scaling and yielding these fluxes is described. The method is tested on the low intensity ionization rates of hydrogen calculated in the Kramers-Henneberger frame. Despite a difference by 7 orders of magnitude of the partial rates for one- and two-photon ionization derived from the same wave function, accurate determination is demonstrated.

1. Introduction

Decay processes are characterized by their total rates, but also by their partial rates when there is more than one route to the final reaction products. Since only very few situations can be treated analytically, much effort has been devoted to the implementation of efficient numerical procedures for the calculation of these rates. For the total rate a familiar approach is the determination of the complex energy of the metastable (or resonant) state, the information being then in the imaginary part of the energy. A

E.J. Brändas and E.S. Kryachko (eds.),
Fundamental World of Quantum Chemistry, Vol. II, 107–115.

great variety of methods have been advanced for the partial rates: Analysis of the elements of the **S** matrix [1, 2]; Stabilization methods [3, 4]; Analysis of the fluxes of the multichannel resonance wave function calculated either with a real reaction coordinate [5-7] or a complex reaction coordinate [8-13]; Propagation of a wavepacket [14]. It is also possible to use the exterior complex scaling method [15, 16] which offers the simplicity of allowing a flux analysis in the real part of the integration axis [17]. This is the method which is reviewed in this note. An application is made to the low intensity one- and two-photon ionization rates of hydrogen, where exact analytical results allow for an assessment of the procedure.

2. Numerical Methodology

Let the dynamics of a system be described by a set of channel functions $U_i(r)$ obeying the coupled equations:

$$[-\frac{\hbar^2}{2m}\frac{\partial}{\partial r^2} + V_{i,i} - E]U_i(r) + \sum_{j\neq i} V_{i,j}U_j(r) = 0 \tag{1}$$

Such equations are generally produced by the elimination of known basis functions of some other coordinates, describing for example rotational motion. The remaining (reaction) coordinate r is the relative distance between two interacting entities. m is the reduced mass of these entities. These differential equations are to be solved with boundary conditions imposed both at short range (zero at the origin) and at long range. If asymptotically the channel functions go to a combination of both ingoing and outgoing waves, the equations describe a collision process, an **S** matrix can be determined and from it transition probabilities can be calculated. If, on the other hand, only outgoing waves are imposed asymptotically (Siegert-like boundary conditions [18]), the equations describe a decay process. Total and partial rates are then the target of the calculation. The technique for the solution is basically the same in the two cases. We need to propagate the multichannel wave function through a discretization procedure. The Fox-Goodwin method [19] is particularly efficient [20]. If r and $r+h$ are two adjacent points on the grid, an inward matrix (with a label i) is defined as:

$$\mathbf{P}^{\mathrm{i}}(\mathrm{r}) = \mathbf{Q}^{\mathrm{i}}(\mathrm{r}+\mathrm{h})[\mathbf{Q}^{\mathrm{i}}(\mathrm{r})]^{-1}. \tag{2}$$

If r and $r-h$ are two adjacent points on the grid, we define the outward matrix (with a label o) as:

$$\mathbf{P}^{\mathrm{o}}(\mathrm{r}) = \mathbf{Q}^{\mathrm{o}}(\mathrm{r}-\mathrm{h})[\mathbf{Q}^{\mathrm{o}}(\mathrm{r})]^{-1}. \tag{3}$$

$\mathbf{Q}^{\mathrm{i}}(\mathbf{Q}^{\mathrm{o}})$ is a matrix of independent vector solutions $\mathbf{U}(r)$ of the coupled-channel equations satisfying the outer(inner) boundary conditions. If $r-h_1$,

r and $r + h_2$ are three adjacent points on the grid, the propagation is done with the two equations:

$$\mathbf{P}^{\mathrm{i}}(\mathrm{r}-\mathrm{h}_1)) = [\boldsymbol{\beta}(\mathrm{r}) - \boldsymbol{\alpha}(\mathrm{r}+\mathrm{h}_2)\mathbf{P}^{\mathrm{i}}(\mathrm{r})]^{-1}\boldsymbol{\gamma}(\mathrm{r}-\mathrm{h}_1), \tag{4}$$

$$\mathbf{P}^{\mathrm{o}}(\mathrm{r}+\mathrm{h}_2) = [\boldsymbol{\beta}(\mathrm{r}) - \boldsymbol{\alpha}(\mathrm{r}-\mathrm{h}_1)\mathbf{P}^{\mathrm{o}}(\mathrm{r})]^{-1}\boldsymbol{\gamma}(\mathrm{r}+\mathrm{h}_2). \tag{5}$$

$\boldsymbol{\alpha}(r)$, $\boldsymbol{\beta}(r)$ and $\boldsymbol{\gamma}(r)$ are the Numerov matrices for a variable step propagation. In an exterior complex scaling procedure, the three adjacent points may be either all real, or all complex (after complex scaling of the reaction coordinate) or mixed (near the transition point between real and complex coordinates), Let r and $r + h$ be the two points chosen for the matching of the inward and outward solutions. The matching condition has the form [21]:

$$det|\mathbf{P}^{\mathrm{i}}(\mathrm{r}) - (\mathbf{P}^{\mathrm{o}}(\mathrm{r}+\mathrm{h}))^{-1}| = 0. \tag{6}$$

For a one-channel case described by a wave function $U(r)$, this relation reduces to a transparent condition for the two functions produced by the inward and outward propagations:

$$[\frac{U^i(r+h)}{U^i(r)} - (\frac{U^o(r)}{U^o(r+h)})^{-1}] = 0 \tag{7}$$

The matching condition generates a complex energy if the inward matrix incorporates outgoing boundary conditions. It is an implicit relation which has to be solved iteratively, starting from a trial value for the energy. Once the complex eigenenergy has been determined, the multichannel wavefunction, written as a vector $\mathbf{U}(\mathrm{r})$, can be calculated at the matching point because its components satisfy the set of homogeneous linear equations:

$$[\mathbf{P}^{\mathrm{i}}(\mathrm{r}) - (\mathbf{P}^{\mathrm{o}}(\mathrm{r}+\mathrm{h}))^{-1}]\mathbf{U}(\mathrm{r}) = 0, \tag{8}$$

The wavefunction at the other grid points can be obtained from:

$$\mathbf{U}(\mathrm{r}+\mathrm{h}) = (\mathbf{P}^{\mathrm{o}}(\mathrm{r}+\mathrm{h}))^{-1}\mathbf{U}(\mathrm{r}) \quad \mathbf{U}(\mathrm{r}-\mathrm{h}) = (\mathbf{P}^{\mathrm{i}}(\mathrm{r}-\mathrm{h}))^{-1}\mathbf{U}(\mathrm{r}) \quad \text{etc...} \tag{9}$$

For the calculation of partial widths in the open channels the Humblet-Rosenfeld formulation [5] is appropriate. We recall briefly the derivation of the partial width expression. After multiplication of Eq. (1) by U_i^*, addition of all equations and subtraction of the complex conjugate of the resulting expression, there is obtained:

$$-\frac{1}{2m}\sum_i[U_i(r)^*U_i^{"}(r) - U_i(r)U_i^{*"}(r)] - (E-E^*)\sum_i|U_i(r)|^2 = 0 \tag{10}$$

For a resonance the energy is written $E = E_R - i\Gamma/2$ and $(E - E^*)$ is $-i\Gamma$. Integration from the origin (where all functions vanish) to some point r_0 in the asymptotic region produces:

$$\frac{1}{2m}\sum_i [U_i^*(r_0)U_i'(r_0) - U_i(r_0)U_i^{*\prime}(r_0)] + i\Gamma \int_0^{r_0} |U_i(r_0)|^2 dr = 0 \qquad (11)$$

Solving this equation for $\Gamma/2$ gives:

$$\frac{\Gamma}{2} = \frac{1}{2m}\frac{\sum_i Im[U_i^*(r_0)U_i'(r_0)]}{\sum_j \int_0^{r_0} |U_j(r)|^2 dr}. \qquad (12)$$

The total half-width is expressed as a normalized sum of fluxes which are identified to the partial half-widths. Only the open channels contribute in the numerator if r_0 is chosen far enough to ensure vanishing of the closed channel functions. However *all* channels contribute to the denominator. Independence with respect to r_0 is easily proven. This analysis is valid only if the channels are asymptotically decoupled. This is the case for the multichannel situation described in the next Section.

3. The multichannel wave function of the hydrogen atom in the Kramers-Henneberger frame.

Most studies of the ionization process of the hydrogen atom have as an objective the determination of the total rate. This is obtained from the complex resonance energy $E_R - i\Gamma/2$. The total rate is then Γ (in atomic units). However the ionization channels are indexed by the number of absorbed photons and the electron angular momentum. The latter index is generally irrelevant and a summation over angular momentum quantum numbers compatible with the number of absorbed photons is performed. If the photon energy is above the ionization threshold, Above Threshold Ionization (ATI) is possible. This means that not only is there the possibility for one photon to be absorbed, but also for two, three, etc.. photons. This process is in principle favoured by an increase of the intensity of the laser. However a striking feature of the high intensity rate at high frequency is that the atom tends to stabilize [22]. To account for the behaviour of atoms exposed to a high-intensity field, many algorithms have been developped. To quote a few, L^2 functions [23, 24] or a finite element basis [25], both with complex rotation, give the energy through a diagonalization procedure. This is also the case with a Sturmian basis [26]. Coupled-channel approaches yielding equations of the type discussed in the previous Section have been formulated. This is done sometimes with a change of gauge or representation as integration along the radial coordinate goes on [27, 28].

The coupled equations can also be written in the Kramers-Henneberger frame [29, 30] which ensures an asymptotic decoupling of the channels. The efficiency of this method has been demonstrated by Faisal and coll. [31, 32]. Low intensity checks of the methods developped for high intensity have been generally limited to the total rate (see however [23, 27, 28]). It is true that there is such a difference in the orders of magnitudes of one- and two- photon rates for an above-threshold case that it is legitimate to compare the total rate with the one-photon rate. However there is more information than this in the wave function. This is the purpose of this note: we will use an efficient method developped for high-intensity studies and apply it to the low intensity case, with the purpose to check if the wave function has information on processes which differ widely in their efficiency. We follow with a few changes the method developped by Faisal and coll. [31, 32]. In the Kramers-Henneberger representation[29, 30], an electron exposed only to a continuous wave laser field has a free wave as its wave function. It is this frame which guarantees the asymptotic decoupling of the channels. We have described elsewhere [33] our own implementation of the method for a high-intensity study of the partial rates. Exterior complex scaling of the radial coordinate and a variable step propagation scheme of the coupled equations were already used. We will sketch very briefly the formalism, since it is developped at length in the papers of Faisal and his group [31, 32]. The Schrödinger equation in the Kramers-Henneberger frame is for hydrogen (in atomic units):

$$[-\frac{1}{2}\Delta - \frac{1}{|\vec{r} - \alpha_0 \hat{z} cos(\omega t)|}]\Psi(\vec{r}, t) = i\frac{\partial}{\partial t}\Psi(\vec{r}, t) \tag{13}$$

α_0 is equal to E_0/ω^2, where E_0 is the peak amplitude of an electric field linearly polarized in the laboratory z direction and ω the field frequency. After Fourier expansion of the potential:

$$-\frac{1}{|\vec{r} - \alpha_0 \hat{z} cos(\omega t)|} = \sum_{n=-\infty}^{n=+\infty} V_n(\vec{r}) exp[in\omega t] \tag{14}$$

introduction of the partial wave expansion of the wave function:

$$\Psi(\vec{r}, t) = \frac{1}{r}\sum_{n,\ell,m} U_{n,\ell,m}(r) Y_{\ell,m}(\hat{r}) exp[-iEt + n\omega t] \tag{15}$$

into the wave equation produces for the unknown functions of the radial coordinate $U_{n,\ell,m}(r)$ the set of coupled equations:

$$[E - n\omega]U_{n,\ell,m}(r) \quad = \quad [-\frac{1}{2}\frac{d^2}{dr^2} + \frac{\ell(\ell+1))}{2r^2}]U_{n,\ell,m}(r)$$

$$+ \sum_{n',\ell',m'} < \ell, m|V_{n-n'}(\vec{r})|\ell', m' > U_{n',\ell',m'}(r)(16)$$

We describe now in more detail the advantage of the variable-step procedure in the present context. We have previously shown on some multichannel models [34] that in an inward propagation the transition from the complex to the real part of the integration grid with the variable step technique produces an inward $\mathbf{P}^{\mathrm{i}}$ matrix on the real part of the grid which is very close to that provided by an entirely real grid and explicit introduction of Siegert asymptotic conditions. Another problem is related to the form of the potential matrix of the coupled equations [32]. Depending on the value of the radial coordinate, the elements of the potential matrix assume two different forms. The cross-over from one form to the other occurs when r reaches the value α_0. In order to reach the low intensity limit, α_0 must take small values. In the present calculations α_0 is given systematically the value 0.001 a.u., corresponding, at the threshold frequency, to an intensity of 2.2×10^9 W/cm^2. We choose to take for the first 100 steps the value $h = 0.00004$ a.u.. In this way the first 25 steps correspond to the internal region (with $r < \alpha_0$). The step is then changed into $h = 0.004$ a.u.. If with this value for h we wanted to have 25 steps in the internal region, we would need to take the value 0.1 for α_0, with for the intensity $I = 2.2 \times 10^{13}$ W/cm^2. This is no longer appropriate for the validity of the perturbative approach. By choosing a variable step, we partition the integration grid into two regions associated to the two forms taken by the potential matrix. These different treatments of the two regions replace, in a sense, the change of gauge which is sometimes performed when going from short to long range [27, 28].

4. Comparaison of analytic perturbative results and numerical coupled channel estimates

For one-photon ionization, analytic perturbative results go back to the early days of quantum mechanics (see the review by Bethe and Salpeter [35]). For two photons there is a rich literature, with a variety of methods which finally agreed to produce reliable results [36-41]. Table 1 shows some results obtained when the photon frequency exceeds the ionization threshold (0.5 a.u. at low intensity). The photon quantum number goes from -1 to $+3$. $n = -1$ and $n = 0$ correspond to two closed channels, the latter describing in fact the initial state. The electron angular quantum number goes up to 3. This choice gives a total of 10 channels. The number of points on the grid is 16000. The step for the first 100 points, as mentiond above, is $h = 0.00004$ a.u.. The step is then changed into $h = 0.004$ a.u.when r reaches α_0. Complex rotation starts after 10000 steps, with an angle

TABLE 1. Low intensity above threshold coupled-channel 1- and 2-photon rates of hydrogen. The 1-photon rate Γ_1 and 2-photon rate Γ_2 are in atomic units. (cc) stands for coupled-channel estimates. (B-S) stands for the Bethe-Salpeter estimate [35]. (K) denotes the 2-photon cross section given by Karule [41]. The unit for σ_2/I is cm^4/W. Γ_1 and Γ_2 which differ by 7 orders of magnitude are obtained from an asymptotic analysis of the same wave function. Parenthesis are powers of 10.

λ (Å)	ω (a.u.)	Γ_1 (cc)	Γ_1 (B-S)	Γ_2 (cc)	σ_2/I (cc)	σ_2/I (K)
20	22.7821	1.28589(-7)	1.28590(-7)	3.4997(-13)	1.606(-44)	1.61(-44)
100	4.556412	1.82829(-7)	1.82830(-7)	1.1328(-13)	4.065(-40)	4.09(-40)
200	2.278206	1.91629(-7)	1.91629(-7)	6.5216(-14)	2.995(-38)	3.02(-38)
300	1.518804	1.89673(-7)	1.89673(-7)	4.6401(-14)	3.642(-37)	3.67(-37)
400	1.139103	1.84653(-7)	1.84653(-7)	3.6320(-14)	2.135(-36)	1.15(-36)
500	0.911282	1.78606(-7)	1.78606(-7)	3.0044(-14)	8.435(-36)	8.49(-36)
600	0.756412	1.72122(-7)	1.72122(-7)	2.5855(-14)	2.665(-35)	2.62(-35)
700	0.650916	1.65979(-7)	1.65979(-7)	2.2912(-14)	6.772(-35)	6.83(-35)
800	0.569552	1.59865(-7)	1.59885(-7)	2.0771(-14)	1.562(-34)	1.581(-34)
900	0.506268	1.53563(-7)	1.54076(-7)	1.7187(-14)	2.950(-34)	3.33(-34)

$\theta = 0.5$ rad.. The matching point has index 1000. We see the very good agreement (six figures) of the one-photon rate with the exact result from the Bethe-Salpeter formula. The agreement with the results of Karule [41] for the two-photon cross sections is very satisfactory, in view of the fact that the flux analysis gives rates for the two-photon process which are 7 orders of magnitude lower that those of the one-photon process. The only line in Table 1 with some significant deviation between our two-photon cross section and that of Karule [41] is for ω approaching the threshold. This is a more difficult case for a coupled-channel approach, since the interaction region between the closed initial channel and the final open channels may be very far from the origin. This is also reflected in the slight deterioration of the one-photon rate.

We recall that the low intensity two-photon cross section σ_2 is related to the two-photon partial rate Γ_2 by the formula:

$$\frac{\sigma_2}{I} = \frac{\Gamma_2 \omega}{I^2} \tag{17}$$

In order to obtain σ_2/I in cm^4/W, Γ_2, which is produced by the code in a unit which is the inverse of the atomic unit of time, has to be converted in sec^{-1} and ω in joules. The peak amplitude of the electric field, E_0, is determined by choosing the parameter $\alpha_0 = E_0/\omega^2$ to be systematically 0.001 au. The intensity I follows from the fact that $E_0 = 1$ a.u. corresponds

TABLE 2. Low intensity below threshold coupled-channel 2-photon rates of hydrogen. Γ_2 is in a.u. and σ_2/I in cm^4/W. (cc) stands for coupled-channel estimates, while column (K) refers to the values given for the 2-photon cross section by Karule [41]. Parenthesis are powers of 10.

λ ($\mathring{A}$)	ω (a.u.)	Γ_2 (cc)	σ_2/I (cc)	σ_2/I (K)
1100	0.414222	5.6219(-15)	3.931(-34)	4.(-34)
1200	0.379703	4.9324(-13)	634.1(-34)	642.(-34)
1300	0.340495	5.6669(-14)	127.6(-34)	127.(-34)
1400	0.325460	2.2350(-14)	84.53(-34)	84.5(-34)
1600	0.284777	9.5311(-15)	91.56(-34)	91.5(-34)
1700	0.268026	9.8872(-15)	101.4(-34)	103.(-34)

to $I = 3.51 \times 10^{16}$ W/cm^2. Efficiency of the variable step algorithm is demonstrated both by the abrupt change imposed in the real part of the integration axis at $r = \alpha_0$ and the change from real to complex coordinate inherent to the exterior complex rotation method.

For sake of completeness we give also in Table 2 some below threshold results and compare them with the numbers provided by Karule [41]. The total width is now practically due only to the two-photon processes, with very small contributions from higher order processes. Satisfactory agreement is again observed, reproducing the rapid oscillations of the cross section.

5. Conclusion

The available analytic above threshold ionization rates of ground state hydrogen in the perturbative regime offer a challenging case for the study of the relation between partial rates and asymptotic fluxes of a multichannel wave function. The main result which emerges is the confirmation of the power of the Humblet-Rosenfeld formalism [5] in such an extreme case where the fluxes in different channels differ by many orders of magnitude.

References

1. Macek, J. Phys Rev A 1970, 2, 1101.
2. Schwenke, D.W.; Truhlar, D.G. J Chem Phys 1987, 87, 1085.
3. Fels, M.F.; Hazi, A.U. Phys Rev A 1972, 5, 1231.
4. Tucker, S.T.; Truhlar, D.G. J Chem Phys 1988, 88, 3667.
5. Humblet, J.; Rosenfeld, L. Nucl Phys 1961, 26, 529.
6. Friedmann, R.S.; Du, M.L.; Dalgarno, A. J Chem Phys 1990, 93, 2375.

7. Alexander, M.H. J Chem Phys 1991, 95, 8931.
8. Yaris, R.; Taylor, H.S. Chem Phys Lett 1979, 66, 505.
9. McCurdy, C.W.; Rescigno, T.N. Phys Rev A 1979, 20, 2346.
10. Noro, T.; Taylor, H.S. J Phys B 1980, 13, L377.
11. Bacic, Z.; Simons, J. Int J Quantum Chem 1982, 21, 727.
12. Peskin, U.; Moiseyev, N.; Lefebvre, R. J Chem Phys 1990, 92, 2902.
13. Moiseyev, N.; Peskin, U. Phys Rev A 1990, 42, 255.
14. Jolicard, G.; Billing, G.D. J Chem Phys 1992, 97, 997.
15. Nicolaides, C.A.; Beck, D.R. Phys Lett A 1978, 65, 11.
16. Simon, B. Phys Lett A 1979, 71, 211.
17. Chrysos, M.; Atabek, O.; Lefebvre, R. Phys Rev A 1993, 48, 3844, 3855.
18. Siegert, A.F.J. Phys Rev 1939, 56, 750.
19. Fox, L. The Numerical Solutions of Two-Point Boundary Value Problem in Ordinary Differential Equations; Oxford Univ. Press: London, 1975.
20. Norcross, D.W.; Seaton, M.J. J Phys B: At Mol Phys 1973, 6, 604.
21. Atabek, O.; Lefebvre, R. Phys Rev A 1980, 18, 17.
22. Gavrila, M. Atomic Structure and Decay in High-Frequency Fields, in "Atoms in Intense Laser Fields" (Editor: M. Gavrila); Academic Press, Boston, 1992, pp. 435-508.
23. Maquet, A.; Chu, S.-I.; Reinhardt, W. P. Phys Rev A 1983, 27, 2946.
24. Plummer, M.; McCann, J.F. 1995 J Phys B: At Mol Opt Phys 1995, 28, 4073.
25. Scrinzi, A.; Elander, N.; Piraux, B. 1993 Phys Rev A 1993, 48, R2527.
26. Dörr, M.; Potvliege, R.M.; Proulx, P.; Shakeshaft, R. 1991 Phys Rev A 1991, 43, 3729.
27. Giusti-Suzor, A.; Zoller P. Phys Rev A 1987, 36, 5178.
28. Dörr, M.; Burke, P.G.; Joachain, C.J.; Noble, C.J.; Purvis, J.; Terao-Dunseath, M. J Phys B: At Mol Opt Phys 1993, 26, L275.
29. Kramers, H.A. Quantum Mechanics; North Holland: Amsterdam, 1956.
30. Henneberger, W.C. Phys Rev A 1968, 21, 838.
31. Dimou, L.; Faisal, F.H.M. Phys Rev Lett. 1987, 59, 872.
32. Lambrecht, U.; Dimou, L.; Faisal, F.H.M. Laser Phys 1998, 3, 440.
33. Lefebvre R.; Stern B. Int J of Quantum Chem 2001, 84, 552.
34. Chrysos, M.; Lefebvre, R. J Phys B: At Mol Opt Phys 1993, 26, 2627.
35. Bethe, H.A.; Salpeter, E.E. Quantum mechanics of one- and two-electron systems, Encyclopedia of Physics, vol XXXV; Springer: Berlin, 1957
36. Zernik, W. Phys Rev A 1964, 135, 51.
37. Chang, T.N.; Tang, C.L. Phys Rev A 1969, 185, 42.
38. Klarsfeld, S. Lett Nuovo Cim 1970, 3, 395.
39. Gontier, Y.; Trahin, M. Phys Rev A 1971, 5, 1896.
40. Khristenko, S.V.; Vetchinkin, S.I. Opt Spektr, 1976, 40, 417.
41. Karule, K.S. J Phys B: At Mol Opt Phys 1978, 11, 441.

GREEN FUNCTION FOR ELASTIC SCATTERING FROM OPEN-SHELL MANY-BODY TARGETS

Ofir E. Alon and L. S. Cederbaum
Theoretische Chemie, Physikalisch-Chemisches Institut, Universität Heidelberg, Im Neuenheimer Feld 229, D-69120 Heidelberg, Germany

ABSTRACT

The standard one-particle Green function (GF) is extended to account for open-shell many-body targets. This extended GF, which we term the open-shell Green function (OSGF), possesses information about the energies and intensities of ionization and attachment processes in degenerate targets. The OSGF is shown to satisfy a generalized Dyson equation and to possess a self-energy (kernel) which is an *exact* optical potential (OP) for elastic scattering from open-shell targets. Moreover, by utilizing general symmetry arguments we show that to *each* scattering process with well-defined quantum numbers there is a *symmetry-irreducible* GF and OP, thus reducing the multichannel scattering problem to symmetry-irreducible ones. The symmetry-irreducible OP's are the analogs of the standard (non-degenerate) OP of Bell and Squires [Phys. Rev. Lett. **3**, 96 (1959)], they exhibit similar analytic properties and, for most atomic and molecular systems, can be evaluated through the usual diagrammatic techniques. Working equations for the symmetry-irreducible optical wavefunctions are constructed. As an explicit example we work out in detail the case of spin-half particles and open-shell targets with spin degeneracy.

I. INTRODUCTION

Elastic scattering is a basic experimental tool to study the internal structure of composite many-body targets. In comparison to closed-shell targets, scattering from open-shell targets is more intricate since for *any* energy of the projectile the scattering process is, in general, a multichannel one. In particular, spin-flips and

E.J. Brändas and E.S. Kryachko (eds.), Fundamental World of Quantum Chemistry, Vol. II, 117–153.

zero-energy excitations are examples of open channels due to the exchange interaction, even if spin-orbit and spin-spin couplings are negligible. There are ample experimental examples of scattering from open-shell targets, including cases where either the projectiles and/or the targets are polarized, thus enabling one to detect the above mentioned quantum effects. We mention here the scattering of protons and neutrons from nuclei (^{3}He [1], ^{12}C, ^{28}Si and others [2]) and the scattering of electrons from atoms (Cs, Mn and others [3,4]), ions (Ar^+ [5]) and molecules (O_2 and NO [4]).

For open-shell atoms and molecules it is possible to compute successfully scattering matrix (S-matrix) elements (see, e.g., Refs. [6,7]). For nuclei, where the underlying potentials are not accurately known, and for larger molecules, where the computational effort is substantial, it is a very attractive concept to describe the scattering from the many-body target by an effective one-particle potential (optical potential (OP)). The construction of OP's for elastic scattering from many-body targets and their application has been a vivid field of research in nuclear, atomic and molecular physics (see, e.g., Refs. [8,9] and references therein). Both Feshbach OP's [10] (constructed by projection operators applied to the coupled-channel equations) and that of Bell and Squires [11] (the self-energy of the one-particle Green function (GF)) are *exact* OP's for elastic scattering from the closed-shell many-body system under investigation. However, for open-shell targets the standard one-particle GF [12,13] does not apply and, independently, Feshbach OP's are not well-behaved potentials [8]. It is instructive to mention that, by neglecting the antisymmetrization between the projectile and the target and by adopting the impulse approximation, Kerman, McManus and Thaler [14] have developed a successful OP for high-energy nucleon-nucleus scattering. To the best of our knowledge, a general OP for elastic scattering from open-shell targets, valid for all projectile's energies, does not appear in the literature (a concise account for our results has been recently published [15]). It is the purpose of this work to construct *exact* OP's for elastic scattering from open-shell many-body targets. Specifically, we generalize the landmark work of Bell and Squires [11] to open-shell targets. It should be emphasized that, although we speak of energetically-elastic scattering, the scattering dynamics is that of a *multichannel* problem, since the target can occupy *different* states before and after the collision.

Having in mind the construction of a theory that generalizes the text-book GF [12,13], we mention that for degenerate targets the statistical average of one-particle GF's (defined with respect to all members of the degenerate manifold) has been considered as a natural generalization of the usual (non-degenerate) one-particle GF [16]. Indeed, the statistical GF admits the correct density, particle-number and energy of the degenerate ground state [16]. Moreover, correct energy splittings and relative spectral intensities in ionization of open-shell targets can be extracted from the statistical GF [17]. Unfortunately, information about *polarized* scattering processes cannot be extracted from the statistical GF, owing to its definition as an averaged quantity (interestingly, the statistical GF has not been discussed in the context of scattering [18]; Bell and Squires OP is valid for non-degenerate targets only [11]).

To describe the various multichannel (or, polarized) scattering processes associated with open-shell targets and, in particular, to construct OP's for such processes we, therefore, have to resort to another line of reasoning. In Refs. [19,20] inelastic propagators have been formally defined. Their self-energies have been shown to be OP's for inelastic scattering from many-body targets. Owing to the complexity characterizing the addition and removal of a particle indistinguishable from those comprising the many-body target, the evaluation of these inelastic propagators is still an open question [19,20]. In particular, several oddities occur when one is trying to accomplish that. If one is to maintain the particle-hole symmetry of the standard GF, that is having the (minus) affinities and (minus) ionization energies as poles, one finds a non-vanishing OP even for non-interacting particles [19,20]. If the latter is to be repaired, one has to break the particle-hole symmetry embedded in the standard GF and increase substantially the number of poles in the denominators of the inelastic propagators. Below we show that if one restricts the discussion of inelastic propagators to degenerate ground states then it is possible to define a generalized GF (hereafter referred to as open-shell GF (OSGF)) which is to open-shell targets what the standard GF is to closed-shell ones. Moreover, the relation between the standard GF and OSGF is to become even more transparent once symmetry analysis is utilized to factorize the latter. While we refer to many-electron targets (e.g., atoms and molecules) throughout this paper, our analysis and results are valid for any type of many-fermionic targets (e.g., nuclei). Adaptation of our treatment and results to bosons is straightforward and will not be considered here.

This paper is organized as follows. In Sec. II we review briefly the theory of the (standard) one-particle GF as far as it is required for our purpose. In Sec. III we introduce the OSGF and derive its self-energy via a generalized Dyson equation. In Sec. IV we utilize symmetry arguments to factorize the OSGF and to discuss briefly the application of the standard Feynman diagrams to evaluate its self-energy. In Sec. V we show that the self-energy of the OSGF is an OP for elastic scattering from open-shell many-body targets. In Sec. VI we discuss the analytical properties of the OSGF and the OP (self-energy) in the context of scattering. A short summary is given in Sec. VII.

II. THEORETICAL BACKGROUND

A. The standard one-particle GF

Let the total Hamiltonian $\boldsymbol{H}$ of the system take on the following appearance in second quantization [12]:

$$\boldsymbol{H} = \sum_p \varepsilon_p a_p^\dagger a_p + \sum_{p,q} W_{pq} a_p^\dagger a_q + \frac{1}{2} \sum_{p,q,k,l} V_{pqkl} a_p^\dagger a_q^\dagger a_l a_k \tag{1}$$

Here, the first sum on the right-hand-side (r.h.s.) is the Hamiltonian $\boldsymbol{H}_0$ of the non-interacting particles. The second sum denotes the interaction $\boldsymbol{W}$ of the electrons with the nuclei of the target. $\boldsymbol{W}$ may also include an external potential, e.g., electric field, if present. The last term on the r.h.s. of (1) is the electron-electron Coulomb interaction $\boldsymbol{V}$ with the matrix elements

$$V_{pqkl} = \int dr_1 dr_2 \varphi_p^*(r_1)\varphi_q^*(r_2)\frac{1}{|r_1 - r_2|}\varphi_k(r_1)\varphi_l(r_2). \tag{2}$$

$\{\varphi_p\}$ are the one-particle functions (orbitals) used to define the destruction operators. We remark that orbitals other than those of free particles (e.g., canonical Hartree-Fock orbitals) can be used to define these operators. This amounts, however, to repartitioning the one-particle part $\boldsymbol{H}_0 + \boldsymbol{W}$ of the Hamiltonian $\boldsymbol{H}$ into a new diagonal $\boldsymbol{H}_0{}'$ and potential $\boldsymbol{W}'$ terms defining the orbital basis (see, e.g., Refs. [21,22]).

Suppose that the system (1) has a *non-degenerate* ground state $|0\rangle$. To investigate those properties of the system (1) associated with adding or removing a particle from the ground state, common text-books define the one-particle GF (also referred to as the 2-point GF) which takes on the following appearance in time space [12,13]:

$$\begin{aligned} g_{pq}(t,t') &= g^{+}_{pq}(t,t') + g^{-}_{pq}(t,t') = \\ &= -i\theta\,(t-t')\left\langle 0\left|a_p(t)a^{\dagger}_q(t')\right|0\right\rangle + \\ &\quad +i\theta\,(t'-t)\left\langle 0\left|a^{\dagger}_q(t')a_p(t)\right|0\right\rangle . \end{aligned} \tag{3}$$

Here $a_p(t) \equiv \exp[i\boldsymbol{H}t]\,a_p \exp[-i\boldsymbol{H}t]$ and $a^{\dagger}_p(t)$ denote the Heisenberg representation of the annihilation and creation operators for projectiles in projectile one-particle states φ_p.

The standard GF possess an appealing physical interpretation in terms of wavefunction autocorrelations. The advanced-particle GF, $\boldsymbol{g}^{+}(t,t')$, can be expressed as the *multiplication* of $(N+1)$-particle column and row vectors

$$\boldsymbol{g}^{+}(t,t') = -i\theta\,(t-t')\left[\ldots\, a^{\dagger}_p(t)|0\rangle \,\ldots\right]^{\dagger} \times \left[\ldots\, a^{\dagger}_q(t')|0\rangle \,\ldots\right], \tag{4}$$

whereas the retarded-hole part, $\boldsymbol{g}^{-}(t,t')$, can be written as (the transpose of) the *multiplication* of $(N-1)$-particle column and row vectors

$$\boldsymbol{g}^{-}(t,t') = i\theta\,(t'-t)\left\{\left[\ldots\, a_q(t')|0\rangle \,\ldots\right]^{\dagger} \times \left[\ldots\, a_p(t)|0\rangle \,\ldots\right]\right\}^{t} . \tag{5}$$

Several properties of interest can be read off the representation of $\boldsymbol{g}(t,t')$ in energy space

$$\boldsymbol{g}(\omega) = \int_{-\infty}^{\infty} e^{i\omega(t-t')}\boldsymbol{g}(t,t')d(t-t'). \tag{6}$$

Let the system under investigation (target) have N_e electrons and ground state energy equal to $E^{[0]}$. The so-called Lehmann representation of $\boldsymbol{g}(t,t')$ then reads [12,13]:

$$g_{pq}(\omega) = \sum_s \frac{\left\langle 0 \left| a_p \right| s(N_e+1) \right\rangle \left\langle s(N_e+1) \left| a_q^\dagger \right| 0 \right\rangle}{\omega + E^{[0]} - E^{[s]}_{N_e+1} + i0^+} + \\ + \sum_r \frac{\left\langle 0 \left| a_q^\dagger \right| r(N_e-1) \right\rangle \left\langle r(N_e-1) \left| a_p \right| 0 \right\rangle}{\omega + E^{[r]}_{N_e-1} - E^{[0]} - i0^+}, \quad (7)$$

where $|s(N_e+1)\rangle$ and $E^{[s]}_{N_e+1}$ are the eigenstates and energies of the target with one additional electron; $|r(N_e-1)\rangle$ and $E^{[r]}_{N_e-1}$ are the corresponding quantities of the target with one electron less, i.e., of the ionized target. Obviously, $\boldsymbol{g}(\omega)$ possesses poles which are directly connected to observables: to electron affinities and ionization potentials of the target system. The residues are products of the transition amplitudes (also known as Feynman-Dyson amplitudes): $\left\langle r(N_e-1) \left| a_p \right| 0 \right\rangle$ and $\left\langle 0 \left| a_p \right| s(N_e+1) \right\rangle$. These quantities are related to the spectral intensities of the ionization and electron-attachment processes, respectively. Finally we note that other measurable quantities, e.g. the ground state energy $E^{[0]}$ and expectation values of one-particle operators, can be computed from the one-particle GF (see, e.g., Refs. [12,13], [23] and references therein).

B. Dyson equation and self-energy

The equation-of-motion (EOM) of the one-particle GF – the well-known Dyson equation – reads as follows in matrix notation [12,13,23]:

$$\begin{aligned} \boldsymbol{g}(\omega) &= \boldsymbol{g}^{(0)}(\omega) + \boldsymbol{g}^{(0)}(\omega)\boldsymbol{\sigma}(\omega)\boldsymbol{g}(\omega), \\ \boldsymbol{\sigma}(\omega) &= \boldsymbol{g}^{(0)^{-1}}(\omega) - \boldsymbol{g}^{-1}(\omega). \end{aligned} \quad (8)$$

$\boldsymbol{g}^0(\omega)$ denotes the free GF computed without particle-particle interaction and is given by

$$g^0_{pq}(\omega) = \delta_{pq}\left[\frac{1-n_p}{\omega - \varepsilon_p + i0^+} + \frac{n_p}{\omega - \varepsilon_p - i0^+}\right]. \quad (9)$$

$n_p = 1$ ($n_p = 0$) for φ_p occupied (unoccupied) in the unperturbed ground state, which is represented by a single Slater determinant.

The self-energy $\boldsymbol{\sigma}(\omega)$ is expressed by a sum of energy-independent (static) and energy-dependent (dynamical) terms:

$$\boldsymbol{\sigma}(\omega) = \boldsymbol{\sigma}(\infty) + \boldsymbol{m}(\omega). \tag{10}$$

The static term is given by

$$\sigma_{pq}(\infty) = W_{pq} + \sum_{n,l} V_{pn[ql]} \cdot \rho_{ln}, \quad V_{pq[kl]} \equiv V_{pqkl} - V_{pqlk}, \tag{11}$$

where

$$\rho_{ln} = \left\langle 0 \left| a_n^\dagger a_l \right| 0 \right\rangle \tag{12}$$

is a one-particle density of the target in the fully-correlated ground state $|0\rangle$ and depends on the target's properties only. The dynamical term takes on the appearance

$$m_{pq}(\omega) = \frac{1}{4} \sum_{n,k,l,u,v,m} V_{pn[kl]} \cdot r_{nlk,uvm}(\omega) \cdot V_{vu[mq]}, \tag{13}$$

where the 6-point response function is given by

$$r_{nlk,uvm}(\omega) = g_{nlk,uvm}(\omega) - \sum_{r,s} g_{nlk,r}(\omega) \left[\boldsymbol{g}(\omega)^{-1}\right]_{rs} g_{s,uvm}(\omega). \tag{14}$$

The 6-point GF appearing in Eq. (14), $g_{nlk,uvm}(\omega)$, is given by the following relation in time space:

$$g_{nlk,uvm}(t,t') = -i\theta(t-t') \left\langle 0 \left| a_n^\dagger(t) a_l(t) a_k(t) a_u^\dagger(t') a_v^\dagger(t') a_m(t') \right| 0 \right\rangle$$
$$+ i\theta(t'-t) \left\langle 0 \left| a_u^\dagger(t') a_v^\dagger(t') a_m(t') a_n^\dagger(t) a_l(t) a_k(t) \right| 0 \right\rangle. \tag{15}$$

The 4-point GF's appearing in Eq. (14), $g_{k,uvm}(t,t')$ and $g_{nlk,u}(t,t')$, are formally obtained from $g_{nlk,uvm}(t,t')$ in Eq. (15) by replacing there, respectively, the quantities $a_n^\dagger(t) a_l(t)$ and $a_v^\dagger(t') a_m(t')$ by 1.

C. The self-energy and elastic scattering

To show that the self-energy $\boldsymbol{\sigma}(\omega)$ is an exact OP governing the elastic scattering processes $a_q^\dagger|0\rangle \to a_p^\dagger|0\rangle$ [11,24] we recall that the corresponding S-matrix is given by $s_{pq} = {}^{-}\langle p | q \rangle^{+}$ [24,25]. The incoming and outgoing scattering states $|p\rangle^{\pm}$ can be expressed using creation operators as $|p\rangle^{\pm} = \lim_{t\to\mp\infty} \exp[-i\varepsilon_p t]\, a_p^\dagger(t)\, |0\rangle$, where $\varepsilon_p = \frac{p^2}{2m}$ is the energy of the projectile [24,25]. It is now possible to relate the elements of the S-matrix to those of the standard GF:

$$s_{pq} = i \lim_{\substack{t\to+\infty \\ t'\to-\infty}} e^{+i\varepsilon_p t} g_{pq}^{[M,N]}(t,t') e^{-i\varepsilon_q t'}. \tag{16}$$

By breaking up the Dyson equation (8) into the two equations

$$\begin{aligned} \boldsymbol{g}(\omega) &= \boldsymbol{g}^{(0)}(\omega) + \boldsymbol{g}^{(0)}(\omega)\, \boldsymbol{t}(\omega)\, \boldsymbol{g}^{(0)}(\omega), \\ \boldsymbol{t}(\omega) &= \boldsymbol{\sigma}(\omega) + \boldsymbol{\sigma}(\omega)\, \boldsymbol{g}^{(0)}(\omega)\, \boldsymbol{t}(\omega), \end{aligned} \tag{17}$$

the final expression for the S-matrix takes on the following familiar appearance:

$$s_{pq}(\varepsilon_p) = \delta_{pq} - 2\pi i\, t_{pq}(\varepsilon_p)\, \delta(\varepsilon_p - \varepsilon_q). \tag{18}$$

In the context of many-body theory, $\boldsymbol{t}(\omega)$ in Eq. (17) plays the role of the improper self-energy [12,13], whereas in the context of scattering theory it is the transition matrix (T-matrix). Following standard scattering theory, relations (17-18) imply that $\boldsymbol{\sigma}(\omega)$ is a one-body effective potential (i.e., an OP) governing the scattering process. To account for the many-body effects, $\boldsymbol{\sigma}(\omega)$ is, in general, a non-local and an energy-dependent potential. For projectile's energies below the first inelastic threshold, $\boldsymbol{\sigma}(\omega)$ is an Hermitian and non-singular OP.

III. THE ONE-PARTICLE OPEN-SHELL GREEN FUNCTION

A. Definition and general properties

Several conditions have to be satisfied by an OSGF, making it a *faithful* generalization of the standard GF. First, it should reduce to the standard one-particle GF for closed-shell targets. Secondly, it should be time-translational invariant, to enable for a Fourier transform from time to energy space. The latter representation should reveal the energies and intensities of ionization and attachment processes in degenerate targets. Thirdly, the matrix representation of the OSGF should be invertible. Only then can it be subject to a generalized Dyson equation with a proper kernel defining a self-energy, in analogy to the closed-shell case. Fourthly, the self-energy of the OSGF should be an OP for *state-specific* (i.e., polarized) scattering processes from the open-shell manifold, thus generalizing the result of Bell and Squires [11] to this *multichannel* case. Of course, one would like also to be able to extract one-particle expectation values, transition moments (between any two members of the degenerate manifold) and the ground state energy from the OSGF.

Let us denote the ground state manifold by $\{|M\rangle, M = 0, \ldots, K\}$. It would be quite appealing to import the interpretation of $\boldsymbol{g}(t,t')$ in terms of autocorrelations of $(N \pm 1)$-particle states (see Eqs. (4-5)) to the structure of the OSGF. Thus, generalizing the advanced-particle part $\boldsymbol{g}^{+}(t,t')$ (Eq. (4)) we define the following super-matrix:

$$\boldsymbol{G}^{+}(t,t') = -i\theta(t-t')\left[\ldots a_p^{\dagger}(t)|0\rangle \ldots a_p^{\dagger}(t)|M\rangle \ldots\right]^{\dagger} \times \left[\ldots a_q^{\dagger}(t')|0\rangle \ldots a_q^{\dagger}(t')|N\rangle \ldots\right],$$
$$G_{pq}^{+[M,N]}(t,t') = -i\theta(t-t')\left\langle M \left| a_p(t) a_q^{\dagger}(t') \right| N \right\rangle. \tag{19}$$

M and N are target indices and range over the entire ground state manifold. It is convenient to view $G_{pq}^{+[M,N]}$ as matrix elements of a super-matrix $\boldsymbol{G}^{+}$ in a double-index space, one characterizing the projectile and the other the target. This super-matrix consists of blocks $\boldsymbol{G}^{+[M,N]} = \left\{G_{pq}^{+[M,N]}\right\}$ which are square matrices of the dimension of the elastic GF. Similarly, to generalize the retarded-hole part $\boldsymbol{g}^{-}(t,t')$ (Eq. (5)) we define the following super-matrix:

$$\widetilde{\boldsymbol{G}}^{-}(t,t') = +i\theta(t'-t)\left\{\left[\ldots a_q(t')|0\rangle \ldots a_q(t')|M\rangle \ldots\right]^{\dagger} \times \left[\ldots a_p(t)|0\rangle \ldots a_p(t)|N\rangle \ldots\right]\right\}^{t},$$
$$\widetilde{G}_{pq}^{-[M,N]}(t,t') = +i\theta(t'-t)\left\langle N \left| a_q^{\dagger}(t') a_p(t) \right| M \right\rangle.$$

It is "tempting" to suggest the super-matrix $\boldsymbol{G}^{+}(t,t') + \widetilde{\boldsymbol{G}}^{-}(t,t')$ as our faithful generalization of the standard GF to the open-shell case. Obviously, it reduces to the standard GF for closed-shell targets and it is time-translational invariant. Unlike the situation for the standard GF, unfortunately, the Fourier transform of

$\boldsymbol{G}^{+}(t,t') + \widetilde{\boldsymbol{G}}^{-}(t,t')$ *is not* invertible, in general. Consequently, it *cannot* be subject to a generalized Dyson equation with a proper kernel defining a self-energy. Therefore, to construct an OSGF we have to employ *either* $\boldsymbol{G}^{+}(t,t')$ or $\widetilde{\boldsymbol{G}}^{-}(t,t')$ along with their corresponding time-ordered partners. Alternatively speaking, to construct an OSGF we have to *depart* from the picture of closed-shell targets where *both* the advanced-particle and the retarded-hole parts can be represented as autocorrelations of $(N \pm 1)$-particle row and column vectors. Since scattering is a process in the course of which $N+1$ particles are involved, we define the OSGF starting from $\boldsymbol{G}^{+}(t,t')$,

$$\begin{aligned} G_{pq}^{[M,N]}(t,t') &= G_{pq}^{+[M,N]}(t,t') + G_{pq}^{-[M,N]}(t,t') = \\ &= -i\theta\,(t-t')\left\langle M \left| a_p(t) a_q^{\dagger}(t') \right| N \right\rangle + \\ &\quad + i\theta\,(t'-t)\left\langle M \left| a_q^{\dagger}(t') a_p(t) \right| N \right\rangle . \end{aligned} \tag{20}$$

Indeed, in definition (20) $\boldsymbol{G}^{-}(t,t')$ (the retarded-hole part) *cannot* be expressed also as a multiplication of column and row vectors, as is stipulated for $\boldsymbol{g}^{-}(t,t')$ in Eq. (5).

Aiming at studying the properties of $\boldsymbol{G}(t,t')$ (Eq. (20)) we first note that for closed-shell targets ($M = N = K = 0$) it boils down to the standard GF in Eq. (3). By construction, the OSGF is time-translational invariant (as is the standard one). Its Fourier transform into energy space takes on the following appearance:

$$\begin{aligned} G_{pq}^{[M,N]}(\omega) &= \sum_{s} \frac{\left\langle M \left| a_p \right| s(N_e+1) \right\rangle \left\langle s(N_e+1) \left| a_q^{\dagger} \right| N \right\rangle}{\omega + E^{[0]} - E_{N_e+1}^{[s]} + i0^{+}} + \\ &+ \sum_{r} \frac{\left\langle M \left| a_q^{\dagger} \right| r(N_e-1) \right\rangle \left\langle r(N_e-1) \left| a_p \right| N \right\rangle}{\omega + E_{N_e-1}^{[r]} - E^{[0]} - i0^{+}} . \end{aligned} \tag{21}$$

In principle, all blocks $\boldsymbol{G}^{[M,N]}(\omega)$ possess the same poles which do not depend explicitly on the projectile's (and target's) indices. By comparing the denominators of $\boldsymbol{G}(\omega)$ to those of $\boldsymbol{g}(\omega)$ (see Eq. (7)) it is obvious that these poles correspond to ionization and affinity energies of the open-shell target in its ground state, what generalizes the case of the standard GF. From a practical point of view this is a very valuable property since no additional poles must be computed for constructing the OSGF. The residues of $\boldsymbol{G}(\omega)$ are products of the generalized transition amplitudes (or, generalized Feynman-Dyson amplitudes): $\left\langle r(N_e-1)\left|a_p\right|N\right\rangle$ and $\left\langle M\left|a_p\right|s(N_e+1)\right\rangle$. These quantities are related to the spectral intensities of *state-specific* ionization and electron-attachment processes, respectively, and will be discussed further elsewhere [18].

To argue that the super-matrix $\boldsymbol{G}(\omega)$ is invertible we calculate its structure at very large energy ω,

$$\lim_{\omega\to\infty}\left[\omega\cdot G_{pq}^{[M,N]}(\omega)\right]=\left\langle M\left|a_p a_q^\dagger\right|N\right\rangle+\left\langle M\left|a_q^\dagger a_p\right|N\right\rangle=\delta_{pq}\delta_{NM}. \tag{22}$$

The matrix $\boldsymbol{G}(\omega)$ becomes diagonal at this limit and is invertible: $\boldsymbol{G}^{-1}\Longrightarrow \mathbf{1}/\omega$. It is well known that the standard GF behaves similarly at very large energies and is invertible, in general. We shall, therefore, assume that the OSGF (see Eq. (21)) exhibits the similar property, meaning that it is, in general, invertible (this property can be verified with some effort for finite-level models).

B. Generalized Dyson equation and self-energy

Aiming at deriving a generalized Dyson equation for the OSGF, we study the EOM of its elements $G_{pq}^{[M,N]}(t,t')$ using Zubarev method [19,20,26]. We start by taking the derivative of a general element $G_{pq}^{[M,N]}(t,t')$ with respect to time t and making use of the EOM of the *projectile's* annihilation operator $a_p(t)$,

$$\begin{aligned} i\frac{d}{dt}a_p(t) &= [a_p(t),\boldsymbol{H}]= \\ &=\varepsilon_p a_p(t)+\sum_q W_{pq}a_q(t)+\sum_{q,k,l}V_{pqkl}a_q^\dagger(t)b_l(t)b_k(t). \end{aligned} \tag{23}$$

The result takes on the appearance

$$\begin{aligned} \left[i\frac{d}{dt}-\varepsilon_p\right]G_{pq}^{[M,N]}(t,t') &= \delta(t-t')\delta_{pq}\delta_{MN}+ \\ &+\sum_l W_{pl}G_{lq}^{[M,N]}(t,t')+\frac{1}{2}\sum_{n,k,l}V_{pn[kl]}G_{nlk,q}^{[M,N]}(t,t'), \end{aligned} \tag{24}$$

where the 4-point OSGF, $G^{[M,N]}_{nlk,q}$, is defined as

$$G^{[M,N]}_{nlk,q}(t,t') = -i\theta(t-t')\left\langle M \left| a^\dagger_n(t)a_l(t)a_k(t)a^\dagger_q(t') \right| N\right\rangle + \\ + i\theta(t'-t)\left\langle M \left| a^\dagger_q(t')a^\dagger_n(t)a_l(t)a_k(t) \right| N\right\rangle . \quad (25)$$

It is convenient to transform Eq. (24) to energy space and to multiply the result with $1/(\omega - \varepsilon_p)$, leading to

$$G^{[M,N]}_{pq}(\omega) = \frac{\delta_{pq}\delta_{MN}}{\omega - \varepsilon_p} + \\ + \frac{1}{\omega - \varepsilon_p}\left\{\sum_l W_{pl}G^{[M,N]}_{lq}(\omega) + \frac{1}{2}\sum_{n,k,l} V_{pn[kl]}G^{[M,N]}_{nlk,q}(\omega)\right\}. \quad (26)$$

Overlooking for the moment the exact infinitesimals $\pm i0^+$ (see Sec. VI), the first term on the r.h.s. is the free OSGF

$$\boldsymbol{G}^{(0)}(\omega) = \mathbf{1} \otimes \boldsymbol{g}^{(0)}(\omega), \quad G^{(0)[M,N]}_{pq}(\omega) = \frac{\delta_{pq}\delta_{MN}}{\omega - \varepsilon_p + i0^+}, \quad (27)$$

computed without particle-particle interaction. As can be seen from Eq. (27), the free OSGF is related to the standard free GF by the direct product relation. The dimension of the unit matrix $\mathbf{1}$ is that of the degenerate manifold. Interestingly, $\boldsymbol{G}^{(0)}(\omega)$ describes both the case of non-interacting particles and the case of a free projectile which is not interacting with a fully correlated target (see also Sec. VI).

It is quite straightforward to cast Eq. (26) in the form of a generalized Dyson equation which reads in matrix notation

$$\boldsymbol{G}(\omega) = \boldsymbol{G}^{(0)}(\omega) + \boldsymbol{G}^{(0)}(\omega)\boldsymbol{\Sigma}(\omega)\boldsymbol{G}(\omega), \quad (28)$$

with the kernel

$$\Sigma^{[M,N]}_{pq}(\omega) = W_{pq}\delta_{MN} + \frac{1}{2}\sum_P^K \sum_m \left[\sum_{n,k,l} V_{pn[kl]}G^{[M,P]}_{nlk,m}\right]\left[\boldsymbol{G}^{-1}(\omega)\right]^{[P,N]}_{mq}. \quad (29)$$

Equivalently, the self-energy $\boldsymbol{\Sigma}(\omega)$ is defined via the formal solution of Eq. (28)

$$\Sigma(\omega) = G^{(0)^{-1}}(\omega) - G^{-1}(\omega). \tag{30}$$

Both Eqs. (29) and (30) make it clear that *all* members of the degenerate manifold are coupled within the self-energy of the OSGF.

Below we show that the multichannel self-energy $\Sigma(\omega)$ can be casted into a *homomorphic* structure to that possessed by the standard self-energy $\sigma(\omega)$ (see Eqs. (10-15)). To this end, we start from the kernel $\Sigma_{pq}^{[M,M]}(\omega)$ in Eq. (29). To proceed, we need the EOM of the 4-point OSGF $G_{nlk,q}^{[M,N]}(t,t')$ (see Eq. (25)) which takes on the appearance

$$\left\{i\frac{d}{dt} - \varepsilon_q\right\} G_{nlk,q}^{[M,N]}(t,t') = \delta(t-t')\left\{\delta_{qk}\rho_{ln}^{[M,N]} - \delta_{ql}\rho_{kn}^{[M,N]}\right\} + \\ + \sum_m G_{nlk,m}^{[M,N]}(t,t')W_{mq} + \frac{1}{2}\sum_{m,u,v} G_{nlk,uvm}^{[M,N]}(t,t')V_{uv[qm]}, \tag{31}$$

where, generalizing the 6-point GF $g_{nlk,uvm}(t,t')$, the 6-point OSGF is given by:

$$G_{nlk,uvm}^{[M,N]}(t,t') = \\ -i\theta(t-t')\left\langle M\left|a_n^\dagger(t)a_l(t)a_k(t)a_u^\dagger(t')a_v^\dagger(t')a_m(t')\right|N\right\rangle + \\ +i\theta(t'-t)\left\langle M\left|a_u^\dagger(t')a_v^\dagger(t')a_m(t')a_n^\dagger(t)a_l(t)a_k(t)\right|N\right\rangle. \tag{32}$$

To arrive at our final result we would also make use of the equivalent form of the self-energy

$$\Sigma_{pq}^{[M,N]}(\omega) = \\ = W_{pq}\delta_{MN} + \frac{1}{2}\sum_P^K\sum_m\left[G^{-1}(\omega)\right]_{pm}^{[M,P]}\left[\sum_{n,k,l} G_{m,kln}^{[P,N]}(\omega)V_{lk[nq]}\right]. \tag{33}$$

Here, $G_{m,kln}^{[P,N]}$ is another 4-point OSGF defined analogously to $G_{nlk,q}^{[M,N]}$ as

$$G_{m,kln}^{[M,N]}(t,t') = -i\theta(t-t')\left\langle M\left|a_m(t)a_k^\dagger(t')a_l^\dagger(t')a_n(t')\right|N\right\rangle + \\ +i\theta(t'-t)\left\langle M\left|a_k^\dagger(t')a_l^\dagger(t')a_n(t')a_m(t)\right|N\right\rangle. \tag{34}$$

Expression (33) for the self-energy can be derived from the generalized Dyson equation in reversed order

$$G(\omega) = G^{(0)}(\omega) + G(\omega)\,\Sigma(\omega)\,G^{(0)}(\omega) \tag{35}$$

in analogy to the derivation of Eq. (29) from the 'original' Dyson equation (28). Both Dyson equations (28) and (35) are easy to determine using the basic relation (30) which serves to define the self-energy $\Sigma(\omega)$.

Next, we transform the EOM of $G^{[M,N]}_{nlk,q}(t,t')$ (Eq. (31)) to energy space and multiply the result from the right by the free GF $G^{(0)}(\omega)$ (see Eq. (27)), leading to

$$G^{[M,F]}_{nlk,f}(\omega) = \sum_{Q}^{K}\sum_{s}\Big\{\Big[\delta_{ks}\rho^{[M,Q]}_{ln} - \delta_{ls}\rho^{[M,Q]}_{kn}\Big] + \\ + \sum_{r} G^{[M,Q]}_{nlk,r}(\omega)W_{rs} + \frac{1}{2}\sum_{u,v,r} G^{[M,Q]}_{nlk,uvr}(\omega)V_{uv[sr]}\Big\}\,G^{(0)[Q,F]}_{sf}(\omega). \tag{36}$$

This equation is now multiplied from the right by the (fF, mP)-element of

$$G^{(0)^{-1}}(\omega)G(\omega) = \mathbf{1} + \Sigma(\omega)G(\omega), \tag{37}$$

where here $\mathbf{1}$ is the unit super-matrix of the $K+1$ channels comprising the degenerate manifold. Eq. (37) follows immediately from the Dyson equation (28). Note that in Eq. (37) we use for $\Sigma(\omega)$ the expression given in Eq. (33). After a slight rearrangement of terms the result for the 4-point OSGF reads

$$G^{[M,P]}_{nlk,f}(\omega) = \sum_{Q}^{K}\sum_{s}\Big\{\Big[\delta_{ks}\rho^{[M,Q]}_{ln} - \delta_{ls}\rho^{[M,Q]}_{kn}\Big] + \\ + \frac{1}{2}\sum_{u,v,r} R^{[M,Q]}_{nlk,uvr}(\omega)V_{uv[sr]}\Big\}G^{[Q,P]}_{sm}(\omega), \tag{38}$$

where

$$R^{[M,N]}_{nlk,uvm}(\omega) = \\ = G^{[M,N]}_{nlk,uvm}(\omega) - \sum_{P,Q}^{K}\sum_{r,s} G^{[M,P]}_{nlk,r}(\omega)\Big[G^{-1}(\omega)\Big]^{[P,Q]}_{rs} G^{[Q,N]}_{s,uvm}(\omega) \tag{39}$$

is a generalized 6-point response function (compare to Eq. (14) for $r_{nlk,uvm}(\omega)$ – the standard 6-point response function). Substituting

Eq. (38) into expression (29) for $\mathbf{\Sigma}(\omega)$ and rearranging terms gives $\mathbf{\Sigma}(\omega)$ a *homomorphic* structure to that of the standard self-energy $\boldsymbol{\sigma}(\omega)$ (see Eqs. (10-15)):

$$\Sigma_{pq}^{[M,N]}(\omega) = \Sigma_{pq}^{[M,N]}(\infty) + M_{pq}^{[M,N]}(\omega). \tag{40}$$

The static (energy-independent) term is given by

$$\Sigma_{pq}^{[M,N]}(\infty) = W_{pq}\delta_{MN} + \sum_{n,l} V_{pn[ql]} \cdot \rho_{ln}^{[M,N]}, \tag{41}$$

where

$$\rho_{ln}^{[M,N]} = \left\langle M \left| a_n^\dagger a_l \right| N \right\rangle \tag{42}$$

is a generalized one-particle density of the degenerate target and depends on the target properties only. The first term in (41) describes the one-body interaction $\boldsymbol{W}$ of the projectile with the target (e.g., nuclei). Hence it is diagonal in the *target's* indices. The second term describe the electrostatic and exchange interactions of the projectile with the exact generalized densities $\rho_{ln}^{[M,N]}$. The dynamical (energy-dependent) term

$$M_{pq}^{[M,N]}(\omega) = \frac{1}{4} \sum_{n,k,l,u,v,m} V_{pn[kl]} R_{nlk,uvm}^{[M,N]}(\omega) V_{vu[mq]}, \tag{43}$$

couples *all* states in the degenerate manifold via the generalized 6-point response function $R_{nlk,uvm}^{[M,N]}(\omega)$.

As is the case with the OSGF itself, the higher-order OSGF's and the self-energy $\mathbf{\Sigma}(\omega)$ reduce to the standard ones for $M = N = K = 0$. The reasons for this property are the EOM's of the n-point OSGF's being linear and homogenous equations, and the free OSGF being diagonal in terms of the target indices (see Eq. (27)). Moreover, for *any* partial subset of members chosen from the degenerate manifold we can construct a generalized Dyson equation the self-energy of which takes on the *same structure* expressed by Eqs. (40-43). Of course, for different subsets chosen the corresponding self-energies differ one from the other. In particular, *any* of the diagonal blocks $\boldsymbol{G}^{[M,M]}(\omega)$ satisfies the (same) Dyson equation fulfilled by the standard GF (Eqs. (10-15)) where the corresponding quantities (density, higher-order GF's) associated with the $[M, M]$ block are substitutes there instead!

Finally, it is important to emphasize that the above construction of the OSGF along with the algebraic formulation of the generalized Dyson equation and the corresponding self-energy poss no restrictions on the *nature* of the degenerate states. In particular, these states need not correlate to *single* Slater determinants in the absence of particle-particle interaction. We remark also that our formulation applies to spatial, spin, or any other type of degeneracy.

IV. SYMMETRY-BASED FACTORIZATION OF THE OSGF

So far, we have used *only* the degeneracy property of the ground state manifold in order to construct the OSGF. Consequently, $\boldsymbol{G}(\omega)$ should account for scattering processes which *a priori* may couple *all* states within the degenerate manifold. Having in mind that the many-body Hamiltonian is invariant under a specific *symmetry group* we can substantially simplify the appearance of the OSGF and gain further physical insight into the scattering of indistinguishable particles from open-shell targets. Since the N-particle states comprising the degenerate manifold span an irreducible representation of the group of $\boldsymbol{H}$ and, therefore, are interrelated by symmetry operations, we can first reduce the number of independent blocks $\boldsymbol{G}^{[M,N]}(\omega)$. To do so, we employ the so-called irreducible tensor method (see, e.g., Ref. [27] for its application to molecular symmetry groups) which, in the more familiar case of spherical symmetry, is commonly known as Wigner-Eckart theorem. Secondly, we project the *combined* projectile–target $(N + 1)$-particle states on the irreducible representations of the group of $\boldsymbol{H}$ and thereby factorize the OSGF and its self-energy into symmetry-irreducible components (channels). We note that the notion of symmetry-irreducible quantities used here relates to the partial and disjoint sets of $(N + 1)$-particle states possessing the same quantum numbers.

Rather than continuing the general presentation of the former section, we chose to study explicitly the case of spin degeneracy and a spin-half particles. These types of symmetry and degeneracy allows for an as transparent treatment as possible of scattering processes. We would like to stress that the symmetry analysis performed does not depend on the symmetry group of the many-body Hamiltonian.

A. Spin-irreducible GF's

Let the spin of the many-body target in its ground state be S. The ground state manifold is denoted by $\{|M_S\rangle, M_S = S, \ldots, -S\}$. In the single-particle annihilation and creation operators we separate spatial and spin degrees of freedom: $a_{p\sigma}(t)$ and $a^\dagger_{p\sigma}(t)$, where σ can assume two values - α (up; $+1/2$) and β (down; $-1/2$).

To reduce the number of independent blocks $\boldsymbol{G}^{[M_S,M'_S]}(\omega)$ we would like to utilize Wigner-Eckart theorem and express the OSGF through the highest-multiplicity diagonal blocks

$$\boldsymbol{G}^{[S,S]} = \begin{pmatrix} \boldsymbol{\mathcal{G}}^\alpha & 0 \\ 0 & \boldsymbol{\mathcal{G}}^\beta \end{pmatrix}, \quad \mathcal{G}^\sigma_{pq} \equiv G^{[S,S]}_{p\sigma,q\sigma}. \tag{44}$$

The dimension of each block $\boldsymbol{\mathcal{G}}^\sigma$ is half of that of the standard GF, since it depends on the *spatial* one-particle indices only. The reason for choosing the highest-multiplicity block, namely $M_S = S$, is to be clarified in the next subsection.

Next, we note that the creation $a^\dagger_{p\sigma}$ and annihilation $(-1)^{1/2+\sigma}a_{p(-\sigma)}$ operators transform with respect to rotations of the spin coordinates as the σ-th component of a tensor operator of the rank $1/2$, $T^{(1/2)}_\sigma$. A product of the latter can be expended as a linear combination of tensor operators [28]

$$T^{(1/2)}_\sigma T^{(1/2)}_{\sigma'} = \sum_{k=0}^{1} \sqrt{2k+1}(-1)^{-\sigma-\sigma'} \begin{pmatrix} \frac{1}{2} & \frac{1}{2} & k \\ \sigma & \sigma' & -\sigma-\sigma' \end{pmatrix} T^{(k)}_{\sigma+\sigma'}, \tag{45}$$

where the $3j$-symbols are related to the Clebsch-Gordon coefficients (Condon and Shortley phase convention is utilized [28]) by:

$$\begin{pmatrix} j_1 & j_2 & j_3 \\ m_1 & m_2 & m_3 \end{pmatrix} = \frac{(-1)^{j_1-j_3-m_3}}{\sqrt{2j_3+1}} \left(j_1 m_1 j_2 m_2 \left| j_1 j_2 j_3 - m_3\right.\right). \tag{46}$$

To proceed we make use of Wigner-Eckart theorem

$$\left\langle M_s \left| T^{(k)}_{\sigma+\sigma'} \right| M'_s \right\rangle = (-1)^{S-M_s} \begin{pmatrix} S & S & k \\ -M_S & M'_S & \sigma+\sigma' \end{pmatrix} \times \\ \times \begin{pmatrix} S & S & k \\ -S & +S & 0 \end{pmatrix}^{-1} \left\langle S \left| T^{(k)}_0 \right| S \right\rangle \tag{47}$$

and the inverse of Eq. (45)

$$T^{(k)}_0 = \sum_{q=-1/2}^{1/2} \sqrt{2k+1} \begin{pmatrix} \frac{1}{2} & \frac{1}{2} & k \\ q & -q & 0 \end{pmatrix} T^{(1/2)}_q T^{(1/2)}_{-q}. \tag{48}$$

Combining Eqs. (45) to (48) gives the generic result

$$\left\langle M_s \left| T^{(1/2)}_\sigma T^{(1/2)}_{\sigma'} \right| M'_s \right\rangle = \\ = (-1)^{-\sigma-\sigma'} (-1)^{S-M_s} \sum_{k=0}^{1} (2k+1) \begin{pmatrix} \frac{1}{2} & \frac{1}{2} & k \\ \sigma & \sigma' & -\sigma-\sigma' \end{pmatrix} \times \\ \times \begin{pmatrix} S & S & k \\ -M_S & M'_S & \sigma+\sigma' \end{pmatrix} \begin{pmatrix} S & S & k \\ -S & +S & 0 \end{pmatrix}^{-1} \times \\ \times \sum_{q=-1/2}^{1/2} \begin{pmatrix} \frac{1}{2} & \frac{1}{2} & k \\ q & -q & 0 \end{pmatrix} \left\langle S \left| T^{(1/2)}_q T^{(1/2)}_{-q} \right| S \right\rangle. \tag{49}$$

From Eq. (49) we learn that only two distinct blocks, $\left\langle S \left| T^{(1/2)}_{1/2} T^{(1/2)}_{-1/2} \right| S \right\rangle$ and $\left\langle S \left| T^{(1/2)}_{-1/2} T^{(1/2)}_{+1/2} \right| S \right\rangle$, contribute to the general term $\left\langle M_s \left| T^{(1/2)}_\sigma T^{(1/2)}_{\sigma'} \right| M'_s \right\rangle$. In addition, the invariance of expression (49) under the transformation $\sigma \leftrightarrow \sigma', q \leftrightarrow -q$ underlines the identical decomposition of both the advanced-particle and retarded-hole terms $\boldsymbol{G}^{\pm[M_S, M'_S]}_{p\sigma, q\sigma'}(\omega)$. In other words, the application of the Wigner-Eckart theorem yields the same result for *any* time-ordering of the annihilation and creation operators (see also the last paragraph of this subsection). By substituting the equivalent expressions for the creation $a^\dagger_{p\sigma'}$ and annihilation $a_{p\sigma}$ operators into Eq. (49) we obtain the compact result

$$\boldsymbol{G}(\omega) = \boldsymbol{A} \otimes \left[\boldsymbol{\mathcal{G}}^\alpha(\omega) - \boldsymbol{\mathcal{G}}^\beta(\omega) \right] + \mathbf{1} \otimes \boldsymbol{\mathcal{G}}^\beta(\omega), \tag{50}$$

where the elements of $\boldsymbol{A}$ are given by

$$A_{\sigma,\sigma'}^{M_S,M_S'} = (-1)^{-\sigma'+S-M_S+\frac{1}{2}} \sum_{\rho=0}^{1} (2\rho+1) \begin{pmatrix} \frac{1}{2} & \frac{1}{2} & \rho \\ -\sigma & \sigma' & \sigma-\sigma' \end{pmatrix} \times$$

$$\times \begin{pmatrix} S & S & \rho \\ S & -S & 0 \end{pmatrix}^{-1} \begin{pmatrix} S & S & \rho \\ M_S' & -M_S & \sigma'-\sigma \end{pmatrix} \begin{pmatrix} \frac{1}{2} & \frac{1}{2} & \rho \\ -\frac{1}{2} & \frac{1}{2} & 0 \end{pmatrix}. \quad (51)$$

Note that for $S = 0$ one has $\mathcal{G}^{\alpha}(\omega) = \mathcal{G}^{\beta}(\omega)$ and the familiar decomposition of the standard GF in case of spin-independent potentials is recovered [12]. The indices of the coefficient matrix $\boldsymbol{A}$ depend on the projectile and target spin variables only, hence its dimension is $2\cdot(2S+1)$. The matrix $\boldsymbol{A}$ possesses several symmetry properties. In particular, it is straightforward to prove that $\boldsymbol{A}$ is symmetric

$$A_{\sigma,\sigma'}^{M_S,M_S'} = A_{\sigma',\sigma}^{M_S',M_S}, \quad (52)$$

possesses the "inversion" symmetry

$$A_{\sigma,\sigma'}^{M_S,M_S'} = A_{-\sigma',-\sigma}^{-M_S',-M_S}, \quad (53)$$

and satisfies the sum rule

$$A_{\alpha,\alpha}^{M_S,M_S} + A_{\beta,\beta}^{M_S,M_S} = 1, \quad \forall M_S. \quad (54)$$

So far we have used symmetry (by employing Wigner-Eckart theorem) to simplify the appearance of $\boldsymbol{G}(\omega)$, namely, reducing the number of its independent blocks. The forthcoming analysis is to enable us to associate a scattering-based interpretation with the OSGF. From Eqs. (51-52) one can verify that the matrix $\boldsymbol{A}$ is real and symmetric, properties that corroborate the spin-independent particle-particle interaction. The simple structure of $\boldsymbol{A}$ allows one to diagonalize it. The resulting eigenvalues are (their multiplicities are indicated in curly brackets):

$$a_1 = 1\,\{2(S+1)\}, \quad a_2 = -\frac{1}{2S}\,\{2S\}. \quad (55)$$

Now we can block diagonalize $\boldsymbol{G}(\omega)$ in Eqs. (50-51) to obtain only two distinct blocks

$$\begin{aligned} \mathcal{G}^{J=S+1/2}(\omega) &= \mathcal{G}^{\alpha}(\omega), \\ \mathcal{G}^{J=S-1/2}(\omega) &= \frac{2S+1}{2S}\mathcal{G}^{\beta}(\omega) - \frac{1}{2S}\mathcal{G}^{\alpha}(\omega), \end{aligned} \quad (56)$$

with the same corresponding multiplicities of $2(S+1)$ and $2S$. This outcome suggests the following physical interpretation: The spin-irreducible GF's $\mathcal{G}^J(\omega)$ describe *scattering* processes in which the $(N+1)$-particle states possess the total spin of $J = S \pm 1/2$ and the corresponding multiplicities are just the usual magnetic multiplicities. In proving this claim we recall the structure of the advanced-particle and retarded-hole parts of $\boldsymbol{G}(t,t')$. Following Eq. (19), $\boldsymbol{G}^+(t,t')$ can be expressed, suppressing a factor of $-i\theta\,(t-t')$, as the *multiplication* of the column and row vectors

$$\Big[\ldots a^\dagger_{p\sigma}(t)|S\rangle \ldots a^\dagger_{p\sigma}(t)|-S\rangle \ldots\Big]^\dagger \times$$
$$\Big[\ldots a^\dagger_{q\sigma'}(t')|S\rangle \ldots a^\dagger_{q\sigma'}(t')|-S\rangle \ldots\Big].$$

The components of these vectors are $(N+1)$-particle states describing the *addition* of two spins: The spin-half projectile and the spin-S composite target. Therefore, block-diagonalizing of $\boldsymbol{G}^+(t,t')$ simply means going from a basis of two independent spins to the basis where their total spin is a good quantum number (we use the term spin-irreducible). To support this interpretation, we construct the following $(N+1)$-particle states:

$$\begin{aligned}|\psi_{q,J,M_J}(t)\rangle \equiv& \sqrt{\frac{S+1/2 \pm M_J}{2S+1}}\,a^\dagger_{q\alpha}(t)\,|M_J-1/2\rangle \\ &\pm \sqrt{\frac{S+1/2 \mp M_J}{2S+1}}\,a^\dagger_{q\beta}(t)\,|M_J+1/2\rangle \\ J = S \pm 1/2,&\quad -J \le M_J \le J.\end{aligned} \tag{57}$$

Eq. (57) is nothing but the text-book formula for angular-momentum addition of spin-half and spin-S particles, however it is adapted to second quantization formalism. Accordingly, it is straightforward to show that application of the z-component of the total spin and the elevating and lowering operators,

$$\hat{S}_z = \frac{1}{2}\sum_q \left(a^\dagger_{q\alpha}a_{q\alpha} - a^\dagger_{q\beta}a_{q\beta}\right), \quad \hat{S}_+ = \sum_q a^\dagger_{q\alpha}a_{q\beta}, \quad \hat{S}_- = \sum_q a^\dagger_{q\beta}a_{q\alpha}, \tag{58}$$

onto $|\Psi_{q,J,M_J}(t)\rangle$ follows the text-book relations of spin (angular-momentum) algebra. Consequently, autocorrelations of the spin-irreducible $(N+1)$-particle states $|\Psi_{q,J,M_J}(t)\rangle$ vanish, unless two states of the same symmetry are matched together

$$-i\theta\,(t-t')\times\left\langle\psi_{p,J,M_J}(t)\,\middle|\psi_{q,J',M'_J}(t')\right\rangle=\delta_{J,J'}\delta_{M_J,M'_J}\mathcal{G}^{+,J}_{pq}(t,t'). \tag{59}$$

Here, $\mathcal{G}^{+,J}_{pq}(t,t')$ is the advanced-particle part of the spin-irreducible GF $\mathcal{G}^{J}_{pq}(t,t')$. This proves our claim that the block-diagonalized OSGF (see Eq. (56)) accounts for scattering processes governing spin-irreducible $(N+1)$-particle states.

Can symmetry-based arguments help us uncover also the structure of retarded-hole part $\mathcal{G}^{-,J}_{pq}(t,t')$? When defining in Eq. (20) of the previous section the OSGF we saw that, unlike the advanced-particle part, the retarded-hole one cannot be expressed as an autocorrelation of column and row vectors. That is, $\mathcal{G}^{-,J}_{pq}(t,t')$ cannot be expressed as autocorrelations of spin-irreducible $(N-1)$-particle states [29]. Fortunately, from Eqs. (50) and (56) we see that the same transformation which block-diagonalizes $\boldsymbol{G}^{+}(t,t')$ also block-diagonalizes $\boldsymbol{G}^{-}(t,t')$ and, hence, $\boldsymbol{G}(t,t')$. The reason is that in evaluating matrix elements with the help of representation theory we ask whether the direct-product representation contains the trivial one. The *order* of the elements in the integrand, however, is not important for determining the *symmetry* of the matrix elements. In our case, interchanging the order of the creation and annihilation operators between $\mathcal{G}^{+,J}_{pq}(t,t')$ and $\mathcal{G}^{-,J}_{pq}(t,t')$, as causal time-ordering dictates, leaves the symmetry of the corresponding matrix elements unchanged. In other words, the selection rules given in Eq. (59) for $\mathcal{G}^{+,J}_{pq}(t,t')$ governs the block structure of $\boldsymbol{G}^{-}(t,t')$ as well (see also the discussion below Eq. (49)).

B. Spin-irreducible self-energies and their evaluation

Having decoupled the OSGF $\boldsymbol{G}(\omega)$ into its spin-irreducible components $\boldsymbol{\mathcal{G}}^{J}(\omega)$ (Eq. (56)), we would like to inquire whether there are self-energies for these GF's *alone*. The answer is positive. Owing to the structure of the generalized Dyson equation (Eqs. (28,30)), one can block diagonalize $\boldsymbol{\Sigma}(\omega)$ to give the spin-irreducible self-energies

$$\mathcal{G}^J(\omega) = \mathcal{G}^{(0)}(\omega) + \mathcal{G}^{(0)}(\omega)\boldsymbol{\Sigma}^J(\omega)\mathcal{G}^J(\omega)$$
$$\boldsymbol{\Sigma}^J(\omega) = \left[\mathcal{G}^{(0)}(\omega)\right]^{-1} - \left[\mathcal{G}^J(\omega)\right]^{-1} \tag{60}$$

by utilizing the same transformation employed to block-diagonalize $\boldsymbol{G}(\omega)$. Here,

$$\mathcal{G}^{(0)}_{pq}(\omega) = \frac{\delta_{pq}}{\omega - \varepsilon_p + i0^+} \tag{61}$$

is computed without particle-particle interaction, where p and q stand for spatial orbitals. In other words, to *each* spin-irreducible $(N+1)$-particle subspace there is a specific self-energy, thus reducing the multichannel problem to spin-irreducible ones (recall that the notion of spin-irreducibility relates to states of the $(N+1)$-particle system, all possessing the same quantum numbers J and M_J).

To arrive at the final expressions for the spin-irreducible self-energies we resort to some of the results of the previous section (see the discussion after Eq. (43)). Specifically, we write the (ordinary) Dyson equation for the $[S, S]$ diagonal block (see also Eq. (44)):

$$\begin{pmatrix} \mathcal{G}^\alpha & \mathbf{0} \\ \mathbf{0} & \mathcal{G}^\beta \end{pmatrix} = \begin{pmatrix} \mathcal{G}^{(0)} & \mathbf{0} \\ \mathbf{0} & \mathcal{G}^{(0)} \end{pmatrix} + \\ + \begin{pmatrix} \mathcal{G}^{(0)} & \mathbf{0} \\ \mathbf{0} & \mathcal{G}^{(0)} \end{pmatrix} \begin{pmatrix} \boldsymbol{\Sigma}^\alpha & \mathbf{0} \\ \mathbf{0} & \boldsymbol{\Sigma}^\beta \end{pmatrix} \begin{pmatrix} \mathcal{G}^\alpha & \mathbf{0} \\ \mathbf{0} & \mathcal{G}^\beta \end{pmatrix}, \tag{62}$$

thus defining the self-energies $\boldsymbol{\Sigma}^\alpha$ and $\boldsymbol{\Sigma}^\beta$. Of course, the explicit expressions for $\boldsymbol{\Sigma}^\sigma$ are given by those for $\boldsymbol{\sigma}(\omega)$ (see Eqs. (10-15)), where the quantities related to the $[S, S]$ block are substitutes there instead. Solving Eq. (60), while using Eqs. (56,62), we arrive at our final expressions for the spin-irreducible self-energies:

$$\boldsymbol{\Sigma}^{J=S+1/2}(\omega) = \boldsymbol{\Sigma}^\alpha, \tag{63}$$

and

$$\boldsymbol{\Sigma}^{J=S-1/2}(\omega) = \mathcal{G}^{(0)^{-1}} - 2S\left(\frac{2S+1}{\mathcal{G}^{(0)^{-1}} - \boldsymbol{\Sigma}^\beta} - \frac{1}{\mathcal{G}^{(0)^{-1}} - \boldsymbol{\Sigma}^\alpha}\right)^{-1}. \tag{64}$$

So far, we have discussed the existence and formal appearance of the spin-irreducible self-energies of the OSGF. We would like to stress that, in obtaining Σ^{σ}, no assumptions have been made as to the nature of the N-particle state $|S\rangle$, but its symmetry. Consequently, to *evaluate* the spin-irreducible self-energies one has to resort to algebraic methods, for instance, by (approximately) solving the set of coupled equations for the n-point OSGF's. Notwithstanding, for some relevant cases Σ^{σ} can be systematically *evaluated* by the *usual* diagrammatic analysis, as utilized for the text-book proper self-energy $\boldsymbol{\sigma}(\omega)$ [12,13].

Generally, in the absence of particle-particle interaction a member $|M\rangle$ of the degenerate manifold correlates to a linear combination of Slater determinants. Consequently, the diagrammatic approach for calculating the corresponding self-energy via Wick theorem [12,13] breaks down. Fortunately, for most atomic and molecular systems there are genuine one-particle configurations within the *unperturbed* degenerate manifold. Due to Hund rule, e.g., the state with maximal/minimal magnetic quantum number, $|M_S = \pm S\rangle$, correspond to a *single* Slater determinant whenever spin-orbit and spin-spin couplings are neglected. For these cases, we can utilize the *usual* diagrammatic analysis to evaluate the self-energy [17]. Here, the self-energies $\Sigma^{\sigma}(\omega)$ (see Eq. (62)) fall into this category and, following our choice to utilize the highest-multiplicity states $|S\rangle$, the spin-irreducible self-energies can be systematically evaluated by the *usual* diagrammatic analysis via relations (63,64).

V. SCATTERING FROM OPEN-SHELL MANY-BODY TARGETS

A. Spin-irreducible optical potentials

So far, we have studied the properties of the OSGF and its decoupling into spin-irreducible channels. These channels are the generalization of the closed-shell single-channel case. Along this line, we would like to resort to the work of Bell and Squires, where the self-energy of the standard GF has been shown to be an optical potential for scattering from closed-shell targets [11,24]. Generalizing their result, we show below that the spin-irreducible self-energies, $\Sigma^{J}(\omega)$, are well-behaved OP's governing the *elastic* scattering from open-shell targets in the $(N+1)$-particle spin-irreducible channels defined in the previous section.

To describe the *various* scattering processes $a^{\dagger}_{q\sigma'}\,|N\rangle \to a^{\dagger}_{p\sigma}\,|M\rangle$ we need the corresponding *multichannel* S-matrix [30]

$$S^{[M,N]}_{p\sigma,q\sigma'} = {}^{-}\langle p;\sigma M \mid q;\sigma' N\rangle^{+}\,, \tag{65}$$

where the incoming and outgoing scattering states $|p;\sigma M\rangle^{\pm}$ can be expressed using creation operators as [25]

$$|p;\sigma M\rangle^{\pm} = \lim_{t\to\mp\infty} \exp[-i\varepsilon_p t]\, a^{\dagger}_{p\sigma}(t)\,|M\rangle\,, \quad \varepsilon_p = \frac{p^2}{2m}. \tag{66}$$

However, following the symmetry analysis of the previous section it is preferential to examine the spin-irreducible scattering process (channel) $|\psi_{q,J,M_J}\rangle \to |\psi_{p,J,M_J}\rangle$, which couples the *minimal* number of (spin-irreducible) $(N+1)$-particle states. Such process leads naturally to the definition of the spin-irreducible S-matrix,

$$\mathcal{S}^{J}_{pq} = {}^{-}\langle \psi_{p,J,M_J} | \psi_{q,J,M_J}\rangle^{+}\,, \tag{67}$$

where the incoming and outgoing spin-irreducible scattering states are given by,

$$|\psi_{p,J,M_J}\rangle^{\pm} = \lim_{t\to\mp\infty} \exp[-i\varepsilon_p t]\,|\psi_{p,J,M_J}(t)\rangle\,. \tag{68}$$

It is now possible to relate $\boldsymbol{\mathcal{S}}^{J}$ to $\boldsymbol{\mathcal{G}}^{J}(t,t')$, namely,

$$\mathcal{S}^{J}_{pq} = i \lim_{\substack{t\to+\infty\\ t'\to-\infty}} e^{+i\varepsilon_p t}\mathcal{G}^{+,J}_{pq}(t,t')e^{-i\varepsilon_q t'} = i \lim_{\substack{t\to+\infty\\ t'\to-\infty}} e^{+i\varepsilon_p t}\mathcal{G}^{J}_{pq}(t,t')e^{-i\varepsilon_q t'}\,. \tag{69}$$

The last step in Eq. (69) is valid since the spin-irreducible retarded-hole parts, $\boldsymbol{\mathcal{G}}^{-,J}(t,t')$, vanish identically for $t > t'$ (see also Sec. VI).

To proceed we break up the spin-irreducible Dyson equation (60) into two equations which take on the following appearance in time space:

$$\begin{aligned}
\boldsymbol{\mathcal{G}}^{J}(t,t') &= \boldsymbol{\mathcal{G}}^{(0)}(t,t') + \\
&\quad + \int_{-\infty}^{\infty} dt_1 \int_{-\infty}^{\infty} dt_2 \boldsymbol{\mathcal{G}}^{(0)}(t,t_1)\,\boldsymbol{\mathcal{T}}^{J}(t_1,t_2)\,\boldsymbol{\mathcal{G}}^{(0)}(t_2,t'),\\
\boldsymbol{\mathcal{T}}^{J}(t,t') &= \boldsymbol{\Sigma}^{J}(t,t') + \\
&\quad + \int_{-\infty}^{\infty} dt_1 \int_{-\infty}^{\infty} dt_2 \boldsymbol{\Sigma}^{J}(t,t_1)\,\boldsymbol{\mathcal{G}}^{(0)}(t_1,t_2)\,\boldsymbol{\mathcal{T}}^{J}(t_2,t').
\end{aligned} \tag{70}$$

The free GF (see Eq. (61)) is given by

$$\mathcal{G}^{(0)}_{pq}(t,t') = -i\delta_{pq}\theta(t-t')\exp[-i\varepsilon_p\,(t-t')]\,. \tag{71}$$

Comparing Eq. (70) to its standard version, Eq. (17), we find that $\mathcal{T}^J(\omega)$ plays the role of a spin-irreducible improper self-energy. Substituting the free GF (71) into the expression for the spin-irreducible S-matrix, Eq. (69), we find the trivial contribution

$$\mathcal{S}^{(0)}_{pq} = \delta_{pq}\,. \tag{72}$$

To obtain the non-trivial contribution to the spin-irreducible S-matrix we recall that $\mathcal{G}^J(t,t')$ (and, hence, also $\mathcal{T}^J(t,t')$) depends only on the time difference of its arguments, (t_1-t_2). Inserting the first relation in Eq. (70) into expression (69), changing the integration variables into (t_1-t_2) and $(t_1+t_2)/2$, and Fourier transforming from time to energy space one arrives at the final expression for the spin-irreducible S-matrix:

$$\mathcal{S}^J_{pq}(\varepsilon_p) = \delta_{pq} - 2\pi i\,\mathcal{T}^J_{pq}(\varepsilon_p)\,\delta(\varepsilon_p-\varepsilon_q). \tag{73}$$

$\mathcal{T}^J_{pq}(\varepsilon_p)$ is an element of the spin-irreducible T-matrix in energy space at $\omega=\varepsilon_p$, the conserved total energy of the elastic scattering process.

Eq. (73) reminds of the text-book equation connecting the S-matrix to the T-matrix [30]. We have, thus, proved that the matrices $\mathcal{T}^J(\varepsilon_p)$ defined in Eq. (70) are nothing but the spin-irreducible components of the (energetically-elastic) T-matrix. Following standard scattering theory we see from relation (70) that $\boldsymbol{\Sigma}^J(\omega)$ are *one-particle* potentials governing the spin-irreducible scattering processes. In other words, $\boldsymbol{\Sigma}^J(\omega)$ are OP's. We note that the spin-irreducible OP's are *independent* of the azimuthal quantum number (within each total spin manifold), as they should be for spin-independent interactions. Since the spin-irreducible scattering channels do not couple due to symmetry restrictions, $\boldsymbol{\Sigma}^J(\omega)$ are Hermitian and non-singular OP's (functions) at projectile energies below the first spin-irreducible inelastic thresholds. The latter correspond to the first $(N+1)$-particle spin-irreducible excited states. This result gives the spin-irreducible OP's of the OSGF the same role played by Bell and Squires OP, $\boldsymbol{\sigma}(\omega)$, for $\boldsymbol{g}(\omega)$. *Our result puts the scattering of electrons from open-shell atoms and molecules on a firm theoretical foundation, similarly to that possessed by scattering from closed-shell targets.*

All elastic scattering differential cross sections can be computed by utilizing the OP's found above. On the one hand, in scattering of unpolarized projectiles from unpolarized targets the (average) cross section depends on the weighted sum of the spin-irreducible transition probabilities. Spin-flip cross sections, on the other hand, can be expressed from the difference between spin-irreducible transition amplitudes [30]. The spin-irreducible transition amplitudes, $\boldsymbol{\mathcal{T}}^J(\omega)$, can be calculated from the spin-irreducible OP's, $\mathbf{\Sigma}^J(\omega)$, through the usual relations in Eq. (70). This allows us to calculate spin-flip and average cross-sections from the spin-irreducible OP's $\mathbf{\Sigma}^J(\omega)$.

It should be emphasized that the connection between the spin-irreducible quantities discussed above holds as well for the (original) *multichannel* ones [15]. Specifically, the multichannel S-matrix (65) can be connected to the OSGF and, upon breaking up the generalized Dyson equation (Eq. (28)) into two equations as we did in Eq. (70), to the multichannel T-matrix. Consequently, $\mathbf{\Sigma}(\omega)$ itself plays a role of a *multichannel* OP accounting for *all* various scattering processes $a^\dagger_{q\sigma'}|N\rangle \to a^\dagger_{p\sigma}|M\rangle$. As its standard analog $\boldsymbol{\sigma}(\omega)$ and its spin-irreducible "components" $\mathbf{\Sigma}^J(\omega)$, $\mathbf{\Sigma}(\omega)$ is Hermitian and non-singular at projectile energies below the first inelastic threshold.

Following this latter conclusion, we argue further on the Hermitian properties of multichannel and single-channel self-energies (OP's) in the context of scattering from open-shell targets. Suppose that one considers only a *partial* subset of states comprising the degenerate manifold. Following the results of the previous section we know that the multichannel GF for this subset satisfies a generalized Dyson equation and possesses a multichannel self-energy. It is not hard to repeat the derivation presented in this section for $\mathbf{\Sigma}^J(\omega)$ (or that performed for $\mathbf{\Sigma}(\omega)$ in Ref. [15]) and prove that such a multichannel self-energy is a multichannel OP describing scattering from the chosen *partial* subset of degenerate target states. Is this multichannel OP Hermitian? In general it is not, to account for losses to those members of the degenerate manifold not taken in the calculation. To demonstrate this point, consider the extreme case where only a single target state is chosen. For convenience, let it be the highest-multiplicity state $|S\rangle$. For this choice, $\boldsymbol{\Sigma}^{\alpha}(\omega) = \boldsymbol{\Sigma}^{J=S+1/2}(\omega)$ is indeed an Hermitian and well-behaved function, because the elastic scattering processes *allowed* by symmetry in this case, $a^{\dagger}_{q\alpha}|S\rangle \to a^{\dagger}_{p\alpha}|S\rangle$, do not couple other members of the degenerate manifold. Unfortunately, the self-energy $\boldsymbol{\Sigma}^{\beta}(\omega)$ is a non-Hermitian potential and on top of that exhibits a cut starting already at zero energy. This can readily be understood by noticing that $\boldsymbol{\Sigma}^{\beta}(\omega)$ accounts for the spin-flip process $a^{\dagger}_{q\beta}|S\rangle \to a^{\dagger}_{p\alpha}|S-1\rangle$, which is an open channel for *any* projectile energy. Consequently, the usage of $\boldsymbol{\Sigma}^{\beta}(\omega)$ as an OP *by itself* is hardly practical. As dictated by symmetry arguments, on the other hand, the spin-irreducible OP's $\boldsymbol{\Sigma}^{J}(\omega)$ are Hermitian and well-behaved functions and therefore are practical tools for scattering calculations. For the spin-irreducible OP $\boldsymbol{\Sigma}^{J=S-1/2}(\omega)$ this is achieved by the coupling of both $\boldsymbol{\Sigma}^{\alpha,\beta}(\omega)$ (see Eq. (64)) which *shifts* the singular properties of the OP $\boldsymbol{\Sigma}^{\beta}(\omega)$ to the (spin-irreducible) *inelastic* threshold.

B. Spin-irreducible optical wavefunctions

In the previous subsection the spin-irreducible self-energies $\boldsymbol{\Sigma}^{J}(\omega)$ have been identified as spin-irreducible OP's. This allows one to transform all equations relevant for the scattering of a projectile from a many-body target, in a spin-irreducible channel, into effective one-particle equations. We follow below a line analogous to that employed in the discussion of elastic scattering from closed-shell targets [25] and inelastic scattering of a projectile distinguishable [31] or indistinguishable [20] from the particles comprising the target.

We define spin-irreducible scattering amplitudes (also referred to as scattering overlaps)

$$f^{J,M_J}_{pq}(t) = f^{J}_{pq}(t) = \langle \psi_{p,J,M_J}(t) \mid \psi_{q,J,M_J} \rangle^{+} . \qquad (74)$$

that are projection of the spin-irreducible outgoing scattering states on the spin-irreducible $(N+1)$-particle states $|\psi_{p,J,M_J}(t)\rangle$ (hence the independence of the matrix element on M_J). This allows us to express the spin-irreducible S-matrix (see Eq. (67)) in terms of the spin-irreducible scattering amplitudes,

$$\mathcal{S}^J_{pq} = \lim_{t\to\infty} \exp[i\varepsilon_p t]\, f^J_{pq}(t). \tag{75}$$

It can be seen that knowledge of the spin-irreducible scattering amplitudes suffices to solve the spin-irreducible scattering problems and, hence, also, the (original) multichannel one.

To make the connection with standard one-body potential scattering more evident, we express all quantities in coordinate space. Defining the spin-irreducible wavefunction

$$\begin{aligned} |\psi_{q,J,M_J}(r,t)\rangle &\equiv \sqrt{\frac{S+1/2\pm M_J}{2S+1}}\Psi^\dagger_\alpha(r,t)\,|M_J-1/2\rangle \\ &\pm \sqrt{\frac{S+1/2\mp M_J}{2S+1}}\Psi^\dagger_\beta(r,t)\,|M_J+1/2\rangle \\ J &= S\pm 1/2, \quad -J\le M_J\le J, \end{aligned} \tag{76}$$

where we have used the definition of the field operator [12]

$$\Psi^\dagger_\sigma(r,t) = \sum_p \varphi^*_p(r) a^\dagger_{p\sigma}(t), \tag{77}$$

the spin-irreducible scattering amplitudes (74) take on the following appearance in coordinate space:

$$f^J_q(r,t) = \sum_p f^J_{pq}(t). \tag{78}$$

The spin-irreducible S-matrix now reads:

$$\mathcal{S}^J_{pq} = \lim_{t\to+\infty} \exp[i\varepsilon_p t] \int \varphi^*_p(r) f^J_q(r,t)dr. \tag{79}$$

To proceed we show that the spin-irreducible scattering amplitudes are intimately related to the spin-irreducible GF's. Noting that in coordinate space the latter functions take on the appearance

$$\mathcal{G}^J(rt,r't') = \sum_{p,q} \varphi_p(r)\mathcal{G}^J_{pq}(t,t')\varphi^*_q(r'), \tag{80}$$

we find using Eqs. (59,74,78) that

$$f_q^J(r,t) = i \lim_{t'\to-\infty} \exp[-i\varepsilon_q t'] \int \mathcal{G}^J(rt, r't')\varphi_q(r')dr', \qquad (81)$$

which is the basic equation to compute the spin-irreducible scattering amplitudes.

The relation between the spin-irreducible scattering amplitudes and spin-irreducible GF's can be further evaluated by applying the generalized Dyson equation (60). In coordinate–time space this equation reads

$$\begin{aligned}\mathcal{G}^J(rt, r't') &= \mathcal{G}^{(0)}(rt, r't') + \\ &+ \int dr_1 dr_2 dt_1 dt_2 \mathcal{G}^{(0)}(rt, r_1t_1)\Sigma^J(r_1t_1, r_2t_2)\mathcal{G}^J(r_2t_2, r't'), \qquad (82)\end{aligned}$$

where the free (spin-irreducible) GF (61) reads explicitly in coordinate–time space

$$\mathcal{G}^{(0)}(rt, r't') = -i\theta\,(t-t')\,\delta(r-r')\exp[-i\boldsymbol{h}_0\cdot(t-t')]\,. \qquad (83)$$

$\boldsymbol{h}_0$ is the Hamiltonian of the free projectile fulfilling

$$\boldsymbol{h}_0(r)\varphi_p(r) = \varepsilon_p\varphi_p(r). \qquad (84)$$

Note that in second quantization it is related to $\boldsymbol{H}_0$ introduced below Eq. (1). We also notice that, because of the spin-irreducible outgoing scattering state $|\psi_{q,J,M_J}\rangle^+$ being an eigenstate of the total Hamiltonian $\boldsymbol{H}$ with the energy $E^{[0]} + \varepsilon_q$, we can easily eliminate the time in the spin-irreducible scattering amplitudes,

$$f_q^J(r,t) = f_q^J(r)\exp[-i\varepsilon_q t]\,. \qquad (85)$$

Inserting Eq. (82) into the basic relation (81), integrating over the intermediate times t_1 and t_2 after defining (t_1-t_2) and $(t_1+t_2)/2$ as new integration variables, and making use of Eq. (85) to eliminate the time, one finds the final result in energy space

$$\begin{aligned}f_q^J(r) &= \varphi_q(r) + \\ &+ \int dr_1 dr_2 \mathcal{G}^{(0)}(r, r_1; \omega = E_q)\Sigma^J(r_1, r_2; \omega = E_q) f_q^J(r_2). \qquad (86)\end{aligned}$$

Note that the spin-irreducible free GF and self-energy appearing in Eq. (86) are evaluated at the energy $\omega = \varepsilon_q$, the total energy of the scattering. *Eq. (86) is a spin-irreducible Lippmann-Schwinger equation where the scattering amplitudes play the role of spin-irreducible optical wavefunctions for the spin-irreducible elastic scattering channels.*

It is convenient to transform the spin-irreducible Lippmann-Schwinger equation into another from with one integration less. This is achieved by multigrating Eq. (86) from the left by ${\mathcal{G}^{(0)}}^{-1}$. The resulting equation takes on the appearance

$$\hat{\boldsymbol{L}}^J(\omega) f_q^J(r) = \omega f_q^J(r), \quad \omega = \varepsilon_q, \tag{87}$$

where

$$\hat{\boldsymbol{L}}^J(\omega) = \boldsymbol{h} + \hat{\boldsymbol{\Sigma}}^J(\omega) \tag{88}$$

is a spin-irreducible Layzer-like operator (see Ref. [16] for more details on the Layzer operator). The "hat" on $\hat{\boldsymbol{\Sigma}}^J$ indicates a non-local operator in coordinate space, i.e.,

$$\hat{\boldsymbol{\Sigma}}^J f_q^J(r) = \int \boldsymbol{\Sigma}^J(r, r') f_q^J(r') dr'. \tag{89}$$

The spatial boundary conditions for different spin-irreducible wave-functions $f_q^J(r)$ and $f_q^{J'}(r)$ are the same. In a general symmetry-irreducible elastic scattering problem, however, the spatial–spin boundary conditions may vary for different symmetry-irreducible channels. Of course, the energy $\omega = \varepsilon_q$ of the scattering process is independent of the quantum numbers (here, J and J') characterizing the different symmetry-irreducible channels, owing to the degeneracy of the ground state manifold.

To summarize, the self-energies $\boldsymbol{\Sigma}^J$ introduced in the previous section have been shown to be OP's for elastic scattering from open-shell targets. The multichannel scattering problem has, therefore, been reduced to a set of one-particle equations for the elastic spin-irreducible optical wavefunctions.

VI. ANALYTICAL CONTINUATION OF THE OSGF

A. Motivation, definition and general properties

In time space the self-energy $\mathbf{\Sigma}(t,t')$ does not vanish at either $t > t'$ or $t < t'$, and, consequently, in energy space $\mathbf{\Sigma}(\omega)$ consists of simple poles both in the upper and lower half of the complex energy plane, as does the OSGF itself (see, e.g., Eq. (21)). This fact does not cause any problem as long as we have a procedure to determine the self-energy. The EOM's we have derived for the n-point OSGF's (see subsection III B) in order to determine the self-energy in a systematic way are initially in time space. While the transformation to energy space is rather straightforward on the real energy axis, it is quite cumbersome if the result should reflect the *correct* analytic properties in the complex energy plane mentioned above. In fact, even the free OSGF, in contrast to the free standard GF (Eq. (9)), possesses non-vanishing non-diagonal contributions in the complex energy plane. To see that, we observe that in the absence of particle-particle interaction ($\boldsymbol{W}$ and $\boldsymbol{V}$ equal to zero) the time dependence of the creation and annihilation operators is particularly simple (see Eq. (23))

$$a_p(t) = e^{iH_0t}a_pe^{-iH_0t} = a_pe^{-i\varepsilon_pt}. \tag{90}$$

Substituting Eq. (90) into $\boldsymbol{G}(t,t')$ (definition (20)) and Fourier transforming of the result into energy space gives the correct analytic structure of the free OSGF (compare to Eq. (27)):

$$G_{pq}^{(0)[M,N]}(\omega) = \frac{\left\langle M \left| a_p a_q^\dagger \right| N \right\rangle}{\omega - \varepsilon_p + i0^+} + \frac{\left\langle M \left| a_q^\dagger a_p \right| N \right\rangle}{\omega - \varepsilon_p - i0^+} = \\ = \frac{\delta_{pq}\delta_{MN}}{\omega - \varepsilon_p + i0^+} + 2\pi i \rho_{pq}^{[M,N]}\delta(\omega - \varepsilon_p). \tag{91}$$

In Eq. (91) the members of the degenerate manifold $\{|M\rangle\}$ correspond to single or linear combinations of Slater determinants. Thus, the generalized density $\rho_{pq}^{[M,N]}$ is not diagonal and, consequently, so is $G_{pq}^{(0)[M,N]}(\omega)$, what complicates the evaluation of the self-energy in the complex energy plane.

For practical reasons, scattering calculations are preferentially carried out in energy space. Therefore, it would be desirable to possess an expression for the OP (self-energy) that has the correct analytic properties, i.e., to know separately the two parts analytic in the upper and lower half of the complex energy plane. To this end, we may either investigate in more detail the transformation of the EOM's of the n-point OSGF's from time to energy space or, alternatively, redefine slightly the retarded-hole part $\boldsymbol{G}^{-}$ to allow for a more straightforward Fourier transformation. Such an alternative approach is discussed below.

The advanced OSGF is defined such that it lives *only* for $t > t'$:

$$
\begin{aligned}
G_{pq}^{[M,N]}(t,t') &= G_{pq}^{+[M,N]}(t,t') + G_{pq}^{-[M,N]}(t,t') = \\
&= -i\theta\,(t-t')\left\langle M \left| a_p(t) a_q^\dagger(t') \right| N \right\rangle \\
&\quad -i\theta\,(t-t')\left\langle M \left| a_q^\dagger(t') a_p(t) \right| N \right\rangle . \qquad (92)
\end{aligned}
$$

For simplicity we refrain from giving the redefined quantities new mathematical notation but add the prefix 'advanced' to distinguish them from the 'causal' ones defined in subsection III A. Fourier transforming of Eq. (92) into energy space reads as follows:

$$
\begin{aligned}
G_{pq}^{[M,N]}(\omega) &= \sum_s \frac{\left\langle M \left| a_p \right| s(N_e+1) \right\rangle \left\langle s(N_e+1) \left| a_q^\dagger \right| N \right\rangle}{\omega + E^{[0]} - E_{N_e+1}^{[s]} + i0^+} + \\
&+ \sum_r \frac{\left\langle M \left| a_q^\dagger \right| r(N_e-1) \right\rangle \left\langle r(N_e-1) \left| a_p \right| N \right\rangle}{\omega + E_{N_e-1}^{[r]} - E^{[0]} + i0^+} . \qquad (93)
\end{aligned}
$$

We see that all the poles are located in the lower half plane (compare to Eq. (21)). Accordingly, the advanced version of the *free* OSGF is now diagonal (compare to Eq. (91)) and its exact analytic structure is given by Eq. (27), both for non-interacting particles as well as for the case of a free projectile which is not interacting with a fully correlated target.

Naturally, the higher-order OSGF's are redefined analogously as advanced functions only. The EOM's of the advanced n-point OSGF's are identical with the ones computed in subsection III B for the causal n-point OSGF's. The transformation of these EOM's to energy space becomes now particularly straightforward. One has just to identify the energy variable ω appearing in any equation discussed thereafter with $\omega + i0^+$. Consequently, the advanced OSGF obeys the generalized Dyson equation as well. The symmetry-based analysis and results of section IV hold equally as well for the advanced $\boldsymbol{G}(\omega)$. Thus, the spin-irreducible quantities of that section, among them are the spin-irreducible self-energies $\boldsymbol{\Sigma}^J(\omega)$, can easily and systematically be analytically continued to the complex energy plane.

B. Application to scattering

The introduction of the advanced OSGF is only of relevance in the context of scattering if we are able to connect its spin-irreducible components to the spin-irreducible S-matrix elements $\mathcal{S}^J_{pq}$. Specifically, the question arises whether Eq. (69) in section V is also valid for the advanced spin-irreducible GF's, since the advanced version of $\mathcal{G}^{-,J}$ no longer vanish for $t > t'$. It is known that for the simpler case of elastic scattering from non-degenerate targets the expression derived for the S-matrix in terms of the standard GF is valid also for the advanced version of the standard GF [32]. Following Ref. [20], we briefly show below that this is also the case for elastic scattering from degenerate targets.

Since the causal and advanced versions of $\mathcal{G}^{+,J}(t,t')$ are identical, to show that Eq. (69) is also valid for the advanced version of $\mathcal{G}^J(t,t')$ we should prove that the contribution coming from the advanced version of $\mathcal{G}^{-,J}(t,t')$ vanishes, namely

$$\mathcal{S}^{-,J}_{pq} \equiv i \lim_{\substack{t\to+\infty \\ t'\to-\infty}} e^{+i\varepsilon_p t}\mathcal{G}^{-,J}_{pq}(t,t')e^{-i\varepsilon_q t'} = 0. \tag{94}$$

If this equation is indeed fulfilled, i.e., if $\mathcal{S}^{-,J}_{pq} = 0$, all the discussion below Eq. (69) applies to the advanced version of $\mathcal{G}^J(t,t')$ as well. In particular, the spin-irreducible self-energies $\boldsymbol{\Sigma}^J(\omega)$ of the advanced version of $\mathcal{G}^J(t,t')$ are exact OP's for elastic scattering in spin-irreducible channels.

To prove that $\boldsymbol{\mathcal{S}}^{-,J} = 0$ we note that the time limits in Eq. (94) can be replaced with Abelian limits:

$$\lim_{t\to+\infty} f(t) = \lim_{\eta\to 0+} \eta \int_0^\infty e^{-\eta t} f(t)dt. \tag{95}$$

Recalling that the retarded-hole parts, $\mathcal{G}^{-,J}$, can be expressed through the (retarded-hole parts of the) highest multiplicity blocks $\mathcal{G}^{\sigma}(\omega)$ (see Eq. (56)) and twice making use of Eq. (95) we arrive at the condition

$$\lim_{\substack{\eta\to 0+\\ \eta'\to 0+}} \left\langle S \left| a^{\dagger}_{p\sigma} \frac{i\eta'}{H+\varepsilon_p - E^{[0]} + i\eta'} \frac{i\eta}{H+\varepsilon_q - E^{[0]} + i\eta} a_{q\sigma} \right| S \right\rangle \stackrel{?}{=} 0. \tag{96}$$

By inserting completeness, $\sum_r |r(N_e-1)\rangle\langle r(N_e-1)|$, we see that $\mathcal{S}^{-,J}_{pq}$ vanishes unless the real parts of the denominators vanish, i.e., if $\varepsilon_p, \varepsilon_q = E^{[0]} - E^{[r]}_{N_e-1}$, where $E^{[r]}_{N_e-1}$ is an eigenenergy of the target less one electron. Since $E^{[r]}_{N_e-1} - E^{[0]}$ (= ionization energy) as well as $\varepsilon_p, \varepsilon_q$ (the projectile kinetic energy) are positive numbers, the denominator cannot vanish and hence

$$\mathcal{S}^{-,J} = 0 \tag{97}$$

for the advanced version of $\mathcal{G}^{-,J}$. To conclude, we have proved that the spin-irreducible self-energies of the *advanced* version of the OSGF are exact OP's for elastic scattering from the degenerate manifold as well.

VII. BRIEF SUMMARY

The OSGF has been introduced in order to tackle the problem of elastic scattering from open-shell many-body targets. To be able to describe state-specific (polarized) scattering processes the OSGF takes the form of a super-matrix in a double index space, one characterizing the one-particle orbitals and the other the members of the ground state manifold. It has been shown that the OSGF satisfies a generalized Dyson equation and possesses a generalized self-energy (kernel). This self-energy couples all members of the degenerate manifold and comprises of static and dynamical terms, similarly to the structure of the self-energy of the standard GF. Taking into account the symmetry of the many-body Hamiltonian, the OSGF and its self-energy have been factorized into their spin-irreducible (in general, symmetry-irreducible) components. The spin-irreducible

GF's and self-energies are the analogs of the standard GF and self-energy. For most atomic and molecular systems these spin-irreducible functions can even be evaluated by the usual Feynman diagrams.

The central result of this work is that the spin-irreducible self-energies of the OSGF are *exact* OP's for elastic scattering in spin-irreducible scattering channels. *Namely, the multichannel problem of scattering of indistinguishable particles from open-shell targets has been reduced to exact one-body potential-scattering problems.* For kinetic energies below the first spin-irreducible inelastic thresholds, these OP's are hermitian and non-singular potentials, what should make them practical tools for calculating scattering cross sections. Our results generalize the classical work of Bell and Squires [11] and open the door to treat the much intricate scattering processes from open-shell targets with tools as attractive and common as is done for closed-shell targets.

ACKNOWLEDGMENTS

The authors wish to thank Jochen Schirmer for fruitful and stimulating discussions. OEA is grateful to the European Community for financial support (Marie-Curie fellowship).

REFERENCES

[1] K. Lee *et. al*, Phys. Rev. Lett. **70**, 738 (1993); M. Viviani *et. al*, Phys. Rev. Lett. **86**, 3739 (2001).

[2] H. Sakaguchi *et. al*, Phys. Rev. C **26**, 944 (1982); S. Kato *et. al*, Phys. Rev. C **31**, 1616 (1985); M. Ibaraki *et. al*, Nucl. Instrum. Methods Phys. Res. A **446**, 538 (2000).

[3] G. Baum *et. al*, Phys. Rev. Lett. **82**, 1128 (1999); R. Meintrup *et. al*, J. Phys. B **33**, L289 (2000).

[4] T. Hegemann *et. al*, Phys. Rev. Lett. **66**, 2968 (1991).

[5] J. B. Greenwood *et. al*, Phys. Rev. Lett. **75**, 1062 (1995).

[6] M.-Rao Lee *et. al*, J. Phys. B **25**, L505 (1992).

[7] F. J. da Paixão *et. al*, Phys. Rev. Lett. **68**, 1698 (1992); Phys. Rev. A **53**, 1400 (1996).

[8] F. Capuzzi and C. Mahaux, Ann. Phys. (NY) **239**, 57 (1995).

[9] F. Michel and G. Reidemeister, Phys. Rev. C **53**, 3032 (1996).

[10] H. Feshbach, Ann. Phys. (NY) **19**, 287 (1962).

[11] J. S. Bell and E. J. Squires, Phys. Rev. Lett. **3**, 96 (1959).

[12] A. L. Fetter and J. D. Walecka, *Quantum Theory of Many-Particle Systems* (McGraw-Hill, NY, 1971).

[13] E. K. U. Gross and E. Runge, *Vielteilchentheorie* (Teubner, Stuttgart, 1986).

[14] A. K. Kerman, H. McManus and R. M. Thaler, Ann. Phys. (NY) **8**, 551 (1959).

[15] O. E. Alon and L. S. Cederbaum, J. Phys. A. **35**, L303 (2002).

[16] A. J. Layzer, Phys. Rev. **129**, 897 (1963).

[17] L. S. Cederbaum and J. Schirmer, Z. Phys. **271**, 221 (1974).

[18] O. E. Alon and L. S. Cederbaum, unpublished.

[19] L. S. Cederbaum, Phys. Rev. Lett. **85**, 3072 (2000).

[20] L. S. Cederbaum, Ann. Phys. (NY) **291**, 169 (2001).

[21] M. Müller and L. S. Cederbaum, Phys. Rev. A **42**, 170 (1990).

[22] H.-G. Weikert and L. S. Cederbaum, Few-Body Syst. **2**, 33 (1987).

[23] L. S. Cederbaum, *The Encyclopedia of Computational Chemistry*, edited by P. v. R. Schleyer, N. L. Allinger, T. Clark, J. Gasteiger, P. A. Kollmann, H. F. Schaefer and P. R. Schreiner (Wiley, Chichester, 1998).

[24] P. Ring and P. Schuck, *The Nuclear Many-Body Problem* (Springer, Berlin, 1980).

[25] G. Csanak *et. al*, Adv. Atom. Mol. Phys. **7**, 287 (1971).
[26] D. N. Zubarev, Sov. Phys. Usp. **3**, 320 (1960).
[27] J. S. Griffith, *The Irreducible Tensor Method for Molecular Symmetry Groups* (Prentice-Hall, London, 1962).
[28] A. de-Shalit and I. Talmi, *Nuclear Shell Theory* (Academic Press, NY, 1963).
[29] For some cases, a much stronger claim holds. Consider, for instance, $S = \frac{1}{2}$. It can be easily verified that even with the most *general* $(N-1)$-particle state,

$$|\psi_q(t)\rangle = \left[C_1 a_{q\alpha}(t) + C_2 a_{q\beta}(t)\right] |1/2\rangle + \\ + \left[C_3 a_{q\alpha}(t) + C_4 a_{q\beta}(t)\right] |-1/2\rangle$$

$(C_1, \ldots, C_4$ are arbitrary constants), one cannot obtain any element of $\boldsymbol{\mathcal{G}}^{-,J=0}(t,t') = 2\boldsymbol{\mathcal{G}}^{\beta}(t,t') - \boldsymbol{\mathcal{G}}^{\alpha}(t,t')$.
[30] B. H. Bransden, *Atomic Collision Theory* (Benjamin, London, 1983).
[31] L. S. Cederbaum, Few-Body Syst. **21**, 211 (1996).
[32] F. Capuzzi and C. Mahaux, Ann. Phys. (NY) **245**, 147 (1996).

UNDERSTANDING THE RATES OF CHEMICAL REACTIONS

John N.Murrell

School of Chemistry, Physics and Environmental Science
University of Sussex
Brighton BN1 9QJ UK

Abstract. A review is given of the key steps that have led to our current understanding of the rates of chemical reactions. These steps include the law of mass action, the Arrhenius equation, catalysis, collision theory, potential energy surfaces, transition state theory, unimolecular reactions, and classical trajectories.

1. Introduction

This paper gives a historical view of our understanding of the rates of chemical reactions. The topic covers approximately 200 years. Almost as soon as people carried out reactions that we would recognise today as chemical, they asked how fast they went. The subject has matured to the point that the foundations on which it has been built are unlikely to be removed. Yet there is still a long way to go before we can predict the rate of a new reaction with confidence. Also, we have pushed forward the questions we ask about a reaction; not just how fast it goes but what are the details of the energy changes that accompany the reaction.

2. The Law of Mass Action

In 1777 Wenzel [1] proposed that as a measure of chemical affinity one could take the rate at which chemical reactions occurred. For the affinity of dilute acids for metals he noted that "If an acid liquid dissolve a drachm of copper or zinc in an hour, a liquid half as strong will require two hours to effect the same, provided that the surfaces and the heats be equal in the two cases". According to the famous 19th century physical chemist Ostwald this was the first statement of the Law of Mass Action, but it took another 50 years before the laws describing the rates of chemical reactions were formulated more precisely.

E.J. Brändas and E.S. Kryachko (eds.),
Fundamental World of Quantum Chemistry, Vol. II, 155–180.

An understanding of chemical formulae, and of the difference between compounds and mixtures was achieved early in the 19th century, and at that time many new chemicals were being synthesised in a manner that we would now recognise as typical of modern chemistry. The importance of chemical mass in determining the rates of chemical reactions (we would today normally talk about concentrations), was first proposed by Berthollet in several papers and books at the beginning of the 19th century. He also made the very important observation that chemical reactions do not always go to completion, and that the position of equilibrium was affected by the quantity of chemicals involved. In his book 'Recherches sur les Lois de Affinité' published in 1801 (translated into English in 1804) [2], he states that 'In opposing the body A to the compound BC, the combination AC can never take place completely, but the body C will be divided between the bodies A and B in proportion to the affinity and quantity of each, or in the ratio of their masses'.

Many experiments were performed to confirm or refute Berthollet's conclusion, and a dynamical view of chemical equilibrium as a balance of two opposing reactions, forward and backward, gradually gained acceptance. In 1850 Williamson [3] said ' it is clear that the relative velocity of interchange must be greatest between the elements of that couple of which the quantity (*at equilibrium*) is least'. More succinctly Malaguti [4], in 1853, said that equilibrium is reached when the velocities of the two opposing reactions are equal.

If we consider a simple chemical equilibrium

$$A + B \leftrightarrow C + D$$

and write the equilibrium constant[1]

$$K = [C][D]/[A][B]$$

then we can take the rate of the forward reaction to be equal to the product k[A][B], and the rate of the backward reaction to be equal to k´[C][D], where k and k´ are called rate constants. If at equilibrium the two rates are equal, then K=k/k´.

[1] What is meant by the square brackets depends on the phase being studied. For example, in solution it would represent concentration, or more exactly, the thermodynamic activity.

A study of the equilibria established in esterification reactions was made by Berthelot and St. Gilles [5], and they deduced that the rate of the forward reaction was proportional to the product of the alcohol and acid concentrations. This work stimulated two Norwegian scientists, Guldberg and Waage [6] to examine a large number of reactions, and they made the first clear and general statement of the law of mass action for chemical reactions. They made two new important points. Firstly, that the quantities entering the rate law could be raised to integer powers other than one, and secondly, for reactions in solution, it is concentration not mass that enters the equations. Thus if a reaction follows the chemical equation

$$nA + mB \rightarrow \text{products}$$

then, according to Guldberg and Waage, the rate of reaction is proportional to the product $[A]^n[B]^m$. They followed this by stating that 'If the number of molecules in unit volume be denoted by p and q, the product pq will represent the frequency of the encounters of these molecules'. In the meantime, van't Hoff [7] had also looked at the work of Berthelot and St. Gilles, and he also recognised the importance of reaction velocities in determining the balance at equilibrium.

One of the most convincing of many later studies which confirmed the law of mass action was that of Lemoine [8] on the reaction

$$H_2 + I_2 \leftrightarrow 2HI$$

He measured the rates of the forward and backward reactions (in the presence of a platinum catalyst), and showed that they approached equality as the reaction approached equilibrium. The rate constants in both directions were later determined through extensive work by Bodenstein [9], although it was much later to be shown that the $H_2 + I_2$ reaction was not a simple bimolecular reaction [10].

3. Rate Equations and Integration

In a typical kinetic experiment the concentration of one or more of the chemical components is followed as a function of time, and the rate of the reaction is then deduced from the rates of change of the reactants. These rates of change are the target of the kinetic theories which will be described later.

In some simple but very important examples the relation between the concentrations of reactants or products and the rates of change of these concentrations can be deduced by simple calculus, and the first such analysis was made by Wilhelmy in 1850 [11]. He used a polarimeter to follow the inversion of cane sugar (sucrose) by dilute acid (it takes up water to give dextrose and levulose), and showed that the initial rate was proportional to the concentrations of both sugar and acid. He set up a differential equation for the rate of loss of sucrose and integrated it to show how the concentration of sugar decreased with time. His analysis agreed with his experimental observations.

A more thorough examination of the kinetics of simple reactions was made by Harcourt (a chemist) and Essen (a mathematician) [12]. They analysed the differential equations for what we now call first order and second order reactions, and for consecutive second order reactions. The term 'order' was first used by Ostwald to describe the powers of the molecular concentrations which enter into the rate equations, although others had earlier implied this term. Incidently, Harcourt and Essen were the first to make a detailed kinetic study of the famous iodine clock reaction (the oxidation of an iodide by hydrogen peroxide in which the iodine formation is followed by a starch-thiosulphate indicator).

4. The Arrhenius Equation

The next important step in understanding the rates of chemical reactions was the interpretation by Arrhenius of the effect of temperature on the rate constant; as we shall see this lead to the idea that many reactions have to surmount an energy barrier to proceed.

The fact that the rates of most chemical reactions increase with a rise in temperature was well known, for example from the work of Berthelot and St Gilles already mentioned [5]. Wilhelmy [11] was the first to propose a specific temperature law for rate constants from his studies on the inversion of cane sugar, mentioned above. He concluded that the rate of inversion had an exponential dependence on the reciprocal temperature (the logarithm of the rate constant is proportional to 1/T). Other early workers who found an exponential temperature law were Hood [13] in his experiments on the oxidation of ferrous sulphate by potassium dichromate, and Bodenstein [9] in studies of the thermal decomposition of HI.

Van't Hoff [14] pointed out that an exponential temperature dependence of rate constants could be anticipated from the fact that equilibrium constants are the ratio of forward and backward rate constants, because it was known from thermodynamics that equilibrium constants followed an exponential temperature law. For a reaction at constant pressure the equilibrium constant is given by

$$\ln K = -\Delta G/RT = -(\Delta H - T\Delta S)/RT \tag{1}$$

where ΔG is the Gibbs free energy change for the reaction. Taking the temperature variations of the enthalpy ΔH and entropy ΔS to be negligible over a small temperature range, this leads to

$$(d \ln K/dT) = \Delta H/ RT^2 \qquad (2)$$

The dependence on temperature of the forward and backward reactions must be such that from their ratio one must arrive at equation (2). This was the starting point for van't Hoff's deduction that the rate constants would vary with temperature according to the law

$$(d \ln k/dT) = B/T^2 \qquad (3)$$

which would integrate to

$$\ln k = -B/T + C \qquad (4)$$

B and C being constants.[2]

When examining the temperature effects on the rates of reactions Arrhenius concluded that they were much too large to be attributed to factors such as the temperature increase in the energy of collision of the molecules, or in the decrease in the viscosity of the solutions [15]. He reached the important conclusion that there was an equilibrium between normal molecules which had the potential to react and those which actually do react, and that B in expression (4) could be identified as the energy needed to create these reactive molecules. He specifically commented on the cane sugar inversion reaction studied by Wilhelmy [11], and used the term 'active cane sugar' for the form that undergoes the reaction. He speculated that the active form was related to the inactive "by displacement of atoms or addition of water". He noted that there must be a 12% increase in the proportion of active cane sugar per degree rise in temperature.

Arrhenius's law is traditionally written

$$k = A \exp(-E_a/RT) \qquad (5)$$

---

[2] One could add an arbitrary function of temperature to equation (3) and recover equation (2) if this function was exactly the same for the forward and backward reactions, but this is an unlikely possibility.

where E_a is called the activation energy (which is not a term that was used by Arrhenius), and A is called the frequency factor; or together they are called the Arrhenius parameters. It is often said that the exponential term in (5) can be interpreted as the fraction of molecules with energy greater than E_a. The basis for this is that the population of energy levels is given by the Boltzmann distribution function

$$P(E) = N\, g(E)\exp(-E/RT) \tag{6}$$

where g(E) is a degeneracy factor for the number of states with energies E, and N is chosen so that the total population is unity. The fraction with energy greater than E_a is obtained by integrating this from E_a to infinity, and this would be equal to the Arrhenius exponential if g(E) was unity. However, this is not true even for collision energies (for which g(E) varies as E^2), so this simple interpretation of the term is incorrect.

A full understanding of the significance of E_a took some time to emerge. There were a number of suggestions that reactions required the formation of intermediate species, but although this is true for certain reactions, there was no evidence for its generality. Marcelin [16] was the first to advance our understanding when he said that molecules in their average state as regards their internal energy were not capable of reacting, and they only became reactive when their internal energy rose above a critical value. Lewis [17] later called the difference between the energy of the average state and the critical state the 'critical increment', and he identified this with E_a. Rice was following a similar line of argument [18]. Tolman [19], in an important paper dealing with the statistical mechanics of chemical reactions, showed that the activation energy could be equated to the difference between the average energy of reacting molecules and the average energy of all molecules.

It is worth emphasising that if a reaction is studied over a short temperature range, then the results can often be fitted by other temperature laws; a straight line relationship between lnk and lnT can, for example, almost always be achieved. The Arrhenius equation is widely accepted because its interpretation leads to the important concept of the barrier to reaction, and the theories which follow from this, rather than because of its superior fit to the data.
The question of how molecules acquire their energy of activation was tackled by several workers, and the most popular early view was that it was provided by infra-red radiation. Trautz [20], Lewis [18,21], and particularly Perrin [22] developed this idea, using Wien's law for the radiation density. This theory was finally rejected (except for reactions which are specifically photochemical) by Langmuir [23] who showed for several examples that the reactants did not absorb light in the frequency region required to reach the activation barrier, and at higher frequencies where they do absorb, there was

insufficient radiation density to produce the experimental reaction rate. The fact that an exponential function appears in both the Arrhenius equation and the Wien radiation law is just a consequence of the statistical basis of both laws; one for molecules and one for photons. It was eventually concluded that for most reactions a sufficient concentration of activated molecules could be produced by molecular collisions alone to give the observed reaction rate. An interesting discussion on this point occurred in a Faraday Society Discussion [24]. It is interesting to note that as late as 1920 Tolman [19] in his paper said that the importance of radiation in producing activated molecules was "unescapable"

One can take van't Hoff's approach to reactions further by noting that expression (2) can be written

$$\ln k - \ln k' = -\Delta G/RT \qquad (7)$$

where k and k′ are the rate constants for the forward and backward reactions respectively. This lead Kohnstamm and Scheffer [25] (following less precise ideas of Marcelin [26]) to write a rate constant as

$$k = \nu \exp(-\Delta G^{\#}/RT) = \nu \exp(-\Delta S^{\#}/R) \exp(-\Delta H^{\#}/RT) \qquad (8)$$

where $\Delta S^{\#}$, and $\Delta H^{\#}$ are the entropy and enthalpy of activation respectively, and ν is a constant for all reactions. By equating $\Delta H^{\#}$ to the activation energy, we get the Arrhenius equation if $\nu\exp(-\Delta S^{\#}/R)$ is identified with the frequency factor A. It should be noted that this approach gives rate constants for the forward and backward reactions which are consistent with their ratio being the equilibrium constant, a property that is not generally held by the collision theories which are described later.

5. Catalysis

The rate of a chemical reaction can be strongly influenced by adding substances to the medium in which it is carried out. Many examples of this had been discovered early in the 19th century, and in 1835 Berzelius had introduced the concept of a catalytic force (katalyska kraft), to cover many of them [27]. 'The nature of the catalytic force seems to consist essentially in the circumstance that substances are able to bring into activity some affinities which are dormant at this particular temperature, and this not by their own affinity, but by their presence alone.' In his catalytic reactions Berzelius included enzyme processes which had been already identified in fermentation reactions, and the role of metals in heterogeneous catalysis which had been known for a long time; Davy [28], in particular, had noted that platinum and palladium wires glowed in mixtures of air and combustible gases. Berzelius was keen to attribute catalysis to some electrical phenomenon as this was his principal interest at the time.

Ostwald took the study of catalysis further, first studying homogeneous reactions which were accelerated by the addition of acids [29]. He was given a Nobel Prize for this work in 1909. Ostwald defined catalysis as the acceleration of a slowly occurring chemical change by the presence of a foreign body, a body which is not necessary for the reaction. He later proposed an alternative definition that a catalyst is any substance that alters the rate of a chemical reaction without appearing in the end product [30]. He also recognised the existence of negative catalysts or inhibitors, which slow reactions by their presence. His most important proposal was that the total energy change in the reaction should be the same with and without the catalyst; it follows that a catalyst will not alter an equilibrium constant, and hence it alters
the rate constants of the forward and backward reactions in the same proportion.

Ostwald did not support Berzelius's idea that there exists a catalytic force, and although he had no specific theory of catalysis of his own, he liked to use analogies such as 'the effect of oil on machinery, or of a whip on a horse'. His views that there is no reaction that cannot be catalysed, and that there is no substance that cannot act as a catalyst are too strong for today, but they do emphasise the point that catalysis is a very widespread phenomenon.

One of the most influential workers in the field of catalysis was Paul Sabatier, who in his book 'La Catalyse en Chemie Organique' [31], first published in 1912, emphasised the importance of understanding the mechanistic or chemical interpretation of catalytic action, and this can be quite specific to the reaction in question; he was awarded a Nobel Prize for this work in 1912. The simplest mechanistic case would be where the catalyst forms a complex with one of the reactants, and it is this complex which reacts. Thus if we start with the uncatalysed reaction

$$A + B \rightarrow \text{products}$$

then the catalytic mechanism might be

$$A + C \leftrightarrow AC$$
$$AC + B \rightarrow \text{products} + C$$

Such a mechanism would alter the energy profile of the reaction; the intermediate AC would be associated with a minimum on the reaction path, but the important point is that the energy barrier from AC + B to products plus C will be lower than the barrier for the uncatalysed reaction. An important point to note is that with the above mechanism the rate of the reaction would depend on the concentration of C; it is not true that catalysts are always effective in very small concentrations.

If the mechanistic role of the catalyst is more complicated than the simple one given above, then it may completely change the route (path) of the reaction. This lead Hinshelwood [32] to liken the catalytic role to the opening of a by-pass road with easier gradients. Given the broad range of catalytic mechanisms, we must conclude that the only statement that embraces all catalytic processes is that catalysts alter the topography of the energy surface leading from reactants to products; we say more about such energy surfaces later in this essay.

6. Collision Theory of Chemical Reactions

Once the atomistic view of chemistry had been established, it was natural to think of chemical reactions as taking place through the collisions of atoms and molecules. In solution this process will be moderated by the solvent; indeed the rate determining step might be the rate of diffusion together of the reactants rather than the energy of their collision. For gas phase reactions, however, an understanding of collision dynamics is essential for the reaction process, and this task began with the work of Maxwell, Clausius and others who developed the kinetic theory of gases in the 19th century. Early theories only treated atoms and molecules as hard spheres.

Two hard spheres with diameters d_A, and d_B, will collide when the distance between their centres is

$$d = (d_A + d_B)/2 \tag{9}$$

and the collision cross section will be

$$\sigma = \pi d^2 \tag{10}$$

The mean relative speed of these molecules at the time of collision can be calculated from the Maxwell-Boltzmann distribution of velocities as

$$v = (8kT/\pi\mu)^{1/2} \tag{11}$$

where μ is the reduced mass of the two molecules, $m_A m_B/(m_A + m_B)$. It follows that the collision frequency per unit concentrations of molecules A and B, which can be equated to the Arrhenius frequency factor, is [33]

$$A = \sigma v = d^2(8\pi kT/\mu)^{1/2} \tag{12}$$

Note that A depends on T, so that in this theory a strict Arrhenius law ($\ln k \propto 1/T$) would not be found. However, one does not usually have rate constants measured over a sufficiently large temperature range to detect any deviation from the strict law.

The first application of this result to a gas reaction was by Lewis [21]. For the thermal gas phase decomposition of HI he found 'satisfactory' agreement with experiment, and for the reverse reaction of H_2 and I_2 , he found 'moderate' agreement with experiment. He concluded that the main error in his theory was in the estimation of d, the distance between the centres of the molecules at collision. Guggenheim and Prue [34] analysed more recent data on the decomposition of HI and found that collision theory gave good agreement with experiment if d was 4.12Å.

A more detailed treatment of the collision theory of reactions was given by Fowler and Guggenheim [35] (see also [36]). By applying the Maxwell-Boltzmann expression to the relative velocity of collisions, it can be shown that the number of collisions per unit volume per unit time in which the relative velocity lies between v and v+dv, and the angle between $\underline{v}$ and the line of centres on impact lies between ϑ and ϑ+dϑ, is

$$n_A n_B (\mu/2\pi kT)^{3/2} 8\pi^2 d^2 \exp(-\mu v^2/2kT)\, v^3 dv\, \sin\vartheta \cos\vartheta\, d\vartheta \qquad (13)$$

If the two molecules are of the same species then this expression has to be divided by a symmetry number of 2.
If we now say that the probability of a reactive collision is $\kappa(v,\vartheta)$, and note that for a bimolecular reaction the number of reactive collisions per unit volume per unit time is $n_A n_B k$, then the second order rate constant is

$$k = (\mu/2\pi kT)^{3/2} 8\pi^2 d^2 \iint \kappa(v,\vartheta) \exp(-\mu v^2/2kT)\, v^3 dv\, \sin\vartheta \cos\vartheta\, d\vartheta \qquad (14)$$

This formula is based on the assumption that either the two reactant molecules are of different species, or that if they are of the same species then both molecules react. It also assumes that only one internal state (rotation, vibration, electronic) is relevant for each species (which is not usually the case).

For the two molecules to be approaching one another ϑ must lie between 0 and $\pi/2$. If the probability of reaction is independent of ϑ (an extreme assumption in most cases), we can integrate over ϑ to give the result

$$k = (\mu/2\pi kT)^{3/2} 4\pi^2 d^2 \int \kappa(v) \exp(-\mu v^2/2kT)\, v^3 dv \qquad (15)$$

The kinetic energy of collision is $E = \mu v^2/2$, and changing the variable in (15) to E gives

$$k = (\mu/2\pi kT)^{3/2}(8\pi^2 d^2/\mu^2)\int \kappa(E) \exp(-E/kT)\, E dE \qquad (16)$$

If we now take $\kappa(E)$ to be zero below the energy barrier E_0, and one above it, then we can integrate (16) to give

$$k= (8\pi kT/\mu)^{1/2}\, d^2\, (1 + E_0/kT) \exp(-E_0/kT) \qquad (17)$$

The energy independent factor is identical to that obtained by the simple collision analysis given above, except for the extra term E_0/kT, which is usually small. In the HI reaction at 556K, for example, this term has the value 0.04.

Although this form of collision theory leads to a broadly satisfactory interpretation of the Arrhenius equation, it requires the adoption of several highly questionable assumptions. In addition, the ratio of the collision theory rates for forward and backward reactions is not equal to the equilibrium constant because there is no explicit treatment of entropy changes in the reaction. However, the Arrhenius form is certainly more robust than is implied by simple collision theory.

To develop collision theory further one has to start by defining a reaction cross section, and determining how this depends on the collision energy and the internal quantum states of the reactants. Some analytical developments have been made along these lines (see Gardiner [36]), but they have been overtaken by the advances that have come along with high speed computers and the work that has emerged from classical trajectory calculations; this will be looked at later.

An interesting application of collision theory was made by Herzfeld [37], who obtained an expression for the rate constant of the unimolecular dissociation process

$$AB \rightarrow A + B$$

This work came before an understanding of the role of pressure in such reactions, which we will discuss later. Herzfeld used the fact that the ratio of the backward to the forward rate constants was equal to the equilibrium constant, and he took the backward rate constant from collision theory. The equilibrium constant could be calculated from statistical mechanics, the case of diatomic molecule dissociation having been analysed by Stern [38]. Herzfeld obtained the result

$$k_u = N(kT/h)(d^2/s^2)(1 - \exp(-h\nu/kT))\exp(-\Delta E/RT) \qquad (18)$$

where d is the collision diameter for the atomic collision, s is the diatomic bond length, ν is the vibration frequency of the diatomic molecule, and ΔE its dissociation energy. Herzfeld went on to equate d and s, which is certainly not acceptable. The most important feature of this work was that by using equilibrium theory he obtained an expression containing the factor (kT/h), which was later to enter the transition state theory developed by Eyring and others.

7. The Potential Energy Surface

From the Arrhenius interpretation of reaction rates we have a picture of a reaction proceeding along a potential energy path and crossing a barrier. The first developments of this idea were by Marcelin [39], and Rodebush [40], who refer specifically to the rate at which the system crosses a critical surface in phase space; a concept used in statistical mechanics in which the phase space variables are the coordinates and momenta of the atoms in the system.

For the collision between two atoms the potential energy is just a function of the distance between the two, but for collisions involving molecules the potential energy will depend on the molecular orientations, and also on the internal geometries of the molecules themselves. This is where the important concept of a potential energy surface comes in (some use the term hypersurface to emphasize that it is multidimensional). This is the energy of the reacting system as a function of a set of coordinates which define completely the instantaneous relative positions of all the atoms in the molecules. If there are N atoms in the reacting molecules there will be 3N-6 relative position coordinates; 3N is the number needed to define the absolute positions of all the atoms in space, and we can subtract from this the 3 coordinates of the overall centre of mass and the 3 which define the overall orientation of the system.

Fritz London [41] was the first to make some conjectures about the nature of such surfaces, indeed he was the first to state explicitly that the majority of chemical reactions could be considered as the motion of atoms on a single surface that extended from reactants to products; what we call an adiabatic process. There are reactions that can only be interpreted by considering jumps between surfaces, but these non-adiabatic processes are a small fraction of the total. London examined the general features of simple gas phase reactions involving only three or four atoms [42], such as

$$D + H_2 \rightarrow DH + H$$

and

$$2HI \rightarrow H_2 + I_2$$

From the new theory of the chemical bond [43], which he had jointly authored, he concluded that for certain three atom reactions of the above type, which are model substitution or abstraction reactions, the lowest energy barrier on the reaction path occurs when the atoms are collinear.

More detailed calculations on such potential energy surfaces, with the collinear restriction, were made by Eyring and Polanyi [44], using a similar Heitler-London model. As well as the hydrogen atom exchange they looked at the two reactions

$$H + HBr \rightarrow H_2 + Br$$

and

$$H + Br_2 \rightarrow HBr + Br$$

The collinear H_3 potential energy surface possesses a col or saddle point in the region where the two H-H distances are equal (both about 1Å), and this leads to an important extension of the one-dimensional Arrhenius view; not all reactions pass exactly through the lowest point on the saddle, but they do pass through the saddle region. The curve that passes from the bottom of the reactant valley, through the saddle point, and emerges at the bottom of the product valley is called the reaction coordinate; this can be defined for any number of coordinates although its precise form depends on the coordinate system used.

Eyring and Polanyi went on to consider the dynamics of motion on their potential energy surfaces according to classical mechanics, and a more thorough analysis was given soon after by Pelzer and Wigner [45] who stressed the importance of the region around the saddle point. For a particular set of initial conditions of the reactants (these will be the initial positions and momenta of all the atoms which according to classical mechanics will determine a particular outcome), it is in principle possible to calculate a trajectory across the potential energy surface. Some of these trajectories will end up back in the region of the reactants, and will be representative of a non reacting collision, and others will pass over the saddle to the product region and be representative of a reactive collision. Pelzer and Wigner derived an expression for the rate of reaction by considering the number of trajectories that passed over the saddle.

For the collinear atom plus diatomic molecule collision of the type examined by Eyring and Polanyi, a trajectory can be pictured as the motion of a ball rolling over a surface, but only if the surface is drawn in skew coordinates; for three atoms of equal mass the appropriate angle is 60°.[3]

We might now take the view that the degree to which we 'understand' a chemical reaction (at least in the gas phase) is the degree to which it can be interpreted by a potential energy surface, using one of the dynamical theories which allow us to calculate the rate expected from that surface. The potential energy surface can, in principle, be calculated from the quantum mechanical Schrodinger equation for the electrons of the reacting molecules, and we can deduce some information from experiments. For reactions in solution, we can usually obtain only a little information about the full surface; for gas reactions it may be possible to obtain quite a lot of information.

8. Transition state theory

In all types of experiment there is some loss of idealised data due to limitations in the measuring equipment; the temperature range may not be as large as one would like, there is always some averaging over the conditions of the reactants or the state of the products. In highly averaged experiments, particularly those in which only the temperature is a specified condition, we expect statistical theories to be very useful, and the most important of these is called transition state theory.

The work of Eyring and Polanyi [44], and of Pelzer and Wigner [45], and indeed most of the people who adopted Arrhenius's view that reactions were due to activated molecules, lead to a thermodynamic view of reactions which we call transition state theory (also called activated complex theory) [46,47,48].

Evans and Polanyi say that their work was forshadowed in several earlier paper [49,50,51].
Evans and Polanyi were the first to use the term transition state to refer to the region of the
potential energy surface in the vicinity of the saddle point, and they and Eyring derived a rate constant for the reaction by assuming that the transition state (or activated complex) was in equilibrium with the reactants. Of course there cannot be a true equilibrium because there can be no uniquely defined

---

3 This was pointed out by Eyring and Polanyi but they had an incorrect sign in their analysis so their axes were at an angle of 120° rather than 60°. The correct result appears to be due to J.O.Hirschfelder in his PhD Dissertation (Princeton University, (!935). The basis of this result is to find orthogonal coordinates which give the classical kinetic energy as a sum of diagonal terms.

area in which the transition state can be said to exist; it is only defined by the coordinate surface that passes through the saddle point and is orthogonal to the reaction coordinate. The transition state is at best a transient species whose lifetime is related to the rate of passage across the barrier, although with modern fast spectroscopic techniques the properties near the barrier can be examined.

If X represents the transition state arising from a bimolecular reaction between A and B, then the equilibrium constant is

$$K = [X]/[A][B] \tag{19}$$

We can obtain an expression for this equilibrium constant using formulae derived from statistical mechanics; it can be written

$$K = (q_X/q_A\, q_B) \exp(-E_0/kT) \tag{20}$$

where the quantities q are called partition functions, and E_0 is the height of the potential energy barrier. There are specific formulae for the partition functions of molecules which take into account their translational, rotational, and vibrational motion, and any electronic degeneracy they may possess. The important question is what to use for that part of the partition function of the transition state which represents the motion along the reaction path. In the original formulation due to Eyring this was taken as the translational partition function of a molecule confined to a one-dimensional box of length δ, and the final expression for the rate constant was shown to be independent of δ. In an alternative treatment, perhaps more difficult to justify, it has been taken as the vibrational partition function for a one-dimensional harmonic oscillator in the limit that the vibrational frequency approaches zero. Both methods lead to the same answer for the rate constant, namely [52]

$$k = (kT/h)\,(q^{\#}/q_A\, q_B) \exp(-E_0/kT) \tag{21}$$

where $q^{\#}$ is the partition function for the transition state except for that part representing motion along the reaction coordinate.

The only modification to the above theory which was in the original work was the introduction of a multiplying factor κ in expression (21) called the transmission coefficient. This took into account the possibility that some systems that passed through the transition state might not end up as products, but would be turned back by other features of the potential energy surface and emerge again as reactants. One would not expect κ to be much less than unity, but a weakness of the original theory was that there was no way of determining its value. In a much later development of the theory it was argued that it was not necessary to define the transition state by a surface that passed through the saddle point. If another surface, orthogonal to the reaction coordinate, was selected so that all trajectories which passed through it went on to emerge as products, then κ would be unity. This led to a modification called variational transition state theory (VTST), in which a surface was chosen to minimise the subsequent calculation of the rate constant [53] This theory is particularly appropriate for reactions which do not usually have a saddle point on the potential energy surface, such as molecular dissociation reactions and the bimolecular addition of free radicals.

The problem with VTST is that the transition state surface is not independent of the energy at which the reactants collide, and hence it must be temperature dependent. It is therefore necessary to formulate the theory in terms of phase space rather than coordinate space, as in the early statistical theory of Marcelin [39], and later of Wigner [54].

The development of VTST is largely due to Keek [55], although there were others having similar ideas at the time, and later there were further generalisations. In his review article Keek shows how his theory relates to several other theories of reaction rates, including conventional transition state theory. He also deals in detail with the problem of three-body recombination.

9. Unimolecular Reactions

A unimolecular reaction is one represented by the process

$$A \rightarrow \text{products}$$

but the rate of such a reactions may not be proportional to the concentration of A, that is be a first order reaction. Indeed in the early history of gas kinetics it was not certain whether first order reactions should exist. If A is a normal stable molecule then there must be some intervention to make it undergo a reaction. This may, for example, be the absorption of light, or, more commonly, it is the effect of molecular collisions, either by some species that is not chemically involved in the reaction, or between molecules A themselves. To understand unimolecular reactions one needs to understand the mechanism through which A first has enough energy to react, and then does so.

A characteristic of unimolecular gas reactions is that their apparent order depends on the pressure of the system. Lindemann [56] was the first to suggest that such reactions could take place by collisions,[4] and Hinshelwood [57] explained why first order behaviour was maintained down to lower pressures than would be expected on the basis of simple collision theory. The so-called Lindemann-Hinshelwood mechanism is described by the following equations

$$A + M \rightarrow A^{*} + M$$
$$M + A^{*} \rightarrow A + M$$
$$A^{*} \rightarrow \text{products}$$

A^{*} is a molecule which has sufficient energy to give products. If the rate constant for collisional excitation is k_1, and for de-excitation is k_2, and k_a is the rate constant for the formation of products from A^{*}, then in the steady state (when the amount of A^{*} is constant), the effective first order rate constant for the disappearance of A is

$$k_u = k_1 k_a / (k_2 + k_a/[M]) \qquad (22)$$

In a gas phase reaction the concentration of M will be proportional to the pressure.
In the so-called high pressure limit ($k_2 > k_a/[M]$), the reaction will be strictly first order, with a rate constant of ($k_1 k_a/k_2$); at low pressures the reaction is second order, the rate being $k_1[A][M]$. The effective first order rate constant k_u falls off with decreasing pressure from $k_1 k_a/k_2$ to $k_1[M]$, and a graph of this against pressure is known as the fall-off graph. The position of the fall-off region in this graph provides a measure of the relative importance of the energizing-deenergizing process and the rate at which A* gives products.

Although the Lindemann-Hinshelwood mechanism explains the passage from high to low pressures, it simplifies the problem by assuming that there is a single species A^{*} which can be assigned a rate constant. But the population of A^{*} cannot be taken to conform to the Boltzmann formula, that is to be in a true equilibrium with A. There are several reasons for this but the main one is that for A^{*} to react its energy must be in particular modes. Molecules that undergo reaction will be those whose extra energy is in vibrations rather than translations or rotations; moreover, not all vibrational modes will be equally effective. Important early work by Chambers and Kistiakowsky [58] showed that for the isomerisation of cyclopropane to propene, only between one half and two thirds of the 21 vibrational modes of cyclopropane were effective in leading to reaction.

---

4 J.A.Christiansen also proposed the Lindemann model in a PhD thesis titled Reaktionskinetisch Studier, of the University of Copenhagen in 1921.

Theories of unimolecular reactions have to tackle the energizing-deenergizing processes, and the rate of reaction of A^*. The first theory to do this is known as the RRK theory after the work of Rice and Ramsperger [59], and of Kassel [60]. Central to their work is the distinction between a molecule which has sufficient energy to undergo reaction (A*, an energised molecule), and one that will pass quickly to products because the energy is in a favourable bond or bonds; this is called the activated complex, $A^{\#}$. The passage from A* to $A^{\#}$ is assigned a rate constant k_a, and the passage of $A^{\#}$ to products a rate constant k_3. However, in this theory it is always assumed that $A^{\#}$ goes very rapidly to products so that the rate of reaction is independent of k_3.

Marcus developed a quantum mechanical version of this theory, which is referred to as RRKM theory [61]. Its main feature is to recognise the importance of zero point vibrational energy as this is unavailable for the reaction process.[5]

In RRKM theory the energy dependence of k_a is calculated by taking the available energy in A* and determining how much of this is in a particular vibrational mode that leads rapidly to reaction. Assuming that the vibrational states can be taken in the harmonic model, then if there are s oscillators having j quanta, the probability that a particular oscillator has m quanta is equal to

$$(j-m+s-1)!j!/(j-m)!(j+s-1)! \qquad (23)$$

and if j is very large this is approximately equal to

$$((j-m)/j)^{s-1} = ((E - E^*)/E)^{s-1} \qquad (24)$$

where E is the total energy and E* is the minimum energy that is necessary for the reaction to occur.

The theory further assumed that energy passes between modes very rapidly so that the rate at which energy appears in the particular mode is proportional to expression (24) (ie k_a is proportional to (24)). The amount of energy needed in the vibrational modes to produce a reaction is usually too large to validate the harmonic model, but in a strictly harmonic model there would be no energy transfer between modes. It is therefore the anharmonic part of the potential

---

5 A molecule does not have to be very large before its zero point energy is bigger than its dissociation energy, although the difference between the zero point energy of the reactant and the products is usually small compared with the dissociation energy.

energy which validates the assumption that energy flows rapidly between modes.[6]

Although the essential element of RRKM theory is that k_a depends on energy, the basis for its success is that s, the number of normal modes relevant to the reaction, is treated as a parameter. As mentioned above, the value of s which gives agreement with the fall-of curve is usually about half the total number of normal modes possessed by the molecule.

The literature on unimolecular reactions is extensive, and is thoroughly treated in several books and reviews (eg the book by Forst [62]).

10. Classical or Quantum mechanics

We know that the dynamics of individual atoms and molecules is governed by the equations of quantum mechanics, yet most practical theories of reaction rates are based on classical mechanics, the mechanics of massive bodies. It is important therefore to know why classical mechanics is successful, and to know when it fails.

In quantum mechanics particles of momentum p have an associated wave whose wavelength λ is given by the de Broglie equation

$$\lambda = h/p \tag{25}$$

A quantum mechanical approach is usually required if this wavelength is large compared with the range over which particles interact; typical classical systems are those for which the wavelength is small even by comparison with the size of the body.

Wave motion is characterised by diffraction and interference, which are easily recognised, and by the phenomenon of tunnelling which is less obvious; none of these is found in classical mechanics. In classical mechanics an initial discrete set of coordinates and momenta of interacting particles will evolve into a specific discrete final set of these variables. In quantum mechanics one cannot have a system with discrete coordinates and momenta; there is a distribution of these variables which can be represented by wave packets, and

---

[6] An alternative approach to the vibrational energy was given by Slater [6] who took a strictly harmonic model (ie no energy transfer between modes) and calculated the rate at which the cooperative motion of the modes led to a critical displacement appearing in a breaking bond. This theory does not seem to be as successful as RRKM theory and is now little used.

under the influence of potentials an initial wave packet will evolve to a final wave packet.

At standard temperatures in the gas phase molecules typically travel at about the speed of sound, say 300m/sec, and a hydrogen atom at these speeds have a wavelength of about 10^{-9}m (10Å). We would therefore expect hydrogen atoms to easily show quantum phenomena, but for atoms heavier than carbon, say, the wavelengths will typically be less than 1Å and quantum behaviour will be difficult to observe. However, even for light atoms one can only observe diffraction and interference in experiments for which the momenta of the particles are highly discrete; one does not observe these wave phenomena for a continuum of wavelengths (eg for white light). It follows that chemical reactions in which the molecules have a range of velocities determined only by the temperature of the system can usually be interpreted by classical dynamics. The only exception to this statement is for the phenomenon of tunnelling. This is the property of quantum systems to pass through a potential barrier even if their total energy is less than the height of the barrier. So, even for a temperature averaged system, light atoms such as hydrogen can be transferred in a reaction if their energy is less than the activation energy. The usual way that this is established is to show that isotope substitution (eg D for H), slows a reaction much more than is expected from a classical picture. Also, if there is tunnelling a reaction will not follow the Arrhenius law for variation of the rate constant with temperature. There is certainly evidence for tunnelling being significant for the H + H_2 reaction, and for several proton transfer reactions in solution [63,64].

11. The Crossed Beam Experiment

In temperature averaged chemical reactions there are usually only two quantities that can be extracted from experimental data, namely the two Arrhenius parameters. For any substantial understanding of the potential energy surface this is clearly inadequate. Other types of experiment must be used to extract such information, and most of these are based on the crossed beam experiment. The origins of beam experiments can be traced back to Rutherford, who showed by passing α particles (He^{2+}) through metal foils that atoms possessed a small positively charged nucleus. The first crossed beam experiments to determine the potentials between atoms (alkali metals and mercury), were by Fraser and Broadway in 1933 [65], and the first experiments to study chemical reactions were by Taylor and Daz on K + HBr in 1955 [66].[7]

---

[7] A slightly earlier crossed beam study of the reaction between Cs and CCl_4 by Bull and Moon did not have velocity resolution in the beams [7].

The normal set-up for the experiment is to have two beams of atoms or molecules directed along lines at right angles to one another, which meet in a collision chamber, and the products of the collision are detected and analysed at various angles in the plane of the incident beams. A laser beam is often applied to the collision region at right angles to this plane.

The density in the collision chamber is sufficiently low that the outcome can be explained by single collisions of the reactants. Ideally both incident beams should have narrowly defined velocities, so that their collision energy is narrowly defined. Again, ideally, the reactants should be in specific rotational, vibrational and electronic quantum states. The products should be analysed to give their velocities, and their quantum states, and this should be done in narrowly collimated beams over a range of angles of the detector.

The experiment leads to reaction cross sections (with the scattering angle as variable) for each value of the collision energy, and for specific quantum states of the reactants, with as wide a range of the collision energy as possible. This is a huge amount of information, which, in principle, would specify in fine detail the features of the potential energy surface. However, any lack of resolution in the energy spread of the reactants or products, or lack of resolution in the scattering angle will ultimately lead to a loss of detail in the potential energy surface. Some loss is inevitable in the experiment, not least because there is a compromise between angular resolution and the need to have sufficient intensity in the exit beam.

The main snag in determining the potential from the experimental data is that there is no direct route for this (we say there is no inversion procedure). The only way to determine the potential is by the indirect route, which is effectively by trial and error. Given a potential we can calculate the reaction cross sections that this would lead to. If these agree with the experimental data we conclude that the potential is accurate within the limits of the data.

12. Classical trajectory calculations

Given a potential energy surface, or more specifically given an algebraic function that describes the energy of the surface with the coordinates as variables, we can calculate the dynamics on that surface by classical or quantum mechanics. We have already seen that quantum features do not emerge in most experiments; putting that another way we can say that if one wants to observe quantum features then special systems have to be studied and special care taken to obtain the highest resolution. Classical mechanics has therefore been used much more extensively than quantum mechanics to interpret chemical reactions, even those arising from crossed beam experiments.

Potential energy functions for molecular reactions are almost always sufficiently complicated that numerical methods have to be used to determine the resulting dynamics, using either classical or quantum theory. This was clear from the earliest studies of potential surfaces by Polanyi, Eyring and others, but the computational facilities were not available at that time. Hirschfelder and co-workers [67] were the first to calculate a trajectory on the collinear $H + H_2$ surface, and showed how it became trapped in a shallow minimum at the top of the surface (a minimum later shown to be an artefact of the method of calculation). Such 'by-hand' calculations were extremely laborious and could not be done in sufficient numbers to explore the dynamics. The first serious numerical studies of classical trajectories were by Wall and co-workers in 1958 [68], using the ILLIAC computer and the Runge-Kutta-Gill method for numerical integration of Hamilton's equations. They calculated several hundred trajectories for the collinear $H + H_2$ surface, and then in 1961[69] calculated 700 trajectories for the three dimensional surface of this reaction. Sufficient trajectories to determine a statistically significant rate constant were calculated by Karplus and co-workers [70], using a more accurate potential surface. The technique has not changed greatly to the present, and is well described in reviews [71]. The first widely available computer programs were for atoms colliding with diatomic molecules, but within about 10 years collisions between atoms and triatomic molecules were being studied [72].

In a classical trajectory study there are a certain number of variables that have to be chosen to specify the initial conditions. Some of these have values determined by the conditions of the crossed beam experiment; the relative velocity of the colliding species, for example, and perhaps the quantum state energies of the reactants. Other quantities are unspecified by the experiment such as the initial orientations of the molecules with respect to the collision axis, and the perpendicular displacement of the initial collision vector from this axis (called the impact parameter). The unspecified parameters are chosen either serially or randomly to cover the full variation they may have, with appropriate statistical weighting, and a set of trajectories is calculated covering these.

The number of trajectories that must be calculated for each set of collision energies and initial quantum states to give statistically significant results depends on the type of analysis that is made of the product states. For example, if one is just interested in whether there has been a chemical reaction or not, then about 1000 trajectories may be enough. If one is interested in knowing the energy states of the product molecules then one could well ask for about 100 trajectories for each possible energy state. If one is interested in the scattering angle that these products and their energy states emerge from the collision, then one would need this number for each band of angles that is covered. In short, the number of trajectories depends on the experimental data being examined or predicted.

By calculating the fraction of all trajectories leading from specified initial conditions to specified final conditions, and multiplying by the area spanned by the impact parameter, one obtains cross sections for specified initial conditions leading to specified final conditions, and these can be summed or averaged to give quantities measurable in experiments. For example, cross sections for scattering into specified angles of emergence (in a relative coordinate system), which are called differential cross sections, can be summed over all angles to give total cross sections; total cross sections can be averaged over initial states and summed over final states to give reaction cross sections, $S(E_T)$, for collision energies E_T. The rate constant for a temperature averaged system is obtained by further averaging of these cross sections over the collision energy, using the Maxwell-Boltzmann formula. The final expression has the form

$$k = (8/\pi\mu(kT)^3)^{1/2} \int E_T\, S(E_T) \exp(-E_T/kT)\, dE_T \qquad (24)$$

However, one might conclude at the end of such calculations that a lot of work has gone in to produce a single quantity, moreover, a quantity that could perhaps be obtained directly by one of the transition state theories. In fact, one is unlikely to embark on trajectory calculations unless one is interested in quantities other than the rate constant; for example the experiment may show the populations of different energy states in the products, and this is most easily obtained from trajectories. Seminal papers along this line were published by John Polanyi and co-workers in 1969, in which they showed how the product energy in an atom-diatomic reaction was disposed according to the position and height of the potential energy saddle point [73,74].

13. Conclusions

I am aware that there are several important topics within the field of chemical reactions which I have not dealt with, or have touched on only briefly: photochemistry (particularly time-resolved), explosions, reactions in condensed phases (particularly atom recombination), and quantum mechanical calculations, are some of the more obvious. Each of these has been associated with key ideas that advanced our understanding of the subject. But finally I should be clear on what is the most important step for understanding the rate of a reaction, and I have no doubt that this is to be able to write down the mechanism in terms of the primary chemical steps which occur. Many chemical reactions are multi-step, but detailed understandings, and particularly numerical calculations, are always made on single step reactions.

I have read many books and reviews specializing in reaction kinetics, and would like to acknowledge particularly the information I received on historical matters from two of these. The first is Partington's epic volumes 'A History of Chemistry' [75], and the second is Laidler's 'The World of Physical Chemistry' [76]. I also acknowledge many helpful discussions with my colleague Tony Stace.

I never worked closely with Per-Olov Lowdin, but I sometimes sent preprints of my papers to him to see if they were nonsense; occasionally they were.

---

1 K.F.Wenzel, Lehr von der Verwandtschaft der Korper, Dresden, (1777)
2 C.L.Berthollet, Recherches sur les lois de Afinite, sm8, Paris IX,(1801)
3 A.W.Williamson, Phil.Mag., 37, 350,(1850)
4 F.J.M.Malaguti, Ann. Chim., 37, 198 (1853)
5 M.Berthelot, and L.P.St.Gilles, Ann.Chim., 65, 385,(1862)
6 C.M.Guldberg, and P.Waage, Forhandlinger I Videnskabs-Selskabet I Christiana 35, (1864-5); 1, (1879)
7 J.H.van't Hoff, Ber., 10, 669, (1877)
8 C.G.Lemoine, Ann.de Chim., 12,145, (1877); 26, 289,(1882)
9 M.Bodenstein, Z.Physik.Chem., **13**, 22 (1894); **22**, 1 (1897); **29**,295 (1899)
10 J.H.Sullivan, J.Chem.Phys., **46**,73 (1967)
11 L.F.Wilhelmy, Ann.Phys., 81, 413, 419,(1850)
12 A.V.Harcourt and W.Essen, Phil. Trans., 156, 193(1866);157, 117 (1867)
13 J.J.Hood, Phil.Mag., 6,371 (1878); 20, 323,(1885)
14 J.H.van't Hoff, Etudes de Dynamique Chimique, 115,127 (1884)
15 S.A.Arrhenius, Z.Phys.Chem., 4,226,(1889)
16 R.Marcelin, Ann.Phys., 3, 120 (1915)
17 W.C.Mc.Lewis, Rep.Brit.Ass.,394 (1915)
18 J.Rice, Rep.Brit.Ass., 397 (1915)
19 R.C.Tolman, J.Amer.Chem.Soc., 42, 2506 (1920)
20 M.Trautz, Z.Anorg.Chem., 102, 81(1918)
21 W.C.McC Lewis, J.Chem.Soc.,113,471,(1918); Phil.Mag., 39,26 (1920)
22 M.W.Perrin, Ann.Phys., 11, 5 (1919)
23 I.Langmuir, J.Amer.Chem.Soc., 42,2190 (1920)
24 Trans.Faraday Soc., 17, 598 (1922)
25 P.Kohnstamm and F.E.C.Scheffer, Ver.Konink.Akad.Wet.Amsterdam, 13,789 (1911)
26 R.Marcelin, Compt.rendu., 151, 1052 (1910)
27 J.J.Berzelius, Jahres-Berichte,15,237,(1835)
28 H.Davy, Phil. Trans., 67, 45,(1817)

29 W.Ostwald, Z.phys.chem., 15,705,(1894)
30 W.Ostwald, Phys.Z., 3, 313,(1902)
31 P.Sabatier, La Catalyse en Chimie Organique, Paris (1912)
32 C.N.Hinshelwood, The Kinetics of Chemical Change in Gaseous Systems, Clarendon Press,(1926)
33 *For a more detailed analysis see, for example, W.Kauzmann, Kinetic Theory of Gases, p175, W.A.Benjamin,(1966)*
34 E.A.Guggenheim and M.A.Prue, Physicochemical Calculations,4th ed., North-Holland, p413,(1959)
35 R.Fowler and E.A.Guggenheim, Statistical thermodynamics, CUP, p501,(1960)
36 W.C.Gardiner, Rates and Mechanisms of Chemical Reactions, W.A.Benjamin, p85 (1969)
37 K.F.Herzfeld, Ann.der Physik, 59,635 (1919)
38 O.Stern, Ann.der Physik, 44, 497(1913)
39 A.Marcelin, Ann.Phys., 3,158 (1915)
40 W.H.Rodebush, Z.fur Physik, 45,606 (1923)
41 F.London, Sommerfeld Festschrift, S 104, S.Hirzel, Leipzig (1928)
42 F.London, Z.Electrochem, 35, 552 (1929)
43 W.Heitler and F.London, Z.Physik., 44, 455 (1927)
44 H.Eyring and M.Polanyi, Z.physik.Chemie, B12, 279 (1931). Part of this is translated in. M.H.Back, and K.J.Laidler, Selected Readings in Chemical Kinetics, Pergammon Press (1967)
45 H.Pelzer and E.Wigner, Z.Physik.Chem., B15,445 (1932)
46 M.G.Evans and M.Polanyi, Trans. Faraday Soc., 31, 875 (1935)
47 H.Eyring, J.Chem.Phys., 3,107 (1935)
48 W.F.K.Wynne-Jones and H.Eyring, J.Chem.Phys., 3,492 (1935)
49 R.C.Tolman, Statistical Mechanics, Chem. Catalogue Co. (1920)
50 M.Polanyi, Z.Physik., 139,439 (1928)
51 K.F.Herzfeld, Kinetisch Theorie der Warme, Muller-Poullets Handbuch der Physik.Z.Aufl., (1925)
52 K.J.Laidler, Chemical Kinetics, p72-79,McGraw-Hill,(1965)
53 J.Horiuti, Bull.Chem.Soc.Japan, 13, 210 (1937)
54 E.Wigner, J.Chem.Phys., 5, 720 (1937)
55 J.C.Keek, J.Chem.Phys., 32,1035 (1960); Adv.Chem.Phys.,13, 85 (1967)
56 F.A.Lindemann, Trans.Faraday Soc., 17, 598 (1922)
57 C.N.Hinshelwood, Proc.Roy.Soc., A113, 230 (1927)
58 T.S.Chambers and G.B.Kistiakowsky, J.Amer.Chem.Soc., 56, 399 (1934)
59 O.K.Rice and H.C.Ramsperger, J.Amer.Chem.Soc., 49, 1617 (1927);50,617 (1928)
60 L.S.Kassel J.Phys.Chem.,32,225,1065 (1928)
61 R.A.Marcus, J.Chem. Phys., 20, 359 (1952)
62 W.Forst, Theory of Unimolecular Reactions, Academic Press (1973)

63 W.R.Schulz, and D.J.Leroy, J.Chem.Phys., **42**, 3869 (1965)
64 I.D.Clark, and R.P.Wayne, Comprehensive Chemical Kinetics, Eds.C.M.Bamford and C.F.H.Tipper, vol 2 p302 (1969)
65 Fraser and Broadway, Proc.Roy.Soc., A141,626 (1933)
66 E.H.Taylor and S.Daz, J.Chem.Phys.,23,1711 (1955)
67 J.O.Hirschfelder, H.Eyring, and B.Topley, J.Chem.Phys., 4, 170 (1936)
68 F.T.Wall, L.A.Hiller, and J.Mazur, J.Chem.Phys., 29, 255 (1958)
69 F.T.Wall, L.A.Hiller, and J.Mazur, J.Chem.Phys., 35, 1284, (1961)
70 M.Karplus, R.N.Porter, and R.D.Sharma, J.Chem.Phys., 40, 2033(1964); 43,3259 (1965)
71 R.N.Porter, and L.M.Raff, Modern Theoretical Chemistry vol 2, Ed. W.H.Miller, Plenum Press, (1976)
72 A.J.Stace and J.N.Murrell, J.Chem.Phys.,68, 3028 (1978)

73 J.C.Polanyi and W.H.Wang, J.Chem.Phys., 51, 1439,(1969)
74 M.H.Mok and J.C.Polanyi, J.Chem.Phys., 51, 1451 (1969)
75 J.R.Partington, A History of Chemistry, vols 3 and 4 ,Macmillan (1966)

76 K.J.Laidler, The World of Physical Chemistry, OUP, (1993)

MULTICHANNEL QUANTUM-CLASSICAL DIFFUSION EQUATIONS

M. V. Basilevsky
Karpov Institute of Physical Chemistry, Vorontsovo Pole 10, 103064 Moscow, Russia

A. I. Voronin
Institute of Problems of Chemical Physics RAS, 142432, Chernogolovka, Moscow, Russia

I. Introduction

A description of chemical reaction dynamics and kinetics in terms of the Fokker-Planck (FP) and Smoluchovsky diffusion theory, as well as the closely related stochastic treatment in terms of generalized Langevin equations (GLE) is now a cornerstone of theoretical research concerned with reactive processes which take place in the condensed phase [1-3]. A generalization of this approach for a wider number of chemically interesting objects seems to be promising. An important branch of such an advanced research is multichannel diffusion kinetics. For a non-classical two-level reactive system immersed in a medium, henceforth referred to as a 'solute', the corresponding dynamical evolution combines a diffusive motion along a classically treated reaction coordinate with non-adiabatic transitions between the quantum states of the remainder solute variables. The idea of using this 'Landau-Zener diffusion problem' in the chemical kinetics of the condensed phase was raised by Burshtein *et al* [4-6] and discussed by others [7-10].The work by Zusman on electron transfer kinetics [11,12] is a familiar successful application of this theory.

The Burshtein-Zusman approach is based on a two-channel counterpart of the Smoluchovsky diffusion equations which was originally formulated on intuitive grounds. A derivation of two-state diffusion equations from the microscopic-level description can be performed in terms of different techniques. The schemes of reasoning reported in the literature include the path-integral analysis of the influence functional approach [13] , the decoupling of the hierarchy of kinetic equations [14,15], the quantum-classical reduction within the Redfield formulation of the quantum relaxation theory [16] and a stochastization of the conventional quasiclassical trajectory approach of the gas phase theory of nonadiabatic transitions [17,18] . In any case, the reduction is

E.J. Brändas and E.S. Kryachko (eds.),
Fundamental World of Quantum Chemistry, Vol. II, 181–209.

not at all rigorous. It actually suggests a set of approximations that trace a guideline along which the coupled diffusion dynamics could emerge from fully quantum-mechanical original model. This situation is in a striking contrast with the one-channel (a single state) case where a rigorous one-to-one correspondence between the underlying physical model and the diffusion (Fokker-Planck) equation is well known [19].

Recently, the validity of this two-state diffusion equation was brought to a doubt [20,21]. The promoting argument referred to an unphysical behavior of the transition element ρ_{12} of the reduced density matrix ρ. It was observed within a short time interval ($0 < t < \hbar/k_B T$) for a specially constructed, rather artificial, model case. The pathologic effect is especially visible when the gap Δ between the energy levels is large in the transition region. This seriously doubts a possibility of a classical diffusion description of quantum variable ρ_{12}. This comment can be relaxed by a counterargument [22] that it is a much longer timescale, governing the evolution of the level populations ρ_{11} and ρ_{22}, which really matters.The fast transition dynamics, i.e. the quantum beats of ρ_{12} on the frequency $\omega_0 = \Delta/\hbar$, is smeared [23] by a standard stationary condition $\partial\rho_{12}/\partial t = 0$ which adjusts the average motion of fast variable ρ_{12} to a evolution of levels populations. The unphysical trends become unobservable on a coarse-grained timescale of the population dynamics.

The problems with multichannel diffusion equations are not, however, fully resolved by this notion. Actually, no satisfactory proofs that fundamental properties of reduced density matrix ρ cannot be violated in a two-state diffusion equation has been presented. The kinetic regime with high-frequency quantum beats ($\hbar\omega_0 >> k_B T$) seems to be especially dangerous. It pertains to transitions proceeding far away from the cross-point of adiabatic energy levels [22] . In a conventional two-state diffusion equation, there appear three parameters having the dimension of frequency, namely, ω_0 , $\omega_{tr} = |V|/\hbar$ (ω_{tr} - the transition frequency, V - dynamic coupling) and η (the friction coefficient which measures the coupling strength between the dynamical system and its environment). The case

$$\eta/\omega_{tr} >> 1;\ \omega_0/\omega_{tr} >> 1;\ \omega_0/\eta >> 1 \tag{1.1}$$

is typical, for instance, for proton transfer system. An inspection whether the quantum-classical diffusion equations are valid within this range of parameters is quite necessary. Several observations signalized that situation is hardly satisfactory. Most obvious is the violation of the detailed balance in two-state diffusion equation [22,24].This effect is immediately recognized for large ω_0 ($\hbar\omega_0 >> k_B T$). Nonadiabatic transitions due to dynamic coupling V are temperature independent. An alternative (complementary) thermal transition mechanism is expected to exist but it has not been revealed so far. Secondly, unphysical (negative) dynamical transition rates cannot be avoided in endothermic electron transfer reactions (a special case of Fig.1b) [21] .

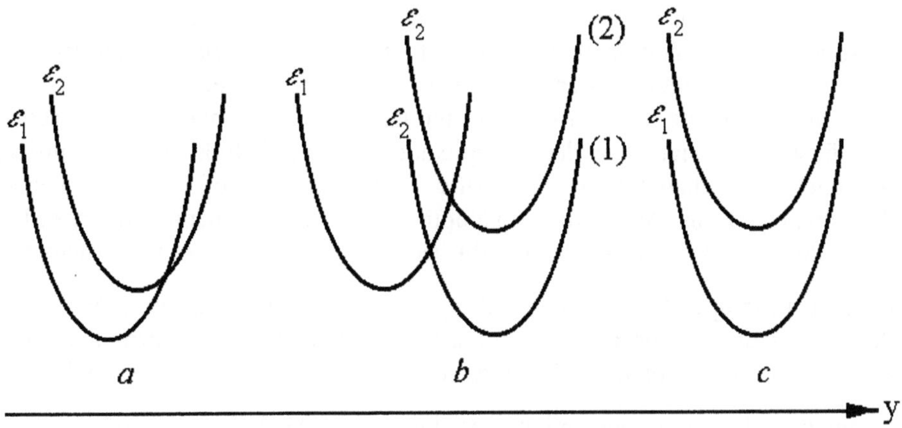

Figure 1. Typical cases of interlevel transitions. The transition occurs between energy profiles ε_1 and ε_2; $\Delta=\varepsilon_2-\varepsilon_1$ is the level separation. a) the "inverted" case, b) the "normal case"; two versions (1) and (2) for ε_2 correspond to exothermic and endothermic reactions, c) the "resonance" case (no relative shift of equilibrium positions of curves ε_1 and ε_2). For harmonic potentials with the same curvatures Δ is constant for case c).

Finally, we add here a third comment concerning a derivation of miltichannel Smoluchovsky equations. They were conventionally deduced [13,16] from the original coupled Fokker-Planck equations in the overdamped kinetic limit, when the friction coefficient η is implied to be the largest quantity having dimension of frequency in a given system [25] .This condition is certainly violated in Eq. (1.1), thus questioning the two-channel Smoluchovsky approximation. All these ambiguities motivated our present study. We reconsidered a derivation of the two-state diffusion equation from the basic equation of the quantum relaxation theory [26-28]

$$\frac{\partial \rho}{\partial t}=\frac{1}{i\hbar}[h,\rho]+\hat{K}\rho \tag{1.2}$$

Here h is the solute Hamiltonian, the brackets mean a commutator and $\hat{K}$ stands for the relaxation integral operator responsible for the solute-medium interaction. Our objective is to take the classical limit in this expression with respect to a single translational solute coordinate but retain quantization for other variables. The corresponding 'quantum-classical' reduction of the dynamical part in Eq. (1.2) (the commutator) was thoroughly discussed [29-

33]. We extend this analysis for the relaxation operator $\hat{K}$ and demonstrate that, within the oscillator medium model together with a standard linear interaction, its classical counterpart turns out to be an FP operator which is non-local in time. In this way a non-Markov master equation emerges quite naturally. Generalizing the present technique for a larger number of classical degrees of freedom is straightforward. This is not considered here in order to preserve the presentation from details of secondary importance. Because our treatment mainly addresses applications to the theory of chemical reactions, the classical coordinate we introduce is thought of as a conventional 'reaction coordinate' although a wider interpretation is not forbidden. We consider separately the diabatic and adiabatic representations of the two-channel diffusion equations. As in other developments in the multichannel theory of Eq. (1.2) [26,27,34-35], we invoke a perturbation theory to analyze the structure of the relaxation operator $\hat{K}$. The resulting constraint does not seem to be rigid in the context of deriving a diffusion master equation. It can be formulated as a short -time approximation for the solute dynamical evolution applied in evaluating $\hat{K}$, thus neglecting correlations between solute and medium dynamics at this stage. Such correlations also appear and are explicitly treated at the stage of solving Eq. (1.2), containing both dynamical and relaxation terms, so their modification of operator $\hat{K}$ itself is a higher-order effect. A similar approximation was implicitly presented in earlier multichannel theories [13,14]. Once this crucial assumption is made, the perturbation expansion arises as its immediate consequence and can be used without special motivation. As compared with alternative approach, working with a formally exact propagator of the oscillator medium model [13,14] , our perturbational reasoning gains more flexibility and suggests the most general form of the so-obtained diffusion-type master equation. For the non-Markov single-channel case, the FP equation obtained by this technique is dynamically equivalent to a GLE which in turn can be derived from a microscopic dynamical model without approximation. This demonstrates how the classical diffusion limiting transition cancels the errors introduced by a perturbational treatment. The cancellation is exact for a single channel case.

II. Quantum-classical reduction of the dynamical operator

Here, we discuss a reduction of a dynamical equation for the density matrix:

$$\frac{\partial \rho}{\partial t} = \frac{1}{i\hbar}[h, \rho] \tag{2.1}$$

where h is the full Hamiltonian of a dynamical subsystem (solute). The total amount of solute variables is divided into quantum and classical subsets and the reduction performs a classical (or quasi-classical) limiting transition for classical variables. The general reduction procedure was described in Refs. [36,29]. Before extending this treatment to a more general relaxation Eq. (1.2), we briefly introduce the basic concepts. The quantum-classical reduction procedure is based on the Weyl representation for the operators and Wigner representation for the density matrix. The Weyl representation for the product of two operators A and B has the form [36]

$$AB \leftrightarrow exp\{ \frac{i\hbar}{2}(\frac{\partial^a}{\partial s}\frac{\partial^b}{\partial p} - \frac{\partial^a}{\partial p}\frac{\partial^b}{\partial s}) \} a(p,s)\, b(p,s) \tag{2.2}$$

where $a(p,s)$, $b(p,s)$ are the classical functions in phase space $(s,\ p)$ corresponding to the operators A and B; indices a and b mean the action of derivatives on function $a(s,p)$ and $b(s,p)$. In turn, the Weyl transformation of the density operator reads

$$\rho\ (p,s,t) = (2\pi\hbar)^{-3} \int dx \exp(ipx/\hbar)\psi(s - x/2,t)\psi^*(s + x/2,t) \tag{2.3}$$

where s-coordinate, p-momentum.The evolution in time of the Wigner function $\rho\,(p,s,t)$ can be derived from Eq. (2.1)-(2.3)

$$\frac{\partial\rho}{\partial t} = \frac{2}{\hbar}\sin\{ \frac{i\hbar}{2}(\frac{\partial^h}{\partial s}\frac{\partial^\rho}{\partial p} - \frac{\partial^h}{\partial p}\frac{\partial^\rho}{\partial s}) \} h(p,s)\,\rho \tag{2.4}$$

where $h(p,s)$ is is the Weyl transformation of the quantum Hamiltonian H

$$h\,(p,s) = (2\pi\hbar)^{-3} \int dx \exp(ipx/\hbar)\psi(s - x/2)H\psi^*(s + x/2) \tag{2.5}$$

For the matrix representation of operators we use indices l, m, n for the internal degrees of freedom. With this specific notation, Eq. (2.1) takes the form adapted for the classical limit reduction

$$\frac{\partial\rho_{lm}}{\partial t} = \frac{1}{i\hbar}[\, h_{ln}\,\rho_{nm} - \rho_{ln}\,h_{nm}\,] \tag{2.6}$$

By generalization of Eq. (2.6) for the Wigner function in the matrix form we obtain

$$\frac{\partial\rho_{lm}}{\partial t} = \frac{1}{\hbar}\sin\{ \frac{\hbar}{2}(\frac{\partial^h}{\partial s}\frac{\partial^\rho}{\partial p} - \frac{\partial^h}{\partial p}\frac{\partial^\rho}{\partial s}) \}\Sigma\,[h_{nl}\,\rho_{nm} + \rho_{ln}\,h_{nm}\,] -$$

$$\frac{i}{\hbar}\cos\left\{\frac{\hbar}{2}\left(\frac{\partial^h}{\partial s}\frac{\partial^\rho}{\partial p}-\frac{\partial^h}{\partial p}\frac{\partial^\rho}{\partial s}\right)\right\}\Sigma\,[h_{ln}\,\rho_{nm}-\rho_{ln}\,h_{nm}] \tag{2.7}$$

where h_{ln} depends on the classical variables *p, s* and ρ_{nm} depends on *p, s, t*. Taking the limit $\hbar \to 0$, we can only retain the leading terms: sin $(x) \to x$, cos $(x) \to 1$. In this way, the following reduction rules for a commutator [...,...] and anticommutator $[...,...]_+$ are obtained

$$[A,B]_{ij} = [a,b]_{ij} - \frac{i\hbar}{2}\sum_k \{a_{ik}, b_{kj}\} + \{a_{kj}, b_{ik}\}$$

$$[A,B]_{+ij} = [a,b]_{+ij} - \frac{i\hbar}{2}\sum_k \{a_{ik}, b_{kj}\} - \{a_{kj}, b_{ik}\} \tag{2.8}$$

where $\{a,b\}$ denotes a Poisson bracket

$$\{a,b\} = \frac{\partial a}{\partial p}\frac{\partial b}{\partial s} - \frac{\partial a}{\partial s}\frac{\partial b}{\partial p} \tag{2.9}$$

a. Diabatic case

A simple non-trivial example of this Weyl technique is given by a two-level problem with the diabatic classical Hamiltonian

$$h^c(p,s) = \begin{pmatrix} p^2/2m + V_{11}, V_{12} \\ V_{21}, p^2/2m + V_{22} \end{pmatrix} \tag{2.10}$$

where mass *m* is associated with the *s*-motion and $V_{12}(s) = V_{21}(s)$. The equation of motion for classical density matrix ρ represented as a Liouville vector

$$\rho = \begin{pmatrix} \rho_{11} \\ \rho_{12} \\ \rho_{21} \\ \rho_{22} \end{pmatrix} \tag{2.11}$$

reads

$$\frac{\partial \rho}{\partial t} = \frac{1}{i\hbar}L[h(s,p)]\rho - u\frac{\partial \rho}{\partial s} + B\frac{\partial \rho}{\partial p} \tag{2.12}$$

Here $u = p/m$ (the velocity), and Liouville matrices $L\,[h^c\,]$ (the commutation with h) and B are

$$L\left[h^c(s,p)\right]=\begin{pmatrix} 0 & -V_{21} & V_{12} & 0 \\ -V_{12} & V_{11}-V_{22} & 0 & V_{12} \\ V_{21} & 0 & V_{22}-V_{11} & -V_{21} \\ 0 & V_{21} & -V_{12} & 0 \end{pmatrix} \tag{2.13}$$

and

$$B=\frac{1}{2}\begin{pmatrix} 2V'_{11} & V'_{21} & V'_{12} & 0 \\ V'_{12} & V'_{11}+V'_{22} & 0 & V'_{12} \\ V'_{21} & 0 & V'_{11}+V'_{22} & V'_{21} \\ 0 & V'_{21} & V'_{12} & 2V'_{22} \end{pmatrix} \tag{2.14}$$

In the last matrix the superscripts mean $\partial/\partial s$. We also present a general prescription for convenient evaluation of the Liouville representation of a commutator $[A,...]$ as a direct matrix product

$$L\,[A]=A \times I - I \times A^{\sim} \tag{2.15}$$

where I is an unit matrix and $A^{\sim}$ is the transpose of A.

b. Adiabatic case

The adiabatic two-state Hamiltonian $h^c(p, s)$ can be represented as

$$h^c(p,s)=\begin{pmatrix} p^2/2m+\varepsilon_1, -i\hbar uJ \\ i\hbar uJ, p^2/2m+\varepsilon_2 \end{pmatrix} \tag{2.16}$$

Here ε_1, ε_2 represent adiabatic energy levels, coupling element J will be defined below.

For the present adiabatic case

$$L\left[h^c\right]=i\hbar\begin{pmatrix} 0 & -uJ & -uJ & 0 \\ uJ & i\omega_0 & 0 & -uJ \\ uJ & 0 & -i\omega_0 & -uJ \\ 0 & uJ & uJ & 0 \end{pmatrix} \tag{2.17}$$

$$B = \begin{pmatrix} \varepsilon_1' & 0 & 0 & 0 \\ 0 & \frac{1}{2}(\varepsilon_1' + \varepsilon_2') & 0 & 0 \\ 0 & 0 & \frac{1}{2}(\varepsilon_1' + \varepsilon_2') & 0 \\ 0 & 0 & 0 & \varepsilon_2' \end{pmatrix} \tag{2.18}$$

Here $\omega_0 = (\varepsilon_2 - \varepsilon_1)/\hbar$ and superscipts mean $\partial/\partial s$.

III. The solute/bath interaction

a. Diabatic case

In the diabatic representation, for the dynamic solute subsystem embedded in a medium we assume a simple form of the interaction term as a direct product

$$W = \Lambda \times \Phi \tag{3.1}$$

where Λ is the function of s and Φ is defined in terms of medium variables. A conventional approach suggests

$$\Lambda = s, \quad \Phi = \sum_\nu c_\nu Q_\nu, \quad W = \Lambda\Phi \tag{3.2}$$

where the medium is treated as an ensemble of oscillator modes $Q_\nu(t)$ with masses m_ν and frequencies ω_ν.

b. Adiabatic case

In the alternative adiabatic representation, one usually deals with the p-dependent solute/bath interaction. We describe below the general scheme of reasoning to receive such an interaction. Usually, the problem of quantum-classical reduction of equation of motion arises in systems with two highly anisotropic dinamical timescales. We consider the Hamiltonian H_{xy} for a combination of a quantum (fast) and classical (slow) variables x and y

$$H_{xy} = h_x + h_y + h_{xy} \tag{3.3}$$

Here h_x and h_y depend only on x, $\partial/\partial x$ and y, $\partial/\partial y$ respectively; h_{xy} represents the interaction. The classical limit transition for y will be made as the next step. The dynamical system (x, y) interacts with the environment composed of harmonic medium modes Q_ν: $Q = \{ Q_\nu\}$ with the bath Hamiltonian H_Q.. The total (system + medium) Hamiltonian is

$$H_{xyQ} = H_{xy} + H_Q + \Lambda_{xy}\,\Phi \tag{3.4}$$

where Φ is a linear combination of medium operators (Q_v or $\partial / \partial Q_v$) as specified below. We separate variable x and define the eigenfunctions φ_i and eigenvectors ε_i

$$(h_x + h_{xy})\, \varphi_i = \varepsilon_i\, \varphi_i \tag{3.5}$$

The Schrodinger equation (3.5) defines the y-dependent adiabatic representation : $\varphi_i = \varphi_i\,(x|y)$, $\varepsilon_i = \varepsilon_i\,(y)$. The first step towards quantum-classical equations suggests averaging (i.e. integration over x) with functions $\varphi_i\,(x|y)$. Consider a two-level model with $i,j = 1,2$ and focus on the operator Λ_{xy} in the coupling term (3.4). After averaging, it converts into 2x2 matrix

$$\Lambda = \begin{pmatrix} \Lambda_{11}, \Lambda_{12} \\ \Lambda_{21}, \Lambda_{22} \end{pmatrix} \tag{3.6}$$

The matrix elements depend on y and/or $\partial / \partial y$: $\Lambda_{ij} = \langle \varphi_i | \Lambda_{xy} | \varphi_j \rangle$, where $\langle\ \rangle$ means integration over x. As will be shown later, a necessary condition that the relaxation operator $\hat{K}$ in Eq. (1.2) contains the terms responsible for the detailed balance can be formulated as

$$\Lambda_{12} \neq 0, \quad \Lambda_{21} \neq 0 \tag{3.7}$$

This conclusion follows from an inspection of the derivation of the thermal equilibrium conditions from the quantum relaxation theory based on Eq. (1.2) [26-28]; the present work suggests an explicit implementation of such an analysis. Matrix elements (3.7) always vanished in earlier studies of multi-channel quantum-classical equations. The structure of relaxation operator $\hat{K}$ in Eq. (1.2) is determined by the form of interaction operator Λ_{xy} . The model, where Λ_{xy} is x - independent, introduces a significant simplification and produces a realistic background for applications. In the diabatic theories $\Lambda_{xy} = y$ (we will use notation s). Here we assume

$$\Lambda_{xy} \rightarrow \Lambda_y = -i\,\hbar\, \frac{\partial}{\partial y} \tag{3.8}$$

The explanation how the expression (3.8) can arise is suggested below. This assumption results in the following structure of matrix Λ :

$$\Lambda = \Lambda_1 + \Lambda_2$$

$$\Lambda_1 = -i\hbar \frac{\partial}{\partial y}\begin{pmatrix} 1 & 0 \\ 0 & 1 \end{pmatrix}; \quad \Lambda_2 = -i\hbar J \begin{pmatrix} 0 & 1 \\ -1 & 0 \end{pmatrix} \tag{3.9}$$

where the coupling term is

$$J(y) = \left\langle \varphi_1 \left| \frac{\partial}{\partial y} \right| \varphi_2 \right\rangle = \frac{\left\langle \varphi_1 \left| \frac{\partial W_{xy}}{\partial y} \right| \varphi_2 \right\rangle}{\varepsilon_2 - \varepsilon_1} \tag{3.10}$$

We imply that the interaction in Eq. (3.4) is a potential term: $h_{xy} = W_{xy} = W(x, y)$. In the diabatic reperesentation (y- independent $\varphi_i(x)$) no coupling can be inserted in matrix Λ. The both necessary conditions, i.e. the form (3.8) of the system-bath interaction and the adiabatic representation have never been satisfied in earlier studies; this was followed by an immediate consequence of a violation of conditions (3.7).

c. The shifted bath oscillators as natural environment variables

Originally, the variables y (the slow dynamical coordinate), q_ν (the bath) and their interaction are defined in terms of the Hamiltonian

$$H_{yq} = h_y + h_q + h_{yq} \tag{3.11}$$

where h_q represents a standart ensemble of purely oscillatory modes q_ν with masses m_ν and freguencies ω_ν .The bilinear interaction is defined as

$$h_{yq} = y\sum_\nu c_\nu q_\nu \tag{3.12}$$

where c_ν are coupling amplitudes. A standard procedure eliminates this interaction by shifting the equilibrium positions of bath oscillatords as

$$Q_\nu = q_\nu - q_{0\nu}$$
$$q_{0\nu} = -y\frac{c_\nu}{m_\nu \omega_\nu^2} \tag{3.13}$$

Usually, this procedure is performed with a harmonic potential for the system coordinate y [1,19,37,38]; then y involved in the transformation (3.13) resulting with the corresponding shift of its equilibrium position (the spin-boson model). The novel element of the present approach is that we do not change y but treat shifted modes Q_ν as modified bath variables adjusted to y. On the other hand, as compared to the one-channel stochastic approach [1,19], the change of variables $q_\nu \rightarrow Q_\nu$ is an essential step. Such a change is unnecessary in one-channel applications, when a transformation of dynamic equation to the GLE form is exact [39,40]. In our multi-channel case, a modified bath formulation is a necessary ingredient of the proper zero order approximation underlying the second-order Redfield Eq. (1.2). The change of variables results in the following modification of Hamiltonian (3.11):

$$H_{yq} \rightarrow H_{yQ} = h_y + h_Q + h_{yQ} \tag{3.14}$$

Here h_y is modified by adding an extra potential term (we do not change the notation for brevity)

$$h_y \rightarrow h_y \tag{3.15}$$

Because the coordinate transformation (3.13) is *y*- dependent, Q_ν are coupled *via* kinetic energy operator. Diagonalization of the pertaining bath Hamiltonian provides the remainder components of Eq. (3.14)

$$h_Q = \sum_\nu -\frac{\hbar^2}{2}\frac{\partial^2}{\partial Q_\nu^2} + \frac{\omega_\nu^2}{2}Q_\nu^2$$

$$h_{yQ} = -i\hbar\frac{\partial}{\partial y}\sum_\nu \gamma_\nu\left(-i\hbar\frac{\partial}{\partial Q_\nu}\right) \tag{3.16}$$

The frequency spectrum is now modified as compared to the original bath; we conserve for brevity the same notations Q_ν and ω_ν . The new interaction amplitudes γ_ν can be explicitly expressed in terms of c_ν and other parameters of the original Hamiltonian (3.11) (see [41] for details). Eqs. (3.14)-(3.16) constitute a background for the basic Hamiltonian (3.4), (3.5). Its most essential feature is the form (3.6) of the solute/bath interaction: just the same as required in order to regenerate the detailed balance.

d. Correlation functions

For the bath Hamiltonian and the interaction as given by Eqs. (3.1), (3.2), (3.16) we will further need the correlation function

$$\mathrm{K}(\tau) = \sum_{\nu\mu} f_{\nu\nu}\langle\nu|\Phi|\mu\rangle\langle\mu|\Phi|\nu\rangle \exp\left[-\frac{i\tau}{\hbar}(\varepsilon_\nu - \varepsilon_\mu)\right]$$

$$f_{\nu\nu} = Z^{-1} \exp\left(-\frac{\varepsilon_\nu}{k_B T} \right) \tag{3.17}$$

where $\varepsilon_\nu = \hbar\omega_\nu$, Z is the partition function, oscillator states are denoted as ν and μ. For the interaction expressed in the coordinate form (3.2), all results can be expressed in terms of the two related correlation functions

$$\begin{aligned} K_+(\tau) &= (1/2)[\,K(\tau) + K(-\tau)\,], \\ K_-(\tau) &= (1/2)[\,K(\tau) - K(-\tau)\,] \end{aligned} \tag{3.18}$$

which are available in a conventional form

$$\begin{aligned} K_+(\tau) &= \sum_\nu c_\nu^2 \frac{\hbar}{2m_\nu\omega_\nu} \coth\frac{\hbar\omega_\nu}{2k_B T} \cos\omega_\nu\tau \\ K_-(\tau) &= \sum_\nu i c_\nu^2 \frac{\mathrm{h}}{2m_\nu\omega_\nu} \sin\omega_\nu\tau \end{aligned} \tag{3.19}$$

For the interaction in the momentum representation (3.16), we should replace

$$c_\nu^2 / m_\nu\omega_\nu \to \gamma_\nu^2 \tag{3.20}$$

IV. Quantum-classical reduction of relaxation operator

When the dynamic solute subsystem described in Section 2 is embedded in a medium, its equation of motion (2.6) has to be modified by adding a relaxation term $\hat{K}\rho$, thus resulting in Eq. (1.2). The expression for the relaxation operator $\hat{K}$ reads [26,27,34,35]

$$\hat{K}\rho = -\frac{1}{\hbar^2}\int_0^t \overline{K}(\tau)\rho(t-\tau)d\tau \tag{4.1}$$

The Liouville representation of kernel $\overline{K}(\tau)$ is described below. In the classical limit for variable *s*, in diabatic representation the dynamical part of Eq. (2.6) reduces to Eq. (2.12). We perform here a similar reduction for the kernel $\overline{K}(\tau)$.

The total quantum-mechanical Hamiltonian can be rewritten as

$$\boldsymbol{H} = \boldsymbol{h} + \boldsymbol{G} + \boldsymbol{W} = \boldsymbol{H}_0 + \boldsymbol{W}$$
$$\boldsymbol{H}_0 = \boldsymbol{h} + \boldsymbol{G} \tag{4.2}$$

where $\boldsymbol{h}$ and $\boldsymbol{G}$ are Hamiltonians of the solute and medium subsystems, respectively, and $\boldsymbol{W}$ represents the corresponding solute-medium interaction. As usually, we consider medium states $|\nu\rangle$, the eigenfunctions of the medium Hamiltonian with energy eigenvalues ε_ν

$$\boldsymbol{G}\,|\nu\rangle = \varepsilon_\nu\,|\nu\rangle \tag{4.3}$$

In a classical limit for variable *s*, the solute Hamiltonian reduces to matrix $\boldsymbol{h}^{c}$ (s,p). By looking at Eq. (2.10) as an example, we infer that in general $\boldsymbol{h}$ is is a non-diagonal matrix

$$\boldsymbol{h} = \|h_{mn}\| \tag{4.4}$$

In a standard relaxation theory both Hamiltonians $\boldsymbol{h}$ and $\boldsymbol{G}$ are considered in their diagonal representations [26,34], but such a limitation for $\boldsymbol{h}$ is too restrictive for the present treatment. Let us consider the expression for the relaxation kernel [26,34,35], defined as a Liouville matrix in the subspace of solute indices *m, l, n, k*,

$$[\overline{K}(\tau)]_{mn,lk} = \sum_{\nu\mu} f_{\nu\nu} K(\tau)^{m\nu n\nu}_{l\mu k\mu}$$

$$f_{\nu\nu} = Z^{-1}\exp(-\frac{\varepsilon_\nu}{kT}) \tag{4.5}$$

where $f_{\nu\nu}$ are medium Boltzmann factors with Z being the medium partition function. The supermatrix $K(\tau)$ is defined in the full eight-index Liouville space including both solute (indices *m, l, n, k*) and medium (indices ν, ν', μ, μ') variables. It is a product of three Liouville matrices [26]

$$K(\tau) = L[W]exp\left(-\frac{i\tau}{\hbar}(1-P)L[H]\right)L[W]$$

$$\approx L[W]\exp\left(-\frac{i\tau}{\hbar}L[H_0]\right)L[W] \tag{4.6}$$

The commutation matrices $L[W]$, $L[H_0]$ and $L[H]$ are introduced according to the Eqs. (2.8), (2.15). The second line of Eq. (4.6), where the full Hamiltonian is substituted by H_0, is a usually applied second-order perturbation theory [26,27,34,35]. The projection operator 1 - *P* of Zwanzig theory [35] can be omitted in this case owing to the special projection properties of $\boldsymbol{H}_0$ and $\boldsymbol{W}$ [26, 27]. For diabatic case we assume a simple form of the interaction term represented by Eqs. (3.1)-(3.2); an explicit evaluation of expression (4.5) is given earlier [16]. This derivation accounts for the non-diagonal form (4.4) of matrix $\boldsymbol{h}$ and results in the final operator expression

$$\overline{K}(\tau) = K_{+}(\tau)L(\Lambda)R(\tau)L[\Lambda] - K_{-}(\tau)L(\Lambda)R(\tau)M[\Lambda] \tag{4.7}$$

where propagator $R(\tau)$ has the form

$$R(\tau) = exp\left(-\frac{i\tau}{\hbar}L[h]\right) \tag{4.8}$$

The commutation matrices $L[\Lambda]$ and $L[h]$ are defined as in Eq. (2.15) and $M[\Lambda]$ is an anticommutation matrix

$$M[\Lambda] = \boldsymbol{\Lambda} \times \boldsymbol{I} + \boldsymbol{I} \times \boldsymbol{\Lambda}^{\sim} \tag{4.9}$$

Being expressed in a closed operator form, the Eq. (4.7) is fully prepared for performing the desired limit transition. Indeed, by applying the reducion rules (2.8), we find

$$L[\Lambda] = -i\hbar\left(\frac{\partial\Lambda}{\partial p}\frac{\partial}{\partial s} - \frac{\partial\Lambda}{\partial s}\frac{\partial}{\partial p}\right)$$

$$M[\Lambda] = 2\Lambda \tag{4.10}$$

The classical limit of the operator $L[h]$ is given in Eq. (2.15). For the two-state diabatic model, the right part of Eq. (2.12) yields

$$L[h] = L[h^{c}] - u\frac{\partial}{\partial s} + B\frac{\partial}{\partial p} \tag{4.11}$$

The insertion of limiting expressions (4.10) and (4.11) in Eq. (4.7) provides an ultimate quantum-classical reduction for the relaxation operator.

For the adiabatic case, the expressions for $L[\Lambda]$ and $M[\Lambda]$ are more complicated. Taking into account Eq. (3.9) and reduction rules (2.8), we obtain

$$L[\Lambda] = L_1 + L_2 + L_3$$
$$M[\Lambda] = M_1 + M_2 + M_3 \tag{4.12}$$

where

$$L_1 = -i\hbar\frac{\partial}{\partial y},\quad L_2 = -i\hbar JP,\quad L_3 = -i\hbar J'Q\frac{\partial}{\partial p}$$

$$M_1 = 2p_y,\quad M_2 = -i\hbar JQ,\quad M_3 = \quad + \tag{4.13}$$

Here two matrices are introduced

$$P=\begin{pmatrix}0 & 1 & 1 & 0\\ -1 & 0 & 0 & 1\\ -1 & 0 & 0 & 1\\ 0 & -1 & -1 & 0\end{pmatrix};\ Q=\begin{pmatrix}0 & -1 & 1 & 0\\ 1 & 0 & 0 & 1\\ -1 & 0 & 0 & -1\\ 0 & -1 & 1 & 0\end{pmatrix} \tag{4.14}$$

V. Two-channel diffusion equations in the diabatic case

a. Diffusion limit

In a diffusion limit we look for multichannel FP-type evolution operators to which the quantum-classical expression (4.7) can be further converted. The result is gained after reducing the solute dynamical evolution in this expression by taking the limit $\tau\to 0$ in the solute propagator. In the diabatic case, we treat ω_0, the frequency of interlevel quantum beats, as a small quantity (only transitions in the vicinity of the crossing points of diabatic levels are taken into account). Thereby, we will accept an approximation $exp\ (i\omega_0\ \tau) \to 1$. An alternative case when $\omega_0\ \tau >> 1$ is considered later for adiabatic case. With this in mind, we assume

$$\exp\left(-\frac{i\tau}{\hbar}L[h]\right)\to 1$$

$$\frac{\partial}{\partial\tau}\exp\left(-\frac{i\tau}{\hbar}L[h]\right)\to\frac{1}{i\hbar}L[h] \tag{5.1}$$

The corresponding kernel

$$\overline{K}(\tau)=-\hbar^2K_+(\tau)\frac{\partial}{\partial p}\exp\left(-\frac{i\tau}{\hbar}L[h]\right)\frac{\partial}{\partial p}-K_-(\tau)i\hbar\frac{\partial}{\partial p}\exp\left(-\frac{i\tau}{\hbar}L[h]\right)(2s)$$

(5.2)

is substituted into the relaxation term (4.1) combined with the reduction results (4.10). The final result for the relaxation kernel is:

$$\frac{\partial b}{\partial p}(\tau)=-\hbar^2\left[K_+(\tau)\frac{\partial^2}{\partial p^2}+J(\tau)\frac{\partial}{\partial p}\left(\frac{p}{m}\right)\right]$$

$$J(\tau)=\sum_\nu\frac{c_\nu^2}{m_\nu\omega_\nu^2}\cos\omega_\nu\tau \tag{5.3}$$

The first and second terms of this expression are interpreted, respectively, as a purely diffusion and a drift contribution to a generalized FP-type equation with a memory. This seems to be obvious for the first term and becomes also so for the second term in the high-temperature limit

$$\frac{\hbar\omega_v}{2kT} << 1; \quad J(\tau) = \frac{1}{kT} K_+(\tau) \tag{5.4}$$

The full relaxation evolution proceeds in this limit according to the expression

$$\hat{K}\rho = \int_0^t K_+(t-\tau)\left[\frac{\partial^2}{\partial p^2} + \frac{1}{kT}\frac{\partial}{\partial p}\left(\frac{p}{m}\right)\right]\rho(\tau)d\tau \tag{5.5}$$

In the Markov limit it reduces to a standard FP evolution with the diffusion coefficient

$$D = \int_0^\infty K_+(\tau)d\tau \tag{5.6}$$

Finally, we compare the present treatment with the earlier derivation of the master equation as suggested by Caldeira and Leggett [39] and repeated in the later work for a multichannel case [15]. Formally, a route from the evolution integral operator of the full quantum theory (1.2), (4.1) to second partial derivative operators appearing in its diffusion-limiting expression proceeds *via* an expansion of the full reduced quantum propagator $\overline{K}(\tau)$ over a small time interval $\delta\tau$ [39,42]. In the short-time limit $\delta\tau \to 0$ for the full propagator, the perturbation theory specified in Eq. (4.6) becomes an exact procedure. This means that the strategy accepted in our work, namely assuming the short-time limit Eq. (5.1) only for the pure solute Hamiltonian plus the perturbation expansion for the full propagator, is equivalent to the short-time approximation taken for the full propagator as accepted in earlier studies [15,39]. This also means that, in the limit $\delta\tau \to 0$, the work with the exact propagator [15,39] has no advantage over the perturbational treatment. On the other hand, a work with a simplified expression of the perturbation theory facilitates formal expressions. In this way, the basic original results of the present study were obtained. These results are:

(a) the expression (5.3) exposing the diffusion limit for the relaxation kernel without the high-temperature constraint Eq. (5.4);
(b) the non-local generalization of the diffusion theory which is transferred without changes from the original non-Markov expression (4.1) to the final Eq. (5.5).

In its general form, with expression (5.3) for the integral kernel, this theory forms a closed approximate model of non-adiabatic transitions in a solute accompanied by dissipation. Its validity is not restricted by a trivial condition that the interaction is small. Actually, it may be large if the timescale specified for the solute evolution by the short-time *ansatz* $\delta\tau \to 0$ becomes sufficiently small. An attempt at circumventing it leads to a sequence of hierarchy equations [14], and the theory becomes more complicated at this advanced level.

b. Fokker-Planck equations

Here we derive and discuss the balance equations describing the kinetic evolution of diagonal elements of the density matrix. The full quantum-classical relaxation equation is transformed to this form after projecting out the evolution of the off-diagonal matrix elements. The balance equations deal with level populations $\rho_{ii}(u,s,t)$, the purely classical quantities. In order to contract the notations, we consider the high-temperature limit Eq. (5.5) and additionally reduce it to a local Markovian case

$$\hat{K}\rho = D\left[\frac{\partial^2}{\partial p^2} + \frac{1}{kT}\frac{\partial}{\partial p}\left(\frac{p}{m}\right)\right]\rho \qquad (5.7)$$

The non-Markov description can be easily regenerated in the final expressions by a simple change of the integral operator

$$\hat{D}(\ldots) = \int_0^t K_+(t-\tau)(\ldots)d\tau \qquad (5.8)$$

for the diffusion coefficient D.

The corresponding set of equations for the two-channel diabatic classical Hamiltonian (2.10) is

$$\frac{\partial\rho_{11}}{\partial t} = \left[-v\frac{\partial}{\partial s} + V_{11}'\frac{\partial}{\partial p} + \hat{K}\right]\rho_{11} + \frac{1}{i\hbar}(V_{12}\rho_{21} - V_{21}\rho_{12})$$

$$\frac{\partial\rho_{12}}{\partial t} = \left[-v\frac{\partial}{\partial s} + (V_{11}' + V_{22}')\frac{\partial}{\partial p} + \hat{K}\right]\rho_{12} + \frac{1}{i\hbar}(V_{11} - V_{22})\rho_{12} -$$

$$\frac{V_{12}}{ih}(\rho_{11} - \rho_{22})$$

$$\frac{\partial\rho_{21}}{\partial t} = \left[-v\frac{\partial}{\partial s} + (V_{11}' + V_{22}')\frac{\partial}{\partial p} + \hat{K}\right]\rho_{21} - \frac{1}{i\hbar}(V_{11} - V_{22})\rho_{21} +$$

$$\frac{V_{12}}{ih}(\rho_{11} - \rho_{22})$$

$$\frac{\partial\rho_{22}}{\partial t}=\left[-v\frac{\partial}{\partial s}+V_{22}^{'}\frac{\partial}{\partial p}+\hat{K}\right]\rho_{22}+\frac{1}{i\hbar}(V_{21}\rho_{12}-V_{12}\rho_{21})$$

(5.9)

Here only leading terms of the expansion in power of $\hbar$ are retained in all non-adiabatic coupling terms and superscipts mean $\partial/\partial s$. Simplifying assumptions regarding the potential-energy matrix (prompted by the Landau-Zener model) are fu rther introduced in order to remove unnecessary details:

$$V_{12}=V_{21}=V=\text{real constant;}$$

$$V_{11}=U_1(s);\quad V_{22}=U_2(s) \tag{5.10}$$

Then Eqs. (5.9) in their local version (Eq. (5.7)) become equivalent to those reported earlier [13,15].

c. Balance equations

It is now expedient to eliminate the off-diagonal matrix elements ρ_{12} and ρ_{21} (the quantum quantities) and obtain a pair of balance equations for the level populations ρ_{11} and ρ_{22}. With simplifications of Eq. (5.10) and contracted operator notations

$$\hat{L}_{11}=-u\frac{\partial}{\partial s}+V_{11}^{'}\frac{\partial}{\partial p},\qquad \hat{L}_{22}=-u\frac{\partial}{\partial s}+V_{22}^{'}\frac{\partial}{\partial p}$$

$$\hat{L}_{12}=\hat{L}_{21}=-u\frac{\partial}{\partial s}+\frac{1}{2}(V_{11}^{'}+V_{22}^{'})\frac{\partial}{\partial p} \tag{5.11}$$

we obtain the exact balance equations as

$$\frac{\partial\rho_{11}}{\partial t}=(\hat{L}_{11}+\hat{K})\rho_{11}-\frac{2V^2}{\hbar^2}\mathrm{Re}\int_0^t\exp\left\{\left[\hat{L}_{12}+\hat{K}-\frac{i}{\hbar}(U_1-U_2)\right](t-\tau)\right\}\times[\rho_{11}-\rho_{22}]d\tau$$

$$\frac{\partial\rho_{22}}{\partial t}=(\hat{L}_{22}+\hat{K})\rho_{22}+\frac{2V^2}{\hbar^2}\mathrm{Re}\int_0^t\exp\left\{\left[\hat{L}_{12}+\hat{K}-\frac{i}{\hbar}(U_1-U_2)\right](t-\tau)\right\}\times[\rho_{11}-\rho_{22}]d\tau \tag{5.12}$$

The three contributions appearing on the right-hand side correspond to the solute dynamical motion $\hat{L}_{ii}$, i=1,2, the relaxation $\hat{K}$ and non-adiabatic transitions. We have assumed ρ_{12} (t=0) = ρ_{21} (t=0) = 0 in the derivation. The operator exponential of the latter contribution accounts for a modulation of non-adiabatic transitions by a classical (dynamical and diffusional motion). In a discussion of useful simplifications, we start with the most reliable assumption and take the local limit of the non-local non-adiabatic term. This results in an approximation

$$\mathrm{Re}\int_0^t \exp\left\{\left[\hat{L}_{12}+\hat{K}-\frac{i}{\hbar}(U_1-U_2)\right](t-\tau)\right\}[\rho_{11}-\rho_{22}]d\tau$$

$$\approx -\mathrm{Re}\frac{1}{L_{12}+K-\frac{i}{\hbar}(U_1-U_2)}[\rho_{11}(t)-\rho_{22}(t)] \qquad (5.13)$$

The next approximation, which cannot be generally rationalized, states that the transition frequency $|(U_1 - U_2)/\hbar| = \omega_{12}$ is important in establishing the timescale for the ρ_{12} kinetics

$$\mathrm{Re}\frac{1}{L_{12}+K-\frac{i}{\hbar}(U_1-U_2)} \approx \hbar\pi\delta(U_1-U_2) \qquad (5.14)$$

This generates the so-called 'local' or 'contact' transition model [5,6,11] which is fit for treating strongly non-adiabatic situations.

d. The Smolukhovsky limit

We finally consider the Smoluchovsky limit leading to Burshtein-Zusman balance equations [5,6,11].The elimination of the velocity variable proceeds in a standard way [25] by expanding the densities over the set of Hermite polynomials. This can be done either in the balance equations (5.12) or in the original set of Eqs. (5.9) with its following reduction to Smoluchovsky balance equations. The first way seems more rigorous but hardly tractable in its full version because of the complicated velocity dependence of the operator integral kernel in Eq. (5.12). In the local approximation (Eq. (5.14)) for this kernel both ways result in the same Burshtein-Zusman theory. We expose below a refined version of Smoluchovsky balance equations. It is obtained by performing the Smoluchovsky limit in Eqs. (5.9) and using an intermediate approximation of the type of Eq. (5.13) for the corresponding integral kernel. After introducing the reduced density matrix

$$\sigma(s,t) = \int \rho(u,s,t)du \qquad (5.15)$$

we obtain the equation for the its off-diagonal element

$$\frac{\partial \sigma_{12}}{\partial t} = \hat{l}_{12}\,\sigma_{12} - \frac{i}{\hbar}(U_1 - U_2)\sigma_{12} + \frac{V}{i\hbar}(\sigma_{22} - \sigma_{11}),$$

$$\hat{l}_{12} = D_s \frac{\partial}{\partial s}\left[\frac{\partial}{\partial s} + \frac{1}{kT}(U_1' + U_2')\right] \tag{5.16}$$

where D_s is the Smoluchovsky diffusion coefficient. The contracted Green's operator, similar to that appearing in Eq. (5.13)

$$\hat{G}_{12} = -\frac{1}{\hat{l}_{12} - \frac{i}{\hbar}(U_1 - U_2)} \tag{5.17}$$

drives the non-adiabatic evolution in the balance equations

$$\frac{\partial \sigma_{11}}{\partial t} = D_s \frac{\partial}{\partial s}\left(\frac{\partial}{\partial s} + \frac{1}{kT}U_1'\right)\sigma_{11} - \frac{2V^2}{\hbar^2}\operatorname{Re}\hat{G}_{12}(\sigma_{11} - \sigma_{22})$$

$$\frac{\partial \sigma_{22}}{\partial t} = D_s \frac{\partial}{\partial s}\left(\frac{\partial}{\partial s} + \frac{1}{kT}U_1'\right)\sigma_{22} + \frac{2V^2}{\hbar^2}\operatorname{Re}\hat{G}_{12}(\sigma_{11} - \sigma_{22})$$

(5.18)

The explicit evaluation of the last transition term in Eq. (5.18) needs a preliminary solution of the equation for the Green's function $\hat{G}_{12}(s, s')$

$$\hat{l}_{12} - \frac{i}{\hbar}(U_1 - U_2)G_{12} = -\delta(s - s') \tag{5.19}$$

Computations with this Green's function are available and have been reported [22]. Eqs. (5.18) are valid when the following conditions are satisfied

$$\frac{kT}{mD_s}\,\sigma^1_{ij} \gg \partial\sigma^1_{ij}/\partial t, \qquad \frac{VmD_s}{kT} \ll 1, \qquad \frac{|U_1 - U_2|mD_s}{kT\hbar} \ll 1 \tag{5.20}$$

where σ^1_{ij} are the next (after σ_{ij}) expansion terms of ρ_{ij} over the Hermite polynomials, that is to say, the first distribution moments. The conditions (5.20) are the specific limitations of the diabatic two-channel theory not discussed before.The set of equations (5.18), (5.19) neglects the solute internal dynamics (it enters the problem only *via* diffusion coefficient D_s) but it dynamically accounts for the combination of a one-dimensional diffusion with non-adiabatic transitions. The merit of this approach is a possibility for treatment of both adiabatic and non-adiabatic transition regimes, so that the limitation of locality of transitions inherent to the Burstein-Zusman model [5,6,11] is circumvented. We recall that the non-Markov counterpart of this approach is recovered by changing the diffusion coefficient D_s by a relevant time-integral operator similar to the expression given by the symbolic Eq. (5.8).

Our final comment is concerned with the scope of the basic Eqs. (5.9). Even without the following approximation, Eq. (5.10), they are limited by the requirement that the channel potentials V_{ii} and interaction potentials V_{ij} are the functions of s. Generally, they may be functions of both p and s but the p-dependence is suppressed in an explicit evaluation of the Poisson bracket term in Eq. (2.10), leading to the expression (2.14) for the matrix B. For practical applications this means that master equations discussed in the present section are well adapted to treat a diabatic representation of the multichannel theory. In the alternative adiabatic representations one usually deals with the p-dependent transition matrix elements and hence the basic master equation must be rederived.

VI. Two-channel diffusion equations in the adiabatic case.

In order to obtain the coupled diffusion equations in the adiabatic case in a tractable form, a set of approximations is required which retain "essential" and drop away "inessential" terms in Eq. (4.7). The relaxation term (4.1) can be modified by taking into account Eq. (4.8) [16,21]

$$\rho(t-\tau) = R(-\tau)\rho(t) \tag{6.1}$$

This approximation implies a purely dynamic evolution without relaxation. It does not violate the accuracy of Eq. (4.1) which was obtained as a second order of perturbation expansion relative to the relaxation operator; any relevant correction added to Eq. (6.1) would contribute to a higher order of the perturbation theory.

By this means, the integral operator (6.1) is transformed to Liouvillian matrix K:

$$K\rho = -\frac{1}{\hbar^2}\int_0^\infty \overline{K}(\tau)R(-\tau)d\tau \tag{6.2}$$

Let us take into account the hierarchy of timescales associated with different kinds of motion

$$\begin{aligned}
\tau_{\text{cl}} &= \omega_{\text{cl}}^{-1} && \text{(motion on classical trajectory)}\\
\tau_{\text{tr}} &= \left(uJ_0\right)^{-1} && \text{(the nonadiabatic transitions)}\\
\tau_{\text{beats}} &= \omega_0^{-1} && \text{(the quantum beats)}
\end{aligned} \tag{6.3}$$

Here ω_{cl} denotes a charasteric frequency for the classical y- motion and coupling $J(y)$ is taken as $J(y) = J_0 + J_1 y + ...$ Provided there exists a hierarchy of timescales

$$\tau_{beats} << \tau_{cl} ,\ \tau_{tr} \tag{6.4}$$

we can conclude that for short time period t satisfying

$$\tau_{beats} << t << \tau_{cl} ,\ \tau_{tr} \tag{6.5}$$

one can assume that

$$R(\tau) \approx R_0(\tau)$$

where the propagator $R_0(\tau)$ has the form

$$R_0(\tau) = \begin{pmatrix} 1 & 0 & 0 & 0 \\ 0 & e^{i\omega_0\tau} & 0 & 0 \\ 0 & 0 & e^{-i\omega_0\tau} & 0 \\ 0 & 0 & 0 & 1 \end{pmatrix} \tag{6.6}$$

The upper limit in Eq. (6.2) $t = \infty$ becomes legitimate after extending the hierarchy of timescales suggested by Eqs. (6.3), (6.4):

$$\tau_{beats} << \tau_{bath} << \tau_{cl} ,\ \tau_{tr} \tag{6.7}$$

Here τ_{bath} denotes the characteristic period of the bath relaxation by the widths of bath correlation functions. The inequality (6.7) is a standart condition eliminating non-Markovian effects from relaxation kinetics. It is expedient to rewrite the integrand of relaxation matrix K in a more explicit form (see Eqs. (4.12)-(4.14))

$$\overline{K(\tau)R(-\tau)} = K_+(\tau)A_+ + K_-(\tau)A_-$$

$$A_+ = (L_1+L_2+L_3)R_0(\tau)(L_1+L_2+L_3)R_0(-\tau)$$

$$A_- = (L_1+L_2+L_3)R_0(\tau)(M_1+M_2+M_3)R_0(-\tau) \tag{6.8}$$

The simplified expression for the relaxation operator is obtained by counting orders of magnitude of its ingredients, withdrawing higher order terms in the Plank constant $\hbar$. Such an analysis separates the leading terms and provides the expression

$$A_{+}= L_1 R_0(\tau)L_1 R_0(-\tau) + L_2 R_0(\tau)L_2R_0(-\tau)$$

$$A_{-}= L_1 R_0(\tau)M_1 R_0(-\tau) + L_2 R_0(\tau)M_2R_0(-\tau) \tag{6.9}$$

Additionally, we accept the "resonance" approximation $\omega'_0=0$ and neglect by the "interference" terms

$$L_1 R_0(t)L_2 R_0(-t),\quad L_2 R_0(t)L_1 R_0(-t),\quad L_2 R_0(t)M_1 R_0(-t) \tag{6.10}$$

a. The relaxation equations

After selecting the most important contributions to the relaxation operator, the quantum-classical relaxation equations acquire the form

$$\frac{\partial\rho_{11}}{\partial t} = N_{11}\rho_{11} + uJ(\rho_{12}+\rho_{21}) - J^2D(\omega_0)\rho_{11} + J^2D(-\omega_0)\rho_{22}$$

$$\frac{\partial\rho_{12}}{\partial t} = N_{12}\rho_{12} + i\omega_0\rho_{12} + uJ(\rho_{11}-\rho_{22}) - 2J^2[D_0\rho_{12} + D_+(2\omega_0)\rho_{21}]$$

$$\frac{\partial\rho_{21}}{\partial t} = N_{21}\rho_{21} + i\omega_0\rho_{21} + uJ(\rho_{11}-\rho_{22}) - 2J^2[D_0\rho_{21} + D_+(-2\omega_0)\rho_{12}]$$

$$\frac{\partial\rho_{22}}{\partial t} = N_{11}\rho_{22} + uJ(\rho_{12}+\rho_{21}) + J^2D(\omega_0)\rho_{11} - J^2D(-\omega_0)\rho_{22}$$

(6.11)

Here the following definitions and notations are accepted

$$N_{ij} = D_0\frac{\partial}{\partial y}\left[\frac{\partial}{\partial y} + \frac{1}{k_BT}\frac{\varepsilon_i' + \varepsilon_j'}{2}\right] - u\frac{\partial}{\partial y} + \frac{\varepsilon_i' + \varepsilon_j'}{2m}\frac{\partial}{\partial u} \tag{6.12}$$

$$D(\omega) = \int_0^\infty d\tau[K_+(\tau)\cos\omega\tau + K_-(\tau)i\sin\omega\tau] = \frac{1}{2}\int_0^\infty d\tau\left[K(\tau)e^{i\omega\tau} + K(-\tau)e^{-i\omega\tau}\right]$$

$$D_+(\omega) = \int_0^\infty d\tau K_+(\tau)e^{i\omega\tau}$$

$$D_0 = D(\omega=0) = D_+(\omega=0) \tag{6.13}$$

The comments given below are aimed at clarifying several novel features of Eqs. (6.11).

1. Operators N_{ij} (i,j = 1,2) are diagonal in the Liouvillian space, i.e. they correspond to an independent evolution of elements ρ_{ij} of density matrix. Their diffusion part originates from matrices L_1, M_1. They generate equations for diagonal elements ρ_{11}, ρ_{22} that reperesent a true adiabatic evolution of a quantum-classical system. A remarkable fact is that the Smolukhovsky diffusion limit is directly reached in terms of the present formulation, without separating the Fokker-Plank equation as an intermediate step. This result is gained due to expressing the solute/environment interaction in terms of momentum operators. Earlier derivations [13,16] started from the bilinear coordinate-dependent interaction and resulted in multi-channel Fokker-Planck equations with the diffusion coefficient $D_p = m\eta k_B T$ proportional to friction coefficient η. As discussed in the Introduction and in Section 5, their reduction to the Smolukhovsky limit by using standart methodologies [23,25] is formally prohibited in the case $(\eta / \omega_0) \leq 1$, when η proves to be not the smallest parameter having dimension of frequency. The approach described in Section 3c suggests a transparent way to circumvent this difficulty.

2. The terms proportional to uJ are responsible for purely dynamical interlevel transitions. Terms of this sort were also present in earlier multi-channel theories [5,6,11,13,16] with a modification due to the diabatic representation of channel basis functions that has been always invoked before. After eliminating off-diagonal elements of the density matrix by using the Green's function technique [16,31,43], the coupled balance equations for ρ_{11} and ρ_{22} have been obtained. We repeat this concluding part of a derivation in the next section.

3. The cross terms containing the correlation functions $D(\pm\omega_0)$ comprise direct coupling of the adiabatic states promoted by the solute/bath interaction. They vanish in the diabatic representation. The partial rates of corresponding nonadiabatic transitions obey the detailed balance, because [26,27]

$$\frac{D(\omega_0)}{D(-\omega_0)} = \exp(-\hbar\omega_0 / k_B T) \tag{6.14}$$

A search of these terms was the main objective of the present adiabatic approach. They arise as a result of multiplications of relevant Liouvillian matrices in Eq. (4.7).

4. We recall two simplifying approximations introduced on the way from the full expression (4.7) for the relaxation operator to Eqs. (6.11). This is, first, the "resonance conditions" $\omega'_0 = 0$ which provides most simple form for the diffusion part of operators N_{ij}. Second, the "interference terms" (6.10) are

neglected in the relaxation operator. The direct multiplication of Liouville matrices shows that they contribute only to nonadiabatic transitions coupling ρ_{ii} and ρ_{jj} and contain quantities proportional to $J\partial/\partial y$. On this background, it is expedient to introduce the third approximation directly in Eqs. (6.11), which would make them more tractable. We shall withdraw the last terms (with square brackets) in the second and third Eqs. (6.11), determining the evolution of ρ_{12} and ρ_{21}. A rational motivation for this step could be given by the relations

$$\omega_0 >> J^2 D_0 >> J^2 | D_+(2\omega_0) | \qquad (6.15)$$

which are expected to be valid for the case of large ω_0 considered here.

b. The Smoluchovsky limit

Our final goal is to derive the Smolukhovsky limit to Eq. (6.11) under conditions providing the following hierarchy of timescales which is in accord with Eqs. (6.3)-(6.5)

$$\omega_0 >> \eta >> \omega_{cl} >> \tau_y^{-1} \qquad (6.16)$$

Here η denotes the friction coefficient (η^{-1} is the velocity relaxation period), ω_{cl} is the characteristic frequency of the classical y-motion and τ_y is the diffusion relaxation period. These quantities are additionally interrelated as

$$D_0 = k_B / m\eta, \quad \eta = \tau_y\, \omega_{cl}^{\,2} \qquad (6.17)$$

where D_0 is the diffusion constant in Eq. (6.12). The y-dependent of ω_0 is suppressed in this simplified treatment.

The first conventional step eliminating off-diagonal elements of the density matrix in Eq. (6.11) is that one assumes

$$\partial\rho_{12}/\partial t = \partial\rho_{21}/\partial t = 0 \qquad (6.18)$$

Provided the last term on the right-hand sides of Eqs. for ρ_{12} and ρ_{21} are negligible, as discussed in item 4, these Eqs. become uncoupled and it suffices to consider only ρ_{12}. Under conditions (6.16), there exists time interval Δt which is much shorter than τ_y but much larger than $\omega_0^{\,-1}$. Eqs. (6.18) are valid for ρ_{12} averaged over such interval. The averaging eliminates the strongly oscillating (with frequency ω_0) component of ρ_{12} but retains its regular component evolving on the diffusion timescale.

With this prelimenary, we introduce the Green's operator resolving Eq. (6.11) for ρ_{12}

$$G_{12} = (i\omega_0 + N_{12})^{-1} \cong (i\omega_0)^{-1}(1 - N_{12}/i\omega_0) \qquad (6.19)$$

According to Eq. (6.12), we separate the diffusion and classical dynamic components of operator $N_{12} = L_{12} + M_{12}$

$$L_{12} = D_0 \frac{\partial}{\partial y}\left[\frac{\partial}{\partial y} + \frac{1}{2k_B T}\left(\varepsilon_1' + \varepsilon_2'\right)\right]$$

$$M_{12} = -u\frac{\partial}{\partial y} + \frac{1}{2m}\left(\varepsilon_1' + \varepsilon_2'\right)\frac{\partial}{\partial u} \tag{6.20}$$

The characteristic frequency of N_{12} is ω_{cl} and and the inequality $\omega_{cl} / \omega_0 << 1$ accounts for the last approximation involved in Eq. (6.19). The resulting expression for the regular component of ρ_{12}

$$\rho_{12}(t) = -(i\omega_0)^{-1}[1-(L_{12} + M_{12})/ i\omega_0]uJ [\rho_{11}(t) - \rho_{22}(t)] \tag{6.21}$$

is now inserted in the first and fourth lines of Eqs. (6.11). Note that $\rho_{11}(t)$ and $\rho_{22}(t)$ are the functions of y and u; on the other hand, we suppress here for simplicity the y-dependence of J, similar to the case of ω_0 as noted above. Provided only the coarse-grained diffusion timescale (Δt no less than the velocity relaxation period) is essential, one introduces spatial distributions $\sigma_{11}(y,t)$ and $\sigma_{22}(y,t)$ by assuming

$$\rho_{11}(y,u,t) = (m/2\pi k_B T)^{1/2} \exp(-mu^2/2k_B T)\, \sigma_{11}(y,t)$$

$$\rho_{22}(y,u,t) = (m/2\pi k_B T)^{1/2} \exp(-mu^2/2k_B T)\, \sigma_{22}(y,t) \tag{6.22}$$

The dynamic components of operators N_{11} and N_{22} vanish after operating with the Maxwell distributions (6.22). The simplified version of balance equations which ultimately arises under approximations (6.18), (6.19), (6.21), (6.22) is expressed as the nonadiabatic Smoluchovsky equations

$$\frac{\partial \sigma_{11}}{\partial t} = D_0 \frac{\partial}{\partial y}\left[\frac{\partial}{\partial y} + \frac{\varepsilon_1'}{k_B T}\right]\sigma_{11} + J^2\left[D(-\omega_0)\sigma_{22} - D(\omega_0)\sigma_{11}\right] -$$

$$\frac{2J^2 k_B T}{\omega_0^2 m} L_{12}\left[\sigma_{11} - \sigma_{22}\right] \tag{6.23}$$

and similar equations for σ_{22} .Here we supposed for simplicity that J does not depend on y. The nonadiabatic transitions generated by the second and third terms of Eq. (6.23) originate, respectivaly, from the thermal solute-bath interaction and from the direct dynamic interlevel interactions. It is remarkable that the dynamic term vanishes when the diffusion relaxation is completed and σ_{11} and σ_{22} become Boltzmann distribution. This is so because diffusion operator L_{12} annihilates Boltzmann distributions. This conclusion is exact provided ω_0 and J are y-independent, as assumed above. It can be stated that the quantities ω_0 and J should be treated like quasiclassical potentials which means that their commutations with operator $\partial/\partial y$ may be neglected. Altogether, we infer that only thermal transitions seem to occur once the intralevel equilibrium is attained. On the other hand, dynamic nonadiabatic transitions are significant in a case of strongly nonequilibrium spatial level distributions.

Finally, let us discuss the different kinetic regime determined by the condition alternative to Eq. (6.16), namely $\omega_0 << \eta$. Under this assumption the velocity relaxation with the period $\tau_u = \eta^{-1}$ provides the shortest timescale as compared to other processes and we can apply the approximation similar to Eq. (6.22) directly to $\rho_{12}(t)$ and $\rho_{21}(t)$ .Then averaging over u with Maxwell distributions annihilates the terms coupling diagonal and off-diagonal elements of the density matrix in Eqs. (6.11). As a result, one obtains diffusion equation (6.23) without dynamic transitions (the last term of Eq. (6.23) is dropped). The adiabatic representation of this sort has been considered earlier by Shushin and Tachiya [17,18].

VII. Conclusion

The earlier works with the two-state diffusion equations (TSDE) [5,6,11,13,21,40,44] were mainly addressed to a "normal" electron transfer processes (ET) (Fig.1b) in diabatic representation. Such approach is constrained by a condition that the transition state must be localized [11,5,19]. The alternative adiabatic treatment met some inherent problems [17,18]. It was revived recently [18,31,43,45] in a context of the "inverted" ET (Fig.1a). The present study has established a natural background for deriving adiabatic quantum-classical TSDE in the case of delocalized interlevel transitions. It covers all examples shown in Fig.1 provided the adiabatic energy gap is not too small ($\bar{u} J / \omega_0$ <<1, here $\bar{u}$ means the mean thermal velocity). We formulated the two conditions that are sufficient for making consistent the quantum-classical limit transition. First, the solute/medium interaction must be expressed in terms of momenta rather the coordinates; this description emerges when the adjusted bath model is accepted as a zero-order approximation for the second-order relaxation Eq. (1.2). Second, an adiabatic rather diabatic representation of the channel (electronic) basis functions must be used. The distinguished role of the adiabatic representation for reproducing the detailed balance effect is a consequence of the interaction model introduced in Section 3. The direct interaction of the fast quantum variable x with thermal bath modes (q) is disregarded. The coupling of the dynamic subsystem to the bath is performed exclusively *via* the slow coordinate y, see Eq. (3.8). As a result, the system-induced factor Λ of the bilinear interaction term becomes a diagonal matrix in adiabatic two-state representation. On the other hand, matrix Λ acquires its off-diagonal coupling term responsible for the detailed balance if the direct *x/q* interaction is introduced, irrespective of whether the basis states are diabatic or adiabatic and whether the interaction is expressed in terms of coordinates or their momenta. We prefered, however, a pure model requiring minimum amount of input information in order to reproduce the desired effect. In fact, it naturally covers electron and proton transfer systems and thereby has a number of applications.

The thermal transitions ensuring detailed balance are supplemented by purely dynamical transitions which can be important in the adiabatic regime until intralevel Boltzmann equilibrium is gained. Such a description seems to be quite relevant for the resonant case (Fig.1c) when the transitions are located around the minima of adiabatic potentials far away from the cross-point region. For an inverted ET case (Fig.1a), a competition between transitions of this sort and those at the cross-point should be investigated.

The present adiabatic Smoluchovsky model applies for two extreme cases when ω_0 is large (ω_0 / η >>1, Eq. (6.23)) or small (ω_0 / η <<1, Eq. (6.23)) without the last "dynamic" term; this approach was originally derived in Ref. (11). Intermediate cases do not allow for a separation of timescales and thus require a complete treatment in terms of Eqs. (6.11) involving both coordinate y and velocity u. The diabatic representation seems to be most relevant for a treatment of avoided crossing region with a small level splitting (($\bar{u} J / \omega_0$ >>1) and conventional procedure [5,6,11, 19] are well adapted for this case.

Acknowledgements

The research was made possible in part by the International Association for the Promotion of Cooperation with Scientists from the New Independent States of the Former Soviet Union (project INTAS-RFBR IR-97-620). The authors acknowledge also financial support from Russian Foundation for Basic Research (RFBR projects No. 00-15-97295 and 99-03-33196).

References

[1] P. Hanggi, P. Talkner and M. Borkovec, Rev. Mod. Phys. **62**, 251 (1990).

[2] J. T. Hynes in *Theory of Chemical Reaction Dynamics* (ed. M. Baer, CRC Press, Florida, vol. IV, 171, 1985)

[3] U. Weiss, *Quantum Dissipative Systems, Series in Modern Condensed Matter Physics, vol.2,* (World Scientific, Singapore, second edition, 1999)

[4] A. I. Burshtein, *Kvantovaya kinetika* (NGY, Novosibirsk, Part II, 1968)

[5] B. J. Jakobson and A. I. Burshtein, Chem. Phys. **49**, 385, (1980).
[6] A. I. Burshtein and A. A. Zharikov, Chem. Phys. **152**, 23 (1991).
[7] Ed. L. T. Muus, P. W. Atkins, K.A. McLauchlin and J. B. Pedersen, *Chemicaly Induced Magnetic Polarization* (Riedel, Dordrecht, 1977).
[8] K. L. Salikhov, Yu. M. Molin, R. Z. Sagdeev and A. L. Buchachenko, *Spin Polarization and Magnetic Field Effects in Radical Reactions* (Elsevier, Amsterdam, 1984).
[9] I. V. Alexandrov, Mol. Phys. **37**, 1949 (1979).
[10] I. V. Alexandrov and R. Gabrielyan, Mol. Phys. **37**, 1963 (1979).
[11] L. D. Zusman, Chem. Phys. **49**, 295 (1980).
[12] L. D. Zusman, Chem. Phys. **80**, 29 (1983).
[13] A. Garg, J. N. Onuchic and V. Ambegaokar, J. Chem. Phys. **83**, 4491 (1985).
[14] Y. Tanimura and R. Kubo, J. Phys. Soc. Jpn **58**, 101 (1989).
[15] Y. Tanimura and S. Mukamel, J. Chem. Phys. **101**, 3049 (1994).
[16] M. V. Basilevsky and A. I. Voronin, J. Chem. Soc. Faraday Trans. **93**, 989 (1997).
[17] A. I. Shushin, Chem. Phys. **60**, 149 (1981).
[18] A. I. Shushin and M. Tachiya, Chem. Phys. **235**, 267 (1998).
[19] E. Pollak in *Dynamics of Molecules and Chemical Reactions* (edited by R.E.Wyatt and J.C.H.Zhang, Marcel Decker, New York, 1996).
[20] P. Frantsuzov, Chem. Phys. Lett. **267**, 427 (1997).
[21] P. Frantsuzov, J. Chem. Phys. **111**, 2075 (1999).
[22] M. V. Basilevsky, A. V. Soudakov and A. I. Voronin, Chem. Phys. **235**, 281 (1998).
[23] C. W. Gardiner, *Handbook of Stochastic Methods* (Springer, Berlin, 1985).
[24] L. D. Zusman, Chem. Phys. **80**, 29 (1983).
[25] H. Risken, *The Fokker-Planck Equation* (Springer, Berlin, 1984).
[26] B. Fain, *Theory of Rate Processes in Condensed Media* (Springer, Berlin, 1984).
[27] R. Kubo, M. Toda and N. Hashitsume, *Statistical Physics II* (Springer, Berlin, 1985).
[28] W. T. Pollard, A. K. Felts and R. A. Friesner, Adv. Chem. Phys. **93**, 77 (1996).
[29] I. V. Alexandrov, Z. Naturforsch. A **36a**, 902 (1981).
[30] W. Boucher and J. Traschen, Phys. Rev. D **37**, 3522 (1988).
[31] D. Y. Yang and R. I. Cukier, J. Chem. Phys. **91**, 281 (1989)
[32] R. Kapral and G. Ciccotti, J. Chem. Phys. **110**, 8919 (1999).
[33] S. Nielsen and R. Kapral, J. Chem. Phys. **115**, 5805 (2001).
[34] A. G. Redfield, Adv. Magn. Reson. **1**, 1 (1965).
[35] R. Zwanzig, J. Chem. Phys. **33**, 1338 (1960).
[36] S. R. De Groot and L. G. Suttorp, *Foundations of Electrodynamics* (North-Holland, Amsterdam, 1972).
[37] J. Ulstrup, *Charge Transfer in Condensed Media* (Springer, Berlin-New York, 1979).
[38] A. M. Kuznetsov and J. Ulstrup, *Electron Transfer in Chemistry and Biology* (Wiley, Chichester, 1999).
[39] A. O. Caldeira and A. Leggett, Ann. Phys. **149**, 374 (1983).
[40] E. Cortes, B. J. West and K. Lindenberg, J. Chem. Phys. **82**, 2708 (1985).
[41] M. V. Basilevsky and A. I. Voronin, J. Phys.: Condens. Matter, in press.
[42] R. P. Feynman and A. R. Hibbs, *Quantum Mechanics and Path Integrals* (McGraw-Hill, New York, 1965).
[43] J. Cao, J. Chem. Phys. **112**, 6719 (2000).
[44] J. Jortner and M. Bixon, Adv. Chem. Phys. **106**, 1 (1999).
[45] P. A. Frantsuzov, S. F. Fisher and A. A. Zharikov, Chem. Phys. **241**, 95 (1999).

Field Energy Density In Chemical Reaction Systems

Akitomo Tachibana
Department of Engineering Physics and Mechanics, Kyoto University, Kyoto 606-8501, JAPAN; Phone & Fax : +81-75-753-5184; Electronic mail : akitomo@kues.kyoto-u.ac.jp

Abstract: Completely new realization of chemical interaction energy density will be presented. The strategy is based upon quantum electrodynamics (QED). "Action through medium" is the basic physics underlying the framework of QED. On the other hand, "action at a distance" is the basis of the Coulomb law, which has long been the guiding principle of chemical interaction. We shall treat chemical interaction from the new picture of "action through medium" in contrast with the alternative conventional picture of "action at a distance." A Hamiltonian density operator has been derived from the non-relativistic limit of QED. It is composed of the density operators of electromagnetic fields and the kinetic energy density operators of electrons and nuclei. The conventional Coulomb interaction energy densities disappear due to the QED variational principle, but it has been rigorously verified that the integrals of the two kinds of the Hamiltonian density operators all over the space should be the same. Although the integral picture should be the same, the new Hamiltonian density carries "action though medium" of the special theory of relativity in the underlying physics, while the conventional ab initio Hamiltonian density carries "action at a distance" of the Coulomb law in the underlying physics. Reflecting the difference in the picture of the underlying physics, the approach presented here in this paper has given new insight on the local picture of interaction of electrons and nuclei in chemical reaction systems. The basic interaction picture of a pair of charged particles has been described by the emergence of the characteristic sphere within which compensation should occur for the primordial interaction in between the charged particles. The local interaction picture has been demonstrated in application to H atom and electrophilic addition reactions of HF to C_2H_4 and C_2H_2.

Key words: field energy density, QED, chemical reaction

E.J. Brändas and E.S. Kryachko (eds.),
Fundamental World of Quantum Chemistry, Vol. II, 211–239.

1. INTRODUCTION

Theory of electromagnetism is invariant under Lorentz transformation and has been formulated by Einstein as special theory of relativity.[1] According to the theory of electromagnetism, moving charged particle modifies the associated electromagnetic field with speed of light. The varying electromagnetic field propagates through space and gives impetus to other moving charged particle according to the Lorentz force. This demonstrates "action through medium" as the field theoretical nature of the electromagnetic interaction of charged particles. Quantum electrodynamics (QED) provides quantum mechanical framework for the field theoretical "action through medium" as the fundamental law of electrons interacting with atomic nuclei and radiation field of photons (see Appendix A).[2] Quantum mechanics gives rise to tension force over and above the Lorentz force.[3] As a matter of fact, it has been proved that for stationary states the tension density exactly cancels the Lorentz force density.[3] Recently, we have demonstrated in several chemical reaction systems that the tension density and the kinetic energy density of electrons moving around nuclei are clearly visualized in real space to participate in transformation of chemical bonds.[3]

For light atoms the speed of electrons is much slower than the speed of light, and hence we rely upon quantum mechanical electrostatic Coulomb law described by conventional ab initio electrostatic Hamiltonian. The Coulomb law demonstrates "action at a distance" as the electrostatic nature of the conventional interaction of charged particles. In contrast to the correct "action through medium" in QED, the conventional electrostatic interaction is instantaneous since the speed of light is eventually infinite for "action at a distance" in the Coulomb law. The interaction of charged particles with radiation field is then treated approximately since Lorentz invariance is lost in the present conventional treatment of the ab initio Hamiltonian.

The difference in between the two viewpoints is remedied for electrostatic phenomena[4]. First, let us start from the QED Hamiltonian density operator $\hat{H}_{\mathrm{QED}}(\vec{r})$ and perform the non-relativistic limit, and then we obtain $\hat{H}_{\text{non-relativistic QED}}(\vec{r})$. Second, for the electrostatic phenomena, we arrive at $\hat{H}_{\text{electrostatic non-relativistic QED}}(\vec{r})$, where the integral in the whole space is transformed into a different integral $\int d^3\vec{r}\hat{H}_{\text{electrostatic non-relativistic QED}}(\vec{r}) = \int d^3\vec{r}\hat{H}_{\mathrm{Coulomb}}(\vec{r})$ (see Appendix B). There emerges the energy density operator $\hat{H}_{\mathrm{Coulomb}}(\vec{r})$ which represents the Coulomb law, obeying the "action at a distance" picture. In this picture, the concept of field disappears but the static potential dictated by the Coulomb law takes place instead. Thus the ab initio Hamiltonian density operator $\hat{H}_{\mathrm{Coulomb}}(\vec{r})$ is "equivalent" to $\hat{H}_{\text{electrostatic non-relativistic QED}}(\vec{r})$ for the electrostatic non-relativistic phenomena of electrons and nuclei provided that it be integrated in the whole space. Although the physical meaning of the energy density is different from each other, both of the energy densities are "equivalent" since the integral in the whole space should be the same.

Using the ab initio electrostatic Hamiltonian, relativistic correction has been invoked, which has been ubiquitous in the field of chemical physics. The relativistic correction is small in the observable quantities and merely exhibits significant numbers only for heavy atoms where speed of core electrons becomes comparable with the speed of light.

In the present paper, however, we shall show that the non-relativistic limit of the QED Hamiltonian density gives rise to completely new Hamiltonian density as compared with the conventional ab initio Hamiltonian density. The observable quantities derived from the new Hamiltonian density should of course be exactly the same as the conventional ab initio Hamiltonian density. But the local nature of physical and chemical interactions of charged particles with each other and with radiation field should be completely new if we use the new Hamiltonian density. This is because the new Hamiltonian density is based on the correct "action through medium" as the field theoretical nature of the electromagnetic interaction in contrast to the approximate "action at a distance" for the conventional ab initio electrostatic Hamiltonian density, where the picture of chemical interactions we have hitherto developed is based on the electrostatic "action at a distance" of the Coulomb law. We are now in a position to get the new correct "action through medium" as the natural proper picture of chemical interactions. Response of charged particles under external electric field in condensed matter physics[5-19] has also been an application of the present treatment.[20] Furthermore, the new Hamiltonian density is essentially advantageous for further treatment of the relativistic phenomena because the theoretical framework has been directly derived from and linked to QED.[21] All the calculations are performed at the HF level of theory using 6-311G** basis set. Unless otherwise stated, we use mathematical conventions in Ref. 3.[3]

2. FIELD ENERGY DENSITY

Conventional QED theory assumes clamped-nuclei Hamiltonian, where the atomic nuclei are clamped in space and are treated as external static source of force for electrons. But in chemical reaction systems, the rearrangement of atomic configuration is of primary interest, and hence the dynamical treatments of atomic nuclei often play an important role.

We shall hence elaborate first the incorporation of the kinetic energy density of atomic nuclei into the general framework of QED and present the general theory of the field energy density in chemical reaction systems. The atomic nucleus is treated as a Schrödinger field. This is an approximate treatment since the atomic nucleus is treated as a composite united particle, not elementary particles such as quarks.

2.1 General settings

The conventional QED Hamiltonian density operator $\hat{H}_{\text{QED}}(\vec{r})$ is composed of the electromagnetic field energy density operator $\hat{H}_{\gamma}(\vec{r})$ and the energy density operator $\hat{H}_{e}(\vec{r})$ of electrons interacting with the electromagnetic field.[3] We may now add the energy density operator $\hat{H}_{\text{atom}}(\vec{r})$ of atomic nuclei interacting through the electromagnetic field and the electron field, leading to the rigged QED Hamiltonian density operator denoted as $\hat{H}_{\text{rigged QED}}(\vec{r})$:

$$\hat{H}_{\text{rigged QED}}(\vec{r}) = \hat{H}_{\gamma}(\vec{r}) + \hat{H}_{e}(\vec{r}) + \hat{H}_{\text{atom}}(\vec{r}). \tag{1}$$

First, the $\hat{H}_{\gamma}(\vec{r})$ is defined as

$$\hat{H}_{\gamma}(\vec{r}) = \frac{1}{8\pi}\left(\hat{\vec{E}}^{2}(\vec{r}) + \hat{\vec{B}}^{2}(\vec{r})\right). \tag{2}$$

The electric field operator $\hat{\vec{E}}(\vec{r})$ and the magnetic field operator $\hat{\vec{B}}(\vec{r})$ constitute the elements of the electromagnetic tensor field density operator $\hat{F}_{\mu\nu}(\vec{r})$ derived from the electromagnetic vector potential operator $\hat{A}_{\mu}(\vec{r})$ under Coulomb gauge condition as follows:

$$\hat{F}_{\mu\nu}(\vec{r}) = \partial_{\mu}\hat{A}_{\nu}(\vec{r}) - \partial_{\nu}\hat{A}_{u}(\vec{r})$$

$$= \begin{pmatrix} 0 & \hat{E}_x(\vec{r}) & \hat{E}_y(\vec{r}) & \hat{E}_z(\vec{r}) \\ -\hat{E}_x(\vec{r}) & 0 & -\hat{B}_z(\vec{r}) & \hat{B}_y(\vec{r}) \\ -\hat{E}_y(\vec{r}) & \hat{B}_z(\vec{r}) & 0 & -\hat{B}_x(\vec{r}) \\ -\hat{E}_z(\vec{r}) & -\hat{B}_y(\vec{r}) & \hat{B}_x(\vec{r}) & 0 \end{pmatrix}, \quad \partial_k \hat{A}^k(\vec{r}) = 0. \tag{3}$$

We use Dirac bracket for the constrained field quantization and get the canonical quantization rule:

$$\hat{A}^i(\vec{r})\hat{A}^j(\vec{s}) - \hat{A}^j(\vec{s})\hat{A}^i(\vec{r}) = 0, \tag{4}$$

$$\hat{E}_i(\vec{r})\hat{E}_j(\vec{s}) - \hat{E}_j(\vec{s})\hat{E}_i(\vec{r}) = 0, \tag{5}$$

$$\frac{1}{4\pi c}\left(\hat{A}^i(\vec{r})\hat{E}_j(\vec{s})-\hat{E}_j(\vec{s})\hat{A}^i(\vec{r})\right)$$

$$=i\hbar\delta^i_j\delta^3(\vec{r}-\vec{s})+i\hbar\frac{\partial}{\partial s^j}\frac{\partial}{\partial r_i}\left(-\frac{1}{4\pi}\cdot\frac{1}{\left|\vec{r}-\vec{s}\right|}\right). \tag{6}$$

Second, the electronic energy density operator $\hat{H}_e(\vec{r})$ is defined as

$$\hat{H}_e(\vec{r})=\frac{c}{2}[\hat{\bar{\psi}}(\vec{r})\left(-i\hbar\gamma^k\hat{D}_{ek}(\vec{r})+m_e c\right)\hat{\psi}(\vec{r})$$

$$+\left(i\hbar\hat{D}_{ek}(\vec{r})\hat{\bar{\psi}}(\vec{r})\gamma^k+m_e c\hat{\bar{\psi}}(\vec{r})\right)\cdot\hat{\psi}(\vec{r})], \tag{7}$$

with

$$\hat{D}_{e\mu}(\vec{r})=\partial_\mu+i\frac{Z_e e}{\hbar c}\hat{A}_\mu(\vec{r}),\ \ Z_e=-1. \tag{8}$$

The m_e and Z_e denotes the mass and charge of electron, respectively. The canonical quentization rule of the Dirac spinor field is

$$\hat{\psi}(\omega)\hat{\psi}(\omega')+\hat{\psi}(\omega')\hat{\psi}(\omega)=0, \tag{9}$$

$$\hat{\psi}^+(\omega)\hat{\psi}^+(\omega')+\hat{\psi}^+(\omega')\hat{\psi}^+(\omega)=0, \tag{10}$$

$$\hat{\psi}(\omega)\hat{\psi}^+(\omega')+\hat{\psi}^+(\omega')\hat{\psi}(\omega)=\delta(\omega-\omega'), \tag{11}$$

where, for the sake of simplicity of presentation, Cartesian and spin variables altogether are represented by ω.

Third, the atomic nuclear energy density operator $\hat{H}_{\text{atom}}(\vec{r})$ is purely the kinetic energy density operator as follows:

$$\hat{H}_{\text{atom}}(\vec{r})=\sum_a\hat{T}_a(\vec{r}), \tag{12}$$

where

$$\hat{T}_a(\vec{r})=-\frac{\hbar^2}{2m_a}\cdot\frac{1}{2}\left(\hat{\chi}^+_a(\vec{r})\hat{D}^2_{ak}(\vec{r})\hat{\chi}_a(\vec{r})+\hat{D}^{+2}_{ak}(\vec{r})\hat{\chi}^+_a(\vec{r})\cdot\hat{\chi}_a(\vec{r})\right), \tag{13}$$

with

$$\hat{D}_{a\mu}(\vec{r})=\partial_\mu+i\frac{Z_a e}{\hbar c}\hat{A}_\mu(\vec{r}). \tag{14}$$

The m_a and Z_a denotes the mass and charge of atomic nucleus a, respectively. The canonical quantization rule of the Schrödinger field is anticommutation relationship for Fermions (+) and commutation relationship for Bosons (-):

$$\hat{\chi}_a(\omega)\hat{\chi}_a(\omega') \pm \hat{\chi}_a(\omega')\hat{\chi}_a(\omega) = 0, \tag{15}$$

$$\hat{\chi}_a^+(\omega)\hat{\chi}_a^+(\omega') \pm \hat{\chi}_a^+(\omega')\hat{\chi}_a^+(\omega) = 0, \tag{16}$$

$$\hat{\chi}_a(\omega)\hat{\chi}_a^+(\omega') \pm \hat{\chi}_a^+(\omega')\hat{\chi}_a(\omega) = \delta(\omega - \omega'), \tag{17}$$

where, for the sake of simplicity of presentation, Cartesian and spin variables altogether are represented by ω.

It should be noted that the present theory is gauge invariant but the presence of the Schrödinger fields $\hat{\chi}_a$ and $\hat{\chi}_a^+$ violates Lorentz invariance in the Lagrangian density, and if we neglect the Schrödinger fields, then we recover the conventional QED theory with the Lorentz invariance as well as the gauge invariance.

The inclusion of the Schrödinger fields $\hat{\chi}_a$ and $\hat{\chi}_a^+$ to the theory in the form of the kinetic energy density $\hat{H}_{\text{atom}}(\vec{r})$ in Eq.(12) enables us to treat the dynamical processes of nuclei in chemical reaction systems.

2.2 Equations of motion of fields

In this subsection, three sets of equations of motion of fields will be presented. First, the Maxwell's equations of motion are found for the electromagnetic fields. Using Eq. (3) rewritten as

$$\hat{\vec{E}}(\vec{r}) = -\text{grad}\hat{A}_0(\vec{r}) - \frac{1}{c}\frac{\partial \hat{\vec{A}}(\vec{r})}{\partial t}, \quad \text{div}\hat{\vec{A}}(\vec{r}) = 0, \tag{18}$$

$$\hat{\vec{B}}(\vec{r}) = \text{rot}\hat{\vec{A}}(\vec{r}), \tag{19}$$

we have

$$\text{rot}\hat{\vec{E}}(\vec{r}) + \frac{1}{c}\frac{\partial \hat{\vec{B}}(\vec{r})}{\partial t} = 0, \tag{20}$$

$$\text{div}\hat{\vec{B}}(\vec{r}) = 0, \tag{21}$$

$$\text{div}\hat{\vec{E}}(\vec{r}) = 4\pi\hat{\rho}(\vec{r}), \tag{22}$$

$$\text{rot}\hat{\vec{B}}(\vec{r}) - \frac{1}{c}\frac{\partial \hat{\vec{E}}(\vec{r})}{\partial t} = \frac{4\pi}{c}\hat{\vec{j}}(\vec{r}), \tag{23}$$

where $\hat{\rho}(\vec{r})$ is the charge density operator and $\hat{\vec{j}}(\vec{r})$ is the charge current density operator. The $\hat{\rho}(\vec{r})$ is decomposed into

$$\hat{\rho}(\vec{r}) = \hat{\rho}_e(\vec{r}) + \sum_a \hat{\rho}_a(\vec{r}), \tag{24}$$

$$\hat{\rho}_e(\vec{r}) = Z_e e \hat{\bar{\psi}}(\vec{r}) \gamma^0 \hat{\psi}(\vec{r}) = Z_e e \hat{N}_e(\vec{r}), \quad \hat{N}_e(\vec{r}) = \hat{\bar{\psi}}(\vec{r}) \gamma^0 \hat{\psi}(\vec{r}), \tag{25}$$

$$\hat{\rho}_a(\vec{r}) = Z_a e \hat{\chi}_a^+(\vec{r}) \hat{\chi}_a(\vec{r}) = Z_a e \hat{N}_a(\vec{r}), \quad \hat{N}_a(\vec{r}) = \hat{\chi}_a^+(\vec{r}) \hat{\chi}_a(\vec{r}), \tag{26}$$

where $\hat{\rho}_e(\vec{r})$ is the electronic charge density operator and $\hat{\rho}_a(\vec{r})$ is the charge density operator of atomic nucleus a, and where $\hat{N}_e(\vec{r})$ and $\hat{N}_a(\vec{r})$ is the position probability density operator of electron and atomic nucleus a, respectively.

The $\hat{\vec{j}}(\vec{r})$ is decomposed into

$$\hat{\vec{j}}(\vec{r}) = \hat{\vec{j}}_e(\vec{r}) + \sum_a \hat{\vec{j}}_a(\vec{r}), \tag{27}$$

$$\hat{\vec{j}}_e(\vec{r}) = c Z_e e \hat{\bar{\psi}}(\vec{r}) \vec{\gamma} \hat{\psi}(\vec{r}), \tag{28}$$

$$\hat{\vec{j}}_a(\vec{r}) = Z_a e \hat{\vec{S}}_a(\vec{r}), \tag{29}$$

$$\hat{S}_a^{\ k}(\vec{r}) = \frac{1}{2m_a}\left(-i\hbar \hat{\chi}_a^+(\vec{r}) \hat{D}_{ak}(\vec{r}) \hat{\chi}_a(\vec{r}) + i\hbar \hat{D}_{ak}^*(\vec{r}) \hat{\chi}_a^+(\vec{r}) \cdot \hat{\chi}_a(\vec{r}) \right), \tag{30}$$

where $\hat{\vec{j}}_e(\vec{r})$ is the electronic charge current density operator and $\hat{\vec{j}}_a(\vec{r})$ is the charge current density operator of atomic nucleus a, and $\hat{\vec{S}}_a(\vec{r})$ is the probability current density operator of atomic nucleus a.

The charge current density operator $\hat{\vec{j}}(\vec{r})$ may further be decomposed into the transversal component $\hat{\vec{j}}_{\mathrm{T}}(\vec{r})$ and the longitudinal component $\hat{\vec{j}}_{\mathrm{L}}(\vec{r})$ as follows:

$$\hat{\vec{j}}(\vec{r}) = \hat{\vec{j}}_{\mathrm{T}}(\vec{r}) + \hat{\vec{j}}_{\mathrm{L}}(\vec{r}), \quad \mathrm{div}\hat{\vec{j}}_{\mathrm{T}}(\vec{r}) = 0, \quad \mathrm{rot}\hat{\vec{j}}_{\mathrm{L}}(\vec{r}) = 0. \tag{31}$$

The equations of motion for the gauge potentials are derived from Eqs. (22) and (23) using Eqs. (18) and (19) as follows:

$$-\Delta \hat{A}_0(\vec{r}) = 4\pi \hat{\rho}(\vec{r}) \tag{32}$$

$$\frac{1}{c}\mathrm{grad}\frac{\partial \hat{A}_0(\vec{r})}{\partial t} + \left(\frac{1}{c^2}\frac{\partial^2}{\partial t^2} - \Delta \right) \hat{\vec{A}}(\vec{r}) = \frac{4\pi}{c} \hat{\vec{j}}(\vec{r}). \tag{33}$$

The solution of Eq. (32) is

$$\hat{A}_0(\vec{r}) = \hat{\phi}_{\text{radiation}}(\vec{r}) + \int d^3\vec{s}\,\frac{\hat{\rho}(\vec{s})}{|\vec{r}-\vec{s}|}, \quad \Delta\hat{\phi}_{\text{radiation}}(\vec{r}) = 0. \tag{34}$$

From canonical quantization of the radiation fields, we put the radiation gauge potential $\hat{\phi}_{\text{radiation}}(\vec{r}) = 0$ in Eq. (34) (see Appendix A):

$$\hat{\phi}_{\text{radiation}}(\vec{r}) = 0. \tag{35}$$

Next, using Eq. (31), Eq. (33) is rewritten in such a way that the transversal component $\hat{\vec{j}}_{\text{T}}(\vec{r})$ of the charge current density operator and the longitudinal component $\hat{\vec{j}}_{\text{L}}(\vec{r})$ satisfy

$$\left(\frac{1}{c^2}\frac{\partial^2}{\partial t^2} - \Delta\right)\vec{A}(\vec{r}) = \frac{4\pi}{c}\hat{\vec{j}}_{\text{T}}(\vec{r}), \tag{36}$$

$$\text{grad}\frac{\partial \hat{A}_0(\vec{r})}{\partial t} = 4\pi\hat{\vec{j}}_{\text{L}}(\vec{r}). \tag{37}$$

The last Eq. (37) gives directly the solution of $\hat{\vec{j}}_{\text{L}}(\vec{r})$. This is further used in Eq. (31) to get $\hat{\vec{j}}_{\text{T}}(\vec{r})$ as

$$\hat{\vec{j}}_{\text{T}}(\vec{r}) = \hat{\vec{j}}(\vec{r}) - \frac{1}{4\pi}\text{grad}\frac{\partial \hat{A}_0(\vec{r})}{\partial t}. \tag{38}$$

On the other hand, Eq. (36) is solved as follows:

$$\hat{\vec{A}}(\vec{r},t) = \hat{\vec{A}}_{\text{radiation}}(\vec{r},t) + \frac{1}{c}\int d^3\vec{s}\,\frac{\hat{\vec{j}}_{\text{T}}(\vec{s},u)}{|\vec{r}-\vec{s}|}, \quad u = t - \frac{|\vec{r}-\vec{s}|}{c}, \tag{39}$$

where $\hat{\vec{j}}_{\text{T}}(\vec{r})$ is given in Eq. (38) and where the radiation gauge potential satisfy (see Appendix A)

$$\left(\frac{1}{c^2}\frac{\partial^2}{\partial t^2} - \Delta\right)\hat{\vec{A}}_{\text{radiation}}(\vec{r}) = 0, \quad \text{div}\hat{\vec{A}}_{\text{radiation}}(\vec{r}) = 0. \tag{40}$$

It should be noted that in Eq. (39) due to causality only the retarded potential is added to the radiation gauge potential $\hat{\vec{A}}_{\text{radiation}}(\vec{r},t)$. Substituting Eq. (39) into Eq. (19), we get

$$\hat{\vec{B}}(\vec{r},t) = \hat{\vec{B}}_{\text{radiation}}(\vec{r},t) + \frac{1}{c}\int d^3\vec{s}\,\frac{\hat{\vec{j}}_{\text{T}}(\vec{s},u)\times(\vec{r}-\vec{s})}{|\vec{r}-\vec{s}|^3}$$

$$+\frac{1}{c}\int d^3\vec{s}\,\frac{1}{|\vec{r}-\vec{s}|}\mathrm{rot}\hat{\vec{j}}_{\mathrm{T}}(\vec{s},u),\quad u=t-\frac{|\vec{r}-\vec{s}|}{c},\tag{41}$$

$$\hat{\vec{B}}_{\text{radiation}}(\vec{r},t)=\mathrm{rot}\hat{\vec{A}}_{\text{radiation}}(\vec{r},t).\tag{42}$$

It should be noted that rot is the differential operation at $\vec{r}$ not $\vec{s}$ in Eq. (41). So that if we observe phenomena in the region $|\vec{r}-\vec{s}| << ct$, then the third term in Eq. (41) is omitted and we get

$$\hat{\vec{B}}(\vec{r},t)\approx\hat{\vec{B}}_{\text{radiation}}(\vec{r},t)+\frac{1}{c}\int d^3\vec{s}\,\frac{\hat{\vec{j}}_{\mathrm{T}}(\vec{s},u)\times(\vec{r}-\vec{s})}{|\vec{r}-\vec{s}|^3},\quad |\vec{r}-\vec{s}|<<ct.\tag{43}$$

The second term may be called the quantum Biot-Savart law.

Second, the Dirac spinor field satisfies

$$i\hbar\gamma^{\mu}\hat{D}_{e\mu}(\vec{r})\hat{\psi}(\vec{r})=m_e c\hat{\psi}(\vec{r}),\tag{44}$$

$$-i\hbar\hat{D}^{+}_{e\mu}(\vec{r})\hat{\bar{\psi}}(\vec{r})\gamma^{\mu}=m_e c\hat{\bar{\psi}}(\vec{r}).\tag{45}$$

Third, the Schrödinger field satisfies

$$i\hbar\frac{\partial}{\partial t}\hat{\chi}_a(\vec{r})=-\frac{\hbar^2}{2m_a}\hat{D}^2_{ak}(\vec{r})\hat{\chi}_a(\vec{r})+Z_a e\hat{A}_0(\vec{r})\hat{\chi}_a(\vec{r}),\tag{46}$$

$$-i\hbar\frac{\partial}{\partial t}\hat{\chi}^{+}_a(\vec{r})=-\frac{\hbar^2}{2m_a}\hat{D}^{*2}_{ak}(\vec{r})\hat{\chi}^{+}_a(\vec{r})+Z_a e\hat{A}_0(\vec{r})\hat{\chi}^{+}_a(\vec{r}).\tag{47}$$

The equations of motion of fields thus obtained will be used in the next subsection to derive equations of motion of momentums.

2.3 Equations of motion of momentums

First, the momentum of the electromagnetic field is represented as the Poynting vector.

The Poynting vector density operator defined as

$$\hat{\vec{G}}(\vec{r})=\frac{1}{8\pi c}\left(\hat{\vec{E}}(\vec{r})\times\hat{\vec{B}}(\vec{r})-\hat{\vec{B}}(\vec{r})\times\hat{\vec{E}}(\vec{r})\right)\tag{48}$$

satisfies the equation of motion

$$\frac{\partial}{\partial t}\hat{\vec{G}}(\vec{r})=-\mathrm{div}\hat{\vec{\sigma}}(\vec{r})-\hat{\vec{L}}(\vec{r}).\tag{49}$$

In this expression, $\hat{\vec{\sigma}}(\vec{r})$ is the Maxwell's stress tensor density operator and $\hat{\vec{L}}(\vec{r})$ is the Lorentz force density operator as follows:

$$\hat{\vec{\sigma}}(\vec{r}) = \hat{\vec{\sigma}}_E(\vec{r}) + \hat{\vec{\sigma}}_M(\vec{r}), \tag{50}$$

$$\hat{\sigma}_E^{ij}(\vec{r}) == \frac{1}{8\pi}\left(\hat{\vec{E}}^2(\vec{r})\delta^{ij} - \left(\hat{E}^i(\vec{r})\hat{E}^j(\vec{r}) + \hat{E}^j(\vec{r})\hat{E}^i(\vec{r})\right)\right), \tag{51}$$

$$\hat{\sigma}_M^{ij}(\vec{r}) == \frac{1}{8\pi}\left(\hat{\vec{B}}^2(\vec{r})\delta^{ij} - \left(\hat{B}^i(\vec{r})\hat{B}^j(\vec{r}) + \hat{B}^j(\vec{r})\hat{B}^i(\vec{r})\right)\right), \tag{52}$$

and

$$\hat{\vec{L}}(\vec{r}) = \frac{1}{2}(\hat{\vec{E}}(\vec{r})\hat{\rho}(\vec{r}) + \hat{\rho}(\vec{r})\hat{\vec{E}}(\vec{r})) + \frac{1}{2c}(\hat{\vec{j}}(\vec{r}) \times \hat{\vec{B}}(\vec{r}) - \hat{\vec{B}}(\vec{r}) \times \hat{\vec{j}}(\vec{r}))$$

$$= \hat{\vec{L}}_e(\vec{r}) + \sum_a \hat{\vec{L}}_a(\vec{r}). \tag{53}$$

$$\hat{\vec{L}}_e(\vec{r}) = \frac{1}{2}(\hat{\vec{E}}(\vec{r})\hat{\rho}_e(\vec{r}) + \hat{\rho}_e(\vec{r})\hat{\vec{E}}(\vec{r})) + \frac{1}{2c}(\hat{\vec{j}}_e(\vec{r}) \times \hat{\vec{B}}(\vec{r}) - \hat{\vec{B}}(\vec{r}) \times \hat{\vec{j}}_e(\vec{r})), \tag{54}$$

$$\hat{\vec{L}}_a(\vec{r}) = \frac{1}{2}(\hat{\vec{E}}(\vec{r})\hat{\rho}_a(\vec{r}) + \hat{\rho}_a(\vec{r})\hat{\vec{E}}(\vec{r})) + \frac{1}{2c}(\hat{\vec{j}}_a(\vec{r}) \times \hat{\vec{B}}(\vec{r}) - \hat{\vec{B}}(\vec{r}) \times \hat{\vec{j}}_a(\vec{r})), \tag{55}$$

where $\hat{\vec{L}}_e(\vec{r})$ is the electronic Lorentz force density operator and $\hat{\vec{L}}_a(\vec{r})$ is the Lorentz force density operator of atomic nucleus a.

Second, the electronic kinetic momentum density operator $\hat{\vec{\Pi}}_e(\vec{r})$ defined as

$$\hat{\Pi}_e^k(\vec{r}) = \frac{1}{2}\left(-i\hbar\hat{\psi}^+(\vec{r})\hat{D}_{ek}(\vec{r})\hat{\psi}(\vec{r}) + i\hbar\hat{D}_{ek}(\vec{r})\hat{\psi}^+(\vec{r})\cdot\hat{\psi}(\vec{r})\right) \tag{56}$$

satisfies the equation of motion

$$\frac{\partial}{\partial t}\hat{\vec{\Pi}}_e(\vec{r}) = \hat{\vec{\tau}}_e^{\Pi}(\vec{r}) + \hat{\vec{L}}_e(\vec{r}). \tag{57}$$

Aside from the electronic Lorentz force density operator $\hat{\vec{L}}_e(\vec{r})$, given in Eq. (54), the $\hat{\vec{\tau}}_e^{\pi}(\vec{r})$ denotes the electronic tension density operator given as the divergence of the electronic stress tensor density operator $\hat{\vec{\tau}}_e^{\Pi}(\vec{r})$ as follows:[3]

$$\hat{\tau}_e^{\Pi k}(\vec{r}) = \partial_l \hat{\tau}_e^{\Pi kl}(\vec{r}), \tag{58}$$

where

$$\begin{aligned}\hat{\tau}_e^{\Pi k}(\vec{r}) = \frac{i\hbar c}{2}[&\hat{D}_{el}^{+}(\vec{r})\hat{\bar{\psi}}(\vec{r})\gamma^{\ell} \cdot \hat{D}_{ek}(\vec{r})\hat{\psi}(\vec{r}) \\ &+ \hat{\bar{\psi}}(\vec{r})\gamma^{\ell}\hat{D}_{ek}(\vec{r})\hat{D}_{el}(\vec{r})\hat{\psi}(\vec{r}) \\ &- \hat{D}_{ek}^{+}(\vec{r})\hat{D}_{el}^{+}(\vec{r})\hat{\bar{\psi}}(\vec{r})\gamma^{\ell} \cdot \hat{\psi}(\vec{r}) \\ &- \hat{D}_{ek}^{+}(\vec{r})\hat{\bar{\psi}}(\vec{r})\gamma^{\ell} \cdot \hat{D}_{el}(\vec{r})\hat{\psi}(\vec{r})] \\ &- \frac{1}{c}(\hat{\vec{j}}_e(\vec{r}) \times \vec{B}(\vec{r}))^k,\end{aligned} \tag{59}$$

and

$$\hat{\tau}_e^{\Pi kl}(\vec{r}) = \frac{i\hbar c}{2}[\hat{\bar{\psi}}(\vec{r})\gamma^{\ell}\hat{D}_{ek}(\vec{r})\hat{\psi}(\vec{r}) - \hat{D}_{ek}^{+}(\vec{r})\hat{\bar{\psi}}(\vec{r})\gamma^{\ell} \cdot \hat{\psi}(\vec{r})]. \tag{60}$$

Third, the kinetic momentum density operator $m_a\hat{\vec{S}}_a(\vec{r})$ of atomic nucleus a satisfies the equation of motion

$$m_a \frac{\partial}{\partial t}\hat{\vec{S}}_a(\vec{r}) = \hat{\vec{\tau}}_a^{S}(\vec{r}) + \hat{\vec{L}}_a(\vec{r}). \tag{61}$$

Aside from the Lorentz force density operator $\hat{\vec{L}}_a(\vec{r})$, given in Eq. (55), the $\hat{\vec{\tau}}_a^{S}(\vec{r})$ denotes the tension density operator given as the divergence of the stress tensor density operator $\hat{\vec{\tau}}_a^{S}(\vec{r})$ as follows:

$$\hat{\tau}_a^{Sk}(\vec{r}) = \partial_l \hat{\tau}_a^{Skl}(\vec{r}), \tag{62}$$

where

$$\begin{aligned}\hat{\tau}_a^{Sk}(\vec{r}) = \frac{\hbar^2}{4m_a}[&\hat{\chi}_a^{+}(\vec{r})\hat{D}_{ak}(\vec{r})\hat{D}_{al}^{2}(\vec{r})\hat{\chi}_a(\vec{r}) + \hat{D}_{ak}^{+}(\vec{r})\hat{D}_{al}^{+2}(\vec{r})\hat{\chi}_a^{+}(\vec{r}) \cdot \hat{\chi}_a(\vec{r}) \\ &- \hat{D}_{ak}^{+}(\vec{r})\hat{\chi}_a^{+}(\vec{r}) \cdot \hat{D}_{al}^{2}(\vec{r})\hat{\chi}_a(\vec{r}) - \hat{D}_{al}^{+2}(\vec{r})\hat{\chi}_a^{+}(\vec{r}) \cdot \hat{D}_{ak}(\vec{r})\hat{\chi}_a(\vec{r})] \\ &- \frac{Z_a e}{c}(\hat{\vec{S}}_a(\vec{r}) \times \hat{\vec{B}}(\vec{r}))^k,\end{aligned} \tag{63}$$

$$\hat{\tau}_a^{Skl}(\vec{r}) = \frac{\hbar^2}{4m_a}[\hat{\chi}_a^+(\vec{r})\hat{D}_{ak}(\vec{r})\hat{D}_{al}(\vec{r})\hat{\chi}_a(\vec{r}) + \hat{D}_{ak}^+(\vec{r})\hat{D}_{al}^+(\vec{r})\hat{\chi}_a^+(\vec{r}) \cdot \hat{\chi}_a(\vec{r})$$

$$- \hat{D}_{ak}^+(\vec{r})\hat{\chi}_a^+(\vec{r}) \cdot \hat{D}_{al}(\vec{r})\hat{\chi}_a(\vec{r}) - \hat{D}_{al}^+(\vec{r})\hat{\chi}_a^+(\vec{r}) \cdot \hat{D}_{ak}(\vec{r})\hat{\chi}_a(\vec{r})]. \quad (64)$$

Combining Eqs. (57) and (61), and using Eq. (53), we obtain

$$\frac{\partial}{\partial t}\hat{\vec{\Pi}}(\vec{r}) = \hat{\vec{\tau}}(\vec{r}) + \hat{\vec{L}}(\vec{r}), \quad (65)$$

where

$$\hat{\vec{\Pi}}(\vec{r}) = \hat{\vec{\Pi}}_e(\vec{r}) + \sum_a m_a \hat{\vec{S}}_a(\vec{r}), \quad (66)$$

$$\hat{\vec{\tau}}(\vec{r}) = \hat{\vec{\tau}}_e^{\Pi}(\vec{r}) + \sum_a \hat{\vec{\tau}}_a^S(\vec{r})$$
$$= \mathrm{div}\hat{\overleftrightarrow{\tau}}(\vec{r}), \quad (67)$$

$$\hat{\overleftrightarrow{\tau}}(\vec{r}) = \hat{\overleftrightarrow{\tau}}_e^{\Pi}(\vec{r}) + \sum_a \hat{\overleftrightarrow{\tau}}_a^S(\vec{r}). \quad (68)$$

2.4 Conservation laws

In this subsection, conservation laws of charges, momentums, and energies will be derived.

First, the charge density conservation laws are obtained as the continuity equations

$$\frac{\partial}{\partial t}\hat{\rho}_e(\vec{r}) = -\mathrm{div}\hat{\vec{j}}_e(\vec{r}) = -\mathrm{div}\hat{\vec{j}}_{e\mathrm{L}}(\vec{r}), \quad (69)$$

for electron, using Eqs. (25),(28),(31) and

$$\frac{\partial}{\partial t}\hat{\rho}_a(\vec{r}) = -\mathrm{div}\hat{\vec{j}}_a(\vec{r}) = -\mathrm{div}\hat{\vec{j}}_{a\mathrm{L}}(\vec{r}), \quad (70)$$

for atomic nucleus a, using Eqs. (26),(29),(31).

Combining Eqs. (69) , (70) and (24) we obtain the continuity equation

$$\frac{\partial}{\partial t}\hat{\rho}(\vec{r}) = -\mathrm{div}\hat{\vec{j}}(\vec{r}) = -\mathrm{div}\hat{\vec{j}}_{\mathrm{L}}(\vec{r}), \quad (71)$$

which is the combined charge density conservation law of the matter fields.

Second, using Eqs. (49) and (65), we obtain

$$\frac{\partial}{\partial t}\left(\hat{\vec{G}}(\vec{r}) + \hat{\vec{\Pi}}(\vec{r})\right) = -\mathrm{div}\left(\hat{\overleftrightarrow{\sigma}}(\vec{r}) - \hat{\overleftrightarrow{\tau}}(\vec{r})\right) \quad (72)$$

which is the momentum conservation law of the rigged QED system.

Third, the energy conservation law is found as follows:

$$\frac{\partial}{\partial t}\hat{H}_{\gamma}(\vec{r}) = -c^2 \mathrm{div}\hat{\vec{G}}(\vec{r}) - \frac{1}{2}\left(\hat{\vec{E}}(\vec{r})\cdot\hat{\vec{j}}(\vec{r}) + \hat{\vec{j}}(\vec{r})\cdot\hat{\vec{E}}(\vec{r})\right), \tag{73}$$

$$\frac{\partial}{\partial t}\hat{H}_{e}(\vec{r}) = -c^2 \mathrm{div}\hat{\vec{\Pi}}_e(\vec{r}) + \frac{1}{2}\left(\hat{\vec{E}}(\vec{r})\cdot\hat{\vec{j}}_e(\vec{r}) + \hat{\vec{j}}_e(\vec{r})\cdot\hat{\vec{E}}(\vec{r})\right), \tag{74}$$

$$\frac{\partial}{\partial t}\hat{H}_{\mathrm{atom}}(\vec{r}) = -\mathrm{div}\sum_a \hat{\vec{s}}_a(\vec{r}) + \frac{1}{2}(\hat{\vec{E}}(\vec{r})\cdot\sum_a \hat{\vec{j}}_a(\vec{r}) + \sum_a \hat{\vec{j}}_a(\vec{r})\cdot\hat{\vec{E}}(\vec{r})), \tag{75}$$

with

$$\begin{aligned}\hat{s}_a^k(\vec{r}) = \frac{1}{2i\hbar}\left(\frac{\hbar^2}{2m_a}\right)^2 \Big(&\hat{D}_{ak}^*(\vec{r})\hat{D}_{al}^{*2}(\vec{r})\hat{\chi}_a^+(\vec{r})\cdot\hat{\chi}_a(\vec{r}) \\ &-\hat{D}_{al}^{*2}(\vec{r})\hat{\chi}_a^+(\vec{r})\cdot\hat{D}_{ak}(\vec{r})\hat{\chi}_a(\vec{r}) \\ &+\hat{D}_{ak}^*(\vec{r})\hat{\chi}_a^+(\vec{r})\cdot\hat{D}_{al}^2(\vec{r})\hat{\chi}_a(\vec{r}) \\ &-\hat{\chi}_a^+(\vec{r})\hat{D}_{ak}(\vec{r})\hat{D}_{al}^2(\vec{r})\hat{\chi}_a(\vec{r})\Big),\end{aligned} \tag{76}$$

leading to

$$\frac{\partial}{\partial t}\hat{H}_{\mathrm{rigged\ QED}}(\vec{r}) = -\mathrm{div}\left(c^2\hat{\vec{G}}(\vec{r}) + c^2\hat{\vec{\Pi}}_e(\vec{r}) + \sum_a \hat{\vec{s}}_a(\vec{r})\right), \tag{77}$$

which is the energy conservation law of the rigged QED system.

2.5 Relationship with conventional QED

Neglecting the Schrödinger fields $\hat{\chi}_a(\vec{r})$ and $\hat{\chi}_a^+(\vec{r})$ totally, we get the conventional QED Hamiltonian density operator as

$$\hat{H}_{\mathrm{QED}}(\vec{r}) = \hat{H}_{\gamma}(\vec{r}) + \hat{H}_e(\vec{r}). \tag{78}$$

The momentum conservation law is reduced from Eq. (72) to

$$\frac{\partial}{\partial t}\left(\hat{\vec{G}}(\vec{r}) + \hat{\vec{\Pi}}_e(\vec{r})\right) = -\mathrm{div}\left(\hat{\vec{\sigma}}(\vec{r}) - \hat{\vec{\tau}}_e(\vec{r})\right). \tag{79}$$

The energy conservation law is reduced from Eq. (77) to

$$\frac{\partial}{\partial t}\hat{H}_{\mathrm{QED}}(\vec{r}) = -\mathrm{div}\left(c^2\hat{\vec{G}}(\vec{r}) + c^2\hat{\vec{\Pi}}_e(\vec{r})\right). \tag{80}$$

2.6 Non-relativistic limit of QED

Taking the non-relativistic limit of $\hat{H}_e(\vec{r})$, the rigged QED energy density operator $\hat{H}_{\text{rigged QED}}(\vec{r})$ reduces to:[3,4]

$$\hat{H}_{\text{non-relativistic QED}}(\vec{r}) = \frac{1}{8\pi}\left(\hat{\vec{E}}^2(\vec{r}) + \hat{\vec{B}}^2(\vec{r})\right) + \hat{T}(\vec{r}). \tag{81}$$

Then the contribution of the matter fields to the energy density operator $\hat{H}_{\text{non-relativistic QED}}(\vec{r})$ should simply be the kinetic energy density operator:

$$\hat{T}(\vec{r}) = \sum_{\alpha} \hat{T}_\alpha(\vec{r}), \tag{82}$$

$$\hat{T}_\alpha(\vec{r}) = -\frac{\hbar^2}{2m_\alpha}\cdot\frac{1}{2}\left(\hat{\chi}_\alpha^+(\vec{r})\hat{D}_{\alpha k}^2(\vec{r})\hat{\chi}_\alpha(\vec{r}) + \hat{D}_{\alpha k}^{+2}(\vec{r})\hat{\chi}_\alpha^+(\vec{r})\cdot\hat{\chi}_\alpha(\vec{r})\right), \tag{83}$$

with

$$\hat{D}_{\alpha\mu}(\vec{r}) = \partial_\mu + i\frac{Z_\alpha e}{\hbar c}\hat{A}_\mu(\vec{r}), \tag{84}$$

where $\alpha = e$ represents the non-relativistic electron with charge $Z_e = -1$, while $\alpha = a$ represents atomic nucleus a with charge Z_a. The $\hat{\chi}_e$ and $\hat{\chi}_e^+$ should of course satisfy Eqs. (15)-(17) with sign +.

The position probability density operator should change from Eq. (25) to

$$\hat{N}_e(\vec{r}) = \hat{\chi}_e^+(\vec{r})\hat{\chi}_e(\vec{r}). \tag{85}$$

Similarly, if not stated explicitly, as in Eq. (85) notation changes in the non-relativistic limit should be obvious.

2.7 Electrostatic limit

We shall apply the general theory thus developed to chemical reaction systems under no external electromagnetic field. The interaction is purely internal in nature and the field operators are of electrostatic type:

$$\hat{\vec{E}}(\vec{r}) = \hat{\vec{E}}_{\text{int}}(\vec{r}),\ \hat{\vec{E}}_{\text{int}}(\vec{r}) = -\text{grad}\hat{\phi}(\vec{r}),\ \hat{\phi}(\vec{r}) = \int d^3\vec{s}\,\frac{1}{|\vec{r}-\vec{s}|}\hat{\rho}(\vec{s}), \tag{86}$$

$$\hat{\vec{B}}(\vec{r}) = 0,\quad \hat{\vec{A}}(\vec{r}) = 0. \tag{87}$$

It should be noted that the electrostatic field operators in Eqs. (86) and (87) define the "electrostatic" Hamiltonian density operator. As a matter of fact, we may substitute Eqs. (86) and (87) into Eq.(81), we get the $\hat{H}_{\text{electrostatic non-relativistic QED}}(\vec{r})$:

$$\hat{H}_{\text{electrostatic non-relativistic QED}}(\vec{r}) = \frac{1}{8\pi}\hat{\vec{E}}^2(\vec{r}) + \hat{T}(\vec{r}). \tag{88}$$

2.8 Adiabatic limit

In the treatment of chemical reaction, we usually rely upon the Born-Oppenheimer adiabatic approximation. The adiabatic Hamiltonian density operator $\hat{H}_{\text{adiabatic}}(\vec{r})$ for any nucleus a clamped at position $\vec{R}_a$ is obtained from Eq. (81) using the adiabatic conditions,

$$\hat{N}_a(\vec{r}) = \delta(\vec{r} - \vec{R}_a), \quad \hat{T}_a(\vec{r}) = 0, \tag{89}$$

as follows:

$$\hat{H}_{\text{adiabatic}}(\vec{r}) = \frac{1}{8\pi}\hat{\vec{E}}^2_{\text{adiabatic}}(\vec{r}) + \hat{T}_e(\vec{r}), \tag{90}$$

with

$$\begin{aligned}\hat{\vec{E}}^2_{\text{adiabatic}}(\vec{r}) = {} & \sum_a \frac{\partial}{\partial\vec{r}}\frac{Z_a e}{|\vec{r}-\vec{R}_a|} \bullet \sum_b \frac{\partial}{\partial\vec{r}}\frac{Z_b e}{|\vec{r}-\vec{R}_b|} \\ & -2\sum_a \frac{\partial}{\partial\vec{r}}\frac{Z_a e}{|\vec{r}-\vec{R}_a|} \bullet \int d^3\vec{s}\, \frac{\partial}{\partial\vec{r}}\frac{e}{|\vec{r}-\vec{s}|}\hat{N}_e(\vec{s}) \\ & +\int d^3\vec{s}\left(\frac{\partial}{\partial\vec{r}}\frac{e}{|\vec{r}-\vec{s}|}\right)^2 \hat{N}_e(\vec{s}) \\ & +2\int d^3\vec{1}\int d^3\vec{2}\, \frac{\partial}{\partial\vec{r}}\frac{e}{|\vec{r}-\vec{1}|} \bullet \frac{\partial}{\partial\vec{r}}\frac{e}{|\vec{r}-\vec{2}|}\hat{\chi}_e^{\,+}(\vec{1})\hat{N}_e(\vec{2})\hat{\chi}_e(\vec{1}),\end{aligned} \tag{91}$$

where • denotes inner product of vectors. In Eq. (91) we have infinite quantities: the $a = b$ terms in the first line represent the infinite self energies of nuclei and the third line represents the infinite self energies of electrons. The self energies are immaterial in analyzing observable phenomena. Excluding the infinite self energies, we finally arrive at the expression

$$\hat{H}_{\text{adiabatic finite}}(\vec{r}) = \frac{1}{8\pi}\hat{\vec{E}}^2_{\text{adiabatic finite}}(\vec{r}) + \hat{T}_e(\vec{r}), \tag{92}$$

where

$$\hat{\vec{E}}^2_{\text{adiabatic finite}}(\vec{r}) = 2\sum_{a<b} \frac{\partial}{\partial\vec{r}}\frac{Z_a e}{|\vec{r}-\vec{R}_a|} \bullet \frac{\partial}{\partial\vec{r}}\frac{Z_b e}{|\vec{r}-\vec{R}_b|}$$

$$-2\sum_{a}\frac{\partial}{\partial\vec{r}}\frac{Z_{a}e}{\left|\vec{r}-\vec{R}_{a}\right|}\bullet\int d^{3}\vec{s}\,\frac{\partial}{\partial\vec{r}}\frac{e}{\left|\vec{r}-\vec{s}\right|}\hat{N}_{e}(\vec{s})$$

$$+2\int d^{3}\vec{1}\int d^{3}\vec{2}\,\frac{\partial}{\partial\vec{r}}\frac{e}{\left|\vec{r}-\vec{1}\right|}\bullet\frac{\partial}{\partial\vec{r}}\frac{e}{\left|\vec{r}-\vec{2}\right|}\hat{\chi}_{e}^{+}(\vec{1})\hat{N}_{e}(\vec{2})\hat{\chi}_{e}(\vec{1}). \quad (93)$$

In Eq. (93), we observe basic interaction scheme for a pair of charged particles. This is brought about through the inner product of a pair of electric fields associated with the charged particle each. In order to clarify the interaction, we shall treat for the sake of simplicity a system composed of only two classical point particles with charge Z_α and Z_β, being situated at $\vec{R}_\alpha$ and $\vec{R}_\beta$, respectively. If a pair of charged particles with charge Z_α and Z_β are united, then the united particle carries charge $Z_\alpha + Z_\beta$. Then, we shall consider the reverse reaction process of charge separation, when the interaction energy density is found to be

$$\frac{Z_\alpha Z_\beta e^2}{4\pi}\frac{\partial}{\partial\vec{r}}\frac{1}{\left|\vec{r}-\vec{R}_\alpha\right|}\bullet\frac{\partial}{\partial\vec{r}}\frac{1}{\left|\vec{r}-\vec{R}_\beta\right|}=\frac{Z_\alpha Z_\beta e^2}{4\pi}\frac{\left(\vec{r}-\vec{R}_\alpha\right)\bullet\left(\vec{r}-\vec{R}_\beta\right)}{\left|\vec{r}-\vec{R}_\alpha\right|^3\left|\vec{r}-\vec{R}_\beta\right|^3}. \quad (94)$$

(a)

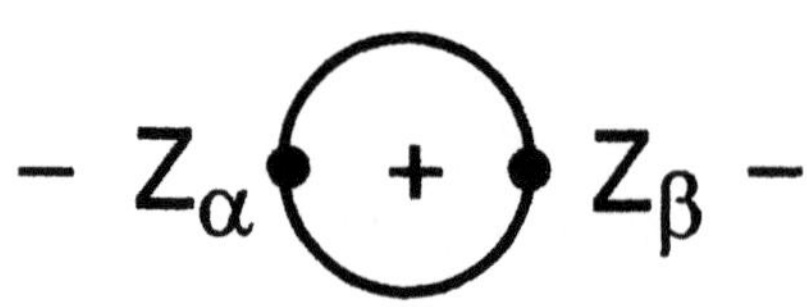

(b)

Figure 1. Basic interaction scheme for a pair of charged particles with charge Z_α situated at $\vec{R}_\alpha$ and charge Z_β at $\vec{R}_\beta$, where (a) if $Z_\alpha Z_\beta > 0$ then the electric field interaction energy is almost everywhere repulsive but attractive within a sphere with poles at $\vec{R}_\alpha$ and at $\vec{R}_\beta$, and (b) if $Z_\alpha Z_\beta < 0$ then the electric field interaction energy is almost everywhere attractive but repulsive within a sphere with poles at $\vec{R}_\alpha$ and at $\vec{R}_\beta$.

As shown in Fig.1, a characteristic sphere with poles at $\vec{R}_\alpha$ and $\vec{R}_\beta$ appears. For equal sign of charge, $Z_\alpha Z_\beta > 0$, the electric field energy density is negative inside the sphere leaving positive outside. Everywhere outside the characteristic sphere is found the intrinsic repulsive interaction for a pair of charged particles carrying equal sign. This is reasonable because the charge separation process should occur by itself. The charge separation process is further enhanced by the emergence of the characteristic attractive sphere, when the repulsive energy density outside is compensated by the formation of negative energy density inside. This compensation is promoted if the charged particle pairs are separated further, accompanying enlarged region of the characteristic attractive sphere. On the contrary, for different sign of charge, $Z_\alpha Z_\beta < 0$, the electric field energy is positive inside the sphere leaving negative outside. This is reasonable again because everywhere outside the characteristic sphere is found the intrinsic attractive interaction for a pair of charged particles carrying different sign. The charge separation process is then inhibited intrinsically. However, the emergence of the characteristic sphere diminishes the binding force. The attractive energy outside is compensated by the formation of positive energy inside. This compensation is enlarged if the charged particle pairs are separated further, accompanigying enlarged region of characteristic repulsive sphere leaving behind. The graph of Eq. (94) is shown in Fig. 2. As found in Fig. 2 explicitly, the magnitude of interaction should of course be significant in the neighborhood of charged particles. In Eq. (93), aside from the internuclear interaction that has exactly the same from as in Eq. (94), electron-nuclear interaction and the interelectronic interaction can be obtained by integrating Eq. (94) over the quantum mechanical density distribution of electrons.

Thus, we have found basic interaction scheme of charged particles in terms of the picture of "action through medium" in the field theory. Although the integral in the whole space should give the same number as that given by the picture of "action at a distance" in the Coulomb law (proof given in Appendix B), the local interaction picture is totally different from each other. The new interaction scheme thus obtained will shed new light on the basic physical interaction scheme underlying electronic processes in chemical reaction systems. Some applications in the next section will demonstrate the new interaction picture.

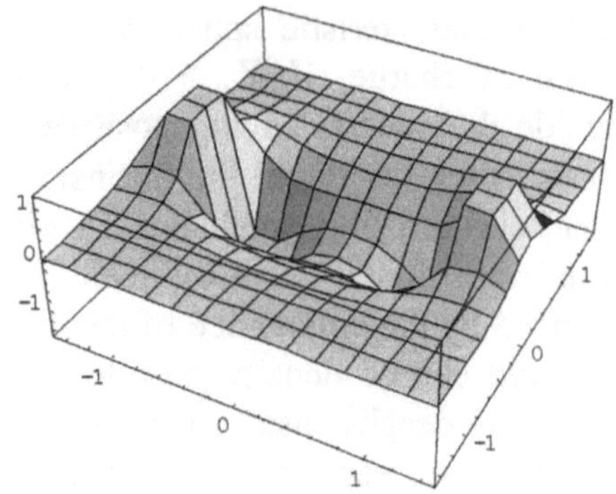

(a)

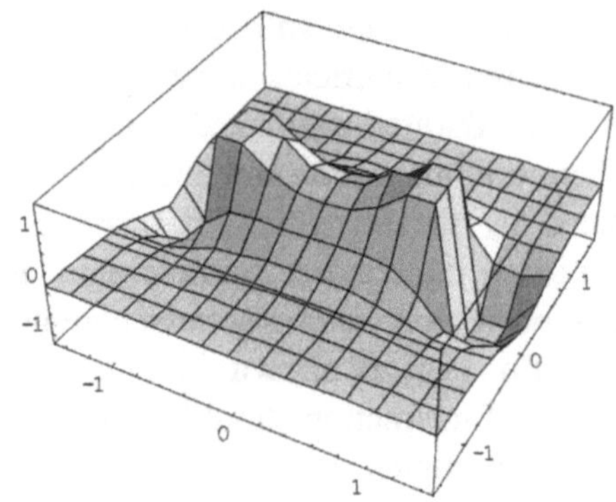

(b)

Figure 2. Graphs of Eq. (94) for (a) $Z_\alpha Z_\beta > 0$ and (b) $Z_\alpha Z_\beta < 0$, where arbitrary unit is used but $\vec{R}_\alpha = (-1,0,0)$ and $\vec{R}_\beta = (1,0,0)$ according to Fig. 1.

3. APPLICATION

Let ensemble average $\langle \hat{A} \rangle$ of field operator $\hat{A}$ over density matrix in the Fock space be performed in order to get observable quantity.

First, by performing ensemble average of Eq. (92), we obtain

$$n_{H_{\text{adiabatic finite}}}(\vec{r}) = \frac{1}{8\pi} E^2_{\text{adiabatic finite}}(\vec{r}) + n_{T_e}(\vec{r}), \tag{95}$$

where

$$n_{H_{\text{adiabatic finite}}}(\vec{r}) = \left\langle \hat{H}_{\text{adiabatic finite}}(\vec{r}) \right\rangle, \tag{96}$$

$$E^2_{\text{adiabatic finite}}(\vec{r}) = \left\langle \hat{\vec{E}}^2_{\text{adiabatic finite}}(\vec{r}) \right\rangle, \tag{97}$$

$$n_{T_e}(\vec{r}) = \left\langle \hat{T}_e(\vec{r}) \right\rangle. \tag{98}$$

3.1 H atom

For hydrogen-like atom in the ground state, analytical formulas of the energy densities are obtained as follows:

$$\frac{1}{8\pi} E^2_{\text{adiabatic finite}}(\vec{r}) = \frac{4Z^5 e^2}{\pi a^4} \frac{1}{x^4} \left(-1 + (1 + x + \frac{1}{2} x^2) e^{-x} \right), \tag{99}$$

$$\xrightarrow{x \to 0} -\frac{2Z^5 e^2}{3\pi a^4} \frac{1}{x}, \tag{100}$$

$$\xrightarrow{x \to \infty} -\frac{4Z^5 e^2}{\pi a^4} \frac{1}{x^4}, \tag{101}$$

$$n_{T_e}(\vec{r}) = \frac{Z^5 e^2}{\pi a^4} \frac{1}{x^4} (\frac{2}{x} - \frac{1}{2}) e^{-x}, \tag{102}$$

$$\xrightarrow{x \to 0} \frac{2Z^5 e^2}{\pi a^4} \frac{1}{x}, \tag{103}$$

$$\xrightarrow{x \to \infty} O(e^{-x}), \tag{104}$$

where

$$x = \frac{2Z}{a} r, \tag{105}$$

$$a = \frac{\hbar^2}{me^2}. \tag{106}$$

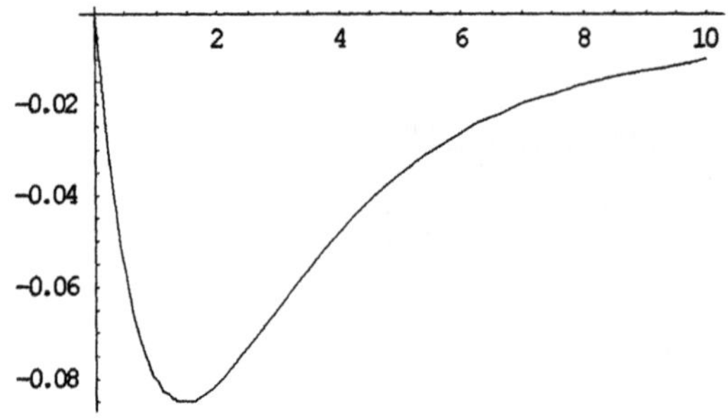

(a)

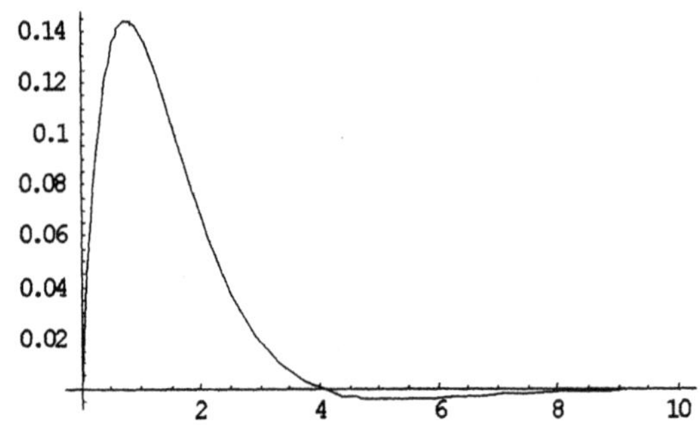

(b)

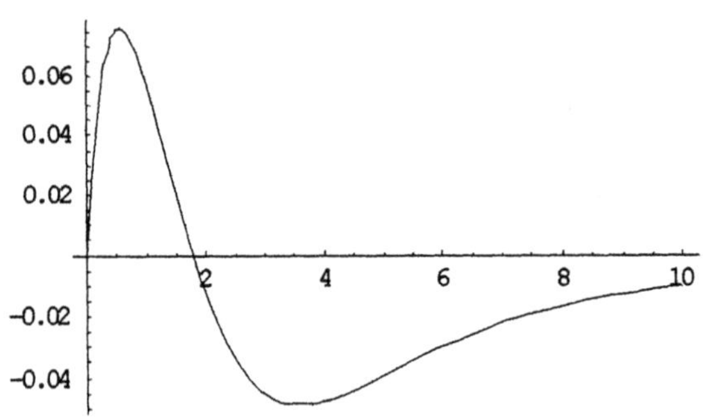

(c)

Figure 3. Graphs of the radial distribution functions $4\pi r^2 f$ for $f = (a)$ Eq. (99), (b) Eq. (102), and (c) Eq. (107), where arbitrary unit is used but abscissa denotes x.

In Fig. 3 are shown the graphs of the radial distribution functions $4\pi r^2 f$ of (a) $f =$ Eq. (99), and (b) $f =$ Eq. (102), respectively. As shown in Fig. 3(a), the electric field energy density is negative definite and is monotone decreasing to minus infinity at the position of proton. The coincidence in between the electron density maximum and the position of nucleus results in this behavior as expected from Fig. 1 (b) and the second line in the right hand side of Eq. (93). This negative most behavior in the electric field energy density should then be used to identify the position of "atom" in real space. On the other hand, as shown in Fig. 3 (b), the kinetic energy density is monotone increasing to plus infinity at the position of proton, which becomes negative beyond $r = 2$ a.u. that is the interface radius of the electronic drop region.[3]

The total energy density with its asymptotic behavior is then found to be as

$$n_{H_{\text{adiabatic finite}}}(\vec{r}) = \frac{4Z^5e^2}{\pi a^4}\frac{1}{x^4}\left(-1+(1+x+\frac{1}{2}x^2+\frac{1}{2}x^3-\frac{1}{8}x^4)e^{-x}\right) \tag{107}$$

$$\xrightarrow{x\to 0} \frac{4Z^5e^2}{3\pi a^4}\frac{1}{x}, \tag{108}$$

$$\xrightarrow{x\to\infty} -\frac{4Z^5e^2}{\pi a^4}\frac{1}{x^4}. \tag{109}$$

At the position of nucleus, the $n_{H_{\text{adiabatic finite}}}(\vec{r})$ diverges to plus infinity that dictates cusp condition.[3] Far away from nucleus, the leading term in $(8\pi)^{-1}E^2_{\text{adiabatic finite}}(\vec{r})$ assures that the $n_{H_{\text{adiabatic finite}}}(\vec{r})$ approaches zero from below. The overall behavior is shown in Fig.3(c). After integration we get

$$\int d^3\vec{r}\, n_{H_{\text{adiabatic finite}}}(\vec{r}) = -\frac{Z^2e^4m}{2\hbar^2}, \tag{110}$$

which is exactly the ground state energy of hydrogen-like atom.

3.2 Electrophilic addition reaction of HF to C_2H_4

Here we examine the electroplilic addition reaction of HF to C_2H_4. The maps of the electric field energy density $(8\pi)^{-1}E^2_{\text{adiabatic finite}}(\vec{r})$ and the energy density $n_{H_{\text{adiabatic finite}}}(\vec{r}) = (8\pi)^{-1}E^2_{\text{adiabatic finite}}(\vec{r}) + n_{T_e}(\vec{r})$ are shown along the IRC of Cs symmetry in Fig. 4 (a)-(c) and Fig. 5 (a)-(c), respectively, in which (a) at the reactant cell, (b) at the transition state (TS) corresponding to the saddle point on the potential energy surface, and (c) at the product complex.

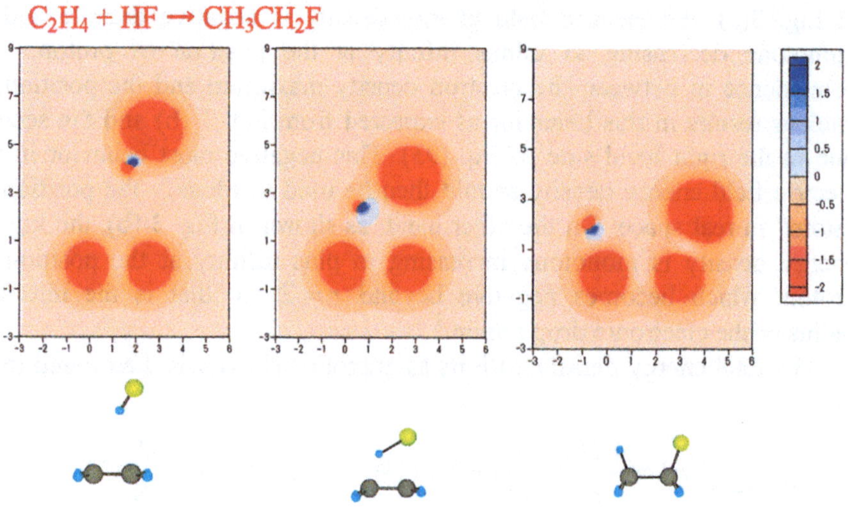

(a) (b) (c)

Figure 4. Graphs of purely the electronic contributions to Eq. (93) multiplied by $(8\pi)^{-1}$ for an electrophilic addition reaction of HF to C_2H_4 along the IRC of C_s symmetry in (a) at the reactant cell, (b) at the transition state (TS) corresponding to the saddle point on the potential energy surface, and (c) at the product complex. The scale of length is a.u. and the scale of density is hartree bohr^{-3}.

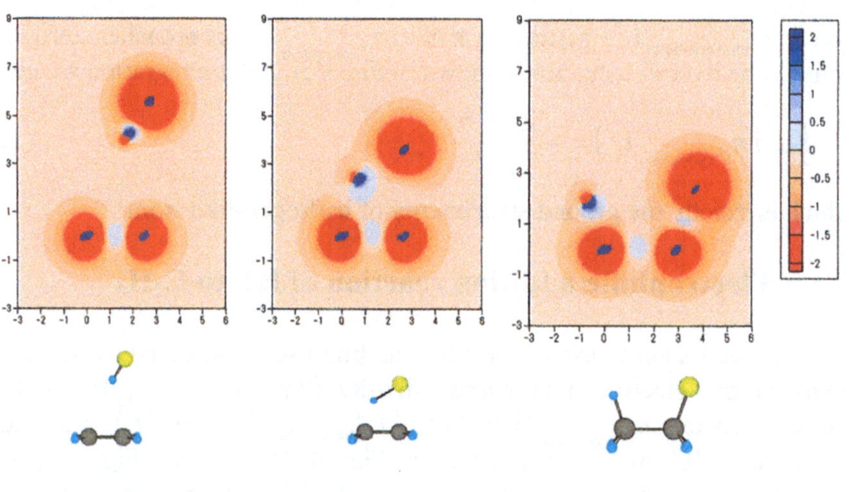

(a) (b) (c)

Figure 5. Graphs of purely the electronic contributions to Eq. (95) for the electrophilic addition reaction of HF to C_2H_4. For other details, see the caption of Fig. 4.

It should be noted that only the electronic contributions are depicted in Figs. 4 and 5, excluding purely classical nuclear contributions as added to the first line in the right hand side of Eq. (93). As shown in Fig. 4, large negative value of the electric field energy density around the position of nucleus is clearly found as expected in Fig. 3(a), where we can identify " atom" in real space. Furthermore in Fig. 4, we can easily observe positive-valued region of the electric field energy density surrounded by negative-valued region in between H and heavier atom A. This dictates chemical bond in between $H^{\delta+}$ and $A^{\delta-}$, because the combination of positive-charged nucleus and negative-charged electron density on the adjacent nucleus yields the basic interaction scheme depicted in Fig. 1 (b). The addition of the kinetic energy density $n_{T_e}(\vec{r})$ changes Fig. 4 into Fig. 5.This is easily performed since we have already published the maps of the kinetic energy density.[3]

3.3 Electrophilic addition reaction of HF to C_2H_2

Here we examine the electrophilic addition reaction of HF to C_2H_2. The maps of the electric field energy density $(8\pi)^{-1} E^2_{\text{adiabatic finite}}(\vec{r})$ and the energy density $n_{H_{\text{adiabatic finite}}}(\vec{r}) = (8\pi)^{-1} E^2_{\text{adiabatic finite}}(\vec{r}) + n_{T_e}(\vec{r})$ are shown along the IRC of Cs symmetry in Fig. 6 (a)–(c) and Fig. 7 (a)-(c), respectively, in which (a) at the reactant cell, (b) at the transition state (TS) corresponding to the saddle point on the potential energy surface, and (c) at the product complex.

It should be noted that only the electronic contributions are depicted in Figs. 6 and 7, excluding purely classical nuclear contributions as added to the first line in the right band side of Eq. (93). The characteristic behaviors found in Figs. 4 and 5 are similarly observed here in Figs. 6 and 7, respectively. It should be noted that the difference in double bond and triple bond is clearly shown in Fig. 4 (a) and Fig. 6 (a), respectively, because more negatively dense value is found around the region remote from the single bond in Fig. 6 (a) over and above that in Fig. 4 (a).

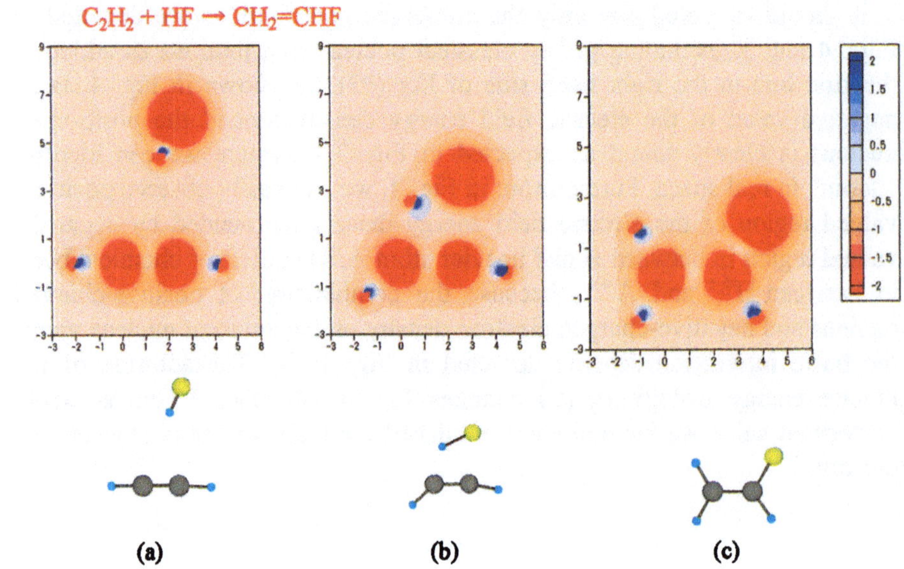

Figure 6. Same as Fig. 4 for an electrophilic addition reaction of HF to C_2H_2.

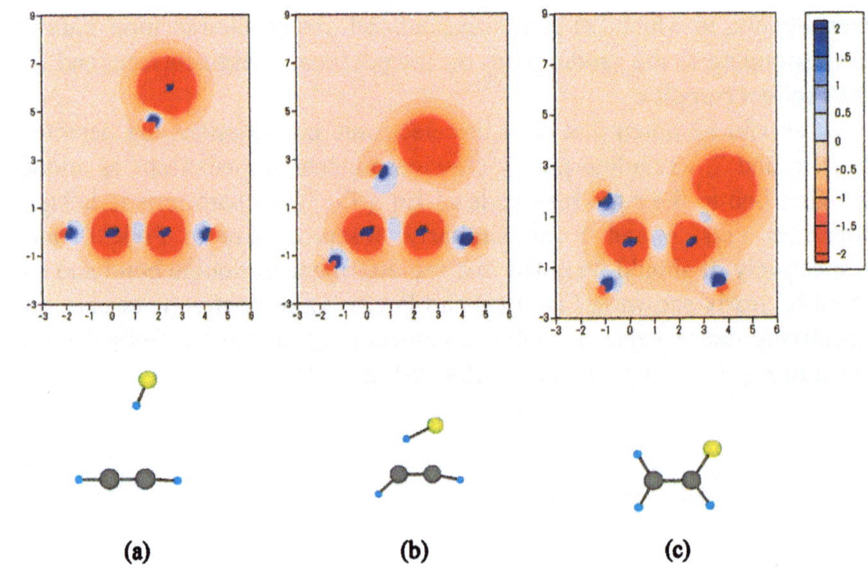

Figure 7. Same as Fig. 5 for the electrophilic addition reaction of HF to C_2H_2.

4. CONCLUSION

The electronic energy density scheme in Ref. 3 has been extended to the field energy density scheme. The field energy density has been derived from the non-relativistic limit of quantum electrodynamics (QED). The QED Hamiltonian energy density operator reduces to Eq. (81) in this limit. It has considerably novel feature as compared with the conventional ab initio electrostatic Hamiltonian energy density operator, but is decomposed into the density operators of the electromagnetic fields and the kinetic energy density operators of electrons and nuclei. The conventional Coulomb interaction energy densities disappear due to the QED variational principle. However, it has been rigorously verified that the integrals of the two kinds of Hamiltonian energy density operators all over the space should be the same. Although the integral picture should be the same, the new Hamiltonian density carries "action though medium" of the special theory of relativity in the underlying physics, while the conventional ab initio electrostatic Hamiltonian density carries "action at a distance" of the Coulomb law in the underlying physics. Reflecting the difference in the picture of the underlying physics, the approach presented here in this paper has given new insight on the local picture of interaction of electrons and nuclei in chemical reaction systems. The basic interaction picture of a pair of charged particles has been described by the emergence of the characteristic sphere within which compensation should occur for the primordial interaction in between the charged particles. The validity of this basic interaction picture has been demonstrated in application to H atom and electrophilic addition reactions of HF to C_2H_4 and C_2H_2.

To conclude, our theory starts from the quantum electro"dynamics" presented in Eq. (1). The electrons and atomic nuclei are "dynamic" entities, not "static" ones!

Only in the electro"static" limit in Sec. 2.7, our theory becomes equivalent to the conventional electro"static" ab initio theory.

Our theory treats observable phenomena: For example, the dynamical motion of atomic nuclei creates gauge potential and exerts electromagnetic interaction on the other entities, electrons and nuclei. This is observable in real phenomena, such as the so called "quantum Biot-Savart law" Eq. (43). Therefore, our theory has the well-founded predictable power of picture that should be observable in electro"dynamics" such as in chemical reaction systems. But this picture is lacking in the conventional treatment of the ectro"static" ab initio Hamiltonian, where, for example, the electro"static" nuclear wave packet dynamics on the potential energy surface averaged over the electro"static" ab initio electronic Hamiltonian does not creat any gauge potential.

ACKNOWLEDGMENTS

This work has been supported by Grant-in-Aid for Scientific Research from the Ministry of Education, Science and Culture of Japan, for which we express our gratitude.

APPENDIX A

The radiation field is characterized by the gauge potential summed over helicity $\sigma = \pm 1$ as

$$\hat{A}^{\mu}_{\text{radiation}}(x) = \frac{K}{\sqrt{(2\pi\hbar)^3}} \sum_{\sigma=\pm 1} \int \frac{d^3\vec{p}}{\sqrt{2p^0}} \cdot \left[\hat{a}(p,\sigma) e^{\mu}(p,\sigma) e^{-ix_\mu p^\mu / \hbar} + \hat{a}^+(p,\sigma) e^{\mu}(p,\sigma)^* e^{+ix_\mu p^\mu / \hbar} \right], \tag{A1}$$

with

$$K = \sqrt{4\pi\hbar^2 c}, \tag{A2}$$

where the field operators $a(p,\sigma)$ and $\hat{a}^+(p,\sigma)$ satisfy

$$\hat{a}(p,\sigma)\hat{a}(\overline{p},\overline{\sigma}) - \hat{a}(\overline{p},\overline{\sigma})\hat{a}(p,\sigma) = 0, \tag{A3}$$

$$\hat{a}^+(p,\sigma)\hat{a}^+(\overline{p},\overline{\sigma}) - \hat{a}^+(\overline{p},\overline{\sigma})\hat{a}^+(p,\sigma) = 0, \tag{A4}$$

$$\hat{a}(p,\sigma)\hat{a}^+(\overline{p},\overline{\sigma}) - \hat{a}^+(\overline{p},\overline{\sigma})\hat{a}(p,\sigma) = \delta^3(\vec{p} - \vec{\overline{p}})\delta_{\sigma\overline{\sigma}}, \tag{A5}$$

and where the polarization vectors $e^{\mu}(p,\sigma)$ defined as

$$e^0(p,\sigma) = 0, \tag{A6}$$

$$p^{\mu} e_{\mu}(p,\sigma) = 0, \tag{A7}$$

should satisfy

$$\sum_{\sigma=\pm 1} e^i(p,\sigma) e^j(p,\sigma)^* = \delta_{ij} - \frac{p^i p^j}{|\vec{p}|^2}, \tag{A8}$$

$$e^k(p,\overline{\sigma}) e^k(p,\sigma)^* = \delta_{\overline{\sigma}\sigma}, \tag{A9}$$

$$e^k(-\vec{p},\overline{\sigma}) e^k(\vec{p},\sigma) = \delta_{\overline{\sigma}\sigma}. \tag{A10}$$

The radiation field is then characterized by the energy density constructed by the transversal electric field

$$\hat{\vec{E}}_{\text{radiation}}(\vec{r}) = \hat{\vec{E}}_{\text{T}}(\vec{r}) = -\frac{1}{c}\frac{\partial}{\partial t}\hat{\vec{A}}_{\text{radiation}}(\vec{r}), \tag{A11}$$

and the magnetic field

$$\hat{\vec{B}}_{\text{radiation}}(\vec{r}) = \text{rot}\hat{\vec{A}}_{\text{radiation}}(\vec{r}), \tag{A12}$$

both derived from the $\hat{\vec{A}}_{\text{radiation}}(\vec{r})$ in Eq. (A1). Then the energy density reduces to

$$\hat{H}_{\gamma\,\text{radiation}}(\vec{r}) = \frac{1}{8\pi}\left(\hat{\vec{E}}^2_{\text{radiation}}(\vec{r}) + \hat{\vec{B}}^2_{\text{radiation}}(\vec{r})\right), \tag{A13}$$

leading to

$$\begin{aligned} &\int d^3\vec{r}\hat{H}_{\gamma\,\text{radiation}}(\vec{r}) \\ &\quad = \sum_{\sigma=\pm 1}\int d^3\vec{p}cp_0\frac{1}{2}\left(\hat{a}(\vec{p},\sigma)\hat{a}^+(\vec{p},\sigma) + \hat{a}^+(\vec{p},\sigma)\hat{a}(\vec{p},\sigma)\right) \\ &\quad = \sum_{\sigma=\pm 1}\int d^3\vec{p}cp_0\left(\hat{a}^+(\vec{p},\sigma)\hat{a}(\vec{p},\sigma) + \frac{1}{2}\delta^3(\vec{p}-\vec{p})\right). \end{aligned} \tag{A14}$$

APPENDIX B

We shall examine the integral of Eq. (88). It is easy to prove

$$\frac{1}{8\pi}\int d^3\vec{r}\hat{\vec{E}}^2(\vec{r}) = \frac{1}{2}\int d^3\vec{r}\,\hat{\phi}(\vec{r})\hat{\rho}(\vec{r}), \tag{B1}$$

and hence

$$\int d^3\vec{r}\hat{H}_{\text{electrostatic non-relativistic QED}}(\vec{r}) = \int d^3\vec{r}\hat{H}_{\text{Coulomb}}(\vec{r}), \tag{B2}$$

where

$$\hat{H}_{\text{Coulomb}}(\vec{r}) = \frac{1}{2}\hat{\phi}(\vec{r})\hat{\rho}(\vec{r}) + \hat{T}(\vec{r}). \tag{B3}$$

The $\hat{H}_{\text{Coulomb}}(\vec{r})$ represents the Coulomb law, where "action at a distance" picture results. It should be noted that although the integrals over the whole space are the same, the alternative densities

$\hat{H}_{\text{electrostatic non-relativistic QED}}(\vec{r})$ and $\hat{H}_{\text{Coulomb}}(\vec{r})$ definitely represent the difference in picture adopted for each. The common ingredient is the kinetic energy density operator $\hat{T}(\vec{r})$.

This kind of transformation should be valid even in the adiabatic limit. Then, we shall examine the integral of Eq. (92), when the ab initio electronic Hamiltonian density $\hat{H}_{\text{ab initio}}(\vec{r})$ is extracted from the $\hat{H}_{\text{adiabatic finite}}(\vec{r})$ in such a way that the integral of the $\hat{H}_{\text{adiabatic finite}}(\vec{r})$ in the whole space is equal to that of the $\hat{H}_{\text{ab initio}}(\vec{r})$ plus a constant that represents the nuclear repulsion energy:

$$\int d^3\vec{r}\hat{H}_{\text{adiabatic finite}}(\vec{r}) = \int d^3\vec{r}\hat{H}_{\text{ab initio}}(\vec{r}) + V_{\text{nuc}}. \tag{B4}$$

In this expression,

$$\hat{H}_{\text{ab initio}}(\vec{r}) = \hat{T}_e(\vec{r}) + \hat{V}_e(\vec{r}) + \hat{W}_e(\vec{r}), \tag{B5}$$

with $\hat{T}_e(\vec{r})$ given in Eq.(92) and

$$\hat{V}_e(\vec{r}) = -\sum_a \frac{Z_a e^2}{\left|\vec{r} - \vec{R}_a\right|}\hat{N}_e(\vec{r}), \tag{B6}$$

$$\hat{W}_e(\vec{r}) = \frac{1}{2}\hat{\chi}_e^+(\vec{r})\int d^3\vec{s}\frac{e^2}{\left|\vec{r} - \vec{s}\right|}\hat{N}_e(\vec{s})\hat{\chi}_e(\vec{r}), \tag{B7}$$

$$V_{\text{nuc}} = \sum_{a<b}\frac{Z_a Z_b e^2}{\left|\vec{R}_a - \vec{R}_b\right|}, \tag{B8}$$

where $\hat{V}_e(\vec{r})$ denotes the nuclear attraction energy density operator of electrons and where $\hat{W}_e(\vec{r})$ denotes the inter-electron potential energy density operator of electrons excluding infinite self energy and where V_{nuc} denotes the nuclear repulsion energy excluding infinite self energy. It should be noted that although the integrals over the whole space are the same, the picture represented should be definitely different with each other.

REFERENCES

[1]A. Einstein, Ann.Phys.**17**, 891 (1905).
[2]S. Weinberg, *The Quantum Theory of Fields* (Cambridge Univ., Cambridge, 1995).
[3]A. Tachibana, J. Chem. Phys. **115**, 3497 (2001).

[4]A. Tachibana, in *Reviews in Modern Quantum Chemistry*, edited by K. D. Sen (World Scientific, Singapore, 2002), in press.
[5]H. B. Huntington and A. R. Grone , J. Phys. Chem. Solids **20**, 76 (1961).
[6]I. A. Blech and E. Kinsbron, Thin Solid Films **25**, 327 (1975).
[7] I. A. Blech, J. Appl. Phys. **47**, 1203 (1976).
[8] J. R. Black, IEEE Trans. Electron Devices **ED-16**, 338 (1969).
[9] P. S. Ho and T. Kwok, Rep. Prog. Phys. **52**, 301 (1989).
[10]C. V. Thompson and J. R. Lloyd, MRS Bulletin, 19 (1993).
[11]H. Kawasaki, M. Gall, D. Jawarani, R. Hernandez, and C. Capasso, Thin Solid Films **320**, 45 (1998).
[12] C. Bosvieux and J. Friedel J. Phys. Chem. Solids **23**, 123 (1962).
[13]P. Kumar and R. S. Sorbello, Thin Solid Films **25**, 25 (1975).
[14]R. S. Sorbello and B. B. Dasgupta, Phys. Rev. B **21**, 2196 (1980).
[15]A. Lodder, J. Phys. F: Met. Phys. **14**, 2943 (1984).
[16]A. Lodder, Solid State Comm. **71**,259 (1989).
[17]R. S. Sorbello, Solid State Phys. **51**, 159 (1998).
[18]A. Lodder and J. P. Dekker, "The Electromigration Force in Metallic Bulk," in *Stress Induced Phenomena in Metallization*, edited by Okabayashi et al ., American Institute of Physics, 1998, pp.315-328.
[19]K. Iguchi and A. Tachibana, Surface Science **159/160**, 167 (2000).
[20]A. Tachibana, in *Stress Induced Phenomena in Metallization*, edited by S. P. Baker (American Institute of Physics, New York, 2002), in press.
[21]A. Tachibana (to be published).

Quantum Chemical Methods applied to Solids

P. Fulde, Max Planck Institute for the Physics of Complex Systems, Nöthnitzer Straße 38, 01187 Dresden (Germany)

Abstract: We discuss the applicability of quantum chemical methods to the computation of many-electron wavefunctions of solids. In addition to the theoretical framework results are presented for the ground state of semiconductors. The effects of correlations on energy bands and on spectral densities are discussed.

Keywords: ground-state wavefunctions, spectral densities, strongly correlated electrons

1 Introduction

One of the most important and active areas of condensed matter theory are electronic structure calculations. Based on breath-taking advances in computer hard- and software and on new theoretical methods, in particular density-functional theory, this field has made progress beyond imaginations 20 or 30 years ago. The dominant role of density-functional theory, more precisely of approximations based on it like the local-density approximation (LDA) is based on its relative simplicity. By deliberately avoiding the problem of computing the many-electron wavefunction of a solid and instead computing directly ground-state properties such as the electronic density distribution or magnetization density the theory reduces to a self-consistent computation of a one-particle Schrödinger equation. Compared with computations based on many-electron Schrödinger equations this seems a tremendous advantage and

E.J. Brändas and E.S. Kryachko (eds.),
Fundamental World of Quantum Chemistry, Vol. II, 241–255.

simplification. Nevertheless, for a good understanding of electron correlations the wavefunction of a solid in a particular state would be highly desirable. It would allow for a study of the correlation hole which an electron is carrying with it. This is a fairly local object except when special correlations like those resulting in superconductivity are considered. Therefore its description should not be more difficult than for an electron in a reasonably seized molecule. In the following we want to show that the ground-state wavefunction of semiconductors and insulators can be calculated with an accuracy which compares with that of molecules. We want also to give examples of calculations of excited state wavefunctions and their associated energies. Here quantum chemical accuracy can be achieved presently only for elemental semiconductors. By using model Hamiltonians also spectral densities can be calculated for strongly correlated electrons.

2 GROUND-STATE: BASIC FORMALISM

We want to concentrate on the computation of the ground-state wavefunction of a solid by means of quantum chemical methods. For that purpose we work with basis sets consisting of Gauss-type orbitals (GTO's). The starting Hamiltonian is

$$H = \sum_{ij\sigma} t_{ij} a^{+}_{i\sigma} a_{j\sigma} + \frac{1}{2} \sum_{\substack{ijkl \\ \sigma\sigma'}} V_{ijkl} a^{+}_{i\sigma} a^{+}_{k\sigma'} a_{l\sigma'} a_{j\sigma} \tag{1}$$

where $a^{+}_{i\sigma}$, $a_{i\sigma}$ are creation and annihilation operators of electrons with spin σ in states related to the GTO functions $f_i(\mathbf{r})$. More precisely, when the overlap matrix between the different GTO's is

$$\hat{S}_{ij} = \int d^3 r f_i^*(\mathbf{r}) f_j(\mathbf{r}) \tag{2}$$

then the operators

$$\hat{a}^{+}_{i\sigma} = \sum_j \hat{S}_{ji} a^{+}_{j\sigma} \tag{3}$$

create electrons in basis states $|f_{i\sigma}\rangle = \hat{a}^{+}_{i\sigma}|0\rangle$. The matrix elements in (1) are given by

$$\begin{aligned} t_{ij} &= \int d^3 r f_i^*(\mathbf{r}) \left(-\frac{1}{2m} \nabla^2 + V(\mathbf{r}) \right) f_j(\mathbf{r}), \\ V_{ijkl} &= e^2 \int d^3 r d^3 r' f_i^*(\mathbf{r}) f_j(\mathbf{r}) \frac{1}{|\mathbf{r} - \mathbf{r}'|} f_k^*(\mathbf{r}') f_l(\mathbf{r}') \quad . \end{aligned} \tag{4}$$

The potential V(**r**) is set up by the nuclei and the inner-shell electrons. We split the Hamiltonian H into

$$H = H_{\text{SCF}} + H_{\text{res}} \tag{5}$$

and assume that the self-consistent field (SCF) part H_{SCF} can be determined with the help of a high-performance program package like CRYSTAL [1] or WANNIER [2]. Our aim is to approximate the exact ground state $|\psi_0\rangle$ with good accuracy, under the assumption that the ground state $|\Phi_{\text{SCF}}\rangle$ of H_{SCF} has been determined and is known. We write the ground-state energy $\mathcal{E}_0$ is the form

$$\mathcal{E}_0 = \langle \Phi_{\text{SCF}} \mid H \mid \psi_0 \rangle^c \tag{6}$$

where the upper script c indicates that the cumulant is taken. Cumulants have been applied to classical mechanics as well as to quantum mechanics, in particular by Kubo [3]. When applied to perturbations theory they correspond to linked diagrams. Formally they are defined by

$$ln\ \langle \phi_1 \mid e^{\lambda A} \mid \phi_2 \rangle = \langle \phi_1 \mid e^{\lambda A} - 1 \mid \phi_2 \rangle^c \quad . \tag{7}$$

This shows their main virtue. Size consistency which is ensured, e.g., for the free energy by taking the logarithm of the partition function can be obtained *without* a logarithm by working instead with cumulants. Equation (6) can be rewritten by introduction of a cumulant wave operator Ω. We recall that the ordinary wave operator $\tilde{\Omega}$ transforms $|\Phi_{\text{SCF}}\rangle$ into $|\psi_0\rangle$, i.e., $|\psi_0\rangle = \tilde{\Omega}|\Phi_{\text{SCF}}\rangle$. The cumulant wave operator does the same when working with cumulants, i.e.,

$$\begin{aligned} \mathcal{E}_0 &= \langle \Phi_{\text{SCF}} \mid H\Omega \mid \Phi_{\text{SCF}} \rangle^c \\ &= (H \mid \Omega) \quad . \end{aligned} \tag{8}$$

Here we have introduced the notation

$$(A \mid B) = \langle \Phi_{\text{SCF}} \mid A^+ B \mid \Phi_{\text{SCF}} \rangle^c \quad . \tag{9}$$

Note that $|\Omega)$ is not uniquely defined. But the difference $\Delta\Omega = \Omega_1 - \Omega_2$ of two different forms of the cumulant wave operator must have the property that $(H|\Delta\Omega) = 0$. Different forms of $|\Omega)$ are

$$\mid \Omega) = \lim_{z \to 0} \left| \ 1 + \frac{1}{z - H}\ H_{\text{res}} \right) \tag{10}$$

or

$$| \Omega) = | e^S) \tag{11}$$

where S is a prime operator, i.e., an operator which is treated as an entity when cumulants are evaluated [4]. The goal is to find a form for $|\Omega)$ which is converging as rapidly as possible when an approximation scheme is applied.

From the properties of cumulants is follows that

$$(A \mid H\Omega) = 0 \tag{12}$$

for any arbitrary operator A, excluding c-numbers [5].

Equations (8) and (12) constitute the basis equations for determining the ground state and its energy of a solid. Note that when in the cumulant notation $|\Omega)$ is known the exact ground-state wavefunction is known. The expectation value of any operator A with respect to $|\psi_0\rangle$ is given by

$$\frac{\langle \psi_0 \mid A \mid \psi_0 \rangle}{\langle \psi_0 \mid \psi_0 \rangle} = (\Omega \mid A\Omega) \quad . \tag{13}$$

Therefore all properties of $|\psi_0\rangle$ can be studied by working with $|\Omega)$ instead.

Note that by setting $S = \sum_i \eta_i S_i$ where the S_i generate an increasing number of electron-hole excitations with $S_i^+|\Phi_{SCF}\rangle = 0$ we recover from (8) and (11) the coupled-cluster (CC) equations [6, 7, 8], i.e.,

$$\mathcal{E}_0 = (H \mid e^S) \quad , \qquad (S_i \mid He^S) = 0 \quad . \tag{14}$$

The coupled-electron-pair approximation (CEPA-0) is recovered by making an ansatz of the form

$$| \Omega) = \ | 1 + \sum_{K\Gamma} \eta_\Gamma^K A_\Gamma^K) \quad . \tag{15}$$

Here

$$A_\Gamma^K = \begin{cases} c_i^+ c_\mu & \\ c_i^+ c_j^+ c_\nu c_\mu & ; \ i < j \text{ and } \mu < \nu \end{cases} \tag{16}$$

where c_i^+ is the creation operator for an electron in the virtual spin orbital i while c_μ destroys an electron in the occupied spin orbital μ. The parameters η_Γ^K follow from $(A_\Gamma^K|H\Omega) = 0$, i.e., by solving the system of linear equations

$$(A_{\Gamma'}^{K'} \mid H_{\text{res}}) + \sum_{K\Gamma} \eta_{\Gamma}^{K} (A_{\Gamma'}^{K'} \mid H\ A_{\Gamma}^{K}) = 0 \quad . \tag{17}$$

The correlation energy is given by

$$E_{\text{corr}} = \sum_{K\Gamma} \eta_{\Gamma}^{K} (H_{\text{res}} \mid A_{\Gamma}^{K}) \quad . \tag{18}$$

3 Reduction to a Few-Electron Problem

When calculating the ground-state wavefunction of correlated electrons in a solid one must use the periodicity of the lattice and the fact mentioned above, that the correlation hole of an electron is a fairly local object. With the help of an appropriate approximation scheme this enables us to reduce the problem of essentially infinitely many electrons to that of a few ones. The crucial question is that of formulating a fast converging computational scheme of that kind.

Determining $|\Omega)$ is equivalent to determining the cumulant scattering matrix $|S)$ of the many-electron system,

$$| S) = \ | \Omega - 1) \quad . \tag{19}$$

We want to construct that matrix step by step within the same spirit as applied by Faddeev for the three-particle problem [9]. He showed how the scattering matrix for three particles can be expressed by using the scattering matrices for the two-particle scattering problem. Similarly we want to express the cumulant scattering-matrix |S) of the N electron problem in terms of scattering matrices involving electrons of increasing number. This is done by the method of increments [10]. By including more and more electrons only the corresponding increments of the scattering matrix are added.

By starting from $|\Phi_{\text{SCF}}\rangle$ we keep all electrons frozen except on atom (or bond) I. The resulting excitations on that atom (bond) due to H_{res} define the one-center scattering matrix $|S_I)$. It can be computed routinely by employing a quantum chemical program package like MOLPRO [11] or GAMESS [12]. Within that approximation

$$| S) = \sum_{I} | S_I) \quad . \tag{20}$$

Next we keep all electrons frozen except on atoms (bonds) I and J and include all excitations on these two centers. This yields the two-center scattering matrix $|S_{IJ})$ and an improved form of $|S)$, i.e.,

$$| S) = \sum_I | S_I) + \sum_{\langle IJ \rangle} | S_{IJ} - S_I - S_J) \quad . \tag{21}$$

In a next step electrons on three different atoms or bonds are released and all excitations on these three centers are taken into account. This adds all increments $|\delta S_{IJK})$ of the different triples $\langle IJK \rangle$ to |S) where

$$| \delta S_{IJK}) = \ | S_{IJK} - S_{IJ} - S_{IK} - S_{JK} - S_I - S_J - S_K) \quad . \tag{22}$$

This can be continued but it turns out that this expansion of $|S)$ is terms of increments is very rapidly convergent. Three-center contributions are the most what is usually required. For a more detailed derivation of this approximation scheme for the cumulant scattering matrix see [13]. The corresponding expansion of the correlation energy is

$$\begin{aligned} E_{\text{corr}} &= \sum_I \ (H \mid S_I) + \sum_{\langle IJ \rangle} \ (H \mid \delta S_{IJ}) + ... \\ &= \sum_I \ \epsilon_I + \sum_{\langle IJ \rangle} \ \epsilon_{IJ} + ... \end{aligned} \tag{23}$$

where $|\delta S_{IJ}) = |S_{IJ} - S_I - S_J)$.

The above theory is very well suited for semiconductors and insulators, where we can define well localized Wannier functions centered at different atoms or bonds. For metals with partially filled conduction bands this is not the case and therefore some modifications are required. We project the atomic orbitals $\{\phi_i\}$ of the valence electrons onto the occupied and the unoccupied (virtual) SCF space. These projections $\{P_{\text{occ}}\phi_i\}$ and $\{P_{\text{virt}}\phi_i\}$ result in non-orthogonal and overcomplete sets of orbitals in the respective spaces. Overcompleteness can be avoided by identifying pairs $\{\phi_i, \phi_j\}$ with large overlap. We eliminate one member of each pair until overcompleteness has been remedied. For example, when considering Li metal with a cubic structure we eliminate from each nearest-neighbor pair $\{P_{\text{occ}}\phi_{2s}\}(i)$, $\{P_{\text{occ}}\phi_{2s}\}(j)$ one member and keep only projected functions on every other site. After a mutual pairwise orthogonalization we can again use the standard quantum-chemical grogram packages. The RPA screening can be taken into account by an ansatz of the form

$$| \Omega) = \ | 1 + \sum_\nu \eta_\nu \omega_\nu + \sum_{\nu\nu'} \eta_\nu \eta_{\nu'} \omega_\nu \omega_{\nu'}) \tag{24}$$

where the ω_ν denote double excitations. When we neglect the last term the Tamm-Dancoff approximation is obtained.

In solids with rare earth or actinide ions we deal with 4f and 5f electrons, respectively, which are strongly correlated. Also in the cuprates with the well known superconductivity properties, electron correlations are strong. Therefore a SCF calculation is a poor starting point for these electrons. Alternatively one may state that the cumulant scattering matrices $|S_I)$ have to be especially well calculated in order to incorporate the strong correlations. This is achieved by expressing not only the occupied but also the unoccupied orbitals of strongly correlated electrons in localized form. The strong correlations are taken into account by Complete Active Space (CASSCF) calculations. When the single-center cumulant scattering matrix $|S_I)$ is determined the active space consists of localized occupied and virtual d or f orbitals of that center. Similarly, when the two-center cumulant scattering matrix $|S_{IJ})$ is calculated, the active space consists of the localized d or f electrons on centers I and J. When CASSCF calculations are performed all electrons are kept frozen except those in the active space. The remaining correlations are weak and can be handled by the usual methods.

4 EXAMPLES

Within the frame work of the above theory a considerable number of ground-state calculations have been informed. They concern semiconductors of groups IV [14], III-V [15], II-VI [16], the oxides MgO [17], CaO [18], NiO [19] and TiO_2 [20], rare-gas crystals [21] and the rare-earth compound GdN with the 4f electrons kept within the core [22]. The SCF part has been done in most cases by applying CRYSTAL together with a localization procedure in order to obtain the Wannier functions. An alternative way of obtaining the SCF ground-state in Wannier form is by performing a cluster calculation. For example, consider silicon. By choosing a cluster cut out from a solid and saturating the dangling bonds with H atoms one can identify the central bond with the Wannier orbitals of the solid. An example is $X_{35}H_{36}$ where X = C, Si, Ge, Sn (see Fig 1).

We find that the central bond agrees very well with the one obtained from CRYSTAL. In order to demonstrate the degree of accuracy which we obtain we list in Table I the results for the cohesive energy of diamond, silicon, germanium and tin. The basis set used for C is (9s4p2d2f)/[3s2p2d2f] while for Si, Sn and Ge a four- valence electron pseudopotential was employed with optimized (4s2p2d1f)/[3s3p2d1f] basis sets. For more details we refer to the original literature.

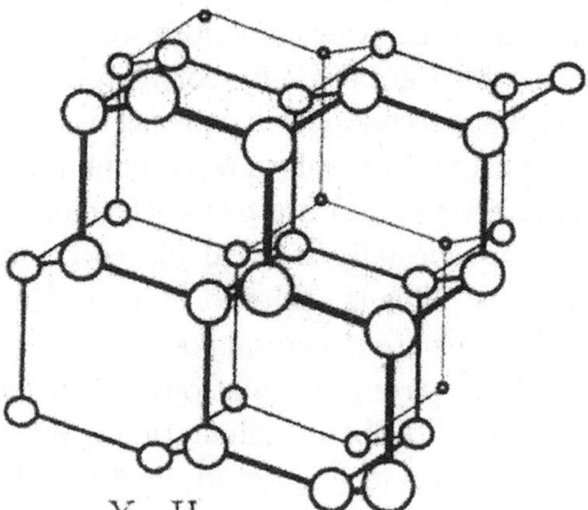

Fig. 1: *$X_{35}H_{36}$ cluster (X = C, Si, Ge, Su) used for the computation of localised SCF orbitals. The H atoms are not shown.*

		C	Si	Ge	Sn
E^{coh}_{SCF}		10.74	6.18	4.25	3.65
E^{coh}_{incr}		14.36	8.83	6.98	6.08
		96 %	94 %	90 %	97 %
Expt.	[23]	15.10	9.39	7.75	6.23
LDA	[24],[25],[26]	17.25	10.58	9.06	–
GGA	[26]	–	8.79	6.83	–
QMC	[26]	14.90	9.76	–	–

Table 1: *Cohesive energy per unit cell (in eV) as obtained from a SCF calculation and including correlations by the method of increments (E^{coh}_{incr}): deviations from experimental values are given in parentheses (in percent). Also shown are LDA, generalised gradient approximation (GGA) and Monte Carlo (QMC) results (From Ref. [14]*

In order to demonstrate the insight which can be gained concerning correlations we consider GaAs [15]. In Fig. 2 we show the correlation contribution to the cohesive energy and its dependence on the quality of the basis set. Basis set A consists of a basis (4s4p)/[3s3p] + 1d for Ga as well as for As while set B consists of a basis (4s4p) + 2d + 1f. For more detailed information see Ref. [15]. Fig. 3 shows a decomposition of the total correlation energy into one-, two-, and three-body increments. Finally, we show in Fig. 4 the fast decrease of the two-body increments as function of distance. The numbers of the different nearest neighbor are included in the diagram so that the contributions add up to the total two-body increment of Fig. 3.

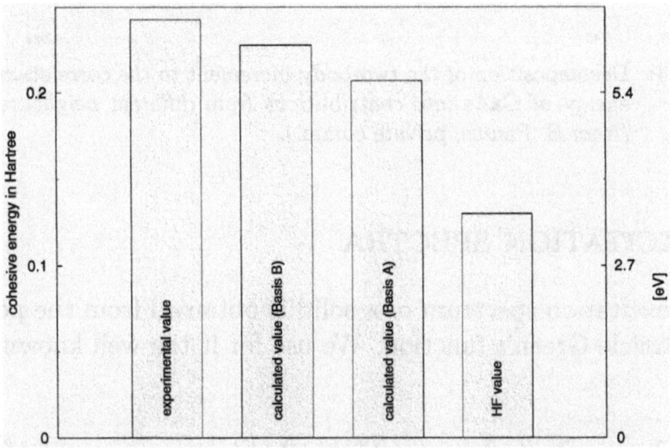

Fig. 2: *Decomposition of the cohesive energy per unit cell of GaAs into a Hartree-Fock or SCF part and a correlation contribution. Shown is the basis-set dependence of the correlation part. The basis sets A and B are described in the text (from B. Paulus, private comm.).*

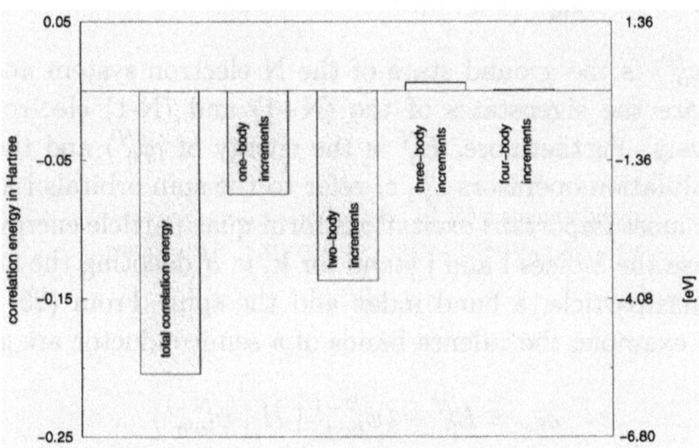

Fig. 3: *Decomposition of the correlation energy per unit cell of GaAs (from B. Paulus, private comm.).*

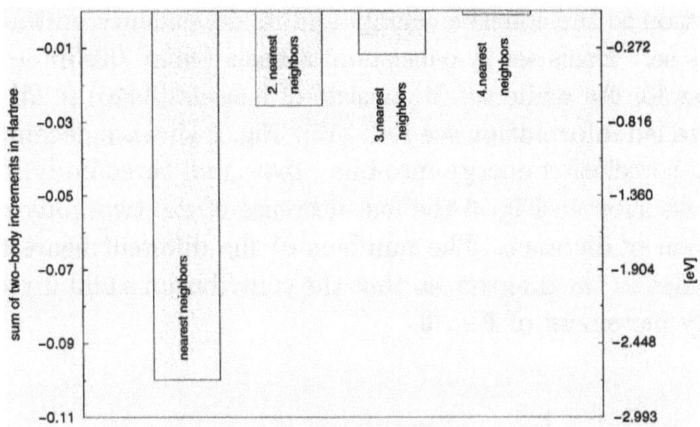

Fig. 4: *Decomposition of the two-body increment to the correlation energy of GaAs into contributions from different neighbors (from B. Paulus, private comm.).*

5 EXCITATION SPECTRA

The excitation spectrum of a solid is obtained from the poles of the single-particle Green's function. We use for it the well known form

$$\begin{aligned} G_{ij}(\omega) \;=\; & \sum_{mm'} \langle \psi_0^N \mid c_i \mid \psi_m^{N+1} \rangle \langle \psi_m^{N+1} \mid \frac{1}{\omega - H + E_0^N} \mid \psi_{m'}^{N+1} \rangle \\ & \langle \psi_{m'}^{N+1} \mid c_j^+ \mid \psi_0^N \rangle \\ & + \sum_{mm'} \langle \psi_0^N \mid c_j^+ \mid \psi_m^{N-1} \rangle \langle \psi_m^{N-1} \mid \frac{1}{\omega + H - E_0^N} \mid \psi_{m'}^{N-1} \rangle \\ & \langle \psi_{m'}^{N-1} \mid c_i \mid \psi_0^N \rangle \end{aligned} \tag{25}$$

where $|\psi_0^N\rangle$ is the ground state of the N electron system and $|\psi_m^{N+1}\rangle$, $|\psi_m^{N-1}\rangle$ are the eigenstates of the (N+1) and (N-1) electron system, respectively. Furthermore, E_0^N is the energy of $|\psi_0^N\rangle$ and the creation and annihilation operators c_j^+, c_i refer to the spin orbitals i and j. In a solid the most important excitations form quasiparticle energy bands in which case the indices i and j stand for $\mathbf{k}$, ν, σ denoting the momentum of the quasiparticle, a band index and the spin. From (25) it follows that, for example, the valence bands of a semiconductor are given by

$$\omega_{\mathbf{k}\nu} = E_0^N - \langle \psi_{\mathbf{k}\nu\sigma}^{N-1} \mid H \mid \psi_{\mathbf{k}\nu\sigma}^{N-1} \rangle \quad . \tag{26}$$

Let us define hole states $|\mathbf{R}n\sigma\}$ via the decomposition

$$| \psi_{\mathbf{k}\nu\sigma}^{N-1} \rangle = \frac{1}{\sqrt{N_0}} \sum_{\mathbf{R}n} \alpha_{\nu n}(\mathbf{k}) e^{i\mathbf{kR}} \mid \mathbf{R}n\sigma \} \tag{27}$$

where the quasihole refers to the localized spin orbital $(\mathbf{R}, n, \sigma)$. Here n denotes the different localized states within a unit cell centered at the lattice vector **R**. With the help of the identity

$$\{\mathbf{0}n'\sigma \mid H \mid \mathbf{R}n\sigma\} = \langle \mathbf{0}n'\sigma \mid H \mid \mathbf{R}n\sigma\}^c \tag{28}$$

where c denotes taking the cumulant we may write

$$\epsilon_{\mathbf{k}\nu} = \sum_{\mathbf{R}} \sum_{nn'} \alpha_{\nu n}(\mathbf{k}) \alpha^*_{\nu n'}(\mathbf{k}) e^{i\mathbf{kR}} \{\mathbf{0}n'\sigma \mid H \mid \mathbf{R}n\sigma\} - E_0^N \quad . \tag{29}$$

Here we have introduced $\epsilon_{\mathbf{k}\nu} = -\omega_{\mathbf{k}\nu}$ in order to follow the conventional notation. The correlation contribution to the valence bands becomes

$$\begin{aligned} \Delta\epsilon_{\mathbf{k}\nu} &= \sum_{\mathbf{R}} \sum_{nn'} \alpha_{\nu n}(\mathbf{k}) \alpha^*_{\nu n'}(\mathbf{k}) e^{i\mathbf{kR}} \langle \mathbf{0}n'\sigma \mid H_{\text{res}}\, S \mid \mathbf{R}n\sigma\rangle^c - E_{\text{corr}} \\ &= \sum_{\mathbf{R}} \sum_{nn'} \alpha_{\nu n}(\mathbf{k}) \alpha^*_{\nu n'}(\mathbf{k}) e^{i\mathbf{kR}} (c_{\mathbf{0}n'\sigma} \mid H_{\text{res}}\, S c_{\mathbf{R}n\sigma}) \end{aligned} \tag{30}$$

where E_{corr} is the correlation energy contribution to E_0^N and $|Rn\sigma\rangle$ is the state where a hole has been generated in the SCF ground state $|\Phi_{\text{SCF}}^N\rangle$. As before S = Ω-1. We can evaluate (30) by applying again the method of increments [27]. In a first step we keep all electrons frozen except those in orbitals (**R**, n) and (**0**, n') which we correlate. This gives us a contribution to the matrix element $\langle \mathbf{0}n'\sigma \mid H_{\text{res}}S \mid \mathbf{R}n\sigma\rangle^c$. After that we do the corresponding calculations for $\langle \Phi_{\text{SCF}}^N \mid H_{\text{res}}\text{S} \mid \Phi_{\text{SCF}}^N\rangle^c$. The difference of the two matrix elements yields $\Delta H^{(0)}_{\mathbf{R}nn'} = (c_{\mathbf{0}n'\sigma} \mid H_{\text{res}}S\ c_{\mathbf{R}n\sigma})$ to lowest order. Next we release electrons also in orbital $(\mathbf{R}_1, n_1)$. This leads to an increment $\Delta H^{(\mathbf{R}_1 n_1)}_{\mathbf{R},nn'}$ to $(c_{\mathbf{0}n'\sigma} \mid H_{\text{res}}S\ c_{\mathbf{R}n\sigma})$. When this is continued we obtain

$$(c_{\mathbf{0}n'\sigma} \mid H_{\text{res}} S c_{\mathbf{R}n\sigma}) = \Delta H^{(0)}_{\mathbf{R}nn'} + \sum_{\mathbf{R}_1 n_1} \Delta H^{(\mathbf{R}_1 n_1)}_{\mathbf{R},nn'} + \ldots \quad . \tag{31}$$

The long-range part of the polarization cloud generated by the hole is taken into account by a continuum approximation. If we neglect the polarization outside a sphere of radius R, the resulting energy ϵ(R) is related to the true polarization energy $\epsilon(\infty)$ through $\epsilon(R) = \epsilon(\infty) - c/R$.

By calculating $\epsilon(\mathrm{R})$ for two different radii R_1 and R_2 we can determine $\epsilon(\infty)$. For more details we refer to [27]. Results for the valence bands of silicon are shown in Fig. 5. One notices considerable changes in the band width due to correlations.

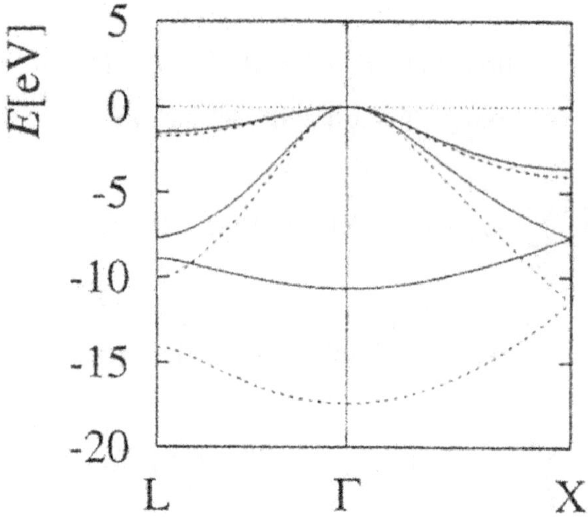

Fig. 5: *The SCF (dashed lines) and correlated (solid lines) band structures for silicon (From [27])*

Of particular interest are the excitations of strongly correlated electrons. Spectral densities are quantities which provide important informations about the latter. For the purpose of determining them we define a retarded Green's function matrix $\mathrm{G}_{mn}(\mathbf{k},\mathrm{t})$ with respect to a set of dynamical variables $\mathrm{A}_m(\mathbf{k})$. One of these variables is the creation operator $c^+_{\mathbf{k}\nu\sigma}$ with indices as explained before. The matrix has the form

$$G_{mn}(\mathbf{k},t) = -i\Theta(t)\ \langle\psi_0 \mid [A^+_m(\mathbf{k},t),\ A_n(\mathbf{k})]_+ \mid \psi_0\rangle \tag{32}$$

where $\Theta(\mathrm{t})$ is the step function and the corresponding spectral functions $\mathrm{A}_{mn}(\mathbf{k},\omega>0)$ are defined as

$$A_{mn}(\mathbf{k},\omega) = -\frac{1}{\pi}\lim_{\eta\to 0} Im\ \{G_{mn}(\mathbf{k},\omega+i\eta)\},\quad \omega \geq 0\quad . \tag{33}$$

Note that for $\omega<0$ the advanced Green's function has to be used instead of the retarded one. We identify with the Fourier transform of the $\mathrm{A}_m(\mathbf{k})$ those local operators, which couple most strongly to an electron $c^+_{\mathbf{k}\nu\sigma}$ (or hole) added to the system. Examples are found, e.g., in [5].

After Fourier transforming (32) and using (13) we can write

$$G_{mn}(\mathbf{k}, z) = (\Omega \mid \left[A_m^+(\mathbf{k}), \left(\frac{1}{z-L}A_n(\mathbf{k})\right)^\bullet\right]_+ \mid \Omega) \quad . \tag{34}$$

Here L is the Liouvillean which acts on arbitrary operators according to LA=[H,A]$_-$. The notation $(AB)^\bullet$ indicates that the operator product has to be treated as an entity when the cumulant is calculated. A formal solution of (34) is

$$\mathcal{G}(\mathbf{k}) = \chi(\mathbf{k})\left[z\chi(\mathbf{k}) - \omega(\mathbf{k})\right]^{-1}\chi(\mathbf{k}) \quad . \tag{35}$$

The susceptibility- and frequency matrices are defined by

$$\begin{aligned} \chi_{mn}(\mathbf{k}) &= \left(\Omega\middle|\left[A_m^+(\mathbf{k}), A_n(\mathbf{k})\right]_+\Omega\right) \\ \omega_{mn}(\mathbf{k}) &= \left(\Omega\middle|\left[A_m^+(\mathbf{k}), (LA_n(\mathbf{k}))^\bullet\right]_+\Omega\right) \quad . \end{aligned} \tag{36}$$

For each **k** point those matrices have to be calculated separately. Their dimension is given by the number of dynamical variables $A_m(\mathbf{k})$. For example, when we describe Ni metal by using a five-band Hubbard Hamiltonian for the d electrons it turns out that the dimension of the two matrices is 66 x 66 [28]. By calculating the spectral density of the d electrons the band narrowing due to correlations can be explained as well as the appearance of a pronounced satellite peak approximately 7eV below the top of the d bands.

It remains a challenging task for the future to calculate spectral densities for strongly correlated electrons with the same accuracy as we can do it at present for weakly correlated electron systems like silicon. Until now calculations have been limited to model Hamiltonians like the five-band Hubbard Hamiltonian for Ni metal, or a three-band Hubbard Hamiltonian for the Cu-O planes of the high-T_c cuprates [29].

CONCLUSIONS

We expect that wavefunction calculations for solids based on quantum chemical techniques will gain increasing attention in the future. Without them our understanding of electron correlations will remain rudimentary. As was shown in the previous sections such calculations are feasible and in fact have been performed for a number of cases. Nevertheless, this field is still at its beginning. The quantum chemical framework has to be properly modified, of course, in order to be applicable to solids. The technique of partitioning and projections is in this respect an important tool. Löwdin has recognized and propagated it many years ago [30, 31], and the Uppsala group has done pioneering work to show its importance (see, e.g., Refs. [32, 33]). It is one of very many ideas and contributions for which Per-Olov Löwdin will be remembered.

References

[1] Program package CRYSTAL, described in Pisani, C., Dovesi, R., Roetti, C. *"Hartree-Fock Ab Initio Treatment of Crystalline Systems"*, Lect. Notes Chem., Vol. 48 (Springer, Berlin, Heidelberg 1988).

[2] Shukla, A., Dolg, M., Stoll, H., Fulde, P., Chem. Phys. Lett. 262, 1996 *and* Phys. Rev. B 57, 1998.

[3] Kubo, R., J. Phys. oc. Jpn. 17, 1962.

[4] Schork, T., Fulde, P., J. Chem. Phys. 97, 1992.

[5] Fulde, Peter *"Electron Correlations in Molecules and Solids"* 3rd ed., Springer, Berlin, Heidelberg, 1995.

[6] Cizek, J., Adv. Chem. Phys. 14, 1969.

[7] Kümmel, H., Lührmann K.H., Zabolitzky, J.G., Phys. Lett. C 36, 1978.

[8] Bishop, R.T., Theor. Chim. Acta 80, 1991.

[9] Faddeev, L.D., Zh. Eksp. Teor. Fiz. 39, 1960, [Engl. transl.: Sov. Phys. - JETP 12, 1961].

[10] Stoll, H., Phys. Rev. B 46, 1992 *and* Chem. Phys. Lett. 191, 1992.

[11] MOLPRO is a package of *ab initio* programs written by Werner H.-J., Knowles, P.J., with contributions from Almlöf, J., Amos, R., Deegan, M.J.O., Elbert, S.T., Hampel, C., Meyer, W., Peterson, K., Pitzer, R., Stone A.J., Taylor, P.R.

[12] GAMESS - Schmidt, M.W., Baldridge, K.K., Boatz, J.A., Elbert, S.T., Gordon, M.S., Jansen, J.J., Koseki, S., Matsunaga, N., Nguyen, K.A., Su, S., Windus, T.L., Dupuis, M., Montgomery, J.A., J. Comput. Chem. 14, 1993.

[13] Fulde, P., Adv. Phys. 51, No. 2, 2002.

[14] Paulus, B., Fulde P., Stoll, H., Phys. Rev. B 51, 1995.

[15] Paulus, B., Fulde P., Stoll, H., Phys. Rev. B 54, 1996.

[16] Albrecht, M., Paulus, B., Stoll, H., Phys. Rev. B 56, 1997.

[17] Doll, K., Dolg, M., Fulde, P., Stoll, H., Phys. Rev. B 52, 1995.

[18] Doll, K., Dolg, M., Stoll, H., Phys. Rev. B 54, 1996.

[19] Doll, K., Dolg, M., Fulde, P., Stoll, H., Phys. Rev. B 55, 1997.

[20] Rośiszewski, K., Doll, K., Paulus, B., Fulde, P., Stoll, H., Phys. Rev. B 57, 1998.

[21] Rośiszewski, K., Paulus, B., Fulde, P., Stoll, H., Phys. Rev. B 60, 1999.

[22] Kalvoda, S., Dolg, M., Flad, H.-J., Fulde, P., Stoll, H., Phys. Rev. B 57, 1998.

[23] Yin, M.T., Cohen, M.L., Phys. Rev. B 24, 1981; Harrison, W.A., *ibid.* 23, 1981, and references cited therein

[24] Fahy, S., Wang, X.W., Louie, S.G., Phys. Rev. B 42, 1990.

[25] Glötzel, D., Segall, B., Andersen, O.K., Solid State Comm. 36, 1980.

[26] Juan, Y.-M., Kaxiras, E., Gordon, R.G., Phys. Rev. B 48, 1993.

[27] Gräfenstein, J., Stoll, H., Fulde, P., Phys. Rev. B 55, 1997.

[28] Unger, P., Igarashi, J., Fulde, P., Phys. Rev. B 50, 1994.

[29] Unger, P., Fulde, P., Phys. Rev. B 47, 1993; *ibid* B 48, 1993.

[30] Löwdin, P.O., J. Mol. Spectrosc. 10, 1963, *ibid* 13, 1964, *ibid* 14, 1964 and Phys. Scr. 21, 1980.

[31] Löwdin, Per-Olov in *"Supercomputer Simulations in Chemistry"* ed. by Dupuis, M., *"Lecture Notes in Chemistry"* Vol. 44, p. 1, Springer, Berlin, Heidelberg, 1986.

[32] Pickup, B.T., Goscinski, O., Mol. Phys. 26, 1973.

[33] Linderberg, T., Öhrn, Y., *Propagation in Quantum Chemistry*, Academic Press, London, 1973.

SUPERCONDUCTING AND SPIN GAPS IN $d_{x^2-y^2}$-WAVE HIGH T_C CUPRATES

Lawrence J. Dunne [a,b] and Erkki J. Brändas [c]

[a] School of Computing, Information Systems and Mathematics, South Bank University, London SE1 0AA, UK
[b] School of Chemistry, Physics and Environmental Science, University of Sussex, Falmer, Brighton, BN1 9QJ, UK
[c] Quantum Chemistry Department, University of Uppsala, Box 518, Uppsala, S-75120, Sweden

Abstract In this paper a theory is presented which accounts in a straightforward way for the existence of the condensate and spin gap in high T_C superconductive cuprates. The synthesis of many types of high T_C superconducting cuprates occurred over a decade ago but the cause of the superconducting condensation and electronic structure of such materials is still a matter of significant deliberation and there is no consensus as to the origin of pairing or the cause of the spin gap. In our study these effects arise as a correlation effect in a weakly doped electronic system within the cuprate environment.

1. INTRODUCTION

Since the finding of high T_c superconductive cuprates the nature of the superconducting condensate and electronic structure of such materials is still a matter of significant argument (1-32). Furthermore, the origin of the spin gap and its relation to the superconductive energy gap is still unclear. In this paper a theory is proposed which accounts in a straightforward way for the existence of the condensate and spin gap in high T_c superconductive cuprates. In our study these gaps arise as a correlation effect in a weakly doped electronic system. We study the properties of a Bohm-Pines type Hamiltonian for Fermions on a lattice embedded in a dielectric medium, with the plasma modes removed (30). The short range electronic interactions may have a repulsive core and at long range exhibit Friedel oscillations or some other long-range attractive tail due perhaps to some other virtual Boson-exchange mechanism. The well-known data which we seek to interpret in a single scheme is shown in Figure 1 which is reproduced from reference (2).

E.J. Brändas and E.S. Kryachko (eds.),
Fundamental World of Quantum Chemistry, Vol. II, 257–270.

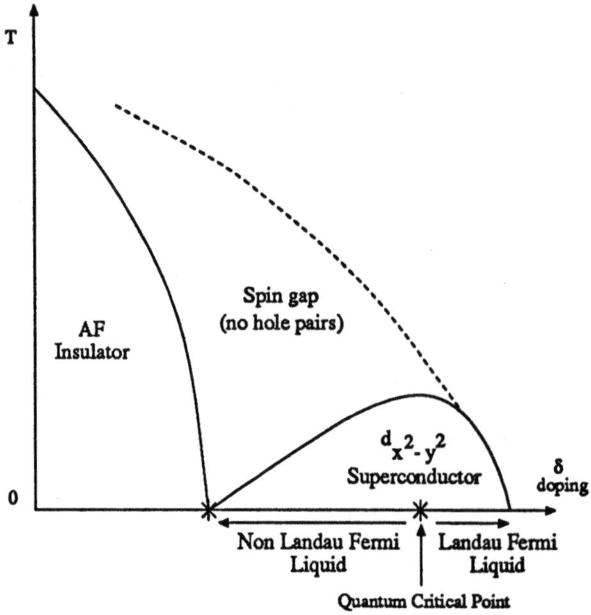

Figure 1 Schematic phase diagram for superconducting cuprates taken from reference (2)

The fundamental structural feature of n or p type copper oxide superconductors are the layers shown in Figure 2 where the quadratic configuration of lattice points is centred on a Cu atom shown in black.

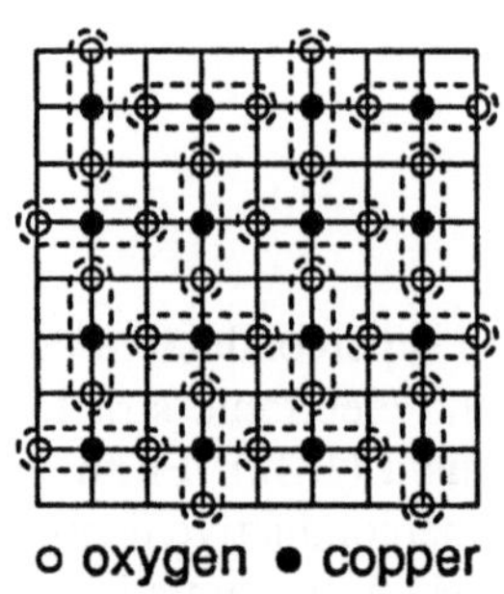

Figure 2 Copper/oxygen layer of Superconducting cuprates. Black and white circles represent copper and oxygen atoms respectively. A pair of E-basis Wannier functions, which are a linear combination of Cu and O atomic orbitals, transforming as (x, y) are localised on every unit cell.

Each cuprate layer is a four-sided arrangement of N/2 lattice points with local C_{4v} point group symmetry about the Cu atoms and with integer index l. The need for pairs of electrons to avoid the short range repulsion and to reside in the longer range attractive region of a potential can enable the system to adopt a low energy superconductive electronic state. Thus our theory presented in this paper is based on the correlation of electrons interacting with a short-range screened Coulomb repulsion and perhaps a longer-range attraction.

We adopt an extended Hubbard Hamiltonian postulated for the active space of the form

$$\mathcal{H}=\sum_{l,m}\langle l|\mathrm{h}|m\rangle a_l^{+}a_m+\frac{1}{2}\sum_{k,l,m,n}\langle kl|\mathrm{v}|mn\rangle a_k^{+}a_l^{+}a_m a_n \tag{1}$$

where h and v are pertinent one and two body parts of the effective Hamiltonian. We will presume that the effective pair interaction has the form $v(i,j) \approx \exp(-\gamma r_{ij})e^2/Dr_{ij}$ plus a long range attractive tail. Because of the short range repulsion v has positive off-diagonal matrix elements for electron pair transfers between nearest neighbour Wannier orbitals on a lattice. v is the same form as discussed in reference (30) but where γ is the Thomas-Fermi screening constant calculated for the carriers induced by doping and D is the background dielectric constant discussed by Mott and Davis (33). This background dielectric constant D is discussed in detail in references (30,33) and has been exhaustively studied in the Mott-Edwards-Sienko relationship $n_c^{1/3}a^* \sim 1/4$ in connection with metal-insulator transitions in doped dielectric media (10-11,30-34). Here n_c is the critical density and $a^* = a_0 D$ is the effective radius of the Fermion bound to a donor centre and where a_0 is the Bohr radius.

Electrons or holes introduced onto the layer must have a counter ion in an electrically neutral system. At low doping electrons or holes will be bound to such a centre and unable to move freely between layers. The origin of the spin gap in this model is simply related to the energy to free a Fermion from its binding centre so that at some stage of doping a metal-insulator transition may occur. In reality this may be anisotropic caused by the differing degrees of orbital overlap in these layer compounds. The spin gap can be interpreted as the energy to excite carriers beyond the mobility shoulders into delocalised or extended states. At low carrier densities it turns out that the spin gap in our theory is the energy to excite carriers across the mobility shoulder into delocalised states extending across the interlayer space. The conductivity above T_c at the point where the spin gap disappears should have a value approximately given by the Mott-Edwards-Sienko minimum metallic conductivity (10-11,31-33). In some respects the theory has similarities and builds on that discussed by Alexandrov (8-9)

2. CUPRATE LAYERS ELECTRONS AND OFF-DIAGONAL LONG-RANGE ORDER

Our theory for the superconductive state arising from repulsive electronic correlations is based on an straightforward mathematical argument given by us previously (18-20) but which is short enough that it bears repetition in this paper. In the usual BCS theory of conventional superconductivity with attractive off-diagonal matrix elements of the phonon-induced electron-electron interaction, it is commonplace to find the existence of the superconductive ground state explained by so-called n-fold stabilisation. Thus, if there are n quasi-degenerate states at energy ε above the Fermi level and these are coupled by an attractive matrix element $-v$ then the Hamiltonian matrix has $(n-1)$ eigenvalues at energy ε and one which turns out to be the superconductive state, which can cross below the Fermi level with an energy $\varepsilon - nV$. However, such a stabilisation can also occur with off-diagonal matrix elements which have the opposite sign and are repulsive. A straightforward demonstration of this is obtained by considering the symmetrical eigen-problem shown below

$$\left(\begin{array}{ccc|ccc} \varepsilon & & \mathbf{0} & & & \\ & \ddots & & & \mathbf{V} & \\ \mathbf{0} & & \varepsilon & & & \\ \hline & & & \varepsilon & & \mathbf{0} \\ & \mathbf{V}^T & & & \ddots & \\ & & & \mathbf{0} & & \varepsilon \end{array}\right) \left(\begin{array}{c} \cdot \\ 1 \\ \cdot \\ \hline \cdot \\ \pm 1 \\ \cdot \end{array}\right) = (\varepsilon \pm \kappa V) \left(\begin{array}{c} \cdot \\ 1 \\ \cdot \\ \hline \cdot \\ \pm 1 \\ \cdot \end{array}\right) \qquad (2)$$

where the sub-blocks have κ rows and columns. The lowest eigenvalue at energy $\varepsilon - kv$ splits off from the other eigenvalues with energy ϵ and, if this eigenvalue is negative then it may become the ground state. The Hamiltonian matrix shown in eqn.2 seems to be directly relevant to doped cuprate sheets. It can be shown in these two cases that, if the basis functions are pairwise occupied as in a BCS wavefunction, then both lowest states have a macroscopic number of electron pairs characteristic of the superconductive state. Following Yang (34) or Coleman (35), or Gor'kov (36) the second order reduced electronic density matrix must, and here does have, one macroscopically large eigenvalue representing a superconductive condensate of pairs. This will be shown below. A review is given in ref.(40).

The experimentally proved (12,14,16) d-wave symmetry of the wavefunction describing the condensed electrons in hole-doped superconductive cuprates can be subjected to a group theoretical analysis (4,20) which shows that such a symmetry cannot originate in a single band description of the copper oxide superconductors. Our preference, argued in references (4,20), is for a pair of localised E-basis functions in the C_{4v} group (37) associated with each unit cell which are symmetry adapted combinations of Cu e-orbitals and in-plane oxygen p_z orbitals. Apical oxygen p_x, p_y orbitals (8) too have such symmetry and one can combine all these orbitals of the correct symmetry to give a pair of E-basis functions on every lattice point. However, recent field effect experiments on simple cuprate layers (7) without apical oxygen show that such p_x and p_y orbitals are not essential for high Tc cuprate superconductivity.

We assume then that there are a pair of localised Wannier type orbitals centred on each cell l of the form $\{\phi_{l,x}(r), \phi_{l,y}(r)\}$ transforming together as the E degenerate irreducible representation in the C_{4v} point group. For mathematical analysis every pair of l, x orbitals is assigned a signature $(-1)^{l+1}$ and $(-1)^l$ for every pair of l, y orbitals so that each pair of e-orbitals on the same unit cell has opposite signatures. Hole doping is supposed to occur from a full valence band and electron doping into a vacant conduction band formed from the appropriate set of Wannier functions. Although we will not dwell on this point here following Alexandrov (8) the important features of the angular resolved photoemission spectra discussed by Shen et al. (6) can be reproduced using the E-orbital assumption. The flatness of the bands which are a widely discussed feature of copper oxide superconductors are shown in reference (20). The band dispersion along the c-axis of the cuprates is supposed to be small.

We adopt an extended Hubbard Hamiltonian postulated for the active space discussed above. The summation in eqn.1 is over cell orbitals and spins and where only on-cell and nearest neighbour matrix elements will be taken into account as significant in a localised basis.

If we suppose that there are N orbitals in the cuprate sheet accessible to $2M$ electrons the degree of doping is given by $\rho = M/N$ where for hole doping a full set of cells with $\rho = 1$, and for electron doping $\rho = 0$, will be important. Cell orbitals in the cuprate sheet can be occupied by a pair of singlet coupled electrons (a dimer ($\uparrow\downarrow$)) or singly occupied by single unpaired electron with down ($\downarrow$) or up ($\uparrow$) spin. Every configuration of electrons in the cuprate layer active orbitals can be represented by a Slater determinant and an approximate many electron wavefunction obtained from diagonalisation of of the Hamiltonian matrix represented in the basis of Slater determinants. Although an exact diagonalisation of such a Hamiltonian matrix seems impossible we have devised a mean-field theory which provides an insight into the behaviour of high T_c cuprate superconductors.

Every Slater determinant basis function, represented in the Hamiltonian matrix, corresponds to a unique configuration of electrons on the cuprate sheet and these can be grouped into $N!/(N-M\uparrow-M\downarrow)!M\uparrow!M\downarrow!$ sets corresponding to the possible ways of combining the spin down ($\downarrow$) and up ($\uparrow$) spin monomers in equal numbers and to sub-classify the grouping of the remaining $(N-M\uparrow-M\downarrow)!/(N-M\uparrow-M\downarrow-M_D!)M_D!$ dimer ($\downarrow\uparrow$) configurations into 2 groups assigned by the $+/-$ sign of the product of the occupied dimer orbital signatures as described in detail in references (18-19). Neglecting the coupling between blocks the Hamiltonian matrix has the form shown in eqn.3.

The off-diagonal matrix elements are most significant for pair transfers between Slater determinant basis functions of opposite signature. It is important to note that every sub-block in the Hamiltonian matrix has the form displayed in eqn.3.

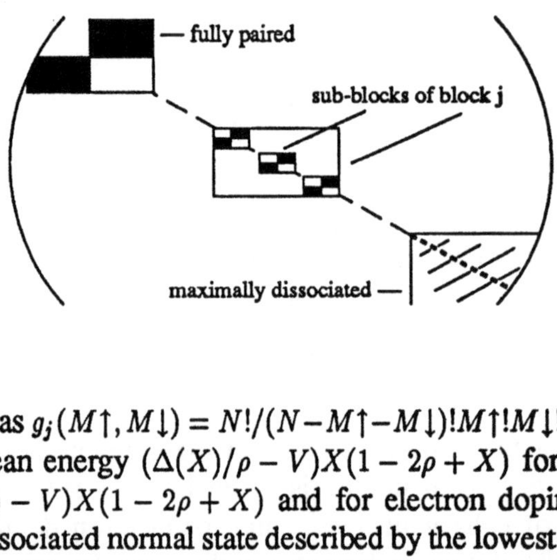

(3)

Each block j has $g_j(M\uparrow,M\downarrow)=N!/(N-M\uparrow-M\downarrow)!M\uparrow!M\downarrow!$ quasi-degenerate states with mean energy $(\Delta(X)/\rho-V)X(1-2\rho+X)$ for hole doping and $(\Delta(X)/(1-\rho)-V)X(1-2\rho+X)$ and for electron doping relative to the completely dissociated normal state described by the lowest block on the right hand side of the Hamiltonian matrix. $\Delta(X)$ is the screened diagonal Coulomb repulsion for bringing electrons together into the same orbital and $X=M_D/N$ and V is equal to the sum of nearest neighbour pair transfer matrix elements and is repulsive. D is the screened diagonal Coulomb repulsion for bringing electrons into the same orbital when its neighbours are fully paired. The linearisation procedure discussed in reference (18) and used above is likely to be reliable at low doping only.

The ground state wavefunction is the geminal power given by

$$(N!/(N-M)!M!)^{-\frac{1}{2}}(1/M!)\left[\sum_l \exp(i\pi l)(a^+_{lx\uparrow}a^+_{lx\downarrow} - a^+_{ly\uparrow}a^+_{ly\downarrow})\right]^M |0\rangle \quad (4)$$

To deduce some thermal properties of the superconductor excited state energies and wavefunctions are necessary which are obtained as follows. Every block j of the Hamiltonian matrix has $g_j(M\uparrow, M\downarrow)$ low-lying states with eigenvectors which are a linear combination of the sub-block degenerate states which may be brought together to give excited state wavefunctions given by

$$(N!/M_D!(N-M\uparrow - M\downarrow - M_D)!M\uparrow!M\downarrow!)^{-\frac{1}{2}}\sum_s c_s\psi_s \quad (5)$$

where $|c_s| = 1$ and $s = 1, 2, .., g_j(M\uparrow, M\downarrow)$ refers to the sub-block s with eigenvector ψ_s of the block j of our concern. The above wavefunction describes a general state of the superconductor for any degree of pair dissociation up to T_c .

The second order electronic density matrix can be constructed from the Hamiltonian matrix eigenvector given in eqn(5) above to give

$$\begin{pmatrix} M_D/N & -\frac{M_D(N-2M+M_D)}{N(N-1)} & \cdots & \cdots \\ -\frac{M_D(N-2M+M_D)}{N(N-1)} & M_D/N & \cdots & \cdots \\ \vdots & \vdots & \ddots & \vdots \\ \cdots & \cdots & \cdots & M_D/N \end{pmatrix} \times \begin{pmatrix} 1 \\ -1 \\ \vdots \\ \cdots \end{pmatrix}$$

$$= [X + M_D(1 - 2\rho + X)] \begin{pmatrix} 1 \\ -1 \\ \vdots \\ \cdots \end{pmatrix} \quad (6)$$

This can be seen to have a macroscopically large eigenvalue signifying a condensate of electron pairs. All $g_j(M\uparrow, M\downarrow)$ low-lying states have the same density matrix eigenvector and eigenvalue.

Consequently the thermally averaged second order reduced electronic density matrix eigenvector is

$$\psi(\mathbf{r}_1\sigma_1, \mathbf{r}_2\sigma_2)$$
$$= \lambda\sum_l(-1)^l[\phi_{lx}(\mathbf{r}_1)\phi_{lx}(\mathbf{r}_2) - \phi_{ly}(\mathbf{r}_2)\phi_{ly}(\mathbf{r}_1)](\uparrow(1)\downarrow(2) - \uparrow(2)\downarrow(1)) \quad (7)$$

which has $d_{x^2-y^2}$ pair symmetry in the C_{4v} group in conformity with determinations of the condensate symmetry in High T_c superconducting cuprates. The function $\psi(\mathbf{r}_1\sigma_1, \mathbf{r}_2\sigma_2)$ transforms as the B_1 irreducible representation in the C_{4v} group. λ is the macroscopically large reduced second order electronic density matrix eigenvalue corresponding to a population $M_D(1-2\rho+X)$ condensed electron pairs where X is calculated from the Helmholtz free energy F defined below.

The distance range over which $\psi(\mathbf{r}_1\sigma_1, \mathbf{r}_2\sigma_2)$ as $\mathbf{r}_1 - \mathbf{r}_2$ increases is a measure of the pair size or coherence length and in this theory is of the order of a few Angströms.

3. T_c FOR SUPERCONDUCTING CUPRATES

Following the reasoning of reference (18) the Helmholtz free energy/orbital, F, is given up to T_c by

$$F = kT[(1-2\rho+2X)\ln(1-2\rho+2X) + 2(\rho - X)\ln(\rho - X)] \\ + [X + \Delta/t(1-2\rho+2X)] \tag{8}$$

where $t = \rho$ when $\rho > 1/2$ and $t = (1-\rho)$ when $\rho < 1/2$. This is derived using the degeneracy factor $g_j(M_\uparrow, M_\downarrow)$ and energy relative to the normal state E_j for each block j given above and the formula $F = -(kT/N)\ln Q$ where $Q = \sum g_j \exp(-E_j/kT)$ is the Canonical partition function.and where the sum is evaluated by the method of the maximum term. Therefore minimisation of F with respect to X gives the equilibrium condition as a solution of the non-linear equation

$$2kT\ln[(1-2\rho+2X)/(\rho - X)] + [-V + \Delta/t)(1-2\rho+2X) = 0 \tag{9}$$

The root of this non-linear equation gives the equilibrium value of X and hence the density of condensed electrons. The transition temperature can be located by noting that at T_c the equilibrium values of X are constrained by $X = 0$ for electron doping and $X = 2\rho - 1$ for hole doping corresponding to zero density of electronic condensate.

Hence the transition temperature T_c is given by

$$kT_c = (V - \Delta/(1-\rho))(1-2\rho)/2\ln[(1-2\rho)/\rho] \tag{10}$$

for electron doping when $\rho < 1/2$ and

$$kT_c = (V - \Delta/\rho)(2\rho - 1)/2\ln[(2\rho - 1)/(1 - \rho)] \qquad (11)$$

for hole doping when $\rho > 1/2$.

The doping phase diagram given by eqns(10-11) is shown in Figure 3 which reproduces the well-known electron-hole symmetry. For reasonable values of matrix elements and background dielectric constant D given in the caption to Figure 3, our theory depicts superconductivity at high temperatures in the same doping range as for copper oxide superconductors.

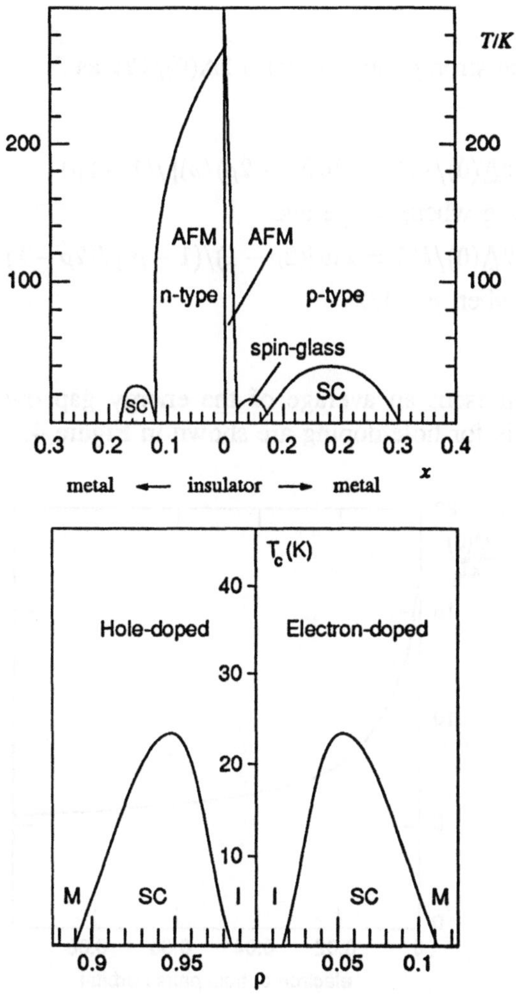

Figure 3 Theoretical phase diagram and doping dependence of the transition temperature predicted by our theory for the parameters $D = 5$, $V = 0.564\text{eV}$ compared with the well known experimental doping T_c curve shown above.

The doping dependence of energy gap and strong coupling ratio $2\Delta(0)/kT_c$ be derived as follows.The energy to remove one electron pair from the condensate at absolute zero is

$$2\Delta(0) = V - \Delta/(1-\rho) \tag{12}$$

for electron doping when $\rho < 1/2$ and

$$2\Delta(0) = V - \Delta/\rho \tag{13}$$

for hole doping when $\rho > 1/2$

This the gives the strong coupling ratio $2\Delta(0)/kT_c$ as

$$2\Delta(0)/kT_c = 2\ln\left[(1-2\rho)/\rho)\right]/(1-2\rho) \tag{14}$$

for electron doping when$\rho < 1/2$ and

$$2\Delta(0)/kT_c = 2\ln\left[(2\rho-1)/(1-\rho)\right]/(2\rho-1) \tag{15}$$

for hole doping when $\rho > 1/2$

These results represent an average of the energy gap over k-space. The predictions of this for hole doping are shown in Figure 4.

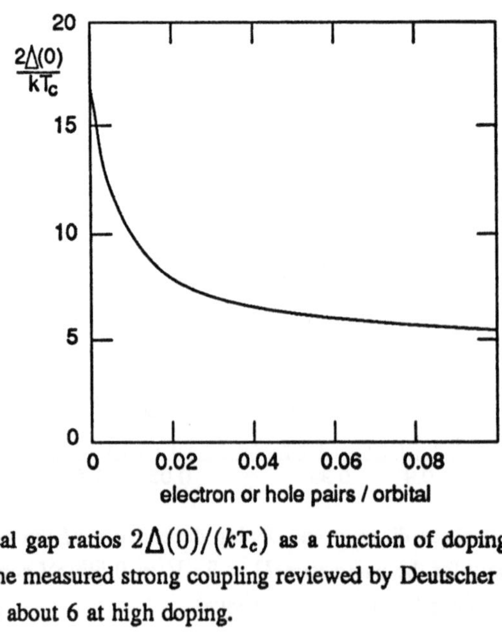

Figure 4 Theoretical gap ratios $2\Delta(0)/(kT_c)$ as a function of doping level for superconducting cuprates. The measured strong coupling reviewed by Deutscher (5) range from about 20 at low doping to about 6 at high doping.

A compilation of gap data given by Deutscher (5) shows the strong coupling ratio in cuprate superconductors varies from about 20 to 6 and falls with rising doping levels as predicted by our model.

4. THE ORIGIN OF THE SPIN GAP AND METAL-INSULATOR TRANSITIONS

If we assume that the band dispersion is small we may interpret the spin gap as the energy to unbind a carrier from its counter ion. We estimate this by numerically solving the radial Schrödinger equation for an electron or hole in an s-state attracted to an oppositely charged centre with the potential $v(r) \approx \exp(-\gamma r)e^2/Dr$. The screening length varies with doping according to the Thomas-Fermi relation given by Mott and Davis (33) but the backgound dielectric constant is fixed. A review of some papers on disorder in high temperature superconductors is given in ref,(39).

We also plot the spin gap in Figure 5 as a function of doping. While we do not take the numerical values derived theoretically too seriously, the behaviour is quite close to that observed experimentally for many cuprates with the spin gap collapsing to zero or a small value near to the optimum doping level. Thus we come close to a metal-insulator transition.

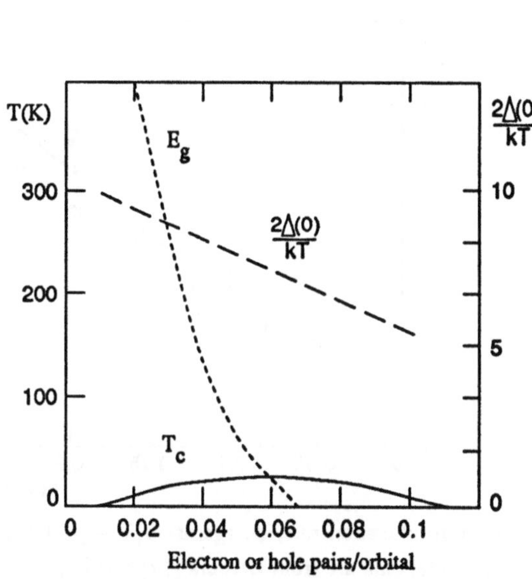

Figure 5 Doping dependence of Tc , spin gap and the strong coupling ratio $2\Delta(0)/kT_c$ for the parameters $D = 5$, $V = 0.564$eV.

Following the Mott-Edwards-Sienko relation $n_c^{1/3}a^* \sim 1/4$ for $D = 5$ this transition should occur when the dopant concentration is about 8×10^{26} carriers/m^3 with the conductivity close to Mott's minimum metallic conductivity (10,11,33) calculated to be roughly 100 Ohm cm^{-1}. The data reviewed by Edwards and Rao (10,11) gives values close to what we expect at the point where the spin gap disappears. When the spin gap is finite we expect the layers to behave as doped semiconductors above T_c .

In planned future studies we hope to compute the magnetic susceptibility and the temperature dependence conductivity on either side of the metal-insulator transition. We are therefore in accord with Loram et al (29) whose view is that the normal state spin gap is not essentially related to the superconductive pairing but is indicative of a metal-insulator transition at the point where the spin gap disappears.

5. DOPING DEPENDENCE OF DENSITY OF THE ELECTRONIC CONDENSATE

The inference that the zero temperature density of condensed electrons should be proportional to $\rho(1-\rho)$ is tested against experiments in Figure 6 where the concentration of holes is estimated from the chemical formula of a series of copper oxide superconductors by valence counting. It can be observed that the prediction is in broad agreement with experiments (39).

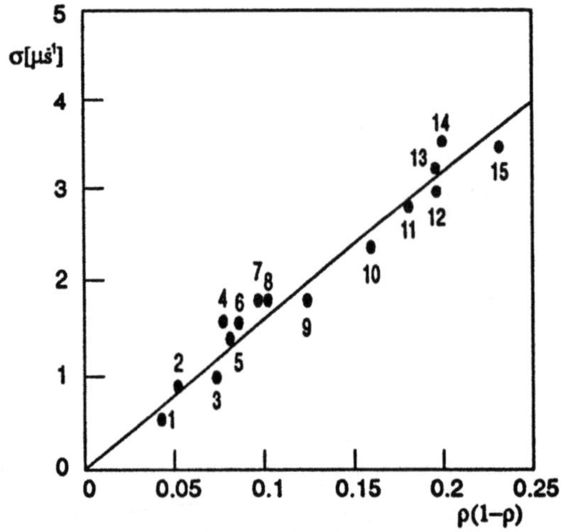

Figure 6 Plot of muon spin relaxation rate, ϕ, against $\rho(1-\rho)$ for $La_{2-x}Sr_xCuO_4$ and $YBa_2Cu_3O_y$. The experimental data are from Uemura [27] and $2\rho(1-\rho)$ is estimated from the chemical formula. There are 14 points starting from the left which correspond to the following x or y values. 1. $x = 0.08$, 2. $x = 0.1$, 3. $x = 0.15$, 4. $x = 0.15$, 5. $y = 6.66$ 6. $y = 6.67$, 7. $x = 0.20$, 8. $x = 0.21$, 9. $y = 6.67$, 10. $y = 6.86$, 11. $y = 6.95$, 12. $y = 6.95$, 13. $y = 7.0$, 14. $y = 7.0$, 15. $Y_{0.7}Ca^{2+}_{0.3}Ba_2Cu_3O_7$.

6. CONCLUSION AND SUMMARY

This paper describes an extension of the repulsive electron correlation model to embrace the d-wave condensate and spin gap. Our conclusion is that high temperature superconductivity is primarily an electron correlation effect possibly supplemented by longer range polaronic attraction of the type discussed by Mott and Alexandrov (see references (8,9) for other references) or Friedel oscillations. The spin gap is associated with the unbinding of holes from their donor centres. We have presumed that the effective pair interaction has the form $v(i,j) \approx \exp(-\gamma r_{ij})e^2/Dr_{ij}$ plus a long range attractive tail with Fiedel oscillations and due to short range repulsion v has positive off-diagonal matrix elements for electron pair transfers between Wannier orbitals. At low carrier densities it turns out that the spin gap in our theory is the donor ionisation energy. The conductivity at the point where the spin gap disappears should be related the Mott-Edwards-Sienko minimum metallic conductivity (10-11,31,32,33). Hence in a single scheme with an idealised Hamiltonian we can account for both the spin gap and superconducting gap as manifestations of strong electronic correlations.

Dedication We dedicate this paper with gratitiude and respect to the memory of our friend and teacher, Professor Per-Olov Löwdin.

REFERENCES

[1] D. Pines (1997), *Physica* C **282-287**, 273-278

[2] T.M Rice (1997), *Physica* C **282-287**, xix-xxiii

[3] B. Batlogg (1997), *Physica* C **282-287**, xxiv-xxx

[4] L J Dunne, E.J Brändas, J.N Murrell, V. Coropceanu (1998), *Solid State Comm.* **108**, 619-623

[5] G. Deutscher (1999), *Nature* **397**, 410-412

[6] Z.X Shen, W.E Spicer, D.M King, D.S Dessau and B.O. Wells (1995), *Science* **267**, 343-350

[7] J.H Schön, M. Dorget, F.C Beuran, X.Z Zu, E. Arushanov, C. Deville Cavellin, and M. Laguës (2001), *Nature* **414**, 434-436

[8] A.S Alexandrov (1996), *Phys Rev.* **53 B**, 2863-2869

[9] A.S Alexandrov (1998), *Phil. Trans. Roy. Soc.* **A 356**, 197-209

[10] P.P Edwards, R.L Johnston, C.N.R Rao, D.P Tunstall and F. Hensel (1998), *Phil. Trans. Roy. Soc.* **A 356**, 5-22

[11] C.N.R Rao (1998), *Phil. Trans. Roy. Soc.* **A 356**, 23-29

[12] D.A Wollman, D.J Van Harlingen, W.C. Lee, D. M. Ginsberg, A.J. Leggett (1993), *Phys. Rev. Lett.* **71**, 2134-2137

[13] S.G Ovchinikov (1997), *Uspekhi (Physics)* **40 (10)**, 993-1017

[14] DJ Van Harlingen (1995), *Rev. Mod. Phys.* **67**, 503-513

[15] M. Sigrist, T.M Rice (1995), *Rev. Mod. Phys.* **67**, 503-513

[16] C.C Tsuei, J.R Kirtley (1997), *Physica C* **282-287**, 4-12

[17] DJ Van Harlingen (1997), *Physica* C **282-287**, 128-131

[18] L.. Dunne (1994), *Physica* C **223**, 291-312

[19] LJ Dunne, J.N Murrell, EJ Brändas (1990), *Physica* C **169**, 501-507

[20] LJ Dunne,.E.J Brändas (2001), *Advances in Quantum Chemistry* **40**, 225-247

[21] C.N Yang (1989), *Phys. Rev. Lett.* **63**, 2144-2148

[22] L J Dunne (1997), *Physica C* **282-287**, 1787-1788

[23] LJ Dunne and T.P Spiller (1993), *J. Phys: Condens. Matter* **4**, L563-L566(1992) and **5** 145 (1993)

[24] J.W Loram, K.A Mirza, J.R Cooper, J.L Tallon (1997), *Physica* C **282-287**, 1405-1406

[25] T. Schneider and H. Keller (1992), *Phys. Rev. Lett.* **69**, 3374-3378

[26] R.S Liu, P.P Edwards (1993) *Synthesis and Characterisation of High Temperature Superconductors (1993)*, Trans. Tech. Publ. Ltd, Switzerland, eds. J.J. Pouch et al.

[27] N. Wada, T. Obana, Y. Nakamura, K. Kumaga (1990), *Physica* **B 165-166**, 1341-1342

[28] S. Uchida (1997), *Physica* C **282-287**, 12-18

[29] J.W Loram, K.A Mirza, J.R Cooper (1998) *Research Review, High Temperature Superconductivity*, University of Cambridge ed. W.Y Liang, 77-97

[30] L J. Dunne (1999), *Int. J. Quant. Chem.* **71**, 111-120

[31] P.P Edwards, T.V Ramakrishnan, and C.N.R Rao (1995), *J. Phys. Chem.* **99**, 5228-5239

[32] P.P Edwards, MJ Sienko (1987), *Phys. Rev.* **B17**, 2575-2581

[33] N.F Mott, E.A Davis (1971) *Electronic Processses in Non-Crystalline Materials*, Clarendon Press, Oxford

[34] C.N Yang (1962), *Rev. Mod. Phys.* **34**, 694-704

[35] A.J Coleman (1967), *Can. J. Phys.* **45**, 1271-1273

[36] L.P Gor'kov (1958), *J. Exp. Teoret. Phys.* **38**, 735-739 (Translation JETP **7** (1958) 505-509)

[37] L. Jansen, M. Boon (1967) *Theory of Finite Groups. Applications in Physics*, North-Holland Publishing Co., Amsterdam

[38] Y.J Uemura et al. (1989), *Phys. Rev. Lett.* **62**, 2317-2321

[39] R.W Hill, C. Proust, L. Tailefer, P. Fournier and R.L Greene (1998), *Nature* **414**, 711-715

[40] EJ Brändas (1995) *Dynamics During Spectroscopic Transitions*, Eds. E. Lippert and J.D Macomber, Springer Verlag, Berlin

Quantum Theory of Disordered Chains

Electronic Structure and Properties

Janos J. Ladik

Chair for Theoretical Chemistry and Laboratory of the National Foundation of Cancer Research Friedrich-Alexander University Erlangen-Nürnberg, Egerlandstr. 3, D-91058 Erlangen, Germany

Abstract: In this paper first we discuss the Single Site Coherent Potential Approximation (CPA) with a momentum and energy-dependent self energy and apply it for the calculation of the density of states of the mixed $(SN)_X$ – $(SNH)_X$ system. Next the Negative Factor Counting (NFC) Method is presented both in its simple (tight binding) and ab initio matrix block form. The latter formalism can be applied also to take into account correlation.
As further step the random walk theory of Lax and coworkers (generalized for the case of arbitrary number of basis functions per site) is discussed in some detail. It is shown that by the knowledge of the Anderson localized one-electron wave functions (computed with the aid of the inverse iteration method) one can easily determine the hopping frequencies in a disordered chain. The latter quantities serve as input into the random walk theory which applying Einstein's relation provides the frequency-dependent complex hopping conductivity.
As applications first the DOS are presented of different aperiodic polymers. It turns out that due to the aperiodicity the DOS histograms broaden strongly as compared to the periodic case if the number of component is at least 3. The calculations have shown as that by a small degree of disorder the CPA DOS gives a reasonable good envelope curve of the NFC histogram. This is, however, not the case of the DOS of larger disorder when the CPA curve gives only a rather poor approximation to the NFC histogram.
The hopping conductivities calculated for different proteins (insulin, lysozyme, subtilisni etc.) are comparatively large ($\sigma(\omega) \approx 10^{-2}\Omega^{-1}\mathrm{cm}^{-1}$ at large frequencies as $\omega = 10^{11}\ sec^{-1}$), and they are usually substantially larger in the biologically active form of the protein, than in the inactive one.
The hopping conductivity of disordered nucleotide base pairs is in the order of $5 \cdot 10^{-1}\Omega^{-1}cm^{-1}$ at $\omega = 10^{11}\ sec^{-1}$. The conductivity increases by 2.5 orders of magnitudes if one introduces a better basis and introduces at the NFC stage also correlation in the MP2 level.
The hopping conductivity of aperiodic polynucleotides decreases

E.J. Brändas and E.S. Kryachko (eds.), Fundamental World of Quantum Chemistry, Vol. II, 271–303.

exponentially with decreasing temperature (until about 60K). Around 60K the hopping conductivity becomes hardly depending on the activation energy in the Boltzmann factor. This makes it very probable that at higher *T*-s variable range hopping is the dominant mechanism of the conductivity, while at low temperatures the conduction happens via multichannel tunneling between the stacked molecules.

All the results obtained for the aperiodic base pair stacks are in excellent agreement (also the *T*-dependence) with the electron energy loss experiments in a hole resonator performed recently on λ-phage DNA.

In the last full point of the paper describes a newly developed intermediate exciton theory for disordered chains. The somewhat involved formalism can give a theoretical description of the excitonic energy levels (both in the singlet or triplet case) and their intensity.

As last point further probable directions of development of the quantum theory of disordered chains (like non-linear optical properties, theory of light emitting devices, exciton transfer, mixtures of polymers with metals) are summarized.

Key words: Coherent Potential Approximation
Disordered Chains
Disordered Polymer
Hopping Conductuvity of Disordered Chains
Intermediate Exciton Theory of Disordered Chains
Negative Factor Counting Method

1. INTRODUCTION

The quantum theory of disordered chains plays a very important role first of all in the case of biopolymers, DNA, RNA, different proteins etc. Further in reality any 1D polymer (or crystal) has some degree of disorder after a few hundred units caused by kinks in the chain (or dislocations in the crystal), impurities, lattice imperfection etc. We shell not discuss these local disorders here (this can be done with the help of a Green's matrix formalism [1]. Also the case of copolymers (in the simplest case of disorder caused by one larger periodic block of one component followed by smaller periodic block of the second component with different lengths etc.) will not be treated here.

The subject of this review paper is the calculation of the electronic structure and some properties (like electronic density of states [DOS] and hopping conductivity) of *substitutionally* disordered chains (different chemical components in the chains following each other in a disordered way) which of course causes also *structural* (geometrical) disorder. Since the applied methods are *ab initio* ones, the two types of disorders will be treated on the same footing.

The *ab initio* Hartree-Fock (HF) (*ab initio* SCF LCAO) treatment of 3D, 2D and 1D periodic systems goes back to Per-Olov Löwdin's famous review paper in Advances in Physics [2].

The formalism given by Löwdin was worked out in more detail in a form suitable for coding in three subsequent papers in 1967 [3,4]. These papers among other things applied Löwdin's method of symmetric orthogonalization [5] to get rid of the overlap matrices.

The quantum theory of the electronic structure of substitutionally disordered chains is based on Dean's negative eigenvalue theorem [6,7], from which the negative factor counting (NFC) method can be derived. This is at present the most successful way to treat this problem. It was first applied for the treatment of vibrational spectra of 3D disordered crystals [7]. M. Seel has first applied it in its simple tight binding form for the calculation of the electronic DOS of a long disordered polypeptide chain [8]. This method was formulated also in its ab initio matrix block form (given the total DOS of all electronic energy levels) by Day and Martino [9].

The present paper will be organised in the following way. In Point 2 the single site Coherent Potential Approximation (CPA) will be shortly described and applied for a mixture of $(SN)_x$ and $(SNH)_x$. In Points 3 and 4 the NFC method will be summarised in its two forms. In Point 5 the calculation of hopping conductivity in the case of disordered chains will be discussed in detail Subsection 5.1 deals with the determination of the hopping frequencies using the Anderson localized wave functions for the electrons and their interactions with acoustic phonons. In 5.2 the generalization of the random walk theory of Lax and coworkers will be presented and it will be shown how this theory-through Einstein's relation can be applied for the calculation of the complex frequency-dependent hopping conductivities.

In subsection 5.3 numerical results will be given for the density of states of disordered polypeptides, for different proteins and for disordered nucleotide base stacks. In subsection 5.4 results will be shown for the hopping conductivities of different proteins and disordered nucleotide base(-stacks). In the case of the base stacks also the basis sets and correlation effects will be discussed and the dependence of the conductivity on the temperature will be presented. Finally it will be shown that the results obtained for the hopping conductivity of disordered base stacks (including the T-dependence) are in excellent agreement with electron energy loss experiments on λ phage DNA in a hole resonator.

In Point 6 a newly dependent Intermediate Exciton Theory for the electronic spectra of disordered chains will be described.

Finally in Point 7 developments expected in the near future on the non-linear optical properties of disordered polymers on disordered light emitting diodes, on exciton transfer in disordered photosynthetic systems etc., will be mentioned.

2. APPLICATION OF THE SINGLE-SITE COHERENT POTENTIAL APPROXIMATION TO $(SN)_X$ WITH HYDROGEN IMPURITIES

The Coherent Potential Approximation (CPA) of the electronic structure of disordered systems assumes that the disordered system can be replaced with such an effective medium that the average of the statistical fluctuations from this potential should be zero. In other words each site but one is replaced by an unknown coherent potential. One then puts on a reference site (see Fig.1.) the A or B component (in a binary disordered system) with probabilities $1\text{-}f$ and f. After then one solves the problem of a single impurity inbedded in the effective medium.

If one assumes that no two subsequent scatterings can happen on the same site, one speaks about single site CPA [10]. If, however, one relaxes this condition (two or more subsequent scattering can happen on the same site before an other scattering happens on other site) one obtains the method of cluster CPA which is more relevant if extended molecules, rather than single atoms or ions are situated at lattice sites. Since, however, the latter method requires a rather elaborate formalism, here only the single site CPA's formalism will be given for a binary system (with components $(SN)_X$ and $(SNH)_X$, where the constituents are small molecules) in the one-band case.

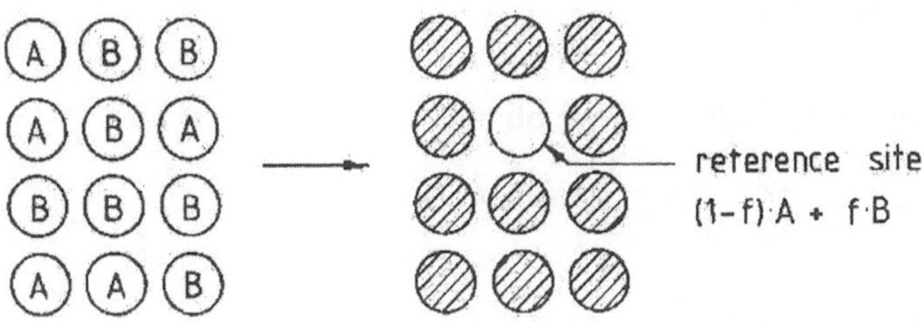

Fig 1. Replacement of a binary system with components A and B by an effective medium (hatched circle). At the reference site we can find component A or B with the probabilities 1-*f* and *f*, respectively.

2.1 FORMULATION OF THE SINGLE-SITE CPA EQUATION

One can start the formulation by writing down the Dyson equation (in the single site, one-orbital per unit cell approximation) in a scalar form

$$G = G^0 + G^0 \Delta G \tag{1}$$

Here

$$G^0(\vec{k},E) = \left[E - \varepsilon^A(\vec{k})\right]^{-1} \tag{2}$$

if A is the host system. Integration of this equation over $\vec{k}$ gives

$$G_0(E) = \Omega^{-1} \sum_{\vec{k}} \left[E - \varepsilon^A(\vec{k})\right]^{-1} \tag{3}$$

(Ω is the volume of the first Brillouin zone). Δ occurring in (3) can be defined (only in the single site approximation) in its Fourier transform from as

$$\Delta(\vec{k}) = \varepsilon^B(\vec{k}) - \varepsilon^A(\vec{k}) \tag{4}$$

The scattering matrix element T can be introduced in expression (1) by the equation

$$G = G^0 + G^0 TG \tag{5}$$

We can rewrite equ. (1) as

$$G = (1 - G^0 \Delta) G^0 \tag{6}$$

Substituting into this equ. (5) we obtain

$$G = G^0 + G^0 \Delta (1 - G^0 \Delta) G^0 \tag{7}$$

Putting into this equ. (7) we can write

$$T = \Delta (1 - G^0 \Delta)^{-1} \tag{8}$$

The introduction of an effective medium makes possible to rewrite the Dyson equ. as

$$G_e = G_e (1 + \Sigma G_e)^{-1} \tag{9}$$

and the self-energy Σ is defined by equ. (9). With the help of equ. (9) we can express also G^0 as a function of G_e:

$$G^0 = G_e (1 + \Sigma G_e)^{-1} \tag{10}$$

If we substitute this back into the original Dyson equ. (1) we obtain

$$(1 + \Sigma G_e) G_e^{-1} G = 1 + \Delta G$$

Multiplying this equation from the left by G_e gives

$$G = G_e \Sigma G = G_e + G_e \Delta G$$

or

$$G = G_e + G_e (\Delta - \Sigma) G = G_e + G_e \Delta_e G \tag{11}$$

with

$$\Delta_e = \Delta - \Sigma \tag{12}$$

At sites A $\Delta = 0$ and therefore $\Delta_e = -\Sigma$, while at sites B $\Delta_e = \Delta - \Sigma$.

Applying equ. (8), substituting instead of Δ, Δ_e and G^0 by G_e and using definition (12) we obtain for T

$$\langle T \rangle = (1-f) - \Sigma[1 + G_e \Sigma]^{-1} + f(\Delta - \Sigma)[1 - G_e(\Delta - \Sigma)]^{-1} \tag{13}$$

From the condition ⟨T⟩ = 0, (eq. (13)) one can derive finally the single-site CPA equations [10]

$$\Sigma(\vec{k},E) = f\Delta(\vec{k})[1 + (\Sigma(\vec{k},E) - \Delta(\vec{k}))G_e(E)]^{-1} \tag{14}$$

One can show from equ. (9)

$$G(\vec{k},E) = [E - \varepsilon^A(\vec{k}) - \Sigma(\vec{k},E)]^{-1} \tag{15}$$

$$G(E) = \Omega^{-1} \sum_{\vec{k}} G_e(\vec{k},E) \tag{16}$$

If the DOS curves calculated for the periodic constitutes have the same shape, but are shifted only by a constant energy value, $\Sigma(k,E) \approx \Sigma$ (constant) [10] and then by comparing equations (2) and (16) one can employ the equation

$$G_e(E) = G^0(E + \Sigma) \tag{17}$$

In contrary to the DOS curves of the constituents of alloys, this is not so in the case of aperiodic polymers as we shall see in the mixed $(SN)_X$ $(SNH)_X$ system. Therefore one had to work in this case with the $\Sigma(k,E)$ function.

Let us recall the expression of the spectral density [12]

$$A(\vec{k},E) = -\pi^{-1} \, Im G_e(\vec{k}, E + i0^+) \tag{18}$$

and for the DOS

$$\rho(E) = \Omega^{-1} \sum_{\vec{k}} A(\vec{k},E) = -\pi^{-1} \, Im G_e(\vec{k}, E + i0^+) \tag{19}$$

Further we have to apply the Kramers-Kronig relation [12] for the Green's function of the effective medium.

$$G_e(E)=\int_{-\infty}^{\infty}\frac{\rho_e(E')}{E-E'}dE' \tag{20}$$

To calculate the expression of the spectral density from (18) we need the well-known ratio of two complex numbers

$$\frac{\alpha_1+i\beta_1}{\alpha_2+i\beta_2}=\frac{\alpha_1\alpha_2+\beta_1\beta_2}{\alpha_2^{\,2}+\beta_2^{\,2}}+i\frac{\alpha_2\beta_1+\alpha_1\beta_2}{\alpha_2^{\,2}+\beta_2^{\,2}} \tag{21}$$

Applying this to equ. (15) ($\alpha_1=1 \quad \beta_1=0$) one obtains for

$$Im G_e(\vec{k},E)=\frac{-Im\left[\Sigma(\vec{k},E)\right]}{\left[E-\varepsilon^A(\vec{k})-Re\Sigma(\vec{k},E)\right]^2+Im\Sigma(\vec{k},E)} \tag{22}$$

Only in a very few case were calculations with $\vec{k}$ and E-dependent self-energy reported. These calculations started with equ. (27) using as zeroth iteration the virtual crystal density value $\rho_{VC}(E)$ which can be obtained from the energy bands of A and B via

$$\rho_{VC}(\vec{k})=(1-f)\varepsilon^A(\vec{k})+f\varepsilon^B(\vec{k}) \tag{23}$$

to determine the DOS $\rho_{VC}(E)$. Putting (23) into (20) one obtains for instance with the help of the method of Kirckpatrick et al already a starting value for $G_e(E)^o$. Substituting this together with $\Sigma(\vec{k})^{(0)}=f\Delta(\vec{k})$ (which is again a virtual crystal approximation for $\Sigma(\vec{k})$ into the CPA equ. (14) leads to a new complex self-energy $\Sigma(\vec{k},E)^{(1)}$. Inserting this to equ. (22) and the resulting expression for $Im G_e(\vec{k},E)$ into equations (18) and (19) we can recalculate the DOS $\rho(E)^{(1)}$ in the next iteration. Substituting $\rho(E)^{(1)}$ back into (20) the procedure can be repeated until self-consisting is reached. This was the way as we have calculated the $(SN)_X$-$(SNH)_X$ mixed polymer [14].

At IBM (San Jose) 4-8 mol % hydrogen was found in $(SN)_X$ with unknown positions of the H impurities [15]. It is interesting to find out the shift of the Fermi level and the change of the density of states if one mixes the S = N- units (3π electrons) ($S=N<_{H}$) units (4π electrons) in a random way. To answer these questions we have applied the above described CPA equation with a $\Sigma = \Sigma(k, E)$. (This was necessary because the DOS curves of $(SN)_X$ and $(SNH)_X$ are completely different).

As SCF criteria of the iterations

$$\max_i \left| \rho_e^{(n)}(E_i) - \rho_e^{(n-1)}(E_i) \right| \leq 10^{-3} \quad \text{a.u.mol}^{-1} \tag{24}$$

was chosen. About 20 iteration steps were needed to reach self consistency in the cases of $0.3 \leq f \leq 0.10$. The mesh points of the calculation were not taken equidistantly, but were chosen more densely in the range where $\rho_{(SNH)_x} \neq 0$ and in regions with numerous peaks (for more detail see [14]).

93 points were chosen in the range $0 \leq E \leq 10\,\text{eV}$ and the same number of k points in the first Brillouin zone have been applied. For each E-value and k-point $\Sigma(\text{k}, \text{E})$ was recalculated in each iteration step, the further details of the rather involved numerical procedure one can find in [14].

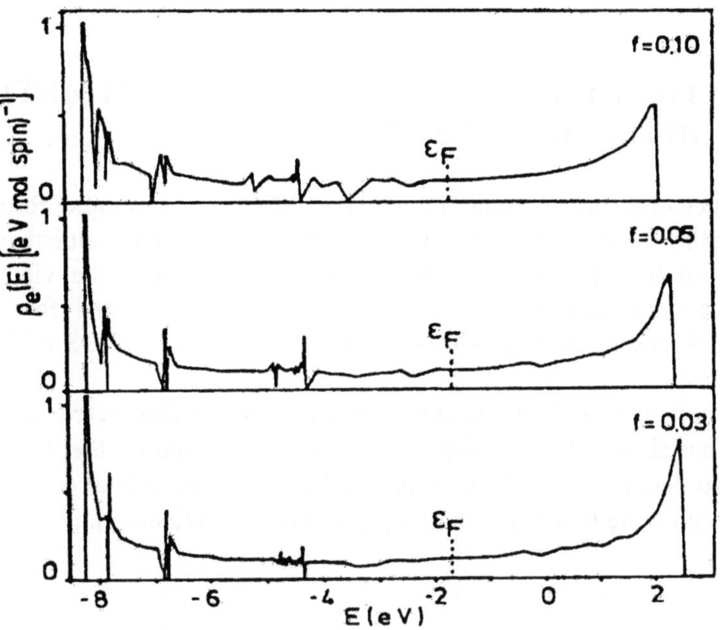

Fig.2 Density of States curves of the $(SN)_X$ and $(SNH)_X$ mixed system obtained with the CPA method with f=0.03, 0.05 and 0.010.

In Fig. 2 the DOS curves for the $(SN)_X$ – $(SNH)_X$ random chain in the cases of f=0.03, 0.05 and 0.10 are shown. One can see that due to the presence of the (SNH) impurities the DOS curves with spikes and dipes to zero) are much more complicated than in the case of the periodic chains [14]. They occur already at f=0.03 which was not described in the oversimplified case of Σ=const. On the other hand one can observe that the Fermi level hardly changes with f.

It is important to notice that with increasing f the DOS at the Fermi level is increasing. This is especially so in the case if one goes to larger f values $0.2 \le f \le 0.5$ (see [14]). As it is well known according to the BCS theory of superconductivity [16] the superconductive transition T_C depends (in a good approximation) exponentially on the electronic density of states at the Fermi level ($\rho(\varepsilon_F)$)

$$T_C = 1.14 k_B \exp[-1/\rho(\varepsilon_F)V] \tag{25}$$

where V is the effective BCS electron-electron interaction parameter.

As it is known $(SN)_X$ is superconducting with $T_C = 0.28\,K$ [17]. Therefore one would expect on the basis of [25] that by increasing the H impurity concentration T_C will increase. As far as the author knows, such experiments have not yet been performed.

3. THE NEGATIVE FACTOR COUNTING METHOD IN ITS SIMPLE FORM

The negative factor counting method (NFC) in its simple form was developed by Dean [6] (based on his so-called negative eigenvalue theorem). It was used to determine the density of states (DOS) in the vibrational spectra of disordered systems. This method cannot be used for simple topological reasons for the determination of the electronic DOS of 2D and 3D systems.

It is, however easily applicable in the case of 1D disordered chains, as Seel [8] has shown it in the following way. Let us suppose that we have a linear disordered chain of N units and we have only one orbital per unit cell. Then we can write in the tight binding (Hückel) approximation.

$$|\mathbf{H}(\lambda)| = \begin{vmatrix} \alpha_1 - \lambda & \beta_2 & 0 & 0 & 0 \\ \beta_2 & \alpha_2 - \lambda & \beta_3 & & \\ 0 & \beta_3 & \alpha_3 - \lambda & \beta_4 & \\ & & & & \\ 0 & & & \beta_N & \alpha_N - \lambda \end{vmatrix} = 0 \tag{26}$$

Here α_i and β_j are usual Hückel parameters and λ denotes the unknown root of the determinant. (In a disordered chain the parameters are different from each other.) The original tridiagonal determinant can be easily transformed into a didiagonal form with the help of a successive Gaussian elimination. To achieve this we must substract from the second row of the determinant the first row multiplied by $\beta_2/(\alpha_1-\lambda)$. This eliminates the element β_2. Continuing this procedure one obtains zero for all the elements of lower diagonal of the originally tridiagonal determinant. After this we can write

$$|\mathbf{H}(\lambda)|=\prod_{i=1}^{N}(\lambda_1-\lambda)=|\mathbf{H}(\lambda)|=\prod_{i=1}^{N}\varepsilon_i(\lambda) \tag{27}$$

(the value of a diagonal determinant is the product of its diagonal elements). The ε_i -s can be given by the simple recursion

$$\varepsilon_i(\lambda)=\alpha_i-\lambda-\beta_i^2/\varepsilon_{i-1}-(\lambda), \tag{28a}$$

$$\varepsilon_i(\lambda)=(\lambda_i-\lambda) \tag{28b}$$

due to the Gaussian elimination procedure. Looking at equ.-s (28) it is clear that for a given trial value of λ the number of eigenvalues $\lambda_i<\lambda$ must be equal to the number negative factors $\varepsilon_i(\lambda)$. (Though with the help of modern computers it is quite possible to determinate all the eigenvalues of a disordered chain of length 10^5-10^6, but the calculations would be quite formidable, especially (see next point) if in the unit cell there are large number of basis functions. On the other hand using equ-s (28a) and (28b) the computation of the factor $\varepsilon_i(\lambda)$ is very rapid.) By giving λ different values, going through the spectrum and taking the difference of the number of negative values of $\varepsilon_i(\lambda)$ belonging to consecutive values of λ, one can obtain a histogram for the distribution of eigenvalues (electronic density of states, DOS) of **H** to any desired accuracy.

In actual calculations one has to calculate the band structure of each component (assuming that the elements of the chains are periodically repeated). The values α_i can be determined from the positions of the bands of the components (the middle point or weighted middle point of the band) and the off-diagonal elements β_i from the widths of the bands (more precisely the algebraic average of the widths of the bands belonging to the consecutive units).

It is clear that the NFC method in this simple form will provide only the DOS of the levels of a single band (for instance valence or conduction band).

Unfortunately the NFC method cannot be applied for simple topological reasons to 2D or 3D disordered systems. Namely the Hückel matrices cannot be expressed in a tridiagonal form. In the 2D case applying again only first neighbors interactions, the matrix would be five-diagonal in the 2D case, and seven-diagonal in the 3D case. To bring a five-diagonal matrix into a three-diagonal form using the Givens-Householder method would require about one third of the CPU time of the complete diagonalization [18]. Therefore in these cases one has to return to the methods well known in solid state physics for the treatment of such systems.

4. THE NEGATIVE FACTOR COUNTING METHOD IN ITS AB INITIO MATRIX BLOCK FORM

The NFC method described in the previous point can be generalized to the case when there are arbitrary number of basis functions in the different units. This can be done both in ab initio HF [9] and different semi empirical forms (which will not be discussed here). One can introduce easily even correlation in the form of many body perturbation theory (MBPT).

In the ab initio HF case one can write instead of the tridiagonal determinant (26) the tri-block diagonal determinant

$$|\mathbf{H}(\lambda)| = |\mathbf{F} - \lambda\mathbf{S}| = \\ = \begin{vmatrix} \mathbf{A}_1 - \lambda\mathbf{S}_1 & \mathbf{B}_2 - \lambda\mathbf{Q}_2 & \mathbf{0} & & \\ \mathbf{B}_2^T - \lambda\mathbf{Q}_2^T & \mathbf{A}_2 - \lambda\mathbf{S}_2 & \mathbf{B}_3 - \lambda\mathbf{Q}_3 & & \\ \mathbf{0} & \mathbf{B}_2^T - \lambda\mathbf{Q}_3^T & \mathbf{A}_3 - \lambda\mathbf{S}_3 & & \\ & & & \ddots & \\ & & & & \mathbf{A}_N - \lambda\mathbf{S}_N \end{vmatrix} = 0 \tag{29}$$

Here $\mathbf{A}_i$ and $\mathbf{B}_{i+1}$ are the diagonal and off-diagonal blocks, respectively, of the Fock matrix and $\mathbf{S}_i$ and $\mathbf{Q}_{i+1}$ are the corresponding blocks of the overlap matrix. The superscript the different diagonal matrix blocks have the order of the different units in the disordered chain, that is $\mathbf{A}_i \neq \mathbf{A}_j$ and $\mathbf{B}_{i+1} \neq \mathbf{B}_{j+1}$,. $\mathbf{S}_i \neq \mathbf{S}_j$ and $\mathbf{Q}_{i+1} \neq \mathbf{Q}_{j+1}$.

We can write $\mathbf{M}(\lambda)$ in the form

$$\mathbf{M}(\lambda) \equiv \mathbf{F} - \lambda\mathbf{S} = \mathbf{S}^{1/2}\left[\mathbf{S}^{-1/2}\mathbf{F}\mathbf{S}^{-1/2} - \lambda\mathbf{1}\right]\mathbf{S}_{\mathrm{F}}^{1/2} \tag{30}$$

and therefore

$$\det \mathbf{M}(\lambda) = \det \mathbf{S}^{1/2} \det\left(\widetilde{\mathbf{F}} - \lambda \mathbf{1}\right) \det \mathbf{S}^{1/2} = \\ = \det \mathbf{S}^{1/2} \det \mathbf{S}^{1/2} \det\left(\widetilde{\mathbf{F}} - \lambda \mathbf{1}\right) = \det \mathbf{S} \det\left(\widetilde{\mathbf{F}} - \lambda \mathbf{1}\right) \tag{31}$$

with $\widetilde{\mathbf{F}} = \mathbf{S}^{-1/2}\mathbf{F}\mathbf{S}^{-1/2}$

The determinant $|\mathbf{F}-\lambda\mathbf{S}|$ can be easily brough again to a diblockdiagonal form with the help of successive Gaussian elimination. One obtains in this way in complete analogy to equ. (28a)

$$\mathbf{U}_i(\lambda) = \mathbf{A}_i - \lambda \mathbf{S}_i - \left(\mathbf{B}_i{}^T - \lambda \mathbf{Q}_i{}^T\right) \mathbf{U}_{i-1}^{-1}(\lambda)\left(\mathbf{B}_i - \lambda \mathbf{Q}_i\right) \tag{32}$$

with $\mathbf{U}_1(\lambda) = \mathbf{A}_1 - \lambda \mathbf{S}_1$.

Because of (31) and (32) we can finally write

$$det(\mathbf{F} - \lambda\mathbf{S}) = \left(\prod_{i=1}^{n} s_i\right)\left[\prod_{j=1}^{n} \lambda_j - \lambda\right] = \prod_{i=1}^{N}\left[\prod_{k=1}^{l_i} u_{ik}(\lambda)\right] \quad \left(n = \sum_{i=N}^{N} l_i\right) \tag{33}$$

Here the s_i-s are the eigenvalues of the $n \times n$ matrix $\mathbf{S}$ and λ_j-s are (the unknown) roots of the generalized eigenvalue equation

$$\mathbf{F}c_j = \lambda_j \mathbf{S} c_j \tag{34}$$

$u_{ik}(\lambda)$ denotes further the *k*-th eigenvalue of the matrix block $\mathbf{U}_i(\lambda)$ of dimension $l_i \times l_i$ which can be easily diagonalized for a chosen value of λ. The number of negative $u_{ik}(\lambda)$ factors must again be equal to the number of eigenvalues λ_j which are smaller than the chosen trial λ-value. Varying λ through the whole spectrum one obtains in this way the whole total electronic density of states histogram to any desired accuracy (see in more detail in point 2.).

The matrix blocks appearing in eq. (28) can be easily constructed performing HF calculations for segments of the disordered chain. In the simplest case of a binary chain with first neighbors' interaction approximation for a chain say

Fig 3 The overlapping dimmers of the chain AABAAABA

one has to perform an ab initio SCF calculation for the dimmers

$$\begin{pmatrix} \mathbf{F}_A & \mathbf{F}_{AA} \\ \mathbf{F}_{AA}^T & \mathbf{F}_A \end{pmatrix}, \begin{pmatrix} \mathbf{F}_A & \mathbf{F}_{AB} \\ \mathbf{F}_{AB}^T & \mathbf{F}_B \end{pmatrix}, \begin{pmatrix} \mathbf{F}_B & \mathbf{F}_{BA} \\ \mathbf{F}_{BA}^T & \mathbf{F}_A \end{pmatrix}, \begin{pmatrix} \mathbf{F}_A & \mathbf{F}_{AA} \\ \mathbf{F}_{AA}^T & \mathbf{F}_A \end{pmatrix}, \begin{pmatrix} \mathbf{F}_A & \mathbf{F}_{AB} \\ \mathbf{F}_{AB}^T & \mathbf{F}_B \end{pmatrix}, \begin{pmatrix} \mathbf{F}_B & \mathbf{F}_{BA} \\ \mathbf{F}_{BA}^T & \mathbf{F}_B \end{pmatrix} \tag{35}$$

One should mention that one can easily introduce also correlation into equ-s (34) and (35). We can write in instead of the HF expression of the dimmer IJ [19, 20]

$$\widetilde{\mathbf{F}}_{\text{dimer}} = \begin{pmatrix} \widetilde{\mathbf{F}}_I & \widetilde{\mathbf{F}}_{IJ} \\ \widetilde{\mathbf{F}}_{IJ}^+ & \widetilde{\mathbf{F}}_J \end{pmatrix} = \mathbf{F}_{\text{dimer}} + \mathbf{S}_{\text{dimer}} \mathbf{U} \boldsymbol{\Sigma} \mathbf{U}^+ \mathbf{S}_{\text{dimer}} \tag{36}$$

Here Σ is the self-energy matrix for which any size consistent correlation expression (for instance MP2 or coupled cluster expression) can be written. The matrix $\mathbf{U}$ is calculated from the eigenvectors u_i of the dimer $\mathbf{F}$

$$\mathbf{F}_{\text{dimer}} u_i = \varepsilon_i \mathbf{S}_{\text{dimer}} u_i \tag{37}$$

If one multiplies (36) from the left by $\mathbf{U}^+$ and from the right by $\mathbf{U}$ and takes into account that $\mathbf{USU}^+ = \mathbf{U}^+\mathbf{SU} = \mathbf{1}$ one obtains the inverse Dyson equation [19]

$$\varepsilon_i^{QP} = \varepsilon_i^{HF} + (\Sigma)_{ii} \tag{38}$$

if one applies the diagonal approximation [21]. Here ε_i^{QP} is the correlation corrected quasi particle energy. As we shall see later this method was successfully applied to aperiodic protein chains.

To close this point one should mention that by knowing the grid used for the NFC calculation one can determine the wave function belonging to any energy level of interest (usually levels near to the gap). For this the standard inverse iteration method can be applied. The results show that until a few hundred levels from the gap (until the mobility edges) both in the valence bands and conduction bands regions, respectively, the wave function are strongly localized on one or two units with some tails (Anderson localization).

5. CALCULATION OF THE HOPPING CONDUCTIVITY OF DISORDERED CHAINS

5.1 DETERMINATION OF THE HOPPING FREQUENCIES

Having the one-electron energy levels and wave functions of a disordered chain one can calculate the hopping frequencies which are due to the interaction of the electrons with the acoustic phonons in the chain. (In the case of proteins these acoustic phonons are the vibrations of the side chains with respect to each other, in nucleotide base stacks the vibrations of the base pairs relative to each other.) For the hopping frequencies one can use in a first approximation the very simple expression given by Mott and Davis [23, 24] for the hopping probability form site A and level i to site B and level j

$$W_{A,i\to B,j} = \nu_{\text{phonon}}\left(\sum_{r\in A}\sum_{s\in B} c_{i,r}^{(A)} c_{j,s}^{(B)} \left\langle \chi_r^{(A)} \middle| \chi_s^{(B)} \right\rangle\right)^2 \exp\left(-\Delta E_{ij}/k_B T\right) \tag{39}$$

Equ. (39) is a generalization for the case when there are arbitrary member of basis functions χ_r say on site A. ν_{phonon} is the acoustic phonon frequency (for its value usually 10^{12} sec^{-1} is taken), the c-s are the LCAO coefficients of the localized one-electron orbitals $\varphi_i^{(A)e}$ and $\varphi_j^{(B)e}$, respectively, and the $\langle\chi_r^{(A)}|\chi_s^{(B)}\rangle$-s are the overlap integrals between AO-s centered on site A and B, respectively. Finally ΔE_{ij} in the Boltzmann factor is the energy difference between $\varepsilon_i^{(A)}$ and $\varepsilon_j^{(B)}$.

One should point out that instead of (39) it would be more correct to take into account directly the interactions of the localized electrons with the localized phonons. In this way one should write

$$\begin{aligned} W_{A,i\to B,j} = {} & \left\langle \varphi_i^{(A)e} \middle| \hat{H}_{el\text{-}ph}^{(A)} \middle| \psi_i^{(A)phonon} \right\rangle \left(\sum_{r\in A}\sum_{s\in B} c_{i,r}^{(A)} c_{j,s}^{(B)} \left\langle \chi_r^{(A)} \middle| \chi_s^{(B)} \right\rangle\right)^2 \times \\ & \times \left\langle \psi_j^{(B)phonon} \middle| \hat{H}_{el\text{-}ph}^{(B)} \middle| \varphi_j^{(B)e} \right\rangle \begin{cases} e^{-\Delta E_{ij}/k_B T} & if \quad \Delta E_{ij} > 0 \\ 1 & if \quad \Delta E_{ij} \le 0 \end{cases} \end{aligned} \tag{40}$$

Here the operators $\hat{\mathbf{H}}_{\text{el-ph}}$ are electron-phonon interaction operator and $\psi_i^{(A)phonon}$ and $\psi_j^{(B)phonon}$ are the localized vibrational wave functions (because of the disorder the phonon wave functions are also Anderson localized) at the two sites.

One could think also on the interactions of the localized electronic wave function with the many vibrational wave functions at each site (which would have also larger frequencies). However, until we are interested in hopping along a polypeptide or polynucleotide stack, certainly the above described acoustic phonons are the interesting ones and from the point of view of conductions along the chains the interactions with the other phonons can be neglected. Finally is should be mentioned that until now no hopping conductivity calculations on any disordered chain have been performed using equ. (40) or its generalization.

5.2 CALCULATION OF THE HOPPING CONDUCTIVITY USING A RANDOM WALK THEORY

After the calculation of the localized one-electron energy levels and wave functions of a disordered chain and having from them the hopping frequencies between the different sites (usually only first or second neighbors hopping) one can use these quantities as input into the random walk theory of Lax and coworkers [25].

There is a more generalized formulation of the random walk theory [26] than the one developed by Lax el al [25], but this could be applied only if such special restrictions would be introduced which are not compatible with the ab initio formalism (see point 4) which we have applied for complicated biopolymers (see point 6). On the other hand the expressions of Lax el al [25] had to be generalized again for arbitrary numbers of basis functions at the different sites.

The starting point of the random walk theory of Lax et al is the master equation

$$\frac{\partial P[(n,j),t|(n_0,i_0),0]}{\partial t} = -\Gamma_{(n,j)}P[(n,j),t|(n_0,t_0),0] + \\ + \sum_{n',j'\neq n,j} h_{n',j'\to n,j}P[(n',j'),t|(n_0,i_0),0] \tag{41}$$

where

$$\Gamma_{(n,j)} = \sum_{n',j'\neq n,j} h_{n,j\to n',j'} \ . \tag{42}$$

Here $P[(n,j),t|(n_0,i_0),0]$ denotes the probability for a charge carrier to be at the energy level j of site n at time t, if it was at the level i_0 of site n_0 at time 0 and $h_{n,j\to n',j'}=W_{n,i\to n',j}$ is the hopping frequency defined before.

The complex ω-frequency-dependent a.c. conductivity is given by the generalized Einstein's relation

$$\sigma(\omega) = \frac{n_V e^2}{k_B T} D(\omega) \tag{43}$$

where n_V is the number of charge carriers in the volume of the disordered chain, and k_B the Boltzmann constant. The diffusion constant of the electrons (holes) $D(\omega)$ can be written with the help of the Laplace transform of $P[(n,j)t|(n_0,i_0)0]$

$$\widetilde{P}[(n,j,i\omega)|(n_0,i)] = \int_0^\infty e^{-i\omega t} P[(n,j)t|(n_0,i_0)0]dt \tag{44}$$

as

$$D(\omega) = -\frac{\omega^2}{2d} \sum_{n_0,i_0} \sum_{n,j} [(n,j) - (n_0,i_0)]^2 \widetilde{P}[(n,j)i\omega|(n_0,i_0) f(E_{n_0},i_0)] \tag{45}$$

Here d is the dimensionality of the system (for chains d=1) and the functions $f(E_n)$ are the equilibrium distribution functions of the localized carriers. They have to fulfill the equation

$$h_{n',j' \to n,j} f_{n',j'} = h_{n,j \to n',j'} f_{n,j} \tag{46}$$

Finally E_{nj} is the j-th level of the charge carrier at site n (if the wave functions are localized on a molecule which contains several atoms, one has to define a center of the charge distribution on a given site and one assumes that the hopping takes place between these different centers).

It can be shown [25] that the Laplace transform $\widetilde{P}$ [] fulfills the equation

$$\left(i\omega - \Gamma_{(n,j)} \widetilde{P}[(n,j)i\omega|(n_0,i_0)]\right) - \sum_{n',j' \neq n,j} h_{n',j' \to n,j} \widetilde{P}[(n',j')i\omega|(n_0,i_0)] = \delta_{n,n_0} \delta_{j,i_0} \tag{47}$$

This leads to a formal solution for $\widetilde{P}$ []

$$\widetilde{P}[(n,j)i\omega|(n_0,i_0)] = [(i\omega)\mathbf{1} - \mathbf{H}]^{-1}_{n,j,n_0,i_0} \tag{48}$$

The elements of the hopping matrix **H** are defined as

$$H_{(n,j),(n,j)} = -\Gamma_{(n,j)} \tag{49}$$

$$H_{(n,j),(n',j')} = h_{n',j' \to n,j} \quad (n',j' \neq n,j) \tag{50}$$

From equ. (50) one can see that the matrix **H** is non-symmetric.

The functions $f_{n,i}$ can be obtained as the components of the normalized right eigenvector of **H** belonging to the non-degenerate eigenvalue O [25]. This eigenvector can be written (after reordering its indices) as

$$\begin{bmatrix} f(E_1) \\ \vdots \\ f(E_m) \end{bmatrix} \tag{51}$$

Here E_1, ..., E_m are those energy levels $E_{n,j}$ which were used in the construction of the matrix **H**.

In an actual calculation one has to know of course the sequence and the conformation of the whole disordered chain. From these the number density n_V can be computed. After this one can construct the matrix **H** with the aid of equ-s (49) and (50). Knowing **H** using equ (48) one can determine the Laplace transforms $\widetilde{P}$ and solving the eigenvalue problem of **H** the distribution functions $f(E_0, i_0)$ can be computed. Substituting the two latter quantities into equ (45) one obtains $D(\omega)$. Finally $D(\omega)$ through equ. (43) provides the hopping conductivity. Schematically the above described procedure can be written as

$$\text{Eqs (49) and (50)} \begin{matrix} \to & \text{Eq (48)} & \to \\ \to & \text{Eq (51)} & \to \end{matrix} \text{Eq (45)} \to \text{Eq (43)}.$$

5.3 DENSITY OF STATES OF DISORDERED POLYPEPTIDES, PROTEINS AND NUCLEOTIDE BASE STACKS

Fig. 4 shows the electronic DOS-s of a two-component (glycine-alanine chain) at 1% (a), 30% (b) and 50% alanine concentration.

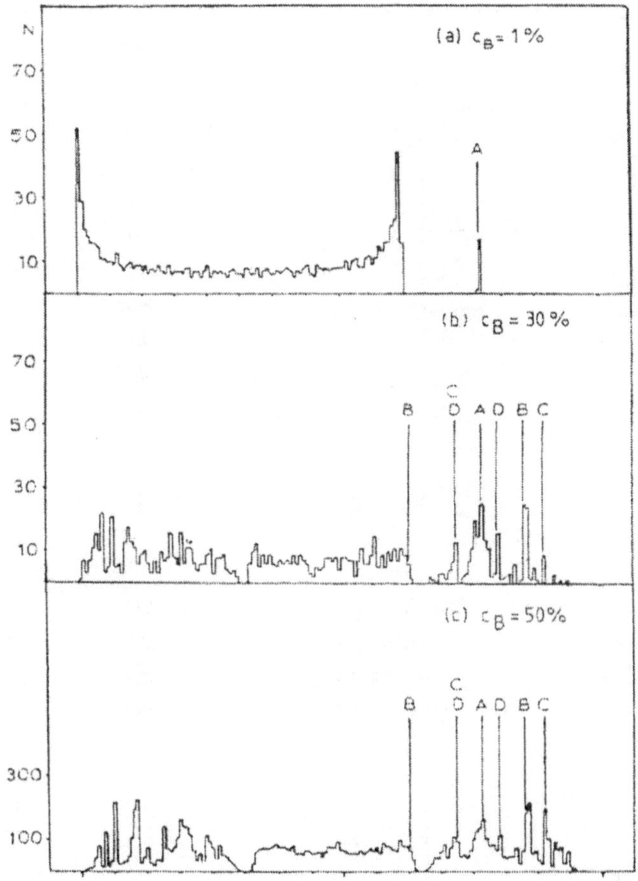

Fig.4 It shows the DOS of a disordered two-component glycine-alanine chain containing 1000 units (a) at 1% glycine, (b) at 30% and (c) at 50% alanine concentration. N is the number of eigenvalues in a histogram interval of 0.005 eV

One can notice that with increasing alanine concentration the DOS curve broadens significantly (at 50% alanine concentration there is hardly any gap). In Fig. 5 the DOS of a binary disordered 1000 units long AT and GC base pair stack calculation is shown.

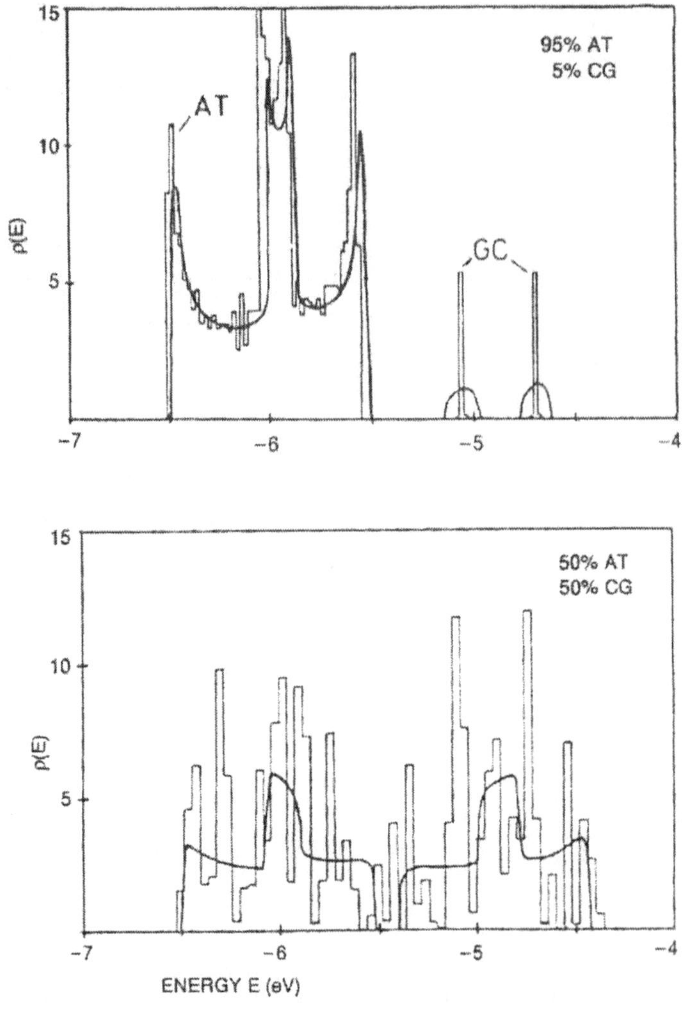

Fig. 5 The DOS for the binary random nucleotide base-pair stack AT and GC. (a) 5% GC, (b) 50% GC with 1000 base pairs. The histogram is the result of a matrix block NFC calculations the smooth lines comes from a CPA calculation.

The smooth curves correspond to a CPA calculation. It is apparent that in the case of small impurity concentration (5% GC) the CPA curves give a good envelope curve for the NFC histogram. Not this is the case for large disorder (5090 GC) when the CPA curve gives a rather poor approximation of the NFC histogram. The reason of this is that the CPA method was developed for a 3D disordered system and therefore gives a poor approximation in the 1D case at larger disorder.

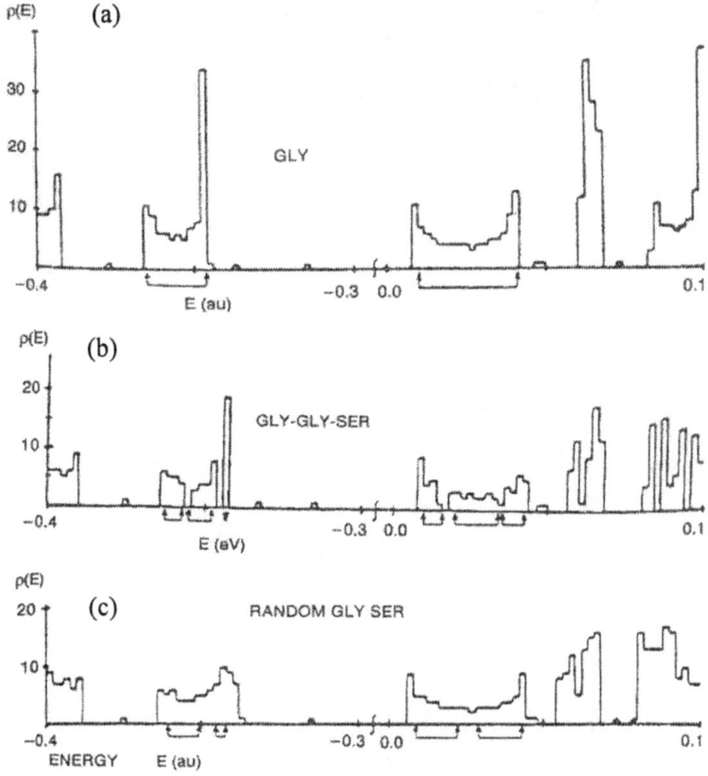

Fig. 6 The DOS curves of the periodic glycine (a), of periodic(gly-ser)$_x$ (b), and of a random gly-ser chain with 38 glycine and 42 serine units (c) calculated with the aid of the matrix block NFC method. In parts (a) and (b) of the Figure the arrows show the bands edges obtained with help of direct ab initio HF crystal orbital calculations.

In Fig. 6 the DOS curves of a periodic glycine chain (a), of a periodic gly-gly-ser chain (20 units) (b) and of a random glycine-serine chain (with 38 and 42 units, respectively) are shown. From the (a) and (b) parts of Fig. 6 one sees that the matrix block NFC method provides in a very good approximation the DOS-s of the periodic chain. On the other hand in the aperiodic case (c) the DOS curve is significantly different from the periodic (gly-ser)$_X$ DOS one. At the same time one can observe again that the disorder broadens the regions where the DOS is not zero. This phenomenon is still much stronger for disordered chains with more than 2 components. To understand the reason of this let us take a disordered nucleotide base stack and let us take only first-neighbors' interactions. One can see from Fig. 7 in a periodic ABCD chain (a)

ABCDABCDABCD ... (a)

AABCBCDABCBA ... (b)

Fig. 7 The sequence of units in a periodic quaternary chain (a) and a disordered four-component chain

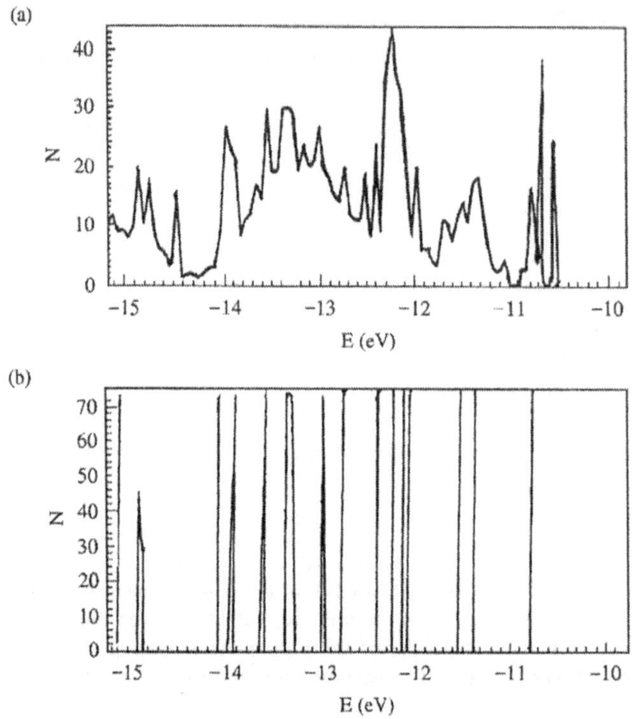

Fig. 8 The DOS curve (a) of the valance bands region of a 300 units long 1:1:1:1:1 gly., ser., cys., asn. polypeptide chain with a random sequence. The histogram interval is 0.05 eV and N is the number of states in this interval. (b) The DOS curve for the same chain but with a periodic sequence (gly., ser., cyst., asn.)

A given unit (for instance D) can have only two next neighbors (C from the left and A from the right). On the other hand in the disordered chain (b) each unit can have four different left and four different right neighbors for instance D can have as left neighbors A, B, C and D and it has the same four possibilities as right neighbors. This means that in an aperiodic chain every unit has 16 different possible next (left + right) neighbors and with it 16 different possible next neighbor interactions, while in the periodic case (independently) from the number of components the number of possible interactions is maximally 2. This very large difference in the possible interactions explains the strong broadening of the DOS curves in multicomponent disordered systems (see Fig 7).

The hopping frequencies of the system for which the DOS histograms are shown in Fig. 6, are in the first neighbors hopping cases $W_{A,i\rightarrow B,j}$ between $5\cdot10^{7}$ and $7\cdot10^{8}\,\sec^{-1}$ (the hopping frequencies are not normalized to one) at 310 K. The second neighbors hopping occur with a frequency of $10^{5}\,\sec^{-1}$. One should mention that these calculations were performed with a minimal basis. If one would use a double ζ basis and take into account also correlation, because of the smaller average level spacing the ΔE_{ij} volumes occurring in the Boltzmann factors of (39) would become also smaller and with them the hopping frequencies larger. As we shall see below this is also the case with the hopping conductivities.

5.4 NUMERICAL RESULTS FOR THE HOPPING CONDUCTIVITIES OF PROTEINS AND DISORDERED NUCLEOTIDE BASE(-PAIR) STACKS

Albert Szent-Györgyi had assumed already in 1941 [28] that there is electronic conduction along the main chain of proteins in the dark and without doping. He has come to this idea on the basis of his studies on oxygen metabolism in the living cell (Szent-Györgyi-Krebs oxygen metabolism cycle [29] in which electrons are transferred through a series of protein molecules to the final acceptors, the $^{3}O_2$ molecules). This idea has been generally neglected by the scientific community, because (1) of the large gap (7-9 eV) between the valence- and conduction bands regions in proteins and (2) because polypeptide chains in proteins are disordered systems. (It was rather assumed that electron transfer in proteins occur via hopping or tunneling between different segments of a folded protein or with the aid of the motion of electrons coupled to protons and ions in hydrated proteins.) Early measurements performed on proteins [30] and DNA [31] have shown a weak semiconduction also in the dark (probably due to electron transport). However, it was not clear whether the conductivity comes from the protein chains themselves or are caused by impurities.

Our frequency-dependent hopping conductivity calculations have shown that there is a rather large hopping conductivity up to $10^{-2}\,\Omega^{-1}\mathrm{cm}^{-1}$ at $\omega=10^{11}\,\sec^{-1}$ at different proteins (pig insulin which has 51 amino acid residues both in its active and inactive form, hen egg white lysozyne with its 139 amino acid residues, subtilisin etc.). The calculations were performed with the known sequences and conformations of the corresponding proteins. It was found that the conductivity was always strongly conformation-dependent, but not too much sequence dependent (the inactive form of a protein had always smaller conductivity than its active form). These results obtained with the help of modern techniques for the treatment of disordered systems have shown that there can be a considerable hopping conductivity along the main chain of disordered protein chain in contradiction to earlier assumptions.

There are no reliable a.c. conductivity measurements on proteins. On the other hand one can compare our results with the hopping conductivities of amorphous glasses like the chalcenogides [32] which were measured in the frequency range $10^4 - 10^8$ sec^{-1}. One finds that $|\sigma(\omega)|$ of pig insulin lies in this frequency range between the those of Te_2AsSi and As_2S_4 and has the same order of magnitude in most frequency regions as $Te_{48}As_{30}S_{12}Ge_{10}$ [32]. (The same is true for $|\sigma(\omega)|$ of hen egg white lysozyme).

Concerning the large gap in proteins one can point out that in vivo there is always a possibility of charge transfer between different macromolecules or between macromolecules and molecules. For instance in a nucleoprotein most probably there is an electron transfer from the negative phosphate groups (which have according to detailed calculation instead of –1.0e – 1.2e negative charge) to the positive argininive residues or to the protonated lysine side chains of the histone molecules.

As next step we have performed hopping conductivity calculations for aperiodic nucleotide base or base pair stacks. We have applied first a minimal basis and two different sequences. The first sequence was a part of an oncogen [33] and the second one was a random sequence generated with the help of a Monte Carlo program. For the geometry the refined data of the Dickerson Group on DNA B [34] have been applied.

According to the results obtained taking a 100 base pair long sequence $|\sigma(\omega)|$ was $5\cdot10^{-1}\Omega^{-1}cm^{-1}$ at $\omega=10^{11}\,sec^{-1}$ [35]. On the other hand if we took into account only a single base stack (in both cases using seq. 1.) $|\sigma(\omega)|$ decreased to $10^{-5}\Omega^{-1}cm^{-1}$ (see Fig. 9). The reason of this is that in the 100 base pair case one has 200 levels in the double stack and only 100 ones in the 100 units case [35]. Since in this way the average level spacing in the double strand case is in the neighborhood of the gap much smaller than in the single strand case, through the Boltzmann factors the hopping frequencies will be larger in the first case. Finally the hopping frequencies because of their Boltzmann factors essentially determine the conductivity values. This explains the large increase of the $|\sigma(\omega)|$ values in the 200 levels case as compared to the 100 levels one.

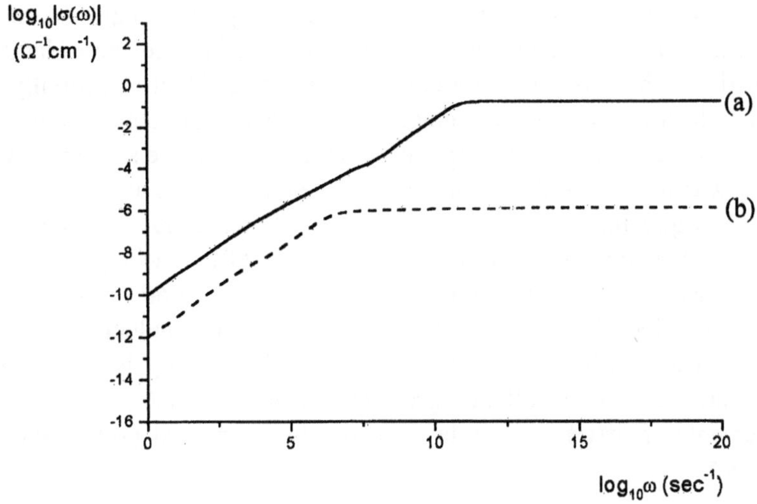

Fig. 9 Logarithm of the absolute value of the complex hopping conductivities ($|\sigma(\omega)|$) as a function of the logarithm of the frequency ω: (a) (-) double strand with sequence 1 in the valence bands region taking into account 200 levels (b) (---) with 100 levels of the same system. The first curve shows a saturation value at $\omega = 10^{12}$ sec^{-1}.

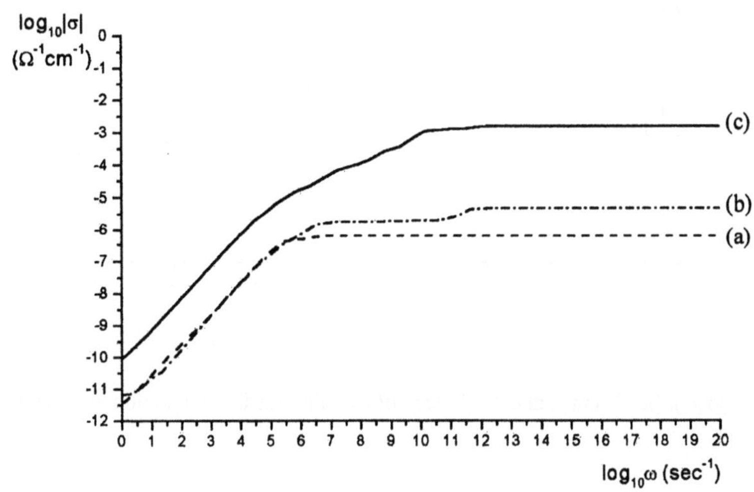

Fig. 10 $log_{10}|\sigma(\omega)|$ for a single stack with sequence 1 and 100 levels in the valence bands region (a) (---) minimal basis, (b) (-·-·-) double ζ basis and (c) (——) double ζ basis and correlation.

Fig. 10 shows the effect of a better basis set and correlation correction at the MP2 level [36]. As one can see the better basis set increases σ by one order of magnitude and the correlation correction by additional 2.5 order of magnitudes. These results are in accordance with the considerations mentioned at the hopping frequencies and mean that essentially the Boltzmann factors determine the magnitude at the hopping conductivities.

In a subsequent paper the temperature dependence of the saturation value of $|\sigma(\omega)|$ was investigated as a function of the temperature at $\omega = 10^{11}\,\mathrm{sec}^{-1}$. In Fig. 11 $log_{10}|\sigma(\omega)|$ as function of $10^3/T$ is shown at $\omega = 10^{11}\,\mathrm{sec}^{-1}$. One can observe that while at higher temperatures ($T > 60\mathrm{K}$) the $\log_{10}\sigma - 10^3/T$ curve decreases exponentially with decreasing T, at $T \approx 60K$ its slope has nearly zero value. In other words at $T \approx 60K$ the conductivity has an activation energy, while at $T > 60\mathrm{K}$ its activation energy is, nearly zero. These results are in excellent agreement with the electron energy loss measurements of $|\sigma(\omega)|$ in a hole resonator by Tran et al performed on λ phage DNA.

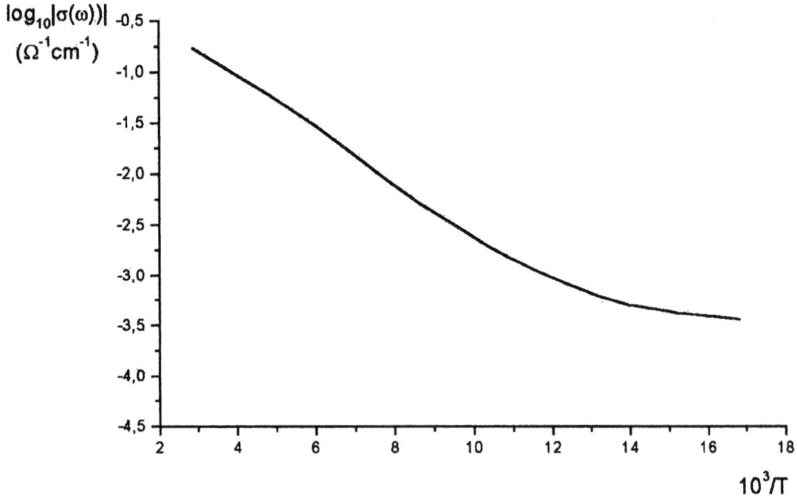

Fig. 11 $\log_{10}|\sigma(\omega)|$ (in $\Omega^{-1}\mathrm{cm}^{-1}$) at $\omega = 10^{11}\,\mathrm{sec}^{-1}$ of a double stack as a function of temperature in $10^3/T$ units.

From Fig. 11 one can see easily calculate ΔE_{ij} in the Boltzmann factor of the hopping frequencies. One finds really that ΔE_{act} is at $T \approx 60\mathrm{K} \approx 0$, while at $T > 60\mathrm{K}$ has a finite value [37]. This means that at higher temperatures the dominant mechanism of the conductivity in aperiodic nucleotide base stacks is variable range hopping, while around 60K the thermal energy is not sufficient to overcome the barriers of hopping. In this case multichannel sufficient to overcome the barriers of hopping. In this case multichannel tunneling remains as the dominant mechanism of the conductivity of the nucleotide base stacks.

6. INTERMEDIATE EXCITON THEORY FOR THE ELECTRONIC SPECTRA OF DISORDERED CHAINS

The most important biopolymers, DNA, RNA, proteins are disordered chains. The determine theoretically their experimentally rather well known electronic absorption spectra one needs a generalization of the intermediate exciton theory [39] which works quite well for periodic 1D organic polymers [40]. One has to emphasize that in the Green's matrix formalism of the intermediate exciton theory one had to apply always correlation corrected (quasi particle (QP)) one-electron energies in the denominators of the Green matrices to obtain realistic results.

An intermediate exciton theory is also important to calculate the observed spectra of other disordered chains (and copolymers which have significant technical importance).

The recently developed theory gives of course only static exciton bands of the aperiodic chains. This is, however, the first step to construct a dynamic theory for exciton transfer in proteins in photosynthetic systems [41]. For the dynamic description of the excitons a generalized master equation was used with some simplification [42]. Besides the other approximations a serious shortcoming of the theory, applied now for exciton transfer in disordered systems is that they have assumed completely localized (Frenkel) excitons, while all our calculations on organic chains have shown that the excitons in the periodic case are spread out to 4-5 neighbors in both direction from the reference cell of excitations. Probably the disorder will localize the excitons better, but without numerical calculations no answer can be given to this question. This was the other main motivation to develop the below described theory.

In Point 4 it was shown how one can determine the DOS (and with it the energy levels) with the help of the ab initio matrix block negative factor counting (NFC) method [9]. In the same point it was described also the way how to obtain correlation corrected QP energies ε_i^{QP} [18, 19] . Having these with the help of the inverse iteration method [22] also the localized one-electron wave function φ_i (near the gap) could be determined. As further ingredient to the new theory one needs the sequence and conformation of the disordered chain for which one wants to perform a calculation. In the case of

Let as start with a disordered chain of N units and 4 components. The calculation should be performed until the 5th neighbors on both sides from a chosen reference cell. Since the sequence is known we can write

$$\chi_1^{(j)}\chi_2^{(j)}\chi_3^{(j)}\chi_4^{(j)}\chi_5^{(j)}\chi_6^{(j)}\ldots\chi_{100}^{(j)} \quad (j=1,2,3,4) \tag{52}$$

The lower index stands for the position in the chain and the superscript indicates which component of the possible four is at the given position.

Let us choose a reference cell at position s with a certain component j_1. Its one-electron wave function in the valance bands region belonging to the i-th QP level ε_i^{QP} is in an LCAO form

$$\varphi_i^{(V)}\left(\vec{R}_s^{(j_1)}\right)=\sum_{t=1}^{m_{j_1}} c_{i,t}^{(V)}\left(\vec{R}_s^{(j_1)}\right)\chi_t^{(j_1)}\left(\vec{R}_s^{(j_1)}\right) \quad \left(t=1,2,\ldots m_{j_1}\right) \tag{53}$$

In (53) $c_{i,t}^{(V)}\left(\vec{R}_s^{(j_1)}\right)$ is the LCAO coefficient belonging to the basis function $\chi_t^{(j_1)}(\vec{R}_s)$ (the t-th basis function of the j_1-th kind of unit at position $\vec{R}_s$ at the level i in the valance band region), m_{j_1} stands for the number of basis functions in the component j_1 and i runs from 1 until $i_{\max}$,

$$i_{\max}=\sum_{j=1}^{4} N_j m_j \tag{54}$$

where N_j is the number of j-th components.
After this one can define one-electron excited state wave functions

$$\varphi_l^{(c)}\left(\vec{R}_s^{(j_1)}+\vec{R}_u^{(j)}\right) \quad \left(u=\pm 1,\pm 2,\pm 3,\pm 4,\pm 5\right)\left(j=1,2,3,4\right) \tag{55}$$

which are localized near to $\vec{R}_s^{(j_1)}$. If within 5 neighbors no such wave functions occur (which in the case of larger unit cells is rather improbable) one has to restart the calculation with an another ground state reference cell $\vec{R}_{s'}^{(j_1')}$.

This makes it possible to write down a Slater determinant of any excited state

$$\left|\Psi_{i\to l}^{V\to C}\left[\vec{R}_s^{(j_1)},\vec{R}_u^{(j)}\right]\right\rangle=\sum_{j_1}^{4}\hat{a}^{+(C)}_{\left[\vec{R}_s^{(j_1)}+\vec{R}_u^{(j_1)}\right],l}\,\hat{a}^{(V)}_{\vec{R}_s^{(j_1)},i}\left|\Phi_0\right\rangle \tag{56}$$

One should notice that if j_1 fixed and the sequence is known, starting from $\vec{R}_s^{(j_1)}$ fixes the value of j in any $\vec{R}_s^{(j_1)}+\vec{R}_u^{(j)}$.

Allowing any ground state reference cell in the disordered chain one can write instead of (54)

$$\left|\Psi_{i\to l}^{V\to C}\left(\vec{R}_u^{(j)}\right)\right\rangle = N^{-1/2}\sum_{\vec{R}_s^{(j_1)}=1}^{N_{j_1}}\sum_{j_1=1}^{4} d_{\vec{R}_s^{(j_1)},i\to l}^{V\to C}\left|\Psi_{i\to l}^{V\to C}\left[\vec{R}_s^{(j_1)},\vec{R}_u^{(j)}\right]\right\rangle \quad (j=1,2,3,4) \tag{57}$$

Finally allowing also for $\vec{R}_u^{(j)}$ to take any u value between –5 and 5 one obtains

$$\left|\Psi_{i\to l}^{V\to C}\right\rangle = \sum_{u=-5}^{5}\Omega_{i\to l}^{V\to C}\vec{R}_u^{(j)}\left|\Psi_{i\to l}^{V\to C}\vec{R}_u^{(j)}\right\rangle \tag{58}$$

For the calculation of the coefficients Ω in (58) one has to know the potential energy between the excited electron and the remaining hole which is the matrix element of the electron-hole interaction operator H_{eh} with the Slater determinants $\Psi_{l-i}^{V\to C}\vec{R}_u^{(j)}$ and $\Psi_{l-i}^{V\to C}\left[\vec{R}_t^{(j_2)}\right]$. Performing the algebra one obtains

$$V^{(V,C)}\left(\vec{R}_s^{(j)},\vec{R}_{t,l-i}^{(j_2)}\right)= \\ =\sum_{j_1=1}^{5}\sum_{u=-5}^{5}\left[\left\langle \varphi^{(C)}_{\left[\vec{R}_s^{(j_1)}+\vec{R}_u^{(j)}\right]}(1)\varphi^{(V)}_{\vec{R}_s^{(j_1)}}(2)\left|\frac{1}{r_{12}}\right|\varphi^{(V)}_{\vec{R}_u^{(j)}}(1)\varphi^{(C)}_{\vec{R}_t^{(j_2)}}(2)\right\rangle - \text{Exch}\right] \tag{59}$$

$$(j, j_1\, j_2 = 1,2,3,4)$$

After this we can write down the Hamiltonian of the system

$$\hat{H} = \hat{F} + \hat{H}_e + \hat{H}_h + \hat{H}_{e-h} \tag{60}$$

which has the eigenvalue equation

$$\hat{H}\left|\Psi_{l-i}^{V\to C}\right\rangle = E_{l-i}^{V\to C}\left|\Psi_{l-i}^{V\to C}\right\rangle \tag{61}$$

Putting $E_{HF}=0$ as the zero point of the energy scale one obtains from (60) and (61) (treating H_{eh} as a perturbation operator)

$$\Psi_{l-i}^{V\to C} = \left[E_{l-i}^{V\to C} - \hat{H}_e - \hat{H}_h\right]^{-1}\hat{H}_{eh}\Psi_{l-i}^{V\to C} \tag{62}$$

Multiplying this from the left by $\langle \Psi_{\vec{R}_r^{(j_3)}}{}_{l-i}|$, inserts the resolution of identity

$$I = \sum_u \left|\Psi_{\vec{R}_u^{(j)} l-i}\right\rangle\left\langle\Psi_{\vec{R}_u^{(j)} l-i}\right| \tag{63}$$

on the right hand side and takes into account equ. (61) one finally obtains

$$\begin{aligned}\Omega_{\vec{R}_r,l-i} = \sum_{\vec{R}_u^{(j)}}\sum_{\vec{R}_t^{(j_2)}} \left\langle\Psi_{\vec{R}_r^{(j_3)},l-i}\right|\left[E_{l-i} - \left(\hat{H}_e + \hat{H}_h\right)\right]^{-1}\left|\Psi_{\vec{R}_u^{(j)},l-i}\right\rangle \times \\ \times \left\langle\Psi^{(j)}_{\vec{R}_u^j,l-i}\right|\hat{H}_{e-h}\left|\Psi_{\vec{R}_t^{(j_2)},l-i}\right\rangle \qquad (j_3 = 1,2,3,4)\end{aligned} \tag{64}$$

Since

$$\left\langle\Psi_{\vec{R}_s^{(j_1)},l-i}\right|\hat{H}_{eh}\left|\Psi_{\vec{R}_t^{(j_2)}}\right\rangle = V^{(V,C)}\left(\vec{R}_s^{(j_1)}, \vec{R}_{t,l-i}^{(j_2)}\right) \tag{65}$$

we have the final result

$$\Omega^{(V,C)}_{\vec{R}_r l-i} = \sum_{\vec{R}_s^{(j_1)}}\sum_{\vec{R}_t^{(j_2)}} G^{V,C}(\vec{R}_r^{(j_3)}, \vec{R}_s^{(j_1)}, E_{l-i}) V^{V,C}(\vec{R}_s^{(j_1)}, \vec{R}_{t,l-i}^{(j_2)}) \Omega^{(V,C)}_{t,l-i} \tag{65}$$

with

$$G^{V,C}(\vec{R}_r^{(j_3)}, \vec{R}_s^{(j_1)}, E_{l-i}) = N^{-1} \frac{\sum_i \left\langle \Psi\left(\vec{R}_r^{(j_3)}\right)_{l-i} \Psi\left(\vec{R}_s^{(j_1)}\right)_{l-i}\right\rangle}{E_{l-i} - \left(\varepsilon_i^{QP(C)} - \varepsilon_{2i-l}^{QP(V)}\right)} \tag{65}$$

In actual calculations one starts from equ. (59) for which one uses the localized wave functions obtained from the inverse iteration (see above). Having $V^{(V,C)}$, one solves the Green's matrix element (65) for E_{l-i} which is the exciton energy. Calculating different E_{l-i} values one can obtain the bunches of excitonic levels which occur now instead of the exciton bands.

Substituting for a given *l-i* value this back into equation (65) together with $V^{(V,C)}$ one can solve the equ. system (65) for the coefficients $\Omega^{(V,C)}_{\vec{R}_r l-i}$. With the help of these coefficients one can obtain from equ. (58) the excitonic wave functions.

One should emphasized that one can perform such calculations for both singlet and triplet excitonic states of a disordered chain. Further the number of units, of the components and the sequences as well the conformation of the chain can freely chosen. Finally it should be mentioned the whole method could be easily generalized to 2D and 3D disordered systems. The main difficulty is that for the quasi particle energies and for the wave functions belonging to them some other (usually more difficult method like CPA) have to be used, because (as was discussed already in Point 3) the NFC method works only for 1D systems.

7. OUTLOOK

Despite of the fast development in the quantum theory of disordered chains which took place in the last twenty years there remained many problems which were not even approached. In the rapidly progressing and technically important field of non-linear optics of polymers nothing happened until now to treat also disordered polymers.

The correlation problem in disordered chains is only poorly solved if the units are large (like in case of large molecules).

In the problem of exciton transfer in disordered chains some work has already been done [41,42], but the solution of the problem is only at its beginning.

Light emitting diodes have drown rather large attention, because of their technical importance, but very little is understood about the mechanism which govern this and related phenomena. It has been recently found [43] that the electroluminescence quantum efficiency of organic light emitting diodes is proportional to their chain length. The phenomenon is not yet properly understood and one has no idea what would happen, if instead of a periodic chain a disordered chain would be investigated.

Polymers can be used as materials coupled with metal or metal oxides. Hardly any theoretical investigation have been done to understand the properties of these non-periodic systems.

Acknowledgements

The author should like to express his deep gratitude to the Late Professor Per-Olov Löwdin, who through his numerous invitations to the Uppsala Quantum Chemistry Department (starting from the early sixties) and the Sanibel Symposia has introduced him to the international scientific community. The interesting and thought provoking discussions with Him has contributed very much to his knowledge and have kept his interest in the problems of quantum biology all the time alive. He feels really very proud that he is a student of Löwdin.

He is further very much indebted to Professor Erkki Brändas for providing the texts of many early papers of Per-Olov and to Dr Y.-J. Ye for performing the sophisticated hopping conductivity calculations.

The author is also very grateful Dr. Ferenc Bogár for his continous discussions and cooperation and his help in preparing this manuscript.

References

[1] See for instance: Ladik, J.J. (1988) *Quantum Theory of Polymers as Solids*, Plenum Press, New York – London, p168.
[2] Löwdin, P.-O. (1956) *Adv. in Physics* **5**, 1.
[3] Del Re, G., Ladik, J.J., and Biczo, G. (1967) *Phys. Rev.* **155**, 997.
[4] André, J.-M., Gouverneur, L. and Leroy, G. (1967) *Int. J. Quant. Chem.* **1**, 427, 451.
[5] Löwdin, P-O. (1951) *J. Chem. Phys.* **18**, 365.
[6] Dean, P. (1960) *Proc. Roy. Soc. London* **A254**, 507, ibid (1961) **A260**, 263.
[7] Dean, P. (1972) *Rev. Mod. Phys.* **44**, 127.
[8] Seel, M. (1979) *Chem. Phys.* **43**, 103.
[9] Day, R.S., and Martino, F. (1981) *Chem. Phys. Lett.* **84**, 86; for more details see ref [1] Point 4.4 (p. 140).
[10] Elliott, J., Krumhansl, J.A., and Leath, P. (1974) *Rev.Mod.Phys*, **46**, 465; Falicov, L.M., Yndurain, F. (1975), *Phys.Rev.*, **B12**, 5664.
[11] See for instance: Martino, F (1974) in André, J-M., Delhalle, J., and Ladik, J. (eds.), *Quantum Theory of Polymers*, Reidel Publ. Comp., Dordrecht-Boston.
[12] See for instance: Fetter, A.L., and Walecka, J.D. (1971) *Quantum Theory of Many Particle Systems*, McGraw Hill, New York–London.
[13] Kirckpatrick, S., Velicky, B., and Ehrenreich, H. (1970) *Phys. Rev.* **B11**, 3250.
[14] Seel, M., Collins, T.C., Martino, F., Rai, D.K., and Ladik J. (1978) *Phys. Rev.* **B18**, 6160.
[15] Street, B. (personal communication)
[16] Bardeen, J., Cooper, L., and Schrieffer, R. (1957) *Phys. Rev.* **108**, 1175.
[17] Hein, R.A., Gibson, J.W., and Blaughter, R.D. (1964) *Rev. Mod. Phys*. **36**, 149.
[18] Ladik, J., and Čížek, J.I. (unpublished)
[19] Liegener, C.-M. (1989) *Chem. Phys.* **133**, 73.
[20] Liegener, C.-M., Sujianto, A., and Ladik, J. (1990) *Chem. Phys.* **145**, 385.
[21] Cederbaum, L.S., and Domcke, W. (1977) *Adv. Chem. Phys.* **36**, 205.
[22] Wilkinson, J.H. (1965) *The Algebraic Eigenvalue Problem*, Calendron Press, Oxford, p. 633.
[23] Mott, M.F., and Davis, E.A. (1971) *Electronic Processes in Non-Crystalline Materials*, Clarendon Press, Oxford.
[24] Grünewald, M., Polenann, B., Mohagran, B., and Wüntz, D. (1985) *Phys. Rev.* **B32**, 8191.
[25] Scher, M., and Lax, M. (1973) *Phys. Rev.* **B7**, 4491; Odagaki, M., and Lax, M. (1981) *Phys. Rev.* **B24**, 5284; *ibid* (1982) *Phys. Rev.* **B26**, 6480; Odagaki, T., Lax, M., and Day, R.S. (1984) *Phys. Rev.* **B30**, 6911.
[26] Haus, K.W., and Kehr, K.W. (1987) *Phys. Rep.* **263**.
[27] Ye, Y.-J., and Ladik, J. (1983) *Phys. Rev.* **B48**, 120.
[28] Szent-Györgyi, A. (1941) *Nature* **148**, 157; (1941) *Science* **93**, 609.
[29] See for instance: Lehninger, A.L. (1975) *Biochemistry*, North Publ., New York, p. 493.
[30] Eley, D.D., Parfitt, G.P., Perry, M.B., and Taytum, O.H. (1953) *Trans. Far. Soc.* **49**, 79; Liang, C.Y. and Scalco, E.G. (1964) *J. Chem. Phys.* **40**, 9199; Pethig, R., and Szent-Györgyi, A. (1977) *Proc. North. Acad. Sci. USA* **74**, 226; Bone, S., Lewis, T.F., Pethig, R., and Szent-Györgyi, A, (1978) *Proc. North. Acad. Sci. USA* **75**, 315.

[31] Eley, P.D., and Spivey, D.J. (1960) *Trans. Far. Soc.* **56**, 1432; ibid (1985) **58**, 411; O'Konski, C.T., and Shirai, M. (1963) *Biopolymers* **1**, 557; O'Konski, C.T. (1963) *Rev. Mod. Phys.* **35**, 721.

[32] See Fig. 7.15 of Mott, M.F., and Davies, E.A. (1971) *Electronic Processes in Non-Crystalline Materials*, Clarendon Press, Oxford.

[33] Bell, G.I., Picklett, R., and Writter, I. (1980) *Nucleic Acad. Research* **8**, 4091.

[34] Dickerson, R.A., Drew, H.R., Counter, B.N., Wing, R.M., Fratini, A.V., and Kopka, M.L. (1982) *Science* **216**, 475.

[35] Ye, Y.-J., Chen, R.-S., Martinez, A., Otto, P., and Ladik, J. (1999) *Solid State Comm* **112**, 139.

[36] Shen, L., Ye, Y.-J., and Ladik, J. (2002) *Solid State Comm.* **121**, 35.

[37] Tran, P., Alavi, B., and Grüner, G. (2000) *Phys. Rev. Lett.* **85**, 1564.

[38] Martinez, A., Čížek, J., and Ladik, J. (2002) *Phys. Lett B* (submitted).

[39] Takeuti, Y. (1957) *Progr. Theor. Phys. (Kyoto)* **18**, 421; ibid (1959) *Suppl.* **12**, 75.

[40] Suhai, S. (1984) *Phys. Rev. B* **29**, 4570; Suhai, S. (1984) *Int. J. Quant. Chem.* **11**, 223; Suhai, S. (1985) *J. Mol. Struct.* **123**, 97; See also further examples in Ladik, J. (1988) *Quantum Theory of Polymers as Solids*, Plenum Press, New York – London, Chapter 8.

[41] Barvik, I. (1991) in Blumen, A. (ed) *Dynamical Processes in Condensed Molecular Systems*, World Scientific, Singapore, p. 275.

[42] Dahlbohm, M., Minami, T., Chernyak, V., Pullerits, T., Sundström, V., and Mukamel, S., (2000) *J. Phys. Chem.* **B104**, 3976.

[43] Wohlgenannt, M., Iiarg, X.M., Vardeny, Z.V., and Janssen, R.A. (2002) *Phys. Rev. Lett.* **88**, 19740.

MOLECULAR ORBITAL THEORY OF THE GASEOUS BOSE-EINSTEIN CONDENSATE:

NATURAL ORBITAL ANALYSIS OF STRONGLY CORRELATED GROUND AND EXCITED STATES OF AN ATOMIC CONDENSATE IN A DOUBLE WELL

William P. Reinhardt and Heidi Perry
University of Washington, Seattle, WA 98121-1700, U.S.A.

Keywords: Gaseous Bose-Einstein condensate in a double well trap, natural orbitals, condensate fragmentation, Josephson effect, macroscopic quantum tunneling, correlations in Bose systems.

Abstract The possibility, envisaged in 1925 by Einstein following the suggestion of Bose, of a dilute gas of atoms being *condensed* into a single quantum state was experimentally achieved in 1995 following decades of research. An avalanche of experiment and theory has followed, leading to the awarding of the 2001 Nobel Prizes in Physics to three of the pioneering experimentalists. Theory, mostly couched in the language and formalism of condensed matter physics, has developed apace. What we point out here is that a condensate in a double well trap may be thought of exactly as a diatomic molecule with delocalized molecular orbitals (MOs) filled with Bosons, rather than electrons. As in quantum chemistry, configuration interaction must be included to supplement the MO picture if dissociation is to be correctly described. Such dissociation is called *condensate fragmentation* in the condensed matter theory literature. Natural orbitals and their occupation numbers provide a clear and succinct way of determining "where the atoms are" in highly correlated states as fragmentation occurs, and are particularly helpful in understanding what is actually happening when energy levels with macroscopic occupancy "cross" as a function of breaking the symmetry between the trap wells. It is seen that only a few particles are actually active in the level crossing. Recent experiments in super-conductive Josephson loops showing tunneling of macroscopic quantum states are easily included in this picture.

E.J. Brändas and E.S. Kryachko (eds.),
Fundamental World of Quantum Chemistry, Vol. II, 305–348.

1. INTRODUCTION

In 1924, even before the advent of modern quantum mechanics, Bose and Einstein posited the possibility of particles, now called Bosons, eschewing the exclusion principle, and Einstein (1925) pointed out the possibility of an ideal gas of such particles achieving a density wherein some *would not participate in the usual thermal agitation.* A composite particle, such as a neutral atom, will be a Boson if it contains an even number of total Fermions, and interacts with its environment in such a manner that its composite nature is not apparent. Thus, for example ^{23}Na is a Boson, with 23 neutrons and protons, and 11 electrons; while a mass 22 or 24 isotope of sodium would be a composite Fermion. In modern terms Einstein's criterion for such a *condensation*, referred to as a Bose-Einstein condensate or BEC, is that the relative deBroglie wavelength would exceed the mean separation between atoms. Approximate ideality requires that both of these length scales be large compared to an atomic size. To condense an actual gas of Bosonic atoms, such as ^{1}H, ^{7}Li, ^{23}Na, or ^{85,87}Rb in a magnetic trap, held via the magnetic moment of the outer unpaired electron, temperatures of the order of tens of nano-Kelvin are required. Such conditions were first achieved, and clear evidence of a quantum phase transition to the condensate observed, in the laboratory in 1995. Good overviews of the pre-and post- 1995 state of experiments and theory appear in Griffin, Snoke, and Stringari (1995) and Pethick and Smith (2002), respectively. Recent theoretical overviews are Dalfovo et. al., (1999), and Leggett (2001).

The Griffin, Snoke, and Stringari (1995) collection of articles contains extended discussions of condensates in more complex condensed matter systems: superfluids and superconductors are Bose condensates. Tinkham (1996) discusses superconductors; and, Tilley and Tilley (1990) overview both superfluids and superconductors. Both, of course, had been observed well before those in the gas phase. What makes the gaseous condensates distinctly different from those of traditional condensed matter physics is that in the gaseous condensate: i) number densities are under experimental control over many orders of magnitude; ii) via optical pumping between hyperfine states, the number of components in a condensate is under experimental control; iii) the effective interaction between condensed atoms is fully tunable in many systems, even with respect to sign; iv) optical manipulations of the original single well magnetic traps allows creation of two or many well condensates, with the tunneling of particles between them also tunable; v) the condensates are *naked*, that is not hidden inside a solid or liquid, a photograph or other image of a gaseous condensate is an image of the condensate particle density itself. The gaseous BEC thus offers an exceptional playground for theory and experiment, and thus a test bed for understanding and testing some of the basic underpinnings of condensed matter quantum physics.

What does all of this have to do with quantum chemistry? We take a condensate in a single well potential to be the analog of an atom, with the trap playing the role of the attractive nucleus, and the electrons being replaced by the condensate atoms, which being uncharged can easily be of variable number, and ask for the ground state to be described by a fully symmetric Hartree wave function. We then enquire as to the importance of correlations. These are typically found to be small in three-dimensional single well traps, but important as one or two of the dimensions are reduced beyond a critical length scale, see, for example, Girardeau et. al. (2001), and references therein.

However, if a single well condensate is the analog of an atom, then multi-well condensates are molecules! Once molecules are introduced, any self-respecting quantum chemist will know that correlations must be properly included if dissociation is to be described. Experimentally in a gaseous BEC such a dissociative process is called *condensate fragmentation*, and may be achieved by raising the tunneling barrier in a double (or multi-) well condensate as discussed in Orzel et. al. (2001) and Greiner et. al. (2002). Extended multi-well fragmentation is a "Mott" transition, familiar in the theory of insulator-conductor transitions in the band theory of electrons in solids, as discussed in, for example, Ashcroft and Mermin (1976).

More recently neutral Fermionic atoms have been condensed in the gas phase, DeMarco and Jin, (1999, 2002), offering the possibility of model atoms and molecules with arbitrary numbers of wells and arbitrary numbers of "uncharged" analogs of electrons. When correlations become important in either the Bose or Fermi cases, a clear opportunity for quantum chemists arises. Per-Olov would be delighted!

We make no attempt to give a complete overview of this material here. Rather, in the style of traditional quantum chemistry we assume that a proper zero order description of the ground state is to simply put all of the Bosons in a single configuration of atomic or molecular orbitals (MOs), and then use the appropriate mean-field or configuration interaction (CI) corrections, for example see McWeeny and Sutcliffe (1969), to achieve the necessary level of description. Once CI is introduced, a natural question is as to what diagnostics are to be used to allow us to describe "what the correlated particles" are doing in terms of an orbital picture. A most natural and convenient manner in which to do this is to use the Bosonic equivalent of Löwdin's natural orbitals (NOs) and their corresponding occupation numbers, Löwdin (1955).

To illustrate the utility of the NO-MO analysis we consider the simple problem of a gaseous BEC in a symmetric double well trap. This "two mode" BEC problem has been discussed many times before in quite different languages, see, for example: Milburn et. al. (1997), Cirac et. al. (1998), Steele and Collett (1998), Gordon et. al. (1999), Raghavan et. al. (1999), Spekkens and Sipe (1999), Franzosi and Penna (2001).

Three problems will be of interest here:

1) understanding the multi-state CI (Born-Oppenheimer like) state correlation diagram as a function of tunneling strength in an MO picture. Alternative, and quite different, discussions of this problem appear in Steele and Collett (1998), and Franzosi and Penna (2001).

2) the description of condensate fragmentation in the NO-MO framework both for the ground and highly excited states; and,

3) the understanding of what particles are actually involved in the tunneling of macroscopic quantum states as these become near degenerate and are only weakly coupled via a tunneling barrier. We will see that the avoided crossings of macroscopic tunneling states can involve a very small number of active particles.

In the course of generating the full correlation diagram, looking at all of the energy levels resulting from a full CI (complete CI in a restricted basis) as a function of barrier height between the double wells, we find immediate evidence for a collective and sharp macroscopic quantum transition. This observation, although made in a novel manner, is not at all new: we are immediately taken back to P. W. Andersons's interpretation of the Josephson effect as a physical pendulum; Anderson (1964, 1966, and 1984). The physical pendulum has two distinctly differing types of motion: oscillations about a single minimum, and hindered rotation. Projection of our CI wave functions into the proper semi-classical phase space, Husimi (1940), shows that these two types of motion are precisely what is implied in the various types of quantum mechanical excited states of the double well condensate. Where hindered or free rotation is at hand both energies and occupation numbers are (exactly or nearly) doubly degenerate, as the rotation may be in either of two directions, and a simple linear combination of the correlated wavefunctions gives fully localized quantum states, and NOs. We thus interpret the correlation diagram in terms of an MO to AO localization process.

The outline of the paper is a follows: in Section 2 the Hartree theory of a BEC localized in a single well is introduced, and its low energy excitations briefly discussed. In particular, it is noted that typical low energy excitations are "collective" motions of all particles, rather than the atomic/molecular picture of "single particle excitations" from a single ground state configuration. The reason for the lack of low-lying single particle excitations is discussed following Huang and Yang (1957). The single particle excitations will, perhaps surprisingly, and pleasingly to quantum chemists who might wonder where they appear, reappear in the excitation spectrum of the double well condensate, and in the description of macroscopic quantum tunneling. A simple highly restricted CI model for a

BEC in a double well potential is then introduced following the work of Spekkens and Sipe (1999), who focus on fragmentation of the ground state. The Spekkens-Sipe Hamiltonian is seen to be easily parameterized, and to then simplify to a Bose-Hubbard site-to-site tunneling model, see, for example Fisher et.al. (1989), used directly in many of the previous studies of two-mode BECs, and indicating the generality of the results obtained here. Finally, in this introductory section, the first order reduced density matrix (1-matrix) is introduced, and factored. This is first done within the framework of the Onsager-Penrose, Penrose (1951), Penrose and Onsager (1956), *ansatz* of ODLRO, or "Off Diagonal Long Range Order." Mathematical implementation of the ODLRO *ansatz* requires introduction of phase-number coherent states, Anderson (1966), Loudon (1973), if accomplished directly as a factorization in 2nd quantization. These coherent states do not correspond to states with a fixed particle number, but are introduced as they will prove useful in the later analysis. As an alternative to the use of coherent states to obtain the ODLRO factorization, the Löwdin NO-occupation number analysis is introduced; see, Löwdin (1955), McWeeny (1959), Davidson, (1976). In Section 3 an overview of numerical results for the full CI spectrum of energy states of the double well BEC is given as a function of barrier height separating the two wells. Condensate fragmentation is observed. The correlation diagram, and occupation number analysis, shows immediate and clear evidence for a macroscopic quantum phase transition, which is the crossover from a delocalized MO picture to degenerate pairs of states which may be easily combined to give a localized orbital description.

The interpretation of this crossover is discussed in Section 4, where contact with the early work of Anderson (1964, 1966) and his interpretation of the Josephson effect as a "physical pendulum" is made. Using an appropriate coherent state basis, a Husimi projection, Husimi (1940), Martens (1989), from individual stationary states for the N particle system is shown to give precisely the semi-classical behavior predicted by Anderson, even for N as small as 20 to 40 particles. The description of the fragmented "rotor" states in terms of localized CI states and corresponding localized NOs is then given. In Section 5 the problem of macroscopic quantum tunneling, recently observed in superconducting current loops, i.e. see van der Wal et. al., (2000), Freidman et. al. (2000), and Yu et. al. (2002), is considered. The symmetry of the Spekkens-Sipe double well is broken by addition of an additional parameter: as a function of this parameter, rotor state degeneracy is broken, and the quantum tunneling splittings studied. Near the crossover region these tunneling splittings are observable. The appropriate NO analysis shows, however, that actually only a few (usually one to three, but variable according to the tuning and magnitude of the splitting) particles, out of many hundreds modeled theoretically, or millions in actual experiments, are actually "moving" from one MO to another in the crossing process. In this situation "few body" excitations actually become the dominant excitation mechanism, in strong contrast to the usual picture of superfluid excitations being collective modes. This also immediately clarifies why a single microwave photon can couple such macroscopic states, an issue not addressed clearly in the experimental papers above. Section 6 contains conclusions and prospects for further work.

2. A Gaseous BEC in a Double Well Potential: Correlations and Their Analysis

In this Section the mean-field equations for a BEC in a single potential well are briefly derived from a wavefunction point of view, and the Fermi-contact interaction is introduced (Section 2.1). This results in the Gross-Pitaevskii (GP), or non-linear Schrödinger equation (NLSE); see Gross (1961), Pitaevskii (1961), Sulem and Sulem (1999), respectively. The single configuration GP level description encompasses much of the observed behavior of trapped BECs. Densities, low-lying collective excitations, e.g. Ruprecht et. al. (1995), Edwards et. al, (1996) and even strongly localized non-linear excitations such as solitons, Reinhardt and Clark (1997), Reinhardt (1998, and references therein) and vortices, Gross (1961), Pitaevskii (1961) are collective modes, in that all N particles remain in an identical single particle orbital. Notably absent are low energy "single particle" excitations of the type common in atoms and molecules where the usual optical transitions take a single electron from one orbital to another. The reason for absence of such single particle excitations appears in first order Rayleigh-Schrödinger perturbation theory, as exposed by Huang and Yang (1957). It may be interpreted as a Bosonic amplification effect for particles to remain in the same single particle state, once that single state has "macroscopic occupancy". This restriction is immediately removed if instead of a single condensate, we have two, or more, weakly coupled condensates. A single particle may then easily jump from one already macroscopically occupied state to another. Tunneling between macroscopically occupied states is a facile single particle process controlled by a one particle tunneling matrix element. We will not, then, be surprised, to see that the NO analysis of excited states of a double well condensate shows that states typically differ by moving small numbers of particles from one NO to another. In Section 2.3 the model Hamiltonian used herein is introduced, and it's relationship to the Bose-Hubbard model of condensed matter physics made clear. Finally, the 1-matrix is introduced in Section 2.4, and factored: i) by fiat; and, ii) more sensibly to a quantum chemist, by following the lead of P.-O. Löwdin, by introduction of a natural orbital analysis, Löwdin (1955).

2.1 Mean Field Theory for a Single Well BEC

The usual derivation of the condensed matter mean field theory follows from factorization of the ground state expectation value of the second quantized Heisenberg equation of motion for a Bosonic field operator. See, for example, Griffin (1996), Fetter and Walecka (1971). Quantum chemists will not be surprised that identical mean field equations may be derived from the *ansatz* that the ground state wavefunction for an N Boson ground state consists of the fully symmetric product of N identical one particle orbitals (MOs in what follows), and then minimizing the energy of this single "perminantal" wavefunction with respect to its single orbital. This is simply the Bosonic analog of, for example, the derivation of the Roothaan restricted Hartree-Fock equations for a closed shell atom or molecule from a single "determinantal" wavefunction, Roothaan (1952). The resulting Hartree (i.e. no exchange) equation is

$$\left(-\frac{\hbar^2}{2m}\nabla_r^2 + V^{trap}(r)\right)\phi(r) + \sum_{i=1}^{N-1}\left[\int \phi_i^*(r')v(r',r)\phi_i(r')d^3r'\right]\phi(r) = \mu\phi(r) \qquad (2.1)$$

where $v(r_i,r_j)$ is the pair potential. Here, a physical assumption has been made. For non-relativistic electrons, for example, there *is* only a spatial pair interaction $-1/r_{ij}$. This is not true for atoms in a condensate. For a gas of cold atoms it must be assumed that the density is low enough so that all three-body effects, including three body contributions to the potential energy, may be neglected in writing Eqn. (2.1). This is consistent with the existence of the condensate itself: if three body interactions become important, actual diatomic molecules or clusters may form, and be lost from the condensate.

Given that the condition for formation of a BEC is that local relative deBroglie wavelengths are long compared to inter-particle distances, the local interaction between any pair of atoms is dominated by the s-wave interaction. Following Fermi, the pair interaction may then be replaced by the contact, or point, interaction, long familiar as the Fermi contact interaction in the description of an s-state electron interacting with an atomic nucleus (see, for example, Flygare, 1978). The contact interaction "pseudo-potential" replacement is

$$v(r_i,r_j) \rightarrow \frac{4\pi\alpha_s\hbar^2}{m}\delta^3(r_i - r_j) \equiv g\delta^3(r_i - r_j) \qquad (2.2)$$

and is thoroughly discussed by Huang, Lee, and Yang (Huang and Lee, 1957), and Lee et. al. (1957), in the context of the perturbation theory of an extended N Boson system, where the ground state orbital is taken as a k = 0 plane wave, namely a constant. The constant α_s in Eqn (2.2) is the s-wave scattering length, and is taken so that a Born approximation to the scattering amplitude gives the "exact" s-wave scattering amplitude. This is thus the lowest order part of a renormalization process, by which a hard core repulsive and short range attractive Born-Oppenheimer molecular pair potential of considerable complexity is replaced by a regularized potential, allowing plane wave matrix elements to be taken in a simple manner. This is a matter of some subtlety, see Huang (1957), above, and also Leggett (2001). Accepting the replacement of the pair potential by the contact interaction of Eqn (2.2), which is certainly fully empirically justified, the Hartree mean field equation for the ground state single particle orbital, ϕ, becomes:

$$\left(-\frac{\hbar^2}{2m}\nabla_r^2 + V^{trap}(r) + g(N-1)|\phi(r)|^2\right)\phi(r) = \mu\phi(r), \qquad (2.3)$$

as all of the ϕ_i are the same for the ground state configuration. Hartree self-consistency follows from the requirement that the same single particle orbital ϕ(r) sits inside the effective potential, g$|\phi(\mathrm{r})|^2$, and also solves the mean field eigenvalue equation (2.3). The eigenvalue μ is, as usual, a chemical potential, or "ionization" energy, i.e. Koopman's Theorem holds. In an extended large N system of neutral atoms, the energy to add or subtract a single particle will be the same, and both equal to μ. Eqn. (2.3) is usually referred to as the non-linear Schrödinger equation (NLSE) or the Gross-Pitaevskii (GP) equation, from its original use in condensed matter theory, see again, Gross (1961), Piteavskii (1961), Sulem and Sulem (1999). Hartree (1928) could have written down Eqn. (2.3) at once, if he had been considering Bosons, and had known to employ the pseudo-potential. Bravery, hubris, and spectacular empirical agreement with many classes of experiment suggest the time dependent generalization

$$i\hbar\frac{\partial}{\partial t}\phi(r,t)=\left(-\frac{\hbar^2}{2m}\nabla_r^2+V^{trap}(r)+g(N-1)\left|\phi(r,t)\right|^2\right)\phi(r,t). \quad (2.4)$$

Eqn. (2.4) has been successfully used to describe the dynamics of phonons, where the time dependent GP equation gives energies equivalent to those of the RPA, Ruprecht et. al. (1995), Edwards et.al. (1996), Thouless (1961). It also well describes dark solitons, Reinhardt and Clark (1997), Reinhardt (1998), Denschlag et. al. (2000); bright solitons, Strecker et.al. (2002), Khaykovich et. al. (2002); and, soliton-vortex interactions, Anderson, B. P., et. al. (2001), and vortex-solitonic structures, Brand and Reinhardt (2002), in properly confined condensates. Leggett (2001) has correctly noted that these seemingly successful applications can take the GP theory way beyond its proven domain of validity. For example, Eqn. (2.4) does not even describe the decay of the excitations it seemingly describes well. Thus its usage is not fully theoretically justified, as the timescales over which it is valid are currently unknown. This, of course, has not prevented many successful applications. The use of the GP, or time dependent GP, equations *does* neglect correlated motions, and thus cannot correctly describe excitations wherein one or more particles "leave" the ground state configuration. Correlations do dominate the dynamics of 1-dimensional condensates, e.g. Girardeau (2001), and references therein. However, in 3 dimensions even creation of strongly localized "high energy" excitations, whether these are kinks in stationary states (Carr et. al., 2000a,b), or time evolving excitations such as solitons or vortices, see above, may not involve correlations, and may thus be described by a single configuration N-Boson wavefunction, wherein all N Bosons have an identical single particle wavefunction, $\phi(r,t)$. As long as this is the case, the time dependent Hartree theory of Eqn. (2.4) can be applied. Where, then, are the "single and double" excitations familiar to quantum chemists in the CI picture of correlations in the electronic structure of atoms and molecules, where they are needed for accurate determination of energetics, and also for proper description of dissociation? Such excitations also form the basis for discussion of the low energy shell model electronic spectroscopy of these same systems. A hint as to their absence is given by first order perturbation theory.

2.2 First Order Perturbation Theory for Single Particle Excitations

Huang and Yang (1957) introduced a first order perturbative description of a dilute, hard sphere Bose gas in a macroscopic uniform volume, V. That is, N Bosons in a box with periodic boundary conditions. This was soon followed by estimates of the 2nd and 3rd order corrections to the energy and density in the same uniform thermodynamic limit, Lee et. al., (1957), and Wu (1959). However, the simple first order result is of exceptional interest, in that, for a condensate of repulsive effective interaction, i.e. positive scattering length, α_s, it is immediately seen that N-Boson systems achieve low energy via multiple occupancy of a single orbital. Consider a uniform Bose system with zero order energies, ε_i, and corresponding occupation numbers, n_i. The zero order energy is

$$E_0 = \sum_{i=0}^{\infty} n_i \varepsilon_i \tag{2.5}$$

where $N = \Sigma_i n_i$. The ground state zero order energy is thus given by the configuration $n_0 = N$, and $n_i = 0$ for all $i > 0$, and is thus $E_0 = N\varepsilon_0$. Huang and Yang (1957) found that the first order correction to this unperturbed energy for a configuration with occupation numbers n_i is

$$E_1 = \left(\frac{g}{V}\right)\left(N^2 - \frac{1}{2}N - \frac{1}{2}\sum_{i=0}^{\infty} n_i^2\right) . \tag{2.6}$$

For the ground state this immediately reduces to (for $n_0 = N$)

$$E_1^{(0)} = \left(\frac{g}{V}\right)\left(\frac{N(N-1)}{2}\right) \tag{2.7}$$

which is simply the number of pairs of Bosons, N(N-1)/2, times the effective pair interaction strength. At first glance the expression of Eqn. (2.6) seems hard to interpret; however, consider the "first" excited configuration, $n_0 = (N-1)$, and $n_1 = 1$. Eqn. (2.6) may then be rearranged as

$$\left(\frac{g}{V}\right)\left(\frac{N(N-1)}{2}\right) + \left(\frac{g}{V}\right)(N-1). \tag{2.8}$$

Comparing Eqns. (2.7) and (2.8) it is evident that the energetic cost of the single particle excitation is

$$\Delta = \left(\frac{g}{V}\right)(N-1). \tag{2.9}$$

As (N-1) ~ N for a macroscopic, or even mesoscopic, system (typical laboratory gaseous BECs contain 10^4 to 10^{11} particles) Eqn. (2.9) shows that there is a *macroscopic* cost for creating a single particle excitation. This situation does not arise for electrons, as, due to the exclusion principle, no Fermionic orbitals have such macroscopic occupancy. This is the origin of the "force" leading towards the Bose condensation. Restating this Huang/Yang idea: dilute, repulsive ($\alpha_s > 0$), Bose condensates condense into a single particle state space, these being plane waves for an extended uniform system, or the ground state orbital, $\phi(r)$, in a trap. Conversely, attractive condensates ($\alpha_s < 0$) condense in configuration space, where the NLSE is actually unstable with respect to collapse of the mean field wavefunction for large enough N. See, Kosmatov, et. al. (1991), Sulem and Sulem (1999), Sackett et. al. (1999), Roberts et. al. (2001), for discussions of the theory and the current experimental situation for the instability of attractive condensates. We restrict our attention to repulsive condensates in what follows. Interestingly, this local stability of repulsive condensates with respect to excitation of a single particle from a single macroscopically occupied configuration does not only apply to the ground state configuration. Macroscopic occupancy of an excited state is also locally stable with respect to single particle excitations, or de-excitations! Low lying excitations for a repulsive condensate are thus "collective" sloshings of the density, wherein all N particles stay in the same, now time dependent, orbital $\phi(r,t)$. This is the Bogoliubov picture of phonon excitations, see for example, Fetter and Walecka (1971), Leggett (2001). This, also, may be part of the explanation of the success of the time dependent GP description of highly energetic and non-linear phenomenon such as solitons, vortices and their interactions, a success surprising to traditional condensed matter physicists (again, e.g. Leggett (2001), page 352).

The argument above should not, however, be construed to imply that BECs are weakly interacting: the same factor of N in Eqn (2.9) implying a high cost of single particle excitations also appears in the effective potential of the GP equation (2.4): mean field effects are often enormous, and, for example can completely out weigh the kinetic energy. In this, the Thomas-Fermi limit, Eqn. (2.4) may be simply solved for the ground state particle density:

$$\rho(r) = N\,|\,\phi(r)\,|^2 = (\mu - V^{trap}(r))/g, \tag{2.10}$$

where $(\mu - V^{trap}(r)) > 0$. This is, in fact, often an excellent approximation to actual ground state densities of trapped (repulsive) BECs.

An interesting question for quantum chemists thus arises: When will correlations involving excitations of a small number of particles from the ground state configuration occur and be observable? The answer is clear: such excitations are needed to properly describe the division, or fragmentation, of a condensate into two (or more) parts. In this case the mean field MO picture does not allow localization of the particles in the separated pieces of the condensate. Said another way, just as in the theory of electronic structure, the MO picture does not allow molecules to dissociate to their atomic ground states: correlation is required. Additionally, for a BEC trapped in a double well, transfer of a single particle from one well to another can be a low energy event, simply because the particle is transferred from one macroscopically occupied state, say in the left well, to another already macroscopically occupied state in the neighboring right well. The perturbative argument of Huang and Yang, seemingly forbidding single particle excitations (or, more generally, few body excitations from the initial configuration), no longer applies as we have both a macroscopic energy cost, and an equivalent macroscopic re-capture of energy, as a single particle is transferred from one condensate of identical particles to another. We now introduce a simple model for such a double well condensate.

2.3 A Simple Model for a BEC in a Double Well

Many recent papers have treated the problem of a double well condensate. For example: Milburn et. al. (1997), Cirac et. al. (1998), Gordon et. al. (1999), Raghavan et. al. (1999), Spekkens and Sipe (1999), Franzosi and Penna (2001), Leggett (2001, for an overview), to name only a few. We introduce the model of Spekkens and Sipe (1999) here. Spekkens and Sipe consider the problem of fragmentation of a condensate ground state as a barrier height is raised. Our analysis goes well beyond this, but begins with the same model Hamiltonian. Approximation, and parameteriztion, of the Spekkens-Sipe model gives rise to the Bosonic Hubbard Hamiltonian (Fisher et. al. 1989), which may be used for double and multi-well tunneling processes.

Figure 1 shows a typical symmetric double well potential, V(x), in a one dimensional co-ordinate space, and we will use this model to typify the correlations which arise when a condensate is trapped in such a situation. Symmetry requires that the only two NOs are the symmetric and anti-symmetric linear combinations of the localized AOs. Neglecting mean field effects, which *is* a serious approximation but we focus on correlation effects here, let us assume that if a single particle is in the left well its localized Schrödinger wavefunction (AO) is $\phi_L(r)$, and if on the right, $\phi_R(r)$.

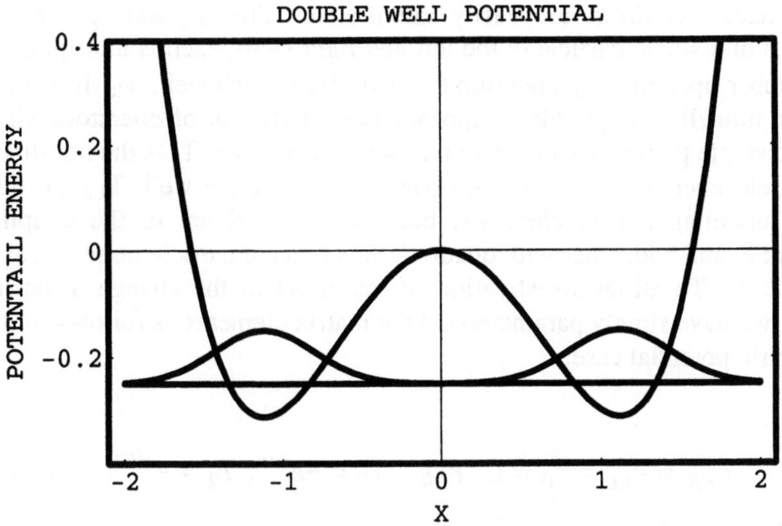

Figure 1. A double well potential, V(x), which might contain a single particle, or many particles. Shown also, are localized "atomic like" (AOs) orbitals on the right, $\phi_R(x)$, and the left, $\phi_L(x)$. Molecular like orbitals may be found as linear combinations of these.

As it is convenient to work in the language of second quantization, let the Fock state |N,M> denote the state with N particles on the left, and M on the right of the barrier, in the single particle states $\phi_L(r)$ and $\phi_R(r)$, respectively. Denoting the corresponding single particle creation-annihilation operators as $C^\dagger_L$, C_L and $C^\dagger_R$, C_R, the Spekkens-Sipe Hamiltonian is:

$$\begin{aligned} H = {} & \varepsilon_{RR}\hat{N}_R + \varepsilon_{LL}\hat{N}_L + \left(\varepsilon_{LR} + gT_1(\hat{N}-1)\right)\left(C^\dagger_L C_R + C^\dagger_R C_L\right) \\ & + g\frac{T_0}{2}(\hat{N}^2_L + \hat{N}^2_R - \hat{N}) \\ & + g\frac{T_2}{2}(C^\dagger_L C^\dagger_L C_R C_R + C^\dagger_R C^\dagger_R C_L C_L + 4\hat{N}_L\hat{N}_R) \qquad (2.11) \end{aligned}$$

where $\hat{N}_L = C^\dagger_L C_L$; $\hat{N}_R = C^\dagger_R C_R$; $\hat{N} = \hat{N}_L + \hat{N}_R$ and

$$\varepsilon_{ij} = \int \phi_i(r)\left(-\frac{\hbar^2}{2m}\nabla^2 + V^{trap}(r)\right)\phi_j(r)d^3r \quad \text{where} \quad i,j = R,L$$

$$T_0 = \int \phi^4_{L,R}(r)d^3r; \quad T_1 = \int \phi^3_L(r)\phi_R(r)d^3r$$

$$T_2 = \int \phi^2_L(r)\phi^2_R(r)d^3r \qquad (2.12)$$

These matrix elements are easily interpreted. The ε_{LL} and ε_{RR} are the energies of a single particle in the left and right wells, each is multiplied by the number operator $N_{L,R}$ counting the particles in each well. ε_{RL} is a single particle tunneling amplitude, it appears as a coefficient of operators which allow a single particle to hop from one well to the other. T_0 is the non-linear mean-field energy associated with particles in a single well. $T_{1,2}$ are non-linear tunneling matrix elements; both are proportional to the coupling constant g, and both enter with quadratic functions of particle numbers, as in Eqn. (2.7). To allow investigation of the effect of the change in barrier height, we have simply parameterized the matrix elements as follows for the symmetric potential case:

$$\varepsilon_{RR} = \varepsilon_{LL} = T_0 = 1; \;\; \varepsilon_{LR} = T_1 = -e^{-\alpha}; \; T_2 = e^{-2\alpha} \qquad (2.13)$$

This parameterization allows a simple study of continuous change in the linear and non-linear tunneling via variation of the single parameter α, not to be confused with the scattering length, α_s, which is part of the effective coupling constant "g". Large α implies small tunneling, leading to fragmentation.

In Section 3 we present results of carrying out full CI in the restricted N particle basis of Fock states $|N-n, n>$, $n = 0,1,\ldots N$ allowing the possibility of all particles being in either well, and all intermediate arrangements. This is carried out as a function of α, yielding a correlation diagram. Such a full CI is, as usual, size consistent, and may thus be applied to large numbers of particles. The simplicity of the two state double well model is that an N particle problem results in a (N+1) x (N+1) dimensional CI matrix, with eigenvector $\{c^i\}$ for the ith eigenvalue of $H(\alpha)$. The occupation representation eigenfunctions are then of the form

$$|\Psi_i\rangle = \sum_{n=0}^{N} c_n^{(i)} \, |N-n, n\rangle \; . \qquad (2.14)$$

Before giving results of computations with in this model, we note that for large N, the ε_{LR} term may be neglected, and if we also ignore the terms T_2 which scales as $e^{-2\alpha}$, rather than $e^{-\alpha}$, the model reduces to

$$H = \hat{N} + g\frac{T_0}{2}(\hat{N}_L^2 + \hat{N}_R^2 - \hat{N}) + gNT_1(C_L^\dagger C_R + C_R^\dagger C_L), \qquad (2.15)$$

which is similar in form to a two well Bosonic-Hubbard model (Fisher, 1989), but differs in that the there is a factor of "N" in the T_1 tunneling term. The usual Bose-Hubbard model contains only the linear Schrödinger-like, tunneling term ε_{LR}. However for fixed N, the two models are simply related by a scaling of parameters, with perhaps an additive constant. The R,L labels may also be extended to an arbitrary number of wells, leading to the widespread use of this model in discussions of BECs in multi-well optical traps.

2.4 Analysis of Correlated Wavefunctions: The 1-Matrix, ODLRO, Coherent States and Natural Orbitals

The key tool in our analysis of the correlated wavefunctions, obtained from the full CI calculations for the double well BEC as modeled in Section 2.3, is the first order reduced density matrix, or 1-matrix. The 1-matrix has played a key role in the theory of super-conductivity, as well as in quantum chemistry. P.-O. Löwdin played a seminal role in this latter development. In second quantized notation, for example Fetter and Walecka (1971), the 1-matrix for an N body ground state (approximate or exact) $|N,0\rangle$ is defined as

$$\gamma(1,1') = \langle N,0 | \psi^{\dagger}(1)\psi(1') | N,0 \rangle , \tag{2.16}$$

where $\psi^{\dagger}(1)$, $\psi(1')$ are the Bosonic field operators, $\psi^{\dagger}(1)$ creating a particle at space-spin point 1, and $\psi(1')$ destroying a particle at 1'. The diagonal element, $\gamma(1,1)$, is, of course, the particle density $\rho(1)$. $\gamma(1,1')$ is normally a non-separable function of the two space-spin co-ordinates 1,1'. Penrose (1951), and Penrose and Onsager (1956), proposed that a key descriptor and indication of a transition to the superfluid or superconducting state would be the factorization:

$$\gamma(1,1') = f(1) f(1') \tag{2.17}$$

for a suitable function $f(1)$. In the superfluid literature a 1-matrix for which this property holds is said to have the property off-diagonal long range order, or ODLRO, namely that the ground state is uniform in the sense that there is no correlation between points at which the system is probed by the operator $\psi^{\dagger}(1)\psi(1')$. We will see below, following Löwdin, that this is simply a statement of macroscopic occupancy of a single quantum state by the vast majority of particles.

But, we first ask, what is the function $f(1)$? P. W. Anderson (1966) suggested that ODLRO resulted from the factorization

$$\langle\psi^{\dagger}(1)\psi(1')\rangle = \langle\psi^{\dagger}(1)\rangle\langle\psi(1')\rangle. \tag{2.18}$$

In Eqn. (2.18) it is immediately evident that the expectation value <,> cannot be taken with respect to the N particle ground state |N,0>, as this is a state with precisely N particles, and the operators $\psi^{\dagger}$, ψ only have matrix elements between states which have particle numbers differing by ±1. Anderson thus suggested that the appropriate expectation value to take to effect the possible ODLRO factorization would be a non-number conserving *coherent state* of a form such as :

$$|\Theta\rangle = \sum_{N} a_N |N,0\rangle, \qquad a_N \cong e^{i\theta N} Exp\left(-\frac{(N-\overline{N})^2}{(\Delta N)^2}\right). \tag{2.19}$$

Use of such states has many positive attributes: in diagrammatic perturbation theory a version of Wick's theorem exists for coherent states, but not for Bosonic Fock states, see Fetter and Walecka (1971). Use of coherent states allows a distinct phase to be associated with the condensate wave function, now defined as

$$\phi(r) = \langle\psi(r)\rangle = \sqrt{\rho(r)}e^{i\theta} \tag{2.20}$$

and identified with the Landau *complex order parameter* empirically used in the phenomenological theory of superfluids, see, for example, Tilley and Tilley (1990). The phase of the condensate wave function is of exceptional importance: phase gradients drive supercurrents, Anderson (1964,1966); and the phase offset, $\Delta\theta$, between two weakly coupled superfluids drives a current proportional to $\sin(\Delta\theta)$. This latter is the Josephson effect of superconductivity. See Josephson (1962), Anderson (1964, 1966), Tinkham 1996). $\Delta\theta$ also controls the velocity of solitons in the BEC: Reinhardt and Clark (1997). We will make use of coherent states, and a projection into a phase space defined by the conjugate variables (N,θ) in Section 4. The fact that the particle number and phase are conjugate quantum mechanical variables implies an uncertainty relationship, the meaning of which will also be illustrated in Section 4.

In the thermodynamic limit $N \to \infty$, at constant density, $\rho = N/V$, the particle number uncertainty to maintain a fixed phase is vanishingly small compared to the overall number of particles, and the use of the Anderson coherent states, and the use of Eqns. (2.19 and 2.20) is perfectly sensible.

The fact that a condensate has some definite phase (out of all possible phases) is then considered a gauge symmetry breaking, analogous to the broken symmetry as a magnet orients its magnetic moment in one particular direction, rather than another. However, it does seem un-natural to assume that in a laboratory experiment with, say, 11,929 particles in a BEC in a magnetic trap, that we should consider the system *not* to have a fixed number of particles, or that even the wavefunction for a single particle might not have a perfectly definite phase, even if the physics is dependent on its relative, rather than absolute value. These statements are not universally accepted, at least partly due to semantic confusion, and the shared view of many condensed matter theorists that no aspects of the gaseous BEC can be simply described by direct use of a wavefunction. To paraphrase Leggett (2001, page 317), see also Reinhardt (1998, pages 281-4): *It should be emphasized that there are no circumstances in which Eqns. (2.19-20) are the physically correct description of the system, or even a part of it...However, the reader should be warned that this opinion is controversial, and there are even those who feel that Eqns. (2.19-20) are not only a possible, but the only legitimate, definition of the order parameter.*

With so much seemingly at stake, it is refreshing to note that if we follow Löwdin (1955), rather than Anderson (1964, 1966), a very simple interpretation of ODLRO for a fixed number of particles emerges immediately. The 1-matrix, considered as an integral kernel

$$\int \gamma(1',1)\varphi_i(1')d1' = \eta_i \varphi_i(1) \tag{2.21}$$

generates a complete set of orthogonal natural orbitals, φ_i ,(NOs hereafter) and corresponding occupation numbers, η_i. In terms of these

$$\gamma(1,1') = \sum \eta_i \varphi_i^*(1)\varphi_i(1'). \tag{2.22}$$

The natural orbitals, if used in a CI calculation, provide optimal convergence for a one-particle basis. The occupation numbers η_i tell us how important a particular NO is in a non-separable correlated wavefunction. The usual normalization of $\gamma(1,1')$ is such that

$$\sum \eta_i = N. \tag{2.23}$$

For a Bosonic system a single quantum state may have macroscopic occupancy, and so, unlike the Fermi case discussed by Löwdin, where the spin NO occupation numbers are restricted, $0 \leq \eta_i \leq 1$, for Bosons this is replaced by $0 \leq \eta_i \leq N$, and only restricted by the fixed total number of particles, N, via Eqn. (2.23). Should a single occupation number dominate the sum, we have

$$\gamma(1,1') = N\varphi_0^*(1)\varphi_0(1'), \qquad (2.24)$$

which immediately gives an interpretation of the ODLRO factorization. Namely, the system has chosen to put essentially all of the particles in a single orbital, which is now seen not to necessarily be homogeneous in space. This is both macroscopic occupancy of a single state and factorization of $\gamma(1,1')$, but without the artifice of introducing a non-number preserving formalism. The condensate phase, θ, is seemingly lost, but is actually recoverable as seen in Section 4, where the Anderson coherent states will be seen to provide a most valuable interpretive insight into the nature of the dynamics of the correlated double well states, even though number conservation is kept exactly for the overall system.

3. Energy Level Correlation Diagram, and Natural Orbital/Occupation Numbers for the BEC Double Well Problem

In this Section the energy correlation diagram resulting from the full CI treatment of the double well problem of Section 2.3 is presented. This is followed by a natural orbital and occupation number analysis. Individual wavefunctions typifying various regions from the correlation diagram are presented. It is seen that both localized and delocalized (fragmented) states can occur for a wide range of values of α, which measures the tunneling strength. For excited fragmented states it is seen that proper linear combinations of CI eigenstates are fully localized, even for tunneling couplings large enough so that the ground state is still delocalized. A physical interpretation of this dual behavior is presented in Section 4.

3.1 An Energy Correlation Diagram Shows a Phase Transition

Choosing the parameters in the Spekkens-Sipe Hamiltonian, Section 2, as $\varepsilon_{LL} = \varepsilon_{RR} = T_0 = 1$; $g = +0.8$; $\varepsilon_{LR} = T_1 = -\exp(-\alpha)$; and, $T_2 = \exp(-2\alpha)$, Figure 2 shows all 21 eigenvalues for a system of 20 particles in the double well potential for α ranging from $\alpha = 0$ to $\alpha = 5$. The tunneling parameters, ε_{LR} and T_1, allowing particles to move through the barrier, thus vary from 1 to 0.0067, taking us from a region where the particles are completely delocalized between the two wells, to a region where most excited states are fragmented, and, if localized, will have definite numbers of particles in each of the two wells.

The correlation diagram of Figure 2 shows a striking set of energy level merging and avoided crossings. Starting at about $\alpha = 1.8$ the highest two energy levels become visually degenerate. As α continues to increase, corresponding to raising the barrier between the wells, subsequent pairs of levels do the same. In the limit of even larger α, only the ground state of the repulsive double well condensate remains non-degenerate, and in that ground state 10 particles are in the state ϕ_R and 10 in ϕ_L: the ground state is thus completely fragmented. Spekkens and Sipe (1999) discuss this limit for the ground state, with an identical conclusion. We focus on the overall picture of all states, and will attempt to understand the origins of the crossover ridge separating the non-degenerate delocalized states on the left of the diagram, from the doubly degenerate pairs on the right.

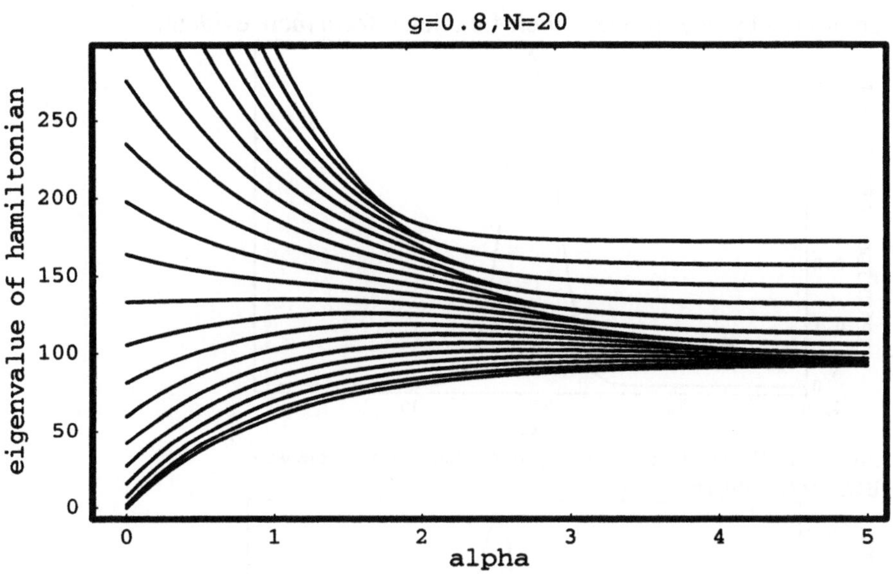

Figure 2. Correlation diagram for all energy levels for 20 Bose particles in a symmetric double well potential, as discussed in Section 2.3. Tunneling is controlled by the parameter α, as $T_1 = -\exp(-\alpha)$, and thus tunneling is strongly allowed at the left, and mostly quenched on the right of the diagram. The most prominent feature of the diagram is the ridge of avoided crossings, and the onset of level degeneracies.

Our analysis of this correlation diagram and the structure of excited state wavefunctions is quite different from that discussed in the work of Steel and Collett (1998), and Franzosi and Penna (2001), who use a transformation to an angular momentum representation, as do Milburn et.al. (1997), in their analysies of wave functions of the form of Eqns. (2.14, 3.1) in discussions of the double well correlation problem. Neither of these groups introduce natural orbitals.

One can ask why we describe the small α orbitals as localized, and those for larger α as delocalized. Examination of a few of the coefficient vectors will make this distinction clear. Figure 3. shows the coefficients of the eigenvectors corresponding to the three lowest energy eigenvalues of the Spekkens-Sipe Hamiltonian, now for 40 particles. This number has been made larger simply to make the coefficient patterns "smoother" to the eye. What is plotted are the coefficients, c^j_n, of the number basis states $|N\text{-}n,n>$ in the expansion for the jth eigenfunction of the Hamiltonian,

$$|\Psi_j\rangle = \sum_{n=0}^{N} c_n^j \,|N-n,n\rangle \tag{3.1}$$

as a function of "n" the number of particles in the "right" well, and these are connected by lines to make their functional form more evident.

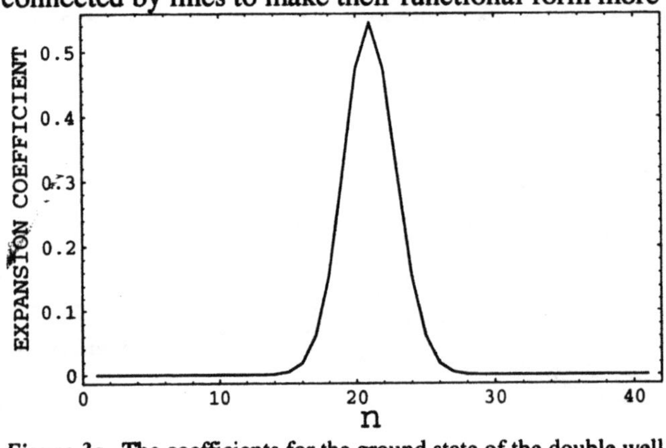

Figure 3a. The coefficients for the ground state of the double well BEC for N = 40, and α = 4.

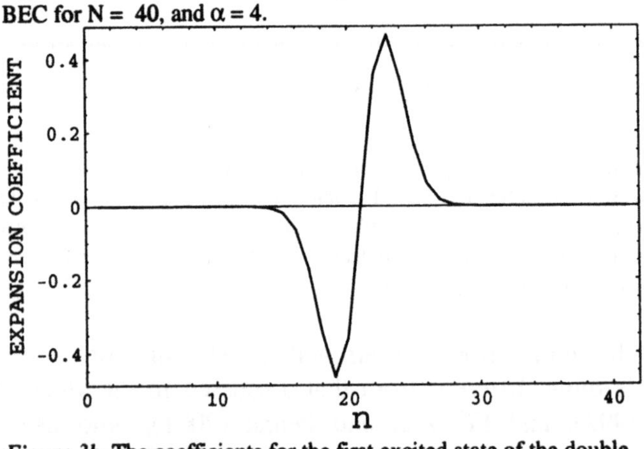

Figure 3b. The coefficients for the first excited state of the double well BEC for N = 40, and α = 4.

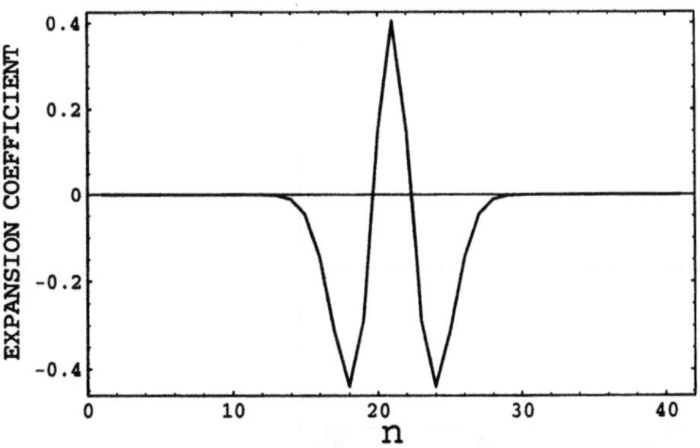

Figure 3c. The coefficients for the second excited state of the Double well BEC for N = 40, and $\alpha = 4$.

These lists of coefficients were calculated for $\alpha = 4$, and appear to be Harmonic Oscillator co-ordinate space wave functions. That this is not a coincidence is the subject of Section 4. Figure 4 shows similar lists of coefficients, again joined to help the eye follow their patterns, for two states on the other side of the ridge separating the doubly degenerate from the non-degenerate states. Note that these do not look like oscillator wave functions. Odd and even linear combinations of degenerate pairs give fully localized orbitals, one of which is shown in Figure 4c.

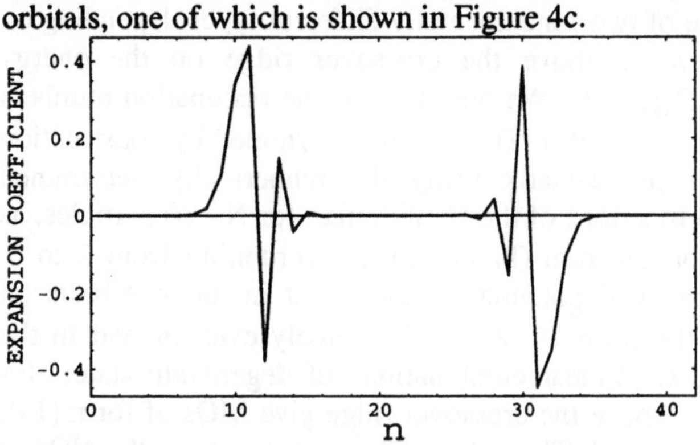

Figure 4a. The coefficients for a fragmented excited odd symmetry state for 40 particles in the double well for $\alpha = 4$.

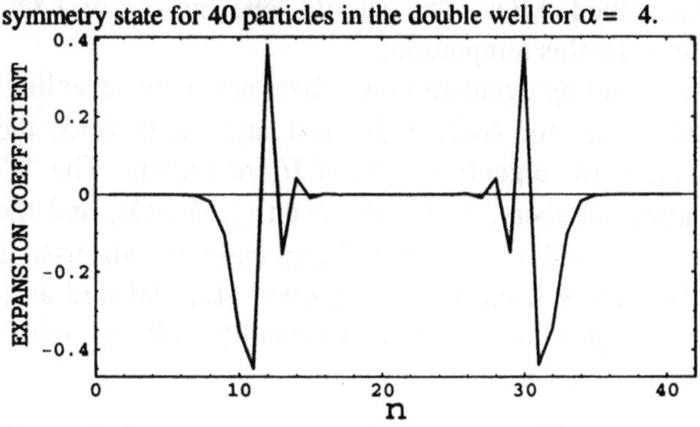

Figure 4b. The even symmetry pair of the state of Fig. 4a.

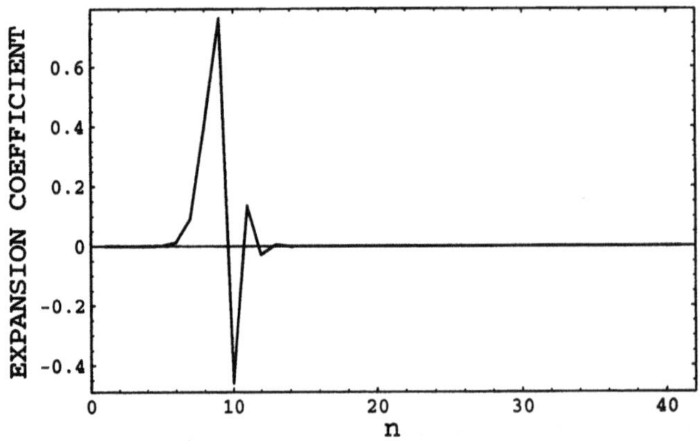

Figure 4c. A localized linear combination of the even and odd states similar to those of Figs. 4a,b, but at an even higher energy. A sign change would give localization in the opposite well.

3.2 Natural Orbitals and Occupation Numbers

The wavefunctions illustrated in the coefficient space of Figures 3 and 4, indicate that the wave functions for the ground and excited states of the double well problem are of two fundamentally different types depending on whether they lie below or above the crossover ridge on the energy correlation diagram of Figure 2. We now look at the occupation numbers for a series of differing values of α. These were determined by construction and diagonalization of the 1-matix using the numerically determined coefficients of the eigenfunctions of the Hamiltonian for N= 20 particles, as in the energy correlation diagram (Figure 2), for α running from 2 to 5. Symmetry of the double well potential requires that in the AO basis all natural orbitals are of the form $1/(\sqrt{2})(1, \pm 1)$, namely even or odd in the symmetric co-ordinate, x. Linear combinations of degenerate states, for example as in Figure 4c, above the crossover ridge give MOs of form (1,0) or (0,1) and are fully localized. Thus, there are no surprises in the NOs as they are symmetry determined MOs. Or, should one now say COs, *condensate orbitals*? We resist this temptation.

However, the corresponding occupation numbers are quite revealing. In Figures 5a,b, for α=2; 6a,b, for α=3; 7a,b, α=4 and, 8a,b, α=5; the occupation numbers for all of the eigenfunctions of H are shown. The "a" figures show the occupation numbers, η_j, for the even (1,1) NOs, and the "b" figures show the η_j for the odd, (1,-1), NOs. In all cases the abscissa is simply the number of the energy level with the ground state labeled as j, running from 1 to 21, the energies increasing monotonically with the index.

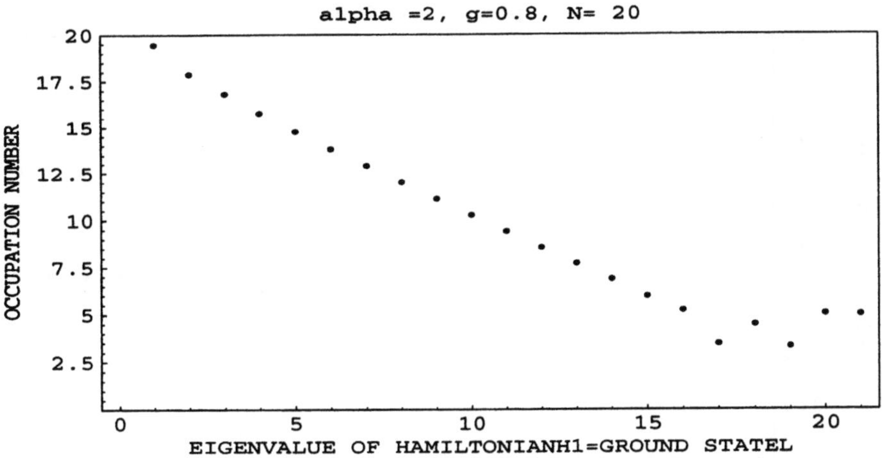

Figure 5a. Occupation numbers, η_j, for the $\alpha = 2$ even NOs. These begin at η_j near 20 for the ground state, j =1 and decrease monotonically until the corresponding energies enter and pass through the crossover region near j = 17.

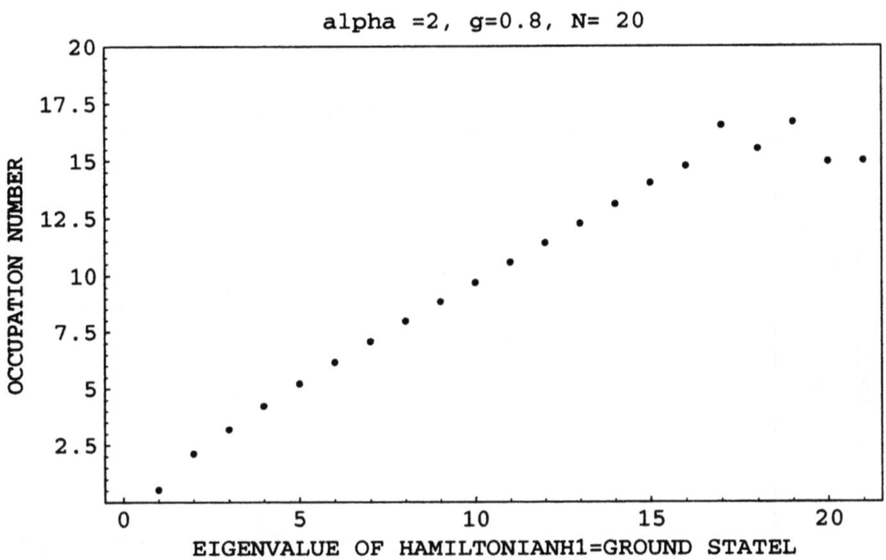

Figure 5b. Occupation numbers, η_j, for the $\alpha = 2$ odd NOs. These begin near 0 for the ground state, j =1 and increase monotonically until the corresponding energies enter and pass through the crossover region near j = 17.

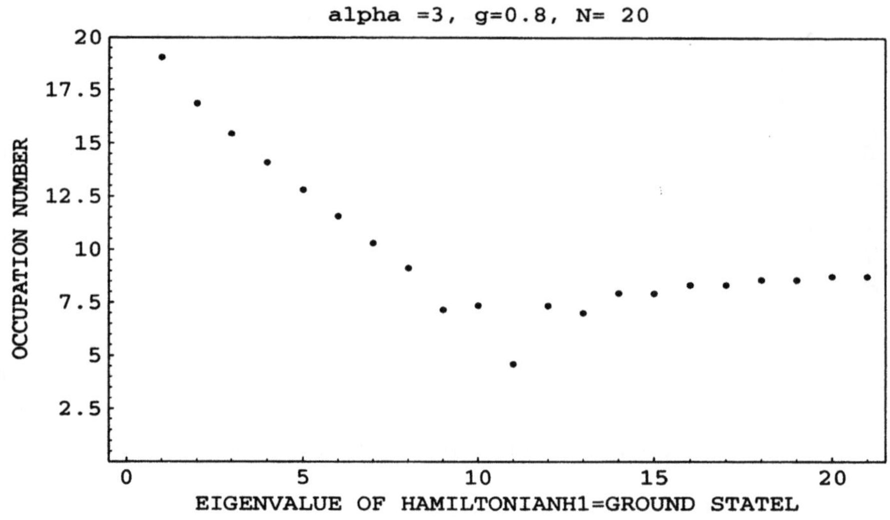

Figure 6a. Occupation numbers, η_j, for the $\alpha = 3$ even NOs. These begin at η_j near 20 for the ground state, j =1 and decrease monotonically until the corresponding energies enter and pass through the cross over region near j = 10.

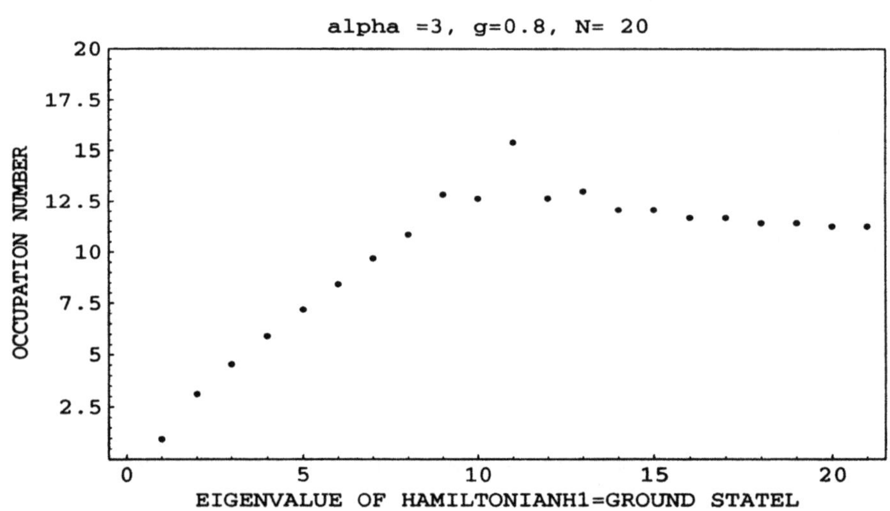

Figure 6b. Occupation numbers, η_i, for the $\alpha = 3$ odd NOs. These begin near 0 for the ground state, j =1 and increase monotonically until the corresponding energies enter and pass through the cross over region near j = 10.

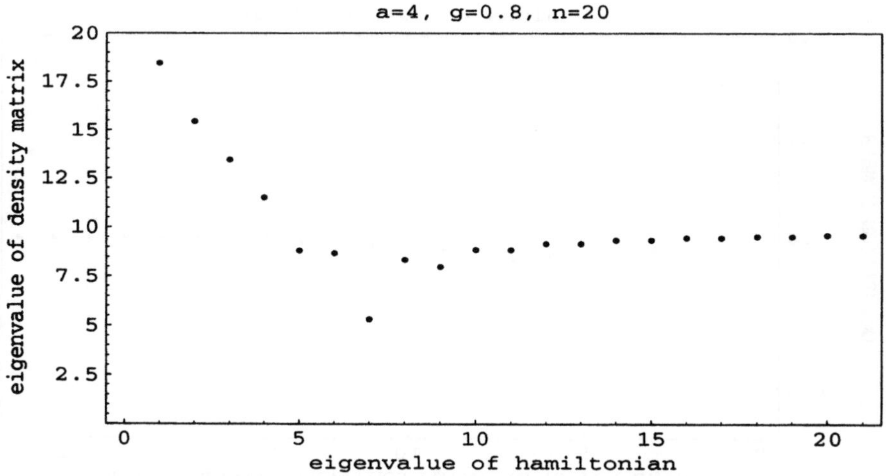

Figure 7a. Occupation numbers, η_j, for the $\alpha = 4$ even NOs. These begin at η_j near 19 for the ground state, j =1 and decrease monotonically until the corresponding energies enter and pass through the cross over region near j = 6.

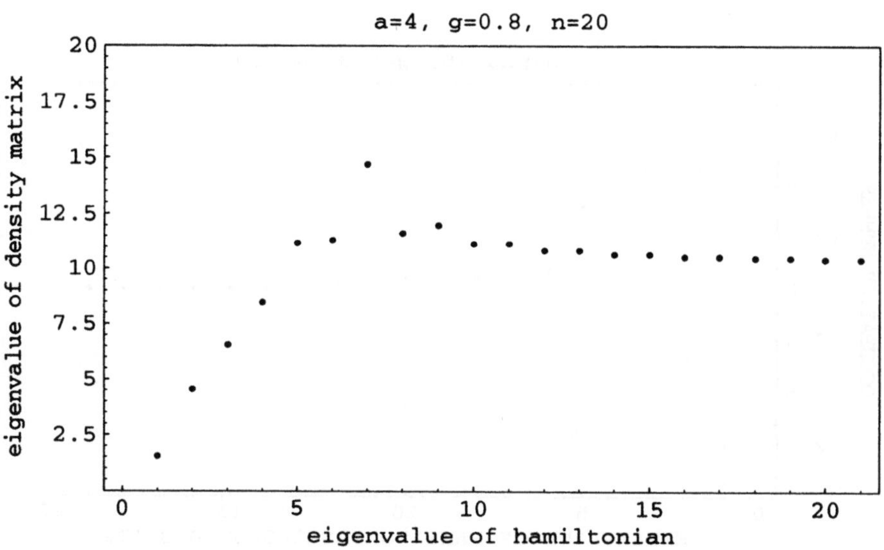

Figure 7b. Occupation numbers, η_j, for the $\alpha = 4$ odd NOs. These begin near 1 for the ground state, j =1 and increase monotonically until the corresponding energies enter and pass through the cross over region near j = 6.

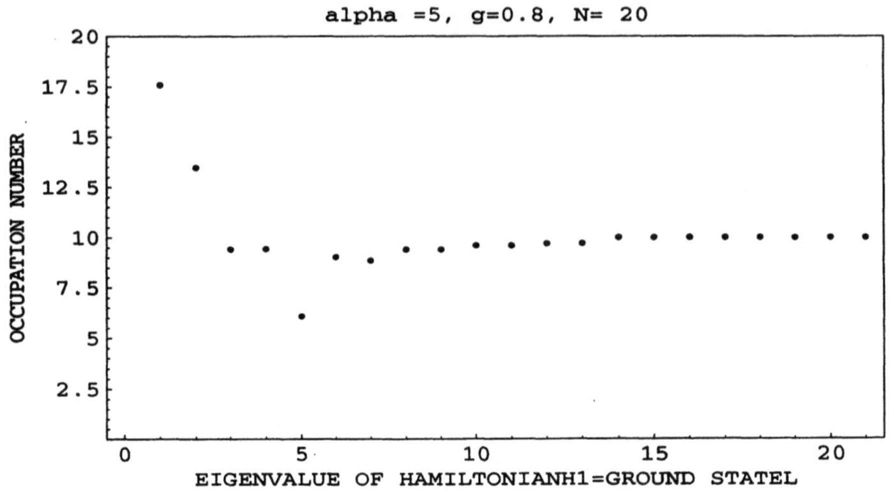

Figure 8a. Occupation numbers, η_j, for the $\alpha = 5$ even NOs. These begin at η_j near 18 for the ground state, j =1 and decrease monotonically until the corresponding energies enter and pass through the cross over region near j = 4. It is evident that all of the states with j > 15 have occupations numbers near 10. In this regime localized AOs provide a better zero order MOs.

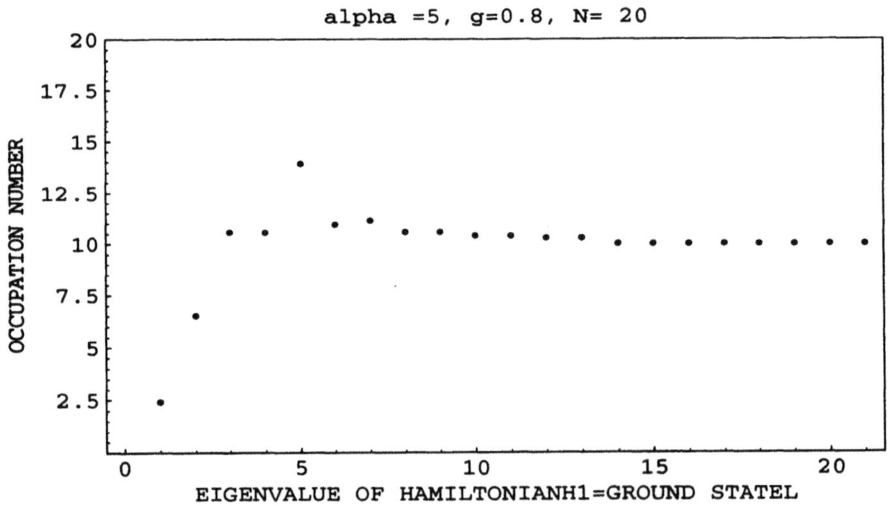

Figure 8b. Occupation numbers, η_j, for the $\alpha = 5$ odd NOs. These begin near 2 for the ground state, j =1 and increase monotonically until the corresponding energies enter and pass through the cross over region near j =4.

For all of the values of α considered, the behavior of the occupation numbers, η_j, is strikingly similar as j runs from 1 to 21, from the lowest to highest energy eigenstate of the full CI. The orbital occupancy for the ground state is close to the expected MO configuration $(\phi_R + \phi_L)^N$. The successively higher states differ from this ground configuration by transfer of atoms from $(\phi_R + \phi_L)$ to $(\phi_R - \phi_L)$. The zero order description of these states are the properly symmetrized versions of $(\phi_R + \phi_L)^{N-n} (\phi_R - \phi_L)^n$, "n" now being the number of electrons transferred from the even to the odd MO. This is precisely what a primitive MO picture would suggest, at least for energies up to the crossover region. Past the crossover energy, odd and even superpositions (which are Schrödinger Cat states) of the localized configurations $(\phi_R)^{N-n}(\phi_L)^n$ are more appropriate. These localized zero order states are *fragmented* in that fixed numbers of atoms are on a given side of the double well. As tunneling is further decreased, as a function of increasing α, the crossover region rapidly approaches the ground state, indicating that as the barrier separating the wells further increases that even the ground state will fragment, with the resulting ground state being $(\phi_R)^{N/2}$ $(\phi_L)^{N/2}$. This is in complete agreement with the Spekkens-Sipe analysis of ground state fragmentation. Note that, in this limit, the case of even and odd N must be considered separately. We only discuss N even, herein.

For intermediate values of α, the NO analysis clearly measures the macroscopic occupancy of the ground state, and the eventually also macroscopic transfer of electrons to the higher energy odd MO, and past the crossover region the transition to localized orbitals. The crossover region is thus the boundary for a localization-delocalization transition. Interestingly, there is another interpretation of this transition, one which requires introduction of a new phase dependent fourier transformed version, $\Psi(\theta)$, of the occupation number wave function $|\Psi>$ of Eqn. (3.1). This new representation will also make clear the origin of the coherent state number-phase uncertainty relationship $\Delta N\, \Delta\theta \geq 2\pi$.

4. Transformation to a Phase Representation: The Tunneling Problem as a Physical Pendulum

A phase representation of the wavefunction is introduced in Section 4.1 via a fourier transform. This purely mathematical step is made *physical* in Section 4.2 where a phase difference in the single particle condensate wave function on opposite sides of the double well is seen to drive a current, leading to Josephson-type oscillations of particles back and forth between the two wells. This suggests pendulum-like oscillatory motion. Both pictures lead to phase-number uncertainty relations. Both such relations are actually the same, although confusion can arise as the language describing them is, at first reading, very different. In Section 4.3 an old trick, the

Husimi transform, is used to project classical phase space information from individual quantum stationary states, and the pendulum interpretations of Section 4.2 are seen to appear automatically, even though there is no explicit time dependence anywhere in the discussion.

4.1 A Transformed Representation of the Wavefunction

In ordinary quantum theory a familiar transformation is between the co-ordinate and momentum representations. A harmonic oscillator wave function in co-ordinate space, Q, $\Psi_n(Q)$, may be fourier transformed to give the momentum space counterpart

$$\Psi_n(P) = \frac{1}{2\pi}\int_{-\infty}^{+\infty} e^{-iPQ}\Psi_n(Q)dQ \ . \tag{4.1}$$

Interestingly, as with proper scaling, the harmonic oscillator Hamiltonian is essentially $H_{osc} = P^2/2 + Q^2/2$, the P and Q space wavefunctions are, not surprisingly given the symmetric role of P and Q in the Hamiltonian, mathematically identical functions, each being a Gaussian times a Hermite polynomial in the appropriate P or Q variables. An equivalent transform of the occupation number representation for the fully correlated double well BEC, may be defined. The discrete vector of coefficients $\{c^j_n\}$, which we refer to as the n-representation may be used to define a θ-representation wave function, via a discrete analog of the fourier transform of Eqn (4.1):

$$\Psi_j(\theta) \equiv \sum_{n=0}^{N} c_n^j e^{-in\theta} \ . \tag{4.2}$$

This is a projection from the coefficient space vectors labeled by the discrete index “n” via the fourier functions exp(-*i*nθ). Figure 9a-c shows the θ-representation wave functions corresponding to the n-representation functions show in Figure 3. From the earlier remark about the P and Q oscillator functions being the same, we should not be overly surprised. The fourier transform of an oscillator function is an oscillator function! But perhaps the reader *is* surprised: what is oscillating in the double well BEC?

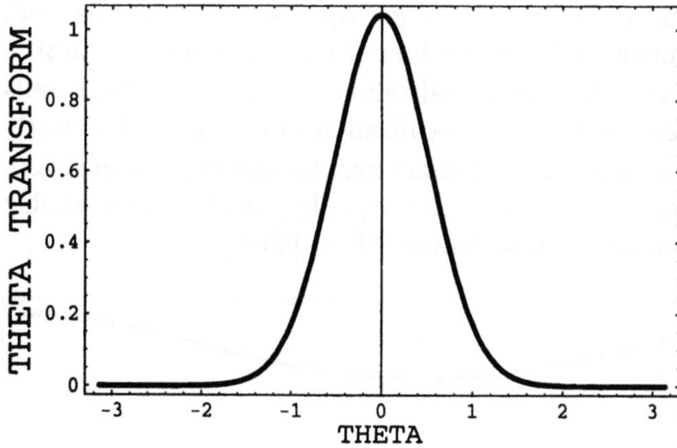

Figure 9a. Ψ(θ) for the α = 4 ground state of the double well BEC, plotted as a function of θ. This corresponds to the coefficient list of Figure 3a. The θ-transform of that Gaussian, is, of course, a Gaussian.

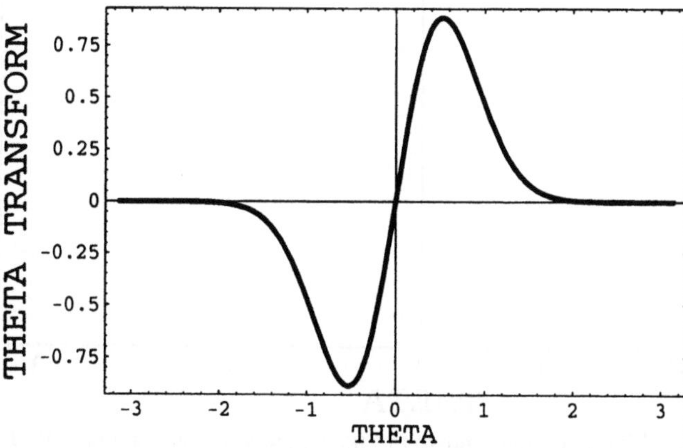

Figure 9b. Ψ(θ) for the α = 4 first excited state of the double well BEC, plotted as a function of θ. This corresponds to the coefficient list in of Figure 3b. The Q-transform of the oscillator function is still an oscillator function.

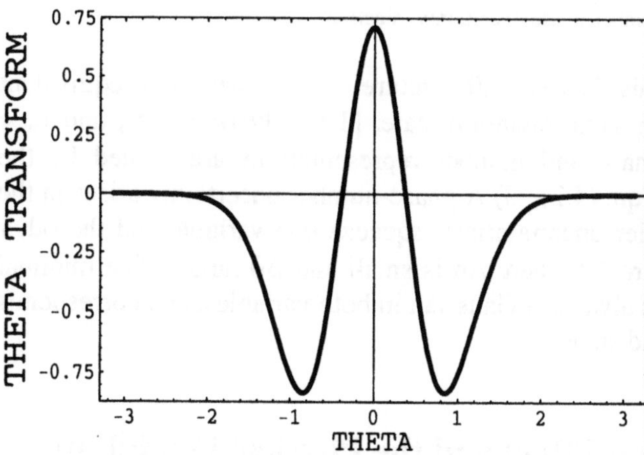

Figure 9c. Ψ(θ) for the α = 4 second excited state of the double well BEC, plotted as a function of θ. This corresponds to the coefficient list of Figure 3c.

Now looking on the opposite side of the crossover ridge, Figure 10 shows the θ-representation of the localized n-representation function of Figure 4c. Just as the localized n-representation function was not an oscillator like function, its θ-transfrom is perhaps even more unexpected. Note at once that we might well expect a fourier type "uncertainty" $\Delta n \Delta\theta \geq 2\pi$ from the definition of Eqn (4.2). Localization in "n", in one of the wells or another, will then correspond to delocalization in θ. Figure 10 clearly indicates this, as $\Psi(\theta)$ is now nearly constant over the full range of angle coordinate, indicating that all angles are almost equally probable, were we able to measure the angle probability distribution $P(\theta) \propto |\Psi(\theta)|^2$.

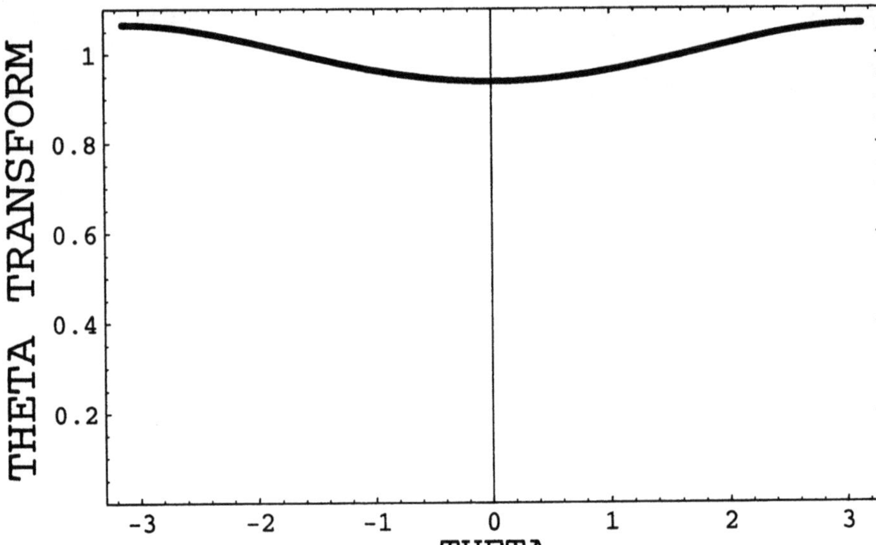

Figure 10. $|\Psi(\theta)|$ for the $\alpha = 4$ localized state shown in Fig. 4c, plotted as a function of θ. In this case neither the $\{c_n\}$ nor θ-representations of the wavefunction are oscillator functions. As clearly illustrated in this θ-plot, the narrow n-distribution of Fig. 4c results in a broad, almost flat $\Psi(\theta)$: this is number-phase uncertainty in action.

Summarizing this fourier (n,θ) picture: 1) we have two equivalent representations of the same quantum state, $|\Psi\rangle$: $\Psi(\theta)=\langle\theta|\Psi\rangle$, and $c_n = \langle n|\Psi\rangle$. 2) These phase and number representations are related by the fourier transform of Eqn. (4.2). 3) A phase-number uncertainty arises in the usual manner of fourier uncertainties: squeeze one variable and the other broadens...as in Figure 10 where Δn is small and $\Delta\theta$ large. The minimal uncertainty state is, as always, a Gaussian in both variables, and corresponds to the oscillator ground state.

4.2 The Josephson Effect and the Physical Pendulum

The transformation between the n- and θ-representations is discussed at length by Anderson and in monographs, Anderson (1984, 1966, 1964),

Tinkham (1996), and Tilley and Tilley (1990) to cite only a few examples. Is this more than a simple change of representations, as in the familiar co-ordinate and momentum representations of ordinary quantum mechanics, which have identical content?

Namely, may we impose, or discover, the physical nature of n and θ and understand their intertwined relationship? This is not a novel question, and it is addressed in the above three Anderson references, and originates in Anderson's elucidation of the Josephson effect. Josephson (1962) envisaged a situation where two (homogeneous) superconductors were separated by a "tunneling junction" (now referred to as a Josephson junction). Were the separate pieces of condensate to have identical densities, their condensate wavefunctions, see Eqn (2.20), might still have a relative phase offset θ, so that |N-n,n> differs in phase $\exp(i(n-m)\theta)$ from |N-m,m>, as a relative phase $\exp(\pm i\theta)$ is introduced for each particle moved between wells, with the sign depending on the direction of such transfer. Josephson then suggested (Josephson, 1962) that a current would flow through the junction, and have a magnitude depending on a tunneling matrix element, E_0, and the phase offset θ:

$$\text{Josephson Current} \propto E_0 \sin(\theta). \tag{4.3}$$

Such a current, see Anderson (1964, 1966), Tinkham (1996), Tilley and Tilley (1990) for fuller discussion, corresponds to a potential energy

$$V(\theta) = E_0(1 - \cos(\theta)). \tag{4.4}$$

Noting that the current causes a flow of particles through the junction, Anderson identified this with $n = N_R - N_L$, the difference between the number of particles on the Left and Right of the junction, and suggested that the dynamics might be well described by the classical Hamiltonian:

$$H(n,q) = \boldsymbol{K}n^2 + V(\theta). \tag{4.5}$$

If V(θ) is that of Eqn (4.5), then a natural interpretation is that the difference variable n is the analog of a momentum, and thus "$\boldsymbol{K}n^2$" is a kinetic energy, $\boldsymbol{K}$ being a constant, and the Hamiltonian of Eqn (4.5) is then that for a physical pendulum with (n,θ) being the conjugate generalized variables. The physical pendulum is a point mass held at a fixed distance from a pivot by a rigid rod. In a gravitational field two types of motion are possible: small oscillations about the potential minimum, where a Taylor expansion of cos(θ) gives

$$V(\theta) = E_0\, \theta^2/2, \tag{4.6}$$

and we see that the potential is harmonic, E_0 plays the role of a spring constant, and we expect small amplitude harmonic oscillations about $\theta = 0$. On the other hand, if the kinetic energy $\boldsymbol{K}n^2$ is greater than E_0, see Eqn (4.4), the classical motion will not be small oscillations about $\theta = 0$, but hindered and then free rotation as $\boldsymbol{K}n^2$ continues to increase. For the Josephson effect the constant $\boldsymbol{K}$ is $e^2/(2C)$, see, Anderson (1964), Tinkham (1996), as charge moves across the barrier, with a restoring propensity depending inversely on the junction capacitance, C. For the Spekkens-Sipe Hamiltonian in our parametrization, $\boldsymbol{K} \propto gT_0$, and $E_0 \propto gT_1$, to within additive constants, and for small n. This identification immediately makes clear that small tunneling (i.e. small T_1) lowers the energy barrier for the pendulum, allowing early onset of free rotation as a function of increasing total energy. Such free rotation well describes the system at energies higher than the crossover energies of Fig. 2. Large α, weak tunneling, then corresponds to a low energy of the crossover, and the apparent double degeneracy of the energy levels is essentially that of the double degeneracy of a particle on a circular ring. Quantization of the physical pendulum of Eqn (4.5) then follows from the correspondence rule

$$n \rightarrow -i\,\partial/\partial\theta, \tag{4.7}$$

making clear the quantum mechanical origin of the $(\Delta n \Delta\theta)$ uncertainty.

Summarizing the physical content of this Josephson-Anderson interpretation, within the context of the double well BEC: 1) A phase offset of θ for a single particle moving from one well to the other will cause a Josephson current to flow....as the condensate consists of repulsive particles, this current will ultimately reverse, and particles will flow back and forth in the frictionless quantum system. This is the pendulum undergoing small oscillations, and as there is a time dependent number of particles on either side of the potential barrier, we have a particle uncertainty. 2) Phase-number uncertainty for a stationary state of the fixed N double well system now takes on a specific physical meaning. Exact knowledge of n, the difference in particle numbers in the two wells implies a fragmented system, and then we have no knowledge of the phase relationship of the wavefunctions across the barrier. Conversely, should we wish to have maximal knowledge of the phase difference between the two wells, the spread in n will be maximal. Minimal (n,θ) uncertainty arises in the ground state, where both the n and θ wave functions are Gaussians, and then it is possible that $\Delta n \Delta\theta = 2\pi$. This is all well illustrated in the comparisons of Figures 3,4 with 9,10.

4.3 Phase-Number Coherent States and the Husimi Projection into Classical Phase space

Schrödinger (1926), reprinted in English translation (1982), in one of the founding papers of wave mechanics noted that quantum time evolution of a displaced harmonic oscillator ground state led to a minimal uncertainty and non-spreading state, whose center followed a classical phase space (q,p) trajectory. Such a state is peculiar to harmonic time evolution, and is nowadays referred to as a coherent state. Such coherent states may be used to project, following Husimi (1940), see also, Gutzwiller (1990), Martens (1989), classical phase space (q,p) behavior from stationary state quantum wavefunctions in either the q or p representations. They also form a starting point for discussions of quantum optics, where the conjugate variables are then N and θ, the number of photons and the phase the electromagnetic wave. This N,θ coherent state representation of the electromagnetic field was introduced by Glauber, see Loudon (1973). Anderson, as in Sections 2.4 and 4.2, introduced it in his early discussions of ODLRO and the Josephson effect. Following Perleman (1986) the (q,p) coherent state is defined as

$$|\alpha\rangle = e^{-|\alpha|^2/2}\sum_{n=0}^{\infty}\frac{\alpha^n}{\sqrt{n!}}|n\rangle, \tag{4.8}$$

where $|n\rangle$ is the nth eigenestate of the scaled variable harmonic oscillator Hamiltonian $H(q,p) = p^2/2 + q^2/2$, and $\alpha = q + ip$, is a complex number, not to be confused with the α of the double well Hamiltonian, whose real and imaginary parts are the phase space variables (q,p). For our purposes, a more useful representation of the state $|q+ip\rangle$ is via its co-ordinate (q') or momentum (p') representations.

$$\langle q'|\alpha = q+ip\rangle = \pi^{-1/4}e^{i\sqrt{2}pq'}e^{-(q'-\sqrt{2}q)^2/2} \tag{4.9}$$

$$\langle p'|\alpha = q+ip\rangle = \pi^{-1/4}e^{-i\sqrt{2}qp'}e^{-(p'-\sqrt{2}p)^2/2} \tag{4.10}$$

These are interpreted as being Harmonic oscillator ground states displaced to the point (q,p) in the classical phase space. Following Schrödinger, their time (harmonic) evolution preserves the shape of the probability distribution, and its center follows the classical phase space orbit starting at the displaced point (q,p). These representations allow projection of classical phase space information from p or q representation wavefunctions: thus for example, as in Husimi (1940)

$$\langle q+ip\,|\,\Psi\rangle = \int_{-\infty}^{+\infty} \Psi^*(q')\langle q'\,|\,p+iq\rangle dq', \tag{4.11}$$

or

$$\langle q+ip\,|\,\Psi\rangle = \int_{-\infty}^{+\infty} \Psi^*(p')\langle p'\,|\,p+iq\rangle dp'. \tag{4.12}$$

These projections are the Husimi transforms of semi-classical quantum mechanics, and may be profitably used in the interpretation of wavefunctions in terms of their underlying phase space classical dynamics. Equations (4.11) or (4.12) may be used to construct phase space probability distributions, $P(q,p) \equiv |\langle q+ip|\Psi\rangle|^2$. The Husimi transform is also a Gaussian averaged Wigner-Weyl transform of the wavefunction, a point of view not developed herein (but see, for example, Gutzwiller, 1990). Note that Steel and Collett (1998) use *Bloch sphere* Wigner-Weyl projections in their analysis of the double well problem. What we do here is rather simpler.

The (n,θ) analogs of Eqns. 4.9, 4.10 are

$$\langle \theta'\,|\,\alpha = \theta + in\rangle = \pi^{-1/4} e^{i\sqrt{2}n\theta'} e^{-(\theta'-\sqrt{2}\theta)^2/2} \tag{4.13}$$

$$\langle n'\,|\,\alpha = \theta + in\rangle = \pi^{-1/4} e^{-i\sqrt{2}\theta n'} e^{-(n'-\sqrt{2}n)^2/2} \tag{4.14}$$

The Husimi transform of an n-representation vector c^j_n is thus:

$$\langle \theta + in\,|\,\Psi_j\rangle = \sum_{n'} c^j_{n'} e^{-i\sqrt{2}\theta n'} e^{-(n'-\sqrt{2}n)^2/2} \tag{4.15}$$

where, following Anderson, $n' = N_R - N_L$, rather than being the simpler Left-Right particle counter of Eqn. (2.14). This is simply a shift in origin. These forms assume that the oscillator Hamiltonian is now $H(n,\theta) = n^2/2 + \theta^2/2$, namely that appropriately scaled variables are used throughout. Figure 11 shows the Husimi probability projection

$$P_j(n,\theta) = \left|\langle \theta + in\,|\,\Psi_j\rangle\right|^2 \tag{4.16}$$

into the classical (n,θ) phase space of two of the number representation eigenstates of the Spekkens-Sipe Hamiltonian, one state being below the crossover ridge, and one above.

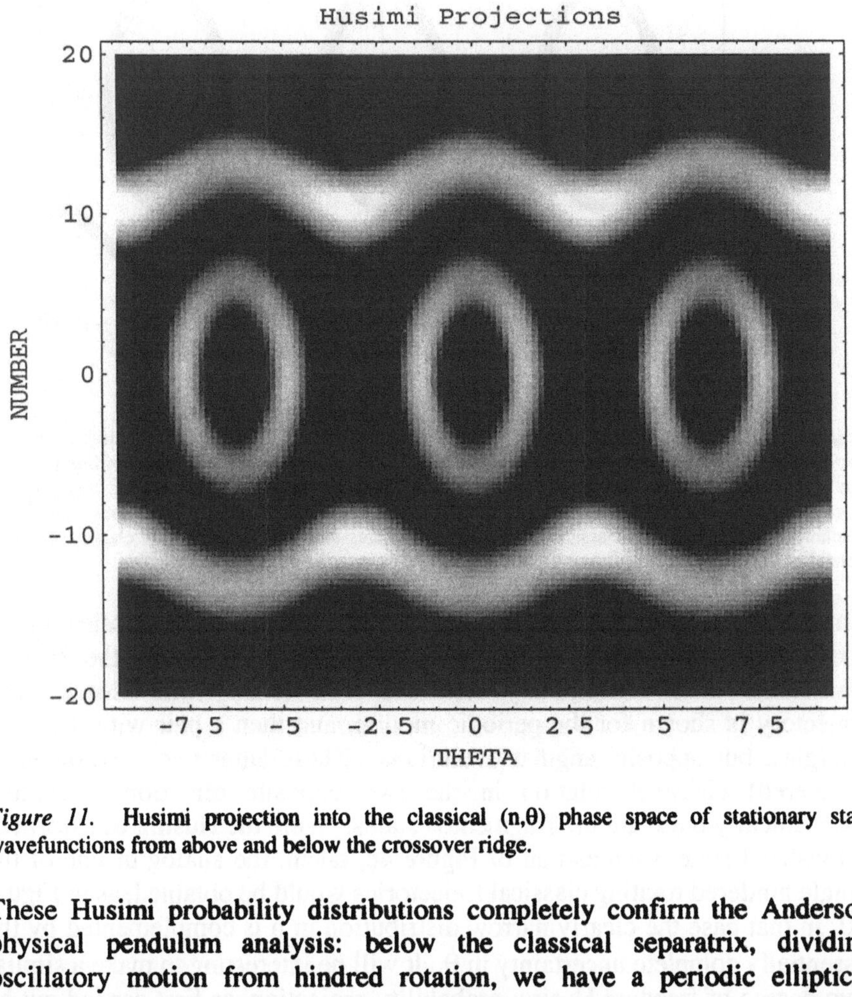

Figure 11. **Husimi projection into the classical (n,θ) phase space of stationary state wavefunctions from above and below the crossover ridge.**

These Husimi probability distributions completely confirm the Anderson physical pendulum analysis: below the classical separatrix, dividing oscillatory motion from hindred rotation, we have a periodic elliptical classical phase space orbit; above the crossover the orbits are no longer periodic, leading, rather, to ever increasing values of θ. Note also that 20 to 40 total particles (in a two basis function Hilbert space) are all that are needed to reach this semi-classical limit, wherein the underlying classical dynamics is evident. Figure 12 shows the classical orbits run at the energies of the eigenstates corresponding to the states shown in Figure 11.

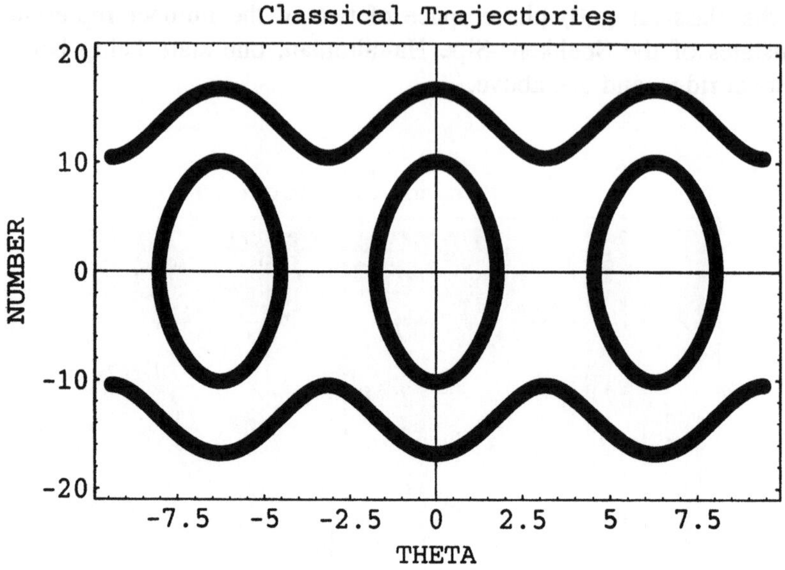

Figure 12. One periodic and oscillatory orbit (replicated three times, as the system is periodic in θ), and two hindered rotational orbits for a physical pendulum, with the classical Hamiltonian of Eqn (4.5). In this simple case these classical phase space orbits are simply constant energy contours. It is evident, from the comparison of Figs. 11 and 12, that the quantum stationary states for the double well BEC contain an underlying semi-classical phase space dynamics identical to that of the classical physical pendulum.

The only difference between Figures 11 and 12 is that three classical phase space orbits are shown in Fig. 12, while Fig. 11 contains the Husimi transforms of two stationary state wavefunctions. One closed orbit trajectory is shown for the periodic motion, and then a pair with identical energies, but opposite angular evolutions. These latter two correspond to (hindered) classical rotation in the two opposite directions; quantum mechanically these are the fragmented states. Were the Husimi transform of a localized state, such as that of Figure 4c, taken, the analog of one of the single hindered rotating classical trajectories would be obtained, as in Figure 10. In that case the clearly narrow distribution in n is complemented by the essentially complete uncertainty in θ. It will be interesting to make a similar two degree of freedom Husimi probability projection, as first carried out by Martens (1989) for a two degree of freedom molecular system, for a three coupled well BEC where, not unexpectedly, Salmond et. al. (2001) and for a 1 and ½ degree of freedom physical pendulum system, and Franzosi et. al., (2002a,b) for a full 2 degree of freedom system, have found evidence for chaotic dynamics. Such would be the expected generic situation for one driven or two coupled non-linear pendula, see for example, Gutzwiller (1996), Blümel and Reinhardt (1997).

5. Macroscopic Quantum Tunneling, a Natural Orbital-Occupation Number Analysis

Recently, there have been several reports of quantum superposition of macroscopically occupied quantum states. In all cases, rather than in gaseous BECs, as discussed here, these have been observed in superconducting current loops, with Josephson junctions inserted. van der Wal et. al., (2000); Freidman et. al. (2000); and Yu et. al. (2002) have observed avoided crossings of θ-space double well quantum states. These systems correspond to three wells of the type discussed above coupled in a circle. In these Josephson systems the effective potential V(θ) has two non-equivalent wells which may be tuned into a symmetric double well via through-the-superconducting-loop magnetic fields. The result is that quantum states initially localized in the two wells become degenerate and mix. What is actually observed in the experiments is the single photon absorption spectrum yielding the splitting between the even and odd superpositions of macroscopic quantum states as a function of the double well asymmetry. In this case the Bosons are Cooper pairs, and it is suggested that superpositions of macroscopic quantum states with millions of such Cooper pairs have been observed. What does this mean? What we see, by analysis of analogous tunneling splttings in our double well model, is that very small number of such Bosons have changed their single particle quantum state, and an NO analysis shows that only a small number of particles are involved in the spectra observed. The observed tunneling splittings are thus the few-body excitation spectrum: seemingly paradoxically the macroscopic quantum superposition of large numbers of particles is actually controlled by a very few of these.

Although the double well BEC discussed here is not precisely equivalent to the Josephson systems mentioned above, the analogous macroscopic superposition states may be indeed observed in the excited states of the double well model discussed above. To find such strong mixing of near degenerate states, one only need look at states at energies just above the cross-over ridge of Fig. 2. Too far above this ridge, such tunneling splittings will be too small of easily observe, but by tuning the tunneling, via α, and choosing the quantum states carefully, double well states with small, but not too small, splittings may be easily found, and their behavior observed by adjusting the one of the two well depth parameters ε_{LL} or ε_{RR}, which we have previously taken to be equal. We have chosen to take $\varepsilon_{LL}(\tau) = \varepsilon_{LL} + 0.1\tau$, and $\varepsilon_{RR}(\tau) = \varepsilon_{LL} - 0.1\tau$, and to let τ vary from -1/2 to 1/2, $\tau = 0$ then being the point of the minimal avoided crossing between the two levels. Figure 13 shows some of the lower lying levels for α =8 plotted as a function of τ for 50 particles. The first noticeable avoided crossing is for the splitting between levels 4 and 5. A magnification of the avoided crossing for these same two levels is shown in Figure 14. A natural-orbital and occupation number analysis of the states of Fig. 14 allows us to ask and answer the question: in a transition between these levels how many particles actually change quantum state? For single photon spectroscopy to successfully observe such a splitting one would guess that the answer must be a fairly small number, or selection rules would make it unlikely that the transition would be observed. In principle the transition

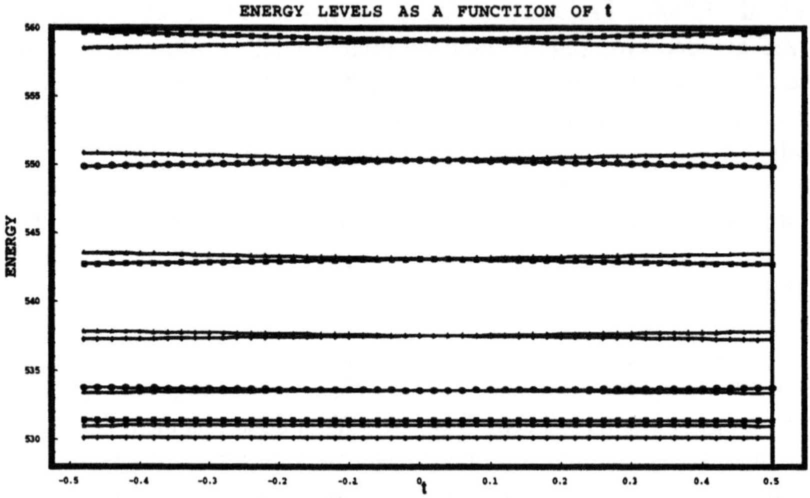

Figure 13. Energy levels for the asymmetric double well, $\varepsilon_{RR} = \varepsilon_{LL} + 0.1\tau$, for τ running from $-1/2$ to $+1/2$. Levels are for 50 particles with a high barrier, $\alpha = 8$, implying that levels 4,5 are already above the crossover ridge, and are near degenerate at $\tau = 0$. A magnification of this lowest energy avoided crossing is shown in the following figure.

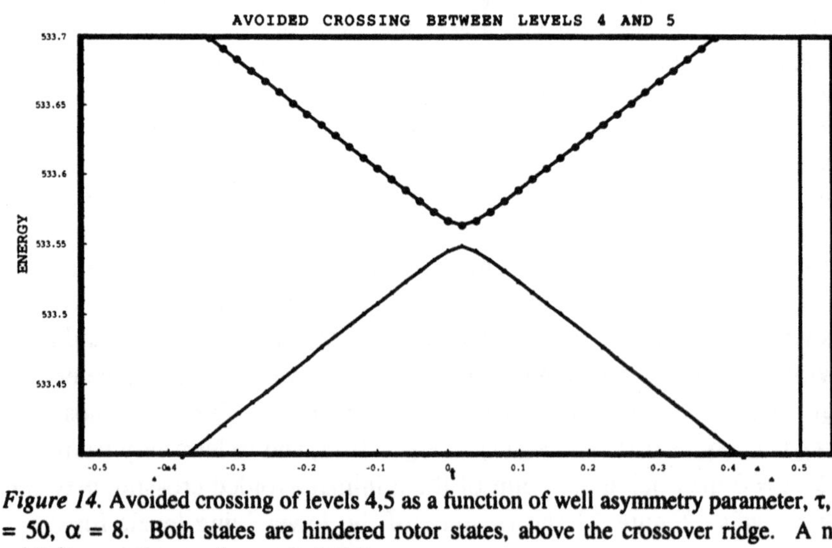

Figure 14. Avoided crossing of levels 4,5 as a function of well asymmetry parameter, τ, for N = 50, $\alpha = 8$. Both states are hindered rotor states, above the crossover ridge. A natural orbital/occupation number analysis follows.

from an even to an odd macroscopic state might only involve the transition of a single particle from an even to an odd orbital, or the reverse.

Thus even though large numbers of particles might be involved in superconducting current loops, the experimentally observed splittings might only involve the transition of a single, or a relatively small number of Bosons. What, then, are the occupation numbers for the states involved in such an avoided crossing as shown in Figure 14? These, and similar occupation numbers for (approximately) equivalent avoided crossings, for varying numbers of particles are shown in Table I.

Table I. Occupation numbers for nearly degenerate states for 26, 50, 100, 200, 400, and 800 particles. In each case $\tau = 0$, $\alpha = 8$, and the occupation numbers are shown for the even (1,1) and odd (1,-1) MOs as labeled by the pair of energy levels involved in the avoided crossing. Although the details do vary over the states shown, it is evident that the occupation numbers generally change by amounts small compared to the total number of particles. For example in the N = 400 case, only "about 3" particles actually change from even to odd MOs in the transition between the states 18 and 19. Thus although macroscopic quantum superpositions are indeed involved, the spectra may involve only few body transitions. As α increases, the number of particles involved in the "split" macroscopic superposition pair approaches 1, and eventually 0 as the condensate is fully fragmented.

Number of particles	State numbers	occupation # (1,1)	occupation # (1,-1)
26	2	13.5185	12.4815
	3	10.6135	15.3865
50	4	23.8434	26.1566
	5	23.0011	26.9989
100	6	45.2315	54.7685
	7	43.6142	56.3858
200	10	84.5568	115.443
	11	81.9966	118.003
400	18	157.629	242.371
	19	154.557	245.443
800	32	277.862	522.138
	33	250.013	549.987

The occupation numbers in Table I. show that each natural orbital is certainly macroscopically occupied, that is both have a major fraction of all particles. The occupation numbers of successive levels in the avoided crossings region give an indication that at the curve crossing, $\tau = 0$, the state mixing may well involve only a few particles changing from even to odd MOs. In fact, as the tunneling barrier is further raised, the number of particles moving from an even to an odd (or vice versa) MO converges zero, representing a fully fragmented state with zero tunneling splitting. This is consistent with the fact that observed spectra in the papers, van der Wal et.

al. (2000), Freidman et. al. (2000), and Yu et. al. (2002), are single photon spectra. One photon will not couple states wherein too many particles move co-operatively. Thus, for example in Freidman et. al. (2000) where the macroscopically mixed states are oppositely propagating supercurrents, it is likely that single photon transitions, and angular momentum conservation, imply that what is being observed is the transition of one or a small number of particles from one macroscopically occupied configuration to another. An occupation number analysis will give further insight into that more complex situation. But, at last, the "single particle excitation spectrum" of a quantum fluid has been identified. This interpretation differs from that of the above experimentalists, who are used to thinking in terms of collective modes of superfluids and superconductors. What we seem to have here, in contrast, are the types of single particle excitations characteristic of those many parts of atomic, molecular, and nuclear physics.

6. Summary and Conclusions

The natural orbitals and occupation numbers, introduced by Löwdin in 1955 are found to provide a most useful diagnostic for the states of a gaseous Bose condensate in a model double well potential. To describe the fragmentation, or dissociation, of such a system correlations are needed, as the Hartree-GP mean field picture does not provide a correct description. Once correlations are introduced the Löwdin occupation numbers provide a clear diagnostic as to what states are occupied by the condensed Bosons, and in doing so make it evident that the transitions observed in recent macroscopic tunneling experiments may be, in fact, part of the few-body, rather than the more usual collective excitations of such systems. Similar techniques should be applicable to looking at fragmentation and number squeezing in multi-well gaseous BECs and in superconducting current loops. This is of particular interest as such systems are possible realizations of the qubits of quantum computers.

7. Acknowledgements

The authors acknowledge helpful conversations with J. Brand and K. Mahmud, and the support of the U.S. National Science Foundation through Grant PHY- 0140091. The paper is dedicated to the memory of P.-O. Löwdin whose enthusiastic and seminal work on natural-orbitals, and the related N-representability problem, inspired, trained, and stimulated many generations of theoretical scientists. It is a pleasure to be able to apply these ideas to illuminate the physics of a novel set of systems.

8. References

Anderson, B. P., Haljan, P. C., Regal, C. A., Feder, D. L., Collins, L. A., Clark, C. W., and Cornell, E. A., Watching Dark Solitons Decay into Vortex Rings in a Bose-Einstein Condensate, Phys. Rev. Letts., 2001; 86,2926.

Anderson, P. W., Special Effects in Superconductivity, in Lectures on the Many-Body Problem, Vol. 2, E. R. Caianeillo, Ed., Academic Press, New York, 1964.

Anderson, P. W., Considerations on the flow of superfluid Helium, Rev. Mod. Phys. 1966; 38, 298.

Anderson, P. W., Basic Notions of Condensed Matter Physics, Addison-Wesley, Reading, MA, 1984.

Ashcroft, N. W. and Mermin, N. D., Solid State Physics, International Thompson Publishers, 1976.

Blümel, R. and Reinhardt, W. P., Chaos in Atomic Physics, Camb. Univ. Press, Camb., 1997.

Brand, J., and Reinhardt, W. P., Svortices and the fundamental modes of the snake instability: Possibility of observation in the gaseous Bose condensate, Phys. Rev. A., 2002; 65,043612.

Carr, L.D., Clark, C. W., and Reinhardt, W. P. Stationary States of the one-dimensional nonlinear Schrödinger equation: I. Case of repulsive non-linearity; II. Case of attractive non-linearity, Physical Review A, 2000a; 62, 63610-1-11; and, 2000b; 62, 63611-1 –11.

Cirac, J. I., Lewenstein, M., Molmer, K., and Zoller, P., Quantum superposition states of Bose-Einstein condensates, Phys. Rev A 1998; 57,1208.

DeMarco, B. and Jin, D. S., Onset of Fermi Degeneracy in a Trapped Atomic Gas, Science, 1999; 285,1703.

DeMarco, B. and Jin, D. S., Spin Exchange in a Gas of Fermi Atoms, Phys. Rev. Letts., 2002; 89,040405.

Dalfovo, F., Giorgini, S., Pitaeviskii, L. P., Stringari, S., Theory of Bose-Einstein condensation in trapped gases, Rev. Mod. Phys. 1999: 71,463.

Davidson, E. R., Reduced Density Matrices in Quantum Chemistry, Academic Press, NY, 1976.

Denschlag, J., Simsarian, J. E. , Feder, D. L. , Clark, C. W., Collins, L. A., Cubizolles, J., Deng L., Hagley, E. W., Helmerson K., Reinhardt, W. P., Rolston, S. L., Schneider, B. I. , and Phillips, W. D., Generating solitons by phase engineering of a Bose-Einstein condensate, Science, 2000; 287,97-101.

Dirac, P. A. M., Principles of Quantum Mechanics, Oxford Univ. Press, Oxford, 1947, pp 131-35.

Edwards, M., Ruprecht, P.A., Burnett, K., Dodd, R. J., and Clark, C. W., Collective excitations of atomic Bose-Einsein condensates, Phys. Rev. Letts., 1996; 77, 1671.

Einstein, A., Quantentheorie des einatomigen idealen Gases, II, Sitz. Der Preuss. Akad. Der Wissenschaften, 1925; K1, 3.

Fetter, A. L., and Walecka, J. D., Quantum Theory of Many-Particle Systems, McGraw-Hill, New York, 1971.

Flygare, W. H., Molecular Structure and Dynamics, Prentice-Hall, Englewood Cliffs, N.J., 1978, p 298.

Fisher, M. P. A., Weichman, B. P., Grinstein, G., and Fisher, D. S. Bosonic localization and the superfluid-insulator transition, Phys. Rev. D 1989; 40,546.

Franzosi, R., and Penna, V., Spectral properties of coupled Bose-Einstein condensates, Phys. Rev. A 2001; 63, 043609.

Franzosi, R., Penna, V., Zecchina, R. Quantum dynamics of coupled Bosonic wells within the Bose-Hubbard Picture, 2002a; arXiv:cond-mat/0205209v1.

Franzosi, R., Penna, V., Collective modes, chaotic behavior and self-trapping in the dynamics of three coupled Bose-Einstein condensates, 2002b; arXiv:cond-mat/0203509v3.

Freidman, J. R., Patel, V., Chen, W., Tolpygo, S. K., Lukens, J. E., Quantum superposition of distinct macroscopic states, Nature 2000; 406,43.

Girardeau, M., Wright, E. M., and Triscari, J. M., Ground state properties of a one-dimensional system of hard-core Bosons in a harmonic trap, Phys. Rev. A 2001; 63,033601.

Gordon, D., and Savage, C. M., Creating macroscopic quantum superpositions with Bose-Einstein condensates, Phys. Rev. A 1999; 59, 4623.

Greiner, M., Mandel, O., Esslinger, T., Hänsch, T. W., Bloch, I., Quantum phase transition for a superfluid to a Mott insulator in a gas of ultracold atoms, Nature 2002; 415, 39.

Griffin, A., Snoke, D. W., and Stringari, S., Bose-Einstein Condensation, Cambridge Univ. Press., Camb., 1995.

Griffin, A., Conserving and gapless approximations for an inhomogeneous Bose gas at finite temperatures, Phys. Rev. A 1996; 53,9341.

Gross, E. P., Structure of a Quantized Vortex in Boson Systems, Il Nuovo Cimento, 1961; 20,454.

Gutzwiller, M. C., Chaos in Classical and Quantum Mechanics, Springer, New York, 1990.

Hartree, D. R., The Wave Mechanics of an Atom in a Non-Coulombic Central Field, Part I, Theory and Methods, Proc. Camb. Phil. Soc., 1928; 24,89.

Hunag, K., and Yang, C. N., Quantum-Mechanical Many-Body Problem with Hard Sphere Interaction, Physical Review, 1957; 105,767.

Husimi, K., Some Formal Properties of the Density Matrix, Proc. Physico- Math. Soc. Japan, 1940; 22,264.

Josephson, B. D., Possible New Effects in Superconductive Tunneling, Phys. Letts, 1962: 1,251.

Khaykovich, L., Schreck, F., Ferrari, G., Bourdel, T., Cubizolles, J., Carr, L. D., Castin, Y., and Salomon, C., Formation of a Matter-Wave Bright Soliton, Science, 2002; 296, 1290-1293.

Kosmatov, N. E., Shvets, V. F., and Zakharov, V. E., Computer simulation of wave collapse in the nonlinear Schrödinger equation, Physica D 1991; 52,16-35.

Lee, T. D., Huang, K., and Yang, C. N., Eigenvalues and eigenfunctions of a Bose system of hard spheres and its low temperature properties, Physical Review, 1957; 106,1135.

Leggett, A. J., Bose-Einstein condensation in the alkali gases: Some fundamental concepts, Rev. Mod. Phys., 2001; 73, 307.

Loudon, R., The Quantum Theory of Light, Oxford, Oxford, (UK), 1973.

Löwdin, P.-O., Physical Interpretation by Means of Density Matrices, Natural Spin-Orbitals, and Convergence Problems in the Method of Configuration Interaction, Physical Review, 1955; 97,1474.

Martens, C. C., Vibration-rotation interaction in the rigid bender: a quantum mechanical phase space view, J. Chem. Phys. 1989: 90,7064.

McWeeny, R. The Density Matrix in Many Electron Quantum Mechanics I., Proc. Roy. Soc. 1959; 253,242.

McWeeny, R., and Sutcliffe, B. T., Methods of Molecular Quantum Mechanics, Academic Press, New York, 1969.

Milburn, G. J., Corney, J., Wright, E. M., Walls, D. F., Quantum dynamics of an atomic Bose-Einstein condensate in a double well potential, Phys. Rev. A 1997; 55, 4318.

Orzel, C., Tuchman, A. K., Fenselau, M. L., Yasuda, M., Kasevich, M. A., Squeezed States in a Bose Einstein Condensate, Science 2001; 291, 2386.

Penrose, O., On the quantum mechanics of Helium II, Phil. Mag. 1951; 42,1373.

Penrose, O., and Onsager, L., Bose-Einstein Condensation and Liquid Helium, Phys. Rev. 1956; 104,576.

Perelomov, A. Generalized Coherent States and Their Applications, Springer, Berlin, 1986.

Pethick, C. J., and Smith, H., Bose-Einstein Condensation in Dilute Gasses, Camb. Univ. Press, Camb., 2002.

Pitaevskii, L. P., Vortex Lines in an Imperfect Bose Gas, Sov. Phys. JETP 1961; 13,451.

Raghavan, S., Smerzi, A., Fantoni, S., and Shenoy, S. R., Coherent oscillations between two weakly coupled Bose-Einstein condensates: Josephson effects, π-oscillations, and macroscopic quantum self-trapping, Phys. Rev. A 1999; 59, 620.

Reinhardt, W.P. and Clark, C.W., Soliton dynamics in the collisions of Bose-Einstein condensates: an analog of the Josephson effect, J. Phys. B 1997; 30,L785.

Reinhardt, W.P., Solitons in the Bose Condensate, in *Tunneling in Complex Systems*, S. Tomsovich, Ed. World Scientific, Singapore; 1998, p 277-326.

Roberts, J. L., Claussen, N. R., Cornish, S. L., Donley, E. A., Cornell, E. A., and Wieman, C. E., Controlled Collapse of a Bose-Einstein Condensate, Phys. Rev. Letts. 2001; 86,4211.

Roothaan, C. C. J., New Developments in Molecular Orbital Theory, Rev. Mod. Phys. 1951; 23,69.

Ruprecht, P. A., Holland, M. J., Burnett, K., Edwards, M., Time-dependent solution of the nonlinear Schrödinger equation for Bose-condensed trapped neutral atoms, Phys. Rev. A 1995; 51, 4704.

Sackett, C. A., Gerton, J. M., Welling, M., and Hulet, R. G., Measurements of Collective Collapse of a Bose-Einstein Condensate with Attractive Interactions, Phys. Rev. Letts. 1999; 82,876.

Salmond, G. L., Holmes, C. A., and Milburn, G. J., The dynamics of a strongly driven two component Bose-Einstein Condensate, Phys. Rev. A 2002: 65,033623.

Schrödinger, E. in Collected Papers on Wave Mechanics, Chelsea, New York, 1982; p 43,44. (Translation from the German, Schrödinger, E., Die Naturwissenschaften, 1926; 28,664-666.)

Spekkens, R. W., and Sipe, J. E., Spatial Fragmentation of a Bose-Einstein condensate in a double well potential, Phys. Rev. A, 1999; 59,3868.

Steel, M. J., and Collett, M. J., Quantum state of two trapped Bose-Einstein condensates with a Josephson coupling, Phys. Rev. A 1998; 75, 2920.

Strecker, K.E., Partridge, G. B., Truscott, A. G., Hulet, R. G., Formation and propagation of matter-wave soliton trains, Nature, 2002; 417,150.

Sulem, C. and Sulem, P.-L., The Nonlinear Schrödinger Equation, Self-Focusing and Wave Collapse, Springer, Berlin, 1999.

Thouless, D. J., The Quantum Mechanics of Many-Body Systems, Academic Press, New York, 1961.

Tilley, D. R., and Tilley, J., Superfluidity and Superconductivity, Institute of Physics, Bristol (UK), 3rd Ed. 1990.

Tinkham, M., Introduction to Superconductivity, McGraw-Hill, New York, 2nd Ed., 1996.

van der Wal, C.H., ter Haar, A. C. J., Wilhelm, F. K., Schouten, R. N. , Harmans, C. J. P. M., Orlando, T. P., Lloyd, S. , and Mooij, J. E., Quantum Superposition of Macroscopic Persistent-Current States, Science 2000; 290,773.

Wu, T. T., Ground state of a Bose system of hard spheres, Physical Review, 1959; 115,1390.

Yu, Y., Han, S., Chu, X., Chu, S.-I., Wang, Z., Coherent Temporal Oscillations of Macroscopic Quantum States in a Josephson Junction, Science 2002; 296, 889.

MODEL STUDIES ON THE ELECTROPHILIC SUBSTITUTION OF METHANE WITH VARIOUS ELECTROPHILES E (E = NO_2^+, F^+, Cl^+, Cl_3^+, HBr_2^+, HCO^+, OH^+, $H_2O–OH^+$, AND Li^+)

Peter R. Schreiner,*[a] Andrey A. Fokin,[b] Paul von Ragué Schleyer,[a] and Henry F. Schaefer III[a]

[a]*Department of Chemistry, University of Georgia, Athens, Georgia 30602-2556, USA* prs@chem.uga.edu, [b]*Department of Organic Chemistry, Kiev Polytechnic Institute, 37 Pobeda Avenue, 03056 Kiev, Ukraine*

Abstract. The reactions of NO_2^+, $^1F^+$, $^1Cl^+$, Cl_3^+, HBr_2^+, HCO^+, $^1OH^+$, $H_2O–OH^+$, and Li^+ with methane were investigated *ab initio* at the MP4SDTQ/6-31G**/MP2/6-31G** + ZPVE level of theory. The more stable electrophiles (NO_2^+, HCO^+, Cl_3^+) attack the methane carbon directly; the first intermediate typically is a substituted, strongly-bound methyl cation-dihydrogen complex (ECH_2^+•••H_2). This either loses H_2 to give the $H_2C–E^+$ cation or rearranges to the thermodynamically most favored *formal insertion* product $H_3C–EH^+$. The generally small activation energies for hydrogen loss (0.4–2.0 kcal mol^{-1}) are lower than the barriers for other reactions (5–50 kcal mol^{-1}). In contrast, less stable, high energy electrophiles (e.g., H^+, R^+, $^1F^+$, $^1Cl^+$, $^1OH^+$) react with methane without a barrier, so that direct C–H insertion or carbon attack pathways are not differentiated. The most favorable binding arrangements involve the electrophile bound to carbon directly, and not via 3c-2e bonding (CHE^+). Dihydrogen attachment (ECH_2^+•••H_2) affords additional stabilization. Initial CH_4•••E^+ complexes were only found for the more stable electrophiles (NO_2^+, HCO^+, Cl_3^+, HBr_2^+, $HO–OH_2^+$, Li^+). However, these CH_4•••E^+ complexes need not be involved in the most favorable pathways. The reactions of $HO–OH_2^+$ and Cl_3^+ are more realistic models for $^1OH^+$ and $^1Cl^+$. Since the charge delocalized ground states involving $HO–OH_2^+$ and Cl_3^+ are more stabilized, the overall activation barriers are 4–7 kcal mol^{-1} higher, but the general features of the reaction are quite similar. Methane activation with charged electrophiles-oxidizers is accompanied by substantial charge transfer from the hydrocarbon to the electrophile. The hydrocarbon moieties resemble the structure of the methane radical cation with two elongated C–H bonds in the lowest energy transition structures. We conclude that no single mechanism describes the electrophilic reactions of methane.

E.J. Brändas and E.S. Kryachko (eds.),
Fundamental World of Quantum Chemistry, Vol. II, 349–375.

INTRODUCTION

One of Olah's most notable contributions to chemistry is his recognition that electrophilic substitution reactions can be carried out on aliphatic as well as aromatic hydrocarbons.[1-5] Attack of an electrophile E^+ was envisaged to occur at a C–H bond (eq. 1), leading to a three center-two electron (3c-2e) species (**1**) as a transition structure or intermediate.

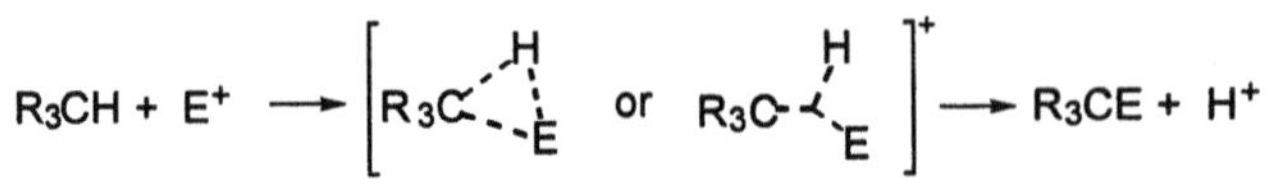

(1)

1

This conceptually attractive postulate was based on the known behavior of protonated alkanes,[6-10] CH_5^+,[9, 11-18] $C_2H_7^+$,[19-21] $C_3H_9^+$,[22-24] etc., where such 3c-2e bonds (**1**, with $E^+ = H^+$ or alkyl$^+$) are implicated. However, the experimental support for Olah's electrophilic substitution mechanism has been based largely on product studies; some secondary H/D kinetic isotope effect results are available, but they are not conclusive.[18]

Inspired by Olah's findings, we have sought to clarify the mechanistic details of such electrophilic substitution reactions.[25, 26] Likewise, Bach and Su[27] investigated the reaction of protonated hydrogen peroxide (H_2OOH^+, as an "OH^+" model), as well as the related reactions of water oxide (H_2OO)[28] and alkyl carbenes with alkanes.[29] Methane activation with highly electrophilic BH_2^+ was also studied by DePuyDamrauer et al.[30] but this report concentrated more on the properties of BH_2^+.

Methane is protonated quite exothermically (PA_O = 130.5 kcal mol^{-1}) in the gas phase.[31] The resulting methonium ion CH_5^+ is exceptionally stable towards dissociation into the methyl cation (CH_3^+) and H_2 (D_O = 45.3 kcal mol^{-1}).[32] H/D exchange occurs when CH_4 reacts with D^+ or CD_4 with H^+.[13, 18] Although never observed in solution directly, the methonium ion also may exist as a short lived intermediate in H/D exchange reactions of methane with DF-SbF_5.[33, 34]

Methane also reacts exothermically in the gas phase with CH_3^+ or with the higher alkyl cations.[6-8] The two isomeric $C_2H_7^+$ cation isomers (**2** and **3**, the bond lengths are shown at MP2/6-31G**)[20, 21] are prototypes for the formal "insertion" of the methyl cation into a C–H bond of an alkane (**2**) or for attack of CH_3^+ at the methane carbon (**3**). The C–C protonated form (**2**) is 4.4 to 9 kcal mol^{-1} more favorable[19] than the C–H protonated one (**3**) at BLYP/6-31G** (8.8 kcal mol^{-1}), MP2/6-31G** (6.1 kcal mol^{-1}), and CCSD(T)/cc-pVTZ (6.4 kcal mol^{-1}).[21] In marked contrast to CH_5^+, the H_2 dissociation energy of **3** is much lower (+4 kcal mol^{-1}).[12, 19-21, 35]

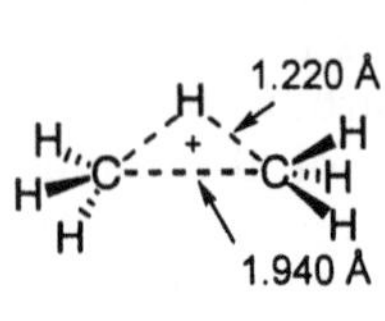

2

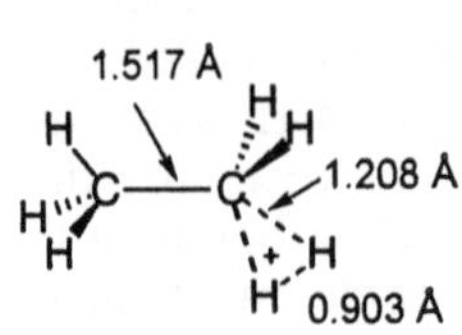

3

Similarly, $C_2H_5^+$ reacts with methane to yield the C–H and C–C protonated forms of the $C_3H_9^+$ cation. The latter is more stable by only 1 kcal mol^{-1},[22, 23] and the i-$Pr^+\cdots H_2$ complex is very weakly bound. Such intermediates are implicated in Olah's reactions of C_2H_5F with CH_4 in SbF_5 solutions.[33, 34]

Methane can be substituted with other electrophiles as well. The reaction of $NO_2^+BF_4^-$ in CH_2Cl_2/HSO_3F gives nitromethane[36, 37], "F^+" (in the form of $N_2F^+AsF_6^-$) yields mixtures of CH_3F, CH_2F_2, and CHF_3;[38] $SbF_5\cdots Cl_2$ in SO_2ClF leads to the dimethylchloronium ion $[H_3C–Cl–CH_3]^+$,[39] and Cl_3^+ gives $[H_3C–Cl–H]^+$ and $[H_3C–Cl–Cl]^+$ in the gas phase.[40] No evidence for radicals was reported in the halonium reactions.

Despite the conceptual attractiveness of the generalized mechanism (eq. 1) for these reactions,[1-4] difficulties arise when the assumed 3c-2e species **1** are subjected to closer scrutiny. Consider **1** as a resonance hybrid:[41]

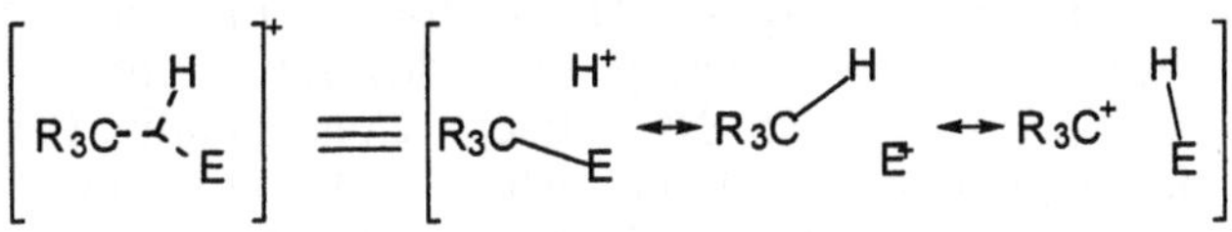

1 **1A** **1B** **1C**

Cyclic three center-two electron bonding can only be effective if the contributions of **1A**, **1B**, and **1C** are *all* significant, i.e., that all have comparable weights as resonance contributors.[42] That this necessary balance[43] is not easily achieved is illustrated by CH_5^+, by **2**, and by **3**. Balance between **1A**, **1B**, and **1C** is even less likely inherently when the electrophile (E^+) is different from H^+ and R_3C^+.[43]

• CH_5^+ has a highly fluctional structure (see below);[17, 44-46] form **1C** (E = H) contributes little.
• The CC bond in **2** is very long; form **1A** (E = CH_3) does not contribute effectively.[47]
• The H_2 moiety in **3** (and in the i-$Pr^+\cdots H_2$ complex) is only weakly bound; form **1C** (E = H) contributes much more than **1A** and **1B** (E = CH_3).

While it was conceptually useful to assume that all electrophiles react like the proton or alkyl cations and that electrophilic aliphatic reaction mechanisms follow the same general pattern (eq. 1),[1-3] these assumptions suppose that a) the methonium ion is a good model for intermediates or transition structures of general type **1**; and b) that C–H attack is favored. In contrast, it was demonstrated[45 48, 49] both computationally and experimentally that CH_5^+ is highly fluctional.[14, 17, 18] Although the C_s complex (**4a**) of the methyl cation and dihydrogen may be the most populated CH_5^+ structure at 0 K,[48] both the second C_s structure (**4b**) and the C_{2v} form (**4c**) are essentially equal in energy at the zero-point vibrational level (within 0.2 kcal mol^{-1}). Complete hydrogen scrambling occurs with unprecedented rapidity (eq. 2).[18] Hence, CH_5^+ is *not* a particularly good model for electrophilic aliphatic substitutions. As noted above, the $CH_3^+ + H_2$ complexation energy also is exceptionally large.

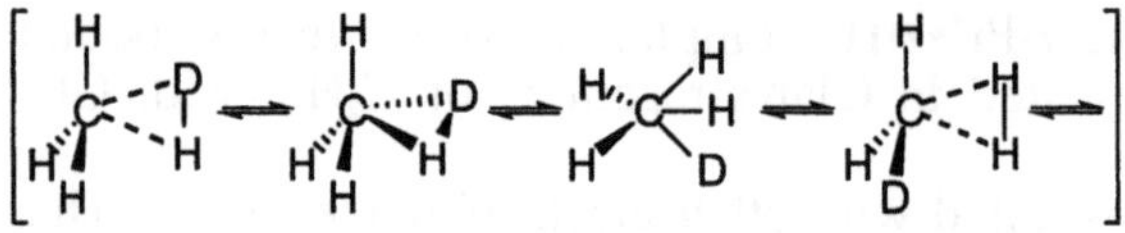

4a **4b** **4c** **4a** (2)

Our earlier investigation of the model reactions of methane with the nitrosonium ion (NO^+),[25] showed that the mechanisms (eq. 1) proposed for the proton ($E^+ = H^+$) and carbenium ions ($E^+ = R^+$) were not applicable to this much more stable electrophile. Carbon rather than hydrogen is attacked by NO^+ (eq. 3, $E^+ = NO^+$) A 3c-2e intermediate **5** (a weak complex between H_2 and CH_2NO^+) is formed eventually, but this does not involve the electrophile ($E^+ = NO^+$) in 3c-2e bonding which has become fully bonded to the carbon in the initial stages. The distortion of methane during the reaction to lower symmetry, leading to a developing lone pair, can be envisaged to be *concomitant* with NO^+ attack. Conceptually, the electrophile binds to the developing lone pair on carbon. Of course, large part of the energy required for the methane distortion is recovered from the new bonding interaction. There is a transition structure which formally corresponds to the direct insertion into a C–H bond, but the C–N bond has formed already and this pathway is 14.4 kcal mol^{-1} less favorable.

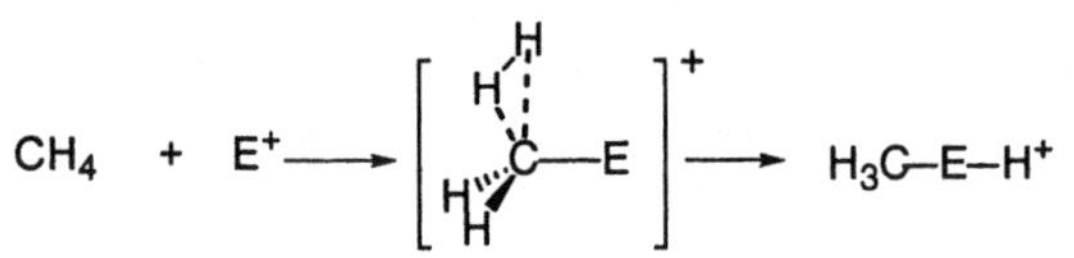

(3)

5

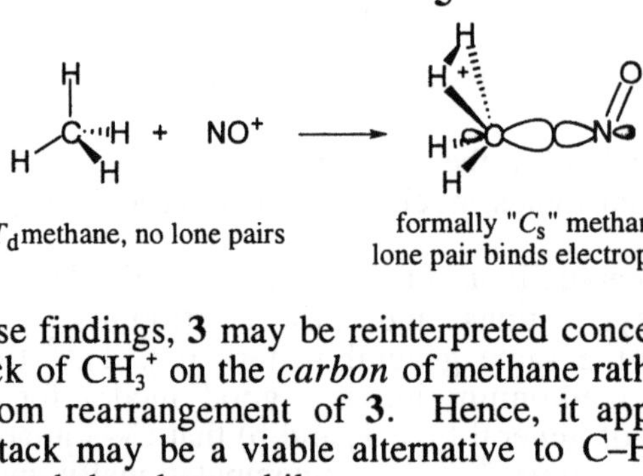

$T_d$methane, no lone pairs

formally "C_s" methane, lone pair binds electrophile

Based on these findings, **3** may be reinterpreted conceptually as resulting from direct attack of CH_3^+ on the *carbon* of methane rather on a C–H bond; **2** may result from rearrangement of **3**. Hence, it appears plausible that direct carbon attack may be a viable alternative to C–H attack, depending upon the alkane and the electrophile.

In their very descriptive frontier molecular orbital (FMO) analysis of the OH^+ and $H_3O_2^+$ reactions with methane, Bach and Su also emphasized a more delocalized view of an electrophile approaching methane.[27] The triply degenerate methane HOMO shares coefficients both form carbon and two hydrogens. As there is no single orbital which describes one separate C–H bond, Bach and Su conclude that "*...the electrophilic oxygen (OH^+) attacks the carbon atom and approaches the hydrocarbon along the approximate atomic 2p carbon orbital that comprises the σ_{CH_2} fragment orbital.*" and that "*...the insertion of an electrophilic reagent into a C–H bond involves the entire CH_2 fragment orbital and not just an isolated C–H σ bonding orbital.*"[27]

Olah *et al.'s* computational reexamination[50] of our earlier work[25] confirmed "...*that the theoretical reaction of CH_4 with NO^+ showed that the reaction path involves attack on carbon instead of C–H bond insertion,*" but attributed this to NO^+ being "...*an ambident and relatively weak electrophile...*" Instead, "*With hard electrophiles, as D^+ (H^+), CH_3^+, $C_2H_5^+$, OH^+, and F^+ methane preferentially reacts via 3c-2e C–H bond insertion path as previously suggested by Olah...*"[51, 52] However, the latter conclusions were not supported computationally. A separate study of the reaction of singlet F^+ with methane failed to reveal any stationary point corresponding to **1** ($E^+ = F^+$); the only 3c-2e species was a $FCH_2^+ \cdots H_2$ complex, corresponding to **5** (see above with $E^+ = F^+$).[26]

We also have investigated the reactions of the next higher alkane, ethane, with the NO^+ model electrophile.[26] Only two viable mechanisms were located although several plausible alternative pathways leading to nitrosomethane, to nitrosoethane, to loss of H_2 (after addition of NO^+ to ethane), as well as insertion of the electrophile into the C–C or a C–H bond of ethane were explored. These two viable mechanisms lead to a) the nitrosomethylene cation and methane through C–C bond cleavage (C_2H_6 + $NO^+ \rightarrow H_2C\text{–}NO^+ + CH_4$) , and to b) protonated nitrosoethane via hydride abstraction ($C_2H_6 + NO^+ \rightarrow C_2H_5^+ + HNO$) followed by addition of HNO to the ethyl cation ($C_2H_5^+ + HNO \rightarrow C_2H_5NHO^+$). The computed activation energies of these two reactions are similar (33.5 and 31.1 kcal mol^{-1}, respectively). We found no evidence for *direct* insertion of the electrophile into a C–H or a C–C bond of ethane.

Our findings for the reactions of methane[25] and ethane[26] with the nitrosonium ion, and Bach's[27-29] as well as Olah's[50] recent discussions of the mechanisms of electrophilic aliphatic substitutions,[53] have stimulated us to investigate the reactions of methane with a larger set of electrophiles from weak to strong (Li^+, HCO^+, NO_2^+, Cl^+, Cl_3^+, HBr_2^+, $H_2O\text{–}OH^+$, OH^+, F^+). The results provide systematic comparisons and refine earlier insights.

METHODS

The *ab initio* programs Gaussian 94[54] and Gaussian 98[55] were employed to optimize geometries fully within the designated symmetry constraints at the restricted Hartree-Fock (HF) level using gradient optimization techniques and the standard 6-31G(d,p) basis set.[56] Optimizations including electron correlation were then performed at the second order Møller-Plesset level (MP2) keeping the core electrons frozen (MP2-fc). MP4SDTQ(fc) single-point energy calculations were carried out on all MP2/6-31G(d,p) optimized structures. This level of theory has successfully been applied to closely related problems.[25, 57] Analytic vibrational frequencies were obtained through the MP2/6-31G(d,p) level of theory to determine the number of imaginary frequencies (NIMAG) and to characterize stationary points (minima have NIMAG=0; transition structures have NIMAG=1). Vibrational frequencies and zero-point vibrational energies (ZPVE) were scaled by the empirical factor 0.91,[58] to correct for anharmonicity and correlation effects. All reaction enthalpies are based on MP4SDTQ/6-31G(d,p)//MP2/6-31G(d,p) + ZPVE energies, unless stated otherwise. Standard notation is used;[56] "//" means "at the stationary point geometry of."

While F^+, Cl^+, and OH^+ have triplet ground states, the (open-shell) singlet states model –in an amplified way–the reactivities of halonium salts (e.g.,

$N_2F^+AsF_6^-$) and halonium salt precursors (e.g., Cl_2•••$AlCl_3$). Our comparison between H_2O–OH^+ (see below and ref.[27]) and OH^+ is instructive in this context. Cl_3^+ and HBr_2^+ (or other positively charged halogen models) were also studied for comparison.

Another complicating feature is that the ionization potentials for F, Cl, and OH are higher than for methane, i.e., oxidation of methane to the radical cation should occur first if the cations of these electrophiles were available freely. However, as noted above, our cationic electrophiles serve as models for the reactions of complexed electrophiles. Association, even with N_2, reduces the ionization potential for the halo- and OH-radicals dramatically (cf. eqs. 4 and 5). It seems therefore justified to use the open-shell singlet cations as non-oxidizing electrophiles computationally.

$$F^+ + CH_4 \rightarrow F^{\bullet} + CH_4^{\bullet +} \qquad \Delta H = -114.8 \text{ kcal mol}^{-1} \quad (4)$$

$$N_2F^+ + CH_4 \rightarrow N_2F^{\bullet} + CH_4^{\bullet +} \qquad \Delta H = +20.4 \text{ kcal mol}^{-1} \quad (5)$$

We also note that the theoretical treatment is only qualitatively reliable for these species since the open-shell singlets are not necessarily described well with a single reference method. A multi-reference treatment, however, is currently not feasible for the large number of species treated in the present work.

RESULTS

In general, the reactions of methane with electrophiles (E^+) might proceed via three different pathways: direct carbon attack, insertion into a C–H bond, and hydrogen abstraction followed by addition of EH to the methyl cation. Our discussion focuses on these three mechanisms (Tables 1 and 2).

CH_4 + NO_2^+ (Fig. 1). Methane is known to react with $NO_2^+BF_4^-$ in CH_2Cl_2/HSO_3F to yield nitromethane, but only with very low yield (0.1% at 25 °C).[37] Although the nitronium cation is known to be a powerful hydride abstraction reagent,[59] eq. 1 (R=H) depicts the reaction of methane with NO_2^+ as an *insertion* with $E^+ = NO_2^+$.

The nitronium ion and methane form an initial complex (**MIN1**, D_O = 3.1 kcal mol^{-1}) which is similar to the CH_4•••NO^+ complex (D_O = 3.6 kcal mol^{-1}).[60] We find computationally that NO_2^+ can react with methane in two different ways. While the energetically favored ($\Delta H^{\ddagger}$ = 49.0 kcal mol^{-1}) hydride abstraction (**TS1**) yields transient CH_3^+ + HONO (and subsequently the formal insertion product H_3C–$ONOH^+$), direct carbon attack either gives H_4C–NO_2^+ (*via* **TS2**, $\Delta H^{\ddagger}$ = 55.9 kcal mol^{-1}) or leads to loss of H_2 (*via* **TS3**, $\Delta H^{\ddagger}$ = 61.6 kcal mol^{-1}). No direct C–H insertion pathway was located. As suggested by Olah, the lowest lying transition state (**TS1**) is non-linear, consistent with a small kinetic primary H/D isotope effect during the nitration of adamantane with $NO_2^+BF_4^-$.[61] The final product in this reaction is O-protonated nitromethane (**MIN3**) which is the most stable structure located on this part of the PES.

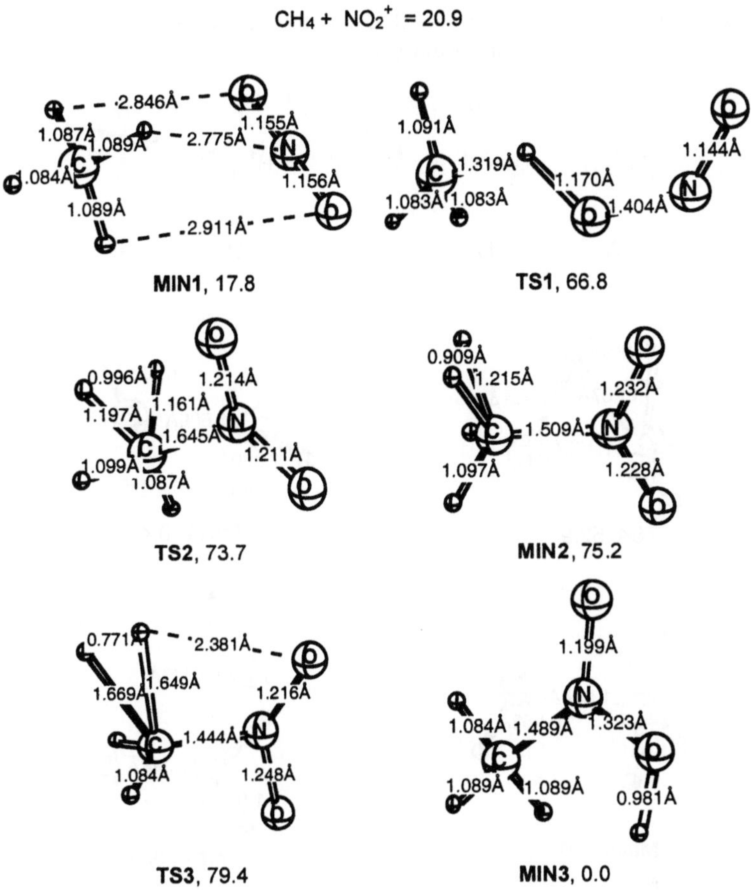

Figure 1. The MP2/6-31G** geometries of the stationary structures for the reaction of methane with the nitronium cation NO_2^+ (relative energies in kcal mol^{-1}).

CH_4 + $^1F^+$ (Fig. 2 and Scheme 1). Methane can be halogenated using a dihalogen-Lewis acid combination (e.g., Cl_2:SbF_5 or Br_2:$AgSbF_6$).[39] Although this approach fails for fluorinations ("fluoronium ions" are unknown in solution chemistry), Olah *et al.* succeeded quite remarkably in using $N_2F^+AsF_6^-$ and $NF_4^+AsF_6^-$ salts as "F^+ equivalents" for the fluorination of methane.[38] Their publication also included QCISD/6-31G* computations of several of the stationary structures (**A–E**) on the CH_4F^+ energy hypersurface.

No C–H insertion transition structure was reported, and "*...attempts to locate the transition structure of the reaction CH_4 + "F^+" leading to formation of 3c-2e structure [E] (involving initial "F^+" insertions directly into the C–H bond) failed. This is due to the extremely high reactivity of electrophilic "F^+" reacting with methane practically without any activation energy barrier.*"[38] Structure **A** was found to be the lowest lying isomer, but the rotation about the C–F bond is virtually free (the barrier is only 0.1 kcal mol^{-1}). Both **C** and **D** are higher order saddle points. Since a transition state for C–H insertion was not located, structure **E** was suggested (but not optimized) as a model for this reaction.

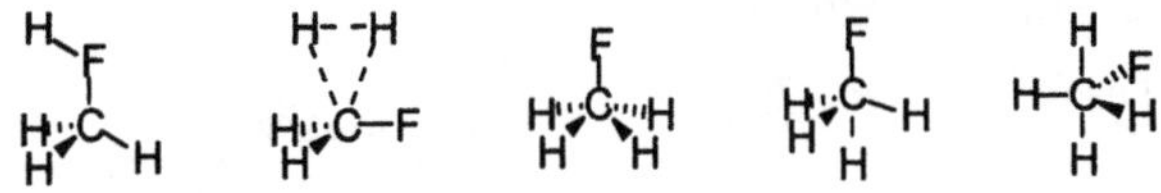

A **B** **C** **D** **E**

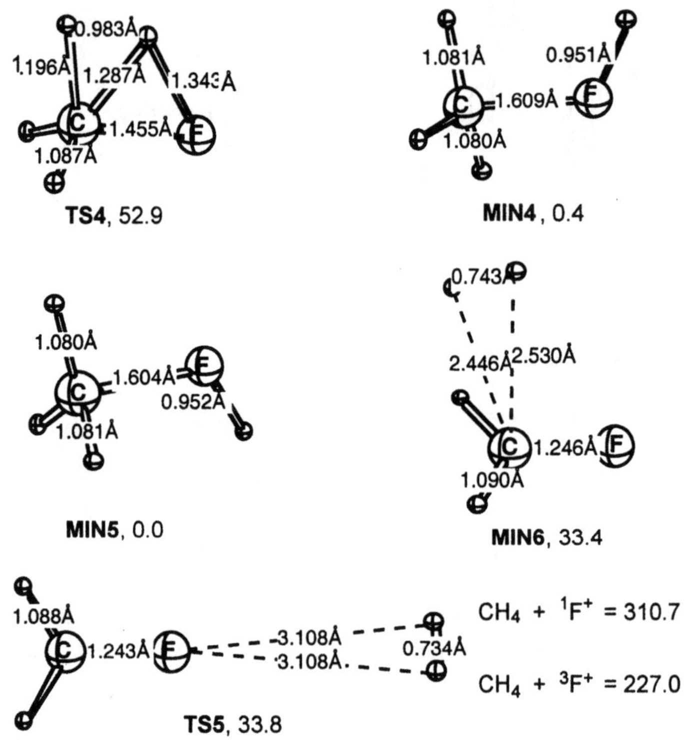

Figure 2. The MP2/6-31G** geometries of the stationary structures for the reaction of methane with singlet state F^+ (relative energies in kcal mol^{-1}).

In addition to reproducing Olah's findings (**MIN4** and **MIN5**, Fig. 2), we were also able to locate a transition structure (**TS4**) which appears to correspond to the C–H insertion step.[62] However, an IRC calculation revealed that this is not the case; surprisingly, **TS4** connects **MIN4** with **MIN6**! Since no other initial transition state (e.g., Olah's structural suggestion **E**) or complex was found, methane and F^+ must yield either a) **MIN6** as an intermediate for fluoronium attacking carbon directly or b) **MIN5** as the direct product of a formal C–H insertion (Fig. 3). The absence of a barrier for either reaction is due the large exothermicity for the CH_4 + F^+ → **MIN6** (–276.6 kcal mol^{-1}) or the CH_4 + F^+ → **MIN5** (–310.7 kcal mol^{-1}) reactions. If **MIN6** is formed first, the activation energy to give **MIN4** (and **MIN5**) is relatively small (19.5 kcal mol^{-1}), and can easily overcome by the large exothermicity of the overall CH_4 + "F^{+}" reaction (−158.6 kcal mol^{-1}, eqs. 6 and 7). Both **MIN4** and **MIN5** are virtually identical in energy, allowing free rotation of the H–F moiety in protonated methyl fluoride. However, no 3c-2e intermediates or transition structures utilizing C–H–F bonding arrangements are apparent. Structure **MIN6** is favored over a 3c-2e binding arrangement including the fluorine atom because $H_2C{-}H^+$ is much less stable than $H_2C{-}F^+$ (eq. 8).

CH_3FH^+ (**MIN5**)	→	$CH_3F + H^+$	ΔH_0 = +152.1 kcal mol^{-1}	(6)
$CH_4 + F^+$	→	CH_3FH^+ (**MIN5**)	ΔH_0 = −310.7 kcal mol^{-1}	(7)
$CH_3^+ + CH_3F$	→	$CH_2F^+ + CH_4$	ΔH_0 = −20.0 kcal mol^{-1}	(8)

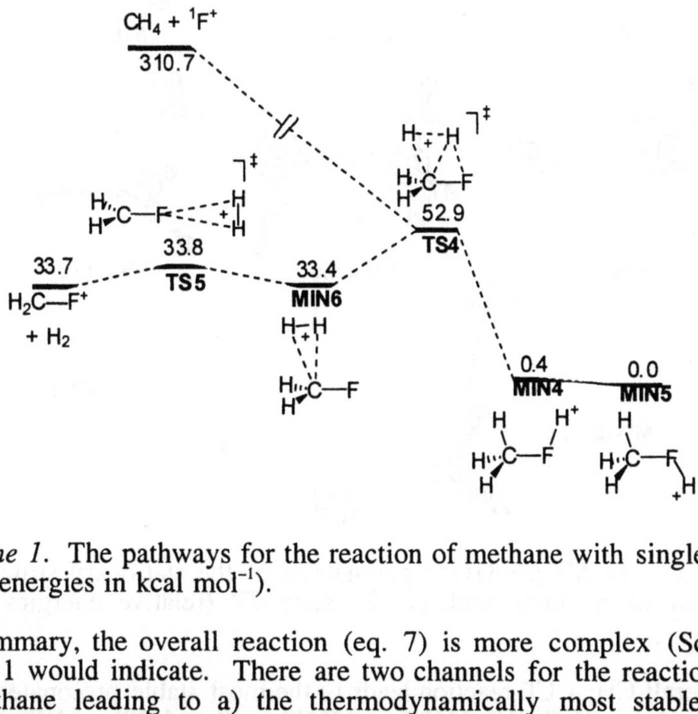

Scheme 1. The pathways for the reaction of methane with singlet state F^+ (relative energies in kcal mol^{-1}).

In summary, the overall reaction (eq. 7) is more complex (Scheme 1) than eq. 1 would indicate. There are two channels for the reaction of "F^{+}" with methane leading to a) the thermodynamically most stable product **MIN5** via **TS4**; and b) loss of H_2 from the first intermediate **MIN6** via **TS5**. Although path b) is favored kinetically, **MIN5** is the thermochemically most stable product. While **MIN5** formally derives from an F^+ insertion into a C–H bond, no direct insertion steps need be involved.

The reactions of methane strongly electrophilic BH^{2+} are quite comparable to our findings for F^+; DePuy, Damrauer et al. also found a hydrogen-shift transition structure similar to our **TS4** for the complext of $H_2B^+{\cdots}H_2CH_2$ going to $[(H_2)HB{-}CH_3]^+$.[30]

CH_4 + $^1Cl^+$ (Fig. 3 and Scheme 2). The reaction of the open-shell singlet Cl^+ with methane involves more stationary points than the $CH_4 + F^+$ reaction, due to the higher stability of Cl^+ (F^+ + HCl —> Cl^+ + HF ΔH_f^o = −125 kcal mol^{-1}). As with $^1F^+ + CH_4$, no activation barriers were located along the pathway to intermediate **MIN8** or the most stable product **MIN9**. In contrast to the $^1F^+$ reaction, structure **MIN8**, formed via direct carbon attack, is a CH_5^+ derivative, but with a 3c-2e C, H, H bond (like in **3**). However, **MIN8** dissociates readily via **TS8** to the very weakly bound $H_2CCl^+{\cdots}H_2$ complex **MIN10** (practically to $H_2C–Cl^+$ and H_2). An alternative pathway leads to the formal C–H insertion products **MIN7** or **MIN9**.

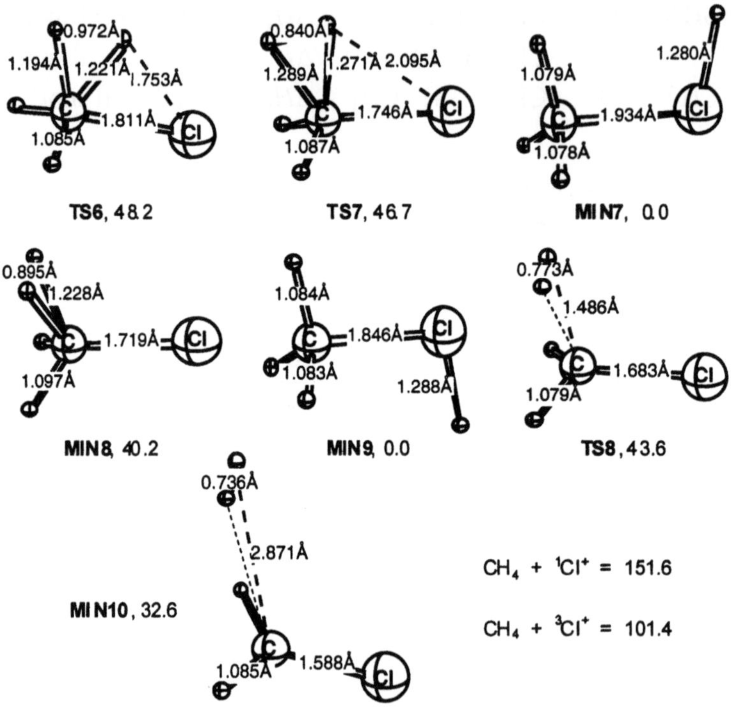

Figure 3. The MP2/6-31G** geometries of the stationary structures for the reaction of methane with singlet state Cl^+ (relative energies in kcal mol^{-1}).

The overall $CH_4 + Cl^+$ reaction leads to the most stable protonated methyl chloride products (H_3CClH^+, **MIN7** or **MIN9**, eq 9 and 10) and is much less exothermic than the reaction of F^+ with methane (cf. eqs. 6 and 7):

$$H_3CClH^+ \text{ (MIN7/9)} \rightarrow CH_3Cl + H^+ \qquad \Delta H_O = +113.0 \text{ kcal mol}^{-1} \qquad (9)$$

$$CH_4 + Cl^+ \rightarrow H_3CClH^+ \text{ (MIN7/9)} \qquad \Delta H_O = -151.6 \text{ kcal mol}^{-1} \qquad (10)$$

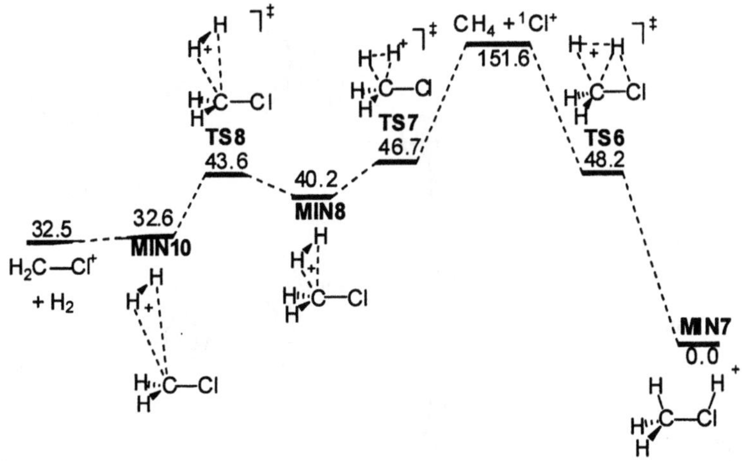

Scheme 2. The pathways for the reaction of methane with singlet state Cl^+ (relative energies in kcal mol^{-1}).

CH_4 + Cl_3^+ (Fig. 4 and Scheme 3). A more realistic alkane halogenation model is the reaction of methane with Cl_3^+, which has singlet ground state[63] and is stable in the gas phase.[64] The gas phase reaction of methane with Cl_3^+ was studied by ion cyclotron resonance spectrometry and yields a mixture of $[H_3C–Cl–H]^+$ and, unexpectedly, $[H_3C–Cl–Cl]^+$, which cannot form *en route* by the electrophilic insertion into a methane C–H bond. Previous computations[40] failed to locate transition structures on the CH_4 + Cl_3^+ PES. We were able to find three transitions structures (Fig. 4) which describe either direct carbon attack (**TS9**), C–H insertion (**TS10**), and C–H abstraction (**TS11**).

Because of insignificant Cl•••Cl_2 interactions, the CH_4•••Cl moieties in **TS9** and **TS10** are very similar to that for **TS7** and **TS6**, respectively. The resulting complex of protonated methyl chloride with Cl_2 (**MIN13**) is the global minumum for this part of the PES; loss of Cl_2 leads to the experimentally observed $[H_3C–Cl–H]^+$. Direct carbon attack and C–H insertions are barrierless, while the H-abstraction path via **TS11** requires substantial activation ($\Delta H^{\ddagger}$ = 32.9 kcal mol^{-1}). However, this path yielding $[H_3C–Cl–Cl]^+$ probably is observed experimentally as a result of an abstraction/recombination reaction to **MIN14**.

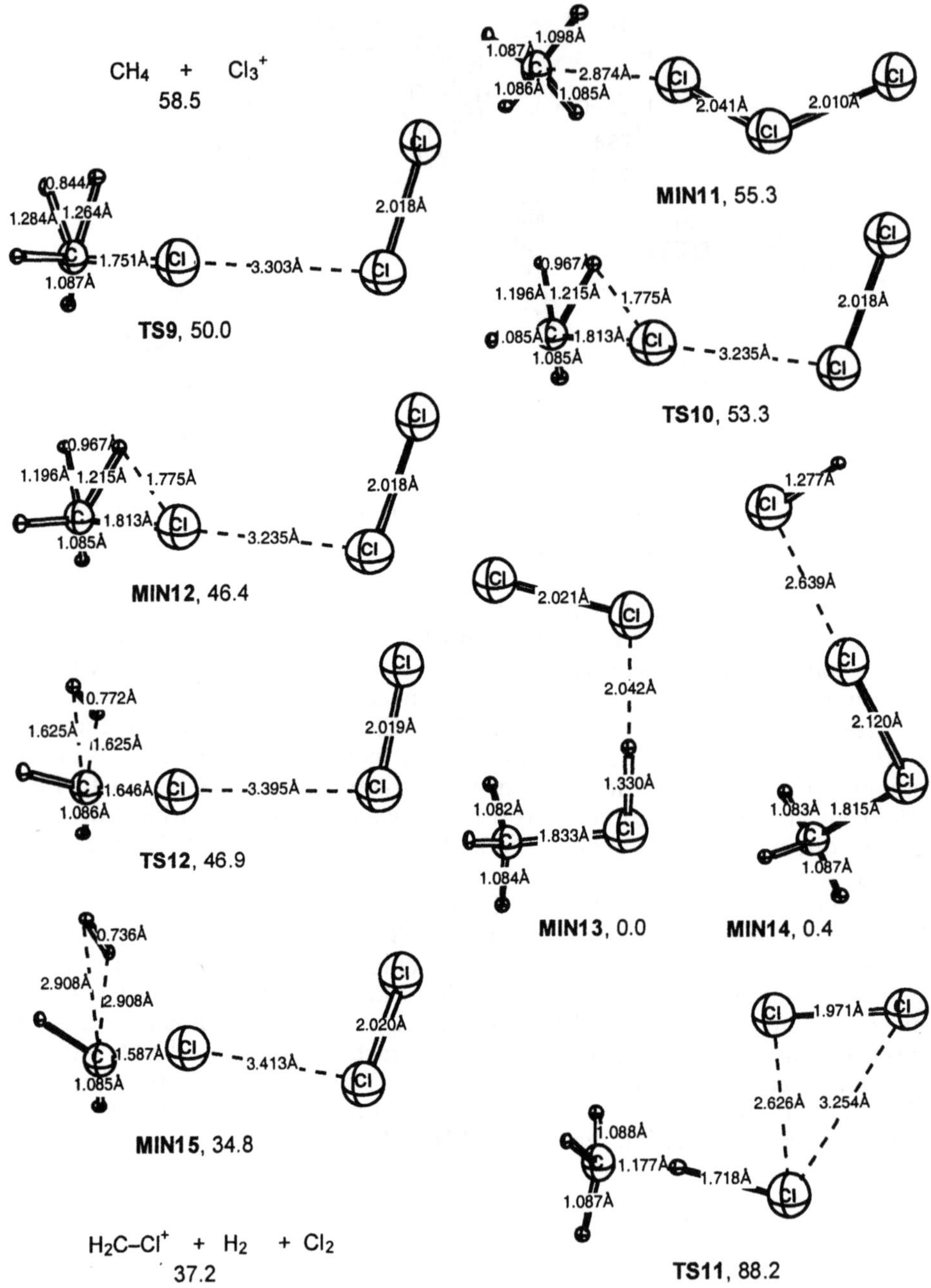

Figure 4. The MP2/6-31G** geometries of the stationary structures for the reaction of methane with Cl_3^+ (relative energies in kcal mol^{-1}).

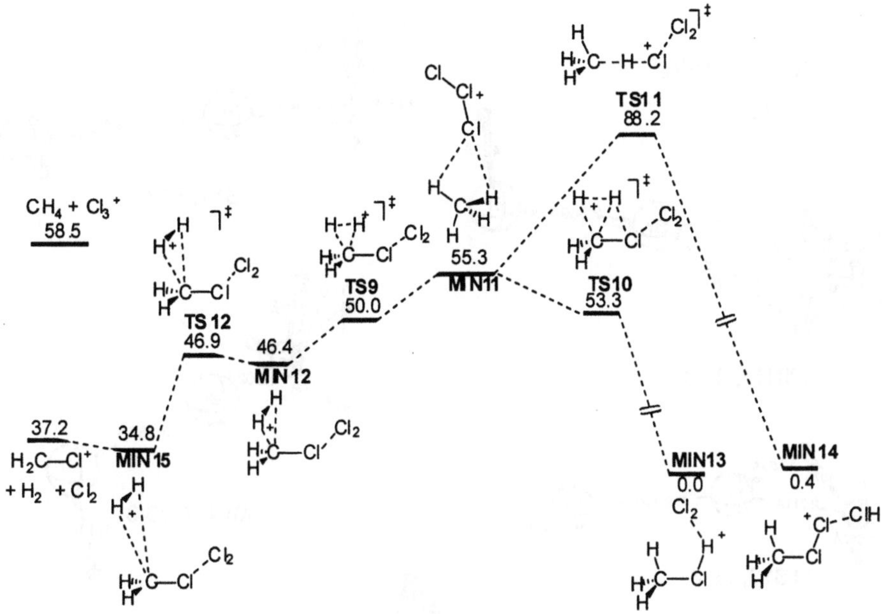

Scheme 3. The pathways for the reaction of methane with Cl_3^+ (relative energies in kcal mol^{-1}).

CH_4 + HBr_2^+ (Fig. 5 and Scheme 4). Positively charged bromine species were generated by Olah in $AgSbF_6/Br_2/CH_2Cl_2$ system and were used for the bromination of a number of alkanes.[65] Generally, the rates for C–H bromination increase in the presence of Lewis acids ("aprotic superacids," e.g., $Br^+AlBr_4^-$).[66] The HBr_2^+ cation seems to be a relevant model for bromine-containing singlet ground state electrophiles. In contrast to the reaction with Cl_3^+, direct attack of HBr_2^+ on the carbon atom of methane (**TS14**, $\Delta H^\ddagger$ = 23.2 kcal mol^{-1}) is much less favorable than the C–H insertion reaction (**TS13**, $\Delta H^\ddagger$ = 8.0 kcal mol^{-1}), which leads to a stable complex of protonated methylbromide with HBr (**MIN18**).

The product of the direct carbon attack (**MIN17**) forms endergonically; while characterized by a low barrier for H_2 loss (**TS15**), the pathway to complex **MIN19** cannot compete with the CH_4 + HBr_2^+ insertion reaction to **MIN18**.

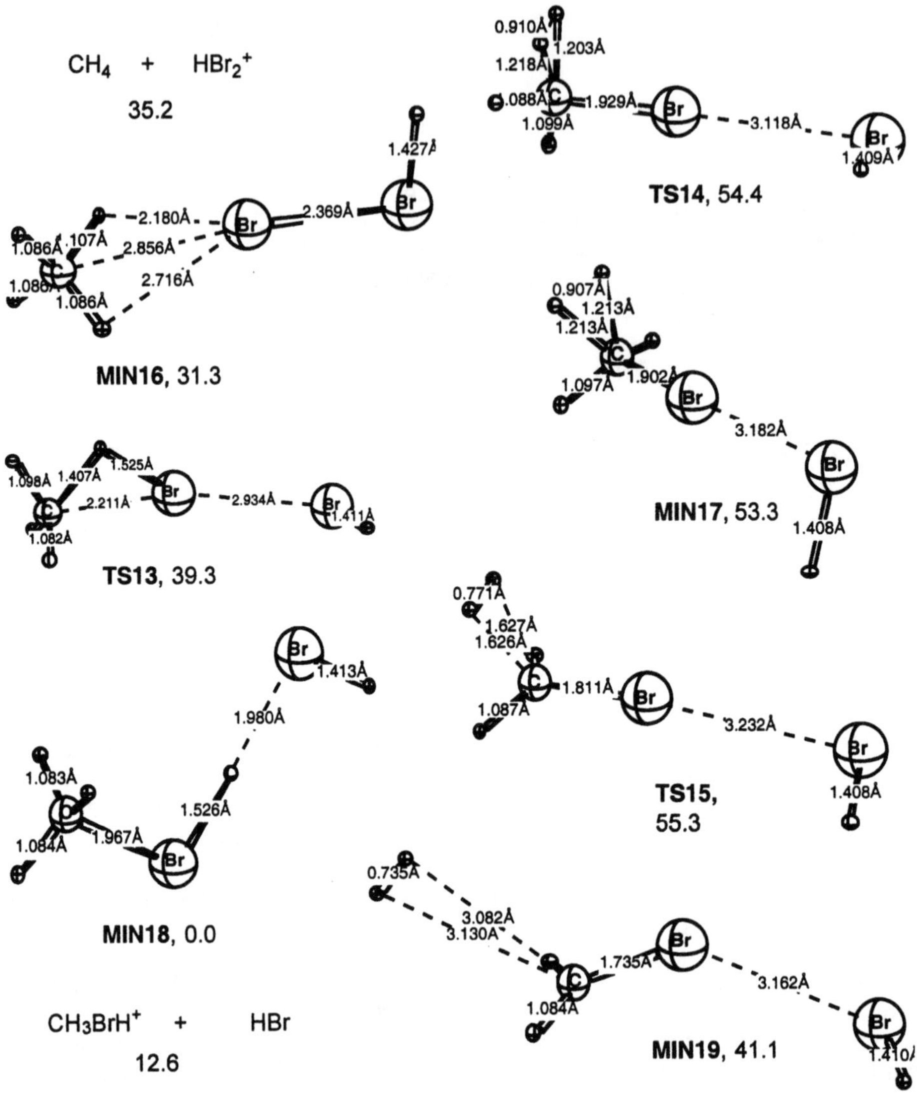

Figure 5. The MP2/6-31G** geometries of the stationary structures for the reaction of methane with HBr_2^+ (relative energies in kcal mol^{-1}).

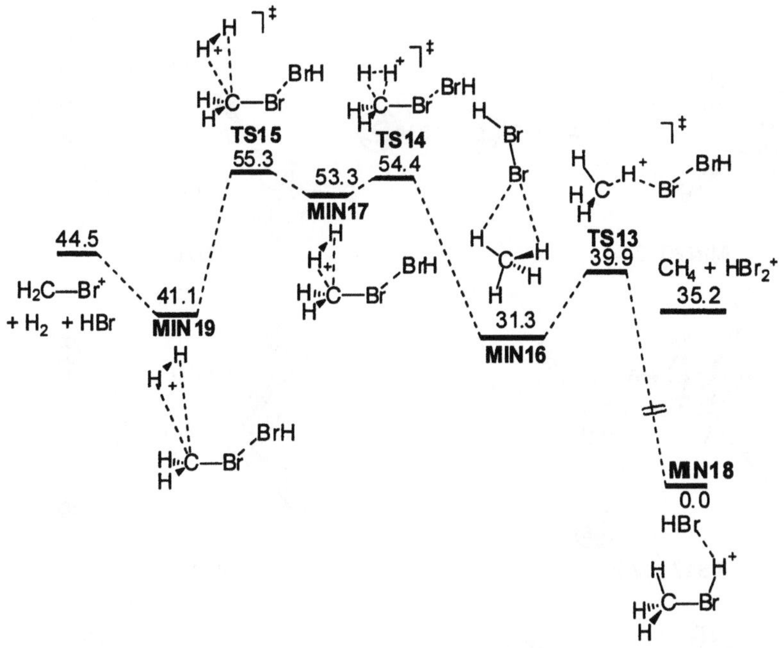

Scheme 4. The pathways for the reaction of methane with HBr_2^+ (relative energies in kcal mol^{-1}).

CH_4 + HCO^+ (Fig. 6). Superacidic (electrophilic) hydroformylations of C_3 to C_6 alkanes in the presence of HCl and stoichiometric amounts of Al_2Cl_6 at 20–150 °C and 100–150 atm gives aldehydes, ketones, and carboxylic acids.[67] However, methane does not react under non-radical conditions; usually, Fe(IV), Ag(I) or Cu(I) salts are involved as single electron transfer (SET) reagents.[68]

The proton transfer reactions of protonated carbon monoxide and methane are thermodynamically unfavorable due to the considerable stability of HCO^+ (eqs. 11 and 12), based on experimental heats of formation).[31] The formation of methylated formaldehyde (eq. 13), protonated vinyl alcohol (eq. 14), and protonated acetaldehyde (eq. 15) is exothermic.

$$CH_4 + HCO^+ \rightarrow CH_5^+ + CO \qquad \Delta H_O = +8.3 \text{ kcal mol}^{-1} \quad (11)$$

$$\rightarrow CH_3^+ + H_2CO \qquad \Delta H_O = +54.0 \text{ kcal mol}^{-1} \quad (12)$$

$$\rightarrow H_2C{=}O{-}CH_3^+ \qquad \Delta H_O = -24.3 \text{ kcal mol}^{-1} \quad (13)$$

$$\rightarrow H_2C{=}CH{-}OH_2^+ \qquad \Delta H_O = -33.3 \text{ kcal mol}^{-1} \quad (14)$$

$$\rightarrow H_3C{-}CH^+{-}OH \qquad \Delta H_O = -42.3 \text{ kcal mol}^{-1} \quad (15)$$

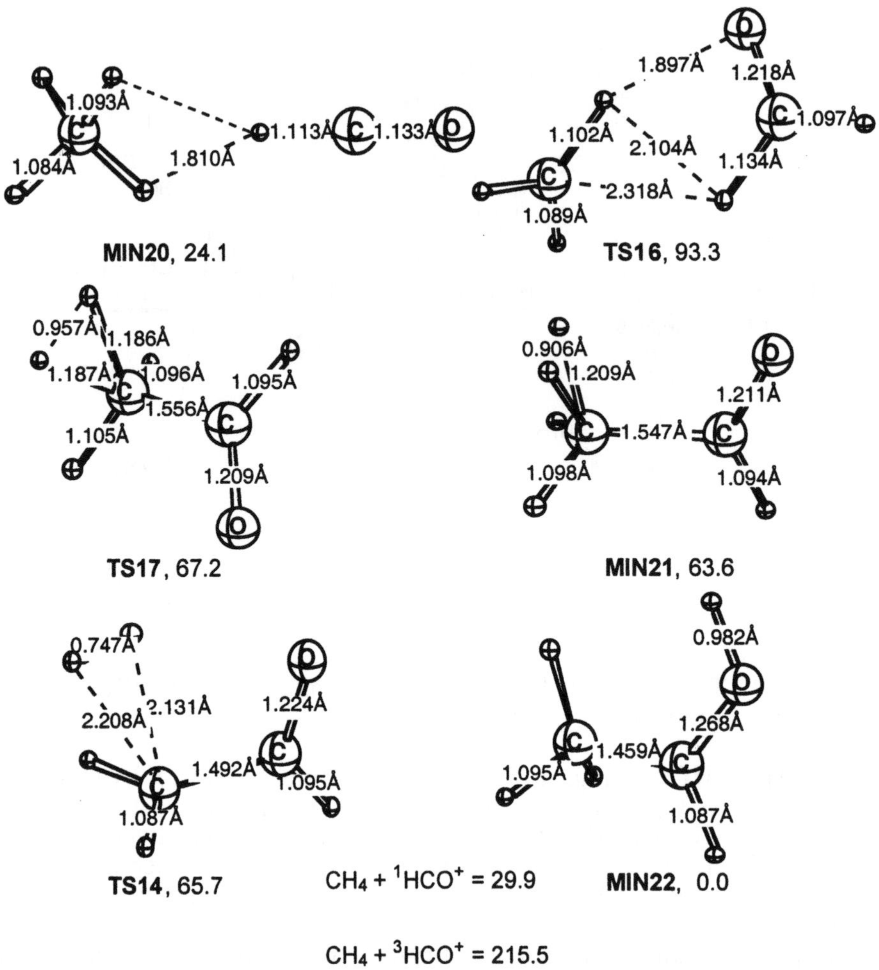

Figure 6. The MP2/6-31G** geometries of the stationary structures for the reaction of methane with HCO^+ (relative energies in kcal mol^{-1}).

As for other stabilized electrophiles (such as NO^+, NO_2^+, Cl_3^+ and HBr_2^+ see above), methane forms an initial complex (**MIN20**, C_{2v}, D_O = 5.8 kcal mol^{-1}) with the hydroformyl cation. We located one transition structure for hydride abstraction (**TS16**) and one for direct carbon attack (**TS17**). No TS for direct C–H insertion was found. However, as expected from the thermodynamics, both transition structures are high in energy. Hydride abstraction is highly unfavorable (eq. 12; $\Delta H^‡$ = 69.2 kcal mol^{-1}) compared to carbon addition ($\Delta H^‡$ = 43.1 kcal mol^{-1}). No TS was located for direct proton transfer from **MIN21** to **MIN22**. Thus, the overall reaction of methane with the hydroformyl cation to yield the protonated acetaldehyde **MIN22** is possible but the barriers are relatively high.

CH_4 + $^1OH^+$ and CH_4 + $HOOH_2^+$ (Figs. 7 and 8, Scheme 5). Hydroxylation of alkanes with peracids (e.g., trifluoroperoxy acetic acid) yields protonated alcohols by formal insertion of "OH^+" into a C–H bond.[69] The electrophilic oxidation of hydrocarbons with hydrogen peroxide and acid has also been studied by Olah *et al.* [70, 71] The active electrophile is assumed to be the hydrated form of the OH^+ ion, $HO–OH_2^+$ (hydroperoxonium ion), i. e., protonated hydrogen peroxide. Bach and Su studied the hydroperoxonium ion reactions computationally with methane, ethane, propane, and isobutane.[27] The activation barrier for the reaction of $HO–OH_2^+$ with methane is only 5.3 kcal mol^{-1} (MP4/6-31G*//MP2/6-31G*). Bach and Su concluded that the entire CH_2 fragment orbital and not just an isolated C–H σ-bonding orbital is attacked by the electrophile. The very long C–O distance (2.462 Å at MP2/6-31G**) in their insertion transition structure (**TS20** in the present study) suggested that *"...the geometry of the hydroxylation transition state has many characteristics resembling those of a free radical hydrogen abstraction"*, and that *"...E^+ attacks the πCH_2 fragment orbital along the axis of the carbon 2p orbital."* [27]

We re-examined these processes in order to compare the reactions of $H_3O_2^+$ with $^1OH^+$ directly. Water interaction stabilizes "OH^+" enormously (eqs. 16–18, based on the experimental ΔH_f^o values[31]), and the $HO–OH_2^+$ ion is much less reactive than $^1OH^+$. Indeed, in contrast to OH^+, $HO–OH_2^+$ forms a weakly bound complex (**MIN25**, D_0 = 4.6 kcal mol^{-1}) with methane (Fig. 8).

$$^1OH^+ + H_2O \rightarrow HO–OH_2^+ \quad \Delta H_0 = -81 \text{ kcal mol}^{-1} \tag{16}$$

$$^1OH^+ + H_2 \rightarrow H_3O^+ \quad \Delta H_0 = -166 \text{ kcal mol}^{-1} \tag{17}$$

$$HO–OH_2^+ + H_2 \rightarrow H_3O^+ + H_2O \quad \Delta H_0 = -85 \text{ kcal mol}^{-1} \tag{18}$$

The $^1OH^+$ ion is highly reactive and lies energetically between $^1F^+$ or $^1Cl^+$ (eqs. 19 and 20).

$$H_2O + {}^1F^+ \rightarrow HF + {}^1OH^+ \quad \Delta H_0 = -119 \text{ kcal mol}^{-1} \tag{19}$$

$$H_2O + {}^1Cl^+ \rightarrow HCl + {}^1OH^+ \quad \Delta H_0 = +16 \text{ kcal mol}^{-1} \tag{20}$$

No initial complex of $^1OH^+$ with methane exists as no activation barrier is involved. Two stable intermediates, the addition product **MIN24** and the formal insertion product **MIN23,** form directly due to the large exothermicities of these reactions (–186.4 and –215.4 kcal mol^{-1}, respectively). No transition state corresponding to Bach and Su's "insertion-TS" for $HO–OH_2^+$ (see below) was located for $^1OH^+$. As observed for the analogous singlet fluoronium and chloronium ions, **MIN6** and **MIN8**, there are two reaction exit channels for **MIN24**, yielding either the thermodynamically most stable product, protonated methanol (**MIN23**) via **TS18** ($\Delta H^‡$ = 38.8 kcal mol^{-1}), or the kinetic products, the $H_2C–OH^+$ cation and H_2.

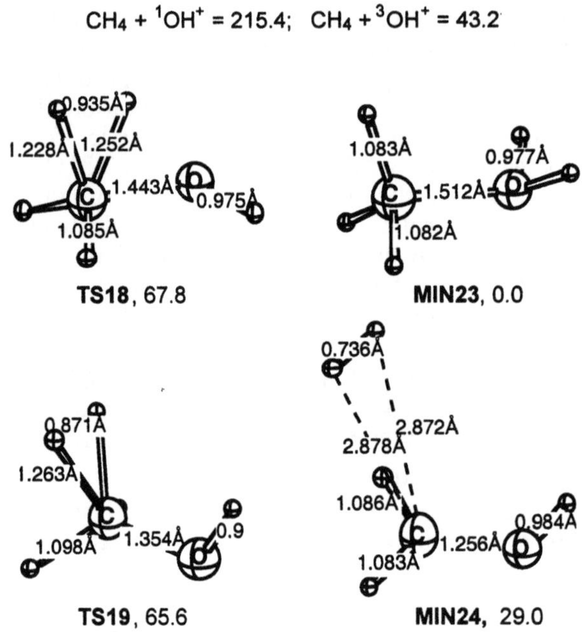

Figure 7. The MP2/6-31G** geometries of the stationary structures for the reaction of methane with $^1HO^+$ (relative energies in kcal mol^{-1}).

We located two transition structures for the $HO{-}OH_2^+ + CH_4$ reaction (Fig. 8). One of these corresponds to Bach and Su's "insertion-TS" (**TS20**) but with a different orientation of the "OH^+" hydrogen relative to the remaining in-plane methane C–H bond. While the orientation of this bond is *anti* in Bach and Su's MP4/6-31G*//MP2/6-31G* study, we find **TS20** to have a *syn* conformation at MP4/6-31G**//MP2/6-31G**. However, this discrepancy does not alter the overall picture. As noted by Bach and Su, **TS20** refers to abstraction of a hydride rather than to C–H insertion. We also located an initial complex between the $H_3O_2^+$ cation and methane, **MIN25** (D_O = 4.6 kcal mol^{-1}), which precedes **TS20** and increases the activation barrier to 7.7 kcal mol^{-1}.

As an alternative to this abstraction/association (formally insertion) process, methane and $H_3O_2^+$ may also form the complex **MIN27** as an intermediate. The activation barrier ($\Delta H^{\ddagger}$ = 49.4 kcal mol^{-1}) for hydride transfer from **MIN27** via **TS21** to the methanol-water complex **MIN26** is higher than for the reaction of $^1OH^+$ ($\Delta H^{\ddagger}$ = 38.8 kcal mol^{-1}). This is due to the highly delocalized nature of **TS15** and **TS21**; the positive charge is distributed over the entire molecule in both.[72] The corresponding ground states also are delocalized, but both **MIN26** and **MIN27** benefit more from the interaction with water (note the increased O–H bond length in the complexed ground states), leading to a higher barrier. Otherwise, the structural features for the solvated vs. unsolvated species are quite comparable.

Thus, although $^1HO{-}OH_2^+$ is much more stable than $^1HO^+$ (eqs. 16–18), the reactions of $^1HO^+$ and $^1HO{-}OH_2^+$ are rather similar. Direct C–H activation requires no (for CH_4 + $^1HO^+$ or the direct attack path to **MIN27** with $^1HO–OH_2^+$) or little energy (7.7 kcal mol^{-1} for the CH_4 + $^1HO–OH_2^+$ abstraction/recombination).

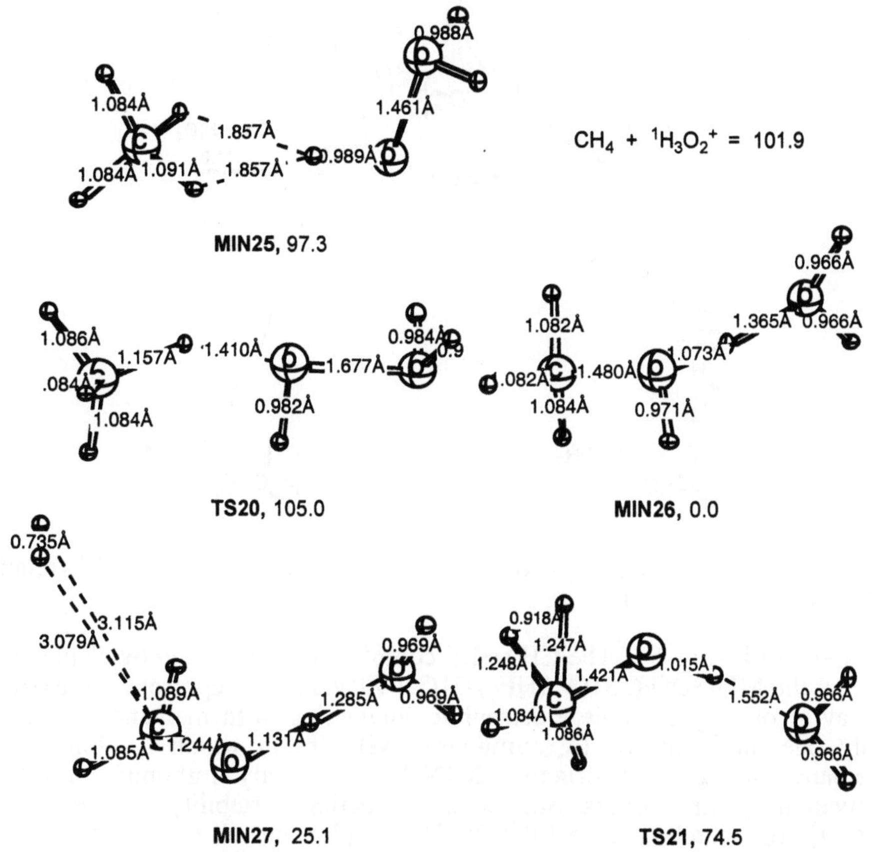

Figure 8. The MP2/6-31G** geometries of the stationary structures for the reaction of methane with $H_3O_2^+$ (relative energies in kcal mol^{-1}).

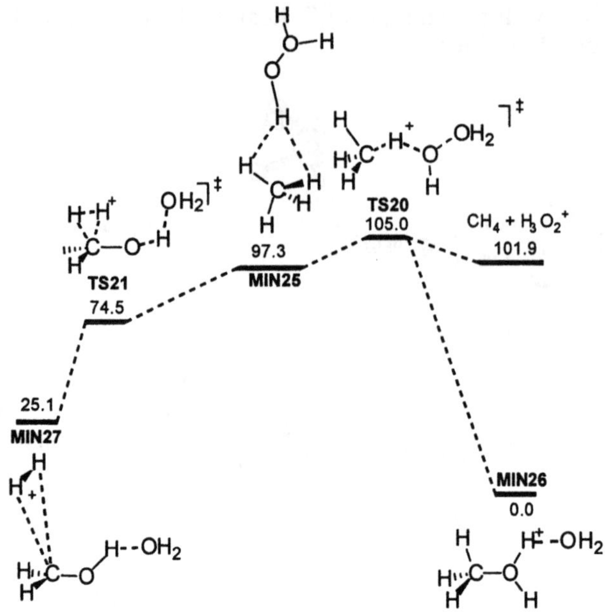

Scheme 5. The pathways for the reaction of methane with $H_3O_2^+$ (relative energies in kcal mol^{-1}).

CH_4 + Li^+ (Fig. 9). The CH_4•••Li^+ complexes, first studied by Schleyer *et al.*[73] at the MP4SDQ/6-31G*//HF/3-21G + ZPVE level, represent the extreme behavior of very a stable electrophile interacting with methane. There is only one minimum of C_{3v} symmetry (**MIN28**) in which Li^+ binds to a methane face. Since structure **MIN28** is the only minimum, no C–H activation by Li^+ occurs, due to the exceptional stability of the lithium cation. We searched for a $LiCH_2^+$•••H_2 complex but all starting geometries led to **MIN28** on optimization (a C_{2v} symmetric H2C–Li^+•••H_2 complex does exist, however). An edge-bound C_{2v} form (**TS23**) serves as the transition state for facile ($\Delta H^{\ddagger}$ = 1.7 kcal mol^{-1}) face-to-face migrations of the lithium cation. Two more stationary points of C_{2v} (**TS24**) and C_{4v} (**TS25**) symmetry have higher Hessian indices (NIMAG = 2 and 3). Additionally, we located a C_s transition structure (**TS22**) for hydrogen scrambling within the methane moiety ($\Delta H^{\ddagger}$ = 64.2 kcal mol^{-1} relative to CH_4 + Li^+). This is particularly interesting as the stereomutation barrier (> 100 kcal mol^{-1} for methane)[74] is significantly reduced by the lithium cation (electrostatic acceleration).[75]

While the qualitative results remain the same, the relative energies of the C_{4v} (**TS25**) and C_{3v} (**TS24**) structures are 50.8 kcal mol^{-1} lower and 17.0 kcal mol^{-1} higher in energy, respectively, than the data previously reported at much lower theoretical levels.[75] However, the 11.0 kcal mol^{-1} CH_4 + Li^+ complexation energy is the same at our reference level and at MP4SDQ/6-31G*//HF/3-21G.[75]

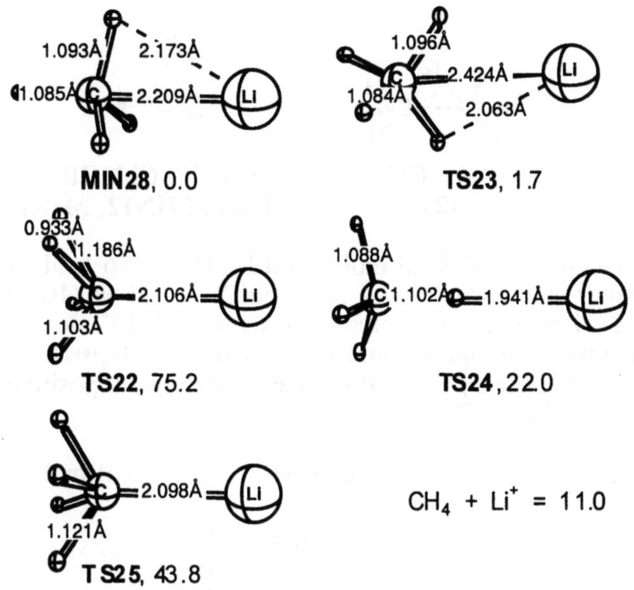

Figure 9. The MP2/6-31G** stationary structures for the CH_4 + Li^+ interaction (in kcal mol^{-1}).

DISCUSSION

The primary goal of the present work was to examine whether a general mechanistic description of the conceptually simple reactions of methane with various electrophiles can be formulated. Using the methonium ion (**1**, R=H) as the prototype (E^+ = H^+), Olah proposed that the C–H bond is attacked by an electrophile E^+ (eq. 1) leading to 3c-2e C, H, E^+ bonding. In contrast, we have suggested that stabilized electrophiles such as the nitrosonium ion may behave differently and attack carbon directly.[25, 26] While a species with 3c-2e binding generally ensues, this does *not* involve the incoming electrophile (eq. 3). In response to this conclusion, Olah suggested that polytopial rearrangements may take place easily, and several forms might coexist (**1A**, **1B**, and **1C**, see introduction). However, the delicate balance required for 3c-2e bonding among C, H, and E^+ with respect to electronegativity and bond strength is difficult to achieve for elements other than carbon, boron, and hydrogen.[43] While multicenter bonding arrangements are common among, e.g., carbocations and carboranes, they are essentially unknown for combinations of carbon and hydrogen with a third element (with the possible exception of boron[30, 76]). Bridging hydrogens are the most commonly observed 3c-2e examples comprised of more than one element. In contrast, 3c-2e CHE^+ arrangements are rare and unfavorable when quite different C–E, C–H, and E–H bond energies are involved. The following progression depicting the most stable multicenter forms emphasizes this point. Irrespective of the nature of E^+, only CCH and CHH 3c-2e bonding not involving the electrophile is found!

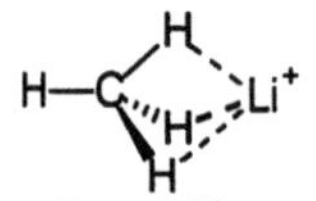

$E = Li^+$
MIN28

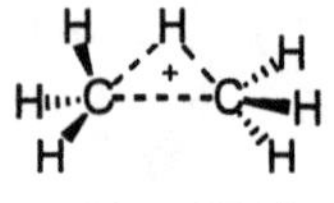

$E = CH_3^+$
(2)

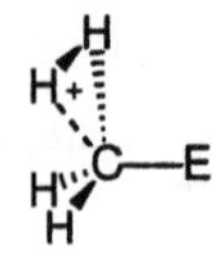

$E = Cl^+, Cl_3^+, HBr_2^+$
MIN10, MIN12, MIN17

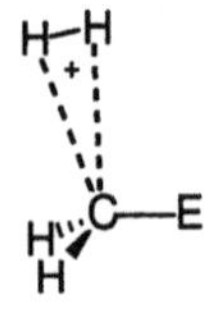

$E = OH^+$
MIN24

Since both Li–C (ca. 42 kcal mol^{-1}) and Li–H (ca. 56 kcal mol^{-1})[31] bonds are much weaker than C–H (ca. 97 kcal mol^{-1})[31] or H–H (103 kcal mol^{-1}),[31] no stable 3c-2e bonding arrangement exists for CH_4Li^+; not even a weakly bound $[LiCH_2{\bullet\bullet\bullet}H_2]^+$ complex could be located. The $H_2{\bullet\bullet\bullet}H_2CE^+$ complexes are favored increasingly by the ability of E to stabilize a positive charge and by the strength of the C–E^+ bonds:

$$CH_3^+ + HE \longrightarrow CH_3E + H^+ \quad \Delta H_r\ (\text{kcal mol}^{-1}) \qquad (21)$$

E =		ΔH_r
	Li	+99[77]
	H	+87
	CH3	+71
	Cl	+63

Except for Li^+ the overall $CH_4 + E^+ \rightarrow CH_3EH^+$ reaction is thermodynamically favorable for all E^+ (Table 2). The stability order of the various electrophiles, E^+, can be evaluated from the overall exothermicity of this reaction (boldface for electrophiles with $\Delta H_0 < 0$) as:

F^+ >> OH^+ > Cl^+ > HO–OH_2^+ >> Cl_3^+ > HBr_2^+ >> HCO^+ > NO_2^+ > NO^+ > Li^+

and from the $CH_4 + E^+ \longrightarrow H_2C{-}E^+ + H_2$ reaction[78] (bold for electrophiles with $\Delta H_0 < 0$) as:

F^+ >> OH^+ >> Cl^+ > HO–OH_2^+ > Cl_3^+ > HBr_2^+>> NO^+ > HCO^+ > Li^+ > NO_2^+

Thus, a change in mechanism can well be expected in going from left to right. All electrophiles except Li^+ react with methane, but with widely different activation barriers. The alternative reaction pathways (C–H insertion or carbon attack) for the highly reactive electrophiles such as H^+, R^+, $^1F^+$, $^1Cl^+$, and $^1OH^+$ cannot be differentiated due to the absence of a barrier. Highly stabilized electrophiles Cl_3^+, HBr_2^+, NO^+ and NO_2^+ are at the borderline where several pathways may be found.

The first reaction intermediate following direct carbon attack can usually be characterized as a complex of a substituted methyl cation and dihydrogen, $H_2{\bullet\bullet\bullet}CH_2E^+$. This finding is consistent with the two well-documented[19] minima found for, e.g., protonated ethane, **2** (with CCH 3c-2e bonding) and **3** (with CHH 3c-2e bonding). No transition states along the pathways leading to these $C_2H_7^+$ isomers have been reported. Due to the similar H–H and methane C–H bond energies, both **2** and **3** can coexist.[6, 7, 19-21] This is not the case for other electrophiles when the binding energies among the different elements are quite different. The large variation in stability of the H_2CE^+ cation (vs. E^+), depending upon C–E vs. C–H and H–H interactions emphasizes this point: H_2CE^+ is stabilized by p (E = F, OH, Cl, Cl_3, HBr_2, HO–OH_2) and by s interactions (E = Li). The interaction of methane with strong electrophiles-oxidizers (E = F, OH, Cl, Cl_3, HBr_2, HO–OH_2) is accompanied by substantial charge transfer from the hydrocarbon to the electrophile (the net charge transfer approaches 1e^-in most cases).[57, 79-81] As a consequence, the CH_4 moiety resembles the structure of the methane radical cation displaying *two* elongated C–H bonds[82-84] (**MIN29**, Fig. 10). Note the structural relationships to the above TSs for carbon attack and C–H insertion! In contrast, the TSs for H-abstractions with *one* elongated C–H bond (like **TS11**) are dissimilar to **MIN29** and typically are highly unfavorable.

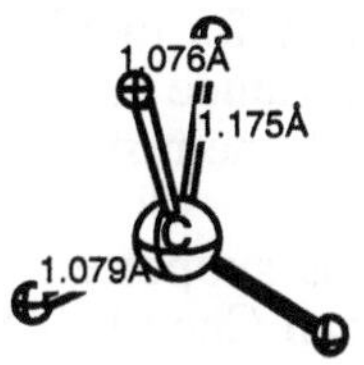

MIN29

Figure 10. The MP2/6-31G** geometry of the methane radical cation (C_{2v}).

Generally there are two exit channels from the dihydrogen complex, ECH_2^+•••H_2. While the loss of H_2 to give a substituted methyl cation (H_2C–E^+) is particularly facile for the most reactive electrophiles, hydride shift leading to a protonated methane derivative is more favorable thermodynamically (to give the overall H_3C–EH^+ product). While the loss of dihydrogen requires little structural reorganization, the excess thermal energy due to the large exothermicity allows the two fragments (H_3C–E^+ and H_2) to overcome the hydride shift barriers and give the most stable H_3C–EH^+ product.

CONCLUSIONS

Three categories of electrophilic reactions with methane can be differentiated. The most reactive electrophiles (e.g., $^1F^+$) attack CH_4 directly without a barrier; there is no transition state which would allow carbon and C–H attack pathways to be distinguished. These electrophiles give H_3C–E^+•••H_2 complexes as intermediates. Three center-two electron bonding is involved in the H_2 complexes, but the E–C bond is fully formed.

Very unreactive electrophiles, like Li^+, merely form complexes with CH_4; substitution is highly endothermic. However, the electrostatic field in such methane complexes reduces the barrier to CH_4 stereomutation considerably.[74]

Electrophiles with intermediate stability, such as NO^+ or NO_2^+, react with methane with moderate overall exothermicity. Three pathways can be differentiated: C–H insertion, direct carbon attack, and hydrogen abstraction. While C–H abstraction is the most favorable reaction pathway for NO_2^+, NO^+ reacts with methane via direct carbon attack. Alternative C–H abstraction transition structures were located for NO^+, Cl_3^+, $H_3O_2^+$, and HBr_2^+, but this reaction is only favorable energetically for the last electrophile in this series.

There is no general evidence for 3c-2e bonding between C, H, and the electrophile. Instead, activation proceeds through charge transfer TS's, resembling the structure of the methane radical cation with two elongated C–H bonds. The latter observation is the key for understanding methane activation with electrophile-oxidizers ($^1F^+$, $^1Cl^+$, $^1OH^+$, Cl_3^+, HBr_2^+, NO^+ and NO_2^+). Solvation (probed by comparison of the reactions of $^1OH^+$ vs. H_2O–OH^+, and $^1Cl^+$ vs. Cl_2–Cl^+ with methane) increases the activation barriers only slightly.

Finally, we wish to stress that our findings do not question Olah's experimental results. Electrophiles attack alkanes, but do so in a variety of ways. While we are unable to support Olah's assumed 3c-2e C–H activation mechanism computationally, at least for methane, this certainly does not detract from the conceptual importance of his ideas which led to the development of a whole new area of chemistry.

Acknowledgments. This work was supported by the Petroleum Research Fund (administrated by the American Chemical Society), the NATO Science Program, and the Fundamental Research Foundation of the Ukraine.

References and Notes

1. G. A. Olah and A. Molnár, in 'Hydrocarbon Chemistry', (John Wiley & Sons, Inc., 1995),
2. G. A. Olah, O. Farooq, and G. K. S. Prakash, in 'Activation and Functionalization of Alkanes', (John Wiley & Sons, Inc., 1989),
3. G. A. Olah, G. K. S. Prakash, R. E. Williams, L. D. Field, and K. Wade, in 'Hypercarbon Chemistry', (Wiley-Interscience, 1987),
4. G. A. Olah, G. K. S. Prakash, and J. Sommer, in 'Superacids', (Wiley-Interscience, 1985),
5. G. A. Olah, Acc. Chem. Res., **20**, 422 (1987).
6. K. Hiraoka and P. Kebarle, Can. J. Chem., **53**, 970 (1975).
7. K. Hiraoka and P. Kebarle, J. Chem. Phys., **63**, 394 (1975).
8. K. Hiraoka and P. Kebarle, Can. J. Chem., **54**, 1739 (1976).
9. K. Hiraoka and P. Kebarle, J. Am. Chem. Soc., **97**, 4179 (1975).
10. K. Hiraoka, T. Mori, and S. Yamabe, Chem. Phys. Lett., **207**, 178 (1993).
11. V. L. Tal'roze and A. K. Lyubimova, Dokl. Akad. Nauk. SSR, **86**, 909 (1952).
12. K. Hiraoka, I. Kidaka, and S. Yamabe, Chem. Phys. Lett., **184**, 271 (1991).
13. A. J. R. Heck, L. J. de Koning, and N. M. M. Nibbering, J. Am. Soc. Mass Spectrom., **2**, 453 (1991).
14. T. Oka and E. T. White, Science, **286**, 1051 (1999).
15. E. T. White, J. Tang, and T. Oka, Science, **284**, 135 (1999).
16. A. L. L. East, M. Kolbuszewski, and P. R. Bunker, J. Phys. Chem. A, **101**, 6746 (1997).
17. P. R. Schreiner, Angew. Chem. Int. Ed. Engl., **39**, 3239 (2000).
18. P. Ahlberg, A. Karlsson, A. Goeppert, S. O. N. Lill, P. Diner, and J. Sommer, Chem. Eur. J., **7**, 1936 (2001).
19. L. I. Yeh, J. M. Price, and Y. T. Lee, J. Am. Chem. Soc., **111**, 5597 (1989).
20. J. W. M. Carneiro, P. v. R. Schleyer, M. Saunders, R. Remington, H. F. Schaefer, A. Rauk, and T. S. Sorensen, J. Am. Chem. Soc., **116**, 3483 (1994).

[21] A. L. L. East, Z. F. Liu, C. McCague, K. Cheng, and J. S. Tse, J. Phys. Chem., **102**, 10903 (1998).

[22] L. Radom, D. Poppinger, and R. C. Haddon, in 'Carbonium Ions', ed. G. A. Olah and P. v. R. Schleyer, (New York, 1976).

[23] G. A. Olah, J. R. De Member, and J. Shen, J. Am. Chem. Soc., **95**, 4952 (1973).

[24] S. J. Collins and P. J. O'Malley, Chem. Phys. Lett., **228**, 246 (1994).

[25] P. R. Schreiner, P. v. R. Schleyer, and H. F. Schaefer III, J. Am. Chem. Soc., **115**, 9659 (1993).

[26] H. F. Schaefer III, P. R. Schreiner, and P. v. R. Schleyer, J. Am. Chem. Soc., **117**, 453 (1995).

[27] R. D. Bach and M.-D. Su, J. Am. Chem. Soc., **116**, 10103 (1994).

[28] R. D. Bach, J. L. Andrés, M.-D. Su, and J. J. W. McDouall, J. Am. Chem. Soc., **115**, 5768 (1993).

[29] R. D. Bach, M.-D. Su, E. Aldabbagh, J. L. Andrés, and H. B. Schlegel, J. Am. Chem. Soc., **115**, 10237 (1993).

[30] C. H. DePuy, R. Gareyev, J. Hankin, G. E. Davico, M. Krempp, and R. Damrauer, J. Am. Chem. Soc., **120**, 5086 (1998).

[31] S. G. Lias, J. E. Bartmess, J. F. Liebman, J. L. Holmes, R. D. Levin, and W. G. Mallard, J. Phys. Chem. Ref. Data, **17 (S1)**, 1 (1988).

[32] Based on the heats of formations: ΔH_f of CH_5^+ = 216.0 kcal mol^{-1} (ref. 4); ΔH_f of CH_3^+ = 261.3 kcal mol^{-1} (Traeger, J. C.; McLoughlin, R. G. *J. Am. Chem. Soc.* **1981**, *103*, 3647).

[33] G. A. Olah, J. Shen, and R. H. Schlosberg, J. Am. Chem. Soc., **92**, 3831 (1970).

[34] G. A. Olah, G. Klopman, and R. H. Schlosberg, J. Am. Chem. Soc., **91**, 3261 (1969).

[35] K. Hiraoka and P. Kebarle, J. Am. Chem. Soc., **98**, 6119 (1976).

[36] G. A. Olah and N. Friedman, J. Am. Chem. Soc., **88**, 5330 (1966).

[37] G. A. Olah and H. C. Lin, J. Am. Chem. Soc., **93**, 1259 (1971).

[38] G. A. Olah, N. Hartz, G. Rasul, Q. Wang, G. K. S. Prakash, J. Casanova, and K. O. Christe, J. Am. Chem. Soc., **116**, 5671 (1994).

[39] G. A. Olah, R. Renner, P. Schilling, and Y. K. Mo, J. Am. Chem. Soc., **95**, 7686 (1973).

[40] F. Cacace, G. de Petris, F. Pepi, M. Rosi, and A. Troiani, Chem. Eur. J., **5**, 2750 (1999).

[41] Note that 3c-2e [C--H--C]$^+$ can also be linear: McMurray, J. E.; Lectka, T. *Acc. Chem. Res.* **1992**, *25*, 47–53).

[42] C. Ramsden, J. Phys. Chem., **106**, 2777–2780 (2002).

[43] T. Kar, J. Mol. Struct. (Theochem), **102**, 313 (1993).

[44] P. v. R. Schleyer and J. W. D. Carneiro, J. Comput. Chem., **13**, 997 (1992).

[45] P. R. Schreiner, S.-J. Kim, H. F. Schaefer III, and P. v. R. Schleyer, J. Chem. Phys., **99**, 3716 (1993).

[46] S.-J. Kim, P. R. Schreiner, P. v. R. Schleyer, and H. F. Schaefer III, J. Phys. Chem., **97**, 12232 (1993).

[47] F. Sun and T. S. Sorensen, J. Am. Chem. Soc., **115**, 77 (1993).

[48] D. Marx and M. Parrinello, Nature, **375**, 216 (1995).

[49] D. Marx and M. Parrinello, Science, **284**, 59 (1999).

[50] G. A. Olah, N. Hartz, G. Rasul, and G. K. S. Prakash, J. Am. Chem. Soc., **117**, 1336 (1995).

[51] An alternative process, preactivation of NO^+ by protonation (to give the protonitrosonium dication HNO_2^+) also was proposed. The authors recognized that the dominant reaction between HNO_2^+ and methane must be proton transfer (to give NO^+ + CH_5^+) rather than the production of diprotonated nitrosomethane ("All attempts to locate a stable isomer of $H_3C\text{–}NH_2O_2^+$ failed."). Although the concept of diprotonated

superelectrophiles has been employed to explain chemical behavior, it is recognized that proton transfers from diprotonated species to hydrocarbons are highly exothermic.

[52] G. A. Olah, Angewandte Chemie-International Edition in English, **32**, 767 (1993).

[53] For a review on CH activations by rare transition metal oxide cations see: Schröder, D.; Schwarz, H. *Angew. Chem. Int. Ed. Engl.* **1995**, *34*, 1973.

[54] M. J. Frisch, G. W. Trucks, H. B. Schlegel, P. M. W. Gill, B. G. Johnson, M. A. Robb, J. R. Cheeseman, T. Keith, G. A. Petersson, J. A. Montgomery, K. Raghavachari, M. A. Al-Laham, V. G. Zakrzewski, J. V. Ortiz, J. B. Foresman, J. Cioslowski, B. B. Stefanov, A. Nanayakkara, M. Challacombe, C. Y. Peng, P. Y. Ayala, W. Chen, M. W. Wong, J. L. Andres, E. S. Replogle, R. Gomperts, R. L. Martin, D. J. Fox, J. S. Binkley, D. J. Defrees, J. Baker, J. P. Stewart, M. Head-Gordon, C. Gonzalez, and J. A. Pople, Gaussian 94, Revision D.3, Pittsburgh PA, 1995.

[55] M. J. Frisch, G. W. Trucks, H. B. Schlegel, G. E. Scuseria, M. A. Robb, J. R. Cheeseman, V. G. Zakrzewski, J. J. A. Montgomery, R. E. Stratmann, J. C. Burant, S. Dapprich, J. M. Millam, A. D. Daniels, K. N. Kudin, M. C. Strain, O. Farkas, J. Tomasi, V. Barone, M. Cossi, R. Cammi, B. Mennucci, C. Pomelli, C. Adamo, S. Clifford, J. Ochterski, G. A. Petersson, P. Y. Ayala, Q. Cui, K. Morokuma, D. K. Malick, A. D. Rabuck, K. Raghavachari, J. B. Foresman, J. Cioslowski, J. V. Ortiz, B. B. Stefanov, G. Liu, A. Liashenko, P. Piskorz, I. Komaromi, R. Gomperts, R. L. Martin, D. J. Fox, T. Keith, M. A. Al-Laham, C. Y. Peng, A. Nanayakkara, C. Gonzalez, M. Challacombe, P. M. W. Gill, B. Johnson, W. Chen, M. W. Wong, J. L. Andres, C. Gonzalez, M. Head-Gordon, E. S. Replogle, and J. A. Pople, Gaussian 98, Revision A.7, Pittsburgh PA, 1999.

[56] W. J. Hehre, L. Radom, P. v. R. Schleyer, and J. A. Pople, in 'Ab Initio Molecular Orbital Theory', (Wiley Interscience, 1986),

[57] A. A. Fokin, P. R. Schreiner, P. A. Gunchenko, S. A. Peleshanko, T. E. Shubina, S. D. Isaev, P. V. Tarasenko, N. I. Kulik, H.-M. Schiebel, and A. G. Yurchenko, J. Am. Chem. Soc., **122**, 7317 (2000).

[58] R. S. Grev, C. L. Janssen, and H. F. Schaefer, J. Chem. Phys., **95**, 5128 (1991).

[59] G. A. Olah, Acc. Chem. Res., **13**, 330 (1980).

[60] However, correction for basis-set superposition errors (BSSE) may reduce all complexation energies significantly.

[61] Olah, G. A.; Ramaiah, P.; Rao, C. B.; Sandford, G.; Golam, R.; Trivedi, N. J.; Olah, J. A. *J. Am. Chem. Soc.* **1993**, *115*, 7246. Note that $NO^+BF_4^-$ also reacts with adamantane. The initial product, 1-nitrosoadamantane, is oxidized into 1-nitroadamantane subsequently.

[62] This transition structure also exists at QCISD/6-31G*; the geometry is very similar to that at our reference level.

[63] Y. Li, X. Wang, F. Jensen, K. N. Houk, and G. A. Olah, J. Am. Chem. Soc., **112**, 3922 (1990).

[64] G. de Petris, F. Pepi, and M. Rosi, Chem. Phys. Lett., **304**, 191 (1999).

[65] G. A. Olah and P. Schilling, J. Am. Chem. Soc., **95**, 7680 (1973).

[66] A. L. Chistyakov, I. V. Stankevich, N. P. Gambaryan, and I. S. Akhrem, Russ. Chem. Bull., **49, 799** (2000).

[67] D. H. R. Barton and D. R. Hill, Tetrahedron Lett., **35**, 1431 (1994).

[68] For a review see: Knaggs, E. A.; Nussbaum, M. L.; Shultz, A. in "Kirk-Othmer Encyclopedia of Chemical Technology," Grayson, M.; Eckroth, D. (Eds.), 3rd ed., Vol 22, Wiley: New York, 1983.

[69] H. J. Schneider and W. Müller, J. Org. Chem., **60**, 4609 (1985).

[70] G. A. Olah, D. G. Parker, N. Yoneda, and F. Pelizza, J. Am. Chem. Soc., **98**, 2245 (1976).

[71] G. A. Olah, D. G. Parker, and N. Yoneda, Angew. Chem. Int. Ed. Engl., **17**, 909 (1978).

[72] P. R. Schreiner, D. L. Severance, W. L. Jorgensen, P. v. R. Schleyer, and H. F. Schaefer III, J. Am. Chem. Soc., **117**, 2663 (1995).

[73] P. v. R. Schleyer, B. Tidor, E. D. Jemmis, J. Chandrasekhar, E. U. Wurthwein, A. J. Kos, B. T. Luke, and J. A. Pople, J. Am. Chem. Soc., **105**, 484 (1983).

[74] M. J. M. Pepper, S. I., P. v. R. Schleyer, M. N. Glukhovtsev, R. Janoschek, and M. Quack, J. Comput. Chem., **16**, 207 (1995).

[75] For the introduction of this electrostatic acceleration on transition structures see: Jiao, H.; Schleyer, P. v. R. *Angew. Chem. Int. Ed. Engl.* **1993**, *32*, 1760.

[76] Z.-X. Wang and M.-B. Huang, J. Am. Chem. Soc., **120**, 6758 (1998).

[77] A value of 29.2 kcal mol^{-1} was taken for the heat of formation for methyl lithium: a) Bickelhaupt, F. M. ; v. Eikema Hommes, N. J. R.; Fonseca Guerra, C.; Baerends, E. J. *Organometallics*, **1996**, *15*, 2923–2931; b) Kaufmann, E.; Raghavachari, K.; Reed, A. E.; Schleyer, P. v. R., *Organometallics* **1988**, *7*, 1597.

[78] Note that this equation can be used to evaluate the overall reaction of a metal atom with methane. If the M–C bond strength in MCH_2^+ exceed 111 kcal mol^{-1}, the activation is exothermic: Heinemann, C.; Schwarz, H.; Koch, W.; Dyall, K. G. *J. Chem. Phys.* **1996**, *104*, 4642.

[79] A. A. Fokin and P. R. Schreiner, Chem. Rev., **102**, in press (2002).

[80] A. A. Fokin, P. A. Gunchenko, S. A. Peleshanko, P. v. R. Schleyer, and P. R. Schreiner, Eur. J. Org. Chem., 855 (1999).

[81] A. A. Fokin, P. R. Schreiner, P. v. R. Schleyer, and P. A. Gunchenko, J. Org. Chem., **62**, 6494 (1998).

[82] M. N. Paddon-Row, D. J. Fox, J. A. Pople, K. N. Houk, and D. W. Pratt, J. Am. Chem. Soc., **107**, 7696 (1985).

[83] M. Matsushita, T. Momose, T. Shida, and L. B. Knight, Jr., J. Chem. Phys., **103**, 3367 (1995).

[84] R. Signorell, M. Sommavilla, and F. Merkt, Chem. Phys. Lett., **312**, 139 (1999).

APPLICATION OF MRD-CI FOR COMPUTING EXCITED STATES OF RADICAL REACTIONS IMPORTANT FOR ATMOSPHERIC CHEMISTRY:THE ELECTRONIC SPECTRUM OF BROMINE NITRATE $BrONO_2$ IN COMPARISON TO ISOVALENT $ClONO_2$ AND PHOTODISSOCIATION OF $BrONO_2$ ALONG O-N AND Br-O CLEAVAGE

Max Mühlhäuser and Sigrid D. Peyerimhoff

Institut fur Physikalische und Theoretische Chemie der Universität Bonn,

Wegelerstrasse 12, 53115 Bonn, FRG

Abstract

Multi-reference configuration interaction calculations are employed to compute the electronic spectrum of bromine nitrate $BrONO_2$ and isovalent $ClONO_2$. The most important transitions of $ClONO_2$ are obtained at 6.10 eV ($3^1A'' \leftarrow X^1A'$), 6.34 eV ($3^1A' \leftarrow X^1A'$) and 6.52 eV ($4^1A' \leftarrow X^1A'$) in reasonable agreement with prior calculations (5.85 eV, 6.47 eV and 6.62 eV) and in accordance with experimental measurements of 5.76 eV (215 nm, $3^1A''$) and 6.52 eV (190 nm, $4^1A'$). The corresponding transitions of $BrONO_2$ are computed at energies of 6.19 eV for $3^1A' \leftarrow X^1A'$ and 6.22 eV for $4^1A' \leftarrow X^1A'$. In addition the first dipole-allowed transition ($1^1A'' \leftarrow X^1A'$) is calculated at 4.23 eV, which is somewhat lower in energy than the corresponding transition of $ClONO_2$ (4.57 eV). This finding is in line with a smaller HOMO-LUMO gap of $BrONO_2$ compared to $ClONO_2$.

Several low-lying singlet and triplet excited states of $BrONO_2$ in the energy range between 3.3 and 7.0 eV are found to be highly repulsive for BrO + NO_2 and Br + ONO_2 cleavage, so that bromine nitrate $BrONO_2$ is expected to be important for atmospheric chemistry as reservoir of Br and BrO radicals.

E.J. Brändas and E.S. Kryachko (eds.),
Fundamental World of Quantum Chemistry, Vol. II, 377–392.

1. Introduction

Quantum chemical calculations are an ideal tool to investigate gas phase reactions of small molecules and radicals, to characterise short-lived intermediates and to detect and identify possible competitive reaction routes. Such calculations have been particularly successful for the investigation of complex reactions which occur in atmospheric chemistry, generally at low pressure and/or under the influence of sun light. Catalytic cycles involving a series of reactions are typical for the mechanism of the stratospheric ozone depletion[1-4]. Among halogens the chlorine compounds are believed to be especially effective in promoting ozone loss in the lower stratosphere, and numerous studies have been undertaken in the past 20 years to detect sources for the chlorine reservoir.

On the other hand, bromine compounds released at the earth's surface could also destroy ozone in the atmosphere[5-8]. Even though bromine compounds are emitted in smaller quantities than corresponding chlorine analogues, a greater percentage of bromine is available in reactive forms because conversion into HBr is negligible while chlorine atoms can be diverted from catalytic cycles through formation of HCl. Bromine is believed to 40 or even 100 times more efficient than chlorine for antarctic ozone loss[9,10].

For this reason considerable work has started in recent years aimed at better characterising bromine compounds that are thought to be prevalent in the stratosphere. In analogy to $ClONO_2$[11-13] which acts as a reservoir for chlorine, $BrONO_2$ could also be a prominent player in the ozone destruction[14,15]. In the lower atmosphere formation of $BrONO_2$ occurs between BrO and NO_2 in a three-body reaction. The main rote for destruction involves presumably photolysis due to ultraviolet sun light during the day, i.e.

$$BrONO_2 + h\nu \rightarrow BrO + NO_2 \qquad (1)$$

$$BrONO_2 + h\nu \rightarrow Br + NO_3 \qquad (1)$$

In the present work these processes will be investigated using high-level quantum chemical methods and comparison will be made with the analogous reactions for the chlorine species.

Such endeavour requires first the calculation of many electronically excited states of $BrONO_2$ and $ClONO_2$ and the probability for populating these states. Multi-reference configuration procedures are required foe a balanced description of such states. The computation of the potential energy surfaces for ground and excited states for BrO + NO_2 cleavage and

Br + NO_3 separation will follow. A qualitative interpretation of the photochemical finding will be undertaken and the results will be summarised in the final section.

2. Computational Techniques

The computations of the electronically excited states were performed with the multi-reference single and double excitation configuration interaction method MRD-CI implemented in the DIESEL program[16]. We used the recently completed version 1.16 in which the selection of reference configurations by a summation threshold can be carried out automatically. We used a summation threshold of 0.85, which means that the sum of the squared coefficients of all reference configurations selected for each state (root) is above 0.85. The set of reference configurations per irreducible representation (IRREP) was in the range between 26 and 50 for $BrONO_2$ and between 24 and 46 for $ClONO_2$. An analysis of the molecular orbitals (MO) involved in this selected reference configurations justified the prior choice of treating the 30 valence electrons active while keeping the remaining electrons in doubly-occupied orbitals (frozen).

From this set of reference configurations (mains) all single and double excitations in the form of configuration state functions (CSFs) are generated. From this set all configurations with an energy contribution $\Delta E(T)$ above a given threshold T were selected, i.e. the contribution of a configuration larger than this value relative to the energy of the reference set is included in the final wavefunction. A selection threshold of $T=10^{-7}$ hartree was used for the singlet states of both molecules, whereas $T=5x10^{-8}$ hartree was used for the corresponding triplet states. The effect of those configurations which contribute less than $T=10^{-7}$ (or $5x10^{-8}$) hartree is accounted for in the energy computation (E(MRD-CI)) by the perturbative λ-extrapolation[17]. The contribution of higher excitations is estimated by applying a generalised Langhoff-Davidson correction formula $E(MRD\text{-}CI{+}Q) = E(MRD\text{-}CI) - (1-c_0^2)\,[E(ref) - E(MRD\text{-}CI)]/\,c_0^2$, where c_0^2 is the sum of squared coefficients of the reference species in the total CI wavefunction and E(ref) is the energy of the reference configurations.

In total we computed 24 states (six singlet and six triplet states in each of the two IRREP A' and A"). In addition we considered up to seven singlet roots of each IRREP in preliminary computations. The number of configuration state functions (CSF) directly included in the energy calculations of the A" states of $BrONO_2$ are as large as 1.2 million (singlet) and 2.1 million (triplet) selected from a total space of 5.6 million (singlet) and 9.6 million (triplet) generated configurations. For the A' states 980.000 (singlet) and 2.14 million (triplet) CSFs are selected from a total space of 10.4 million (singlet) and 12.9 million (triplet) generated configurations. For the A' states of $ClONO_2$ 1.4 million (singlet) and 2.6 million (triplet) CSFs were selected from a total generated space of 14.2 million (singlet) and 15.4 million (triplet) CSFs. For the states with A" symmetry 1.7 million (singlet) and 2.8 million (triplet) CSFs were selected from 8.4 million (singlet) and (12.5 million) triplet generated configurations.

For the calculations of excited states we used the correlation consistent AO basis set of Dunning of double zeta quality cc-pVDZ[18]. In addition the basis set was enlarged by s-Rydberg functions located at the nitrogen and a negative ion function for the chlorine and bromine centres (cc-pVDZ+SP). The exponents taken are α_s(N)= 0.028, α_p(Cl)= 0.049 and α_p(Br)= 0.032. From our prior studies[19,20] it is known that changes between the results obtained with cc-pVDZ+SP double and cc-pVTZ+SP triple basis sets are small. The cc-pVDZ+SP basis is flexible with respect to polarisation and electron correlation and is considered to be fairly balanced for all states treated so that transition energies of the interesting region below 7.5 eV are expected to be within an error margin below 0.3 eV.

Two models were adopted for the potential energy surfaces: One was changing the BrO-NO_2 bond length stepwise in the range from 1.30 Å to 10 Å, while all other geometrical variables were optimised. For this relaxation we used the economic density functional method (DFT) with the B3LYP functional as implemented in the GAUSSIAN 98[22] program package. This is justified since we obtained a fairly reliable equilibrium geometry of $BrONO_2$ already at the B3LYP/6-31G* level (as can be seen from fig.1) and since it is known from various investigations that DFT/B3LYP calculations can be used to predict overall molecular structures quite successfully.

In the second model the potential energy curve is obtained by elongating the Br-ONO_2 bond length stepwise from 1.80 Å to 10 Å. Again all other bond lengths and angles were optimised at the DFT-B3LYP/6-31G* level.

We examined twenty low-lying excited states (five singlet and five triplet states of each IRREP) upon BrO-NO_2 and Br-ONO_2 cleavage. For the computation of the excited states potential energy curves only 14 electrons were treated active while the remaining electrons were kept in doubly occupied orbitals (frozen). This is justified from an analysis of the

MOs involved in the selected reference configurations and by comparison with the spectrum computed with all 30 valence electrons kept active. The changes in transition energies are below 0.3 eV.

3. Results and Discussion

As a starting point of our study of excited states we adopted the equilibrium geometries of bromine nitrate $BrONO_2$ and chlorine nitrate $ClONO_2$ from ref.'s 7 and 11. They are given for completeness in figure 1. The structure of $BrONO_2$ was determined experimentally [7], whereas the $ClONO_2$ equilibrium structure results from QCISD optimisations[11]. The global minimum of both isovalent molecules is characterised by C_S

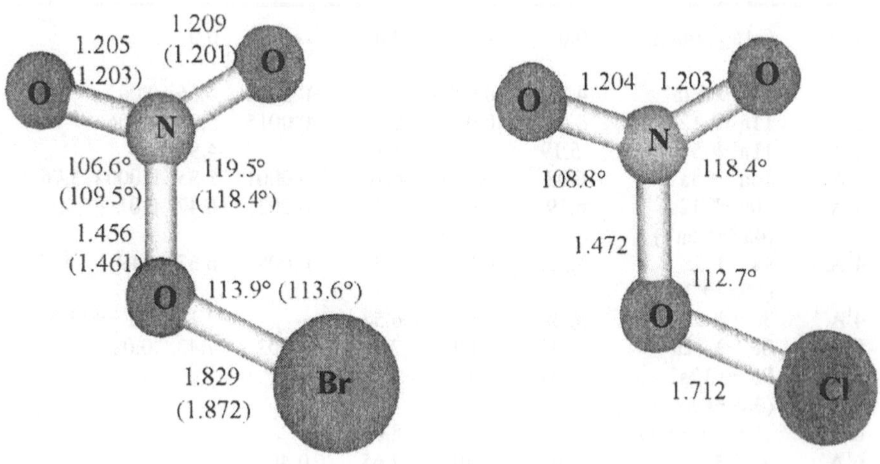

Figure 1: Equilibrium geometries of bromine nitrate $BrONO_2$ and Chlorine nitrate $ClONO_2$ as adopted from ref.'s 7 and 11. The lengths are given in Å, angles in degrees. The values for $BrONO_2$ result from experimental work [7], whereas the $ClONO_2$ structures was obtained by QCISD/6-31G* optimisations of ref. 11. The values in parentheses result from DFT-B3LYP/6-31G* optimisations of the present work

symmetry and besides the somewhat longer OBr bond their equilibrium structures are quite similar. Furthermore it can be seen from fig.1, in which we also included the results of DFT-optimisations of $BrONO_2$, that a reasonably reliable equilibrium geometry of $BrONO_2$ is already obtained at the DFT-B3LYP/6-31G* level.

3.1 The electronic spectrum of $BrONO_2$ in comparison to $ClONO_2$:

In tables 1 and 2 we present the comparison of the computed excitation energies between $BrONO_2$ and $ClONO_2$. The ground states of both molecules are singlet states.

Table 1: Calculated electronic transition energies ΔE(eV) and oscillator strengths f from the ground state X^1A' of C_s symmetric $BrONO_2$ to its low-lying electronic excited singlet states and comparison with the corresponding transitions in $ClONO_2$. For comparison the results of a recent calculation of Graña, Lee and Head-Gordon [11] and experimental values [12,13] available for $ClONO_2$ have been included. The excitation energies are given with respect to the lowest state of each molecule. The ground state configuration for the valence electrons is $(11a')^2\,(4a'')^2$. The values have been obtained at the MRD-CI+Q level with the use of a cc-pVDZ+SP basis as explained in the computational techniques.

State	$BrONO_2$ Excitation	$ClONO_2$ $\Delta E^{(1)}$	$f_1^{(1)}$	$ClONO_2$ $\Delta E^{(1)}$	$f_2^{(1)}$	$\Delta E^{(2)}$	$f_3^{(2)}$	$\Delta E_{(exptl.)}^{(3)}$
X^1A'	$(11a')^2\,(4a'')^2$	0.0	--	0.0	--	0.0	--	
$1^1A''$	4a''→ 12a'	4.23	0.0003	4.57	0.0008	4.36	0.001	
$2^1A'$	11a'→ 12a'	4.97	0.001	5.44	0.0015	5.21	0.006	
$2^1A''$	11a'→ 5a''	5.19	--	5.17	--	4.93	--	
$3^1A''$	10a'→ 5a''	6.08	0.0001	6.10	0.0001	5.85	0.0004	5.76
$3^1A'$	10a'→ 12a' (4a''→ 5a'')	6.19	0.01	6.34	0.003	6.47	0.01	
$4^1A'$	4a''→ 5a'' (10a'→ 12a')	6.22	0.02	6.52	0.008	6.62	0.19	6.52
$4^1A''$	3a''→ 12a'	6.30	--	6.52	--	7.10	(3a''→13a')	
$5^1A''$	2a''→ 12a'	6.88	0.003	7.49	0.003	7.43	0.002	
$5^1A'$	9a'→ 12a' (3a''→ 5a'')	7.04	0.14	7.15	0.12			
$6^1A''$	10a'11a'→ 12a'5a''	7.42	--	8.04	--			
$6^1A'$	3a''→ 5a'' (9a'→ 12a')	7.62	0.40	7.65	0.40			

(1) this work.
(2) Vertical CCSD/6-311(2+2+)G** excitation energies and CIS/6-31G* oscillator strengths as calculated by Graña, Lee and Head-Gordon[11].
(3) Experimental data taken from ref.'s 12, 13.

Because of spin conservation transitions to singlet excited states are most likely, but recent experimental photodissociation studies have shown that spin-orbit coupling has a non-negligible influence for some chlorine species like HOCl so that weak transitions to triplet states are possible[22-24]. Because of this we have also studied triplet states as shown in table 2.

Table 2: Calculated electronic transition energies ΔE(eV) and oscillator strengths f from the ground state X^1A' of C_s symmetric $BrONO_2$ to its low-lying electronic excited triplet singlet states and comparison with the corresponding transitions in $ClONO_2$. For comparison the results of a recent calculation of Graña, Lee and Head-Gordon [11] for $ClONO_2$ have been included. The excitation energies are given with respect to the singlet ground state of each molecule. The ground state configuration for the valence electrons is $(11a')^2\,(4a'')^2$. The values have been obtained at the MRD-CI+Q level with the use of a cc-pVDZ+SP basis as explained in the computational techniques.

State	Excitation	$BrONO_2$ $\Delta E^{(1)}$	$ClONO_2$ $\Delta E^{(1)}$	$ClONO_2$ $\Delta E^{(2)}$
X^1A'	$(11a')^2\,(4a'')^2$	0.0	0.0	0.0
$1^3A''$	$4a'' \rightarrow 12a'$	3.35	3.53	3.43
$1^3A'$	$11a' \rightarrow 12a'$	4.09	4.47	4.30
$2^3A'$	$4a'' \rightarrow 5a''$ $(3a'' \rightarrow 5a'')$	4.86	4.72	4.56
$2^3A''$	$11a' \rightarrow 5a''$	4.90	4.78	4.71
$3^3A'$	$10a' \rightarrow 12a'$	5.50	5.73	5.77
$3^3A''$	$10a' \rightarrow 5a''$	5.79	5.88	
$4^3A'$	$3a'' \rightarrow 5a''$ $(4a'' \rightarrow 5a'')$	6.01	6.02	
$5^3A'$	$9a' \rightarrow 12a'$	6.09	6.08	
$4^3A''$	$3a'' \rightarrow 12a'$	6.05	6.22	
$5^3A''$	$2a'' \rightarrow 12a'$	6.37	6.75	
$6^3A'$	$8a' \rightarrow 12a'$	7.04	7.35	
$6^3A'$	$10a'\,11a' \rightarrow 12a'5a''$	7.30	7.79	

(1) this work.
(2) Vertical CCSD/6-311(2+2+)G** excitation energies as calculated by Graña, Lee and Head-Gordon[11].

The electronic spectrum of $ClONO_2$ has been the subject of extensive prior experimental[12,13] and theoretical[11] research. Therefore we will concentrate on the electronic spectrum of $BrONO_2$ and restrict the present discussion of $ClONO_2$ to those points which show interesting similarities to $BrONO_2$. For $ClONO_2$ two transitions are known from experiments [12,13]. The transition at 215 nm (5.76 eV) can be assigned to the transition 10a' –> 5a" ($3^1A''$) and the transition at 190 nm (6.52 eV) corresponds to the $4^1A'$ state[11]. It can be seen from table 1 that our values for the excited states of $ClONO_2$ up to 7 eV are generally in good agreement with the experimental measurements[12,13] and the values computed by Graña, Lee and Head-Gordon[11]. These computations were performed using coupled cluster methods (CCSD) so that higher excitations (above 7eV) are somewhat dubitative. Instead of 3a"–> 13a' we find 3a"–> 12a' to be the most important configuration of the $4^1A"$ state which explains that we compute the $4^1A"$ state about 0.62 eV lower than Graña, Lee and Head-Gordon. The use of the rather poor CIS/6-31G* method in this early study[11] can also explain the different magnitude in various computed oscillator strengths, although the general trends are fully consistent between our MRD-CI values and the study of Graña, Lee and Head-Gordon.

The SCF-MO energy scheme of $ClONO_2$ and $BrONO_2$ can be seen from figure 2. With respect to the 30 valence electrons treated active in the CI calculation the ground state configuration of both molecules is $(11a')^2$ $(4a")^2$. As can be seen from figure 2, beside the HOMO-LUMO excitation 4a" –> 12a' also transitions originating from the outer valence orbitals 11a', 3a", 10a', 9a' and 2a" can be expected to be in the low-energy region. Furthermore it can be seen from fig. 2 that the valence MOs of $ClONO_2$ are somewhat lowered compared to the valence MOs of $BrONO_2$ in line with the shorter (and stronger) Cl-O bond compared to Br-O. This leads to a larger HOMO-LUMO gap in $ClONO_2$ ($\Delta\varepsilon_{SCF}$ (HOMO-LUMO) = 1.57 eV). As a consequence the first dipole-allowed transition ($1^1A"$ <– X^1A') of $BrONO_2$ is computed with somewhat smaller excitation energy (4.23 eV) than the corresponding transition of $ClONO_2$ (4.57 eV).

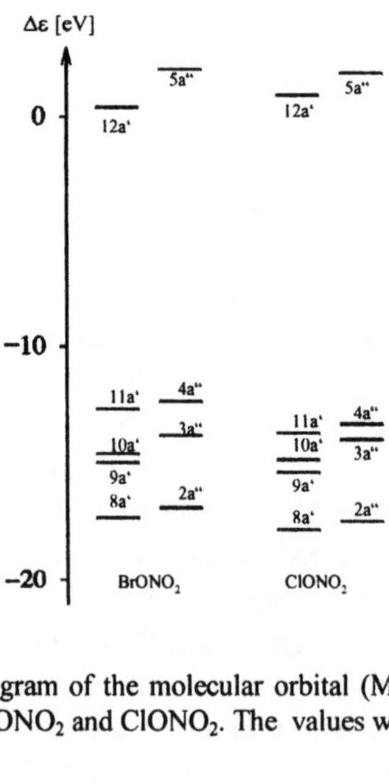

Figure 2: Schematic diagram of the molecular orbital (MO) energy spectrum of ground state configuration of $BrONO_2$ and $ClONO_2$. The values were obtained at the SCF level.

The next singlet excited states $2^1A'$ and $2^1A''$ of $BrONO_2$ are also lowered compared to the corresponding transitions in $ClONO_2$. Both states origin from the valence MO 11a'. Since in $BrONO_2$ the LUMO 12a' is lowered somewhat more than the virtual MO 5a" the energetic ordering of $2^1A'$ and $2^1A''$ is reversed in $BrONO_2$. However this should not be overestimated since both states lie energetically very close.

Generally the spectra of $ClONO_2$ and $BrONO_2$ are very similar. Like in $ClONO_2$ the most important transitions of $BrONO_2$ are $3^1A' \leftarrow X^1A'$ and $4^1A' \leftarrow X^1A'$. They are obtained at 6.19 eV and 6.22 eV respectively. Both states correspond to a linear combination of 10a'->12a' and 4a"->5a" excitation, jugded on the basis of the squared coefficient c^2 of the corresponding wavefunction in the leading configurations: c^2(10a'->12a')=0.50 and c^2(4a"->5a")=0.21 in $3^1A'$, whereas c^2(10a'->12a')=0.21 and c^2(4a"-> 5a")=0.34 in $4^1A'$. This multi-reference character of the states $3^1A'$ and $4^1A'$is in line with almost equal $\Delta\varepsilon$(SCF) values for 10a'->12a' and 4a"->5a".

As can be seen from figure 3, in which contour diagrams of selected MOs are given, 10a' possesses mainly lone pair character at the oxygen centres, whereas 12a' is strongly σ^*(Br-O) antibonding. Therefore 10a'->12a' can be considered as n -> σ^* type excitation along the Br-O distance.

Further strong transitions are computed at 7.04 eV ($5^1A'$) and 7.62 eV ($6^1A'$). Again both states show multi-reference character. The linear combinations involved are 9a'→12a' and 3a''→5a''. Like 10a'->12a' also 9a'-> 12 a' can be considered as n -> $\sigma*$ type excitation along the BrO distance. MO 9a' is composed of lone pair p_x atomic orbitals (AOs), whereas MO 10a' corresponds to p_y lone pair AOs located at the oxygen centres. The large oscillator strength of these two transitions can be understood from the 3a''→ 5a'' transition, which corresponds to π -> $\pi*$ type excitation localised on the NO_2 group of the molecule.

Since the oscillator strength of the transitions to $3^1A'$, $4^1A'$, $5^1A'$ and $6^1A'$ of $BrONO_2$ is of the same magnitude (or even stronger) than the experimentally observed transitions of $ClONO_2$ it can be expected that these states of $BrONO_2$ should also be experimentally discernible.

Due to spin conservation only transitions to singlet states are obtained with a non-zero oscillator strength. The location of the corresponding triplet states of $BrONO_2$ are summarised in table 2. Due to spin conservation only the singlet state is obtained with a non-zero oscillator strength. In table 2 again comparison is made to the triplet states of

$ClONO_2$ computed with our method (MRD-CI) and the values obtained by Graña, Lee and Head-Gordon[11]. Again we find relatively small differences between CCSD/6-311(2+2+)G**[11] and MRD-CI/cc-pVDZ+SP (this work) excitation energies also for the triplet states of $ClONO_2$. Furthermore the vertical excitation energies from the singlet ground state to the excited triplet states of $BrONO_2$ are of similar magnitude as in $ClONO_2$. The excited triplet states of both molecules involve the same valence excitations.

The singlet triplet splitting of single-configurational $BrONO_2$ states is in the order of 0.3 to 0.9 eV as can be expected for typical valence transitions. Contrary to the corresponding singlet states the states $3^3A'$ and $4^3A'$ are characterised by a single excitation and no mixing between 10a'–> 12a' and 4a"–>5a" is found in these states.

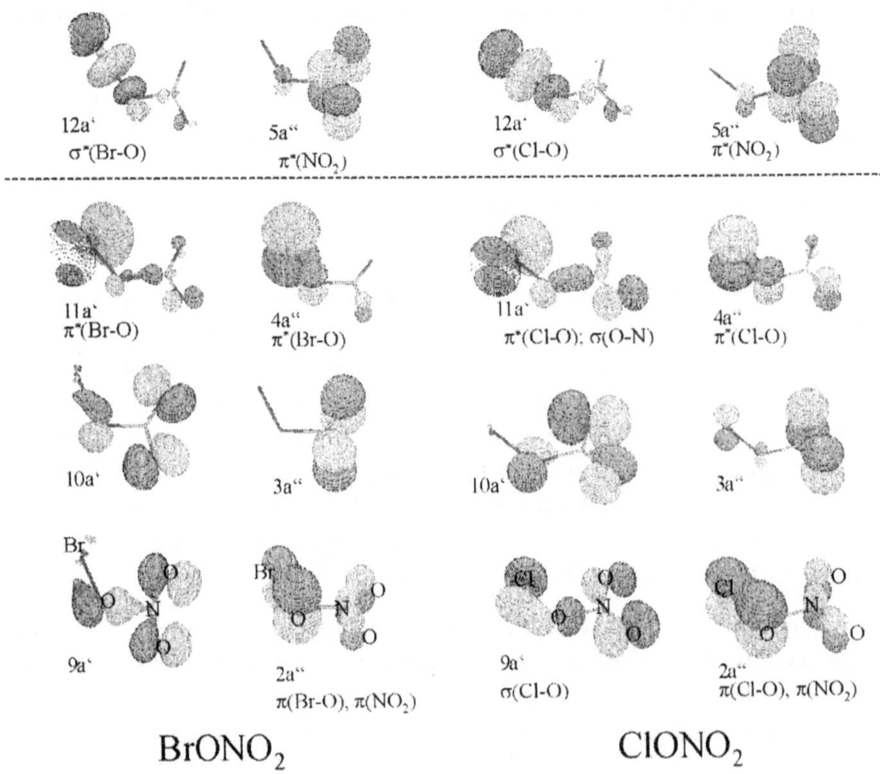

Figure 3: Charge density contours of characteristic valence (9a' to 11a' and 2a" to 4 a") and low-lying virtual (12a' and 5a") orbitals of $BrONO_2$ and $ClONO_2$. The numbering of MOs is with respect to the 30 valence MOs.

3.2. Photodissociation of $BrONO_2$ along O-N and Br-O cleavage

Since isovalent $ClONO_2$ is known to be an important reservoir of Cl and ClO radicals[1-4] photofragmentation processes such as

$$BrONO_2 \rightarrow BrO + NO_2 \qquad (1)$$

and

$$BrONO_2 \rightarrow Br + ONO_2 \qquad (2)$$

can also be expected to have great influence on the concentrations of Br and BrO_x[5-8]. Thus the fragmentations (1) and (2) are expected to be strongly related with the atmospheric ozone reduction. Therefore these two photofragmentation reaction pathways are investigated in the present study and multi-reference configuration interaction calculations are carried out to compute potential energy curves for O-N and Br-O elongation leading to the fragments of (1) and (2). The values obtained for the fragmentation energies of the reactions (1) and (2) at the E(MRD-CI+Q)/cc-pVDZ+SP level, namely 34.0 kcal/mol for (1) and 34.4 kcal/mol for (2), are in reasonable agreement with bond energies of 28 kcal/mol and 32 kacl/mol obtained by Parthibian and Lee[14] combining theoretical $\Delta H^0_{f,298}$ values of $BrONO_2$ with experimental heats of formation.

The MRD-CI potential energy curves for the pathways (1) and (2), i.e. the fragmentation along N-O (1) and Br-O (2) bonds, are presented in figures 4 and 5. It can be seen that the formation of bromine nitrate $BrONO_2$ in its ground state from the fragments BrO and NO_2 as well as from the fragments Br and ONO_2 is an energetically favoured process. On both cuts through the ground state potential energy surface we have not found a barrier. In other words the fragmentation processes according to (1) or (2) on the ground state surface requires more than 30 kcal/mol and is therefore very unlikely.

Along pathway (2) the O-N bond in the ONO_2 fragment is somewhat shortened to 1.203 Å. This observation is in line with three equivalent ON bonds in NO_3 (D_{3h}). The experimental BrO ($X^2\Pi$) distance is 1.717 Å[25] in reasonable agreement to our calculated value of 1.742 Å using DFT-B3LYP/6-31G* optimisation.

In figure 4 we present the calculated MRD-CI potential energy curves of twenty (five singlet and five triplet states of each IRREP) low-lying excited states of $BrONO_2$ for BrO + NO_2 cleavage following pathway (1). It is seen that the first three excited states with vertical excitation energies of $1^3A''$=3.35 eV, $1^3A'$=4.09 eV and $1^1A''$=4.23 eV are repulsive. Together with the ground state they correlate with the dissociation channel calculated at 1.47 eV (34.0 kcal/mol) which corresponds to the ground state of NO_2 (X^2A_1) and BrO ($^2\Pi$). All three excited states are populated by excitations into the LUMO 12a' which possesses not only Br-O but also O-N antibonding character. Their origins are the valence MOs 11a' ($^3A'$) and 4a" ($^1A''$, $^3A''$). The third dissociation channel calculated at 4.98 eV corresponds to BrO in its excited $A^2\Pi$ state. The calculated energy difference $\Delta\varepsilon$ of 3.51 eV is in good agreement with the 3.437 eV reported in the literature[25].

Figure 4: Calculated MRD-CI potential energy curves of twenty (five singlet and five

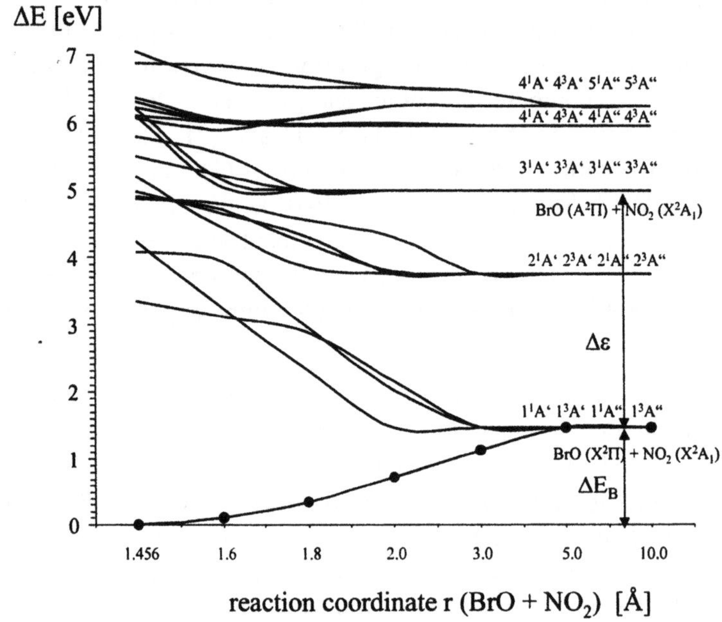

triplet states of each IRREP) low-lying excited states of $BrONO_2$ for BrO + NO_2 cleavage. The points indicate actually computed distances and resemble relative stabilities (in eV) with respect to the equilibrium structure given in figure 1. The fragmentation energy ΔE_B and the energy difference $\Delta\varepsilon$ between two important dissociation channels are illustrated.

The results for the second model adopted for the potential energy curves, i.e. the elongation of the Br + ONO_2 bond as reaction coordinate are displayed in figure 5. Again MRD-CI potential energy curves of twenty (five singlet and five triplet states of each IRREP) low-lying excited states of $BrONO_2$ are computed for Br + NO_2 cleavage following pathway (2). Five low-lying excited states with vertical excitation energies of $1^3A''$=3.35 eV, $1^3A'$=4.09 eV, $2^3A'$=4.86 eV, $1^1A''$=4.23 eV and $2^1A'$= 4.97 eV are found to be highly repulsive for Br + ONO_2

elongation. The energy surfaces of the ground state and of these five low-lying repulsive states end up in the same dissociation channel at calculated 1.49 eV (34.4 kcal/mol) which corresponds to the X^2A_2' ground state of NO_3 and a separated bromine atom in its ^2P ground state. The repulsive character of these states can be understood on the basis of qualitative MO considerations already. The contour diagrams of the MOs involved, which are illustrated in figure 3, show that the LUMO (12a'), which is populated in these states shows Br-O antibonding character. Due to a nodal plane perpendicular to the Br-O axis this LUMO 12a' is σ*(Br-O)-type. It is populated in the 1^3A" and 1^1A" states by 4a"->12a' excitation and in the states 1^3A' and 2^1A' by 11a'->12a' transition. The repulsive 2^3A' results from 4a"->5a" excitation as can be seen from table 2 in conjunction with figure 3. The dipole allowed singlet transition 11a'->12a' is obtained with considerable oscillator strength so that photo-fragmentation along the elongation of the Br-O bond involving this excitation is very likely.

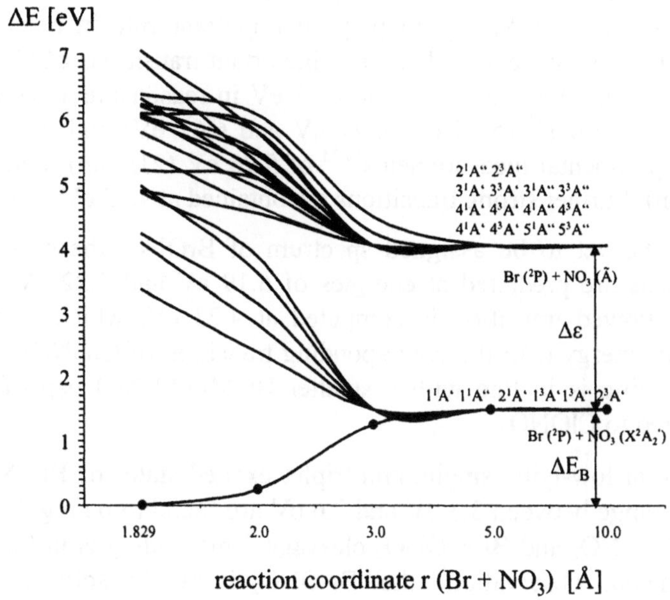

Figure 5: Calculated MRD-CI potential energy curves of twenty (five singlet and five triplet states of each IRREP) low-lying excited states of $BrONO_2$ for Br + NO_2 cleavage. The points indicate actually computed distances and resemble relative stabilities (in eV) with respect to the equilibrium structure given in figure 1. The fragmentation energy ΔE_B and the energy difference Δε between the two lowest dissociation channels are illustrated.

The second dissociation channel that can be seen from figure 5 corresponds to the first excited state of NO_3 (Ã). This second channel is calculated at 4.05 eV, so that an energy difference of $\Delta\varepsilon$=2.56 eV is obtained. This value should correspond to the first excited states of NO_3. The NO_3 spectrum is characterised by diffuse bands in the range between 1.8 eV and 2.5 eV [25]. Again the MOs 12a' and 5a" are populated in these cases so that these excited states are also repulsive for Br-O elongation.

Since all excited states considered are found to be repulsive for BrO-NO_2 and Br- ONO_2 elongation, photofragmentation along BrO + NO_2 and Br + ONO_2 cleavage is very likely. Such photofragmentation will either lead to BrO and NO_2 in their ground states or involve excited states of NO_2 and BrO, which are likely to interact with various reaction cycles of ozone destruction.

4. Summary and Conclusion

We performed MRD-CI calculations for ground and excited states of bromine nitrate $BrONO_2$ in comparison to isovalent chlorine nitrate $ClONO_2$, which is known to play an important role in reactions with ozone in the stratosphere. The most important transitions of $ClONO_2$ are obtained at 6.10 eV, 6.34 eV and 6.52 eV in reasonable agreement with prior calculations[11] (5.85 eV, 6.47 eV and 6.62 eV) and in accordance with experimental measurements[12,13] of 5.76 eV (215 nm) and 6.52 eV (190 nm). Further strong transitions are obtained at 7.15 eV and 7.65 eV.

For the yet to be assigned spectrum of $BrONO_2$ the corresponding transitions are predicted at energies of 6.19 eV and 6.22 eV. The first dipole-allowed transition is computed at 4.23 eV, which is somewhat lower in energy than the corresponding transition of $ClONO_2$ (4.57 eV). This finding is in line with a smaller HOMO-LUMO gap of $BrONO_2$ compared to $ClONO_2$.

Several low-lying singlet and triplet excited states of $BrONO_2$ in the energy range between 3.3 eV and 7.6 eV are found to be highly repulsive for BrO + NO_2 and Br + ONO_2 cleavage. This finding is in line with the experimental observation[8], that Br-ONO_2 is rapidly splits into Br and ONO_2 between 208 nm (5.36 eV) and 250 nm (4.96 eV). Although due to the outcome of our computations a number of excited states of $BrONO_2$ fall into this energy region, we expect the $2^1A' \leftarrow X^1A'$ transition computed at 4.97 eV to be mainly responsible for such a dissociation, because the corresponding excitation 11a'–>12 a' can be considered n(Br)–>σ*(BrO) type and is obtained with a (moderate) oscillator strength of 0.001.

In summary the present computations of excited states, the vertical electronic absorption spectrum and photodissociation along $BrO\text{-}NO_2$ and $Br\text{-}ONO_2$ cleavage can provide characteristic fingerprints as a guideline for further experimental search and lead to a better understanding of radical reactions of $BrONO_2$ significant for atmospheric chemistry to clarify the role of $BrONO_2$ as a possible reservoir of Br and BrO radicals.

Acknowledgement

The present study is part of a NATO science project "Study of elementary steps of radical reactions in atmospheric chemistry". The financial support from the NATO collaborative linkage grant "Environmental Earth Science and Technology" EST.CLG.977083 is gratefully acknowledged. M. Schnell is thanked for helpful discussions and M. Hanrath for various improvements of the DIESEL program package.

References:

1 J. G. Anderson, D. W. Tooley, W. H. Brune, Science, **251**, 39 (1991).
2 S. Solomon, Nature, **347**, 347 (1990)
3 .P. J. Crutzen, R. Müller, C. Brühl, T. Peter, Geophys. Res. Letters, **19**, 1113 (1992).
4 J. N. Crowley, P. Campuzano-Jost, G. K. Moortgat, J. Phys. Chem., **1000**, 3601 (1996).
5 S. C. Wofsy, M. B. Mc Elroy, Y. L. Yung, Geophys. Res. Lett. **2**, 215 (1975).
6 J. E. Spencer, F. S. Rowland J. Chem. Phys. **82**, 7 (1978).
7 B. Casper, P. Lambotte, R. Mimkwitz, H. Oberhammer, J. Phys. Chem. **97**, 9992 (1993).
8 Deters, J. P. Burrows, J. Orphal, J. Geophys. Res. **103**, 3563 (1998).
9 S. Solomon, M.Mills, L.E. Heidt, W.H. Pollock, A.F. Tuck, J. Geophys. Res. **97**, 825 (1992).
10 R. Garcia, S. Solomon J. Geophys. Res. **99**, 12937 (1994).
11 A. M. Graña, T. J. Lee, M. Head-Gordon, J. Phys. Chem. **99**, 3493 (1995).
12 F. S. Rowland, J. E. Spencer, J. M. Molina, J. Chem. Phys. **80**, 2711 (1976).
13 L. T. Molina, J. M. Molina J. Photochem. **11**, 139 (1979).
14 S. Parthiban, T. J. Lee, J. Chem. Phys. **109**, 525 (1998).
15 S. Parthiban, T. J. Lee, J. Chem. Phys. **113**, 145 (2000).
16 M. Hanrath and B. Engels, Chem. Phys. **225**, 197, (1997).
17 R. J. Buenker, S. D. Peyerimhoff, Theor. Chim. Acta **35**, 33 (1974); **39**, 217 (1975).
18 T. H. Dunning, Jr., J. Chem. Phys. **90**, 1007 (1989).
19 M. Mühlhäuser, M. Schnell, S. D. Peyerimhoff, Chem. Phys. Lett **334**, 519 (2001).
20 M. Schnell, M. Mühlhäuser, S. D. Peyerimhoff, Mol. Phys. **100**, 509 (2001).
21 GAUSSIAN 98: M. J. Frisch, G. W. Trucks, H. B. Schlegel, P. M. W. Gill, B. G. Johnson, M. A. Robb, J. R. Cheeseman, T. A. Keith, G. A. Peterson, J. A. Motgomery, K. Raghavachari, M. A. Al-Laham, V. G. Zakrewski, J. V. Ortiz, J. B. Foresman, J. Cioslowski, B. B. Stefanov, A. Nanayakkara, M. Challacombe, C. Y. Peng, P. Y. Ayala, W. Chen, M. W. Wong, J. L. Andres, E. S. Replogle, R.

L. Matrin, D. J. Fox, J. S. Binkley, D. J. Defrees, J. Baker, J. P. Stewart, M. Head-Gordon, C. Gonzales and J. A. Pople, Gaussian Inc., Pittsburgh PA, 1999
22 Y. Li, J. S. Francisco, J. Chem. Phys., **111**, 8384, 1999
23 R. J. Barnes, M. Lock, J. Coleman, A. Sinha, J. Phys. Chem., **100**, 453, 1996
24 J. S. Francisco, M. R. Hand, I.H.Williams, J. Phys. Chem. A, **100**, 9250, 1996.
25 K. P. Huber and G. Herzberg, *Constants of Diatomic Molecules,* Molecular Spectra and Molecular Structure Vol. IV, Van Nostrand Reinhold, New York, 1979.

NITROUS OXIDE: ELECTRON ATTACHMENT and POSSIBLE SCENARIO for ITS REACTION with *ns* METAL ATOMS

Oksana Tishchenko[&], Eugene S. Kryachko[&,#], and Minh Tho Nguyen[&]
[&] *Department of Chemistry, University of Leuven, Celestijnenlaan 200F, B-3001 Leuven, Belgium*
[#] *Bogoliubov Institute for Theoretical Physics, Kiev, Ukraine 03143*
[#] *Corresponding author: eugene@bohr.quantchem.kuleuven.ac.be*

Abstract Nitrous oxide (N_2O) is known as the "laughing gas" having many beneficial actions but also an important "greenhouse gas" whose emission in the atmosphere should be strongly reduced and strictly controlled. The present review discusses the molecular mechanisms of some chemical processes relevant to its reduction, including the dissociative electron attachment as well as its reactions with lithium and copper atoms, as the representative *ns* metal atoms. Novel structural and mechanistic aspects are revealed using high level ab initio quantum chemical calculations.

Keywords: Nitrous oxide, electron attachment, metal atoms, reaction mechanisms, *ab initio* calculations

1. Prologue: Nitrous Oxide, a Molecule for All Seasons

The year 1772 witnessed the identification, by some great chemists, of two natural and long-lived gases whose intensive use has ever since made primordial contributions to the advances of chemical sciences: molecular nitrogen (N_2) and nitrous oxide (N_2O). While N_2 was independently isolated in pure form from air by Rutherford, Scheele and Cavendish [1], N_2O was actually discovered by Priestley from reduction of NO with iron or iron/sulphur mixtures and dubbed it "diminished nitrous air" [2]. Sir Humphrey Davy inhaled N_2O during his medical training in 1799 as a part of the experimental study on the psychological effect of

E.J. Brändas and E.S. Kryachko (eds.), Fundamental World of Quantum Chemistry, Vol. II, 393–423.

a variety of gases and found its ability to promote lightheadness and called it as a "laughing gas" [3] (see also Ref.[4]). The dentists Horace Wells and Gardner Colton showed its anaestethic properties in 1844-1846. Nowadays, N_2O has a panoply of both positive applications and negative consequences in our daily life. Let us mention a few of them.

In medicine, N_2O is widely used as both anesthetic and analgesic agents mainly in obstetric medicine or as an anesthetic carrier gas on anesthetic machines [5]. It is also injected as a tracer gas to measure the blood flow [6] and lung volume [7] owing to its biological inertness when present in low concentration.

In radiation chemistry, N_2O is usually present in aqueous media as a scavenger for hydrated electrons [8]. Moreover, it is commonly served as a propellant for pressurized food containers [9] in the food industry or as a possible direct product of nitric oxide (NO) synthetic enzyme in biology [10]. Nitrous oxide behaves as an oxidant in catalysis to yield highly reactive oxygen anion (O^-) upon dissociative adsorption on a catalyst surface [11]. Decomposition of N_2O is a mean of evaluating the catalytic activity of various oxide surfaces [12]. N_2O plays the role of an indicator for electron availability at semiconductor surfaces [13] and for the determination of the surface area of highly dispersed supported metal particles, using adsorption measurements [14]. It is also used in the diary industry as a mixing and foaming agent since it is non-flammable, bacteriostatic, colorless (under ambient conditions) and exhibit a slightly sweet odor and taste. It also presents in the interstellar medium [15, 16] and in fact only one known interstellar compound where all five valence electrons of the nitrogen atom are engaged [17].

Recently, there has however been considerable concern about the less laughing facet of nitrous oxide due to its negative impact on the troposphere and stratosphere. In the "Kyoto Protocol" which was signed in 1997 in Japan and confirmed in 2002 in Marrakech, Marocco, and already ratified by more than 170 countries around the world with the aim to organise the common fight against the climate changes, nitrous oxide was indeed listed as one of the six strong greenhouse gases, along with carbon dioxide, methane, fluorohydrocarbons, perfluorohydrocarbons and hexafluorosulfur [18]. Production of these gases should be significantly reduced and strictly controlled by participating countries. Nitrous oxide is a minor component of the atmosphere. It is a potent and harmful greenhouse gas whose global warming potential is larger than that of CO_2 by a factor of ca. 310 [18, 19, 20, 21]. N_2O is characterized by a long lifetime: its half-life in the atmosphere is estimated as 110 -168 years [22]. Thus, although the former gas exists only at the trace level in the troposphere, its effect on the atmospheric chemistry is expected to be by far larger than its weights [23]. Due to the existence of a gap between the Shumann-Runge bands of molecular oxygen and the Hartley band of ozone, and the Herzberg continuum of O_2, energetic solar radiation easily reaches the lower stratosphere (at ≈ 20 km, the UV window covers the range between 197 and 214 nm [24]). It photolyses N_2O as

$$N_2O + h\nu \rightarrow N_2 + O(^1D). \qquad (1)$$

Characterized by the unit quantum yield $D_o = 351$ kJ/mol [25], the reaction (1) involves about 90 % of atmospheric N_2O. The remainder reacts with electronically excited oxygen atoms:

$$N_2O + O(^1D) \rightarrow 2NO \;\; (6\%) \qquad (2)$$

and

$$N_2O + O(^1D) \rightarrow N_2 + O_2 \;\; (4\% \;\; [26]). \qquad (3)$$

An instantaneous lifetime of 123 years is determined by the reactions (1) and (2) [27]. (3) is actually the dominant source of reactive odd nitrogen to the stratosphere [28] whose cycle plays a key role in the depletion of nonpolar ozone [29, 30]. The reaction (2) yields the nitric oxide radical (NO) [31]. The latter is well known for its role as a destroyer of the ozone layer which protects the Earth from harmful ultraviolet radiation [32, 33, 34, 35, 36, 37]. As a consequence, N_2O concentration is expected to affect substantially the global warming and stratospheric ozone loss. Kim and Gordon [38] suggested that the N_2O isotopomer concentration in the troposphere results from the light N_2O, produced by soil, N_2O from oceans, and a back flux of isotopically heavy N_2O from the stratosphere. The reactions (1) - (3) are treated as the sources of the stratospheric enrichment, except the photolysis of N_2O at 185 nm [39] whose understanding has been significantly improved thanks to Yung-Miller model, which takes the zero-point energy [40] and its current generalization [41] including the transition dipole moment surface, upper-state dynamics and bending excitation.

The concentration of N_2O in the atmosphere amounts 315 ppb and it continuously increases at an annual rate of 0.8 ppbv [42, 43]. Measurements of N_2O in polar ice show that its global concentration is higher now than any time during the past 45000 years [44]. After the late ice age, the concentration of N_2O was grown and then remained at the level of ≈ 275 ppb for about 10000 years untill the 19th century. Since then, its concentration has been significantly increased. Atmospheric nitrous oxide is derived from natural and anthropogenic sources (anthropogenic nitrate [45]), but the latter are the main and only ones that can be minimized by improvements in technology or changes in land use. Several anthropogenic sources of N_2O emissions have been identified, including biomass burning, fossil fuel combustion, industrial production of adipic and nitric acids, nylon manufactures and vehicular traffics [46]. Other well-known N_2O sources are waste water treatment plants, fertilized soils [47], nitrogen-enriched rivers, estuaries and ground water [48].

In this global context, considerable effort has thus been devoted to find out the ways of reducing the emission of nitrous oxide gas. Numerous studies were focused on modelling and understanding of favorable and workable mechanisms regulating the N_2O production and balancing it with its destruction. The latter task has opened a wide area of experimental and theoretical studies. An interesting proposal for a solution is to selectively recover N_2O on an adsorbent to produce a concentrated stream of N_2O during desorption and to use it as an oxidant for organic chemistry [49]. Another possible process is oxidizing phosphines (PR_3) by nitrous oxide at or below 100 C [50]. Oxidations of most organic compounds by nitrous oxide, giving N_2 as the only by-product, are thermodynamically favorable but are found to be extremely slow except at high temperatures. Within this approach, let us however mention a dramatic procedure put forward by Solutia (former Monsanto) which consists in recovering and using N_2O as a selective oxidant to convert benzene to phenol, and then hydrogenate phenol to cyclohexanone. The latter is then oxidized by nitric oxide to adipic acid, returning N_2O as a by-product and thus closing the N_2O-cycle [49]. Reactions of N_2O with radicals [51] and light atoms, such as nitrogen, oxygen, etc. [52] producing N_2, were also studied but the corresponding reactions do not appear to be efficient for practical purpose.

Due to the fact that the three-atomic N_2O species is isoelectronic with fulminic acid (HCNO), diazomethane (H_2CNN) and hydrazoic acid (HNNN), it belongs to the important class of 1,3-dipoles [53, 54]. Thus, the possibility to use acetylenes and olefins that might exist in different natural environments, as dipolarophiles to trap N_2O within the framework of 1,3-dipolar cycloaddition reactions, has recently been investigated [55]. Addition and subsequent elimination of N_2 from the primary five-membered rings have been found to be facile processes to achieve.

2. Metal Atoms as Reducing Agents

Of the various available proposals, the addition of metals to combustion systems has often been considered as an efficient mean of destroying N_2O, thanks to the large exothermicity and rapidity of the corresponding reactions [56]. As for an alternative, the catalytic decomposition on various surfaces involving a dissociative electron attachment to N_2O has been put forward [57]. Apparently, an optimum procedure is the use of heterogeneous catalysts including transition metals, copper in particular [58, 59, 60, 37, 61, 62, 63, 64, 65]. As well known, the oxidation reaction of metal atom with N_2O proceeds via the following scheme (M = metal):

$$N_2O(X^1\Sigma^+) + M \rightarrow MO + N_2. \quad (4)$$

It is worthwhile mentioning that such reactions are quite interesting *per se* owing to the fact that metal oxide product (MO) is often formed on energetically accessible lower-lying excited-state potential energy surfaces (PESs) that gives rise to a strong chemiluminiscence. Moreover, some of these reactions exhibit a rather striked non-Arrhenius behavior at higher temperatures which could be considered as a clue to its mechanistic complexity, indicating that either non-adiabatic transitions take place between different PESs, and/or other reaction pathway(s) on the ground-state PES becomes opened [67].

In this review, we will consider the molecular mechanisms of the reaction (4), confining our discussion to metals with *ns* valence electrons: lithium and copper atoms will be taken as representative examples of alkali and transition metals, respectively. For mechanistic features of the reaction (4) with other metals see theoretical studies [68, 69, 70, 71].

Let us first briefly outline the key features of these reactions observed experimentally. First of all, it has been revealed that the metal atom, participating in (4), cleaves the N-O bond. The reaction (4) is characterized by a rather large exothermicity of 40.4 kcal/mol (M = Li) [72, 73] and 27.1 kcal/mol (M = Cu) [104] thanks to the oxygen atom which, on the one hand, strongly binds the metal and, on the other hand, is not sufficiently strong bound to N_2 in nitrous oxide. Moreover, experiments clearly demonstrate the existence of two different kinetic regimes [67, 75, 76]. One is a typical Arrhenius with the rate constant $k_A(T)$ determined as

$$k_A(T) = k_A^{(o)} exp(-E_A/RT) \quad (5)$$

where, for M = Cu, the pre-exponential factor $k_A^{(o)} = 1.70 \times 10^{-10}$ $cm^3 \cdot mol^{-1} \cdot s^{-1}$ (470 K < T < 1190 K) [75] or $2.4 \pm 0.6 \times 10^{-10}$ $cm^3 \cdot mol^{-1} \cdot s^{-1}$ (458 K < T < 980 K) [76] and the activation enthalpy amounts to E_A = 5129 K = 10.2 kcal/mol [75] or 11.6 ± 0.3 kcal/mol [76]. For the reaction (4) with M = Li, the corresponding parameters are as follows: $k_A^{(o)} = 4.8 \times 10^{-10}$ $cm^3 \cdot mol^{-1} \cdot s^{-1}$ (T < 500 K) and $E_A = 3.4 \pm 0.1$ kcal/mol [67].

Another regime shows a non-Arrhenius behavior and takes place at higher temperatures: above 500 K for Li, and above 1190 K for Cu. Its rate constant obeys the following empirical fit:

$$k_{non-A}(T) = k^{(o)}_{non-A} T^{\alpha} exp(-E_{non-A}/RT) \tag{6}$$

where for Cu, $k^{(o)}_{non-A}$ = 3.04 × 10^{-20} $cm^3 \cdot mol^{-1} \cdot s^{-1}$, α = 2.97, and E_{non-A} = 6.1 kcal/mol [76], and for Li, $k^{(o)}_{non-A}$ = 6.1 ± 0.3 × 10^{-16} $cm^3 \cdot mol^{-1} \cdot s^{-1}$, α = 2.11 ± 0.03, and E_{non-A} = 2.2 ± 0.1 kcal/mol [77]. In this context, two legitimate questions could be posed: what are then the mechanisms underlying reaction (4) and how do structurally Arrhenius and non-Arrhenius reaction channels look like?

In order to elucidate the mechanisms of the reaction (4), three theoretical models have been proposed. Within the electron transfer (ET) model [78, 79, 80, 81, 82, 83, 84], the electron transfer from M to N_2O, which occurs at a definite M-N_2O distance via the harpoon mechanism [85], predetermines (4). The activation energy E_A then depends on the ionization energy IE_M, being equal to 5.392 eV for Li and 7.72 eV for Cu [86], although such dependence has recently been questioned [82, 83, 84, 87]. This is on one hand. On the other one, the electron affinity (EA) of the ground state N_2O molecule is firmly negative and equal to ca. -6 kcal/mol (*vide infra*) [88] (see Section 3.2). This implies that the electron hopping from M to the ground state N_2O has to be questioned too. The model of the direct abstraction (DA) [78, 79] is based on the suggestion that M directly abstracts the oxygen from N_2O.

The third model is the resonance interaction (RI) model proposed by Fontijn and co-workers [89, 90, 91, 92, 93, 94] who examined empirical correlations between the reaction rate constant and activation energy E_A, on the one hand, and, on the other, the sums of ionization energy and *s-p* promotion energy of M, PE_M (for instance, the PE_{Cu} = 87.2 kcal/mol). The RI model assumes that the reaction of any of the aforementioned metal atoms M with N_2O is governed by an activated complex (*AC*) composed of three resonating structures S_1, S_2, and S_3 described by the corresponding wave functions ψ_1, ψ_2, and ψ_3. S_1 is primarily formed by the covalent interaction between the oxygen atom of N_2O and the *s* bonding orbital of M. The structure S_2 is analogous to S_1 where the covalent interaction is realized throughout the *p* bonding orbital of M whereas the third one, S_1, is suggested to be ionic and composed of the anion N_2O^- interacting with cation M^+. Therefore, the total wavefunction Ψ_{AC} of the activated complex is expressed as a linear combination of these three wave functions

$$\Psi_{AC} = \sum_{i=1}^{3} c_i \psi_i \tag{7}$$

with real variational parameters $c_i, i = 1, 2, 3$ subject to the normalization condition $\sum_{i=1}^{3} c_i^2 = 1$, chosen under the assumption of the mutual orthogonality of the wave functions corresponding to the different resonating structures. Ψ_{AC} is then defined as the eigenfunction corresponding to the lowest eigenenergy of the 3×3 Hamiltonian matrix

$$\begin{pmatrix} H_{11} & H_{12} & H_{13} \\ H_{21} & H_{22} & H_{23} \\ H_{31} & H_{32} & H_{33} \end{pmatrix}$$

where the diagonal elements are the following: $H_{11} = V$ is the barrier of the formation of S_1 treated as a model parameter, $H_{22} = \mathrm{PE} + V/2$, and H_{33} is approximated as

$$H_{33}^{(1)} = \mathrm{IE_M} - \mathrm{E}A_{\mathrm{N_2O}} - \frac{e^2}{R_o} - \frac{\alpha e^2}{2R_o^4} + \frac{B}{R_o^\delta} \tag{8}$$

or

$$H_{33}^{(2)} = \mathrm{IE_M} - \mathrm{EA_{N_2O}} - \frac{e^2(\delta - 1)}{\delta R_o} \tag{9}$$

including particularly the attractive Coulomb and polarization potentials and the repulsive Born-type potential. In Eqs.(8) and (9), e is the proton charge, R_o is the sum of the univalent radii of M^+ and O^-, α is the sum of their polarizabilities, and finally, δ is the averaged Born exponent.

3. The Ground State N_2O Molecule

3.1 Structure and key properties

The molecule of nitrous oxide is formed by three second-period atoms and contains 16 valence electrons. It is therefore isoelectronic to CO_2. In its ground electronic state, N_2O is a linear asymmetrical molecule N-N-O, characterized by the point symmetry group $C_{\infty v}$ and thus assigned to its $X^1\Sigma^+$ irreducible representation [95-101]. It can be also characterized by two Lewis resonance structures [99] (see also Ref.[101]):

The bond lengths r(N-N) and r(N-O) of the ground-state N_2O molecule are collected in Table 1. $N_2O(X^1\Sigma^+)$ has a relatively small dipole moment of 0.161 D [119, 120]. Its electronic configuration is characterized by the following molecular orbital (MO) pattern: $(1\sigma)^2(2\sigma)^2(3\sigma)^2(4\sigma)^2$

Table 1. The experimental and theoretical bond lengths (in Å) of the ground-state nitrous oxide.

	Reference	r(N-N)	r(N-O)
Experimental data			
Rotational spectroscopy	[102-104]	1.1282	1.1842
Cotton and Wilkinson	[105]	1.129	1.188
Brüning *et al.*	[57]	1.126	1.186
Teffo and Chédin	[107]	1.127292	1.185089
CRC book	[108]	1.1284	1.1841
Computational data			
MP2/6-31G(d)	[109]	1.172	1.193
MP2/6-311++G(d)	[110]	1.163	1.181
MP2/6-311+G(2d)	[109]	1.157	1.186
B3LYP/6-311+G(2d)	[109]	1.126	1.184
B3LYP/6-311+G(2d,2p)	[111]	1.122	1.188
B3P86/6-311+G(2d,2p)	[111]	1.120	1.181
SVWN/6-311+G(2d,2p)	[111]	1.129	1.178
BLYP/6-311+G(2d,2p)	[111]	1.140	1.202
BP86/6-311+G(2d,2p)	[111]	1.1392	1.196
MP2/6-311+G(2d,2p)	[111]	1.157	1.186
BP/TZP	[112, 113]	1.145	1.120
VWN/TZVP	[112]	1.1375	1.1855
BP/TZVP	[[113]	1.145	1.197
GGA-BP/TZVP	[112]	1.1453	1.2004
B3LYP/DZP^{++}	[118]	1.141	1.202
B3P86/DZP^{++}	[118]	1.138	1.195
BHLYP/DZP^{++}	[118]	1.119	1.190
BLYP/DZP^{++}	[118]	1.161	1.218
BP86/DZP^{++}	[118]	1.158	1.210
LSDA/DZP^{++}	[118]	1.150	1.193
CISD/TZ2P	[114]	1.1073	1.1809
CISD(T)/TZ2P	[115]	1.15053	1.2068
CCSD(T)/TZVP	[106]	1.137	1.193
QCISD/aug-cc-pVDZ	[116]	1.138	1.201
QCISD(T)/aug-cc-pVDZ	[116]	1.150	1.204
CCSD(T)/cc-pVTZ	[115]	1.13285	1.18960
QCISD/aug-cc-pVTZ	[116]	1.121	1.189
QCISD(T)/aug-cc-pVTZ	[116]	1.134	1.193
CCSD(T)/cc-pVQZ	[116]	1.12909	1.18703
CCSD(T)/ANO[5421]	[117]	1.1311	1.1874
CCSD(T)/ANO	[112]	1.1303	1.1896

Table 2. Fundamental and harmonic (in parentheses) vibrational frequencies of the $X^1\Sigma^+$ N_2O (in cm^{-1}). Calculated frequencies are referred to the harmonic unless the otherwise is noticed.

	ν_1	ν_2	ν_3
Experimental data			
Herzberg [126]	1284.9	588.78	2223.76
Brüning *et al.* [57]	1277	589	2224
Teffo and Chédin [107]	1271.4 (1298.3)	588.8 (596.3)	2223.8 (2282.1)
Computational data			
MP2/6-311++G(d) [110]	1286	538	2243
CISD/DZP [138]	1352		2402
CCSD/DZP [138]	1290		2296
CCSD(T)/DZP [138]	1269		2222
CISD/DZP^{++} [138]	1345		2400
CCSD/DZP^{++} [138]	1281		2292
CISD/TZ2P [138]	1354 (1345.9) [114])	645.5 [114]	2417 (2419.5) [114]
CCSD/TZ2P [138]	1288		2307
CCSD(T)/TZ2P [115]	1256	579	2221
VWN/TZVP [112]	1339.72	618.67	2338.12
BP/TZVP [113]	1280.67	589.48	2253.07
GGA-BP/TZVP [112]	1280.65	589.70	2253.07
QCISD/aug-cc-pVDZ [116]	1282	586	2311
QCISD(T)/aug-cc-pVDZ [116]	1265	564	2233
CCSD(T)/cc-pVTZ [115]	1271.0 (1297.2)	594.4 (601.1)	2224.9 (2282.8)
QCISD/aug-cc-pVTZ [116]	1297	615	2340
QCISD(T)/aug-cc-pVTZ [116]	1274	590	2258
CCSD(T)/cc-pVQZ [115]	1303.5	602.1	2287.9
CCSD(T)/ANO[5421] [117]	1272.0 (1299.1)	590.2 (604.3)	2227.3 (2285.9)

$(5\sigma)^2(6\sigma)^2(1\pi)^4(7\sigma)^2(2\pi)^4$ [121]. The highest occupied 1π and 2π *MOs* consist of π-type orbitals lying orthogonal to the molecular axis [122]. The former has a strong bonding character whereas the latter is of a dominant non-bonding character though, precisely, there exists a positive and constructive interference between the nitrogen atoms and the destructive one with the oxygen that gives an overall non-bonding character [123, 124]. The highest σ *MO*, 7σ, has a predominant N-O bonding character [121, 125]. The lowest unoccupied *MO* (*LUMO*) of N_2O ($X^1\Sigma^+$) is the antibonding 3π.

The infrared spectrum of the ground-state N_2O molecule, surveyed in Table 2, exhibits three strong fundamental absorption bands: ν_1 (asym-

Table 3. Spectroscopic constants of the $X\,^1\Sigma^+$ state of N_2O (in Å and cm^{-1}).

	Expt. [107]	CCSD(T)/cc-pVTZ [115]	CISD/TZ2P [114]
X_{11}	-15.138	-15.069	-14.14
X_{12}	-27.207	-26.858	-32.57
X_{13}	-14.328	-14.264	-12.88
X_{22}	-4.319	-3.766	-4.05
X_{23}	-5.374	-5.204	-6.10
X_{33}	1.112	1.211	1.30
G_{33}	-0.575	-0.631	-0.54
B_e	0.42112	0.417488	0.42988
$10^6 D_e$	0.17610 (= D_o)	0.1691	0.16743
$10^{12} H_e$	-0.0156(13) (= H_o)	-0.01417	-0.0278
α_1	0.003444	0.003437	0.00308
α_2	0.001925	0.001836	0.00211
α_3	-0.0005697	-0.0005655	-0.000541

metric stretch, Σ^+), ν_2 (bending, Π) and ν_3 (symmetric stretch, Σ^+), which reveal a superimposed rotational fine structure [126]. What is important for a further discussion in Section 4 is that the bending mode ν_2 is doubly degenerate (the Π irreducible representation of the $C_{\infty v}$ group is two-dimensional) since, according to the known rule that a linear molecule, consisting of N atoms has $3N - 5$ internal degrees of freedom, N_2O possesses 4 internal degrees of freedom. This degeneracy of the ν_2 vibrational mode can be lifted upon interaction of N_2O with surrounding, paricularly with a metal atom M (see Section 4).

The dipole-moment derivatives of N_2O have recently been accurately determined in Ref.[127]. Spectroscopic and force constants are correspondingly shown in Tables 3 and 4. The recent NMR shielding constants of the title molecule are collected in Table 5.

The PESs of the lower-lying electronic states of N_2O were extensively studied by Peyerimhoff and Buenker [133], Hopper *et al.* [125], Hopper [121], Chutjian and Segal [134], and recently, by Balint-Kurti and coworkers [135] and Hwang and Mebel [109]. Notice that in addition to the ground-state NNO molecule, there exist two N_2O isomers. One is a cyclic isomer with the symmetry C_{2v} which was obtained by Klapötke and Schulz [136, 137] in their MP2/6-31+G(d) study of the dissociation of the N_4O molecule and later was thoroughly investigated by Galbraith and Schaefer [138]. It will be demonstrated in Section 3.2 that this

Table 4. Force constants of the ground-state N_2O molecule (in Jacobi internal coordinates, aJ; Å, rad).

	Experimental data				Computational data			
					CCSD(T)/ TZ2P	CCSD(T)/ cc-pVTZ	CCSD(T)/ ANO[5421]	DVR
	[107]	[128]	[129]	[130]	[114]	[115]	[117]	[131]
f_{rr}(NN)	18.2512	18.2362	18.588	18.1904	21.914	18.12625	18.14833	18.653
f_{rR}	.0276	1.0286	1.134	1.0240	1.635	0.96047	0.95093	0.718
f_{RR}(NO)	11.95962	11.6662	11.802	12.0308	12.096	12.02091	12.07914	11.693
$f_{\alpha\alpha}$	.6659	0.6664	0.66550	0.6660	0.763	0.68257	0.68766	0.661
f_{rrr}	133.5984	-132.4326	-150.48	-134.6650	-161.1	-137.96941	-136.188	-151.262
f_{Rrr}	6.8725	-9.8418	-4.54	-1.9200	-1.069	-2.53165	-3.5356	-2.548
f_{RRr}	.4979	2.4512	-5.90	-0.4728	-0.928	-0.29464	-0.37687	2.906
f_{RRR}	98.8309	-96.3270	-94.74	-102.9000	-104.3	-98.00215	-99.719	-94.746
$f_{\alpha\alpha r}$	-1.5796	-2.5674	-2.918	-1.6812	-1.722	-1.69056		-5.944
$f_{\alpha\alpha R}$	-1.5374	-1.1012	-0.784	-1.5744	-1.655	-1.44902		0.755
f_{rrrr}	691.3727	674.7480	52.56	745.2912	931.1	811.36004	796.70	949.184
f_{Rrrr}	46.4463	65.6076	-51.60	10.4292	5.699	10.11118	25.232	-12.017
f_{RRrr}	-3.4847	8.7712	171.96	-8.6808	5.077	0.0797	-2.0384	-12.719
f_{RRRr}	-7.6910	-20.7588	9.84	-12.8936	12.17	17.77822	24.276	-27.299
f_{RRRR}	634.9213	590.3304	623.28	614.9256	614.0	587.16909	584.63	571.393
$f_{\alpha\alpha rr}$	1.8083	12.6700	-1.44	3.0024	0.237	1.10549		30.755
$f_{\alpha\alpha Rr}$	5.1053	7.2216	0.20	5.0150	4.117	4.0413		4.131
$f_{\alpha\alpha RR}$	1.4348	-6.1652	3.90	3.1852	2.491	2.39417		-18.765
$f_{\alpha\alpha\alpha\alpha}$	1.8947	2.2824	2.6520	1.8120	2.414	1.98497	1.82216	8.159

Table 5. NMR shielding constants (in ppm) of the ground-state N_2O molecule (N_1-N_2-O), calculated with the qz2p basis set at the experimental equilibrium geometry [132].

	HF	CCSD	CISD(T)
$^{14}N_1$	-33.7	7.6	15.7
$^{14}N_2$	62.4	102.7	110.5
^{17}O	172.3	200.3	201.3

isomer has a spectacular implication on the electron affinity of N_2O. The most recent CCSD/TZ-ANO [112, 113] calculations place the C_{2v} isomer of N_2O 64.8 kcal/mol above the NNO global minimum and predict its bond lengths r(N-N) = 1.188 Å and r(N-O) = 1.536 Å and the bond angle $\angle$NNO= 67.3^o. The magnetic properties of the cyclic N_2O isomer have been studied by Wüllenand Kutzelnigg [139]. The other isomer N-O-N with the $D_{\infty h}$ symmetry has recently been found by Crawford and Klapötke [110] and Wang and Harcourt [112] who located it at 110.7 kcal/mol above the global minimum and calculated the N-O bond length at 1.202 Å [112].

3.2 Electron attachment to N_2O

Due to a possibility of the electron transfer from metal to N_2O in the reaction (4), a knowledge of the electron affinity of N_2O and the PES of its anion is a necessary accessary to elucidate its mechanism. A retrospective look at this problem reveals that the EA of N_2O was a controversial issue - this is clearly seen from Table 6 collecting reported experimental and theoretical data. Let us briefly summarize the main available results that disagree with each other on this quantity. The first experimental estimations dated back to the early 1970s, when a value of 0.27 ± 0.17 eV was derived from the electron capture detector experiments [140] and a lower limit for the EA of -0.15 ± 0.1 eV was obtained from collisional dissociation or collisional ionization spectra [141]. The positive value of 0.22 eV currently recommended in, among other compilations, the CRC-Handbook [142] was taken from a 1976 estimation of Hopper *et al.* [125] which was based on results of molecular beam experiments. A negative vertical electron affinity of -2.23 eV and a dissociation limit of 0.42 eV relative to the $N_2 + O^-$ asymptote were also derived in this study. This implies that the anionic system gains a stabilization of 2.45 eV following geometry relaxation. These authors [125] also carried out *ab initio* MCSCF-CI calculations using a double-

Table 6. Experimental and Theoretical Electron Affinities (in eV) of N_2O.[a,b]

EA	Method	Ref.
0.27 ± 0.17	electron capture detector	[140]
$> -0.15 \pm 0.10$	neutral beam ionization potentials	[141]
0.22 ± 0.1 (-2.23 ± 0.20)	from electron affinity of radical / enthalpy of formation of anion relationship	[125]
0.22 ± 0.1	collision induced dissociation threshold	[143]
$< 0.76 \pm 0.10$ (ca. 1.5)	laser photoelectron spectroscopy	[144]
-1.7 (-4.1)	*ab initio* MCSCF-CI calculations	[125]
0.39	*ab initio* Hartree-Fock calculations	[68]
-0.13	*ab initio* calculations using G2 approach	[145]
0.3 - 0.8	density functional calculations	[118]
-0.0205 -0.149	*ab initio* QCISD and QCISD(T) calculations	[116]
-0.03 ± 0.1	*ab initio* calculations up to CCSD(T) level	[88]

[a]Vertical EAs are given in parentheses.
[b]Partly reproduced from Ref.[142].

zeta quality basis set augmented by diffuse functions to determine the adiabatic EA. Nevertheless, these calculations, quite extensive by the standard of that time, indicated that both vertical and adiabatic EAs are negative by -4.1 and -1.7 eV, respectively, even though the difference between these values seems to be comparable with the experimental estimate. Accordingly, the N_2O^- anion has a bent form with NNO angle of 132.7^o and the N-O and N-N distances significantly stretched as compared to those in the neutral species. In a 1983 theoretical study, Yarkony [68] reported that the $^2A'$ state of N_2O^- is actually stable by 0.39 eV with respect to ground-state product limit. Such a positive EA came no doubt from the fact that only Hartree-Fock calculations were used in these calculations.

In the later 1990s, the electron affinity of N_2O was further evaluated in three theoretical works [145, 118, 116]. Yu *et al.* [145] considered in 1992 this quantity for a series of triatomic species and paid a special attention to the N_2O/N_2O^- case. Electronic energies of ions and neutrals were calculated by the G2 approach using geometry optimization at the (U)MP2/6-31(d) level. With the N-N and N-O bond lengths of 1.128 Å and 1.674 Å, and the NNO angle of 134.6^o, a negative value of 0.13 eV was derived. The N-O distance of 1.674 Å, which is rather long with respect to the standard N-O distances, was no doubt due to the absence of diffuse functions in the basis set inherent in the geometry optimization step of the G2 procedure. In 1997, a density functional theory study by Tschumper and Schaefer [118] indicated that nitrous oxide has positive electron affinity ranging from +0.3 to +0.8 eV. Nevertheless, the negative EA value was recently confirmed in a 1999 study by McCarthy *et al.* [116] using essentially quadratic configuration interaction QCISD and QCISD(T) methods with aug-cc-pVDZ and aug-cc-pVTZ basis sets. Indeed, these calculations led to values for EA in the range of -0.021 and -0.149 eV.

Yu *et al.* [145] also attempted to figure out the reason for the disagreement between experiment and theory and suggested that it apparently came from the presense of an energy barrier of about 0.15 eV to the decomposition of N_2O^- into N_2 and O^- which was not taken into account by Hopper *et al.* [125] in the interpretation of their experimental results for the threshold energy of the collisionally induced reactions of the anions.

Seeking to unravel the mystery of the electron attachment to N_2O, the authors of Ref.[88] have recently undertaken a thorough study of the ground state PES of N_2O^-, which allowed a resolution of this long-standing controversy and a new look on the dissociative electron attachment to N_2O. A novel entity - a T-shaped ground state anion **A2** (Figure 1) has been revealed and fully characterized. This cyclic anion turns out to be more stable than the open-chain **A1** form by 5.3 kcal/mol, lying, at most, 0.7 kcal/mol above the ground-state neutral **N1** molecule, and can thus be regarded as the adiabatic anion. In view of these results, the adiabatic electron affinity of nitrous oxide, evaluated as the energy difference between **A2** and **N1** forms, amounts to only -0.03 eV.

Let us consider the properties of **A2** in some detail. First of all, it is characterized by the $N \cdots O$ distances of 2.7 - 2.8 Å and can be considered as a weakly-bound complex of N_2 and O^-. The population analysis shows that in fact the negative charge is almost entirely located on the oxygen atom. Asymptotically, this ground-state **A2** anion correlates with a higher-lying cyclic $^1A_1$ isomer of neutral nitrous oxide (see Section 3.1). Since in C_s symmetry, both **A1** and **A2** forms belong to the representation $^2A'$, they are smoothly connected without surface crossing via a transition state structure **ATS** (Figure 1). The latter represents a stationary point on the ground-state PES with a single imaginary frequency describing the normal mode of the predominant N-O stretching character mixed with the NNO bending vibration. It is placed by 6.0 and 12.0 kcal/mol above the anion **A1** and neutral **N1**, respectively, and characterized by the following geometrical parameters: r(N-N) = 1.12 - 1.16 Å (cf. 1.17 - 1.21 Å in **A1**), r(N-O) = 1.63 - 1.69 Å (cf. 1.30 - 1.33 Å in **A1**), $\angle$NNO = 131 - 140° (cf. 129 - 135° in **A1**). Although this **ATS** structure was earlier reported by Yu *et al.* [145],

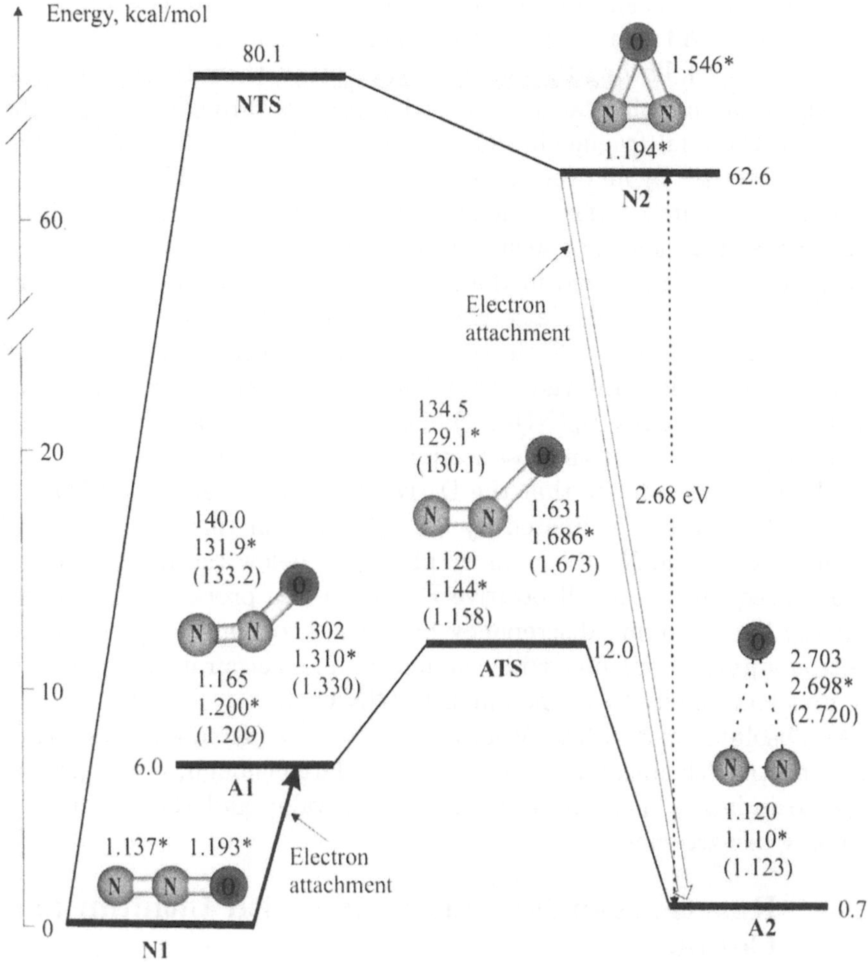

Figure 1. The potential energy profile of the dissociative electron attachment to NNO, calculated at a variety of computational levels (reading from the top down): MP2/TZV, CCSD(T)/TZVP (indicated by the asterisk) and QCISD(T)/aug-cc-pVDZ (in parentheses) [88, 106]. Bond lengths are given in Å, bond angles in ^{o}.

it has in contrast been interpreted as the transition state governing the dissociation of **A1** to the product fragments N_2 and O^-.

In summary, following the electron attachment, the nitrous oxide system undergoes a rather facile dissociation process throughout the cyclic isomer to the final fragments **N1** $\rightarrow$ **A1** $\rightarrow$ **ATS** $\rightarrow$ **A2** $\rightarrow$ $\mathbf{N_2 + O^-}$ and characterizes by an overall barrier of 12.0 kcal/mol. The presence of **ATS** allows a number of experimental findings to be rationalized, either in gas phase reactions or transformations at metal surfaces [88].

Hopper *et al.* [125] used in their study the following thermodynamic cycle to estimate the adiabatic EA of N_2O: $EA(N_2O) = -D_o(NN\text{-}O) + EA(O) + D_o(NN\text{-}O^-)$. While the dissociation energy D_o(NN-O) and the electron affinity of oxygen EA(O) were accurately known, the value for dissociation energy $D_o(NN\text{-}O^-)$ of 0.43 ± 0.1 eV was only estimated from fitting of the reaction cross sections, that led to an error in D_o. The calculations [88] indicate that the $D_o(NN\text{-}O^-)$ value derived by Hopper *et al.* [125] was overestimated by 0.2 eV which in turn overestimated the stability of the N_2O^- anion by placing it below the linear neutral. As a consequence, a small positive $EA(N_2O)$ was proposed. It is clear that the long-standing discrepancy between theory and experiment on electron affinity of nitrous oxide came from a less accurate evaluation of the dissociation energy of the anion $D_o(NN\text{-}O^-)$.

An absolute error of 0.2 eV appears quite reasonable and lies within the expected "chemical accuracy", but it is large enough, in the present case, to induce a qualitative change of the value and thereby a long-standing disagreement.

4. Reactions of Nitrous Oxide with Lithium and Copper

On the basis of the mechanism of the dissociative electron attachement to N_2O, outlined in Section 3.2, it would be reasonable to suggest that it might play an important role in predetermining the pathways of chemical reactions of the type (4) of metals with N_2O, where the latter acts as an electron acceptor in the initial stage. Two remarkable examples of the reaction (4) with M = Li and Cu will be thoroughly examined in the present Section [106, 146]. Their PES profiles are vividly displayed in Figures 2 and 3 (Refs.[106] and [146]).

4.1 Entrance channel and transition states

The entrance channel of the reaction (4) with M = Li and Cu is represented by complex **I**, which is bound by 2.8 kcal/mol for the former and 0.24 kcal/mol for the latter (see Figures 2 and 3). **I** is characterized

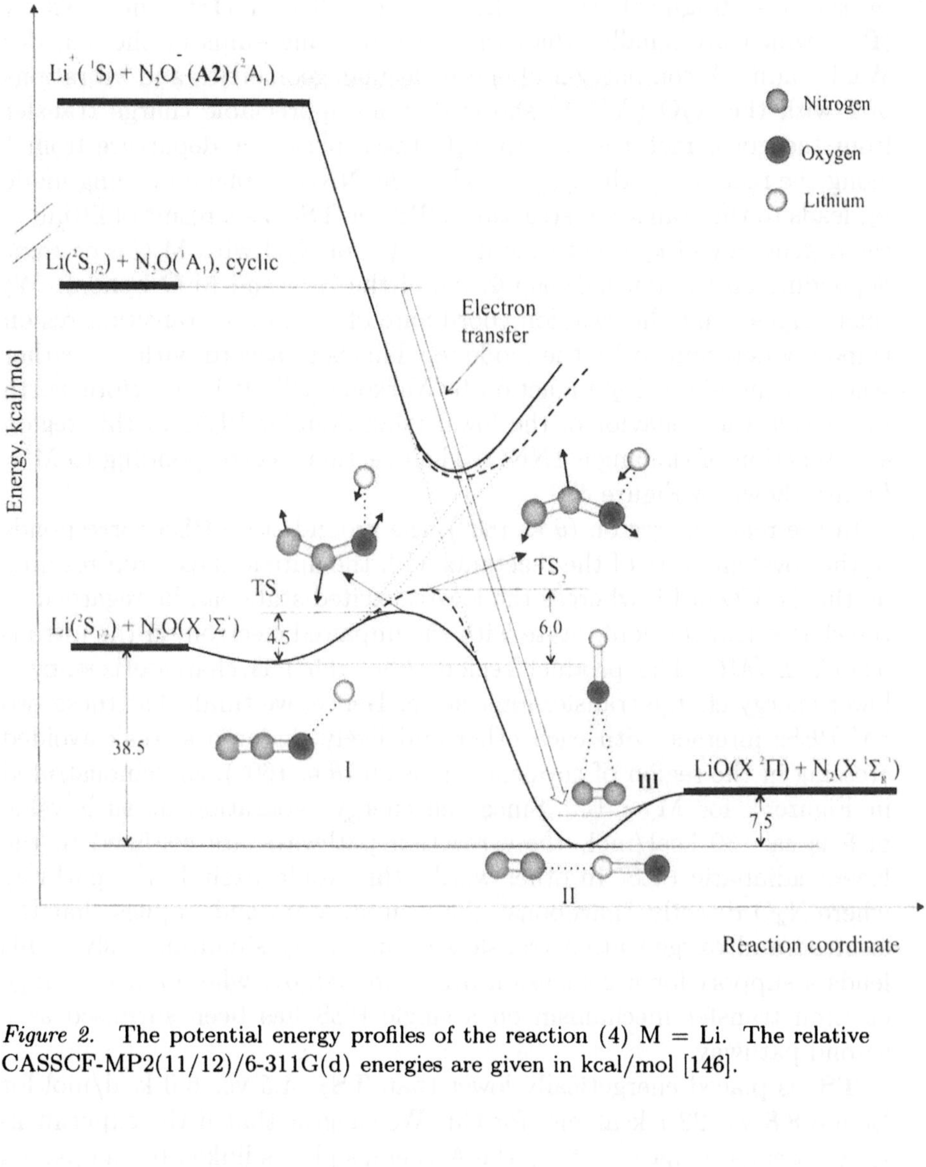

Figure 2. The potential energy profiles of the reaction (4) M = Li. The relative CASSCF-MP2(11/12)/6-311G(d) energies are given in kcal/mol [146].

by the inter-fragment M$\cdots$O distances of 2.098 Å ($\mathbf{I}^{Li}$) and 3.695 Å ($\mathbf{I}^{Cu}$) which are smaller than the corresponding sums of the van der Waals radii. A comparison of the molecular geometries and vibrations of **I** with the N_2O ($X^1\Sigma^+$) shows that no appreciable charge transfer from the given metal atoms to N_2O takes place. A departure from **I** along the reaction pathway, parallel to the NNO in-plane bending mode ν_2, leads to the transition structures, $\mathbf{TS}_1$ or $\mathbf{TS}_2$, as a result of lifting of the degeneracy of ν_2 due to the interaction of N_2O with M (*cis* or *trans* depending on the mutual disposition of the N-N and M-O bonds). We thus suggest that the reaction coordinate of (4) in the transition region is mainly determined by the mode ν_2. This is in accord with the earlier study on the Mg + N_2O reaction by Yarkony [68]. It is therefore worth to analyze the behavior of the lower-lying doublet PESs in this region as a function of the angle ∠NNO. Their sections, corresponding to M = Li, are shown in Figure 4.

In the reactant region ($\theta \approx 180^o$), the ground-state PES corresponds to the covalent state of the reactants with the unpaired electron residing on the 2*s* *MO* of Li, whereas the lowest excited state can be regarded as the charge-transfer ionic state with the unpaired electron on the nitrous oxide's *LUMO*. The product region ($\theta < 155^o$) is characterized by a lower-energy charge-transfer ionic state. Hence, we think that these two ^{2}A$'$ PESs interact with each other and likely undergo a weak avoided crossing in the region of covalent repulsion ($\theta \approx 160^o$), as demonstrated in Figure 2 for M = Li. Since the energy separation at such value of θ is ca. 30 kcal/mol, these reaction pathways are confined to the lowest adiabatic PES. In other words, this would exclude the pathway where N_2O directly "harpoons" the lithium atom and implies that the electron and oxygen atom transfers occur nearly simultaneously. This lends a support for a conclusion drawn in Ref.[67] where a short-range electron transfer mechanism on a single PES has been surmised as a second pathway.

$\mathbf{TS}_1$ is placed energetically lower than $\mathbf{TS}_2$: 4.5 vs. 6.0 kcal/mol for Li and 8.8 vs. 22.1 kcal/mol for Cu. We suggest that if the experimentally observed departure from the Arrhenius plot is linked to an opening of a higher-energy channel, associated with a *trans*-$\mathbf{TS}_2$, then such energy differences may pinpoint the reason why the reaction with lithium exhibits a non-Arrhenius behavior starting from 600 K, whereas the reaction with copper shows a similar regime above 1190 K. In the latter case, the alternative rection pathway with the activation barrier of 22.1 kcal/mol (for the reaction enthalpy being 16.3 kcal/mol) is much less accessible and embarks on to interfere perceptibly with the lower-energy pathway at higher temperatures.

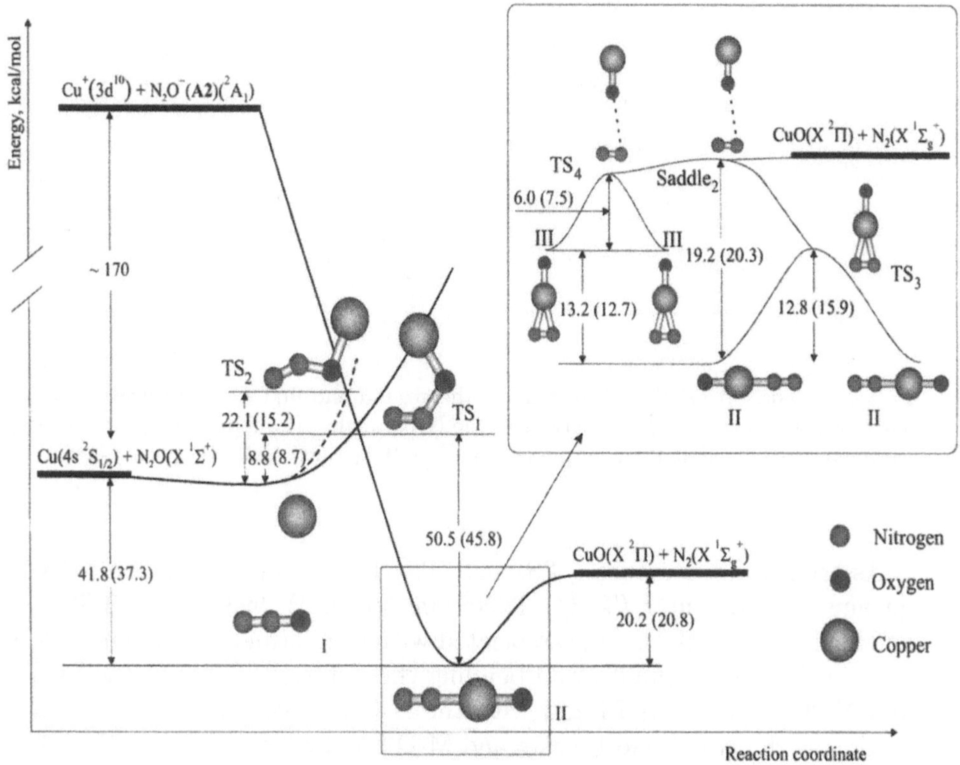

Figure 3. Potential energy profiles of the reaction (4) with M = Cu. The relative CCSD(T)/TZVP and B3LYP/TZVP (in parentheses) energies (in kcal/mol), taken at the B3LYP/TZVP geometries, are indicated [106].

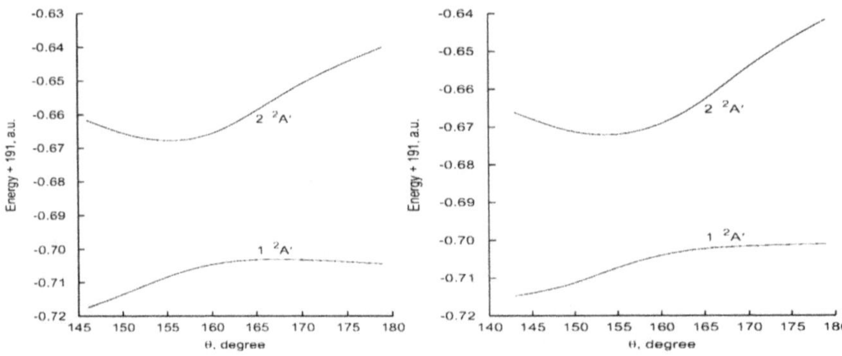

Figure 4. The CASSCF-MP2 sections of the ground and first excited PESs of the reaction (1) with M = Li as functions of the NNO angle. The remaining geometrical parameters are fixed at their optimized values in $\mathbf{TS}_1$ (left) and $\mathbf{TS}_2$ (right) [146].

As transition structures, $\mathbf{TS}_1$ and $\mathbf{TS}_2$ have a single imaginary frequency at 1481i cm^{-1} ($\mathbf{TS}_1^{Li}$), 2090i cm^{-1} ($\mathbf{TS}_2^{Li}$), 593i cm^{-1} ($\mathbf{TS}_1^{Cu}$), and 567i cm^{-1} ($\mathbf{TS}_2^{Cu}$). It is associated with the normal mode characterized by the predominant NNO bending character, mixed with the Li-O and N-O stretches for Li and with the N-O stretch for Cu, and functioning as the N-O bond-breaker and M-O bond-maker. It is worthwhile mentioning an existence of a third reaction pathway for M = Li which begins at the $^1A_1$ cyclic form of nitrous oxide and correlates with the T-shaped complex **III**. Due to the large electron affinity of 2.68 eV of the cyclic form [88], the electron transfer along this pathway is likely to occur non-adiabatically, resembling to a certain extent the "harpoon" mechanism. Since the cyclic form of N_2O lies, as estimated in earlier works, by 2.55 - 2.72 eV [121, 88] above its linear isomer, such pathway seems to be rather hardly accessible under conditions examined by Plane [67].

4.2 Product channels

The product (ionic) portion of PESs describing the reaction (4) comprises of two energy-minimum structures: one is the linear structure **II** residing in its global minimum, and the other, in case of M = Li, is a T-shaped structure **III** placed 6.0 kcal/mol higher. In **II**, MO is bound to N_2 on the metal side with r(M-N) = 2.202 Å ($\mathbf{II}^{Li}$) and 1.844 Å ($\mathbf{II}^{Cu}$). Therefore, the metal-nitrogen bond is appreciably stronger for M = Cu, that is also reflected by the frequency of the M-N stretch of 164 cm^{-1} ($\mathbf{II}^{Li}$) and 352 cm^{-1} ($\mathbf{II}^{Cu}$). There obviously exist two identical linear structures **II**, linked to each other via the T-shaped transition state $\mathbf{TS}_3$

which is placed 4.4 kcal/mol (**TS**$_3{}^{Li}$) and 12.8 kcal/mol (**TS**$_3{}^{Cu}$) above **II**.

The T-shaped complex **III**^{Li} mimics the structure of the ground-state NNO^- **A2** interacting with the metal cation. This is apparently a consequence of that the molecular orbitals of the lithium oxide are mainly built from those of the oxygen anion. Therein, the lithium oxide is bound to N_2 mainly by the oxygen atom, being separated from both nitrogens by a longer distance of 3.027 Å. If M = Cu, the corresponding T-shaped structure represents either the transition state **TS**$_4$ or the saddle point **Saddle**$_2$, depending on the N···O distance. The former links two identical T-shaped complexes **III**^{Cu}, where CuO is bonded to the molecular nitrogen mainly by the copper atom. The latter, **Saddle**$_2$, connects transition structures **TS**$_3{}^{Cu}$ and **TS**$_4{}^{Cu}$. These subtle patterns of the ionic part of PES of the reaction (4) with copper are displayed in insert of Figure 3. In fact, the reaction pathway, passing through the structures **II**^{Cu} → **TS**$_3{}^{Cu}$ → **TS**$_4{}^{Cu}$ → **III**^{Cu}, establishes another channel of the reaction (4) which might also contribute to its non-Arrhenius behavior. Notice that the structures on the ionic part of the PESs in both Li and Cu cases can be partitioned into two classes depending on whether the metal atom pervades into the weak O-N_2 bond in **A**$_2$ or it approaches O of **A**$_2$ perpendicular to N-N bond. After all, the studied reaction pathways barrierlessly exit to the product limits CuO (X $^2\Pi$) + N_2 (X $^1\Sigma_g^+$) or LiO(X $^2\Pi$) + N_2(X $^1\Sigma_g^+$).

We suggest that the existence of the linear complexes **II** is nicely confirmed by the earlier experimental photoelectron spectrum by Wright *et al.* [147] who studied the product channel of the reaction (4) with Li. These authors observed two broad bands with the corresponding vertical ionization energies of 8.8 ± 0.2 and 9.4 ± 0.1 eV, respectively. Based on the present results, it is reasonable to suggest that the observed spectrum is associated with ionization of a mixure containing **II** and separated LiO, rather than solely of the latter. Indeed, **II** is bound by 7.5 (CASSCF-MP2/6-311G(d)) and 6.3 kcal/mol (CASSCF-MP2/6-311+G(d)) with respect to the ground-state separated products. Hence, the weak onset of the lower-energy band at 7.6 eV [147] should likely be interpreted as the first adiabatic energy of **II**, which, due to the stabilization of the positive charge within this complex, is expected to be lower than that of LiO (8.44 - 9.0 eV) [148, 149, 150, 151, 152, 153, 154, 155]). This is affirmatively demonstrated by (U)MP2/6-311G(d) and (U)CCSD(T)/6-311G(d) calculations yielding the first adiabatic ionization energy for **II** of 7.6 eV perfectly matching the experimental value [147]. The equilibrium ground-state geometry of the cation **II**$^+$ (Figure 5) markedly differs from that of the neutral, exhibiting the Li···O bond

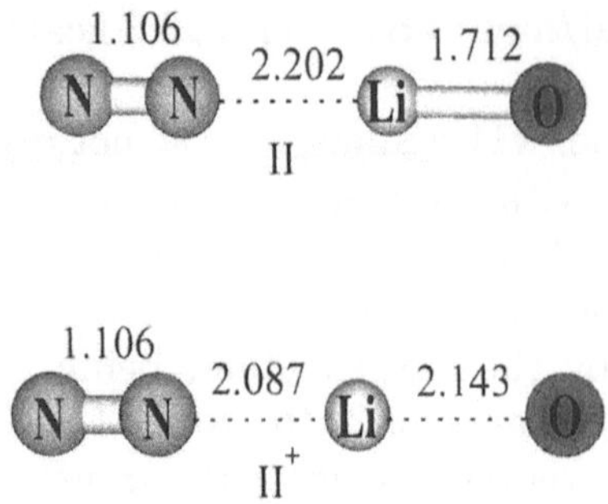

Figure 5. The CASSCF(11/12)/6-311G(d) equilibrium structures of the global-minimum complex **II** and its cation **II**$^+$. Bond lengths are given in Å.

length of 2.143 Å (vs. 1.712 Å in **II**) and the Li···N bond length of 2.087 Å (vs. 2.202 Å in **II**); the lithium atom is now closer to nitrogen. This causes significant changes in normal vibrations. In particular, while the vibration at 453 cm^{-1} in cation describes a migration of Li between O and N, the mode at 175 cm^{-1} corresponds to a N···O stretch. These features are likely relevant to a rather broad nature of the corresponding bands in experimental photoelectron spectrum.

5. Epilogue

A thorough theoretical investigation of the lower-lying PES for dissociative electron attachment of nitrous oxide and its reaction with lithium and copper atoms provides us with a clear view on the molecular mechanisms of such processes, whose key features can be summarized as follows:

Addition of an electron to the linear NNO ground state leads first to the open-chain bound anion lying 6.0 kcal/mol above the neutral, which is further transformed upon cyclization with the activation barrier of 6.0 kcal/mol to the T-shaped anionic complex **A2**. The latter is more stable than the open isomer and can be thus treated as the adiabatic anion, lying at most 250 cm^{-1} above the neutral ground state. This process of the dissociative electron attachment plays an important role in predetermining the mechanistic features for the reaction (4) which, as we have shown for Li and Cu, exhibits a rather similar landscape of the ground-state PES. Starting from the reactant complex **I**, which foreshadows the reactant channel of this reaction, excursions along the NNO in-plane bending mode ν_2 lead to two transition-state linkers $\mathbf{TS}_1$ and $\mathbf{TS}_2$, and hence, to the two-channel reaction pathway, that brings reactants to such molecular arrangement which promotes a charge transfer. Although the nature of both transition structures is rather similar, they differ from each other energetically. At lower temperatures, the former activation barrier plays a dominant role in determining the reaction kinetics. It starts to interplay appreciably with the higher-energy activation barrier at higher temperatures, where the latter channel mainly controls the curvature of the 1/T plot. The energy separation between two lower-lying PESs ($^2A'$), calculated for M = Li to be equal to $\approx$ 30 kcal/mol in both transition regions, allows us to rule out the dominance of a process by non-adiabatic electron transfer.

The lower-energy product portion of the PES of the reaction (4) is found to be far from an exhaustion due to the presence of solely N_2 and MO, but instead, it is populated by a variety of structures which can be considered as originating from a weakly bound T-shaped anion interacting with metal. In both lithium and copper cases, the linear structure **II** attains the global minimum of the total PES, being 7.5 and 20.2 kcal/mol, respectively, more stable than the corresponding product limits. For M = Li, the existence of such structure also allows us to explain the origin of the broad bands and relatively low ionization energies in the earlier experimental photoelectron spectrum [147].

Needless to say again that both Li and Cu atoms have *ns* unpaired electrons implying their doublet electronic ground states. The scenario for the reaction (4), involving alkaline earth metals (Mg, Ca, K, etc.) and transition metals, interacting in higher spin states, is expected to be different. Much still remains to be done to elucidate the mechanisms behind this chemically fundamental and socially pertinent reaction (4). Let us hope that in a foreseeable future, the nitrous oxide molecule could fully recover its primary status as "the laughing gas" ("no laughing matter" [8]), as a gas for all seasons!

Acknowledgments

The authors are indebted to the government of the Flemish Community of Belgium, the Fund for Scientific Research (FWO-Vlaanderen) and the KULeuven Research Council for continuing financial support. We also thank Chris Vinckier for valuable discussions and support.

Dedication

We dedicate this work to the unforgettable memory of Per-Olov Löwdin who [156], in 1957, introduced and defined Quantum Chemistry in terms of its goals, one of which, consisting in finding "the correct electronic interpretation of the ordinary chemical phenomena", has been particularly guided us in this study of the reaction (4).

References

[1] N. N. Greenwood and E. Earnshaw, *Chemistry of the Elements* (Pergamon, Oxford, 1984), p. 466.

[2] J. W. Mellor, *Comprehensive Treatise on Inorganic and Theoretical Chemistry* (Wiley, New York, 1922), Vol. VIII.

[3] A. V. Leont'ev, O. A. Fomicheva, M. V. Proskurnina and N. S. Zefirov, Russ. Chem. Rev. **70**, 91 (2001).

[4] B. Wang and X.-Y. Li, Anal. Chem. **70**, 2181 (1998).

[5] W. J. Grant, *Medical Gases: Their Properties and Uses* (HM+M Publishers, Aylesbury, 1978).

[6] E. Zeidifard, S. Godfrey and E. E. Davies, J. Appl. Physiol. **41**, 433 (1976).

[7] C. E. W. Hahn, A. M. S. Black, S. A. Barton and I. Scott, J. Appl. Physiol. **75**, 1863 (1993).

[8] A. N. Frumkin, O. A. Petrii and B. B. Damaskin, in *Comprehensive Treatise in Electrochemistry*, edited by J. O. Bockris, B. E. Conway and E. Yeager (Plenum Press, New York, 1980), Vol. 1.

[9] C. A. Simpler and R. C. Webster, in *Industrial Use of Nitrous Oxide - No Laughing Matter*, in *Nitrous Oxide*, edited by D. W. Eastwood (Davies, Philadelphia, 1964), Vol 1.

[10] H. H. H. W. Schmid, H. Hofmann, U. Schindler, Z. S. Shutenko, D. D. Cunningham and M. Feelisch, Proc. Natl. Acad. Sci. USA **93**, 14492 (1996).

[11] Z. Sojka and M. Che, J. Phys. Chem. **100**, 14776 (1996).

[12] G. Delahay, B. Coq, S. Kieger and B. Neveu, Cat. Today **54**, 431 (1999).

[13] J. Cunningham, J. J. Kelly and A. Penny, J. Phys. Chem. **74**, 1992 (1970).

[14] G. C. Chinchen, C. M. Hay, G. D. Vandervell and K. C. Waugh, J. Catal. **103**, 79 (1987).

[15] B. E. Turner, Symp Int. Astron. Union **150**, 181 (1992).

[16] L. M. Ziurys, A. J. Apponi, J. M. Hollis and L. E. Snyder, Astrophys. J. **436**, L181 (1994).

[17] W. J. Wilson and L. E. Snyder, Astrophys. J. **246**, 86 (1981).

[18] J. Desne, Le Monde (France), 14 February (2002).

[19] Y. L. Yung, W. C. Wang and A. A. Lacis, Geophys. Res. Lett. **3**, 619 (1976).

[20] O. Badr and S. D. Probert, Appl. Energy **44**, 197 (1993).

[21] H. B. Bange, Nature **408**, 301 (2000).

[22] R. Prinn, D. Cunnold, R. Rasmussen, P. Simmonis, F. Alyea, A. Crawford, P. Frase and R. Rosen, J. Geophys. Res. Atm. **95**, 18369 (1990) and references therein.

[23] M. Prather, in *Climate Change 1995: The Science of Climate Change, Contribution of Working Group I to the Second Assessment of the Intergovernmental Panel on Climate Change*, edited by J. T. Houghton (Cambridge University Press, Cambridge 1996), pp. 86 - 103.

[24] W. B. De More, S. P. Sander, C. J. Howard, A. R. Ravishankara, D. M. Golden, C. E. Kolb, R. F. Hampson and M. J. Molina, *Chemical Kinetics and Photochemical Data for Use in Stratospheric Modelling, Evaluation Number 12* (JPL Publication 97-4, Jet Propulsion Laboratory, Caltech, Pasadena, 1997).

[25] H. Okabe, *Photochemistry of Small Molecules* (Wiley, New York, 1978), Ch. 6.

[26] C. A. Cantrell, R. E. Shetter and J. G. Calvert, J. Geophys. Res. **99**, 3739 (1994).

[27] K. Minschwander, R. J. Salawitch and M.B. McElroy, J. Geophys. Res. **98(D6)**, 10543 (1993).

[28] D. W. Fahey, S. Solomon, S. R. Kawa, M. Loewenstein, J. R. Podolske, S. E. Strahan and K. R. Chan, Nature **345**, 698 (1990).

[29] P. J. Cruzen, Q. J. Roy. Meteorol. Soc. **96**, 320 (1970); J. Geophys. Res. **76**, 7311 (1971).

[30] H. S. Johnston, Science **173**, 517 (1971).

[31] R. P. Wayne, *Chemistry of Atmospheres* (Clarendon Press, Oxford, 1991), pp. 116 - 150.

[32] W. Seiler and R. Conrad, J. Air Pollut. Control **31**, 767 (1981).

[33] T. E. Graedel and P. J. Crutzen, *Atmospheric Change: an Earth System Perspective* (Freeman, New York, 1993).

[34] R. P. Wayne, *Chemistry of Atmospheres* (Clarendon, Oxford, 1996).

[35] P. J. Crutzen, Annu. Rev. Earth Planet Sci. **7**, 443 (1979).

[36] H. Rodhe, Science **248**, 1217 (1990).

[37] G. Centi, S. Perathoner and F. Vanazza, CHEMTECH **12**, 48 (1999).

[38] K.-R. Kim and H. Craig, Science **262**, 1855 (1993).

[39] J. C. Johnston, S. S. Cliff and M. N. Thiemens, J. Geophys. Res. **100(D8)**, 16801 (1995).

[40] Y. L. Yung and C. E. Miller, Science **278**, 1778 (1997).

[41] M. S. Johnson, G. D. Billing, A. Gruodis and M. H. Janssen, J. Phys. Chem. A **105**, 8672 (2001).

[42] J. T. Houghton, *Climate Change 1996: The Science of Climate Change* (Cambridge University Press, Cambridge, 1996).

[43] P. B. McMahon, B. W. Bruce, M. F. Becker, L. M. Pope and K. F. Dennehy, Environ. Sci. Technol. **34**, 4873 (2000).

[44] M. Leuenberger and U. Siegenthaler, Nature **360**, 449 (1992).

[45] W. C. Trogler, J. Chem. Ed. **72**, 973 (1995).

[46] *The Draft 1998 Inventory of U.S. Greenhouse Gas Emissions and Sinks (1990-1996)* (U.S. Environmental Protection Agency 1998).

[47] P. Czepiel, P. Crill and R. Harriss, Environ. Sci. Technol. **29**, 2352 (1995).

[48] J. J. Cole and N. F. Caraco, Environ. Sci. Technol. **35**, 991 (2001).

[49] G. Centi, P. Gerenali, L. dall'Olio, S. Perathoner and Z. Rak, Ind. Eng. Chem. Res. **39**, 131 (2000).

[50] S. Poh, R. Hernandez, M. Inagaki and P. G. Jessop, Org. Lett. **1**, 583 (1999).

[51] I. Lein and P. Potzinger, Organomet. **19**, 4701 (2000).

[52] A. Fernendez and A. Fontijn, Int. J. Chem. Kin. **33**, 387 (2001).

[53] G. Leroy, M. Sana, L. A. Burke and M. T. Nguyen, in *Quantum Theory of Chemical Reactions*, edited by R. Daudel, A. Pullman, L. Salem and A. Veillard (Reidel, Dordrecht, 1979), Vol. 1, p. 91.

[54] A. Padwa, *1,3-Dipolar Cycloaddition Chemistry* (Wiley, New York, 1984).

[55] T. L. Nguyen, F. De Proft, A. K. Chandra, T. Uchimaru, M. T. Nguyen and P. Geerlings, J. Org. Chem. **66**, 6096 (2001).

[56] R. A. Perry and J. A. Miller, Int. J. Chem. Kinet. **28**, 217 (1996).

[57] F. Brüning, S. Matejcik, E. Illenberger, Y. Chu, G. Senn, D. Muigg, G. Denifl and T. D. Märk, Chem. Phys. Lett. **292**, 177 (1998).

[58] K. Narita, N. Takezawa, H. Kobayashi and I. Toyoshima, React. Kinet. Catal. Lett. **19**, 91 (1982).

[59] J. W. Evans, M. S. Wainwright, A. J. Bridgewater and D. Young, J. Appl. Catal. A **75**, 7 (1983).

[60] Y. J. Li and J. N. Armor, Appl. Catal. B - Environm. **1**, L21 (1992).

[61] F. Kaptijn, J. Rodrigues-Mirasol and J. A. Moulijn, Catal. B - Environm. **9**, 25 (1996).

[62] L. Chen, H. Y. Chen, J. Lin and K. L. Tan, Surf. Interf. Anal. **28**, 115 (1999).

[63] S. Sato, R. Takahashi, T. Sodesawa, K. Yuma and Y. Obata, J. Catal. **196**, 195 (2000).

[64] G. Ceballos, H. Wende, K. Baberschke and D. Arvanitis, Surf. Sci. **482**, 15 (2001).

[65] W. F. Schneider, K. C. Hass, R. Ramprasad and J. B. Adams, J. Phys. Chem. B **102**, 3692 (1998).

[67] J. M. C. Plane, J. Phys. Chem. **91**, 6552 (1987).

[68] D. R. Yarkony, J. Chem. Phys. **78**, 6763 (1983).

[69] B. Kiran, C. Vinckier and M. T. Nguyen, Chem. Phys. Lett. **344**, 312 (2001).

[70] A. Stirling, J. Phys. Chem. **102**, 6565 (1998).

[71] A. Stirling, J. Am. Chem. Soc. **124**, 4059 (2002).

[72] K. P. Huber and G. Herzberg, *Molecular Spectra and Molecular Structure. IV. Constants of Diatomic Molecules* (Van Nostrand, New York, 1989).

[73] *Handbook of Physics and Chemistry*, 65th ed., edited by R.C. Weast (CRC Press, Boca Raton, FL, 1985).

[74] M. W. Chase, C. H. Davis, J. R. Downey and D. Y. Frurip, J. Phys. Chem. Ref. Data **14**, Suppl No. 1 (1985).

[75] A. S. Narayan, P. M. Futerko and A. Fontijn, J. Phys. Chem. **96**, 290 (1992).

[76] C. Vinckier, T. Verhaeghe and I. Vanhees, J. Chem. Soc. Faraday. Trans. **90**, 2003 (1994).

[77] J. M. C. Plane, in *Gas Phase Metal Reactions*, edited by A. Fontijn (Elsevier, Amsterdam 1992), p. 39.

[78] D. Ritter and J. C. Weisshaar, J. Phys. Chem. **93**, 1576 (1989).

[79] D. Ritter and J. C. Weisshaar, J. Phys. Chem. **94**, 4907 (1990).

[80] M. L. Campbell and R. E. McClean, J. Phys. Chem. **97**, 7942 (1993).

[81] M. L. Campbell and R. E. McClean, J. Chem. Soc. Faraday. Trans. **91**, 3787 (1995).

[82] M. L. Campbell, J. Chem. Phys. **104**, 7515 (1996).

[83] M. L. Campbell and J. R. Metzger, Chem. Phys. Lett. **253**, 158 (1996).

[84] M. L. Campbell, E. J. Kölsch and K. L. Hooper, J. Phys. Chem. A **104**, 11147 (2000).

[85] D. R. Herschbach, Adv. Chem. Phys. **10**, 319 (1966).

[86] C. E. Moore, *Ionization Potentials and Ionization Limits Derived from the Analyses of Optical Spectra* (NIST, Washington, 1971).

[87] A. Delabie, C. Vinckier, M. Flock and K. Pierloot, J. Phys. Chem. A **105**, 5479 (2001).

[88] E. S. Kryachko, C. Vinckier and M. T. Nguyen, J. Chem. Phys. **114**, 7911 (2001).

[89] P. M. Futerko and A. Fontijn, J. Phys. Chem. **95**, 8065 (1991).

[90] P. M. Futerko and A. Fontijn, J. Phys. Chem. **97**, 3861 (1992).

[91] P. M. Futerko and A. Fontijn, J. Phys. Chem. **98**, 7004 (1993).

[92] D. P. Belyung, P. M. Futerko and A. Fontijn, J. Phys. Chem. **102**, 155 (1995).

[93] A. S. Blue, D. P. Belyung and A. Fontijn, J. Phys. Chem. **107**, 3791 (1997).

[94] A. Fontijn and P. M. Futerko, in *Gas Phase Metal Reactions*, edited by A. Fontijn (Elsevier, Amsterdam 1992), p. 93.

[95] R. Mecke, Z. Phys. **64**, 173 (1930).

[96] P. C. Mahanti, Phys. Z. **34**, 108 (1931).

[97] L. Pauling and S. B. Hendricks, J. Am. Chem. Soc. **48**, 641 (1926).

[98] E. K. Plyler, Elisha Mitchell Sci. Soc. **48**, 33 (1932).

[99] L. Pauling, Proc. Natl. Acad. Sci. USA **18**, 498 (1932).

[100] R. Wierl, Ann. Phys. **8**, 521 (1931); Phys. Z. **31**, 1028 (1930).

[101] L. R. Maxwell, V. M. Mosley and L. S. Deming, J Chem. Phys. **2**, 331 (1934).

[102] J. L. Griggs, K. N. Rao, L. H. Jones and R. M. Potter, J. Mol. Spectrosc. **25**, 24 (1968).

[103] J. H. Callomon and F. Creutzberg, Phil. Trans. **277A**, 157 (1974).

[104] M. V. Chase, C. A. Davies, J. Downey, Jr., D. J. Frurip, R. A. McDonald and A. N. Syverud, J. Phys. Chem. Ref. Data, Suppl. 1 **4**, 1554 (1985).

[105] F. A. Cotton and G. Wilkinson, *Advanced Inorganic Chemistry* 5th ed. (Wiley, New York, 1998), p. 321.

[106] E. S. Kryachko, O. Tishchenko and M. T. Nguyen, Int. J. Quantum Chem. **00**, 000 (2002).

[107] J.-L. Teffo and A. Chédin, J. Mol. Spectrosc. **135**, 389 (1989).

[108] *CRC-Handbook Book of Chemistry and Physics*, 74th ed., edited by D. R. Lide (Chemical Rubber, Boca Raton, FL, 1993).

[109] D.-Y. Hwang and A. M. Mebel, Chem. Phys. **259**, 89 (2000).

[110] M.-J. Crawford and T. M. Klapötke, Inorg. Chem. **38**, 3009 (1999).

[111] B. S. Jursic, Int. J. Quantum Chem. **58**, 41 (1996).

[112] F. Wang and R. D. Harcourt, J. Phys. Chem. A **104**, 1304 (2000).

[113] F. Wang, F. P. Larkins, M. J. Brunger, M. T. Michalewicz and D. A. Winkler, Spectrochim. Acta, Part A **57**, 9 (2001).

[114] W. D. Allen, Y. Yamaguchi, A. G. Csaszar, D. A. Clabo, Jr., R. B. Remington and H. F. Schaefer III, Chem. Phys. **145**, 427 (1990).

[115] J. M. L. Martin, P. R. Taylor and T. J. Lee, Chem. Phys. Lett. **205**, 535 (1993).

[116] M. C. McCarthy, J. W. R. Allington and K. O. Sullivan, Mol. Phys. **96**, 1735 (1999).

[117] A. T. Wong and G. B. Bacskay, Chem. Phys. Lett. **207**, 360 (1993).

[118] G. S. Tschumper and H. F. Schaefer III, J. Chem. Phys. **107**, 2529 (1997).

[119] K. Mogi, T. Komine and K. Hirao, J. Chem. Phys. **95**, 8999 (1991).

[120] D. Bellert, T. Buthelezei, V. Lewis, K. Dezfulian and P. J. Brucat, Chem. Phys. Lett. **240**, 495 (1995).

[121] D. G. Hopper, J. Chem. Phys. **80**, 4290 (1984).

[122] J. W. Rabalais, J. M. McDonald, V. Scherr and S. P. McGlynn, Chem. Rev. **71**, 73 (1971).

[123] D. A. L. Kilcoyne, S. Nordholm and N. S. Hush, Chem. Phys. **107**, 225 (1986).

[124] K. Jug, J. Am. Chem. Soc. **100**, 6581 (1978).

[125] D. G. Hopper, A. C. Wahl, R. L. C. Wu and T. O. Tiernan, J. Chem. Phys. **65**, 5474 (1976).

[126] G. Herzberg, *Molecular Spectra and Molecular Structure. II. Infrared and Raman Spectra of Polyatomic Molecules* (Van Nostrand Reinhold, New York, 1945).

[127] E. I. Lobodenko, V. I. Perevalov, O.M. Lyulin and J.-L. Teffo, J. Mol. Spectrosc. **205**, 239 (2001).

[128] M. Kobayashi and I. Suzuki, J. Mol. Spectrosc. **125**, 24 (1987).

[129] M. Lacy and D. H. Whiffen, Mol. Phys. **45**, 241 (1982).

[130] I. Suzuki, J. Mol. Spectrosc. **25**, 479 (1968).

[131] G. Yan, H. Xian and D. Xie, Chem. Phys. Lett. **271**, 157 (1997).

[132] J. Gauss, J. Chem. Phys. **116**, 4773 (2002).

[133] S. D. Peyerimhoff and R. J. Buenker, J. Chem. Phys. **49**, 2473 (1968).

[134] A. Chutjian and G. A. Segal, J. Chem. Phys. **57**, 3069 (1972).

[135] A. Brown, P. Himeno and G. G. Balint-Kurti, J. Phys. Chem. A **103**, 11089 (1999).

[136] T. M. Kläpotke and A. Schulz, Chem. Ber. **128**, 201 (1995).

[137] T. M. Kläpotke, A. Schulz and R. D. Harcourt, *Quantum Mechanical Method in Main Group Chemistry* (Wiley, New York, 1998).

[138] J. M. Galbraith and H. F. Schaefer III, J. Am. Chem. Soc. **118**, 4860 (1996).

[139] C. v. Wüllen and W. Kutzelnigg, J. Chem. Phys. **104**, 2330 (1996).

[140] W. E. Wentworth, E. Chen and R. Freeman, J. Chem. Phys. **55**, 2075 (1971).

[141] S. J. Nalley, R. N. Compton, H. C. Schweinler and V. E. Andersson, J. Chem. Phys. **59**, 4125 (1973).

[142] *CRC-Handbook of Chemistry and Physics* (Web Edition, 1999), pp. 10-150.

[143] T. O. Tiernan and R. L. C. Wu, Adv. Mass. Spectrom. **7A**, 136 (1978).

[144] J. V. Coe, J. T. Snodgrass, C. B. Freidhoff, K. M. McHugh and K. H. Bowen, Chem. Phys. Lett. **124**, 274 (1986).

[145] D. Yu, A. Rauk and D. A. Armstrong, J. Phys. Chem. **96**, 6013 (1992).

[146] O. Tishchenko, E. S. Kryachko, C. Vinckier and M. T. Nguyen, Chem. Phys. Lett. (submitted).

[147] T. G. Wright, A. M. Ellis and J. M. Dyke, J. Chem. Phys. **98**, 2891 (1993).

[148] J. Berkowitz, W. A. Chupka, G. D. Blue and J. L Margrave, J. Phys. Chem. **63**, 644 (1959).

[149] D. White, K. S. Seshadri, D. F. Dever, D. E. Mann and M. J. Linevski, J. Chem. Phys. **39**, 2463 (1963).

[150] K. F. Zmbov, P. Ficarola and J. L. Margrave, J. Inorg. Nucl. Chem. **30**, 2059 (1968).

[151] R. Yamdagni, C. Pupp and R. F. Porter, J. Inorg. Nucl. Chem. **32**, 3509 (1970).

[152] D. L. Hildenbrand, J. Chem. Phys. **57**, 4556 (1972).

[153] E. N. Verkhoturov, PhD Thesis, Moscow State University, 1977, reported in: M. F. Butman, L. S. Kudin and K. S. Krasnov, Russ. J. Inorg. Chem. **29**, 1228 (1984).

[154] C. H. Wu, H. Kudo and H. R. Ihle, J. Chem. Phys. **70**, 1815 (1979).

[155] H. Nakagawa, M. Asano and K. Kubo, J. Nucl. Mater. **102**, 292 (1981).

[156] P.-O. Löwdin, J. Phys. Chem. **61**, 55 (1957).

HEAVY ELEMENT QUANTUM CHEMISTRY—THE MULTICONFIGURATIONAL APPROACH

Björn O. Roos and Per-Åke Malmqvist
Department of Theoretical Chemistry, Chemical Center, P.O.Box 124, S-221 00 LUND, Sweden

Laura Gagliardi
Dipartimento di Chimica G. Ciamician, Università di Bologna, Via F. Selmi 2, 40126, Bologna, Italy

Results from some illustrative calculations performed with the quantum chemistry software MOLCAS-5 are presented. They all involve compounds that contain an actinide ion. The CUO molecule is used to illustrate Löwdin's natural orbital concept and also to show the difficulty of determining the correct ground state for actinide compounds with covalent bonds. Calculations of the ionization energies of ThF_4 are reported, where the ionized states are first computed in a localized approach, which is then symmetry adapted using the non-orthogonal CI approach of the MOLCAS software (CASSCF state interaction). This calculation also yields the spin-orbit splitting of the ionized states. Finally we describe the coordination between uranyl and neptunyl ions with water and carbonate ions. All these results have been obtained using multiconfigurational SCF methods (CASSCF) with dynamic correlation added using second order perturbation theory (CASPT2).

1 INTRODUCTION

This paper is written in honor of Professor Per-Olof Löwdin. We shall give some examples of how the Complete Active Space (CAS) SCF approach can be used to study compounds formed with actinide atoms, in particular uranium. It may not be generally known that a prelude to modern multiconfigurational SCF theory, in particular CASSCF, was given by Löwdin already in the famous 1955 Physical Review article [1]. In this article he defines the *Extended Hartree—Fock Equations*. A limited number of MOs are used to build a CI wave function and the condition for optimal orbitals is written down. This condition is nothing but what has later been known as the extended Brillouin's theorem. Thus, the basic formulation was already there in 1955. It was to take another 20 years until this was fully utilized in effective and general implementations.

Quantum Chemistry for compounds containing atoms from the lower part of the periodic system faces problems that we are not used to for lighter elements. Relativistic effects have to be included already at the stage where the wave

E.J. Brändas and E.S. Kryachko (eds.), Fundamental World of Quantum Chemistry, Vol. II, 425–441.

function is determined. Large effects of spin-orbit coupling can be expected for some elements (in particular the main group atoms). The basis sets have to be adapted to the relativistic treatment.

Atomic levels are densely spaced and it is not always clear what the valence state will be. For example: the uranium atom has the ground state electronic configuration $(5f)^3(6d)^1(7s)^2$. This is not an obvious choice and other electronic configurations are very close in energy. In most compounds uranium will be highly oxidized, normally as U(IV) or U(VI). The latter ion has lost all its valence electrons and is therefore reasonably simple to treat. U(IV) has two valence electrons and it is not obvious which they will be. Actually in a molecule like OUO it is $(5f)(7s)$ [2], while in UF_4 it is $(5f)^2$ (see below). It is clear that a flexible wave function approach is needed to deal with all these possibilities.

The MOLCAS quantum chemistry software is presently being extended to be able to deal with all atoms of the periodic system at the multiconfigurational level [3]. A fully relativistic treatment has not been implemented. Instead the new features are based on the unrelativistic treatment. Here we shall illustrate the use of the CASSCF method [4] with dynamic correlation added using second order perturbation theory (CASPT2) [5, 6]. Scalar relativistic effects are included either by perturbation theory at the CASSCF level, adding the mass-velocity and Darwin terms to the one-electron Hamiltonian, or by using the Douglas-Kroll (DK) Hamiltonian [7, 8]. In the latter case special relativistic AO basis sets have to be used, which have been contracted using the DK Hamiltonian. Such basis sets are presently under development at this department [9]. An alternative is to use effective core potentials and a variety of such basis sets are available in MOLCAS.

Spin orbit coupling effects are included a posteriori at the CASSCF level using the CASSCF State Interaction method [10], which has recently been extended to include such effects [11]. A set of CASSCF calculations are first performed for the electronic states that are assumed to couple under the influence of the spin-orbit Hamiltonian. The CASSI program is then used to set up and diagonalize the interaction matrix. Eventually, dynamic correlation effects may be included by shifting the diagonal elements of this Hamiltonian to energies obtained, for example, in a CASPT2 calculation. Experience shows that this method gives accurate results for almost all heavy atoms, including actinides. The largest errors are found for the heaviest main group elements.

We shall below present some examples of calculations performed with this software. They will all involve actinide compounds. The results of the study of the CUO and ThF_4 molecules in the gas phase will be reported. Some actinide carbonate complexes in water will then be discussed.

2 THE CUO MOLECULE

As a first illustration of a chemical bond involving an actinide we choose the simple triatomic CUO. This, and similar molecules can be created in a matrix and their IR spectrum have recently been recorded [12]. These data may be used to test the theoretical models used for actinide compounds. A number of molecules with the general formula XUY (X,Y=C,N,O) were recently studied using the CASSCF approach with dynamic correlation effects calculated at second order in perturbation theory with a CASSCF reference function (CASPT2) [13]. The computed IR frequencies were found to be in agreement with experiment and it was concluded that the approach is valid for these type of compounds.

Here we shall not report these results in detail but only use the CUO molecule to study in some detail the electronic structure. For the analysis we shall use Löwdins natural orbitals. CUO is a closed shell molecule, where uranium is in the oxidation state VI. Formally we can write it as $C^{4-}U^{6+}O^{2-}$, which of course is far from the real charge distribution but helps to understand the binding. The $2p$-orbitals form bonds with hybrid orbitals of uranium formed from the $5f$, $6d$, $7s$, and $7p$ orbitals. In principle we have six bonding orbitals, two of σ- and four of π-type. These bonds can, however, be expected to be rather weak and we anticipate rather large occupations also of the antibonding orbitals.The strength of a chemical bond depends on the occupation of both the bonding and the antibonding orbital. The bonding orbital has an occupation number close to two in a fully developed bond, while both orbitals have occupation number one when the bond is broken.

It is then quite clear how a CASSCF calculation on a the CUO molecules shall be performed. One needs in all 12 active orbitals, six bonding and six antibonding, with 12 active electrons. Such a calculation was carried out at the computed equilibrium geometry: r(CU)= 1.773 Å, r(OU)=1.808 Å. [13]. The Stuttgart relativistic effective core potential (ECP) and the corresponding valence basis set was used for the uranium atom including 32 valence electrons (small core) [14]. The ANO-L basis set of the MOLCAS library was used for C and O contracted to $4s3p2d$. It was shown in our earlier study that these basis sets are not large enough to reproduce the vibrational frequencies accurately, but they will do for our present purpose of illustrating the chemical bond.

The resulting natural orbitals are shown in Fig. 1. They have been paired

Figure 1: The natural orbitals in CUO. From left to right and starting with the lowest row, the orbitals (occupation numbers) are: $\sigma_{OU}(1.97)$, $\sigma^*_{OU}(0.02)$, $\sigma_{CU}(1.96)$, $\sigma^*_{CU}(0.06)$, $\pi_{OU}(1.97)$, $\pi^*_{OU}(0.03)$, $\pi_{CU}(1.94)$, $\pi^*_{CU}(0.06)$. Only one component of the degenerate π-orbitals is shown.

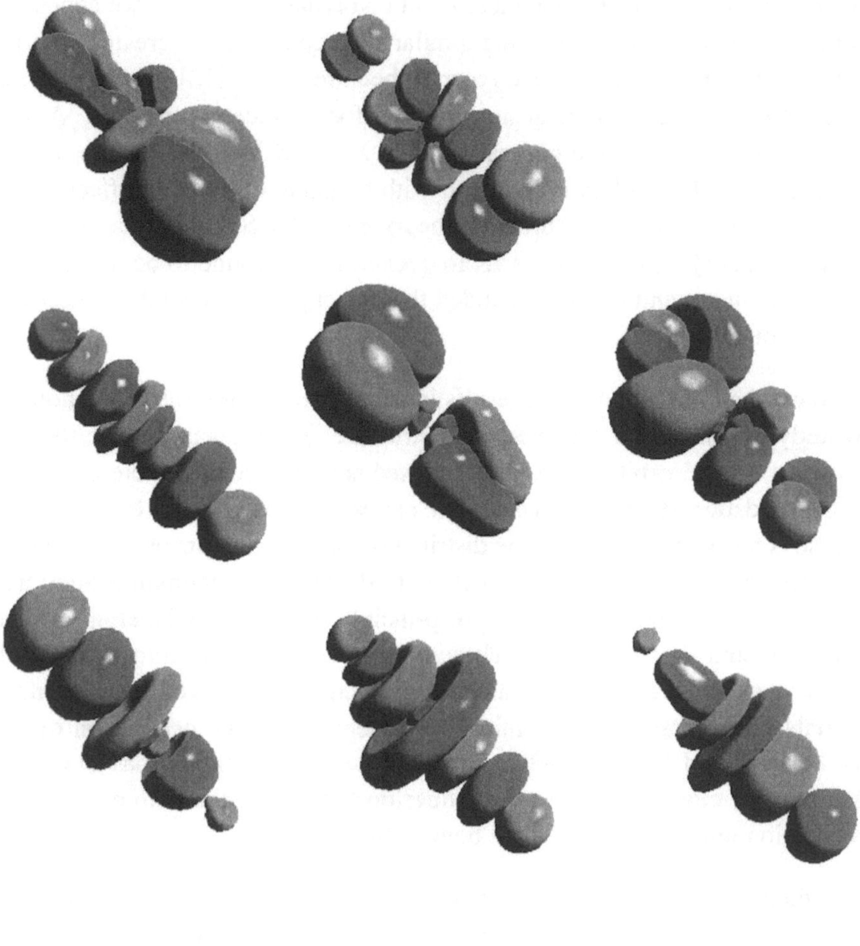

such that each bonding orbital is followed by the antibonding counterpart. The orbitals are in the figure ordered from left to right, starting with the bottom row. The first orbital gives mainly the σ-bond between U and O. It is composed of oxygen $2p$ plus a hybrid of $7s$-, $6p$-, $6d$-, and $5f$-orbitals on U with a slight dominance of $6d\sigma$. Notice that we have written $6p$ and not $7p$. The total population of this shell in the molecule is 5.66, showing that these orbitals should be considered as valence and not belonging to the core. The occupation number for σ_{OU} is 1.97. The corresponding antibonding orbital, σ^*_{OU}, has an occupation number of 0.02. The next pair of orbitals gives the CU σ-bond. The bonding orbital is somewhat less occupied (1.96) electron and the contribution from uranium is here dominated by $5f\sigma$. The next pair gives one of the OU π-bonds. Again, the contribution from uranium is dominated by $6d\pi$, while the CU π-bond contains mostly $5f\pi$. This is clearly illustrated in the picture of the orbitals.

The total occupation number for the antibonding orbitals is 0.27, which gives a total bond order of 5.73 (we define the bond order for a bond i as $(\eta_i - \eta_i^*)/2$, where η_i and η_i^* are the occupation numbers of the bonding and antibonding orbital, respectively.

We thus see two triple bonds formed. They are not very ionic. The total charge on uranium is +1.57, which is quite far from the formal charge 6. The negative charge is equally distributed between oxygen and carbon. The population on uranium is: $7s$, 0.12; $6p$, 5.66; $6d$, 1.79, $5f$, 2.85. It is clear that uranium uses all its valence shells to form the chemical bonds.

The above bonding situation is typical for U(VI) compounds of the structure XUY. An important example is the uranyl ion OUO^{2+}. If we add one electron to the system, as for example in NUO, it will go into the $5f$-shell and the ground state will be $^2\Phi$ with the odd electron in a non-bonding orbital. The structure of the bonding is unaltered. The situation becomes more complicated if we add one more electron as for example in OUO. An intuitive solution would be to occupy a second $5f$-orbital with the electronic configuration ^3H_g. However, the system prefers the electronic configuration $(5f)(7s)$. A recent study actually found five states with this configuration below $^3H_g(\Omega = 4)$ at 0.5 eV above the ground state $^3\Phi_u(\Omega = 2)$ [2]. We refer to this article and [13] for a more detailed discussion. The iso-electronic NpO_2^+ species on the other hand has a ^3H_g ground state, and this is due to the stabilization of the 5f orbitals along the actinide series.

So, we think we understand the CUO molecule. However, recently Li *et al.* reported on the complexation of CUO molecule with Ar, Kr, and Xe in a noble gas matrix [15]. They showed that the frequencies for CU and OU stretch changes from 1047 and 872 in an Ne matrix to 852 and 804 in an Ar matrix. The same shift is obtained with heavier noble gas atoms and also when 1% Ar is added to an Ne matrix. These large shifts cannot be explained as an increased matrix effect or by complexation of noble gas atoms to the CUO moiety. By performing DFT calculations on the Ar-CUO complex they find that the ground state changes from $^1\Sigma^+$ to $^3\Phi$. Even without the Ar atom they find this state only 0.03 eV above the ground state. This must clearly be wrong, because if spin-orbit effects are added it would make $^3\Phi(\Omega = 4)$ the ground state. Still, it seems clear that complexation will change the ground state from a singlet to a triplet.

In order to check whether this result holds true also at the *ab initio* level of theory we have computed the properties of the $^3\Phi$ state. A first set of calculations used an the Stuttgart ECP basis set for uranium [14] combined with the MOLCAS basis set ANO-L,4s3p2d for C and O. The active space described above was used with the two $5f\phi$-orbitals added (12 electrons in 14 orbitals). The geometries for the ground state and the excited states were optimized at the CASPT2 level.

The resulting bond distances were 1.77 (CU), 1.81 (OU) Å for the ground state and 1.89 (CU), 1.84 (OU) Å for the excited state. The excited electron comes mainly from the CU bond. The adiabatic excitation energy was at this level of theory 0.11 eV. A calculation of the vibrational frequencies gave 858 and 796 cm^{-1} for the excited state to be compared to the experimental values 853 and 804 cm^{-1} for CUO in the Ar matrix [15]. The nature of the electronic state thus seems clear. It might be added that the frequencies computed by Li *et al.* using DFT are 902 and 843 cm^{-1}, thus giving a less clear indication of the nature of the electronic state [15]. For the ground state we have earlier reported the frequencies: 1077 and 870 cm^{-1}, respectively. The DFT results are 1049 and 874 cm^{-1}. These numbers may be compared to the experimental results obtained in a Ne matrix: 1047 and 872 cm^{-1}. The DFT results are for this closed shell structure much closer to experiment.

In order to add spin-orbit coupling to these result, the calculations were redone using an all-electron basis set for U. A newly developed relativistic basis set of the ANO type was used with the contraction: 9s8p7d5f2g [9]. The adiabatic excitation energy was now found to be 0.03 eV in complete agreement with the DFT result [15]. The zero-field splitting of the $^3\Phi$ state is, however, computed to be 0.35 eV. Thus, the $^3\Phi(\Omega = 4)$ component is stabilized this amount compared to the spin-free calculation and becomes the ground state. The vibrational spectrum discussed above shows that this is unlikely to be the case for the free molecule. Thus, both DFT and the wave function based approach predict the wrong ground state for CUO. At present we do not have any explanation for this result, but will continue to investigate it.

3 ON THE IONIZATION ENERGIES OF ThF_4

One problem in actinide chemistry is to find reliable experimental results, which can be used to calibrate the theoretical methods. The research group in Southampton under the leadership of Professor John Dyke has for some time been measuring ionization energies of actinide compounds in the gas phase. These are excellent quantities for testing purposes, because they are sensitive measures of the electronic structure. In 1981 the spectra for a number of actinide halides were measured [16] and preliminary assignments were made of the different bands. A reassignment was made in 1988 with the aid of relativistic Hartree-Fock-Slater calculations [17]. The agreement with experiment was, however, not impressive. An upward shift of all IPs of about 2 eV was made in order to use the data for assignment of the spectrum.

We shall address this problem using the multiconfigurational approach. The example we shall use is the ThF_4 molecule, which is a closed shell system with Th in the oxidation state IV. The first set of calculations were performed using and ECP basis set for Th with a small core (30 electrons explicitly treated) and the valence basis set 9s8p7d5f [14]. For F we used the ANO-S basis set of MOLCAS contracted to 4s3p2d. No spin-orbit effects were included at this level of theory. First the ground state geometry was optimized at the SCF/MP2 level of theory. This gives a bond distance of 2.127 Å to be compared to the experimental value 2.14 [18].

This geometry was used to compute the vertical IPs. These calculations were made without symmetry (the symmetry of the molecule is T_d) such that degenerate states are treated equivalently. The active orbitals comprised the 12 $2p$-orbitals of the fluorine atoms with 23 active electrons. One state average CASSCF calculation was performed for all the 12 ionized states followed by a multi-state (MS) CASPT2 calculation [19] where all valence electrons are correlated.

The results of this first calculation is presented in Table 1. We notice that the computed IPs at the MS-CASPT2 level are too low and that the deviation from experiment increases with the value of the IP from 0.8 eV to 1.71 eV. This seems to indicate that the relaxation effects are not well described at the CASSCF level, which in turn is over-compensated at the CASPT2 level of theory. The IPs obtained at the CASSCF level are all too high, while one would

Table 1: Ionization enrgies in ThF_4 computed with an ECP basis set (in eV). Delocalized (deloc.) and localized(loc.) orbitals. No spin orbit coupling.

Orbital	CAS(deloc.)	PT2(deloc.)	CAS(loc.)	PT2(loc.)	Exp.[a]
$4t_2$	15.36	12.72	12.50	13.97	13.56[b]
$1t_1$	16.43	13.58	13.13	14.60	14.52
$1e$	16.87	13.99	13.59	15.06	15.43
$3a_1$	17.01	14.07	13.65	15.12	16.11
$3t_2$	17.22	14.40	13.95	15.42	16.11

[a]from Ref. [16].

[b]Average over two spin-components.

expect them to be too low if relaxation was properly described. If the interaction between the four fluorine atoms is weak one might expect the ionization to be localized and it would be difficult to describe the relaxation effects in a delocalized picture. Is it possible to perform the calculation in a localized model?

Yes, it is. The CASSCF State interaction code (CASSI) in MOLCAS [10] can be used for this purpose. The following procedure is applied. First, a state average CASSCF/CASPT2 calculation is performed for the three local ionizations on each fluorine atom. An active space of three orbitals is used with five active electrons (F2p). Secondly, the 12 CASSCF wave functions (three for each F atom) are allowed to interact using the CASSI method. The result is twelve symmetry adapted wave functions. Note that the basic CASSCF wave functions are non-orthogonal as are the four different orbital sets. Thus, a 12*12 non-orthogonal CI is performed to obtain the final result. The localized CASPT2 correlation energy is finally added to obtain the final ionization energies.

A small trick had to be used in order to make the orbitals localize. A small positive charge was placed outside one fluorine atom. The calculation then directly localized the orbitals. This charge is then removed and the calculation is repeated using these localized input as starting orbitals. The resulting orbitals remain localized.

The results are also presented in Table 1. We notice that the CASSCF IPs are now all too low, as one would expect. The CASPT2 correction (1.47 eV) brings them in reasonably good agreement with experiment. The errors vary between 0.08 and 0.99 eV. The largest error is found for the $^2A_1$ state. The theory at all levels predict a difference in the IP of about 0.3 eV between the

$^2A_1$, $^2T_2$ ionized states, while experimentally they are accidentally degenerate. We believe that part of the error obtained in this region of the spectrum is due to the low resolution of the experimental spectrum.

However, one thing is still missing, the spin-orbit coupling. It arises through the an interaction of the F2p orbitals with the Th 6p electrons, which have a large spin-orbit coupling constant. Thus, we expect to see the largest effects in the $^2T_2$ states. The spin-orbit coupling is carried out using the CASSI technique as described earlier. It is necessary to use an all-electron basis set, because the approach has not been implemented with ECPs. A relativistic ANO-type basis set developed by U. Wahlgren was used [20]. It was contracted: 9s8p7d5f. The same basis set as described above was used for F. The calculations were performed using the localized approach, but the active space was extended to 6 orbitals per F atom, 2p and 3p. Thus, the most important correlating natural orbitals were optimized at the CASSCF level of theory. No CASPT2 correction was made to the total energies, since it is not clear how much double counting of the correlation energy this would lead to. The CASPT2 correction to the IPs were anyway small with this extended active space.

Table 2: Ionization enrgies in ThF_4 computed with an all-electrons basis set (in eV). Localized orbitals were used. Spin orbit coupling included. Relative values within parentheses.

Orbital	IP(theory)	Exp.[a]
$4t_2(6u_{3/2})$	15.21(0.00)	13.36(0.00)
$4t_2(4e_{5/2})$	15.60(0.39)	13.77(0.41)
$1t_1(5u_{3/2})$	16.41(1.20)	14.52(1.16)
$1t_1(4e_{1/2})$	16.42(1.21)	14.52(1.16)
$3a_1(3e_{1/2})$	17.07(1.86)	15.43(2.07)
$1e(4u_{3/2})$	17.15(1.94)	16.11(2.75)
$3t_2(3u_{3/2})$	17.61(2.40)	16.11(2.75)
$3t_2(3e_{5/2})$	17.66(2.45)	16.11(2.75)

[a]from Ref. [16].

The results of this calculation are shown in Table 2. Clearly we see a problem with the approach: all ionization energies are between 1 and 2 eV too large. The reason for this behavior is most probably the large effect of the delocalization on the ground state energy. When the four localized ground state wave functions are mixed a stabilization of 2.07 eV is obtained. It seems that this use of CASSI may lead to an unbalanced description of the correlation energy between the ground and ionized states. It should, however, be pointed out that the basis set is rather limited and it is possible that a more extended calculation may lead to improved results.

However, the relative energies and the spin-orbit splitting of the levels is described well. The lowest IP (from the $4t_2$ orbital) is split into two components: $6u_{3/2}$ and $4e_{5/2}$: The splitting is 0.39 eV to be compared to the experimental value 0.41 eV. The second $^2T_2$ state is split 0.05 eV. Here the resolution of the experimental spectrum does not allow a more detailed analysis and only one value is reported [16]. The other states have negligible spin-orbit splittings.

4 THE COORDINATION OF ACTINYL IONS WITH CARBONATE AND WATER

Actinide chemistry mainly occurs in solution and in the following we shall present some examples of the study of actinide complexes in water, carried out with the CASSCF/CASPT2 method. We will concentrate on uranyl and neptunyl complexes of carbonate ions. Uranium and neptunium are mainly present as uranyl and neptunyl and they are highly soluble under typical ground-water conditions [21]. Carbonate is an important natural ligand because of its relatively high concentration and strong actinide-complexing ability [22, 23]. Therefore carbonato complexes of U(VI) and Np(V) are expected to play an important role in the fate and transport of U and Np in natural aquatic environment. We have considered the carbonate system of uranyl (VI) and (V), $[UO_2(CO_3)_3]^{4-}$ and $[UO_2(CO_3)_3]^{5-}$ [24], Figure 2.

Both solid-state structures and EXAFS information are available for complexes of this type. Docrat et al. [25] claim from their cyclic voltammetry study that the U(VI)/U(V) couple is not reversible. We investigated the conformation changes between the two carbonate complexes that eventually occur. Such a change could arise from a twist about the U-C direction which moves the coordinated O atoms out of the equatorial plane, as has been observed in the solid state for $[UO_2(CO_3)_3]^{4-}$ compounds [26, 27, 28].

The structure of $[UO_2(CO_3)_3]^{4-}$ has previously been studied by Pyykkö *et al.* [29] at HF level of theory. We initially optimized the geometry of the two compounds at second order perturbation level of theory, (MBPT2 and CASPT2), by imposing D_{3h} symmetry. We then relaxed the symmetry of the U(V) compound towards a twisted structure, and investigated if this implied extra-stabilization. Negative ions are rarely stable in gas phase, while they are stabilized by the interaction with a surrounding medium, for example a po-

Figure 2: The structure of $[UO_2(CO_3)_3]^{4-/5-}$ and $[NpO_2(CO_3)_3]^{5-}$.

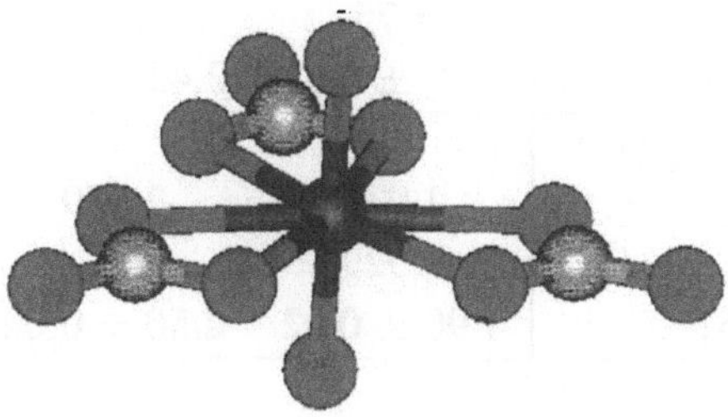

lar solvent like water. In quantum chemistry one often uses a solvent model. In this study the aqueous environment has been modeled with a reaction-field Hamiltonian throughout all the calculations. The choice of the active space for these species has been made in analogy with what described in the previous section for CUO.

In Table 3 the typical bond distances and angles are reported. The theoretical values are in agreement with experiment. The (5-) compound was successively distorted from D_{3h} symmetry to a C_3 twisted structure. A slight distortion of ca. 5 degrees from planarity produces a structure which is ca. 6 kcal mol^{-1} higher in energy than the D_{3h} structure. This indicates that the D_{3h} structure is a real minimum on the energy surface. In agreement with experiment, the reduction of $[UO_2(CO_3)]^{4-}$ to $[UO_2(CO_3)]^{5-}$ causes a U-O equatorial increase of ca 0.12 Å and a U-O uranyl increase of ca 0.08 Å, while no substantial rearrangement of the structure occurs.

The neptunyl carbonate complexes were also considered [30]. Clark *et al.* [31, 32] investigated the carbonate complexation of neptunyl. Solutions containing $[NpO_2(CO_3)]^-$ (Figure 3), $[NpO_2(CO_3)_2]^{3-}$ (Figure 4), and $[NpO_2(CO_3)_3]^{5-}$ (Figure 2) were prepared and the solution species were studied using various spectroscopic techniques. These studies made it possible to determine the molecular structure of the carbonate complexes.

Table 3: Typical bond distances, Å, in $[UO_2(CO_3)_3]^{4-}$ and $[UO_2(CO_3)_3]^{5-}$, in the presence of the reaction field Hamiltonian.

System	$R_{UO_{ax}}$	$R_{UO_{eq}}$	R_{UC}
$[UO_2(CO_3)_3]^{4-}$			
MBPT2	1.881	2.407	2.841
CASPT2 (12/12)	1.845		
HF [29]	1.75	2.55	
Exp. [25]	1.80 ± 0.02	2.43 ± 0.02	2.89 ± 0.04
$[UO_2(CO_3)_3]^{5-}$			
MBPT2	1.933	2.529	2.947
CASPT2 (13/13)	1.929		
Exp. [25]	1.90 ± 0.02	2.50 ± 0.02	2.94 ± 0.04

We were mainly interested in the molecular structure of the complexes and the explicit coordination of water molecules. Clark *et al.* [31] consider almost definitive the structure of $[NpO_2(CO_3)_3]^{5-}$, while they seem more uncertain about the structures of $[NpO_2(CO_3)]^{-}$ and $[NpO_2(CO_3)_2]^{3-}$. This uncertainty concerns the number of water molecules bound to the complexes. They suggest that *the structural information reported in their work may enable more accurate modeling of neptunyl carbonate reactivity*. From now on they will be indicated as NpMono, NpBis, and NpTris respectively. The structure of these systems has been determined and the water coordination has been investigated.

Table 4: Typical bond distances, Å, and angles, degrees, in $[NpO_2(CO_3)_2]^{3-}$, obtained at CASPT2 level with different active spaces (2/7 and 10/12), and basis sets (BS-1 and BS-2). All results obtained in the presence of the reaction field Hamiltonian.

	2/7	10/12	10/12	2 H_2O^a; 10/12	Exp.[b]
Bond distance	BS-1	BS-1	BS-2	BS-1	
$R_{NpO_{ax}}$	1.854	1.850	1.817	1.854	1.85 ± 0.02
$R_{NpO_{eq}}$	2.532	2.548	2.565	2.548	2.48 ± 0.03
R_{NpC}	2.954	2.970	2.986	2.891	2.93 ± 0.03
$\angle O_{eq}NpO_{eq}$	52	51	52	52	
$\angle O_{eq}CO_{eq}$	117	118	117	123	

[a] Np-water distance is 2.585 Å.
[b] Ref [31].

Figure 3: The structure of $[NpO_2(CO_3)]^- \cdot 3\ H_2O$.

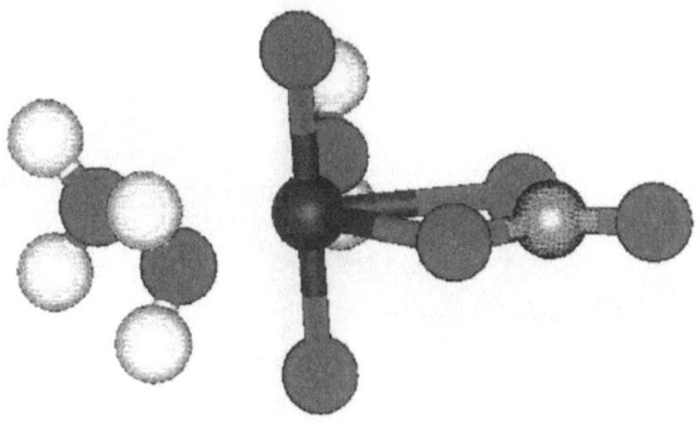

As an example, we will report the structure of NpBis in Table 4 obtained with different active spaces and basis sets. For the details one should see the original paper [30]. The results are in agreement with experiment. The inclusion of two explicit water molecules (Figure 4) in the 10/12 calculation does not seem to affect the Np-O_{ax} and Np-O_{eq} bonds. However, some energetic considerations show that NpBis $\cdot$ 2 H_2O is ca. 50 kcal/mol lower in energy than NpBis with no explicit water. The binding energy for one water molecule is thus estimated to be about 25 kcal/mol.

The system that we found most problematic to describe was NpTris (Figure 3). The (5-) anion turned out to be unbound even in the presence of the cavity. With such a highly negative charge, the rather simplified solvent model used was probably not able to keep the complex bound. Actually a bound complex could only be obtained by explicitly including counterions in the complex. This can only in an approximate way describe the complex dynamic situation in the solvent and the deviation from experiment is therefore not surprising.

From a comparison between uranyl and neptunyl, it seems that the neptunyl ion has less complexing strength than uranyl. An indication of that is that the UO bond in UO_2^+ is lengthened 0.17 Å in the complex, while the corresponding Np-O bond only becomes 0.11 Å longer in forming the complex. Also, the perturbation of the CO_3 units is larger in neptunyl than in uranyl carbonate

Figure 4: The structure of $[NpO_2(CO_3)_2]^{3-}$.

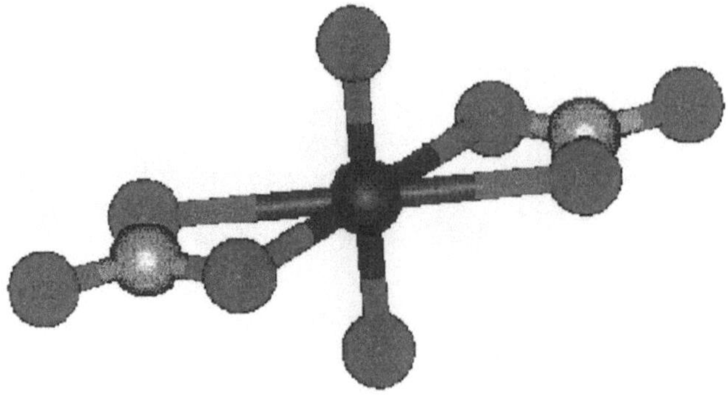

complexes.

These calculations do not give a definitive answer on the difficult issue of determining the coordination of actinyl ions in water. They should on the other hand be considered as initial tests to see if our quantum models reproduce the structures experimentally available. The coordination of water can not be addressed by adding an explicit number of water molecules at the quantum level, but it needs to be simulated at the macroscopic scale. An investigation has recently started, where the intermolecular forces obtained from quantum mechanical calculations are used in statistic mechanical simulations of actinyl ions in solutions.

5 SUMMARY

We have in this paper given some illustrations of how the MOLCAS-5 software can be used in studies of actinide compounds. The main emphasis has been on the multiconfigurational approach (CASSCF), which is vital for the description of the highly degenerate electronic structures that are typical for these systems. Dynamic correlation effects have been added using second order perturbation theory. For the ionization energies of ThF_4 we have also illustrated how the CASSCF state interaction method can be used to go from a localized picture to a delocalized and symmetry adapted one at the same time as the effect of spin-orbit coupling is included in the calculation. The conclusion from this, and several other similar applications, is that CASSCF/CASPT2 is a powerful technique for these applications. The illustrations show that also for these heavy element compounds is it possible to start from a non-relativistic description of the wave function with scalar relativistic effects included using the Douglas-Kroll Hamiltonian (or relativistic effective core potentials) and add spin-orbit effects a posteriori. It should be emphasized, however, that it is necessary to develop special AO basis sets with the scalar relativistic effects included. Other applications have shown that the same approach may be used all over the periodic system, with the possible exception of the heavier main group elements, but even there are the errors within the limits of the general accuracy of the approach.

6 ACKNOWLEDGMENTS

This work has been supported by the Italian Ministero dell'Istruzione dell'Università e della Ricerca and the Swedish Research council.

References

[1] Löwdin, P.-O. *Phys. Rev.* **1955**, *97*, 1474.

[2] Gagliardi, L.; Roos, B. O.; Mamqvist, P.-A.; Dyke, J. M. *J. Phys. Chem. A* **2001**, *105*, 10602.

[3] Andersson, K.; Baryz, M.; Bernhardsson, A.; Blomberg, M. R. A.; Carissan, Y.; Cooper, D. L.; Cossi, M.; Fleig, T.; FIscher, M. P.; Gagliardi, L.; de Graaf, C.; Hess, B. A.; Karlstrm, G.; Lindh, R.; Malmqvist, P.-.; Neogrdy, P.; Olsen, J.; Roos, B. O.; Schimmelpfennig, B.; Schtz, M.; Seijo, L.; Serrano-Andrs, L.; Siegbahn, P. E. M.; Stlring, J.; Thorsteinsson, T.; Veryazov, V.; Wierzbowska, M.; ; Widmark, P.-O. MOLCAS *Version 5.2.* Dept. of Theor. Chem., Chem. Center, Univ. of Lund, P.O.B. 124, S-221 00 Lund, Sweden, Lund **2001**.

[4] Roos, B. O. In Lawley, K. P., ed., *Advances in Chemical Physics; Ab Initio Methods in Quantum Chemistry - II.* John Wiley & Sons Ltd., Chichester, England, 1987 399.

[5] Andersson, K.; Malmqvist, P.-Å.; Roos, B. O.; Sadlej, A. J.; Wolinski, K. *J. Phys. Chem.* **1990**, *94*, 5483.

[6] Roos, B. O.; Andersson, K.; Fülscher, M. P.; Malmqvist, P.-Å.; Serrano-Andrés, L.; Pierloot, K.; Merchán, M. In Prigogine, I.; Rice, S. A., eds., *Advances in Chemical Physics: New Methods in Computational Quantum Mechanics, Vol. XCIII:219–331*. John Wiley & Sons, New York, 1996 219–332.

[7] Douglas, N.; Kroll, N. M. *Ann.Phys.* **1974**, *82*, 89.

[8] Hess, B. *Phys.Rev. A* **1986**, *33*, 3742.

[9] Roos, B. O.; Widmark, P.-O.To be published.

[10] Malmqvist, P. Å.; Roos, B. O. *Chem. Phys. Letters* **1989**, *155*, 189.

[11] Malmqvist, P.-Å.; Roos, B. O.; Schimmelpfennig, B. *Chem. Phys. Letters* **2002**, *in press*, xxx.

[12] Zhou, M.; Andrews, L.; Li, J.; Bursten, B. E. *J. Am. Chem. Soc.* **2000**, *121*, 9712.

[13] Gagliardi, L.; Roos, B. O. *Chem. Phys. Letters* **2000**, *331*, 229.

[14] Institute für Theoretische Chemie, Universität Stuttgart. ECP and corresponding valence basis sets. http://www.theochem.uni-stuttgart.de.

[15] Li, J.; Bursten, B. E.; Liang, B.; Andrews, L. *Science* **2002**, *295*, 2242.

[16] Dyke, J. M.; Josland, G. D.; Morris, A.; Tucker, P. M.; Tyler, J. W. *J. Chem. Soc. Faraday Trans. 2* **1981**, *77*, 1273.

[17] Boerrigter, P. M.; Snijders, J. G.; Dyke, J. M. *J. Elec. Spectrosc. Rel. Phenom.* **1988**, *46*, 43.

[18] Girichev, G. V.; Petrov, V. M.; Giricheva, N. I.; Zasorin, E. Z.; Krasnov, K. S.; Kiselev, Y. M. *Zh. Strukt. Khim.* **1983**, *24*, 61.

[19] Finley, J.; Malmqvist, P.-Å.; Roos, B. O.; Serrano-Andrés, L. *Chem. Phys. Letters* **1998**, *288*, 299.

[20] Wahlgren, U.Private Communication.

[21] Nitsche, H.; Lee, S. C.; Gatti, R. C. *J. Radioanal. Nucl. Chem.* **1988**, *124*, 171.

[22] Clark, D. L.; Hobart, D. E.; Neu, M. P. *Chem. Rev.* **1995**, *95*, 25.

[23] Nash, K. L.; Cleveland, J. M.; Rees, T. F. *J. Environ. Radioact.* **1988**, *7*, 131.

[24] Gagliardi, L.; Grenthe, I.; Roos, B. O. *Inorg. Chem.* **2001**, *40*, 2976.

[25] Docrat, T. I.; Mosselmans, J. F. W.; Charnock, J. M.; Whiteley, M. W.; Collison, D.; Livens, F. R.; Jones, C.; J., E. M. *Inorg. Chem.* **1999**, *38*, 1879.

[26] Coda, A.; Della Giusta, A.; Tazzoli, V. *Acta Crystallogr.* **1981**, *B37*, 1496.

[27] Mereiter, K. *Acta Crystallogr.* **1986**, *C42*, 1682.

[28] Effenberger, H.; Mereiter, K. *Acta Crystallogr.* **1988**, *C44*, 1172.

[29] Pyykko, P.; Li, J.; Runeberg, N. *J. Phys. Chem.* **1994**, *98*, 4809.

[30] Gagliardi, L.; Roos, B. O. *Inorg. Chem.* **2002**, *41*, 1315.

[31] Clark, D. L.; an S. A. Ekberg. N. J. Hess, S. D. C.; Neu, M. P.; Palmer, P. D.; Runde, W.; Tait, C. D. *J. Am. Chem. Soc.* **1996**, *118*, 2089.

[32] Clark, D. L.; an S. A. Ekberg. N. J. Hess, S. D. C.; ; Janecky, D. R.; Neu, M. P.; Palmer, P. D.; Tait, C. D. *New J. Chem.* **1996**, *20*, 211.

DENSITY FUNCTIONAL THEORY PERFORMANCE IN METAL CONTAINING SYSTEMS:

Some Representative Cases

M.C. MICHELINI, T. MARINO, N. RUSSO, E. SICILIA and M. TOSCANO
Dipartimento di Chimica, Università della Calabria, I-87030 Arcavacata di Rende (CS), Italy. Email: russo@unical.it

Abstract: The performance of density functional theory in determining the physico-chemical properties of a series of metal containing systems is presented. The use of different exchange-correlation functionals and basis sets is proposed to determine geometrical structures, coordination modes, binding energies and potential energy surfaces. The comparison with the experimental and the high level ab-initio data evidences the reliability and the defninciency of density functional based methods and underlines their potentiality in the study of large systems.

1. Introduction

Metal containing systems are important in a wide series of fundamental and applied sciences including catalysis, materials and surfaces, biology and medicine, conductors and superconductors, membranes and nano-technologies.

For these reasons their investigation has attracted the attention of many researchers and the development in synthetic strategies, experimental techniques and theoretical methods to obtain and characterize them is grown rapidly.

From a theoretical point of view, the metal containing systems are fascinating because their core, the metal, is not easy to treat at quantum-mechanical level. In fact, metals and in particular transition metals, are characterized by several electronic spin states that fall in a narrow range of energy, often present open shell configurations and require the treatment of relativistic effects. Furthermore, the great number of valence electrons makes indispensable the computation of correlation effects by powerful computational tools. So, also the seemingly simple determination of the ground state of a transition metal atom is not a trial operation. It is well known that traditional Hartree-Fock methods fail in predicting correctly the properties of the transition metal containing systems and the post-Hartree-Fock methods are difficult to be applied in absence of powerful and fast computers. In these last 30 years, the density functional theory, in the Kohn-Sham formulation [1], established itself as an economic and rigorous alternative to describe the complex molecular systems. Several books, reviews and thousand of scientific papers have been produced that demonstrate their reliability [2-8].

E.J. Brändas and E.S. Kryachko (eds.),
Fundamental World of Quantum Chemistry, Vol. II, 443–465.

In this article, we will report a summary of our recent works on the density functional treatment of inorganic systems including examples of interest in bio-inorganic, organometallic, surface sciences and catalysis fields.

The properties studied here involve geometrical structures, coordination modes, binding energies or metal ion affinities and potential energy hypersurfaces.

2. Method and computational details

Computations were performed using the density functional (DF) method in the Kohn-Sham formulation selecting the results obtained employing two exchange-correlation potentials in order to underline their respective performance:

- the three parameter hybrid functional of Becke [9] and Lee, Yang and Parr [10] (B3LYP) for the exchange and correlation parts, respectively;
- the gradient corrected functional composed by the exchange of Perdew and Wang [11] and the correlation of Perdew [12] (PWP).

Orbital basis sets of double-zeta quality (DZ) were generally used, except in some cases, explicitly indicated in the text.

deMon [13] and Gaussian 94 [14] codes were employed for the gradient corrected and for the hybrid computations, respectively.

3. Structural parameters

3.1 PLATINUM-IMINO ETHER COMPLEXES

In the last two decades considerable attention has been devoted to the reactions of nucleophilic attack upon coordinated nitriles in metal complexes [15]. In fact, some platinum-imino ether compound arising from this process has antitumor activity [16].

Platinum nitriles (*cis*- and *trans*- $[PtCl_2(NCR)_2]$, R=Me, Ph) react with methanol, in the presence of a base, to give imino ether derivatives, $[PtCl_2\{HN{=}C(OMe)R\}_2]$, with different configuration at the imino ether ligands (E and Z) and at the C=N double bond (cis and trans). The structures of some of these molecules were assigned on the basis of X-ray [17] data, so we had the possibility to compare our computed B3LYP/6-311+G(2df,2p) geometry with the experimental counterpart. The *trans*- EE $[PtCl_2\{HN{=}C(OMe)Me\}_2]$ complex shown in the figure 1 was the object of our study.

Theoretical geometrical parameters are listed in the table 1 together with the available experimental ones.

The agreement between theoretical and experimental bond distances can be considered satisfactory if we take into account that X-ray data are effected by the crystal environment and that in the theoretical data the relativistic effect of platinum atom was not considered. In fact, the maximum deviations (0.04-0.07 Å) occur in the bond lengths involving the metal.

The valence angles are well reproduced except in the case of N_2-Pt-N_1 and C_{11}-N_1-Pt for whose we found differences of about 6 degrees. A deeper glance at the X-ray data reveals that the molecule is not symmetric as we should expect. This evidences the role played by the other molecules present in the crystal.

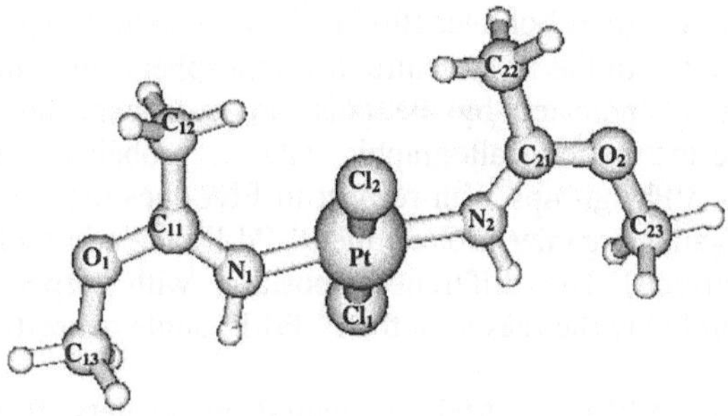

Figure 1. Equilibrium conformation of *trans*- EE $[PtCl_2\{HN=C(OMe)Me\}_2]$ complex.

TABLE 1. Bond distances (Å) and valence angles (degrees) for *trans*-EE $[PtCl_2\{HN=C(OMe)Me\}_2$ complex. X-ray values in parentheses represent estimated standard deviations.

Distances	Theory	X-ray	Angles	Theory	X-ray
Cl_1-Pt	2.367	2.298(2)	Cl_2-Pt-Cl_1	179.9	179.4(1)
Cl_2-Pt	2.367	2.301(2)	N_1-Pt-Cl_1	87.0	87.8(2)
N_1-Pt	2.029	1.989(7)	N_1-Pt-Cl_2	92.9	91.7(2)
N_2-Pt	2.029	2.009(8)	N_2-Pt-N_1	174.0	179.1(2)
C_{11}-O_1	1.339	1.315(11)	C_{13}-O_1-C_{11}	119.8	119.4(8)
C_{21}-O_2	1.339	1.306(11)	C_{11}-N_1-Pt	134.2	125.6(6)
C_{11}-N_1	1.280	1.307(11)	N_1- C_{11}-O_1	123.3	124.4(8)
C_{21}-N_2	1.280	1.294(12)	C_{12}-C_{11}-O_1	112.1	113.4(8)
C_{13}-O_1	1.430	1.411(14)	C_{12}-C_{11}-N_1	124.6	122.2(8)
C_{23}-O_2	1.430	1.455(14)	N_2-Pt-Cl_1	87.0	92.8(2)
C_{12}-C_{11}	1.488	1.502(12)	N_2-Pt-Cl_2	92.9	87.6(2)
C_{22}-C_{21}	1.488	1.519(14)	C_{23}-O_2-C_{21}	119.8	117.9(8)
			C_{21}-N_2-Pt	134.2	126.7(7)
			N_2- C_{21}-O_2	123.3	125.1(9)
			C_{22}-C_{21}-O_2	112.1	111.1(8)
			C_{22}-C_{21}-N_2	124.6	123.7(8)

3.2 PALLADIUM DIPHOSPHENE COMPLEXES

The diphosphene ligands show an interesting coordination chemistry because of the different bonding modes to the metal center. In the monometallic complexes diphosphene can coordinate *via* the lone pair of one phosphorus (η^1) or *via* the P=P π system (η^2) as in the metal-olefin interaction [18]. Of course, the coordination mode depends on the electronic configuration of the metal fragment. In the case of palladium-ML_2 fragment, diphosphene can assume both η^1 and η^2 conformation, but the former is the most stable one as shown by the X-ray structure of the $(PPh_3)_2Pd(CF_3PPCF_3)$ [19].

Our density functional study on the $(PH_3)_2Pd(HPPH)$ model complex, performed using the gradient corrected PWP exchange-correlation potentials, confirms the experimental coordination mode giving structural data close to the X-ray counterpart. Geometrical parameters are reported in table 2.

The calculation gives bond lengths and valence angles in good agreement with the experimental results. In the η^2 structure the diphosphene and phosphines ligands lie in the same plane as indicated by the twist angle value. The slight deviation from planarity, suggested by crystallographic data, is probably due to the larger steric hindrance of the PPh_3 groups with respect to PH_3 ones used in the model system. In fact, the PPh_3 ligands organize around the P'-Pd-P' angle in such a manner to alleviate the repulsion effects. The different disposition with respect to the simpler PH_3 substituents is probably the reason of the P'-Pd-P' angle overestimation (about 11°).

TABLE 2. Main structural parameters for the $(PH_3)_2Pd(HPPH)$ model complex. X-ray are referred to the $(PPh_3)_2Pd(CF_3PPCF_3)$ system. Distances are in Å and angle in degrees.

Parameters	Theory	X-Ray
Pd-P	2.380	2.364
Pd-P'	2.323	2.354
P-P	2.125	2.121
P-Pd-P	53.0	53.3
P'-Pd-P'	122.2	109.2
Twist angle	0.0	9.0

P and P' refer to diphosphene and to phosphine moiety, respectively. X-ray data are taken from ref. [19].

3.3 COBALT AND CHROMIUM COMPLEXES

The nitrosyl and carbonyl complexes with the first row transition metals have been widely studied for their importance in the organometallic and inorganic chemistry.

We report here, as representative cases, the optimized structures of a mixed NO and CO cobalt $[Co(NO)(CO)_3]$ and an esacarbonyl chromium $[Cr(CO)_6]$ complexes for whose experimental investigations are available [20].

Computations for $[Co(NO)(CO)_3]$ system have been performed using the PWP exchange-correlation functionals imposing the C_3v symmetry. The computed Co-N and Co-C bond lengths are 1.676 and 1.839 Å, respectively, while the corresponding experimental distances are 1.671 and 1.842 Å [20]. The N-Co-C valence angle is about 8° larger than measured one (116.1° versus 108.7°) because theoretical computations take not into account the intramolecular interactions that occur in the crystal.

For chromium complex, the PWP Cr-C distance, obtained optimizing in Oh symmetry, is in very good agreement with the experimental one (1.915 versus 1.918 Å) [21].

3.4 ADSORPTION DISTANCES

Adsorption of oxygen and nitrogen atoms on copper surfaces has been the subject of many theoretical [22,23] and experimental [24,25] studies because its importance in heterogeneous catalysis and corrosion. In the framework of cluster model approach, we have performed a density functional investigation of the $O\text{-}Cu_5$ and $N\text{-}Cu_5$ systems.

Computations have been done using the PWP exchange-correlation functionals in conjunction with the (7111/411/1*) and (4,4;4,4) orbital and auxiliary basis sets for oxygen and nitrogen. A relativistic extended (Cu^{+17}) model core potential that allows the explicit treatment of $3p^6$ $3d^{10}$ $4s^1$ valence electrons has been used for copper [26]. The Cu_5 cluster model (see figure 2), consisting of a first layer of four and a second of one atoms, has been fixed in its bulk geometry and the position of adsorbed atoms optimized in the fourfold open site as experimentally indicated.

Results are reported in table 3 together with other theoretical and experimental data.

For the O-Cu_5 system, the DF distances appear to be overestimated with respect to both ab-initio and experimental values [26], while a better agreement is obtained in the case of N-Cu_5. Although the discrepancies can be justified invoking several factors (poor description of the surface, lack of surface reconstruction and relaxation upon adsorption, coverage effects), the computational tool seems to be not sufficiently accurate to describe reliably the geometric aspect of this interaction.

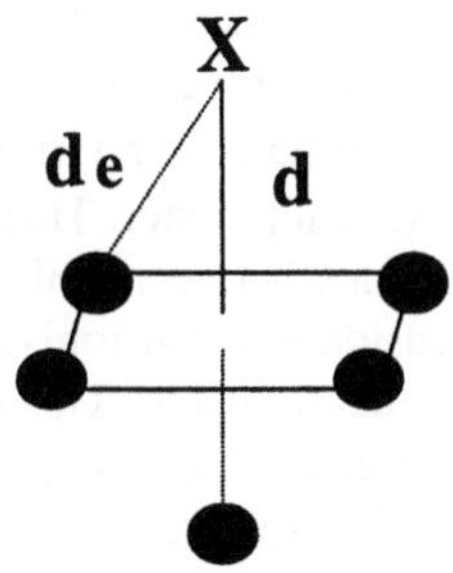

Figure 2. Schematic drawing of Cu_5 cluster. d and de represent the distance from the adsorbate (X=O, N) to its next neighbours surface metal atoms and the perpendicular bond length to the surface, respectively.

TABLE 3. Theoretical and experimental adsorption distances (Å) for O-Cu_5 and N-Cu_5 systems.

Method	O-Cu_5		N-Cu_5	
	de (Å)	d (Å)	de (Å)	d (Å)
HF	0.90	2.02	0.49	1.87
MP2	0.92	2.03	0.32	1.83
DFT	1.03	2.08	0.48	1.87
EXP	0.80	1.88	0.4-0.6	/

4. Coordination sites

4.1 INTERACTION OF CARBON DIOXIDE WITH Sc, Ti, Ni AND Cu ATOMS

The coordination of carbon dioxide on a metallic centre seems to be the most important steep in its reduction [27]. A number of experimental [28,29] and theoretical [29-36] studies have been devoted to this subject with the aim to highlight the initial steep of the metal-catalysed conversion of CO_2.

From these studies it emerged that some first row transition metal atoms form M-CO_2 complexes whereas other elements insert spontaneously into C-O bond, yielding OMCO systems. The coordination modes are difficult to be determined experimentally, but the theoretical works have considered also this aspect.

Five different coordination modes have been take into account in our study (see figure 3) for each metal [27]. Computations have been performed using the gradient corrected exchange-correlation PWP and the hybrid B3LYP functionals.

For Sc-CO_2 interaction, only two minima are possible. The PWP calculations indicates the η^2C,O in the $^2A'$ electronic state as the most stable form but the $^2A_1$ η^2O,O coordination mode is favoured at B3LYP level. Previous theoretical MCPF computations [30b] gives the same result obtained by B3LYP ones although the energy differences between the two forms is always very small and it reduces still taking into account the ZPE and BSSE corrections.

Ti-CO_2 complex can assume the η^2C,O, η^2O,O and η^1O forms. With titanium atom, in its electronic state of triplet, the preferred coordination mode, at B3LYP level, is η^2C,O followed by the η^2O,O. The η^1O lies at 26.5 kcal/mol above the absolute minimum. Gradient corrected computations give, instead, different results. In particular η^2C,O complex does not appear as a minimum in the triplet potential energy surface and the η^2O,O coordination is the most stable one. This same indication arises from previous HF results [34]. The discrepancy between B3LYP and both gradient corrected and HF data can be explained by considering that ionisation potentials obtained by the latter methods are far from the experimental value. This error in the ionisation potential evaluation can significantly affect the amount of charge transfer from the metal to the ligands with serious consequences on the preferred coordination modes.

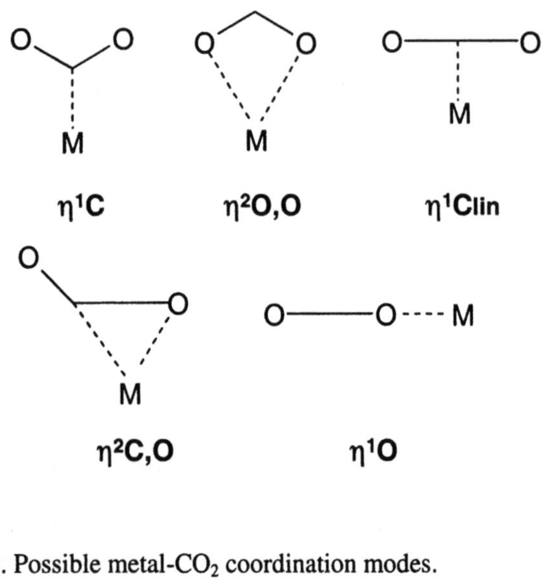

Figure 3. Possible metal-CO_2 coordination modes.

Both PWP and B3LYP functionals suggest that of the five considered coordination modes, only the η^2C,O and η^2O,O structures are minima for Ni-CO_2 system in the singlet ($^1A_1$) and triplet ($^3A_1$) states, respectively, in agreement with previous Becke-Perdew investigation [29]. Although all computations and experiment [27,29] agree on the fact that the Ni-CO_2 is a very weakly bond complex, the relative energy value of η^2O,O form is very different at the various levels of theory. The differences can be ascribed to the documented major difficulty of the gradient corrected with respect to hybrid methods to describe such a type of interaction [36].

IR spectra on the Cu-CO_2 system have suggested that it is a thermo labile and weakly bond complex [28]. We have considered for this system the lowest doublet spin state. Full geometry optimisation of the five coordination modes carried out at the PWP and B3LYP levels, give only a minimum in the potential energy surface, that corresponds almost to the η^1O coordination. The geometrical parameters of the minimum obtained with the two functionals differ considerably underlying again the difficulty of the gradient corrected methods to treat the weakly bond systems. No comparison is possible with previous CI computations [32] because, in this last study, the lack of the vibrational analysis does not allow to establish if the three proposed structures (η^2O,O, η^2O and η^1C) are real minima on the potential energy surface.

4.2 COORDINATION MODES OF AMMONIA AND ETHENE ON Mo_2.

The experimental studies [37,38] on the interactions between molybdenum and different small organic molecules have shown that the reactivity of Mo clusters changes drastically with the cluster size.

These results are very significant because of the importance of molybdenum in biological and interfacial catalysis [39,40]. For these reasons, the reactivity of Mo_n systems has been the subject of a series of theoretical works performed at different level of theory including HF [41], MCPF [42,43] and DFT [44-48]. The theoretical investigations are important because the experiments in gas-phase do not give information about the structure and coordination modes although they are able to provide reliable binding energies (BE).

In a work recently published by some of us [49], different coordination modes arising from the interaction Mo_2 fragments with NH_3 and C_2H_4 have been analyzed using the PWP exchange-correlation functional. In the case of the interaction of ammonia with the Mo_2 dimer, we have found that only through the coordination mode (b) it is possible to obtain a stable complex. All the other considered coordination modes (see figure 4) give unbounded situations or geometries that collapse, during the optimization, in the (b) form. The computed BE value is close to the experimental indications [37] confirming indirectly the preferred coordination mode.

In the Mo_2NH_3 complex, a value of about 115° is found for the Mo-Mo-N angle (α).

This bent geometry arises from the interaction between the N atom in NH_3 and a delta molecular orbital of Mo_2. In previous theoretical studies on Cu_2NH_3 [50] and Ag_2NH_3 [51] systems, Fournier [50,51] found a linear structure because the donation in both systems involves the sigma antibonding molecular orbital arising from the dz^2 atomic orbitals of the metal. Mulliken, natural bonding and the Mayer bond order analyses have been performed with the aim to analyze the nature of the bonding in these systems. From these data we can deduce that only a bonding molecular orbital (41A), formed by the nitrogen p (80%) and molybdenum d_{xz} orbital (10%) at -9.2 eV, is present. The second order perturbation theory analysis shows that in the formation of Mo_2NH_3, a stabilizing effect comes from the interactions between molybdenum and ammonia orbitals including the nitrogen lone pair.

The complexes between molybdenum dimer and ethane have not observed in the experimental study of Lian et al. [37] due to the repulsive interactions occurring in the entrance channel. The only stable structures found for the $Mo_2C_2H_4$ complex are those having the C=C bond perpendicular to the Mo_2 moiety (d) (see figure 4).

The structural analysis of the Mo_2-ethane system shows an elongation of the C=C bond (i.e. 1.35 Å for the free versus 1.48 Å for the bonded ligand).

For this complex, it is easy to hypothesize a bond formation as described by the general scheme given by Chatt, Dewar and Ducanson for the interactions between a transition metal and an olefin. In this model, the ethene donates π-electrons to an empty orbital of the metal. This donation causes a weaken of the π-bond of ethane, a lowering of the π^*-orbital energy and consequently an electron back-donation from the d orbitals of the metal. The orbital energy diagram reveals a gap that allows the donation from ethene to the transition metal. The Mulliken net charges as well as the natural net charges show a partial transfer of d metal electrons to the π^*-system of the C_2H_4. The amount of charge transfer is not high enough to allow a strong donation back-donation as occurs in other second row metal-C_2H_4 complexes [43]. The low BE values account for this phenomena.

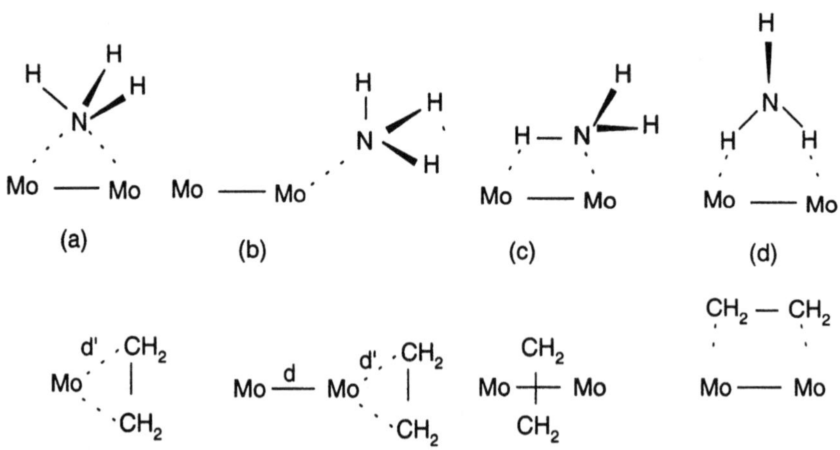

Figure 4. Coordination modes of ammonia and ethane on Mo_2 cluster.

4.3 COORDINATION SITES OF METAL IONS ON GUANINE

The reactions between atomic metal ions and organic molecules in gas-phase have aroused considerable interest. In the last years, the studies in this field have grown in the same proportion of the possible practical implications. In fact, the metal ions play a significant role in several biological processes. They affect syntheses, replication, structural integrity and cleavage of nucleic acids.

Experimental gas-phase studies on this matter have been performed essentially by using different mass spectrometry methods. The information coming from these studies are precious from the energetic point of view, but does not yield evidence on the coordination mode of metal ion and on the role of various nucleic acid bases tautomers.

Indirect conclusions about the coordination sites from experiment can be however drawn on the basis of metal ion affinity values that depend strongly on the particular metal coordination.

Computations have been performed using the hybrid B3LYP correlation potential, in connection with the 6-311+G(2df,2p) orbital basis set.

Full geometry optimization of five low lying guanine tautomers and their complexes with lithium, sodium, potassium and manganese ions has been mandatory to determine the possible and favoured coordination sites.

The interaction of a metal ion with a nucleic acid is controlled by the bond strength between the metal ion and the possible donor centers on the bases. The known sites for cation coordination of guanine are mainly the N7, N9 and O6 atoms (see figure 5). The existence of metal-N7 binding in guanine has been confirmed by various spectroscopic methods [52-54] and in some cases involves essentially divalent cations.

The metalation process leads to complexes in which the metal cation bicoordinated gives rise to further cycles (four- or five- membered) in the structures. The five-membered ring formation is, in principle, favoured with respect to that of four-membered one because of the minor annular strain, and the interaction of M^+ is better with the nitrogen or oxygen atoms having a noticeable negative charge. In particular, the carbonyl oxygen is preferred over the hydroxyl one and the imino nitrogen over the amino one.

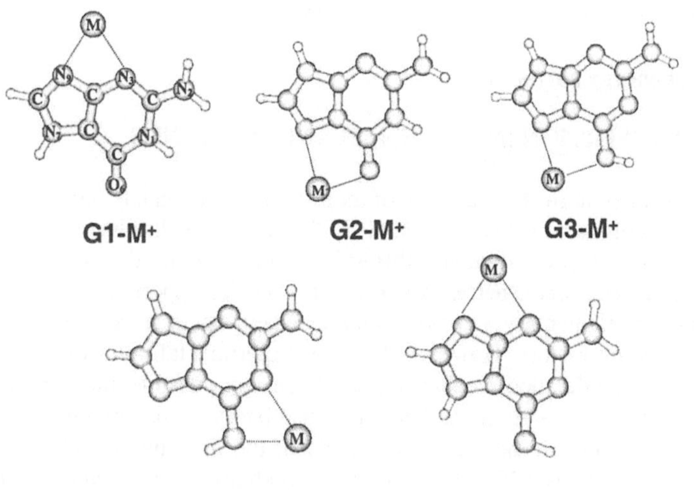

Figure 5. Possible coordination modes of a metal ion on the five lowest lying guanine tautomers.

The stability order of complexes of guanine, is consistent with all these propositions and arises from a balance of the two effects.

In the case of lithium ion, the most stable species is G2-Li^+ followed by G1-Li^+, G5-Li^+ and G3-Li^+ at 10.73, 12.44 and 14.58 kcal/mol, respectively. The less stable complex is G4-Li^+ (at 21.30 kcal/mol) because, other than a four-membered ring formation, the metalation reaction involves oxygen atom of the -OH group [55].

The computed MIA values closer to the experimental measurement (57.18 kcal/mol)[56] are those referred to the G3-Li^+ (60.28 kcal/mol) and the G4-Li^+ (54.50 kcal/mol) systems.

For both the Na^+ and K^+ ions the most stable complex is obtained starting from the G2 tautomer and is characterized by a further five-membered ring formation involving the metal species. The remaining complexes follow the same stability order irrespective of the cation considered: G1-Na^+ (G1-K^+) > G5-Na^+ (G5-K^+) > G3-Na^+ (G3-K^+) > G4-Na^+ (G4-K^+) at 9.1 (8.6), 11.0 (10.4), 12.9 (11.8) and 18.5 (17.0) kcal/mol, respectively.

The best agreement between the theoretical and experimental [56] sodium ion affinity is obtained for the G3-Na^+ complex (42.8 vs 43.5 ± 1.0 kcal/mol) but the value obtained for the G1-Na^+ compound (44.9 kcal/mol) can be also taken into account.

For the potassium ion, the G4-K^+ complex yields the MIA value (26.7 kcal/mol) closest to the experimental determination (28.0 ± 1.0 kcal/mol) [57].

The complexes with manganese cation have the following stability order: Mn^{2+} -G1 > Mn^{2+} -G5 > Mn^{2+} -G2 > Mn^{2+} -G3 > Mn^{2+} -G4 [58]. The results show as the N, N coordination is preferred over the N, O one and the N, O coordination is favoured when the oxygen atom is not involved in the O-H bond. The manganese ion affinity for guanine goes from 207.1 kcal/mol (for the G1 isomer) to 166.5 kcal/mol (for the G4 isomer). No experimental or previous theoretical determination has been available in this case.

5. Potential energy surfaces.

5.1 METAL INSERTION INTO C-H, N-H AND O-H BONDS

The growing interest in the reactions of atomic transition metals with small molecules containing prototype bonds (e.g. C-H, C-C, N-H and O-H) has been prompted by their importance in catalytic processes [59-84]. Among the used approaches gas-phase experiments, mass spectrometric in particular, give the opportunity to obtain valuable information on elementary reactions and mechanisms in a controlled environment without disturbing factors. However, from an experimental point of view a complete elucidation of the detailed mechanisms is, often, not possible and a close interaction between theory and experiment becomes mandatory. This interaction has already begun to yield results in the case of the study of the first-row transition metal cations with water interactions [85-88] and of the scandium ion with water, ammonia and methane [89].

Experimentally it has been shown that the H_2 elimination process is the most thermodynamically favoured in the reaction of early (Sc^+-V^+) and middle (Cr^+) first-row transition metal cations with water, ammonia and methane, while at higher energies other reaction products became accessible [63,65-67,71,73,76,77].

$$M^+ + XH_n \rightarrow MXH_{n-2}^+ + H_2 \quad (M=Sc, Ti, V) \quad (X=O, N, C)$$

It has been suggested that the studied reactions proceed *via* oxidative addition, i.e. insertion of M^+ (M=Sc, Ti, V) into the H-X (X=O,N,C) bond, to form an insertion intermediate that may lie either lower or higher in energy than the reactants.

Comparison between the proposed paths (see figure 6) shows that, while the mechanisms of reaction with H_2O, NH_3 and CH_4 are similar, the energetics are significantly different.

The observed similarities have been explained as a result of the fact that water, ammonia and methane are isoelectronic in the sense that the central heavy atom has the same number of valence electrons with the same sp^3 hybridization [90]. The differences in energetics have been attributed [76,77] to the different numbers of lone pairs: that is, two on the oxygen atom of water, one on the nitrogen atom of ammonia and none on the carbon atom of methane. On the other hand, differences in the behaviour of the considered cations are induced by the different atomic orbital occupancies, which dictate the accessibility of metal orbitals to electron donation from the ligand.

A very important topic of interest in these reactions, is the influence of the electronic state of the metal cation upon the reactivity and reaction path. The hypothesis of a crossing between high and low-spin surfaces, probably occurring in the region of the ion-molecule complex, M^+-XH_n (X=N, C), has been proposed [65-67,75,76] to explain the nature of the final products with respect to the reactants and the efficiency of the reaction. For this reason, both high- and low-spin surfaces have been examined for the insertion reactions of the Sc^+ (^3D, ^1D), Ti^+ (^4F, ^2F) and V^+ (^5D, ^3F) cations.

According to experimental observations the first step of the reaction is the formation of a stable ion-dipole complex. The reasonable mechanism for the next H-X bond breaking step is M^+ insertion into an H-X bond to form a H-M^+-XH_{n-1} intermediate.

This process, indeed, is thermodynamically favourable since the broken H-X bond energy is replaced by the energy of the H- M^+ and M^+-XH_{n-1} bonds formed.

From the insertion intermediate a molecular hydrogen complex is obtained *via* a tight four-center transition state. Then, without an energy barrier, the complex decomposes to give the reaction products directly.

The key reaction intermediates, H-M^+-XH_{n-1}, surmised by the experimentalists have been localized and characterized. Because of the excited nature of this intermediate with respect to the ground state of the reactants, the reactions are presumed to occur *via* a crossing from the high-spin to the low-spin surface.

The thermodynamical stability, with respect to ground state asymptotes, of the stationary points is very different along the three considered paths. The interaction between M^+ and the ligand in the ion-molecule complex is strongest in the case of ammonia and decreases in going to water and methane. The magnitude of the bonding between the metal ions and the simple non inserted ligand is not only electrostatic in nature; the magnitude of the binding energy is determined by a balance between the electrostatic interaction and Pauli repulsion. Indeed, the decreasing values of the dipole moments for water, ammonia and methane are inconsistent with the observed trends.

The available mechanisms adopted by the metal cation to reduce Pauli repulsion, therefore, are responsible for the reported order of the values.

The next minimum along the surfaces is the low-spin H-M-XH_{n-1}complex, whose formation is the key step for all the considered processes.

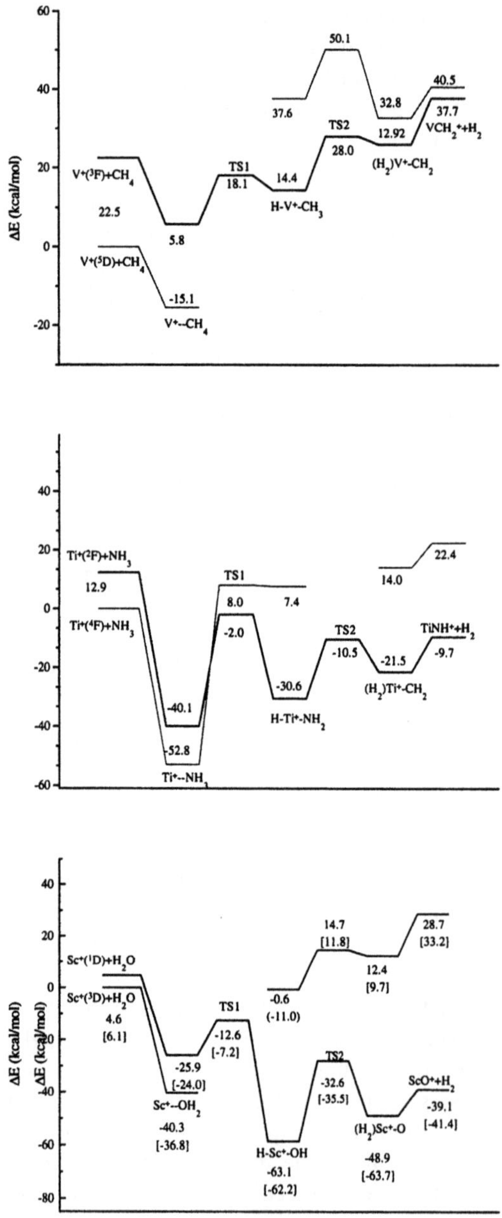

Figure 6. Potential energy surfaces for the insertion mechanism of M^+ (M=Sc, Ti, V) in XH_n bonds (X=C, N, O).

The formation of two M-L σ bonds is the outcome of the pairing of two metal electrons with two electrons originally occupying the X-H σ bonds and explains the enhanced stability of the low-spin hydrido-complexes. Several factors can be invoked to explain the observed stability trends, as the M^+ $4s^1 3d^{n-1}$ configuration and the spin-pair of the appropriate orbitals of the metals with those of the ligands to form the two σ bonds. When the ground state of the cation is $3d^n$ the effect of the promotion to an excited state is a weakening of the bond. Moreover, there is a change in the exchange terms of the energy because of the spin pairing of the electrons of the metal and, again, a weakening of the bond.

Also in this case the presence of different numbers of lone pairs on the oxygen, nitrogen and carbon atoms can be considered responsible of the decrease, from water, to ammonia and to methane, of the stability of the intermediate complexes observed along the paths for the three ligands. Indeed, the possibility of oxygen to form two dative bonds, one in the case of nitrogen and none for methane corresponds to a gradual weakening of the bond strength.

The barrier heights for TS1 and TS2 formations also change significantly. While, along the path for the insertion of Sc^+ into H-O bond the energy barriers are both well below the energy of the ground state asymptote, the position of the TS1 barrier eight is only slightly below for Ti^+ and, finally, V^+ ion inserts into the H-C bond with a barrier in excess of reactants.

To the nature of the metal-oxygen, nitrogen or carbon bond, can be ascribed the different stability of dehydrogenation products. Indeed, the strength of the bond between the metal with the unpaired electrons of oxygen and nitrogen is enhanced by the donation of electron density from the lone pairs of *2p* electrons into empty *d* orbitals. Unlike oxygen and nitrogen, carbon has no lone pair to share with the metal, thus accounting for the weaker M-C bond strength.

5.2 INTERACTION OF ALKALI METAL IONS WITH α-ALANINE .

The alkali metal ions are involved in different biological phenomena including enzyme regulation, stabilization of structural elements and transport of amino acids to proteins through membranes [91,92]. The elementary mechanism that regulates these processes are not yet well understood and continuous research work are devoted on this field.

The potential energy surfaces for the Li^+-, Na^+- and K^+- α-alanine interactions are reported in figure 7. Table 4 collects the relative energies of the minima and transition states (TS) (see figure 8).

The isomerization sequence reported in the PESs involves only one-step processes, thus, the various minima can be obtained only from a particular starting complex. The general shape of the three paths is very similar except for the height of the energetic barriers between the different minima that decreases in going from lithium to potassium complexes (see table 4). The three PESs are characterized by two separate interconversion channels connecting two different sets of minima. The first channel at high energy, not reported here [93], starts from minimum **1** that via TS6 should arrive at minimum **5**. The isomerization processes include the cleavage of the O2-M^+ bond in **1** to form a monodentate complex and the rotation about the C-Cα bond with the formation of a new bond between O1 and M^+ as in structure **5**. Transition states TS6 are characterized by the following relative energy values: 23.3 (for Li^+), 17.0 (for Na^+) and 16.7 kcal/mol (for K^+). The alternative path, in which the in plane H-shift could be hypothesized, is energetically unfavourable (in fact the transition states TS6 lie, in this case, at 51.2, 41.3 and 40.5 kcal/mol for Li^+, Na^+ and K^+, respectively).

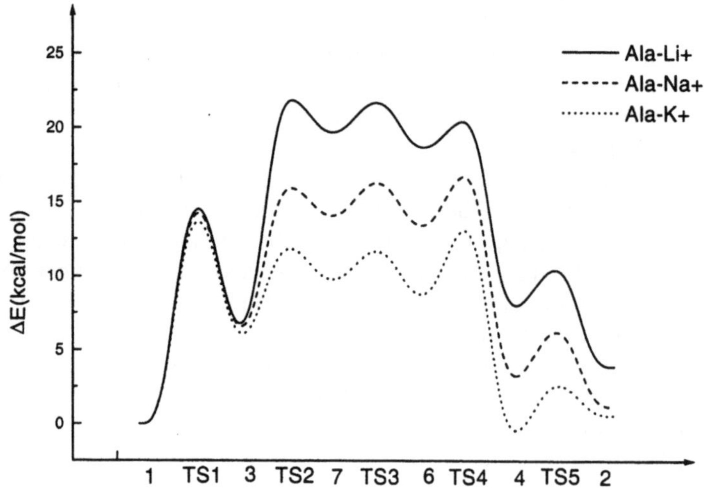

Figure 7. Potential energy surfaces for the α-alanine-M^+ (M= Li, Na, K) interactions.

The barrier for the next transformation **5-8**, consisting of a simple migration of M^+ toward the amine nitrogen (see TS7 in figure 8), is slightly higher with respect to the previous one, for lithium and sodium, but becomes lower for potassium cation. The formation of complex **8** starting from minimum **1** via species **5** is, in all the three cases, an endothermic process.

In the second channel (see figure 7) the step **1-3** is a cis-trans conversion that involves the hydrogen atom of the carboxyl group. In TS1 the hydrogen is placed at 90° with respect to the plane defined by O1-C-O2.

This isomerization occurs without particular problems as indicated by the values of the energetic barriers that fall in the range 14.2-14.9 kcal/mol for the three alkali metals.

In complex **3,** the shift of the bicoordinated metal towards the O2 that yields the monocoordinated species **7**, implies TS2 formation in which the cation appears to be out of the plane containing the O2, C, Cα and N atoms. This transition state presents a slightly different geometry depending on the particular metal ion. The conversion **3-7** requires 15.8, 9.9 and 6.2 kcal/mol for Li^+, Na^+ and K^+, respectively. The height of the barriers for the **7-6** isomerization is practically the same for the three metal species (about 2 kcal/mol). The transition state is obtained as a result of a torsion around the C-Cα axis and a rearrangement of the hydroxyl hydrogen that occur simultaneously. TS3 are product-like species, so, the very low energetic cost, can be easily justified. The **7-6** isomerization is the first exothermic process along the three paths.

From complex **6** it is possible to arrive at minimum **4** through TS4. The **6-4** energetic barriers are of 1.9, 4.0 and 4.8 kcal/mol for Li^+, Na^+ and K^+, respectively.

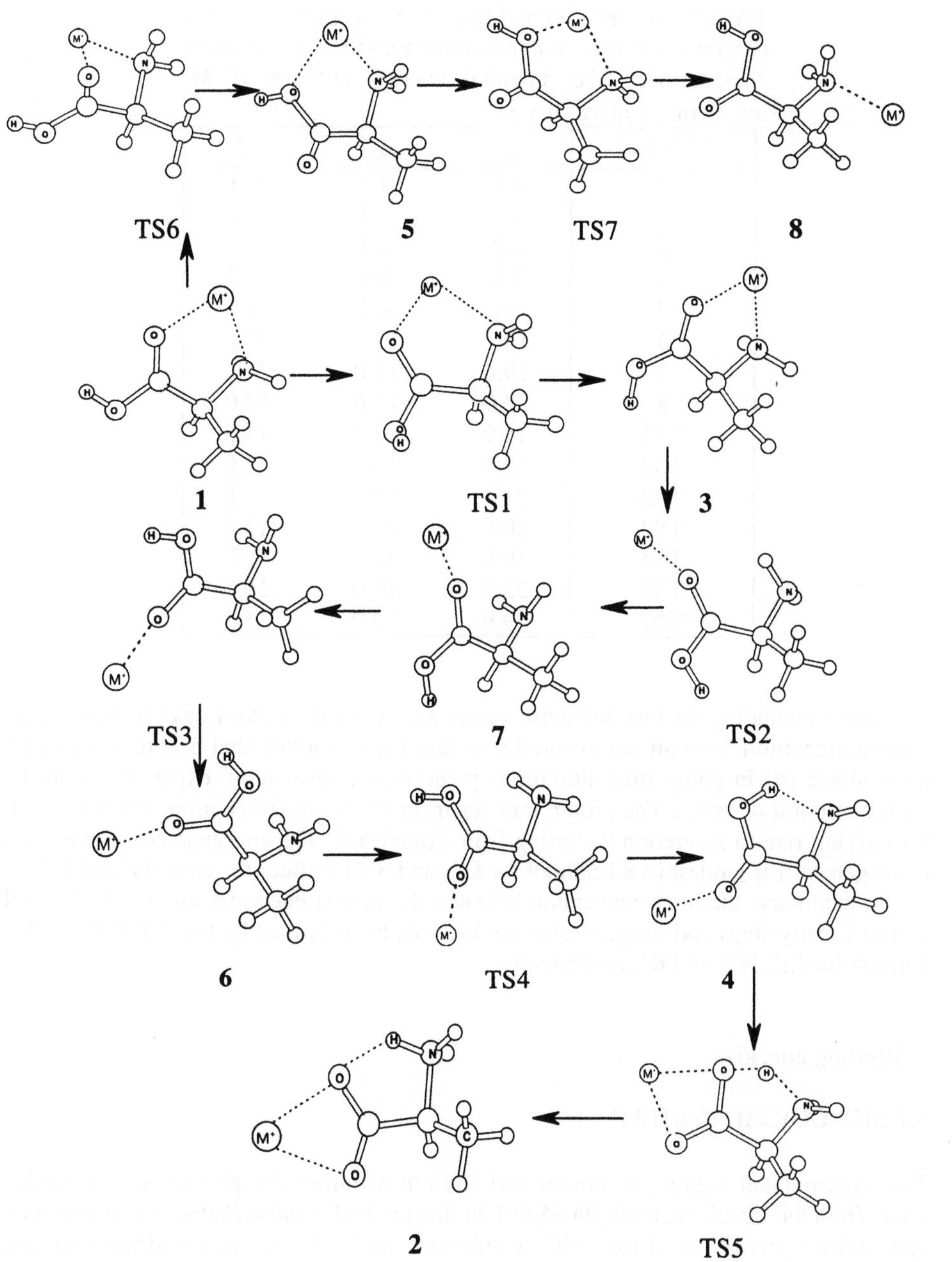

Figure 8. Minima and transition states on potential energy surfaces of the α-alanine-alkali ions interactions.

TABLE 4. B3LYP/6-311++G** relative energies (kcal/mol) at 0 K, for the minima and transition states characterizing the potential energy surfaces of M^+ interacting with α-alanine.

Structure	Li^+	Na^+	K^+
1	0.0	0.0	0.0
2	3.9	1.2	0.6
3	6.3	6.2	5.8
4	7.8	3.0	-0.7
5	9.2	8.6	8.1
6	18.7	13.3	8.6
7	19.6	14.0	9.7
8	23.9	17.6	14.0
TS1	14.9	14.7	14.5
TS2	22.1	16.1	12.0
TS3	21.8	16.4	11.8
TS4	20.6	17.0	13.4
TS5	10.6	6.4	2.7
TS6	23.3	17.0	16.7
TS7	29.9	19.0	15.3

Notwithstanding the fact that these values are very similar, they nevertheless reflect that the transition from an unflavoured structure (**6**) to another that becomes gradually more stable (**4**) in going from lithium to potassium, is ever more expensive from the energetic point of view . The proton transfer from the O1 to the N atom occurs through TS5 giving rise to zwitterion **2** starting from complex **4**. The energetic cost for the 1,4 hydrogen shift is modest (2.8 kcal/mol for Li^+, and 3.4 kcal/mol for both Na^+ and K^+).

In summary, the inter-conversion between the lowest energetic minima **1**, **2** and **4** requires many steps and the processes are kinetically controlled by the **3-7**, **6-4** and **1-3** barriers for Li^+, Na^+ and K^+, respectively.

6. Binding energies

6.1 BIOLOGICAL SYSTEMS

The experimental values of lithium and sodium affinities for glycine and α-alanine come from the kinetic method [94,95,96]. In this method metal affinities are obtained by generating a cationized dimer with a reference molecule of known affinity for the cation.

The binding characteristics of the two ligands in the dimer should be sufficiently similar in order to neglect relative dissociation entropy. These assumptions can be a source of errors and for this reason many authors [56,97,98] believe it necessary to revise the experimental data. The reference molecule used by Bojesen et al.[94] is N,N-dimethylformamide (DMF) for which an indirect measurement of MIA has been taken into account.

Our MIA B3LYP values at 298 K and corrected by BSSE, are reported in table 5.

The observation that more than one isomer can give MIA values in agreement with the experimental data, suggests another factor that must be taken into account in the theoretical evaluation of MIA and in the interpretation of the experimental measure. For this reason all the isomers of glycine and alanine have been complexed with the cations but only the result concerning the most stable system has been reported in the table 5 [93,99].

Excellent agreement exists for experimental MIA absolute values in the case of Na^+ cation. On the other hand, the MIA for Li^+ appear overestimated by about 7 kcal/mol.

Because the used computational tool has been widely tested with very good results for a great number of systems, the discrepancies found for the MIA values of glycine and alanine complexes with lithium, can be ascribed to the scarce accuracy in the experimental measurements obtained by the kinetic method.

The lithium, sodium and potassium ions affinities for the DNA and RNA nucleobases have been determined by the density functional method using the hybrid B3LYP exchange correlation potential and the 6-311+G(2df,2p) basis set (see table 5).

TABLE 5. MIA values (kcal/mol) of DNA, RNA bases and some selected amino-acids.

System	Theory	Experiment
Li^+-uracil	51.5	50.4
Li^+-thymine	51.6	51.4
Li^+-cytosine	53.8	55.4
Li^+-guanine	54.5	57.2
Li^+-adenine	49.0	54.1
Li^+-glycine	58.3	51.0
Li^+-alanine	58.8	52.6
Na^+-uracil	34.1	33.7
Na^+-thymine	34.2	34.4
Na^+-cytosine	42.2	42.3
Na^+-guanine	42.8	43.5
Na^+-adenine	29.7	41.1
Na^+-glycine	39.1	38.0
Na^+-alanine	40.5	39.4
K^+-uracil	25.3	24.1
K^+-thymine	25.0	24.4
K^+-cytosine	26.0	26.3
K^+-guanine	26.7	28.0
K^+-adenine	18.2	25.3

All the molecular complexes, obtained by the interaction between several low lying tautomers of nucleic acid bases and the alkali ions on the different binding sites, have been considered because the metal ion affinity value can depend on both the particular tautomer and the coordination mode of the cation to the base. Despite the fact that the small energy differences between the free bases tautomers have only a modest influence on the values of MIA, the different stable complexes that they produce can, on the contrary, play a decisive role. Results in these cases show that there is always only one tautomer to which it corresponds the metal affinity value closer to the experimental measurements. This tautomer is not necessarily the most stable one.

According to the experimental interpretation [100,101] for uracil, thymine and adenine, the MIA value can be associated to the most stable tautomer of the corresponding free bases. In fact, the high energy difference between their tautomers allows to the most stable ones to predominate also in the experimental conditions and to be easily metalated.

For cytosine, adenine and guanine the values of MIA closest to experimental data are obtained in correspondence of complexes different by those considered in the interpretation of the experimental work. For this reason the metal ion affinity value closest to experiment must be determined taking into account the role played by the different tautomers of the free bases with similar energy and all the possible complexes obtained by them.

The B-M^+ bond energies decrease as the size of the metal ion becomes larger. The affinities for nucleobases increase in the order adenine < uracil < thymine < cytosine < guanine for lithium and sodium and adenine < thymine < uracil < cytosine < guanine in the cases potassium cation.

The most important differences with the experimental trends concern mainly the adenine bases for which we obtain metal affinity values that appear quite a lot lower than all the measured values obtained by various techniques.

The level of theory used for the computations of metal ion affinity allows reliable predictions of this property with absolute mean deviations of 2.3 kcal/mol with respect to the values obtained by the kinetic method.

6.2 FIRST AND SECOND ROW TRANSITION METAL COMPLEXES

In this section we report some gas-phase metal-ligand bond strengths useful in the elucidation of bonding in transition-metal systems as well as in understanding the mechanism of a variety of catalytic reactions.

The binding energies calculated at B3LYP/DZVP level together with the experimental counterparts are reported in table 6.

The following trends can be evidenced:

The binding energy of ammonia and water complexes, at both experimental [63,79, 101-105] and theoretical levels, decrease in going from titanium to manganese cations, whereas the methylene ones have a similar behaviour with the exception of manganese for which the value increases. This behaviour can be attributed [106] to the loss of exchange energy and to the required promotion energy, if responsible of the formation of the bond is considered to be the spin-pairing of 4s orbital of M^+ in its $4s3d^{n-1}$ configuration.

The reactivity of Mo and Mo_2 with ammonia, and ethene molecules has been investigated by using B3LYP functional coupled with LANL2DZ effective potential [49] for molybdenum and DZVP all electron basis set for C, N and H .

The results obtained show that the molybdenum atom is able to react with C_2H_4 and binds weakly with NH_3. The dimer Mo_2 gives stable complexes with ammonia and ethene. For the Mo_2NH_3 complex, all the employed levels of theory give binding energies in good agreement with the experimental value while in the case of the MoC_2H_4 system, the use of model core potentials coupled with gradient-corrected exchange-correlation functionals overestimates the binding energies.

TABLE 6. Theoretical and experimental binding energies at 0 K for some selected transition metal containing systems. All values are in kcal/mol.

System	Theory	Experiment
Ti^+-NH_3	52.9	46.6± 1.6
V^+-NH_3	47.8	44.5± 1.8
Cr^+-NH_3	45.0	43.6±2.3
Mn^+-NH_3	37.2	39.0± 1.9
Sc^+-CH_3	57.5	55.8± 2.3
Ti^+-CH_3	54.0	51.0± 0.7
V^+-CH_3	45.4	46.1± 1.6
Cr^+-CH_3	34.2	26.3± 0.9
Mn^+-CH_3	54.5	49.1± 0.9
Ti^+-H_2O	38.2	37.7± 1.4
V^+-H_2O	38.4	35.8± 1.2
Cr^+-H_2O	35.1	31.7± 2.1
Mn^+-H_2O	30.3	29.1± 1.4
Mo-C_2H_4	16.7	17.0± 4.3
Mo_2-NH_3	15.1	14.0± 2.0

For MoC_2H_4 and Mo_2NH_3 we obtain binding energy values of 16.70 and 15.1 kcal/mol, respectively. The general agreement with experimental data is very satisfactory.

4. Conclusions

In this work we have shown the performance of the density functional method in the study of metal containing systems. The explored properties for a selected number of cases include the geometrical structures, the metal coordination modes, potential energy hypersurfaces, binding energies and metal ion affinities.

Our results confirm that this theoretical approach is able to correctly reproduce many important physico-chemical parameters, so it can be used for their predictions in systems for which the experiment is not easy to perform.

The computational efforts necessary to carry out the investigations are less demanding with respect to the methods for which the correlation effects must be treated at post-Hartree-Fock level. This aspect is very important in view of the possibility to study large systems with hundred-thousand atoms systems.

Of course, all that glitters is not gold, and other work is needed to improve the method and the relative computer codes. This work is in progress in many laboratories.

Acknowledgements

This work was supported by Università della Calabria, MIUR and progetto Memobiomar.

References

1.Kohn, W. and Sham, L.J. (1965) *Phys. Rev.* A**140**, 1133.

2.Andzelm, J. (1991) in J.K. Labanovski and J. Andzelm, Eds., *Density Functional Methods in Chemistry*, Springer Verlag, NewYork, p.195.

3.Salahub, D.R. and Zerner M.C. (1989) *The Challenge of d and f Electrons*, American Chemical Society, Washington, DC.

4.Russo, N. (1992) Density Functional Model Calculations for Homogeneous and Heterogeneous Catalysis, in N. Russo and D.R. Salahub (eds.), *Metal-ligand interactions. From Atoms, Clusters and Surfaces,* Kluwer, Dordrecht, pp.341-366.

5.Russo, N., Toscano, M. and Mineva T. (1995) Metal-ligand interactions in gas-phase and in solution. The density functional point of view, in N. Russo and D.R. Salahub (eds.), *Metal-ligand interactions. Structure and Reactivity*, Kluwer, Dordrecht, pp.453-472.

6.Marino T., Mineva T., Russo N., Sicilia E. and Toscano M. (2000) Potential energy surfaces of metal-ligand interactions obtained using density functional theory, in N. Russo and D. R. Salahub (Eds), *Metal-ligand interactions in chemistry, physics and biology,* Kluwer, Dordrecht, pp. 207-226.

7.Russo N., Abashkin Y., Calaminici P., Mineva T., Sicilia E. and Toscano M. (1995) Gaussian density functional method. An alternative tool for the prediction physico-chemical properties in D.P. Chong, (ed), *Recent advances in density functional methods* (Part I.), World Scientific, Singapore, pg. 200-250; De Luca, G., Mineva, T., Russo, N., Sicilia E. and Toscano M. (1997) Continuum dielectric models for the solvent and density functional theory: the state-of-the-art, in D.P. Chong (ed), *Recent Advances in Density Functional Methods* (part II),World Scientific, Singapore, pp. 41-59.

8.Abashkin, Y., Russo, N., Sicilia E. and Toscano M. (1995) Constrained optimization procedure for finding transition states and reaction pathways in the framework of gaussian based density functional method : the case of isomerization reactions, *Theoret. Comput. Chem.*, **2**, 255; Belcastro, M., Marino, T., Mineva, T., Russo, N., Sicilia E. and Toscano M. (1996) Density functional theory as a tool for the prediction of the properties in molecules with biological and pharmacological significance, *Theoret. Comput. Chem.,* **4**, 743.

9.Becke, A.D. (1993) *J. Chem. Phys.* **98,** 5648.

10.Lee, C., Yang, W. and Parr, R.G. (1988) *Phys. Rev. B* **37,** 785.

11.Perdew, J. P. and Wang, Y. (1986) *Phys. Rev. B* **33**, 8800.

12.Perdew, J. P. (1986) *Phys. Rev. B* **33**, 8822.

13.St-Amant, A., (1992) Ph.D Thesis, Université de Montréal.

14.Frisch, M.J, Trucks, G.W., Schlegel, H.B., Gill, P.M.W., Johnson, B.G., Robb, M.A., Cheesman, J.R., Keith, T.A., Petersson, G.A., Montgomery, J.A., Raghavachari, K., Al-Laham, M.A., Zakrzewski, V.G., Ortiz,, J.V., Foresman, J.B., Croslowski, J., Stefanov, B.B., Nanayakkara, A., Challacombe, M., Peng, C.Y., Ayala, P.Y., Chen, W.; Wong, M.W., Andres, J.L., Peplogle, E.S., Gomperts, R., Martin, R.L., Fox, D.J.; Binkley, J.S., Defrees, D.J., Baker, J., Stewart, J.P., Head-Gordon, M., Gonzales, C. and Pople, J.A. (1995) *Gaussian 94* (Revision A.1) Gaussian, Inc. Pittsburgh, PA.

15.Storhoff, B.N. and Lewis, H.C. Jr. (1977) *Coord. Chem.* **23**, 1.

16.Coluccia, M., Nassi, A., Loseto, F., Boccarelli, A., Mariggiò, M.A., Giordano, D., Intini, F. P., Caputo, P. and Natile, G. (1993) *J. Med. Chem.* **36**, 510.

17.Cini, R., Caputo, P.A., Intini, F.P. and Natile, G. (1995) *Inorg. Chem.* **34**, 1130.
18.Chatt, J. and Duncanson, L.A. (1953) *J. Chem. Soc.* 2239, and references therein.
19.Phillips, I.G., Ball, R.G. and Cavell R.G. (1992) *Inorg. Chem.* **31**,1633.
20.Hedberg, K., Hedberg, L., Hagen, K., Ryan, R.R. and Jones, L.H. (1985) *Inorg. Chem.* **24**, 2771.
21.Jost, A., Rees B. and Yelon W.B. (1975) *Acta Cryst.* **B31**, 2649.
22.Madhavan, P.V. and Newton, M.B. (1987) *J. Chem. Phys.* **87**, 4030.
23.Illas, F., Bachs, M., Rubio, J. and Ricart, J.M. (1989) *J. Chem. Phys.* **91**, 5466.
24.Lederer, T., Arvanitis, D., Comelli, G., Troger, L. and Barbershke, K. (1993) *Phys. Rev. B* **48**, 15390.
25.Lederer, T., Arvanitis, D., Tischer, M., Comelli, G., Troger, L. and Barbershke, K. (1993) *Phys. Rev. B* **48**, 11277.
26.Torras, J., Toscano, M., Ricart, J.M. and Russo, N. (1997) *J. Mol. Cat. A: Chem.* **119**, 387.
27.Braunstein, P., Matt, D. and Nobel, D. (1988) *Chem. Rev.* **88**, 747.
28.Mascetti, J. and Tranquille, M. (1988) *J. Phys. Chem.* **92**, 2177.
29.Galan, F., Fouassier, M., Tranquille, M. and Mascetti, J. (1997) *J. Phys. Chem.* **101**, 2626.
30.a) Rodriguez-Santiago, L., Sodupe, M. and Brachandell V. (1996) *J. Chem. Phys.* **105**, 9966; b) Sodupe, M., Branchadell and Oliva, A. (1995) *J. Phys. Chem.* **92**,2177.
31.Jeung, G.H. (1988) *Mol. Phys.* **65**, 669.
32.Caballol, R., Sanchez Marcos E. and Barthelat, J. (1987) *J. Phys. Chem.* **91**, 1328.
33.Sirois, S., Castro, M. and Salahub, D. (1994) *Int. J. Quantum Chem.* **S28**, 645.
34.Jeung, G.H. (1989) *Mol. Phys.* **67**, 747.
35.Mele, F., Russo, N. and Illas, F. (1998) *Chem. Phys. Lett.* **295**, 409.
36.Martinez, A., Koster, A. M. and Salahub, D.R. (1997) *J. Phys. Chem. A* **101**, 1532.
37.Lian L. , Mitchell S.A. and Rayner D.M. (1994) *J. Phys. Chem* **98**, 11637.
38.Carroll J.J. , Haug K.L. and Weisshaar J.C. (1993) *J. Am. Chem. Soc*, **115**, 6962.
39.Holm R.H. (1990) *Coord. Chem. Rev.* **100**, 183.
40.Oku H. , Ueyama N. and Nakamura A. (1995) *Inorg. Chem.* **34**, 3667.
41.Jeung G.H.(1994) *Chem. Phys. Lett.* **221**, 237.
42.Bauschlicher C.W. Jr , Partridge H. , Sheeky J.A. , Langhoff S.R. and Rosi M. (1992) *J. Phys. Chem*, **96**, 6969.
43.Blomberg M.R.A. ; Siegbahn P.E.M. and Svensson M. (1992) *J. Phys. Chem.* **96**, 9794.
44.Diez R.P. (2000) *Int. J. Quantum Chem.* **76**, 105.
45.Mineva T. , Russo N. and Sicilia E. (1999) *Bulg. Chem. Ind.* **70**, 22.
46.Berces A. , Mitchell S.A. and Zgierski M.Z. (1998) *J. Phys. Chem. A* **102**, 6340.
47.Martinez A. ; Koster A.M. and Salahub D.R. (1997) *J Phys. Chem. A* **101**, 1532.
48.Torrent M., Gili P., Duran M. and Solà M. (1997) *Int. J. Quantum Chem.*, **61**, 405.
49.Lacaze-Dufaure, C., Mineva, T. and Russo, N. (2001) *J. Computat. Chem.* **22**,1557.
50.Fournier R. (1995) *J. Chem. Phys.*, **102**, 5396.
51.Rayner D.M., Lian L., Fournier R., Mitchell S.A. and Hatckett P.A. (1995) *Phys. Rev. Lett.*, 2070.
52.(a) Jia, X.; Zon, G.; Marzilli, L. G. (1991) *Inorg. Chem.* **30**, 228. (b) Froystein, N. A., Davis, J. T., Reid, B. R. and Sletten, E. (1993) *Acta Chem. Scand.* **47**, 649. (c) Yang, D., van Boom, S.S.G.E., Reedijk, J., van Boom, J.H., Wang, A.H.-J. (1995)

Biochemistry, 34, 12912. (d) Marzilli, L.G., Ano, S.O., Intini, F. P. and Natile, G. (1999) *J. Am. Chem. Soc.*, **121,** 9133.

53.Duguid, J., Bloomfield, V. A., Benevides, J. and Thomas, G. J. (1993) *Biophys. J.*, **65**, 1916.

54.(a) Abrescia, N. G. A.; Malinina, L.; Fernandez, L. G.; Huynh-Dinh, T.; Neidle, S.; Subirana, J. A. *Nucleic Acids Res.* **1999**, 27, 1593. (b) Egli, M.; Williams, L. D.; Fredericks, C. A.; Rich, A. *Biochemistry* **1991**, 30, 1364. (c) Coste, J.; Maligne, J.-M.; Serre, L.; Shepard, W.; Roth, M.; Leng, M.; Zelwer, C. *Nucleic Acids Res.* **1999**, 27, 1837. (d) Takahara, P. M.; Fredericks, C. A.; Lippard, S. J. *J. Am. Chem. Soc.* **1996**, 118, 12309. (e) McFail-Isom, L.; Shui, X.; Williams, L. D.; *Biochemistry* **1998**, 47, 17105. (f) Harper, A.; Brannigan, J. A.; Buck, M.; Hewitt, L.; Lewis, R. J.; Moore, M. H.; Schneider, B. *Acta Crystallogr. D* **1998**, 54, 1273. (g) Correl, C. C.; Freeborn, B.; Moore, P. B.; Steitz, T. (1997)*Cell* **91,** 705. (h) Wang, A. H.-J.; Hakoshima, T.; van der Marel, G.; van Boom, J. H.; Rich, A. (1984) *Cell* **37**, 321.

55.Russo, N., Toscano, M. and Grand, A. (2001) *J. Phys. Chem.* **105**, 4735.

56.Cerda B. A. and Wesdemiotis C. (1996) *J. Am. Chem. Soc.* **118**, 11884.

57.Russo, N., Toscano, M. and Grand, A. (2001) *J. Am. Chem. Soc.* **123**, 10272.

58.Russo N., Sicilia E., Toscano M. and Grand A. (2002) *Int. J. Quantum Chem.* In press

59.Freas, R. B.; Ridge D. P. (1980) *J. Am. Chem. Soc.*, **102**, 7129.

60.Aristov, A.; Armentrout, P.B (1984) *J. Am. Chem. Soc.* **106**, 4065.

61.Tolbert, M.A. ; Beauchamp, J.L. B (1984) *J. Am. Chem. Soc.* **106**, 8117.

62.Reents, W.L., Strobel, F., Freas, R.B., Wronka, J. and Ridge D.P. (1985) *J. Phys. Chem.* **89**, 5666.

63.Aristov, N. and Armentrout, P.B. (1986) *J. Phys. Chem.* **108**, 1806.

64.Kang, N. and Beauchamp, J.L. (1986) *J. Am. Chem. Soc.* **108**, 7502.

65.Aristov, N. and Armentrout, P.B. (1987) *J. Phys. Chem.* **91**, 6178.

66.Sunderlin, L.S. and Armentrout P.B. (1988) *J. Phys. Chem.* **92**, 1209.

67.Georgiadis, R. and Armentrout, P.B (1988) *J. Phys. Chem.* **92**, 7067.

68.Tonkyn, R., Ronan, M. and Weisshaar, J.C. (1988) *J. Phys. Chem.* **92**, 92.

69.Buckner, S.W. , Gord, J.R. and Freiser, B.S. (1988) *J. Am. Chem. Soc.* **110**, 6606.

70.Irikura, K.K. and Beauchamp, J.L. (1989) *J. Am. Chem. Soc.* **111**, 75.

71.Sunderlin, L.S. and Armentrout P.B. (1989)*J. Am. Chem. Soc.* **111**, 3845.

72.Magnera, T.F., David, D.E. and Michl, J. (1989) *J. Am. Chem. Soc. **111***, 4100.

73.Armentrout, P.B and Beauchamp, J.L. (1989) *Acc. Chem. Res.* **22**, 315.

74.Russel, D.H, Ed., (1989) *Gas Phase Inorganic Chemistry*, Plenum, New York, p.412.

75.Armentrout, P.B, (1990) *Annu. Rev. Phys. Chem.* **41**, 313.

76.Clemmer, D.E., Sunderlin, L.S. and Armentrout, P.B. (1990) *J. Phys. Chem.* **94,** 208.

77.Clemmer, D.E., Sunderlin, L.S. and Armentrout, P.B. (1990) *J. Phys. Chem.* **94**, 3008.

78.Armentrout, P.B, (1991) *Science* **41**, 175.

79.Fisher, E.R.. and Armentrout, P.B (1992) *J. Am. Chem. Soc.* **114**, 2049.

80.Guo, B.C., Kerns, K.P. and Castleman, A.W. (1992) *J. Phys. Chem.* **96**, 4879.

81.Clemmer, D.E., Aristov, N. and Armentrout, P.B. (1993) *J. Phys. Chem.* **97**, 544.

82.Chen, Y., Clemmer, D.E. and Armentrout, P.B. (1994) *J. Phys. Chem.* **98**, 11490.

83.Chen, Y. and Armentrout, P.B. (1995) *J. Phys. Chem.* **99**, 10775.

84.Haynes, C.L.; Chen, Y.; Armentrout, P.B. (1995) *J. Phys. Chem.* **99**, 9110.
85.Irigoras, A., Fowler, J.E. and Ugalde, J.M. (1998) *J. Phys. Chem.* **102**, 293.
86.Irigoras, A., Fowler, J.E. and Ugalde, J.M. (1999) *J. Am. Chem. Soc.* **121**, 574.
87.Irigoras, A., Fowler, J.E. and Ugalde, J.M. (1999) *J. Am. Chem. Soc.* **121**, 8549.
88.Irigoras, A., Elizalde, O., Silanes, I., Fowler, J.E. and Ugalde, J.M. (2000) *J. Am. Chem. Soc.* **122**, 114.
89.Russo, N. and Sicilia, E. (2001) *J. Am. Chem. Soc.* **123**, 2588.
90.Bent, H.A. (1966) *J. Chem. Educ.* **43**, 170.
91.Hoyau, S., Norrman, K., McMahon, T.B. and Ohanessian G. (1999) *J. Am. Chem. Soc.*, **121**, 8864.
92. Stryer, L. (1988) *Biochemistry*, W. H. Freeman, New York.
93.Marino T., Russo N. and Toscano M. (2001) *Inorg. Chem.* **40**, 6439.
94.Bojesen G., Breindhal T. and Andersen U. (1993) *Org. Mass Spectrom.* **28**, 1448.
95.Cooks R.G. , Patrick J.S. , Kohaho T. and McLuckey S. A. (1994) *Mass Spectrom. Rev.* **13**, 287.
96.McLuckey S. A., Cameron D. and Cooks R. G. (1981) *J. Am. Chem. Soc.* **103**, 1313.
97.Hoyau S. and Ohanessian G. (1998) *Chem. Eur. J.* **4**, 1561.
98.Rodgers M. T. and Armentrout P. B., (2000) *J. Am. Chem. Soc.* **122**, 8548.
99.Marino, T., Russo, N. and Toscano, M. (2000) *J. Inorg. Biochem.* **79**, 179.
100.Rodgers, M.T. and Armentrout, P.B. (2000) *J. Am. Chem. Soc.* **122**, 8548.
101.Derek, W. and Armentrout, P.B. (1998) *J. Am. Chem. Soc.* **120,** 3176.
102. Armentrout, P.B. (1990) *ACS Symp. Ser.* **4281,** 18.
103.Sunderlin, L.S. and Armentrout, P.B. (1989) *Int. J. Mass Spectrom. Ion Processes*, **94**, 149.
104.Georgiadis, R. and Armentrout, P.B. (1989) *Int. J. Mass Spectrom. Ion Processes* **89**, 227.
105.Dalleska, N.F., Honma, K., Sunderlin, L.S. and Armentrout P.B. (1994) *J. Am. Chem. Soc.* **116,** 3519.
106.Sicilia, E. and Russo, N. (2002) *J. Am. Chem. Soc.* **124**,1471.

REVIVAL OF PARAMETRIZED THEORETICAL METHODS FOR APPLIED SCIENCE STUDIES: THE EXAMPLE OF TRANSITION-METAL CLUSTERS

G. Berthier [1], M. Defranceschi [2], C. Le Bris [3]

(1) LETMEX, Université de Nice-Sophia-Antipolis et Observatoire de la Côte d'Azur, UMR 6103, Parc Valrose, 06108 Nice Cedex 2, France

(2) CEA/DEN/DSOE/RB, Centre de Saclay, Bât. 121, 91191 Gif-sur-Yvette Cedex, France

(3) CERMICS, Ecole Nationale des Ponts et Chaussées, 6 & 8, avenue Blaise Pascal, Cité Descartes, Champs-sur-Marne, 77455 Marne-La-Vallée Cedex 2, France

Abstract : The application of parametrized molecular orbital method to transition metal clusters is surveyed. Theoretical tools for the interpretation of their chemical reactivity in terms of electronic structures are presented. Problems of catalytic reactivity in mixed Pd-Rh clusters and changes connected to the formation of mixed Tc-Ru clusters by spontaneous transmutation of ^{99}Tc are examined.

1 Foreword

The endless increase of computer possibilities during the last decades has resulted in an unexpected feedback in Theoretical Chemistry. Parallely to the improvement of the accuracy of *ab initio* calculations involving large basis sets, big configuration interactions for rather small molecules, etc., there was a revival of more or less parametrized procedures in the field of applications concerning realistic systems. This is mainly due to the excellent ratio quality/cost (in material and human resources) of these methods for those who are interested in reliable predictions rather than by subtle details of electronic structures.

Striking examples of the present situation are given by the modelling of problems of biological interest, using Molecular Dynamics methods through well-known commercial codes. Basically, such approaches are lying on mechanical potentials, which have been fitted carefully for the study of molecular conformations in Organic Chemistry [1]. As concerns condensed matter and its environment (i.e., solvation problems), there is apparently a tendency now to use methods derived from Density Functional Theory, which include - in addition to the parametrization of the functionals themselves- that of the thermostats added in order to avoid blow-up risks [2, 3].

A second class of parametrized treatments proceeds from the concept of pseudopotentials borrowed from Solid State theories. They are

E.J. Brändas and E.S. Kryachko (eds.),
Fundamental World of Quantum Chemistry, Vol. II, 467–482.

methods which enable us to eliminate particular electrons from a system by arguing that their consideration is burdening the computation time without physical necessities; for instance, inner-shell electrons for the description of the chemical bonds. The interaction between the remaining electrons and the cores formed by the former ones and the cores (i.e., ions in the language of Solid State Physics) are described by effective potentials (the ECP's), that is algebraic expressions fitted, most often by least-square procedures, on appropriate theoretical or experimental data [4, 5]. Spin Hamiltonian theory [6] or Molecular Mechanics [1] are limiting cases of such techniques, where the electronic variables are reduced to their spin components or there are no electrons at all.

Finally, it is still possible to have recourse to more traditional semi-empirical treatments culminating in the nineteen fifties and sixties with the extended Hückel method EXTHUC [7] provided a few improvements suggested by modern *ab initio* methods are taken into account. In this respect, the authoritative surveys published by Per-Olov Löwdin in Philosophical Magazine [8] or Advances in Quantum Chemistry [9] are irreplaceable especially with all the problems connected to overlap. Our work on transition-metal clusters belongs to this category of parametrized methods.

Using the keyword "clusters" in order to access to electronic data files yields information on a large variety of molecular species such as polynuclear coordination complexes, molecular crystals, glasses, aggregates of carbon, silicon, noble gases [10], etc... besides metallic clusters. If we agree to consider as "genuine clusters" compounds formed by only one or more kinds of atoms very close chemically speaking (i.e., "pure" and "mixed" clusters), there are many metals that are known as giving molecules of the cluster type : alkalines, transition metals, lanthanides, ... The simplest metal clusters for theoretical studies are those formed by alkalines : Li, Na, K, etc., because it is possible to model their electronic structure on that of an one-orbital, one-electron system [11, 12], that is a network looking like a π-electron system in aromatic molecules, except we have *s*-orbitals instead of π ones and, consequently, more overlap. This opens the door to very useful graph theory arguments [13] as those presented below.

In two recent papers [14, 15], we were concerned with second-row transition metal clusters (from Tc to Pd) in connection with their chemical reactivity and their catalytic activity. The presence of more or less fully occupied *d*-shells was the main difficulty we had to overcome, in order to understand their electronic structure. Let us mention some important points : First of all, the presence of *d*-electrons not assigned to definite chemical bonds gives rise, when the size of the cluster increases, to incompletely filled shells prefiguring the conduction band of metallic solids. Consequently, monodeterminantal *ab initio* models (or, alternatively, pure states v-representable with integers as occupation numbers of molecular orbitals in the DFT terminology) are most defective. To this former factor of complexity, a second one is added by the effective electronic structure of the atomic components of the clusters, i.e. for a transition metal M=[core]$nd^q(n+1)s$ rather than a configuration of the type nd^{q+1} or $nd^{q-1}(n+1)s^2$, because the promotion of an *s* valence electron is necessary to achieve the chemical bonding of two neighbouring atoms [16].

As a matter of fact, the post Hartree-Fock calculations performed up to now for dimers and small metal particles involving transition elements (for instance for Pd [17-21] and Rh clusters [22]) indicate that these molecular systems have a huge number of very close almost degenerated electronic states, with different spin multiplicities and various geometries, in the vicinity of the expected ground state, the latter being subjected to Jahn-Teller distortion effect in its valence shells [23, 24]. The same could be said, *mutatis mutandis*, for DFT studies [25, 26]. In any case, use of sophisticated *ab initio* methods, taking into account, possibly, relativistic effects [27] are necessary.

To circumvent these intricacies, leading to prohibitive computational costs for big clusters a come-back to traditional semi-empirical methods is hardly avoidable. Our initial idea was the possibility of simulating, by an appropriate choice of parameters, the main features of the *ab initio* treatments by means of tight-binding/EXTHUC calculations leading to clear-cut chemical interpretations. So, we postpone the description of our method after that of the chemical questions which it has been designed for.

2 Chemical aspects of cluster computations

Given their semi-empirical background, an important criterion for the validity of the parametrized calculation methods is their ability of depicting the trend of a series of closely related compounds as regards the variation of different properties, for instance binding energies. The study of metal particles formed by second-row transition elements has a special advantage. The physical observables are more smoothly connected with the atomic structure of their components [28], without some complications arising from the poor overlap of the 3*d*-orbitals on nearest neighbours in the first row series [29] or the relativistic 6*s*, 6*p*-shell contraction and the 5*d*-shell expansion in the third series [30].

The first information given by a quantum mechanical calculation on a pure cluster P_n is its atomization energy (AE), that is the balance of the dissociation process:

$$P_n \rightarrow nP \quad + (AE) \qquad (1)$$

where atom P may be taken in a convenient valence state rather than in its spectroscopic ground state if some electron (i.e., for transition elements, one of the outer shell s^2, has been decoupled for chemical purposes). To compare the AE values for clusters of different size, it is customary to divide the energy differences obtained in this way either by the number n of atoms P or the number m of atom pairs P-P directly connected on the graph of the cluster [26]. This defines atomization energies per atom or per bond:

$$D_P = AE / n, \quad D_{P-P} = AE / m \quad (n \leq m \leq n(n-1)) \qquad (2)$$

which may have some relationship with the bulk formation energies of solids. Such a procedure, however, implies that the geometrical arrangement of the atoms in the cluster, in other words its point group, has something to do with the structure of the infinite lattice corresponding to the bulk [31], i.e. a face-centered cubic one (fcc) in our series of transition metals. Clearly, the preceding definitions are meaningless for mixed clusters formed of different atoms P and Q, thus including different types of bonds P-P, Q-Q, P-Q. Our approach of the problem was the following : Consider, for simplicity, a pair of monosubstituted clusters $P_{n-1}Q$ and $Q_{n-1}P$ whose parents, P_n and Q_n, have the same number m of bonds. We can calculate atomization energies $AE_P/n = D_P$ and $AE_Q/n = D_Q$ or, alternatively, atomization energies per bond $AE_P/m=E_{PP}$ and $AE_Q/m=E_{QQ}$. From these basic quantities as reference data, we can try to express the atomization energies AE_{PQ} of the mixed clusters in terms of atom increments δ_a and bond increments δ_b, that is for $P_{n-1}Q$:

$$AE_{PQ} - (n-1)D_P - D_Q = \delta_a \qquad (3)$$

$$AE_{PQ} - (m - m_Q)E_{PP} - m_Q E_{PQ} = \delta_b \qquad (4)$$

m_Q being the number of P-Q bonds. The second procedure requires the evaluation of the bond parameter E_{PQ}, for which we have chosen the Pauling arithmetic rule for bond energies :

$$E_{PQ} = \tfrac{1}{2}(E_{PP} + E_{QQ}) \qquad (5)$$

a relationship working well for dissociation energies of heteronuclear diatomics with transition metals [32]. We get:

$$(AE_{PQ} - \delta_b) = (m - \frac{m_Q}{2})E_{PP} + \frac{m_Q}{2}E_{QQ} \qquad (6)$$

Extension of this formula to polysubstituted systems is straightforward. Even if in some cases, e. g. the tetrahedral and the octahedral clusters, both increments coincide for algebraic reasons, bond parameters δ_b have to be preferred because the quotient $\varepsilon_{PQ} = \delta_b / m_Q$ has a formal structural meaning, namely that of a positive or negative energy increment (deficit or excess of energy respectively), localized on the mixed bonds of the $P_{n-1}Q$ molecule. Atomization energies as defined above are connected to the magnitude of the energy stored in a cluster; if the latter becomes available for chemical reactions, we are led to interpret ε_{PQ} as a reactivity index with respect to reactions involving bonds specifically.

Rough, but significant predictions concerning the attack of specific positions in a molecule by various reagents (ions or radicals) may be often made by inspecting the distribution of charge and spin densities among the various atoms, as given by conventional Mulliken population analysis for total and unpaired-electron densities including overlap. In addition to charges, the spin distribution is relevant for probing the reactivity of the different sites of clusters formed by transition metals having incomplete outer shells of electrons. The corresponding quantities given in the form of unpaired-electron distributions normalized to unity and, for simplicity, averaged for similarly located atomic centers, are useful as indices for reactions involving radicalic species, but also for reactions with cations and anions in the frame of the frontier-electron theory; they indeed coincide with the so-called condensed Fukui functions [33], whose values are identical for electrophilic, nucleophilic and radicalic reactions in the case of HOMO and LUMO molecular orbitals belonging to the same set of energy levels.

Let us finish this section by a warning on the atomic charges obtained by one-orbital, one-electron models (or their avatars) for a 2n-electron system distributed among n identical centers. Using well-known arguments of the π-electron theory, it is found that the resulting charge distribution is uniform or in other words the treatment is self-consistent [34] if we have an alternant (bichromatic) chemical graph. On the other hand, there are exaggerated charge transfers every time the graph includes odd-membered rings, but the matrix elements of the corresponding molecular orbital model can be easily corrected [35]. Before using them for chemical purposes, similar improvements are to be made on cluster calculations involving odd cycles, as it is the case for the icosahedral structures considered below.

3 Parameterization

The EXTHUC method in the primitive form suggested by Wolfsberg and Helmholz for the study of transition metal ions [36] has been used as the core of our simulation program: The molecular orbitals are computed in a non-iterative way where the diagonal elements of the matrix Hamiltonian for the *s*, *p*, *d* basis functions of the valence electrons are identified to corresponding atomic terms and off-diagonal elements are evaluated by means of the standard formula:

$$H_{PQ} = K(H_{PP} + H_{QQ}).S_{PQ}^{-1} = \frac{k}{2}(\omega_P + \omega_Q)S_{PQ}. \tag{7}$$

In the second-quantization language, this means that we start with an effective one-electron model Hamiltonian of the form:

$$H = \sum_S \sum_{P,Q} \left[(1-k)\delta_{PQ} + \frac{k}{2}(\omega_P + \omega_Q) \right] . S_{PQ} a_{PS}^{+} a_{QS} \tag{8}$$

where S is an index denoting spin-up and spin-down values [37]. A semi-empirical formalism of this kind for molecules is basically equivalent to a tight-binding approach for solids, if overlap is included in both theories [38]. Its application to a regularly arranged chain of atoms has been proved to generate the expected band structure when the number of atoms goes to infinity [39].

Diagonalizing H yields the total electronic energy of the system as a sum of energy orbitals e_i multiplied by occupation numbers ν_i appropriate to its most plausible configuration, namely:

$$E_1 = \sum_i \nu_i e_i \tag{9}$$

(ν_i = 2 or 0 for closed-shell, fractional numbers for open shells). To get bonding however, it is necessary to introduce repulsion terms, generally in parametrized exponential form, between the various atomic cores P and Q to which the valence orbitals are attached [40]. For clusters, the Born-Mayer formula [41, 42] :

$$E_C = \sum_{k'} \sum_{k \neq k'} \alpha_{PQ} e^{-\beta_{PQ}(r_{k,k'} - d_{PQ})} \tag{10}$$

with a specific energy prefactor α_{PQ} for the pair of atoms P Q located at centers k and k' and a dimensionless damping factor $\beta_{PQ(r_{k,k'} - d_{PQ})}$, is particularly convenient, for these parameters can be evaluated algebraically from reference data (at least for pure dimers and tetramers). The constant d_{PQ} can be identified to be an effective bulk interatomic distance. The preceding formula is valid in so far the interaction potential inside the molecule is isotropic; this is the case of the most symmetrical structures P_4, P_6 and P_{13} we have studied. Otherwise, extended Born-Mayer repulsion formulas with three-dimensional α_{PQ} nd β_{PQ} parameters should be considered.

Now, let us give numerical details on the simulation of the *ab initio* results by our EXTHUC method in the Tc to Pd series of clusters.

i) The values given by Alvarez for Pd and Rh in a comprehensive survey [43] have been selected (Tab. 1) and the overlap integrals, S_{PQ}, have been calculated from Slater integrals with the exponents recommended by Pietro and Hehre for the first and second-row transition metals [44]. The proportionality constant was fixed to K=3, a rather high value that has the advantage of isolating the block of the molecular orbitals of dominant *d*-character from the others in the one-electron spectra of all the clusters considered. This choice, indeed, leads to a clear-cut distribution of the occupied molecular orbitals into two families, those having a dominant *s*-character, which can be associated to the skeleton of the cluster, and those having a dominant *d*-character, which mimic the *d*-band of solids in the case of finite-size system. The orbital energies e_d, forming an almost closed-shell are distributed around the H_{dd} matrix elements in a rather small interval, whereas the energies of the former look like the eigenvalues of the topological graph formed by a network

Table 1. Résumé of EXTHUC parameters used for second-row transition metal clusters

P	orbital	H_{PP}	ζ	d_{PP} tetra	hexa	13-atom
Tc	5s-5s 5p-5p 4s-4s	-0.370 -0.198 -0.471	1.85 1.85 3.00	4.8545	4.9468	5.0415
Ru	5s-5s 5p-5p 4s-4s	-0.330 -0.184 -0.466	1.75 1.75 3.20	4.7500	4.9210	4.9330
Pd	5s-5s 5p-5p 4s-4s	-0.2690 -0.1380 -0.4420	1.80 1.80 3.60	5.0758	5.1280	5.1695
Rh	5s-5s 5p-5p 4s-4s	-0.2970 -0.1680 -0.4590	1.75 1.75 3.55	4.8150	4.9210	5.0153

including n equivalent vertices [13]. If, for simplicity, the matrix element H_{pq} associated to every atom pair of the graph is supposed to have a common value γ, one finds a lowest eigenvalue e_i equal to $(n-1)\gamma$ and (n-1) highest eigenvalues equal to $-\gamma$. Therefore, it is possible to easily recognize the pseudo *d*-band of the cluster by taking a not too small K value ; increasing K, however, does not imply a corresponding variation of the binding energy of the cluster because the core repulsion terms included in the treatment has to be readjusted accordingly.

ii) The total electronic energy E_1, computed as a sum of occupied orbital energies e_i, involve occupation numbers ν_i. They are parameters to be determined by taking into account the specificity of the cluster.

For the orbitals having a *d*-dominant character, ν_i has been fixed to the average population of the *d*-orbitals of the isolated atom in its chemical significant $d^q s^1$ valence state, i.e. $\nu_d = q/5$ for a pure cluster P_n and $(nq + nq')/5(n + n')$ for a mixed cluster $P_nQ_{n'}$. This choice gets some support in the values computed by *ab initio* calculations of very simple clusters concerning the populations of the *d* natural orbitals [17]. Furthermore, the *d* pseudo-core built in this way does not contribute very much to the cluster binding energies, just because it includes the same number of bonding and antibonding molecular orbitals with the same populations as in the isolated atoms.

The total number of remaining electrons, exactly equal to the number of atomic centers, is distributed among the *s*-dominant orbitals with a double occupancy or an equidistribution for those located below or above the *d*-block. The decoupling of the outer electrons in our model is reminiscent of the local spin density picture of metallic solids [45]. This neglects the degeneracy loss of the outer orbitals occurring -even in the graph theory- as soon as we distinguish between neighbor and remote atomic centers of the molecular skeletron. Let us add that for sufficient large clusters (e.g., n=13), low *s*-dominant energy levels may be interposed between the lowest one and the *d*-block; for them, a double occupancy is postulated.

iii) All the calculations have been carried out with frozen geometries, in the hypothesis they correspond to interatomic distances at equilibrium. Trial computations were performed on tetrahedral clusters Pd_4 and Rh_4, for which precise *ab initio* results are available [20, 22]. The bond lengths necessary to a similar study for Ru_4 and Tc_4 were obtained by linearly extrapolating those of Rh_4 from bulk. To supply the lack of data for larger clusters, a three-point interpolation formula has been used, where the variables are the bond orders given by the graph theory and the reference data corresponding to dimers, tetramers and bulk. Bond lengths between different metals in mixed clusters were taken as arithmetic averages of their homologues in pure clusters of the same size, and the overall geometries of these systems have been constructed accordingly.

iv) Regarding the determination of the α and β repulsion parameters, our simulation technique has its starting point in the results of preliminary calculations made on reference clusters, i.e. they have to reproduce the *ab initio* atomization energies of tetrahedral pure clusters. In the EXTHUC approach, the atomization energies (binding energies) are obtained by substracting the atomic energies, corrected possibly by the contribution of the core repulsion term E_C. Separating the part A included in E_1 as the sum of the $(4d^9 5s^1)$ matrix elements from their molecular counterpart B and writing that the forces acting on the cores are zero at the equilibrium, we have the two conditions:

$$B + C = E_{B+C} \tag{11}$$

$$\partial B/\partial r_k + \partial C/\partial r_k = 0 \qquad r = (x, y, z) \tag{12}$$

where the first equation involves the value E_{B+C} assigned to the atomization energy of the reference compound, and the second equation the Hellmann-Feynman derivatives of orbital energies e_i of E_1 and pair repulsion terms of E_C with respect to the positions r_k of nuclei.
The preceding calculations implemented in our EXTHUC program with fixed prefactors α_{PQ} and damping exponents β_{PQ} simultaneously are straightforward for the Pd_4 and Rh_4 clusters, *ab initio* atomization energies being available. It is not the case for Ru_4 and Tc_4 so that an extrapolation procedure from the Rh_4 value to the other two metals has been used to have the necessary data, assuming that the ratio λ between the partial atomization energy B and the *ab initio* result is transferable. The β_{PQ} parameters computed in this way for tetrahedral geometries have been kept for larger clusters, but the α_{PQ} prefactors have been reoptimized for the 6- and 13-atom compounds, using the condition of zero Hellmann-Feynman forces (see Fig.1 for values). This indeed, is required for ensuring the size consistency of the whole procedure.
The α_{PQ} and β_{PQ} parameters and the bulk distances d_{PQ} in mixed clusters have been estimated by taking the geometrical means of the α_{PQ}'s and the arithmetic means of the β_{PQ}'s and d_{PQ}'s used for corresponding pure clusters. The geometrical means for a prefactor was suggested by Born-Mayer paper in 1932.
v) Modelling large clusters with nearly equivalent atoms being the main goal of this study, the charge corrections in 13-atom compounds have been achieved by postulating zero net charge transfers between the

	Symmetry	n m	α	β	d
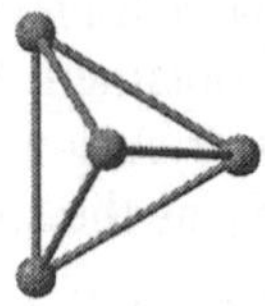	*Td*	4 6	Tc 0.054	0.697	5.11
			Ru 0.055	0.526	5.01
			Pd 0.060	0.522	5.2
			Rh 0.041	0.618	5.08
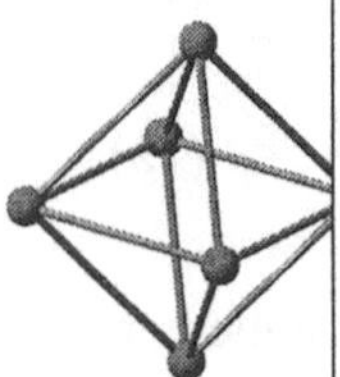	*Oh*	6 12	Tc 0.038	0.697	5.11
			Ru 0.031	0.526	5.01
			Pd 0.023	0.522	5.2
			Rh 0.045	0.618	5.08
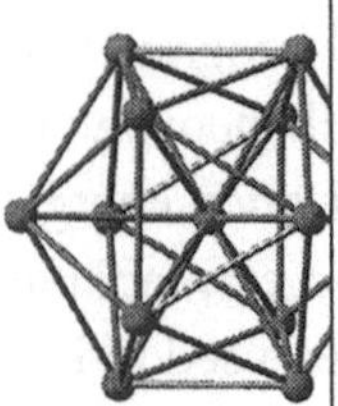	*Ih*	13 42	Tc 0.022	0.697	5.11
			Ru 0.017	0.526	5.01
			Pd 0.030	0.522	5.2
			Rh 0.017	0.618	5.08

Figure. 1 Born Mayer parameters for second-row transition metal clusters

surface and center atoms of the octahedron. The neutrality of pure clusters of this family is recovered by small positive increments of some thousands (in a.u.) for the former and, of course, a negative correction twelve times greater for the latter.

4 Some results

Three points illustrating the possibilities of the preceding methodology for the study of problems concerning the electronic structure of transition-metal clusters will be briefly discussed in this final section: The variation of atomization energies of pure clusters from Tc to Pd with respect to their size and shape; the origin of the increased catalytic reactivity of mixed Pd-Rh clusters; the change of electronic structure of Tc clusters arising from the spontaneous transmutation of the ^{99}Tc nucleus into ^{99}Ru.

Table 2. Atomization energies per atom in pure clusters (in a.u) with respect to $4d^{q}5s$ configurations; a: without charge corrections and b) with charge corrections.

-AE/n		Tc	Ru	Pd	Rh
n=4		0.1668	0.1594	0.139	0.074
n=6		0.1791	0.1886	0.154	0.085
n=13	a	0.2658	0.2601	0.216	0.141
	b	0.2707	0.2688	0.221	0.145
Bulk formation enthalpies (kcal/mol)		7.03	6.66	5.77	3.92

The quantities reported in Table 2 are atomization energies per atom, as defined in Section 2, for pure clusters whose chemical nature (Tc, Ru, Rh, Pd), size (n= 4, 6, 13) and geometry (symmetry point group: Td, Oh, Ih) were chosen both for convenience reasons and a possible relationship with bulk properties accessible to experiment. The general tendency exhibited by these values is a regular increase with the size of the cluster and a decrease when going from the left to the right of the second row of transition metals. We have no qualitative explanation concerning the local maximum of $|AE|/n$ for Ru_6, except it displays the influence of the geometry for a cluster of finite size on the magnitude of its binding energy, a fact already noted by various authors (see e. g. [10, 31]). In spite of evident reserves from the physical and chemical points of view, the theoretical trend to a decrease of atomization energy per atom from Tc to Pd can be compared to the enthalpies for bulk at 298K [46], and the overall agreement of theory with experiment may be considered as a good mark for the present EXTHUC parametrization (especially for the Alvarez H_{pp} values).

In Table 3 presenting the results of investigations on mixed Pd-Rh clusters in connection with catalysis, it is seen that the accidental insertion of a foreign atom, for instance Rh, increases the atomization enrgy of a pure cluster, i.e. Pd_n. If we identify the resulting increase of binding energy to an additional storage of chemically available energy we are led to think to a rise of reactivity due to the heteronuclear bonds Pd-Rh. The theoretical ingredient employed [14] to quantify this idea is the

Table 3. Atomization energies (in a.u.) for pure and mixed Pd/Rh clusters (without charge corrections for 13-atom pure and surface-monosubstituted clusters).

Clusters	-AE	$m_{Pd\text{-}Rh}$	$-\varepsilon_{Pd\text{-}Rh}$
Pd_4	0.296	0	-
Pd_3Rh	0.383	3	0.007
Pd_2Rh_2	0.455	4	0.007
Rh_3Pd	0.513	3	0.007
Rh_4	0.556	0	-
Pd_6	0.510	0	-
Pd_5Rh	0.604	4	0.006
Rh_5Pd	0.882	4	0.005
Rh_6	0.924	0	-
Pd_{13}	1.833	0	-
$Pd_{12}Rh$	1.938	16	0.005
$Rh_{12}Pd$	2.738	16	0.006
Rh_{13}	2.795	0	-

increment $\varepsilon_{Pd\text{-}Rh}$ expressing the variation of the atomization energy per mixed bond with respect to its mean value for pure clusters Pd_n and Rh_n. The values of the $\varepsilon_{Pd\text{-}Rh}$ reported in column 3 are not only consistent with the well-known charcteristics of palladium catalysors including rhodium impurities, but also they suggest a better efficiency of very small mixed clusters probably because of a lowering of the energy barrier for the formation of adducts on the Pd-Rh bonds. This could be a clue to understand the special properties of colloidal dispersions of Pd/Rh bimetallic clusters [47].

Our last study summarized by Table 4 is a typical example for the use of Quantum Chemistry in Applied Sciences, namely a reactivity problem in Nuclear Chemistry. It involves Tc clusters we have taken as models [15] for investigating the consequences of the radioactive transformation of the Tc long-lived isomer (2.10^5 years):

$$^{99}Tc \rightarrow {}^{99}Ru + \beta + 0.292\,MeV$$

in Tc wastes coming from spent fuels of nuclear industry [48]. All the types of chemical reactions, bond attacks or atomic site attacks, have been considered, since we have no *a priori* reasons to pay attention to some of them. In such a case, the evolution of the various theoretical indices associated to the electronic structure as Tc atoms are replaced by Ru ones in technetium clusters should give a lot of information on their properties.

Table 4. Atomization enrgies (in a.u.) and Fukui indices for pure and mixed Tc/Ru clusters. With charge corrections for 13-atom pure and surface-monosubstituted.

	-AE	$-\varepsilon_{Tc\text{-}Ru}$	Fukui indices	
			f_{Tc}	f_{Ru}
Tc_4	0.6672	-	1/4	-
Tc_3Ru	0.6675	0.0026	0.260	0.219
Ru_3Tc	0.6528	0.0025	0.279	0.240
Ru_4	0.6378	-	-	1/4
Tc_6	1.0743	-	1/6	-
Tc_5Ru	1.0962	0.0031	0.1701	0.1495
Ru_5Tc	1.1335	0.0029	0.1837	0.1632
Ru_6	1.1315	-		1/6
Tc_{13}	3.5194	-	0.0616	-
$Tc_{12}Ru$	3.5296	0.0019	0.0621	0.0583
$Ru_{12}Tc$	3.5110	0.0024	0.0752	0.0680
Ru_{13}	3.4949	-	-	0.0690

First of all, the (negative) values of the $\varepsilon_{Tc\text{-}Ru}$ increments indicate that two neighbour atoms Tc and Ru accumulate energy, as in the case of Pd-Rh pairs previously discussed. In metallic clusters, the stored atomization energy plays a role somewhat similar to the conjugation energy of aromatic hydrocarbons: It is possible to distribute this excess of energy between the various C-C bonds according to the values of their bond orders and to predict in this way whether an addition reaction is easier or not with respect to a reference compound, for instance benzene [49]. In a similar way, $\varepsilon_{Tc\text{-}Ru} < 0$ suggests an increase affinity of the corresponding Tc-Ru bond with regard to possible chemical additions in mixed clusters.

Second, the substitution of a Tc nucleus by a Ru in a technetium cluster induces important charge transfers between atoms, the magnitude of which may be revealed by a standard Mulliken population analysis giving atomic charges, spin densities and so on ... The combined effects of the delocalization of the electron cloud between centers equivalent by symmetry (as in pure clusters) and the changes of nuclear charges resulting from the Tc-Ru transmutations are at the origin of these intermolecular charge transfers. The theoretical indices most significant for reactivity purposes are the spin densities, just because in our EXTHUC approximation for transition metal clusters they numerically coincide with the common value taken by the condensed Fukui indices f [33], valid here for all types of site reactions, i.e. electrophilic, nucleophilic or radicalic chemical attacks on definite atoms. In table 4 summarizing the results obtained for monosubstituted mixed clusters, it is seen that admitting Ru atoms in Tc frameworks produces a significant increase of the Fukui index computed for the remaining Tc atoms. A similar effect is observed in polysubstituted Tc clusters (see [15] for details).

From this simple Hückel treatment conveniently parametrized for second-row transition elements, we conclude that the unescapable decay of technetium nuclei may cause an increasing reactivity of remaining Tc atoms, whatever the type of chemical reaction may be: Bond additions or atom attacks. This could lead to the dissemination of undesirable compounds containing β-decaying Tc nuclei.

Acknowledgments: This paper is dedicated to the memory of Professor P.O. Löwdin whose research works and teaching during Sweden summer schools and Florida winter schools were of paramount importance for the evolution of Quantum Chemistry in the XXth century. The authors are indebted to Professor O. Salvetti (University of Pisa) for having initiated this work.

References:

[1] N.F. Allinger, J. Amer. Chem. Soc., 99 (1977) 8127.
[2] R. Car, M. Parrinello, Phys. Rev. Lett., 55 (1985) 2471
[3] P.E. Blöchl, M. Parrinello, Phys. Rev., B45 (1992) 9413
[4] J.C. Barthelat, Ph. Durand, A. Serafini, Molec. Phys., 19 (1976) 159
[5] S. Topiol, J.W. Moskowitz, C. Melius, J. Chem. Phys., 68 (1978) 2364
[6] D. Maynau, M. Said, J.P. Malrieu, J. Amer. Chem. Soc., 105 (1983) 5244
[7] R. Hoffmann, J. Chem. Phys., 39 (1963) 397
[8] P.O. Löwdin, Phil. Mag. Supp., 5 (1956) 1
[9] P.O. Löwdin, Adv. Quant. Chem., 12 (1980) 263
[10] A. Julg, *Crystals as Giant Molecules* in Lecture Notes in Chemistry, Vol. 8, Springer-Verlag, 1978
[11] A. Julg, M. Bénard, M. Bourg, M. Gillet, E. Gillet, Phys. Rev., B9 (1974) 3248
[12] A. Julg, G. Del Re, V. Barone, Phil. Mag., 35 (1977) 517
[13] A.T. Balaban, *Chemical Application of Graph Theory*, A.T. Balaban, ed., Acad.Press, London, 1976, p. 14
[14] G. Berthier, M. Defranceschi, R. Montagnani, O. Salvetti, Int. J. Quant. Chem., 82 (2001) 26
[15] M. Defranceschi, G. Berthier, J. Nucl. Mat., xx (2002) XXXX
[16] C.F. Melius, Chem. Phys. Lett., 39 (1976) 287
[17] I. Shim, K.A. Gingerich, J. Chem. Phys., 80 (1980) 5107

[18] F. Illas, J. Rubio, J. Conellas, T.M. Ricart, J. Chem. Phys., 93 (1990) 2603
[19] M. A. Blomberg, P.E.M. Siegbahn, M. Svensson, J. Phys. Chem., 96 (1992) 5783
[20] a) K. Balasubramanian, J. Chem. Phys., 89 (1988) 6310;
b) K. Balasubramanian, J. Chem. Phys., 91 (1989) 307;
c) D. Dai, K. Balasubramanian, J. Chem. Phys., 103 (1995) 648;
d) D. Dai, K. Balasubramanian, Chem. Phys. Lett., 310 (1999) 303.
[21] Q. Cui, D.G. Musaev, K. Morokuma, J. Chem. Phys., 108 (1998) 8418
[22] a) K. Balasubramanian, D.W. Liao, J. Phys. Chem., 93 (1989) 3989;
b) K.K. Das, K. Balasubramanian, J. Chem. Phys., 93 (1990) 625;
c) D. Dai, K. Balasubramanian, Chem. Phys. Lett., 195 (1992) 207;
d) D. Dai, K. Balasubramanian, Chem. Phys. Lett., 263 (1996) 703;
e) D. Majumdar, K. Balasubramanian, J. Chem. Phys., 106 (1997) 4053;
f) D. Majumdar, K. Balasubramanian, J. Chem. Phys., 108 (1998) 2495
[23] J. R. Letelier, A. Toro-Labbe, Bol. Soc. Chil. Quim., 44 (1999) 87
[24] J. R. Letelier, A. Toro-Labbe, Ying-Nan Chin, J. Chin. Chem. Soc., 46 (1999) 101
[25] A. Goursot, I. Papai, D.R. Salahub, J. Amer. Chem. Soc., 114 (1992) 7452
[26] G. Valerio, H. Toulhoat, J. Phys. Chem., 100 (1996) 10827
[27] C. Xiao, S. Krüger, T. Belling, M. Mayer, N. Rösch, Int. J. Quant. Chem., 74 (1999) 405
[28] R. L. DeKock, H.B. Gray, *Chemical Structure and Bonding*, p. 425, Benjamin Cummings, Menlo Park, Calif., 1980
[29] L. Brewer, Science, 161 (1968) 115
[30] P. Pyykkö, J.P. Desclaux, Acc. Chem. Res., 12 (1979) 276
[31] D.E. Ellis, Theory of Electronic Properties of Metal Clusters and Particles. In *Physics and Chemistry of Metal Cluster Compounds*, L.J. de Jongh ed., p. 135, Kluwer Academic Publishers, 1994
[32] K.S. Gupta, K. Gingerich, J. Chem. Phys., 70 (1979) 5350
[33] W. Yang, W. J. Mortier, J. Am. Chem. Soc., 108 (1986) 5708
[34] C.A. Coulson, G.S. Rushbrooke, Proc. Camb. Phil. Soc., 36 (1940) 193
[35] G. W. Wheland, D.E. Mann, J. Chem. Phys., 17 (1949) 264
[36] M. Wolfsberg, L. Helmholz, J. Chem. Phys., 20 (1993) 837
[37] G. Del Re, A. Julg, A. Lami, J. Phys., 38 (1977) C2-33
[38] G.R. Baldock, Proc. Phys. Soc. (London), A66 (1953) 2
[39] R.P. Messmer, Phys. Rev., B15 (1977) 1811
[40] A. Julg, in *Growth and Properties of Metal Clusters*, J. Bourdon Ed., p. 256, Elsevier, Amsterdam (1980)
[41] M. Born, J. E. Mayer, Zeitsch. Phys., 75 (1932) 1
[42] N. Lathiotakis, A.N. Andrichs, M. Menon, J. Connot, J. Chem. Phys., 104 (1996) 992
[43] S. Alvarez, *Tables of parameters for extended Hückel calculations*, Report of the University of Barcelona (1989)
[44] J. Pietro, W.J. Hehre, J. Comput. Chem., 4 (1983) 241
[45] R.O. Jones, O. Gunnarson, Rev. Mod. Phys., 61 (1989) 689
[46] *Handbook for Chemistry*, 78th Edition, pp.5.5, CRC Press, Boca Raton, New York, 1997
[47] M. Harada, R. Asakawa, Y. Ueji, N. Toshima, J. Phys. Chem., 97 (1993) 10742

[48] R. Mills et.al.; A. Tobias, J. Blachot et. al. - *JEF-2.2 Radioactive decay data - JEF Report 13*. AEN/OCDE - 328 pages. Août 1994.

[49] B. Pullman, A. Pullman, *Les théories électroniques de la chimie organique*, Masson, Paris, 1952

QUANTUM CHEMICAL CALCULATIONS OF MOLECULAR G-TENSORS OF BIOLOGICAL RADICALS

Olav Vahtras,
Maria Engström,
and Hans Ågren
Theoretical Chemistry, SCFAB,
Royal Institute of Technology,
S-106 91 Stockholm, Sweden
agren@theochem.kth.se

Abstract

We review state-of-the art quantum chemical methodology to calculate g-tensors that appear in electron paramagnetic resonance (EPR) spectra with applications to radicals. The most important EPR parameters are the g-tensor and the hyperfine coupling A-tensor. While the latter tensor has been much used and simulated in computational models, the calculation of g-tensors is a relatively new theoretical activity, restricted in time to the last few years, a situation which is somewhat ironic from the point of view of the important role g-tensors play in EPR analysis, in particular in high-field experiments. In this paper we first review g-tensor calculations employing a response theory algorithm derived at our laboratory. The theory goes ultimately back to Löwdins inner projections and the formulation of response functions by Olsen and Jørgensen. We review response theory results on g-tensors calculated with the restricted open-shell Hartree-Fock (ROHF) and multiconfiguration self-consistent field (MCSCF) reference wave functions. The recent focus on development of a concomitant density functional theory (DFT) of molecular g-tensors has lead to a widening of applications to include biologically interesting radicals. This development is commented on in this paper and illustrated by some DFT results for property-to-structure relations of amino acid radicals and for spin labels in proteins (ribonucleotide reductase and galactose oxidase). Calculations on a number of substituted benzene radicals as a test bed show that ROHF wave functions provide quantitative results for the g-tensor for a majority of such cases. Some radicals call for a more sophisticated approach, and the multi-configurational method can in most cases provide accurate results where ROHF fails. The role of correlation can

E.J. Brändas and E.S. Kryachko (eds.),
Fundamental World of Quantum Chemistry, Vol. II, 483–524.

qualitatively be traced to the organization of the few lowest excited states of the radical, while quantitatively the contributions from all excited states must be accounted for, as obtained by means of the response theory method. The g-tensor is a sensitive probe of molecular interactions in the close vicinity of the radical, and calculations reviewed in this work confirm such sensitive dependence of g-values on hydrogen bonding of radicals for a few selected spin labels. For example, MCSCF calculations on tyrosyl radical models result in the same g-shift as observed experimentally in ribonucleotide reductase. Furthermore, a sulfur substitution, modeling the tyrosyl radical in galactose oxidase, does unexpectedly not influence the g-tensor or the spin density distribution. Calculations on sulfur centered radicals with different structures and charges show that the g-tensor pattern is significantly different among these radicals. The calculations provide means to distinguish different radical structures from each other. The examples indicate that simulations of the kinds reviewed here may have significant implications for the interpretation of experimental EPR spectra of transient radicals, such as the intermediate cysteine radicals in RNR.

1. Introduction

During the last two decades one has witnessed an intense development in theoretical chemistry modeling. Extensive research has led to a situation where calculations provide exquisite information on chemical reaction mechanisms, molecular interactions, properties and structures, and where they can confirm experimental data and provide new information and interpretations of experimental results. One early fundamental task for applied quantum chemistry was to optimize total energies and determine electronic structures and chemical bonds. In more recent times geometric structure determinations have come at focus, and form now standard practice at many levels of theory. Another chemically interesting, but arduous, field is given by calculations of transition states which are gateways of reactions. As an example, one can now model active sites in enzymes and enzymatic mechanisms in photosystem II (Siegbahn and Crabtree, 1999), ribonucleotide reductase (Siegbahn et al., 1998a; Siegbahn et al., 1998b) and galactose oxidase (Himo et al., 2000) with considerable success. These and many other examples (Siegbahn and Blomberg, 2000) show that calculations of molecular energies, conformations and transition states provide information on the possible reaction pathways. An alternative way to explore a molecular system is given by calculations of properties. This means calculations of the molecular response upon perturbations from external and/or internal electromagnetic fields. One purpose with such calculations is to explore the basic properties *per se*, another is to derive structure-to-property and function-to-property relationships. These relationships can then be used together with experimental data to derive structural and functional information. Calculations of optical (Norman et al., 1997; Pierloot et al., 1998), vibrational (Plashkevych et al., 2000; Qin and Wheeler, 1995), electronic (Carravetta et al., 2000; Gel'mukhanov et al., 2000; Luo et al., 1999), and magnetic (Helgaker et al., 2000; Schreckenbach and Ziegler, 1998) properties may thus provide additional and complementary information about structures, functionality, and inter- and intramolecular interaction.

In this review, we present state-of-the-art quantum chemical methodology for calculations of electron paramagnetic resonance (EPR) g-tensors with applications to radicals in biomolecular systems. Special emphasis is put on structure-to-property relations of amino acid radicals and spin labels in proteins. EPR spectra are commonly interpreted from simulations, where experimental results are matched with parameters in the theoretical energy expression, the spin Hamiltonian. The most important parameters are the g-tensor and the hyperfine coupling tensor, A, which define the spectral shift and line splitting due to the interaction between magnetic moments and magnetic fields. These parameters are sensitive to geometric structure and also to solvent and intermolecular interaction in the close vicinity of the radicals. The A-tensor has been the most important EPR parameter for organic radicals. The hyperfine splitting pattern is often well resolved and provides a fingerprint of the radical in the actual sample. The g-value of carbon, nitrogen and oxygen centered radicals are close to the free electron value and the different components of the g-tensor can often not be resolved in ordinary low field (9 GHz) EPR. The interest in g-tensor calculations has been limited by these reasons. However, with the introduction of high field ($>$ 90 GHz) EPR, it is possible to resolve all the tensor components. Owing to these high field techniques it has been discovered that even the g-tensor provides significant patterns which can separate different radicals from each other. Following this technical development the interest in accurate calculations has increased.

Until a few years ago g-tensors had only been little attended with first principle methods which is somewhat surprising considering the role they play for the proper interpretation of EPR spectroscopy of radicals. The reason for this can probably be referred to the complex quantum expression, involving spin-orbit interaction and a summation of matrix elements over a complete set of excited states. We solved this problem through an implementation of the g-tensor expression using response functions (Vahtras et al., 1997; Engström et al., 1998), thereby opening the possibility to make accurate g-tensor calculations for a variety of systems. The g-tensor response functions have mainly been calculated from restricted open-shell Hartree–Fock (ROHF), representing the one-electron picture, and multi-configuration self-consistent field (MCSCF) wave functions, representing many-electron approaches. The poor scaling of MCSCF methods makes calculations on large and asymmetrical molecular systems computationally demanding. Thus, for computational reasons it is advantageous to use ROHF wave functions since they involve lower computational cost with respect to molecular size. This is exemplified with ROHF linear response calculations of the g-tensor of substituted benzene radicals. ROHF results are in good to excellent agreement with experimental data in many cases, however, a few examples have been found where this situation does not hold. For instance, the g-tensor of the phenoxyl radical is incorrect by ROHF. A problem like that can often be traced to a complex electronic structure and might be solved by using a multi-configuration, correlated, approach. Thus, although ROHF is an accurate approach for several radicals it can not be used for molecules which are characterized by strong electron correlation.

Density functional theory (DFT) is another approach, which recently has become available for g-tensor calculations. This method opens up possibilities for calculations on larger systems of molecules which demands electron correlation methods. Malkina et al., van Lenthe et al. and Ziegler et al. all show that DFT calculations on g-tensors of bio-radicals constitute a fully realistic proposition with rewarding results (Malkina et al., 2000; van Lenthe et al., 1997; Schreckenbach and Ziegler, 1997), see also recent PhD theses in this area produced by Engström (Engström, 2001) and Stein (Stein, 2001).

Evaluation of recent g-tensor calculations give two main conclusions: It is possible to make accurate quantum chemistry calculations of EPR parameters with now available computational methods, and: With relevant choices of molecular models, these calculations can be of great significance for the interpretations of experimental EPR results. There is thus optimism for the further development of this field, and for the possibilities of computational modeling to assist the experiments. With this work we thus like to review some important landmarks in this field, with special emphasis on response theory.

The review is organized as follows. In the next section (Section 2), we briefly recapitulate some basic facts of the EPR experiment, in particular the role of g-tensors in these measurements. In section 3 we review theory for current calculations of g-tensors, including basic definitions of measurable quantities and of operators relevant for the spin Hamiltonian. The theory part covers the ab initio response theory implementation for g-tensors developed and gives also an example of the DFT method used in recent applications by Malkina et al (Malkina et al., 2000) and in our laboratory. Section 3 also describes some practical aspects of efficient evaluations of ingoing matrix elements. In section 4 we evaluate the theory and computational methods for some sample systems before entering the application of the biological radicals. The latter comprise amino acids in section 5 and special spin labeled proteins in Section 6.

2. Electron Paramagnetic Resonance (EPR)

Electron Paramagnetic Resonance (EPR; or Electron Spin Resonance, ESR) is a spectroscopic method for studies of species with unpaired electrons, such as free radicals and transition metal complexes. The EPR technique takes advantage of the interaction between the electron spin and external magnetic fields to obtain information about the structure and dynamics of radicals.

To prepare for an EPR measurement, the sample is inserted in a cavity in a homogeneous magnetic field (Figure 1). This static field causes an alignment of the magnetic dipole moments and a separation of degenerate energy levels, i.e. the Zeeman effect. The energy E of a magnetic dipole with moment $\vec{\mu}$ in the presence of a magnetic field $\vec{B}$ is given by

$$E = -\vec{\mu} \cdot \vec{B} \tag{1}$$

The magnetic field direction defines the z-axis in the laboratory coordinate system. The magnetic moments of electrons are related to the electron spin:

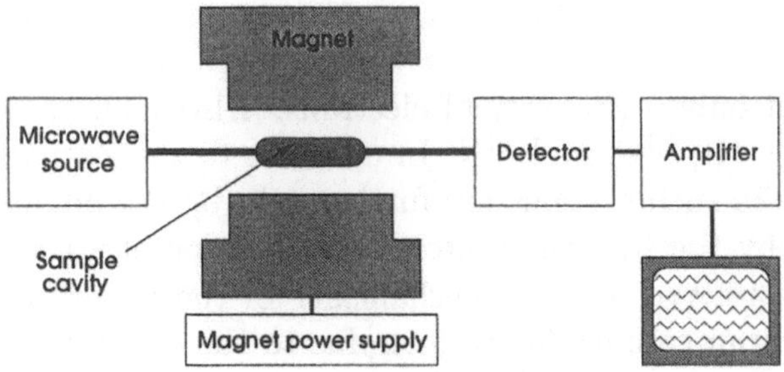

Figure 1. Block diagram of a simple EPR spectrometer.

$$\mu_z = -g_e \mu_B m_S \tag{2}$$

where g_e is the electronic g-factor, $\mu_B = e\hbar/2m_e = 9.2740 \times 10^{-24}$ J/T is the Bohr magneton, and m_S is the electron spin quantum number. The energy separation ΔE between the two states of an unpaired electron is obtained by combining Equations 1 and 2:

$$\Delta E = g \mu_B B \tag{3}$$

The magnitude of the separation depends on the g-value and the strength of the magnetic field. For a free electron the g-value is g_e=2.002319. In molecules, however, the g-value is directionally dependent. That is, the g-value is anisotropic and will have different values depending on the orientation of the radical in the magnetic field. The g-values for organic radicals are commonly close to the free electron value and the different g-tensor components (g_{xx}, g_{yy}, and g_{zz}) are not resolved in low-field (X-band) EPR. As the separation of energy levels depends on B, high-field (W-band) EPR are used for better resolution of the g-tensor components (Figure 2).

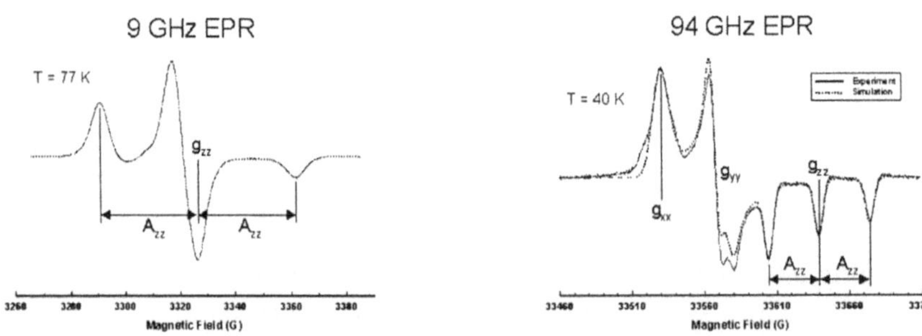

Figure 2. Low (9 GHz) and high (94 GHz) field EPR spectra of the spin label MTSSL. The threefold splitting of the z component is due to $I = 1$ of ^{14}N. The hyperfine splittings of the x and y components are too small to be visible in the spectrum.

Spin is not only a property of electrons. Also some nuclei possess a net spin, e.g. ^{1}H, ^{13}C, and ^{14}N. In addition to the Zeeman effect, the energy levels in such systems are further split by the intrinsic magnetic field caused by the hyperfine interaction between electron and nuclear spins. There are two distinct mechanisms for the hyperfine interaction. One is the Fermi contact interaction due to the electron density at the nucleus. The other is the dipole-dipole interaction between electron and nuclear magnetic moments.

3. Theory and Calculations of EPR Parameters

3.1 Magnetic fields

The unperturbed Hamiltonian in the Schrödinger equation describes the internal interaction between electrons and nuclei. This Hamiltonian is sufficient for the description of a molecule in the absence of external electromagnetic fields. However, the presence of such fields contribute to additional terms; the kinetic momentum that enters the kinetic energy operator will then have a field dependent correction

$$\vec{\pi} = \vec{p} + e\vec{A} \tag{4}$$

where $\vec{\pi}$ is the kinetic momentum, $\vec{p}$ the canonical momentum, and $\vec{A}$ the magnetic vector potential. In addition, there will be an interaction between the electron spin magnetic moments and the external magnetic field

$$H = \frac{e}{m_e}\vec{s}\cdot\vec{B} \tag{5}$$

The field sources are either external to the molecule

$$\vec{A}_{\text{field}} = \frac{1}{2}\vec{B}_{\text{ext}} \times \vec{r} \tag{6}$$

which holds for a uniform magnetic field, or nuclear magnetic dipoles,

$$\vec{A}_{\text{nucleus}} = \frac{\mu_0\gamma_N}{4\pi r^3}\vec{I}\times\vec{r}, \tag{7}$$

where γ_N is the magnetogyric ratio for the nucleus, μ_0 is the vacuum permeability and $\mathbf{I}$ is the nuclear spin. The magnetic fields associated with the nuclear dipoles are thus,

$$\vec{B}_{\text{dip}} = -\frac{\mu_0\gamma_N}{4\pi r^3}(\vec{I} - \frac{3\vec{r}(r\cdot\vec{I})}{r^2}) \tag{8}$$

$$\vec{B}_{\text{FC}} = \frac{2}{3}\mu_0\gamma_N\vec{I}\delta(\vec{r_N}) \tag{9}$$

Eq. 8 describes the dipolar magnetic field from a nucleus at a certain distance $\mathbf{r}$. The Fermi contact term arises from the assumption that the nuclei are ideal point particles (c.f. Eq. 7). The Hamiltonian of an electron in a magnetic field given by

$$H = \frac{(\vec{p}+e\vec{A})^2}{2m_e} + V + \frac{e}{m_e}\vec{s}\cdot\vec{B} + H_{\text{so}}\ , \tag{10}$$

where we in addition have the now field dependent spin-orbit interaction

$$H_{\text{so}} = \frac{\alpha^2}{2}\frac{Z_N\vec{r}\times(\vec{p}+e\vec{A})}{r^3}\ , \tag{11}$$

is sufficient to determine all relevant EPR parameters.

3.2 Relativistic quantum mechanical operators

The several operators crucial for the calculations of EPR parameters can be viewed in a relativistic context. Three space and one time coordinates are required to describe the 4-coordinate system for a particle in a relativistic framework (Helgaker et al., 1998). The Schrödinger equation is not Lorentz invariant, that is not invariant with respect to transformations between different time/space coordinate systems, since it contains second derivatives with respect to space coordinates together with a first order time derivative. Dirac's relativistic equation is based upon the classical Hamiltonian for a relativistic electron in an electromagnetic field

$$\mathcal{H} = c\sqrt{m_e^2c^2 + \vec{\pi}^2} - e\phi \tag{12}$$

where m_e is the electron mass, c the velocity of light in vacuum, e the electron charge, ϕ the scalar electromagnetic potential, and π is the kinetic momentum (Eq. 4). Using the following definition for the expression under the square root symbol in the Hamiltonian (Eq. 12)

$$m_e^2c^2 + \vec{\pi}^2 = (\vec{\alpha} \cdot \vec{\pi} + \beta mc)^2 \tag{13}$$

the Dirac equation is obtained from the now linear energy operator

$$H_D\Psi = E\Psi \tag{14}$$

The Dirac equation has a four-component solution:

$$\Psi = \begin{bmatrix} \psi_{L\alpha} \\ \psi_{L\beta} \\ \psi_{S\alpha} \\ \psi_{S\beta} \end{bmatrix} \tag{15}$$

The large components ψ_L are associated with electrons and the small components ψ_S with positrons. In the relativistic framework, spin is included in the definition of the wave function ψ, while in non-relativistic quantum mechanics, spin is explicitly added to the wave function $\psi(\vec{x}) = \theta(\vec{r})\eta(s)$.

The four-component Dirac equation can be reduced to a two-component expression, represented by a Pauli-type Hamiltonian. With this representation, the electronic states can be retrieved without considering the positronic solutions. Additional relativistic approximations for many-particle systems are obtained by the Breit-Pauli Hamiltonian. With this approach, it is possible to describe the interaction due to spin and orbital motion in the non-relativistic limit. The Breit-Pauli Hamiltonian can be written as a sum of one- and two-electron operators:

$$H_{\mathrm{BP}} = \sum_i h_i + \frac{1}{2}\sum_{ij} h_{ij} \tag{16}$$

Several of these operators are crucial for the calculations of EPR parameters:

- The **Zeeman** operator describes the interaction between magnetic fields and magnetic moments which are generated by electron and nuclear spins.

- The **spin-orbit** operator couples particle spin and electronic orbital motion. The electronic spin-orbit operator provides the largest contribution to the g-tensor. It also provides a mechanism for intersystem crossings and phosphorescence in molecules. The nuclear spin-orbit operator makes a small contribution to the hyperfine coupling constants. This contribution is not considered in most theoretical applications.

- **Spin-spin** operators couple electronic and nuclear spins. The coupling between electrons and nuclei contribute to the hyperfine coupling constant via the dipolar and Fermi contact terms. The electron spin-spin interaction causes zero field splittings in EPR spectra.

- **Diamagnetic** operators describe the interaction due to the vector potential $\vec{A}$ included in the kinetic momentum (Eq. 4). The gauge correction to the g-tensor is obtained from the diamagnetic contribution.

- The **mass velocity operator** makes a contribution to the kinetic energy due to the relativistic dependence of mass on velocity.

Modern computational techniques for calculations of EPR parameters use the Breit-Pauli Hamiltonian as a starting-point. In parallel to this development, Nørager and Jensen (Nørager, 2000) have implemented a method to calculate the g- and A-tensors from the Dirac equation (Nørager, 2000). In this approach, the tensors are calculated as expectation values of the relativistic four-component wave function (Eq. 15). Two-component relativistic approaches using the so-called ZORA formalism has been developed in a DFT framework by van Lenthe et al. (van Lenthe et al., 1997).

3.3 Spin Hamiltonian

EPR spectra are commonly interpreted from simulations, where experimental results are matched with theoretical parameters. For radicals exposed to magnetic fields, the most dominating interaction will be found between the spins of the system. The energy operator for such systems can therefore be expressed as a spin Hamiltonian

$$\begin{aligned} H_{\text{spin}} &= H_{\text{Zeeman}} + H_{\text{hyperfine}} + H_{\text{zfs}} \\ &= \mu_B \vec{S} \cdot \vec{\vec{g}} \cdot \vec{B} + \vec{S} \cdot \vec{\vec{A}} \cdot \vec{I} + \vec{S} \cdot \vec{\vec{D}} \cdot \vec{S} \end{aligned} \tag{17}$$

The Zeeman term describes the interaction between the electron spin $\vec{S}$ and the external magnetic field $\vec{B}$, parameterized by the g-tensor. $\mu_B = e\hbar/2m_e$ is the Bohr magneton. The hyperfine interaction between the electron spin $\vec{S}$ and the nuclear spin $\vec{I}$ is described by the A-tensor. Calculations of the g- and A-tensors are based on the Breit-Pauli Hamiltonian (Eq. 16) which is derived from relativistic many-body theory. The zero-field splitting tensor parameterizes the internal interaction between the electron spins and is only relevant for systems with more than one open shell. A detailed description of this formalism can be found in *Theoretical foundations of electron spin resonance* by Harriman (Harriman, 1978).

3.3.1 *A*-tensor. The A-tensor is calculated for magnetic nuclei, i.e. nuclei with spin $I \neq 0$. The hyperfine interaction includes a sum of two first-order terms

$$\vec{\vec{A}} = A_{\text{iso}} \vec{\vec{1}} + \vec{\vec{A}}_{\text{dip}} \tag{18}$$

These terms are derived from the expressions for the magnetic field arising from the nucleus (Eqs. 8 and 9). The isotropic term A_{iso} is often called the Fermi contact interaction and it is proportional to the spin density at the nucleus in question. The Fermi contact term is calculated from the isotropic hyperfine Hamiltonian

$$H_{\text{iso}} = -\frac{2}{3} g_e \gamma_e \gamma_N \mu_0 \delta(\vec{r}_N) \vec{S} \cdot \vec{I} \tag{19}$$

where γ_e and γ_N are the magnetogyric ratios for the electron and nucleus, respectively, μ_0 is the vacuum permeability and $\delta(\vec{r}_N)$ is the delta function which extracts the spin density at the nucleus. A_{dip} in Eq. 18 corresponds to the classical dipolar coupling between magnetic dipoles.

$$H_{\rm dip} = \frac{\mu_0 g_e \gamma_e \gamma_N}{4\pi r^3}[\vec{S}\cdot\vec{I} - 3\frac{(\vec{S}\cdot\vec{r})(\vec{r}\cdot\vec{I})}{r^2}] \tag{20}$$

where r is the electron-nucleus distance. There are also second order contributions to the A-tensor, derived from the Breit-Pauli Hamiltonian (Eq. 16). These contributions include cross terms between the orbital hyperfine operator and spin-orbit coupling operators (see the book by Harriman (Harriman, 1978) for further details). The second order contributions are expected to be negligible for organic radicals, but they contribute significantly to the hyperfine coupling of radicals in metal complexes (Belanzoni et al., 2000).

During the last decade, several DFT (Himo et al., 1997; Rega et al., 1996) and MCSCF (Fernández et al., 1992; Fernández et al., 1996) calculations on A tensors have been reported, for example, Martell et al. (Martell et al., 1997) have calculated the hyperfine coupling constants (hfcc) of sulfur containing radicals and Eriksson et al. (Himo and Eriksson, 1997; Eriksson et al., 1997) have calculated hfcc for several biological important radicals. Investigations reported in references (Rega et al., 1996; Fernández et al., 1996) deal with solvent effects based on polarizable continuum dielectric models. In reference (Rega et al., 1996) hydrogen bonds are explicitly included in the molecular model to mimic hydrogen bonding solvents. All these calculations consider the dominating first order contributions to the hyperfine interaction, i.e. Eq. 18.

3.3.2 *g*-tensor. The g-tensor which enters the spin Hamiltonian (Eq. 17) is calculated as a correction to the free electron value, g_e = 2.002319

$$\vec{\vec{g}} = g_e \vec{\vec{1}} + \Delta\vec{\vec{g}} \tag{21}$$

The diagonal elements of the g-tensor, g_{xx}, g_{yy}, and g_{zz}, correspond to the experimental values obtained in anisotropic systems, i.e. crystals and frozen solutions. At room temperature the paramagnetic species are exposed to rapid tumbling and random reorientation, resulting in an isotropic g-value averaged over all orientations. The isotropic g-value is thus calculated as the average value of the three components,

$$g_{\rm iso} = \frac{g_{xx} + g_{yy} + g_{zz}}{3} \tag{22}$$

Several terms, derived from relativistic quantum mechanics, contribute to the g-shift $\Delta\mathbf{g}$; 1) the relativistic mass contribution (RMC); 2) the gauge corrections (GC); and 3) the spin-orbit coupling (SOC) and orbital Zeeman (OZ) one- and two-electron cross terms.

$$\begin{aligned}\Delta g^{ab} &= \Delta g_{\mathrm{RMC}} + \Delta g_{\mathrm{GC}}(1e) + \Delta g_{\mathrm{GC}}(2e) \\ &+ \Delta g_{\mathrm{SOC/OZ}}(1e) + \Delta g_{\mathrm{SOC/OZ}}(2e)\end{aligned} \tag{23}$$

These terms are obtained from a perturbation expansion of the molecular energy as exhaustively described by Harriman (Harriman, 1978).

Although the knowledge of the magnitude and the anisotropy of g-tensors often is indispensable for interpretations of EPR experiments, comparatively little has so far been accomplished by computational methods. Stone developed a theory for calculations of g-tensors for organic radicals (Stone, 1963; Stone, 1964). This theory is based upon the dominating second order contribution to the g-tensor (Carrington and McLachlan, 1979).

$$g_{aa} = g_e - 2\zeta \sum_n \frac{\langle \Psi_0 | L_z | \Psi_n \rangle \langle \Psi_n | L_z | \Psi_0 \rangle}{E_n - E_0} \tag{24}$$

g_{aa} denotes here the diagonal elements of the g-tensor, ζ is the spin-orbit coupling constant, and L_z is the angular momentum operator. Several semi-empirical calculations based on this simplified treatment have been performed (Minaev, 1989; Ishii et al., 1989). For radicals in biological systems, semi-empirical results have been reported with excellent agreement with experiments (Burghaus et al., 1993; Knüpling et al., 1997; Un et al., 1995). A disadvantage with these methods is the difficulty in describing molecular systems which are little investigated. Therefore, the interest in developments of ab initio and DFT methods has increased during the last decade. Jayatilaka (Jayatilaka, 1998) made calculations of the g-tensor as a derivative of the energy using a general Hartree–Fock method. Lushington et al. (Lushington et al., 1995; Lushington and Grein, 1997) developed a multireference configuration interaction (MRCI) method including all relevant Breit-Pauli contributions to the g-tensor. They use a truncated sum-over-state (SOS) approach in calculating the spin-orbit coupling (SOC) and orbital-Zeeman (OZ) cross terms. The truncated SOS procedures could be questionable for second order properties due to the very slow convergence of such sums (Vahtras et al., 1992a), and choosing the truncation limit of the SOS expansion becomes a delicate operation. This problem is circumvented in the method developed by Vahtras et al. (Vahtras et al., 1997) where the matrix elements and excitation energies are calculated according to linear response theory. With this approach, excited states are calculated implicitly and the error in having to truncate the summation over excited states is thereby completely avoided.

3.4 Response theory

There are different approaches for calculating physical and chemical properties from Rayleigh-Schrödinger perturbation theory. A straightforward formulation, one that is usually presented in textbooks, is to express the results of perturbation theory in terms of summations over excited states. Consider *e.g.* a system described by the Hamiltonian H_0, with the energy eigenvalue E_0 corresponding the ground state $| \, 0\rangle$. The second-order energy correction due to a perturbation V is

$$E^{(2)} = \sum_{k>0} \frac{|\langle 0 \mid V \mid k \rangle|^2}{E_0 - E_k} \, . \tag{25}$$

The main difficulty with this formulation is that it requires complete knowledge of the excited states, so one is forced to pray for rapid convergence and truncate the series after the first few excited states. A more general form of Eq. 25 was given by Löwdin (Löwdin, 1971)

$$E^{(2)} = \langle 0 \mid V R(E_0) V \mid 0 \rangle \tag{26}$$

where R is a resolvent operator involving the inverse of the Hamiltonian,

$$R(E) = (E - H_0)^{-1} \, . \tag{27}$$

The resolvent operator as a function of E has poles at the eigenvalues of H_0. Strictly speaking, for the expression in Eq. 26 to be meaningful, the singular component of the resolvent must be removed. We note that Eq. 25 can be derived from Eq. 26 if we insert a resolution of the identity (and removing the ground state) on each side of the resolvent,

$$R(E) = \sum_k \mid k \rangle \langle k | (E - H_0)^{-1} \sum_l \mid l \rangle \langle l | = \sum_k \frac{\mid k \rangle \langle k |}{E - E_k} \, . \tag{28}$$

If the sum over states is incomplete Eq. 28 is an *outer projection* of the resolvent R onto the subspace spanned by the states $| \, k\rangle$.

Another approach to perturbation theory is response functions, which is the subject of this paper. The basis for all response function methods is an *inner projection* of the resolvent, see Löwdin (Löwdin, 1971) and Oddershede (Oddershede, 1987)

$$R(E) = \mid \mathbf{k} \rangle \langle \mathbf{k} | E - H_0 \mid \mathbf{k} \rangle^{-1} \langle \mathbf{k} | \tag{29}$$

where $| \, \mathbf{k}\rangle$ are the states $| \, k\rangle$ collected in a row matrix and $\langle \mathbf{k} |$ the corresponding column matrix. If the functions k indeed are the eigenstates of H_0, then Eqs. 28 and 29 are identical. There is a clear advantage with the inner projection scheme. We can let $\mid \mathbf{k}\rangle$ be some other basis than the one formed by the energy eigenfunctions. In that case the evaluation of the second order energy is a two-step procedure; solve the linear system of equation

$$\langle \mathbf{k}|E_0 - H_0 \mid \mathbf{k}\rangle \mathbf{N} = \langle \mathbf{k}|V \mid 0\rangle \tag{30}$$

and form

$$E^{(2)} = \langle 0 \mid V \mid \mathbf{k}\rangle \mathbf{N} \ . \tag{31}$$

The summation over excited states is effectively replaced by solving a linear system of equations, and this is done without prior knowledge of the excited states.

These introductory comments leads to the response theory formalism. This formalism is first of all a way of formulating time-dependent perturbation theory. In fact, time-dependent and time-independent perturbation theory are treated on equal footing, the latter being a special case of the former. As the name implies, response functions describe how a property of a system responds to an external perturbation. If initially we have a system in the state $\mid 0\rangle$ (the reference state), as a weak perturbation $V(t)$ is turned on, the average value of an operator A will develop in time according to

$$\langle A\rangle = \langle A\rangle_0 + \langle A\rangle_1 + \langle A\rangle_2 \cdots \ , \tag{32}$$

where $\langle A\rangle_0$ is the average value with respect to the reference state, and $\langle A\rangle_1$ and $\langle A\rangle_2$ the first and second-order responses, respectively. These may be written in their Fourier representations,

$$\langle A\rangle_1 = \int d\omega_1 \langle\langle A; V\rangle\rangle_{\omega_1} e^{-i\omega_1 t + \epsilon t} \ , \tag{33}$$

and

$$\langle A\rangle_2 = \frac{1}{2}\int\int d\omega_1 d\omega_2 \langle\langle A; V, V\rangle\rangle_{\omega_1,\omega_2} e^{-i\omega_1 t - i\omega_2 t + 2\epsilon t} \ . \tag{34}$$

Eq. 33 defines the linear response function

$$\langle\langle A; V\rangle\rangle_{\omega_1} = \langle \frac{AV_{\omega_1}}{\omega_1 - \hat{\omega} + i\epsilon} - \frac{V_{\omega_1}A}{\omega_1 + \hat{\omega} + i\epsilon}\rangle \tag{35}$$

where $\hat{\omega} = H_0 - E_0$ and we use the notation

$$\frac{AV}{\omega - \hat{\omega}} \equiv A(\omega - \hat{\omega})^{-1} V \tag{36}$$

The linear response function $\langle\langle A; V^{\omega}\rangle\rangle_{\omega}$ can in the basis of eigenfunctions $|\Psi_n\rangle$ of H_0 be written

$$\langle\langle A; V^{\omega}\rangle\rangle_{\omega} = \sum_{n\neq 0} \frac{\langle\Psi_0|A|\Psi_n\rangle\langle\psi_n|V^{\omega}|\Psi_0\rangle}{\omega - (E_n - E_0)} - \sum_{n\neq 0} \frac{\langle\Psi_0|V^{\omega}|\Psi_n\rangle\langle\psi_n|A|\Psi_0\rangle}{\omega + (E_n - E_0)} \tag{37}$$

The response functions describe the response of the system to a frequency-dependent perturbation and do inherently contain information about the manifold of excited states. They have poles at the excitations energies of the system and residues related to transition moments between ground and excited states. The time-dependent response function is identical to the time-independent function in the limit of a static perturbation, e.g. a static magnetic field. Thus the time-independent expression is obtained for $\omega = 0$.

The linear response function in Eq. 35 has the same structure as the second-order energy expression in Eq. 26 and we note that for $A = V$ and $\omega_1 = 0$ they are identical, except for a factor of two. Similarly, Eq. 34 defines the quadratic response function

$$\begin{aligned}\langle\langle A; V, V\rangle\rangle_{\omega_1,\omega_2} &= \langle\frac{AV_{\omega_1}V_{\omega_2}}{(\omega_1+\omega_2-\hat{\omega}+2i\epsilon)(\omega_2-\hat{\omega}+i\epsilon)}\rangle \\ &+ \langle\frac{V_{\omega_2}V_{\omega_1}A}{(\omega_2+\hat{\omega}+i\epsilon)(\omega_1+\omega_2+\hat{\omega}+2i\epsilon)}\rangle \\ &- \langle\frac{V_{\omega_1}AV_{\omega_2}}{(\omega_1+\hat{\omega}+i\epsilon)(\omega_2-\hat{\omega}+i\epsilon)}\rangle + P_{12}\end{aligned} \tag{38}$$

where P_{12} is a permutation that denotes symmetrization with respect to the frequencies ω_1 and ω_2. We note that by taking the residue of the quadratic response function at an excitation energy ω_k, one obtains a second-order transition moment, *i.e.* a transition matrix element of an operator A between the ground state and the excited state $|\, k\rangle$ that is induced by the perturbation V. The double residue of the quadratic response function at two excitation energies, ω_f and $-\omega_g$, gives a matrix element between the corresponding excited states, $\langle g|A|k\rangle$.

A major advantage with the response function technology is the relative computational simplicity. In fact, explicit wave function optimization is only necessary for the reference state (preferably the molecular ground state). Hence, the reference state can be either an SCF, an MC-SCF or a coupled cluster (CC) state. Either choice of reference state automatically defines an accompanying response function, and a set of molecular properties are thereby indirectly being defined. The linear response function may be used for calculations of excitation energies, transition moments, and various properties like the spin-orbit coupling matrix elements needed for g-tensor calculations.

The response function formalism used in this paper dates back to Zubarev (Zubarev, 1974). Linear and quadratic response functions were subsequently derived and implemented by several authors (Yeager and Jørgensen, 1979; Dalgaard, 1980; Dalgaard, 1982). A general formulation of linear, quadratic, and cubic response functions for exact and approximate reference states were given in 1985 by Olsen and Jørgensen (Olsen and Jørgensen, 1985). The derivation we use for g-tensors follow a similar outline. The last 15 years have witnessed a large increase in quantum chemical methods applicable in the response theory context and an explosion of molecular properties calculated by these methods. For example, response theory can now be used in parallel SCF calculations for large-scale systems (Norman et al., 1996; Jonsson et al., 1998; Ruud et al., 1998), and CC methods allow highly electron correlated calculations for small and medium-sized systems. A hierarchy of coupled cluster models has thus recently been developed for accurate calculations of linear and nonlinear response properties (Christiansen et al., 1995a; Christiansen et al., 1995b).

3.5 Spin-orbit response theory

Before describing the actual implementation for g-tensors we give some basic expressions and examples of spin-orbit response theory. Common for these is the evaluation of the electronic spin-orbit interaction operator, referred to as the Breit-Pauli spin-orbit Hamiltonian, given by

$$H_{\text{so}} = \frac{\alpha^2}{2}\left[\sum_{iA} Z_A \frac{\vec{l}_{iA}\cdot\vec{si}}{r_{iA}^3} - \sum_{ij}\frac{\vec{l}_{ij}\cdot(\vec{s}_i + 2\vec{s}_j)}{r_{ij}^3}\right], \tag{39}$$

where i, j refer to electrons and A to nuclei. Note that the relative orbital angular momentum between two particles, *e.g.* $\vec{l}_{ij} = (\vec{r}_i - \vec{r}_j)\times\vec{p}_i$ is asymmetric ($\vec{l}_{ij} \neq \vec{l}_{ji}$). This means that the two-electron integrals over this operator do not have the permutation symmetry as do the electron repulsion integrals. They could have been symmetrized, had it not been for the for the asymmetry in the spin part, which has its origin in the Thomas precession (Thomas, 1927).

The results of response theory are most conveniently cast into a formalism based on second quantization. In a second quantized representation of (39), the spin-operators will transfer to triplet excitation operators which are weighted by the integrals over the orbital parts. The

z-component will have the form

$$H_{\rm so}^{z} = \sum_{ij} (\widehat{ij}|Z) E_{ij}^{-} + \sum_{ijkl} (\widehat{ij}|kl) \left(e_{ijkl}^{-+} + 2e_{ijkl}^{+-} \right) \tag{40}$$

where the one-particle excitations are given in by

$$E_{rs}^{\pm} = a_{r\alpha}^{\dagger} a_{s\alpha} \pm a_{r\beta}^{\dagger} a_{s\beta} \tag{41}$$

the two-particle excitations by

$$e_{ijkl}^{+-} = E_{ij}^{+} E_{kl}^{-} - E_{il}^{-} \delta_{kj} \tag{42}$$

and the integrals, written in a Mulliken-like notation, by

$$(\widehat{ij}|Z) = \frac{\alpha^2}{2} \sum_{A} Z_A \int d\vec{r} \phi_i(\vec{r}) \frac{l_A^z}{r_A^3} \phi_j(\vec{r}) \tag{43}$$

$$(\widehat{ij}|kl) = -\frac{\alpha^2}{2} \int d\vec{r}_1 d\vec{r}_2 \phi_i(\vec{r}_1) \phi_k(\vec{r}_2) \frac{l_{12}^z}{r_{12}^3} \phi_j(\vec{r}_1) \phi_l(\vec{r}_2) \tag{44}$$

The Mulliken notation is being used because Eq. 44 can be seen as the electron repulsion integrals between a charge distribution kl and a differentiated charge distribution $\widehat{ij}$. Due to the nature of the orbital angular momentum, these integrals are symmetric in kl and real anti-symmetric in ij. The other components of the spin-orbit operator are obtained by assigning the proper integrals and tensor components to Eq. 40. In practice, only the operators of Eqs. 41 and 42 are needed, since matrix elements involving other components can easily be written in this form using the Wigner-Eckart theorem (Vahtras et al., 1992b).

A large number of spin-orbit properties can now be derived from the response functions. From the linear response function we can deduce the second-order energy correction due to SOC

$$E^{(2)} = \frac{1}{2} \langle\langle H_{\rm so}; H_{\rm so} \rangle\rangle_0 = \frac{1}{2} H_{\rm so}^{[1]\dagger} (E^{[2]})^{-1} H_{\rm so}^{[1]} \tag{45}$$

where $H_{\rm so}^{[1]}$ is the spin-orbit gradient and the SOC constants are derived from the residue

$$\lim_{\omega \to \omega_f} (\omega - \omega_f) \langle\langle H_{\rm so}; H_{\rm so} \rangle\rangle_\omega \quad \to \quad \langle 0 \mid H_{\rm so} | f \rangle \tag{46}$$

and which is formed by solving the eigenvalue equation

$$(E^{[2]} - \omega_f S^{[2]}) X_f = 0 \tag{47}$$

and forming

$$\langle 0|H_{\rm so}|f\rangle = H_{\rm so}^{[1]\dagger} X_f \tag{48}$$

One of the most important applications of quadratic response theory, pertaining to spin-orbit properties, is the calculation of the spin-orbit induced dipole moment which can be derived from the residue

$$\lim_{\omega\to\omega_f} (\omega - \omega_f)\langle\langle \vec{r}; H_{\rm so}, C\rangle\rangle_{0,\omega} \quad \to \quad \langle {}^1 0 \mid \vec{r} \mid {}^3 f\rangle \tag{49}$$

where C is an arbitrary triplet operator, and $\mid {}^3 f\rangle$ is a triplet state. The matrix element to the right is zero in the absence of perturbations. The residue to the left corresponds to inserting the first-order corrections to the singlet state $\mid {}^1 0\rangle$ and the triplet state $\mid {}^3 f\rangle$, due to SOC. In this case we solve the eigenvalue equation (47) for X_f and the excitation frequency ω_f and the linear response equations

$$(E^{[2]})N^{\rm so} = H_{\rm so}^{[1]} \tag{50}$$

$$N^r(E^{[2]} - \omega_f S^{[2]}) = \vec{r}^{[1]} \tag{51}$$

to form

$$\begin{aligned}\langle {}^1 0 \mid \vec{r} \mid {}^3 f\rangle = & -N_j^r H_{{\rm so}_{jl}}^{[2]} X_{lf} - N_j^{\rm so}(\vec{r}_{jl}^{[2]} + \vec{r}_{lj}^{[2]}) X_{lf} + \\ & + N_j^r (E_{jml}^{[3]} + E_{jlm}^{[3]} - \omega_f S_{jlm}^{[3]}) N_m^{\rm so} X_{lf}\end{aligned} \tag{52}$$

The matrix element $\langle l|H_{\rm so} \mid k\rangle$, where $\mid l\rangle$ and $\mid k\rangle$ are two arbitrary excited states, can be derived from the double residue

$$\lim_{\omega_1\to\omega_k} (\omega - \omega_k) \lim_{\omega_2\to -\omega_l} (\omega + \omega_k)\langle\langle H_{\rm so}, B, C\rangle\rangle_{\omega_1,\omega_2} \tag{53}$$

where are B and C are two arbitrary operators, one singlet and one triplet. In this case we solve two eigenvalue equations of the form (47) one for $X_,\omega_f$ and one for $X_{-g}, -\omega_g$ and the linear response equation

$$N^{\rm so}(E^{[2]} - (\omega_f - \omega_g)S^{[2]}) = H_{\rm so}^{[1]\dagger} \tag{54}$$

The matrix element is in this case

$$\begin{aligned}\langle g|H_{\rm so} \mid f\rangle = N_j^{\rm so}(& E_{jkl}^{[3]}(X_{k-g}X_{lf} + X_{kf}X_{l-g}) + \omega_g S_{jkl}^{[3]} X_{k-g} X_{lf} - \\ & -\omega_f S_{jkl}^{[3]} X_{kf} X_{l-g}) - A_{jk}^{[2]}(X_{j-g}X_{kf} + X_{jf}X_{k-g})\end{aligned} \tag{55}$$

Thus, from a single calculation we may calculate a complete spin-orbit matrix over a manifold of eigenstates of H_0.

3.6 First-order corrections to the g-tensor

The first-order contributions are expectation values of the terms in Eq. (10) and (11), that are bilinear in electron spin and magnetic field. These are the relativistic mass correction and the diamagnetic spin-orbit terms. The first correction can be derived from the ordinary spin-Zeeman term by expanding the mass in Eq.(5) to first order in v^2/c^2,

$$\Delta g^{ab}_{\mathrm{RMC}} = -\frac{\alpha^2}{S}\langle\Psi_0|\sum_i p_i^2 s_{z,i}|\Psi_0\rangle\delta^{ab} \tag{56}$$

where the sum runs over all electrons i, δ^{ab} is the Kronecker delta symbol, α is the fine structure constant, and p is the canonical momentum. We assume that Ψ_0 refers to the ground state component with maximum spin projection $M_s = S$. Furthermore, we do not make distinction between g_e and 2 in higher-order expressions.

The other term involves the spin-orbit operator alone, by considering the magnetic part of the kinetic momentum. This contribution is either called gauge-correction (e.g. in Harriman) or diamagnetic (in analogy with the NMR shielding tensor). The field will depend on an arbitrary gauge origin $\vec{r}_C$

$$\vec{A} = \frac{1}{2}\vec{B}\times\vec{r}_C \tag{57}$$

Substituting this into the spin-orbit operator yields the one- and two-electron gauge corrections.

$$\Delta g^{ab}_{\mathrm{GC}}(1e) = \frac{\alpha^2}{4S}\langle\Psi_0|\sum_i\sum_N\frac{Z_N}{r_{iN}^3}[(\vec{r}_{iN}\cdot\vec{r}_{iC})\delta^{ab} - \vec{r}^{\,a}_{iC}\cdot\vec{r}^{\,b}_{iN}]s_{z,i}|\Psi_0\rangle \tag{58}$$

$$\begin{aligned}\Delta g^{ab}_{\mathrm{GC}}(2e) = \frac{\alpha^2}{4S}\langle\Psi_0|\sum_i\sum_{j\neq i}\frac{1}{r_{ij}^3}[\vec{r}_{ij}\cdot(2\vec{r}_{jC}-\vec{r}_{iC})\delta^{ab}\\ - (2\vec{r}^{\,a}_{jC}-\vec{r}^{\,a}_{iC})\cdot\vec{r}^{\,b}_{ij}]s_{z,j}|\Psi_0\rangle\end{aligned} \tag{59}$$

where the sums run over all electrons i, j and nuclear centers N. The origin r_C is chosen as the electronic charge centroid (ECC) defined by

$$\vec{r}_C = \frac{1}{K}\sum_{i=1}^{K}\vec{r}_i \tag{60}$$

Previous calculations (Lushington et al., 1995; Lushington and Grein, 1997) showed that the choice of gauge origin in general has a modest effect on the g-value if the basis set includes polarization functions. The basis sets used in the work reviewed here include such functions and the calculations are thereby expected to be practically independent of the choice of gauge origin.

In the articles referenced herein, the two-electron gauge correction is not included. This contribution is not expected to influence the g-values significantly, except in those cases where the g-shift is close to zero. In most cases, this correction is one order of magnitude smaller than the SOC/OZ contribution and does not influence the result appreciably. However, in those cases when the cross terms are small, exclusion of the two-electron correction may indeed cause a pronounced deviation in the calculated g-shift.

3.7 Second-order corrections to the g-tensor: linear response

The main contributions to the g-tensor arise from the SOC operator which enters the second order cross-term between the field independent spin-orbit operator and the orbital Zeeman interaction. The Breit-Pauli spin-orbit Hamiltonian contains both one-electron and two-electron contributions:

$$H_{\text{so}} = \frac{\alpha^2}{2}\left[\sum_{iN} \frac{Z_N \mathbf{l}_{iN}}{r_{iN}^3} \cdot \vec{s}_i - \sum_{ij} \frac{\mathbf{l}_{ij}}{r_{ij}^3} \cdot (\vec{s}_i + 2\vec{s}_j)\right] \tag{61}$$

where i, j and N refer to the electrons and the nuclei, respectively. Thus $\mathbf{l}_{iN}$ and $\mathbf{l}_{ij}$ define the angular momenta of electron i with respect to the position of nucleus N or electron j, $\vec{r}_{ij}$ are the corresponding position vectors, and $\vec{s}_i$ the electron spin operators. α is the fine-structure constant. The SOC Hamiltonian couples with the OZ interaction

$$H_{OZ} = \frac{1}{2}\sum_i \vec{l}_{iC} \cdot \vec{B}_0 \tag{62}$$

in the $\Delta g_{\text{SOC/OZ}}$ one- and two-electron cross terms

$$\Delta g^{ab}_{\text{SOC/OZ}}(1e) = \frac{\alpha^2}{S}\sum_n \frac{\langle\Psi_0|\sum_i\sum_N \frac{Z_N}{r_{iN}^3} l^a_{iN} s_{z,i}|\Psi_n\rangle\ \langle\Psi_n|\sum_i l^b_{iC}|\Psi_0\rangle}{E_0 - E_n} \tag{63}$$

$$\Delta g^{ab}_{\mathrm{SOC/OZ}}(2e) = \\ -\frac{\alpha^2}{S}\sum_n \frac{\langle\Psi_0|\sum_i\sum_{j\neq i}(\frac{l^a_{ij}}{r^3_{ij}} + 2\frac{l^a_{ji}}{r^3_{ij}})s_{z,i}|\Psi_n\rangle\,\langle\Psi_n|\sum_i l^b_{iC}|\Psi_0\rangle}{E_0 - E_n} \tag{64}$$

where the sum runs over excited states Ψ_n of same spin as the ground state Ψ_0 and l_{iN}, l_{iC}, and l_{ij} denote angular momentum elements. Excited state properties, such as energies, spin-orbit coupling and angular momentum matrix elements, are retrieved from the response equations. In a response formulation this cross-term may be written as

$$\Delta g^{ab}_{\mathrm{SOC/OZ}} = \langle\langle H^a_{\mathrm{so}}; L^b_C\rangle\rangle \tag{65}$$

where the response function in a stationary picture is identical to the energy derivative formula

$$\langle\langle H^a_{\mathrm{so}}; L^b_C\rangle\rangle = \frac{\partial\langle H^a_{\mathrm{so}}\rangle}{\partial\lambda}\left(\frac{\partial^2 E}{\partial\lambda^2}\right)^{-1}\frac{\partial\langle L^b_{\mathrm{so}}\rangle}{\partial\lambda} \tag{66}$$

if the wave function has been parameterized by λ. Excited states are thereby calculated only implicitly. In this way, the error in having to truncate the summation over excited states is avoided, which is the major advantage with the response function technology.

3.8 DFT method for g-tensor calculations

As the interest for quantum chemical calculations of larger molecular systems is growing, DFT methods have become important for g-tensor calculations. Schreckenbach and Ziegler (Schreckenbach and Ziegler, 1997) developed a DFT algorithm using gauge-including atomic orbitals (GIAOs) to avoid the gauge origin problem of magnetic property operators. This method has recently been applied for calculations on transition metal complexes (Patchkovskii and Ziegler, 1999; Patchkovskii and Ziegler, 2000). van Lenthe et al. (van Lenthe et al., 1997) implemented the zero-order regular approximation (ZORA) for relativistic effects. Malkina et al. (Malkina et al., 2000) adopt a different DFT approach by using spin-orbit pseudopotentials and the atomic mean field approximation (AMFI) with the IGLO basis set (Individual Gauge of Localized Orbitals).

The description of g-tensor calculations in this section refers to the recent DFT implementation within the deMon code (Salahub et al., 1991; Malkina et al., 2000). In this approach the SOC (Eq. 61) and the OZ (Eq. 62) operators are coupled to the SOS expression for the $\Delta g_{\mathrm{SOC/OZ}}$ term

$$\Delta g_{\mathrm{SOC/OZ},\mu\nu} = \frac{\alpha^2}{2} g_e \left[\sum_k^{\mathrm{occ}(\alpha)} \sum_a^{\mathrm{virt}(\alpha)} \frac{\langle \psi_k^\alpha | H_{\mathrm{SOC},\nu} | \psi_a^\alpha \rangle \langle \psi_a^\alpha | l_{O,\mu} | \psi_k^\alpha \rangle}{\varepsilon_k^0 - \varepsilon_a^0 - \Delta E_{k\to a}^{\mathrm{xc}}} - \sum_k^{\mathrm{occ}(\beta)} \sum_a^{\mathrm{virt}(\beta)} \frac{\langle \psi_k^\beta | H_{\mathrm{SOC},\nu} | \psi_a^\beta \rangle \langle \psi_a^\beta | l_{O,\mu} | \psi_k^\beta \rangle}{\varepsilon_k^0 - \varepsilon_a^0} \right] \tag{67}$$

$\psi_k^{\alpha/\beta}$ and $\psi_a^{\alpha/\beta}$ refer to unperturbed occupied and virtual α/β MOs, respectively, and ε_k and ε_a are the corresponding Kohn–Sham eigenvalues. $H_{\mathrm{SOC},\nu}$ denotes the ν-component of the spatial part of the field-free SOC Hamiltonian. See the paper by Malkina et al. (Malkina et al., 2000) for further details.

3.9 Atomic mean field approximation

The lack of particle symmetry in the spin-orbit operator, (Eq. 61), brings about considerable computational costs and disk space requirements for storing the integrals. Rather severe bottlenecks for the g-tensor calculations are thus given by computing and storing the spin-orbit integrals in the full Breit–Pauli form. The calculations are costly even for light elements. This computational cost is due to low symmetry of the full spin-orbit operator, which includes two-electron terms with contributions from spin-own-orbit and spin-other-orbit interactions. The full spin-orbit integrals can be approximated by the one-electron part of the operator and/or the one-center contributions (Turro, 1979; Minaev, 1982). The latter approximation is often well motivated, however, the complete neglect of the two-electron terms can sometimes fail (Knuts et al., 1994). An alternative solution is to treat the two-electron interaction in an average way, screening the nuclear charge for the effective one-electron spin-orbit interaction, by an atomic mean-field approximation (AMFI) to the spin-orbit operator. Heß et al. (Heß et al., 1996) describe a Fock-type generalization of the spin-orbit operator which accounts for the interaction between single valence-valence excited state determinants, i.e. the mean field spin-orbit operator. The computational cost is further reduced by neglecting all multi-center contributions to both the one- and two-electron spin-orbit interaction (Marian and Wahlgren, 1996). The screening of the nuclear charge and the calculations of the spin-orbit integrals is thereby reduced to individual atoms with full atomic symmetry imposed.

AMFI is thus efficient in reducing the computational cost in g-tensor calculations and it forms a rewarding balance between efficiency and accuracy (Heß et al., 1996; Marian and Wahlgren, 1996; Vahtras et al., 2002). It has been shown also in connection with response theory g-tensor calculations that AMFI provides substantial time-savings in the calculations of the integrals and that the disk space requirement is greatly alleviated (Vahtras et al., 2002). It is found that the accuracy using AMFI for g-tensors is remarkable for systems containing atoms beyond the second row, in fact the errors are found completely negligible to other errors which arise from the wave function parameterization. In particular, the correlation error is an order of magnitude larger than the errors associated with AMFI.

4. Method evaluation

The description of g-tensor calculations in this section refers to the linear response method derived in (Vahtras et al., 1997; Engström et al., 1998) and implemented in the DALTON quantum chemistry program (Helgaker et al., 2001). The full Breit-Pauli Hamiltonian as implemented by Vahtras et al. (Vahtras et al., 1992b) has been employed in the majority of these studies. The comparing DFT calculations have been made with the DeMon code (Malkina et al., 2000).

4.1 MCSCF: diatomic molecules

The capability of the response theory approach for g-tensor calculations was examined by applications to a series of diatomic molecules with a triplet ground state in (Engström et al., 1998). In that preparatory study a sequence of molecules was selected for response calculations including the NH, NF, PH, PF, NCl, O_2, SO and S_2 molecules, which earlier were used by Wayne and Colbourn (Wayne and Colbourn, 1977) for calculations of spin-spin interaction and singlet-triplet transition intensities. In Ref. (Engström et al., 1998) the two ions OH^+ and SH^+ were also considered, thereby forming a series with related electronic structures.

Table 1 shows calculated contributions to the g-shift of diatomic molecules in the selected set. The dominating contribution originates from the second order spin-orbit coupling and orbital-Zeeman cross term. The first order term contributes with only a fraction of the total g-shift. The two-electron gauge correction, not considered here, is expected to be of the same magnitude as the one-electron correction, but with opposite sign. Results from different sized complete active spaces (CAS) are also presented in Table 1. Electron correlation effects on first order correc-

Table 1. Contributions to the g-shift. Δg_{RMC} is the relativistic mass contribution, $\Delta g_{\mathrm{GC}}(1e)$ the one electron gauge correction, and $\Delta g_{\mathrm{SOC/OZ}}$ the second order spin-orbit/orbital-Zeeman contribution. All values are given in ppm (parts per million). Data from Reference (Engström et al., 1998).

Molecule	Δg_{RMC}			$\Delta g_{\mathrm{GC}}(1e)$			$\Delta g_{\mathrm{SOC/OZ}}$		
	ROHF[a]	*CAS1*[b]	*CAS2*[c]	*ROHF*[a]	*CAS1*[b]	*CAS2*[c]	*ROHF*[a]	*CAS1*[b]	*CAS2*[c]
O_2	-322.	-326.	-367.	208.	204.	216.	3470.	2774.	2957.
SO	-232.	-254.	-259.	254.	235.	255.	4321.	4187.	4068.
S_2	-211.	-216.	-222.	251.	239.	263.	12622.	11424.	11177.
NH	-192.	-212.	-206.	106.	111.	109.	1637.	1537.	1536.
NF	-238.	-253.	-271.	133.	137.	143.	1413.	1897.	2056.
NCl	-219.	-221.	-240.	142.	142.	174.	4026.	4226.	4141.
PH	-145.	-151.	-148.	137.	139.	136.	4090.	4506.	4045.
PF	-169.	-177.	-184.	158.	157.	161.	-1251.	-2679.	-602.
OH^+	-305.	-335.	-328.	152.	160.	157.	4623.	4154.	4197.
SH^+	-220.	-227.	-226.	176.	168.	168.	8178.	8744.	7901.

a. ROHF like wave functions with 2 electrons in a small π-orbital active space.
b. MCSCF with complete active space (CAS) which in C_{2v} symmetry notation $(A_1B_1B_2A_2)$ reads 4220.
c. MCSCF with complete active space (CAS) which in C_{2v} symmetry notation $(A_1B_1B_2A_2)$ reads 6331.

tions are small. The largest variation among the first order terms is less than 2% of the total correction. However, the difference between ROHF and MCSCF second order contributions makes a considerable portion of the total g-shift, indicating that correlated wave functions are essential for accurate calculations of the second order contributions to the g-shift of these diatomic molecules, whereas the first order corrections are less sensitive.

Table 2 shows the total correction of the transversal g-values $\Delta g_{\perp}$. The spin-orbit coupling is increased for heavier elements and a distinct trend can be recognized. The g-shifts increase when going from lower to higher row of atoms: $\Delta g_{\perp}(\mathrm{NH}) < \Delta g_{\perp}(\mathrm{PH})$, $\Delta g_{\perp}(\mathrm{NF}) < \Delta g_{\perp}(\mathrm{NCl})$ and $\Delta g_{\perp}(\mathrm{O_2}) < \Delta g_{\perp}(\mathrm{SO}) < \Delta g_{\perp}(\mathrm{S_2})$. All results are in good agreement with experimental data, concerning both the trends and the magnitudes of the total values. In particular, the anomalous sign and the magnitude for PF is reproduced. The calculations of the PF g-tensor indicate the need to include the complete summation of the excited-state contributions, since upper excited states provide a delicate balance in sign and magnitude of their different contributions.

4.2 ROHF: substituted benzene radicals

In the previous section it was stated that MCSCF wave functions are essential for accurate calculations of second order contributions of diatomic triplets. However, the size and and in many cases the lack of

Table 2. Total correction to g-factors. $\Delta g_{\perp}$ values of diatomic molecules with the ground triplet $^3\Sigma^-$ state. The g-shifts are given in ppm. Data from (Engström et al., 1998).

		$\Delta g_{\perp}$		
Molecule	*Distance (Å)*[a]	*ROHF*	*CAS1*	*CAS2*
O_2	1.207	3355.	2652.	2806.
SO	1.481	4343.	4168.	4064.
S_2	1.889	12662.	11448.	11218.
NH	1.036	1552.	1436.	1438.
NF	1.317	1308.	1781.	1928.
NCl	1.614	3949.	4147.	4074.
PH	1.422	4082.	4494.	4033.
PF	1.590	-1263.	-2698.	-625.
OH^+	1.029	4470.	3980.	4026.
SH^+	1.374	8134.	8686.	7842.

a Reference (Huber and Herzberg, 1979)

Table 3. Isotropic g-values, g_{iso}, of substituted benzene radicals. The molecules are oriented in the xy-plane with the substituted groups along the x-axis.

	Calculation	*Experiment*	*Ref.*
Benzene cation	2.00266	2.00242	(Carter and Vincow, 1967)
Benzene anion	2.00285	2.00285	(Segal et al., 1965)
Dihydropyrazine cation	2.00319	2.00325	(Zeldes and Livingston, 1974)
Hydropyrazine	2.00343	2.00340	(Zeldes and Livingston, 1972)
Aniline	2.00439	2.00331	(Neta and Fessenden, 1974)
Phenoxyl	2.02652	2.00462	(Bresgunov et al., 1992)
Benzoquinone anion	2.00383	2.00468	(Hales, 1975)
Nitropyridine anion	2.00481	2.00492	(Meisel and Neta, 1975)
Nitrobenzene anion	2.00478	2.0048	(Flockhart et al., 1970)

symmetry make this approach impractical for g-tensor calculations of biological radicals. ROHF wave functions, on the other hand, have a good computational performance with respect to the size of the system. Thus, it is a computational advantage if the ROHF linear response method can be used to provide qualitatively acceptable results. The applicability of the ROHF method was therefore examined by g-tensor calculations on a group of substituted benzene radicals with similar structure.

In Table 3 the calculated and experimental isotropic values are presented. The isotropic g-values are calculated as the average value of the individual components, $g_{iso} = (g_{xx} + g_{yy} + g_{zz})/3$. The results are in excellent agreement with experiment for six out of the nine radicals in the chosen set. Results for the *p*-benzoquinone anion and aniline radicals showed slight deviation, while the *g*-tensor of the phenoxyl radical was incorrect. The largest error in the ROHF phenoxyl radical *g*-tensor appears in the component along the CO-bond, g_{xx}. This problem was traced to a complex electronic structure with strong correlation, and to the presence of very low-lying states with strong coupling. Accordingly, the connection between the *g*-value and the optical spectrum gives a possibility to anticipate difficult cases for which the applicability of Hartree–Fock is limited. MCSCF calculations of the phenoxyl radical *g*-tensor resulted in *g*-values consistent with experiment, $g_{iso} = 2.0048$. The majority of cases, however, confirm the ROHF response theory as a viable approach to calculate *g*-values of organic radicals.

4.3 MCSCF: sulfur centered radicals

It is sometimes difficult to experimentally assign the radical giving rise to the EPR spectrum of a short-lived intermediate, as for instance for the intermediate radicals important for the enzymatic function of ribonucleotide reductase (RNR). In those cases calculations provide means to distinguish different radical structures from each other. A transient radical at the substrate active site of RNR of unknown origin with an isotropic *g*-value, g_{iso}=2.01, was observed with EPR (Persson et al., 1998). This rather high *g*-value contributed to the interpretation of a sulfur centered radical on a cysteine moiety. The radical was later assigned as a disulfide anion (Lawrence et al., 1999) which also was confirmed by MCSCF calculations of disulfide radicals of different conformations and charges (Engström et al., 2000c).

The *g*-tensor of the RS·radical is highly anisotropic and the large g_1 value separates the monosulfide radicals from the disulfide radicals (Table 4). The *g*-tensor component directed along the CS bond, g_1=2.428, is the largest among the radicals in Table 4. This value is substantially decreased to g_1=2.277 in a simple hydrogen bond model. Hence, solvent dependency of thiyl radicals may explain the range of experimental values, g_1=2.13-2.32 (Nelson et al., 1977).

The calculated *g*-tensor of the neutral disulfide radical is rhombic with a rather high isotropic value, g_{iso}=2.03 (Table 4). Several EPR experiments on disulfide radicals have been described in the literature (Hadley and Gordy, 1974a; Nelson et al., 1977). These report *g*-tensors similar to the calculated values, giving the conclusion that the disulfide radical model reproduced the experimental results. The *g*-values for the cation radicals were higher (g_{iso}=2.02) compared with those for the

Table 4. Experimental (E) and calculated (C) g-tensors of thiyl radicals.

		g_1	g_2	g_3	*Ref.*
Monosulfide radicals					
RCH_2-S·	E	2.158			(Nelson et al., 1977)
RH_2-S·H-bond	E	2.164	2.012	2.003	(Matsuki et al., 1993)
CH_3-S·	C	2.428	2.017	2.002	(Engström et al., 2000c)
CH_3-S·H-bond	C	2.277	2.018	2.002	(Engström et al., 2000c)
Disulfide radicals					
RCH_2-SS·	E	2.053	2.026	2.000	(Hadley and Gordy, 1974a)
	E	2.057	2.023	2.000	(Parast et al., 1995)
RCH_2-S·S-$(R'CH_2)_2$	E	2.058	2.025	2.001	(Nelson et al., 1977)
CH_3-SS·	C	2.063	2.028	2.002	(Engström et al., 2000c)
Disulfide cations					
RH_2-S·S-RCH_2^+	E	2.032	2.019	2.004	(Gillbro, 1974)
	E	2.055	2.026	2.002	(Hadley and Gordy, 1974b)
CH_3-S·S-CH_3^+ (1)[a]	C	2.047	2.017	2.002	(Engström et al., 2000c)
CH_3-S·S-CH_3^+ (2)[b]	C	2.031	2.020	2.002	(Engström et al., 2000c)
Disulfide anions					
RCH_2-S·S-RCH_2^-	E	2.020	2.020	2.000	(Gillbro, 1974)
	E	2.018	2.018	2.000	(Hadley and Gordy, 1974b)
	E	2.023	2.015	2.002	(Lawrence et al., 1999)
CH_3-S·S-CH_3^- (1)[c]	C	2.028	2.009	2.002	(Engström et al., 2000c)
CH_3-S·S-CH_3^- (2)[d]	C	2.023	2.021	2.002	(Engström et al., 2000c)

[a] Planar *trans* configuration.
[b] Planar *cis* (∠(CSSC)=1.3°) configuration.
[c] Planar *trans* configuration.
[d] Angled (∠(CSSC)=85°) configuration.

anion radicals. Among the investigated models, only the disulfide anion radical induced the isotropic value g_{iso}=2.01 as observed in RNR. The g-tensor of the disulfide anion radicals in cysteine crystals is axial (Hadley and Gordy, 1974b). The angled configuration of the anion radical is also axial whereas the planar configuration is rhombic. In contrast to the disulfide anion observed in cysteine crystals, the radical anion in RNR was suggested to have a more planar configuration. Thus it is possible to obtain structural information about radicals by examining their g-tensors. Complementary information is obtained from the A-tensors. Calculations of these EPR parameters provide distinct patterns which separate different radical structures from each other. These results may be important for the interpretation of experimental EPR spectra of transient radicals.

4.4 ROHF/MCSCF versus DFT: nitroxide radical

Density functional theory (DFT) has recently been demonstrated as an effective approach for g-tensor calculations of large molecular systems which require correlated methods. The work by Malkina and co-workers (Malkina et al., 2000) thus proves DFT to be a promising method in this context. For example, they calculated g-tensors of the phenoxyl and tyrosyl radicals with magnitudes comparable to experimental results.

The g-tensor of a nitroxide radical (See Figure 8) were computed in (Engström et al., 2001) with different computational methods, see Table 5. The results show that the ROHF values are exaggerated in comparison with high field EPR experiments. However, it must be pointed out that that the experiment referred to in Table 5 was performed with the nitroxide radical in ethanol solution. A protic solvent, such as ethanol, reduces the tensor component along the NO bond, g_{xx}, as is discussed in Section 6. Despite eventual solvent effects, it is interesting to note that the DFT method turns out well in comparison with the experimental results. The time saving with DFT was in these calculations considerable compared with the MCSCF method.

5. Sample applications: Amino acid radicals

Chemical and biological processes are frequently mediated by radicals which usually are reactive and short-lived compounds with open shells. The radicals react with other molecules in a way that leads to pairing of the unpaired electrons. Protein radicals have important functions in life processes such as enzymatic reactions. Some of these radicals occur as reaction intermediates. They are transients and appear

Table 5. g-tensor of a nitroxide radical (Figure 8) at different levels of theory. All calculations were performed with the same B3LYP/6-31G** optimized geometry and with the atomic mean field approximation (AMFI).

	g_{xx}	g_{yy}	g_{zz}	*Ref.*
ROHF/cc-pVDZ	2.01155	2.00705	2.00231	(Engström et al., 2001)
MCSCF/cc-pVDZ[a]	2.00841	2.00659	2.00216	(Engström et al., 2001)
BP86/IGLO-II	2.00832	2.00568	2.00214	(Owenius et al., 2001)
Experiment[b]	2.00835	2.00593	2.00200	(Owenius et al., 2001)

a. Calculated with 9 electrons in 8 active orbitals which in c_{2v} symmetry notation reads 4320.
b. W-band (95 GHz) EPR measurements of MTSSL spin label in ethanol solution.

as elements in chemical reactions. Other radicals function as mediators of electron/hydrogen atom transfer and are stabilized by the protein framework, e.g. the tyrosyl radical in ribonucleotide reductase (RNR), galactose oxidase (GO), and photosystem II.

The tyrosyl radical in RNR was the first stable radical detected in biological systems (Ehrenberg and Reichard, 1972; Sjöberg et al., 1977). This radical has a vital part in the catalytic activity of the enzyme. It is kept stable within the protein until the enzymatic reaction is initiated. Various intermediate amino acid radicals are also important for the enzymatic function of RNR and other enzymes. The disulfide anion radical at the substrate active site (Lawrence et al., 1999) together with tryptophan, cysteine and glycine radicals (Gräslund and Sahlin, 1996; Persson et al., 1998) are examples of such radicals.

For most enzymes little is known about the reaction mechanisms. A better understanding of these mechanisms may enhance the possibilities to produce synthetic enzymes and well-designed pharmaceuticals, and research on protein radical properties would therefore contribute to yet another piece of knowledge in this field.

5.1 Tyrosyl radical

5.1.1 Ribonucleotide reductase (RNR). RNR participates in the formation of DNA by catalyzing the reduction of ribonucleotides to deoxyribonucleotides. Three classes of RNR with different protein structures are known. These enzymes use different protein radicals for the enzymatic function. Class I enzymes occur in all higher organisms and in some prokaryotes and have been widely investigated by EPR ever since the detection of the stable tyrosyl radical in the 1970*s* (Ehrenberg

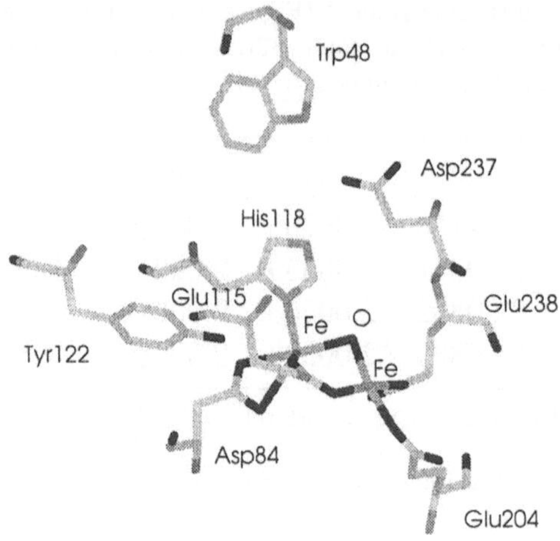

Figure 3. Catalytic site of *E. coli* RNR. The tyrosyl radical (Tyr122) is stabilized and maintained within the protein framework by a diferric oxygen center.

and Reichard, 1972; Sjöberg et al., 1977). The tyrosyl radical is stabilized by the protein framework and a neighboring diferric oxygen center which participates in the formation and maintenance of the radical (Gräslund and Sahlin, 1996; Reichard, 1993) (Figure 3). The distance between the radical at the catalytic site of the enzyme and the substrate binding site is remarkably large. These two sites are connected by a network of hydrogen bonded amino acids.

One way to explore the mechanism of the tyrosyl radical essential for the catalytic activity of class I RNR is to investigate the protein structure in the vicinity of the radical. In this context, it is of importance to explore possible hydrogen bond donors, since the communication between the radical and the enzyme active site is mediated by the hydrogen bonded network. The g-tensor of the tyrosyl radical provides a different pattern whether there is a hydrogen bond to the oxygen atom of the radical or not. This means that the g-tensor component directed along the tyrosyl CO bond g_{xx} is significantly decreased in those species where a hydrogen atom is detected close to the phenolic oxygen (van Dam et al., 1998).

MCSCF calculations of different hydrogen bonding models (Figure 4) resulted in the same shift as observed experimentally in RNR. The linear model (Figure 4a) corresponds to the tyrosyl H-bond structure in RNR suggested by van Dam et al. (van Dam et al., 1998). A water molecule at the hydrogen bonding distance $R(\mathrm{O} \cdots \mathrm{H}) = 1.95$ Å reduces

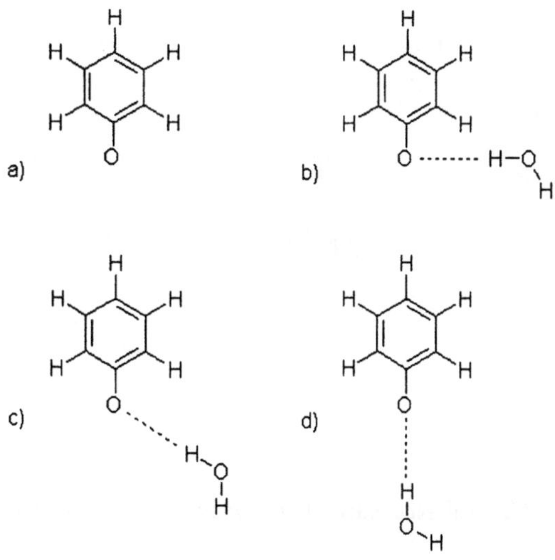

Figure 4. (a) Phenoxyl radical. (b) Phenoxyl radical with a water molecule as a model of a hydrogen bond at 90° with respect to the phenoxyl CO bond. (c) Phenoxyl radical with the water molecule in 135° orientation. (d) Phenoxyl radical and water molecule in linear configuration.

Table 6. Experimental g-tensor of tyrosyl radical in RNR from different sources compared with calculated values for phenoxyl + water complexes. Data from Reference (Engström et al., 2000b).

	hydrogen bond	g_{xx}	g_{yy}	g_{zz}
E. coli RNR (Un et al., 1995)	-	2.0087	2.0042	2.0020
S. typhimurium RNR (Allard et al., 1996)	-	2.0089	2.0043	2.0021
Phenoxyl	-	2.0087	2.0050	2.0025
HSV1 RNR (van Dam et al., 1998)	+	2.0076	2.0043	2.0022
Mouse RNR (van Dam et al., 1998)	+	2.0076	2.0043	2.0022
Phenoxyl + water R(O···H) = 1.95	+	2.0081	2.0049	2.0025
Phenoxyl + water R(O···H) = 1.70	+	2.0079	2.0048	2.0025
Phenoxyl + water R(O···H) = 1.50	+	2.0076	2.0047	2.0025
Phenoxyl + water 180°[a]	+	2.0081	2.0049	2.0025
Phenoxyl + water 135°[a]	+	2.0078	2.0049	2.0025
Phenoxyl + water 90°[a]	+	2.0074	2.0049	2.0025

[a] R(O···H) = 1.95

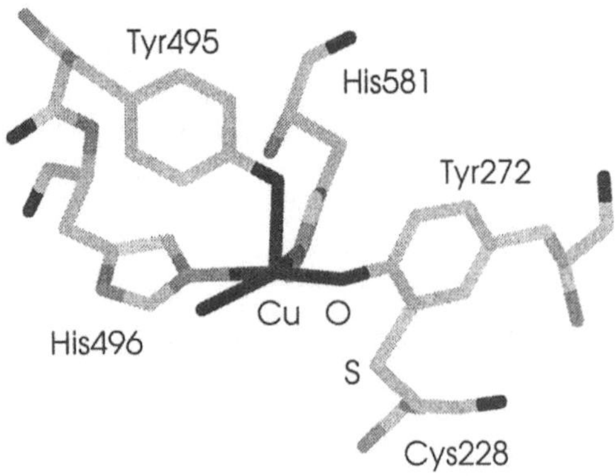

Figure 5. Crystal structure of the active site of galactose oxidase.

the g_{xx} value from 2.0087 to 2.0081 (Table 6). This shift depends on the lower angular momentum and spin-orbit coupling matrix elements for the excitation between the oxygen lone-pair n and the singly occupied π orbitals ($n \rightarrow \pi_{SOMO}$ excitation). The g_{xx} value is further decreased by hydrogen bonding in bent CO$\cdots$H configurations due to a blue shift of the $n \rightarrow \pi_{SOMO}$ excitation. The different performance between the linear and bent models can be explained by the overlap between phenoxyl lone-pair and water orbitals which is found in a phenoxyl-water complex in bent configuration. This overlap produces a decreased orbital energy and a following increased $n \rightarrow \pi_{SOMO}$ excitation energy.

5.1.2 Galactose oxidase (GO). Galactose oxidase (GO) is a funghial enzyme that catalyzes the oxidation of galactose to the corresponding aldehyde. The catalytic site of GO consists of a tyrosyl radical which is antiferromagnetically coupled to a metal site composed of a single copper atom (Figure 5). The coupling to the copper site makes the tyrosyl radical EPR silent. However, tyrosyl EPR spectra have been recorded in the apo-enzyme where the copper atom has been removed (Babcock et al., 1992).

The tyrosyl radical in GO is covalently bound to a cysteine. A heavy atom, such as the cysteine sulfur, close to the radical, is expected to substantially influence the magnetic properties of the radical. In contrast to these expectations, no major shifts in the g-tensor were detected by EPR (Babcock et al., 1992). MCSCF calculations of a sulfur substituted

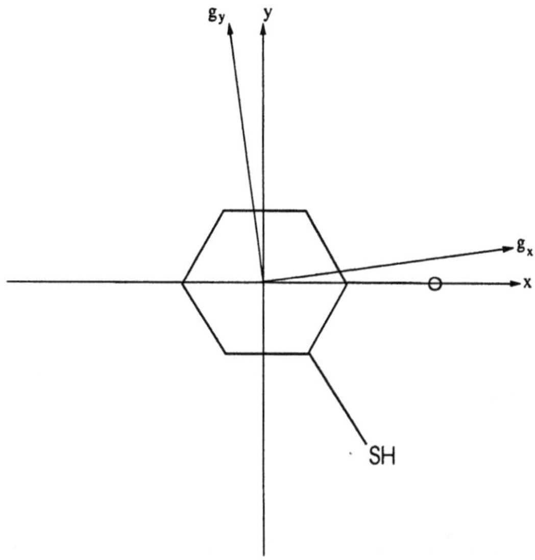

Figure 6. The principal axes of the g-tensor in the sulfur substituted phenoxyl radical compared with the molecular coordinate system. The g_{xx} component of the phenoxyl radical is aligned with the CO bond whereas g_{xx} of phenoxyl-S is rotated 8° away from the CO bond.

phenoxyl radical as a model for the tyrosyl radical in GO show that the spin density pattern is basically the same as in the unsubstituted radical (Engström et al., 2000a). The g-tensor of the phenoxyl is essentially rhombic with g_{xx} directed along the CO bond, g_{yy} in the molecular plane, and g_{zz} out of the plane (Table 6). The sulfur substituted tyrosyl radical induces only a small shift of the g-tensor components towards less rhombic values compared with other phenoxy type radicals: g_{xx}=2.0079, g_{yy}=2.0052, and g_{zz}=2.0025. The principal axes of the sulfur substituted phenoxyl g-tensor is rotated by 8° in the molecular plane (Figure 6). The x-axis is shifted away from the sulfur atom and is no longer collinear with the C-O bond.

6. Sample applications: Spin labels

Nitroxide spin labels are widely used together with EPR to investigate structure and mobility of proteins (Berliner and Reuben, 1989). The labels are artificially introduced at specific sites of the protein by site directed mutagenesis. By this technique the nucleotide sequence of the protein gene is altered producing a mutated protein structure. In this structure one amino acid is replaced with a cysteine that can be linked to the spin label. The EPR spectrum of a nitroxide spin label is highly

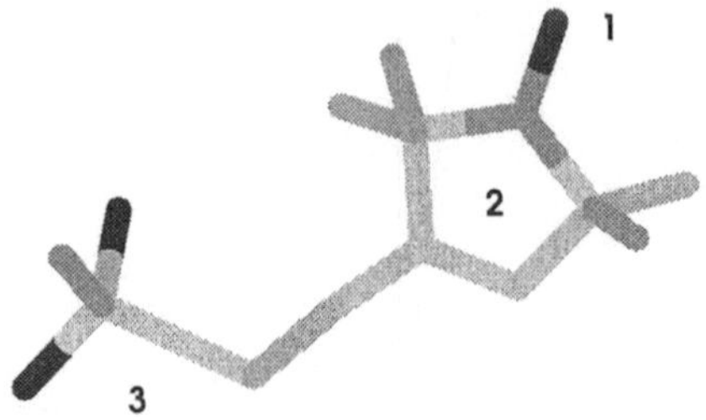

Figure 7. Methane Thio Sulfonate Spin Label (MTSSL). 1. Paramagnetic nitroxide group 2. Bridge 3. Linker.

sensitive to the mobility of the label and information about the local environment around the label can thereby be obtained.

The methanethiosulfonate spin label (MTSSL) is commonly used in studies of protein structure and dynamics (Hubbell et al., 1998; Steinhoff et al., 2000). This label consists of three parts: 1) the paramagnetic group (NO); 2) the bridge with methyl groups that stabilize the radical; and 3) the linker to the protein (Figure 7). The spin density of the label is mainly concentrated to the nitrogen and oxygen in almost equal parts. Interactions with solvent molecules involve a redistribution of the spin density with accompanying changes in the EPR parameters. The exposed lone-pair orbital of the oxygen atom makes the nitroxide sensitive to intermolecular interactions in terms of hydrogen bonding. Thus apart from the sensitivity to the motion of the protein-label system, the EPR spectrum of a nitroxide spin label is sensitive to solvent interactions through the EPR parameters.

The influence of protic solvents on the g-tensor was studied with the ROHF linear response method in (Engström et al., 2001). In that study MTSSL was simulated with a truncated model of a pyrrolidine radical without the linker to the protein (Figure 8). This choice of molecular model was based upon the observation that removal of the linker structure did not appreciably alter the spin density distribution of the radical. The results indicated that the g-tensor component directed along the NO bond of the nitroxide decreases upon hydrogen bonding to the oxygen atom. However, the ROHF method clearly overestimates the influence of hydrogen bonding. By DFT, however, the hydrogen bonding effect was found to correspond to experimental values for MTSSL in various solvents (Table 7).

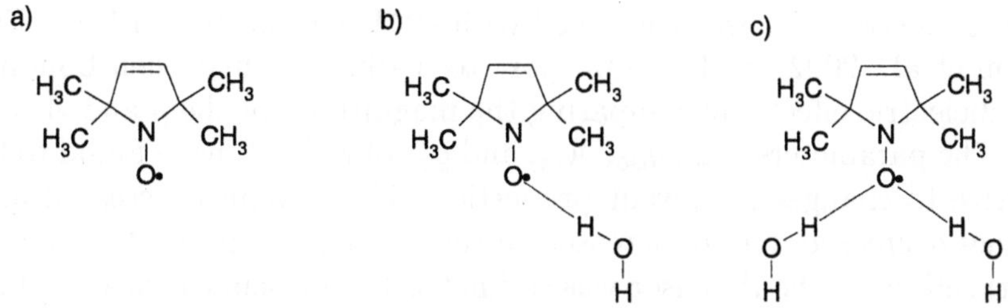

Figure 8. Molecular models describing three possible ways that MSL (model spin label) could interact with its local environment. The number of hydrogen bonded solvent molecules to MSL varies for the models. (a) model A, no H-bonds; (b) model B, one H-bond; and (c) model C, two H-bonds.

Table 7. Calculated and experimental isotropic and anisotropic EPR parameters of the MSL spin label model in some media with different properties.

	g_{xx}	g_{yy}	g_{zz}
Calculation			
Model A[a]	2.00832	2.00568	2.00214
Model B[a]	2.00788	2.00561	2.00214
Model C[a]	2.00750	2.00554	2.00213
Experiment			
Ethanol	2.00835	2.00593	2.00200
Methanol	2.00812	2.00578	2.00189
Water	2.00800	2.00586	2.00199

[a] Applied structural model according to Figure 8.

EPR spectroscopy was combined with DFT calculations in Ref. (Engström et al., 2002) with the purpose to distinguish hydrogen bonding from dielectric effects by comparing the magnitudes of the g and A tensors. The parameters A_{iso}, g_{iso}, A_{zz}, and g_{xx} of MTSSL were found to be sensitive to changes in solvent properties. The A-values increased and g-values decreased due to increased solvent polarity or proticity. The dielectric effect on the A-tensor was estimated from calculations according to the polarizable continuum model (PCM). In this model the solvent is considered as a uniform medium with dielectric constant ε and the solute molecule is inserted in a cavity formed from van der Waals surfaces of the radical. From a combined experimental and theoretical qualitative analysis it was possible to specify the causes of the parameter shifts in detail. At low dielectric constants ($\varepsilon < 10$), the sensitivity of A_{iso} and A_{zz} to ε was large. However, at large ε ($\varepsilon > 30$), the sensitivity to the polarity of the continuum was very small. The most dominating shifts in the EPR parameters were instead determined by specific interactions with solvent molecules, such as water.

Considering the dominating effect on the EPR parameters by specific intermolecular interactions, further investigations on this issue are appropriate. MTSSL in a surrounding protein structure was investigated in Ref. (Engström et al., 2002). The complex of two proteins, tissue factor (TF) and factor VIIa (FVIIa) involved in blood coagulation, was chosen for that purpose. A reasonable model of MTSSL in the sTF/FVIIa complex obtained from a molecular mechanics structure (Owenius et al., 1999) was selected for DFT calculations on the EPR parameters of the spin label. The aim was to investigate the influence on the EPR parameters from the amino acids in the nearest environment of MTSSL in the protein complex. This was accomplished by DFT calculations. The fact that this method can be applied for EPR properties of such a large system has wide ramifications and brings optimism for further applications on spin labels at different positions in proteins.

References

Allard, P., Barra, A. L., Andersson, K. K., Schmidt, P. P., Atta, M., and Gräslund, A. (1996). Characterization of a new tyrosyl free radical in *Salmonella typhimurium* ribonucleotide reductase with EPR at 9.45 and 245 GHz. *J. Am. Chem. Soc.*, 118:895–896.

Babcock, G. T., El-Deeb, M. K., Sandusky, P. O., Whittaker, M. M., and Whittaker, J. W. (1992). Electron paramagnetic resonance and electron nuclear double resonance spectroscopies of the radical site in galactose oxidase and of thioether-substituted phenol model compounds. *J. Am. Chem. Soc.*, 114:3727–3734.

Belanzoni, P., van Lenthe, E., and Baerends, E. J. (2000). An evaluation of the density functional approach in the zeroth order regular approximation, ZORA, for relativistic effects: magnetic interactions in small metal complexes. DFT 2000, Menton, France.

Berliner, L. J. and Reuben, J. (1989). *Biological Magnetic Resonance*, volume 8, Spin Labeling - Theory and Applications. Plenum Press, New York.

Bresgunov, A. Y., Dubinsky, A. A., Poluektov, O. G., and Lebedev, Y. S. (1992). The structure of phenoxyl radicals as studied by 2 mm band ESR. *Mol. Phys.*, 75:1123–1130.

Burghaus, O., Plato, M., Rohrer, M., Möbius, K., MacMillan, F., and Lubitz, W. (1993). 3-mm high-field EPR on semiquinone radical anions $Q^{\cdot -}$ related to photosynthesis and on the primary donor $P^{\cdot +}$ and acceptor $Q_a^{\cdot -}$ in reaction centers of *Rhodobacter sphaeroides* R-26. *J. Phys. Chem.*, 97:7639–7647.

Carravetta, V., Ågren, H., Vahtras, O., and Jensen, H. J. A. (2000). Ab initio calculations of molecular resonant photoemission spectra. *J. Chem. Phys.*, 113:7790–7798.

Carrington, A. and McLachlan, A. D. (1979). *Introduction to magnetic resonance*. Chapman and Hall, London.

Carter, M. K. and Vincow, G. (1967). Electron spin resonance of the benzene positive-ion radical. *J. Chem. Phys.*, 47:292–302.

Christiansen, O., Koch, H., and Jørgensen, P. (1995a). The 2nd-order approximate coupled-cluster singles and doubles model CC2. *Chem. Phys. Lett.*, 243:409–418.

Christiansen, O., Koch, H., and Jørgensen, P. (1995b). Response functions in the CC3 iterative triple excitation model. *J. Chem. Phys.*, 103:7429–7441.

Dalgaard, E. (1980). *J. Chem. Phys.*, 72:816.

Dalgaard, E. (1982). *Phys. Rev. A*, 26:42.

Ehrenberg, A. and Reichard, P. (1972). Electron spin resonance of the iron-containing protein B2 from ribonucleotide reductase. *J. Biol. Chem.*, 247:3485–3488.

Engström, M., Himo, F., and Ågren, H. (2000a). Ab initio g-tensor calculations of the thioether substituted tyrosyl radical in galactose oxidase. *Chem. Phys. Lett.*, 319:191–196.

Engström, M., Himo, F., Gräslund, A., Vahtras, O., Minaev, B., and Ågren, H. (2000b). Hydrogen bonding to the tyrosyl radical analyzed by *ab initio* g-tensor calculations. *J. Phys. Chem.*, 104:5149–5133.

Engström, M., Minaev, B., Vahtras, O., and Ågren, H. (1998). Linear response calculations of electronic g-factors and spin-rotational coupling constants for diatomic molecules with a triplet ground state. *Chem. Phys.*, 237:149–158.

Engström, M., Owenius, R., and Vahtras, O. (2001). *Ab initio* g-tensor calculations of hydrogen bond effects on a nitroxide spin label. *Chem. Phys. Lett.*, 338:407–413.

Engström, M., Vaara, J., and Schimmelpfennig, B. (2002). Density functional theory calculations of epr parameters of a nitroxide spin label in tissue factor and factor viia protein complex. Submitted to Phys. Chem. Chem. Phys.

Engström, M., Vahtras, O., and Ågren, H. (2000c). Mcscf and dft calculations of epr parameters of sulfur centered radicals. *Chem. Phys. Lett.*, 328:483–491.

Engström, M. (2001). *Quantum Chemical Calculations of Electron Paramagnetic Resonance of Biological Radicals.* PhD Thesis, Linköping University, Linköping, Sweden,, 2001.

Eriksson, L. A., Himo, F., Siegbahn, P. E. M., and Babcock, G. T. (1997). Electronic and magnetic properties of neutral and charged quinone and plastoquinone radicals. *J. Phys. Chem.*, 101:9496–9504.

Fernández, B., Christiansen, O., Bludsky, O., Jørgensen, P., and Mikkelsen, K. V. (1996). Theory of hyperfine coupling constants of solvated molecules: Applications involving methyl and ClO_2 radicals in different solvents. *J. Chem. Phys.*, 104:629–635.

Fernández, B., Jørgensen, P., Byberg, J., Olsen, J., Helgaker, T., and Jensen, H. J. A. (1992). Spin polarization in restricted electronic structure theory: Multiconfiguration self-consistent-field calculations of hyperfine coupling constants. *J. Chem. Phys.*, 97:3412–3419.

Flockhart, B. D., Leith, I. R., and Pink, R. C. (1970). Electron-transfer at alumina surfaces. part 3.-reduction of aromatic nitro-compounds. *Trans. Faraday Soc.*, 66:469–476.

Gel'mukhanov, F., Salek, P., Shalagin, A., and Ågren, H. (2000). X-ray raman scattering under pulsed excitation. *J. Chem. Phys.*, 112:5593–5603.

Gillbro, T. (1974). ESR and structures of sulfur-centered radicals and radical ions. γ-irradiated dimethyl disulfide and methane thiol single crystals at 77°K. *Chem. Phys.*, 4:476–482.

Gräslund, A. and Sahlin, M. (1996). Electron paramagnetic resonance and nuclear magnetic resonance studies of class I ribonucleotide reductase. *Annu. Rev. Biophys. Biomol. Struct.*, 25:259–286.

Hadley, J. H. and Gordy, W. (1974a). Nuclear coupling of ^{33}S and the nature of free radicals in irradiated crystals of cystine dihydrochloride. *Proc. Nat. Acad. Sci., USA*, 71:3106–3110.

Hadley, J. H. and Gordy, W. (1974b). Nuclear coupling of ^{33}S and the nature of free radicals in irradiated crystals of cystine dihydrochloride: Part II, charged radicals. *Proc. Nat. Acad. Sci, USA*, 71:4409–4413.

Hales, B. J. (1975). Immobilized radicals. I. Principal electron spin resonance parameters of the benzosemiquinone radical. *J. Am. Chem. Soc.*, 97:5993–5997.

Harriman, J. E. (1978). *Theoretical foundations of electron spin resonance.* Academic Press, New York.

Helgaker, T., Jensen, H. J. A., Jørgensen, P., Olsen, J., Ruud, K., Ågren, H., Auer, A. A., Bak, K. L., Bakken, V., Christiansen, O., Coriani, S., Dahle, P., Dalskov, E. K., Enevoldsen, T., Fernandez, B., Hättig, C., Hald, K., Halkier, A., Heiberg, H., Hettema, H., Jonsson, D., Kirpekar, S., Kobayashi, R., Koch, H., Mikkelsen, K. V., Norman, P., Packer, M. J., Pedersen, T. B., A.Ruden, T., Sanchez, A., Saue, T., Sauer, S. P. A., Schimmelpfennig, B., Sylvester-Hvid, K., Taylor, P. R., and Vahtras, O. (2001). *Dalton, an ab initio electronic structure program, Release 1.2. See http://www.kjemi.uio.no/software/dalton/dalton.html.*

Helgaker, T., Jørgensen, P., and Olsen, J. (1998). Quantum chemistry and molecular properties. The fifth sostrup summer school.

Helgaker, T., Watson, M., and Handy, N. C. (2000). Analytical calculation of nuclear magnetic resonance indirect spin-spin coupling constants at the generalized gradient approximation and hybrid levels of density-functional theory. *J. Chem. Phys.*, 113:9402–9409.

Heß, B. A., Marian, C. M., Wahlgren, U., and Gropen, O. (1996). A mean-field spin-orbit method applicable to correlated wavefunctions. *Chem. Phys. Lett.*, 251:365–371.

Himo, F. and Eriksson, L. A. (1997). Theoretical study of model tryptophan radicals and radical cations: Comparisons with experimental data of DNA photolyase, cytochrome c peroxidase, and ribonucleotide reductase. *J. Phys. Chem.*, 101:9811–9819.

Himo, F., Eriksson, L. A., Maseras, F., and Siegbahn, P. E. M. (2000). Catalytic mechanism of galactose oxidase. A theoretical study. *J. Am. Chem. Soc.*, 122:8031–8036.

Himo, F., Gräslund, A., and Eriksson, L. A. (1997). Density functional calculations on model tyrosyl radicals. *Biophysical Journal*, 72:1556–1567.

Hubbell, W. L., A, G., R, L., and A, L. M. (1998). Recent advances in site-directed spin labeling of proteins. *Curr. Opin. Struct. Biol.*, 8:649–656.

Huber, K. P. and Herzberg, G. (1979). *Molecular spectra and molecular structure. Constants of diatomic molecules.* Van Nostrand Reinhold Company, New York.

Ishii, M., Morihashi, K., and Kikuchi, O. (1989). Calculations of g-values of free radicals by finite perturbation theory. *J. Mol. Struct. (Theochem)*, 187:209–218.

Jayatilaka, D. (1998). Electron spin resonance g tensors from general Hartree-Fock calculations. *J. Chem. Phys.*, 108:7587–2498.

Jonsson, D., Norman, P., Ruud, K., Ågren, H., and Helgaker, T. (1998). Electric and magnetic properties of fullerenes. *J. Chem. Phys.*, 109:572–577.

Knüpling, M., Törring, J. T., and Un, S. (1997). The relationship between the molecular structure of semiquinone radicals and their g-values. *Chem. Phys.*, 219:291–304.

Knuts, S., Minaev, B. F., Ågren, H., and Vahtras, O. (1994). The phosphorescence of benzene obtained by ab initio and semiempirical calculations. *Theor. Chim. Acta*, 87:343–371.

Lawrence, C. C., Bennati, M., Obias, H. V., Bar, G., Griffin, R. G., and Stubbe, J. (1999). High-field EPR detection of a disulfide radical anion in the reduction of cytidine 5'-diphosphate by the E441Q R1 mutant of *Escherichia coli* ribonucleotide reductase. *Proc. Natl. Acad. Sci. USA*, 96:8979–8984.

Löwdin, P. (1971). *Int. J. Quant. Chem.*, 4:31.

Luo, Y., Norman, P., Macak, P., and Ågren, H. (1999). Solvent effects on the polarizabilities and hyperpolarizabilities of conjugated polymers. *J. Chem. Phys.*, 111:9853–9858.

Lushington, G. H., Bündgen, P., and Grein, F. (1995). Ab initio studies of molecular g-tensors. *Int. J. Quant. Chem.*, 55:377–392.

Lushington, G. H. and Grein, F. (1997). Multireference configuration interaction calculations of electronic g-tensors for NO_2, H_2O^+ and CO^+. *J. Chem. Phys.*, 106:3292–3300.

Malkina, O. L., Vaara, J., Schimmelpfennig, B., Munzarova, M., Malkin, V. G., and Kaupp, M. (2000). Density functional calculations of electronic g-tensors using spin-orbit pseudopotentials and mean-field all-electron spin-orbit operators. *J. Am. Chem. Soc.*, 122:9206–9218.

Marian, C. M. and Wahlgren, U. (1996). A new mean-field and ecp-based spin-orbit method. applications to pt and pth. *Chem. Phys. Lett.*, 251:357–364.

Martell, J. M., Eriksson, L. A., and Goddard, J. D. (1997). A density functional study of the hyperfine properties of sulfur-containing radicals and radical ions. *Acta Chem. Scand.*, 51:229–232.

Matsuki, K., Hadley, J. H., Nelson, W. H., and Yang, C.-Y. (1993). ESR studies of monosulfide radicals in irradiated *N*-acetyl-*L*-cysteine single crystals. *J. Magn. Res.*, 103:196–202.

Meisel, D. and Neta, P. (1975). One-electron redox potentials of nitro compounds and radiosensitizers. corelation with spin densities of their radical anions. *J. Am. Chem. Soc.*, 97:5198–5203.

Minaev, B. F. (1982). *Zh. Struct. Khim.*, 23:7.

Minaev, B. F. (1989). Solvent induced emission of molecular $^1\delta_g$ oxygen. *J. Mol. Struct. (Theochem)*, 183:207–214.

Nelson, D. J., Petersen, R. L., and Symons, M. C. R. (1977). Unstable intermediates. Part 178. The structure of intermediates formed in the radiolysis of thiols. *J. C. S. Perkin II*, 0:2005–2015.

Neta, P. and Fessenden, W. (1974). Hydroxyl radical reactions with phenols and anilines as studied by electron spin resonance. *J. Phys. Chem.*, 78:523–529.

Nørager, A. (2000). *Udvikling af en relativistisk model til beregning af ESR-parametre*. PhD thesis, Odense Universitet.

Norman, P., Jonsson, D., Ågren, H., Dahle, P., Ruud, K., Helgaker, T., and Koch, H. (1996). Efficient parallel implementation of response theory: Calculations of the second hyperpolarizability of polyacenes. *Chem. Phys. Lett.*, 253:1–7.

Norman, P., Luo, Y., Jonsson, D., and Ågren, H. (1997). Ab initio calculations of the polarizability and hyperpolarizability of c_{60}. *J. Chem. Phys.*, 106:8788.

Oddershede, J. (1987). *Adv. Chem. Phys.*, 69:201.

Olsen, J. and Jørgensen, P. (1985). *J.Chem. Phys.*, 82:3235.

Owenius, R., Engström, M., Lindgren, M., and Huber, M. (2001). Influence of solvent polarity and hydrogen bonding on the EPR parameters of a nitroxide spin label studied by 9-ghz and 95-ghz epr spectroscopy and dft calculations. *J. Phys. Chem. A*, 105:10967–10977.

Owenius, R., Österlund, M., Lindgren, M., Svensson, M., Olsen, O. H., Persson, E., Freskgard, P.-O., and Carlsson, U. (1999). Properties of spin and fluorescent labels at a receptor-ligand interface. *Biophys. J.*, 77:2237–2250.

Parast, C., Wong, K. K., and Kozarich, J. W. (1995). Electron paramagnetic resonance evidence for a cysteine-based radical in pyruvate formate-lyase inactivated with mercaptopyruvate. *Biochemistry*, 34:5712–5717.

Patchkovskii, S. and Ziegler, T. (1999). Prediction of electron paramagnetic resonance g-tensors of transition metal complexes using density functional theory: First applications to some axial d^1MEX_4 systems. *J. Chem. Phys.*, 111:5730–5739.

Patchkovskii, S. and Ziegler, T. (2000). Prediction of EPR g tensors in simple d^1 metal porphyrins with density functional theory. *J. Am. Chem. Soc.*, 122:3506–3516.

Persson, A. L., Sahlin, M., and Sjöberg, B.-M. (1998). Cysteinyl and substrate radical formation in active site mutant E441Q of *Escherichia coli* class I ribonucleotide reductase. *J. Biol. Chem.*, 273:31016–31020.

Pierloot, K., De Kerpel, J. O. A., Ryde, U., Olsson, M. H. M., and Roos, B. O. (1998). Relation between the structure and spectroscopic properties of blue copper proteins. *J. Am. Chem. Soc.*, 120:13156–13164.

Plashkevych, O., Ågren, H., Carravetta, V., Contini, G., and Polzonetti, G. (2000). Vibronic structure fingerprints in nexafs: a theoretical study of 2-mercaptobenzoxazole. *Chem. Phys. Lett.*, 327:7–12.

Qin, Y. and Wheeler, R. A. (1995). Similarities and differences between phenoxyl and tyrosine phenoxyl radical structures, vibrational frequencies and spin densities. *J. Am. Chem. Soc.*, 117:6083–6092.

Rega, N., Cossi, M., and Barone, V. (1996). Development and validation of reliable quantum mechanical approaches for the study of free radicals in solution. *J. Chem. Phys.*, 105:11060–11067.

Reichard, P. (1993). From RNA to DNA, why so many ribonucleotide reductases. *Science*, 260:1773–1777.

Ruud, K., Jonsson, D., Norman, P., Ågren, H., Saue, T., Jensen, H. J. A., Dahle, P., Helgaker, T., and H, K. (1998). Generalized integral-screening for efficient calculations of nonlinear optical properties of large molecules. *J. Chem. Phys.*, 108:7973–7979.

Salahub, D. R., Fournier, R., Mlynarski, P., Papai, P., St-Amant, A., and Ushio, J. (1991). In Labanowski, J. and Andzelm, J., editors, *Density Functional Methods in Chemistry*, New York. Springer.

Schreckenbach, G. and Ziegler, T. (1997). Calculation of the G-tensor of electron paramagnetic resonance spectroscopy using gauge-including atomic orbitals and density functional theory. *J. Phys. Chem.*, 101:3388–3399.

Schreckenbach, G. and Ziegler, T. (1998). Density functional calculations of NMR chemical shifts and ESR g-tensors. *Theor. Chem. Acc.*, 99:71–82.

Segal, B. G., Kaplan, M., and Fraenkel, G. K. (1965). Measurement of g-values in the electron spin resonance spectra of free radicals. *J. Chem. Phys.*, 43:4191–4200.

Siegbahn, P. E. M., Blomberg, M., and Pavlov, M. (1998a). A comparison of electron transfer in ribonucleotide reductase and the bacterial photosynthetic reaction center. *Chem. Phys. Lett.*, 292:421–430.

Siegbahn, P. E. M. and Blomberg, M. R. A. (2000). Transition-metal systems in biochemistry studied by high-accuracy quantum chemical methods. *Chem. Rev.*, 100:421–438.

Siegbahn, P. E. M. and Crabtree, R. H. (1999). Manganese oxyl radical intermediates and O-O bond formation in photosynthetic oxygen evolution and proposed role for the calcium cofactor in photosystem II. *J. Am. Chem. Soc.*, 121:117–127.

Siegbahn, P. E. M., Eriksson, L., Himo, F., and Pavlov, M. (1998b). Hydrogen atom transfer in ribonucleotide reductase (RNR). *J. Phys. Chem.*, 102:10622–10629.

Sjöberg, B.-M., Reichard, P., Gräslund, A., and Ehrenberg, A. (1977). Nature of the free radical in ribonucleotide reductase from *Escherichia coli*. *J. Biol. Chem.*, 252:536–541.

Stein, M. (2001). *Insight into the Mechanism of [NiFe] Hydrogenase by means of Magnetic Resonance Experiments and DFT Calcualtions*. PhD Thesis, Technische Universität Berlin, Berlin, Germany, 2001.

Steinhoff, H. J., Savitsky, A., Wegener, C., Pfeiffer, M., Plato, M., and Möbius, K. (2000). High-field epr studies of the structure and conformational changes of site-directed spin labeled bacteriorhodopsin. *Biochim. Biophys. Acta*, 1457:253.

Stone, A. J. (1963). *g* factors of aromatic free radicals. *Mol. Phys.*, 6:509–515.

Stone, A. J. (1964). *g* tensors of aromatic hydrocarbons. *Mol. Phys.*, 7:311–316.

Thomas, L. (1927). *Phil. Mag.*, 3:1.

Turro, N. J. (1979). *Modern molecular photochemistry*. Benjamin/Cumming, Menlo Park, Calif.

Un, S., Atta, M., Fontecave, M., and Rutherford, A. W. (1995). *g*-values as a probe of the local protein environment: High-field EPR of tyrosyl radicals in ribonucleotide reductase and photosystem II. *J. Am. Chem. Soc.*, 117:10713–10719.

Vahtras, O., Ågren, H., Jørgensen, P., H.J.Aa. Jensen, Helgaker, T., and Olsen, J. (1992a). Multiconfigurational quadratic response functions for singlet and triplet perturbations – the phosphorecence lifetime of formaldehyde. *J. Chem. Phys.*, 97:9178–9187.

Vahtras, O., Ågren, H., Jørgensen, P., Jensen, H. J. A., Helgaker, T., and Olsen, J. (1992b). Spin-orbit coupling constants in a multiconfiguration linear response approach. *J. Chem. Phys.*, 96:2118–2126.

Vahtras, O., Engström, M., and Schimmelpfennig, B. (2002). Electronic g-tensors obtained with the mean-field spin-orbit hamiltonian. *Chem. Phys. Lett.*, 351:424–430.

Vahtras, O., Minaev, B., and Ågren, H. (1997). Ab initio calculations of electronic *g*-factors by means of multiconfiguration response theory. *Chem. Phys. Lett.*, 281:186–192.

van Dam, P. J., Willems, J.-P., Schmidt, P. P., Pötsch, S., Barra, A.-L., Hagen, W. R., Hoffman, B. M., Andersson, K. K., and Gräslund, A. (1998). High-frequency EPR and pulsed Q-band ENDOR studies on the origin of the hydrogen bond in tyrosyl radicals of ribonucleotide reductase R2 proteins from mouse and herpes simplex virus type 1. *J. Am. Chem. Soc.*, 120:5080–5085.

van Lenthe, E., Wormer, P. E. S., and van der Avoird, A. (1997). Density functional calculations of molecular *g*-tensors in the zero-order regular approximation for relativistic effects. *J. Chem. Phys.*, 107:2488–2498.

Wayne, F. D. and Colbourn, E. A. (1977). Effective spin-spin interactions and singlet-triplet transition intensities in a class of open shell diatomic molecules. *Mol. Phys.*, 34:1141–1155.

Yeager, D. L. and Jørgensen, P. (1979). *Chem. Phys. Lett.*, 65:77.

Zeldes, H. and Livingston, R. (1972). Paramagnetic resonance study of liquids during photolysis. XIV Pyridine, pyrazine, pyrimidine and pyridazine. *J. Phys. Chem.*, 76:3348–3355.

Zeldes, H. and Livingston, R. (1974). Electron spin resonance of photolytically reduced pyrazine. *Mol. Phys.*, 27:261–264.

Zubarev, D. N. (1974). In *Nonlinear Statistical Thermodynamics*, New York. Plenum.

Electron Propagator Calculations on the Ionization Energies of Nucleic Acid Bases, Base–Water Complexes and Base Dimers

O. Dolgounitcheva, V. G. Zakrzewski and J. V. Ortiz [1]
Department of Chemistry
Kansas State University
Manhattan, KS 66506–3701

Introduction

Among the principal patrons of quantum biology, the field where quantum mechanics, especially molecular electronic structure theory, is applied to systems of biological significance, was our friend and teacher, Per–Olov Löwdin. Quantum mechanical phenomena in the generation and expression of genetic information were a special interest of his. He also provided guidance and encouragement to those who develop and apply propagator methods in quantum chemistry. Our interest in the electronic structure of nucleic acids is, to a significant extent, a consequence of conversations, symposia and informal discussions that occurred as a result of Professor Löwdin's sponsorship.

Chemical reactivity associated with nucleotides, nucleosides and H–bonded and stacked dimers of the purine and pyrimidine bases known as nucleobases is a central concern in molecular biology and biochemistry [1, 2]. Since these patterns of reactivity may be related to molecular ionization energies (IEs), precise data on these

[1] ortiz@ksu.edu

E.J. Brändas and E.S. Kryachko (eds.),
Fundamental World of Quantum Chemistry, Vol. II, 525–555.

energy differences are needed by specialists in several fields of biological chemistry. However, the instability of nucleic acid fragments impedes experimental measurements of their IEs in the gas phase. Accurate theoretical techniques therefore can perform an important service in supplementing and interpreting spectral data. We presently employ electron propagator methods that are computationally feasible for calculations on systems of this size.

Methods

Electron propagator methods have become routine tools for prediction and interpretation of molecular ionization spectra [3]–[8]. Quasiparticle approximations, where the off–diagonal elements of the self–energy matrix in the canonical, Hartree–Fock orbital basis are neglected, have proven especially useful in the study of large molecules [9]. The recently proposed partial third order approximation (P3) [10] and the older Outer Valence Green Function approximations (OVGF) [6] have been implemented efficiently in the Gaussian–98 suite of programs. P3 and OVGF are applicable when the one–electron picture of ionization embodied in Koopmans's theorem remains qualitatively valid and these methods therefore are used chiefly for outer valence ionization energies and electron affinities. Previous implementations included in the Gaussian–98 code consisted of two, separate codes: one for symmetry–adapted calculations with closed–shell reference states and another, more general, but less efficient, code for UHF reference states without symmetry adaptation.

Algorithms

The present implementation of P3 and OVGF approximations represents a generalization of previous versions in several respects. The present code is designed to use electron repulsion integrals in the molecular orbital (MO) basis set as generated by the Gaussian code. Antisymmetrized matrix elements are created for various subsets according to spin cases and the occupancy of the spin–orbital indices. (Let i,j,k,... be occupied spin–orbital indices and a,b,c,... be virtual spin–orbital indices.) These subsets include

- all occupied indices <ij||kl>
- three occupied indices, one virtual index <ij||ka>
- two occupied, two virtual indices <ij||ab>, <ia||jb>
- one occupied, three virtual indices <ia||bc>
- all virtual indices <ab||cd>

The double–bar notation stands here for antisymmetrization over the last two indices: <pq||rs> = <pq|rs> - <pq|sr>. Abelian group symmetry is taken into account. The direct product decomposition scheme of integral storage employed in several kinds of post–Hartree–Fock calculations [11] has been employed. It assumes that two–electron matrix elements are stored such that

$$\begin{array}{ccccc} \Gamma_p \otimes \Gamma_q & = & \Gamma_r \otimes \Gamma_s & = & 1 \\ \Gamma_p \otimes \Gamma_q & = & \Gamma_r \otimes \Gamma_s & = & 2 \\ \Gamma_p \otimes \Gamma_q & = & \Gamma_r \otimes \Gamma_s & = & 3 \\ \vdots & & \vdots & & \vdots \end{array}, \tag{1}$$

where Γ_p is the index of the irreducible representation of orbital p. Irreducible representations of Abelian groups may be labelled by the natural numbers from 1 to 8. Every set of matrix elements consists of a set of rectangular arrays which obey the previous equation. This scheme also may be used for cases where there are restrictions on indices such as $i < k$ or $i \leq k$. Elements that are non–zero by symmetry are ordered in (2,2) (that is, pairwise) symmetry order.

This storage technique is especially convenient for contractions over a pair of indices, for they may be executed as matrix multiplications with pair superindices. It has been implemented in Gaussian–98 for post–Hartree–Fock methods. This manner of storing integrals, while useful, is only one of many alternatives and it entails performance of various manipulations. Such manipulations over symmetry–compressed entities include

- expansion of four–dimensional entities with restrictions on indices to their full sizes, such as <ij|ab> with i<j,a<b into all i,j,a,b

- changes in the order of indices, such as (i,j,k,l) into (i,k,j,l)
- straightforward contraction over two indices, or somewhat more involved contractions over one or three indices.

These manipulations can be done in various ways, some of which are discussed in Ref. [11]. Molecular orbitals are assumed to be ordered according to irreducible representations within the occupied and unoccupied subspaces for both spin cases. The only information necessary to describe symmetry–related properties, and therefore to perform the operations mentioned above, is the distribution of MOs between representations within each of the subspaces.

Quasiparticle propagator methods include contractions involving matrix elements such as <pi||ab> or <pa||ij>, where index p is restricted to just one value, and consequently to a certain irreducible representation. This case was treated by introducing formal distributions with all entries set equal to zero except for one particular representation. Contractions with matrix elements like this lead to intermediates with the same formal size of o^2v, or ov^2, where o is the number of occupied spin–orbitals and v is the number of virtual spin–orbitals. Computations involving these terms are arranged with the goal of reducing the amount of I/O by keeping intermediates of o^2v and v^2o size in core.

Special options exist to carry out all types of quasiparticle calculations with, or without, transformed integrals with all four virtual indices. Contributions from integrals with four virtual indices can be taken into account optionally in a direct way. Computation of P3 and OVGF ionization energies (electron detachment from occupied MOs) can be also done in the case when contributions from integrals with four and three virtual indices can be done in direct way. These features make the present code rather flexible with respect to usage of disk space.

The current code was applied to a number of systems with at least one element of symmetry. Two of these, hydrogen–bonded and stacked uracil dimers, are included into this review.

Details of Calculations

All calculations were performed with the development version of the Gaussian–99 suite of programs [12]. Molecular geometries of free bases and the uracil, hydrogen–bonded dimer (HB) were optimized with MP2 [13] and 6–311G** bases [14]. The most stable HB uracil dimer of Ref. [15] was taken. For the uracil stack dimer (SB), the MP2–optimized geometry of uracil was used and a 3.4 Å distance between stacked monomers was adopted, as recommended in Ref. [16]. Head–to–head orientation of monomers was taken into consideration. Four isomers of uracil–water complexes were optimized with MP2 and the 6–311++G** basis [27]. Molecular and orbitals were obtained with the MOLDEN graphing program [17].

Results and Discussion

Calculated IEs of free bases together with available experimental data are summarized in Tables 1 through 5. IEs of uracil–water complexes are summarized in Table 6. Calculated IEs of HB and SB uracil dimers are gathered in Table 7. Corresponding Dyson orbital (DO) diagrams are presented in Fig.s 3, 5, 8, 10, 12, 16 and 17. The numbering scheme for the purine core is given in Fig. 1 below.

Purines

Adenine

Optimized with MP2 and the 6–311G** basis, the *9H* isomer of adenine (Fig. 2) is about 8 kcal/mol lower than any other tautomer. The dependence of IEs upon the basis sets and propagator methods applied was described in detail in Ref. [19]. P3 gives excellent agreement with available experimental data of Ref. [18]. (See Table 1.) The position of π_1 coincides with the experimental value of 8.48 eV [18]. The corresponding DO (Fig. 3) is delocalized over the entire molecular framework. Two and three–center binding π lobes on both rings contribute to this orbital in an antibonding mode with each other. Significant participation of amino nitrogen is noted.

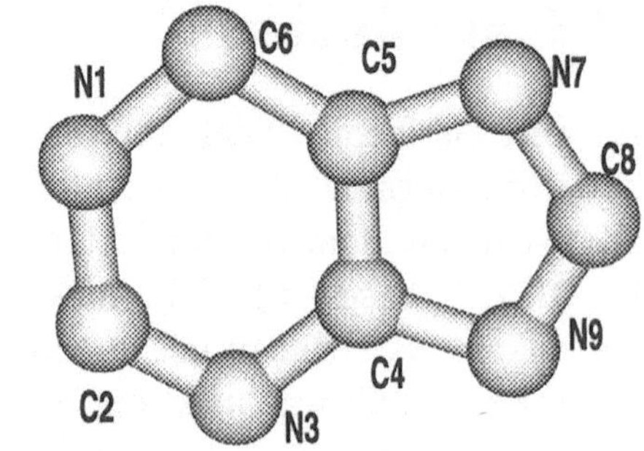

Purine core

Figure 1: Numbering Scheme in Purines.

A very small gap of ~0.1 eV is predicted for the σN_- and π_2 energy levels. This splitting is caused by a very large Koopmans defect for the first "lone pair" σN_- and explains the unresolved experimental band at ~9.6 eV. The σN_- DO, although dominated by out–of–phase interaction of lone electron pairs of pyridinic nitrogens, is rather delocalized. There is also a significant contribution from N_7 of the imidazolic ring.

Table 1: *9H*–Adenine Ionization Energies, eV

Orbital	KT	P3	PES [18]
π_1	8.40	8.49	8.48
σN_-	11.13	9.45	~9.6
π_2	10.14	9.56	~9.6
π_3	11.56	10.50	~10.5
σN_+	12.30	10.48	~10.5
σN_-	13.51	11.54	11.39
π_4	13.49	12.15	12.10
π_5	15.09	13.46	13.21

The π_2 DO is also delocalized. Virtually no splitting is predicted by P3 for the next two levels, π_3 and σN_+. This result is consistent with the

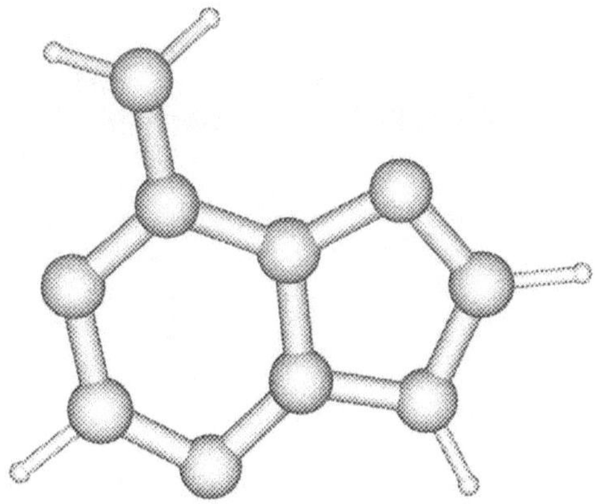

Figure 2: *9H*–Adenine

experimental data: both energies occur at ~10.5 eV. Contributions from a three–center binding lobe at the imidazolic ring are out of phase with respect to those from N_7 and N_{amino} atomic orbitals in the π_3 DO. The σN_+ DO is characterized by the in–phase arrangement of all N lone–pair lobes, but there is significant delocalization into the inner space of both heterocyclic rings. The ionization energy of the next, σN_- DO is slightly overestimated. The DO is dominated by out–of–phase interaction of N_3 and N_7 lone pairs.
The position of a π_4 level is predicted within 0.05 eV of experiment. The electron density distribution in the corresponding DO is characterized mostly by the antibonding arrangement of two large π–bonding fragments in the pyrimidinic moiety. The last P3 IE is 0.24 eV higher than the experimental maximum, but shake–up effects might take place for this ionization. The presence of a four–center, binding π

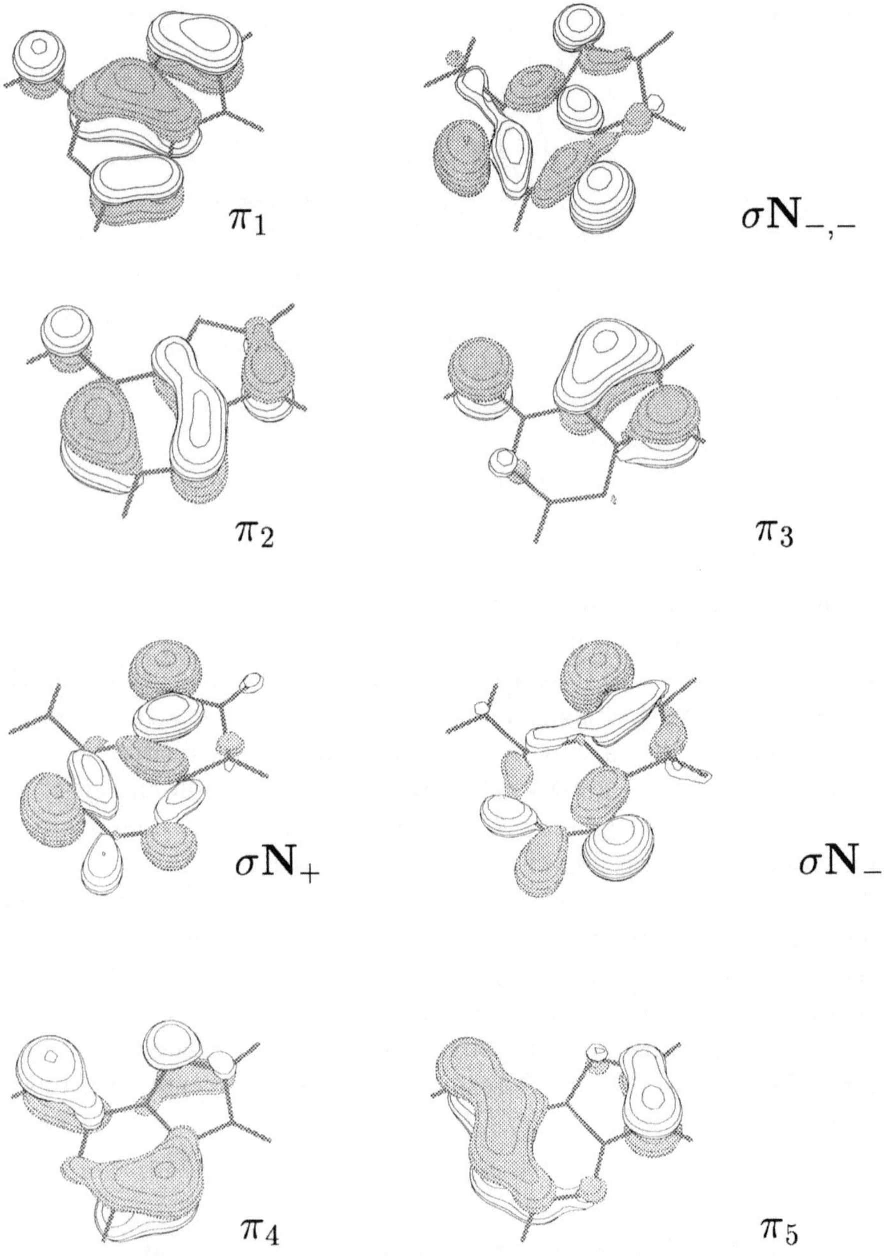

Figure 3: DOs of Adenine.

lobe in the corresponding DO is also an indicator of a possible shake–up.

Guanine.

Guanine is one of the two nucleic acid bases with a pronounced trend to tautomerism (another is cytosine). The relationship between tautomerism in guanine and photoelecton spectra was discussed in our earlier work [21]. Here we present the P3 results on *9H*–guanine which is the major form of guanine in nature (see Fig. 4).

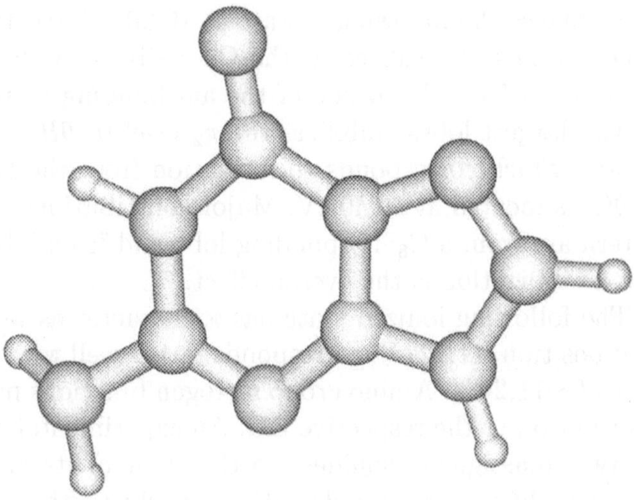

Figure 4: *9H*–Guanine

The P3 IEs of *9H*–guanine are summarized in Table 2 together with the experimental data of Ref. [20]. (IEs up to π_7 can be found in Ref. [21].) All P3 IEs are in very good agreement with the experimental peak positions. Corrections to KT nearly vanish for the first ionization energy, which occurs at 8.13 eV in our P3 calculations. This value is well within the first ionization band in the PES of guanine of Ref. [20]. There is a strong resemblance between the π_1 DOs of *9H*–guanine (see Fig. 5) and *9H*–adenine. Additional anti–bonding conjugation of the carbonyl oxygen and the ring system provides for a decrease in the first ionizaton energy of guanine. The next experimental peak at ~9.9 eV is not resolved and was assigned to ionizations from π_2 and n_1 orbitals. P3 calculations place two final states at 9.82 and 10.02 eV and these are assigned to ionizations from σ DOs with large O, N_7 and N_3 contributions. There remains extensive delocalization into the molecular core. Koopmans defects are large (about 2 eV) in both cases. A considerable flow of electron density toward the N_7 position is noticed in both DOs. For the higher ionization energy, σ antibonding in the C_5–N_7 region is accentuated. Both DOs are destabilized.

P3 places the following π state at 10.20 eV and the corresponding π_2 DO is mostly localized at the C_5–N_7 bond with a notable contribution from N_9. Alleviation of the antibonding relationship between the two largest lobes stabilizes the π_2 level in *9H*–guanine. The next π state, which corresponds to ionization from the highly delocalized π_3 DO, is located at 10.40 eV. Major contributions to the delocalized pattern are from a C_8–N_7 bonding lobe and from lobes at N_1, N_3 and O. Destabilization is the overall effect.

The following ionized state has π character as well, and its calculated position, 11.12 eV, corresponds rather well with the experimental value of ~11.2 eV. Amino group nitrogen functions provide the largest contribution to the respective DO. An experimental peak at ~11.2 eV has two unassigned shoulders to the right of its maximum. One of these shoulders corresponds rather closely to the P3 value of 11.50 eV. The corresponding DO is the first almost pure nonbonding DO in 9H–guanine and it describes the out–of–phase interaction of two, nonadjacent, nitrogen nonbonding functions at N_3 and N_7.

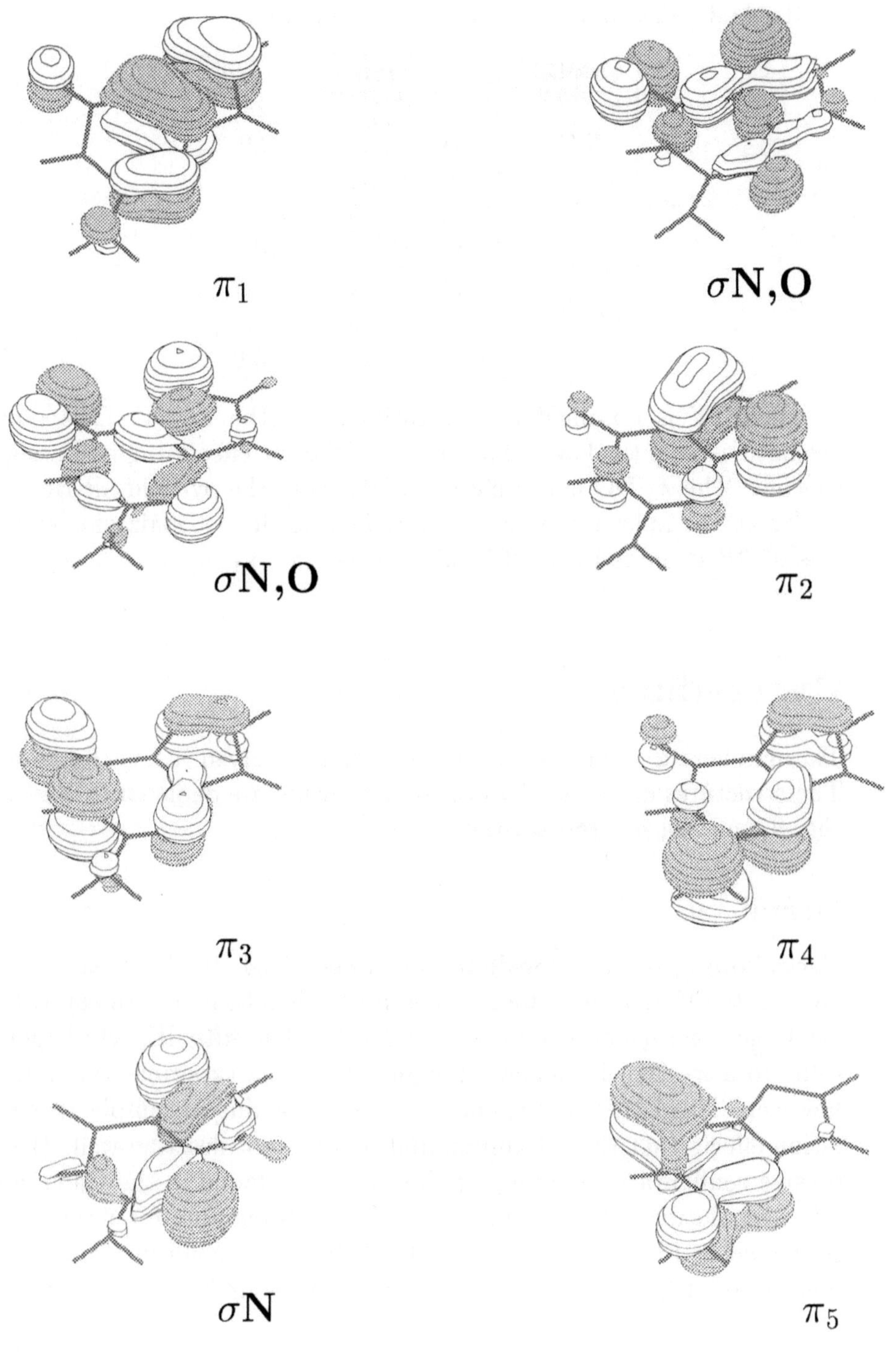

Figure 5: DOs of *9H*–Guanine.

Table 2: *9H*-Guanine Ionization Energies, eV

Orbital	KT	P3	PES [20]
π_1	8.12	8.13	8.28
σN,O	11.68	9.82	~9.9
σN,O	12.00	10.02	~9.9
π_2	11.10	10.20	~10.4
π_3	11.71	10.40	~10.4
π_4	12.49	11.12	~11.2
σN	13.51	11.50	~11.2
π_5	14.89	13.28	~13.0

13.28 eV is obtained for ionization from the π_5 DO. The pole strength is slightly lower here and correlation effects might be significant. The π_5 DO is completely localized in the pyrimidinic ring.

Thus, the following order was obtained for lowest ionization energies of *9H*-guanine by the P3 calculations: π_1, n_1, n_2, π_2, π_3, π_4, n_3, and π_5.

Pyrimidines

The numbering scheme for pyrimidine rings is given in Fig. 6 below. The structures of uracil, thymine and cytosine are depicted in Fig. 7, Fig. 9, and Fig. 11, respectively.

Uracil

The following IEs were predicted for uracil: 9.54, 10.15, 10.52, 11.12 and 12.91 eV. The first four values are in excellent agreement with available experimental data (see Table 3). The fifth IE, which pertains to a π_3 DO, is somewhat higher than the experimental value, however. This larger discrepancy is likely to be due to complex correlation effects that the P3 approximation does not describe well. (Extensive configuration mixing for this state was found in the Gaussian lobe calculations of Ref. [23] as well.) DO diagrams for uracil are presented in Fig. 8. π_1 and π_2 DOs qualitatively confirm simple depictions of atomic orbital contributions to MOs in Ref. [22] that were

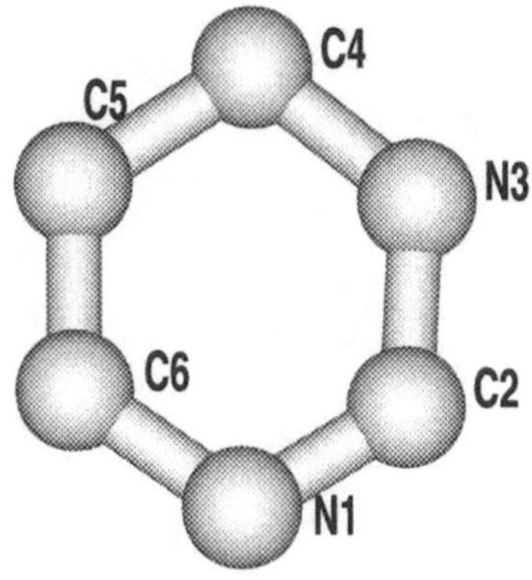

Figure 6: Numbering Scheme in Pyrimidines.

based on semiempirical, CNDO calculations.

Table 3: Uracil Ionization Energies, eV

Orbital	KT	P3	PES [22]
π_1	9.95	9.54	9.5
σO_-	12.14	10.15	10.1
π_2	11.78	10.52	10.6
σO_+	13.10	11.12	11.2
π_3	14.22	12.91	12.63

The π_1 density is fairly delocalized over most heavy atoms with the largest contributions from a C_5C_6 bonding lobe and an N_1 atomic orbital with the opposite phase. In addition, there are two π lobes localized on both oxygens. The DO of the n_1 level, although dominated by oxygen contributions, is widely delocalized into the ring. Three lobes with alternating phases on the oxygens and N_3 obtain in the π_2 DO. Oxygen contributions predominate in the n_2, σO_+ orbital, but with the opposite phase relationship between the oxygen–centered lobes. For the π_3 DO, two, three–center π lobes have opposite phases. One of these lobes is delocalized over the $N_1C_6C_5$ fragment, while the other settles on the OCC_2N_3 region. C_4 participation is not impor-

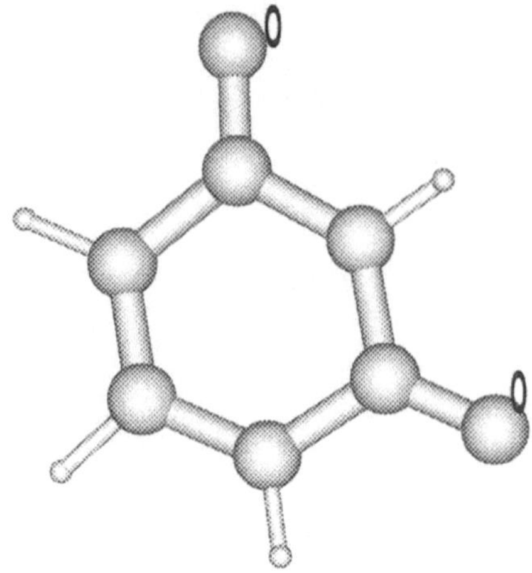

Figure 7: Uracil

tant. The following order of P3 energies was obtained: π_1, n_1, π_2, n_2, and π_3.

Thymine

Only the *cis* form of thymine (Fig. 9) is a rotational minimum. Since the rotational barrier is only 1.43 kcal/mol, methyl group rotation is almost free.

IEs of two rotamers of thymine are almost identical with the exception of the ionization from the σO_- orbital, for which a difference of 0.14 eV obtains. For the first four ionizations, very good correspondence with experiment is achieved. (See Table 4.) The position of the fifth, π_3, band is predicted at higher energy than observed; larger discrepancies between theory and experiment are typical for higher, π–hole IEs [19, 21]. The trends observed for experimental ionization energy shifts [22] are well reproduced in our P3 calculations. Maximum shifts of 0.4 eV were obtained for ionizations from π_1 and π_3 levels.

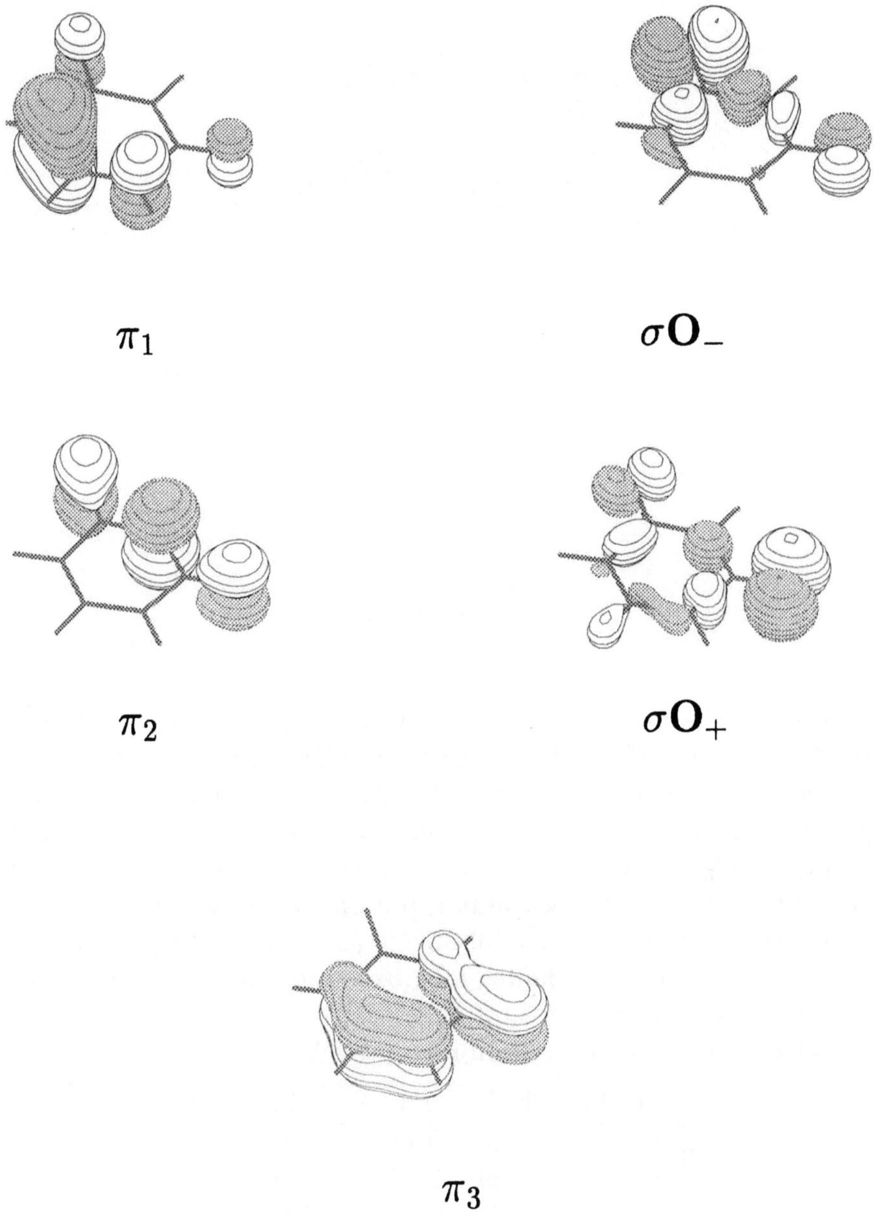

Figure 8: DOs of Uracil.

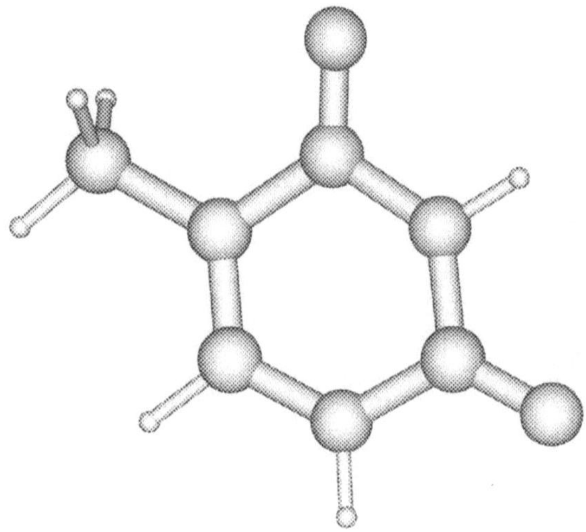

Figure 9: Thymine

The IE of a π_2 level changes only slightly, by 0.1 eV (0.2 eV in Ref. 9), while the IEs of two σ levels are affected by 0.06–0.20 eV. A comparison of the corresponding DO diagrams of uracil and thymine gives an explanation for this phenomenon (see Fig. 10). All thymine DOs qualitatively resemble their uracil counterparts. The only two DOs with appreciable contribution from a methyl group are π_1 and π_3. In both cases, a pseudo–π lobe of CH_3 is in antiboding conjugation with the nearest C–C π binding lobe. No such interaction takes place for the σO or π_2 DOs.

Table 4: Thymine Ionization Energies, eV

Orbital	KT	P3	[22]
π_1	9.53	9.14	~9.1
σO_-	12.09	9.95	~10
π_2	11.73	10.43	10.40
σO_+	12.98	10.99	~10.8–11.0
π_3	13.86	12.52	12.30

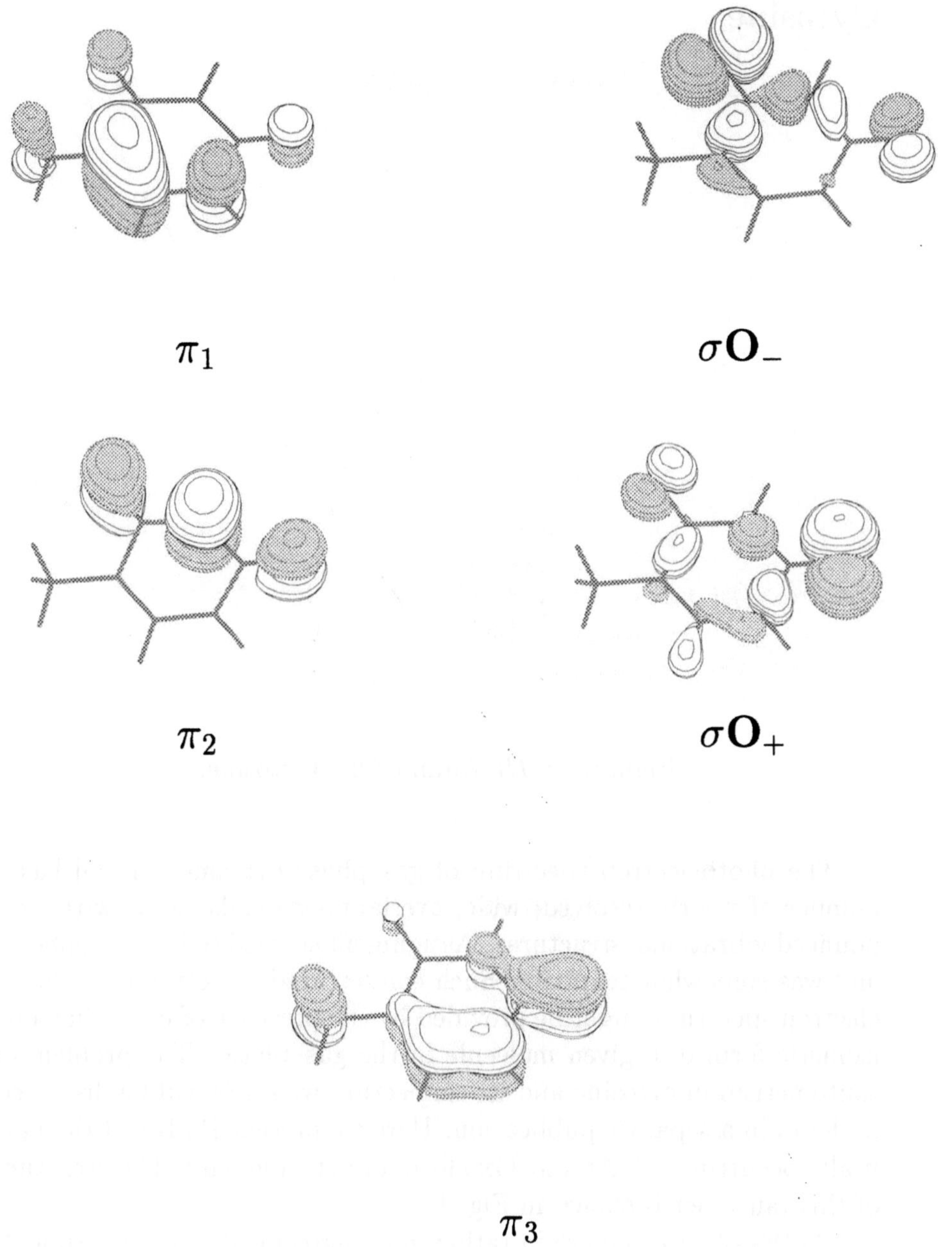

Figure 10: DOs of Thymine.

Cytosine

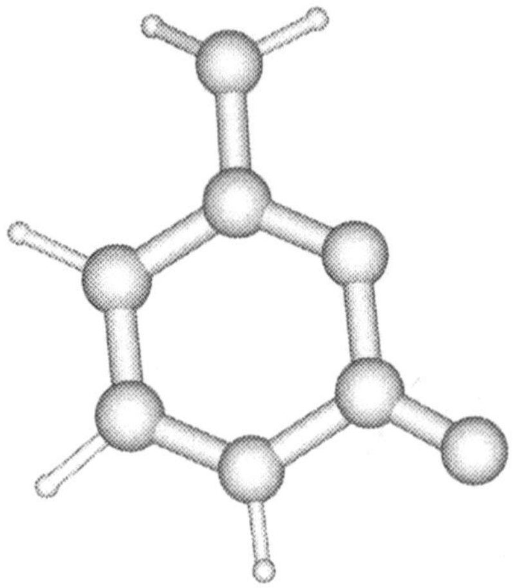

Figure 11: *1H*–Amino–Oxo Cytosine.

The photoelectron spectrum of gas–phase cytosine [24, 25] has a number of poorly resolved, wide, overlapping bands, some with pronounced vibrational structures. Assignment seemed to be very difficult and was somewhat tentative. Such complicated structure in a photoelectron spectrum might be provided by the presence of more than one isomeric form of a given molecule in the gas–phase. The problem of tautomerism in cytosine and its connection with IEs will be discussed in detail in a separate publication. Here we present P3 IEs of the naturally occurring 1H–Amino–Oxo form of cytosine only. The structure of this tautomer is shown in Fig. 11.

P3 IEs of cytosine are in rather good agreement with experiment. The following order of P3 energies was obtained: π_1, π_2, n_1, n_2, π_3, and π_4. (See Table 5.) For the first ionization, the calculated value and the experimental maximum practically coincide. Electron density distribution in the corresponding DO (see Fig. 12) is to some extent similar to that of the π_1 DO in uracil. The orbital is delocalized over most heavy atoms in the ring and the carbonyl oxygen. No participation of the amino group is revealed. The second ionization occurs also from a π DO which presents a delocalized pattern of out–of–phase–lobes with the highest amplitudes on the amino–group nitrogen and the adjacent C–N bond in the ring. The energy gap between the π_2 and n_1 (σN,O$_+$) is only 0.1 eV at the P3 level. The σN,O$_+$ DO is dominated by the in–phase n lobes of pyrimidinic nitrogen and carbonyl oxygen. The relative position of these two levels is different from that of uracil. A very simple explanation for this may be found: the π_2 DO in cytosine is dominated by the anti–bonding conjugation of the amino group nitrogen lone pair and an adjacent three–center bond in the ring. Such an interaction leads to a significant shift in the energy of this level. The next ionization energy is predicted at 9.96 eV and pertains to yet another σN,O DO. The three close–lying energies explain the complicated structure of the second ionization band in the experimental PES of cytosine [24]. Positions of the next two levels are predicted at 12.08 eV and 13.02 eV, while assignments made on the basis of comparison with the PES of N_1–methyl cytosine in Ref. [25] gives 11.76 eV and 13.12 eV, respectively. Two factors might be responsible for such a discrepancy: (a) the experimental bands cover ionization events from other cytosine tautomers; (b) shake–up effects take place. The latter possibility has already been discussed for other bases and heterocycles [26].

Table 5: Cytosine Ionization Energies, eV

Orbital	KT	P3	[25]
π_1	9.30	8.79	8.80
π_2	10.55	9.54	~9.8–10.3
σN,O$_+$	11.35	9.64	9.6–10.1
σN,O$_-$	12.05	9.96	9.6–10.1
π_3	13.35	12.08	11.76
π_4	14.53	13.02	13.12

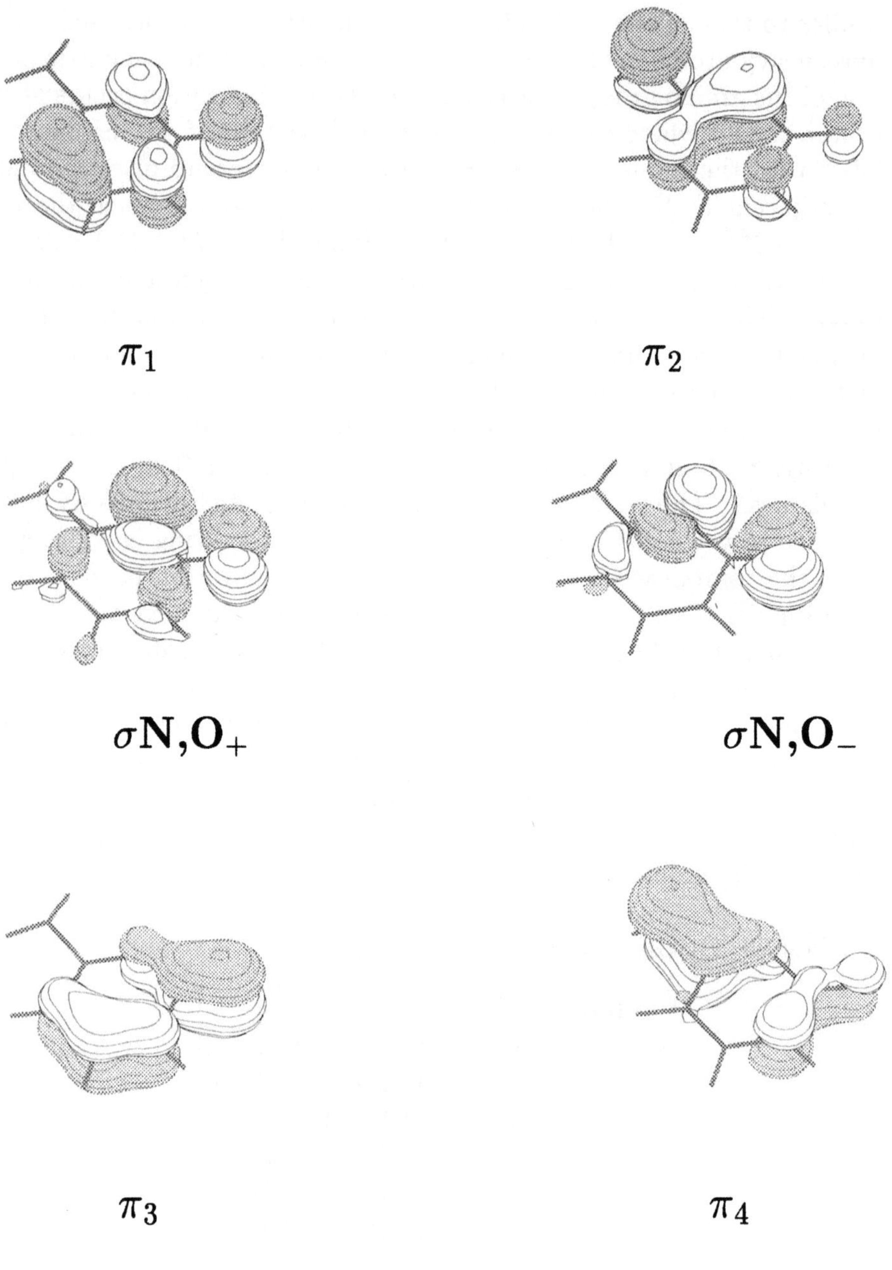

Figure 12: DOs of *1H*–Amino–Oxo Cytosine.

Uracil–Water Complexes

In our earlier work on valence–bound anions of monohydrated uracil tautomers, four stable neutral tautomeric structures were found [27]. This work showed that hydration plays a crucial role in the formation of stable anionic complexes, for uracil itself can form only a metastable valence anion [28]. It seemed interesting to know whether solvation by one water molecule would have any influence upon the IEs of uracil. Four uracil–water complexes are depicted in Fig. 13. They are very close in energy, complex C being the most stable [27]. Hydrogen bond energies are within 7.3–9.2 kcal/mol.

Ionization energies of all four isomers are summarized in Table 6. In all cases, the changes in IE values are not really significant and the shifts do not exceed ±0.24 eV. For isomers A, C and D all shifts are upward, while the opposite is true for isomer B. The largest change in the first IE, 0.14 eV, is obtained for the most stable isomer C. In all cases with the exception of complex B, ionizations from a water fragment occur as separate events that could be very easily distinguished from uracil ionizations.

Table 6: Ionization Energies of Uracil–H_2O Complexes, eV

	P3 Ionization Energies				
Orbital	A	B	C	D	Uracil
π_1	9.47	9.59	9.40	9.49	9.54
n_1	10.16	10.40	10.10	9.98	10.15
π_2	10.33	10.66	10.49	10.32	10.52
n_2	10.93	11.26	11.15	11.07	11.12
H_2O	11.80	11.23	12.30	11.90	
π_3	12.78	12.93	12.69	12.80	12.91
H_2O	13.85	13.46	13.83	13.76	

Hydrogen–bonded Uracil Dimer

The optimized structure of the most stable HB uracil dimer is shown in Fig. 14. Bond lengths and bond angles in both uracil units in HB dimer are practically unchanged compared with the free uracil molecule. The $N_1H...O$ hydrogen bonds are 1.785 Å.

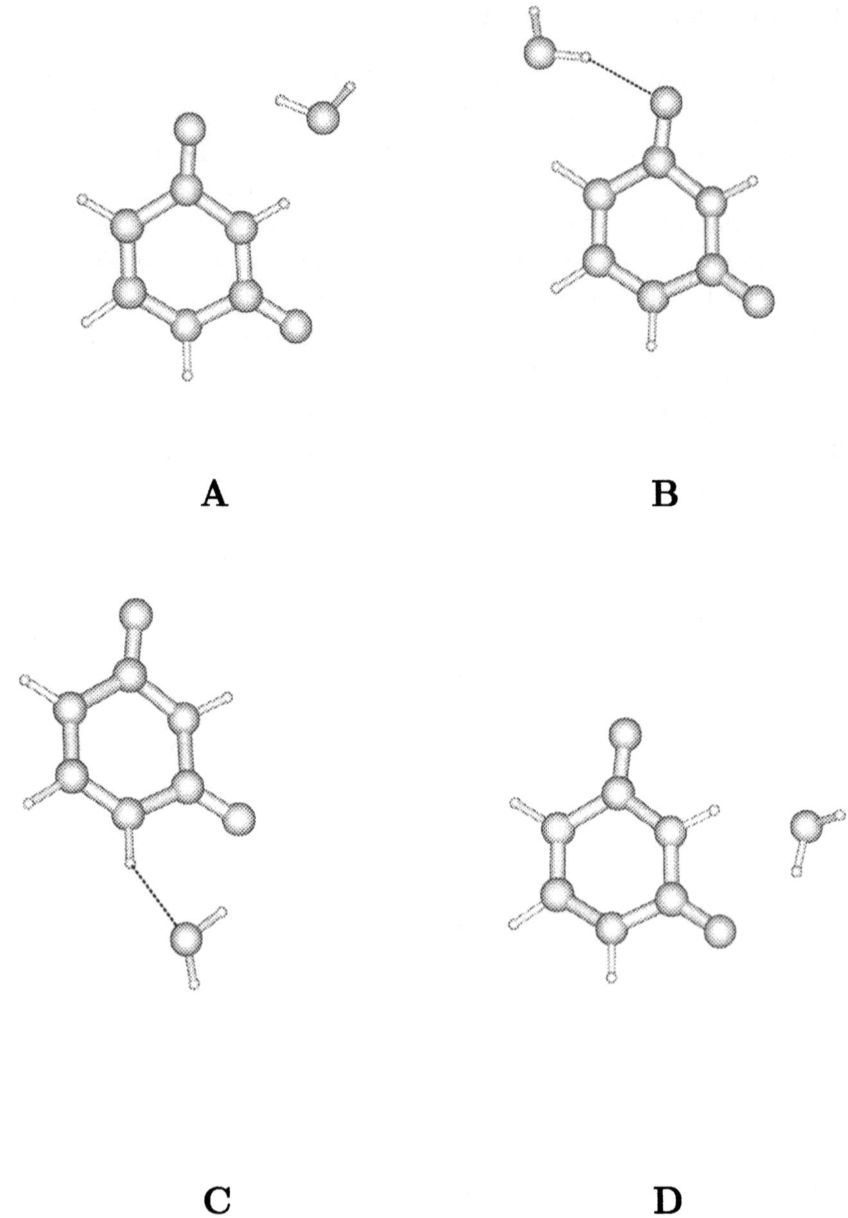

Figure 13: Uracil Complexes with Water.

Ionization energies in this system appear in pairs, the order of which completely corresponds to that of free uracil: π_1, n_1, π_2, n_2, π_3.

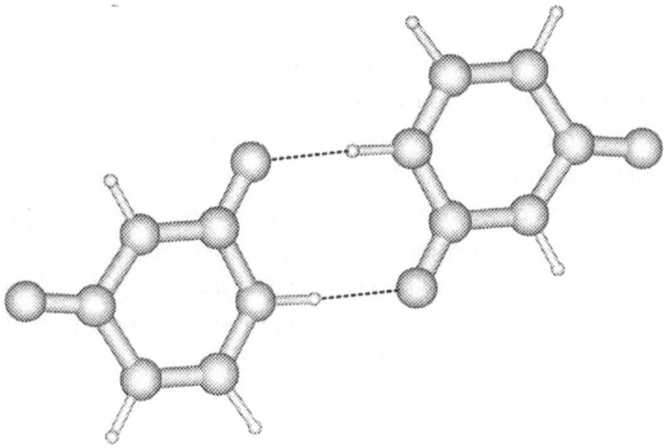

Figure 14: Hydrogen–Bonded Uracil Dimer

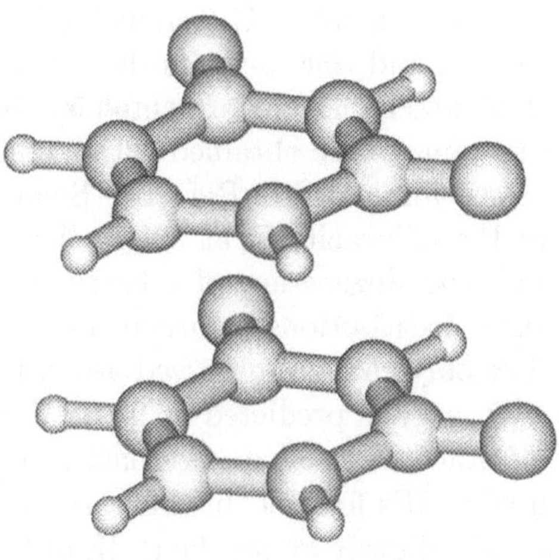

Figure 15: Uracil Stack Dimer

The first pair, $\pi_1 - \pi_1'$, is split by 0.19 eV. Although the corresponding DO (Fig. 16) shows no disturbance of electron density in either unit, the upper counterpart is shifted 0.1 eV from the IE of U. The π_1' level is essentially at the same energy as π_1 in uracil and the only difference between the DOs is in the inverted phases between the monomers.

The n_1 level is very little influenced by dimerization. The energies of both in–phase (n_1) and out–of–phase (n_1') levels go down just a little, and the splitting is insignificant. The downshift of the following, π_2 level is about 0.2 eV for either counterpart which makes this level doubly degenerate. Significant splitting of 0.35 eV is predicted for the next pair of n_2 energies. Although both energies generally decrease, close arrangement of carbonyl oxygen lone pairs leads to this relatively large splitting. An upward shift is predicted by P3 for the last ionization under consideration, while the gap between the π_3 and π_3' is only 0.1 eV.

Stacked Uracil Dimer.

Base stacking is among the most important phenomena in the folding and unfolding of DNA and RNA [16]. Uracil presents a very convenient system for initial studies of IEs in stack dimers. The planarity of uracil permits employment of C_s symmetry in a "head–to–head" oriented SD (Fig. 15) and thus permits the use of the full set of valence orbitals. IEs of uracil SD are presented in the last two columns of Table 7. These values were obtained with a distance of 3.4 Å between monomers, as recommended in Ref. [16]. Some general trends are observed for all IEs of uracil SD at 3.4 Å. These are: large splittings of π levels and near degeneracy of n levels. In all DOs (see Fig. 17), electron density distributions in uracil units remain the same as in free uracil. The only changes observed are in phase relations between units. The π_1 energy is predicted at 9.16 eV which is 0.38 eV above the first IE of uracil. The π_1' counterpart energy is predicted at 9.89 eV. The pair of n_1 IEs follows. In this case, n_1 and n_1' are almost degenerate and only slightly above the n_1 IE of free uracil. The splitting pattern repeats in the π_2 pair. The upper π_2 is 0.33 eV above the π_2 level of uracil. The π_2' is only 0.13 eV lower. The trends observed for the n_1 energies hold for the n_2 levels as well. Both n_2 energies are above that of free uracil and the gap is only 0.06 eV.

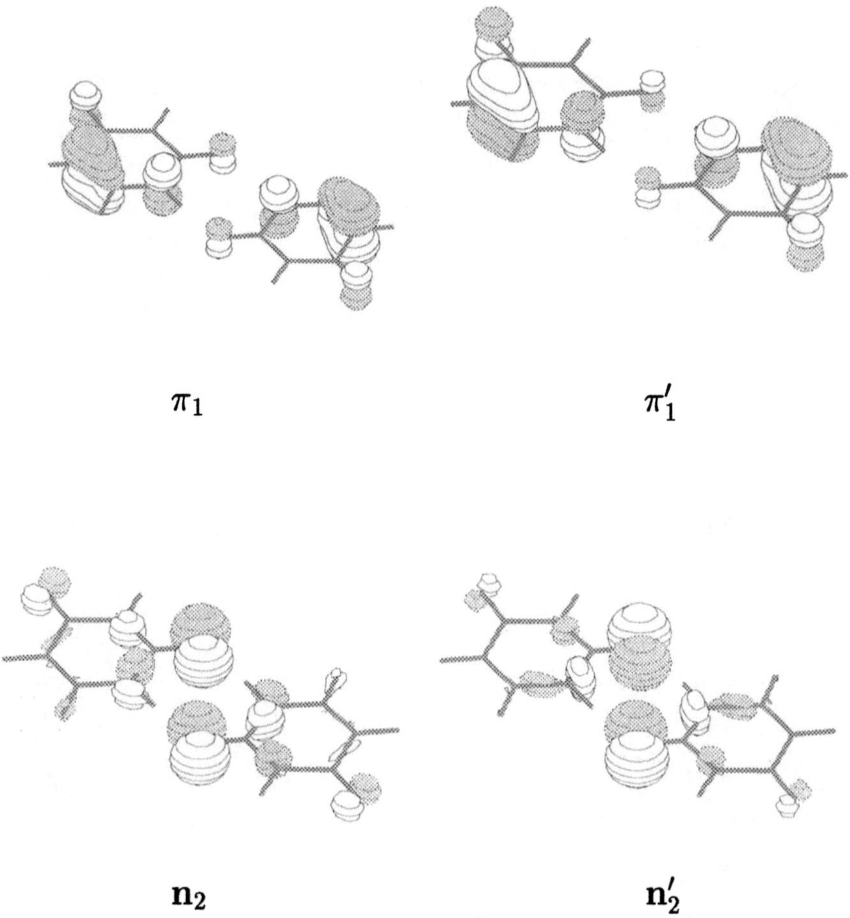

Figure 16: DOs of Hydrogen–Bonded Uracil Dimer.

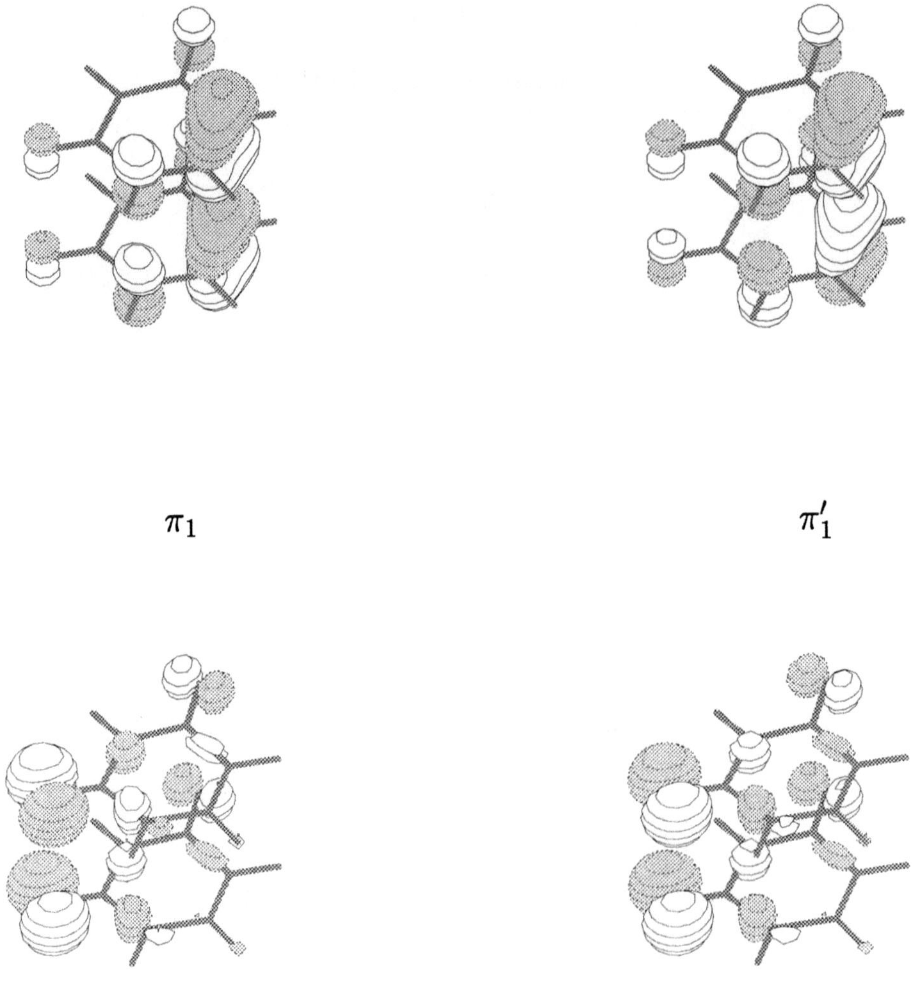

Figure 17: DOs of Stacked Uracil Dimer.

Table 7: Ionization Energies (eV) of Uracil and its Dimers

Uracil			Uracil H–Dimer			Uracil S–Dimer		
MO	KT	P3	MO	KT	P3	MO	KT	P3
			π_1	9.92	9.44	π_1	9.56	9.16
π_1	9.95	9.54						
			π_1'	10.00	9.53	π_1'	10.46	9.89
			n_1	12.25	10.22	n_1	12.09	10.03
σ_1	12.14	10.15						
			n_1'	12.29	10.26	n_1'	12.14	10.08
			π_2	12.08	10.71	π_2	11.49	10.19
π_2	11.78	10.52						
			π_2'	12.10	10.72	π_2'	12.00	10.65
			n_2	13.30	11.26	n_2	13.06	11.00
σ_2	13.10	11.12						
			σ_2'	13.74	11.61	σ_2'	13.12	11.06
			π_3	14.18	12.73	π_3	13.92	12.54
π_3	14.22	12.91						
			π_3'	14.34	12.84	π_3'	14.66	13.27

The splitting is largest in case of the π_3 levels and is mostly provided by the increase in the π_3' IE.

To discover the dependence of IEs in an SB dimer with respect to the normal distance between stacked units, P3 energies of uracil SB dimer were calculated at a number of inner stack distances different from the recommended equilibrium value. At distances larger than the equilibrium value, the order of energy levels does not differ from that in free uracil. The same order pertains at the equilibrium distance, while the splittings of the original levels become larger. When the inner stack distances are smaller than the equilibrium value, crossing of different energy levels occurs, and the first ionization event is facilitated.

Conclusions

An electron propagator approximation that has been designed for applications on large molecules, P3, has been successfully used to predict and assign the photoelectron spectra of the five nucleobases found in DNA and RNA. Efficient algorithms also permit the use of this method in the study of hydrogen–bonded dimers and dimer stacks of nucleobases. With the advent of more powerful computational technology, applications to larger fragments of genetic matter are at hand.

Acknowledgments

This work was supported by the National Science Foundation under grants CHE–9873897 and CHE-0135823.

References

[1] Kim,N. S.; Zhu, Q.; LeBreton, P. R. *J. Am. Chem. Soc.* **1999,** *121*, 11516 and references therein.

[2] Sponer, J.; Hobza, P.; Leszczynski, J. in *Computational Molecular Biology*, (Elsevier, Amsterdam, 1999), Vol. 8, p. 85.

[3] J. Linderberg, J. Öhrn, Y. *Propagators in Quantum Chemistry* (Academic Press, New York, 1973).

[4] P. Jorgensen, P.; J. Simons, J. *Second Quantization–Based Methods in Quantum Chemistry* (Academic Press, New York, 1981).

[5] Öhrn, Y.; Born, G. in *Advances in Quantum Chemistry*, Löwdin, P.-O., Editor (Academic Press, New York, 1981), Vol. 13, p. 1.

[6] Niessen, W. von; Schirmer, J.; Cederbaum, L. S. *Comput. Phys. Rep.* **1984,** *1*, 57.

[7] Ortiz, J. V. in *Computational Chemistry, Reviews of Current Trends*, Leszczynski, J., Editor (World Scientific, Singapore, 1997), Vol. 2, p. 1.

[8] Ortiz, J. V.; Zakrzewski, V. G.; Dolgounitcheva, O. in *Conceptual Trends in Quantum Chemistry*, Kryachko, E. S., Editor (Kluwer, Dordrecht, 1997), Vol. 3, p. 465.

[9] Ferreira, A. M.; Seabra, G.; Dolgounitcheva, O.; Zakrzewski, V. G.; Ortiz, J. V. in *Quantum–Mechanical Prediction of Thermochemical Data*, Cioslowski, J., Editor (Kluwer, Dordrecht, 2001), p. 131.

[10] Ortiz, J. V. *J. Chem. Phys.* **1996**, *104*, 7599.

[11] Stanton, J. F.; J. Gauss, J.; Watts, J. H.; Bartlett, R. J. *J. Chem. Phys.* **1991**, *94*, 4334.

[12] Frisch, M. J.; Trucks, G. W.; Schlegel, H. B.; Scuseria, G. E.; Robb, M. A.; Cheeseman, J. R.; Zakrzewski, V. G.; Montgomery, Jr. , J. A.; Stratmann, R. E.; Burant, J. C.; Dapprich, S.; Millam, J. M.; Daniels, A. D.; Kudin, K. N.; Strain, M. C.; Farkas, O.; Tomasi, J.; Barone, V.; Mennucci, B.; Cossi, M.; Adamo, C.; Jarmillo, J.; Cammi, R.; Pomelli, C.; Ochterski, J.; Peterson, G. A.; Ayala, P. Y.; Morokuma, K.; Malick, D. K.; Rabuck, A. D.; Raghavachari, K.; Foresman, J. B.; Ortiz, J. V.; Cui, Q.; Baboul, A. G.; Clifford, S.; Cioslowski, J.; Stefanov, B. B. Liu, G.; Liashenko, A.; Piskorz, P.; Komaromi, I.; Gomperts, R.; Martin, R. L.; Fox, D. J.; Keith, T.; Al-Laham, M. A. Peng, C. Y. Nanayakkara, A.; Challacombe, M.; Gill, P. M. W.; Johnson, B.; Chen, W.; Wong, M. W.; Andres, J. L.; Gonzalez, C.; Head-Gordon, M.; Replogle, E. S.; Pople, J. A. Gaussian 99, Development Version (Revision B.06+) Gaussian, Inc., Pittsburg, PA, 1998

[13] Bartlett, R. J. *Annu. Rev. Phys. Chem.* **1981**, *32*, 359 and references therein.

[14] (a) Krishnan, R.; Binkley, J. S.; Seeger, R.; Pople, J. A. *J. Chem. Phys.* **1980**, *72*, 650; (b) Petersson, G. A.; Bennett, A.; Tensfeldt, T. G.; Al–Laham, M. A.; Shirley, W. A.; Mantzaris, J. *J. Chem. Phys.* **1988**, *89*, 2193; (c) Petersson, G. A.; Al–Laham, M. A. *J. Chem. Phys.* **1991**, *94*, 6081.

[15] Kratochvil, M.; Engkvist, O.; Šponer, J.; Jungwirth, P.; Hobza, P. *J. Phys. Chem. A.* **1998**, *102*, 6971.

[16] Šponer, J.; Leszczynski, J.; Hobza, P. *J. Phys. Chem.* **1996**, *100*, 5590.

[17] Schaftenaar, G. MOLDEN 3.4, CAOS/CAMM Center, The Netherlands, 1998.

[18] (a) Peng, S.; Padva, A.; LeBreton, P. R. *Proc. Natl. Acad. Sci. USA* **1976**, *73*, 2966; (b) Lin, J.; Yu, C.; Peng, S.; Akiyama, I.; Li, K.; Lee, L. K.; LeBreton, P. R. *J. Am. Chem. Soc.* **1980**, *102*, 4627.

[19] Dolgounitcheva, O.; Zakrzewski, V, G.; Ortiz, J. V. *Int. J. Quantum Chem.* **2000**, *80*, 831.

[20] Lin, J.; Yu, C.; Peng, S.; Akiyama, I.; Li, K.; Lee, L. K.; LeBreton, P. R. *J. Am. Chem. Soc.* **1980**, *102*, 4627.

[21] Dolgounitcheva, O.; Zakrzewski, V, G.; Ortiz, J. V. *J. Am. Chem. Soc.* **2000**, *122*, 12304.

[22] Padva, A.; O'Donnell, T. J.; LeBreton, P. R. *Chem. Phys. Lett.* **1976**, *41*, 278.

[23] O'Donnell, T. J,; LeBreton, P. R.; Petke, J, D.; Shipman, L. L. *J. Phys. Chem.* **1980**, *84*, 1975.

[24] (a) Dougherty, D.; McGlynn, S. P. *J. Chem. Phys.* **1977**, *67*, 1289; (b) Dougherty, D.; Younathan, E. S.; Voll, R.; Abdulnur, S.; McGlynn, S. P. *J. Electron Spectrosc. Relat. Phenom.* **1978**, *13*, 379; (c) Yu, C.; Peng, S.; Akiyama, I.; Lin, J.; LeBreton, P. R. *J. Am. Chem. Soc.* **1978**, *100*, 2303.

[25] Urano, S.; Yang, X.; LeBreton, P. R. *J. Mol. Struct.* **1989**, *214*, 315.

[26] (a) Zakrzewski, V. G.; Dolgounitcheva, O.; Ortiz, J. V. *J. Chem. Phys.* **1996**, *105*, 8748; (b) Dolgounitcheva, O.; Zakrzewski, V. G.; Ortiz, J. V. *J. Phys. Chem. A* **1997**, *101*, 8554; (c) Dolgounitcheva, O.; Zakrzewski, V. G.; Ortiz, J. V. *J. Phys. Chem.*

A **2000**, *104*, 10032; (d) Zakrzewski, V. G.; Dolgounitcheva, O.; Ortiz, J. V. *Int. J. Quantum Chem.* **2000**, *80*, 836.

[27] Dolgounitcheva, O.; Zakrzewski, V. G.; Ortiz, J. V. *J. Phys. Chem.* **1999**, *103*, 7912.

[28] Dolgounitcheva, O.; Zakrzewski, V. G.; Ortiz, J. V. *Chem. Phys. Lett.* **1999**, *307*, 220.

EFFECTS OF IONIZATION AND CATIONIZATION ON INTERMOLECULAR PROTON TRANSFER REACTIONS IN DNA BASE PAIRS.

Juan Bertran*, Marc Noguera, Mariona Sodupe
Departament de Química, Universitat Autònoma de Barcelona
Bellaterra, 08193 SPAIN

1. INTRODUCTION

In a pioneering work in 1963, Löwdin[1-2] introduced the hypothesis that the double proton transfer reaction in DNA base pairs could be a possible source of spontaneous mutations, since rare tautomers could be formed which might disturb the genetic code. Since then, many theoretical studies have been devoted to check Löwdin's hypothesis. First studies,[3-9] considered the double proton transfer process on the ground state of neutral pairs. However, because of the size of the Adenine-Thymine(AT) and Guanine-Cytosine(GC) base pairs, low computational levels of theory were used. In addition, these studies, using semiempirical[3-6] and ab initio[7-9] methods, were performed using fixed geometries for the monomers during the proton transfer processes. Consequently, both the single and double proton transfer reactions were found to be too unfavorable. More recent studies, in which geometries were fully optimized, found smaller energy barriers.[10-12] All studies agree with the fact that the single proton transfer reaction is less favorable than the double proton transfer one, because the single transfer process implies a charge separation in the formation of the resulting ion-pair complex, while in the double proton transfer process the electroneutrality is maintained. Nevertheless, the computed energy barrier of the double proton transfer process was always high.[12]

In the last years, the double proton transfer process, induced by photoexcitaction, in DNA base pairs or model systems has attracted a great

E.J. Brändas and E.S. Kryachko (eds.),
Fundamental World of Quantum Chemistry, Vol. II, 557–581.

interest, both from an experimental[13-20] and a theoretical point of view.[21-25] In particular, Zewail et. al.[13-14] have studied the double proton transfer reaction in the 7-azaindole dimer, (a model system of DNA base pairs) by means of femtosecond dynamic experiments. Other authors have also considered the same system by using similar techniques.[15-20] The central points of the discussion have been whether the mechanism of this reaction is step-wise or concerted and the nature of the intermediate. Recent theoretical studies[21-22] on the first excited state of this system, using the CIS method, have shown that the double proton transfer process is a two-step reaction with an important energy barrier, in contrast to what was obtained by Catalán et al. using fixed geometries derived from ground state optimizations.[23,24] Moreover, results show[22] that the intermediate corresponding to this stepwise mechanism presents a zwitterionic character. In addition, calculations provide another much more stable stationary point in the region of the intermediate that does not belong to the same potential energy surface of the other stationary points. This structure arises from a different electronic state and has a neutral character since the transfer of a single proton is compensated by a charge transfer electronic excitation. Such intermediate could be responsible for the stepwise reaction observed in experiments through a nonadiabatic transition from the initially accessed electronic state to the one of the neutral intermediate.[22] Similar results have been obtained in a theoretical study of the same authors[25] for Guanine-Cytosine and Adenine-Thymine Watson-Crick base pairs. That is, the energy barrier of the double proton transfer reaction in the first singlet excited state (although smaller than that corresponding to the ground state) is significant, whereas a charge transfer excited electronic state was found to be at low energies in the region corresponding to a single proton transfer that eventually neutralizes the charge separation induced by the electronic transition. This state can be populated through internal conversion from another electronic state initially obtained upon photoexcitation. The derived structures may play a key role in the DNA misspairing.

In the limit, a photoexcitation becomes a photoionization process. One-electron oxidations in DNA have recently received considerable attention due to their connection with DNA damage caused by ionizing radiation,[26-27] oxidizing agents[28-29] and photoirradation using endogenous photosensitizers.[30-33] The initial ionization of DNA by the 193-nm light is predicted to occur mainly at the guanine residue, which has the lowest ionization potential.[34,35] Moreover, initially oxidized radical species on other fragments can migrate to the most easily oxidized nucleobase guanine. Thus, the DNA damage is predicted to be produced at this site.[36]

One proposed pathway[36] to strand breakage in DNA goes through deprotonated species of the guanine radical cation that produce specific hydrogen atom abstraction reactions from the sugar moiety, causing the heterolytic elimination of the phosphate ester bond.[37-38] Moreover, proton transfer reactions between base pair ion radicals or to surrounding hydrogen bonded water molecules, can be important determinants of ion radical stabilization and migration in DNA.[28,39] Recently, several works have discussed how proton transfer reactions can influence (stop) the migration of holes or electrons in DNA.[40-43] So, it is not surprising that different proton transfer reactions between base pair radical cations have been studied from a theoretical point of view.[44-48]

Another possible way to induce intermolecular proton transfer reactions in Watson-crick base pairs is by the interaction of metal cations with the purine bases. Although the primary influence of metal cations is to neutralize the negative charges on the backbone phosphate groups, they can also interact, to a lesser extent, with the DNA bases, modifying the strength and structure of the H-bonded pairs. Several theoretical and experimental works have determined the metal cation affinity of the common DNA and RNA nucleobases.[49-55] Among the various binding sites, the N_7 of Adenine and the N_7 and O_6 of Guanine are the preferred ones. The influence of the metal cation (naked and hydrated) on the structure and binding energy of the base pair has also been considered from a theoretical point of view.[56-64] It is shown that the nucleobase pairing, especially in the case of Guanine-Cytosine pair, is enhanced by the presence of the metal cation, mainly due to polarization effects. However, only one work has considered the influence of the metal cation on intermolecular proton transfer reactions. It is found that the presence of the metal cation stabilizes the proton transferred structure but no kinetical study of this process has yet been reported.[62]

In this chapter we will focus on the single and double proton transfer reactions in Watson Crick DNA base pairs induced by ionization or by interaction of metal cations. Both the ionization and cationization processes can be related when the metal cation considered presents not only electrostatic but also oxidant effects.

2. IONIZATION OF WATSON-CRICK BASE PAIRS

Radical cations, generated by ionization, exhibit a very rich and varied chemistry.[65] This is due to the changes produced in the charge and spin when an electron is removed from a system. On one hand, ionization creates a positive charge that introduces large electrostatic effects whereas, on the other hand, the deficit of one electron modifies, directly and indirectly, the strength of covalent bonds. Such effects result in unusual structures and reactivities, which are often significantly different from those of their neutral precursors.

One of the processes that is strongly affected by ionization is the proton transfer reaction, which in many cases occurs spontaneously leading to the so-called distonic radical cations. In the last years, the effect of ionization in several hydrogen bonded systems has been studied in our laboratory.[66-72] For intermolecular H-bonded complexes it is observed that when ionization is produced on the proton donor molecule, the hydrogen bond is strengthened and the proton transfer reaction is favored. This is due to an increase of acidity of the donor monomer. In contrast if ionization is produced in acceptor monomer the hydrogen bond is weakened and the proton transfer reaction becomes very unfavorable, in such a way that other rearrangements are frequently observed.

Guanine-Cytosine (GC) and Adenine-Thymine (AT) Watson Crick base pairs are systems with multiple hydrogen bonds (three and two, respectively) with the interacting monomers acting both as proton donor and proton acceptor simultaneously. Such systems are more complex than the ones with a single hydrogen bond because once the system is ionized many different proton transfer reactions can take place. That is, the proton transfer reaction can either be produced from any of the two monomers and the existence of more than one hydrogen bond opens the possibility of having not only single proton transfer reactions but also double proton transfer processes.

All calculations have been performed using the B3LYP density functional approach[73-74] with a double-ζ plus polarization function basis set. Calibration calculations for the formamidine-formamide dimer, which can be taken as a model system for the Adenine-Thymine base pair, showed that the B3LYP method behaves better than MP2 and reasonably well compared to the highly correlated CCSD(T) method.[48]

2.1 Equilibrium geometries, Dimerization energies and Ionization Potentials

Figure 1 shows the optimized hydrogen bond distances of the neutral and cationic species of GC and AT base pairs. For comparison we have also included in parentheses the experimental values. Although the isolated monomers with an NH_2 group are non planar,[12,75-77] the Watson-Crick GC and AT dimers have C_s symmetry and are planar due to the formation of the hydrogen bonds. The lowest out-of-plane frequencies are, however, very small, which indicates that those dimers are very flexible. The comparison between the computed and experimental values for the neutral systems show that whereas for AT the agreement is very good, for GC the obtained values show larger differences. In particular, the theoretical results show that the O_6-N_4 distance is shorter than the N_2-O_2 one, in contrast to the experimental values.[78] Part of this difference may arise from the fact that the experimental results have been obtained from crystallographic data and not from gas phase studies. In fact, a very recent study has shown that the discrepancies between theory and experiment arise from the molecular environment in the crystals.[63]

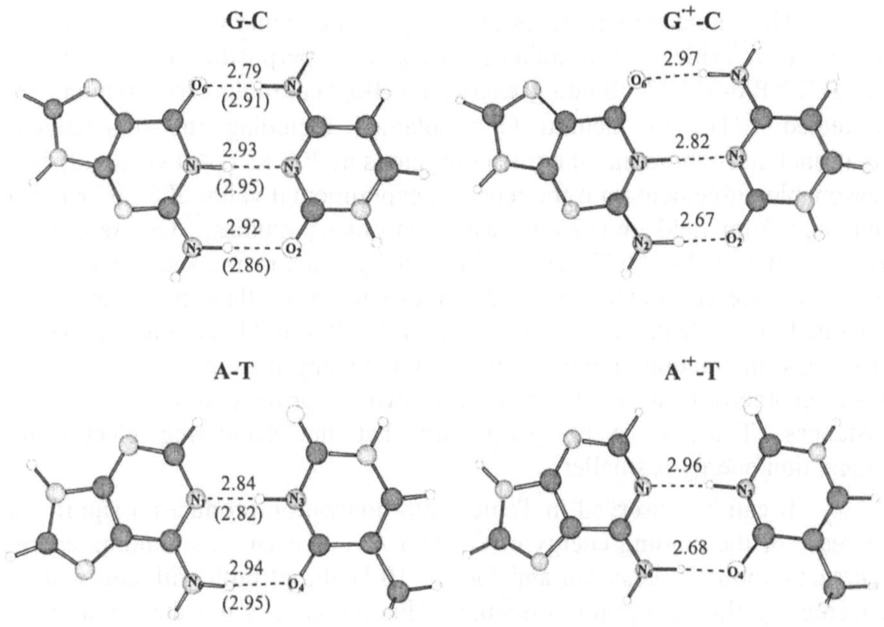

Figure 1. B3LYP hydrogen bond distances (in Å) of the neutral and ionized GC and AT dimers. In parentheses experimental values.

It can also be observed in Figure 1 that the H-bonds distances change dramatically upon ionization. This can be understood considering the nature of the ionized species. The lowest ionic state of GC and AT is a $^2A''$ state. Population analysis indicates that the ionization of GC and AT is mainly localized at the Guanine and Adenine monomers, respectively. For

GC, the charge on Guanine is 0.82 and the spin density 1.00 whereas for AT, the charge on Adenine is 0.79 and the spin density 0.86. This was to be expected considering that Guanine and Adenine have a lower ionization potential than Cytosine and Thymine, respectively.[34,35] Since Guanine and Adenine are the two monomers that lose the electron and thus, become more acid, those hydrogen bonds in which these two monomers act as proton donor become stronger in the ionized system. This implies a shortening of the distance between the two heavy atoms and a lengthening of the H-X bond involved. In contrast, those H-bonds in which Guanine and Adenine act as proton acceptor become weaker. That is, the N_1-N_3 and N_2-O_2 H-bond distances of GC decrease after ionization while the O_6-N_4 increases. For AT, the N_6-O_4 is the bond that gets shorter while the N_1-N_3 becomes longer.

The interaction energies of neutral and cationic GC and AT dimers are given in Table 1. Correcting for basis set superposition error decreases the B3LYP/6-31G** binding energy by 3-5 kcal/mol. The counterpoise corrected $-\Delta H^0_f$ for neutral GC, obtained including the translational, rotational and vibrational thermic corrections at 298 K, (24.0 kcal/mol) is in reasonable agreement with the reported experimental value of 21.0 kcal/mol obtained from field ionization mass-spectrometry studies.[79] Our results are very similar to the DFT values obtained by Santamaría and coworkers,[80] using a different functional, and somewhat larger than previous results obtained at the MP2 level of calculation.[81,82] It should be noted, however, that those interaction energies were obtained using the equilibrium Hartree-Fock geometries, which have been shown to provide too large H-bond distances. Thus, it is not surprising that the stabilizing electrostatic interaction energy is smaller.

It can be observed in Table 1 that ionization produces a significant increase of the binding energy of GC and AT. For GC, the binding energy increases about 19 kcal/mol and for AT 10 kcal/mol both with and without considering the zero point correction. This increase is not due to an equal strengthening of the three hydrogen bonds of GC or the two hydrogen bonds of AT. As it has already been mentioned, ionization strengthens those hydrogen bonds in which the ionized monomer acts as proton donor and weakens those in which it acts as acceptor. Thus, the N_1-N_3 and N_2-O_2 hydrogen bonds of GC and the N_6-O_4 of AT would become strong hydrogen bonds according to the energetical classification (12 to 24 kcal/mol).[83] Moreover, the optimized H-bond distances of these strong bonds are close to the sum of the Van der Waal radii, 2.65 for O-N and 2.75 for N-N, and so, could be denoted as short-strong hydrogen bonds (SSHB).[84]

Table 1. B3LYP interaction energies of Watson-Crick neutral and cationic GC and AT base pairs (in kcal/mol). In parentheses are counterpoise corrected values.

	GC		AT	
	neutral	cation	neutral	cation
D_e	30.3(25.5)	48.4(44.3)	16.4(12.3)	25.4(22.3)
D_0 [a]	28.8(24.0)	46.8(42.7)	15.3(11.2)	25.1(22.0)
ΔH^0_{298K} [b]	28.8(24.0)	47.1(43.0)	15.0(10.9)	24.8(21.7)
ΔG^0_{298K} [b]	17.4(12.6)	34.1(30.0)	3.8(−0.3)	13.6(10.5)

[a] Includes zero point energy computed from the unscaled harmonic B3LYP frequencies.
[b] After correction for translational, rotational and vibrational energies determined at the B3LYP level.

A simple thermodynamical cycle shows that the decrease of the ionization potential of the dimers compared to that of the free monomers is just the increase of the binding energy produced by ionization. That is, the lowering of the adiabatic ionization potential, including the zero point energy and correcting for BSSE, of GC compared to G, is 0.81eV, and that of AT compared to A is 0.47 eV. We expect the lowering of the IP to be more accurate than the computed IP values. Thus, our best estimate of the adiabatic ionization potential of GC(6.96eV) and AT(7.79eV) are obtained by subtracting the computed lowering to the experimental IP values[85] of the monomers G and A, respectively. A similar value for GC (6.90 eV) has been obtained recently by Sevilla et al.[46]

2.2 Single and Double proton transfer reactions

The following scheme shows the single and the double proton transfer processes considered. The nature of the H-bonds in each structure has been indicated with the letters s (strong) and w (weak).

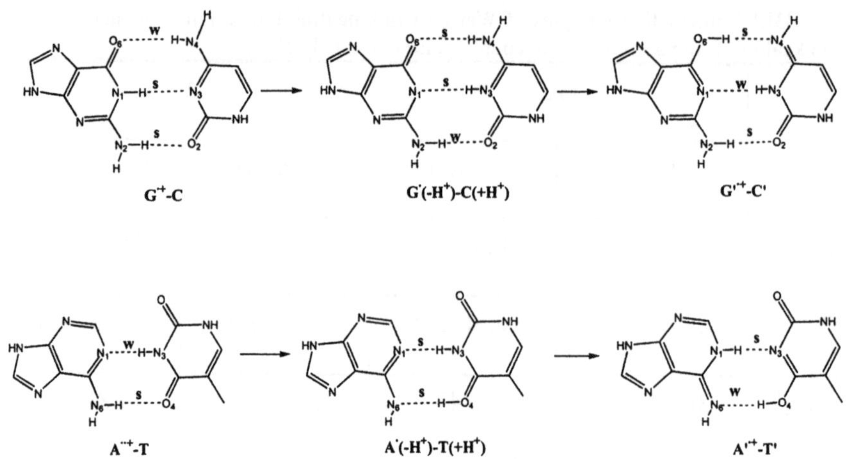

As mentioned, ionized GC presents one weak and two neighbor strong H-bonds, that is, a (w-s-s) situation. Any of the two strong H-bonds can be involved in the single proton transfer reaction. Results show that the transfer from N_1 to N_3 is the most favorable process. After the proton is transferred the charge lies on protonated Cytosine and the radical character on deprotonated guanine and so, the obtained structure corresponds to a distonic radical cation.[86] Population analysis confirms that the nature of the $G^{+}C \rightarrow G^{\cdot}(-H^{+})\text{-}C(+H^{+})$ process is that of a proton transfer reaction since the charge on $C(+H^{+})$ is 0.82 whereas the spin density of $G^{\cdot}(-H^{+})$ is 1.0. Because the charge moves to Cytosine, $C(+H^{+})$, those hydrogen bonds in which this monomer acts as proton donor become strong, whereas the one in which it acts as proton acceptor becomes weak. The resulting structure, $G^{\cdot}(-H^{+})\text{–}C(+H^{+})$, shows a (s-s-w) pattern, similar to the reactant. The two strong hydrogen bonds can benefit of the enhanced electrostatic interaction by decreasing simultaneously the N_1-N_3 and N_4-O_6 distances, while the N_2-O_2 distance increases to reduce repulsion. The main changes occur on the terminal hydrogen bonds; that is, the N_4-O_6 distance decrease from 2.97 Å to 2.70 Å whereas the N_2-O_2 one increases from 2.67 Å to 2.92 Å. It must be noted that the transfer from N_2 to O_2 would produce an isomer with an alternated (s-w-s) situation. Such pattern is not favorable due to the presence of a central weak bond, which does not allow the dimer to get a geometrical compromise that benefit the two terminal strong short H-bonds, without introducing the central H-bond into a repulsive region. Because of that, this s-w-s structure has not been localized as a minimum on the potential energy surface. The double proton transfer reaction leads to $G'^{+\cdot}\text{–}C'$, with also an alternated (s-w-s) situation. Thus, it is not surprising that any attempt to localize this energy minimum collapsed to the single proton transferred structure.

For the $A^{+\cdot}$-T dimer, we have one weak and one strong H-bond (w-s) before any proton transfer is produced. Once the proton is transferred, the positive charge moves to protonated Thymine and thus, both H-bonds become strong (s-s). As for GC the $A^{+}T \rightarrow A^{\cdot}(-H^{+})$-$T(+H^{+})$ reaction is that of a proton transfer process. Population analysis shows that the charge on $T(+H^{+})$ is 0.76 whereas the spin density on $A^{\cdot}(-H^{+})$ is 1.0. Thus, this structure also corresponds to a distonic radical cation. The values of the optimized H-bond distances confirm the expected changes. That is, the N_1-N_3 distance decreases from 2.96 Å to 2.81 Å whereas the N_6-O_4 bond becomes stronger, the distance being particularly small (2.58 Å). The double proton transfer reaction leads to a situation similar to the initial one, but with the two H-bonds reversed (s-w). The H-bond distances in this structure (2.61 Å and 2.84 Å) are smaller than in the initial $A^{+}T$ one. For this $A'^{+\cdot}$-T' system, the charge and spin density of Adenine are 0.84 and 1.00, respectively. Charge and spin density are now in the same monomer as a consequence of the second proton transfer reaction. It has to be mentioned that in addition to the variations of H-bond distances during the proton transfer reactions, important geometry changes take also place in the rings of the purine and pirimidine monomers. This indicates that the studied processes have a high multidimensional character.

The relative energies of the single and double proton transferred structures with respect to the ground state asymptote are given in Figure 2. For comparison we have also included the relative energies of the single and double proton transfer asymptotes. It can be observed that for GC there is not a minimum corresponding to the double proton transfer reaction. As mentioned, any attempt to localize such stationary point collapsed to the single proton transferred isomer. However, at a value of 1.05 Å for the O_6-H distance, which could be taken as the double proton transfer structure, the energy is about 10 kcal/mol higher than the initial $G^{+\cdot}$-C. On the other hand, for Adenine-Thymine it is found that the double proton transfer reaction shows a very shallow minimum, the transition state being very close in energy.

Two factors determine the stability of the different species: a) the relative proton affinities of the centers involved in the proton transfer reaction or in other words, the relative stability of the asymptotes from which they derive, and b) the number and sequence of strong and weak H-bonds formed. For the same number of hydrogen bonds, the most stable structures are those in which the two strong H-bonds are neighbors.

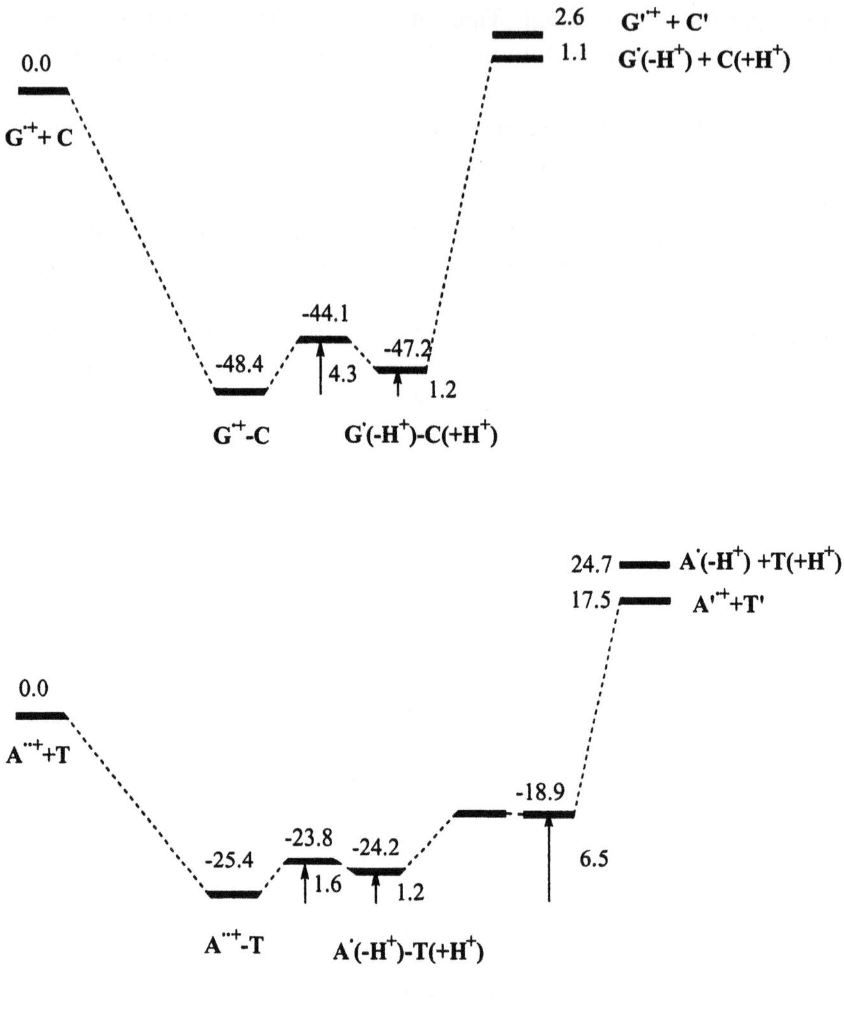

Figure 2. Energy profiles for the single and double proton transfer processes in cationic Guanine-Cytosine and Adenine Thymine base pairs. Energies (in kcal/mol) are relative to the ground state asymptote.

For example, for GC, the non-proton-transfer, the single proton transfer G˙(-H$^+$) + C(+H$^+$) and the double proton transfer, G'$^{\cdot+}$ + C' asymptotes lye more or less at the same energy (0.0, 1.1 and 2.6 kcal/mol, respectively). However, only the non-proton transfer and single proton transfer structures are stable. On the other hand, it can be observed in Figure 2, that the energy difference between the non transferred, G$^{+\cdot}$-C, and the single proton transferred, G˙(-H$^+$)-C(+H$^+$), structures (1.2 kcal/mol) is almost the same than that found for the corresponding asymptotes thus, indicating that the strength of the H-bond interaction is very similar in the two structures. These results are not surprising considering that both G-C and G(-H$^+$)-C(+H$^+$) show the same (s-s-w) pattern, with two neighbors strong H-bonds, whereas the double proton transferred structure presents an unfavorable (s-w-s) situation.

Similar arguments hold for the AT cationic system. The single proton transferred asymptote lies much higher in energy (24.7 kcal/mol) than the ground state one. However, the derived proton-transferred A(-H$^+$)-T(+H$^+$) dimer is almost degenerate with the non-proton transferred A$^{+\cdot}$-T one. This is due to the fact that in A$^{+\cdot}$-T we have one weak and one strong H-bonds (w-s) while in A(-H$^+$)-T(+H$^+$) the two H-bonds are strong (s-s). Thus, the larger H-bond interaction compensates for the relative energy of the two asymptotes. The double proton transfer structure has the same pattern than the initial structure but it lies higher because the asymptote from which it derives is also less stable. The energy difference is, however, reduced from 17.5 to 6.5 kcal/mol, because the short H-bond in A'$^{+\cdot}$-T' is stronger than in the initial reactant.

Both for GC and AT radical cations the single proton transfer reaction presents a low energy barrier (4.3 kcal/mol for GC and 1.6 kcal/mol for AT), the AT system behaving more as a strong short low-barrier hydrogen bond than GC. In contrast, the double proton transfer process is not a favorable process. It has to be mentioned that recently an experimental gas phase study using enhanced multiphoton ionization (REMPI) has found evidences for proton transfer in GC pair.[87]

The behavior of the radical cation species is very different to that observed for the neutral base pairs, for which the single proton transfer reaction has been found to be very unfavorable due to the formation of an ion-pair complex.[12] The facility of radical cation species to produce the single proton transfer reactions is due to the increased acidity of the ionized monomer and to the fact that the proton transfer does not imply a creation of charges but a transfer of a positive charge.

3. CATIONIZATION OF WATSON-CRICK BASE PAIRS

As mentioned, another possible way to induce proton transfer reactions in Watson-Crick base pairs is by the cationization of the purine bases. However, most of the theoretical studies performed till now have considered the cationization of mono and dipositive closed shell metal cations to the N_7 of purines and focused their attention to the changes produced on the dimerization energies due to cation binding. To our knowledge only one work[62] has considered the interaction of metal cations (hydrated Mg^{+2} and Zn^{+2}) to the structure derived from the proton transfer process from N_1 to N_3 in Guanine-Cytosine base pair, but no energy barrier for the reaction has been reported.

In this chapter we will analyze the effect of cationization of GC by Cu^{+2}, an open shell metal cation (d^9) for which both electrostatic and oxidant effects are important. The changes produced on the structure and dimerization energy of GC will be compared with those obtained for the closed shell Cu^+ (d^{10}) and Ca^{+2} for which the metal-GC interaction is mainly due to electrostatic and polarization effects. Comparison with previous results will also be reported.

Secondly, we will present the energy profiles (minima and transition states) corresponding to different intermolecular proton transfer processes. The case of Cu^{+2} is especially interesting since its interaction with GC can lead to an Cu^+-$GC^{+\cdot}$ complex due the oxidant character of this metal cation. Therefore, this system is expected to show certain analogies with the ionized GC system previously reported. Moreover, the comparison of this cation with Cu^+ and Ca^{+2} will allow us to analyze the importance of each effect (electrostatic or oxidant).

All calculations have been performed using the B3LYP density functional method and the same basis set used for the ionized systems. For Cu we have used the all electron Wachters-Hay basis set [88-89] and for Ca the basis set of Blaudeau et al.[90] Previous theoretical calculations have shown that B3LYP approach is a cost-effective method for studying transition metal ligand systems.[91]

3.1 Equilibrium geometries, Dimerization energies and Cation Affinities

The optimized H-bond distances of GC cationized by Ca^{+2}, Cu^+ and Cu^{+2} are sown in Figure 3. For comparison we have also included those of isolated GC. All systems show C_s symmetry, the lowest electronic state being a $^1A'$ for Ca^{+2} and Cu^+ and a $^2A''$ for Cu^{+2}. First, it can be observed that, in all cases, the metal cation interacts both with the N_7 and O_6 of Guanine; that is, the complex shows a bidentate coordination. The largest metal-ligand distances corresponds to Ca^{+2}. This was to be expected considering that Ca^{+2} is, among the three cations, the one with a larger ionic radius. However, in contrast to what one would initially expect the metal-ligand distances for Cu^{+2} are larger than for Cu^+. This is due to the fact that the interaction of Cu^{+2} induces an oxidation in the base pair; that is, there is an electron transfer from GC to the metal cation. Thus the final situation can be viewed as the interaction of Cu^{+} with GC radical cation, which produces an increase of the metal-guanine distances due to the repulsive electrostatic term between the two fragments. This is confirmed by the population analysis which shows that the metal charge in Cu^{+}-GC (0.89) and Cu^{+2}-GC (0.92) is very similar. Moreover, the population analysis shows that the spin density in the Cu^{+2} system mainly lies on the Guanine fragment (0.98) and not in the metal cation.

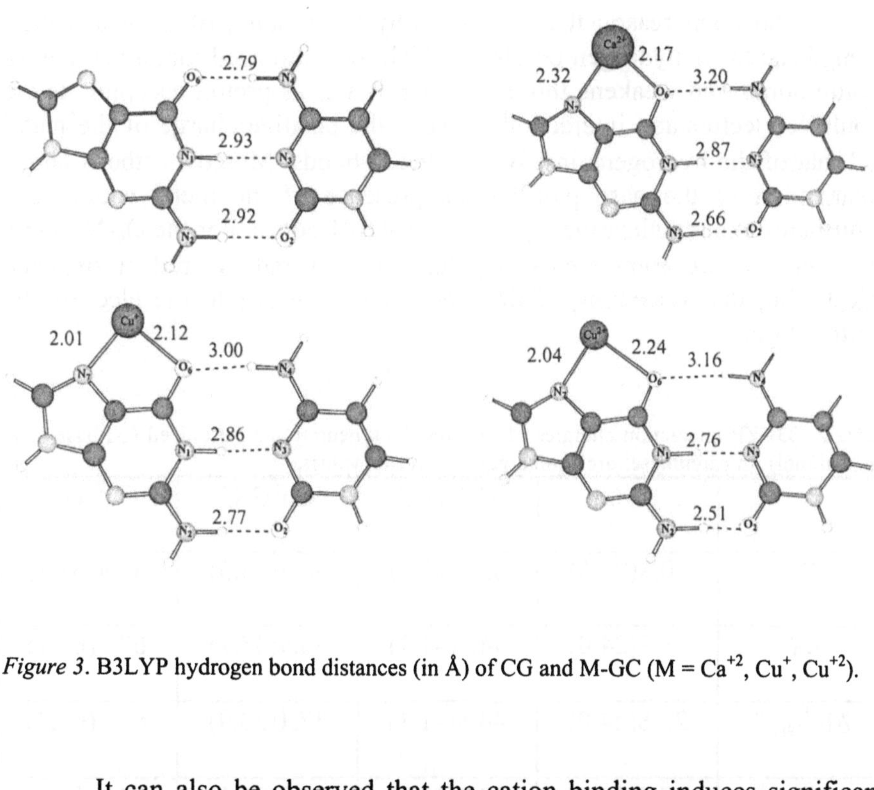

Figure 3. B3LYP hydrogen bond distances (in Å) of CG and M-GC (M = Ca^{+2}, Cu^{+}, Cu^{+2}).

It can also be observed that the cation binding induces significant changes on the H-bond distances. In all cases, the O_6-N_4 hydrogen bond increases whereas the other two, N_1-N_3 and N_2-O_2, decrease, especially the latter one. Such changes follow the same trend observed for ionized GC. That is, cationization strengthens those hydrogen bonds in which the cationized monomer (Guanine) acts as proton donor and weakens those in which it acts as proton acceptor. Similar trends have been observed in previous theoretical studies.[56] As expected, the changes on the hydrogen bond distances are more pronounced for the divalent cations than for the monovalent ones. On the other hand, the comparison between the two divalent cations shows that the strengthening of the H bonds is significantly larger for Cu^{+2}. In fact the differences between Ca^{+2} and Cu^{+2} are more important than between the monovalent Cu^{+} and divalent Ca^{+2}. This is due to the oxidant character of Cu^{+2} which leads to the formation of the Cu^{+}-$GC^{+\cdot}$ complex, for which both the electrostatic and oxidant effects contribute to the changes observed. The comparison between Figures 1 and 3 shows that the effect of oxidation of GC is more important than that produced by the binding of a monovalent metal cation. On the other hand, if one adds the changes observed for $GC^{+\cdot}$ and for $Cu^{+}GC$, the obtained values are very similar to those of $Cu^{+}GC^{+\cdot}$.

The main reason that justifies why the binding of a metal cation strengthens those hydrogen bonds in which the cationized monomer acts as proton donor and weakens those in which it acts as proton acceptor is the repulsive electrostatic interaction between the positive charge of the metal and that of the hydrogens involved in the H-bonds. Moreover, the σ and π polarization of the base pair by the presence of the metal cation also contributes to strengthen the N_1-N_3 and N_2-O_2 H-bonds. For the O_6-N_4 bond, however, the two components of polarization (σ and π) produce opposite effects. So, the weakening of this bond is mainly due to the electrostatic contribution.

Table 2. B3LYP interaction energies of Watson-Crick neutral and cationized GC base pairs (in kcal/mol). In parentheses are counterpoise corrected values.

	G-C	Ca^{2+}G-C	Cu^{+}G-C	Cu^{2+}G-C
D_e	30.3(25.5)	45.8(42.1)	38.7(35.5)	66.4(63.3)
D_0[a]	28.8(24.0)	44.8(41.1)	36.9(33.7)	67.2(64.1)
ΔH^0_{298K}[b]	28.8(24.0)	44.8(41.1)	37.1(33.9)	67.3(64.2)
ΔG^0_{298K}[b]	17.4(12.6)	33.4(29.7)	24.3(21.1)	54.4(51.3)

[a]Includes zero point energy computed from the unscaled harmonic B3LYP frequencies.
[b]After correction for translational, rotational and vibrational energies determined at the B3LYP level.

The hydrogen bond interaction of GC base pair isolated and cationized by Ca^{+2}, Cu^{+} and Cu^{+2} is shown in Table 2. First, it can be observed that cationization strengthens the G-C base pair interaction. The increase induced by the metal cation is 15.5 kcal/mol for Ca^{+2}, 8.4 kcal/mol for Cu^{+} and 36.1 kcal/mol for Cu^{+2}. As expected the divalent cations produce a larger effect than the monovalent one. However, it is surprising the important increase induced by Cu^{+2} compared to Ca^{+2}. As mentioned this is due to the fact that in the case of Cu^{+2}, the metal cation oxidizes the base pair and so, part of this important increase results from the formation of GC radical cation. It can be observed in Table 1 that ionization of GC implies an increase of the H-bond stabilization of 18 kcal/mol. Considering that the interaction of Cu^{+} produces an increase of about 8 kcal/mol and assuming that both effects, ionization and cationization, are additive the resulting stabilization should be around 26 kcal/mol. This value is about 10 kcal/mol smaller than what is observed indicating important cooperative effects. These changes follow the geometrical variations observed on the H-bonds and discussed above; that is, the larger the decrease of the H-bond distances in which Guanine act as proton donor, the larger the increase of the interaction energy.

Basis set superposition error has been corrected using the counterpoise procedure. The values (in parentheses) show that the correction for the cationized systems is very similar for the three metals and smaller (about 1-2 kcal/mol) than the value obtained for the isolated base pair. Therefore from counterpoise corrected values the increases in the interaction energies induced by the metal cations would be slightly larger. The results obtained for Ca^{+2} (42.1 kcal/mol) and Cu^{+} (35.5 kcal/mol) are very similar to those previously reported at the MP2 level using HF geometries and pseudopotentials for the metal cation (38.8 and 33.1 kcal/mol, respectively).[56]

Table 3. B3LYP Metal-GC interaction energies (in kcal/mol). In parentheses are counterpoise corrected values.

	Ca^{2+}-GC	Cu^{+}-GC	Cu^{2+}-GC
D_e	177.3(172.4)	104.0(98.4)	343.9(340.3)
D_0^{a}	176.2(171.3)	102.6(97.0)	345.2(341.6)
$\Delta H^0_{298K}{}^{b}$	176.8(171.9)	103.1(97.5)	345.8(342.2)
$\Delta G^0_{298K}{}^{b}$	169.7(164.8)	97.7(92.1)	336.3(332.7)

[a]Includes zero point energy computed from the unscaled harmonic B3LYP frequencies.
[b]After correction for translational, rotational and vibrational energies determined at the B3LYP level.

The metal-GC interaction energies are shown in Table 3. The counterpoise corrected values are shown in parentheses. It should be mentioned that for $Cu^{+2}GC$ the counterpoise correction has been computed assuming the Cu^{+} and $GC^{+\cdot}$ fragments since that is the real situation observed in the complex. As expected, the metal cation-base pair interaction energy is larger for the divalent metals (Ca^{+2} and Cu^{+2}) than for the monovalent one (Cu^{+}). However, the difference between the metal ion affinities of Ca^{+2} and Cu^{+2} is too large even if the differences between the metal-ligand distances are considered. As mentioned above, differences between Ca^{+2} and Cu^{+2} arise from the oxidant character of Cu^{+2} which leads to the formation of GC radical cation. Such oxidation is not surprising considering the large ionization potential of Cu^{+} (20.85 eV at the B3LYP level) compared to that of GC (6.53 eV). Therefore, the electron transfer from GC to Cu^{+2} implies an extra stabilization of 330.2 kcal/mol. The interaction energy with respect to the lowest Cu^{+} (^{1}S) + $GC^{+\cdot}$($^{2}A''$) asymptote is thus, 13.8(10.2) kcal/mol. This binding energy is much smaller than the one computed for Cu^{+}-GC due to the repulsive electrostatic interaction between the Cu^{+} and $GC^{+\cdot}$ radical cation. In fact the 13.7(10.1) kcal/mol binding energy arises mainly from the polarization of the ionized base pair induced by the positive charge of Cu^{+}.

The binding energies computed in the present work for Ca^{+2} and Cu^{+} are somewhat larger than those reported in previous studies[52,56] This is mainly due to the fact that previous calculations have been performed at the Hartree-Fock optimized geometries which provides too large metal–ligand distances and so, the electrostatic contribution to the binding energy is underestimated.

3.2 Single and Double proton transfer reactions

The following scheme shows the single and the double proton transfer processes considered.

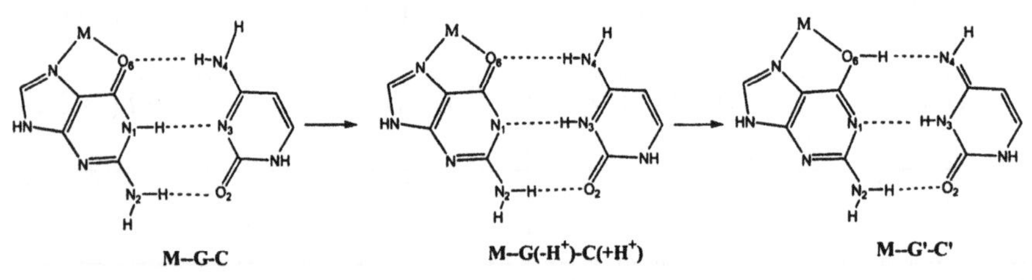

As mentioned, cationization of GC strenghtens those H-bonds in which Guanine acts as proton donor (N_1-N_3 and N_2-O_2) and weakens the one in which it acts as proton acceptor (O_6-N_4). Thus any of the two strengthened H-bonds can be involved in the single proton transfer reaction. Results show that the transfer from N_1 to N_3 is the only favorable process for Ca^{+2} and for Cu^+ since all attempts to optimize a N_2-O_2 proton transferred structure collapsed to the initial one. For Cu^{+2} a minimum structure was obtained although less stable than the one arising from the N_1-N_3 transfer. Although the analysis previously performed for the ionized systems cannot be applied here in the same way due to the presence of many different factors acting in opposite directions, the preferred situations are again those in which two similar in strength H-bonds are neighbors. This is due to the rigidity of the bases involved in the three hydrogen bonds, which do not benefit alternate situations, with a different central H-bond. The double proton transfer reaction also leads to an unfavorable alternate situation. Moreover, the acceptor oxygen O_6 is interacting with the metal cation, decreasing significantly its basicity. As a matter of fact, all optimizations starting from a double proton transferred structure have collapsed to the single proton transferred one. Finally, it should be mentioned that the population analysis confirms that the reaction corresponds to a proton transfer process since the charge of Cytosine in the M-G(-H^+)-C(H^+) structure for M = Ca^{+2}, Cu^+ and Cu^{+2} is 0.93, 0.86 and 0.98, respectively.

Table 4. B3LYP hydrogen bond distances (in Å) for the non-proton transferred M-GC and single proton transferred M-G(-H^+)C(+H^+)

	Ca^{2+}		Cu^+		Cu^{2+}	
	GC	G(-H^+)C(+H^+)	GC	G(-H^+)C(+H^+)	GC	G(-H^+)C(+H^+)
M-N_7	2.32	2.29	2.01	2.00	2.04	2.06
M-O_6	2.17	2.15	2.12	2.07	2.24	2.14
O_6–N_4	3.20	2.95	3.00	2.70	3.16	3.04
N_1–N_3	2.87	2.92	2.86	2.79	2.76	2.92
N_2–O_2	2.66	2.92	2.77	2.95	2.51	2.85

Table 4 shows the metal-ligand and hydrogen bond distances of the three cationized systems before and after the proton transfer process N_1—N_3 occurs. First, it can be observed that, in general, metal Guanine distances decrease in the proton transferred structure. For Ca^{+2} and Cu^+ this is consistent with the fact that the Guanine fragment G(-H+) acquires a negative charge which enhances the Metal-Guanine binding. For Cu^{+2}, the decrease of the M-O6 distance is attributed to the fact that the positive charge of ionized Guanine moves to protonated Cytosine diminishing its electrostatic repulsion with Cu^+.

With respect to the hydrogen bond distances, it can be observed that in all cases the shortest one corresponds to one of the extreme hydrogen bonds (N_4-O_6 or N_2-O_2) in agreement with the above mentioned rigidity of the base pair. For the monovalent Cu^+, the proton transferred structure shows shorter H-bonds than the non-transferred one indicating a stronger interaction between the two fragments, G(-H^+) and C(+H^+). In contrast for Ca^{+2} and Cu^{+2} it is observed a clear weakening of the hydrogen bonds.

The relative energies of the single proton transferred structure with respect to the MG + C asymptote are given in Figure 4. For comparison we have also included the relative energies of the single proton transfer MG(-H^+) + C(+H^+) asymptote.

Let us first discuss the case of the monovalent cation Cu^+. It can be observed that in this case the single proton transfer process is slightly endothermic (4.6 kcal/mol), the energy barrier being 6.2 kcal/mol. The process for this system is much more favorable than the one in absence of the metal cation, which has been computed to be 24 kcal/mol.[12] This is due to an increase of the Guanine acidity induced by the metal binding. On the other hand, it is worth noting that whereas the energy difference between the single proton transferred and the non proton transferred structure is only 4.6 kcal/mol, the energy difference of the respective asymptotes is 13.8 kcal/mol, which indicates that the H-bond interaction in MG(-H^+)-C(+H^+) (48.0 kcal/mol) is significantly larger than in MG-C (38.7 kcal/mol). Such strengthening of the hydrogen bonds in MG(-H^+)-C(+H^+) arise from the formation of an ion pair which is stabilized by the positive charge of the metal cation. Consistently, the hydrogen bonds in the proton transferred structure are shorter than in the initial species (see above). Thus, as noted for the ionized systems two factors determine the stability of the different species: a) the relative proton affinities of the centers involved in the proton transfer reaction or in other words, the relative stability of the asymptotes from which they derive, and b) the strength of the hydrogen bond interaction.

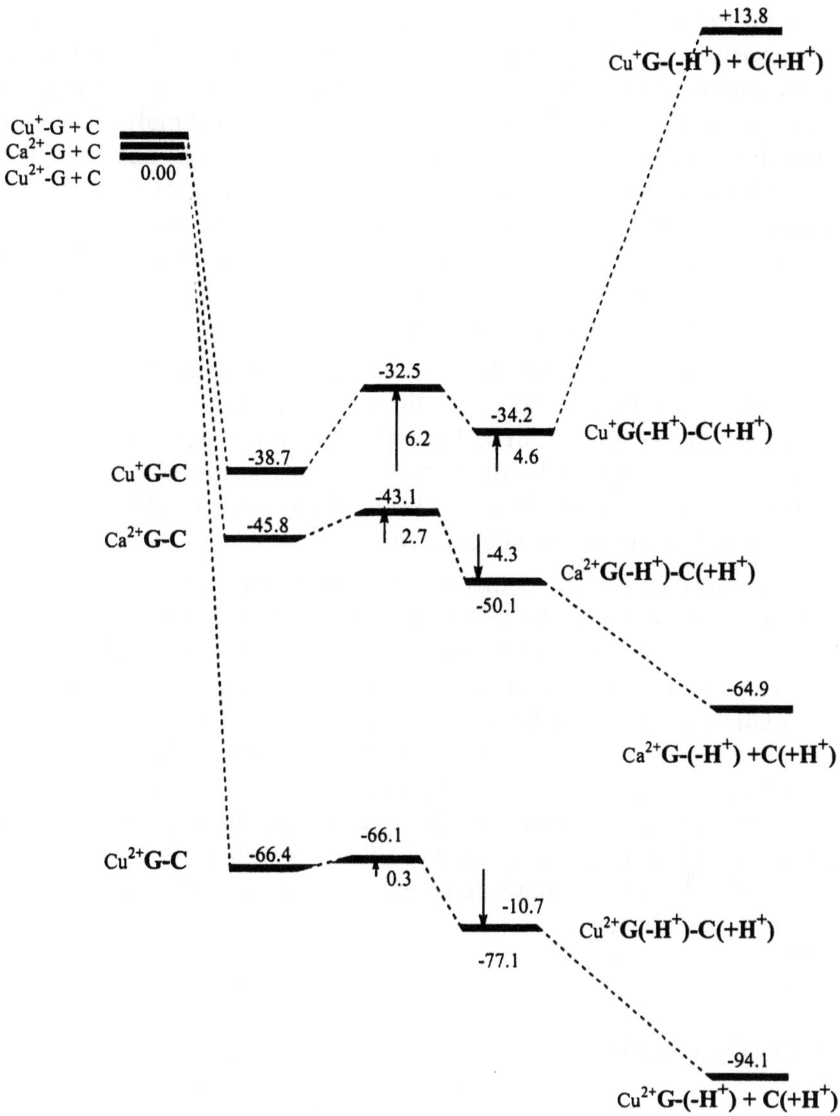

Figure 4. Energy profiles for the single proton transfer processes in cationized Guanine-Cytosine base pairs. Energies (in kcal/mol) are relative to the ground state asymptote

Divalent cations provide a completely different picture. The main change is that the proton transferred asymptote lies below the proton transferred minima MG(-H^+)-C(+H^+). This different behavior is due to the electrostatic repulsion between the two fragments, C(+H^+) and [MG(-H^+)]$^+$. The weakening of the hydrogen bond interaction in such proton transferred species for the divalent cations results in an increase of the H-bond distances as previously seen. The proton transfer reaction for the Ca^{+2} and Cu^{+2} are now exothermic (-4.3 and –10.7 kcal/mol, respectively) and consequently present smaller energy barriers (2.7 and 0.3 kcal/mol). Therefore, considering the small energy barrier of the reaction and the large exothermicity associated to the fragmentation of MG(-H^+)-C(+H^+), the interaction of this metal cation to Guanine is expected to induce a spontaneous proton transfer process followed by the separation of the two fragments. Interestingly, broken GC base pairs and protonated Cytosine were recently observed in a platinated duplex.[92]

Even though both divalents cations provide a similar picture there are several differences to point out. First, Cu^{+2} leads to a stronger MG-C complex compared to Ca^{+2}. In addition the proton transfer reaction is more exothermic and the energy barrier very small. This seems to suggest that the presence of both electrostatic and oxidant effects is more efficient for this process than the electrostatic effects alone.

It must be remarked that the present results correspond to a gas phase situation, which can be very different from that in real living systems, since in these cases the metal cation is solvated by water molecules and interacting with the negatively charged backbone. Therefore, the electrostatic effects will be largely screened, which will make less favorable the studied reaction. Single point calculations for pentahydrated Mg^{+2} have shown that both minima are nearly degenerate.[62] Although the changes induced by cation solvation on the electrostatic interactions are to be expected, the changes on the oxidant character as well as the contribution of the two effects can be more complex. Studies on this direction are in progress in our laboratory.

4. CONCLUSIONS

The intermolecular proton transfer processes in DNA base pairs, as a possible source of spontaneous mutations leading to cancer and aging was treated by P. O. Löwdin in many publications.[1,2,9,93-102] In this chapter we have tried to develop the posterior evolution of this bright intuition that Löwdin had forty years ago. We have mainly focused on the works performed in our laboratory concerning the effects of ionization and cationization on these reactions. During all these years, the developments in quantum chemistry methods and the increase in computational resources had allowed to perform gas phase quantum mechanical studies on these processes at a reasonably accurate level of calculation. The main conclusions of these studies are the following:

a) For neutral systems all studies agree that the single proton transfer reaction is less favorable than the double proton transfer one, because the single transfer process implies a charge separation due to the formation of an ion-pair complex, whereas in the double proton transfer process the electroneutrality is maintained. Nevertheless, the energy barrier of the double proton transfer process is high.

b) For the excited states the energy barrier of the double proton transfer reaction (although smaller than that corresponding to the ground state) is also significant. However, a charge transfer excited electronic state has been found to be quite stable in the region corresponding to a single proton transfer, which eventually neutralizes the charge separation induced by the electronic transition. This state could be populated through a crossing with the state initially obtained by phtotexcitation.

c) For the ionized GC and AT systems, the single proton transfer reaction appears to be very favorable since ionization of the purine bases increases their acidity and the reaction does not imply a creation of charges, but the transfer of a positive charge. The double proton transfer reaction, however, is not so favorable. The energy profile is determined by two factors: the proton affinity of the two centers involved in the proton transfer and the number and sequence of strong and weak hydrogen bonds.

d) Cationization of Guanine at the N_7 and O_6 positions favors the single proton transfer reaction, especially for the divalent cations for which the process becomes exothermic and the energy barrier is very small. Moreover, for these cations the fragmentation of the proton transferred structure is exothermic due to the electrostatic repulsion between the two fragments. So, cationization of GC by divalent metal cations is expected to induce spontaneous proton transfer followed by separation of the two fragments, changing the genetic information. These effects are more important for Cu^{+2} due to the oxidant character of this metal cation.

Although these results may suggest interesting gas phase experiments to test the theoretical predictions and analyze the intrinsic reactivity (in absence of solvent) of these systems, a step forward for understanding these processes in real living systems requires the introduction of the environment. For instance, the formation of an ion-pair by a single proton transfer process in neutral systems will be strongly favored by a polar medium. On the other hand, solvation of metal cations will screen the electrostatic effects described. Although some initial steps have been performed in this direction, it still remains many work to be done to fully understand these complex biological processes. Finally, protons are not classical particles but "wave packets" obeying laws of quantum theory and so, tunneling can occur readily if fluctuations of the environment change the energy profile to a symmetrical double well situation. Therefore, as Löwdin emphasized in many occasions, tunnel effects can also play a very important role in these mutagenic reactions.

REFERENCES

1. Löwdin, P.O. *Rev. Mod. Phys.* **1963**, *35*, 724;
2. Löwdin, P.O. *Adv. Quantum Chem.* **1965**, *2*, 213.
3. Rein, R.; Harris, F.E. *J. Chem. Phys.* **1964**, *41*, 3393.
4. Lunell, S.; Sperber, G. *J. Chem. Phys.* **1967**, *46*, 2119.
5. Scheiner, S.; Kern, C.W. *J. Am. Chem. Soc.* **1979**, *101*, 4081.
6. Scheiner, S.; Kern, C.W. *Chem. Phys. Lett.* **1978**, *57*, 331.
7. Clementi, E.; Mehl, J.; von Niessen, W. *J. Chem. Phys.* **1971**, *54*, 508.
8. Clementi, E. *Proc. Natl. Acad. Sci. U.S.A.* **1972**, *69*, 2942.
9. Kong, Y.S.; John, M. S.; Löwdin, P.O. *Int. J. Quantum Chem. Quantum Biol. Symp.* **1987**, *14*, 189.
10. Hrouda, V.; Florián, J.; Hobza, P. *J. Phys. Chem.* **1993**, *97*, 1542.
11. Florián, J.; Hrouda, V.; Hobza, P. *J. Am. Chem. Soc.* **1994**, *116*, 1457.
12. Florián, J.; Leszczynski, J. *J. Am. Chem. Soc.* **1996**, *118*, 3010.
13. Douhal, A.; Kim, S. K.; Zewail, A. H. *Nature* **1995**, *378*, 260.
14. Chachisvilis, M.; Fiebig, T.; Douhal, A.; Zewail, A. H. *J. Phys. Chem. A* **1998**, *102*, 669.
15. Nakahima, A.; Hirano, M.; Hasumi, R.; Kaya, K.; Watanabe, H.; Carter, C. C.; Williamson, J. M.; Miller, T. A. *J. Phys. Chem. A* **1997**, *101*, 392.
16. López-Martens, R.; Long, P.; Sogaldi, D.; Soep, B.; Syage, J.; Millie, Ph. *Chem. Phys. Lett.* **1997**, *273*, 219.
17. Takeuchi, S.; Tahara, T. *Chem. Phys. Lett.* **1997**, *227*, 340.
18. Takeuchi, S.; Tahara, T. *J. Phys. Chem. A* **1998**, *102*, 7740.
19. Folmer, D. E.; Poth, L.; Wisniewski, E. S.; Castleman, A. W. Jr. *Chem. Phys. Lett.* **1998**, *287*, 1.
20. Folmer, D. E.; Wisniewski, E. S.; Castleman, A. W. Jr. *Chem. Phys. Lett.* **2000**, *318*, 637.
21. Douhal, A.; Guallar, V.; Moreno, M.; Lluch, J. M. *Chem. Phys. Lett.* **1996**, *256*, 370.
22. Moreno, M.; Douhal, A.; Lluch, J. M.; Castaño, O.; Frutos, L. M. *J. Phys. Chem. A* **2001**, *105*, 3887.
23. Catalán, J.; del Valle, J. C.; Kasha, M. *Proc. Natl. Acad. Sci. U.S.A.* **1999**, *96*, 8338.
24. Catalán, J.; del Valle, J. C.; Kasha, M. *Chem. Phys. Lett.* **2000**, *318*, 629.
25. Guallar, V.; Douhal, A.; Moreno, M.; Lluch, J. M. *J. Phys. Chem. A* **1999**, *103*, 6251.
26. Sevilla, M.D.; Becker, D.; Yan, M.; Summerfield, S.R *J. Phys. Chem.* **1991**, *95*, 3410.

27. Candeias, L.P.; O'Neill, P.; Jones, G.D.D.; Steenken, S. *Int. J. Radiat. Biol.* **1992**, *61*, 15.

28. Steenken, S. *Chem. Rev.* **1989**, *89*, 503.

29. Johnston, D.H.; Glasgow, K.C.; Thorp, H.H. *J. Am. Chem. Soc.* **1995**, *117*, 8933.

30. Hall, D.B.; Holmlin, R.E.; Barton, J.K. *Nature* **1996**, *382*, 731.

31. Stemp, E.D.A.; Arkin, M.R.; Barton, J.K. *J. Am. Chem. Soc.* **1997**, *119*, 2921.

32. Murphy, C.J.; Arkin, M.R.; Ghatlia, N.D.; Bossmann, S.; Turro, N.J.; Barton, J.K. *Proc. Nat. Acad. Sci. U.S.A.* **1994**, *91*, 5315.

33. Arkin, M.R.; Stemp, E.D.A.; Holmlin, R.E.; Barton, J.K.; Hörmann, A.; Olson, E.J.C.; Barbara, P.F. *Science* **1996**, *273*, 475.

34. Sevilla, M.D.; Besler, B.; Colson, A.O. *J. Phys. Chem.* **1995**, *99*, 1060.

35. Steenken, S.; Jovanovic, S. V. *J. Am. Chem. Soc.* **1997**, *119*, 617.

36. Melvin, T.; Botchway, S.W.; Parker, A.W.; O'Neill, P.O. *J. Am. Chem. Soc.* **1996**, *118*, 10031.

37. Breen, A.P.; Murphy, J.A. *Free Radical Biol. Med.* **1995**, *18*, 1033.

38. Gugger, A.; Batra, R.; Rzadek, P.; Rist, G.; Giese, B. *J. Am. Chem. Soc.* **1997**, *119*, 8740.

39. Steenken, S.; Telo, J.P.; Novais, H.M.; Caudeias, L.P. *J. Am. Chem. Soc.* **1992**, *114*, 4701.

40. Steenken, S. *Biol. Chem.* **1997**, *378*, 1293.

41. Murphy, C.; Arkin, M.; Jenkins, Y.; Ghatlia, N.; Bossmann, S.; Turro, N.; Barton, J. K. *Science*, **1993**, *262*, 1025.

42. Kelley, S. O.; Barton, J. K. *Science* **1999**, *283*, 375.

43. Bhattacharya, P. K.; Barton, J. K. *J. Am. Chem. Soc.* **2001**, *123*, 8649.

44. Colson, A.O.; Besler, B.; Sevilla, M.D. *J. Phys. Chem.* **1992**, *96*, 9787.

45. Colson, A.O.; Sevilla, M.D. *Int. J. Radiat. Biol.* **1995**, *67*, 627.

46. Li, X.; Zhongli, C.; Sevilla, M. D. *J. Phys. Chem. B* **2001**, *105*, 10115.

47. Hutter, M.; Clark, T. *J. Am. Chem. Soc.* **1996**, *118*, 7574.

48. Bertran, J.; Oliva, A.; Rodríguez-Santiago, L.; Sodupe, M. *J. Am. Chem. Soc.* **1998**, *120*, 8159.

49. Anwander, E. H. S.; Probst, M. M.; Rode, B. M. *Biopolymers* **1990**, *29A*, 757.

50. Burda, J. V.; Šponer, J.; Hobza, P. *J. Phys. Chem.* **1996**, *100*, 7250.

51. Cerda, B. A.; Wesdemiotis, C. *J. Am. Chem. Soc.* **1996**, *118*, 11884.

52. Gadre, S. R.; Pundlik, S. S.; Limaye, A. C.; Rendell, A. P. *Chem. Comm.* **1998**, 573.

53. Rodgers, M. T.; Armentrout, P. B. *J. Am. Chem. Soc.* **2000**, *122*, 8548.

54. Risso, N.; Toscano, M.; Grand, A. *J. Am. Chem. Soc.* **2001**, *123*, 10272.

55. Csaszar, K.; Špaková, N.; Štefl, R.; Šponer, J.; Leontis, N. B. *J. Mol. Biol.* **2001**, *313*, 1073.

56. Burda, J. V.; Šponer, J.; Leszczynski, J.; Hobza, P. *J. Phys. Chem. B* **1997**, *101*, 9670.

57. Šponer, J.; Burda, J. V.; Sabat, M.; Leszczynski, J.; Hobza, P. *J. Phys. Chem. A* **1998**, *102*, 5951.

58. Šponer, J.; Sabat, M.; Burda, J. V.; Leszczynski, J.; Hobza, P. *J. Phys. Chem. B* **1999**, *103*, 2528.

59. Šponer, J.; Burda, J. V.; Leszczynski, J.; Hobza, P. *J. Biomol. Struct. Dyn.* **1999**, *17*, 61.
60. Gresh, N.; Šponer, J. *J. Phys. Chem. B* **1999**, *103*, 11415.
61. Pelmenschikov, A.; Ziberberg, I.; Leszczynski, J.; Famulari, A.; Sironi, M.; Raimondi, M. *Chem. Phys. Lett.* **1999**, *314*, 496.
62. Šponer, J.; Sabat, M.; Gorb, L.; Leszczynski, J.; Lippert, B.; Hobza, P. *J. Phys. Chem. B* **2000**, *104*, 7535.
63. Fonseca Guerra, C.; Bickelhaupt, F. M.; Snijders, J. G.; Baerends, E. J. *J. Am. Chem. Soc.* **2000**, *122*, 4117.
64. Muñoz, J.; Šponer, J.; Hobza, P., Orozco, M., Luque, F. J. *J. Phys. Chem. B* **2001**, *105*, 6051.
65. Gebicki, J.; Bally, T. *Acc. Chem. Res.* **1997**, *30*, 477.
66. Sodupe, M.; Oliva, A.; Bertran, J. *J. Am. Chem. Soc.* **1994**, *116*, 8249.
67. Sodupe, M.; Oliva, A.; Bertran, M. *J. Am. Chem. Soc.* **1995**, *117*, 8416.
68. Sodupe, M.; Oliva, A.; Bertran, J. *J. Phys. Chem. A.* **1997**, *101*, 9142.
69. Sodupe, M.; Bertran, J.; Rodríguez-Santiago, L.; Baerends, E. J. *J. Phys. Chem. A.* **1999**, *103*, 166.
70. Rodríguez-Santiago, L.; Sodupe, M.; Oliva, A.; Bertran, J. *J. Am. Chem. Soc.* **1999**, *121*, 8882.
71. Rodríguez-Santiago, L.; Sodupe, M.; Oliva, A.; Bertran, J. *J. Phys. Chem. A.* **2000**, *104*, 1256.
72. Rodríguez-Santiago, L.; Vendrell, O.; Tejero, I.; Sodupe, M.; Bertran, J. *Chem. Phys. Lett.* **2001**, *334*, 112.
73. Becke, A. D. *J. Chem. Phys.* **1993**, *98*, 5648.
74. Lee, C.; Yang, W.; Parr, R. G. *Phys. Rev. B* **1988**, *37*, 785.
75. Šponer, J.; Hobza, P. *J. Am. Chem. Soc.* **1994**, *116*, 709.
76. Šponer, J.; Leszczynski, J.; Hobza, P. *J. Phys. Chem.* **1996**, *100*, 5590.
77. Šponer, J.; Florián, J.; Hobza, P.; Leszczynski, J. *J. Biomol. Struct. Dyn.* **1996**, *13*, 827.
78. Saenger,W. Principles of Nucleic Acid Structure; Springer-Verlag **1984** New York; pp 123 and 124 and references therein.
79. Yanson, I.K.; Teplitsky, A. B.; Sukhodub, L.F. *Biopolymers* **1979**, *18*, 1149.
80. Santamaría, R.; Vázquez, A. *J. Comput. Chem.* **1994**, *9*, 981.
81. Gould, I.R.; Kollman, P.A. *J. Am. Chem. Soc.* **1994**, *116*, 2493.
82. Šponer, J.; Leszczynski, J.; Hobza, P. *J. Phys. Chem.* **1996**, *100*, 1965.
83. Frey, P.A.; Whitt, S.A.; Tobin, J.B. *Science* **1994**, *264*, 1927.
84. Pan, Y.; McAllister, M.A. *J. Org. Chem.* **1997**, *62*, 8171.
85. Hush, N.S.; Cheung, A.S. *Chem. Phys. Lett.* **1975**, *34*, 11.
86. Yates, B.F.; Bouma, W.; Radom, L. *J. Am. Chem. Soc.* **1984**, *106*, 5805.
87. Nir, E.; Kleinermanns, K.; de Vries, M. S. *Nature* **2000**, *408*, 949.
88. Wachters, A. J. H. *J. Chem. Phys.* **1970**, *52*, 1033.
89. Hay, P. J. *J. Chem. Phys.* **1977**, *66*, 4377.
90. Blaudeau, J. P., McGrath, M. P., Curtiss , L. A.; Radom, L. *J. Chem. Phys.* **1997**, *107*, 5016.

91. Koch, W.; Kolthausen, M. C. "*A Chemist Guide to Density Functional Theory*", WILEY-VCH Weinheim **2001**.

92. Coste, J.; Maligne, J.-M.; Serre, L.; Shepard, W.; Roth, M.; Leng, M.; Zelwer, C. *Nucleic Acid Res*. **1999**, *27*, 1837.

93. Löwdin, P. O. *Laekartidningen* **1977**, *74*, 3419.

94. Löwdin, P. O. *Int. J. Quantum. Chem., Quantum. Biol. Symp*. **1977**, *4*, 185.

95. Löwdin, P. O. *Phys. –Chem. Mech. Carcinog.* **1969**, 203.

96. Löwdin, P. O. *Mol. Ass. Biol.* **1968**, 539.

97. Löwdin, P. O. *Ann. N. Y. Acad. Sci*. **1969**, *158*, 86.

98. Löwdin, P. O. *Int. J. Quantum. Chem., Symp*. **1968**, *2*, 207.

99. Löwdin, P. O. *Study Week Mol. Forces* **1967**, 637.

100. Löwdin, P. O. *Adv. Chem. Phys*. **1965**, *8*, 177.

101. Löwdin, P. O. *Electron. Aspects Biochem*. **1964**, 167.

102. Löwdin, P. O. *Mut. Res*. **1965**, *2*, 218.

The Origin of Spontaneous Point Tautomeric Mutations in DNA:

Löwdin's Mechanism of Proton Tunneling in DNA Base Pairs

Eugene S. Kryachko
Bogoliubov Institute for Theoretical Physics, Kiev, 03143 Ukraine
and
Department of Chemistry, University of Leuven
Celestijnenlaan 200 F, B-3001 Leuven, Belgium
E-mail: eugene@bohr.quantchem.kuleuven.ac.be

Abstract This review is about the hydrogen-bonded patterns in the guanine-cytosine and adenine-thymine DNA base pairs and their role in the origin of spontaneous point tautomeric mutations (mismatched pairs), whose mechanism via proton tunneling along particular hydrogen bond(s) was proposed by Löwdin, about preopening of these base pairs via a partial breaking of some hydrogen bonds, a role of water in facilitating this process and about of cure of DNA base pairs against appearance of mismatches via invoking so-called covalent and hydrophobic base pairs.

1. Introductory Excursion in the DNA Kingdom

If you ask Nature the right question,
she will give you the right answer.

Richard Feynman

E.J. Brändas and E.S. Kryachko (eds.),
Fundamental World of Quantum Chemistry, Vol. II, 583–629.

In the DNA and RNA molecules, nature has found an excellent mechanism for storing long-time information in the form of a proton-electron pair code of the type
 *H.*

Per-Olov Löwdin ([1], p. 347)

The structure of **d***eoxyribo***n***ucleic* **a***cid*, (DNA), originally proposed by Watson and Crick [2-4] and refined by Langridge et al. [5], furnishes the double helix. Two types of interaction, hydrogen bonding and stacking, maintain the DNA double helix structural motif. The hydrogen bonds orchestrate the base pairing between the adenine A and thymine T (or 5-methyl-uracil) nucleic acids, on the one hand, and on the other one, between the guanine G and cytosine C nucleic acids in the Watson-Crick double helix of DNA [2] (see also Ref.[6] for review and references therein) linking together that way two DNA strands. Stacking interaction primarily determines the helix architecture [7]. The nucleic acid base pairs A·T and G·C of DNA, referring usually to the Watson-Crick complementary or canonical ones, are pictured in Figure 1. The canonical A·T base pair is formed between A and T due to the two hydrogen bonds: N(A)-H···O(T) and N(T)-H···N(A) [8] while the G·C one between G and C via three hydrogen bonds: N(C)-H···O(G), N(G)-H···N(C), and N(G)-H···O(C) where the base is indicated in parentheses.

What is the very place of hydrogen bonding in maintaining the DNA double helix architecture comparing with the interstrand base-stacking and interstrand cross-

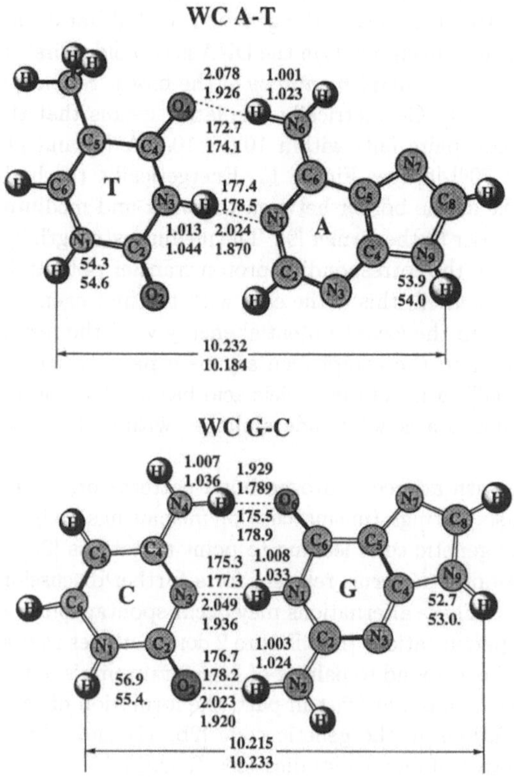

Figure 1. The correct Watson-Crick (WC) A·T and G·C base pairs of DNA. The optimized geometrical parameters of the A·T pair, including the C′ to C′ distance R, defined as the distance between $H_9(A)$ and $H_1(T)$, and the glycosyl bond angles, $\alpha_1 \equiv \angle N_9H_9(A)H_1(T)$ and $\alpha_2 \equiv \angle N_1H_1(T)H_9(A)$, correspond to the HF (upper entry; reading from the top down) and B3LYP (lower entry) computational levels in conjunction with 6-31+G(d) basis set. Similarly for the G·C base pair. The distances are given in Å, angles in deg.

stacking interactions? How much does this hydrogen bonding stabilize the DNA? Is that due to the hydrogen bonding that adenine "likes" thymine and guanine "likes" cytosine? Whether it actually looks like, loosely speaking, as a well-known double-head Janus: on the one side, the hydrogen bonding masters to hold together the DNA base pairs, and on the other, it "misbehaves" by allowing a proton transfer to occur and to lead to mutations in the genetic code? What does it actually mean?

The role of the hydrogen bonds in the DNA is twofold. First, they are responsible for a so called *electronic complementarity* of the bases. Second, they hold the DNA *shape complementarity*. Geometrically, the latter means that the C′ to C′ distance R of the DNA base pairs falls within 10.6 - 10.8 Å [6] and glycosyl bond angles α_1 and $\alpha_2 \in [52^o, 70^o]$ [6] (see Figure 1). Energetically, the hydrogen bonds in the DNA base pairs lie at the border between the weak and medium-strength hydrogen bonds, for being closer to the former [9]. The medium-strength hydrogen bonds often have double wells on the corresponding proton transfer potential energy surface (PT PES). As was anticipated, this is the case with the hydrogen bonds in the DNA. If each proton resides in the lowest potential energy well, the resulting hydrogen-bond patterns correspond to the complementary base pairs which are composed of the "major" or "normal" forms of the nucleic acid bases. This specific hydrogen-bonded pattern predetermines a genetic code alphabet within the Watson-Crick model of DNA [2, 5].

Another case with altered hydrogen-bond patterns or, in the other words, non-Watson-Crick base pairings (mismatches or mispairings) might therefore cause an alternation of the genetic code leading to point mutations [2b]. A couple of typical altered hydrogen-bond patterns, relevant for a further discussion, are schematically shown in Figure 2. These alternations may occur spontaneously or due to a presence of some external perturbations [10]. Figure 2 demonstrates that such hydrogen-bond patterns actually correspond to pairing of "rare" tautomers of the DNA bases which, in turn, as believed, cause, at least in part, the formation of mismatches, producing spontaneous mutations in the genetic code [2b, 11] and which, in fact, were and still are a subject of numerous studies (see Refs.[11-15] for review and references therein).

It seems to deserve to be mentioned, at the end of the present Section, that mutations were discovered by de Vries in 1902. In their classical work, dated by 1943, Luria and Delbrück [16] assumed that there was a "fixed small chance per time unit ... to undergo a mutation" and that the time unit should be a division of cycles of the bacteria, that is, a constant probability of mutation per cell division. It is a long belief of biology that mutations occur randomly, non-directly, continuously, and irrelatively to any advantage, offered by the surrounding enviroment [11]. Estimates of spontaneous mutation frequency fall between $> 1 \times 10^{-7}$ and $\approx 5 \times 10^{-11} s^{-1}$ per

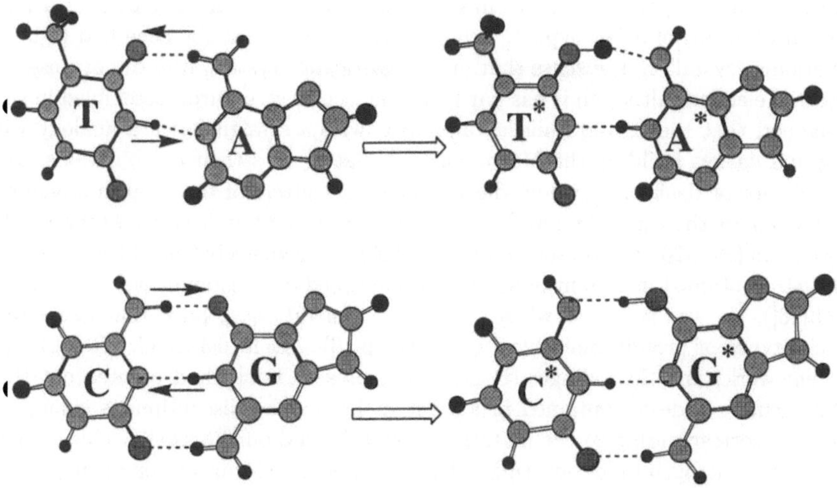

Figure 2. The double-proton transfers in the canonical A·T and G·C base pairs of DNA.

a base pair [11n-o].

2. Löwdin's Mechanism of Spontaneous Point Tautomeric Mutations in DNA

... the genetic code is essentially a "proton code".

Per-Olov Löwdin ([1], p. 215)

In 1963, Per-Olov Löwdin [1, 17-24] proposed a model of proton transfer as the origin of spontaneous point mutations in the DNA base pairs. His work "Some Aspects on DNA Replication; Incorporation Errors and Proton Transfer" [18], Löwdin started with quoting Schrödinger's book (Ref.[25], the present work): "Schrödinger ... also pointed to the fact that there may be a parallelism between the discontinuous changes in the genetic materials leading to 'mutations' and the quantum jumps ... . According to Schrödinger, it seems plausible that the genetic information was contained in an aperiodic crystal in the form of a special arrangement of a large number of atoms of a few types". Schrödinger believed that the gene had to be an "aperiodic crystal" in the sense that, unlike inorganic crystals, it does not possess a repeated element, although it has particular molecular structures, assembled in such a manner, that they could build an alphabet, with a straightforward analogy with dots and dashes building the Morse code (the latter uses their combinations, from a single dot or dash to four dots/dashes for the 26 letters of the English alphabet).

"At about the same time as Schrödinger gave his Dublin lecture, Avery and his co-workers (Ref.[26], the present work) could show conclusively that the genetic message was contained in a giant organic molecule called deoxyribonucleic acid (DNA)" (Ref.[18]). It was the time when "there was general acceptance that genes were special types of protein molecules" (Ref.[2d], p.13). "Delbrück et al. (Ref.[27], the present work) and Schrödinger (Ref.[25], the present work) had pointed out that, if the genetic code is contained in a giant molecule, the discontinuous changes in the code corresponding to the mutations could depend on the existence of isomeric forms. The nitrogen bases occuring in DNA show simple forms of isomerism, namely the keto-enol tautomerism and the amino-imine tautomerism, which are forms which have been studied since 1911. Watson and Crick assumed that the nitrogen bases in DNA normally occurred in the keto and amino forms, but they also considered the existence of the "rare" tautomeric forms. ... This implies that, if a tautomeric base for some reason occurs in the original DNA molecule or in the nutrition material involved in the replication process, there will be an error introduced in the copying procedure which changes a G-C pair into an A-T pair, and vice versa" (Ref.[18]).

A scenario of Löwdin's model of the spontaneous mutations in DNA is the following. Let us, first, suggest that the genetic alphabet, stored in the DNA double helix, depends upon the stability of the hydrogen-bonded base pairs and the motion of the proton(s) along these hydrogen bonds as well. Each proton, belonging to a given hydrogen bond between the A·T or G·C base pairs, may be transferred via quantum mechanical tunneling or "quantum jumping" from the position, corresponding to the normal base pair, to the position corresponding to the rare one (if it does exist). The latter implies a (partial) loss or, in other words, a damage of the genetic alphabet. It was Löwdin's suggestion that such transfer is significantly facilitated if the simultaneous double proton tunneling takes place. Its tunneling rate determines the probability of the occurrence of spontaneous point tautomeric mutations.

Since 1963, when Löwdin offered his model of the origin of spontaneous point mutations in DNA base pairs for the first time [17], much theoretical efforts have been devoted to the study of his mechanism, and particularly, to the accurate description of the landscape of the PT PESs in the DNA base pairs [11c, 28-32]. Using semi-empirical calculations, Rein and Ladik [28b] and Rein and Harris [28c-e] have shown that the PT PES of the single proton transfer in the base pair G·C has a double-well character. At the same time, similar calculation has been performed for the A·T base pair by Lunell and Sperber [28f]. Further ab initio calculations have been conducted by Clementi and co-workers [29] and by Scheiner and Kern [30]. Analyzing the single well proton potential, similar to that found by Colson et al. [32], Clementi and co-workers [29b] have led to the conclusion that an enlargement of employed basis set may result in a double well PT PES.

The crucial steps in this area have recently been made due to the works by Hobza and co-workers [31a-b] and Florián and Leszczynski [31c]. Hobza and co-workers [31a-b] have studied the double proton transfer in the A·T base pair via using the HF approach in conjunction with MINI-1 and MIDI-1 basis sets. Within the geometry relaxation framework, they have shown, first, that the tautomeric conversion A·T $\Rightarrow$ A*·T*, shown in Figure 2, is controlled by the activation barrier of the 9.7 kcal/mol height relative to the normal A·T pair and second, that the rare tautomer pair A*·T* lies below the barrier by only 0.2 kcal/mol. The latter means its metastability (see also Ref.[31c]). Florián and Leszczynski [31c] have investigated the whole PT PES in the G·C base pair and concluded that the tautomeric transition G·C $\Rightarrow$ G*·C*, shown in Figure 2 too, is energetically accessible and its equilibrium constant falls within the range of the measured DNA replication fidelity, and therefore, their findings support Löwdin's mechanism. This will be discussed in detail in Section 3 (see also Chapter "Effects of Ionization and Cationization on Intermolecular Proton Transfer" by J. Bertran, M. Noguera and M. Sodupe in this Volume).

We end our introductory excursion to the DNA kingdom with mentioning other possible mechanisms which might also contribute to mutations. DNA bases can be damaged in many ways. These are, for example, the isomerization of bases [11a-c], an existence of pyrimidine dimers [33], effect of ionizing radiation and UV irradiation [10b, 34], formation of excimers [35], mutagenicity of metals [36], double-proton transfer in the electronically excited base pairs [37]. In particular, Catalán and Kasha [37b] have demonstrated, as experimentally as well as computationally, by employing the B3LYP/6-31G(d,p) method, that the tautomerization mechanism in the double-hydrogen bonded dimer of 7-azaindole, as a model DNA base pair, involves a quantum tunneling throughout the excited-state potential barrier. The dynamics of the double proton transfer in the 7-azaindole dimer has recently been investigated using femtosecond transient absorption and fluorescence upconversion techniques by Zewail and co-workers [37d].

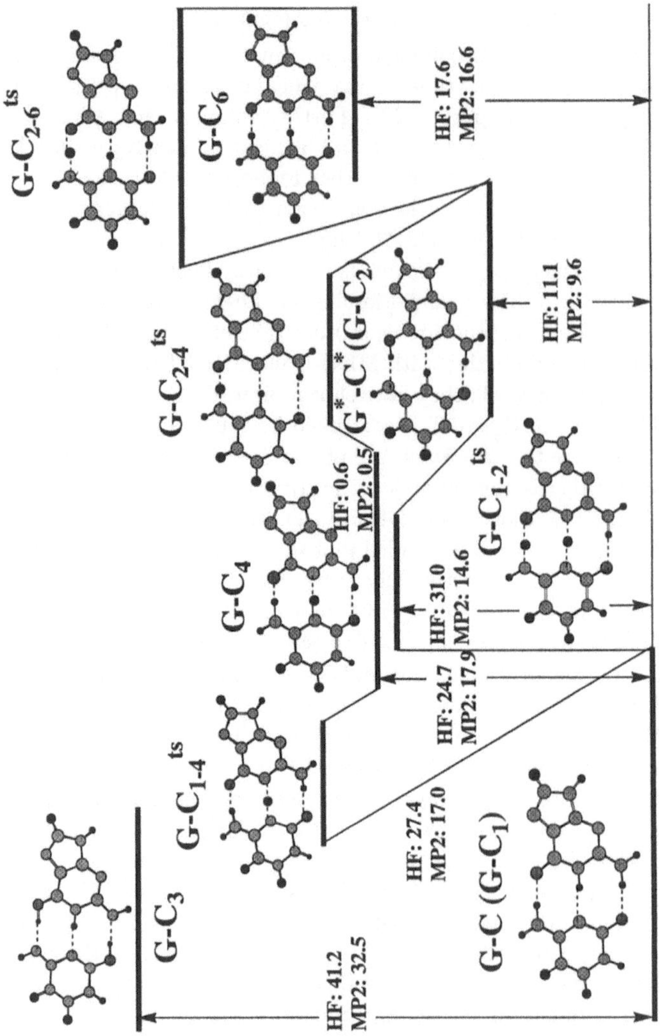

Figure 3. The proton transfer PES of the G·C base pair [31c] (see Section 3).

3. Tautomeric Misparings in the G·C Base Pair

The Watson-Crick G·C base pair of DNA is formed thanks to the three hydrogen bonds: N_4(C)-H···O_6(G), N_3(C)···H-N_1(G) and O_2(C)···H-N_2(G). Generally speaking, there are nine possible and different arrangements of the protons, sharing these hydrogen bonds, which correspond to the normal and rare tautomeric base pairs, shown in Figure 3. Among them, there are only two base pairs, the normal G·C ($\equiv$ G·C_1) and the most stable rare, imino-keto/amino-enol G*·C* ($\equiv$ G·C_2), which lie at the bottom of the PES of all proton rearrangements: the corresponding energy offset amounts to 9.6 kcal/mol at the MP2(fc)/6-31G(d,p)//HF/6-31G(d) computational level [31c]. Solvent effects place the mismatched base pair G·C_2 10 $\pm$ 2 kcal/mol less stable than the normal. This implies that the corresponding equilibrium constant of the tautomeric transition G·$C_1 \Rightarrow$ G·C_2 falls into $\approx 10^{-6} - 10^{-9}$ range, that nicely fits the range of the measured DNA replication fidelity.

Structurally, the mismatched base pair G·C_2 has the following hydrogen bonds: N_4(C)···H-O_6(G), N_3(C)-H···N_1(G) and O_2(C)···H-N_2(G). A straight comparison of its pattern with that inherent for G·C_1 shows that they only differ in the position of the protons in the two, N_4(C)-H···O_6(G) and N_3(C)···H-N_1(G), hydrogen bonds. This reflect, as shown by Florián and Leszczynsky [31c], two different energetic pathways on the lower energy portion of the PES, governing proton rearrangements within the G·C base pair and linking G·C_1 to G·C_2. The higher-energy pathway is mediated by the mismatched ionic base pair-linker G·C_4, located 17.9 kcal/mol above the WC pair (at the same computational level [31c]). Actually, this pathway consists of two transition structures, G·C^{ts}_{1-4} and G·C^{ts}_{2-4}, as the linkers between G·C_1 and G·C_4, and G·C_4 and G·C_2, respectively. They both are placed by $\approx$ 17 kcal/mol above the WC base pair.

The lower-energy pathway passes throughout the transition pair-linker G·C^{ts}_{1-2} which governs the double proton transfer along the N_3(C)···H···N_1(G) and N_4(C)···H···O_6(G) hydrogen bonds and which is placed by only 14.6 kcal/mol above the WC pair and by 5.0 kcal/mol above G·C_2 [31c]. It is in fact a saddle point of the third order, reflecting the planarity constraint imposed in Ref.[31c]. It is interestingly to mention the geometrical changes in the aforementioned hydrogen bonds, accompanied by the double proton transfer. The bond length r(N_3(C)-N_1(G)) and the bond angle $\angle N_3$(C)HN_1(G) undergo changes from 3.043 to 2.608 and to 3.056 Å and from 176.4^o to 176.8^o and to 176.0^o under the double proton transfer G·$C_1 \Rightarrow$ G·$C^{ts}_{1-2} \Rightarrow$ G·C_2, respectively. The similar changes, particularly, a contraction of the hydrogen bond at the transition structure, are predicted for the bond length r(N_4(C)-O_6(G)) and the bond angle $\angle N_4$(C)HO_6(G): from 2.929 to 2.442 and to 2.872 Å and from 177.0^o to 177.5^o and to 171.5^o, respectively. The third hydrogen bond, O_2(C)-H···N_2(G), not really participated in the double proton transfer, undergoes a less contraction, namely, from 3.018 to 2.867 and to 3.134 Å. The rate constant of the double proton transfer reaction $G{\cdot}C_1 \Rightarrow G{\cdot}C_2$ falls into the $0-10^4 s^{-1}$ range at $300\,K$ [31c].

4. Mechanism of the Origin of Misparings in the A·T Base Pair

We have already mentioned in Section 2 that Hobza and co-workers [31a-b] have shown that the tautomeric transition A·T ⇒A*·T* is not energetically reliable (see also Ref.[31c]). What does it actually imply? An absence of any proton transfer and therefore, a non-occurrence of the spontaneous point tautomeric mutations within the A·T base pair? Or there exists a completely different mechanism of proton transfer that leads to them? This paradigm has recently been resolved by Kryachko and Sabin [38] who proposed a novel and physically sound model of the A·T tautomeric rearrangements. Before discussing this model, let us first notice what is the difference in the hydrogen-bonded patterns of the A·T and G·C pairs. As shown in Figure 1, guanine is linked to cytosine via three hydrogen bonds, two belonging to the N-H···O type and the third is of the N-H···N type [1, 6]. According to the general theory of hydrogen bonding [9], the former two are usually stronger than the latter. Since A·T contains one N-H···O and one N-H···N, it is obviously a less strongly bonded pair compared to G·C. Such observation is also supported by the mass spectrometric measurements [39] of the enthalpies of formation of such base pairs as 9-methyl-A·1-methyl-T and 9-methyl-G·1-methyl-C amounting, respectively, to -13.0 and -21.0 kcal/mol. These values are consistent with quantum chemical estimations [40, 15a, 15c]. Summarizing, we conclude that the A·T pair can more easily undergo some tautomeric rearrangements than G·C. This is the important conjecture in view of the well-known biological relevance of A·T-rich regions of DNA in molecular processes [41]. For instance, as known, the stretched A·T base pairs, partly in alternating order, compose the promoter sequences of many eukaryotic structural genes and A·T-rich regions separate functional domains in eukaryotic DNA as, e. g., long A·T-rich regions are found between the clustered yeast genome coding sequences. The A·T pair also frequenly occurs in transposable elements.

The lower-energy portion of the PT PES for the A·T base pair, excluding the single-step double-PT tautomeric conversion A·T ⇒ A*·T*, already shown in Figure 2 and discussed at length in Section 2, is schematically displayed in Figure 4 [38]. It clearly demonstrates that the novel mechanism for the tautomeric conversion as a cause of mutation in the hydrogen-bonded pattern of the A·T pair of DNA, consists of four principal stages [38].

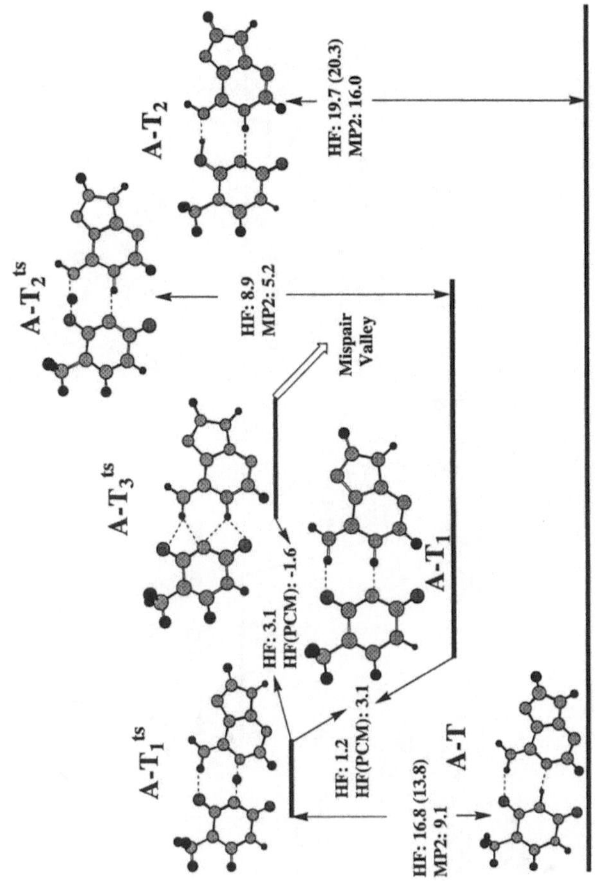

Figure 4. The portion of the PES of the A·T base pair demonstrating the novel model of the rearrangements of its hydrogen-bonded patterns. Energies are given in kcal/mol; HF corresponds to the HF/6-31+G(d) while the MP2 to the MP2(fc)//HF/6-31+G(d); HF(PCM) refers to the polarizable continuum model [38].

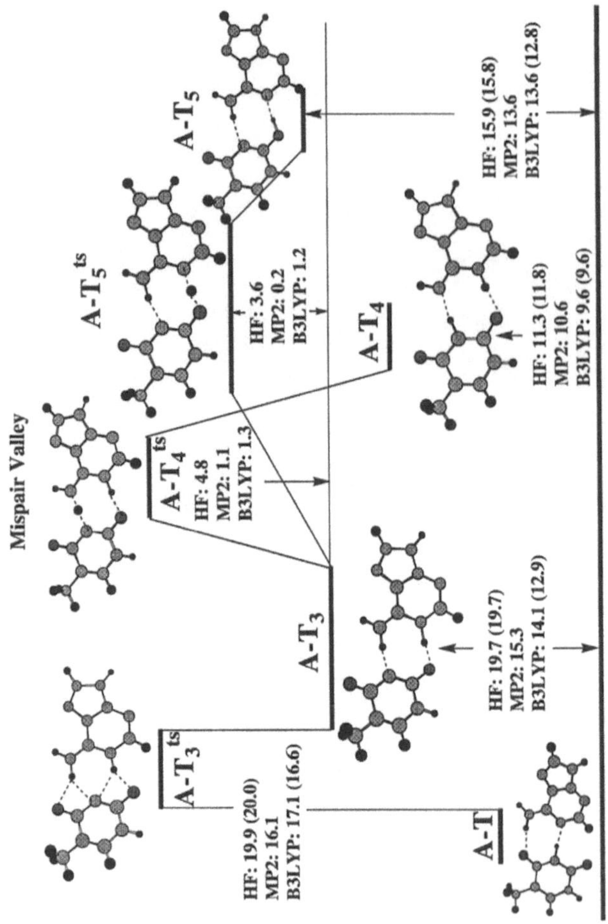

Figure 5. The mispairing valley in the PES of the A·T base pair. Energies in kcal/mol; the upper entry corresponds to the HF/6-31+G(d) while the lower one to the B3LYP/6-31+G(d); MP2 refers to MP2//HF/6-31+G(d) [38].

4.1. Proton Transfer Along the N_3(T)-H$\cdots$$N_1$(A) Hydrogen Bond

At the first, preliminary stage, the proton migrates along the weaker hydrogen bond N_3(T)-H$\cdots$$N_1$(A) of A·T. This is shown in Figure 4. Such protonic motion is coupled to the vibrational stretching mode $\nu_{N_3(T)H}$. The proton passes the transition state A·T_1^{ts} and arrives at the ionic pair A·T_1. The corresponding energetic barrier amounts to 9.1 kcal/mol (MP2/6-31+G(d)//HF/6-31+G(d); 16.8 without and 13.8 kcal/mol with ZPVE at the HF/6-31+G(d) level) [38]. Interestingly, it is smaller than the analogous barrier in the G·C base pair [31c]. This proton migration is accompanied by a substantial contraction of the hydrogen bridge N_3(T)$\cdots$H$\cdots$$N_1$(A) by 0.46 Å, compared to that in the WC A·T base pair. The neighboring bridge O_4(T)$\cdots$H$\cdots$$N_6$(A) undergoes a smaller contraction of 0.22 Å.

The resulting ionic pair structure A·T_1 with the total dipole moment of 5.4 D is formed by two alligned hydrogen bonds N_1(A)-H$\cdots$$N_3$(T) and N_6(A)-H$\cdots$$O_4$(T). Energetically, at the HF/6-31+G(d) computational level, A·T_1 lies 1.2 kcal/mol lower the transition structure A·T_1^{ts} that implies its relative metastability, although it however appears more stable than A·T_2. Such picture is valid in the gas phase. A solvent, particularly the water environment inherent to the DNA structure, actually favors its further stabilization due to a larger dipole moment (5.4 D in A·T_1 vs. 3.0 in A·T_1^{ts}). Modelling water environment by means of the polarizable continuum approach (PCM) results in a difference of the electronic energies of the A·T_1 and A·T_1^{ts} base pairs of 3.1 kcal/mol (see Figure 4) [38].

There however appears a nuance in estimating a relative metastability of A·T_1 due to a richness of the landscape of the PES of the A·T base pair. It arises due to the existence of another transition state in a vicinity of A·T_1 strongly affecting its metastability. As demonstrated in Figure 4, this is A·T_3^{ts} energetically placed rather close to A·T_1^{ts}. Its very existence is likely attributed to the low-frequency bending mode at 82 cm^{-1} ($\approx$ 0.2 kcal/mol) assigned to the relative motion of A and T within their pair. Notice that the A·T_1^{ts} is linked to the higher-frequency stretch at 2612 cm^{-1} of the hydrogen bridge N_3(T)$\cdots$H$\cdots$$N_1$(A) [38]. Therefore, since the low-frequency mode obviously undergoes excitation much easily at ambient temperatures than does its higher-frequency counterpart, A·T_3^{ts} seems to be easily accessible compared to A·T_1^{ts}. Furthermore, as shown in Ref.[38], the PCM lowers A·T_3^{ts} compared to A·T_1^{ts} by 1.6 kcal/mol due to particularly the larger dipole model of the former (7.7 D vs. 3.0 D). In overall, A·T_1 $\Rightarrow$ A·T_3^{ts} constitutes a more preferable pathway than A·T_1 $\Rightarrow$ A·T_1^{ts}. Interestingly, the former pathway is more smooth : its slope comprises of 57 $\imath$ cm^{-1} whereas in the latter, it is equal to 1333 $\imath$ cm^{-1} [38].

Two intermediate stages, second and third, both start at A·T_1 and are distinct from each other by that, at the former, the proton moves along the N_6(A)-H$\cdots$$O_4$(T) bridge. In other words, there occurs a bending of A and T relative to each other

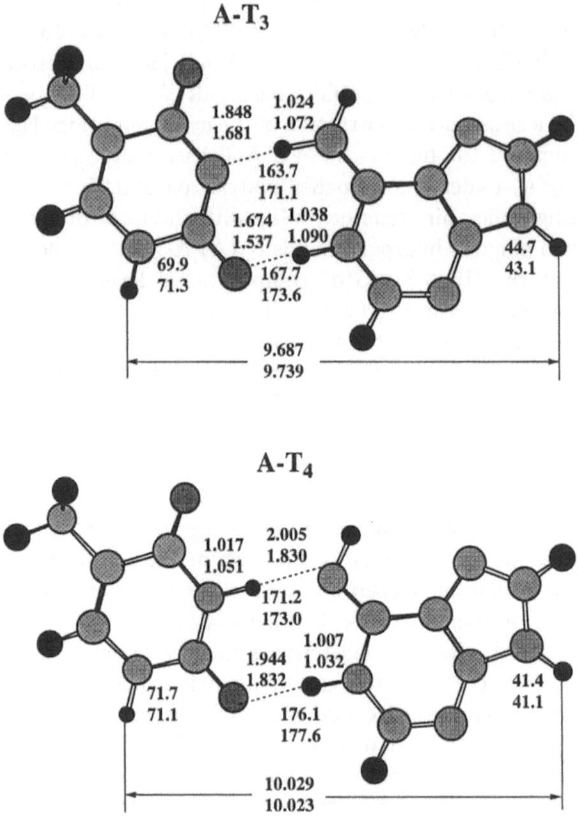

Figure 6. The mismatched A·T base pairs belonging to the mispairing valley. Distances are given in Å, angles in deg.; the order in the optimized geometrical parameters is the following: the upper entry corresponds to the HF/6-31+G(d) while the lower to the B3LYP/6-31+G(d) [38].

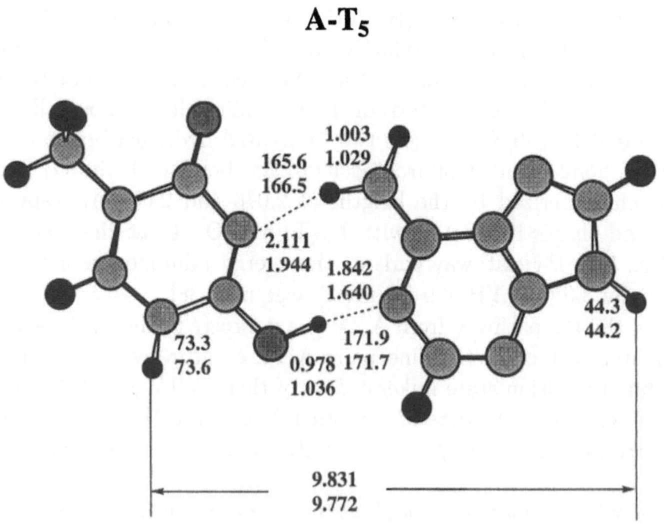

Figure 7. The A·T_5 mismatched base pair from the mispairing valley of the PES of the A·T base pair. Distances are given in Å, angles in deg.; the order in the optimized geometrical parameters is the following: the upper entry corresponds to the HF/6-31+G(d) while the lower to the B3LYP/6-31+G(d) [38].

that switches both the N_6(A)-H and N_1(A)-H bonds partly to N_3(T) and O_2(T), respectively, as shown in Figure 4. The second stage consists in that the proton surmounts the transition structure A·T_2^{ts} via migration along the N_6(A)···H···O_4(T) hydrogen bond with the activation barrier of 8.9 kcal/mol at the HF/6-31+G(d) level substantially reduced to 5.2 kcal/mol at the MP2//HF/6-31+G(d). These two stages lead to the rare tautomeric pair A·T_2 placed 19.7 (20.3), 16.0, and 14.9 (14.0) kcal/mol above the normal at the HF, MP2//HF, and B3LYP (after ZPVE) computational levels in conjunction with 6-31+G(d) basis set, respectively [38]. Notice that the HF magnitude agrees fairly well with the value of 20.1 (19.7 for ZPVE) calculated in Ref.[31b] using of HF/MIDI-1 (ZPVEs refer to the HF/MINI-1; see Table 2 in Ref.[31b]). The Gibbs free energy difference between these pairs amounts to 21.1 and 14.6 kcal/mol (HF and B3LYP, respectively) which is also in a good agreement with the value of 16.6 kcal/mol, estimated in Ref.[31c].

4.2. Mispairing Valley

The reaction pathway, covering the third and fourth stages, is the most spectacular in this scenario [38]. It starts at A·T_3^{ts} whose activation barrier amounts to 19.9 [(HF); 16.1 (MP2); 16.6 (B3LYP)] kcal/mol (without ZPVE) with respect to A·T and 4.3 (HF) kcal/mol relative to A·T_1. Its HF (B3LYP) transition frequency of 57 (114) $\imath$ cm^{-1} describes the concerted, in-plane, hindered rotations of both intramolecular hydrogen bonds of the A·T pair, which, in overall, results in the interbase shift [38]. A·T_3^{ts} possesses two bifurcated hydrogen bonds (see Figure 5). The N_6(A)-H bond forms a bifurcated hydrogen bond with the O_4 and N_3 atoms of thymine characterized by the lengths of 2.076 and 2.051 Å, respectively. The N_1(A)-H bond shares its proton with N_3(T) and O_2(T) at the distances of 2.119 and 1.837 Å. The PT pathway ends at the energy minimum structure A·T_3 lying below A·T_3^{ts} by 3.0 (B3LYP; 0.9 for MP2) kcal/mol [38].

Actually, the PT pathway from A·T_3^{ts} is a doorway to the valley which embodies three mismatched adenine-thymine pairs A·T_{3-5} naturally interconnected via the corresponding transition state linkers. Two of them A·T_4^{ts} and A·T_5^{ts}, characterized by the small activation barriers of 1.3 and 1.2 kcal/mol (B3LYP) relative to A·T_3, govern, correspondingly, the proton migrations A·T_3 $\Rightarrow$ A·T_4 and A·T_3 $\Rightarrow$ A·T_5 [38].

Let us now have a more close look at the mispairing valley. The most stable mismatched pair in this valley is A·T_4. The energy difference between the mismatched A·T_4 and the correct WC A·T pair amounts to 10.6 kcal/mol at the MP2//HF computational level and 9.6 (9.6) kcal/mol at the B3LYP one (without and with ZPVE). In terms of the Gibbs free energy, taken at room T = 298.15 K, it is less stable by 9.7 $\pm$ 2 kcal/mol than the WC pair (within the accuracy of the calculation). Two other base pairs, A·T_3 and A·T_5, are energetically higher and placed above the WC by 14.10 (12.9) and 13.6 (12.8) kcal/mol, respectively (B3LYP) [38].

Interestingly, A·T_4 is rather polar (4.4 D) compared to the WC pair (2.1 D). Its hydrogen-bond pattern is composed of the two hydrogen bonds, N_3(T)-H$\cdots$$N_6$(A) and N_1(A)-H$\cdots$$O_2$(T) with the bond lengths of 2.876 and 2.863 Å, respectively (see Figure 6). In other words, it is slightly more compact than the WC pair: the A·T_4 mispair is characterized by $R = 10.023$ Å and $\alpha_1 = 41.1^o$ and $\alpha_2 = 71.1^o$ (cf. Figure 1). We interpret this as that such mispairing minimally distorts the double-helix structural motif.

It follows from Figure 4 that A·T_5 and A·T_3 are nearly isoenergetic, although the former is neutral with the total dipole moment of 4.0 D and, in contrast, the latter is ionic (7.1 D). This explains rather short hydrogen bond lengths, 2.745 and 2.623 Å, of N_6(A)-H$\cdots$$N_3$(T) and N_1(A)-H$\cdots$$O_2$(T) in A·$T_3$, respectively (see Figures 6 and 7), and the B3LYP red shift of the N_6(A)H stretch, equal to 713 cm^{-1}, between A·T_5 and A·T_3. The proton migration along the former bond, accompanied by its further contraction by 0.16 Å, governs the ionic-to-neutral transition A·T_3 $\Rightarrow$ A·T_4. A large change of the N_6(A)$\cdots$$N_3$(T) distance, occuring during the proton transfer, demonstrates a strong coupling between the motions of the heavier subsystem (precisely, the $N_6(A)\cdots N_3(T)$) and the lighter one. Therefore, the proton transfer is actually two-dimensional [38].

4.3. Terse Intermediate Conclusions

Summarizing, we have briefly delineated two theoretical scenarios of the rearrangements of the hydrogen-bonded patterns within the G·C and A·T base pairs of DNA leading to its mismatches and thus supporting the Löwdin's mechanism of quantum tunneling of the protons. In the latter case, it mainly takes place in the valley formed by the mispairs $A{\cdot}T_{3-5}$ (see Figure 5-7). Comparing these models, it is worth noticing that the proton rearrangements in the A·T pair is actually a multi-stage process which is primarily governed by the transition state $A{\cdot}T_3^{ts}$. On the one hand, the latter is not directly related to proton transfer, rather to switching of the hydrogen-bond pattern in the ionic $A{\cdot}T_1$ pair, originating from the proton transfer along the weaker hydrogen bond $N_3(T)\text{-}H\cdots N_1(A)$ of the WC A·T pair. On the other one, this transition state separates the A·T from the ionic pair-structure $A{\cdot}T_3$ which can be stabilized in the DNA polar environment. This is in fact a feature emphasizing that the discussed mechanism of spontaneous point mutations in the A·T pair is not purely tautomeric. Interestingly, the formation of such ionic pair opens an access to the valley in the lower-energy portion of the A·T PES where mispairings occur rather easily due to the lower-barrier single proton transfers. Following Ref.[38], we demonstrate that the double proton transfer $A{\cdot}T_4 \Leftrightarrow A{\cdot}T_5$ is sequential via $A{\cdot}T_3$, and that the concerted route is much higher in energy. In terms of the Gibbs free energy taken at T = 298.15 K, the most stable mispair, $A{\cdot}T_4$, is less stable by 9.7 ± 2 kcal/mol than the WC pair, thus implying that the equilibrium constant of the tautomeric conversion $A{\cdot}T \Rightarrow A{\cdot}T_4$ falls into the interval of $10^{-6} - 10^{-9}$ [11j-k]. The latter estimate nicely matches the range of the measured DNA replication fidelity.

It is worth also mentioning the tautomeric composition of the $A{\cdot}T_{3-5}$ mispairs. It has been long believed that the favorable mispair should consist of the most preferential rare tautomers (excluding the amino-imino tautomerization of adenine which is less favorable compared to the H_9-H_7 rearrangement [42]). This is, for instance, the case of the $A^*{\cdot}T^*$ mispair (Figure 2), where thymine transforms to its enol O_4 tautomer which is the most energetically favorable within the present tautomerization scheme and energetically 11.8 kcal/mol above the normal thymine (the B3LYP/6-31+G(d,p) computational level, Ref. [15f-g]). However, the most stable $A{\cdot}T_4$ possesses the normal thymine base while $A{\cdot}T_5$, the less minor one whose B3LYP/6-31+G(d,p) relative stability amounts to 19.0 kcal/mol (see Ref. [15f-g] for the complete list of the uracil tautomers). As mentioned, the $A{\cdot}T_3$ mispair is ionic and composes of the di-deprotonated thymine.

In conclusion, it is worth juxtaposing these two mechanisms of the tautomeric conversions in G·C and A·T by presenting them schematically:

$$G \cdot C \Rightarrow G \cdot C^{ts} \Rightarrow G^* \cdot C^* \quad (1)$$

and

$$A \cdot T \Rightarrow A \cdot T_1^{ts} \Rightarrow A \cdot T_1 \Rightarrow A \cdot T_3^{ts} \Rightarrow A \cdot T_3 \Rightarrow A \cdot T_4^{ts} \Rightarrow A \cdot T_4. \quad (2)$$

Comparing these schemes, we conclude [38] that, in contrast to the single-step concerted quantum tunneling of two protons in the G·C pair, the mechanism of the hydrogen-bonded mispairing in the A·T is sequential and includes at least three stages. Following Ref. [38], we may suggest that further studies on both experimental and theoretical sides, better in concert, are highly demanding to resolve the paradigm of the existence of "tauterogens" (Ref. [13d]) in DNA and to show, if they do exist, which of them, that originating from the G·T or that from A·T tautomeric conversions, would be preferential.

5. Preopening of the A·T Base Pair

Water is the natural biological solvent and the water environment is inherent to the DNA structure (see Figure 8). How water affects the mechanism of the origin of spontaneous point tautomeric mutations in the DNA base pairs? Whether water molecules might play a role as a mediator in such processes in a sense that proton transfer from one base to another takes place along the water-mediated hydrogen bridge? Obviously, such water molecules actually "intrude" into the inter-base area. How they might be placed therein [43]? This is precisely what the concept of *preopening* of the DNA base pairs is about.

5.1. Short Introduction

Geometries of WC base pairs are known from crystallographic experiments [44, 6]. At physiological conditions, the mutual disposition of nucleobases in pairs undergoes structural fluctuations implying that base pairs in the DNA double helix may occupy a large variety of different conformational structures that indeed means that the total potential energy surface of the DNA double helix has an extremely rich landscape. Some of them may be particularly referred to the WC configuration, whose inter-base hydrogen bonds are completely formed ("closed"), and to opened

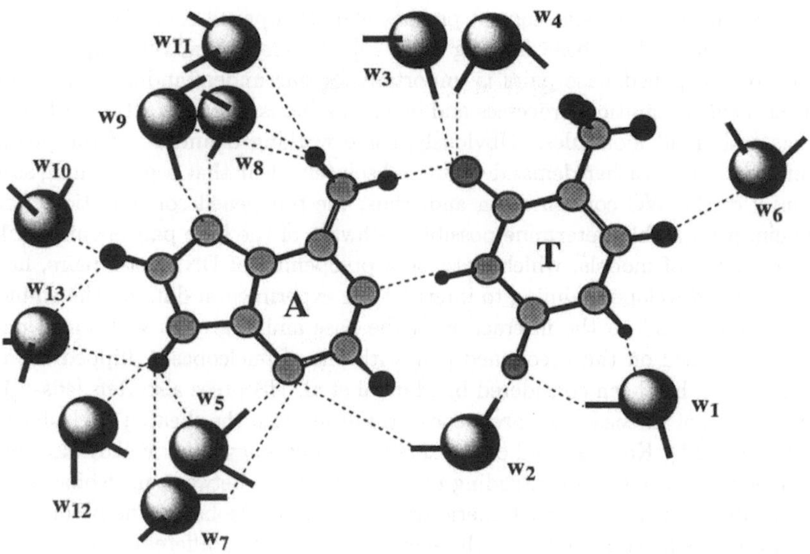

Figure 8. The Watson-Crick A·T base pair with the preferential hydration sites.

configurations with partly distorted, partly broken hydrogen bonds ("opened") [45, 46]. As well known, transription, replication, and recombination of the genome often cause significant deviations of the DNA structure and, in particular, of the hydrogen-bonded patterns of the constituent base pairs [46].

Structures of the opened configurations of DNA have not been clearly identified so far. We are aware, on the one hand, of existence of a complete opening of the base pair when all its inter-base hydrogen bonds are ruptured and the nucleobases are not stacked conventionally in the oligonucleotide strands [45]. An occurrence of such complete opening has been verified in the experiments on the interaction of the base imino groups with formaldehyde and the imino proton exchange [45, 47]. On the other hand, there is a strong evidence of that there exist partly opened (preopened) configurations of the base pair preceding its complete opening and somewhat intermediate between the WC and completely opened configurations. In the latter case, the corresponding structural changes are less frustrated and disfavor the complete rupture of the hydrogen-bonded pattern and base unstacking [45, 48]. An appearance of preopened configurations has been manifested in the experimental studies on the interaction of the base amino groups with formaldehyde and on the exchange of the amino protons with solvent protons [48]. Preopened configurations may also result in a so called "base flipping" [6c, 46]. Summarizing, this implies that the study of preopened base pairs is important for our understanding of the protein-nucleic acid recognition processes and of the mechanism of interaction of DNA with drugs and small molecules. Obviously, some working definition of the preopened configurations is rather demanded. It is absolutely clear that a preopening starts in vicinity of the WC configuration and, thus, the preopened configurations, located therein, presumably determine possible pathways of the base pair opening in DNA.

A variety of models, which describe a preopening of DNA base pairs, have recently been developed, aiming to interpret the experimental data on the amino proton exchange and on the interaction of the base amino groups with formaldehyde. A model, based on the preopened pair with one of nucleobases, flipped out of the double helix, has been considered by Mandal et al. [48c] (see also Refs.[46a-d]). Another model of the base pair preopeness, resulting from the sheared nucleobases, has been studied by Kollman and co-workers [49a]. Pechenaya and Volkov [49b] treated preopening as a result of intruding of water molecule between nucleobases. Based on possible occurrence of tautomeric transitions in nucleobases, the preopened configuration has been modelled by the sheared bases taking different tautomeric forms [49c]. Base pairs with two nucleobases, linked by a single or bifurcated hydrogen bond, have recently been treated by Leontis and Westhof [6e].

A location of the preopened base-pair configurations as stationary points on the DNA PES, characterized by a certain hydrogen-bonded pattern, is actually an unfeasible task due to the enormous number of the degrees of freedom of DNA. A typical way to resolve this problem consists in choosing pathways of a particular conformational rearrangement [50]. Some pathways, which lead to the base-pair opening, have been determined via using the empirical force field models [51]. However, this approach was unsuccessful in locating any local minima in the neighborhood of the WC configuration. In particular, such pathways have been recently examined by Lavery and co-workers [50b, 51d-f] using the AMBER potential energy function. The A·T and G·C base pairs, surrounded by water molecules, have been separated along their principal hydrogen bond axis. Lavery and co-workers revealed two minima on the PES profiles of these base pairs along the chosen coordinate. One of them corresponds to the WC configuration whereas the other, located e. g. at 5.6 Å for the A·T pair, describes actually the configuration A·w·T where water mediates the base pairing [50b]. They particularly demonstrated that the angle of base pair opening varies in the range of $34^o - 50^o$ depending on a chosen pathway and the DNA pair base bending. Interestingly, without water, such PESs profiles show a single WC minimum.

5.2. Preopening Structures of A·T Pair Mediated by Water Molecule

In Figures 9a-b we display the key, most remarkable structures of our recent works [52a-b] on the preopening of the A·T base pair mediated by water molecule. The latter intrudes into the base pair interior on the major or minor groove side and recover "loose" hydrogen bonds. In the two former structures, P-WC-w_1 and P-WC-w_2, optimized at the B3LYP/6-31+G(d) computational level, water molecule penetrates between the N_6(A) and O_4(T) atoms. Its formation is related to that, at the preliminary stage, water molecule is attached to the base pair on the side of the major groove of the double helix and forms the corresponding hydrogen bonds (see Figure 9a). Looking at both these configurations, we may conclude that two hydrogen bonds, N_3(T)-H$\cdots N_1$(A) and C_2(A)-H$\cdots O_2$(T), for the latter being shorter by ≈ 0.76 Å compared to that in the WC base pair, play the key role in its stabilization.

The third structure P-WC-w_3, displayed in Figure 9b, is formed via bonding of water molecule to the N_1(A) and N_3(T) atoms. It is somewhat buckled. Notice that it is similar to the base pair configuration reported by Poltev and Steinberg [52c] and characterized by the following geometrical parameters: $R = 12.758$ Å, $\alpha_1 = 26.4^o$, and $\alpha_2 = 30.2^o$. The formation of such configuration is indeed a two-stage process. At the first stage, water molecule attaches the double helix on the minor groove side (see Figure 8). An elongation and a further breaking of the C_2(A)-H$\cdots O_2$(T) bond take place at the second stage when water molecule intrudes into the interior of this bond, then moves to the N_3(T)-H$\cdots N_1$(A) bond, and mediates it by donating its proton to the N_1 atom of adenine and accepting the N_3-H bond of thymine.

The P-WC-w_3 configuration possesses relatively large binding energy of 8.6 kcal/mol (6.0 kcal/mol after ZPVE), comparable with that inherent to the most stable complexes of the WC A·T base pair with water [43, 52a-c]. However, comparing its geometrical parameters with the canonical WC base pair, we may notice that their formations cause larger perturbations of the sugar-phosphate backbone. We may thus conclude that the appearance of such preopened configuration is less probable. The binding energies of both P-WC-w_1 and P-WC-w_2 configurations are equal to 6.0 kcal/mol which reduces to 3.6 kcal/mol after taking ZPVE into account.

Therefore, in summary, we show the existence of three structures on the lower-energy portion of the PES of the A·T-w describing its preopeness due to the intruded water molecule. Among them, the most favorable structure is with water molecule being attached to the A·T base pair on the minor groove side of the double helix. However, less energetically configurations P-WC-w_1 and P-WC-w_2 where water molecules intrudes into the A·T interior on the major groove side resemble those inherent for the W-C configuration (see Figure 8). Hence, the insertion of water

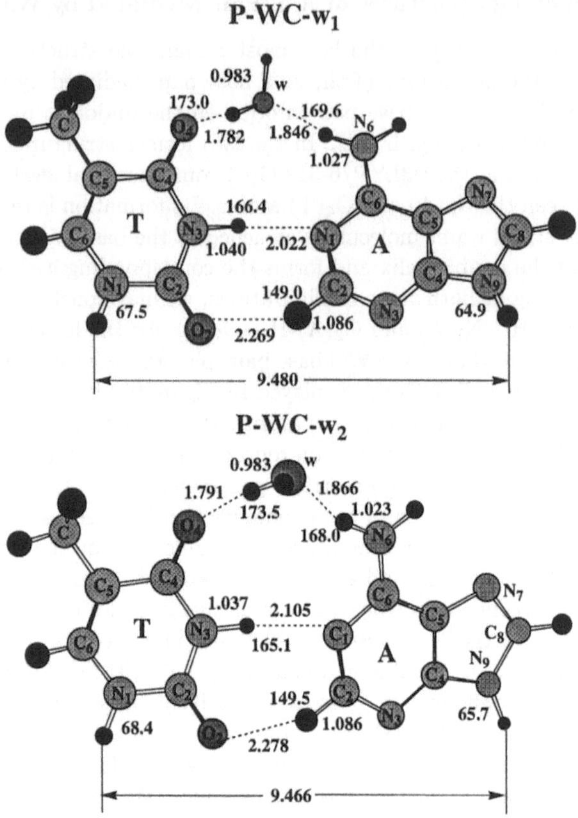

Figure 9a. Two remarkable structures of the preopening of the A·T base pair mediated by water molecule on the major groove side.

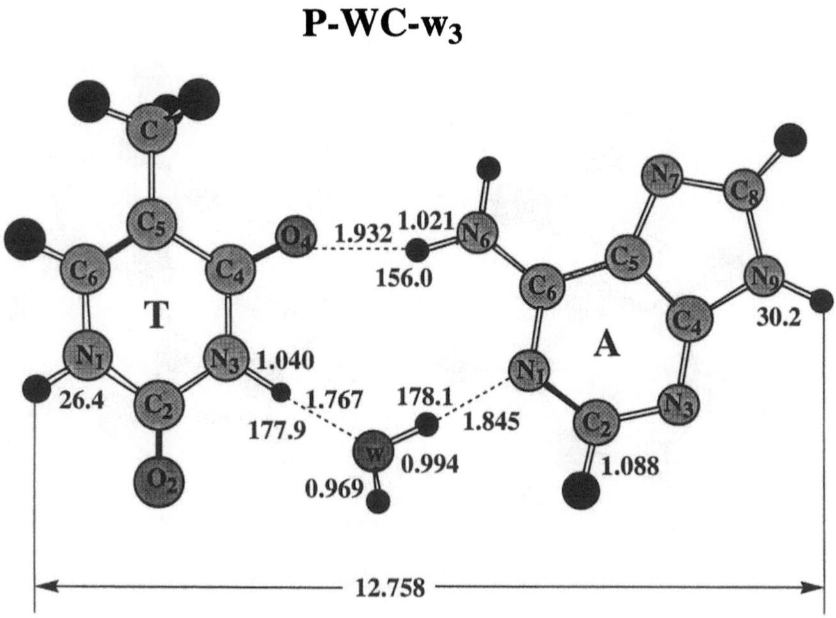

Figure 9b. The structure of the preopening of the A·T base pair mediated by water molecule on the minor groove side.

molecule into the A·T pair on the major groove side is quite probable. Interestingly, the formation of such preopened configurations is accompanied by the displacement of the amino group of adenine toward to the major groove of the double helix. We suggest that this can be tested in experiments on the interaction of probe molecules such as, e. g., formaldehyde, with the adenine amino group and on the amino proton exchange [52a].

5.3. A "Base Flipping" Pathway of A·T Pair

We have noticed in Subsection 5.1 that particular preopened configurations may result in a so called "base flipping" [6c, 46a, 53]. Flipped-out bases has been observed in the duplex DNA in a crystal structure [46b] and in NMR experiments of *B* form of DNA containing unpaired bases [46c-d]. The base flippings also occur in some damage-specific DNA repair enzymes [53]. What is the mechanism of flipping of the A·T base pair? This problem has recently been addressed in Ref. [38].

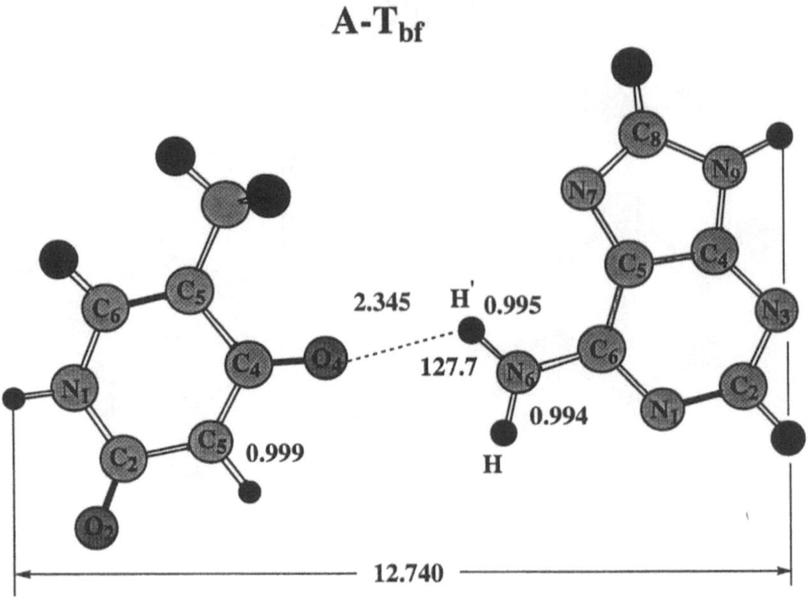

Figure 10. The "base flipping" A·T pair (HF/6-31+G(d)) [38]. The distances are given in Å, angles in deg.

As known, the WC A·T base pair is characterized by the low-frequency mode centered at 95 cm^{-1}, assigned to the bending motion of the N_6(A)-H bond and leading to its closer contact with the N_3(T)-H bond [38]. Their mutual repulsion results in flipping of the adenine base around the C_4-O_4 bond of thymine, as shown in Figure 10 where the corresponding minimum energy and planar (neglecting the known non-planarity of the amino group of adenine) "base flipping" structure A·T_{bf} is displayed. Energetically, A·T_{bf} is 6.2 kcal/mol above the WC pair at the HF/6-31+G(d) computational level, after taking the ZPVE correction into account [38]. Geometrically, the hydrogen atoms of the amine group of adenine are located by about 2.3 - 3.1 Å from the O_4 atom of thymine, whereas the actual hydrogen bond arises between the H′ atom of the amino group of adenine, with the bond length of 2.345 Å and bond angle of 127.7°.

6. Hydrogen Bonding Detour to DNArt

Would you tell me, please, which way I ought to go from here?
That depends a good deal on where you want to go to, said the Cat.
I don't much care where - said Alice
Then it does not matter which way you go, said the Cat.

Lewis Carroll, Alice's Adventures in Wonderland

We could now partly answer on the question of what is the key place of the hydrogen bonding in maintaining the DNA double helix architecture, comparing with the interstrand base-stacking and interstrand cross-stacking interactions? This fundamental paradigm, arisen since the discovery of the DNA structure, led in the mid 1980s to the concept of a "covalent base pair", introduced by Devadas and Leonard [54]. During the last two decades, this concept greatly inspired an appearance of a variety of covalently linked base pairs [54-58]. Nowadays, it has been reconsidered mainly thanks to the studies by Kool, Romesberg, and Schultz [59, 60, 6f] (see also Ref. [61]). The current work in this area has been particularly focused on the base pairing between adenine and difluorotoluene and on a design of new covalently bonded base pairs by Gao and Orgel [57] and Kishi and co-workers [58].

The concept of a covalent base pairing has been recently extended to other types of the pairing of the unnatural bases, particularly those, proposed by Schultz and Romesberg and co-workers, which are formed via hydrophobic interactions [60]. An advantage of the hydrophobic bases is first of all in that they are incapable of tautomerization and therefore, an insertion of unnatural base pairs into the DNA liaisons might significantly disfavor a mispairing. One of the most important questions in this area is how the unnatural base pairs accommodate the standard Watson-Crick architecture, without steric clashes [60, 63] or, put in other words, a *shape mimicry* based on the geometric effect or a shape complementarity, which has long been thought to be the important factor in the DNA replication fidelity [10c, 6c-d, 61a, 60]. The hypothesis of shape mimicry is the following: whether it is possible to devise such mimic pair comprising of two non-hydrogen-bonding nucleoside shape mimic for, say, adenine and thymine? It has been recently shown that the aromatic molecule difluorotoluene structurally mimics thymine whereas 4-methylbenzimidazole less perfectly mimics adenine [59g-j, 59c]. Our recent work [64] reports a high-level performance B3LYP/6-31+G(d) theoretical study of the covalently bonded base pairs, focusing on the Gao-Orgel and Kishi models, with the emphasize on the analysis of their shape complementarity within the DNA double helix architecture. It particularly shows and discusses their optimized geometries and IR spectra and demonstrates some subtle features of these base pairs, related particularly to their tautomerism, which might have an impact on a further development of the covalently bonded base-pair architecture. Finally, it offers new covalent base pairs as the analogues of the Gao-Orgel one and elaborates their properties.

6.1. Gao-Orgel Covalent Base Pair and Its Tautomers

The Gao-Orgel covalently bonded base pair GO-CP_o with the formyl-type hydrogen bond, shown in Figure 11, has been synthesized in the reaction of two nucleoside analogues, the dialdehyde and N^4-aminocytidine. The semiempirical PM3 calculation, conducted by Gao and Orgel, has demonstrated that, first, the base pair GO-CP_o closely resembles geometrically the Watson-Crick base pair, second, it is in fact planar, and third, it minimally distorts the double-helix structure motif (see Figure 1 for the definition): its C$'$ to C$'$ distance R equal to 10.5 Å and glycosyl bond angles $\alpha_1 = 48.5^o$ and $\alpha_2 = 57.1^o$ are very close to those pertaining to the double-helix DNA ($R = 10.8$ Å and both $\alpha_1 = \alpha_2 = 52^o$ [6b-c] or $\alpha_1 = \alpha_2 \approx 56^o$ [63a]; $R = 10.60 \pm 0.15$ and $\alpha_1 = \alpha_2 = 68 \pm 2^o$ [6]; $R = 10.60 \pm 0.15$ Å [6]; see also Refs. [44, 63b-c]).

The B3LYP/6-31+G(d,p) optimized structure of the Gao-Orgel covalently bonded base pair GO-CP_o is displayed in Figure 11 [64]. It is perfectly planar and characterized by the electronic energy of -979.429674 hartree. The total dipole moment of 8.5 D exceeds that of the A·U base pair by a factor of 4.7. The geometrical parameters R^{GO-CP_o}, $\alpha_1^{GO-CP_o}$, and $\alpha_2^{GO-CP_o}$ of the pair GO-CP_o are equal 10.03 Å, 42.9^o, and 58.7^o, respectively. Comparing them, for instance, with the corresponding values of R^{AU}, α_1^{AU}, and α_2^{AU} of the A·U base pair, calculated at the same level [15d], we conclude that R^{GO-CP_o} is smaller than R^{AU} by 0.09 Å and $\alpha_1^{GO-CP_o}$ by 11.5^o than α_1^{AU} whereas $\alpha_2^{GO-CP_o}$ is slightly larger (by 3.2^o) compared to the corresponding α_2^{AU}.

GO-CP_o is characterized by two intramolecular hydrogen bonds (see Figure 11): the O_1-$H_2 \cdots N_3$ with the bond length of 1.532 Å and bond angle of 148.3^o between the formyl group C_4=O_1 and the N_3 atom of the dialdehyde ring and the C_5-$H_6 \cdots N_7$ hydrogen bond whose bond length is equal to 2.089 Å and bond angle to 124.6^o. Since the hydrogen atom H_2 is bonded to the oxygen O_1, the former refers to the formyl-type hydrogen bond and is manifested in the appearance of the highly IR active vibrational mode at 2412 cm^{-1} describing the O_1-H_2 stretch.

A proton transfer along the hydrogen bond O_1-$H_2 \cdots N_3$ results in the amide tautomeric form GO-CP_1 of the covalent base pair GO-CP_o, which appears to be more stable by 5.9 kcal/mol (see Figure 11). Due to the entropy excess, equal to 6.4 cal/mol·T of the GO-CP_1 base pair over the original GO-CP_o one, ambient temperatures enhance its stabilization to 7.1 kcal/mol. The tautomerization process

GO-$CP_1 \Leftrightarrow$ GO-CP_o is accompanied by substantial changes in the geometry in the

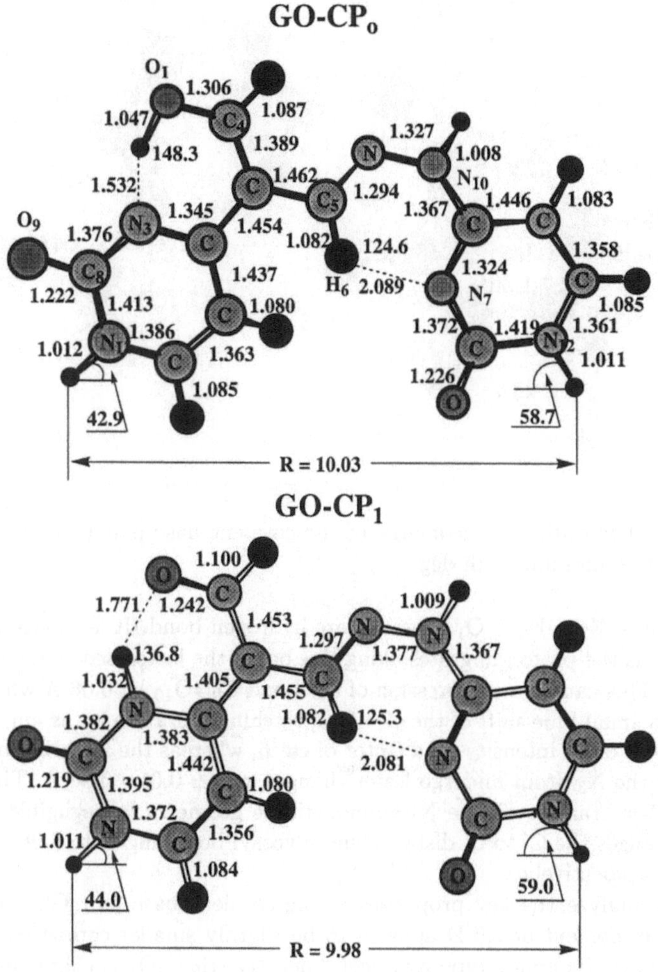

Figure 11. The optimized geometries of the covalent base pairs GO-CP_{o-1}. Distances are given in Å and angles in deg.

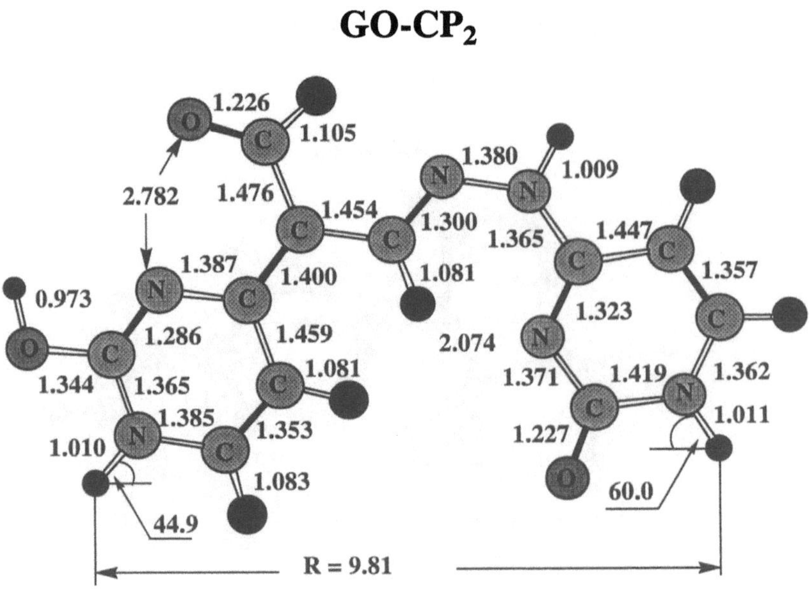

Figure 12. The optimized geometry of the covalent base pair GO-CP_2. Distances are given in Å and angles in deg.

vicinity of the $N_3 \cdots H_2 \cdots O_1$ proton-share hydrogen bond. It is shown in Figure 11 that as far as the proton migrates along this bond, the latter becomes contracted by ~ 0.14 Å. This causes a compression of the bond $C_9{=}O_{10}$ by 0.06 Å which, in turn, results in a small blue shift of the $C_4{=}O_1$ stretching vibration by 19 cm^{-1} and in an enhancement of IR intensity by a factor of ca. 5, whereas the intraring bonds in the vicinity of the N_3 atom undergo lesser changes, ca. $\pm$ 0.01 - 0.04 Å. The net effect of the proton transfer on the N_4-aminocytidine geometry is negligible although it slightly changes the C′ to C′ distance and glycosyl bond angles to 9.98 Å, and 44.0^o and 59.0^o, respectively.

Let us analyze the key properties of the covalent base pair GO-CP_1 [64]. Its total dipole moment of 7.9 D appears to be sligthly smaller compared to the GO-CP_o base pair. Its amide-type hydrogen bond $N_3{-}H_2 \cdots O_1$ is characterized by the bond length equal to 1.771 Å, the bond angle to 136.8^o, and the N-H stretching vibration predicted at 3266 cm^{-1}.

A further proton transfer from the N_3 atom to the carbonyl oxygen atom O_7 leads to the enol-type tautomer GO-CP_2 displayed in Figure 12 [64]. It is less stable by 18.8 kcal/mol and less polar by 3.7 D than the covalent base pair GO-CP_o. The GO-CP_2 tautomer has a slightly smaller C′ to C′ distance of 9.81 Å. A comparison of the geometries of the dialdehyde ring in the GO-CP_1 and GO-CP_2 base pairs demonstrates that such keto-enol tautomerism produces a substantial elongation of the C_4-O_1 bond by 0.12 Å and compression of the N_3-C_8 bond by ca. 0.1 Å.

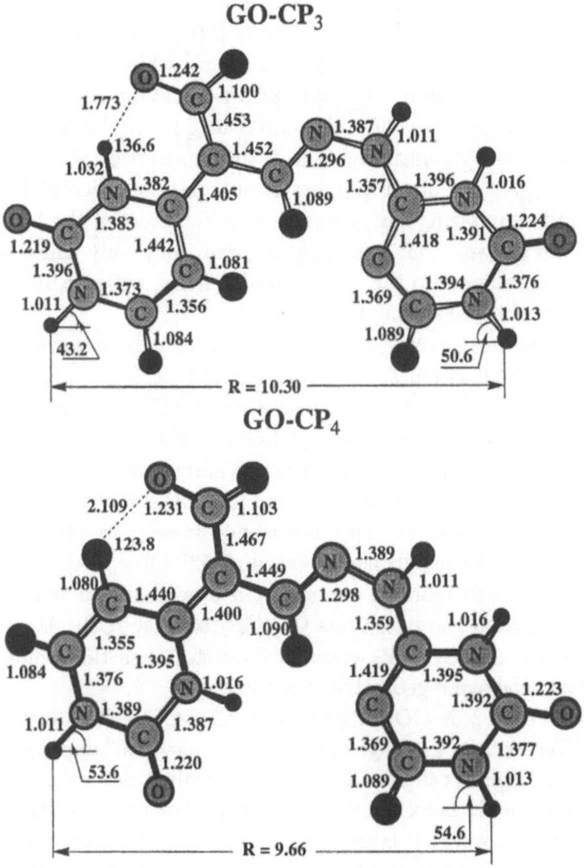

Figure 13. The optimized geometries of the covalent base pairs GO-CP_{3-4}. Distances are given in Å and angles in deg.

Let us now estimate the occurrence of the GO-CP_2 base pair using the expression for the equilibrium constant $k_{GO-CP_1 \Leftrightarrow GO-CP_2}$ of the tautomerization process GO-

$CP_1 \Leftrightarrow$ GO-CP_2: $k_{GO-CP_1 \Leftrightarrow GO-CP_2} = exp[-(\Delta H - T\Delta S)/k_B T]$ where ΔH is the difference in enthalpies of GO-CP_1 and GO-CP_2, ΔS is their corresponding entropy difference, and k_B is Boltzmann constant equal to $198.72156 \cdot 10^{-2}$ cal/mol·T. We then obtain $k_{GO-CP_1 \Leftrightarrow GO-CP_2} = 6.0 \cdot 10^{-19}$ which is extremely lower, by the eleven orders, than the estimated misinsertion frequency of the spontaneous point mutations in the conventional DNA base pairs [11n-o, 65]. Therefore, the covalently bonded base pair GO-CP_1 will not suffer of the possible GO-CP_2-type mismatches. Although, on the other hand, the equilibrium constant $k_{GO-CP_1 \Leftrightarrow GO-CP_2}$ of the tautomerization GO-$CP_1 \Leftrightarrow$ GO-CP_o, equal to $6.0 \cdot 10^{-6}$, demonstrates that the occurrence of the GO-CP_o mismatches is of the same order as that in the DNA base pairs. However, this type of mismatches is not precisely identical to that occurred in DNA [66] and, comparing the geometrical parameters of the base pairs GO-CP_1 and GO-CP_o, we suggest that such tautomerism may only cause negligibly minor perturbations with respect to the overall double helix, without, likely, any extension beyond the mismatched covalent base pair.

In our work [64], we have offered two new covalent base pairs GO-CP_3 and GO-CP_4. Their B3LYP/6-31+G(d,p) optimized and perfectly planar geometries are shown in Figure 13. GO-CP_3 is characterized by the C′ to C′ interglycosidic distance R equal to 10.30 Å and by the glycosyl bond angles $\alpha_1 = 43.2^o$ and $\alpha_2 = 50.6^o$. We may thus notice that its distance R is certainly in a better agreement with the experimental value than that of GO-CP_o and GO-CP_1. Furthermore, the bond angle α_1 is nearly the same as in the aforementioned base pairs GO-CP_o and GO-CP_1, while α_2 is closer to the experimental value of 52° [2b-c]. Altogether, this implies that GO-CP_3 has much more in common (in a geometrical sense) with the DNA double helix motif than the base pairs GO-CP_o and GO-CP_1. By analogy with the base pair GO-CP_1, the GO-CP_3 covalent base pair has the amide type hydrogen bond with nearly the same geometrical parameters.

The covalent base pair GO-CP_4 has a smaller C′ to C′ interglycosidic distance R of 9.66 Å. However, its glycosyl bond angles $\alpha_1 = 53.6^o$ and $\alpha_2 = 54.6^o$ perfectly match their experimental range and thus, such covalent pair excellently accomodates within the DNA double helix among all covalent base pairs of the Gao-Orgel type treated in the present work. It is also characterized by the lower total dipole moment equal to 2.5 D which is comparable to that of the conventional WC base pairs.

6.2. Kishi Covalently Cross-Linked Base Pair

Kishi and co-workers [58] have recently proposed the CH_2-bridged base pairs (see Figure 14) and reported their synthesis. Comparing them, for instance, with the WC A·T base pair, we notice that they possess the single hydrogen bond of the N-H···O type which, as known, is characterized by the proton-transfer barrier of more than 12 kcal/mol, and therefore, the resulting tautomeric mispair is unlikely

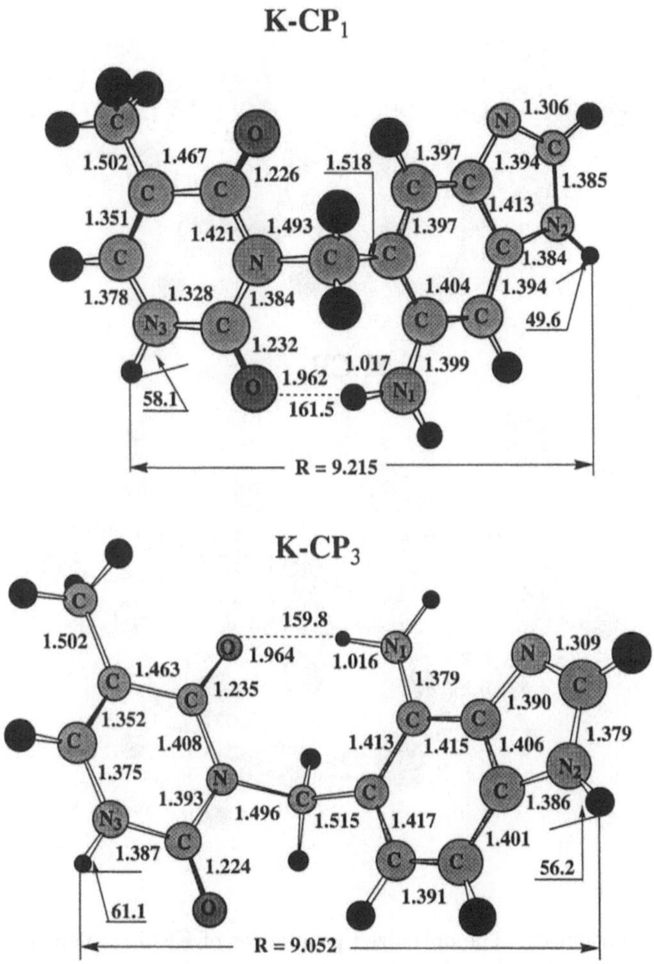

Figure 14. The optimized geometries of Kishi covalent base pairs (distances are indicated in Å and angles in deg).

K-CP$_2$

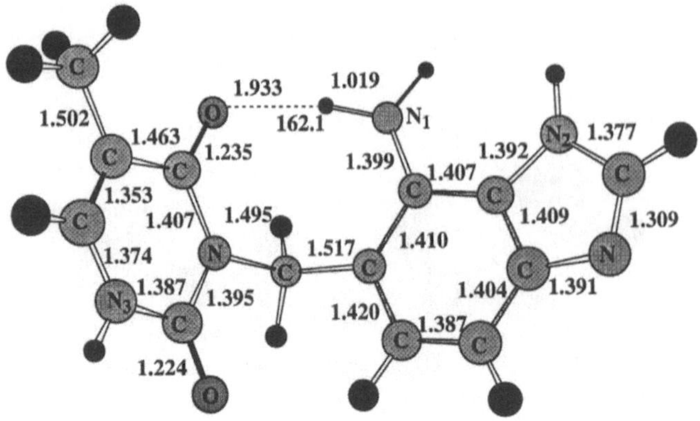

K-CP$_4$

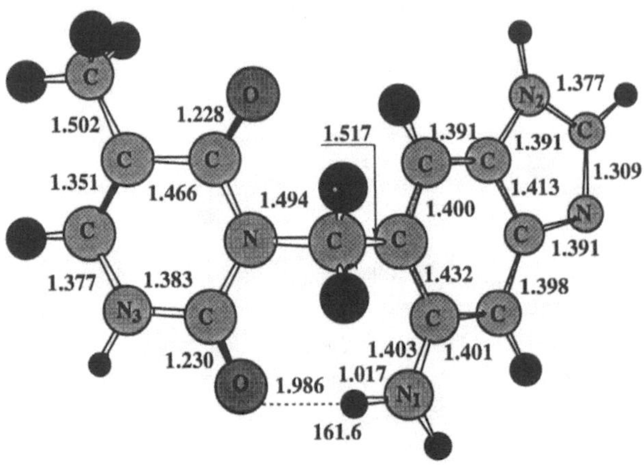

Figure 14 (continued). The optimized geometries of Kishi covalent base pairs (distances are indicated in Å and angles in deg).

to occur. Following the order of Ref. [58], Figure 14 displays the B3LYP/6-31+G(d,p) optimized geometries of four CH_2-bridged base pairs: two of type I, K-CP_1 of the WC type and K-CP_2 of the reverse WC type, and two of type II, K-CP_3 of the WC type and K-CP_4 of the reverse WC type. It is worth first of all noticing that all these pairs are non-planar, as already mentioned in Ref. [58a]. They are nearly iso-energetic, although K-CP_3 is sligthly favored over the others: by 4.5 (4.0) kcal/mol over K-CP_1 and K-CP_4 and 4.4 (3.8) kcal/mol over K-CP_2 (after ZPVE evaluated at the B3LYP/6-31G computational level) [64].

In K-CP_3, the C′ to C′ distance R is equal to 9.052 Å and glycosyl bond angles $\alpha_1 = 56.2^o$ and $\alpha_2 = 61.1^o$, that implies that a distortion of the DNA double helix motif is not so substantial. However, K-CP_1, the less energetically favorable covalent base pair, fits it better: $R = 9.215$ Å, $\alpha_1 = 49.6^o, \alpha_2 = 58.1^o$. R correlates rather well with the X-ray structural analyses reported by Qiao and Kishi [58a].

Kishi-type covalent base pairs K-CP_1 and K-CP_2 have an interesting feature which distinguishes them from Gao-Orgel pair. This is their ability to form the reverse WC base pairing which is energetically equivalent to the normal one. One of these reverse pairs, namely the K-CP_2, has the smallest hydrogen bond length equal to 1.933 Å and correspondingly, the lowest $\nu_{N-H\cdots O}$ stretch at 3502 cm^{-1} that, in overall, might likely lead to the feasible formation of its tautomeric mispairing counterpart.

6.3. Summary of the Detour

Summarizing the presented theoretical results, we conclude that the amide covalently bonded base pair GO-CP_1 of the Gao-Orgel type is more stable compared to the formyl one although, due to a larger dipole moment of the latter one, polar solvents like e. g. water favor its stronger stabilization. We also reveal that there exist some preferences of the covalent base pair GO-CP_1 over the canonical A·T and G·C pairs which are expressed particularly in that the former less suffers of possible fluctuational rotations since its rotational constant A exceeds, for instance, that of the A·U pair by a factor of 3.4. Moreover, GO-CP_1 is much stronger stabilized by a polar environment and likely favors a stacking due to a larger mean polarizability and total dipole moment, which, as shown, is superior that of the A·U base pair by a factor of nearly 4.5. Its C′ to C′ interglycosidic distance R equal to 9.98 Å and glycosyl bond angle $\alpha_2 = 59.0^o$ nearly perfectly coincide with the corresponding calculated values of the A·U pair. Nevertheless, the agreement in glycosyl bond angle $\alpha_1 = 44.0^o$, indicating a slight shift of the dialdehyde towards the minor groove, is less satisfactory. Hovewer, they fall within the range of the polymerase clamp [2, 12, 14] and, therefore, GO-CP_1 may be incorporated into the double helix without

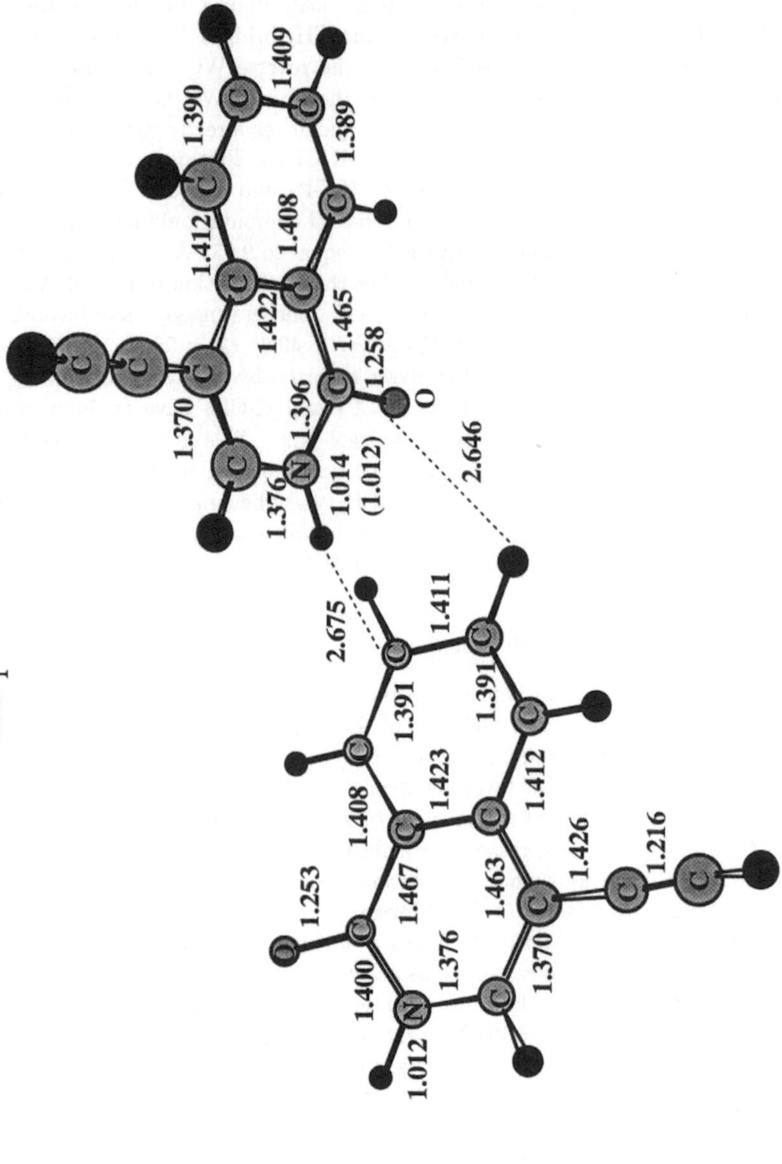
1.409
1.390
1.389
1.412
1.408
1.422
1.465
1.258
O
1.370
1.396
N
1.376
1.014
(1.012)
2.646
2.675
1.411
1.391
1.391
1.423
1.408
1.412
1.253
1.467
1.463
1.426
1.216
1.400
1.376
1.370
1.012

HB_1

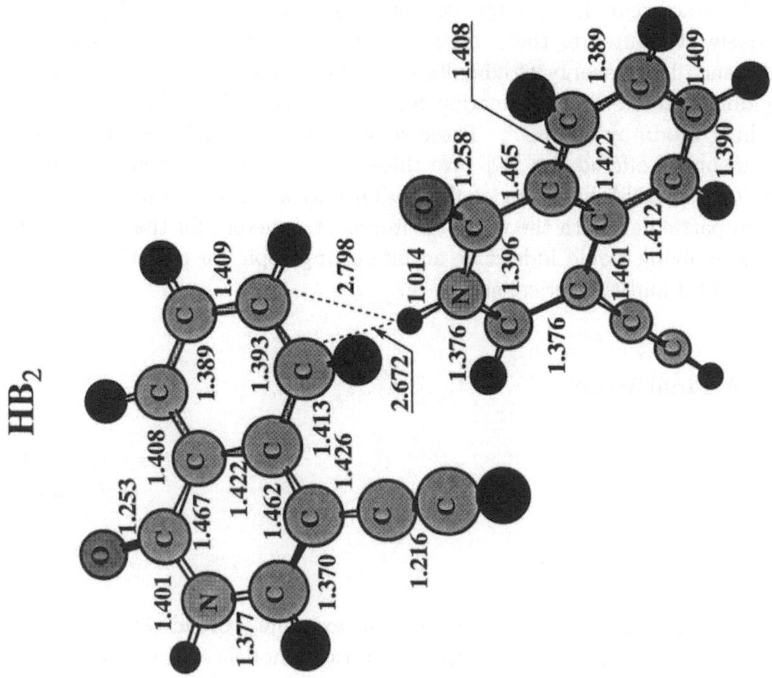

Figure 15. The B3LYP/6-31G hydrophobic base pairs. Distances are given in Å.

causing major perturbations. Also, the calculated frequencies of the covalently bonded base pairs can be used as its "fingeprints" in future experimental studies. Altogether, we indeed demonstrate a useful consistency of the Gao-Orgel and Kishi et al. experiments and a high-level computational approach toward a further exploiting of the concept of covalently bonded base pairing in the DNA double helix.

Finally, it would be worth mentioning the hydrophobic analogue of the base pair which has recently been proposed by Schultz, Romesberg and co-workers [60]. Figure 15 displays the two B3LYP/6-31G optimized structures, HB_1 and HB_2 of the 7-ethynyl isocarbostyril base pair [64] (a lower analogue of the 7-propynyl isocarbostyril base pair considered in Ref. [60]), excluding a very weak and planar van der Waals pair, characterized by the large distance $R = 11.586$ Å and by a head-to-head arrangement of the C-C≡C-H bonds. The energies of the pair formation of HB_1

and HB_2 amount to 4.1 and 4.0 kcal/mol, respectively, after ZPVE correction, that it is likely attributed to the π hydrogen bonding N-H$\cdots\pi$ and two weak C-H$\cdots$O bonds (such large energy of complexation is not consistent with the suggested hydrophobic character). Its formation results in a red shift of the corresponding ν_{NH} stretching mode by 38 cm^{-1}. These structures are actually inconsistent with the experimentally offered one [60]. We think that such inconsistency in the retention of the typical DNA base pair configuration may arise due to neglecting solvent effects, in particular, with the water environment, inherent for the DNA double helix, and its resolving would indeed be an interesting topic for a further study, on both experimental and theoretical sides.

7. A Final Word

What is must be changeable
if it is not to be all.

Theodor Adorno[1]

The present review partly "touches" the extremely broad DNA kingdom though it chiefly focuses on discussing the theoretical models of the tautomerization of the A·T and G·C base pairs which lead to mismatches in their hydrogen-bonded patterns and thereby support Löwdin's proton-tunneling mechanism of the origin of the spontaneous point tautomeric mutations. In this relation, we think that it would be premature to lead to an affirmative conclusion that the long-standing perception of that the Watson-Crick hydrogen-bonded pattern is very important for an efficient and highly selective replication [10c] has to be changed in view of a new (covalent and hydrophobic) base pair mimicry and the fact that the mismatched DNA base pairs are also hydrogen bonded, and also the results of new experiments showing a rather efficient replication of non-hydrogen-bonded DNA base pairs [59k, 60] particularly including one of the promising type of unnatural DNA base pair as 7-azaindole dimer. *How do we know that we know all what we have to know to make such a firm conclusion?* The reason is that, as we have already mentioned in Section 1, the Watson-Crick hydrogen-bonded pattern definitively means a rather specific hydrogen bonding and it is not yet clear how such pattern influences the replication

[1]Th. W. Adorno, *Negative Dialectics* (Continuum, New York, 1973), p. 398. Translated from German: "Nur wenn, was ist sich ändern lässt, ist das, was ist, nicht alles" (*Negative Dialektik* (Suhrkamp Verlag, Frankfurt am Main, 1966). I thank Christian L. Hart Nibbrig for his kind consultation.

process. On the other hand, *do we really need absolutely "clean" DNA base pairs?* Put in other words, *are the mutations always a wrong matter?*

Furthermore. It is even not absolutely clear a role of the C_2-H(A)$\cdots O_2$(T) weak hydrogen bond in the A·T base pair which we have excluded from the present treatment (recall Ref. [8]). Nevertheless, it is remarkable in a sense that it should be referred to a so-called weak "blue-shifting" hydrogen bonds [67]. The B3LYP/6-31+G(d) computational level, used in the present work, demonstrates that the C_2-H bond of adenine undergoes a contraction by 0.2 mÅ while forming the WC A·T base pair. The corresponding stretching vibration ν_{CH} becomes then blue-shifted by 6 cm^{-1} that is smaller by a factor of two than the blue shift of the C-H stretching vibration in the $CH_4 \cdot H_2O$ complex [67b, d]. Notice finally that the C_2-H(A)·O_2(T) bond length is equal to 2.937 Å that is about 0.4 Å longer than in the $CH_4 \cdot H_2O$ complex [67d]. It seems that this is actually the first computational observation of a presence of the "blue-shifting" C_2-H(A)$\cdots O_2$(T) bond in the Watson-Crick A·T base pair of DNA (cf. Ref. [67c]).

"Quantum biology" is a new field
which describes the life processes
and the functioning of the cell
on a molecular and submolecular level.

Per-Olov Löwdin [68]

Acknowledgments

I gratefully acknowledge many years of fruitful and inspiring interactions with Viktor Ya. Antonchenko, Erkki J. Brändas, Jean-Louis Calais, Enrico Clementi, Viktor I. Danilov, Alexander S. Davydov, Valeri Poltev, Jack Sabin, Steve Scheiner, Serge Volkov, and Georg Zundel. Finally, I thank Pavel Hobza and Jerzy Leszczynski for providing reprints of their related publications and Thérèse Zeegers-Huyskens and Minh Tho Nguyen for hospitality. This work was partly supported by the Research Grant of University of Leuven.

Dedication

I dedicate this work to the memory of Per-Olov Löwdin whom I met for the first time in 1978 in Kiev and since then, the subsequent 22 years were the years of our rather close "bonding", full of exciting, fruitful, and warm discussions about almost everything: the N-representability problem of reduced density matrices, density functional theory, the molecular structure, the Born-Oppenheimer approach and non-adiabaticity, the hydrogen bonding, particularly the base pairing vs. mispairing in DNA and Löwdin's mechanism of the spontaneous point mutations.

♡

References

[1] P.-O. Löwdin, Adv. Quantum. Chem. **2**, 213 (1965).

[2] (a) J. D. Watson and F. H. C. Crick, Nature **171**, 737 (1953) (reprinted in Ann. N. Y. Acad. Sci. **758**, 737 (1995), 758, 737). (b) J. D. Watson and F. H. C. Crick, Nature **171**, 964 (1953). (c) F. H. C. Crick and J. D. Watson, Proc. Roy. Soc. London **A 223**, 80 (1954). (d) J. D. Watson, *The Double Helix* (Atheneum, New York, 1968). (e) F. H. C. Crick, *What Mad Pursuit: A Personal View of Scientific Discovery* (Basic Books, New York, 1988). (f) J. D. Watson and F. H. C. Crick, J. Am. Med. Ass. **269**, 1966, 1967 (1993). (g) J. D. Watson, *Genes, Girls and Gamow* (Oxford University Press, Cambridge, 2001) (see below Ref.[4]).

[3] For their outstanding work in the discovery of the double helix structure of DNA, James D. Watson and Francis H. C. Crick shared the 1962 Nobel Prize for Physiology and Medicine with Maurice Wilkins. Rosalind Franklin, who greatly contributed to this discovery, died in 1958, and the Nobel Prize rules did not allow a Nobel Prize to be awarded posthumously.
The account of the events surrounding the discovery of DNA has just been released in the Rockefeller archive. This is the memorandum from the diary of

Gerard Roland Pomerat, who was the assistant director of the natural sciences programme at the Rockefeller Foundation, and at that time, he was at the Cavendish Laboratory in Cambridge, visiting its director Sir Lawrence Bragg. Pomerat tells that "Elizabeth, Watson's sister had typed the final version of the famous *Nature* paper on Saturday 28 March, and it was this manuscript that "Sir L. (Lawrence Bragg) today authorizes them to send." Watson says that the manuscript was mailed on Thursday 2 April; a later letter from Wilkins tells Crick that "all is safe in the hands of Gale (editor of *Nature*) who may be able to get it in April 25." (for other comments J. A. Witkowski, Nature **415**, 473 (2002)).

[4] The new book "Genes, Girls and Gamow" [2g] has been recently reviewed by H. F. Judson (Nature **413**, 775 (2001)) who tells also a story of publishing of "The Double Helix" [2d] by Atheneum although the former contract on this book was assigned with Harvard University Press.
Interestingly, Jeff Goldblum, who plays J. D. Watson in the movie, based on Watson's book [2d] and which in fact was called "Life Story" in the U.K. and "Double Helix" in the U.S., has played the science characters: The Fly in "The Fly" and Ian Malcolm in "Jurassic Park" and in "The Lost World".
Another exciting area in the DNA kingdom is DNA sequences and the origin of long-range correlations therein if they do exist (see, e. g., B. Audit, C. Thermes, C. Vaillant, Y. d'Aubenton-Carafa, J. F. Muzy and A. Arneodo, Phys. Rev. Lett. **86**, 2471 (2001); W. Li, Ibid. **86**, 5815 (2001); I. Grosse, H. Herzel, S. V. Buldyrev and H. E. Stanley, Phys. Rev. E **61**, 5624 (2000); S. Guharay, B. R. Hunt, J. A. Yorke and O. R. White, Physica D **146**, 388 (2000); E. Yeramian, Gene **255**, 139 (2000); M. Y. Azbel, Phys. Rev. Lett. **75**, 168 (1995) and references therein). Usually, DNA sequences are inhomogeneous, i. e., regions, enriched by the G and C (so-called G*p*C islands), alternate with those which are poor by G and C (known as "isochores"; see R. Román-Roldán, P. Bernaola-Galván and J. L. Oliver, Phys. Rev. Lett. **80**, 1344 (1998) and references therein). Defining A, T, G, and C bases as "words" and using the standard analysis of frequency occurrence of "words" in DNA sequences, one obtains that the G·C pairs show the largest standard deviation $\sigma = 1.43$ (M. Ortuño, P. Carpena, P. Bernaola-Galván, E. Muñoz and A. M. Somoza, Europhys. Lett. **57**, 759 (2000)). DNA sequences generate *DNA music* (Susumu Ohno; see also WWW sites: http://www.algoart.com/dnamusic/; http://wildcat.arizona.edu//papers/89/89/06_1_m.html, http://linkage.rockefeller.edu/wli/dna_corr/music.html). For a current review see J. Maruani, R. Lefebvre and M. Rantanen, *Science and Music, A Unifying Concept: From the Music of the Depths to the Music of the*

Spheres, presented at the Annual General Meeting of the European Academy in Brussels, Belgium, September 2000 and at the ICTCP-IV Congress in Marly-le-Roi, France, July 2002.

[5] R. Langridge, H. R. Wilson, C. W. Hooper and M. H. F. Wilkins, J. Mol. Biol. **2**, 19 (1960).

[6] (a) W. Saenger, *Principles of Nucleic Acid Structure* (Springer, Berlin, 1984). (b) R. R. Sinden, *DNA Structure and Function* (Academic, San Diego, 1994). (c) *Nucleic Acids in Chemistry and Biology*, G. M. Blackburn and M. J. Gait (Eds.) (Oxford University Press, Oxford, 1996), p. 21. (d) A. Kornberg and T. A. Baker, *DNA Replication* (Freeman, New York, 1992). (e) N. Leontis and E. Westhof, Quart. Rev. Biophysics **31**, 399 (1998). (f) L. Wang and P. G. Schultz, Chem. Commun. 1 (2002).

[7] P. Schuster and P. Wolschann, Monatsh. Chem. **130**, 947 (1999) and references therein.

[8] The C(A)-H···O(T) hydrogen bond is rather weak: (a) R. D. Green, *Hydrogen Bonding by C-H Groups* (MacMillan, London, 1974). (b) G. R. Desiraju and T. Steiner, *The Weak Hydrogen Bond in Structural Chemistry and Biology* (Oxford University Press, New York, 1999). See discussion in Section 7.

[9] (a) G. C. Pimentel and A. L. McClellan, *The Hydrogen Bond* (Freeman, San Francisco, 1960). (b) *The Hydrogen Bond. Recent Developments in Theory and Experiments*, P. Schuster, G. Zundel and C. Sandorfy (Eds.) (North-Holland, Amsterdam, 1976). (c) S. Scheiner, *Hydrogen Bonding - A Theoretical Perspective* (Oxford University Press, New York, 1997). (d) J. E. Del Bene, in *The Encyclopedia of Computational Chemistry*, P. v. R. Schleyer, N. L. Allinger, T. Clark, J. Gasteiger, P. A. Kollman, H. F. Schaefer III and P. R. Schreiner (Eds.) (Wiley, Chichester, 1998), Vol. 2, p. 1263. (e) J. E. Del Bene and M. J. T. Jordan, *Int. Rev. Phys. Chem.* **18**, 119 (1999). (f) J. E. Del Bene, in *Recent Theoretical and Experimental Advances in Hydrogen Bonded Clusters, NATO ASI Series C, Vol. 561*, S. S. Xantheas (Ed.) (Kluwer, Dordrecht, 2000), p. 309.

[10] (a) J. W. Drake, *The Molecular Basis of Mutation* (Holden-Day, San Francisco, 1970). (b) C. Auerbach, *Mutations Research. Problems, Results and Perspectives* (Chapman and Hall, London, 1976). (c) J. D. Watson, N. H. Hopkins, J. W. Roberts, J. A. Steitz and A. M. Weiner, *Molecular Biology of the Gene* (Benjamin/Cummings, Menlo Park, 1987).

[11] (a) P. D. Lawley and P. Brookes, Nature **192**, 1081 (1961) (b) P. D. Lawley and P. Brookes, P. J. Mol. Biol. **4**, 216 (1962). (c) J. Ladik, *Quantenbiochemie für*

Chemiker und Biologen (Akadémiai Kiadó, Budapest, 1972). (d) L. C. Sowers, B. R. Shaw, M. L. Veigl and W. D. Sedwick, Mutation Res. **177**, 201 (1987). (e) M. D. Topal and J. R. Fresco, Nature **263**, 285 (1986). (f) M. D. Topal and J. R. Fresco, Nature **263**, 289 (1986). (g) V. I. Danilov and G. F. Kventsel, *Electronic Representation in the Point Mutation Theory* (Naukova Dumka, Kiev, 1971). (h) R. Rein, M. Shibata, R. Garduno-Juarez and T. Kieber-Emmons, in *Structure and Dynamics: Nucleic Acids and Proteins*, E. Clementi and R. H. Sarma (Eds.) (Adenine Press, New York, 1983), p. 269. (i) M. J. Lutz, J. Horlacher and S. A. Benner, Bioorg. Med. Chem. Lett. **8**, 1149 (1998). (j) B. G. Hall, Genetics **126**, 5 (1990). (k) B. G. Hall, FEMS Microbiol. Lett. **117**, 237 (1994). (l) P. L. Foster and J. Cairns, Genetics **131**, 783 (1992). (m) For a hypothetical quantum model of direct mutations see V. V. Ogryzko, BioSystems **43**, 83 (1997) (see also A. Goswami and D. Todd, Integrative Physiol. Behav. Sci. **32**, 132 (1997) and referencies therein). (n) J. Cole, and T. R. Skopek, Mutation Res. **304**, 364 (1996). (o) B. L. Parsons and R. H. Heflich, Ibid. **387**, 97 (1997).

[12] (a) J. S. Kwiatkowski, T. J. Zielinski and R. Rein, Adv. Quantum Chem. **18**, 85 (1986). (b) J. S. Kwiatkowski and W. B. Person, in *Theoretical Biochemistry and Molecular Biophysics. Vol. 1: DNA*, D. Beveridge and R. Lavery (Eds.) (Adenine Press, New York, 1990), p. 153. (c) H. Echols and M. F. Goodman, Annu. Rev. Biochem. **60**, 477 (1991).

[13] (a) J. S. Kwiatkowski and B. Pullman, Adv. Heterocycl. Chem. **18**, 199 (1975). (b) W. B. Person, K. Szczepaniak, M. Szczepaniak, J. S. Kwiatkowski and R. J. Czerminski, J. Mol. Struct. **194**, 239 (1989). (c) J. Šponer, J. Leszczynski and P. Hobza, J. Biomol. Struct. Dyn. **14**, 117 (1996). (d) P. Strazéwski and C. Tamm, Angew. Chem. Int. Ed. Engl. **29**, 36 (1990). (e) S. Steenken, Chem. Biol. **378**, 1293 (1997).

[14] (a) J. S. Kwiatkowski and J. Leszczynski, Int. J. Quantum Chem. **61**, 453 (1997). (b) V. I. Poltev, A. V. Teplukhin and J. S. Kwiatkowski, J. Biomol. Struct. Dyn. **9**, 747 (1992).

[15] (a) P. Hobza and J. Šponer, Chem. Rev. **99**, 3247 (1999). (b) S. Morpurgo, M. Bossa and G. O. Morpurgo, Adv. Quantum Chem. **36**, 169 (2000). (c) J. Leszczynski, Adv. Mol. Struct. Res. **6**, 209 (2000). (d) E. S. Kryachko and M. T. Nguyen, Adv. Quantum Chem. **40**, 79 (2001). (e) E. S. Kryachko, M. T. Nguyen and Th. Zeegers-Huyskens, J. Phys. Chem. A **105**, 3379 (2001). (f) E. S. Kryachko, M. T. Nguyen and Th. Zeegers-Huyskens, J. Phys. Chem. A **105**, 1288 (2001). (g) E. S. Kryachko, M. T. Nguyen and Th. Zeegers-Huyskens,

J. Phys. Chem. A **105**, 1934 (2001). (h) E. S. Kryachko, M. T. Nguyen and Th. Zeegers-Huyskens, Chem. Phys. **264**, 21 (2001). (i) S.-I. Kawahara, T. Uchimaru and M. Sekine, J. Mol. Struct. (Theochem) **530**, 109 (2000).

[16] S. F. Luria and M. Delbrück, Genetics **28**, 491 (1943).

[17] P.-O. Löwdin, Rev. Mod. Phys. **35**, 724 (1963).

[18] P.-O. Löwdin, in *Electronic Aspects of Biochemistry*, B. Pullman (Ed.) (Academic, New York, 1964), p. 167.

[19] P.-O. Löwdin, Biopolymers Symp. **1**, 161 (1964).

[20] P.-O. Löwdin, Biopolymers Symp. **1**, 293 (1964).

[21] P.-O. Löwdin, Mutation Res. **2**, 218 (1965).

[22] P.-O. Löwdin, Pont. Acad. Sci. Scripta Varia **31**, 323 (1967).

[23] P.-O. Löwdin, Int. J. Quantum Chem.: Quantum Biol. Symp. **13**, 311 (1986).

[24] Y. S. Kong, M. S. Jhon and P.-O. Löwdin, Int. J. Quantum Chem.: Quantum Biol. Symp. **14**, 189 (1987).

[25] E. Schrödinger, *What Is Life? The Physical Aspect of the Living Cell* (Cambridge University Press, London, 1945).

[26] O. T. Avery, C. M. McLeod and M. McCarty, J. Exptl. Med. **79**, 137 (1944).

[27] M. Delbrück, N. W. Timoféëff and K. G. Zimmer, Nachr. Akad. Wiss. Göttingen Biol.-Physiol.-Chem. Abt. **1**, 189 (1935).

[28] (a) J. Ladik, J. Theor. Biol. **6**, 201 (1964). (b) R. Rein and J. Ladik, J. Chem. Phys. **40**, 2466 (1964). (c) R. Rein and F. E. Harris, Ibid. **41**, 3393 (1964). (d) R. Rein and F. E. Harris, Ibid. **42**, 2177 (1965). (e) R. Rein and F. E. Harris, Ibid. **43**, 4415 (1965). (f) S. Lunell and G. Sperber, Ibid. **46**, 2119 (1967).

[29] (a) E. Clementi, J. Mehl and W. von Niessen, J. Chem. Phys. **54**, 508 (1971). (b) E. Clementi, G. Corongiu, J. Detrich, S. Chin and L. Domingo, Int. J. Quantum Chem.: Quantum Chem. Symp. **18**, 601 (1984).

[30] (a) S. Scheiner and C. W. Kern, Chem. Phys. Lett. **57**, 331 (1978). (b) S. Scheiner and C. W. Kern, J. Am. Chem. Soc. **101**, 4081 (1979).

[31] (a) V. Hrouda, J. Florián and P. Hobza, J. Phys. Chem. **97**, 1542 (1993). (b) J. Florián, V. Hrouda and P. Hobza, J. Am. Chem. Soc. **116**, 1457 (1994). (c) J. Florián and J. Leszczynski, J. Am. Chem. Soc. **118**, 3010 (1996). (d) D. Barsky and M. E. Colvin, J. Phys. Chem. A **104**, 8570 (2000).

[32] (a) A. O. Colson and M. D. Sevilla, Int. J. Radiat. Biol. **67**, 627 (1995). (b) A. O. Colson, B. Besler and M. D. Sevilla, J. Phys. Chem. **96**, 9787 (1992).

[33] P. Rosen, Int. J. Quantum Chem. **9**, 473 (1975).

[34] (a) W. G. Cooper, Int. J. Quantum Chem. **14**, 71 (1978). (b) W. G. Cooper, Int. J. Quantum Chem.: Quantum Biol. Symp. **5**, 463 (1978). (c) M. D. Sevilla, D. Becker, M. Yan and S. R. Summerfield, J. Phys. Chem. **95**, 3409 (1991). (d) D. Becker and M. D. Sevilla, Adv. Radiat. Biol. **17**, 121 (1993). (e) E. D. A. Stemp, M. R. Arkin and J. K. Barton, J. Am. Chem. Soc. **119**, 2921 (1997). (f) R. Itoh, Progr. Theor. Phys. Suppl. **17**, 69 (1961).

[35] (a) J. Eisinger, Photochem. Photobiol. **7**, 597 (1968). (b) A. Lamola, Ibid. **7**, 619 (1968). (c) M. Kasha, P. Horowitz and M. El-Bayoumi, in *Molecular Spectroscopy, Modern Research*, K. N. Rao and C. W. Mathews (Eds.) (Academic, New York, 1972), Chapter 6.

[36] (a) J. Šponer, M. Sabat, J. V. Burda, J. Leszczynski, P. Hobza and B. Lippert, J. Biol. Inorg. Chem. **4**, 537 (1999). (b) J. Müller, R. K. O. Sigel and B. Lippert, J. Inorg. Biochem. **79**, 261 (2000) and references therein. (c) J. Šponer, J. Leszczynski and P. Hobza, Biopolymers (Nucleic Acid Sci.) **61**, 3 (2002) and references therein.

[37] (a) C. A. Taylor, M. A. El-Bayoumi and M. Kasha, Proc. Natl. Acad. Sci. USA. **63**, 253 (1969). (b) J. Catalán and M. Kasha, J. Phys. Chem. A **104**, 10812 (2000). (c) D. E. Folmer, L. Poth, E. S. Wisniewski and A. W. Castleman, Jr. Chem. Phys. Lett. **287**, 1 (1998). (d) T. Fiebig, M. Chachisvilis, M. Manger, A. H. Zewail, A. Douhal, I. Garcia-Ochoa and A. de La Hoz Ayuso, J. Phys. Chem. A **103**, 7419 (1999) and references therein.

[38] E. S. Kryachko and J. R. Sabin, Int. J. Quantum Chem. (submitted).

[39] (a) I. Yanson, A. Teplitsky, and L. Sukhodub, Biopolymers **18**, 1149 (1979). (b) W. L. Jorgensen and co-workers (W. L. Jorgensen and J. Pranata, J. Am. Chem. Soc. **112**, 2008 (1990); J. Pranata, S. G. Wierschke and W. L. Jorgensen, Ibid. **113**, 2810 (1991)) showed that the 9-methyl-G·1-methyl-C is stronger by 10.7 kcal/mol than 1-methyl-uracil·2,6-diaminopyridine.

[40] (a) I. R. Gould and P. A. Kollman, J. Am. Chem. Soc. **116**, 2493 (1994). (b) K. Brameld, S. Dasgupta and W. A. Goddard III, J. Phys. Chem. B **101**, 4851 (1997) and references therein.

[41] (a) G. A. Patikoglou, J. L. Kim, L. Sun, S.-H. Yang, T Kodadek and S. K. Burley, Genes Dev. **13**, 3217 (1999). (b) Z. S. Juo, T. K. Chiu, P. M. Leiberman, I. Baikalov, A. J. Berk and R. E. Dickerson, J. Mol. Biol. **261**, 239 (1996). (c) J. Moreau, M. F. Maschat, J. Kejlarova-Lepesant, J.-A. Lepesant and K. Scherrer, Nature **295**, 239 (1982). (d) T. Oosumi, B. Garlick and W. R. Belknap, Proc. Natl. Acad. Sci. USA **92**, 8886 (1995). (e) N. G. A. Abrescia, A. Thompson, T. Huynh-Dinh and J. A. Subirana, Ibid. **99**, 2806 (2002).

[42] J. Gu and J. Leszczynski, J. Phys. Chem. A **103**, 2744 (1999).

[43] (a) E. Clementi and G. Corongiu, J. Chem. Phys. **72**, 3979 (1980) and references therein. (b) M. Mezei, J. Am. Chem. Soc. **118**, 8493 (1996) and references therein. (c) D. L. Beveridge and K. J. McConnell, Curr. Opin. Struct. Biol. **10**, 182 (2000) and references therein. (d) E. T. Kool, Annu. Rev. Biophys. Biomol. Struct. **30**, 1 (2001), p. 5*ff* and references therein.

[44] (a) M. Spencer, Acta Cryst. **12**, 66 (1959). (b) S. Arnott, Progr. Biophys. Mol. Biol. **21**, 267 (1970). (c) N. C. Seeman, J. Rosenberg, F. L. Suddath, J. J. P. Kim and A. Rich, J. Mol. Biol. **104**, 145 (1976).

[45] (a) C. R. Cantor and P. R. Schimmel, *Biophysical Chemistry* (Freeman, New York, 1980), Ch. 22.4. (b) M. D. Frank-Kamenetskii, Molek. Biol. (USSR) **17**, 639 (1983). (c) S. W. Englander and N. R. Kallenbach, Quart. Rev. Biophys. **16**, 521 (1984).

[46] (a) D. M. F. van Aalten, D. A. Erlanson, G. L. Verdine and L. Joshua-Tor, Proc. Natl. Acad. Sci. USA **96**, 11809 (1999). (b) L. Joshua-Tor, F. Frolow, E. Appella, H. Hope, D. Rabinovich and J. L. Sussman, J. Mol. Biol. **225**, 397 (1992). (c) K. M. Morden, Y. G. Chu, F. H. Martin and I. Tinoco, Jr., Biochemistry **22**, 5557 (1983). (d) M. W. Kalnik, D. G. Norman, M. G. Zagorski, P. F. Swann and D. J. Patel, Biochemistry **28**, 294 (1989). (e) J. Pérez-Martín and M. Espinosa, Science **260**, 805 (1993). (f) D. M. Crothers, M. R. Gartenberg and T. E. Shrader, Methods Enzym. **208**, 118 (1991). (g) S. A. Wolfe and G. L. Verdine, J. Am. Chem. Soc. **115**, 12585 (1993) and references therein.

[47] M. Gueron, M. Kochoyan and J.-L. Leroy, Nature **328**, 89 (1987).

[48] (a) H. Teitelbaum and S. W. Englander, J. Mol. Biol. **92**, 55 (1975). (b) A. V. Lukashin, A. V. Vologodskii, M. D. Frank-Kamenetskii and Yu. L. Lyubchenko,

J. Mol. Biol. **108**, 665 (1976). (c) C. Mandal, N. R. Kallenbach and S. W. Englander, J. Mol. Biol. **135**, 391 (1979). (d) V. V. Demidov and Yu. S. Lazurkin, Molek. Biol. (USSR) **14**, 448 (1980). (e) V. V. Demidov, Dokl. Akad. Nauk USSR **251**, 1268 (1980). (f) Y. Hayashi, M. Nakanishi, M. Tsuboi, I. Tazawa and Y. Inoke, Biochim. Biophys. Acta **656**, 120 (1981). (g) R. S. Preisler, C. Mandal, S. W. Englander, N. R. Kallenbach, J. Frazier, N. T. Miles and F. B. Howard, Biopolymers **23**, 2099 (1984).

[49] (a) J. W. Keepers, P. A. Kollman, P. K. Weiner and T. L. James, Proc. Natl. Acad. Sci. USA **79**, 5537 (1982). (b) V. I. Pechenaya and S. N. Volkov, Molek. Biol. (USSR) **18**, 1134 (1984). (c) S. N. Volkov, Ibid. **29**, 1086 (1995).

[50] (a) R. Lavery, Adv. Comput. Biol. **1**, 69 (1994). (b) E. Stofer, C. Chipot and R. Lavery, J. Am. Chem. Soc. **121**, 9503 (1999). (c) E. Giudice, P. Várnal and R. Lavery, CHEMPHYSCHEM **2**, 673 (2001).

[51] (a) J. W. Keepers, P. A. Kollman and T. L. James, Biopolymers **23**, 2499 (1984). (b) V. I. Poltev and N. V. Shulyupina, Molek. Biol. (USSR) **18**, 1549 (1984). (c) V. I. Poltev and N. V. Shulyupina, J. Biomol. Struct. Dyn. **3**, 739 (1986). (d) J. Ramstein and R. Lavery, Proc. Natl. Acad. Sci. USA **85**, 7231 (1988). (e) J. Ramstein and R. Lavery, J. Biomol. Struct. Dyn. **7**, 915 (1990). (f) J. Bernet and R. Lavery, J. Mol. Struct. (Theochem) **398-399**, 473 (1997). (g) P. Cieplak and P. A. Kollman, J. Am. Chem. Soc. **110**, 3734 (1988). (h) L. X. Dang and P. A. Kollman, Ibid. **112**, 503 (1990).

[52] (a) E. S. Kryachko and S. N. Volkov, Int. J. Quantum Chem. **82**, 193 (2001). (b) E. S. Kryachko, J. Phys. Chem. B **106**, 000 (2002). (c) V. I. Poltev and S. V. Steinberg, Molek. Biol. **21**, 704 (1987).

[53] C. D. Mol, S. S. Parikh, C. D. Putnam, T. P. Lo and J. A. Tainer, Annu. Rev. Biophys. Biomol. Struct. **28**, 101 (1999).

[54] (a) B Devadas and N. J. Leonard, J. Am. Chem. Soc. **108**, 5012 (1986). (b) B. Devadas and N. J. Leonard, Ibid. **112**, 3125 (1990). (c) see also B. Bhat, N. J. Leonard, H. Robinson and A. H.-J. Wang, Ibid. **118**, 10744 (1996).

[55] (a) T. R. Webb and M. D. Matteucci, J. Am. Chem. Soc. **108**, 2764 (1986). (b) A. L. Weber, Origins Life Evol. Biosphere **19**, 179 (1989). (c) A. E. Ferentz and G. L. Verdine, J. Am. Chem. Soc. **113**, 4000 (1991). (d) M. Cowart and S. J. Benkovic, Biochemistry **30**, 788 (1991). (e) M. A. Mitchell, R. C. Kelly, N. A. Wicnienski, N. T. Hatzenbuhler, M. G. Williams, G. L. Petzold, J. L. Slightom, and D. R. Siemieniak, J. Am. Chem. Soc. **113**, 8994 (1991). (f) I. Dieter-Wurm,

M. Sabat and B. Lippert, Ibid. **114**, 357 (1992). (g) R. W. Armstrong, M. E. Salvati and M. Nguyen, Ibid. **114**, 3144 (1992). (h) T. Ohtsuki, M. Kimoto, M. Ishikawa, T. Mitsui, I. Hirao and S. Yokoyama, Proc. Natl. Acad. Sci. USA **98**, 4922 (2001).

[56] (a) J. J. Kirchner, S. T. Sigurdsson and P. B. Hopkins, J. Am. Chem. Soc. **114**, 4021 (1992). (b) H. Huang, M. S. Solomon and P. B. Hopkins, Ibid. **114**, 9240 (1992). (c) F. C. Seaman and L. Hurley, Biochemistry **32**, 12577 (1993). (d) S. E. Osborne, R. J. Cain and G. D. Glick, J. Am. Chem. Soc. **119**, 1172 (1997).

[57] K. Gao and L. E. Orgel, Proc. Natl. Acad. Sci. USA **96**, 14837 (1999).

[58] (a) X. Qiao and Y. Kishi, Angew. Chem. **111**, 977 (1999); Angew. Chem. Int. Ed. **38**, 928 (1999). (b) Y.-L. Qiu, H.-Y. Li, G. Topalov and Y. Kishi, Tetrahedron Lett. **41**, 9425 (2000). (c) H.-Y. Li, Y.-L. Qiu, E. Moyroud and Y. Kishi, Angew. Chem. Int. Ed. **40**, 1471 (2001).

[59] (a) S. Moran, X.-F. R. Rex, S. Rumney and E. T. Kool, J. Am. Chem. Soc. **119**, 2056 (1997). (b) J. C. Morales and E. T. Kool, Nature Struct. Biol. **5**, 950 (1998). (c) K. M. Guckian, T. K. Krugh and E. T. Kool, Ibid. **5**, 954 (1998). (d) D. Barsky, E. T. Kool and M. E. Colvin, J. Biomol. Struct. Dyn. **16**, 1119 (1999). (e) E. T. Kool, J. C. Morales and K. M. Guckian, Angew. Chem. Int. Ed. **39**, 990 (2000). (f) K. M. Guckian, T. R. Krugh and E. T. Kool, J. Am. Chem. Soc. **122**, 6841 (2000). (g) E. T. Kool, Annu. Rev. Biophys. Biomol. Struct. **30**, 1 (2001) and references therein. (h) K. M. Guckian, B. A. Schweitzer, R. X. F. Ren, C. J. Sheils, P. L. Paris, D. C. Tahmassebi, and E. T. Kool, J. Am. Chem. Soc. **118**, 8182 (1996). (i) B. A. Schweitzer and E. T. Kool, J. Org. Chem. **59**, 7238 (1994). (j) B. A. Schweitzer and E. T. Kool, J. Am. Chem. Soc. **117**, 1863 (1995). (k) T. J. Matray and E. T. Kool, Nature **399**, 704 (1999). See also: M. Meyer and J. Sühnel, J. Biomol. Struct. Dyn. **15**, 619 (1997); X. Wang and K. N. Houk, Chem. Commun. 2631 (1998); E. Cubero, E. C. Sherer, F. J. Luque, M. Orozco and C. A. Laughton, J. Am. Chem. Soc. **121**, 8653 (1999); E. C. Sherer, S. J. Bono and G. C. Shields, J. Phys. Chem. B **105**, 8445 (2001); C&EN March 3, 9 (1997); Ibid. November 9, 69 (1998).

[60] (a) D. L. McMinn, A. K. Ogawa, Y. Wu, J. Liu, P. G. Schultz and F. E. Romesberg, J. Am. Chem. Soc. **121**, 11585 (1999). (b) A. K. Ogawa, Y. Wu, D. L. McMinn, J. Liu, P. G. Shultz and F. E. Romesberg, Ibid. **122**, 3274 (2000). (c) E. L. Tae, Y. Wu, G. Xia, P. G. Schultz and F. E. Romesberg, Ibid. **123**, 7439 (2001). See also: R. F. Service, Science **289**, 232 (2000); C&EN August 6, 32 (2001).

[61] (a) M. F. Goodman, Proc. Natl. Acad. Sci. USA **94**, 10493 (1997). (b) U. Diederichsen, Angew. Chem. Int. Ed. **37**, 1655 (1998).

[62] (a) M. J. Lutz, H. A. Held, M. Hottinger, U. Hübscher and S. A. Benner, Nucleic Acid Res. **24**, 1308 (1996). (b) E. T. Kool, Biopolymers **48**, 3 (1998).

[63] (a) R. E. Dickerson, Methods Enzymol. **211**, 67 (1992). (b) A. Voet and A. Rich, Progr. Nucl. Acid Res. Mol. Biol. **10**, 183 (1970). (c) J. M. Rosenberg, N. C. Seeman, R. O. Day and A. Rich, J. Mol. Biol. **104**, 145 (1976).

[64] E. S. Kryachko and M. T. Nguyen, J. Phys. Chem. B **106**, 000 (2002).

[65] L. V. Mendelman, M. S. Boosalis, J. Petruska and M. F. Goodman, J. Biol. Chem. **264**, 14415 (1989).

[66] (a) A. J. Lomant and J. R. Fresco, in *Progress in Nucleic Acid Research Molecular Biology*, A. J. Lomant and J. R. Fresco, (Eds.) (Academic, New York, 1975), Vol. 15, p. 185. (b) T. Lindhal, Annu. Rev. Biochem. **51**, 61 (1982). (c) D. L. Sloane, M. F. Goodman and H. Echols, Nucleic Acid Res. **26**, 6465 (1998). (d) S. Moran, R. X. Ren, C. J. Sheils, S. Rumney and E. T. Kool, Ibid. **24**, 2044 (1996).

[67] (a) P. Hobza and Z. Havlas, Chem. Rev. **100**, 4253 (2000). (b) Y. Gu, T. Kar and S. Scheiner, J. Am. Chem. Soc. **121**, 9411 (1999). (c) P. Hobza, J. Šponer, E. Cubero, M. Orozco and F. J. Luque, J. Phys. Chem. B **104**, 6286 (2000). (d) E. S. Kryachko and Th. Zeegers-Huyskens, J. Phys. Chem. A **105**, 7118 (2001). (e) E. S. Kryachko and Th. Zeegers-Huyskens, Ibid. A **106**, 000 (2002). (f) E. S. Kryachko and Th. Zeegers-Huyskens, Ibid. A **106**, 000 (2002).

[68] P.-O. Löwdin, Adv. Quantum Chem. **1**, vii (1964).

Modern Quantum Look at the Periodic Table of Elements

V. N. Ostrovsky
V.A.Fock Institute of Physics, The University of St Petersburg,
198904 St Petersburg, Russia;
E-mail: Valentin.Ostrovsky@pobox.spbu.ru

Abstract. The famous *Aufbau principle* is operative for all systems of fermions. Various Periodic Laws are specified by different lengths of periods, or, in other terms, different sets of *magic numbers*. For the Chemical Periodic Table the period lengths are described by the $(n + l, n)$-rule. The origin of this rule lies in the specifics of effective one-electron potential in an atom. The fascinating properties of this potential are exposed. Alternative approaches to the Periodic Table based on group theory are discussed.

> It is perhaps remarkable that, in axiomatic quantum theory, the simple energy rule $(n+l, n)$ has not yet been derived from first principles.
>
> P. O. Löwdin, Some Comments on the Periodic System of Elements,
>
> *International Journal of Quantum Chemistry (Symposium)* IIIS: 331, 1969.

1. Introduction

The foundation for quantum interpretation of the Periodic System of Elements was laid down in the famous series of papers by Niels Bohr [see Collected Papers, Bohr (1977)]. In modern terms it consists of three general principles[1]:

– • Many-electron atom can be characterized by set of one-electron quantum numbers n and l that label occupied one-electron orbitals (in a simplistic wording each electron is characterized by some n and l);

– • No two electrons in an atom may have the same four quantum numbers n, l, m, m_s (Pauli exclusion principle);

– • One-electron orbitals are filled consequently in order of increasing energy (*Aufbau principle*).

Due to spherical symmetry, in non-relativistic case the energy levels depend on principal n and orbital l quantum numbers, but not on

E.J. Brändas and E.S. Kryachko (eds.),
Fundamental World of Quantum Chemistry, Vol. II, 631–674.

azimuthal m and spin m_s numbers; hence each nl orbital could be occupied by $2(2l + 1)$ electrons.

These principles are inevitably formulated in all exposures of quantum theory of the Periodic System. Clearly, to apply in practice the last principle one needs to know *ordering in energy* of one-electron orbitals. In fact the three principles are operative for *any system composed of fermions* and lead to *some* Periodic Law, i.e. (quasi)periodic pattern of properties depending on the number of particles in the system. The list, as recently reviewed by Ostrovsky (2001), in addition to the well known 'chemical' Periodic Table, includes its modifications for ionized, compressed, or excited atoms, but also the Periodic Laws for atomic nuclei, clusters of atoms or molecules and particles in the traps. Each such system is characterized by a particular set of *magic numbers* closely related period lengths. On the contrary, for bosons the Pauli exclusion principle is not operative and hence (quasi)periodicity does not appear. Instead, for a system of boson one finds famous phenomenon of Bose-Einstein condensation: all the particles occupy single (the lowest in energy) one-particle orbital. Properties of such a system depend smoothly and monotonously on the number of particles. One can argue that fermions play crucial role in diversity of our world which would be much more dull if the principal constituent particles (electrons and nucleons) were bosons.

Various Periodic Laws appear exclusively due to different ordering of one-particle orbitals in energy. In other words, three general principles allow one to anticipate that *some* Periodic Table exist, but its specific form is governed by actual ordering of orbital energies. In this sense the type of ordering stands as a crucial issue of the theory. The ordering pattern is defined by the shape of effective one-particle potential $V(r)$.

Often (sometimes implicitly) exposures of the Periodic Law presume that the one-electron levels are ordered according to the distorted Hydrogen atom (n, l) scheme: the levels with higher principal quantum number have lower binding energy; among the levels with the same n (degenerate in non-relativistic H atom) these with higher orbital quantum number l have lower binding energy. The simple justification of this splitting pattern follows from the fact that at large distances from the atomic nucleus the valence electron feels Coulomb attraction to the core with unit positive charge. At smaller distances stronger attraction to the bare nucleus prevails. This potential deepening influences primarily electrons with smaller orbital momentum l, whereas higher l electrons do not penetrate extended centrifugal barrier[2].

Comparison with the detailed numerical quantum calculations for multielectron systems and with the empirical data shows that for the *highly ionized* atoms such (n, l) ordering is operative indeed (Goudsmith and Richards, 1964). Its heuristic derivation given above presumes that the deviation of the effective one-electron potential from $-1/r$ behaviour can be treated as a *small* perturbation.

However the structure of *neutral atoms* is of primary importance when the Periodic System is concerned. Here the (n, l) rule fails (although some authors consider it as an 'ideal' form of the Periodic Table). It is subject to a large number of violations for heavier atoms as clearly seen from Fig. 2 in the paper by Demkov and Ostrovsky (1971c). This implies that the effective potential exhibits *strong* deviation from the Coulomb shape, leading to the substantial rearrangement of the spectrum. The overlap appears between the groups of the energy levels with different principal quantum numbers n. The n-grouping of levels disappears[3], but the new type of regularities emerge.

Quite long ago it was noticed that the $(n + l, n)$ *ordering rule* describes much better the structure of the Periodic Table of Neutral Atoms. Being purposely given in two parts, it reads:

As the nuclear charge Z increases in neutral atom,

- **(i) the one-electron orbitals are filled in the order of increasing sum $N \equiv n + l$, and**
- **(ii) for the fixed $(n + l)$ the orbitals are filled in the order of increasing n.**

The simple meaning of the sum $(n + l)$ is that it represents ordinal number of period in the Table (see, however, discussion below). Some authors expresses critical attitude towards $(n + l, n)$ rule (Scerri 1991, 1997a, 1997b, 1998a) since it has 'too much' exceptions: for 18 atoms one occupied orbital differs from the rule prediction; in 2 more cases two orbitals are different. However it is sufficient to compare Fig. 1 and Fig. 2 presented in the paper by Demkov and Ostrovsky (1971c) to see that the exceptions from $(n + l, n)$ rule (Fig. 1) represent only a weak distortions of well manifested general trend, whereas arrangement according to (n, l) ordering (Fig. 2) looks fairly irregular for heavier elements, i.e. for larger nuclear charges Z.

The periods within (n, l) ordering scheme correspond to the hydrogenic n-shells. Taking into account the electron spin, one obtains the period lengths

$$2n^2 = 2,\ 8,\ 18,\ 32,\ 50,\ \ldots\ . \tag{1}$$

Although the $(n + l, n)$-rule provides mostly correct ordering of elements in the Periodic Table, the dimensions of N-groups ($N = n+l$),

$$2,\ 2,\ 8,\ 8,\ 18,\ 18,\ 32,\ 32,\ \ldots\ , \tag{2}$$

differ from the period lengths in the conventional form of the Periodic Table,

$$2,\ 8,\ 8,\ 18,\ 18,\ 32,\ 32,\ \dots\ . \tag{3}$$

The difference is shown in more detail in the following scheme

$$\overbrace{\underbrace{1s}_{\text{dim}=2}}^{n+l=1} \ll \overbrace{\underbrace{2s}_{\text{dim}=2}}^{n+l=2} < \overbrace{\underbrace{2p \ll 3s}_{\text{dim}=8}}^{n+l=3} < \overbrace{\underbrace{3p \ll 4s}_{\text{dim}=8}}^{n+l=4} < \overbrace{\underbrace{3d < 4p \ll 5s}_{\text{dim}=18}}^{n+l=5} <$$

$$< \overbrace{\underbrace{4d < 5p \ll 6s}_{\text{dim}=18}}^{n+l=6} < \overbrace{\underbrace{4f < 5d < 6p \ll 7s}_{\text{dim}=32}}^{n+l=7} < \overbrace{\underbrace{5f < 6d < 7p \ll 8s}_{\text{dim}=32}}^{n+l=8} < \dots \tag{4}$$

where dimensions of $(n + l)$-shells are indicated with account for the electron spin. We denote by symbol $\ll$ large energy gaps between the one-electron levels. These gaps (dividing the periods in the Periodic Table) *do not coincide* with the borders between the $(n + l)$-groups of levels.

Comparing sequence of numbers (2) with results of the hydrogenic $(n,\ l)$-rule, set (1), one sees that the period lengths met are the same in both cases, being equal to $2\mathcal{N}^2$ with integer $\mathcal{N} = 0, 1, 2, 3\dots$. However in case of $(n+l,\ n)$-rule (2) each length appears twice. Sometimes such pairs of periods are referred to as *diads.* These *length doubling* is a very important feature to be discussed below (section 6).

The early history of the $(n + l,\ n)$ rule is traced in the section 2 below. At that time the rule was considered as purely empirical one that created natural dissatisfaction among researchers. It was expressed by Per-Olov Löwdin in 1969 in the words cited as epigraph of the present paper. He felt the situation much better than many other people who then (and now) think that physics long ago had told the final word about the Periodic Table and nothing could be added. Löwdin pinpointed the object for research very timely: in two or three years the outburst of activity followed from a number of groups all over the world that used very different approaches. The objective of the present paper is to provide a review of these studies and characterize the current state of the problem. In section 3 we demonstrate that $(n + l)$ rule leads to special pattern of classical electron trajectories, namely the trajectories close after two rotations around the force center. The interaction potential $V(r)$ might be constructed from this trajectory property, if some additional constraints are imposed (section 4). These constraints could be chosen to obtain a potential with additional fascinating features: focusing properties, special pattern of degeneracy for zero-energy levels, symmetry under operation of *inversion in a sphere.* This potential has

Coulomb singularity at the force center and reproduces very well effective one-electron potential in atom as given by Thomas-Fermi theory. It provides firm non-empirical basis for interpretation and explanation of the $(n + l,\, n)$ rule (section 5). The symmetry aspects of the Periodic Table and their analysis in terms of mathematical group theory are considered in section 6. The section 7 contains some concluding remarks and in particular, briefly some discusses closely-lying issues that remained beyond the mainstream of the present paper.

2. Early history: $(n + l,\, n)$ as empirical rule

The early history of the $(n + l)$ ordering rule looks rather obscure[4]. Without giving references Löwdin (1969) ascribes this rule to Bohr, but remarks that "Bohr himself was never too explicit about his 'Aufbau'-principle and $(n+l,\, n)$ rule is sometimes referred to as Goudsmith-rule or Bose-rule" (no references to Goudsmith or Bose are given either). Regretfully, this leaves in obscurity and controversy important historical details. For instance, Goudsmith himself (Goudsmith and Richards, 1964) says nothing about his own role in discovery of the rule, but recalls that he received private communication from Madelung about it in December, 1926. Note, however, that this testimony appeared as late as in 1964. Goudsmith did not mention the rule when he discussed electronic configurations of the heaviest elements (Wu and Goudsmith, 1933) although Karapetoff (1930) did this earlier with some success as discussed further in more detail. Most authors state (incorrectly, as demonstrated below) that the first publication of the rule belongs to Madelung (1936). In his *Mathematischen Hilfsmittel des Physikers* the sum $n + l$ appears in entries of the adopted form of the Periodic Table. In a brief comment Madelung indicates (i) that the table is arranged according to $(n + l,\, n)$ rule, (ii) that the rule is empirical (*'lexicographic'*) and its theoretical background is unknown; (iii) that several deviations from this rule are observed. All these points deserve some closer look.

In the Madelung's chart each chemical element is labeled by four quantum numbers: $\{n + l,\, l,\, m,\, m_s\}$ (we employ modern notations; Madelung uses s instead of our m_s). For a given $N = n + l$ the quantum number n varies by unit steps from $\left[\frac{1}{2}(N+1)\right]$ to N, i.e. l varies from $\left[\frac{1}{2}(N-1)\right]$ to 0 (here $[z]$ stands for integer part of z). For a given l the quantum number m varies by unit steps from $-l$ to l, and, for all other quantum numbers fixed, m_s takes two values, $-\frac{1}{2}$ and $\frac{1}{2}$. The Madelung's term *lexicographic order* reminds us the alphabetic words ordering in vocabularies that does not depend on the word meaning. However, in fact his approach is not that formal, but more physical, since Madelung ascribes the four quantum numbers $\{n+l,\ l,\ m,\ m_s\}$ to 'the last electron' added to neutral atom when the charge of the atomic nucleus receives unit increment, from $(Z-1)$ to Z; it is supposed that quantum numbers of all other electrons, that were already present in $(Z-1)$ atom, are not changed. It should be stressed that within the modern description that employs language of atomic configurations (not used by Madelung) only n and l quantum numbers have direct physical meaning being related to the number of nodes in radial ($n_r = n - l - 1$) and angular (l) one-electron wave functions (orbitals). The orbitals enter Slater determinants that are used to approximately construct wave function of a multielectron atom. Thus n and l correspond to physical quantities that are approximately conserved in an atom, i.e. to one-electron integrals of motion. On the contrary, the quantum numbers m and m_s are not good even approximately, since the one-electron angular momenta are to be coupled, for instance, according to LS scheme. The related wave function is generally a sum of Slater determinants with different (m, m_s)-orbitals. This means that the labels m and m_s loose sense. The atomic wave function is eigenfunction of operators L and S (total orbital momentum and total spin of an atom) that are exact integrals of motion within non-relativistic approach. Projections of $\vec{L}$ and $\vec{S}$ vectors, M_L and M_S, are good quantum numbers in this approximation, but become invalid in relativistic theory where exact integrals of motion are J ($\vec{J} = \vec{L} + \vec{S}$) and M_J. Thus the Madelung's numbers m and m_s are just formal non-physical labels. However, they can be simply related to physical magnitude: occupation number ν_i for $n_i l_i$ orbital. It is easy to see that the occupation number of the *outer* $n_i l_i$ orbital in Madelung's labeling is

$$\nu_i = l_i + m + m_s + \frac{3}{2}\,. \tag{5}$$

In much later development within group-theoretical approach Fet [1980, 1992, see also chapter 2 in the book by Hefferlin (1989)] understood the numbers $\{n+l,\ l,\ m,\ m_s\}$ as pure *lexicographic* labels not related to electronic structure of an atom. He ascribed the origin of such interpretation to Madelung and suggested the name *Madelung numbers* to the quantum numbers interpreted in this way. As discussed above, this attribution is to some extent applicable to m ands m_s, but not to $(n+l)$, n and l. Even in the former case the *lexicographic* labels are related to electronic structure by virtue of formula (5).

The distinction between physical and formal labeling becomes very prominent when we consider exceptions to the $(n+l, n)$ rule. Within the formal approach the exceptions simply do not exist, or, in other words, remain unnoticed. Indeed, the sequence of natural numbers (the element ordinal numbers $Z = 1, 2, 3, \ldots$) always can be labeled according to the formal $\{n+l, l, m, m_s\}$ ordering. The exceptions are recovered only when one relates the labels to electron configuration of an atom. This is why the exceptions are never discussed in the formal group-theoretical treatments (see discussion in section 6). Madelung himself mentions that the exceptions do exist, but he does not provide any detail. He also stresses that his lexicographic rule is empirical with unknown theoretical background. The entire piece of his text concerning the Periodic Table is very terse and was not changed in subsequent editions of his book (the last one was in 1964).

The first reference to Madelung could be found in the text of the book by Bauer (1938) in the form 'Madelung, 1936', although not included in the reference list. Bauer reiterates Madelung's table providing more information on atomic structure; the exceptions to the rule are not mentioned. The next reference is contained in the short note by Auwers (1948) who discussed the rule and the form of the periodic chart and referred also to Bauer.

It is only rarely acknowledged in the literature (Romanovskaya, 1986) that in fact *the first published formulation* of the $(n+l, n)$ rule belongs to Vladimir Karapetoff (1930). Apparently atomic physics was kind of a hobby for this author being characterized by Brittain (1991) as a 'pioneer electrical engineering educator'. He published a single paper on the subject acknowledging support by the grant on research of some insulators. Karapetoff stressed empirical character of the rule. Importantly, he clearly formulates both its first and the second parts. Karapetoff also makes a practical use of the rule by predicting configurations of transuranian elements – the issue which is not fully settled until now. He describes his attitude towards these predictions in such terms

> All the known elements seem to obey this rule; for this reason hypothetical unknown elements or mono-nuclear combinations, up to the atomic number 124, are also shown in the chart at the extreme right, in accordance with this rule .... This extension of the chart is not necessarily a prophecy, since even the heaviest known elements are quite unstable (radioactive). The chart has been extended partly to test that the rule does not give any mathematically absurd results and partly to indicate that some kinds of orbits ... [5f and 6d orbits in the modern notations], which do not occur in any of the known elements, might be realisable with

> heavier nuclei. In fact all the possible kinds of orbits indicated along the axis of ordinates, are covered by the proposed rule. Perhaps knowing which orbit to expect may help in a search for new mono-nuclear systems beyond uranium.

It seems that the paper by Karapetoff is not widely known. For instance, $(n+l,\, n)$ rule was applied to discussion of electron configuration in yet unknown high-Z elements by Gol'danskii (1970), Neubert (1970), Abrahamson (1971), and Rau *et al* (1972), but these authors did not refer to Karapetoff's paper. However Seaborg (1949) and Cabezas *et al* (1961) were aware of his work. Some other authors (Mazurs, 1974, Dockx, 1959) do refer, but do not notice formulation of the rule contained in his paper. In particular, Dockx (1959) in a book devoted to 'fundamental theory of the Periodic System of Elements' remains unaware of $(n+l,\, n)$ rule and refers to Karapetoff, but only in connection to electron configurations of heavy elements.

Sometimes in the literature $(n+l,\, n)$ rule is related to the papers where form and structure of the Periodic Table was discussed, but the sum $n+l$ does not appear explicitly. However we are interested here in explicit formulation and use of this rule.

Apparently a brief exposure of $(n+l,\, n)$ rule in the handbook by Madelung (1936) at first did not attract much attention, particularly among chemists. The rediscoveries of $n+l$ rule started to appear in mid-forties in this community. The authors did not refer to Madelung or Karapetoff. Wiswesser (1945a,b,c) published a series of tutorial papers about quantum interpretation of the Periodic Table. In the first of them he says: " ...the patterns will be filled in increasing order of $(n+l)$ rather than n alone. A better approximation, completely specific, is given by $n+l-l/(l+1)$." Some not very convincing handwaving arguments in favour of the sum $n+l$ are presented based on the nodal structure of wave functions. After this the idea of $n+l$ ordering is dropped to appear again in a short remark in the last paper of the series: "The empirical sequence measure, $n+l-l/(l+1)$, distorts the correct order of subgroups, but not the pattern sequence within any subgroup". Apparently this author still believes in 'ideal' $(n,\, l)$ filling sequence that is 'distorted' by empirical $(n+l,\, n)$ ordering.

Yeou Ta (1946) names the sum $n+l$ the *mixed quantum number of transition* and gives full formulation of the $(n+l,\, n)$ rule [the preceding short note by this author (Yeou Ta, 1945) does not mention the rule]. Simmons (1947, 1948) develops an 'arithmetical' construction of the Periodic Table. In the middle of the first paper the sum $n+l$ appears without any motivation and discussion being dropped almost immediately. In the second paper, when discussing a role of the sum $(n+l)$, Simmons (1948) refers to earlier paper by Yeou Ta and calls $(n+l,\, n)$

rule *Yeou Ta rule* [the paper contains also references to Pao Fang Yi (1947) and Hakala (1948)]. From this we see that Simmons does not claim priority in formulating the rule. Probably he did not consider the rule as an important issue; at least in a subsequent short note on the periodic classification of elements (Simmons, 1960) the rule is not mentioned at all.

In his letter to the editor Hakala (1948), commenting on previous letter by Pao Fang Yi (1947), discusses ordering of elements without formulating $(n + l)$ rule, albeit this specific combination of quantum numbers appears (without comments) in suggested formula[5] for the subshell energy sequence:

$$E_{\text{ord}} = (n + l)(n + l + 2)/4 + \left[1 - (-1)^{n+l}\right]/8 - l\ . \qquad (6)$$

In the extended publication Hakala (1952) applies mathematical difference scheme to analyze the empirical structure of the Periodic Table. Along these lines he finds that the sum $n + l$ emerges in quite a natural way. A number of formulae similar to (6) is obtained. Normally formulae are derived in pairs, being different for odd and even $n+l$, although each pair could be combined in a single formula using (quite artificial) term $(-1)^{n+l}$, as in Eq. (6). Among previous authors cited above Hakala refers only to Simmons (1947).

The paper by Carroll and Lehrman (1948) follows that by Simmons (1948) in the magazine. It formulates both parts of $(n + l,\, n)$ rule without mentioning any predecessors.

Independently the $(n + l,\, n)$ rule was derived by Klechkovskii who studied systematically its various aspects in much detail starting from 1951 (1951; 1952a; 1954; 1952c; 1953a; 1953b; 1954; 1954b; 1956; 1960; 1961; 1962). Klechkovskii summarized his work in a book (Klechkovskii, 1968).

It seems that since sixties the $(n + l,\, n)$ rule became quite broadly known although some rediscoveries occurred much later (Mills, 1972, Ausubel, 1976, Singh and Dikshit, 1991). Keller (1956) and Balaban (1963) popularized $(n + l,\, n)$ rule with references to predecessors. The rule was employed and/or discussed by a number of authors, see, for instance, papers by Dash (1969), Gol'danskii (1970), Neubert (1970), Odabasi (1973), Bosi (1983) and monographs on atomic structure by Rubinowicz (1954), Condon and Odabasi (1980) and Veselov and Labzovsky (1986). Only rarely it is mentioned in textbooks on chemistry. The books by Cooper (1960), Sisler (1963), Puddephatt and Monaghan (1986) devoted especially to the Periodic Table of Elements do not mention the rule at all. The fundamental *Chemistry of Elements* by Greenwood and Earnshaw (1984) spends literally three lines of text to expose the rule: "it is found that the sequence of increasing electron energy levels in an atom follows the sequence of values $n + l$; if 2 electrons have the same $n + l$ then one with smaller n is more tightly bound". Very often till now the rule is considered as an empirical one. As shown below, the physicists tried to rationalize the rule and reveal its dynamic origin, whereas many chemists treated the rule as purely mnemonic device or numerological exercise that does not have any serious background and implications. The extremist manifestation of this attitude could be exemplified by a recent note by Ludwig (1992). The situation did not change much from 1969 when Löwdin wrote:

> "It is ... a startling fact that the simple energy rule $(n + l,\, n)$ has not entered any major chemistry textbook, as far as I know, and it is still this rule that gives the first explanation of the occurrence of the transition metals, the rare earth metals, and over-all structure of the electronic shell of atoms" (Löwdin, 1969).

3. Classical orbits of valence electrons in atoms

For a long time *dynamical* origin of the sum of principal n and orbital l quantum numbers remained mysterious. For better understanding of the problem we compare situation with that for Hydrogen atom. Here 'primordial' quantum numbers are radial, n_r, and orbital, l, numbers that appear from counting the nodes, respectively, in radial and angular wave functions. The sum $n = n_r + l + 1$ emerges in a natural way when the Schrödinger equation is solved: the level energy depends only on this linear combination of quantum numbers, but not on n_r and l separately. This mathematical observation finds deep roots within the group-theoretical approach (Fock, 1935). On the contrary, the combination $n + l$ never appeared as a result of solution of any Schrödinger equation[6].

It is well known that the Thomas-Fermi theory was successfully employed to evaluate nuclear charge Z when the first electron with given orbital momentum l appears in neutral atom (Fermi, 1927, 1928, Jensen and Luttinger, 1952, Ivanenko and Larin, 1953, Oliphant, 1956, Theis, 1955, Tietz, 1958, 1960, Klechkovskii, 1961, Essén, 1982, Bosi, 1983, ten Hoor, 1986). Kónya (1964a, 1964b) used the Thomas-Fermi scheme to analyze the first appearance of an electrons with a given radial quantum number n_r. However in this approach the sum $(n + l)$ also does not appear in a natural way. The general systematic trends in properties of atomic potentials along the Periodic Table were studied, for instance, by Göppert-Mayer (1941), Latter (1955), Rao and Fano (1968), Marson (1969), Kaldor (1977), however without relation to $n+l$

rule [although Rao and Fano refer to the paper by Goudsmith and Richards (1964)].

The first part of $(n+l, n)$ rule is of primary importance here, since it implies grouping of one-electron energy levels with the same sum of quantum numbers $(n+l)$. Within a crude approximation such grouping may be considered as an (approximate) degeneracy of one-electron energy levels with the same $(n+l)$. Importantly, the concept of degeneracy does not belong solely to quantum mechanics. It appears also in classical dynamics when the systems with two or more degrees of freedom are considered. If dynamical variables are separable, then each (j-th) degree of freedom is characterized by some frequency ω_j. If one makes transition to action-angle variables, then the Hamiltonian function H depends only on the action variables I_j, and the frequencies are obtained as derivatives (Landau and Lifshitz, 1976)

$$\omega_j = \frac{\partial H}{\partial I_j} \ . \tag{7}$$

Consider for simplicity the case of two degrees of freedom. *The classical degeneracy* appears when H depends only on linear combination of action variables with some integer coefficients ν_1 and ν_2: $H = H(\nu_1 I_1 + \nu_2 I_2)$; then $\nu_2\,\omega_1 = \nu_1\,\omega_2$, i.e. ratio of frequencies is rational number and classical trajectory is closed (periodic).

Within the semiclassical approximation the Bohr-Sommerfeld quantization rule

$$I_j = n_j + \gamma_j \tag{8}$$

allows one to express action variables via integer quantum numbers n_j and some constants $0 \le \gamma_j \le 1$ (we use hereafter atomic units $e = m_e = \hbar = 1$). Thus semiclassical counterpart of (7) is

$$\omega_j = \frac{\partial E_{n_1\,n_2,\ldots}}{\partial n_j} \ , \tag{9}$$

where the bound state energy $E_{n_1\,n_2,\ldots}$ is considered as a continuous function of quantum numbers. Degeneracy implies that $E_{n_1\,n_2} = E(\nu_1 n_1 + \nu_2 n_2)$. Thus the classical and quantum degeneracies are closely related.

In our particular case the degeneracy $E_{nl} = E(n+l) = E(n_r + 2l + 1)$ implies that $2\omega_r = \omega_\varphi$, where ω_r is frequency of radial oscillations, ω_φ is frequency of rotation. In terms of corresponding periods T_r and T_φ one finds relation

$$T_r = 2T_\varphi \ . \tag{10}$$

This means that the trajectory of valence electron (approximately) closes after rotation over angle 4π around the force center. This corollary of $n + l$ filling rule within classical mechanics framework was realized independently and simultaneously by Wheeler (1971) and Demkov and Ostrovsky (1971c). Wheeler (1971) coined the name *double necklace* for such orbits and called them also *chemical orbits.* He drew these orbits schematically, without knowing related interaction potential $V(r)$ that is effective one-electron potential in atom. Demkov and Ostrovsky (1971c) plotted the double-necklace trajectories as they were calculated for specific potential (30) that provides a good approximation for effective one-electron potential in atoms (see discussion in section 5).

Powers (1971) was inspired by Wheeler to seek for the actual potential. However from the very beginning of his analysis he made quite a restrictive, unjustified and in fact unnecessary presumption, namely, a power law for the potential, $V(r) \sim r^s$. Powers concluded that for atoms the index s lies somewhere in the interval $-7/4 \le s \le -3/2$. It seems that this result testified only that the power-law potential form is not applicable.

The oscillation frequencies (9) generally depend on the values of action variables I_j, and on the total energy E. Generally the degeneracy condition is satisfied only for some values of these parameters. In case of a central potential this means that degeneracy occurs only for some values of integrals of motion E and l. According to well known Bertrand (1873) theorem [see also discussion by Berrondo *et al* (1974)], only for two potentials, Coulomb potential $V(r) = -Z/r$ and 3D oscillator potential $V(r) = kr^2$, a degeneracy is operative for all E and L. Hereafter we denote by L classical orbital momentum that varies continuously, to distinguish it from integer quantum number l.

Our task is to describe potentials which generate closed classical trajectories for all L and single value of energy E. It does not contradict Bertrand theorem which considers trajectories for *all* L and E. Indeed, only valence electrons are essential when the structure of the Periodic Table is discussed. The binding energies of these electrons are small compared with energies of inner electrons. Approximately we can say that the energy of interest is $E = 0$. This formulation casts our task as a *classical inverse problem.*

4. Classical inverse problem

In this section we describe all the central potentials which, for the fixed energy E, give closed periodic orbits with the prescribed ratio $\tilde{\mu}$

of angular and radial periods,

$$\tilde{\mu} = T_\varphi / T_r \tag{11}$$

The subsequent development is based on the paper by Ostrovsky (1997) which, however, contains a misprint: $\tilde{\mu}$ is defined as inverse of (11). The case of double necklace orbit (10) corresponds to $\tilde{\mu} = \frac{1}{2}$. In distinction to Powers (1971) no particular functional form of potential is presupposed.

It is universally known that for the finite motion in a central field the radial variable r oscillates between the inner $[r_{\min}(L)]$ and outer $[r_{\max}(L)]$ turning points as the particle vector rotates. Half of the oscillation [i.e. variation of r from $r_{\min}(L)$ to $r_{\max}(L)$] corresponds to the vector rotation over the angle

$$\varphi_0(L) = \int_{r_{\min}(L)}^{r_{\max}(L)} \frac{(L/r^2)\,dr}{\sqrt{2\mathcal{M}\,[E - V(r)] - (L^2/r^2)}}\,, \tag{12}$$

where $\mathcal{M}$ is particle mass [see, for instance, book by Landau and Lifshitz (1977)]. The closed trajectories are obtained when $\varphi_0 \equiv \pi/\tilde{\mu}$ does not depend on the orbital momentum L, and $\tilde{\mu}$ is a rational number.

We consider at first more general problem when $\varphi_0(L)$ is any given function. Then (12) could be considered as an integral equation for the interaction potential $V(r)$.

In this formulation our task is similar to the one-dimensional inverse classical problem: restoration of the potential function $U(x)$ if the oscillation period $T(E)$ is a known function of the energy E. Solution of this problem is known to be non-unique (Landau and Lifshitz, 1977, Section 12]: if $x_1(E)$ and $x_2(E)$ are the classical turning points ($x_2 > x_1$), then only the function $x_2(U) - x_1(U)$ is restored. This means that the function $x_1(U)$ $[x_2(U)]$ [i.e., the potential to the left (right) of its minimum] could be chosen arbitrary; then the function $x_2(U)$ $[x_1(U)]$ [i.e., the potential to the right (left) of its minimum] is uniquely fixed by the explicit solution of the inverse problem.

The other closely related situation appears in the classical inverse scattering problem, i.e. when potential is restored from the scattering angle known as a function of impact parameter for a given collision energy. This inverse problem was solved by Firsov (1953) [7]; his solution is also reproduced in the well-known textbook (Landau and Lifshitz, 1977). Some modification of this procedure is applicable to the problem under consideration [8]; the final result could be presented as:

$$\frac{r_{\max}(\mathcal{L})}{r_{\min}(\mathcal{L})} = \exp\left\{\frac{2}{\pi}\int_{\mathcal{L}}^{L_m} \frac{\varphi_0(L)\,dL}{\sqrt{L^2 - \mathcal{L}^2}}\right\}\,, \tag{13}$$

$$r_{\min}\sqrt{2\mathcal{M}\left[E - V(r_{\min})\right]} = \mathcal{L}\ , \tag{14}$$

$$r_{\max}\sqrt{2\mathcal{M}\left[E - V(r_{\max})\right]} = \mathcal{L}\ . \tag{15}$$

We have four functions defined parametrically via the parameter $\mathcal{L}$:

$$\begin{aligned} r_{\min}(\mathcal{L})\ , \qquad & r_{\max}(\mathcal{L})\ , \\ w_{-}(\mathcal{L}) &\equiv r_{\min}\sqrt{2\mathcal{M}\left[E - V(r_{\min})\right]} \quad \text{and} \\ w_{+}(\mathcal{L}) &\equiv r_{\max}\sqrt{2\mathcal{M}\left[E - V(r_{\max})\right]}\ . \end{aligned} \tag{16}$$

If these functions are known, the potential $V(r)$ is defined automatically. However there are only three relations (13)-(15) defining four functions. This means that the solution is not unique. For instance, just as in the one-dimensional problem, the potential $V(r)$ can be chosen arbitrarily for $r > R_m$ ($r < R_m$). Then the relations (13)-(15) allow us to restore the potential for $r < R_m$ ($r > R_m$). The radius R_m of circular orbit for given energy E corresponds to the position of the bottom in the effective potential well (the latter includes centrifugal potential term). One could impose also some alternative restrictions in order to make the solution unique. Of course, only the potential in the classically accessible domain [where $V(r) < E$] can be restored.

The function $\varphi_0(L)$ is defined at some interval $0 \leq L \leq L_m$ where the classical motion is possible (for higher orbital the classical finite motion with the given energy E becomes forbidden because of the effective centrifugal barrier). The trajectory with the largest possible orbital momentum L_m is a circle with some radius R_m. If R_m is chosen, then the potential $V(R_m) \equiv V_m$ is defined from the relation

$$L_m = R_m\sqrt{2\mathcal{M}(E - V_m)}\,. \tag{17}$$

The turning points satisfy inequality $r_{\min}(L) \leq R_m \leq r_{\max}(L)$. As L decreases from its maximum value L_m, the domain of classically allowed radial motion extends in both outer and inner directions from R_m. Hence the classical motion probes the potential $V(r)$ in more broad region around R_m. The information on the potential behaviour in the expanding region is enclosed in the function $\varphi_0(L)$; it is extracted with the help of equations (13)-(15). This is the essence of the classical inversion procedure.

As an example we turn to the case of interest in the present context with L-independent angle $\varphi_0 = \pi/\tilde{\mu}$. Then the integral in Eq. (13) is calculated analytically:

$$\frac{r_{\max}(\mathcal{L})}{r_{\min}(\mathcal{L})} = \left[\frac{L_m}{\mathcal{L}} + \sqrt{\left(\frac{L_m}{\mathcal{L}}\right)^2 - 1}\,\right]^{2/\tilde{\mu}} . \tag{18}$$

Now we consider three forms of additional constraint which makes the solution unique. At first we impose the following relation

$$r_{\min}(L)\, r_{\max}(L) = R_m^2 , \tag{19}$$

which means that the inner turning point of radial motion $r_{\min}$ can be obtained from the outer turning point $r_{\max}$ (or vice versa) by *inversion in the sphere* of radius R_m. Then using formulae (14), (15), (18) we obtain the explicit expression for the potential

$$V_{\tilde{\mu}}(r) = E - \frac{2L_m^2}{\mathcal{M} r^2 \left[(r/R_m)^{\tilde{\mu}} + (R_m/r)^{\tilde{\mu}}\right]^2} . \tag{20}$$

If we are interested in the potentials with the zero-energy asymptote $[V(\infty) = 0]$, then we have to put $E = 0$ in (20). After this the potential reads

$$U_\mu(r) = -\frac{2v}{r^2 R^2 \left[(r/R)^\mu + (R/r)^\mu\right]^2} , \tag{21}$$

where $\mu = \tilde{\mu}$, $R = R_m$ and $v = R_m^2 L_m^2/\mathcal{M}$. The original derivation by Demkov and Ostrovsky (1971c) was somewhat different. Its extensive discussion was given by Wheeler (1976) in a broad context of classical and quantum inverse problems. For the potential (21) the classical trajectories $r(\varphi)$ (φ is the azimuthal angle in the orbit plane) are defined by the formula

$$\sin\left[2\mu\,(\varphi - \phi)\right] = \frac{\beta + 4 - 2\zeta(r)}{\beta - 4} , \tag{22}$$

$$\zeta(r) = \left[\left(\frac{r}{R}\right)^\mu + \left(\frac{R}{r}\right)^\mu\right]^2 , \qquad \beta = \frac{2\mathcal{M}v}{R^2L^2} , \tag{23}$$

where ϕ is a constant of integration. Various trajectories in the family differ by the orbital momentum L, i.e. by the parameter β. Each individual trajectory is invariant under the operation $\{r \Rightarrow R^2/r,\ \varphi \Rightarrow \varphi + \pi/\mu\}$. Note that the transformation of the radial coordinate here corresponds to the inversion in the sphere $r = R$ (19).

The trajectory always crosses the sphere $r = R$. For the part of the trajectory lying inside the sphere the particle vector rotates on the angle π/μ; and the same rotation angle corresponds to the part of trajectory outside the sphere. Applying aforementioned operation twice, we see that trajectories are invariant under rotation over angle $2\pi/\mu$.

Within well known analogy between classical mechanics and geometrical optics, the potential (21) corresponds to continuous distribution of refractive index

$$n_\mu(r) = \frac{2n_0}{r\left[(r/R)^\mu + (R/r)^\mu\right]} \,. \tag{24}$$

with some unessential normalization factor n_0. This family of refractive indices was found originally by Lenz (1928) in his paper on theory of optical images. The particular case $\mu = 1$ was discovered long before this by Maxwell (1854) [see also Maxwell (1952), Collected Papers, p. 76] who had demonstrated that such a distribution of refractive index possesses unusual properties: all rays (trajectories) coming from some arbitrary point in space (source) are *exactly* focussed in another point (image). Maxwell gave name "fish-eye" to this *perfect optical system.* The related wave (quantum) problem was studied by Demkov and Ostrovsky (1971b) and later by Buchdahl (1978) and Frank *et al* (1990). Lenz (1928) succeeded in generalizing Maxwell's result in geometrical optics and found that refractive indices (24) also ensure perfect focusing. Contrary to the fish-eye special case, generally the rays outgoing from the point source form caustics of circular lying on $r = R_m$ sphere.

Thus an extra presumption (19) makes solution of the classical inverse problem unique and, in addition, produces smooth potentials with focusing properties. As in some sense opposite limiting case we replace the presumption (19) by the condition that for all L the outer turning point lies at the same position, i.e.

$$r_{\max}(L) = R_m \,. \tag{25}$$

This corresponds to the spherical potential $V(r)$ with the vertical *outer* reflective wall for $r = R_m$. Note that this implies that just at the position of the wall ($r = R_m$) the potential is not defined uniquely and the equation (15) becomes non-informative. Using (18) and (14) we obtain the potential

$$V_{\tilde{\mu}/2}(r) = E - \frac{2L_m^2}{\mathcal{M}r^2\left[(r/R_m)^{\tilde{\mu}/2} + (R_m/r)^{\tilde{\mu}/2}\right]^2} \,, \tag{26}$$

for $r < R_m$. The classical trajectories do not penetrate the complementary domain, $r > R_m$, hence the potential $V(r)$ remains arbitrary here [only the limiting condition $V(R_m + 0) > E$ is to be satisfied]. The expression (26) differs from (21) by the replacement $\tilde{\mu} \Rightarrow \tilde{\mu}/2$. The meaning of this result is clear: when the particle (moving inside the sphere $r = R$) hits the vertical wall, it is scattered elastically and reflected. After this reflection it continues motion inside the sphere along the same trajectory as before the scattering, but rotated over the angle $\pi/\tilde{\mu}$. Therefore the same symmetry of the classical trajectories is achieved with the parameter $\tilde{\mu}/2$ instead of $\tilde{\mu}$.

The other simple choice of an additional constraint in the classical inverse problem is to look for the spherical potential $V(r)$ with the vertical *inner* wall at $r = R_m$. This means that instead of condition (25) we have now

$$r_{\min}(L) = R_m \, . \tag{27}$$

It is easy to see that in this case potential is again defined by the analytical expression (26) but now it is to be applied for $r > R_m$. In the complementary domain, $r < R_m$, the classical trajectories do not probe the potential that implies limiting condition $V(R_m - 0) > E$. Note that if, as usual, an additive constant in the potential is fixed by condition $V(r \to \infty) = 0$, then we have to put $E = 0$ in formula (26) in case of potential with vertical inner wall, whereas in case of outer wall E remains free parameter not related to potential behaviour at $r \to \infty$.

5. Effective one-electron potential in atoms

As emphasized in the beginning if preceding section, in the context of the Periodic Table the interesting case is that of $\tilde{\mu} = \frac{1}{2}$. Here we obtain effective one-electron potential in atoms in the form

$$U_{1/2}(r) = -\frac{2v}{r\,R\,(r+R)^2} \, . \tag{28}$$

The related classical double-necklace orbits are

$$\sin(\varphi - \phi) = \frac{\beta - 2r/R - 2R/r}{\beta - 4} \, . \tag{29}$$

An interesting feature of this potential is $\sim -1/r$ singularity at the origin that for the one-electron potential corresponds to Coulomb attraction to the atomic nucleus. In order to get proper normalization we relate potential strength parameter v to the nuclear charge Z by formula $2v/R^3 = Z$ and rewrite (28) as

$$U_{1/2}(r) = -\frac{Z}{r\,(1 + r/R)^2} \, . \tag{30}$$

Note that such a physical singularity is obtained exclusively from the degeneracy pattern and that for general potential $U_\mu(r)$ (21) the proper singularity is obtained only in case $\mu = \frac{1}{2}$.

Actually potential of the form (30) was suggested in atomic physics some 15 years before it appeared in the paper by Demkov and Ostrovsky (1971c). It is well known that Thomas-Fermi statistical theory provides approximate expression for effective one-electron potential in atom (Slater, 1947; Gombas, 1949; Landau and Lifshitz, 1977)

$$U_a^{TF}(r) = -\frac{Z}{r}\chi(kr) , \qquad k = \left(\frac{8\sqrt{2}}{3\pi}\right)^{2/3} Z^{1/3} . \tag{31}$$

The universal function $\chi(x)$ is found as a solution of non-linear differential equation but does not have analytic expression. This was considered as serious drawback in pre-computer era and some people, among them T. Tietz looked for simple analytical approximations for $\chi(x)$. The approximation suggested by him (Tietz, 1954, 1955) had form

$$\chi(x) = \frac{1}{(1+\alpha x)^2} . \tag{32}$$

It is easy to see that expression (31) combined with formula (32) gives the same result as (30) provided parameters α and R are related as

$$R = \alpha^{-1} Z^{-1/3} \left(\frac{3\pi}{8\sqrt{2}} .\right)^{2/3} \tag{33}$$

More complete bibliography of numerous works by Tietz on Thomas-Fermi theory of atom and its applications could be found in the paper by Demkov and Ostrovsky (1971c). Tietz used two different methods to choose parameter α: normalization condition in momentum space gave $\alpha = (\pi/8)^{2/3} \approx 0.569$ (Tietz 1954, 1955) and variational principle (Tietz, 1955) gave $\alpha = 2^{-4}(35\pi/4)^{2/3} \approx 0.538$. Both these values ensure a good approximation for exact Thomas-Fermi function [its quality is quite robust against variation of α; Demkov and Ostrovsky (1971c) suggested to use numerologically attractive value $\alpha = \frac{1}{2}$ which also gave a good result].

In the present context the most interesting is zero-energy solution of the Schrödinger equation with potential (30). Indeed, as nucleus charge Z increases, new levels appear in this potential [note that the parameter R is Z-dependent, according to formula (33)]. Generally this occurs at some $Z = Z_{nl}$ when the electron states have energy $E = 0$. As Z increases further, the binding energy becomes larger and eventually these levels are filled by atomic electrons according to the Aufbau scheme. Thus solution of zero-energy Schrödinger equation allows one to trace *the genesis of the atomic orbitals*. Demkov and Ostrovsky (1971c) found exact zero-energy solutions of the Schrödinger equation with the potential (30). This gave the critical values Z_{nl} in a simple closed form

$$Z_N = \frac{4}{3\pi}\left[\alpha N(N+1)\right]^{3/2} \ . \qquad (34)$$

They depend not on the quantum numbers n and l separately, as one could expect for a potential of general form, but only on the sum $N = n+l$. Hence as Z increases, the new bound levels appear simultaneously in groups with the same N, i.e. the levels are ordered according to the first part of the $(n + l,\, l)$ rule. At the moment of appearance (i.e. at $E = 0$) the levels exhibit an additional degeneracy of the type that clearly correlates with a closed pattern of the double necklace orbits discussed above.

To deduce the second part of the rule one has to consider how the levels within the same N-group are ordered for a deeper potential, i.e. for $Z > Z_N$, when the level energies differ from zero, N-degeneracy is lifted and the double-necklace orbits start to precess. Such analysis was carried out within perturbation theory by Demkov and Ostrovsky (1971c) and the second part of the rule was confirmed.

It could be said that we consider here 'potential curves' for the Periodic Table, i.e. dependence of one-electron orbital energies in neutral atoms E_{nl} on element ordinal number Z. This approach helps to interpret one more feature in the Periodic Table: the difference between the Table periods and $(n+l)$ group shown in scheme (4). The conventionally accepted period boundaries [see however discussions on the subject in the book by Mazurs (1974)] correspond to the symbol $\ll$ in the scheme. The situation looks as if ns electrons, lying highest in their $(n + l)$ group, leave $(N = n)$-group to which they formally belong and join the next (adjacent) $(N = n + 1)$-group. Namely this regrouping of $n + l$ groups produces the chemical periods. The regrouping was interpreted by Demkov (1975) in terms of special features in the potential curves behaviour in the vicinity of $E = 0$ point. Namely, for all $l > 0$ the curves are well approximated by straight lines that cross $E = 0$ horizontal line in the plot as illustrated by Fig. 1 in the paper by Ostrovsky (1981). However, for $l = 0$ the potential curves behave parabolically and are tangent to $E = 0$ line. This implies formation of large energy gap between s-curve and lower-lying curves with $l > 0$ (with the same quantum number N).

When continued to $Z < Z_n$ domain, the 'potential curves' become complex-valued that means shape resonances in electron-atom scattering. Due to N-degeneracy of zero-energy levels, the resonances in different partial waves are anticipated to be clustered (Berezina and Demkov, 1975) – this theoretical conclusion is yet to be tested experimentally.

Numerical values of the critical parameters Z_N evaluated according to formula (34) were found to be in reasonable agreement with empirical values of nuclear charge $Z^{(N)}$ when the first electron with the quantum number N appear in the Periodic Table (Demkov and Ostrovsky, 1971c). It is worth mention here yet another application of the potential (30) to the problem of subshell filling along the Periodic Table. Abrahamson (1971) used it to evaluate critical charge Z_l when the first electron with given L appears. This author looked for minimum of effective radial potential [obtained by adding to (30) the centrifugal repulsion term $(l+1/2)^2/(2r^2)$] and presumed that the energy level appears as soon as potential at minimum is negative. This simple approach gave good agreement with empirical data. Abrahamson (1971) did not mention $(n+l,\, n)$ rule.

Tietz looked also for solution of Schrödinger equation with the potential (30) (see, for instance, Tietz, 1964, 1968), but in general case this could not be achieved analytically. He did not notice possibility of exact solution for $E = 0$ and N-clustering of levels, although in the case $l = 0$ he found zero-energy solution for somewhat more general potential than (30) and used it to estimate scattering lengths for electron-atom collisions (Tietz, 1966).

The relation between the potential (30) and more accurate description of electron motion in an atom is quite peculiar. For large binding energies E_b the levels are close to that for Coulomb potential of atomic nucleus $-Z/r$, with relatively small screening by other electrons that could be considered as a perturbation. Here the model does not provide an essentially new picture. At large separations r the electron in a neutral atom feels Coulomb attraction $-1/r$ to the residual atomic core that is not reproduced by the model (30). This attraction is crucial at very small binding energies E_b leading to formation of Rydberg series of excited atomic states. Only in the intermediate and relatively narrow domain of E_b starting from few eV the model provides realistic and essentially new picture. However, namely this domain is the most essential one for the Aufbau ordering and for chemistry.

The more general potentials (21) also allow exact solution at $E = 0$. The zero-energy bound states exist for particular values of potential strength

$$v_{N_\mu} = R^2\mu^2\left(N_\mu + \frac{1}{2\mu}\right)\left(N_\mu + \frac{1}{2\mu} - 1\right), \qquad (35)$$

$$N_\mu = n + \left(\frac{1}{\mu} - 1\right) l \, . \tag{36}$$

Generally such a formulation is known as Sturmian problem: the energy E is kept fixed and some other parameter (in our case the potential strength v) is quantized. Formula (35) shows that that with the focusing potentials (21) one can obtain degeneracy patterns of zero-energy levels characterized by an arbitrary linear combination N_μ of quantum numbers n and l. If one chooses μ somewhat different from our value $\mu = \frac{1}{2}$, then the ordering of one-electron levels is changed, but the difference in comparison with $(n+l, n)$ rule becomes noticeable only for large n and l. The lower part of thus obtained Periodic Table remains the same. Such an exercise was done by Purdela (1988) who promoted $n + \frac{1}{2}l$ ordering that corresponds to $\mu = \frac{2}{3}$. The feasible advantage of this choice remains awkward, but the disadvantage is clear: unphysical $\sim r^{-2/3}$ singularity in the potential (21) compared to the physical Coulomb behaviour $\sim r^{-1}$ for $\mu = \frac{1}{2}$ discussed in this section. It is worthwhile to note here that $n + \frac{1}{2}l$ appeared earlier in the paper by Katriel and Jorgensen (1982) as a part of their diagram correlating $(n+l)$ ordering for neutral atoms and n-ordering for positive ions.

6. Group Theoretical Approaches

6.1. Classification, symmetry and group theory

The symmetry considerations are natural when the Periodic System of Elements is discussed. One feasible way of reasoning is to notice that the *double necklace* orbits are obviously symmetrical[9]. Another way to symmetrical point of view stems from the general observation that as a problem of *classification*, the structure of the Periodic System is subject of the Group Theory that is ultimately related to symmetry analysis.

In 1971–1972, when after Demkov and Ostrovsky (1971c) found approximate one-electron potential in atom (30) that explains appearance of $(n+l)$ quantum number and Wheeler (1971) suggested the name *double-necklace orbits* for classical electron trajectories in atoms, three groups of researchers[10] independently and unaware of cited papers applied group-theoretical approach to the Periodic Table of Elements: (i) Rumer and Fet (1971) and Konopel'chenko (1972) in Novosibirsk, USSR; (ii) Barut (1972) (University of Colorado, USA) who published his paper in New Zealand, in Proceedings of Rutherford Centennial Symposium; (iii) Novaro and Wolf (1971) and Novaro and

Berrondo (1972) who worked in Mexico. The methodology was apparently inspired by the success of the group-theoretical methods in hadron physics. Below we call this Elementary Particle Approach (EPA) in the theory of the Periodic System of Elements.

The *symmetry* of a physical system implies that there are some operations that do not change the system (more exactly, leave invariant its Hamiltonian operator)[11]. These operations comprise the *symmetry group*. They could have geometrical meaning (such as rotations in the case of spherically symmetrical potential), but also can evade direct geometrical interpretation (so called hidden symmetries). By applying to eigenstate an operation belonging to the symmetry group we obtain another (degenerate) eigenstate with the same energy. The symmetry group operators (generators) commute with the system Hamiltonian.

Another important notion is the *dynamical group* which includes the symmetry group, but contains also operators allowing one to construct a complete set of eigenstates starting from any particular one. This means that all possible states of the physical system are contained in a single irreducible representation (IR) of its dynamical group. Hence the generators of dynamical group generally do not commute with the Hamiltonian.

If the group is known, then mathematical (group-theoretical) techniques allows one to produce full set of labels necessary for classification of states; the dynamical group provides complete classification whereas the symmetry group is able to predict degeneracy patterns met in the system. For instance, symmetry of central potentials is described by the three-dimensional rotation groups designated as O(3). This group provides l and m labels and predicts $(2l + 1)$-fold degeneracies of energy levels, i.e., independence of energies on the azimuthal quantum number m. The methodology of construction of the symmetry group from the degeneracy pattern was considered in the enlightening paper by Demkov (1963).

Both symmetry and dynamical groups could be *derived* from analysis of the Hamiltonian operator of the physical system under consideration. In this way one can find not only rather obvious geometrical symmetries, but also hidden symmetries, such as four-dimensional rotation group O(4) for the hydrogen atom revealed by Fock (1935) via quite sophisticated construction including transition to the momentum space and its subsequent stereographic projection on a sphere in four-dimensional space. This group provided a symmetry explanation of degeneracy of hydrogen energy levels in orbital quantum number l, that otherwise has been regarded as an accidental one.

For the hydrogen atom the dynamical group was found by Malkin and Man'ko (1965, 1966) as the group[12] O(4,2) [see also Barut and

Kleinert (1967)]. The IRs of this group were analyzed by Barut and Böhm (1970) who labeled them by two numbers S and k_0. The IR ($S = 0$, $k_0 = 0$) is related to the H atom, comprising all its discrete spectrum. Its reduction over the chain of subgroups

$$O(4,2) \supset O(4,1) \supset O(4) \supset O(3) \supset O(2) \tag{37}$$

contains Fock's symmetry group of H atom O(4). This IR remains irreducible for O(4,1) subgroup of O(4,2). Hence both O(4,2) and O(4,1) could be considered as dynamic group for the discrete states of the hydrogen atom.

When the Periodic System is considered from a group-theoretical viewpoint, the natural task is to analyze the symmetry properties of the one-electron potential in multielectron atoms, just like the symmetry properties of the Coulomb potential were analyzed in case of hydrogen atom. This line of analysis is referred to below as Atomic Physics Approach (APA), see section 6.3. However at first we discuss its counterpart, EPA in more detail.

6.2. Elementary Particle Approach to the Periodic Table

For some physical systems it might occur that the Hamiltonian is not known (or even does not exist, at least in the conventional sense), but the underlying symmetry or dynamical group could be somehow *guessed.* This was the background for successful applications of group theory in elementary particle theory. Its great achievement was prediction of some new particles (resonances). These particles were related to some missing members in the adopted group-theoretical scheme. Such a conclusion is possible only if a correspondence between mathematical (IRs) and physical (particles) objects is established. In case of application to the Periodic System such a correspondence is a weak point of many EPA studies.

EPA, contrary to APA, implies that the knowledge of an atom consisting of electrons and atomic nucleus is dropped. Instead, the chemical elements are considered formally as various states of some imagined artificial object: *'atomic matter'* (Barut, 1972) or *'structureless particle with inner degrees of freedom'* (Rumer and Fet, 1971). Various states of such a system can be labeled by the quantum numbers. For example, if the symmetry group of the object is O(4), then the integer quantum numbers n, l, m identify the states (IRs members); the range of variation for these numbers follows from the group theory, being the same as for the hydrogen atom ($0 \le l \le n-1$, $-l \le m \le l$). However these quantum numbers *are not related* to the principal, orbital and azimuthal quantum numbers of any electron state: as emphasized above, the approach simply does not rely on the fact that the atoms are built from the electrons. In order to distinguish the 'internal' quantum numbers in EPA sense we will denote them by the tilde sign: $\tilde{n}$, $\tilde{l}$, $\tilde{m}$. One can find some roots for this interpretation in the early book by Madelung (1936), as discussed in section 2.

In case of (n, l) rule (actually applicable to highly ionized atoms, see section 1) the order Q of the element in the chart can be evaluated via these quantum numbers as

$$Q(\tilde{n},\tilde{l},\tilde{m},\tilde{\sigma}) = \frac{1}{12}2\tilde{n}(2\tilde{n}-1)(2\tilde{n}-2) + 2\tilde{l}(\tilde{l}+1) + 2\tilde{m} + \tilde{\sigma} + \frac{3}{2}\ . \tag{38}$$

In atomic physics Q is interpreted as the number of electrons, or ordinal number of the element[13]. The variable $\tilde{\sigma}$ adopts two values: $\pm\frac{1}{2}$. The neighboring states with the same $\tilde{n}$-values comprise the multiplets of O(4) as in the non-relativistic hydrogen atom. The dynamic group O(4,2) contains operators which change one element into another; hence they change $Q(\tilde{n},\tilde{l},\tilde{m},\tilde{\sigma})$, so that integer number $\tilde{n}$ varies from 1 to ∞.

The algebraic formulation of the $(n+l, n)$ rule similar to (38) takes the form (Barut, 1972)

$$\begin{aligned} Q &= \frac{1}{6}\left[(\tilde{n}+\tilde{l})^3 + 3(\tilde{n}+\tilde{l})^2 + 5(\tilde{n}+\tilde{l}) + 3\right] - 2\left[\tilde{l}(\tilde{l}+1) + \tilde{m}\right] + \tilde{\sigma} - \frac{1}{2} \\ &\quad (\text{for odd } \tilde{n}+\tilde{l}); \qquad (39) \\ Q &= \frac{1}{6}(\tilde{n}+\tilde{l})(\tilde{n}+\tilde{l}+1)(\tilde{n}+\tilde{l}+2) - 2\left[\tilde{l}(\tilde{l}+1) + \tilde{m}\right] + \tilde{\sigma} - \frac{1}{2} \\ &\quad (\text{for even } \tilde{n}+\tilde{l})\ . \qquad (40) \end{aligned}$$

Thus one has two different analytical expressions, one for odd and the other for even values of $\tilde{n}+\tilde{l}$. This peculiarity, apparently unavoidable, correlates with the doubling of periods ultimately related to the $(n+l, n)$ rule, see discussion above. One can always write down such a pair of formulae as a single expression by using term $(-1)^{n+l}$, cf. (6). A large number of similar formulae was suggested by Klechkovskii (1968) without using group-theoretical technique [see also book by Mazurs (1974)].

Since EPA is necessarily a phenomenological approach, it is particularly important (i) to specify exactly the empirical basis for the choice of the group and (ii) to outline how the formal mathematical scheme can be interrelated to the observable physical objects and quantities. The authors who apply EPA to the Periodic System, as anticipated, choose to forget about structure of atom (even about such basic ideas that an atom consists of electrons and nucleus). Nevertheless, the choice of group is in fact based on some information outside Group Theory, and there should be a possibility of comparing the results with empirical data. It is highly desirable to demonstrate that the output suitable for comparison exceeds the empirical input. From this point of view the statement that "Within this approach we have to give up all available chemical and spectroscopic information" (Konopel'chenko and Rumer, 1979) looks as unfounded extremism, since information of this kind is necessary just for choice of the particular group among the infinite number of options provided by pure mathematics.

While some of the works in the field fail to satisfy criteria (i)-(ii)[14], there are a number of papers that admit the goal of finding the group which allows the degeneracy pattern coinciding with the empirically known period lengths in the empirical Periodic Table. Let us start with more simple case of (n, l) rule applicable to ionized atoms. The period lengths in this case are $2\mathcal{N}^2$, $\mathcal{N} = 1, 2, 3, \ldots$, i.e. the same as degeneracies met in hydrogen atom. This fact allows one to conclude that the Periodic Table with (n, l) rule is described by O(4) (broken) symmetry group and O(4,2) dynamical group. In other words EPA message is like this: "You can forget that an atom consists of electron and nucleus, but *if you believe* that it possesses (broken) O(4) symmetry and O(4,2) dynamical groups, then you can conclude that the Periodic System of Elements is operative, with the period lengths $2\mathcal{N}^2$, as found in Nature".

Imagine analogous situation for hydrogen atom: the knowledge of $2\mathcal{N}^2$ degeneracy pattern (say, from analysis of spectroscopic observations) is retained, but information about H atom consisting of electron and proton is dropped. Then following EPA methodology one can say that hydrogen atom has O(4) symmetry group and O(4,2) dynamical group. Hardly anything else could be added to this if one remains at EPA standpoint. An actual historical development was different and in present terminology could be described as APA treatment. Fock (1935) revealed O(4) group of the Coulomb problem by stereographic projection of 3D momentum space on a sphere in 4D space. This *dynamic treatment* allowed him to relate hydrogenic wave functions to spherical harmonics in four-dimensional space. In particular, this resulted in new and useful sum rules for H atom (see, for instance, a paper by Chibisov *et al* (2000) and references therein). The generators of O(4) group are components of two vectors: angular momentum of electron and its Runge-Lenz vector. Physical interpretation of this fact led to solution of important physical problems, such as energy levels of hydrogen atom in crossed electric and magnetic fields (Demkov *et al*, 1969) mixing of Rydberg atom states by collisions with charged particle (Kazansky and Ostrovsky, 1996, Vrinceanu and Flannery, 2001); atomic pseudospin resonance (Fregenal *et al*, 2001) and so on. Dynamic group was used less intensely, although Barut and Kleinert (1967) based on it elegant derivation of electromagnetic transition probabilities in H atom. In all these applications appealing and important analytical results were obtained. This comparison is to illustrate the difference between the outputs that that could be anticipated from EPA and APA.

The case of $(n+l,\ l)$ rule differs by doubling of periods, as compared to $(n,\ l)$ rule, see discussion in section 5. Hence the task is to modify the Coulomb field group so as to incorporate this doubling. Solution of this problem proves to be non-unique, the specific groups suggested by different authors[15] (Barut, 1972; Rumer and Fet, 1971; Konopel'chenko, 1972; Novaro and Wolf, 1971; Novaro and Berrondo, 1972; Berrondo and Novaro, 1973; Novaro, 1973; Fet, 1979; Fet, 1980; Kalloniatis, 1988; Novaro, 1989) have been critically reviewed by Ostrovsky (1996). Within EPA approach there is no reason to prefer any of these formal constructions.

Here we briefly mention that Barut (1972) had suggested SO(4,2) as the dynamic group of the Periodic Table whereas Novaro and Berrondo (1972) had chosen SU(2)⊗SU(2)⊗SU(2). The scheme by Fet evolved: at first this author was not aware of the period doubling and suggested dynamic group as $O(4,2)\otimes SU(2)_S$ (Fet 1974, 1975, Byakov *et al*, 1976-1977) where the subscript S indicates that this group is responsible for the 'spin doubling', i.e. to the fact that according to Pauli exclusion principle every nlm orbital could be occupied by two electrons with opposite projections of spins. In other words SUS(2) group is used to explain doubling of period lengths from $\mathcal{N}^2$ to $2\mathcal{N}^2$. Later (Fet 1979, 1980) the group was changed to

$$\mathrm{G_F} = \mathrm{O}(4,2) \otimes \mathrm{SU}(2)_S \otimes \mathrm{Z}_2\ , \tag{41}$$

where the cyclic group (i.e. permutation group with two elements) Z_2 was introduced to account for the period doubling. Note that two kinds of doubling (spin- and period-) are described by very different mathematical means [$\mathrm{SU}(2)_S$ and Z_2 groups] that has no justification and esthetically does not look attractive.

The most simple and uniform way to achieve period doubling is to choose dynamical group for the Periodic System in form of direct product

$$\mathrm{G_{PS}} = \mathrm{O}(4,2) \otimes \mathrm{SU}(2)_S \otimes \mathrm{SU}(2)_T \tag{42}$$

where $SU(2)_T$ is group of 'atomic isospin' responsible for period doubling. When one chooses two-dimensional IR of the latter group (with pseudospin $T = \frac{1}{2}$), each IR of $O(4,2)$ is met twice in IR of the product group $O(4,2) \times SU(2)_T$: once with $T_3 = \frac{1}{2}$, and for the second time with

$T_3 = -\frac{1}{2}$. The group (42) was put forward at first in the conference proceedings (Ostrovsky, 1980) in the same volume as suggestion of the group (41) (Fet, 1980); more accessible is the journal paper (Ostrovsky, 1981). Subsequently the group (42) was accepted by Fet (1992) [see also Chapter 2 written by this author for the book by Hefferlin (1989)].

The group (42) was suggested and developed (Ostrovsky 1980, 1981) as a phenomenological construction within EPA spirit. Most importantly, simultaneously it was related along APA lines to the particular symmetry of potential (30), as discussed in section 6.3.

Ideally one could vision how EPA predicts existence of the group of noble gases in the Periodic Table. The basis could be the fact that without these elements the period lengths is $2\mathcal{N}^2 - 1$. Since no suitable group with such IRs dimensions is found, a brave group-theoretician chooses to stick to O(4) symmetry group, declares that one group of elements is missing in the chart and suggests experimental search for it. Such a prediction would be even more difficult than prediction of new elements by Mendeleev, since the latter author used interpolation between properties of already known elements, but in case of noble gases productive interpolation is hardly possible. Nobody knows if such a group-theoretical prediction could occur in reality, since the noble gases are already known for about one hundred years. "A chemist at that time could not wait until quantum mechanics was completed" (Göppert-Mayer and Jensen, 1955).

Unfortunately further perspectives of EPA remain unclear. Its current achievements look like translation of empirical information about period lengths to the mathematical language which is probably more appealing for a certain part of scientific community, but nothing more. The empirical input (period lengths) is cast in *mathematical* terms of the dynamical group, but hardly anything more than period lengths return back to the interested researcher who could aspire for some new results and theoretical predictions to be compared with experiment, or additional *physical* insight. For most workers in the field such translation into specialized mathematical language does not justify an accompanying sacrifice: giving up all references to electronic structure of atoms.

6.3. Atomic Physics Approach and Secondary Periodicity

Contrary to EPA, the APA is not based on empirical information but seeks to find symmetry properties of exact or approximate Hamiltonians that are already available from *ab initio* quantum theory. The indications of deep symmetry of the effective atomic potential (28) were revealed by Demkov and Ostrovsky (1971c). Namely, as discussed above, all the classical trajectories are closed for $E = 0$ and, moreover, possess focusing properties. The quantum levels for $E = 0$ appear simultaneously for different values of orbital momentum l.

The relevant *quantum* symmetry analysis was developed by Ostrovsky (1981) who used stereographic projection to map the three-dimensional quantum problem with potential (28) onto four-dimensional sphere. This could be compared with Fock's (1935) construction when the *momentum space* was mapped on 4D sphere. Ostrovsky carried out his transformation in conventional coordinate space and cast problem in 4D coordinates ξ_1, ξ_2, ξ_3, ξ_4 with the constraint $\xi_1^2 + \xi_2^2 + \xi_3^2 + \xi_4^2 = R^2$, where R is radius of the sphere. The coordinates ξ_1, ξ_2, ξ_3 lie in conventional 3D space whereas ξ_4 is orthogonal to it. Denote as T_3 a *discrete* operation of reflection $\xi_4 \Rightarrow -\xi_4$. Ostrovsky (1981) explicitly demonstrated that the approximate atomic Hamiltonian with effective potential (28) is *invariant* under this transformation and thus revealed an additional (hidden) integral of motion. In conventional 3D coordinate space T_3 generates the following *discrete* transformation of the wave function [see also Demkov and Ostrovsky (1971b)]:

$$T_3\,\psi(\vec{r}) = \frac{R}{2r}\,\psi\left(\vec{r}\,\frac{R^2}{r^2}\right) . \tag{43}$$

In geometry the transformation $\vec{r} \Rightarrow \vec{r}\,R^2/r^2$ is known as *inversion in the sphere* of radius R. It conformally maps inner parts of the sphere $r = R$ in 3D space onto its outer part, and vice versa. In quantum mechanics this operation was considered by Demkov and Ostrovsky (1971b) and Demkov and Semenova (1984).

The discrete operator T_3 has only two eigenvalues $\frac{1}{2}\tau$ with $\tau = \pm 1$. This situation is similar to the electron spin projection that takes values $m_s = \pm\frac{1}{2}$. In analogy with the terminology of nuclear physics the quantum number τ could be named *atomic isospin*. The isospin vector operator has three components $\{T_1, T_2, T_3\}$ in some abstract space. The conventional commutation relations for these three operators generate algebra of $SU(2)_T$ group that enters (42). Importantly, T_3 operator has simple meaning in the coordinate representation, as expressed by formula (43). The zero-energy eigenstates of potential (30) are also eigenstates of the operator T_3. The odd and even values of $(n+l)$ correspond to $\tau = 1$ and $\tau = -1$ respectively. Thus an additional integral of motion is revealed, providing an additional classifying label τ. It allows one to distinguish $2\mathcal{N}^2$-dimensional periods met in pairs as discussed above.

Thus the dynamical group (42) for the Periodic System suggested by Ostrovsky (1981) incorporates operator T_3 with a clear geometrical meaning. This is its advantage as compared with the groups suggested within EPA. The additional atomic isospin classification also has physical (or chemical) meaning.

For a rather long time the so called 'secondary' periodicity effect was discussed in chemistry. Originally Biron (1915) noticed that for the elements in a given group some chemical and physical properties are reproduced most completely not in the adjacent periods, but in every second period. As one of examples he considered tendency of N, As and Bi to be trivalent while P and Sb are pentavalent. According to Smith (1924) "it is ... a general observation that alternate members of a valence group in the periodic table show the greatest chemical resemblance"[16]. In chemical literature the secondary periodicity is also referred to as an 'alternation effect', manifested, for instance, for electronegativity (Sanderson 1952, 1960), or for heats of formation of oxides, and also for ionization potentials and sizes of ions (Phillips and Williams, 1965). Some additional bibliography can be found in the papers by Kapustinskii (1951a), Ostrovsky (1981) and Pyykkö (1988). The latter reference interprets secondary periodicity in terms of properties of Hartree-Fock orbitals. The convincing graphical illustration of the secondary periodicity was provided by Odabasi (1973) who depicted variation of ionization potentials and mean value of r^{-2} along some groups of elements in the Periodic Table. These properties exhibit saw-like modulation of general smooth trend along the Table column.

The elements within the group that belong to every second period in the Table correspond to the same quantum number τ [17]. The atomic model gives here a new insight *relating the secondary periodicity to the properties of the atomic orbitals*, namely, the isospin, or T_3-symmetry. From this point of view it is particularly important that the T_3-symmetry is stable with respect to the variation of the atomic potential and the energy of the electron. Indeed, as shown by Ostrovsky (1981), the *realistic atomic orbitals* [calculated within a simplified version of the Hartree-Fock approximation (Herman and Skillman, 1963)] in a good approximation possess a definite parity under the transformation T_3 (43). The possible applications of this symmetry for the calculation of some matrix elements are also discussed in this paper.

Kitagawara and Barut (1983; 1984) modified scheme by Ostrovsky (1981) to consider mapping of (unphysical) two-dimensional problem[18] on the three-dimensional sphere. In this case the treatment is much easier due to possibility to use well developed mathematical theory of complex variables. Although these authors claimed to correct some drawback of treatment by Ostrovsky (1981), theirs own scheme constructed for three-dimensional problem suffers from very serious deficiencies (Ostrovsky, 1996). Note also that Kitagawara and Barut misinterpret meaning of the potential (28) by fully limiting it to EPA framework: "The Demkov-Ostrovsky equation is just a mathematical model providing the quantum number $n + l$ and its degeneracy. The coordinates appearing in this equation do not have a direct physical meaning such as the spatial coordinates of the valence electron in an atom". In the same spirit Scerri *et al* (1998b) call this potential "heuristic" [in the sense *ad hoc*] that "leaves us with necessity to explain where this particular potential came from".

7. Conclusion

The Periodic Law embraces vast amount of knowledge. Only part of it could be cast in physical terms and expressed in quantitative measure (such as atomic radii or ionization potentials, and densities, values of specific heat etc; electronic configurations of atoms also belong here). The other part of information, probably more important for chemists, belongs to pure chemistry and indicates a variety of facts and trends concerning valences, oxidation states, chemical compounds and reactivities, and so on. These chemical properties are often difficult to formalize and express numerically. From these vast piles of information we have chosen to consider mostly the Periodic Table of Neutral Atoms that is well formalized in quantum language of atomic configurations. The notion of configuration is well known to be an approximation and the most sophisticated and numerically accurate calculations describe atomic wave function by superposition of atomic configurations[19]. However the fact that the configurations are used in *overall* quantum analysis of the Periodic Table of Atoms is not only necessary and inevitable but at the same time advantageous since approximations are gnosiologically crucial in providing interpretations and explanations [see more discussion by Ostrovsky (2001)].

As shown in section 4, the phenomenological $(n + l, n))$ rule alone does not allow to uniquely restore effective one-electron potential in atom. The choice of a particular potential form (28) is seminal, since it provides a simple analytical approximation to the Thomas-Fermi one-electron potential, with a proper Coulomb singularity at the force center. This potential has focusing properties, allows exact solution of Schrödinger equation and ensures realization of both parts of $(n + l, n))$ rule. Moreover, it possesses additional (hidden) symmetry under operation (43) that provides a physical quantum interpretation for periodic doubling and secondary periodicity[20]. A combination of these features as well as applicability of $(n + l, n)$ rule in a very broad range of atomic numbers could be considered as something above expectations. Therefore situation urges for deeper understanding. For instance, it remains unclear how the special properties of the potential (28) are related to the well known high symmetry of the Coulomb interaction between the electrons and nuclei. A number of paper was devoted to the search for O(4) and other symmetries in atoms (Novaro, 1971; Ellis and Higginbotham, 1975; Herrick and Kellman, 1978; Herrick, 1980; Wulfman, 1979; Wulfman, 1983; Wulfman and Levine, 1984), and references therein, see also book by Judd (1967). However, the relation between these studies and the symmetry of the Periodic Table was not established.

Sometimes the question is asked, if quantum mechanics had fully explained the Periodic Law (see, for instance, Scerri, 1997a). To our opinion, this question is similar to: "Did physics explain what is an electron ?". The answer is both 'yes' and 'no'. We know a lot about properties of electron and can predict and govern its behaviour in various situations. However the story is not finished yet. The explanations of deeper levels on what is the nature of electric charge or how the electron mass could be precalculated are not achieved yet. The deep problems, to which the Periodic Law of course belongs, are never closed, but change their formulations.

Our *overall description* of the Periodic Table does not embrace more subtle effects and conclusions that could be drawn from quantum analysis of the periodicity phenomenon. To exemplify, we cite only part of the related lists (Greenwood and Earnshaw, 1984, p. 32, and Pyykkö 1991, 2001, Fig. 1): the 'anomalous' properties of elements in the first period (from lithium to fluorine); partial screening effects: d-shell contraction, lanthanoid contraction; inert pair effect; relativistic contraction and stabilization of s and p electrons, relativistic expansion and destablization of d and f electron, and so on. These features till now evade description based on symmetry principles that led Pyykkö (1991, 2001) to quite pessimistic conclusion: "No deep group-theoretical principle". Some authors go further and argue that the connection between electronic configuration and the chemical behaviour is loose for heavy elements: "There is no simple and clear-cut relation between the chemistry of given elements and the electronic structure of its monoatomic entities" (Jorgensen, 1973, 1979). The question, however, is if a similar relation exists between chemistry of an element and its position in the Periodic Table.

Although the present discussion of the group theoretical results has strong critical flavour, one potential advantage of this approach should not be forgotten: it could hope for description of the Periodic Law from more broad point of view than the electronic structure of an atom and even more broad than physics by incorporating entire wealth of chemical information referred to above. Indeed, the group theoretical schemes within EPA require so little of quantum mechanics that probably could cut the links at all and emerge as some universal methods of description. The future would show to what extent this ideal goal could be achieved. The quantum mechanics is not necessary in this pursuit; moreover, the objective could be achieved only within some more broad theory or mathematical discipline. The group theory as a mathematical device is, in this sense, already broader than quantum mechanics. It seems that research in Periodic Tables of Molecules generally follow along these lines (Hefferlin *et al*, 1979a; Hefferlin *et al*, 1979b; Hefferlin and Kuhlman, 1980; Hefferlin and Kutzner, 1980; Kong, 1982; Zhuvikin and Hefferlin, 1983; Hefferlin *et al*, 1984; Hefferlin, 1989; Carlson *et al*, 1996; Karlson *et al*, 1995; Hefferlin *et al*, 1996; Cavanaugh *et al*, 1996; Hefferlin and Knoll, 2000; Hefferlin, 2000; Hefferlin and Matus, 2001). Some novel non-quantum mathematical techniques were probed recently in application to the Periodic Law (Gorban', 1992; Horia *et al*, 1996; Scerri *et al*, 1998b; Khramov *et al*, 1998; Khramov *et al*, 2000; Sneath, 2000; Tarbeev *et al*, 2001). The future studies will show how productive are these approaches.

Acknowledgements

The fruitful discussions with Yu. N. Demkov are highly appreciated. This work was partially supported by the National Science Foundation through a grant for the Institute for Theoretical Atomic and Molecular Physics at Harvard University and Smithsonian Astrophysical Observatory.

Notes

[1] The first of these principles is an approximation that is not universal but legitimate and productive for ground states of atoms, and also for one-electron excitations. For the doubly excited states it could be invalid. The Pauli exclusion principle has well known general formulation in terms of wave function symmetry that does not resort to one-electron quantum numbers. The minimum energy of multielectron system does not always correspond to filling of one-electron orbitals according to Aufbau principle, that could cause some minor deviations from general scheme.

[2] As Barut (1972) remarks, relativistic effects in H atom lead to the same (n, l) ordering. Of course, non-Coulomb core effects are far more important for valence electrons in neutral atoms.

[3] Perelomov *et al* (1966) discussed condition under which the n-grouping remains valid.

[4] A valuable bibliographic source is book by Mazurs (1974). However being interested mostly in graphical representations of the Periodic System, this author does not present complete historical survey of the $(n+l)$ rule. For instance, although the paper by Karapetoff (1930) is referred to, its role in establishing $(n + l)$ rule is not mentioned. The book by van Spronsen (1969) is devoted to pre- or non- quantum aspects of the Periodic Law; for instance, such notion as *quantum number* practically does not play a role in the exposure. The new bibliographical source was provided recently by Scerri and Edwards (2001).

[5] Later Klechkovskii (1968) derived this formula as a corollary of $(n + l, n)$ rule.

[6] The attempt by Neubert (1970) to correlate $(n + l)$ rule and pure Coulomb potential does not look convincing.

[7] For some further bibliographical remarks see paper by Demkov *et al* (1971a), where uniqueness of the solution is also analyzed.

[8] The semiclassical version of this problem was considered by Miller (1969); see also the book by Child (1991).

[9] However, the entire issue of symmetry in classical mechanics is not simple. There is a known statement (Mukunda, 1967) that *all* classical systems involving three degrees of freedom automatically possess both SO(4) and SU(3) symmetries. From this point of view all the classical systems are "equally symmetrical". However, obviously some hierarchy with respect to the symmetry properties exists even in classical mechanics, and there is a need for some tool to distinguish "more and less symmetrical" systems. The appeal to the periodic orbits seems to provide such a possibility.

[10] In the most general terms a suggestion to apply group theory to the Periodic System could be found in the papers by Neubert (1970).

[11] We present here only a brief qualitative discussion of group theoretical approach bearing in mind its applications to the Periodic Table reviewed by Ostrovsky (1996). From a different standpoint group theoretical approach was reviewed by Wulfman (1979) and Fet [1992; see also chapter 2 in the book by Hefferlin (1989)].

[12] The group O(4,2) differs from SO(4,2) only by the presence of the discrete parity operator; the relation between IRs of these groups is very simple (Barut and Böhm, 1970). Various authors consider alternatively O(4,2) and SO(4,2). For uniformity of discussion in this paper we refer everywhere to the groups with the parity operator included.

[13] Barut (1972) and subsequent authors (Odabasi, 1973) equate the right-hand-side of this formula to 'the number of nuclear protons' Z which for the ions differs from Q. However, the Aufbau rule is related to the electronic shell, i.e. to Q. For instance, for the highly ionized atoms with the fixed Q the nuclear charge Z can be varied (under condition $Z \gg Q$), but the structure of electronic shell remains the same.

[14] Rumer and Fet (1971) and Fet (1974, 1975) establish correspondence between formal IRs states and chemical elements in order of increasing atomic weights which, as shown by Mendeleev, are not related directly to the properties classified by the Periodic Table. Rumer and Fet (1971), Konopel'chenko (1972), Fet (1974, 1975) do not even mention $(n + l, n)$ rule that describes real structure of the Periodic Table.

[15] It is worthwhile to indicate that Novaro and Berrondo (Novaro and Berrondo, 1972; Berrondo and Novaro, 1973; Novaro, 1973; Novaro, 1989) looked for the group which describes the chemical periods, whereas other authors have concentrated on the $(n + l)$ grouping. The phenomenological difference between these patterns expressed by the scheme (4) was discussed in more detail in section 5. It seems

that the interpretation of this difference due to Demkov (1975) essentially lifts the problem.

[16] It is worthwhile to give here extended citation from the book by Smith (1924) that might be not easily available: "It is, however, probable that radium is more closely allied to strontium (both give intensely red flame coloration), just as thorium is most closely allied to zirconium, and uranium most closely allied to molybdenum. It is, in fact, a general observation that alternate members of a valence group in the periodic table show the greatest chemical resemblance, for example, iodine and chlorine; bromine and fluorine; bismuth, arsenic and nitrogen; the antimony and phosphorus."

[17] The secondary periodicity was considered by by Barut (1972) [and also by some other authors, (Odabasi, 1973, Condon and Odabasi, 1980)] as a direct confirmation of his EPA scheme. However this effect is a mere manifestation of the period doubling which is intrinsic for $(n+l, n)$ rule adopted by Barut (1972) as a phenomenological foundation of his group-theoretical scheme.

[18] The Periodic Table in a two-dimensional world was considered also by Asturias and Aragon (1985), Négadi and Kibler (1996) and Négadi (2000).

[19] The statement that "The electronic configurations ... cannot be derived using quantum mechanics ... because the fundamental equation of quantum mechanics, the Schrödinger equation, cannot be solved exactly for atoms other than hydrogen" (Scerri, 1998b) is not exact. In modern quantum mechanics there is no other way to *derive* electronic configurations than starting from exact Schrödinger equation and developing scheme for its approximate solution. Therefore it looks too strong to say that "quantum mechanics forbids any talk of electrons in orbitals and hence electronic configuration" (Scerri, 1997a). Equally, it cannot be said that the atomic orbitals stem from approximations that logically contradict the theory [H. Post, as cited by Scerri (1989)]. The orbitals are simply the one-electron building blocks routinely used in quantum mechanics to construct multi-electron wave function. The Hartree-Fock method allows one to choose these blocks in an optimal way for each particular atom.

[20] Note that recent quantum formulation of inverse quantum problem in application to the Periodic Table (Tarbeev *et al*, 1997) leads either to potential (28) or to other versions which do not have symmetry properties discussed here.

References

A. A. Abrahamson, Onset of Atomic-Subshell Filling in Ordinary and Superheavy Elements. *Physical Review A* 4: 454-456, 1971.

F. J. Asturias and S. R. Aragon, The Hydrogen Atom and the Periodic Table in Two Spatial Dimensions. *American Journal of Physics* 53: 893-899, 1985.

R. Ausubel, A Simple Method for Determining the Order in which Electrons Fill Shells. *Journal of Chemical Education* 53: 645, 1976.

O. v. Auwers, Periodisches System auch Quantenzahlen. *Physikalische Blätter* 4: 423-425, 1948.

A. T. Balaban, Numere Magice de Electroni si de Nucleoni. *Revista de Chemie* 14: 158-160, 1963.

A. O. Barut, Group Structure of the Periodic System. In *The Structure of Matter, Proceedings of Rutherford Centennial Symposium*, edited by B. G. Wybourne,

Christchurch, New Zealand: University of Canterbury Publications, Bascands, 1972, pp. 126-136.

A. O. Barut and A. Böhm, Reduction of a Class of O(4,2) Representations with Respect to SO(4,1) and SO(3,2). *Journal of Mathematical Physics* 11: 2938-2945, 1970.

A. O. Barut and H. Kleinert, Transition Probabilities of the Hydrogen Atom from Noncompact Dynamical Groups. *Physical Review* 156: 1541-1545, 1967.

H. Bauer, *Grundlagen der Atomphysik*, Wien, Springer 1938.

N. B. Berezina and Yu. N. Demkov, Low-Energy Resonance Scattering of Electrons by Thomas-Fermi Atom. *Zhurnal Eksperimental'noi i Teoreticheskoi Fiziki* 68: 848-854, 1975 [*Soviet Physics – Journal of Experimental and Theoretical Physics* 41: 421-424, 1975].

M. Berrondo and O. Novaro, On a Geometrical Realization of the Aufbau Scheme. *Journal of Physics B* 6: 761-769, 1973.

M. Berrondo, J. Flores, and O. Novaro, La Théorème de Bertrand a Cent Ans de sa Formulation. *Compte Rendus* 278: B455-B457, 1974.

J. Bertrand, Theóreme Relatif au Mouvement d'un Point Attire vers un Centre Fixe. *Comptes Rendus* 77: 849, 1873.

E. V. Biron, Yavleniya Vtorichnoi Periodichnosti [In Russian: Phenomena of Secondary Periodicity]. *Zhurnal Russkogo Fiziko-Khimicheskogo Obschestva, Chast' Khimicheskaya* 47: 964-988, 1915.

N. Bohr, *Collected Works*, vol. 4 *[Periodic System (1920-1923)]*, edited by J. Rud Nielsen, Amsterdam: North Holland, 1977.

L. Bosi, The Aufbau Principle Revisited. *Nuovo Cimento B* 76: 89-96 (1983).

J. E. Brittain. Vladimir Karapetoff: A Pioneer Electrical Engineering Educator. *Proceedings of the Institute of Electrical and Electronic Engineers* 85: 1662-1663, 1997.

H. A. Buchdahl. Kepler Problem and Maxwell Fish-Eye. *American Journal of Physics* 46: 840-843, 1978.

V. M. Byakov, V. I. Kulakov, Y. B. Rumer, and A. I. Fet, *Group-Theoretical Classification of Chemical Elements. I. Physical Foundations, Preprint ITEP-26*, Moscow: Institute of Theoretical and Experimental Physics, 1976; *Group-Theoretical Classification of Chemical Elements. II. Description of Applied Groups, Preprint ITEP-90*, Moscow: Institute of Theoretical and Experimental Physics, 1976; *Group-Theoretical Classification of Chemical Elements. III. Comparison with the Properties of Elements, Preprint ITEP-7*, Moscow: Institute of Theoretical and Experimental Physics, 1977.

A. Y. Carlson, I. Lingren, and R. Marrus, Atomic-Beam Investigations of Electronic and Nuclear Ground States in the Rare-Earth Region. *Physical Review* 122: 1796-1801, 1961.

C. M. Carlson, R. J. Cavanaugh, R. A. Hefferlin, and G. V. Zhuvikin, Periodic Systems of Molecular States from the Boson Group Dynamics of $SO(3)\times SU(2)_s$. *Journal of Chemical Information and Computational Science* 36: 396-398, 1996.

B. Carroll and A. Lehrman, The Electron Configuration of the Ground State of Elements. *Journal of Chemical Education* 25: 662-666, 1948.

R. Cavanaugh, R. Marsa, J. Robertson, and R. A. Hefferlin, Adjacent DIM Isoelectronic Molecules and Chemical Similarity among Triatomics. *Journal of Molecular Structure* 382: 137-145, 1996.

M. Chibisov, A. M. Ermolaev, F. Brouillard, and M. H. Cherkani, New Summation Rules For Coulomb Wave Functions. *Physical Review Letters* 84: 451-454, 2000.

M. S. Child, *Semiclassical Mechanics with Molecular Applications*, Oxford: Clarendon Press, 1991.

E. U. Condon and H. Odabasi, *Atomic Structure*, Cambridge: Cambridge University Press, 1980.

D. G. Cooper, *The Periodic Table*, London: Buttenworths, 1960.

H. H. Dash, A Quantum Table of the Periodic System of Elements. *International Journal of Quantum Chemistry (Symposium)* IIIS: 335-340, 1969.

Yu. N. Demkov, The Definition of the Symmetry Group of a Quantum System. The Anisotropic Oscillator. *Zhurnal Eksperimental'noi i Teoreticheskoi Fiziki* 44: 2007-2010, 1963 [*Soviet Physics – Journal of Experimental and Theoretical Physics* 17: 1349-1351, 1963].

Yu. N. Demkov, B. S. Monozon, and V. N. Ostrovsky, Energy Levels of a Hydrogen Atom in Crossed Electric and Magnetic Fields. *Zhurnal Eksperimental'noi i Teoreticheskoi Fiziki* 57: 1431-1434, 1969 [*Soviet Physics – Journal of Experimental and Theoretical Physics* 30: 775-776, 1970].

Yu. N. Demkov, V. N. Ostrovsky, and N. B. Berezina, Uniqueness of the Firsov Inversion Method and Focusing Potentials. *Zhurnal Eksperimental'noi i Teoreticheskoi Fiziki* 60: 1604-1610, 1971a [*Soviet Physics – Journal of Experimental and Theoretical Physics* 33: 867-870, 1971].

Yu. N. Demkov and V. N. Ostrovsky, Internal symmetry of the Maxwell "fish-eye" problem and the Fock group for the hydrogen atom. *Zhurnal Eksperimental'noi i Teoreticheskoi Fiziki* 60: 2011-2018, 1971b [*Soviet Physics – Journal of Experimental and Theoretical Physics* 33: 1083-1087, 1971].

Yu. N. Demkov and V. N. Ostrovsky, $n + l$ Filling Rule in the Periodic System and Focusing Potentials. *Zhurnal Eksperimental'noi i Teoreticheskoi Fiziki* 62: 125-132, 1971c; *Errata* 63: 2376, 1972 [*Soviet Physics – Journal of Experimental and Theoretical Physics* 35: 66-69, 1972].

Yu. N. Demkov, *unpublished*, 1975.

Yu. N. Demkov and N. V. Semenova, *Teoreticheskaya i Matematicheskaya Fizika* 60: 423-431, 1984.

D. Dockx, *Theorie Fondamentale du Systeme Periodique des Elements, Archives de l'Institut International des Sciences Theorique* # 10, Bruxelles: Office International de Librairie, 1959.

D. G. Ellis and J. Higginbotham, Broken O_4 Symmetry in Alkali-Like Atoms. *Journal of Physics B* 8, 685-694 (1981).

H. Essén, Periodic Table of the Elements and the Thomas-Fermi Atom. *International Journal of Quantum Chemistry* XXI: 717-726, 1982.

E. Fermi, *Rend. Accad. Lincei* 6: 602, 1927.

E. Fermi, Eine Statistische Methode zur Bestimmung einiger Eigenshaften des Atoms und ihre Anwendung auf die Theorie des Periodischen Systems der Elemente. *Zeitschrift für Physik* 48: 73-79, 1928.

A. I. Fet, Conformal Group and Chemical Affinity. *Pis'ma v ZhETF* **20**, 24-26 (1974) [*JETP Letters.* **20** 10-11 (1974)].

A. I. Fet, Conformal symmetry of chemical elements. *Teoreticheskaya i Matematicheskaya Fizika* 22: 323-334, 1975 [*Theoretical and Mathematical Physics* 22: 227-235, 1975].

A. I. Fet, *The System of Elements from the Group-Theoretic Viewpoint*, Preprint 1, Novosibirsk: Institute of Inorganic Chemistry, 1979.

A. I. Fet, The Madelung Numbers and the System of Chemical Elements. In *Group Theoretical Methods in Physics, Proceedings of International Symposium*, edited by M. A. Markov, Moscow: Nauka, 1980, vol.1, pp. 327-336.

A. I. Fet, Gruppa Simmetrii Khimicheskikh Elementov [In Russian: Symmetry Group of Chemical Elements]. In *Matematicheskoe Modelirovanie v Biologii i Khimii [In Russian: Mathematical Modelling in Biology and Chemistry]*, edited by R. G. Khlebopros, Novosibirsk: Nauka, 1992, pp. 118-203.

O. B. Firsov, *Zhurnal Eksperimental'noi i Teoreticheskoi Fiziki* 24: 279, 1953.

V. Fock, Zur Theorie des Wasserstoffatoms. *Zeitschrift für Physik* 98: 145-154, 1935.

A. Frank, F. Leyvraz, and K. B. Wolf, Hidden Symmetry and Potential Group of the Maxwell-Fish-Eye. *Journal of Mathematical Physics* 31: 2757-2768, 1990.

D. Fregenal, T. Ehrenreich, B. Henningsen, E. Horsdal-Pedersen, L. Nyvang, and V. N. Ostrovsky, Atomic pseudospin resonance. *Physical Review Letters* 87: 223201 (1-4), 2001.

V. I. Gol'danskii, The Periodic System of D. I. Mendeleev and Problems of Nuclear Chemistry. *Journal of Chemical Education* 47: 406-416, 430, 1970.

P. Gombas, *Die Statistische Theorie des Atoms und ihre Anwendungen*, Vienna: Springer, 1949.

M. Göppert-Mayer, Rare-Earth and Transuranian Elements. *Physical Review* 60: 184-187, 1941.

M. Goeppert-Mayer and J. H. D. Jensen, *Elementary Theory of Nuclear Shell Structure*, New York, J. Wiley and Sons, 1955.

A. N. Gorban', E. M. Mirkes, A. P. Svitin, Poluempiricheskii Metod Klassifikatzii Atomov i Interpolyatziya ikh Svoistv [In Russian: Semiempirical Method of Classification of Atoms and Interpolation of Their Properties]. In *Matematicheskoe Modelirovanie v Biologii i Khimii [In Russian: Mathematical Modelling in Biology and Chemistry]*, edited by R. G. Khlebopros, Novosibirsk: Nauka, 1992, pp. 204-220.

S. A. Goudsmith and P. I. Richards, The Order of Electron Shells in Ionized Atoms. *Proceedings of National Academy of Science USA* 51: 664-671, 1964.

N. N. Greenwood and A. Earnshaw, *Chemistry of the Elements*, Pergamon, Oxford, 1984.

R. Hakala, Letter to the Editor. *Journal of Chemical Education* 25: 229, 1948.

R. Hakala, The Periodic Law in Mathematical Form. *Journal of Physical Chemistry* 56: 178-181, 1952.

R. Hefferlin, R. Campbell, and H. Kuhlman, The Periodic Table of Diatomic Molecules - I. An Algorithm for Retrieval and Prediction of Spectroscopical Properties. *Journal of Quantitative Spectroscopy and Radiation Transfer* 21: 315-336, 1979a.

R. Hefferlin, R. Campbell, and H. Kuhlman, The Periodic Table of Diatomic Molecules - II. Predicted Internuclear Separations from Curve-Fitted Data. *Journal of Quantitative Spectroscopy and Radiation Transfer* 21: 337-354, 1979b.

R. Hefferlin and H. Kuhlman, The Periodic Table of Diatomic Molecules - III. Theoretical Articulation. *Journal of Quantitative Spectroscopy and Radiation Transfer* 24: 379-383, 1980.

R. Hefferlin and M. Kutzner, Systematics of Ground-State Potential Minima between Two Main-Group Atoms or Ions. *Journal of Chemical Physics* 75: 1035-1036, 1981.

R. A. Hefferlin, G. V. Zhuvikin, K. E. Caviness, and P. J. Duerksen, Periodic Table of N-Atom Molecules. *Journal of Quantitative Spectroscopy and Radiation Transfer* 32: 257-268, 1984.

R. A. Hefferlin, *Periodic Systems and Their Relation to the Systematic Analysis of Molecular Data*, Lewiston: The Edwin Mellen Press, 1989.

R. A. Hefferlin, A. V. Sepman, and G. V. Zhuvikin, Molecular Multiplets of Alkaline Atoms. *Journal of Chemical Information and Computational Science* 36: 399-401, 1996.

R. A. Hefferlin and J. Knoll, On the Use of Laplace's Equation for Global Prediction of Internuclear Separation and Dissociation Energy. *Journal of Mathematical Chemistry* 28: 169-192, 2000.

R. Hefferlin, The Textures of Chemical Spaces. *Journal of Molecular Structure (Theochem)* 506: 71-86, 2000.

R. A. Hefferlin and M. T. Matus, Molecular Similarity for Small Species: Refining the Isoelectronic Index. *Journal of Chemical Information and Computational Science* 41: 484-494, 2001.

F. Herman and S. Skillman, *Atomic Structure Calculations*, New York: Prentice-Hall, 1963.

D. R. Herrick and M. E. Kellman, New Interpretation, Anomalous Degeneracies, and SO(4) Theory for Electron Correlation in First-Row Atoms and Multiply-Excited States. *Physical Review A* 18: 1770-1780 (1978).

D. R. Herrick, Pairing Model for Doubly Excited Atoms. *Physical Review A* 22: 1346-1357 (1980).

Y. Ho and C. Wulfman, Configuration Mixing within $3s - 3p$ Subshell. *Chemical Physics Letters* 103: 35-39, 1983.

M. J. ten Hoor, Atomic Thomas-Fermi Theory and Electron Subshell Filling. *International Journal of Quantum Chemistry* 30, 831-844 (1986).

D. Ivanenko and S. Larin, *Doklady Akademii Nauk* 88: 45-48, 1953.

J. H. D. Jensen and J. M. Luttinger, Angular Momentum Distribution in the Thomas-Fermi Model. *Physical Review* 86: 907-910, 1952.

C. K. Jorgensen, The Loose Connection between Electronic Configuration and the Chemical Behaviour of the Heavy Elements. *Angewandte Chemie, International Edition in English* 12: 12-19, 1973.

C. Jorgensen, The Periodical Table and Induction as Basis of Chemistry. *Journal de Chemie Physique* 76: 630-635, 1979.

B. R. Judd, *Second Quantization and Atomic Spectroscopy*, Baltimore: Johns Hoppkins Press, 1967.

U. Kaldor, A Simple Screening Potential Explaining the Structure of the Periodic Table. *Chemical Physics Letters* 49: 384-385, 1977.

A. C. Kalloniatis, SO(4) Group Generators for k-Ordered Atomic Systems. *Journal of Physics A* 21: L573-L576, 1988.

A. F. Kapustinskii, *Doklady Akademii Nauk* 80: 365-368, 1951a.

V. Karapetoff, A Chart of Consecutive Sets of Electronic Orbits within Atoms of Chemical Elements. *Journal of Franklin Institute* 210: 609-614, 1930.

S. M. Karlson, R. J. Cavanaugh, R. A. Hefferlin, and G. V. Zhuvikin, in *Group Theoretical Methods in Physics*, edited by A. Arima, T. Eguchi, N. Nakamishi, Singapore: World Scientific, 1995, pp.211-214.

J. Katriel and C. K. Jorgensen, Possible Broken Supersymmetry behind the Periodic Table. *Chemical Physics Letters* 87: 315-319, 1982.

A. K. Kazansky and V. N. Ostrovsky, Rydberg Atom-Ion Collision: Quantum Theory of Intrashell Transitions. *Physical Review Letters* 77: 3094-3097, 1996.

J. Keller, Configuration Electronica de los Atomos. *Ciencia* 16: 86-88, 1956.

N. N. Khramov, S. A. Bartenev, G. S. Markov, and V. N. Romanovsky, Chemometric Modelling of Periodicity of Properties for Chemical Elements and Elementary Substances. In *XVI Mendeleev Congress on General and Applied Chemistry*, Proceedings, vol. 1, pp. 107-108, Moscow, 1998.

N. N. Khramov, S. A. Bartenev, G. S. Markov, and V. N. Romanovsky, Forecast of Heavy Element Properties on the Basis of Harmonic Analysis. In *Matematicheskie Metody v Teknike i Tekhnologiyakh [In Russian: Mathematical Methods in Technique and Technologies]*, Proceedings, vol. 3, pp. 175-176, St Petersburg, 2000.

M. Kibler, The Periodic System of Chemical Elements: Old and New Developments. *Journal of Molecular Structure (Theochem)* 187: 83-93, 1989.

Y. Kitagawara and A. O. Barut, Period Doubling in the $n + l$ Filling Rule and Dynamical Symmetry of the Demkov-Ostrovsky Atomic Model. *Journal of Physics B* 16: 3305-3327, 1983.

Y. Kitagawara and A. O. Barut, On the Dynamical Symmetry of the Periodic Table: II. Modified Demkov-Ostrovsky Atomic Model. *Journal of Physics B* 17: 4251-4259, 1984.

V. M. Klechkovskii, *Doklady Akademii Nauk* 80: 603-606, 1951.

V. M. Klechkovskii, *Doklady Akademii Nauk* 83: 411-414, 1952a.

V. M. Klechkovskii, *Zhurnal Eksperimental'noi i Teoreticheskoi Fiziki* 23: 115-122, 1952b.

V. M. Klechkovskii, *Doklady Akademii Nauk* 86: 691-694, 1952c.

V. M. Klechkovskii, *Doklady Akademii Nauk* 92: 923-926, 1953a.

V. M. Klechkovskii, *Zhurnal Eksperimental'noi i Teoreticheskoi Fiziki* 25: 179-187, 1953b.

V. M. Klechkovskii, *Zhurnal Eksperimental'noi i Teoreticheskoi Fiziki* 26: 760-761, 1954a.

V. M. Klechkovskii, *Doklady Akademii Nauk* 95: 1173-1176, 1954b.

V. M. Klechkovskii, On the First Appearance of Atomic Electrons with l, n, n_r i $n+l$ Given, *Zhurnal Eksperimental'noi i Teoreticheskoi Fiziki* 30: 199-201, 1956 [*Soviet Physics – Journal of Experimental and Theoretical Physics* 3: 125-127, 1956].

V. M. Klechkovskii, *Doklady Akademii Nauk* 135: 655-658, 1960.

V. M. Klechkovskii, Justification of the Rule for Successive Filling of $n + l$ Groups. *Zhurnal Eksperimental'noi i Teoreticheskoi Fiziki* 41: 465-466, 1961 [*Soviet Physics – Journal of Experimental and Theoretical Physics* 14: 334-335, 1962].

V. M. Klechkovskii, $n + l$ Groups in the Thomas-Fermi Model. *Optika i Spektroskopiya* 12: 434-437, 1962 [*Optics & Spectroscopy* 12: 238-240, 1962].

V. M. Klechkovskii, The Consequtive Filling Rule for $n + l$ Groups and the Multiplicity of the Ground Terms of Atoms. *Optika i Spektroskopiya* 19: 441-443, 1965 [*Optics & Spectroscopy* 19: 245-246, 1965].

V. M. Klechkovskii, *Raspredelenie Atomnyh Elektronov i Pravilo Posledovatel'nogo Zapolnenya $(n + l)$-Grupp; [In Russian: The Distribution of Atomic Electrons and the Rule of Successive Filling of $(n + l)$-Groups]*, Moscow: Atomizdat, 1968.

F. Kong, The Periodicity of Diatomic Molecules. *Journal of Molecular Structure (Theochem)* 90: 17-28, 1982.

B. G. Konopel'chenko, *Gruppa SO(2,4) + R i Tablitza Mendeleeva* [In Russian: *SO(2,4) + R Group and Mendeleev Table*], Preprint IYaF 40-72, Novosibirsk: Institute of Nuclear Physics, 1972.

B. G. Konopel'chenko and Yu. B. Rumer, Atoms and Hadrons (Classification Problem). *Uspekhi Fizicheskikh Nauk* 129: 339-345, 1979 [*Soviet Physics – Uspekhi* 22: 837-840, 1979].

A. Kónya, Quantum Numbers and Energy Levels in the Thomas-Fermi Atom. *Acta Academiae Scientiarum Hungaricae* 18: 39-53, 1964a.

A. Kónya, Theoretical Interpretation of Some Properties of the Periodic System by the Thomas-Fermi Model. *Acta Academiae Scientiarum Hungaricae* 18: 129-130, 1964b.

L. D. Landau and E. M. Lifshits, *Mechanics*, Oxford: Pergamon, 1976.

L. D. Landau and E. M. Lifshits, *Quantum Mechanics: Non-Relativistic Theory*, Oxford: Pergamon, 1977.

R. Latter, Atomic Energy Levels for the Thomas-Fermi and Thomas-Fermi-Dirac Potentials. *Physical Review* 99: 510-519, 1955.

W. Lenz, Zur Theorie der Optischen Abbildung. In *Probleme der Modernen Physik*, edited by P. Debye, Leipzig: Hirsel, 1928, p. 198-207.

P. O. Löwdin, Some Comments on the Periodic System of Elements. *International Journal of Quantum Chemistry (Symposium)* IIIS: 331-334, 1969.

O. G. Ludwig, The Best Aufbau Mnemonic: The Periodic Table. *Journal of Chemical Education* 69: 430, 1992.

E. Madelung, *Die Mathematischen Hilfsmittel des Physikers*, 3rd edn., Berlin: Springer, 1936, p.359 [7th edn., Berlin, 1964].

I. A. Malkin and V. I. Man'ko, Symmetry of the Hydrogen Atom. *Pis'ma v ZETF* 2: 230-233, 1965 [*JETP Letters* 2:, 146-148, 1966];

I. A. Malkin and V. I. Man'ko, *Yadernaya Fizika* **3**, 372-382, 1966 [*Sov.J.Nucl.Phys.* **3**, 267-274, 1966].

S. T. Manson, Dependence of the Phase Shift on Energy and Atomic Number for Electron Scattering By Atomic Fields. *Physical Review* 182: 97-103, 1969.

J. C. Maxwell, Solution of Problems. *Cambridge and Dublin Mathematical Journal* 8: 188, 1854.

J. C. Maxwell, *The Scientific Papers*, New York: Dover, 1952, pp. 74-79.

E. G. Mazurs, *Graphic Representations of the Periodic System during One Hundred Years*, The University of Alabama Press, 1974.

W. H. Miller, WKB Solution of Inversion Problem for Potential Scattering. *Journal of Chemical Physics* 51: 3631-3638, 1969.

R. Mills, Systematic Table of Elements. *American Journal of Physics* 40: 1169-1170 1972.

N. Mukunda, Dynamical Symmetries and Classical Mechanics. *Physical Review* 155: 1383-1386, 1967.

T. Négadi and M. Kibler, The Periodic Table in Flatland. *International Journal of Quantum Chemistry* 57: 53-61, 1996.

T. Négadi, On the Planar Periodic Table. *International Journal of Quantum Chemistry* 78: 206-211, 2000.

D. Neubert, Double Shell Structure of the Periodic System of the Elements. *Zeitschrift für Naturforschung* 25a: 210-217, 1970.

O. Novaro and A. Freyre, O(4) and U(3) Symmetry Breaking in Second Row of Periodic Table. *Molecular Physics* 20: 861-871, 1971.

O. Novaro, Comment on the Group Theoretical Justification of the Aufbau Scheme. *International Journal of Quantum Chemistry (Symposium)* N° 7: 53-56, 1973.

O. Novaro, Group Theoretical Aspects of the Periodic Table of the Elements. *Journal of Molecular Structure (Teochem)* 199: 103-118, 1989.

O. Novaro and M. Berrondo, Approximate Symmetry of the Periodic Table. *Journal of Physics B* 5: 1104-1110, 1972.

O. Novaro and K. B. Wolf, A Model Hamiltonian for the Periodic Table. *Revista Mexicana de Fisica* 20: 265-268, 1971.

H. Odabasi, Some Evidence about the Dynamic Group SO(4,2) Symmetries of the Periodic Table of Elements. *International Journal of Quantum Chemistry (Symposium)* N° 7: 23-33, 1973.

T. A. Oliphant, Angular Momentum Distributions in the Thomas-Fermi-Dirac Model of the Atom. *Physical Review* 104: 954-958, 1956.

V. N. Ostrovsky, Teoretiko-Gruppovye Aspekty Periodicheskoi Sistemy Elementov. in *Group Theoretical Methods in Physics, Proceedings of International Symposium*, edited by M. A. Markov, Moscow: Nauka, 1980, vol.1, pp. 181-188.

V. N. Ostrovsky, Dynamic Symmetry of Atomic Potential. *Journal of Physics B* 14: 4425-4439, 1981.

V. N. Ostrovsky, Group theory and Periodic System of Elements. In *Latin-American School of Physics XXX ELAF*, edited by O. Castaños, R. Lópes-Peña, J. G. Hirsch and K. B. Wolf, AIP Conference Proceedings 365, 1996, pp.191-216.

V. N. Ostrovsky, $3n_r + l$ Grouping of Energy Levels and Clusters. *Physical Review A* 56: 626-631, 1997.

V. N. Ostrovsky, What and how Physics contributes to understanding the Periodic Law ? *Foundations of Chemistry* 3: 145-182, 2001.

Pao Fang Yi, *Journal of Chemical Education* 24: 567, 1947.

A. M. Perelomov, V. S. Popov, and M. V. Terent'ev, "Hidden" Symmetry of Highly Excited Atomic Levels. *Zhurnal Eksperimental'noi i Teoreticheskoi Fiziki* 51: 600-605, 1966 [*Soviet Physics – Journal of Experimental and Theoretical Physics* 24: 401-403, 1967].

C. S. G. Phillips and R. J. P. Williams, *Inorganic Chemistry*, Oxford: Oxford University Press, 1965, Chapters 18.4, 20.2, 30.2.

H. F. Pop, C. Sarbu, O. Horowitz and D. Dmitriesu, A Fuzzy Classification of the Chemical Elements. *Journal of Chemical Information and Computer Science* 36: 465-482, 1996.

R. T. Powers, Frequencies of Radial Oscillation and Revolution as Affected by Features of a Central Potential. In *Atti dei Convegno Mendeleeviano*, Edited by M. Verde, Academia delle Scienze di Torino, Torino, 1971, pp. 235-242.

R. J. Puddephatt and P. K. Monaghan, *The Periodic Table of the Elements*, Oxford: Clarendon Press, 1986.

D. Purdela, A Regular Periodic Table of the Elements and its Quantum Mechanical Requirement. *International Journal of Quantum Chemistry* XXXIV: 107-119, 1988.

P. Pyykkö, Relativistic Effects in Structural Chemistry. *Chemical Reviews* 88: 563-594, 1988.

P. Pyykkö, Relativistic Effects on Periodic Trends. In *The Effects of Relativity in Atoms, Molecules, and the Solid State*, edited by S. Wilson, I. P. Grant, and B. L. Gyorffy, New York, Plenum Press, 1991, pp. 1-13.

P. Pyykkö and Y. Zhao, The Elements of Flatland: Hartree-Fock Atomic Ground States in Two Dimensions for $Z = 1 - 24$. *International Journal of Quantum Chemistry* XI: 527-544, 1991.

P. Pyykkö, A Note on Nodal Structures, Partial Screening, and Periodic Trends among Alkali Metals and Alkaline Earths. *International Journal of Quantum Chemistry* 85: 18-21, 2001.

A. R. P. Rau and U. Fano, Atomic Potential Wells and the Periodic Table. *Physical Review* 167: 7-10, 1968.

A. R. P. Rau, R. O. Mueller, and L. Spruch, Where Along the Periodic Table will a g Electron First Appear ? *Comments on Atomic and Molecular Physics* 3: 87-93 (1972).

T. B. Romanovskaya, *Istoriya Kvantovo-Mekhanicheskoi Interpretatzii Periodichnosti [In Russian: History of Quantum Mechanical Interpretation of Periodicity]*, Moscow: Nauka, 1986.

W. Rubinowicz, *Kwantowa teoria atomu*, Warsaw, Panstw. Wyd. Naukowe, 1954.

Yu. B. Rumer and A. I. Fet, Spin(4) Group and Mendeleev's Table. *Teoreticheskaya i Matematicheskaya Fizika* 9: 203-210, 1971 [*Theoretical and Mathematical Physics* 9: 1081-1085, 1972].

R. T. Sanderson, An Explanation of Chemical Variations within Periodic Major Groups. *Journal of American Chemical Society* 74: 4792-4794, 1952.

R. T. Sanderson, *Chemical Periodicity*, Reinhold Publishing Corporation: New York, 1960, Chapter 2.

E. R. Scerri, Transition Metal Configurations and Limitations of the Orbital Approximation. *Journal of Chemical Education* 66: 481-483, 1989.

E. R. Scerri, Chemistry, Spectroscopy and the Question of Reduction. *Journal of Chemical Education* 68: 122-126, 1991.

E. R. Scerri, Has the Periodic Table been Successfully Axiomatized ? *Erkenntnis* 47: 229-243 1997a.

E. R. Scerri, The Periodic Table and the Electron. *American Scientist* 85: 546-553, 1997b.

E. R. Scerri, Periodicity, Chemical. In *MacMillan's Encyclopaedia of Chemistry*, edited by J. J. Lagowski, New York: Macmillan Reference, Vol. III, 1998a, pp. 22-32.

E. R. Scerri, The Evolution of the Periodic System. *Scientific American* 279: 56-61, 1998b.

E. R. Scerri, V. Kreinovich, P. Wojciechowski, and R. R. Yager, Ordinal Explanation of the Periodic System of Chemical Elements. *International Journal of Uncertainty, Fuzziness, and Knowledge-Based Systems* 6: 387-399, 1998b.

E. R. Scerri and J. Edwards, Bibliography of Secondary Sources on the Periodic System of the Chemical Elements. *Foundations of Chemistry* 3: 183-196, 2001.

G. T. Seaborg, Place in Periodic System and Electronic Structure of the Heaviest Elements. *Nucleonics* 5: #5, 16-36, 1949.

L. M. Simmons, A Modification of the Periodic Table. *Journal of Chemical Education* 24: 588-591, 1947.

L. M. Simmons, Display of Electronic Configurations by a Periodic Table. *Journal of Chemical Education* 25: 658-661, 1948.

L. M. Simmons, Letter to Editor. *Nature* 186: 154, 1960.

R. Singh and S. K. Dikshit, Aufbau Rewritten. *Journal of Chemical Education* 68: 396, 1991.

H. H. Sisler, *Electronic Structure, Properties and the Periodic Law*, Reinhold Publishing, New York, 1963.

J. C. Slater, *Quantum Theory of Atomic Structure*, New York: McGraw Hill, 1960.

J. D. M. Smith, *Chemistry and Atomic Structure*, London: Ernest Benn, Ltd, 1924, p. 126.

P. H. A. Sneath, Numerical Classification of the Chemical Elements and its Relation to the Periodic System. *Foundations of Chemistry* 2: 237-263, 2000.

J. van Spronsen, *The Periodic System of Chemical Elements. A History of the First Hundred Years*, Elsevier, Amsterdam, 1969.

Yu. V. Tarbeev, N. N. Trunov, A. A. Lobashev, and V. V. Kukhar', Extraction of Information from the Periodic System of Elements as an Inverse Problem. *Zhurnal Eksperimental'noi i Teoreticheskoi Fiziki* 112: 1226-1238, 1997 [*Journal of Experimental and Theoretical Physics* 85: 666-72, 1997].

Yu. V. Tarbeev, N. N. Trunov, A. A. Lobashev, and V. V. Kukhar', The Periodic System and Metrology of Mesoscopic Quantum Systems. *Izmeritel'naya Tekhnika* #9, 53-58, 2001 [*Measurement Techniques* 44: 944-952, 2001].

W. R. Theis, Struktur des Phasenraumes und ein Zusammenhang zwischen Bahndrehimpulsverteilung und Dichteverteilung der Nukleonen im Atomkern. *Zeitschrift für Physik* 140: 1-15, 1955.

T. Tietz, Approximate Analytic Solution of the Thomas-Fermi Equation for Atoms. *Journal of Chemical Physics* 22: 2094, 1954.

T. Tietz, Über eine Approximation der Fermischen Verteilungsfunktion. *Annalen der Physik* 15: 186-188, 1955.

T. Tietz, Über eine Approximation der Fermischen Verteilungsfunktion fur Positive Freie Ionen. *Nuovo Cimento* 1: 968-971, 1955.

T. Tietz, Eine Analitische Formel für Theorie der Bildung der Elektrongruppen im periodischen System der Elemente. *Acta Academiae Scientiarum Hungaricae* 9: 73, 1958.

T. Tietz, Electronengruppen im periodischen System der Elemente in der Statistischen Theorie Des Atomes. *Annalen der Physik* 5: 237-240, 1960.

T. Tietz, Über Eigenverte und Eigenfunctionen in der Statistischen Theorie des Atoms. *Annalen der Physik* 12: 373-376, 1964.

T. Tietz, Low Energy Electron-Atom Scattering Cross Section in Thomas-Fermi Theory. *Zeischrift für Natruforschung* 21a: 360-361, 1966.

T. Tietz, Contribution to the Thomas-Fermi Theory. *Journal of Chemical Physics* 49: 4391-4393, 1968.

M. G. Veselov and L. N. Labzovsky, *Teoriya Atoma [In Russian: Theory of an Atom]*, Moscow: Nauka, 1986.

D. Vrinceanu and M. R. Flannery, Analytical Quantal Collisional Stark Mixing Probabilities. *Journal of Physics B* 34: L1-L8, 2001.

J. A. Wheeler, From Mendeleev's Atom to the Collapsing Star. In *Atti dei Convegno Mendeleeviano*, Edited by M. Verde, p. 189-233, Academia delle Scienze di Torino, Torino, 1971.

J. A. Wheeler, Semiclassical Analysis Illuminates the Connection between Potential and Bound States and Scattering. In *Studies in Mathematical Physics*, Edited by E. H. Lieb, B. Simon and A. S. Wightman, Princeton Series in Physics, Princeton University Press, Princeton, 1976, pp. 351-422.

W. J. Wiswesser, The Periodic System and Atomic Structure. I. An Elementary Physical Approach. *Journal of Chemical Education* 22: 314-322, 1945a.

W. J. Wiswesser, The Periodic System and Atomic Structure. II. Detailed Introduction to Wave Mechanical Approach. *Journal of Chemical Education* 22: 370-379, 1945b.

W. J. Wiswesser, The Periodic System and Atomic Structure. III. The Wave Mechanical Interpretation. *Journal of Chemical Education* 22: 418-426, 1945c.

C. E. Wulfman, Dynamical Groups in Atomic and Molecular Theory. In *Recent Advances in Group Theory and Their Application to Spectroscopy*, edited by J. C. Donini, Plenum, New York, 1979, pp.329-403.

C. Wulfman, Configuration Mixing within $2s-2p$ Shell. *Physical Review Letters* 51: 1159-1162, 1983.

C. Wulfman and R. D. Levine, Transition from Collective to Independent-Particle Motion within the Valence Shell of the Second- and Third-Row Atoms. *Physical Review Letters* 53: 238-241, 1984.

T. Y. Wu and S. Goudsmith, The Order of Electron Shells in Ionized Atoms. *Physical Review* 43: 496, 1933.

Yeou Ta, Une nouvelle représentation du Tableau Periodique Basé sur la Structure Electronique des Éléments. *Compt. Rend. (Paris)* 221: 441-442, 1945.

Yeou Ta, Une nouvelle représentation du Tableau Periodique des Éléements. *Annales de Physique (Paris)* 1: 88-99, 1946.

G. V. Zhuvikin and R. Hefferlin, Periodicheskaya Sistema Dvukhatomnykh Molekul: Theoretiko-Gruppovoi Podkhod [In Russian: Periodic System of Diatomic Molecules: Group-Theoretical Approach]. *Vestnik Leningradskogo Universiteta* N° 16: 10-16, 1983.

LOWDIN'S REMARKS ON THE AUFBAU PRINCIPLE AND A PHILOSOPHER'S VIEW OF AB INITIO QUANTUM CHEMISTRY.

Eric R. Scerri,
Department of Chemistry & Biochemistry,
UCLA,
Los Angeles,
CA 90095,
USA
scerri@chem.ulca.edu

1. Introduction.

It is indeed a great honor to be invited to contribute to this memorial volume. I should say from the outset that I never met Löwdin but nevertheless feel rather familiar with at least part of his wide-ranging writing. In 1986 I undertook what I believe may have been the first PhD thesis in the new field of philosophy of chemistry. My topic was the question of the reduction of chemistry to quantum mechanics. Not surprisingly this interest very soon brought me to the work of Löwdin and in particular his analysis of rigorous error bounds in ab initio calculations (Löwdin, 1965).
I later discovered a short article in which Löwdin made some interesting remarks that resonated with me (Löwdin, 1969).

> The energy rule for the neutral atoms was obviously in contradiction to Bohr's calculation on the hydrogen atom, which indicated that the energies should be increasing with increasing *n*. It is typical of the nature of "frontier-research" that Bohr abandoned this rule for the higher atoms, since it led to the wrong structure of the periodic system, and the modified rule $[(n + \ell, n)]$ seems to have been obtained in a more intuitive way. Bohr himself was never too explicit about his "Aufbau" -principle, and [the rule] is sometimes referred to as the Goudsmit-rule or the Bose-rule. It is perhaps remarkable that, in axiomatic quantum

E.J. Brändas and E.S. Kryachko (eds.),
Fundamental World of Quantum Chemistry, Vol. II, 675–694.

> theory, the simple energy rule has not yet been derived from first principles. (p.332).
>
> Bohr's quantum numbers (n, ℓ, m) have fully entered chemistry, and every chemistry student learns about the symbols *1s, 2s, 2p, 3s, 3p, 3d* etc. It is hence a startling fact that the simple energy rule has not entered any major chemistry textbooks, as far as I know, and it is still this rule which gives the first explanation of the occurrence of the transition metals, the rare-earth metals, and the over-all structure of the electronic shells of atoms. (p.334).
>
> It would certainly be worth while to study the energy rule from first principles, i.e. on the basis of the many-electron Schrödinger equation. (p.334).

In the case of the second excerpt I think I can safely say that Löwdin is wrong. The simple energy rule regarding the order of filling of orbitals in neutral atoms has now entered *every* textbook of chemistry, although his statement may have been partly true in 1969 when he wrote his article.[1] Although Löwdin càn be excused for not knowing what was in chemistry textbooks I think it is also safe to assume that he is correct in his main claim that this important rule has not been derived. Nor as I have claimed in a number of brief articles has the rule been derived to this day (Scerri, 1998).

The examination of this idea has subsequently formed an integral part of my research in the philosophy of chemistry. This has also led to a certain amount of disagreement with other authors who appear to interpret Löwdin's remark in a somewhat different manner (Ostrovsky, 2001). I now deeply regret not having contacted Löwdin directly in order to seek his own clarification. In the present contribution I intend to revisit this question and to take the opportunity to respond to some critics as well as hopefully injecting some new ideas into the discussion.

2. General approach of this study and ab initio calculations in science generally.

Quantum Mechanics has been the most spectacularly successful theory in the history of science. As is often mentioned the accuracy to which the anomalous magnetic moment of the electron can be calculated is a staggering nine decimal places. Quantum Mechanics has revolutionized the study of radiation and matter since its inception just over one hundred years ago. The impact of the theory has been felt in such fields as solid state physics, biochemistry, astrophysics, materials science and electronic engineering not to mention chemistry.

Quantum Mechanics offers the most comprehensive and most successful explanation of many chemical phenomena such as the nature of valency and bonding as well as chemical reactivity. It has also provided a fundamental explanation of the periodic system of the elements which summarizes a vast amount of empirical chemical knowledge. Quantum Mechanics has become increasingly important in the education of chemistry students. The general principles provided by the theory mean that students can now spend less time memorizing chemical facts and more time in actually thinking about chemistry.

My project is not to critique of the power of quantum chemistry that I regard to be a self-evident fact. But with the triumph of quantum mechanics I believe there has been some tendency to exaggerate its success, especially on the part of some practicing quantum chemists and physicists. As a philosopher of chemistry I have the luxury of being able to examine the field as an outsider and of asking the kinds of questions which true practitioners might not even contemplate. The approach I take in this article is a philosophical one in the sense that I am concerned with principles and not just with technical details, although I try to be as accurate as possible with the latter.

Quantum mechanics is part of the reductionist tradition in modern science, and the general claim, often just made implicitly as in any branch of reduction, is that the highest ideal one can aspire to is to derive everything from the theoretical principles. The less experimental data one needs to appeal to, the less one is introducing measured parameters the purer the calculation and the closer it approaches to the ideal of Ockham's razor of being as economical as possible (Hoffman, Minkin, Carpenter, 1996).[2]

Of course there is no such thing as a completely *ab initio* calculation and if one looks far enough back at the history of any scientific theory one finds that it began with the assumption of at least some experimental data. But it is also fair to say that once the basic principles of a theory have been arrived at the theorist may 'kick away' the historical-experimental scaffolding. The modern student of quantum mechanics, for example, is not obliged to follow the tortuous route taken by Planck, Einstein, De Broglie, Schrödinger and others. She can go directly to the postulates of quantum mechanics where she will find a procedure for doing all kinds of calculations and she can safely ignore the historical heritage of the theory. Indeed many argue, and correctly in my view, that it is actually a hindrance for the practitioner to get too involved in the historical aspects of the theory although it may of course be culturally enriching to do so.

The epitome of the ab initio approach in science is something like Euclidean geometry where one begins with a number of axioms and one derives everything from this starting point without any recourse whatsoever to empirical data. Needless to say geometry, Euclidean or otherwise, has its origins in the dim distant past when agrarian people needed to think about lines and angles and areas of land. But once the concepts of line, angle and distance had been sufficiently abstracted the agrarian heritage could be completely forgotten.

In a similar way, my question in this article will be to be to ask to what extent the periodic table of the elements can be explained strictly from first principles of quantum mechanics without assuming any experimental data whatsoever. I suspect that some readers and fellow contributors to this volume might well experience some irritation at the almost perverse demands which I will make on what should be derivable from the current theory. If so, then I apologize in advance.

By adopting a perspective from the philosophy of science I will attempt to cross levels of complexity from the most elementary chemical explanations based on electron shells to those based on ab initio methods. Such a juxtaposition is seldom contemplated in the chemical literature. Textbooks provide elementary explanations which necessarily distort the full details but allow for a more conceptual or qualitative grasp of the main ideas. Meanwhile the research literature focuses on the minute details of particular methods or particular chemical systems and does not typically examine the kind of explanation that is being provided. To give a satisfactory discussion of explanation in the context of the periodic table we need to consider both elementary and deeper explanations within a common framework.

One of the virtues of philosophy of science is that it can bridge different levels in this way since it primarily seeks the 'big picture' rather than the technical details. In fact supposedly elementary explanations often provide this big picture in a more direct manner. Of course what is also needed is to connect the elementary explanation to the technical details in the deeper theories.

The question of whether or not different levels of explanation for any particular scientific phenomenon are in fact consistent and whether they form a seamless continuum has been the subject of some debate. For example the philosopher of science Nancy Cartwright goes to some lengths to argue that many different explanations can be found for the action of lasers and suggests that these explanations are not necessarily consistent with each other (Cartwright, 1983). In other writings she has expressed some support for the thesis that the various special sciences are dis-unified (Galison, Stump, 1996).

My own view differs from Cartwright's in that I think that the sciences are unified and that explanations given for the same scientific phenomenon at different levels are essentially consistent, although the connection if frequently difficult to elaborate in full (Scerri, 2000). In this paper I will attempt to draw such connections for the various explanations of the periodic table given at different levels of sophistication. Another way of regarding the present project is to consider typical 'chemical explanations', full of visualizations and sometimes naïve realism, and contrast them with the more abstract mathematical explanations favored by the theorist.

3. What do the terms ab initio and first principles really mean in current quantum chemistry?

It is interesting to consider the meanings of the terms of "ab initio" calculations as well as the closely related term "first principles calculations". How are these terms currently used by the computational chemistry community? Do these terms mean the same thing?

The answer to this question as well as the question of the precise meaning of the term ab initio itself in the context of quantum chemistry seems to differ considerably according to the particular researcher that one might consult.[3] Some authors I have questioned claim that the two terms are used interchangeably to mean calculations performed without recourse to any experimental measurement. This would include Hartree-Fock, and many of the DFT functionals, along with quantum Monte Carlo and CI methods.

Others report that the term first principles is being increasingly used because the Latin term "ab initio" can been seen as unnecessarily elitist and that there is no scientific significance in using one term rather than the other one. Others disagree, suggesting that first principles is favored by the DFT community since in this kind of work it is quite clear that some semi-empirical elements are introduced whereas the wavefunction or traditional ab initio approaches are less semi-empirical.

Turning to the question of exactly what is implied by the term ab initio, I have encountered an even larger variety of opinions. According to some sources the origin of the term is purely accidental. They claim that the term was originally applied to the Roothaan-Hall approach through an amusing accident. Robert Parr was apparently collaborating in some work of this kind with a group in England and in reporting one of his calculations, is said to have described it as "ab initio", implying that the whole of that particular project had been carried out from the beginning in his laboratory. Very soon the term was being used for all kinds of accurate theoretical work which, at least at first sight, did not involve any fixing of paramenters.

Regarding current ab initio calculations it is probably fair to say that they are not really ab initio in every respect since they incorporate many empirical parameters. For example, a standard HF/6-31G* calculation would generally be called "ab initio", but all the exponents and contraction coefficients in the basis set are selected by fitting to experimental data. Some say that this feature is one of the main reasons for the success of the Pople basis sets. Because they have been fit to real data these basis sets, not surprisingly, are good at reproducing real data. This is said to occur because the basis set incorporates systematical errors that to a large extent cancel the systematical errors in the Hartree-Fock approach. These features are of course not limited to the Pople sets. Any basis set with fixed exponent and/or contraction coefficients have at some point been adjusted to fit some data. Clearly it becomes rather difficult to demarcate sharply between so-called ab initio and semi-empirical methods.[4]

To some other experts the meaning of the term ab initio is rather clear cut. Their response is that "ab initio" simply means that all atomic/molecular integrals are computed analytically, without recourse to empirical parametrization. They insist that it does not mean that the method is exact nor that the basis set contraction coefficients were obtained without recourse to parametrization. Yet others point out that even the integrals need not be evaluated exactly for a method to be called ab initio, given that, for instance, Gaussian employs several asymptotic and other cutoffs to approximate integral evaluation.

But regardless of these disagreements as to what exactly the term ab initio means I want to presume to define a new term that I shall call "super ab initio". In the rest of this article I would like to suggest the kind of explanations which might be possible when we arrive at the point of being able to perform super-ab initio calculations in which no feature whatsoever is introduced by recourse to experimental data. This will be done especially regarding attempts to explain, or reduce, the periodic system of the elements by using quantum mechanics.

4. First the elementary approach.

Let us start at an elementary level or with a typically "chemical" view. Suppose we ask an undergraduate chemistry student how quantum mechanics explains the periodic table. If the student has been going to classes and reading her book she will respond that the number of outer-shell electrons determines, broadly speaking, which elements share a common group in the periodic table. The student might possibly also add that the number of outer-shell electrons causes elements to behave in a particular manner.

Suppose we get a little more sophisticated about our question. The more advanced student might respond that the periodic table can be explained in terms of the relationship between the quantum numbers which themselves emerge from the solutions to the Schrödinger equation for the hydrogen atom.[5]

But does the fact that the third shell can contain 18 electrons, for example, which emerges from the relationships among the quantum numbers, also explain why some of the periods in the periodic system contain eighteen places? Actually not exactly. If electron shells were filled in a strictly sequential manner there would be no problem and the explanation would in fact be complete. But as everyone is aware, the electron shells do not fill in the expected sequential manner. The configuration of element number 18, or argon is,

$$1s^2, 2s^2, 2p^6, 3s^2, 3p^6$$

This might lead one to think that the configuration for the subsequent element, number 19, or potassium, would be

$$1s^2, 2s^2, 2p^6, 3s^2, 3p^6, 3d^1$$

since, up to this point, the pattern has been to add the new electron to the next available orbital in the sequence of orbitals at increasing distances from the nucleus. However spectroscopic evidence shows quite clearly that the configuration of potassium should be denoted as,

$$1s^2, 2s^2, 2p^6, 3s^2, 3p^6, 4s^1$$

As many textbooks state this can be explained from the fact that the 4s orbital has a lower energy than the 3d orbital. In the case of element 20 or calcium the new electron also enters the 4s orbital and for the same reason.

5. Transition metal configurations.

The interesting part is what happens next. In the case of the following element, number 21, or scandium, the orbital energies have reversed so that the 3d orbital has a lower energy. Textbooks almost invariably claim that since the 4s orbital is already full there is no choice but to begin to occupy the 3d orbital. This pattern is supposed to continue across the first transition series of elements, apart from the elements Cr and Cu where further slight anomalies are believed to occur.

In fact this explanation for the configuration of the scandium atom and most other first transition elements is inconsistent. If the 3d orbital has a lower energy than 4s starting at scandium then, if one were really filling the orbitals in order of increasing energy, one would expect that all three of the final electrons would enter 3d orbitals. The argument which most textbooks present is incorrect since it should be possible to predict the configuration of an element from a knowledge of the order of its own orbital energies (Scerri, 1989, Vanquickenborne, 1994). It is incorrect to consider the configuration of the previous element and assume that this configuration is carried over intact on moving to the next element, especially in cases where orbital energies cross over each other as they do in this case. It should be possible to predict the order of orbital filling for the scandium atom on its own terms. If one tries to do so however one predicts a configuration ending in $3d^3$ contrary to the experimental facts.

The full explanation of why the $4s^2 3d^1$ configuration is adopted in scandium, even though the 3d level has a lower energy, emerges from the peculiarities of the way in which orbital energies are defined in the Hartree-Fock procedure. The details are tedious but have been worked out and I refer anyone who is interested in pursuing this aspect to the literature (Melrose, Scerri, 1996).[6, 7]

6. How are configurations derived from the theory?

But let me return to the question of whether the periodic table is fully and deductively explained by quantum mechanics. In the usually encountered explanation one assumes that at certain places in the periodic table unexpected orbital begins to fill as in the case of potassium and calcium where the 4s orbital begins to fill before the 3d shell has been completely filled. This information itself is not derived from first principles. It is justified *post facto* and by some tricky calculations (Melrose, Scerri, 1996; Vanquickenborne, Pierloot, Devoghel, 1994).

But if we ignore the conceptual paradox of why 4s fills preferentially even though it has a higher energy than 3d we can just concentrate on calculations aimed at determining the ground state configuration. The Hartree-Fock method[8] can be used to compare the energies of the scandium atom with two alternative configurations, [Ar] $4s^2\ 3d^1$ and [Ar] $4s^1\ 3d^2$.

configuration $4s^2\ 3d^1$

Non-Relativistic	-759.73571776	(in Hartrees)
Relativistic	-763.17110138	

configuration $4s^1\ 3d^2$

Non-Relativistic	-759.66328045
Relativistic	-763.09426510

It would appear that both non-relativistic and relativistic ab initio calculations correctly compute that the $4s^2$ configuration has the lowest energy in accordance with experimental data. But these calculations, including the ones for subsequent elements must be done on a case-by-case basis. As Löwdin has pointed out, there is not yet a general derivation of the formula which governs the order of filling, that is the $n + \ell$, or Madelung rule, which states that given a choice of filling any two orbitals the order of filling goes according to increasing values of $n + \ell$.

But similar calculations do not fare as well in other atoms. Consider the case of the chromium atom for example.[9]

configuration $4s^1\ 3d^5$

Non-Relativistic	-1043.14175537
Relativistic	-1049.24406264

configuration $4s^2\ 3d^4$

Non-Relativistic	-1043.17611655
Relativistic	-1049.28622286

It appears that both non-relativistic and relativistic calculations at this level of accuracy fail to predict the experimentally observed ground state which is the $4s^1\ 3d^5$ configuration. Of course I do not deny that if one goes far enough in a more elaborate calculation then eventually the correct ground state will be recovered. But in doing so one knows what one is driving at, namely the experimentally observed result. This is not the same as strictly predicting the configuration in the absence of experimental information.

In addition, if one goes beyond the Hartree-Fock approximation to something like the configuration interaction approach there is an important sense in which one has gone beyond the picture of a certain number of electrons into a set of orbitals.[10] If one insists on picturing this, then rather than just every electron being in every possible orbital in the ground state configuration, every electron is now in every one of the orbitals used to express the thousands or even millions of contributing configurations.

But I want to return to my claim that quantum mechanics does not really explain the fact that the third row contains 18 elements to take one example. The development of the first of the period from potassium to krypton is not due to the successive filling of 3s, 3p and 3d electrons but due to the filling of 4s, 3d and 4p. It just so happens that both of these sets of orbitals are filled by a total of 18 electrons. This coincidence is what gives the common explanation its apparent credence in this and later periods of the periodic table. As a consequence the explanation for the form of the periodic system in terms of how the quantum numbers are related is semi-empirical, since the order of orbital filling is obtained form experimental data. This is really the essence of Löwdin's quoted remark about the $(n + \ell, n)$ rule.

Closing of shells,
Occurs at Z = 2, 10, 28, 60, 110, etc.
Closing of periods,
Occurs at Z = 2, 10, 18, 36, 54, etc.

It is the second sequence of Z values shown above that really embodies the periodic system and not the first.[11] For all we know electron shells or orbitals may not even exist or may be replaced by some other concept in a future theory. But the fact that chemical repetitions occur at Z = 2, 10 18, 36, for example, are chemical facts which will never be superceded.

Only if shells filled sequentially, which they do not, would the theoretical relationship between the quantum numbers provide a purely deductive explanation of the periodic system. The fact the 4s orbital fills in preference to the 3d orbitals is not predicted in general for the transition metals but only rationalized on a case by case basis as I have argued. Again, I would like to stress that whether or not more elaborate calculations finally succeed in justifying the experimentally observed ground state does not fundamentally alter the overall situation.[12]

To sum-up, we can to some extent recover the order of filling by calculating the ground state configurations of a sequence of atoms but nobody has yet deduced the $n + \ell$ rule from the principles of quantum mechanics.[13]

7. Choice of basis set.

There is another general problem which mars any hope of claiming that electronic configurations can be fully predicted theoretically and that quantum mechanics thus provides a purely deductive explanation of what was previously only obtained from experiments. In most of the configurations we have considered, with the exception of cases mentioned above, it has been possible to use a quantum mechanical method to calculate that this particular configuration does indeed represent the lowest energy possibility. However, in performing such calculations the candidate configurations which are subjected to a variation procedure are themselves obtained from the aufbau principle and other rules of thumb such as Hund's principle or by straightforward appeal to experimental data. There is a very simple reason for this state of affairs. The quantum mechanical calculations on ground state energies involve the initial selection of a basis set, which in its simplest form is the electronic configuration of the atom in question.

Quantum mechanical calculations are not capable of actually generating their own basis sets that must therefore be put in "by hand". So whereas the correct ground state electronic configurations can in many cases be selected among a number of plausible options, the options themselves are not provided by the theory. I suggest this is another weakness of the present claims to the effect that quantum mechanics explains the periodic system and it is an aspect that might conceivably corrected by future developments.[14] In addition the commonly used basis sets are almost invariably constructed in a semi-empirical manner by reference to experimental data on some particular chemical system. This is rather undeniable given that the exponents and contraction coefficients in the basis set are generally selected by fitting to experimental data as mentioned above in section 3.[15]

8. Qualitative explanation of PT in terms of electrons in shells.

I will now attempt to take stock of the various senses of the claim that the periodic system is reduced, or fully explained, by quantum mechanics and to extend the scope of this work to more elaborate theoretical approaches. As any student of chemistry knows, the approximate recurrence of elements after certain regular intervals is explained by the possession of a certain number of outer-shell electrons. This form of explanation appears to be quantitative to some people because it deals in number of electrons but in fact turns out to be rather qualitative. It cannot of course be used to predict quantitative data on a particular atom with any degree of accuracy.

Whereas the crude notion of a particular number of electrons in shells or orbitals does not produce very accurate calculations the process can be refined in several well known ways. The first refinement is perhaps the use of the Hartree method of calculating self-consistent orbitals while at the same time minimizing the energy of the atom.[16] The next refinement lies in making the method consistent with the notion that electrons are indistinguishable. By appealing to the Hartree-Fock method and by performing a permutation of all the electrons in the atom each electron finds itself simultaneously in all occupied orbitals at once to put it in pictorial terms. A third refinement might be to include any number of excited state configurations for the atom, in a configuration interaction calculation. Having reached this level of abstraction we have really left behind the homely picture of electrons in particular shells. If one still insists on visualization, each electron is now in every orbital of every single configuration which we choose to consider.

There is still a connection with the elementary homely model but it is also fair to say that the move towards greater abstraction has somewhat invalidated the naïve model. This now raises the question as to whether the elementary model really does have explanatory power. I would argue that it does not. It may have led historically to these more sophisticated approaches but it has been rendered vastly more abstract in the process. But if we are considering the general question of explanation it is not essential to retain the homely picture that can be grasped by the general chemist of the beginning student of physical chemistry. We must move on to enquire about how the more abstract approaches actually fare. The short answer I believe is much better but still not in strictly, or super, ab initio fashion.

9. Ab initio calculations based on wavefunctions.

Of course the Hartree-Fock method and the configuration interaction approaches fare much better and are a serious contenders for the claim of a full explanation of the periodic system than the elementary and qualitative notion of electrons in shells.

Within these ab initio approaches the fact that certain elements fall into the same group of the periodic table is not explained by recourse to the number of outer-shell electrons. The explanation lies in calculating the magnitude of a property such as the first ionization energy and seeing whether the expected periodicity is recovered in the calculations. For example Clementi shows a diagram of the experimental ionization energies for the first 53 elements in the periodic table, along with the values calculated using ab initio quantum mechanical methods (Clementi, 1980, p. 12). The periodicity is captured remarkably well, even down to the details of the anomalous sections of the graph occurring between elements in groups II and III in each period of the table where ionization energies show a decrease with increasing Z. Clearly the accurate calculation of atomic properties can be achieved by the theory. The quantum mechanical explanation of the periodic system within this approach represents a far more impressive achievement than merely claiming that elements fall into similar groups because they share the same number of outer-electrons.

And yet in spite of these remarkable successes such an ab initio approach may still be considered to be semi-empirical in a rather specific sense. In order to obtain calculated points shown in the diagram the Schrödinger equation must be solved separately for each of the 53 atoms concerned in this study. The approach therefore represents a form of "empirical mathematics" where one calculates 53 individual Schrödinger equations in order to reproduce the well known pattern in the periodicities of ionization energies. It is as if one had performed 53 individual experiments, although the "experiments" in this case are all iterative mathematical computations. This is still therefore not a general solution to the problem of the electronic structure of atoms.

10. Density Functional Approach.

In 1926 Llewellyn Thomas proposed treating the electrons in an atom by analogy to a statistical gas of particles. Electron-shells are not envisaged in this model, which was independently rediscovered by Enrico Fermi two years later. For many years the Thomas-Fermi method was regarded as a mathematical curiosity without much hope of application since the results it yielded were inferior to those obtained by the method based on electron orbitals.[17]

On the other hand the Thomas-Fermi method, which treats the electrons around the nucleus as a perfectly homogeneous electron gas, yields a mathematical solution that is universal, meaning that it can be solved once and for all. This feature already represents an improvement over the method which seeks to solve Schrödinger equation for every atom separately. This was one of the features that made people go back to the Thomas-Fermi approach in the hope of making progress. As is well known much progress has now been achieved and the current density functional theories, that are modern descendants of the Thomas-Fermi method have become as powerful as methods based on orbitals and wavefunctions and in many cases can outstrip the wavefunction approaches in terms of computational accuracy.

An important conceptual, or even philosophical, difference between the orbital/wavefunction methods and the density functional methods is that, at least in principle, the density functional methods do not appeal to orbitals. In the former case the theoretical entities are completely unobservable whereas electron density invoked by density functional theories is a genuine observable. Experiments to observe electron densities have been routinely conducted since the development of X-ray and other diffraction techniques (Coppens, 2001).[18]

Meanwhile orbitals cannot be observed either directly, indirectly since they have no physical reality contrary to the recent claims in Nature magazine and other journals to the effect that some d orbitals in copper oxide had been directly imaged (Scerri, 2000). Orbitals as used in ab initio calculations are mathematical figments that exist, if anything, in a multi-dimensional Hilbert space.[19] Electron density is altogether different since it is a well-defined observable and exists in real three-dimensional space, a feature which some theorists point to as a virtue of density functional methods.

11. Density Functional Theory in Practice.

But alas most of what has been described so far concerning density theory applies in theory rather than in practice. The fact that the Thomas-Fermi method is capable of yielding a universal solution for all atoms in the periodic table is a potentially attractive feature but is generally not realized in practice. The attempts to implement the ideas originally due to Thomas and Fermi have not quite materialized. This has meant a return to the need to solve a number of equations separately for each individual atom as one does in the Hartree-Fock method and other ab initio methods using atomic orbitals.

In addition most of the more tractable approaches in density functional theory also involve a return to the use of atomic orbitals in carrying out quantum mechanical calculations since there is no known means of directly obtaining the functional that captures electron density exactly.[20] The work almost invariably falls back on using basis sets of atomic orbitals which means that conceptually we are back to square one and that the promise of density functional methods to work with observable electron density, has not materialized.

To make matters worse, the use of a uniform gas model for electron density does not enable one to carry out good calculations. Instead a density gradient must be introduced into the uniform electron gas distribution. The way in which this has been implemented has typically been in a semi-empirical manner by working backwards from the known results on a particular atom, usually the helium atom (Gill, 1998). It has thus been possible to obtain an approximate set of functions which often serve to give successful approximations in other atoms and molecules. As far as I know, there is no known way of yet calculating, in an ab initio manner, the required density gradient which must be introduced into the calculations.

By carrying out this combination of semi-empirical procedures and retreating from the pure Thomas-Fermi notion of a uniform electron gas it has actually been possible, somewhat surprisingly, to obtain computationally *better* results in many cases of interest than with conventional ab initio methods. True enough, calculations have become increasingly accurate but if one examines them more closely one realizes that they include considerable semi-empirical elements at various levels. From the purist philosophical point of view, or what I call "super - ab initio" this means that not everything is being explained from first principles.

As readers of this volume are also aware, the best of both approaches have been blended together with the result that many computations are now performed by a careful mixture of wavefunction and density approaches within the same computations (Hehre et al., 1986). But the unfortunate fact is that, as yet, there is really no such thing as a pure density functional method for performing calculations. The philosophical appeal of a universal solution for all the atoms in the periodic table based on observable electron density, rather than fictional orbitals, has not yet borne fruit.[21, 22]

12. Conclusions.

My aim has not been one of trying to decide whether or not the periodic system is explained by quantum mechanics. Of course broadly speaking quantum mechanics does provide an excellent explanation and certainly one better than was available using only classical mechanics. But the situation is more subtle.

Whereas most chemists and educators seem to believe that all is well, I think that there is some benefit in pursuing the question of how much is strictly explained from the theory. It is indeed something of a miracle that quantum mechanics explains the periodic table to the extent that it does at present. But we should not let this fact seduce us into believing that it is a completely ab initio explanation. We have not yet arrived at the super – ab initio phase of quantum chemistry and nor are we even close. If anything, the compromises that have been struck with the acceptance of parametrization as well as the mixing of wavefunction and DFT approaches begin to question the earlier promise of ab initio quantum chemistry.

One thing that is sure is that the attempt to explain the details of the periodic table continues to challenge the ingenuity of quantum physicists and quantum chemists. For example, a number of physicists are trying to explain the periodic table by recourse to group theoretical symmetries in combination with quantum mechanics (Ostrovsky, 2001). Meanwhile Dudley Herschbach and colleagues have worked on a number of approaches which aim at obtaining a global solution to the energies of the atoms in the periodic table by drawing analogies with the work of physicists on second-order phase transitions (Kais, Sung, Herschbach, 1994)

But to return to the general question of explaining the periodic table, perhaps philosophers of chemistry have a role to play here. Unconstrained by what can presently be achieved, or even what might be achieved in the foreseeable future, one can point out the limitations of the current state of the art. One may thereby place the research in the wider context of scientific reductionism in general and what it might mean for a calculation to be really ab initio. This is not a denial of the progress achieved in quantum chemistry but is more of an unrestrained look at what more could conceivably be done. Of course this might require a deeper theory than quantum mechanics or maybe a cleverer use of the existing theory. There is really no way of telling until the next significant advance has been achieved.

NOTES

1 I doubt whether Löwdin's claim was even true of textbooks in 1969, although I have not conducted any survey on this point.

2 Although actually in this article Hoffman et al. argue that using Ockham's razor may not always be the ideal approach to take in chemistry.

3 I would like to thank the many subscribers to the Computational Chemistry Listserver (CCL) who responded to my questions on this topic.

4 Meanwhile others object to the suggestion that the optimization of basis sets are carried out by reference to experimental data. While accepting that the exponents and contraction coefficients are generally optimized in atomic calculations, they insist that these optimizations are in themselves ab initio.

5 In fact the fourth quantum number does not emerge from solving Schrödinger's equation. It was initially introduced for experimental reasons by Pauli, as a fourth degree of freedom possessed by each electron. In the later treatment by Dirac the fourth quantum number emerges in a natural manner but the extension of this theory to many-electron system has proved to be difficult. The Schrödinger approach is still the most commonly used one. Spin is thus generally added to the functions derived from the Schrödinger equation.

6 It is gratifying to see that this article has now been cited by about ten chemistry textbooks including those by Atkins, Huheey, Levine, Zumdahl etc.

7 Very briefly, the issue is resolved by recognizing that the energies of the 4s and 3d orbitals vary depending on the configuration in question (Melrose, Scerri, 1996).

8 In a recent paper Ostrovsky has criticized my claiming that electrons cannot strictly have quantum numbers assigned to them in a many-electron system (Ostrovsky, 2001). His point is that the Hartree-Fock procedure assigns all the quantum numbers to all the electrons because of the permutation procedure. However this procedure still fails to overcome the basic fact that quantum numbers for individual electrons such as ℓ in a many-electron system fail to commute with the Hamiltonian of the system. As aresult the assignment is approximate. In reality only the atom as a whole can be said to have associated quantum numbers, whereas individual electrons cannot.

9 These calculations were carried out using the Internet web pages constructed by Charlotte Froese-Fischer, http://hf5.vuse.vanderbilt.edu/hf.html

10 Broadly speaking it is still an orbital based method of course but not one that corresponds to the elementary concept of a particular number of electrons in the shells of an atom.

11 I thank Professor Henry Bent for pointing out an error in an earlier draft of this paper.

12 Although the C.I. approach it is capable of calculating the energy of the entire atom, given that the calculation involves a mixture of so many different configurations, it could be argued that it does not calculate the ground state configuration as such. In any case this will depend on the weighting coefficients which result from each calculation.

13 Of course the calculations can be improved by adding extra terms until this failure is eventually corrected. However, these additional measures are only taken after the facts are known. In addition, the lengths to which theoreticians are forced to go to in order to obtain the correct experimental ordering of terms

does not give one too much confidence in the strictly predictive power of quantum mechanical calculations in the context of the periodic table. For example, the very accurate calculations on nickel include the use of basis sets which extend up to 14s, 9p, 5d as well as f orbitals (Raghavachari, Trucks, 1989).

14 It is not inconceivable that this procedure could be automated but I am not aware that it has yet.

15 Details can be found in chapter 4 of the book by Hehre et al. as cited in the references. As a further aside, it is worth noting that despite of the title of this book which is "Ab Initio Molecular Orbital Theory", the authors admit on the very first page that most of the book is devoted to calibrated work!

16 I am doing a certain amount of back-tracking given that this method was mentioned above when some results were quoted for tranition metals.

17 Edward Teller showed that the Thomas-Fermi method cannot predict binding in atoms.

18 This is why I and some others have been agitating about the recent reports, starting in Nature magazine in September 1999, that atomic orbitals had been directly observed. This is simply impossible unless one is using the word "orbital" rather perversely to mean charge density (Scerri, 2000).

19 I have tried to stress the educational implications of the claims for the observation of orbitals in other articles and will not dwell on the issue here (Scerri, 2000 in Science and Education).

20 The fact that an exact density functional exists is known from a theorem proved by Hohenberg and Sham and Kohn. However, this is a non-constructive proof since it does not actually give the form of the exact functional. DFT theorists must try to approximate this functional as well as they can.

21 Some preliminary work aimed at developing pure density methods has been carried out (Wang, Carter, 2000).

21 Peter Gill has recently given a more accurate as well as more witty account of what he regards as the demise of the DFT approach (Gill, 2001).

REFERENCES

Cartwright, Nancy. C., *How the Laws of Physics Lie*. Oxford: Clarendon Press, 1983.

Clementi, Enrico, *Computational Aspects for Large Chemical Systems.* Berlin: Springer-Verlag, 1980.

Galison, Peter, Stump, David, (eds.), *The Disunity of Science.* Stanford University Press: Stanford, California, 1996.

Gill, P.M.W., In *Encyclopedia of Computational Chemistry*, Vol 1, P. von Ragué Schleyer (ed.), Chichester: Wiley, 1998.

Gill, P.M.W., Obituary: Density Functional Theory (1927-1993), Australian Journal of Chemistry, 2001; 54: 661-662.

Hehre, Warren .J., Radom, Leo, Schleyer, Paul, v.R., Pople, John, *Ab Initio Molecular Orbital Theory*. New York: John Wiley, 1986.

R. Hoffman, Minkin, B. Carpenter, Ocham's Razor and Chemistry, Bulletin de la Societée Chimique Française, 1996; 133: 117-130.

Kais, S., Sung, S.M., Herschbach, D.R., Large-Z and –N Dependence of Atomic Energies of the Large-Dimension Limit, International Journal of Quantum Chemistry, 1994; 49: 657-674.

Löwdin, P-O., Studies in Perturbation Theory. X. Bounds to Energy Eigenvalues in Perturbation Theory Ground State, Physical Review, 1965; 139A: 357-364.

Löwdin, P-O., Some Comments on the Periodic System of the Elements, International Journal of Quantum Chemistry, 1969; IIIS: 331-334.

Melrose, M.P.; Scerri, E.R., Why the 4s Orbital is Occupied Before the 3d. Journal of Chemical Education, 1996; 73: 498-503.

Melrose, M.P.; Scerri, E.R., The authors Reply to "Why the 4s Orbital Is Occupied before the 3d", Journal of Chemical Education, 1997; 74: 616-616.

Ovstrovsky, V.N., How and What Physics Contributes to Understanding the Periodic Law, Foundations of Chemistry, 2001; 3: 145-182.

Raghavachari, K., Trucks, G.W., Highly Correlated Systems. Ionization Energies of First Row Transition Metals Sc-Zn, Journal of Chemical Physics, 1989; 91: 2457-2460.

Scerri, Eric R. Transition Metal Configurations and Limitations of the Orbital Approximation, Journal of Chemical Education, 1989; 66: 481-483.

Scerri, Eric R. How Good Is the Quantum Mechanical Explanation of the Periodic System? Journal of Chemical Education, 1998; 75: 1384-1385.

Scerri, E.R., The Failure of Reduction and How to Resist the Disunity of Science in Chemical Education", Science and Education, 2000; 9: 405-425.

Scerri, E.R. Have Orbitals Really Been Observed? Journal of Chemical Eduation, 2000; 77: 1492-1494.

Vanquickenborne, L. G.; Pierloot, K.; Devoghel, D., Electronic Configurations and Orbital Energies, Inorganic Chemistry, 1989; 28: 1805-1813, 1989.

Vanquickenborne, L. G.; Pierloot, K.; Devoghel, D.

Transition Metals and the Aufbau Principle, Journal of Chemical Education, 1994; 71: 469-471.

Wang, A., Carter, E.A., In *Theoretical Methods in Condensed Phase Chemistry*, Schwartz, S.D. (ed.), Dordrecht: Kluwer, 2000, pp. 117-184.

INDEX

MIX
Papier aus verantwortungsvollen Quellen
Paper from responsible sources
FSC® C105338

If you have any concerns about our products,
you can contact us on
ProductSafety@springernature.com

In case Publisher is established outside the EU,
the EU authorized representative is:
Springer Nature Customer Service Center GmbH
Europaplatz 3, 69115 Heidelberg, Germany

Printed by Libri Plureos GmbH
in Hamburg, Germany

MIX
Papier aus verantwortungsvollen Quellen
Paper from responsible sources
FSC® C105338

If you have any concerns about our products,
you can contact us on
ProductSafety@springernature.com

In case Publisher is established outside the EU,
the EU authorized representative is:
Springer Nature Customer Service Center GmbH
Europaplatz 3, 69115 Heidelberg, Germany

Printed by Libri Plureos GmbH
in Hamburg, Germany